Jane's

ARMOUR AND ARTILLERY

Edited by Christopher F Foss

Twenty-third Edition

2002-2003

Total Number of Entries 907	New and Updated Entries 709
Total Number of Images 2,125	New Images 345

ISBN 0 7106 2425 5

In the USA and its dependencies
Jane's Information Group Inc, 1340 Braddock Place, Suite 300, Alexandria, Virginia 22314-1651, USA

roketsan
We offer you our Artillery Experience
122 mm calibre Extended Range Multi Barrel Rocket System
+40 km range
with superior dispersion.
The system is in the inventory of Turkish Armed Forces.
TR-122 rocket family can be fired from T-122 and other standard 122 mm rocket launchers.
T-122 is capable of firing classical short range 122 mm Grad type rockets as well as TR-122 Rocket family.
TR-122, 122 mm Extended Range Rocket Family is grouped into three as TR-122, TRB-122 and TRK-122 having three kinds of warhead; High Explosive, Steel Ball and Submunition warheads respectively.
The system has a NATO Stock Code Number.
Roketsan is certified to AQAP 110 Quality System and complies with MIL-STD-973 Configuration Management System.
TR - 122
T-122 Multi Barrel Rocket Launcher in firing position of TR-122, 122 mm Extended Range Artillery Rocket
See you at
IDEX 2003 (Abu Dhabi/UAE)
and IDEF 2003 (Ankara/Turkey)
T-122
Ankara - Samsun Karayolu 40.km Elmadağ, 06780 Ankara - TURKEY Phone: + 90 312 863 4200 (Pbx) Fax: + 90 312 863 4208 e-mail: rs@rs.com.tr
Design by. Serkan Mumcuoglu e-mail: serkanmumcu@hotmail.com

Contents

ASHOT ASHKELON
Your one stop partner for new production
and upgrades of Automotive Systems
Transmissions, Gear Systems, Final Drives, Shafts, Suspension
Systems, Track Parts, Gears and Assemblies for:
• Tanks
• Armored Personnel Carriers
• Tracked Vehicles
• Wheeled Heavy Vehicles
MERKAVA MBT
UPGRADED M-60
Ashot-Ashkelon Industries Ltd.
P.O.B. 21 Ashkelon 78100, Israel
Tel. +972-8-6721505, Fax. +972-8-6728167
E-mail: boaz.tadmor@ashot.co.il
IDF Combat Proven Hardware
Meets and Mil-Q-9858A
www.ashot.co.il

Jane's

Users' Charter

This publication is brought to you by Jane's Information Group, a global company with more than 100 years of innovation and an unrivalled reputation for impartiality, accuracy and authority.

Our collection and output of information and images is not dictated by any political or commercial affiliation. Our reportage is undertaken without fear of, or favour from, any government, alliance, state or corporation.

We publish information that is collected overtly from unclassified sources, although much could be regarded as extremely sensitive or not publicly accessible.

Our validation and analysis aims to eradicate misinformation or disinformation as well as factual errors; our objective is always to produce the most accurate and authoritative data.

In the event of any significant inaccuracies, we undertake to draw these to the readers' attention to preserve the highly valued relationship of trust and credibility with our customers worldwide.

If you believe that these policies have been breached by this title, you are invited to contact the editor.

A copy of Jane's Information Group's Code of Conduct for its editorial teams is available from the publisher.

INVESTOR IN PEOPLE

How to use *Jane's Armour and Artillery*

In order to help our readers we have included reference pages at the beginning of this publication, the first being this page on How to use *Jane's Armour and Artillery* which gives an overview of what is contained in each section and an explanation of the Entry tags we have added.

These pages and the Foreword and Contents are printed on coloured paper to make it easier for readers to find the page they need.

For ease of reference *Jane's Armour and Artillery* is divided up into the following sections:

Tanks
MBTs and medium tanks
Light tanks
Reconnaissance vehicles
Armoured personnel carriers (tracked)
Armoured personnel carriers (wheeled)
Armoured fighting vehicle families
Tank destroyers
Self-propelled guns and howitzers (tracked)
Self-propelled guns and howitzers (wheeled)
Towed anti-tank guns, guns and howitzers
Multiple rocket launchers
Coastal artillery guns and missiles
(including fixed-fortification weapons)
Armour and artillery in service

Each entry normally has detailed sections covering development, description, variants, specifications, status and manufacturer. For some entries we have added the total number of vehicles ordered or delivered. These figures do not imply that all of these vehicles or weapons are fully operational, as most armies usually allocate significant quantities of vehicles for training and war reserve stocks.

All the entries are arranged under country of origin. In the case of armoured vehicles, the newest equipment is normally given first with the oldest systems last.

This arrangement enables the reader to trace the development of vehicles in a particular country.

In the UK, for example, the latest Challenger 2 MBT is given first, followed by the Challenger 1, Khalid, Chieftain and Centurion. In the USA, the M1 is followed by the older M60 and M48.

In the artillery and multiple rocket launcher sections, the weapons are arranged under calibre with the largest calibre weapons first and the smallest calibre last.

If there are a number of weapons with the same calibre then the latest ones are always given first and the oldest last.

To help users of this title evaluate the published data, Jane's Information Group has divided entries into three categories. A full list of all entries indicating their current status is provided in the index.

● ***VERIFIED*** The editor has made a detailed examination of the entry's content and checked it's relevancy and accuracy for publication in the new edition to the best of his ability.

● ***UPDATED*** During the verification process, significant changes to content have been made to reflect the latest position known to Jane's at the time of publication.

● ***NEW ENTRY*** Information on new equipment and/or appearing for the first time in the title.

All new pictures are dated with the year of publication. New pictures this year are dated 2002. Some are followed by a seven digit number for ease of identification by our image library.

Total Number of Entries 907 New and Updated Entries 709
Total Number of Images 2,125 New Images 345

Copyright enquiries
Contact: Keith Faulkner, Tel/Fax: +44 (0) 1342 305032, e-mail: keith.faulkner@janes.co.uk

British Library Cataloguing-in-Publication Data.
A catalogue record for this book is available from the British Library.

Printed and bound in Great Britain by Bath Press, Bath and Glasgow

Alphabetical list of advertisers

Abbreviations

AA	anti-aircraft
AAAV	Advanced Amphibious Assault Vehicle
AAD	Armoured Amphibious Dozer
AAG	anti-aircraft gun
AAK	Appliqué Armor Kit (US)
AAPC	Advanced Armoured Personnel Carrier (Turkey)
AAV	Assault Amphibian Vehicle
ABIT	Advanced Built-In Test
ABS	Anti-skid Braking System
ABSV	Armoured Battlegroup Support Vehicle
ACAVP	Advanced Composite Armoured Vehicle Platform (UK)
ACCE	Abrams/Crusader Common Engine
ACCV	Armored Cavalry Cannon Vehicle (US)
ACE	Armored Combat Earthmover (US)
ACLOS	Automatic Command Line Of Sight
ACRV	Armoured Command and Reconnaissance Vehicle
ACS	Artillery Communications System
ACV	Armored Cannon Vehicle (US)
ACV	Armoured Combat Vehicle
ACVT	Armored Combat Vehicle Technology Program (US)
ADAM	Area Denial Artillery Munition
ADAMS	Air Defense Advanced Mobile System (US)
ADATS	Air Defence Anti-Tank System
ADEA	Army Development and Employment Agency
AEV	armoured engineer vehicle
AF	air force
AFARV	Armored, Forward Area, Re-arm Vehicle (US)
AFAS	Advanced Field Artillery System
AFCS	Automatic Fire-Control Systems
AFD	automatic feeding device
AFSV	Armoured Fire Support Vehicle
AFV	Armored Family of Vehicles (US)
AFV	armoured fighting vehicle
AGF	Army Ground Forces
AGL	above ground level
AGLS	Automatic Gun Laying System
AGS	Armored Gun System (US)
AGV	Assault Gun Vehicle
AGVT	Advanced Ground Vehicle Technology
Ah	ampère hour
AIFS	Advanced Indirect Fire System
AIFV	Armored Infantry Fighting Vehicle (US)
AIFV	armoured infantry fighting vehicle
AIPS	Advanced Integrated Propulsion System (US)
ALAAVS	Advanced Light Armored/Amphibious Vehicle System (US)
ALC	Advanced Land Combat (US)
ALS	Advanced Laying System
ALSV	Armoured Logistics Support Vehicle
ALT	Armoured Launching Turret
AMC	Advanced Mortar Carrier (Turkey)
AMC	Armoured Mortar Carrier
AMC	Army Matériel Command
AMCCOM	Armament Munitions and Chemical Command
AMDS	Anti-Missile Discarding Sabot
AMF	*Amphibische Mehrzweck-Fahrzeuge* (multipurpose amphibious vehicle)
AML	*Automitrailleuse Légère* (light armoured car)
AMOS	Advanced Mortar System
AMR	*Automitrailleuse de Reconnaissance*
AMS	Armoured Mortar System
AMX	*Atelier de Construction d'Issy-les-Moulineaux*
ANAD	Anniston Army Depot
AOI	Arab Organisation for Industrialisation
AOS	add-on stabilisation
AP	armour-piercing
APAM	anti-personnel, anti-matériel
APC	armoured personnel carrier
APC	armour-piercing capped
APC-T	armour-piercing capped – tracer
APCT-BF	armour-piercing capped tracer – base fuze
APDS	armour-piercing discarding sabot
APDS-T	armour-piercing discarding sabot – tracer
APE	*Amphibisches Pionier-Erkundungsfahrzeug* (amphibious front-line reconnaissance vehicle)
APERS	anti-personnel
APERS-T	anti-personnel tracer
APFIDS	armour-piercing fragmentation incendiary discarding sabot
APFSDS	armour-piercing fin-stabilised discarding sabot
APFSDS(P)	armour-piercing fin-stabilised discarding sabot (practice)
APFSDS-T	armour-piercing fin-stabilised discarding sabot – tracer
APG	Aberdeen Proving Grounds
APGM	Autonomous Precision-Guided Munition
APHE	armour-piercing high explosive
API	armour-piercing incendiary
APIT	armour-piercing incendiary tracer
APM	anti-personnel mine
APS	Advanced Propulsion System
APS	Artillery Pointing System
APSE	armour-piercing secondary effect
APSE-T	armour-piercing secondary effect – tracer
APT	armour-piercing tracer
APTE	Abrams Power Train Evolution (US)
APU	auxiliary power unit
APV	armoured patrol vehicle
AR/AAV	Armored Reconnaissance/Airborne Assault Vehicle (US)
ARDEC	Armament Research Development and Engineering Center (US)
ARDNOT	Automatic Day/Night Optical Tracker
ARE	*Atelier de Construction Roanne*
ARETS	Armour Remote Target System
ARMAD	Armoured and Mechanised Unit Air Defence
ARMSCOR	Armament Manufacturing Corporation (South Africa)
ARMVAL	Anti-Armor Vehicle Evaluation (US)
ARP	anti-radiation projectile
ARP	Armoured Repair Plates
ARRADCOM	Armament Research and Development Command
ARRV	Armoured Repair and Recovery Vehicle
ARSV	Armored Reconnaissance Scout Vehicle (US)
ARV	armoured recovery vehicle
ASA	Advanced Security Agency
ASARC	Army Systems Acquisition Review Council
ASCOD	Austrian Spanish Co-operative Development
ASM	Armored Systems Modernisation (US)
ASTROS	Artillery Saturation Rocket System (Brazil)
ASV	ammunition supply vehicle
ASV	Armored Security Vehicle (US)
AT	anti-tank
ATACS	Advanced Tank Cannon System (US)
ATCAS	Advanced Towed Cannon System
ATD	Advanced Technology Demonstration
ATD	Advanced Technology Demonstrator (US)
ATD	Automatic Target Detection
ATDT	Automatic Target Detection and Tracking
ATDU	Armoured Trials and Development Unit (UK)
ATFCS	Automated Targeting and Fire Control System
ATG	anti-tank gun
ATGM	Anti-Tank Guided Missile
ATGW	anti-tank guided weapon
ATLAS	Advanced Technology Light Artillery System
ATLV	Artillery Target Location Vehicle
ATM	anti-tank mine
ATR	Automotive Test Rig
ATR – EDS	Automotive Test Rig – Electric Drive System
ATS	*Atelier de Construction de Tarbes*
ATTC	All Terrain Tracked Carrier
ATTS	Air Transportable Towed System
ATV	Armoured TOW Vehicle (Turkey)
AVGP	Armoured Vehicle General Purpose (Canada)
AVH	Armoured Vehicle Heavy
AVL	Armoured Vehicle Light
AVLB	armoured vehicle-launched bridge
AVM	Armoured Vehicle Medium
AVR	armoured vehicle reconnaissance
AVRE	Armoured Vehicle Royal Engineers (UK)
AVRE	Assault Vehicle Royal Engineers (UK)
AVT	Advanced Vehicle Technologies (US)
AWE	Advanced Warfighting Experiment
BAe	British Aerospace
BARV	Beach Armoured Recovery Vehicle
BASE	British Aerospace Systems and Equipment
BATES	Battlefield Artillery Target Engagement System
BB	base bleed
BCC	battery control centre
BCP	Battery Command Post
BCT	Brigade Combat Team (US)
BCV	Battle Command Vehicle
BD	base detonating
BE	base ejection
BFV	Bradley Fighting Vehicle
bhp	brake horsepower
BIFF	Battlefield Identification Friend or Foe
BITE	built-in test equipment
BL	Blank
BLR	*Blindado Ligero de Ruedas*
BL-T	blank tracer
BMF	Belgian Mechanical Fabrication
BMR	*Blindado Medio de Ruedas*
BMS	Battlefield Management System
BNS	Bill Night Sight
BOCV	Battery Operations Centre Vehicle
BOFI	Bofors Optronic Fire-control Instrument
BOL	Bearing Only Launch
BOLTS	Bolt On Loading Tray System
BTA	best technical approach
CAD	computer-assisted design
CAL	Canadian Arsenals Limited
CAP	Combustible Augmented Plasma
CARRV	Challenger Armoured Repair and Recovery Vehicle
CAT/FCS	Command Adjusted Trajectory/Fire-Control System (US)
CAT/LCV	Combined Arms Team/Lightweight Combat Vehicle (US)
CATT-B	Component Advanced Technology TestBed (US)
CAV	Composite Armored Vehicle (US)
CAWS	Cannon Artillery Weapons Systems (US)
CCC	combustible cartridge case
CCD	Charge Coupled Device
CCP	Computer Control Panel
CCR	Crown Copyright Reserved
CCTV	closed-circuit TV
CCU	Central Control Unit
CCV	Close Combat Vehicle
CCV	Command and Control Vehicle (US)
CCV-L	Close Combat Vehicle – Light
CDM	Coastal Defence Missile
CDU	Command Display Unit
CDU	Computer Display Unit
CE	Chemical Energy
CECOM	Communications-Electronics Command
CENTCOM	Central Command
CEP	Circular Error of Probability
CET	Combat Engineer Tractor
CEU	Computer Electronics Unit
CEV	combat engineer vehicle
CF	controlled fragmentation
CFE	Conventional Forces Europe
CFV	Cavalry Fighting Vehicle (US)
CGS	Crew Gunnery Simulator
CHARM	Challenger Chieftain Armament
CHIP	Challenger Improvement Programme
CIFV	Composite Infantry Fighting Vehicle
CILAS	*Compagnie Industrielle des Lasers*
CIS	Chartered Industries of Singapore
CIS	Commonwealth of Independent States
CITV	Commander's Independent Thermal Viewer (US)
CIWS	Close-In Weapons System
CLAMS	Clear Lane Marking System (US)
CLAWS	Close Combat Light Armor Weapon System (US)
CLGP	Cannon-Launched Guided Projectile (US)
CMS	Compact Modular Sight
CMV	Combat Mobility Vehicle (US)
COG	Course Over Ground
Comp B	Composition B
COMVAT	Combat Vehicles Armament Technology (US)
COS	Chief of Section
COTAC	*Conduite de Tir Automatique pour Char* (tank automatic fire)
COTS	Commercial-off-the-shelf
COV	Counter Obstacle Vehicle (US)
CP	concrete-piercing
CPS	Cardinal Points Specification (UK)
CPV	command post vehicle
CR	Capability Requirement
CRR	*Carro de Reconhecimento Sobre Rodas* (reconnaissance tracking scout car)
CRT	cathode ray tube
CRU	Cable Reel Unit
CS	Communications Subsystem
CSB	Combat Support Boat
CSF	Combined Service Forces
CSI	computer-synthesised image
CSS	Computer Sighting System
CTA	Case Telescoped Ammunition
CTI	Central Tyre Inflation
CTIS	Central Tyre Inflation System
CTRA	*Carro de Transporte Sobre Rodas Anfibo* (amphibious tracking scout car)
CTT	Challenger Training Tank

CV90	Combat Vehicle 90
CVAST	Combat Vehicle Armament System Technology (US)
CVRDE	Combat Vehicle Research and Development Establishment (India)
CVR(T)	Combat Vehicle Reconnaissance (Tracked) (UK)
CVR(W)	Combat Vehicle Reconnaissance (Wheeled) (UK)
CVT	controlled variable time
CVTTS	Combat Vehicle Targeting System
CWR	continuous wave radar
CWS	cupola weapon station
DAHA	dual axis head assembly
DAREOD	Damaged Airfield Reconnaissance Explosive Ordnance Disposal
DARCOM	US Army Matériel Development and Readiness Command
DARPA	Defense Advanced Research Projects Agency
DAS	Defensive Aids Suite
DAS	Defensive Aid System
DASP	Demountable Artillery Surviellance Pod
DCA	*Défense Contre Avions* (anti-aircraft)
DD	Detroit Diesel
DDA	Detroit Diesel Allison
DDS	Department of Defense Support
DDU	digital display unit
DECA	Digital Electronic Control Assembly
DEFA	*Direction des Études et Fabrications d'Armement*
DERA	Defence Evaluation and Research Agency
DESO	Defence Export Sales Organisation
DFCS	Digital Fire-Control System
DFSV	Direct Fire Support Vehicle
DFV	Desert Fighting Vehicle
DHSS	Data Handling SubSystem
DIA	Defense Intelligence Agency (US)
DIP	driver instrument panel
DMU	Distance Measurement Unit
DNRS	Day/Night Range Sight
DoD	Department of Defense (US)
DOP	Department of Productivity
DPA	Defence Procurement Agency (UK)
DPICM	Dual-Purpose Improved Conventional Munition (US)
DRA	Defence Research Agency (UK)
DROPS	Demountable Rack Offloading and Pick-up System
DSACS	Direct Support Armored Cannon System (US)
DSETS	Direct Support Electrical Test System
DSO	Defence Sales Organisation
DS/T	practice discarding sabot/tracer
DSWS	Division Support Weapon System
DTAT	*Direction Technique des Armements Terrestres*
DTT	Driver Training Tank
DU	depleted uranium
DV	Deming Vehicle
DVE	Driver's Vision Enhancer
DWFK	Deep Water Fording Kit
EAAK	Enhanced Appliqué Armor Kit (US)
EAOS	Enhanced Artillery Observation System
EBG	*Engin Blindé Génie* (armoured combat vehicle)
EBR	*Engin Blindé de Reconnaissance* (armoured reconnaissance vehicle)
EBRC	*Engin Blindé a roues de Contact*
EC	European Community
ECS	Environmental Control Subsystem
ECV	Enhanced Capacity Vehicle
EDS	Electric Drive System
EFAB	*Etablissement d'Etudes et de Fabrications d'Armement de Bourges*
EFC	equivalent full charge
EFCR	equivalent full charge rounds
EFM	Explosives Factory Maribyrnong
EFP	Expanded Feasibility Phase
EFP	Explosively Formed Penetrator
EFVS	Electronic Fighting Vehicle System (US)
EG	External Gun
ELKE	Elevated Kinetic Energy Weapon (US)
ELS	Electromagnetic Launcher System
ELSAP	*Elektronische Schiessanlage für Panzer* (electronic fire system for tanks)
EM	electromagnetic
EMC	Executive Management Committee
EMD	Engineering and Manufacturing Development
EMDG	Euromissile Dynamics Group
EMG	externally mounted gun
EMP	electromagnetic pulse
ENGESA	*Engesa Engenheiros Especializados* (Brazil)
EOD	explosive ordnance disposal
EOR	Explosive Ordnance Reconnaissance
EOTS	Electro-Optical Tracking System
EPC	electronic plane conversion
EPC	*Engin Principal de Combat* (future main battle tank)
EPG	European Production Group
EPG	European Programme Group
EPS	Electrical Power Subsystem
EPU	Electronics Processing Unit
ER	enhanced radiation
ER	extended range
ERA	explosive reactive armour
ERA	extended range ammunition
ERC	*Engin de Reconnaissance Canon*
ERFB	extended range full-bore
ERFB-BB	extended range full-bore – base bleed
ERGFCDS	Extended Range Gunnery Fire-Control Demonstration System
ERGP	extended range guided projectile
ERMIS	Extended Range Modification Integration System
ERP	Extended Range Projectile
ERSC	extended range subcalibre
ERV	Engineer Reconnaissance Vehicle
ERV	emergency rescue vehicle
ESD	*Electronique Serge Dassault*
ESPAWS	Enhanced Self-Propelled Artillery Weapon System (US)
ESRS	electro-slag refined steel
ETA	Estimated Time of Arrival
ETC	ElectroThermal Cannon
ETM	Electronic Technical Manuals
ETS	Elevated TOW System
ETS	Engineer Tank System
EW	electronic warfare
EWK	*Eisenwerke Kaiserslautern Göppner*
EWS	external weapon station
FAAD	Forward Area Air Defense (US)
FAAR	Forward Area Alerting Radar (US)
FAASV	Field Artillery Ammunition Support Vehicle (US)
FACE	Field Artillery Computer Equipment
FARS	field artillery rocket system
FARV-A	Future Armored Resupply Vehicle – Artillery (US)
FAST	Forward Area Support Team
FAV	Fast Attack Vehicle (US)
FBRV	Future Beach Recovery Vehicle (UK)
FCC	Fire Command Centre
FCC	Fire-Control Computer
FCCVS	Future Close-Combat Vehicle System (US)
FCE	fire-control equipment
FCLV	Future Command and Liaison Vehicle (UK)
FCS	fire-control system
FCS	Future Combat System
FCU	Fire Control Unit
FDC	fire direction centre
FDCV	fire direction centre vehicle
FDSWS	Future Direct Support Weapon System
FEBA	Forward Edge of the Battlefield Area
FEP	Firepower Enhancement Programme
FET	Future Engineer Tank
FFAR	Folding Fin Aerial Rocket
FG	field gun
FH	field howitzer
FIFV	Future Infantry Fighting Vehicle (US)
FIS	Fuze Interface System
FISTV	Fire Support Team Vehicle (US)
FITOW	Further Improved TOW (US)
FLIR	forward-looking infra-red
FLOT	forward line of own troops
FMC	Food Machinery Corporation
FMS	Foreign Military Sales
FMTV	Family of Medium Tactical Vehicles
FN	*Fabrique Nationale*
FOG	Fibre Optic Gyro
FOM	Fibre Optic Missile
FOO	forward observation officer
FOTT	Follow On To TOW
FOV	field of view
FRAG	fragmentation
FROG	free rocket over ground
FRES	Future Rapid Effects System
FSCV	fire support combat vehicle
FSED	full-scale engineering development (US)
FST	Future Soviet Tank/Follow-on Soviet Tank
FSV	fire support vehicle
FSV	Future Scout Vehicle (US)
FTMA	Future Tank Main Armament
FTS	Future Tank Study
FUE	First Unit Equipped
FUG	*Felderitö Usó Gépkosci*
FV	fighting vehicle
FVDD	Fighting Vehicle Development Division
FV/GCE	fighting vehicle gun control equipment
FVRDE	Fighting Vehicle Research and Development Establishment (UK)
FVS	Fighting Vehicle System (US)
FVSC	Fighting Vehicle Systems Carrier (US)
FY	fiscal year
G	gendarmerie
GAO	General Accounting Office
GAP	Gun Aiming Post
GCE	gun control equipment
GCT	*Grande Cadence de Tir* (high rate of fire)
GCW	Gross Combat Weight
GD	General Dynamics
GDLS	General Dynamics Land Systems
GDU	Gun Display Unit
GFE	Government Furnished Equipment
GH	gun-howitzer
GIAT	*Groupement Industriel des Armements Terrestres*
GLC	gun lay computer
GLH-H	Ground-Launched Hellfire-Heavy (US)
GLLD	Ground Laser Locator Designator (US)
GLS	*Gesellschaft für Logistischen Service*
GMC	General Motors Corporation
GMLRS	Guided Multiple Launch Rocket System
GM, MVO	General Motors, Military Vehicle Operations
GMS	Gun Management System
GOCO	government-owned, contractor-operated (US)
GP	General Purpose
GP	guided projectile
GPMG	General Purpose Machine Gun
GPO	gun position officer
GPS	Global Positioning System
GPS	gunner's primary sight
GPSS	Gunner's Primary Sight Subsystem
GSR	General Staff Requirement (US)
GSRS	General Support Rocket System
GST	General Staff Target
GST	*Gesellschaft für System-Technik*
GTCS	Gun Test and Control System
GTI	German Tank Improvement
GVW	Gross Vehicle Weight
GW	guided weapon
HAB	Heavy Assault Bridge (US)
HAWK	Homing-All-the-Way-Killer (US)
HB	heavy barrel
HC	high capacity
HCDS	Harpoon Coast Defence Unit
HCER	high-capacity extended range
HCHE	high-capacity high explosive
HCT	HOT Compact Turret
HE	high explosive
HE-APERS	high-explosive anti-personnel
HEAP-T	high-explosive anti-personnel – tracer
HEAT	high-explosive anti-tank
HEAT-FS	high-explosive anti-tank fin-stabilised
HEAT-MP	high-explosive anti-tank multipurpose
HEAT-MP(P)	high-explosive anti-tank multipurpose (practice)
HEAT-T	high-explosive anti-tank – tracer
HEAT-T-HVY	high-explosive anti-tank – tracer – heavy
HEAT-T-MP	high-explosive anti-tank – tracer – multipurpose
HEAT-TP-T	high-explosive anti-tank – target practice – tracer
HED-D	Hybrid Electric Drive-Demonstrator
HEDP	high-explosive dual-purpose
HEER	High-Explosive Extended Range (US)
HE-FRAG	high-explosive fragmentation
HE-FRAG-FS	high-explosive fragmentation – fin-stabilised
HE-FS	high-explosive – fin-stabilised
HEI	high-explosive incendiary
HEIT	high-explosive incendiary tracer
HEL	High-Energy Laser (US)
HEL	Human Engineering Laboratory (US)
HELP	Howitzer Extended Life Program (US)
HEMAT	Heavy Expanded Mobility Ammunition Trailer (US)
HEMTT	Heavy Expanded Mobility Tactical Truck (US)
HEP	high-explosive plastic
HEPD	high-explosive point detonating
HE/PR	high-explosive practice
HEP-T	high-explosive practice – tracer
HERA	high-explosive rocket-assisted
HE-S	high-explosive spotting
HESH	high-explosive squash head
HESH-T	high-explosive squash head – tracer
HE-T	high-explosive tracer
HET-PF	high-explosive tracer – percussion fuze
HE-T SD	high-explosive tracer – self-destruct
HETS	Heavy Equipment Transport Semi-trailer
HFCC	Howitzer Fire-Control Computer (US)
HFHTB	Human Factors Howitzer TestBed (US)
HFM	Heavy Force Modernisation (US)
HIFV	Heavy Infantry Fighting Vehicle
HIMAG	High-Mobility Agility Test Vehicle (US)
HIMARS	High Mobility Artillery Rocket System (US)
HIP	Howitzer Improvement Program (US)
HIRE	Hughes Infra-Red Equipment
HIU	Heading Indicator Unit

HMC	Howitzer Motor Carriage
HMD	Helmet-Mounted Display
HMLC	High-Mobility Load Carrier
HMMWV	High-Mobility MultiPurpose Wheeled Vehicle
How	howitzer
HOE	Holographic Optical Element
hp	horsepower
HPFP	High-Performance Fragmentation Projectile
HPS	Helmet Pointing System (or Sight)
HPT	high-pressure test
HRU	Heading Reference Unit
HSCLCS	Harpoon Ship Command Launch Control Set
HSDS	Hellfire Shore Defence System
HSS	Hunter Sensor Suite
HSTV(L)	High-Survivability Test Vehicle (Lightweight) (US)
HTTB	High-Technology TestBed (US)
HVAP	high-velocity armour-piercing
HVAPDS-T	high-velocity armour-piercing discarding sabot – tracer
HVAPFSDS	high-velocity armour-piercing fin-stabilised discarding sabot
HVAP-T	high-velocity armour-piercing tracer
HVM	Hypervelocity Missile
HVSS	horizontal volute spring suspension
HVSW	Hypervelocity Support Weapon
HVTP-T	high-velocity target practice – tracer
HWSTD	High Water Speed Technology Demonstrator
HYPAK	Hydraulic Power Assist Kit
IAFV	infantry armoured fighting vehicle
IAI	Israel Aircraft Industries
IASD	Instant Ammunition Selection Device
IBAS	Improved Bradley Acquisition System
ICC	information co-ordination centre
ICM	improved conventional munition
ICM BB	improved conventional munition base bleed
ICV	Infantry Combat Vehicle
IDF	Israel Defence Forces
IEPG	Independent European Program Group
IFCS	Improved Fire-Control System
IFCS	Integrated Fire-Control System
IFF	Identification Friend or Foe
IFV	infantry fighting vehicle
IFVwCM	Infantry Fighting Vehicle with Integrated Countermeasures (US)
II	image intensification
ILL	illuminating
ILMS	Improved Launcher Mechanical System
ILS	Integrated Logistic Support
IMMLC	Improved Medium Mobility Load Carrier
IOC	initial operational capability
IPF	Initial Production Facility
IPO	International Programme Office
IPPD	Integrated Product Process Development
IR	infra-red
IRBM	intermediate-range ballistic missile
IRU	Inertial Reference Unit
IS	internal security
ISD	In Service Date
ISU	Integrated Sight Unit
ISV	Internal Security Vehicle
ITV	Improved TOW Vehicle (US)
IVPDL	Inter-Vehicle Positioning and Data Link (US)
IWS	Improved Weapon System
JBMoU	Joint Ballistic Memorandum of Understanding
JGSDF	Japanese Ground Self-Defence Force
JPO	Joint Project Office
JSC	Joint Steering Committee
JSDFA	Japanese Self-Defence Force Agency
KAAV	Korean Armoured Amphibious Vehicle
KE	kinetic energy
KEC	Kaman Electromagnetics Corporation
KEM	Kinetic Energy Missile (US)
KIFV	Korean Infantry Fighting Vehicle
KPA	Korea People's Army
LAAG	light anti-aircraft gun
LAD	light aid detachment
LADS	light air defense system (US)
LAPES	Low-Altitude Parachute Extraction System
LASIP	Light Artillery System Improvement Plan
LATS	Light Armoured Turret System
LAU	Light Armoured Unit
LAV	Light Armored Vehicle (US)
LAV	Light Assault Vehicle (US)
LAV-AD	Light Armored Vehicle – Air Defense
LAW	light anti-tank weapon
LCS	Loader Control System
LCU	Landing Craft Utility
LCV	Light Contingency Vehicle (LCV) (US)
LED	light emitting diode
LEP	Life Extension Programme
LEU	Launcher Erector Unit
LIW	Lyttleton Engineering Works
LIA	Linear Induction Accelerator
LIMAWS	Lightweight Mobile Artillery Weapon System
LKP	Loader Keyboard Panel
LLAD	low-level air defence
LLLTV	low light level television
LLM	Launcher Loader Module
LMG	light machine gun
LNS	Land Navigation System
LNS	Laying and Navigation System
LOAL	Lock-On After Launch
LOBL	Lock-On Before Launch
LOS	Line Of Sight
LOSAT	Line Of Sight Anti-Tank
LP	Liquid Propellant
LPC	Launcher Pod Carrier
LPC	Launch Pod Container
LPT	Low-Profile Turret
LPTS	Lightweight Protected Turret System
LRAR	Long Range Artillery Rocket
LRASS	Long-Range Advanced Scout System
LRAT	long-range anti-tank
LRBB	long-range base bleed
LRF	laser range-finder
LRF	Low Recoil Force
LRHB	long-range hollow base
LRN	Low Recoil NORICUM
LRU	Launcher Replenishment Unit
LRU	line-replaceable unit
LTD	laser target designator
LTFCS	Laser Tank Fire-Control System
LVA	landing vehicle assault (US)
LVT	landing vehicle tracked (US)
LVTC	landing vehicle tracked command (US)
LVTE	landing vehicle tracked engineer (US)
LVTH	landing vehicle tracked howitzer (US)
LVTP	landing vehicle tracked personnel (US)
LVTR	landing vehicle tracked recovery (US)
LWML	LightWeight Multiple Launcher
LWMS	LightWeight Modular Sight
LWS	laser warning system
LWT	light weapon turret
MAC	Medium Armored Car (US)
MACS	Modular Artillery Charge System
MADLS	Mobile Air Defence Launching System
MAMBA	Mobile Artillery Monitoring Battlefield Radar
MAOV	Mobile Artillery Observation Vehicle
MAP	Military Aid Programme
MAPS	Modular Azimuth Position System
MARDI	Mobile Advanced Robotics Defence Initiative (UK)
MARS	Multiple Artillery Rocket System
MAV	maintenance assist vehicle
MAVD	MLRS Aim Verification Device
MBA	main battle area
MBB	*Messerschmitt-Bölkow-Blohm*
MBT	main battle tank
MCCS	Marconi Command and Control Systems
MCRV	Mechanised Combat Repair Vehicle
MCS	Microclimate Conditioning System
MCS	modular charge system
MCSK	Mine Clearance System Kit (US)
MCT	Medium Combat Tractor (US)
MCT	MILAN Compact Turret
MCT	Mercury Cadmium Telluride
MCV	Mechanised Combat Vehicle
MCWL	Marine Corps Warfighting Laboratory
MCWS	Minor Calibre Weapons Station (US)
MDU	Map Display Unit
MENS	Mission Element Need Statement (US)
MEV	medical evacuation vehicle (US)
MEWS	Mobile Electronic Warfare System
MEWSS	Mobile Electronic Warfare Support System
MF	multifunction
MFC	mortar fire controller
MFCV	Missile Fire-Control Vehicle
MFF	munition filling factory
MGTS	MultiGun Turret System
MG	machine gun
MGB	Medium Girder Bridge
MGU	Mid-course Guidance Unit
MICOM	Missile Command (US)
MICV	mechanised infantry combat vehicle
MILAN	*Missile d'Infantrie Léger Antichar* (light infantry anti-tank missile)
MILES	Multiple Integrated Laser Engagement System (US)
MIPS	Medium Integrated Propulsion System (US)
MLA	Manufacturing Licence Agreement
MLC	Military Load Class
MLC	Modular Load Carrier
MLI	Mid-Life Improvement
MLRS	Multiple Launch Rocket System (US)
MMBF	mean miles between failures
MMS	mast-mounted sight
MoA	Memorandum of Agreement
MoD	Ministry of Defence
MODA	Ministry of Defence and Aviation
MOLF	Modular Laser Fire Control (Germany)
MoU	Memorandum of Understanding
MPBAV	MultiPurpose Base Armoured Vehicle
MPC	MultiPurpose Carrier (Netherlands)
MPGS	Mobile Protected Gun System (US)
MPS	Maritime Prepositioning Ships (US)
MPV	Multi-Purpose Vehicle
MPWS	Mobile Protected Weapon System (US)
MRAV	MultiRole Armoured Vehicle
MRBF	mean rounds before failure
MRCS	Marconi Radar and Control Systems
MRS	multiple rocket system
MRS	muzzle reference system
MRSI	Multiple Round Simultaneous Impact
MRV(R)	Mechanised Recovery Vehicle (Repair)
MRVR	Mechanised Repair and Recovery Vehicle (US)
MSTAR	Man-portable Surveillance and Target Acquisition Radar
MSV	Modular Support Vehicle
MT	mechanical time
MTB	Mobility TestBed
MTBF	Mean Time Between Failure
MTI	moving target indication
MTL	Materials Technology Laboratory (US)
MTR	Mobile Test Rig (US)
MTSQ	mechanical time and super-quick
MTU	*Motoren- und Turbinen-Union*
MTVL	mobile tactical vehicle light
MULE	Modular Universal Laser Equipment (US)
MV	muzzle velocity
MVEE	Military Vehicles and Engineering Establishment
MVRS	Muzzle Velocity Radar System
MWS	Manned Weapon Station (US)
NATO	North Atlantic Treaty Organisation
NBC	nuclear, biological, chemical
NBMR	NATO Basic Military Requirement
NDU	Navigation Display Unit
NTC	National Training Center (USA)
NUGP	nominal unit ground pressure
NVE	Night Vision Equipment
NVESD	Night Vision and Electronic Sensors Directorate
OBR	Optical Beam-Riding
OCC	*Obus à Charge Creusé* (shaped-charge shell)
OCDS	OTOMAT Coastal Defence System
OCU	Operational Centre Unit
ODE	Ordnance Development and Engineering (Singapore)
OEO	optical and electro-optical
OP	observation post
OP	Operators Panel
OPV	Observation Post Vehicle
OT	operational test
OTA	overflight top attack
OTEA	Operational Test and Evaluation Agency (US)
P^3I	Pre-Planned Product Improvements
PAR	pulse acquisition radar
PAT	power-assisted traverse
PCB	printed circuit board
PD	point detonating
PDNA	Positioning Determining/Navigation Unit
PDRR	Program Definition and Risk Reduction
PE	Procurement Executive
PEC	Printed Electronic Circuits
PFD	proximity fuze disconnector
PFHE	prefragmented high explosive
PFPX	prefragmented proximity fuzed
P How	pack howitzer
PIE	pyrotechnically initiated explosive
PIP	Product Improvement Programme
PLA	People's Liberation Army
PLARS	Position Location And Reporting System (US)
PLC	Programmable Logic Control
PLO	Palestinian Liberation Army
PLS	Palletised Load System (US)
PM	*porte mortier* (mortar carrier)
PMO	Program Management Office (US)
PNU	Position Navigation Unit
POF	Pakistan Ordnance Factories
PMS	Pedestal-Mounted Stinger
POS	*Postes Optiques de Surveillance*
PPI	plan position indicator
PPS	Precise Positioning Service
PRAC	practice
PRAC-T	practice tracer
PRI	projector reticle image
PSD	Propulsion System Demonstrator

PSS	Primary Sight System
PTO	power take-off
PVP	*Petit Véhicule Protégé*
PWI-SR(GR)	*Panser Wagen Infanterie-Standaard (Groep)*
QCB	Quick Change Barrel
QE	quadrant elevation
RA	Royal Artillery
RAAMS	Remote Anti-Armor Mine System (US)
RAC	Royal Armoured Corps
RADIRS	Rapid Deployment Multiple Rocket System (US)
RAM-D	reliability, availability, maintainability and durability
RAO	Rear Area Operations
RAP	rocket-assisted projectile
RAPI	Reactive Armour Protection
RARDE	Royal Armament Research and Development Establishment (UK)
RATAC	*Radar de Tir pour L'Artillerie de Campagne* (radar for field artillery fire)
RBL	Range and Bearing Launch
RCAAS	Remote-Controlled Anti-Armor System (US)
RCC	Roland Co-ordination Center (US)
RCDU	Remote-Controlled Defence Unit
RC/MAS	Reserve Component/Modified Armament System
RCS	Radar Cross Section
RCT	Royal Corps of Transport
RCV	Robotic Command Vehicle
RDF	Rapid Deployment Forces
RDF/LT	Rapid Deployment Force Light Tank (US)
RDJTF	Rapid Deployment Joint Task Force (US)
RDT&E	Research Development Test and Evaluation
REME	Royal Electrical and Mechanical Engineers
RFAS	Russian Federation and Associated States
RFI	request for information
RFP	request for proposals
RFPI	Rapid Force Projection Initiative
RFQ	Request For Quotations
RHA	rolled homogeneous armour
RHA	Royal Horse Artillery
RISE	Reliability Improved Selected Equipment
RLC	Royal Logistic Corps
RLG	Ring Laser Gyro
RMG	ranging machine gun
RO	Royal Ordnance
ROBAT	Robotic Counter-Obstacle Vehicle (US)
RoC	Republic of China
ROC	required operational characteristics
ROF	rate of fire
RoK	Republic of Korea
RoKIT	Republic of Korea Indigenous Tank
ROR	range only radar
RP	rocket-propelled
RPC	Rocket Pod Container
RPV	remotely piloted vehicle
RR	Recoilless Rifle
RRPR	Reduced Range Practice Rocket
RSS	Rosette Scanning Seeker
RTT	*Roues Transporteur de Troupes*
RUC	Royal Ulster Constabulary
SABCA	*Société Anonyme Belge de Constructions Aéronautiques*
SADARM	Sense And Destroy Armor (US)
SADF	South African Defence Force
SAE	Society of Automotive Engineers
SAL	semi-active laser
SAM	surface-to-air missile
SAMM	*Société d'Applications des Machines Motrices*
SANDF	South African National Defence Force
SANG	Saudi Arabian National Guard
SAPI	semi-armour-piercing incendiary
SATCP	*Système Anti-aérien à Très Courte Portée* (very short-range anti-aircraft system)
SAVA	Standard Army Vectronics Architecture
SAW	Squad Automatic Weapon
SCORE	Stratified Charge Omnivorous Rotary Engine
SD	self-destruct
SEME	School of Electrical and Mechanical Engineering
SEN	shell extended range NORICUM
SFCS	Simplified Fire-Control System
SFIM	*Société de Fabrication d'Instrument de Mesure*
SFIRR	solid fuel integral rocket/ramjet
SFM	Sensor Fuzed Munitions
SGTS	Second-Generation Tank Sight
SHORAD	Short-Range Air Defence System
SH/PRAC	squash head practice
shp	shaft horsepower
SICPS	Standard Integrated Command Post Systems
SIP	System Improvement Plan
SIPS	Small Integrated Propulsion System
SKOT	*Sredni Kolowy Opancerzny Transporter* (armoured personnel carrier)
SLAP	saboted light armor penetrator
SLEP	Service Life Extension Program (US)
SLR	Super Low Recoil
SM	smoke
SMG	sub-machine gun
Smoke BE	smoke base ejection
Smoke WP	smoke white phosphorus
SMP	Surface Mine Plough
SOG	Speed Over Ground
SP	self-propelled
SPAAG	self-propelled anti-aircraft gun
SPAAM	self-propelled anti-aircraft missile
SPAG	self-propelled assault gun
SPARK	solid propellant advanced ramjet kinetic energy missile
SPATG	self-propelled anti-tank gun
SPAW	self-propelled artillery weapon
SPG	self-propelled gun
SPH	self-propelled howitzer
SPL	self-propelled launcher
SPLL	Self-propelled Loader Launcher
SPM	self-propelled mortar
SPS	Standard Positioning Service
SPSM	*Sensorgezundete Panzerabwehr SubMunition*
SR	Staff Requirement
SRC	Space Research Corporation
SRG	shell replenishment gear
SRU	Shop Replaceable Unit
SRU	slip ring unit
SRV	Surrogate Research Vehicle (US)
SSA	Special Spaced Armour
SSG	Single Shot Gun
SSK	Single Shot Kill
ST	Staff Target
STA	shell transfer arm
STAFF	Small Target Activated Fire-and-Forget (US)
STARTLE	Surveillance and Target Acquisition Radar for Tank Location and Engagement (US)
STE/ICE	Simplified Test Equipment/Internal Combustion Engine (US)
STUP	spinning tubular projectile
TACMS	Army Tactical Missile System (US)
TACOM	Tank-automotive and Armaments Command (US)
TADDS	Target Alert Display Data Set (US)
TADS	Target Acquisition and Designation System (US)
TAM	*Tanque Argentino Mediano*
TAS	tracking adjunct system
TAS	Turet Attitude Sensor
TBAT	TOW/Bushmaster Armored Turret (US)
TCO	Tactial Combat Operations
TCU	tactical control unit
TD	tank destroyer
TDCS	Tank Driver Command System
TDR	Target Data Receiver
TEL	Transporter-Erector-Launcher
TELAR	Transporter-Erector-Launcher And Radar
TES	target engagement system
TGMTS	Tank Gunnery Missile Tracking System
TGP	Terminally Guided Projectile
TGS	Tank Gun Sight
TGSM	terminally guided submunition
TGTS	tank gunnery training simulator
TI	thermal imaging
TICM	thermal imaging common modules
TIIPS	Thermal Imaging and Integrated Position System
TIPU	Thermal Image Processing Units
TIRE	Tank Infra-Red Elbow
TIS	thermal imaging system
TLC	Transport Launching Container
TLD	Top Level Demonstrations
TLE	Treaty Limited Equipment
TLR	Tank Laser Range-finder
TLS	Tank Laser Sight
TM&LS	Textron Marine & Land Systems
TMC	Turret Management Ballistic Computer
TMS	Turret Modernisation System
TMUAS	Turreted Mortar Under Armor System (US)
TOE	table of organisation and equipment
TOGS	Thermal Observation and Gunnery System (UK)
TOPAS	*Transporter Obrneny Pasovy*
TOTE	Tracker, Optical Thermally Enhanced
TOW	Tube-launched Optically tracked Wire-guided (US)
TP	target practice
TP-FL	target practice flash
TPFSDS-T	target practice fin-stabilised discarding sabot – tracer
TP-SM	target practice – smoke
TP-SP	target practice – spotting
TP-T	target practice – tracer
TRACER	Tactical Reconnaissance Armoured Combat Equipment Requirement (UK)
TRACKSTAR	Tracked Search and Target Acquisition Radar System (US)
TRADOC	Training and Doctrine Command (US)
TSFCS	Tank Simplified Fire-Control System
TSQ	time and super-quick
TT	*transport de troupes* (troop transporter)
TTB	Tank TestBed (US)
TTG	Time To Go
TTS	Tank Thermal Sight
TU	Terminal Unit
TUA	TOW Under Armor (US)
TUR	*Tiefflieger-Überwachungs-Radar* (low-level surveillance radar)
TWD	Thermal Warning Device
TWMP	Track Width Mine Plough
TYDP	Ten Year Defence Programme (Australia)
UBLE	Universal Bridge Launching Equipment
UDR	Ulster Defence Regiment
UET	Universal Engineer Tractor (US)
UGV	Unmanned Ground Vehicle
ULC	Unit Load Container
USAADS	US Army Air Defense School
USMC	United States Marine Corps
USN	United States Navy
UTS	Universal Turret System
VAB	*Véhicule de l'Avant Blindé* (front armoured car)
VAB	Vickers Armoured Bridgelayer
VADS	Vulcan Air Defense System (US)
VAE	*Vehículo Armado Exploracion*
VAPE	*Vehículo Apoyo y Exploracion*
VARRV	Vickers Armoured Repair and Recovery Vehicle
VARV	Vickers Armoured Recovery Vehicle
VBC	*Véhicule Blindé de Combat* (armoured combat vehicle)
VBCI	*Véhicule Blindé de Combat d'Infanterie*
VBL	*Véhicule Blindé Léger* (light armoured vehicle)
VBM	*Véhicules Blindés Modulaires*
VCA	*Véhicule Chenillé d'Accompagnement* (tracked support vehicle)
VCC	*Veicolo Corazzato de Combattimento*
VCG	*Véhicule de Combat du Genie* (armoured engineer vehicle)
VCI	*Véhicule de Combat d'Infanterie; Vehículo combate infanteria* (infantry combat vehicle)
VCR	variable compression ratio
VCR	*Véhicule de Combat à Roues* (wheeled combat vehicle)
VCR/AT	*Véhicule de Combat à Roues/Atelier Véhicule*
VCR/IS	*Véhicule de Combat à Roues/Intervention Sanitaire*
VCR/PC	*Véhicule de Combat à Roues/Poste de Commandement*
VCR/TH	*Véhicule de Combat à Roues/Tourelle HOT*
VCR/TT	*Véhicule de Combat à Roues/Transport de Troupes*
VCTIS	Vehicle Command and Tactical Information System
VCTM	*Vehículo de Combate Transporte de Mortero* (armoured mortar carrier)
VCTP	*Vehículo de Combate Transporte de Personal* (armoured personnel carrier)
VDA	*Véhicule de Défense Anti-aérienne* (anti-aircraft defence vehicle)
VDAA	*Véhicule d'Auto-Défense Anti-aérienne*
VDM	viscous damped mount
VDSL	Vickers Defence Systems Ltd
VDU	visual display unit
VEC	*Vehículo de Exploración de Caballerie*
VEDES	Vehicle Exhaust Dust Ejection System (US)
VERDI	Vehicle Electronics Research Defence Initiative (UK)
VHIS	visual hit indicator system
VIB	*Véhicule d'Intervention du Base*
VIDS	Vehicle Integrated Defence System
VINACS	Vehicle Integrated Navigation and Command System
VIRSS	Visual and InfraRed Smoke Screening System
VITS	Video Image Tracking Systems
VLAP	Velocity-enhanced Long-range Artillery Projectile
VLC	*Véhicule Léger de Combat* (light armoured car)
VLSMS	Vehicle-Launched Scatterable Mine System
VMBT	Vickers Main Battle Tank
VMS	Vehicle Motion Sensor

VNAS	Vehicle Navigation Air System
VRL	*Véhicule Reconnaissance Léger* (light reconnaissance vehicle)
VSEL	Vickers Shipbuilding and Engineering Ltd
VT	variable time
VTP	*véhicule transport de personnel* (personnel carrier)
VTT	*véhicule transport de troupe* (troop transporter)
VVSS	vertical volute spring suspension
VXB	*Véhicule Blindé à Vocations Multiples* (multipurpose armoured car)
WAM	Wide Area Mines (US)
WAPC	Wheeled Armoured Personnel Carrier (Canada)
WASAD	Wide Angle Surveillance and Automatic Detection Device
WAV	Wide Angle Viewing
WFSV	Wheeled Fire Support Vehicle (Canada)
WHA	Weapons Head Assembly
WHSA	Weapons Head Support Assembly
WLR	Weapon Locating Radar
WMRV	Wheeled Maintenance and Recovery Vehicle (Canada)
WP	white phosphorus
WP-T	white phosphorus – tracer
YMRS	Yugoslav multiple rocket system

The 120S MBT has been developed as a private venture by General Dynamics Land Systems for the export market. It essentially consists of an upgraded M60 series MBT chassis fitted with the complete combat proven turret of the M1A1 MBT armed with a 120 mm M256 smoothbore gun (General Dynamics Land Systems) **2002**/0132281

Foreword

Early in 2002, the US Army took delivery from GM GDLS Defense Group (General Motors/General Dynamics Land Systems) of the first of at least 2,131 Light Armoured Vehicle – III (LAV-III) (8 × 8) models, which will be issued to the Brigade Combat Teams (BCTs) that are now being formed.

The first vehicle to enter service is the basic Infantry Carrier Vehicle (ICV), which has recently been named the Stryker. This is fitted with a Norwegian Kongsberg Protech AS Remote Weapon Station (RWS), armed with a 12.7 mm M2 machine gun or a 40 mm Mk 19 automatic grenade launcher.

By the end of this year almost 300 vehicles are expected to have been delivered to the first BCT, the 3rd Brigade, 2nd Infantry Division and expected to be ready for full-scale operations by May 2003.

One of the key requirements for this family of 8 × 8 vehicles is that they must be fully airportable in a Lockheed Martin C-130 Hercules transport aircraft. For some of the variants, especially the 105 mm armed Mobile Gun System (MGS), this will not be easy and efforts are now underway in an effort to reduce the overall weight of the system without the removal of essential onboard equipment.

Today, the US Army has a mix of heavy units, airborne and light units. The latter two can be fairly rapidly deployed anywhere in the world but they currently lack armoured vehicles to enable personnel to be rapidly transported from the drop zone/air head to the area where they are required.

The forming of the BCTs by the US Army will allow the US to rapidly deploy highly mobile combat units anywhere in the world. These units will arrive in a ready-to-fight condition and the LAV-III will provide the troops with at least a measure of protection and firepower.

When all of the LAV-III variants are deployed the BCT will have considerable firepower, being armed with TOW and Javelin anti-tank missiles, mortars and 105 mm rifled guns. The main role of the latter is not to engage any threat of armoured units, but to provide direct fire support to the infantry using a variety of 105 mm munitions. It should be noted that the BCT will not have the offensive fire capability of the current heavy armoured units equipped with M1A1/M1A2 Abrams Main Battle Tanks (MBTs) and M2 Bradley Infantry Fighting Vehicles (IFVs).

It is worth pointing out that some 20 years ago there was a joint US Army/Marine Corps programme to procure earlier generation LAV (8 × 8) vehicles, but in the end, the US Army pulled out and concentrated only on tracked vehicles.

The US Marine Corps subsequently purchased 758 vehicles in six basic configurations and these have seen widespread service in various trouble spots around the world; a number of vehicles were deployed to Afghanistan at an early stage. These US Marine Corps LAVs are now being upgraded to extend their operational lives well into the 21st century.

As this edition of *Jane's Armour and Artillery* was being prepared, the US Army selected Boeing/Science Applications International Corporation (SAIC) as the lead systems integrator for the very ambitious Future Combat Systems (FCS). The FCS is a complex network of sensors, weapons platforms and manned and unmanned

Patria Vehicles XA-180 (6 × 6) series armoured personnel carrier of the Finnish Army during operations in the Balkans (Richard Stickland) **2002**/0132435

Early in 2002 the US Army took delivery of its first LAV-III (8 × 8) Infantry Carrier Vehicle (called Stryker), armed with a remote-controlled 12.7 mm M2 machine gun **2002**/0096841

robotic vehicles which will form the heart of the US Army's Objective Force.

The Objective Force will be much lighter and more mobile than today's units, the US Army requiring the ability to put a combat-capable brigade anywhere in the world within 96 hours, a full division in 120 hours and five divisions on the ground within 30 days.

Until the Objective Force is ready, the US Army will rely on a mix of the new BCTs and the so-called Legacy Force. The Legacy Force is built around units equipped with older and heavier systems such as the General Dynamics Land Systems M1A1/M1A2 MBT MBT and the United Defense LP M2 series Bradley IFV.

The US Army is hoping that its FCS first unit equipped (FUE) will be ready for operations by 2010 which, given the complexity of the system, is a very tight schedule – it remains to be seen if this goal can be achieved.

Over the next five years, the US Army is expected to spend US$4 billion on the FCS programme, with even more going to production systems. The Defense Advanced Research Projects Agency (DARPA), working with the Army, is the contracting agency.

It is of note that an Armoured Fighting Vehicle (AFV) contractor was not selected to head up the FCS programme and General Dynamics Land Systems and United Defense LP are currently not involved, although this is expected to change as the programme moves ahead. At this stage it is still not clear as to whether the FCS platform will be tracked or wheeled, or a mix of these types.

In order to fund the FCS, the US Army has had to make some hard choices, with a number of key programmes having recently been cancelled. These included further upgrades to the widely deployed M113 series troop carrier and the cancellation of the Wolverine Heavy Assault Bridge (HAB) and Grizzly Counter Obstacle Vehicle (COV), both of which were previously regarded as key elements for supporting the heavy force. Indeed, the withdrawal of the M728 Combat Engineer Vehicle (CEV), without there being a replacement, has left a significant gap in the US Army combat engineer capability.

In the longer term, the US Army will not be investing any further in its heavy units but for many other countries around the world, including some in Europe and the Middle East, investments continue to be earmarked for heavier forces.

Many countries have no intention of becoming involved in costly and extended peacekeeping operations far from the borders of their own countries. For some countries, involvement in United Nations (UN) peacekeeping-type operations has been a very time consuming and expensive process. Units often have to be retrained or re-roled for peacekeeping-type operations and considerable additional funding is often required for deployment, spare parts and maintenance.

Experience has shown that vehicles engaged on almost continuous peacekeeping operations, such as constant patrolling on roads and

Giat Industries Leclerc MBT of the French Army in Kosovo, with turret traversed right (Richard Stickland) ***2002***/0132436

over extended distances, have much higher utilisation rates when compared to the normal peacetime training cycle. For some countries engaged in peacekeeping operations, there has been a clear move to a more balanced fleet of tracked and wheeled armoured vehicles. Not only do the latter have significantly lower operating costs than their tracked counterparts, but they are also regarded as appearing less threatening.

In the past, the foreword of *Jane's Armour and Artillery* has covered virtually every type of vehicle and artillery system within its content; this year the main emphasis is on tanks, reconnaissance vehicles, armoured personnel carriers and experimental vehicles.

Main Battle Tanks

Some countries in NATO will start phasing out their current MBT fleets from around 2015, but the majority of the others have decided for the time being at least that they will retain their current MBTs in service until 2025 or later.

The French MoD has now awarded Giat Industries a contract for the final batch of 54 Leclerc MBTs which will be delivered to the French Army between late 2003 and late 2005. This will bring the French Army Leclerc MBT fleet up to a total of 406 units. These will be issued to eight units, each with 40 vehicles, with the remainder being used for training, war reserves and undergoing maintenance. Since the first Leclerc rolled off the production line at Roanne in 1992, the vehicle has continued to be improved. The latest so-called Series 2 Leclerc vehicles incorporate an independent air conditioning system, as well as a track tensioning system operated by the driver and a rear view camera for reversing.

Even more enhancements are expected for the future, including an improved armour package and a defensive aids suite to detect and decoy incoming threats such as anti-tank guided missiles.

Until now there has been only one export sale of the Leclerc MBT – the United Arab Emirates ordered 390 MBTs and 46 DNG Armoured Recovery Vehicles (ARVs) and deliveries of these have been underway for some years.

On 1st March 2002, the Hellenic government finally decided in favour of purchasing the German Leopard 2 MBT to meet its future MBT requirements. Following detailed contract negotiations an order is expected to be placed for up to 170 Leopard 2 MBTs and variants which will include ARVs as well as armoured vehicle-launched bridges.

Krauss-Maffei Wegmann Leopard 2A6 armed with a Rheinmetall 120 mm L/55 smoothbore gun (Christopher F Foss) ***2002***/0132437

At this stage it is not known which variant of the Leopard 2 the Hellenic Army will select, the 2A5, 2A5 plus or the latest 2A6 with the 120 mm L/55 Rheinmetall smoothbore gun, which offers a significant increase in combat effectiveness.

The Leopard 2 MBT has now been selected by 10 countries in Europe, with the obvious training and logistic advantages that this entails. These are Austria, Denmark, Finland, Germany, Netherlands, Norway, Poland, Spain, Sweden and Switzerland. The first Leopard 2 came off the production line in 1979 but since then it has been continuously upgraded and improved, with the current production model being the Leopard 2A6.

Krauss-Maffei Wegmann is, however, studying many more improvements, so that the Leopard 2 will remain an effective weapon system for some years to come. In addition to new types of 120 mm ammunition, a new MTU power pack and a defensive aids suite could also be fitted.

India has taken delivery of the first batch of Russian T-90 series MBTs which were recently ordered, pending the possible introduction of the Arjun MBT. The order is for a total of 310 T-90 MBTs, of which 186 will be assembled in India.

Early in 2002 India revealed the prototype of a hybrid MBT called the Tank EX. This is essentially the chassis of the locally manufactured T-72M1 MBT fitted with the complete three-person turret of the Arjun MBT armed with a 120 mm rifled tank gun. Arjun has been under development for over 20 years but the programme has been plagued with problems.

Tank EX has been developed by the Indian Defence Research and Development Organisation (DRDO) and if adopted would recoup some of the considerable investment made in the Arjun MBT over the years. The Arjun chassis could well be fitted with the South African Denel Ordnance 155 mm/52 calibre T6 artillery turret that has already undergone extensive tests in India.

India is also upgrading 250 models of its current fleet of locally built T-72M1 series MBTs, with a new computerised fire-control system developed by PCO of Poland; this is the new SKO-1Y Drawa-T Thermal Imaging Fire Control System (TIFCS). First deliveries will be made later this year and they will be completed by 2004. This is a direct sale and involves no local production.

In Israel, the Merkava Mk IV has now replaced the early Mk III on the production line. This model should have made its first public appearance by the time this edition of *Jane's Armour and Artillery* has been published.

Hägglunds Vehicle of Sweden has released additional details of the CV90120-T tank that it has developed as a private venture. This is essentially a modified CV90 chassis and it has now been selected by four countries for the IFV role, fitted with an advanced new three-person turret armed with a 120 mm smoothbore gun, computerised fire-control system and an advanced optronics suite.

CV90120-T has the firepower of much heavier vehicles, such as the Leopard 2 and Challenger 2, but it is much lighter and is fully air transportable in the European A400M transport aircraft.

While the CV90120-T will not have the same level of protection as heavier MBTs, it has a high level of combat survivability owing to its to compact shape, which incorporates stealth characteristics and a

Shown for the first time early in 2002 was the Indian Tank EX, which is essentially the Russian T-72M1 chassis fitted with the complete turret of the locally developed Arjun MBT ***2002***/0096866

Hägglunds Vehicle CV90120-T tank weighs less than 30 tonnes but has the firepower of MBTs weighing twice this amount **2002**/0132438

very advanced defensive aids suite which can counter a variety of battlefield threats.

Early in 2002, Vickers Defence Systems completed the last of 386 Challenger 2 MBTs for the British Army and unless export orders are placed, these may well be the last vehicles of their type ever to be produced in the UK.

For the export market, Vickers Defence Systems developed the Challenger 2E, which in many respects is more advanced than the earlier Challenger 2 operated by the British Army, especially in the areas of target acquisition and power pack. Challenger 2E was one of the contenders for the Greek MBT competition which was recently won by Krauss-Maffei Wegmann of Germany.

The Challenger 2E power pack consists of the German MTU 1,500 hp EuroPowerPack that has also been installed in the Leclerc MBTs being delivered to the UAE and for trials purposes in a US General Dynamics Land Systems M1A2 series MBT.

Unlike some other countries, the British Army is investing in its key MBT support vehicles and has recently awarded Vickers Defence Systems a contract for 66 Engineer Tank Systems (ETS); the first of these should be completed late in 2002. The ETS order comprises 33 Trojan engineer vehicles and 33 Titan armoured vehicle-launched bridges which will be based on a new chassis incorporating Challenger 2 components.

In the US, General Dynamics Land Systems is continuing to upgrade at least part of the Abrams fleet. By the end of this decade, the front line fleet is expected to consist of 966 M1A2SEP models in

Early in 2002 Vickers Defence Systems delivered the last of 386 Challenger 2 MBTs to the British Army (Richard Stickland) **2002**/0132439

the Counter Attack Corps and 790 M1A1AIM models in the Containment Force.

The Abrams is the second largest consumer of sustainment dollars in the US Army (the first being the Apache attack helicopter). For this reason, significant efforts are being made to reduce operation and support costs, the major part being the installation of the Abrams Crusader Common Engine (ACCE).

All M1A1/M1A2 series MBTs are powered by the AGT 1500 gas turbine engine, production of this having finished in 1992 after some 12,163 engines had been built. This is now regarded as 1960s technology and the installation of the ACCE, which will also be a turbine, is expected to offer the user a number of significant advantages, including improved reliability and reduced fuel consumption.

Some countries with larger armies are going for a mix of new and upgraded vehicles. Turkey, for example, has evaluated a number of foreign MBT designs and recently decided to upgrade at least part of its older MBT fleet.

Israel Military Industries (IMI) has been selected to upgrade 170 US-supplied General Dynamics Land Systems M60A1 series MBTs, under a contract reported to be worth up to US$668 million. This will be based on the latest IMI Sabra III design and will incorporate a

The General Dynamics Land Systems M1A2 will be upgraded by the installation of the new ACCE gas turbine power pack **2002**/0132440

Alvis Vehicles Scarab (4 × 4) fitted with Thales AFV Systems SWARM during FCLV trials (Christopher F Foss) **2002**/0132441

Vickers Defence Systems RG-32M fitted with Thales SWARM Lite during FCLV trials (Christopher F Foss) **2002**/0132442

120 mm smoothbore gun, new fire control and sighting system, new 1,000 hp power pack (consisting of an MTU diesel coupled to a Renk automatic transmission) and a new armour package. Also bidding for this contract was General Dynamics Land Systems of the US (with the 120S MBT) and FNSS Defence Systems of Turkey.

Turkey has also awarded the local company of ASELSAN Electronic Industries a contract for the modernisation of 162 Leopard 1 series MBTs, featuring a new computer-based fire-control system called Volkan. It is expected that if the programme runs to plan the first vehicles will be completed within 24 months. They will provide the Leopard 1 with a significantly increased first-round hit probability under day and night conditions, while the target is either stationary or on the move.

General Dynamics Land Systems has also developed the 120S MBT for the export market and this has already been offered to a number of countries. It was shown for the first time in Turkey late in 2001. The 120S is essentially an upgraded M60 series MBT chassis, featuring many improvements and fitted with the complete combat-proven turret of the M1A1 Abrams MBT armed with a 120 mm M256 smoothbore gun. This was offered to Turkey and it is not known if marketing of this vehicle continues.

Light tanks

There is little new to report in the Light tanks section. With the increased emphasis on rapid deployment forces it would have been expected that there would have been more movement in this area.

Reconnaissance vehicles

One of the most advanced reconnaissance platforms ever to be developed is the UK/US Tactical Reconnaissance Armoured Combat Equipment Requirement (TRACER)/Future Scout Cavalry System (FSCS). Two consortiums, Lancer and SIKA International, have recently completed Advanced Technology Demonstrators (ATDs), which are now being evaluated in the UK and US. For a variety of reasons, funding for this very advanced sensor-dominated platform will cease in June 2002, but the UK and the US are hoping to use the technology in other platforms.

In the case of the UK, the technology could well be used in the projected Future Rapid Effects System (FRES), which includes a complete family of light armoured vehicles to be fully airportable in a Lockheed Martin C-130 Hercules transport aircraft. Other elements will include a 155 mm lightweight artillery system, rocket launcher and aviation elements. At this stage, it is not certain as to whether this will be a tracked or wheeled armoured vehicle but a number of variants are projected, including armoured personnel carrier, mortar carrier, command post, air defence, engineer and direct fire. FRES could enter service as early as 2008, with the first operational variant being the armoured personnel carrier.

The ARGE Fennek consortium has been awarded a contract worth about €500 million for the supply of 612 Fennek (4 × 4) reconnaissance vehicles for the German and Netherlands armies. If the programme runs to schedule then it is expected that the first Fennek production vehicles will be delivered in 2003 and final deliveries will be made in 2007. The German Army will take delivery of 202 vehicles, while the Netherlands will take delivery of 410 vehicles.

For several years the British Army has been working on its Future Command and Liaison vehicle (FCLV) requirement. This has now been reduced to 422 vehicles, with the heavier requirement moving over to the FRES programme. It is expected that a production contract will be placed in 2003, the vehicle having a projected in-service date of 2006.

The three companies that submitted vehicles for the trials phase, which recently took place at the Armoured Trials and Development Unit (ATDU) at Bovington Camp, were Alvis Vehicles (with the Scarab and IVECO Multirole Light Vehicle), Insys (with the latest ACMAT TCM-based vehicle) and Vickers Defence Systems (with the RG-31M and RG-32M). When fielded, the FCLV will be used to undertake a wide range of battlefield roles and will be issued to both the army and the RAF Regiment.

A number of countries, including the Czech Republic and Poland, are also upgrading their older reconnaissance vehicles, such as the Russian BRDM-2 (4 × 4) amphibious scout car.

Armoured Personnel Carriers (tracked)

As mentioned before, a number of countries are now placing greater emphasis on the procurement of wheeled armoured vehicles, especially armoured personnel carriers. France has decided to eventually replace its fleet of Giat Industries AMX-10P full tracked IFVs with the new Satory Military Vehicles VBCI (8 × 8).

Some countries, however, are still procuring or at least upgrading their fully tracked armoured vehicles as well. One of the best-selling vehicles of the type in Europe is the Swedish CV9040 series developed by Bofors Defence and Hägglunds Vehicle. The Swedish Army model is armed with a 40 mm cannon.

For the export market, Hägglunds Vehicle developed the CV9030. This is fitted with a two-person turret armed with a Boeing 30 mm Bushmaster cannon. A total of 104 have been delivered to Norway, while Finland has ordered 57 and Switzerland 186.

In order to win these key export contracts, Hägglunds Vehicle has had to provide as much work to local industry as possible. In the case of Switzerland, for example, the chassis will come from Sweden while production of the turret and integration of the turret and chassis

The Hägglunds Vehicle CV9030 has been selected by Finland, Norway and Switzerland **2002**/0132443

Alvis Vehicles Warrior fitted with an RO Defence/CTA International two-person turret armed with a 40 mm Case Telescoped Weapon System (CTWS) **2002**/0096862

The Krauss-Maffei Wegmann Dingo (4 × 4) APCV is now in service with the German Army – it is shown here without armament **2002**/0132445

will be undertaken in Switzerland. A number of other Swiss companies will also be involved at the subsystems level in the areas of optronics and weapons.

The standard Infantry Combat Vehicle (ICV) of the German Army is the Marder 1, the design of which is now well over 30 years old. Later this decade the German Army hopes to field a new vehicle called the Schutzenpanzer 3. The baseline vehicle is expected to weigh around 32 tonnes, in order for it to be transportable in the future European A400M strategic transport aircraft. But when fitted with its full armour package it is expected to weigh up to 45 tonnes, with a high level of protection being provided against direct and indirect fire weapons.

The UK is currently studying a number of options to upgrade at least part of its fleet of Alvis Vehicles Warrior ICVs. The heart of this upgrade would include the installation of a new weapon system that could be fitted to a one- or two-person turret or possibly a remote-controlled weapon system.

In March 2002 an Alvis Vehicles Warrior was shown fitted with a two-person turret armed with a 40 mm Case Telescoped Weapon System (CTWS) developed by CTA International, a joint venture company formed some years ago by Giat Industries of France and RO Defence of the UK.

In the US, upgrades to the United Defense LP Bradley continue and for trials purposes a vehicle fitted with a Boeing 30 mm MK 44 cannon has recently been tested. This weapon system has already been selected for the US Marine Corps Advanced Amphibious Assault Vehicle (AAAV). At present, however, there is no requirement for the current Boeing 25 mm cannon to be replaced in the Bradley.

IFVs/ICVs usually have to operate in close co-operation with much heavier and well-protected MBTs. For this reason a number of countries, for example Israel, Russia and Ukraine, have developed very heavy infantry combat or infantry support vehicles which have been designed to operate in the direct fire zone in close co-operation with MBTs. These are often based on MBT chassis and have a very high level of protection.

Combat experience with Israel and Russia has shown that in many battlefield situations speed and mobility is no substitute for adequate armour protection, especially against infantry weapons such as the highly effective Russian RPG-7 system which can now fire unguided rockets with a variety of warhead types.

Armoured Personnel Carriers (wheeled)

There have been continuing problems with the ADI Bushmaster (4 × 4) wheeled APC infantry mobility vehicle. This was to have entered service in 2001, but its introduction has been delayed due to reliability problems in a number of key areas. These are now being addressed and it is expected that the vehicle will enter initial production in 2003.

The Austrian company of Steyr-Daimler-Puch has recently completed prototypes of its new generation Pandur II 6 × 6 and 8 × 8 series of vehicles. These incorporate many improvements over the original Pandur (6 × 6) vehicle that is now in service with Austria, Belgium, Kuwait, Slovenia and the US.

The Pandur II 8 × 8 vehicle can be fitted with a much wider range of weapon systems and the first example is fitted with a Steyr-Daimler-Puch two-person turret armed with a 30 mm cannon. This model retains its full amphibious capability, being propelled in the water by two water jets mounted on either side at the rear of the hull.

Early in 2002, the Finnish Defence Force ordered two examples of the Patria Vehicles Armoured Modular Vehicle (AMV) and these should be delivered in 2003 for an extensive series of trials. This vehicle has a significant increase in payload and mobility over the current Patria Vehicles 6 × 6 XA series, of which over 1,000 have now been built.

The German BWB has now placed a total of three contracts with Krauss-Maffei Wegmann for its Dingo All Protected Carrier Vehicle (APCV) and some from the first order have already been deployed to Afghanistan and Kosovo. The latest contract, placed in March 2002, is for 34 vehicles that are due to be delivered to the German Army between June and October 2002.

Orders continue to flow for the Piranha family of 8 × 8 APCs. The Piranha was originally designed by MOWAG in Switzerland, which is now owned by General Motors Defense of Canada. The US Army has ordered over 2,131 models for its new BCTs.

Austrian Steyr-Daimler-Puch Pandur II (8 × 8) fitted with a two-person turret armed with a 30 mm cannon **2002**/0132444

The Piranha IV (8 × 8) is the latest in a long line of Piranha vehicles developed by the MOWAG company of Switzerland **2002**/0132446

The SEP technology demonstrator has a number of advanced features including electric drive and a rubber band track (Hägglunds Vehicle) **2002**/0132447

From the Swiss production line deliveries have recently been made to Denmark, Ireland and Switzerland. The Spanish Marines recently placed an order for 18 Piranha (8 × 8) III series vehicles that will be delivered from 2003 through to 2004. These will be supplied in three versions – APC, command post and ambulance and will be the latest production standard. MOWAG has also completed the prototype of the first of its latest Piranha IV (8 × 8) series, which has increased protection and payload over the earlier Generation III vehicle, although this will continued to be marketed.

Vickers Defence Systems has a licence to manufacture and market the Piranha IV for a number of markets including South Africa and the UK. If the UK did trim its MRAV order back, then some believe that the Piranha IV would be a very cost-effective alternative.

On the international front, development of the German/Netherlands/UK MRAV (8 × 8) Multi-Role Armoured Vehicle continues, although the programme is already behind schedule. The first prototype should have been rolled out in August 2001, but this was delayed by more than six months.

Industry sources believe that MRAV will enter service, although it will not be procured in the numbers originally anticipated. At one time the UK was considering purchasing well over 1,000 vehicles but figures of 600 to 700 are now being mentioned, with the main anticipated users being the British Army's armoured brigades.

Experimental and trials AFVs

This section was deleted from *Jane's Armour and Artillery* several years ago to allow for additional coverage elsewhere. Experimental and trials armoured vehicles do, however, play an important role as these provide the springboard to the next generation of AFVs.

In the UK, QinetiQ has completed trials of the Advanced Composite Armoured Vehicle Platform (ACAVP), which makes use of a Desert Warrior power pack and suspension. This was fitted with a 30 mm Fox turret to simulate the weight of a weapon system. According to QinetiQ, these trials have shown that it is possible to design a vehicle with a composite hull and this would offer the user a number of advantages over a traditional hull of welded aluminium or steel construction.

The Swedish SAT Mark stealth technology demonstrator is based on a modified Hägglunds CV90 series full-tracked chassis (Hägglunds Vehicle) **2002**/0132448

Sweden has recently released some details of two of its experimental vehicles, the SEP and a stealth vehicle based on the CV90 currently in service with the Swedish Army. The SEP has been developed by Hägglunds Vehicle and features a decoupled running gear, electric transmission, rubber band tracks, databus, built-in test equipment and advanced mine protection. In the SEP vehicle the front-mounted power pack is coupled to a generator to provide power for the electric motors, which in turn drive the tracks. The SEP has a combat weight of 13.5 tonnes and a maximum road speed of 85 km/h. It is envisioned that production vehicles would be fitted with a mission module which would be fully interchangeable with the SEP 6 × 6 wheeled vehicle which is also under development.

The SEP is probably the most advanced vehicle of its type and Sweden is now looking for other countries to share development costs in order that the development of the tracked and wheeled family can be completed.

The SAT/Mark stealth vehicle was funded by the Swedish Defence Material Administration (FMV) and developed by a consortium of Swedish defence contractors. Within this consortium, Hägglunds Vehicle is responsible for the tracked chassis, electro-optical outlooks and the overall project co-ordination; Bofors Defence (now part of United Defense of the US) is responsible for the turret and weapon; Saab Baracuda was responsible for the surface treatment and unique rapid camouflage system; and Saab Tech Systems managed the sensor pack. Saab Bofors Dynamics has also been involved in the project.

For its stealth version, CV90 has been extensively modified to achieve low signatures in all wavelengths. This has been achieved by the vehicle's unique shape and by incorporating special materials in the hull and turret. The upper part of the hull and turret has been sloped and special paint applied. The tracks and suspension have also been covered.

Land-Based Air Defence

Details of self-propelled and towed anti-aircraft guns are given in *Jane's Land-Based Air Defence*. This publication also gives details of man-portable surface-to-air missile systems, self-propelled surface-to-air missile systems, towed anti-aircraft gun sights, static and towed surface-to-air missile systems, shelter- and container-based surface-to-air missile systems, anti-aircraft control systems and a comprehensive inventory listing.

When an armoured vehicle chassis is used as a basis for an air defence system, such as the German Krauss-Maffei Wegmann Gepard twin 35 mm self-propelled anti-aircraft gun system, which is based on a modified Leopard 1 MBT chassis, the Euromissile Roland surface-to-air missile system based on a French Giat Industries AMX-30 chassis and the German Rheinmetall Landsystem Marder 1 mechanised combat vehicle chassis, brief details and a photograph will usually be given in *Jane's Armour and Artillery* under their respective entries. There is also a cross-reference to *Jane's Land-Based Air Defence*.

Subsystems

Details of AFV armament, anti-tank guided weapons, turrets and cupolas, engines, transmissions and power packs, are given in *Jane's Armour and Artillery Upgrades*.

This volume also contains detailed sections covering automatic loaders, land navigation systems, suspensions and tracks, armour systems, fire detection and suppression systems, laser detectors, smoke dischargers and decoys, weapon control and stabilisation systems, fire-control systems and commander's, gunner's and driver's day and night vision equipment currently in service or under development. There is also a section entitled Towed and self-propelled artillery upgrades, most of which is also covered in *Jane's Armour and Artillery*.

Many of the more recent AFVs are now being offered with a variety of subsystems such as power packs, fire-control systems, armament and day/night vision equipment.

This has allowed the AFV manufacturer to tailor a vehicle to meet the specific operational requirements of the user. The German 1,500 hp MTU EuroPowerPack, for example, was developed as a private venture for the export market. The EuroPowerPack is installed in all of the 436 Giat Industries Leclerc MBTs currently being built for the United Arab Emirates and is also installed in the export Vickers Defence Systems Challenger 2E. For trials the EuroPowerPack has also been installed in a US General Dynamics Land Systems M1 series MBT and more recently has been selected by the French Army for its Leclerc ARVs, first examples of which have now been delivered.

Some countries are now upgrading their armoured vehicles, including MBTs, IFVs and APCs, as a cost-effective way to increase their operational effectiveness. Other countries have elected to invest money in new systems rather than upgrade systems whose original designs are over 30 years old.

The publication of *Jane's Armour and Artillery Upgrades* has enabled the many essential subsystems that go into a modern armoured vehicle to be covered in much more detail than was previously possible. For additional information on current and future upgrade programmes the reader is also referred to *Jane's Defence Upgrades*, which is published 24 times a year by Jane's Information Group.

Ammunition

Brief details of ammunition are normally given in each entry, but full details are given in *Jane's Ammunition Handbook*.

In recent years the conventional natures of artillery ammunition, high explosive, smoke and illuminating, have been supplemented by carrier projectiles which contain submunitions to attack the vulnerable upper surfaces of tanks and armoured vehicles. To enable targets to be engaged at longer ranges, Rocket Assisted Projectiles (RAPs) and Extended Range Full Bore Base Bleed (ERFB-BB) artillery projectiles have been developed and placed in production. More recently a number of countries have developed 155 mm long-range artillery projectiles which have base bleed and are rocket assisted. Even more advanced and longer range projectiles are now under development, most notably in Europe and the US.

Now entering production, or in the final stages of development, are a number of 155 mm smart munitions which have been designed to attack formations of armoured vehicles at long range with a very high hit probability. The first of these was the US Aerojet SADARM, which has recently been followed by the German 155 mm SMArt and the French/Swedish Bonus.

Military Logistics

Jane's Armour and Artillery contains brief details and photographs of specialised variants of AFVs such as ARVs, engineering and specialised vehicles. Full details of these vehicles, and the countless other vehicles, tracked and wheeled, armoured and unarmoured, that support the front line AFVs will be found in a companion yearbook, *Jane's Military Vehicles and Logistics*.

Acknowledgements

One of the last tasks of the editor is to thank everyone who has been involved, however small the part, in the preparation and production of this yearbook. It is very much a team effort.

The editor would like to take this opportunity to thank the many manufacturers, defence forces and individuals that have contributed to the new edition of *Jane's Armour and Artillery*.

Special thanks are due to Paul Beaver, Raymond Cheung, Desmond Dempsey, Kensuke Ebata, Adrian English, Robin Fletcher, Terry Gander, Marsh Gelbert, Michael Green, David C Isby, Michael Jerchel, Ian Kemp, Franz Kosar, Francisco A Marin, David Markov, Barry Marriot, Denis J McCarthy, Julio Montes, Richard M Ogorkiewcz, Rupert Pengelley, Richard Stickland, E Q Tee, Dr Arthur Volz, Steven Zaloga and C R Zwart for their most valuable assistance.

The drawings used in *Jane's Armour and Artillery* have been prepared by a number of people. Steven Zaloga has allowed me to reproduce some of his excellent drawings, while others have been provided by Vasco Barbic and Henry Morshead.

My wife Sheila has also been of invaluable assistance in the preparation of this edition of *Jane's Armour and Artillery*. Mike Bryant has given the editor maximum possible support throughout the preparation of this edition of *Jane's Armour and Artillery*. Special thanks are also due to the production team, headed by Neil Grace, who have processed and checked all of the editor's copy.

Additional information and photographs for the next edition of *Jane's Armour and Artillery* should be sent as soon as possible. The address for all correspondence is:

Jane's Information Group, Sentinel House, 163 Brighton Road, Coulsdon, Surrey, CR5 2YH, UK
Tel: (+44 20) 700 37 00
Fax: (+44 20) 700 37 88
E-mail: jaa@janes.co.uk

Christopher F Foss
March 2001

CHRISTOPHER F FOSS

Chris Foss is the Editor of *Jane's Armour and Artillery*, co-editor of *Jane's Armour and Artillery Upgrades* and *Jane's Military Vehicles and Logistics* and Associate Editor of *Jane's Land-Based Air Defence*; he also acts as the Land Forces Editor of *Jane's Defence Weekly*.

Chris Foss began his association with Jane's as early as 1970, when he began writing for *Jane's Weapon Systems*. Since 1979 he has acted as a self-employed contractor with the company. His achievements have included conceiving and editing *Jane's Defence Review*, the first magazine ever produced by Jane's Information Group and proposing and launching *Jane's Missiles and Rockets*.

He has written numerous other books relating to armoured vehicles and weapons systems both for Jane's and for other publishers. He has also lectured in a number of countries, chaired many conferences and acted as a defence consultant for numerous private sector contractors and government agencies.

2002/0129064

EDITORIAL AND ADMINISTRATION

Publishing Director: Alan Condron, e-mail: Alan.Condron@janes.co.uk

Managing Editor: Mike Bryant, e-mail: Mike.Bryant@janes.co.uk

Global Content Manager: Anita Slade, e-mail: Anita.Slade@janes.co.uk

Content Editing Manager: Jo Fenwick, e-mail: Jo.Fenwick@janes.co.uk

Pre-Press Manager: Christopher Morris, e-mail: Christopher.Morris@janes.co.uk

Team Leaders: Sharon Marshall, e-mail: Sharon.Marshall@janes.co.uk
Neil Grace, e-mail: Neil.Grace@janes.co.uk

Production Editor: Mick Durkin, e-mail: Mick.Durkin@janes.co.uk

Production Controller: Victoria Powell, e-mail: Victoria.Powell@janes.co.uk

Content Update: Jacqui Beard, Information Collection Co-ordinator
Tel: (+44 20) 87 00 38 08 Fax: (+44 20) 87 00 39 59
e-mail: yearbook@janes.co.uk

Jane's Information Group Limited, Sentinel House, 163 Brighton Road, Coulsdon, Surrey CR5 2YH, UK
Tel: (+44 20) 87 00 37 00 Fax: (+44 20) 87 00 37 88
e-mail: jaa@janes.co.uk

SALES OFFICE

Send EMEA enquiries to: *Group Sales Manager*
Jane's Information Group Limited, Sentinel House, 163 Brighton Road, Coulsdon, Surrey CR5 2YH, UK
Tel: (+44 20) 87 00 37 00 Fax: (+44 20) 87 63 10 06
e-mail: info@janes.co.uk

Send USA enquiries to: *Robert Loughman – Vice-President Product Sales*
Jane's Information Group Inc, 1340 Braddock Place, Suite 300, Alexandria, Virginia 22314-1651, USA
Tel: (+1 703) 683 37 00 Fax: (+1 703) 836 02 97 Telex: 6819193
Tel: (+1 800) 824 07 68 Fax: (+1 800) 836 02 971
e-mail: info@janes.com

Send Asia enquiries to: *David Fisher – Group Sales Manager*
Jane's Information Group Asia, 60 Albert Street, #15-01 Albert Complex, Singapore 189969
Tel: (+65) 331 62 80 Fax: (+65) 336 99 21
e-mail: info@janes.com.sg

Send Australia/New Zealand enquiries to: *David Moden – Business Manager*
Jane's Information Group, PO Box 3502, Rozelle Delivery Centre, New South Wales 2039, Australia
Tel: (+61 2) 85 87 79 00 Fax: (+61 2) 85 87 79 01
e-mail: info@janes.thomson.com.au

ADVERTISEMENT SALES OFFICES

(Head Office)
Jane's Information Group
Sentinel House, 163 Brighton Road,
Coulsdon, Surrey CR5 2YH, UK
Tel: (+44 20) 87 00 37 00
Fax: (+44 20) 87 00 38 59/37 44
e-mail: defadsales@janes.co.uk

Richard West, Senior Key Accounts Manager
Tel: (+44 1892) 72 55 80 Fax: (+44 1892) 72 55 81
e-mail: richard.west@janes.co.uk

Kate Hamlin, Advertising Sales Manager
Tel: (+44 20) 87 00 38 53 Fax: (+44 20) 87 00 38 59/37 44
e-mail: kate.hamlin@janes.co.uk

Joni Beeden, Advertising Sales Executive
Tel: (+44 20) 87 00 39 63 Fax: (+44 20) 87 00 38 59/37 44
e-mail: joni.beeden@janes.co.uk

Steve Soffe, Advertising Sales Executive
Tel: (+44 20) 87 00 39 43 Fax: (+44 20) 87 00 38 59/37 44
e-mail: steven.soffe@janes.co.uk

(USA/Canada office)
Jane's Information Group
1340 Braddock Place, Suite 300,
Alexandria, Virginia 22314-1651, USA
Tel: (+1 703) 683 37 00
Fax: (+1 703) 836 55 37
e-mail: defadsales@janes.com

USA and Canada
Katie Taplett, US Advertising Sales Director
Tel: (+1 703) 683 37 00 Fax: (+1 703) 836 55 37
e-mail: katie.taplett@janes.com

Northern USA and Eastern Canada
Harry Carter, Northeast Region Advertising Sales Manager
Tel: (+1 703) 683 37 00 Fax: (+1 703) 836 55 37
e-mail: harry.carter@janes.com

South Eastern USA
Kristin D Schulze, Advertising Sales Manager
PO Box 270190, Tampa, Florida 33688-0190
Tel: (+1 813) 961 81 32 Fax: (+1 813) 961 96 42
e-mail: kristin@intnet.net

Western USA and West Canada
Richard L Ayer
127 Avenida Del Mar, Suite 2A, San Clemente, California 92672
Tel: (+1 949) 366 84 55 Fax: (+1 949) 366 92 89
e-mail: ayercomm@earthlink.com

Australia: *Richard West* (see UK Head Office)

Benelux: *Steve Soffe* (see UK Head Office)

Brazil: *Katie Taplett* (see USA address)

Eastern Europe: MCW Media & Consulting Wehrstedt
Dr. Uwe H. Wehrstedt
Hagenbreite 9, D-06463 Ermsleben, Germany
Tel: (+49) 0700/WEHRSTEDT / (+49) 03 47 43/620 90
Fax: (+49) 03 47 43/620 91
e-mail: info@Wehrstedt.org

France: Patrice Février
BP 418, 35 avenue MacMahon,
F-75824 Paris Cedex 17, France
Tel: (+33 1) 45 72 33 11 Fax: (+33 1) 45 72 17 95
e-mail: patrice.fevrier@wandadoo.fr

Germany and Austria: *MCW Media & Consulting Wehrstedt* (see Eastern Europe)

Greece: *Steve Soffe* (see UK Head Office)

Hong Kong: *Joni Beeden* (see UK Head Office)

India: *Joni Beeden* (see UK Head Office)

Israel: Oreet – International Media
15 Kinneret Street, IL-51201 Bene Berak, Israel
Tel: (+972 3) 570 65 27 Fax: (+972 3) 570 65 27
e-mail: admin@oreet-marcom.com
Defence: Liat Shaham
e-mail: liat_s@oreet-marcom.com

Italy and Switzerland: Ediconsult Internazionale Srl
Tel: (+39 010) 58 36 59 Fax: (+39 010) 56 65 78
e-mail: genova@ediconsult.com

Japan: Skynet Media, Inc
748, 1-7 Akasaka 9-chome, Minato-ku, Tokyo 107-0052, Japan
Contact: Mr Osamu Yoneda
Tel: (+81 3) 54 74 78 35
Fax: (+81 3) 54 74 78 37
e-mail: skynetme@wonder.ocn.ne.jp

Middle East: *Steve Soffe* (see UK Head Office)

Pakistan: *Joni Beeden* (see UK Head Office)

Russian Federation: Simon Kay
33 St John's Street, Crowthorne, Berkshire RG45 7NQ, UK
Tel: (+44 1344) 77 71 23 Mobile: (+44 7702) 54 96 84
Fax: (+44 1344) 77 58 85
e-mail: crowkay@msn.com/crowkay@yahoo.com

Scandinavia: The Falsten Partnership
PO Box 21175, London N16 6ZG, UK
Tel: (+44 20) 88 06 23 01 Fax: (+ 44 20) 88 06 81 37
e-mail: sales@falsten.com

Singapore: *Richard West/Joni Beeden* (see UK Head Office)

South Africa: *Richard West* (see UK Head Office)

South Korea: JES Media Inc
2nd Floor, ANA Building, 257-1 Myungil-Dong, Kandong-Gu, Seoul 134-070, Korea
Contact: Mr Young-Seoh Chinn, President
Tel: (+82 2) 481 34 11/34 13
Fax: (+82 2) 481 34 14
e-mail: jesmedia@unitel.co.kr

Spain: Via Exclusivas SL
Contact: Julio de Andres
Viriato 69SC, E-28010 Madrid, Spain
Tel: (+34 91) 448 76 22 Fax: (+34 91) 446 02 14
e-mail: j.a.deandres@viaexclusivas.com

Turkey: *Richard West* (see UK Head Office)

ADVERTISING COPY
Delphine Gandelin (Jane's UK Head Office)
Tel: (+44 20) 87 00 37 42 Fax: (+44 20) 87 00 38 59/37 44
e-mail: delphine.gandelin@janes.co.uk

For North America, South America and Caribbean only:
Shanee Johnson (Jane's USA address)
Alexandria, Virginia 22314-1651, USA
Tel: (+1 703) 683 37 00 Fax: (+1 703) 836 55 37
e-mail: shanee.johnson@janes.com

TANKS

MBTs and medium tanks
Light tanks

MBTs AND MEDIUM TANKS

ARGENTINA

TAMSE TAM medium tank

Development

The German company of Rheinmetall Landsysteme (then Henschel Wehrtechnik) developed a medium tank called the TAM *(Tanque Argentino Mediano)* to meet the requirements of the Argentine Army. The first prototype of the TAM was completed in Germany in 1976 and the second prototype the following year. Chassis construction and final assembly took place in Buenos Aires with turret construction and integration of the 105 mm rifled gun and fire-control system at Rio Tercero.

Since the first TAM prototypes were delivered to Argentina, development was continued by TAMSE and, over a period of time, more than 1,450 engineering modifications were introduced into production vehicles. Additional information on the TAM vehicle is given in the entry for the Rheinmetall Landsysteme R 301 (previously TH 301) tank under Germany.

An ICV called the VCTP has been developed to work with the TAM, details of which are given in the Armoured personnel carriers (tracked) section.

The total Argentine requirement was for 512 105 mm armed TAM medium tanks and VCTP infantry combat vehicles. It is understood that for financial reasons production was completed after about 350 vehicles had been built.

In May 1994, TAMSE stated that the delivery of the remainder of the initial production of 120 vehicles to the Argentine Army was to be completed by August 1994 by which time the total production of the TAM family would have reached about 400 vehicles of all types. Recent information has indicated that this objective was not in fact achieved. Final deliveries of the TAM and VCTP to the Argentinian Army are understood still to take place.

Variants

VCA 155

Full details of the VCA 155, a 155 mm self-propelled artillery system based on a lengthened TAM chassis, are given in the Self-propelled guns and howitzers (tracked) section under Argentina. This uses the same turret as installed on the Italian Otobreda Palmaria 155 mm self-propelled artillery system. This is now in service with Argentina. The Palmeria is used by Libya and Nigeria.

VCRT

In late 1986, it was confirmed that the Argentine Army had developed a new armoured recovery vehicle based on the VCTP infantry combat vehicle and known as the VCRT. It has a new superstructure and recovery rig supplied by MaK (today Rheinmetall Landsysteme) of Germany.

The VCRT has a rear-mounted dozer blade to provide a secure anchor for the main 30 tonne winch which is located within the sloping hull front. Mounted on the right side of the superstructure rear is a crane jib which can be traversed through 213.5° and lift a maximum load of 22 tonnes.

The first prototype of the VCRT was completed in mid-1987, but by early 2002 no production of this version of the TAM had been undertaken.

Rocket launchers

The Israel Military Industries LAR 160 multiple rocket system has been trialled on an Argentine TAM chassis with the original turret replaced by two 18-round launchers.

The multiple rocket launcher version is called the VCLC (Vehiculo de Combate Lanza Cohetes). In addition to being installed with the LAR 160 system, also known as the CAL, it can also be installed with a new 350 mm system called the CAM. A small number of these systems have been delivered to the Argentine Army.

TAMSE TAM medium tank with roof-mounted 7.62 mm machine gun 0018579

Status

Production complete. In service with Argentina. No longer being marketed.

Manufacturer

TAMSE.
This company is no longer trading.

UPDATED

Specifications

Crew: 4
Combat weight: 30,000 kg
Power-to-weight ratio: 24 hp/t
Ground pressure: 0.788 kg/cm²
Length:
(gun forward) 8.23 m
(hull) 6.775 m
Width: 3.29 m
Height:
(overall) 2.71 m
(hull top) 2.43 m
Ground clearance: 0.45 m
Track: 2.62 m
Max road speed: 75 km/h
Range: 590 km
Range with long-range tanks: 940 km
Fuel capacity: 640 + 400 litres
Fording:
(without preparation) 1.5 m
(with preparation which takes 20 min) 4 m
Gradient: 60%
Side slope: 30%
Vertical obstacle: 1 m
Trench: 2.5 m
Engine: MTU MB 833 Ka-500 6-cylinder diesel developing 720 hp at 2,400 rpm
Transmission: Renk HSWL-204 planetary with 4 speeds forward and reverse
Steering: double differential with hydrostatic steering control
Suspension: torsion bar
Electrical system: 24 V
Batteries: 8 × 12 V, 100 Ah
Armament:
(main) 1 × 105 mm FRT L51 rifled gun
(coaxial) 1 × 7.62 mm MG
(anti-aircraft) 1 × 7.62 mm MG
Smoke grenade dischargers: 2 × 4 76 mm
Ammunition:
(main) 50
(machine gun) 6,000
Gun control equipment
Turret power control: electrohydraulic/manual
(by commander) yes
(by gunner) yes
(commander's override) yes
Turret traverse: 360°
Gun elevation/depression: +18°/−7°
Gun stabiliser:
(vertical) yes
(horizontal) yes
Commander's periscope stabilised: yes
NBC system: yes
Night vision equipment: yes

CANADA

Canadian Leopard 1 MBTs

To replace its British-supplied 105 mm armed Centurion MBTs, Canada selected a modified version of the German Krauss-Maffei (which on 1 January 1999 became Krauss-Maffei Wegman) Leopard 1A3 fitted with the Belgian SABCA computerised fire-control system.

The Canadian order comprised 114 MBTs, six bridgelayers and eight armoured recovery vehicles and was valued at C$187 million. The first Leopard 1A3s, called the C1 by Canada, were handed over in June 1978 by Krauss-Maffei and, under the terms of the contract, Krauss-Maffei took back the entire fleet of Canadian Centurion MBTs. Some of the Centurion turrets, complete with 105 mm L7 guns, were subsequently purchased by Austria for use in the static defence role.

The Canadian C1 Leopard MBTs and variants were deployed in Canada and in Germany, but in 1993 the latter were returned to Canada following the withdrawal of all Canadian Forces from Europe.

The first Canadian Leopard C2 which is essentially a Leopard 1A5 turret installed on the original Leopard 1A3 chassis (Canadian Armed Forces) **2000**/0034403

Leopard AVLB of the Canadian Forces in travelling configuration (Canadian Forces) 0002404

The Canadian Army subsequently ordered nine of the latest version of the Leopard 1 armoured engineer vehicle called Badger. These were built by MaK (today Rheinmetall Landsysteme) of Kiel with first vehicles being handed over in 1990.

Late in 1996 it was announced that the Canadian Forces were to carry out a major update on their fleet of C1 MBTs which involved the replacement of the existing turret with the complete turret of the German Leopard 1A5; this is covered in detail under Germany.

The Leopard 1A5 turret features the STN ATLAS Elektronik EMES-18 computerised fire-control system which incorporates a Carl Zeiss thermal imager. A total of 123 Leopard 1A5 turrets were purchased which has enabled all of the 114 Canadian C1 MBTs to be fitted with the new turret.

Of the nine remaining turrets, five will be used as training turrets, two as complete spares for the EMES-18 fire-control system and two maintenance testbeds.

The 105 mm L7 rifled guns in the Leopard 1A5 turrets will not be retained but will be replaced by 105 mm guns from the existing Canadian Leopard C1 MBTs. The computers have been reprogrammed to match 105 mm ammunition used by Canada.

GLS refurbished the turret, removed the 105 mm gun, modify the turret where required, including the installation of the new radios ordered under the Tactical Command, Control and Communications System project.

The turret was shipped to Canada where a subcontractor installed the 105 mm barrel and mounted the turret on the existing chassis for final delivery to the Canadian Forces. The first upgraded Leopard 1 fitted with a new turret system was handed over to the Canadian Armed Forces in November 1999. These are designated the Leopard C2 MBT. Final deliveries have now been made.

Leopard 1 MBT of the Canadian Armed Forces fitted with a new passive armour package over its frontal arc (Michael Jerchel) **2001**/0105856

Prime contractor is the German company GLS, a subsidiary of Krauss-Maffei Wegmann which built almost all of the Leopard 1 series MBTs for the home and export markets. Total value of the contract is estimated to be C$145 million. The surplus Leopard 1A5 chassis will be retained by GLS and sold as spares.

For operations in the Balkans a number of Leopard 1 MBT's of the Canadian Armed Forces were fitted with a new passive armour package developed by IBD of Germany. This was fitted over the frontal arc of the Leopard 1 to provide a higher level of battlefield protection.

Canadian Armoured Combat Vehicle

It is expected that in the future the Canadian Leopard 1 MBTs will be replaced by the future Armoured Combat Vehicle, available details are given under Canada in the Light Tanks section.

UPDATED

CHINA, PEOPLE'S REPUBLIC

NORINCO Type 98 Main Battle Tank

Development

One of the many new weapons shown for the first time in the huge military parade that took place in Beijing late in 1999 was the Type 98 Main Battle Tank (MBT) which is understood to have entered service with the PLA some time in 1998.

The Type 98 is the most advanced MBT to have entered service with the PLA to date and features significant improvements in the three key areas of MBT design, armour, mobility and firepower.

The Type 98 is probably related to the Type 90 series MBT but has a different and longer hull with some sources calling it the Type 90-III MBT.

It is believed that the Type 98 has the industrial designation of the WZ 123 with the first prototypes being called Type 96. The latter differed in a number of areas including having different skirts.

Some sources have stated that the export versions of the Type 98 are in fact the Type 90-IIM or Type 90-III.

Description

The layout of the Type 98 MBT is conventional, with the drivers compartment at the front, fighting compartment and turret in the centre and the power pack at the rear.

The driver is seated at the front of the vehicle in the centre and has a single piece roof hatch that lifts and opens to the right. In front of this are day periscopes, the centre one of which can be replaced by a passive night vision periscope.

The turret is of a new design and well protected over the frontal arc. The armour package on the turret front either side of the 125 mm smoothbore gun is of an advanced modular design. This allows it to be changed if it is hit or requires to be replaced by a more advanced armour package in the future. To enable the armour package to be rapidly changed in the field lifting eyes are provided.

The gunner is seated on the left side of the turret and has a single piece hatch cover that opens to the front, and in front of this is the stabilised day/night sight that also includes a laser range-finder. The latter feeds information to the computerised fire-control system that provides the Type 98 MBT with a high first round hit probability against stationary and moving targets.

The commander is seated on the right side of the turret and has a roof mounted stabilised day/night and also operates the 12.7 mm anti-aircraft machine gun. Mounted either side of the turret is a bank of five electrically operated smoke grenade launchers firing forwards.

The Type 98 MBT is armed with a 125 mm smoothbore gun fed by an automatic loader mounted below the turret which first loads the projectile and then the charge. The 125 mm smoothbore gun is fitted with a fume extractor and a thermal sleeve. A 7.62 mm machine gun is mounted coaxial with the main armament.

A computerised fire-control system is installed and commander and gunners stabilised day/thermal sights enable the Type 98 to carry out hunter killer type target engagements.

The chassis is similar in some respects to the Russian T-72 MBT with a well sloped glacis plate complete with V type splash board, horizontal ribs and

The new Chinese Type 98 MBT clearly showing the 125 mm smoothbore gun but with roof mounted laser detector and associated laser dazzle device covered up ***2001***/0105857

a dozer/entrenching blade that is mounted under the nose of the vehicle.

Unlike the Russian T-72 and earlier T-62, T-55 and T-54 MBTs the Type 98 MBT does not have the characteristic exhaust outlet on the left side of the hull.

This indicates that a new diesel power pack may be fitted that may not be of Chinese origin. The air inlet and air outlet louvres are in the hull roof.

Some sources have stated that this is powered by a 8-cylinder turbocharged diesel developing 1,200 hp while other sources have indicated a 1,500 hp diesel engine.

Suspension is probably of the torsion bar type with either side consisting of six large dual rubber tyred road wheels with the drive sprocket at the rear, idler at the front and track return rollers.

The Type 98 MBTs shown in Beijing were fitted with simple side skirts but it is probable that more advanced armoured skirts would be fitted for combat.

Mounted on the right side of the turret is what appears to be a pole type laser detector, while mounted on the left side is what is believed to be a laser dazzle device that would be used to neutralise any ground or helicopter based electro-optical sighting systems. It is believed that the laser counter measure device may be designated the ZM-87 (laser disturber). A cage type stowage basket runs around the turret and this provides some protection against attack from HEAT projectiles and rockets. Mounted on the turret rear is a snorkel for deep fording operations.

Standard equipment includes an NBC system and some reports have stated that it is fitted with a digital command, control and communications system. To extend the operational range of the Type 98 MBT additional diesel fuel drums can be rapidly mounted at the rear of the hull.

No detailed specification of the Type 98 MBT has yet been released, but speculation is that it has a combat weight of at least 50 tonnes with some sources stating a weight of between 52 and 54 tonnes.

Status

Production. In service with Peoples Liberation Army.

Manufacturer

China North Industries Corporation

UPDATED

NORINCO Type 90-II MBT (MBT 2000)

Development

The existence of the NORINCO (China North Industries Corporation) Type 90 MBT was first revealed in late 1991 and, compared to earlier Chinese MBTs, it has significant improvements in the three key areas of MBT design: armour, mobility and firepower.

Under a contract signed in May 1990, the Type 90-II is also to be manufactured under licence in Pakistan and further development has resulted in the Khalid MBT which is tailored to meet the specific operational requirements of the Pakistani Armoured Corps.

Additional details of the Khalid, the first prototype of which was completed in June 1991 and which is also referred to as the MBT-2000 or P-90, are given under Pakistan. Late in 1998, it was revealed that China, Pakistan and the Ukraine were working together on MBTs and available details are given in the entry on the MBT-2000 under Pakistan.

In many respects the NORINCO Type 90-II is similar to the Type 85-IIM although the former is heavier with improved protection and is powered by a 1,200 hp diesel engine.

Early in 1997, additional information on the Type 90-II MBT was released by NORINCO and at the same time it was revealed that it was now being offered fitted with explosive reactive armour to its hull and turret for improved battlefield survivability.

By early 2002 it is believed that volume production of the Type 90-II MBT had yet to commence. It is understood that this vehicle is for the export market and not for the People's Liberation Army.

Description

The overall layout of the Type 90-II MBT is conventional with the driver's compartment at the front, fighting compartment in the centre and power pack at the rear.

The hull is of welded steel construction with an additional layer of composite armour at the front. Explosive reactive armour has been added to the glacis plate and nose of the vehicle.

The driver is seated in the centre and has a single-piece hatch cover that lifts and swings to the right and in front of this are three day periscopes, the centre one of which can be replaced by a passive periscope for driving at night.

The turret is in the centre and is of all-welded steel armour construction to which a layer of composite armour has been added over the frontal arc. In addition to the turret front, packs of explosive reactive armour have also been added to the sides and forward roof.

The commander is seated on the right and the gunner is seated on the left with both provided with single-piece roof hatches and vision devices.

NORINCO Type 90-II MBT without explosive reactive armour fitted over frontal arc ***2000***/0064628

NORINCO Type 90-II MBT from the rear showing stowage basket which runs around the turret rear 0018580

Specifications
Crew: 3
Combat weight: 48,000 kg
Power-to-weight ratio: 25 hp/t
Length:
(main armament forwards): 10.067 m
(main armament rear): 9.687 m
Width
(without skirts): 3.40 m
(with skirts): 3.50 m
Height (turret roof): 2.37 m
Firing height: n/avail
Ground clearance: 0.45 to 0.51 m
Track: 2.79 m
Track width: n/avail
Length of track on ground: 4.78 m
Maximum speed:
(1st gear, low): 9 km/h
(1st gear): 15 km/h
(2nd gear): 22 km/h
(3rd gear): 39 km/h
(4th gear): 62.3 km/h
Acceleration (0 to 32 km/h): 10 s
Maximum range: 450 km
Fording:
(without preparation): 1.40 m
(with preparation): 5 m
Gradient: 60%
Side slope: 40%
Vertical obstacle: 0.85 m
Trench: 2.7 m
Engine: Perkins Engines Company CV12-1200 TCA 12-cylinder, water-cooled, electronic-controlled diesel developing 1,200 hp at 2,300 rpm
Transmission: ESM 500 hydromechanic with infinitely variable hydraulic steering, 4 forward and 2 reverse gears
Brakes: n/avail
Suspension: torsion bar
Electrical system: n/avail
Batteries: n/avail
Armament:
(main): 1 × 125 mm smooth bore gun
(coaxial): 1 × 7.62 mm MG
(anti-aircraft): 1 × 12.7 mm MG
Smoke grenade dischargers: 2 × 6, can also lay smoke screen by injecting diesel into the exhaust outlets at the rear
Ammunition:
(main): 39
(coaxial): 3,000
(anti-aircraft): 500
Gun control equipment
Turret power control: powered/manual
(by commander): yes
(by gunner): yes
Turret traverse: 360°
Gun elevation/depression: +14/–6°
Gun stabiliser: 2-axis
NBC: yes
Night vision equipment: yes
More recent data released by NORINCO has stated that the vehicle has a combat weight of 46 tonnes and a hull length of 6.487 m and an overall length of 9.865 m with turret traversed to the front.

Main armament consists of a 125 mm smoothbore gun fitted with a thermal sleeve and fume extractor. It is fed by an automatic loader which holds a total of 22 projectiles and charges which can be loaded at the rate of six to eight per minute. After the 125 mm smooth bore gun has fired it returns to an angle of 4° 15′ for loading and then returns to the firing position. The 125 mm smoothbore gun is stated to have a life of 500 rounds.

Types of separate loading ammunition that can be fired by the 125 mm smoothbore gun include APFSDS (muzzle velocity 1760 m/s), HEAT (muzzle velocity 850 m/s) and HE-FRAG (muzzle velocity 950 m/s). According to NORINCO, the APFSDS-T projectile will penetrate 460 mm of conventional steel armour at a range of 2,000 m. A total of 39 projectiles are carried in the normal ratio of four APFSDS-T, three HEAT and three HE-FRAG.

Mounted coaxial to the right of the main armament is a 7.62 mm machine gun, while mounted on the commander's cupola is a 12.7 mm machine gun, for engaging ground and aerial targets.

The 7.62 mm coaxial machine gun has a cyclic rate of fire of 700 to 800 rds/min with a practical rate of fire of 250 rds/min. Effective range is quoted as 1,000 m.

The 12.7 mm machine gun has a cyclic rate of fire of 650 to 700 rds/min with a practical rate of fire of 80 to 100 rds/min. Traverse is 360° with elevation from –4.5 to +70°. Effective range is quoted as 1,600 m.

The commander is provided with a two-axis stabilised panoramic sight while the gunner has a two-axis stabilised fixed roof-mounted sight.

The Image Stabilised Fire-Control System (ISFCS) includes a thermal imager module, laser range-finder integrated into the gunner's sight, control panel, ballistic computer and a roof-mounted sensor for crosswind, tilt and angular velocity.

This, together with the stabilised 125 mm gun and roof-mounted stabilised sights, enables the Type 90-II to engage stationary and moving targets while it is stationary or moving. If required, the tank commander can take over and lay and fire the main armament.

According to NORINCO, Type 90-II has a 71 per cent hit probability on a moving target while the vehicle itself is moving. Target engagement time, stationary to stationary target is quoted as 7 seconds, while on the move 10 seconds.

Mounted either side of the turret is a bank of six electrically operated smoke grenade launchers which fire forwards and to the rear of this is a large stowage basket. When not required, the snorkel is normally carried in this stowage basket.

The power pack consists of the engine, transmission and cooling system and can be removed as a complete unit in 30 minutes. The engine is a British Perkins Engines Company Shrewsbury CV12-1200 TCA diesel developing 1,200 hp coupled to a French SESM ESM 500 automatic transmission.

The V-12 diesel engine is similar to that installed in the Vickers Defence Systems Challenger 2 (used by the British Army) and the Challenger 1 (used by Jordan under the local name of Al Hussein) while the ESM 500 transmission is installed in the French Giat Industries Leclerc MBT used by the French Army.

Suspension is of the torsion bar type with hydraulic shock absorbers and either side consists of six large dual rubber-tyred roadwheels with the drive sprocket at the front, idler at the rear and track return rollers. The upper part of the suspension is covered by a rubber skirt with the forward part on either side being covered by explosive reactive armour.

Standard equipment includes an anti-neutron liner for the crew compartments, collective NBC system, explosion/fire detection and suppression system and infrared reflecting paint. It can also lay a smoke screen by injecting diesel fuel into the exhaust outlets at the rear of the hull.

To extend the operational range of the Type 90-II MBT, two additional diesel fuel drums can be mounted externally at the rear.

Variants
None announced so far.

Status
Prototypes. Not yet in production or service.

Manufacturer
Chinese state factories.
Enquiries to China North Industries Corporation.

UPDATED

NORINCO Type 85-III MBT

Development
Early in 1995, NORINCO (China North Industries Corporation) revealed that it was marketing a new version of its Type 85 MBT called Type 85-III.

This is a further development of the earlier Type 85-II (armed with a 105 mm rifled gun and a crew of four) and the Type 85-IIM (armed with a 125 mm smoothbore gun and a crew of three). The latter has been manufactured in Pakistan as the Type 85-IIAP, which was first observed in March 1993.

It is understood that some subsystems of this tank are also used in the Type 90-II (or MBT-2000 as it is also referred to) covered in the previous entry.

Description
The overall layout of the Type 85-III MBT is virtually the same as that of the Type 85-II, covered in the following entry, with the main improvements being summarised as follows:

China North Industries Corporation Type 85-III MBT showing 125 mm smoothbore gun fed by automatic loader
0018581

(1) Installation of a new power pack consisting of a transversely mounted V-type diesel engine developing 1,000 hp
(2) The complete power pack is integrated and can be removed and replaced in the field in less than 40 minutes
(3) The transmission has been improved and now has hydraulic control and the driver can select either automatic, semi-automatic or manual modes of operation
(4) The composite armour panels on the turret and hull front provide a very high level of protection against both kinetic and chemical energy attack. The hull front armour is fixed while the turret armour can be replaced with a different armour package if required
(5) Commander and gunner have roof-mounted stabilised (in two planes) day/image

Specifications
Crew: 3
Combat weight: 42,500 kg
Power-to-weight ratio: 23.52 hp/tonne
Length:
(gun forward) 10.428 m*
(gun rear) 9.508 m*
Width: 3.40 m
Height: 2.20 m
Max road speed: 65 km/h
Max range: 600 km (approx)
Fording: 5 m
Engine: V-type diesel developing 1,000 hp
Transmission: planetary with 7 forward and 1 reverse gears
Armament:
(main) 1 × 125 mm smooth bore gun
(coaxial) 1 × 7.62 mm MG
(anti-aircraft) 1 × 12.7 mm MG
Smoke grenade dischargers: 2 × 6
Smoke laying capability: yes
Ammunition:
(main) 42 × 125 mm
Turret traverse: 360°, powered/manual
Gun stabiliser:
(vertical) yes
(horizontal) yes
Rangefinding device: yes, laser
NBC system: yes
Night vision equipment: yes
*Including long-range fuel tanks at hull rear.

intensification (second-generation) night sights and the computerised fire-control system allows stationary or moving targets to be engaged with a high first round hit probability when the vehicle is moving

(6) Main armament comprises a 125 mm smoothbore gun fed by an automatic loader. This is believed to be similar to that fitted to the Russian T-72/T-80 MBTs and has allowed the crew to be reduced to three. Maximum quoted rate of fire is 8 rds/min and a total of 42 rounds of separate loading (projectile is loaded first followed by the charge with its semi-combustible cartridge case) ammunition are carried

(7) The 125 mm gun fires three types of 125 mm ammunition, APFSDS, HEAT and HE-FRAG. The 125 mm gun can fire at least two types of 125 mm APFSDS-T ammunition developed by NORINCO, the 125-I and 125-II. This ammunition can also be fired by the Russian T-72 series MBT and the Chinese Type 85-IIAP MBT.

Tungsten alloy with variable density is used as the core material with the 125-I projectile having a muzzle velocity of 1,730 m/s which will penetrate 460 mm of RHA at a range of 2,000 m. The projectile weighs 7.37 kg and is 28 mm in diameter.

The 125-II projectile has a muzzle velocity of 1,740 m/s and will penetrate 600 mm of RHA at a range of 2,000 m. The projectile weighs 7.44 kg and has a diameter of 26 mm.

Standard equipment includes a collective NBC system, fire/explosion detection/suppression system, bank of six forward-firing smoke grenade launchers mounted either side of the turret and the ability to lay a smoke screen by injecting diesel fuel into the exhaust.

The Type 85-III MBT has more recently been offered fitted with a Global Positioning System.

To extend the operational range of the Type 85-III MBT, drum-type long-range fuel tanks can be fitted at the rear.

Status
Development complete. Ready for production.

Manufacturer
Chinese state factories.
Enquiries to China North Industries Corporation.

UPDATED

NORINCO Type 85-II MBT

Development
Early in 1989, NORINCO (China North Industries Corporation) stated that it had developed two new MBTs based on the Type 80 MBT chassis (qv), the Type 85-II and the Type 85-IIA.

These MBTs were developed to meet the operational requirements of the People's Liberation Army (PLA) with firing trials carried out in mid-1989.

All previous Chinese MBTs had a cast steel turret but the Type 85 has a welded steel turret with compound armour for improved battlefield survivability. In addition it has an enhanced computerised fire-control system which, when used with the independently stabilised sighting system, allows the gunner to engage moving targets when the tank is in motion.

The actual difference between the Type 85-II and Type 85-IIA has not been disclosed although the latter is slightly heavier at 39.5 tonnes and slightly shorter in length.

Main armament of the Type 85-II and Type 85-IIA is the same as that of the earlier Type 80, a 105 mm rifled tank gun with the Type 85-II carrying 46 rounds and the Type 85-IIA carrying 44 rounds of ammunition.

The original Type 80 MBT has an individual NBC system while Types 80-II, 85-II and 85-IIA have a collective system. The Type 80-II, 85-II and 85-IIA also have a Type 889B radio system.

In mid-1991, it was announced that the Type 85-II MBT was to be co-produced in Pakistan (qv). The Pakistani designation for this vehicle is the Type 85-IIAP.

Late in 1992, NORINCO revealed it was manufacturing another version of the Type 85 called the Type 85-IIM; this is similar to the earlier models but is heavier and is armed with a 125 mm rifled tank gun which is fed by an automatic loader which feeds the projectile and then the charge into the weapon.

Recent information has indicated that the version of the Type 85 selected by Pakistan is designated the Type 85-IIAP. This made its first appearance during a parade held in Islamabad in March 1993.

According to information released by the United Nations, China exported the following quantities of Type 85 series MBTs in recent years to Pakistan:

Year	Quantity	Comment
1992	97	
1993	35	
1994	82	
1995	51	plus 20 ARV
1996	3	
1997	nil	
1998	nil	
1999	nil	
2000	nil	

NORINCO Type 85-IIM MBT which is armed with a 125 mm smoothbore gun fed by an automatic loader

In 1996, Pakistan placed a contract with Ukraine for 320 T-80UD MBTs with first deliveries taking place in early 1997 and final deliveries being made in late 1999. It is of note that deliveries from China ran down before the commencement of deliveries of the T-80UD to Pakistan.

Description
The layout of the Type 85-IIM MBT is conventional with the driver's compartment at the front, fighting compartment in the centre and power pack at the rear. The basic hull and turret is of welded steel armour with an additional layer of composite armour added to the glacis plate and the turret front and sides for improved battlefield survivability.

The driver is seated at the front left and has a single-piece hatch cover and day periscopes for forward observation, one of which can be replaced by a passive periscope for driving at night.

Specifications
Type 85-IIM
Crew: 3
Combat weight: 41,000 kg
Power-to-weight ratio: 18.5 hp/t
Length: (gun forward) 10.28 m
Width: (over skirts) 3.45 m
Height: (turret top) 2.30 m
Ground clearance: 480 mm
Max road speed: 57.25 km/h
Vertical obstacle: 0.8 m
Trench: 2.7 m
Fording: 1.4 m
Gradient: 60%
Engine: V-12 supercharged diesel developing 730 hp
Armament:
(main) 1 × 125 mm smooth bore gun
(coaxial) 1 × 7.62 mm MG
(anti-aircraft) 1 × 12.7 mm MG
Smoke grenade dischargers: 2 × 6
Ammunition:
(main) 40 × 125 mm
(coaxial) 2,000 × 7.62 mm
(anti-aircraft) 500 × 12.7 mm
Gun elevation/depression: +14/–6°, powered/manual
Turret traverse: 360°, powered/manual
Gun stabiliser:
(vertical) yes
(horizontal) yes
Range-finding device: yes, laser
NBC system: yes
Night vision equipment: yes

The gunner is seated on the left of the turret with the commander on the right. The gunner has a single-piece hatch cover which opens forwards. It has a circular opening for the installation of a snorkel for deep fording as well as an observation periscope. The gunner's sight is mounted in the turret roof forwards of the hatch cover.

The commander has a cupola and a periscope that can be traversed a full 360°. Mounted on the turret roof is a 12.7 mm anti-aircraft machine gun which can be used by the commander or gunner. This weapon is believed to be the Chinese NORINCO 12.7 mm tank anti-aircraft machine gun Type 59 which has been made under licence by Pakistan Ordnance Factories.

Main armament of the Type 85-II comprises a 125 mm smoothbore gun which is fitted with a fume extractor and thermal sleeve. This is fed by an automatic loader which has enabled the crew to be reduced to three; it uses three types of separate loading (projectile and charge) ammunition, APFSDS, HE and HEAT.

The 125 mm gun can fire at least two types of 125 mm APFSDS-T ammunition developed by NORINCO, the 125-I and 125-II. This ammunition can also be fired by the Russian T-72 series MBTs.

Tungsten alloy with variable density is used as the core material with the 125-I projectile having a muzzle velocity of 1,730 m/s which will penetrate 460 mm of RHA at a range of 2,000 m. The projectile weighs 7.37 kg and is 28 mm in diameter.

The 125-II projectile has a muzzle velocity of 1,740 m/s and will penetrate 600 mm of RHA at a range of 2,000 m. The projectile weighs 7.44 kg and has a diameter of 26 mm.

According to NORINCO, this round can also be fired from the 125 mm smoothbore gun installed in the Russian T-72 MBT.

A 7.62 mm machine gun is mounted coaxially with the main armament and a bank of six electrically operated smoke grenade dischargers is mounted either side of the turret. In addition the vehicle can lay its own smoke screen by injecting diesel fuel into the exhaust.

The Image-Stabilised Fire-Control System (ISFCS) includes a laser range-finder integrated into the stabilised gunner's sight, control panels, ballistic computer and sensors which give a high first round hit probability against both stationary and moving targets while the Type 85-IIM is stationary or moving. Second-generation image intensification night vision equipment is standard for the commander, gunner and driver.

The supercharged V-12 diesel engine develops 730 hp and is coupled to a transmission of an undisclosed type. It is possible that the power pack is the same as that installed in the Type 80 MBT. The earlier NORINCO Types 59, 69 and 79 MBTs have their exhaust outlets located on the left side of the hull, just above the last roadwheel station as on the Soviet T-54, T-55 and T-62 MBTs. The hull of the Type 85-IIM is a new design and does not appear to be based on the Type 80 MBT.

NORINCO Type 85-IIAP MBT of the Pakistan Army showing roof-mounted 12.7 mm anti-aircraft machine gun 0018582

The suspension either side consists of six dual rubber-tyred roadwheels with torsion bars, hydraulic shock absorbers, an idler at the front, drive sprocket at the rear and track-return rollers. The upper part of the suspension is covered by a skirt. The steel tracks have replaceable rubber pads.

Standard equipment includes a Type 889B radio system, NBC system and explosion detection/ suppression system.

Variants

Type 88C MBT

It is understood that the People's Liberation Army designation for the Type 85-IIM MBT is the Type 88C and was first seen late in 1999. This weighs 41.5 tonnes and has an overall length of 10.65 m (turret traversed to the front), width of 3.3 m and a height of 2.3 m. It has a crew of three and is armed with a 105 mm rifled tank gun (although some sources have claimed it to be a 125 mm smooth boregun) fed by an automatic loader. The ISFCS-212 computerised fire-control system is fitted.

Some Type 88C MBTs have been observed fitted with a laser jammer type device on the left side of the turret roof. This is operated by remote control from within the turret.

It is possible that there are variants of the Type 85 series chassis under development, including an armoured recovery vehicle.

Type 85 IIAP MBT

This is the version of the Type 85-II MBT built in Pakistan.

Type 85-III MBT

Details of this version that features a 125 mm smooth bore gun fed by an automatic loader are given in the previous entry.

Status

Production as required. In service with:

Country	Quantity	Comment
China	600	estimate
Pakistan	300+	estimate

Manufacturer

Chinese state factories.
Enquiries to China North Industries Corporation.

UPDATED

NORINCO Type 80 MBT

Development

NORINCO (China North Industries Corporation) started concept studies for a second-generation MBT in 1978 with the first prototypes being completed in 1985. Recent information has indicated that the Type 80 MBT was developed at Factory No 617. Development was completed in 1988 and it was designated the Type 80 MBT, although it has also been referred to as the Type 88 and the Type 69-III.

Main improvements to the Type 80 compared with the earlier Type 69 MBT include the installation of a new computerised fire-control system for increased first range hit probability, laser range-finder, stabilisation system, 105 mm rifled tank gun, new torsion bar suspension with six roadwheels for increased cross-country mobility, more powerful diesel engine for increased power-to-weight ratio, passive night vision equipment and the ability to be fitted with a snorkel for deep fording.

The Type 80 MBT can engage moving targets from a stationary firing position.

Production of the Type 80 MBT was completed by NORINCO some time ago and export marketing is now being concentrated on its more recently developed MBTs such as the Type 85 and Type 90 and their variants. According to United Nations sources, China did not export any Type 80 MBT series vehicles between 1992 and 2000.

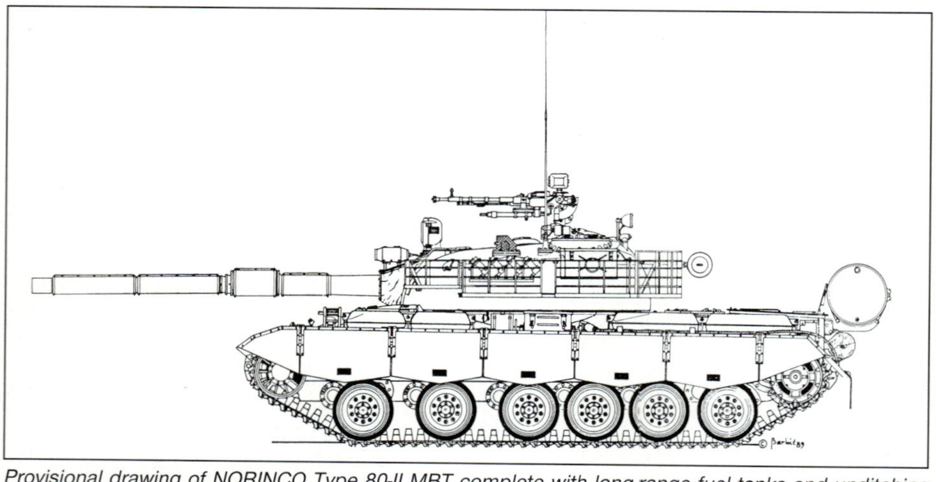

Provisional drawing of NORINCO Type 80-II MBT complete with long-range fuel tanks and unditching beam at rear (Vasco Barbic)

Description

The hull of the Type 80 MBT is of all-welded steel construction and has the driver's compartment at the front, the fighting compartment in the centre and the engine and transmission at the rear.

The driver sits at the front of the hull on the left side and has a single-piece hatch cover that lifts and swings to the left. In front of this are two day periscopes, one of which can be replaced by an infra-red or passive periscope for driving at night.

Mounted on the well-sloped glacis plate is a splashboard to stop water rushing up and into the driver's compartment while the tank is fording.

To further improve the Type 80's protection a panel of composite armour can be added at the front.

The turret is in the centre of the vehicle with the commander on the left, the gunner forward and below the commander and the loader on the right. Composite

Specifications
Type 80
Crew: 4
Combat weight: 38,000 kg
Power-to-weight ratio: 19.2 hp/t
Length:
(gun forward) 9.328 m
(gun rear) 8.611 m
(hull) 6.325 m
Width:
(over tracks) 3.354 m
(over skirts) 3.372 m
Height:
(turret roof) 2.290 m
(including 12.7 mm AA MG) 2.874 m
(firing) 1.784 m – 1.798 m
Ground clearance: 0.48 m
Track: 2.7 m
Track width: 580 mm
Length of track on ground: 4.064 m
Max road speed: 65 km/h
Max road range: 430 km
Fuel capacity: 1,400 litres
Vertical obstacle: 0.8 m
Trench: 2.7 m
Fording: 1.4 m
Fording with snorkel: 5 m
Gradient: 60%
Side slope: 40%
Engine: VR36 diesel developing 730 hp at 2,000 rpm (early versions had 700 hp engine)
Suspension: torsion bar
Armament:
(main) 1 × 105 mm rifled gun
(coaxial) 1 × 7.62 mm MG
(anti-aircraft) 1 × 12.7 mm MG
Smoke grenade dischargers: 2 × 4
Ammunition:
(main) 44
(coaxial) 2,250
(anti-aircraft) 500
Gun elevation/depression: +18°/–4.5°, powered/manual
Turret traverse: 360°, powered/manual
Gun stabiliser:
(vertical) yes
(horizontal) yes
Range-finding device: yes, laser
NBC system: yes
Night vision equipment: yes

NORINCO Type 80-II MBT which is armed with a 105 mm rifled gun fitted with a fume extractor and a light alloy thermal sleeve
0018583

armour can also be added to the turret for improved protection.

The commander has a cupola which can be traversed manually through a full 360° and a single forward-opening hatch cover with two integral periscopes for observation to the sides. Forward of this hatch cover are a further three day periscopes for observation over the frontal arc. The loader has a circular hatch cover that opens to the left.

Main armament comprises a stabilised 105 mm rifled gun which fires Western- or Chinese-produced ammunition. This has powered elevation from –4.5 to +18° with the turret having powered traverse through 360°. The 105 mm rifled tank gun, which is similar in appearance to the Western L7/M68, is fitted with a fume extractor and a thermal sleeve. The 105 mm gun fires the following NORINCO-developed rounds: APFSDS-T (muzzle velocity 1,455 m/s), HEAT-T (1,173.5 m/s), HESH (731.5 m/s) and HE (850 m/s).

More recently two new 105 mm rounds have been developed by NORINCO for use with the Type 80 series MBT and other MBTs armed with a 105 mm rifled tank gun.

These two new 105 mm rounds are designated the Type 105-I and 105-II and both use a brass cartridge case and have a dispersion at a range of 1,000 m of less than 0.3 mils horizontal by 0.3 mils vertical.

The Type 105-I has a muzzle velocity of 1,455 m/s and will penetrate 150 mm of RHA at an angle of 60° at a range of 2,000 m. This is equal to 300 mm of RHA at 0°.

The Type 105-II has a muzzle velocity of 1,530 m/s and will penetrate 150 mm of RHA at an angle of 71° at a range of 2,000 m. This is equal to 460 mm of RHA at 0°.

A 7.62 mm machine gun is mounted coaxially with the 105 mm gun and there is a 12.7 mm machine gun on the loader's hatch for anti-aircraft use.

Mounted either side of the turret are four forward-firing electrically operated smoke grenade dischargers. A stowage basket runs around the sides and rear of the turret. The stowage basket provides some protection against HEAT projectiles as these would be detonated away from the main armour of the turret.

The fire-control system includes a ballistic computer, a laser range-finder integrated into a stabilised sight, sensors and a gun stabilisation system. Night vision equipment fitted is of the second-generation image intensification type.

The Type 80 MBT is powered by a Model VR36 diesel developing 700 hp at 2,000 rpm but late production engines develop 730 hp at 2,000 rpm. Type 80 has two circular exhaust outlets on the right side of the hull. Diesel fuel can be injected so that the tank can lay its own smoke screen while moving.

To extend the operational range of the tank two large drum fuel tanks can be mounted at the rear; these can be jettisoned if required. An unditching beam is normally carried at the rear of the hull under the long-range fuel tanks.

The torsion bar suspension is new with six dual rubber-tyred roadwheels, drive sprocket at the rear, idler at the front and three track-return rollers. The first, second, fifth and sixth roadwheel stations have hydraulic shock-absorbers.

The upper part of the track is covered either side by skirts with a wavy lower edge which are hinged up to allow access to the suspension for maintenance.

Standard equipment includes an overpressure NBC system, night vision equipment for the commander, gunner and driver and provision to install a snorkel in the roof forward of the loader's hatch; this allows the tank to submerge to a depth of 5 m and cross rivers up to 600 m wide. An unditching beam is carried at the rear of the hull under the long-range fuel tanks.

Communications equipment installed includes a YRC-83 radio and VIC-8 intercom system. To further increase the tank's survivability on the battlefield, an automatic explosion suppression and fire extinguishing system is fitted.

Variants

Type 80-II MBT

This is virtually identical to the basic Type 80 MBT except that it has a combat weight of 38.5 tonnes and a power-to-weight ratio of 18.96 hp/t. Specifications are identical apart from the overall length of 9.336 m. The Type 80-II has a Type VRC-83 radio. It may also have a semi-automatic transmission in place of the standard manual transmission and both the Type 80 and the Type 80-II can be fitted with a different fire-control system consisting of a projecting sight, laser range-finder, ballistic computer and a bidirectional stabiliser. The photograph of the Type 80-II MBT does not have the stowage basket that is usually fitted to the NORINCO Type 80 series MBTs. The type 80-II is also known in the PLA as the Type 88B. This is armed with a 105 mm rifled tank gun.

Status

Production complete. In service with:

Country	Quantity	Comment
China	500	estimate
Myanmar	20+	estimate

Manufacturer

Chinese state factories.
Enquiries to China North Industries Corporation.

UPDATED

NORINCO Type 79 MBT

Development

The NORINCO (China North Industries Corporation) Type 79 MBT made a brief appearance in the October 1984 parade in Beijing when it was referred to simply as a Main Battle Tank. Its chassis is similar to the Type 59/Type 69 but has the towing eyes under the nose and a headlamp arrangement similar to the Type 80 MBT.

The Type 79 MBT is a further development of the Type 69-II MBT with work being carried out by No 617 Factory. According to Chinese sources, the main improvements of the Type 79 MBT are in the areas of firepower, fire control, night vision equipment and communications equipment. As far as is known, the Type 79 MBT has not been exported and is no longer being marketed.

Some sources have also given the Type 79 MBT the designation of the Type 69-III with the industrial designation being the WZ121D. Some sources have stated that the total production of the Type 79 was 800 vehicles or less with one source quoting a figure of just 519 units built.

Some reports have also mentioned an improved version of the Type 79 called the Type 79-II. This is understood to be fitted with a fire-control system which has the laser range-finder integrated into the gunner's sight, rather than mounted externally over the main armament where it is highly vulnerable to small arms fire and shell splinters.

Description

The turret is of cast armour and is slightly different from the Type 59/Type 69 as there is a new large rectangular

Chinese Type 79 MBT with 105 mm gun and smoke grenade dischargers mounted either side of the main armament

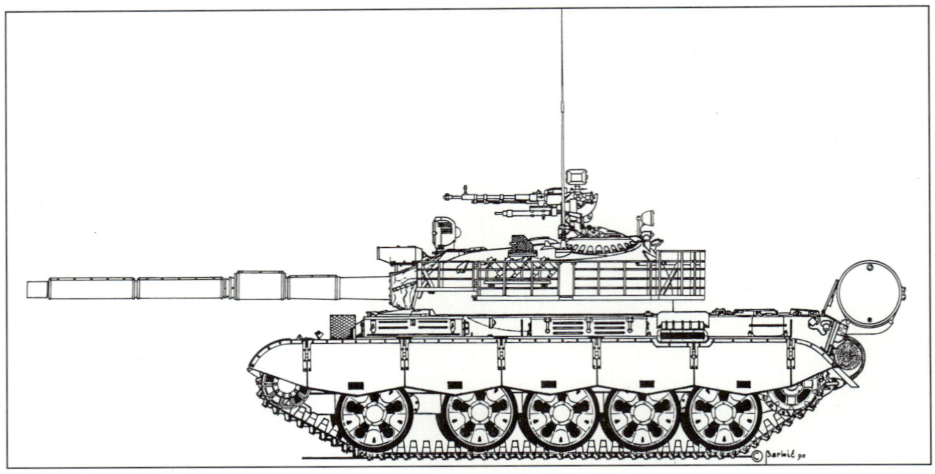

NORINCO Type 79 MBT armed with 105 mm gun (Vasco Barbic)

window located ahead of the commander's cupola which could be an internally mounted laser range-finder.

Mounted either side of the turret is a bank of four electrical smoke grenade dischargers, which are very similar to the German Krauss-Maffei Wegmann 76 mm type and, to the immediate rear of these, is a metal stowage box which is believed to house additional grenades.

Main armament comprises a 105 mm rifled tank gun which is provided with a fume extractor and a thermal sleeve.

A 7.62 mm machine gun is mounted coaxially with the main armament and a 12.7 mm machine gun is mounted on the loader's cupola for local and anti-aircraft defence. There are slight differences in the rear of the hull which indicate a different power pack.

Chinese sources state that the Type 79 MBT weighs 37.5 tonnes compared with the 36 tonnes of the Type 59 MBT and has a maximum road speed of 50 km/h. Standard equipment includes a simplified fire-control system and night vision equipment for commander, gunner and driver.

The upper part of the suspension of the Type 79 MBT is covered by five removable skirts. The tank has also been observed fitted with new tracks with removable rubber pads in place of the more usual all-steel track fitted to other older Chinese MBTs.

A stowage basket runs around the rear half of the turret from just forward of the smoke grenade launchers and it is claimed to provide some protection against HEAT projectiles by detonating their warheads before they reach the turret.

Standard equipment on the NORINCO Type 79 MBT includes NBC system, unditching beam and long-range diesel fuel tanks which are carried at the rear.

The Type 79 MBT can also lay its own smoke screen by injecting diesel fuel into the exhaust outlet on the left side of the hull.

Status

Production complete. In service with the Chinese Army. No longer being marketed.

Manufacturer

Chinese state factories.
Enquiries to China North Industries Corporation.

UPDATED

NORINCO Type 69 MBT

Development

The NORINCO (China North Industries Corporation) Type 69 MBT made its first public appearance during a parade near Zhangiakou outside Beijing in September 1982.

Iraq placed orders with China for between 100 and 200 Type 69-I tanks armed with the 100 mm gun and first deliveries of these were made in 1983 via Saudi Arabia. Total deliveries of the Type 69 MBT to the Gulf combatants are estimated at between 1,800 and 2,500 over a three-year period up to late 1987. The Type 69-I MBT has the industrial designation of the WZ121 with the later Type 69-II having the industrial designation of the WZ121A.

In early 1987, Thailand placed orders for quantities of Chinese military equipment including Type 69-II MBTs. Type 69-II is called the Type 30 MBT by the Royal Thai Army. Recent information has indicated that Thailand may have only taken delivery of 50 of these vehicles. Pakistan is now manufacturing the 105 mm Type 69-II MBT under licence. Additional details are given under Pakistan.

Pakistan has carried out a major upgrade on its Type 69 MBT with the latest Phase III having a 125 mm smoothbore gun fitted. This is called the Al Zarra and additional details are given later in this section under Pakistan.

There has been no recent production of the Type 69 and it is no longer being marketed by NORINCO.

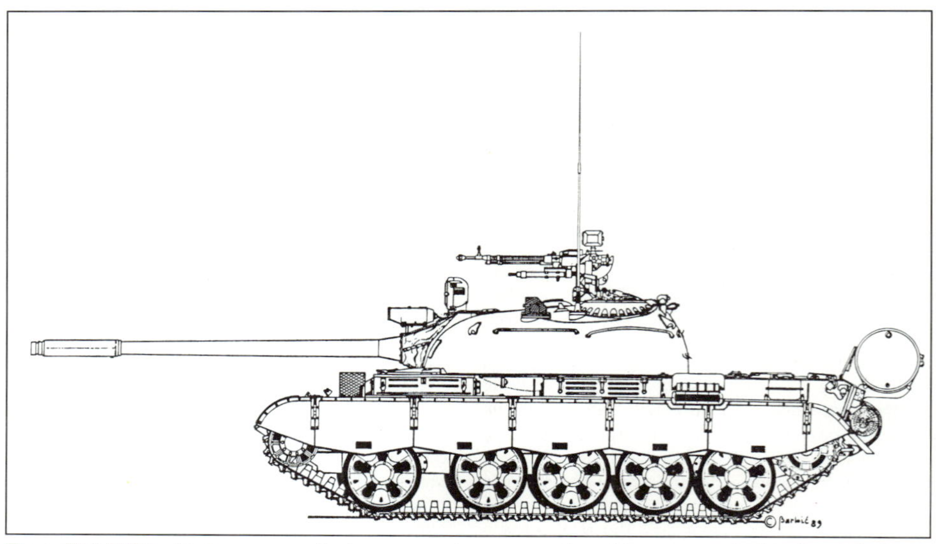

Type 69-II MBT with laser range-finder over main armament and long-range fuel tanks at hull rear (Vasco Barbic)

Description

First production Type 69s were produced with rifled and smoothbore guns but extensive trials demonstrated that the rifled gun was more accurate and had greater armour penetration characteristics than the smoothbore weapon. After 150 Type 69s had been built with smoothbore guns it was decided to concentrate on the rifled gun which was used on all subsequent production Type 69 series MBTs for home and export markets. The Type 69-I has the smoothbore gun; the Type 69-II has the rifled gun and a different fire-control system.

The Type 69-I MBT is a further development of the Type 59. It differs mainly in the areas of armament, fire control and night vision equipment. The actual layout of both vehicles is virtually identical.

The Type 69-I MBT is armed with a 100 mm smoothbore gun as distinct from the 100 mm rifled gun installed in the earlier Type 59 MBT. This is slightly longer than the rifled gun and has a bore evacuator near the end of the muzzle. Types of 105 mm ammunition that can be fired by the smoothbore gun include anti-personnel, high-explosive, high-explosive anti-tank and high-velocity armour-piercing discarding sabot.

A laser range-finder, similar to that on the Type 59 MBT described in the following entry, is mounted externally over the main armament in front of the mantlet.

Night vision equipment includes an infra-red searchlight on the commander's cupola, an infra-red sight for the gunner, an infra-red searchlight above and

NORINCO Type 69-II command tank which is externally recognisable by its two radio antennas 0018586

Twin 57 mm Type 80 self-propelled anti-aircraft gun system 0018585

Type 84 AVLB in travelling configuration 0018587

to the immediate right of the 100 mm main armament and an infra-red driving light on each running board.

Mounted externally on the loader's cupola is a 12.7 mm Type 54 anti-aircraft machine gun. Mounted coaxially with the main armament is a 7.62 mm Type 59T machine gun and there is a similar weapon in the bow of the tank.

The Type 69 MBT has a complete NBC system, semi-automatic fire extinguishing system and, like Russian T-series tanks, can lay a smoke screen by injecting diesel fuel into the exhaust system on the left side of the hull. The vehicle is also coated with infra-red reflecting paint.

Variants

Type 69-II MBT

Further development in 1983 of the Type 69-I MBT by No 617 Factory resulted in the Type 69-II MBT.

Main improvements include the installation of the Tank Simplified Fire-Control System with laser range-finder, rubber shielding for the side skirts, turret grid armour, hydraulic booster for the steering mechanism and main clutch, double pin rubber tracks, NBC system, automatic fire detection and extinguishing system, engine low-pressure alarm and smoke grenade launchers either side of the turret. The turret grid armour, which is also referred to as 'Boom Shields' has been designed to detonate projectiles fitted with a HEAT warhead before they impact the main amour of the Type 69 series MBT.

Main armament consists of a 100 mm rifled gun which is stabilised in both elevation and traverse. This fires Chinese-developed HEAT, HE, APHE and APFSDS ammunition.

At least three types of APFSDS ammunition have been developed including one with a semi-combustible cartridge case.

The TSFCS consists of three main components, the TLR1A tank laser range-finder, a 1A ballistic computer and the TGS-A tank gunsight. The TLR1A measures the distance to the target, and this information is automatically fed into the ballistic computer which supplies information to control the automatic range-setting mechanism of the sight and simultaneously the gun control system and sets the 100 mm gun elevation required in order to hit the target.

The laser range-finder is mounted externally above the 100 mm gun and is vulnerable to shell splinters and small arms fire. The TSFCS-L, which is an option for the Type 69-II, has the laser range-finder combined with the gunsight in a single unit mounted inside the tank.

The gunner's day sight has a magnification of ×3.5 (18° field of view) and ×7 (9° field of view) while the gunner's night sight has a magnification of ×7 (6° field of view) and a maximum effective range of 800 m.

The commander's day sight has a magnification of ×5 (12° field of view) while the night sight has a magnification of 6 (8° field of view) and a maximum effective range of 500 m.

The laser range-finder has a range from 300 to 3,000 m and can calculate the correct aiming point for the 7.62 mm coaxial machine gun and the 100 mm main armament.

The Type 69 MBT can also be fitted with the TSFCS-C, which can also be retrofitted into existing T-series tanks of Chinese and Soviet origin. This system consists of the gunner's sight, laser range-finder, fire-control computer, control panel and sensors for elevation and azimuth rates and cant, crosswind, air temperature and ammunition charge temperature. The computer calculates target range, direction and lead angle and also interfaces with the weapon stabilisation system.

The driver's day periscope can be replaced by a night periscope with a range of 60 m and a 30° field of view.

The Type 69-II MBT has a Type 889 radio and a Model 803 intercom for the commander, gunner, loader and driver. There are two command versions of the MBT. The Type 69-II Mk B command tank has one Type 889 radio and one Type 892 radio while the Type 69-II Mk C has two Type 889 radios.

Some Type 69 MBTs have been observed fitted with an external stowage bin on the turret rear and armour protection for the anti-aircraft machine gun. Thai Army Type 69-IIs MBTs have had their Chinese 12.7 mm weapons replaced by US 12.7 mm M2 MGs.

Type 69-IIB MBT command

This has the industrial designation of the WZ121B and has two radios, whereas the standard Type 69-II has only one radio.

Type 69-IIC MBT command

This uses two examples of the Type 889 radio as fitted to the standard Type 69-II MBT which normally has just one.

Specifications
Type 69-II MBT
Crew: 4
Combat weight: 36,700 kg
Power-to-weight ratio: 15.8 hp/t
Ground pressure: 0.823 kg/cm²
Length:
(gun forward) 8.589 m
(gun rear) 8.498 m
(hull) 6.243 m
Width:
(over skirts) 3.307 m
(over hull) 3.27 m
Height:
(to axis of AA MG) 2.807 m
(with AA MG at max elevation) 3.909 m
Axis of fire: 1.75 m
Ground clearance: 425 mm
Track: 2.64 m
Track width: 580 mm
Length of track on ground: 3.845 m
Max road speed: 50 km/h
Fuel capacity: 935 litres
Max road range: 420-440 km
Vertical obstacle: 0.8 m
Trench: 2.7 m
Fording: 1.4 m
Gradient: 60%
Side slope: 40%
Engine: Type 12150L-7BW V-12 diesel developing 580 hp at 2,000 rpm
Suspension: torsion bar
Armament:
(main) 1 × 100 mm gun
(coaxial) 1 × 7.62 mm MG
(bow) 1 × 7.62 mm MG
(anti-aircraft) 1 × 12.7 mm MG
Ammunition:
(main) 44
(7.62 mm) 3,000
(12.7 mm) 500
Gun elevation/depression: +18/–5°
Turret traverse: 360°
NBC system: yes
Night vision equipment: yes

Type 653 armoured recovery vehicle of the Iraqi Army captured in Kuwait (Christopher F Foss) 0018589

Chinese Type 69-II MBT used by the Iraqi Army and captured in Kuwait ***2002***/0130605

Chinese Type 69-II MBT captured in the Middle East and showing laser range-finder over mantlet of 100 mm gun and fitted with front-mounted mineclearing system 0011552

Twin 37 mm SPAAG

This is essentially the chassis of the Type 69 MBT fitted with a power-operated two-man turret and armed with twin 37 mm cannon. As far as it is known this system remains at the prototype stage. Details are given in *Jane's Land-Based Air Defence 2002-2003.*

Twin 37 mm SPAAG

During a defence exhibition held in Beijing late in 1988, another twin 37 mm self-propelled anti-aircraft gun system was shown for the first time.

As far as it is known, this system has yet to enter production. Full details of this are given in *Jane's Land-Based Air Defence 2002-2003.*

Twin 57 mm Type 80 SPAAG

This is the Chinese equivalent of the Soviet-designed ZSU-57-2 twin 57 mm SPAAG based on a modified T-54 tank chassis. The Type 80 is a modified Type 69-II MBT chassis fitted with an open-topped turret armed with twin 57 mm cannon for which a total of 300 rounds of ammunition is carried.

The Type 80 has a crew of six and a combat weight of 31 tonnes. Full details of this are given in *Jane's Land-Based Air Defence 2002-2003.*

Type 84 AVLB

The Type 84 armoured vehicle-launched bridge is essentially a Type 69 MBT with its turret removed, fitted with a bridge-launching mechanism similiar to that on the German Leopard 1 Biber system. When opened out, the bridge, which is launched over the front of the vehicle, is 18 m long, can span gaps of up to 16 m and take tracked and wheeled vehicles weighing up to 40 tonnes. Full details of the Type 84 are given in *Jane's Military Vehicles and Logistics 2002-2003.*

Type 653 ARV

This is based on the chassis of the Type 69 MBT and is designed not only to recover this and other vehicles on the battlefield but also to undertake major repairs such as changing complete power packs, clearing obstacles and preparing fire positions. Standard equipment includes a front-mounted hydraulically operated dozer blade, a hydraulic crane on the right side of the hull, a main winch with a capacity of 70 tonnes, auxiliary winch, tools, towbars and cables. Full details are given in *Jane's Military Vehicles and Logistics 2002-2003.* In addition to being used by the Chinese Army, small quantities of these have been sold to Bangladesh, Iraq, Pakistan and Thailand as part of the Type 69 MBT sale. An upgraded version is also available and this has been sold to Kuwait to support its NORINCO 155 mm/45 calibre self-propelled artillery systems. The upgraded model is referred to as the Type 653A.

Type 84 armoured mineclearing vehicle

This carries rocket propelled mineclearing equipment on the rear deck. There is an unidentified minesweeping version based on the Type 69 series MBT chassis

Status

Production complete. In service with the following countries.

Country	Quantity	Comment
Bangladesh	80	includes Type 59
China	200	provisional figure
Iran	200	estimate
Iraq	200	estimate
Myanmar	50-80	Type 69-II
Pakistan	250	see Pakistan (some being upgraded)
Thailand	50	estimate, believed to be now in store
Zimbabwe	10	

Manufacturer

Chinese state factories.
Enquiries to China North Industries Corporation.

UPDATED

NORINCO Type 59 MBT

Development/Description

In the early 1950s the Soviets supplied China with a quantity of T-54 MBTs, production of which was subsequently undertaken in China by NORINCO (China North Industries Corporation) under the designation Type 59. The first production models were similar to the Soviet T-54 and not fitted with infra-red night vision equipment or a stabiliser for the 100 mm main armament. Later models were fitted with a fume extractor as on the T-54A. Late production Type 59s have been fitted with an infra-red searchlight for both the commander and gunner and a larger infra-red searchlight has been mounted above the 100 mm main armament and moves in elevation with it. Some Type 59 vehicles have been seen fitted with a laser range-finder to the right of the infra-red searchlight mounted over the main armament.

Some years ago British company MEL (which has since been taken over) supplied China with 30 sets of passive night vision equipment for the Type 59 MBT. These include the commander's image-intensified periscope Type DC 1026/00, the gunner's image-intensified periscope Type DC 1024/00 and the driver's image-intensified periscope Type DC 1028/00.

The Chinese have developed and placed in production a 100 mm APFSDS-T projectile for the Type 59 MBT. This is designated the AP100-2 and has a long rod penetrator and a semi-combustible cartridge case. The round has a saddle-type sabot and eight fins. All that remains after firing is the stub case which is ejected when the breech opens. NORINCO has also developed a 100 mm APFSDS-T round with a conventional brass cartridge case. This has a muzzle velocity of 1,480 m/s and will penetrate 150 mm of conventional steel armour at an angle of 65° at a range of 2,400 m.

It is believed that production of the Type 59 tank began in 1957 to 1960 at Baotou in the Beijing military region. The industrial designation of the Type 59 MBT is now known to be the WZ120.

More recently, some Type 59 MBTs have been observed fitted with a bank of four smoke grenade dischargers on the side of the turret.

The Type 59 series was followed by the very similar Type 69. Production of both vehicles was completed many years ago and neither is now marketed by NORINCO.

Variants

Type 59-I MBT

Factory No 617 carried out a number of improvements to the Type 59 MBT which resulted in the Type 59-I. This featured hydraulic booster technology to improve the opening and closing mechanisms for the safety door cover and cupola door cover, installation of a simplified fire-control system, laser range-finder, engine low-pressure alarm and automatic fire extinguisher.

Type 59-II MBT

Further development resulted in the Type 59-II which has the 105 mm rifled tank gun that can fire APFSDS,

Type 59-I MBT fitted with side skirts and showing laser range-finder over 100 mm gun

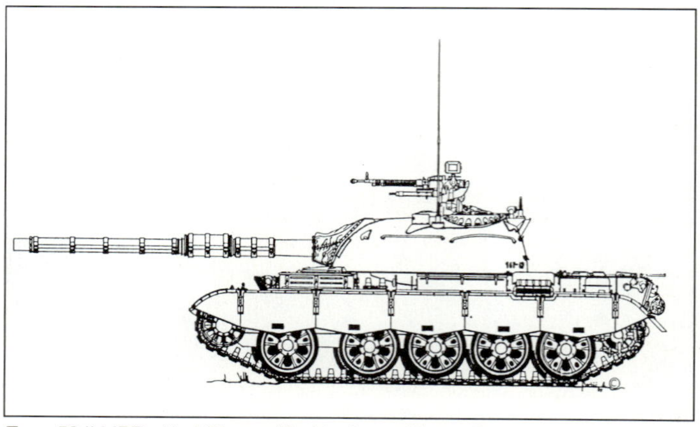

Type 59-II MBT with 105 mm rifled tank gun (Vasco Barbic)

Specifications
Crew: 4
Combat weight: 36,000 kg
Power-to-weight ratio: 14.44 hp/t
Ground pressure: 0.8 kg/cm²
Length:
(gun forward) 9 m
(hull) 6.04 m
Width: 3.27 m
Height: 2.59 m
Ground clearance: 0.425 m
Track: 2.855 m
Track width: 580 mm
Length of track on ground: 3.84 m
Max road speed: 40-50 km/h
Fuel capacity:
(main) 815 litres
(external) 400 litres
Range:
(main fuel supply) 420-440 km
(main and external supply) 600 km
Fording:
(without preparation) 1.4 m
(with preparation) 5.5 m
Gradient: 60%
Vertical obstacle: 0.79 m
Trench: 2.7 m
Engine: Model 12150L V-12 liquid-cooled diesel developing 520 hp at 2,000 rpm
Transmission: manual, 5 forward and 1 reverse gears
Suspension: torsion bar
Electrical system: 24 V
Armament: (see text)
(main) 1 × 100 mm; Type 59 rifled gun
(coaxial) 1 × 7.62 mm; Type 59T MG
(bow) 1 × 7.62 mm; Type 59T MG
(anti-aircraft) 1 × 12.7 mm; Type 54 MG
Ammunition:
(main) 34
(coaxial/hull) 3,500
(anti-aircraft) 200
Gun elevation/depression: +17/–4°
Turret traverse: 360°
Armour
Hull front upper: 97 mm at 58°
Hull front lower: 99 mm at 55°
Hull sides upper: 79 mm at 0°
Hull sides lower: 20 mm at 0°
Hull rear upper: 46 mm at 0°
Hull rear lower: 46 mm at 0°
Hull top: 33 mm
Hull floor – front: 20 mm
Hull floor – rear: 20 mm
Turret front: 203 mm at 0°
Turret sides: 150 mm at 0°
Turret rear: 64 mm at 0°
Turret roof: 39 mm at 79°
Mantlet: incl in above figures
NBC system: yes
Night vision equipment: yes

APDS, HEAT and HESH ammunition, two-axis stabilisation system for the main armament, light spot fire-control system, automatic fire extinguishing and explosion suppression system, 580 hp diesel engine, new radios and communications system.

The 105 mm rifled tank gun is similar to the Western L7 (UK)/M68 (US) and is fitted with a fume extractor and thermal sleeve. This has been referred to by US forces as the Model 1984 as this was the first year that it was identified.

The Chinese 105 mm rifled tank gun fires Chinese-developed ammunition and this gun is also installed in a number of other Chinese MBTs including the Type 79 (qv), Type 80 (qv) and Type 85 (qv).

NORINCO Type 59 retrofit package

NORINCO has offered an upgrade package for the Type 59 MBT. The original 520 hp diesel will be replaced by a new 730 hp diesel giving a much higher power-to-weight ratio and therefore greater acceleration on the battlefield.

Other improvements include APFSDS-T ammunition for the 100 mm gun, rubber-bushed tracks, NBC system, friction shock-absorbers, gun stabilisation in both elevation and traverse and a new fire-control system incorporating spot injection in the gunner's sight.

Chinese Type 59 MBT fitted with bank of four smoke grenade dischargers on side of turret 0018590

Chinese Type 59 MBT armed with NORINCO 120 mm smoothbore gun which is fitted with a fume extractor and thermal sleeve 0018591

Status

Production complete. In service with the countries listed in the table.

Country	Quantity	Comment
Albania	721	
Bangladesh	80	incl Type 69
Cambodia	200	incl some T-54 and T-55
China	6,000	estimate, also used by Marines
Congo, Democratic Republic	20	estimate
Iran	220	estimate
Iraq	200	estimate
Korea, North	175	
Pakistan	1,200	some upgraded with 105 mm gun
Sudan	10	Type 59
Tanzania	30	about 15 are operational
Vietnam	350	
Zambia	20	
Zimbabwe	22	

Type 59 with 105 mm L7 gun

As a private venture the now BAE Systems, RO Defence of the UK has fitted a Type 59 MBT with the 105 mm L7A3 rifled tank gun. This, together with three other upgraded Type 59s, was tested in Pakistan in 1987. Additional details are given under Pakistan in this section. So far no country has adopted the 105 mm L7 upgrade for the Type 59 MBT, although Egypt has adopted this upgrade for its Russian T-54/T-55 series MBTs.

NORINCO 120 mm T-54/Type 59 Upgrade

China North Industries Corporation (NORINCO) has developed an upgrade package for the widely deployed Chinese Type 59 and Russian T-54/T-55 MBT that includes the replacement of the current 100 mm rifled tank gun by a 120 mm smoothbore gun. As of early 2002 there are no known orders for the NORINCO 120 mm T-54/Type 59 MBT upgrade. This has also been referred to in Chinese sources as the Type 59 Kai.

The exact origin of the Chinese 120 mm smoothbore is not known as in the past, with the exception of the Western 105 mm rifled tank gun, all Chinese tank guns have been of Eastern design.

Some years ago China did develop and place in production the 120 mm self-propelled anti-tank gun Type 89 (details of which are given in the Tank destroyers section) which has a similar 120 mm smoothbore gun and the 120 mm weapon for the Type 59 upgrade is probably identical. This was seen in public for the first time late in 1999 during a major parade held in China.

The 120 mm smoothbore gun is 6 m long and weighs 2,600 kg complete with breech mechanism. It is fitted with a fume extractor and thermal sleeve but has no muzzle reference system. To give an increase in barrel life the interior is chrome plated.

As well as firing Western natures of armour-piercing fin stabilised discarding sabot (APFSDS) ammunition, such as the German Rheinmetall DM23 and DM33 and HEAT multipurpose, it can also fire 120 mm ammunition developed by NORINCO.

Like the Western 120 mm smoothbore ammunition, this has a semi-combustible cartridge case and all that remains after firing is the stub cartridge case. The penetrator has a respectable length-to-diameter ratio of 25.2/1 and complete with saddle weighs 7.33 kg. When in flight the penetrator weighs 4.1 kg.

The NORINCO APFSDS round has a muzzle velocity of 1,660 m/s and is claimed to have a similar armour penetration performance to NATO natures of ammunition, for example around 550 mm of rolled homogenous armour (RHA) at a range of 1,500 m.

Turret traverse is 360° with weapon elevation from –5 to +18°. A total of 28 rounds of 120 mm ammunition are carried compared to 34 rounds of 100 mm.

To improve first round hit probability a simplified fire-control system has been fitted which includes a laser range-finder. This enables the upgraded tank to engage moving targets when it is stationary, quoted engagement times are around 7 seconds for a stationary target and 10 seconds for a moving target.

Range, target horizontal angular speed and tilt angle of the 120 mm gun is input automatically while charge temperature, ambient air temperature, muzzle velocity and crosswind are entered manually as required.

While the weapon is normally laid on the target by the gunner, if required the commander can also take over, aim and fire. Various types of night vision equipment can be fitted as required by the user.

The secondary armament remains the same and consists of a roof-mounted 12.7 mm machine gun, 7.62 mm coaxial machine gun and a 7.62 mm machine gun which is fixed in the glacis plate and operated by the driver.

Basic ARV

This may well be a local modification and is essentially a Type 59 with its turret removed, armed with a single 12.7 mm machine gun. As this vehicle is not believed to have a winch, it is limited to towing operations.

Manufacturer

Between 1950 and 1951 three factories in China were converted into tank repair facilities. These were in Beijing, Changchun and Harbin. It is possible that these were subsequently upgraded to tank production facilities.

Enquiries to China North Industries Corporation.

UPDATED

CROATIA

RH-ALAN Degman MBT

Development

In mid-1999, Croatia finally revealed its new Degman Main Battle Tank (MBT), which is an enhanced version of the Yugoslav M-84A MBT that has been manufactured in Croatia for several years.

The M-84A is in turn an improved version of the original Yugoslav M-84 which in turn is based on the Russian T-72 MBT. Full details are given later in this section under Yugoslavia.

Under current plans, production of the M-84A was expected to be completed in Croatia in the near future, after which it will be followed on the production lines by the improved Degman.

For trials purposes, the new computerised fire-control system has been tested in the current M-84A and it is expected that in the future the whole Croatian M-84A/Degman MBT fleet will be brought up to a common fire-control system standard. Details of the M-84A/B are given in the following entry.

Description

In overall layout the Degman, which has sometimes been referred to as the M-95, is identical to the M-84/M84A with the driver at the front, two-person turret in the centre and with the power pack at the rear.

Degman is armed with a 125 mm D-81 smoothbore gun fed by an automatic loader installed in a new all-welded steel turret, whereas the earlier M-84/M-84A had a cast turret.

To provide a higher level of battlefield survivability, Degman is fitted with explosive reactive armour panels of a new design. These are installed on the turret front and sides, glacis plate and side skirts and provide protection against HEAT attack.

Degman has a new power pack consisting of a 1,000 hp diesel engine coupled to a locally developed automatic transmission.

The fire-control system of the Degman is an improved version of the existing Omega system developed by Fotona of Slovenia. Fotona has combined with Duro Dakovic of Slavonski Brod to develop the Degman.

Other electronic enhancements to the Degman include an advanced NBC system, fire detection and suppression system and computer-based driver's control panel that replaces the existing panel to integrate drive train monitoring with a navigation system and other functions. The radio communications package is provided by the now Thales of the UK.

Specifications

Not available.

Status

Trials. Expected to enter production in the near future.

Manufacturer

RH-ALAN.

UPDATED

Degman MBT fitted with ERA **2000**/0048566

Degman MBT fitted with ERA **2000**/0064629

Degman MBT from the rear showing additional fuel drums **2000**/0064630

Degman MBT from the front clearly showing ERA on the turret and hull **2001**/0105858

RH-ALAN M-84AB MBT

Development/Description

The M-84 series of MBT is essentially an upgraded version of the Russian T-72 MBT manufactured in the former Yugoslavia and full details of this are given under Yugoslavia later in this section. First production vehicles were called the M-84, followed by the improved M-84A.

At the time of the locally called Homeland War, Croatia only produced about 21 per cent of the M-84, the remainder came from other parts of the country.

Assembly of the M-84 was, however, undertaken in Croatia at the Duro Dakovic Workshops in Slavonski Brod and in 1996 RH-ALAN of Croatia stated that the latest version of the M-84, the M-84B, was now in production in Croatia.

Under current plans, production of the M-84A series of MBT was due to be completed in 2001, after which it was to be followed by the improved Degman MBT, covered in detail earlier in this section.

Standard M-84AB MBT built in Croatia **2001**/0105859

Croatian built M-84AB MBT **2001**/0105860

The M-84A was the version developed for Kuwait which ordered a total of 200 vehicles in 1989, including 15 command tanks and 15 armoured recovery vehicles.

By the time of the Iraqi invasion of Kuwait in 1990, about 15 vehicles had been delivered and more were delivered to the remains of the Kuwait Army in Saudi Arabia. Some of these vehicles took part in the recapture of Kuwait in early 1991.

It has been confirmed that the Kuwaiti vehicles have the SUV-M-84 computerised fire-control system with the gunner having a DNNS-2 day (×3 and ×7) and night (×8.5) sight with independent stabilisation in two planes and with integral laser range-finder.

Other optical devices include TNP-160 commander's periscope (two), TNPO-168 V driver's periscope, TNPA-65 auxilliary periscopes (two for commander, two for driver and one for gunner), DNKS-2 day/night periscope for commander and PPV-2 passive persicope for driver.

The M-84AB is also fitted with a GPK-59 gyrocompass, VRQ 316 communications equipment and BCC 600 intercom system.

More recently, RH-ALAN have stated that the M-84A MBT is available with two different fire-control options, the FCS-84 or the FCS Omega-84.

The FCS-84 is believed to be the SUV-M-84 which was fitted to the M-84A MBTs built for Kuwait while the FCS Omega-84 is also fitted to the M-84A4 Snajper (Sniper).

Key components of the FCS Omega-84 include the DNZN-2 commander's day/night sight, UMZ commander's control module, SCN-84 stabilised gunner's sight, UMC computer control board, UIR control and regulator, INZ commander's indicator, UPC computer control board, ZB gyro unit, elevation and traverse mechanisms, RUB distribution control unit, LIRD laser irradiation indicator and warner, intercom device and smoke grenade launchers. The heart of the FCS Omega-84 is the DBR-84 fire-control computer.

The SCS-84 gunner's sight has a day channel with a magnification of ×10 and a 6° field of view, while the night channel has a magnification of ×7.5 and a 7.5° field of view. The laser range-finder has a maximum range of 10,000 m. The night channel can be second- or third-generation image intensification.

The DNZN-2 commander's sight has a day channel with a magnification of ×4.9 and a 10° field of view while the night channel has a magnification of ×4.3 and a 10.9° field of view. The night channel can be second- or third-generation image intensification. The commander also has two TNP-60 and two TNPA-65 periscopes.

Variants

M-84ABN

This is the M-84AB fitted with navigation equipment.

M-84ABK

This is the M-84AB fitted with extensive communications equipment, land navigation system and generator for the command role.

M-84ABI

This is the armoured recovery version of the M-84 series MBT and fitted with specialised recovery equipment. In appearance, the M-84ABI ARV is virtually identical to the Polish WZT-3 armoured recovery vehicle.

M-84A4 Snajper (Sniper)

This is understood to have a number of improvements including a new SCS-84 stabilised day/night sight, DBR-84 ballistic computer and improved elevation and traverse sensors.

The SCS-84 sight uses an integrated three-channel optical system, with its day sight being fed into the right monocular and its third-generation image intensifier night sight is fed into the left monocular.

The day sight has a ×10 magnification with a ballistic graticule to a range of 6,000 m. The sight's Nd:Yag laser range-finder has a maximum range of 9,900 m with an error of ±7.5 m.

The DBR-84 ballistic computer is designed to operate with all of the upgraded sensors, including a roof-mounted meteorological sensor. With automatic and manual data entry modes, the computer is said to be able to make calculations for APFSDS and HE-FRAG projectiles.

The M-84AB is fitted with a laser irradiation system with smoke grenade launchers for the jamming of guided missiles.

M-95 Degman MBT

This is a further development of the M-84 by Croatia and is covered in detail in a separate entry. In the long term, the Croatian M-84A and the Degman will have a common computerised fire-control system.

Specifications

Same as Yugoslav M-84 covered later in this section.

Status

Production as required. In service with Croatia.

Manufacturer

RH-ALAN.

UPDATED

CZECH REPUBLIC

T-72 CZ MBT

Development

In 1979-80 the then Czechoslovakia commenced licensed production of the Russian T-72 MBT with ZTS Martin undertaking production of the complete chassis and ZTS Dubnica nad Vahom responsible for the complete T-72 MBT turret.

As both of these companies are now in Slovakia there is a separate entry under Slovakia for this T-72 MBT and its many variants, some of which are in service with the Czech Republic.

Following a competition in 1995 the VOP 025 (Military Repair Plant 025) was selected to be the prime contractor for the Czech Republic T-72M1 upgrade programme.

Two prototypes of an upgraded T-72M1 MBT were completed by VOP 025, these being designated the T-72CZ M3 and the T-72CZ M4. The only major difference between the two vehicles is that the T-72CZ M4 has a new Israeli NIMDA power pack.

From an automotive aspect, the T72CZ M4 was by far superior as it had a higher power-to-weight ratio, greater acceleration and slightly higher top speed.

Trials of the upgraded T-72CZ M3/M4 have now been completed following problems encountered with integrating the Italian fire-control system in the turret.

T-72CZ M4 upgraded under the leadership of VOP 025 to meet the requirements of the Czech Army **2000**/0064633

Specifications
(Where different from standard T-72 MBT)

T-72CZ M3
Crew: 3
Combat weight: 46,000 kg
Power-to-weight ratio: 18.65 hp/t
Ground pressure: 0.902 kg/cm²
Length gun forwards: 9.55 m
Width: 3.755 m
Height to turret roof: 2.185 m
Maximum road speed: 60 km/h
Armament:
(main): 1 × 125 mm 2A46M smoothbore gun
(coaxial): 1 × 7.62 mm PKT MG
(anti-aircraft): 1 × 12.7 mm NSVT MG
Ammunition:
(main): 37
(coaxial): 2,000
(anti-aircraft): 720
Engine: V-46 TC diesel developing 858 hp at 2,000 rpm
Transmission: original
NBC system: yes
Night vision equipment: yes

T-72CZ M4
Crew: 3
Combat weight: 48,000 kg
Power-to-weight ratio: 20.83 hp/t
Ground pressure: 0.941 kg/cm³
Length gun forwards: 9.88 m
Width: 3.755 m
Height to turret roof: 2.180 m
Maximum road speed:
(forward) 61 km/h
(reverse) 14 km/h
Acceleration: 0 to 32 km/h in 8.5 s
Range:
(road) 600+ km
(cross-country) 400+ km
Armament:
(main): 1 × 125 mm 2A46M smoothbore gun
(coaxial): 1 × 7.62 m PKT MG
(anti-aircraft): 1 × 12.7 mm NSVT MG
Ammunition:
(main): 37
(coaxial): 2000
(anti-aircraft): 720
Engine: Perkins Engines Company CV-12 1000 diesel developing 1,000 hp at 2,300 rpm
Transmission: Allison XTG-411-6 fully automatic with 4 forward and 2 reverse gears
NBC system: yes
Night vision equipment: yes

It was expected that as many as 300 T-72 series MBTs will be upgraded to the enhanced standard but recent information has indicated that as few as 140 will be upgraded due to budgetary problems. As of January 2002, no orders had been placed for the T-72 CZ series of MBTs by the Czech Republic.

The Czech Republic T-72 CZ MBT is also referred to as the T-72 M40Z.

Description

Each T-72M1 has been upgraded in a number of key areas. To reduce target engagement times and improve first round hit probability of the 125 mm smoothbore gun under day and night conditions, a new Italian Officine Galileo TURMS-T (Tank Universal Reconfigurable Modular System) fire-control system has been installed.

This fire-control system is also installed in the Italian Ariete MBT and Centauro (8 × 8) tank destroyer and is also in quantity production for other armoured vehicles.

The commander and gunner have primary stabilised day/thermal sights with the gunner also having a laser range-finder. The commander's sight is the roof-mounted periscope type and allows the commander to find targets and then designate them to the gunner.

This enables hunter/killer target engagements to be carried out. Under night conditions, targets can be typically recognised at a range of 4,200 m and identified at a range of 2,100 m.

The heart of the fire-control system is the Turret Management ballistic Computer (TMC) which combines the functions of the ballistic computer and gun control system within a single unit. This replaces the existing K1 box and has seven function blocks which include analogue interface, resolver interface, digital and relay logic interface, communications interface, hydraulic driver power switching and power supply.

The Russian 125 mm 2A46M smoothbore gun fitted with a fume extractor and thermal shield is retained but a muzzle reference system has been installed.

The existing T-72M1 hydraulic gun control system is retained but digital control and filtering has been incorporated as has control and compensation of the existing mechanical backlashers of the MBT.

New tank commander's and gunner's joysticks have been fitted as have new sensors including turret position sensor, turret traverse accelerometer, meteorological sensor, ammunition powder temperature sensor, elevation transducer and turret gyro, vertical sensor and weapon gyro.

The commander is provided with a new fixed cupola on the forward part of which is mounted the stabilised panoramic sight with a new 12.7 mm NSV machine gun support on the right side.

The turret has been fitted with a Polish PCO SSC-1 laser warning system while battlefield survivability has been further enhanced by the installation of Explosive Reactive Armour (ERA) over the frontal arc. A German Kidde Deugra fire detection and suppression system has also been fitted.

T-72CZ M4 showing the new explosive reactive armour package fitted to the hull and turret 0045690

The ERA, called DYNA, has been designed in Poland and is claimed to give protection against both kinetic energy (for example armour piercing fin-stabilised discarding sabot) and chemical energy (High Explosive Anti-Tank) attack.

In addition to VOP 025, many local Czech Republic companies have also been involved in the T-72CZ M3 and M4 upgrade. Synthesia has developed enhanced 125 mm APFSDS-T ammunition while Letecke Pristroje Praha SRO has developed the NBV-97 inertial/Global Positioning System land navigation system and the DITA-97 diagnostic system.

Mounted on the front of the hull is the locally designed Metra Blankso SP system which has been designed to neutralise anti-tank mines fitted with a magnetic fuze.

Mesit Uherske Hradiste SRO has developed the internal communications system while Meopta Prerov AS has developed the driver's TKN-3P passive night vision periscope and DICOM has developed the external communications equipment.

As previously mentioned, there are two versions of the Czech Republic upgraded T-72CZ, the M3 and M4. The former retains the original power pack with the addition of two turbo-blowers for the engine and a modified transmission.

The T-72CZ M4 has the new NIMDA power pack which has been designed specifically for the upgraded T-72 MBT. This consists of a British Perkins Engines Company Condor CV-12 rated at 1,000 hp (the 1,200 hp version is standard on the Challenger 1 and 2 MBTs), coupled to a US-supplied Allison Transmission XTG-411-6 fully automatic transmission with four forward and two reverse gears, thermostat cooling system and a 650 A alternator.

This is the 23rd power pack to have been designed by NIMDA and can be removed in about 60 minutes. It is envisioned that if selected CKD local assembly of the engine while Praga would carry out local assembly of the transmission with VOP 025 being responsible for the complete power pack.

If required, the T-72M3 CZ and T-72M4 CZ can be fitted with the NBZ-90 articulated front-mounted dozer system to enabling the vehicle to prepare fire positions and clear battlefield obstacles. It can also be fitted with a device on the front of the hull to activate anti-tank mines fitted with a magnetic fuze.

New crew heaters have also been installed on the upgraded vehicle.

PSP T-72 MP MBT upgrade

Details of this upgrade, a joint development between PSP Bohemia AS of the Czech Republic, SAGEM of France and the Ukraine were given in *Jane's Armour and Artillery 2000-2001,* page 16. It never passed the prototype stage.

Status

Prototypes. Not yet in production or service.

Manufacturer

VOP 025 Nový Jičín.

UPDATED

T-55AM2 MBT

Development/Description

Caliber Prague Limited of the Czech Republic is offering for export surplus quantities of the upgraded T-55 series MBT, designated the T-55AM2.

This upgrade incorporates most of the improvements carried out some years ago to some T-55 MBTs operated by the former Warsaw Pact (see RFAS T-54/T-55 entry) with the exception of the ability to fire the Russian-designed and built 100 mm laser-guided projectile.

This upgrade retains the 100 mm rifled tank gun but has been improved in the key areas of armour, mobility and firepower.

Additional passive armour has been fitted to the hull and turret to provide for increased protection against mines and anti-tank guided missiles.

T-55A MBT fitted with explosive reactive armour package developed in the Czech Republic, this also has a new fire-control system with laser range-finder over 100 mm gun (Terry J Gander) 0045693

Specifications
T-55AM2
(Where different from standard T-55 MBT)
Crew: 4
Combat weight: 38,500 kg
Combat weight with additional diesel fuel drums: 38,900 kg
Length overall: 9.00 m
Height to met mast: 3.03 m
Width overall: 3.76 m
Max engine output at 2,000 rpm: 610 hp
Max driving speed: 50 km/h
Fording depth: 1.4 m
Width of fording: up to 1,000 m
Max fording depth: 5 m
Cross-country range with additional fuel drums: 410-450 km
Road range with fuel drums: 650-715 km
Range of 100 mm gun: 5,000 m max
Range of 7.62 mm MG: 1,800 m max

Other modifications include improvements allowing for more accurate firing, in the dynamic and driving qualities of the vehicle, the engine cooling and an increase in protection of the crew from the effects of mines, anti-tank guided missiles and missiles launched from low-flying aircraft and helicopters.

The key areas of modification can be summarised as follows:

TOH — thermal protection of the gun barrel
The installation of a thermal sleeve enhances shooting accuracy and protects the gun barrel against temperature changes which may cause warping or bending of the barrel.

DOK — improved protection of the chassis
Mounted on the glacis plate is a block of special passive armour that provides an increase in armour protection against attack by anti-tank guided missiles with a HEAT (High Explosive Anti-Tank) warhead.

DOV — improved protection for the turret
The turret front and sides have been strengthened by mounting semicircular blocks of passive armour of a special mass which protects against missiles fitted with a HEAT warhead.

ZOD — improved bottom resistance
The area below the driver's position has been fitted with additional passive armour protection against mines.

BOK — lateral chassis protection
The fuel tanks and maintenance equipment mounted on the running board on either side of the chassis, just above the suspension, have been fitted with a vertical flap for increased protection against missiles.

SODA — protection against the effects of napalm-type inflammable warfare agents
The exterior of the vehicle has been improved against inflammable agents. This refers to numerous modifications in the design of vulnerable external parts of the vehicle.

TUCA — electric firing of grenades
Mounted on either side of the turret rear is a bank of 81 mm electrically operated smoke grenade launchers that fire over the frontal arc of the T-55AM2.

ZVM — enhanced engine output
Modifications have been made to the diesel engine, for example changes in the angle of pre-injection and changes in aspiration and exhaust piping have enhanced engine output up to 610 hp.

ZVJ — improvement of suspension
To improve cross-country mobility, modifications have been made to the suspension including adjustments of the torsion bars, the rocker arms and the shock-absorber levers. In addition new torsion bars have been fitted.

KPS — tracks with rubber-metallic joint
The life of the tracks has been almost doubled by replacing the original all-metal tracks by tracks with rubber-metallic joint. Track guidance has been improved by the addition of a guiding pulley on the rear driving sprocket.

Barrels — additional fuel drums
To increase the operational range of the T-55AM2, permanent connection of additional diesel fuel drums to the fuel delivery system has been made.

Export T-55AM2 MBTs

In September 2000 Georgia took delivery of the first batch of 11 upgraded T-55AM2 Main Battle Tanks (MBTs) out of the 120 recently ordered from the Czech Republic.

The sale of the surplus T-55AM2 MBTs by the Czech Republic brings a number of advantages to the country: first, the tanks do not have to be broken up, which costs 67,000 crowns per vehicle; second, it brings in foreign exchange; and third, this currency helps pay for army modernisation.

The current Georgian MBT fleet consists of 48 T-55 and 31 T-72 series MBTs and this is the first time that the country has purchased significant quantities of weapons itself. In the past these have been gifts from Russia.

According to United Nations sources, between 1992 and 2000 the following quantities of T-55 were exported by the Czech Republic:

From	To	Quantity	Comment
Czech Rep	Cambodia	40	T-55 in 1994
Czech Rep	Georgia	10	T-55 in 2000
Czech Rep	Latvia	3	T-55 in 2000
Czech Rep	Sri Lanka	15	T-55 in 1996
Czech Rep	Sri Lanka	3	T-55 in 1997
Czech Rep	Sri Lanka	11	T-55 in 2000
Czech Rep	Yemen	97	T-55 in 2000

MT-55A Bridgelayer

This is based on a modified T-55 MBT chassis. Details are given in *Jane's Military Vehicles and Logistics 2002-2003.*

VT-55A ARV

This is also referred to as the MT-55A ARV and is fitted with a main winch with a capacity of 25,000 kg, an auxiliary winch with a capacity of 800 kg, a crane with a maximum lifting capacity of 1,500 kg and a front-mounted dozer/stabiliser blade. Details are given in *Jane's Military Vehicles and Logistics 2002-2003.*

VOP 25 Cancer combat engineer vehicle

VOP 25, the leading upgrade facility in the Czech Republic, is now marketing an upgrade package for the T-54/T-55 which is called Cancer 55 (Zenijni Tank 2 Generace — Pioneer Tank of the Second Generation).

The 100 mm gun has been removed and the position plated over. On the right side of the turret roof has been fitted a hydraulically operated crane.

The crane can be fitted with various attachments to carry out specific operations, for example, a shovel with a maximum capacity of 0.6 m^3.

To the immediate left of the crane has been fitted a raised commander's cupola with observation devices, single-piece roof hatch that opens to the rear and a roof mounted 12.7 mm machine gun.

Mounted at the front of the vehicle is a hydraulically operated dozer blade designated NBZ 90 which is controlled by the driver and has a maximum capacity, depending on soil conditions, of up to 220 m^3/h.

Full details are given in *Jane's Military Vehicles and Logistics 2002-2003.*

Civil conversions

Prototypes of a number of specialised civilian conversions of T-54/T-55 MBT chassis have been completed including the SDS-214 special construction machine with crane and various firefighting vehicles.

Modernised T-55 MBT

Some years ago the then Czechoslovak Army started to field modernised T-55s in significant numbers as

Modernised T-55AM2 showing additional passive armour installed on hull and turret for added protection
2002/0130606

insufficient T-72s were being built to replace the older tanks on a one-for-one basis.

The modifications included full-length track skirts which cover the upper part of the track, a thermal sleeve for the 100 mm gun, additional passive armour protection covering the frontal 180° of this turret (this being very similar to that fitted to Soviet T-62 tanks), 12.7 mm ammunition boxes stowed externally on the right side of the turret and a bank of forward-firing 81 mm smoke grenade dischargers installed on the right side of the turret to the immediate rear of the turret armour. Lateral protection is also provided for the diesel fuel tanks on either side of the hull above the roadwheels.

To improve the first round hit probability of the 100 mm gun, a new fire-control system was fitted which includes a laser range-finder mounted externally over the 100 mm gun, a ballistic computer and a meteorological sensor which is mounted on the turret roof at the rear. Most of these improvements are identical to those currently being offered on the export upgraded T-55AM2 MBT.

T-55 Armour upgrade

To improve the battlefield survivability of the widely deployed T-54/T-55 MBTs, a Czech Republic consortium is now marketing an Explosive Reactive Armour (ERA) package.

This consortium includes VOP 025 Nový Jičín armoured vehicle facility, Vyzkumny Ř Zkusebni Ustav PV 010 and Synthesia who supply the actual explosive element. This Czech-designed ERA package is quite different to that fitted to some Russian T54/T-55 series MBTs.

The ERA armour system for the T-55 comprises ERA boxes and panels attached to the glacis plate, both sides of the tank turret, left and right mudguards and the upper part forward and between the commander's and loader's roof hatches.

Extensive trials in the Czech Republic of a T-55A series tank fitted with this ERA package showed the following:

(1) A 100 to 200 per cent increase in protection against hand-held anti-tank weapons and missiles whose penetration effect is more than 650 mm
(2) Increase of 250 to 300 per cent in protection against shaped charge sub-munitions of the top attack type
(3) Increase of 100 to 140 per cent against artillery shaped charge projectiles
(4) It was also evident that there was some increase in ballistic protection against sub-calibre armour piercing projectiles.

Status

Production complete. The T-55AM2 is in service with a number of countries and quantities of vehicles are available for immediate delivery. Known recent customers include Cambodia (40), (Georgia will eventually have 120), Latvia (3), Sri Lanka (29) and the Yemen (97).

Manufacturer

State factories.
Enquiries to Calibre Prague Limited.

UPDATED

EGYPT

Egyptian production of the M1A1 MBT

Late in 1988 the USA and Egypt signed a Memorandum of Understanding (MoU) allowing Egypt to co-produce up to 555 General Dynamics Land Systems M1A1 MBTs. Under the terms of this MoU some 40 per cent of the tank's components would be manufactured in Egypt with the remainder being produced in the USA and shipped to Egypt for final assembly.

The first 25 M1A1 MBTs were shipped direct from the USA to Egypt and they were followed by partly assembled tanks and then kits.

The first co-produced M1A1 MBT rolled off the production line at the Egyptian Tank Plant in June 1992. The Egyptian Army M1A1 is identical to that built for the US Army except that it does not have the special heavy armour package as fitted to late production M1A1 MBTs.

Egypt will not be allowed to export any of the tanks without the permission of the USA and in the final stages of the project a maximum 35 per cent of the M1A1 was made in Egypt.

In April 1989, the Us DoD announced that General Dynamics Land Systems had been awarded a contract for Manufacturing Technical Assistance (MTA) to support the co-production of the M1A1 MBTs in Egypt.

Although the original intention was for a total of 555 M1A1s, this was recently reduced to 524, with the first 25 coming direct from General Dynamics Land Systems and the remaining 499 being assembled in Egypt and divided up into six increments:

(1) 75 produced from July 1992 through to August 1993 with the US providing the turret, welded hull and all hull components in kits. The Egyptians added brackets and harnesses to the engines and did assembly work on the power pack. In addition they also assembled the hull components, ground the race ring that holds the turret to the hull and did some initial painting
(2) 100 produced from September 1993 to September 1994 with additional work being performed in Egypt including welding 200 accessories to the hull and installing the rear grill door
(3) 92 vehicles produced from September 1994 through to August 1995 with additional tasks including hull machining, welding additional accessories, torsion bar tunnel welding, sponson installation, top deck fabrication, turret assembly and prime paint. During this period Factory 200 began to receive the 120 mm main armament and gun mount from Factory 100. The latter facility, also known as the Abu Zaabal Engineering Industrial Company, is the sole source in Egypt for artillery and tank barrels
(4) Covered 100 vehicles produced from September 1995 through to August 1996 with additional tasks in Egypt including turret machining, all accessory welding, ammunition door installation and turret prime and paint
(5) This covered 100 vehicles produced from August 1996 through to September 1997 with additional tasks performed including hull fabrication, plate flame cutting, plate straightening, all edge preparation, component subassembly for the hull and hull structures fabrication
(6) This was the final batch of 32 vehicles and was produced from September 1997 through to late 1998.

Assembly of the General Dynamics Land Systems M1A1 Abrams MBT underway at the Egyptian Tank Plant.
0036329

Early in 1996 Egypt decided to restore 31 M1A1s to the programme, so the total is now 555 vehicles.

In mid-1999, through the US government's Foreign Military Sales (FMS) programme, the Egyptian government requested a co-production programme with the US for the possible sale of 100 General Dynamics Land Systems M1A1 Main Battle Tanks (MBTs) at an estimated cost of US$564 million.

The request also includes 100 120mm M256 smoothbore guns, 100 12.7 mm M2 machine guns, 200 M240 7.62 mm machine guns, 12 5.56 mm M16A2 rifles plus spare and repair parts, support and test equipment, personnel training and training equipment.

It was expected that another 200 M1A1s would be built in Egypt but for budgetary reasons this has been cut back to 100 vehicles with another 100 vehicles to be procured in the future.

Early in 2000, General Dynamics Land Systems was awarded a US$156 million firm fixed contract for 100 M1A1 MBT kits for Egypt with deliveries running from January 2001 through to 2003.

General Dynamics Land Systems have confirmed that as the production line for the Textron Lycoming AGT 1500 1,500 hp gas turbine engine is now closed, these Egyptian M1A1s will be fitted with new engines from US Army stocks.

In the short term, part of the slack taken up by the ending of co-production of the M1A1 MBT at the ETP was taken up by co-production of 50 United Defense LP M88A2 Hercules Armoured Recovery Vehicles (ARVs).

Final assembly of these takes place in Egypt using components supplied by the US. The M88A2 ARV will be used to recover the M1A1 MBT.

It was revealed in 1999 that Egypt will be the launch customer for the Primex Technologies 120 mm KEW-A1 APFSDS-T (Armour Piercing Fin-Stabilised Discarding Sabot - Tracer) round.

Production commenced late in 1999 and final load, assembly and pack will be undertaken at the US Army Ammunition Plant at Middletown, Iowa. The actual projectile will be made at Primex Technologies Red

Lion facility in Pennsylvania. A total of 10,800 rounds were delivered by 2001.

In 2002 it was stated that the US Department of Defense was considering a request from Egypt to add an additional 100 General Dynamics Land Systems M1A1 Main Battle Tanks (MBTs) to its co-production programme.

If the deal goes ahead, it would be through the Foreign Military Sales (FMS) programme and be worth an estimated US$590 million. In addition to the 100 M1A1 platforms, Egypt also requested 100 120 mm M256 smoothbore tank guns (made by Watervliet Arsenal), 100 0.50 calibre M2 machine guns, 200 7.62 mm M240 machine guns, 12 5.56 mm M16A2 rifles plus other associated spare parts and training equipment.

When completed, the second batch of 100 M1A1 MBTs will bring the total Egyptian M1A1 fleet up to 755 vehicles.

Status

Production. A total of 555 M1A1 have been delivered to Egypt with another 100 now being built/assembled. These are expected to be followed by another batch of 100, making a grand total of 755 M1A1 MBT.

Manufacturer

Egyptian Tank Plant.

UPDATED

Egyptian tank fleet

As of early 2002, the Egyptian front-line tank strength is believed to consist of 1,000 T-54/T-55s, 500 T-62s, 1,459 M60 series MBTs (759 M60A3 supplied brand new and 700 M60A1 delivered in 1991/1992) and 555 M1A1 MBTs.

In the 1970s, several hundred Egyptian T-54/T-55/T-62 MBTs were fitted with German AEG/Telefunken white/infra-red searchlights to the right of the 100 mm main armament and some were also fitted with the Iskra laser range-finder. Some Egyptian T-62s have also been fitted with two launchers either side of the turret rear for surface-to-surface unguided smoke rockets.

Late in 1984, General Dynamics Land Systems was awarded a US$150 million contract to build a new tank plant outside Cairo. The facility is called Factory 200 and also overhauls US-supplied M60A3 MBT and M88A1 armoured recovery vehicles.

Factory 200 is also referred to as the Egyptian Tank Plant and this is where production of the M1A1 MBT was undertaken for the Egyptian Army.

Early in 1990, Egypt announced that it had accepted an offer from the United States for 700 M60A1 MBTs free of charge, with Egypt only having to pay for the shipping costs. All of these had been delivered by early 1992.

Egypt would like to upgrade the M60A1 MBTs to the more recent M60A3 standard, enabling its whole M60 MBT fleet to be of the same configuration.

T-55 fitted with BAE Systems, RO Defence 105 mm L7 rifled tank gun **2002**/0130607

General Dynamics Land Systems 120S

This is an upgraded M60 chassis fitted with the combat proved M1A1 120 mm turret. This has been developed by General Dynamics Land Systems and first shown in 2001. The 120S is aimed at the export market and a number of countries have been briefed on the vehicle, including Egypt and Turkey. Full details are given in this section under the US.

Egyptian production of M1A1 Abrams MBT

Details of this are given in the previous entry.

Ramses II MBT

In November 1984, Teledyne Continental Motors (taken over by General Dynamics Land Systems) of the USA was awarded a contract to upgrade the firepower and mobility of a single T-54 MBT. This was originally called the T-54E but was subsequently renamed Ramses II. The first prototype of the Ramses II was sent to Egypt for extensive firepower and mobility trials in January 1987 and these were completed in late 1987.

Late in 1989, Egypt signed a technical assistance agreement with TCM to support continued Egyptian testing of the Ramses II, with testing commencing in the summer of 1990. By early 2002 Ramses II MBT upgraded MBT had not entered production/upgrade for the Egyptian Army.

The rear hull has been modified to accommodate the new power pack consisting of a TCM AVDS-1790-5A turbocharged diesel developing 908 bhp (which has 80 per cent commonality with the engine installed in the M60A3 MBT) coupled to a Renk RK-304 transmission. Two new exhaust pipes, one either side of the hull rear, replace the single exhaust outlet in the left side of the hull.

The original suspension has been replaced by new General Dynamics Land Systems, Model 2880 in-arm hydropneumatic suspension units each fitted with an M48-type roadwheel, idler at the front, large drive sprocket at the rear, two new track-return rollers and US pattern tracks replacing the original Russian tracks.

A SABCA Titan Mk 1 laser fire-control system has been installed which includes a modified Avimo TL10-T sight incorporating the laser range-finder and an integrated in-eyepiece CRT alphanumeric graphic display. The fire-control system also includes an original SABCA double digital processor, an image intensification periscopic night sight, automatic attitude and atmospheric sensors and associated control boxes.

The gun and turret stabilisation system are provided by HR Textron Incorporated of the USA.

Other modifications include a fire detection and suppression system, new final drives, new fuel tanks, new air filtration system, British William Cook Defence (at the time called Blair Catton) tracks, new communications system and a new turret basket.

The original 100 mm gun has been replaced by the 105 mm M68 ordnance which is already fitted in Egyptian M60A3 MBTs, the original breech of the 100 mm DT-10T gun has been retained and modified and the recoil system has also been modified. A muzzle reference system is fitted as standard and an M60 day/night searchlight is mounted over the 105 mm gun.

Mounted either side of the turret is a bank of four electrically operated smoke dischargers. A collective type NBC system has been installed.

Egypt continues to purchase spare parts for its large M60 fleet from the United States and, early in 1997, purchased 30 M60 series engines from General Dynamics Land Systems, valued at US$5.943 million, with deliveries taking place by the end of 1997.

German modernised T-54

Under a contract awarded early in 1988, the German company Jung Jungenthal has modernised an Egyptian Army T-54. There is a separate entry for this MBT under Germany later in this section. As of early 2002 there were no known plans for this modernisation programme to move into the production phase.

Egyptian T-62 with 105 mm gun

The Egyptian Army has evaluated a T-62 MBT with its original 115 mm gun replaced by a 105 mm L7A1 gun modified by NORICUM of Austria and designated the TG N-105. As of early 2002 there were no known plans

Ramses II MBT undergoing mobility trials in the USA before shipment to Egypt 0018597

T-55 tank upgraded with a 105 mm L7 series gun supplied by BAE Systems, RO Defence; an infra-red/white light searchlight is mounted coaxially with the 105 mm gun 0018598

for this modernisation programme to move into the production phase.

Egyptian modernised T-34/85 tank

The Egyptian Army Number 5 Tank Workshop has modernised one example of the T-34/85 tank which it is now offering for export, or existing T-34s could be brought up to the new configuration in the user's own facilities. The first prototype of the upgraded T-34/85 tank was shown in late 1987.

According to the Egyptian Army, the 480 hp diesel of the T-34/85 lasted only 150 km before requiring an overhaul and gave a power-to-weight ratio of 15 hp/t.

In the upgraded T-34/85, the original engine has been replaced by a T-54 diesel developing 520 hp which gives a power-to-weight ratio of 16.25 hp/t and lasts 4,500 km before requiring an overhaul.

Other automotive improvements include modified fuel, air and cooling systems; the rear engine deck cover now has torsion bars to assist in opening; a modified final drive; control linkage and rubber dampers; and rubber side skirts have been fitted to reduce the amount of sand thrown up by the tracks.

An explosion and fire detection and suppression system has been installed in the crew compartment that can suppress a fire in 80 ms. The modified T-34/85 can also lay its own smoke screen by injecting diesel fuel into the rear exhaust pipes.

Egyptian M88A2 ARV co-production

In September 1998, the Ground Systems Division of United Defense LP was awarded a US$12.9 million Foreign Military Sales (FMS) contract by the US Army Tank-automotive & Armaments Command (TACOM) to provide Manufacturing Technical Assistance (MTA) to Egypt.

This contract was the first of two contracts which will lead to the co-production of 50 M88A2 Hercules heavy Armoured Recovery Vehicles (ARVs) at the Egyptian Tank Plant.

The second M88A2 contract provided the necessary parts and kits for co-production as well as the manufacturing and support required to carry out hull machining and welding work at the United Defence LP, Ground Systems Division in York.

This facility, which is currently building M88A2s for the US Army, will send hulls to the ETP who will undertake final assembly and delivery to the Egyptian Land Forces with final deliveries to take place in 2002. Total value of the Egyptian contract, including spares, is around US$200 million.

Manufacturer

Egyptian Tank Plant (ETP).

UPDATED

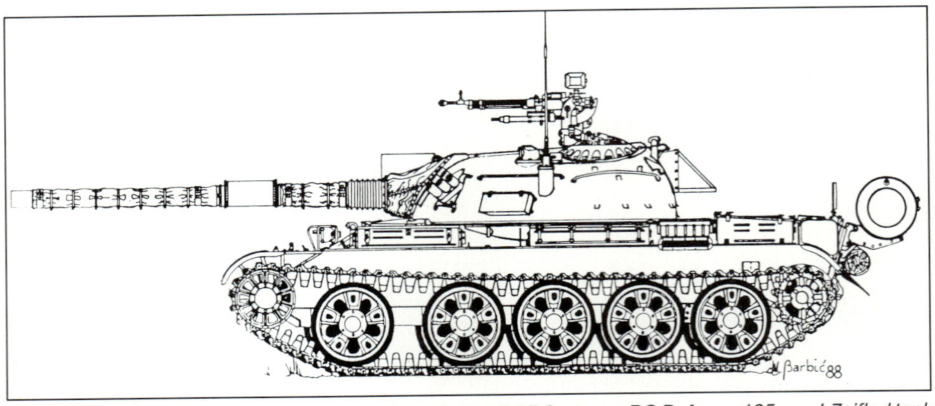

Upgraded T-55 MBT of the Egyptian Army armed with a BAE Systems, RO Defence 105 mm L7 rifled tank gun fitted with a fume extractor and thermal sleeve, smoke grenade launchers on turret sides, smoke grenade boxes, US-type radios and aerials and German AEG/Telefunken searchlight (Vasco Barbic)

FRANCE

Giat Industries Leclerc MBT

Development

Project definition on the Engin Principal de Combat (EPC) was completed by the Atelier de Construction d'Issy-les-Moulineaux in 1985 and by that time five test rigs were already being used in component development, one for the suspension, three for automotive trials and the fifth for the weapon system.

On 30 January 1986 it was announced that the EPC would be named the Leclerc.

Project definition for the Leclerc was completed in mid-1986 and in the same year it was decided to build six complete prototypes of the vehicle; the first of these was completed late in 1989. The Leclerc made its first public appearance at the 1990 Satory exhibition of defence equipment for the ground forces.

The first production Leclerc MBT was completed by Giat Industries in December 1991 and handed over to the French Army on 14 January 1992.

A total of six Giat Industries facilities is involved in the production of the Leclerc MBT: Toulouse, Saint-Chamond, Tulle, Bourges, Tarbes and Roanne.

Turret construction and integration are undertaken at Tarbes while Roanne builds the complete chassis and then integrates this with the turret. The completed Leclerc is then tested and delivered.

Each French Army Leclerc MBT regiment has a total of 40 vehicles, one at regimental HQ and three squadrons each with 13 vehicles, one at squadron HQ and three troops each with four vehicles.

The first regiment equipped with the new Leclerc MBT is the 501-503Rd RCC.

In 2001, the French DGA placed a contract with Giat Industries for a batch of 52 Leclerc Main Battle Tanks (MBTs) for the French Army. This new order will extend production of the Leclerc at Roanne through to 2006, assuming that no additional export orders are placed.

This is the last order to be placed by the DGA for the French Army. Under current contracts Giat Industries is building 354 Leclerc MBTs with final deliveries due in 2003. This brought the total number of Leclerc MBTs, for the French Army up to 406.

The last batch of 52 Leclerc MBTs will be to the latest build standard known as the Block II+ (or Block III internally at Giat Industries). At one time over 1,000 Leclerc MBTs were to have been built for the French Army, but the end of the Cold War and restructuring have reduced the total to 406 MBTs, plus the armoured recovery vehicles.

Tropicalised Leopard of the United Arab Emirates deployed in Kosovo **2000**/0064637

Giat Industries Leclerc MBTs of the French Army during training exercises in Qatar **2002**/0130610

Giat Industries Tropicalised Leclerc MBT during trials in the Middle East **2002**/0130609

United Arab Emirates

Following an international competition, early in 1993 the United Arab Emirates (Abu Dhabi) selected the Giat Industries Leclerc MBT to meet its future requirements and placed an order for a total of 436 vehicles for delivery from 1994.

This order comprises 388 Leclerc MBTs, two driver training tanks and 46 armoured recovery vehicles. In addition the contract includes ammunition, spares, training equipment and logistic support.

The first two production Leclerc MBTs for the United Arab Emirates were flown direct to the country late in 1994 in an Antonov 124 transport aircraft while the remaining three were delivered in December 1994 to the French armour training school at Saumur where they were used for tactical training.

According to the United Nations, deliveries of the Leclerc MBT to the UAE from Giat Industries were for the period 1994 to 2000 as follows:

Quantity	Year
5	1994
25	1995
58	1996
63	1997
64	1998
62	1999
32	2000

This version, or Tropicalised Leclerc as it is also referred to, has a number of modifications including the installation of a EuroPowerPack consisting of an MTU 883 1,500 hp V-12 diesel engine coupled to a Renk HSWL 295 TM automatic transmission with five forward and three reverse gears, new cooling and air filtration system, additional armoured side skirts, air conditioning system, diesel auxiliary power unit, modified turret stowage, commander's sight with laser range-finder, thermal channel and installation of a battle management system, which is now called FINDERS (Fast Information, Navigation, Decision and Reporting System).

The hull of the Leclerc has been slightly extended at the rear to enable the MTU power pack to be installed and this has enabled the fuel capacity of the vehicle to be increased. An air conditioning system has also been fitted. As the German engine develops 1,500 hp at 2,700 rpm in this application, the final drives of the Leclerc have been modified, as have the vehicle's electronics.

FINDERS has a colour map display giving the position of the Leclerc MBT as well as other friendly MBTs and can also be used for message handling, route planning and the designation of enemy targets.

Description

The Leclerc MBT has a conventional layout with the driver sitting at the front, the turret in the centre and the power pack at the rear.

The driver is seated at the front left of the vehicle with reserve 120 mm ammunition in a drum to his right. He has a single-piece hatch cover that opens to the left which is provided with three day periscopes, the centre one of which is a TTD OB-60 day/night periscope. The driver steers the vehicle with a conventional steering wheel.

The gunner of the Leclerc is seated on the right of the turret with the commander on the left, both being provided with a single-piece hatch cover. The commander has eight day periscopes and the gunner is also provided with day periscopes.

Main armament comprises a Giat 120 mm smoothbore gun designated the CN 120-26 (or F1) which has a chamber the same size as that of the 120 mm smoothbore gun fitted to the Leopard 2, M1A1 and M1A2.

The French gun, however, has a calibre length of 52 compared with the 44 of the other weapons and therefore a higher muzzle velocity of 1,790 m/s when firing APFSDS projectiles. The gun is fitted with a thermal sleeve and a compressed air system for blowing out propellant fumes. A dynamic muzzle reference system is fitted.

Mounted in the turret bustle is the automatic loader which contains 22 rounds of ready-use ammunition normally loaded from outside through a small hatch in the back of the turret bustle. The autoloader can distinguish between six different types of ammunition and will automatically recognise and respond to ammunition selection commands.

The 120 mm gun is automatically depressed to −1.8° for reloading and the automatic loader is separated from the turret by bulkheads with blowout panels being provided in the roof.

In addition, it can be loaded by the gunner using the 18 rounds stowed in the drum located to the right of the driver.

The installation of the automatic loader enables a cyclic rate of fire of 10 rds/min to be achieved with 10 to 12 rds/min being the effective rate of fire.

There are two main types of 120 mm ammunition fired, APFSDS with a muzzle velocity of 1,790 m/s and HEAT with a muzzle velocity of 1,100 m/s, both having a semi-combustible cartridge case containing a dual-base solid propellant. Its primer is also combustible. A DU 120 mm APFSDS round called OFL 120 F2 is now in production for the French Army. Giat Industries is also developing a 120 mm anti-helicopter round. The APFSDS round is called the OFL 120 F1 and has a tungsten penetrator with a length-to-diameter ratio of 20:1. It is understood that a new DU 120 mm round is to be introduced in the future called PROCIPAC (Projectile Cinetique a Penetration Accrue).

A 12.7 mm machine gun is mounted coaxially with the main armament and a 7.62 mm machine gun is mounted externally on the turret roof. The 7.62 mm anti-aircraft machine gun can be aimed and fired under complete armour protection.

Mounted along either side of the turret roof towards the rear are seven launcher tubes which can launch various grenades, for example smoke or anti-personnel. In addition, four vertical launchers are dedicated to the launching of infra-red decoys.

The fire-control system includes a modular thermal imaging camera, Avimo HL-58 laser range-finder and stabilised sights for the commander and gunner.

The SAGEM HL 60 gyrostabilised gunner's sight is an essential part of the fire-control system and is linked mechanically to the gun.

It combines low-drift gyroscopes, high-precision digital angular sensors, digital processors and advanced sensors, an Avimo laser range-finder and thermal camera. These components are arranged in three separate modules, the stabilised sight head, the acquisition system and an electronics unit.

The stabilised head of the sight supplies the attitude and vertical references necessary for on-the-move firing calculations and the cardinal reference for north bearing navigational purposes.

It incorporates a large mirror made of beryllium-based alloy which has a lower inertia and better mechanical strength than conventional glass mirrors and can transmit good images in the visible region as well as in the near and far infra-red. The mirror is stabilised about two axes by a gyroscope and an accelerometer. A second gyroscope and two accelerometers are used to provide the vertical and navigational references without recourse to magnetic north.

The acquisition system incorporates a direct day channel with magnifications of ×3.3 and ×10, video day channel with magnification of ×10, thermal infra-red channel which allows for detection at 5,000 m, recognition at 2,500 m and identification at 2,000 m, laser range-finder channel, boresighting channel and an emergency mode with operation of the mirror.

The modular, fully digital electronic unit is connected to the databus and the unit calculates and provides the angular position and speed of the target with a precision of 0.1 mrad.

The gunner's HL 60 sight is also linked to visual displays at the commander's station so enabling the tank commander to fire the main armament using his duplicate controls.

The commander has a SAGEM roof-mounted stabilised panoramic HL-70 sight which can incorporate a laser range-finder, eight day periscopes and three control panels. The first panel is a master control and monitoring unit and has five systems: start-up switches for the turret services, fire-control system, sights, thermal imager and automatic loader. Using automatic built-in test equipment, all the systems are operational 1 minute after switch-on over the full operational temperature range.

The second panel contains the main 120 mm armament controls and warning lights and controls the use of the automatic data transmission system. This allows the burst transmission of standardised format messages which can be modified using a keyboard entry system.

The third panel contains the secondary armament controls and the sensor inputs. These include the meteorological sensor, ammunition selector (up to six types can be programmed) and a repeat fire selector which reloads the same ammunition type. This panel also contains the vehicle screening smoke and self-defence grenade firing buttons.

The VDU, in addition to displaying the thermal picture from the gunner's sight, is the crew interface with the vectronics.

Giat Industries Leclerc MBT showing fire on the move capability **2000**/0064636

The tank commander's SAGEM HL-70 gyrostabilised panoramic sight has an intermediate unit, an aiming sight and an electronic unit. The sight head includes a Galilean system for changing the field of view (20° or 5°) for magnifications of ×2.5 and ×10. It also includes a second-generation image-intensifying night channel with a magnification of ×2.5 and a 13° field of view. The sight can recognise a target at 4,000 m and identify it at 2,500 m. A thermal camera can be fitted if required.

The tank commander also has a 15 cm TV monitor that repeats the image seen in the gunner's sight and, in particular, the image from the thermal camera.

Each of the commander's eight day periscopes has a switch which, when activated, brings the gun onto its line of sight.

The gunner has two control panels, a VDU, three observation periscopes and his main sight. The controls provide all of the functions for the tank commander with the exception of the turret services start-up switches.

The fire-control system enables six different targets to be engaged in 35 seconds and allows the tank to engage both stationary and moving targets while moving across country itself. First round hit probability is being quoted as 95 per cent. Typical engagement cycle time will be 4 to 6 seconds.

A battle management system is installed together with a multiplexed numeric datalink with three computers. The systematic use of digital electronics in all subsystems of the Leclerc and turret means that the tank uses some 30 central processing units with 8-, 16- or 32-bit microprocessors. The crew will have access to all the subsystems of the tank. For example, the tank commander can find out the amount of fuel remaining in the fuel tanks just as easily as the driver, the number of rounds left in the automatic loader or where the tank is at any given time.

Positioning data is upgraded as the Leclerc moves and rendezvous points for replenishment can be anticipated. The vehicle can automatically transmit combat status, location data and logistical data to field command vehicles.

The shell of the hull and turret of the Leclerc MBT is of welded steel construction to which the composite armour will be added; this provides a very high degree of protection against both KE and CE attack.

The armour system is of a modular design so that it can be easily replaced or upgraded allowing new armour developments to be incorporated in the future when this becomes available. The high power-to-weight ratio of the Leclerc will enable the tank to be up-armoured to meet new threats without substantially lowering its power-to-weight ratio.

Giat Industries 120 mm ammunition for Leclerc MBT, from left to right HEAT-MP (training), APFSDS, HEAT-MP and APFSDS (DU). The bands around the projectile are bar codes for the automatic loader 0018601

Turret traverse and gun elevation are electric and both the commander and gunner are able to lay and fire the main and secondary armament. Manual turret controls are provided for emergency use.

Prime contractor for the gun power equipment is Matra Defense and the turret can be slewed through 180° in less than 6 seconds.

The French Army Leclerc is powered by an SACM V8X-1500 Hyperbar diesel developing 1,500 hp coupled to an SESM ESM 500 fully automatic transmission with five forward and two reverse gears and a hydrostatic transmission.

The Suralmo-Hyperbar process involves adding an auxiliary combustion chamber to the engine together with a bypass circuit of variable cross-section. The result is the equivalent of a gas turbine.

The engine is managed by an electronic system supplied by SAGEM with the computer determining the fuel requirements, starting procedure sequences and monitoring the engine.

A Turbomeca TM-307B acts as an auxiliary power unit when the main engine is shut down to conserve fuel.

After evaluating both hydropneumatic and torsion bar with rotary damper suspension systems the former was selected for the Leclerc with SAMM being the prime contractor. The Leclerc has six roadwheels, the drive sprocket at the rear, idler at the front and track-return rollers.

Standard equipment on the Leclerc is a new design of NBC system, deep fording, fire detection and suppression system for both crew and engine compartments, and a communications system from Thales.

Variants

Leclerc Block I

The first batch of 132 Leclerc MBTs for the French Army were called Block I and delivered between 1992 and 1996. These were optimised for a Central European battlefield environment.

Leclerc Block II

A total of 178 Block II versions are being delivered to the French Army, with deliveries running from 1997 through to 2003. These have a number of enhancements including a climate-control system for worldwide deployment, enhanced software, add-on armour plates to the hull sides and an oil cooler for the final drive.

Five Leclerc Block II MBTs undertook an extensive series of trials in Qatar in mid-1989 in temperatures of up to +54°C.

Leclerc Block II+

Within Giat Industries this is also referred to as Leclerc Block III and will have a number of improvements, including improved passive armour over the turret front and sides. This means that the turret stowage arrangements will be changed.

The commander and gunner's sights will incorporate the SAGEM second-generation FLIR camera, called Iris, for enhanced target acquisition under almost all weather conditions.

Artist's impression of Leclerc PTG armoured vehicle-launched bridge launching its scissors bridge over the front 0018602

Giat Industries Leclerc ARV ***2000***/0064639

Giat Industries Leclerc ARV fitted with the K2D mineclearing kit 0018603

In addition it will also be fitted with a battle management system called Icone. This is a further development of the SIT-V1 system ordered by the French Army in 2000 for its AMX-10RC (6 × 6) armoured cars and Panhard VBL (4 × 4) scout cars.

It will also be fitted to receive the Thales Battlefield Identification Friend or Foe (BIFF) system. Giat Industries hope that, funding permitting, the French Army will decide to upgrade earlier vehicles to the enhanced Block II+ standard.

Future improvements to Leclerc MBTs of the French Army

These could include:

2002	enhanced command and control with battle management system and fire detection and suppression system
2006	battle identification-friend-or-foe, defensive aids suite and new thermal imager
2008	enhanced armour, automatic target tracking and new integrated command and control system

140 mm turret

Giat has designed a turret for the Leclerc that could accommodate a 140 mm smoothbore gun fed by a bustle mounted automatic loader.

Leclerc ARV

The first prototype of the Giat Industries Leclerc ARV was completed in the spring of 1994 and shown for the first time at the EuroSatory 94 exhibition in June 1994.

Giat Industries built two prototypes of the Leclerc ARV, or DNG (Depanneur Nouvelle Generation) as it is also referred to. The first customer is the United Arab Emirates which has ordered 46. The French Army has placed an initial order for 15 Leclerc ARVs with the total requirement being for 30 (15 plus 15 with the last batch being 5 + 5 + 5) vehicles as the current AMX-30D ARV cannot recover the heavier Leclerc MBT. Production of this vehicle is now under way and first vehicles have been delivered to the UAE.

The Leclerc ARV has been designed to carry out three missions: recover a disabled tank, tow a broken down or damaged tank and carry out repairs in the forward area, for example change a complete power pack.

The Leclerc ARV is based on automotive components of the export version of the Leclerc MBT, including the power pack, but has a longer chassis with seven roadwheels either side (the Leclerc MBT has six) and a new superstructure. French Army Leclerc ARVs have the same powerpack as those of the United Arab Emirates.

The SAMM hydropneumatic suspension system is retained, as is the track tensioning system that allows the driver to adjust the track tension without leaving his position.

Mounted at the front of the Leclerc ARV is a hydraulically operated blade which can be used as a stabiliser when the crane or winch is being used, or as a dozer blade to clear battlefield obstacles.

The hydraulic crane has a maximum capacity of 30,000 kg while the main winch has a maximum capacity on a single line pull of 35,000 kg and is provided with 180 m of cable. An auxiliary winch is fitted with a capacity of 1,500 kg.

The recovery equipment used in the Leclerc ARV has been supplied by the now Rheinmetall Landsysteme of Germany and is the same as that in the Büffel ARV in service or on order for Germany, Netherlands, Spain, Sweden and Switzerland.

Full details of the Leclerc ARV are given in *Jane's Military Vehicles and Logistics 2002-2003.*

Leclerc ARV with K2D mineclearing kit

In 1997, Giat Industries of France promoted the chassis of the Leclerc MBT for a much wider range of roles under the designation E-Force, with the E standing for Engineer.

This covers the Leclerc ARV covered previously, K2D mineclearing kit, EPG Armoured Engineering Vehicle and the PTG armoured bridge launching vehicle.

The K2D mineclearing kit can quickly be fitted to the Leclerc ARV. Mounted at the front of the hull is a British Pearson Engineering Full Width Mineplough (FWMP) which incorporates the latest DEMETER electromagnetic signature duplicator from Giat Industries.

This is designed to trigger magnetic mines ahead of the vehicle. The vehicle also has the Pearson Engineering Pathfinder dual minefield marking system which has been specially designed for this application. Each unit has 100 reusable markers which are fired into the ground using compressed air and allow the following units to determine the cleared path through the minefield.

Mounted either side at the rear of the hull is a Polish Pronit rocket propelled mineclearing system which fires a rocket to which is attached an explosive charge across the minefield.

This falls to the ground and is then detonated so clearing any mines by creating an overpressure. Each

Giat Industries Leclerc of the French Army (Pierre Touzin) ***2002***/0130611

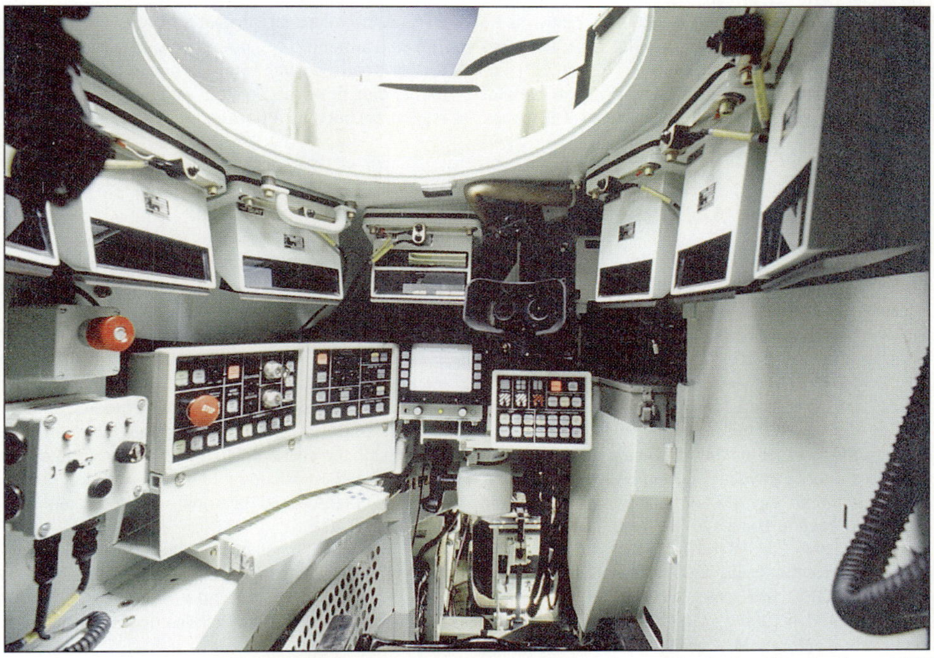

Giat Industries FINDERS battle management system installed in a Giat Industries Leclerc MBT ***2002***/0130608

Specifications

French Army model
Crew: 3
Combat weight: 56,500 kg
Power-to-weight ratio: 26.54 hp/t
Ground pressure: 0.9 kg/cm^2
Length:
(gun forward) 9.87 m
(hull) 6.88 m
Width: 3.71 m
Height:
(turret roof) 2.53 m
Ground clearance: 0.5 m
Length of track on ground: 4.318 m
Max road speed: 72 km/h
Max cross-country speed: 50 km/h
Max reverse speed: 38 km/h
Acceleration: 0/32 km/h in 5.5 s
Fuel capacity: 1,300 litres (1,700 litres with external fuel)
Max road range:
(with internal fuel) 450 km
(with external fuel) 550 km
Vertical obstacle:
(forwards): 1.25 m
(reverse) 0.6 m
Trench: 3 m
Fording: 1 m
Fording with preparation: 4 m
Engine: SACM V8X-1500 8-cylinder Hyperbar diesel developing 1,500 hp at 2,500 rpm
Transmission: SESM ESM 500 automatic with 5 forward and 2 reverse gears, microprocessor-controlled hydraulic multiple plate clutch
APU: Turbomeca TM-307B gas turbine
Suspension: hydropneumatic
Vertical roadwheel travel: 300 mm
Roadwheel rebound: 135 mm
Armament:
(main) 1 × 120 mm F1 smoothbore gun
(coaxial) 1 × 12.7 mm MG
(anti-aircraft) 1 × 7.62 mm MG
Smoke-laying equipment: 2 × 7 either side of turret
Ammunition:
(main) 40 (22 ready use)
(12.7 mm MG) 950
Gun control equipment
Turret power control: electric/manual
(by commander) yes
(by gunner) yes
Turret traverse: 360°
Max turret traverse speed: 40°/s
Max gun elevation speed: 30°/s
Commander's fire-control override: yes
Gun elevation/depression:
(main armament) +15/–8°
Gun stabiliser:
(vertical) yes
(horizontal) yes
Range-finder type: yes, laser
Elevation quadrant: yes
Traverse indicator: yes
NBC system: yes
Night vision equipment: yes

Giat Industries Leclerc MBT of the French Army ***2002***/0130612

Pronit clears a path about 110 m long and between 4 and 8 m wide depending on the type of mine.

The Pathfinder and Pronit are mounted on a subframe which can be quickly fitted or removed from the Leclerc ARV; it can also be fitted to the Leclerc EPG engineering vehicle. Another possible alternative is the installation of a module carrying the Giat Industries Minotaur mine scattering system.

These modules are exchanged using the vehicles onboard crane which can lift a maximum load of 30 tonnes.

The K2D can quickly be installed on the Leclerc ARV after the removal of the front-mounted dozer blade. The crane and winches are retained so allowing the vehicle to recover damaged and disabled vehicles.

EPG armoured engineer vehicle

The EPG armoured engineering vehicle is similar to the ARV and has the same hydraulic arm and winches. The former can, however, be fitted with specialist engineer equipment such as a bucket, auger, grapple, hammer and hook. The front-mounted dozer blade can be used for a variety of roles. The vehicle can also carry fascines to lay in ditches and other obstacles.

The first prototype of the EPG, built using company funds, was completed late in 2001.

Optional equipment includes explosive reactive armour, remote-control capability, thermal viewing device for driver and the FINDERS (Fast Information Navigation DEcision and Reporting System). Full details of this are given in *Jane's Military Vehicles and Logistics 2002-2003.*

PTG armoured vehicle launched bridge

The PTG armoured bridge laying vehicle consists of the Leclerc MBT with turret removed and fitted with a UBLE (Universal Bridge Launching Equipment) which is a hydraulically driven mechanical bridge laying system developed in co-operation with Vickers Bridging of the UK.

This can launch No 10 (26 m) and No 12 (13.5 m) bridges over the front of the vehicles which are already in production for the British Army as part of the BR 90 programme. These bridges are to MLC 70 (Military Load Class 70) standard.

Status

Production. In service with:

Country	Quantity	Comment
France	406	delivery by 2006
UAE	436	including ARV

Prime contractor

Giat Industries.
The Leclerc turret is manufactured at Tarbes while the hull is manufactured at Roanne where integration of hull and chassis takes place.

UPDATED

Giat Industries AMX-30 MBT

Development

Design work was carried out on the AMX-30 MBT at the Atelier de Construction d'Issy-les-Moulineaux, with the first two prototypes being completed in 1960. A further seven prototypes built with the Hispano-Suiza HS 110 12-cylinder multi-fuel engine, which was also fitted to production tanks, were completed early in 1963 and in July of that year the tank was adopted by the French Army as the replacement for its US-supplied M47 tanks. Prior to production two preproduction tanks were built.

Production of the AMX-30 began in 1966 at the Centre de Roanne and the AMX-30 was also built under licence in Spain (qv). In 1987, Cyprus placed an order with France for the supply of 15 AMX-30 B2 MBTs and a single AMX-30D armoured recovery vehicle; these were delivered in 1988. Late in 1989 Cyprus placed an order with Giat for the supply of a further 35 AMX-30 B2 MBTs and another AMX-30D armoured recovery vehicle.

AMX-30 series MBTs were used by France, Saudi Arabia and Qatar during operations conducted early in 1991 to retake Kuwait following its invasion by Iraq in 1990.

Production of the AMX-30 series was completed late in 1993 when the last AMX-30 EBGs were completed for the French Army. Late in 1994 the chassis went back in production as the French Army ordered another batch of 20 GCT 155 mm self-propelled howitzers.

Although the Giat Industries 120 mm armed Leclerc MBT is being introduced into service with the French Army the older AMX-30 B2 is now expected to remain in service until 2010.

For this reason it has been upgraded in a number of key areas including the installation of Giat Industries explosive reactive armour and a new RVI Mack diesel engine. Additional details of these upgrades are given later in this entry.

Giat Industries completed production of the AMX-30 series of MBTs some years ago and according to the United Nations arms transfer list, between 1992 and 2000 the following quantities were exported:

From	Country	Quantity	Date
France	Chile	21	1998
France	Chile	10	1999
France	Qatar	10	1998
Greece	Cyprus	15	1996
Greece	Cyprus	27	1997
Greece	Cyprus	10	1997

In 2000 and 2001 it was reported that Saudi Arabia was considering passing on some of its fleet of AMX-30 MBTs to other countries. Tunisia was to receive 30 vehicles with the Lebanon also expected to receive some vehicles.

Description

The hull of the AMX-30 is made of rolled steel plates welded together. It is divided into three compartments:

driver's at the front, fighting in the centre and the power pack at the rear.

The driver is seated at the front of the vehicle on the left with a single-piece hatch cover opening to the left and three day periscopes. The centre periscope, depending on the model of the tank, can be either a day periscope which can be replaced by an image intensification night periscope (Thales TH 9478), or a SOPELEM OB-16A periscope. This has a binocular system for infra-red light by night and a monocular system for day driving. The infra-red system has a magnification of ×1 and a 35° field of view and the day system has a similar magnification and a 24° field of view.

The other three crew members are seated in the turret, with the commander and gunner on the right and the loader, who also operates the radio, on the left. The commander's cupola is a TOP 7 with 10 day periscopes for all-round vision and a single-piece hatch cover opening to the rear. Mounted on the forward part of the commander's cupola is a SOPELEM M 270 prism head. This comprises an armoured housing and a thick glass behind which is a prism, swivelling in elevation and reflecting the image of the terrain towards the M 267 day sight or the OB-23A infra-red telescope (magnification ×4, 9° field of view), the object lenses of which fit into its lower part. On the right of the prism is a swivelling arm which supports the 7.62 mm machine gun and its PH-9-A infra-red searchlight, it has a 500 m range when being used in the infra-red mode and a 700 m range in the white light mode. The weapon can be elevated from -10 to +45° by a handwheel in the turret roof. The prism may also be used to aim the coaxial 20 mm cannon, in which case the head (of the prism) is electrically servo-controlled by the elevation swivel of the 20 mm cannon. The M 270 has a magnification of ×10 and allows the tank commander to locate and identify targets and bring the turret to bear onto the target. The tank commander also operates the SOPELEM M 208 range-finder with a magnification of ×6 and a range of 600 to 3,500 m.

Mounted on the left side of the turret, coaxial with the main armament, is a SOPELEM PH-8-B searchlight, which has a maximum range of 2,000 m when used in the white light mode and a range of 1,000 m in the infra-red mode.

The gunner, who is seated below and in front of the commander, has an M 271 day sight with a magnification of ×8 which can be changed for an OB-17A night sight. This is mounted in the roof of the turret and has a luminous graticule, magnification of ×5.4 and a 7° field of view. When used with the PH-8-B infra-red searchlight it has a range of 800 m. The gunner also has two periscopes.

The loader is seated on the left of the turret and is provided with two periscopes and a single-piece hatch cover that opens to the rear. A small circular hatch on the left side of the turret is used for replenishing ammunition and ejecting spent cartridge cases manually.

The engine is to the immediate rear of the bulkhead separating the engine compartment from the fighting compartment. The Hispano-Suiza HS 110 engine was manufactured by Renault and operates on oil, petrol or paraffin. The engine is water-cooled and air is drawn in above the chassis in the rear part of the tank and blown down through the radiator by a fan driven by the engine.

The complete power pack, comprising the engine, combined gearbox and steering unit, and clutch assembly, can be removed in 45 minutes by a team of three. The power pack can also be run outside the tank before installation.

The transmission consists of an automatic clutch, combined gearbox and steering unit, brakes and two final drives. In the basic version the centrifugal-type clutch is activated electrically by the gear shift lever; a non-synchronised reverser enables the same number of gears to be engaged in reverse as forwards. The combined gearbox and steering mechanism contains the mechanically operated gearbox giving five speeds both forward and in reverse and a triple differential steering system.

The brakes are hydraulically operated and are used as both service and parking brakes. Each final drive comprises spur-type right-angle gears and an epicyclic gear train.

The torsion bar suspension either side consists of five rubber-tyred twin roadwheels with the idler at the front and the drive sprocket at the rear. There are five track-return rollers which support the inside of the track only. The first, second, fourth and fifth roadwheels are mounted on bogies and the first and fifth roadwheels are provided with hydraulic shock-absorbers. The centrally guided steel track has removable rubber pads and each track weighs 1,580 kg. When new it has 83 links.

The AMX-30 can ford to a maximum depth of 1.3 m without preparation and 2 m with preparation. When deeper rivers are encountered a snorkel is erected over the loader's hatch. Two types of snorkel are available, a wide one for training and a much thinner one for operations, similar in concept to the snorkels fitted to the Russian T-54/T-55 and T-62 MBTs. Before entering the water a ring is inflated around the turret, mantlet and cupola using the electrically driven compressor and, two blanking plates are fitted over the engine compartment louvres; these are carried on the right side of the glacis plate when not in use.

Standard equipment on all AMX-30s includes a battery-operated electric pump for refuelling and a lubricating pump using the tank's compressed air circuit. The power receptacle in the driver's compartment can be used for recharging the tank's batteries from another tank and also for supplying the electric pump used for refuelling. The tank is also equipped with an NBC system, heater, automatic fire alarm system, radios, crew intercom system and an infantry telephone at the rear.

Main armament of the AMX-30 is a Giat Industries 105 mm smoothbore gun designated the CN-105-F1 with a length of 56 calibres. The gun does not have a muzzle brake or a fume extractor but is fitted with a magnesium alloy thermal sleeve. A compressed air system evacuates any fumes from the barrel. The recoil system consists of two diametrically opposed hydraulic brakes and an oleo-pneumatic recuperator for counter-recoil of the barrel.

The 105 mm gun can fire APFSDS, HEAT, HE, phosphorus smoke or illuminating rounds of a French design. It can also fire standard 105 mm ammunition as used with the L7 series of weapons mounted in the Leopard 1 and M60 series of MBTs. A 105 mm DU round has been developed by Giat Industries. This is now in service with the French Army but has not been released for export.

In total 47 rounds of 105 mm ammunition are carried, 19 of which are in the turret (18 in the bustle) and 28 in the hull to the right of the driver.

Mounted to the left of the main armament is a Giat Industries 20 mm Model F2 (Type M693) cannon, which can be elevated with the main armament and also on its own to a maximum of +40° for use against slow-flying aircraft and helicopters. The cannon has a maximum effective range of 1,500 m and can be fired by the gunner or tank commander. The 20 mm cannon can be either dual feed (with HEI rounds with a muzzle velocity of 1,050 m/s and armour-piercing rounds with a muzzle velocity of 1,250 m/s), or single feed firing American M56 type ammunition. In total 480 rounds of ready use ammunition are carried, with a further 550 rounds held in reserve. When originally introduced into service a 12.7 mm M2 machine gun was mounted to the left of the main armament. This was provided with 600 rounds of ammunition.

Mounted to the right of the commander's cupola is a 7.62 mm Model F1 machine gun which can be aimed and fired from within the cupola. The weapon has an elevation of +45° and a depression of −10°. In total 2,050 rounds of 7.62 mm ammunition are carried, of which 550 are ready for immediate use. The empty cartridge cases are automatically ejected outside the tank. Maximum effective range of the 7.62 mm machine gun is quoted as 700 m.

There are two smoke grenade dischargers mounted either side of the turret which can lay a smoke screen covering the tank in 8 seconds.

The hydraulic aiming control system of the AMX-30 is a SAMM Model CH 27-1S with elevation controlled through a hydraulic actuating cylinder/shock-absorber and traverse controlled through a hydraulic motor. The gunner is provided with dual-handle controls and the tank commander has a single handle control which can also be used to override the gunner.

A simplified version of the AMX-30 was developed for export with no night vision or NBC system. The commander's cupola was replaced by an S 470 cupola with an externally mounted 12.7 mm machine gun, rather than a 7.62 mm machine gun as installed on French Army AMX-30s, and the coaxial weapon with a 7.62 mm machine gun rather than a 12.7 mm machine gun or a 20 mm cannon.

For Operation Desert Storm in the Middle East in 1991, most of the French Army AMX-30 B2 MBTs deployed to Saudi Arabia were modified with the installation of an infra-red decoy system mounted on the forward part of the turret and smoke/decoy launchers mounted either side of the turret.

AMX-30S

For desert operations the AMX-30S has been developed with the following modifications: the fitting of sand shields, reduction in the ratios of the gearbox which limits its speed to 60 km/h and an engine developing only 620 hp at 2,400 rpm. At least one foreign army (Saudi Arabia) has adopted the AMX-30S and its tanks are fitted with a CILAS/SOPELEM M 409 sight. This has a day sight with a magnification of ×8 and an 8° field of view, infra-red night sight with a magnification of ×4.5 and a 10° field of view and a laser range-finder with a range of between 400 and 10,000 m. This development allows the tank commander to range onto a target without having to traverse the turret. In addition the laser range-finder increases the first round hit probability.

AMX-30 B2

The AMX-30 B2, announced in June 1979, is essentially an AMX-30 with an integrated fire-control system based

Giat Industries AMX-30 B2 DT mineclearing tank in travelling configuration (Stefan Marx) 0011632

Giat Industries AMX-30B MBT (Michael Jerchel) 0018606

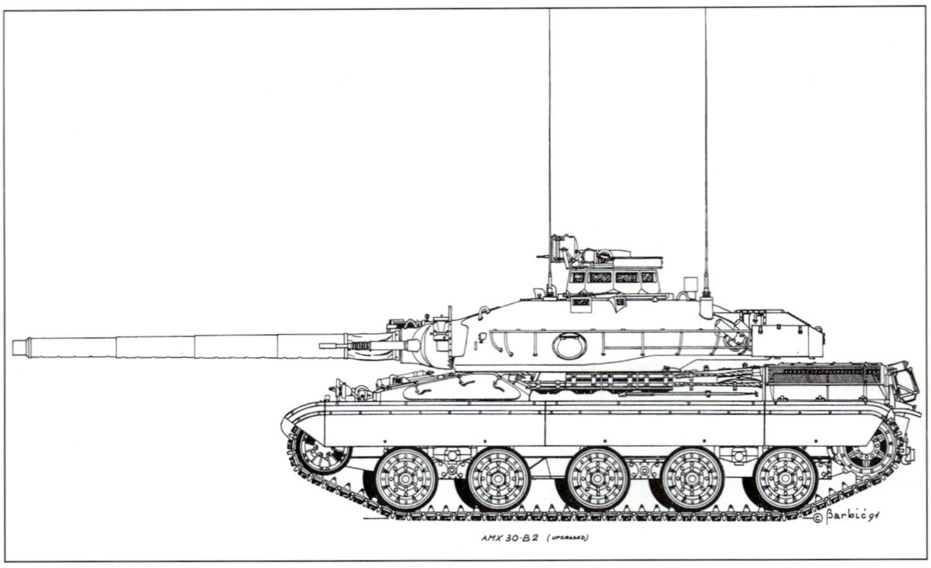

Giat Industries upgraded AMX-30 B2 MBT (Vasco Barbic)

on a laser range-finder and a thermal system, new gearbox and other improvements. The French Army ordered 50 vehicles in both the 1981 and 1982 defence budgets and first deliveries were made to the 503rd Regiment at Mourmelon in January 1982. The French Army took delivery of 166 brand new AMX-30 B2 MBTs with 493 existing AMX-30s upgraded to the AMX-30 B2 standard.

The COTAC FCS, officially designated the APX M581, has been designed by the AMX-APX in collaboration with the Ateliers de Construction de Tarbes. Its main components are the gunner's APX M544 telescopic (×10 magnification) sight which is combined with an APX M579 electronic control system and an APX M421 optical module containing a computer-controlled graticule. The laser range-finder is the APX M550 and is made by CILAS under the designation TCV 80. It has a maximum range of 10,000 m and is accurate to ±5 m. The SOPELEM telescopic sight has a magnification of ×10 and is directly coupled to the main armament. In addition to the M581 range-finder telescope the gunner has a rotatable M282 periscope, a fixed M223 periscope and a TV monitor associated with a thermal monitor.

The COTAC FCS also incorporates an accelerometer-type trunnion tilt sensor and a control panel. The tank commander manually inserts information via the control panel concerning the type of ammunition being fired, drift and jump angles, crosswind velocity, altitude and ambient temperature. Apart from the ammunition details, most of this information has to be inserted only once a day. To determine lead angles for the engagement of moving targets, a two-axis gyroscope is mounted on the sight to measure target tracking rates in both azimuth and elevation.

Other parameters are automatically acquired by the gunner's range-finder telescope, including target distance, its displacement speed in elevation and azimuth and turret slant. These parameters are processed by a lead computer built into the telescope which transmits the elevation and azimuth corrections to the gunner's day sight by means of an optical compensator and the graticule of the thermal camera for engaging targets at night.

The tank commander is provided with an M496 sight (×8 magnification) for firing the 105 mm gun and the 20 mm cannon (land targets).

In the cupola, the commander is provided with an M591.02 prism head and an OB-49 periscope for firing the 20 mm cannon (air targets) and the 7.62 mm machine gun.

Mounted externally on the right side of the turret is the CASTOR thermal camera with both the tank commander and gunner being provided with a TV monitor screen. The screens display an aiming mark which enables the commander or gunner to engage targets at ranges up to 4,000 m.

The loader has a rotatable M282 periscope and two fixed M223 periscopes, with the driver having a central M223 periscope (or an OB-31A image intensification periscope) and two fixed M223 periscopes.

The engine of the AMX-30 B2 has been developed from the AMX-30 and develops 700 hp at 2,600 rpm.

The HS-110-2 is fitted with new and more powerful turbo-compressors which provide a 45 hp increase as well as improving the operational life of the engine and its torque characteristics.

A new hydrostatically controlled engine system now proposed by Giat incorporates oil heat exchangers cooled by air instead of engine coolant. The air filters now have removable covers to facilitate field maintenance.

To allow the AMX-30 B2 to fire the newer APFSDS ammunition a new gunner's sight graticule is fitted to match the ballistics of the round while the turret and ammunition racks have been modified to accept the differently shaped APFSDS round.

The SESM ENC 200 is a gearbox of the lock-up torque converter type with five forward and reverse speeds and hydrostatic steering controlled by a conventional steering wheel instead of sticks. This system allows on-the-spot turning and gear changing in bends.

The suspension uses new torsion bars which give improved cross-country mobility. As an option it is possible to fit the AMX-30 B2 with connector tracks which are quieter and offer less resistance than conventional tracks.

The AMX-30 B2 is also fitted with a new collective pressurisation system with two circuits, one with complete filtration for improved NBC protection.

In addition to these improvements, Giat offers a number of other options for the AMX-30 B2. French Army AMX-30 B2s have the OB-49 roof-mounted sight with two day and one night channels. This can be replaced by the M427 with day and image intensification night channels and a laser range-finder or a roof-mounted SAGEM M527 gyrostabilised sight. The gunner has an M581 range-finder telescope but the driver's infra-red periscope can be replaced by an OB-31 or CN2-516 image intensification periscope.

Other options include new 125 Ah batteries with a longer life and needing less maintenance, new filters, DUK DUR 440 radiation counter and different land navigation systems.

French Army AMX-30 B2s have a thicker gun shield; additional add-on armour protection can be provided for the turret sides and armoured skirts could be fitted.

In place of the standard bank of two electrically operated smoke dischargers either side of the turret, the Giat/Lacroix GALIX AFV protection system could be fitted. This can fire a wide range of different types of grenade to meet various battlefield threats. An exhaust

AMX-30 B2 MBT fitted with Giat Industries G2 explosive reactive armour modules (C R Zwart) **2001**/0105865

AMX-30D ARV **2000**/0064640

AMX-30 bridgelayer with bridge starting laying sequence **2000**/0064641

Giat Industries AMX-30 MBT fitted with Russian KMT-5 plough and roller mine clearing system **2000**/0064647

AMX-30 DCA with twin 30 mm cannon which is in service only with Saudi Arabia 0018608

system smoke generator can also be fitted to the AMX-30 B2 if required.

Apart from France, the only known user of the AMX-30 B2 is Cyprus.

Variants

AMX-30 with ERA

Under contract to the French Army, Giat Industries has developed an ERA package for part of the French Army's fleet of AMX-30 B2 MBTs.

The ERA package uses the Giat Industries BS (Brique de Surblindage) G2 module which is claimed to be resistant to small arms fire up to 20 mm in calibre but when hit by a HEAT warhead the module provides protection equivalent to 400 mm of conventional steel armour at an angle of 60°.

Two AMX-30 B2 battalions of the FAR (Rapid Deployment Force) have been equipped with the ERA package while an additional two battalions will be fitted for, but not, with the ERA package.

This ERA package is being offered for installation on other MBTs by Giat Industries.

French Army re-power of AMX-30 B2 MBT

The AMX-30 B2 was the final production model of the AMX-30 MBT and is powered by a Hispano-Suiza HS-110 12-cylinder water-cooled multi-fuel engine built by Renault developing 720 hp at 2,600 rpm coupled to a SESM ENC 200 transmission.

The HS-110 has never been considered a satisfactory engine and following trials with a number of engines the French Army selected the Renault Vehicules Industriels (RVI) Mack E9 diesel and placed an order for 500 engines.

The first Mack E9 engines were delivered to the French Army early in 1998 as they will be carrying out the actual conversion work although the design authority remains Giat Industries.

The Mack E9 engines will first be installed in the AMX-30 B2 tank followed by some of the more specialised versions, including the 155 mm GCT and the AMX-30D armoured recovery vehicle.

For the AMX-30 B2 application the E9 is rated at 552 kW (750 hp) and according to RVI is almost a direct replacement for the current Hispano-Suiza HS-110 engine as no modificatons are required to be carried out for the actual hull. The SESM ENC 200 transmission is retained as this was fitted on all AMX-30 B2 series tanks.

The specialised versions of the AMX-30 had older manual transmissions which are expected to be replaced by the SESM ENC 200 in which the driver steers the vehicle with a steering wheel rather than the old tillers as originally installed.

The military E9 features double stage turbocharging on either side (2 × 2 turbochargers of the Garrett type), a Bosch direct injection system and a dry sump. The latter is a specific military requirement as it allows the AMX-30 B2 to operate on steep side slopes and gradients which are not required for the civilian Mack E9 engine.

The Mack E9 engines for the AMX-30 B2 are manufactured in the USA and are then sent to the RVI facility in Limoges where they are modified for the military role. When compared to the existing engine, the Mack E9 is more reliable and offers lower life cycle costs.

AMX-30 Upgrades

Giat Industries is also offering a number of subsystem upgrades for the AMX-30 MBT including the following:

(1) AMX-30 B2 suspension for improved mobility
(2) Installation of microprocessor control box for the transmission
(3) Replacement of the existing Renault engine by the new MACK E9-750 diesel which develops 750 hp at 2,400 rpm
(4) Installation of an ENC 200 transmission
(5) Modification of ammunition racks and sights to enable the AMX-30 to fire the latest 105 mm APFSDS projectiles
(6) Installation of a DIVT thermal camera with monitors for commander and gunner
(7) Installation of a fire/explosion detection/suppression system for the crew compartment
(8) Installation of a new smoke/decoy launching system either side of the turret
(9) Installation of an anti-tank missile decoy system on the turret roof.

Giat Industries AMX-30 EBG combat engineer vehicle in travelling configuration **2000**/0064642

Spanish AMX-30 upgrade

In January 1987, after studying a number of proposals and trials with prototype vehicles, Spain approved a plan to upgrade 150 of its AMX-30 MBTs in two stages, first the installation of a new power pack and second a new fire-control system. Details of this plan are given under Spain.

AMX-30 Venezuela

In 1985, SABCA of Belgium was awarded a contract to modernise Venezuelan AMX-30 MBTs with a new fire-control system. Teledyne Continental Motors (now General Dynamics Land Systems) of the USA has also modernised at least one Venezuelan AMX-30 with a 908 hp version of the AVDS-1790 series diesel engine coupled to a fully automatic transmission; speed is increased to 65 km/h and the operating range with increased fuel capacity is up from 535 km to 721 km. The AMX-30 was also fitted with a new fire-control system incorporating a laser range-finder, weapon stabilisation system and sensors for wind, temperature and humidity and a modernised commander's station. The driver's station has also been modernised, a fire detection and suppression system installed and the vehicle can now lay its own smoke screen by injecting diesel fuel into the exhaust system.

Other options include appliqué armour protection and the replacement of the torsion bar suspension with Model 2880 hydropneumatic suspension units for improved cross-country mobility.

AMX-30D armoured recovery vehicle

This is designated the AMX-30D and is designed to carry out three major tasks: the recovery of disabled and damaged vehicles, replacement of major components such as engines and engineer work. Equipment provided includes a hydraulically operated dozer blade at the front of the hull which is used to stabilise the vehicle while the crane is being used or during winching or for dozing. Mounted on the right side of the vehicle is a hydraulic crane which can lift 12,000 kg through 240° or 15,000 kg when being used to the front of the vehicle. The main winch has a maximum capacity of 35,000 kg and is provided with 90 m of 34 mm diameter cable while the auxiliary winch has 3,500 kg capacity and is provided with 120 m of 11.2 mm diameter cable. The AMX-30D has a loaded weight of 36,000 kg and normally carries a spare engine at the rear. Details of this version are given in *Jane's Military Vehicles and Logistics 2002-2003.*

AMX-30 bridgelayer

The AMX-30 bridgelayer is not used by the French Army, although a few are used by Saudi Arabia. It is basically an AMX-30 MBT with its turret removed and replaced with a scissors-type bridge which is launched over the rear of the hull in 5 minutes. The bridge is a class 50 and when opened out is 22 m long and will span a gap up to 20 m. The bridge is 3.1 m wide or 3.92 m when fitted with widening panels. Weight is 42,500 kg with the bridge and 34,000 kg without. Full details of the AMX-30 armoured bridgelayer are given in *Jane's Military Vehicles and Logistics 2002-2003.*

AMX-30 Combat Engineer Tractor

The AMX-30 Combat Engineer Tractor, or Engin Blindé de Génie (EBG), was developed by Giat Industries to meet the requirements of the French Army, with a total of 71 vehicles being built, the last of which was completed late in 1993.

Specifications
(Specifications in square brackets relate to the AMX-30 B2 where this differs from the AMX-30)
Crew: 4
Combat weight: 36,000 [37,000] kg
Unloaded weight: 34,000 [35,300] kg
Power-to-weight ratio: 20 [18.91 hp/t
Ground pressure: 0.77 [0.90] kg/cm²
Length:
(gun forward) 9.48 m
(gun rear) 8.73 m
(hull) 6.59 m
Width: 3.1 m
Height:
(to hull top) 1.5 m
(to turret top) 2.29 m
(overall, including searchlight) 2.86 m
Firing height: 1.81 m
Ground clearance: 0.44 m
Track: 2.53 m
Track width: 570 mm
Length of track on ground: 4.12 m
Max speed:
(1st gear) 7 km/h
(2nd gear) 15 km/h
(3rd gear) 26 km/h
(4th gear) 43 km/h
(5th gear) 65 km/h
Average speed:
(road) 50 km/h
(cross-country) 35-40 km/h
Fuel capacity: 970 [900 litres
Max road range: 500-600 [400-450] km
Fording: 1.3 m
(with preparation) 2.2 m
(with snorkel) 4 m
Gradient: 60%
Side slope: 30%
Vertical obstacle: 0.93 m
Trench: 2.9 m
Engine (see text): Hispano-Suiza HS 110 12-cylinder, water-cooled supercharged multifuel developing 720 hp at 2,000 rpm [700 hp at 2,600 rpm]
Transmission (see text): mechanical with 5 gears in both directions
Steering (see text): triple differential
Clutch: centrifugal
Suspension: torsion bar
Electrical system: 28 V
Batteries: 8 × 12 V, 100 Ah
Armament:
(main) 1 × 105 mm smooth bore gun
(coaxial) 1 × 20 mm cannon; (or 1 × 12.7 mm MG)
(anti-aircraft) 1 × 7.62 mm MG
Smoke-laying equipment: 2 smoke grenade dischargers either side of turret
Ammunition:
(main) 47
(coaxial) 1,050 [480]
(anti-aircraft) 2,050 [2,070]
Gun control equipment
Turret power control: electrohydraulic/manual
(by commander) yes
(by gunner) yes
(commander's override) yes
Commander's fire-control override: yes
Max rate power traverse: 360° in 12-13 s
Max rate power elevation: 5.5°/1 s
Main armament elevation/depression: +20/–8°
Secondary armament elevation/depression: +40/–8°
7.62 mm armament elevation/depression: +45/–10°
Gun stabiliser:
(vertical) no
(horizontal) no
Elevation quadrant: yes
Traverse indicator: yes
Armour (estimated)
Hull front: 79 mm
Hull sides front: 57 mm
Hull sides rear: 30 mm
Hull top: 15 mm
Hull bottom: 15 mm
Hull rear: 30 mm
Turret front: 80.8 mm
Turret sides: 41.5 mm
Turret rear: 50 mm
Turret top: 20 mm
NBC system: yes
Night vision equipment: yes

The EBG is designed for use in forward areas and is equipped with a front-mounted hydraulically operated dozer blade which is fitted with six scarifying teeth, a hydraulic winch with a maximum capacity of 20,000 kg, a hydraulic PTO and a working arm pivoted at the front of the hull on the right side. The arm is used to lift obstacles out of the way and can be fitted with pincers to remove tree trunks or an auger which can bore 220 mm diameter holes to a depth of 3 m.

Armament consists of a 7.62 mm machine gun, four smoke grenade dischargers, a launching tube for demolition charges and four mine-launching tubes. The 142 mm demolition charge has a maximum range of 300 m. Each of the four mine-launching arms has a tube containing six 139 mm diameter anti-tank mines. The vehicle's crew of three consist of the commander, a sapper and a driver. It has a maximum road speed of 65 km/h and can ford to a depth of 2.5 m without preparation or 4 m with preparation. Details of this vehicle are given in *Jane's Military Vehicles and Logistics 2002-2003.*

The French Army EBGs are to be provided with an MLC 65 (Military Load Class) folding assault bridge (details of which are given in *Jane's Military Vehicles and Logistics 2002-2003.*

The TCP *(Travure Courte Portée)* folding assault bridge has been developed by CNIM which is currently supplying the French Army with the *Pont Flottant Motorisé* (PFM) motorised pontoon bridge system.

This bridge is fitted onto the EBG from a single-axle trailer using only the EBG's crane. Once in position, the scissors bridge is opened by a combination of cables and gravity and can be used to span a gap up to 12 m wide. When opened out the bridge is 14 m long, 4 m wide and weighs a total of 4,800 kg.

AMX-30 B2 as used by the French Army in Operation Desert Storm (Pierre Touzin) **2002**/0130613

AMX-30 MDR Flail Tank

Development of this system has been cancelled. For trials purposes an AMX-30 MBT has been fitted with Israeli mineclearing rollers at the front of the hull. For Operation Desert Storm, some French Army AMX-30 B2 MBTs were fitted with Russian KMT-5 plough/roller type mineclearing systems supplied by Germany from ex-East German Army stocks.

AMX-30 B DT/AMX-30 B2 DT mineclearing tank

Under contract to the French Army, Giat Industries has developed a mineclearing version of the AMX-30 MBT and this is now in service with the French Army engineers.

Shahine firing unit as delivered to Saudi Arabia 0018609

Roland 2 SAM system of the French Army on a modified AMX-30 chassis in travelling configuration (Pierre Touzin) 0011634

Status
Production complete. In service with the countries in the table.

Model	AMX-30	AMX-30S	AMX-30 B2	AMX-30	AMX-30	AMX-30	GCT	Shahine	DCA	Roland
Type	MBT	MBT	MBT	AVLB	ARV	CET	SPG	SPAAM	SPAAG	SPAAM
Bosnia	36	0	0	0	0	0	0	0	0	0
Chile	50	0	0	0	0	0	0	0	0	0
Croatia	42	0	0	0	0	0	0	0	0	0
Cyprus	52	0	102	0	2	0	0	0	0	0
France	387	0	659	0	134	71	251	0	0	181*****
Greece	154***	0	0	0	14	0	0	0	0	0
Iraq	0	0	0	0	0	0	85	0	0	13
Kuwait	0	0	0	0	0	0	18	0	0	0
Nigeria	0	0	0	0	0	0	0	0	0	16
Qatar	0	34******	0	0	1	0	0	0	0	3
Saudi Arabia	0	290	0	12	57	0	51	109*	53	0
Spain	299**	0	0	0	10	0	0	0	0	18
UAE (Abu Dhabi)****	64	0	0	0	4	0	0	0	0	0
Venezuela	81	0	0	0	4	0	0	0	0	0

* *36 acquisition and 73 firing units*
** *210 now operational*
*** *in reserve (52 passed on to Cyprus)*
**** *additional 36 transferred to Bosnia in 1997*
***** *being reduced*
****** *+10*
The figures for Iraq are for the total number delivered and few of these are believed to remain operational

When based on an AMX-30 B MBT chassis it is called the AMX-30 B DT, and when based on the more recent AMX-30 B2 MBT chassis it is called the AMX-30 B2 DT, with the DT standing for Demineur Tank (Mineclearing Tank).

The vehicle is based on a standard AMX-30 B or AMX-30 B2 MBT chassis with the 105 mm main armament and secondary armament removed and the position plated over.

Mounted at the front of the hull is an Israeli RAMTA or UK Pearson Engineering plough-type mine clearing system and a Giat Industries DEMETER magnetic field generator.

Mounted on either side of the hull rear is a Pearson Engineering Pathfinder marking system which dispenses marker poles into the ground as the system moves forward.

As an alternative to mineclearing devices of the plough type it can also be fitted with roller-type mineclearing devices, for example those developed by Urdan Industries of Israel.

The AMX-30 B DT/AMX-30 B2 DT mineclearing tanks can be operated with a normal crew, or under full remote control with the aid of an optional kit.

Shahine acquisition unit as delivered to Saudi Arabia 0018610

Roland anti-aircraft missile system
This is a modified AMX-30 MBT chassis with two Euromissile Roland surface-to-air missiles in the ready to launch position and a further eight rounds in reserve in the hull, four either side, ready for automatic reloading. The surveillance radar is mounted on the turret rear with the tracking radar being mounted on the turret front. This version is in service with France (181), Iraq (13), Qatar (3), Spain (18) and Nigeria (16). Full details of this system are given in *Jane's Land-Based Air Defence 2002-2003*. The exact number of Iraqi systems operational is uncertain. The French figure has recently been reduced, as 20 shelter-mounted models have been built by Euromissile using some key parts from the French Army systems based on the AMX-30 MBT chassis.

Shahine anti-aircraft missile system
This was developed by the now Thales Air Defence (at that time Thomson-CSF) specifically to meet the requirements of Saudi Arabia. It consists of a firing unit with six missiles in the ready to fire position on a modified AMX-30 chassis and an acquisition unit with a large surveillance radar on a similar chassis. This is an all-weather system. The firing unit does not carry any reserve missiles, these being brought up by a cross-country truck fitted with a crane. Full details of this system are given in *Jane's Land-Based Air Defence 2002-2003.*

Twin 30 mm anti-aircraft gun system
This is an updated version of the turret installed on the AMX-13 light tank chassis, but installed on an AMX-30 chassis specifically to meet the requirements of Saudi Arabia which took delivery of a total of 53 vehicles. The chassis is fitted with a power-operated turret armed with twin 30 mm cannon for which a total of 600 rounds of ready use ammunition is carried (300 for each gun) with a further 900 rounds being carried in reserve. Mounted on the turret rear is the radar system. Full details of this system are given in *Jane's Land-Based Air Defence 2002-2003.*

155 mm GCT self-propelled Howitzer
Details of this system are given in the Self-propelled guns and howitzers (tracked) section. In addition to being in service with the French Army, the GCT is also used by Iraq, Kuwait and Saudi Arabia. This turret, on a T-72M1 MBT, has been evaluated in India.

Driver Training Tank
This is the AMX-30 MBT with its turret replaced by an observation-type turret and is used for driver training. This is referred to as the AMX-30 écolage by the French Army.

Manufacturer
Giat Industries.

UPDATED

GERMANY

Krauss-Maffei Wegmann Leopard 2A5/2A6 MBT

Development
There are two separate combat improvement programmes planned for the Leopard 2 MBT which has now been in service with the German Army for over 20 years.

KWS I covers the installation of a new Rheinmetall 120 mm 55 calibre smoothbore gun as the replacement for the current Rheinmetall 120 mm 44 calibre smoothbore gun.

In addition to firing the current range of ammunition, it will also fire the improved 120 mm kinetic energy ammunition (DM53) with increased armour penetration characteristics.

The KWS II combat improvement programme is already underway for the German and Royal Netherlands armies under the designation of the Leopard 2A5 and these and other improvements were incorporated into the latest production Leopard 2 MBTs for the Swedish Army.

The first of three prototypes of the Leopard 2A5 was completed by Krauss-Maffei (on 1 January 1999 Krauss-Maffei and Wegmann merged to become Krauss-Maffei Wegmann GmbH and Co KG) late in 1990 and following trials, was accepted for service. Development of the Leopard 2A5 was funded by

Germany, Netherlands and Switzerland. Krauss-Maffei was awarded the contract to upgrade the Leopard 2 to the Leopard 2A5 standard in January 1994.

In September 1995, the German Federal Office for Defence Technology and Procurement took delivery from Krauss-Maffei of the first of 225 Leopard 2 MBTs (covered in detail in the following entry) which are being upgraded to the Leopard 2A5 configuration, or Leopard 2 (Improved) as it was previously referred to.

The 225 vehicles were upgraded by Krauss-Maffei, the prime contractor for the Leopard 2 MBT. The turrets were upgraded in Kassel which delivers them to Munich which then integrates the turret with the chassis and delivers the completed vehicle to the German Army.

Krauss-Maffei delivered 16 Leopard 2A5s in 1995 with production running at six vehicles a month from January 1996, production for the German Army has now been completed. Krauss-Maffei Wegmann is now working on the second batch of 125 vehicles for the German Army.

The German Army has 20 active and 14 reserve tank battalions. Of these, 26 have the Leopard 2 and the remainder the Leopard 1. Of the 26 Leopard 2 battalions, three have the latest Leopard 2A5 while the remainder have older Leopard 2 MBTs. These three battalions are earmarked for the Crisis Reaction Forces which will be increased to 6 Leopard 2A5 battalions in the future.

In March 1994 the Royal Netherlands Army decided to upgrade 180 of its Leopard 2s to the Leopard 2A5 standard, with an option on a further 150 vehicles. First deliveries were made to the Netherlands in March 1996 and ran through to September 2000.

Switzerland took delivery of the last of its Leopard 2 MBTs as recently as 1993. As of January 2001 it has not decided when to upgrade any of its Leopard 2 MBTs to this standard.

The Leopard 2A5, with a number of further improvements, especially in the areas of armour and command and control, has been ordered by the Swedish Army (120 delivered plus option on 90 which has not been exercised) with Spain (219) also selecting this version. Additional details are given under Spain and Sweden later in this entry.

In March 2002, the Hellenic Army selected the Leopard 2 to meet its future requirements and will place an order for a total of 170 vehicles following contract negotiations. It is not known which version of the Leopard 2 series MBT will be ordered.

Description

The layout of the Leopard 2A5 is virtually identical to that of the Leopard 2 with the main areas of improvement being:

(1) The commander's roof-mounted periscope now has a thermal sight whose image is transmitted to a monitor inside the turret. This allows the commander and gunner to engage targets under all weather conditions and also allows them to operate as a hunter (commander) and killer (gunner) team. The commander's monitor can also show the gunner's day and night vision images while the commander can see his own day vision through his eyepiece

(2) Installation of a new all-electric gun control equipment as a replacement for the earlier hydraulic system. This new all-electric system is not only quieter but is also easier to maintain and consumes less electrical energy

(3) Improved armour protection over the frontal arc with the turret front having a distinctive arrowhead shape. This provides a significant increase in protection against both kinetic and chemical energy attack. The turret interior has been fitted with spall liners. The side skirts also incorporate composite improved armour. The new externally mounted armour panels can be removed with onboard equipment and can be replaced by armour modules with a higher level of protection

(4) New driver's hatch that slides to the right

(5) Installation of a TV camera at the rear of the hull with monitor for the driver to allow for safer reversing of the tank. This has a 65° field of view in horizontal and vertical planes

(6) Hybrid navigation system based on fibre-gyro technology with GPS support enables the tank commander to navigate in any operational environment and represents a key element in command and control system which can be retrofitted at a future date

(7) The laser range data processor is modified. Until now, only the second (more distant) echo of range measurement has been accepted as valid. It will now also be possible to evaluate the first echo if required. The first echo is required for shorter reaction times when engaging helicopters and for engaging these targets APFSDS projectiles would be used as they have such a flat trajectory.

The Swedish Leopard 2 is called the MBT 122 (Strv 122) and has some additional improvements which increases the combat weight of the vehicle from the 59.7 tonnes of the German Leopard 2A5 to 62 tonnes.

Additional improvements required by Sweden include:

(1) Improved protection of the front and sides of the chassis similar to that of the turret, including spall liner

(2) Improved protection against the effects of bomblets on the turret roof

(3) A modular Tank Command and Control System (TCCS) connected via a databus to the subsystems of the tank and including both the navigation system and a new generation of Swedish radio equipment

(4) Eye-safe laser (Raman shifted Nd-YAG) range-finder.

Additional details of the Swedish Leopard 2 programme are given under Sweden.

Krauss-Maffei Wegmann Leopard 2A6 MBT armed with Rheinmetall 120 mm L/55 weapon **2000**/0064643

Leopard 2A5 MBT of the Royal Netherlands Army fitted with Rheinmetall 120 mm 55 calibre smoothbore gun **2001**/0105866

Specifications

Crew: 4
Combat weight: 59,700 kg
Power-to-weight ratio: 25.12 hp/t
Ground pressure: 0.89 kg/cm²
Length:
(gun forward) 9.97 m
(gun rear) 8.49 m
Width with skirts: 3.74 m
Width railway transport: 3.55 m
Height over turret roof: 2.64 m
Height over commander's periscope: 3.00 m
Firing height: 2.01 m
Ground clearance: 0.50 m
Length of track on ground: 4.95 m
Track width: 635 mm
Max road speed:
(forwards) 72 km/h
(reverse) 31 km/h
Cruising range road: 500 km
Fuel capacity: 1,200 litres
Fording: 1 m
(with preparation) 2.35 m
(with snorkel) 4 m
Gradient: 60%
Side slope: 30%
Vertical obstacle: 1.1 m
Trench: 3.00 m
Engine: MTU MB 873 Ka-501 4-stroke, 12-cylinder diesel, exhaust turbocharged, liquid-cooled developing 1,500 hp at 2,600 rpm
Transmission: Renk HSWL 354 hydrokinetic planetary gear shift, 4 forward and 2 reverse gears
Clutch: torque converter
Suspension: torsion bar, friction damper and hydraulic bump stops
Electrical system: 24 V
Batteries: 8 × 12 V, 125 Ah
Armament:
(main) 1 × 120 mm smoothbore gun
(coaxial) 1 × 7.62 mm MG3
(anti-aircraft) 1 × 7.62 mm MG3
Smoke-laying equipment: 8 × 76 mm smoke grenade dischargers either side of turret
Ammunition:
(main) 42
(MG) 4,750
Gun control equipment
Turret power control: electric/manual
(by commander) yes
(by gunner) yes
(commander's override) yes
Gun elevation/depression: +20/–9°
Turret traverse: 360°
Gun stabiliser:
(vertical) yes
(horizontal) yes
Range setting device: yes, laser
Elevation quadrant: yes
Traverse indicator: yes
NBC system: yes
Night vision equipment: yes

The Leopard 2A6 EX MBT armed with the Rheinmetall 120 mm L/55 smoothbore gun and additional armour **2001**/0105867

Variants

In March 2001, Krauss-Maffei Wegmann handed over to the German Army the first upgraded Leopard 2A6 series Main Battle Tank (MBT) armed with the new Rheinmetall 120 mm L/55 smoothbore tank gun.

Under current plans, the German Army is to take delivery of 350 Leopard 2A6 MBTs of which 225 have been funded so far with an additional contract for the remaining 125 expected in the future.

The installation of the Rheinmetall 120 mm L/55 smoothbore tank gun, when used with the latest Rheinmetall 120 mm DM53 Armour-Piercing Fin-Stabilised Discarding Sabot - Tracer (APFSDS-T) ammunition, using a non-depleted uranium penetrator, gives the Leopard 2A6 an additional combat range of 1,600 m and much greater armour penetration characteristics.

Rheinmetall are also developing a new modular High Explosive - Tracer (HE-T) round that will enable targets to be engaged out to 5,000 m, with more accuracy and greater target effectiveness than the current HEAT-MP-T.

Rheinmetall supply the 120 mm L/55 smoothbore gun kit to Krauss-Maffei Wegmann who actually fit the weapon. The kit includes the 120 mm barrel with muzzle reference system, steel shell case holder, breech and breech guard and thermal shield for the barrel. The existing breech mechanism is retained. The recoil breaks and trunnion bearings are also strengthened. The current 120 mm L/44 barrel weighs 1,190 kg while the new 120 mm L/55 weighs 1,347 kg.

Modifications made to the fire-control computer include an additional ballistic circuit board, modification of diagnostic sequences and new circuit boards for stabilisation electronics.

A modification is also made to the reticle of the FERO Z18 gunners sight with modifications also being carried out to the acceleration sensor, elevation drive and gun control equipment.

The full potential of the Leopard 2A6 MBT is obtained with the use of the 5th generation DM53 APFSDS-T round. The first customer for this was Switzerland who took delivery of their first batches in 1999 and recently the DM53 has been ordered by Germany and the Netherlands.

Netherlands Leopard 2A6

The Royal Netherlands Army is to upgrade 180 of its existing Leopard 2 MBTs to the Leopard 2A6 standard. The first of these was delivered in late 2001 under the designation of the Leopard 2 A6 NL. The first vehicle was converted in the facilities of Krauss-Maffei Wegmann in Germany, with the remainder being upgraded in the Netherlands at MCW Mechanische Central Werkplaats.

Spanish Leopard 2A6

Spain is the first country to order brand new Leopard 2A6 MBTs with the 120 mm L/55 smoothbore gun with the first of 219 to be handed over to the Spanish Army by Krauss-Maffei Wegmann in Munich late in 2002. The first 30 will be built in Germany with some major components being provided by Santa Barbara Sistemas in Spain with the remainder being built in Seville, Spain. These will have the Spanish Army designation of the Leopard 2A6E.

Leopard 2A6 EX

A further development of the Leopard 2A6, the private venture Leopard 2A6 EX has been developed for the export market and has a number of additional improvements including an auxiliary power unit, air conditioning unit, and improved armour protection. The Leopard 2A6 EX took part in the recent Turkish MBT trials during which it traveled over 3,000 km and fired 400 rounds.

Swedish Strv 122

Additional details of this are given later in this section under Sweden.

Further Improvments

Krauss-Maffei Wegmann has leased two Leopard 2 MBTs from the German Army for use as technology demonstrators. The first of these is called the Leopard 2A6 Ex and is covered above and was completed in 2000.

The second technology demonstrator will be completed in 2002 and will feature the German MTU EuroPowerPack developing 1,500 hp which is already installed in the Giat Industries Leclerc MBT and variants for the United Arab Emirates. This is lighter and more compact than the current power pack and will provide additional space inside of the hull and restore the power-to-weight of the Leopard 2 which has been degraded over the years with the installation of additional armour.

Other additional improvements could include installation of a battlefield identification friend or foe system, improved sensors, defensive aids suite and the installation of a more powerful engine.

Status

In production. In service with the following countries:

Country	Quantity	Comment
Germany	225	Deliveries complete, second batch of 125 vehicles now being upgraded
Greece	170	Selected March 2002
Netherlands	180	Upgrade complete, now to be upgraded to Leopard 2A6
Spain	219	First deliveries late 2001
Sweden	120	Deliveries complete

Manufacturer

Krauss-Maffei Wegmann GmbH & Co KG.

UPDATED

Leopard 2A5 of the German Army deployed to the Balkans (Michael Jerchel) **2002**/0130614

Krauss-Maffei Wegmann Leopard 2 MBT

Development

In 1970, Germany took the decision to proceed with a new tank called the Leopard 2 and 16 hulls and 17 turrets were integrated by Krauss-Maffei (which on 1 January 1999 became Krauss-Maffei Wegmann with facilities in Munich and Kassel) between 1972 and 1974. All of the prototypes used the Renk transmission and MTU diesel from the MBT-70.

Ten of the Leopard 2 prototypes were equipped with a Rheinmetall 105 mm smoothbore gun and the remainder were fitted with a Rheinmetall 120 mm smoothbore gun. Two of the chassis were fitted with a hydropneumatic suspension but the advanced torsion bar suspension with integrated friction dampers was subsequently adopted.

In 1974, the USA and Germany signed a Memorandum of Understanding (MoU) under which both countries affirmed their intention to make all

Leopard 2A4 of German Army with roof mounted 7.62 mm MG3 machine gun (Stefan Marx) ***2001***/0105868

reasonable efforts to standardise their tank programmes. In 1977, the MoU was amended to include efforts to standardise some tank components between the two countries. These included engine, transmission, gunner's telescope, night vision equipment, fire-control system, tracks and main armament.

To meet the requirements of the USA, Germany built another model called the Leopard 2(AV) (Austere Version). It had a different fire-control system and a turret and hull which incorporated a spaced multilayer armour, as well as many other detail improvements influenced by the conflict in the Middle East in 1973. Two Leopard 2(AV) hulls and three turrets were built. One complete Leopard 2(AV) tank and one chassis were delivered to the USA for trials in September 1976 but, as expected, the USA chose one of the two competing American designs, the Chrysler XM1. The USA did, however, adopt the Rheinmetall 120 mm smoothbore gun under the designation M256.

In 1977, the German Army selected Krauss-Maffei as prime contractor for series production of the Leopard 2 and placed an order, with options, for 1,800 MBTs, 990 of which were to be built by Krauss-Maffei and the remaining 810 by MaK (which is today Rheinmetall Landsysteme).

The first preproduction Leopard 2 MBT was delivered to the German Army late in 1978 for training. Krauss-Maffei and MaK delivered three preproduction Leopard 2 MBTs in early 1979. The first production Leopard 2 was handed over by Krauss-Maffei in October 1979 in Munich. Six tanks were delivered in 1979, 100 in 1980, 220 in 1981 and by 1982 production was running at 300 a year (25 a month). In 1982 it was announced that the cost of the Leopard 2 programme for the German Army was DM5,100 million, with the first production lots being for 380, 450 and 300 tanks. The last two production lots comprised 300 and 370 tanks.

In June 1987, an order for the sixth production batch of 150 Leopard 2s was placed by the German government. Of these, 55 per cent were built by Krauss-Maffei with the remaining 45 per cent by MaK. In 1988, the contract was placed for the seventh production batch of 100 Leopard 2s, with the contract for the eighth production batch of 75 vehicles being placed in 1990. Final deliveries of the Leopard 2 to the German Army were made in 1992.

There was a total of eight production lots of Leopard 2 MBTs for the German Army and details of these are listed here.

1st production lot

Has circular protection ring over optical blocks at the tank commander's station and small circular plate cover on the turret roof covering the base of the former crosswind velocity sensor, which is no longer installed.

As originally built they also had the interim PZB 200 image intensification night sight which was subsequently replaced by an integral thermal sight which was fitted as standard from the second production lot onwards.

1st to 4th production lot

Has circular ammunition resupply hatch in left side of the turret, this was subsequently welded over.

2nd production lot

In addition to elimination of the crosswind velocity sensor this lot had relocation of the tank filler openings, improved exhaust gratings and modified tool stowage.

5th production lot

From this production lot the ammunition resupply hatch in the left side of the turret was deleted, a new three-colour paint scheme was incorporated, there was a digital core for the fire-control computer (also retrofitted to lots 1 to 4) and installation of a fire suppression system.

6th production lot

Similar to the 5th production lot.

7th production lot

From this production lot reshaped forward sections of the side skirts were produced.

8th production lot

In addition to the above new side skirts were fitted.

Austrian Leopard

The Austrian Army has taken delivery of 114 Leopard 2 MBTs from the Netherlands with 35 being delivered in 1997 and 79 in 1998.

Danish Leopard 2

Under a contract signed in 1998 the Danish Army has taken delivery of 51 Leopard 2 MBTs from the German Army with final deliveries in 2000. Krauss-Maffei Wegmann will upgrade these to the Leopard 2A5 plus standard.

Late in 2000, it was stated that Elop of Israel had signed a contract worth US$3 million to supply Thermal Imaging Module (TIM) to the Danish Army for its 51 Leopard 2 MBTs over a two year period. These will be installed in the tank commander's sights and enable 24-hour observation and surveillance capability under harsh weather conditions. At that time the company stated that it had supplied about 700 TIM systems for use on Leopard 2 series MBTs used by Germany, Netherlands and Sweden.

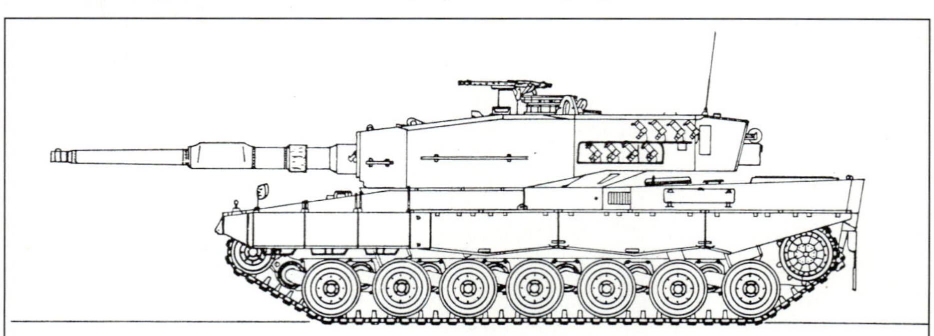

Leopard 2 MBT armed with 120 mm smoothbore gun

Finnish Leopard 2 programme

Early in 2002 it was stated that Finland was having discussions with Germany to acquire up to 100 Leopard 2A4 MBTs.

Netherlands Leopard 2 programme

In March 1979, the Netherlands placed an order for 445 Leopard 2s for delivery from 1982 to 1986, Krauss-Maffei built 278 and MaK (now Rheinmetall Landsysteme) the remaining 167.

The first four Leopard 2 MBTs for the Royal Netherlands Army were completed in mid-1981. Main contract deliveries began in July 1982 and by November that year 10 vehicles a month were being delivered, with final deliveries taking place in July 1986. They have different 7.62 mm machine guns (FN MAG in place of Rheinmetall MG3s), smoke dischargers, passive night periscope for the driver, radios and intercom equipment from the German version.

In 1983, the Royal Netherlands Army's 41st Armoured Brigade, stationed in Germany, became the first in the RNLA to be equipped with the Leopard 2. The RNLA also ordered 20 driver-training versions of the Leopard 2 MBT.

In January 1993, the Royal Netherlands Army stated that it would be phasing out 115 of its 445 Leopard 2s. Early in 1999 it was announced that an additional 130 Leopard 2A4s would be phased out of service and sold.

A total of 180 Royal Netherlands Army Leopard 2 MBTs have been upgraded to the Leopard 2A5 standard. In early 2001 Krauss-Maffei Wegmann was awarded the contract to upgrade these vehicles to the Leopard 2A6 standard. The first one was upgraded in Germany with the remainder being upgraded in the Netherlands. The first upgraded Leopard 2A6 was handed over to the Netherlands Army last 2001.

The Austrian Army has taken delivery of 114 Leopard 2A4 MBTs from the Royal Netherlands Army.

Norwegian Leopard 2

Norway is taking delivery of 52 Leopard 2A4 series MBTs from Germany and the first of these have now been delivered. These have been fitted with Norwegian standard radios, smoke grenade launchers and machine gun mounts. These replace the 59 Leopard 1A1Nos MBTs.

Polish Leopard 2

In the future, Poland is expected to lease 110 Leopard 2A4 series MBTs from Germany plus Biber AVLBs on the Leopard 1 chassis, four Keiler mineclearing vehicles, eight Büffel armoured recovery vehicles as well as other specialised vehicles and equipment.

Spanish Leopard 2 programme

In mid-1995, it was announced that Spain was to manufacture under licence 219 Leopard 2A5 MBTs for the Spanish Army. This contract was finally signed in December 1998 and also included 16 Büffel armoured recovery vehicles.

Spain has already taken delivery of one Leopard 2 MBT from the German Army for training purposes and as an interim measure Spain has leased 108 Leopard 2s from the German Army for five years. These were delivered between November 1995 and June 1996.

Additional details of the Spanish programme are given in the entry for Spanish MBT Fleet under Spain. The Spanish Leopard 2s will have the Rheinmetall 120 mm L/55 gun.

Büffel armoured recovery vehicle of the German Army armed with 7.62 mm MG3 machine gun (Stefan Marx) ***2001***/0105869

Leopard 2 MBT of the German Army fitted with snorkel for deep fording ***2001***/0105871

Swedish Leopard 2 programme

Following a competition between the Krauss-Maffei Leopard 2A5, Giat Industries Leclerc and General Dynamics Land Systems M1A2 for a new MBT to replace Sweden's older Centurions and S-tanks, the Leopard 2A5 was selected early in 1994. Additional details are given under Sweden.

Swiss Leopard 2 programme

In August 1983, the Swiss Army announced that after evaluating the Leopard 2 and M1 MBTs, the Leopard 2 had been selected. A total of 380 was ordered in one lot, the first 35 coming direct from Krauss-Maffei in Germany and the remainder being produced under licence in Switzerland. The first 35 were delivered in 1987 and used for training. The first battalion was fully operational with Leopard 2 MBTs in January 1988.

Prime contractor for the licence production in Switzerland was Contraves with final assembly taking place at the Federal Construction Works (now known as Swiss Ordnance Enterprise) at Thun. Production of the Leopard 2 in Switzerland started in 1987, with the first Swiss-built vehicles delivered in December 1987 and production running at six per month through to March 1993. The Swiss Leopard 2 is similar to the German vehicles but with Swiss radios and intercoms and Swiss coaxial and anti-aircraft machine guns. Switzerland has also decided to install the British Kidde-Graviner Crew Bay fire and explosion suppression system as did the German Army for the fifth production batch of Leopard 2 MBTs, beginning late in 1985.

Description

The hull of the Leopard 2 has a spaced multilayer armour and is divided into three compartments: driver at the front, fighting in the centre and power pack at the rear.

The driver is seated at the front of the hull on the right side and is provided with a single-piece hatch cover that opens to the right and three day observation periscopes. The centre periscope can be replaced by a passive night periscope. Some of the ammunition supply is stowed to the left of the driver.

The turret is in the centre of the vehicle with the commander and gunner on the right and the loader on the left. The commander is provided with a circular hatch cover that opens to the rear and periscopes for all-round observation. A PERI-R17 primary stabilised panoramic periscope with magnification of ×2 and ×8 is mounted in front of the commander's hatch and can be traversed through 360°, enabling the commander to observe the terrain and lay the 120 mm main armament.

The gunner has a dual-magnification stabilised EMES-15 sight with integrated laser range-finder and thermal image unit (WBG) which are linked to the fire-control computer. He also has an auxiliary sighting telescope FERO-Z18 with a magnification of ×8. The tank commander uses the thermal sight, which is integrated with the gunner's EMES-15, to observe the battlefield. The picture is transmitted to the commander's PERI-R17 so he can see the same frame as the gunner. Furthermore the commander can control all functions of the fire-control system and the weapon slave system by a computer-controlled testing board called the RPP 1-8.

The EMES-15 is a primary stabilised binocular sighting instrument for the gunner. The mirror head of the periscope is stabilised around two axes; the day path has a magnification of ×12 and a 5° field of view.

The laser range-finder has a range of 9,990 m and is accurate to ±10 m with measuring distances shown to three digits, together with the fire preparation and selected type of 120 mm ammunition in the lower part of the gunner's sight.

The fire-control computer successively calculates the angle of sight and lateral angular lead for the main armament. The following parameters are taken into account: target distance (from laser range-finder), angle of tilt of the vehicle, direction of motion in regard to the target and ballistic data of ammunition. The data calculated by the fire-control computer are fed into the weapon slave system which guides the weapon to the line of sight of the EMES-15 or PERI-R17. The thermal image unit integrated in the EMES-15 enables the armament to be aimed and fired at night, in bad weather or at camouflaged targets.

The gunner also has a roof-mounted observation periscope. The loader is seated on the left side of the turret and has a single-piece hatch cover that opens to the rear and a single day observation periscope. An ammunition resupply hatch is provided in the left side of the turret and there is a stowage basket at the turret rear. The first has been eliminated on later production batches. This has been changed with the introduction of the seventh production batch.

The engine compartment at the rear of the hull is separated from the fighting compartment by a fireproof bulkhead. The MTU MB 873 engine is coupled to a Renk HSWL 354 hydrokinetic planetary gear shift with an integral service brake.

The suspension with torsion bars consists each side of seven dual rubber-tyred roadwheels with the idler at the front, the drive sprocket at the rear and four track-support rollers. Advanced friction dampers are provided at the first, second, third, sixth and seventh roadwheel stations. The Diehl end connector tracks have rubber-bushed pins and removable rubber pads, which can be replaced by snow grousers. The rear two-thirds of the top of the tracks are covered by steel-reinforced rubber skirts (those of the 8th production lot use a composite material) which can be folded up to allow access to the suspension for maintenance. The front third of the top of the track is covered by special armoured boxes which can be rotated upwards both for maintenance and to reduce the overall width of the vehicle for rail travel.

Standard equipment on the Leopard 2 includes a collective NBC system, power pack preheating, crew compartment heater, a fire extinguishing system, electric bilge pumps and an escape hatch in the hull floor behind the driver.

The 120 mm smoothbore gun has been developed by Rheinmetall and fires two types of ammunition, APFSDS-T (DM-13) and HEAT-MP-T (DM-12), each of which has a practice version. The APFSDS-T has an effective range of over 2,000 m and the HEAT-MP-T has a high degree of effectiveness against both soft and hard targets. Both fin-stabilised rounds have a semi-combustible cartridge case with a metal base stub which is ejected into a box under the breech.

In 1983, the second-generation APFSDS-T round was introduced under the designation DM-23. This has a new monobloc tungsten penetrator. A third-generation APFSDS-T round is the DM-33. This has a much greater length-to-diameter ratio. This was followed by the DM-44 which was not adopted by the German Army. The latest round is designated the DM-53 which is now in production with the first customer being Switzerland. All of the German 120 mm APFSDS-T projectiles have non-depleated uranium penetrators. Early in 2001 the DM53 round was ordered by Germany and the Netherlands.

Rheinmetall are also developing as a private venture new modular High Explosive-Tracer (HE-T) round. It will enable targets to be engaged out to 5,000 m with more accuracy and greater target effectiveness than the current HEAT-MP-T.

Of the 42 rounds of ammunition carried, 27 are stored to the left of the driver, 15 in the left side of the turret bustle and separated from the fighting compartment by an electrically driven door. If the bustle is hit it will explode upwards due to blow-out panels.

A 7.62 mm Rheinmetall MG3 machine gun is mounted coaxially to the left of the main armament and a similar weapon can be mounted on the loader's hatch. Mounted on either side of the turret are eight smoke grenade dischargers.

Leopard 2 2A4 DK MBT of the Danish Army ***2001***/0105870

Leopard 2 MBT of the Royal Netherlands Army (Richard Stickland) ***2002***/0130615

Leopard 2 driver training tank with simulated 120 mm gun ***2002***/0130616

Status

Production. In service with the following countries:

Country	Quantity	Comment
Austria	114	from Netherlands, now delivered
Denmark	51	from Germany, deliveries completed in 2000. To be upgraded to Leopard 2A5 plus standard
Germany	2,125	225 upgraded to Leopard 2A5, 108 leased to Spain, 160 sold to Sweden in 1994, 51 sold to Denmark. Additional 125 being upgraded to Leopard 2A5 standard and first 225 now being upgraded to Leopard 2A6 standard.
Netherlands	445	180 upgraded to Leopard 2A5 standard, 114 sold to Austria, remainer for sale. RNLA is upgrading 180 Leopard 2A5 to 2A6 standard
Norway	52	ex-German Army, delivered from 2001
Spain	108	leased from Germany, delivery 1995-96
	219	to be built in Spain
Sweden	120	deliveries completed 2001
	160	ex-German Army, delivered 1995
Switzerland	380	

Variants

Leopard 2A5

This is the latest version of the Leopard 2 and details are given in a separate earlier entry.

Leopard 2 with 120 mm/L55 gun

Rheinmetall has developed a 120 mm/L55 smoothbore tank gun which is a direct replacement for the 120 mm/L44 smoothbore tank gun currently fitted to the Leopard 2 MBT used by the German Army.

The new 120 mm/L55 smoothbore gun is fitted with a thermal sleeve, fume extractor and a muzzle reference system.

This 120 mm barrel is 1.30 m longer than the current barrel which leads to a significant increase in muzzle velocity, especially with the latest APFSDS-T rounds which have more propellant and heavier projectiles.

The geometry of the chamber and the pressure limits are defined in the quadrilateral Interface Control Document (ICD) for the 120 mm tank main armament systems.

In accordance with the ICD the Rheinmetall 120 mm L55 has the identical chamber geometry as the current fielded barrel and complies with the pressure limits. This means that all current 120 mm projectiles can be fired with the new 120 mm L55 barrel.

In order to minimise the extent of changes, the interface measures of the 120 mm L55 to the breech and to the cradle are unchanged.

According to Rheinmetall, two further improvements are possible with the 120 mm L55 barrel. First, the breech could be made from the same steel as the barrel and this would lift the pressure load of the gun by 500 bar compared to the current weapon.

Second, the design of the 120 mm L55 barrel permits a higher pressure load over a longer range of the barrel than is known from the fielded barrel. So this new barrel is prepared for future interior ballistics which are expected to show a wider maximum pressure instead of the now usual pressure peak.

The Leopard 2A5 when fitted with the 120 mm L55 Barrel will become the Leopard 2A6. An illustration of this appears in the previous entry.

Leopard 2 with 140 mm gun

Under contract to the BWB, Rheinmetall has been developing a 140 mm smoothbore gun. This could be fitted into the turret of the current Leopard 2 and would be fed by a bustle-mounted automatic loader which would reduce the turret crew to two. As of early 2002, there are no plans for the 140 mm weapon to enter production.

Specifications

Crew: 4
Combat weight: 55,150 kg
Power-to-weight ratio: 27 hp/t
Ground pressure: 0.83 kg/cm^2
Length:
(gun forward) 9.668 m
(gun rear) 8.49 m
(hull) 7.722 m
Width:
(over skirts) 3.7 m
Height:
(to turret top) 2.48 m
(to commander's periscope) 2.787 m
(hull top) 1.769 m
(firing height) 2.01 m
Ground clearance:
(front) 0.537 m
(rear) 0.487 m
Track: 2.785 m
Track width: 635 mm
Length of track on ground: 4.945 m
Max road speed:
(forwards) 72 km/h
(reverse) 31 km/h
Fuel capacity: 1,200 litres
Max road range: 550 km
Fording: 1 m
(with preparation) 2.25 m
(with snorkel) 4 m
Gradient: 60%
Side slope: 30%
Vertical obstacle: 1.1 m
Trench: 3 m
Engine: MTU MB 873 Ka-501 4-stroke, 12-cylinder diesel, exhaust turbocharged, liquid-cooled developing 1,500 hp at 2,600 rpm
Transmission: Renk HSWL 354 hydrokinetic planetary gear shift, 4 forward and 2 reverse gears
Clutch: torque converter
Suspension: torsion bar, friction damper and hydraulic bump stops
Electrical system: 24 V
Batteries: 8 × 12 V, 125 Ah
Armament:
(main) 1 × 120 mm smooth bore gun
(coaxial) 1 × 7.62 mm MG3 MG
(anti-aircraft) 1 × 7.62 mm MG3 MG
Smoke-laying equipment: 8 × 76 mm smoke grenade dischargers either side of turret
Ammunition:
(main) 42
(MG) 4,750

Gun control equipment
Turret power control: electrohydraulic/manual
(by commander) yes
(by gunner) yes
(commander's override) yes
Commander's fire-control override: yes
Gun elevation/depression: +20/–9°
Turret traverse: 360°
Gun stabiliser:
(vertical) yes
(horizontal) yes
Range setting device: yes, laser
Elevation quadrant: yes
Traverse indicator: yes
NBC system: yes
Night vision system: yes

Leopard 2 AVLB

The German and Royal Netherlands armies have a requirement for a new armoured vehicle launched bridge on the Leopard 2 chassis. Following a competition, MAN won with its modular bridge system. Krauss-Maffei Wegmann will build new chassis for these AVLB's

Büffel Armoured Recovery Vehicle

In August 1990, MaK (now Rheinmetall Landsysteme) announced that it had been selected by the BWB as the main contractor for series production of the Büffel Armoured Recovery Vehicle (ARV) based on automotive components of the Leopard 2 MBT.

In total 100 Büffel ARVs have been built, 75 for the German Army and 25 for the Royal Netherlands Army, with first production vehicles completed late in 1991. Of the 100 vehicles, 55 were built by MaK and 45 by Krauss-Maffei.

To carry out its recovery role, the Büffel ARV is fitted with a hydraulic crane which is pivoted at the front of the hull on the right side, winch, dozer/stabilising blade and other specialised equipment. Additional details of the vehicle are given in *Jane's Military Vehicles and Logistics 2002-2003*.

Spain has ordered 16 Büffel ARVs of which four will be built in Germany and the remainder under licence in Spain. Sweden and Switzerland have also ordered the Buffel ARV.

Leopard 2 driver-training vehicle

The German Army has taken delivery of 22 driver-training versions of the Leopard 2 while the Royal Netherlands Army has taken delivery of 20 vehicles. These are essentially a Leopard 2 with its turret replaced by an observation-type turret and weighted to give the vehicle similar handling characteristics to the standard Leopard 2. Also provided are a dummy 120 mm gun barrel, seats for the instructor, two trainees and override facilities for the instructor.

Other variants

The chassis of the Leopard 2 could also be used to mount the turret of the Gepard anti-aircraft tank.

A Leopard 2 prototype was fitted with the US Textron Lycoming AGT 1500 gas turbine in 1978 for trials purposes by MaK (now Rheinmetall Landsysteme) at Kiel.

For trials purposes the MTU 883 V-12 diesel developing 1,475 hp at 3,000 rpm has been installed in a Leopard 2 MBT. This produces the same output as the MTU 873 Ka-501 currently installed in the Leopard 2 but is lighter, more compact and more fuel efficient.

Manufacturers

Krauss-Maffei Wegmann GmbH & Co KG.

Rheinmetall Landsysteme GmbH (except for Netherlands upgrades and Sweden).

UPDATED

Krauss-Maffei Wegmann Leopard 1 MBT

Development

The Leopard 1 MBT was developed to meet the requirements of the Federal German Army and, following trials with prototype vehicles, was accepted for service in 1963. Full details of the development history of the Leopard 1 MBT was given in *Jane's Armour and Artillery 1997-98* page 35.

In July 1963, Krauss-Maffei of Munich (which in January 1999 became Krauss-Maffei Wegmann with facilities in Munich and Kassel) was nominated prime contractor for the Leopard 1 MBT and MaK of Kiel was selected as general contractor for production of the armoured recovery vehicle, armoured engineer vehicle and the armoured bridgelayer. MaK also built a small number of Leopard 1 MBTs under subcontract to Krauss-Maffei. MaK became Rheinmetall Landsysteme in 2000.

The first production Leopard was handed over to the Federal German Army in September 1965. Production of the Leopard 1 was completed by Krauss-Maffei in 1979 but resumed by Krauss-Maffei and MaK early in 1981 to meet the requirements of Greece and Turkey.

By late 1999, the Leopard 1 MBT was no longer in front-line service with the German Army although it was still used by reserve units.

Description

Leopard 1A1

The all-welded steel hull is divided into two compartments, the crew compartment at the front and engine compartment at the rear.

The driver is seated at the front of the hull on the right and is provided with a single-piece hatch cover that opens to the left. In front of the hatch are three day periscopes, the centre one of which can be replaced by an image intensification periscope for night driving.

The all-cast steel armour turret is mounted in the centre of the hull with the commander and gunner seated on the right and the loader on the left. Both the commander and the loader have a single-piece hatch cover that opens to the rear. The commander is provided with eight day periscopes for all-round observation, one of which can be replaced by an image intensification periscope for night observation. The gunner's station has the sights and one day periscope, the loader has two day periscopes for observation of a designated battlefield sector. Mounted in the turret roof, in front of the commander's hatch, is a TRP 2A zoom periscope with a magnification of ×4 up to ×20. This is swivel-mounted and can also be moved in elevation by hand within the gun elevation range. When the commander actuates a switch, the periscope remains in the given position, while the turret is being traversed. For night observation the commander can replace the zoom periscope with an active infra-red sight.

The gunner is seated in front of and below the commander and is provided with a TEM 2A range-finder which can be used in either the coincidence or stereoscopic modes. It has a magnification of ×16 and is mechanically linked to the gun and provided with superelevation cams to compute the superelevation angles for two types of ammunition. The gunner also has a TZF 1A telescope mounted coaxially with the main armament which has a magnification of ×8 and is provided with a movable scaled graticule to set the superelevation for the different types of ammunition. There is an ammunition resupply hatch in the left side of the turret and a stowage basket at the turret rear. Mounted over the main armament is an XSW-30-U infra-red/white searchlight, which can be removed and stowed at the rear of the turret when not required. In the infra-red mode, in conjunction with an infra-red sight which can be mounted in place of the commander's TRP 2A periscope, it has a range up to 1,200 m depending on conditions and in the white light mode a range of 1,500 m. The Leopard 1A1A2 and A1A4 has an LLLTV observation and sighting system.

The engine compartment at the rear is separated from the fighting compartment by a fireproof bulkhead. The complete power pack, consisting of the engine, transmission and cooling system, is provided with quick-disconnect couplings allowing the entire power pack to be replaced in the field within 20 minutes.

The steer/shift transmission is coupled directly to the engine and has four forward and two reverse gears as well as a pivot turn capability, a torque converter and lock-up clutch. The gears can be shifted electrohydraulically without interruption of tractive forces. The gear changing is dependent on the type of gear selector, either automatic or manual. The speed range can be preselected manually with the automatic gear selector.

The running gear includes seven dual light metal rubber-tyred roadwheels with the drive sprocket at the rear, idler at the front and four track-return rollers on either side. The first, second, third, sixth and seventh roadwheel stations are provided with hydraulic shock-absorbers. The Diehl tracks are provided with rubber-bushed track pins and the rubberised track pads can be replaced by snow grousers. Steel-reinforced rubber skirts for the top of the tracks can be folded vertically for maintenance. The track skirts perform two functions: they increase the ballistic protection and also reduce the whirl-up of dust.

The NBC system is installed at the front of the hull and produces an overpressure in the crew compartment. In normal use the system provides the crew with fresh air and in the NBC mode, filters contaminated particles by separate purification processes.

Standard equipment includes an automatic fire detection and extinguishing system, heater, hull escape hatch and infra-red reflection suppressing decontamination paint.

For underwater operations there are two fording shafts available: one for fording to a depth of 2.25 m and an extra shaft for fording to a depth of 4 m. Either shaft is to be fitted over the commander's hatch. The combustion air is drawn down through the shaft and the exhaust leaves via the exhaust pipes in the normal manner. Before entering the water all openings are sealed, the turret ring by an inflatable rubber ring. A hydraulic system controls the engine combustion air intakes, dust ejection blower valves of the combustion air cleaners, exhaust flaps of the engine and heater, pressure ventilation in the engine compartment as well as the dust ejection valves and the air intake of the NBC system. Two bilge pumps remove any water that enters the vehicle.

Main armament of the Leopard 1 is a now BAE Systems, RO Defence 105 mm L7A3 rifled tank gun which consists of a single-piece barrel with a screwed-on breech-ring and a bore evacuator. The barrel can be changed in the field in about 20 minutes. The semi-automatic breech mechanism automatically opens after each round is fired, ejecting the empty cartridge case into the spent cartridge container under the breech.

The Leopard 1 can fire all the standard 105 mm rounds manufactured by Canada, France, Germany,

Leopard 1A2 with thermal sleeve for 105 mm L7A3 gun and infra-red/white light searchlight on mantlet

Upgraded Leopard 1A5 MBT

Israel, UK and USA. Of the 60 rounds carried, 42 are in the hull and 18 in the turret.

Krauss-Maffei Wegmann states that the first round hit probability of the Leopard 1 weapon system, even when on the move and firing at moving targets, was significantly increased by the incorporation of the gun stabilisation system and modern fire-control systems, the latest of which has a primary stabilised line of sight, laser range-finder and an integral thermal imaging system. From 1986 to 1993 1,225 German Leopard 1A1A1/1A1A2 tanks were retrofitted and designated Leopard 1A5.

Mounted coaxially with the main armament is a 7.62 mm Rheinmetall MG3 machine gun which is provided with 1,250 rounds of ready use ammunition. The coaxial machine gun is operated by the gunner but can also be operated by the commander in the override mode. A second 7.62 mm MG3 machine gun is mounted at the commander's or loader's station for anti-aircraft defence. A total of 5,500 rounds of 7.62 mm machine gun ammunition is carried for the coaxial and anti-aircraft machine guns. Four electrically-operated 76 mm smoke grenade dischargers are mounted on either side of the turret.

German Army Leopard 1A4 MBT showing welded turret 0018616

Variants

New Leopard 1 Designators

Since the Leopard 1 MBT entered service with the German Army in 1965, it has gone through a number of upgrades including the installation of new radios and in some cases, the PZB 200 low light level TV (LLLTV) camera which is mounted over the 105 mm rifled tank gun. Most of the PZB 200s were transferred from Leopard 2 MBTs when the latter were fitted with a thermal sighting system.

Original designator	Equipment modification	New designator
Leopard 1A1A1	SEM 80/90 digital radios	Leopard 1A1A3
Leopard 1A1A2	SEM 80/90 digital radios and PZB 200 LLLTV camera	Leopard 1A1A4
Leopard 1A2	SEM 80/90 digital radios	Leopard 1A2A2
Leopard 1A2A1	SEM 80/90 digital radios and PZB 200 LLLTV camera	Leopard 1A2A3
Leopard 1A3	SEM 80/90 digital radios	Leopard 1A3A2
Leopard 1A3A1	SEM 80/90 digital radios and PZB 200 LLLTV camera	Leopard 1A3A3
Leopard 1A4	phased out of German Army service before receiving SEM 80/90 digital radios and/or PZB 2000 LLLTV camera	
Leopard 1A5	SEM 80/90 digital radios	Leopard 1A5A1

Leopard 1A1A1

The first model of the Leopard to enter service with the Federal German Army was called the Leopard 1 of which 1,845 were built in the first, second, third and fourth batches. From 1971 they were fitted with a thermal sleeve for the main armament, gun stabilisation system, new tracks and lateral skirts. The stabilisation system controls the gun in elevation and traverse so that the gunner is able to observe the terrain, acquire and then engage the target while moving across country, with a high degree of first round hit probability. With these modifications the Leopard 1 became the Leopard 1A1. These tanks have since been fitted with Blohm+Voss passive add-on armour on the turret and gun shield and are now known as the Leopard 1A1A1. The add-on armour consists of flexibly mounted, screwed-on armour steel plates with two-faced rubber lining. The armour also covers the turret bustle back and the gun shield is reinforced with armour steel plates. In addition, armour steel plates are welded on to the sloped front roof section. The extra weight caused by the fitting of this additional armour is about 760 kg, which makes the vehicle approximately the same weight as the Leopard 1A4. Another improvement on the Leopard 1A1A1 is the significantly enhanced combustion air intake filters. Leopard 1A1A1s with the PZB 200 LLLTV system are designated the Leopard 1A1A2.

Leopard 1A2

The fifth production batch of Leopards for the Federal German Army consisted of 342 tanks of which 232 were built with the modifications of the Leopard 1A1 (but without the spaced armour on turret and mantlet) and in addition had a turret of thicker cast steel, an improved NBC system and passive image intensification night vision equipment for the commander and driver.

Leopard 1A3

A total of 110 Leopard 1A3 tanks out of the fifth production batch were delivered to the Federal German Army, incorporating all the improvements of the Leopard 1A1 and 1A2 but with a new welded turret of spaced armour with a wedge-shaped mantlet. The turret rear stowage bin has been incorporated into the contour of the turret which accommodates the searchlight when it is not mounted over the main armament. In addition, the loader's periscope is movable in both elevation and azimuth. In 1981, the Greek government ordered 106 Leopard 1A3 MBTs fitted with the EMES 12A3 computerised fire-control systems, plus an option for an additional 101 vehicles for delivery from February 1983 through to April 1984. Krauss-Maffei of Munich built 73 MBTs and MaK (now Rheinmetall Landsysteme) of Kiel 33 MBTs and four ARVs. The Turkish order for 81 Leopard 1s comprised four ARVs built by MaK and 77 Leopard 1A3 MBTs fitted with the EMES-12A3 fire-control system (54 built by Krauss-Maffei and 23 by MaK).

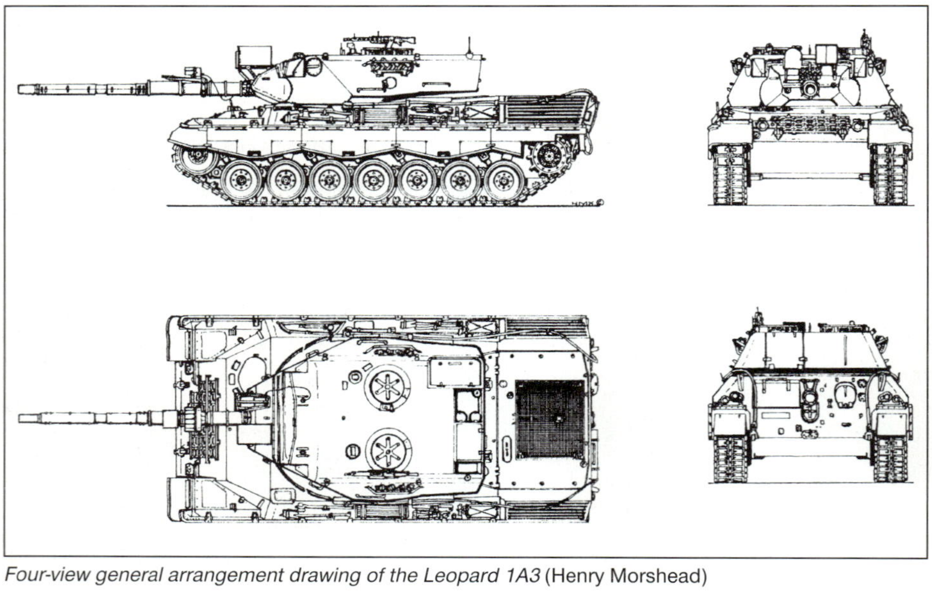

Four-view general arrangement drawing of the Leopard 1A3 (Henry Morshead)

Leopard 1A4

This was the final production model of the Leopard 1 of which 250 were built for the Federal German Army, 215 by Krauss-Maffei and 35 by MaK (now Rheinmetall Landsysteme). It is similar to the Leopard 1A3 but has an integrated fire-control system consisting of a stabilised panoramic telescope for the tank commander, gunner's primary sight with integral stereoscopic range-finder coupled to the fully stabilised main armament and controlled by a ballistic computer.

Late in 1985 it was stated that Turkey was to receive 150 German Army Leopard 1A4s. Before delivery to Turkey, 150 tanks were converted to the T1 configuration of the 77 Leopard 1s already in use by the Turkish Army. The first four Leopard 1 MBTs were handed over to the Turkish Army in mid-1990 and the final vehicles were delivered late in 1991.

Leopard 1A5

In 1982-83, six German Army Leopard 1 MBTs were fitted with different fire-control systems for comparative trials. These were the AEG/Telefunken Lemstar M/EMES-17, STN Atlas Elektronik FLP-10/EMES-18, and the Zeiss AFS-L/EMES-12A4, all being completed by a thermal imaging system for application under poor sighting and night combat conditions.

The German Army finally accepted the STN Atlas Elektronik GmbH EMES-18 combined with the thermal imaging system in 1984. First converted tanks were delivered to the German Army in December 1986 and a total of 1,300 was delivered through to 1992. Conversion work was undertaken by Wegmann of Kassel under a contract worth DM829 million; the major subcontractors were STN Atlas Elektronik for the

Leopard AEV with its auger being used to the front of the vehicle 0018617

Combat Improved Netherlands Gepard twin 35 mm SPAAG which has different tracking and surveillance radars (C R Zwart) ***2001**/0100282*

EMES-18 fire-control system and Carl Zeiss for the thermal imaging system. Of the 1,300 vehicles upgraded for the German Army, a batch of 75 was subsequently diverted to the Greek Army with deliveries being carried out by Krauss-Maffei until early 1994.

A recently developed modification kit for the suspension, which greatly increases the Leopard 1's cross-country mobility, has been delivered to some of the current users of the Leopard 1 by Krauss-Maffei Wegmann.

Late in 1996, it was confirmed that the Canadian Forces were to purchase 123 Leopard 1A5s. These turrets have now been fitted to the existing Canadian Forces C1 Leopard MBTs. Additional details are given earlier in this section under Canada. First vehicles were delivered to the Canadian Forces in late 1999.

In 1998, Germany and Greece signed a contract covering the sale of a further 170 surplus Leopard 1A5 MBTs.

Optional equipment for Leopard 1 MBT

Krauss-Maffei Wegmann offers a large number of modification kits for the Leopard 1 MBT. These include additional armour for the turret and mantlet, armoured skirts, automatic transmission and gear shift, the STN Atlas Elektronik fire-control system EMES-18 that has an integral thermal imaging system based on the American Common Modules, passive LLLTV sighting system PZB 200, brackets on glacis plate for snow grousers, dozer blade (already adopted by Australia, Canada, Denmark), cleaning system for driver's periscopes, improved turret and trunnion bearings, modifications to FCS to allow a 105 mm gun to fire APFSDS rounds, improved combustion air cleaners, external stowage boxes on hull sides (already adopted by Australia, Belgium, Denmark and the Netherlands), snorkel, passive periscope for commander and driver, track with replaceable track pads, stabilisation system for the main armament and a tropical kit. It can also be fitted with a fire (explosion) suppression system and fuel cell explosion suppressants.

The PZB 200 system consists of a combined low light level TV camera, for terrain surveillance and sighting of targets, which is mounted on the gun mantlet and so moves in elevation with the main armament, a monitor that can be seen by the commander and gunner, a control panel with on/off switch, a dimmer control for graticule illumination, a focusing switch and a selector switch for extreme illumination levels and simple built-in test equipment.

Denmark and Norway have already awarded contracts for the retrofit of the EMES-18 fire-control system.

Under contract to the Canadian Department of National Defense, Krauss-Maffei has developed and integrated into the Leopard C1 a gun/turret electric drive unit which replaces the standard hydraulic drive. ESW EXTEL is responsible for the electric drives, while Krauss-Maffei Wegmann is responsible for integrating it into the vehicle.

Norway has adopted the turret electric drive system for fitting into its existing Leopard/MBTs in conjunction with the EMES-18 fire-control system. These have now been delivered.

Other more recent options include a reverse camera system and a modification kit to convert the fire extinguishing system from Halon to nitrogen as the fire extinguishant, the latter having been developed by Krauss-Maffei Wegmann. These kits, except the camera, are currently being delivered to Norway and Denmark and will be implemented in the 170 Leopard 1A5s purchased by Greece.

Leopard 1 dozer blade unit

The complete dozer blade unit consists of the dozer blade, two pivot-mounted push arms and a hydraulic unit for lifting and lowering the blade. The complete dozer blade can be attached or removed in approximately 10 minutes without the need for a crane or an ARV and can be locked mechanically for travelling. All control elements are packaged into a watertight unit so that earth moving is possible as well as fording operations. A utility outlet for power supply is standard equipment on the vehicle. A simple control panel is attached to the driver's station and the power supply and control cable passes through the open driver's hatch or, alternatively, one of the slots normally occupied by one of the vision blocks.

To enable the Canadian Leopard C1 MBTs to operate the dozer blade as well as the mine plough and mine roller systems, an Engineer Equipment Interface Modification Kit (EEJM) has been developed by Krauss-Maffei. This allows the tank to operate the dozer blade or one of two mineclearing systems, permitting operation with the driver's hatch closed. A scoop welded on the turret roof plate of the chassis supplies power and control for the three adaptable engineer equipments at two receptacles. The exterior cable routed from the armour scoop to the add-on equipment is protected by armoured covers.

The dozer blade is used by Australia, Canada and Denmark. Details of this are given in *Jane's Military Vehicles and Logistics 2002-2003*. The EEJM kit is used by Australia and Canada.

Leopard 1 with 120 mm gun

For trials purposes, Rheinmetall has fitted one German Army Leopard 1 series MBT with the same 120 mm smoothbore gun as installed in the Leopard 2 and US M1A1/M1A2 MBTs. The said MBT and an additional MBT fitted with the Rheinmetall 120 mm gun, dummies of the latest state-of-the-art add-on armour and other improvements, such as infra-red suppressing measures, were subjected to compatibility trials in the first half of 1987. Krauss-Maffei Wegmann was the prime contractor for this programme.

Both prototypes were based on a Leopard 1A1A1 tank; a third prototype had the welded turret. All prototypes were fitted with the EMES-18 fire-control system.

When fitted with the 120 mm smoothbore gun, 42 rounds are carried, the same as in the Leopard 2, with 13 in the turret, two ready for use and 27 in the hull.

In early 2002 this upgrade remained at the prototype stage.

Leopard 1 with 105 mm IWS

Early in 1991, a Leopard C1 MBT of the Canadian Forces was fitted with the new private venture BAE Systems, RO Defence 105 mm Improved Weapon System (IWS).

The objective was to design a weapon system that would increase the firepower of existing in-service MBTs armed with the 105 mm L7/M68 rifled tank gun such as the Leopard 1 and M60, having performance levels comparable to current in-service 120 mm gun systems but not using a Depleted Uranium (DU) tungsten alloy penetrator.

The 105 mm IWS fits directly into the existing vehicle mounting and utilises the existing gun cradle and recoil system.

It is also capable of firing all existing NATO-approved 105 mm rounds (excluding APDS) as well as existing training ammunition. Built-in safety features ensure that the new high-pressure APFSDS round cannot be incorporated into standard 105 mm L7/M68 guns or into an IWS incorporating L7 components.

The new high-pressure ordnance weighs 1,334 kg and is made of Electro-Slag-Refined (ESR) steel to withstand increased pressures, with the barrel being chrome-plated to increase barrel life which is expected to be a minimum of 450 EFC (new ammunition).

It has a vertical sliding breech mechanism, fume extractor, concentric thermal sleeve and a screw-on muzzle brake with an efficiency of 25 per cent incorporated to reduce recoil loads.

The IWS has a recoil distance of 279 mm and can also be fitted with a muzzle reference system.

The new 105 mm APFSDS round has been designed to fit into existing ammunition stowage arrangements without any reduction in the number of rounds being carried by the MBT.

The new APFSDS round weighs a total of 19 kg and is 1.03 m long, with a muzzle velocity of 1,420 m/s.

The tungsten-nickel-iron, long, rod penetrator, manufactured by BAE Systems, RO Defence Speciality Metals, is located in a three-segment sabot which has been designed to reduce parasitic weight to a minimum. The length-to-diameter ratio is 23:1.

According to BAE Systems, RO Defence, the new penetrator has a mass approximately 40 per cent greater than that of current APFSDS rounds. No actual details of penetration have been revealed but at least 520 mm of RHA at a range of 2,000 m is probable.

The propelling charge comprises a stub brass cartridge case and a combustible sleeve which together house the multibase granular propellant and primer.

As of early 2002, 105 mm Improved Weapon System had not entered production.

Leopard 1 artillery observation vehicle

Following a competition, early in 1997 the German BWB awarded Krauss-Maffei Wegmann a contract for the design, development and construction of two prototypes of a new Armoured Artillery Observation Vehicle Leopard 1A5 (Beob Pz Art Leop).

These two vehicles have now been completed and are currently undergoing extensive trials by the German Army. It is expected that there will be another competition for the full-scale production phase with the original requirement being for 333 vehicles.

Krauss-Maffei Wegmann is the overall prime contractor for the Beob Pz Art Leop which is based on a modified Leopard 1A5 Main Battle Tank (MBT) chassis.

Krauss-Maffei is responsible for the overall system as well as for fitting and integration services regarding the chassis and the artillery installation kit. Wegmann, under contract to Krauss-Maffei, is responsible for all modifications to the turret. On 1 January 1999

Leopard 1 Biber AVLB laying bridge in position (C R Zwart) **2001**/0105872

Krauss-Maffei and Wegmann merged to form Krauss-Maffei Wegmann.

The modifications to the turret have been extensive and include the removal of the 105 mm gun and mount, loader's control panel, computer control panel and all ammunition racks.

The 7.62 mm MG3 machine gun mounted co-axial in the left side of the gun mantlet is retained as is the bank of four electronically operated 76 mm forward-firing grenade launchers either side of the turret.

The new Honeywell MAPS (Modular Azimuth Position System) navigation system (also fitted to the PzH 2000) is installed on the still freely moving mantlet.

Artillery specific equipment installed in the vehicle includes the navigation system, command and control computer, auxiliary display for the crew commander, interface electronic fire-control system and an operator's control panel.

The observer uses the existing stabilised HZF sight telescope to observe the terrain with the navigation system being adjusted so that it is parallel to the line of sight of the gunner's telescope. Target range-finding is accomplished using the laser range-finder integrated into the existing EMES-18 computerised fire-control system.

The range from the laser range-finder is transmitted as a digital signal to the new coupling electronic fire-control system (KEFLA) where they are linked to other data required for target acquisition. Integrated into the HZF sight is a WGB-X thermal image system. The observer can also lay the 7.62 mm MG3 machine gun using a TZF 3A sight.

The operator is seated in a new position to the rear of where the 105 mm gun was and has a control panel, work table and seat. The operator's control panel includes a command and control computer, distribution box, crew commander's auxiliary display, GPS receiver and radio. The command and control computer has a Windows interface with optional graphic map presentation and is the link between the operator and the artillery platoon fire direction centre or the fire support officer.

The command and control computer is the heart of the Beob Pz Art Leop and carries out a number of functions including operation and display of data of the navigation system and target acquisition, determination of own postion and target co-ordinates, connection to and from the fire-control system with an option for graphic position presentation.

The vehicle commander, who is seated at the tank commander's station, has a stabilised TRP 5A panoramic sight and vision blocks.

For reasons of cost it has not been possible to bring the engine, steering and exhaust system up to Euro II standard so an exception permit will be required.

Modifications to the chassis include a new turret/ mantlet lock, a new interface card integrated into the automatic gear selector unit, new rear view mirrors, improved crew station ergonomics, stowage and lighting. The existing single circuit brake system has been modified into a more efficient dual circuit system.

Australian Leopards

Between 1976 and 1978, the Australian Army took delivery of 103 new Leopard 1 MBTs and variants in the following configurations:

Designation	Quantity
Leopard AS1 MBT	90
Leopard AS1 ARV Medium	8
Leopard AS1 AVLB	5

The Australian company Tenix Defence Systems has installed an air conditioning system in the Australian Army Leopard 1 series MBTs. This was installed at the army's Darwin facilities.

Australia has recently obtained 55 Leopard 1 turrets from Canada for use as spare parts.

Australia is now considering a number of options to extend the life of its Leopard AS1 MBT fleet. These include installing a thermal night sight and a new or upgraded fire-control system. It is also possible that the existing turret could be replaced by a Leopard 1A5 turret with its more advanced fire-control system and thermal sighting system.

Belgian Leopards

After evaluating a number of tanks, Belgium placed its first Leopard 1 order in 1967, the first NATO country (apart from Germany) to order the tank. The first production Leopard 1 for Belgium was completed in February 1968. The Belgian vehicles had their 7.62 mm MG3 machine guns replaced by FN 7.62 mm MAG weapons plus minor stowage changes. From 1975 the tanks were fitted with stowage boxes similar to those on the Netherlands Leopards, a thermal sleeve for the main armament and the HR Textron Incorporated stabilisation system. The tanks were also fitted with the SABCA FCS which has since been adopted by Australia and Canada.

The SABCA FCS basically consists of an optical sight with an integral laser range-finder, seven sensors and an analogue computer. The computer determines the angles between the line of sight and the gun axis from the information it receives about the range of the target and other variables. The output of the computer is transformed, through a 2° of freedom gimballed mirror system, with torque DC motor drives and a compensated resolver feedback network, into a displacement of cross-hairs in the gunner's sight and to the gun via the gun stabilisation. The cross-hairs are brought back on to the target when the gun is automatically laid with the correct target elevation (or superelevation) and azimuth. The sensors measure ambient temperature, air pressure, temperature in the ammunition stowage area, gun wear, crosswind, trunnion cant and rate of turret traverse. Following trials with a prototype Leopard 1 fitted with the German Blohm+Voss add-on armour, as already installed on German and Netherlands Leopard 1 MBTs, the Belgian Army has refitted 132 tanks with this system. Conversion work was carried out at the Rocourt Arsenal in Belgium.

Early in 1987, the Belgian MoD awarded SABCA and OIP a BFr177 million contract for the development of a modification kit for the fire-control system, incorporating a thermal imaging system, which has now been installed in the 132 Leopard 1 MBTs of the Belgian Army originally delivered by Krauss-Maffei from 1968.

SABCA and OIP were jointly responsible for the integration and factory test of the new sight (under the leadership of OIP) and integration and factory test of the new complete TFCS (under the leadership of SABCA).

There were four phases of the programme. Phase 1 covered system definition and was completed in mid-1986. Phase 2 covered development and evaluation, Phase 3 covered industrialisation and preseries production and Phase 4 covered series production.

Under Phase 2, integration and factory test of the sight was carried out in October 1987 with the integrated and factory tested TFCS being handed over to the Belgian MoD at the end of April 1988 and final MoD review taking place in late July 1988. The first prototype modernised Leopard 1 MBT was delivered to the Belgian authorities on 6 October 1988.

The Phase 3 industrialisation phase was completed by mid-1989 with five preproduction systems completed by the end of 1990 and production systems being delivered through to 1997.

The installation of the new fire-control system provides the Leopard 1 with a high-performance thermal imaging device in a day/night sight with integrated laser range-finder.

OIP-Instrubel was created by Oldelft of the Netherlands in order to take over the former OIP-Optics which went bankrupt in 1987.

The Belgian Army decided to upgrade just 132 Leopard 1 MBTs to Leopard 1A5(BE) configuration with the remaining 202 to be disposed of by sale. Of these, 87 have been sold to Brazil with first deliveries taking place late in 1997.

One of the Leopard 1A5(BE)s has been fitted with additional armour on the turret and gun shield and is now known as the Leopard 1A6(BE). It is not certain if the remainder of the Leopard 1A5(BE)s will be upgraded to the Leopard 1A6 (BE) standard. The Belgium Army expects to purchase a new MBT between 2008 and 2015.

Brazilian Leopards

The Brazilian Army has purchased 87 unmodified Leopard 1 MBTs from the Belgian Army with the first batch of 33 vehicles arriving in Brazil late in 1997.

The second batch of Leopard 1 MBTs was delivered in mid-1998 with the remaining batch following early in 1999.

Canadian Leopards

Details of these are given under Canada.

Chilean Leopard 1 MBTs

In January 1999 the Chilean Army took delivery of the first batch of 14 Leopard 1-V MBTs.

A total of 200 used Leopard 1-V MBTs armed with a 105 mm L7 rifled tank gun were ordered from the Netherlands in 1998 and these have now replaced older vehicles, some of which date back to 1944.

The Royal Netherlands Army had phased out of service all of its 468 Leopard 1-V MBTs and Greece has already taken delivery of 170 vehicles.

Before delivery in Chile, the 200 Leopard 1-V MBTs have been fitted with the PZB 200 image intensification night vision system which is being supplied by GLS of Germany.

The PZB 200 was originally fitted to the Leopard 2 MBT as an interim solution, pending the installation of a thermal night vision system. They were then removed from the Leopard 2 and fitted to some Leopard 1s of the German Army.

When they have all been delivered, the Chilean MBT fleet will consist of 200 Leopard 1 MBTs and 60 French Giat Industries AMX-30 MBTs. Chile took delivery of 19 AMX-30s many years ago but has recently taken delivery of additional vehicles.

It is understood that GLS, a subsidiary of Krauss-Maffei Wegmann, submitted a letter of offer for the supply of an additional 170 Leopard 1 MBTs from German Army stocks. This batch would comprise 70 Leopard 1A5s and 100 earlier Leopard 1A1/1A2 MBTs. So far Chile has not exercised this option.

Chilean Leopard 1 variants

In 2001 RDM Technology of the Netherlands was awarded a contract by Chile for the supply of 11 specialised variants of the German built Leopard 1 series Main Battle Tank (MBT) from Royal Netherlands Army (RNLA) stocks.

The order comprised three BrPz Biber Armoured Vehicle Launched Bridges (AVLB), three PiPz-1 Armoured Engineer Vehicles (AEV) and two mineclearing breaching MBTs.

The latter will be based on a Leopard 1-V MBT chassis and are fitted with the Pearson Engineering Trackwidth Mine Plough (TWMP) and a device for carrying fascines to drop into anti-tank ditches and other battlefield obstacles.

In addition, another three Leopard 1-V MBTs have had their turrets removed and converted into bridge carrying vehicles to carry replacement bridges for the Biber AVLB. Each of the latter carries one bridge, which when opened out, is 22 m long and can span a gap of up to 20 m.

First Leopard 1 variants were delivered to Chile late in 2001 and all were delivered by early 2002.

Denmark

Denmark ordered 120 Leopard 1A3s in June 1974, the first of which were delivered in March 1976 and the last in November 1978.

Denmark has modernised the existing Leopard 1s to the 1A5 configuration and obtained a further 110 Leopard 1A3s from German Army stocks.

Late in 1989, Denmark awarded STN Atlas Elektronik a contract for the supply of 230 EMES-18 fire-control systems for installation in its original Leopard 1 MBTs plus those supplied by Germany. The whole tank fleet is now converted to the Leopard 1A5 configuration.

Greek Leopards

In 1981, Greece placed an order for the supply of 106 Leopard 1A3 MBTs plus four armoured recovery vehicles and these were delivered between 1983 and 1984.

The Greek Leopard 1A3 MBTs have the EMES-12A3 computerised fire-control system and the PZB 200 LLLTV system.

More recently the Netherlands has supplied Greece with 170 Leopard 1-V MBTs while Germany has supplied two batches of ex-German Army Leopard 1s; the first batch consisted of 75 upgraded Leopard 1A5s while the second batch consisted of a further batch of 170 Leopard 1A5s. The latter agreement was signed in 1998.

It is expected that the Greek Army will upgrade another batch of 170 Leopard 1 series MBTs with a new fire-control system, probably the STN Atlas Elektronik MOLF 2000, which is a further development of the MOLF installed in the upgraded M48A5 MBTs of the Greek Army.

The Greek Army is expected to upgrade its fleet of Leopard 1A3GR and 1(V) MBTs to the Leopard 1A5 plus configuration which will have either an STN ATLAS MOLF or EMES 18 computerised fire-control system installed.

It is expected that 105 Leopard 1A3GRs and 120 Leopard 1(V)s will be upgraded, with the remaining 50 Leopard 1(V)s being converted into specialist vehicles such as ARV and AEV.

A first batch of 10 Leopard 1(V)s have been converted for use in the mineclearing role having been fitted with the UK's Pearson Engineering Full Width Mine Plough system.

It is also said that consideration is also being given to convert some Leopard 1A5 MBTs into artillery observation post vehicles.

Italian Leopards

Although Italy was one of the three countries originally involved in the search for a European tank, it was not until 1970 that the Italians ordered the Leopard 1. The first 200, 92 of which were delivered in 1971 and the remaining 108 in 1972, were bought direct from Krauss-Maffei as it took several years for Otobreda to tool up for production of the tank. The first Italian Leopard was completed in 1974 and the last of the original order for 600 was completed early in 1978. The Italian Army then ordered a further 120 Leopard 1 MBTs which have now been delivered. MaK (today Rheinmetall Landsysteme) also delivered 69 ARVs and 12 AEVs to Italy.

Leopard 1A5(BE) of the Belgian Army from the rear clearly showing the large roof-mounted gunner's sight (C R Zwart) **2002**/0130617

Otobreda built 160 specialised versions of the Leopard 1 for the Italian Army, 68 ARVs, 28 AEVs and 64 AVLBs. For trials purposes Italy ordered four sets of the Blohm+Voss add-on armour system as installed on German and Netherlands Leopard 1 MBTs.

In 1993, a prototype Leopard 1 MBT was retrofitted into the Leopard 1A5IT configuration. Following trials with this prototype, 60 Leopard 1s of the Italian Army were upgraded to this standard in 1995, followed by a second lot of 60 vehicles in 1996.

Bofors demining vehicle

In late 1995, the now Bofors Defence of Sweden, teamed with Swedish DeMining and Toonstad Maskinfabrik of Norway, completed and demonstrated the prototype of a Demining Vehicle (DV) based on the chassis of the Leopard 1 MBT.

The DV can be operated from the fragment-protected cab mounted on top of the hull or remotely controlled and TV supervised from up to 500 m away with the crew under armour protection.

Mounted at the front of the vehicle is the demining roller which has its own engine. The actual roller is made up of a series of tungsten carbide teeth that either ignite the mines or chew them into small harmless pieces.

The DV weighs around 45 tonnes and can carry out demining to a maximum depth of 500 mm with maximum demining width being 4 m. The demining roller is powered by a 900 hp engine and the complete roller can be replaced in 10 minutes.

To enable the roller to follow the contour of the ground, it is tiltable and is also adjustable for depth. The roller's hydraulically operated supports are used when replacing the roller and for loading the roller on and off a transport vehicle.

For clearing flat surfaces, such as roads, a special tool can be fitted to the demining roller. Additional details of this vehicle are given in *Jane's Military Vehicles and Logistics 2002-2003*.

Minebreaker 2000

This has been developed by the German company of FFG and is based on a Leopard 1 MBT chassis and is used for humanitarian mineclearing operations.

76 mm OTOMATIC SPAAG

In 1991, for trials purposes, a Leopard 1 MBT of the Italian Army was fitted with a private venture Otobreda 76 mm OTOMATIC air defence turret and this was tested by the Italian Army and Air Force.

The two-man power-operated turret is fitted with a 76 mm gun that can fire a variety of ammunition types including a special APFSDS round. To engage aircraft or attack helicopters a five-round burst would typically be used. The turret is fitted with tracking and surveillance radars. Details of the 76 mm OTOMATIC system are given in *Jane's Land Based Air Defence 2002-2003*. This system remains at the prototype stage.

Netherlands Leopards

After evaluating the Leopard, Chieftain and MBT-70 in 1967-68, the Netherlands ordered 468 Leopard 1 MBTs between 1968 and 1970. The Leopards were delivered between 1969 and 1972.

The Netherlands Leopards have different radios, Netherlands-type smoke grenade dischargers and three stowage panniers. At a later date the tanks were fitted with a Honeywell gun stabilisation system, modified optical sights for British L52 APDS ammunition, a passive periscope for the driver and 7.62 mm FN MAG coaxial machine guns rather than German 7.62 mm MG3 weapons.

All 468 Leopard 1s of the Royal Netherlands Army were modified by 574 Tankwerkplaats at Leusden and fitted with appliqué turret armour supplied by Blohm+Voss of Germany. This is the same as that fitted to the Leopard 1A1A1 of the German Army. When modified they are designated the Leopard 1-V, V standing for *Verbeterd* or Improved. The EMES-12A3 AFSL-2 (Netherlands configuration) gunner's fire-control system was installed at the same time as the new armour. The EMES-12A3 was supplied by

Combat Improved Gepard twin 35 mm self-propelled anti-aircraft gun **2000**/0045459

Pionierpanzer 2 Dachs Armoured Combat Engineer Vehicle in travelling configuration **2001**/0105874

Leopard 1 artillery observation vehicle which has 105 mm gun removed
2000/0069523

Leopard Armoured Recovery Vehicle (Bergepanzer 2) in travelling configuration (Richard Stickland)
0011563

Specifications

Model	**Leopard A1A1, A2, A3, A4**
Crew	4
Combat weight	42,400 kg
Unloaded weight	40,400 kg
Power-to-weight ratio	19.57 hp/t
Ground pressure	0.88 kg/cm²
Length:	
(gun forward)	9.543 m
(hull)	7.09 m
Width: (overall)	3.41 m (with skirts)
Height	
(to top of commander's periscope)	2.764 m
Firing height	1.88 m
Ground clearance	0.44 m
Track	2.7 m
Track width	550 mm
Length of track on ground	4.236 m
Max road speed	65 km/h
Fuel capacity	985 litres
Max range	
(road)	600 km
(cross-country)	450 km
Fording	2.25 m
(with preparation)	4 m
Gradient	60%
Side slope	30%
Vertical obstacle	1.15 m

Trench	3 m
Engine	MTU MB 838 Ca M-500, 10-cylinder multifuel developing 830 hp at 2,200 rpm
Transmission	ZF 4 HP 250 planetary-gear shift with hydraulic torque converter, 4 forward and 2 reverse gears
Steering	double differential
Clutch	torque converter with mechanical interlock
Suspension	torsion bar
Electrical system	24 V
Batteries	8 × 12 V, 100 Ah (each)
Armament	
(main)	1 × 105 mm L7A3 rifled gun
(coaxial)	1 × 7.62 mm MG3 MG
(anti-aircraft)	1 × 7.62 mm MG3 MG
Smoke-laying equipment	4 × 76 mm smoke grenade dischargers either side of turret
Ammunition	
(main)	60
(7.62 mm)	5,500
Gun control equipment	
Turret power control	Electrohydraulic/manual
(by commander)	Yes
(by gunner)	Yes
(commander's override)	Yes

Commander's fire-control override	Yes
Max rate of power traverse	360° in 15 s
Max rate of power elevation	5.4°/s
Gun elevation/ depression	+20/–9°
Gun stabiliser	
(vertical)	Yes
(horizontal)	Yes
Range setting device	Yes
Elevation quadrant	Yes
Traverse indicator	Yes
Armour	
Hull nose	70 mm at 55°
Hull glacis	70 mm at 60°
Hull glacis top	25 mm at 83°
Hull sides upper	35 mm at 50°
Hull sides lower	25 mm at 0°
Hull top	10 mm
Hull floor	15 mm
Hull rear	25 mm at 12°
Turret mantlet	n/avail
Turret front	n/avail
Turret sides	n/avail
Turret rear	n/avail
NBC system	Yes
Night vision equipment	Yes

Honeywell and Zeiss with Oldelft as the Netherlands subcontractor.

In addition the Leopard 1-V has new track skirts fitted as standard. All Leopard 1-Vs of the Royal Netherlands Army were phased out of service by 1995 and Greece has already taken delivery of 170 vehicles free of charge. More recently 200 have been sold to Chile with first deliveries in January 1999.

Norwegian Leopards

Details of these and the specialised Norwegian models of the Leopard 1 are given under Norway later in this section.

Turkey

Under an agreement signed in Bonn in November 1980, Turkey received 77 Leopard 1A3 MBTs (54 built by Krauss-Maffei and 23 by MaK) and four ARVs (built by MaK) which were delivered between September 1982 and December 1983. These were fitted with the EMES-12A3 fire-control system and the PZB 200 LLLTV system.

In 1990-91, Turkey took delivery of 150 Leopard 1s from Krauss-Maffei which had been upgraded to the Leopard T1 configuration with tank fire-control system EMES-12A3. A further 80 Leopard 1A1A1s have already been delivered by the German government.

Some of these are to be fitted with a new fire-control system developed by the Turkish company of ASELSAN.

Leopard 1A5 MBT of the Danish Army from the rear with turret traversed to the right (Richard Stickland)
2002/0130618

Leopard armoured recovery vehicle (Bergepanzer 2)

The Leopard ARV was designed by Porsche and the first production vehicle was completed in September 1966 by MaK (Now Rheinmetall Landsysteme) of Kiel. The ARV is based on the chassis of the MBT but has a new hull. Standard equipment includes a dozer blade at the front of the hull, hydraulically operated crane on the right side of the hull which is pivoted at the front, main winch with a capacity of 35 tonnes (which can be increased to 70 tonnes with a guide pulley), electric wrench and a welding system. The vehicle, which weighs 39,800 kg and has a crew of four, has been designed for the recovery of damaged vehicles, towing disabled vehicles, changing vehicle components and lifting complete vehicles weighing up to 20 tonnes, carrying a spare Leopard MBT power pack, dozing and refuelling and defuelling other vehicles. Armament consists of a bow-mounted 7.62 mm MG3 machine gun, 7.62 mm MG3 anti-aircraft machine gun and smoke grenade dischargers. In 1978, the Federal German Army took delivery, from MaK (now Rheinmetall Landsysteme), of 100 product-improved Leopard ARVs which have hydraulic jacks mounted at the rear of the hull on the right side and a more powerful crane (Bergepanzer 2 LS). Details of the Leopard ARV are given in *Jane's Military Vehicles and Logistics 2002-2003.* The designation of the Bergepanzer 1 was allocated to the US-supplied M88 which was supplied in small numbers to the German Army.

Leopard Armoured Engineer Vehicle

This is based on the Leopard ARV and entered production in 1968. The only differences between the ARV and the AEV are that a heat exchanger has been installed, explosives are carried for demolition work, an auger is carried in place of the spare power pack on the rear of the hull and the dozer blade can be fitted with scarifiers to rip up the surface of roads. Full details of the Leopard AEV are given in *Jane's Military Vehicles and Logistics 2002-2003.*

Leopard Bridgelayer (Biber)

After trials with two competing designs, Model B was selected for production and the first production vehicles were completed by MaK (now Rheinmetall Landsystems) in 1975. The aluminium bridge is launched over the front of the vehicle as follows: on arrival at the obstacle the vehicle lowers the support blade at the front of the hull, the lower half of the bridge slides forwards until its end is lined up with the end of the upper half, the two sections are locked together, extended over the gap and lowered into position, the cantilever arm is withdrawn, the support blade raised and the vehicle can then be driven away. The bridge is 22 m long when opened out and can span a gap up to 20 m. Weight of the bridge and vehicle is 45,300 kg. Full details of the Leopard bridgelayer are given in *Jane's Military Vehicles and Logistics 2002-2003.*

MAN Leguan AVLBs

Details of these new AVLB on Leopard 1 chassis for the Norwegian Army are given later in this section under Norway. Further details will be found in *Jane's Military Vehicles and Logistics 2002-2003.*

Leopard driver training tank

This is essentially an MBT with its turret removed and replaced by an observation cabin with seats for the instructor and two trainee drivers with the third trainee driver seated in the normal position at the front of the hull. If required, the instructor can take over control of the vehicle at any time. The driver training tanks used by Belgium and the Netherlands do not have the dummy gun barrel as fitted to the German Leopard driver training tanks.

Pionierpanzer 2 Dachs Armoured Combat Engineer vehicle

Under contract to the German MoD, MaK (now Rheinmetall Landsysteme) developed a new Leopard Armoured Engineer Tank based on the chassis of the Leopard 1. This has a front-mounted dozer blade and an extendable hydraulic arm with a bucket on one side of the hull. In 1987, MaK was awarded a contract for the conversion of 104 Leopard 1 ARVs and the 36 existing Leopard 1 AEVs into the Pionierpanzer 2 Armoured Combat Engineer Vehicle. At the same time, the Canadian Forces placed an order for nine vehicles, the first of which was handed over in March 1990.

Main tasks of this vehicle are the preparing of river crossing sites and the preparation and removal of obstacles and blockages on the battlefield. Additional tasks include the recovery of vehicles and equipment, lifting of loads, towing, recovering vehicles using the main winch, recovery of tanks and other vehicles during deep fording operations and carrying out engineer work.

Full details of this vehicle are given in *Jane's Military Vehicles and Logistics 2002-2003.*

Leopard 1 new production orders as of 1 January 1992

	MBT	ARV	AEV	Bridgelayer	Gepard	Training	Total
Australia	90	8	nil	5	nil	nil	103
Belgium	334	36	6	nil	55	12	443
Canada	114	8	9	9	nil	nil	137
Denmark	120	nil	nil	nil	nil	nil	120
Germany	2,437	544*	36	105	420	60	3,602
Greece	106	4	nil	nil	nil	nil	110
Italy	920**	137	40	64	nil	nil	1,161
Netherlands	468	52	25	25	95	12	677
Norway	78	6	nil	nil	nil	nil	84
Turkey	77	12	nil	nil	nil	nil	89

* 444 standard plus 100 product improved
** 200 from Germany, remainder built under licence in Italy
Brazil: 87 vehicles from Belgium
Chile: 200 delivered from Netherlands
Germany: now has about 500 after passing vehicles to other NATO countries including Denmark (110), Greece (245), Norway (92) and Turkey (230)
Belgium: 132 have been upgraded to Leopard 1A5 (BE) with remaining 202 up for sale with 87 of these being sold to Brazil
Denmark: an additional 110 have been received from Germany and all are now Leopard 1A5 standard
Greece: total is now 525 following transfer of vehicles from Germany and the Netherlands
Netherlands: all are now for sale and 170 have been transferred to Greece and 200 to Chile
Norway: total is now 170 following transfer of 92 from Germany
Turkey: total is now 307 following transfer of 230 from Germany

Gepard Anti-aircraft Vehicle

This is used by Germany and the Netherlands (this last version having different tracking and surveillance radars) and consists essentially of a modified Leopard 1 MBT fitted with a power-operated two-man turret armed with twin 35 mm Oerlikon Contraves 35 mm cannon, tracking radar on the turret front and surveillance radar on the turret rear. Each 35 mm cannon has a total of 310 rounds of ready use ammunition for use in the air defence role and 20 rounds of APDS-T for use in engaging ground armoured targets.

For details of the Gepard self-propelled anti-aircraft gun system, see *Jane's Land-Based Air Defence 2002-2003.* Part of the Gepard fleet used by Germany and the Netherlands is being upgraded and first vehicles were completed in 1998. The Belgian systems have been phased out of service. Romania has taken delivery of a batch of ex-German Army Gepard SPAAG's.

76 mm anti-aircraft tank

MaK (now Rheinmetall Landsysteme) and Otobreda of Italy have suggested that the Leopard 1 MBT or Palmaria chassis could be fitted with the same turret armed with the 76 mm gun as installed on the OTO 76/62 76 mm self-propelled anti-aircraft gun. This system has been proposed as an anti-helicopter system and is designated the HEFAS (Hubschrauber-Erdziel, Flugabwehr-System — Helicopter Ground Target and Air Defence System). For trials purposes, Otobreda has fitted the 76 mm two-man power-operated turret of the OTOMATIC air defence system on a modified Leopard 1 chassis of the Italian Army with trials carried out between 1991 and 1992. As of early 2002 the OTOMATIC remains at the prototype stage.

Leopard 155 mm Self-propelled Gun

This was developed as a private venture by Krauss-Maffei Wegmann and Giat of France. It basically consisted of a Leopard chassis with the turret of the French 155 mm GCT self-propelled gun which is normally mounted on the AMX-30 chassis. This has not so far been adopted by any country.

Netherlands BARV

The Royal Netherlands Marine Corps has deployed a Beach Armoured Recovery Vehicle (BARV) based on a modified Leopard 1 MBT chassis with its turret removed. This was converted by the Dutch Army depot at Leusden.

Future Beach Recovery Vehicle

Late in 1999, the UK Defence Procurement Agency awarded Hugglunds Moelv of Norway a contract covering the supply of four Future Beach Recovery Vehicles (FBRVs) based on a modified Leopard 1 MBT chassis. These will replace the current Centurion Beach Armoured Recovery Vehicle (BARV). The first prototype was completed in 2001 and additional details are given under Norway.

NFW AVLB

As a private venture, the German company of NFW have built the prototype of a Leopard 1 AVLB which consists of a modified Leopard 1 MBT chassis fitted with bridge and launching system of the BLG-60 AVLB which was normally installed on a T-54/T-55 MBT chassis.

Firefighting vehicle

GLS and Pietzch have developed the heavy-duty firefighting vehicle called Jumbo - Tank PHF 20T which is based on a Leopard 1 MBT chassis and four of these have been sold to Argentina.

Status

Production complete. In service with Australia, Belgium, Brazil, Canada, Chile, Denmark, Germany, Greece, Italy, Norway and Turkey.

Manufacturers

Krauss-Maffei Wegmann GmbH & Co KG.
Rheinmetall Landsysteme GmbH.

UPDATED

Rheinmetall Landsysteme - R 301 (TAM) tank

Development

In mid-1974, the German company now known as Rheinmetall Landsysteme (at that time Thyssen Henschel) was awarded a contract for the design and development of a new medium tank for the Argentine Army under the designation of the TAM (Tanque Argentino Mediano), as well as an infantry combat vehicle under the designation of the VCI (Véhiculo Combate Infanteria), subsequently renamed VCTP. The contract covered the design, development and construction of three prototypes of both the TAM and the VCI.

The first prototype of the TAM was completed in September 1976 with the second and third prototypes finished the following year. On completion of company trials in Germany the TAMs were shipped to Argentina for both operational and firing trials. Chassis construction and final assembly were carried out in Buenos Aires while construction of the turret and the integration of the gun and fire-control system took place at Rio Tercero. The total Argentine requirement was for 512 TAMs and VCTP infantry fighting vehicles, but production was completed after about 350 vehicles had been built. Additional information on the TAM programme is given in this section under Argentina.

In 1981, Malaysia signed a contract for 102 members of this family in three versions: MBT (Lion), armoured personnel carrier (Tiger) and armoured recovery vehicle (Elephant), but no deliveries were made.

The R 301 (or TH 301 as it was originally referred to) was tested at the Thai Cavalry Centre in 1985-86, this being the first time that this vehicle had been tested in a non-NATO country. The vehicle was fitted with an AEG/

Telefunken fire-control system, FWM electrohydraulic gun-control system and with the main 105 mm armament stabilised in both elevation and traverse, gunner's sight with laser range-finder, gyrostabilised panoramic telescope for the tank commander and PZB 200 passive LLLTV system over the main armament with screens for the commander and gunner.

In early 1988, a TAM was trialled in Ecuador and a R 301 was demonstrated in Malaysia, but by 2002 none of these trials had resulted in firm orders being placed.

Rheinmetall Landsysteme R 301 with 105 mm gun showing the PZB 200 LLLTV camera which moves in elevation with the main armament mounted on the right side of the mantlet 0045706

Description

The chassis of the vehicle is based on the one used for the Marder 1 ICV used in large numbers by the German Army, but strengthened to take account of the increased weight of the vehicle and the stresses caused by firing the main armament.

The hull of the R 301 is all-welded steel with the well-sloped glacis plate providing protection against small arms fire, shell splinters and armour-piercing rounds up to 40 mm in calibre.

The driver is seated at the front of the hull on the left side and steers the vehicle with a conventional steering wheel. He is provided with three day periscopes and a single-piece hatch cover that opens to the right. The centre one can be replaced by a passive periscope for night driving. Two escape hatches are provided, one in the floor of the vehicle and one at the hull rear.

The power pack is mounted to the right of the driver with the glacis plate hinged on the right to allow access to the engine for maintenance. On the prototype, the MTU engine was identical to the one installed in the Marder 1 ICV and developed 600 hp at 2,200 rpm but production vehicles have a supercharged engine that develops 720 hp at 2,400 rpm. The engine is coupled to a Renk HSWL 204 planetary gearbox which gives four gears in both directions. The steering system is of the stepless hydrostatic type.

The all-welded turret is mounted to the rear of the vehicle with the commander and gunner on the right and the loader on the left. The commander is provided with a single-piece hatch cover and eight day periscopes. Mounted in front of the hatch is a non-stabilised Steinheil Lear Siegler TRP-2A panoramic sight with a magnification of between ×6 and ×20, identical to that fitted to the Leopard 1 MBT and which can be replaced by a night sight. The commander also operates the coincidence range-finder. The gunner is seated forward and below the commander and is provided with a Zeiss TZF sight with a magnification of ×8, combined with a swivelling and tilting day periscope. The loader is provided with a single-piece hatch cover that opens to the rear with a tilting periscope mounted in front of him. An ammunition loading port is provided in the left side of the turret. Mounted in the hull rear is a hatch that can be used either as an emergency exit or to reload the vehicle rapidly with ammunition.

Rheinmetall Landsysteme R 301 from the rear showing long-range fuel tanks **2002**

The torsion bar suspension system either side consists of six rubber-tyred roadwheels with the drive sprocket at the front, idler at the rear and three track-return rollers. The first, second, fifth and sixth roadwheel stations have hydraulic shock-absorbers. The tracks are by Diehl with removable rubber pads. The tops of the tracks are covered by detachable rubber skirts.

To extend the R 301's range, two 200 litre long-range diesel fuel tanks can be fitted at the rear. The basic vehicle can ford to a depth of 1.4 m but with preparation 2.25 m, or 4 m with a snorkel. Standard equipment includes an NBC system, a combustion heater for cold engine starts and the crew compartment, electric bilge pumps and a fire extinguishing system that can be operated manually or automatically.

Main armament is a 105 mm rifled gun developed in Argentina, although the Rheinmetall Rh 105-30 can also be fitted. Both weapons fire all standard 105 mm tank ammunition. It is provided with a bore evacuator, a thermal sleeve and is fully stabilised in both planes.

Specifications

(TAM, data in square brackets refer to the R 301, where different)
Crew: 4
Combat weight: 31,600 [31,000] kg
Unloaded weight: 29,200 kg
Power-to-weight ratio: 22.78 [24.19 hp/t
Ground pressure: 0.74 [0.79] kg/cm²
Length:
(gun forward) 8.17 [8.455 m
(hull) 6.775 [6.72 m
Width: 3.306 m
Height:
(to turret top) 2.436 m
Ground clearance: 0.45 m
Track: 2.62 m
Track width: 500 mm
Length of track on ground: 3.9 m
Max road speed:
(1st gear) 16 km/h
(2nd gear) 31 km/h
(3rd gear) 47 km/h
(4th gear) 76 [72] km/h
Fuel capacity: 650 litres
Max road range: 550 [500] km
(with auxiliary tanks) 900 [750] km
Fording: 1.4 m
(with preparation) 2.25 m
(with snorkel) 4 m
Gradient: 60%
Side slope: 30%
Vertical obstacle: 0.9 m [1.0 m]
Trench: 2.9 m
Turning radius:
(1st gear) 6.5 m
(2nd gear) 13.5 m
(3rd gear) 20 m
(4th gear) 30 m
Engine: supercharged MTU MB 833 Ka-500 6-cylinder diesel developing 720 hp at 2,200 rpm [750 hp at 2,400 rpm]
Transmission: Renk HSWL 204 planetary with 4 speeds both forward and reverse
Steering: double differential with hydrostatic steering control
Suspension: torsion bar
Electrical system: 24 V
Batteries: 8 × 12 V, 100 Ah
Armament:
(main) 1 × 105 mm rifled gun
(coaxial) 1 × 7.62 mm MG
(anti-aircraft) 1 × 7.62 mm MG
Smoke-laying equipment: 4 × 76 mm smoke grenade dischargers either side of turret
Ammunition:
(main) 50
(machine gun) 6,000
Gun control equipment
Turret power control: electrohydraulic/manual
(by commander) yes
(by gunner) yes
(commander's override) yes
Turret traverse: 360°
Commander's fire-control override: yes
Gun elevation/depression: +18/–7°
Gun stabiliser:
(vertical) yes
(horizontal) yes
NBC system: yes
Night vision equipment: yes

There are 50 rounds of 105 mm ammunition carried, 20 of which are in the turret. Mounted coaxially with the main armament is a 7.62 mm machine gun. A similar weapon is mounted on the roof of the tank for anti-aircraft defence. Four Krauss-Maffei Wegmann 76 mm smoke grenade dischargers are mounted on either side of the turret.

Optional equipment includes additional armour protection, radios and an intercom system.

Variants

R 301

Using company funds, Rheinmetall Landsysteme has built a fourth prototype of the TAM called the R 301 (this was originally called the TAM-4), which was completed late in 1978. Externally it is very similar to the model designed for Argentina but internally has significant differences in its power pack and fire-control system.

In addition to eight periscopes for all-round observation, the commander has a PERI-R12 sight as fitted to the final production model of the Leopard 1, the Leopard 1A4. This stabilised sight has an elevation of +60°, a depression of −10° and can be traversed through a full 360°. It has a magnification of ×2 or ×8 and is fitted with an external wiper.

The gunner has a fully stabilised periscope with a magnification of ×8 and an integral laser range-finder. Additional vision equipment includes one day periscope for the gunner and one day periscope for the loader.

Its main armament is a Rheinmetall-developed 105 mm gun designated the Rh 105-30 firing all standard 105 mm tank rounds and with an elevation of +18° and depression of −7°. The turret can be traversed through a full 360° in 15 seconds and the main armament is fully stabilised in both planes.

Information on target range is fed to the computer manually by the gunner or automatically by the laser range-finder, which has a range of 400 to 5,000 m. Angle of cant and lead angle are also fed automatically and the type of ammunition to be fired is fed in manually.

For engaging targets at night there is an LLLTV camera mounted on the mantlet which moves in elevation with the main armament. Both the commander and gunner have one TV monitor screen and, in addition to the optical cross-hairs being displayed, a second set of electronically generated cross-hairs is superimposed on the TV picture. The latter changes in accordance with the variable angle of aim (independent of ammunition type and target range) and is also corrected for vehicle cant. The two sets of cross-hairs are then brought together and the main armament fired.

The R 301 is powered by an MTU 833 Ka-500 diesel developing 750 hp coupled to a Renk HSWL 204 transmission which is fitted with a hydrodynamic retarder which is used to augment the steering effort and to brake the vehicle.

155 mm SPG

Argentina placed an order with Otobreda of Italy for 25 155 mm Palmaria turrets which will be installed on lengthened TAM chassis, with seven roadwheels either side instead of six. The turrets were delivered in 1986-87 with the first chassis being completed in Argentina in 1987. Full details of this are given in the Self-propelled guns and howitzers (tracked) section under Argentina.

Status

Production complete in Argentina but can be undertaken in Germany or Argentina upon receipt of orders.

Manufacturer

Rheinmetall Landsysteme GmbH.

UPDATED

Jung Jungenthal upgraded T-54/ T-55 MBT

Development/Description

Early in 1991 Jung Jungenthal completed the first prototype of an upgraded Russian T-54/T-55 MBT that it developed to meet the requirements of the Egyptian government.

Although the prototype is based on the T-54/T-55 chassis, it is also applicable to the Chinese equivalent of the vehicle, the NORINCO (China North Industries Corporation) Type 59 MBT.

The T-54/T-55 has been upgraded in the key areas of armour, mobility and firepower.

Battlefield survivability has been considerably improved by the fitting of additional passive armour onto the glacis plate of the vehicle while the frontal 180° of the turret has a horseshoe-type armour array which is similar to that installed on some Russian T-54/T-55/ T-62 MBTs. This armour provides both Chemical Energy (CE) and Kinetic Energy (KE) protection.

The original T-54/T-55 is not fitted with side skirts but the modernised T-54/T-55 is provided with these, with those either side of the driver's compartment being armoured. These hinge up vertically to allow access for maintenance purposes.

The engine and crew compartments are fitted with a fire detection/suppression system provided by Deutsche Graviner.

The original 100 mm D-10 series rifled tank gun is retained although a 105 mm BAE Systems, RO Defence L7 or US M68 rifled tank gun can be installed if required. A 7.62 mm machine gun is mounted coaxially with the main armament and a 12.7 mm machine gun is mounted on the turret roof for local and air defence purposes.

A bank of four Krauss-Maffei Wegmann electrically operated 76 mm smoke grenade dischargers can be mounted either side of the turret firing to the front of the vehicle.

The original gun control equipment has been replaced by a new all-electric gun control and stabilisation system developed by Curtiss-Wright Antriebstechnik (previously SIG) of Switzerland which gives a higher first round hit probability. A new turret race ring has also been fitted.

The torsion bars for the T-54/T-55 have been retained but new GLS hydraulic bump stops have been installed and the original tracks and roadwheels have been replaced by those fitted to the T-62. New ZF final drives have been fitted.

The hull now has a different shape, with new Superflexit bag-type fuel tanks installed under armour protection along the sides of the vehicle. These contain 1,200 litres of diesel fuel, which is an increase in capacity over those fitted in the standard T-54/T-55 MBT.

The original V-12 water cooled diesel engine has been retained but the manual transmission has been replaced by the German Renk LSG 3000 series fully automatic power shift transmission which has been in volume production for some years.

The power pack is fitted with the improved Behr cooling system and new combustion air filters. Like the original T-54/T-55 MBT, the upgraded vehicle can lay its own smoke screen by injecting diesel fuel into the exhaust outlet.

The company is now considering a number of upgrades, including the engine, for the Russian T-72 MBT used by many armies.

Specifications
T-54/T-55
Crew: 4
Combat weight: 41,000 kg
Length:
(with 100 mm gun) 9 m
(hull) 6.54 m
Height: 2.4 m
Ground clearance: 0.5 m
Max road speed: 48 km/h
Range: 400 km
Fuel capacity: 1,200 litres
Engine: V-55 V-12 water-cooled diesel developing 580 hp at 2,000 rpm
Transmission: Renk LSG 3000 fully automatic with 4 forward and 2 reverse gears
Armament:
(main) 1 × 100 mm D-10 series rifled gun
(coaxial) 1 × 7.62 mm MG
(anti-aircraft) 1 × 12.7 mm MG
(smoke grenade dischargers) 2 × 4 76 mm
NBC system: yes
Night vision equipment: yes

Other programmes

In addition to upgrading a single prototype of the T-54/ T-55 MBT, Jung Jungenthal has also been involved in a number of programmes to upgrade MBTs into Armoured Recovery Vehicles (ARVs) or Armoured Vehicle-Launched Bridges (AVLBs):

M48 MBT into ARV

This uses recovery equipment from the Leopard 1 ARV and is in service with Germany (eight), Greece (three) and Turkey (four). Details are

Jung Jungenthal upgraded T-54/T-55 MBT showing additional passive armour protection on the hull and turret 0045707

The power pack for the Jung Jungenthal upgraded T-54/T-55 consists of the original diesel engine coupled to a new ZF LSG 3000 automatic transmission 0045708

given in *Jane's Military Vehicles and Logistics 2002-2003.*

M48 MBT into AVLB
One prototype of this has been completed and details are given in *Jane's Military Vehicles and Logistics 2002-2003.*

Leopard 1 MBT into ARV
One prototype has been completed by the company and series production is now under way. This uses the same recovery equipment used in the Leopard 1 ARV and is covered in *Jane's Military Vehicles and Logistics 2002-2003.*

Leopard 1 MBT into AVLB
The first prototype of this was completed in May 1994. This has been developed to meet the requirements of the export market.

Status
Prototype. In early 2002, the upgraded T-54/T-55 MBT remained at the prototype stage. So far all the trials of this vehicle have taken place in Germany.

Manufacturer
Jung Jungenthal GmbH.

UPDATED

GREECE

Greek Main Battle Tank Competition

Late in 1998, the Greek Army completed its extensive series of trials of six MBTs with the detailed trials report being submitted to the Tank Division of the Greek Ministry of Defence in January 1999.

The six MBTs tested between October and December were General Dynamics Land Systems M1A2 Abrams (USA); Giat Industries Leclerc (France); Krauss-Maffei Wegmann Leopard 2A5 (in Swedish Strv 122 version) (Germany); Malyshev Plant T-84 (Ukraine); Omsk Machine Construction Plant T-80U (Russia); Vickers Defence Systems Challenger 2E (UK).

These trials lasted nine weeks and included firepower and mobility trials as well as road and cross-country trials. All the tanks were manned by a Greek crew for some of the trials.

The initial Greek Army requirement is for 246 MBTs, 24 ARV's, 12 AVLBs, 12 driver training tanks 12 turret maintenance systems plus training and spares. Most of these will be assembled/manufactured in Greece.

The first 250 new MBTs will probably be used to equip three independent armoured brigades (21st, 23rd and 35th) based in northeast Greece.

There is only one armoured fighting vehicle manufactured in Greece, Hellenic Vehicle Industry (ELBO), currently building the Leonidas full-tracked armoured personnel carrier for the home and export market. This company is expected to play a major role in the production of whichever MBT is selected by the Greek Army.

On 1st March, it stated that the Greek Tank Competition had been won by Krauss-Maffei Wegmann of Germany with the Leopard 2. Following negotiations, an order is expected to be placed for 170 vehicles rather than the 250 originally expected.

UPDATED

INDIA

Arjun MBT

Development/Description
In 1972, the Indian Army issued a requirement for a new MBT to replace the Vijayanta in production and work on this began at the Combat Vehicle Research and Development Establishment in 1974.

By the time the first prototype of the Arjun was unveiled in April 1984 Rs300 million had already been spent on the project.

A number of foreign companies provided technical advice and/or assistance in the development of the Arjun MBT including Krauss-Maffei (prime contractor for the Leopard 1 and Leopard 2 MBTs for the German Army), MTU (engine), Renk (automatic transmission) and Diehl (tracks) of Germany and Oldelft (Netherlands).

Suspension is of the hydropneumatic type. The aluminium alloy rubberised track blocks have rubber-bushed pins, the track tensioners have a built-in overload protection.

Initially six prototypes of the Arjun were built, fitted with a German MTU MB 838 Ka-501 diesel developing 1,400 hp coupled to a Renk automatic transmission. It is reported that none of these was armoured but had hulls and turrets of mild steel.

Production vehicles have a new composite armour, developed by the Indian Defence Metallurgical Laboratory, called Kanchan armour. This will be produced by Mishra Dhatu Nigam. The thermal night vision equipment has been developed by the DRDO laboratory at Dehra Dun.

Between 1983 and 1989, India is reported to have imported 42 foreign engines and transmissions for the prototypes of the Arjun at a total cost of US$15 million.

By late 1987, 10 prototypes of the Arjun, or MBT 90 as it is sometimes referred to, had been completed under the designation of the Mark I. Of these, six were handed over to the Indian Army for extensive trials and the remaining four have been retained for further development work and trials at the Combat Vehicle Research and Development Establishment (CVRDE).

The integrated fire-control system installed in the Arjun consists of a laser range-finder, ballistic computer, thermal imaging night sight, stabilised panoramic sight for the tank commander, standby (articulated) sighting telescope and an electronic gate to ensure a high first round hit probability.

According to the CVRDE, the third-generation director-type fire-control system with a 120 mm rifled gun electronically slaved to the sights, enables the gunner to locate, identify, track, range and engage moving targets while the Arjun is itself moving. This gives a high first round hit probability.

One of the prototypes of the Arjun MBT clearly showing the 120 mm rifled gun which is fitted with a fume extractor and thermal sleeve
0011559

The gunner's main sight consists of a day sight, thermal sight, laser range-finder and stabilised head common to all three channels. The common sighting head mirror is stabilised in elevation and azimuth.

The day sight has dual magnification while the thermal imager provides a night vision facility to the gunner and the commander to observe and engage targets under conditions of total darkness and smoke. The laser range-finder is integral with the gunner's sight.

The panoramic sight enables the commander to effect an all-round surveillance on the battlefield without moving his eyes from the sight and without being disturbed by the motion of the turret. The field of view is stabilised with the help of a two-axis rate gyro mounted on the platform of the head mirror. The sight has dual magnification.

The ballistic computer computes the gunlaying data from information provided by a number of sensors mounted in the vehicle and on manual inputs and provides the signals corresponding to the tangent elevation and azimuth offset.

To increase the accuracy of the firing, the fire-control system has a coincidence window that allows firing of the main 120 mm armament only when it reaches the desired position as dictated by the computer.

The 120 mm rifled gun will fire APFSDS, HE, HEAT, HESH and smoke rounds developed by the Indian Explosive Research and Development Establishment at Pune. The 120 mm gun is of ESR steel and fitted with a thermal sleeve and fume extractor. A 7.62 mm machine gun is mounted coaxial with the main armament and a 12.7 mm anti-aircraft machine gun is fitted for anti-aircraft defence.

The Explosives Research and Development Laboratory at Pune in Maharashtra has developed a new high-energy propellant used in the 120 mm rifled gun to give a higher muzzle velocity and therefore greater penetration characteristics.

In addition to the previously mentioned 120 mm rounds, a special anti-helicopter round is believed to be

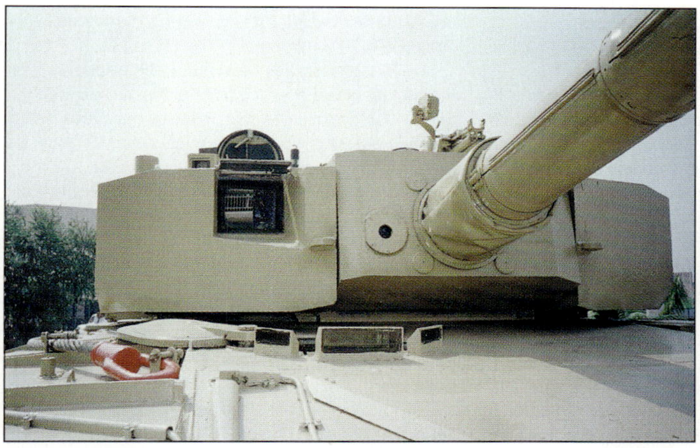

Close up of Arjun MBT turret showing gunner's sight **2000**/0069524

Arjun MBT from the rear showing additional fuel tanks **2000**/0069525

under development. All 120 mm rounds use a semi-combustible cartridge case.

Turret traverse and weapon elevation are all-electric, with the prototype systems being provided by FWM of Germany.

Mounted either side of the turret is a bank of nine forward-firing smoke grenade dischargers, the upper layer having five tubes and the lower layer four.

It was expected that production Arjun MBTs could have a locally-designed engine developing 1,500 hp coupled to a locally designed semi-automatic transmission with four forward and two reverse gears, hydrodynamic torque converter, retarder and integral system. The service brake consists of a hydraulically operated high-performance brake disc that is incorporated into the final drive.

The Arjun has an NBC system designed and built by the Bhabha Atomic Research Centre (BARC).

To further enhance battlefield survivability, the Arjun has an automatic fire detection and suppression system. Ammunition is also stowed in watertight containers to reduce the risk of fire.

In March 1993, it was reported that the Arjun MBT had sucessfully completed its firing trials.

In mid-1998, it was stated that a total of 32 Arjun MBTs had been built. This comprises 12 prototypes, 32 pre-production series (PPS) tanks, two torsion bar tanks, one test vehicle, one recovery tank and one Mk II Arjun.

The first batch of 125 Arjun MBT's have been ordered and the first of these have been delivered to the 2nd Independent Armoured Brigade.

In late 2001, MTU of Germany stated that it had overall responsibility for the power pack for the Arjun MBT and that India had placed a contract with the company for the supply of 124 power packs for installation in the first batch of production vehicles.

In addition, a further 31 replacement power packs had been ordered to bring the total up to 155 units.

This power pack consists of an MTU 10-cylinder MB 838 Ka-501 diesel with an output of 1,030 kW at 2,400 rpm, coupled to a Renk RK 304 S transmission and a cooling system.

Specifications

Crew: 4
Combat weight: 58,500 kg
Power-to-weight ratio: 23.93hp/t
Ground pressure: 0.84 kg/cm^2
Length: 10.194 m
Width: 3.847 m
Height: 2.32 m
Ground clearance: 0.45 m
Max road speed: 72 km/h
Max cross-country speed: 40 km/h
Range: 450 km (estimate)
Fuel capacity: 1,610 litres
Gradient: 77%
Side slope: 40%
Vertical obstacle: 0.90 m
Trench: 2.43 m
Engine: MTU 838 Ka 501 10-cylinder liquid-cooled diesel developing 1,400 hp at 2,500 rpm
Transmission: Renk RK 304 S epicyclic with hydrodynamic torque converter, 4 forward and 2 reverse gears
Suspension: hydropneumatic
Armament:
(main) 1 × 120 mm rifled gun
(coaxial) 1 × 7.62 mm MG
(anti-aircraft) 1 × 12.7 mm MG
Smoke laying equipment: 2 × 9 smoke grenade launchers
Ammunition:
(main) 39
(coaxial) 3,000 (estimate)
(anti-aircraft) 1,000 (estimate)
Gun control equipment:
Turret power control: electrohydraulic/manual
(by commander) yes
(by gunner) yes
(commander's override) yes
Gun elevation/depression: +20/–10°
Gun stabiliser:
(vertical) yes
(horizontal) yes
Range setting device: yes (laser range-finder)
NBC system: yes
Night vision equipment: yes

Variants

Projected variants include an armoured recovery vehicle, engineer tank, armoured vehicle-launched bridge and air defence (gun or missile) and artillery support, including an observation post vehicle.

For trials purposes, one of the prototype Arjun MBT chassis has been fitted with the South African LIW 155 mm/52 calibre T-6 artillery turret system (qv Self-propelled Guns and Howitzers - Tracked). As of early 2002 the Indian Army had not placed any orders for this system.

Tank EX

In February 2002, India revealed the Tank EX which has been developed by the Heavy Vehicles Factory as a private venture. This is essentially an T-72M1 series chassis fitted with the complete turret of the Arjun MBT.

Status

Low-rate production with 125 vehicles expected to be produced by the year 2002/2003.

Manufacturer

Heavy Vehicles Factory, Avadi.

UPDATED

Indian T-72M1 MBT production

The original Indian intention was to order only a small number of export T-72M1 MBTs from Russia pending the production of the locally-designed Arjun MBT which is covered in the previous entry.

It was then decided to undertake local production of the Russian T-72M1 MBT at Avadi and the first vehicles were completed in 1987 with first vehicles being delivered to the Indian Army the following year.

The first 175 tanks were produced with kits supplied by Russia. This was followed by progressive local manufacture. The eventual aim was to produce as much of the tank as possible in India with the target being 97 per cent. It is understood that this target was never achieved.

Production of the T-72M1 in India was running at about 70 vehicles a year with final vehicles being delivered in March 1994. In Indian Army service the T-72M1 is known as the Ajeya and by 1995, the Indian Army had around 1,100 vehicles in service.

Recent information has stated that India has 35 armoured regiments equipped with the T-72 M1 MBT Each regiment has 55 T-72M1 MBTs. This would give a total Indian T-72 MBT fleet of about 1,900 vehicles.

Under the T-72M1 upgrade programme, which is known in India as 'Operation Rhino', it is expected that

Indian-built T-72M1 MBT showing smoke grenade dischargers either side of the turret but without a 12.7 mm anti-aircraft gun fitted

the T-72M1 MBTs will be upgraded in a number of key areas including the installation of a new fire-control system, new diesel power pack, explosive reactive armour, land navigation system, laser warning system frequency-hopping radios and a new NBC system.

It is expected that around 180 to 200 T-72M1 MBTs will be upgraded to the full standard with a larger number only having part of the upgrade package.

Early in 1997, it was reported that Russia had offered India its Arena defensive aid suite as a possible counter to the recent acquisition by Pakistan from the Ukraine of the T-80UD MBT. It is superior in many respects to the T-72M1 which is currently the latest MBT in service with the Indian Army. Details of the Arena system are given in the entry for the Russian T-80 MBT later in this section.

T-90S in India

India has ordered 310 Russian T-90S MBTs which will be sufficient to equip at least five armoured regiments assuming that each is equipped with 55 T-90S MBTs. The first 124 will come from the Russian production line, with the remaining vehicles to be manufactured under licence at the Heavy Vehicles Factory. First T-90S MBT's were delivered to India in December 2001.

Variants

155 mm SPG

The Indian Army has a requirement for a 155 mm self-propelled artillery system mounted on a locally built T-72M1 MBT chassis. Four foreign 155 mm turrets were tested on a T-72 chassis in India with the South African LIW Division of Denel T-6 eventually being selected as the winner.

As of early 2002, it is understood that no production order had been placed for the T-6 turret system which is covered in detail in the Self-propelled guns and howitzers (tracked) section.

ARV

The Indian Army has taken delivery of 113 VT-72B armoured recovery vehicles based on the T-72 MBT chassis from ZTS of Slovakia and more recently Bharat Heavy Electricals Ltd (BHEL) has obtained a licence to undertake production of this vehicle in India. It is understood that the Indian Army has a requirement for up to 400 ARVs.

AVLB

India is known to have built at least one prototype of an armoured vehicle-launched bridge on the T-72 MBT chassis under the local designation of the Bridge Layer Tank (BLT) T-72.

The T-72 chassis has been fitted with a scissors-type bridge that is laid over the front of the vehicle with the bridge being similar to that fitted to the Kartik AVLB which is based on a stretched Vijayanta MBT chassis (qv).

The bridge itself is very similar to the former East German BLG-60 series. The system can lay a bridge 20 or 22 m long when opened out.

Details of the BLT T-72 are given in *Jane's Military Vehicles and Logistics 2002-2003*, as are details of the new Tank-mounted Multi Hop Assault Bridge (MHAB) and the Extended Span Assault Bridge (ESAB). For the trials programme, old Centurion MBT chassis have been used.

Manufacturer

Heavy Vehicles Factory, Avadi.

UPDATED

Vijayanta MBT production

In August 1961, following the evaluation of competing British and German MBT designs to meet an Indian requirement for a new MBT to be manufactured in India, an agreement was signed between Vickers (now Vickers Defence Systems) Limited of the UK and the Indian government. This agreement covered building prototypes in the UK, supplying 90 production tanks and building a new facility at Avadi, near Madras to undertake production of the Vickers MBT. The Indians call the tank the Vijayanta, or Victorious.

The first two prototypes were completed in 1963 and one was sent to India and the other retained in the UK for development work. The first production tanks were delivered from Vickers' Elswick works in 1965. The first Indian Vijayanta, which was mainly built from components supplied by the UK, rolled off the production line at Avadi in January 1965. India progressively undertook the production of more and more of the tank and eventually almost all of the tank was produced in India.

Production of the Vijayanta was completed some years ago after about 1,800 (although some sources have given figures varying from 1,200 to 2,200 units) vehicles were built. Full details of the Vijayanta tank are given in the UK section under the Vickers MBT.

Early in 1981, the then Marconi Radar Systems was awarded a US$6 million contract for its SFCS 600 tank fire-control system for the Vijayanta MBT with a US$6 million option for additional systems. Each contract was for 70 systems making a total of 140, but the option on the additional 70 systems was never exercised.

Indian Kartik AVLB on a stretched Vijayanta MBT chassis with bridge in travelling position 0002505

The Tank Electronics Support Centre at Madras has developed the Bharat Electronics Tank Fire-Control System Mk 1A (AL 4420) for the Vijayanta.

The Mk 1A features an improved sight mount and a fire-control linkage designed to minimise the play between the sight mount and linkage and between the linkage and the 105 mm rifled gun. A muzzle reference system is also provided to check and correct misalignment between the gun and sight axes caused by thermal deformation.

Bharat has also developed the much more sophisticated Mk 1B system (AL 4421) which incorporates a British Barr & Stroud (now Thales Optronics) Tank Laser Sight and a computer to increase first round hit probability.

Extensive trials have been conducted on two repowered Vijayanta MBTs and the configuration has been accepted by the Indian Army for introduction into service.

In mid-1993, Indian sources stated that, as the Arjun had been delayed, the programme to upgrade part of the Vijayanta MBT fleet, which was originally proposed as far back as the early 1980s under the name Bison, was to go ahead with up to 1,100 vehicles to be upgraded.

It is uncertain as to whether all of these will be upgraded to the same standard. The modifications are said to include:

(1) Installation of the T-72 M1 MBT diesel engine
(2) New fire-control system
(3) Additional passive armour
(4) Passive night vision equipment, including thermal sights
(5) Land navigation system.

The fire-control system is understood to be the Yugoslav SUV-T55A which was originally developed for the Russian-designed T-54/T-55/T-62 MBTs.

The armour to be fitted to the Vijayanta is believed to be the advanced Kanchan composite armour fitted to the Arjun which was designed by the Defence Metallurgical Research Laboratory at Hyderabad.

Indian-built Vijayanta armoured recovery vehicle with A-frame in the operating position 0002504

Specifications
Weight: 40,400 kg
Length:
(gun forward) 9.788 m
Width:
(with skirts) 3.168 m
Height:
(cupola doors shut) 2.711 m
Max road speed: 48.3 km/h
Gradient: 60%
Vertical obstacle: 0.914 m
Fording: 1.3 m
Operating range: 354 km
Commander's sight: periscope binocular and sighting periscope
Loader's periscope: No 23 Mk 1
Gunner's sight: sighting periscope No 30 Mk 1
Day vision:
(commander) episcope No 7 Mk 1 (M1)
(driver) periscope AFV 101A
Night vision:
(commander) periscope, armoured vehicle, IR 101A
(gunner) sight periscope, armoured vehicle, IR 102C
(driver) binocular, armoured vehicle, IR 101A
Engine: Leyland L60 developing 535 hp at 2,375 rpm

Indian-built Vijayanta MBT armed with a 105 mm gun which is not fitted with a thermal sleeve

Although the Vijayanta is essentially the Vickers Mk 1 MBT, fully described under the UK, Indian specifications differ from those of the latter in some areas and Indian sources have provided the following information.

In mid-2001, it was stated that the Indian Army will retire its fleet of 2,141 Vijayanta Main Battle Tank's (MBTs), replacing them with the newly acquired Russian T-90S MBTs that started to arrive by the end of 2001, and the Ajeya, the locally built Russian T-72M1 MBT.

A senior officer stated that the Vijayantas, first inducted into service in 1966, had 'outlived their expected life and are in the process of being phased out'.

Today 37 of India's 58 armoured regiments are equipped with around 1,700 T-72s and T-72M1 Ajeyas. 12 with some 540 Vijayanta and nine with over 400 T-55s, locally retrofitted with a 105 mm rifled gun.

In the late 1980s, the Defence Research and Development Organisation (DRDO) abandoned Operation Bison after upgrading 130 of a proposed 1,100 Vijayantas for around Rs10 million (US$222,222) each by equipping them with the then Marconi Radar and Control Systems SFCS 600 fire-control system and better sights. This made way for assembling and later building Ajeyas and to develop the indigenous Arjun MBT.

Operation Rhino, to upgrade around 1,500 Ajeyas was launched in the mid-1990s when the DRDO faced technical problems with Arjun, but this too was scaled down a few years later, when the purchase of the T-90S became imminent.

Only around 200 Ajeyas will now be fully retrofitted with new fire-control systems (FCSs) and power packs, explosive reactive armour, land navigation and laser warning systems, frequency hopping radios and nuclear, biological and chemical equipment for around Rs15 million each. The rest would be partially modernised.

Kartik AVLB

Seen for the first time early in 1989 was an armoured vehicle-launched bridge. It is based on the lengthened Vijayanta chassis used for the 130 mm Catapult self-propelled artillery system. This has been fitted with a scissors-type bridge of Eastern European design and is launched and recovered over the front of the chassis. Details of this are given in *Jane's Military Vehicles and Logistics 2002-2003*.

Vijayanta ARV

An armoured recovery vehicle based on a modified Vijayanta MBT has been designed and built in production quantities and details of this are given in *Jane's Military Vehicles and Logistics 2002-2003.*

Vijayanta 130 mm SPG

The Indian Army is believed to have about 100 Catapult self-propelled guns in service. This is a Vijayanta chassis with an additional roadwheel either side. The Russian designed and built turret has been removed and the chassis fitted with a 130 mm M-46 field gun firing over the rear. The weapon has limited traverse and about 30 projectiles and their associated charges are carried. Available details of this system are given in the Self-propelled guns and howitzers (tracked) section later in this volume.

Manufacturer

Heavy Vehicles Factory, Avadi.

UPDATED

IRAN

Defence Industries Organisation Zulfiqar MBT

Development/Description

Early in 1994, Iran unveiled a new MBT called the Zulfiqar which has been developed by 'Construction Crusade', an arm of the Islamic Revolutionary Guards Corps.

Russia is known to have provided Iran with a quantity of T-72 MBTs and recent reports have indicated that the Zulfiqar uses some components of the Russian T-72 MBT including the 125 mm smoothbore gun and automatic loader. The T-72S, which is an export version of the T-72 and is fitted with ERA, is now being manufactured in Iran under licence.

According to United Nations sources, between 1992 and 2000, Iran took delivery of significant quantities of Russian designed T-72 series MBTs from a number of countries, these included the following:

Country	Quantity	Delivery
Belarus	8	2000
Poland	34	1994
Poland	70	1995
Russia	100	1993
	20	1994
	2	1996
	3	1997

A detailed analysis of available information and photographs of the Zulfiqar reveal that the hull and turret are of welded steel construction and bear little resemblance to the T-72 at all.

The Russian T-72 MBT, like the earlier T-54/T-55/T-62, is powered by a diesel engine with the air intakes/outlets in the hull roof and the single exhaust outlet on the left side of the hull towards the rear.

The Iranian developed Zulfiqar MBT is armed with a 125 mm smoothbore gun fed by an automatic loader
2000/0069526

The new Iranian MBT is powered by a diesel engine with two exhaust outlets in the rear of the hull; this could indicate the engine is a V-type.

The layout of the Zulfiqar MBT is conventional with the driver front left, turret in the centre and the power pack at the rear. The driver has a single-piece hatch cover that opens to the right and three day periscopes for driving when closed up. The centre day periscope can be replaced by a passive periscope for driving at night.

Suspension either side appears to be of the torsion bar type with either side having six dual rubber-tyred roadwheels, idler at the front and large drive sprocket at the rear; there are five return rollers.

The roadwheels and other parts of the suspension appear to be very similar to those of the US M60 series MBT which has been in service with Iran for many years. The hull of the Iranian MBT is of the box shape rather than the boat shape of the M48/M60 series.

The front of the turret is well sloped and is angled to the rear to provide the maximum possible level of protection. There is a domed ventilator in the turret roof

Iranian Zulfiqar MBT with turret traversed to rear and clearly showing suspension 0018624

Iranian Zulfiqar MBT firing its 125 mm smoothbore gun 0018625

on the right side and this is similar to that fitted to the US M48/M60 series MBTs, as are the day/night driving lights mounted on the glacis plate.

The commander is seated on the right with the gunner on the left. The two examples of the Zulfiqar do have a number of minor differences in the commander's cupola. The first one is similar to that used in the T-72 MBT while the second one has a cupola that is similar to that developed in Israel and has an externally mounted 12.7 mm MG.

Main armament of the Zulfiqar comprises a 125 mm smoothbore gun which is fitted with a fume extractor and is understood to be fed by an automatic loader. The 125 mm gun is positioned in a very narrow mantlet and there does not appear to be a coaxial machine gun fitted.

Iran may well have the capability to assemble an MBT, but it must be considered very doubtful if every single component used in the Zulfiqar is produced in Iran. Some key subsystems must still be imported.

No detailed specifications of the Zulfiqar have been released but Iran sources have stated that it has a combat weight of 40 tonnes, is powered by a 1,000 hp diesel and has a maximum road speed of 70 km/h.

Iranian sources also claim that the Zulfiqar is fitted with a weapon stabilisation system and a computerised fire-control system which includes a laser range-finder. Night vision equipment is also fitted.

Late in 1999, it was stated that development of the third-generation Zulfiqar MBT had been completed and volume production for the Iranian ground forces had commenced.

Since the existence of the Zulfiqar MBT was first revealed in 1994, further development has taken place with the latest version claimed to have fundamental differences, especially in the turret.

According to Iran, the Zulfiqar MBT features an NBC system, good cross-country mobility, advanced fire-control system (possibly of European origin) and laser range-finder for improved first round hit capability and reinforced passive armour.

It is believed that the 125 mm smoothbore gun and its associated automatic loading system is the same as that installed in the T-72S MBT which is now being manufactured in Iran under a deal signed with Russia several years ago. This feeds the 125 mm projectile and then the charge into the 125 mm gun. A 7.62 mm machine gun is mounted coaxial with the main armament.

In many respects the Zulfiqar MBT is very similar in appearance to the now defunct Brazilian ENGESA Osorio MBT but with suspension similar to that of the 105 mm armed M60 MBT already in service with Iran. Prototypes of the Osorio were armed with a 105 mm rifled tank gun or a 120 mm smoothbore gun, both of which were manually loaded.

It is possible that the power pack of the Zulfiqar MBT may also be used in the upgraded Iranian T-54/T-55/ Type 59 MBT, which is also called the Type 72Z and covered in a separate entry. This consists of a Russian V-46-6 V-12 diesel developing 780 hp coupled to a SPAT 1200 transmission. This gives Zulfiqar a maximum road speed of 65 km/h.

FCS for T-72 MBT

The Islamic Republic News Agency has reported that Iran has now developed a 'firing-control system' for the Russian developed T-72 series Main Battle Tank (MBT), armed with a 125 mm smoothbore gun fed by an automatic loader.

During the unveiling of this system in Tehran, Defence Minister Ali Shamkhani stated that this control system could well enhance the capability of the T-72 MBT during operations carried out in darkness.

He added that the system costs about half of the current world market price and added that "it can compete technically" with similar European products.

It is assumed that Iran has developed a complete T-72 fire-control system that includes, not only the fire-control computer, but also gun-laying equipment, sensors, laser range-finder and day/night vision devices for commander and gunner.

In the past most Russian MBTs have used image intensification night sights, although more recently thermal devices have been developed which have greatly increased capabilities. Often this has been done in association with European defence companies trying to break into Eastern European markets.

Iran also recently revealed that it was manufacturing 125 mm ammunition for the Russian designed T-72 MBT which has been manufactured in Iran for several years. It is assumed that this T-72 MBT licence agreement also covered the complete fire-control system.

Iran has a large fleet of older MBTs that are in urgent need of being upgraded in a number of key areas, especially fire control. This new fire-control system for the T-72 may also be fitted to the local built Zulfiqar MBT which is now in production. Slovenia has supplied its EFCS-3 computerised fire-control system for installation in the Iranian Type 72Z, that is based on a Russian T-54/T-55 or Chinese Type 59 MBT.

Status

In production. In service with the Iranian Army. There are no known exports of the Zulfiqar.

Manufacturers

Defence Industries Organisation.
Shahid Kolah Dooz Industrial Complex.

UPDATED

Defence Industries Organisation Type 72Z MBT

Development

The Type 72Z MBT is essentially a Russian T-54/T-55 or Chinese NORINCO (China North Industries Corporation) Type 59 MBT upgraded in the key areas of mobility and firepower by the Shahid Kolah Dooz Industrial Complex of the Tank Modernising Industry.

The existence of the Type 72Z was first revealed late in 1996 by which time it was already in service with the Iranian Army using T-54/T-55 and Type 59 MBTs captured from Iraq as the basis for the upgrade. The designation T-72Z is derived from the date in the Iranian calendar and not to the Russian T-72 MBT design.

Description

The overall layout of the Type 72Z MBT is identical to the Russian T-54/T-55, covered in detail later in this volume, with the main changes being in the diesel power pack, armament and fire-control system.

The standard 100 mm rifled gun has been removed and replaced by a Western 105 mm M68 rifled tank gun fitted with a fume extractor and a thermal sleeve.

To improve first round hit probability against stationary and moving targets a Slovenian Fontona EFCS-3 computerised fire-control system has been fitted. This includes an Nd:YAG laser range-finder with a maximum range of 10,000 m and is accurate to plus or minus 5° and a ballistic computer which can be

Iranian T-72Z MBT which is armed with a 105 mm M68 rifled tank gun fitted with fume extractor and thermal sleeve 0018627

Specifications
Crew: 4
Combat weight: 36,000 kg
Power-to-weight ratio: 21.66 hp/t
Ground pressure: n/avail
Length:
(gun forward) 8.90 m
(hull) 6.54 m
Width: 3.35 m
Height (turret top): 2.37 m
Track width: 585 mm
Length of track on ground: 3.85 m
Max road speed:
(forward) 65 km/h
(reverse) 4.63 km/h
Fuel consumption:
(road) 240 litres/100 km
(cross-country) 260-450 litres/100 km
Fording: 1.4 m
Gradient: 60%
Side slope: 40%
Vertical obstacle: 0.85 m
Trench: 2.7 m
Engine: modified V46-6 12-cylinder four-stroke water-cooled diesel developing 780 hp at 2,000 rpm
Transmission: SPAT 1200 with 7 forward and 1 reverse gears
Armament:
(main) 1 × 105 mm M68 rifled gun
(coaxial) 1 × 7.62 mm MG
(anti-aircraft) 1 × 12.7 mm MG
Smoke laying equipment: 2 × 4 smoke grenade launchers
Ammunition:
(main) 38
(coaxial) 2,000 (estimate)
(anti-aircraft) 500 (estimate)
Gun control equipment:
Turret power control: powered/manual
(by commander) yes
(by gunner) yes
Turret traverse: 360°
Gun elevation/depression: +18 to –5°
Range setting device: yes (laser)
NBC system: yes
Night vision equipment: yes
Armour:
(side) 80 mm
(top) 20 mm
(rear) 40 mm

programmed for APDS, HEAT, HESH and APFSDS natures of 105 mm ammunition as well as the 7.62 mm coaxial machine gun.

The computer processes information from a number of sensors, including the gunner's laser range-finder and roof-mounted mast met system and is claimed to give a hit probability of more than 80 per cent.

The gunner's day sight has a magnification of ×10 and a 6° field of view while the second-generation image intensification night sight has a magnification of ×7 and a 6° field of view.

Key components of the EFCS-3 include gunner's day/night sight with laser range-finder, gun elevation sensor, met sensor, commander's panel, powder temperature sensor, digital ballistic computer, remote display, gunner's control block kit, gun triggering box, distribution box, servo electronic box and top cover assembly.

According to Iran, the EFCS-3 computerised fire-control system has the following operational advantages:

(1) High hit probability and very low reaction time
(2) Automatic and manual stabilisation and positioning of the gun on the target (within firing gate)
(3) Possibility of engaging stationary and moving targets when the Type 72Z is stationary or moving
(4) Automatic alignment test and correction capability at all ranges as well as independent line of sight
(5) Automatic or manual input of meteorological and ballistic parameters into computer calculations
(6) Software can be programmed for installation in different tanks and with different weapons
(7) Fitted with high-resolution, second-generation night vision device

The 7.62 mm coaxial and 12.7 mm anti-aircraft machine guns have been retained and a bank of four electrically operated smoke grenade launchers are mounted either side of the turret firing forwards.

Like most Russian MBTs, the Type 72Z MBT has the ability to lay a smoke screen by injecting diesel fuel into the exhaust outlet on the left side of the hull.

At least one example of the Type 72Z has been fitted with a roof-mounted laser warning device which is probably coupled to a commander's display and the electrically-operated smoke grenade launchers either side of the turret.

To keep down dust, new track skirts have been fitted as has a new live double pin track with replaceable rubber pads, which allow the Type 72Z to be used on roads without damaging the surface.

The original Russian T-54 is powered by a V-54 V-12 diesel developing 520 hp while the T-55 is powered by the V-55 V-12 diesel developing 580 hp.

Iranian sources state that the upgraded Type 72Z is powered by the V-46-6 V-12 diesel engine developing 780 hp which has been integrated into a new power pack which also includes the SPAT 1200 transmission which can be used in automatic or semi-automatic modes. The transmission also includes the brakes, steering, final drives and transfer case.

The V-46 V-12 diesel engine is also installed in early production T-72 series MBTs (for example T-72 and T-72A) and Iran could obtain these from a variety of sources apart from Russian.

The power pack consists of the engine, transmission and cooling system with the radiators being mounted over the transmission. It is reported that the power pack of the Type 72Z is also installed in the Iranian Zulfiqar MBT (covered in a separate entry) which is now being manufactured in Iran.

Iran has developed an explosive reactive armour system, details of which are given in a separate entry and this could be fitted to the Type 72Z to further improve its survivability on the battlefield.

Status

Production. In service with the Iranian Army. There are no known exports of this MBT.

Manufacturers

Defence Industries Organisation.
Shahid Kolah Dooz Industrial Complex.

UPDATED

Defence Industries Organisation Safir-74 MBT

Development/Description

Early in 1996, it was announced that the Islamic Revolutionary Guards Corps had completed trials of an upgraded T-54 MBT under the local name of the Safir-74 (Messenger 74).

These Russian-designed 100 mm armed T-54 MBTs were captured from Iraq during the 1980 to 1988 Gulf War and some sources have estimated that up to 190 T-54/T-55 series MBTs may be in service with the Iranian Army.

Iran is also understood to have 220 Chinese NORINCO (China North Industries Corporation) Type 59 MBTs in service (these being the Chinese version of the T-54), as well as 200 of the improved Chinese Type 69.

Recent information has stated that Iran took delivery of 300 Type 59 MBT's between 1982 and 1984 with another 200 being delivered between 1985 and 1986. These were followed by 240 Type 69 MBTs that were delivered between 1987 to 1988. According to United Nations sources, China did not export any MBTs of any type to Iran between 1992 and 2000.

The only information on the Safir-74 MBT is that it has major improvements to the output of the engine, transmission and firepower and that a fire extinguishing system has been fitted.

A new fire-control system has been installed (which may include a laser range-finder) which allows the Safir-74 to engage both stationary and moving targets while it is stationary or moving itself. The armour is also understood to have been upgraded.

It is possible that the original 100 mm gun has been replaced by a Western 105 mm L7/M68 rifled tank gun of the type fitted to the M60 and M48A5 MBTs supplied to Iran many years ago. For some years Iran has been manufacturing 105 mm tank ammunition.

Type 72Z MBT

Late in 1996, it was revealed that Iran had developed an upgraded T-54/T-55 MBT under the local name of Type 72Z. Full details of this are given in the previous entry. It is possible that the Safir-74 and Type 72Z are related, or could even be the same vehicle.

The upgrade work on the Type 72Z has been carried out by the Shahid Kolah Dooz Industrial Complex of the Tank Modernising Industry. This in turn is part of the Defence Industries Organisation.

T-54/T-55 MBT fitted with Iranian-developed explosive reactive armour package to its hull and turret
0011560

Iranian Explosive Reactive Armour

Iran has developed an Explosive Reactive Armour (ERA) system which could be fitted to new and existing armoured fighting vehicles, including MBT's, improving their battlefield survivability against attack by Chemical Energy (CE) and Kinetic Energy (KE) projectiles over the frontal arc.

The Iranian ERA package is very similar to that currently manufactured for the home and export market by Russia and this also provides protection from both CE and KE attack. First-generation Russian ERA only provided protection against CE attack.

This Iranian ERA system, which is understood not to have entered volume production so far, is manufactured by the Shahid Kolah Dooz Industrial Complex and during trials, stopped the main armour of an undisclosed MBT being penetrated by an Anti-Tank Guided Missile (ATGM) fitted with a tandem HEAT warhead.

The Iranian armour system comprises one composite layer that provides protection against KE and CE projectiles and an additional energetic material layer (explosive) that has the ability to protect against CE projectiles.

The complete ERA package consists of a number of small blocks which are bolted onto the MBT, typically on the glacis plate, nose, frontal part of the turret and skirting.

While most of the tank has a single layer of explosive reactive armour, some parts, for example the turret, have a double layer.

The T-72 ERA turret package weighs a total of 416 kg while the hull front (glacis plate and nose) package weighs 281 kg and the hull side (for example, skirts) package weighs 494 kg. This gives a total weight of 1,191 kg.

When allowing for spaces for the sighting systems, main and secondary armament, between 60 and 65 per cent of the turret front is protected by the ERA while for the hull front it is between 70 and 80 per cent and for the hull sides 60 to 62 per cent.

According to the Iranian manufacturer, this ERA package is applicable to any type of MBT, for example T-55, T-72 and T-80 with some modifications of the attachment assembly.

Main advantages have been summarised as low weight, low cost, high level of protection, easy to install and maintain, waterproof and cannot be activated by gas arc welding and cutting.

The ERA packages will not detonate if dropped from a height of 5 m, complete safety from small arms fire up to 30 mm in calibre, cannot be activated by grenades, resistant to attack by napalm incendiary mixtures and safety from attack by 125 mm HE projectile fragments at ranges of less than 10 m.

Status

Modernisation programme under way. In service with the Iranian Army. As far as it is known, there have been no exports of any MBTs by Iran.

Manufacturers

Defence Industries Organisation.
Shahid Kolah Dooz Industrial Complex.

UPDATED

IRAQ

Modified Iraqi MBTs

Development

As a result of combat experience during eight years of fighting against Iran, Iraq modified many Chinese and Russian supplied armoured fighting vehicles to meet its own specific requirements and some of these were shown during an international defence exhibition held in Baghdad in April 1989.

In 1989, Iraq also stated that it was assembling the Russian T-72M1 MBT in Iraq under the local name of the Assad Babyle (Babylon Lion), with local industry manufacturing the 125 mm ordnance, breech mechanism, ammunition and some of the electronics.

A small number of Iraqi T-72M1 MBTs was fitted with an ATGW decoy system which was mounted on the turret roof forward of the gunner's hatch. This was designed to decoy Western type TOW, MILAN and HOT-type ATGWs.

Following the 1991 Gulf War, it is understood that there has been no further local production/assembly of the T-72M1 in Iraq.

During the Baghdad International Exhibition for Military Production held in the spring of 1989, Iraq exhibited a number of upgraded MBTs.

Although the Iraqi Army referred to these as T-54/T-55 conversions, some of the chassis/turrets were in fact Chinese-supplied NORINCO (China North Industries Corporation) Type 59/Type 69 chassis, large numbers of which were supplied to Iraq during the 1980s conflict with Iran.

As far as is known, the only upgraded T-54/T-55/Type 59/69 MBTs encountered by the Allied Forces during Operation Desert Storm were a few with additional passive armour over their frontal arc which gave a high degree of protection against HEAT projectiles.

Before the invasion of Kuwait, the Iraqi Army's MBT fleet is believed to have consisted of some 5,500 vehicles of all types. By early 2002, it is estimated that Iraq had about 2,700 MBTs of all types, but many of these were non-operational due to a shortage of spare parts following the 1990-91 Gulf War.

Recent information has stated that China supplied Iraq with at least 1,300 Type 59 MBTs which were delivered between 1982 and 1988 and 1,500 Type 69 MBTs which were delivered between 1983 and 1987.

T-55 with additional armour

Some T-55s have been fitted with appliqué multilayer passive armour to the glacis, hull sides and turret front, side and rear. Inside, the armour package is understood to consist of layers of steel and thin rubber.

In addition the vehicle was fitted with new night vision equipment and a bank of four electrically operated

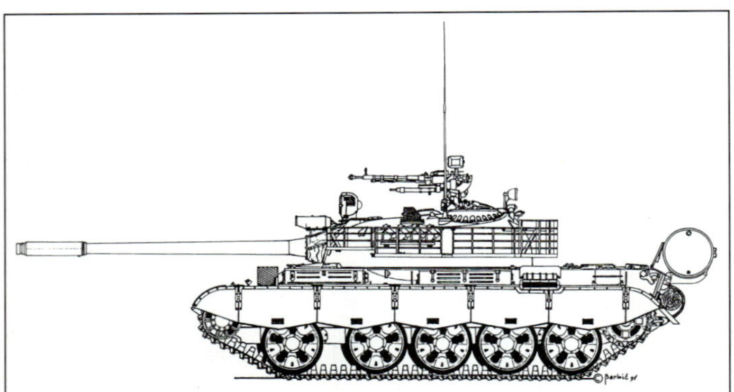

Iraqi NORINCO Type 69-II MBT with a laser range-finder over the 100 mm gun (Vasco Barbic)

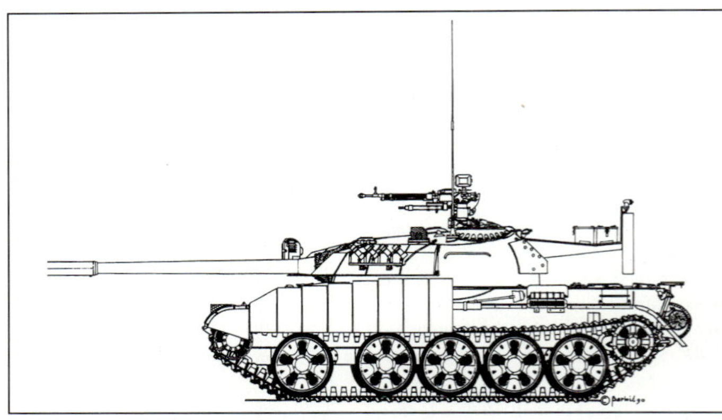

Iraqi NORINCO Type 69-II MBT with additional armour fitted to the turret and hull (Vasco Barbic)

T-55 fitted with a 125 mm smoothbore gun as installed in the T-72M1 MBT
0011558

Iraqi T-72M1 MBT showing ATGW decoy system mounted on the turret roof and with the circular shutter in the closed position
0011557

smoke grenade dischargers either side of the turret, firing forwards.

These improvements have, however, increased the weight of the vehicle so the power-to-weight ratio has been reduced from 16 to 14.7 hp/t.

Other modified Iraqi MBTs

These include a T-55 series (or its Chinese equivalent) with a 125 mm smoothbore gun, artillery observation post vehicle with obsevation mast and 160 mm self-propelled mortar. As far as it is known none of these ever entered service with the Iraqi Army. Details were given in *Jane's Armour and Artillery 1998-99*.

Status

The T-55 with additional armour entered service with the Iraqi Army in quantity. As far as it is known, Iraq has never exported any MBTs.

UPDATED

T-55 with additional armour from the front
0011555

ISRAEL

Merkava MBT

Development

Israeli experience in the 1967 Middle East conflict, campaign proved that mobility was no substitute for armour protection. It therefore decided at an early stage that the main emphasis would be placed on armour, with firepower and mobility second and third priorities. Design work on the tank started as early as 1967 but detailed design work, under the direction of Gen Israel Tal, did not begin until August 1970. Before the construction of the first prototypes of the Merkava MBT a number of test rigs based on M48 and Centurion tank chassis was completed to prove the basic concept. The first prototype of the Merkava was completed in 1974.

In May 1977, Israel finally announced that it had developed a new MBT called the Merkava (or Chariot) to the prototype stage and that a series production run of 40 tanks was being built. First production tanks were delivered to the Israel Defence Force 7th Armoured Brigade in 1979. The Merkava was first used in action during fighting in the Lebanon in the summer of 1982. First production Merkava Mk 2s were delivered to the Israel Defence Force in December 1983.

No production figures of the Merkava have been released, but it is estimated that by early 2002 total production amounted to over 1,000 vehicles, the current production version being the Mk 3 which became operational with the Israeli Armour Corps in early 1990.

As the Mk 1 and Mk 2 vehicles come back for base overhaul they are being upgraded with some of the components of the Mk 3, but this excludes the 120 mm smoothbore gun.

Merkava Mk 1 MBT showing turret profile and 105 mm M68 gun fitted with fume extractor and thermal sleeve. Note the exhaust port at the front of the hull above drive sprocket (Israeli Ministry of Defence)
0018628

An outline comparison of the three Marks of the Merkava MBT is shown in the accompanying Table 1.

In 1996, it was revealed that the more advanced fire-control system of the Merkava Mk 2 had been backfitted to the earlier Mk 1 as they are overhauled.

Description

Merkava Mk 1

The hull of the Merkava is made of cast and welded armour with a well-shaped glacis plate with the right

Specifications
Table 1

Model	**Mk 1**	**Mk 2**	**Mk 3**
Weight	60,000 kg	61,000 kg	65,000 kg
Main armament	105 mm rifled gun	105 mm rifled gun	120 mm smoothbore gun
Main gun ammunition	62	62	48
60 mm mortar	external	Internal	internal
Engine	900 hp diesel	900 hp diesel	1,200 hp diesel
Transmission	semi-automatic	automatic, electronic control	automatic, electronic control
Suspension system	1st generation	2nd generation	2nd generation
Fire-control system	new electronic computer, new laser range-finder, new night vision equipment	new electronic computer, new laser range-finder, new night vision equipment	new electronic computer, new, laser range-finder, new night vision equipment, stabilised vertical and horizontal line of sight and automatic target tracker
Turret-control system	electrohydraulic	Electrohydraulic	electric/electronic
Armour	spaced armour	combined spaced and add-on special armour	modular special armour
Ammunition survivability	heat-resistant	heat-resistant	heat-resistant and resistant to chain detonation
NBC system	overpressure	Overpressure	overpressure, central crew air conditioning
Electromagnetic warning system	elementary	Elementary	improved

The current production model is the Mk 3.

Merkava Mk 3 MBT fitted with a more powerful engine, a 120 mm smoothbore gun and new special armour (Marsh Gelbart) 0045709

side higher than the left. Behind the first layer of cast armour is a space filled with diesel fuel and then another layer of armour. This spaced armour gives the tank protection from HEAT projectiles and ATGWs.

The layout of the Merkava is unconventional, with the turret and fighting compartment at the rear of the vehicle. The driver is seated on the left side of the hull, forward of the turret, with the engine compartment to his right. The driver is provided with a one-piece hatch cover that opens to the left and three day observation periscopes for driving with the hatch closed; the centre one can be replaced by a passive one for night driving. The driver can reach his compartment through the main crew compartment as the backrest of the driver's seat folds forwards.

The Teledyne Continental (now General Dynamics Land Systems) AVDS-1790-6A V-12 diesel develops 900 hp and is coupled to an Allison Transmission Division of General Motors CD-850-6BX transmission. The power pack is in fact a more powerful version of the engine fitted to the M60 and M60A1 MBTs and the transmission is also similar to that installed in these tanks. Access to the engine compartment is via two flaps which are opened by springs after the locks are released. The engine can be replaced in the field in about 60 minutes. The air-cooling filter vent is positioned in the upper part of the hull, forward of the driver's seat, with the outlet being located on the opposite side. The exhaust outlet is on the right side of the hull, above the skirting plates over the second roadwheel.

The wedge-shaped turret, which has been designed to accept either a 105 mm (M68) or Israel Military Industries' 120 mm smoothbore gun, is cast with a welded front. It has a small cross-section and a large overhang at the rear. The radios and hydraulics are mounted in the turret bustle. The commander is seated on the right side of the turret with the gunner seated forward and below the commander. The commander has no cupola but is provided with a hatch cover that opens to the rear and five day periscopes for all-round observation. The commander's hatch can be raised manually giving direct all-round observation while retaining full overhead protection. Mounted forward of this hatch cover in the roof of the tank is a sight that can be traversed through 360°, with a zoom magnification of from ×4 to ×20. The rotating head of the commander's periscope is linked to the turret traverse system by a counter-rotating device. The gunner's optics are in the forward part of the turret roof and right-angled ribs in front of the optics stop shell splinters and small arms fire from damaging them. The gunner's periscope has magnifications of ×1 and ×8 and incorporates a laser range-finder. The loader is seated on the left rear side of the turret and is provided with a single-piece hatch cover opening to the rear and a single periscope.

There are three hatch covers in the rear of the hull: the left one gives access to the batteries and the right one gives access to the NBC pack. The centre one is a two-part door, the upper part opening upwards and the lower part downwards, through which ammunition or wounded can be loaded. This hatch can be opened from the outside but locked from the inside. A 60 litre water tank is provided above the rear hatch. The infantry telephone is mounted at the rear of the hull on the left side. Many Merkava MBTs have been fitted with closely spaced chains with ball ends around the lower part of the turret bustle. This has been designed to detonate HEAT projectiles before they can hit the turret ring.

The Mk 1 Merkava has six 790 mm Centurion-type rubber-tyred roadwheels either side with the drive sprocket at the front, idler at the rear and four track-return rollers. Each roadwheel is suspended by a separate helical spring with suspension arms for two roadwheels, each caged in a housing. The first, second and fourth track-return rollers support the inside of the track only.

A white light and an infra-red driving light mounted on either side of the glacis plate can be folded down in action to avoid damage from shell splinters and small arms fire. The 1 kW searchlight is mounted vertically in the bustle under armour protection to the rear of the loader's position and is controlled by the tank commander. The Merkava is equipped with an NBC system and a Spectronix explosion suppression system. This was not fitted to original Merkava MBTs but is now standard on all new production vehicles and has been backfitted to older vehicles as they return for overhaul.

Main armament is a standard 105 mm M68 rifled tank gun fitted with a thermal sleeve, manufactured in Israel by Israel Military Industries and also fitted to most other Israeli tanks. It has an elevation of +20° and a depression of −8.5°. A travelling lock is provided for the 105 mm gun on the right side of the glacis plate. The Merkava carries a total of 62 rounds of 105 mm ammunition of which six are stowed below the turret ring for ready use and the remainder in the hull rear, 12 in two-round containers and 44 in four-round containers. No ammunition is stowed above the turret ring or in the hull front and all ammunition is in special containers. In addition to the standard 105 mm HEAT and HESH rounds the gun will fire an APFSDS-T round called the M111 developed by Israel Military Industries, which is claimed to be superior to the American M735. More recently, Israel Military Industries has introduced a new 105 mm APFSDS-T round called the M413 which has a muzzle velocity of 1,455 m/s when fired from an L7/M68 gun with a maximum effective range of 6,000 m.

The 105 mm gun is fitted with a thermal sleeve designed and built by Vidco Industries and this is also fitted to Israeli M48, M60 and Centurion series MBTs. This eliminates most of the error caused by barrel bending due to sun, wind, rain and other effects and improves first round hit probability.

A 7.62 mm machine gun is mounted coaxially to the left of the main armament and a similar weapon is mounted at the commander's and loader's station; these weapons can be lowered to reduce the profile of the Merkava.

The 7.62 mm coaxial machine gun is fed from a 2,000 round continuous belt which is between the plates of the spaced armour. The machine guns are 7.62 mm MAGs, manufactured under licence from FN Herstal. Some Merkavas have been observed with a 12.7 mm M2 HB machine gun over the gun barrel. A 60 mm Soltam Systems Commando mortar is carried by the Merkava for which 30 mortar bombs are stored in a compartment in the turret rear. This mortar fires high-explosive, smoke and illumination bombs and helps conserve 105 mm ammunition.

During the 1973 Middle East war, many Israeli tanks ran out of ammunition and so the Merkava has been designed to carry a large supply in the rear of the hull. The Merkava can also be used as a command post with the ammunition supply containers removed. By reducing the ammunition load the Merkava can also carry troops, for example 10 infantry can be carried by reducing the ammunition load by 45 rounds or a commando squad of three together with their radios if the ammunition load is reduced by 25 rounds. It must be emphasised that the ability to carry infantry is only an option for use in special circumstances as the infantry have no vision devices at all.

The Merkava has a digital fire-control system designed by Elbit Computers Limited of Haifa and a laser range-finder which can be used by the commander or gunner. This fire-control system is called the Matador Mk 1. The laser range-finder is manufactured by ELOP Electro-Optics Industries and feeds information into the computer. The system is built

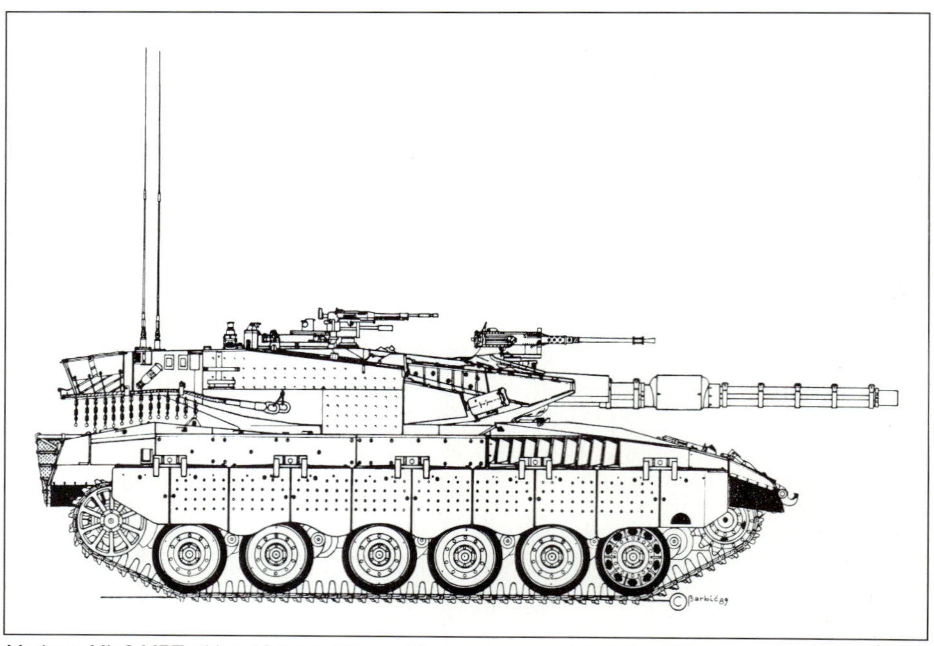

Merkava Mk 3 MBT with a 12.7 mm M2 machine gun mounted over the 120 mm smoothbore gun and showing chains under the turret bustle

around a central processing unit and consists of operation units, control and feedback servo loops and sensors. The system includes three operation units, gunner's, commander's and loader's. The gunner's is the main unit and provides all manual inputs necessary for ballistic computation. It also includes a logistic panel which enables system boresighting and system BITE, as well as the display of preselected inputs. The unit also includes the following manual inputs: type of ammunition (six types) and recoil compensation insertion for each type of ammunition, in elevation and deflection. The commander's unit provides a readout of the system's display, range and ammunition inputs. The loader's unit provides ammunition inputs.

The control loop transfers the computer superelevation information to the hydraulic gun elevation drive and the ballistic move drive. In addition, deflection data is transferred to the moving graticule. The feedback loop ensures that the actual superelevation and graticule deflection data are identical to the computed data and will correct the error accordingly.

The system includes automatic sensors, a laser range-finder, a turret cant angle indicator and a target angular velocity sensor. Optional sensors include crosswind velocity, charge temperature, barrel bend and ambient air density.

Manual operation of the electrical system, if the computer fails, is made possible with a handwheel drum and mounted scales, connected to the mechanical gearbox, to give elevation angle compensation and range to target for all types of projectiles. The commander can, if required, take over control of the main armament and fire the gun. The main armament is stabilised in both planes as US HR Textron Incorporated stabilisation system is fitted as standard. This is manufactured under licence in Israel by PML Precision Mechanism Ltd.

Specifications

Merkava Mk 1
(specifications in square brackets relate to the Mk 3 where different)
Crew: 4
Combat weight: 60,000 kg [65,000 kg]
Unloaded weight: 58,000 kg [62,000 kg]
Power-to-weight ratio: 15 hp/t [18.46 hp/t]
Ground pressure: 0.9 kg/cm² [0.96 kg/cm²]
Length:
(gun forward) 8.63 m [9.04 m]
(hull) 7.45 m [7.97 m]
Width: 3.7 m [3.72 m]
Height:
(to commander's cupola) 2.75 m [2.80 m]
(to turret roof) 2.64 m [2.66 m]
Firing height: 2.15 m
Ground clearance: 0.47 m [0.49 m]
Track width: 640 mm [660 mm]
Length of track on ground: 4.78 m
Max road speed: 46 km/h [60 km/h]
Acceleration:
(0 to 32 km/h) 13 s
Range: 400 km [500 km]
Fording
(without preparation) 1.38 m [1.40 m]
(with preparation) 2 m [2.4 m]
Fuel capacity: 1,250 litres [1,400 litres]
Gradient: 70%
Side slope: 38% [40%]
Vertical obstacle: 0.95 m [1.05 m]
Trench: 3 m [3.55 m]
Engine: General Dynamics Land Systems AVDS-1790-6A V-12 diesel developing 900 hp [General Dynamics Land Systems AVDS 1790-9AR V-12 turbocharged diesel developing 1,200 hp at 2,400 rpm]
Transmission: Allison Transmission CD-850-6BX semi-automatic [Ashot hydromechanic with 4 forward and 3 reverse gears]
Suspension: independent, helical spring and volute bumper spring [Independent trailing arm, coil springs]
Electrical system: 24 V
Batteries: 8 × 12 V, 500 Ah [10 × 12 V, 625 Ah]
Armament:
(main) 1 × 105 mm rifled gun [1 × 120 mm smoothbore gun]
(coaxial) 1 × 7.62 mm MG
(anti-aircraft) 2 × 7.62 mm MG
(mortar) 1 × 60 mm
Ammunition:
(main) 62 (nominal, can be as high as 85) [48]
(coaxial/anti-aircraft) 10,000
Gun control equipment
Turret power control: electrohydraulic [electric/manual
manual
(by commander) yes
(by gunner) yes
Turret traverse: 360°
Gun elevation/depression: +20/–8.5° [+20/–7°]
Gun stabiliser
(vertical) yes
(horizontal) yes
Range setting device: yes (laser)
NBC system: yes
Night vision equipment: yes

Variants

Dozer/Mineclearing Equipment

Some Merkava MBTs have attachment plates under the hull front to enable them to be fitted with Israeli-designed and produced mineclearing roller equipment and dozer blades.

Platoon communications versions of the Merkava are also in service; these have additional communications equipment and an additional antenna.

Details of the Urdan RKM bulldozer attachment are given in *Jane's Military Vehicles and Logistics 2002-2003*. Details of the various Israeli mineclearing systems, some of which can be fitted to the Merkava and other Israeli MBTs, are given in the same volume.

Late production Merkava Mk 2 MBTs were built with bolts and fastners for the attachment of a dozer blade or mineclearing systems. Early vehicles were backfitted with these attachments.

Merkava Mark 2

The Mk 2 entered production in 1982 with first production vehicle being completed in August 1983. Main improvements over the original production model include a layer of special armour on the front and sides of the turret, special armour on the hull front, the 60 mm mortar is mounted in the left side of the turret roof and can be loaded and fired from within the turret, and the steel skirts that protect the suspension are backed by special armour. The fire-control system, called the Matador Mk 2, has a number of improvements including an Nd:YAG laser range-finder in place of the original Neodymium-glass laser range-finder and a more advanced computer. The new side skirts are attached by heavy spring hinges so that the skirts do not become detached while crossing rough terrain. The original 900 hp General Dynamics Land Systems diesel engine has been retained but the new Israeli-designed and -built transmission is much more efficient. It has increased the range of the Merkava by 25 per cent to 500 km with only a slight increase in fuel capacity. Production of the Merkava Mk 2 was completed late in 1989.

Merkava Mark 3

Design work on the Merkava Mk 3 commenced in August 1983 shortly after the first production Merkava Mk 2 was handed over. First production Mk 3 was delivered to the 188th Armoured Brigade in the spring of 1989.

Merkava Mk 3 Baz (Hawk) with modular armour suite and improved suspension. Note the integrated thermal sleeve on the 120 mm gun (IDF) **2001**/0067239

Merkava Mk 3 Baz (Hawk) MBT fitted with a more powerful engine, a 120 mm smoothbore gun and new special armour (Christian Dumont) **2002**/0130960

Rear view of Merkava Mk 3 Baz (Hawk) MBT showing external stowage and chains hanging below the turret bustle (Christian Dumont) **2002**/0130959

While the Merkava Mk 2 was an improvement of the original Merkava Mk 1, the Merkava Mk 3 is essentially a new design. The basic hull has been lengthened by 457 mm and additional fuel cells have been added to the hull rear.

The basic turret casting has been lengthened by 230 mm and the complete turret armour array around the crew redesigned to a modular configuration which can be changed or upgraded in the future.

The Merkava Mk 3 is built in production blocks and by mid-1996 at least three production blocks had been built. Block I and Block II differences were minor and mainly internal.

Block III changes were major and mainly external. These included turret armour components changed all round exterior, armour section on loader's side increased by 125 mm, 60 mm mortar port area significantly changed along with deflector area in front of the gunner's sight aperture.

The key improvements of the Merkava Mk 3 can be summarised as:

(1) Main armament is an Israel Military Industries 120 mm smoothbore gun fitted with a thermal sleeve. The 120 mm smoothbore gun is ballistically identical to that of the 120 mm smoothbore gun installed in the US M1A1 and M1A2 MBTs and German Leopard 2 MBTs. The 120 mm gun has an elevation of +20° and a depression of –7° with a total of 48 rounds of ammunition being carried. A 7.62 mm MG is mounted coaxially with the main armament and there are also two 7.62 mm roof-mounted machine guns; a total of 10,000 rounds of 7.62 mm ammunition is carried. Some of the 48 rounds of 120 mm ready use ammunition are in a drum on the floor of the turret compartment which provides the same level of protection as the individual containers. The 60 mm mortar is aimed, loaded and fired from within the turret and, in addition to firing HE bombs against anti-tank teams, it can also fire illuminating rounds.

(2) Mounted either side of the turret is the Israel Military Industries CL-3030 instantaneous self-screening system for combat vehicles

(3) Turret controls are now all-electric with controls for both gunner and commander. The commander's panoramic sight has a magnification of ×4.8 and ×12, with an optical relay to the gunner's sight which is a two-axis stabilised day/night sight with a magnification of ×5 (thermal) and ×12 (day). The gunner's sight incorporates an Nd:YAG laser range-finder and automatic target tracker. The advanced fire-control system has been developed by ELOP and Elbit. The fire-control system is integrated with the turret- and gun-control equipment and allows line of sight stabilisation for the tank commander and gunner. It allows the Merkava Mk 3 to engage moving targets while it is also moving

(4) An advanced threat warning system developed by Amcoram is fitted, with threat warnings being displayed on a small panel at the commander's station

(5) The armour is of a modular special type and covers both the turret and hull. This armour provides a much higher level of protection than that of the Merkava Mk 1 and 2 and can also be changed in the field. The new modular armour is attached to the hull and turret of the Merkava by bolts and as new armour systems become available they can replace the earlier armour packages

(6) The suspension is new consisting of 12 roadwheels, six either side, independently mounted on trailing arms and sprung by pairs of concentric coil springs. The first two and last two roadwheel stations either side have a hydraulic rotary damper developed by SHL. Ashot Ashkelon provides the final drives and suspension system for the Merkava Mk 3. Total roadwheel travel is 604 mm, 300 mm bump and 304 mm rebound

(7) It is powered by a General Dynamics Land Systems (previously Teledyne Continental Motors) AVDS-1790-9AR V-12 air-cooled diesel developing 1,200 hp and this is coupled to a new Ashot transmission

(8) The NBC system is still of the overpressure type but with a central filter and an air conditioning capability.

First round hit capability against moving targets is further enhanced by the installation of an Elbit Automatic Target Tracker (ATT) which automatically positions the aiming point accurately onto the target without operator interaction.

The ATT is based on image processing techniques implemented in the existing video source which can either be the thermal system or TV channel in the day sight.

The auto-tracker is of the centroid type and maintains the sight on the target once it has been laid by the tank gunner. The line of sight locks irrespective of the motion of the own tank and the target, and it brings the sight back on track even when the latter has been obscured for a short time such as by a terrain feature.

It has been confirmed that the Merkava Mk III is fitted with the Elbit Knight Mk III advanced computerised fire-control system.

The 120 mm smoothbore gun, which fires ammunition developed by Israel Military Industries, has been retained but improvements have been made to make the loader's job easier.

The forward part of the turret roof, to the immediate front of the commander's and loader's hatches, has been provided with an additional layer of passive armour to provide an increased level of protection against top attack weapons such as bomblets.

Merkava Mk 3 Baz

The Merkava Mk 3 was the first MBT to be protected by a modular layered armoured suite. The armour modules are fitted with special attachment gear that enables the entire system to be replaced in the field, as well as to withstand the high dynamic loads occurring from high speeds or ballistic shock.

The armour fitted to the latest Mk 3 Baz (Hawk) around the turret front and sides has a distinct elliptical shape and provides a higher level of ballisitc protection than the armour fitted to earlier vehicles.

The Merkava Mk 3 Baz (Hawk) is fitted with an auto-tracking system that locks onto targets at a range of several kilometres with great accuracy. Based on the video output from either a TV camera (daylight channel) or a thermal imaging camera (night/adverse weather channel), the auto-tracker is of the centroid type and maintains the sight on the target once it has been laid by the gunner.

The line of sight locks on irrespective of the motion of either the target or the tank, and brings the sight back on track even when the target has been obscured temporarily by screening terrain.

It is claimed that the auto-tracker is extremely effective against fleeting targets such as helicopters or fast moving enemy tanks. The device is incorporated into the fire-control system as an integral part.

The Merkava Mk 3 has a new suspension system of a spring and rotary coil-spring design, differing from the double spring system of the previous models of the Merkava.

Either side consists of six road wheels with the drive sprocket at the front, idler at the rear and five track return rollers. The first and last return rollers are of the double type with the centre three of the single type.

The crew of the Merkava Mk 3 is protected against NBC threats and can fight without NBC protective clothing and harness gear. The system operates on the principle of a pressure chamber inside the fighting compartment and as an option, filtered air can be directed to flow into the protective masks and personal gear when worn by the crew when operating with open hatches.

The vehicle also incorporates a modern air conditioning system which inserts fresh filtered air into the tank and crew clothing.

Some Merkava Mk 3 series MBTs have been fitted with a device either side of the 120 mm smoothbore gun which is similar in some respects to the Russian TshU 1-7 Shtora-1 countermeasure system, which has been fitted to T-90 and other Russian tanks. This has been designed to decoy ATGW before they impact the MBT. This version of the Merkava Mk 3 was also fitted with additional grenade launchers either side of the turret.

Merkava Mark 4

In mid-1991, it was stated that Israel had started to develop a new generation of MBT which will enter service in the next decade. No detailed information is available, although speculation is that it will have an integrated vehicle protection system and perhaps a new weapon system, which may consist of a larger calibre gun or an electrothermal weapon. Late in 1992, Israel Military Industries revealed that it was working on a 140 mm APFSDS-T round fired from a smoothbore gun.

No technical details were released, although it is estimated that the tungsten rod penetrator has a length-to-diameter ratio of about 25:1. The three-part sabot has a plastic driving band on the rear with the penetrator having eight fins.

It is possible that this 140 mm smoothbore gun is for the Merkava Mk 4 MBT. It is expected that the Mk 4 model will be 100 per cent Israeli built and incorporate no foreign-built components.

Israel tank ammunition developments

Israel Military Industries has continuously developed 105 and 120 mm types of ammunition for the various models of the Merkava MBT as well as a number of export customers.

Known types of Israel Military Industries 120 mm MBT ammunition include APFSDS-T M711 (CL 3254), HEAT-MP-T M325 (CL 3105) and TPCSDS-T M324 (CL 3139).

Israel Aircraft Industries, MBT Division, is developing the 105 and 120 mm LAHAT (Laser Homing Anti-Tank) gun launched projectile which has a tandem HEAT warhead to defeat targets fitted with explosive reactive armour.

It has a conventional cartridge case with the booster charge launching the projectile from the 105 mm gun tube at low velocity after which the four fins/control surfaces unfold and the main sustainer motor cuts in and accelerates the missile until target impact.

LAHAT is a semi-active laser guided missile and while it would normally be fired with the missile being designated by the launch platform it can also be fired with a lock on after launch capability with target designation being carried out by another platform.

Test firings of the LAHAT were carried out in mid-1999 from a 105 mm M68 armed M60A3 and Merkava Mk II of the Israel Defence Force.

Israel Military Industries is also developing a new top attack fire-and-forget extended range tank projectile called Excalibur in 105 and 120 mm versions, both of which are fitted with a tandem HEAT warhead.

Mounted in the nose of Excalibur is an advanced seeker that acquires, tracks and homes onto the target. The onboard computer controls all of the projectile sequences.

155 mm Self-propelled Gun

Details of the Soltam Systems Slammer 155 mm self-propelled howitzer based on a Merkava chassis are given in the Self-propelled guns and howitzers (tracked) section under Israel. This system is still at the prototype stage with two vehicles built for trials and development work.

Status

In production. In service with the Israel Defence Force.

Manufacturer

Israeli Ordnance Corps, facility at Tel a Shomer, near Tel Aviv, Israel.

Enquiries to SIBAT-Foreign Defence Assistance and Defence Export.

UPDATED

Israel Military Industries Sabra main battle tank

Development

The Slavin Plant of Israel Military Industries has developed, as a private venture, a comprehensive upgrade package for the US M60A3 series MBT called Sabra.

The M60A3 has been phased out of service with the US Army and large numbers of these, and the later and improved M60A3 TTS (Tank Thermal Sight), have been passed to other countries.

The new M60A3 Sabra upgrade draws on experience obtained in the extensive series of upgrades to the M48/M60 MBT carried out in Israel which have culminated in the MAGACH-7 and MAGACH-8 as well as the Merkava series of MBT.

Israel Military Industries has already offered the upgraded M60A3 package to Turkey who has received over 900 M60 series MBTs. The design of the upgrade is modular so that the customer can pick only those parts he requires.

For example if they have a M60A3 with TTS, they may well opt for only the 120 mm smoothbore gun, additional armour package and perhaps some automotive improvements.

Turkey is expected to upgrade 170 M60A1 MBTs, but as of December 2001, no firm contract had been placed with Israel Military Industries owing to the overall price submitted.

Specifications

Crew: 4
Combat weight: 55,000 kg*
Power-to-weight ratio: 16.50 hp/t
Length:
(gun forward) 9.40 m
(hull) 8.26 m
Width: (without skirts) 3.63 m
Height: (to commanders cupola) 3.05 m
Ground clearance: 0.45 m
Track width: 610 m
Max road speed: 48 km/h
Acceleration: 9.6 s (0 to 32 km/h)
Range: 450 km
Fording:
(without preparation) 1.40 m
(with preparation) 2.40 m
Gradient: 60%
Side slope: 30%
Vertical obstacle: 0.91 m
Trench: 2.60 m
Engine: General Dynamics Land Systems AVDS 1790-5A four stroke V-12 diesel with two turbochargers developing 908 hp at 2,400 rpm
Transmission: automatic
Suspension: independent trailing arm with three dampers and three bumpers either side, total road wheel travel is 220 mm
Electrical system: 24 V
Armament:
(main) 1 × 120 mm smoothbore gun
(coaxial) 1 × 7.62 mm MG
(anti-aircraft) 2 × 7.62 mm (or 5.56 mm) MG
(mortar) 1 × 60 mm
Ammunition: (main) 44
Gun control equipment: powered/manual:
(by commander) yes
(by gunner) yes
Turret traverse: 360°
Gun elevation/depression:
(vertical) yes
(horizontal) yes
Range setting device: yes (laser)
NBC system: yes
Night vision equipment: yes
• depends on level of ballistic protection.

Description

The overall layout of the Sabra MBT is identical to that of the M60 series covered later in this section under the United States but has significant improvements in the key areas of armour, mobility and firepower.

The M60A3 is armed with a 105 mm M68 rifled tank gun which is a US version of the now BAE Systems, RO Defence 105 mm L7 rifled tank gun but with a US designed breech.

In the Sabra upgrade this has been replaced by an Israeli Military Industries 120 mm smoothbore gun which is already fitted to the Merkava Mk 3 MBT.

This 120 mm smoothbore gun is fitted with a fume extractor and thermal sleeve and fires standard 120 mm smoothbore ammunition which is available from a variety of sources.

The 120 mm smoothbore gun firing armour-piercing fin-stabilised discarding sabot (APFSDS) ammunition has greater armour penetration and longer range than the older 105 mm gun.

A 7.62 machine gun is mounted coaxial with the 120 mm smoothbore gun and two 7.62 mm machine guns are pintle mounted on the roof of the vehicle, one for the commander and one for the gunner. Like the Merkava MBT, the upgraded M60A3 is fitted with a Soltam Systems 60 mm mortar for infantry suppression.

Sabra upgraded M60A3 MBT by Israel Military Industries armed with 120 mm smoothbore gun fitted with fume extractor and thermal sleeve 0045719

To enable moving targets to be engaged with a high first round hit probability while the vehicle is stationary or moving, an Elbit/Elop Knight series computerised fire-control system has been fitted.

The gunner has a new two-axis stabilised day (magnification of ×8) and thermal (magnification of ×5.3) sight with a laser range-finder. The tank commander has a day sight with an optical relay of the gunner's thermal picture.

A hybrid turret and gun control system is fitted which consists of an all electric traverse control system and hydraulic elevation and depression system. Either commander or gunner can lay and fire the main armament while the vehicle is stationary or moving.

A new passive armour package has been installed which gives a high protection level over the frontal arc. This armour package is of modular construction so that it can be exchanged or be upgraded as the threat changes or as new armour development takes place. If required, Israeli explosive reactive armour could also be installed on the Sabra.

A locally-developed threat warning system has been fitted and mounted either side of the turret. It is an Israel Military Industries CL-3030 instantaneous self-screening system which is fitted as standard on all Israel MBTs as well as an increasing number of support vehicles. This can be activated manually or automatically.

The M60A3 has a combat weight of 52.6 tonnes but the upgraded M60A3 can weigh, depending on the armour package selected by the user, over 54 tonnes.

For this reason the automotive aspects of the M60A3 have been improved. The original General Dynamics Land Systems AVDS-1790-2C series diesel engine developing 750 hp has been replaced by the improved AVDS-1790-5A diesel with two turbochargers which develops 908 hp at 2,400 rpm. With a combat weight of 55 tonnes this gives a power-to-weight ratio of 16.50 hp/tonne.

For improved mobility across country, the suspension has been upgraded and new torsion bars, three piston type dampers and heavy duty dampers have been installed. Total road wheel travel is now

220 mm. This upgraded suspension gives a better ride for the crew and a more stable firing platform. New single pin steel tracks have also been fitted.

Other improvements include an automatic fire detection and explosion suppression system for higher crew survivability.

Sabra Mk II MBT

More recently, the Slavin Plant of Israel Military Industries has confirmed that it has recently completed the first prototype of its Sabra Mk II MBT.

The latest Sabra Mk II retains the 120 mm MG251 smoothbore gun of the earlier Mk I, which was first revealed several years ago and is also installed in the Merkava Mk III MBT used by the Israel Defence Force (IDF).

A total of 40 rounds of 120 mm ammunition are carried. A 7.62 mm or 5.56 mm Machine Gun (MG) is mounted coaxial with the main armament. Two similar weapons are mounted on the turret roof and a Soltam Systems 60 mm mortar is mounted externally.

The passive armour package of the Sabra Mk I was optimised to provide the crew with protection from attack by kinetic energy (KE) projectiles. The Sabra Mk II bolted hybrid (explosive reactive and passive) armour package provides protection from KE and chemical energy (CE) attack.

For enhanced hit probability against stationary and moving targets under day and night conditions, the Sabra Mk II is fitted with the Elbit/Elop Knight III computerised fire-control system with a new all electric gun control and stabilisation system, which is similar to that installed in the Merkava Mk III MBT.

The gunner's sight is a two-axis stabilised day/thermal sight with eye-safe laser range-finder and automatic target tracker, with the commander having a display of the gunner's sight.

Israel Military Industries Sabra Mk II MBT armed with a 120 mm smoothbore gun and new hybrid armour package **2001**/0089832

The Sabra Mk II is being offered with two different diesel engines, a German MTU diesel developing 1,000 hp or a General Dynamics Land Systems AVDS 1790 series V-12 diesel developing 1,200 hp with modified final drives. The combat weight depends on the armour package but is between 56 to 59 tons.

For improved cross-country mobility, advanced torsion bars and three rotary dampers have been installed either side with maximum road wheel travel now being 265 mm. Two track options are now available, steel single dry pin or rubber bushed double pin with replaceable pads.

Other standard equipment includes a defensive aids suite which includes instantaneous grenade launchers and laser detectors and an automatic fire and explosion detection and suppression system.

Status

Development complete. Ready for production. Has been proposed to the Turkish Land Forces Command.

Manufacturers

Israel Military Industries, Slavin facility.

UPDATED

Israeli upgraded M60 series MBT (MAGACH)

Development

The first US designed and built M48 series MBT came from Germany and at the time of the 1967 war 250 were in service of which about 20 had been fitted with a 105 mm gun, the rest retained their original 90 mm guns.

Of these 250 re-engined vehicles, 40 were ex-German Army M48s, 100 were ex-US Army M48A1s and the remaining 110 were ex-US Army M48A2Cs.

MAGACH is actually a Hebrew acronym (Ma-Ga-Ch) from the Hebrew letters Mem and Chaf which stand for 48 and the middle letter of Gimel standing for Germany, as this was the first source of Israeli M48s.

For many years Israel has had a policy of upgrading its MBTs as the threat evolves and as new technology becomes available.

There is insufficient money in the IDF budget to replace all existing tanks with the more recent Merkava and today about 75 per cent of the IDF's armour units still have older tanks.

Initial upgrade

From the 1970s, the M60s were upgraded in a number of areas including the addition of an Israeli-developed thermal sleeve for the 105 mm gun, the addition of Israeli-developed explosive reactive armour, installation of a new Urdan commander's cupola (also adopted by the US Army for its M48A5 tanks), two roof-mounted 7.62 mm machine guns and the Israel Military Industries CL-3030 instantaneous self-screening system either side of the main armament. Details of the Blazer ERA and CL-3030 system are given in the Centurion entry.

It is believed that the M60 series fitted with ERA were called the MAGACH-6. It is understood that these have now been replaced by later and enhanced versions.

Latest upgrade

The key part of the latest upgrade is the new passive armour, while other changes include a new engine, tracks and fire-control system.

The passive armour package has been fitted to the glacis, nose, turret front and sides and the forward part of the roof. Armoured skirts have also been fitted to provide lateral protection, something that earlier Israeli upgrade kits failed to provide. The forward part of the skirt, which covers the crew compartment, is of a different design from the rear.

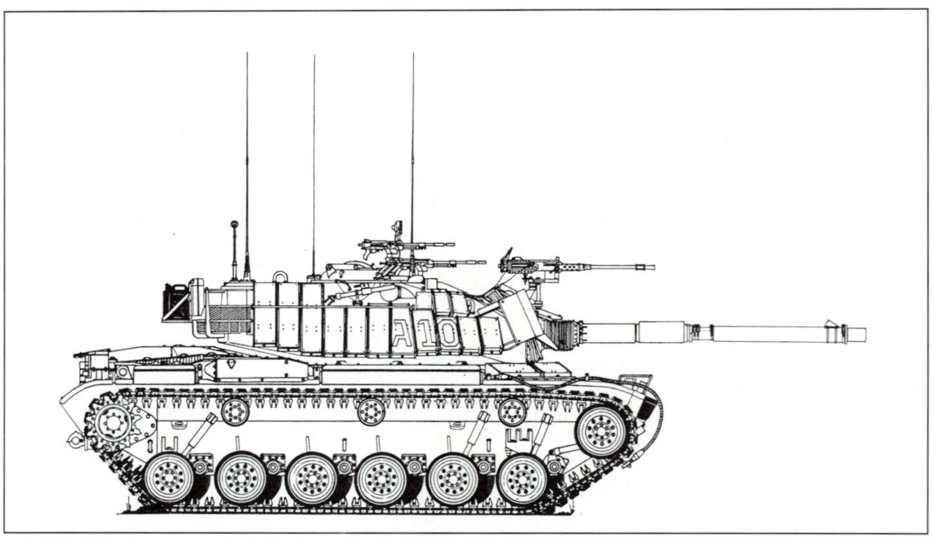

M60A1 MBT (command) of the Israel Defence Force fitted with a 12.7 mm M2 machine gun over the 105 mm M68 gun and Blazer explosive reactive armour system (Vasco Barbic)

MAGACH-8 MBT with 105 mm gun and turret traversed left (Marsh Gelbert) 0045710

M60A1 of the Israel Defence Force fitted with Blazer ERA and showing the positions of two 7.62 mm machine guns and one 12.7 mm machine gun

MAGACH-7 is fitted with additional passive armour to its hull and turret for improved battlefield survivability 0018788

The new armour package also provides some additional protection against top attack, at least on the forward part of the roof.

Openings have been left in the right side of the armour package for the gunner's sight and in the left side of the armour package for the 7.62 mm coaxial machine gun.

Passive armour will give the M60 protection against Kinetic Energy (KE) and Chemical Energy (CE) attack, unlike the earlier explosive reactive armour used on M60 upgrades which only gave protection against CE like the Soviet AT-3 'Sagger' anti-tank guided weapon used by Arab armies.

The installation of additional armour and other improvements has led to the installation of a new engine and tracks so that the vehicle retains its cross-country mobility. The new tracks, produced by Urdan Industries, are the same as those used on the Merkava.

The Israeli Armoured Corps has standardised on the US Teledyne Continental (now General Dynamics Land Systems) AVDS-1790 series diesel engine for many years and this has been retrofitted to all M60s and Centurions. It was already fitted to the M60s when delivered.

The upgraded MAGACH-7 MBT is fitted with the Teledyne Continental (now General Dynamics Land Systems) AVDS-1790-5A V-12 diesel engine developing 908 hp which has also been fitted with an automatic starter.

The Matador computerised fire-control system developed by Elbit and ELOP Electro-Optics Industries Limited is also being installed. The system is already in service on Merkava and includes a laser range-finder. The M60A3 version of the M60s fielded by the IDF already has a modern fire-control system including a laser range-finder and this may point to the new upgrade being limited to earlier M60/M60A1 versions.

The 105 mm M68 gun, manufactured locally by Israel Military Industries and installed in all Israeli M60, Centurion and Merkava Mk 1 and 2 vehicles, remains unchanged.

In addition to firing the first-generation Israel Military Industries M111 APFSDS round, it also fires the second-generation APFSDS-T M413 round which has greater armour penetration characteristics.

A further development of the MAGACH-7 is called, sometimes incorrectly, the MAGACH-8 and this has similar capabilities but a different and probably more advanced armour package. The turret armour is different with the turret front having an arrowhead type configuration. The armour package used on the MAGACH-8 is similar to that used on the latest production Merkava Mk 3 MBT.

In addition to the 7.62 mm coaxial machine gun there is a 12.7 mm machine gun mounted above the 105 mm M68 tank gun and two roof-mounted 7.62 mm machine guns on pintle mountings. The CL-3030 screening system is fitted and these are mounted on either side of the turret.

MAGACH MBT designations

Recent information has indicated that there are the following known models of the MAGACH MBT:

MAGACH 6 - Blazer ERA and Urdan type 7.62 mm machine gun (initial upgrade)

MAGACH 6A - Steel tracks as fitted to Merkava MBT

MAGACH 6B Gal - fitted with Gal (Wave) computerised fire-control system

MAGACH 7A - similar to MAGACH 6B but with new type of armour and skirts

MAGACH 7B - fitted with additional applique armour for turret sides

MAGACH 7C Gimel - fitted with General Dynamics Land Systems AVDS-1790-6A diesel developing 900 hp and again new type of modular turret armour with a pointed front. Gimel is the third letter of the Hebrew alphabet

MAGACH 7D - has improved hull and saucer turret armour package and is also sometimes incorrectly referred to as the MAGACH 8.

Sabra MBT

For the export market Israel Military Industries is marketing a further development of the MAGACH-7/8 fitted with a 120 mm smoothbore gun. This is called the Sabra and is covered in detail in a separate entry. More recently the improved Sabra II has been revealed. As of early 2002, both of these systems remain at the prototype stage.

Pele Robotic M60A1

Details of this remote-controlled M60 series MBT, which can be used for a variety of battlefield missions such as a remote-controlled mineclearing system, were given in *Jane's Armour and Artillery 2000-2001*.

Nagmachon APC

This armoured personnel carrier is based on the M48 chassis and details are in the Armoured personnel carriers (tracked) section later in this volume.

M60 MBT with SMCD

Late in 1991, RAMTA Structures and Systems Division of Israel Aircraft Industries, announced that it had developed an add-on Scatterable MineClearing Device (SMCD) that can be quickly installed on tracked and wheeled armoured vehicles, with the former including the M60A1 MBT.

The system can be quickly installed at the front of the vehicle and has been developed as an effective and low-cost solution to the problem of clearing scatterable mines laid on the surface of roads and runways.

The SMCD can be installed without any structural modification or degradation of the vehicle and when not in use it is locked up in the travel position.

The system is operated by the vehicle driver. When a button is pressed, the locking device is released and the system drops down ready for use. The system weighs approximately 500 kg and is about 4 m wide.

Status

Upgrade programme under way. In service with Israel Defence Force.

Manufacturer

Israeli Ordnance Corps.

The upgraded M60 programme is carried out at Israeli Ordnance Corps Workshops at Tel a Shomer, near Tel Aviv, Israel.

UPDATED

Israeli upgraded Centurion (Sho't) MBT

Development/Description

The first order placed by Israel with the UK was for the supply of 16 used Centurion Mk 5 and 14 new Mk 8 vehicles armed with a 20-pounder (83.8 mm) gun. First deliveries were made to Israel in December 1958.

In May 1960, a further 60 used Mk 5 Centurions were purchased from the UK. In October 1961, the UK agreed to upgun 90 Centurion MBTs with the BAE Systems, RO Defence 105 mm L7 series rifled tank gun. In 1962, the UK supplied 45 Mk 5s, already fitted with 105 mm L7 guns and by the 1967 war, a total of 385 Centurions were in service.

One of the main drawbacks of the Centurion was its Meteor petrol engine which gave the tank a relatively low power-to-weight ratio and a very high fuel consumption limiting its operational range. The cooling system and air cleaners also gave a lot of trouble.

In 1967, a decision was taken to replace the Meteor petrol engine with a diesel. After trials with three different engines, the Teledyne Continental (now General Dynamics Land Systems) AVDS-1790-2A diesel was selected, as it was also being installed in the Israeli Armoured Corps' M48A2 tanks which at that time were being modernised. The original Merritt-Brown Z51R manual gearbox was replaced by the Allison Transmission CD-850-6 automatic, which greatly reduced driver fatigue as well as simplifying training.

The first contract, signed in May 1967, was for the supply of 400 AVDS-1790 series diesel engines for the Centurion and the first upgraded Centurion vehicles were completed in May 1970.

To accommodate the new power pack, the rear of the hull had to be enlarged, but even so, the engine had to be installed at an inclination of 3.5° front side up. This meant that elevated top decks had to be fitted to accommodate the cooling air outlets that vented upwards through louvres above the transmission compartment. Externally, the modified Israeli Centurion, officially designated the Upgraded Centurion is recognisable by its raised engine decks and air filter boxes on the trackguards.

The Upgraded Centurion, which entered service with the Israeli Armoured Corps in May 1970 and was shown for the first time in public in February 1973, also has an increased fuel capacity, more efficient air-cooled braking system, new fire extinguishing system, improved ammunition layout with 72 rounds of 105 mm ammunition and a new electrical system.

Some Centurions have been observed with a 12.7 mm M2 machine gun over the gun barrel.

The Upgraded Centurion has a maximum road speed of 50 km/h and twice the cruising range of the Centurion Mk 5 on which it is based.

Recent information is that the Centurion MBT is no longer in front line service with the IDF. SIBAT, part of the Israel Ministry of Defence, is now offering upgraded Centurion MBTs for sale as well as upgraded M48A3/M48A5, T-55 and T-62 series MBTs.

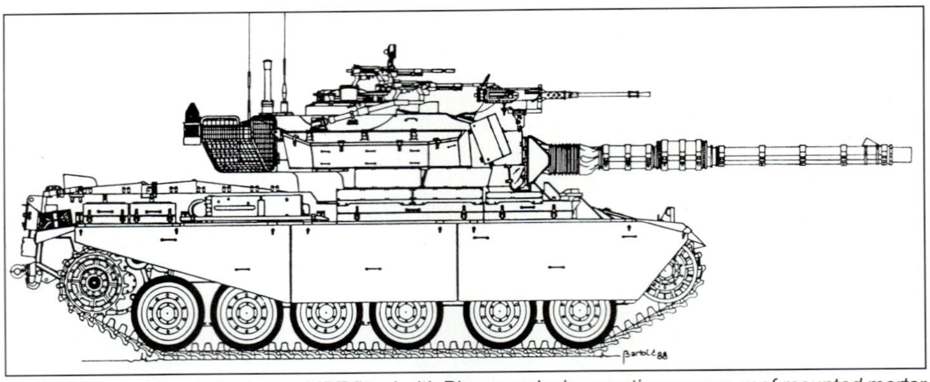

Sho't Upgraded Israeli Centurion MBT fitted with Blazer explosive reactive armour, roof-mounted mortar, two 7.62 mm machine guns on the roof, a 12.7 mm M2 machine gun over the 105 mm gun, a thermal sleeve for the 105 mm gun, new low-visibility aerials, new exhaust ports, a different stowage basket at the hull rear and the Matador computerised fire-control system (Vasco Barbic)

Variants

All of the Centurion, M48 and M60 series MBTs of the Israel Defence Force were fitted with the Israel Military Industries Blazer explosive reactive armour package, details of which are given later in this entry. Most were fitted with the Israel Military Industries CL-3030 instantaneous self-screening system for combat vehicles.

This system is fitted to many other Israeli armoured vehicles including the upgraded M48/M60, Merkava and more recently the upgraded M113 series armoured personnel carrier.

The two basic components of the system are the two launch containers, mounted one either side of the main armament and, the IS-10 system which holds 10 smoke projectiles in two cassettes of five projectiles each. The smoke projectiles contain red phosphorus which is non-toxic and, when fired, create a screen at a distance of 40 to 50 m from the tank that is initially 60 m wide and 4 to 8 m high.

The Upgraded Centurion and other Israeli MBTs can also be fitted with the Ramta Structures and Systems Track Width Mine Plough and the Urdan RKM mineclearing roller system (*Jane's Military Vehicles and Logistics 2002-2003*). It can also be fitted with the RKM bulldozer attachment.

The Israel Defence Force also uses the Centurion ARV Mk 2 and some tanks have had their turrets removed for use as tugs.

Specifications

Upgraded Israeli Centurion MBT
Crew: 4
Combat weight: 53,500 kg
Power-to-weight ratio: 14.01 hp/t
Length (gun forwards): 7.55 m or 7.84 m (depending on model)
Width: 3.38 m
Height: 2.94 m
Ground clearance: 0.42 m
Track: 2.641 m
Track width: 610 mm
Length of track on ground: 5.572 m
Max road speed: 50 km/h
Fuel capacity: 1,090 litres
Range (cruising): 500 km
Fording: 1.45 m
Gradient: 60%
Vertical obstacle: 0.914 m
Trench: 3.352 m
Engine: General Dynamics Land Systems AVDS-1790-2AC air cooled diesel developing 750 hp at 2,400 rpm
Transmission: Allison CD-850-6A1 fully automatic
Suspension: Horstmann
Electrical system: 24 V
Armament:
(main) 1 × 105 mm L7 series gun
(co axial) 1 × 7.62 mm MG
(anti-aircraft) 2 × 7.62 mm MG
Ammunition:
(105 mm) 72
(7.62 mm) 7,500
Gun control equipment: electric/manual
Max rate power traverse: 360° in 26 s
Gun elevation/depression: +20/–10°
Commanders override: yes
Gun stabiliser:
(vertical) yes
(horizontal) yes
Range setting device: yes
Elevation quadrant: yes
Traverse indicator: yes
NBC system: yes
Night vision equipment: yes

Puma APC

The Puma armoured personnel carrier is based on a modified Centurion chassis and was developed to meet the requirements of the Israel Defence Force Engineering Corps. Available details are given in the Armoured personnel carriers (tracked) section under Israel.

Nagmashot APC

The Nagmashot armoured personnel carrier is based on a modified Centurion chassis and was designed to meet the requirements of the Infantry Corps. Available details are given in the Armoured personnel carriers (tracked) section under Israel.

Note on Israeli MBTs

During the fighting in Lebanon in the summer of 1982, many Israeli M48, M60 and Centurion series MBTs were observed to be fitted with appliqué armour. In late 1983, it was confirmed that this was explosive reactive armour and had been developed by RAFAEL Armament Development Authority with production and marketing now being undertaken by Israel Military Industries. This armour, called Blazer, is suitable for installation on all types of MBT and adds less than one tonne to the vehicle weight. It consists of small panels of armour bolted onto the hull and turret, reacting to HEAT attack but not activated by small arms ammunition, fire or artillery fragments. Once a panel has been hit it no longer provides protection, but the manufacturer believes that the chance of a panel being hit twice is remote. On an M60A1 type MBT about 10 m² is protected including the turret roof. Blazer provides a high level of protection against attack from HEAT projectiles such as those fitted to RPG-7 and 'Sagger' type weapons. An illustration of an M60A1 with Blazer explosive reactive armour appears in this section in the M60 MBT entry.

Blazer is not initiated by small arms ammunition up to 23 mm API at all angles of attack from zero range, artillery and mortar fragments from 2 m and arc welding and autogen cutting. Blazer provides complete arrest of the AT-3 'Sagger' at all angles of attack over 70 per cent of the frontal arc of the hull. On the front of the hull Blazer provides complete arrest of RPGs at all angles of attack (0° incidence) and, complete arrest of AT-3 'Sagger' at 30° (NATO 60°) incidence on the side of the turret.

In early 1984, it was announced that the Israeli Ordnance Corps had developed a low-cost fire-control system which would initially be fitted to Israeli M48 and Centurion MBTs. The fire-control system can be installed in under a day and includes a computer developed by Elbit Computers and electronics and optics by ELOP Electro-Optics Industries Limited.

Manufacturer

Israeli Ordnance Corps.
The Upgraded Centurion programme was carried out at Israeli Ordnance Corps Workshops at Tel a Shomer, near Tel Aviv, Israel.

UPDATED

Israeli Upgraded Centurion MBT fitted with the Israel Aircraft Industries, Ramta Structures and Systems, Track Width Mine Plough (TWMP) system

Upgraded Centurion MBT of the IDF from the rear with 105 mm gun depressed and showing extensive external turret stowage (IDF) **2000**/0069527

Israeli T-54/T-55 and T-62 series MBT

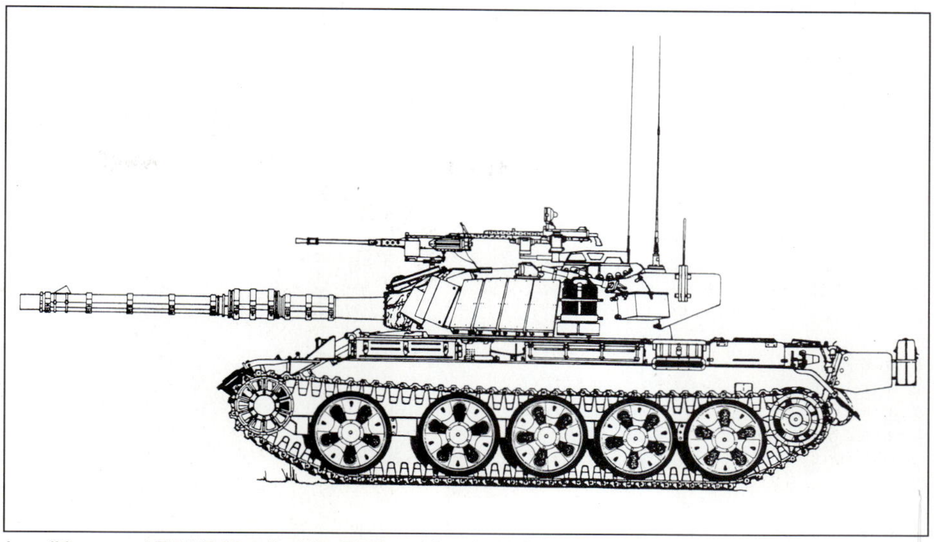

Israeli improved T-54/T-55 called the Ti-67 and fitted with Blazer ERA (Vasco Barbic)

Israeli T-54 and T-55

The Israel Defence Force captured large numbers of Russian T-54 and T-55 tanks during the 1967 and 1973 conflicts, many of which were subsequently modified for their own use. These modifications, which have not been carried out on every Israeli T-54 or T-55, include replacing the 100 mm gun with a 105 mm M68 rifled gun; modification of the ammunition racks to accept 105 mm ammunition; new communications equipment; modification of the commander's seat; replacing the gunner's seat; installation of an azimuth indicator; the driver's hatch can now be opened from the outside; installation of a commander's traverse control; modification of the sighting system for 105 mm ammunition; replacing the coaxial machine gun with a 7.62 mm (0.30) Browning machine gun and the 12.7 mm DShKM roof-mounted anti-aircraft machine gun with a 12.7 mm (0.50) Browning M2 HB machine gun; a new fire-control and electrical system; an air conditioning system; new radio mounts on the turret rear; a US infantry telephone on the hull rear; a Browning 0.30 machine gun at the loader's station; upward angling of the exhaust outlet; additional track stowage; installation of a fire extinguishing system; and new night vision equipment. These tanks are designated the Ti-67. None of these is understood to be in front-line service with Israel, some have been converted to other roles. These include the Achzarit heavy troop carrier covered in the Armoured personnel carriers (tracked) section.

Israeli T-54/T-55/T-62 Model S

This was announced in 1984 and, in addition to all the previous modifications such as replacement of the 100 mm gun with the 105 mm rifled gun, it has many other improvements including the American Detroit Diesel 8V-71T engine developing 609 hp; new semi-automatic hydromechanical transmission equipped with a torque converter; new air cleaners as fitted to the Israeli M60 and Merkava tanks; Blazer explosive reactive armour added to the hull and turret; an HR Textron Incorporated weapon stabilisation system; installation of an ELOP computerised Matador fire-control system; new low-profile commander's cupola; infra-red detectors; passive night vision equipment for the commander, gunner and driver; Spectronix fire detection and suppression system; new turret basket and extensive external stowage; modernised driver's station including replacement of sticks with a steering wheel; new final drives; new fuel system with all fuel now internal and improved suspension.

As far as is known, none of these upgraded T-54/T-55 MBTs is now in operational use with the Israel Defence Force, some have been sold while others were passed on to Lebanon.

A number of T-54/T-55 chassis have recently been upgraded to specialised armoured personnel carrier configurations and armoured engineer vehicles with the Israeli NIMDA company being involved in part of this programme. These are called the Achzarit. Details of this are given in the Armoured personnel carriers (tracked) section later in this book.

These upgraded T-55 series MBTs armed with the 105 mm gun are also known as the TIRAN-5. These, together with slightly modified T-62 series MBTs retaining their 115 mm smoothbore guns are currently being offered for export sale by SIBAT, the Foreign Defence Assistance and Defence Export division of the Israel Ministry of Defence.

According to United Nations sources, between 1992 and 2000 Israel only exported MBTs to one country. This was Uruguay, who took delivery of 15 T-55 MBTs from Israel in 1997.

Specifications
(Tiran-5)
Crew: 4
Weight (combat): 36,000 kg
Power-to-weight ratio: 16.11 hp/t
Ground pressure: 0.86 kg/cm^2
Length: 6.30 m
Width: 3.27 m
Height: 2.40 m
Ground clearance: 0.34 m
Max road speed: 50 km/h
Cruising range: 450 km (650 km with external tanks)
Fuel capacity: 675 litres + 285 litres external
Gradient: 60%
Engine: Model V-55, V-12 diesel developing 580 hp at 2,000 rpm
Transmission: manual, 5 forward and 1 reverse gears
Electrical system: 24 V
Armament:
(main) 1 × 105 mm rifled tank gun
(coaxial) 1 × 7.62 mm MG
(anti-aircraft) 1 × 7.62 mm MG
Ammunition:
(main) 40
(coaxial) 3,500

Manufacturer

Israel Ordnance Corps.
This upgrade programme was carried out at the Israel Ordnance Corps Workshops at Tel a Shomer near Tel Aviv, Israel.

UPDATED

Tiran-5 upgraded T-55 MBT with 105 mm rifled tank gun and 12.7 mm machine gun over 105 mm gun (IDF) ***2000***/0069528

Modified T-62 MBT being offered for export by SIBAT and retaining 115 mm gun (IDF) ***2000***/0069529

ITALY

Ariete MBT

Development

In 1982, the Italian Army issued a broad specification for a new main battle tank to be designed and built in Italy. By 1984, the specifications had been agreed with industry and design work on key subsystems such as power pack and armament was well under way.

In the same year Otobreda and IVECO formed a consortium, with headquarters in Rome, to handle the Italian Army's new generation of tracked and wheeled armoured vehicles. For the Ariete MBT, Otobreda has overall responsibility with IVECO being responsible for the power pack and suspension system. For the Centauro 105 mm (8 × 8) tank destroyer, IVECO is responsible for the chassis while Otobreda has the responsibility for the turret and 120 mm smoothbore main armament.

The first prototype of the Ariete was completed in 1986 and by 1988 all six prototypes had been completed and trials by the Italian Army and the manufacturer were well under way. The six prototype Ariete MBTs fired more than 3,000 rounds of 120 mm ammunition and underwent more than 450 days of

Ariete MBT showing the 120 mm smoothbore gun fitted with fume extractor, thermal sleeve and muzzle reference system
2001/0105877

combat tests. The Italian Army has placed an order for 200 Ariete MBTs with production undertaken at the Otobreda facility at La Spezia with IVECO at Bolzano supplying the complete power pack (engine, transmission and cooling system) ready for installation in the vehicle.

The first production Ariete MBT was delivered late in 1995 and final production vehicles will be delivered to Italian Army in 2002.

The total price of the 200 Ariete MBTs for the Italian Army will be around, US$970 million including the total ammunition, spare parts and logistics. In addition to IVECO and Otobreda there are almost 248 subcontractors on the Ariete MBT programme.

Studies for the so-called Ariete Mk 2 are already under way and this would have a more powerful 1,500 hp turbocharged diesel engine, hydropneumatic suspension, automatic loading system for the 120 mm smoothbore gun, more advanced fire-control system and increased protection for improved battlefield survivability.

As of early 2002 there were no firm plans for the Ariete Mk 2 to enter production.

Description

The Ariete's hull and turret are of all-welded steel construction with an additional layer of advanced armour over the frontal arc, including the nose, glacis plate and turret front and sides.

Exact protection levels are classified, but the 54 tonne weight limit allows for protection against 105 mm APFSDS attack at least and against the HEAT warheads of most light anti-tank weapons.

Layout is similar to that of most other current generation tanks, with the driver's compartment front right, the turret in the centre and the power pack at the rear. The driver's single-piece hatch cover lifts and swings to the right. It has three integral forward-facing day periscopes, the central one of which can be replaced by an MES VG/DIL 100 passive periscope for night driving. The driver's seat is hydraulically adjustable up or down and he directs the vehicle using a steering wheel rather than tillers. To his left are stowed 27 rounds of 120 mm ammunition and in the floor immediately behind him is the hull escape hatch. The fighting compartment is fitted with a fire and explosion detection and suppression system.

The turret is in the centre of the hull with the commander on the right, the gunner ahead of and below him and the loader on the left. The sides and rear of the turret are vertical but the front is well sloped to provide the best possible protection. The commander has a rear-opening single-piece hatch cover and eight day periscopes for all-round observation.

The loader has a rear-opening single-piece hatch cover and two roof-mounted day periscopes looking forward and to the left of the vehicle. Forward of this hatch is a Marconi Defence Systems Division laser warning sensor with the display at the commander's station.

There are blow-out panels in the turret roof to vent any ammunition explosion upward and away from the turret crew, plus an ammunition resupply hatch in the left side and a large stowage basket at the rear of the bustle.

The commander's and layer's periscopes (TURMS - Tank Universal Modular System), manufactured by Officine Galileo, are common to the Centauro 8 × 8 105 mm armed tank destroyer of which 400 have been delivered to the Italian Army. In 1999 the Spanish Army placed a contract for 22 new build Centauro vehicles which have now been delivered. More recently another 62 vehicles have been ordered by Spain.

The firing systems main components are the commander's primary stabilised panoramic day/night (image intensification) sight; the gunner's primary stabilised periscope laser sight; a ballistic computer; sensors; muzzle reference system; and commander's, gunner's and loader's control panel. The commander's sight has a magnification of ×2.5 and ×10 and a thermal image from the gunner's sight is displayed on a separate TV monitor at the commander's station. The roof-mounted sight can be traversed through 360° and elevated from −10 to +60°. The commander's head remains stationary while the sight head moves.

The gunner's roof-mounted sight integrates four main modules-primary stabilised head mirror, visual unit, laser and thermal image unit in a single housing protected by armoured shutters which open left and right. The daylight ×5 magnification and thermal night vision with wide and narrow fields of view are all routed via the common head mirror.

The ballistic computer performs all fire computations and controls and manages the optical sight, laser range-finder and servos as well as the sensors, built-in test equipment and training. It also enables the system to be reconfigured from the normal operating level to back-up modes in the event of partial failure. The key

Specifications

Crew: 4
Combat weight: 54,000 kg
Power-to-weight ratio: 24 hp/t
Ground pressure: 0.90 kg/cm²
Length:
(including main armament) 9.67 m
(hull) 7.59 m
Width:
(overall) 3.601 m
(over tracks) 3.42 m
Height:
(turret roof) 2.5 m
(hull top) 1.82 m
Firing height: 1.98 m
Ground clearance: 0.48 m
Track: 2.802 m
Track width: 618 mm
Max speed (cruising): over 65 km/h
Acceleration:
(0 to 32 km/h) 6 s
Max range (cruising): over 550 km
Fording:
(with preparation) 2.1 m
(without preparation) 1.2 m
(deep fording with preparation) 4 m
Gradient: 60%
Side slope: 30%
Vertical obstacle: 1 m
Trench: 3 m
Engine: IVECO V-12 MTCA, turbocharged, intercooled, 12-cylinder diesel developing 1,300 hp at 2,300 rpm
Transmission: Renk LSG 3000 automatic with 4 forward and 2 reverse gears. This allows 3 ranges of steering radii and pivot
Brakes: hydraulic actuated with transmission retarder
Suspension: torsion bar
Electrical system: 24 V
Batteries: 8 × 12 V, 100 Ah
Armament:
(main) 1 × 120 mm smoothbore gun
(coaxial) 1 × 7.62 mm MG
(anti-aircraft) 1 × 7.62 mm MG
Smoke grenade dischargers: 2 × 4 80 mm
Ammunition:
(main) 42
(MG) 2,400
Gun control equipment
Turret power control: electrohydraulic/manual
(by commander) yes
(by gunner) yes
Turret traverse: 360°
Commander's fire-control override: yes
Gun elevation/depression: +20/−9°
Gun stabiliser:
(2-axis stabilisation) yes
NBC system: yes
Night vision equipment: yes

Ariete MBT showing roof-mounted 7.62 mm MG
2001/0105878

TURMS sensors provide meteorological, tank attitude and barrel wear data. The meteorological and wind sensors are mounted on the turret roof.

The Ariete can engage both stationary and moving targets while either stationary or mobile itself. The tank commander would normally acquire the first target and hand it over to the gunner for engagement while he searches for additional targets. The gunner's back-up sight is an Officine Galileo C-102 coaxial telescope with a magnification of ×8 and with three manually selectable aiming graticules.

Main armament is an Otobreda 120 mm 44 calibre auto-frettaged smoothbore gun fitted with a muzzle reference system, thermal sleeve and fume extractor. The coaxial recoil system consists of a recoil buffer and a nitrogen-actuated recuperator, with an internal reservoir installed on the gun cradle. The vertical sliding wedge breech block remains open after the weapon has recoiled and the stub of the cartridge case has been ejected.

The gun chamber is identical in size to that of the 120 mm smoothbore gun installed in the Leopard 2 and the M1A1/M1A2 so their ammunition is interoperable. The Italian company, SIMMEL DIFESA, is prime contractor for the APFSDS and HEAT-MP 120 mm ammunition. A total of 42 rounds of 120 mm ammunition is carried, 15 in the turret bustle and the remaining 27 in the hull.

The fixed mantlet has vertical slots for the auxiliary gunner's telescope, the 120 mm gun and the 7.62 mm coaxial machine gun on its left. The coaxial machine gun is fired electrically by the gunner, who also has a mechanical back-up. Another 7.62 mm machine gun is mounted on the turret roof for local and anti-aircraft defence. Operated by the tank commander or by the loader and mounted on a spring-balanced cradle, it can be traversed rapidly through 360° and elevated from −9 to +65°.

Turret traverse and weapon elevation is electrohydraulic with manual back-up. An all-electric gun control and stabilisation system is now offered for the Ariete MBT. The commander and gunner have dual-control handles.

A bank of four 80 mm electrically operated, forward-firing smoke grenade dischargers, identical to those fitted to the Centauro and Dardo HITFIST infantry fighting vehicle, is mounted on each side of the turret firing towards the front. The smoke grenade launchers can be operated automatically by the laser warning system.

The complete power pack, consisting of engine, transmission and cooling system, is delivered to Otobreda by IVECO ready for installation in the vehicle and can be replaced in less than an hour. The IVECO V-12 MTCA engine, a 12-cylinder, four-stroke, direct-injected, water-cooled, supercharged, after-cooled diesel, develops 1,300 hp and gives a power-to-weight ratio of 24 hp/t. Supercharging is by two turbochargers, one for each bank, arranged at the rear of the engine.

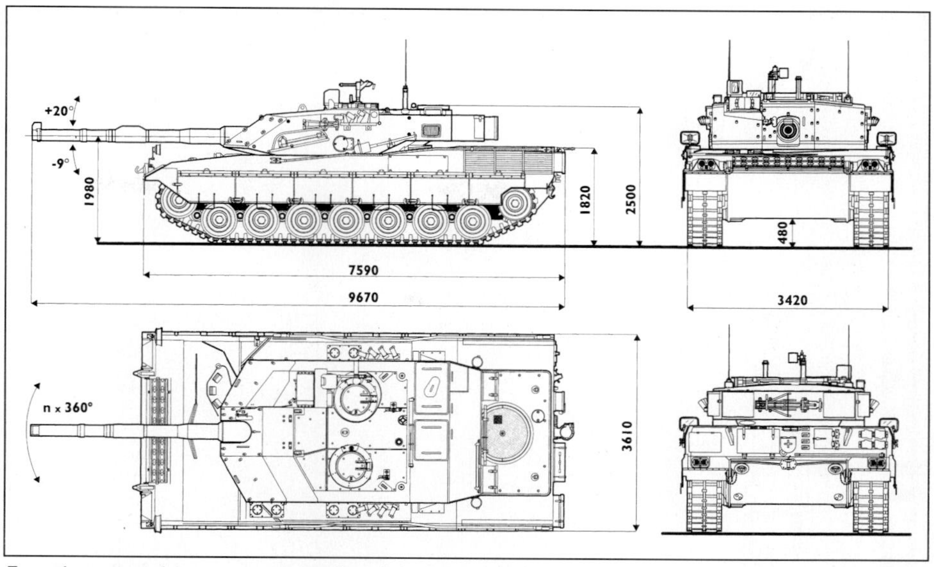

Four view general arrangement drawing of the Ariete MBT ***2001****/0105879*

Their inlets are cooled by air-water heat exchangers. The air intake is above the engine compartment and there are exhaust outlets either side of the hull at the rear.

The engine is coupled directly to a German Renk LSG 3000 fully automatic transmission (manufactured under licence by IVECO in Italy) which provides four forward and two reverse speeds, three turning radii and pivot. Forward and reverse gears are selected through a power-operated shift control, with emergency mechanical shifting of the second forward and second reverse speeds available as a back-up.

The transmission incorporates an auxiliary retarder to absorb about 75 per cent of the complete braking energy according to an integration logic with the mechanical brakes. The hydraulically operated main and emergency brakes act in conjunction with the automatic transmission retarder and the mechanical disc brakes on the final drive. The parking brake is of the mechanical type and acts on both disc brakes.

The two GRP main fuel tanks, located either side of the rear of the fighting compartment are selectable through a two-way valve actuated by the driver. An auxiliary tank filled from the main tank pipes ensures that fuel is available even when the tank is going up slopes and the tanks are partially empty. The fuel is forced into the engine by an electrically operated pump.

The torsion bar suspension either side consists of seven dual rubber-tyred roadwheels, idler at the front, drive sprocket at the rear and four track-return rollers. The first, second, third, sixth and seventh roadwheels have hydraulic shock-absorbers and all seven suspension arms either side have hydraulic bumpers to limit excessive roadwheel travel. The Type 840 track was designed by Diehl of Germany and is manufactured under licence in Italy. It has rubber-bushed connectors, with the individual track shoes being joined by end connectors and centre guiding teeth.

The upper part of the suspension is covered by skirts that can be quickly hinged upward allowing access to the suspension for maintenance.

The 24 V electrical system incorporates a total of eight 12 V, 100 Ah batteries to give a total of 400 Ah at 24 V. There are four batteries located on each side of the hull for ease of access via a panel above the fifth and sixth roadwheel stations.

The SP-180 NBC pack, manufactured by Sekur, has unprotected and protected modes. Under normal unprotected conditions it provides forced air ventilation, sucking in air from outside through a cyclone filter. In the protected mode, the forced air from the cyclone filter enters a battery of composite filters containing the final filter and the carbon filter. The outlet is in the left side of the turret toward the rear.

Status

Production complete. In service with the Italian Army which has ordered a total of 200 vehicles.

Prime contractor

ConsorzioIveco Oto (which is also prime contractor for the Centauro (8 × 8) 105 mm tank destroyer and the DARDO/HITFIST infantry combat vehicle recently ordered by the Italian Army.

UPDATED

Otobreda OF-40 MBT

Development

The OF-40 MBT was designed from 1977 by Otobreda and FIAT specifically for the export market. Otobreda was responsible for overall design and production and FIAT the power pack. In appearance the OF-40 resembles the late-production German Krauss-Maffei Wegmann Leopard 1A4 and it uses some components of the Leopard 1 MBT. The first prototype of the OF-40 (O for Otobreda, F for FIAT and 40 for the approximate weight in tons) was completed in 1980.

The United Arab Emirates placed an order for 18 OF-40 Mk 1s and the first of these was delivered in 1981. Since then, the United Arab Emirates has taken delivery of a further 18 OF-40 Mk 2 MBTs and three armoured recovery vehicles, while the original 18 OF-40 Mk 1 MBTs have been brought up to Mk 2 standard.

The OF-40 MBT has also been tested in Thailand, demonstrated in Egypt and offered for local production in Spain and Greece. The Otobreda OF-40 MBT is no longer being marketed and the only customer to date is the United Arab Emirates (Dubai).

Description

The all-welded steel hull of the OF-40 MBT is divided into three compartments: driving at the front, fighting in the centre and engine and transmission at the rear. The driver is seated at the front on the right side and has a single-piece hatch cover that lifts and swings to the left. There are three unity periscopes in front of him, the centre one of which can be replaced by an Alenia image intensification periscope for night driving. To the left of the driver is the NBC pack and 42 rounds of 105 mm ammunition and behind him there is an oval emergency hull escape hatch.

The all-welded steel turret is in the centre of the hull with the commander and gunner on the right and the loader on the left. The commander has a circular hatch cover that opens to the rear and eight unity periscopes for all-round observation, one of which can be replaced by an Alenia image intensification periscope. The commander's seat can be elevated by a manually operated hydraulic system and a foot-operated valve will enable the commander to lower the seat quickly in an emergency.

The commander does not have a cupola, but mounted in the roof forward of the hatch is a day/night sight from Officine Galileo, developed in a joint venture with SAGEM of France. This is stabilised and fitted with image intensification night vision equipment. This enables the tank commander to detect and identify the target which can then either be handed over to the gunner or engage the target personally.

The gunner is seated in front of and below the commander and has one forward-facing roof-mounted periscope and an optical sight, the Alenia C215 articulated telescope with a 7.5° field of view and a magnification of ×8. It is mounted coaxially with the 105 mm gun in an M114 telescope mount.

The loader sits on the left of the turret and has a circular hatch cover that opens to the rear, in front of which are two roof-mounted day periscopes that give observation to the front and left side.

The main armament of the OF-40 is an Otobreda-designed and -manufactured 105 mm 52 calibre rifled tank gun with falling wedge breech block, concentric buffer and spring recuperator. The gun is fired electrically but can also be fired by a manually actuated impulse generator. During counter-recoil the automatic breech block opens and ejects the spent cartridge case into a bag under the breech.

The barrel is provided with a thermal sleeve and a bore evacuator. The gun fires all standard NATO 105 mm ammunition including APDS, canister, HEAT, HESH and smoke and a well-trained crew can achieve a rate of fire of 9 rds/min. As an option, the 105 mm main armament can be stabilised in both elevation and traverse.

A total of 57 rounds of 105 mm ammunition is carried, 42 to the left of the driver and 15 in the turret for ready use.

A 7.62 or 12.7 mm machine gun is mounted coaxially to the left of the main armament and a similar weapon is mounted on the turret roof for anti-aircraft use.

OF-40 Mk 2 MBT 0045713

Mounted externally on either side of the turret are four 76 mm electrically operated smoke grenade dischargers.

The fire-control system has been designed by Officine Galileo and consists of the gunner's fire-control system, gunner's telescope which moves in elevation with the main armament, commander's stabilised day/night roof-mounted sight and the turret/gun electrohydraulic drive system.

The gunner's fire-control system consists of an Officine Galileo OG14 sight with a magnification of ×8 and range scale for APDS, HEAT and HESH 105 mm projectiles and 7.62 mm machine gun ammunition; Alenia VAQ-33 laser range-finder with a range of between 400 and 9,995 m; laser feeding and control equipment; fire-control computer; control device and readout for the tank commander, gunner's controls and elevation sensor.

The commander's roof-mounted VS 580-B sight can be traversed through a full 360°, has a magnification of ×8, a stadiametric range-finding capability and range scales for APDS, HEAT and HESH ammunition.

Turret traverse and gun elevation are electrohydraulic, with turret traverse at a rate of 360° in 17 seconds and gun elevation from –9 to +20° at 7°/s. The turret/gun drive system is controlled from the gunner's handle but an override system allows the commander to take over the laying and firing of the main armament if required and manual controls are provided for emergency use.

The engine, transmission and cooling system are assembled to form the power pack which can be removed by four personnel with a crane in under 45 minutes. The cables and wires are provided with quick disconnect couplings and, if required, the power pack can be run outside the vehicle for testing.

The engine is an MTU 10-cylinder, four-stroke, precombustion chamber, supercharged multifuel developing 830 hp at 2,200 rpm and is fitted with a tropicalisation kit to control the fuel supply and prevent the engine from overheating.

Two fuel tanks, with a total capacity of 1,000 litres, are mounted each side of the engine compartment and fuel can be drawn from one or both tanks.

Combustion air is taken in by two supercharged blowers through the two air cleaners. The two dry air cleaners are fitted in the side recess in front of the engine compartment and air enters through the intake scoops on the hull deck. Screens mounted in front of the scoops prevent large particles such as leaves entering the vehicle. A cyclone filter battery cleans the air of coarse dust and large-surfaced micro-top filters retain the fine dust. The coarse dust ejected by the cyclone filters is continuously blown outside by a dust blower installed in each air cleaner. The exhaust gases are taken from each cylinder liner by flexible corrugated tube compensators through an exhaust manifold into the exhaust mufflers in the side recesses of the engine compartment.

The engine of the OF-40 is liquid cooled through a self-contained cooling system with internal pressure controlled by a pressure relief and underpressure valve. On either side of the transmission is a cooler with a control fan on the vertical axis between them to suck the necessary cooling air from the top through the cooling air intake grille and force it through the two coolers. When deep fording, this duct is filled with water and the fan and coolers are therefore completely flooded.

Engine torque is transmitted to a torque converter which is followed by a four-speed planetary gearbox with two reverse speeds. The individual speeds are shifted electrohydraulically by a shift lever alongside the driver. There is a separate clutch between the engine and Renk transmission which can be operated manually from the crew compartment. This provides easy starting at low temperatures as the starter has to crank the engine only and not the transmission.

The steering works through a longitudinally superimposed gearbox, so steering movements do not cause essential power losses in the gearbox. Steering movements are mechanically transmitted from the control handle at the driver's station to the control valve.

The final drives are on either side at the rear of the vehicle and are of the planetary type. The transmission is connected to the final drives by an internally toothed sleeve which can be loosened from the outside when the power pack has to be removed.

The suspension is of the torsion bar type with seven dual rubber-tyred roadwheels on each side. The first three and last two roadwheels have hydraulic shock-absorbers and bump stops to prevent excessive wheel travel. The adjustable idler is at the front, drive sprocket at the rear and there are five track support return rollers on each side.

The track is the rubber-bushed connector type with the individual track shoes joined by end connectors and centre guiding teeth. The upper part of the track and suspension is covered by a steel-reinforced rubber skirt which keeps dust down and provides some level of protection against attack by HEAT projectiles.

The OF-40 has three independent braking systems; service, parking and emergency systems.

The tank can ford to a depth of 1.2 m without preparation and an optional semi-deep fording hydraulic system enables it to cross streams to a maximum depth of 2.25 m. A snorkel can be fitted which enables the OF-40 to ford to a depth of 4 m. An electrically operated bilge pump is fitted in both the engine and crew compartment, each with a capacity of 120 litres/min.

The overpressure NBC system is mounted at the front of the hull to the left of the driver. Fresh air is drawn into the vehicle by a blower and is then separated from the coarse dust in a cyclone filter. The NBC protection filter is downstream of the coarse dust filter and retains all known NBC agents. The dust collection compartment of the cyclone filter is continuously kept free of dust by a dust ejector blower. Pressure in the tank is monitored by a pressure gauge in the crew compartment.

An automatic fire extinguishing system in the engine compartment has spray tubes and nozzles connected to a battery of bottles in the crew compartment containing the fire extinguishing agent. This system is activated at 180°C through a firewire in the engine compartment.

Specifications
Mk 1
Crew: 4
Combat weight: 45,500 kg
Unloaded weight: 43,100 kg
Power-to-weight ratio: 18.24 hp/t
Ground pressure: 0.92 kg/cm²
Length:
(gun forward) 9.222 m
(gun rear) 8.114 m
(hull) 6.893 m
Width:
(without skirts) 3.35 m
(over skirts) 3.51 m
Height:
(turret roof) 2.45 m
(top of commander's sight) 2.68 m
(top rear hull deck) 1.76 m
Firing height: 1.89 m
Ground clearance: 0.44 m
Track width: 584 mm
Length of track on ground: 4.25 m
Max road speed:
(forward) 60 km/h
(reverse) 25 km/h
Fuel capacity: 1,000 litres
Max road range: 600 km
Fording:
(without preparation) 1.2 m
(with preparation) 2.25 m
(with snorkel) 4 m
Gradient: 60%
Side slope: 30%
Vertical obstacle: 1.1 m
Trench: 3 m
Turning radius: pivot turns
Engine: MTU 90° 10-cylinder, 4-stroke, precombustion chamber, supercharged multifuel developing 830 hp at 2,200 rpm*
Transmission: Renk power shifting 4 forward and 2 reverse planetary type with hydraulic torque converter
Suspension: torsion bar
Electrical system: 24 V
Batteries: 8 × 12 V, 400 Ah
Armament:
(main) 1 × 105 mm rifled gun
(coaxial) 1 × 7.62 mm MG
(AA) 1 × 7.62 mm MG (or optional 12.7 mm MG)
Smoke grenade dischargers: 4 × 76 mm either side of turret
Ammunition:
(main) 57
(MG) 5,700
Gun control equipment
Turret power control: electrohydraulic/manual
(by commander) yes
(by gunner) yes
(commander's override) yes
Commander's fire-control override: yes
Max rate power traverse: 360°/17 s
Max rate power elevation: 7°/s
Gun elevation/depression: +20°/–9°
Gun stabiliser:
(vertical) no (optional)
(horizontal) no (optional)
Range setting device: yes
Elevation quadrant: yes
Traverse indicator: yes
NBC system: yes
Night vision equipment: yes
* (950 hp diesel option)

Variants

OF-40 Mk 2 MBT
The OF-40 Mk 2 MBT is essentially the Mk 1 fitted with the Galileo OG14L2A fire-control system which includes a stabilisation system for the 105 mm gun and sensors for wind velocity, powder temperature, ambient temperature and type of ammunition. The OG14L2B is similar but has a stabilised line of sight.

In the OF-40 Mk 2 the tank commander has a day/night periscope that is self-stabilised during panoramic search and slaved to the gun stabilisation when used for firing.

The gunner has a sight with a magnification of ×7 and ×14 that incorporates a laser range-finder as well

OF-40 ARV in the travelling configuration carrying replacement power pack 0018790

as a telescopic sight, with a magnification of ×8, coaxial with the 105 mm gun.

Mounted externally on the right side of the mantlet is an LLLTV camera with the monitor in the turret.

Deliveries of the second batch of 18 OF-40 Mk 2 tanks began in 1984 and were completed in 1985. The Mk 1 tanks have now been brought up to Mk 2 standard.

OF-40 ARV

Three OF-40 armoured recovery vehicles have been delivered by Otobreda. These have a front-mounted winch with a tractive effort of 35,000 kg in first gear; this can be increased with the aid of a tackle. Mounted on the left side of the hull at the front is a hydraulic jib with a maximum lifting capacity of 18,000 kg with the front-mounted stabiliser blade in position. The blade can also be used as a dozer blade to clear away obstacles and fill in trenches. The OF-40 ARV has a crew of four and basic specifications are: length 7.68 m, height 2.35 m, ground clearance 0.44 m and combat weight 45,000 kg. If required, a replacement OF-40 power pack can be carried on the rear decking. Full details of this vehicle are given in *Jane's Military Vehicles and Logistics 2002-2003*.

Systems on OF-40 Chassis

The hull of the OF-40 can also be used to mount other weapon systems such as a twin 35 mm anti-aircraft turret similar to the one on the Gepard anti-aircraft gun system. The OF-40 chassis, with a different power pack, is also used as the basis for the Otobreda Palmaria 155 mm self-propelled howitzer described later in this book. The Palmaria chassis also forms the basis for the 76 mm OTOMATIC self-propelled air defence system. Details of the OTOMATIC are given in *Jane's Land-Based Air Defence 2002-2003*. As of early 2002, the OTOMATIC system remained at the prototype stage.

Status

Production complete. In service with Dubai (part of the United Arab Emirates). The OF-40 MBT is no longer being marketed.

Manufacturer

Otobreda.

UPDATED

JAPAN

Mitsubishi Type 90 MBT

Development

Research and development of a new Japanese MBT began in 1976 under the designation of the STC, subsequently called the TK-X, by the Technical Research Headquarters of the Japanese Self-Defence Agency.

Funding for the diesel engine, gun, gun ammunition and fire-control system started in Fiscal Year 1977 with funding for the suspension and new armour in FY78.

Prime contractor for the Type 90 MBT is Mitsubishi Heavy Industries, which has also been prime contractor for all Japanese post-Second World War MBTs including the Type 74 and the Type 61.

With the exception of the 120 mm smoothbore gun, which is made under licence from Rheinmetall of Germany, the Type 90 MBT and its subsystems are all designed and built in Japan.

Major subcontractors include Japan Steel Works, Mitsubishi Electric Corporation, Fujitsu Company and the NEC Corporation.

Requirements of the Type 90 were completed in 1980 with two prototypes, both armed with a Japanese 120 mm gun firing Japanese ammunition, completed in 1982 to 1984. These were used in extensive trials that continued through to 1986.

A second series of four prototypes was built between 1986 and 1988 which incorporated changes as a result of trials with the first two prototypes. These were armed with the Rheinmetall 120 mm smoothbore gun also fitted to the German Leopard 2 and, in a modified version, in the US M1A1/M1A2 Abrams MBT.

These prototypes were used for development and then user trials, all of which were completed by 1989.

The Type 90 was type classified in 1991, as of late 1999 a total of 172 were in service with production still under way. The Mitsubishi Type 90 MBT and its variants have not been offered on the export market. Recent production has been at the rate of about 17 Type 90 MBTs per year.

Description

No firm details of the Type 90 MBT armour type used in the construction of the Type 90 MBT have been released, but it is understood that extensive use of composite armour is made in both hull and turret over the frontal area of the tank. The Mitsubishi Steel Works and Kyoto Ceramic Company have been responsible for much of the armour development.

The layout of the Type 90 MBT is conventional with the driver's compartment at the front, turret in the centre and engine and transmission at the rear. The driver is seated at the front left and has a single-piece hatch cover with three day periscopes being provided for forward observation; the centre one can be replaced by a passive periscope for night driving.

A quantity of 120 mm ammunition is stowed to the right of the driver's position.

The turret is similar to that of the German Leopard 2 MBT with vertical front, sides and rear and a bustle that extends well over the rear engine compartment.

The commander is seated on the right of the turret with the gunner on the left and the automatic loader being mounted in the turret bustle. Blow-out panels are provided in the turret bustle roof. The commander is provided with day vision blocks for all-round observation and both commander and gunner have roof hatches.

Main armament comprises a 120 mm Rheinmetall smoothbore gun made under licence from Germany, which is fitted with a thermal sleeve, fume extractor and muzzle reference system. The ordnance and breech are identical to the German 120 mm smoothbore gun but the recoil system and gun mount were designed and built in Japan.

Type 90 MBT (Paul Beaver) 0018792

Type 90 ARV in travelling configuration (Japan Military Review)

Specifications
Crew: 3
Weight: 50,000 kg
Power-to-weight ratio: 30 hp/t
Ground pressure: 0.89 kg/cm^2
Length:
(overall) 9.755 m
(hull) 7.5 m
Width:
(overall) 3.43 m
(over tracks) 3.33 m
Height:
(overall) 3.045 m
(hull roof) 2.335 m
Ground clearance: 0.45 m (variable between 0.2 and 0.6 m front and rear)
Track: 2.71 m
Length of track on ground: 4.55 m
Track width: 600 mm
Max road speed:
(forward) 70 km/h
(reverse) 42 km/h
Fuel capacity: 1,100 litres
Max range: 400 km
Fording: 2 m
Gradient: 60%
Side slope: 40%
Vertical obstacle: 1 m
Trench: 2.7 m
Engine: Mitsubishi 10ZG 10-cylinder water-cooled diesel developing 1,500 hp at 2,400 rpm
Transmission: automatic
Suspension: torsion bar/hydropneumatic
Armament:
(main) 1 × 120 mm smoothbore gun
(coaxial) 1 × 7.62 mm MG
(anti-aircraft) 1 × 12.7 mm MG
Smoke-laying equipment: 2 × 3 smoke grenade dischargers
Gun control equipment
Turret power control: electric/manual
(by commander) yes
(by gunner) yes
Gun stabiliser:
(vertical) yes
(horizontal) yes
Range setting device: yes (laser)
NBC system: yes
Night vision equipment: yes

Type 90 MBT moving through snow 0011615

No details of the gun elevation/depression have been made available but it is probably similar to the Type 74 which has an elevation of +15° and a depression of –10° by pitching the hull forwards or backwards. Turret traverse and weapon elevation are all-electric with manual controls for emergency use.

The 120 mm smoothbore gun is fed by an automatic bustle-mounted loader designed by Mitsubishi Heavy Industries. This is believed to hold a total of 16 rounds of ready use 120 mm ammunition of the APFSDS-T or HEAT-MP type and can be loaded from inside the Type 90 MBT or through a hatch in the turret roof. The ammunition has a semi-combustible cartridge case and all that remains after the weapon has fired is the stub cartridge case.

When the gun has fired, the ordnance returns to 0° elevation where it is loaded by the automatic loader, the gun then automatically returns to the firing position. Manual back-up of the automatic loader is possible.

A 7.62 mm machine gun is mounted coaxial to the left of the main armament and a 12.7 mm machine gun is mounted on the roof for anti-aircraft defence. This is mounted on the forward part of the turret roof so that it can be used by the commander or gunner. Mounted either side of the turret is a bank of three electrically operated smoke grenade dischargers firing forwards.

The fire-control system of the Type 90 includes a gunner's periscopic sight which is stabilised in azimuth, commander's periscopic sight with 180° traverse with dual-axis stabilisation and a digital fire-control computer.

The gunner's sight has a daylight channel, thermal sensor and an Nd:YAG laser range-finder.

The gunner's sight is provided by Nikon Corporation, commander's sight by Fuji Photo Optical Company and the thermal unit by Fujitsu Company.

The fire-control system includes an advanced auto-tracking capability that is based on the output of the thermal imager.

The automatic tracker is effective against ground targets and can be used when the Type 90 MBT is stationary or on the move.

When engaging without the auto-tracker, the gunner or commander uses his manual controls to track the target. To engage the auto-tracker, the commander or gunner only has to press the lock-on switch as soon as he has acquired the target and it falls within the lock-on gate in the sight.

If the target disappears temporarily, for example when it moves behind cover, the sight continues tracking at the same rate so that when the target reappears, the gunner can quickly lock-on again.

The commander's dual-axis stabilised sight, which only has a day capability, not only permits him to acquire and engage targets directly, but also provides a hunter killer capability. By pushing a button on his override control handle, the commander is able to hand-off a target that he has acquired to the gunner. The commander can then search for new targets while the gunner engages the first target.

In addition, the gunner's sight incorporates a thermal unit which is fed into two displays, one at the gunner's station and the other at the commander's. The tank commander also has unity vision blocks arranged to give 360° vision for surveillance.

The complete fire-control system provides a range of viewing options for both the commander and gunner, with redundancy provided by a back-up coaxial telescope mounted alongside the gun for use by the gunner.

The digital fire-control computer is the heart of the fire-control system. When engaging static and moving targets, calculations are made by the computer to compensate for range, wind (with the wind sensor being mounted on the turret bustle), temperature, gun tube bend (information from the muzzle reference system which is located on the left of the muzzle and is fed into the gunner's sight and fire-control computer), and cant angle (trunnion tilt).

Suspension of the Type 90 is of the hybrid type with torsion bars for the centre two roadwheels and hydropneumatic units for the first two and last two roadwheel stations, with six dual rubber-tyred roadwheels, drive sprocket at the rear, idler at the front and return rollers. The upper part of the track is covered by a lightweight skirt. The suspension enables the tank to change attitude in longitudinal direction only, for example on reverse slopes. Unlike the current Type 74, it will not be able to compensate for transverse changes.

The Mitsubishi 90° 10ZG is a two-stroke water-cooled direct injection diesel engine that uses an electric fuel-control system. It is supercharged by a Rootes blower, the engine having two exhaust turbochargers and an intercooler. Its maximum power of 1,500 hp is achieved at 2,400 rpm.

The engine is coupled to an automatic transmission which incorporates a torque converter with an automatic lock-up clutch in series with a four-speed planetary range box. Two speeds are available in reverse.

Hydrostatically controlled regenerative differential steering is utilised and oil-cooled multiple disc brakes are incorporated into the transmission.

The three engine cooling radiators are mounted above the transmission and are used together with three mixed flow fans. The fans are driven by hydraulic motors and are understood to be of the variable speed type depending on the engine and transmission temperatures. Air cleaners are located either side of the transmission.

Mounted on the forward part of the turret roof is a laser detector which provides an audible signal, as well as a direction indication, at the commander's station. This system can be used to automatically launch smoke grenades to counter heat-seeking missiles.

An NBC system is fitted as standard and a dozer blade can be mounted at the front of the vehicle if required.

Type 90 mineclearing MBT

Mounted at the front of the hull is the Type 92 mineclearing roller system. Development of this commenced in 1998 and it was type classified in 1992. Mounted in front of the rollers are magnetic force emitting devices for the detonation of magnetic mines.

Type 90 ARV

The Type 90 armoured recovery vehicle is based on the chassis of the Type 90 MBT and has a new superstructure with a crane mounted at front right, dozer/stabiliser blade at the front of the hull and a hydraulically operated winch. Additional details are given in *Jane's Military Vehicles and Logistics 2002-2003*. Recent production has been at the rate of two vehicles per year.

Type 91 AVLB

This is the chassis of the Type 90 MBT fitted with a scissors-type bridge that is launched over the front of the vehicle. The bridge is class 60 and can span a gap up to 20 m. Additional details are given in *Jane's Military Vehicles and Logistics 2002-2003.*

Status

Production. In service with Japanese Ground Self-Defence Force.

Manufacturer

Mitsubishi Heavy Industries, Sagamihara Limited.

UPDATED

Mitsubishi Type 74 MBT

Development

Mitsubishi Heavy Industries and the Japanese Ground Self-Defence Force began project definition studies for a new MBT to succeed the Type 61 in 1962. To prove the basic concept, a number of test rigs were built and tested between 1964 and 1967. Construction of the first two prototypes began at the Maruko works of Mitsubishi Heavy Industries late in 1968 and was completed in September 1969. They were called the STB-1 and contained many features of other tanks under development at that time.

The STB-1 was followed in 1971 by the STB-3 which had the automatic loader removed as it proved both too complex and too expensive. The remote-controlled M2 12.7 mm (0.50) machine gun was also replaced by a simpler mount and the turret was slightly different in shape having a much longer bustle.

The first production contract was placed before the final model, the STB-6 and appeared in 1973. The first Type 74 MBT was completed in September 1975 and by the time production was complete a total of 873 Type 74 MBTs had been built by Mitsubishi Heavy Industries. It has been followed in production by the Type 90 MBT armed with a 120 mm smoothbore tank gun. Details of the Type 90 MBT are given in the previous entry.

The Type 74 replaced the 90 mm armed Type 61 MBT. About 560 of these were built, but by the end of 1999, only 40 remained in front line service.

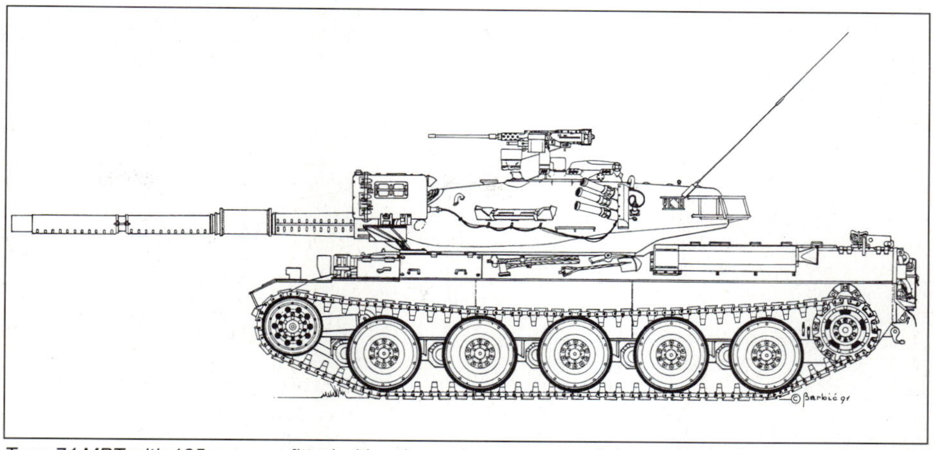

Type 74 MBT with 105 mm gun fitted with a thermal sleeve and an infra-red/white light searchlight to the left of the main armament (Vasco Barbic)

Specifications

Crew: 4
Combat weight: 38,000 kg
Unloaded weight: 36,300 kg
Power-to-weight ratio: 18.94 hp/t
Ground pressure: 0.86 kg/cm²
Length:
(gun forward) 9.42 m
(gun rear) 7.84 m
(hull) 6.7 m
Width: 3.18 m
Height:
(including AA MG) 2.67 m (with 0.65 m ground clearance)
(to turret top) 2.48 m (with 0.65 m ground clearance)
(to turret top) 2.03 m (with 0.2 m ground clearance)
Ground clearance: adjustable from 0.2 to 0.65 m
Track width: 550 mm
Length of track on ground: 4 m
Max road speed: 60 km/h
Track width: 550 mm
Fuel capacity: 950 litres
Max range: 400 km
Fording: 1 m
(with preparation) 2 m
Gradient: 60%
Side slope: 40%
Vertical obstacle: 1 m
Trench: 2.7 m
Engine: Mitsubishi 10ZF Type 22 WT 10-cylinder air-cooled diesel developing 720 hp at 2,200 rpm
Transmission: Mitsubishi MT75A manual with 6 forward and 1 reverse gears
Steering: double-differential
Suspension: hydropneumatic, variable
Electrical system: 24 V
Armament
(main) 1 × 105 mm rifled gun
(coaxial) 1 × 7.62 mm MG
(anti-aircraft) 1 × 12.7 mm MG
Smoke-laying equipment: 3 smoke grenade dischargers either side of turret
Ammunition:
(main) 55
(coaxial) 4,500
(anti-aircraft) 660
Gun control equipment
Turret power control: electric/manual
(by commander) yes
(by gunner) yes
Turret traverse: 360°
Gun elevation: +9° (+15° using suspension)
Gun depression: –6° (–12° using suspension)
Gun stabiliser:
(vertical) yes
(horizontal) yes
Elevation quadrant: yes
Traverse indicator: yes
Range setting device: yes (laser)
NBC system: yes
Night vision equipment: yes

Description

The hull of the Type 74 MBT is of all-welded steel and is divided into three compartments: driver's at the front, fighting in the centre and power pack at the rear.

The driver is seated at the front of the vehicle on the left side and is provided with a single-piece hatch cover that opens to the left. There are three JM 17 Mod 2 day periscopes mounted forward of this hatch cover and an infra-red or passive night periscope can be mounted in the centre of the hatch cover and used in conjunction with the infra-red driving lights.

The turret is all-cast steel with the commander and gunner seated on the right and the loader on the left. The commander's cupola can be traversed through 360° and has a single-piece hatch cover that opens to the rear. There is a J3 infra-red periscope sight with an integral laser range-finder and a magnification of ×1 and ×8 in the forward part of the commander's cupola. He also has five day periscopes, two on either side and one to the rear. The gunner is seated forward of the commander and is provided with a J2 IR periscopic sight, with a magnification of ×1 and ×8, in the turret roof and a telescope linked to the main armament. The loader has a single-piece hatch cover that opens to the rear and a day periscope which can be traversed through a full 360° mounted forward of his hatch. There is an external stowage basket at the rear of the turret.

The engine and transmission are mounted at the rear of the tank with the exhaust pipes and silencers mounted on the running boards to the rear of the hull. The maximum quoted road speed of the Type 74 is 53 km/h but reliable sources have indicated that its top speed is at least 60 km/h.

The suspension is hydropneumatic and can be adjusted to suit the type of terrain being crossed. Either side are five dual rubber-tyred roadwheels with the drive sprocket at the rear and the idler at the front. There are no track-return rollers. The hydropneumatic suspension can be operated either by the commander or the driver and enables the tank to be inclined 6° forwards or backwards, 9° left or right and raised or lowered giving a minimum ground clearance of 200 mm and a maximum ground clearance of 650 mm.

Standard equipment includes infra-red driving lights, an infra-red searchlight to the left of the main armament and an NBC system. In the white light mode the searchlight has a maximum range of 3,000 m and in the infra-red mode a range of 1,000 m. One Type 74 MBT in each company is fitted with a dozer blade.

Main armament consists of a 105 mm rifled tank gun which fires the standard range of ammunition including APDS-T and HESH-T. Early in 1984, it was announced that the Japanese Ground Self-Defence Force was to introduce the American 105 mm M735 APFSDS-T round for the Type 74 MBT, with prime contractor in Japan to be Daikin Industry. There are 14 rounds of ready use ammunition carried in the turret. The 105 mm

Type 78 ARV with hydraulic crane in operating position

Type 87 twin 35 mm self-propelled anti-aircraft gun system with radars erected
2002/0130961

gun has a drop block-breech mechanism and a new concentric recoil mechanism to reduce the volume of the upper part of the gun as well as reducing the frontal area of the turret. The fire-control system includes a Nippon Electric laser range-finder mounted in the commander's sight, which also provides inputs to the Mitsubishi Electric ballistic computer connected to the gunner's sight. Target range is fed to the computer automatically but trunnion tilt, barrel wear, ammunition type and temperature are fed in manually. The 105 mm main armament is stabilised in both the vertical and horizontal planes.

A 7.62 mm Type 74 machine gun is mounted coaxially with the main armament and a 12.7 mm (0.50) Browning M2 HB machine gun is pintle-mounted in the centre of the turret forward of the commander's and loader's positions. This has an elevation of +60° and a depression of –10°. Three smoke grenade dischargers are mounted either side of the turret. A stowage basket extends around the rear of the turret.

Late in 1988 it was observed that a number of Type 74 MBTs had been fitted with a thermal sleeve for their 105 mm guns, thermal night vision equipment and armour on the top and forward parts of their turrets.

More recently a number of other modifications have been carried out to the Type 74 MBT. These include the installation of a laser warning system on the roof (similar to that fitted to the Type 90 MBT), modifications to the fire-control system, modified rear drive sprocket and the fitting of track skirts.

With preparation the Type 74 can ford to a depth of 2 m. This kit consists of a drum type extension collar above the commander's hatch and two thin extensions to the exhaust pipes, one either side.

Type 61 MBT

The type 61 MBT was the first post Second World War tank to be designed and built in production quantities. It has now almost been phased out of service with the Japanese Ground Self-Defence Force.

Details of the Type 61 MBT and its variants, the Type 67 AVLB, Type 70 ARV and Type 67 AEV, were given in *Jane's Armour and Artillery 1999-2000*. The Type 61 MBT was never exported. By the end of 1999 only 40 remained in service.

Type 74 MBT fitted with a thermal sleeve for its 105 mm rifled tank gun (Kensuke Ebata) 0018793

Variants

Type 78 Armoured Recovery Vehicle

The prototype of an ARV based on the chassis of the Type 74 MBT was completed in 1974 and standardised as the Type 78 ARV in 1978. In appearance it is very similar to the French AMX-30D and German Leopard 1 ARVs, with the crew compartment at the front, hydraulically operated dozer blade at the front of the hull and a hydraulic crane on the right side of the hull. Details of the Type 78 ARV are given in *Jane's Military Vehicles and Logistics 2002-2003*. At least 50 of these vehicles have been built.

Type 87 Twin 35 mm SPAAG

This is essentially a Type 74 MBT chassis fitted with a new two-person power-operated turret armed with twin 35 mm Oerlikon Contraves 35 mm KDA cannon, tracking and surveillance radars. Prototype systems were designated the AWX but production systems are known as the Type 87 twin 35 mm self-propelled anti-aircraft gun system. Details of this system are given in *Jane's Land-based Air Defence 2002-2003*. By the end of 1999, a total of 45 of these systems have been delivered.

Status

Production complete. In service with the Japanese Ground Self-Defence Force.

Manufacturer

Mitsubishi Heavy Industries, Sagamihara Limited.

UPDATED

JORDAN

King Abdullah II Design and Development Bureau MBT programmes

The King Abdullah II Design and Development Bureau (KADDB) was established in August 1999 as an independent civilian structure within the Jordanian Armed Forces reporting direct to His Majesty King Abdullah II.

Since 1996 a nucleus of engineers has been working at the King Hussein Main Workshops on a range of design and development activities. These include the already revealed AB1 M47 armoured repair and recovery vehicle, AB2 armoured troop carrier, AB3 Black Iris utility vehicle, AB4 Al Akrup Scorpion upgrade, AB5 special forces Land Rover and the upgraded M60A1 Main Battle Tank (MBT) the AB9. As of early 2002 none of the tank based systems was in quantity production.

AB9 MBT

This was first shown in 1999 and has recently been further enhanced and now called the AB9B1. It is an M60A1 MBT with its 105 mm gun replaced by a 120 mm smoothbore L/50 Compact Tank Gun supplied by SW Ordnanace Enterprise Corp and a new Raytheon Integrated day/night fire-control system which features second-generation FLIR (Forward Looking Infra-Red) imaging, eye-safe laser range-finder and digital computer.

The AB9B1 is also fitted with a Goodrich 218S laser warning receiver which is used in conjunction with the electrically operated smoke grenade launchers mounted either side of the turret. Trials with this upgrade are still underway.

AB91 MBT with turret traversed to rear to show 120 mm Compact Tank Gun **2002**/0067714

AB9C4 MBT

With the introduction of up to 288 British Army Vickers Defence Systems Challenger I MBTs into the Jordanian Army under the local name of Al Hussein, some 300 upgraded Centurion MBTs, called the Tariq, have now become surplus to requirements.

Under project AB9C4 a new compact turret called Falcon is being designed for the Tariq which will mount the same 120 mm Compact Tank Gun as the AB9B1 which will be fed by a bustle mounted automatic loader with the commander, gunner and driver all seated below the turret ring. This will make for a lighter and more survivable platform.

Falcon will feature an optronics suite from Thales Optronics of the UK and the vehicle will also incorporate the Trilom signature management system, Maske multispectral obscurants and the Thales AFV Systems Cerebus defensive aid system which includes a Thales Optronics (Taunton) LWD2 laser warning device.

AB13 heavy infantry fighting vehicle

Project AB13 is a complete rebuild of the Tariq MBT in association with the Malyshev Tank Factory in the Ukraine (home of the T-80UD and T-84 MBTs) into a heavy infantry fighting vehicle.

This will have a crew of three and carry seven infantry and fitted with a Ukrainian turret type 99A armed with twin 30 mm cannon, 7.62 mm machine gun, 30 mm AGS-19 grenade launcher and 81 mm smoke grenade launcher system.

The driver and one crew member is at the front, turret in the centre and with the engine compartment at the rear. The latter has been modified and a raised troop compartment is provided at the rear complete with entry hatches and firing ports.

AB14 Temsah Heavy Infantry Fighting Vehicle

The AB14 Temsah (Arabic for crocodile) is based on a much modified Tariq MBT chassis with the power pack at the front, behind which is the commander and gunner. The troop compartment is at the rear and the 10 troops enter and leave via a large power operated ramp.

Full details of this vehicle, which remains at the prototype stage, are given in the Armoured personnel carriers (tracked) section.

The AB14 has been developed by the KADDB in association with the Mechanology Design Bureau of South Africa.

Status

Development. None of these vehicles is currently in quantity production.

Manufacturer

King Abdullah II Design and Development Bureau.

UPDATED

Project AB13 heavy infantry fighting vehicle on modified Centurion MBT chassis and fitted with new turret armed with 30 mm cannon, 30 mm grenade launcher and 7.62 mm machine gun
2001/0067716

KOREA, SOUTH

Hyundai K1 MBT

Development

Following proposals from a number of armoured fighting vehicle manufacturers, in 1980 the South Korean government selected the now General Dynamics Land Systems of the United States to design and build two prototypes of a new MBT to meet its own specific requirements.

The first of two prototypes of the XK-1 MBT was completed in 1983, the Automotive Test Rig (ATR). It was shipped to Aberdeen Proving Grounds in November 1983 for automotive performance, endurance and reliability testing. The ATR was a fully payloaded MBT chassis fitted with a non-operational turret.

The second prototype, called the Fire-Control Test Rig (FCTR), was rolled out at a ceremony at Selfridge Air National Guard base in December 1983 and shipped to Aberdeen Proving Grounds in February 1984 to begin fire-control tests.

Production of the XK-1 began in South Korea in 1984 and first production vehicles were completed in 1985. The vehicle was subsequently type-classified as the K1 MBT and the tank made its first appearance in September 1987, by which time several battalions had been equipped with it. The first production batch of 210 vehicles was completed in 1987 with the second batch consisting of 325 vehicles.

Many key components of the K1, such as the Computing Devices Company (now owned by General Dynamics Land Systems) ballistic computer, MTU diesel engine, Renk transmission and SAGEM roof-mounted sight are now being manufactured in-country.

Hyundai K1A1 MBT armed with a 120 mm M256 smoothbore tank gun 0018796

By 2002, total production of the K1 MBTs and variants amounted to just over 1,000 units.

As of early 2002, the K1 MBT and its variants had not been exported. A variant of the vehicle, designated the K1-M, has been designed and manufactured to meet the future Malaysian MBT requirement.

This has a number of modifications to satisfy the Malaysian Army requirement including the installation of an air conditioning system, NBC system, laser warning system and explosive reactive armour as well as a reduction in the combat weight of the vehicle to less than 50 tonnes

As one of its marketing efforts, Hyundai exhibited the K1 MBT, K1 ARV and K1 AVLB at the DSA '96 show in Malaysia in 1996. Following the exhibition Hyundai carried out an operation and mobility performance in front of the Malaysian Army officers at the Malaysian Armour School located on Port Dickson.

Hyundai K1 MBT armed with 105 mm M68A1 rifled tank gun 0018795

Description

The layout of the K1 MBT is conventional, with the driver's compartment at the front, fighting compartment in the centre and engine and transmission at the rear.

Over the frontal arc of the K1 MBT, including the nose, turret front and sides, advanced armour of the Chobham type is fitted, which is manufactured in the US; this provides protection from both kinetic and chemical energy attack.

The driver is seated front left and is provided with a single-piece hatch cover pivoted on the left that lifts upwards to open. This has three integral day periscopes, the centre one of which can be replaced by a passive night driving periscope. The commander is seated on the right of the turret with the gunner below and forward of his position and the loader on the left.

The commander has a French SAGEM (previously SFIM) two-axis independent stabilised panoramic sight which has a magnification of ×3 and ×10, periscopes for all-round observation and a single-piece hatch cover that opens to the rear. This is manufactured in South Korea by Samsung Electronics Co Ltd, under licence

Hyundai K1 armoured vehicle-launched bridge 0018797

from SAGEM (previously SFIM). This has three modes of operation, align with main armament, independent target tracking and gunner override.

The gunner's two-axis stabilised day/night sight device incorporates a laser range-finder and thermal imaging system which is similar to that installed in the General Dynamics Land Systems M1A1 MBT and has magnifications of ×1 and ×10 (day) and ×3 and ×10 (night). The gunner's articulated auxiliary sight is provided by the Electro-Optical Division of the Kollmorgen Corporation and Opto Mechanik Inc (OMI). This has a magnification of ×8.

The latest Texas Instruments (now part of the Raytheon Systems Company) Gunner's Primary Tank Thermal Sight incorporates an eye-safe carbon dioxide laser range-finder with final integration and testing taking place in South Korea.

Turret drive and weapon elevation is electrohydraulic with manual controls for emergency use. Stabilisation is provided in both elevation and traverse.

The fire-control system includes a Computing Devices Company digital ballistic computer with a number of sensors including crosswind and allows the tank to engage both stationary and moving targets while it is stationary or moving itself.

Main armament comprises a US-designed 105 mm M68A1 (which is based on the BAE Systems, RO Defence L7 rifled tank gun) rifled gun which is also fitted on many South Korean M48A5 MBTs and for which ammunition is made in South Korea, including APFSDS. The barrel has a fume extractor, thermal sleeve and a muzzle reference system.

A 7.62 mm M60E2 machine gun is mounted coaxially with the main armament. The loader has a 7.62 mm M60D roof-mounted machine gun while the commander has a 12.7 mm K6 roof-mounted machine gun. Mounted either side of the forward part of the turret is a bank of six electrically operated smoke grenade dischargers.

An unusual feature of the K1 is the hybrid suspension system with improved torsion bars in the centre and hydropneumatic suspension units at each end. This allows the tank to kneel so that the main armament can be depressed to −10°. The remote track adjusting system has two mode settings.

The drive sprocket is at the rear, idler at the front, with six dual rubber-tyred roadwheels, track-return rollers with the upper part of the track being covered by an armoured skirt. For the track, either the replaceable or integral rubber pad can be fitted.

The rear-mounted power pack consists of a German MTU MB 871 Ka-501 1,200 hp diesel coupled to a German Renk LSG 3000 fully automatic transmission with acceleration from 0 to 32 km/h in 9.4 seconds.

Standard equipment includes a passive image intensification periscope for the driver, hydraulic bilge pump, heater, automatic Halon fire detection and suppression system for both crew and engine compartments and a VRC-947K and/or VRC-964K and a VIC-7K for intercom system.

The NBC system consists of an M8A1 alarm system and an M13A1 gas particle filter.

K1A1 MBT

In 1996, Hyundai completed two prototypes of the K1A1 MBT based on the proven mobility and fire-control system of the earlier M1 MBT.

These two prototype vehicles successfully passed all severe testing conducted through February 1997. These trials were carried out by the Republic of Korea government under a variety of weather and terrain conditions.

The K1A1 MBT has an increased firepower performance over the earlier K1 MBT owing to the adoption of a Korean Commander's Panoramic Sight (KCPS) which has been locally developed.

This has a greatly improved performance and function compared with the existing Commander's Panoramic Sight (CPS) for the K1. The K1A1 is also armed with the 120 mm M256 smoothbore gun which is also installed in the General Dynamics Land Systems M1A1 and M1A2 Abrams MBTs.

The main characteristics of the K1A1 MBT have been summarised by the manufacturer as follows:

(1) The significantly enhanced armour penetration power and combat firing range due to the installation of a 120 mm M256 smoothbore gun instead of the 105 mm M68A1 rifled gun of the earlier K1
(2) Improved performance in the driving and stabilisation of the gun and turret by improving the gun/turret drive system for the 120 mm gun
(3) Enhanced processing speed and capability of the ballistic computer
(4) Enhanced night combat and operation capability by adding a thermal imaging capability to the commander's panoramic sight
(5) Improved sealing capability in fording operations and the turret slewing characteristics during moving by improvement of the race ring (or turret bearing) as well as enhanced survivability for the crew by adoption of the improved fire suppression system including thermal wire sensor for the engine bay
(6) Easier maintainability and lowered maintenance cost by utilising a track with replaceable pads.

Future MBT

It is understood that based upon the capability and experience of research and development for the development of the K1 and K1A1 MBTs, Korea is now developing a new concept MBT taking into account the 21st century battlefield environment.

While no characteristics of this new MBT have been officially disclosed, it is assumed that the following factors have been considered:

(1) Significantly enhanced firepower with the main gun utilising the latest gun technology
(2) Significantly enhanced crew survivability by adoption of special armour utilising new materials and active defence system
(3) Automated and intelligent fire-control system and other control systems
(4) Maximised combat efficiency by adoption of vetronics and battlefield management system.

It is understood that ADD, an agency of the Korean government and Hyundai, prime contractor of the M1 and M1A1 MBTs, undertook the concept study and funding for the Future Main Battle Tank (FMBT) for the Korean government.

Armoured vehicle-launched bridge

Early in 1989, Vickers Defence Systems of the UK was awarded a contract by Hyundai Precision and Industry Co Ltd to design and build the prototype of an Armoured Vehicle-Launched Bridge (AVLB) system and bridge.

The AVLB will launch a scissors-type bridge similar to the British No 8 tank bridge which is carried and launched over the front of a Chieftain. This bridge has an overall length of 22 m and can span a gap up to 20.5 m depending on the firmness of the bank.

The bridge and its associated launching system was built at Vickers Defence Systems' Newcastle-upon-Tyne facility and shipped to South Korea in 1990 where it was integrated with the chassis, based on the K1 MBT, designed and built by Hyundai Precision and Industry Co Ltd.

Late in 1993, Hyundai awarded Vickers Defence Systems a contract worth £23 million to supply the bridges for the K1 AVLB.

Specifications

K1
Crew: 4
Combat weight: 51,100 kg
Power-to-weight ratio: 23.48 hp/t
Ground pressure: 0.87 kg/cm²
Length:
(gun forward) 9.672 m
(hull) 7.477 m
Width: (over skirts) 3.594 m
Height: (to turret top) 2.248 m
Ground clearance: 0.46 m
Max road speed: 65 km/h
Average cross-country speed: 40 km/h
Speed on 60% gradient: 8 km/h
Cruising range: 437 km
Fording:
(without kit) 1.2 m
(with kit) 2.20 m
Gradient: 60%
Side slope: 30%
Vertical obstacle: 1 m
Trench: 2.74 m
Engine: MTU MB 871 Ka-501 diesel developing 1,200 hp at 2,600 rpm
Transmission: Renk LSG 3000 automatic with mechanical lock-up, 4 forward and 2 reverse gears
Braking: hydraulic and mechanical, two circuits
Suspension: torsion bar/hydropneumatic
Electrical system: 24 V
Batteries: 6 × 12 V, 100 Ah
Generator: 23 hp
Armament:
(main) 1 × 105 mm M68A1 rifled gun
(coaxial) 1 × 7.62 mm M60E2 MG
(commander) 1 × 12.7 mm K6 MG
(loader) 1 × 7.62 mm M60D MG
Smoke-laying equipment: 2 × 6 smoke grenade dischargers
Ammunition:
(105 mm) 47
(12.7 mm) 1,000
(7.62 mm coaxial) 7,200
(7.62 mm loaders) 1,400
Gun control equipment
Turret power control: electrohydraulic/manual
(by commander) yes
(by gunner) yes
Commander's fire-control override: yes
Turret traverse: 360°
Gun elevation/depression: +20/−10°
Gun stabiliser:
(vertical) yes
(horizontal) yes
NBC system: yes
Night vision equipment: yes

The contract covered the manufacture of eight bridges and 41 launching mechanisms in the UK which were installed on the K1 AVLB chassis by Hyundai. The total South Korea requirement is for 56 K1 AVLBs and the balance of the bridges and launching mechanisms will be built in South Korea by Hyundai under a technology transfer agreement.

The bridge takes 3 minutes to be launched and 10 minutes to be recovered, with maximum bank height differences being 2.4 m.

The K1 AVLB has a total weight of 54.70 tonnes and in travelling configuration is 12.56 m long, 4.0 m wide and 4.0 m high. The bridge itself weighs 12.9 tonnes and is MLC 66 (this means that it can take AFVs to a maximum combat weight of 60 tonnes). It has a crew of two and is armed with a 7.62 mm M60D machine gun. Full details of this are given in *Jane's Military Vehicles and Logistics 2002-2003.*

Armoured recovery vehicle

Hyundai Precision and Industry Co Ltd has developed an ARV based on the chassis of the K1 in co-operation with the German company Rheinmetall Landsysteme (then MaK). In appearance it is similar to the ARV based on the Leopard 1 and Leopard 2 MBT chassis which the company developed to meet the requirements of the German Army.

Standard equipment includes a front-mounted dozer/stabiliser blade, main winch with a maximum capacity of 35 tonnes (70 tonnes with pulley) and 150 m of cable, with a crane being mounted at the front right side of the hull. When travelling this is stowed alongside the right side of the hull. It can be traversed through 270° and has a maximum lifting weight of 25 tonnes. An auxiliary winch, auxiliary power unit, electric impact wrench and welding equipment are also provided.

In November 1991, Hyundai of South Korea, prime contractor for the K1 family of MBTs, awarded the then MaK company a production contract for subsystems for the K1 ARV.

Production of these is undertaken at the now Rheinmetall Landsysteme Kiel facility, with first deliveries completed late in 1993. This vehicle is now in service, with the first order being for 90 vehicles followed by a second order for 59. Since then a third order has been placed.

The recovery equipment is integrated into the chassis by Hyundai which then delivers the complete vehicle to the South Korean Army. Hyundai is building the remainder of the recovery equipment such as dozer, winch and crane system in South Korea as part of a technology transfer from the now Rheinmetall Landsysteme.

The K1 ARV has a crew of four and weighs 52 tonnes, or 57 tonnes when carrying a replacement power pack.

Hyundai K1 armoured recovery vehicle with crane traversed to front and stabiliser/dozer blade lowered and lifting complete K1 MBT turret 0011641

Table 1	**K1**	**K1-M**
Combat weight	51,100 kg	49,700 kg
Ground pressure	0.87 kg/cm²	0.84 kg/cm²
Width (over skirts)	3.595 m	3.5 m
Ammunition		
(105 mm)	47	41
(12.7 mm)	1,000	2,000
Laser range-finder	Nd:YAG/CO_2	Raman shift Nd:YAG
Air conditioning system	no	Yes
Laser warning system	no	Yes

It is armed with a 12.7 mm machine gun operated by the vehicle commander. Full details of this are given in *Jane's Military Vehicles and Logistics 2002-2003.*

K1 mineclearing vehicle

For trials purposes the K1 MBT has been fitted with a roller type mineclearing system at the front of the hull. When clearing mines this has a maximum speed of 16 km/h.

K1 Combat mobility vehicle

Studies are already under way on a combat mobility version of the K1 which would use a similar hull to that of the K1 ARV previously described but would carry special equipment to undertake the engineer role.

K1-M MBT

This is the version being proposed to Malaysia and key differences in the specifications listed in Table 1.

Status

In production. In service only with the South Korean Army.

Manufacturer

Hyundai Precision & Industry Co Ltd.

UPDATED

NORWAY

Norwegian Army MBT fleet

Following comparative trials between the German Leopard 1 and the Swedish S-tank, Norway selected the Leopard 1 and the first order was placed in November 1968 with deliveries running between 1970 and 1971. This order comprised 78 Leopard 1 MBTs and six Leopard 1 armoured recovery vehicles.

Subsequently, all Norwegian Leopard 1 MBTs were upgraded to the latest Leopard 1A5 configuration but retaining the cast turret. To enable this upgrade to take place, in 1988, Norway ordered 78 EMES-18 computerised fire-control systems from STN Atlas Elektronik of Germany and at the same time the original hydraulic gun control equipment was replaced by an all-electric system.

In addition to the original 78 Leopard 1 MBTs which were upgraded to the Leopard 1A5 standard, an additional 92 Leopard 1s were obtained from Germany of which 33 were upgraded to the Leopard 1A5 configuration which in Norwegian Army service is called the Leopard 1A5NO with the remaining unmodified Leopards being known as the Leopard 1A1NO.

As of early 2001, the Norwegian Army MBT fleet consisted of 111 Leopard 1A5NO and 59 Leopard 1A1NO, all armed with a 105 mm gun.

Early in 2001, the Norwegian Army Material Command signed a contract with Netherlands for the supply of 52 surplus Leopard 2A4 MBTs under a contract worth US$168 million. There is also an option on another four Leopard 2A4 MBTs for training purposes.

These Leopard 2A4 MBTs will replace the 59 Leopard 1A1NO MBTs which will be phased out of service. The 111 Leopard 1A5NO MBTs will remain in service.

The Leopard 2A4 MBTs will be modified by installing Norwegian standard radios, smoke grenade launchers and machine gun mounts. The first Leopard 2A4 MBTs were delivered to Norway in 2001.

Leopard 1 AVLB

Early in 1996, the German company MAN GHH was awarded a contract by the Norwegian Army Matériel Command for the conversion of nine Leopard 1 MBT chassis into armoured vehicle-launched bridges with the Leguan systems.

While MAN was the prime contractor and supplied the bridge and associated launching system, the actual conversion and integration of the laying equipment was carried out by Krauss-Maffei (which on 1 January merged with Wegmann to become Krauss-Maffei Wegmann) in Munich. The Leopard 1 chassis were purchased from surplus German Army stocks by Norway and have been supplied by Krauss-Maffei.

The Leopard 1 MBT chassis has had its turret removed and hull reinforced to accept the bridgelaying mechanism which is launched over the front of the vehicle. In addition, a new reinforced two-circuit brake system has been fitted. Mounted at the front of the hull is a tiltable dozer blade which can also be used to stabilise the vehicle when the bridge is being layed in position.

In addition to the nine AVLB systems, a total of 13 MAN Leguan bridges which, when opened out are 26 m long and are MLC 70 (Military Load Class), have been supplied. The bridges are fitted with special adaptors to allow them to be incorporated into ferries.

The first of these Leopard 1 AVLBs were handed over to the Norwegian Army Matériel Command in December 1997 and final deliveries taking place in 1999.

Leopard 1 AEV

In 1995, Hägglunds Moelv was awarded a contract by the Norwegian Army for the conversion of 22 Leopard 1 MBT chassis into Armoured Engineer Vehicles.

Main roles of this vehicle can be summarised as:

(1) Repair and re-establishment of axis of advance
(2) Rapid construction of slit trenches for armoured vehicles and assault personnel
(3) Construction of field locations and fire emplacements for MBTs and armoured infantry fighting vehicles

Leopard 1A5NO of the Norwegian Army with turret traversed to the left (Svein Wiiger Olsen) **2002**/0130962

AMCV developed by Hägglunds Moelv and based on a much modified Leopard 1 series MBT chassis with flail system deployed ready for use (HV) **2000**/0069530

Hägglunds FBRV being put through its paces (UK MoD/CCR) **2002**/0096165

Hägglunds Vehicle Leopard 1 AEV of the Norwegian Army with dozer blade raised and hydraulic arm with bucket traversed to front (Richard Stickland) **2002**/0130965

(4) The preparation of access and egress sites/entry and exit ramps for water crossing and to make river beds trafficable
(5) Load gravel, packaged cargo, timber and building material onto trucks
(6) Construct and remove obstacles and barriers on the battlefield
(7) The destruction of roads.

The conversion includes the removal of the complete turret and the installation of a new all-welded superstructure for the crew of three and new engineer systems.

Mounted at the front of the vehicle is a hydraulically operated dozer blade. The primary mission of this is to clear and destroy roads, with the secondary mission being to use the dozer blade as a soil anchor when operating the winch or excavator boom.

Pivoted at the front of the hull is a hydraulically operated arm which is traversed to the rear when not required. This arm can be traversed through 180° and is fitted with a bucket with a capacity of 0.7 m^3 for digging purposes. The arm has a maximum reach of 8 m. If required this can be replaced by a soil drill.

Two hydraulic capstan winches are provided for recovery purposes and these would normally be used in conjunction with the dozer blade. One winch has a towing force of 90 kN with a total theoretical force of 720 kN.

Options include an overhead weapon station (Krauss-Maffei Wegmann Type 2038 or RAFAEL 12.7D), larger capacity (1 m^3) excavator bucket, additional armour protection and a soil drill.

The first prototype of the Leopard 1 AEV was deployed to Kosovo by the Norwegian Army in late 1999 with the first two production vehicles being completed in October and November 1999. All 22 vehicles have now been delivered to the Norwegian Army. Within the Norwegian Army this Leopard 1 AEV vehicle is called the Leopard 1 INGPV (Ingeniorpanservagn).

Leopard 1 mineclearing vehicle

Late in 1996, Hägglunds Moelv was awarded a contract worth SKr130 million from the Norwegian MoD for the supply of seven flail type heavy armoured mineclearing vehicles based on a Leopard 1 MBT series chassis.

It is expected that the Hägglunds Moelv Armoured Mine Clearing Vehicle (AMCV) will complete its final batch of extensive trials with the Norwegian Army in early 2002, with the main production run of six vehicles being completed between late 2002 and late 2003. The prototype will then be rebuilt to the full production standard and delivered back to the Norwegian Army.

For this application the turret of the Leopard 1 has been removed and replaced by a new all-welded steel-armour superstructure that has the same level of armour protection as the modified chassis. The chassis has been provided with additional protection against mines and internally there are spall liners.

The vehicle has a crew of two, consisting of a driver who also operates the mineclearing system and the vehicle commander who operates the Krauss-Maffei Wegmann overhead weapon station armed with a 12.7 mm M2 machine gun. The latter can also be used in the Explosive Ordnance Disposal (EOD) role.

The flail based mineclearing system is mounted on a turntable at the front of the vehicle. When travelling, is rotated and raised it so that it is stowed on the forward part of the vehicle roof to reduce the overall width of the vehicle for travelling.

When deployed at the front of the vehicle, the flail system is powered by an external drive system that connects to the rear drive sprockets, located one either side at the rear of the hull.

The flails rotate at a speed of 300 rpm and clear a path 4 m wide to a maximum depth of 0.3 m with speed being a maximum of 7 km/h, but this does depend on the nature of the soil.

The AMCV has been designed to clear surface and buried anti-tank and anti-personnel mines under all-weather conditions. It can also be operated by remote control. It has a maximum weight of 47 tonnes and has similar mobility to the Leopard 1 MBT on which it is based.

Mounted on the roof of the AMCV, at the rear, are the two units of the United States Mine Clearing Line Charge (MICLIC) each of which will clear a path about 100 m long and 12 m wide. These are fired over the minefield and then fall to the ground where the explosive charge is detonated. This creates an over-pressure which activates the mines.

These are supplied as Government Furnish Equipment (GFE) as is the Israel Military Industries Clear Lane Marking System (CLAMS). The latter is mounted at the rear of the AMCV and deploys markers on the ground. These indicate the centre of a cleared path so that following vehicles do not stray into the mine field.

Leopard 1 AVLB of the Norwegian Army in travelling configuration complete with bridge (Richard Stickland) **2002**/0130963

Leopard 1 AVLB of the Norwegian Army in travelling configuration after it has laid its bridge in position (Richard Stickland) **2002**/0130964

In a typical battlefield mission, the MICLIC would normally be used first, then the AMCV would enter the minefield to clear any remaining mines. If required the AMCV could also be fitted with an electronic mineclearing and detection system. This would detect and activate mines fitted with a magnetic or electronic fuze.

Exported specialised vehicles

Leopard 2 AEV

To meet the requirements of Sweden and Switzerland, Hägglunds Moelv has submitted a proposal for the development of a new Armoured Engineer Vehicle (AEV) based on the chassis of the Leopard 2 MBT. This would have similar equipment to that fitted to the Leopard 1 AEV developed and manufactured by the company for the Norwegian Army.

Future Beach Recovery

Following a competiton, Hägglunds Moelv was awarded a contract by the UK Defence Procurement Agency for the supply of four (with an option on a fifth) Future Beach Recovery Vehicle (FBRV) to replace the current Centurion BARV. The first vehicle was delivered for trials in 2001 and with the main production run being carried out between 2002 and 2003.

The FBRV is based on a much modified Leopard 1 MBT chassis with its turret removed and replaced by a new all-welded steel superstructure.

Some of the Leopard 1 series subsystems have been modified to enable the FBRV to operate in salt water. The air intakes for the main engine and crew compartment are located in the top rear end of the crew compartment. The FBRV will be able to operate in water up to 2.95 m in depth.

UPDATED

PAKISTAN

Pakistan MBT modernisation plans

Until fairly recently, the Pakistan tank fleet consisted of approximately 1,200 Chinese Type 59s (100 mm), 40/50 Russian T-54/T-55 (100 mm), 50 US M47M (90 mm) and 345 US supplied M48A5 (105 mm) vehicles.

Recent deliveries of brand new vehicles from China North Industries Corporation (NORINCO) include at least 250 Type 69-IIP and at least 250 Type 85-IIAP MBTs with the P in the designation standing for Pakistan.

The Type 85-IIAP is armed with a 125 mm smoothbore gun fed by an automatic loader.

It is believed that first deliveries of the Type 85-IIAP MBTs were made to Pakistan in mid-1992 with the following quantities of Type 85-IIAP being delivered to Pakistan:

Year	Quantity	Comment
1992	97	–
1993	35	–
1994	82	–
1995	51	plus 20 ARV
1996	3	–
1997	nil	–
1998	nil	–
1999	nil	–
2000	nil	–

NORINCO Type 69 series MBT of the Pakistan Army upgraded with 105 mm rifled tank gun and a bank of six smoke grenade launchers mounted either side of the turret (Christopher F Foss) ***2002***/0130966

With Chinese assistance, Pakistan has established the Heavy Rebuild Factory at Taxila. The facility covers 48 acres and is capable of rebuilding tanks and power packs as well as manufacturing individual optical and electrical components.

The Engine Rebuild Plant was completed at Taxila first, the plant for the rebuilding of the hull was completed shortly afterwards. The facility was fully operational by 1979-80 and normally rebuilds approximately 100 Type 59 tanks and 250 Type 59 tank engines a year. Of the 11,000 components used in the rebuild of the Type 59 some 8,000 are now manufactured locally.

More recently the Heavy Rebuild Factory (HRF) at Taxila was renamed Heavy Industries Taxila.

The Military Vehicle Research and Development Establishment has developed to the prototype stage the M47M armoured vehicle-launched bridge system. This lays a scissors bridge with a maximum span of 21.4 m. Full details are given in *Jane's Military Vehicles and Logistics 2002-2003*.

In August 1989, an agreement was signed between Pakistan and General Dynamics Land Systems, to establish an overhaul facility in Pakistan for M47 and M48 tanks, M109 and M110 self-propelled howitzers, M113 series APCs and M88A1 armoured recovery vehicles. Pakistan has also manufactured the M113A2 APC under licence and may in the future build the United Defense Armored Infantry Fighting Vehicle under licence. Details of the Pakistan M113 production are given in the Armoured personnel carriers (tracked) section.

In 1988, two 120 mm armed M1A1s were trialled in Pakistan but, in the short term at least, no purchase or local manufacture of the M1A1 is expected to take place in Pakistan.

Early in 1994 it was disclosed that Perkins Engines (Company) of the UK had completed tests of a new integrated power pack for an MBT in the 50 tonne range and this was believed to be for the MBT 2000 project.

This power pack consists of a Perkins Engines (Company) CV-12 diesel developing 1,200 hp (as installed in the Challenger 1 and Challenger 2 MBTs) coupled to a French SESM ESM 500 automatic transmission (as installed in the Leclerc MBT used by the French Army).

In late 1996, it was revealed that the Dr A Q Khan Research Laboratories in Pakistan had developed at least three types (A, B and C) of explosive reactive armour (ERA) kits for installation on MBTs and other armoured vehicles, tracked and wheeled.

It is probable that this ERA will be fitted to current and future Pakistan MBTs to improve their battlefield survivability against HEAT projectiles.

In August 1996, Pakistan placed an order with the Ukraine for the supply of 320 T-80UD MBTs under a deal worth about US$580 million. The first 15 were supplied in February 1997 with final deliveries taking place in late 1999. Full details of the T-80UD are given later in this section under Russia.

UPDATED

M48A5 MBT of the Pakistan Army with a 105 mm rifled tank gun and a roof-mounted 12.7 mm anti-aircraft MG of Chinese origin 0018798

The Pakistan Army has now taken delivery of 350 T-80UD MBT's from the Ukraine 0011612

MBT 2000 (Al-Khalid)

Development

On 1 October 1988, it was announced that Pakistan would manufacture a new MBT (MBT 2000) with the assistance of the People's Republic of China and with the now Heavy Industries Taxila in Pakistan carrying out the programme.

The project was approved by the Defence Committee of the Cabinet on 16 January 1990 and the contract with China North Industries Corporation (NORINCO) was signed in May 1990.

Rather than build a new MBT from scratch, Pakistan elected a four-phased approach:

(1) Upgrade the existing Type 59 MBT. Details of the three phase Type 59 MBT upgrade are given in the following entry.
(2) Assembly/manufacture of the Type 69-II MBT; this has 50 per cent of its components from the Type 59 and 50 per cent new components. Production of Type 69-II hulls commenced in Pakistan in 1993 with the turrets still coming from China. The Pakistan designation is the Type 69-IIMP with the engine developing 580 hp. Details of the Type 69 are given under China
(3) Co-production and manufacture of the Type 85 MBT; this has 20 per cent components from the Type 59, 30 per cent components from the Type 69-II MBT and 50 per cent new components. Co-production of the Type 85 has now commenced and a batch of these vehicles has already been supplied from China. In Pakistan service these are designated the Type 85-IIAP. Details of the Type 85 are given under China
(4) Production of the MBT 2000 with 45 per cent of its components taken from existing vehicles (10 per cent Type 59, 15 per cent Type 69-II and 20 per cent Type 85) and 55 per cent new components.

The Heavy Industries Taxila, also known as P-711, has been established and running for some time. It was then decided to build additional facilities for specific components:

P-882	All types of AFV hull
P-883	All types of turret
P-884	All types of engine from 520 to 1,500 hp
P-885	Progressive manufacture of guns from 105 to 203 mm
P-886	Flexible machining centre

The first prototype of MBT 2000, which is also referred to as the Al-Khalid or P-90, was completed on schedule in June 1991, with trials being carried out in 1992.

It is expected that initially 65 to 70 per cent of the key components will be made in Pakistan, 20 to 25 per cent will be supplied by China and the remainder will come from other countries.

During the development phase of the Al Khalid a number of prototypes of the vehicle were built and tested in various configurations:

Configuration 1 was based entirely on the Chinese systems viz the Chinese 125 mm smoothbore gun and automatic loader, fire-control system, gun-control system and diesel power pack. The latter comprises a 1,200 hp diesel engine manufactured in China and based on the MTU 396 series engine manufactured under licence from MTU of Germany coupled to a Renk LSG 3000 automatic transmission.

Configuration 2 used a Chinese 125 mm smooth bore gun and automatic loader, fire-control system and gun-control equipment with a Western power pack consisting of a British Perkins Engine Company 1,200 hp Condor diesel coupled to a French SESM-500 automatic transmission.

Configuration 3 was based on a 1,200 hp power pack, 125 mm smoothbore gun and automatic loading system from Ukraine with the fire-control system and gun-control system being of East European design.

Configuration 4 would have been based on Western armament and power pack with the latter consisting of an MTU 871 or General Dynamics Land Systems AVDS-1790 series diesel engine coupled to a Renk LSG 3000 automatic transmission. This option did not materialise owing to difficulties in the acquisition of these systems due to trade embargoes.

During the planning stage, a thorough technical evaluation of hydrogas suspension and a Renk 304 automatic transmission was carried out. In the end these were not considered to be feasible due to problems associated with maintainability, reliability and other limitations.

Due to high ambient temperatures and dust in areas where the Al Khalid MBT was due to operate, special emphasis was placed on the design and installation of high-capacity cooling and efficient air clearing systems.

Trials with the prototype Al Khalid MBT have been undertaken in ambient temperatures of up to 55°C and with very fine sand/dust particles.

According to the prime contractor, the users and technical experts were fully satisfied with the trials conducted in the summer of 1998 in the southern deserts of Pakistan and subsequently in different regions of the country presenting different terrain environments.

Late in 2000 it was stated that a preproduction batch of 15 Al Khalid MBTs would be built to enable trials to be undertaken at squadron level before volume production starts. The first of these 15 pre-production Al Khalid MBTs was handed over to the Pakistan Army for extensive trials in July 2001 and all have now been delivered.

The Heavy Industries Taxila facility has successfully undertaken manufacture of Chinese Type 69-II and Type 85II AP MBTs and the US M113 series APC.

It is understood that a major upgrading and rearranging of production lines has been under way at the HIT for several years to allow for production of the new Al Khalid.

This also includes the establishment of CNC machining centres for precision machining of the hull, turret and other critical parts to undertake production of the Al Khalid.

Pakistan is also understood to have been making simultaneous efforts towards the development of special armour for use in tank production. The HIT has been successful is developing special steel for tank guns and composite armour materials for use during series production.

During successive trials, armour protection of the hull and turret of the tank have been tested through live firing which defeated all types of 120 and 125 mm tank projectiles and selected other tank ammunition.

Specifications

Crew: 3
Combat weight: 46,000 kg
Power-to-weight ratio: 26.08 hp/t
Length: 6.9 m
Width: 3.4 m
Height: 2.3 m
Max road speed: 70 km/h
Acceleration: 0 to 32 km/h: 10 s
Range (road): 400 km
Vertical obstacle: 0.85 m
Trench: 2.7 m
Gradient: 60%
Side slope: 40%
Engine: 8-cylinder, 4-stroke, water-cooled, turbocharged diesel developing 1,200 hp
Transmission: semi-automatic hydromechanical power shift with 4 forward and 2 reverse gears
Suspension: torsion bars, shock-absorbers and hydraulic bump stops
Armament:
(main) 1 × 125 mm smoothbore gun
(coaxial) 1 × 7.62 mm MG
(anti-aircraft) 1 × 12.7 mm MG
(smoke grenade dischargers) 2 × 4
Ammunition:
(125 mm) 39 (22 in automatic loader)
(12.7 mm) 500
(7.62 mm) 2,000
(grenades) 16 (12 smoke, 4 HE)
Gun control equipment
Turret power control: powered/manual
(by commander) yes
(by gunner) yes
Gun stabiliser:
(vertical) yes
(horizontal) yes
Range setting device: yes (laser)
NBC system: yes
Night vision equipment: yes

Description

The layout of the MBT 2000 is conventional, with the driver's compartment at the front, turret in the centre and the power pack at the rear. In many respects the MBT 2000 appears to be almost identical to the Chinese NORINCO Type 90-II MBT.

The turret and hull are of all-welded steel armour construction and an additional layer of composite armour has been added over the frontal arc, to which explosive reactive armour can be added if required. Turret thickness at the front is estimated to be 600 mm with the glacis/nose estimated to be 450 to 470 mm.

The armour is of modular design enabling the user to change the damaged modules or replace the existing models with new armour packages as the threat evolves or as new technology becomes available.

Main armament is a 125 mm smoothbore gun fitted with a thermal sleeve and a fume extractor. The 125 mm gun is autofrettaged and chrome plated. This can fire APFSDS (muzzle velocity 1,760 m/s), HEAT (muzzle velocity 850 m/s) and HE-FRAG (muzzle velocity 950 m/s). It can also fire a laser guided projectile fitted with a HEAT warhead. The gun is fed by

MBT 2000 (Al-Khalid) MBT which is very similar to the Chinese Type 90-II MBT and is shown here fitted with explosive reactive armour to its turret and chassis (HIT) **2000**/0069532

Chinese supplied Type 85-IIAP MBT (Christopher F Foss) **2001**/0105975

First preproduction MBT 2000 (Al Khalid) MBT was handed over to the Pakistan Army in 2001 and is now undergoing extensive trials **2002**/0095825

an automatic loader enabling the crew to be reduced to three – commander, gunner and driver.

A 7.62 mm machine gun is mounted coaxially with the main armament and a 12.7 mm machine gun is mounted on the roof for anti-aircraft and local defence. Mounted either side of the turret, towards the rear, is a bank of four electrically operated forward-firing smoke grenade dischargers. A turret basket is provided at the rear.

The computerised fire-control system includes a bi-axis stabilised dual magnification gunners sight, bi-axis stabilised commanders sight with hunter killer capability, computer, commander's control panel, laser range-finder, crosswind sensor, tilt sensor and angle velocity sensor. This allows the MBT 2000 to engage moving targets under day and night conditions.

The gunner's integrated biaxis stabilised sight incorporates day/night channels, thermal imager and a laser range-finder. It has dual magnification of ×3 and ×10 and two fields of view, 20°and 6°.

The commander's roof-mounted panoramic sight is also biaxis stabilised and has day, second-generation image intensification and a laser range-finder. The sight has a magnification of ×7.5 and a 7.5° field of view.

The laser range-finder has a range of 200 to 5,000 m and interfaced with the gunner's sight is an auto tracker.

The complete power pack, which consists of the engine, transmission and cooling system, can be removed from the vehicle in 30 minutes to facilitate field replacement.

Suspension is of the torsion bar type with six large dual rubber-tyred roadwheels either side, idler at the front, drive sprocket at the rear and track-return rollers. The upper part of the suspension is covered by a track skirt with a wavy lower part.

Variants

There are no known variants of the MBT 2000.

Status

Entering production for the Pakistan Army who have placed a contract for the first batch of 15 preproduction vehicles which have now been delivered.

Manufacturer

Heavy Industries Taxila.

UPDATED

Pakistan Type 59 MBT Upgrades

Development/Description

Outside China, Pakistan is the largest user of the China North Industries Corporation (NORINCO) Type 59 (or T-59 as it is also referred to) Main Battle Tank (MBT) with an estimated 1,200 vehicles in service.

The Type 59 is itself a further development of the Russian T-54 MBT and is almost identical in appearance and retains the 100 mm rifled gun as its main armament. A 7.62 mm machine gun is mounted coaxially with the main armament and a 12.7 mm machine gun is mounted on the turret roof for air defence purposes.

While more recent MBTs have been introduced into service with the Pakistani Army, including the NORINCO Type 69-II and Type 85-IIAP, a decision was made to upgrade a significant part of the Type 59 MBT fleet at the Heavy Industries Taxila facility in three phases.

In 1993, the Phase I upgrade of the Type 59 MBT started, during which some 28 improvements were carried out to improve the combat effectiveness of the Type 59 MBT under day and night conditions. The most significant of these was the installation of a 105 mm rifled tank gun which would fire standard natures of ammunition, computerised fire-control system for improved first round hit capability, two-axis stabilisation system for the 105 mm main armament and image intensification sights for the tank commander, gunner and driver. A total of 500 Type 59 MBTs were upgraded to the Type 59 Phase I configuration.

The new gunner's thermal sight installed in the Phase II and Phase III could be the locally developed by Al-Technique Corporation of Pakistan (ATCOP). Two versions of this thermal sight have been developed, uncooled and cooled. Both these include the gunner's thermal sighting system, associated electronics and a monitor for the tank commander so that he can see the same thermal picture of the target as the gunner.

Earlier optical systems developed by ATCOP include the TR-2 and TR-3 laser range-finders, GNS 1 image intensification gunner's night sight and the DNS-3 drivers image intensification night periscope. All these have been developed for the T-series MBTs. The company has also developed the LTS laser threat warning system.

From 1998, a further 300 Type 59 MBTs were upgraded to the Phase II standard. This included all the features of the Phase I with the addition of the installation of a more powerful diesel engine and thermal night vision equipment rather than image intensification night vision devices.

Type 59 MBT upgraded to the latest Phase III standard clearly showing the additional passive armour on the turret and hull and the 125 mm smoothbore gun (HIT) **2000**/0069533

The Phase III Al Zarrar is the most ambitious upgrade ever attempted on the Type 59 MBT or its Russian equivalent, the T-54. It has now been disclosed that three different prototypes of this were built for extensive user trials:

Prototype 1

This is fitted with the 125 mm smoothbore tank gun as installed in the T-80U MBT supplied to Pakistan by the Ukraine. For this application the turret has been raised 100 mm off the chassis. A new power pack has been installed that includes the Ukrainian 5TDF five-cylinder diesel, developing 700 hp. It shares many common components with the 6TD-1 six-cylinder diesel developing 1,000 hp which is already installed in the 320 T-80UD/T-84 MBTs, recently supplied by the Ukraine to Pakistan.

The suspension has been upgraded and reinforced to take into account the higher combat weight of the vehicle. This includes the installation of stronger torsion bars. The vehicle is now fitted with track return rollers and new double-pin tracks with rubber pads have been installed.

Battlefield survivability is enhanced by the introduction of additional passive armour to the turret with a fully enclosed stowage box now being installed on the turret rear. Explosive Reactive Armour (ERA) has been installed over the frontal arc. A new computerised fire-control and sighting system has also been installed.

Prototype 2

This retains the 100 mm gun of the Type 59 but is fitted with the V46-5M 12-cylinder water-cooled diesel engine that develops 690 hp coupled to a modified manual transmission. The suspension has been improved. For increased survivability an anti-mine bottom plate has been installed and ERA can be fitted.

Prototype 3

This features the power pack of the NORINCO Type 85-IIAP MBT with the suspension upgraded. The 125 mm smoothbore gun of the Type 85 has been installed. The fire-control system consists of image stabilised sights and an upgraded computerised fire-control system. Armour protection is the same as Prototype 1.

Specifications
(Type 69IIMP)
Crew: 4
Combat weight: 36,700 kg
Power-to-weight ratio: 15.80 hp/tonne
Ground pressure: 0.823 kg/cm²
Length:
(gun forward) 8.589 m
(hull) 6.243 m
Width: 3.27 m
Height: (to axis of AA MG) 2.807 m
Max road speed: 50 km/h
Range: 300 km
Verticle obstacle: 0.50 m
Trench: 2.50 m
Gradient: 32%
Side slope: 30%
Fording: 1.4 m
Engine: water-cooled diesel developing 580 hp
Transmission: manual
Armament:
(main) 1 × 105 mm rifled gun
(coaxial) 1 × 7.62 mm MG
(bow) 1 × 7.62 mm MG
(anti-aircraft) 1 × 12.7 mm MG
(smoke grenade launchers) 2 × 6
Ammunition:
(105 mm) 44
(12.7 mm) 500
(7.62 mm) 3,000

Gun control equipment
Turret power control: powered/manual
(by commander) yes
(by gunner) yes
Gun stabiliser:
(vertical) yes
(horizontal) yes
Range setting device: yes (laser)
NBC system: yes
Night vision equipment: yes

The 125 mm smoothbore gun is not fed by a fully automatic loader as installed in the Type 85 and Al Khalid MBTs so the upgraded Al Zarrar MBT retains its four-man crew consisting of commander, gunner, loader and driver. The 125 mm ammunition, already manufactured by the Pakistan Ordnance Factories (POF) is of the separate loading type, projectile and charge with its semi-combustible cartridge case.

All of the upgraded vehicles retain the 7.62 mm coaxial and 12.7 mm anti-aircraft machine guns and have a bank of electrically operated smoke grenade launchers either side of the turret, firing forwards.

The different Al Zarrar configurations were built to give the user the maximum possible flexibility in making a decision on which systems to install in production vehicles.

Different sources of key subsystems also allow for increased competition. The ERA, for example, could be supplied by NORINCO or from two sources in Pakistan, the Dr A Q Khan Research Laboratories or the National Development Complex (NDC).

The final configuration of production Al Zarrar is now being decided.

It has also been disclosed that the HIT built a batch of 45 NORINCO Type WZ 653 armoured recovery vehicles and proposals have been made to upgrade this vehicle enabling it to recover the heavier Al Khalid MBT. The Type 653 ARV was originally developed to recover the Type 59 and Type 69 MBTs but the Type 85-IIAP and Al Khalid are heavier, hence the requirement for an improved recovery vehicle.

A key part of these upgrades has not only been to improve the combat capabilities of the Type 59 but also to improve the overall realiability of the vehicle and reduce maintenance.

Most of the new subsystems installed in the Type 59 MBT upgrade are also used in the Type 69 and/or Type 85 series MBTs used by Pakistan. This reduces the number of spare parts in the supply chain and has a significant impact on logistics and training as so many parts are common across the whole fleet.

Fire-control upgrade

In 2001, it was stated that the Al Technique Corporation of Pakistan (ATCOP) was to improve the combat capability of about 200 NORINCO Type 85-IIAP series MBTs in service with the Pakistani Armoured Corps.

It is estimated that Pakistan took delivery of almost 500 of these vehicles from China between 1991 and 1996, with additional vehicles being built/assembled at the Heavy Industries Taxila (HIT) facility under a co-production agreement.

The Type 85-IIAP gunner currently has an image intensification night sight with limited night/poor weather capabilities and a UK BAE Systems, Avionics (previously GEC-Marconi) thermal imaging camera has been installed in one example of the Type 85-IIAP MBTs for trials purposes.

The BAE Systems, Avionics thermal imager has been designed for installation in MBTs such as the Type 85-IIAP and is claimed to have a detecting range of 7,000 m and a recognition range of 3,000 m.

The installation of a thermal imaging gunners sight will enable targets to be engaged with greater accuracy and at a longer range than the current image intensification sight. The tank commander will retain an image intensification sight and the driver can replace his standard day periscope by an image intensification periscope for driving at night.

Status

In service with Pakistan Army. Upgrades are continuing.

Manufacturer

Heavy Industries Taxila.

UPDATED

Prototype 1 of the Al Zarrar MBT clearly showing the 125 mm gun and turret with additional external stowage at rear and ERA fitted to the hull and turret (Christopher F Foss) ***2001***/0105976

Prototype 3 of the Al Zarrar MBT with turret armed with 125 mm smoothbore gun fitted with ERA to the hull and turret (Christopher F Foss) ***2001***/0105977

POLAND

PT-91 MBT

Development

The PT-91 MBT, which is also referred to as the Twardy (Hard), is a further development by the Zaklady Mechaniczne Bumar-Labedy SA version of the Russian-designed T-72M1 (covered in detail in a separate entry) which has been produced under licence in Poland for many years.

The first prototype of the PT-91 was completed late in 1992 and underwent trials with the Polish Army.

It was originally understood that 78 PT-91 series MBTs were built but recent information has stated that a total of 226 were built, although some of these could have been conversions of older T-71M1 series MBTs. The manufacturer has the capability to build up to 30 PT-91 MBTs a year. Although the production line is now closed it could be restarted again if significant orders were placed.

In addition to offering new build PT-91 MBTs, Poland is also offering to upgrade users' existing T-72M1 MBTs to the PT-91 standard with the aid of kits.

Description

The PT-91 MBT has a number of improvements over the earlier T-72M1 in the key areas of armour, mobility and firepower. In addition to producing new vehicles to the PT-91 standard, most of these modifications can be backfitted to existing T-72M1s while others, for example the armour package, can be backfitted to other vehicles.

The main improvements of the PT-91 over the original Polish-built T-72M1 can be summarised as follows:

(1) Installation of the Polish-developed explosive reactive armour Type ERAWA-1 and ERAWA-2 to the hull and turret, details of this are given in the entry for the Polish T-72M1 MBT
(2) Four laser warning antennas are fitted around the turret which alerts the crew if it is being targeted

by a laser range-finder or designator. This system is known as the OBRA laser warning system and works in automatic and manual modes. The OBRA laser warning system was developed and produced by the Polish company PCO.

(3) Mounted either side of the turret are 12 Tellur smoke grenade dischargers which cover the frontal arc of the vehicle. These are operated in conjunction with the OBRA laser warning system
(4) The skirts are steel with the forward part covered by ERAWA-1 explosive reactive armour blocks
(5) New computerised fire-control system installed, with the gunner's passive image intensification sight having the option of being changed for a Drawa thermal sight. A laser range-finder is fitted as standard and the 125 mm smoothbore main armament is stabilised in both elevation and traverse. The fire-control system has been developed by PCO of Poland.
(6) The commander (POD-72) and driver (PNK-72) have new passive night observation devices
(7) Improved NBC protection system
(8) Installation of hull escape hatch
(9) Driver's seat now provides some protection against mines
(10) The standard T-72M1 MBT 12-cylinder water-cooled supercharged diesel develops 760 hp but for the PT-91 this has been uprated to 850 hp and the cooling system improved. In 1994 it was stated that a 1,000 hp 12-cylinder water-cooled supercharged engine was being tested
(11) The engine and crew compartment have been fitted with a more modern fire detection and suppression system
(12) Tracks are fitted with rubber pads which reduce surface noise as well as improving traction on roads
(13) New communications system installed
(14) The Polish developed rocket-propelled mineclearing system Type PW-LWD can be installed on the rear of the hull.

In addition, the manufacturer also states that parts of the hull of the PT-91 can be fitted with radar-absorbent material.

Recent photographs of the PT-91 released by the manufacturer show a number of differences from the original photographs shown in *Jane's Armour and Artillery 1994-95*.

The latest vehicle has a higher pole-type met sensor mounted on the left side of the turret roof while the commander's cupola has been redesigned and has a periscopic sight to the front. This may well be stabilised and allows the tank commander full 360° observation without moving his head.

If required the PT-91 can be fitted with the LIW division of Denel Tiger computerised fire-control system which has already been successfully demonstrated in the Polish T-71M1Z MBT, covered in a separate entry.

Enhanced PT-91 for export

Zaklady Mechaniczne Bumar Labedy have developed for the export market an enhanced version of their PT-91 MBT, that is currently in service with the Polish Army.

The original locally produced T-72M1 was powered by a 760 hp 12-cylinder diesel, but the PT-91 is powered by a locally produced 860 hp diesel which gives a power-to-weight ratio of 18.76 hp/tonne. The latest export version of the PT-91 has a computerised fire-control/sighting system and a more powerful engine.

The former is the French SAGEM SAVAN-15 system which is a member of a complete family of computerised fire-control systems, marketed under the Stabilised Aiming Vertical sensing and Navigation (SAVAN) gunner's multichannel stabilised sights.

The SAVAN 20 system is installed in the Giat Industries Leclerc MBT, while the SAVAN 10 is fitted with the British Army's Vickers Defence Systems Challenger 2 MBT.

The tank commander has the SAGEM VIGY 15 stabilised commander's periscope sight, while the gunner has a stabilised day/thermal sighting system incorporating a laser range-finder. This new fire-control system gives the PT-91 the ability to engage stationary and moving targets with a high first-round-hit probability under almost all-weather conditions.

The vehicle is fitted with the locally developed WOLA S-1000 four-stroke turbocharged and after-cooled water-cooled diesel engine developing 1,000 hp at 2,000 rpm. This has already completed some 2,000 km of extensive reliability trials.

Improvements have also been made to the transmission system as well as other detailed upgrades. These include the installation of Diehl Type

PT-91 MBT with roof-mounted 12.7 mm anti-aircraft machine gun traversed to front. This is the final configuration after implementation of all solutions resulting from evaluations carried out by the Polish Army **2001**/0014437

Upgraded Polish PT-91 series MBT that is fitted with a more powerfull diesel engine and a new computerised SAGEM day/night fire-control system (Christopher F Foss) **2002**/0130967

Specifications

Crew: 3
Combat weight: 45,300 kg
Power-to-weight ratio: 18.76 hp/t
Ground pressure: 0.94 kg/cm^2
Length:
(overall, with long-range fuel tanks) 9.67 m
(overall, without long-range fuel tanks) 9.53 m
Width:
(overall) 3.59 m
(over tracks) 3.37 m
Height: (turret roof) 2.19 m
Ground clearance: 0.43 m
Track: 2.79 m
Length of track on ground: 4.27 m
Max road speed: 60 km/h
Max range: 650 km
Gradient: 60%
Side slope: 50%
Vertical obstacle: 0.85 m
Trench: 2.8 m
Fording: (with snorkel) 5 m
Electrical system: 27 V
Batteries: 4 × 12 V, 180 Ah
Engine: Type S-12U four-stroke, multifuel, supercharged, water-cooled diesel developing 850 hp at 2,300 rpm
Transmission: manual, 7 forward and 1 reverse gears
Armament:
(main) 1 × 125 mm smoothbore D81T (2A46) gun
(coaxial) 1 × 7.62 mm PKT MG
(anti-aircraft) 1 × 12.7 mm NSV MG
Smoke-laying equipment: 2 × 12 smoke grenade launchers plus system integral with exhaust
Ammunition:
(main) 42 (22 ready use)
(7.62 mm) 2,000
(12.7 mm) 300
Turret traverse: 360°
Gun elevation/depression: +13° 47′/–6° 13′
Gun stabiliser:
(vertical) yes
(horizontal) yes
Optics:
(commander's day periscopes) 2 × TNPO-160
(driver's day periscope) 1 × TNPO-168W
(gunner's day periscope) 1 × TNP-165A
(commander's image intensification night periscope) POD-72
(driver's image intensification night periscope) PNK-72
NBC system: yes
Night vision equipment: yes

840 l tracks that meet western standards and new RPC 9500 communications equipment. Cross-country mobility is improved by the installation of Elastomeric bumpers.

One potential customer for this upgraded PT-91 MBT could be Malaysia. The earlier T-72M1Z-fitted with a South African Denel Tiger fire-control system, was demonstrated several years ago with no sales being made. The Polish company is now concentrating marketing of the latest PT-91 with the French SAGEM system for this and other markets.

This latest PT-91 is regarded as a stepping stone to yet another version Phase C, or PT-91/120 which would feature a NATO calibre 120 mm smoothbore gun. The current PT-91 has a 125 mm smoothbore gun fed by an automatic loader, which first loads the projectile and then the charge.

The NATO 120 mm smoothbore gun fires one-piece ammunition which is loaded manually. The only company to have built a 120 mm automatic loader for this type of ammunition is Giat Industries of France for their Leclerc MBT, of which, over 500 have been built for the home and export market.

Future Polish MBT Fleet

Early in 2002, it was stated that Poland expected to lease 124 ex-German Army Leopard 2 A4 MBTs. This was parallel with a programme to upgrade at least part of their existing T-72 series fleet, to achieve greater interoperability with Poland's NATO allies.

In 2001, Poland withdrew from service 815 T-55A/T-55AM MBTs. This left the Polish Army with 662 T-72 MBTs of which 287 were T-72M/M1D and 226 PT-91 series.

It is expected that by 2006 to 2008 the Polish MBT fleet will consist of about 400 vehicles which is expected to include the 124 German Leopard 2A4s, plus some upgraded T-72/PT-91 series vehicles. The upgrade could well include the installation of a 120 mm smoothbore gun.

A number of Polish companies have submitted proposals for this, including the manufacturer of the PT-91, OBRUM and HSW.

Gorilla MBT

This was being developed as a follow-on to the PT-91 but work on this has now stopped. Details were given in *Jane's Armour and Artillery 1995-96*.

Status

Production complete. In service with the Polish Army.

Manufacturer

Zaklady Mechaniczne Bumar-Labedy SA.

UPDATED

Polish T-72M1 MBT

Development/Description

For some years, Poland has manufactured the Russian T-72M1 MBT under licence for both the home and export markets. It is similar to the Russian vehicle but there are some minor modifications. The full specifications, from Polish sources, are provided as there are some differences between Russian- and Polish-built vehicles.

Standard equipment on Polish built T-72M1 MBTs includes the 125 mm smoothbore main armament stabilised in both planes with the 2E28M electrohydraulic stabilisation system, an NBC protection system, deep fording equipment, integral smoke screening system, smoke grenade launchers, fire detection and suppression system, radios, vehicle intercom, self-entrenching blade mounted under the nose of the vehicle and the ability to be fitted with mineclearing systems at the front of the hull.

The range-finder sight is designated the TPD-K1 and incorporates a laser range-finder with a stabilised field of view in the vertical plane; it has a range-measuring capability from 500 to 3,000 m.

The fire-control system of the T-72M1 enables targets to be engaged under both day and night conditions, it lays the 125 mm gun on the target using information supplied by the computer which takes in information regarding distance and speed of the target, turret rotation speed and side inclination of the tank. Data on the type of ammunition to be used is fed into the fire-control system manually.

Analytical ranges for engaging targets are 5,000 m for HE projectiles, 4,000 m for APFSDS projectiles, 4,000 m for HEAT projectiles and 1,800 m for a 7.62 mm coaxial machine gun. As far as it is known, Poland never obtained the technology required to launch a 125 mm laser-guided projectile through the 125 mm smoothbore gun barrel.

The T-72M1 driver has a TNPO-168 day vision periscope that can be replaced by a TWNE-4B night device and the tank commander has a TKN-3 day/night device.

Poland has supplied Iran with a total of 104 T-72M1 MBTs. 34 were delivered in 1994 and the remaining 70 in 1995. Since then Poland has undertaken not to supply any additional vehicles to Iran.

T-72M1Z MBT from the side showing the explosive reactive armour array on the glacis and turret front
2001/0105980

Specifications

Crew: 3
Combat weight: 41,500 kg
Length: (gun forward) 9.53 m
Width:
(overall) 3.59 m
(over tracks) 3.37 m
Height: (turret roof) 2.19 m
Ground clearance: 0.449 m
Length of track on ground: 4.27 m
Max speed:
(cross-country) 35/45 km/h
(road) 60 km/h
Range: 460-700 km
Fording: 1.8 m
(with snorkel) 5 m
Gradient: 60%
Side slope: 30%
Engine: V-46.6 12-cylinder, 4-stroke, multifuel, water-cooled, supercharged diesel developing 780 hp at 2,000 rpm
Transmission: mechanical power transmission consists of an intermediate gear and two gearboxes with side drives set up in centreline of gearboxes. The planetary gearboxes have friction clutches and hydraulic steering, 7 forward and 1 reverse gears
Suspension: torsion bar
Armament:
(main) 1 × 125 mm smoothbore D81T gun
(coaxial) 1 × 7.62 mm PKT MG
(anti-aircraft) 1 × 12.7 mm NSV MG
Smoke-laying equipment: 12 smoke grenade launchers plus system integral with exhaust
Range:
(max of 125 mm gun) 10,000 m
(direct shot at 2 m high target) 2,100 m
Rate of fire: (automatic loading) 8 rds/min; (manual loading) 2 rds/min
Ammunition:
(125 mm) 44 (22 ready use)
Turret traverse: 360°
Gun elevation/depression: +13° 47′/–6° 13′
Gun stabiliser:
(vertical) yes
(horizontal) yes
NBC system: yes
Night vision equipment: yes

Variants

New armour package

The Polish Military Institute of Armament Technology (WITU) has developed two types of Explosive Reactive Armour (ERA) which can be fitted to the T-72M1 MBT and also installed on the more recent PT-91 MBT covered in a separate entry. Details of the Polish MBT defensive aids system are also given in the PT-91 entry.

The turret of the T-72M1 is protected by 108 blocks of ERA, the hull by 118 blocks and the side skirts by 84 blocks, giving the vehicle a total of 394 blocks of ERA.

These ERA blocks are of a different shape and design from those in the Russian Federation which have a much tighter fit.

There are two types of explosive reactive armour blocks, ERAWA-1 (with one layer of explosive inside) and ERAWA-2 (with two layers of explosive inside).

According to the design authority, these reduce the penetration capability of advanced shaped charge rounds by between 50 and 70 per cent, with the actual figure depending on the incidence angle of the shaped charge jet on the armour surface.

The armour package is insensitive to 7.62, 12.7 and 14.5 mm small arms fire as well as cannon fragments between 20 and 30 mm. Only the ERA block hit by a HEAT round goes off with the adjacent blocks being damaged or torn off.

Although designed for the T-72M1, this armour package can also be fitted to infantry fighting vehicles, for example.

Wolf MBT

The Wolf (or *Wilk* in Polish) is the locally built T-72M1 MBT fitted with a PCO day/night passive gunner's sight, a PCO passive night driver's periscope, a new communications system, two-layer explosive reactive armour, new side skirts, an uprated diesel engine, fast reacting fire extinguishers and a laser warning device coupled with smoke grenade launchers on either side of the turret.

T-72M1Z MBT

This vehicle is basically the Polish T-72M1 MBT fitted with the South African LIW, a Division of Denel, Tiger computerised fire-control system which enables stationary and moving targets to be engaged with a high probability of first round hit while the T-72M1Z is moving.

Other improvements, which are also incorporated in the latest Polish PT-91 MBT, include the installation of a more powerful S12U 850 hp diesel engine, laser warning system and the Polish-developed ERAWA-1 and ERAWA-2 explosive reactive armour package.

The T-72M1Z MBT has been tested in Malaysia, Poland and South Africa but as of early 2002 no orders had been placed for the T-72M1Z MBT.

PT-91 MBT

Further development of the T-72M1 in Poland has resulted in the more recent PT-91 MBT covered in a separate entry.

T-72M1K commander's tank

This is the T-72M1 fitted with a second short-wave radio set with a maximum range up to 50 km, a combined stub antenna 3 m high, a navigational device serving for continuous automatic elaboration of geographic co-ordinates of points and a generating set.

PMC-90 armoured vehicle-launched bridge

This is the Polish-built T-72 MBT chassis fitted with a launching system and bridge similar to that fitted to the Slovakian ZTS MT-72 armoured bridgelayer which is described in detail in *Jane's Military Vehicles and Logistics 2002-2003*. The scissors type bridge has a total length of 20 m and can span gaps up to 19 m.

Polish-built T-72M1 MBT fitted with the Polish-designed explosive reactive armour package 0018805

WZT-3 armoured recovery vehicle

This has been based on the T-72M1 MBT chassis and is armed with a 12.7 mm machine gun which is fitted to the tank commander's hatch.

Standard equipment includes a TD-50 crane with a telescopic jib that can lift a maximum load of 15 tonnes, front-mounted stabilising/dozer blade, main and secondary winches, fire detection and suppression system, smoke grenade dischargers, the capability to lay its own smoke screen and an NBC system. It normally has a crew of four and weighs 42 tonnes. The WZT-3 is in service with the Polish Army and a quantity has also been built for the Indian Army. Details of the WZT-3 ARV are given in *Jane's Military Vehicles and Logistics 2002-2003*. Polish vehicles are to be retrofitted with the S12U 850 hp engine fitted to the PT-91 MBT.

MID armoured engineering vehicle

This is based on the WZT-3 ARV but has a large hydraulically powered heavy jib crane arm located on the right side of the superstructure. This can be fitted with either gripper claws to enable battlefield obstacles to be removed or with a digger bucket. Maximum capacity of the jib crane is 7,000 kg. Located at the front of the vehicle is a V-shaped dozer blade and other equipment includes two winches and welding equipment. So far three prototypes of this vehicle have been completed. Details of the vehicle are given in *Jane's Military Vehicles and Logistics 2002-2003.*

Training vehicle

This is also referred to as the SJ 09 Training Vehicle and is used to train T-72M1 driver/mechanics in how to drive and operate the vehicle as well as carrying out maintenance.

The turret has been replaced by a fully enclosed rectangular observation cabin with seats for the instructor and other trainee drivers who can then observe the driver being trained. A simulated 125 mm gun is provided. Two versions are available. Version 1 has full equipment and version 2 enables existing T-72M1 MBTs to be refitted to the SJ 09 standard.

LOARA-G twin 35 mm SPAAG

This has been developed under the leadership of RADWAR to meet the requirements of the Polish Army. It consists of a T-72 series chassis fitted with a new turret armed with twin 35 mm cannon, tracking and surveillance radars. Two prototypes are now being tested.

Status

Production as required. In service with the Polish Army and other armed forces including Iran.

Manufacturer

Zaklady Mechaniczne Bumar-Labedy SA.

UPDATED

T-72M1 of the Polish Army fitted with plough type mineclearing equipment at the front of the hull (Jaroslaw Cislak) 0011610

PMC-90 armoured vehicle launched bridge laying its scissors bridge in position over front of hull 0011609

ROMANIA

TM-800 medium tank

Development

Late in 1994, Romania released details of the TM-800 MBT which, in many respects, is similar to the TR-580 MBT also developed in Romania and covered in a separate entry and may well be an export version of this vehicle.

Like the TR-580 MBT, the TM-800 has six roadwheels either side with the upper part of the suspension being covered by a light steel skirt.

Recent information has indicated that this vehicle remains at the prototype stage and has not yet entered volume production. According to United Nations sources, there were no exports of MBTs from Romania between 1992 and 2000.

Description

The overall layout of the TM-800 is similar to that of the TR-580 and TR-85 MBTs with the driver seated at the front left with some ammunition stowed to his right. The driver has a single-piece hatch cover with day periscopes for forward observation. The centre one can be replaced by a passive periscope for driving at night.

The cast turret is in the centre of the vehicle with the commander and gunner on the left and loader on the right as in the Russian T-54/T-55 MBT on which this vehicle is based.

Both commander and loader have a cupola with a single-piece hatch cover with the loader manning the 12.7 mm anti-aircraft machine gun. Boxes of ready use 12.7 mm machine gun ammunition are stowed on either side of the turret.

The engine compartment is at the rear of the hull. On the Soviet T-54/T-55/T-62/T-72 MBTs there is a distinct exhaust outlet on the left side of the hull above the last roadwheel station. The TM-800 MBT lacks this exhaust outlet. Mounted at the rear of the hull is an unditching beam.

Main armament comprises a 100 mm D-10 series rifled gun fitted with a fume extractor which is stabilised in both elevation and traverse. In addition, the gunner has an independently stabilised telescope. A 7.62 mm machine gun is mounted coaxially with the main armament.

The integrated fire-control system includes a ballistic computer and a laser range-finder to enable targets to be engaged at ranges from 150 to 4,000 m.

Standard equipment includes infra-red night vision equipment for the commander, gunner and driver, an NBC system, a fire detection and suppression system, the ability to mount mineclearing systems at the front of the hull, an anti-radiation lining, radios, internal communications equipment and a deep fording capability.

Although the TM-800 is similar to the older Russian T-54/T-55 MBTs it is some 9,000 kg heavier. According to the Romanians the hull and turret incorporate sandwich armour for improved battlefield survivability and this armour, together with the longer engine compartment, could account for the overall weight increase of the vehicle.

If required, a dozer blade or mineclearing device can be installed on the front of the hull.

Specifications

Crew: 4
Combat weight: 45,000 kg
Power-to-weight ratio: 18.45 hp/t
Ground pressure: 0.895 kg/cm²
Length:
(overall) 9.00 m
(hull) 6.74 m
Width: (overall) 3.30 m
Height: (turret top) 2.35 m
Ground clearance: 0.425 m
Track: 2.64 m
Track width: 580 mm
Length of track on ground: 4.25 m
Max speed:
(road) 64 km/h
(cross-country) 40 km/h
Fuel capacity: 1,100 litres
Max range:
(without external fuel tanks) 500 km
(with additional fuel tanks) 700 km
Fording:
(without preparation) 1.4 m
(with preparation) 5 m
Gradient: 60%
Side slope: 40%
Vertical obstacle: 0.9 m
Trench: 2.8 m
Engine: 830 hp diesel
Transmission: manual
Armament:
(main) 1 × 100 mm D-10 series rifled gun
(coaxial) 1 × 7.62 mm MG
(anti-aircraft) 1 × 12.7 mm MG
Smoke-laying equipment: 2 × 5 smoke grenade dischargers either side of turret
Ammunition:
(main) 43
(7.62 mm) 3,500
(12.7 mm) 500
(smoke grenades) 20
Gun stabiliser:
(vertical) yes
(horizontal) yes
NBC system: yes
Night vision equipment: yes

Variants

TER-800 ARV

The TER-800 (Pentru Evacuare Si Reparatti) is based on the chassis of the TM-800 MBT but has a new superstructure enabling it to carry out different missions. Standard equipment includes a winch, crane and specialised tools and specialist equipment. Loaded weight is quoted at 45.50 tonnes.

TR-125 MBT

This is similar in some respects to the Russian T-72 but never entered volume production. Details were given in *Jane's Armour and Artillery 1995-96,* page 102.

Status

Development complete. Ready for production. As of early 2002, there are no known export customers for the TM-800 MBT.

Manufacturer

Arsenalul Armatei.

UPDATED

TM-800 MBT from the left side showing the side skirts that cover the upper part of the suspension 0018809

TM-800 MBT from the front with commander's and gunner's hatches in open position 0018808

TR-85 MBT

Development

The TR-85 is a local Romanian design and is similar in appearance to the Russian T-55 series of MBT.

In 1993, the Romanian Army stated that it had 632 TR-85 MBTs in service. Since then the figure has dropped to 314.

The TR-85 is based on the Russian T-55 but with a new suspension system and a new German diesel engine. The T-800 covered in a separate entry is the export version of the TR-85. The turret of the TR-85 is said to be new. It is believed that the TR-85 was originally built at the state run Brashove Factory.

Description

This is similar to the Russian T-55 with the 100 mm main armament being fitted with a fume extractor near the muzzle and a thermal sleeve. Mounted above the mantlet is a laser range-finder similar to that fitted to some Chinese MBTs including the NORINCO Type 69.

Mounted externally on the turret sides and to the rear are boxes of 12.7 mm ammunition while on the forward left side of the turret is a rectangular stowage box similar to that fitted to Polish-built T-55 MBTs.

Romanian TR-85N MBT (Paul Beaver) 0011564

Specifications
(TR-85M1)
Crew:4
Combat weight: 50,000 kg
Power-to-weight ratio: 17.2 hp/t
Ground pressure: 1 kg/cm²
Length:
(gun forwards) 9.96 m
(gun rear) 8.75 m
Width: 3.435 m
Height: 3.10 m
Track: 2.67 m
Max road speed: 60 km/h
Road range: 310 km
Vertical obstacle: 0.9 m
Trench: 2.80 m
Engine: 8-cylinder 4-stroke diesel developing 860 hp at 2,300 rpm
Transmission: hydromechanical
Suspension: torsion bar with telescopic hydro-gas shock absorbers
Armament:
(main) 1 × 100 mm rifled tank gun
(coaxial) 1 × 7.62 mm MG
(anti-aircraft) 1 × 12.7 mm MG
(smoke grenade launchers) 2 × 10
Ammunition:
(main) 41
(7.62 mm) 4,500
(12.7 mm) 750
Gun stabiliser:
(horizontal) yes
(vertical) yes
NBC system: yes
Night vision equipment: yes

The chassis of the TR-85 is similar to the T-55, but each side has six roadwheels (the T-55 has five) with a distinct gap between the first/second, second/third and fifth/sixth roadwheels with the idler at the front and drive sprocket at the rear.

The upper part of the running gear is covered by a light sheet steel skirting of local design which is ribbed for increased strength and has a number of footholds in the lower part.

The driver is seated at the front left and to his immediate front are ribs that run across the glacis plate and a low-profile splashboard going two-thirds of the way across the glacis plate. On the right side of the glacis plate are two headlamps and low down are several spare track links.

The TR-85 also has a different power pack, or at the least a different engine, which is believed to be a 600 hp diesel, as the top of the engine compartment is different from the T-54/T-55 series. The TR-85 does not have the common exhaust outlet on the left side that is a standard feature of T-54/T-55 series vehicles.

The larger circular inlet in the centre of the engine deck is believed to be the air inlet with the air outlets being positioned on either side. Due to the installation of a new power pack, the hull of the TR-85 extends more to the rear. It is believed that the engine is of German origin.

The TR-85 has a combat weight of 43.3 tonnes and is fitted with a full range of infra-red night vision equipment for the commander, gunner and driver, an NBC system, an unditching beam and a snorkel at the hull rear. In addition, it can be fitted with long-range fuel tanks at the hull rear. Mounted under the long-range fuel tanks at the rear of the hull is a wooden unditching beam.

The upgraded Romanian TR-85M1 MBT showing additional armour protection for the turret, turret bustle and smoke grenade launchers 0043999

Variants

TR-85N MBT
This is understood to be an upgraded TR-85 although the exact details of the modifications are not known. The Romanian defence industries are now offering the DE-830CP eight-cylinder diesel upgrade pack for the TR-85 series MBT.

TR-85M1 MBT
Early in 2000, the Romanian Rompres News Agency stated that the Romanian Army was carrying out final tests on 13 upgraded TR-85 MVTs.

Following these user tests a decision will be taken on how many of the Romanian Army fleet of 300 plus TR-85 MBT, will be upgraded to the enhanced TR-85 M1 standard.

The TR-85 M1 is an upgraded version of the TR-85 designed and built in Romania in the 1970s using German chassis technology but retaining some features of the Russian T-54/T-55 MBT.

The upgraded TR-85M1 MBT has a number of improvements including additional passive armour over the turret front and new all-electric gun control and stabilisation system.

A new fire-control system has been fitted with ballistic computer and this together with the new sights and laser range-finder, increase the first round hit probability.

To retain turret balance, a new welded steel turret bustle has been added which extends over the rear engine compartment. New electrically operated 81 mm smoke grenade launchers are mounted each side of the turret and these are integrated with a laser warning system.

The 100 mm rifled gun is retained but a fume extractor has been added and it now fires an Armour Piercing Fin-Stabilised Discarding Sabot (APFSDS) round with increased armour penetration characteristics.

A 7.62 mm machine gun is mounted coaxial with the main armament and a 12.7 mm machine gun is fitted on the turret roof for anti-aircraft defence.

The TR-85M1 MBT weighs 50 tonnes and is powered by an 8-cylinder diesel which develops 860 hp, this gives the vehicle a maximum road speed of 60 km/h and a road range of 310 km.

Prime contractor for the upgraded TR-85 M1 is Arsenalul Armetei with a number of foreign companies being involved including MBDA, SAGEM, Kollmorgen and Racal.

The German company of Krauss-Maffei Wegmann have had discussions on possible future MBT collaboration with Germany, but no contracts have yet been signed.

T-55 upgrade
Arsenalul Armatei is offering an upgrade for the Russian T-55 MBT that includes many of the features of the TR-85M1 MBT upgrade. This also includes a new power pack consisting of a Perkins Engines Company CV-8 800 hp diesel coupled to an Allison XTG-411-5 fully automatic transmission. As of 2002, this had not been sold.

New Romanian MBT
Romania has had discussions with the German company of Krauss-Maffei Wegmann, prime contractor for the Leopard 1 and Leopard 2 MBTs, concerning the joint development of a new Romanian MBT.

The prime contractor in Romania would be Arsenalul Armetei which is also the prime contractor for the TR-85 and other Romanian MBTs

Status
Production complete. In service with the Romanian Army. The TR-85 fleet is expected to be upgraded to the enhanced TR-85M1 configuration.

Manufacturer
Arsenalul Armatei.

UPDATED

TR-580 MBT

Development
This is another Romanian MBT development which has a similar hull and turret to the TR-85 (qv) and is believed to have preceded this in production. In 1993, the Romanian Army stated that it has 398 TR-580 MBTs in service. Since then quantities of the TR-580 MBT in service with the Romanian Army have been further reduced and by early 1999 about 88 vehicles remained in service. It is believed that the Romanian Naval Infantry has 120 TR-580 MBTs in service.

The designation TR-580 means 'Romanian tank with 580 hp engine' (Tank Romanist). The TR-580 has also been referred to as the TR-77 or the M1977. It is possible that the TR-580 was probably used as a testbed for the TR-85M1 MBT upgrade programme (qv) as some tanks were observed with a similar modified turret.

Description
The cast armoured steel turret lacks the rectangular stowage box on the left and the 100 mm rifled tank gun only has a fume extractor, no thermal sleeve or laser range-finder being installed. Ammunition boxes for the 12.7 mm anti-aircraft machine gun are fitted either side of the turret.

The hull is similar to that of the TR-85 but the TR-580 has six unique spoked roadwheels with a gap between the first/second roadwheels.

The upper part of the running gear is covered by a light sheet steel skirt that, like the TR-85, extends right to the front and covers the front idler wheel.

The T-54/T-55 type exhaust outlet is retained above the last two roadwheel stations on the left side and the rear engine decking appears to be similar to the standard T-55 MBT. The TR-580 has a combat weight of 38.2 tonnes.

Mounted at the rear of the hull are two long-range fuel tanks underneath which is an unditching beam.

Variants

TCZ-580 ARV
This is an armoured recovery version of the TR-580.

Romanian T-55AM MBTs
There are at least three different versions of this tank in Romanian series, all of which have the same designator.

The first is similar to the Russian version with the Volna computerised fire-control system, the second is the Czech made T-55AM2 with the Kladivo computerised fire-control system and the last one is equipped with a laser range-finder similar to that used by the TR-85. This is probably the Ciclop computerised fire-control system.

Status
Production complete. In service with the Romanian Army. There has been no recent production of the TR-580 MBT. According to United Nations sources, Romania did not export any MBTs between 1992 and 2000.

Manufacturer
Arsenalul Armatei.

UPDATED

TR-580 MBT from the front

TR-580 MBT from the side

RUSSIAN FEDERATION AND ASSOCIATED STATES (CIS)

Black Eagle MBT

Development/Description
Late in 1997, during a defence equipment exhibition held in Omsk, Russia briefly showed the first example of a new MBT called Black Eagle (Chiorny Oriol).

This was essentially a T-80U MBT chassis fitted with a new two-person turret. It was subsequently revealed that this turret was in fact a mockup.

Some sources have indicated that Black Eagle is being developed at the Omsk Machine Construction Plant for the export market and may well also include some foreign components in its design, especially in the area of optics and fire control.

At a subsequent defence equipment exhibition held in Omsk in June 1999 the Black Eagle was again shown. This time however it featured a new chassis and a complete turret. It is believed that this vehicle has the development designation of Obiekt 640.

The standard production T-80 series MBT has six road wheels either side with the drive sprocket at the rear, idler at the front and track return rollers.

The latest Black Eagle MBT features a new and longer chassis with seven road wheels either side with drivers compartment at the front, turret and fighting compartment in the centre and power pack at the rear.

The longer chassis of the Black Eagle has enabled the armour protection at the front of the hull to be considerable increased.

The turret is a new design which was previously offered for installation of a T-80U chassis called the T-80UM2, although this designation has since been adopted for another version of the standard T-80U MBT.

The new turret is armed with the standard two-axis stabilised 125 mm 2A46M smoothbore tank gun which is fed from a bustle-mounted automatic loader. This is separated from the turret crew by an armoured bulkhead. Blow out panels are located in the turret roof.

The installation of the bustle-mounted automatic loader not only increases the combat survivability of the Black Eagle MBT but would also enable the 125 mm gun to fire more advanced ammunition with greater armour penetration characteristics.

The T-64, T-72, T-80 and T-90 all feature a 125 mm smoothbore gun fed by an automatic loader mounted under the turret. This has enabled the turret crew to be reduced to two, commander and gunner.

The 125 mm automatic loading system first loads the projectile (or in some cases a guided missile) and then the charge.

Combat experience has shown that this design of automatic loading system makes the MBT highly vulnerable when projectiles penetrate the hull. This usually causes the ammunition to ignite so causing the turret to be blown off.

The use of a bustle-mounted automatic loader will increase the combat survivability of the Black Eagle as well as allowing the gun to use new generation ammunition with enhanced armour penetration characteristics. It is also probable that the installation of an automatic loader will enable a higher rate of fire to be achieved.

A computerised fire-control system is fitted to enable stationary and moving targets to be engaged with a very high first round hit probability. Mounted on the roof is the DV-EBS wind sensor which feeds information to the computer.

The gunners stabilised sight has day and thermal channels and an integrated laser range-finder which feeds target information into the fire-control system.

In addition, the commander also has a roof-mounted day/thermal sighting system so allowing the Black Eagle MBT to carry out hunter/killer target engagements.

An Ainet automatic fuze setting system is installed which allows the crew to set the 125 mm high-explosive fragmentation projectiles to explode over the target for maximum effect.

In Black Eagle the gunner is seated on the right and the commander on the left, both being provided with roof hatches.

A 7.62 mm PKT machine gun is mounted coaxial with the main armament and a 12.7 mm NSVT machine gun is mounted above the right side of the turret, this can be aimed and fired under complete armour protection.

For a higher level of battlefield survivability, the Black Eagle MBT is fitted with the latest Kaktus Explosive Reactive Armour (ERA) which provides a higher level of protection against both kinetic energy and chemical energy attack over the frontal arc.

The latest Drozd-2 active protection defensive aids suite is also fitted although as an alternative the Arena system could also be installed.

No firm details of the Black Eagle have been released although it is estimated that it has a combat weight of about 50 tonnes and is powered by a 16-cylinder turbocharged diesel engine developing 1,200 hp.

Standard equipment includes day and night vision equipment, NBC system, front mounted dozer blade, fire detection and suppression system and a battle management system.

Future Russian MBT
There have been persistant reports that Russia is developing a new MBT, perhaps armed with a 152 mm smoothbore gun fed by an automatic loader. It is possible that Black Eagle could be back fitted with this new 152 mm smoothbore gun.

T-95 MBT
Early in 2000 it was stated that a new MBT called the T-95 had been developed and available details of this are given in a separate entry. It is understood that this has yet to enter volume production.

Status
Black Eagle remains at the prototype stage.

Manufacturer
Omsk Machine Construction Plant.

UPDATED

Black Eagle development tank

Development
Late in 1998 Russia released details of an evolutionary version of the T-80 MBT called the T-80UM2. This has a new turret which is probably related to the more recent Black Eagle MBT which is covered in a separate entry.

More recently the designation T-80M2 has been allocated to another version of the T-80 MBT which is covered under variants in the T-80 MBT entry.

It is believed that Russia is probably offering this turret system for installation on new vehicles as well as being retrofitted to older chassis.

This is essentially an upgraded T-80U MBT chassis fitted with a new all-welded cast steel turret incorporating advanced armour with additional armour protection being provided for the hull and turret.

The overall chassis layout of the vehicle is almost identical to the standard T-80U MBT covered in a separate entry.

On the hull front and turret frontal arc, Kaktus explosive reactive armour (ERA) is installed which gives protection against both chemical energy (High Explosive Anti-Tank) and kinetic energy (KE) attack. The track skirts are also of the Kaktus type with anti-fragmentation screens being provided around the lower part of the turret front.

The turret crew positions have been reversed when compared to the current production T-80U as on the Black Eagle development tank the gunner is seated on

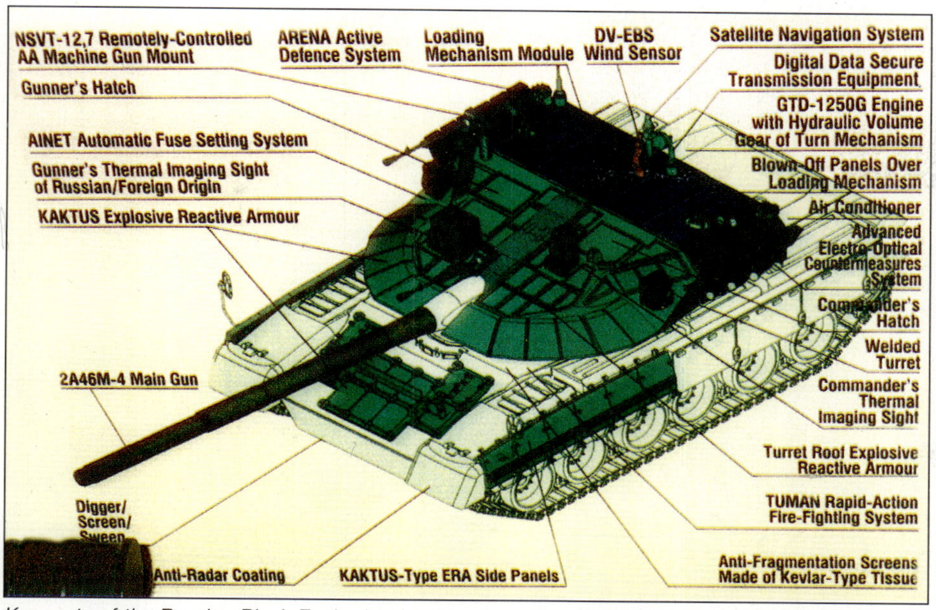

Key parts of the Russian Black Eagle development tank Main Battle Tank features a bustle-mounted automatic loader
0045792

the right and the commander on the left, both are provided with new roof hatches.

The Black Eagle development tank retains the 125 mm 2A45M smoothbore gun of earlier T-80 series MBTs but a new automatic loading system has been installed in the turret bustle with blow out panels above.

By moving the ammunition stowage from below the turret to the turret bustle the survivability of the Black Eagle development tank is considerably improved. During the fighting in the Middle East (T-72) and more recently, Chechnya (T-72 and T-80), many MBTs were lost when the ammunition below the turret exploded, normally removing the complete turret.

In addition, the rate of fire of the 125 mm gun is probably higher, because as with the previous automatic loading system, the loader had to first load the projectile and then the semi-combustible charge. All that remains after firing is the sub-cartridge case.

The installation of the bustle-mounted automatic loader will also enable the 125 mm gun to fire more advanced fixed ammunition with higher armour-piercing capabilities than current types. These, when compared with the latest Western types, have shorter length to diameter ratios.

So far there is no hard evidence that Russia has developed a new generation of 125 mm ammunition of the complete round type.

A 7.62 mm PKT machine gun is mounted coaxial with the 125 mm smoothbore main armament and the roof-mounted 12.7 mm NSVT machine gun can be laid onto the target and fired from under complete armour protection.

A computerised fire-control system is fitted to the Black Eagle development tank which allows stationary and moving targets to be engaged with a high first round hit probability while the vehicle is stationary or moving.

The gunner's stabilised sight has day and thermal channels and a laser range-finder that feeds information to the fire-control computer. The commander has a roof mounted day/thermal sighting system so allowing the vehicle to carry out hunter/killer target engagements.

The Ainet automatic fuze setting system for the 125 mm fragmentation ammunition is also installed and this allows the crew to set the fuze of a 125 mm high-explosive fragmentation projectile so that it detonates above the target for maximum impact.

Mounted on the turret roof is the DV-EBS wind sensor which feeds information to the fire-control computer.

The Black Eagle development tank is powered by a GTD-1250G turbine developing 1,250 hp coupled to a modified transmission. It is understood that in future this turbine may be replaced by a more fuel efficient diesel engine.

According to Russian sources, the Black Eagle development tank MBT is 58 per cent more effective than the standard production T-80 MBT.

As well as being fitted with additional armour protection, the Black Eagle development tank is also fitted with the roof-mounted Arena active defence system. This can detect, track and neutralise incoming anti-tank weapons before they come in contact with the vehicle. This has also been installed on a number of Russian armoured fighting vehicles including the T-80 and BMP-3. An illustration of the Arena system fitted onto the T-80 MBT is given in the entry for the T-80 MBT later in this section.

Standard equipment for the Black Eagle development tank includes a Tuman rapid action fire detection and suppression system, RPz-86M anti-radar coating, advanced electro-optical countermeasure equipment, air conditioning system, secure digital data communications system and satellite navigation system.

Long range drum type fuel tanks can be fitted at the rear to increase the operational range of the Black Eagle development tank and an unditching beam can also be carried at the rear. A self-entrench device is mounted under the nose of the vehicle and various type of plough and roller type mineclearing systems can be quickly fitted at the front of the hull.

Status
Prototype. Not yet in production or service.

Manufacturer
Omsk Machine Construction Plant.

UPDATED

Black Eagle development MBT with mockup turret and chassis with six roadwheel stations (A Chirvalnikov)
2001/0018516

T-95 MBT

Development
In early 2000, the then Russian Defence Minister, Marshal Igor Sergeyev announced that a radically new Main Battle Tank (MBT) designated the T-95 had been developed in Russia.

Sergeyev made his statement after visiting defence industry facilities in Nizhnii Tagil and Yekaterinburg in the Ural region. During the visit to the Uralvagonzavod plant the Minister looked over a full-scale prototype of the new vehicle.

The fact that he named it T-95 suggests possible fielding of the new tank. It is known that such designations are given to operational vehicles such as the T-80 and T-90. Russian pilot and developmental vehicles are usually designated by the word 'object' with a number given to it.

Description
Full technical data, outline and configuration features of the new T-95 MBT are not available. At the same time, a representative of Uralvagonzavod said the tank has a whole new design, rather than a modification of an existing vehicle.

It is claimed that the new T-95 MBT weighs 50 tons, its length and width would be likely about the same as the current T-72, T-80 and T-90 MBTs.

But the main feature of the new tank lies in its radically new configuration. Its gun is in a small unmanned turret. A recently designed automatic loader, typical of the Russian MBTs, is located below the turret which first loads the projectile and then the charge.

The three-man crew consisting of driver mechanic, gunner and commander are seated in a special armoured capsule, separated by an armoured bulkhead from the automatic loader and turret with externally mounted main armament.

This design feature makes it possible not only to reduce the silhouette of the MBT and therefore make it less observable on the battlefield, but also considerably enhance crew safety and survivability. Combat experience has shown that one of the main weaknesses of Russian MBTs has been that when the hull is penetrated near the 125 mm automatic loader the ammunition often explodes, so blowing off the complete turret and destroying the whole vehicle.

It is believed that the T-95 MBT is armed with a 135 mm gun which is believed to be of the smoothbore type and is fitted with a new Fire-Control System (FCS).

Target information comes via optical, thermal imaging and IR channels. The FCS will also include a laser range-finder and possibly a radar. It is expected that it will feature a hunter/killer capability. In this role the target is first detected by the commander who then hands over to the gunner to carry out the actual target engagement.

It should be noted that the new configuration places very strict demands to the FCS, as the crew is prevented from using traditional optical devices.

In the past one of the major weakness of Russian MBTs, when compared to their western counterparts, has been in the area of fire control and night vision optics and it is only recently that Russian MBTs have started to be fitted with thermal vision devices.

The new T-95 MBT is not the sole domestic new-generation MBT. In 1999, at the arms exposition in Omsk, Siberia, there was displayed the 'object 640' (named Black Eagle), developed at the Omsk-based Design Bureau of Transport Machine-building who also build the T-80.

Black Eagle also features a wholly new chassis and turret. Available details are given in a latter entry. Black Eagle retains a conventional turret with seats for the commander and gunner with the automatic loader and a part of the 125 mm ammunition load being placed in a spacious housing located in a rear part of turret.

As of early 2002, it is not clear as to when the new T-95 MBT will enter service as there is no assurance that it will be brought to a production phase within an acceptable time period.

Nizhnii Tagil is where the T-72 and T-90 MBTs are built with Omsk producing the T-80 MBT. Both facilities have been struggling to keep their research, development and production facilities alive against the background of reduced orders from the Russian Army and small export orders. The Omsk facility recently had a major boost when India ordered 310 of the export T-90S series MBTs early in 2001. The first of these were delivered late in 2001 and some will also be manufactured under licence in India.

It is known that OAO Spetsmash Design Bureau in St Petersburg has been working on a two-person MBT fitted with an externally mounted large calibre gun.

Status

Prototype. No firm details of this have yet been released.

Manufacturer

Uralvagonzavod (Nizhnii Tagil).

UPDATED

T-90 MBT

Development

The T-90 MBT was developed at the Kartsev/Venediktov Bureau, 'Vagonka' at Nizhnyi Tagil under the designation of Obiekt 188. The chief designer of T-90 MBT was V Potkin.

The existence of the T-90 was first revealed in 1993 and it is believed to have entered low-rate production in 1994 for the Russian Army.

The T-90 is a further development of the T-72BM MBT. It was also designed and built at Nizhnyi Tagil and incorporates some of the advanced features of the late production T-80 MBT, especially in the areas of defensive aids systems, fire control and Kontakt-5 Explosive Reactive Armour (ERA) system.

The T-90 MBT is also being offered for export under the designation of the T-90S and T-90E.

The T-90 MBT was shown for the first time outside Russia in March 1997 when it was demonstrated in Abu Dhabi. It is being offered for the export market alongside the turbine-powered T-80U with the latter being the more mobile of the two vehicles.

As previously stated, the T-90 is a further development of the T-72BM but it has the latest armour package and a new fire-control system. A comparison of the fire-control system fitted to the T-72BM, T-80U and T-90 MBTs is given in Table 1.

Several years ago the Russian Army standardised on the T-90 MBT to meet its future needs as this was cheaper than the turbine-powered T-80U, but recent information has indicated that no decision has been taken as to which MBT will be procured on a regular basis. Speculation is that the Russian Army cannot afford to keep two MBT production facilities with current budget constraints.

Following several years of negotiation and trials with a number of vehicles, early in 2001, India signed a contract for the supply of 310 T-90S MBTs, with a total value of about US$700 million.

The first 124 vehicles will come direct from Russia with the remainder being assembled at the Heavy Vehicle Factory at Avadi in Southern India where licence production of the older T-72M1 MBT has been undertaken in the past (qv India).

First vehicles were delivered to India late in 2001. It is expected that first locally assembled vehicles will be completed in 2003, followed by local production.

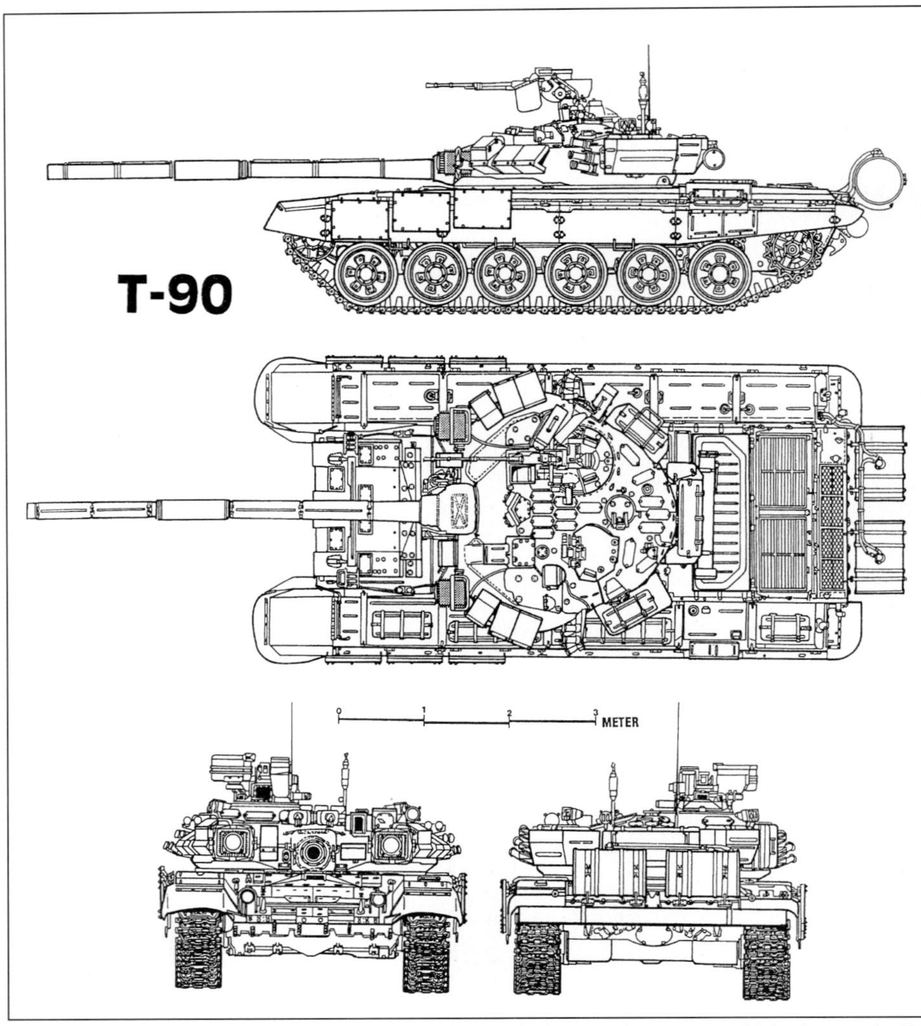

T-90 MBT which is fitted with the latest generation explosive reactive armour which gives protection against both chemical energy and kinetic energy attack (Steven Zaloga) 0002532

Table 1	**T-72BM**	**T-80U**	**T-90**
Fire control	1A40	1A45	1A45T
Gun stabilisation	2E42-2	2E42	2E42-4
Gunner's range-finder Sight	1K13-49	1A42	1A43
Ballistic computer	1V528	1V528	1V528-1
Wind sensor	crosswind	DVE-BS	DVE-BS
Guided missile	Svir	Reflecks	Reflecks

Description

The layout of the T-90 MBT is almost identical to that of the T-72 MBT, with the driver's compartment at the front, turret in the centre and engine compartment at the rear.

The hull and turret of the T-90 over the forward arc is fitted with the latest generation Kontakt-5 ERA which provides protection against APFSDS and HEAT-type projectiles.

In addition to being fitted to the hull and turret, ERA panels are also fitted either side of the hull front to provide lateral protection to each side of the driver's compartment.

The driver is seated at the front of the hull in the centre and has a single day periscope that gives observation through the frontal arc and a single-piece hatch cover that lifts and opens to the right. For driving at night the day periscope can be replaced by a TVN-5 night vision device.

The other two members are seated in the turret with the gunner on the left and the commander on the right.

The tank commander's contrarotating cupola has a single-piece hatch cover that opens forwards with two rear-facing TNPA day vision blocks. In the forward part of the cupola is the TKN-4S Agat-S stabilised day/image intensification sight with a TNP-160 day periscope either side.

The gunner's hatch opens forwards and has a circular opening for mounting the snorkel for deep fording operations. In front of the gunner's hatch is the TNPA-65 day vision block while a TNPA-65 day vision block is fitted in the hatch cover itself.

The gunner of the T-90 is provided with a day and thermal sighting system with the tank commander being provided with a screen to monitor the thermal view seen by the gunner.

The T-90 has an integrated computerised fire-control system designated as 1A45T which allows the tank commander and gunner to lay and fire the 125 mm smoothbore main armament while the vehicle is stationary or moving under day and night conditions.

The gunner's sighting system includes the 1A43 day sight with stabilised field of view in two planes and laser range-finder, IG46 day sight range-finder with missile guidance channel, 1V528-1 digital ballistic computer, DVE-BS wind gauge, gunner's T01-K01 infra-red vision equipment and TPN4-49-23 sight Buran-PA. The last can be replaced by the Agava-2 roof-mounted stabilised thermal sight.

Main armament comprises a 125 mm 2A46M1 (or D-81TM) smoothbore gun fitted with a fume extractor and a thermal sleeve. This weapon is stabilised by the 2E42-4 system in both planes and fed by an automatic loader.

The 125 mm gun fires ammunition of the separate loading type and details are given in a separate entry for the T-80 MBT. In addition it can also fire a special high-explosive fragmentation projectile that can be detonated over the target using the tank's fire-control system. Russian sources state that the T-90 has a maximum rate of fire of 7 rds/min.

Specifications

Crew: 3
Combat weight: 46,500 kg
Power-to-weight ratio: 18.06 hp/t
Ground pressure: 0.91 kg/cm²
Length:
(gun forward) 9.53 m
(hull) 6.86 m
Width:
(without skirts) 3.37 m
(over skirts) 3.78 m
Height: (without AA MG): 2.226 m
Ground clearance: 0.47 m
Track width: 580 mm
Length of track on ground: 4.278 m
Max road speed: 60 km/h
Range:
(road) 550 km
(dirt road) 450 km
Fuel capacity:
(main) 1,200 litres
(auxiliary) 400 litres
Fording:
(without preparation) 1.8 m
(with preparation) 5.0 m
Gradient: 60%
Side slope: 40%
Vertical obstacle: 0.85 m
Trench: 2.8 m
Engine: Model V-84MS 4-stroke 12-cylinder multifuel diesel developing 840 hp
Transmission: manual, 7 forward and 1 reverse gears
Steering: clutch and brake
Suspension: torsion bar
Electrical system: 24 V
Armament:
(main) 1 × 125 mm smoothbore 2A46M gun
(coaxial) 1 × 7.62 mm PKT MG
(anti-aircraft) 1 × 12.7 mm NSVT MG
Ammunition:
(main) 43 (22 ready use)
(coaxial) 2,000
(anti-aircraft) 300
Smoke grenade launchers: 8 × 81 mm
Gun control equipment
Turret power control: electric/manual
Turret traverse: 360°
Gun elevation/depression: +14/–6°
Gun stabiliser:
(horizontal) yes
(vertical) yes
Range setting device: yes (laser)
NBC system: yes
Night vision equipment: yes

In addition, the 125 mm gun can also fire the 9M119 Refleks laser-guided projectile out to a range of 5,000 m. This has the US/NATO designation of the AT-11 'Sniper'. The T-90 normally carries six AT-11 'Sniper' missiles.

The AT-11 'Sniper' weighs 17.2 kg at launch and has four wraparound fins at the rear for stability when the missile leaves the launch tube, and two towards the front for steering.

The complete missile system is called the 9K119 with the complete round, for example missile, reduced charge propellant and a space plug, being called the 9UBK14.

Key parts of the 9S515 missile control system include the gunner's sighting, aiming and range-finder pack, information field generator, voltage charge and control unit.

As well as the 9M119 laser-guided missile, the T-90 can also fire the more recent 9M119M laser-guided missile, with the complete round being designated as 3UBK20. The 9M119M missile has a tandem warhead to defeat vehicles fitted with explosive reactive armour. The first HEAT warhead activated the ERA, so clearing a patch for the main HEAT charge. Only the gunner can launch the Refleks guided missile. The 5,000 m range Refleks is launched by the T-90, T-80U and the T-80UD while the shorter range (4,000 m) Svir is launched from the T-72B and T-72S.

A 7.62 mm PKT machine gun is mounted coaxially to the right of the main armament and a 12.7 mm NVST machine gun is mounted on the commander's cupola.

The T-90 MBT made its first appearance overseas in March 1997 in Abu Dhabi (Christopher F Foss)
0011803

The latter can be aimed and fired under complete armour protection; on the T-72 the commander has to expose the upper part of his body to fire this weapon. The 12.7 mm NSVT machine gun has a PZU-7 optical sight and is fitted with a 1ETs29 vertical stabilisation system.

Mounted either side of the turret is a bank of six electrically operated 81 mm smoke grenade launchers which are in a new low-angle configuration compared to those fitted to earlier Russian MBTs.

The quick forming aerosol screening system comprises the four laser radiation sensors (two coarse and two fine receiving heads), Type 902A Aerosol Forming Grenade Launch System dispensing 81 mm 3D17 aerosol grenades and associated controls.

The aerosol screening system detects laser illumination, determines its direction and type (laser range-finder or designator), generates warning signals, both audio and visual, and lays in automatic or semi-automatic modes, quick forming aerosol screens within 3 seconds at a distance of 50 to 80 m from the tank.

In addition, the T-90 can also lay its own smoke screen by injecting diesel fuel into the exhaust outlet located on the left side of the hull.

To improve its battlefield survivability the T-90 is fitted with the TShU1-7 Shtora-1 countermeasures system which is also fitted to some models of the T-80UD and T-84 (Ukraine) MBTs.

The T-90 also has a laser warning system and an aerosol generating system that uses 3D17 grenades of 902A Tucha (Cloud) designation. These are launched from the standard 81 mm launchers either side of the turret.

The TShU1-7 Shtora (which means shutter or blind) consists of an infra-red source, power supply and the control panel. The T-90 has two infra-red sources, one mounted either side of the 125 mm smoothbore gun. These introduce a spurious signal into the guidance circuitry of the incoming missile through continuously generated coded pulsed infra-red jamming signals.

The V-84MS diesel engine, which is fitted with a preheater for use in cold weather, is coupled to a mechanical transmission which consists of a primary reduction gear, two planetary final gearboxes and two planetary final drives. The engine is fitted with an effective two-stage cleaning system and a temperature warning device.

Although a diesel engine, the V-84MS V-12 will also run on other fuels including gasoline, kerosene and benzine fuels, alone or unblended.

Suspension is of the torsion bar type and either side consists of six dual rubber-tyred roadwheels with the drive sprocket at the rear, idler at the front and three track-return rollers. Hydraulic shock-absorbers are provided at the first, second and sixth roadwheel stations.

Standard equipment includes NBC protection, fire detection and suppression system, nose-mounted dozer blade and a deep fording kit. To increase the operational range of the T-90, two diesel fuel drums can be carried under the hull rear.

Mineclearing equipment of the KMT-7 or KMT-8 type can be mounted at the front of the hull and communications equipment includes an R-173 UHF radio and an R-173P radio.

Variants

For trials purposes, T-90 MBT has been fitted with other and more powerful engines including the V-92 diesel developing 950 hp and the V-96 developing 1,100 hp. For trials purposes a turbine has also been fitted to the T-90 along similar lines to that fitted to the T-80U MBT.

T-90S MBT

The Russian Uralvagonzavod State United Enterprise based in Nizhnyi Tagil, has developed an enhanced version of the T-90S Main Battle Tank (MBT) to meet the operational requirements of countries in Asia. The T-90S has been ordered by India but this is understood not to have all of the features listed below.

The latest version of the T-90S MBT features an air conditioning system for operations in high ambient temperatures. A French Thales Optronics second-generation thermal camera has been installed with the commander and gunner both being provided with a screen.

T-90 MBT from the rear without long-range fuel tanks fitted and showing snorkel attached to turret rear
2000/0069542

The day/night sighting system is called the ESSA (thermal vision and sighting system) and has been developed by Thales, Uralvagonzavod and the Peleng Joint Stock Company of the Belarus.

This enables either the commander or gunner to lay and fire the 125 mm main armament under poor weather conditions with a high first-round-hit probability.

The sighting system may be a development of the Booklet stabilised multichannel sight also developed by these three companies for retrofit into the more widely deployed and much older T-72 MBT.

Booklet features a day channel (two fields of view), Thales Catherine thermal channel (two fields of view), laser range-finder and laser missile guidance channel.

The standard T-90 is powered by a Model V-84MS 4-stroke 12-cylinder multifuel engine developing 840 hp coupled to a manual transmission. With a combat weight of 46.5 tonnes, this gives the T-90S a power-to-weight ratio of 18.06 hp/tonne and a max road speed of 60 km/h.

The latest enhanced T-90S is powered by a B92C2 V-12 diesel developing 1,000 hp, coupled to a manual transmission with seven forward and one reverse gears. This gives the vehicle a maximum road speed of 65 km/h and a power-to-weight ratio of 21.50 hp/tonne.

A turbo charged version of the diesel engine is also available which gives a further increase in power.

A new track has been developed and tested for the T-90S that not only has a longer life but also has replaceable rubber pads that can be quickly removed.

Main armament of the T-90S is still the 125 mm 2A46M smoothbore gun which fires conventional natures of separate loading ammunition (projectile and charge) including HE-FRAG (high-explosive fragmentation) and APFSDS-T (Armour Piercing Fin Stabilised Discarding Sabot - Tracer).

It can also fire a 9M119M laser guided projectile fitted with a tandem HEAT (High Explosive Anti-Tank) warhead out to a maximum range of 5,000 m. A 7.62 mm 6P7K machine gun is mounted coaxial with the main armament and a 12.7 mm 6P49 machine gun is mounted on the turret roof and can be laid and fired from within the turret.

Standard equipment on the T-90S includes the latest generation Kontakt-5 explosive reactive armour which provides a high level of protection against both kinetic and chemical energy attack, an NBC system, fire detection and suppression system, front mounted dozer/blade, long-range fuel tanks and the TshU1-7 Shtora-1 countermeasures system.

This is also fitted to the Ukrainian T-84 MBT, although Russia has stated that it is the only source for the TshU1-7 Shtora-1 system and would not be exported to the Ukraine.

BMR-3M Armoured Demining Vehicle

The BMR-3M armoured demining vehicle is based on a much modified T-72 or the more recent T-90 MBT chassis and has its turret removed and replaced by a fully enclosed raised superstructure for the crew.

Explosive reactive armour is fitted to not only the superstructure but also the glacis plate and covers the upper part of the suspension as well.

Mounted at the front of the vehicle is the KMT-7 mineclearing roller system fitted with an electromagnetic device to activate anti-tank mines fitted with a magnetic fuze.

Status

Production. In service with the Russian Army. Early in 2001, India ordered 310 T-90S MBTs of which 124 will come from Russia and the remainder assembled in India. First vehicles were delivered late in 2001.

Manufacturer

Nizhnyi Tagil.

UPDATED

T-80 MBT

Development

The T-80 MBT was developed at the Leningrad Kirov Plant (LKZ) under the direction of Nikolai S Popov, with the prototype being designated the Obiekt 219 and fitted with a GTD-1000T turbine engine. This was accepted for service in 1976.

Initial production of the T-80 was undertaken in Leningrad although the first version was not built in large numbers. This was soon followed by the much improved T-80B MBT which had the development designation of Obiekt 219R.

Since 1976 the T-80 MBT has been improved as new technology became available, with the latest versions having a more powerful engine, better armour protection and a defensive aids suite.

In 1996, Cyprus ordered a batch of 41 T-80 series MBTs for its National Guard with all vehicles delivered by early 1997. These were delivered from stock as the only remaining T-80 plant in Russia, at Omsk, has a significant number of T-80U MBTs built for the Russian Army but not delivered because of a shortage of funds.

It has been confirmed that the T-80 series MBTs delivered to Cyprus were fitted with the Brod-1 deep fording equipment and the Shtora-1 defensive aids suite.

The Republic of Korea took delivery of 33 T-80U MBTs in 1996/1997.

Production of the T-80 has been completed at Leningrad and the other T-80 plant, Kharkov, is now in Ukraine. This facility has built 320 T-80UD MBTs for Pakistan which were delivered between 1997 and late 1999.

Further development of the T-80UD by the Ukraine has resulted in the T-84 MBT for which there is a separate entry under Ukraine. This features a new all-welded steel armour turret with additional layers of armour and explosive reactive armour.

Description

The overall layout of the T-80 is similar to that of the T-64 MBT series with the driver's compartment at the front, two-man turret in the centre and engine and transmission at the rear. There are, however, many detailed differences.

The glacis plate is of the laminate type for improved protection against kinetic energy and HEAT (High Explosive Anti-Tank) attack and there is a dozer blade carried retracted under the nose of the vehicle.

The driver is seated in the centre and has a single-piece hatch cover that lifts and swings to the right. In front of this are three forward-facing day periscopes, the centre one of which can be replaced by a passive night driving device.

The turret is armoured steel with an inner layer of special armour; the gunner sits on the left and the tank commander on the right.

Whereas the T-64 is powered by a five-cylinder, opposed piston, liquid-cooled diesel engine developing 750 hp coupled to a manual transmission with seven forward and one reverse gears, the T-80 has a SG-1000 gas-turbine engine developing 1,000 hp coupled to a manual transmission with five forward gears and one reverse.

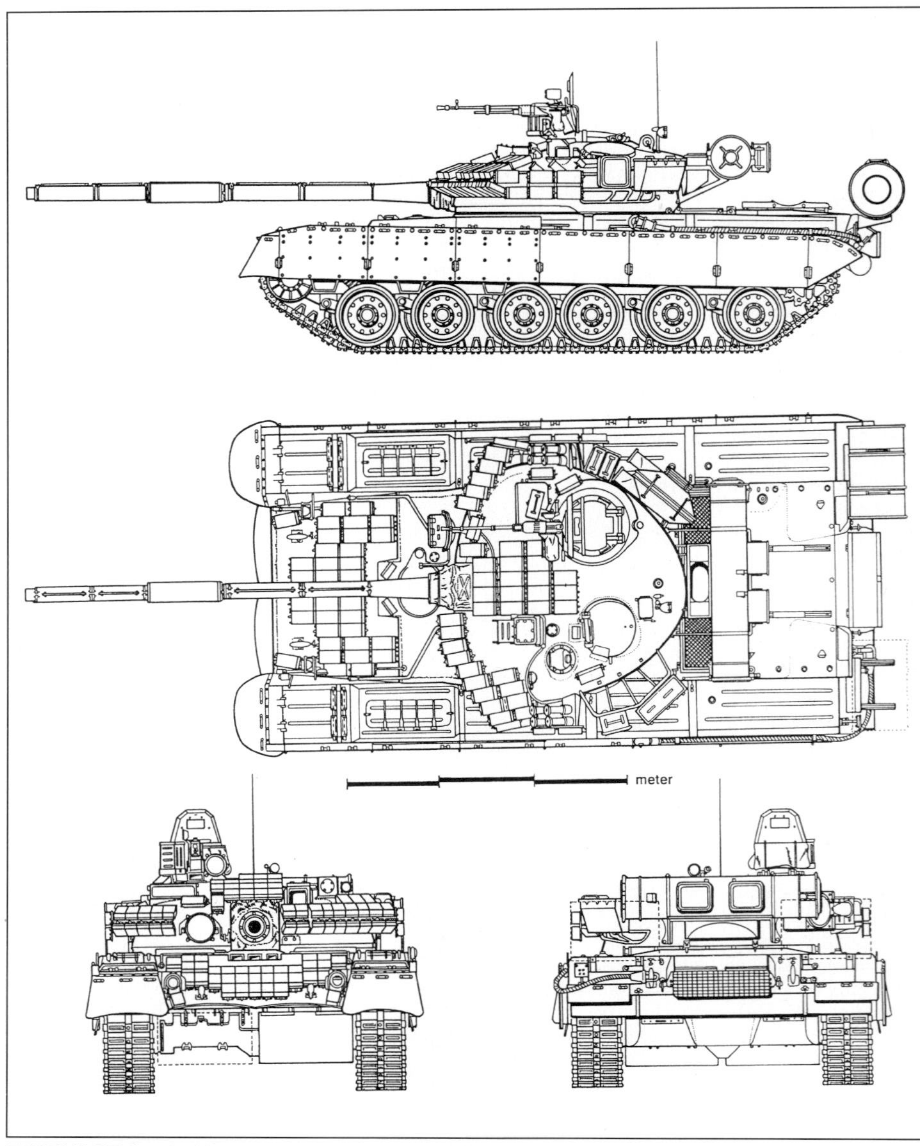

T-80BV MBT fitted with explosive reactive armour (Steven Zaloga)

The T-80's rear hull top is different from the T-64's in that it has a distinct oblong exhaust outlet in the hull rear.

The T-80 reverted back to torsion bar suspension with each side consisting of six forged steel-aluminium rubber-tyred roadwheels, drive sprocket at the rear, idler at the front and five return roadwheels. The rubber-tyred roadwheels are in two halves and bolted together.

As the track of the T-80 is slightly wider than the T-64's and has a longer length in contact with the ground, the ground pressure of the T-80 is improved, as is its power-to-weight ratio. The new track design has end connectors and central guide horns.

The roadwheel spacing is not identical and there are distinct gaps between the second and third, fourth and fifth, and fifth and sixth roadwheels. The side skirt covers the return rollers.

The T-80 has the same 125 mm 2A46 smoothbore gun as the T-72 with a horizontal ammunition stowage system. This can fire either the AT-8 'Songster' ATGW

T-80UD MBT with commander manning 12.7 mm anti-aircraft machine gun and four 81 mm smoke grenade dischargers either side of turret 0018814

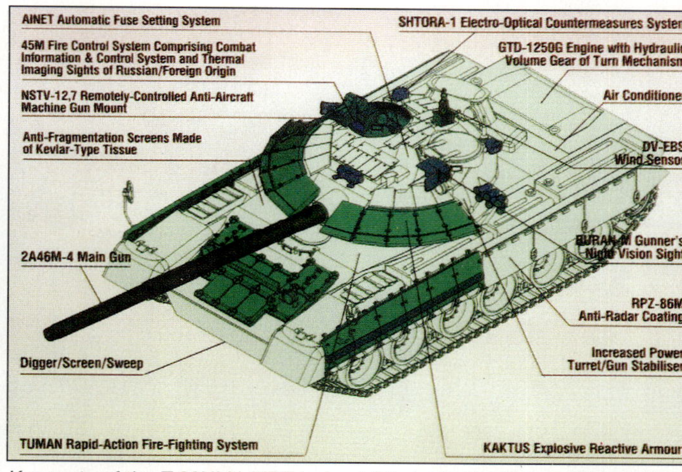

Key parts of the T-80UM1 MBT but not fitted with the Arena active defence system 0036296

or 125 mm ammunition of the separate loading type. With the latter, the projectile is loaded first, followed by the semi-combustible cartridge case; all that remains after firing is the stub base which is ejected. This 125 mm ammunition is common to the T-64, T-72, T-80, T-84 and T-90 MBTs and known types are shown in Table 1. The 125 mm smoothbore gun is stabilised in both elevation and traverse.

The HEAT training round is the BP-5 while the current APFSDS-T training round is the BP-6.

A 7.62 mm PKT machine gun is mounted coaxially to the right of the main armament and a 12.7 mm NSVT machine gun is mounted on the commander's cupola. Full details of the 125 mm gun/missile launcher are given in the entry for the T-64B.

The 125 mm gun/missile launcher tube is fitted with a thermal sleeve and a fume extractor.

Banks of 81 mm electrically operated smoke grenade dischargers are mounted either side of the 125 mm gun/missile launcher, normally five on the left and four on the right.

A total of four AT-8 'Songster' ATGWs is carried and these are identical to those launched by the T-64B MBT deployed many years ago. The missile guidance box is mounted on the right side of the turret roof in front of the commander's cupola and can be removed and stowed inside the vehicle if required.

Standard equipment includes snorkels for deep fording operations, which are carried on the turret rear when not required, an overpressure-type NBC protection system, night vision equipment for all three crew members, unditching beam carried across the hull rear and a laser warning device activated by laser range-finders, laser designators or precision-guided munitions fitted with a laser guidance device. Mounted on the turret rear is a large circular container which carries two snorkels. The larger one is the snorkel for the gas turbine, with another one being fitted onto the radiator grille by means of two adaptors. This provides an air intake for the gas turbine.

To extend the operational range of the T-80, additional drum-type fuel tanks can be mounted at the hull rear. These can be quickly jettisoned if necessary. Each of these fuel drums holds 300 litres of fuel and is connected to the main fuel supply.

Each of the two main fuel drums at the hull rear holds an additional 200 litres of fuel. An additional third fuel tank may be mounted crosswise on the engine hatch cover. This does not affect the combustion air intake which is directly behind the turret ring. The air intake is covered by two hinged rectangular grilles which open forward, uncovering a filter inserted underneath.

In the past, there has been considerable confusion on the correct designations for Russian MBTs, including the T-80 series. The designations listed are from Russian sources.

Variants

T-80 (1976)

This was the initial production T-80 and used many components of the earlier T-64A MBT including the complete turret. It was powered by the SD-1000 three-shaft turbine engine developing 800 to 1,000 hp but was not built in large numbers.

This model was not fitted with a 12.7 mm NSVT anti-aircraft machine gun, the infra-red searchlight was on the left of the 125 mm gun and the gunner had the

Table 1

Round	**Projectile**	**Type**	**Round weight**	**Projectile weight**	**Armour penetration at 2,000 m, 60°**
3VBM3	3BM9	APFSDS	19.5 kg	3.6 kg	
3VBM6	3BM12	APFSDS	19.7 kg	3.8 kg	
3VBM7	3BM15	APFSDS	19.7 kg	3.8 kg	150 mm
3VBM8	3BM17	APFSDS	19.7 kg	3.8 kg	
3VBM9	3BM22	APFSDS	20.2 kg	6.9 kg	
3VBM13	3BM32	APFSDS	20.4 kg	7.1 kg	250 mm *
3VBM17	3BM42	APFSDS	20.4 kg	7.1 kg	
3VP6	3P6	Training	18.5 kg	5.2 kg	
3VBK7	3BK12	HEAT	28.5 kg	19.0 kg	
3VBK7	3BK12M	HEAT	28.5 kg	19.0 kg	
3VBK10	3BK14M	HEAT	28.5 kg	19.0 kg	220 mm
3VBK16	3BK18M	HEAT	28.5 kg	19.0 kg	260 mm
3VBK17	3BK21B	HEAT	29.0 kg	19.5 kg	
	3BK29	HEAT	28.4 kg	18.9 kg	ERA + 300 mm
3VP5	3P11	Training	28.5 kg	19.0 kg	
	9M112	Missile			
3VOF22	3OF19	HE-Frag	33.0 kg	23.0 kg	
3VOF36	3OF26	HE-Frag	33.2 kg	23.2 kg	

* This projectile has a depleted uranium penetrator

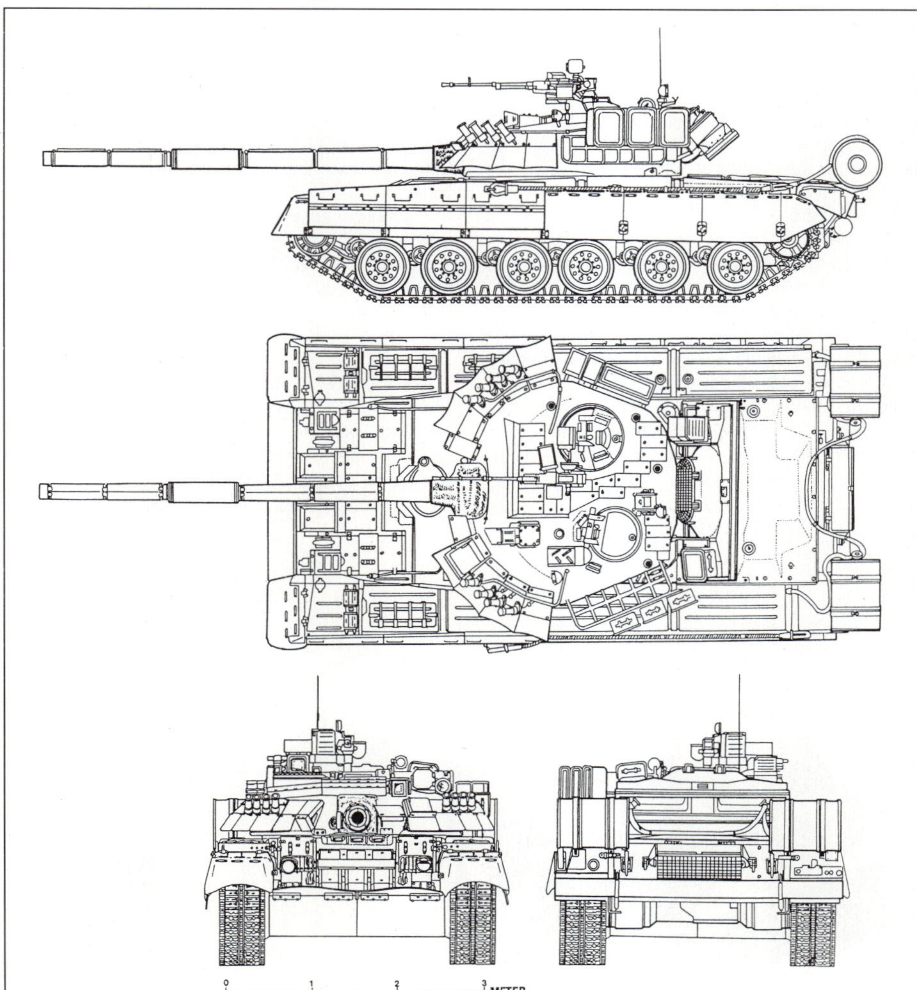

T-80UM Model 1993 MBT fitted with the latest generation explosive reactive armour which gives protection against both CE and KE attack (Steven Zaloga)

T-80UM MBT from the rear showing the large exhaust outlet, unditching beam below and APU on left side (Christopher F Foss) **2001**/0105972

TPD-2-49 optical sight/range-finder with one of the optical ports on the right side of the turret in front of the commander's position.

T-80B (1978)

After two years of production, the T-80 was replaced by the T-80B (development designation Obiekt 219R) which had the 9K112 Kobra missile system which fired the 9M112 Kobra (US/NATO designation AT-8 'Songster') radio-controlled missile, which was also fired by the T-64B MBT. It also had improvements to the fire-control system including the installation of a laser range-finder.

The IA33 fire-control system includes the IG42 range-finder sight with an electronic control panel, 1V517 tank ballistic computer, 1G43 fire selector panel, 2Eh46M weapons stabiliser and input sensors for windspeed, tank speed, cant and bearing, all of which are updated at regular periods.

The T-80B also has a modified turret incorporating a new-generation composite K ceramic armour which provides improved protection against HEAT and APFSDS penetrators.

The T-80B was powered by the SG-1000 gas turbine developing 1,000 hp but from 1978 this was replaced by the GTD-1000TF turbine engine rated at 1,100 hp.

T-80BK (1978)

This is the commander's model of the T-80B with additional communications equipment and it is not fitted with the Kobra radio guided missile system.

T-80BV (1985)

Further development of the T-80B, under the designation of Obiekt 219RV, resulted in the T-80BV which externally is recognisable by the addition of first-generation Kontakt explosive reactive armour to the hull and turret for improved battlefield survivability. The design and function of the explosive reactive armour is identical to that of the T-64BV MBT.

The installation of the explosive reactive armour gave the T-80BV a very high degree of battlefield survivability against anti-tank weapons fitted with a single HEAT warhead.

The T-80BV has a combat weight of 42.5 tonnes, maximum road speed of 70 km/h and a road range of 370 km.

T-80BVK (1985)

This is the commander's model of the T-80BV with additional communications equipment.

T-80U (1985)

This was originally called the SMT (Soviet Medium Tank) M1989 by NATO until its correct Russian designation became known.

This is essentially the second-generation T-80 series MBT with the U in the designation standing for uluchsheniye in Russian, or improved in English.

This had the development designation of Obiekt 219AS with the Kirov plant being responsible for the hull and the Kharkov plant being responsible for the turret and armament.

This has improved armour protection, the updated 125 mm gun and a new fire-control system that allows the tank commander or gunner to aim and fire the main armament.

Main armament is the latest 125 mm 2A46M-1 smoothbore gun with stowage for an additional seven rounds of ammunition when compared to the T-80BV. A built-in muzzle reference system is provided which enables the crew to boresight the gun without leaving the vehicle. The 125 mm gun can fire the 9M119 Refleks (US/NATO designation AT-11 'Sniper') semi-automatic laser beam-riding missile out to a maximum range of 5,000 m.

The fire-control system is designated the Type 1A45 with the stabiliser being the 2Eh42 and the ballistic computer the 1V528. The gunner has a stabilised Irtysh (IG46) day sight stabilised in two planes with a magnification of ×2.7 and ×12 which also includes a laser range-finder and missile guidance channel.

The gunner also has a roof-mounted Buran-PA thermal sight stabilised in two planes with a separate eyepiece, with the tank commander being provided with a monitor so that he can see the thermal image of the gunner's sight.

The tank commander has a roof-mounted, stabilised in vertical plane PNK-4S day/night sight which incorporates the TKN-4S night vision device. The sight has a day magnification of ×5.1 and a night magnification of ×7.5. The commander also has five day periscopes/vision blocks.

All of the T-80U and T-80UM series have the 12.7 mm NSVT machine gun installed on a pintle mount, the T-80UD had the same weapon in the commander's cupola that allowed him to aim and fire the weapon under complete armour protection.

The original production T-80U was powered by the GTD-1000TF gas-turbine engine developing 1,100 hp, although this was replaced on later production vehicles by the GTD-1250 gas turbine developing 1,250 hp. In addition, a GTD-18A gas-turbine auxiliary power unit is mounted at the rear of the hull on the left side and this allows the vehicle to run key subsystems without the main turbine engine running.

The driver's seat is now attached to the roof of the tank rather than the floor and to the left of the driver's seat are pillars. Both these features strengthen the hull in the event of the vehicle running over a mine.

The Kontakt explosive reactive armour of the T-80U provides protection against both APFSDS and HEAT attack over the frontal arc.

The turret roof between the commander's and gunner's hatches has been provided with additional protection against top attack weapons and a collar of rubber skirts hangs from the turret front which is believed to reduce the signature of the T-80U and deflect top attack bomblets.

The upper part of the suspension is covered by side skirts with the forward part either side of the driver's position providing additional armour protection.

The capacity of the internal fuel tanks is 1,090 litres with an additional 680 litres being stowed in five fuel tanks above the tracks. The vehicle can also be fitted with three auxiliary drum-type tanks at the rear of the hull carrying another 200 litres each, bringing the total capacity to 2,370 litres.

Like all members of the T-80 MBT family, the T-80U is fitted with an NBC system, a fire detection and suppression system, smoke grenade dischargers, the ability to lay a smoke screen by injecting fuel into the exhaust, a front-mounted dozer blade, an unditching beam at the rear, internal and external communications equipment and the ability to be fitted with mineclearing equipment, for example the front-mounted KMT-6.

For the export market an air conditioning system is offered as an optional extra.

The T-80U series MBTs produced from 1990 were fitted with a more powerful gas turbine engine, Brod-1 deep fording kit and the GTA-18 auxiliary power unit.

T-80U-1

This is understood to have been a preproduction model of the T-80U and was still equipped with the 9K112 Kobra radio guided ATGW system.

T-80UD (1988)

This was developed at the Malayshev Plant in Kharkov and has the turbine replaced by the 6TF two-stroke diesel developing 1,100 hp, a different transmission and driver controls and built-in second-generation explosive reactive armour. The D in the designation

Artist's impression of T-80U command staff vehicle **2000**/0069535

Artist's impression of T-80U armoured transporting and loading vehicle **2000**/0069536

T-80UM1 Bars (Snow Leopard) fitted with Arena active defensive aids system (Steven Zaloga) **2000**/0069538

T-80UM2 fitted with Drozd active defensive aids system (Steven Zaloga) **2000**/0069539

stands for diesel and this version retains the ability to fire the AT-11 laser-guided missile. This version had the development designation of Obiekt 478. It is estimated that only about 200 of these were delivered to the Russian Army with 300 still in the Ukraine when the USSR broke up. Production of the T-80UD continued in the Ukraine for Pakistan.

T-80UK Command Tank

This is the command version of the T-80U MBT and has the following additional features:

(1) Communications equipment consisting of UHF station, UHF receiver, HF station, UHF and HF aerials and an 11 m telescopic mast, the latter increases the range of the tank when it is deployed in the static role. Range is up to 40 km for the R-163-50U radio and 350 km for the R-163-50K radio. An AB-1-P28 1 kW generator is provided to power the communications equipment when the tank is stationary
(2) Navigation equipment consisting of TNA 4-3 position indicator, plotting board, gyro course indicator, control panel and aiming circle
(3) 1 kW auxiliary generating unit
(4) Installation of TShU1-7 Shtora-1 countermeasures system which is covered in the entry for the T-90 MBT
(5) Internally the T-80UK also features a modernised fire-control system, aiming sight, ballistic computer, commander's hatch control unit, loading mechanism and banks of smoke grenade launchers either side of the turret.

Because of the additional amount of communications equipment carried, this version only carries 30 rounds of 125 mm ammunition. A total of 750 rounds of 7.62 mm and 500 rounds of 12.7 mm machine gun ammunition is also carried.

T-80UE tank

This is a T-80UK with some of the command elements removed while retaining the TshU1-7 Shtora-1 counter measures system.

T-80UM

This is powered by the 1,250 hp gas turbine and has a number of other improvements including the Brod-M deep fording kit and Agava M1 computerised fire-control system which incorporates a thermal imaging sight for the gunner, with the tank commander being provided with a screen to see the gunner's target image. The new gunner's sight is the Agava-M1 and the weapon system can now use the latest 9M119M guided missile.

Some T-80UM vehicles have the 12.7 mm NSVT machine gun removed from the commander's cupola and repositioned on a pintle-type mount on the roof to the left of the commander's hatch. The US Army calls this the T-80UM Model 1993.

T-80UM1

This is also referred to as the T-80UM1 Bars (Snow Leopard) and is fitted with the more advanced Arena (or Kazt) Defensive Aid Suite (DAS).

The Arena DAS includes a mast-mounted multidirectional millimetre-wave radar system mounted on the roof of the turret towards the rear which senses incoming missiles.

This information is passed to the onboard computer which then activates one of the explosive panels mounted around the lower part of the turret so knocking out the approaching missile or rocket. The munitions are arranged in an arc around the front, sides and towards the rear of the turret.

Arena has been designed to increase the battlefield survivability of MBTs against attack from missiles and rockets fitted with a HEAT warhead and is claimed to have a reaction time of around 0.05 seconds.

Specifications

Model	**T-80B**	**T-80U**
Crew	3	3
Combat weight	42,500 kg	46,000 kg
Power-to-weight ratio	25.90 hp/t	27.20 hp/t
Ground pressure	0.86 kg/cm²	0.92 kg/cm²
Length		
(gun forward)	9.9 m	9.656 m
(hull)	7.4 m	7.00 m
Width	3.4 m	3.589 m
Height		
(without AA MG)	2.202 m	2.202 m
Ground clearance	0.38 m	0.446 m
Length of track on ground	4.4 m	4.3 m
Track width	580 mm	580 mm
Max road speed	70 km/h	70 km/h
Cross-country speed	48 km/h	48 km/h
Road range		
(without long-range fuel tanks)	335 km	335 km
(with long-range fuel tanks)	440 km	440 km
Fuel capacity		
(internal)	1,100 litres	1,090 litres
(external)	740 litres	680 litres
Fording		
(without preparation)	1.8 m	1.8 m
(with preparation)	5 m	5 m
Gradient	63%	63%
Side slope	46%	46%
Vertical obstacle	1 m	1 m
Trench	2.85 m	2.85 m
Engine	1,000 hp SG-1000 turbine	1,250 hp GTD-1250 turbine
Transmission	manual, 5 forward, 1 reverse gears	manual, 4 forward, 1 reverse gears
Steering	clutch and brake	clutch and brake
Suspension	torsion bar	torsion bar
Electrical system	27 V	27 V
Armament		
(main)	1 × 125 mm smoothbore gun/missile launcher	1 × 125 mm smoothbore 2A46M-1 gun/missile launcher
(coaxial)	1 × 7.62 mm PKT MG	1 × 7.62 mm PKT MG
(anti-aircraft)	1 × 12.7 mm NSVT MG	1 × 12.7 mm NSVT MG
Smoke grenade dischargers	8-12 × 81 mm	8 × 81 mm
Ammunition		
(main)	36	45 (28 in automatic loader)
(coaxial)	1,250	1,250
(anti-aircraft)	500	500
ATGW	4	6
Gun control equipment		
Turret power control	electric/manual	electric/manual
Turret traverse	360°	360°
Gun elevation/depression	+14/–5°	+14/–5°
Gun stabiliser		
(vertical)	yes	yes
(horizontal)	yes	yes
Range-finder	laser	laser
NBC system	yes	yes
Night vision equipment	yes	yes

T-80UK MBT fitted with explosive reactive armour and TShU-1-7 Shtora-1 counter measures system (Steven Zaloga) **2000**/0069540

BREM-80U ARV in travelling configuration (Steven Zaloga) **2002**/0069308

The system is fully automatic and is said to provide a high degree of protection through about 300° with a dead area to the turret rear.

The T-80UM1 is also fitted with a GTD-1250-G engine, air conditioning system, roof mounted DV-EBS wind sensor, Buran-M stabilised gunner's day/night sight, RPZ-86M anti-radar coating, increased power stabilisation system for the 125 mm 2A46M-4 main armament, KAKTUS explosive reactive armour system, Tuman rapid action fire detection and suppression system, anti-fragmentation screens made of Kevlar type material, NSTV 12.7 mm remote-controlled machine gun, Ainet automatic fuze setting system for 125 mm high-explosive fragmentation ammunition and the 45M fire-control system which includes a day/thermal sighting system of Russian or foreign origin.

The fire-control system is designated the 1A45 and includes a laser range-finder, wind sensor, tank speed indicator, target speed indicator, roll angle sensor, ammunition and surrounding temperature sensors and a ballistic computer. The commander can also lay and fire the 125 mm main armament.

A GTA-18 gas turbine is installed which allows the tank to run all of its systems with the main engine shut down.

T-80UM2

This designation was originally applied to a T-80 chassis fitted with a new turret armed with a 125 mm gun fed by a bustle-mounted automatic loader (*Jane's Armour and Artillery* 1999-2000).

More recently this designation has been applied to the T-80U fitted with the Drozd-1 (Thrush) defensive aids suite.

This was first installed on T-55 series MBTs of the Russian naval infantry and is covered in this entry.

The first-generation Drozd-1 only covered the frontal arc but the latest Drozd-2 uses smaller munitions, improved sensors and can cover a full 360° arc.

T-84

Further development of the T-80 in Ukraine has resulted in the T-84 which was shown for the first time in the Middle East in March 1995. Details of this are given under Ukraine. The T-84 is now in low-rate production and is in service with the Ukraine.

BREM-80U armoured recovery vehicle

To support the T-80U MBT, the Transport Machine Building Design Bureau has developed the BREM-80U ARV with the prototype being built by the Plant of Transport Machine Building.

The baseline T-80 turbine powered chassis is retained in the BREM-80U ARV but a new all-welded steel armoured superstructure has been added at the front for the crew and the winch equipment. This provides protection from small arms fire and shell splinters.

The main winch has a capacity of 35 tonnes but using snatch blocks this can be increased to 140 tonnes. An auxiliary winch is also provided.

The crew of four consists of the commander, mechanic/driver, fitter and wedger/rigger, a fifth seat is provided for an additional crew member.

Mounted at the front of the hull is a hydraulically operated blade. This can be used as a dozer blade or an anchor blade when the winch or crane is being used.

The hydraulically operated crane is pivoted at the left side of the hull and folds back along the right side of the hull when not required. The jib crane is of the telescopic type and has a maximum lifting capacity of 18 tonnes.

Armament consists of a 12.7 mm machine gun with a bank of eight 81 mm electrically operated smoke grenade launchers being mounted at the front of the vehicle on the left side.

Command Staff Vehicle

This proposed vehicle is based on a T-80U chassis and is designed to provide control and communication functions within motorised rifle and tank divisions as well as at brigade, regiment and battalion level.

The chassis is very similar to that used for the BREM-80U ARV and features a new raised superstructure towards the front which provides space for the commander and driver plus two to four staff members. Additional communications equipment is fitted.

Armament consists of a roof mounted 12.7 mm machine gun while mounted either side are banks of 81 mm electrically operated smoke grenade launchers firing forwards. An NBC system is fitted as is a front mounted dozer blade.

Armoured Transporting Loading Vehicle

This has a similar chassis to the command staff vehicle and has been designed to resupply tanks with 125 mm ammunition (projectile and charge) with a total of 135 rounds being carried. The MBTs would draw up alongside of the armoured transporting and loading vehicle with the ammunition being transferred to the MBTs via a chute.

It has a combat weight of 46 tonnes, a crew of two and a maximum road speed of 70 km/h. It is fitted with an NBC system and a front mounted dozer blade.

152 mm 2S19 self-propelled gun

The 152 mm 2S19 self-propelled gun, fully covered in the Self-propelled Guns and Howitzers (Tracked) sections is based on a modified T-80 chassis but with a T-72 series diesel engine. As of early 2002 this had not been exported.

Status

Production. In service with the following countries:

Country	Quantity	Comment
Cyprus	41	delivered 1996-97
Pakistan	320	delivered from Ukraine between 1997 and 1999
Russia	3,500	estimate
South Korea	33	
Ukraine	273	

Manufacturers

Kirov Works, Leningrad (production completed 1990).
Kharkov, Ukraine (completed production of T-80UD in 1990 but production was subsequently started again for home market and Pakistan).
Omsk Machine Production Plant (still in low-rate production of T-80U).

UPDATED

T-72 MBT

Development

The Morozov Design Bureau, based at the Ural Tank Plant in Nizhnyi Tagil and headed by Valeriy Venediktov, developed the T-72 MBT as an alternative to the highly complicated and expensive T-64 MBT (qv).

Before the formal acceptance of the T-72 in 1973 a whole series of developmental vehicles was built and tested. This included the Obiekt 167 built in 1961 using a T-62 MBT hull with six roadwheels. Mounted on the turret rear was a three-round launcher for the AT-3 'Sagger' ATGW and the rear of the hull was also modified.

Obiekt 167GTD followed in 1965 and featured a gas-turbine engine. In 1968, the Obiekt 172 was built which featured a T-64-type hull fitted with a small turret with cross-turret optical range-finder, a searchlight mounted to the left of the gun and no commander's cupola.

The actual T-72 prototype was completed in 1970 and designated as Obiekt 172M. It was fitted with a V-46 diesel engine with supercharger developing 780 hp, a searchlight to the right of the 125 mm gun, a commander's roof-mounted machine gun, a cross-

MTU-90 three part bridge based on T-72 MBT in the travelling configuration 0011800

turret optical range-finder and gill- or flipper-type armoured panels covering the forward part of the hull sides. The armament and fire-control system were virtually identical to those of the T-64A.

In designing the T-72, the hull and turret layout of the T-64 MBT was utilised, together with some of the components from the latter. The engine is an improved version of that installed in the T-62 while the cooling system is similar to the one fitted in the T-55/T-62. The drive train is similar to that utilised in the T-64.

The T-72 is understood to have entered production in 1974 with the tank becoming fully operational by 1975. The T-72 was first seen in public in October 1977 when the French Minister of Defence, Yvon Bourges, paid an official visit to the Tamanskaya Guards Division. It made its first appearance in numbers during the parade held in Red Square, Moscow, in November 1977.

The T-72 was used in combat by the Syrian Army against the Israel Defence Force in fighting in Lebanon in 1982 and by the Iraqi Army during Operation Desert Storm in early 1991. Combat experience has shown that the 105 mm/M68 rifled tank gun, firing APFSDS projectiles, is capable of penetrating the early T-72 MBT over its frontal arc.

Late production vehicles have a higher level of armour protection. Those with the Kontakt-5 explosive reactive armour have a high level of battlefield survivability as this provides protection from both chemical and kinetic energy attack.

Since 1971, the T-72 and its numerous variants have been manufactured in substantial numbers for both the home and export markets. The price of the T-72S, for export, was US$1.2 million early in 1993. It is understood that there has been no recent production of the T-72 series MBT although the production line at Nizhnyi Tagil can be restarted and that quantities of the latest production T-72 are available for immediate delivery.

According to the United Nations, the quantities in the table of T-72 MBTs were exported by various countries from 1992 through to 2000.

From	To	Quantity	Date	Comment
Belarus	Hungary	100	1996	
Belarus	Iran	8	2000	
Belarus	Morocco	70	2000	
Belarus	Yemen	27	2000	
Czech Republic	Syria	81	1992	T-72M
Germany	Finland	97	1992	
Germany	Sweden	5	1992	
Germany	USA	27	1993	
Kazakstan	Belarus	53	2000	
Poland	Iran	34	1994	
Poland	Iran	70	1994	
Russia	Bulgaria	100	1999	
Russia	India	100	1992	
Russia	Iran	100	1993	
Russia	Iran	20	1994	
Russia	Iran	2	1996	
Russia	Iran	4	1998	
Slovakia	Syria	58	1993	
Turkmenistan	Russia	530	1996	
Ukraine	Algeria	27	1998	
Ukraine	Algeria	27	1999	
Ukraine	Sierra Leone	2	1994	

Description

(T-72)

The hull of the T-72 MBT is divided into three compartments: driver's at the front, fighting in the centre and the power pack at the rear. It is believed that the turret has conventional cast armour with a maximum thickness of 280 mm, the nose is about 80 mm thick and the glacis is of a new laminate armour 200 mm thick, which when inclined gives between 500 and 600 mm of protection. Late production T-72s do, however, incorporate advanced armour protection in their turrets.

The glacis is well sloped, transversely ribbed and has a deep V splashboard. The driver is seated at the front of the hull and is provided with a single-piece hatch cover that opens to the right, in front of which is a single wide-angle TVNE-4E day observation periscope. The driver's headlamps are designated FG 125.

The other two crew members are seated in the turret, with the gunner on the left and the commander on the right. The commander's contrarotating cupola has a single-piece hatch cover that opens forward with two rear-facing TNPA day vision blocks. In the forward part of the cupola is a combined TKN-3 day/night sight with an OU-3 infra-red searchlight mounted over the top and, to either side of the combined day/night sight, is another TNP-160 day periscope. Forward and slightly below the commander's cupola is an optical stadiametric range-finder.

The gunner's hatch opens forward and has a circular opening for mounting the snorkel for deep fording operations. In front of the gunner's hatch is a TNP-160 day periscope, while a TNPA-65 day vision block is fitted in the hatch cover itself. In front and to the left of the gunner's hatch is a panoramic day/night sight which is used in conjunction with the infra-red searchlight mounted to the left and in front of the sight. The gunner's sight is the TPD-2-49 while his night sight is the TPN-1-49-23.

According to former Soviet sources, the TPD-2-49 monocular eyepiece stereoscopic range-finder sight enables targets to be engaged out to a range of 4,000 m. It has a field of view stabiliser for its vertical axis. In addition the sight has a ballistic computer that performs calculations based on input data for firing a specific type of round, the temperature and density of the air and crosswind speed.

Using this sighting system, maximum effective target ranges are 4,000 m using APFSDS and HEAT rounds, 5,000 m using HE fragmentation rounds and 1,800 m using the 7.62 mm PKT coaxial machine gun.

Using the gunner's quadrant fitted as standard on the T-72, indirect firing can be accomplished out to 9,400 m.

The L-2 infra-red searchlight is mounted on the right side of the 125 mm main armament rather than the left as in the case of the earlier T-64, although it is occasionally mounted on the left side (initial production vehicles).

Two light steel stowage boxes are mounted on the turret, one at the rear and the other on the right slightly behind the commander's position. The snorkel is carried on the left side of the turret to the rear. Whereas the turret on the T-64 is roughly circular, that of the T-72 has a distinct bulge and external stowage on the two tanks is also different.

Fuel cells extend along the right side of the hull top; on the left hull top are stowage boxes and a single oil cell. There are four internal oil tanks in the T-72, one to the left of the driver, two to his right and one on the floor in the space between the rear of the ammunition carousel and the fire wall/engine bulkhead. An unditching beam is normally carried at the rear and there is also provision for carrying two 200 litre diesel fuel drums at the rear to increase operational range. These can be quickly jettisoned if required by the tactical situation.

The T-72 is powered by a V-12 piston V-46-6 multifuel air-cooled engine which develops 780 hp. The engine will run on three fuels, with the driver being provided with a dial to set the engine for the type of fuel being carried. The three settings are D for Diesel, B for Benzine and K for Kerosene. In the T-72 MBT the diesel engine is mounted transversely. Late production vehicles such as the T-72S have the more powerful V-84 MS diesel developing 840 hp.

The suspension each side consists of six roadwheels with the idler at the front, drive sprocket at the rear and three return rollers supporting the inside of the track only. Shock-absorbers are fitted at the first, second and sixth roadwheel stations. The track of the T-72 is of the single pin type with rubber bushes and this has been back-fitted to some older T-55/T-62 MBTs. Total roadwheel travel on the T-72A is 285 mm but this was increased to 325 mm on the later T-72B.

There are four removable spring-loaded skirt plates fitted over the forward part of the track, which are unclipped in action and spring forward at an angle of 60° from the side of the vehicle, to give a measure of protection against HEAT projectiles.

A dozer blade mounted under the nose of the tank is used for clearing obstacles and preparing fire positions and, like most other Russian tanks, the T-72 can be fitted with mineclearing equipment such as the KMT-5, KMT-6 and KMT-6 M2. The dozer blade can be brought into the operating position in 1 or 2 minutes and enables the T-72 tank to prepare its own defilade position without calling on engineer support. A typical position for a T-72 MBT would be 10 m long, 4.5 m wide and 1.2 m deep, with a total volume of 54 m³. The T-72 is provided with an NBC system and can be fitted with a snorkel for deep fording. The tank takes about 20 minutes to prepare for amphibious use and is ready for action within 2 minutes of leaving the water.

Main armament is a 125 mm (2A46) smoothbore gun fitted with a light-alloy thermal sleeve and a bore evacuator. The thermal sleeve has four removable

Close-up of turret of T-72B1 showing additional appliqué roof armour and top armour either side of driver's position. Note 81 mm smoke grenade launchers on left side of turret

Close-up of T-72M1 turret showing bank of 81 mm smoke grenade launchers (Christopher F Foss) **2000**/0069543

Close up of T-72B turret showing boxes of explosive reactive armour fitted to forward sides and turret roof for added protection (Richard Stickland) ***2000***/0069544

T-72B tank of the Russian Army fitted with explosive reactive armour to its hull and turret (Stefan Marx) 0018816

sections, two in front and two to the rear of the bore evacuator. The 125 mm gun is stabilised in both planes. The gun stabilisation system is designated the 2Eh28M. The 125 mm gun fires three main types of separate loading ammunition, APFSDS with a maximum range of 2,100 m, HEAT-FS with a maximum direct fire range of 4,000 m and HE-FRAG(FS) with a maximum indirect fire range of 9,400 m.

The 125 mm ammunition is of the separate loading type, that is, the projectile is loaded first followed by the semi-combustible cartridge case, then all that remains after firing is the stub base which is ejected. This 125 mm ammunition is common to the T-64, T-72, T-80, T-84 (Ukraine) and the T-90. Details of this ammunition are given in the entry for T-80 MBT earlier in this section.

Of 39 rounds of ammunition carried, 12 are APFSDS-T, 21 HE-FRAG(FS) and the remaining six HEAT-FS. The additional rounds of ammunition are stowed in racks behind the turret basket and in indentations in the rear floor fuel cell and second forward right cell near the driver. The carousel automatic loader is mounted on the turret floor and also on the rear wall of the turret. The projectile is loaded in the lower half of a carrier, the cartridge and propellant in the upper half. The carousel carries 24 ready use projectiles. When the 125 mm gun loads it must pick up the carrier and ram both the projectile and powder charge. This enables a rate of fire of 8 rds/min to be achieved but there are some doubts as to the reliability of the automatic loader. The main armament is stabilised and enables the T-72 to shoot on the move with a high probability of a first-round hit.

A 7.62 mm PKT machine gun is mounted coaxially to the right of the main armament and has 250 rounds of ready use ammunition and a new design 12.7 mm NSV machine gun is mounted on the commander's cupola. The 12.7 mm NSV machine gun can, however, only be used with the commander exposing the upper part of his body. Maximum sight range in the ground role is 2,000 m; maximum sight range in the anti-aircraft role is 1,500 m.

T-72s built for export have a slightly different fire-control system and automatic loader and do not have the internal lining that is standard on Russian T-72s, consisting of a layer of synthetic material that contains lead and provides some degree of protection against the effects of neutron radiation and electromagnetic pulses.

Variants

Since the T-72 MBT entered production in 1971 a large number of variants have been produced both for the home and export markets and many earlier vehicles have been subsequently brought up to more recent production standards.

In recent years there has been considerable confusion concerning the exact designation of members of the T-72 MBT family. The designations used within the RFAS Ground Forces have now been declassified and are listed in the following text.

The date after the designation is the year that it was accepted for service with the Russian Army.

T-72 (1973)

This was the first model to enter service and has the infra-red searchlight mounted on the left of the 125 mm gun, commander's 12.7 mm anti-aircraft machine gun, TPD-2-49 optical cross-turret range-finder and gill or flipper-type armoured panels either side of the forward part of the hull. It has four evenly spaced ribs on the upper glacis plate.

The preseries T-72 (Obiekt 172-2M) also has no thermal sleeve for the 125 mm gun and has a single radio aerial on the forward part of the turret roof.

The standard production T-72 (Obiekt 172M) had the Luna L2AG infra-red searchlight mounted on the right side of the turret, thermal sleeve for the 125 mm gun and single radio aerial on the turret roof towards the rear. Some vehicles were subsequently fitted with additional armour on the turret roof and had the TPD-K1 laser range-finder installed.

T-72K (1973)

Commander's version of T-72 with additional communications equipment. Company commander's vehicles have two R-123M or R-173 radios while battalion and regiment vehicles have one R-123M/

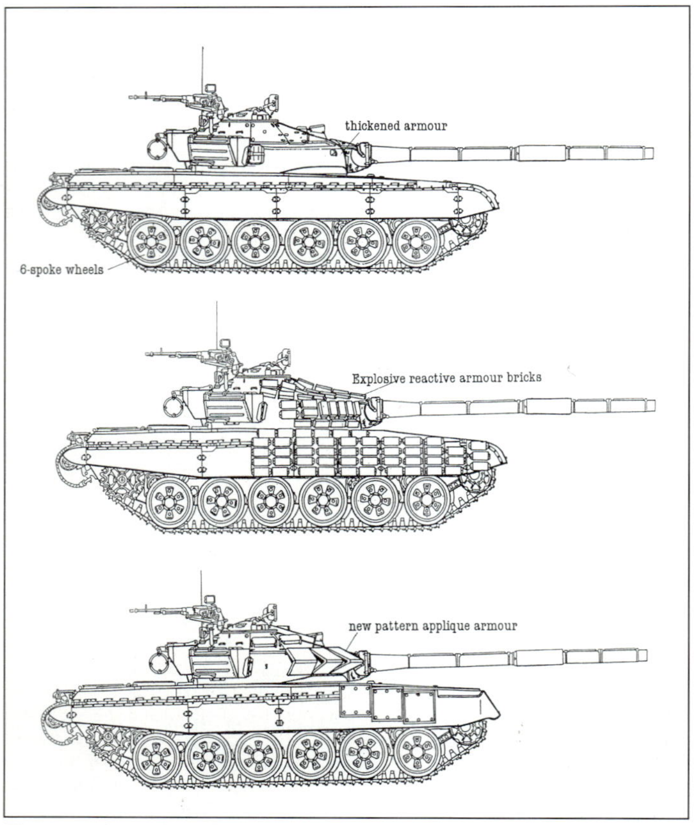

Three versions of the T-72 MBT, from top to bottom, T-72B1, T-72B (with Kontakt ERA) and T-72BM (Steven Zaloga)

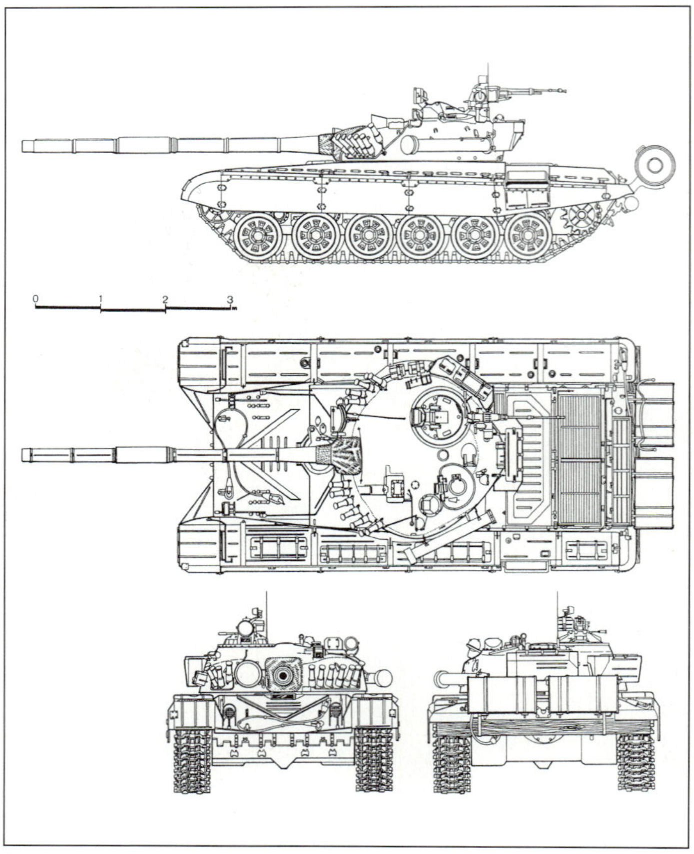

T-72A (Export T-72M1) MBT (Steven Zaloga)

R-173 and one R-130M which uses a 10 m antenna when the vehicle is static.

T-72 (1975)
This was the first export version and differed from Soviet vehicles in having a different armour package over the frontal arc of the turret, as well as differences to the PAZ collective NBC system and the amount of ammunition carried. This version is sometimes referred to as the T-72 Model 1975.

T-72A (1979)
This was accepted for service in June 1979 and is a further development of the T-72, with the optical rangefinder replaced by the TPD-K1 laser range-finder sight for improved first round hit probability, TPN-3-49 gunner's night sight, TPN-3 searchlight, side skirts, 125 mm 2A46 gun (earlier versions had 2A26M), Type 902B 81 mm smoke grenade launcher system, napalm protection system, turn signals, TVNE-4B driver night observation device, upgraded suspension for improved cross-country mobility and installation of a V-46-6 diesel engine. It also has different plastic side skirts covering the upper part of the suspension and separate panels protecting the sides of the fuel and stowage panniers. Three evenly spaced ribs are on the upper glacis plate. This version also had a significant increase in armour protection, especially over the frontal arc of the turret.

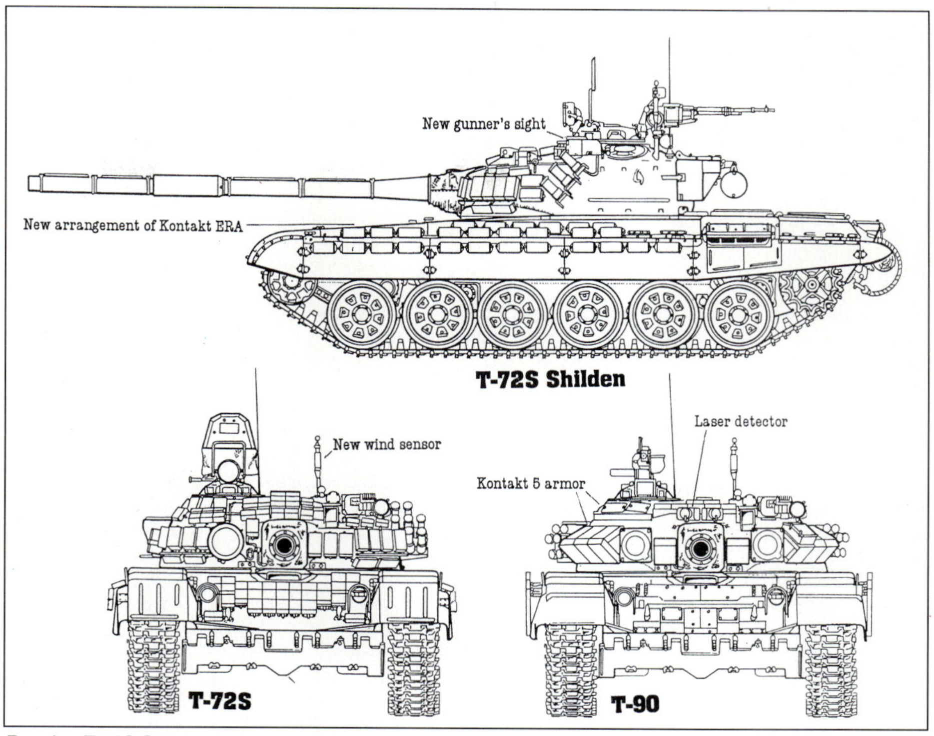

Russian T-72S Shilden is the latest export version of the T-72 series MBT (Steven Zaloga) 0002540

T-72AK (1975)
This is the commander's model of the T-72A with additional communications equipment and a total of two antennas.

T-72M (1980)
This is the export model of the T-72A and has a different armour protection level, carries a different amount of ammunition and has a different type of NBC system fitted. There are four evenly spaced ribs on the upper glacis plate.

T-72M1 (1982)
This is a modernised T-72M and has an additional layer of 16 mm armour plate on the glacis plate and combination armour in the turret with pelletised filler agent. There are two ribs on the upper glacis plate.

T-72AV (1985)
T-72A fitted with an explosive reactive armour array consisting of 227 boxes.

T-72B (1985)
This version was developed under the designation of the Obiekt 182 and later the Obiekt 184 and was first seen in 1986. Its turret has a new armour package that provides a much higher degree of protection than any previous model of the T-72 MBT. In addition the glacis plate is provided with 20 mm of appliqué armour.

Late production vehicles also have an anti-radiation layer on the hull roof which is also fitted to some T-72A vehicles.

Mounted either side of the turret is a bank of 81 mm forward-firing electrically operated smoke dischargers.

The rear of the vehicle has been redesigned as a new fan cooling system has been installed.

Main armament is the 125 mm 2A46M gun which can be removed from the vehicle without removing the turret. This can also fire the 9M119 AT-11 'Svir' laser beam-riding missile out to a range of 5,000 m. A muzzle reference system is fitted as standard. A total of 45 rounds of 125 mm ammunition is carried, 22 in the carousel automatic loader and 23 stowed in the turret and hull.

The fire-control system consists of the 1A40-1 sight complex which is based on the TPD-K1 laser range-finder and sight used in the T-72A with the field of view being stabilised in the vertical plane. Firing at night is achieved with the 1K13-49 sight, which is part of the 9K120 guided weapons complex, and can be used in the active or passive modes.

The 125 mm main armament is stabilised by the 2Eh42-2 system with hydroelectric drive in the vertical plane and electric drive in the horizontal plane. The T-72B is powered by the 840 hp V-84-1 diesel engine.

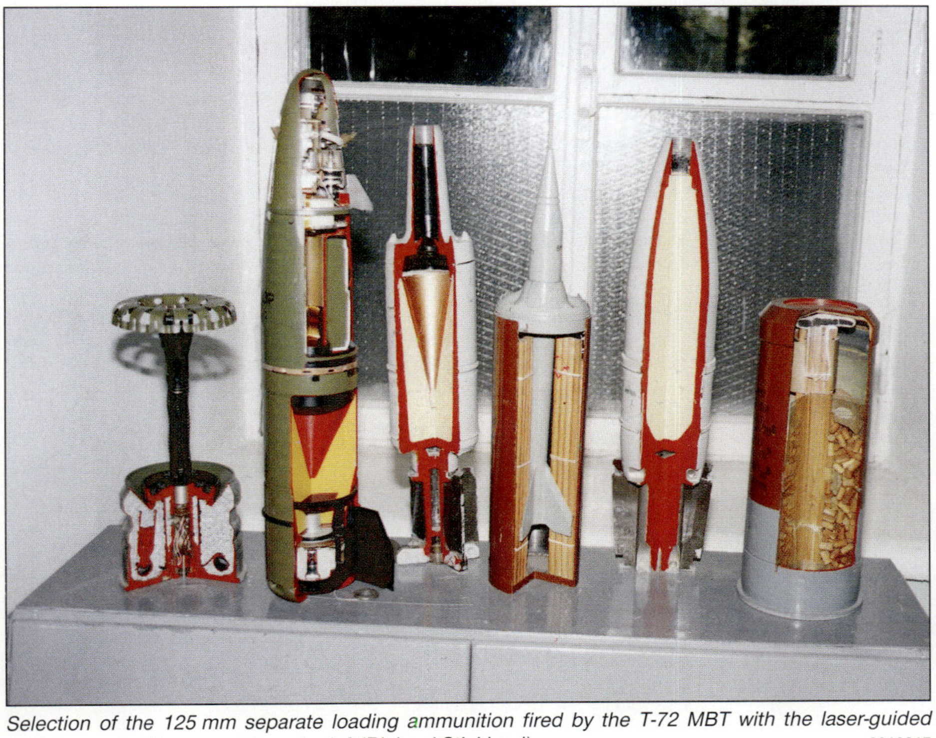

Selection of the 125 mm separate loading ammunition fired by the T-72 MBT with the laser-guided 125 mm projectile second from the left (Richard Stickland) 0018817

T-72BK (1985)
Commander's model of the T-72B with additional communications equipment fitted and two antennas.

T-72B1 (1985)
T-72 MBT without the capability to fire the AT-11 'Svir' laser-guided projectile. The T-72B and T-72B1 are sometimes fitted with the Kontakt explosive reactive armour.

T-72S (1987)
This is the export version of the T-72B and was initially referred to as the T-72M1M. The ERA package consists of 155 boxes, with the hull and turret being identical to those of the T-72M1. There is a different selection of ammunition options for the 125 mm gun and it can fire the AT-11 'Svir' laser-guided missile. This is also referred to as the T-72S 'Shilden' rocket tank for the export market. The T-72S is now being manufactured under licence in Iran.

T-72S1 (1987)
Export version of the T-72B1 with 155 boxes of ERA, the hull and turret are identical to the T-72M1's. There is a different selection of ammunition options for the 125 mm gun.

T-72BM (1992)
Improved T-72B with built-in second-generation explosive reactive armour that provides protection against APFSDS and HEAT attack. The explosive reactive armour fitted to the T-72M is the Kontakt-5. It is powered by an SV-84 diesel engine developing 840 hp.

Export designations
Russia has recently released details of another batch of designations that apply to export models of the T-72 series MBT.

These are T-72 (Eh and Eh-1 export models), T-72M (export models Eh-2, Eh-3 and Eh-4) which corresponded to the domestic T-72A but with different turret armour, ammunition and collective protection system.

T-72M1 (export models Eh-5 and Eh-6) which again corresponds to the domestic T-72A but with different armour in hull and turret. Finally, there is the T-72S which differs from the T-72B in the areas of armour, NBC system and types of ammunition used.

T-72M1 Upgrade Package
Nizhnyi Tagil, prime contractor for the T-72, is now offering an upgrade package for the T-72M1 which is claimed to improve the characteristics of the vehicle by 1.7 times. Many of these improvements are also fitted to the T-72S and T-90 MBTs. Key improvements have been summarised as:

(1) New 125 mm smoothbore gun with bore sighting system

Specifications
T-72S
Crew: 3
Combat weight: 46,500 kg
Power-to-weight ratio: 18.06 hp/t
Ground pressure: 0.90 kg/cm²
Length:
(gun forward) 9.533 m
(hull) 6.95 m
Width:
(without skirts) 3.37 m
(over skirts) 3.59 m
Height: (without AA MG): 2.228 m
Ground clearance: 0.49 m
Track width: 580 mm
Length of track on ground: 4.278 m
Max road speed: 60 km/h
Range:
(road, without long-range fuel tanks) 480 km
(road, with long-range fuel tanks) 550 km
Fuel capacity: 1,000 litres
Fording:
(without preparation) 1.8 m
(with preparation) 5.0 m
Gradient: 60%
Side slope: 40%
Vertical obstacle: 0.85 m
Trench: 2.8 m
Engine: V-12 multifuel (V-84) 840 hp at 2,000 rpm
Transmission: synchromesh, hydraulically assisted with 7 forward and 1 reverse gears
Steering: clutch and brake
Suspension: torsion bar
Electrical system: 24 V
Armament:
(main) 1 × 125 mm 2A46 smoothbore gun
(coaxial) 1 × 7.62 mm PKT MG
(anti-aircraft) 1 × 12.7 mm NSVT MG
Ammunition:
(main) 45 (incl 6 ATGW)
(coaxial) 2,000
(anti-aircraft) 300
Smoke grenade launchers: 8 × 81 mm
Gun control equipment
Turret power control: electric/manual
Turret traverse: 360°
Gun elevation/depression: +14/-6°
Gun stabiliser:
(vertical) yes
(horizontal) yes
NBC system: yes
Night vision equipment: yes

IMR-2 Combat Engineer Vehicle in action (Jane's Intelligence Review)

(2) New fire-control system
(3) Thermal night sight
(4) Stabilisation system for main armament
(5) Installation of laser-guided missile system
(6) 12.7 mm anti-aircraft machine gun that can be aimed and fired under complete armour protection
(7) Installation of explosive reactive armour to hull and turret
(8) Installation of Shtora TShU1-7 jamming system
(9) Improved NBC system
(10) Improved fire detection and suppression system
(11) New communications system
(12) More powerful 840 hp engine
(13) New suspension

T-72 series Upgrade Package

In addition to the comphrensive upgrade described above, Russia is also offering an upgrade package for the T-72K, T-72M, T-72MK, T-72M1 and T-72M1K MBTs in the following key areas:

(1) Improved armour protection by the installation of explosive reactive armour
(2) Special paint and screens to make it more difficult to detect in the visible, near and far infra-red bands
(3) Improved interior protection from radiation
(4) Installation of new fire detection and suppression system
(5) Upgraded suspension system.

Latest Russian T-72 MBT upgrade

In late 1998, Russia formed a new consortium of the leading Russian MBT contractors and sub-contractors to carry out upgrades to the T-72 series MBT.

This consortium includes the Steel Research Institute (Moscow), Heavy Machine Building Research Institute (St Petersburg), Heavy Machine Building Research Bureau (Nizhniy Tagil), Motvilikhskije Factories (Perm), Industrial Union UVZ (Nizhniy Tagil), Machine Building Union Belomo (Minsk) and Barnaultransmach (Barnaul).

These are all key specialists in their respective fields with the Nizhniy Tagil being MBT specialists, Perm being involved in gun design and production and Barnaul with engine design and production.

In December 1998, Russia presented its proposals to NATO in Brussels to upgrade T-72 MBTs used by members of the former Warsaw Pact to enable them to remain effective well into the 21st century. According to Russian sources their modernisation proposal would increase combat effectiveness by between 50 and 70 per cent.

As with most proposed MBT upgrades, the Russian T-72 upgrade covers three key areas of the MBT, armour, mobility and firepower.

The baseline T-72 MBT is very well protected over its frontal arc by various combinations of cast steel armour and laminate armour. Its survivability in the upgrade is further enhanced by the installation of the latest generation explosive reactive armour which gives protection against chemical energy and kinetic energy attack.

A defensive aids suite is also provided to reduce the risk of the upgraded T-72 being attacked by anti-tank guided missiles.

The battlefield survivability of the MBT has also been significantly increased by other means including the installation of a high speed fire detection and suppression system, special paint to reduce the thermal signature of the tank, providing additional protection for the ammunition and filling the fuel tanks with special material.

It is proposed that the current diesel engine be replaced by a KD-34 or B-92C2 (or V-92S2 as it is referred to by Russia) diesel engine developing 1,000 hp or a more powerful 1,200 hp diesel engine with hydromechanical transmission.

The installation of a more powerful engine is of key importance as the additional armour fitted increases the overall weight of the T-72 and therefore lowers its power-to-weight ratio.

Cross-country mobility is improved by the installation of new hydraulic shock absorbers and giving a higher road wheel travel of 320 mm

It is also proposed that a recently developed automated control system for the engine and transmission be installed. To allow the main engine to be shut down an auxiliary power unit (APU) with an output of 19 kW is fitted as well as an air conditioning system.

As far as it is known the existing 125 mm smoothbore gun is retained which fires two part ammunition, projectile and charge, which is loaded using an automatic loader.

The latest version of the 125 mm smoothbore gun is claimed to be 15 per cent more accurate and have a similar increase in range. It is also easier to replace under battlefield conditions.

As well as firing standard natures of 125 mm ammunition such as Armour-Piercing Fin Stabilised Discarding Sabot - Tracer (APFSDS-T) and high-explosive fragmentation, it can also fire a laser-guided High Explosive Anti-Tank (HEAT) missile out to a maximum range of 5,000 m. The latest version has a tandem HEAT warhead to defeat targets with explosive reactive armour.

A new computerised fire-control system is installed which includes a ballistic computer and various sensors, this gives a higher first round hit probability and either the gunner or the tank commander can aim and fire the main armament.

The installation of an electrohydraulic gun stabiliser is said to increase the accuracy of the gun by 50 per cent.

New optics include a new thermal sight which will identify targets at a range of 2,300 to 2,800 m depending on ambient conditions.

Russia is also proposing joint development of other systems which could improve the combat effectiveness of the T-72 even further. These include a computerised fire-control system, data management system, new communications, navigation system as well as a diagnostics capability.

Upgraded T-72M1 MBT

In 2001, the Russian United Enterprise Uralvagnozavod revealed that it had completed the prototype of an upgraded version of the T-72M1 MBT for the export market. The T-72M1 is the export version of the Russian Army T-72A MBT.

Some of the subsystem components of the upgraded T-72 MBT are also used in the more recent T-90 MBT, which is also produced at Uralvagnozavod.

The standard T-72M1 weighs around 43 tonnes and is powered by the V-46-6 diesel developing 780 hp giving a power to weight ratio of 18.1 hp/tonne. This gives a maximum road speed of 60 km/h, with speed on a dirt road being 35 to 40 km/h.

MTU-72 AVLB in travelling configuration

BREM-1 ARV in travelling configuration (Steven Zaloga) ***2000***/0069545

Well camouflaged T-72M1 MBT of the Polish Army (Richard Stickland) ***2000***/0069537

T-72 series MBT with turret traversed to rear, side skirts removed and long-range fuel tanks installed at the rear (C R Zwart) ***2001***/0105981

The upgraded T-72M1 MBT weighs 45 tonnes and is currently offered with a choice of two different diesel engines. The first is a V-92S2 diesel developing 1,000 hp which gives a power-to-weight ratio of 22.2 hp/tonne.

The second is the V-84MS which develops 840 hp and gives a power-to-weight ratio of 18.6 hp/tonne. The installation of the more powerful engine is claimed to increase its road speed to 65 km/h and its dirt road speed to 40 to 45 km/h.

For improved firepower the standard 125 mm 2A46M smoothbore tank gun has been replaced by the more recent 125 mm 2A46M. The improved fire-control system now allows the upgraded T-72M1 to fire the 9M119 series laser-guided missile out to a maximum range of 5,000 m. The latest versions of this missile have a tandem HEAT warhead to defeat MBTs fitted with explosive reactive armour.

First-round-hit probability is further enhanced by the upgraded fire-control system and the installation of the French Thales Optronique Catherine thermal imager which has a claimed detection range of 3,000 to 3,500 m.

The gunner has a two-axis stabilised sight that combines day/thermal and laser guidance channels, in addition he has a TPD-K1 standby laser sight as a backup. The tank commander has a PNK-4M day/night sight that is stabilised in the vertical plane only.

According to Russian sources, the daytime identification range for the tank commander's PNK-4M sight is 4,000 m, while under night conditions this is reduced to 1,200 m. The standard production T-72M1 is equipped with a TKN-3 day/night sight which has a daytime range of 2,000 m and a night-time detection range of only 300 m. In addition, the upgraded T-72M1 is fitted with an automatic target tracking device.

For improved battlefield survivability, the upgraded T-72M1 MBT is fitted with part of the TShU1-7 Shtora active jamming device which is already fitted to a number of other Russian MBTs, including the T-80 and T-90.

To reduce costs, the baseline upgraded T-72M1 MBT is only fitted with the SPZ laser jamming device, but if required by the customer, it can be fitted with the complete system. This includes the SOEP optical jamming system and the 81 mm smoke grenade launching system that can jam ground and air launched laser anti-tank guided missiles.

The standard Russian GPK-59 navigation system has been replaced by a Global Positioning System (GPS) and the standard Russian R-123 radio system has been replaced by the more recent R-163-50U radio set and R-163UP receiver.

French power pack for Russian T-72 MBT

The French companies, Cummins Wartsila Diesel and SESM, are now developing a new power pack as a private venture for the widely deployed Russian T-72 MBT.

This power pack consists of the WARTSILA SACM V8X 1000 diesel engine coupled to an SESM ESM 350 fully automatic powershift transmission and a new cooling system.

For this T-72 application the engine is mounted in a transverse configuration (also referred to as U-type) with the cooling system on the right.

The power pack is installed into the T-72 MBT hull on a three point mounting without any extensive modifications to the actual hull.

It has been designed to be quickly removed for ease of maintenance or replacement and a complete power pack can be exchanged in under 60 minutes.

The V8X 1000 diesel engine is a member of the X-family and has been specifically designed for the T-72 application. It is fitted with two turbochargers and develops 1,000 hp at 2,500 rpm with a maximum torque of 3,500 mN at 1,800 rpm.

The V8X 1000 is basically the same engine as the V8X 1500 currently in full-scale production for the Giat Industries Leclerc MBT for the French Army (those built for the United Arab Emirates have a German 1,500 hp MTU EuroPowerPack) but fitted with conventional turbochargers. The original V8X 1500 has a self-sustained supercharging system.

In the standard T-72 the driver steers the vehicle using tillers but in this upgrade he is provided with a steering wheel as in the latest MBTs such as the Giat Industries Leclerc. This makes the vehicle much easier to handle as well as reducing driver fatigue.

The standard Russian T-72 is powered by a V-12 diesel engine developing 740 hp so the new French power pack offers a considerable increase in power-to-weight ratio.

It is expected that a first maintenance inspection would be required at 400 hours with mean time between overhaul of 1,000 to 1,200 hours.

Booklet gunners sight

The Joint Stock Company Pelang has developed the Booklet multichannel gunners sight for installation in the T-72 series MBT.

This comprises a day TV channel, thermal imaging channel, missile guidance channel and a laser range-finder channel.

KBP Upgrade for T-72 MBT

The Russian KBP Instrument Design Bureau, prime contractor for the gun launched laser-guided missiles (100, 115 and 125 mm), are now marketing new missiles which have a maximum range of 5,000 m as well as a tandem HEAT warhead to defeat targets fitted with explosive reactive armour.

This latest version uses a new sighting and guidance unit which was originally developed for the Russian BMP-3 infantry fighting vehicle which includes a two-axis line of sight stabiliser, optical channel, image intensification night channel and a laser guidance channel. A further enhancement is the replacement of the image intensification channel by a thermal channel and an autotracker.

Czech Republic T-72 CZ Upgrade

Details of this upgrade are given under the entry for this vehicle in the Czech Republic. As of 2002 this remained at the prototype stage.

T-72 MP Upgrade

Details of this upgrade were given in *Jane's Armour and Artillery 2000-2001*. It never passed the prototype stage.

T-72AG Upgrade

This upgrade package has been developed in the Ukraine and details are given later in this section under Ukraine. The Ukraine has also developed a 120 mm version of the T-72 series MBT firing NATO type ammunition. Details of this are also given under Ukraine as are details of the T-84-120 MBT armed with a similar gun fed by a bustle mounted automatic loader.

T-90

This is a further development of the T-72BM and is covered in a separate entry.

M-84

This is a Yugoslav version, fully covered under Yugoslavia. RH ALAN of Croatia now manufacture a version of the M-84 and details are given under Croatia earlier in this section.

T-72 with mineclearing equipment

All members of the T-72 MBT family, with the exception of command tanks, can be fitted with mineclearing equipment at the front of the hull, for example the KMT-6. All vehicles have a dozer blade which folds back under the nose when not required.

Close-up of T-72B turret clearly showing the extra armour bulge forward of the bank of 81 mm smoke grenade launchers 0045798

T-72 with PW-LWD mine clearance system

For some years the Polish Army has deployed a tank-mounted rocket-propelled mineclearing system on a T-54/T-55 chassis called the PW-LWD.

This consists of a rocket which is attached to a 170 m explosive filled hose. The complete system is carried in a boat-shaped container which slides onto a special mount on the hull top at the rear of the vehicle.

More recently the system has appeared on a Polish T-72 MBT which can also be fitted with other types of nose-mounted mine clearance systems.

The vehicle halts on arrival at the minefield, and the PW-LWD system is fired. The explosive hose falls to the ground and is detonated, setting off any mines in its path. This Polish mine clearance system has also been installed on the French Leclerc ARV as part of a complete mineclearing system.

T-72 with 155 mm turret

To meet an Indian Army requirement for up to 600 155 mm self-propelled artillery systems based on a locally built T-72 MBT chassis, four foreign 155 mm turret systems underwent extensive firepower and mobility demonstrations. As of early 2002 no production orders had been placed for any of these turrets although the South African T-6 turret systems was the preferred option.

MTU-72 AVLB

The armies of the RFAS (CIS) have recently started to deploy an armoured vehicle-launched bridge system based on the T-72 MBT or T-90 chassis which is similar in appearance and concept to the older MTU-20, which is based on the T-55 chassis (qv). The MTU-72 AVLB weighs 40,000 kg complete with bridge. The bridge weighs 6,400 kg and when opened out is 20 m long and can be used to span a gap of up to 18 m. Full details are given in *Jane's Military Vehicles and Logistics 2002-2003*.

MTU-90 AVLB

This is the latest AVLB to be developed based on the T-72 or T-90 MBT chassis and has a total of three sections which when opened out, span a gap of 24 m. Available details are given in *Jane's Military Vehicles and Logistics 2002-2003*.

BREM-1 Armoured Recovery and Repair Vehicle

This is based on the chassis of the T-72 and mounted at the front of the hull on the left side is a hydraulic crane which can lift 12 tonnes. Also fitted are a main winch with a capacity of 25 tonnes which can be increased to 100 tonnes, auxiliary winch, hydraulically operated dozer/stabilising blade at the front of the hull, towing equipment and a complete range of tools and recovery equipment. Full details are given in *Jane's Military Vehicles and Logistics 2002-2003*.

IMR-2 Combat Engineer Vehicle

A T-72 MBT chassis is used as the basis for the new IMR-2 combat engineer vehicle.

The crane can be fitted with a number of attachments including pincers for uprooting trees. Pivoted at the front of the vehicle is a dozer blade that can be used in a V-configuration or as a straight dozer blade. When not required it is raised clear of the ground. Details of this are given in *Jane's Military Vehicles and Logistics 2002-2003*.

Table 1

Country	Quantity	Comment
Algeria	350	
Angola	50	Estimate
Armenia	102	
Azerbaijan	147	
Belarus	1,225	
Bulgaria	433	including 100 from Russia in 1996
Croatia	30	Estimate
Czech Republic	541	some built locally
Finland	162	from Russian Federation and Germany
Georgia	31	
Hungary	238	
India	1,100	was manufactured under licence, some sources quote figures of between 800 and 1,980 vehicles
Iran	250	local production of T-72S is now underway
Iraq	500	approx number
Kazakhstan	600	approx number
Kyrgyzstan	210	
Libya	260	
Morocco	70	delivered 2000
Poland	772	some built locally
Romania	30	
Russia	7,000	approx number
Sierra Leone	2	
Slovakia	272	some built locally
Syria	1,500	
Tajikistan	40	
Turkmenistan	570	
Ukraine	1,305	
Uzbekistan	100	Estimate
Yemen	27	delivered 2000
Yugoslavia	304	some built locally as M-84

BMR-3M Armoured Demining Vehicle

The BMR-3M armoured demining vehicle is based on a much modified T-72 or the more recent T-90 MBT chassis and has had its turret removed and replaced by a fully enclosed raised superstructure for the crew.

For a higher level of battlefield survivability, explosive reactive armour is fitted to not only to the superstructure but also the glacis plate and track skirts which cover the upper part of the suspension.

Mounted at the front of the vehicle is the KMT-7 mine clearing roller system fitted with an electromagnetic device to activate anti-tank mines fitted with a magnetic fuze.

Heavy infantry fighting vehicle

Details of this, essentially a modified T-72 MBT chassis fitted with a much modified BMP-3 turret armed with a 100 mm gun, 30 mm coaxial cannon and 7.62 mm machine gun, are given in the Armoured Personnel Carriers (Tracked) section. This remains at the prototype stage.

BMPT tank support vehicle

This was first shown in mid-2000 and has been developed by the Nizhni Tagil State Unitary Enterprise and is called the BMPT Combat Vehicle for Tank Support.

This is based on a modified T-72 MBT chassis with the turret removed. To the rear of the driver's compartment, at the front of the vehicle, the hull has been raised to provide greater internal volume. On top of this is mounted a new turret armed with an externally mounted 30 mm cannon and a 30 mm grenade launcher.

Mounted on the left side of the turret is a four-round launcher for the 9M113 Konkors (NATO AT-5 Spandrel). Mounted one either side of the forward part of the hull is a 7.62 mm machine gun or a 30 mm grenade launcher.

The BMPT has a combat weight of 47 tonnes and a crew of five. It is powered by a B92C2 4-stroke diesel which develops 1,000 hp and this gives a power-to-weight ratio of 21.2 hp/tonne.

The latest generation explosive reactive armour is fitted for a higher level of battlefield survivability.

TOS-1 rocket system

Details of this, a T-72 MBT chassis with its original turret removed and replaced by a new turret with 30 × 220 mm rocket tubes to launch rockets with fuel air explosives, are given in the Multiple Rocket Launchers section. Details of this system, which were only built in small numbers, are given in the Multiple rocket launcher's section.

Status

Production complete but could be resumed again for export market. It has been manufactured under licence in Croatia, Czechoslovakia, India, Iran, Iraq, Poland and Yugoslavia. Additional details are given under their respective country headings (see Table 1).

Manufacturers

Chelybinsk, Nizhnyi Tagil and Kirov (Russian).

UPDATED

T-64 MBT

Development

The T-64 MBT was developed at the Kharkov Transport Machine Plant under the direction of A A Morozov, and was the first Russian MBT to have a three-man crew as a result of the installation of an automatic loading system.

The first prototype vehicles were completed in 1960 under the designation Obiekt 430 and this was followed by Obiekt 432 in 1962, and then 434b in 1967.

To power the T-64, the new 5TDF two-stroke opposing cylinder multifuel engine was developed, this is fitted with a governor to limit excessive wear.

The lightweight suspension system of the T-64 consisted of shortened torsion bars with the first, second and sixth roadwheels being provided with a telescopic hydraulic shock-absorber.

The hull and turret were of cast and welded steel armour which incorporated a mixture of conventional steel and ceramic inserts called Combination K, with the ceramic inserts providing superior protection against HEAT attack. The sides of the hull and upper part of the suspension were provided with gill-type armour panels which sprang outward during combat to reduce the effectiveness of HEAT projectiles.

In 1963, a preproduction batch of T-64 MBT's was authorised and by late 1964, a total of 54 vehicles had been built after which production rapidly increased and by December 1965, 218 had been completed. It was accepted for service in December 1966. Production continued to 1969.

Early production vehicles had the 115 mm D-68 (2A21) smoothbore gun fed by an automatic loader. There were many problems with this first version of the T-64, especially with the automatic loader, power pack (especially the transmission) and the suspension.

Further developments resulted in the T-64A which had many improvements including the 2A26M2 125 mm smoothbore gun fed by an automatic loader. Trials with this version commenced in 1967 and it was accepted for service two years later.

Total production of the T-64 MBT series amounted to around 8,000 but by 2001 the numbers in service with the armies of the RFAS (CIS) were rapidly declining. The T-64 was never offered on the export market as at the time of its introduction it incorporated many advanced features.

The United Nations list of MBT exports for the years 1992 through to 2000, revealed that there were no exports of the T-64 series MBT during this period from either Russia or the Ukraine.

Description

The T-64 MBT has a similar layout to the T-72 but is armed with a 125 mm 2A26M2 smoothbore gun with a vertical ammunition stowage system, whereas the T-72 and T-80 are armed with a 125 mm 2A46 gun with a horizontal ammunition feed system. It has the same NBC system, but narrower tracks, a five-cylinder opposed diesel engine and a slightly different turret.

The suspension either side consists of six small dual roadwheels with the drive sprocket at the rear, idler at the front and four track-return rollers which support the inside of the double-pin track only. The first, second, fifth and sixth roadwheel stations are provided with a hydraulic shock-absorber. Over the top of the suspension, which slopes downward towards the rear, is a rail on which panels of additional armour can be attached.

The infra-red searchlight is mounted on the left rather than the right of the 125 mm main armament. There are either two or three boxes of 12.7 mm ammunition mounted on the left side of the turret, a snorkel is carried on the top of the turret and at the very rear of the turret is a detachable stowage box. The T-64 has two snorkels for deep fording, one fitted to the turret and the other over the engine compartment.

The 125 mm 2A26M2 gun is stabilised in both elevation and traverse with the ordnance being fitted with a thermal sleeve and fume extractor. The commander's sight is designated the TKN-3V. The gunners sight and coincidence optical range-finder is designated the TPD-43B. The 12.7 mm anti-aircraft machine gun has electric elevation from –5 to +70° with manual controls being provided for emergency use. The 12.7 mm anti-aircraft sight is designated the PZU-5.

Unlike the T-72's, the 12.7 mm NSVT anti-aircraft machine gun of the T-64 can be aimed and fired from within the tank. The main 125 mm armament can be laid and fired while the T-64 is moving across country and the commander can override the gunner if required. Standard equipment includes an NBC system, infra-red night vision equipment for the commander, gunner and driver, ability to be fitted with snorkels for deep fording and a laser warning device of a similar type to that fitted on the T-80 MBT.

Variants

T-64

This was the initial production model armed with a 115 mm D-68 (2A21) smoothbore gun for which 40 rounds of ammunition were carried; of this total, 30 rounds were in the automatic loader for ready use. It is believed that 600 were built, but none now remain in service. Most were eventually modified to T-64A/T-64B standards as the T-64R.

T-64A

This was developed under the designation Obiekt 437 and was accepted for service in 1969 with production being undertaken in Kharkov and Omsk.

Major differences include installation of the 125 mm 2A26M2 smoothbore gun with thermal sleeve and a modified automatic loader, a new fire-control system, a TPD-43D sight for the gunner with an enlarged opening, a bump stop for the fourth roadwheel and 81 mm smoke grenade launchers either side of the 125 mm 2A26M2 gun. The commander could aim and fire the 12.7 mm machine gun from within the turret (the cupola is electrically operated with the mount being designated PZU-5), with increased armour protection for the turret, a new PPO fire detection and suppression system, a dozer blade mounted under the nose, fittings for the KMT-6 mineclearing system, a multifuel diesel engine, a deep fording capability, turning signals and the ability to operate at higher altitudes.

T-64AK

This is the command version of the T-64 and was accepted for service in 1973. It has an additional command radio set operating in the HF band, a demountable antenna, navigation equipment and an auxiliary generator.

The handrails on either side of the turret front were omitted and the forward stowage boxes on the right-hand sponson were replaced by additional external fuel tanks as on the left-hand sponson. When deployed in the stationary position, a 10 m high telescopic mast can be erected over the turret and this is held in position by stays that are pegged to the ground. The T-64AK command tank is not fitted with the 12.7 mm roof-mounted machine gun.

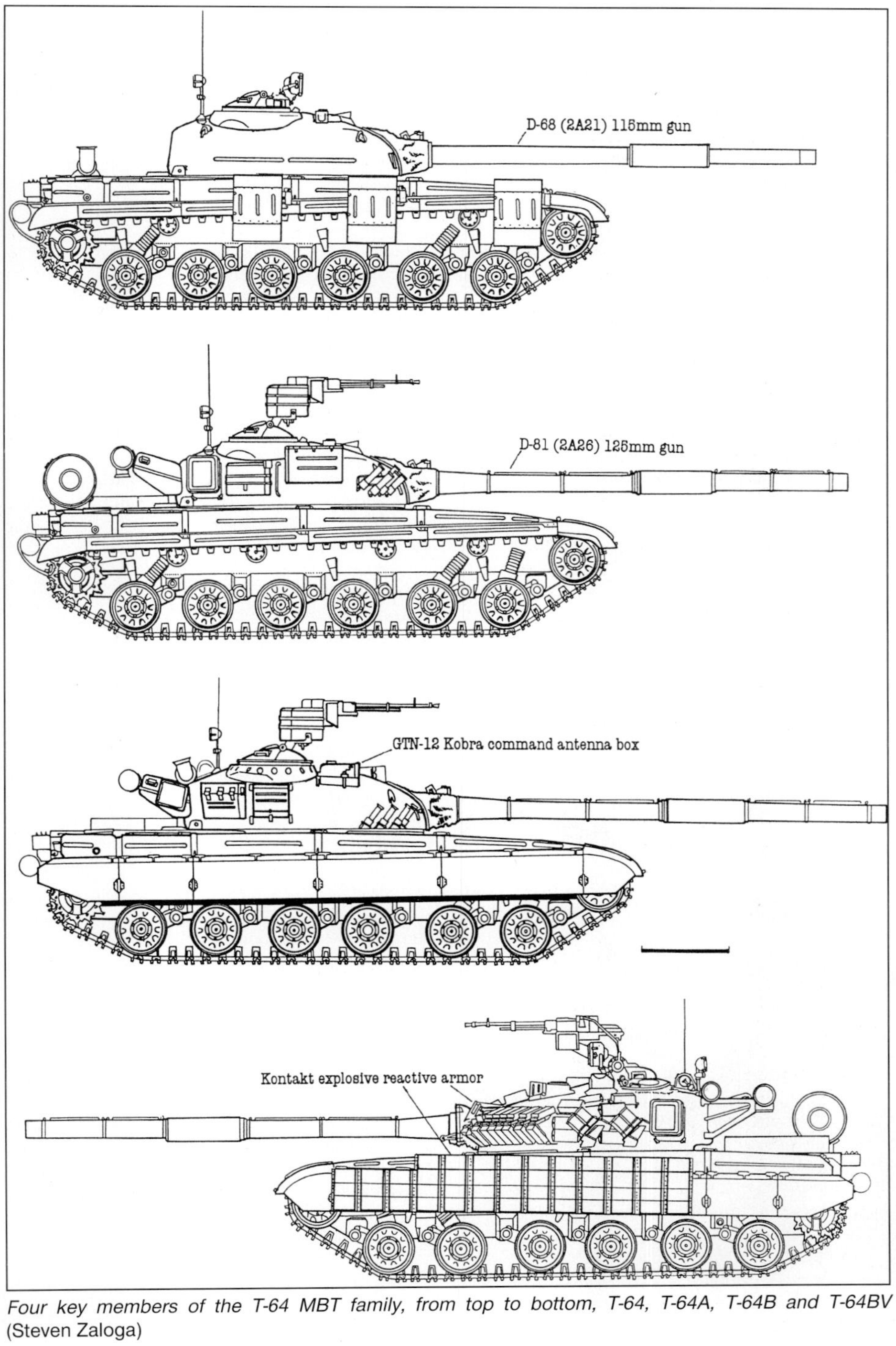

Four key members of the T-64 MBT family, from top to bottom, T-64, T-64A, T-64B and T-64BV (Steven Zaloga)

T-64B

This was a major redesign of the T-64A, under the designation Obiekt 447, to incorporate a new hull and turret armour which was not as bulky as the first-generation Combination K armour on the T-64A, but which offered the same or better protection.

The T-64B also has a new fire-control system, napalm-resistant defence system, smoke grenade launcher system, quick disconnect system for barrel and breech assemblies, side skirts and increased suspension travel.

The tank commander has a TKN-3V combination day/night binocular periscopic sight with a magnification of ×5 in the day mode and ×4.2 in the night mode, plus the ability to work with infra-red night vision equipment. He also has day vision blocks and a 12.7 mm anti-aircraft machine gun mount is fitted with a PZU-5 monocular periscopic sight with panoramic head that provides a 50° field of view.

The gunner has an optical monocular sight with laser range-finder that has a two-axis stabilised field of view and magnification of ×3.9 and ×9; in addition he has unity power day prismatic view blocks and a TPN1-49-23 night sight that operates in conjunction with the L-4A infra-red searchlight.

The fire-control system is designated the 1A33 and enables the T-64B to engage stationary and moving targets while the vehicle itself is stationary or moving.

The fire-control system includes the two-axis stabilised IG42 laser range-finder sight, the 1V517 tank ballistic computer with data inputs being automatic (from the range-finder or manual), the 2Eh36M armament stabiliser, the IG43 fire-control panel and some other elements.

The T-64 gunner can select the type of round he wishes to fire by pushing a button and a rate of fire of 6 to 8 rds/min can be achieved. The 125 mm smoothbore 2A26M2 gun has a sighted range out to 4,000 m using the day sight and 800 m using the night sight. Maximum effective range of the APFSDS-T round is 2,500 m. The 125 mm ammunition is of the separate loading type, that is, the projectile is loaded first followed by the semi-combustible cartridge case, then all that remains after firing is the stub base which is ejected. This 125 mm ammunition is common to the T-64, T-72, T-80, T-84 (Ukraine) and the more recent T-90, details of this ammunition are given in the entry for the T-80 MBT earlier in this section.

The T-64B's 125 mm ordnance also fires the radio command guided AT-8 'Songster' ATGW, which is kept in the automatic loader in two separate parts like standard APFSDS or HEAT-FS rounds and loaded using the automatic loader.

There are at least two versions of the T-64B: new construction T-64Bs, which appear to lack the usual forward gunner's sight, and rebuilt T-64Bs, which are modifications of earlier T-64s.

The AT-8 'Songster' could be used to engage targets such as MBTs and ATGW systems armed with HOT, TOW and Swingfire systems out to approximately

T-64 series MBT from the rear showing the long-range fuel drums mounted above the rear engine deck 0045800

T-64BV fitted for, but not with, explosive reactive armour 0018822

4,000 m and can also engage helicopters. A boost motor launches the AT-8 'Songster' from the gun barrel and the main motor then cuts in and carries it all the way to the target. It takes 9 or 10 seconds to reach a range of 4,000 m.

The complete missile system of the T-64B has the Russian designation 9K112-1 with the actual Kobra missile being designated 9M112M. The US/NATO designation for the 9M112M is the AT-8 'Songster'. The T-64B has a laser range-finder in place of the coincidence range-finder fitted to the earlier T-64A MBT.

In addition to the 9M112M missile, other elements of the 9K112-1 armament complex include the 9V387 control panel relay, 9S461 tank control apparatus and the PO-900 converter. This system is also installed in the T-64BV and T-80BV MBTs.

The Kobra missile system has three operating modes:

(1) Primary, firing with the 125 mm 2A26M2 gun elevated at an angle of 3° higher in relation to the line of sight to the target and programmed return of the missile to the line of sight immediately after firing
(2) Firing from above the target (additional mode), firing with an angle of elevation of around 3° above the line of sight, so that the missile flies 3 to 5 m from (above) the line of sight, and drops back to the line of sight just in front of the target. This mode is used on a dusty battlefield or at night
(3) Firing at ranges of less than 1,000 m (auxiliary mode), firing with an angle of elevation of around 40 minutes of arc relative to the line of sight to the target.

The missile has a muzzle velocity of 125 m/s, which increases to 500 m/s and it weighs 25 kg at launch. The radio transmitter for the AT-8 'Songster' is mounted in a removable steel box in front of the right commander's cupola although the missile is guided to the target by the gunner. All the gunner has to do is to keep his sight on the target to ensure a hit. The box in front of the commander's cupola can be moved and stowed elsewhere.

The AT-8 'Songster' has a single HEAT warhead and can probably penetrate at least 600 mm of steel armour but cannot penetrate Chobham-type and explosive reactive armours . It has a maximum range of 4,000 m and the T-64B can fire a maximum of four AT-8 'Songsters' per minute. A T-64B normally carries six AT-8 'Songsters' plus 36 rounds (projectile and charge) of 125 mm ammunition.

The T-64B has a laser range-finder for improved first round hit probability, explosive reactive armour and strengthened turret hatch protection. It has a different explosive reactive armour package from the T-80.

The reactive armour on the T-64B covers the glacis plate, the forward part of the turret front, sides and roof and hull sides extending to the five roadwheels. When fitted with explosive reactive armour, the 81 mm smoke grenade dischargers are moved from either side of the main armament to either side of the turret rear, roughly in line with the fourth and fifth roadwheels.

T-64BK

This is the command version of the T-64B with additional communications equipment and a land navigation system.

T-64BM

This is a variant of the T-64B with the 6TD engine which is a six-cylinder version of the 5TDF two-cycle multifuel diesel developing 1,000 hp.

T-64BV

This is the designation for the T-64B fitted with explosive reactive armour. It was first seen in 1984.

T-64B1

This was first fielded in 1981 and is the T-64B without the AT-8 'Songster' (9M112) ATGW system installed.

T-64B1K

This is the command tank version of the T-64B1 and has additional communications equipment installed.

T-64BV1K

This is the T-64B fitted with explosive reactive armour and fitted with additional communications equipment for use in the command role.

The T-64BM2 (T-64U) MBT has been developed in the Ukraine and has a number of improvements including new armour (Jim Kinnear) ***2002***/0118055

T-64R

This is an early T-64 upgraded to almost the T-64B standard with the more recent 125 mm 2A46-2 gun, Kobra AT-8 missile system, laser range-finder and other detailed improvements. The R in the designation stands for *Remontirovanniy* or rebuilt.

Driver instruction vehicle

It is known that a driver instruction version of the T-64 was developed. This was a standard T-64 with its turret removed and replaced by a new superstructure, which is part of the upper crew compartment of the BTR-60 series (8 × 8) APC.

Specifications

T-64B
Crew: 3
Combat weight: 39,500 kg
Power-to-weight ratio: 17.7 hp/t
Ground pressure: 0.86 kg/cm²
Length:
(gun forward) 9.9 m
(hull) 7.4 m
Width:
(without skirts) 3.38 m
(with skirts) 4.64 m
Height: (without AA MG) 2.2 m
Ground clearance: 0.377 m
Track width: 580 mm
Length of track on ground: 4.4 m
Max road speed: 75 km/h
Range:
(road, without long-range fuel tanks) 400 km
(road, with long-range fuel tanks) 550 km
Fuel capacity: 1,000 litres
Fording:
(without preparation) 1.8 m
(with preparation) 5.0 m
Gradient: 60%
Side slope: 40%
Vertical obstacle: 0.8 m
Trench: 2.28 m
Engine: 5DTF 5-cylinder opposed piston, diesel, liquid-cooled, 700 hp at 3,000 rpm
Transmission: synchromesh, hydraulically assisted with 7 forward and 1 reverse gears
Steering: clutch and brake
Suspension: torsion bar
Electrical system: 24 V
Armament:
(main) 1 × 125 mm 2A26M2 smoothbore gun
(coaxial) 1 × 7.62 mm PKT MG
(anti-aircraft) 1 × 12.7 mm NSVT MG
Smoke grenade launchers: 12 × 81 mm (in two banks of 6)
Ammunition:
(main) 36*
(coaxial) 1,250
(anti-aircraft) 300

Gun control equipment
Turret power control: electric/manual
Turret traverse: 360°
Gun elevation/depression: +14/–6°
Gun stabiliser:
(vertical) yes
(horizontal) yes
NBC system: yes
Night vision equipment: yes
*+6 ATGW

Ukrainian T-64 upgrades (T-64U)

The Ukraine has started to upgrade at least part of its T-64 MBT fleet to extend its operational life well into the 21st Century.

The T-64 MBT was originally developed at the Kharkov Transport Machine Plant in the Ukraine with production being undertaken from the early 1960s. About 8,000 T-64 series MBTs were built in the Ukraine, but the vehicle was never offered on the export market.

Some years ago at least part of the T-64 fleet was upgraded, with the T-64BV being fitted with Explosive Reactive Armour (ERA) and the T-64B1 with the capability to launch the 4,000 m range AT-8 Songster radio command guided missile.

This early generation ERA gave the T-64 a very high level of additional protection over the frontal arc against HEAT warheads.

The latest T-64 upgrade includes features of the locally built T-80UD and more recent T-84 MBTs. Whereas Russia produced the turbine powered T-80U, the Ukraine manufactured the diesel powered T-80UD and 320 of these were delivered to Pakistan between 1997 and 1999.

This was a major export contract for the Ukraine and brought in valuable foreign exchange as well as keeping their production line going. It also allowed them to invest in new technologies.

The T-64B and T-64B1 were selected to be upgraded to the new T-64U standard, with the U standing for Usovershenniy or improved. The T-64U brings the earlier vehicles up to almost the latest T-80UD/T-84 MBT standard in the areas of armour protection and fire-control system.

In the first phase, T-64BV and T-64BV1 have been equipped with a locally developed ERA which is similar to the Russian Kontakt-5.

This is installed on the hull and turret of the vehicle when they undergo a major maintence and the armour gives protection against kinetic energy attack (APFSD) and chemical energy attack (for example HEAT) projectiles. New composite armour has also been added. This version is called the T-64BM2.

The second model is called the Obiekt 447AM1 and is fitted with the 1A42 aiming system. This includes the IG46 laser range-finder and 1V528-2 fire-control computer, the TO1-KO1 optical system includes the thermal TPN-4E Buran-E and L-4 searchlight. It also has the 9K117 Reflecks barrel-launched 125 mm laser-guided missile, the 6ETs-43 automatic loader and the roof-mounted DVE-BS wind sensor (this replaced the original Russian IB11 system).

The third model and the most advanced, is the Obiekt 447AM2 and has the 1A45 Irtish fire-control system, which includes the 1A42 laser range-finder, TO1-KO1 (without the L-4 searchlight) and PNK-4 with the TKN-4S Agat sight, PZU-7 anti-aircraft sight for the 12.7 mm anti-aircraft machine gun and the 1Ets29 control system.

It is of note that some of the subsystems of the latest T-64 MBT upgrade are also used in Russian MBTs, for example the 1V528 series fire-control computer is used in the later models of the T-72 vehicles as well as the T-80U and T-90. The roof-mounted wind sensor is used in the T-80U and T-90.

It has also been disclosed that the Ukraine has developed its own 125 mm laser-guided projectile, called the Combat, which can be fired from the locally produced T-80UD/T-84 series MBTs. It can also be fired from some versions of the locally upgraded T-64 MBT.

T-64 with mineclearing equipment

All T-64s are fitted with a dozer blade that folds back under the nose of the vehicle when not required and all have provision for the installation of mineclearing equipment such as the KMT-4 mineclearing roller.

BREM-64

An armoured recovery version of the T-64 was developed under the designation BREM-64. This is fitted with a front mounted dozer blade, winches and a crane which is mounted on the left side of the hull.

Status

Production complete. In service only with the armies of the Russian Federation and associated states, of which the Ukraine Army is known to have around 2,200 and Uzbekistan around 100.

Manufacturers

State facilities at Kharkov (KB-60) and Omsk.

UPDATED

T-62 MBT

Development

The T-62 MBT was designed by the Kartsev Bureau, 'Vagonka' at Nizhnyi Tagil from the earlier T-55 MBT. It incorporates a number of components from that vehicle but has a longer and wider hull and a new turret with a diameter of 2,245 mm, as opposed to 1,845 mm.

The development designation of the T-62 was Obiekt 166 which was accepted for service with the Soviet Army in mid-1961. Late in 1961, a preproduction batch of 25 T-62 MBTs was built for trials purposes with volume production starting at Nizhnyi Tagil in mid-1962.

The engine and transmission of the T-55 were retained but the cooling system was improved by the provision of a larger diameter fan. The suspension is identical to the T-55's but the mounts have been rearranged as the hull is longer. The nuclear protection, fire detection and suppression and water fording systems are identical to those installed in the T-54B and the T-55. The T-62 was first seen in public during a parade held in Moscow in May 1965. Production of the T-62 continued in the Soviet Union until 1975 by which time about 20,000 tanks had been completed. It was produced for export in Czechoslovakia between 1973 and 1978 with approximately 1,500 built. It was also produced in North Korea for home and export markets.

Main recognition features of the T-62 compared with the earlier T-54/T-55 are a longer and wider hull, different spacing of the roadwheels as the T-62 has a distinct gap between the third and fourth and fourth and fifth roadwheels, shape of the turret, and the longer and fatter gun barrel with a fume extractor towards its muzzle.

A complete family of variants was developed on the early T-54/T-55 MBT chassis but there were very few variants on the T-62 MBT chassis. As for specialised roles, for example, the armoured recovery vehicle and bridgelayer, the less expensive T-54/T-55 chassis was more than adequate.

According to United Nations sources, Russia did not export any T-62 series MBTs between 1992 and 2000. Some were however exported by Bulgaria. This country supplied 24 to Angola in 1993 and 56 to the Yemen in 1994.

Description

The hull of the T-62 is divided into three compartments, driver's at the front, fighting in the centre and engine and transmission at the rear. The driver is seated at the front of the vehicle on the left side and is provided with a single-piece hatch cover that opens to the left. There are two day vision blocks mounted forward of this hatch cover each with an integral defrosting element. The left vision block can be replaced by a TVN-2 infra-red periscope which has a 30° field of view and a maximum range of 60 m. A hull escape hatch behind the driver's seat opens to the inside of the vehicle. Mounted on the glacis plate is a wave deflector and, mounted to the rear of this on the right side are a white light and an infra-red headlamp.

T-62M MBT with 12.7 mm DShKM machine gun on loader's cupola (Michael Green/US Army)

The cast turret is in the centre of the tank with the commander and gunner seated on the left and the loader on the right. Both are provided with a single-piece hatch cover that opens to the rear and can be locked vertically. Rails outside the turret can be used by infantry or for stowing personal equipment.

The commander's cupola has four day periscopes, two mounted in the hatch cover and two in the forward part of his cupola. The commander's sight, designated TKN-3, is a day/night binocular periscope with an integral infra-red capability, mounted in the forward part of his cupola. For day use it has a magnification of ×5 and a 10° field of view and for night it has a magnification of ×4.2 and an 8° field of view. Effective range when used in conjunction with the OU-3GK searchlight is 400 m. The handles of the sight are used to rotate the commander's cupola and operate the searchlight, target designation equipment and other systems.

The gunner has a TSh2B-41u telescope with a rotating graticule for superelevation required for the different types of 115 mm ammunition and dual magnification, ×3.5 with an 18° field of view and ×7 with a 9° field of view, filter capabilities, stadiametric range-finder (at the bottom of the graticule) and an integral wiper. The maximum sighting ranges are 4,000 m for APFSDS, 3,700 m for HEAT, 4,800 m for HE 18 and 3,600 m for HE 11. The gunner's infra-red sight is the TPN1-41-11 periscope which is used in conjunction with the main L-2G searchlight mounted coaxially to the right of the main armament and has an effective range of 800 m. The TPN1-41-11 has a magnification of ×5.5 and a 6° field of view. The gunner is also provided with a Type TNP-165 periscope with a magnification of ×1. The loader on the right side of the turret is provided with a single TNP-165 day periscope which can be used to the front or rear of the vehicle.

Mounted at the rear of the turret, to the left of the spent cartridge ejection door, is an electrically operated blower worked by the driver using the KUV-3 ventilator control box.

To the rear of the turret, over the engine compartment, is a large rectangular sheet steel plate that covers the engine louvres when the tank is snorkelling. The engine is equipped with a preheater and is normally started by compressed air, although an electrical auxiliary system is also provided. The engine is coupled to the manual transmission and changing up or down is accomplished by double declutching. The two-stage planetary steering system, which also serves as brakes, transmits torque to the final drives, each of which is a two-step, step-down reduction gear.

The torsion bar suspension consists of five dual rubber-tyred roadwheels with the drive sprocket at the rear and the idler at the front. A hydraulic shock-

Specifications

Crew: 4
Combat weight: 40,000 kg
Unloaded weight: 38,000 kg
Power-to-weight ratio: 14.5 hp/t
Ground pressure: 0.77 kg/cm^2
Length:
(gun forward) 9.335 m
(gun rear) 9.068 m
(hull) 6.63 m
Width: 3.3 m
Height: 2.395 m
Ground clearance: 0.43 m
Track: 2.64 m
Track width: 580 mm
Length of track on ground: 4.15 m
Max speed:
(1st gear) 14.5 km/h
(2nd gear) 20 km/h
(3rd gear) 29 km/h
(4th gear) 45.5 km/h
(5th gear) 50 km/h
(reverse) 7 km/h
Fuel capacity:
(internal) 675 litres
(external) 285 litres
(supplementary external) 400 litres
Max range:
(paved road) 450 km
(dirt road) 320 km
(paved road with additional fuel tanks) 650 km
(dirt road with additional fuel tanks) 450 km
Fuel consumption:
(paved road) 3-3.3 litres/km
(dirt road) 1.9-2.1 litres/km
Fording: 1.4 m
(with preparation) 5 m
Gradient: 60%
Vertical obstacle: 0.8 m
Trench: 2.85 m
Turning radius: skid turns
Engine: Model V-55-5 V-12 water-cooled diesel with injection pump fuel system, pressure lubricated, developing 580 hp at 2,000 rpm
Auxiliary engine: none
Transmission: manual with 5 forward and 1 reverse gears
Steering: 2-stage planetary
Clutch: 10 driving, 9 driven steel discs and a release mechanism
Suspension: torsion bar
Electrical system: 24 V
Batteries: 4 × 12 V, 150 Ah
Armament:
(main) 1 × 115 mm 2A20 smoothbore gun
(coaxial) 1 × 7.62 mm PKT MG
(anti-aircraft) 1 × 12.7 mm DShKM MG (on T-62M only)
Smoke-laying equipment: diesel fuel injected into exhaust system
Ammunition:
(main) 40
(coaxial) 2,500
(anti-aircraft) 300
Gun control equipment
Turret power control: electrohydraulic/manual
(by commander) yes (powered only)
(by gunner) yes (electrohydraulic/manual)
Turret traverse: 360°
Gun elevation/depression: +16°/–6°
(commander's override) yes
Gun stabiliser:
(vertical) yes
(horizontal) yes
Elevation quadrant: yes
Traverse indicator: yes

Armour
Hull front upper: 102 mm at 60°
Hull front lower: 102 mm at 54°
Hull sides upper: 79 mm at 0°
Hull sides lower: 15 mm at 0°
Hull rear upper: 46 mm at 0°
Hull rear lower: 46 mm at 0°
Hull top: 31 mm
Hull floor front: 20 mm
Hull floor rear: 20 mm
Turret front: 242 mm
Turret sides: 153 mm
Turret rear: 97 mm
Turret top: 40 mm
Turret mantlet: included in above figures
Turret hatches: 30/31 mm
NBC system: yes
Night vision equipment: yes

absorber is provided at the first and last roadwheel stations. The all-steel track has steel pins that are not secured at the outer end and are free to travel towards the hull. A raised piece of metal welded to the hull, just forward of the sprocket, drives the track pins back into position each time they pass. Each track has 96 links and weighs 1,386 kg when new.

A centralised ethylene-bromide fire extinguishing system is automatically activated by heat sensors of which there are eight in the engine, transmission and fighting compartments, or activated manually by the tank commander or driver.

The T-62 MBT has a PAZ nuclear collective protection system which consists of a radiation detector/actuator (RBZ-1m), five separate explosive squib mechanisms and a blower/dust separator. The box-like radiation detector/actuator (a radiation threshold detector) is mounted on the right side of the turret compartment behind the compressed air tanks.

The detector/actuator senses the initial pulse of radiation (gamma or neutron pulse) which precedes the blast wave and then activates the explosive squib mechanism. This is a system of spring-loaded shutters, dampers or louvres that are held by a detent pin. When the squib mechanism is activated, an explosive charge detonates and forces the detent pin out of place, thus allowing the shutter, damper or louvre to close. The explosive squib mechanism's function is to close the engine louvres, sight aperture, bulkhead ventilation fan, air baffles to the transfer case and the air intake to the blower/dust separator. The blower/dust separator is in the turret below the shell ejection port and is an electric motor mounted with a set of fan blades which draw air into the vehicle and spins it at approximately 7,000 rpm.

It should be noted that the blower/dust separator removes nuclear fall-out only: it does not protect crew from chemical or biological contaminants as the air is not passed through a chemical filter. The tank must pass through the contaminated area as quickly as possible and then be decontaminated before becoming fully operational again. Late production models of the T-62 are fitted with a chemical filter.

Mounted at the rear of the hull is an unditching beam and two diesel drum fuel tanks can be installed at the rear to increase operational range. All vehicles have three external fuel cells on the right side of the vehicle for diesel fuel while the single tank on the left side is for auxiliary oil. The driver can select which fuel to use first, normally the two rear drums, then the three external cells and finally the main fuel tank.

The T-62 can, like other Russian tanks, lay its own smoke screen by spraying diesel oil into the exhaust manifold when it is sufficiently hot, creating thick white smoke which exits from the exhaust ports on the left side of the tank. When laying the smoke screen the tank is always in second or third gear. This consumes about 10 litres of fuel per minute and produces a smoke screen 250 to 400 m long which lasts for 4 minutes, depending on the wind strength.

The tank can ford to a maximum depth of 1.4 m without preparation. It can also ford deep water when fitted with a snorkel. It takes up to 8 hours to prepare the tank for deep fording and two types of snorkel are available; a large diameter one for training and a thinner one for operational use which is normally carried in sections on the rear of the turret. The tank is usually driven across the river bed in first gear and navigated by its GPK-59 gyro compass and a radio link to the far bank. Once ashore, it takes only 1 or 2 minutes to prepare the tank for action again.

Main armament of the T-62 MBT is a U-5TS (2A20) 115 mm smoothbore two-axis stabilised gun fitted with a bore evacuator, with a maximum rate of fire of 4 rds/min when at a standstill. After firing, the main armament of the T-62 automatically elevates to an angle of +3° 30' for loading. The turret cannot be traversed while the weapon is being loaded. The 115 mm U-5TS (2A20) gun fires fixed types of 115 mm ammunition, see Table 1.

An integral spent shell ejection system, activated by the recoil of the gun, ejects the empty cartridge case out of the turret through a trapdoor in the turret rear. Of the 40 rounds (37 in command tanks) of 115 mm ammunition carried, two ready rounds are kept in the turret, one round by the gunner's feet, one by the loader's feet, 16 in the forward part of the tank to the right of the driver and 20 in the rear of the fighting compartment.

A 7.62 mm PKT machine gun mounted coaxially to the right of the main armament has a practical rate of fire of 200 to 250 rds/min and is fed by a belt containing 250 rounds.

The gun elevating and traversing mechanism consists of electric, hydraulic and manual controls. The

T-62 MBT fitted with a laser range-finder over the 115 mm gun mantlet

Russian supplied T-62 (Model 1972) of the Iraqi Army clearly showing additional protection fitted to the searchlight mounted coaxially with the 115 mm gun and the infa-red commander's searchlight **2002**/0130957

gunner can elevate or depress the gun (electric/hydraulic) and both the commander and gunner can traverse the turret by electric controls through a full 360°. Only the gunner can traverse the turret and elevate the gun manually. The gun is fully stabilised in both horizontal and vertical planes.

Variants

Since the T-62 was accepted for service in 1962 it has been updated on a number of occasions. Listed below, from Russian sources, are the key versions of the T-62 MBT with the date in brackets being the year it was accepted for service.

T-62 (Model 1962)

Improved model of the first production version with a new engine deck and other detailed improvements.

T-62 (Model 1972)

T-62 with 12.7 mm DShKM anti-aircraft machine gun mount over loader's position and some other detailed improvements.

T-62 (Model 1975)

This was the final production model and was fitted with the KTD-1 laser range-finder over the 115 mm U-5TS (2A20) gun mantlet to increase first round hit probability.

T-62D

T-62 fitted with the Drozd anti-tank missile defence system, passive armour protection, a V-55U engine and the R-173 communications system. Details of this first-generation Drozd system, now called Drozd-1 are given in the following T-54/T-55 MBT.

T-62D-1

T-62 fitted with the Drozd anti-tank missile defence system, passive armour protection, a V-46-5M engine and the R-173 communications system.

T-62M

T-62 upgrade with the Sheksna (US designation AT-10) laser beam-riding missile system, passive armour protection, a V-55U engine and the R-173 communications system.

Table 1

Round	Projectile	Type	Round weight	Projectile weight	Armour penetration at 1,000 m, 0/60°
3UBM3	3BM3	APFSDS	22.3 kg	4.0 kg	
3UBM4	3BM4	APFSDS	22.5 kg	4.2 kg	
3UBM5	3BM6	APFSDS	22.5 kg	4.0 kg	228/199 mm
3UBM9	3BM21	APFSDS	23.5 kg	3.9 kg	
3UBM13	3BM28	APFSDS	24.0 kg	5.4 kg	
3UBK3	3BK4	HEAT	26.6 kg	13.1 kg	495/248 mm
3UBK3	3BK4M	HEAT	26.6 kg	13.1 kg	
	3BK15	HEAT			
3UBK10-1	9M117	Missile	28.0 kg	17.6 kg	550/275 mm
3UOF1	30F11	HE-Frag	28.1 kg	14.7 kg	
3UOF6	30F18	HE-Frag	30.75 kg	17.7 kg	
3UOF37	30F27	HE-Frag	30.75 kg	17.7 kg	

The 9K116-1 Sheksna fitted to the T-62M is virtually identical to the 9K116 Bastion system installed in the T-55M. The main differences between the munitions and the fire-control system is that one (the T-55) is fired from a 100 mm rifled gun and the other (the T-62) is fired from a 115 mm U5-TS (2A20) smoothbore gun.

The actual round is designated the 3UBK10 with the missile being designated the 9M117. The latter is fitted with a single HEAT (High Explosive Anti-Tank) warhead.

The latest round for the 115 mm gun is designated the 3UBK10M-2 and is fitted with a tandem HEAT warhead which will penetrate at least 600 mm of steel armour protected by explosive reactive armour. The complete round weighs 30.4 kg with the missile having a maximum range of 4,000 m and a minimum range of 100 m. The missile in this version is designated the 9M117M.

The onboard systems and fire controls have been improved only by replacing those elements necessary to support the launching of the missile.

Key parts are the 1K13-1 combined sight and control element, converter, control panel and an electronic distribution panel. The Volna fire-control system includes the KDT-2 laser range-finder mounted over the main armament, which has a maximum range of 4,000 m, TShSM-41U sight, Meteor M1 stabiliser and the BV62 ballistic computer.

The combat propelling charge provides the missile with an initial velocity of 400 to 500 m/s after which the main missile motor takes over. The actual missile weighs 28 kg and has a maximum range of 4,000 m and, in addition to being used to engage other MBTs, it can also engage hovering helicopters.

Method of guidance is semi-automatic laser beam-riding and all the operator has to do to ensure a hit is to keep the cross-hairs of the sight on the target.

The additional armour fitted to the glacis plate and around the forward part of the turret is essentially steel with an additional layer of phenopolyurethane inside.

Additional armour is fitted under the forward part of the hull and a steel frame is provided between the floor and roof of the driver's compartment to help absorb the shock of a mine exploding under the vehicle.

The upper parts of the suspension, either side, are covered with a skirting made of woven rubber material that is 10 mm thick and the forward torsion bars are covered by rubber sheeting.

The crew of the vehicle is provided with 1PZh-1 anti-radiation jackets and protection is provided against attack from napalm.

A bank of eight Type 902B 81 mm smoke grenade launchers is mounted on the right side of the turret and these are operated by the gunner; the ability to inject diesel fuel into the exhaust is retained.

The V-55U diesel engine has an integral supercharger and develops 620 hp; for improved cross-country mobility, new torsion bars with increased road travel have been fitted. The new tracks with rubber-bushed hinges are standard and two auxiliary shock-absorbers are fitted to the second roadwheel station on each side.

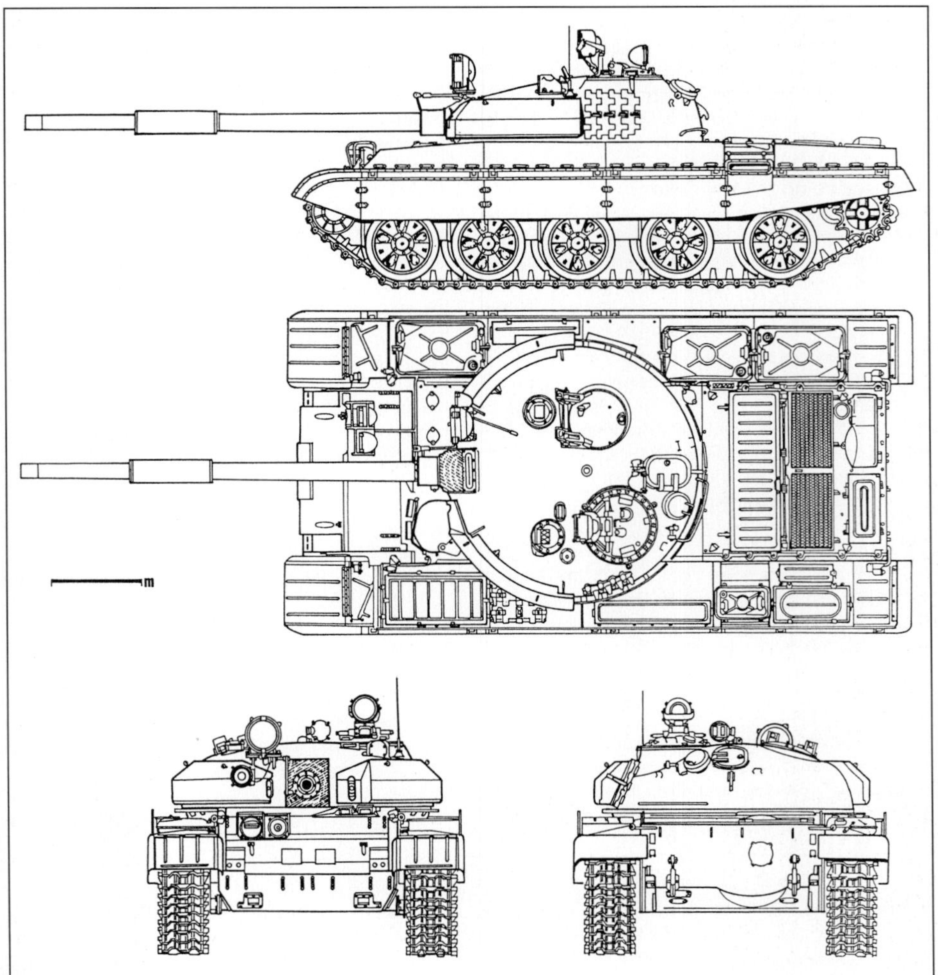

T-62M MBT (with Bra Armour) (Steven Zaloga)

T-62M-1

T-62 upgrade with the Sheksna (US designation AT-10) laser beam-riding missile system, passive armour protection, a V-46-5M engine and the R-173 communications system.

T-62M1

T-62M upgrade without the Sheksna missile system, the passive armour protection, the V-55U engine and the R-173 communications system.

T-62M1-1

T-62 upgrade without the Sheksna guided missile system, the passive armour protection, the V-46-5M engine and the R-173 communications system.

T-62M1-2

T-62M1 upgrade without the Sheksna guided missile system, passive armour protection on turret and belly only, the V-55U engine and the R-173 communications system.

T-62M1-2-1

T-62M1 upgrade without the Sheksna guided missile system, passive armour protection in turret and belly only, and with the V-46-5M engine and the R-173 communications system.

T-62MV

T-62 upgrade with the Sheksna (US designation AT-10) laser-guided missile system, explosive reactive armour system, a V-55U engine and the R-173 communications system.

T-62MV-1

T-62 upgrade with the Sheksna (US designation AT-10) laser-guided missile system, explosive reactive armour protection, a V-46-5M engine and the R-173 communications system.

Listed in Table 2 is a summary of these modifications.

T-62 MBT from rear showing long-range fuel tanks on hull rear (Michael Green/US Army)

T-62M modified for use in Afghanistan and showing additional armour and tank track 0018825

Table 2

Modifications	Tank Designation T-62M	T-62M1	T-62M1-2	T-62MV	T-62D
Sheksna missile system	yes	no	no	yes	no
Volna fire-control system	yes	yes	yes	yes	no
12.7 mm AA machine gun	yes	yes	yes	yes	no
Barrel thermal sleeve	yes	yes	yes	yes	no
Auxiliary armour					
Hull	yes	yes	no	no	yes
Turret	yes	yes	yes	no	yes
Belly	yes	yes	yes	yes	yes
Explosive reactive armour	no	no	no	yes	no
Side skirts	yes	yes	yes	yes	yes
Napalm protection system	yes	yes	yes	yes	yes
Smoke launchers	yes	yes	yes	yes	no
V-55U engine*	yes	yes	yes	yes	yes
Modernised suspension	yes	yes	yes	yes	yes
R-173 radio system	yes	yes	yes	yes	no

*T-62s with the V-46-5M diesel engine (developing 690 hp) are designated T-62M-1, T-62M1-2, T-62M1-2-1, T-62MV-1 and T-62D-1

T-62K command tank

In addition to having extra communications equipment, the T-62K command tank has the TNA-3 land navigation system. This appeared in 1973 and consists of six components: gyroscopic compass, one or two compass indicators, latitude correction device, an odometer, power converter and a calculator.

The gyroscopic compass indicates on what azimuth the vehicle is travelling, is mounted in the left rear of the turret compartment and can determine direction regardless of its position within its mounting. The compass indicator is mounted in the driver's compartment or in the tank commander's cupola; some T-62Ks have two compass indicators, one at each position. The T-62K also has an electric charging system but carries four rounds less of 115 mm ammunition.

The command version of the T-62M is designated the T-62MK by Russia.

T-62 flame-thrower

A flame-thrower version of the T-62 is in service under the designation of TO-62. This has a flame gun mounted coaxially with the 115 mm gun with an effective range of 100 m.

Egyptian T-62s

Some Egyptian T-62s have been observed fitted with twin launchers on either side of the turret rear for indigenously developed Sakr surface-to-surface smoke rockets.

T-62 Iraqi Army

Some of the T-62 series MBTs used by the Iraqi Army were modified in a number of areas including the addition of covers to the turret-mounted searchlights. As far as is known, the Iraqi Army never received any of the T-62 MBTs that were fitted to fire the laser-guided missile through the 115 mm gun barrel.

RO 115 mm tank barrel

BAE Systems, RO Defence is now offering a 115 mm tank barrel as the replacement for the existing 115 mm U-5TS fitted in the T-62, substantial quantities of which have been provided for Egypt as it can no longer obtain barrels from the RFAS.

T-62 with Giat 120 mm smoothbore gun

In mid-1986, Giat proposed the replacement of the Russian T-62's 115 mm U-5TS (2A20) smoothbore gun with the Giat 120 mm smoothbore gun. In this conversion the original elevation and traverse mechanism is retained and the existing 115 mm gun is replaced by the 120 mm smoothbore gun with loading from the right instead of the left, the existing recoil system also being modified. This remains at the concept stage.

KBP upgrade for T-62 MBT

The Russian KBP Instrument Design Bureau, prime contractor for the gun launched laser-guided missiles (100, 115 and 125 mm), are now marketing new missiles which have a maximum range of 5,500 m when fired from the T-62 MBT. The new 115 mm missile also has a tandem HEAT (High Explosive Anti-Tank) warhead to defeat targets fitted with explosive reactive armour.

This latest version uses a new sighting and guidance unit which was originally developed for the BMP-3 infantry fighting vehicle which includes a two-axis line of sight stabiliser, optical channel, image intensification night channel and a laser guidance channel.

A further optional enhancement is the replacement of the image intensification channel by a thermal channel and an autotracker.

T-62 series upgrade package

Russia is now offering a number of upgrades to the T-62 series MBT to improve its battlefield survivability:

(1) Improved armour protection by the installation of two different configurations of passive armour and one configuration of explosive reactive armour
(2) Special paint and screens to make it more difficult to detect in the visible, near and far infra-red bands
(3) Improved interior protection from radiation
(4) Installation of new fire detection and suppression system
(5) Upgraded suspension system.

TV-62M ARV

In 1988, Bulgaria took delivery of 150 T-62M MBTs armed with a 115 mm gun from Russia. Due to a number of problems, understood to be concerned with the 115 mm weapon system, these were soon withdrawn from service.

Some of these T-62 MBTs were subsequently converted into the recovery role and designated the TV-62M.

The modifications included the removal of the three-person turret which was replaced by the fixed upper half of a T-55 or T-55A tank turret which for this role had the 100 mm weapon removed and the turret facing the rear.

The two cupola hatch covers have been retained and a 12.7 mm DShK machine gun is mounted on the cupola. On the left rear of the hull, above the exhaust outlet, has been located a small winch. Tow bars and other equipment is carried.

Status

Production complete. In service with the following countries:

Country	Quantity	Comment
Afghanistan	170	current status uncertain
Algeria	330	
Angola	150	including 24 from Bulgaria in 1993
Belarus	142	
Cuba	400	
Egypt	500	UK supplied new barrels
Ethiopia	100	status uncertain
Iran	200	from North Korea
Iraq	200	approx number
Israel	70	reserve/training
Korea, North	1,800	local production
Libya	350	approx number
Mongolia	250	
Russia	2,000	approx number
Syria	1,000	
Uzbekistan	179	
Vietnam	80	approx number
Yemen	250	approx number including 56 from Bulgaria in 1994

Manufacturer

Nizhnyi Tagil.

The T-62 was also produced in the former Czechoslovakia for export markets. Production of the T-62 was once undertaken in North Korea and about 200 of these have been exported to Iran.

UPDATED

T-54 and T-55 MBTs

Development

In 1944, the Soviet Union designed a new medium tank called the T-44. This was produced in small numbers between 1945 and 1949, but proved unreliable in service. Main improvements over the earlier T-34/85 were its torsion bar suspension, transverse-mounted engine and transmission, well-shaped hull and a turret with a similar shape except that it did not have the thick turret neck. The tank had a crew of four as the bow machine gunner was eliminated in favour of a fixed 7.62 mm machine gun firing forwards, a feature later adopted for the T-54. The turret of the T-44 was a modified T-34/85 turret without the prominent collar at the turret base. The T-44 was used in combat towards the end of the Second World War and again during the Hungarian uprising of 1956.

The T-44 was followed by the T-54, the first prototype of which was completed in 1945 with production following in 1947 in Kharkov. No production figures for the T-54 and T-55 MBTs have been released but it is estimated that over 50,000 were built in the Soviet Union with production also being undertaken in China (as the Type 59), Czechoslovakia and Poland. Production of the T-55 is believed to have continued at the Omsk tank plant until as late as 1981, long after the more well armed T-62 had gone out of production.

Whereas the T-54 MBT was designed by the Morozov Bureau at the Malyshev Plant at Kharkov in Ukraine under the designation Obiekt 137 (or B-40), the T-55 series was developed by the Kartsev Bureau at Nizhnyi Tagil which later went on to design the T-62 and subsequently the T-72 and T-90 MBTs.

With the downsizing of many Eastern European armies quantities of T-54/T-55 MBTs have been sold off in recent years.

According to United Nations sources, between 1992 and 2000 the following quantites of T-54/T-55 were exported by various countries as shown in table 1.

T-54 series MBT without a 12.7 mm anti-aircraft machine gun and narrow board on glacis plate (US Army/Michael Green)

Table 1

From	To	Quantity	Comment
Belarus	Ethiopia	40	T-55 in 1998
Belarus	Sudan	9	T-55 in 1996
Belarus	Uganda	10	T-55 in 2000
Bulgaria	Angola	31	T-55 in 1999
Bulgaria	Ethiopia	50	T-55 in 1998
Bulgaria	Ethiopia	100	T-55 in 1999
Bulgaria	Macedonia	94	T-55 in 1999
Bulgaria	Uganda	90	T-55 in 1998
Bulgaria	Yemen	6	T-55 in 1994
Czech Republic	Cambodia	40	T-55 in 1994
Czech Republic	Georgia	10	T-55 in 2000
Czech Republic	Latvia	3	T-55 in 2000
Czech Republic	Sri Lanka	15	T-55 in 1996
Czech Republic	Sri Lanka	3	T-55 in 1997
Czech Republic	Sri Lanka	11	T-55 in 2000
Czech Republic	Yemen	97	T-55 in 2000
Israel	Uruguay	15	T-55 in 1997
Poland	Cambodia	50	T-55 in 1994
Poland	Yemen	20	T-55 in 1999
Slovakia	Angola	150	T-55 in 1999
Slovakia	Angola	55	T-55 in 2000
Ukraine	Azerbaijan	100	T-55 in 1993
Ukraine	Azerbaijan	50	T-55 in 1994

Description

The all-welded steel hull of the T-54 is divided into three main compartments: driver's at the front, fighting in the centre and the engine and transmission at the rear.

The driver is seated at the front of the tank on the left and is provided with a single-piece hatch cover that swings to the left. There are two day periscopes provided forward of this hatch, one of which can be replaced by an infra-red periscope which is used in conjunction with the infra-red light mounted on the right side of the glacis plate. The driver's infra-red system is known as the TVN-2. A narrow board mounted at right angles to the glacis plate stops water rushing up the glacis plate when the tank is fording shallow rivers. To the right of the driver is space for 100 mm ammunition stowage, batteries and a small fuel tank. A hull escape hatch is provided behind the driver.

The turret is a one-piece casting with the top, which consists of two D-shaped pieces of armour welded together down the centre, welded into position. The commander is seated on the left of the turret, with the gunner on the same side but in a more forward position. The commander has a cupola which can be traversed through 360°, with a single-piece hatch cover that opens forwards with a single day periscope each side. A TPK-1 sight with a single periscope either side is mounted in the forward part of the cupola top. When in action the commander is constantly rotating his cupola for possible targets. On sighting a target he lines it up with his TPK-1 sight and then slews the turret onto the target. The gunner then lays the gun accurately onto the target and fires the main armament. The gunner has a TSh 2-22 sight with a magnification of ×3.5 or ×7. The loader is seated on the right of the turret and is provided with a periscope and a single-piece hatch cover that opens to the rear. The T-54 does not have a rotating turret floor, but the later T-55 has a partial floor, as does the T-62.

The engine is mounted transversely at the rear of the hull. The T-54 uses an electrical start-up system for the engine as its primary means, with a compressed air system for back-up in cold weather. The T-55 uses a compressed air engine start-up system primarily, with an electrical back-up. This is due to the fact that the T-55 has an AK-150 air compressor to refill the air pressure cylinders, unlike the T-54.

The torsion bar suspension, either side, consists of five single rubber-tyred roadwheels with a distinct gap between the first and second roadwheels. The drive sprocket is at the rear and the idler at the front. There are no track-return rollers. The first and fifth roadwheel stations are provided with a hydraulic shock-absorber. The all-steel track has steel pins that are not secured at the outer end and are free to travel towards the hull. A raised piece of metal welded to the hull just forward of the sprocket drives the track pins back into position each time they pass.

The T-54 and T-55 can be fitted with a snorkel for deep fording. Two types of snorkel are available, a thin one for operational use and a thick one for training. These snorkels take between 15 and 30 minutes to fit and on reaching the far bank are blown off. The operational snorkel is mounted over the loader's periscope and, when not fitted, is carried disassembled at the rear of the turret, or at the rear of the hull. The thicker snorkel is mounted over the loader's hatch cover. The tank normally crosses the river in first gear and navigates with the aid of an onboard gyro-compass.

T-55A MBT fitted with a 12.7 mm DShKM anti-aircraft machine gun at the US Army's National Training Center, Fort Irwin, California (Michael Green/US Army)

The T-54 was not originally fitted with an NBC system; it was installed on later production tanks and subsequently refitted to earlier production models. The tank can lay its own smoke screen by injecting diesel fuel into the exhaust on the left side, producing a cloud of smoke some 300 m long which lasts for approximately 2 minutes.

To extend the tank's operational range, two drum-type fuel tanks can be fitted at the rear. On the right running board are four flat light steel tanks. The first, second and fourth tanks each contain 93 litres of fuel, while the third contains lubricating oil. The driver can select which fuel tanks to use first, for example the drums at the rear, followed by the three external tanks and finally the main fuel tanks. There are three stowage boxes on the left running board, the first one for tools and other maintenance equipment, the second for gun-cleaning equipment and the third for infra-red equipment. Mounted at the very rear of the hull is an unditching beam.

When the T-54 entered service it was not fitted with night vision equipment. The first production tank to have this fitted was the T-54B and early tanks were then refitted with it. The tank is normally fitted with the following equipment: an infra-red driving light which is used in conjunction with the driver's infra-red periscope and has a range of 40 to 60 m; an infra-red searchlight with a range of 400 m mounted on the forward part of the commander's cupola and another infra-red searchlight to the right of the main armament with a range of 800 to 1,000 m.

Main armament of the T-54 is a 100 mm D-10T (originally called the M1944) rifled tank gun which was originally developed around naval gun ammunition and also used in a modified form in the SU-100 assault gun. The D-10T has a horizontal sliding wedge breech block with the recoil system consisting of a hydraulic buffer and a hydropneumatic recuperator. The weapon has a maximum range in the indirect fire role of 14,600 m and can fire the types of 100 mm ammunition shown in Table 2.

An average rate of fire for a T-54 is 4 rds/min. The D-10T gun is not stabilised but the D-10TG, first fitted to the T-54A, is stabilised in the vertical plane only, while the D-10T2S, which is stabilised in both planes, is fitted in the T-54 and later production vehicles and has been refitted in earlier tanks.

The turret traversing and gun elevating system was manual on early tanks but later models have full power traverse and elevation with manual controls for use in an emergency.

Mounted to the right of the 100 mm main armament is a 7.62 mm SGMT machine gun and a similar weapon fixed in the centre of the glacis plate fires forward and is operated by the driver by pressing a button on the right steering lever. Mounted at the loader's position is a 12.7 mm DShKM anti-aircraft machine gun.

Variants

Since the T-54 was accepted for service, it has been upgraded many times with further development resulting in the T-55. Listed, from Russian sources, are the key versions of the T-54 and T-55 MBT with the date in brackets being the year it was accepted for service.

T-54 (1949)

First low-rate production model with wide mantlet and turret undercut at the rear. The model is also referred to as the T-54 Model 1949.

T-54 (1951)

Second low-rate production model with narrow so-called 'pig snout' mantlet and with turret undercut at the rear.

T-54 (1953)

Initial full-rate production model with narrow mantlet and hemispherical turret not undercut at the rear. This turret was used for all subsequent models and modifications of the T-54/T-55.

T-54A (1955)

Modified T-54 with fume extractor and vertical axis stabilisation for 100 mm D-10TG main armament. Other improvements include power elevation for the main armament, an electric oil pump, a bilge pump, a modified air filter and an automatic fire extinguishing system. Some were subsequently fitted with infra-red night vision equipment.

This was also the first version of the T-54 to be fitted with the OPVT river fording equipment that allows crossing of rivers up to 5 m deep and 700 m wide.

The T-54A was built in the former Czechoslovakia, China and Poland with modifications. The model is also referred to as the T-54A Model 1951.

T-54B (1957)

T-54A with a two-axis stabiliser called the Tsiklon (Cyclone), an improved 100 mm D-10T2S gun, turret basket, infra-red night vision equipment for the commander, gunner and driver and three external fuel tanks.

Mounted over the main armament is the L-2 infra-red searchlight; the commander's smaller searchlight is designated the OU-3 while the gunner's standard MK-4 observation device has been replaced by the TPN-1 night observation device.

The Polish equivalent of the T-54B was the T-54AM, a designation sometimes used in the West for this type, whether Polish or Russian.

T-54AK

First standard command version of the T-54A with additional communications equipment and therefore reduced ammunition supply.

T-54AD

Polish command version of the T-54A with the D indicating *dowodca* or command. This version has a modified turret with a slight extension on the turret rear to provide space for the command radios. This vehicle

Table 2

Round	Projectile	Type	Round weight	Projectile Weight	Armour penetration at 1,000 m, 0° (60°)
53-UBR-412B	53-BR-412B	AP-T	30.1 kg	15.9 kg	185 mm
3UBR3	52-BR-412B	AP-T	30.6 kg	15.9 kg	185 mm
53-UBR-412D	53-BR-412D	AP-T	30.6 kg	16.0 kg	185 mm
	3BM8	AP-T		4.13 kg	
	3BM25	APFSDS			
3UBK4	3BK5	HEAT	26.6 kg	12.37 kg	380 (190) mm
3UBK4	3BK5M	HEAT	22.6 kg	4.4 kg	380 (190) mm
	3BK17	HEAT			380 (190) mm
3UBK10-1	9M117	Missile	24.5 kg	17.6 kg	550* (255) mm
53UOF-412	53-OF-412	HE-Frag	30.3 kg	15.6 kg	
53-UOF-412U	53-OF-412	HE-Frag	26.7 kg	15.6 kg	
53-UOF-412ZhU	53-OF-412Zh	HE-Frag	26.7 kg	15.5 kg	
53-UO-415	53-O-415	HE-Frag	30.1 kg	15.4 kg	
3UOF10	3OF32	HE-Frag	30.4 kg	15.4 kg	
3UOF11	3OF32	HE-Frag	26.8 kg	15.4 kg	

*Some Russian promotional material claims the penetration to be 660 mm

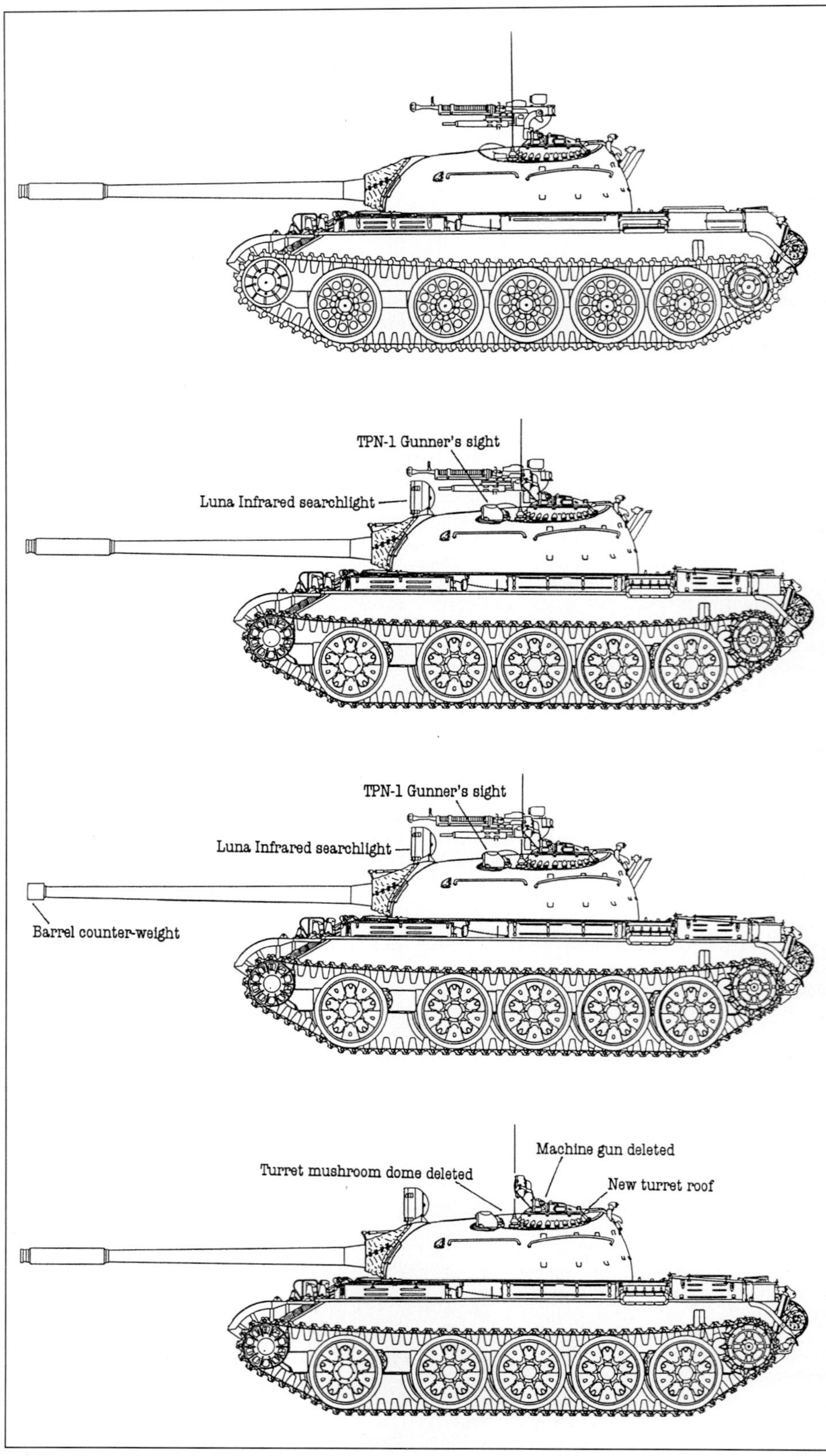

From top to bottom, T-54A Model 1951, T-54B Model 1952, T-54B (T-54 Model 1951 rebuild) and T-55 Model 1958 (Steven Zaloga)

T-55AM2B MBT showing additional armour protection on the turret and a laser range-finder over the 100 mm gun ***2000***/0069546

T-55AM MBT of the Polish Army with 12.7 mm machine gun covered up (Jaroslaw Cislak) 0011793

was designed for use by regimental commanders and regimental chiefs of staff, with a similar T-55 derivative also being built.

The official Polish Army designations are the T-55AD1 (similar to the T-55AK3) and the T-55AD2 (similar to the T-55AK1 or 2). Some have been converted to the T-55AMD1 and D2 configuration. All command tanks have their 100 mm ammunition supply reduced to 37 or 38 rounds.

Upgraded vehicles, for example the T-55AM2K1 and K2 are fitted with R-173 and R-173P radios. The T-55AM2K3 has one R-173 radio, a R-143T2 radio and a 2.2 kW generator set type NS1250B.

T-54M (1983/1988)

T-54A/B upgraded to T-55M standards with new tracks, an upgraded suspension, internal improvements, additional armour, a V-55U engine and an R-173 radio. This model had the development designation Obiekt 140.

T-55 (1958)

Developed by Nizhnyi Tagil under the designation Obiekt 155, this is a T-54 with a new turret without the rooftop mushroom ventilator dome. It has a new stabiliser, an ammunition load increased to 43 rounds, new running gear and a more powerful V-55 diesel engine.

T-55A (1961)

T-55 with radiation shielding added for protection against nuclear weapons. The 7.62 mm SGMT machine guns of the earlier versions were replaced by a 7.62 mm PKT machine gun, the fixed bow-mounted machine gun was also removed which allowed another six rounds of 100 mm ammunition to be carried.

Other improvements include the ability to lay a smoke screen through the exhaust outlet, an NBC system, an air compressor for starting, redesigned front fuel tanks and ammunition compartment, night vision equipment for the commander, gunner and driver, an improved deep fording capability, improved shock-absorbers, an improved fire detection and suppression system and an anti-radiation lining.

T-55 (1970)

T-55A with a 12.7 mm anti-aircraft machine gun mount over the loader's hatch (which was also fitted to the T-62) and other detailed improvements.

T-55A (1970)

T-55 with a 12.7 mm anti-aircraft machine gun mount over the loader's hatch and other detailed improvements.

T-55K Command Vehicles

There are at least three command versions of the T-55 which have additional communications equipment, have a generator and carry less ammunition. The T-55K1 and T-55K2 both carry two R-123s or R-123Ms and an R-124. The T-55K3 carries one R-130M, an R-123M, an R-124 and a 10 m antenna. Variants based on the latter chassis are designated the T-55MK and so on.

T-55M (1983/1988)

T-55 upgrade with Bastion (known as the AT-10 'Stabber' in the West) laser beam-riding missile system, added passive armour protection, a V-55U engine and the R-173 radio system.

The Bastion guided missile complex is designated the 9K116 and uses the 3UBK10-1 round with the 9M117 missile and its associated guidance system. The latter includes the gunner's 1K13 BOM sight and observation device (which replaces the TPN-1 night sight), the 9S831 transformer, a control panel and an electronic panel.

The 3UBK10-1 round is similar in size and shape to the standard 100 mm rounds but is fitted with an ejector device that propels the missile out of the barrel at a speed of 400 to 500 m/s; the sustainer motor of the missile then cuts in and propels the missile to the target. All the operator has to do to ensure a hit is to keep the sight on the target. Missile control is semi-automatic via a laser beam which has high resistance to jamming.

The latest round for the 100 mm rifled gun is designated the 3UBK10M-1 and is fitted with a tandem HEAT warhead which will penetrate at least 600 mm of steel armour protected by explosive reactive armour. The complete round weighs 30.4 kg with the missile having a maximum range of 4,000 m and a minimum range of 100 m. The missile in this version is designated the 9M117M.

The same 9M117/9M117M missile is used for the 100 mm T-12 towed anti-tank gun, 115 mm smoothbore gun of the T-62 MBT and the 100 mm rifled gun of the BMP-3 infantry combat vehicle.

The T-55M also has the Volna fire-control system which includes the KTD-2 laser range-finder with a maximum range of 4,000 m, the BV-55 ballistic computer, the TShSM-32PV vertically stabilised sight and the Tsiklon M1 stabiliser system. The tank commander has the PNK sight.

Fitted to the glacis plate and either side of the turret is additional armour protection which consists of layers of armour plate with the space between them filled with phenopolyurethane.

The armoured side skirts are formed of steel-reinforced rubber sections that are 10 mm thick.

The crew members are also provided with IPZh-1 individual anti-radiation protection vests and additional shielding is provided inside the hull and turret against the effects of nuclear weapons.

Under the belly of the tank is an additional layer of armour to protect the driver if the tank runs over a mine, and the driver's emergency hatch cover has been

T-54 of the Iraqi Army (Christopher F Foss) 0011798

T-55AM2B MBT from the rear fitted with turret-mounted mast-type met sensor (Michael Jerchel) 0018826

T-55 MBT with extensive external stowage and additional armour in service in Croatia (Stefan Marx) **2000**/0069547

strengthened and provided with an armoured cover. Pillars have been installed between the floor and the roof to the right of the driver for increased survivability.

A Type 902B 81 mm smoke grenade launcher system has also been installed which can lay down a smoke screen at ranges of 200 to 350 m from the tank.

The V-55U diesel engine develops 620 hp with inertial supercharging and gives the vehicle a higher power-to-weight ratio. Roadwheel travel has been increased from 135/149 mm to 162/182 mm and the torsion bars are now made of electroslag refined steel. The tracks are improved with a new pattern and new rubber and steel bushings.

The T-55M has a combat loaded weight of 40.5 tonnes and a range of 450 km.

The Czechoslovak tanks have the Kladivo fire-control system with a laser range-finder mounted over the main armament while the Polish tanks have the Merida fire-control system with a laser range-finder integrated into the gunner's sight.

T-55M-1 (1983/1988)

T-55 upgrade with the Bastion (AT-10) laser beam-riding missile system, added passive armour protection, the V-46-5M engine (as installed in the T-62 MBT) and the R-173 radio system.

T-55MV (1983/1988)

T-55 upgrade with the Bastion (AT-10) laser beam-riding missile system, explosive reactive armour protection, a V-55U engine and the R-173 radio system. The explosive reactive armour was developed at the NIIBT Research Institute of the Main Armour Directorate at Kubinka with the V in the designation standing for *vzryvnoi* or explosive. When fitted with the ERA package the T-55AM became the T-55AMZ.

In 1997, the Syrian Army took delivery from Ukraine of its first batch of upgraded T-55MV tanks and these have been forward deployed towards the Golan Heights which have been occupied by Israel since 1967.

It is understood that Syria ordered a total of 200 of these upgraded T-55MV tanks although it is not clear as to whether these upgrades are being carried out in Syria or Ukraine, or the latter is supplying complete vehicles from stock.

For some years, the No 7 Armour Repair Plant at Kyiv and No 17 Armour Repair Plant in Lviv, both in Ukraine, have been offering modernised T-55MV tanks on the export market.

If the T-55BV is fitted with the latest generation explosive reactive armour (Kontakt 5), it affords a high level of protection against both kinetic energy and chemical energy attack.

T-55AM-1 (1983/1988)

T-55A upgrade with the Bastion (AT-10) laser beam-riding missile system, additional passive armour protection, a V-46-5M engine developing 690 hp (or a V-55U developing 620 hp) and an R-173 radio.

Some sources have also quoted the following designations for the version capable of firing the AT-10:

Czech-built	T55-AM2B
Polish-built	T-55AM2P
Russian-built	T-55AM2PB

The T-55AM also has the integrated laser range-finder and sight. Thermal shield for 100 mm gun and the Tsiklon-M stabilisation system.

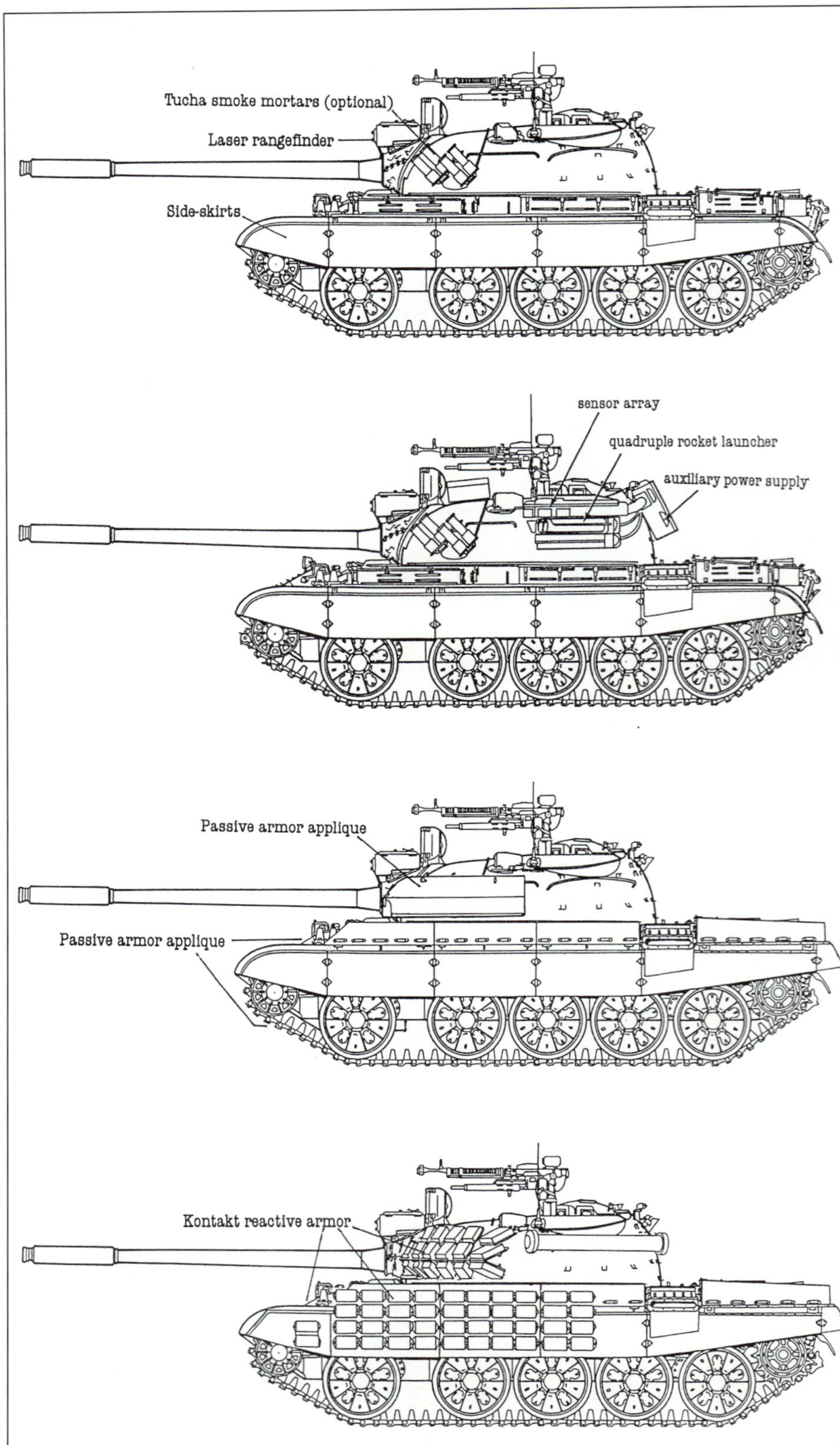

Key members of the T-55 family from top to bottom, T-55AM, T-55AMD Drozd, T-55AM (with Bra Armour) and T-55AMV (Steven Zaloga)

Russian Naval Infantry vehicles with a PT-76 light amphibious tank in the foreground and a modernised T-55MV with explosive reactive armour and a laser range-finder over the mantlet in the background

WZT-2 armoured recovery vehicle of the Polish Army in travelling configuration (Jaroslaw Cislak) 0011796

T-55AD (1983/1988)

T-55M upgrade with the Drozd anti-tank missile system. A motion sensor triggers the incoming missile and this in turn triggers one of the four tubes mounted on either side of the turret firing a shotgun blast of small projectiles. As far as is known, this version is no longer in front-line service with the Russian Army. Some were used by the Russian Naval Infantry. The Drozd system has also been observed fitted to the more recent T-62 and T-80 series MBTs and has been offered on the export market. Illustrations of the Drozd anti-tank missile system installed on the T-80 are given in the entry for this vehicle elsewhere in this section.

T-55AD-1 (1983/1988)

T-55M upgrade with the Drozd anti-tank missile system and a V-46-5M engine.

Table 3 is a summary of recent modifications to the T-55 MBT.

Table 3

Modifications	**Tank Designation**		
	T-55M	**T-55AD**	**T-55MV**
Bastion missile system	Yes	yes	yes
Volna fire-control system	Yes	yes	yes
12.7 mm anti-aircraft MG	Yes	yes	yes
Thermal sleeve for barrel	Yes	yes	yes
Auxiliary armour			
Hull	Yes	yes	no
Turret	Yes	yes	yes
Belly	Yes	yes	yes
Explosive reactive armour	No	no	yes
Active protection measures	No	yes	no
Side skirts	Yes	yes	yes
Napalm protection system	Yes	yes	yes
Smoke grenade launchers	Yes	yes	yes
V-55U diesel engine*	Yes	yes	yes
Modernised suspension	Yes	yes	yes
R-173 radio system	Yes	yes	yes

*Tanks with the V-46-5M engine are designated T-55M-1, T-55AM-1, T-55AD-1 and T-55MV-1.

Russian T-55 MBT upgrades

In 2001, it was revealed that the Russian Omsk Transport Machine Building Bureau had completed tests of at least two upgraded T-55 series MBT, that have been developed for the export market.

The first T-55 upgrade retains the 100 mm gun and is fitted with a new computerised fire-control system. Battlefield survivability has been enhanced by the installation of an Explosive Reactive Armour (ERA) package to its hull and turret. This provides a high level of protection against both chemical energy and kinetic energy attack over the frontal arc. New and improved side skirts and screens are also fitted.

The computerised fire-control system is said to be similar to that installed in the T-80U MBT (also produced at Omsk) and the BMP-3.

T-55 (6) MBT upgrade

The second example is based on a stretched T-55 chassis with an additional roadwheel either side, which also uses some suspension parts of the T-80U for increased cross-country mobility.

The additional roadwheel is mounted at the very front, which has allowed the front of the hull to be extended with a higher level of armour protection.

The standard V-12 water-cooled diesel developing 580 hp of the T-55 has been replaced by a more powerful 690 hp diesel engine.

The original turret has been removed and replaced by a complete T-72 series turret armed with a 125 mm smoothbore gun with a 7.62 mm machine gun being mounted coaxial and a 12.7 mm machine gun on the roof. This turret has an automatic loader that first loads the 125 mm projectile and then the charge. This has allowed the crew to be reduced to three.

The installation of the 125 mm smoothbore gun and new computerised fire-control system, enables the upgraded T-55 to engage targets at a longer range and with a higher hit probability than the current 100 mm armed T-55. The 125 mm gun can also fire a laser-guided missile out to a range of 4,000 to 5,000 m.

Other improvements include the installation of new 81 mm smoke grenade launchers, fire detection and suppression system and a new enhanced paint scheme. The TVK-3 combined tank driver system and the TKKh-1SM commander's system is also installed.

According to the Omsk Transport Machine Building Design Bureau, this T-55 upgrade package is modular, so that the customer can select only those parts which meet their exact operational requirements, or as is more likely, their budget.

Russian T-55M(6) MBT upgrade clearly showing extended hull front with additional roadwheel either side and new armour package (Jim Kinnear) ***2002***/0130968

KBP Upgrade for T-55 MBT

The Russian KBP Instrument Design Bureau, prime contractor for the gun launched laser-guided missiles (100, 115 and 125 mm), are now marketing new missiles which have a maximum range of 5,500 m when fired from the T-55 MBT. The new 100 mm missile also has a tandem HEAT (High Explosive Anti-Tank) warhead to defeat targets fitted with explosive reactive armour.

This latest version uses a new sighting and guidance unit which was originally developed for the now Russian Kurgan BMP-3 infantry fighting vehicle which includes a two-axis line of sight stabiliser, optical channel, image intensification night channel and a laser guidance channel.

A further optional enhancement is the replacement of the image intensification channel by a thermal channel and an autotracker.

OT-54 Flame-thrower Tank

This tank substituted an ATO-1 automatic flame-thrower for the usual 7.62 mm coaxial machine gun. The bow ammunition stowage was modified to permit the carrying of 460 litres of flammable liquid using compressed air for propulsion. Maximum range was 160 m and the system could fire 15 to 20 bursts/min.

Engineer Tanks

In 1959, a quantity of T-55 MBTs was modified to accept the PT-55 mine roller system. These versions are externally identifiable by the attachment fittings on the hull front. The PT-55 mine roller weighed 6/7 tonnes and was installed in the vehicle only when in mineclearing operations. Shortly afterwards, additional fittings were developed for the BTU and BTU-55 bulldozer blades. The BTU-55 was an improved and lightened type (1.4 tonnes versus 2.3 tonnes) and was eventually the more common of the two.

Jung Jungenthal Upgraded T-54/T-55

The German company, Jung Jungenthal, has upgraded a single example of an Egyptian T-54/T-55 MBT and details of this vehicle, which remains at the prototype stage, are given under Germany earlier in this section.

Czech T-55AM2 MBT

Details of the T-55AM2 which is currently being offered for sale by Calibre Prague Limited, are given earlier in this section under the Czech Republic.

T-55 series Upgrade Package

Russia is now offering a number of upgrades to the T-55 series MBT to improve its battlefield survivability:

(1) Improved armour protection by the installation of two different configurations of passive armour and one configuration of explosive reactive armour
(2) Special paint and screens to make it more difficult to detect in the visible, near and far infra-red bands
(3) Improved interior protection from radiation
(4) Installation of new fire detection and suppression system
(5) Upgraded suspension system.

Egyptian T-54/T-55

Details of the modifications carried out and proposed for the Egyptian T-54/T-55 tanks are given under Egypt.

BTS-2 armoured recovery vehicle in travelling configuration and clearly showing snorkel for deep fording operations, this vehicle is in service with Egypt (Paul Beaver) ***2001***/0105982

Polish Army BLG-67 armoured vehicle launched bridge in travelling configuration (Jaroslaw Cislak) 0011791

Finnish T-55 MBTs

Finland has modified 70 T-55 tanks to the T-55M and T-55MK standard. In addition 10 tanks were fitted with BAE Systems, RO Defence, Marksman twin 35 mm turret and these are designated the T-55AM-Marksman (military designation Itpsv 90). They also have 12 ZSU-57-2s which have been upgraded to the ZSU-57-2M configuration by bringing the original chassis up to the T-55A standard.

Vietnamese T-54/T-55

Some Vietnamese T-54/T-55 tanks have been fitted with additional bolt-on armour protection.

Differences between production tanks

Polish-produced tanks often have different stowage arrangements which include the mounting of a rectangular box on the left side of the turret, a smaller square stowage box on the left side of the turret rear and slightly different rear decking.

BTS-1 Armoured Recovery Vehicle

This was the first version of the T-55 to be used as an ARV and is simply a T-54A with its turret removed. It is called the BTS-1 (*Bronetankoviy Tyagach Srdniy* – Medium Armoured Tractor) by the Russian Federation.

BTS-2 Armoured Recovery Vehicle

This was the second version to enter service and has its turret removed and is fitted with a winch and a large spade at the hull rear for vehicle recovery. It has tool and specialised equipment stowage and a small tripod jib crane can quickly be erected. This version was also built in Czechoslovakia as the WZT-1.

BTS-3 Armoured Recovery Vehicle

This was fielded in the 1960s and is also known as the SPK-12G. It is fitted with a hydraulic crane on the front left of the hull which can be used to the front or rear. It has a telescopic jib and can be used to lift complete tank turrets as well as power packs. Other equipment includes a front-mounted BTU-55 dozer blade, a rear-mounted spade, a winch, tools and other specialised equipment. The Czech equivalent was the WZT-2.

BTR-4 Armoured Recovery Vehicle

This is similar to the BTS-2 but has a crane pivoted on the left side of the hull with a telescopic jib which can lift a maximum weight of 1,500 kg. It also has a winch, a rear-mounted spade and extensive external stowage.

Recent information has stated that the BTS-4 and some of the earlier ARV's were in fact based on a T-44 chassis, not a T-54 chassis.

Bosnia T-55/M18

Bosnia has modified some of its SO 76 M-18 (US M18 Hellcat tank destroyer developed in the Second World War) by mounting these turrets on a T-55A chassis. These are then designated the SP 76 M-18 Mod.

Czech VT-55 ARV

This Czech-designed and built ARV has a crane pivoted at the hull rear on the right side and when travelling the telescopic jib is traversed forwards. A cargo-carrying area similar to that installed on Russian ARVs is provided and the vehicle commander has a distinctive cupola on the forward right side of the hull. Additional details are given in *Jane's Military Vehicles and Logistics 2002-2003*.

T-55 series MBT in service with Croatia fitted with rocket propelled mineclearing equipment on either side of hull rear 0011792

Polish T-54/T-55 ARVs

Poland has developed at least two ARVs based on T-54 or T-55 MBT chassis, designated WZT-1 and WZT-2. The WZT-1 is the Polish equivalent of the BTS-2 with some differences while the WZT-2 is a new design and equivalent to the Russian BTS-3 ARV.

Czech MT-55 Bridgelayer

This was developed by Czechoslovakia as the replacement for the earlier MT-34 which was based on the T-34 tank chassis. The MT-55 is fitted with a scissors bridge which is launched hydraulically over the front of the vehicle. When opened out it is 18 m long and will span a gap of 16 m; it will take a maximum load of 50,000 kg. Weight is 37,000 kg and it has a crew of two. Full details of the MT-55 armoured bridgelayer are given in *Jane's Military Vehicles and Logistics 2002-2003*.

German BLG-60 Bridgelayer

This was developed by East Germany and Poland and was used by them in place of the Czech MT-55 bridgelayer. The scissors bridge is launched over the front of the vehicle; it is 21.6 m long and will span a gap of 20 m. Maximum capacity is 50,000 kg. The BLG-60 weighs 37,000 kg and has a crew of two or three. More recently an improved version of the BLG-60 bridgelayer has been introduced under the designation BLG-60 M2. This features improved reliability of the hydraulic system, increased capacity and safety for vehicles using the bridge, use of a number of BLG-60 M2 bridges for multiple crossings of gaps and is more reliable in the field. Details of the BLG-60 are given in *Jane's Military Vehicles and Logistics 2002-2003*. Recently this has been fitted to a Leopard 1 MBT chassis.

Soviet MTU-20 Bridgelayer

The MTU-20 was developed in the late 1960s as the replacement for the older MTU and is launched in a similar manner except that, when travelling, the ends of the bridge are folded on top to reduce the overall length of the equipment. When opened out the bridge is 20 m long and will span a gap of up to 18 m; maximum capacity is 60,000 kg. The MTU-20 weighs 37,000 kg and has a crew of two. Full details of the MTU-20 armoured bridgelayer are given in *Jane's Military Vehicles and Logistics 2002-2003*.

Soviet MTU-1 Bridgelayer

The MTU-1 bridgelayer entered service with the Soviet Army in the late 1950s and is basically a T-54 with its turret removed and replaced by a launching system for a bridge 12.3 m long. The bridge will span a gap of up to 11 m and has a maximum capacity of 50,000 kg. The MTU-1 weighs 34,000 kg and has a crew of two. Full details of the MTU-1 armoured bridgelayer are given in *Jane's Military Vehicles and Logistics 2002-2003*.

Combat Engineer Vehicle (IMR)

This is essentially a T-55 MBT with its turret removed and replaced by a hydraulically operated crane that can be traversed through a full 360°. This is fitted with a pair of pincer-type grabs which are used to remove trees and other obstacles, and which can be replaced by a small bucket. Mounted at the front of the hull is a hydraulically operated dozer blade which can be used in a straight or V configuration, but cannot angle doze. Full details of the IMR are given in *Jane's Military Vehicles and Logistics 2002-2003*.

Dozer T-54

The T-54 and T-55 can be fitted with either the BTU bulldozer blade for clearing soil and obstacles or the STU blade for clearing snow.

Flame-thrower TO-55

A flame-thrower version of the T-55 is known to be in service with the former Soviet Army and Naval Infantry under the designation TO-55. This has its 7.62 mm coaxial machine gun replaced by a flame-thrower for which 460 litres of liquid are carried. Maximum range of the flame gun is 200 m. The 100 mm gun is retained. Because of the amount of space taken up internally by the flame gun and its associated equipment there has been a reduction in the amount of onboard ammunition carried by this version.

SU-122 Tank Destroyer

The 122 mm SU-122 tank destroyer was introduced in 1949 and was a T-54-type chassis with a superstructure similar to that of the SU-100 100 mm assault gun. Mounted in the front of the hull was a 122 mm gun with limited elevation/depression and traverse. The SU-122 tank destroyer was also referred to as the IT-122. These were withdrawn from the late 1950s and converted into armoured recovery vehicles. They have the NATO designation of M1977.

ZSU-57-2 SPAAG

The ZSU-57-2 twin 57 mm self-propelled anti-aircraft gun system was developed to meet the requirements of the Soviet Army and was first seen in 1957. It is a chassis based on T-54 components but with only four roadwheels each side and fitted with an open-topped turret armed with twin 57 mm automatic cannon. Turret traverse is a full 360° with weapon elevation from –5 to +85° and a total of 316 rounds of ready use ammunition is carried in clips of four rounds. This is a clear weather system and has been replaced in front-line service by the more capable ZSU-23-4. It does however remain in service with some other countries. Full details of the ZSU-57-2 are given in *Jane's Land-Based Air Defence 2000-2001*.

Indian T-54 and T-55

Some Indian T-54s have had their Soviet 12.7 mm DShKM anti-aircraft machine guns replaced by a standard 12.7 mm (0.50) Browning M2 HB machine gun as fitted to other Indian AFVs. India fitted sheet steel tubes to the 100 mm barrels to resemble the bore evacuator of the L7 gun to serve as a means of distinguishing Indian T-54/T-55 tanks from Pakistani Chinese-supplied Type 59 MBTs.

Iraqi T-54/T-55 MBTs

Details of some of the Iraqi conversions to their T-54/55 MBTs are given under Iraq.

Israeli T-54/T-55 Upgrades

Details of these are given under Israel. As far as it is known these are no longer in service with the Israel Defence Force but are held in reserve. The Achzarit heavy APC is based on a rebuilt T-54/T-55 chassis and full details of this are given in the Armoured personnel Carriers (tracked) section under Israel.

RO 105 mm L7 Conversion

BAE Systems, RO Defence of the United Kingdom has developed a 105 mm L7A3 conversion package for the T-54/T-55 tank and this is now being offered on the world market. It is also applicable to the Chinese Type 59 MBT.

The original 100 mm gun is replaced by the well-known 105 mm L7A3 which has been turned through 180° as Russian tanks are loaded from the right; a revised recoil system has been designed which includes new buffer components and modified recuperator; there are two new hull fuel tanks to incorporate 105 mm ammunition; the main and emergency firing circuits have been modified; there is a new telescopic graticule to suit new ammunition and a 7.62 mm coaxial machine gun. Following trials in Egypt the Egyptian government placed a production order for this conversion.

Specifications

Model	**T-54**	**T-55**
Crew	4	4
Combat weight	36,000 kg	36,000 kg
Unloaded weight	34,000 kg	33,700 kg
Power-to-weight ratio	14.44 hp/t	16.11 hp/t
Ground pressure	0.81 kg/cm²	0.81 kg/cm²
Length		
(gun forward)	9 m	9 m
(gun rear)	8.485 m	8.485 m
(hull)	6.04 m	6.20 m
Width	3.27 m	3.27 m
Height		
(cupola)	2.4 m	2.35 m
(with 12.7 mm AA MG)	2.75 m	2.70 m
Firing height	1.75 m	1.75 m
Ground clearance	0.425 m	0.425 m
Track	2.64 m	2.64 m
Track width	580 mm	580 mm
Length of track on ground	3.84 m	3.84 m
Max road speed	50 km/h	50 km/h
Total fuel capacity	812 litres	960 litres
Max range		
(road)	510 km	460 km
(with long-range fuel tanks)	720 km	650 km
Fuel consumption	1.9 litres/km	1.9 litres/km
Fording	1.4 m	1.4 m
(with preparation)	5 m	5 m
Gradient	60%	60%
Vertical obstacle	0.8 m	0.8 m
Trench	2.7 m	2.7 m
Engine model	V-54	V-55
Engine type	V-12 water-cooled diesel	V-12 water-cooled diesel
Engine	520 hp/2,000 rpm	580 hp/2,000 rpm
Auxiliary engine	none	None
Transmission	manual with 5 forward and 1 reverse gears	
Steering	geared turn clutch and brake	
Clutch	multiplate	Multiplate
Suspension	torsion bar	Torsion bar
Electrical system	28 V	28 V
Batteries	4 × 12 V, 280 Ah	4 × 12 V, 280 Ah
Model	**T-54**	**T-55**
Armament		
(main)	1 × 100 mm D-10 rifled gun	1 × 100 mm D-10T2S rifled gun
(coaxial)	1 × 7.62 mm SGMT MG	1 × 7.62 mm SGMT MG
(bow)	1 × 7.62 mm SGMT MG	None
(anti-aircraft)	1 × 12.7 mm DShKM MG	None
Smoke-laying equipment	diesel fuel injected into exhaust system	
Ammunition		
(main)	34	43
(7.62 mm)	3,000	3,500
(12.7 mm)	500	None
Gun control equipment		
Turret power control	electrohydraulic with manual controls for emergency use	
(by commander)	yes	Yes
(by gunner)	yes	Yes
Max power traverse rate	360° in 21 s	360° in 21 s
Gun elevation/depression	+17/–5°	+18/–5°
Gun stabiliser		
(vertical)	no	Yes
(horizontal)	no	Yes
Range-finder type	stadiametric	Stadiametric
Elevation quadrant	yes	Yes
Traverse indicator	yes	Yes
Armour		
Hull front upper	97 mm at 58°	97 mm at 58°
Hull front lower	99 mm at 55°	99 mm at 55°
Hull sides upper	79 mm at 0°	79 mm at 0°
Hull sides lower	20 mm at 0°	20 mm at 0°
Hull rear upper	46 mm at 0°	46 mm at 0°
Hull rear lower	46 mm at 0°	46 mm at 0°
Hull top	33 mm	33 mm
Hull floor front	20 mm	20 mm
Hull floor rear	20 mm	20 mm
Turret front	203 mm at 0°	203 mm at 0°
Turret sides	150 mm at 0°	150 mm at 0°
Turret rear	64 mm at 0°	64 mm at 0°
Turret roof	39 mm at 79°	39 mm at 79°
Mantlet	incl in above figures	Incl in above figures
NBC system	yes	Yes
Night vision equipment	yes	Yes

Civilian Conversions

The RFAS is now marketing a number of systems based on the T-54/T-55 MBT for civilian applications and brief details of these are given as follows:

GPM-54M: Firefighting vehicle with a two-man crew.

GPMU-54: Firefighting vehicle with a four-man crew which can carry 9,000 litres of water, foam or powder fire suppressant.

Bronya: Fire reconnaissance vehicle designed for use in conditions of radioactive contamination, suspended

Close up of T-55AD showing turret detail with the Drozd anti-tank missile system installed (Richard Stickland) ***2000***/0038084

T-55M MBT showing laser range-finder over mantlet of 100 mm gun 0011794

Table 4

Country	Quantity	Comment
Afghanistan	700	some Type 59, many were destroyed in recent fighting
Albania	721	Type 59
Algeria	324	
Angola	300	
Azerbaijan	173	including 150 from Ukraine in 1993 and 1994
Bangladesh	140	incl Type 59/Type 69 and ARV
Belarus	381	
Bosnia-Herzegovina	10	approx number
Bulgaria	1,042	
Cambodia	200	some Type 59, Czech Republic supplied 40 T-55 in 1994 and Poland supplied 50 the same year
Central African Republic	4	
Chad	60	
China	Small number	plus many Type 59
Congo	25	plus 15 Type 59
Croatia	210	approx number
Cuba	1,100	
Czech Republic	397	
Ecuador	3	from Nicaragua
Egypt	1,100	some with 105 mm gun
Eritrea	n/a	T-54/T-55
Ethiopia	300+	status uncertain, including 40 from Belarus and 150 from Bulgaria
Finland	70	upgraded, T-55M
Georgia	58	
Guinea	8	
Hungary	597	some modernised
India	700	
Iran	110	
Iraq	1,000	plus some Type 59
Israel	300	reserve
Korea, North	1,175	estimate incl 175 Type 59
Laos	30	
Latvia	3	from Czech Republic in 2000
Lebanon	205	
Libya	1,600	
Macedonia	94	from Bulgaria in 1999
Malawi	n/avail	Unconfirmed user
Mauritania	35	
Mongolia	350	
Mozambique	80	
Namibia	n/avail	Unconfirmed user
Nicaragua	127	
Nigeria	50	
Pakistan	50	plus 1,200 Type 59
Peru	300	about 50 operational
Poland	839	most locally built, some modernised
Romania	821	
Russia	3,000	
Rwanda	12	T-54
Slovakia	206	
Slovenia	55	qv Slovenia
Somalia	n/avail	status uncertain
Sri Lanka	40	from Czechoslovakia from December 1991
Sudan	250	plus 9 from Belarus in 1996
Syria	2,100	
Tanzania	35	all believed to be Type 59
Togo	2	
Uganda	150	including 90 from Bulgaria in 1998 and 60 from Ukraine in 1995
Ukraine	154	
Uruguay	15	T-55 delivered 1997 from Israel
Uzbekistan	80	
Vietnam	900 approx	plus 350 Type 59
Yemen	700 approx	Bulgaria supplied 6 T-55 in 1994
Yugoslavia	785 approx	
Zambia	10	

or aerosol flammable materials. Equipment fitted includes thermal night vision equipment, dosimeter and chemical sampling equipment, land navigation system, communications and individual crew protection for the three-man crew. The Bronya is based on a T-55 MBT chassis.

Shchit: Multipurpose firefighting mount for use in forest areas, it has a three-man crew and is based on a T-55 MBT chassis.

Irtysh: Mobile protected command post based on a T-55 MBT chassis with a crew of four or five men.

Slavutich: Firefighting system based on a T-54/T-55 MBT chassis. On the front of the chassis is a rocket launcher-type device that fires salvos of flame suppressant materials out to a maximum range of 100 m.

Sojka: Based on a T-55 MBT chassis and designed to extinguish fires under conditions of radioactive contamination.

Ladoga: Telescopic hoist vehicle based on a T-55 MBT chassis; the hoist has a maximum height of 26 m which can be reached in 12 seconds.

Fire engine: Based on a T-54 MBT chassis, it has a two-man crew and is fitted with extendable ladders and foam-type firefighting equipment.

GTU-1 Universal Tracked Tractor: Based on a T-54/T-55 chassis, and fitted with a 25 tonne and 0.5 tonne winch, a scraper/dozer blade, a cargo platform, a crane with a maximum lifting capacity of 3,000 kg, a self-recovery unditching beam and specialised equipment including an electronic welding system.

Mobile power station: Based on a T-55 MBT chassis and fitted with a generator.

T-55 MTU-20 bridgelayer in travelling configuration

BMR mineclearing vehicle fitted with a KMT M1988 mineclearing system. In the background is a T-54 fitted with the older KMT-type system

T-54/T-55 Retrofits

Details of the Teledyne Continental Motors (which in 1996 was taken over by General Dynamics Land Systems) rebuilt T-54 for the Egyptian Army, called Ramses II, are given under Egypt. This remains at the prototype stage.

Romanian T-54/T-55

Available details of the Romanian T-54/T-55 are given in this section under Romania. These are designated the TR-800; TR-85 and TR-580. They have never been exported.

BMR mineclearing vehicle

This was first observed in Afghanistan and uses a T-54/T-55 chassis with a new superstructure with a very well sloped glacis plate. The driver is seated at the front left with the vehicle commander to his right. The latter is provided with a cupola on which is mounted a 12.7 mm machine gun. On the right side of the hull is a bank of smoke grenade dischargers firing forwards. This vehicle has been observed to be fitted with the KMT M1988 mineclearing rollers as well as the KMT-7 system. Details of the vehicle, also referred to as the M1977, are given in *Jane's Military Vehicles and Logistics 2002-2003.*

Mineclearing T-54 and T-55 tanks

These tanks can be fitted with a wide range of mineclearing equipment including Czech roller and plough type systems, Russian roller system models PT-54, PT-54M and PT-55, Russian plough system type KMT-4, and the Russian plough and roller-type systems called the KMT-5, KMT-5M, KMT-6 and KMT-6 M2. Details of these systems are given in *Jane's Military Vehicles and Logistics 2002-2003.* A more recent development is the mounting on the rear of the tank of rocket-propelled charges which are launched across the minefield and once on the ground are detonated.

BTR-T armoured personnel carrier

Details of this Russian full-tracked armoured personnel carrier, based on a much modified T-55 series MBT chassis and first shown late in 1997, are given in the Armoured personnel carriers (tracked section). As of early 2002-2003 the BTR-T remained at the prototype stage.

Czech MT-55A AVLB in the travelling position from the front (PB) **2002**/0130958

Achzarit infantry armoured vehicle

Details of this Israeli conversion of a T-54/T-55 chassis into a well armoured infantry fighting vehicle are given in the Armoured personnel carriers (tracked) section under Israel. Large numbers of these vehicles are currently used by the Israel Defence Force.

Status

Production complete. In service with the following countries: (see Table 4).

In addition they are used by a number of irregular units in the Middle East and elsewhere and some countries also use them for training. Under the continuing CFE talks many of the figures are being substantially reduced.

Manufacturers

Three former Soviet tank plants plus Labedy in Poland and Martin in Czechoslovakia. Poland produced the T-54 from 1956 to 1964 and the T-55 from 1964 onward. Was manufactured in China as the NORINCO Type 59 (qv).

UPDATED

SLOVAKIA

ZTS T-72 series MBT

Development/Description

From 1979-80, the then Czechoslovakia undertook licensed production of the Russian-designed T-72M1 MBT for the Czech Army and for the export market.

ZTS Martin was responsible for the chassis and ZTS Dubnica nad Vahom was responsible for the turret. Following the split of Czechoslovakia into the Czech Republic and Slovakia, the former is no longer involved in the production of MBTs and this is now concentrated at the ZTS Tees Martin facility in Slovakia.

This facility is responsible for marketing the Russian-designed T-72M1, full details of which are given under the T-72 entry in the Russian Federation and Associated States (RFAS). According to United Nations sources, the only Slovakian MBT exports that took place between 1992 and 1998 was the supply of 58 T-72 series MBTs to Syria in 1973.

Slovakia has also exported surplus quantities of older T-55 series MBTs. According to the United Nations, between 1992 and 2000 exports were made to just one country, Angola. This country took delivery of 150 vehicles in 1999 and 55 in 2000.

ZTS T-72M1-A MBT clearly showing explosive reactive armour array on turret and chassis (Stefan Marx) **2001**/0105969

More recently, ZTS Tees Martin has developed more specialised versions of the T-72 MBT for the export market, as well as a number of variants based on this chassis and details of these are given below.

Variants

T-72M1 MBT

Standard production vehicle, same as Russian-built T-72M1 MBT.

T-72M1 Moderna MBT

Developed to the prototype stage and fitted with locally developed Explosive Reactive Armour (ERA), two Oerlikon Contraves 20 mm KAA-001 cannon mounted externally one either side of the turret and a new computerised fire-control system.

The gunner has the TPD-K1 × 8 day sight with laser range-finder and a SABCA VEGA thermal night sight with ×1.8 and ×5.5 magnification. The commander has a roof-mounted stabilised SAGEM (previously SFIM) VS580 series day panoramic sight. This version is no longer being marketed.

T-72M1-A MBT

This is a further development of the T-72M1 and features the installation of the Slovenian EFCS3-72A computerised fire-control system which includes the commander's SGS-72A stabilised passive sight with two magnifications, laser detector warning system, installation of more powerful S12U diesel engine developing 850 hp, modified transmission, new driver's PNK-72 night device, improved protection for hull floor with driver's seat now suspended from roof, new DSM 16.1 combined instrument panel for driver, new Explosive Reactive Armour (ERA) package which also provides protection against tandem HEAT warheads, new fire detection and suppression system and modified electrical harness. One prototype of this version has been completed and tested. Main

Specifications
(T-72M2)
Crew: 3
Combat weight: 46,250 kg
Power-to weight ratio: 18.11 hp/tonne
Ground pressure: 0.89 kg/cm^2
Length:
(gun forwards) 9.53 m
(hull) 6.86 m
Width: 3.75 m
Height:
(turret top) 2.19 m
Ground clearance: 0.450 m
Track width: 580 mm
Length of track on ground: 4.27 m
Max road speed:
(road) 60 km/h
(cross-country) 35 - 45 km
Fuel capacity: 1,750 litres (with external fuel tanks)
Max range: 620 km (road)
Fording:
(without preparation) 1.20 m
(with preparation) 5 m
Gradient: 63%
Side slope: 46%
Vertical obstacle: 0.85 m
Trench: 2.80 m
Engine: S12U 12-cylinder, water-cooled multifuel developing 838 hp at 2,000 rpm
Transmission: epicyclic with 7 forward and 1 reverse gears
Clutch: disc
Suspension: torsion bars and hydraulic shock-absorbers
Electrical system: 27 V
Batteries: 4 × 12 V
Ammunition:
(main) 1 × 125 mm 2A46M2 smoothbore gun
(coaxial) 1 × 7.62 mm TK model 95 MG
(anti-aircraft) 1 × 30 mm 2A42 cannon
(smoke laying equipment) 11 × 2 × 80 mm
Gun control equipment:
(by commander) yes
(by gunner) yes
(commanders override) yes
Gun elevation/depression: +13.4/–6°
Turret traverse: 360°
Gun stabiliser:
(vertical) yes
(horizontal) yes
Range setting device: yes, laser (TPD-K1)
Elevation quadrant: yes
Traverse indicator: yes
NBC system: yes
Night vision equipment: yes

emphasis by ZTS is on the T-72M2 covered in detail following.

T-72M2 MBT

This is a further development of the above with the following additional modifications, new fire-control system with a ballistic computer, new intercom and radio set, modernised 125 mm smoothbore gun 2A46M2 and removal of two Oerlikon Contraves 20 mm cannon and installation of 30 mm 2A42 cannon on right side of turret only. The primary role of the latter is to engage low-flying aircraft and helicopters.

The tank commander has a stabilised SAGEM (previously SFIM) MVS 580 day/thermal panoramic sight incorporating a laser range-finder while the gunner has a Fontana TIGS day/thermal night sight with integrated electronic module with ballistic computer, control panel and CRT display. The gunner retains the standard TPD-K1 stabilised day sight which incorporates a laser range-finder.

One prototype of this version has been completed and tested.

Other features of the upgraded T-72M2 MBT include a new Turret Management System (TMS) which includes the ballistic computer, NC regulator to control the movement of the turret and guns and servo system for the 30 mm 2A42 cannon mounted on the right side of the turret.

The TMC also integrates the gunner's original TPD-K1 sight and the roof-mounted commander's MVS 580 stabilised panoramic sight. The latter is normally used to

ZTS VT-72B armoured recovery vehicle with crane elevated 0018831

Upgraded T-72M2 MBT showing explosive reactive armour package installed over frontal arc **2002**/0131036

Close-up of the T-72M2 turret showing 30 mm 2A42 cannon and GALIX grenade launchers **2002**/0131037

control the 30 mm 2A42 cannon and can also control the 125 mm smoothbore gun.

The latest explosive reactive armour is installed over the frontal arc of the vehicle and a new MASK painting system provides masking against radar, infra-red and visible wavelengths.

Combat survivability is further enhanced by the installation of an LIRD-4D laser irradiation detector system which is used in conjunction with the French Galix vehicle protection system. This combination enables a variety of grenades to be launched under manual or automatic modes.

The diesel engine has been upgraded and coupled to a modified transmission. A German Deugra fire and explosion suppression system is fitted as standard. Raychem have provided the cabling system.

One prototype of this version has been completed and tested.

Zuzana T-72M1 SPG

This is essentially the chassis of the T-72M1 MBT fitted with the complete turret of the Zuzana 155 mm self-propelled gun-howitzer which is fully described in the Self-propelled guns and howitzers (wheeled) section. This combination has been evaluated in India to meet the requirements of the Indian Army. As of early 2002 India had not selected any self-propelled artillery system to meet its future requirements.

VT-72B ARV

This is based on a T-72 chassis with a new superstructure. Pivoted at the right side of the hull is a hydraulically operated crane with a telescopic jib which is traversed to the rear when not required.

Mounted on the forward part of the hull is the main winch, while mounted at the front of the hull is a hydraulically operated dozer/stabiliser blade.

Full details of the VT-72B ARV are given in *Jane's Military Vehicles and Logistics 2002-2003*.

India has purchased 113 of these vehicles out of a total requirement of around 400 units. These will be used to support the locally manufactured T-72M1 and new Arjun MBTs.

MT-72 AVLB

This is currently being offered for the export market and consists of a T-72 MBT chassis with its turret removed and replaced by a launching mechanism for a scissors-type bridge which is launched over the front of the vehicle.

When opened out, this bridge is 20 m long and will typically span a gap of 18 m and take tracked and wheeled vehicles weighing up to 50,000 kg. Full details of this vehicle, which has yet to enter production, are given in *Jane's Military Vehicles and Logistics 2002-2003*.

Status

Production as required.

Manufacturer

ZTS Tees Martin.

UPDATED

SLOVENIA

Slovenian M-55 S1 MBT

Development

In the mid-1990's Slovenia started a programme to upgrade its old 100 mm armed T-55 MBTs to a new standard locally known as the M-55 S1.

While some of the sub-systems have been designed and manufactured in Slovenia, especially in the area of optics, other parts have been supplied by foreign countries, including Israel.

Prime contractor in Slovenia for this T-55 MBT upgrade was SZ-Stroji in Tehnoloska Oprema doo who is also the local manufacturer of the Austrian Steyr-Daimler-Puch Pandur (6 × 6) armoured personnel carrier for the Slovenian Army.

It is understood that the first example of the M-55 S1 MBT was completed in 1997 with final deliveries being made to the Slovenian Army in 1999. Israel provided considerable assistance to Slovenia with this upgrade, especially Israel Military Industries, Elbit and possibly Elop.

It is believed that about 30 of the 55 T-55 MBTs in service in Slovenia were upgraded to the enhanced M-55 S1 MBT standard. It is expected that this company will now upgrade about 40 M-84 MBTs which Slovenia retained when it declared independence from Yugoslavia in 1991. Full details of the M-84 series MBT, a further development of the Russian T-72, are given later in this section under Yugoslavia.

Description

The overall layout of the upgraded T-55 S1 MBT is virtually identical to the Russian T-54/T-55 MBT covered in detail in an earlier entry.

The upgrade carried out in Slovenia covers the three main areas of the vehicle, armour, mobility and firepower.

The standard 100 mm D-10T rifled gun has been replaced by a 105 mm L7 type rifled tank gun which can fire a wide range of ammunition types including High Explosive Anti-Tank - Tracer (HEAT-T), Armour Piercing Fin Stabilised Discarding Sabot - Tracer (APFSDS-T) and High Explosive Squash Head - Tracer (HESH-T).

The 105 mm L7 gun is fired electrically as is the 7.62 mm machine gun mounted coaxially with the main armament. The roof mounted 12.7 mm machine gun is retained.

The 105 mm barrel has been fitted with a three part thermal sleeve which is said to reduce barrel bending by over 70 per cent so helping to increase accuracy of the weapon.

A new computerised fire-control system has been installed in the M-55 S1 MBT which is claimed to allow stationary and moving targets to be engaged with a high first round hit probability while the tank itself is stationary or moving.

The fire-control system has three modes of operation, automatic, semi-automatic and manual. In the first mode all the crew have to do is manually insert the type of 105 mm ammunition to be used.

The gunner is provided with the Slovenian Fontana SGS-55 two-axis stabilised day/image intensification sight with a laser range-finder which feeds information to the fire-control computer. As an option, a thermal version of this system is also available.

The vehicle is fitted with a Slovenian COMTOS-55 commander's take over set which enables him to override the gunner and carry out target tracking, ranging, target selection and to fire the main gun/ coaxial machine gun. This is mounted under the existing TPN-10 sight which is retained with the picture of the target coming from the gunner's sight by a charge coupled device.

The driver is provided with a Fontana CODRIS combined day/night driver's periscope which allows day (80° field of view) or night (78° field of view) simply by flicking a switch.

The standard T-55 has a turret and hull of welded and cast steel armour but for added battlefield survivability the upgraded M-55 S1 has been fitted with Explosive Reactive Armour (ERA) to the frontal arc of the turret, glacis plate and upper hull sides.

This armour is believed to be the latest Israeli Super Blazer system which gives a high level of protection against High Explosive Anti-Tank (HEAT) attack but cannot be activated by small arms fire.

According to Slovenia this ERA provides the following protection levels:

(1) Over an arc of 50° left and right from RPG-7 HEAT warheads
(2) Over an arc of 40° against HOT HEAT missile warheads
(3) Over an arc of 30° against TOW HEAT missile warheads

The upper part of the suspension is protected by a five part rubber skirt which provides two functions. The first is to help reduce dust and the second to activate High Explosive Anti-Tank (HEAT) warheads before they hit the main armour of the vehicle.

For improved cross-country mobility the suspension of the vehicle has been replaced and now each side consists of five rubber-tyred road wheels, drive sprocket at the rear, idler at the front and four return rollers. The new 580 mm wide track has been fitted with replaceable rubber pads and to reduce driver fatigue a hydraulic steering system has been installed.

The standard T-55 MBT is fitted with a V-12 water cooled diesel developing 580 hp but this has been replaced by a 600 hp diesel engine. As an option a new power pack that includes a MAN 850 hp diesel can also be installed. With a combat weight of 40 tonnes, this power pack would give a good power-to-weight ratio of just over 21 hp/tonne.

The upgraded M-55 S1 does not only rely on its improved armour for battlefield survivability and is also fitted with Fontana LIRD (Laser IRradiation Detector and warner) system. This consists of a roof mounted detector head unit, indicator unit and harness.

The system has a full 360° coverage in azimuth and from –20 to +60 ° in the vertical plane. It can alert the crew with a pulsed laser from a laser range-finder or a laser illuminator/designator. This allows the crew to take action such as moving to a new position or activating onboard devices such as the grenade launchers.

Mounted either side of the turret is a bank of six Israeli Military Industries IS-6 CL-3030 instantaneous self screening system for combat vehicles — installed

T-55 upgraded in Slovenia to the enhanced M-55 S1 standard which includes the installation of explosive reactive armour and a 105 mm rifled tank gun ***2002***/0131042

on Israeli armoured vehicles for many years. This launches grenades to a range of 40 m and a 60 m wide smoke screen is laid which lasts for about 2 minutes, depending on wind conditions. The tank can also lay a smoke screen by injecting diesel fuel into the exhaust outlet on the left hand side of the hull.

The engine and crew compartments are each fitted with a fire-detection and suppression system. Each of these systems comprises optical detectors, cylinders filled with Halon 1301 gas and a control device which can be operated automatically or manually.

A brand new communication system is installed in the M55 S1 which includes two new antennas, new radios, VIC intercom system and an externally mounted telephone. The telephone is fitted at the rear of the vehicle on the right side and allows the accompanying infantry to talk to the tank crew.

The older Russian-type lighting system has been replaced by a Western system including new headlamps, side lights, turning indicators and a roof-mounted flashing warning lamp for peacetime use.

Mounted on the turret rear is a large stowage box where the items that have been displaced by the installation of vertical hull side armour have been relocated. Equipment such as kit, rations and a camouflage net can be stowed in this basket.

Status

Upgrade completed for Slovenian Army.

Manufacturer

SZ - Stroji in Tehnoloska Oprema doo.

UPDATED

SOUTH AFRICA

Vickers OMC Olifant MBT

Development

It has been revealed that before the introduction of the Olifant Mk 1A there were a number of other Centurion MBT upgrades in South Africa.

In 1972, the Centurion was fitted with a V-12 fuel-injected petrol engine developing 810 hp coupled to a new three-speed (two forward and one reverse) automatic transmission. This project was called the Skokiaan but only eight conversions were made.

This was followed by the Semel project in 1974 which involved fitting the eight Skokiaan vehicles and some unconverted Centurions with a modified engine and some other improvements and these were called the Centurion Mk 5A or Semel. A total of 35 of these vehicles was produced and some were used in the then southwest Africa.

These vehicles were followed by the Olifant Mk 1 whose development commenced in 1976, with the first prototype being completed the same year followed by the second in 1977 and the third in 1978. This was then accepted for service with the South African Armoured Corps.

The Olifant Mk 1 has the same power pack as the Semel but also incorporates many other modifications, especially to the subsystems, including the suspension, turret drives and vision equipment. The tank commander was also provided with a hand-held laser range-finder.

This was followed by the Olifant Mk 1A which entered production in 1983 and entered service two years later and subsequently saw combat service in Angola.

This has all the improvements of the Olifant Mk 1 but also has a number of other improvements. Modifications over the original Centurion can be summarised as the installation of a locally built version of the BAE Systems, RO Defence 105 mm L7-type rifled tank gun, a bank of eight 81 mm electrically operated smoke grenade dischargers either side of the turret and a new power pack consisting of a V-12 diesel engine developing 750 hp coupled to a semi-automatic transmission. Passive night vision equipment has been installed as has a new commander's cupola.

The Olifant Mk 1A does not have a computerised fire-control system but the commander is provided with an AFV No 4 sight and the gunner with an AFV No 18 sight with magnifications of ×1 and ×6, range drum, elevation indicator and a traverse indicator.

Vickers OMC Olifant Mk 1A showing white light searchlight mounted over 105 mm gun **2002**/0131041

The gunner is provided with an Eloptro MSZ-2 two-channel sight with an integral laser range-finder working in both channels. The right-hand channel has a ×8 magnification and an 8° field of view while the left-hand channel houses the night elbow which has an image intensifier with a ×7.2 magnification and a 7.5° field of view. If required, the night elbow can be replaced by a ×1 magnification observation prism.

The driver can replace his standard day periscope with a locally designed passive night vision device.

The installation of the new power pack greatly increases the operational range of the Olifant 1A compared with the original Centurion as the diesel engine is much more fuel efficient; in addition, fuel tanks with a significant increase in capacity have been fitted.

Development work on the Olifant Mk 1B commenced in 1983, with the first prototype being completed in 1985. Following trials with prototype vehicles by the South African Armoured Corps, the first production Olifant Mk 1B vehicles were completed in early 1991. It is believed that the first production batch of Olifant Mk 1B amounted to just under 50 vehicles. This upgrade programme has now been completed.

The South African Olifant MBT programme was originally carried out by OMC Engineering (Pty) Limited with the OMC standing for the Olifant Manufacturing Company. This company eventually became Reumech OMC and was taken over by Vickers Defence Systems of the UK late in 1999 and then renamed Vickers OMC.

Description

The overall layout of the Olifant Mk 1A and Mk 1B is virtually identical to that of the basic Centurion MBT described under the UK.

Whereas the Olifant Mk 1A is based on the original Centurion MBT, the Mk 1B is a complete rebuild and incorporates many features not previously fitted to the Centurion. These can be summarised as:

(1) Replacement of the bogie-type suspension by torsion bars which greatly improve cross-country mobility. Total wheel travel is now 435 mm. The

Olifant Mk 1B with turret traversed to the left. Side skirts are not fitted (T J Gander)

Olifant Mk 1B with 105 mm gun depressed and with side skirts fitted (Christopher F Foss) 0018832

Specifications

Olifant Mk 1A
Crew: 4
Combat weight: 56,000 kg
Power-to-weight ratio: 13.39 hp/t
Length: 8.29 m
Width: 3.39 m
Height: 2.94 m
Max road speed: 45 km/h
Fuel capacity: 1,240 litres
Max range:
(road) 500 km
(cross-country) 240 km
Gradient: 60%
Trench: 3.45 m
Engine: V-12 diesel developing 750 hp at 2,300 rpm
Transmission: semi-automatic, 2 forward and 1 reverse gears
Armament:
(main) 1 × 105 mm L7 gun
(coaxial) 1 × 7.62 mm MG
(anti-aircraft) 1 × 7.62 mm MG
(smoke grenade dischargers) 2 × 4 81 mm
Ammunition:
(main) 72
(7.62 mm) 5,600
NBC system: no
Night vision equipment: yes

Olifant Mk 1B
Crew: 4
Combat weight: 58,000 kg
Power-to-weight ratio: 16.38 hp/t
Length: 8.61 m
Width: 3.42 m
Height: 3.55 m
Ground clearance: 0.345 m
Max road speed: 58 km/h
Fuel capacity: 1,468 litres
Max range:
(road) 500 km
(cross-country) 260 km
Fording: 1.5 m
Gradient: 60%
Vertical obstacle: 0.98 m
Trench: 3.45 m
Engine: V-12 diesel developing 950 hp at 2,400 rpm
Transmission: automatic with 4 forward and 2 reverse gears
Armament:
(main) 1 × 105 mm L7 gun
(coaxial) 1 × 7.62 mm MG
(anti-aircraft) 1 × 7.62 mm MG
Smoke-laying equipment: 2 × 4 81 mm smoke grenade launchers
Ammunition:
(main) 68
(7.62 mm) 5,000
NBC system: no
Night vision equipment: yes

first and last two roadwheel stations either side have a hydraulic damper with hydraulic bump stops at each wheel station
(2) The hull now has a double floor with the torsion bars between the two floor plates; the double floor also provides increased protection against mines
(3) New power pack consisting of a V-12 diesel developing 950 hp coupled to a new AMTRA III automatic transmission with four forward and two reverse gears, which also incorporates a double-differential steering system with a two-speed mechanical steering drive and hydraulic retarder
(4) Fitting of new final drives with a ratio of 6.167:1
(5) Installation of additional passive armour to the glacis plate, nose, forward part of the turret roof and turret front and sides. Turret front and side armour is placed well away from the actual turret
(6) Fitting of modular side skirts that provide some protection against HEAT attack as well as helping to keep dust down
(7) Redesign of the driver's compartment with new single-piece lift and swing-type hatch cover and three new day periscopes, one of which can be replaced by a passive night vision device. A hull escape hatch is fitted. The driver has a new instrument panel and the tillers have been replaced by a yoke-type system
(8) A turret bustle is fitted for increased stowage
(9) Two new unity-powered roof-mounted day periscopes are fitted for the loader, who also has a new one-piece hatch cover
(10) The 105 mm gun L7 is now fitted with a thermal sleeve and a fume extractor
(11) New solid-state gun control equipment
(12) Gunner's sight now has day/night capability and incorporates a laser range-finder for improved first round hit capability. Commander has new day/night sight
(13) The headlamps are now armoured to enhance bush bashability
(14) Mounted over the 105 mm main armament is an infra-red/white light searchlight that moves in elevation with the main armament
(15) Crew compartment (driver's and fighting) is fitted with a fire detection/suppression system.

The Olifant Mk 1B has a combat weight of 58 tonnes and a maximum speed of 58 km/h and a total of 68 rounds of 105 mm ammunition is carried.

For operations in dense bush the Olifant can be fitted with a V-shape bush basher bar which is mounted on the front of the tank and enables the vehicle to push its way through dense bush without sustaining damage.

The nose of the Olifant can be fitted to take an electrohydraulic dozer blade or mechanical mine rollers. The 3.5 m wide dozer blade weighs 1,500 kg and takes 7 minutes to install using a crane and 15 minutes without a crane.

Variants

Olifant 2 turret

The LIW Division of Denel have developed to the prototype stage a new turret called the Olifant 2. This is claimed to be a state-of-the-art MBT turret featuring hunter-killer, fire-on-the-move and day/night target engagement capabilities.

Main armament can either be the LIW 105 mm GT-8 rifled tank gun or a LIW developed 120 mm smooth bore gun. A 7.62 mm machine gun is mounted coaxial with the main armament and banks of 81 mm electrically operated smoke grenade launchers are mounted either side of the turret towards the rear.

The gunner, seated on the right hand side and foward of the commander, is provided with a periscope stabilised day/thermal sighting system that incorporates a laser range-finder. The tank commander has an independent panoramic stabilised day/passive night sight.

In a typical target engagement the target would first be acquired by the tank commander who would then hand over the target to the gunner. The tank commander would then start to search for another target to be engaged.

The turret is also fitted with a full solution fire-control system with automatic sensors for meteorological parameters and dynamic tilt together with a state-of-the-art gun and turret drive system. According to LIW, the turret has a high hit probability when the vehicle or the target is moving or stationary.

Protection for the crew of three is provided by means of modular armour packages that are attached to the all-welded steel shell. The front of the turret is shaped like an arrow head in a similar manner to that of the latest German Leopard 2A5 MBT. A turret bustle-mounted ammunition carousel is suitable for the stowage of kinetic ammunition.

Crew protection from ammunition ignition is enhanced by blow off panels on the turret roof in a similar manner to those of the US M1/M1A1/M1A2 Abrams series of MBT.

Ready rounds of 105 mm ammunition are located in the turret basket for ease of access by the gunner positioned on the left hand side of the turret. According to LIW a comfortable rate of fire of 10 rds/min can be achieved.

The turret basket and ring gear are fully compatible with the British developed Centurion and Chieftain MBTs.

Future South African MBT

It was expected that the current Olifant Mk 1 series of MBT would have been replaced by a new MBT in the early part of the 21st century.

There were two contenders to meet this requirement, the French Giat Industries Tropicalised Leclerc and the British Vickers Defence Systems Challenger 2E.

The requirement was for 96 new MBTs, six armoured recovery vehicles and four armoured vehicle launched bridges on a similar chassis.

Late in 1998, South Africa announced that it was to order a major package of new defence equipment but MBTs were not part of this package.

It is now possible that South Africa will upgrade its Olifant 1 series MBTs to extend their operational lives.

Mineclearing Vehicle

The Olifant Mk 1A and Mk 1B can both be fitted with plough- or roller-type mineclearing systems.

Armoured Recovery Vehicle

An armoured recovery vehicle based on the chassis of the Olifant Mk 1A has been completed and is now in service with the South African Armoured Corps. Details are given in *Jane's Military Vehicles and Logistics 2002-2003.*

Armoured Vehicle-Launched Bridge

South Africa has developed an armoured vehicle-launched bridge based on a Centurion (Olifant) MBT chassis. This bridge is not of the scissors type and is extended over the front of the vehicle.

Status

Conversions to Mk 1A and Mk 1B have been completed. Late in 1996, South Africa stated that it had a total of 224 Olifant Mk 1A and Mk 1B MBTs in service. In addition there were 28 bridge-laying tanks, 10 bridge-laying vehicles, 20 bridge carrying vehicles and 30 26 m bridges.

Prime contractor

Vickers OMC.

UPDATED

SPAIN

Spanish MBT fleet

Until the mid-1990's, the Spanish Army and Marines fielded a total of 854 MBTs:
375 M47E1 (now scrapped)
164 M48A5E2 (retained)
299 AMX-30E (89 scrapped)
16 M48A3 (Marines, now replaced by M60A3)

Under the Conventional Forces Europe Treaty, Spain is allowed to have a total of 794 MBTs. Spain has already agreed to lower this figure to a ceiling of 684 vehicles.

In the spring of 1995, Spain scrapped 118 M47 and 16 M48A3 MBTs in the final phase of force reduction obligations under the Conventional Forces Europe Treaty.

As of early 2002, the front-line Spanish MBT fleet consisted of the following vehicles:

Type	Quantity
AMX-30 ER1	60
AMX-30 EM2	150
M48A5E2	164
M60A1E	50
M60A3E	154
M60A3E TTS	106
Leopard 2A4	108

All of these above are used by the Spanish Army with the exception of 16 M60A3 MBTs which are used by the Spanish Marines.

Figures issued by the United Nations show that the US delivered 96 M60 series MBTs to Spain in 1992 followed by 311 in 1993. This is a total of 407. The difference could be that some were used as spare parts or for conversions to more specialised roles.

Following extensive trials with the prototype of an M60A3E, the remaining 153 may be upgraded to the M60A3E1 standard. These will have a number of improvements including an upgraded diesel power pack developing 900 to 950 hp, explosion/fire detection and suppression system for the crew compartment, new fire-control system similar to that of the M48A5E2 and probabaly a new Explosive Reactive Armour (ERA) package from Santa Barbara Sistemas.

Another 94 M60A3 MBTs will be upgraded in the area of mobility only, for example fitted with the more powerful diesel engine.

Spanish M60A1/M60A3/M60A3 TTS have a bank of four 76 mm smoke grenade launchers mounted either side of the turret. Two prototypes have been provided with ERA for their hulls, turrets and side skirts.

Late in 1996, the Spanish MoD stated that 36 US-supplied M60 series MBTs would be converted for mine clearance operations at a cost of around US$26.6 million.

Early in 1997, the Spanish Ministry of Defence announced that Peugeot Talbot SA had been awarded a contract for the conversion of 38 M60 MBTs into the M60 CZ-10/25E Alacran for delivery from 1997 through to 1999.

M60A3 TTS MBT with German-type 76 mm smoke grenade launchers (Francisco A Marin) 0018833

Additional details are given in the entry for Peugeot Talbot MBT conversions later in this section. Also competing for this contract was GAMESA of Spain which also built a prototype vehicle for trials by the Spanish Army.

In the future the current fleet of MBTs used by the Spanish Army will be supplemented by a Spanish version of the German Krauss-Maffei (which became Krauss-Maffei Wegmann on 1 January 1999) Leopard 2A5E. Details of this programme are given in a separate entry.

UPDATED

Spanish Leopard 2A5 MBT

In mid-1995, it was announced that Spain was to manufacture under licence 200 German Krauss-Maffei Leopard 2A5 MBTs for the Spanish Army with deliveries to run at the rate of 40 vehicles a year.

The government-to-government Memorandum of Understanding (MoU) was signed in Brussels between Germany and Spain on 5 June 1995 covering this and other defence projects.

Spain has already taken delivery from Germany of 108 Leopard 2A4 MBTs from the German Army for a period of five years. These were delivered to Spain between November 1995 and June 1996 and will cost US$6.6 million a year to lease. This lease has now been extended.

The first Leopard 2s went to the two armour battalions which Spain has committed to the Eurocorps and some 50 are to be used to equip the two cavalry regiments based in the Spanish North African enclaves of Ceuta and Melilla.

In Spanish Army service, the Leopard 2A4 is designated the Leopard 2A4E, while the new production Leopard 2A5 will become the Leopard 2A5E.

Prime contractor for the Leopard 2A5E is Santa Barbara Sistemas (now part of the US General Dynamics Group), which is also the prime contractor for the Spanish-built Pizarro infantry combat vehicle. Spanish content of the Leopard 2A5E is expected to be between 60 and 70 per cent.

Early in March 1998, negotiations finally started between Krauss-Maffei (Krauss-Maffei Wegmann) and Santa Barbara for the production of 219 Leopard 2A5 MBT's and 16 armoured recovery vehicles.

The contract for production of the Leopard 2A5 in Spain was finally signed late in December 1998. The first vehicles are expected to be completed in late 2002 with the main production run commencing the following year.

While Santa Barbara Sistemas will be overall prime contractor for the Spanish Leopard 2A5E programme, many other Spanish companies will be involved including Bazan (MTU V-12 diesel engine), ENSB (turret and 120 mm smoothbore gun), and Plasencia de las Armas for the tracks and suspension. Amper will be responsible for the communications system. The fire-control system will be manufactured by EWS.

Leopard 2A4E MBT of the Spanish Army (Francisco A Marin) 0011786

According to Spanish sources, the Leopard 2A5E contract will provide work for Santa Barbara Sistemas for a period of seven years.

Late in 1999, it was announced that Spain had decided to arm its new Leopard 2 MBTs with the latest Rheinmetall DeTec AG 120 mm L/55 calibre smoothbore tank gun.

The 120 mm L/55 calibre smoothbore tank gun was developed by Rheinmetall DeTec under contract to the German MoD as the replacement for the current 120 mm L/44 smoothbore weapon installed in all production Leopard 2 series MBTs built since 1979.

The 120 mm L/55 smoothbore gun has now been ordered by Germany and the Netherlands, with the first German Army Leopard 2A6 being handed over early in 2001. The US has also purchased three 120 mm L/55 weapons for trials in an M1 series Abrams MBT.

The new 120 mm L/55 is 1.3 m longer than the L/44 which leads to a significant increase in muzzle velocity and effective combat range. It also allows the weapon to fire future 120 mm rounds with increased armour penetration characteristics.

The 120 mm L/55 weapons will be manufactured in Spain by Santa Barbara Sistemas with the assistance of Rheinmetall DeTec under a deal worth DM155 million.

Santa Barbara Sistemas has awarded the German company of STN ATLAS Elektronik GmbH and the Spanish Indra EWS company a contract for the delivery of 219 fire-control systems for the Spanish built Leopard 2 MBT. This covers the supply of not only the fire-control system but also panoramic periscope, testing equipment and a battlefield command system.

In addition it covers the supply of 16 battlefield command systems for installation in the Büffel Armoured Recovery Vehicle's (ARVs) also to be supplied to the Spanish Army.

Amper Programas of Madrid will supply the communication systems for the Spanish Leopard 2 MBTs and variants with the BAE Systems ROVIS digital intercommunication system being selected to meet part of this requirement.

Of these 16 ARVs, four will be supplied by the now Rheinmetall Landsysteme (previously MaK) in Germany and the remainder manufactured under licence in Spain.

Under the Conventional Forces Europe (CFE) treaty, Spain is allowed a total of 794 MBTs, although for financial reasons Spain intends to reduce its total MBT fleet to 684, and this figure could well be reduced even further.

UPDATED

Spanish AMX-30 MBT production

Development

Following delivery of the first 19 vehicles from France, Spain was building the AMX-30 MBT under licence from France from 1974, initially from components supplied by France, but gradually the Spanish content increased until about 85 per cent of the tank was built in Spain. A total of 19 tanks were supplied by the now Giat Industries direct from France plus 10 AMX-30D ARVs, with 280 built under licence in Spain with final deliveries in 1983-84.

The Spanish tanks are almost identical to the French models but have an MG3S 7.62 mm machine gun for the commander and other minor differences. On Spanish vehicles the coaxial weapon is a 12.7 mm M2 machine gun. Production of the AMX-30 was undertaken by the Sevilla Artillery Factory of Santa

Barbara, Empresa Nacional de Optica SA (ENOSA) providing the optical equipment, and Experiencias Industriales SA (EISA) the electrical and hydraulic equipment.

In 1980, Spain purchased the entire licence from Giat in order to continue development itself.

With the introduction of the German-supplied Leopard 2A4 and US-supplied M60 series MBTs no further enhancements are now planned for the AMX-30 ER1/EM2 MBTs of the Spanish Army.

With the introduction of Spanish built Leopard 2A5E MBTs it is expected that the AMX-30 ER1 and EM2 will eventually be phased out of service.

Prime contractor for the AMX-30 and Leopard 2A5 MBT in Spain is now Santa Barbara Sistemas which is part of the General Dynamics Group.

AMX-30R Roland SAM system

In 1984, Spain signed a contract with Euromissile for the supply of 18 Roland 2 SAM systems and 414 missiles. Much of this work was carried out in Spain including manufacture of the AMX-30R launcher vehicle. The first Spanish Roland SAM system on an AMX-30 chassis was handed over to the Spanish Army late in 1988. Deliveries are now complete and the AMX-30 production line in Spain was closed down some years ago.

AMX-30 ER1

This upgrade was carried out by the Spanish Army and included the following:

(1) Replacement of the original transmission by a US Allison CD-850-6A transmission made under licence in Spain
(2) Slightly modified cooling system and overhauled engine
(3) Modified rear engine compartment with the top deck now 15 cm higher, modified hull floor and modified hull rear
(4) Driver has a new seat, a new instrument panel and a steering wheel instead of the original two tillers.

The Spanish AMX-30D ARVs and four AMX-30 Driver Training Tanks have also been modified in a similar manner, as have the 18 Roland SAM systems which have now been designated the AMX-30RE. There are 60 AMX-30 ER1s in service.

AMX-30 EM2

This is the final upgrade and has the new power pack, the new Hughes (now Raytheon Systems Company) computerised fire-control system and additional armour. The actual conversion was carried out by Santa Barbara with the engine and transmission being manufactured under licence in Spain by Empresa Nacional Bazan.

The new power pack consists of the MTU MB 833 Ka-501 six-cylinder water-cooled, supercharged diesel engine developing 850 hp at 2,450 rpm coupled to a ZF LSG 3000 automatic transmission with four forward and two reverse gears. To accommodate the new power pack the rear engine top deck and exhaust have been modified.

The Hughes (Raytheon Systems Company) Mk 9 fire-control system includes a computerised ballistic computer, a mast-mounted sensor, a muzzle reference system, a laser range-finder and a HIRE thermal camera. The 105 mm main armament is not stabilised in elevation or traverse. The tank commander now has a 12.7 mm machine gun.

The vehicle has also been fitted with a Halon fire extinguishing system in the engine compartment and a new explosion detection/suppression system in the crew compartment.

The suspension has been upgraded with new torsion bars and shock-absorbers and new side skirts fitted.

Explosive reactive armour developed by Santa Barbara will also be fitted to improve battlefield survivability.

The AMX-30 EM2 has a similar specification to the French Army AMX-30 except for the different power pack, a combat weight of 36.5 tonnes, a width of 3.22 m, a maximum road speed of 64 km/h and a road range of 400 km. There are 150 of these remaining in service.

AMX-30I

The Spanish Army has four of these vehicles which are used for driver training purposes. The turret has been removed and replaced by an observation cabin.

Status

Production complete. In service with the Spanish Army. With the introduction of the Leopard 2 and M60 series MBTs into the Spanish Army, no further upgrades of the AMX-30 ER1 or AMX-30 EM2 MBTs are expected to be made.

Manufacturer

Santa Barbara Sistemas (part of the US General Dynamics Group).

UPDATED

Upgraded AMX-30 EM2 MBT (Francisco A Marin) 0018834

Upgraded AMX-30 EM2 MBT from the rear (Francisco A Marin)

Peugeot-Talbot MBT conversions

In 1951, Barreiros Diesel SA was formed for the conversion of petrol engines to diesel operation. It later began manufacturing its own diesel engines, followed by trucks and other vehicles. In 1965, it formed a partnership with Chrysler Corporation (now General Dynamics Land Systems) of the USA and the defence division was manufacturing M60 MBTs as well as supplying spare parts for the M48 MBT.

In the 1960s, the company built trucks for the Spanish Army and in the 1970s started to overhaul and modernise M47 and M48 tanks and M113 series armoured personnel carriers for the Spanish Army. In 1978, the company went into partnership with PSA and its name was changed to Talbot SA.

MBT conversions currently offered by Talbot are listed below with a resumé of army/marine conversions. More recently the company has become Peugeot-Talbot. It is considered unlikely that there will be any further conversions of the M47/M48 series MBTs in Spain. As far has it is known, Spain has never exported any of its M47/M48 upgrades.

M47 Upgrades

Talbot upgraded US-supplied M47 MBTs to the M47E1 and M47E2 standard but these have been phased out of service.

M48A5E2 MBT of the Spanish Army with turret traversed to the rear (Francisco A Marin) 0018835

Specifications
M48A5E2
Crew: 4
Combat weight: 49,000 kg
Unloaded weight: 44,000 kg
Power-to-weight ratio: 15.51 hp/t
Ground pressure: 0.86 kg/cm²
Length:
(gun forward) 9.3 m
(gun rear) 8.04 m
(hull) 6.88 m
Width: 3.63 m
Height:
(without AA MG) 2.87 m
(with AA MG) 3.28 m
Ground clearance: 0.42 m
Track: 2.92 m
Track width: 710 mm
Length of track on ground: 4 m
Max road speed: (forward) 50 km/h
Fuel capacity: 1,500 litres
Max road range: 550 km
Fording: 1.2 m
Gradient: 60%
Vertical obstacle: 0.91 m
Trench: 2.59 m
Engine: Continental (now General Dynamics Land Systems) V-12, AVDS-1790-2D developing 760 hp at 2,400 rpm
Transmission: Allison CD series 850-6A
Suspension: torsion bar
Electrical system: 24 V
Batteries: 6 × 12 V
Armament:
(main) 1 × 105 mm rifled gun
(coaxial) 1 × 7.62 mm
(anti-aircraft) 1 × 12.7 mm
Ammunition:
(main) 54
(coaxial) 8,000
(anti-aircraft) 600
NBC system: yes
Night vision equipment: yes

M48A3E/M48A5E Upgrades
These have now been phased out of service.

M48A5E2
A total of 164 of these remain in service. This is fitted with a Hughes (now Raytheon Systems Company) computerised fire-control system with many components being manufactured under licence in Spain by EISA/ENOSA and about 70 per cent of the fire-control system being the same as that fitted to the M60A3 MBT.

M48A5E1
As the M48A5E but it has the Teledyne Continental Motors (now General Dynamics Land Systems) AVDS-1790-2D 12-cylinder diesel and a new fire-control system. The gunner has an M35 periscope, a laser range-finder, a passive night vision viewer, a control panel, an output unit, an ammunition selection unit, a graticule projection unit, handgrips with switches for laser/lead/battle range, a cant sensor and an analogue electronic computer. The commander has an M17 range-finder and an ammunition selection unit, the loader has an ammunition selection display unit, and the driver has three M27 periscopes, the centre one of which can be replaced by an AN/VVS-2 passive night driving periscope. The vehicle has a TEESS engine exhaust smoke generating system and smoke grenade dischargers installed.

In 1992, the prototype of the M48A5E3 was built which is based on the M48A5E2 but also has a gun stabilisation system, a thermal camera and a new fire/explosion detection system in the crew compartment.

GAMESA M47VR armoured recovery vehicle in travelling configuration 0018836

M47E2I Engineer Tank
Only one of these was built and no further conversions to this standard are planned.

M47E2R Recovery Tank
Only one of these was built and no further conversions to this configuration are planned.

M47E2VLP Bridgelayer
Only one of these was built and no further conversions to this configuration are planned.

M47E2Z Engineer Tank
No production examples of this vehicle have been completed and in the end an M60 chassis was selected for this conversion.

M47VR Recovery Tank
Following a competition, in January 1994 the GAMESA company was selected to build 22 ARVs under this designation. These vehicles have the same power pack as the M47 and M48E consisting of an AVDS-1790-2A diesel developing 760 hp coupled to a CD-850-6A transmission. Equipment includes a winch with a capacity of 35 tonnes, a hydraulic crane with a lifting capacity of 5.6 tonnes, welding equipment and a platform to carry a complete power pack of the M60 series MBT. Delivery of the last of 22 production M47VR ARVs was made in 1996. Full details of the M47VR are given in *Jane's Military Vehicles and Logistics 2002-2003*.

M47 Leguan Bridgelayer
A Spanish Army M47 chassis has been fitted with a German MAN 26 m Leguan bridgelaying system for trials purposes. This bridge is launched horizontally over the front of the vehicle. Mounted at the front of the M47 is a dozer/stabilising blade. The weight of the system with the bridge is 53 tonnes and without the bridge is 48 tonnes.

This is now in service with the Spanish Army and details are given in *Jane's Military Vehicles and Logistics* 2002-2003.

M60 Mineclearing Tanks
Late in 1996, the Spanish MoD stated that 36 US-supplied M60 series MBTs would be converted for mine clearance operations at a cost of around US$26.6 million.

M60 CZ-10/25E Alacran Engineer Tanks
Early in 1997, the Spanish Ministry of Defence announced that Peugeot Talbot SA had been awarded a contract for the conversion of 38 M60 MBTs into the M60 CZ-10/25E Alacran for delivery from 1997 through to 1999. Also competing for this contract was GAMESA of Spain which also built a prototype vehicle under the designation M60 VZ Engineer Tank.

The M60 CZ-10/25E retains the hull and turret of the M60A1 MBT but mounted at the front of the hull is a dozer/stabiliser blade. Mounted on the front of the turret is the Case Poclain mechanical excavator which can be replaced by an NPK hammer or a Verachtert VTC-30 cutting tool.

It also has a rear-mounted hydraulic winch with a direct traction force of 25,000 kg which can be increased to 50,000 kg with a return block. Armament consists of a 12.7 mm M2 machine gun and two banks of four electrically operated 76 mm smoke grenade launchers. Full details of this are given in *Jane's Military Vehicles and Logistics* 2002-2003.

Status
Production complete. In service with the Spanish Army.

Manufacturer
Peugeot-Talbot Espana SA.

UPDATED

SWEDEN

Swedish Leopard 2 MBT programme

Development/Description
Following extensive trials in Sweden between the French Giat Industries Leclerc, the German Krauss-Maffei (which became Krauss-Maffei Wegmann on 1 January 1999) Leopard 2A5 and the General Dynamics Land Systems M1A2 Abrams MBT, early in 1994 Sweden selected the Leopard 2A5 to meet its future MBT requirements.

In June 1994, the Swedish Defence Matériel Administration (FMV) signed a contract with Krauss-Maffei of Germany for the manufacture and delivery of 120 new Leopard 2 (S) (Sweden) MBTs for the Swedish Army. Total value of the contract is DM1,020 million.

The agreement involved extensive industrial co-operation between Germany and Sweden, with the Swedish companies, Bofors Defense (at the time Bofors Weapon Systems) and Hägglunds Vehicle having the largest part of the contract in Sweden.

Krauss-Maffei has undertaken to ensure that the committed industrial co-operation of up to 100 per cent of the total order within a nine year period is achieved following the signing of the contract. In addition, Krauss-Maffei has voluntarily undertaken to accomplish industrial co-operation up to a further approximately 20 per cent of the contract value.

The 2A5 is called the Leopard 2 (S) by Sweden with the Swedish Army designation being the Stridsvagn 122 (Strv 122).

The Leopard 2 is now the only MBT deployed by Sweden with the S-tank and Centurion having now been phased out of service.

In addition, the contract covers the supply of training, simulators, spare parts, maintenance, documentation, an option for the procurement of additional vehicles and Büffel armoured recovery vehicles as well as interfaces for equipment already used by the Swedish Army such as the BT46 gunnery simulator.

By late 1994, Sweden had taken delivery of 160 ex-German Army Leopard 2 MBTs which are designated the Strv 121 in Swedish Army service.

The first firm order was for 120 vehicles placed direct with Krauss-Maffei by the FMV in June 1994. There is an option for an additional 90 vehicles which would bring the Swedish Leopard 2 fleet up to a total of 370 vehicles. At present there are no plans that this option will be exercised.

In addition to being overall prime contractor, Krauss-Maffei subcontracted the chassis to Hägglunds Vehicle in northern Sweden. For the turret, Wegmann has subcontracted the work to Bofors Defence while for the fire-control system STN ATLAS Elektronik subcontracted work to Saab.

Within Sweden, Bofors Defence was responsible for the turret and armament while Hägglunds Vehicle was

Swedish Army Stridsvagn 122 MBT is a further development of the Leopard 2A5 **2000**/0069551

Swedish Army Stridsvagn 122 MBT has a Swedish-developed armour package **2000**/0069552

responsible for the chassis as well as integrating the chassis with the turret and delivering the complete vehicle to the Swedish Army.

Bofors Defence completed the first turret shells and from numbers 1 to 29 were sent to Wegmann in Germany which integrated the turret and returned it to Sweden. From turret number 30 to 120 Bofors Defence integrated the turrets. A similar arrangement exists for the chassis. Hägglunds Vehicle delivered the first Leopard 2 hull to Krauss-Maffei in Germany in October 1995, on schedule.

Bofors Defence manufactured 50 per cent of the Rheinmetall 120 mm smoothbore gun with Rheinmetall of Germany doing a similar amount.

Ammunition was not a part of the contract and, following a competition, Israel Military Industries was awarded a contract worth more than US$20 million for a large quantity of 120 mm Armour-Piercing Fin-Stabilised Discarding Sabot (APFSDS) ammunition and its associated training round. The APFSDS round will have a conventional penetrator, not a depleted uranium one as favoured by the US.

The other Leopard 2 users have the HEAT-MP-T (High Explosive Anti-Tank - MultiPurpose - Tracer) as the secondary type of ammunition but Sweden will field an HE-T (High Explosive-Tracer). This has been developed by NAMMO (Nordic Ammunition Company), which is prime contractor and responsible for the projectile, and Israel Military Industries, which is responsible for the charge and fins. The projectile is fitted with a point detonating fuze which is armed 50 m after it leaves the 120 mm smoothbore gun.

The Swedish-built Leopard 2 (S) MBTs differ from their German counterparts in a number of areas including the installation of the French Giat Industries Galix vehicle protection system, which is also fitted to the Combat Vehicle 90 and a new command and control system developed by Saab.

In addition, the Leopard 2 (S) has a new passive armour system developed in Sweden by Akers, which will provide a very high degree of protection from both kinetic energy and chemical energy attack.

Even when compared to the latest German Army Leopard 2A5, the Swedish Strv 122 will be the best armoured Leopard 2 in service with any army in the world.

This is because it will be the first Leopard 2 to have a full command and control system. It will also have a much improved battlefield survivability as it has additional armour protection over the frontal arc of the vehicle as well as on the upper surfaces to protect against top attack munitions.

The following additional modules have been incorporated into the Swedish Leopard 2:

Chassis–This has been fitted with add-on armour modules consisting of frontal modules and modules on the upper glacis plate. The crew compartment in the hull and the driver's station are protected by a spall liner. For greater lateral protection, the hull is fitted with new heavy track skirts.

Turret–The roof features add-on armour modules for overhead protection against bomblets in the area of the crew compartment with sliding hatches in the turret making it easier to handle the heavy weight of the armour.

The power pack has been modified with the integration of suction ventilation improved service brake, electronic brake wear sensor and integration of a radiator shut-off device.

The suspension has been upgraded to take into account the heavier weight of the vehicle. Other modifications include the integration of six Swedish batteries, seat belts at each crew station, emergency switch in the driver's compartment and Swedish radios and intercom.

The following production schedule was agreed:

- Assembly of tanks in Germany started in the second quarter of 1996
- Assembly of tanks in Sweden started in the first quarter of 1997
- Deliveries of the first German-built tank started in the fourth quarter of 1996
- The first Swedish-built tank was delivered by Hägglunds Vehicle early in 1998.

The Swedish Defence Matériel Administration handed over the first two of 120 Leopard 2 (S) MBTs to the Swedish Army on 22 May 1997. One of each vehicle was handed over to Skaraborgsbrigaden (MekB 9) in Skovde and Norrbottensbrigaden (MekB 19) in Boden, which are the two mechanised brigades to be equipped with the Leopard 2 (S) MBT.

The three other mechanised brigades MekB 7, 8 and 10 are equipped with the Leopard 2A4 MBTs purchased from the German Army.

The Command and Control System of the Leopard 2 (S) was developed and delivered by STN ATLAS Elektronik GmbH, Saab (of Sweden) and IBP Pietzsch.

The system is based on the IFIS system which has been developed by IBP Pietzsch under contract to the German MoD.

The Leopard 2 (S) will be the first MBT in Europe with an advanced command and control system that will include, for example, map functions, presentation of position, decision support and handling system.

Specifications

(Strv 122)
Crew: 4
Combat weight: 62,000 kg
Power-to-weight ratio: 24.19 hp/t
Ground pressure: 0.94 kg/cm^2
Length:
(gun forward) 9.97 m
(gun rear) 8.74 m
Width:
(over skirts) 3.81 m
(travelling) 3.55 m
Height:
(to turret top) 2.64 m
(to commander's periscope) 3.00 m
(firing height) 2.01 m
Ground clearance: 0.50 m
Track: 2.785 m
Track width: 635 mm
Length of track on ground: 4.95 m
Max road speed:
(forward) 72 km/h
(reverse) 31 km/h
Fuel capacity: 1,200 litres
Max road range: 470 km
Fording: 1.40 m
Gradient: 60%
Side slope: 30%
Vertical obstacle: 1.10 m
Trench: 3.10 m
Engine: MTU MB 873 Ka-501 4-stroke, 12-cylinder diesel, exhaust turbocharged, liquid cooled, developing 1,500 hp at 2,600 rpm
Transmission: Renk HSWL 354 hydrokinetic planetary gear shift, 4 forward and 2 reverse gears, integrated service brake
Clutch: torque converter
Suspension: torsion bar, friction damper and hydraulic bump stops
Electrical system: 24 V
Batteries: 8 × 12 V, 125 Ah
Armament:
(main) 1 × 120 mm smoothbore gun
(coaxial) 1 × 7.62 mm MG3 MG
(anti-aircraft) 1 × 7.62 mm MG3 MG
Smoke laying equipment: 2 × 4 Galix
Ammunition:
(main) 42
(MG) 4,750
Gun control equipment:
(by commander) yes
(by gunner) yes
(commander's override) yes
Gun elevation/depression: +20/–9°
Turret traverse: 360°
Gun stabiliser:
(vertical) yes
(horizontal) yes
Range setting device: yes, laser
Elevation quadrant: yes
Traverse indicator: yes
NBC system: yes
Night vision equipment: yes

The Swedish TCCS allows for the following functions:

- Command and control of own MBT
- Command and control of other MBTs at the platoon and company level
- Command post for battalion commander.

General system functions include tactical and logistics information exchange via data communication, positioning and navigation, situation information, topographical map display, generation and exchange of orders and situation reports and automatic target location.

Saab has also signed a co-operation agreement with STN Atlas Elektronik regarding development and production of the countermeasures system for the Leopard 2 (S). In addition, the used Leopard 2s that Sweden received may be upgraded with the warning system and countermeasures.

Status
Production complete. In service with Swedish Army.

Manufacturers
Bofors Defence (part of united Defense Group of the USA) (turret)
Hägglunds Vehicle (chassis)

UPDATED

Hägglunds Vehicle CV90120-T Tank

Development
The CV90120-T Tank has been developed by Hägglunds Vehicle as a private venture to meet emerging requirements for a tank with the firepower of current in service Main Battle Tanks (MBT) such as the Leopard 2, but with greater tactical and strategic mobility as well as having a high level of combat survivability.

Within the designation CV90120-T, the CV90 is for Combat Vehicle 90 chassis on which the system is based, 120 is for the calibre of the main armament in mm and T is for Tank.

Current MBTs such as the Leopard 2, Challenger 2 and the M1A2 all weigh 60 tonnes or more, and this limits their operational deployment. These heavy MBTs can only be transported by the largest transport aircraft such as the Lockheed Martin C-5A Galaxy.

In addition, in many countries the existing infrastructure of bridges and roads cannot take these heavy vehicles. The CV90120-T will be airportable in the common European A400M transport aircraft.

The first prototype of the CV90120-T was completed in mid-1998 and commenced its static firing trials in Sweden late in 1998. These were followed by moving firing trials in 1990 and 2000.

Development of the vehicle was completed late in 2001, although further development in a number of areas continues, especially electronics, as new technology becomes available.

CV90120-T tank from the front and clearly showing the well sloped turret front with the SaabTech Vetronics Panoramic Low Signature Sight in the centre **2002**/0131038

Description
The chassis of the CV90120-T is a further development of that used for the CV 9040 infantry fighting vehicle (in service with the Swedish Army) and the CV 9030 in service with Norway and ordered by Finland (CV9030FIN) and Switzerland (CV9030CH). Full details of these vehicles are given in the Armoured Personnel Carriers (Tracked) section under Sweden.

The CV90 chassis has been modified to take into account the larger and heavier three person turret and additional passive armour has been fitted over the frontal arc.

The armour package of the CV90120-T is modular so that if a customer requires a higher level of protection then it could be fitted. The basic hull and turret are of all welded armoured steel with internal spall liners.

The additional hull and turret appliqué armour provides protection against medium calibre attack of an undisclosed type, but probably around 30 mm Armour Piercing Discarding Sabot (APDS) through the frontal arc.

Through the remainder of the vehicle protection is provided against 14.5 mm armour piercing attack. If required, Explosive Reactive Armour (ERA) can be fitted to the hull and turret for a higher level of protection against HEAT type warheads fitted to anti-tank weapons.

The power pack is situated at the front right side of the vehicle consists of the latest Scania V-8 4-cycle diesel developing 670 hp at 2,200 rpm coupled to an Perkins Engines Company/Allison X-300 series automatic transmission with four forward and two reverse gears.

With a combat weight of 26 tonnes this gives the CV90120-T a high power-to-weight ratio of 25.76 hp/t, maximum road speed of 70 km/h and a cruising range of 600 km.

The driver is seated at the front left side of the hull and has a single piece hatch cover that is hinged on the left side. To his immediate front are three periscopes, the centre one of which can be replaced by an AN/VVS-CV90 passive periscope for driving at night.

Suspension is of the torsion bar type. Either side has seven rubber tyred road wheels with the drive sprocket at the front, track return rollers and idler at the rear.

The latter is adjustable by the driver without leaving the vehicle. Rotary dampers are provided for the 1st, 2nd and 7th road wheel station for improved ride across country. The upper part of the suspension is covered by a skirt.

CV90120-T tank firing its Swiss RUAG Land Systems high pressure 120 mm smoothbore L/50 Compact Tank Gun **2002**/0131035

The turret is in the centre of the hull with the commander seated on the right, gunner forward and below the commander and the loader on the left. Both the commander and loader have single piece hatch covers that open to the rear.

Main armament of the CV90120-T comprises a fully stabilised Swiss RUAG Land Systems high pressure 120 mm L/50 smooth bore Compact Tank Gun (CTG) which is fitted with a thermal sleeve, fume extractor and a pepperpot muzzle brake to reduce recoil forces. It has a vertical sliding breech mechanism.

This can fire all types of current and projected in service 120 mm ammunition which is available from a number of sources including the latest 5th generation Rheinmetall DeTec DM53 Armour Piercing Fin Stabilised Discarding Sabot - Tracer (APFSDS-T) round.

The 120 mm gun can also fire new types of High Explosive (HE) rounds from Rheinmetall DeTec and the NAMMO 120 mm High Explosive - Tracer (HE-T) round which has been in service with the Swedish Army for several years as an alternative to the older High Explosive Anti-Tank - Multi-Purpose -Tracer (HEAT-MP-T) round. All of these 120 mm rounds have a semi-combustible cartridge case and all that remains after firing is the sub cartridge case.

A load assist device is fitted as standard which, according to Hägglunds Vehicle, allows a maximum rate of fire of 12 to 14 rds/min to be achieved. A 12.7 mm machine gun is mounted coaxial with the main armament.

A total of 45 rounds of 120 mm ammunition are carried of which 12 are located in the semi-automatic magazine located in the turret bustle. An additional 33 120 mm rounds are stowed in the hull rear with ammunition rapidly being reloaded via the large door in the hull rear. The latter can also be used by the crew to exit the vehicle in an emergency without exposing themselves to small arms fire and shell splinters.

An all electric gun control and stabilisation system is fitted as standard with turret traverse through a full 360° and weapon elevation from −8 to +22° with both commander and gunner having a full fire-control system.

The advanced fire-control system enables stationary and moving targets to be engaged while the CV90120-T is stationary or moving. The gunner is provided with a roof mounted stabilised SaabTech Vetronics UTAAS (Universal Tank and Anti-Aircraft System) sight with day/thermal channels and an eye safe laser range-finder which feeds information to the fire-control computer. UTAAS can be elevated from −10 to +25° and in silent watch mode has a traverse of 18° left and right.

Specifications

Crew: 4
Combat weight: 26,000 - 27,700 kg
Power-to-weight ratio: 25.76 hp/t
Ground pressure: 0.55 to 0.66 kg/cm²
Length:
(gun forward) 8.95 m
(chassis) 6.45 m
Width: 3.19 m
Height:
(panoramic sight) 2.80 m
(roof top) 2.40 m
Ground clearance: 450 mm
Track width: 533 mm
Length of track on ground: 3.980 m
Max speed:
(forwards) 70 km/h
(reverse) 43 km/h
Acceleration: 0 to 30 km/h 8 s
Gradient: 60%
Side slope: 40%
Trench: 2.4 m
Vertical obstacle: 1 m
Fording: 1.5 m
Range: 600 km
Engine: Scania Scania V-8 4-stroke diesel developing at 670 hp at 2,200 rpm
Transmission: Perkins Engines Company/Allison X300 series automatic with 4 forward and 2 reverse gears
Suspension: torsion bar with rotary dampers
Electrical system: 24 V
Armament:
(main) 1 × 120 mm L/50 smoothbore
(coaxial) 1 × 12.7 mm MG
Ammunition:
(120 mm) 45 (12 ready plus 33 stowed)
(7.62 mm) 3,000
Gun control equipment: electric/manual
Gun elevation/depression: +22/-8°
Turret traverse: 360°
(by commander) yes
(by gunner) yes
(commander's override) yes
Range setting device: yes (laser)
NBC system: yes
Night vision equipment: yes

The CV90120-T is based on the chassis of the CV90 series of infantry fighting vehicles which are in service with Norway and Sweden and have been ordered by Finland and Switzerland **2002**/0131039

The commander's Thales Optronics (Taunton) DNGS stabilised sighting system has a day channel with magnifications of ×1.5, ×6 and ×24 with a remote display of the gunner's thermal sight with option of a separate image intensification channel if required. He also has a laser range-finder. The commanders cupola can be traversed 110° left and right.

Mounted on the turret roof is the new SaabTech Vetronics Panoramic Low Signature Sight (PLSS). This stabilised armoured modular sight can be fitted with various sensors including day/thermal cameras and a eyesafe laser range finder. The PLSS can be operated by the commander or loader.

The installation of the PLSS allows the CV90120-T to undertake hunter/killer target engagements in which the target is first acquired by the vehicle commander, tracked and then handed over to the gunner who carries out the actual target engagement.

The use of three independent sights is claimed to provide the crew of the CV90120-T with superior battlefield awareness.

Each crew member, commander, gunner, loader (second commander) and driver have a display which are connected via an internal data/video base.

This allows each crew member to be provided with a wealth of information if required. This can include navigation, fire control, video, intercom, radio, threat warning and vehicle support system information.

According to Hägglunds Vehicle, the CV90120-T has increased functionality through the extensive use of sensors and actuators, database systems, advanced displays and controls with built in test equipment as standard. In addition, it has significant growth potential that allows new technology to be inserted as it becomes available.

The integrated, scalable and open electronic architecture of the CV90120-T comprises BIT (Built In Test), DAS (Defensive Aid Suite) and a complete video network with displays at each crew station.

The BIT can also be used as a Computer Based Trainer (CBT) to allow training to take place wherever the vehicle is deloyed. A Battle Management System (BMS) is available as an option.

The combat survivability of the CV90120-T is further enhanced by the installation of a modular defensive aid suite (DAS) which includes laser, radar and missile approach warners.

Mounted on each corner of the turret is an Thales Optronics (Taunton) LD-2 laser warning system which indicates the direction of the laser as well as its type, for example laser rangefinder or laser designator.

Mounted on each corner below the laser warner is a Grintek Avitronics Missile Approach Warner (MAW) that identifies a missile launch so that the appropriate counter measure can be deployed before the missile impacts on the CV90120-T.

In addition to the front, sides and rear of the turret is a Grintek Avitronics radar warner to warn the crew of any active radar, for example missiles or battlefield surveillance radar.

Mounted on the turret roof is a top attack radar which can identify smart indirect munitions. All of these sensors feed information into a central processor that informs the crew as to the location of the threat, classification of threat and response selection. It can be operated in the manual or fully automatic modes with information displayed on all screens if required.

In addition, the CV90120-T survivability is further enhanced by the installation of the Multi Spectral Aerosol (MSA) active counter measure system that includes 12 grenade launchers mounted either side of the turret rear.

The CV90 chassis was originally designed with a low signature but for the CV90120-T application has been further enhanced by the installation of radar absorbing track skirts.

The turret, with its carefully shaped angles also incorporates stealth characteristics. The acoustic and thermal signature of the vehicle is further reduced by ventilated double shell principle and by routing the exhaust pipe to the rear of the vehicle on the right side.

Standard equipment on the CV90120-T includes a hybrid overpressure NBC protection system and a fire detection and suppression system for the engine compartment.

Optional equipment includes an air conditioning system and various intercom/radio systems according to the users specific operational requirements.

Status

Development complete. Ready for production.

Manufacturer

Hägglunds Vehicle AB (taken over by Alvis plc in 1997)

NEW ENTRY

SWITZERLAND

Leopard 2 (Pz 87 Leo) MBT

Development

Between August 1981 and June 1982, two German Leopard 2 (120 mm smoothbore gun) and two American M1 (105 mm rifled gun) tanks underwent technical and user trials in Switzerland.

In August 1983, Switzerland selected the Leopard 2 and late in 1984 the Swiss Parliament authorised the purchase of 380 Leopard 2s. The first 35 Leopard 2s came direct from Krauss-Maffei (which on 1 January 1999 became Krauss-Maffei Wegmann) in Munich; the remainder were manufactured under licence in Switzerland with Contraves (today Oerlikon Contraves) as prime contractor. Final assembly took place at the now SW Swiss Ordnance Enterprise Thun (previously k+w Thun) where the production and overhaul of the Pz 61 and Pz 68 were undertaken. Swiss Ordnance

Swiss Army Leopard 2 MBT from the front (Andreas Kirchhoff via Michael Jerchel)

Enterprise was also responsible for assembling the turrets and power units. The Rheinmetall 120 mm smoothbore gun was also built under licence.

Between 60 and 70 per cent of the Leopard 2s were built in Switzerland. Swiss Leopard 2s are similar to the German vehicles but have Swiss radios, antennas and machine guns, an improved driver's hatch, a digital computer in place of the analogue computer, a Deugra fire/explosion detection and suppression system for the crew compartment, improved NBC protection, hydraulic track tensioning units, a Baird passive night driving periscope, optical master warning for the driver when driving with the hatch open and some other minor modifications.

The first 35 Leopard 2s, or Panzer 87s as they are designated by the Swiss Army, were delivered in 1987, with the first battalion being formed in 1988. The first two Swiss-built Pz 87s were delivered to the Defence Procurement Agency (GR) on 17 December 1987, as originally scheduled. Late in 1984, Hughes Electro-Optical and Data Systems Group (which is now part of the Raytheon Group) signed a manufacturing and technical assistance agreement with Wild Heerbrugg Ltd of Switzerland, under which the latter company would build the laser tank fire-control system for the Swiss Panzer 87 Leopard, or Pz 87 Leo as it is usually known.

Production was undertaken at Swiss Ordnance Enterprise Thun at the rate of 72 vehicles a year, with final deliveries taking place in 1993.

It is expected that at least part of the Leopard 2 MBT fleet of the Swiss Army will be upgraded to the Leopard 2A5 configuration in the future.

As of January 2002, no contracts had been placed for this upgrade work.

The Swiss Army has selected the new German Rheinmetall 120 mm DM53 APFSDS-T round and placed an order for 20,000 rounds. First deliveries were made in 1999 and the Swiss Army was the first customer for this advanced APFSDS-T round.

Variants

Buffel

Late in 2001, Switzerland ordered 25 Buffel ARVs under the designation BPz Buffel CH.

Leopard 2 with 140 mm gun

For trials purposes, a Swiss Army Leopard 2 has been fitted with a 140 mm smoothbore gun. There are no plans for this to be retrofitted to the existing in-service Leopard 2s. Details of the 140 mm Leopard 2 were given in *Jane's Armour and Artillery 1997-98.*

Manufacturer

SW Swiss Ordnance Enterprise Corp.

UPDATED

Pz 68 MBT

Development

In 1951, the Technical Section of the Swiss General Staff carried out a study to see if an MBT could be designed in Switzerland. This study concluded that it was possible and two years later design work on the first Swiss MBT began at the now SW Swiss Ordnance Enterprise Corp.

The first prototype of the Swiss MBT, designated the KW30, was completed in 1958 and armed with a Swiss-designed 90 mm gun. The second prototype was completed the following year. These two prototypes were followed by 10 preproduction tanks, Pz 58 (Panzer 58), built between 1960 and 1961 and armed with a British 20 pounder (83.4 mm) gun as installed in the Pz 55. In March 1961, a production order for 150 Pz 61s (armed with a British Royal Ordnance (now BAE Systems, RO Defence) 105 mm rifled gun) was placed, with deliveries to the Swiss Army taking place between January 1965 and December 1966.

The Pz 61 MBTs were all phased out of service with the Swiss Army by late 1996.

Further development of the Pz 61 resulted in the Pz 68. The first prototype of this was completed in 1968 with production being authorised the same year.

Between January 1971 and July 1974, a total of 170 Pz 68 Mk 1s was delivered to the Swiss Army. Main differences between the Pz 61 and Pz 68 can be summarised as the installation of a stabilisation system for the 105 mm gun to allow the Pz 68 to engage targets accurately while moving across country, 105 mm ammunition supply hatch in the left side of the turret, more powerful MTU engine coupled to a modified transmission, wider tracks with replaceable pads and greater length of track in contact with the ground.

The Pz 68 has been upgraded on a number of occasions including the installation of the thermal sleeve for the 105 mm gun and the installation of the Bofors Lyran system (Swiss Army designation 7.1 cm Le GW 74) on the turret roof between the commander's and loader's positions. This can launch an illuminating rocket to a maximum range of 1,300 m.

A total of 390 Pz 68 series MBTs was built for the Swiss Army. In 1999 a total of 186 Pz 68 was in service together with 186 of the upgraded Pz 68/88 covered in detail later in this entry.

In January 1998 it was stated that the unmodified Pz 68 MBTs would be phased out of service by 2000.

Description

The hull of the Pz 68, like its turret, is a one-piece casting supplied to Thun by Georg Fischer of Schaffhausen, and is divided into three compartments: driver's at the front, fighting in the centre and the power pack at the rear.

The driver is seated at the front of the hull in the centre and is provided with a one-piece hatch cover hinged at the rear. Three day periscopes are mounted forward of the hatch. The centre one can be replaced by a passive periscope for driving at night.

The turret has no bustle, is in the centre of the hull with the commander and gunner seated on the right and the loader on the left. The commander's cupola has eight day periscopes and a single-piece hatch cover that opens to the rear. The commander also operates the split image coincidence range-finder which has a magnification of ×8. The gunner is seated forward and below the commander and has a day periscope with a magnification of ×8 and ×2.7. The loader's cupola is provided with six day vision blocks and split hatch covers that open left and right. The loader's cupola is a little higher than the commander's whose area of observation is therefore restricted.

The engine compartment is at the rear, separated from the fighting compartment by a fireproof bulkhead. The complete power pack, consisting of the engine, auxiliary engine, transmission, cooling and exhaust system, can be removed from the vehicle in about 1 hour. The MTU engine was supplied by Germany and is coupled to a Swiss SLM transmission consisting of a semi-automatic gearbox with six forward and two reverse gears and a double-differential steering system with a hydrostatic steering drive providing progressive and continuous steering control.

The suspension of the Pz 68 consists of six dual rubber-tyred roadwheels, each independently located and sprung by Belleville washers. The idler is at the front and the drive sprocket at the rear. There are three track-return rollers. The tracks of the Pz 68 are of cast manganese and have no rubber pads.

Main armament of the Pz 68 is a British-designed Royal Ordnance (now BAE Systems, RO Defence) 105 mm L7 series gun manufactured in Switzerland with a number of modifications, known as the Pz Kan 61 and firing a Swiss-designed HE round (muzzle velocity 600 m/s) in addition to the normal APDS (muzzle velocity 1,470 m/s), HESH (muzzle velocity 730 m/s) and smoke rounds. Switzerland adopted the Israel Military Industries M111 APFSDS projectile. The gun control system for the Pz 68 is a French SAMM Type CH 25 electrohydraulic system similar to that developed for the French Giat Industries AMX-30 MBT.

Mounted on the forward part of the loader's hatch is a 7.5 mm MG 51 machine gun for anti-aircraft defence which can be elevated from –4 to +77°, and mounted on either side of the turret are three L Pz 51 (80.5 mm) smoke grenade dischargers.

Standard equipment on the Pz 68 includes an NBC system, a hull escape hatch and a drinking water tank. There is no provision for deep fording.

Upgraded Pz 68/88 MBT showing 105 mm gun which is fitted with a fume extractor, thermal sleeve and muzzle reference system ***2000***/0069553

Variants

Pz 68 Mk 2

This was approved for production in 1974 at a cost of SFr146.3 million, and is basically the Pz 68 Mk 1 with an alternator, a thermal sleeve for the 105 mm main armament, and a system for extracting carbon monoxide. Between March and December 1977, 50 Pz 68 Mk 2 tanks were delivered to the Swiss Army.

Pz 68 Mk 3

This has all the improvements of the Mks 1 and 2 but also has a larger turret. Production was authorised in 1975 at a cost of SFr447 million and 110 tanks were delivered between 1978 and September 1979.

Pz 68 Mk 4

This is similar to the Mk 3 and 60 were ordered in 1978 for delivery between July 1983 and April 1984. The Pz 68 Mk 4 cost SFr207 million. In general the terms Mks 1, 2, 3 and 4 are not used in Switzerland. They are normally called the 1st, 2nd, 3rd (or Pz 68/75) and 4th series.

Pz 68 Upgrade

It was announced in early 1988 that 195 Pz 68 MBTs were to be upgraded by the installation of a new fire-control system supplied by Honeywell's German subsidiary and licence-built in Switzerland by Contraves (today Oerlikon Contraves). The fire-suppression system would also be upgraded with all work being carried out at the then Swiss Ordnance Enterprise Thun where the Leopard 2 (Pz 87) was built. Of the total of 195 Pz 68 MBTs upgraded, 25 were from the second batch (Pz 68 Mk 2) with the remainder all of the third (Pz 68 Mk 3) and fourth (Pz 68 Mk 4) batches. When upgraded the Pz 68 was redesignated the Pz 68/88.

Additional modifications included a muzzle reference system and a primary stabilised sighting system for the gunner with an eyepiece for the tank commander. Suspension was upgraded with hydraulic bump stops and an improved collective NBC system was fitted. New paint was introduced and the steel diesel fuel tanks were replaced by GRP tanks.

120 mm gun for Pz 68

Some years ago the Swiss Army tasked the now SW Swiss Ordnance Enterprise Corp to design, develop and test a 120 mm smoothbore gun that would be almost a direct replacement for the current 105 mm gun installed in the Pz 68.

This would give the Pz 68 a significant increase in firepower while also giving greater ammunition operability across the tank fleet.

This Swiss 120 mm smoothbore tank gun was developed and type classified as the 120 mm Compact Tank Gun (CTG) and successfully tested in the Pz 68 MBT.

The end of the Cold War and significant reductions in the Swiss Army MBT fleet meant that the expected up gunning of the Pz 68 MBT with the CTG did not take place. SW Swiss Ordnance Enterprise Corp are now offering their 120 mm CTG for other applications.

The 120 mm CTG is available with a calibre length of 50 or 44 with the former being the version currently being marketed. With the addition of a pepper pot type muzzle brake the weapon can be installed on combat vehicles in the 25 to 30 tonne class. It is fitted with a fume extractor and a thermal sleeve.

The design of the weapon is flexible and the design of the cradle and muzzle brake as well as the length of the recoil travel may be adopted to meet the specific requirements of the application.

For trials purposes the 120 mm CTG has also recently been installed in a General Dynamics Land Systems M60A1 MBT of the Jordanian Army. It is also fitted in the private venture Hägglunds Vehicle CV 90120 Tank covered in detail elsewhere in this section under Sweden.

SW Swiss Ordnance Enterprise Corporation are now marketing surplus unmodified Pz 68 MBTs for the export market as well as offering a major upgrade package. The Swiss Army has a reputation of having well maintained vehicles and the surplus Pz 68s are ready for almost immediate use.

This Pz 68 upgrade is modular so that the user can adopt only those parts required. It can include the installation of the 120 mm CTG, new Explosive Reactive Armour (ERA) for increased battlefield survivability and a new computerised fire-control system, for example the same as that installed in the Pz 68/88 (or Pz 68A6 as it is sometimes referred) for improved first round hit probability. The ERA proposed for the upgraded Pz 68 was originally developed for use on the Swiss Army 155 mm/52 calibre Bison fortress guns.

This Pz 68 upgrade work could be carried out at the SW Swiss Ordnance Enterprise Corp facility in Switzerland or carried out in the end users own country using kits and expertise supplied from Switzerland.

Specifications

Crew: 4
Combat weight: 39,700 kg
Unloaded weight: 38,700 kg
Power-to-weight ratio: 16.62 hp/t
Ground pressure: 0.86 kg/cm^2
Length:
(gun forward) 9.49 m
(gun rear) 8.6 m
(hull) 6.88 m
Width: 3.14 m
Height:
(including AA MG) 2.88 m
(commander's cupola) 2.75 m
Firing height: 1.96 m
Ground clearance: 0.41 m
Track: 2.59 m
Track width: 520 mm
Length of track on ground: 4.43 m
Max road speed: 55 km/h
Fuel capacity: 710 litres
Max range:
(road) 350 km
(cross-country) 160 km
Fording: 1.1 m
Gradient: 60%
Vertical obstacle: 0.75 m
Trench: 2.6 m
Engine: MTU MB 837 Ba-500 8-cylinder diesel developing 660 hp at 2,200 rpm
Auxiliary engine: OM 636 diesel developing 35 hp at 2,600 rpm
Transmission: SLM semi-automatic, 6 forward and 6 reverse gears
Steering: double differential with hydrostatic steering control
Suspension: Belleville washers
Electrical system: 24 V
Batteries: 4 × 6 V, 200 Ah
Armament:
(main) 1 × 105 mm rifled gun
(coaxial) 1 × 7.5 mm MG
(anti-aircraft) 1 × 7.5 mm MG
Smoke-laying equipment: 3 smoke grenade dischargers either side of turret
Ammunition:
(main) 56
(7.5 mm) 5,200
Gun control equipment
Turret power control: electrohydraulic/manual
(by commander) yes
(by gunner) yes
(commander's override) yes
Commander's fire-control override: yes
Max rate power traverse: 360° in 9.5 s
Max rate power elevation: 20.8°/s
Gun elevation/depression: +21/–10°
Gun stabiliser:
(vertical) yes
(horizontal) yes
Range setting device: yes
Elevation quadrant: yes
Traverse indicator: yes
Armour: 15-120 mm
NBC system: no
Night vision equipment: yes

Armoured Recovery Vehicle

Development of an ARV based on the chassis of the Pz 61 MBT began at Thun in 1961 with the first prototype being completed in 1967-68. After trials with this prototype a modified version based on the chassis of the Pz 68 was placed in production. First production vehicles, known as the Entpannungspanzer 65 (or Entp Pz 65) were completed in 1970. The vehicle is fitted with a main winch with 120 m of cable and has a maximum capacity of 25,000 kg, which can be increased to 75,000 kg with the use of snatch blocks. The auxiliary winch used to pull out the main cable has 240 m of cable. Mounted at the front of the vehicle is a hydraulically operated dozer blade which is used to stabilise the vehicle or for dozing. Pivoted at the front of the hull is an A-frame with a maximum lifting capacity of 15,000 kg. A full range of tools and cutting equipment is carried. The Entp Pz 65 has a crew of five and a loaded weight of 38,000 kg. Armament consists of a single 7.5 mm machine gun and eight smoke dischargers. Full details of the Entp Pz 65 are given in *Jane's Military Vehicles and Logistics 2002-2003*. The last production Entp Pz 65 was handed over in May 1985. Upgraded Entp Pz 65 vehicles are designated the Entp Pz 68/88.

Armoured Bridgelayer

The prototype of the bridgelayer was based on the chassis of a Pz 61 tank but production vehicles, designated the Brückenlegepanzer 68, were based on the Pz 68 tank chassis. The bridge is launched as follows: the vehicle stops short of the obstacle and the bridge is tilted forwards, a beam is slid over to the far bank and the bridge is then advanced along the beam. Once the bridge has reached the far bank, the beam is retracted. The bridge has an overall length of 18.23 m and a maximum capacity of 60,000 kg, its normal capacity being 50,000 kg. The bridgelayer has a crew of three and weighs 44,600 kg complete with the bridge. Full details of the Brü Pz 68 are given in *Jane's Military Vehicles and Logistics 2002-2003*. More recently these have been fitted with a fully automatic hydraulic bridgelaying system and redesignated the Brü Pz 68/88. The bridge and bridgelaying system used with this vehicle have also been utilised by Jung Jungenthal of Germany for its M47 and M48 tank chassis conversion to the armoured bridgelayer role. Details of these are given in *Jane's Military Vehicles and Logistics 2002-2003*.

Target Tank

A special version of the Pz 68, called the Panzerzielfahrzeug or Pz Zielfz 68, was developed for training anti-tank crews. It has a heavily armoured turret with a dummy gun and additional armour is provided for the hull. The tops of the tracks are covered by armoured skirting. The Pz Zielfz 68 weighs 38,000 kg and has a crew of two.

Status

Production complete. In service only with the Swiss Army.

Manufacturer

SW Swiss Ordnance Enterprise Corp.

UPDATED

Entp Pz 65/88 ARV in travelling configuration 0011780

Brü Pz 68/88 armoured bridgelayer crossing a bridge that it has just laid in position over a gap 0011779

TAIWAN

M48H MBT

Development/Description

Late in 1988, the Taiwanese Army announced that it had successfully fired a home-produced version of the US-built 105 mm M68 rifled tank gun from a modified M48 tank.

The 105 mm M68 is the US version of the British Royal Ordnance (now BAE Systems, RO Defence) L7 weapon but with a US-designed breech. It has been used for many years, not only in the US M60/M60A1/M60A3, but also the M1, M48A5 and South Korean K1, as well as Israeli tanks including the Merkava Mk 1 and Mk 2.

The United Services Institute of Taiwan was responsible for producing and testing the 105 mm rifled tank gun that is fitted to the M48H.

The M48H, also referred to as the Brave Tiger MBT, was shown for the first time in 1990. This is also known as the CM-11, Courageous Tiger.

The M48H is a hybrid vehicle which uses new M60A3 hulls purchased from General Dynamics Land Systems, before the production line closed down in 1987, fitted with modified M48 turrets. This has an advanced fire-control system incorporating a ballistic computer and thermal imaging sight which was developed by the Chung Shan Institute of Science and Technology for the vehicle.

It is understood that the roof-mounted stabilised sight, which incorporates a thermal imaging channel, is similar to that installed on the United States General Dynamics Land Systems M1 Abrams MBT and is slaved to the 105 mm gun. This sight has dual-axis stabilisation which allows improved target tracking on the move. The laser range-finder is believed to be of the Nd:YAG type.

Late in 1988, Urdan Industries of Southfield, Michigan, USA, was awarded a US$3.8 million fixed price contract for 520 low-profile commander's cupolas for installation on M48 tanks by the US Army Tank Automotive Command. This order was a Foreign Military Sale (FMS) to Taiwan with the actual work being carried out at Netanya, Israel. Production was completed late in 1989.

The M48H is powered by a General Dynamics Land Systems AVDS-1790-2C diesel engine which also powered the final production model of the M60 series, the M60A3. This develops 750 hp at 2,400 rpm and is coupled to a CD-850 series automatic transmission.

From a firepower and mobility aspect, the M48H is on a par with the US Army M60A3, but its turret armour is inferior, although this could easily be improved by the addition of appliqué passive or Explosive Reactive Armour (ERA).

It is believed that Taiwan has recently purchased from abroad (possibly from SNPE of France) an ERA system for its M48H series MBTs to improve their battlefield survivability against anti-tank weapons fitted with a high-explosive anti-tank warhead.

An explosion/fire detection/suppression system is fitted to the M48H MBT, this was provided by Santa Barbara Dual Spectrum of the USA, with the initial order being for 171 systems.

Until the introduction of the M48H, the most powerful tank in the Taiwanese tank fleet was the M48 series of MBT. These were originally armed with the 90 mm gun but it is believed that these have all now been regunned with the 105 mm M68 weapon plus other modifications bringing them up to the M48A5 configuration. It is believed that 100 M48A5s are in service.

Taiwanese M48H Brave Tiger MBT which is based on the modified M60 MBT chassis and M48 series MBT turret (L J Lamb)
0018839

In addition to the M48A5 and M48H MBTs, the remainder of the tank fleet is understood to consist of 230 to 325 upgraded M24 Chaffee light tanks with 90 mm guns and a new NAPCO diesel power pack and 675 M41 light tanks (of which perhaps some 650 are fully operational), some of which have been upgraded to the Type 64 configuration. Additional details are given in the Light tanks section under Taiwan.

Exact numbers of M48H tanks built are not known although, early in 1985, orders were placed with General Dynamics Land Systems for 450 M60 chassis and 265 assembly ready hulls for Taiwan.

It is believed that a total of 450 M48H tanks were built in Taiwan with final deliveries being made in early 1995.

Taiwan has already acquired from surplus US Army stocks 360 M60A3 TTS (Tank Thermal Sights) MBTs.

Recent figures released by the United Nations show the following exports of M60 series to Taiwan in recent years:

1992	nil
1993	nil
1994	nil
1995	21
1996	107
1997	30
1998	120
1999	98
2000	27

Late in 1996, it was revealed that Taiwan was planning to develop its own third-generation MBT which is expected to become operational in 10 to 15 years, according to the Army Chief of Staff.

This would replace the older MBTs in service with Taiwan, first followed by the M48H and the more recently acquired M60 series, both of which are armed with a 105 mm rifled gun.

No firm details of this future MBT have been released apart from 'It would be streamlined, better to operate in mountainous terrain, with upgraded combat capabilities and protective armour.'

Taiwan is also considering the acquisition of an initial batch of 105 mm armed light tanks with known contenders including the United Defense M8 Armored Gun System and the Textron Marine & Land Systems Stingray II. There is also at least one European contender for this requirement. As of January 2002, no firm contracts are known to have been placed by Taiwan.

Status

Production completed in early 1995. In service in Taiwan.

Manufacturer

Fighting Vehicles Development Centre (prime contractor).

UPDATED

Specifications

M48H
Crew: 4
Combat weight: 50,000 kg
Power-to-weight ratio: 15 hp/t
Length: (overall) 9.306 m
Width: (overall) 3.631 m
Height: 3.086 m
Armament:
(main) 1 × 105 mm rifled gun
(coaxial) 1 × 7.62 mm MG
(anti-aircraft) 1 × 12.7 mm MG; 1 × 7.62 mm MG
Smoke dischargers: 2 × 6
(Performance is similar to M60A1 MBT)
NBC system: yes
Night vision equipment: yes

TURKEY

New Turkish MBT

Late in 1999, following an international competition, the Turkish Land Forces Command (TLFC) short listed the following MBTs to take part in extensive series of fire power and mobility trials in Turkey early in 2000.
Alenia Otobreda of Italy with Ariete
General Dynamics Land Systems of the USA with M1A2
Giat Industries of France with Leclerc
Krauss-Maffei Wegmann of Germany with Leopard 2A6EX
Ukrspetexport of the Ukraine with the T-84

In the end the Ariete MBT did not take part in the trials in Turkey. General Dynamics Land Systems sent a standard M1A2 for the trials and this was followed by an M1A2 MBT fitted with a MTU EuroPowerPack. The Krauss-Maffei Wegmann Leopard 2A6EX is the latest version and features a number of improvements including the 120 mm L/55 smoothbore gun.

As of late January 2002, the Turkish Land Forces Command had not selected a winner for their new MBT competition and the total number of vehicles to be produced, under licence in Turkey, had not been decided, but could be at least 1,000 units.

The Turkish Land Forces Command is also considering upgrading 1,000 of its M60 MBTs for which there are proposals from General Dynamics Land Systems and Israel Military Industries.

The General Dynamics Land Systems proposal is called the 120S and is essentially an upgraded M60 series chassis with an M1A1 Abrams turret armed with a 120 mm M256 smoothbore gun.

The Israel Military Industries vehicle is called the Sabra and retains the existing hull and chassis with many modifications and is fitted with the Israel Military Industries 120 mm smoothbore gun as fitted to the Merkava MBT which is also being offered to Turkey.

As any new MBT selected by Turkey will not be in service in significant numbers for at least five years, Turkey may well opt to upgrade at least part of its M60 fleet with a 120 mm gun as an interim measure.

Details of the Sabra and 120S are given in this section under Israel and the United States of America.

As of January 2002, the Turkish Land Forces Command had not awarded an contracts for the proposed M60 MBT upgrade programme.

UPDATED

Turkish MBT fleet

Early in 2002, it was estimated that the Turkish MBT fleet consisted of around 2,876 M48, 932 M60 (658 M60A3 TTS and 274 M60A1) and 307 Leopard 1.

According to United Nations sources, between 1992 and 2000, Turkey took delivery of the following:

From	Quantity	Type	Delivery
Germany	11	Leopard 1	1992
Germany	85	Leopard 1	1993
USA	577	M60	1992
USA	356	M60	1993

M48 with the now Krauss-Maffei Wegmann retrofit kit as supplied to the Turkish Army showing the thermal sleeve for the 105 mm L7 gun
0018840

The now Krauss-Maffei Wegmann of Germany rebuilt five Turkish M48 tanks to a similar standard as the 650 M48A2GA2s it supplied to the German Army (see entry on M48 in this section). In addition, the Turkish vehicles have been fitted wth an MTU MB 837 Ea-500 diesel engine with all peripheral equipment (for example control units and cooling system), a water preheating system, fuel tanks with pumps and controls, an exhaust system and electrical equipment including a control panel and batteries. There are also modifications to the chassis which include ammunition stowage, dust shields, a heating system and a fire detection and extinguishing system.

The now Krauss-Maffei Wegmann was then awarded a contract by the German government to provide 165 kits, plus personnel support, enabling Turkey to carry out the work itself.

Under an agreement signed in Bonn in November 1980, Turkey received 77 Leopard 1A3 MBTs (54 built by the then Krauss-Maffei and 23 by the then MaK) and four ARVs (built by MaK). The first of these was delivered in 1982 and final deliveries were made in 1984.

In 1990-91, Turkey took delivery of 150 Leopard 1 MBTs from the then Krauss-Maffei that had been upgraded to the Leopard 1A5 configuration as well as 80 Leopard 1A1A1 MBTs and a further eight Leopard 1 ARVs from German Army stocks.

Following trials with two prototype vehicles fitted with the locally developed Aselsan Volkan fire-control system installed in two Leopard 1 series MBTs of the Turkish Land Forces Command, a contract was placed for a total of 162 systems.

The Volkan computerised fire-control system includes day/thermal sights, laser range-finder, computer and a number of sensors. According to Aselsan, the installation of the Volkan computerised fire-control system allows the upgraded Leopard 1 to engage moving targets under adverse conditions with a high first round hit probability when the platform is stationary or moving.

In late 1979, a team of Turkish officers visited the Anniston Army Depot in the USA where the M48A5 tank conversion programme was being carried out. It was determined that Turkey would be able to convert its ageing fleet of M48 series tanks into the M48A5 configuration, with a target date of 1983 for first conversion vehicles to be completed.

In 1982, the USA leased heavy industrial equipment to Turkey for installation in the depot at Kayseri which was to become the first M48 tank conversion plant. Turkey ordered its first two conversion kits from the USA in 1982 with a further 400 being ordered later the same year.

Between August 1982 and May 1983, the industrial plant arrived in Turkey, was installed and the operators were trained. Conversion work on the first five M48s started early in May 1983 and by September the same year initial trials of these five vehicles, designated the M48A5T1, had been completed.

Upgraded M48A5T2 of the Turkish Army clearly showing the 105 mm rifled tank gun fitted with a thermal sleeve and fume extractor
0018841

The second M48A5 conversion facility at Arifiye started work in January 1984 and by the summer of that year the first Turkish M48 battalion had converted to the M48A5T1 MBT.

It is understood that the final conversion kits were shipped by the United States late in 1990 with the final batch of M48A5T1s delivered in 1993.

The M48A5T1 is the equivalent of the US Army M48A5 and is described in detail in the entry on the M48 tank later in this section.

Late in 1985, the US Department of Defense notified Congress of a letter of offer to Turkey, valued at US$206 million, for the supply of 760 tank conversion kits, comprising tank thermal sights, ballistic computers and add-on stabilisation systems. Prime contractors for this package were the then Texas Instruments (thermal sight), Computer Devices Company (ballistic computer) and HR Textron Incorporated (stabilisation system).

When fitted with the thermal sight, ballistic computer and stabilisation system, the vehicle is known as the M48A5T2.

In August 1986, HR Textron Incorporated announced that it had been awarded a US$30 million contract by the US Army Armament, Munitions and Chemical Command for 760 tank stabilisation systems. First deliveries were made in 1987 with follow-on orders bringing the total to more than 2,000 units through 1995.

Further improvements under consideration include additional armour protection and improved suspension.

For some years Turkey has been making the US 105 mm FP105 APFSDS rounds for the 105 mm armed Leopard 1 and M48A5T1 MBTs.

It is understood that the M48 upgrade programme in Turkey is now complete, with a total of 1,369 M48A5T1 and 751 M48A5T2 MBTs in service. These serve alongside 658 M48A3 and 183 M48T5 MBTs.

Pending the introduction of a new Turkish MBT (covered in a separate entry), Israel Military Industries (Sabra) and General Dynamics Land Systems (120S) are marketing upgraded M60 MBTs to the Turkish Army.

Both of these are armed with a 120 mm smoothbore gun and full details are given under Israel and the United States of America. Other contenders include FNSS Defense Systems of Turkey and the Malyshev Plant of the Ukraine. As of February 2002, no contract had been placed for the M60 upgrade by the Turkish Land Forces Command.

UPDATED

UKRAINE

T-84-120 Oplot Main Battle Tank

Ukrainian T-84-120 MBT, shown here with a defensive aids suite and explosive reactive armour, is armed with a 120 mm smoothbore gun fed by a bustle mounted automatic loader (KhMDB) **2001**/0073602

Specifications
Crew: 3
Combat weight: 48,000 kg
Power-to-weight ratio: 25 hp/t
Ground pressure: 0.93 kg/cm^2
Length:
(gun forwards) 9.72 m
(hull) 7.985 m
Width:
(over tracks) 3.56 m
(over hull) 3.775 m
Height:
(turret roof) 2.74 m
(over 12.7 mm MG) 2.74 m
Ground clearance: 0.515 m
Track width: 580 mm
Track: 2.8 m
Length of track on ground: 4.29 m
Max road speed:
(forwards) 70 km/h (road)
45 km/h (cross-country)
(reverse) 32 km/h
Fuel capacity: 1,300 litres
Range:
(road) 500 km
(cross-country) 400 km
Fording:
(without preparation) 1.8 m
(with preparation) 5 m
Gradient: 63%
Side slope: 36%
Vertical obstacle: 1 m
Trench: 2.85 m
Engine: Type 6TD-2 twin-stroke, multifuel, liquid-cooled 6-cylinder diesel developing 1,200 hp
Transmission: mechanical, epicycle train with 7 forward and 1 reverse gears
Steering: clutch and brake
Suspension: torsion bar with hydraulic shock-absorbers on 1st, 2nd and 5th road wheel stations
Armament:
(main) 1 × 120 mm smoothbore gun
(coaxial) 1 × 7.62 mm machine gun
(anti-aircraft) 1 × 12.7 mm machine gun
Smoke grenade launchers: 2 × 6 81 mm
Ammunition:
(main) 40 (of which 22 are ready use)
(coaxial) 1,250
(anti-aircraft) 450

Gun-control equipment
Turret-power control:electric/manual
Turret traverse: 360°
Gun elevation/depression: +13° 30'/-5° 40'
Gun stabiliser:
(vertical) yes
(horizontal) yes
Range-finder: yes, laser
NBC system: yes (overpressure)
Night vision equipment: yes

Development

In late 2000, the Khar'kov Morozov Machine-building Design Bureau (KhMDB) based in the Ukraine, released details of the 120 mm smoothbore gun armed T-84 Main Battle Tank (MBT). It was demonstrated in Turkey early in 2001 to meet the operational requirements of the Turkish Land Forces Command (TLFC). This is designated the T-84-120 MBT, or Oplot. Development of the T-84-120 is complete and production can commence when orders are placed.

Description

The overall layout of the T-84-120 Oplot is almost identical to the standard production T-84 covered in detail in the following entry. The major difference is its 120 mm NATO calibre gun and associated ammunition handling system.

The basic T-84 MBT, a further development of the T-80UD MBT of which 320 have been supplied to Pakistan, is armed with a standard 125 mm smoothbore gun with the automatic loader being positioned below the turret. This feeds the projectile and then the charge with its semi-combustible cartridge case, to the 125 mm smoothbore gun.

Combat experience has shown that the position of this type of automatic loader below the turret makes the vehicle very vulnerable and, if the ammunition is ignited the turret is blown off and the vehicle totally destroyed.

The 125 mm smoothbore gun has been replaced by a NATO standard 120 mm smoothbore gun from an undisclosed source. It is fully stabilised and fitted with a fume extractor, thermal sleeve and muzzle reference system.

The bustle mounted automatic loader contains 22 rounds of ready use 120 mm ammunition with a further 18 rounds stowed in a mechanised stowage position in the hull. The automatic loader is separated from the turret by armoured bulkheads with blow out panels in the roof of the turret. According to the manufacturer, the 120 mm smoothbore gun can be changed under field conditions.

A 7.62 mm machine gun is mounted co-axial with the main armament and there is also a 12.7 mm roof mounted machine gun. A computerised fire-control system is fitted. The commander and gunner have day/night passive sights with an integrated laser range-finder with the gunner also having a thermal sight with a monitor for the commander. The gunner also controls the laser guided 120 mm gun launched missile.

The T-84-120 MBT can engage moving targets when it is stationary or moving and the stabilised sighting system allows for hunter/killer type target engagements to be carried out. Either the commander or gunner can lay and fire the 125 mm main armament as well as the 7.62 and 12.7 mm machine guns.

The T-84-120 has a two axis stabilised 120 mm smoothbore gun that fires standard NATO-type one-piece ammunition with a semi-combustible cartridge case and all that remains after firing is the stub cartridge case.

The manufacturer claims a rate of fire of 8 to 10 rounds a minute with the system having two other modes of operation, semi-automatic and manual gun loading. The KhMDB also state that it fires a laser-guided projectile. This has been developed in the Ukraine and is called Combat, additional details are given in the entry for the T-84 MBT under the Ukraine.

The T-84-120 also features a new turret of all welded armour construction and enhanced levels of protection for hull and turret. Explosive reactive armour is provided over the hull and turret that provides protection against kinetic and chemical energy attack.

A Shtora defensive aids suite is fitted as standard. Laying smoke and aerosol screens create interference to decoy away laser beam riding guided missiles and infra-red guided missiles. Mounted either side of the turret is a bank of six 81 mm grenade launchers.

It is powered by a very compact 1,200 hp 6TD-2 diesel engine which, with a combat weight of 48 tonnes, gives a power-to-weight ratio of 25 hp/tonne. Maximum road speed is quoted as 70 km/h.

A Vehicle Movement Control System is fitted which ensures control of the engine operation modes and automated gear shifting in order to reduce fuel consumption and driver fatigue. It also controls engine starting modes and displays current information about the power pack operation mode.

Other features include an NBC system, auxiliary power unit, navigation system that incorporates a Global Positioning System, new driver's panel, new driver's controls, auxiliary power unit and a fire detection and suppression system.

The installation of the 8 kW under-armour APU provides electrical power for all tank systems while the engine is shut down. In addition, the APU can also be used for starting the main diesel engine.

Status

Development complete. Ready for production.

Manufacturer

Malyshev Plant.

UPDATED

T-84 MBT

Development

The T-84 MBT was developed by the Kharkov Morozov Design Bureau (previously the A A Morozov Design Bureau) which is in the Ukraine, with production being undertaken by the Malyshev Plant, also in the Ukraine.

This facility developed the T-64 MBT and produced some 8,000 vehicles for the Soviet Army. This was never exported and is not often seen in public.

The T-64 was followed by the improved T-64A and then the T-64B (1976). Subsequently the Malyshev Factory also undertook production of the diesel-powered T-80UD (1986) MBT but, with the break-up of the Soviet Union, production of this ceased around 1989/1991 as about 70 per cent of the MBT was imported from other parts of the country such as the turret, hull, gun and some key electronic components.

In 1993, the Ukraine took a decision to develop the T-80UD further and this resulted in the T-84 MBT.

The T-80 was designed jointly by the Leningrad Kirov Plant and the Kharkov Malyshev Plant. Production of the T-80UD powered by a 1,250 hp gas turbine continues at Omsk but has ceased at St Petersburg (formerly Leningrad). At Kharkov, only the T-80UD powered by the 6TD diesel developing 1,000 hp had been produced.

According to the manufacturer, some 98 per cent of the T-84 is now manufactured in the Ukraine and it is planned to increase this to the full 100 per cent.

The T-84 is based on a T-80UD (this being powered by a more fuel-efficent diesel engine rather than a gas-turbine engine) fitted with a new turret of all-welded construction rather than the cast steel turret which was previously imported from Russia.

While under development the T-84 had a number of different designations including Object 478B, Object 478DU and Object 478DU2. The last two were powered by the 1,000 hp 6TD diesel engine and had different road wheels.

In mid-2001, the Ukraine Army took delivery from the local Malyshev Plant of two new T-84 series MBTs. These were the first new MBTs to be delivered to the Ukraine Army since the country became independent from Russia.

The late production vehicles supplied to Pakistan (see below) have been built to the T-84 standard but were powered by the 1,000 hp 6TD engine.

The T-84 has been shown and demonstrated in a number of countries including Pakistan, Turkey and the United Arab Emirates.

It is understood that a new MBT is being designed at the Kharkov Morozov Design Bureau which, like the tank and engine facilities, is still owned by the government. No details of this are available, although it could well be an unconventional design with a large calibre gun.

In August 1996, Pakistan placed an order with the Ukraine for the supply of 320 T-80UD MBTs with the first batch of 15 vehicles being delivered in March 1997 and with the second batch of 35 following in mid-1997.

Final deliveries were made to Pakistan late in 1999. Late production vehicles for Pakistan have now been confirmed as being almost to the T-84 configuration including the new all-welded turret which incorporates advanced armour.

The Kharkov Morozov Design Bureau/Malyshev Plant has developed an upgrade package for the T-72 called the T-72AG and details of this are given in a separate entry. This vehicle was shown for the first time in early 1997.

T-84 MBT which is armed with a 125 mm smoothbore gun fed by an automatic loader **2001**/0103436

Description

In many respects the T-84 is similar in layout to the T-80 MBT covered in this section under the Russian Federation, with the driver's compartment at the front, turret in the centre and the power pack at the rear.

The manufacturer claims that the armour protection of the T-84, which includes Explosive Reactive Armour (ERA) for the turret and chassis, is superior to that fitted to the T-72 and T-80. The ERA gives protection against both chemical and kinetic energy attack.

The driver is seated at the front of the hull in the centre and is provided with a single-piece hatch cover and three day periscopes for forward observation. The centre one can be replaced by a passive periscope for driving at night.

The gunner is seated on the left of the new turret and commander on the right, both being provided with a roof hatch. It is understood that the turret consists of multiple layers of steel and non-metallic armour with six layers on the front and five on the sides.

Main armament comprises a 125 mm KBA-3 (this is the local built version of the Russian 125 mm 2A46M-1) smoothbore gun fitted with a thermal sleeve and fume extractor. This is fed by an automatic loader which is similar to that installed in the T-64 in that the separate loading ammunition is stowed vertically whereas that of the T-72 and T-80 is stowed horizontally.

The T-84 has a total of 28 rounds (projectile and charge) of ready use ammunition which can be of up to six different types including APFSDS, HE-FRAG, HEAT, laser-guided projectile and two natures of HE with a timed fuze (delayed detonation).

It is claimed that the 125 mm KBA-3 gun can fire 7 to 9 rds/min, two to three guided missiles or four delayed detonation protectiles. Once the gun has been fired it returns to its pre-assigned loading position where the projectile and then the charge is loaded. It then returns to its previously laid position. It takes between 7 and 12.5 seconds to load a new projectile, charge and return the gun to the firing position.

The 125 mm KBA-3 smoothbore gun barrel can be changed without the need to remove the gun from the tank. It is also provided with a special clearance device for the barrel in the gun cradle. In addition, servicing of the recoil system has been simplified by installing indicators on the recoil absorbers and recuperator.

The T-84 fires the same types of 125 mm ammunition as the Russian T-72 and T-80, including the laser beam-riding guided missile, details of which are given in their respective entries earlier in this section.

In 2000, the Ukrainian State Enterprise "Specialised Foreign Trade Firm Progress" confirmed that in 1999 the Ukrainian Army took delivery of the first batch of Combat laser guided missiles that are launched from the 125 mm smoothbore guns installed in the locally built T-80UD and T-84 series Main Battle Tanks (MBT).

All of the T-80UD/T-84 series MBTs delivered to Pakistan were fitted with an advanced fire-control system that enables the 125 mm smoothbore gun to fire a laser-guided projectile.

The latter were developed in Russia by the Tula KBP Design Bureau for Instrument Building (Refleks with a maximum range of 5,000 m) and the older KBP design Bureau for Instrument Building (Svir with a maximum range of 4,000 m).

The Ukraine decided to develop its own laser-guided missile for use with the T-80UD/T-84 MBT and this entered production as the Combat in 1999.

Combat consists of two parts, the pusher and the actual missile, both of which are stowed in the automatic loader of the T-80UD/T-84 located below the turret. This first loads the missile and then the pusher.

The gunner first acquires the target and when aim has been established, fires the 125 mm gun. The pusher ejects the missile from the gun and its rocket motor then cuts in. It then rides the laser beam until it impacts the target. The pusher is ejected out of the gun internally.

According to Progress, the new Combat laser guided missile can be fired from the T-72, T-80UD and T-84 series MBTs, provided that they have the correct fire-control system installed while the platform and target are moving.

Although the primary role of the Combat laser guided projectiles is against MBTs at ranges beyond that of the 125 mm tank gun firing conventional ammunition, it can also be fired against other battlefield targets such as hovering helicopters and pill boxes.

The Combat missile has a maximum range of 5,000 m and is fitted with a tandem HEAT (High-Explosive Anti-Tank) warhead to defeat targets fitted with explosive reactive armour.

Flight time to 5,000 m is quoted as 16 seconds with the projectile having four wraparound fins at the rear. Total weight of the round, for example pusher and missile, is 31 kg, which is considerably more than its Russian equivalent. Although the method of guidance is similar, the Ukrainian missile is a new design and has a diffrent shape and much larger pusher.

A 7.62 mm PKT machine gun is mounted coaxially with the main armament and the commander is provided with a 12.7 mm NSVT machine gun for use in the surface/air and surface/surface roles which can be laid and fired under complete armour protection. The 12.7 mm machine gun has an elevation of +70° and a depression of –5°.

Mounted on either side of the turret is a bank of six electrically operated 81 mm smoke grenade dischargers which, unlike the Russian T-72/T-80, are provided with a cowl. The T-84 can also lay its own smoke screen by injecting diesel fuel in the exhaust.

The T-84 is also fitted with the Shtora-1 vehicle protection system which has also been fitted to the

The T-84 MBT is a further development of the T-80UD MBT but has many improvements and features a new all-welded turret armed with a 125 mm smoothbore gun **2002**/0095478

T-84 series MBT with skirts removed to show suspension detail being loaded onto a semi-trailer **2002**/0131040

Russian T-90 MBT and some versions of the T-80. This increases the survivability of the T-84 by decoying away incoming ATGWs.

The T-84 is fitted with the same 1A45 fire-control system as the T-80U and T-80UD and either the commander or gunner can lay the 125 mm KBA-3 main armament while the tank is stationary or moving with a high first round hit probability.

The commander has a TKN-4S Agat day/night vertically stabilised sight while the gunner has an IG46 Irtysh two-plane (vertical and horizontal) day sight that also incorporates a laser range-finder and a missile guidance capability. The gunner's Buran-E night sight is stabilised in the vertical plane only. Thermal night vision equipment is optional on the T-84.

The commander's and gunner's day/night system is an integrated part of the fire-control system and, according to the manufacturer, provides the following degree of accuracy:

- Laser guided missiles out to 5,000 m (day)
- Conventional tank rounds out to 2,500 m (day)
- Night out to 1,200 m (Buran-E system)
- Night out to 3,000 m (thermal imager).

The computerised analogue/digital fire-control system lays the gun onto the target after taking into account all the inputs from the sensors including tank and target speed, cant, windspeed, ambient temperature, projectile and charge temperature, barrel wear and so on. It also computes the time when the HE projectile with controlled detonation should be detonated over the target. The ballistic computer is designated the 1V528 while the gun stabilisation system is designated the 2E42.

The power pack is at the rear and consists of a model 6TD-2 diesel developing 1,200 hp with a 1,500 hp version currently under development, coupled to a new mechanical transmission with seven forward and one reverse gears. An APU is mounted at the rear of the hull on the left side.

The air inlet allows air to be ducted from the least dusty quarter and enables water obstacles to be crossed to a water depth of 1.8 m without preparation. With preparation the T-84 can ford to a depth of 5 m.

There are two parts to the air filtration system, the centrifugal precleaners and the air cleaner casing. This enables the tank to be operated in hot and dusty conditions for up to 1,000 km without a change of filters and to carry out combat under radioactive conditions.

The 1,500 hp engine version, currently under development, is designated the 6TD-3 and features intermediate cooling of the supercharged air. The engine compartment is provided with thermal protection.

The suspension is of the torsion bar type with each side having six dual rubber-tyred roadwheels with the idler at the front, drive sprocket at the rear and track-return rollers. Hydraulic shock absorbers are provided for the first, second and sixth road wheel stations either side.

The upper part of the suspension is covered by a skirt, the forward part of which is armoured. A rubber mat hangs at the front of the vehicle and this helps to keep down dust.

Standard equipment includes an NBC system, night vision equipment for commander, gunner and driver, provision for deep fording, fire detection/suppression system, radiation shielding and a dozer blade mounted under the front of the hull.

It can also be fitted with various types of mineclearing system at the front of the hull including the KMT-6 plough-type system. Mounted at the rear of the hull is an unditching beam and two long-range fuel tanks. One of the options being offered for the export market is an air conditioning system.

T-84U MBT

In 1999, the Ukraine started marketing an improved version of the T-84 MBT. This is a further development of the T-80U MBT of which 320 have been supplied to Pakistan under a US$580 million contract placed in 1996.

The Ukrainian Kharkov Morozov Machine Building Design Bureau has been responsible for the T-84U design aspects while the State Enterprise Malyshev Plant will be responsible for production aspects.

Main features of the T-84U can be summarised as an improved Explosive Reactive Armour (ERA) package, new R-163-50K radio, new-generation 1V528-2 ballistic computer, new 1 KRNA land navigation system, new auxiliary power unit, installation of a command and control system and the option of being fitted with two types of track.

The R-163-50K is a short-wave two-way radio set operating in the 2,000 to 30,000 kHz frequency band with 16 pre-selected frequencies. Maximum range using a vertical positioned rod aerial is said to be 50 km. The range can be increased up to 250 km using a declined rod aerial or up to 350 km if a symmetric vibrator aerial is used.

One type of track is fitted with un-removable rubber pads from both sides of the track links, this is normally used for road use. The second is a multipurpose track with replaceable rubber pads and removable metal grousers that give improved traction in poor soil conditions.

The 1 KRNA land navigation system is of the satellite radio navigation type and based on the Global Positioning System (GPS) NAVSTAR and GLONASS system.

The new ERA, which is installed over the frontal arc of the T-84U, is claimed to provide a higher level of protection against both kinetic and chemical energy attack.

To further improve its combat survivability, the T-84U can be fitted with additional side screens, camouflage net and dazzle painting. The paint also helps to ensure a low level of reflection of laser and infra-red irradiation.

The installation of an auxiliary power unit allows the main systems of the T-84U to be switched on and in use with the main engine being shut down to conserve fuel. The diesel electric power unit has an output of 8 kW and has sufficient fuel for 24 hours of continuous operations.

As well as offering brand new vehicles to this T-80U standard, the Ukraine will also offer upgrade kits to enable future customers of theT-84 to upgrade them to the latest T-84 standard in their own facilities.

Specifications

Crew: 3	**Engine:** Model 6TD-2 twin-stroke, multifuel, liquid-cooled 6-cylinder diesel, fuel injected, developing 1,200 hp
Combat weight: 46,000 kg	**Transmission:** mechanical, epicycle train with 7 forward and 1 reverse gears
Power-to-weight ratio: 26.08 hp/t	**Steering:** clutch and brake
Ground pressure: 0.93 kg/cm^2	**Suspension:** torsion bar with hydraulic shock absorbers on 1st, 2nd and 6th road wheel stations
Length: (gun forward) 9.72 m; (hull) 7.085 m	**Electrical system:** 27 V
Width: (over tracks) 3.56 m; (with skirts) 3.775 m	**Armament:** (main) 1 × 125 mm KBA-3 smoothbore gun; (coaxial) 1 × 7.62 mm PKT MG; (anti-aircraft) 1 × 12.7 mm NSVT MG
Height: (turret roof) 2.215 m; (including 12.7 mm MG) 2.74 m	**Ammunition:** (main) 43 (28 in automatic loader); (coaxial) 1,250; (anti-aircraft) 450
Ground clearance: 0.515 m	**Smoke grenade launchers:** 2 × 6 81 mm
Track width: 580 mm	**Gun control equipment**
Track: 2.8 m	**Turret power control:** electric/manual
Length of track on ground: 4.29 m	**Turret traverse:** 360°
Max road speed: 65 km/h	**Gun elevation/depression:** (turret front) +13° 30'/–5° 40'; (turret rear) +16° 40'/–2° 20'
Average cross-country speed: 45-50 km/h	**Gun stabiliser:** (vertical) yes; (horizontal) yes
Fuel capacity: 1,300 litres	**Range-finder:** yes, laser
Range: (road) 540 km; (cross-country) 350-400 km	**NBC system:** yes
Fording: (without preparation) 1.8 m; (with preparation) 5.0 m	**Night vision equipment:** yes
Gradient: 63%	
Side slope: 36%	
Vertical obstacle: 1.0 m	
Trench: 2.85 m	

T-84 MBT from the rear with skirts removed and showing suspension in detail **2001**/0103437

Variants

T-84-120 MBT

Details of the T-84, which is armed with a NATO standard 120 mm smoothbore gun fed by a bustle mounted automatic loader, are given in the previous entry.

BTMP-84 HIFV

It has been disclosed that the Kharkov Morozov Machine Building Design Bureau in the Ukraine has built the prototype of a new Heavy Infantry Fighting Vehicle (HIFV), based on the T-84 MBT chassis called the BTMP-84. Details of this are given in the Armoured personnel carriers (tracked) section under Ukraine. This has yet to enter production.

BREM-84 ARV

The Ukraine has proposed that a new-generation armoured recovery vehicle could be developed on the T-84 chassis called the BREM-84.

New 125 and 140 mm guns

In 1998, the Ukraine's State Scientific and Technical Center of Artillery and Rifle Arms (SSTC ARA) revealed that it had developed to the prototype stage new 125 and 140 mm smoothbore tank guns.

The 125 mm gun, known as the Vitiaz is 6 m long with a maximum recoil length of 310 mm. When firing an armour-piercing fin-stabilised discarding sabot (APFSDS) projectile, a muzzle velocity of 2,030 m/s is attained. This will penetrate 380 mm of steel armour at an angle of 60°.

The 140 mm gun, the Bagira, is 7 m long and also has a maximum recoil length of 310 mm. When firing an APFSDS projectile a muzzle velocity of 870 m/s is attained and this will penetrate 450 mm of armour at an angle of 60°.

According to the SSTC ARA, both smoothbore weapons can be fitted into new MBTs as well as being retrofitted into other MBTs such as the T-64, T-72, T-80UD and the T-84.

T-84 SPAAG

A T-84/T-80UD chassis has been used for trials with the Donets air defence system. This consists of a modified ZSU-23-4 turret armed with two SA-13 Gopher type SAM each side while retaining its four 23 mm cannon. The turret is fitted on a raised superstructure that is in turn installed on the T-84/T-80UD chassis. The Donets remains at the prototype stage. Details of this system are given in *Jane's Land-Based Air Defence 2000-2001 pages 96 to 97.*

Chinese/Pakistan/Ukraine Co-operation

Late in 1998 it was stated that the Malyshev Plant has started work on technical aspects of a joint Chinese/Pakistan/Ukraine MBT project.

It is possible that an MBT turret armed with a 125 mm smoothbore tank gun fed by an automatic loader supplied by NORINCO of China could be integrated in Pakistan onto a T-80/T-84 chassis supplied by the Malyshev Plant.

T-64 MBT upgrade

The Ukraine has started to upgrade part of its T-64 series MBTs and available details are given in the entry for the T-64 under Russia.

Status

Production as required. In service with Ukraine and Pakistan

Manufacturer

Malyshev Plant.

UPDATED

T-72AG MBT

Development

Based on its extensive experience in the development and production of the T-64 and T-80UD MBTs, and development of the T-84 MBT, which is fully covered in the previous entry, the Kharkov Morozov Design Bureau and the Malyshev Plant have developed an upgrade package for the widely deployed Russian T-72 series MBT.

This upgrade covers the three key areas of the T-72, armour, mobility and firepower with the upgraded vehicle being called the T-72AG. Many of the key components of the upgraded T-72AG are also used in the locally produced T-80UD or the T-84 MBT.

The T-72AG is a further development of the T-72 Banan upgrade which was first revealed in 1995.

Description

Overall layout of the upgraded T-72AG MBT is identical to that of the Russian T-72 MBT covered in detail earlier in this section.

The Model V46-6, four-stroke 12-cylinder diesel of the T-72 develops 780 hp and has been replaced by the locally built Model 6TD-1 two-stroke, 6-cylinder liquid-cooled diesel engine which develops 1,000 hp. This engine is installed in the 320 T-80UD MBTs supplied by the Ukraine to Pakistan with final deliveries taking place in late 1999.

T-72 MBT upgraded to the T-72AG configuration (Christopher F Foss) 0018842

With a combat weight of 45.5 tonnes, the upgraded T-72AG has a maximum road speed of 65 km/h and an acceleration of 0 to 40 km/h in 15 seconds. The air cleaners only need to be replaced after 1,000 km compared to between 300 and 500 km for those fitted to the original T-72 MBT. The specific fuel consumption of the Model 6TD diesel engine is also lower than the Model V46-6 engine of the T-72 MBT.

Main armament comprises the 125 mm 2A46M smoothbore gun fitted with a thermal sleeve and fume extractor which is fed from an automatic loader.

A 7.62 mm PKT machine gun is mounted coaxial with the main armament and the commander's 12.7 mm NSVT machine gun can be aimed and fired from within the turret using the TNK-4S sight.

The T-72 tank commander's unstabilised TKN-3 day/image intensification sighting device has been replaced in the T-72AG upgrade by the more recent PNK-4S day/image intensification system which is stabilised in one plane. The day channel has a magnification of ×7.5 with the night channel having a magnification of ×5. The commander also has a monitor so he can see the same thermal image as the gunner.

The gunner's original TPD-K1 ×8 day sight is stabilised in one plane and has been replaced by the latest IG46 day sight which is stabilised in two planes and has zoom magnification from 2.7 to ×12. The gunner's TPN-3 night sight has been replaced by a French SAGEM SANOET thermal sight which has a maximum effective range of 3,000 m, as an alternative the TPN-4 night sight can be fitted.

The manual ballistic computer of the T-72 has been replaced by a new 1V528 computer with automatic inputs for angle of elevation and lateral lead angle with inputs for crosswind, target angular and linear velocities, angle of position, air temperature and pressure, charge temperature, gun tube wear, course angle and tank speed.

According to the manufacturer, the improvements to the sighting and fire-control systems give a hit probability of up to 85 per cent on a target at a range of 2,000 m with target engagement time reduced to between 10 and 15 seconds.

Battlefield survivability is also improved by the installation of the latest Explosive Reactive Armour (ERA) system which provides protection from both chemical and kinetic energy attack over the frontal arc. This ERA is believed to be the same as the Russian Kontakt-5. The side screens are also of advanced armour.

The latest fire detection and suppression, radiation protection and NBC system has been fitted to the T-72AG MBT. In addition to the bank of six 81 mm smoke grenade launchers fitted either side of the turret, the T-72AG can lay its own smoke screen by injecting diesel fuel into the exhaust outlet at the rear.

Variants

The T-72AG upgrade package is modular so the customer can adopt only those systems required to meet specific operational requirements. As an option, the 6TD-1 1,000 hp engine can be replaced by the 6TD-2 diesel engine which develops 1,200 hp and would improve the power-to-weight ratio even further. The T-72AG can also be fitted with the Russian designed Shtora-1 vehicle protection system.

T-72-120 MBT fitted with a 120 mm smoothbore tank gun fed by a bustle-mounted automatic loader (Christopher F Foss) ***2000***/0069557

T-72 with 120 mm smoothbore gun

In early 1999, the Ukraine revealed that it had developed to the prototype stage a version of the Russian designed T-72 MBT armed with a NATO standard 120 mm smoothbore gun fed by a bustle-mounted automatic loader. This has the Ukrainian designation of the T-72-120.

The standard T-72 MBT has a 125 mm 2A46 series smoothbore gun fed by an automatic loader mounted under the turret which first feeds the projectile and then the charge. The breech is then closed and the gun fired.

Combat experience has shown that this arrangement of ammunition stowage is a major drawback in that in the event of the ammunition exploding the complete turret is normally blown off and the tank destroyed.

By installing the bustle-mounted automatic loader, any explosion is hopefully vented through blow out panels in the turret roof. When fitted with the 120 mm smoothbore gun the T-72 carries a total of 44 rounds of separate loading (projectile and charge) 120 mm ammunition of which 22 rounds are in the turret mounted bustle.

The 120 mm smoothbore gun fires standard NATO types of single piece 120 mm ammunition, all that remains after firing is the stub cartridge case. Additional 120 mm ammunition is stowed at the rear of the fighting compartment, well below the turret.

According to the Ukraine, a modified 125 mm 2A46M breech mechanism is used which can accept 125 and 140 mm smoothbore barrels also developed in the Ukraine.

It is believed that the 120 mm smoothbore gun has been developed by the State Scientific and Technical Center for Artillery and Rifle Arms (SSTC ARA) who have also recently developed the 140 mm Bagira and 125 mm Vitiaz smoothbore tank guns.

Mounted coaxial with the main armament is a 7.62 mm machine gun while mounted on the commanders cupola is a 12.7 mm machine gun which can be aimed and fired from under complete armour protection.

The T-72-120 MBT has been developed by the Kharkov Morozov Machine Building Design Bureau and the State Enterprise Malyshev Plant.

As well as installing the 120 mm smoothbore gun, the T-72-120 can be upgraded in a number of other areas to improve its overall battlefield capabilities.

Two different diesel engine options are available for the T-72-120, both of which have been developed in the Ukraine. These are the 6TD-1 diesel developing 1,000 hp and the 6TD-2 developing 1,200 hp.

In the standard T-72-120 MBT the gunner has an Irtysh day sight and an Alis-Buran night sight but as an option the French SAGEM-15 day/thermal sight can be fitted.

The tank commander has a TKH-4S1 Agat day/night sight or as an option the French SAGEM (previously SFIM) VS 580 stabilised day/night sight which is also available fitted with a laser range-finder.

The installation of this VS 580 sight enables hunter/killer target engagements to take place. A computerised fire-control system is installed which uses a 1V528-3 computer.

As well as the installation of the 120 mm gun, automatic loader and improved sighting systems, the T-72-120 can also be fitted with the Russian designed Shtora optronic countermeasures system and additional armour of the explosvie reactive type.

The additional armour is installed over the frontal arc and gives protection against both chemical energy (CE) and kinetic energy (KE) attack. It can also be fitted with a satellite supported navigation system.

With a combat weight of 48 tonnes the T-72-120 has a maximum road speed of 69 km/h forwards and 32 km/h in reverse. By late 2000, it was understood that five prototypes of this vehicle had been built. More recently the Ukraine has developed a new version of the T-84 fitted with a NATO standard smoothbore gun. This is called the T-84-120 and there is a separate entry on this vehicle.

New 125 mm and 140 mm guns

In 1998, the Ukraine's State Scientific and Technical Center of Artillery and Rifle Arms (SSTC ARA) revealed that it had developed to the prototype stage new 125 and 140 mm smoothbore tank guns.

The 125 mm gun, known as the Vitiaz is 6 m long with a maximum recoil length of 310 mm. When firing an Armour-Piercing Fin-Stabilised Discarding Sabot (APFSDS) projectile, a muzzle velocity of 2,030 m/s is attained. This will penetrate 380 mm of steel armour at an angle of 60°.

The 140 mm gun, the Bagira, is 7 m long and also has a maximum recoil length of 310 mm. When firing an APFSDS projectile a muzzle velocity of 870 m/s is attained and this will penetrate 450 mm of conventional steel armour at an angle of 60°.

According to the SSTC ARA, both smoothbore weapons can be fitted into new MBTs as well as being retrofitted into other MBTs such as the T-64, T-72, T-80UD and the T-84.

PSP T-72 MP

The T-72 MP MBT upgrade programme was developed under the leadership of the PSP Bohemia AS company of the Czech Republic. As far as it is known this is no longer being marketed. Details were given in *Jane's Armour and Artillery 2000-2001*.

Other T-72 upgrades

The No 7 Armour Repair Plant, Kyiv and No 17 Armour Repair Plant, Lviv, both in the Ukraine, are also offering upgrades for the T-72 series MBT.

This includes the installation of the R-173 radio system with R-174 intercom, V-84 or V-46-6 diesel engine, Soda napalm protection system, turret and hull fitted with ERA for improved battlefield survivability, new night vision devices, upgraded suspension for improved cross-country mobility, new A-4 headlights, Tucha 81 mm smoke grenade launchers mounted either side of the turret and TPDK-1 1A40 laser range-finder day and TPN-1 night sight.

The No 7 and No 17 Armour Repair Plants also offer upgrade packages for the older T-55 to bring them up to the T-55MV standard and in 1997 it was reported that Syria was upgrading 200 of its existing T-55 MBTs to this standard which includes the ability to fire a laser guided projectile from its 100 mm gun and the installation of an ERA package to the hull and turret for improved battlefield survivability.

Status

Development complete. Ready for production on receipt of orders.

Manufacturer

Malyshev Plant.

UPDATED

M60A3 MBT Upgrade

Development

Late in 2001, the Kharkov Morozov Machine Building Design Bureau, who is prime contractor for the T-80UD and more recent T-84 series Main Battle Tanks (MBTs), revealed that it had developed to the concept stage an extensive upgrade package for the widely deployed US General Dynamics Land Systems M60A3 MBT.

The main aim of this upgrade is to increase the operational life of the M60 series MBT, whose design is now over 40 years old, as well as to improve the overall combat efficiency of the the vehicle.

One early potential customer for the upgraded M60A3 MBT is the Turkish Land Forces Command. As of early 2002, the front runner for this contract was considered to be the Israel Military Industries Sabra II upgrade package.

There are now at least three other contenders, these are the Morozov Machine Building Plant with this design, the US company of General Dynamics Land Systems with the 120S and a proposal from the local company of FNSS Defense Systems who are the largest manufacturer of tracked armoured vehicles in Turkey.

Description

The overall layout of the upgraded M60A3 from the Ukraine is identical to that of the standard production vehicle with the driver's compartment at the front, turret and fighting compartment in the centre and the very compact diesel power pack at the rear.

As usual with most upgrades, to reduce overall life cycle costs, proven subsystems are incorporated in the upgraded M60A3 wherever possible. The three key areas upgraded are firepower, mobility and protection.

According to the Kharkov Morozov Machine Building Design Bureau, a customer can upgrade the M60A3 in all three of these areas or select just one or two depending on the actual operational requirements.

The current M60A3 MBT is armed with a 105 mm M68 series rifled tank gun which is loaded manually and has a crew of four, commander, gunner, loader and driver.

Scale model of M60A3 MBT upgraded by the Ukraine ***2002***/0096202

The M60A3 upgraded in the Ukraine has a crew of three, commander, gunner and driver as the 105 mm gun has been replaced by a 120 mm smoothbore gun which is fed by a bustle mounted automatic loader which was originally developed for the export T-84-120 Oplot MBT.

A maximum of 42 rounds of 120 mm ammunition are carried including Armour Piercing Fin Stabilised Discarding Sabot - Tracer (APFSDS-T), High Explosive Anti-Tank -MultiPurpose - Tracer (HEAT-MP-T) and High Explosive Fragmentation HE-FRAG). Blow out panels are provided in the turret roof and a bulkhead separates the ammunition from the crew.

In addition a laser guided anti-tank round can also be fired out to a range of 5,000 m. The latter is based on the locally produced Combat projectile originally developed in 125 mm calibre.

A 7.62 mm machine gun is mounted coaxial with the main armament and a 12.7 mm machine gun is mounted on the turret roof. Banks of 81 mm electrically operated grenade launchers are mounted either side of the turret firing forwards.

An advanced computerised fire-control system is installed in the upgraded M60A3 MBT that provides a high first round hit probability under day and night conditions when either the platform or the target are moving.

The original AVDS-1790 series 750 hp powerpack has been replaced by a new powerpack built around the very compact 6TD-2 diesel engine developing 1,200 hp. With a combat weight of 56.4 tonnes this gives a power-to-weight ratio of 21.15 hp/tonne and a maximum road speed of 55 to 60 km/h compared to that the standard M60A3. Cross-country speed is quoted as 35 to 40 km/h.

Ground pressure of the upgraded M60A3 is quoted as 0.95 kg/cm^2 with the overall physical characteristics being similar to the standard M60A3 except for an overall width of 3.85 m.

As the new engine is much more compact than the current AVDS-1790 series diesel engine an additional 400 litres of fuel are carried which increases the operation range of the upgraded vehicle.

The upgraded tank can ford to a depth of 1.8 m but a deep fording kit has also been developed which allows the upgraded M60A3 to ford to a maximum depth of 5 m. When not required the snorkel is carried in a horizontal position on the turret rear.

For improved battlefield survivability the turret and chassis can be fitted with a new armour package that includes a mix of passive and advanced explosive reactive armour systems. This provides a high level of protection against Chemical Energy (CE) and Kinetic Energy attack (KE) over the frontal arc.

Status
Proposal.

Manufacturer
Malyshev Plant

NEW ENTRY

UNITED KINGDOM

MObile DIrect FIre Equipment Requirement (MODIFIER)

Development/Description
Although the Vickers Defence Systems Challenger 2 MBT (covered in detail in a separate entry) has only recently entered service with the Royal Armoured Corps, the UK MoD is already studying a follow-on armoured vehicle which is currently called the MObile DIrect FIre Equipment Requirement (MODIFIER).

MODIFIER is being studied against the background of the lack of a clearly defined threat while at the same time realising that following the end of the Cold War there is considerable instability in many parts of the world.

It is envisaged that future operations will be of an official nature and be carried out on a 24 hour basis under all weather conditions.

Operations will also take place over a much wider area with greater dispersion of the units taking part. Units will only be concentrated at the point of effort.

Synchronisation will be achieved through digitisation and there will be increased strategic and operational mobility. Compared with in-service systems, there will also be reduced logistic burden and the system will be easily supportable.

Applications for MODIFIER are expected to include national force projection as well as forming a key part of any joint and combined forces. It will also be strategically deployable with the ability to operate in a wide range of climatic conditions.

Its primary task will be to destroy ground based anti-armour systems at longer ranges than today and be employable in small formations or small packages across the spectrum of warfare.

Compared with the current Vickers Defence Systems Challenger 2 MBT, MODIFIER will be much lighter, less than 50 tonnes and ideally 40 tonnes. It will be transportable by sea (roll on/roll off); the ability to be transported by C-17 and C-5 transport aircraft is desired.

Operational mobility by road or rail is required to be significantly better than current MBTs.

No timescale for MODIFIER has been determined but it is anticipated that the earliest In-Service Date (ISD) would be around 2020. Germany has a requirement for a new MBT with an ISD of around 2007 while the USA's vehicle, the Future Combat System, has a possible ISD of 2008.

Given the cost of developing such a complex system as MODIFIER, it is possible that a collaboration programme similar to the UK/USA Tactical Reconnaissance Armoured Combat Equipment Requirement (TRACER)/Future Scout Cavalry Vehicle (FSCV). This was a joint UK/USA programme and available details are given in the Reconnaissance vehicles section. For a variety of reasons this programme will not proceed past the demonstrator programme stage.

The Defence Procurement Agency (DPA) Armour Systems Future Projects Group (FPG) has started conceptual work on MODIFIER.

Considerations being taken into account by the FPG include:

(1) Procurement risk reduction activities on what promises to be a very high technology project.
(2) Location and capability of industrial partners in the long term project time for MODIFIER.
(3) International collaboration opportunities for joint development and procurement.
(4) The potential for multinational deployment and operation.

Late in 1999, the Defence Procurement Agency stated that MODIFIER was at the pre User Requirement Document Phase.

Key technologies are seen to lie in the areas of advanced material sciences; propulsion and power generation; fire power; defensive and evasive aid systems; mission support systems and crew ergonomics.

Both tracked and wheeled (8 × 8) solutions are being studied with manned and unmanned weapon stations of various types.

Status
Studies.

Manufacturer
Not selected.

UPDATED

Vickers Defence Systems Challenger 2E MBT

Development
In the early 1990s as a private venture, Vickers Defence Systems started development of an enhanced export model of the Challenger 2 MBT.

The latter is now the only MBT deployed by the British Army who took delivery of 386 vehicles with another 38 being built for Oman. Final deliveries of Challenger 2 MBT were made to the British Army in February 2002.

This export model eventually became the Challenger 2E and ran for the first time in 1994 and since then development has continued as new technology has become available.

By early 2002, the prototype of the Challenger 2E had completed well over 14,000 km of trials in Europe and the Middle East.

Following over six years of development of the Challenger 2E MBT, Vickers Defence Systems is now building a production standard vehicle which should be completed in 2003.

The additional features and technology embodied in the latest Challenger 2E, combined with its extensive growth potential, take it a generation ahead of current in-service MBTs according to Vickers Defence Systems.

Cutaway drawing of production standard Challenger 2E MBT showing position of main subsystems
2002/0095687

Description
The overall layout of the Challenger 2E is almost identical to the Challenger 2 MBT which is covered in detail in the following entry with the main difference being in key subsystems.

The first prototype of the Challenger 2E is essentially an ex-British Army Challenger 2 MBT with a number of enhanced capabilities in the key areas of mobility, situation awareness, target acquisition and survivability.

The Challenger 2E is fitted with the the proven German MTU EuroPowerPack which consists of the MTU 883 V-12 diesel developing 1,500 hp (with further growth potential) coupled to a transversely mounted Renk HSWL 295 TM automatic transmission with five forward and three reverse gears.

The MTU EuroPowerPack is much more compact than that 1,200 hp diesel power pack currently installed in the British Army Challenger 2 MBT and this has freed up additional space.

Some of this has been used to increase the amount of diesel fuel being carried that has allowed the operational range to be increased to beyond 550 km.

With the new MTU EuroPowerPack the Challenger 2E has not only a greater power-to-weight ratio but also higher speed and improved cross-country mobility.

It also features second-generation hydrogas suspension, an improved hydraulic track tensioner operated by the driver, rubber tyred idler and a second-generation double-pin track with rubber pads more suited to global use.

While the British Army Challenger 2 has a good suite of day/thermal night observation and gunnery equipment, Challenger 2E has greatly enhanced target acquisition.

The commander has the latest French SAGEM MVS 580 IRIS roof-mounted panoramic day/Thermal Imaging (TI) sight which also incorporates a TI contour image enhancement facility and an eyesafe Laser Range-Finder (LRF). The TI channel has ×5 and ×15 magnification plus ×30 (zoom).

The gunner has a SAGEM SAVAN 15 IRIS roof-mounted gyro-stabilised sight with similar capabilities except that it is not panoramic.

The Vickers Defence Systems Challenger 2E is a further development of the Challenger 2 in service with Oman and the UK **2002**/0131054

Both of these sights have common 2nd generation thermal imaging systems and allow either the commander or gunner to detect and engage targets under almost all weather conditions at long range with greater accuracy.

A Battle Management System (BMS) developed by Vickers Defence Systems has been fitted which incorporates US TRW FBCB2 software. This provides the tank commander with own vehicle position/ bearing, friendly and enemy force location and enhances battlefield awarness as well as reducing commander's workload. In addition it has a weapon system interface and provides information such as steering direction to the driver and a surveillance mode based on map co-ordinates of target range on map display.

Main armament comprises the BAE Systems, RO Defence 120 mm L30A1 rifled tank gun which fires Armour Piercing Fin Stabilised Discarding Sabot (APFSDS), L31 High Explosive Squash Head (HESH) and L34 smoke ammunition of the separate loading type, for example projectile and charge.

While the British Army uses the L27A1 APFSDS Depleated Uranium (DU) round, for the export market the L28 APFSDS has been developed which uses a tungsten penetrator. The L28 has a maximum stated range of 3,500 m and a total of 45 projectiles and associated propelling charges are carried in Challenger 2E.

A 7.62 mm MGA3A1 machine gun is mounted coaxial with the 120 mm L30A1 gun and a 7.62 or 12.7 mm machine gun is mounted at the loaders station.

The fire-control system is an enhanced version of that installed in the British Army Challenger 2. A new combined commander's control panel and fire-control panel has been installed which combines the functions of the existing fire- control panel, commander's control panel and relaxed view monitor. The gunner also has a new control and display panel.

The British Army Challenger 2 can ford to a depth of 1.07 m but Challenger 2E can deep ford to 2 m using an onboard kit. This consists of a bellows type air intake mounted to the turret rear which draws in air for the engine. Two bilge pumps are fitted which operate independent of the main engine. While fording the Challenger 2E retains a full NBC capability.

The driver's compartment of Challenger 2E has been redesigned and new controls have been fitted together with a brand new display panel to his left with a databus link to the power pack and to the BMS. Finally it has a combined APU and air conditioning system.

Unlike the British Army Challenger 2 MBT, the Challenger 2E is fitted with a fully automatic fire/ explosion detection and suppression system in the crew and engine compartments.

The Challenger 2E MBT has been developed for the export market **2002**/0131053

Survivability is further enhanced as there is no explosive material above the turret ring with the 120 mm charges stowed in armoured charge bins. Turret drives are all electric so there is no danger from hydraulic fluid in the vehicle.

The overall signature of the Challenger 2E has also been reduced and the hull and turret incorporate an unspecified new advanced armour technology that incorporates new materials for improved protection against both chemical energy and kinetic energy attack.

The new production standard Challenger 2E has been designed to be easier and quicker to manufacture. By careful design the number of welds have been reduced and the whole electrical system in the chassis redesigned.

There are now 50 less harnesses and this will be more reliable as well as being quicker and cheaper to build. In addition, more castings have been introduced to reduce costs and simplify manufacture.

Combat identification, direct fire weapons effect simulator and live fire monitoring equipment have already been integrated in the Challenger 2E for trials.

As an alternative to the pintle mounted 12.7 mm machine gun a servo-controlled overhead weapon platform can also be fitted and this would be slaved to the commander's roof-mounted stabilised periscopic sight.

Specifications

Crew: 4
Combat weight: 62,500 kg
Power-to-weight ratio: 24 hp/tonne
Ground pressure: 1.015 kg/cm^2
Length:
(gun forward) 11.35 m
(gun rear) 9.80 m
(hull) 8.13 m
Width:
(overall) 3.55 m
(over tracks) 3.42 m
Height (turret roof): 2.86 m
Ground clearance: 0.50 m
Length of track on ground: 4.79 m
Track width: 650 mm
Max road speed: 65 plus km/h
Usable fuel capacity: 1,962 litres
Max range:
(road) 550+ km
(cross-country) 350+ km
Vertical obstacle: 1.0 m
Gradient: 60%
Side slope: 30%
Trench: 2.5 m
Fording:
(standard) 1.07 m
(with preparation) 2 m
Engine: MTU 883 V-12 diesel developing 1,500 hp (with further growth potential)
Transmission: Renk HSWL 295 TM automatic with five forward and three reverse gears
Suspension: improved hydropneumatic
Electrical system: 24 V
Batteries: 6
Armament:
(main) 1 × 120mm L30A1 rifled tank gun
(coaxial) 1 × 7.62 mm L94A1 Chain Gun
(anti-aircraft) 1 × 7.62 mm or 12.7 mm MG
Smoke grenade launchers: 2 × 5 66 mm
Ammunition:
(main) up to 45 projectile stowage positions
(7.62 mm) 4,000
Gun control equipment: electric/manual
(by commander) yes
(by gunner) yes
Turret traverse: 360°
Gun elevation/depression: +20/-10°
(commander's override) yes
Gun stabiliser:
(vertical) yes
(horizontal) yes
Range setting device: yes, laser
Elevation quadrant: yes
Traverse indicator: yes
NBC system: yes
Night vision equipment: yes

Variants

Vickers Defence Systems can offer a complete family of vehicles including not only the Challenger 2E MBT but also the key support vehicles including the Driver Training Tank (DDT), Armoured Repair and Recovery Vehicle (ARRV) and an Armoured Vehicle Launched Bridge (AVLB). Under a £250 million contract the company is building 66 Engineer Tank Systems (ETS) for the British Army that will be based on Challenger 2 components.

Status

Development complete. Ready for production.

Manufacturer

Vickers Defence Systems

NEW ENTRY

Vickers Defence Systems Challenger 2 MBT

Development

Work on the Challenger 2 started in November 1986 as a private venture and, in March 1987, Vickers Defence Systems made its first presentation to the UK MoD. Throughout 1987 the Challenger 2 design was further refined and in February 1988, Vickers Defence Systems submitted its formal proposal to the UK MoD following the issue of the Staff Requirement.

Early in 1988, Vickers Defence Systems took the decision to build nine turrets using company funding and the first turret shell was completed in the Autumn of 1988.

Vickers Defence Systems and General Dynamics Land Systems both put their firm production proposals to the UK MoD in mid-1988 and in December of that year it was announced that Vickers Defence Systems was to be awarded a £90 million fixed-price contract to undertake a demonstration phase (also referred to as the proof-of-principle phase) which lasted until the end of September 1990.

An important part of the demonstration contract awarded to Vickers Defence Systems was to show that a new depleted uranium projectile, together with a new charge system for the now BAE Systems, RO Defence

Vickers Defence Systems Challenger 2 MBT of the British Army firing 120 mm L30 gun 0045810

120 mm CHARM gun, could be successfully developed to the standard and in the timescale required.

Design of the Challenger 2 was essentially completed by August 1989 with the last bought-out items being delivered in February 1990. Of the nine Challenger 2 prototypes built, seven were built at Leeds and two at Newcastle-upon-Tyne. One additional Challenger 2 turret was built at Leeds.

Vickers Defence Systems completed the demonstration phase of Challenger 2 on schedule at the end of September 1990. In June 1991, the UK government selected the Challenger 2 and placed an order worth around £520 million for 127 Challenger 2 MBTs and 13 Driver Training Tanks. Production commenced in 1993 with first production vehicles completed at Vickers Defence Systems' Leeds facility in July 1994.

The first Challenger 2 Driver Training Tanks were completed in 1993 to allow training to start ahead of deliveries of the Challenger 2 MBT to the British Army.

The Challenger 2 is the first British Army tank since the Second World War to be designed, developed and produced exclusively by a contractor, Vickers Defence Systems, with set reliability goals laid down in the fixed-price contract.

The Challenger 2 MBT was formally accepted for service with the British Army on 16 May 1994.

In July 1994, Vickers Defence Systems was awarded a second contract by the UK MoD for the supply of an additional 259 Challenger 2 MBTs and nine Driver Training Tanks plus training and logistic support.

Under the Strategic Defence Review the Royal Armoured Corps will have a total of six Challenger 2 MBT regiments, each of which will be equipped with 38 Challenger 2 MBTs. The first regiment to be equipped with the Challenger 2 was the Royal Scots Dragoon Guards who achieved their in-service date of June 2000.

So far the only export customer for the Challenger 2 MBT is Oman, which placed a contract worth around £140 million in July 1993.

This contract covered the supply of 18 Challenger 2 MBTs, four Challenger Armoured Repair and Recovery Vehicles, two Challenger Driver Training Tanks, four Alvis Stormer command post vehicles, nine UNIPOWER M series (6 × 6) tank transporters and semi-trailers plus spares and training. Oman took delivery of its first Challenger 2 MBTs in 1995 with final deliveries taking place in 1996.

In November 1997, Vickers Defence Systems was awarded a contract worth more than £100 million for the supply of a further 20 Challenger 2 MBTs for Oman. The final batch of 10 vehicles were shipped to Oman late in 2000 and this enabled the Royal Oman 1st Main Battle Tank Regiment, based at Shaffa, to be fully equipped.

These are to the latest Challenger 2 (Oman) Phase II standard. The first vehicles were delivered to Oman in the Phase I configuration and these have now been brought up to the latest Phase II standard under a rolling programme so that the complete Omani Challenger 2 MBT fleet is now a common build standard.

When compared to the British Army Challenger 2 MBTs, the Omani vehicles have a number of modifications to enhance their operational performance in the high-ambient temperatures encountered in the Middle East.

The Omani Challenger 2 MBTs have a modified power pack consisting of a Perkins CV12 diesel developing 1,200 hp, coupled to a David Brown Engineering TN54 transmission. The water cooling system and air flow have been modified to enable full engine power to be maintained in ambient temperatures up to +52°C.

Other modifications include a 12.7 mm M2 machine gun for the loader, single-pin tracks, improved air conditioning system, new communications equipment and Magellan Global positioning System.

Early in 1999, the UK MoD confirmed that the Vickers Defence Systems Challenger 2 MBT had exceeded the most rigorous reliability targets ever set anywhere, during a demanding series of trials under battlefield conditions.

A squadron of 12 Challenger 2 MBTs manned by experienced British Army crews, completed the highly successful battle testing at the Armoured Trials and Development Unit (ATDU) at the Royal Armoured Corps Centre at Bovington, Dorset.

All 12 Challenger 2 MBTs used during the trials out-performed the UK MoD's exacting reliability targets, according to Vickers Defence Systems.

At the end of February 2002, the British Army took delivery of the last of 386 Challenger 2 MBTs from Vickers Defence Systems at its Newcastle upon Tyne facility.

Of the 386 Challenger 2 MBTs, 167 were built at the Newcastle upon Tyne facility with the remainder being built at Leeds.

Although the production site at Leeds has been closed, VDS still have over 100 design and engineering staff at this facility who are engaged in a number of current and future armoured fighting vehicle programmes.

The final 10th batch of British Army Challenger 2 MBTs passed their acceptance trials late in 2001. Under batch testing, four production Challenger 2 MBTs are taken from the production line and put through four clearly defined 'battlefield days'.

Each of these 'battlefield days' consists of 16 hours of weapon system operation, 34 rounds of 120 mm ammunition firing, 1,000 rounds of secondary armament firing and 13.42 hours of engine running.

Description

The turret is a new design, incorporating second-generation Chobham armour with a significant increase in protection against both KE and CE attack.

During peacetime training the armoured side skirts are not normally fitted to the Challenger 2 MBT but are rapidly fitted when the vehicle deploys on operations.

The Challenger 2 MBTs deployed by the British Army to the Balkans were also fitted with the Explosive Reactive Armour (ERA) package to the front of the hull.

This was originally developed and fitted to the older Challenger 1 deployed to the Middle East in 1990/91 for Operation Desert Storm, the recapture Kuwait.

The turret crew positions on Challenger 2 are similar to Challenger 1, with the commander seated on the right, the gunner forward and below the commander, and the loader on the left side. Each of them is provided with a lap-type seat belt.

Both the commander and the loader have a single-piece hatch cover that opens to the rear.

The tank commander has eight day unity magnification day periscopes for all-round observation. Under each periscope is a red button, when this is pressed the commander's sight slews so that it lines up with the periscope.

The loader has a single roof-mounted day periscope that can be traversed.

The NBC system is mounted in the turret bustle and fully meets all known threats. The crew environmental control system provides both heating and cooling for the crew compartment. This is the first time such a system has been installed in a British MBT.

Main armament of the Challenger 2 MBT consists of a BAE Systems, RO Defence 120 mm rifled tank gun designated the L30A1 which is of ESR steel, autofrettaged, chrome-lined and fitted with a thermal sleeve, fume extractor and muzzle reference system. It has a split-block sliding breech mechanism with an elastomeric obturating pad.

The 55 calibre L30A1 is the first British tank gun to be chrome-lined which gives the barrel a longer life as well as giving more consistent accuracy.

In addition to firing all current 120 mm ammunition, with the exception of the obsolete APDS round, the L30A1 also fires a new DU round L27A1 with a new BAE Systems, RO Defence L17A1 stick charge propellant system.

The first DU round is part of the CHARM 1 system (gun, charge and projectile) with the actual projectile being designated the L26. The CHARM 3 projectile is also of the DU type but has a greater length-to-diameter ratio and greater penetration characteristics. CHARM 3 is the main wartime round for the Challenger 2 although it will not normally be fired in peacetime.

Late in 2000 it was revealed that BAE Systems, RO Defence had developed, as a private venture, a new APFSDS-T projectile with a conventional penetrator which has the unofficial designation of the L28. This is

Challenger 2 MBTs in service with the British Army and fitted with peacetime training skirts **2000**/0069560

used with the L16 charge system. This is being offered for export customers of the Challenger 2E MBT who do not want to use the DU projectile.

Ammunition is of the separate loading type, that is projectile and charge, with all explosive ammunition being stowed below the turret ring for increased battlefield survivability. The charges are stowed in armoured bins.

Mounted coaxially, to the left of the main armament, is The Boeing Company 7.62 mm Chain Gun which is already in service with the British Army and installed in the Alvis Vehicles Warrior mechanised combat vehicle. This weapon has the British Army designation of the L94A1 and was manufactured by BAE Systems, RO Defence under licence in the UK. The loader has a Thales AFV Systems mount fitted with an externally mounted 7.62 mm L37A2 machine gun.

Mounted either side of the turret front is a bank of five 66 mm Thales AFV Systems electrically operated smoke grenade dischargers that cover the forward arc of the vehicle.

The heart of the fire-control system of the Challenger 2 is the latest generation digital computer from Computing Devices Canada, which is an improved version of that installed in the M1A1 Abrams. It also has growth capacity for future enhancement such as a Battle Management System (BMS) and navigation aids. It will also be capable of being used for training purposes.

Computer generated impression of the Vickers Defence Systems Trojan ETS based on a modified Challenger 2 MBT chassis carrying a fascine at the rear, Pearson Engineering Pathfinder lane marking system, hydraulically operated shovel and Pearson Engineering Full Width Mine Plough **2001**/0073410

The commander has a roof-mounted SAGEM (previously SFIM) stabilised sight which is similar to that installed on the French Giat Industries Leclerc MBT; it gives full 360° capability without having to move the head. This has a magnification of ×3.2 and ×10.5 in the day mode and also has an Nd:YAG laser range-finder.

A thermal image with a magnification of ×4 and ×11.5 can be injected into the sight with a separate thermal image relaxed viewing monitor provided to the right of the sight for surveillance purposes.

As with all British MBTs since the Second World War, Challenger 2 has an all-electric gun control and stabilisation system from the now BAE Systems, RO Defence. In addition it incorporates a MIL-STD-1553 databus, this being the first to be installed in a British AFV. Turret traverse is 360° with weapon elevation from −10 to +20°.

The gunner's roof-mounted stabilised sight has been developed by Thales Optronics (previously Pilkington Optronics) with SAGEM of France being the major subcontractor; it includes day optics, thermal image input and Nd:YAG laser range-finder. The gunner's sight has a magnification of ×3 and ×10 in the day mode and also has a traverse of 7° left and right. A thermal image with the same magnification as that of the commander can be injected at the flick of a switch.

Like the tank commander, the gunner has a separate relaxed viewing thermal image monitor to the left of his sight for surveillance purposes.

A typical target engagement for Challenger 2 would take place as follows. The tank commander first spots the target and then brings his roof-mounted sight to bear on the target. He does this by laying the dot in the centre of his sight onto the enemy vehicle.

By pressing his align switches, the tank commander is able to bring the gun onto the target. At this point he hands over to the gunner using a standard fire order giving the ammunition nature 'FIN' and the nature of the target 'TANK'.

The gunner then reports that he has identified the target by stating 'ON'. From this point on, the gunner becomes entirely responsible for the target, allowing the tank commander the freedom to find another target with his own roof-mounted stabilised sight.

Having found a second target, the tank commander is able to determine the range using the laser range-finder in his own sight. The computer is able to store and process two separate sets of target data at the same time.

By this time, the gunner should be on the verge of completing the first engagement. The commander waits for the gunner to fire and, having confirmed that the first target has been destroyed, he immediately presses his align switches followed by his firing switch.

By pressing his align switches, the tank commander causes the turret to traverse automatically until the gun comes into alignment with his sight.

The moment that the gun achieves coincidence with the commander's sight, the computer fires the gun. According to the MoD, this is a very powerful routine which, in reality, allows the Challenger 2 to take on two targets at the same time.

Mounted over the 120 mm L30A1 gun is the Thales Optronics (previously Pilkington Optronics) thermal imaging sight head that moves in elevation with the main armament. This is the same sight head as that fitted to the Challenger 1 but the thermal imaging system has fewer components and improved reliability.

The gunner also has a BAE Systems L30 reversionary mode telescope mounted coaxially with the main armament with an etched graticule, this is also used for muzzle reference purposes.

According to Vickers Defence Systems, Challenger 2 has plenty of growth potential including the installation of a 140 mm gun, this made possible by the width of the mantlet; improved protection including Explosive Reactive Armour (ERA) arrays covering the sides of the vehicle; automatic muzzle reference system; automatic target detection and tracking systems and various defensive aids that could detect and counter a variety of threats.

The existence of the 1553 databus and the Computing Devices Canada computer makes many of the electronics improvements much easier.

The shape of the hull of the Challenger 2 is virtually identical to that on the current Challenger 1 and like the turret, incorporates stealth technology to minimise the radar signature.

Many improvements have been carried out on the chassis, for example the driver's steering controls are more rugged, fire retardant bag-type fuel tanks have been fitted, new lights have been installed and the engine decks have been redesigned for ease of operation. In all, the Challenger 2 chassis has 156 improvements over the original Challenger 1 MBT chassis.

Challenger 2 is powered by the Perkins Engines Company 12-cylinder diesel developing 1,200 hp at 2,300 rpm coupled to the David Brown Engineering TN54 transmission which is already installed in the Challenger Armoured Repair and Recovery Vehicle.

The TN54 epicyclic transmission has six forward and two reverse gears with automatic control embodying double-differential regenerative steering with hydrostatically transmitted power giving infinitely variable output. The TN54 transmission also gives increased manoeuvrability in confined spaces.

Challenger 2 is also fitted with the Dowty Defence and Air Systems Digital Automotive System Control Unit (DASCU) and a new APU which is also installed in the Challenger Armoured Repair and Recovery Vehicle.

The hydropneumatic suspension system, fitted to all members of the Challenger 1 family, has been upgraded and new seals have been incorporated. Total wheel travel is 450 mm.

Challenger 2 is also fitted with hydraulic track tensioners at the front, enabling the track to be correctly tensioned from under armour protection by the driver. This facility reduces track wear and tear and improves hull stability by ensuring that track tension is constantly optimised.

Following extensive trials of two competing new double-pin track designs, the British-designed William Cook Defence track was selected for production Challenger 2 MBTs.

Challenger 2 can lay its own smoke screen by injecting diesel fuel into the exhaust outlets on either side of the hull top towards the rear.

All Challenger 2s are fitted to carry two standard 45 gallon (Imperial) drums at the rear of the hull, the racks being provided with a quick-release mechanism. These are not connected to the vehicle's fuel supply. An electric pump is fitted to allow fuel to be quickly transferred from these drums to the on-board fuel tanks or to/from another vehicle.

The Challenger 2 is also fitted to accept the Pearson Engineering Combat Dozer Blade adopted by the British Army in 1990.

Vickers Defence Systems Challenger 2 MBT in service with Oman showing roof-mounted 12.7 mm M2 machine gun **2002**/0073305

Vickers Defence Systems Challenger 2 MBT in the Balkans and fitted with ERA on front of hull and armoured side skirts (Richard Stickland) ***2002***/0131047

Vickers Defence Systems Challenger 2 MBT in service with the British Army in the Balkans and fitted with Pearson Engineering dozer blade (Richard Stickland) ***2002***/0131048

Specifications
(British Army Challenger 2)
Crew: 4
Combat weight: 62,500 kg
Power-to-weight ratio: 19.2 hp/t
Ground pressure: 0.9 kg/cm²
Length:
(gun forward) 11.55 m
(gun rear) 9.86 m
(hull) 8.327 m
Width:
(overall) 3.52 m
(over tracks) 3.42 m
Height: (to turret roof) 2.49 m
Ground clearance: 0.50 m
Track: 2.77 m
Length of track on ground: 4.79 m
Track width: 650 mm
Max road speed: 56 km/h
Average cross-country speed: 40 km/h
Usable fuel capacity: 1,592 litres
Max range:
(road) 450 km
(cross-country) 250 km
Vertical obstacle: 0.9 m
Gradient: 60%
Trench: 2.34 m
Fording: 1.07 m
Engine: Perkins Engines Company CV-12 TCA Condor V-12 12-cylinder 26.1 litre diesel developing 1,200 bhp at 2,300 rpm
Auxiliary engine: Plessey
Transmission: David Brown Engineering TN54 epicyclic transmission with 6 forward and 2 reverse gears
Steering: Commercial Hydraulics double differential, hydrostatic control
Suspension: hydropneumatic
Electrical system: 24 V
Batteries: 6
Armament:
(main) 1 × 120 mm L30A1 rifled gun
(coaxial) 1 × 7.62 mm L94A1 Chain Gun MG
(anti-aircraft) 1 × 7.62 mm L37A2 MG
Smoke grenade dischargers: 2 × 5 66 mm
Ammunition:
(main) up to 50 projectile stowage positions
(7.62 mm) 4,000
Gun control equipment
Turret power control: electric/manual
(by commander) yes
(by gunner) yes
Turret traverse: 360°
Gun elevation/depression: +20/–10°
(commander's override) yes
Commander's fire-control override: yes
Gun stabiliser:
(vertical) yes
(horizontal) yes
Range setting device: yes, laser
Elevation quadrant: yes
Traverse indicator: yes
NBC system: yes
Night vision equipment: yes

Variants

Challenger 2 Control
Used by brigade commanders.

Challenger 2 Command
Used by squadron commanders.

Omani Challenger 2
These have a number of modifications to suit operations in the Middle East, especially maintaining full engine power of 1,200 hp in temperatures of up to +52°C.

The core engine and transmission have not been changed but the water cooling system and airflow have been modified. Larger radiators have been installed and larger fans positioned at the hull rear; air enters through the top louvres and exits through the new louvres in the hull rear, improving airflow and reducing the vehicle's thermal signature.

Other modifications include the replacement of the loader's 7.62 mm machine gun with a pintle-mounted 12.7 mm M2 machine gun. The vehicle also has single pin tracks, a different communications fit, an improved air conditioning system and provision for installation of a Magellan GPS.

Future Challenger 2 improvements
There are three possible key areas for future British Army Challenger 2 improvements, spend to save, globalisation and Technology Insertion Programmes (TIPS), all of which are dependent on UK MoD funding.

The current Digital Automotive Systems Control Unit (DASCU) was developed many years ago for the Challenger 1 MBT. This is now considered obsolete and will be replaced by a more compact Vickers Integrated Control System (VICS).

Prototypes of this have already been trialled in a Challenger 2 with first production application expected to be the British Army's new Engineering Tank System (ETS).

The current Auxiliary Power Unit (APU) could be replaced by the new Vickers Specialist Engines APU 2000 that has already been selected for the VDS ETS.

Late in 2001, Vickers Specialist Engines were awarded a £2.7 million contract by Vickers Defence Systems for the supply of four prototypes and 66 production APU 2000 systems. These will be delivered from June 2003, with series production units being delivered from October 2004.

The current Challenger 2 MBT was designed for operations in Europe, while the 38 Challenger 2 MBTs built for Oman had a number of modifications to allow them to operate in the high-ambient temperatures encountered in the Middle East.

For the Omani Challenger 2 MBT and variants, the core engine and transmission have not been changed but the cooling system and airflow have been modified.

Larger radiators have been installed and larger fans positioned at the rear. The air now enters through the top louvres and exits through the new louvres in the almost vertical hull rear.

Even prior to British Army Challenger 2s going to Oman for Exercise 'Saif Sareea II' in late 2001, Vickers Defence Systems had already completed a number of design studies to improve the performance of the vehicle in very high temperatures.

Responsibility for the Challenger 2 has now passed from the Defence Procurement Agency to the Defence Logistics Organisation (DLO) and the UK MoD is now considering future improvements to the vehicle as a result of the Omani exercise in 2001.

But as with all UK programmes, this is dependent on funding but is expected to include a redesigned air cleaner system.

TIPS could well leverage off the considerable investment made by VDS in the export Challenger 2E. This has many improvements including the German MTU 1,500 hp EuroPowerPack, commander's and gunner's sights with second-generation thermal imager, redesigned driver's compartment and battle management system to name just a few.

It is understood that one of the earliest TIPS could well include providing the commander with a panoramic thermal sight. Today the commander has a roof-mounted stabilised SAGEM ×2 magnification day sight with laser range-finder.

The commander also has a monitor that provides a thermal view of the gunner's TI sight. The incorporation of a commander's stabilised day/TI sight would enable faster target acquisition under day and night conditions and permit full hunter/killer target engagements.

There are however a number of other emerging technologies that could be incorporated into the Challenger 2 such as a defensive aids suite and automatic target tracking.

Challenger 2E
This has been developed for the export market and has a number of improvements over the current Challenger 2. Details are given in the previous entry.

Challenger 2 of British Army from rear showing additional fuel drums and 120 mm gun in travelling lock (Richard Stickland) ***2002***/0131052

Challenger 2 of British Army with turret traversed to rear and showing ERA on hull front and armoured side skirts (Richard Stickland) ***2002***/0131049

Driver Training Tank

In addition to ordering 386 Challenger MBTs, the UK MoD also ordered 22 new Driver Training Tanks (DTTs) which are similar to the Challenger Training Tank (CTT) covered in the following entry but with improved instrumentation for the instructor and mounted on a Challenger 2 chassis. First production DTTs were completed late in 1993. Deliveries are now complete.

Engineer Tank System

In August 2000, following a competition, the UK Defence Procurement Agency (DPA) selected Vickers Defence Systems to supply the British Army with a new Engineer Tank System (ETS) which was previously called the Future Engineer Tank (FET).

This will cover the supply of 66 ETS against the original requirement for 102 vehicles to replace the current in service Armoured Vehicle Royal Engineer (AVRE) and Armoured Vehicle Launched Bridge (AVLB) systems, based on the now obsolete Chieftain MBT chassis.

This order comprises 33 Titan bridgelayers and 33 Trojan engineer vehicles. A total of four prototype ETS are now being built, two Titan and two Trojan and the first of these will be completed late in 2002. Following trials, these will be brought up to full production standard.

There will be two versions of the ETS, the Titan bridgelaying vehicle and the Trojan flexible obstacle/mine clearance vehicle. Titan will be able to carry and lay the current in-service Vickers 26 m and 13.5 m long aluminium bridges with the crew under complete armour and NBC protection.

Trojan will be able to clear battlefield obstacles and clear a path through minefields. Standard equipment will include a winch and shovel and a Pearson Engineering Track Width Mine Plough can be mounted at the front to clear mines.

A Pearson Engineering Pathfinder clear lane marking system will also be fitted. It can also carry fascines to drop into ditches and tow a trailer-mounted BAE Systems, RO Defence Python rocket-propelled mineclearing system.

The introduction of the Vickers Defence System ETS will enable the older Chieftain AVRE and AVLB systems to be phased out of service and in the future the British Army will have a common heavy fleet chassis based on the Challenger 2 MBT chassis and subsystems.

It was expected that Vickers Defence Systems would have been awarded a contract for the supply of 32 Challenger 2 based Heavy Armoured Repair and Recovery Vehicles (HARRV) to enable the remaining Chieftain ARVs to be replaced. This order has now been moved to the right but may be placed when sufficient funds are available.

Status

Production complete. Production can commence again if additional orders are placed.

Country	Quantity	Comment
Oman	38	deliveries complete
UK	386	deliveries completed February 2002

Manufacturer

Vickers Defence Systems (Newcastle-upon-Tyne).

UPDATED

Vickers Defence Systems Challenger 1 MBT

Development

In September 1978, the British Ministry of Defence placed an order with the Royal Ordnance Factory Leeds for the supply of 243 Challenger 1s, essentially a further development of the Shir 2 originally developed for Iran. This order was valued at £300 million.

The Challenger 1 was accepted by the General Staff in December 1982 by which time production was already well under way at Royal Ordnance Leeds.

The first British Army order for Challenger 1 was for 243 tanks, sufficient to equip four regiments. The first of these was handed over to the British Army by Royal Ordnance Leeds in March 1983. In June 1984, the MoD placed an order for an additional 64 Challenger 1 MBTs, sufficient to equip a fifth regiment.

The Royal Tank Regiment received its first vehicles late in 1984 and by May 1985 the Royal Hussars were completely equipped with the Challenger 1 MBT. A further 18 tanks were ordered in June 1985.

In June 1986, Royal Ordnance Weapons and Fighting Vehicle Division took over responsibility for design authority and future post-design services work on Challenger 1. This was previously undertaken by the Royal Armament Research and Development Establishment at Chertsey.

In July 1986, it was announced that Vickers Defence Systems was to buy the Royal Ordnance Leeds tank factory for approximately £11 million and that a new facility, identical to Vickers Defence Systems Armstrong Works at Newcastle-upon-Tyne, would be built at a cost of £14 million. The new Vickers Defence Systems Leeds factory was fully operational by late 1987. At the same time, the UK MoD placed a further order for 76 Challenger 1 MBTs at a cost of £100 million. The last of 420 Challenger 1s was scheduled for delivery to the British Army in September 1989 but this was subsequently stretched until mid-1990.

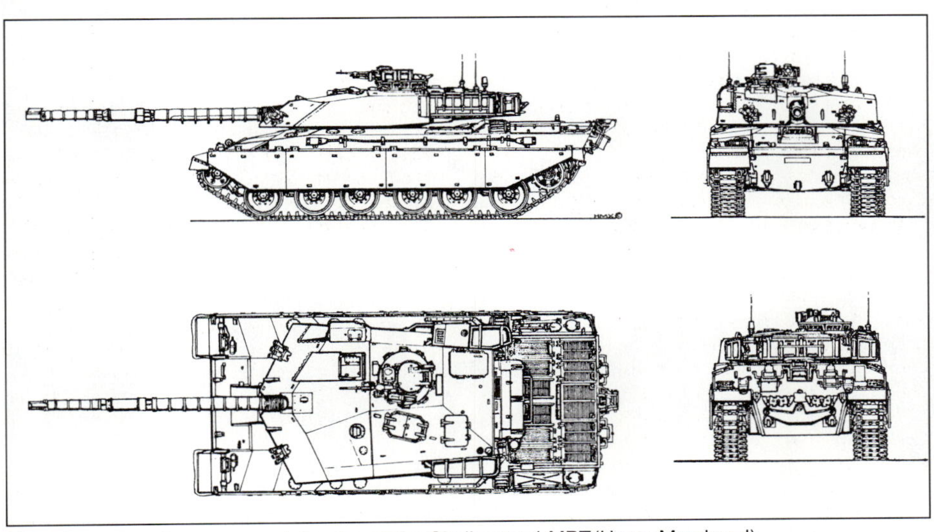

Four-view general arrangement drawing of the Challenger 1 MBT (Henry Morshead)

The Challenger 1 MBT was finally phased out of service with the British Army late in 2000 following the introduction of the Vickers Defence Systems Challenger 2 MBT.

In March 1999, the UK Minister for Defence Procurement, Lord Gilbert, said that the UK was considering the supply of an unspecified quantity of Challenger 1 MBTs to Jordan as a replacement for the current fleet of 293 Tariq MBTs.

It was subsequently disclosed that a total of 288 Challenger 1 MBTs will be supplied to Jordan with the first batch of 24 being delivered late in 1999.

Before delivery, the initial batch of Challenger 1 MBTs, to be renamed Al Hussein in Jordanian service, were brought up field standard by the Army Base Repair Organisation (ABRO) facility at Bovington Camp, Dorset, home of the Royal Armoured Corps.

Officials declined to disclose the armour package for the Jordanian vehicles but it is believed that the standard British Army baseline Chobham armour package and other systems will be retained. The latest British Army Challenger 2 has a more advanced armour system called Dorchester.

The first battalion of Al Hussein MBTs was fully equipped in 2000 and it is understood that the UK will supply sufficient vehicles to enable all of the Tariq fleet to be replaced.

Late in 2000 the UK shipped another batch of 30 ex-British Army Challenger I MBTs) from Marchwood Military Port, Southern England, to Jordan.

The first batch of 14 Challenger 1 MBTs were shipped to Jordan in 1999 and this delivery enabled the first Jordanian Challenger 1 battalion to be fully equipped with 44 vehicles. This unit is the 7th Battalion of the 99th Armoured Brigade.

Prior to their shipment to Jordan, 18 of the Challenger 1 MBTs were inspected and overhauled at the Army Base Repair Organisation (ABRO) facility at Bovington, Dorset. The remaining 12 came from British Army stacks.

The UK has also provided a considerable amount of Challenger 1 spare parts, including power packs to Jordan as electrical and mechanical training rigs and publications. The latest delivery also included five Challenger 1 MBTs in running condition for spares and three Driver Training Tanks.

As well as giving the Jordanian Army a fleet of vehicles with increased armour, mobility and firepower over the current Tariq, the introduction of the Al Hussein will also help to standardise its MBT fleet that also inlcuded 274 Khalid MBTs also built at the Royal Ordnance Factory Leeds and delivered from 1982.

Khalid was basically a Chieftain MBT armed with a 120 mm L11 rifled gun, computerised fire-control system, uprated suspension and new power pack consisting of a Perkins CV12 1200 hp diesel engine coupled to a David Brown Engineering TN37 transmission.

The provision of the Challenger 1 to Jordan is a government to government deal and on the UK side is being handled by the Disposal Sales Agency (DSA). A support package is also being provided.

Challenger 1 MBT being prepared for Jordan at the ABRO facility at Bovington, Dorset (Christopher F Foss) **2001**/0103439

Description

Challenger 1 is essentially the FV4030/3 (Shir 2) modified to suit the operational requirements of the British Army. The layout of Challenger 1 is similar to that of the Chieftain, with the driver's compartment at the front, turret and fighting compartment in the centre, and engine and transmission at the rear. The turret and hull incorporate Chobham armour over the frontal arc for increased battlefield survivability.

The driver has a single-piece hatch cover that lifts and swings forward horizontally allowing him to drive in the head-out position. To the rear of this is a single wide-angle day periscope which can be replaced by a Thales Optronics (previously Pilkington Optronics) Badger passive periscope for driving at night. The driver can also leave the vehicle via the fighting compartment.

The commander is seated on the right of the turret with the gunner forward and below the commander and the loader on the left side of the turret.

The commander originally had a modified No 15 cupola that was designated the No 32 cupola. This has a No 37 day sight which is capable of being quickly replaced by a Rank Pullin image intensification swap sight. The tank commander also has nine day periscopes for all-round observation.

The No 37 swap sight was an interim solution as in July 1981 the MoD awarded a development contract to Barr & Stroud (today Thales Optronics) for the Thermal Imaging Surveillance and Gun Sighting Sights (also known as TOGS - Thermal Observation and Gunnery Sight). The turret had been designed to accept this system when it was available. A single thermal imager in an armoured box on the right of the turret provides separate outputs for both commander and gunner in the relaxed observation mode or as a gunsight.

In April 1987, the UK MoD announced that orders had been placed for the final production batches of the Thermal Observation and Gunnery Sights worth about £35 million. This order completed the buy of TOGS for the British Army's Challenger 1 and Chieftain fleets. Half the order was placed with Barr & Stroud, the original producer of the system, with half going to Avimo. This was the third order for TOGS and Barr & Stroud supplied 85 per cent of the MoD's total TOGS requirement.

The loader has a roof-mounted ×1 periscope swivel-mounted forward of the two-piece hatch cover that opens front and rear.

Challenger 1s are armed with the standard BAE Systems, RO Defence 120 mm L11A5 rifled tank gun fitted with a thermal sleeve, a fume extractor and a muzzle reference system. Types of ammunition available are APDS-T (L15A4), DS-T (L20A1), HESH (L31), HESH practice (L32A5), Smoke WP (L34) and APFSDS-T (L23A1). There are up to 42 charge stowage and 64 projectile stowage positions. Each charge location takes either one DS charge or two HESH/Smoke charges. A typical mix would be 20 DS and 44 HESH/Smoke with all charges stowed in special containers below the turret ring. A fire suppressant fluid in these containers reduces the risk of fire and, on later vehicles, these containers are armoured.

A 7.62 mm L8A2 machine gun is mounted coaxially with the main armament and a 7.62 mm L37A2 machine gun is mounted at the commander's cupola. Mounted either side at the front of the turret is a cluster of five electrically operated 66 mm smoke grenade dischargers, each with a coverage of 100°.

The Challenger 1 has the former GEC-Marconi Radar and Defence Systems, Defence Control Systems Division, Improved Fire-Control System (IFCS), which is fully described in the entry for the Chieftain MBT and has a target tracking rate of 30 mils/s in traverse and 10 mils/s in elevation. Slewing rate in traverse is 0.2 to 480 mils/s and 0.2 to 200 mils/s in elevation. The gun control system is also similar to the Chieftain's but a number of units have been redesigned to replace the thermionic valve equipment with solid-state equivalents.

The gunner has a periscope with a Thales Optronics Tank Laser Sight No 10 Mark 1 with a magnification of ×1 and ×10 with an 8.5° field of view. The laser rangefinder is an Nd:YAG unit with an operating range of 300 to 10,000 m and an accuracy of ±10 m for 90 per cent of shots. The gunner also has an emergency No 87 periscopic sight. This is stowed under armour but can quickly be deployed and collimated with the gun. It projects through an armoured hatch in the turret roof, rather than a conventional telescope as used in Chieftain, to avoid detracting from the frontal integrity of the armour.

Mounted at the rear of the turret is an NBC environmental control system.

Challenger 1 is powered by a Perkins Engines Company Condor V-12 1200 diesel developing 1,200 bhp at 2,300 rpm, which is fitted with two Garrett-AiResearch turbochargers. The transmission is a David Brown Engineering TN37. Torque converter is a Borg-Warner with lock-up clutch giving four forward and three reverse gears. Steering is a Commercial Hydraulics STN37 double differential with hydrostatic, infinitely variable control.

The hydropneumatic suspension of Challenger 1 was designed and developed by the then MVEE in conjunction with the now Hostman Defence Systems, with production being undertaken by Vickers Defence Systems. Challenger 1 has six aluminium roadwheels with the drive sprocket at the rear, idler at the front and two track-return rollers either side. The upper part of the track is covered by conventional aluminium skirts similar to those fitted to the Chieftain. In times of crisis or war these would be replaced by skirts incorporating Chobham armour which were first deployed during the 1990-91 Gulf War. The tracks of the Challenger 1 are of a new design and are not interchangeable with those fitted to the Chieftain.

Variants

Challenger Mk 1 to Mk 4

First production Challenger 1 MBTs were not fitted with TOGS and were called the Mk 1. As soon as TOGS was installed they became the Mk 2. in 1987 these were reworked with a quick-release mechanism for the external fuel drums at the rear. The Challenger Mk 3 has a safer interior with reworked charge stowage and these are called Challenger Mk 2 (ACB). The late production model was the Mk 4.

Challenger 1 MBT without appliqué armour installed (Richard Stickland) **2001**/0103440

Al Hussein (Challenger 1) MBT for Jordan at the ABRO facility Bovington, Dorset, prior to shipment to Jordan (Christopher F Foss) **2002**/0131051

Challenger Training Tank (CTT) (Terry J Gander) 0011767

Al Hussein (Challenger 1) MBT for Jordan with turret traversed to rear (Christopher F Foss) **2002**/0131050

Challenger 1 Control

This is a Challenger 1 command tank for brigade commander with only 44 rounds for the main gun.

Challenger 1 Command

Command tank for squadron commander, also has 44 rounds for main gun.

Challenger 1 in the Middle East

Challenger 1 MBTs were deployed to Saudi Arabia late in 1990 and subsequently took part in Operation Desert Storm. The 7th Armoured Brigade deployed two regiments of Challenger 1 MBTs each with 57 MBTs (four squadrons) while the fourth brigade deployed one regiment of Challenger 1 MBTs with 43 MBTs (three squadrons). Challenger Armoured Repair and Recovery Vehicles were also deployed, as were additional Challenger 1 MBTs, as battlefield replacement vehicles.

In early 1991, these were improved in a number of key areas including the installation of BAE Systems, RO Defence Explosive Reactive Armour (ERA) on the nose and glacis plate and Vickers Defence Systems passive armour along the sides of the hull.

BAE Systems, RO Defence also improved the firepower of the Challenger 1 120 mm L11 rifled tank gun with a new projectile and charge system. The new APFSDS projectile gave greater accuracy and penetration and incorporated features of the new system developed for the L30A1 rifled tank gun installed in the Challenger 2. The charge was designated the L14A1 and the depleted uranium projectile the XL26E1.

During Operation Desert Storm not a single Challenger 1 MBT was knocked out by Iraqi forces but they destroyed some 300 Iraqi MBTs.

Challenger 2 MBT

Details of this are given in the previous entry.

Challenger Training Tanks

In February 1988, the British Army placed an order with Vickers Defence Systems at Leeds for the supply of 17 Challenger Training Tanks valued at £18 million. The first Challenger Training Tank (CTT) was completed in August 1989 with the main production run being carried out between May and September 1990. The vehicles were used to provide realistic driver and maintenance training at Royal Armoured Corps and

Specifications

Crew: 4
Combat weight: 62,000 kg
Unloaded weight: 60,000 kg
Power-to-weight ratio: 19.35 bhp/t
Ground pressure: 0.97 kg/cm²
Length:
(from front mudguard to gun crutch) 8.327 m
(from gun muzzle to gun crutch – gun forward) 11.56 m
(from front mudguard to gun muzzle – gun rearward) 9.8 m
Width:
(overall) 3.518 m
(over tracks) 3.42 m
Height:
(to top of commander's sight) 2.95 m
(turret roof) 2.5 m
Ground clearance: 0.5 m
Track: 2.77 m
Track width: 650 mm
Length of track on ground: 4.79 m
Max road speed:
(1st gear) 13 km/h
(2nd gear) 22 km/h
(3rd gear) 34 km/h
(4th gear) 56 km/h
(reverse 1st gear) 14 km/h
(reverse 2nd gear) 22 km/h
(reverse 3rd gear) 36 km/h
Min turning radius:
(1st gear) 6.8 m
(2nd gear) 11.1 m
(3rd gear) 17.4 m
(4th gear) 28.7 m
(reverse 1st gear) 7.3 m
(reverse 2nd gear) 11.1 m
(reverse 3rd gear) 18.6 m
Usable fuel capacity: 1,592 litres
Max range:
(road) 450 km
(cross-country) 250 km
Fording: 1.07 m
Gradient: 58%
Vertical obstacle: 0.9 m
Trench: 2.8 m
Engine: Perkins Engines Company Condor V-12 1200 12-cylinder diesel, water-based cooled, developing 1,200 bhp at 2,300 rpm
Auxiliary engine: Coventry Climax H30 diesel, 3-cylinder, developing 37 bhp at 3,000 rpm*
Transmission: David Brown Engineering TN37. Borg-Warner torque converter with lock-up clutch giving 4 forward and 3 reverse gears
Steering: Commercial Hydraulics STN 37 double-differential, hydrostatic control
Suspension: hydropneumatic, 450 mm total wheel travel
Electrical system: 24 V
Batteries: 6 × 6TN (4 in hull, 2 in turret)
Armament:
(main) 1 × 120 mm L11A5 rifled gun
(coaxial) 1 × 7.62 mm L8A2 MG
(anti-aircraft) 1 × 7.62 mm L37A2 MG
Smoke-laying equipment: 2 × 5 66 mm smoke grenade dischargers
Ammunition:
(main) up to 64 projectile stowage positions (see text)
(7.62 mm) 4,000

Gun control equipment
Turret power control: electric/manual
(by commander) yes
(by gunner) yes
Turret traverse: 360°
Gun elevation/depression: +20/–10°
Max rate power traverse: 480 mils/s
Max rate power elevation: 200 mils/s
(commander's override) yes
Commander's fire-control override: yes
Gun stabiliser:
(vertical) yes
(horizontal) yes
Range setting device: yes, laser
Elevation quadrant: yes
Traverse indicator: yes
NBC system: yes
Night vision equipment: yes
* On later vehicles, Perkins diesel Type 4108 with a capacity of 1.76 litres giving 37 bhp at 2,850 rpm

Challenger Training Tank from the rear (Christopher F Foss) **2002**/0002567

Challenger Armoured Repair and Recovery Vehicle with stabiliser blade lowered and main winch being used (Richard Stickland) **2002**/0131034

Royal Electrical and Mechanical Engineer Establishments.

The CTT was the first vehicle of its type to be used by the British Army and replaced Challenger 1 gun tanks then being used for this role.

As the CTT weighs the same as the Challenger 1 MBT the automotive performance of the two vehicles is identical. In addition to being used for driver training, the CTT can be used for maintenance training, MBT recovery or as a dozer vehicle.

The CTT is essentially a Challenger 1 MBT with its turret replaced with a new non-rotating turret that accommodates an instructor and up to four students. The new turret provides roll-over protection and the instructor has duplicate instrumentation and controls that allow him to control turret equipment, monitor the driver's performance and inject automotive faults into the driver's instrument panel. In an emergency the instructor can also stop the vehicle.

Combat Dozer Blade

Early in 1990, the British Army selected the Pearson Combat Dozer Blade (UDK1) for service on its Challenger 1 MBTs and every Challenger 1 in the British Army has been modified to accept it.

Typical roles of the UDK1 include digging hull down fighting positions, crossing anti-tank ditches and similar obstacles, construction of defensive earthworks, removing obstacles and creating roadways for follow-on vehicles through wood, bush and difficult terrain.

The dozer can be mounted on or removed from the Challenger 1 MBT in under 15 minutes and it is fully interchangeable with other equipment such as the Pearson mine plough. The dozer unit contains its own power pack with only electrical power being required from the vehicle.

The UDK1 is also fitted to the Chieftain Armoured Vehicle Royal Engineers developed by Vickers Defence Systems for the British Army. Full details of this are given in the Chieftain entry later in this section.

Challenger Armoured Repair and Recovery Vehicle (CR ARRV)

In June 1985, it was announced that Vickers Defence Systems of Newcastle-upon-Tyne had been awarded a contract for the development and initial production of 30 Challenger Armoured Repair and Recovery Vehicles for the British Army.

The first of six preproduction CR ARRVs was completed in August 1987 and all six were completed by 1988.

Late in 1989, an additional 47 CR ARRVs were ordered from Vickers Defence Systems and in 1990, a further four were ordered, bringing the total order for the British Army to 81 vehicles.

Challenger 1 MBT without appliqué armour installed on front and side (Richard Stickland) ***2002***/0131033

Production was undertaken at both Newcastle-upon-Tyne and Leeds and the first production vehicles were completed at Newcastle-upon-Tyne in April 1990 with the first vehicles being completed at Leeds in November 1990. In total, 12 Challenger Armoured Repair and Recovery Vehicles were deployed to Saudi Arabia in late 1990 and used in Operation Desert Storm.

Standard equipment on the vehicle includes a 52 tonne hydraulic winch, an independent auxiliary winch, a hydraulically operated crane that can lift a complete Challenger 1 or 2 power pack and a front-mounted multipurpose blade which can be used as an earth anchor, dozer blade or crane stabiliser.

Full details of the Challenger Armoured Repair and Recovery Vehicle are given in *Jane's Military Vehicles and Logistics 2002-2003.* In 1993, Oman ordered four Challenger Armoured Repair and Recovery Vehicles and these were built at Newcastle-upon-Tyne.

Omani CR ARRV

These have been modified in such a way as to meet the customer's specific demands and needs. The vehicle's specification was developed for operation in the climatic conditions of the Middle East. The power pack cooling group was modified to allow operation without power cutback in ambient temperatures above 50°C and an air conditioning system has been installed.

Challenger 1 with Marksman Anti-aircraft Turret

In June 1985, the now BAE Systems, RO Defence Group Marksman twin 35 mm anti-aircraft turret was taken to Leeds mounted on a Centurion tank chassis which had been used for firing trials.

So far the only customer for the Marksman anti-aircraft turret is Finland, who fitted it to a number of T-55 tank chassis.

Status

Production complete. Withdrawn from the front line service with British Army in late 2000. Jordan is to take delivery of 288 Challenger 1 MBTs with the first batch of 14 delivered late in 1999.

Manufacturer

Vickers Defence Systems.

UPDATED

Vickers Defence Systems Khalid MBT

Development

In December 1974, Iran ordered 125 Shir 1 (FV4030/2) and 1,225 Shir 2 (FV4030/3) MBTs but the order was cancelled by the new Iranian government in July 1979. The first three FV4030/2 prototypes were completed by January 1977.

By this time, production of the FV4030/2 was under way at the Royal Ordnance Factory Leeds (which was subsequently taken over by Vickers Defence Systems and now closed down as an AFV production facility) with first production tanks scheduled for delivery in 1980.

In November 1979, Jordan placed an order with the UK for 274 Khalid MBTs worth £266 million for delivery from 1981.

Further development of the Shir 2 (FV4030/3) resulted in the Challenger 1 MBT which entered service with the British Army in 1983 and for which there is a separate entry. The Challenger 1 MBT was withdrawn from service with the British Army late in 2000 and most of these will be transferred to Jordan.

Modifications carried out since the Khalid MBT entered service with the Jordanian Army have included modifications of sights and stowage to allow for the carrying and firing of the BAE Systems, RO Defence 120 mm APFSDS-T ammunition and the upgrading of the David Brown Engineering TN37 Mk 2 transmission to TN37 Mk 2A standard.

Vickers Defence Systems Khalid MBT 0011765

Description

The Khalid is essentially the FV4030/2 MBT with minor modifications to suit Jordanian requirements; it is based on a late production Chieftain with major changes in the fire-control system and new power pack.

The power pack consists of the Perkins Engines Company Condor V-12 1200 diesel, the David Brown Engineering TN37 transmission and a cooling system by Howden Aircontrol.

The power packs for the Khalid (FV4030/2) and Challenger 1 are almost identical and both have twin electric starters.

The cooling group essentially consists of two air-to-water radiators, two air-cooled charge air coolers and three 380 mm mixed flow fans, mounted on top of the vehicle drive transmission. Cooling air enters through armoured louvres, passes through the heat exchangers to the fans and discharges through armoured louvres.

Specifications
Crew: 4
Combat weight: 58,000 kg
Unloaded weight: 56,300 kg
Power-to-weight ratio: 20.68 hp/t
Ground pressure: 0.90 kg/cm^2
Length:
(gun forward) 11.55 m
(gun rear) 9.86 m
(hull) 8.39 m
Width:
(over skirts) 3.518 m
(over tracks) 3.42 m
Height:
(overall) 2.975 m
(turret roof) 2.435 m
Ground clearance: 0.508 m
Track: 2.77 m
Track width: 650 mm
Length of track on ground: 4.965 m
Max road speed: 56 km/h
Fuel capacity: 950 litres
Max range:
(road) 400 km (estimated)
(cross-country) 250 km (estimated)
Fording: 1.066 m
Gradient: 60%
Side slope: 30%
Vertical obstacle: 0.914 m
Trench: 3.149 m
Engine: Perkins Engines Company Condor V-12 1200 12-cylinder diesel, water-cooled developing 1,200 hp at 2,300 rpm
Transmission: David Brown Engineering TN37. Borg-Warner torque converter with lock-up clutch giving 4 forward and 3 reverse gears
Steering: Commercial Hydraulics STN 37 double differential, hydrostatic control
Suspension: hydropneumatic
Electrical system: 24 V
Batteries: 6 × 6TN (4 in hull, 2 in turret)
Armament:
(main) 1 × 120 mm L11A5 rifled gun
(coaxial) 1 × 7.62 mm L8A2 MG
(anti-aircraft) 1 × 7.62 mm L37A2 MG
Smoke-laying equipment: 2 × 6 66 mm smoke grenade dischargers mounted either side of turret
Ammunition:
(main) 64
(7.62 mm) 6,000
(smoke grenades) 24
Gun control equipment
Turret power control: electric/manual
(by commander) yes
(by gunner) yes
Max rate of power:
(traverse) 360° in 16 s
Gun elevation/depression: +20/–10°
(commander's override) yes
Commander's fire-control override: yes
Gun stabiliser:
(vertical) yes
(horizontal) yes
Range setting device: yes, laser
Elevation quadrant: yes
Traverse indicator: yes
NBC system: yes
Night vision equipment: yes

A separate fan was required for cooling the auxiliary power unit when used during the vehicle 'silent watch' situation.

The Khalid has a bogie-type suspension which is a further development of that fitted to Chieftain with nearly twice the suspension travel.

The fire-control system of the Khalid is the Computer Sighting System which is similar to the Chieftain Improved Fire-Control System described in the entry for the Chieftain. The Thales Optronics (previously Pilkington Optronics) Tank Laser Sight is also fitted to the Khalid.

The commander's cupola is a No 15 which was modified to accept the No 84 sight. The No 84 sight has been developed by Thales Optronics and is a combined day/passive night sight plus projector reticle image unit, and provides the commander with a 24 hour vision and firing capability. The No 84 sight incorporates two independent channels for day and night use and interfaces with the 120 mm L11A5 main armament via the projector reticle image unit which injects optical graticule information into the sight and also enables spot injection for the CSS. It has a fully armoured hood, can be elevated from –10 to +35° and is also provided with a wiper blade.

The main armament consists of a standard 120 mm L11A5 rifled tank gun, a 7.62 mm L8A2 machine gun mounted coaxially with the main armament, a 7.62 mm L37A2 machine gun which can be aimed and fired from inside the commander's cupola and six electrically operated 66 mm smoke grenade dischargers either side of the turret.

The driver can exchange the day driving periscope for a Thales Optronics image intensification night periscope.

Early in 1987, Jordan placed a multimillion pound order with Kidde-Graviner of the UK for its Crew Bay fire and explosion suppression system. These systems were retrofitted to all 274 Khalid MBTs, 293 Tariq modernised Centurions, M60A1/M60A3 MBTs and the ENGESA EE-11 Urutu (6 × 6) APC ordered for police use.

Vickers Defence Systems Khalid MBT 0011764

Jordan has also taken delivery of a quantity of Royal Ordnance L23A1 APFSDS rounds and is considering a number of improvements for the Khalid including night vision equipment and an additional armour protection package, but funding is a problem at the present time.

The Khalid MBTs are now to be supplemented by 288 Challenger 1 MBTs supplied by the UK. These are called the Al Hussein with the first 14 being delivered late in 1999.

Status
Production complete. In service with the Jordanian Army. Iraq has also supplied Jordan with one Chieftain Mk 3/3(P) and 89 Mk 5/5(P) MBTs plus 60 M47s and 20 CVR(T) Scorpion vehicles. As far as is known, none of these armoured vehicles has been taken in service with the Jordanian Army.

Manufacturer
Vickers Defence Systems.

UPDATED

Vickers Defence Systems Chieftain MBT

Development
In 1956, Leyland Motors, which was the design parent for the Centurion Mk 7, built three prototypes of an MBT called the FV4202. This was similar in some respects to the earlier Centurion but had only five roadwheels either side and a new turret without a mantlet, and the driver was seated reclining at the front of the hull allowing the hull's height to be reduced. Both the last two details were subsequently adopted for the Chieftain MBT (FV4201), the characteristics of which were issued in 1958. The first mockup was completed in early 1959 and the first prototype later that year. The Chieftain was first shown to the public in 1961.

A further six prototypes were built between July 1961 and April 1962 and in May the following year the Chieftain was accepted for service with the British Army. Two production lines for the Chieftain were established, one at Royal Ordnance Leeds (taken over by Vickers Defence Systems in 1986) and the other at Vickers' plant at Elswick.

About 900 Chieftain MBTs were built for the British Army, with production completed early in the 1970s. In 1971, Iran placed an order for approximately 707 Chieftain MBTs: the Mk 3/3(P) and Mk 5/3(P) plus a quantity of ARVs and bridgelayers, all of which were delivered by 1978.

Iran also took delivery of 187 improved Chieftains called the FV4030/1. These carried more fuel than the Mk 5/5P, had improved mine protection and additional shock-absorbers on the rear station. They also incorporated electronic control of the now David Brown Engineering TN12 transmission to permit automatic operation and easier driver training.

In 1974, Iran ordered 125 Shir 1 and 1,225 Shir 2 MBTs for delivery from 1980 but the order was cancelled in 1979 before deliveries could begin from Royal Ordnance Leeds.

The Shir 1 (FV4030/2) was subsequently ordered by Jordan under the name Khalid and the Shir 2 (FV4030/3) was the basis for the Challenger 1 MBT for the British Army. There are separate entries for both vehicles earlier in this section.

The Challenger 1 MBT was withdrawn from British Army service late in 2000 and most are now being passed onto Jordan.

During fighting in the Middle East, Iraq captured between 200 and 300 Chieftain MBTs from Iran, many of which were undamaged. Late in 1988, Iraq gave Jordan a large amount of military equipment captured during fighting with Iran. This included 60 M47 tanks, 90 Chieftains, 20 Scorpions and 35 armoured personnel carriers.

In August 1981, Oman took delivery of 12 British Army Chieftain Mk 7/2C MBTs. These were on loan but subsequently purchased by Oman together with 15 new Chieftains called the Qayd Al Ardh, or Mk 15 by the manufacturer, delivered between 1984 and 1985. These Chieftains, as well as the Scorpions used by Oman, are fitted with the BAE Systems L20 sight which incorporates the Type 520 laser range-finder.

Chieftain Mk 5/2K showing the 120 mm L11 rifled tank gun fitted with a thermal sleeve and muzzle reference system 0018847

Chieftains of the Kuwaiti Army engaged Iraqi MBTs, including T-72s, during the Iraqi invasion of Kuwait in the summer of 1990.

The Chieftain MBT was finally phased out of front-line service with the British Army in March 1996. Specialised versions of the Chieftain remain in service in the support role with the British Army.

Description (Mk 5)

The hull of the Chieftain is of cast and rolled steel sections welded together and is divided into three compartments with driving at the front, fighting in the centre and engine at the rear.

The driver is seated at the front of the hull and is provided with a single-piece hatch cover that lifts and opens to the right. To the rear of the hatch cover is a single wide-angle AFV No 36 Mk 1 periscope which can be replaced by a passive periscope for night driving.

The turret is made of cast and rolled steel sections with the loader seated on the left and the commander and gunner on the right. The commander's cupola can be traversed through 360° by hand and has a single-piece hatch cover that opens to the rear, nine AFV No 40 Mk 2 observation periscopes and a single AFV No 37 Mk 4 sighting periscope. Mounted on the right side of the commander's cupola is a spotlight which is mounted coaxially with the cupola-mounted 7.62 mm L37A1 machine gun.

The gunner is seated in front of and below the commander and has a sight periscope No 59 with a magnification of ×1 and ×8 or a Thales Optronics (previously Pilkington Optronics) Tank Laser Sight Unit. The gunner also has a No 70 telescopic sight. Both the commander and gunner can replace their day sights with an infra-red sight with a magnification of ×3: the commander's night sight is the L1A1 and the gunner's the L3A1.

The loader has a two-piece hatch cover that opens to the front and rear of his position and a folding rotatable periscope AFV No 30 Mk 1.

Mounted on the left side of the turret is a former GEC-Marconi Radar and Defence Systems, Defence Control Systems Division, infra-red/white light searchlight. The very high-intensity light operates as a target illuminator and is mounted directly onto the gun turret so that it remains in line with the gun during traverse. In elevation a servo-control system on the mirror assembly ensures that the light beam remains aligned automatically onto the target. The searchlight has a range in the infra-red mode of at least 1,000 m and a range in the white light mode of at least 1,500 m. The searchlight overhangs the side of the turret and has an armoured cover.

British Army Chieftains were fitted with the Thales Optronics Thermal Observation and Gunnery Sight which is also installed on the Challenger 1 MBT.

The Leyland L60 engine is at the rear of the tank and is coupled to a TN12 gearbox.

The TN12 epicyclic gearbox incorporates a Merritt-Wilson differential steering system and an electrohydraulic gear selector. It has six forward and two reverse gears plus an emergency mechanical selector of second gear forward and low reverse.

Of five portable fire extinguishers provided, two large ones are mounted in the engine compartment. A FIREWIRE automatic fire detection system actuates a warning horn and internal flashing indicators and creates an audible signal in the crew's headphones.

The Horstmann suspension consists of three bogies each side, each with two dual roadwheels and a set of three horizontal springs. The drive spocket is at the rear, idler at the front and there are three track-return rollers. The first roadwheel station has a hydraulic shock-absorber. The tracks are of cast steel with dry pins and replaceable rubber pads. Each track has 96 links when new and the total weight of the pair of tracks on the Chieftain is 4,719 kg. The top half of the track is covered by an armoured skirt which can be removed for maintenance.

An air ventilation and filtration system is mounted in the rear of the turret which provides the crew with clean air for normal ventilation and works in conjunction with an NBC filter pack which does not require the crew to wear respirators inside the tank.

Mounted on the turret roof is an infra-red detector which has three silicon photovoltaic cells covering 360°. Any infra-red light source detected can be localised to within an arc of about 62°.

If required, a dozer blade could be mounted at the front of the hull. This consists of an electrohydraulic power pack which is fitted in place of the right-hand front stowage bins. The blade itself is aluminium and is operated by the driver who is provided with a joystick control unit.

Main armament is a BAE Systems, RO Defence 120 mm L11A5 rifled tank gun which is provided with a fume extractor, a muzzle reference system, and a thermal sleeve. It has a vertical sliding breech block and a maximum rate of fire of 8 to 10 rounds for the first minute and 6 rds/min thereafter.

The ammunition is of the separate loading type, that is, projectile and charge. The projectiles are stowed alongside the driver, under the gun and in the turret and the charges are stored in pressurised bins below the turret ring. The gun can fire the following types of ammunition manufactured by BAE Systems, RO Defence: APDS-T (L15A4), DS-T (L20A1), HESH (L31), SH/Practice (L32A5), Smoke-WP (L34) and APFSDS-T (L23A1).

Mounted coaxially to the left of the main armament is a 7.62 mm L8A1 machine gun. The commander's cupola is provided with a 7.62 mm L37A1 machine gun which can be aimed and fired from within the tank. A 12.7 mm (0.50) L21A1 ranging machine gun can be mounted over the main armament. This was removed from all British Army Chieftain MBTs and ranging was achieved using the Tank Laser Sight. Mounted either side of the turret is a bank of six electrically operated smoke grenade dischargers.

The gun control system is a Marconi Fighting Vehicle Gun Control Equipment (FV/GCE) No 7 Mk 4 which is a further development of the system used on the Centurion. This has the following modes of operation: stabilised power control, power control, emergency battery control and emergency hand control.

When originally introduced into service, the Chieftain was provided with a 12.7 mm Ranging Machine Gun (RMG) as the main means of aiming its 120 mm gun. For an APDS-T engagement the gunner fires short bursts from the RMG using four predetermined graticule marks. The burst which lands over the target indicates which main armament graticule mark should be used. Using this method the target can be engaged quite rapidly. For the lower-velocity HESH ammunition up to 2,000 m the RMG is used for obtaining the close bracket required for this round, which has a higher trajectory. The main armament is not fired until a hit is a virtual certainty.

In the early 1970s, the Thales Optronics Tank Laser Sight Unit was introduced. This houses the laser transmitter module and optical system, the receiver system and the optical sight system. The line of sight in elevation is synchronised with the axis of the gun by means of a precision parallel linkage and in azimuth by coincident turret mounting. Final boresighting is achieved by instrument-mounted controls. Gunlaying is achieved through a ballistic graticule. Laser ranging may be initiated by the gunner or remotely by the tank commander; the range is displayed in the left eyepiece and remotely displayed at the commander's station. In a smoke situation, target range uncertainty is effectively diminished by selection of first and last range logic. The laser has an operating range of 500 to 10,000 m and is accurate to ±10 m for 90 per cent of shots.

The British Army's Chieftain fleet (except Mk 1s) was retrofitted with the fully integrated IFCS (Improved Fire-Control System) developed and manufactured by the then GEC-Marconi Radar and Defence Systems, Defence Control Systems Division. IFCS provides a significantly higher first round hit probability on stationary tank targets to at least 3,000 m and on moving targets out to more than 2,000 m. The IFCS has been installed in the Leopard 1 (with successful firings in 1979), and has been adopted for the Khalid and Challenger 1 as the Computer Sighting System (CSS).

The IFCS uses the Marconi 12-12P digital computer, which is programmed for the specific combination of tank, gun and ammunition carried. It automatically gathers and updates information from various sensors, calculates the ballistic solution and correct laying offsets for each target engagement, and controls the

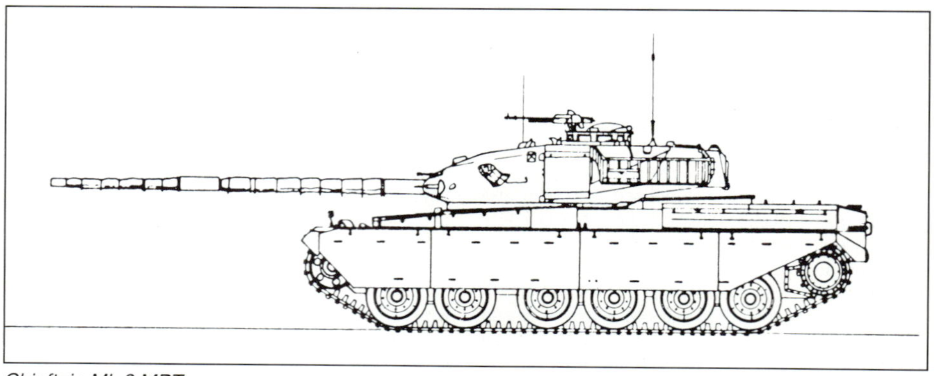

Chieftain Mk 3 MBT

automatic laying of the gun in line and elevation (with tracking of moving targets) in readiness to fire. Both gunner and commander have full fire controls, the commander also has override control and can manually insert data whenever desired.

The system comprises four subsystems; the Data Handling SubSystem (DHSS) with the 12-12P digital computer, commander's control and monitor unit, and the firing handles; the sighting subsystem with the Tank Laser Sight (TLS) and aiming ellipse electronics; the sensor subsystem of various sensors located about the tank: for example, wind direction and velocity, air temperature and pressure, trunnion tilt, angle of sight, charge temperature, barrel wear, target displacement and ammunition type, and processing electronics; and the Gun Control Equipment (GCE) which has been uprated for IFCS as the FV/GCE No 10 and No 11. The meteorological sensor was removed from British Army Chieftain MBTs.

Marks of the basic Chieftain MBT

Mk 1 of which 40 were built and issued for training in 1965-66, 585 bhp engine.
Mk 1/2 is Mk 1 brought up to Mk 2 standard and used for training.
Mk 1/3 is Mk 1 with a new power pack and used for training.
Mk 1/4 is Mk 1 with a new power pack and modified RMG, used for training.
Mk 2 was the first model to enter service, first issued in 1967, with a 650 bhp engine.
Mk 3 entered service in 1969, has improved auxiliary generator, 650 bhp engine, dry-air cleaner element, modified No 15 Mk 2 cupola with L37A1 7.62 mm machine gun, oil-filled top rollers, axle arms and track tensioner.
Mk 3/G was a prototype only with turret air-breathing for engine aspiration.
Mk 3/2 is Mk 3/G modified.
Mk 3/S is production model of Mk 3/G with turret air-breathing and commander's firing switch.
Mk 3/3 is Mk 3 with extended range RMG, fitted to accept the Thales Optronics laser range-finder, modified 720 bhp engine, new low-loss air cleaner system, turret aspiration and modified NBC pack.
Mk 3/3P is Mk 3/3 export version for Iran.
Mk 4, of which only two were built, had increased fuel capacity and other minor modifications.
Mk 5 is a development of the Mk 3/3 with uprated 720 bhp engine and strengthened gearbox, gunner's telescope and commander's collimator fitted with graticule for use with extended range 12.7 mm RMG ammunition, modified exhaust system, air cleaning filters for generator, battery heating and lagging, new type gun clamp to accept new thermal sleeve, modified charge bins, improved No 43 gunner's telescope mounting, improved stowage, 50 per cent reduction in RMG ammunition-carrying capacity, increase in APDS-T projectiles carried, No 37 Mk 4 commander's sight, stabilised infra-red detector, commander's machine gun elevation to +90° and new NBC pack.
Mk 5/2K is Mk 5 for Kuwait, produced by Royal Ordnance Leeds and Vickers at Elswick with final vehicles completed late in 1979. These are no longer in service with Kuwait.
Mk 5/3P is a Mk 5 for Iranian Army.
Mk 6 is Mk 2 with a new power pack and modified RMG.
Mk 7 is Mk 3 and Mk 3/S with an improved engine and modified RMG.
Mk 8 is Mk 3/3 with above modifications.
Mk 9 is Mk 6 with IFCS.
Mk 10 is Mk 7 with IFCS.
Mk 11 is Mk 8 with IFCS, Stillbrew, TOGS and No 11 NBC system.
Mk 12 is Mk 5 with IFCS, Stillbrew, TOGS and No 11 NBC system.

Variants

Chieftain with Stillbrew Armour

This passive armour package was only fitted to British Army Chieftain MBTs which have now been withdrawn from front-line service.

Chieftain with Hydrostrut Suspension

For trials purposes a Chieftain MBT was fitted with the Hydrostrut suspension system developed by Vickers Defence Systems Leeds and Air-Log Limited. This has not been adopted by any Chieftain users.

Chieftain Armoured Repair and Recovery Vehicle (T J Gander) **2000**/0069565

Chieftain Armoured Vehicle Royal Engineer carrying pipe fascine and call on rolled up trackway (Richard Stickland) **2000**/0069567

Specifications

Model Mk 5
Crew: 4
Combat weight: 55,000 kg
Unloaded weight: 53,500 kg
Power-to-weight ratio: 13.63 bhp/t
Ground pressure: 0.9 kg/cm²
Length:
(gun forward) 10.795 m
(gun rear) 9.87 m
(hull) 7.518 m
Width:
(incl searchlight) 3.657 m
(over skirts) 3.504 m
(over tracks) 3.327 m
Height: (overall) 2.895 m
Ground clearance: 0.508 m
Track: 2.718 m
Track width: 610 mm
Length of track on ground: 4.8 m
Max road speed: 48 km/h
Fuel capacity: 950 litres
Max range:
(road) 400-500 km
(cross-country) 200-300 km
Fording: 1.066 m
Gradient: 60%
Vertical obstacle: 0.914 m
Trench: 3.149 m
Engine: Leyland L60, 2-stroke, compression ignition, 6-cylinder (12 opposed pistons) multifuel
Engine model: No 4 Mk 8A
Output: 750 bhp at 2,100 rpm
Auxiliary engine: Coventry Climax H30 No 4 3-cylinder, vertically opposed compression ignition developing 23 bhp at 2,000 rpm
Transmission: David Brown Engineering TN12 with 6 forward and 2 reverse gears plus emergency mechanical selection for 2nd gear forward and low reverse
Steering: triple differential incorporated in TN12
Clutch: centrifugal
Suspension: Horstmann
Electrical system: 28.5 V
Batteries: 200 Ah (4 × 12 V) batteries in hull for engine starting and general electrical services. 100 Ah (2 × 12 V) in turret for radio load and emergency power supply to fighting equipment
Armament:
(main) 1 × 120 mm L11A5 rifled gun
(coaxial) 1 × 7.62 mm L8A1 MG
(anti-aircraft) 1 × 7.62 mm L37A1 MG
(RMG) * 1 × 12.7 mm L21A1 MG
Smoke-laying equipment: 2 × 6 66 mm smoke grenade dischargers mounted on each side of turret
Ammunition:
(main) 64
(7.62 mm) 6,000
(12.7 mm)* 300
Gun control equipment
Turret power control: electric/manual
(by commander) yes
(by gunner) yes
Max rate power:
(traverse) 360° in 16 s
Gun elevation/depression: +20/–10°
(commander's override) yes
Commander's fire-control override: yes
Gun stabiliser:
(vertical) yes
(horizontal) yes
Range setting device: yes**
Elevation quadrant: yes
Traverse indicator: yes
NBC system: yes
Night vision equipment: yes
* When fitted with the IFCS the RMG and its ammunition is removed. IFCS was only in British Army Chieftains and not in Iranian or Kuwaiti vehicles.
** Retrofitted British Army vehicles had Tank Laser Sight fitted.

Chieftain AVLB launches the first of its pair of Close Support No 12 13.5 m bridges, retaining the other No 12 bridge for deployment elsewhere 0011722

Chieftain AVRE carrying pipe fascine and using front mounted dozer blade. This vehicle is also fitted with appliqué armoured skirts (Richard Stickland) ***2002***/0131043

Chieftain AVRE

Although there was to be a Chieftain Armoured Vehicle Royal Engineer (FV4203) this never entered service. In 1986, the Engineer Workshops of 40 Army Engineer Support Group at Willich, Germany, converted 12 Chieftain gun tanks into Chieftain AVREs for use in BAOR.

The Chieftain AVREs produced at Willich were, however, considered to be interim vehicles and late in 1989 Vickers Defence Systems at Newcastle-upon-Tyne was awarded a contract by the UK MoD for the design, development and conversion of 48 Chieftain gun tanks into Chieftain Armoured Vehicle Royal Engineers. The 48 vehicle contract covered two prototype vehicles and 46 production vehicles (one batch of 30 and one of 16).

First two prototypes were completed in 1991 with production running through to late 1994. Conversion work was undertaken by Vickers Defence Systems at Leeds using Chieftain MBT chassis provided from Army stocks.

Production Chieftain AVREs have a new superstructure, a winch with a capacity of 10 tonnes at the hull rear, an ATLAS hydraulic crane and top-mounted hampers on which the engineer equipment is carried.

Main roles of the Chieftain AVRE can be summarised as:

(1) Mineclearing using the Track Width Mine Plough or Giant Viper rocket-propelled system
(2) Laying Class 60 aluminium trackway
(3) Removing battlefield obstacles using a front-mounted Bulldozer Earth Moving Attachment (BEMA) Blade or crane
(4) Filling in ditches using BEMA Blade or plastic pipe fascines.

Details of the Chieftain AVRE are given in *Jane's Military Vehicles and Logistics 2002-2003*. It will eventually be replaced by the Vickers Defence Systems Challenger 2 Based Trojan.

Armoured Repair and Recovery Vehicle

This is essentially a Chieftain ARV fitted with a hydraulic crane that can lift a complete Challenger 1 MBT power pack. ARRVs were conversions from existing ARVs. Vickers Defence Systems has supplied Hydraulic Track Tensioners for the British Army Chieftain AVLB and ARRV.

Chieftain Armoured Recovery Vehicle (FV4204)

The Chieftain ARV was manufactured by Vickers at Elswick and was based on the chassis of the Mk 5 Chieftain MBT. The main double-capstan winch with electrohydraulic controls is provided with 122 m of 28 mm diameter cable. The similar auxiliary winch is hydraulically operated and provided with 260 m of 11 mm diameter cable. Power for both winches is taken from a PTO on the main engine. Mounted at the front of the ARV is a hydraulically operated earth anchor which, when lowered, allows the vehicle to exert a pull of up to 90,000 kg. The Chieftains delivered to Iran had an ATLAS AK 6000M crane which can lift a maximum load of 5,803 kg. The Chieftain ARV has a crew of four and a loaded weight of 56,000 kg. Armament consists of a cupola-mounted 7.62 mm machine gun and 66 mm smoke grenade dischargers. Full details of the Chieftain ARV are given in *Jane's Military Vehicles and Logistics 2002-2003*. In total, 257 ARVs were built at Newcastle-upon-Tyne, of this figure Iran ordered 73 (of which 41 were delivered) with Jordan taking a further 30 vehicles.

The Chieftain ARV will remain in service with the British Army to at least 2003. It is used to support units equipped with the AS90 self-propelled artillery system and engineer units equipped with the Chieftain AVLB and ARV.

The Chieftain with the hydraulic crane is called the Chieftain Armoured Repair and Recovery Vehicle (ARRV) by the British Army. The original ARV is no longer in service.

Chieftain Mk 6 AVLB

In March 1984, Vickers Defence Systems was awarded a £4.5 million contract by the UK MoD to convert 11 Chieftain Mk 1/4 gun tanks into Chieftain Mk 6 AVLBs. User trials were completed in November 1985 and the first three vehicles were handed over in June 1986 with final deliveries in September 1986.

The conversion work included removing the complete turret and replacing it with an armoured roof plate with a commander's hatch; fitting armoured lugs for mounting the launching mechanism on the hull and roof plate to maintain the same geometrical position as on production AVLBs; installing the hydraulic bridge-launching mechanism; stripping the vehicle and refitting the driver's/crew compartment and external parts to accommodate the AVLB components; modifying the hull to accept the HD1400 constant running pump installation; manufacturing the maraging steel launch structure; and providing mounting points at the front of the hull for the Pearson Engineering Track Width Mine Plough system, which was already in service on Centurion Assault Vehicle Royal Engineers.

For Operation Desert Shield/Desert Storm, a number of the Chieftain AVLBs of the Royal Engineers were used in the mineclearing role. In addition to the Pearson Engineering plough system at the front of the hull they were also fitted with a device to detect and activate mines fitted with a magnetic fuze. The vehicles also towed a trailer-mounted BAE Systems, RO Defence Giant Viper mineclearing system.

To improve battlefield survivability, additional Chobham-type armour was fitted to the sides of the vehicle similar to that fitted to the Challenger 1 MBT.

Chieftain Armoured Vehicle-Launched Bridge (FV4205)

The Chieftain AVLB is basically a Chieftain MBT chassis without a turret and fitted with a hydraulic system for laying and recovering a bridge. The vehicle can carry and lay either a No 8 or a No 9 Tank Bridge. The No 8 Tank Bridge is carried folded and launched over the front of the vehicle. It has an overall length of 24.384 m and can span a gap of up to 22.86 m (hard banks) or 22.25 m (soft banks). The No 9 Tank Bridge is carried horizontally and is swung vertically through 180° and laid in position over the front of the vehicle. The No 9 Tank Bridge is 13.411 m long and can span a gap up to 12.192 m (firm banks). The Chieftain AVLB (with No 8 Tank Bridge) weighs 53,300 kg and has a crew of three. Full details of the Chieftain AVLB are given in *Jane's Military Vehicles and Logistics 2002-2003*. In total, 51 AVLBs were built by Leeds, 37 for the UK and 14 for Iran. The Chieftain AVLB chassis used with the British Army's new Bridging for the 1990s programme covered in *Jane's Military Vehicles and Logistics 2002-2003*. The Chieftain AVLB will be replaced by the Vickers Defence Systems Challenger 2 based Titan.

Chieftain AVLB carrying No 10 tank bridge which is of the scissors type and launched over front of vehicle. This vehicle is also fitted with appliqué armoured skirts (Richard Stickland) ***2002***/0131044

Close Support Bridge System

The British Army's current Chieftain AVLB carried a No 8 or No 9 Tank Bridge. This is now being replaced by a new Close Support Bridge (CSB) system that is a key part of the overall Bridging for the 1990s (BR90) system which has been developed by Vickers Bridging under contract to the UK MoD.

The Close Support Bridge system consists of a set of three tank-launched bridges, a modification kit for the in-service Chieftain bridgelayer tank and a Tank Bridge Transporter (TBT) truck which is based on a new high-mobility Alvis UNIPOWER (8 × 8) vehicle.

All of these bridges are launched with the crew under complete armour protection and take 3 minutes to lay and 5 minutes to recover and can be laid and recovered from either end.

The scissors No 10 Tank Bridge is 26 m long when opened out and can span a gap of 24.5 m. The No 11

Tank Bridge is of the up-and-over type and is 16 m long and can span a gap of 14.5 m. The No 12 Tank Bridge is also of the up-and-over type, is 13.5 m long and can span a gap of up to 12 m. A Chieftain can carry two of these bridges.

Full details of this are given in *Jane's Military Vehicles and Logistics 2002-2003.*

Chieftain with Marksman turret

For demonstration purposes a Chieftain has been fitted with a BAE Systems, RO Defence, Marksman twin 35 mm air defence turret. So far the Marksman twin 35 mm air defence turret has been adopted by Finland which has fitted it to a T-55 MBT chassis.

Full details of the Marksman are given in *Jane's Land-Based Air Defence 2002-2003.*

Status

Production complete. In service with:

Country	Quantity	Comment
Iran	200	probably not all operational, also ARV and AVLB
Iraq	100	status uncertain, some 90 have been given to Jordan
Jordan	90	status uncertain, plus some Chieftain ARVs
Oman	27	replaced by Challenger 2 MBTs, now held in reserve
UK	nil	following phase out of MBT version only now uses AVLB, AVRE, and ARRV

Manufacturer

Vickers Defence Systems.

UPDATED

Vickers Defence Systems Centurion MBT

Development

In 1943, the Department of Tank Design was asked to start design work on a new heavy cruiser tank under the designation A41. This tank was required to have good armour protection, be armed with a 17-pounder (76.2 mm) gun and have a good cross-country performance: a high road speed was not considered essential at that time. The first mockup of the A41 was completed in 1944 and six prototype tanks were completed in early 1945. These were sent to Germany but arrived too late to see any combat. The A41 later became known as the Centurion Mk 1 and the up-armoured A41A became the Centurion Mk 2, both armed with the 17-pounder gun.

Throughout its life the Centurion has proved capable of being up-armoured and up-gunned. It was originally armed with the 17-pounder which was replaced first by the 20-pounder and finally by the BAE Systems, RO Defence 105 mm L7 series rifled gun which was subsequently adopted by many other countries. Other improvements carried out during its life included increased fuel capacity, contrarotating commander's cupola and improved stowage. All British models of the TN12 tank used the same basic Meteor petrol engine and transmission.

Production of the Centurion was undertaken by the Royal Ordnance Factories at Leeds and Woolwich, Vickers at Elswick and Leyland Motors at Leyland. Most were built at Leeds and Elswick. Production of the Centurion was finally completed in 1962, by which time 4,423 tanks had been built, of which at least 2,500 were exported.

In the British Army the Centurion was replaced from the late 1960s by the Chieftain MBT. The last user of the Centurion MBT in Europe was Sweden and this has now been phased out of service and replaced by the German Leopard 2 series MBT.

Description (Mk 13)

The all-welded steel hull of the Centurion is divided into three compartments: driving at the front, fighting in the centre, and engine and transmission at the rear.

The driver sits at the front of the hull on the right side and has two hatch covers that open either side of his position, each with a day periscope. To the left of the driver a quantity of ammunition is carried.

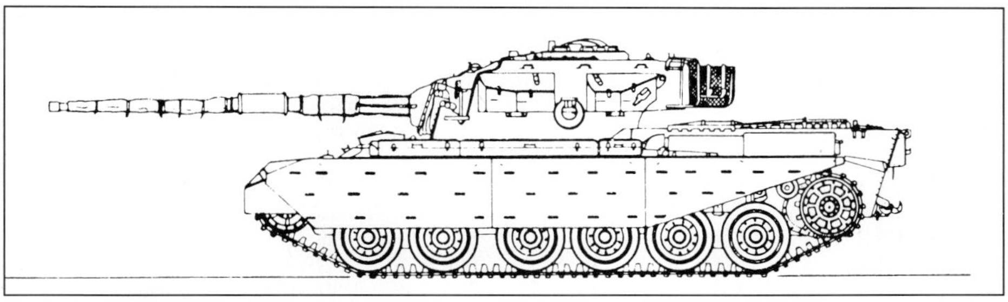

Centurion Mk 10 MBT

The turret in the centre of the hull is of cast construction with the roof welded into position. There is an ammunition resupply hatch mounted externally in the left side of the turret and stowage boxes on either side of the turret. Most vehicles have a wire stowage rack at the rear of the turret. The loader is seated on the left of the turret, the commander on the right, and the gunner in front of and below the commander.

The commander's cupola can be manually traversed through 360° by hand and is provided with split hatches, a periscopic day sight with a ballistic pattern and seven day periscopes for all-round observation. A searchlight mounted at the commander's station can be fitted with an infra-red filter. The gunner has a periscopic day sight with ballistic pattern graticule, linked to a range drum for targets between 3,000 and 8,000 m. The commander's and gunner's sights are linked by a heat-compensated bar.

The loader is provided with twin hatch covers that open front and rear and a single day observation periscope.

The engine compartment at the rear is separated from the fighting compartment by a fireproof bulkhead. Air is drawn in through the engine deck louvres by two fans. Power is transmitted from the engine through the clutch to the transversely mounted Merritt-Brown Z51R transmission. This is a combined change-speed and steering mechanism and incorporates a differential. Drive from the output epicyclic gears at each end of the gearbox is transmitted through an internally toothed coupling ring and an externally toothed driving shaft to the final drive.

The Horstmann suspension consists of three units on each side, each carrying two pairs of roadwheels sprung by one set of concentric springs. The first and last roadwheel units are fitted with shock-absorbers. The drive sprocket is at the rear, idler at the front and there are six track-return rollers, the four dual roller centre ones and the single front and rear ones which support the inside of the track only. The tops of the cast manganese steel tracks are covered by removable armour skirts which provide protection against HEAT projectiles.

As originally built the Centurion was not fitted with any infra-red night vision equipment, but in the 1960s many British Army Centurions were fitted with infra-red driving lights, an infra-red searchlight to the left of the main armament and infra-red sights for both the commander and gunner. The tank has no NBC system or amphibious capability although a deep fording kit was developed. A dozer blade can be mounted at the front of the hull for clearing obstacles and preparing fire positions.

One of the shortcomings of the Centurion was its short operational range, although later production tanks had increased fuel capacity and the Mk 5 could tow a monowheel trailer full of fuel to increase its operational range.

Main armament of the Centurion Mk 13 is the now BAE Systems, RO Defence 105 mm L7A2 rifled tank gun which is provided with a fume extractor, and many countries have fitted a thermal sleeve. It has an effective range of 1,800 m with APDS-T rounds or between 3,000 and 4,000 m using HESH. A well-trained crew can fire 8 rds/min.

The 105 mm weapon is normally aimed using a 12.7 mm (0.50) ranging machine gun mounted coaxially with the main armament which has a maximum range of 1,800 m and fires in three-round bursts using tracer ammunition. The gun is fully stabilised and the gunner can select any one of the following modes of operation: manual elevation and traverse, non-stabilised power traverse, stabilised powered elevation and traverse or emergency, single-speed power traverse.

The 105 mm gun fires the following types of ammunition, all of which are of the fixed type with the projectile securely attached to a brass cartridge case which contains the propellant and initiated by an electric primer: APDS-T (L28A1), APDS-T (L52A1), APFSDS-T (L64 and later developments), DS/T (L45A1), HESH (L37) and Smoke (L39), all manufactured by BAE Systems, RO Defence.

Mounted coaxially to the left of the main armament is a 7.62 mm (0.30) machine gun and there is a similar weapon on the commander's cupola for anti-aircraft use. Six electrically operated 66 mm smoke grenade dischargers are mounted either side of the turret.

There have probably been more variants of the Centurion than any other post-Second World War vehicle in the MBT class. A full list of the gun tank models and variants of the vehicle still in service follows; experimental vehicles have been excluded.

Centurion Mk 2 ARV in travelling configuration (T J Gander) 0011721

Jordanian Army Tariq MBT

Centurion Beach Armoured Recovery Vehicle (BARV) deployed by the Royal Marines to the Middle East with British amphibious forces (Richard Stickland)
2002/0131032

Specifications
Model Mk 13
Crew: 4
Combat weight: 51,820 kg
Power-to-weight ratio: 12.54 hp/t
Ground pressure: 0.95 kg/cm²
Length:
(gun forward) 9.854 m
(hull) 7.823 m
Width: 3.39 m
Height: (without AA MG) 3.009 m
Ground clearance: 0.51 m
Track: 2.641 m
Track width: 610 mm
Length of track on ground: 4.572 m
Max road speed: 34.6 km/h
Fuel capacity: 1,037 litres
Max road range: 190 km
Fording: 1.45 m
(with preparation) 2.74 m
Gradient: 60%
Vertical obstacle: 0.914 m
Trench: 3.352 m
Engine: Rolls-Royce Mk IVB 12-cylinder liquid-cooled petrol developing 650 bhp at 2,550 rpm
Auxiliary engine: Morris USHNM 4-cylinder petrol, 20 bhp at 2,500 rpm
Transmission: Merritt-Brown Z51R manual with 5 forward and 2 reverse gears
Steering: triple differential
Clutch: triple dry plate
Suspension: Horstmann
Electrical system: 24 V
Batteries: 4 × 6 V, 115 Ah
Armament:
(main) 1 × 105 mm L7A2 rifled gun
(coaxial) 1 × 7.62 mm MG
(anti-aircraft) 1 × 7.62 mm MG
(RMG) 1 × 12.7 mm MG
Smoke-laying equipment: 2 × 6 66 mm smoke grenade dischargers either side of turret
Ammunition:
(main) 64
(7.62 mm) 4,750
(12.7 mm) 600
Gun control equipment: electric/manual
Max rate power traverse: 360° in 26 s
Gun elevation/depression: +20/–10°
(commander's override) yes
Gun stabiliser:
(vertical) yes
(horizontal) yes
Range setting device: yes
Elevation quadrant: yes
Traverse indicator: yes

Armour
Hull glacis: 118 mm*
Hull nose: 76 mm*
Hull sides front: 51 mm*
Hull sides:
(rear upper) 38 mm*
(rear lower) 20 mm*
Hull floor: 17 mm*
Turret front: 152 mm*
NBC system: yes
Night vision equipment: yes
* estimated

Centurion Gun Tank Models

Centurion Mk 1 armed with 17-pounder (76.2 mm) gun; none remains in service.
Centurion Mk 2 armed with 17-pounder gun; none remains in service.
Centurion Mk 3 armed with 20-pounder (83.8 mm) gun; none remains in service, most brought up to Mk 5 standard.
Centurion Mk 4 was to have been a close support model armed with 95 mm howitzer but was not placed in production.
Centurion Mk 5 was designed by Vickers at Elswick and armed with a 20-pounder gun.
Centurion Mk 5/1 is Mk 5 up-armoured.
Centurion Mk 5/2 is Mk 5 with a 105 mm gun.
Centurion Mk 6 is a Mk 5 up-armoured, with additional fuel capacity at rear of hull and 105 mm gun.
Centurion Mk 6/1 is a Mk 6 with infra-red night vision equipment and stowage basket at rear of turret.
Centurion Mk 6/2 is Mk 6 with 12.7 mm ranging machine gun for 105 mm gun.
Centurion Mk 7 was designed by Leyland and is designated the FV4007. Armed with a 20-pounder gun with fume extractor and carries 61 rounds of 20-pounder ammunition.
Centurion Mk 7/1 is a Mk 7 up-armoured and designated FV4012.
Centurion Mk 7/2 is a Mk 7 with a 105 mm gun.
Centurion Mk 8 was developed from the Mk 7 and has a resiliently mounted gun mantlet with no canvas cover and the commander's cupola is contrarotating. The commander can also raise his twin hatch covers in an umbrella fashion for improved visibility without revealing his position.
Centurion Mk 8/1 is Mk 8 up-armoured.
Centurion Mk 8/2 is Mk 8 with 105 mm gun.
Centurion Mk 9 is Mk 7 up-armoured and up-gunned with 105 mm gun, designated the FV4015.
Centurion Mk 9/1 is Mk 9 with infra-red night vision equipment and stowage basket on rear of turret.
Centurion Mk 9/2 is Mk 9 with 12.7 mm ranging machine gun.
Centurion Mk 10 is Mk 8 up-armoured, up-gunned with 105 mm gun and with an ammunition capacity of 70 rounds, designated the FV4017.
Centurion Mk 10/1 is Mk 10 with infra-red night vision equipment and stowage basket on rear of turret.
Centurion Mk 10/2 is Mk 10 with 12.7 mm ranging machine gun.
Centurion Mk 11 is Mk 6 with ranging machine gun, infra-red night vision equipment and stowage basket on turret rear.
Centurion Mk 12 is Mk 9 with infra-red night vision equipment, 12.7 mm ranging machine gun and stowage basket on turret rear.
Centurion Mk 13 is Mk 10 with 12.7 mm ranging machine gun and infra-red night vision equipment.

Variants

Centurion (Mk 2) ARV (FV4006)

The Mk 2 was preceded by the Mk 1 ARV but none remains in service. The ARV Mk 2 is essentially a Centurion MBT with its turret removed and replaced by an all-welded superstructure behind the driver's position. The commander's cupola can be traversed through 360° and is fitted with a 7.62 mm (0.30) machine gun. Mounted at the rear of the hull are spades which are used to stabilise the vehicle when the winch is being used. The winch's 31,000 kg capacity can be increased with the aid of snatch blocks to a maximum of 90,000 kg. The vehicle has a loaded weight of 50,295 kg and a crew of four. Full details of the Centurion ARV Mk 2 were given in *Jane's Military Vehicles and Logistics 1995-96*. All Centurion Mk 2 ARVs were built by Vickers Defence Systems at Newcastle-upon-Tyne, with 345 being built between 1954 and 1959.

Centurion BARV (FV4018)

This is used only by the British Army and only 12 were built for use by amphibious forces. It is basically a Centurion tank with its turret replaced by a superstructure enabling it to operate in water up to 2.895 m deep. It has a crew of four, one of whom is a trained diver. Loaded weight is 40,643 kg. Full details of the Centurion BARV were given in *Jane's Military Vehicles and Ground Support Equipment 1987*. There are a total of five BARVs still operational. The Centurion BARV will be replaced by a new BARV based on a Leopard 1 MBT chassis for which the Norwegian company of Hägglunds Moelv was awarded a contract in December 1999. The first four vehicles were delivered to the UK in 2001.

Israeli Centurions

Details of Israeli Centurion MBTs are given in this section under Israel. Details of their armoured personnel carrier versions are given in the Armoured personnel carriers (tracked) section under Israel. As far as it is known the Centurion gun tank has now been phased out of service with Israel.

Jordanian Centurions

Centurions of the Jordanian Army have been refitted with the American General Dynamics Land Systems, AVDS-1970 series diesel engine in place of the original Meteor petrol engine. These upgraded vehicles are called the Tariq. They have also been fitted with a Belgian SABCA fire-control system which incorporates a laser range-finder manufactured under licence from the then Hughes company of the USA. This system is similar to the SABCA fire-control system installed in the Leopard 1 MBTs used by Australia, Belgium and

Status

Production complete. Centurion MBTs and variants are still in service with the following countries:

Country	Quantity	Comment
Israel	1,000	most converted for specialist roles
Jordan	293	all 105 mm, called Tariq to be replaced by Challenger 1 MBTs (first vehicles delivered in late 1999)
Singapore	12+	based in Brunei and Taiwan, some sources state 63 vehicles
South Africa	224	all 105 mm, called Olifant, plus ARV

Canada. They have also been fitted with a US HR Textron turret drive and stabilisation system and hydropneumatic suspension. The last of 293 conversions to the Tariq standard was completed in 1985.

In early 1986, Kidde-Graviner of the UK was awarded a contract by Jordan to supply its Crew Bay fire and explosion suppression system for installation in all Jordanian Army Khalid, Tariq and M60A1/M60A3 MBTs plus the ENGESA EE-11 Urutu (6 × 6) APCs which were delivered in 1987.

These are now rapidly being replaced by ex-British Army Challenger 1 MBTs under a deal signed early in 1999. A total of 288 Challenger 1 MBTs are being supplied to Jordan under the local name of Al Hussein. The first 14 vehicles were delivered to Jordan by sea late in 1999. The first regiment became operational in the second half of 2000.

Singaporean Centurions

It is believed that Singapore has a minimum of 12 Centurions, six of which are based in Brunei and six in Taiwan. All of these have 105 mm guns and new diesel engines. Some reports indicate that Singapore may have as many as 63 Centurion MBTs.

South African Centurions

Details of the extensive improvements carried out by South Africa to its Centurion MBTs, now called the Olifant, are given in this section under South Africa.

Manufacturers

Leyland Motors, Leyland.
Royal Ordnance Factory, Leeds (taken over by Vickers Defence Systems in 1986).
Vickers Limited, Elswick.

UPDATED

Vickers Defence Systems MBT

Development

The Chieftain MBT (covered in detail in an earlier entry) was designed to meet the requirements of the British Army during the late 1950s. Realising that several countries would not buy such a heavy and expensive tank, Vickers (today called Vickers Defence Systems) designed a new 37 tonne MBT armed with the now BAE Systems, RO Defence 105 mm rifled gun from the Centurion and the engine, transmission, brakes, steering and fire-control system of the Chieftain. The use of these Chieftain components in a tank weighing only 37 tonnes improved its reliability and durability.

In January 1961, a team of Indian defence experts visited Germany and the United Kingdom to examine tank designs that could be produced in India. The Vickers proposal was accepted and in August 1961 an agreement was signed between Vickers and the Indian government under which the company would establish a factory in India for the production of the Vickers MBT.

The first two prototypes of the Vickers MBT were completed in 1963; one was retained in the United Kingdom and the other sent to India. Vickers began production at its Elswick works in 1964 and the first production tank was delivered to India in 1965. The Indian factory at Avadi, near Madras, was built and the first Vijayanta (the Indian name for the Vickers tank) rolled off the production line in January 1965. Some 2,200 Vijayantas have been built in India although some reports have quoted a lower figure of 1,600; production has been phased out. These are expected to be phased out of service with the Indian Army from about 2007 to 2008.

In 1968, Kuwait placed an order for 70 Vickers Mk 1 MBTs which were delivered between 1970 and 1972. These are no longer in service.

In 1975, Vickers replaced the original Leyland L60 engine with the Detroit Diesel 12V-71T turbocharged diesel engine developing 720 bhp at 2,500 rpm, but retained the TN12 transmission. This increased the power-to-weight ratio to over 18 bhp/t, as well as increasing the maximum road speed to over 50 km/h.

In 1977, Kenya placed an order for 38 Vickers Mk 3 MBTs plus three ARVs, which were all completed by late 1980. In December 1978, Kenya placed a second order for 42 Vickers Mk 3 MBTs plus four ARVs which were delivered from 1981 to 1982.

In August 1981, the Nigerian government placed an order worth some £60 million with Vickers Defence Systems for 36 Vickers Mk 3 MBTs, five ARVs and six bridgelayers. The MBTs are fitted with the Detroit Diesel 12V-71T diesel engine, the David Brown Engineering TN12 V5 transmission, the BAE Systems Avionics L23 gunner's sight which has a SIMRAD LV352 laser rangefinder, the Thales Optronics (previously Pilkington Optronics) Condor commander's day/night sight and the then GEC-Marconi Radar and Defence Systems, Defence Control Systems Division, SFCS 600 fire-control system. First production vehicles for Nigeria were completed in mid-1983. Early in 1985, Nigeria placed a repeat order for Vickers Mk 3 MBTs; first deliveries were made late in 1985 and final deliveries in late 1986. Included in this order was a further batch of five ARVs and six AVLBs.

In 1989, Vickers Defence Systems supplied Tanzania with four Mk 3 ARVs to support its Chinese-supplied NORINCO Type 59 MBTs.

Late in 1991, it was disclosed that Vickers Defence Systems had received a contract worth £150 million for the supply of a significant number of MBTs and specialised support vehicles. These were for Nigeria, with deliveries taking place from late 1992 through to late 1995.

Description (Mk 1)

The all-welded rolled steel hull of the Vickers MBT is divided into three compartments: driver's at the front, fighting in the centre and the engine and transmission at the rear.

The driver's compartment is on the right, with a single-piece hatch cover opening to the right. Forward of the cover is a single AFV No 44 Mk 2 wide-angle day periscope for closed-down driving. This can be replaced by a passive periscope for driving at night. To the left of the driver are stowed 25 rounds of 105 mm ammunition.

The all-welded steel turret has an ammunition reloading hatch in the left side and a stowage basket at the rear. The loader is seated on the left of the turret and the commander and gunner on the right.

The commander is provided with a single-piece hatch cover that opens to the rear, a sight with a magnification of ×10 and six day periscopes for all-round observation. The gunner is seated in front of and below the commander and is provided with a single day sighting telescope with a ballistic graticule. The loader has a two-piece hatch cover that opens front and rear and is provided with a day observation periscope in front of his position.

The engine, transmission, steering system and brakes are at the rear of the hull. The complete power

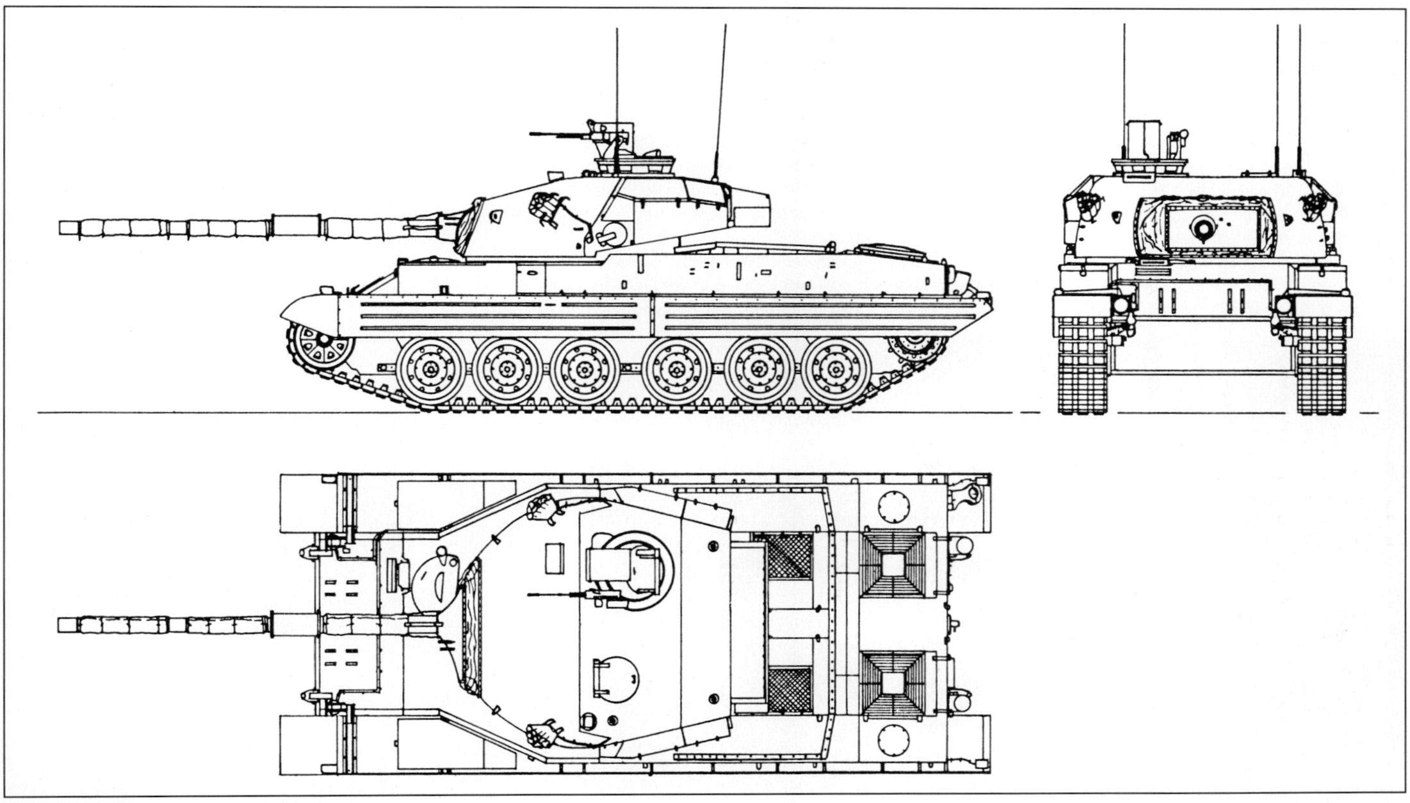

Vickers Mk 3 MBT as built for the Kenyan Army

Specifications

Model	Mk 1	Mk 3
Crew	4	4
Combat weight	38,600 kg	38,700 kg
Unloaded weight	36,000 kg	36,100 kg
Power-to-weight ratio	16.83 bhp/t	18.60 bhp/t
Ground pressure	0.87 kg/cm²	0.88 kg/cm²
Length		
(gun forwards)	9.728 m	9.788 m
(hull)	7.92 m	7.65 m
Width	3.168 m	3.168 m
Height		
(overall without AA MG)	2.64 m	3.099 m
(to turret roof)	2.438 m	2.476 m
Ground clearance	0.406 m	0.406 m
Track	2.533 m	2.533 m
Track width	520 mm	520 mm
Length of track on ground	4.28 m	4.29 m
Max speed		
(road, forward)	48 km/h	50 km/h
Fuel capacity	1,000 litres	1,000 litres
Max road range	480 km	490 km
Fording	1.143 m	1.1 m
Gradient	60%	60%
Side slope	30%	30%
Vertical obstacle	0.914 m	0.914 m
Trench	2.438 m	2.4 m
Engine	Leyland L60 Mk 4B 6-cylinder water-cooled multifuel developing 650 bhp at 2,670 rpm	Detroit Diesel 12V-71T turbocharged diesel developing 720 bhp at 2,500 rpm
Transmission	David Brown Engineering TN12 semi-automatic, 6 forward and 2 reverse speeds	
Steering	triple-differential	triple-differential
Suspension	torsion bar	torsion bar
Electrical system	24 V	24 V
Batteries	4 × 6 V, 115 Ah	4 × 12 V, 200 Ah
Armament		
(main)	1 × 105 mm L7 rifled gun	1 × 105 mm L7 rifled gun
(coaxial)	1 × 7.62 mm MG	1 × 7.62 mm MG
(anti-aircraft)	1 × 7.62 mm MG	1 × 7.62 mm MG
(ranging machine gun)	1 × 12.7 mm MG	1 × 12.7 mm MG

Model	Mk 1	Mk 3
Smoke-laying equipment	6 smoke grenade dischargers mounted either side of turret	
Ammunition		
(main)	44	50
(7.62 mm)	3,000	2,400
(12.7 mm)	600	700
Gun control equipment		
Turret power control	electric/manual	electric/manual
(by commander)	yes	yes
(by gunner)	yes	yes
Max rate of power		
(traverse)	360° in 13 s	360° in 15 s
(commander's override)	yes	yes
Gun elevation/ depression	+20/–7°	+20/–10°
Commander's fire-control override	no	yes
Gun stabiliser		
(vertical)	yes	yes
(horizontal)	yes	yes
Range setting device	yes	yes (laser range-finder)
Elevation quadrant	yes	yes
Traverse indicator	yes	yes
Armour		
Hull nose	80 mm	80 mm
Hull glacis	60 mm	60 mm
Hull sides front and intermediate	40 mm	40 mm
Hull sides rear	30 mm	30 mm
Hull top	25 mm	25 mm
Hull floor	17 mm	17 mm
Hull rear	20 mm	20 mm
Turret front	80 mm	n/a
Turret sides	40-60 mm	n/a
Turret rear	40 mm	40 mm
Turret top	25 mm	25 mm
NBC system	yes	yes
Night vision equipment	yes	yes

pack, consisting of the L60 engine, transmission, radiators, fans, coolant and oil filter, can be removed from the tank as a complete unit. Cooling air is drawn by way of the louvres, through the radiators and engine compartment and is discharged by fans through outlet louvres over the gearbox compartment.

The TN12 gearbox combines the Wilson epicyclic gear change principle with the Merritt steering system and provides six forward and two reverse speeds. Input to the transmission incorporates a centrifugal clutch and an input-driven pump provides oil pressure for gear engagement. Steering is controlled by hydraulically applied disc brakes with a mechanical interlock to prevent simultaneous engagement.

In addition to the belt-driven 24 V generator, an auxiliary three-cylinder engine drives a second generator to provide turret power and heat as well as charging the batteries when the main engine is not running.

The torsion bar suspension system either side consists of six dual rubber-tyred roadwheels with the idler at the front, drive sprocket at the rear and three track-return rollers. Suspension units one, two and six mount secondary torsion bars which are brought into action by stops mounted on the hull. The first, second and sixth roadwheel stations are provided with a hydraulic shock-absorber. The tracks are of cast manganese with removable rubber pads.

Standard equipment for the Mk 1 includes fire warning and internal firefighting systems. Optional equipment includes an NBC system, night vision equipment and a flotation screen carried collapsed around the top of the hull which would take between 15 and 30 minutes to erect. The tank would then be steered and propelled in the water by its tracks at a speed of 6.4 km/h. Some Indian vehicles have an infra-red/white searchlight mounted to the left of the main armament.

Main armament consists of a BAE Systems, RO Defence 105 mm L7A1 gun which is normally aimed using a 12.7 mm (0.50) ranging machine gun mounted coaxially with the main armament. This has a maximum range of 1,800 m and fires three-round bursts of tracer ammunition. The 105 mm gun fires the following types of BAE Systems, RO Defence ammunition: APDS-T (L28A1), APDS-T (L52A1), APFSDS-T (L64), DS/T (L45A1), HESH (L37) and Smoke (L39), or any of the other makes of 105 mm ammunition available from many sources.

The tank is fitted with the then GEC-Marconi Radar and Defence Systems, Defence Control Systems Division, EC517 gun control and stabilisation system which has three modes of operation: non-stabilised, stabilised and emergency. A coaxial 7.62 mm machine gun is mounted to the left of the main armament, and a second 7.62 mm gun is mounted on the commander's cupola. On each side of the turret are six electrically operated smoke grenade dischargers.

Additional details of Indian production of the Vickers Mk 1 and its variants are given under India earlier in this section.

Vickers Armoured Repair and Recovery Vehicle in travelling configuration carrying spare power pack

Vickers Armoured Bridgelayer in travelling configuration

Vickers Mk 1 without a thermal sleeve for the 105 mm gun and fitted with a 12.7 mm RMG and 7.62 mm coaxial and 7.62 mm anti-aircraft machine guns

Vickers Mk 3 MBT as built for the Nigerian Army **2000**/0069568

Variants

Vickers MBT Mk 2

This was a project only and basically a Mk 1 with two integral missile launchers either side of the turret rear for the Swingfire ATGW missile. One example of the Mk 2 was built.

Vickers MBT Mk 3

The all-welded rolled steel hull of the Vickers Mk 3 MBT is divided into three compartments: driver's at the front, fighting in the centre and the engine and transmission in the rear.

The driver's compartment is on the right, with a single-piece hatch cover opening to the right. Forward of the cover is a single AFV No 44 Mk 2 wide-angle day periscope for closed-down driving. This can be replaced by a passive periscope for driving at night. To the left of the driver 25 rounds of 105 mm ammunition are stowed.

The steel turret has a cast front welded to armour plate to give improved ballistic protection. It has an ammunition reloading hatch in the left side and a stowage basket on the rear. The loader sits on the left of the turret and the commander and gunner on the right.

The commander's cupola has 360° hand traverse and has a rear-opening single-piece hatch cover. The commander has a Thales Optronics (previously Pilkington Optronics) Condor combined day/night sight; this has day magnifications of ×1 and ×10 and a night magnification of ×4. Using the Condor, the commander can aim and fire the main 105 mm armament at night or in poor light. The sight has an injected ballistic graticule from the collimator, a range readout from the laser range-finder, and controls for operating the laser and for laying and firing the main armament. The commander also has six day periscopes for observation.

The gunner has a BAE Systems Avionics L23 periscopic sight, with magnifications of ×1 and ×10, incorporating an Nd:YAG laser range-finder and a ballistic graticule. The gunner's sight is linked to the gun by a temperature-compensated link bar and to a collimator in the commander's cupola. The collimator projects an illuminated ballistic graticule image into the field of view of the commander's sight when the cupola and the turret are lined up.

The loader has a single-piece hatch cover that opens forward and an AFV No 30 Mk 1 day observation periscope.

The main armament is the BAE Systems, RO Defence 105 mm L7A1 gun which fires APFSDS, APDS, HEAT, HESH, HE, smoke and canister rounds.

The tank is fitted with the then GEC-Marconi Radar and Control Systems, Defence Control Systems Division, EC620 gun control and stabilising system which has three modes of operation: non-stabilised, stabilised and emergency.

The 12.7 mm ranging machine gun is retained. It is a very effective heavy machine gun for use against lightly armoured and soft-skinned vehicles. It also provides a back-up in the event of failure of the laser range-finder or fire-control computer.

The secondary armament is a coaxially mounted 7.62 mm machine gun. A further 7.62 mm machine gun is provided on the commander's cupola, in front of and to the left of the hatch. This weapon can be elevated from −10° to the vertical, and can be mechanically cocked, aimed and electrically fired from under armour. A spotlight is fitted to the cross shaft of the machine gun mounting. Elevation is achieved by the commander's sight elevation gear; thus the machine gun and the spotlight follow the commander's line of sight through all angles of elevation.

There are 50 rounds of 105 mm ammunition carried, 18 rounds in the turret below the ring, 25 stowed horizontally in the front of the hull and seven stowed vertically in the hull centre section.

Electrically operated smoke grenade dischargers are fitted on each side of the turret.

The engine, transmission, steering system and brakes are at the rear of the hull. The power pack consisting of the engine, radiators, coolant and oil filter can be removed as a complete unit. All connections to the power pack are by means of self-sealing couplings, plugs and sockets so that the power pack can be readily removed from the vehicle for major overhauls. Power is provided by a Detroit Diesel 12V-71T, two-stroke, turbocharged diesel developing 720 bhp at 2,500 rpm. If required the complete power pack can be run outside the vehicle.

The now David Brown Engineering TN12 V5 automatic gearbox combines the Wilson epicyclic gear change principle with the Merritt steering system and provides six forward and two reverse gears. The driver can initiate an emergency sequential downchange to first and also engage one forward and one reverse gear manually. Transmission is by a centrifugal clutch; an input shaft-driven pump provides oil pressure for gear engagement. Steering is controlled by hydraulically applied disc brakes with a mechanical interlock to prevent simultaneous engagement. The final drive gear ratio has been raised to take advantage of the extra power available.

Cooling air is drawn in through inlet louvres over the engine compartment, through the radiators and engine compartment and is discharged by fans through outlet louvres over the gearbox compartment.

The torsion bar suspension system consists of six dual rubber-tyred roadwheels with the idler at the front, drive sprocket at the rear and three track-return rollers. Suspension units one, two and six incorporate secondary torsion bars which are brought into action by stops mounted on the hull. The first, second and sixth roadwheel stations have hydraulic shock-absorbers. The tracks are of cast manganese steel with removable rubber pads.

Mk 3(M)

This vehicle was developed for Malaysia and had many improvements including explosive reactive armour. It is no longer being marketed. Details were given in *Jane's Armour and Artillery 2001-2002.*

Status

Production as required. In service with the countries in the accompanying table.

Country	Model	Quantity	Comment
India	Mk 1	2,200	most locally built, L60 engine to be phased out by 2007-2008
Kenya	Mk 3	80	Detroit Diesel engine, plus 7 ARVs
Kuwait	Mk 1	70	L60 engine. None now believed to be operational
Nigeria	Mk 3	136	Detroit Diesel engine, plus 12 ARVs and 26 AVLBs
Tanzania	n/a	n/a	four ARVs with Detroit Diesel engine

Vickers Armoured Bridgelayer

The Nigerian orders for Vickers Mark 3 MBTs placed in 1981, 1985 and 1991 each included six bridgelayers. The Vickers Armoured Bridgelayer (VAB) is designed to transport, launch and recover a tank bridge which is 13.41 m long and to provide a clear span of military bridge of Class 60/70. The bridge is launched over the front of the vehicle, with the bridge launching equipment hydraulically operated with power supplied by a pump driven by a PTO from the main engine. Full details of the vehicle are given in *Jane's Military Vehicles and Logistics 2002-2003.* Alternatively, the bridgelayer can incorporate a horizontally launched bridge.

Vickers Armoured Repair and Recovery Vehicle

The Vickers Armoured Repair and Recovery Vehicle is based on the Vickers MBT Mk 3 hull. Three (two with a crane and one without) were delivered to Kenya in 1980 and a further four (two with a crane and two without) in 1981-82. Nigeria has taken delivery of 26 ARRVs since 1983. In 1989, a small number, believed to be four, of Vickers ARRVs was delivered to Tanzania. The vehicle has three compartments: winch compartment at the front with the driver to the left of the winch, the other three crew members in the centre of the vehicle and the engine and transmission at the rear.

The main winch, as developed for Chieftain, is of the capstan type with a nominal direct line pull of 25,000 kg working in conjunction with the anchor spade fitted to the front of the vehicle. Where necessary the line pull can be increased to a nominal 65 tonnes by multireeving of the rope using recovery equipment provided with the vehicle. The vehicle is provided with a full range of recovery equipment including pulleys, cables and towbars and is also equipped with a crane capable of lifting 4,000 kg to 3.62 m, which is primarily intended for removing/replacing the V800 power pack, or a TN12 transmission. The vehicle has a weight of 36,800 kg and is armed with a 7.62 mm machine gun and two six-barrelled smoke grenade dischargers. Full details of the vehicle are given in *Jane's Military Vehicles and Logistics 2002-2003.*

Vickers Anti-aircraft Tank

This is the Vickers Mk 3 chassis fitted with the now BAE Systems, RO Defence, Marksman twin 35 mm turret. The Marksman turret has only been installed for trials purposes. The only customer so far is Finland which is fitting it to T-55 tank chassis. Details of the Marksman twin 35 mm turret are given in *Jane's Land-Based Air Defence 2002-2003.*

Manufacturer

Vickers Defence Systems.

UPDATED

UNITED STATES OF AMERICA

Future Combat Systems (FCS)

Development/Description

Today, the US Army operates a fleet of General Dynamics Land Systems M1A1 and M1A2 MBTs while the US Marine Corps operates a much smaller fleet of M1A1 MBTs, all of which are armed with a 120 mm M256 smoothbore gun which is manually loaded.

Until recently, the US Army was studying five tracked vehicles to meet its possible future operational requirements.

These included a 57 ton (US) MBT armed with an 120 mm/140 mm external gun, 55 ton (US) MBT armed with an advanced weapon system, 40 ton (US) tank armed with a 120 mm smoothbore gun and two versions of a 10 ton (US) anti-tank vehicle.

Details and illustrations of these vehicles, together with other background information, was given in *Jane's Armour and Artillery 2000-2001* pages.

With the increased emphasis on rapid deployment forces, the US Army in the short term has selected the GM Defense Light Armoured Vehicle III (8 × 8) to meet its medium Interim Armoured Vehicle (IAV) requirement for its Brigade Combat Teams (BCT).

For the longer term the US Army is placing its main emphasis on the Future Combat Systems (FCS) which is expected to eventually replace a number of current US AFVs, including the General Dynamics Land Systems M1A1/M1A2 MBT.

In May 2000, the US Defence Advanced Research Projects Agency (DARPA) and the US Army selected four contractor teams for the first phase of the FCS programme.

Each of these contractor teams entered into a 'Section 845/804 Other Transaction for Prototypes' agreement for the initial 24 month conceptual design which was completed in late 2001.

These four teams are:

(1) The Boeing Co, Phantom Works, Seattle, Washington. Agreement amount (cost-shared with company) US$23,299,998. Government share US$10 million.

(2) Science Applications International Corp, McLean, Virginia. Agreement amount (cost-shared with company) US$12,830,470. Government share US$10 million.

(3) TEAM FoCuS Vision Consortium, c/o Raytheon Co, Plano, Texas. Agreement amount (cost shared with consortium) US$14 million. Government share US$10 million.

(4) Team Gladiator comprising: TRW Inc, Carson, California; Lockheed Martin Inc; Lockheed Martin Missiles and Fire Control - Dallas, Texas; CSC/Nichols Research, Huntsville, Alabama; Carnegie Mellon Research Institute, Pittsburg, Pennsylvania; Battelle Memorial Institute, Colombus, Ohio; and IITRI/AB Tech Group, Alexandria. Agreement (cost shared with consortium) US$15,461,499. Government share US$10 million.

The FCS demonstration programme is a collaborative DARPA/US Army programme to design and demonstrate future combat systems that can be the centrepiece of the Army's Objective Force vision.

During the first phase of the FCS programme, contractor teams studied two design concepts. Each of these must provide the deployability, agility, versatility, lethality, survivability and sustainability envisioned for the Objective Force.

The first concept will be for a network centric distributed force that will include a manned command and control element/personnel carrier, a robotic direct fire system, a robotic non-line of sight system, an all weather robotic sensor system, coupled with other layered sensors.

The second concept will be the teams own design approach for a system or systems.

According to DARPA, the envisioned Objective Force must provide the US Army with a significant combat overmatch against all foreseeable enemies extending through the 2025 timeframe.

It should provide for rapid unit deployment and successful offensive, defensive and stability and support operations while using smaller combat formations capable of very-high operational tempo while requiring a significant lower logistic support structure.

In addition, it should provide unrestricted transportability and deployability that would enable a brigade sized force to be deployed within 96 hours, a division within 120 hours and five divisions within 30 days.

The contractors started with a clean sheet of paper' and it has yet to be decided if the FCS will be tracked, wheeled or even on an airborne platform of some type.

The FCS will carry out a number of missions including:

- Direct fire
- Indirect fire
- Air defence
- Situational understanding
- Troop transport
- Non-lethal
- Mobility/counter mobility
- C4I
- Combat service support
- Combat support
- Support

The FCS must be deployable with current transport aircraft including the Lockheed Martin C-130 and will weigh about 20 tonnes.

It is also required to have a 50 per cent decrease in consumption, have a 100 km/h burst speed and a sustained 60 km/h cross-country speed.

On March 8th 2002, the US Department of Defense announced that it had selected The Boeing Company and Science Applications International Corporation (SAIC) as the Lead System Integrator (LSI) for the US Army's FCSs.

According to Boeing, the win also represented a key component in Boeing's vision of the integrated battlespace of the future where networked information and communications systems provide a competitive edge to soldiers in the field and commanders in the control room.

The FCS agreement, subject to negotiations, has an estimated value of US$154 million that represents a portion of the programme which is estimated at US$ 4 billion over the next five years. DAPRA, working for the US Army, is the contracting agency.

FCS is expected to draw on the considerable investment made by the UK/US in the Tactical Reconnaissance Armoured Combat Equipment Requirement (TRACER)/Future Scout Cavalry System (FSCS) for which funding ceases in mid-2002.

US Marine Corps MEFFV programme

The US Marine Corps is currently formulating its requirements for the Marine Air Ground Task Force Expeditionary Family of Fighting Vehicles (MEFFV) programme.

Technology funding is expected to commence in FY 2004 and some technologies are expected to come from the US Army's Future Combat Systems (FCS) programme.

MEFFV is expected to replace the Marine Corps current M1A1 MBTs and LAV (8 × 8) vehicles. MEFFV may well comprise seven to nine vehicles which would range in size from 30 tonne assault vehicles to 10 tonne platforms that can be transported by helicopters. A possible fielding date for the MEFFV is 2018-2020.

UPDATED

General Dynamics Land Systems M1/M1A1/M1A2 Abrams MBT

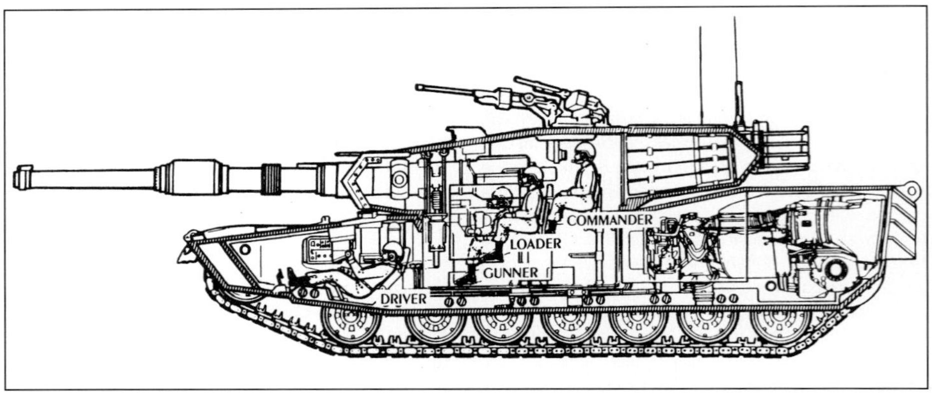

Cutaway drawing of M1A1 MBT showing position of main components. This is not representative of armour thickness

Development

The XM1 Programme was established in December 1971 and in February the following year the US Army activated a task force with the user, trainer and developer participating to formulate the concept for the new battle tank. This report was published in August 1972 and the proposed characteristics were then reviewed to eliminate unnecessary features and reduce costs to the minimum. The final programme was contained in a Development Concept Paper, as amended, which was approved by the Deputy Secretary of Defense in January 1973.

In June 1973, contracts for the prototype development validation phase of the new tank, called the XM1, were awarded to the two prime contractors. Each contractor was required to develop a tank which met the material need requirements while remaining within an average 'design-to-unit hardware' cost of US$507,790 in 1972 dollars for production tanks. Each contractor was required to deliver to the army one prototype tank, one automotive test rig and one hull and turret for ballistic tests. The Defense Division of the Chrysler Corporation (now General Dynamics Land Systems), which also then manufactured the M60 series, was awarded a contract worth US$68.1 million, and the Detroit Diesel Allison Division of the General Motors Corporation was awarded one worth US$88 million.

In February 1976, the army accepted the prototype vehicles from both US contractors and operational and engineering testing was conducted through to April 1976.

In November 1976, the Secretary of the Army announced that the Chrysler Corporation prototype had been selected to enter Full-Scale Engineering Development (FSED). Both final bids included the AGT 1500 turbine engine, in the initial offer General Motors only bid with the diesel engine.

Chrysler was awarded a three-year contract worth US$196.2 million for the FSED phase, during which 11 XM1 pilot vehicles with their associated spares were produced at the then Chrysler-run Detroit Arsenal Tank Plant. The first pilot was completed in February 1978 and the last in July 1978.

In March 1982, Chrysler sold its tank-building subsidiary (Chrysler Defense Incorporated) to General Dynamics for US$348.5 million.

Production of the M1, M1A1 and M1A2 MBT was undertaken at the Lima Army Tank Plant in Ohio, where M1s are now upgraded to the M1A2 configuration. The first production M1 MBT was completed in February 1980, and M1 assembly began at the Detroit Arsenal

Tank Plant in 1982. The original intention was to procure 3,312 M1s at a cost of US$4,900 million, but in 1978 this figure was increased to 7,058 by the end of FY88; in 1984, the figure was further increased to 7,467. Production increased from 30 vehicles a month in 1981 to 60 a month in 1982 and this was maintained until January 1984. At that time production was increased to 70 vehicles a month.

Total production of the M1 was 2,374 tanks, with last vehicles being delivered in February 1985 when production switched to the Improved M1 which was completed in May 1986.

The Improved M1 was followed by the M1A1, with first production example being completed in August 1985 and continued until early 1993.

General Dynamics Land Systems began delivering the first of 299 FY90 M1A1 MBTs in April 1991. Delivery was completed by September 1991, after which tank assembly ceased at Detroit Arsenal Tank Plant. At this point, 7,467 tanks had been delivered. The Detroit Arsenal Tank Plant closed down at the end of 1996. This was operated by General Dynamics Land Systems.

Meanwhile, a second increment of FY90 tanks began delivery in July 1991 at Lima Army Tank Plant. By August 1992, Lima delivered all 401 for this increment.

The first of five pilot M1A2 MBTs was delivered in March 1992 and in the same month the US Army approved Low-Rate Initial Production (LRIP) of 62 vehicles, the first of which was handed over to the US Army in November 1992 with the last being delivered in the Spring of 1993.

The upgrade of the M1 Abrams MBT to the enhanced M1A2 configuration was confirmed on 18 December 1992 when the Office of the Secretary of Defense authorised Phase I to upgrade 210 MBTs. The first contract, which covered production planning and the conversion of four M1 MBTs into pilot vehicles, was issued to General Dynamics Land Systems Division in February 1993.

In April 1994, the US Army Systems Acquisition Review Council approved plans to move forward with low-rate production of the M1 to the M1A2 upgraded programme.

Under Phase I of the upgrade General Dynamics delivered four prototype vehicles and 206 production vehicles, with deliveries running from October 1994 through to September 1996.

In mid-1996, General Dynamics Land Systems was awarded a US$1.5 billion multiyear contract for the upgrade of a further 580 M1 to M1A2 MBTs for the US Army to run from October 1996 through to July 2001.

It is expected that the US Army will order a total of 1,079 M1 upgrades to the M1A2 standard, this includes the two batches of 206 and 580 already ordered. These would be added to a fleet of 62 new M1A2 MBTs and 19 deployable M1A2 prototypes and pilots.

The first export customer for the M1 series MBT was Egypt, which selected the M1A1. The sale of 555 M1A1 MBTs to Egypt has been agreed and final deliveries were made in 1998. Since then another agreement has been signed with Egypt to produce another 100 M1A1 MBTs at the Egyptian Tank Plant. Additional details are given in this section under Egypt.

Saudi Arabia has a requirement for 700 new MBTs and has already ordered 315 M1A2 MBTs which have now all been delivered.

In October 1992, Kuwait selected the M1A2 over the British Vickers Defence Systems Challenger 2 MBT and a total of 218 vehicles was delivered from October 1994 through to 1996.

M1/M1A1 in Southwest Asia

In March 1991, the US DoD released provisional information on the performance of a number of key US weapon systems deployed during Operation Desert Storm. The following is an extract on the M1/M1A1 MBT.

"After 100 hours of offensive operations, the operational readiness rates for both the VII Corps and XVIII Airborne Corps exceeded the Army's 90 per cent standard. Especially noteworthy was a night move by the 3rd Armored Division covering 200 km (120 miles). None of the more than 300 tanks in the division broke down.

"Seven separate M1A1 crews reported being hit by T-72 tank rounds. These M1A1s sustained no damage, attesting to the effectiveness of our heavy armour. Other crews reported that the M1A1 thermal sight allowed them to acquire Iraqi T-72s through the smoke from oil well fires and other obscurants. The T-72 did not have the same advantage. This situation gave the Abrams a significant edge in survivability, engagement range and night manoeuvre. Additionally, tank crews reported that the M829A1 tank round was extremely effective against the T-72. In sum, the combined performance of the Abrams armour, thermal sight and ammunition attest to the systems' exceptional lethality and survivability.

"Of the 1,955 M1A1 Abrams tanks in theatre, four were disabled and four were damaged but are repairable. No M1A1 crew members were killed by enemy fire in the many tank engagements. Overall M1A1 operational readiness rates exceeded 90 per cent before and during combat."

Description (M1A1)

The hull and turret of the M1 is of advanced armour construction similar to the Chobham armour developed in the UK and gives protection against ATGWs and other battlefield weapons.

The driver is seated at the front of the vehicle in the centre and operates the vehicle from a semi-reclining position when driving with the hatch closed. Steering is accomplished by rotating a motorcycle-type T-bar which actuates the steering lever on the transmission to produce the steering speed bias of the track. At both ends of the T-bar are twist grip controls which serve as the throttle for the electronic fuel management system. The condition of fluid levels, filters, batteries, electrical connectors and circuit breakers are displayed on the driver's maintenance monitoring panel. The driver is provided with a single hatch that opens to the right with three integral day periscopes for observation when the hatch is closed. The centre day periscope can be replaced by an image intensification periscope for night driving. The driver has a 120° field of view and his night driving periscope will fit into the loader's periscope housing for night-time surveillance.

The commander and gunner are seated on the right of the turret and the loader on the left. The commander is provided with six day periscopes which cover 360°, as well as a day sight with a magnification of ×3 for the 12.7 mm M2 machine gun mounted over his position and an optical extension of the gunner's primary sight. The gunner has a primary sight (GPS) with dual day optics with a magnification of ×10 (narrow field of view), magnification of ×3 (wide field of view), close-in surveillance magnification of ×1 and an 18° field of view, thermal imaging night vision optics with a magnification of ×10 (narrow field of view), magnification of ×3 (wide field of view), sight stabilisation in elevation and a Raytheon Systems Company laser range-finder. The turret is stabilised in azimuth with a gyro reticle compensation (gyroscope) drive to keep the aim point on target in deflection.

The gunner's auxiliary day sight (a Kollmorgen Model 939) has a magnification of ×8 and an 8° field of view. The loader is provided with a periscope with a magnification of ×1 which can be traversed through 360°.

The fire-control system includes the laser range-finder, full-solution solid-state digital computer and stabilised day/thermal night sight. The stabilisation system permits accurate firing on the move and the gunner merely places his reticle in GPS (a graticule is used in the gunner's auxiliary sight) on the target and uses the laser range-finder (Nd:YAG) to determine the range. The computer then determines and applies the weapon sight offset angles necessary to obtain a target hit and the gunner opens fire. The main 120 mm smoothbore M256 armament is equipped with a muzzle reference system to measure the bend of the gun tube. Information from a wind sensor mounted on the turret roof and a pendulum static cant sensor at the turret roof centre is fed automatically to the computer together with inputs from the laser range-finder and the lead angle. The following data are manually set: battle sight range, ammunition type, barrel wear, muzzle reference compensation, barometric pressure and ammunition temperature.

The infra-red Thermal Imaging System (TIS) has been developed by the Raytheon Systems Company and produces an image by sensing the small difference in heat radiated by the objects in view. The detected energy is converted into electrical signals which are displayed on a cathode ray tube, similar to a TV picture, and the image displayed is projected into the eyepiece of the gunner's sight. In addition, the sight displays target range information and indicates if the laser range-finder has received more than one return. The operator can select a first or last return mode for lasing the target before firing. Ready to fire indication and confirmation that the systems are working properly are also provided.

The thermal imaging system generates a reticle pattern boresighted to the day graticule and to the laser range-finder. This allows the gunner to operate the TIS just as he would the day sight. The infra-red sight is based on use of common modules, components standardised to specifications of the US Army Night Vision and Electro-Optics Center.

The digital fire-control computer is produced in Canada by Computing Devices Canada (now owned by General Dynamics). The fire-control computer hardware consists of an electronics unit and a separate data entry and test panel. The electronic unit contains the computing element, the power regulators and interfaces with other elements of the fire-control system. The entry and test panel contain the keyboard, control switches and indicators, and a numeric display.

General Dynamics Land Systems M1A1 fitted with RAMTA Structures and Systems Track Width Mine Plough System (Michael Jerchel) ***2001***/0103444

General Dynamics Land Systems M1A2 Abrams fitted with German MTU EuroPowerPack during trials in the United States 0011717

The fire-control computer carries out a continuous monitoring of its internal function and memory, and provides a visual display of any malfunction. A manually initiated self-test facility gives fault diagnosis in either unit of the system to the replaceable subassembly level.

Power for the electrohydraulic gun and turret drive system is provided by an engine-driven pump through a slip-ring in the turret/hull interface, to a power valve in the manifold beneath the main armament.

The crew compartment is separated from the fuel tanks by armour bulkheads. Sliding armour doors and armour-protected boxes isolate the crew from onboard main armament ammunition explosion. An automatic Halon fire extinguishing system in the tank reacts to the outbreak of a fire in 2 ms and extinguishes fire in less than 250 ms. Ready use 120 mm ammunition is stowed in the turret bustle and in the event of penetration by a HEAT projectile, the explosion would blow off the top panels with the crew being protected by the access doors normally kept in the closed position. The loader holds the switch closed with his knee to keep the doors open. The doors close automatically when the pressure switch is relieved. In addition to venting upwards, the turret bustle magazine vents to the rear.

The M1 is powered by a Lycoming Textron AGT 1500 gas turbine. The engine operates primarily on diesel or kerosene-based fuel, but can operate on petrol during emergencies. Approximately 70 per cent of the engine accessories and components can be removed without removing the power pack from the tank. The complete power pack can be removed and replaced in less than an hour compared with 4 hours for the M60 series. The gas turbine delivers more horsepower to the sprocket than a comparable diesel engine because of the low cooling requirement. The exhaust for the gas turbine is at the rear of the hull with the air inlet on the hull top.

The engine is coupled to an Allison Transmission X-1100-3B fully automatic transmission with four forward and two reverse speeds. The transmission also provides integral brakes, variable hydrostatic steering and pivot steering.

The improved torsion bar suspension has rotary shock-absorbers at the first, second and seventh roadwheel stations with 381 mm of roadwheel travel compared with 162 mm in the M60 series. The top of the suspension is protected by vertical armoured skirts which hinge outwards to allow access to the suspension for maintenance. The drive sprocket is at the rear with the idler at the front and there are two return rollers. Standard equipment includes an AN/VDR-1 (RADIAC) Radiological Warning Device, a chemical agent detector kit, a collective NBC protection unit and personnel heaters.

The turret has been designed to accept the standard 105 mm M68 series rifled gun (which is the barrel of the now BAE Systems, RO Defence 105 mm L7 with a US developed breech mechanism) or the German Rheinmetall 120 mm smoothbore gun which has the US designation M256.

The 105 mm M68E1 gun will fire the standard ammunition as used for the M60 as well as the more recent M735 APFSDS-T round, M774 APFSDS-T and M833 APFSDS-T depleted uranium round and the M737 TPDS training round. Of the 55 rounds of 105 mm ammunition carried, 52 are compartmentalised, 44 rounds stowed in the turret bustle, 22 on the left side and 22 on the right, three stowed horizontally in spall-proof containers on the turret basket and the remaining eight stowed in the hull in a compartmentalised hull box.

General Dynamics Land Systems M1A1 fitted with deep fording kit 0018857

The USA decided to adopt the 120 mm Rheinmetall smoothbore gun for the M1. The final production decision was taken in December 1984 and the first production M1 with the 120 mm gun, designated M1A1, was completed in August 1985.

Gun/turret integration was carried out by General Dynamics. The new gun mount is similar to that employed on the basic 105 mm M1 MBT. Software changes to the computer were made on completion of firing table data collection. The ammunition racks, blow-off panels and crew compartment sliding door have been redesigned.

US-manufactured ammunition has been type-classified to support 1985 M1A1 production. Types of ammunition fired by the 120 mm gun include:

(a) M829 APFSDS-T, which has a DU penetrator. Since then the M829A1 (development designation M829E1) and M829A2 (development designation M829E2) have been placed in production
(b) M830 HEAT-MP-T (the latest round to enter production is the M830A1 which is fitted with a new projectile and fuze)
(c) M831 TP-T training round
(d) M865 TPCSDS-T training round.

In August 1998, Alliant Techsystems was awarded a US$30 million contract from the US Army for engineering and manufacturing development of the new 120 mm APFSDS-T round designated XM829E3.

This has been type classified as the M829A3 and would replace the earlier M829A1 and the M829A2 rounds. The contract also includes two options for production quantities which would bring the total value of the contract to US$127 million through to the year 2004.

These rounds are interoperable with the 120 mm smoothbore gun installed in the Leopard 2 and Leclerc MBTs, with the round consisting of the projectile and semi-combustible cartridge case.

At least two new rounds are now under development for the 120 mm armed M1A1 and M1A2 MBTs, the XM943 STAFF and the XM1007 TERM-KE.

The XM943 STAFF (Smart Target-Activated Fire and Forget) round is of the top attack type and is effective against stationary and moving targets. It is understood that work on this round has now ceased.

Late in 1998, the US Army awarded competitive contracts to Alliant Techsystems (US$37 million potential), Boeing (US$35 million potential) and Raytheon Systems Company (US$49 million potential) for initial design, development and testing of the XM1007 Tank Extended Range Munition – Kinetic Energy (TERM-KE).

TERM-KE will enable the M1A2 MBT to engage threat systems at longer battlefield ranges than is possible with current 120 mm tank rounds.

The US Army hopes to complete an entire TERM-KE mission, from target engagement to target impact, in less than 30 seconds.

The inclusion of a designation mode in the guidance system allows a helicopter or other platform to spot the intended target platform just before impact.

Provisional details released of the Alliant Techsystems XM1007 proposal shows a projectile with eight folding fins at the rear, composite projectile case and nose mounted manoeuvre mechanism, battery, guidance electronics and a MMW sensor. Late in 2001, it was stated that development of this would cease to fund other US Army programmes.

The US Army has fielded a special bunker defeat round for US Army units deployed in South Korea following the withdrawal of the M60 based M728 Combat Engineer Vehicle. The round is called the XM908 High-Explosive Obstacle-Reducing Tracer (HE-OR-T) and could be used against obstacles like the 'Dragon Teeth' in case of armed conflict.

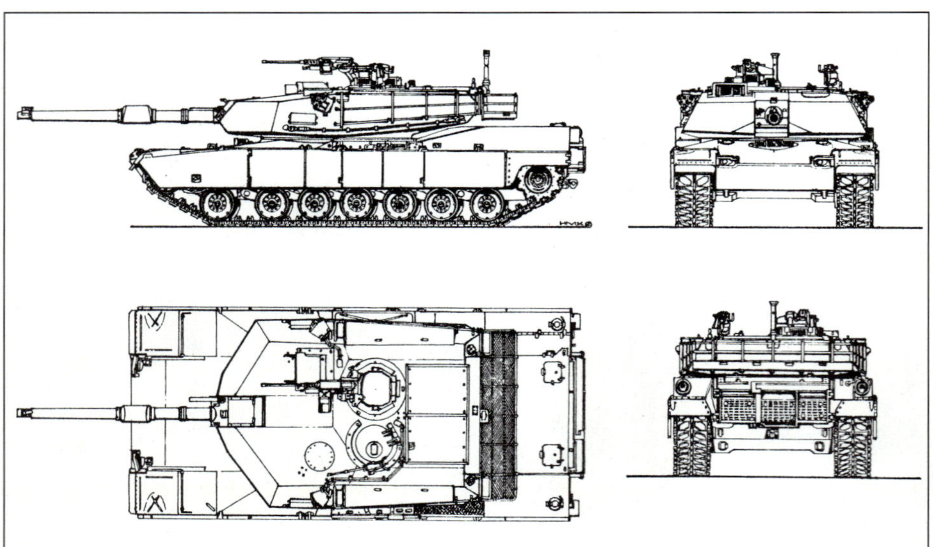

Four-view general arrangement drawing of the General Dynamics Land Systems M1A1 Abrams MBT (Henry Morshead)

Mounted coaxially to the right of the main armament is a 7.62 mm M240 machine gun, and a similar weapon skate-mounted on the left side of the turret for the loader can be elevated from -30 to +65°, total traverse being 265°. A total of 11,400 rounds of 7.62 mm machine gun ammunition is carried. Mounted at the commander's station is a standard 12.7 mm (0.50) Browning M2 HB machine gun which can be elevated from -10 to +65° and traversed through 360°. This weapon has powered and manual controls for traverse and manual controls for elevation. The 12.7 mm machine gun has electric power traverse and can be aimed and fired from within the turret. A total of 1,000 rounds of 12.7 mm machine gun ammunition is carried. Mounted on either side of the turret is a British-designed (L8A1) six-barrelled smoke grenade discharger designated the M250.

Improved M1 MBT

The first Improved M1 MBT was completed in October 1984 and final deliveries were made in May 1986. The Improved M1 is essentially the basic M1 with improved armour protection. In total, 894 Improved M1 MBTs were built before production switched solely to the M1A1 with the 120 mm gun.

General Dynamics Land Systems M1A2 Abrams MBT **2000**/0069569

M1A1 MBT

The first production M1A1 was completed in August 1985 and of the US Army buy of 8,064 tanks, 4,796 are M1A1s. In addition to the upgraded armour of the Improved M1 MBT, the M1A1 has the 120 mm M256 smoothbore gun and an integrated NBC system. This NBC system provides the crew with conditioned air for breathing and also supplies cooling or heating for the crew as required while they are wearing their protective suits and face masks. Other improvements on the M1A1 include re-indexed torsion bars, increased damping rate of shock-absorbers, strengthened compensating idler arms, modified transmission, roadwheels and final drive, redesigned loader's seat, redesigned stowage under loader's seat, new loader's shoulder guard, new tank commander's panel, electrical harness re-routed and fold-up turret bustle rack.

The M1 has three blow-off panels in the turret roof while the M1A1 has two. Of the 40 rounds of 120 mm ammunition carried by the M1A1, 34 are in the turret bustle and six in a rear hull box. In addition to the smoke grenade dischargers the M1A1 also has an engine-operated smoke-laying system.

In May 1987, General Dynamics Land Systems received a multiyear contract worth US$3.5 billion from TACOM for 3,299 M1A1 tanks to be delivered over a four-year period. The first tank under this contract was delivered in May 1987 and the last in September 1991.

Through 1991, General Dynamics delivered 7,602 Abrams MBTs to the US Army and Marine Corps. This figure includes the 25 shipped to Egypt as part of the co-production contract.

April 1993 marked the end of M1A1 MBT production with a total of 4,796 vehicles being built at the Lima, Ohio and Detroit, Michigan, Tank Plants.

M1A1 with DU armour

On 14 March 1988, the US Department of Defense announced that a new version of the M1A1 MBT (with significantly improved armour) was due to enter production in the near future.

The armour design modification to the M1A1 incorporates steel-encased depleted uranium which is two and a half times the density of steel and is already used in a broad spectrum of civilian applications. Sealed within the tank, depleted uranium has a very low level of natural radiation which is within the acceptable range established by the US Nuclear Regulatory Commission.

First production M1A1s with the new armour were completed in 1988. These tanks, which were shipped to US units in Germany, weighed about 65 tons. The heavy armour package deployed in Europe includes DU in the turret. This is also referred to as the M1A1 HA with the latter standing for heavy armour.

M1A2 MBT

The M1A2 MBT provides enhanced operational fightability over that of the M1A1 MBT. The improvements of the M1A2 include Improved Commander's Weapon Station (ICWS), Commander's Independent Thermal Viewer (CITV), an Inter-Vehicular Information System (IVIS), Position/Navigation System (POS/NAV) and several survivability initiatives.

Late in 1988, the US Army awarded General Dynamics Land Systems a contract worth over US$350 million to carry out Full-Scale Engineering Development (FSED) of the M1A2 programme.

The first production M1A2s were completed late in 1992 with IOC achieved in 1993.

In 1989, General Dynamics Land Systems awarded subcontracts for the following M1A2 subsystems:

(1) Commander's Independent Thermal Viewer (CITV) to Texas Instruments (now Raytheon Systems Company)
(2) Hull/Turret Electronics Unit (H/TEU) to Texas Instruments (now Raytheon Systems Company)
(3) Line of Sight/Dual-Axis Head Assembly (LOS/DAHA) to Hughes Aircraft (now Raytheon Systems Company)
(4) Position/Navigation Unit to Smiths Industries
(5) Direct Support Electrical System Test Set (DSESTS) to Chrysler Pentastar.

General Dynamics Land Systems manufactures the following M1A2 subsystems:

(1) Commander's Integrated Display (CID)
(2) Driver's Integrated Display (DID)
(3) Fire Control Electronics Unit (FCEU)
(4) Gunner's Control and Display Panel (GCDP)
(5) Hull Power Distribution Unit (HPDU)
(6) Hull/Turret Position Sensor (HTPS)
(7) Improved Commander's Weapon Station (ICWS)
(8) Radio Interface Unit (RIU)
(9) Improved suspension.

Under the full-scale development contract, General Dynamics built 10 M1A2 MBTs for test. In January 1990, the US government authorised the manufacturing planning Production Pilot Vehicle options. The first of these five pilot vehicles was delivered in March 1992.

Long-lead low-rate initial production funding was awarded for 62 M1A2 MBTs in April 1990 and the first of these was completed late in 1992.

The US Army M1A2 is the baseline for export versions of the M1A2 although there are differences in the armour package and communications.

By mid-1996, production of new M1A2 MBTs for Kuwait and Saudi Arabia was complete. The Lima facility is currently upgrading M1s to M1A2 standard.

The M1A2 is still being marketed, especially in the Middle East and if further orders are placed the production line can be reopened.

It should be noted that the upgrade of M1 to the M1A2 configuration includes the installation of a new turret on an upgraded chassis.

M1A2 Technology Transfer

Late in 1991, the US Army provided General Dynamics Land Systems with a Bradley Fighting Vehicle (BFV) for the purpose of transferring the M1A2 IVIS technology to this vehicle. The company delivered a BFV demonstration vehicle late in 1992 with an M1A2 IVIS system integrated.

In March 1993, General Dynamics Land Systems demonstrated for the US Army a horizontally integrated battlefield command and control system for combined weapons.

Using the General Dynamics Land Systems Inter-Vehicular Information System (IVIS), five M1A2 MBTs, one Bradley Fighting Vehicle and one OH-58D helicopter were able to maintain constant digital communication between each other and a base communication system.

The US Army awarded General Dynamics Land Systems a contract to install the IVIS system into six additional Bradleys for use with the M1A2 MBTs during IOTE (Initial Operational Training and Evaluation).

M1A2 SEP

In September 1994, General Dynamics Land Systems was awarded a contract valued at US$5 million to conduct preliminary design of the M1A2 System Enhancement Package (SEP).

This programme is designed to introduce the M1A2 SEP into the M1 to M1A2 upgrade production and refit all previously produced M1A2 MBTs to a single enhanced M1A2 configuration.

In February 1995, the requirements for the M1A2 SEP were established at the System Requirements Review/Functional Design Review. Major improvements to the M1A2 SEP tank include:

(1) Colour tactical display
(2) Keyboard
(3) Voice synthesis
(4) Digital terrain maps
(5) Increased memory
(6) Improved processors
(7) Force 21 command and control
(8) EPLARS
(9) Global Positioning System
(10) Improved intercom
(11) Standard army architecture
(12) Under armour auxiliary power unit
(13) Crew compartment cooling/conditioning

General Dynamics Land Systems M1A2 System Enhancement Package MBT **2000**/0069570

(14) Improved power distribution
(15) Improved component reliability.

In August 1995, General Dynamics Land Systems was awarded a contract valued at US$108 million to perform the final design phase of the M1A2 SEP programme.

The M1A2 SEP is also being designed to accommodate new systems being developed as part of the US Army's Horizontal Technology Integration (HTI) programme, these items include:
(1) CITV second-generation FLIR
(2) GPS second-generation FLIR
(3) Battle combat identification system
(4) Multipurpose Integrated Chemical Agent Detector (MICAD)
(5) Eye-safe laser range-finder.

The M1A2 SEP prototype MBTs were delivered for formal government testing in the summer of 1998.

General Dynamics Land Systems was awarded a long-lead production contract by the US Army in December 1997. Work on the first of 240 SEP M1A2 MBTs commenced late in 1998 and 10 M1A2 MBTs with the SEP will be delivered to the US Army each month starting in August 1999 and running through to July 2001. It is expected that the US Army will procure 1,150 SEPs by 2004.

Late in 2001, TACOM awarded General Dynamics Land Systems a contract to upgrade a further 14 M1A2 to the M1A2 SEP configuration. In addition there was an option for another 11 vehicles. Deliveries under this contract start in December 2002. This contract picked up where the 307 multiyear contract left off.

M1A1 for the US Marine Corps

The US Marine Corps originally intended to replace its then current fleet of 716 M60A1 MBTs with 476 M1A1 MBTs at a total cost of US$1,357 million.

The US Marine Corps decided to procure only 221 M1A1s, 66 in FY89 and 155 in FY90. The first M1A1 for the US Marine Corps was delivered in November 1990. Since then the US Army and US Marine Corps tanks have been identical. The tank has a Deep Water Fording Kit (DWFK), interface hardware to accept the Position Location Reporting System (PLRS) and additional tie-down points for stowage on board ship and transportation on LCACs of the US Navy.

The DWFK was developed by General Dynamics Land Systems for the US Marine Corps and was procured by TACOM. Key parts of the DWFK are two vertical air intake tubes located on the left side of the hull and a larger exhaust tube pipe at the rear. The kit can be installed by three men in less than 60 minutes and is quickly removed by traversing the turret. When fitted it allows the M1A1 to ford to a depth of 1.981 m with its turret forward.

Every US Marine Corps M1A1 is issued with a DWFK as part of its onboard equipment.

The US Marine Corps took delivery of 221 new M1A1 MBTs but in the spring of 1995 took delivery of the first of 50 M1A1 MBTs from the US Army at Anniston Army Depot, Anniston, Alabama.

Each M1A1 underwent some 62 modifications to bring the vehicles up to the 12th year 'common' configuration which is almost the same as the M1A1 common tanks that the US Marine Corps already operates.

These modifications include:
Battlefield override
Deep water fording kit
Position Locating Reporting System (PLRS)
External auxiliary power unit
Missile countermeasure device mounting kit
Digital electronic control unit.

These 50 M1A1 MBTs were all operational with the US Marine Corps by late 1995.

An additional 132 M1A1s have been transferred from the US Army to the Marine Corps, of which 84 were signed for in June 1995 and 24 in July 1995. The last consisted of a further 24 vehicles.

The US Marine Corps is considering a number of other potential upgrades to its fleet of M1A1 MBTs including thermal sights, target detection and automatic tracking, new eye-safe laser range-finder.

In September 2000, DRS Technologies announced that it had been awarded a contract by the US Marine Corps for the first phase of the M1A1 Abrams MBT Firepower Enhancement Programme (FEP).

According to the company, successful completion of this competitive integration and development phase contract could lead to a final integration and prove out phase award; this could be followed by a production contract for over 400 units to be retrofitted into their fleet of M1A1 series MBTs.

Under the terms of the US$2.2 million award, DRS will build and integrate prototypes of a second-generation FLIR thermal imaging resolution, target range, detection capability and reliability according to DRS.

Heavy Assault Bridge with 26 m bridge being extended over front of vehicle 0011709

USMC Assault Breaching Vehicle

This demonstrator for the US Marine Corps consists of an M1 chassis with its turret removed and plated over and fitted with a full width mine plough at the front, MICLIC mineclearing system on top of the hull and a lane marking system which dispenses rods into the ground either side of the hull.

It can be operated by remote control and is fitted with a commanders weapon station armed with a 12.7 mm M2 machine gun. An NBC overpressure system is not provided for the demonstrator.

Abrams AIM XXI

The first Abrams Integrated Management M1A1 MBT for the 21st century (AIM XXI) was accepted by the US Army in a joint industry and depot ceremony at the Lima Army Tank Plant in December 1996.

This was just six months after General Dynamics Land Systems was awarded a US$4.1 million firm fixed-price contract to refurbish 18 M1A1 MBTs in a Proof of Principle programme by the US Army in a teaming arrangement with the Anniston Army Depot (ANAD).

The AIM XXI programme is designed economically to rebuild and maintain the M1A1 MBT to a 'like new' condition. This will improve fleet readiness and reduce sustainment costs through a Service Life Extension Programme (SLEP). AIM XXI integrates the original equipment manufacturer and depot support of the M1A1 MBT fleet in the areas of vehicle restoration, field support and information management.

In August 1998, the US Army awarded General Dynamics a contract worth US$20.7 million to refurbish 45 M1A1 MBTs to the AIM XXI by July 2000.

In the AIM XXI programme M1A1 MBTs were completely disassembled at the Anniston Army Depot and overhauled at the Lima Army Tank Plant.

The US Army's 1st/66th Armor Battalion of the 4th Infantry Division, Food Hood, Texas, received their first eight refurbished AIM XXI M1A1 MBTs in July 2000. The first 45 AIM XXI MBTs are sufficient for one battalion and these vehicles are also referred to as M1A1D with the latter meaning digitised.

It was announced in January 2000 that TACOM had awarded contracts to General Dynamics Land Systems for 70 Abrams Integrated Management AIM XXI MBTs and procure long-lead items for an additional 135 MBTs.

The US Army fielded its first AIM XXI MBT battalions in Europe in 2001. These are described as 'digitally capable' and have factory installed wiring that will permit the bolt on of necessary components when the US Army in Europe digitises it forces. There are six M1A1 battalions in Germany.

In addition, the US Army National Guard has funded a pilot programme for five Abrams AIM XXI MBTs for delivery in FY2001 with the expectation of adding two National Guard battalions each year starting in FY2003.

Diesel-powered M1A2

Late in 1997, an M1A2 MBT fitted with a German EuroPowerPack diesel power pack, successfully completed a series of trials in the USA.

The diesel-powered M1A2 Abrams has been developed specifically for the export market.

The first series of trials consisted of four weeks of shake-down testing at the Sterling Heights and General Motors Milford test track. These trials validated the automotive integration of the EuroPowerPack into the M1A2 Abrams MBT.

The installation of the MTU EuroPowerPack is in direct response to market requirements as potential customers believe that the diesel engine is more fuel efficient than the turbine and will give an increased operating range.

The MTU EuroPowerPack consists of the MTU 883 V-12 diesel developing 1,500 hp coupled to a Renk HSWL 295 TM automatic transmission with five forward and three reverse gears together with a cooling and air filtration system.

The first production application for the MTU EuroPowerPack, which was designed by the company

Grizzly vehicle with Power-Driven Arm in travelling configuration and blade with teeth, lowered at front of hull
2001/0103445

General Dynamics Land Systems M1A1 HA deployed with KFOR (Richard Stickland) **2002**/0131045

General Dynamics Land Systems M1A1 HA deployed with KFOR (Richard Stickland) **2002**/0131046

as a private venture, is for the French Giat Industries Tropicalised Leclerc MBT which is already in service with the United Arab Emirates. The total Tropicalised Leclerc order comprises 388 Leclerc MBTs, two driver training tanks and 46 armoured recovery vehicles. For this application, a total of 480 EuroPowerPacks were being provided.

For trials purposes, the EuroPowerPack has also been installed in the export Vickers Defence Systems Challenger 2E.

Abrams/Crusader Common Engine (ACCE) programme

Late in 2000, the US Army awarded an industry partnership of Honeywell and General Electric the contract for the Abrams/Crusader Common Engine (ACCE) programme.

This three-year contract is worth about US$196.6 million and covers the development, production and testing of 24 LV100-5 gas-turbine engines, for use as a retrofit engines for the M1 series MBT as well as the new Crusader 155 mm self-propelled artillery system.

Of the US$196 million, US$144.8 million is for the common engine design, testing and development, US$28.3 million for Abrams unique design, testing and prototype and US$22.5 million for Crusader unique design, testing and prototype.

There are two phases to the ACCE. Phase I covers integration and trials, with Phase II covering production and logistics support.

The projected requirement is to re-engine 2,845 M1s over a period of eight years commencing in FY2004.

According to the US Army the main reasons for the ACCE are:

(1) Reduced operating and support costs
(2) Assisting in the weight reduction of the Crusader
(3) Optimises commonality between the two systems
(4) Leverages on commercial/military technology

Other potential money saving upgrades to the M1 series could include replacing over older analogue systems with digital systems and using commercial-off-the-shelf technology where applicable.

ACCE integration

In July 2001, US Army TACOM awarded General Dynamics Land Systems a US$92 million Abrams/Crusader Common Engine (ACCE) integration contract.

Under the terms of this contract, General Dynamics will develop hardware and software to enable the integration of the ACCE engine and adaptor kit into the Abrams M1A2 SEP and M1A1 Abrams Integrated Management (AIM) MBTs.

The contract calls for the delivery of nine prototype tanks (four M1A2 SEP and five M1A1 AIM) tanks and one M1A2 SEP Systems Integration Laboratory (SIL) by December 2004. It also includes funding for training and testing.

Other M1 developments

Late in 1986, General Dynamics and others were awarded a two-year contract from TACOM to build and demonstrate a transversely mounted engine propulsion system for the M1 MBT. This uses an advanced AGT 1500 engine which achieves 15 per cent fuel savings and develops 1,540 hp. The system was completed and successfully demonstrated in 1990. The engine is mounted transversely and aft against the transmission so saving 35 per cent of the engine compartment. Prime contractor is General Dynamics; Allison Transmission and Textron Lycoming are subcontractors.

FMC (now United Defense LP), General Motors and General Dynamics Land Systems each completed full-scale wooden mockups of a 120 mm Manned Weapon Station (MWS) for the M1 in mid-1986.

In July 1987, TACOM issued an RFP for a new track for the M1/M1A1 to replace the current T156 track which has a short life.

In July 1988, it was announced that the then Steel Products Division of the then FMC Corporation had won the competition to produce the new T158 track for the M1/M1A1 MBT. The T158 is a joint development between the US Army Tank Automotive Command, FMC Steel Products and Goodyear. The first year of a five-year contract was worth US$9.6 million to FMC with the second worth US$20.9 million. Production of the T158 was undertaken by FMC at Anniston, with Goodyear the main subcontractor for rubber components.

The new T158 track is of the double-pin type with replaceable rubber pads and is designed to last at least 3,400 km before replacement, a 300 per cent increase over the current T156 track. No modifications are required to install the T158 track on the M1/M1A1. All new production M1A1/M1A2 MBTs are fitted with the new T158 track.

An M1 with a complete in-arm hydropneumatic suspension system was completed in October 1985. Following company trials in 1986 it was sent to the Waterways Experimental Station (WES) where its test results over a course were compared to those of a standard M1 fitted with torsion bar suspension.

The US Army has trialled an M1 series MBT fitted with a hydropneumatic suspension system but, at present, there is no funding for this to move into the production phase.

M1 with APU

In September 1993, General Dynamics Land Systems was awarded a contract valued at US$9.37 million for the installation of 1,500 External Auxiliary Power Units for the M1A1 MBT.

Saudi Arabian M1A2 MBTs

The 315 M1A2 Abrams MBTs supplied to Saudi Arabia have a number of differences including the installation of Jaguar radios, an external auxiliary power unit, an advanced IVIS and a mineclearing capability.

Visually Modified M1A1

The National Training Center, Fort Irwin, California, has 28 Visually Modified M1A1 MBTs to resemble Russian designed T-80 for training purposes.

M1 Abrams Bulldozer Kit

In 1978, the Defense Division of the Chrysler Corporation (now General Dynamics Land Systems) studied the possibility of adapting the M9 bulldozer kit to fit the M1 Abrams MBT. The result of this study was a recommendation that a new kit be designed with a revised mould-board geometry to improve the driver's vision and the overall system performance and yet take advantage of the low profile of the M1 Abrams. It was also recommended that improved hydraulic components be used, allied to a rapid coupling and release mechanism.

Design of the new bulldozer kit was undertaken by the US Army Mobility Equipment Research And Development Command (MERADCOM), now the Belvoir Research and Development Center at Fort Belvoir, Virginia. In 1981, the centre awarded a production contract to Barnes and Reinecke Inc to build a prototype kit and this prototype was tested at Fort Knox during 1982.

The M1 Abrams bulldozer kit fits onto the MBT's lifting eyes and towing lugs and is powered by the tank's 24 V electrical system.

M1 AVLB

Following a competition, early in 1994, the US Army Tank Automotive Command awarded General Dynamics Land Systems a US$26 million, 39 month Engineering, Manufacturing and Development Phase contract for the Heavy Assault Bridge (HAB). Under this phase an additional two systems, now called the Wolverine, were built, with trials starting in June 1996.

Late in 1996, General Dynamics Land Systems was awarded a contract by the US Army for an additional six Heavy Assault Bridge (HAB) systems based on a modified M1 MBT chassis.

Two prototype HAB have already completed extensive user trials under the Engineering Manufacturing Development (EMD II) Phase. The major subcontractor for the HAB is MAN of Germany which developed the Leguan bridge and its associated laying system.

All HAB are conversions of existing M1 series MBTs and not new-build chassis. The latter will however be brought up to the latest M1A2 digital standard for fleet commonality by US Army depots which will then deliver these to the Lima Army Tank Plant.

The HAB is also known as the Wolverine and consists of an MLC 70 (Military Load Class) horizontally launched bridge integrated onto a modified M1 MBT chassis. When opened out, the bridge is 26 m long and can be used to span gaps of up to 24 m.

The new HAB has a number of advantages over the current US Army scissors-type Armoured Vehicle Launched Bridge (AVLB) based on the old M48/M60 MBT chassis.

Not only does it have a higher road speed, improved cross-country mobility and greater armour protection but as the bridge is laid horizontally it is more difficult to detect and the complete system is more survivable.

The computer-controlled automatic launch system allows the driver/operator or commander to launch or recover the bridge in within 5 minutes under complete armour protection using a single joystick.

Mounted at the front of the M1 chassis is a blade which is lowered to the ground to stabilise the system while the bridge is being launched. The blade can also be used for bridge cleaning.

There are no hydraulics or electronics on the bridge. A separate Launch Power Unit (LPU) is fitted to supply hydraulic power for spade and actuators on the launch mechanism with back-up capability from the chassis power pack-driven pump.

The HAB is unarmed but is fitted with a multispectrum smoke grenade system and a fog-oil-based obscuration reinforcing system.

An NBC system is fitted as standard and there are also separate automatic fire extinguishing systems for the crew, chassis power pack and launch power unit. Full details are given in *Jane's Military Vehicles and Logistics 2002-2003*.

In mid-1998, the US Army awarded General Dynamics Land Systems a US$106 million low-rate initial production contract for 29 HAB to be delivered between August 1999 and December 2001.

The HAB programme was terminated, but in mid-2000, General Dynamics was awarded another contract for 10 systems for delivery from 2001.

Grizzly (Breacher) armoured vehicle

In January 1997, the US Army Tank Automotive Command (TACOM) awarded United Defense LP, Ground Systems Division, a US$129 million contract for the Engineering and Manufacturing Development (EMD) phase of the Grizzly (Breacher) armoured vehicle.

This has been designed to clear minefields, neutralise obstacles, demolish berms and fill in anti-tank ditches for highly mobile forces.

Under the terms of this 52 month contract, the two prototypes of the Grizzly was modified in a number of key areas including the existing architecture and automatic depth control. Main emphasis will be on cost reduction, performance and production readiness.

The two modified Grizzly vehicles were delivered back to the US Army for further evaluation and testing in 1999.

The Grizzly is based on a General Dynamics Land Systems M1 MBT chassis with the same power pack and suspension (latest M1A2 standard) but has a new armoured superstructure for the crew of two that provides the same level of protection as the M1 MBT.

To carry out its battlefield role it is fitted with specialised equipment. Mounted at the front of the vehicle on the right side is the Power-Driven Arm (PDA). This is telescopic and has a maximum reach of 9.14 m and can lift 1,814 kg when fully extended. This is normally fitted with a bucket which can dig to a maximum depth of 5.27 m. When travelling, this lays back along the right side of the hull.

Mounted at the front of the hull is the MineClearing Blade (MCB) which clears surface and buried anti-tank and anti-personnel mines. For increased accuracy when ploughing it has an automatic depth control.

Armament consists of banks of electrically operated smoke grenade dischargers at the front of the hull and a remote controlled 12.7 mm machine gun on the roof.

The Grizzly is fitted with an advanced open systems vehicle architecture with 1553 databus and provision for digital battlefield command and control. An overpressure NBC system is fitted as standard.

The commander's station allows him to control the MineClearing Blade, Power-Driven Arm and Weapon Mount System using a joystick with day and night vision cameras being provided.

As of the late 2000 two examples of the Grizzly had been completed but there is no current production funding for the system. All work on this programme has now ceased.

M1/M1A1 mineclearing attachments

The Mineclearing Roller System developed by Chrysler, now General Dynamics Land Systems Division, for the M60 MBT along with an adaptor kit enabling the roller to be mounted on M1/M1A1 MBTs, has been type-classified for US Army use.

In January 1989, it was announced that RAMTA Structures and Systems of Israel was to supply 400 mineclearing blades to the US Army for use on the M1/M1A1 MBT. This design evolved from an M60/M728 Track Width Mine Plough (TWMP) design developed in the early 1970s for the US Army by Chrysler.

Full details of this system are given in *Jane's Military Vehicles and Logistics 2002-2003*.

Both of these mineclearing attachments were used by the US Army during Operation Desert Storm in early 1991.

The US Army began fielding M1 Battalion Countermine Sets in the autumn of 1990. Each set consists of four mineclearing rollers, six mounting kits, four Cleared LAne Marking Systems (CLAMS), four CLAMS mounting kits and 12 mineclearing blades.

Abrams Recovery Vehicle (ARV)

A private venture, General Dynamics Land Systems has built the prototype of the Abrams Recovery Vehicle (ARV). This competed against the BMY M88A1E1 ARV for a US Army requirement for an Improved Recovery Vehicle and, following bid evaluations, the M88A1E1 was selected. General Dynamics Land Systems is, however, still marketing the Abrams Recovery Vehicle for the export market. ARV is based on the chassis of the M1A1 and has a new superstructure. Pivoted at the front left side is a 35 ton hydraulic crane and mounted at the front of the hull is a hydraulically operated dozer/stabilising blade. The main winch has a capacity of 70 tonnes and the auxiliary winch has a capacity of 4 tonnes. Details of the vehicle are given in *Jane's Military Vehicles and Logistics 2002-2003*.

Specifications

M1A2 Abrams
Crew: 4
Combat weight: 63,086 kg
Power-to-weight ratio: 23.77 hp/t
Ground pressure: 1.08 kg/cm²
Length:
(gun forward) 9.83 m
(gun rear) 9.033 m
(hull) 7.92 m
Width:
(overall) 3.657 m
(tracks) 3.479 m
Height:
(over 12.7 mm MG) 2.885 m
(turret roof) 2.375 m
(hull) 1.727 m
Firing height: 1.89 m
Ground clearance:
(hull centre) 0.483 m
(hull sides) 0.432 m
Track width: 635 mm
Length of track on ground: 6.65 m
Max speed:
(road) 67.6 km/h
(cross-country) 48.28 km/h
(on 10% gradient) 27.36 km/h
(on 60% gradient) 6.6 km/h
Acceleration: (0-32 km/h) 7.2 s
Fuel capacity: 1,907.6 litres
Max cruising range: 426 km
Fording: 1.219 m
(with preparation) 1.98 m
Gradient: 60%
Side slope: 40%
Vertical obstacle: 1.067 m
Trench: 2.743 m
Turning radius: pivot to infinitely variable
Engine: Textron Lycoming AGT 1500 gas turbine developing 1,500 hp at 30,000 rpm (to be replaced by Abrams/Crusader Common Engine)
Transmission: Allison Transmission X-1100-3B automatic with 4 forward and 2 reverse gears
Final reduction ratio: 4.67:1
Braking system: hydromechanical
Suspension: advanced torsion bar
Electrical system: 24 V
Batteries: 6 × 12 V
Armament:
(main) 1 × 120 mm M256 smoothbore gun
(coaxial) 1 × 7.62 mm M240 MG
(anti-aircraft commander) 1 × 12.7 mm M2 MG
(anti-aircraft loader) 1 × 7.62 mm M240 MG
Smoke-laying equipment: 2 × 6 smoke grenade launchers plus integral engine smoke generator
Ammunition:
(main) 40
(12.7 mm) 1,000
(7.62 mm) 12,400
(smoke grenades) 24

Gun control equipment
Turret power control: electrohydraulic/manual
(by commander) yes
(by gunner) yes
Turret traverse: 360°
Max rate of power traverse:
(tracking) 4.2°/s
(slew rate with stabilisation) 42°/s
Max rate of power elevation:
(tracking) 1.4°/s
(slew rate with control handles) 22.5°/s
(slew rate with stabilisation commands) 42°/s
Gun elevation/depression: +20/-10°
(commander's override) yes
Commander's fire-control override: yes
Gun stabiliser:
(vertical) yes
(horizontal) yes
Range setting device: yes (laser)
Elevation quadrant: yes
Traverse indicator: yes
Armour: classified
NBC system: yes
Night vision equipment: yes

Status

Production of the M1, M1A1 and M1A2 is complete for home and export markets. Upgrade of M1 to M1A2 standard for the US Army continues. If further export orders for the M1A2 are obtained the production line can be restarted.

Manufacturers

Egyptian Tank Plant (M1A1).
General Dynamics Land Systems (M1A2 and upgrades).

UPDATED

Country	Model	Quantity	Comment
Egypt	M1A1	555	co-production programme (see Egypt)
Kuwait	M1A2	218	production completed 1995
Saudi Arabia	M1A2	315	production completed 1994
US Army	M1	2,374	production complete, number reduces as M1 to M1A2 programme gets under way
US Army	M1 to M1A2	750	upgraded plan under way
US Army	M1 Improved	894	production complete
US Army	M1A1	4,796	production complete
US Army	M1A2	77	production complete (includes prototypes)
US Marines	M1A1	403	production complete

General Dynamics Land Systems 120S Main Battle Tank

Development

The 120S MBT has been developed by General Dynamics Land Systems as a private venture for the export market and is a follow on to the earlier M60-2000 MBT *(Jane's Armour and Artillery 2001-2002)*.

Development of the 120S commenced in December 2000 and in August 2001 the company rolled out a fully functional prototype of the 120S MBT at their Detroit, Michigan, facility.

This made its first public appearance at the IDEF Exhibition held in Turkey in October 2001, after which it was shown to senior Turkish Land Forces Command (TLFC) officers before returning to the United States. Other countries such as Egypt have also been briefed on the 120S programme.

In the designation 120S the 120 represents the 120 mm smoothbore gun and the S represents speed and survivability. The M60 part of the designation is no longer considered relevant as so much of the tank is new and in appearance it is very similar to the M1A1 Abrams.

General Dynamics Land Systems is the overall design authority of the M60 series MBT and this company has considerable experience in the upgrading of this and other MBTs for the export market as well as the transfer of tank technology abroad.

General Dynamics Land Systems 120S MBT clearly showing the M1A1 turret **2002**/0095936

Rear view of General Dynamics Land Systems M120S MBT showing the cleaned up chassis and new ballistic side skirts **2002**/0095938

A good example is the co-production programme with Egypt for the M1A1 MBT where the Egyptian Tank Plant not only assembles the M1A1 but also makes some of the subsystems.

For the 120S programme the M1A1 turret and M60 series chassis have been leased from the US Army. To reduce overall life cycle costs, many of the subsystems are already well proven and in many cases in volume production. In the case of the chassis, for example, the tranmission, final drive, track and torsion bars are from the M1A1 MBT.

Description

The 120S MBT is essentially a combat proven M1A1 turret installed on an upgraded M60 series chassis and a detailed description of each of these is provided elswehere in this section. It has been stressed that there is no DU armour in the 120S hull or turret.

While the existing M60 turret could have been upgraded a decision was taken to fit the combat proven M1A1 turret for a number of reasons. This includes the higher level of armour protection and the fact that the 120 mm ammunition is separated in the turret bustle.

The 120S turret is armed with a 120 mm M256 smoothbore gun with a 7.62 mm M240 coaxial machine gun. The commander is provided with a 12.7 mm M2 machine gun while the loader has a 7.62 mm M240 machine gun.

A bank of six electrically operated smoke grenade launchers is mounted either side of the turret firing forwards. In addition, the 120S can also lay its own smoke screen by injecting diesel fuel into the exhaust outlet at the hull rear.

The combat proven 120 mm M256 smoothbore gun can fire a complete range of ammunition from a variety of sources including the US, France, Germany and Israel, all of which has a semi-combustible cartridge case.

The computerised fire-control system enables targets to be engaged while the 120S is moving under day and night conditions with a high first-round-hit probability.

The gunner is provided with day/FLIR stabilised sight with an eye-safe laser range-finder and 240 × 4 or 480 × 2 second-generation thermals although the former is not cleared for export to all countries at the present time.

The turret is installed on the existing M60 chassis using a simple adapter ring that allows the use of the M1A1 wire race ring with no turret modifications.

The M60 chassis has been fitted with the enhanced torsion bar system of the M1 series MBT to take account of the additional weight of the M1A1 turret as well as the additional armour that would be fitted to production vehicles. The track is the T-158 light weight.

The upgraded suspension coupled with the improved power-to-weight ratio gives a faster speed across country and a more stable firing platform as well as a better ride for the crew.

If required by the customer, the torsion bars could be replaced by hydropneumatic units for improved cross-country ride as well as saving some space inside of the hull.

The upper part of the suspension is fitted with new ballistic skirts and new sponsons are provided for additional storage such as batteries and, if required by the user, additional diesel fuel. These sponsons give a much smoother outline.

The fully functional prototype of the120S retains the standard M60 series power pack consisting of a General Dyanimics Land Systems AVDS-1790 series diesel engine developing 750 bhp coupled to an Allison X-1100-5 series automatic transmission. The final drives are from the M1A1 MBT.

Specifications

Crew: 4
Combat weight: 56,250 kg
Power-to-weight ratio: 21.2 hp/tonne
Ground pressure: 1.04 kg/cm^2
Length:
(gun forward) 9.547 m
(hull) 6.947 m
Width: 3.772 m
Height:
(gunner's sight) 2.894 m
(12.7 mm MG) 3.249 m
Ground clearance: 0.50 m
Length of track on ground: 4.262 m
Max speed:
(road) 51.6 km/h
(cross-country) 40.2 km/h
(60% gradient) 6.1 km/h
Acceleration: 0 to 32 km/h in 9.2 sec
Max road range: 443 km
Fording: 1.2 m
Gradient: 60%
Side slope: 30%
Vertical obstacle: 0.9 m
Trench: 2.6 m
Engine: General Dynamics Land Systems AVDS-1790-9 V-12 diesel developing 1,200 hp
Transmission: Allison XT1100-5 automatic with 4 forward and 1 reverse gears
Suspension: advanced torsion bar and rotary shock-absorbers
Armament:
(main) 1 × 120 mm M256 smoothbore gun
(coaxial) 1 × 7.62 mm M240 MG
(commander)1 × 12.7 mm M2 MG
(loader) 1 × 7.62 mm M240 MG
Smoke laying equipment: 2 × 6 M250 smoke grenade launchers either side of turret plus diesel fuel injected into the exhaust
Ammunition:
(120 mm) 42 (36 in turret)
(12.7 mm) 1,000
(7.62 mm) 11,400
Gun control equipment:
Turret power control:electro hydraulic/manual
(by commander) yes
(by gunner) yes
Turret traverse: 360°
Gun elevation/depression: +20/–10°
Commander's fire-control override: yes
Gun stabiliser:
(vertical) yes
(horizontal) yes
Range setting device: yes (laser)
Elevation quadrant: yes
Traverse indicator: yes
NBC system: yes
Night vision equipment: yes

It is envisioned that production vehicles would have the more powerful General Dynamics Land Systems AVDS-1790-9 diesel developing 1,200 hp.

This engine has already been produced in production quantities for the United Defense M88A2 armoured recovery vehicle and the Merkava Mk 3 MBT. General Dynamics Land Systems will offer the AVDS series engine with or without the digital electronic fuel-control system.

The existing M60 chassis has been fitted with a mock up armour as the armour type and position depends on the user's operational requirement.

General Dynamics Land Systems are offering a number of options for the 120S including alternative power packs (for example MTU diesel coupled to automatic transmission), external auxiliary power unit, appliqué digital command and control computer, thermal driver's viewer, commander's independent thermal view with first- or second-generation FLIR, thermal management system and on board Built In Test Equipment (BITE) to name but a few.

Status

Functional prototype completed in 2001.

Manufacturer

General Dynamics Land Systems

NEW ENTRY

General Dynamics Land Systems M60 series of MBT

Development

Early in 1956, it was decided that the current M48 tanks armed with a 90 mm gun should be further developed to produce an improved tank with increased operational range and mobility which would require a minimum of refuelling and servicing as well as incorporating an improved main armament.

In November 1956, a decision was taken to install an AVDS-1790-P compression ignition engine in an M48 tank, which was subsequently tested at Yuma Test Station during the summer of 1957. As a result of a meeting in February 1958, three M48 tanks (designated the M48A2E1) were rebuilt to incorporate the new power pack contemplated for use in the new tank, which had been designated the XM60.

During October and November 1958, several main armament candidates were tested at Aberdeen Proving Ground and, based on the test results, the British 105 mm L7A1 barrel, with the US T254E2 breech, was selected as the main armament for the XM60 and designated the M68 cannon.

In February 1957, the M60 (XM60) vehicle characteristics and production planning schedules were established and, in March 1959, the M60 was classified as Standard A.

The initial bid package for the production of M60 tanks was released in April 1959 and in June of the same year a production contract was awarded to Chrysler Corporation, Delaware Defense Plant, for the production of 180 M60 MBTs. In August 1959, an engineering release bid package was released for the second production buy of M60s to be built at Delaware. Subsequent production buys, beginning with October 1960 production, were made from the Chrysler Corporation Detroit Tank Plant, with production of the M60A3 completed in September 1987.

The M60 entered service with the US Army in 1960 and from October 1962 the tank was succeeded in production by the M60A1 (development designation M60E1). The first M60s had M48-type turrets with 105 mm guns. The M60A1 has the same basic chassis as the M60 but a redesigned turret with greater ballistic protection and other modifications. It carries 63 rounds of 105 mm ammunition rather than 57 as carried by the older M60.

For most of the 1960s and early 1970s production of the M60A1 was maintained at a very low rate, the minimum necessary to sustain the production base.

As a result of the Middle East War of 1973 a major effort was made to increase production of the M60A1 for two reasons: to replace M60A1 tanks supplied to Israel and to increase war reserve stocks which were then very low. It took some time to build up production of the M60A1 owing, in the main, to a shortage of hull and turret castings. By 1975, production had been boosted to 48 tanks per month, which increased to 72 a month in 1977 and 104 a month in December 1977. Peak production rate was achieved in October 1978 when 129 vehicles were completed. Production continued at a high rate until April 1979 when it started to drop and by the summer of 1980 it was running at 50 vehicles a month. The last M60A1 was completed in May 1980 and all production after this date was of the M60A3. Production of the M60A3 MBT was expected to be completed in May 1985 after over 15,000 M60 series MBTs had been built but, in May 1985, the DoD notified Congress of a letter of offer for 94 M60A3 MBTs for Egypt at a cost of US$165 million. M60A3s being produced early in 1985 were for Saudi Arabia. Taiwan also purchased M60A3 hulls from General Dynamics Land Systems. These are used in the M48H MBT programme in Taiwan, additional details of which are given under Taiwan earlier in this section.

The FY80 request was for 116 M60A3s for the US Army and a further 444 for foreign military sales, which were built between April 1981 and July 1982. There was no US Army funding for M60A3 production after FY80 but in FY81, between July and September 1982, 167 M60A3s were built for foreign military sales.

In 1988, it was stated that all M60 series tanks, including those in service with the National Guard, will be replaced by 1997.

M60 series MBTs took part in Operation Desert Storm in the spring of 1991 when they were used by Egypt (M60A1), Saudi Arabia (M60A3) and the United States Marine Corps (M60A1) with explosive reactive armour packages originally developed for the US Army.

Brazil has recently leased 91 M60A3 MBTs from the USA for a period of five years at a total cost, including training, transport and limited spare parts, of US$15 million.

In recent years the United States has supplied substantial quantities of M60 MBTs to friendly foreign governments at little or no charge other than the cost of transportation.

According to United Nations sources, the United States delivered the following quantities of M60 MBTs between 1992 and 2000 (see Table 1).

Table 1

Country	1992	1993	1994	1995	1996	1997	1998	1999	2000
Bosnia	nil	nil	nil	nil	45	nil	nil	nil	nil
Egypt	nil	nil	374	214	nil	nil	nil	nil	nil
Greece	492	671	nil	nil	nil	nil	nil	nil	nil
Jordan	nil	nil	nil	nil	50	38	nil	nil	nil
Morocco	nil	120	120	nil	nil	nil	nil	nil	nil
Portugal	nil	80	nil	13	nil	nil	nil	nil	nil
Saudi Arabia	nil	nil	148	89	124	nil	nil	nil	nil
Spain	96	311	nil	nil	nil	nil	nil	nil	nil
Taiwan	nil	nil	nil	21	107	30	120	nil	nil
Thailand	nil	nil	nil	24	nil	101	nil	98	27
Turkey	577	356	nil	nil	nil	nil	nil	nil	nil

United States Government support for the M60 series MBT ceased on 1st January 1999 and since then General Dynamics Land Systems, Customer Service & Support Company, has positioned itself to provide timely fleet management support to M60 MBT tank owners.

This includes the supply of spare parts, upgrade packages and a Logistics Automated Maintenance and Supply System (LAMSS).

Description (M60A1)

The hull of the M60A1 is made of cast-steel sections and forged floor plates welded together. It is divided into three compartments: driver's at the front, fighting in the centre and the engine and transmission at the rear.

The driver is seated at the front of the vehicle and is provided with a single-piece hatch cover that opens to the right. Three M27 day periscopes are mounted forward of his hatch and an M24 infra-red periscope can be installed in a mount in the centre of his hatch cover for driving at night. The M24 has now been replaced by the AN/VVS-2 night viewer, which is of the passive rather than infra-red type. A hull escape hatch is provided near the driver's position.

The all-cast-steel turret is in the centre of the vehicle with the loader on the left and the commander and gunner on the right. There is an external stowage basket at the rear of the turret. The loader is provided with a single-piece hatch cover that opens to the rear with an integral M37 day periscope which can be traversed through 360°.

The commander has a cupola that can be traversed through 360° by hand, a single-piece hatch cover that opens and swings to the rear, an M28C sight in the forward part and eight vision blocks for all-round observation. The M28C can be replaced by an M36 infra-red periscope or an M36E1 passive periscope for night vision. The gunner is seated in front of and below the commander and is provided with an M31 day periscope with a magnification of ×8 and an M105D day telescope with a magnification of ×8 and a 7.5° field of view. The M31 periscope can be replaced by an M32 infra-red periscope or an M35E1 passive periscope for night engagement of targets. The M17A1 or M17C range-finder has a magnification of ×10, a 4° field of view and a range of between 500 and 4,400 m.

The engine compartment at the rear of the hull is separated from the fighting compartment by a fireproof bulkhead, and is equipped with a fire extinguishing system.

The torsion bar suspension system either side consists of six dual rubber-tyred roadwheels with the idler at the front, drive sprocket at the rear and three track-return rollers. The first, second and sixth roadwheel stations are provided with a hydraulic shock-absorber.

The NBC system of the M60 is of the central air filtration type which pipes fresh air to each crew member via a tube. A full range of night vision equipment is fitted as standard including an infra-red searchlight over the 105 mm M68 main armament. The latter is either the AN/VSS-1 or the more recent AN/VSS-3A. The AN/VSS-3A can be used in both the visible or infra-red modes with three types of beam; compact, spread or variable width.

General Dynamics Land Systems Division M60A1 from above clearly showing turret and commander's cupola **2000**/0069571

M60A1 MBT fitted with dozer blade (US Army)

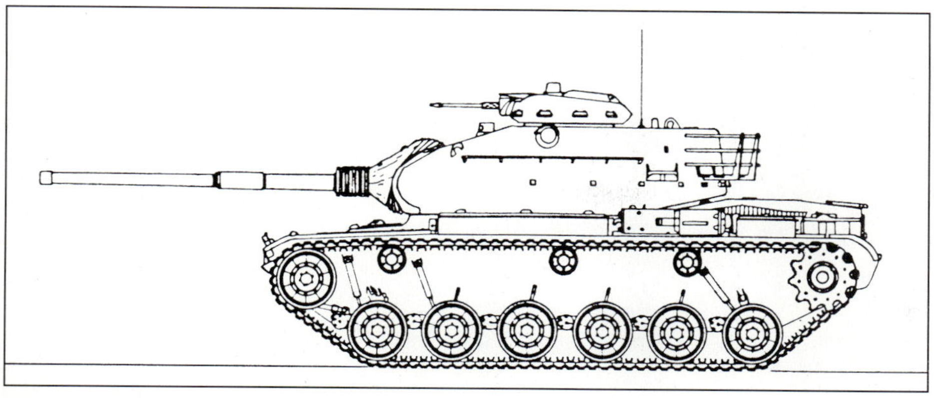

M60A1 MBT armed with a 105 mm gun

The crew compartment is provided with a heater and a RADIAC NBC detector can be fitted if required. The tank can ford to a depth of 1.219 m without preparation and with preparation to a depth of 2.438 m. It can also be fitted with an M9 bulldozer blade kit on the front of the hull for preparing fire positions and clearing obstacles.

Main armament of the M60, M60A1 and M60A3 tanks is a 105 mm M68 rifled tank gun with a bore evacuator. A well-trained crew can fire between 6 and 8 rds/min. Of the 63 rounds of ammunition carried, 26 are carried in the forward part of the hull, to the left and right of the driver's position, 13 in the turret for ready use, 21 in the turret bustle and the remaining three under the gun.

The 105 mm M68 rifled gun can fire the following types of fixed ammunition:
APDS-T (M728), APFSDS-T (M735/M735A1), APFSDS-T (M774), APDS-T (M392A2), APERS-T (M494), HEAT-T (M456 series), TP-T (M467), TP-T (M490), TPDS-T (M724), TPDS (M737), HEP-T (M393A1/M393A2), Dummy (M457), TP-T (M393A1), and Smoke WP-T (M416).

Mounted in the commander's cupola is an M85 12.7 mm (0.50) machine gun with an elevation of +60° and a depression of –15°. Mounted coaxially to the left of the main 105 mm M68 rifled armament is a 7.62 mm M73 machine gun, replaced by the 7.62 mm M240 weapon.

A number of M60A1s have been updated with the RISE engine, the main armament fully stabilised in both elevation and traverse, the fitting of a top-loading air cleaner, new T142 tracks and improved night vision equipment. Additional details of this programme are given in the entry for the M60A3.

M60A3

The M60A3 (development designation M60A1E3) is a product-improved M60A1 and some of the improvements in the tank, for example the add-on stabilisation system, the RISE engine and the smoke grenade launchers, were first fitted to the M60A1 some years ago.

The first M60A3s completed at the Detroit Arsenal Tank Plant in February 1978 were the first of a low-rate of initial production quantity of 296 M60A3s which were funded in the FY76 transitional quarter and FY77. In 1978, the M60A1 monthly production rate was 116 tanks at Detroit. The army had 5,400 M60A3 TTS tanks, 1,686 of which are new production M60A3s, 114 are M60A1 passive tanks field-retrofitted to M60A3 TTS configuration, and 3,600 are conversions carried out by the Mainz Army Depot in Germany and Anniston Army Depot in the USA. In January 1983, Saudi Arabia placed an order for 100 M60A3 MBTs at a total cost, including spares and other support equipment, of US$176 million. In 1987, it ordered 150 conversion kits to convert M60A1 tanks into M60A3s at a cost of US$120 million. These conversions included the installation of a Tank Thermal Sight (TTS).

The main improvements are in the fire-control system. A Raytheon Systems Company laser range-finder with a maximum range of 5,000 m replaces the optical range-finder and a solid-state computer replaces the mechanical computer. Following successful trials with 20 prototype units, Raytheon Systems Company was awarded its first production contract for the laser range-finder in 1976. The contract was for 69 units at a total cost of US$11 million. Range data are fed to the computer and, with other data such as crosswind velocity, air temperature, gun trunnion tilt, air density, altitude, target tracking rate and ammunition ballistics, provide the correct azimuth and elevation firing commands to the tank gunner and commander. The modifications to the fire-control system, combined with the passive night sight, improve the night-fighting capabilities of the tank.

The Raytheon Systems Company fire-control system is composed of two major subsystems, the AN/VVG-2 laser range-finder and the M21 ballistic computer system. The laser/sight subsystem is composed of two main units: a commander's integrated laser/sight with control unit and a laser electronics unit. Range may be fed automatically to the computer or, if more than one return is received, the commander may select the range return based on his assessment of the situation. Since the gunner's sight is boresighted with the laser range-finder, either the commander or the gunner can fire the laser and/or the gun.

The computer subsystem includes the ammunition selector sensors. The commander or gunner may select one of the four basic types of ammunition to be used. The computer stores solutions for six 105 mm ammunition types with the four basic types to accommodate ammunition stored in different locations. The switch at the bottom of the ammunition select unit is used to set up conditions according to whether the tank is moving or stationary. The computer is a solid-state hybrid and processes all the input data and commands the sightlines for laying the gun.

The top half of the gunner's control unit panel contains three manual inputs: air temperature, pressure altitude (slowly varying inputs, not required to be very accurate) and manual range in case the laser fails. The rate unit, crosswind sensor and the cant unit are automatic. The rate unit is used against moving targets from a stationary tank. When the tank is moving it is switched out and replaced by input from the gun stabilisation system gyros, since inertial tracking rates are required. The crosswind sensor can be stowed horizontally and is spring-mounted to prevent damage when the tank encounters low branches. Since it is the only unit not under armour, a circuit in the computer continuously monitors its output and can switch to manual input in the event of a failure. The cant unit senses gun trunnion roll.

A row of lights on the top of the gunner's control unit is for self-test. Either a green 'System OK' light is lit, or a light comes on indicating which unit is faulty. A red light indicates failure of the system and an orange light indicates that the system will continue to function, but less accurately. This self-test system removes the troubleshooting burden from the tank crew, and replacing system units in the field is simple. The bottom half of the panel, which is normally covered, is used for boresighting, zeroing, manual crosswind input (if desired) and the switch to select the four basic ammunition types.

At the beginning of FY80, AN/VGS-2 Tank Thermal Sights (TTSs) were added, which not only improved the night fighting capability of the tank but also enabled the tank to see through smoke and ground cover. The AN/VSG-2 thermal imaging equipment replaced the gunner's passive night vision periscope.

Major M60A3 improvements are as follows:

(1) Main armament fully stabilised in both elevation and traverse*
(2) Top-loading air cleaner installed*
(3) AN/VSS-1 searchlight replaced by AN/VSS-3A on passive tanks
(4) Thermal gunner's sight
(5) T97 tracks replaced by T142 tracks with removable pads*
(6) AVDS-1790-2C RISE (Reliability Improved Selected Equipment) engine fitted*
(7) Thermal sleeve for 105 mm main armament
(8) Laser range-finder
(9) 650 A oil-cooled alternator
(10) Solid-state computer
(11) Passive night vision devices, followed by thermal sights
(12) Coaxial machine gun replaced by 7.62 mm M240 weapon
(13) British-style six-barrelled smoke grenade dischargers fitted to either side of turret
(14) Engine smoke generator
(15) Automatic Halon fire extinguishing system.

* also fitted to late production M60A1s.

New production M60A3s were fitted with an engine exhaust smoke system and older tanks retrofitted. Fuel is sprayed into the exhaust manifold, creating smoke.

TACOM and General Dynamics developed an improved air filtration system for the M60 series called the Vehicle Exhaust Dust Ejector System (VEDES) which was retrofitted to the fleet starting in late 1984.

M60 MBT showing searchlight over 105 mm gun 0045820

US Army M60A3 MBT with its 105 mm M68 gun fitted with a thermal sleeve and fume extractor **2001**/0103448

Specifications

Model	M60	M60A1	M60A3
Crew	4	4	4
Combat weight	49,714 kg	52,617 kg	52,617 kg
Unloaded weight	45,631 kg	48,684 kg	48,684 kg
Power-to-weight ratio	15.08 bhp/t	14.24 bhp/t	14.24 bhp/t
Ground pressure	0.8 kg/cm²	0.87 kg/cm²	0.87 kg/cm²
Length			
(gun forward)	9.309 m	9.436 m	9.436 m
(hull)	6.946 m	6.946 m	6.946 m
Width	3.631 m	3.631 m	3.631 m
Height	3.213 m	3.27 m	3.27 m
Firing height	2.095 m	2.095 m	2.095 m
Ground clearance	0.463 m	0.463 m	0.45 m
Track	2.921 m	2.921 m	2.921 m
Track width	711 mm	711 mm	711 mm
Track adjustment	mechanical	mechanical	hydraulic
Track type	T97/T142	T97/T142	T142 replaceable pads
Length of track on ground	4.235 m	4.235 m	4.235 m
Max road speed	48.28 km/h	48.28 km/h	48.28 km/h
Fuel capacity	1,457 litres	1,420 litres	1,420 litres
Max road range	500 km	500 km	480 km
Fording	1.219 m	1.219 m	1.22 m
(with preparation)	2.438 m	2.438 m	2.4 m
Gradient	60%	60%	60%
Side slope	30%	30%	30%
Vertical obstacle	0.914 m	0.914 m	0.914 m
Trench	2.59 m	2.59 m	2.59 m
Turning radius	pivot to infinity on all models		
Engine	Continental* AVDS-1790-2A		Continental* AVDS-1790-2C
	12-cylinder air-cooled diesel developing 750 bhp at 2,400 rpm		
Transmission	General Motors Corporation, cross-drive, single-stage with 2 forward and 1 reverse ranges		
Suspension	torsion bar	torsion bar	torsion bar
Electrical system	24 V	24 V	24 V
Batteries	6 × 12 V, 100 Ah	6 × 12 V, 100 Ah	6 × 12 V, 100 Ah
Armament			
(main)	1 × 105 mm M68 rifled gun	1 × 105 mm M68 rifled gun	1 × 105 mm M68 rifled gun
(coaxial)	1 × 7.62 mm MG	1 × 7.62 mm MG	1 × 7.62 mm MG
(anti-aircraft)	1 × 12.7 mm MG	1 × 12.7 mm MG	1 × 12.7 mm MG
Smoke-laying equipment	nil	2 × 6 smoke grenade launchers either side of turret	
Ammunition			
(main)	57	63	63
(7.62 mm)	6,000	6,000	6,000
(12.7 mm)	900	900	900
Gun control equipment			
Turret power control	electrohydraulic/manual in all models		
(by commander)	yes	yes	yes
(by gunner)	yes	yes	yes
Max rate power traverse	360° in 15 s	360° in 15 s	360° in 15 s
Gun elevation/depression	+20/–9°	+20/–10°	+20/–10°
(commander's override)	yes	yes	yes
Commander's fire-control override	yes	yes	yes
Gun stabiliser			
(vertical)	no	yes**	yes
(horizontal)	no	yes**	yes
Range setting device	yes	yes	yes
Elevation quadrant	yes	yes	yes
Traverse indicator	yes	yes	yes
Armour	no details released, but armour protection is considered an improvement over the M48 series, especially over frontal area and turret		
NBC system	yes	yes	yes
Night vision equipment	yes	yes	yes

* Now General Dynamics Land Systems.
** As originally built, the M60A1 was not fitted with a gun stabilisation system but these are now retrofitted to some models.

M60A4

This was never type-classified and was a proposal to further improve the M60A1/M60A3 for the National Guard.

M60A1 with ERA

The US Army was to have fielded an Explosive Reactive Armour (ERA) package for its M60A3 MBTs but, although production did start and quantities were produced, it was decided not to field the system. The US Marine Corps fitted some of its M60A1s with the ERA package, and these were used during Operation Desert Storm early in 1991. By late 1991 however, all M60A1 MBTs had been phased out of service with the US Marine Corps.

In 1993, 10 of these were shipped from the USA to Somalia for use by Italian troops operating as part of United Nations forces.

120S MBT

Details of this vehicle, developed as a private venture by General Dynamics Land Systems. Details are given in an earlier entry. Details of the previous M60-2000 were given in *Jane's Armour and Artillery 2001-2002.*

Austrian Army M60A1 MBT Upgrade

Late in 1986, an Austrian joint venture (M60 ARGE) of Steyr-Daimler-Puch and NORICUM, was awarded a contract by the Austrian Army to upgrade 118 M60A1 MBTs to the M60A3 (Passive) configuration, to include a new fire-control system incorporating a laser range-finder system produced by the now Raytheon Systems Company and a Honeywell stabilisation system.

The new AVDS-1790-2C engine was supplied to the Austrian Army to the ARGE and purchased from the US Army's Mainz depot. The engine was upgraded to the AVDS-1790-2CA version.

In addition to upgrading the M60A1 to the M60A3 (Passive), the ARGE performed the complete overhaul of these vehicles. The ARGE purchased the conversion kits and spares for the overhaul from the FMS Corporation (now Marvin Land Systems) of Los Angeles, California, at a total cost of US$36 million.

With all of these modifications the upgraded M60A1 is equivalent to the US M60A3. The first M60A3s were handed over to the Austrian Army in mid-1988 and the last ones in November 1989. The Austrian Army already has 50 new M60A3s. With the introduction of the Leopard 2 MBT the completed Austrian fleet of M60A3 MBTs is now up for disposal.

Israeli M60 series

Outside the United States, the Israel Defence Force is the largest user of the M60 series. There is a separate entry for the M60 under Israel, as this country has carried out so many improvements to the vehicle.

M60A1 with 120 mm gun

In early 1999, the Jordanian King Hussein Main Workshops and SW Swiss Ordnance Enterprise Corp completed the prototype of a US supplied M60A1 MBT with its original 105 mm M68 rifled tank gun replaced by the SW Swiss Ordnance Enterprise Corp 120 mm smoothbore Compact Tank Gun (CTG).

The M60A1 turret was sent to the SW Swiss Ordnance Enterprise Corp facility in Thun where the new 120 mm CTG gun was fitted. The complete turret was then sent back to Jordan and integrated onto the M60A1 MBT chassis at the King Hussein Main Workshops.

The first example was completed early in 1999 with the second vehicle being completed late in 1999. According to the SW Swiss Enterprise Corp the programme was still in its early stages with no funded production programme yet established.

The 120 mm CTG is fitted with a semi-automatic vertically sliding wedge breech block with an electrical firing mechanism. For the M60A1 application the 50 calibre ordnance is not fitted with a muzzle brake.

It is fitted with two recoil brakes and one recuperator and a filament wound glass fibre composite fume extractor. If required a muzzle reference system can be fitted.

It can fire all current and projected natures of 120 mm smoothbore ammunition which are available from a wide range of sources. The 120 mm Armour-Piercing Fin-Stabilised Discarding Sabot (APFSDS) projectile has a significant increase in penetration characteristics over current 105 mm APFSDS rounds.

The 120 mm CTG was originally developed to meet requirements of the Swiss Army for a 120 mm smoothbore gun which could be retrofitted its current 105 mm armed Pz 68 MBTs with the minimum of modifications.

There are no longer any plans to retrofit the 120 mm CTG into the Pz 68 MBTs of the Swiss Army and SW Swiss Ordnance Enterprise Corp are now looking at other applications for this weapon.

The first non-Swiss application was for the Swedish Hägglunds Vehicles Vehicle CV90120-T which is based on the chassis of the CV90 series of infantry combat vehicle. Details of the CV90120-T are given earlier in the section under Sweden.

Saudi Arabian M60A1s

In early 1988, General Dynamics Services Company was awarded a contract by Saudi Arabia for the upgrading of 150 M60A1 MBTs to the M60A3 standard, with the work carried out in existing Saudi Arabian facilities over a three-year period. The first vehicle was handed over in December 1989 and final deliveries were made in May 1991. Modifications to the M60A1 included thermal sights, a new fire-control system, a ballistic computer, a laser range-finder and a thermal sleeve for the 105 mm M68 tank gun.

Mine roller system

General Dynamics Land Systems has developed a mine roller system that can be fitted to the M60A1 and M60A3 MBTs.

M9 bulldozer kit

This was developed as a depot retrofit package and is used to give already fielded M60s bulldozing capabilities similar to those of the M728 Combat Engineer Tractor. Full details of this

M60A3 MBT of the US Army fitted with attachments for a dozer blade (Peter Siebert)

M728 Combat Engineer Vehicle of the Singapore Army fitted with Pearson Engineering Track Width Mine Plough (David Boey) 0018861

bulldozer are given in *Jane's Military Vehicles and Logistics 2002-2003.*

M60 AVLB

The M60 Armoured Vehicle-Launched Bridge (AVLB) is an M60 fitted with a hydraulic launching mechanism and an aluminium scissors bridge which is launched over the front of the vehicle and weighs 14,470 kg. The bridge takes 2 minutes to lay and when open has an overall length of 19.202 m and can span a gap of up to 18.288 m. Anniston Army Depot started a programme in 1985 to convert M60A2 tanks into M60A1 AVLBs. The M60 AVLB weighs 55,746 kg with the bridge and 41,685 kg without it. Details of the M48/M60 AVLB are given in *Jane's Military Vehicles and Logistics 2002-2003.* In addition to being used by the US Army, M60 AVLBs are used by Israel, Singapore (8) and Spain. In US Army service the M48/M60 AVLB will be supplemented by the Heavy Assault Bridge (HAB) based on the M1 MBT chassis.

Upgraded M60 (AVLB-70)

Late in 1994, it was disclosed that an international industrial team had developed to the prototype stage a new mobility upgrade for not only the M48/M60 AVLB but also for the M60 series of MBTs which are still being used by many countries.

A contractor team was assembled to fund the development of an AVLB-70 prototype system with team members including Teledyne Vehicle Systems (now General Dynamics Land Systems), Allison Transmission and the Israeli company NIMDA, which have extensive experience in the upgrading of tracked armoured vehicles.

The prototype AVLB-70 was completed at the Anniston Army Depot early in 1994 and has since undergone extensive trials.

Teledyne Vehicle Systems provided the AVLB-70 package with the up-powered version of the AVDS-1790 diesel engine that had already been developed for the Israeli Merkava Mk 3 MBT (the new 1,200 hp AVDS-1790-9A replaced the 900 hp AVDS-1790-6A used in the Merkava Mk 1).

A dieselised version of the Allison Transmission X-1100-3B transmission used in the M1 MBT, designated the X-1100-5, was coupled to the new AVDS-1790 engine to form a new power pack.

The original torsion bar suspension system has been replaced by a General Dynamics Land Systems, hydropneumatic system which has been sold to a number of customers.

NIMDA provided the upgrade package with the integrated hardware in the vehicles driver's compartment.

The prototype AVLB-70 was completed in early 1994 and has already demonstrated a top speed of 64 km/h with an acceleration of 0 to 32 km/h in under eight seconds. Details of this version are given in *Jane's Military Vehicles and Logistics 2002-2003.*

M728 Combat Engineer Vehicle

The Combat Engineer Vehicle was developed under the designation T118E1 and was standardised in 1963 as the M728. It entered production in 1965 and entered service with the US Army in 1968. By 1983, over 300 vehicles had been completed, with no subsequent production. The M728 is also used by Saudi Arabia and Singapore (12).

The M728 is armed with a short-barrelled 165 mm M135 demolition gun which fires the M123A1 HEP (or HESH) round. A 7.62 mm machine gun is mounted coaxially with the main armament and a 12.7 mm machine gun is mounted in the commander's cupola for ground and anti-aircraft use.

Mounted at the front of the hull is a hydraulically operated dozer blade which is used for clearing obstacles and preparing fire positions. Pivoted towards the front of the turret is an A-frame which can be folded back over the rear of the turret when not required. The two-speed winch at the rear of the turret has a maximum capacity of 11,340 kg. An

Status

Production complete. In service with the following countries in the accompanying table.

Country	Model	Quantity	Comment
Austria	M60A3	168	for sale or disposal
Bahrain	M60A3	180	includes 20 delivered in 1995
Bosnia	M60A3	45	delivered 1996
Brazil	M60A3	91	deliveries completed
Egypt	M60A3	759	final deliveries 1987 (new)
	M60A1	700	ex-US Army, delivery 1990-91
Greece	M60A1/M60A3	669	from US Army 357 M60A1 and 312 M60A3
Iran	M60A1	355	approx 160 left
Israel	M60/M60A1/ M60A3	1,350	many local modifications (see Israel)
Jordan	M60A1/M60A3	354	
Morocco	M60A1/M60A3	300	60 M60A1 and 240 M60A3
Oman	M60A1/M60A3	79	6 M60A1 and 73 M60A3 plus some M728 CEVs
Portugal	M60A3	93	being upgraded with new fire-control system
Saudi Arabia	M60A3	460	
Singapore	n/a	n/a	has 8 M728 CEVs and 12 M60 AVLBs
Spain	M60A3	260	154 + 106 TTS
	M60A1	50	
Sudan	M60A3	20	delivered 1982, current status uncertain
Taiwan	M60A3	403	delivered between 1995 and 2000
Thailand	M60A3	125	
	M60A1	53	
Tunisia	M60A3/M60A1	84	54 M60A3 and 30 M60A1
Turkey	M60A1	274	from US Army
	M60A3	658	from US Army (with TTS)
USA	M60/M60A1/ M60A3, M728, M60 AVLB	192	approx number, no longer in front-line service. USAF has some M60A3s for EOD duties upgraded from M60A1s
Yemen	M60A1	64	supplied 1979, current status uncertain

Note: See also introduction list of M60 series MBTs recently supplied by the United Nations as the two lists do not always agree. Some of the M60s may well be used for spare parts.

US supplied M06A1 MBT fitted with SW Swiss Ordnance Enterprise Corp 120 mm Compact Tank Gun (Terry J Gander) ***2000***/0069572

infra-red searchlight similar to that fitted to the M60A1 is mounted over the main armament. The M728 weighs 52,163 kg loaded and has a crew of four: commander, gunner, loader and driver. Full details of the M728 Combat Engineer Vehicle are given in *Jane's Military Vehicles and Logistics 2002-2003.* The M728 is no longer used by the US Army. Recent information indicates that it is also used by Oman.

Manufacturer

Detroit Arsenal Tank Plantwhich was operated by General Dynamics Land Systems but closed down in December 1996.

UPDATED

General Dynamics Land Systems M48 series of MBTs

M48A5 AVLB with scissors bridge being unfolded over front of vehicle 0018864

Development

In October 1950, Detroit Arsenal Tank Plant began design work on a new tank armed with a 90 mm gun and, in December the same year, the Chrysler Corporation was awarded a letter of intent to design the new tank under the designation T48. Design work began in late December and six prototypes were built, the first being completed in December 1951. In March 1951, before the prototypes had been completed, the Fisher Body Division of the General Motors Corporation and the Ford Motor Company at Livonia were both given a production contract for the tank. Chrysler's first production M48 rolled out of the Delaware Tank Plant and was christened by Mrs George S Patton Junior on 1 July 1952.

The first production M48s were completed in 1952 and production of the M48 series amounted to 11,703 units, about 6,000 of which were built by the Chrysler Corporation at the Delaware Tank Plant which continued production until 1959. Alco Products of Schenectady also built a number of M48A2 tanks.

Many components of the M48 are also used in the M88 ARV as well as the M53 SPG, and further development of the tank resulted in the M60 series, production of which was finally completed at the Detroit Tank Plant in September 1987. Initial production of the M60 MBT was in fact undertaken at the Delaware Tank Plant but in 1959 a decision was taken that all future production would be undertaken at the Detroit Arsenal Tank Plant. The latter facility was run by General Dynamics Land Systems but this has now been closed down.

United Nations sources state that there were no exports of surplus M48 tanks by any country between 1992 and 2000.

Description (M48A3)

The cast hull of the M48 is boat-shaped with additional sections welded into position. A hull escape hatch is provided in the floor of the tank. The turret is a one-piece casting.

The driver is seated at the front of the hull in the centre and is provided with a single-piece hatch cover that opens to the right, forward of which are three M27 day periscopes. For driving at night, an M24 infra-red periscope is placed in the turntable located in the driver's hatch.

The other three crew members are in the turret with the commander and gunner on the right and the loader on the left. The commander is provided with an M1 cupola which he can traverse by hand through 360°. This cupola is equipped with five day vision blocks and an M28C day sight for controlling the 12.7 mm (0.50) machine gun. The M28C has a magnification of ×1.5 and a 48° field of view. The coincidence range-finder is operated by the tank commander and has a maximum range of 4,400 m and a magnification of ×10.

The gunner is seated forward and below the commander and has a roof-mounted periscope day sight with a ×8 magnification and a day telescope with a similar magnification linked to the main armament. The ballistic computer is an electromechanical device which has been designed to compute superelevation angles for the main armament. It receives range information in the form of shaft rotation from the range-finder. The data are then applied to the ammunition data and ballistic corrections which have been entered into the computer by the gunner. The end product is the superelevation angle which is transmitted through the ballistic drive to the periscope. The ballistic drive also transmits the same superelevation angle to the range-finder which is operated by the tank commander. The loader is provided with a single-piece hatch cover that opens to the rear. Mounted to the top rear of the turret is a dome-shaped ventilator cover and a stowage basket is provided at the rear of the turret.

The engine is mounted to the immediate rear of the bulkhead separating the engine and fighting compartments and the transmission is at the rear of the vehicle. The engine compartment is equipped with fire extinguishers but there is not a fire warning system. Power is transmitted to the final drives through the cross-drive transmission, which is a combined transmission, differential, steering and braking unit.

M48A2 showing the 90 mm gun which is fitted with a fume extractor and a 'T'-type muzzle brake ***2001***/0103449

The torsion bar suspension consists of six dual rubber-tyred roadwheels with the idler at the front, drive sprocket at the rear and five return rollers on each side. Some earlier versions of the M48 tank were fitted with a tensioning idler between the drive sprocket and the sixth roadwheel station. Hydraulic shock-absorbers are provided for the first, second and sixth roadwheel stations.

Infra-red driving lights are fitted as standard and most models have an infra-red/white light searchlight mounted over the main armament, which has a maximum range of 2,000 m. Standard equipment includes an NBC system, heaters, an external infantry telephone and provision for installing a dozer blade on the front of the hull. The M48, M48C and M48A1 can be fitted with the M8 blade weighing 3,980 kg and the M48A2, M48A3 and M48A5 with the M8A1 blade which weighs 3,810 kg.

The tank can ford to a depth of 1.219 m without preparation and to 4.438 m with a deep fording kit. Before deep fording, all the openings are sealed and an exhaust extension is fitted vertically to the right rear engine grill, the bilge pump is switched on.

Main armament of the M48A3 is an M41 (T139) 90 mm M41 gun, which consists of the barrel, evacuator chamber, blast deflector and breech mechanism assembly. The breech block is of the vertical sliding type with an inertia percussion firing mechanism. The barrel has a life of 700 equivalent full charge rounds. The following types of fixed rounds can be fired, but it should be noted that not all are currently in use: APERS-T (M580), APC-T (M82), AP-T (M77), AP-T (M318), Canister (M336), Canister (M377), HE-T (M71), HEAT-T (M431), HEAT (M348/M348A1), HVAP-T (M332A1), TP-T (M353) and Smoke WP (M313).

A total of 19 rounds of 90 mm ammunition is stowed to the left of the driver with a further 11 rounds to his right, eight horizontally on the turret floor, 16 stowed vertically around the turret ring and the remaining eight for ready use in the turret.

Mounted coaxially to the left of the main armament is a 7.62 mm M73 machine gun (earlier models have the 7.62 mm (0.30) M1919A4E1 weapon), and mounted in the commander's cupola is a 12.7 mm (0.50) Browning M2 HB machine gun which can be elevated from –10 to +60°, and can be aimed and fired from within the cupola.

Variants

M48

This was the first production model of the series and has the following distinguishing features: small driver's hatch, five track-return rollers, no tensioning idler, no dust shields on the fenders, 'T' or cylindrical-type blast

Royal Thai Army M48A5 MBTs with two roof-mounted 7.62 mm MGs
0018863

M48A5 MBT showing two roof-mounted 7.62 mm M60 series machine guns (US Army)
0002589

deflector on the barrel, and the commander's cupola has the 12.7 mm machine gun on an open mount rather than in a fully enclosed cupola as later production tanks.

M48C
This is identical to the M48 but has a hull of mild steel and is therefore unsuitable for combat. This model was used for training only and has the letter C embossed on the right front of the hull meaning that the tank is non-ballistic.

M48A1
This model has a larger driver's hatch, a fully enclosed commander's cupola, fender dust shields, a rear track idler wheel, five track-return rollers and a 'T'-type blast deflector.

M48A2
Development of this model began in 1954 under the designation T48E2. The first production order was awarded to Alco Products Incorporated of Schenectady, New York, in 1955, followed two years later by an order to the Chrysler Delaware Tank Plant (Lenape Ordnance Modification Center) at Newark, Delaware. Major differences between this and earlier models can be summarised as a fuel injection system for the engine, larger fuel tanks, an improved engine deck to minimise infra-red detection, a constant-pressure turret control system, an improved fire-control system, a modified commander's cupola, a stowage basket mounted at turret rear and the main armament fitted with a 'T'-type blast deflector. The suspension was also modified and jettisonable long-range fuel tanks could be fitted at the rear.

M48A2C
This is almost identical to the M48A2 apart from slight differences in the optical and fire-control equipment. Most models do not have the track tensioner wheel.

M48A3
This had the development designation of M48A1E2 and is basically a rebuild of the earlier M48A1 and M48A2 tanks. Major differences are the replacement of the petrol engine by the same diesel engine as installed in the M60A1 MBT, an improved fire-control system and the modification of the commander's cupola by mounting a circular ring with vision blocks between the roof of the turret and the base of the commander's cupola. Most M48A3s have only three track-return rollers and no rear idler. In addition, they have 'T'-type blast deflectors and fender dust shields.

In 1967, Bowen-McLaughlin-York was awarded a contract to remanufacture and modify 578 M48A1 tanks to M48A3 configuration. Under this programme, the company rebuilt the vehicles and performed extensive modifications in accordance with 74 US government-funded, Chrysler-designed kits. The principal change to this vehicle was the replacement of the petrol engine with the Continental (now General Dynamics Land Systems) AVDS-1790-2A diesel engine. Both remanufacture and modification were carried out in York, Pennsylvania.

The Red River Army Depot converted 400 M48A1s to M48A3 standard while the Anniston Army Depot converted 800 M48A1s to M48A3 standard.

M48A4
It was intended that turrets removed from M60s after they had been fitted with the new turret mounting the laser guided Shillelagh missile system would be fitted to older M48 chassis to be designated the M48A4. This project was cancelled after six prototypes had been completed.

M48A5
The Anniston Army Depot converted 2,064 M48A1/M48A3s to M48A5 configuration, with final deliveries taking place in December 1979. Major modifications required to convert an M48A1 to the M48A5 standard were top-loading air cleaner, top deck grill, engine and transmission shroud, gun travel lock, exhaust grills, power pack (new engine fitted), final drives, tow pintle, engine and transmission mounts, hull turret seal, torsion bar knockout, bulkhead, drain valves, fuel tank and lines, track support rollers and shield, turret basket, modified turret ammunition stowage, T142 track, double bump spring and forward arm, driver's controls, driver's escape hatch, modified hull ammunition stowage, fire extinguisher, hull armour, heater, stowage boxes, driver's periscope, gun shield and cover, turret and gun control, 105 mm M68 gun (as in M60 series), composite headlamp, 2.2 kW searchlight, nylon ballistics shield, M114 mount and M105 telescope and graticule kit, M13B1 quadrant elevation, M10A6 ballistics drive, M32 and M118 periscope mount, M13B1C computer cam kit, M17B1C range-finder and M28C periscope, M104A1 periscope mount, turret electrical kit, cupola adaptor ring and retaining ring, turret manual drive and traverse gearbox, turret stowage, searchlight stowage and cargo rack screen. Major modifications to convert the M48A3 to the M48A5 are the turret electrical kit, turret stowage, top-loading air cleaner, gun travel lock, solid-state regulator, turret basket, T142 track, ammunition stowage, 105 mm M68 gun, gun shield and cover, M87 gun mount, graticule kit and computer cam kit. The fully enclosed commander's cupola was replaced by an Israeli-designed cupola which was manufactured by Associated Steel Foundries in Israel. First US-produced cupolas were delivered in mid-1976. A 7.62 mm M60D machine gun is mounted externally at the commander's station and there are two mounting positions provided at the loader's hatch for a 7.62 mm M60D machine gun to provide suppressive fire against enemy troop and ATGW positions. These M48A5 have now been phased out of service with the National Guard and some passed on to other countries.

German M48s
The German company now called Krauss-Maffei Wegmann rebuilt 650 M48 series tanks to a new configuration known as the M48A2GA2, first deliveries of which were made in mid-1978. The modifications include the replacement of the 90 mm gun with a 105 mm L7A3 rifled gun as installed in the Leopard 1, a thermal sleeve for the main armament, a new gun travelling lock, a new commander's cupola, modified ammunition stowage, passive night vision equipment for the driver, commander and gunner and modifications to the fire-control system. Basic data are as follows: loaded weight 47,800 kg, length gun forwards 9.35 m, length gun rear 8.1 m, height 2.9 m, ammunition stowage 46 rounds of 105 mm and 4,750 rounds of 7.62 mm (2,250 rounds in turret and 2,500 rounds in the hull). German M48s also have four German 76 mm smoke grenade dischargers either side of the turret. The driver has the Elektro image intensifier periscope, while mounted over the 105 mm gun is the PZB 200 passive TV aiming and observation unit from AEG-Telefunken with a screen that can be seen by both the commander and gunner. The first M48A2GA2 was handed over in June 1978, with final deliveries being made in November 1980. GLS, a subsidiary company of Krauss-Maffei (which in January 1999 became Krauss-Maffei Wegmann), has supplied 170 of these kits to Turkey (see Turkey in this section). In addition to the modifications carried out for the German and Turkish M48s, Wegmann has suggested a number of further improvements including add-on armour for the turret, a stabilised periscopic sight for the gunner, a laser range-finder, a thermal imager, a fire-control computer, the fitting of hydraulic shock-absorbers to improve ride, new torsion bars and an NBC system.

The M48A2GA2 MBT is no longer in service with the German Army. A total of 24 M48 series tanks of the German Army have now been converted into the Keiler mineclearing system by MaK System Gesellschaft mbH (now Rheinmetall Landsysteme). Details are given later in this entry.

GLS suspension improvements
The German company GLS, a subsidiary of Krauss-Maffei (now Krauss-Maffei Wegmann), has produced a

Rheinmetall Landsysteme Keiler mineclearing tank based on an upgraded M48 tank chassis in operating configuration
2000/0044207

suspension system retrofit package to improve the cross-country mobility of M48 series MBTs. All the components envisaged for retrofitting are already used on German armoured vehicles such as the Leopard 1, Leopard 2 or Marder 1, or have been derived from such components.

The retrofit package consists of hydraulic bump stops at all roadwheel positions, hydraulic telescopic shock-absorbers at first, second and sixth roadwheel stations and torsion bars of high-strength material.

Greek M48s
The Greek Army has upgraded 400 of its M48 MBTs with the German STN ATLAS Elektronik MOLF computerised fire-control system. These were M48A5s with the 105 mm M68 rifled tank gun and diesel engine.

South Korean M48 rebuild
Many of the South Korean M48 series MBTs have been brought up to the 105 mm M48A5 standard under the leadership of Hyundai Precision & Industry Co Ltd who is also prime contractor for the South Korean K1 MBT. It is understood that a total of 597 were upgraded. Additional details were given in *Jane's Armour and Artillery 1997-98,* page 72.

Spanish M48s
Details of these vehicles are given in the Peugeot Talbot MBT conversions entry under Spain.

Taiwanese M48H
This is fully covered under Taiwan and is an M60A1 chassis with an M48 series turret and a new computerised fire-control system incorporating a laser range-finder.

Turkish M48s
Details of these vehicles and the modernisation programmes, are given under Turkey.

Iranian M48s
M48s of the Iranian Army were brought up to M48A5 standard between November 1970 and February 1972 at a vehicle manufacturing plant built in Iran by Bowen-McLaughlin-York.

Israeli M48s
Many Israeli M48s have been brought up to M60 standard by the installation of a now General Dynamics Land Systems 12-cylinder AVDS 1790 series diesel engine, a 105 mm gun and the Israeli-designed and manufactured tank commander's cupola. Israel has fitted mineclearing ploughs to both the M48 and M60 tanks and provision has also been made for fitting appliqué armour to the hull and turret. Details of this Israeli-developed Blazer explosive reactive armour and other modifications carried out to the M60s of the Israel Defence Force are given in the M60 entry under Israel.

M67/M67A1/M67A2 flame-thrower tanks
None of these is now in service.

General Dynamics M48/M60 upgrades
General Dynamics Land Systems, has developed a complete series of upgrades for the M48 and M60 series of MBTs, both of which are powered by their AVDS 1790 series diesel engines. Details of these and other upgrades offered by this company were given in *Jane's Armour and Artillery 1997-98,* pages 155 to 156.

M48 Armoured Vehicle-Launched Bridge
This is an M48 with its turret removed and fitted with a scissors bridge launched hydraulically over the front of the vehicle in three minutes. The bridge weighs 14,470 kg, has an overall length when opened out of 19.202 m, can span a gap up to 18.288 m and has a maximum capacity of 60,000 kg. The M48 AVLB has a crew of two and weighs 55,746 kg with the bridge and 41,685 kg without it. Details of the M48/M60 AVLB are given in *Jane's Military Vehicles and Logistics 2002-2003.*

In 1990, the Taiwan (Republic of China) Army purchased 12 M48A5 AVLB systems from Marvin Land Systems, Los Angeles, California. The system consisted of a converted M48A5 tank chassis with the CD-850-6A1 transmission and the AVDS-1790-2D diesel engine, a launching mechanism and the span, Class 60 aluminium scissors bridge.

Marvin Land Systems conducted engineering and chassis overhaul and conversion, bridge construction and the assembly and testing of the complete system at its Los Angeles facility.

Specifications

Model	**M48A3**	**M48A5**
Crew	4	4
Combat weight	47,173 kg	48,987 kg
Unloaded weight	44,452 kg	46,287 kg
Power-to-weight ratio	15.89 hp/t	15.89 hp/t
Ground pressure	0.83 kg/cm²	0.88 kg/cm²
Length		
(gun forward)	8.686 m	9.306 m
(hull)	6.882 m	6.419 m
Width	3.631 m	3.631 m
Height		
(overall)	3.124 m	3.086 m
Ground clearance	0.406 m	0.419 m
Track	2.921 m	2.921 m
Track width	711 mm	711 mm
Length of track on ground	4 m	4 m
Max road speed forwards	48.2 km/h	48.2 km/h
Fuel capacity	1,420 litres	1,420 litres
Max road range		
(without external tanks which could be fitted to M48A1 and M48A2)	463 km	499 km
Fording	1.219 m	1.219 m
(with preparation)	2.438 m	2.438 m
Gradient	60%	60%
Vertical obstacle	0.915 m	0.915 m
Trench	2.59 m	2.59 m
Engine (all 12-cylinder air-cooled)	AVDS-1790-2 A	AVDS-1790-2D
(type)	diesel	diesel
(output)	750/2,400 hp/rpm	750/2,400 hp/rpm
Auxiliary engine	none	none
Transmission		
(model)	CD-850-6	CD-850-6A
(type)	planetary gear shift with hydraulic torque converter with 2 forward and 1 reverse gears (low, high and reverse)	
Steering	cross-drive differential in all models	
Electrical system	24 V	24 V
Batteries	6 × 12 V	6 × 12 V
Armament		
(main)	1 × 90 mm M41 rifled gun	1 × 105 mm M68 rifled gun
(coaxial)	1 × 7.62 mm MG	1 × 7.62 mm MG
(anti-aircraft)	1 × 12.7 mm MG	2 × 7.62 mm MG
Smoke-laying equipment	none	M239 smoke grenade launchers and engine smoke-laying system
Ammunition		
(main)	62	54
(7.62 mm)	6,000	10,000
(12.7 mm)	630	none
Gun control equipment		
Turret power control	electrohydraulic with manual for emergency use	
(by commander)	yes	yes
(by gunner)	yes	yes
Rate of power traverse	max/360° in 15 s for all models	
Gun elevation/depression	+19/–9°	+19/–9°
(commander's override)	yes	yes
Commander's fire-control override	yes	yes
Gun stabiliser		
(vertical)	no	no
(horizontal)	no	no
Range-finder	M17B1C	M17A1 or M17B1C
Gunner's telescope	M105	M105D
Elevation quadrant	M13A3	M13B1
Traverse indicator	yes	yes
Armour		
Hull front	101/120 mm	101/120 mm
Hull sides front	76 mm	76 mm
Hull sides rear	51 mm	51 mm
Hull top	57 mm	57 mm
Hull floor	12.7/63 mm	12.7/63 mm
Hull rear	44 mm	44 mm
Turret front	110 mm	110 mm
Turret sides	76 mm	76 mm
Turret rear	50 mm	50 mm
Turret top	25 mm	25 mm
NBC system	yes	yes
Night vision equipment	yes	yes

M48 ARV
The German companies GLS and Jung Jungenthal offer an armoured recovery vehicle based on the M47 or M48 chassis. Details are given in *Jane's Military Vehicles and Logistics 2002-2003.*

M48 mineclearing vehicle
To meet the requirements of the German Army for a mineclearing tank based on an M48 chassis, MaK (now Rheinmetall Landsysteme) designed and built a system called the Keiler armoured mineclearing vehicle and,

following extensive trials, this was adopted by the German Army. MaK of Kiel was awarded a contract to convert an initial batch of 24 M48 chassis into the Keiler mineclearing vehicle with first vehicles being delivered in 1996. Mineclearing is carried out at a speed of 0.6 to 4.2 km/h with the clearing path being 4.7 m wide and the clearing depth 25 cm. Details of the Keiler are given in *Jane's Military Vehicles and Logistics 2002-2003*. Final deliveries were made to the German Army in March 1998.

Manufacturers

Chrysler Corporation, Delaware.
Ford Motor Company, Michigan.
Fisher Body Division of General Motors Corporation, Michigan.
Alco Products of Schenectady.
Although no longer in production, the Systems Technical Support (engineering) work for the M48A5 is still being done by General Dynamics Land Systems (formerly Chrysler Defense Incorporated).

Status

Production complete. In service with the countries in the accompanying table.

Country	Model	Quantity	Comment
Greece	see comment	714	18 M48A3, 303 M48A5 and 400 M48A5 with MOLF fire-control system fitted
Iran	M48A5	80	all 105 mm, some non-operational
Israel	M48A5	500	all 105 mm and locally modified
Jordan	M48A5	approx 200	not in front-line use
Korea, South	various	850	597 to M48A5 standard
Lebanon	M48A5	104	delivered in 1983, also 40 M48A1s in reserve
Morocco	M48A5	224	M48A3 upgraded
Pakistan	M48A5	345	all 105 mm
Portugal	M48A5	86	some from US, some from FGR
Spain	M48A5E2	164	
Taiwan	M48A1	100	up-gunned to 105 mm (this figure excludes the M48H)
Thailand	M48A5	75	some sources state 150
Tunisia	M48A3/M48A5	28	14 of each
Turkey	M48A3	2,876	2,120 upgraded to 105 mm M48A5T1/M48A5T2 in Turkey, includes 170 upgraded tanks from FGR
Vietnam	M48A5	not known	called RS48

Note: Additional details of the Spanish, Taiwanese and Turkish M48 modernisation programmes are given under their respective countries' entries earlier in this section.

UPDATED

YUGOSLAVIA, FEDERAL REPUBLIC

M-84 MBT

Development

In 1977, Yugoslavia decided to build the Russian T-72 MBT under licence and by 1979 all of the documents and drawings required to build the tank had been received.

The next two years were spent carrying out detailed changes to the drawings to suit local manufacture and incorporating some Yugoslav equipment, especially in the turret. There were some 240 main contractors involved in producing the M-84 in Yugoslavia.

Yugoslav-built prototypes were completed in 1982-83 and production got under way in 1983-84, with first production vehicles being completed late in 1984. By early 1991, it was estimated that some 500 had been built in Yugoslavia under the designation of the M-84. Before the commencement of local production, Yugoslavia took delivery of approximately 50 T-72s from the then Soviet Union enabling troop training to get under way.

Although the M-84 is based on the Russian T-72 MBT, many improvements over the original were incorporated to meet the requirements of the Yugoslav Army.

Production of the M-84A, which contains a number of significant improvements over the original production M-84, commenced in 1988.

The Yugoslav M-84 resembles the T-72 only in outward appearance. There are minor differences in the external features of the turret, with a bank of six electrically operated 81 mm smoke grenade dischargers either side of the turret and a meteorology sensor pole on the forward part of the turret roof to the immediate rear of the 125 mm gun; the absence of infra-red searchlights and the sighting device case indicate significant differences inside the turret, particularly in respect of the fire-control system.

M-84AB MBT showing additional armour on the glacis plate and the dozer blade under the nose of the vehicle
0018865

The M-84A has a computerised fire-control system, ballistic data sensors and a day/night sighting device incorporating a laser range-finder and second-generation image intensifier.

In addition to the new fire-control system, the M-84A MBT has better armour protection and numerous other improvements including a V-46TK V-12 diesel developing 1,000 hp which gives a better power-to-weight ratio than that of the T-72 and therefore superior mobility.

Close-up of the Kuwaiti M-84AB MBT

M-84 from the front, clearly showing the mast-mounted sensor on the forward part of the turret roof

Specifications

M-84A

Crew: 3

Combat weight: 42,000 kg

Power-to-weight ratio: 23.8 hp/t

Ground pressure: 0.81 kg/cm²

Length:
(gun forward) 9.53 m
(gun rear) 9.67 m
(hull) 6.86 m

Width:
(overall) 3.57 m
(over tracks) 3.37 m

Height:
(turret roof) 2.19 m
(to centre of front idler) 0.835 m
(to centre of rear drive sprocket) 0.731 m

Firing height: 1.651 m

Ground clearance: 0.426 m

Track: 2.79 m

Track width: 580 mm

Length of track on ground: 4.27 m

Max road speed:
(1st gear) 7.93 km/h
(2nd gear) 14.72 km/h
(3rd gear) 18.59 km/h
(4th gear) 23.26 km/h
(5th gear) 31.97 km/h
(6th gear) 44.21 km/h
(7th gear) 65 km/h
(reverse) 4.53 km/h

Fuel capacity: 1,450 litres (including 200 litre drums)

Range road: 500-700 km

Fording
(without preparation) 1.2 m
(with 5 min preparation) 1.8 m
(with special equipment taking 20 min to install) 5 m
(max water speed 1.5 m/s)

Gradient: 58%

Side slope: 47%

Vertical obstacle: 0.85 m

Trench: 2.8 m

Engine: V-46TK 4-stroke, 12-cylinder multifuel liquid-cooled diesel, with two turbo-compressors for supercharging, developing 1,000 hp at 2,000 rpm

Transmission: two side epicyclic gearbox with 7 forward and 1 reverse gears, hydromechanical control

Suspension: torsion bar

Electrical system: 27 V (for starter mode 48 V)

Batteries: 4 × 12 V, 140 Ah

Armament:
(main) 1 × 125 mm 2A46 smoothbore gun
(coaxial) 1 × 7.62 mm M86 MG
(anti-aircraft) 1 × 12.7 mm M87 MG

Smoke-laying equipment: 12 81 mm smoke grenade launchers (8 on left side of turret and 4 on right side)

Ammunition:
(main) 42 total of which 22 are for ready use in automatic loader
(12.7 mm) 300
(7.62 mm) 2,000

Rate of fire:
(automatic) 6 to 9 rds/min
(manual) 2 rds/min

Maximum traverse speed: 70°/s

Gun control equipment

Turret power control: power/manual

Turret traverse: 360°

Gun elevation/depression: +13° 47'/–6° 13'

Stabilisation system: yes, 2 planes

Stabilisation modes: stabilised, semi-stabilised and mechanical (manual)

Target tracking: acquisition day and night with independent line of sight

Firing data: computed automatically from following sensors, target range, target speed, turret turn rate, tank tilt, air temperature, air pressure, windspeed

Gunner's DNNS-2 day/night sight type: monocular
(measuring range) 200 to 9,995 m

Field of vision:
(day) 9 and 20°
(night) 5°

Magnification:
(day) ×3 and ×7
(night) ×8.5

Computing range: 200 to 6,000 m

Muzzle velocity variation: ±50 m/s

Side wind: ±25 m/s

Tank lateral tilt: 15°

Commander's DNKS-2 day/night sight type: day/night, passive, binocular

Day magnification: ×4.9

Night magnification: ×4.3

Field of view:
(day) 10°
(night) 10.9°

Standard equipment: intercom, radio (20 km range), NBC system, automatic fire extinguishing system, front-mounted dozer blade, camouflage, mounting points for mineclearing kit

Recent information has indicated that production of the M-84 may once again be under way in Yugoslavia and that, in addition to the known sale to Kuwait, sales may have been made to Libya and Syria. The first production M-84 ARV, designated the M-84 ABI, was completed in December 1994.

Listed below is the percentage of M-84 components that were produced in each of the republics of the former Yugoslavia:

Bosnia and Herzegovina	34
Serbia	21
Croatia	21
Slovenia	19
Macedonia	3
Montenegro	2

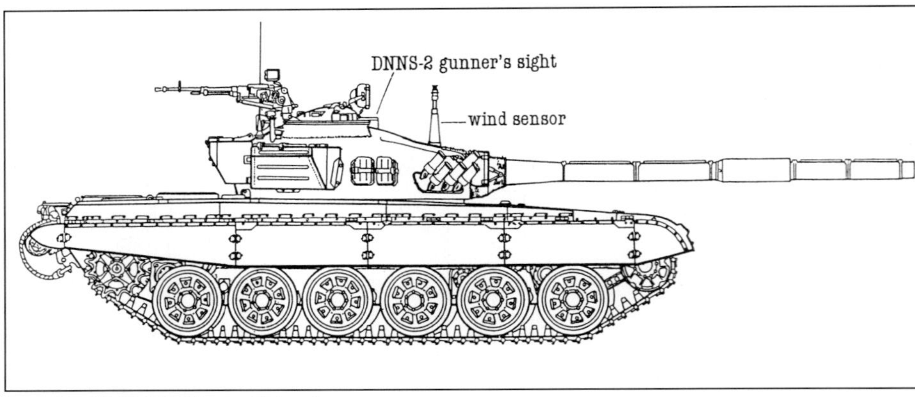

Yugoslav M-84 MBT (Steve Zaloga)

Variants

A command version is known to be in service and an armoured recovery version is in service.

M-84AB MBT

This was the version developed for Kuwait which ordered 200 vehicles in 1989, including 15 command tanks and 15 armoured recovery vehicles. At the time of the Iraqi invasion, about 15 had been delivered and more were delivered to the remains of the Kuwait Army in Saudi Arabia. Some of these vehicles took part in the recapture of Kuwait in early 1991.

It has been confirmed that the Kuwaiti vehicles have the SUV-M-84 computerised fire-control system with the gunner having a DNNS-2 day (×3 and ×7) and night (×8.5) sight with independent stabilisation in two planes and with integral laser range-finder.

Other optical devices include TNP-160 commander's periscope (two), TNPO-168 V driver's periscope, TNPA-65 auxiliary periscopes (two for commander, two for driver and one for gunner), DNKS-2 day/night periscope for commander and PPV-2 passive periscope for driver.

The M-84AB is also fitted with a GPK-59 gyro-compass, VRQ 316 HE/HG communications equipment and BCC 600 intercom system.

M-84AB MBT

This is currently being manufactured in Croatia by RH-ALAN and full details are given earlier in this section under Croatia with the latest version being the M-84A Snajper (Sniper).

Degman MBT

This is a further development of the M-84 by Croatia and details are given earlier in this section under Croatia. The first full up Degman was completed in 1999 and this is due to replace the M-84AB on the production line in the future. As of early 2002, the current status of the Degman MBT was uncertain.

Status

Believed to be in low-rate production. In service with:

Country	Quantity	Comment
Croatia	45	estimate (qv)
Kuwait	150	50% in reserve
Libya	n/k	unconfirmed user
Slovenia	44	
Syria	n/k	unconfirmed user
Yugoslavia	239	

Contractor

Yugoimport SDPR. Production is understood to be undertaken by a consortium led by the Krusevac-based company of 14 October.

UPDATED

TANKS - LIGHT TANKS (under 25,000 kg)

AUSTRIA

Steyr-Daimler-Puch SK 105 light tank

Development

The Steyr SK 105, nicknamed the Kürassier, was developed from 1965 by Saurer-Werke which was taken over in 1970 by the 6Steyr-Daimler-Puch company; it is based on a redesigned chassis of the Saurer APC. Full details of this are given in the Armoured personnel carriers (tracked) section.

The SK 105 was developed to provide the Austrian Army with an independent anti-tank weapon for difficult terrain.

The first prototype of the SK 105 light tank was completed in 1967 with the second following in 1969. Five preproduction vehicles were completed in 1971 and by mid-2001 some 600 vehicles had been built for home and export markets.

Over the last few years, Steyr-Daimler-Puch has carried out a major upgrade to the SK 105 vehicles used by the Austrian Army which refers to the vehicles as the JaPz K (Jagdpanzer Kurassier).

Some 170 vehicles have been fitted with an improved 105 mm gun and fire suppression system and are called the JaPz K A1. Between 1998 and 2000, 120 of these were fitted with a modern fire-control computer and Lansadot thermal imager and designated the JaPz K A2.

In 2000, after a gap of several years, Steyr-Daimler-Puch restarted production of its SK 105 light tank/tank destroyer for a further two export customers.

These customers are understood to be the Brazilian Marines and the Botswana Defence Force with first or follow-up deliveries taking place in late 2000.

The Brazilian Marines placed an order for 17 new SK 105 light tanks and one armoured recovery vehicle (ARV) which have been used to equip its first tank unit.

The Brazilian Marines' SK 105 will be the most advanced version of the SK 105 light tank ever built and will feature a stabilised 105 mm rifled tank gun, stabilised day/night sights for the commander and gunner and laser range-finder integrated into the gunner's sight.

The vehicle will also have an Israeli Elbit computerised fire-control system that will enable stationary and moving targets to be engaged with a high first round hit probability under day and night conditions.

In 1999, Botswana took delivery of 30 SK 105 light tanks and has more recently taken delivery of an additional 20 plus two armoured recovery vehicles and two command post vehicles. This is the first time that Steyr-Daimler-Puch has built a command post version of the SK 105 light tank.

Latest production vehicles are fitted with a more powerful 320 hp diesel engine coupled to a fully automatic ZF 6 HP 600 transmission with six forward and one reverse gears.

The first-generation infra-red night vision equipment fitted to the SK 105 has now given way to image intensification or thermal night vision equipment which is capable of detecting and identifying targets at longer ranges.

Description

The hull of the SK 105 is made of welded steel and is divided into three compartments: driver's at the front, fighting in the centre and the power pack at the rear. The SK 105 is immune to 20 mm armour-piercing rounds over its frontal arc.

Add-on armour on the front part of the chassis and turret is available as an option which provides protection against 35 mm APDS projectiles at a flank arc of ±20°.

The driver sits at the front of the vehicle on the left and is provided with a single-piece hatch cover that opens to the left. There are three day periscopes forward of this hatch cover and, in wet weather, a small windscreen with a wiper can be fitted. The driver's centre day driving periscope can be replaced by a passive night vision periscope if required. Ammunition and the vehicle's batteries are stowed to the right of the driver.

The Steyr JT 1 oscillating turret is similar to that fitted to some members of the French Giat Industries AMX-13 light tank family.

The commander is seated on the left of the turret and the gunner on the right. The commander is provided with seven day periscopes, a periscopic sight with ×1.6 magnification and a 28° field of view or alternatively ×7.5 magnification and a 9° field of view and a single-piece hatch cover that lifts and swivels to open. The commander's infra-red night sight has a magnification of ×6. The gunner has two day observation periscopes, a telescopic sight with a magnification of ×8 and an 8.5° field of view and a one-piece lifting and swivelling hatch cover. Due to the design of the oscillating turret all sights are always linked to the main and secondary armament. For engaging targets at night an infra-red periscopic sight with a magnification of ×6 and a 7° field of view is provided for the commander. A CILAS TCV 29 laser range-finder (range of 400 to 9,995 m) is mounted on the roof of the turret. The XSW-30-U 950 W infra-red/white light searchlight is fitted on the left front plate of the oscillating part of the turret. A fixed fan in the turret draws out fumes when the main or secondary armament is fired.

The engine and transmission are at the rear of the vehicle and the power pack compartment is fitted with a fire extinguisher which can be operated by hand or automatically. Combustion air is drawn in via a cyclone filter system.

The engine drives an oil pump with a variable output linked to a hydraulic motor acting on the differential which controls the tracks. By acting on the steering gear and the speed of the engine the driver can

Greif armoured recovery vehicle in travelling configuration (Stefan Marx)
0045896

Steyr SK 105 light tank in service with Argentina and in United Nations markings (Dirk Caemerlynck)
2002/0114262

Engineer Tank 4KH7FA-AVE in travelling configuration (Stefan Marx)
2001/0093241

Upgraded Steyr SK 105 light tank of the Austrian Army
0045893

Specifications
SK 105/A1
Crew: 3
Combat weight: 17,700 kg
Power-to-weight ratio: 18.1 hp/t
Ground pressure: 0.67 kg/cm²
Length
(gun forward) 7.735 m
(hull) 5.582 m
Width:
(overall) 2.5 m
(over tracks) 2.5 m
Height:
(to top of commander's cupola) 2.529 m
(to hull top) 1.413 m
Firing height: 1.965 m
Ground clearance: 0.4 m
Track: 2.12 m
Track width: 380 mm
Length of track on ground: 3.037 m
Max speed:
(forward) 70 km/h
(reverse) 7.36 km/h
Fuel capacity: 420 litres
Max range: 500 km
Fording: 1 m
Gradient: 75%
Side slope: 40%
Vertical obstacle: 0.8 m
Trench: 2.41 m
Turning radius: 8.5 m in 1st gear
Engine: Steyr 7FA 6-cylinder liquid-cooled 4-stroke turbocharged diesel developing 320 hp at 2,300 rpm
Auxiliary engine: none
Transmission: ZF 6 HP 600 automatic with 6 forward and 1 reverse gears
Steering: hydrostatic split torque type
Suspension: torsion bar
Electrical system: 24 V
Batteries: 2 × 12 V, 180 Ah
Armament:
(main) 1 × 105 mm 105 G1 rifled gun
(coaxial) 1 × 7.62 mm MG
(cupola (optional)) 1 × 7.62 mm MG
Smoke-laying equipment: 3 smoke grenade dischargers either side of turret
Ammunition:
(main) 42
(coaxial) 2,000

Gun control equipment
Turret power control: hydraulic/manual
(by commander) yes
(by gunner) yes
Max rate power; (traverse) 360° in 12-15 s
Gun elevation/depression: +12/–8°
Max power elevation/depression: 4.5-5°/s
(commander's override) yes
(commander's fire-control override) yes
Gun stabiliser:
(vertical) no
(horizontal) no
Armour:
Hull front: 20 mm
Hull sides: 14 mm
Hull top: 8 mm
Hull rear: 12 mm
Turret front: 40 mm
Turret sides: 20 mm
Turret top: 10 mm
NBC system: yes
Night vision equipment: yes

continuously adjust the speed ratio of both tracks. All turning radii can thus be obtained until pivoting on the spot occurs, with both tracks turning in opposite directions at the same speed. Since this system needs no action from the brakes, the whole power produced by the engine is constantly available at the tracks. The main brakes of the SK 105 are hydraulic and foot-operated disc brakes.

The torsion bar suspension either side consists of five dual rubber-tyred roadwheels with the drive sprocket at the rear and the idler at the front. There are three track-return rollers and the first and last roadwheel stations have hydraulic shock-absorbers. The vehicle has tracks with rubber blocks, each track having 78 links. Steel spikes can be fitted for operations in ice and snow.

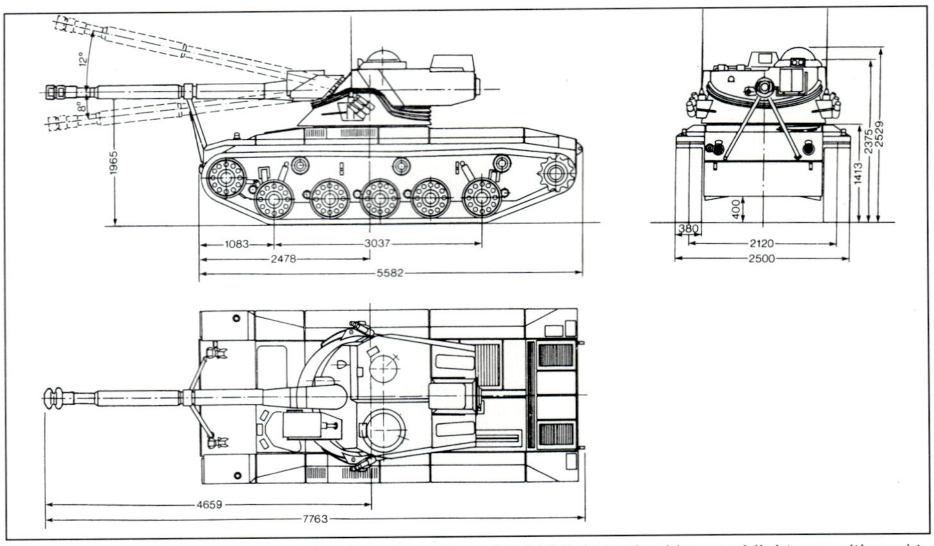

Three-view general arrangement drawing of the Steyr SK 105 light tank with searchlight repositioned to left of the 105 mm gun

The SK 105 has individual NBC protection by means of face masks and a diesel-operated heater is fitted as standard on all models. The SK 105 is not amphibious but can ford to a depth of 1 m.

All production versions of the SK 105 have a 105 mm Giat Industries rifled gun designated the 105 G1 which fires the following types of spin-stabilised fixed ammunition: HE with complete round weighing 18.5 kg and a muzzle velocity of 700 m/s; HEAT with complete round weighing 17.3 kg, a muzzle velocity of 800 m/s and a maximum effective range of 2,700 m which will penetrate 360 mm of conventional steel armour at an incidence of 0° or 150 mm of conventional steel armour at an incidence of 65°; and smoke with complete round weighing 19.1 kg and having a muzzle velocity of 695 m/s.

The 105 mm 105 G1 enables APFSDS rounds to be fired, giving a greater armour-piercing capability than the HEAT projectile.

The modifications to the SK 105 include a single-stage muzzle brake, lower extractor, an ammunition loading chute, six-round revolving magazines in the rear of the turret and the ammunition stowage racks in the hull and turret basket, as APFSDS projectiles are longer.

The APFSDS round has been developed by Giat of France and is designated the OFL 105 G1.

The OFL 105 G1 has a muzzle velocity of 1,460 m/s and will penetrate a NATO triple target at a range of 1,000 m and a NATO heavy single target at a range of 1,200 m. The projectile itself weighs 3.14 kg, 1.84 kg of which is accounted for by the penetrator.

The 105 mm gun is fed from two revolver-type magazines in the turret bustle, one on either side and each holding six rounds of 105 mm ammunition. Once the gun is fired the empty cartridge cases are ejected out of the rear of the turret through a trapdoor hinged on the left. A rate of fire of one round every 5 seconds can be achieved until the two magazines are empty. Once empty the magazines have to be refilled by hand from outside the tank. A gun barrel travelling lock is fitted at the front of the hull and this folds back onto the glacis plate when not in use.

Mounted coaxially to the right of the main armament is a 7.62 mm MG 74 (Steyr) machine gun and mounted on either side of the turret are three smoke grenade dischargers. A cupola-mounted 7.62 mm or 12.7 mm machine gun for close in and anti-aircraft defence, positioned in front of the commander's cupola, can be fitted as an option.

Steyr SK 105/A1 Light Tank

The SK 105/A1 is a further development of the SK 105 but differs in modifications of the power train and the fire-control equipment.

The manual shift gear is replaced by a ZF 6 HP 600 automatic six-speed transmission with torque converter and lock-up clutch.

The fire-control equipment consists of an electronic fire-control system including a digital fire-control computer and the respective observation and sighting system. The fire-control system enables the gunner of the SK 105/A1 to observe independently, range and hit the target by a simple automated firing process. The system allows for immediate growth capabilities, such as introducing and storing parameters for additional types of ammunition.

The major components are digital fire-control computer unit including commander's operating panel and cant angle sensor, single reticle gunner's sight including laser range-finder, day elbow with ×8 magnification, passive night elbow with ×7.2 magnification and interchangeable with a daylight unity prism, moving-view rotating wedges module in elevation and deflection and periscopic mirror head and gunner's control unit. In order to compute the compensation angle in elevation and deflection, the computer is fed with data from several sources, according to the type of firing: range data from the laser range-finder, side cant of the tank from the built-in cant angle sensor, target velocity from angular velocity and displacement sensor for tracking moving targets. The computer enables automatic computed firing corrections for firing in conformance with the type of ammunition and target range, in order to improve second hit probability.

The commander is provided with a passive night sight, interchangeable with the day sight.

As of mid-2001 this remained at the prototype stage.

Steyr SK 105/A2 Light Tank

In 1981, Steyr-Daimler-Puch completed the prototype of the 105/A2, an improved version of the basic SK 105. It has a fully stabilised turret, oscillating part including the turret crew seats, an electrical laying system, a Digital Fire-Control Computer (DFCC) and a fully automatic 105 mm ammunition loading system.

The DFCC processes the following automatic input data, cant (from sensor) and range (from laser range-finder) as well as the manual input data for crosswind, atmospheric density (altitude above sea level) and barrel wear. The ballistic data of four types of ammunition can be stored and calculated. The gunner's sight with integrated laser range-finder, together with the DFCC and armament stabilisation system, provides high first round hit probability.

An integrated passive night sight for the gunner and a changeable passive night sight for the commander ensure full night fighting capability.

The redesigned turret front gives upgraded ballistic protection. The SK 105/A1 and SK 105/A2 both retain the French 105 mm 105 G1 guns. The SK 105A/2 has an all-electric gun control system.

It is envisioned that any new production models would be built to this standard.

Upgrade Package

Steyr-Daimler-Puch is now marketing an upgrade package to enable existing SK 105 users to upgrade their vehicles to the SK 105/A1 configuration.

The modifications include moving the position of the searchlight from the turret top to the left side of the 105 mm gun; installation of a cupola-mounted 7.62 or 12.7 mm machine gun, additional armour protection over the frontal arc (20° to the longitudinal axis) to provide protection against APDS rounds up to 35 mm;

modification of the weapon to fire 105 mm APFSDS rounds; a fire-control computer and a gunner's sight with integrated laser range-finder; thermal sight for gunner with commander's monitor: hydraulic bump stops for improved cross-country mobility; 250 Ah dry batteries; a fire extinguishing system using Halon; and a fully automatic ZF 6 HP 600 transmission with torque converter and lock-up clutch.

For desert operations, improvements include sand shields, an improved cooling system and a cyclone-type air cleaner. The night fighting capability can be improved by image intensification devices for the driver.

Variants

Greif Armoured Recovery Vehicle

The first prototype of the Greif ARV was completed in 1974 with production beginning in 1976-77. The chassis is identical to the SK 105 but a new superstructure has been built at the front of the vehicle. To the right of the vehicle is mounted a hydraulic crane with a maximum lifting capacity of 6,000 kg. The crane is provided with 42 m of cable and its jib can be extended from its normal length of 3 to 3.9 m. The main winch, with a capacity of 20,000 kg, is mounted in the lower part of the hull and is led out through the front. Mounted at the front of the hull is a hydraulically operated blade which can be used for dozing or stabilising the vehicle when the winch is being used. Full designation of this vehicle is the Armoured Recovery Vehicle 4K-7FA, SB 20 (Bergepanzer Greif). Loaded weight is 19,800 kg and it has a crew of four: driver, commander and two mechanics. Full details of the Greif ARV are given in *Jane's Military Vehicles and Logistics 2001-2002*.

Pionier Vehicle/Engineer Vehicle

The Pionier vehicle is designated the Engineer Tank 4KH7FA-AVE and has the same hull and chassis as the Greif ARV. It has a new winch with a capacity of 8,000 kg and is fitted with a larger dozer blade at the front of the hull. On the right side of the hull, where the crane on the ARV is situated, is a hydraulically operated excavator. Loaded weight is 19,000 kg and it has a crew of four. Full details are given in *Jane's Military Vehicles and Logistics 2001-2002*.

Driver Training Vehicle

This is called the 4KH7FA-FA (Fahrschulpanzer) and is based on the chassis of the SK 105. Each SK 105 can be converted into a driver training vehicle within 2 hours.

Status

In production as required. In service with:

Country	Quantity	Comment
Argentina	112	plus 6 ARVs
Austria	286	plus 39 ARVs and 19 AEVs
Bolivia	34	plus 2 ARVs
Botswana	50	plus 2 ARVs and 2 CPVs
Brazil	17	plus 1 ARV
Morocco	111	plus 10 ARVs
Tunisia	42	plus 3 ARVs and 2 AEVs

Manufacturer

Steyr-Daimler-Puch Spezialfahrzeug AG & Co KG.

UPDATED

CHINA, PEOPLE'S REPUBLIC

NORINCO Type 99 light amphibious tank

Development

NORINCO (China North Industries Corporation) has developed a new light amphibious tank called the Type 99. This is understood to have been developed as a replacement for the older Type 63 light amphibious tank that has been in service with the People's Liberation Army for some years.

At this stage it is not clear whether the Type 99 light amphibious tank is a new design or is possibly based on parts of the now obsolete Type 63 light amphibious tank.

Some sources have indicated that it is an upgraded Type 63 light amphibious tank chassis fitted with a new turret.

It is known that at least three prototypes/pre-production models of the Type 99 light amphibious tank were completed by 1997 at the latest. Recent information has stated that the Type 99 light amphibious tank is now operational with the PLA.

Description

The hull and turret of the Type 99 are of all-welded steel armour construction that provides the occupants with protection from small arms fire and shell splinters.

The driver's compartment is at the front, fighting compartment in the centre and the power pack at the rear. The driver is seated at the front left and has a single-piece hatch cover and three periscopes.

The turret is of a new design with a large bustle at the rear and with the sides sloping inwards towards the roof. There is a wire stowage basket that extends from the sides of the turret to the turret rear.

The turret fitted to the Type 99 may be related to the turret fitted to the new NORICO (6 × 6) armoured car, revealed early in 2001.

Main armament comprises a 105 mm rifled gun that is provided with a fume extractor. It is believed that the 105 mm gun is fully stabilised and is probably coupled to a computerised fire-control system with the commander and gunner having day/night sights. It is probable that the gunner's sight incorporates a laser range-finder that feeds information to the onboard ballistic computer.

A 7.62 mm machine gun is mounted co-axial with the main armament and a 12.7 mm machine gun is mounted on the turret roof for air defence purposes. A bank of electrically operated smoke grenade launchers is mounted either side of the turret firing forwards.

Suspension appears to be similar to that of the Type 63 light amphibious tank with either side having six large roadwheels and with the idler at the front and the drive sprocket at the rear.

It is believed that there are no track-return rollers. The upper part of the track and suspension is covered by a skirt that may help reduce drag and therefore improve water speed.

The vehicle is fully amphibious with the minimum of preparation apart from activating the bilge pumps and the trim vane that is mounted at the front of the vehicle.

It is probable that, like the earlier Type 63 light amphibious tank, the Type 99 is propelled in the water by two water-jets positioned one either side at the rear of the hull.

The Type 99 is probably fitted with NBC systems of the overpressure type and a fire detection and suppression system for the engine compartment at least.

Scale model of Type 99 light amphibious tank (A Pinkov) ***2002***/0073744

Specifications

Not available.

Status

Production. In service with the People's Liberation Army.

Manufacturer

Chinese state factories. Enquiries to China North Industries Corporation.

UPDATED

NORINCO Type 63 light amphibious tank

Development

Initial development work on the NORINCO (China North Industries Corporation) Type 63 light amphibious tank was carried out at the Chinese Military Engineering Institute (MEU) and the No 60 Research Institute for the Fifth MMB. This was produced in small numbers and is understood to be called the Type 60 light amphibious tank.

Further development and prototype construction was carried out by the Chinese Armoured Force Science and Technology Establishment and the No 615 Factory.

Final development was carried out by No 256 Factory and after trials the vehicle was accepted for service by the People's Liberation Army as the Type 63 light amphibious tank in 1963.

China is known to have received some Russian PT-76 light amphibious tanks which could well have been used in the development phase of the Chinese Type 63 light amphibious tank.

From a recognition point of view, the major difference between the Chinese Type 63 light amphibious tank and the Russian PT-76 light amphibious tank is that the former has a different turret of cast rather than welded steel construction.

The Type 63 light amphibious tank is heavier than the Russian PT-76 but has a greater road speed and power-to-weight ratio because of its more powerful diesel engine.

As well as being used by the Chinese Army the Type 63 is also used by the Chinese marines who come under naval command.

As far as is known there has been no recent production of the Type 63 light amphibious tank by NORINCO.

More recently China has developed a new light amphibious tank believed to be armed with a 105 mm gun. This may be designated the Type 99. Available details are given in an earlier entry.

It is considered that the Type 99 light tank is the longer term replacement for the old Type 63 light amphibious tank.

NORINCO Type 63 light amphibious tank with a laser range-finder over the 85 mm gun and a crewman aiming a type HN-5 surface-to-air missile 0045894

NORINCO Type 63 light amphibious tank with trim vane folded back on to glacis plate 0045898

Description

The hull of the Type 63 is welded rolled steel and divided into three compartments: driver's at the front, fighting in the centre and engine at the rear. The glacis plate of the Type 63 is much shallower than the Type 60 or PT-76. The driver is seated at the front of the hull on the left and is provided with three day periscopes forward and a single-piece hatch cover that swings to the left. There is a single infra-red driving light on the right side of the hull front.

The turret is in cast steel sections welded together, with the commander and gunner seated on the left of the turret and the loader on the right. The commander's hatch opens forward and the loader's to the rear. A dome-shaped ventilator is mounted in the turret roof to the rear of the commander's and gunner's hatches. The engine and transmission are at the rear of the hull and additional fuel tanks can be fitted on top of the hull to increase operational range.

The torsion bar suspension consists of six large rubber-tyred roadwheels similar to the Soviet PT-76, with the drive sprocket at the rear and the idler at the front. There are no track-return rollers.

Main armament of the Type 63 is an 85 mm rifled gun, probably identical to that installed in the Type 62 light tank, which fires ammunition of the fixed type including APDS (muzzle velocity 792 m/s), HEAT (muzzle velocity 845 m/s) and HE (muzzl velocity 785 m/s). The 85 mm gun has a normal recoil length of between 280 and 320 mm with a maximum recoil length of 330 mm. Maximum rate of fire is 8 rds/min. Maximum range in the direct fire role is quoted as 5,200 m while in the indirect fire role it is 12,900 m.

Some Type 63 light amphibious tanks have been fitted with a Chinese-designed laser range-finder over the rear part of the 85 mm gun. This is identical to that on some Chinese-built Type 59 MBTs. This feature may indicate that this version has a computerised fire-control system for enhanced first round hit probability.

A 7.62 mm machine gun is mounted coaxially to the right of the main armament and a 12.7 mm Type 54 heavy machine gun is mounted at the loader's station for anti-aircraft defence. The Type 54 is the Russian M1938/46 DShKM manufactured in China.

The tank is fully amphibious, being propelled by two water-jets of Russian design. Before entering the water a trim board is erected at the front of the hull; when travelling this folds back on to the glacis plate. As far as is known the Type 63 has no NBC system or night vision equipment for the commander and gunner.

Communications equipment installed includes a Type A-220A radio with a maximum range of 16 km and a vehicle intercom set.

Specifications

Crew: 4
Combat weight: 18,400 kg
Power-to-weight ratio: 21.74 hp/t
Ground pressure: 0.576 kg/cm^2
Length:
(gun forwards) 8.435 m
(gun rear) 7.733 m
(hull) 7.15 m
Width: 3.2 m
Height to turret top:
(excl AA MG) 2.522 m
(incl AA MG) 3.122 m
Axis of fire: 1.96 m
Ground clearance: 0.4 m
Track: 2.82 m
Track width: 360 mm
Length of track on ground: 4.44 m
Max speed:
(road) 64 km/h
(water) 12 km/h
Fuel capacity: 403 litres
Max range: 370 km
Max water range: 120 km
Fording: amphibious
Gradient: 60%
Side slope: 30%
Vertical obstacle: 0.87 m
Trench: 2.9 m
Engine: Model 12150-L 12-cylinder water-cooled diesel developing 400 hp at 2,000 rpm
Transmission: manual, 5 forward and 1 reverse gears
Electrical system: 24 V
Armament:
(main) 1 × 85 mm rifled gun
(coaxial) 1 × 7.62 mm Type 59T MG
(anti-aircraft) 1 × 12.7 mm Type 54 MG
Ammunition:
(main) 47
(12.7 mm) 500
(7.62 mm) 2,000
Turret traverse: 360°
Gun elevation/depression: +18/–4°
Armour:
Hull front (max): 11 mm at 80°
Hull sides upper: 14 mm
Hull top: 10 mm
Hull floor: 10 mm
Hull rear: 10 mm
Turret mantlet: 11 mm
NBC system: no
Night vision equipment: no

Variants

These are understood to include the Article 76 (WZ211) light tank recovery vehicle. A turretless Type 63 has been observed in use during Chinese space recovery activities. The SY-1 coastal defence missile system was also observed on this chassis some years ago. Components of this Type 63 are also used in the Type 77 APC which is covered in the Armoured Personnel Carriers (Tracked section). This chassis is understood to be called the WZ211 chassis. Some Chinese sources have also mentioned Type 63A and this may well relate to a version with the laser range-finder installed over the 85 mm gun.

Status

Production complete. In service with:

Country	Quantity	Comment
China	1,200	Estimate (Army and Marines)
Korea, North	n/k	
Myanmar	105	
Vietnam	150	Estimate

Manufacturer

Chinese state factories. Enquiries to China North Industries Corporation.

UPDATED

NORINCO Type 62 light tank

Development

This NORINCO (China North Industries Corporation) Type 62 light tank was developed to meet the requirements of the Chinese Army from 1958 at No 674 Factory and was the first complete tank to be built in China as the heavier Type 59 was based on the Russian T-54 series MBT.

Following an extensive series of trials, the vehicle was classified as the Type 62 light tank in 1962. It has been designed for use in China's southern river network areas and hilly terrain where heavier vehicles such as the Type 59 MBT are difficult to deploy.

The Type 62 light tank is virtually a scaled-down Type 59 MBT and has a lower ground pressure.

As well as being used by the Chinese Army the Type 62 is also used by the Chinese Marines who come under naval command. As far as is known, there has been no recent production of the NORINCO Type 62 light tank.

More recently China has developed a new light amphibious tank which is believed to be armed with a 105 mm gun. This vehicle may be designated the Type 99 amphibious light tank. Available details are given in a separate entry.

Description

The layout of the Type 62 light tank is identical to the Type 59, with the driver seated at the front of the hull on the left and the other three crew members in the turret.

The commander and gunner are seated on the left of the turret with the loader on the right. The engine and transmission are at the rear of the hull. The torsion bar suspension either side consists of five roadwheels with a distinct gap between the first and second wheels. The drive sprocket is at the rear and the idler at the front. The initial roadwheel was of a stamped design, similar to that of the Russian PT-76 but was subsequently replaced by a new design similar in appearance to that used on the larger T-54A/Type 59 MBT.

Main armament is an 85 mm gun, probably identical to that on the Type 63 light amphibious tank, which fires fixed ammunition including APDS (muzzle velocity 792 m/s), HEAT (muzzle velocity 845 m/s) and HE (muzzle velocity 785 m/s).

Specifications

Crew: 4
Combat weight: 21,000 kg
Power-to-weight ratio: 20.47 hp/t
Ground pressure: 0.71 kg/cm^2
Length:
(gun forward) 7.9 m
(hull) 5.55 m
Width: 2.86 m
Height: 2.25 m
Ground clearance: 0.42 m
Track: 2.39 m
Track width: 380 mm
Length of track on ground: 3.53 m
Max road speed: 60 km/h
Fuel capacity: 730 litres
Max road range: 500 km
Fording: 1.3 m
Gradient: 60%
Vertical obstacle: 0.7 m
Trench: 2.55 m
Engine: liquid-cooled diesel developing 430 hp at 1,800 rpm
Transmission: manual, 4 forward and 1 reverse gears
Suspension: torsion bar
Electrical system: 24 V
Armament:
(main) 1 × 85 mm rifled gun
(coaxial) 1 × 7.62 mm Type 59T MG
(bow) 1 × 7.62 mm Type 59T MG
(anti-aircraft) 1 × 12.7 mm Type 54 MG
Ammunition:
(main) 47
(coaxial/bow) 1,750
(anti-aircraft) 1,250
Turret traverse: 360°
Gun elevation/depression: +20/–4°
Gun stabiliser:
(vertical) no
(horizontal) no
Armour:
(glacis) 25 mm
(lower glacis) 20 mm
(hull sides) 16 mm
(roof and belly) 12.5 mm
(turret) 32 to 50 mm
NBC system: no
Night vision equipment: no

NORINCO Type 62 light tank
0045899

A 7.62 mm machine gun is mounted coaxially to the right of the 85 mm main armament and a 12.7 mm Type 54 heavy machine gun is mounted at the loader's position for anti-aircraft use. The Type 54 is the Russian M1938/46 DShKM manufactured in China.

The Type 62 is not amphibious, has no NBC system and has not been observed with any infra-red night vision equipment.

Recent information has stated that the day and night vision equipment of the Type 62 is identical to that of the Type 59 MBT.

Variant

It is understood that there was a tropicalised version of the Type 62 light tank. There are at least two other variants, the Article 60 light tank armoured recovery vehicle and the Article 82 light engineering chassis.

Status

Production complete. In service with:

Country	Quantity	Comment
Albania	35	
Bangladesh	40	
China	800	estimate
Congo, Democratic Republic	40	
Korea, North	n/k	
Mali	18	
Sudan	70	
Tanzania	30	
Vietnam	200	estimate

Manufacturer

Chinese state factories. Enquiries to China North Industries Corporation.

UPDATED

FRANCE

Giat Industries AMX-13 light tank

Development

Design work on the AMX-13 light tank began in 1946 at the Atelier de Construction d'Issy-les-Moulineaux and the first prototype was completed two years later. Production was undertaken at the Atelier de Construction Roanne (ARE) from 1952, with the first production tanks completed the following year.

The basic AMX-13 light tank chassis, usually in a modified form, has been used for a wide range of vehicles including the AMX VCI mechanised infantry combat vehicle, the 105 mm self-propelled howitzer, the 155 mm self-propelled gun Mk F3 and the twin 30 mm DCA self-propelled anti-aircraft gun system. Details of all these vehicles are given later in the volume with the exception of the 30 mm DCA and 105 mm SPH, which is no longer in service.

Without changing the basic design, many improvements were announced at the 1985 Satory Exhibition of Military Equipment.

These included a new power pack consisting of a more fuel efficient diesel engine coupled to a fully automatic transmission and the replacement of the torsion bar suspension by a new hydropneumatic suspension for improved cross-country mobility.

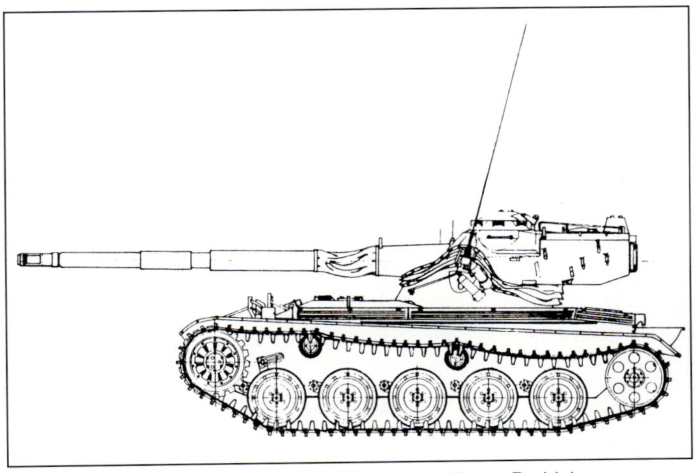

Early hull type AMX-13 armed with a 90 mm gun (Vasco Barbic)

AMX-13 rearmed with a Cockerill 90 mm Mk IVA3 gun with its ordnance fitted with a muzzle brake and a fume extractor
0045902

AMX-13 light tank upgraded by Giat Industries and fitted with a Baudouin 6F 11 SRY diesel engine and an upgraded turret 0045903

AMX-13 light tank 0045904

The armoured vehicle division of Mécanique Creusot-Loire is now part of Giat Industries and is responsible for after sales support for all members of the AMX-13 family of light armoured vehicles, as well as for the provision of upgrade packages. The AMX-13 is no longer being marketed by Giat Industries.

It is estimated that total production of the AMX-13 family of light tracked vehicles, including the light tank, amounted to 7,700 units, of which around 3,400 were exported. The AMX-13 light tank was phased out of service with the French Army many years ago.

Description (basic vehicle)

The hull of the AMX-13 is of all-welded steel and divided into three compartments, with the driver and engine compartments at the front and the turret mounted at the rear. The driver is seated at the front of the vehicle on the left side with the engine compartment to his right and the differential in front of him. The driver is provided with a single-piece hatch cover that opens to the left and three day periscopes, the centre one of which can be replaced by an image intensification periscope for night driving.

The turret is mounted at the rear of the hull and the type of turret depends on the model of the tank and its armament. All turrets are oscillating. In all models the commander is seated on the left of the turret and the gunner on the right. The commander is provided with eight day periscopes and a domed hatch cover that opens to the rear. The gunner has two day periscopes and a single-piece hatch cover that opens to the rear.

The torsion bar suspension either side consists of five rubber-tyred roadwheels with the drive sprocket at the front and the idler at the rear. Trials have been completed with the torsion bars replaced by hydropneumatic suspension units. There are three (or in some cases two) track-return rollers which support the inside of the track only. The first and last roadwheel stations have hydraulic shock-absorbers. The steel tracks have 85 links per side and can be fitted with rubber pads if required.

The AMX-13 does not have an NBC system, cannot be fitted for deep wading and, as built, was not fitted with any night fighting equipment, although several armies have fitted their vehicles with such systems: for example, an infra-red searchlight to the rear of the gunner's position and an infra-red sight for the gunner.

More recently, the AMX-13 has been offered fitted with passive or thermal night firing and night driving equipment, a laser range-finder and an automatic display of the battle sight.

Variants

Model 51 with 75 mm Gun

This was the first model of the AMX-13 to enter service and is fitted with the FL-10 turret armed with a 75 mm gun with a single-baffle muzzle brake. The weapon is fed from two revolver-type magazines in the turret bustle, one either side, each magazine holding six rounds. Once the gun has been fired, the empty cartridge case is ejected from the rear of the turret through a trapdoor hinged on the left. A rate of fire of one round every 5 seconds can be achieved until the two magazines are empty. Once empty, the magazines have to be refilled by hand from outside the tank. The gun fires fixed APC and HE rounds.

A 7.5 or 7.62 mm machine gun is mounted coaxially to the right of the main armament and a similar weapon is often mounted externally at the commander's position. There are 37 rounds of 75 mm and 3,600 rounds (in belts of 200) of machine gun ammunition carried. In addition there are two electrically operated grenade dischargers either side of the turret.

AMX-13 with 90 mm Gun

This entered service in the early 1960s. It is armed with a 90 mm gun in an FL-10 two-person oscillating turret and the barrel is fitted with a thermal sleeve and a single-baffle muzzle brake. The gun fires fixed ammunition including APFSDS, canister, HE, HEAT and smoke.

The turret also has a 7.62 mm coaxial machine gun and a similar 7.62 mm weapon can be mounted at the commander's position. There are 32 rounds of 90 mm ammunition carried, 21 of which are in the turret (12 in the magazines) and 11 in the hull, plus 3,600 rounds of machine gun ammunition.

Existing 75 mm armed AMX-13s can be retrofitted with the 90 mm gun and this was carried out by the French Army from 1970.

AMX-13 with 105 mm Gun

This was developed specifically for export and has an FL-12 two-person osillating turret armed with a 105 mm gun which fires the same smoothbore rounds as the AMX-30 MBT but with a smaller and therefore lighter propellant. This turret is also fitted to the Austrian SK 105 used by Argentina, Austria, Bolivia, Botswana, Brazil, Morocco and Tunisia. Details of this are given earlier in this section under Austria.

AMX-13 with FL-15 Turret

This is very similar to the older FL-12 but observation equipment includes seven M554 day periscopes and a pair of OB-44 night observation binoculars for the tank commander and two M556 day periscopes for the gunner.

Sighting equipment consists of an M212 periscope/telescope with magnifications of ×1.6 and ×6.5 for the tank commander and a telescope with magnifications of ×8 (day) and ×6 (night) for the gunner. A fire-control system will facilitate target acquisition, reduce target engagement time and increase first round hit probability.

Turret traverse is 360° and the top part of the two-person oscillating turret can be elevated from –8 to +12° by the commander or gunner with the former having overriding control.

Types of ammunition fired by the 105 mm gun include Giat Industries APFSDS, HEAT-T, HE, smoke, illuminating, HEAT (training) and HE (training).

Giat Industries Add-on Armour

Giat Industries has developed a range of add-on passive armour packages suitable for its AMX-13 range of light tanks as well as the US M113 APC series.

The add-on package for the AMX-13 light tank weighs under 650 kg with no single unit weighing over 50 kg, enabling it to be installed or removed by the crew using onboard tools.

In the case of the AMX-13 light tank the add-on armour package is installed on the turret front and sides and on the nose and glacis plate of the hull. When fitted, it provides protection for both areas against penetration by 20 mm armour-piercing projectiles fired from a range of 100 m over an arc of approximately 180°. The AMX-13 armour is made of high-hardness steel.

In addition, Giat Industries offers a wide range of kits enabling users to upgrade their vehicles to the latest production standard including improvements in armour, mobility (for example new engine, transmission and suspension) and firepower (for example new armament and fire-control system).

Cockerill 90 mm Regunning Packaging

Cockerill Mechanical Industries offered a 90 mm Mk IVA3 regunning package for the AMX-13 light tank as a replacement for the existing 75 or 105 mm ordnance. The new ordnance incorporates a bore evacuator, a muzzle brake and breech ring accessories with turret modifications including changes to the two six-round ammunition drums, the turret basket ammunition racks and the existing automatic loading system. A total of 22 of these 90 mm regunning packages had been produced for export. This upgrade package is no longer being marketed by Cockerill Mechanical Industries.

Netherlands AMX-13s

The Royal Netherlands Army was one of the largest users of the AMX-13 family of vehicles but all of these have now been phased out of service. Some, including AMX-13 light tanks, AMX VCI infantry fighting vehicles and AMX 105 mm Mk 61 self-propelled howitzers, have been overhauled by RDM Technology of the Netherlands and passed on to Indonesia. This company has also upgraded members of the AMX-13 light tank for other countries, including some in the Middle East.

AMX-13 GTI

GLS of Munich, Germany, a subsidiary of Krauss-Maffei (which became Krauss-Maffei Wegmann on 1 January 1999), has proposed improving the mobility of the AMX-13 light tank by improving its suspension system.

The original suspension system has been replaced by a new connector-type track with centre guide and replaceable track pads. The original roadwheels will be replaced by M113 roadwheels and the original idler will be replaced by an M113 idler. The existing sprockets and final drive will be modified, as will the existing suspension arms. The existing track-return rollers will be modified and telescopic shock-absorbers replaced by dual-action hydraulic telescopic shock-absorbers on the first, second, fourth and fifth roadwheel stations.

GLS also offers new torsion bars for the AMX-13 and installs hydraulic bump stops.

Argentine AMX-13s

In addition to purchasing vehicles from France, Argentina assembled a number of vehicles from components supplied directly from France.

It is believed that most of these have had their original SOFAM petrol engines replaced by KHD V-8 diesels developing 260 hp. Peru is understood to have refitted

Specifications

90 mm version
Crew: 3
Combat weight: 15,000 kg
Unloaded weight: 13,000 kg
Power-to-weight ratio: 16.66 bhp/t
Ground pressure: 0.76 kg/cm²
Length:
(gun forward) 6.36 m
(hull) 4.88 m
Width: 2.51 m
Height: (to top of commander's hatch) 2.3 m
Ground clearance: 0.37 m
Track: 2.159 m
Track width: 350 mm
Length of track on ground: 2.997 m
Max road speed: 60 km/h
Fuel capacity: 480 litres
Max road range: 350-400 km
Fording: 0.6 m
Gradient: 60%
Side slope: 60%
Vertical obstacle:
(forwards) 0.65 m
(reverse) 0.45 m
Trench: 1.6 m
Turning radius: skid turns
Engine: Sofam Model 8Gxb 8-cylinder water-cooled petrol developing 250 hp at 3,200 rpm
Transmission: manual with 5 forward and 1 reverse gears
Steering: Cleveland (Cletrac) controlled differential
Clutch: single-disc
Suspension: torsion bar
Electrical system: 24 V
Batteries: 4 × 12 V, 190 Ah
Armament:
(main) 1 × 90 mm rifled gun
(coaxial) 1 × 7.5 mm (or 7.62 mm MG)
(anti-aircraft) 1 × 7.5 mm or (7.62 mm MG) optional
Smoke-laying equipment: 2 smoke grenade dischargers either side of turret
Ammunition:
(main) 32
(MG) 3,600
Smoke canisters: 12
Gun control equipment
Turret power control: hydraulic/manual
(by commander) yes
(by gunner) yes
Max rate of power:
(traverse) 360°/12 s
(elevation) 5° in 1s
Gun elevation/depression: +12.5/–5°
(commander's override) yes
Gun stabiliser:
(vertical) no
(horizontal) no
Commander's sight: telescope L961 with ×1.5 and ×6 magnification
Gunner's sight: telescope L862 with ×7.5 magnification
Armour:
Hull front: 15 mm at 55°
(equal to 40 mm)
Hull sides: 20 mm
Hull top: 10 mm
Hull rear: 15 mm
Turret front: 25 mm at 45°
(equal to 40 mm)
Turret sides: 25 mm
Turret top: 10 mm
NBC system: no
Night vision equipment: yes

its AMX-13 light tanks with this engine and Venezuela refitted its vehicles in batches of 10.

AMX-13 Model 1987 with 105 mm Gun

Late production Model AMX-13 light tanks have many improvements over earlier vehicles and significant improvements in armour, mobility and firepower.

Externally the appearance of the vehicle has changed and a new hull front with improved ballistic protection is installed, the old torsion bars have been replaced by new hydropneumatic units and sand guards help to keep down the dust, while a 105 mm gun is fitted as standard.

The petrol-engined AMX-13 had a range of 350 to 400 km on roads but when fitted with a diesel engine this is increased to between 550 and 600 km. There are two diesel engines currently offered, both with an output of 280 hp. These are the Detroit Diesel Model 6V-53T or the French Baudouin 6F 11 SRY. They give a road speed of 65 km/h and a new cooling system enables the engine to operate in high ambient temperatures.

An automatic transmission from Rockford Powertrain is proposed.

The installation of the new engine, transmission and suspension system gives a much improved cross-country mobility and provides a better ride for the crew, as well as a more stable firing platform.

The turret has the 105 mm Giat Industries gun which can fire many types of ammunition including APFSDS. An extensive range of modular firing controls with integrated range-finder can be fitted without any modification.

Ecuadorean AMX-13

In 1988, Ecuador awarded a contract for the upgrading of its fleet of 108 AMX-13 light tanks armed with 105 mm guns. SOPELEM supplied the basic day version of its SOPTAC 18 fire-control system which comprises a laser range-finder and its control unit, a digital computer that calculates the firing elevation and a module that injects an aiming graticule into the gunner's original sight. The graticule is positioned automatically to correspond to the gun elevation calculated by the computer, which takes into account the range obtained by the laser range-finder, atmospheric pressure, gun chamber pressure and the tank's horizontal attitude, all of which are obtained by integrated sensors, while the side-wind, type of ammunition and barrel wear information is fed in manually.

Fives-Cail Babcock provided kits to modify the 105 mm gun and its ammunition system to allow APFSDS rounds to be fired.

Work on upgrading the AMX-13s started in Ecuador late in 1988 and was completed in 1990.

In the future Ecuador may replace the current petrol engines with a diesel engine and also give the vehicles a night fighting capability.

Singapore AMX-13 SM1

Details of this version of the AMX-13 are given later in this section under Singapore. Indonesia has upgraded some of its AMX-13 to the AMX-13 SM1 standard. Prime contractor for the SM1 upgrade is now Singapore Technologies Kinetics.

Venezuelan AMX-13s

In addition to the 36 AMX-13 light tanks delivered by France in the mid-1950s, late in 1988, Mécanique Creusot-Loire won a contract from Venezuela for the supply of 31 90 mm armed, overhauled and modernised AMX-13 light tanks. These were ex-French Army vehicles stripped down by the company and refurbished. They now have the US Detroit Diesel Model 6V-53T engine developing 280 hp coupled to a Borg Warner three-speed fully automatic transmission with torque converter, a new Chausson air-water/oil cooler with a thermostatically controlled centrifugal fan, an alternator or 200 A generator and NATO 6TN batteries. The original torsion bar suspension has been removed and replaced by hydropneumatic units for increased cross-country mobility.

A SOPELEM 18-02 SOPTAC fire-control system incorporating an M213 day sight and a TCV-107 laser range-finder has been fitted.

Main armament comprises a 90 mm gun with a 7.62 mm NF1 coaxial machine gun. An Israeli Tadiran communications system is fitted.

Final deliveries of the upgraded AMX-13 light tanks to Venezuela were made in late 1990.

Venezuelan AMX-13 Rocket Launchers

Venezuela has approximately 25 AMX-13 light tank chassis fitted with the Detroit Diesel 6V-53T diesel engine, with the turret replaced by the Israel Military Industries 160 mm LAR multiple rocket launcher system, details of which are given in the Multiple rocket launchers section under Israel.

NIMDA Upgrade Package for AMX-13

The Israeli company NIMDA offers a complete retrofit package for the AMX-13 light tank including a new Detroit Diesel Model 6V-53T developing 275 hp at 2,800 rpm coupled to a new NIMDA N303 automatic transmission, new armament, a computerised fire-control system, additional armour protection and a fire/explosion detection and suppression system.

AMX-13 Armoured Recovery Vehicle

The AMX-13 ARV (Char de Dépannage Model 55) is used to recover other members of the AMX-13 family as well as changing major components such as turrets and engines. Equipment fitted includes a front-mounted A-frame, a 15,000 kg capacity winch with 50 m of 25 mm diameter cable, a secondary winch with 120 m of 6 mm diameter cable, four spades at the rear of the hull and tools and other equipment. The vehicle has a crew of three consisting of commander, driver and winch operator and weighs 15,300 kg. Armament consists of an externally mounted 7.5 mm or 7.62 mm machine gun and grenade dischargers.

AMX-13 Bridgelayer

The AMX-13 bridgelayer (Char Poseur de Pont AMX-13) is fitted with a folding Class 25 bridge which has an unfolded length of 14.01 m. The bridge is launched over the rear of the vehicle and two stabilisers steady it when the bridge is being positioned. The vehicle has a loaded weight of 19,700 kg and without the bridge weighs 15,000 kg.

Manufacturer

Mécanique Creusot-Loire (now Giat Industries).

Status

Production complete. In service with (see table):

Country	Quantity	Comment
Argentina	58	was also assembled in Argentina; also ARV and AVLB
Dominican Republic	2	75 mm model, some sources have stated that 12 vehicles are in service
Ecuador	108	105 mm, plus ARV upgraded
Indonesia	275	75 mm and 105 mm, plus ARV and AVLB. Some upgraded to AMX-13 SM1 standard.
Ivory Coast	5	75 mm model
Lebanon	35	90 mm and 105 mm, out of 100+
Peru	110	80 105 mm model, 30 75 mm model
Singapore	350	75 mm model, most modified, SM1 (qv Singapore)
Venezuela	36	armed with 75 mm guns, 31 90 mm upgraded delivered, MRS

Note: The AMX-13 is no longer used by Algeria, Cambodia, Chile, Djibouti, Egypt, France, India, Israel, Morocco, Nepal, the Netherlands or Tunisia. While the vehicle is no longer in production, Giat Industries can still supply upgraded AMX-13s or kits to upgrade existing vehicles.

UPDATED

INTERNATIONAL

ASCOD 105 light tank

Development

The ASCOD 105 light tank has been developed as a private venture for the export market by ASCOD AIE which is jointly owned by SANTA BARBARA of Spain and Steyr-Daimler-Puch of Austria.

The ASCOD 105 light tank consists of a modified ASCOD armoured infantry fighting vehicle chassis which is already in production in Austria by Steyr-Daimler-Puch (for the Austrian Army who call it the Ulan) and in Spain by SANTA BARBARA (for the Spanish Army who call it the Pizarro).

It is fitted with the complete South African LIW turret installed on the private venture South African Vickers OMC Rooikat 105 mm armoured car (8 × 8) (covered in detail in the Reconnaissance vehicles section).

The first prototype of the ASCOD 105 light tank was completed late in 1996 and it has been demonstrated in Austria to a number of potential export customers, including some in Asia.

Late in 1999, following an international competition, the Royal Thai Marine Corps selected the ASCOD 105 light tank and following contract negotiations an order will be placed with the Austrian company, Steyr-Daimler-Puch for 15 vehicles plus one command post vehicle and one repair and recovery vehicle. This is the first export order for the ASCOD 105 light tank.

As of mid-2001, no production contracts are known to have been placed for any versions of the ASCOD 105 light tank, although marketing continues.

ASCOD 105 light tank fitted with Otobreda 105LRF (Low Recoil Force) turret for trials **2002**/0095359

Specifications (with South African 105 mm turret)

Crew: 4
Combat weight: 28,500 kg
Power-to-weight ratio: 21.05 hp/t
Ground pressure: 0.65 kg/cm^2
Length:
(gun forward) 7.63 m
(hull) 6.61 m
Width:
(overall) 3.150 m
(tracks) 2.958 m
Height:
(hull roof) 1.775 m
(commander's cupola) 2.76 m
(7.62 mm AA MG) 3.03 m
Firing height: n/avail
Ground clearance: 0.45 m
Track: 2.434 m
Track width: 534 mm
Length of track on ground: 3.99 m
Max road speed:
(forward) 70 km/h
(reverse) 33 km/h
Acceleration: 0 to 50 km/h 11 s
Fuel capacity: 650 litres
Cruising range: (road) 500 km
Fording:
(without preparation) 1.2 m
(with preparation) 1.5 m
Gradient: 75%
Side slope: 40%
Vertical obstacle: 0.95 m
Trench: 2.3 m
Turning radius: pivot
Engine: MTU 8V 183 TE22 8 V-90 developing 600 hp at 2,300 rpm
Transmission: Renk HSWL 106C hydromechanical with integrated reversing and steering gear, automatic 6 speed with lock up clutch
Suspension: torsion bar with trailing arm for each roadwheel, rotary dampers for first and sixth roadwheel
Electrical system: 24 V
Batteries: 6 × 24 V, 300 Ah
Armament:
(main) 1 × 105 mm rifled GT 7 gun
(coaxial) 1 × 7.62 mm MG
(anti-aircraft) 1 × 7.62 mm MG
Ammunition:
(main) 40 (16 + 24)
(MG) 4,600 (2,800 + 1,800)
Smoke grenade launchers: 2 × 4 81 mm
Gun control equipment: electric/manual
(by commander) yes
(by gunner) yes
Gun elevation/depression: +17/–8°
Turret traverse: 360°
Gun stabiliser:
(vertical) yes
(horizontal) yes
NBC system: yes
Night vision system: yes

Description

The hull of the ASCOD 105 light tank is all-welded steel armour which provides protection from small arms fire and shell splinters.

The driver is seated front left with the power pack to the right and above his position is a single-piece hatch cover that lifts and opens to the rear. In front of this are three day periscopes, the centre one of which can be replaced by a passive periscope for driving at night.

The all-welded steel turret is in the centre with the commander and gunner on the right and the loader on the left. The commander has a single-piece hatch cover that opens to the rear and eight day periscopes for all-round observation and a roof-mounted stabilised day sight. He also has a monitor of the gunner's main thermal sight.

The gunner has a stabilised day/thermal sight incorporating a laser range-finder and an observation periscope. The loader has a single-piece hatch cover that opens to the rear plus two day periscopes and a vision block. An ammunition resupply hatch is provided in the left side of the turret.

Main armament is a fully stabilised 105 mm LIW GT 7 gun that fires standard natures of 105 mm ammunition with a 7.62 mm machine gun mounted coaxial with the main armament and a roof-mounted 7.62 mm anti-aircraft machine gun.

Suspension either side is of the torsion bar type with rotary dampers for the first and last roadwheel stations. The suspension either side consists of seven dual rubber-tyred roadwheels with the drive sprocket at the front, idler at the rear and track-return rollers.

The rear hull door of the ASCOD armoured infantry fighting vehicle is retained and for a detailed description of the chassis and turret the reader is referred to the detailed entries previously mentioned.

If required, additional armour can be fitted to provide a higher level of battlefield survivability.

Standard equipment for the ASCOD 105 includes an overpressure NBC system, fire detection and suppression system for engine compartment and a diesel-operated heater for driver and crew compartment.

Low-Profile Turret on ASCOD

First firing trials of the General Dynamics Land Systems' 105 mm Low-Profile Turret (LPT) installed on an ASCOD infantry fighting vehicle (IFV) chassis were successfully carried out in the US late in 1998.

The LPT was originally developed in the early 1980s and first installed on a tracked chassis called the Armored Gun System and then the Direct Fire Support Vehicle.

The latest-generation LPT has a number of significant improvements and features a new turret structure that is lighter and stiffer. The 105 mm M68A1 rifled tank gun is fed by an automatic loader mounted below the turret ring and between the commander (seated on the right) and the gunner (seated on the left).

The 105 mm M68A1 gun has been fitted with a cam-operated drop-block breech assembly, bore evacuator, muzzle reference mounting collar and an integral muzzle brake. Low recoil is made possible by the installation of a four-cylinder recoil assembly.

By mounting the 105 mm M68A1 gun externally the LPT is not only much lighter than a conventional turret, allowing it to be installed on much lighter tracked and wheeled armour vehicles, but is also much smaller and therefore much more difficult to hit.

The 7.62 mm machine gun is mounted coaxial with the 105 mm M68A1 gun and a 7.62 or 12.7 mm machine gun or a 40 mm grenade launcher can be mounted at the commander's station.

The commander and gunner have day vision periscopes while the gunner's primary sight has a dual-axis stabilised head mirror, second-generation FLIR, eye-safe laser range-finder and eight direct viewing options.

The commander has a panoramic sight which allows the vehicle to undertake hunter/killer target engagements. Turret traverse and weapon control is all electric and stabilised and either commander or gunner can aim and fire the armament.

It was expected that the LPT would first be tested on a General Motors Defense LAV-III (8 × 8) chassis. As this was not available, an offer to test it on the ASCOD was accepted by General Dynamics. Over 70 rounds of ammunition were fired while the vehicle was stationary and moving with the target also moving.

The aim of the trials was to validate the improvements carried out to the latest-generation LPT system. A further development of the LPT installed on the latest LAV-III chassis has been selected by the US Army to meet the Mobile Gun System requirement of its

ASCOD 105 light tank with South African LIW turret ***2002***/0114263

ASCOD 105 light tank fitted with General Dynamics Land Systems' Low-Profile Turret armed with an externally mounted 105 mm rifled tank gun ***2000***/0069576

new Brigade Combat Team's. It is expected that 204 vehicles will be procured.

Centauro Turret on ASCOD

Early in 2000, the ASCOD chassis was fitted with the Italian Otobreda 105LRF turret which is standard on the 400 Centauro (8 × 8) vehicles delivered to the Italian Army and more recently the 22 Centauro (8 × 8) vehicles delivered to Spain. Full details of this turret are given for the entry of the Centauro in the Reconnaissance vehicles section.

Firing trials were conducted with the vehicle stationary and moving with good results according to Steyr-Daimler-Puch.

Combat weight of this version is 29.49 tonnes.

Status

Development complete. Ready for production. Selected by Royal Thai Marine Corps (see development).

Manufacturers

LIW, division of Denel, (turret).
ASCOD AIE, (chassis).

UPDATED

IRAN

Tosan light tank

Development/Description

Late in 1997, Iran stated that it had developed a new light tank called the Tosan (Fury) armed with a 90 mm gun and that this was expected to enter production in the near future.

No information has been released on the Tosan, apart from that it is capable of 'rapid response and built for strategic missions.'

It is also said to be very light so that special trucks will not be required and it is also claimed to have a superior firing and targeting system.

The Tosan could be a local design, a foreign vehicle manufactured under licence or a mix of new and existing vehicles and equipment.

If the Tosan has not been completely designed and built in Iran, possible existing modified vehicles could include:

(1) Alvis Scorpion with the 76 mm gun replaced by a 90 mm gun. Before the fall of the Shah of Iran the country took delivery of a large number of 76 mm Alvis Scorpion Combat Vehicle Reconnaissance (Tracked) vehicles. All recent export Alvis Scorpion vehicles have been fitted with a Belgian Cockerill 90 mm gun. This is also fitted to the Brazilian ENGESA EE-9 (6 × 6) Cascavel armoured cars supplied in large numbers to Iraq, some of which have been captured by Iran. It is possible that Iran has taken these 90 mm guns, or reverse engineered them, and installed them in the existing Alvis Scorpion turrets. Another alternative would be to fit a new turret with a 90 mm gun.

(2) Locally built Boragh, which is covered in detail in the Armoured personnel carriers (tracked) section, fitted with a two-person turret armed with a 90 mm rifled gun as fitted to the ENGESA EE-9 (6 × 6) Cascavel armoured car. Another alternative could be fitting the complete ENGESA EE-9 (6 × 6) Cascavel turret. This vehicle was designed and manufactured by the Brazilian company ENGESA which ceased trading some years ago. Large numbers of EE-9s were supplied to Iraq and some of these were subsequently captured by Iran.

Specifications

Not available.

Status

Low-rate production. In service with Iran.

Manufacturer

Tosan is believed to be built at the Iranian Revolutionary Guards Corps' facility at Khorranambad.

UPDATED

KOREA, NORTH

M1985 light tank

Development/Description

The M1985 light tank was developed and manufactured in North Korea. It is understood to be based on the chassis of the locally manufactured VTT-323 full-tracked armoured personnel carrier, which in turn is based on the Chinese NORINCO Type 63 full-tracked APC. It was first seen in public in 1985, hence its designation M1985.

The hull of the new light amphibious tank is of all-welded steel armour construction with the driver seated front left, ammunition stowage to the right, turret and fighting compartment in the centre and engine and transmission at the rear.

The turret is similar to that fitted to the Russian PT-76 light amphibious tank with grab rails on the outside.

Specifications
Provisional
Crew: 3-4
Combat weight: 22,000 kg
Power-to-weight ratio: 14.54 hp/tonne
Length: 9.4 m
Width: 3.06 m
Height: 2.8 m
Engine: 320 hp diesel
Armament:
(main) 1 × 85 mm rifled gun
(coaxial) 1 × 7.62 mm MG
(anti-aircraft) 1 × 12.7 mm MG
(ATGW) 1 × AT-3 Sagger
Turret power-control: powered/manual
Gun stabiliser:
(vertical) probably not
(horizontal) probably not
NBC system: probable
Night vision equipment: probable

M1985 85 mm armed North Korean light tanks on parade 0045906

The glacis plate is well sloped and the hull sides above the suspension slope inwards.

Suspension either side consists of six large rubber-tyred roadwheels, drive sprocket at the rear and idler at the front. There are no track-return rollers.

Main armament is believed to comprise an 85 mm gun which is fitted with a fume extractor towards its muzzle. A 7.62 mm machine gun is mounted coaxially with the main armament and, mounted over the 73 mm gun is an ATGW launcher, possibly for a wire-guided AT-3 'Sagger'. A 12.7 mm machine gun can be mounted on the turret roof for anti-aircraft defence.

In the absence of any official North Korean designation, the US Army has given this vehicle the designation M1985, as this was the year it was first seen in public.

Status

Production probably complete. In service with the North Korean Army. There are no known export sales of this light tank. In 2001, it was estimated that the total North Korean light tank fleet consisted of about 500 M1985 and PT-76 light tanks.

Manufacturer

North Korean state factories.

UPDATED

RUSSIAN FEDERATION AND ASSOCIATED STATES (CIS)

PT-76 light amphibious tank

Development

Shortly after the end of the Second World War the Russian Army issued a requirement for a light amphibious tank and an amphibious armoured personnel carrier (APC) which would all share the same automotive components.

After trials with prototype vehicles designated the P-39 (amphibious light tank) and P-40 (amphibious APC), it was decided that the Higher Scientific Research Institute for Transport Machinery Construction (VNII TransMash) and the Chelyabinsk Kirov factory would design and build prototypes of a new light amphibious tank under the leadership of Zh Ya Kotin.

Prototypes of the light amphibious tank were designated the Object 740, while that of the amphibious APC was designated the Object 750 with the first prototypes completed in 1950.

Following trials and further modifications the vehicles were accepted for service in 1951 under the designation of the PT-76 (amphibious tank with 76 mm gun).

Production was undertaken at the Volgograd Tractor Factory from 1958 to 1967 and a total of 12,000 vehicles was built for the home and export markets. Total exports amounted to approximately 2,000 vehicles.

Automotive components of the PT-76 were also used in the Russian BTR-50 full-tracked armoured personnel carrier that has now been phased out of front-line service with Russian forces.

The PT-76 has been used in combat in Africa, the Middle East and during the Indo-Pakistan conflict and was widely used in Vietnam by the North Vietnamese Army. The main drawbacks of the PT-76 are its large size because the original requirement said amphibious; its lack of NBC or night fighting equipment; (although some vehicles were subsequently fitted with infra-red night vision equipment) and its very thin armour which can even be penetrated by heavy machine gun fire. It does have an excellent amphibious capability and is also used by the Russian Naval Infantry.

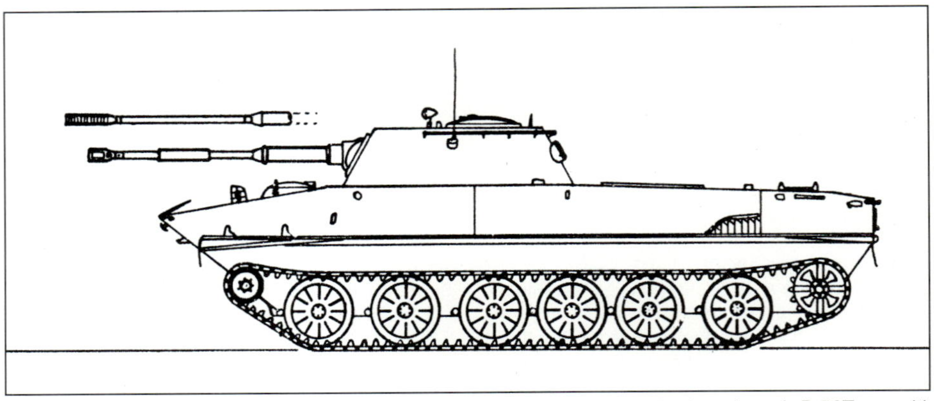

PT-76 light amphibious tank with D-56T gun with double-baffle muzzle brake. Inset is early D-56T gun with multislotted muzzle brake

Specifications

PT-76B
Crew: 3
Combat weight: 14,600 kg
Power-to-weight ratio: 16.4 hp/t
Ground pressure: 0.50 kg/cm^2
Length
(gun forwards) 7.625 m
(hull) 6.91 m
Width: 3.14 m
Height: (overall) 2.255 m
Firing height: 1.82 m
Ground clearance: 0.37 m
Track: 2.74 m
Track width: 360 mm
Length of track on ground: 4.08 m
Max speed:
(road) 44 km/h
(water) 8-9 km/h
Fuel capacity:
(main) 380 litres
(auxiliary) 190 litres
Max range:
(road) 370 km
(road, with auxiliary tanks) 480 km
(water) 120 km
Fuel consumption: 0.96 litres/km
Fording: amphibious
Gradient: 70%
Vertical obstacle: 1.1 m
Trench: 2.8 m
Turning radius: skid turns
Engine: model V-6B 6-cylinder in-line water-cooled diesel developing 240 hp at 1,800 rpm
Auxiliary engine: none
Transmission: manual with 4 forward and 1 reverse gears
Steering: clutch and brake
Suspension: torsion bar
Electrical system: 24 V
Batteries: 2 × 12 V, 100 Ah
Armament:
(main) 1 × 76.2 mm rifled D-56TS gun
(coaxial) 1 × 7.62 mm SGMT MG
(anti-aircraft) 1 × 12.7 mm DShKM MG (optional)
Smoke-laying equipment: diesel fuel injected into exhaust system
Ammunition:
(main) 40
(MG) 1,000
Gun control equipment
Turret power control: electric/manual
(by commander) yes
Max rate of power: (traverse) 360°/21 s
Gun elevation/depression: +30/–4°
Gun stabiliser:
(vertical) yes
(horizontal) yes
Armour
Hull front upper: 11 mm at 80°
Hull front lower: 14 mm at 45°
Hull side upper: 14 mm at 0°
Hull side lower: 14 mm at 0°
Hull rear upper: 7 mm at 0°
Hull rear lower: 7 mm at 45°
Hull top: 7 mm
Hull belly front: 5 mm
Hull belly rear: 5 mm
Turret front: 17 mm at 35°
Turret sides: 16 mm at 35°
Turret rear: 11 mm at 35°
Turret top: 8 mm
Mantlet: 11 mm at 33°
NBC system: yes
Night vision equipment: yes

Description

The hull of the PT-76 is of welded steel and divided into three compartments, with the driver at the front, the fighting compartment in the centre and the power pack at the rear. The driver is seated in the centre of the hull and is provided with a single-piece hatch cover that swings to the right. Three TNP day periscopes are mounted forward of the hatch cover; the centre one can be replaced by a PER-17A day periscope, raised, enabling the driver to see over the front of the vehicle when the trim vane is erected. The centre periscope can be replaced by a TVN-2B infra-red periscope with a 60 m range.

The driver is also provided with a GPK-48 gyrocompass. In the floor of the vehicle, under the driver's seat, is an emergency hatch which can be used by all crew members.

The turret is of all-welded steel with the commander, who also acts as the gunner, seated on the left and the loader on the right. The turret has an oval-shaped hatch cover that hinges forward and can be locked vertical. Mounted in the left side of the hatch cover is a circular cupola for the commander which houses three integral day periscopes and can be traversed through 360° by hand.

The commander, who also acts as the gunner, has a TPKU-1 day sight with a magnification of ×5 which was replaced on the PT-76B by the TPKU-2 and two TNP day periscopes. The loader has a MK-4 observation device.

The engine compartment is at the rear and the Model V-6 engine used in the PT-76 is one bank of that fitted to the T-54 MBT. As the tank operates in a very cold climate an engine preheater is fitted as standard.

The manual gearbox has five forward and one reverse gears and steering is of the clutch and brake type.

The torsion bar suspension either side consists of six roadwheels with the idler at the front and the drive sprocket at the rear. The first and sixth roadwheel stations have hydraulic shock-absorbers and the steel tracks each have 96 links when new. Tracks are steel with a single dry pin.

The PT-76 is fully amphibious, being propelled by two water-jets mounted at the rear of the hull. The only preparation required before entering the water is to erect the trim vane at the front of the hull and switch on the two electric bilge pumps. A manual bilge pump is provided for emergency use. Steering is accomplished by opening and closing the two hatches over the rear water-jets: to go left the left water-jet is covered up and

PT-76 light amphibious tank upgraded with 90 mm CMI gun and new fire-control system as used in Indonesia 0045907

to turn left through 180° the left water-jet sucks in water while the right water-jet pushes it out.

The PT-76 is not fitted with an NBC system although there is a ventilator mounted in the turret rear. Many vehicles used by marines have been fitted with a snorkel extension piece to this ventilator, which often sucked exhaust gases into the fighting compartment. To increase the PT-76's operational range, additional fuel tanks can be fitted; these can be of the drum type or flat type as fitted to one side of the T-54/T-55 MBT. A white light searchlight is mounted on the right side of the turret roof (some models have an infra-red searchlight on the turret) and one of the driving lights may be infra-red. When being used by the Naval Infantry the vehicle is often fitted with a complete set of navigation lights.

The main armament of the PT-76 is the 76.2 mm rifled D-56T gun which has a semi-automatic vertical sliding wedge breech block, a hydraulic buffer and a hydropneumatic recuperator. The gun is 42 calibres long and has an overall length of 3.455 m and weight of 1,150 kg. The gun has a maximum rate of fire of between 6 and 8 rds/min, a maximum range in the indirect fire role of between 12,000 and 13,290 m and can fire the following types of fixed ammunition: AP-T (BR-350 series), API-T (BZR-350B), HE-FRAG (0-350A), HEAT-FS (BK-354M) and HVAP-T (BR-354P).

Other types of projectile are also available, some with cartridges with reduced propellant charges, for HE-FRAG and AP-T projectiles.

A 7.62 mm SGMT machine gun is mounted coaxially to the right of the main armament. Many PT-76s have been fitted with a roof-mounted 12.7 mm DShKM anti-aircraft machine gun.

The turret of the PT-76 light amphibious tank has also been fitted on to a number of river patrol craft such as the Shmel class.

Variants

The first production version of the PT-76 light tank was armed with the D-56T gun and fitted with a multislotted muzzle brake. This version is rarely seen today. The most common version of the PT-76 is fitted with the D-56TM (2A16) gun with a double-baffle muzzle brake and a bore evacuator towards the muzzle. For training, the muzzle brake is sometimes removed.

The PT-76B (development designation Object 740B) is fitted with a 76.2 mm D-56TS gun which has a two-axis stabiliser system. This was accepted for service in 1962 and was also fitted with two auxiliary fuel tanks, each of which holds 95 litres. This increased the overall weight of the vehicle by 200 kg.

For improved buoyancy, the height of the hull was raised by 130 mm and the bow lengthened and the stern was given a slightly steeper reverse angle.

The PT-76B also features an air filtration and ventilation system (FVU) and anti-nuclear protection system (PAZ) to allow it to operate under NBC conditions.

As vehicles came back for overhaul they were often upgraded, for example in 1967 the 7.62 mm SGMT machine guns were replaced by the latest 7.62 mm PKT machine guns.

The driver was provided with a TVN-2E night vision device and the earlier FG-100 and FG-102 headlamps were replaced by the FG-125 and FG-126 versions.

Communications were also upgraded with the older R-113 being replaced by the R-123 and the R-120 intercom replaced by the later R-124.

Chinese NORINCO Type 63

This has a similar hull to the PT-76 but has a different turret; details are given in this section under China.

NIMDA Upgrade Package for the PT-76

The Israeli company, NIMDA, offers a complete retrofit package for the PT-76 light amphibious tank. This includes a new power pack incorporating a Detroit Diesel 6V-92T diesel developing 300 hp at 2,100 rpm coupled to the original transmission with a new clutch assembly. The kit also includes a new alternator (24 V, 200 Ah), a new cooling and electrical system and fuel, exhaust, air inlet, hull and top deck modifications. Another option is the replacement of the original transmission by a fully automatic transmission.

The PT-76 prototype has had its original 76.2 mm gun replaced by a new 90 mm Cockerill Mk III gun that can fire a wide range of ammunition including APFSDS.

The Russian 7.62 mm machine gun has been replaced by a Western 7.62 mm machine gun and a similar weapon can be mounted on the turret roof for air defence purposes. Electrically operated grenade dischargers can be mounted either side of the turret.

To improve the first round hit probability, a new fire-control system has been installed, together with a new day/night sight for the gunner which incorporates a laser range-finder.

A new solid-state all-electric gun/turret stabilisation and power control system has replaced the original Russian system and a new fire extinguishing system has been installed.

Recent information has indicated that part of the Indonesian PT-76 fleet may have been upgraded to this standard. Some sources have indicated that between 50 and 60 vehicles were upgraded.

It is also understood that some of the Indonesian fleet of BTR-50 full-tracked armoured personnel carriers may have been upgraded to a similar automotive standard.

Status

Production complete. In service with:

Country	Quantity	Comment
Afghanistan	60	
Benin	20	status uncertain
Cambodia	10	
Congo	3	
Croatia	5	
Cuba	50	estimate
Guinea	20	
Guinea-Bissau	20	
Indonesia	130	some upgraded 30 (army), 100 (marines)
Iraq	100	status uncertain
Korea, North	500	including M1985
Laos	25	
Madagascar	12	
Nicaragua	10	probably now in reserve
Russia	200	approx number, army and naval infantry
Uganda	20	
Vietnam	300	
Yugoslavia	5	
Zambia	30	

Manufacturer

Volgograd Tractor Plant.

UPDATED

PT-76 Model 2 light amphibious tank

PT-76 Model 2 armed with a 76.2 mm (D-56T) gun fitted with a double-baffle muzzle brake and a fume extractor (Richard Stickland) 0045908

SINGAPORE

Singapore Armed Forces new light tank

Development/Description

The Singapore Ministry of Defence (MINDEF) is understood to be drawing up requirements for a new light tank for the Singapore Armed Forces (SAF), which will enter service early in the 21st century.

The French designed and built Giat Industries (previously Mécanique Creusot-Loire) AMX-13, covered in detail in the following entry and armed with a 75 mm gun fed by a bustle-mounted automatic loader, is the standard light tank of the Singapore Armed Forces.

From 1988, Singapore Technologies Kinetics Ltd upgraded the entire fleet of AMX-13 (believed to consist of 350 vehicles), to the latest AMX-13 SM1 standard.

The 75 mm gun was retained but a new power pack, electrical system and hydropneumatic suspension system were installed giving the vehicle a greater operational range, improved reliability and better mobility over rough terrain.

By today's standard, the 75 mm rifled gun is obsolete and lacks sufficient armour penetration to defeat any of the more modern tanks over their frontal arc.

It is understood that Singapore is to develop a new light tank, having a crew of three with the main armament, which will be at least 105 mm in calibre, fed by an automatic loader. Some sources have indicated that Singapore may opt for a 120 mm smoothbore gun, also fed by a bustle-mounted automatic loader.

It is possible that the Singapore light tank will be based on automotive components, or perhaps even the chassis, possibly with extensive modifications of the new Bionix Singapore Infantry Fighting Vehicle (IFV) which is now in production by Singapore Technologies Kinetics in two baseline versions.

The IFV25 is fitted with an electrically powered two-person turret armed with a US-supplied The Boeing Company 25 mm M242 Chain Gun and 7.62 mm co-axial machine gun. The second version, IFV 40/50 has a 40/50 cupola armed with a 40 mm grenade launcher and a 7.62 mm machine gun. Full details are given in the Armoured Personnel Carriers (Tracked) section.

It is believed that the first production order is for 300 units to be followed by a second batch of 200 units. These would then be followed by the new light tank.

Status

Studies.

Manufacturer

Not selected but would probably be Singapore Technologies Kinetics Ltd.

UPDATED

Singapore Technologies Kinetics Ltd AMX-13 SM1 light tank

Development

Singapore has the largest fleet of French Giat Industries (previously Mécanique Creusot-Loire) built AMX-13 light tanks in the world and, for many years, Singapore Technologies Kinetics (previously Singapore Automotive Engineering) has been overhauling these and other tracked and wheeled armoured vehicles for the Singapore Armed Forces.

More recently, Singapore Technologies Kinetics Ltd has developed a complete chassis refurbishment programme for the AMX-13 with the first production vehicles handed over to the Singapore Armed Forces in June 1988 under the designation of the AMX-13 SM1. It is understood that most of the Singapore fleet of 350 AMX-13 light tanks has been upgraded to the AMX-13 SM1 configuration.

The main area of improvement carried out on the chassis which has been given a total automotive refit with a new Detroit Diesel 6V-53T engine, a new ZF 5WG-180 fully automatic transmission, a new electrical system and a hydropneumatic suspension system.

These improvements not only make the AMX-13 SM1 more reliable but also increase the operational range of the vehicle, giving a higher speed, better acceleration and improved cross-country mobility.

Singapore Technologies Kinetics Ltd is now offering this retrofit kit for other users of the AMX-13. It is believed that Singapore Technologies Kinetics has supplied two batches of upgrade kits to Indonesia which has now upgraded part of its fleet to the AMX-13 SM1 standard.

AMX-13 SM1 light tank modernised by Singapore Technologies Kinetics Ltd **2000**/0069577

Performance comparison

Model	**AMX-13**	**AMX-13 SM1**
Power-to-weight ratio	16.6 hp/t	19 hp/t
Gradient	60%	60%
Side slope	40%	40%
Max road speed	61 km/h	64 km/h
Fuel capacity	480 litres	480 litres
Fuel consumption		
(road (litres/100 km))	110/130	100
(cross-country (litres/h))	50/60	25/30
Range		
(road)	350/400 km	450/500 km
(cross-country)	8/9 h	16/19 h
Wheel travel		
(static to bump stop)	115 mm	275 mm
Cross-country obstacle clearance	115 mm	275 mm
Suspension wheel load capacity (g)	2.25	4.5

Singapore has a requirement for a new light tank and available details of this are given in the Singapore Armed Forces new light tank entry.

As of mid-2001, no official information on a possible replacement for the AMX-13 SM1 light tank had been released.

Description

The original petrol engine has been replaced by a more fuel efficient water-cooled, turbocharged Detroit Diesel Model 6V-53T which develops 290 bhp at 2,800 rpm. This can be replaced in 50 minutes.

The engine cooling system for the upgraded AMX-13 SM1 light tank has been designed, developed and produced by the UK company, Gallay.

The AMX-13 SM1 cooling system consists of three air-cooled heat exchangers which are mounted in series. Air passes from front to back through the assembly, cooling in turn the oil from the steering brakes and differential, the engine cooling water and automatic transmission fluid.

The engine is coupled to a ZF 5WG-180 fully automatic transmission with five forward and two reverse gears. This is controlled by an Electrical Control Unit (ECU) which is programmed to protect against unsafe shifts. In case of an electrical failure there is a mechanical fallback.

A new electrical system has been fitted which is more reliable and is easier to maintain and repair. This is based on four 12 V batteries connected both in series and in parallel. If a single battery fails, the others can still supply power. In addition there is a new 28 V 160 A fully suppressed alternator. This is more reliable than the older dynamo and is easier to maintain.

A new electrical slip-ring has been fitted, as have new instruments for the driver. The new Driver Instrument Panel (DIP) has been ergonomically designed and is user friendly; it instantly displays information to the driver. A new radio harness is sealed in shrinkable polyolefin tubing which provides protection against oil, fuel and moisture. It is also heat retardant and in the event of fire does not give off toxic fumes. All sections can be quickly disconnected and replaced for repairs in the field.

The original torsion bar suspension has been replaced by a new hydropneumatic suspension system which increases total wheel travel, absorbs greater shock loads and permits greater mobility over rough terrain.

The danger of shredding tracks is reduced as the suspension system maintains a more consistent tension on the track. The pressure in each unit strut can be individually adjusted.

The AMX-13 SM1 retains the 75 mm main armament and 7.62 mm coaxial machine gun of the original AMX-13 but an additional 7.62 mm machine gun is pintle mounted on the forward part of the turret and this can be used by the commander or gunner. A bank of two electrically operated grenade launchers is mounted either side of the turret firing forwards. It is understood that the main types of ammunition fired by the 75 mm gun include HE, HEAT and a new locally produced APFSDS-T type.

Variants

STA VCI Upgrade

In February 1994, Singapore Technologies Kinetics Ltd revealed that as a private venture it had completed the prototype of an AMX VCI infantry fighting vehicle (fully covered in the Armoured personnel carriers (tracked) section) fitted with the same power pack and suspension improvements as the AMX-13 SM1 light tank.

As Singapore does not operate any AMX VCI infantry fighting vehicles, this upgrade is aimed specifically at the export market.

In addition, the upgraded AMX VCI was fitted with a new turret designed by the then Chartered Industries of Singapore (CIS) fitted with a CIS 40 mm 40 AGL grenade launcher and a CIS 12.7 mm machine gun. This is also fitted to upgraded M113 series APCs and details are given in the Armoured personnel carriers (tracked) section under Singapore.

Status

Upgrade for Singapore Armed Forces is complete. At least one foreign country has purchased this upgrade package and this is understood to be Indonesia.

Manufacturer

Singapore Technologies Kinetics Ltd.

UPDATED

TAIWAN

Type 64 light tank

Development/Description

The Type 64 light tank is similar to the US-designed and -built M41A2/M41A3 (covered in detail later in this section) in automotive and mechanical details and has the same dimensions, chassis, engine and transmission but an improved electrical system.

The hull is of all-welded steel construction as in the M41 but incorporates advanced welding techniques in its manufacture and uses a new high-hardness alloy steel.

The all-cast three-person turret has the same layout as the original M41 although slight alterations have been made to suit the medium stature of the Chinese tank crews. An additional layer of laminated high-hardness alloy steel armour plate is bolted/welded onto the hull and turret sides to provide greater protection against HEAT and APDS projectiles.

The Type 64 is armed with a 76 mm rifled gun as in the M41 but it is made in Taiwan and claimed to have a performance superior to the original 76 mm M32 gun. A 7.62 mm machine gun is mounted coaxially with the main armament and a 12.7 mm machine gun is mounted at the commander's station for use in the anti-aircraft role. Both these weapons are made in Taiwan and the coaxial weapon probably resembles a modified version of the US M60.

Chinese sources state the Type 64 has an improved fire-control system. Ammunition stowage is probably identical to the M41A2/M41A3.

Later Type 64 tanks may incorporate laser range-finders, ballistic computers and passive night vision equipment. All of these have been developed by the Sun Yat-Sen Scientific Research Institute of Taiwan.

The Type 64 has the same mobility, agility and range characteristics as the M41 but, in view of its additional armour, is probably heavier. Like the M41 it can be fitted for deep fording, but it is not amphibious and has no NBC system. It is fitted with passive night driving lights and improved Taiwan-built communications equipment.

Variant

M41D

Following extensive trials between M41 light tanks upgraded by Techmotiv of Canada (using a US Cummins VTA 903-TR diesel engine) and NAPCO International of the US (using a US Detroit Diesel 8V-71T engine) the latter was selected by the Taiwanese Army early in 1997.

Following contract negotiations, it is understood that an initial order was placed with NAPCO International for 50 M41D upgrade kits with the actual upgrade work to be carried out in Taiwan.

The replacement of the current petrol engine by the more fuel efficient Detroit Diesel 8V-71T V-8 diesel engine developing 405 hp increases the operational range of the tank from 160 to 450 km, as well as reducing the risk of fire. Maximum road speed is 72 km/h and a maximum combat weight is 25 tonnes.

It is believed that by late 1999 the first batch of vehicles had been upgraded to the M41D standard at a unit cost of NT$20 million. The upgrade will extend the life of the M41 light tank by at least 15 years.

It is estimated that Taiwan has a fleet of around 400 (although some sources have stated 650) M41 light tanks which were supplied by the US many years ago.

In the M41D upgrade the 76 mm gun has been retained as new Armour-Piercing Fin-Stabilised Discarding Sabot - Tracer (APFSDS-T) ammunition has already been purchased from AAI Corporation of the United States. The round is designated the M464 and has a muzzle velocity of 1,433 m/s.

The 76 mm gun is designated the M32K1 with 65 rounds of 76 mm ammunition carried. The machine gun mounted coaxial is the T74 with 5,000 rounds carried. A total of 2,175 rounds of 12.7 mm ammunition are carried for the roof mounted 12.7 mm M2 HB weapon.

Other improvements to the M41D light tank include the installation of side skirts to help keep down dust, a bank of six electrically operated 66 mm T85 smoke grenade launchers mounted either side of the turret and the ability to lay a smoke screen by injecting diesel fuel into the exhaust outlets.

The 76 mm rifled gun has been fitted with a fume extractor and a pepperpot-type muzzle brake. The 7.62 mm coaxial machine gun is retained and the commander has a roof-mounted 12.7 mm machine gun for anti-aircraft and local defence. It is understood that the commander also has a day/thermal sighting system. It is believed that this has been supplied by Raytheon Systems Company of the United States.

Taiwan also has a requirement for a new light tank armed with a 105 mm rifled gun and contenders for this requirement include Textron Marine & Land Systems with the Stingray II and United Defense LP with the M8 Armored Gun System. It is probable that first vehicles would be purchased from abroad with gradual licensed production being undertaken in Taiwan. As of mid 2001, no official announcement had been made concerning the selection of a new light tank for the Republic of China Army.

Status

In service with Taiwan. Production complete. Upgrade to M41D under way.

Manufacturer

Fighting Vehicles Command, Republic of China Army, Taichung, Taiwan.

Development by Fighting Vehicles Development Center Republic of China Army, Taichung, Taiwan 400.

UPDATED

UNITED KINGDOM

Alvis Scorpion light tank

Details of the Alvis Scorpion Combat Vehicle Reconnaissance (Tracked), which is also referred to as a light tank, are given in the Armoured fighting vehicle families section. Original production versions were armed with a now BAE Systems, RO Defence 76 mm gun and powered by a Jaguar petrol engine while current production models have a Cockerill 90 mm Mk III gun and are powered by a diesel engine.

The 76 mm armed Scorpion has been phased out of service with the British Army, but it remains in service with many other countries. The current production model, for export, is armed with a Cockerill 90 mm Mk III gun.

Full details of the latest Alvis Stormer 30 armoured reconnaissance vehicle are given in the following section. This has been developed as a private venture and was first shown in 1997.

Alvis Vehicles has now merged with GKN Defence and the Alvis Vehicles facility at Coventry was shut down in the second half of 1999. Production of the Alvis Vehicles Scorpion family of full-tracked vehicles is now undertaken at the previous GKN Defence facility at Telford, Shropshire. There has been no recent production of the basic Alvis Vehicles Scorpion, although upgrade packages have been provided to Oman and the UK.

UPDATED

UNITED STATES OF AMERICA

United Defense LP M8 Armored Gun System (AGS)

Development

Following an international competition, in mid-1992 the US Army selected the FMC Corporation, Ground Systems Division (now United Defense LP), to design and build its Armored Gun System (AGS) to replace the M551 Sheridan light tanks used by the 3rd Battalion, 73rd Armor of the 82nd Airborne Division.

The M551 has now been retired from the US Army without replacement. A number of these vehicles is still, however, used in the training role, especially at the National Training Center. Details of the M551 were given in *Jane's Armour and Artillery 1996-97* pages 174-175.

The XM8 was Type Classified as the M8 Armored Gun System in October 1995 by the US Army and declared production ready. However, on 5 February, 1996 the US Army issued a Partial Stop Work notice to United Defense for the final phase of the EMD contract.

This was followed by formal cancellation of the M8 AGS.

United Defense continues to market the M8 AGS overseas, especially Asia, the Middle East and Europe and one of the six prototypes of the vehicle was shown in Europe in June 1996.

Late in 1997, FNSS Defense Systems of Turkey and United Defense LP, teamed to offer the M8 AGS to the Turkish Land Forces Command (TLFC).

In September 1997, the second of the six prototype M8 AGS built with the highest protection Level III was demonstrated in Turkey during which it undertook firepower and mobility demonstrations for the TLFC.

According to United Defense LP, the M8 AGS is well suited to peacekeeping operations, such as Bosnia, where the TLFC has been deployed, as well as being used in traditional roles of reconnaissance.

As well as being less expensive to procure than an MBT, the M8 AGS has a much lower life cycle cost. According to United Defense, the fuel cost of the M8 AGS is US$0.60 per km compared to US$5.35 per km of an MBT with spare parts per km being US$35 and US$137 respectively.

The M8 is armed with a 105 mm M35 rifled gun, fed by an automatic loader, which fires the same natures of ammunition as all the MBTs (Leopard 1, M48 and M60 series) currently in service with the TLFC.

If selected by Turkey, the M8 would be manufactured under licence by FNSS Defense Systems which is a joint venture company established to build Armoured Infantry Fighting Vehicles (details of this programme are given in the Armoured personnel carriers (tracked) section under Turkey.

The existing Turkish AIFV machinery, tooling and fixtures could be reconfigured for the M8 AGS with a minimum of investment, according to United Defense.

As of mid 2001, there are no plans for the TLFC to adopt the M8 AGS as the acquisition of a new MBT and the upgrading of at least part of the existing fleet of M60 series MBTs with a 120 mm smoothbore gun has a higher priority.

In March 1999, the M8 AGS was demonstrated to officials of the Japan Ground Self-Defence Force which is understood to be formulating its requirements for a new light tank. Japanese officials have also visited companies in France, Germany and Italy.

The M8 AGS was evaluated by the US Army for use by its new Brigade Combat Teams but in the end it was decided to purchase an all-wheeled fleet of armoured vehicles based around the General Motors Defense LAV-III (8 × 8).

Description

The basic hull and turret of the M8 AGS is of all-welded aluminium armour construction which provides protection from small arms fire and shell splinters and to this is fitted a modular armour package.

The modular armour package permits the M8 AGS to be equipped with three increasing levels of external armour protection to meet differing threats.

The level one package, which protects against small arms fire and artillery splinters, is designed for rapid deployment forces and gives a gross vehicle weight of 18,052 kg.

The level two armour package provides light forces with a higher level of protection and a gross combat weight of 20,820 kg.

Contingency forces expecting to face higher threats can equip the M8 AGS with level three armour package resulting in a gross weight of about 23,586 kg.

The heavier armour package will provide the M8 AGS with protection against threats up to and including hand-held anti-tank weapons.

The driver is seated at the front of the vehicle in the centre and has a single-piece hatch cover that lifts and then opens to the rear. This has five integral wide-angle day periscopes for driving when closed down and, in addition, provisions are made for mounting a driver's passive night driving periscope.

Suspension is of the torsion bar type with six dual rubber-tyred roadwheels, idler at the front, drive sprocket at the rear and no track-return rollers. High-temperature shock-absorbers are provided at the first, second, third, fifth and sixth roadwheel positions. The suspension spring system provides a 250 mm jounce.

The front-mounted idler assembly is composed of standard United Defense, Ground Systems Division M113A3 and M2 parts already in the US Army inventory. The track is of the steel double pin T150 type with detachable rubber pads, 152 mm pitch and 381 mm wide and has been developed by United Defense LP, Steel Products Division.

The M8 AGS is powered by a six-cylinder Detroit Diesel Model 6V-92TA developing 550 hp at 2,400 rpm on JP8 fuel and 580 hp on DF2 diesel fuel. This uses 65 per cent common components with the 8V-92TA engine installed in the Oshkosh Heavy Expanded Mobility Tactical Truck (HEMTT) M977 (8 × 8) already in service with the US Army in significant numbers.

The engine is coupled to a General Dynamics Land Systems HMPT-500-3EC hydromechanical transmission, which is already installed in the M2 Bradley IFV, with a final drive ratio of 4.4:1.

The rear engine compartment roof is raised well above the turret ring so the 105 mm gun depression is

Specifications

Crew: 3
Airdrop weight: (level 1) 16,730 kg
Combat weight:
(level 1) 18,052 kg
(level 3) 23,590 kg
Roll-on/roll-off weight:
(level 2 combat) 20,410 kg
Power-to-weight ratio: (level 1 combat) 30.5 hp/t
Ground pressure: (level 2 combat weight) 0.66 kg/cm^2
Length:
(gun forward) 8.97 m
(hull) 6.1 m
Width:
(overall) 2.69 m
(over tracks) 2.54 m
Height:
(commander's cupola) 2.55 m
(turret roof) 2.34 m
(hull top rear) 1.77 m
(hull top forward) 1.49 m
Ground clearance: 0.41 m
Track width: 381 mm (steel, double pin type with detachable rubber pads)
Length of track on ground: 3.61 m
Max road speed: 72.42 km/h
Acceleration: (0 to 32 km/h, level 3 combat weight) 7.5 s
Range: (at 40 km/h) 451 km
Fuel capacity: 568 litres
Fording: 1.02 m
Gradient: 60%
Side slope: 40%
Vertical obstacle: 0.812 m
Trench: 2.06 m
Engine: Detroit Diesel Model 6V-92TIA developing 550 hp at 2,400 rpm, 580 hp on DF2 diesel fuel
Transmission: General Dynamics Land Systems HMPT-500-3EC hydromechanical
Steering: hydrostatic
Brakes: multidisc, oil-cooled
Final drive ratio: 4.4:1
Suspension: torsion bar
Road wheel travel: 254 mm (max)
Electrical system: 24 V
Batteries: 4 × 6TN
Generator: 300 Ah
Armament:
(main) 1 × 105 mm M35 soft recoil rifled gun
(coaxial) 1 × 7.62 mm M240 MG
(commander) 1 × 12.7 mm M2 MG or 1 × 7.62 mm MG or 1 × 40 mm Mk 19 grenade launcher
Smoke-laying equipment: 2 × 8 smoke grenade launchers
Ammunition:
(main) 30 (21 of which are ready use)
(coaxial) 4,500 (1,000 of which are ready use)
(commander) 600 (100 of which are ready use)
(grenades) 32 (16 of which are in launcher)

Gun control equipment
Turret power control: hydraulic/manual
(by commander) yes
(by gunner) yes
Gun elevation/depression: +20/–10°
Turret traverse: 360°
Gun stabiliser: yes
Sights stabilised: yes
NBC system: yes
Night vision equipment: yes

United Defense M8 Armored Gun System (AGS) in Level 1 configuration showing 105 mm M35 gun. The 12.7 mm M2 machine gun is not fitted
2002/0114264

United Defense M8 Armored Gun System (AGS) in Level 2 configuration showing explosive reactive armour package and fitted with roof-mounted 12.7 mm M2 machine gun
2002/0114265

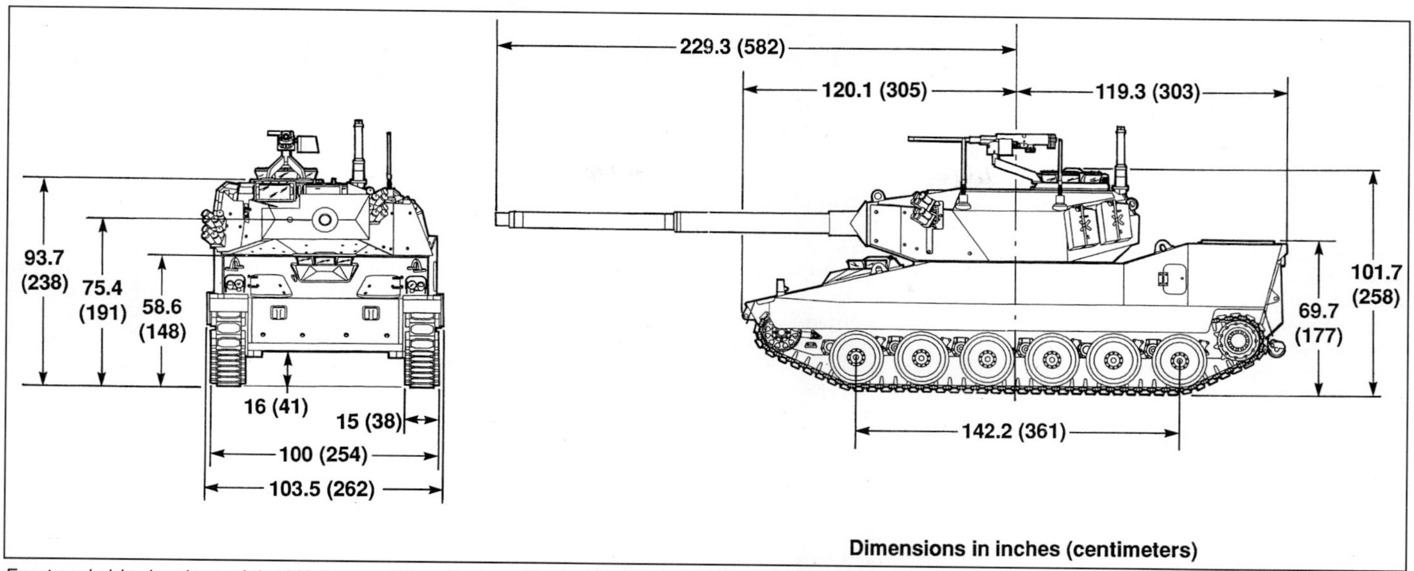

Front and side drawings of the M8 Armored Gun System fitted with a 12.7 mm M2 anti-aircraft machine gun

limited to 0° over the rear arc. The air inlets/outlets are in the roof of the engine compartment with the exhaust at the left.

At the rear of the hull is a hydraulically actuated access ramp which, when horizontal, provides a work platform for ease of maintenance.

The engine, transmission and cooling system are mounted on rails and can be easily cranked out on to the ramp. Trained personnel can remove the complete power pack to the ramp platform where it can be ground hopped in less than 5 minutes. The M2 Bradley fan is belt-driven from a transmission-mounted PTO.

The turret is of all-welded aluminium with the automatic loader on the left and the commander and gunner on the right.

The gunner is seated forward of the commander and has a single-piece hatch cover that opens forwards. Forward of this is the Raytheon Systems Company two-axis stabilised day/thermal night sight with laser range-finder. Both commander and gunner have displays enabling either crew member to use the day or thermal channel. As a back-up, the gunner also has a fibre optic auxiliary sight mounted coaxially to the right of the main armament.

The commander is seated to the rear of the gunner and has a rotatable cupola with seven day periscopes identical to the driver's for all-round observation (unity vision) and a single-piece hatch cover that opens to the rear.

The Raytheon Systems Company gunner's sight is based on a company-developed sight selected for the US Marine Corps LAV-105 assault gun which includes proven components from a number of Raytheon Systems Company products. The sight includes the Raytheon Systems Company HIRE, this is an Infra-Red Equipment (HIRE) which is a high-performance thermal imaging system consisting of three modular components, the sensor and gunner's display, commander's remote display and the power supply/ electronics. The sight also incorporates the laser range-finder from the M1 MBT and the line of sight stabilisation platform based on the South Korean K-1 MBT.

Turret and gun drives have been developed by HR Textron Incorporated and are based on those installed in the General Dynamics Land Systems M1/M1A1/ M1A2 MBT in service with the US Army. The gunner has twin control handles while the commander has a single control handle.

The 105 mm M35 gun is fully stabilised in both axes with an elevation of +20° and a depression of −10° through a frontal 270° arc. Turret traverse is a full 360°. Turret traverse and weapon elevation are hydraulic with manual controls for emergency use.

The Computing Devices Company of Canada digital fire-control computer is a modified version of that installed in the Challenger 2 MBT in service with Oman and the UK. Mounted on the turret roof at the rear is the wind sensor which feeds information to the fire-control computer.

Main armament comprises a 105 mm M35 rifled tank gun which includes a low-force gun mount. The six M8 AGS built so far have a pepperpot muzzle brake but these would probably not be fitted to production vehicles.

Mounted in the left side of the turret is an automatic loader which enables a rate of fire of 12 rds/min to be achieved. This has been designed and built by United Defense, Armament Systems Division.

Once the target has been designated, the gunner selects either automatic (for multiple shots) or single-shot modes as well as the type of round desired. The ammunition is automatically loaded at the rate of one round every 5 seconds and the system can be operated on the move. A back-up design feature allows manual loading of the main gun from under armour should it become necessary.

The automatic loader holds 21 rounds of 105 mm ammunition stowed vertically base downward. A chain drive moves the rounds along a crescent-shaped pathway to the loading position. A loading arm with a ram tray assembly then takes the round and inserts it in the breech of the gun in one upward and forward movement.

Once the gun has fired, it recoils 533 mm to the rear and returns to battery. The 105 mm gun returns to 0° for case ejection and reloading. The breech is hydraulically opened and the empty case is ejected through the loading port in the turret bustle. When the gun is depressed below 7°, a linkage-operated gun cover in the roof lifts to clear the recoiling breech.

The automatic loader is separated from the crew compartment by a bulkhead containing an access panel for manual loading. Blow-out panels are provided in the turret roof.

The 105 mm M35 gun fires all standard US NATO tank ammunition including APFSDS-T, APERS, WP, HEAT and HESH and their respective training rounds.

A standard 7.62 mm M240 machine gun is mounted coaxially to the right of the main armament, provided with 1,000 rounds of ready use ammunition and a further 3,500 rounds of ammunition carried in reserve in the turret.

A number of weapons can be mounted externally on the commander's cupola including a 12.7 mm M2 machine gun, 7.62 mm machine gun or a 40 mm Mk 19 grenade launcher.

Mounted either side of the turret is a bank of eight electrically operated grenade launchers which can fire a wide variety of grenades over the frontal arc and a missile countermeasure package can be fitted to the M8.

The M8 AGS is fully transportable in a Lockheed Martin C-130 Hercules transport aircraft. With the level one armour package it can be air-dropped, with the level two armour package it can be driven on and off a C-130, while with the heaviest level three armour package it is air-transportable in a C-17 transport. The C-17 can carry three M8s while the C-5 can carry five M8 vehicles.

Status

Development complete. Type Classified by US Army as M8 Armored Gun System but not funded for production.

Manufacturer

United Defense LP, Ground Systems Division.

UPDATED

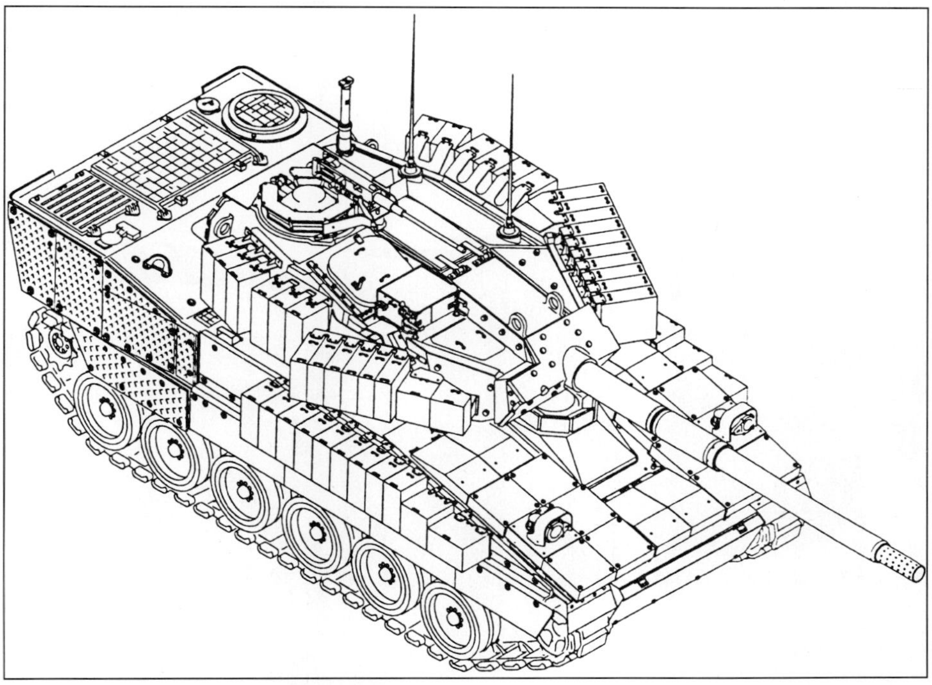

M8 Armored Gun System in Level 3 configuration

Cadillac Gage Stingray II light tank

Development

Late in 1996, Textron Marine & Land Systems completed the first preproduction example of the Stingray II light tank which has been developed by the company as a private venture specifically for the export market.

Stingray II is a further evolutionary development of the Stingray I light tank (covered in detail in a separate entry) which is currently in service with the Royal Thai Army (RTA) which took delivery of 106 vehicles between 1986 and 1990.

Like the earlier vehicle, the Stingray II uses proven and in-production subsystems wherever possible to reduce procurement and total life cycle costs.

According to Textron Marine & Land Systems, key features of the latest Stingray II can be summarised as maximised survivability, superior mobility and agility, improved target engagement capabilities, is production ready and uses proven and in-production automotive, turret and weapons. As of mid-2001, no production contracts for the Stingray II light tank had been announced, although marketing continues.

Specifications

Crew: 4
Combat weight: 22,600 kg
Power-to-weight ratio: 24.43 hp/t
Ground pressure: 0.80 kg/cm²
Length:
(gun forward) 9.35 m
(hull) 6.44 m
Width: 2.71 m
Height: (overall) 2.55 m
Ground clearance: 0.46 m
Track: 2.31 m
Track width: 380 mm (see text)
Length of track on ground: 3.63 m
Max road speed: 70 km/h
Max range (at 48 km/h): 525 km
Fording: 1.07 m
Gradient: 60%
Side slope: 40%
Vertical obstacle: 0.84 m
Trench: 2.13 m
Engine: Detroit Diesel Model 8V-92TA developing 550 hp at 2,300 rpm
Transmission: Allison Transmission XTG-411 fully automatic
Steer unit: integrated into transmission
Final drive: planetary 4:1 ratio
Suspension system: independent trailing arm
Electrical system: 24 V
Batteries: 4 × 6TL
Armament:
(main) 1 × 105 mm LRF rifled gun
(coaxial) 1 × 7.62 mm M240 MG
(anti-aircraft) 1 × 12.7 mm M2 MG
Smoke-laying equipment: 2 × 4 electrically operated grenade launchers
Ammunition:
(main) 32 (8 ready, 24 stowed)
(7.62 mm) 2,400 (400 ready, 2,000 stowed)
(12.7 mm) 1,100 (100 ready, 1,000 stowed)
Gun control equipment
Turret power control: electrohydraulic/manual
(by commander) yes
(by gunner) yes
Gun elevation/depression: +20/–8°
Gun stabiliser: optional
Range setting device: yes (laser)
NBC system: yes
Night vision equipment: yes

Description

The overall layout of the Stingray II is identical to the Stingray I with driver in the front, three-man power-operated turret in the centre and power pack at the rear.

For improved battlefield survivability over the frontal arc, Stingray II is provided with the recently developed '2001' special high-hardness steel armour which provides protection against small arms fire up to 23 mm calibre. To protect the upper part of the suspension, side skirts with a similar level of protection are fitted.

These can be quickly removed allowing access to the suspension for maintenance purposes.

With this armour package fitted, Stingray II weighs 22,600 kg. If the user requires a higher level of protection then an enhanced appliqué armour kit can be provided.

This can be fitted by the user using standard tools in between 2 and 4 hours. When fitted with the enhanced armour package Stingray II weighs 26,308 kg, with protection over the frontal arc being provided against RPG-7-type hand-held anti-tank weapons.

To provide a high degree of hit probability against stationary and moving targets the latest M1A1 digital fire-control system is fitted to the Stingray II.

At present the stabilised gun control system is of the electrohydraulic type but, if required by the user, an all-electric system can be provided. An automatic deck clearance system is fitted as standard on the Stingray II.

The digital fire-control system includes the gunner's day/night range sight (DNRS), the electrohydraulic weapon/turret drive and stabilisation system (WTDS), Digital Fire-Control Computer (from Computing Devices Canada), computer control panel and the gunner's auxiliary sight (GAS) back-up telescopic sight.

The DFCC is based on that used in the M1 series MBT and incorporates a serial data interface with the CCP and facilitates the manual entry of all parameters affecting the ballistic offsets (range, wind, lead, cant, barometric pressure, air temperatures, ammunition temperature, barrel wear, ammunition type and zeroing data), display of these parameters, display of the ballistic and final positional offsets, manual request for system self-test and display of associated results.

The cant sensor is a simple and rugged pendulum device located on the ceiling above the gunner. The pendulum is magnetically dampened and a rugged potentiometer is used to measure pendulum position.

Mounted above the turret bustle is the crosswind sensor which measures the crosswind component of the wind at the vehicle.

The gunner has the latest proven stabilised Raytheon Systems Company HIRE day/thermal night sight with integrated laser range-finder which is currently in volume production for a number of other applications. The gunner also has an auxiliary day telescope as a back-up with a magnification of × 6.2.

The commander has an M36E1 day/image intensification night sight and is provided with a monitor which enables him to have the same thermal picture of the target as the gunner. The commander also has seven day periscopes for all round observation.

The power pack consists of a Detroit Diesel Model 8V-92TA diesel developing 550 hp at 2,300 rpm coupled to an Allison XTG-411 fully automatic transmission with four forward and two reverse gears and Textron Marine & Land Systems final drives.

The XTG-411 features a high-efficiency torque converter with automatic lockup clutch, clutch brake/geared steer and mechanical brakes for service and parking.

The upgraded independent trailing arm suspension either side consists of six dual rubber-tyred roadwheels with the idler at the front, drive sprocket at the rear and three track-return rollers. The standard tracks are 380 mm wide but these can be replaced by 460 mm wide tracks when the appliqué armour package is fitted to enable ground pressure to be maintained.

The main armament of the Stingray II is still the stabilised BAE Systems, RO Defence 105 mm Low-Recoil Force (LRF) gun with a 7.62 mm M240 coaxial machine gun and a 12.7 mm M2 anti-aircraft machine gun. The 105 mm LRF gun is fully stabilised in two axes.

Optional equipment includes various radio and communications systems, a driver's passive night viewer, a GPS navigation system and a ventilated face mask NBC system.

Typical communications fit would be an AN/VRC – 90 (V) high-power SINCGARS radio, RT7000 high-frequency radio (option) and an AN/VIC-3 intercom system.

It can also be fitted with a laser warning system and a system that enables Stingray II to lay its own smoke screen by injecting diesel fuel into the exhaust outlet.

Variants

Stingray II with LAV-105 turret

Studies by Textron Marine & Land Systems have shown that the Stingray II can be fitted with the complete turret of the LAV-105 (8 × 8) assault gun covered in the following section.

The LAV-105 turret was developed by Textron Marine & Land Systems and is armed with a 105 mm M35 gun fed by an automatic loader developed by FHL of the UK.

This version of Stingray II would have a crew of three consisting of commander, gunner and driver and have a combat weight of 21,100 kg.

To improve battlefield survivability, the Stingray II with the LAV-105 turret can also be fitted with a laminate armour package that covers the frontal arc of the hull and turret. This increases the combat weight of the vehicle to 25,500 kg.

Status

Development complete. Ready for production.

Manufacturer

Textron Marine & Land Systems.

UPDATED

Stingray II light tank showing BAE Systems, RO Defence 105 mm Low-Recoil Force gun fitted with muzzle brake, thermal sleeve and fume extractor
0045911

Cadillac Gage Stingray light tank

Development

The Stingray light tank was developed for the export market developed by Cadillac Gage Textron (now Textron Marine & Land Systems) as a private venture.

The concept phase of Stingray development started in January 1983 with actual design work beginning in September the same year. Prototype construction began in February 1984 and the vehicle, without its turret, ran for the first time in August 1984. Design work on the turret began in mid-1983, the prototype turret being completed in June 1984 and installed on an M551 Sheridan light tank chassis for trials.

The Stingray light tank was shown in public for the first time in October 1984.

The prototype Stingray light tank went to Thailand for trials early in 1986 and the second prototype Stingray, which was built to production drawings, was completed in mid-1986. Early in 1987, the first prototype of the Stingray was demonstrated in Malaysia.

In October 1987, Cadillac Gage Textron signed an agreement with Thailand for the supply of 106 Stingray light tanks to the Royal Thai Army at a total cost of US$150 million. First production Stingray light tanks were completed in December 1988 and shipped direct to Thailand. Final deliveries were made early in 1990.

Further development by Textron Marine & Land Systems has resulted in the Stingray II which is covered in detail in the previous entry and marketing is now being concentrated on this vehicle, especially in Asia. As of mid-2001, the Stingray II remained at the prototype stage.

Description

The hull of the Stingray is of all-welded Cadloy steel armour with protection over the frontal arc against penetration by 14.5 mm armour-piercing rounds fired from the Russian KPVT heavy machine gun and protection against 7.62 mm armour-piercing rounds over the remainder of the vehicle.

The layout of the Stingray is conventional, with the driver's compartment at the front, fighting compartment in the centre and engine and transmission at the rear.

The driver sits in the centre of the hull at the front and has a single-piece hatch cover that is hinged at the rear and can be locked open. The hatch cover has three integral day periscopes which give good visibility over 120° of the vehicle's front. The centre periscope can be replaced by a passive periscope for night driving. The driver has an adjustable seat and would normally enter his position via the turret. With the turret traversed fully left or right or with the 105 mm gun in its travel lock, the driver can easily enter via his hatch.

The driver steers the Stingray with an oval steering wheel rather than the more usual sticks. Either side of the driver's position are 14 rounds of 105 mm ammunition. When in a combat area, an optional spall blanket is hung to the sides and rear of the driver's position for added protection. The Stingray has additional internal stowage space above the tracks with an access hatch in the glacis plate either side of the driver.

The Stingray suspension is of the torsion bar type and based on that fitted to the United Defense, Ground Systems Division, M109 155 mm self-propelled howitzer.

The suspension either side consists of six dual rubber-tyred roadwheels, drive sprocket at the rear, idler at the front and three track-return rollers, the last from the M41 light tank. The torsion bars are longer than those of the M109 and are of 166,000 psi steel with a 203.2 mm jounce, 50.8 mm rebound and bottom at 2.5 *g* loading. The first and last torsion bars are stiffer than the remainder and the first, second and sixth roadwheel stations have a hydraulic shock-absorber.

The road arms were originally designed to carry 1,814 kg but in the Stingray carry only 1,588 kg, therefore leaving plenty of scope for additional weight, for example appliqué armour should this be required, without overloading the suspension.

The tracks are 380 mm wide and are of the double-pin type with detachable pads.

The Stingray is powered by the Detroit Diesel, Model 8V-92TA eight-cylinder diesel developing 535 bhp at 2,300 rpm. The engine is mounted crossways to make maximum use of the available space. The engine has twin turbochargers, a Roots blower and an aftercooler.

The engine is coupled to an Allison Transmission XTG-411-2A automatic transmission which has an automatic lock-up torque converter with a torque multiplication of 2.4, geared steer and clutch and brake steering. The 4:1 planetary final drive gearboxes have been designed and built by the company specifically for the Stingray.

Cooling is by three-vane axial fans each of which moves 9,000 cfm. The radiators and oil coolers are sized to provide adequate cooling to 120°C. The air inlets are above the engine compartment, with the air outlet at the left rear of the hull and the exhaust outlet on the right side of the hull. Special attention has been given to making the infra-red signature of the vehicle as small as possible. An extra heavy-duty air cleaner scavenges heavier dirt into the engine exhaust.

Electrical power is provided by a 650 A oil-cooled generator and stored in six batteries in the left side of the hull. Access to these batteries for daily checks is via a door in the side of the hull. The batteries are capable of furnishing start-up current in temperatures as low as −25°F.

The rear engine compartment panels fold down for access to the engine while those on top fold rearwards. Daily checks can easily be carried out on the engine.

The 757 litre fuel tank is integral and located between the crew and engine compartments.

The now BAE Systems, RO Defence 105 mm Low-Recoil Force (LRF) gun has a muzzle brake added, a redesigned fume extractor, a cradle design and a new recoil system. The new cradle design is a monocoque structure that combines high rigidity with light weight.

The standard 105 mm L7A3/M68 gun has a recoil of 300 mm which takes 70 ms and achieves a maximum trunnion force of 34,020 kg. The LRF gun has a longer recoil and achieves a maximum trunnion force of 13,608 kg. In addition to the new recoil system with its longer recoil, the muzzle brake also helps to reduce both trunnion pull and recoil energy.

There are 32 rounds of 105 mm ammunition carried, all of which are stowed below the turret ring. Eight rounds are stowed in the turret, three of which are for immediate use, stowed vertically to the left of the breech and the remainder under the spent case ejection bag. The three ready rounds are kept in special individual racks which are configured to accept all types of 105 mm ammunition. Once the loader releases the catch the first ready round swings forward as it is pivoted at the base. The spent case ejection bag holds five cases which can be disposed of through the ammunition resupply hatch in the left side of the turret.

Following initial trials with the prototype Stingray, a muzzle reference system was fitted to the 105 mm LRF gun and a collapsing breech guard fitted to allow for improved crew movement in the turret.

A 7.62 mm M240 machine gun is mounted coaxially to the left of the main armament and is provided with 400 rounds of ready use ammunition. A further 2,000 rounds of 7.62 mm ammunition are stowed in the turret area in 200-round boxes. It takes the loader 7 seconds to replenish the ready use ammunition for the coaxial machine gun.

A 7.62 mm M240 (with 200 rounds of ready use ammunition) or a 12.7 mm M2 machine gun (with 100 rounds of ready use ammunition) is mounted externally at the commander's station for anti-aircraft use. A total of 1,000 rounds of machine gun ammunition is carried for the M2 anti-aircraft weapon.

A bank of four electrically operated grenade dischargers is mounted either side of the turret, for which 16 grenades are carried.

The three-man turret of the Stingray has been designed for installation on other chassis as well as on Stingray, including the Cadillac Gage LAV-600 (6 × 6),

Cadillac Gage Stingray light tank from the rear ***2002***/0114266

Specifications

Crew: 4
Combat weight: 21,205 kg
Unloaded weight: 19,387 kg
Power-to-weight ratio: 25.9 hp/t
Ground pressure: 0.72 kg/cm^2
Length:
(gun forward) 9.30 m
(hull) 6.448 m
Width: 2.71 m
Height:
(overall) 2.55 m
(reduced) 2.4 m
Ground clearance: 0.46 m
Track: 2.31 m
Track width: 380 mm
Length of track on ground: 3.632 m
Max speed: 67 km/h
(road, on 10% slope) 29 km/h
(road, on 60% slope) 6 km/h
Acceleration: (0 to 32 km/h) 7 s
Fuel capacity: 757 litres
Max range: 483 km (at 40 km/h)
Fording: 1.07 m
Gradient: 60%
Side slope: 40%
Vertical obstacle: 0.76 m
Trench: 2.13 m
Engine: Detroit Diesel Model 8V-92TA developing 535 hp at 2,300 rpm
Transmission: Allison Transmission XTG-411-2A automatic
Steer unit: integrated into XTG-411 transmission
Final drive: planetary, 4:1 ratio
Suspension: torsion bar
Electrical system: 24 V
Batteries: 4 × 6TN
Armament:
(main) 1 × 105 mm LRF rifled gun
(coaxial) 1 × 7.62 mm M240 MG
(anti-aircraft) 1 × 12.7 mm M2 MG
Smoke-laying equipment: 2 × 4 smoke grenade launchers
Ammunition:
(main) 32 (8 ready, 24 stowed)
(7.62 mm) 2,400
(12.7 mm) 1,100

Gun control equipment
Turret power control: electrohydraulic/manual
(by commander) yes
(by gunner) yes
Max rate of turret:
(traverse) 360° at 30°/s (stabilised)
(elevation) 8°/s, 40°/s (stabilised)
Gun elevation/depression: +20 to 7.5°
Gun stabiliser: optional
Range setting device: yes (laser)
NBC system: optional
Night vision equipment: yes

M41 and M551 tanks and heavier chassis such as the M47, M48 and T-series, or any vehicle with a chassis weight exceeding 13 tonnes.

The turret is of all-welded Cadloy steel armour construction providing the same level of protection as the hull. The use of flat plates in the construction of the turret enables the user to up-armour the turret at a later date without extensive rework.

The commander sits on the right of the turret with the gunner below and to his front and the loader on the left. The gunner has a roof-mounted Optic-Electronic Corporation M36E1 day/night sight. As an option this can be replaced by an M36E1 SIRE day/night sight incorporating a laser range-finder or a thermal sight. The commander has an NV-52 day/night sight, seven periscopes for all-round observation and a single-piece hatch cover that opens to the rear.

The loader has a single-piece hatch cover that opens to the rear and a single forward-facing periscope for observation.

Cadillac Gage Stingray light tank from the front ***2002***/0114267

Turret traverse and weapon elevation are via an HR Textron Incorporated electrohydraulic system with a manual back-up. Controls are provided for both the commander and gunner and turret traverse is a full 360° at 40°/s, with weapon elevation from –7.5 to +20° at a similar speed. An automatic deck clearance system is fitted as standard and as an option, a two-axis electrohydraulic stabilisation system can be fitted, enabling the 105 mm gun to be laid and fired while Stingray is moving across country.

Prototypes and production Stingray light tanks have been fitted with the then Marconi Electronic Systems, Land and Naval Systems Group, Digital Fire-Control System (DFCS). This is now part of BAE Systems, RO Defence.

The Stingray can be fitted with the standard M13A1 ventilated face mask system for NBC protection and the vehicle can be painted in chemical-resistant paint.

As an option Stingray can be fitted with a land navigation system and an engine smoke generator which injects diesel oil into the exhaust to create a smoke-screen behind the tank. The prototype is fitted with an HTL engine fire warning and suppression system as well as a fire detection/suppression system for the crew compartment. Radios and crew intercom depend on the user's requirements.

Status

Production complete. In service with Thailand (106). Marketing is now being concentrated on the latest Stingray II covered in a previous entry.

Manufacturer

Textron Marine & Land Systems.

UPDATED

M41 light tank

Development

Production of the M41 light tank was undertaken by the Cadillac Motor Car Division of General Motors Corporation at Cleveland Tank Arsenal.

First production M41s were completed in mid-1951 and 1,802 were built before production was switched to the later M41A1, followed by the M41A2 and finally the M41A3 which differ only in minor details. Total production of the vehicle amounted to about 5,500 units. Its full US Army designation is Tank, Combat, Full-Tracked: 76 mm Gun, M41. It was replaced in the US Army by the M551 Sheridan but is still used by some other armies. Many of the components of the M41 are also used in the M42 twin 40 mm self-propelled anti-aircraft gun system (also built at Cleveland), and the M44 155 mm and M52 105 mm self-propelled howitzers.

Description

The all-welded steel hull of the M41 light tank is divided into three, with the driver's compartment at the front, the fighting compartment in the centre and the power pack at the rear.

The driver is seated at the front of the vehicle on the left side and is provided with a single-piece hatch cover that opens to the right. There are three M17 day periscopes mounted forward of this position and a single one to the left. A hull escape hatch is provided beneath the driver's seat.

The turret, which is provided with a basket, is of cast and welded construction with the commander and gunner seated on the right and the loader on the left. The commander's cupola has a single-piece hatch cover that opens forwards and five vision blocks and an M20A1 day periscope that can be traversed through 360°. The gunner has an M97A1 day telescopic sight for aiming the 76 mm M32 gun and an M20A1 day periscope that can be traversed through 360°. The loader is provided with a single-piece hatch cover that opens forwards and a single M13 periscope. Mounted at the rear of the turret is a light sheet metal stowage box and there is a dome-shaped ventilator in the turret roof towards the rear. The ventilator blower is operated from the driver's position.

The power pack compartment is at the rear of the hull and is separated from the fighting compartment by a fireproof bulkhead. It is equipped with a fire extinguisher that is operated by the driver. The engine is mounted towards the front of the engine compartment with the transmission at the rear.

The torsion bar suspension consists of five dual rubber-tyred roadwheels each side with the drive sprocket at the rear, idler at the front and three track-return rollers. The first, second and fifth roadwheel stations are provided with a hydraulic shock-absorber. The steel tracks, which have 75 or 76 links when new, have detachable rubber pads.

Standard equipment on all tanks includes a heater, deep fording equipment and electric bilge pumps. It is not fitted with an NBC system. The basic model was not fitted with night vision equipment although the final production model did have provision for an infra-red searchlight over the main armament.

Main armament of the M41 consists of a 76 mm gun M32 (T91E3) in a mount M76 (T138E1); the M41A1 has the M32A1 gun in mount M76A1. The gun has a vertical sliding breech block, a spring actuator and an inertia percussion firing mechanism. The recoil system is of the concentric hydrospring type.

A 7.62 mm (0.30) Browning M1919A4E1 machine gun is mounted coaxially to the right of the main armament and an anti-aircraft 12.7 mm (0.50) Browning M2 HB machine gun is mounted at the commander's position. This has a traverse of 360°, elevation limits being from –10 to +65°.

The 76 mm gun fires the following types of fixed ammunition: AP-T (M339), Blank (M355A2), Canister (M363), HE (M352), HEAT-T (M496), HVAP-DS-T (M331A1/M331A2), HVAP-T (M319), TP-T (M340/M340A1) and WP Smoke (M361/M361A1).

In 1982, it was revealed that AAI Corporation had developed a new 76 mm APFSDS-T round for the

M41 light tank from rear with roof-mounted 12.7 mm M2 HB machine gun covered up (US Army) ***2002***/0114268

M41 light tank used by Guatemala (DIDE via Julio Montes)

Specifications
Crew: 4
Combat weight: 23,495 kg
Unloaded weight: 18,457 kg
Power-to-weight ratio: 21.26 bhp/t
Ground pressure: 0.72 kg/cm²
Length:
(gun forward) 8.212 m
(hull) 5.819 m
Width: 3.198 m
Height:
(to cupola) 2.726 m
(including MG) 3.075 m
Ground clearance: 0.45 m
Track: 2.602 m
Track width: 533 mm
Length of track on ground: 3.251 m
Max road speed: 72 km/h
Fuel capacity: 530 litres
Max range: 161 km
Fuel consumption: 3.29 litres/km
Fording: 1.016 m
(with preparation) 2.44 m
Gradient: 60%
Side slope: 30%
Vertical obstacle: 0.711 m
Trench: 1.828 m
Turning radius: skid turns
Engine: M41 and M41A1, Continental AOS-895-3, 6-cylinder air-cooled supercharged petrol developing 500 bhp at 2,800 rpm. M41A2 and M41A3 Continental or Lycoming AOSI-895-5 with fuel injection also developing 500 bhp at 2,800 rpm
Auxiliary engine: GMC Model A41-1
Transmission: GMC Allison Division cross-drive Model CD-500-3 with 1 forward and 1 reverse ranges
Final reduction ratio: 4.25:1
Suspension: torsion bar
Electrical system: 24 V
Batteries: 4 × 6TN, 100 Ah
Armament:
(main) 1 × 76 mm M32 rifled gun
(coaxial) 1 × 7.62 mm M1919A4E1 MG
(anti-aircraft) 1 × 12.7 mm M2 MG
Ammunition:
(main) 57 rounds M41, 65 rounds M41A1, M41A2 and M41A3
(coaxial) 5,000
(anti-aircraft) 2,175
Gun control equipment
Turret power control: hydraulic/electric with manual controls
(by commander) yes (only on M41A1 and later models)
(by gunner) yes
Max rate power: (traverse) 360° in 10 s
Gun elevation/depression: +19° 45′/–9° 45′
Gun stabiliser:
(vertical) no
(horizontal) no
Range-finder: none
Elevation quadrant: Type M9
Gunner's sight: M97 telescope with ×8 magnification and 7° 24′ field of view
Both the commander and the gunner have an M20A1 periscope with 2 optical systems: ×1 system for observation and ×6 for ranging on targets. Illumination of graticule patterns are provided for by a Light Instrument M36
Armour
Hull front glacis: 25.4 mm at 30°
Hull front nose: 31.75 mm at 45°
Hull top: 12/15 mm
Hull floor: 9.25/31.75 mm
Hull rear: 19 mm
Turret mantlet: 38 mm
Turret front: 25.4 mm
Turret sides: 25 mm
Turret rear: 25 mm
Turret roof: 12.7 mm
NBC system: no
Night vision equipment: yes

M32 cannon. This was developed as a private venture for the export market and has been sold to Denmark and Taiwan.

The M41A2 is almost identical to the M41A1 except that it has a simplified turret and a gun control system developed by Cadillac. This consists of manual and hydraulic power traverse for the gunner, with direct mechanically linked control of oil gear pump in lieu of the electrical control in the M41; dual-power traverse by the commander; and manual mechanical rack and pinion-type elevation for the gunner, with slewing elevation control for the gun. This more compact system enabled the tank to carry 65 rounds of 76 mm ammunition compared with 57 on the earlier model. These modifications were incorporated in the M41A2 and the M41A3, both of which also have the fuel injection petrol engine fitted.

Variants

Brazilian M41 Programme

The Bernardini company of São Paulo, has now converted all Brazilian Army M41 light tanks into the M41B/M41C configuration.

The original 500 hp petrol engine has been replaced by a Brazilian-built Saab-Scania DS-14A 04 eight-cylinder diesel developing 405 hp, which is coupled to the standard Allison CD-500-3 transmission. To install this diesel engine the rear hull has been enlarged and a new cooling system, consisting of a radiator and two 12-blade fans, has been installed. Other improvements include a modified electrical system that has two 12 V batteries and four 60 Ah alternators (eliminating the need for the generator previously fitted), a new instrument panel and redesigned fuel tanks.

The first 20 M41Bs retain the original 76 mm gun M32. For trials purposes one prototype was fitted with the standard Cockerill 90 mm Mk III gun. The following 120 to 200 M41Bs were fitted with the M32 gun bored out to fire the same 90 mm ammunition as the Cockerill 90 mm Mk III gun used in the Brazilian Army's ENGESA EE-9 armoured cars. This gun is designated the Ca 76/90 M32 BR1. The ordnance is shorter than the original and has a counterbalance at the forward end of the barrel giving the impression of being a bore evacuator. The original muzzle brake is retained and a torsion bar compensator is fitted at the breech.

From 1984, all rebuilt M41s were fitted with the newer Ca 76/90 M32 BR2 gun which retains the basic length of the M32 and has a bore evacuator and muzzle brake and it fires an APFSDS round. This model is known as the Carro Padrão (Standard Tank), has a thermal sleeve for the 90 mm gun, side skirts and additional spaced armour for the forward part of the hull, glacis plate and turret. Either side of the turret are four smoke grenade dischargers. The electro-optic fit can include night vision equipment and a laser range-finder coupled to the gunsight. The transmission has been upgraded

Brazilian Army M41C light tank fitted with additional armour protection to turret, glacis plate, nose and hull sides and a 90 mm gun
0045915

M41 light tank fitted with the Cockerill 90 mm gun Mk IV
0045916

which improves acceleration and gives a maximum road speed of 70 km/h.

By March 1985, 386 M41 tanks had been upgraded by Bernardini for the Brazilian Army. The Brazilian Marine Corps ordered 35 M41C tanks, these being M41Bs with an up-to-date fire-control system and the 90 mm Ca 76/90 M32 BR3 gun. The army and marine conversion programme is now complete, with final deliveries in 1990. In 1997, Brazil stated that it had a total of 287 M41 light tanks in service. In 1991, Bernardini completed the modernisation of 22 M41 light tanks for Uruguay along similar lines to that carried out for the Brazilian Army. These have a Cockerill 90 mm gun.

Danish M41s

Details of this upgraded vehicle were given in *Jane's Armour and Artillery 2001-2002*. It has now been phased out of service.

Taiwanese M41s

Details of this are given in the earlier entry, under Taiwan, for the Type 64 light tank. This has the local designation of the M41D, with 'D' standing for diesel.

M41 with New Power Pack

In addition to the Brazilian repowered M41s previously mentioned there are at least three other repower packages were offered.

NAPCO of the USA has designed a power pack for the M41 using the Detroit Diesel 8V-71T engine, developing 450 hp at 2,500 rpm, coupled to the original transmission, giving an operating range of over 450 km. This engine has been selected for the Taiwanese M41 upgrade programme.

FFG of Germany has designed a power pack for the M41 that uses the British Perkins Engines Company Condor CV-8 diesel engine which powers the Warrior. This also retains the original transmission.

More recently, AF Budge (Sales) Limited of the UK has developed and trialled an M41 light tank powered by a Perkins CV-8 diesel engine coupled to the original automatic transmission.

M41 with Cockerill 90 mm Gun

The Belgian company Cockerill has replaced the 76 mm gun of the M41 with its 90 mm gun Mk IV which fires the following types of ammunition: HEAT-T, HESH-T, HE-T, Smoke-WP-T and canister. This model has already been adopted by Uruguay. This upgrade package is no longer being marketed by Cockerill.

NIMDA Upgrade Package for M41

The Israeli company NIMDA offers a complete retrofit package for the M41 light tank which includes some or all of the following: replacement of the 76 mm gun by an Israel Military Industries 60 mm HVWS (High-Velocity Weapons System); a new power pack incorporating a Detroit Diesel engine; advanced fire-control systems and optics; a new suspension; and improved protection and survivability. There are two power pack options available; the 8V-71T developing 470 hp coupled to the existing CD-500-3 transmission; or a 6V-92T also developing 470 hp but coupled to an X-300 or the CD-500-3 transmission.

M41 with 105 mm Turret

For trials purposes the M41 light tank chassis has been fitted with the private venture Cadillac Gage 105 mm Low-Recoil Force Turret, which is also fitted to the Cadillac Gage Stingray light tank and the LAV-600 (6 × 6) armoured car. Full details of this turret, which has been produced in production quantities for the Stingray light tank, are given in the entry for the Stingray light tank earlier in this section.

Manufacturer

Cadillac Motor Car Division of General Motors Corporation.

Status

Production complete. In service with:

Country	Quantity	Comment
Brazil	287	Army and Marines with number reducing as Leopard 1 and M60 MBTs enter service
Chile	60	
Denmark	53	upgraded to DK-1 standard. No longer in front line service.
Dominican Republic	12	M41A1 model
Guatemala	10	delivered 1982
Taiwan	675	some sources state 400, some upgraded plus further upgrade under way, see Taiwan
Thailand	250	most now in store
Uruguay	22	Fitted with Cockerill 90 mm gun and all upgraded by Bernardini by 1991

Note: M41 light tanks are no longer used by Argentina, Austria, Ethiopia, Germany, Greece, Japan, the Philippines, Somalia, Spain, Sudan, Tunisia or the USA.

UPDATED

RECONNAISSANCE VEHICLES

RECONNAISSANCE VEHICLES

AUSTRALIA

Tenix Defence Systems Shorland S52 armoured patrol car

Specifications
Crew: 3
Configuration: 4 × 4
Combat weight: 3,600 kg
Power-to-weight ratio:
(petrol) 37.8 bhp/t
(diesel) 30.5 bhp/t
Length: 4.49 m
Width: 1.8 m
Height:
(roof) 1.8 m
Ground clearance: 324 mm
Track: 1.5 m
Wheelbase: 2.795 m
Max road speed: 120 km/h
Fuel capacity: 136 litres
Max range:
(road) over 630 km
Fording: 0.50 m
Vertical obstacle: 0.23 m
Turning radius: 6.6 m
Engine: 3.5 litre Land Rover V8 water-cooled developing 134 bhp at 5,000 rpm or 2.5 litre Land Rover TDi water-cooled 4-stroke turbocharged diesel developing 107 bhp at 3,800 rpm
Transmission: manual with 5 forward and 1 reverse gears
Transfer case: 2 speed
Steering: power assisted
Clutch: single dry plate
Suspension: coil springs, anti-roll bars fitted front and rear
Tyres: 9.00 × 16 (run-flat)
Brakes:
(main) hydraulic, dual circuit front and rear, disc front, drum rear
(parking) mechanical drum operation on transmission
Electrical system: 12 V (24 V optional)
Batteries: 1 × 12 V, 58 Ah
Armament: 1 × 7.62 mm MG
Ammunition: 1,600
Smoke-laying equipment: 4 × 66 mm smoke grenade dischargers either side of turret (optional)

Development

The Shorland armoured patrol car was originally designed in 1965 by the then Short Brothers and Harland, specifically to meet the requirements of the Royal Ulster Constabulary for use in Northern Ireland.

First production vehicles were completed at Short Brothers and Harland's facility at Newtownards in 1965. By the end of 1982, over 1,000 vehicles had been produced and by early 2000, it had been purchased by at least 45 countries.

The first production models were powered by a four-cylinder Rover petrol engine which developed 67 hp at 4,000 rpm. The Mk 2 was based on the Series 2 Land Rover chassis and was powered by a four-cylinder petrol engine which developed 77 bhp at 4,000 rpm. The Mk 3 was introduced in 1972 and fitted with a 2.6 litre six-cylinder engine which developed 91 bhp at 4,500 rpm. The Mk 4, employing the new V-8 engine, went into production in 1980.

The latest Series 5 version was designed to take advantage of the Land Rover Defender 110 in (4 × 4) chassis with its improved coil spring suspension and wider wheel track. Other improvements introduced with the new chassis were improved performance, a tighter turning circle and power steering to reduce driver fatigue.

Today, the complete Series 5 family consists of the S52 Armoured Patrol Car, the S53 Air Defence Vehicle, the S54 Anti-hijack Vehicle and the S55 Armoured Personnel Carrier. Full details of the S55 are given in the Armoured personnel carriers (wheeled) section.

Late in 1996, Shorts sold its complete line of Shorland (4 × 4) light armoured vehicles and the more recently developed S600 (4 × 4) armoured personnel carrier, to British Aerospace Australia Limited.

Early in 1999, Tenix Defence Systems, prime contractor for the Australian Army M113 series upgrade programme, acquired the military vehicle design and production capability of British Aerospace Australia.

In future, depending on the quantity and numbers of vehicles involved, production will be undertaken in Australia or Northern Ireland.

By mid-2001, it is understood that no production of the Shorland S52 or its variants had been undertaken in Australia.

Description

The Shorland S52 armoured patrol car is based on the 110 in (4 × 4) heavy-duty Land Rover Defender chassis which has permanent four-wheel drive and a lockable central differential. Normal power is the 3.5 litre Rover V8 petrol engine with the Rover 2.5 litre Turbo diesel available as an alternative.

The Land Rover chassis is strengthened to take the all-welded armoured body and fitted with stronger axles and suspension and a different final drive ratio. Nevertheless, some 85 per cent of the automotive and mechanical components are stock Land Rover parts which reduces the spares and maintenance problem.

The vehicle hull provides protection against right angle fire by NATO 7.62 × 51 mm ball ammunition down to 25 m range while the bullet-resistant glass provides the same degree of protection as the steel and meets the test requirements of BS 5051 part 1 Class R2.

The engine and radiator are fully armour protected and the floor is made from reinforced glass fibre to give protection against mines and improvised explosive devices. The fuel fillers are under armoured hatches which can also be released from inside the vehicle. Vehicles are trimmed internally with a fire-retardant material which provides thermal insulation.

The manually operated turret has full 360° rotation and is fitted with an adjustable seat for the gunner. A mechanical lock prevents rotation when not in use. An optical day periscope sight is fitted, this being boresighted to the armament. A searchlight is provided and this is also linked to the day periscope sight. The most common armament is the FN HERSTAL SA 7.62 mm GPMG but most other light machine guns can be fitted. Two four-barrelled grenade launchers can be fitted, one to each side of the turret. These fire 66 mm electrically initiated smoke/illumination/CS gas grenades. In the observation role the turret is operated with the rear part folding forwards and upwards and the turret back folding downwards.

The Shorland armoured patrol vehicle has a crew of three, driver and commander seated side by side in the front and the gunner in the turret. Alternative seating for the gunner is fitted behind the commander. The hull also contains the spare wheel and ammunition and there is ample room for other equipment such as radios. An escape hatch is provided in the rear panel.

The large bullet-resistant windscreen and side windows provide excellent visibility for the driver and commander. Two extraction fans are fitted, one mounted on the sloping rear roof is controlled from the dashboard while the second one in the turret is controlled by the gunner and is specifically intended for fume extraction when the armament is being fired.

Standard equipment normally includes seat belts and air conditioning, optional equipment includes drop down visors with vision blocks in place of the bullet-resistant windscreen, run-flat tyres, a loudhailer, automatic fire protection and various radio installations. Right- or left-hand steering can be specified.

Anti-hijack version of the Shorland armoured patrol car to the latest production standard with additional windows for improved all-round visibility 0088009

Shorland S52 armoured patrol car 0088010

Shorland S53 Mobile Air Defence Vehicle

This has been designed specifically for the defence of airfields and other key installations and supply routes against low-level air attack.

Based on the S52 Armoured Patrol Car, the S53 is fitted with a variant of the Thales Air Defence (previously Shorts Missile Systems) Starburst/Blowpipe/Javelin lightweight multiple launcher in place of the turret.

The system uses the standard shoulder launch aiming unit as clip-on equipment and mounts three canistered missiles. The turret ring has an integral hatch cover and periscope view block for surveillance when the hatch is closed. The ring carries a pintle mounting for the lightweight multiple launcher. This has a traverse head assembly with a sealed bearing unit which fits over the pintle. The traverse freedom relative to the turret ring is ±40° and the ring has handgrips so that the aimer can slew it to the approximate target bearing and then track with the lightweight multiple launcher. The system is manually operated.

Six missiles in their ammunition containers are stowed in racks on either side at the rear of the vehicle. The aiming unit is between the missile storage racks during transport.

Shorland S54 Anti-Hijack Vehicle

This has been developed specifically for anti-hijack patrols at airports and other high-risk areas and in addition to the driver's and commander's bullet-resistant windscreens and side windows, additional vision blocks can be fitted in the hull and turret.

The specially designed turret has a vision block and mount which permits an HK33 rifle to be sighted discreetly on a target while both the rifle and the marksman remain concealed.

Status

Production as required. In service with over 45 countries including the following:

Country	Quantity	Comment
Argentina	3	Mk 2, delivered in 1972 for Argentinean Federal Police
Bahrain	8	
Botswana	10	
Burundi	7	
Cyprus	n/avail	police
Guyana	3	
Kenya	8	
Lebanon	n/avail	police
Lesotho	8	
Libya	n/avail	may be non-operational
Malaysia	40	police
Mali	9	
Portugal	n/avail	
Syria	100	
Thailand	32	
UAE	n/avail	
UK	n/avail	

As of mid-2001 there has been no production of this vehicle in Australia.

Manufacturer

Tenix Defence Systems.

UPDATED

BELARUS

2-T Stalker reconnaissance vehicle

Development

Early in 1999, it was revealed that the Minotor Service Company of the Republic of Belarus was developing as a private venture a new highly mobile full-tracked reconnaissance vehicle called the 2-T Stalker.

This has been designed to undertake deep reconnaissance missions on the battlefield and is fitted with an advanced electro-optic computerised fire-control system which is said to allow ground and aerial targets to be rapidly detected, tracked and engaged under day and night conditions.

Although the prime contractor is the Minotor Service Company, it has been assisted by the Peleng Joint Stock Company of Belarus (sighting systems) and KBM of Russia (surface-to-air missile systems and anti-tank guided weapons). Export marketing of the 2-T Stalker is being carried out by Bel Tech Export.

The first prototype of the 2-T Stalker reconnaissance vehicle was completed in the second half of 2000 and was shown for the first time in public during the IDEX 2001 defence equipment exhibition held in Abu Dhabi in March 2001.

It should be noted that the prototype vehicle differs in a number of key areas from that of the original drawings issued in 1999 which appeared in *Jane's Armour and Artillery 2001-2002*, especially in the area of weapons installation.

It is understood that the Army of Belarus has a requirement for 30 vehicles and that the vehicle also carries the nickname of Minokhod.

2-T Stalker clearly showing the launchers for the anti-tank guided missiles and the fire-and-forget surface-to-air missiles in the raised position (Christopher F Foss) **2002**/0059294

Description

The 2-T Stalker reconnaissance vehicle uses some automotive components that are also used in the 2S6M Tunguska self-propelled air defence gun/missile system (*Jane's Land-Based Air Defence 2001-2002*) which is in service with India and Russia.

The layout of the 2-T Stalker reconnaissance vehicle is conventional, with the driver's compartment at the front, power-operated turret in the centre and power pack at the rear.

The driver is seated in the centre of the vehicle and has a single-piece hatch cover and three day periscopes which give observation to the front of the vehicle. The centre day periscope can be quickly replaced by a passive night vision periscope.

The all-welded steel hull has been redesigned and up armoured to provide a higher level of battlefield survivability. Over the frontal 80° of arc (40° left and right) it is claimed to provide protection against penetration from 30 mm armour piercing projectiles fired from a range of 300 m.

The turret has seats for the commander and gunner and it is believed that the commander is seated on the right and the gunner on the left, with both provided with hatch covers that open forwards and can be locked in the vertical position.

Either the commander or gunner can lay and fire the main armament which consists of a fully stabilised 30 mm cannon and 7.62 mm coaxial machine gun. Between 500 and 700 rounds of 30 mm ammunition are carried plus 2,000 rounds of 7.62 mm ammunition.

The commander and gunner have roof-mounted stabilised day/night sights to the immediate front of their hatch covers with another two stabilised day/night sights being positioned one either side towards the front of the turret. These could well be used to control and launch the missiles. It is believed that smoke/decoy grenade launchers are located within the turret of the 2-T Stalker.

According to the prime contractor, the 2-T Stalker features a brand new multichannel optronic suite that ensures day/night surveillance and automatic discrimination and tracking of moving targets as well as laser and optical range- finding. The data obtained is automatically relayed to a higher command post.

Mounted on the right side of the turret is a pod of two fire-and-forget Igla (SA-18 Grouse) or Igla-I (SA-16 Gimlet) surface-to-air missiles (SAM). When travelling the pod of SAM missiles is retracted under armour protection alongside the turret and when required for action are rapidly raised.

Mounted on the left side of the turret is a two-round launcher for the Russian Kolomna KBM 9K114 Kokon (NATO AT-6 Spiral) anti-tank guided missile (ATGM). This is retracted under full armour protection alongside the turret and when required for action is rapidly raised.

Specifications
Crew: 3
Combat weight: 27,400 kg
Length: 7.70 m
Width: 3.386 m
Height: 2.51 m
Ground clearance: 400 mm
Max road speed: 95 km/h
Range: 1,000 km
Armament:
(main) 1 × 30 mm cannon
(coaxial) 1 × 7.62 mm machine gun
(grenade launcher) 1 × 30 mm
Missiles:
(anti-armour) 2 AT-6 (ready to launch)
(air defence) 2 surface-to-air (ready to launch)
NBC system: yes

The 9K114 Kokon is a radio command guided ATGM with a maximum range of 5,000 m and is fitted with a HEAT (High Explosive Anti-Tank) warhead which will penetrate 600 mm of conventional steel armour.

More recent versions of the 9K114 Kokon ATGM have a tandem HEAT warhead for defeating explosive reactive armour as well as a thermobaric war head which is highly effective against softer targets.

The 2-T Stalker reconnaissance vehicle is powered by a diesel engine coupled to a hydromechanical

2-T Stalker from the front with missile launchers retracted **2002**/0114319

2-T Stalker from the rear **2002**/0114320

transmission. The power pack is located at the rear of the hull on the left side with air inlet and air outlet louvres being located in the roof and the exhaust outlet on the left side of the hull.

The suspension is of the hydropneumatic type and allows the driver to adjust the ground clearance from 130 to 570 mm depending on the terrain being crossed.

The 2-T Stalker has six rubber-tyred road wheels either side with the drive sprocket at the rear, idler at the front and track return rollers. The upper part of the track is covered by a skirt.

Standard equipment on the 2-T Stalker includes passive day/night devices for commander, gunner and driver and a full NBC system. A total of 12 anti-tank mines are carried, which are laid manually. It also features a laser warning system and a self-entrenching device is mounted under the nose of the vehicle.

Status

Prototypes. Not yet in production or service.

Manufacturer

Minotor Service Company.

UPDATED

BRAZIL

ENGESA EE-9 Cascavel armoured car

Development

The EE-9 (6 × 6) Cascavel armoured car was designed by ENGESA (Engenheiros Especializados SA) to meet the requirements of the Brazilian Army. Design work began in July 1970 after the company had completed the designing of the EE-11 Urutu (6 × 6) armoured personnel carrier, with the first prototype being completed in November of that year.

Following trials with prototype vehicles, a preproduction order for 10 EE-9 Cascavel vehicles was placed by the Brazilian Army, which named the vehicle the CRR (*Carro de Reconhecimento Sobre Rodas*). These were delivered between 1972 and 1973.

Production of the EE-9 began at a new plant at São José dos Campos in 1974 with the first production vehicles being delivered the same year. The production vehicles were slightly longer and wider than the prototypes and had a different wheelbase.

First production vehicles delivered to the Brazilian Army were fitted with a turret armed with 37 mm guns removed from old American-supplied M3 Stuart light tanks, but first production vehicles for export were fitted with a French CNMP-Berthiez (now Hispano-Suiza) 90 mm H 90 turret. This company is no longer involved in the design, development or production of turret systems.

The EE-9 Cascavel shares many components with the EE-11 Urutu (6 × 6) armoured personnel carrier and many automotive components used in both vehicles are standard commercial parts available worldwide. These include the Detroit Diesel and Mercedes-Benz diesel engines.

The ENGESA company is no longer trading but large numbers of EE-9 vehicles remain in service in many parts of the world.

Details of the ENGESA EE-11 Urutu (6 × 6) armoured personnel carrier are given in the Armoured personnel carriers (wheeled) section.

Specifications

Crew: 3
Configuration: 6 × 6
Combat weight: 13,400 kg
Unloaded weight: 10,900 kg
Power-to-weight ratio: 15.82 hp/t
Length:
(gun forward) 6.2 m
(hull) 5.2 m
Width: 2.64 m
Height:
(top of commander's cupola) 2.68 m
(to turret roof) 2.28 m
(to hull top) 1.75 m
Ground clearance:
(rear axle) 0.34 m
(hull centre) 0.5 m
Track: 2.1 m
Wheelbase: 2.343 m + 1.414 m
Angle of approach/departure: 72°/80°
Max road speed: 100 km/h
Fuel capacity: 390 litres
Max cruising range: 880 km
Fording: 1 m
Gradient: 60%
Side slope: 30%
Vertical obstacle: 0.6 m
Engine: Detroit Diesel 6V-53N 6-cylinder water-cooled diesel developing 212 hp at 2,800 rpm (other engines could also be installed)
Transmission: Allison Transmission MT-643 automatic with 4 forward and 1 reverse gears
Transfer case: ENGESA 2-speed
Steering: ZF integral hydraulic
Suspension:
(front) independent with helical springs and double-action shock-absorbers
(rear) ENGESA Boomerang type with walking beams, leaf springs and double-action shock-absorbers
Differential locking system
Tyres: 12.00 × 20 (with run-flat capability)
Brakes:
(main) air over hydraulic, dual-circuit, disc brakes
(parking) mechanical on transfer case output
Electrical system: 24 V (4 batteries)
Armament:
(main) 1 × 90 mm rifled gun
(coaxial) 1 × 7.62 mm MG
(anti-aircraft) 1 × 7.62 mm or 12.7 mm MG (optional)
Smoke-laying equipment: 3 smoke grenade dischargers either side of turret
Ammunition:
(main) 44
(coaxial) 2,200
Gun control equipment
Turret power control: electric/manual
(by gunner) yes
Max rate of traverse: 360° in 15 s (with power traverse)
Gun elevation/depression: +15° to -8°
Gun stabiliser:
(vertical) no
(horizontal) no
Armour: 8.5-16 mm
NBC system: no
Night vision equipment: optional

Description

The all-welded hull of the EE-9 has dual-hardness armour developed by ENGESA's Research Department with the co-operation of the University of São Paulo. It consists of an outer layer of hard steel and an inner layer of softer steel roll-bonded and heat-treated to give maximum ballistic protection. Increased protection is given over the frontal arc and special emphasis has been given to protecting the vehicle against attack from booby traps, grenades and Molotov cocktails.

The driver is seated at the front of the hull on the left side and is provided with a hatch cover opening to the right. There are three day periscopes in the top part of the glacis plate giving the driver a 120° view. The driver is also provided with a small windscreen and wiper, which folds forward onto the glacis plate when not in use. The driver's seat and steering wheel are both adjustable.

Initial production vehicles were fitted with a French H 90 turret armed with a French Giat Industries 90 mm D 921 gun but final production vehicles have an ENGESA ET-90 turret armed with a 90 mm ENGESA EC-90 gun, which is essentially a locally built Cockerill gun. The commander is seated on the left side of the turret and the gunner on the right, both with a single-piece hatch cover that opens to the rear and four day periscopes. The gunner has a day telescopic sight linked to the main armament with a magnification of ×5.9. EE-9s have electric-powered traverse and elevation as an option for the turret. It is believed that the EE-9s supplied to Tunisia were fitted with an OIP fire-control system, probably the LRS-5.

The engine compartment is at the rear of the vehicle and access to the engine is via two hatches in the hull roof. Power is transmitted from the engine to the gearbox (an Allison Transmission MT-643 automatic with four forward and one reverse gears) to the two-speed ENGESA transfer box, which splits the drive

Early production ENGESA EE-9 Cascavel armoured car 0088011

ENGESA EE-9 Cascavel (6 × 6) armoured cars loaded onto railway flatcars 0088013

Late production ENGESA EE-9 with a mount for an external machine gun on the commander's cupola, a front-mounted winch and a laser range-finder integral with the gunner's sight (Christopher F Foss)

ENGESA EE-9 Cascavel (6 × 6) armoured car with an ENGESA ET-90 turret armed with an ENGESA EC-90 90 mm gun and a laser range-finder over the main armament 0088012

between the front and rear differentials. The front differential is bolted to the hull with the power being transmitted to the front axle by shafts, and the rear differential forms part of the ENGESA Boomerang walking beam suspension. The latter, which is also used on ENGESA 6 × 6 trucks, consists of a rigid axle, connected to the hull by double leaf springs and telescopic dampers, which holds two lateral walking beams. Power is taken from the drive shaft to the four rear wheels by gears inside the walking beam, which can rotate freely around the central hub. The design of the rear suspension allows up to 0.87 m of vertical wheel travel and ensures that all four rear wheels are in contact with the ground at all times. The run-flat tyres enable the vehicle to travel for at least 40 km at a speed of 56 km/h in consistent ground after being completely deflated. The Cascavel has a double 24 V electrical system; a battery set is used for vehicle and turret operations, while a second set is used purely for engine start to avoid lack of power after long radio-watch operations. The Cascavel is also equipped with disc brakes and an automatic tyre inflation system.

Optional equipment included an air conditioning system, heater, laser range-finder, automatic fire extinguisher and passive night vision equipment. The Cascavel could be fitted with various types of fire-control system.

The 90 mm gun is called the EC-90-III and is mounted in an ET-90 turret. The rifled gun fires the following types of ammunition: HE, HEAT, HESH, Smoke White Phosphorus (Smoke WP-T) and HEAT Target Practice. More recently an APFSDS round has been developed. Of the 44 90 mm rounds carried, 24 are in the turret and 20 in the hull. ENGESA also developed a commander's cupola for the EE-9 which can be traversed through 360°, has day vision blocks for all-round observation, and a 12.7 mm (0.50) machine gun mounted externally on a bracket forward of the commander's single-piece hatch cover that opens to the rear.

A 7.62 mm machine gun is mounted to the left of the 90 mm main armament and a similar weapon, or a 12.7 mm (0.50) Browning M2 HB machine gun, can be mounted on the turret roof for anti-aircraft defence. Three smoke grenade dischargers are mounted either side of the turret and are electrically fired from within the vehicle. Late production vehicles for Iraq have a traverse indicator, a clinometer, improved seating, a laser range-finder integrated into the sight which is protected by a shutter when the main armament is fired, and a 12.7 mm M2 HB machine gun on an anti-aircraft mount similar in design to that used by the former Soviet Union with its 12.7 mm DShKM machine guns.

A summary of the various models of the ENGESA Cascavel armoured car follows:

Mk I
Powered by a Perkins 6357 V engine coupled to a manual Clark transmission, no central tyre-pressure regulation system. Did not enter production.

Mk II
First production model armed with a 37 mm gun, Mercedes-Benz OM 352A 170 hp diesel engine, Clark manual transmission, 12 V (hull) and 24 V (turret) electrical system, no central tyre-pressure regulation system. These were built only for the Brazilian Army and most were subsequently rebuilt with the ENGESA ET-90 turret armed with a locally built 90 mm EC-90 gun based on the Belgian 90 mm MECAR weapon.

Mk III
First production model for export, Mercedes-Benz OM 352A 170 hp diesel, Allison Transmission AT-540 automatic transmission, 12 V (hull) and 24 V (turret) electrical system, central tyre-pressure regulation system and fitted with a French Hispano-Suiza H 90 turret with a Giat Industries 90 mm gun.

Mk IV
First production models completed in 1977 and fitted with an ENGESA ET-90 turret with an ENGESA EC-90 gun, Detroit Diesel Model 6V-53 developing 212 hp coupled to an Allison Transmission fully automatic transmission Model MT-643, 24 V (turret) electrical system and a central tyre-pressure regulation system.

Mk V
Powered by Mercedes-Benz OM 352A diesel developing 190 hp coupled to a fully automatic Allison Transmission AT-540 or AT-545 automatic transmission, ENGESA ET-90 turret 24 V (hull and turret) electrical system, disc brakes and a central tyre-pressure regulation system. This system enables the driver to change the tyre pressure from his seat to suit the type of ground being crossed. For example, when travelling on roads the tyre pressure is 4 kg/cm^2, cross-country 3 kg/cm^2 and for mud, sand and snow 2 kg/cm^2.

Mk VI
As Mk V but powered by a Mercedes-Benz OM 352A diesel developing 190 hp at 2,800 rpm.

Mk VII
As Mk VI but fitted with Allison Transmission MT-643 fully automatic transmission.

Variants
There are no variants of the EE-9 apart from the different armament installations and variations between different production batches.

Status
Production complete, no longer marketed. In service with the following:

Country	Quantity	Comment
Bolivia	24	
Brazil	415	army 409, marines 6
Burkina Faso	24	
Chad	small number	
Chile	50	
Colombia	127	also used by marines
Cyprus	124	
Ecuador	22	
Gabon	14	
Ghana	3	
Iran	40	quantity not confirmed (captured)
Iraq	250	quantity not confirmed
Nigeria	75	
Libya	380	
Paraguay	30	
Suriname	6	
Togo	36	
Tunisia	24	
Uruguay	10	
Zimbabwe	90	

Manufacturer
ENGESA Engenheiros Especializados SA.
This company is no longer trading.

UPDATED

ENGESA EE-3 Jararaca scout car

Development

The ENGESA EE-3 (4 × 4) Jararaca scout car was designed specifically for the export market and was not adopted by the Brazilian Army.

As the ENGESA company is no longer trading, the EE-3 Jararaca is no longer being marketed or built. Unconfirmed reports indicate that Iraq took delivery of a small quantity of these vehicles.

Description

The all-welded hull of the EE-3 Jararaca consists of an outer layer of hard steel and an inner layer of softer steel roll bonded and heat treated to give the maximum ballistic protection. The armour is similar to that used on other members of the ENGESA family of 6 × 6 wheeled armoured vehicles, the EE-9 Cascavel armoured car and the EE-11 armoured personnel carrier.

The driver is seated towards the front of the hull, slightly to the left of the vehicle's centreline and has a single-piece hatch cover that lifts and swings to the right to open, although on late production models it opens to the left. There are three day periscopes for forward observation immediately in front of the driver. This can be replaced by a passive periscope for driving at night. A windscreen and wiper, which can be fitted in front of the driver's position for use in wet weather, folds forward on to the glacis plate when not in use. The machine gunner is seated behind the driver on the right side of the hull and has a circular mounting which can be traversed through 360° and a single-piece hatch cover that opens to the rear. A standard 12.7 mm (0.50) M2 HB machine gun is pintle mounted on the forward part of the circular mounting, but there is no provision for aiming the weapon from inside the vehicle. There are standard day vision blocks in both sides of the hull.

The crew of three enters and leaves the EE-3 Jararaca by a door in the right side of the hull that opens forward. Immediately behind the door is a single vision block under which is a firing port. The commander is seated on the left side of the hull to the rear and has a hatch cover that opens to the rear. On late production models, the commander has a hatch that opens to the left with an integral roof-mounted, forward-facing periscope. There is an observation block in the left side of the hull.

The Mercedes-Benz OM 314A turbocharged engine is at the rear of the hull and access to it is by two large hatches in the roof. The suspension, front and rear, consists of semi-elliptical springs and hydraulic shock-absorbers. The axles have hypoid gears. The wheels of the EE-3 Jararaca are not within the armoured part of the vehicle as the light sheet steel covering over the top of each wheel will blow away if the vehicle hits a mine. The EE-3 Jararaca is fitted with an automatic tyre-inflation system as standard. The central tyre-pressure regulation system enables the driver to adjust the tyre pressure to suit the type of ground being crossed without leaving his seat. For example, when travelling on roads the tyre pressure is 2.5 kg/cm², cross-country 2.1 kg/cm² and for mud, sand and snow 1.4 kg/cm².

The tyres of the EE-3 Jararaca, like those of the EE-9 Cascavel and EE-11 Urutu, have the ENGESA run-flat system which consists of an alveolar core inside the pneumatic tyre. When the tyre deflates, it contacts the inner core so ensuring continued mobility.

The standard 12.7 mm (0.50) M2 HB pintle mounted machine gun can be replaced by a 7.62 mm machine gun or a 20 mm cannon. Other armament installations offered for the EE-3 Jararaca included a 60 mm breech-loaded mortar, a Euromissile 2,000 m range MILAN launcher and a cupola with internally operated 7.62 mm machine guns.

ENGESA developed two new turrets that could be fitted to the EE-3 Jararaca: the ENGESA ET-MD one-man turret with a 7.62 mm NATO machine gun and a 12.7 mm M2 HB machine gun, both internally operated; and the ENGESA ET-20 one-man turret with a 20 mm cannon and a 7.62 mm NATO machine gun.

The EE-3 Jararaca has no amphibious capability. Optional equipment included passive night vision equipment and various radios with an intercom system. The EE-3 Jararaca could be fitted with an NBC system.

Variants

Apart from the NBC reconnaissance version and the different armament installations there are no variants of the EE-3 Jararaca. Cyprus uses the EE-3 Jararaca fitted with the 2,000 m range Euromissile MILAN ATGW system. It has not been confirmed that the EE-3 NBC version ever entered production.

Status

Production complete. No longer being marketed. In service with the following countries:

Country	Quantity	Comment
Cyprus	40	some sources say 15 vehicles
Ecuador	10	
Gabon	12	
Uruguay	16	

Manufacturer

ENGESA Engenheiros Especializados SA.
This company is no longer trading.

UPDATED

Specifications

Crew: 3
Configuration: 4 × 4
Combat weight: 5,800 kg
Power-to-weight ratio: 20.7 hp/t
Length: 4.163 m
Width: 2.235 m
Height:
(without armament) 1.56 m
(top of 12.7 mm MG mount) 2.3 m
Ground clearance: 0.335 m
Track: 1.71 m
Wheelbase: 2.6 m
Angle of approach/departure: 58°/54°
Max road speed: 100 km/h
Fuel capacity: 140 litres
Max road range: 700 km
Fording: 0.6 m
Gradient: 60%
Side slope: 30%
Vertical obstacle: 0.4 m
Turning radius: 7.73 m
Engine: Mercedes-Benz OM 314A turbocharged, 4-cylinder water-cooled diesel developing 120 hp at 2,800 rpm
Transmission: Clark 240 V manual with 5 forward and 1 reverse gears
Transfer case: ENGESA 2-speed
Steering: ZF integral hydraulic
Clutch: single dry plate hydraulically operated
Suspension: semi-elliptical springs and hydraulic shock-absorbers, both have hypoid gears and tapered gear, single reduction differential
Tyres: 11.00 × 20
Brakes:
(main) drum, air-assisted
(parking) drum, acting on transfer case
Electrical system: double 24 V system
Batteries: 4 × 12 V, 95 Ah
Armament: 1 × 12.7 mm M2 MG
NBC system: optional
Night vision equipment: optional

ENGESA EE-3 Jararaca (4 × 4) scout car as used by Cyprus and armed with Euromissile MILAN ATGW launcher on one-person turret ***2000***/0064869

ENGESA EE-3 Jararaca (4 × 4) scout car armed with pintle-mounted 12.7 mm M2 machine gun 0088014

BULGARIA

BRM-23 reconnaissance vehicle

Development/Description

The Bulgarian BRM-23 reconnaissance vehicle is a further development of the BMP-23 infantry combat vehicle fully covered in the Armoured personnel carriers (tracked) section.

Production of this vehicle, as well as the BMP-23 and BMP-30, was undertaken by the TERM Joint Stock Company branch of Targovishte whose company headquarters is in Sofia. It is believed that total production of the BMP-23/BMP-30 vehicles and their variants amounted to 114 units. There are no known exports. As far as it is known there has been no recent production of the BRM-23 reconnaissance vehicle.

The vehicle has been designed to carry out battlefield reconnaissance up to 100 km ahead of the main force. The hull and turret are very similar to the BMP-23 infantry combat vehicle but it has a reduced number of firing ports (two on the right side of the hull and one on the left).

Its five-man crew consists of commander, gunner, navigator, radio operator and driver. Of these, two are at the front of the vehicle, two in the turret and one at the rear of the hull on the left side.

The BRM-23 reconnaissance vehicle carries a wide range of specialised equipment to enable it to carry out its mission, including:

(1) LBD-1 hand-held laser range-finder with a maximum range of 9,000 m and an accuracy of ±5 m. If required, the device can also be mounted on a tripod
(2) PGN-9M hand-held passive night observation device which can also be mounted on a tripod
(3) Land navigation system
(4) Orientation equipment consisting of a periscopic vision device model PV-1
(5) PAB-2A artillery aiming circle
(6) RM-130M HF radio with the antenna-type frame in the vertical position at the rear. The radio has a range of up to 120 km and is used to communicate with higher elements in the command chain
(7) R-123M VHF radio with a maximum range of 20 km for vehicle to vehicle communications
(8) R-31M radio which is used for communications between the dismounted members of the vehicle and the vehicle itself
(9) MT-66 man-portable mine detector
(10) VPHR and DP-5V radiation and chemical detectors.

The armament of the BMP-23 is retained in the BRM-23 reconnaissance version. The two-person turret is armed with a 23 mm 2A14 cannon (as used in the ZSU-23-2 light anti-aircraft gun system), 7.62 mm PKT machine gun and launcher for AT-3 'Sagger' wire-guided ATGW.

Status

Production as required. In service with the Bulgarian Army. There are no known export sales of the BRM-23 reconnaissance vehicle.

Manufacturer

TERM Joint Stock Company.

UPDATED

BRM-23 reconnaissance vehicle with rear antenna erected. The slotted device is fitted above the second to fourth roadwheels when afloat, to improve the amphibious characteristics

Specifications

Crew: 5
Combat weight: 14,800 kg
Power-to-weight ratio: 21 hp/t
Ground pressure: 0.485 kg/cm²
Length:
(overall) 7.285 m
Width:
(overall) 3.055 m
(over tracks) 2.85 m
Height:
(overall) 2.53 m
(hull top) 1.71 m
Ground clearance: 0.4 m
Track width: 350 mm
Max road speed: 61.5 km/h
Max water speed: 5 km/h
Range: 500 km
Fuel capacity: 500-560 litres
Gradient: 60%
Side slope: 30%
Vertical obstacle: 0.8 m
Trench: 2.5 m
Engine: diesel developing 315 hp
Transmission: manual
Suspension: torsion bar
Electrical system: 24 V
Armament:
(main) 1 × 23 mm 2A14 cannon
(coaxial) 1 × 7.62 mm PKT MG
(ATGW) 1 × 'Sagger' ATGW launcher
Ammunition:
(23 mm) 600
(7.62 mm) 2,000
(ATGW) 4
NBC system: yes
Night vision equipment: yes

CANADA

Armoured Combat Vehicle Programme

Development/Description

The Canadian Forces Mobile Command is now formulating its requirements for an Armoured Combat Vehicle Programme (ACVP) which will supplement and eventually replace the current Cougar (6 × 6) vehicles armed with a 76 mm gun.

The ACVP requirement is understood to be for up to 200 vehicles and could be worth C$980 million with first vehicles completed in the early 21st century. In the longer term, the ACVP could also replace the Leopard 1 MBTs currently used by the Canadian Forces Mobile Command. The Canadian fleet of Leopard 1 MBTs has recently been upgraded to the C2 standard which includes the installation of an ex-German Army Leopard 1A5 turret with a more advanced computerised fire-control system and day/thermal sights.

Late in 1997, General Dynamics Land Systems of the United States signed a Memorandum of Understanding (MoU) with the now General Motors Defense (then Diesel Division, General Motors of Canada) to co-operate on the ACVP.

Under the terms of this MoU, General Motors Defense would be the prime contractor and provide the Light Armoured Vehicle (LAV) (8 × 8) chassis with General Dynamics Land Systems providing its Low-Profile Turret (LPT) armed with a 105 mm rifled gun fed from an automatic loader. Computing Devices Canada would provide the turret electronics and fire-control software.

The first prototype of this vehicle, called the LAV 105 mm, was completed in 1999 and has already undergone extensive company trials. Details of the LAV 105 mm are given in a separate entry. This was selected by the US Army late in 2000.

The LAV 105 mm has now been adopted by the US Army to meet its requirements for the Mobile Gun System (MGS) part of its new Brigade Combat Teams. A total of 204 will be supplied at a unit cost of US$4.2 million.

Status

Future programme.

Manufacturer

Not selected.

UPDATED

General Motors Defense 105 mm LPT Assault Gun

Development

The 105 mm LPT (Low Profile Target) Assault Gun has been developed by the now General Motors Defense (at that time Diesel Division, General Motors of Canada) as a private venture to meet the possible future operational requirements of Australia, Canada and, more recently, the US Army.

The Canadian Army has a requirement for 164 systems to replace its current Cougar (6 × 6) 76 mm armed light armoured vehicles and supplement its Leopard 1 105 mm armed Main Battle Tanks (MBT).

The first prototype of the 105 mm LPT Assault Gun was completed in 1999.

The system essentially consists of the latest production standard General Motors Defense LAV-III (8 × 8) chassis fitted with the private venture General Dynamics Land Systems Low-Profile Turret (LPT), which has already been installed on a number of other platforms for trials purposes.

The vehicle could be air portable in a number of transport aircraft including the C-5A, C-17, C-130 and C-141. Like more recent versions of the LAV on which it is based, the 105 mm LPT is not amphibious, although it can ford to a depth of 1.7 m.

General Motors Defense, 105 mm LPT Assault Gun (Scott Gourley)
2001/0073184

It should be noted that the description given below relates to the prototype system 105 mm LTP Assault Gun which may well be different from future production versions.

Following a competition, late in 2000 the US Army selected a version of this system based on a LAV-III (8 × 8) chassis to meet its requirement for a Mobile Gun System (MGS) for use with its new Brigade Combat Teams.

Under current plans a total of 204 systems are to be ordered at a unit cost of $4.2 million. These will be issued on the scale of 27 per brigade.

Following trials with 10 preproduction vehicles, full production is expected to start in 2004/2005.

Description

The hull of the 105 mm LPT Assault Gun is of all welded steel armour construction to which an additional layer of passive armour has been added to provide a higher level of protection.

The passive armour package depends on the user but typically would provide protection against 12.7 mm armour piercing (AP) at specified ranges.

For crew protection against ballistic overmatch situations, the vehicle is also equipped with a liner of composite material applied to selected portions of the interior of the vehicle. This liner minimises spalling in the event of hull penetration.

The driver is seated at the front of the vehicle on the left side and can enter his position via a single-piece hatch cover that opens to the rear, or via the crew compartment.

In the forward part of the driver's hatch cover are three M17 day periscopes, the centre one of which can be replaced by an image intensifier for night driving. The hatch provides single eye-level viewing for the driver for both the hatch open and closed positions. A thermal viewing aid can also be inserted up through the hull ahead of the driver's hatch for driving at night.

The engine compartment is to the right of the driver and is provided with an automatic fire detection and suppression system. The prototype vehicle is equipped with a Caterpillar 3126A diesel engine is coupled to an Allison MD 3560 P automatic transmission with six forward and one reverse gears and a two speed transfer case. A larger 400 hp power pack is planned for the future.

Steering is power-assisted on the front two axles and when driven across country, all eight wheels are engaged. When being driven on roads the front four wheels are normally disengaged.

A central tyre pressure regulation system is fitted as standard and this allows the driver to adjust the tyre pressure to suit the type of terrain being crossed.

The power-operated LPT is mounted in the rear section of the hull with the gunner being seated on the left and the vehicle commander on the right.

The LPT is not only much lighter than a conventional turret armed with a 105 mm rifled tank gun, but also presents a much smaller turret to the enemy. This makes it more difficult to locate and therefore more difficult to successfully engage.

Turret traverse is all solid-state digital electric with manual controls as a back up. Turret traverse is a full 360° with weapon elevation from –8 to +18°. Either the commander or gunner can lay and fire the main armament both having identical controls.

The gunner has a roof-mounted stabilised day/thermal sighting system which includes a laser rangefinder.

In addition, the gunner has four M27 day periscopes that give observation to the front, sides and rear and a single-piece hatch cover. There is also an auxiliary day sight with a magnification of ×7 that can be used by the commander or gunner, which is mounted to the right of the 105 mm M68A1 gun and linked mechanically to the gun.

The commander has seven M27 wide field of view day periscopes that give observation to the front, sides and rear and a two-axis stabilised panoramic viewer (planned for production) that is mounted on the right side of the LPT. This has a low-light level television and variable field of view day optics.

This has been developed from the system installed on the Advanced Amphibious Assault Vehicle (AAAV). In addition, the commander has a screen on which he can see a picture from his own sighting system as well as that of the gunner.

Specifications
(provisional)

Crew:	3
Configuration:	8 × 8
Weight (for production):	
(combat)	19,100 kg
(unloaded)	16,016 kg
Power-to-weight ratio:	18.3 hp/t (prototype) 20.9 hp/t (production)
Length (hull):	7.391 m
Width:	2.66 m
Height:	
(overall)	2.819 m
(hull top)	1.995 m
Ground clearance:	0.425 m
Max road speed:	100 km/h
Range:	450 km
Fuel capacity:	215 litres
Gradient:	60%
Side slope:	30%
Vertical obstacle:	0.6 m
Trench:	2.00 m
Engine:	Caterpillar 3126A diesel developing 350 hp (for prototype) (Future plans 400 hp for production)
Transmission:	Allison MD 3560 P automatic, 6 forward and 1 reverse gears
Transfer case:	2 speed
Steering:	power-assisted on front two axles
Turning radius:	8.5 m
Suspension:	independent (hydropneumatic strut on front axles and hydropneumatic strut on rear axles)
Brakes:	air brakes
Tyres:	1200 R20 XML with central tyre inflation system
Electrical system:	24 V
Batteries:	6 (prototype) 4 (production)
Armament:	
(main)	1 × 105 mm M68A1 rifled tank gun
(coaxial)	1 × 7.62 mm MG
(air defence)	1 × 7.62 mm or 12.7 mm MG
(smoke grenade dischargers)	2 × 4 66 mm
Ammunition:	
(main)	22-32 × 105 mm
(coaxial)	1,760
Gun control equipment	
Turret power control:	electric/manual
Elevation/depression:	–8/+18°
Turret traverse:	360°
Gun stabiliser:	
(vertical)	yes
(horizontal)	yes
NBC system:	yes
Night vision equipment	yes

General Motors Defense 105 mm LPT Assault Gun **2000**/0064870

The LPT is armed with a 105 mm M68A1 rifled tank gun which for this application is fitted with a pepper pot type muzzle brake, fume extractor and thermal sleeve. This fires all types of 105 mm standard ammunition including APFSDS, HEAT and smoke.

The 105 mm M68A1 gun is fed by a hydraulically operated loader mounted in the turret basket which contains eight rounds of ready-to-use 105 mm ammunition. The empty cartridge cases are ejected out of the rear of the LTP bustle.

A load cycle takes 6 seconds and the loader can download ammunition as well as uploading under power or manual control. A typical ammunition load would be 24 to 32 rounds. There is also the option of revolver type automatic 105 mm ammunition reloading system which would feed the ready use magazine.

A 7.62 mm machine gun is mounted coaxial to the left of the 105 mm M68A1 gun and a 7.62 or 12.7 mm machine gun can be mounted on a skate mount at the commander's station. Mounted on either side of the

LPT, at the rear, is a bank of four 66 mm electrically operated smoke grenade launchers.

The crew can also enter their positions via the twin doors in the hull rear, or via hatches in the roof.

The heart of the computerised fire-control system is the Computing Device Canada (CDC) fire-control computer. The identification of the target is primarily done by the commander looking through the periscopes around his hatch, using the panoramic camera, or by viewing the same image that the gunner is seeing though his/her day sight or thermal imager.

The usual combat scenario is that the commander will locate the target and then hand off to the gunner; the commander is then free to locate another target.

The commander can also override the gunner's controls and engage targets. As the sights and 105 mm gun are fully stabilised, the system has a high hit probability against stationary and moving targets while the vehicle is stationary or moving.

Typical equipment on the base vehicle includes an anti-tank guided missile wire cutter for the hull and turret, a climate control system comprising a heating and air conditioning system, vehicle and external communications systems, crew compartment fire detection and suppression system, a slave receptacle and an inter-linked global positioning system.

If required, the 105 mm LPT Assault Gun can be fitted with an Auxiliary Power Unit (APU) and/or a hydraulically operated winch mounted on the left side of the vehicle behind the driver's hatch that leads out through the front or rear of the vehicle. This can be used for self-recovery or to recover other moved or disabled vehicles.

Status

Development. The US Army has a requirement for 204 systems to meet its Mobile Gun System requirement.

Manufacturer

General Motors Defense.

UPDATED

General Motors Defense LAV-25 (Reconnaissance) Coyote

Development

To replace the United Defense full-tracked Lynx Command and Reconnaissance Vehicles in service with the Canadian Forces, in April 1992, a modified version of the LAV-25 (8 × 8) light armoured vehicle was selected and, in 1993, an order was placed with the now General Motors Defense (at that time, Diesel Division, General Motors of Canada), for 203 vehicles.

The first four vehicles were completed in January 1996 and production has now been completed.

Three versions have been purchased, one with a mast-mounted sensor suite, one with two tripod-mounted sensor suites and the third built to receive, but not fitted with, any surveillance suites. Details are given in the following description.

The vehicle is fully air-transportable in the following aircraft:

Type	Quantity
C-130	1
C-141	2
C-5A	8

Description

The hull of the LAV-25 (Reconnaissance) is very similar to the LAV-25 used by the US Marine Corps for some years but incorporates many detailed improvements.

The all-welded steel hull provides the crew with protection from small arms fire and shell splinters, with the driver being seated at the front of the hull on the left with the engine compartment to his right.

Add-on armour, when fitted, will provide protection against projectiles up to 14.5 mm all round and against 30 mm over the frontal arc. Unlike earlier versions of the LAV-25, the LAV-25 (Reconnaissance) is not amphibious.

The driver has a single-piece hatch cover that opens to the left and three M17 day periscopes, the centre one of which can be replaced by a thermal vision device.

The driver's periscopes have two levels of laser protection and are upgradable to three levels of protection. There is a hull escape hatch in the left side of the vehicle.

The Delco Defense LAV-25 turret upgraded to full electric drive is in the centre of the hull, with the commander seated on the right and the gunner on the left, with both having a single-piece hatch cover that opens to the rear.

The commander has a total of seven day periscopes with two levels of laser protection which are upgradable to three layers of protection. He also has a DIM 36 target acquisition sight mounted in the forward part of the turret roof which incorporates a Generation III image intensification sight. The commander also has a HIRE thermal sight remote display from the gunner's sight.

The gunner has a sight observation periscope in the left side of the turret roof with two levels of laser protection which is upgradable to three levels of protection. In the forward part of the roof the gunner has a DIM 36 target acquisition sight, HIRE thermal sight with through the sight laser range-finder.

Main armament comprises a fully stabilised The Boeing Company 25 mm M242 Chain Gun with a 7.62 mm C6 machine gun being mounted coaxial to the right and provision for a roof-mounted 7.62 mm C6 machine gun.

The surveillance equipment compartment is at the rear of the hull with two doors that open outwards to the left and right. Each of these has a day vision block with laser protection.

Over the surveillance compartment are two roof hatches that open left and right with the sensor package being elevated when the left hatch is open. The surveillance operator is seated in the rear on the left side with the display to his right.

There are two versions of the LAV-25 (Reconnaissance), battle group kit and fixed brigade kit. Prime contractor for the Reconnaissance Vehicle Surveillance System (RVSS) is Computing Devices Canada Ltd (a General Dynamics Company).

This consists of the sensor module, telescopic mast (if fitted) and the operator's console. The last gives the operator full control of all sensors with automatic target detection and tracking, full use of multisensor synergy and 2 hours of VCR recording of all electro-optical data.

The battle group kit consists of man-portable dismounted surveillance sensors mounted on tripods with a lightweight copper/fibre optic composite cable allowing for the sensor package to be positioned up to 200 m from the operator's station in the vehicle.

LAV-25 (Reconnaissance) with mast-mounted sensor pod in raised position ***2001***/0093524

One tripod has the battlefield surveillance radar with 3,200 mil azimuth coverage while the second tripod has 6,400 mil azimuth coverage with the following sensors:

(1) long-range TV camera
(2) eye-safe laser range-finder
(3) thermal imager.

The fixed brigade kit has an electrically powered extendable mast which projects the surveillance sensors 10 m above the ground. On this mast are mounted:

(1) battle surveillance radar with 3,200 mil azimith coverage which will detect moving targets out to 24 km
(2) electro-optical package with long-range TV camera giving day/night recognition/detection out to 18 km, eye-safe laser range-finder with a range of 10 km and accurate to ± 5 m, and high-performance FLIR with all-weather recognition/identification of targets out to 12 km.

LAV-25 (Reconnaissance) with sensor suite dismounted and installed on two tripods 0088015

LAV-25 (Reconnaissance) in travelling configuration and turret traversed forwards ***2001***/0093523

Specifications
Crew: 4
Configuration: 8 × 8
Combat weight: 13,400 kg
Unloaded weight: n/avail
Power-to-weight ratio: 20.52 hp/tonne
Length: 6.39 m
Width: 2.50 m
Height: 2.692 m
Ground clearance: n/avail
Max road speed: 100 km/h
Fuel capacity: n/avail
Max range: 660 km
Gradient: 60%
Side slope: 30%
Vertical obstacle: n/avail
Trench: 2.057 m
Turning radius: 7.77 m
Engine: Detroit Diesel Model 6V-53T diesel developing 275 hp
Transmission: Allison MT653 fully automatic with 5 forward and 1 reverse gears
Transfer case: single speed
Steering: power assisted, front four wheels
Suspension: independent
Brakes: power brakes
Tyres: n/avail
Electrical system: 24 V
Batteries: 8
Alternator: 300 A
Armament:
(main) 1 × 25 mm M242 cannon
(coaxial) 1 × 7.62 mm C6 MG
(anti-aircraft) 1 × 7.62 mm C6 MG
(smoke grenade dischargers) 2 × 4, 76 mm
Ammunition:
(main) 210 × 25 mm (ready use); 210 × 25 mm (stowed)
(coaxial) 440 × 7.62 mm (ready use); 660 × 7.62 mm (stowed)
(anti-aircraft) 220 × 7.62 mm (ready use)
Gun control equipment:
Turret power control: electric/manual
Elevation/depression: +60° to –10°
Turret traverse: 360°
Gun stabiliser:
(vertical) yes
(horizontal) yes
NBC system: yes
Night vision equipment: yes

The battlefield surveillance radar used in the LAV-25 is the UK Thales (formerly Thomson Racal Defence) MSTAR (Man-portable Surveillance and Target Acquisition Radar) manufactured under licence by Systems and Electronics Inc (SEI), a subsidiary of ESCO Electronic Corporation of the USA.

Navigation equipment fitted includes Digital Compass System (DCS) with driver's and commander's display, DCS interface with PLGR, dead reckoning back-up to PLGR and laser range-finder available on commander's display.

The vehicle has full-time four-wheel drive on the rear wheels with selectable full eight-wheel drive.

The 24 V electrical system is waterproof and radio suppressed; two 500 A slave receptacles are provided.

Threat detection equipment from DRS Technologies includes:

(1) Laser Warning Receiver (LWR)
(2) commander's LWR display that shows threat direction in relation to turret
(3) driver's LWR display that shows threat direction in relation to vehicle.

Standard equipment includes an automatic fire detection and suppression system in the engine and crew compartments; front-mounted hydraulic winch with a capacity of 6,804 kg; NBC system of the ventilated face mask type; GID-3 chemical detector; radiac set AN/VDR 2 radiation monitor; ATGM wire cutter for commander and driver; and a complete communications system.

Hot/cold weather equipment fitted includes dual hydraulically driven air conditioning system, fuel-fired high-capacity crew heater, engine coolant preheater for –40°C cold start and engine glow plugs for cold start.

Variants
None.

Status
Production complete. In service with Canadian Forces (203 vehicles).

Manufacturer
General Motors Defense.

UPDATED

CHINA, PEOPLE'S REPUBLIC

NORINCO 120 mm (8 × 8) Tank Destroyer

Development
Early in 2001 the Chinese defence industry revealed that it had developed to the prototype stage a new 8 x 8 tank destroyer armed with a turret-mounted 120 mm smoothbore gun.

The 8 × 8 chassis of this as yet undesignated vehicle could be a further development of the NORINCO WZ 551 (6 × 6) armoured personnel carrier (APC) which first appeared in public in 1986. Full details of this vehicle are given in the Armoured Personnel Carriers (Wheeled) section.

This entered production in the late 1980s and is now in service in significant numbers with the People's Liberation Army (PLA) and is being adopted for an increasing number of roles.

In many respects the WZ 551 (6 × 6) APC is very similar to the French Giat Industries family of VAB 4 × 4 and 6 × 6 series of APC that have been built in France in large numbers for the home and export markets.

First photograph released of the new Chinese 8 × 8 tank destroyer armed with a 120 mm smoothbore gun

Description
The new Chinese 8 × 8 120 mm tank destroyer is similar in concept to the Italian IVECO/Otobreda Centauro 105 mm (8 × 8) armoured car/tank destroyer but is much higher and has a different layout.

Centauro has the driver front left with the power pack to his right, turret in the centre and ammunition/small troop compartment at the rear.

It is estimated that the Chinese vehicle weighs between 26 and 34 tonnes.

Its chassis could be a further development of the WZ 551 (6 × 6) but it is more likely to be a new chassis with an increased payload to enable it to undertake a wider range of battlefield missions.

The driver and one other person are at the front with the turret in the centre and the power pack at the rear. Steering is power assisted on the front four wheels.

To the immediate front and either side of the driver and one other crew person is a bulletproof window. When in a combat area these are covered by armoured shutters with observation then being via roof-mounted periscopes that cover observation over the frontal arc. These two crew members enter and leave the vehicle via a hatch above their position.

Main armament comprises a 120 mm smoothbore gun fitted with a fume extractor, thermal sleeve and muzzle brake. It is probable that the 120 mm smoothbore gun is similar to that installed on the Type 89 full-tracked tank destroyer and fires APFSDS (armour-piercing fin-stabilised discarding sabot) and other types of ammunition with a semi-combustible cartridge case. The Type 89 has an automatic loader for the 120 mm gun but it is not known at this stage if the new 8 × 8 vehicle has this feature.

A computerised fire-control system is fitted, with the commander being seated on the left and gunner on the right; both have a single-piece hatch cover that open towards the front. There also appears to be a hull entry door in the side of the hull between the second and third roadwheels.

The fire-control system and stabilised day/night sights are probably the same as those installed in the recently revealed NORINCO 105 mm (6 × 6) tank destroyer (covered in a separate entry). It is possible that these new 8 × 8 and 6 × 6 vehicles are related or at the least share a number of common components.

A 12.7 mm machine gun is mounted on the right side of the turret roof for air defence purposes and it is probable that a 7.62 mm machine gun is mounted co-axial with the 120 mm weapon.

Standard equipment probably includes an NBC system. Other equipment may well include a winch and a central tyre-pressure regulation system. It is not amphibious.

Specifications
Not available.

Status
Prototype. Not yet in production or service.

Manufacturer
NORINCO (China North Industries Corporation).

NEW ENTRY

NORINCO 105 mm (6 × 6) Tank Destroyer

Development

Early in 2001 NORINCO (China North Industries Corporation) revealed that it had developed to the prototype stage a new 105 mm 6 × 6 tank destroyer/armoured car.

It is believed that the first prototype of this was completed several years ago but that it has not yet entered volume production for the People's Liberation Army (PLA).

Description

In appearance, this new 6 × 6 tank destroyer/armoured car is similar to the French Panhard ERC Sagaie (6 × 6) armoured car armed with a 90 mm gun that entered service with the French Army in 1984, with the driver's compartment at the front, turret in the centre and power pack at the rear.

The turret of the new Chinese vehicle is armed with a 105 mm rifled gun fitted with a fume extractor, thermal sleeve and a distinctive muzzle brake to reduce recoil forces when firing armour-piercing fin-stabilised discarding sabot - tracer (APFSDS-T) types of ammunition. In many respects the 105 mm gun is very similar to that fitted to the Italian Otobreda/IVECO Centauro (8 × 8) tank destroyer currently in service with Italy (400) and Spain (22).

A 7.62 mm machine gun is fitted coaxial with the 105 mm main armament and there is also a 12.7 mm machine gun mounted on the turret roof for local and air defence purposes. Banks of four electrically operated smoke grenade launchers are mounted either side of the turret, firing forwards.

It is possible that the new, and as yet undesignated, 105 mm tank destroyer/armoured car is fitted with the same turret as that fitted to the recently revealed Type 99 light amphibious tank (qv in Light Tanks section).

This is believed to have a fully stabilised 105 mm gun which is used in conjunction with a computerised fire-control system, with the commander and gunner having day/night sights and the latter also having a laser range-finder.

It is stated that the new 6 × 6 vehicle uses existing components wherever possible to reduce operating and procurement costs and, like all Chinese armoured vehicles, it would be powered by a diesel engine. Standard equipment includes an NBC system. It is possible that the new 105 mm 6 × 6 vehicle is related to the new NORINCO 120 mm (8 × 8) vehicle covered in a separate entry.

The new 105 mm (6 × 6) tank destroyer is fully amphibious with some preparation, which includes erecting a trim vane at the front of the hull and switching on the bilge pumps. When afloat, it is propelled in the water by two water jets located one either side at the rear of the hull.

At this stage no firm details of the vehicle have been released, apart from it having a combat weight of 18.5 tonnes. It is possible that the new vehicle is based on automotive components of the latest version of the WZ 551 (6 × 6) series of armoured personnel carrier (APC) that has been in service with the PLA for some years.

In the past the PLA has been mainly equipped with full-tracked APCs but in recent years has, like many other countries, started fielding wheeled APCs such as the amphibious WZ 551. It is possible that the new 105 mm armoured car/tank destroyer has been developed by NORINCO to provide a complementary tank destroyer/armoured car with similar strategic mobility.

First photograph released of the new NORINCO 105 mm tank destroyer/armoured car (6 × 6) with trim vane folded back on glacis plate ***2002***/0095260

Specifications

Not available.

Status

Prototype. Not yet in production or service.

Manufacturer

NORINCO (China North Industries Corporation).

NEW ENTRY

CZECH REPUBLIC

Snezka reconnaissance vehicle

Development

The Snezka full-tracked reconnaissance system has been developed by the Military Institute for Weapon and Ammunition Technology to meet the operational requirements of the Czech Army and, following trials with two prototype systems, a production order has been placed. The stretched chassis is manufactured by the VOP 026 facility. The official designation of this vehicle is the BVP-1PPK Snezka.

The main role of the Snezka reconnaissance vehicle is to provide a real-time target acquisition capability for self-propelled artillery units including the following:

(1) Detection, recognition and tracking of stationary and moving targets and groups of targets
(2) Observation of fall of shot of artillery projectiles
(3) Measuring co-ordinates of above targets, processing target information and passing this onto the next chain of command so that the targets can be engaged.

Description

Snezka is based on a stretched and modified BMP-1 full-tracked infantry fighting vehicle chassis which has a total of seven roadwheels either side and a raised superstructure towards the rear. The normal BMP-1/BMP-2 has only six roadwheels either side.

Snezka reconnaissance vehicle with hydraulic arm retracted for travelling purposes 0045841

The hull of the vehicle is of all-welded steel armour which provides protection from small arms fire and shell splinters. The driver is seated at the front left with the power pack compartment to his right.

The engine air inlet and outlet louvres are in the roof of the engine compartment with the shrouded exhaust outlet on the right side of the hull. The vehicle can lay its own smoke screen by injecting diesel fuel into the exhaust outlet.

The driver has a single-piece hatch cover that opens to the right and to the front of this and to the immediate left is a total of four TNPO-170A day periscopes, one of which can be replaced by a passive TVNE-1PA periscope for driving at night.

To the rear of the driver is another crew position which is provided with a single-piece hatch cover and three day periscopes for forward observation. In addition, there is another vision device located in the actual hatch cover.

The raised superstructure has two hatches with the one on the left side being provided with a circular cupola which can be traversed through a full 360° and is armed with a 12.7 mm machine gun which can be used for surface/surface and surface/air roles. The hatch on the right opens to the right and is not provided with any observation devices.

Normal means of entry and exit is via two doors in the hull rear with the left door having a single firing port. Mounted on either side of the hull is a bank of three electrically operated 81 mm smoke grenade launchers which fire to the front of the vehicle.

The suspension is of the torsion bar type with either side having seven single rubber-tyred roadwheels with the drive sprocket at the front, idler at the rear and

Snezka reconnaissance vehicle with hydraulic arm raised in operating position 0045840

track-return rollers. The upper part of the suspension is covered by a light steel skirt and shock-absorbers are fitted on the first two and last two roadwheel stations.

The Snezka system is fully amphibious being propelled in the water by its tracks. Before entering the water the trim vane is erected at the front of the vehicle and the bilge pumps are switched on. Standard equipment includes an NBC system and passive night vision devices.

A laser detector system is fitted as standard as are crew compartment heaters. To provide additional power for the wide range of electronic devices installed on the Snezka, a portable generator is carried as part of its standard equipment.

Mounted on the roof is a hydraulically operated arm on the top of which is the sensor package.

To enable the system to carry out its battlefield mission, the following equipment has been installed on the mast: battlefield radar whose antenna can be traversed through 360°, day/night TV system, thermal camera, laser range-finder and wind velocity measuring unit.

The radar can detect a person at a range of 9,000 m and a tank at a range of 33,000 m with radar range for an artillery explosion being 15,000 m. It can measure artillery explosions with an accuracy of 20 m.

The electro-optical module incorporates a laser range-finder, thermal camera and three charge-coupled device TV cameras for day and night operation.

The latter includes a JPJ-II-200 night camera with a second-generation image intensification tube. The near infra-red cameras include a JPJ-S-300 unit with a 10 km tank detection range and a JPJ-0-78 unit with a 5 km tank detection range.

The vehicle is fitted with a TALIN navigation system from Honeywell. This consists of an internal navigation unit, vehicle motion sensor, control and display unit and driver display unit. The inertial navigation unit receives data from three sensors (three digital laser gyroscopes and three accelerometeres), evaluate direction, acceleration and speed of the vehicle by means of navigational algorithms provide the actual position of the vehicle. The vehicle motion sensor is an additional sensing device which improved navigational accuracy. All of the navigation data are shown on the control and display unit and the driver's display units. Any inputs are realised on the control and display unit. Internal functions such as aiming marks, electronic bore sighting and focus adjustment are controlled by an RS-485 serial link.

The mast takes 90 seconds to raise and 60 seconds to return to the travelling position. The overall height of the system is 14 m. The system can be deployed on any terrain with a side slope of up to 7°.

The mast is referred to as the Manipulator of Position (MP) on which the pedestal of the Sensors System Manipulator (SSM) is mounted. The SSM contains the sensor head together with the reconnaissance sensors and the control and propulsion modules.

Internally the Snezka is provided with positions for vehicle commander who is provided with a PC, radar system operator and the operator of all the other surveillance systems.

Internally, the vehicle is fitted with a land navigation system and additional communications equipment which allows target information to be sent to a fire-control unit up to 24,000 m away. Parts of the sensor package can be used while the vehicle is travelling up to a maximum speed of 5 km/h.

Specifications

Crew: 4
Combat weight: 17,440 kg
Power-to-weight ratio: 17.20 hp/t
Ground pressure: n/avail
Length:
(travelling) 7.534 m
Width:
(travelling) 3.150 m
Height:
(travelling) 3.470 m
(combat) 14.77 m
Ground clearance: 0.42 m (estimate)
Track: 2.55 m
Track width: 300 mm
Length of track on ground: n/avail
Max speed:
(road) 55 km/h
(cross-country) 30 km/h
(water) 7 km/h
Fuel capacity: 462 litres
Max road range: 600 km
Fording: amphibious
Gradient: 60% (estimate)
Side slope: 30% (estimate)
Vertical obstacle: 0.7 m (estimate)
Trench: 2.5 m (estimate)
Engine: 4-stroke liquid-cooled, direct injection, UTD-20 6-cylinder diesel developing 300 hp at 2,600 rpm
Transmission: manual, 5 forward and 1 reverse gears
Clutch: multiplate, dry, constant engagement
Suspension: torsion bar, hydraulic shock-absorbers on 1st, 2nd, 6th and 7th roadwheels
Electrical system: 24 V
Armament: 1 × 12.7 mm MG
Smoke-laying equipment: 2 × 3 81 mm smoke grenade dischargers, diesel fuel injected into the exhaust
NBC system: yes
Night vision equipment: yes

Sensor pod
Detection range of radar:
(people) 9,000 m
(tank) 33,000 m
(artillery explosion) 15,000 m
(accuracy) 20 m
JPJ-II-200 camera
(detection range of tank) 1,600 m
JPJ-S-300 unit
(detection range of tank) 10,000 m
JPH-0-78 unit
(detection range of tank) 5,000 m
Laser range-finder
(maximum range) 30,000 m
Mast
(height fully raised) 14 m
(traverse) 190° left and right
(depression) –30/+60°

Status
Production. In service with Czech Republic.

Manufacturer
State factories. Design authority is Military Institute for Weapon and Ammunition Technology. Chassis will be built by VOP 026 facility.

UPDATED

Light Observation System on BMP chassis

Development
The Light Observation System (LOS) was developed and integrated under the leadership of the Military Technical Institute for Weapons and Ammunition to meet the operational requirements of the Czech Republic. The modified chassis is the responsibility of VOP 026 with Omnipol responsible for export sales and marketing.

Its main role is to detect, recognise and identify targets for subsequent engagement by artillery systems under day and night conditions. It can also be used to monitor shell impact, observe the terrain and determine target co-ordinates.

LOS is typically deployed more forwards than the more sophisticated Snezka reconnaissance vehicle based on a stretched BMP series chassis that is covered in a separate entry.

Within the Czech Republic Army LOS forms a key part of the ASPRO artillery fire-control system. LOS has been in service with the Czech Republic for several years on a modified BMP-2 infantry combat vehicle, although it can be integrated into a wide range of other chassis, tracked and wheeled.

Light Observation System (LOS) on modified BMP-2 chassis with sensor pod in raised position
2002/0114321

Description

As previously stated, the Czech Army has integrated the LOS system into a modified BMP-2 infantry combat vehicle chassis. The two-person turret is retained but is now fixed and the 30 mm armament is a mock-up to confuse enemy observers.

The mast-mounted sensor system is integrated into the rear part of the turret and when lowered is under full armour protection. The sensor head is mounted on a telescopic arm that can be raised to a maximum height of 4.3 m. The sensor head can be traversed 200° left and 200° right with a depression of –40° and an elevation of +40°.

The sensor head includes a day/night (infra-red) TV system, eye-safe laser range-finder, laser target marker/designator, north-finding gyro and inertial navigation and GPS. The design of the sensor pod is such that it can be integrated with other types of sensor according to customer's specific operational requirements.

According to Omnipol, the target detection range of the day observation camera is 5 km, while that of the night camera is 10 km. Target detection by the infra-red camera is quoted as being 6 km while the laser range-finder has a minimum range of 10 km and a maximum range of 20 km. The laser designator has a maximum range of 15 km. Travel distance accuracy is 0.35 per cent, while range of the data transmission is 20 km.

The operators are seated within the hull of the vehicle under full armour protection and their equipment includes a special tactical computer with sophisticated software, colour flat-panel display, black/white monitor and a keyboard with mouse and joystick.

On the flat-panel display can be the digital map or the image seen by the day/night sensors with a panoramic view above. In addition, the operator can have a combined panoramic view with a digital map. Control and other information is shown to the right of the display.

LOS is claimed to be simple to operate and requires a very short preparation time. The system can be operated by one person if required. The complete system has a crew of four, including the driver and commander.

Specifications

Similar to Russian BMP-2.

Status

Production as required. In service with Czech Republic.

Manufacturer

See Development.

NEW ENTRY

VOP 026 Upgraded BRDM-2 Amphibious Scout Car

Development

Early in 2001 the Czech Army completed trials of two Russian designed and built BRDM-2 (4 × 4) amphibious scout cars which have been upgraded by the local VOP 026 overhaul facility with the assistance and co-operation of a number of other companies and government establishments in the Czech Republic.

These two prototype vehicles are designated the BRDM-B and the BRDM-VR, with the former being a reconnaissance vehicle and the latter a command and control vehicle with additional communications equipment.

As a result of trials with the prototype vehicles, the user has requested a number of modifications which will be incorporated into a run of preproduction vehicles which are expected to be delivered in 2002. It is understood that the first production batch will consist of 50 units that will be earmarked for use by the Czech rapid reaction units.

Specifications

Crew: 6 (max)
Configuration: 4 × 4
Weight:
(combat) 8,000 kg
(unloaded) 7,190 kg
Power-to-weight ratio: 20.25 hp/t
Length: 5.845 m
Width: 2.35 m
Height:
(turret roof) 2.40 m
(hull roof) 1.92 m
Ground clearance: 0.325 m
Track: 1.84 m
Wheelbase: 3.10 m
Max speed:
(road) 101 km/h
(water) 8 km/h
Fuel capacity: 290 litres
Max road range: 800 km
Fording: amphibious
Gradient: 60%
Side slope: 30%
Vertical obstacle: 0.4 m
Trench crossing: n/available
Engine: Renault turbocharged 5-stroke diesel developing 162 hp
Transmission: ZF manual 5-speed
Clutch: single plate, dry friction
Suspension: semi-elliptical springs and hydraulic shock-absorbers
Tyres: 13.00 × 16
Brakes: hydraulic, dual circuit, air booster
Electrical system: 24 V
Armament:
(main) 1 × 12.7 mm NSVT machine gun
(co-axial) 1 × 7.62 mm PKT machine gun
NBC system: yes
Night vision equipment: yes

Description

The overall layout of the VOP O26 upgraded BRDM-2 (4 × 4) amphibious scout car is almost identical to the original vehicle developed well over 30 years ago, with large numbers being built at the Molotov GAZ Plant in Gorkiy, Russia.

The normal BRDM-2 is fitted with a one-person manually operated turret armed with a 14.5 mm KPVT and a 7.62 mm PKT coaxial machine gun and a crew of four.

On the upgraded BRDM-2 the turret has been modified and is armed with a 12.7 mm NVST and a 7.62 mm PKT machine gun, which are laid by the commander/gunner by remote control from within the hull. The turret is fitted with a TV system with integrated daytime Charge Coupled Device (CCD) television camera and an infra-red camera.

The onboard navigation system consists of a fully integrated Global Positioning System (GPS) receiver and an integrated digital map. In addition, this can be integrated with a higher level information system. All of the information is displayed on the commander's/gunner's flat display screen. A portable laser range-finder is carried as standard.

On the original BRDM-2, two belly wheels were mounted either side of the hull between the front and rear wheels, which were lowered by the driver when rough terrain was being crossed. On the upgraded BRDM-2 these have been removed to increase internal volume as well as allowing a front-opening entry door to be provided in each side of the hull. The hull floor protection has also been improved against anti-personnel mines. The basic armour of the BRDM-2 provided protection against 7.62 mm small arms fire.

BRDM-2 (4 × 4) amphibious scout car upgraded in the Czech Republic to the enhanced BRDM-B standard
2002/0095642

The original BRDM-2 was powered by a GAZ-41 V-8 petrol engine developing 140 hp coupled to a manual transmission. In the upgraded version this has been replaced by a new Renault turbocharged diesel developing 162 hp coupled to a new ZF manual transmission and a new modified transmission for the propeller drive. The cooling and lubrication system has also been improved. A new braking system has been installed as has a new lighting system.

Vision has been enhanced by the installation of bulletproof windows around the forward part of the hull. The large main windscreen is fitted with wipers and washers. A maximum of six people are now carried including the commander/gunner and driver. There is also a bulletproof observation device above the door in either side of the hull.

Like the original BRDM-2, the upgraded vehicle is fully amphibious, being propelled when afloat by a single water-jet mounted at the hull rear. Before entering the water a trim vane is erected at the front of the vehicle and the bilge pumps are switched on.

Other modifications include new brakes, upgraded chassis, new tyres, communications equipment to NATO standards, new NBC system and modern hot water heater. The engine compartment has been fitted with a semi-automatic fire detection and suppression system.

Variants

BRDM-VR command post

The BRDM-VR is similar to the BRDM-B but has been optimised for the command post role and has a crew of four and additional communications equipment installed. It is also provided with a dismounted generator to power the additional communications equipment when the vehicle is used in the static role. It would typically be used in the company command post role.

Other applications
The hull modifications of this upgrade are also applicable to the BRDM-2 tank destroyer armed with Spandel anti-tank guided missiles and the SA-9 Gaskin surface-to-air missile system, both of which are still widely deployed.

Status
Prototypes. Not yet in production or service.

Manufacturer
VOP 026.

NEW ENTRY

FRANCE

Engin Blinde a Roues de Contact

Development/Description
Early in 2000, it was revealed that the French Delegation General pour l'Armement (DGA) had awarded a 16 month sole source contract to Giat Industries for the study phase of a new 8 × 8 Armoured Fighting Vehicle (AFV) for the French Army called the Engin Blinde a Roues de Contact (Wheeled Armoured Fighting System).

It is anticipated that the EBRC could replace a number of tracked and wheeled vehicles currently in service with the French Army.

This would include two battalions of Giat Industries AMX-30B2 Main Battle Tanks (MBT) (80 vehicles), Giat Industries AMX-10RC (6 × 6) armoured car armed with a 105 mm gun (337 delivered to the French Army) and the Panhard ERC Sagaie (6 × 6) armoured car armed with a 90 mm gun (192 delivered to the French Army).

Projected missions of the EBRC would include reconnaissance, early entry and fire support for the Giat Industries Leclerc MBT of which 406 are expected to be in service by 2006.

The replacement of three vehicle types by a common vehicle would offer the French Army significant logistical and training advantages. During the study phase, a large number of EBRC configurations are being studied by Giat Industries.

The DGA and the French Army are closely involved at every stage with the aim to have a small number of EBRC concepts, perhaps three, at the end of the 16 month period.

A number of other companies are already involved at the study phase, the main one being Matra BAe Dynamics as it is expected that some versions of the EBRC will be armed with missiles.

One possible solution is a common turret that will be configured for the mission being undertaken. For example, the reconnaissance version would have a mast-mounted sensor pod, early entry would have a longer range anti-tank guided missile, while the escort vehicle for the Leclerc MBT would have a countermeasures system, radar and perhaps a Matra BAe Dynamics Mistral surface-to-air missile system.

A number of turret configurations are also being studied for the EBRC including a conventional turret, another with the main armament mounted above and third with the armament remote controlled from within the vehicle.

One possible weapon for the EBRC could be the CTA International 40 mm Case Telescoped Weapon System (CTWS) currently under advanced development.

Larger calibre versions of the CTWS are being studied including 90 and 105 mm. Another alternative would be a low-recoil 120 mm smooth gun fed by an automatic loader.

The user has already specified a number of EBRC key requirements including fully air transportable in the projected A400M aircraft, armour protection to be against 14.5 mm small arms fire and RPG-7 type anti-tank weapons, special protection against mines and incorporate stealth characteristics and battlefield digitilisation.

The crew of the EBRC would also depend on the mission but the baseline vehicle would probably have a crew of three, commander, gunner and driver and carry a number of dismounted FELIN (Fantassins a Liaisons Integrees) Land Warrior infantrymen. The reconnaissance version of the EBRC would carry two, early intervention two or three and the escort four or five.

The vehicle will have full 8 × 8 drive and, to reduce development and procurement costs, the projected EBRC would use automotive components from existing vehicles such as the VCI/Vextra with possible technology insertions including electric drive, energy storage and advanced suspension. Details of the Vextra (8 × 8) vehicle are given later in this section.

If all goes to plan, following the completion of the study phase early in 2001, it is expected that the feasibility phase would run from mid-2001 through to late 2003.

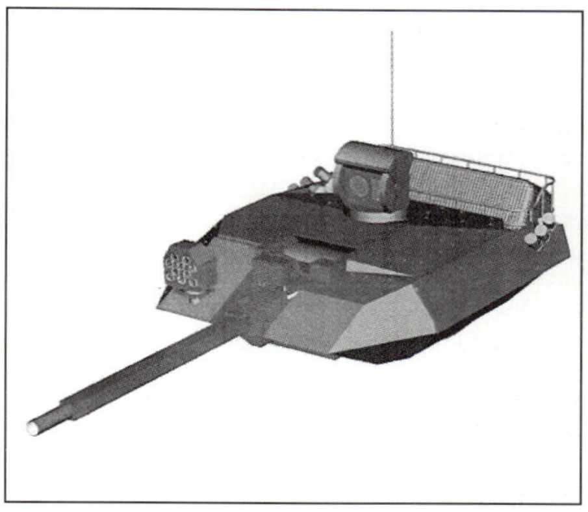

Giat Industries concept for a modular unmanned turret to equip the EBRC (8 × 8) includes the baseline version armed with a 40 mm CTA cannon, infra-red jammer mounted to its right and an IFF interrogator on the mantlet **2001**/0083033

This would be followed by full-scale development which would include trials with prototype vehicles and this would be complete by 2007. Production could start in 2008 with the vehicle entering service the following year. By this time the AMX-10RC would have been in service with the French Army for well over 25 years.

Status
Study phase.

Manufacturer
Not selected.

UPDATED

Giat Industries VEXTRA 105 mm armoured vehicle

Development
The VEXTRA (8 × 8) vehicle was developed as a private venture by Giat Industries for the export market. It was first shown in 1994 fitted with a Giat Industries Dragar one-man turret armed with a 25 mm cannon and a 7.62 mm coaxial machine gun.

In 1997, the VEXTRA (8 × 8) high-mobility concept demonstrator was fitted with the latest version of the Giat Industries three-man TML 105 105 mm turret system and completed a successful series of firing trials in France.

The Giat Industries TML 105 105 mm turret has been developed in two versions, both of which are armed with a Giat Industries 105 mm G2 rifled gun, firing standard NATO 105 mm ammunition.

The first version has an unstabilised 105 mm gun and sights and for trials purposes has been fitted to the Giat Industries AMX-10RC (6 × 6) armoured car as well as the Swiss MOWAG Piranha Armoured Combat Vehicle (10 × 10) and the Hägglunds Vehicle CV 90 full-tracked chassis.

The latest version has a fully stabilised Giat Industries 105 mm G2 gun, stabilised commander's sight and stabilised gunner's day/thermal sight which incorporates a laser range-finder. This enables the system (stationary or moving) to engage stationary and moving targets with a high first round hit probability.

Late in 1997, the VEXTRA fitted with the latest TML 105 105 mm turret was flown to the United Arab Emirates for an extensive series of firepower and

VEXTRA (8 × 8) concept demonstrator fitted with Giat Industries TML 105 105 mm turret **2002**/0095364

VEXTRA (8 × 8) concept demonstrator fitted with Giat Industries Dragar 25 mm turret (Christopher F Foss) **2002**/0114322

mobility trials manned by Giat Industries and UAE crews.

During these trials almost 2,500 km were clocked up on various types of terrain including 500 km in one day to demonstrate the strategic mobility of VEXTRA 105. A total of 115 rounds of 105 mm ammunition was fired at ranges of up to 2,500 m while the vehicle was stationary and moving.

As of mid-2001 there were no orders for the Giat Industries VEXTRA 105 mm armoured vehicle or for any of its variants.

Description

The hull of the VEXTRA (8 × 8) is of all-welded aluminium armour construction with an additional layer of passive armour for added protection. The driver is seated at the front left and the power pack is to the right. The driver has a single-piece hatch cover that opens to the left with three integral day periscopes, the centre one of which can be replaced by a passive periscope for driving at night.

The Giat Industries TML 105 105 mm turret is mounted in the centre of the hull with the commander being seated on the right with the gunner forward and below. The commander has a raised cupola with a single-piece hatch cover that opens to the rear, panoramic stabilised day periscope and observation day periscopes. The gunner has a roof-mounted stabilised day/thermal sighting system.

The loader is seated on the left and has a single-piece hatch cover that opens to the rear and day periscopes that give observation to the front and sides of the turret.

Main armament comprises a stabilised Giat Industries 105 mm G2 gun, 7.62 mm coaxial machine gun and the option for a roof-mounted 7.62 mm machine gun for air defence purposes. A bank of electrically operated Galix grenade launchers is mounted at each corner of the turret so that a full 360° coverage is obtained.

At the rear of the hull is a power-operated ramp with an integral door. This rear area can be configured to carry ammunition or personnel.

To reduce the thermal signature of the vehicle the exhaust outlet is at the left hull rear with the cooling outlet at the right hull rear.

An H-type drive line is fitted with the two front steered axles being designed for optimal mobility/internal volume trade off.

The vehicle has full 8 × 8 drive with power steering on the front four wheels. A central tyre pressure regulation system is fitted which allows the driver to adjust the tyre pressure to suit the terrain. Forward/rear lockable differentials are provided as are left/right lockable differentials.

The VEXTRA is mine protected and provides multi-impact protection against medium calibre threats fired at short range. When fitted with ERA the vehicle provides protection against small calibre infantry weapons fitted with a HEAT warhead. Maximum combat weight with highest level of protection is quoted as 34 tonnes.

VEXTRA (8 × 8) concept demonstrator fitted with Giat Industries TML 105 105 mm turret during cross-country trials in France
2000/0064872

Specifications

Crew: 4
Combat weight: 28,000 kg
Power-to-weight ratio: 25 hp/tonne
Configuration: 8 × 8
Length (chassis): 7.40 m
Width: 3.0 m
Height (hull top): 2.00 m
Wheelbase: 1.7 + 1.8 + 1.50 m
Track: 2.554 m
Maximum speed: 120 km/h
Acceleration: 0 to 30 km/h in 3.4 s
0 to 48 km/h in 7.5 s
Range (road): 800 km
Fording: 1.2 m
Gradient: 60%
Side slope: 30%
Vertical obstacle: 1 m
Trench: 2 m
Armament:
(main) 1 × 105 mm G2 gun
(coaxial) 1 × 7.62 mm MG
Ammunition (105 mm): 35
Engine: Scania diesel developing 700 hp
Transmission: SESM automatic with 7 forward and 2 reverse gears and torque converter
Suspension: independent oleopneumatic, front wheels double-wishbone, rear trailing arms
Steering: power assisted, front four wheels
Turning radius: 7 m
Tyres: 405/95 R22 fitted with tyre pressure regulation system and run-flat insert
Brakes: hydraulic assisted
Electrical system: 24 V
NBC system: optional
Night vision equipment: yes

Variants

The VEXTRA 105 mm is being marketed by Giat Industries as a highly mobile reconnaissance/tank destroyer with a combat weight of 28 tonnes.

It can also be fitted with other turrets, for example the Dragar 25 mm, or used as a base design for other battlefield missions for the export and domestic markets. When fitted with a 25 mm turret, for example, four personnel can be carried in the rear.

According to Giat Industries, the projected infantry fighting vehicle model of the VEXTRA would have a crew of three and carry eight fully equipped troops.

Status

Prototype. Not yet in production or service.

Manufacturer

Giat Industries.

UPDATED

VEXTRA (8 × 8) concept demonstrator fitted with Giat Industries TML 105 105 mm turret during during mobility demonstrations conducted in the United Arab Emirates
2001/0093525

VEXTRA (8 × 8) concept demonstrator fitted with Giat Industries Dragar 25 mm turret (Christopher F Foss)
2001/0093526

Giat Industries AMX-10RC reconnaissance vehicle

Development

The AMX-10RC reconnaissance vehicle was designed from September 1970 at the Atelier de Construction d'Issy-les-Moulineaux (AMX) to meet a French Army requirement for a vehicle to replace the Panhard EBR (8 × 8) heavy armoured car. The first of three chassis was completed in June 1971.

Many of the automotive components of the AMX-10RC are identical to those used in the tracked Giat Industries AMX-10P IFV.

The AMX-10RC equips the reconnaissance regiments of the corps of the French Army, the armoured regiments of the rapid deployment forces and the cavalry regiments of the infantry divisions. The first unit to be equipped with the vehicle was the 2ème Régiment de Hussards based at Sourdun, which received its complement of vehicles in 1981-82.

The original French Army requirement was for 525 AMX-10RC vehicles, with the first production order being placed in 1976, but for a number of reasons only 337 vehicles were ordered. The 12 vehicles for Qatar were from the French Army order so another batch of 12 vehicles was built for the French Army, with final deliveries completed in 1994.

Total production of the AMX-10RC for the home and export markets amounted to 457 vehicles. Production of the turret was undertaken at Tarbes while chassis production and final assembly was undertaken at Roanne. Both these facilities are part of Giat Industries.

Morocco placed an order for 108 AMX-10RC vehicles modified to meet local operating conditions in 1978 and first vehicles were delivered in 1981 for training.

The AMX-10RC was deployed by the French Army in Chad during Operation Manta between August 1983 and November 1984.

The AMX-10RC was also deployed by the French Army in the Persian Gulf during Operation Desert Storm. Two AMX-10RC regiments of the 6th Light Armoured Division made it possible for the US Army XVIIIth Corps to entrust the French Daguet Division with the very important west coverage mission in Iraq for the entire allied forces.

Since it was first introduced into service with the French Army in 1979, the AMX-10RC has been constantly improved. A new diesel engine, improved 105 mm ammunition and additional passive armour protection have been introduced.

For Operation Desert Storm, the recapture of Kuwait in 1991, some AMX-10RC vehicles were fitted with additional passive armour to their frontal arc with an anti-tank guided weapon decoy system being mounted on the front left side of the turret.

In 1994 it was revealed that a batch of improvements was under development for the French Army including:

(1) Installation of a Giat Industries 105 mm G2 gun firing NATO standard ammunition
(2) Installation of a thermal camera
(3) Fitting of a decoy system
(4) Additional armour protection
(5) Installation of a central tyre inflation system
(6) Installation of an electronic control system for the gearbox
(7) Installation of a Land Battle Management System (LBMS).

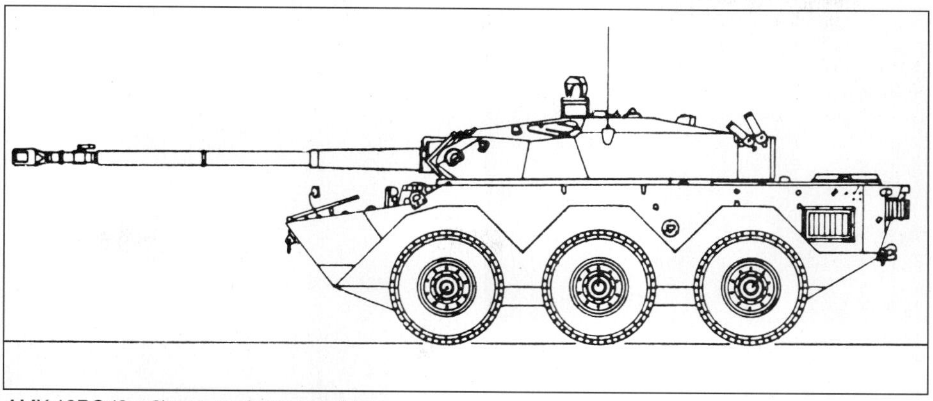

AMX-10RC (6 × 6) reconnaissance vehicle

It has recently been confirmed that the 105 mm F2 gun will be retained and that consideration is being given to the installation of thermal cameras removed from AMX-30B2 MBTs to replace the current low light level TV camera.

The major upgrade previously mentioned was not implemented on cost grounds and in mid-2000, following a competition between SAGEM, Thomson-CSF and Giat Industies teamed with Matra Systemes et Information (MSI), the French Delegation Generale pour l'Armement (DGA) selected the latter team to provide the AMX-10RC (6 × 6) armoured car and the Panhard VBL (4 × 4) scout car with a Terminal Information System (TIS) type Battlefield Management System (BMS). The French Army designation for the system is SIT-V1.

TIS is an evolution of the Giat Industries FINDERS (Fast Information Navigation Decision and Reporting System) which is fitted as standard on the 388 Leclerc Tropicalised Main Battle Tanks (MBT) being delivered by Giat Industries to the United Arab Emirates.

Giat Industries is the TIS prime contractor and will provide the software, integrate elements of the information system and integrate the complete system in the vehicles. MSI, as a subcontractor, will produce the computer and graphics terminal.

Installation of the TIS will allow the reconnaissance vehicles to exchange digitised data such as the tactical situation and have this information displayed on a touchscreen digital map that can be seen by the vehicle commander. In the case of the AMX-10RC, the screen will also show the video image from the thermal camera mounted co-axial with the 105 mm gun.

One squadron of 12 AMX-10RC armoured cars and 12 Panhard VBL scout cars will be equipped with the TIS for trials after which more than 600 vehicles will be retrofitted with the system.

The TIS will also be fitted to the French Army's future Satory Military Vehicles VBCI (8 × 8) infantry fighting vehicle. The first batch will consist of 700 vehicles, 550 in the infantry fighting vehicle role and the remaining 150 in the command post role.

The AMX-10RC will also be fitted with the CEPM (microprocessor electropneumatic command) system that takes into account the driver's decision to change speed and automatically performs the operation at the appropriate time. First CEPM kits were ordered in 1999 and deliveries will run through to 2002.

The turret will be modernised by the installation of the TIS, Galix combat vehicle protection system, Thomson-CSF PR4G radio system and in the future a laser warning system.

Turret ergonomics will be improved and the existing workload redistributed with the loader now operating the radio.

Giat Industries is responsible for the turret while French Army facilities will upgrade the chassis with Giat Industries integrating the turret with the chassis. First series production upgraded AMX-10RC vehicles will be delivered to the French Army in 2002.

Giat Industries is no longer marketing this vehicle but is now concentrating its efforts on the VEXTRA (8 × 8) fitted with the TML 105 105 mm turret. Details of this are given in the previous entry.

Description

The all-welded aluminium hull and turret provides the crew with protection from small arms fire, flash burns and shell splinters. The driver is seated at the front of the vehicle on the left side and is provided with an adjustable seat and a large single-piece hatch cover, that opens to the right. There are three day periscopes, the centre one of which can be replaced by a passive OB-31A periscope for night driving.

The turret is in the centre of the vehicle with the commander and gunner seated on the right and the loader on the left. Both the commander and the loader are provided with a single-piece hatch cover that opens to the rear and the loader has three day periscopes, one to the front, one to the left and one to the rear. The vehicle commander has six day periscopes for all-round observation and a day periscope and panoramic M389 telescope with magnifications of ×2 and ×8 which has 360° full rotation and enables the commander to observe the target regardless of the position of the turret. The M389 can be elevated from −12 to +24° and has automatic projection of graticules, ×2 for the 7.62 mm machine gun and ×8 for

Specifications
(with 0.3 m ground clearance)
Crew: 4
Configuration: 6 × 6
Combat weight: 15,880 kg
Unloaded weight: 14,900 kg
Power-to-weight ratio: 16.45 hp/t
Length:
(gun forward) 9.15 m
(hull) 6.357 m
Width: 2.95 m
Height:
(overall) 2.66 m
(to turret top) 2.29 m
(to hull top) 1.565 m
Ground clearance:
(normal) 0.35 m (adjustable from 0.2-0.6 m)
Track: 2.425 m
Wheelbase: 1.55 + 1.55 m
Max speed:
(road) 85 km/h
(water) 7.2 km/h
Max road range: 1,000 km
Fording: amphibious
Gradient: 50%
Side slope: 30%
Vertical obstacle: 0.8 m
Trench: 1.65 m
Engine: Baudouin Model 6F 11 SRX diesel engine developing 280 hp at 3,000 rpm.
Transmission: preselective with 4 forward and 4 reverse gears
Suspension: hydropneumatic
Tyres: 14.00 × 20 XL
Electrical system: 24 V
Batteries: 6 × 12 V, 100 Ah
Armament:
(main) 1 × 105 mm F2 rifled gun
(coaxial) 1 × 7.62 mm MG
(anti-aircraft) 1 × 7.62 mm MG (French Army vehicles)
Smoke-laying equipment: 2 × 80 mm smoke grenade dischargers either side of turret
Ammunition:
(main) 38
(coaxial) 4,000
(smoke grenades) 16
Gun control equipment
Turret power control: electrohydraulic/manual
(by commander) yes
(by gunner) yes
Turret traverse: 360°
Gun elevation/depression: +20°/−8° (elevation can be increased to +40° using suspension)
Gun stabiliser:
(vertical) no
(horizontal) no
NBC system: yes
Night vision equipment: yes

Terminal Information System installed in a Giat Industries AMX-10RC (6 × 6) armoured car (DGA)
2001/0073149

Giat Industries AMX-10RC (6 × 6) reconnaissance vehicle with additional armour, revised external stowage and an ATGW decoy system mounted on the left side of the turret (BP) **2001**/0093527

the 105 mm gun. The commander also has overriding controls enabling him to lay the gun onto the target. The gunner has three periscopes and an M504 telescope manufactured by SOPELEM and Compagnie Industrielle des Lasers (CILAS) which is the main part of the fire-control system. The M504 has a magnification of ×10 and is combined with an optical compensator with control electronics for automatic input of fire corrections, an M550 laser range-finder and an M553 boresight unit which corrects for any abnormal droop or accidental misalignment of the main armament.

The complete COTAC fire-control system comprises various sensors which provide the computer with the following data: target distance as measured by the laser range-finder with an accuracy of ±5 m between 400 and 10,000 m; horizontal and vertical speed of the target; angle of cant of the vehicle and windspeed, altitude, outside and gun temperatures, which are fed into the computer manually.

In a typical target engagement the gunner aligns the sight with the target, tracks the target and depresses the measuring button for two or three seconds, the time taken for the target information to be fed into the computer. The introduction of fire correction data takes only 1.5 seconds and the gunner opens fire.

There are three modes of engagement, COTAC, tank laser sight and graticule judgement.

For night engagements the AMX-10RC is fitted with a Thales Model DIVT 13 LLLTV. The system consists of a television camera mounted in the mantlet to the left of the main armament, an electronics unit, a control unit and two television screens, one for the commander and the other for the gunner. An aiming graticule is generated electronically and superimposes itself on the monitor screen. The deviation computer of the COTAC fire-control system then introduces fire correction data and shifts this aiming graticule by the required value under the same conditions as for the day graticule.

For Operation Desert Storm, Giat Industries supplied the French Army with the Thales DIVT 16 thermal camera as a replacement for the DIVT 13. This option enabled observation and identification to take place at ranges of up to 4,000 m.

The engine at the rear of the hull is coupled to the gearbox through a hydraulic torque converter with a cut out clutch. The gearbox has three functions:

(1) Gear shifting with four ratios in both directions
(2) Auxiliary drive for the two water-jets
(3) Steering with epicyclic gears and hydraulically actuated brakes.

The AMX-10RC is skid-steered in a similar manner to the tracked AMX-10P.

In 1983, it was confirmed that the last production batches of AMX-10RC vehicles for the French Army would be fitted with the Baudouin Model 6F 11 SRX diesel engine which develops 280 hp, in place of the standard Renault-built HS 115 diesel that develops 260 hp at 3,000 rpm. In addition to being a more powerful engine the Baudouin is more fuel efficient, which will give the vehicle a larger operating range of 1,000 km.

By 1995, all the French Army vehicles were fitted with the Baudouin 6F 11 SRX diesel engine.

The hydropneumatic suspension system, designed and manufactured by Messier Auto-Industrie, can be quickly adjusted by the driver to suit the type of ground being crossed, this being accomplished without the driver leaving his seat.

The minimum ground clearance is 210 mm, but for road travel 350 mm is normal, with 470 mm for cross-country travel and 600 mm for amphibious operations.

The suspension at each wheel station consists of a roadwheel arm and suspension assembly (link rod, level and cylinder with the cylinder acting as spring and shock-absorber and allowing the height to be adjusted). The suspension cylinders are supplied by a hydraulic pump. The tyre pressure can be adjusted to suit the type of ground being crossed, this being accomplished without the driver leaving his seat.

Giat Industries AMX-10RC (6 × 6) armoured car turret fitted with Kit Basique de Contre Measures (KBCM)
2002/0063380

The AMX-10RC is fully amphibious, being propelled in the water by two water-jets either side of the rear of the hull. Before entering the water a trim vane is erected at the front of the vehicle, with a transparent window in the left side to allow the driver to see ahead when afloat. The vehicles delivered to Morocco were not fitted with water-jets.

The NBC system is mounted in the rear of the turret and maintains a slight overpressure in the fighting compartment. For use in cold climates an auxiliary heater is installed.

Main armament of the AMX-10RC is a Giat Industries 105 mm F2 semi-automatic gun with a vertical wedge breech block and a barrel with a thermal jacket and a two-stage muzzle brake. The gun itself is 48 calibres long without the muzzle brake, has a recoil length of 600 mm and a recoil force of about 13 tonnes. The recoil mechanism comprises a brake on the left side and a counter-recoil mechanism on the right side of the gun. The gun fires four types of Giat Industries fixed ammunition, APFSDS, HEAT, HE and smoke. The APFSDS round uses the penetrator of the 90 mm Giat Industries APFSDS round and has a muzzle velocity of 1,400 m/s. This will penetrate a NATO triple heavy tank target at a range of 2,000 m. The new APFSDS round weighs 13.1 kg complete, the projectile weighing 3.8 kg.

Of 38 rounds of 105 mm ammunition carried, 12 are for ready use in the turret. A 7.62 mm machine gun is mounted coaxially to the left of the main armament. Two 80 mm smoke grenade dischargers mounted either side of the turret towards the rear are electrically fired from within the vehicle. French Army AMX-10RC vehicles have been fitted with a roof-mounted 7.62 mm machine gun.

The turret is equipped with a SAMM CH49 electrohydraulic control system consisting of a servo-controlled cylinder, a flow dividing unit, a servo motor, a hydraulic power source, an electronics box, twin handles for the gunner and a single handle for the commander.

If required, the AMX-10RC could have been delivered without the air conditioning system, heater, NBC system, night fire-control equipment, navigation system and amphibious capability. The vehicle can also be fitted with a thermal camera to extend observation and identification out to 4,000 m in any weather conditions and a front wheel-mounted self-recovery device.

Variants

AMX-10RC driver training vehicle

This is an AMX-10RC vehicle with its turret removed and fitted with a fully enclosed cabin for the instructor and several other students.

AMX-10RC 105 TML

This turret, of which two versions have been developed, has been installed on an AMX-10RC (6 × 6) vehicle for trials purposes. This combination is no longer being marketed by Giat Industries.

AMX-10RC with countermeasures kit

For trials purposes an AMX-10RC has been fitted with the Kit Basique de Countre Measures (KBCM). This has a number of subsystems to neutralise a variety of battlefield threats. These sensors include laser warning device, missile warning system, infra-red jammers and the GALIX screening smoke system. Details are given in *Jane's Armour and Artillery Upgrades 2000-2001.*

Status

Production complete. No longer being marketed. In service with the following countries:

Country	Quantity
France	337
Morocco	108
Qatar	12

Manufacturer

Giat Industries.

UPDATED

Renault VBC 90 armoured car

Development

The VBC 90 (*Véhicule Blindé de Combat*) was designed by Renault Véhicules Industriels (RVI) specifically for the export market and was shown for the first time at the 1979 Satory Exhibition of Military Equipment.

The first production VBC 90 armoured car was completed in September 1981. It uses many automotive components of the VAB (4 × 4) which is in service with the French Army. The first of 28 VBC 90 armoured cars ordered by the French Gendarmerie were delivered in 1983 and are fitted with an SFIM SOPTAC 11 fire-control system incorporating a laser range-finder. Oman placed an initial order for four VBC-90 armoured cars with SOPTAC 11 fire-control systems. A further two vehicles were delivered in early 1985.

The Renault VBC 90 (6 × 6) armoured car is no longer being marketed.

Early in 2001, Renault Vehicles Industrials and Giat Industries formed a new joint venture company called Satory Military Vehicles. This will not only handle the French Army's new VBCI (8 × 8) vehicle but also current vehicles such as the VAB and AMX-10 series (tracked and wheeled).

Renault VBC 90 (6 × 6) armoured car as delivered to the French Gendarmerie, with the driver's hatch closed and showing integral bulletproof windows 0088018

Description

The all-welded steel hull protects the crew from small arms fire and shell splinters and is divided into three compartments: driving at the front, fighting in the centre and engine at the rear. The driver is seated at the front of the vehicle on the left side and has a single-piece hatch cover that opens to the right with three bulletproof windows for greater all-round observation, the centre one of which is fitted with a wiper.

The Giat Industries TS 90 turret is of all-welded steel armour with the commander on the left and the gunner on the right, both with a single-piece hatch cover that opens to the rear and an adjustable seat. An extractor fan is mounted in the forward part of the turret roof. Main armament consists of a long-barrelled Giat Industries 90 mm F1 gun with a 35° oblique wedge breech block, a hydropneumatic recoil system, a thermal sleeve and a muzzle brake. There are 20 rounds of ready to use 90 mm ammunition carried, four in the turret basket and 16 in the turret bustle. The empty 90 mm cartridge cases are ejected into a bag under the breech and the commander has an ammunition resupply hatch in the left side of the turret. A further 25 rounds of 90 mm ammunition are carried in the hull.

Types of Giat Industries 90 mm ammunition that can be fired include canister, HE, HE long-range, HEAT, smoke and APFSDS.

Mounted coaxially to the left of the main armament is a 7.62 mm machine gun and there are two electrically operated 80 mm smoke grenade dischargers mounted either side of the turret towards the rear.

A PH 9A white light searchlight is mounted coaxially to the left of the main armament and a second white light searchlight is mounted in the forward part of the turret and operated by the commander. The commander has seven (three day M556, four day M554) periscopes and the gunner five (three day M556, two day M554). The gunner also has a Giat Industries M563 telescopic day sight with a magnification of ×5.9 for laying the main armament.

The gunner can also have a SOPELEM TJ N2 90A sight with a magnification of ×6 (day) and ×6 (night). The SFIM SOPTAC 10A is the TJ N2 90A with a TCV 107 laser range-finder with range displayed to the vehicle commander with automatic display of target elevation in the gunner's sight graticule. SFIM SOPTAC 11A is the TJ N2 90A with a TCV 107 laser range-finder with automatic computing and display of range to the vehicle commander, and automatic display of target elevation in the gunner's sight graticule and lateral deflection (target movement speed) for firing at moving targets. All SFIM SOPTAC sights can be fitted so that the gunner's field of view is also supplied to the commander's sight.

The engine, transmission and fuel tank are at the rear of the hull, with a large door opening to the left giving easy access for maintenance.

Power is transmitted from the engine to the wheels by a torque converter and a gearbox with five forward and one reverse gears. Gears are shifted by a small electrically operated lever which also operates the clutch. The axles have differential reduction gears with double reduction and differential locking. The wheels are independently suspended by torsion bars and hydraulic shock-absorbers. Steering is hydraulically assisted on the two front axles and the tyre pressures can be adjusted to suit the type of ground being crossed. Unlike the VAB family of APCs, the VBC 90 does not have any amphibious capability.

Standard equipment on all vehicles includes a heater and a power-operated winch mounted internally at the front of the hull on the right side. This is provided with

Specifications

Crew: 3
Configuration: 6 × 6
Power-to-weight ratio: 16 hp/t
Combat weight: 13,500 kg
Weight unladen: 12,250 kg
Length:
(gun forward) 8.085 m
(hull) 5.63 m
Width: 2.5 m
Height:
(to turret top) 2.552 m
(to hull top) 1.737 m
Track: 2.035 m
Wheelbase: 1.5 + 1.5 m
Angle of approach/departure: 45°/45°
Max road speed:
(5th gear) 92 km/h
(4th gear) 58 km/h
(3rd gear) 34 km/h
(2nd gear) 20 km/h
(1st gear) 12 km/h
(reverse) 13 km/h
Max road range:
(at 60 km/h) 1,000 km
Fording: 1.2 m
Gradient: 50%
Side slope: 30%
Vertical obstacle: 0.5 m
Trench: 1 m
Engine: Renault MIDS 06.20.45 in-line water-cooled turbocharged 6-cylinder diesel developing 220 bhp at 2,200 rpm. Previous engine MAN D.2356 HM 72 in-line water-cooled 6-cylinder diesel developing 220 hp at 2,200 rpm (standard on VBC 90 until 1983 but no longer available)
Transmission: Renault Transfluid 376 with 5 forward and 1 reverse gears
Steering recirculating ball, hydraulically assisted on first and second axles
Tyres: 14.00 × 20 (run-flat)
Brakes: disc
Suspension: independent (torsion bar and telescopic shock-absorbers)
Electrical system: 24 V
Batteries: 4 × 12 V, 6TN
Armament:
(main) 1 × 90 mm F1 rifled gun
(coaxial) 1 × 7.62 mm MG
(anti-aircraft) 1 × 7.62 mm MG (optional)
Smoke-laying equipment: 2 × 80 mm smoke grenade dischargers either side of turret
Ammunition:
(main) 45
(coaxial) 4,000
Gun control equipment
Turret power control: electrically assisted manual (traverse only)
Max rate of power traverse: 360°/20 s
Max rate of manual elevation: 23°/15 s
Gun elevation/depression: +15°/−8° (except over rear)
Gun stabiliser:
(vertical) no
(horizontal) no
NBC system: optional
Night vision equipment: optional

60 m of cable and has a capacity of 6,000 kg. Optional equipment includes an NBC system and independent heating as well as the various fire-control systems previously mentioned.

Status

In service with France (Gendarmerie, 28 delivered from 1983) and Oman (6). Production complete. The Renault VBC 90 is no longer being marketed. There are no known plans for these vehicles to be upgraded in the near future.

Manufacturer

Renault Véhicules Industriels
Mecanique Creusot-Loire
Production of the VBC 90 was undertaken by Renault Véhicules Industriels with Mecanique Creusot-Loire as co-contractor under the leadership of Renault Véhicules Industriels.

UPDATED

Panhard ERC Sagaie armoured car

Development

The ERC (Engin de Reconnaissance Canon) range of armoured cars was developed by Panhard as a private venture from 1975 and was aimed specifically at the export market.

The range was first shown in 1977 at the Satory Exhibition of Military Equipment and entered production the following year at Panhard's new factory at Marolles. First production vehicles were completed in 1979.

The ERC range of armoured cars share many automotive components with the Panhard VCR range of armoured personnel carriers also shown for the first time in 1977. Full details of these are given in the Armoured personnel carriers (wheeled) section. There has been no recent production of the Panhard VCR (6 × 6) armoured personnel carrier.

The Panhard VCR (6 × 6) was originally developed to meet the specific operational requirements of Iraq as the platform for the Euromissile UTM-800 HOT anti-tank turret. Using automotive components from the VCR (6 × 6) the ERC armoured car was created but with a different wheelbase on account of the turret basket of the new Giat Industries TS90 turret.

The French Army evaluated the ERC 90 F4 Sagaie between 1978-80 and accepted the vehicle in December 1980 although further trials, including hot weather trials in Djibouti in 1982, were subsequently carried out. First production vehicles were delivered to the French Army in 1984 and the total French Army order was for 192 ERC Sagaies with final deliveries in 1990.

The French Rapid Deployment Force uses this vehicle as its standard armoured car. The 1st Regiment Infanterie de Marine of the 9th Marine Division has converted to a reconnaissance regiment with ERC 90 vehicles. The Regiment d'Infanterie de Chars de Marine of the 9th Marine Division received all of its vehicles in 1984, the 1st Regiment of the Parachute Hussards of the 11th Parachute Division received their vehicles in 1985-86 while the 4th Chasseurs Regiment of the 27th Alpine Division received their vehicles in 1988-89. In each case they replaced Panhard AML 60 and AML 90 (4 × 4) armoured cars. Each regiment has three squadrons with each of these having four platoons each with three ERC 90 F4 Sagaies and a Euromissile MILAN ATGW squad.

A total of 42 ERC 90 F4 Sagaie armoured cars is forward deployed overseas, 24 with the 13th Foreign Legion in Djibouti, six with the 6th Marine Corps Battalion in Gabon, six with the 43rd BIMa in the Ivory Coast and six with the 23rd BIMa at Dakar in Senegal.

In the longer term, the Panhard ERC (6 × 6) armoured cars used by the French Army could be replaced by the projected Engin Blinde a Roues de Contact (EBRC) (8 × 8) for which Giat Industries have been awarded a 16 month sole source study contract. Available details of this vehicle are given earlier in this section.

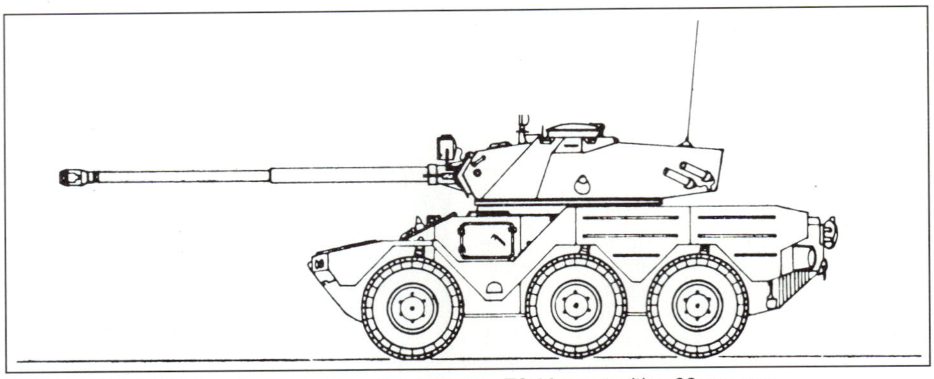

Panhard ERC 90 F4 Sagaie fitted with a Giat Industries TS 90 turret with a 90 mm gun

Description

The all-welded steel hull provides the crew with protection against small arms fire and shell splinters. The floor of the hull consists of two plates welded together to form a flat V which helps the vehicle to slide off obstacles and stiffens the floor against damage from mines.

The driver is seated at the front of the hull, slightly to the left of the vehicle's centreline. This hatch is in two parts: one part, which contains the forward-facing day periscope, folds forward onto the glacis plate and the second part folds upwards. The forward day periscope can be replaced by an image intensification periscope for night driving, and there is an additional day periscope either side of the driver's hatch cover.

The turret is mounted in the centre of the vehicle and as each version has a different turret they are described in detail under Variants. All vehicles in the range are powered by a militarised version of a Peugeot V-6 petrol engine which develops 145 hp at 5,500 rpm. Power is taken from the engine through a single-plate clutch to a transversely mounted Panhard gearbox with six forward and one reverse gears and incorporating a Panhard cam-type limited slip differential from which the drive is taken to either side of the vehicle. Drive is then taken by half-shafts to bevel gears at the centre wheel stations and then shafts take power to the bevel gears at the front wheel stations. The rear wheel bevel gears are driven by shafts directly from the gearbox. The drive from the centre to the front wheel on each side incorporates another cam-type limited slip differential. Power is transmitted from the bevel boxes to the wheels by a train of gears housed within trailing arms which independently locate the wheels. The front and rear wheels are sprung by single coil springs with concentric telescopic hydraulic dampers. The centre wheels, which are raised off the ground when on roads, are sprung by hydropneumatic units. The centre wheels are powered even when raised clear of the road surface.

Only the front wheels are steered, and steering is hydraulically assisted. The low-pressure tyres are fitted with Hutchinson cellular inner tubes, which enable the vehicle to travel for at least 100 km at 30 km/h when the tyres have been punctured by bullets.

The basic vehicle can ford to a depth of 1.2 m without preparation but two different amphibious models of the ERC have been developed. Both models have a trim vane mounted at the front of the hull which is folded back onto the glacis plate when not in use and which has a transparent panel in its centre to enable the driver to have a clear view ahead, and sheet metal floats over the wheel stations filled with closed-cell foam. A rear opening door is provided in the left side of the hull. Before entering the water the trim vane is erected and the two tubular ducts at the sides of the engine compartment are hydraulically swivelled into the vertical position to prevent water entering the engine compartment. The radiator compartment is allowed to flood, as when the vehicle is afloat cooling is by water-to-water heat transfer. The first amphibious model is propelled in the water by its wheels and the second by two Dowty-Messier hydrojets.

All models have a turret-mounted ventilator and internal lights as standard. Optional equipment includes an NBC system, a ground navigation compass, air conditioning equipment and a front-mounted winch with 60 m of cable and a capacity of 3,500 kg.

Variants

ERC 90 F4 Sagaie (ERC-1 Sagaie)

This was the first model to enter production and is fitted with a Giat Industries TS 90 turret armed with a 90 mm gun which can fire canister, HE, HE long-range, HEAT, smoke and APFSDS ammunition.

Panhard ERC 90 F4 Sagaie of the French Army (Christian Dumont)

2000/0064875

Panhard ERC 90 F4 Sagaie 2 in service in Gabon

0045850

A 7.62 mm machine gun is mounted coaxially to the left of the main armament and two 80 mm smoke grenade dischargers are mounted either side of the turret at the rear. The commander is seated on the left side of the turret and the gunner on the right, both with a single-piece hatch cover that opens to the rear. The commander is provided with seven day periscopes and the gunner five; the gunner also has an M563 day telescopic sight with a magnification of ×5.9. Of the 20 90 mm rounds carried in the turret, 12 are HEAT and 8 HE.

Panhard offers six options for this vehicle, each with a weight index of 1 or 2, and the user can select options up to an index of 4. The options are: amphibious kit propelled in the water by wheels (1), amphibious kit propelled in the water by hydrojets (2), NBC system (1), air conditioning system/heater (1), an additional 10 rounds of 90 mm and 1,000 rounds of 7.62 mm ammunition (1) and a ground navigation system (1).

ERC 90 F4 Sagaie 2 (ERC-2 Sagaie)

This model has a longer, wider hull and is powered by two Peugeot XD 3T four-cylinder turbocharged diesel engines each developing 98 hp at 4,150 rpm. These engines are also used in the Panhard VBL light scout car, the Panhard VCR TT 2 APC (two engines) and the upgraded Panhard AML/ Buffalo 4 × 4 vehicles. The engines are coupled to a ZF 4 HP 22 automatic transmission with four forward and one reverse gears and a Panhard two-speed transfer box, one for road travel and the other for cross-country.

As an alternative to the two Peugeot XD 3T engines the Sagaie 2 can also be powered by two Peugeot PVR six-cylinder water-cooled petrol engines developing 145 hp each.

The Sagaie 2 has larger 13.00 × 16 tyres compared to the Sagaie 1 which has 11.00 × 16 tyres, and the central pair of roadwheels has been moved further back.

By 1985, two vehicles had been built, one with two petrol engines and one with two diesel engines. The vehicle can be made amphibious by filling the side bins with foam; this also provides some degree of protection against HEAT attack.

The prototypes have been fitted with the SAMM TTB 190 turret which is the model selected by Gabon, but other turrets can be fitted including TS 90.

The desert version has no water-jets for amphibious propulsion and the buoyancy boxes are replaced by boxes for increased stowage of ammunition, fuel and water. Optional equipment includes an NBC system and an air conditioning unit.

The only known customer for this vehicle, which is also referred to as the Panhard ERC-2, is Gabon. SAMM is no longer building or marketing the TTB 190 turret system.

Specifications

Model	**Sagaie 1**	**Lynx**
Crew	3	3
Configuration	6 × 6	6 × 6
Combat weight	8,300 kg	7,600 kg
Power-to-weight ratio	17.5 hp/t	19.07 hp/tonne
Length		
(gun forward)	7.693 m	6.115 m
(hull)	5.098 m	5.235 m
Width	2.495 m	2.495 m
Height		
(overall)	2.254 m	2.147 m
(to hull top)	1.502 m	1.502 m
Ground clearance		
(road)	0.294 m	0.294 m
(cross-country)	0.344 m	0.344 m
Track	2.135 m	2.135 m
Wheelbase	1.63 + 1.22 m	1.63 + 1.22 m
Angle of approach/departure	48°/45°	48°/45°
Max speed		
(road)	95 km/h	95 km/h
(water, wheels)	4.5 km/h	4.5 km/h
(water, hydrojets)	7.2 km/h	7.2 km/h
Max road range	700 km	700 km
Fuel capacity	242 litres	242 litres
Fording	1.2 m	1.2 m
Gradient	60%	60%
Side slope	30%	30%
Vertical obstacle	0.8 m	0.8 m
Trench	1.1 m	1.1 m
Engine	Peugeot V-6 petrol developing 155 hp at 5,250 rpm	
Transmission	Panhard with 6 forward and 1 reverse gears	
Steering	power-assisted	Power-assisted
Clutch	hydraulic-controlled disc	Hydraulic-controlled disc
Tyres	11.00 × 16	11.00 × 16
Electrical system	24 V	24 V
Armament		
(main)	1 × 90 mm F1 rifled gun	1 × 90 mm F1 rifled gun
(coaxial)	1 × 7.62 mm MG	1 × 7.62 mm MG
(anti-aircraft)	1 × 7.62 mm MG (optional)	1 × 7.62 mm MG (optional)
Ammunition		
(main)	20	30
(coaxial)	2,000	3,000
(anti-aircraft)	included in above	Included in above
Turret power traverse	manual, 360°	Manual, 360°
Gun elevation/depression	+15°/–8°	+35°/–8°
Gun stabiliser		
(vertical)	no	No
(horizontal)	no	No
NBC system	optional	Optional
Night vision equipment	optional	Optional

Overall height depends on the ground clearance; for roads it is normally 294 mm and for cross-country operations it is 344 mm.

ERC 90 F1 Lynx (ERC-1 Lynx)

This is fitted with a Lynx 90 turret designed and manufactured by Hispano-Suiza (which is no longer involved in the design, development and production of armoured fighting vehicle turrets) and armed with a 90 mm F1 gun, a 7.62 mm coaxial machine gun and two 80 mm smoke grenade dischargers on either side of the turret at the rear. The gun fires HEAT, HE, smoke and canister rounds.

The commander is seated on the left of the turret with the gunner on the right. The commander is provided with a cupola with a single-piece hatch cover that opens to the rear and eight day periscopes for all-round observation. The gunner is provided with a single-piece hatch cover that opens to the rear. Vision equipment consists of a combined day and night passive periscope Model TJN 2.90 (SOPELEM), TCV 107 laser range-finder (CILAS/SOPELEM) and 12 day periscopes (nine Type L794B and three Type L794D). A total of 21 rounds of 90 mm ammunition is carried in the turret and, if required, a 7.62 mm machine gun can be fitted for anti-aircraft defence.

Other versions of the Panhard ERC

The following versions of this vehicle are no longer being marketed by Panhard:

ERC 60/12 Mangouste

Panhard ERC 90 F4 Sagaie of the French Army in Somalia **2001**/0093528

Panhard ERC 90 F4 Sagaie of the French Army in Bosnia (Richard Stickland) **2000**/0064876

Panhard ERC 90 F1 Lynx armoured cars in service with the Mexican Army

Panhard ERC 20 Kriss with SAMM TAB 220 20 mm twin turret in service with Gabon 0088019

EMC 81 mm Mortar Gun Carrier
ERC with SAMM TTB 125 turret
ERC with SAMM TTB 190 turret
ERC with 60-20 Serval turret
ERC with SAMM S 530 A 20 mm turret
ERC with TG 120 turret
ERC with TTB 140 40 mm turret
ERC with TA-20 turret
ERC with SAMM TAB 220 20 mm turret (this has a two-man turret with twin 20 mm anti-aircraft cannon and four have been sold to Gabon)

Sagaie upgrade

In 1998, Panhard submitted a proposal to the French DGA for the upgrade of the French Army fleet of Panhard ERC 90 F4 Sagaie (6 × 6) armoured cars fitted with Giat Industries 90 mm TS-90 turret.

The Panhard ERC 90 F4 Sagaie is powered by a Peugeot V-6 petrol engine developing 155 hp coupled to a manual transmission. This gives the vehicle a maximum road speed of 95 km/h with a maximum operational road range of 700 km.

Today these are the only petrol-engined armoured fighting vehicles in service with the French Army which, like most other armies, is now moving to a common diesel fuel policy with the obvious logistical and increased operational range advantages.

The main part of the proposed Panhard upgrade is the replacement of the existing power pack by a new power pack consisting of a more fuel-efficient MTU 4-cylinder diesel developing 170 hp coupled to a Renk fully automatic transmission with torque converter. This maintains road speed at 95 km/h but operational range is increased to 900 km.

The current vehicle weighs 8.3 tonnes and is fully amphibious being propelled in the water by water-jets mounted at the rear of the hull. The upgraded vehicle, with add-on armour, weighs 9.5 tonnes but is not amphibious.

Once it has been type classified by the French Army, the repower upgrade will also be offered by Panhard on the export market.

Since it has entered service with the French Army, the Panhard ERC 90 F4 Sagaie has already been upgraded in a number of areas including additional turret armour, increased external stowage and roof skate mount for a 7.62 mm machine gun. Some have also been fitted with a roof-mounted anti-tank missile decoy system.

Manufacturer

Société de Constructions Mécaniques Panhard et Levassor.

Status

Production as required. In service with:

Country	Quantity	Comment
Argentina	12	Lynx 90 turret, used by marines
Chad	4	Lynx 90 turret
France	192	TS 90 turret
Gabon	10	6 with TTB 120 turret and four ERC 20 Kriss twin 20 mm turret
Côte d' Ivoire (Ivory Coast)	7	TS 90 turret
Mexico	120	Lynx 90 turret
Nigeria	46	TS 90 turret

UPDATED

Panhard AML light armoured car

Development

In 1956, following the successful use of the British Ferret (4 × 4) scout car (200 vehicles were purchased) by the French Army in North Africa, the French Army issued a requirement for a similar vehicle but with a more powerful armament as the Ferret was armed only with a single 7.62 mm machine gun which, in the case of the Mk 1, was pintle mounted.

Panhard, which was then building the EBR (8 × 8) heavy armoured car at the time, was awarded a contract to build the prototype which was completed in 1959 under the Panhard designation Model 245. Saviem and DEFA-AMX also built prototype vehicles but, after trials, the Panhard model was accepted for service with the French Army as the AML (*Automitrailleuse Légère*) and first production vehicles were delivered to the French Army in 1961. The total French Army buy was 210 AML 90s and 425 AML 60-7s; the French Gendarmerie bought 70 AML 60-7s and 45 AML 90s. The Panhard AML light armoured car was phased out of front-line service with the French Army early in 1990.

The AML has also been manufactured under licence in South Africa by the then Sandock-Austral Limited (which in late 1999 became Vickers OMC) as the Eland. There is a separate entry for this vehicle under South Africa. The Panhard M3 armoured personnel carrier, for which there is a separate entry, uses 95 per cent of the

AML 90 with H 90 turret in service with El Salvador (Julio Montes)
2000/0064877

Panhard AML 90 retrofitted with a Peugeot XD 3T diesel engine 0088021

automotive components of the AML, and many armies have bought complete fleets of AML armoured cars and M3 armoured personnel carriers.

Late production models of the AML have had the original Panhard petrol engine replaced by the Peugeot XD 3T turbocharged diesel developing 98 hp and this is also offered as a retrofit package.

In June 1994, Panhard stated that a total of 6,035 Panhard AML (4 × 4) armoured cars and M3/Buffalo (4 × 4) armoured personnel carriers had been made in France with an additional 1,100 being made under licence.

There has been no recent production of the Panhard AML although production can start again if significant additional orders are received. It should be noted that many versions of the Panhard AML armoured car were built with a two-person turret designed and built by Hispano-Suiza. This company no longer manufactures these turrets.

Description

(AML 90)

The all-welded steel hull of the AML is divided into three compartments: driving at the front, fighting in the centre and the power pack at the rear.

The driver is seated at the front of the hull and is provided with a single-piece hatch cover that opens to the right with three integral day periscopes, the centre one of which can be replaced by an image intensification periscope for night driving.

The all-welded turret was then manufactured by Hispano-Suiza. The commander is seated on the left of the turret and the gunner on the right, both with a single-piece hatch cover that opens to the rear. The commander and gunner each have four L794B day periscopes and the gunner also has a Type M262 or M37 day sighting periscope. The M37 has a magnification of ×6. A searchlight is mounted forward of the commander's hatch. Either side of the hull below the turret ring is a door. The left door, on which the spare wheel and tyre are mounted, opens to the rear, the right door opens to the front.

The engine is at the rear of the hull and two access panels are provided over the engine compartment. The gearbox is crosswise and consists of two gearboxes in one (high and low), coupled on both sides of the bevel pinion. The low-range box is for cross-country use and comprises two low gears, a top gear and one reverse gear. When the low-range box is in normal drive the four ratios of the high-range box command the four upper gears of the range: sixth, fifth, fourth and third. The high-range box is for normal road use and has three low gears and one overdrive. There are Panhard-type ball differentials in the gearbox and in each rear transfer box which automatically prevent gear slip.

Drive is transmitted from the gearbox to two lateral transfer boxes via pinions to the rear wheels and via drive shafts that run along the inside of the hull to the front wheels.

The independent suspension at each wheel station consists of coil springs and hydropneumatic shock-absorbers acting on the trailing arms of the wheel mechanism. The tyres are fitted with unpuncturable Hutchinson inner tubes.

The basic vehicle has no NBC system or night fighting aids, but they can be fitted as optional extras, as can an air conditioning system and an amphibious kit. As far as is known the amphibious kit for the AML series was never placed in quantity production.

Specifications

AML with H 90 turret
Crew: 3
Configuration: 4 × 4
Combat weight: 5,500 kg
Power-to-weight ratio: 16.36 hp/t
Length:
(gun forward) 5.11 m
(hull) 3.79 m
Width: 1.97 m
Height:
(overall) 2.07 m
(to hull top) 1.385 m
Ground clearance: 0.33 m
Track: 1.62 m
Wheelbase: 2.5 m
Max road speed:
(6th gear, high range) 90 km/h
(5th gear, high range) 61 km/h
(4th gear, high range) 35 km/h
(3rd gear, high range) 18.8 km/h
(3rd gear, low range) 18.8 km/h
(2nd gear, low range) 9.3 km/h
(1st gear, low range) 4.5 km/h
(reverse) 5.5 km/h
Fuel capacity: 156 litres
Max road range: 600 km
Fording: 1.1 m
(with kit) amphibious
Gradient: 60%
Side slope: 30%
Vertical obstacle: 0.3 m
Trench:
(with 1 channel) 0.8 m
(with 4 channels) 3.1 m
Turning radius: 6 m
Engine: Panhard Model 4 HD 4-cylinder air-cooled petrol developing 90 hp at 4,700 rpm
Transmission: manual with 6 forward and 1 reverse gears
Clutch: centrifugal with electromagnetic automatic control
Suspension: independent, coil spring and hydropneumatic shock-absorbers acting on suspension trailing arms of wheel mechanism
Tyres: 11.00 × 16
Brakes:
(main) hydraulic, dual circuit
(parking) hand-brake operating on gearbox output shaft
Electrical system: 24 V
Batteries: 2 × 12 V
Armament:
(main) 1 × 90 mm gun
(coaxial) 1 × 7.62 mm MG
Smoke-laying equipment: 2 × 2 smoke grenade dischargers
Ammunition:
(main) 20
(coaxial) 2,000
(smoke grenades) 12
Gun control equipment
Turret power control: manual
Max rate of traverse: 360° in 20 s
Gun elevation/depression: +15°/–8°
Gun stabiliser:
(vertical) no
(horizontal) no
Armour: 8-12 mm
NBC system: optional
Night vision equipment: optional

Note: Final production models of the AML are powered by the Peugeot XD 3T diesel engine developing 98 hp coupled to a standard Panhard transmission which gives an increase in operating range of 700 to 800 km depending on the weight of the vehicle. The Peugeot XD 3T diesel engine is also installed in the Panhard VBL (4 × 4) scout car which is used in large numbers by the French Army.

The Panhard AML 90 F1 diesel weighs 5,900 kg combat loaded and has similar specifications to the H 90 version except length overall is 5.48 m, hull length is 4.15 m, height to top of cupola is 2.19 m and height to turret roof is 2.15 m.

Variants

AML with H 90 turret

This model is fitted with an H 90 turret manufactured by Hispano-Suiza and is armed with a Giat Industries 90 mm D 921 F1 gun which fires HEAT, HE, smoke and canister rounds.

A 7.62 mm coaxial machine gun is mounted to the left of the main armament and a similar weapon, or a 12.7 mm (0.50) M2 machine gun, can be mounted on the roof for anti-aircraft defence. Two 80 mm smoke grenade dischargers mounted on either side of the turret are electrically fired from within the vehicle.

To provide more mobile fire support, Ethiopia has removed the turrets from some of its AML-90 armoured cars and fitted these onto the Russian BTR-60 (8 × 8) series of armoured personnel carrier.

AML with Lynx 90 turret

This replaced the AML with H 90 turret in production and has the same armament as the H 90 turret but can

Panhard AML with HE 60-7 turret in service with the Irish Army and armed with a 12.7 mm M2 MG (Denis J McCarthy)

Panhard AML with HE 60-20 Serval turret

0045846

be fitted with passive night vision equipment, a laser range-finder and powered traverse.

AML with HE 60-7 turret

This model is fitted with an H 60 turret armed with twin 7.62 mm machine guns on the left and a single 60 mm mortar on the right. The machine guns can be elevated from −15 to +60° and the mortar from −15 to +80°. The TDA Model HB 60 can be used in the direct role up to a range of 300 m or in the indirect role up to a range of 2,600 m. In the basic model, 53 mortar bombs and 3,800 rounds of 7.62 mm machine gun ammunition are carried; if additional radios are installed for the command role, ammunition is reduced to 32 mortar bombs and 3,200 rounds of machine gun ammunition. Types of mortar bomb include HE, canister, smoke and illuminating.

The turret is provided with a large two-piece hatch cover that opens to the front and rear of the vehicle, seven day periscopes (Model L794B) and a day M112/M113 periscope with a magnification of ×5 for aiming.

In 1992, Manroy Engineering of the UK completed a contract to replace the 60 mm mortars in Irish Army AMLs with the Manroy 0.5 in (12.7 mm) M2 HB QCB (Quick Change Barrel) machine gun.

This entailed removing the mortar and mantlet and replacing them with a new mantlet and cradle for the machine gun.

Modifications are also carried out to the sights and ammunition feed systems as well as the ammunition and storage systems. An electric firing system was installed for the 0.50 in (12.7 mm) machine gun.

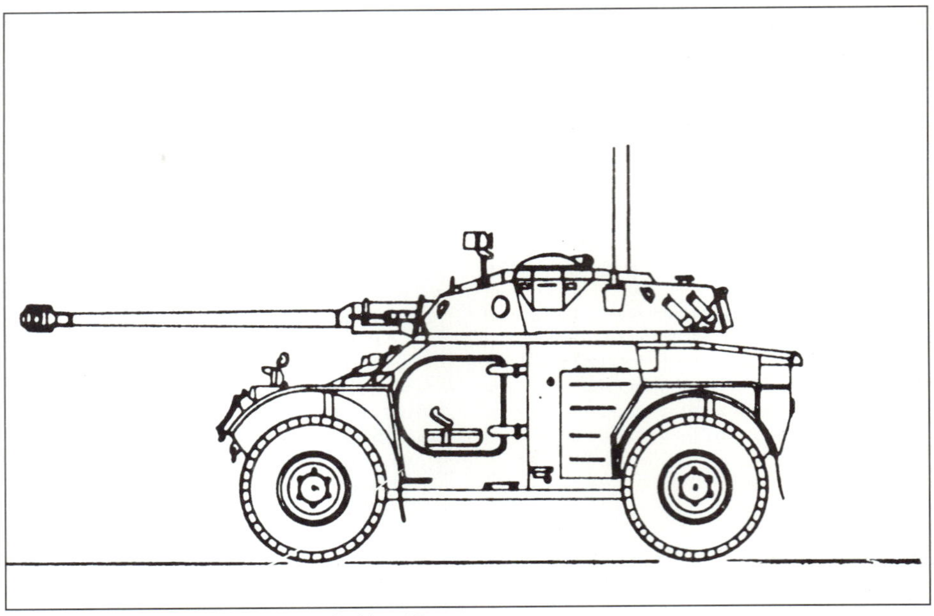

Panhard AML with H 90 turret armed with a 90 mm gun

AML with HE 60-12 turret

This is almost identical to the HE 60-7 turret but is armed with a 12.7 mm (0.50 in) machine gun and a TDA 60 mm HB 60 mortar. The machine gun can be elevated from −15 to +60° and the mortar from −15 to +80°. A total of 43 mortar bombs and 1,300 rounds of 12.7 mm machine gun ammunition is carried. If additional radios are carried, ammunition load is reduced to 31 rounds of 60 mm and 900 rounds of 12.7 mm machine gun ammunition. Loaded weight of an AML with the HE 60-12 turret is 4,800 kg.

AML with HE 60-20 turret

This is almost identical to the HE 60-7 turret but is armed with a 20 mm cannon and a 60 mm DTAT Model CS or a TDA HB 60 mortar. The cannon can be elevated from −8 to +50° and the mortar from −15 to +80°. When fitted with the DTAT mortar 53 mortar bombs and 500 rounds of 20 mm ammunition are carried; if the TDA mortar is installed, 43 mortar bombs are carried and 20 mm ammunition supply remains the same. If additional radios are installed for command vehicle use, the DTAT mortar is provided with 32 mortar bombs and the TDA with 31, and in both cases 500 rounds of 20 mm ammunition are carried. If required, a 7.62 mm machine gun can be mounted on the roof of the turret for anti-aircraft defence (this applies to all HE 60 turrets). This turret is no longer in production. Loaded weight of an AML with the HE 60-20 turret is 4,800 kg.

AML with HE 60-20 Serval turret

This version of the AML is fitted with an HE 60-20 Serval turret designed and manufactured by Hispano-Suiza (which is no longer engaged in the design, development and production of turrets) and is armed with a 60 mm breech-loaded mortar which can be elevated from −8 to +80°, and a 20 mm cannon mounted on the right side of the turret which can be elevated from −8 to +50°. The 60 mm mortar can be either a TDA 60 mm HB 60 (maximum range 2,600 m with M72 bomb) or a TDA 60 mm HB 60 LP long mortar (maximum range with LP bomb being 5,000 m). The 20 mm cannon can be either a French M693 dual-feed cannon with 300 rounds of HE and 37 rounds of AP ammunition for ready use or an HS-820 Oerlikon Contraves KAD-B16 single-feed cannon with 300 rounds of ready to use ammunition. If required, a 7.62 mm machine gun can be mounted coaxially to the right of the 20 mm cannon. Vision equipment includes a day sighting periscope and nine day observation periscopes, with the option of fitting a night periscope.

Panhard AML with HE 60-12 turret armed with a 60 mm mortar and a 12.7 mm machine gun 0088020

AML with S 530 turret

This was developed specifically for export and is a basic AML fitted with a turret manufactured by SAMM (Société d'Applications des Machines Motrices), armed with twin 20 mm M621 cannon with an elevation of +75°, depression of −10°, and turret traverse of 360°.

A total of 600 rounds of 20 mm ammunition is carried and the gunner can select either single shots, bursts or full automatic. Cyclic rate of fire is 740 rds/min/barrel. The only known user of this version is Venezuela, which has 12 vehicles in service. Full details of this are given in *Jane's Land-Based Air Defence 2001-2002*, pages 58 to 59. This system is no longer being marketed.

AML re-engining

As the 4HD petrol engine installed in the Panhard AML and the M3 is no longer produced, Panhard is offering a conversion kit to enable users of these vehicles to retrofit the new Peugeot XD3T liquid-cooled diesel engine in their own facilities with the minimum of specialised equipment.

The modifications can be summarised as:

Hull	
(cutting, drilling and welding)	new rear engine bonnet
Circuits	
(fuel (diesel))	modified
(dashboard)	modified
(main electrical control box)	modified
(alternator)	new
Power Plant	
(XD 3T turbocharged diesel)	new
(hydraulic control)	new
(mechanical clutch)	new
(reinforced gearbox)	new
(2nd gear)	synchronised
(hydraulic control cooling system)	new
(3 pedals unit)	new
(rear suspension)	reinforced
(4 shock-absorbers)	changed
(rear mudguards adapted to body)	modified
(exhaust system)	new

In addition, a more powerful air compressor can be fitted, as can Citröen brakes with disc brakes on front wheels.

Early in 1993, Panhard announced that Ireland had become the latest country to order its upgrade package to replace the existing petrol engine with a more fuel-efficient diesel engine.

In total, eight countries have now selected the modernisation package and the number of vehicles upgraded has now exceeded 300.

Senegal, Rwanda, Ireland, Morocco and four undisclosed countries are the other clients for the upgrades. Morocco is the largest customer with over 200 vehicles converted.

AML weapon modernisation

Panhard can upgrade the AML 90 in three key areas to improve its combat effectiveness:

(1) replacement of current infra-red night vision equipment used by the commander, gunner and driver by new image intensification equipment

(2) the installation of a laser range-finder above the 90 mm gun to increase first round hit capability

(3) use of Giat Industries APFSDS ammunition. In order to fire this ammunition the muzzle brake and recoil brake have to be modified but the same number of rounds of 90 mm ammunition can be carried in the same ready racks.

AML with 40 mm grenade launcher (AML 40)

This was first shown in 1990 and is essentially a Panhard AML 60-7 with the turret modified to accept a 7.62 mm machine gun (right) and the Spanish SANTA BARBARA 40 mm automatic grenade launcher which has a maximum range of 2,200 m, a cyclic rate of fire of 200 rds/min and a muzzle velocity of 240 m/s. A total of 225 rounds of 40 mm and 1,250 rounds of 7.62 mm ammunition is carried. By 2001 this model had not yet entered production.

FDS for Irish AML armoured cars

Building on their experience in the design, development and production of a wide range of turrets for tracked and wheeled armoured fighting vehicles, the LIW division of Denel, South Africa, have developed the Fire Detecting System (FDS) for light turret systems.

The FDS has been developed by LIW as a private venture and is aimed at the export market. It can be rapidly installed on one or two person turrets armed with a weapon system ranging from a 7.62 mm machine gun up to a 90 mm gun.

The FDS consists of two main components, the passive night sight assembly and second the laser range-finder assembly.

Externally mounted night sight elements consist of the night elbow, armour hood, window, armour flap with image intensification iris, wiper and associated water pipe. The internally mounted elements include the remote control at the gunners station, video display unit and the electronic unit.

Externally mounted laser range-finder elements are the laser range-finder, armoured hood, window, armour flap, wiper, water pipe and boresight with mechanical adjustment. Internally mounted elements include laser range-finder remote-control unit with range display for the gunner, range display for the commander and an electronic unit.

The external elements of the FDS are mounted co-axial with the main armament and move in elevation with this weapon. The FDS can be rapidly installed on light armoured vehicles with electrical power being received from the existing turret power supply.

For trials purposes the LIW FDS has been fitted to a South African Eland 20 mm (4 × 4) light armoured car which is a licence built and further developed version of the French Panhard AML series of vehicle.

When fitted with the EDS, a light armoured vehicle can identify and engage targets at a longer range and with a higher first round hit probability under day and night conditions.

The first customer for the FDS is Ireland which placed an order with LIW for 16 complete 20 mm Eland two-person turrets fitted with the FDS which have been installed on some of their existing French supplied Panhard AML (4 × 4) light armoured cars. The first four turrets were accepted from LIW late in 1999.

An additional 16 EDS systems have been supplied to Ireland for installation on existing Panhard AML 90 vehicles currently in service.

Status

Production complete. In service with the following countries:

Country	Quantity	Comment
Algeria	44	AML 60, gendarmerie and republican guard
Argentina	48	AML 90, 60 were purchased but 12 lost in the Falklands campaign
Bahrain	22	AML 90
Benin	12	AML 90 (6), AML 60 (6), all diesel
Bosnia-Herzegovina	44	AML 90 believed supplied by a country in the Middle East
Burkina Faso	15	AML 90
Burundi	18	AML 90 Lynx (12), AML 60 (6)
Chad	50	AML 90 and AML 60
Congo, Democratic Republic	60	AML 90 (30), AML 60-7 (30)
Djibouti	24	AML 90 (20), AML 60-7 (4) (status uncertain)
Ecuador	27	AML 90, AML 60-7
El Salvador	10	AML 90
France	121	AML 90, AML 60-7, gendarmarie
	588	AML 90, AML 60, Army, in reserve
Gabon	24	AML 90, AML 60-7
Guinea	2	AML 90
Iraq	n/avail	AML 90, AML 60*
Ireland	47	AML 90 (15), AML 60 (32), diesel (some upgraded further; see text under Ireland)
Ivory Coast	16	AML 90, AML 60-7
Kenya	72	AML 90, AML 60-7
Lesoto	n/k	AML 90
Malaysia	140	AML 90, AML 60
Malawi	13	from South Africa
Mauritania	60	AML 90 (40), AML 60-7 (20)
Morocco	230	AML 90 (190), AML 60-7 (40), most converted to diesel
Niger	125	most AML 60-7, but also uses AML 90 and Serval 60/20, some in diesel configuration
Nigeria	180	AML 90 (120), AML 60-7 (60)
Portugal	40	AML 60-7, reserve
Rwanda	12	AML 90, AML 60-7*
Saudi Arabia	235	AML 90, AML 60
Senegal	57	AML 90 (27), AML 60-7 (30) (diesel)
Somalia	n/avail	AML 90, AML 60-7*
South Africa	118	see separate entry; plus many for sale
Sudan	6	AML 90
Togo	10	AML 90 (70), AML 60-7 (3)
Tunisia	35	AML 90
UAE	50	AML 90
Venezuela	22	AML 90 and AML 60 (10), AML twin 20 mm SPAAG (12)
Yemen	185	AML 90 (125), AML 60 (60)*
Zimbabwe	20	from South Africa (Eland version)

As far as is known, it is no longer used by Angola, Cambodia, Ethiopia, France, Israel, Lebanon, Pakistan or Spain.

*In view of the recent conflicts in these countries, status of vehicles is uncertain.

Manufacturer

Société de Constructions Mécaniques Panhard et Levassor.

UPDATED

Panhard VBL

Development

In 1978, the French Army issued a requirement for a vehicle weighing under 3,500 kg for two basic roles, anti-tank armed with the Euromissile MILAN ATGW, and intelligence/scout armed with machine guns.

After proposals from five companies were studied, Panhard and Renault were each awarded contracts for the design and construction of three vehicles which were delivered to the French Army for trials in 1983. The official French Army designation for the vehicle is the *Véhicule Blindé Léger* (VBL).

In February 1985, the French Army accepted the Panhard vehicle to meet its requirements for the VBL.

Panhard VBL in LWB version for Kuwait fitted with Raytheon Systems Company TOW ATGW and 7.62 mm MG (Christian Dumont) ***2001***/0093531

Panhard VBL in LWB version for Oman fitted with Raytheon Systems Company TOW ATGW and 7.62 mm MG (Christian Dumont) ***2002***/0073360

Panhard VBL in service with Kuwait fitted with Panhard PL127 cupola armed with Panhard 12.7 mm M2 machine gun (Christian Dumont) **2002**/0114324

The Long Wheelbase version of the Panhard VBL normally has a crew of four **2001**/0093532

Panhard VBL with HOT/ATM installation in retracted position 0088022

Panhard VBL in service with Oman and showing amphibious capabilities (Christian Dumont) **2002**/0114325

Between June and October 1988, Panhard delivered 15 preproduction vehicles to the French Army for a wide range of troop and specialist trials under a variety of environmental conditions. Before this, Panhard delivered three VBLs to the French Army for use in Lebanon.

Under the French Army's first programme for the VBL, a total of 1,585 VBLs was required in three batches.

The first batch of Panhard VBL vehicles for the French Army consisted of 569 vehicles, the first of which was completed late in 1990. All vehicles have now been delivered. This order comprised 201 reconnaissance and 368 anti-tank armed with Giat Industries APILAS anti-tank systems.

Panhard expected that the next order would be for a total of 330 vehicles per year but this was cut back to 110 vehicles a year. The first batch of 110 vehicles (TC 1 - *Tranche Conditionnelle No 1*) has been completed.

Deliveries of the second batch of 110 (TC 2) commenced in June 1995 and were completed by June 1996. In May 1995, Panhard was notified of the third batch of 110 vehicles (TC 3) which were delivered between June 1996 and April 1997.

Normal production rate for the French Army is 10 vehicles a month but there is plenty of capacity at Marolles for production to be increased if additional VBL orders are received.

By December 1995, the French Army had ordered 922 VBL vehicles, this includes prototypes, preproduction vehicles and trials vehicles and includes two long wheelbase models. The standard version of the VBL has a wheelbase of 2.45 m and the long wheelbase version has a wheelbase of 2.7 m.

By late 1998, the French Army had taken delivery of over 900 VBLs with another 350 to be delivered by 2003.

Some French Army VBL vehicles have been fitted with a 12.7 mm M2 HB machine gun with the gunner being provided with additional armour protection from small arms fire and shell splinters.

The 1,000th VBL was completed in late 1995; of these, 983 are in the standard configuration and 17 in the stretched configuration.

Most sales have been made to existing Panhard customers with the VBLs being used as scouts for the larger and better-armed Panhard AML (4 × 4) and ERC (6 × 6) armoured cars.

VBLs have been used by the French Army operating in the former Yugoslavia and some vehicles have clocked between 80,000 and 100,000 km.

Wherever possible, standard and proven commercial components have been used in the design, for example the diesel engine is the same as that installed in the Peugeot 505 and 605 cars, the Peugeot P4 (4 × 4) light vehicle in service with the French Army and the Talbot Tagora car. The German ZF automatic transmission is also used in many European cars.

In 1996, Oman ordered a total of 51 vehicles for delivery from 1997; of these 43 were in the standard wheelbase version and eight in the long wheelbase version and armed with the Raytheon Systems Company TOW 3,750 m ATGW system.

The same year, Indonesia placed an order for an initial batch of 18 VBL vehicles which is expected to be followed by a further 46 vehicles. Kuwait placed an order for 20 VBLs and the first of these were delivered in 1997.

Description

The hull of the Panhard VBL is of all-welded THD steel that varies in thickness from 5 to 11 mm with the engine at the front and the crew compartment at the rear. The driver sits at the front of the crew compartment on the left with the vehicle commander to the right. Both have a door in the side that opens to the front, with a bulletproof window in the upper part. To the front of each crew member is a bulletproof window with an electric wiper and the driver has an emergency periscope. As an option, the normal single-piece bulletproof windscreen for the commander and driver can be replaced by an oblong windscreen that opens upwards for improved ventilation. When in a combat area, this would be covered by an armoured shutter. Over the top of the driver's position is a single-piece hatch cover that opens to the front. The commander has a single-piece circular hatch cover that opens to the rear. To the upper left of the driver's rear and upper right of the commander's rear is a small bulletproof window that gives observation to the rear, although this is not a feature of French Army production vehicles.

The upper sides of the rear crew compartment slope inward. There is a large door in the rear that opens to the left and has a bulletproof window. In the roof at the rear is a single-piece hatch cover that opens to the left side. All of the bulletproof windows have a thickness of 33.5 to 49 mm.

The Michelin combat tyres enable the vehicle to cover 50 km at a speed of 30 km/h after a puncture or being hit by bullets. A tyre-pressure regulation system is fitted as standard. This allows the driver to adjust the tyre pressure to suit the terrain being crossed.

The VBL is fully amphibious after two minutes of preparation and is propelled in the water by a single propeller mounted under the hull rear. Before entering the water a trim vane is erected manually at the front of the vehicle.

For export vehicles the amphibious kit is optional, other options include an NBC system (standard on French Army vehicles), an air conditioning system, a passive night driving periscope, a heater and power-assisted steering.

Variants

The French Army uses the VBL in two basic roles, combat/anti-tank and intelligence.

The combat version has a crew of three: commander, driver and MILAN operator, and is armed with a 7.62 mm machine gun (with 3,000 rounds of ammunition) and a Euromissile MILAN ATGW launcher (with six missiles). The MILAN launcher can be fitted on the front or rear hatches of the VBL or deployed away from the vehicle. This version has a radio, three 5.56 mm FAMAS rifles and nine hand grenades.

The intelligence version has a crew of two, commander/radio operator and a driver. Armament consists of one 7.62 mm machine gun (with 3,000 rounds) or one 12.7 mm machine gun (with 1,200 rounds). The weapon is fitted to a ring mount on the roof of the vehicle to the right of the driver and has an elevation from –10 to +50° and a traverse of 180° left and right. It also has two radios, which can be used by

Specifications

Crew: 2 or 3
Configuration: 4 × 4
Combat weight:
(without NBC system) 3,550 kg
(with NBC system) 3,590 kg
Unloaded weight:
(without NBC system) 2,850 kg
(with NBC system) 2,890 kg
Power-to-weight ratio: 29.57 hp/t
Length:
(with extra fuel can) 3.87 m
(without extra fuel can) 3.7 m
Width: 2.02 m
Height:
(to hull top) 1.7 m
(with 7.62 mm MG) 2.14 m
Ground clearance: 0.37 m
Track: 1.69 m
Wheelbase: 2.45 m
Max speed:
(road) 95 km/h
(water) 4.5 km/h
Max road range: 600 km
(with onboard fuel cans) 800 km
Fording: 0.9 m
Gradient: 50%
Side slope: 30%
Trench: 0.50 m
Turning radius: 5.77 m
Engine: Peugeot XD 3T, 4-cylinder turbocharged diesel developing 105 hp at 4,150 rpm
Transmission: ZF automatic, 3 forward and 1 reverse gears
Transfer case: 2-speed
Suspension:
(front) double-triangle, helical springs, double-effect shock-absorbers
(rear) torsion bars, double-effect shock-absorbers
Tyres: 9.00 × 16
Brakes (main): disc on all wheels
Electrical system: 24 V
Batteries: 2 × 12 V, 125 Ah
Armament: see text
Armour: 5 mm to 11.5 mm
NBC system: yes
Night vision equipment: yes

Panhard VBL of the French Army with roof-mounted 7.62 mm machine gun (Richard Stickland) **2002**/0114326

Long wheelbase Panhard VBL showing three roof hatches 0045866

Panhard VBL fitted with PL 40 cupola armed with 40 mm grenade launcher (Christian Dumont) **2002**/0114327

Panhard VBL (4 × 4) scout car fitted with E6 II A1 turret armed with 20 mm cannon **2002**/0114323

either crew member, two 5.56 mm FAMAS rifles and six hand grenades. An LRAC anti-tank launcher and 12 rockets can also be carried.

French Army TIS for VBL

In mid-2000, the French Delegation Generale pour l'Armement (DGA) selected Giat Industries and Matra Systems et Information to provide the AMX-10RC (6 × 6) armoured car and the Panhard VBL (4 × 4) scout car with a Terminal Information System (TIS) type Battle Management System (BMS).

TIS is an evolution of the Giat Industries FINDERS (Fast Information Navigation Decision Reporting System) which is fitted as standard on the 388 Leclerc Tropicalised Main Battle Tanks (MBT) being delivered by Giat Industries to the United Arab Emirates.

The installation of the TIS will allow the reconnaissance vehicles to exchange digitised data such as the tactical situation and have this information displayed on a digital map that can be seen by the vehicle commander.

Under the first phase one, squadron of vehicles consisting of 12 AMX-10RC armoured cars and 12 Panhard VBL scout cars will be equipped with the TIS for extensive trials after which more than 600 vehicles will be retrofitted with the system.

Export variants

In addition to the two basic models required by the French Army, Panhard has proposed at least 22 other versions including a command vehicle, an armoured personnel carrier, an anti-aircraft vehicle with two Matra BAe Dynamics Mistral surface-to-air missiles, an anti-tank vehicle with Euromissile HOT or Raytheon Systems Company TOW ATGWs, a radio vehicle, a radar vehicle with RASIT battlefield surveillance system, an electronic warfare vehicle, an ammunition carrier and an internal security vehicle. Other armament options are also available including 20 mm cannon. However, some of these are only suggested for the LWB version of the VBL.

Panhard offers two basic chassis, the standard model as used by the French Army, this being the model described, and the ULTRAV M11 which has a stretched chassis. These can be fitted with a wide range of equipment and weapons depending on their operational role, and some of these are listed following.

LWB (4 × 4) VBL

The standard production Panhard VBL has wheelbase of 2.45 m and a crew of two or three depending on the mission and equipment fit.

As a private venture Panhard has built a stretched version with a wheelbase of 2.7 m which can have a crew of four depending on equipment fit.

This stretched version has a total of four roof hatches, each of which swings back through 180° towards the rear with the driver's at front left, commander's front right and two rectangular hatches at the very rear.

Panhard has suggested the following roles for the stretched VBL:

(1) Patrol vehicle with four crew and 7.62 mm roof-mounted machine gun
(2) Squadron command vehicle with 7.62 mm roof-mounted machine gun and additional communications equipment
(3) Scout vehicle with various armament installations

Panhard VBL fitted with Panhard PL 127 cupola armed with Panhard 12.7 mm M2 machine gun (Christian Dumont) **2001**/0093530

Panhard VBL in service with Oman and showing amphibious capabilities (Christian Dumont) **2001**/0064880

including turret armed with 12.7 mm machine gun
(4) Surveillance vehicle with RASIT battlefield surveillance radar system mounted on roof and operator at rear
(5) Air defence surveillance vehicle with roof-mounted radar system and operator at rear
(6) Air defence vehicle with Matra BAe Dynamics ALBI system and two Mistral surface-to-air missiles in the ready to launch position. The first customer for this vehicle was Oman, which placed an order for an undisclosed quantity of systems late in 2000.
(7) Fitted with Raytheon Systems Company TOW anti-tank guided missile system with a maximum range of 3,750 m. This version has already been adopted by Oman which took delivery of eight vehicles in 1997.

As mentioned under Development, this version of the VBL is now expected to be adopted by the French Army for a number of specialised roles.

IS role

The Panhard VBL is also being marketed for use in the Internal Security (IS) role, carrying five men.

VBL with surveillance radar

For the export market a Panhard VBL has been fitted with an air defence surveillance radar system which would be used to designate targets for the VCR (6 × 6) vehicle fitted with the SAAB Bofors Dynamics RBS 70 surface-to-air missile system and the ERC Kriss twin 20 mm (6 × 6) self-propelled anti-aircraft gun system. This version is used by Gabon.

The version has a crew of three and is fitted with a radar system with a range of 16 to 20 km for aircraft and 8 to 10 km for helicopters. The antenna is mounted on a telescopic mast and is raised above the vehicle when operating.

Inside the rear of the VBL are the radar transceiver display console, land navigation system and a radio communications system. This system is used by Gabon.

Panhard VBL with HOT/ATM

This is a Panhard VBL (4 × 4) fitted with a Euromissile HOT/ATM turret with a 4,000 m range HOT anti-tank missile either side of the turret. In the middle of the turret, between the missiles, is an electro-optical package which includes a thermal camera, TV camera and laser range-finder. This can be elevated and used for battlefield surveillance or for aiming the HOT missiles.

While Euromissile is responsible for the turret and missile system, Thales Optronique is responsible for the elevating mast. Additional details are given in the entry for the Wiesel vehicle later in this section.

VBL with 20 mm cannon

In response to the operational requirements of a southeastern European country, Panhard has now completed firepower and mobility trials of a VBL fitted with the German Rheinmetall Landsysteme (previously KUKA) one-person turret E6 II A1.

Key features of this turret have been summarised as: no gases can enter the crew compartment; all-welded steel armour structure provides protection from 7.62 mm armour-piercing ammunition; and gunner is under full armour protection.

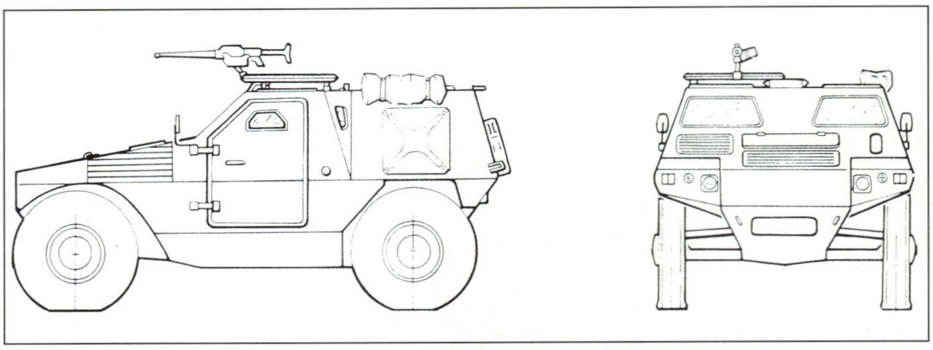

Outline drawing of production Panhard VBL (4 × 4) light vehicle with a roof-mounted 7.62 mm machine gun

Panhard VBL fitted with PL 40 cupola armed with 40 mm grenade launcher (Christian Dumont) **2001**/0064883

Panhard VBL of theFrench Army without armament installed (BP) **2001**/0093529

The turret is mounted on the roof of the VBL and is armed with a 20 mm MK 20 Rh 202 automatic cannon with a cyclic rate of fire of 1,000 rds/min. It has electrical and mechanical modes of operation and the gunner can select single, burst or continuous firing modes.

The 20 mm dual-feed cannon fires standard types of 20 mm × 139 ammunition, with two magazines mounted either side holding ready use ammunition.

One of the magazines would typically hold 100 rounds of high explosive (HE) ammunition, with the other holding 60 rounds of armour-piercing (AP) ammunition.

The turret has manual traverse through 360° with a manual elevation from –10 to +45°. The gunner is provided with a PERI Z 16 day sight with magnification of ×2 and ×6 and five observation periscopes. Total weight of the turret without the gunner is only 404 kg, with 24 V power being run off the vehicle's power supply.

The design of the E6 II A1 turret is modular and optional equipment currently includes an image intensification night sight, various types of electrically operated smoke grenade launchers, searchlight, internal spall liner, additional external armour, electric drive system, intercom and laser protection for the observation and sighting devices. An electric turret drive system is also possible.

In addition to being able to engage ground targets out to a maximum range of 2,000 m, the high elevation of 45° also enables the 20 mm weapon to engage slow-flying aircraft or targets such as high buildings.

The E6 series turret has been designed for installation on a variety of light tracked and wheeled armoured vehicles, with the first production application being the now Rheinmetall Landsysteme Wiesel 1 Armoured Weapon Carrier (AWC). This was developed to meet the requirements of the German airborne forces who took delivery of 133 units fitted with this turret.

Status

In production. In service with:

Country	Quantity	Comment
Cameroon	5	delivered
Djibouti	7	delivered
France	1,250	1,000+ delivered
Gabon	12	delivered
Greece	73	delivery from 1997
Indonesia	18	delivery from 1997
Kuwait	20	delivery from 1997
Mexico	40	delivered, 8 have MILAN ATGW
Niger	7	delivered
Nigeria	72	delivered
Oman	51	delivery from 1997. Additional order placed late 2000 for other variants including armed with twin ALBI launcher
Portugal	25	Delivered
Qatar	16	Delivered
Rwanda	16	Delivered, status Uncertain
Togo	2	Delivered

Manufacturer

Société de Constructions Mécaniques Panhard et Levassor.

UPDATED

SOFRAME RPX 6000 scout car

Development

The RPX 6000 (4 × 4) *Véhicule Blindé Léger à Roues* (VBLR) has been developed as a private venture by SOFRAME (previously the LOHR company) for the export market and was shown for the first time at the 1981 Satory Exhibition of Military Equipment. By mid-2001 five prototypes of the RPX 6000 had been completed, but volume production had yet to commence.

Description

The RPX 6000 scout car vehicle has an all-welded steel hull with the three-man crew compartment at the front and the engine compartment at the rear. The driver is seated at the front of the hull and has a bulletproof windscreen to his front and sides which, in action, are covered by hinged flaps raised by the driver from inside the vehicle. There is a single-piece hatch cover over the driver's position.

In either side of the hull is a two-part entry hatch, the upper part folding to the rear and the lower part folding downwards to form a step. There is a circular roof hatch in the centre of the hull roof on which a variety of armament installations can be mounted including a SAMM BTM 208 one-man turret (former designation S 365-1) armed with a 12.7 mm and a 7.62 mm machine gun, a Thales AFV Systems turret with a 12.7 mm machine gun and external grenade launchers, turret-mounted 20 mm cannon, 60 mm breech-loaded mortar and other light turrets not exceeding a diameter of 1.28 m. It can also be fitted with the Giat Industries one-person turret armed with a 20 mm cannon (500 rounds) and a 7.62 mm machine gun (2,000 rounds). Optional equipment includes an NBC system, night vision equipment and an amphibious capability.

Variants

Variants of the RPX 6000 suggested by the manufacturer include a command vehicle, a personnel carrier, a radar vehicle, a command/radio vehicle and a recovery vehicle. The RPX 6000 D is powered by a BMW turbocharged diesel developing 125 hp at 4,800 rpm giving a maximum road speed of 90 km/h and range of 800 km.

More recently, one example of the RPX 6000 scout car has been modified. This now has a single oval-shaped roof hatch that opens to the rear above the driver's position.

In either side of the hull, to the rear of the driver's position, are three rectangular bulletproof windows. In the hull side, between the front and rear roadwheels, is a firing port with an associated vision block above.

Latest version of SOFRAME RPX 6000 scout car with single-piece roof hatch and windows in hull sides
2001/0093533

Specifications

Crew: 3 (or 2 to 4)
Configuration: 4 × 4
Weight:
(loaded) 6,800 kg
(empty) 5,200 kg
Power-to-weight ratio: 26.47 hp/t
Length: 4.585 m
Width: 2.164 m
Height:
(without armament) 1.65 m
Ground clearance: 0.4 m
Track: 1.828 m
Wheelbase: 2.7 m
Max road speed: 110 km/h
Range: 600 km or 10 h
Fording: 0.95 m
Gradient: 60%
Side slope: 30%
Vertical obstacle: 0.4 m
Engine: BMW 6-cylinder petrol developing 180 hp (the RPX 6000 D has a BMW diesel engine, see text)
Transmission: automatic
Transfer case: two-speed manual with differential locks
Front suspension: transversal parallelogram, coil spring, hydraulic shock-absorber and stabilisation bar
Rear suspension: trailing arm type, coil spring, hydraulic shock-absorber and stabilisation device
Steering: hydraulically assisted
Tyres: 12.5 × 20 XS (run-flat)
Brakes: hydraulic drum on all wheels
NBC system: optional
Night vision equipment: optional

SOFRAME RPX 90 (4 × 4 armoured car)

A single prototype of this was first shown in 1983 but it never entered production and is no longer being marketed. Details were given in *Jane's Armour and Artillery 1990-91*.

SOFRAME VPX 5000 light vehicle

Details of this light tracked vehicle, first shown in 1981, were given in *Jane's Armour and Artillery 1995-96*. This vehicle never entered production.

SOFRAME RPX 3000 light armoured vehicle

Details of this 4 × 4 light armoured vehicle, the prototype of which was completed in 1985, were given in *Jane's Armour and Artillery 1997-98*. This vehicle never entered production or service.

SOFRAME 120 mm mortar

The SOFRAME RPX 40 M 120 mm mortar system on a 4 × 4 chassis is no longer marketed.

Status

Development complete. Ready for production.

Manufacturer

SOFRAME.

UPDATED

GERMANY

Rheinmetall Landsysteme R-400 armoured combat vehicle

Development

Some years ago, the then Thyssen Henschel developed a complete family of wheeled armoured fighting vehicles as a private venture called the TH-200 (4 × 4), TH-600 (6 × 6) and the TH-800 (8 × 8). Details of these were given in *Jane's Armour and Artillery 1997-98*.

Development work on the TH-200 and TH-800 stopped some years ago and Thyssen Henschel subsequently became Henschel Wehrtechnik and development work on the TH-400 continued.

With the acquisition of Thyssen Henschel by Rheinmetall DeTec, the vehicle became the H-400 and, in early 2000, the former company became known as Rheinmetall Landsysteme GmbH.

More recently the designation of the vehicle has been changed yet again to the R-400.

Description

The hull of the R-400 is of all-welded steel armour construction with the driver's compartment at the front, fighting compartment in the centre and the power pack at the rear.

The steel armour provides protection from small arms fire, shell splinters and medium machine gun fire. For improved protection passive and explosive reactive armour is available as an option.

The driver is seated in the centre and has a single-piece hatch cover that opens to the right. In front of this are three day periscopes, the centre one of which can be replaced by a passive periscope for driving at night.

The power-operated turret is in the centre with the commander and gunner on the right and the loader on the left. The commander has a raised cupola with a single-piece hatch cover and observation day periscopes. In front of the cupola is the commander's stabilised panoramic sight.

The gunner is seated in front of and below the commander and has a roof-mounted sight and a sighting system coaxial with the 105 mm main armament. The loader has a single-piece hatch cover that opens to the rear and a roof-mounted periscope.

Main armament comprises a 105 mm rifled gun which is fitted with a fume extractor and thermal sleeve with a 7.62 mm machine gun being mounted coaxially to the left. Mounted either side of the turret is a bank of four 76 mm electrically operated smoke grenade launchers that fire forwards.

The rear-mounted power pack can be quickly removed as a complete unit for field replacement.

The R-400 incorporates a new type of suspension, consisting of a double trailing arm independent wheel suspension, the kinetic properties of which have considerably improved mobility in rough terrain.

The negative lead angle of the springing reduces the vertical and horizontal acceleration when driving rapidly across obstacles so that the physical load on the crew is lower when compared to vehicles equipped with rigid axles.

Standard equipment includes power steering and a central tyre pressure regulation system that allows the driver to adjust the tyre pressure to suit the type of terrain being crossed, even when the vehicle is moving.

Variants

The R-400 is now being promoted as a modular concept capable of taking advantage of the wide range of turrets, weapon systems, fire-control systems and other subsystems available within the Rheinmetall DeTec Group of which Rheinmetall Landsysteme is now a key part.

Rheinmetall Landsysteme GmbH R-400 (6 × 6) vehicle fitted with three-man turret armed with 105 mm rifled gun
2001/0088023

Specifications

Crew: 4
Configuration: 6 × 6
Combat weight: 24,500 kg
Chassis weight: 17,300 kg
Power-to-weight ratio: 17.75 hp/t
Length: 6.36 m
Width: 2.98 m
Height:
(without turret) 2.083 m
(with turret) 2.735 m
Ground clearance: 0.45 m
Max road speed: 105 km/h
Range on road: 1,000 km
Gradient: 60%
Vertical obstacle: 0.60 m
Trench: 1.10 m
Turning radius: 8.8 m
Fording: 1.2 m
Armament:
(main) 1 × 105 mm smooth bore gun
(secondary) 2 ×7.62 mm MG
Smoke laying equipment: 4 × 76 mm
Ammunition: (main) 27 × 105 mm
Engine
(diesel) 435 hp
NBC system: optional
Night vision equipment: optional

Turrets now being offered include:

(1) Rheinmetal Landsysteme turret armed with 105 mm smoothbore gun. Prototypes of the R-400 had a 105 mm rifled tank gun
(2) Rheinmetall Landsysteme E4 two-person turret with internally mounted, fully stabilised 20 to 30 mm automatic cannon
(3) Rheinmetall Landsysteme E8 one-person turret with internally mounted cannon up to 30 mm in calibre. There are various stabilisation and fire-control options.

Weapons now being offered include:

(1) 105 mm smoothbore gun from Rheinmetall Weapons & Munitions with 105 mm ammunition being provided by Rheinmetall Weapons & Munitions and Eurometaal. This weapon has already been test fired and draws on Rheinmetall Landsysteme's experience with the design and development of 120 mm smoothbore tank guns
(2) Mauser 30 mm MK automatic cannon
(3) Mauser/Oerlikon Contraves 35/50 mm weapons systems

Fire-control systems being offered include:

(1) STN ATLAS Elektronik AOZ
(2) STN ATLAS Elektronik SEOSS
(3) STN ATLAS Elektronik FAUST
(4) STN ATLAS Elektronik MOLF

Command and control systems include:

(1) STN ATLAS ACE (Army Command and Control).

Self-protection systems include:

(1) Modular self-protection system MASKE from Buck Neue Technologien.

Status

Development of the R-400 (6 × 6) vehicle is complete and production can start on receipt of orders.

Manufacturer

Rheinmetall Landsystem GmbH (Kassel).

UPDATED

Rheinmetall Landsysteme Spähpanzer Luchs armoured amphibious reconnaissance vehicle

Development

Following extensive trials with 8 × 8 armoured amphibious reconnaissance vehicles built by the Joint Project Office and Daimler-Benz, the latter was selected to meet the requirements of the German Army.

In December 1973, Rheinstahl Wehrtechnik, now Rheinmetall Landsysteme, was awarded a contract worth DM300 million for 408 vehicles. First production vehicles were completed in May 1975 and the first vehicle was officially handed over to the German Army in September 1975; production continued until early 1978.

In the future, the Spähpanzer Luchs is expected to be supplemented in the German Army by Fennek (4 × 4) currently under development to meet the requirements of the German and Royal Netherlands armies. Details of the Fennek are given later in this section under Netherlands. As of July 2001, no production orders had been placed for the Fennek.

Specifications

Crew: 4
Configuration: 8 × 8
Combat weight: 20,000 kg
Power-to-weight ratio: 20 hp/t
Length: 7.743 m
Width: 2.98 m
Height:
(to top of machine gun rail) 2.905 m
(to top of hull top) 2.125 m
Firing height: 2.342 m
Ground clearance:
(axles) 0.44 m
(hull) 0.58 m
Track: 2.54 m
Wheelbase: 1.4 + 2.365 + 1.4 m
Angle of approach/departure: 66°/53°
Max speed:
(road, forward) 90 km/h
(road, reverse) 90 km/h
(water) 10 km/h
Acceleration:
(0-80 km/h) 65 s
Fuel capacity: 500 litres
Max road range: 730 km
Fording: amphibious
Gradient: 60%
Side slope: 30%
Vertical obstacle: 0.6 m
Trench: 1.9 m
Engine: Daimler-Benz OM 403 A 10-cylinder 90° V 4-stroke multifuel with 2 exhaust-driven turbochargers developing 390 hp at 2,500 rpm (diesel fuel) or 320 hp at 2,500 rpm (petrol)
Transmission: ZF 4 PW 95 H 1 4-speed with torque converter
Suspension: coil springs and hydraulic shock-absorbers
Steering: recirculating ball power-assisted
Turning radius:
(all wheel steering) 5.75 m
(front wheel steering) 9.7 m
Tyres: 14.00R x 20
Brakes:
(main) hydraulic, dual circuit
(parking) mechanical operating on transmission
Electrical system: 24 V
Batteries: 6 × 12 V, 100 Ah
Armament:
(main) 1 × 20 mm Rheinmetall MK 20 Rh 202 cannon
(anti-aircraft) 1 × 7.62 mm Rheinmetall MG3 MG
Smoke-laying equipment: 2 × 4 76 mm smoke grenade dischargers
Ammunition in turret:
(main) 375 (300 HE + 75 AP)
(anti-aircraft) 100-round box
Gun control equipment
Turret power control: electrohydraulic/manual
Turret traverse: 360°
Gun elevation/depression: +69°/–15°
Gun stabiliser:
(vertical) no
(horizontal) no
NBC system: yes
Night vision equipment: yes

Description

The all-welded steel hull of the Luchs provides the crew with protection from small arms fire and shell splinters. The hull and turret are immune, over their frontal area, to penetration from 20 mm projectiles.

The driver is seated at the front of the hull on the left side and is provided with a single-piece hatch cover that opens to the right. There are three day periscopes in front of the driver's hatch cover, the centre one of which can be replaced by a passive periscope for night operations.

The Rheinmetall TS-7 two-person power-operated turret is located in the centre of the hull and has spaced armour for improved protection. The commander is seated on the left and the gunner on the right, both with a single-piece hatch cover that opens to the rear. Both the commander and gunner are provided with a PERI-Z-11 A-1 periscope for aiming the 20 mm cannon, and a total of 12 day periscopes. Turret traverse and weapon elevation/depression are electrohydraulic and can be operated by the commander or gunner. The fire-control system includes an azimuth indicator.

An entry door is provided in the left side of the hull. The fourth crew member, the radio operator/rear driver, is seated to the rear of the turret on the left side facing the rear and is provided with a single-piece hatch cover that opens to the right. There are three day periscopes in front of the cover, the centre one of which can be replaced by a passive periscope for night operations.

The engine compartment is behind the turret on the right side and access is by a large oblong hatch that opens to the right. The engine compartment, cooling and exhaust systems are separated from the fighting compartment by gastight welded bulkheads. The engine compartment is fitted with an automatic fire extinguishing system. The complete power pack, consisting of the engine, transmission, air filter, oil cooler and parking brake, can be removed as a complete unit and run outside the vehicle. The cooling system is at the back of the hull and the exhaust outlet is at the very rear of the hull.

Power is transmitted from the engine to a four-speed hydraulic torque converter and then via a distributor gear with differential locks to the wheels. The Luchs has the same speed in both directions.

The suspension consists of four Daimler-Benz rigid axles with differential locks which are supported by longitudinal bars, with bogies for all axles. Each wheel station has a vertical coil spring and a hydraulic shock-absorber. All eight wheels are steered but for road use the front four wheels only are normally used.

The Luchs is fully amphibious, propelled in the water by two Schottel steerable propellers located under the rear of the hull, one on each side. These are powered by a PTO on the main transmission via a bevel-type distributor gear. Before entering the water a trim vane is hydraulically erected at the front of the hull and the bilge pumps, two in the fighting compartment and one in the engine compartment, are switched on.

In 1982, it was announced that a total of 423 sets of thermal night vision equipment would be purchased for

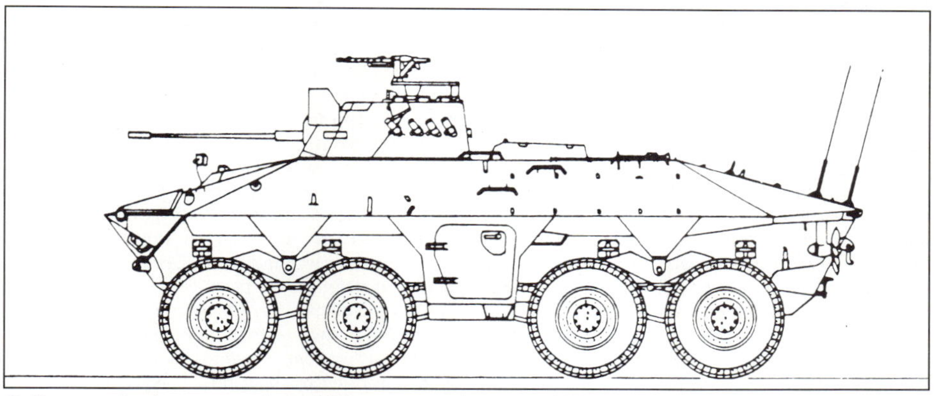

Spähpanzer Luchs armoured amphibious reconnaissance vehicle

Spähpanzer Luchs A2 armoured reconnaissance vehicle showing steering on all eight wheels (Michael Jerchel) **2002**/0114328

Spähpanzer Luchs A2 armoured reconnaissance vehicle with roof-mounted 7.62 mm MG3 machine gun (Michael Jerchel) **2002**/0114329

the Luchs, 63 sets in 1984 and 120 per year in 1985, 1986 and 1987. With the introduction of the thermal night vision equipment the infra-red searchlight was removed from the left side of the turret.

When upgraded with SEM 80/90 radios, the vehicle became the Luchs A1 and, when fitted with thermal night vision equipment, it became the Luchs A2.

For operations in Bosnia, the Luchs has been fitted with additional equipment including a GPS navigation system, gyrocompass and an additional circular antenna.

Standard equipment includes a preheater for the cooling liquid, engine oil and transmission oil and a preheater for the batteries. The NBC system can also be used for ventilating the vehicle when all the hatches are closed.

The Luchs is armed with a 20 mm Rheinmetall MK 20 Rh 202 cannon which is also installed in the Rheinmetall Landsysteme Marder 1 ICV. This cannon has dual feed and the empty cartridge cases and belt links are ejected externally to the right of the turret.

A Rheinmetall 7.62 mm MG3 machine gun mounted on a skate mount over the commander's hatch can be used for both ground and anti-aircraft defence. Four 76 mm smoke grenade dischargers are mounted either side of the turret.

Status

Production complete. In service with the German Army.

Manufacturer

Rheinmetall Landsysteme GmbH (Kassel).

UPDATED

Rheinmetall Landsysteme armoured weapon carrier Wiesel 1, air-portable

Development

The Wiesel 1 air-transportable armoured weapon carrier was developed to meet the requirements of the German Army by Porsche, with production and marketing being carried out by MaK System Gesellschaft mbH of Kiel as general contractor. In early 2000 this became Rheinmetall Landsysteme GmbH.

Between 1989 and 1992, the German Army took delivery of a total of 345 Wiesel 1 vehicles armed with a Raytheon Systems Company TOW ATGW launcher or a Rheinmetall 20 mm cannon. In addition, another seven vehicles were delivered to the US armed forces and four of these were used for robotic trials purposes.

A variety of different turrets and weapons systems could be installed on the Wiesel 1, which would enable it to undertake a wide range of roles on the battlefield. In addition to being used by airborne units, the manufacturer has also suggested that it can be used by attack forces, scouts, military police and border guards as well as rapid reaction forces.

The Wiesel 1 has also been tested and demonstrated in a number of other countries including Greece, Indonesia, Malaysia, Norway, Singapore, Thailand, the United Arab Emirates, the United Kingdom and the USA.

Several Wiesel 1 armoured weapon carriers, in both 20 mm and TOW configuration, were deployed by Germany to Somalia as part of United Nations forces in 1993.

In addition to the Wiesel 1, the now Rheinmetall Landsysteme has more recently developed the multipurpose carrier Wiesel 2, which has an extended base and nearly twice as much internal volume. Due to this, it offers a much wider range of roles and missions. Details of the Wiesel 2 are given in the Armoured personnel carriers (tracked) section.

The Wiesel production has now opened again for the German Army which has selected the Wiesel 2 as the platform for its LeFlaSys Stinger air defence system for which STN ATLAS Elektronik is prime contractor.

Although the Wiesel 1 was originally designed by Porsche, Rheinmetall Landsysteme GmbH is now the design authority and is currently studying a number of additional variants for possible use by the German Army.

Although the German Army took delivery of 345 Wiesel 1 armoured weapon carriers, in early 1997 it had a total of 343 vehicles on strength, 210 armed with the Raytheon Systems Company TOW anti-tank guided missile system and 133 armed with the Rheinmetall 20 mm cannon.

Description

The hull of the Wiesel 1 is of all-welded steel construction providing the crew with protection from 7.62 mm small arms fire and shell splinters. The driver sits at the front of the vehicle on the right and has three day periscopes for observation to the front and sides; the centre one can be replaced by a passive periscope for night driving.

The engine, which is a standard VW production model, is to the left of the driver with an access hatch in the glacis plate. The engine drives via a three-speed ZF automatic transmission and a Porsche two-speed reduction gear with integrated Cletrac steering gear. The complete power pack can be removed and replaced in 10 minutes. The steering gear is operated by the driver via a steering wheel and allows a variable turning radius from infinite to 2.3 m. Pivoting on one track is possible by engaging the parking brake on one side.

The engine exhaust system is located in the cooling air exhaust flow on the left side of the vehicle. The exhaust gases are passed to the rear and cooled in the process. The closed crew compartment is at the rear of the hull and entry is by climbing over the sides of the vehicle. The flexible self-sealing rubber fuel tank has an explosion-retarding polyurethane liner and is mounted externally on the rear of the hull.

Suspension either side consists of three dual roadwheels, drive sprocket at the front, idler at the rear and one track-return roller. A Belleville spring assembly automatically provides track tension when the vehicle is travelling. The track consists of an endless rubber band with wire reinforcements. More recently, the German company Diehl has developed a new track called the Type 622 for the Rheinmetall Landsysteme Wiesel 1 light armoured vehicle.

Following extensive trials the Type 622 track has been accepted for service with the German Army. The track is of the double-pin type and has a life of around 6,000 km.

Airlift capability

According to Rheinmetall Landsysteme GmbH the Wiesel 1 is air-transportable by the following aircraft:

Type	Quantity
CH-53 helicopter	2 internally
CH-47 helicopter	2 internally
C-130 aircraft	3 internally
C-160 aircraft	4 internally
C-141B aircraft	6 internally
Boeing 747 aircraft	24 internally
UH-60 helicopter	1 slung externally
Super Puma helicopter	1 slung externally

Variants

The front-driven AWC Wiesel 1 was designed for a variety of missions and deployments. Based on the same chassis, the integration of various turrets and weapon systems enables the production of a complete family of vehicles according to the requirements of the users.

In addition, the development of the extended base vehicle of the Wiesel offers the possibility of a second family of the Wiesel vehicle. Following the same principle as the short versions, various turrets and weapons systems can be integrated in the extended base vehicle chassis which has an additional roadwheel and is approximately 60 cm longer than the short version.

Short version variants

AWC Wiesel 1 reconnaissance vehicle

Under contract to the German Army, Rheinmetall Landsysteme has converted a batch of 10 TOW versions of the Wiesel 1 into reconnaissance vehicles.

AWC Wiesel 1 with 20 mm cannon (MK 20 A1)

This was built to meet the requirements of the German Army and is fitted with a one-man Rheinmetall Landsysteme (previously KUKA) turret E6-II-A1 armed with a 20 mm Rheinmetall MK 20 Rh 202 dual-feed cannon with 160 rounds of ready use ammunition, 60

Euromissile HOT/ATM on Rheinmetall Landsysteme Wiesel 1 light tracked armoured vehicle with sensor pod raised and armed with HOT ATGM and 20 mm cannon **2000**/0064833

AWC Wiesel 1 MK 20 A1 armed with a 20 mm cannon (Michael Jerchel) **2002**/0114330

rounds on the left side and 100 rounds on the right side, allowing the gunner to select the type of ammunition required to engage different targets. The weapon can be elevated from −10 to +45° and traversed 55° left and right. Elevation and traverse are manual and the gunner aims the weapon via a Zeiss PERI-Z-16 periscopic sight; a passive night sight is also available. This has two crew consisting of commander/gunner and driver.

AWC Wiesel 1 with TOW 2 (TOW A1)

This was built to meet the requirements of the German Army and is fitted with a Raytheon Systems Company TOW ATGW launcher on an elevating pedestal with a traverse of 45° left and right and an elevation and depression of 10°. Seven TOW missiles are carried of which two are for ready use. This model has a crew of three; driver, gunner and loader.

AWC Wiesel 1 BTM 208, reconnaissance vehicle

This is fitted with the French SAMM BTM 208 turret armed with one 7.62 mm machine gun (with 200 rounds of ready use ammunition and 300 rounds in reserve) and one 12.7 mm M2 HB machine gun (with 100 rounds of ready use ammunition and 400 rounds in reserve). This remains at the prototype stage.

AWC Wiesel 1 BTM 263, mortar system

This version has a SAMM BTM 263 turret and is armed with a 60 mm mortar and a 7.62 mm machine gun.

AWC Wiesel 1 with 30 mm RMK 30 cannon

Early in 1997, Mauser-Werke Oberndorf Defense Systems released brief details of a new 30 mm cannon, designated the RMK 30, which it has developed under contract to the German BWB.

The RMK 30 is claimed to be a revolutionary new recoilless automatic cannon which has no recoil forces during single shots or burst rates of fire.

This feature enables the RMK 30 to be installed on helicopters and light armoured vehicles. The first production application for this weapon is expected to be the German Army version of the Eurocopter Tiger helicopter.

At a later date the RMK 30 could also be fitted to the Rheinmetall Landsysteme Wiesel 1 light air-portable armoured vehicles, used by the German Army, which are at present armed with a Rheinmetall 20 mm MK 20 Rh 202 cannon.

Late in 1996, the RMK 30 automatic cannon was demonstrated to high-ranking German Army officers during a live firing exercise at the Heuberg/Baden-Wurttemberg troop testing grounds.

For this trial the weapon was installed on the Rheinmetall Landsysteme Wiesel 1 light air-portable armoured vehicle with the weapon being fired by remote control. Accurate rapid single shots and bursts of four and five rounds were fired on a target at a range of 1,500 m.

The new RMK 30 automatic cannon fires new 30 mm ammunition also developed by Mauser; this is of the telescoped type.

AWC Wiesel 1 with HOT/ATM

In 1997, Euromissile, teamed with Thales Optronique, completed final development and integration of a new weapon system called HOT/ATM which on the same platform combines advanced observation and firing capabilities. In the designation HOT/ATM the latter stands for Anti-Tank Modular. For trials purposes, HOT/ATM has been installed on the Wiesel 1 chassis.

The HOT/ATM system has been developed to meet the requirements of the French and German armies for a system capable of being installed on light tracked and wheeled armoured vehicles which can carry out advanced observation as well as engaging targets with HOT ATGM out to a maximum range of 4,000 m.

The turret is fitted with a mast-mounted stabilised sensor pod which features a TV camera, thermal camera (10 µm), EBO (1 µm electro-optical block) and a laser range-finder.

Roles of the sensor pod include observation, detection, localisation, display, automatic target tracking and missile tracking.

The monitoring and display module is fitted inside the vehicle and allows the operator to carry out battlefield surveillance, accurate target localisation on a digital terrain map, automatic target tracking, target selection prior to engagement and HOT missile launching with a very short reaction time.

The turret is unmanned with the operator being seated in the vehicle on the right with the vehicle commander to his left.

Mounted either side of the turret are one or two HOT ATGMs in the ready to launch position. Other weapon combinations are possible such as 7.62 or 12.7 mm machine gun or the new Mauser RMK 30 recoilless 30 mm cannon which has already been tested during recent unmanned firings. Other ATGM can also be fitted.

In addition, the HOT/UTM can exchange surveillance and firing data with neighbouring systems as well as upper level C³I systems.

In addition to the Wiesel 1, the HOT/ATM system can be fitted on to a wide range of tracked and wheeled

Specifications

Model	**Wiesel MK 20**	**Wiesel TOW**
Crew	2	3
Power-to-weight ratio	30.7 hp/t	30.7 hp/t
Ground pressure	0.36 kg/cm²	0.36 kg/cm²
Weight	2,800 kg	2,800 kg
Length		
(overall)	3.545 m	3.31 m
(hull)	3.31 m	3.31 m
Width	1.825 m	1.827 m
Height		
(overall)	1.825 m	1.897 m
Axis of fire	1.64 m	1.795 m
Ground clearance	0.302 m	0.302 m
Track	1.62 m	1.62 m
Track width	200 mm	200 mm
Length of track on ground	1.82 m	1.82 m
Max road speed	75 km/h	75 km/h
Min cruising speed	4 km/h	4 km/h
Gradient	60%	60%
Side slope	30%	30%
Vertical obstacle	0.4 m	0.4 m
Trench	1.2 m	1.2 m
Fuel capacity	80 litres	80 litres
Cruising range (at 55 km/h)*	300 km	300 km
Track	endless rubber track	
Suspension	torsion bars, hydraulic vibration dampers	
Steering	hydraulic steering brakes on Cletrac steering transmission	
Brakes		
(service)	fully hydraulic dual-circuit disc brake system	
(parking)	mechanical	
Engine	VW 5-cylinder diesel with exhaust turbocharger developing 86 hp at 4,500 rpm	
Transmission	ZF 3 HP 22 3-speed automatic, 2-gear splinter group, total of 6 forward and 2 reverse gears	
Electrical system	24 V	24 V
Batteries	2 × 12 V, 45 Ah	2 × 12 V, 45 Ah
Armament	1 × 20 mm Rheinmetall MK 20 Rh 202 cannon	1 × TOW ATGW launcher
Ammunition	400 × 20 mm	7 × TOW ATGWs
Weapon traverse	110° left/right	45° left/right
Weapon elevation	−10°/+45°	−10°/+10°
NBC system	no	no
Night vision equipment	yes	yes

AWC Wiesel 1 TOW A1 fitted with a Raytheon Systems Company TOW ATGW being unloaded from CH-53G helicopter ***2000***/0064835

Mauser-Werke Oberndorf Defense Systems 30 mm RMK 30 automatic cannon installed for trials purposes on a Wiesel 1 light armoured vehicle 0088025

vehicles such as the French Panhard VBL (4 × 4), Giat Industries VAB (4 × 4 or 6 × 6), Steyr-Daimler-Puch Pandur (6 × 6), Patria Vehicles Oy (formerly SISU) XA-185 series (6 × 6), Russian BMP-1/2, and the US AM General HMMMWV (4 × 4).

AWC Wiesel 2 Stinger, anti-aircraft system

The German Army has a requirement for an air-portable self-propelled air defence system called the *Leichtes Flugabwehrsystem* (LeFlaSys).

In early 1995, the German MoD selected the STN Atlas Elektronik Atlas Short-Range Air Defence (ASRAD) System to meet the LeFlaSys requirement.

This version is also based on the stretched Wiesel 2 chassis covered in detail in the Armoured personnel carriers (tracked) section. First production vehicles were completed late in 2000 and the system became operational with the German Army early in 2001.

SYRANO robotic vehicle

The chassis of the Wiesel 2 is used as the platform for the French SYRANO robotic system which is currently at the prototype stage.

Other variants

Additional proposed variants based on the short version of the Wiesel 1 include the AWC Wiesel 1 MILAN anti-tank system, the AWC Wiesel 1 RATAC-S for surveillance and target acquisition.

Rheinmetall Landsysteme has also completed the prototype of an experimental Wiesel 1 with an electric drive system and this has already completed its company tests.

Wiesel extended base vehicle

The prototype of this version was completed by Rheinmetall Landsysteme early in 1994, details are given in the Armoured personnel carriers (tracked) section. It is now known as the Wiesel 2.

Status

Production of the Wiesel 1 for the German Army was completed in 1992 but further production has commenced again as the Wiesel 2 chassis has been selected for the STN ATLAS Elektronik ATLAS Short-Range Air Defence (ASRAD) System. In 2000, the Wiesel 1 was in service with the following countries:

Country	Quantity	Comment
Germany	343	delivered 1989/1992
USA	7	trials work

Manufacturer

Rheinmetall Landsysteme GmbH.

UPDATED

HUNGARY

FUG (or OT-65) amphibious scout car

Development

The FUG (*Felderito Úszó Gépkocsi*) fulfils a similar role in the Hungarian Army to the Russian BRDM-1 amphibious scout car but has its engine at the rear rather than at the front and two instead of one water-jets for amphibious operations.

The vehicle entered service with the Hungarian Army in 1964 and with the Polish and Czech armies in 1966. The Czechs call the vehicle the OT-65. There are no known plans for the upgrading or modernisation of these vehicles.

It is considered that these vehicles will be phased out of service in the next few years and replaced by more modern vehicles.

In Slovak service its replacement could be the Aligator (4 × 4) light armoured reconnaissance vehicle covered later in this section under Slovakia. First production Aligator vehicles will be completed late in 2001, with final deliveries in 2002.

Specifications

OT-65
Crew: 2 + 4
Configuration: 4 × 4
Combat weight: 7,000 kg
Power-to-weight ratio: 15.87 hp/t
Length: 5.79 m
Width: 2.5 m
Height:
(to hull top) 1.91 m
(with turret OT-65A) 2.25 m
Ground clearance: 0.34 m
Track: 1.9 m
Wheelbase: 3.3 m
Max speed:
(road) 87 km/h
(water) 9 km/h
Fuel capacity: 200 litres
Max cruising range:
(road) 600 km
Fuel consumption: 0.28 litres/km
Fording: amphibious
Gradient: 60%
Side slope: 30%
Vertical obstacle: 0.4 m
Trench: 1.2 m
Engine: Csepel D.414.44 4-cylinder in-line water-cooled diesel developing 100 hp at 2,300 rpm
Transmission: manual with 5 forward and 1 reverse gears
Tyres: 12.00 × 18
Suspension:
(front) helical springs
(rear) semi-elliptical springs
Main armament: 1 × 7.62 mm SGMB MG
Ammunition (main): 1,250
Armour: (max) 13 mm
NBC system: no
Night vision equipment: yes

Description

The hull of the FUG amphibious scout car is of all-welded steel armour, with the driver seated at the front on the left and the vehicle commander on the right. Both are provided with a hatch cover to the front, which is hinged at the top and opens upwards, and a windscreen with a wiper. The hatch is equipped with an integral vision block for use when closed.

The crew compartment is behind the driver and the only means of entry is by the hatch in the roof that opens either side of the vehicle. The roof hatches can be locked vertical to give a measure of protection when the standard 7.62 mm SGMB machine gun is being used. A ventilator is mounted in the roof of the vehicle over the commander's position. Six circular firing ports are provided, two in each side and two in the rear. A vision slit is provided to the left of the driver and to the right of the commander.

The engine compartment is at the rear of the hull with the air inlet immediately behind the crew compartment, the air outlets either side and the exhaust pipe on the right side.

On either side of the vehicle, between the front and rear wheels, are two belly wheels, which are lowered by the driver and driven from a PTO. This gives the FUG improved cross-country ability and allows it to cross trenches. When the belly wheels are lowered the vehicle is driven in first gear.

The FUG is fully amphibious, being propelled in the water by two water-jets at the rear of the hull. Before entering the water a trim board is erected at the front of the vehicle which is stowed under the nose when not in use, and the bilge pumps are switched on. When not required, the water-jets are covered by a hinged shutter. Standard equipment includes infra-red driving lights. Some models are also fitted with an infra-red searchlight and a central tyre-pressure system that allows the driver to adjust the tyre pressure to suit the type of ground being crossed. It has no NBC system.

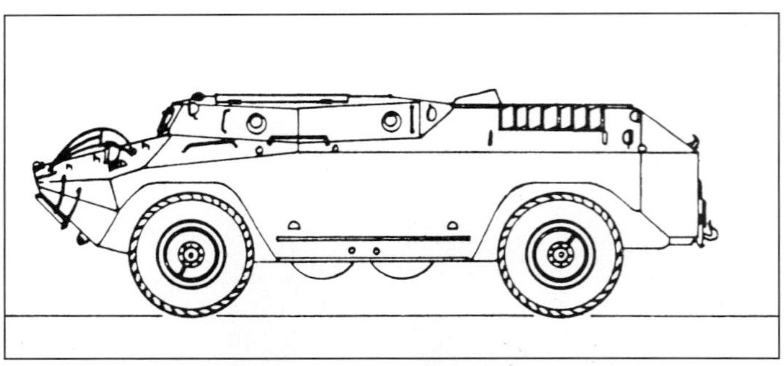

FUG (or OT-65) amphibious scout car without fitted armament

FUG (4 × 4) amphibious scout car with roof hatches in the open position

Main armament comprises a pintle-mounted 7.62 mm SGMB machine gun with limited traverse, which is aimed and fired by one of the crew members with the two hatches open.

Variants

Hungarian versions

D-442VS or FUG-US is an NBC reconnaissance vehicle.
D-422MRP is a radio communications vehicle probably used by forward air control teams

Czech versions

OT-65DP and OT-65DPP are specialised reconnaissance versions.
OT-65Ch and OT-65ChV are NBC reconnaissance vehicles.
OT-65R-2 and OT-65R-112 are radio and command vehicles.
OT-65RL has battlefield surveillance radar.
OT-65M or T-65A is covered below.
OT-65 M1992 is an ambulance version.

NBC reconnaissance version

This performs a similar role to the Russian BRDM-1 RKh vehicle and is used to mark lanes through contaminated areas. Mounted on either side of the hull rear is a rack that carries lane marking poles. When required, the racks are swung vertically through 90° so that they are over the rear of the hull and lane marking poles can then be dispensed into the ground from inside the vehicle.

OT-65A

This is a Czech modification and is the basic vehicle with the turret of the OT-62B tracked armoured personnel carrier mounted on the roof. The turret is armed with a 7.62 mm M59T machine gun with an elevation of +20° and a depression of –10°. The turret can be traversed manually through 360°. Mounted externally on the right side of the turret is an 82 mm T-21 Tarasnice recoilless gun which can be aimed and fired from inside the turret but can be reloaded only from the outside. The T-21 fires a HEAT projectile weighing 2.13 kg with a muzzle velocity of 250 m/s which will penetrate 230 mm of conventional steel armour. Maximum range is 2,500 m, effective range 450 m and about 300 m against moving targets. The crew of the OT-65A enters and leaves the vehicle by a small roof hatch that opens either side of the roof immediately behind the turret. This has also been referred to as the FUGa in Eastern literature.

PSZH-IV APC

There is a separate entry in the Armoured personnel carriers (wheeled) section for the Hungarian PSZH-IV (4 × 4) armoured personnel carrier. This was originally called the FUG-66 and later the FUG-70 until it was established that it was an armoured personnel carrier and not an amphibious scout car. This has also been referred to as the FUG-2 in Eastern literature.

Status

Production complete. In service with the following countries:

Country	Quantity
Czech Republic	182
Hungary	132
Slovakia	90

Manufacturer

Hungarian state factories.

UPDATED

INTERNATIONAL

Tactical Reconnaissance Armoured Combat Equipment Requirement/Future Scout Cavalry System

Development/Description

The British Army had a requirement for a new vehicle called TRACER (Tactical Reconnaissance Armoured Combat Equipment Requirement) (*Jane's Armour and Artillery 1996-97*) while the US Army had a requirement for a vehicle called the Future Scout Cavalry System (ibid page 245).

These were to have been separate development programmes but, late in 1996, the UK and the USA signed a letter of intent jointly to develop a new reconnaissance vehicle for the 21st century to save both development and procurement costs.

In the British Army the new vehicle would replace the current Alvis Scimitar and Sabre reconnaissance vehicles and the Striker anti-tank vehicle which were to have been replaced by TRACER.

The UK would have fielded three versions of TRACER, formation reconnaissance scout, formation reconnaissance guided weapon and close reconnaissance scout.

In the US Army it would have replaced the M3 Bradley and the HMMWV (4 × 4) currently used which were to have been replaced by the Future Scout Cavalry System.

The MoU between the UK and the US was finally signed on 7 July 1998 with the Invitation To Tender (ITT) being issued on 9 July with the bids from the two consortia being submitted early in October 1998.

In January 1999, the UK MoD, also acting on behalf of the US DoD, awarded the two international consortia, Lancer and SIKA International, each a £90 million 42 month contract for the Phase I Project Definition (PD)/ Advanced Technology Demonstrator (ATD) phase of the Armoured Scout and Reconnaissance Vehicle (ASRV).

The Lancer team is lead by the now BAE Systems, RO Defence and includes United Defense and Raytheon Systems Company of the US and Alvis Vehicles of the UK.

SIKA International is a joint British Aerospace/ Lockheed Martin company and also includes Computing Devices, Thales Optronics, Thales Air Defence UK, Smiths Industries and Vickers Defence Systems of the UK and General Dynamics Land Systems/Armament Systems and Northrop Grumman of the US. In January 2000 British Aerospace became BAE Systems.

SIKA International TRACER/FSCS ATD features a crew of three and is armed with an externally mounted 40 mm CTA International cannon **2002**/0096492

Lockheed Martin includes its Electronics & Missiles, Federal Systems, Information Systems, Sanders and Missions Systems divisions.

BAE Systems includes CTA International, Matra BAe Dynamics, British Aerospace Systems and Equipment and Military Aircraft.

The ASRV, which will be tracked, will operate deep inside enemy lines and make extensive use of stealth technology in order to survive.

It will have an extensive suite of day/night sensors in order to carry out its mission and will be armed with a 40 mm CTA International cannon mounted externally. The UK was also seeking an over watch version armed with anti-tank guided missiles to replace the current Striker vehicle.

Priority user requirements include information gathering using a multi-spectrum sensor system and aided target recognition, survivability and C4I. Survivability includes a low profile, signature management stealth, physical protection and a defensive aid system. C4I includes data management/ compression, crew stations with advanced interfaces, suitable bandwidth and fully linked to digitised battlespace.

Secondary priority user requirements cover mobility, lethality and growth. Mobility includes a high power density engine, semi-active suspension and possibly a hybrid electric drive.

Lethality includes a 40 mm weapon using case telescoped ammunition and, for the UK only, an anti-tank guided weapon. Finally, growth includes an open system architecture and P3I options.

According to UK sources, there have been recent significant advances in a number of key areas including stealth, armour and sensor technologies. Advances in sensor technologies include second-generation thermal imagers and advanced battlefield surveillance radars.

It is expected that the TRACER/FSCS will have a crew of three (commander, driver and scout) and will weigh around 18 tonnes and, therefore, be airportable in a Lockheed Martin C-130 Hercules transport aircraft.

In the first nine months of the 42-month period, each consortium will further develop its concept options and reduce them to three. These will be cost constrained, best value and fully compliant.

Between 10 and 22 months each consortium's preferred production system will be further developed and tested while in months 20 to 33 production system definition would be completed and its specification adapted to take account of an updated Technical Specification Requirement (TSR).

During months 33 to 42, the consortia would have received the Invitation to Tender (ITT) for the Phase II Full Development (FD)/Engineering and Manufacturing Development (EMD) phase.

By this time each consortium's demonstrator vehicle will also be running. The Lancer vehicle is expected to feature an electric drive system. It is understood that both of the consortiums have selected the CTA International 40 mm Cased Telescoped Weapon System (CTWS) as the main armament of their technology demonstrator.

The goal of the demonstrator is to meet the requirements of the TSR but, where the requirements cannot be met, the contractor is required to develop a plan on how these requirements may be met in the future.

Late in 2001, the UK MoD stated it would not be funding TRACER after mid-2002 and was studying other ways to meet the requirement.

In mid-2001 it was generally accepted that TRACER/FSCS would not move into the next phase and that the technology would probably be used in other programmes.

In parallel to the TRACER/FSCS, the UK is conducting studies to examine the potential use of unmanned air vehicles. The outcome of these studies will be used to inform the UK decision on the optimum mix of platforms, technology and sensors and the role that TRACER would play in the UK Army's future battlefield information gathering system.

Phase III would have covered Low-Rate Initial Production (LRIP) while Phase IV would have covered full-rate production. It was expected that first production vehicles would have been completed in 2007 and an in service date of 2008.

The US Army had a requirement for about 1,239 vehicles while the British Army had a requirement for about 335 vehicles. The production contract was expected to be worth up to $6.6. billion.

The provisional estimated programme value of the Tracer/FSCS was:

	UK	US
PD/ATD	£90 million	£90 million
FD/EMD	£400 million	$640 million
Production	£1 billion	$5 billion
Support	£1.2 billion	$6.1 billion

In May 1999, Vickers Defence Systems and General Dynamics Land Systems formed a new joint venture company called Vehicle Armour and Armament Ltd, based in Newcastle-upon-Tyne, Northern England.

The company will be the focus of Vickers Defence Systems and General Dynamics Land Systems input into the SIKA International Team.

Vehicle Armour and Armament will be responsible for the design and construction of the demonstrator vehicle as well as providing significant design and management support activities.

As well as designing the full tracked chassis, Vehicle Armour and Armament Ltd will also be responsible for its track and suspension, power pack, auxiliary automotive systems, driver's controls and displays, core ventronics, NBC/climate control, armour, driving vision, situational awareness/direct vision, fire control, weapon station drives, main armament and secondary weapon.

General Dynamics Land Systems is the prime contractor for the M1 series of MBT and the Advanced Amphibious Assault Vehicle currently under development for the US Marine Corps. Vickers Defence Systems are prime contractor for the Challenger 2 MBT now in service with the British Army.

Status

Project Definition/Advanced Technology Demonstrator. No UK funding after mid-2001.

Manufacturers

Lancer Team, SIKA International.

UPDATED

ISRAEL

RAMTA RAM family of light AFVs

Development

The RAM family of light armoured fighting vehicles was developed by the RAMTA Division of Israel Aircraft Industries Ltd, as a successor to the company's earlier RBY Mk 1 light armoured reconnaissance vehicle (fully described in the following entry) and was announced at the Paris Air Show in June 1979.

The RAM family has now been expanded into two basic groups, RAM V-1 (Open), which has an open configuration and RAM V-2 (Closed) which has a closed configuration.

Description

The hull of the RAM vehicle is of all-welded thick steel armour with the bottom of the vehicle having been designed to provide maximum possible protection against mines.

Its very low centre of gravity prevents the RAM from being overturned by mine blasts. The hull provides protection against shell splinters, shrapnel and small arms fire up to 7.62 mm armour piercing.

Battlefield survivability has further been enhanced by the separation of the crew compartment from the

RAMTA RAM V-1 (Open) combat vehicle 0045860

RAMTA RAM V-1 fitted with Israel Aircraft Industries TCM-20 LAAG system 0045859

RAMTA RAM V-2 (closed) armed with 12.7 mm and 7.62 machine guns 0045858

RAMTA RAM V-2 (closed) armed with 12.7 mm and 7.62 machine guns and side door open 0045857

wheels. All critical onboard systems such as fuel, brakes and steering have redundant back-up.

The RAM V-1 (Open) has an open roof troop compartment making it easily accessible and easy to leave. It provides 360° view from all positions and is claimed to give excellent observation. There is also an optional canvas roof that can be fitted if required by the customer.

The front of the vehicle is normally used for equipment including a spare wheel and tyre, with the open topped crew compartment in the center and the engine compartment at the rear. The commander and driver are seated at the front of the compartment with the remaining crew members sitting facing outwards in pairs. The engine is located at the rear of the vehicle.

The RAM V-1 carries a 106 mm M40 series recoilless rifle which can be traversed through a full 360°. Eighteen rounds of 106 mm ammunition are carried in the vehicle as well as night vision equipment and a multichannel transceiver.

It can also carry a Raytheon Systems Company TOW (maximum range 3,750 m) or Euromissile MILAN (maximum range 2,000 m) ATGW on a telescopic pedestal. In the case of the former, 16 TOW ATGW can be carried plus a 7.62 mm machine gun together with 2,500 rounds of ammunition.

Variants

RAM V-2 (Closed)

This has a fully enclosed troop compartment that carries a crew of nine and their personal equipment. It has one door each side for entry and exit and it also has folding armour plate hatches which can be fully or partially opened, this providing all-round observation while affording protection against armour-piercing small arms fire and shell fragments. The personnel can easily fire in all directions.

Various types of armament can be fitted according to mission requirements including one 12.7 mm machine gun or 40 mm grenade launcher, 1,500 rounds of 12.7 mm ammunition or six boxes of ammunition for the 40 mm grenade launcher, three 7.62 mm machine guns plus 5,500 rounds of ammunition, one rocket launcher plus six rounds, 24 fragmentation grenades, nine smoke grenades and two smoke flares. In addition, the vehicle carries night vision equipment and a multichannel receiver.

Specifications

Model	**RAM V-1 (Open)**	**RAM V-2 (Closed)**
Max crew	2 + 7	2 + 7
Configuration	4 × 4	4 × 4
Combat weight	5,750 kg	6,000 kg
Unloaded weight	4,300 kg	4,550 kg
Payload	1,700 kg	1,400 kg
Power-to-weight ratio:	22.95 hp/t	22.95 hp/t
Length	5.52 m	5.52 m
Width	2.03 m	2.03 m
Height		
(overall)	1.72 m	2.2 m
Ground clearance		
(hull)	0.575 m	0.575 m
(axles)	0.31 m	0.31 m
Track	1.726 m	1.726 m
Wheelbase	3.9 m	3.9 m
Angle of approach/ departure	60°/43°	60°/43°
Max road speed	96 km/h	96 km/h
Fuel capacity	160 litres	160 litres
Road range	800 km	800 km
Fording	1 m	1 m
Gradient	64%	60%
Side slope	35%	35%
Vertical obstacle	0.8 m	0.8 m
Engine	Deutz air-cooled, 6-cylinder diesel developing 132 bhp	Deutz air-cooled, 6-cylinder diesel developing 132 bhp
Transmission	Allison Transmission AT-540 automatic with 4 forward and 1 reverse gears	Allison Transmission AT-540 automatic with 4 forward and 1 reverse gears
Transfer case	Harwaythorn 2-speed	Harwaythorn 2-speed
Steering	power-assisted ZF	power-assisted ZF
Turning radius	7.5 m	7.5 m
Suspension	semi-elliptical springs and hydraulic shock-absorbers	semi-elliptical springs and hydraulic shock-absorbers
Tyres	9.00 × 20	9.00 × 20
Brakes		
(main)	hydraulic, power assisted	hydraulic, power assisted
(parking)	mechanical	mechanical
Electrical system	24 V	24 V
Batteries	2 × 12 V	2 × 12 V
Armour	8-10 mm	8-10 mm
NBC system	optional	optional
Night vision equipment	optional	optional

Note: As an option, run-flat tyres can be fitted.

TCM-20 AA

This is the RAM V-1 (Open) fitted with the RAMTA TCM-20 twin 20 mm light anti-aircraft gun system for which a total of 360 rounds of 20 mm ammunition is carried. Full details of the trailer-mounted TCM-20 twin 20 mm light anti-aircraft gun system are given in *Jane's Land-Based Air Defence 2001-2002* pages 208 to 209.

In addition, one 7.62 mm machine gun, night vision equipment and a multichannel transceiver are fitted.

This model, which is only based on the chassis of the RAM V-1 has a crew of four consisting of driver/loader, commander, gunner and loader. It is possible to fire against both ground and air targets while the vehicle is travelling.

Status

Production as required. The RAM series and the earlier RBY are in service with many countries including Botswana, Cameroon, Gabon, Lesotho, Morocco (including some with a 106 mm recoilless rifle) and Democratic Republic of Congo.

Manufacturer

RAMTA Division, Israel Aircraft Industries Ltd.

UPDATED

RAMTA RBY Mk 1 light armoured reconnaissance vehicle

Development

The RBY Mk 1 light armoured reconnaissance vehicle was developed by RAMTA Division of Israel Aircraft Industries Ltd, and was first announced at the Paris Air Show in May 1975.

The RBY Mk 1 is suitable for a wide range of roles including reconnaissance, long-range patrols and commando-type operations, internal security operations and the evacuation of wounded in forward battlefield areas.

Its low weight enables it to be carried by light aircraft and helicopters such as the Sikorsky CH-53 used by the Israeli Air Force.

Upgraded Guatemalan RBY Mk 1 which has a number of improvements including a new power pack and overhead protection (Julio A Montes)
0088026

Specifications
Crew: 2 + 6
Configuration: 4 × 4
Unloaded weight: 3,600 kg
Power-to-weight ratio: 33.33 hp/t
Length: 5.023 m
Width: 2.03 m
Height:
(empty, without armament) 1.66 m
(loaded, without armament) 1.54 m
Ground clearance:
(hull) 0.48 m
(axles) 0.27 m
(transfer case) 0.375 m
Track: 1.665 m
Wheelbase: 3.4 m
Angle of approach/departure: 60°/50°
Max speed:
(road) 100 km/h
(cross-country) 50 km/h
Fuel capacity: 140 litres
Max range:
(road) 550 km
(cross-country) 400 km
Fording: 0.75 m
Gradient: 60%
Side slope: 30%
Engine: Chrysler 225-2 6-cylinder water-cooled petrol developing 120 hp
Transmission: Model 435 manual with 4 forward and 1 reverse gears
Transfer case: 2-speed
Steering: worm and sector, power assisted
Clutch: single dry plate
Suspension: semi-elliptical springs and hydraulic shock-absorbers
Brakes:
(main) hydraulic
(parking) mechanical mounted on drive shaft
Electrical system: 24 V
Batteries: 2 × 12 V
Armour: 8-10 mm
NBC system: no
Night vision equipment: optional

Production of the RAMTA RBY Mk1 light armoured vehicle was completed some years ago and it is no longer being marketed. It was replaced in production by the enhanced RAMTA RAM family of 4 × 4 armoured vehicles which has a number of improvements including the installation of a diesel engine coupled to a fully automatic transmission. Full details of this vehicle are given in the previous entry.

Description

The all-welded hull of the RBY Mk 1 has 8 mm thick armour steel (MIL-A-46100) sides inclined at an angle, and a floor of 10 mm thick plain carbon steel (SAE-1020).

The shape of the hull has been designed to give the crew maximum possible protection from mines. The axles and wheels are at the extreme ends of the vehicle and the effect of a mine explosion when touched off by one of the wheels is that the respective axle is damaged or cut, the glass fibre bumpers disintegrate, but the hull stays intact. The ability of the vehicle to survive mine explosions was proved in a test programme using different types of mine.

The driver is seated at the front of the vehicle on the left with the commander to his right. The front part of the crew compartment can be folded down into the horizontal for increased visibility and is provided with two integral observation hatches that are hinged at the top and open outwards. To the left of the driver and to the right of the commander is another observation hatch, also hinged at the top and opening outwards.

The six passengers are seated three each side, facing outwards, back to back for all-round observation. There are no doors in the hull as this would weaken the structure and the crew enter and leave the vehicle by climbing over the sides.

The petrol engine is at the rear of the hull and has a glass fibre cover and side doors for ease of access. The automotive components of the RBY Mk 1 are of a standard commercial design that meets current military requirements. Two spare wheels are carried, one at the front of the vehicle and one at the rear. Night driving lights and a front-mounted winch with a capacity of 2,722 kg can be fitted if required. Up to five 7.62 mm or 12.7 mm machine guns can be pintle mounted around the top of the hull to provide suppressive fire.

RBY Mk 1 in service with the Guatemalan Army (via Julio Montes)

Variants

Anti-tank armed with 106 mm M40 recoilless rifle

This model has the 106 mm M40 recoilless rifle manufactured by Israel Military Industries, with 16 anti-tank rounds. Secondary armament consists of two 7.62 mm machine guns with 5,000 rounds and anti-tank missiles. A long-range anti-tank model was offered.

Communications

The RBY Mk 1 could be fitted with a variety of communications equipment.

Guatemalan upgraded RBY Mk 1 vehicles

A number of the vehicles in service with Guatemala have been rebuilt with a 6.8 litre V-8 diesel developing 145 hp at 3,600 rpm coupled to an automatic transmission. This power pack came from the Israeli M426 (4 × 4) truck.

Kevlar overhead protection has been fitted and a rear-opening side door added either side. It is believed that a total of five vehicles was upgraded early in 1997.

Status

Production complete. No longer marketed. Replaced in production by the RAM family of 4 × 4 vehicles covered in detail in the previous entry. In service with the following countries:

Country	Quantity
Guatemala	9
Honduras	13
Israel	n/avail
Lesotho	10

Manufacturer

RAMTA Division, Israel Aircraft Industries Ltd.

UPDATED

ITALY

Centauro (8 × 8) tank destroyer

Development

In the 1980s, the Italian Army drew up its requirements for a complete new family of tracked and wheeled armoured fighting vehicles to meet its future requirements for the 1990s.

Four vehicles have been developed, the Centauro (8 × 8) tank destroyer, the Ariete MBT, the Dardo HITFIST infantry fighting vehicle and the Puma 4 × 4 and 6 × 6 family of multirole light armoured vehicles.

These four vehicles have been developed by the Consortium IVECO – Otobreda based in Rome with IVECO at Bolzano being responsible for the wheeled vehicles and the power packs of all the vehicles and Otobreda at La Spezia for the tracked vehicles and the armament of all the vehicles. The Otobreda Division, Alenia Difesa, also markets this turret for other applications.

Early in 2000, the Centauro 105 mm turret was successfully tested on the chassis of the ASCOD used in the light tank role. Additional details of this combination, being offered on the export market, are given in the Light tanks section under International.

Early in 1984, the Italian Army formulated its requirements for a new highly mobile wheeled tank destroyer that was required to have a standard NATO 105 mm rifled tank gun.

It was also required to have high road speed for strategic mobility, long range, good cross-country mobility and a computerised fire-control system which would share many common components with those being developed by Officine Galileo for the Ariete MBT.

The first complete prototype was finished in January 1987 with the second following in mid-1987 and, by December, four complete vehicles were involved in the trials programme. Prototypes 5 to 9 were completed in 1988 and, in addition, a ballistic hull was built.

The prototypes were followed by a preproduction batch of 10 vehicles which were completed from late 1989.

Production of the Centauro commenced late in 1990 with first production vehicles being completed in 1991. The original Italian Army requirement was for a total of 450 Centauro vehicles but this was subsequently cut to 400.

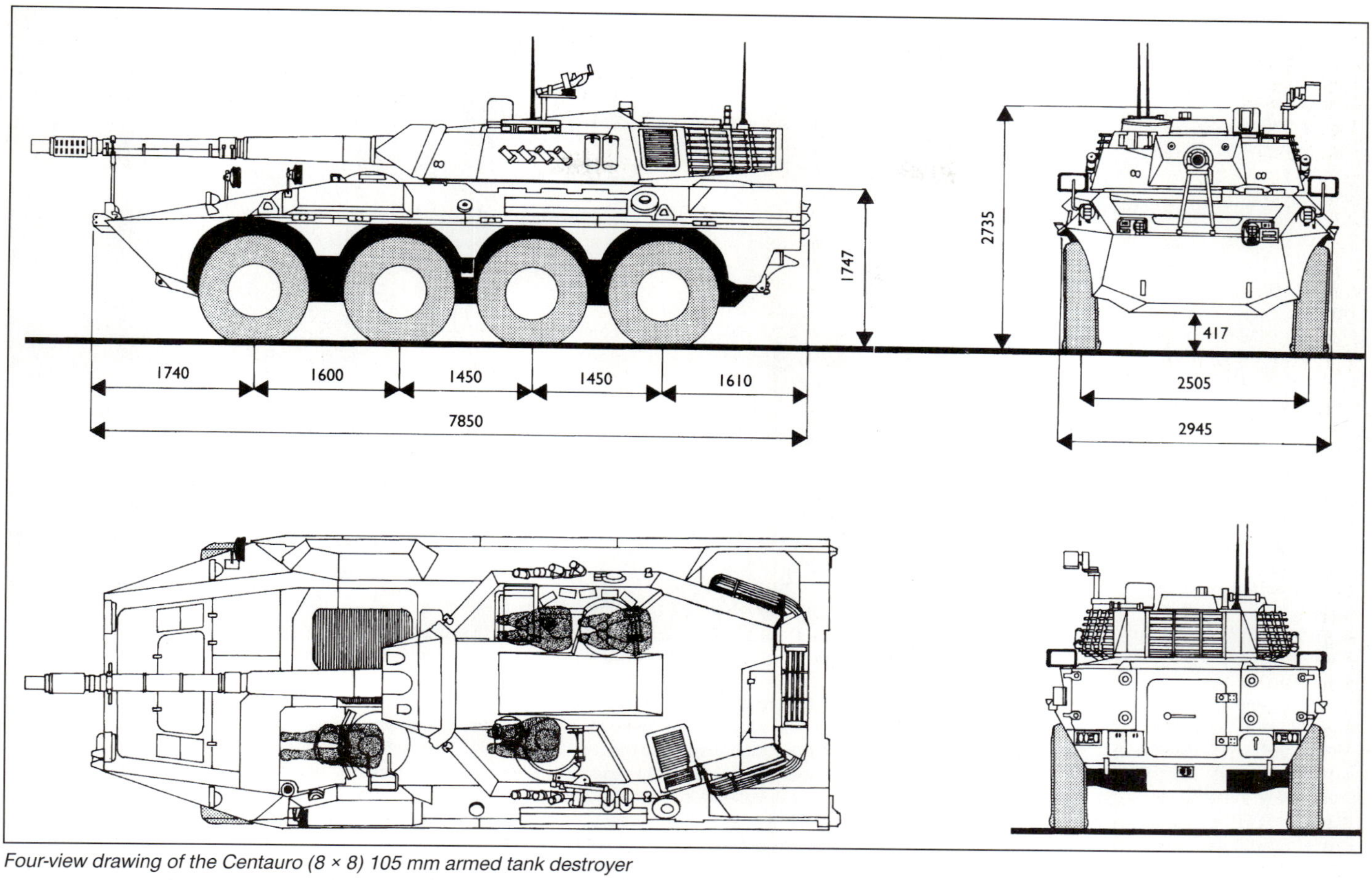

Four-view drawing of the Centauro (8 × 8) 105 mm armed tank destroyer

Centauro (8 × 8) tank destroyer fitted with additional passive armour on the turret and showing mounting attachments on hull for additional armour (Richard Stickland) ***2002***/0114331

Centauro (8 × 8) fitted with Otobreda OTO T60/70A turret armed with 60 mm cannon (Christopher F Foss) ***2001***/0093537

Production Centauro vehicles differ from the prototypes in a number of key areas including improved stowage, a modified turret rear, a door in the hull rear rather than a ramp, improved layout for the crew positions and improvements to the weapon system.

Within IVECO, the Defence Vehicles Division in Bolzano in northern Italy is responsible for the Centauro. IVECO assembles the complete power pack (engine, transmission and cooling system) not only for the Centauro, but also for the Ariete, Dardo HITFIST and Puma.

Late in 1992, eight Centauro (8 × 8) 105 mm armed tank destroyers were deployed to Somalia and during the first four months these averaged 8,400 km without any serious problems.

In 1993, following an international competition, the now BAE Systems, RO Defence of the UK designed, developed and built 20 ROMOR-A explosive reactive armour packages for the Centauro vehicles to increase their battlefield survivability against penetration from anti-tank weapons fitted with a High Explosive Anti-Tank (HEAT) warhead. Ten kits were supplied direct to Italy with the other 10 being supplied direct to Somalia where they were installed on the vehicles.

In addition, Consortium IVECO - Otobreda has developed a new passive armour package for the Centauro to improve its survivability against small arms fire. This armour package is fitted to the hull and turret.

Final deliveries of the Centauro (8 × 8) vehicle were made to the Italian Army late in 1996.

In mid-1999, the Spanish MoD awarded the IVECO Otobreda consortium a contract worth US$70 million for the supply of 22 Centauro (8 × 8) armoured vehicles plus a training and support package.

These were brand new vehicles and were delivered to the Spanish Army late in 2000. These vehicles are referred to as the VRC-105 and equip the 8 Lusitania Light Cavalry Regiment of the Spanish Army Rapid Reaction Force. There is a possibility that additional orders will be placed by Spain in the future.

As far as it is known, the Spanish vehicles are the same as those supplied to the Italian Army but will be capable of being fitted with an explosive reactive armour package developed in Spain by Santa Barbara. This will provide additional protection to the Centauro over the frontal arc against weapons fitted with a HEAT (High Explosive Anti-Tank) warhead.

Early in 2000, it was stated that the Italian Army was to loan the US Army a total of 16 Centauro vehicles for trials purposes for its medium weight interim brigade requirement.

Of these 16 vehicles, 12 will be in the standard configuration with the remaining four in the close defence section configuration. These are expected to remain in the US until at least 2001.

Description

The hull and turret of the Centauro are of all-welded steel armour construction which provides the crew with protection from small arms fire and shell splinters. Over the frontal arc, protection is provided against penetration from 20 mm projectiles and against 12.7 mm attack all round.

The driver is seated at the front of the hull offset to the left with the power pack to his right and separated from the remainder of the vehicle by a fireproof bulkhead. The driver is provided with an adjustable seat, a single-piece hatch cover that opens to the left and three day periscopes for forward observation, the centre one of which can be replaced by an MES VG/DIL passive periscope for night driving.

The turret is mounted on the roof of the hull slightly to the rear of the vehicle's centreline, with the commander on the left, gunner on the right and loader forward and below the gunner.

The commander is provided with four day periscopes to give observation to the front, sides and rear. The stabilised day sight is mounted forward of the commander's hatch and gives full 360° observation without head movement.

Specifications

Crew: 4
Configuration: 8 × 8
Power-to-weight ratio: 20.8 hp/t
Combat weight: 25,000 kg
Length:
(hull) 7.85 m
(incl 105 mm gun) 8.555 m
Width:
(over tyres) 2.945 m
(over hull) 3.05 m
Height:
(to hull top) 1.747 m
(to turret roof) 2.438 m
(overall) 2.735 m
Ground clearance: 0.417 m
Wheelbase: 1.6 + 1.45 + 1.45 m
Track: 2.505 m
Angle of approach/departure: 45°/60°
Max road speed: 105 km/h
Range: 800 km
Fording: 1.5 m
Gradient: 60%
Side slope: 30%
Engine: IVECO VTCA V-6 turbocharged diesel developing 520 hp at 2,300 rpm
Transmission: ZF 5 HP 1500 automatic with 5 forward and 2 reverse gears
Steering: power assisted (first, second and fourth axles)
Turning radius: 9 m
Suspension: independent hydropneumatic
Tyres: 14.00 × 20 (run-flat) (central tyre pressure regulation system fitted)
Electrical system: 24 V
Armament:
(main) 1 × 105 mm gun
(coaxial) 1 × 7.62 mm MG 42/59
(anti-aircraft) 1 × 7.62 mm MG 42/59
Smoke-laying equipment: 2 × 4 76 mm smoke grenade dischargers
Ammunition:
(105 mm) 40 (14 turret, 26 hull)
(coaxial) 1,400
(smoke grenades) 16
Gun control equipment
Turret power control: electrohydraulic/manual
Turret traverse: 360°
Gun elevation/depression: +15°/–6°
NBC system: yes
Night vision equipment: yes

Centauro (8 × 8) tank destroyer fitted with reactive armour to hull and turret (BP) **2001**/0093536

The gunner leaves the vehicle via the loader's single-piece hatch that opens to the rear with five day periscopes being provided in the right side of the turret roof for use by the gunner and loader.

The turret of the Centauro is the complete responsibility of Otobreda at La Spezia and is delivered fully integrated and tested and ready for installation on the chassis.

Main armament comprises an Otobreda-designed 105 mm 52 calibre rifled gun with a long recoil system that fires standard NATO tank ammunition as used in the L7 and M68 guns, including APFSDS. The weapon has a vertical sliding breech mechanism that remains open after the empty cartridge case has been ejected and is fitted with a multislotted muzzle brake, a thermal sleeve with an integral fume extractor and a muzzle reference system. The ordnance is autofrettaged and has a maximum recoil of 750 mm.

There are 40 rounds of 105 mm ammunition carried, 14 of which are in the turret and the remainder in the hull. A 7.62 mm MG 42/59 machine gun is mounted coaxially to the left of the main armament with a similar weapon being mounted on the turret roof for anti-aircraft defence. Mounted either side of the turret is a bank of four electrically operated 76 mm smoke grenade dischargers which fire over the frontal arc of the vehicle.

Turret traverse and weapon elevation are electrohydraulic with manual controls for emergency use with elevation being from –6 to +15°, slightly less than MBTs because of the low turret profile. The centre of the turret roof is raised along the axis of the gun to allow the weapon to have a depression of –6°.

The Centauro is fitted with the same OfficineGalileo TURMS (Tank Universal Reconfigurable Modular System) fire-control system installed in the Ariete MBT.

Main components of the TURMS (Tank Universal Reconfigurable Modular fire-control System) are the commander's stabilised panoramic day sight, stabilised periscopic day/night laser sight for the gunner, digital ballistic computer, sensors, muzzle reference system and commander's, gunner's and loader's display panels.

The commander's stabilised panoramic sight has magnifications of ×2.5 and ×10 and for night vision the thermal image from the gunner's sight is displayed at the commander's station on a separate TV monitor. Traverse is a full 360° with elevation from –10 to +60°.

The gunner's roof-mounted sight integrates the four main modules (the primary stabilised head mirror, visual unit, laser transceiver and thermal image unit) in a single housing. The daylight (×5 magnification) and thermal night vision sights with two fields of view, wide and narrow, are all routed via the common head mirror.

The ballistic digital computer performs all the fire computations, controls and manages the operation of all ballistic units (optical sight, laser range-finder and servos) as well as the sensors, BITE (Built-In Test Equipment) and training. It also permits the reconfiguration of system operation from the normal level to back-up modes in case of partial failures.

The three key sensors are meteorological, vehicle attitude and gun wear.

The gunner also has an Officine Galileo OG C-102 coaxial day telescope with a magnification of ×8 and three aiming graticules which are selectable manually.

Although the 105 mm main armament is stabilised, the Italian Army would normally come to the halt to open fire.

The Centauro is powered by an IVECO VTCA V-6 turbocharged diesel developing 520 hp at 2,300 rpm which, with a combat weight of 25 tonnes, gives a good power-to-weight ratio of 20.8 hp/t.

The engine is coupled to a German ZF 5 HP 1500 fully automatic transmission with five forward and two reverse gears which in turn provides power to a ZF transfer case. The ZF transmission and transfer case are manufactured under licence in Bolzano. The complete Centauro power pack can be replaced in 20 minutes.

From the transmission, power goes to the differential which distributes power to either side of the vehicle via a propeller shaft. Each of the latter then engages a series of shafts and conic couplings (one for each wheel).

Suspension is of the hydropneumatic type with powered steering on the front four and last two roadwheels, with the latter only operating at speeds of under 20 km/h.

A central tyre-pressure regulation system is fitted as standard which allows the driver to adjust the tyre pressure to suit the ground being crossed while the vehicle is still moving. The tyres are of the run-flat type.

Mounted in the turret bustle is the Sekur NBC system which is of a similar design to that installed in the Ariete MBT and is of the overpressure type. An integrated air conditioning system is installed allowing the crew to operate the vehicle with the outside temperature ranging from –30 to +44°C.

Standard equipment includes a front-mounted winch, a fire detection and suppression system for the engine compartment and a fire and explosion detection and suppression system for the crew compartment. A winch is located in the forward part of the hull.

A roof-mounted laser warning device may also be installed to warn the crew of any laser range-finders or weapons.

The standard production Centauro vehicles weigh 25 tonnes combat ready, but additional armour protection could be provided for improved battlefield survivability and in this case maximum combat weight could rise to 28 tonnes.

Variants

IFOR vehicles

The Italian Army deployed a number of IVECO – Otobreda Centauro (8 × 8) tank destroyers to Bosnia as part of IFOR (Peace Implementation Force) and these were fitted with the passive armour package as shown in some of the illustrations.

In addition, roof-mounted 7.62 mm MG 42/59 machine guns with shields for added protection from small arms fire, were supplied to the right and left respectively of the commander's and loader's hatch covers.

Centauro with close defence section

Of the total 400 Centauro vehicles delivered to the Italian Army, the last batch of 150 vehicles have the capability of carrying a close defence section of four men and using the rear door.

These vehicles were delivered to the Italian Army with four seats which are part of the Centauro kit of tools and accessories. These seats are installed at unit level as required after the removal of the two rearmost ammunition racks. When fitted with these seats this version of the Centauro carries a total of 16 rounds of 105 mm ammunition of which 14 are in the turret and two in the hull.

The area for the section of four men carried is fitted with spall liners while the NBC air conditioning system is slightly modified to provide purified air to the increased crew.

Centauro with 60 mm turret

For trials purposes the new Otobreda T60/70A turret, armed with a 60 mm gun, has been installed on the latest version of the Centauro chassis which is longer and wider with wheel base now being 1.605 m +1.60 m + 1.60 m. It has a combat weight of 24 tonnes and can carry six men plus its crew of three. The same chassis is used for the VBC. Details of this turret are given in the entry for the Italian Dardo HITFIST ICV covered under Italy in the Armoured personnel carriers (tracked) section.

Centauro VBC APC
This is a Centauro with its three-man turret removed and fitted with a two-man power-operated turret armed with a 25 mm gun and 7.62 mm coaxial machine gun. It was first shown in public in 1996. Full details are given in the Armoured personnel carriers (wheeled) section.

Recovery vehicle
This is currently being studied.

NBC reconnaissance
Proposal.

Air defence
Gun and missile air defence versions have been proposed.

Armoured bridgelayer
Proposal.

155 mm self-propelled gun
Under a programme funded by the Italian Army, one of the prototypes of the Centauro has been rebuilt as a testbed for a 155 mm self-propelled artillery system with the elevating mass and ordnance being provided by the Italian Army. It is not expected that production systems will use the Centauro (8 × 8) chassis on cost grounds. It is expected that an 8 × 8 truck chassis will be used.

Status
Production as required. A total of 400 vehicles was delivered to the Italian Army between 1991 and 1996. Late in 2000, Spain took delivery of 22 vehicles. The US Army has loaned 16 vehicles for trials purposes.

Manufacturer
Consorzio Iveco Oto.

UPDATED

FIAT - Otobreda Type 6616 armoured car

Development
The Type 6616 (4 × 4) armoured car was a joint development between FIAT and Otobreda, with the former responsible for the hull and automotive components and the latter for the turret and armament installation.

The vehicle shares many common components with the Type 6614 (4 × 4) armoured personnel carrier which was manufactured under licence in South Korea for the Korean Army.

The first prototype of the Type 6616 armoured car was completed in 1972 and after trials a production order was placed by the Italian government for 40 vehicles for the Carabinieri.

Production of the FIAT - Otobreda Type 6616 armoured car has been completed with 300 vehicles built. Production of the vehicle was undertaken in the FIAT (later IVECO Defence Vehicles Division) facilities at Bolzano with the complete turret, ready for installation on the vehicle, being provided by Otobreda. This vehicle, and its numerous variants, are no longer marketed.

Description
The all-welded steel armour hull provides the occupants with protection from small arms fire and shell splinters. In each side of the hull is an escape hatch.

The driver is seated at the front of the vehicle on the right side and is provided with a single-piece hatch cover that opens to the right with an opening in the roof for a passive night driving periscope. To the front and side of the driver are five vision blocks which cover an arc of 200°, the centre three with a wiper blade.

The all-welded steel turret is in the centre of the hull with the commander on the left and the gunner on the right. Their seats, which can be adjusted for height, are mounted on the turret basket. The commander has nine day periscopes for all-round observation and a single-piece hatch cover that opens to the rear. The gunner has a single-piece hatch cover that opens to the rear and a day periscope sight mounted in the turret roof forward of the hatch cover. The Alenia P204 gunner's periscope has a magnification of ×1 for general observation and ×8 for aiming purposes in daylight (with a 9° field of view). It can be replaced by a P194 day image intensifier with a magnification of ×8 and a 7.9° field of view for engaging targets at night. The periscopic head has an elevation of +60° and a depression of −10°. The radio is mounted in the turret bustle.

The engine compartment at the rear of the hull is separated from the fighting compartment by a fireproof bulkhead. The engine is at the very rear of the hull and is

Specifications
Crew: 3
Configuration: 4 × 4
Combat weight: 8,000 kg
Power-to-weight ratio: 20.20 hp/t
Length:
(hull) 5.37 m
Width: 2.5 m
Height:
(to turret roof) 2.035 m
Firing height: 1.76 m
Ground clearance:
(hull front) 0.485 m
(hull centre) 0.44 m
(hull rear) 0.506 m
(axle) 0.37 m
Track: 1.96 m
Wheelbase: 2.75 m
Angle of approach/departure: 45°/45°
Max speed:
(road) 100 km/h
(water) 5 km/h
Fuel capacity: 150 litres
Max range:
(road at 70 km/h) 700 km
(water at 5 km/h) 20 km
Fording: amphibious
Gradient: 60%
Side slope: 30%
Vertical obstacle: 0.45 m
Turning radius:
(land) 7 m
(water) 7.5 m
Engine: Model 8062.24 supercharged liquid-cooled in-line diesel developing 160 hp at 3,200 rpm
Transmission: manual with 5 forward and 1 reverse gears
Transfer case: 2-speed
Steering: assisted
Clutch: single dry plate, hydraulically operated
Suspension: independent
Tyres: 14.50 × 20
Brakes:
(main) drum, air over hydraulic (dual circuit)
(parking) drum, hand-operated, on transfer front output shaft
Electrical system: 24 V
Batteries: 4 × 12 V, 100 Ah
Armament:
(main) 1 × 20 mm MK 20 Rh 202 cannon
(coaxial) 1 × 7.62 mm MG 42/59
Smoke-laying equipment: 3 × 76 mm smoke grenade dischargers either side of turret
Ammunition:
(main) 400
(coaxial) 1,000
(smoke grenades) 39
Gun control equipment
Turret power control: electric/manual
Max elevation speed: 10°/s
Max turret traverse speed: 10°/s
Traverse speed (commander controlled): 30°/s
Gun elevation/depression: +35°/−5°
Gun stabiliser:
(vertical) no
(horizontal) no
Armour:
(hull and turret) 6-8 mm
NBC system: optional
Night vision equipment: optional

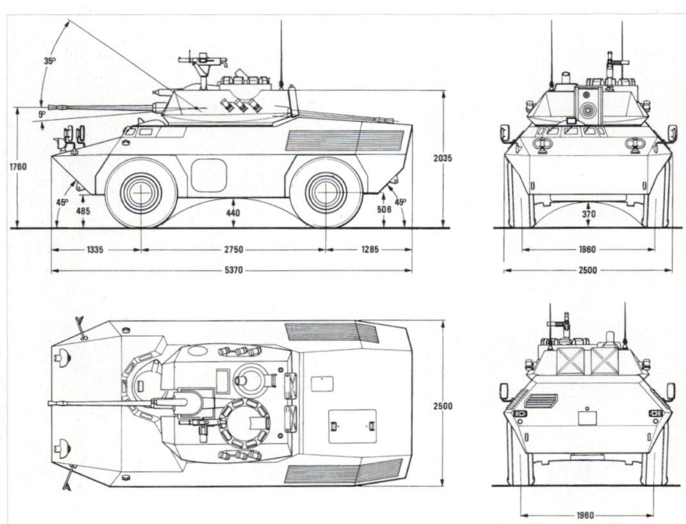

FIAT - Otobreda Type 6616 armoured car

FIAT - Otobreda Type 6616 armoured car with a two-man turret armed with a 20 mm cannon and a 7.62 mm coaxial machine gun
0088028

coupled to a manual five-speed transmission from which power is transmitted to the two-speed transfer case and then to the front and rear axles by propeller shafts. Both front and rear differentials are mechanically lockable. The planetary final drives, front and rear, consist of epicyclic gear trains in the wheel hubs.

The suspension, front and rear, is of the independent strut and link type with a helical spring and coaxial rubber bump stop. Each wheel station is provided with hydraulic shock-absorbers and anti-roll bars are fitted as standard. The tyres are of the run-flat type and the steering is power assisted.

Optional equipment included an NBC system, an air conditioning system, a fire extinguishing system and a front-mounted winch powered by a PTO with a maximum capacity of 4,500 kg and provided with 40 m of 12 mm diameter cable.

The Type 6616 armoured car is fully amphibious, propelled in the water by its wheels. There are four bilge pumps fitted, each with a capacity of 180 litres per minute and it is possible to pressurise the submerged mechanical components.

The main armament consists of a Rheinmetall 20 mm MK 20 Rh 202 cannon. The empty cartridge cases are automatically ejected outside the vehicle. A total of 250 rounds of ready use 20 mm ammunition is carried with a further 150 rounds in reserve. Mounted coaxially over the main armament is a 7.62 mm MG 42/59 machine gun with 300 rounds of ready use ammunition, and a further 700 rounds carried in reserve. Mounted on either side of the turret are three 76 mm smoke grenade dischargers which are electrically fired from within the vehicle. If required a 106 mm M40-type recoilless rifle could be mounted forward of the commander's position on the turret roof.

Status
Production complete. In service with the Italian Carabinieri (40), Peru (15), Somalia (total of 30 delivered between 1978 and 1979 but current status is uncertain) and other undisclosed countries.

Manufacturer
IVECO SpA, Defence Vehicles Division.

UPDATED

JAPAN

Komatsu Type 87 reconnaissance and patrol vehicle

Development
The Type 87 (6 × 6) reconnaissance and patrol vehicle was developed by Komatsu Limited to meet the operational requirements of the Japanese Ground Self-Defence Force and uses the same automotive components as the Mitsubishi Type 82 (6 × 6) command and communications vehicle covered in the following entry.

By mid-2001, it is estimated that the Japanese Ground Self-Defence Force had taken delivery of at least 100 Komatsu Type 87 reconnaissance and patrol vehicles.

Description
The Type 87 reconnaissance and patrol vehicle has a new redesigned hull of all-welded steel construction that provides protection from small arms fire and shell splinters.

The five-man crew of the vehicle consists of the driver, assistant driver/radio operator, commander, gunner and observer.

The driver sits at the front of the vehicle on the right and has a single-piece hatch cover that opens to the right with an integral day periscope in the centre and three fixed forward-facing day periscopes. A passive night driving device can replace one of the centre day periscopes.

The assistant driver/radio operator is to the left of the driver and has a single-piece hatch cover that opens to the left with a single forward-facing day periscope.

The two-man power-operated turret has been designed and built by Komatsu Limited and is in the centre of the hull slightly forward of the second axle.

The gunner is seated on the left with a single-piece circular hatch cover that opens to the rear. Forward of this is a day/night sight and to the left is a single observation periscope.

The commander is seated on the right of the turret and has a single-piece circular hatch cover that opens to the rear, roof-mounted sight and six observation periscopes. Mounted on the turret rear is a stowage box.

Main armament comprises an Oerlikon Contraves 25 mm KBA cannon with a 7.62 mm machine gun mounted coaxially to the right. Mounted either side of the turret is a bank of four forward-firing electrically operated smoke grenade dischargers. The 25 mm Oerlikon Contraves cannon is manufactured under licence by Nippon Seiko Jyo (Japan Steel Works).

The engine compartment is at the rear of the hull on the right with the air inlet/outlet louvres in the roof and the exhaust outlet on the right side of the hull at the rear.

There is an aisle for the observer on the left and he is provided with a single rearward-facing periscope for observation purposes. A small door is provided in the rear of the hull on the left side and this opens to the left.

In the left side of the hull between the second and third axles is an entry door that opens forward, this being provided with a single observation block. A similar door is provided in the right side of the hull that also opens forwards.

Standard equipment includes power steering and passive night vision equipment. The Type 87 does not have an NBC system and is not amphibious.

Variants
None. Components of the Type 87 reconnaissance vehicle are also used in the Type 82 command and communications vehicle which is used in larger numbers by Japan.

Status
Production. In serivce with the Japanese Ground Self-Defence Force. This vehicle has not been offered in the export market.

Specifications
Crew: 5
Configuration: 6 × 6
Combat weight: 15,000 kg
Power-to-weight ratio: 20.33 hp/t
Length:
(hull) 5.99 m
Width: 2.48 m
Height: 2.8 m
Ground clearance: 0.45 m
Maximum road speed: 100 km/h
Range: 500 km
Vertical obstacle: 0.6 m
Trench: 1.5 m
Fording: 1 m
Engine: Isuzu 10PBI 10-cylinder water-cooled diesel developing 305 hp at 2,700 rpm
Transmission: manual; (6 forward and 1 reverse gears)
Steering: power assisted
Suspension: coil springs
Armament:
(main) 1 × 25 mm Oerlikon Contraves 25 mm KBA cannon
(coaxial) 1 × 7.62 mm MG
Smoke-laying equipment: 2 × 4 smoke grenade launchers
NBC system: no
Night vision equipment: yes

Manufacturer
Komatsu Limited.

UPDATED

Type 87 reconnaissance and patrol vehicle (Kensuke Ebata) 0088029

Type 87 reconnaissance and patrol vehicle (Paul Beaver) 0088030

Mitsubishi Type 82 command and communications vehicle

Development

In 1974, two wheeled reconnaissance vehicles were designed to meet the requirements of the Japanese Ground Self-Defence Force.

The 4 × 4 vehicle was designed and built by Komatsu Seisaku-Jyo while the 6 × 6 vehicle was designed and built by Mitsubishi Heavy Industries. After trials it was decided to concentrate on the 6 × 6 model.

There have been two models of the 6 × 6 vehicle developed, a reconnaissance/patrol vehicle and a command and communications vehicle. The reconnaissance patrol vehicle is being built by Komatsu Limited and ¥342 million was allocated in the 1978 budget for the construction of two prototype vehicles. The reconnaissance and patrol vehicle was type classified as the Type 87 and is covered in the previous entry.

The command and communications vehicle was developed by Mitsubishi Heavy Industries. A total of ¥456 million was allocated in the 1978 budget for the construction of four prototypes. The first prototypes were completed in 1980. A total of 250 vehicles will be produced at a unit cost of ¥58 million. Eighty per cent of its components will be identical to those used in the 6 × 6 reconnaissance patrol vehicle. This was subsequently standardised as the Type 82 command and communications vehicle.

By mid-2001, it is estimated that the Japanese Ground Self-Defence Force had taken delivery of at least 240 Mitsubishi Heavy Industries Type 82 command and communications vehicles.

Specifications

Crew: 8
Configuration: 6 × 6
Combat weight: 13,500 kg
Power-to-weight ratio: 22.59 hp/t
Length: 5.72 m
Width: 2.48 m
Height: 2.37 m
Ground clearance: 0.45 m
Wheelbase: 1.5 + 1.5 m
Max road speed: 100 km/h
Range: 500 km
Fording: 1 m
Gradient: 60%
Vertical obstacle: 0.6 m
Trench: 1.5 m
Engine: Isuzu 10PBI 10-cylinder water-cooled diesel developing 305 hp at 2,700 rpm
Transmission: automatic 6 forward and 1 reverse gears
Turning radius: 8.4 m (with 1st and 3rd axles)
Tyres: 14.00 × 20
Armament:
(main) 1 × 12.7 mm M2 MG
(secondary) 1 × 7.62 mm MG
NBC system: no
Night vision equipment: yes

Description

The hull of the Type 82 command and communications vehicle is of all-welded construction, with the driver sitting at the front of the hull on the right side. Above his position is a single-piece hatch cover with an integral day periscope that can be traversed left and right. A second crew member sits to the left of the driver and over his position is a single-piece hatch cover; forward of this is a pintle-mounted 7.62 mm machine gun. The windscreens to the front and either side of this position are covered, in combat areas, by armoured shutters hinged at the top, the forward shutter being provided with two vision slits.

The engine compartment is to the rear of the front two crew members, with the air inlet and outlet louvres in the roof. There is a gangway in the right side of the hull which connects the front and rear crew compartments.

The rear crew compartment, which also houses most of the communications equipment, is raised above the roof of the vehicle and has two circular hatch covers. The left one is provided with six observation day periscopes and a single piece hatch cover that opens to the rear and the right with an externally pintle-mounted 12.7 mm machine gun with a shield. It has a two-piece hatch cover that opens left and right. Late production vehicles have been observed fitted with a larger cupola on the roof with an externally mounted 12.7 mm M2 HB machine gun.

In the hull rear is a large door opening to the right on which is a single firing/observation port that opens upwards. In the left side of the hull, between the second and third axle, is a single forward-opening door again with a single firing/observation hatch opening upwards. To the rear of this door is another single firing/observation port opening upwards. In the right side of the hull, between the first and second axle, is a forward-opening door with an integral firing/observation port that opens upwards and to the rear of this a further two firing/observation ports.

Some Type 82 command and communications vehicles have been observed without the door in the left side of the hull.

Variants

NBC reconnaissance

An NBC reconnaissance version of the Type 82 command and communications vehicle is now in service. This has devices mounted at the rear for the collection of samples which can then be analysed

Status

In production. In service only with Japanese Ground Self-Defence Force. This vehicle has not been offered in the export market.

Manufacturer

Mitsubishi Heavy Industries.

UPDATED

Type 82 (6 × 6) command and communications vehicle from the rear (K Nogi)

Type 82 command and communications vehicle fitted with a 12.7 mm M2 HB machine gun (Kensuke Ebata)

NETHERLANDS

SP aerospace and vehicle systems MultiPurpose Carrier (MPC) (Fennek)

Development

The MultiPurpose Carrier (MPC) (4 × 4) light armoured vehicle has been under development by SP aerospace and vehicle systems bv as a private venture following a market survey which indicated a gap in the market for a vehicle of this type.

The first preseries prototype was completed in late 1992 and was used for extensive company trials.

SP aerospace and vehicle systems has designed the MPC to bridge the gap between Jeep-type vehicles and armoured personnel carriers but the modular design is such that it can easily be adapted to undertake a wide range of missions on the battlefield as well as a number of paramilitary and civil roles.

Typical projected military missions include command and communications, reconnaissance, anti-tank (for example fitted with the Euromissile MILAN ATGW system), support roles, security, surveillance, intelligence-gathering when fitted with various sensors and engineer.

Wherever possible, proven and standardised production components have been used in the design of the MPC, for example the diesel engine is derived from a standard diesel and will be produced in large numbers. Proven components have helped to reduce the life cycle costs of the MPC and the design is such that local production is also possible.

The vehicle has plenty of room for specialised equipment as well as its normal three-man crew of commander, radio operator and driver. If required, the MPC can carry a maximum of five people.

The overall MPC concept allows various adaptations according to customer requirements.

In 1998, RDM Technology who have extensive experience in upgrading armoured fighting vehicles and artillery systems, took a 100 per cent share in SP aerospace and vehicle systems.

Description

The hull of the MPC allows flexible armour protection concepts and is typically of all-welded aluminium armour with additional armour modules providing

Front view of Fennek (4 × 4) reconnaissance vehicle with sensor pod in operating position ***2001***/0093538

Specifications

Fennek – provisional
Crew: 3
Configuration: 4 × 4
Combat weight: 10,200 kg
Empty weight: 8,500 kg
Payload: 1,700 kg
Power-to-weight ratio: 23.52 hp/t
Length: 5.58 m
Width: 2.55 m
Height:
(top of roof) 1.79 m
(overall) 2.29 m
Ground clearance: 0.40 m
Wheelbase: 3.00 m
Fording: 1 m
Max road speed:
(forwards) 112 km/h
(reverse) 23 km/h
Range:
(road) 1,000 km
(cross-country) 460 km
Gradient: 60%
Side slope: 35%
Engine: Deutz BF6M 2013C 6-cylinder turbocharged inter-cooled diesel developing 240 hp
Transmission: automatic six speed with torque converter and lock up
Steering: power-assisted
Turning radius: 6.43 m
Suspension: independent
Electrical system: 24 V
Batteries: 4 × 24 V, 120 Ah
Armament: 1 × 40 mm grenade launcher or 1 × 12.7 mm MG or 1 × 7.62 mm MG
NBC system: yes
Night vision equipment: yes
Gun control equipment
Turret power control: powered/manual
Traverse: 360°
Gun elevation/depression: +40/–10°

protection from 7.62 mm AP small arms fire and shell splinters. The actual armour composition thickness depends on the customer's operational requirement, as does the layout of such subsystems as hatches/doors, the vision system, armament and the communications fit.

The driver is seated at the front of the vehicle with excellent vision to his front via three large bulletproof windows, each of which is provided with an electric wiper.

The driver's layout has been designed for ease of use with power steering being fitted as standard. Driver fatigue is further reduced by the installation of a fully automatic transmission and all-wheel drive.

The internal layout of the MPC depends on the mission. The Fennek, for example, is a member of the MPC vehicle family for the reconnaissance and support role task. In this configuration, to the rear of the driver there is room for the operators of the mission equipment and access is provided by a large door in the side of the hull, above which is a long bulletproof window. In addition, there is a rearward-opening roof hatch.

The flat roof of the MPC has been designed so that a one-man turret can be fitted if required by the user.

The engine compartment is at the rear of the MPC and the power pack can be removed from the vehicle as a single unit to facilitate field replacement.

Standard equipment includes a fire detection and suppression system in the engine compartment, a heating and cooling system and a wash/wipe system. A wide range of optional equipment can be fitted including an NBC system, various communications systems and two water-jets to give a full amphibious capability.

The prototype has a wheelbase of 3 m, but eventually this could be stretched if required by the user, with no changes to the power pack or automotive components apart from using longer drive shafts.

Joint Royal Netherlands/German Army programme

For the Royal Netherlands and German army requirement, SP aerospace and vehicle systems and Krauss-Maffei Wegmann developed a specific derivative of the MPC called the Fennek vehicle.

Following a competition, the Royal Netherlands Army and the German Army selected the Fennek for their planned reconnaissance fleet replacement programme.

The company was awarded a contract to build a total of four vehicles, two for Germany and two for the Netherlands. The first of these was rolled out late in March 1996 and then delivered to Krauss-Maffei Wegmann in Germany to install the sensors and other mission-related equipment.

By late 1998, a total of five prototypes had been built, two for the German Army, two for the Royal Netherlands Army and one to be retained by SP aerospace and vehicle systems and Krauss-Maffei Wegmann for development work and as a company demonstrator.

While SP aerospace and vehicle systems will be responsible for the vehicle and Krauss-Maffei Wegmann for systems integration, the German company, STN ATLAS Elektronik, will be responsible for the sensor pod.

Following trials with the prototype vehicles, a number of problems became apparent and this has led to a delay of several years in the programme for the German and Dutch armies.

Main problem areas discovered during the 1996 trials included armour, performance on and off the road and the braking system.

In mid-1998 the two companies submitted a package of technical options which were selected and one of the prototypes with all the improvements and options were tested.

The main role of the Fennek is that of reconnaissance and mounted towards the rear of the hull on the left side is a retractable mast which is fitted with a sensor pod containing thermal camera, day camera and laser range-finder.

The mast can be raised 1.5 m above the roof of the vehicle and can be controlled in azimuth and elevation with the operator being seated under full armour protection in the hull of the vehicle.

If required, the sensor head can be removed from the vehicle, placed on a tripod and operated at up to 40 m from the vehicle.

The sensor suite is operated from a swivel-mounted operator's console which is also provided with a monitor, enabling the commander or observer to operate the equipment.

Distance and angle measurements are used to survey the location of the target and the vehicle's location can be quickly determined using the onboard hybrid navigation system (GPS and inertial).

In addition to the mast-mounted sensor suite, the Fennek is fitted with the Tactical Command and Control System (TCCS) as well as HF and VHF radio systems. The crew members communicate through an intercom system.

The TCCS enables the reconnaissance vehicles to be integrated into the reconnaissance data information, command and control system.

This is accomplished by means of digitised maps showing the position of their own vehicle and that of the target as well as the current battlefield environment. This data is updated through HF and VHF radio data transmissions through the vehicle and command post.

The Fennek is fitted with an electrically operated gun mount which can be armed with a 7.62 mm or 12.7 mm machine gun or a 40 mm grenade launcher.

The sighting system is derived from the PERI Z17 which is already in service in large numbers. As an option, the periscope can be upgraded with additional plug-in modules, for example an image intensifier.

The gun mount can be traversed through a full 360° and weapon elevated from –10 to +40°. When fitted with a 7.62 mm machine gun, a total of 250 rounds of ready use ammunition are carried or 32 rounds when fitted with a 40 mm grenade launcher.

An amphibious version is also being marketed as are different armour packages. It is expected that the German vehicles may be armed with a new 40 mm grenade launcher developed by Heckler & Koch.

Standard equipment will include an NBC protection system, winch, rear view TV camera system as well as heating, ventilation and air conditioning systems.

As of July 2001, no firm production contracts had been placed for the Fennek vehicle by either Germany or the Netherlands.

The German radio fit will consist of one SEM 80/90 VHF and one HRM 7400 HF.

The STN Elektronik mast-mounted sensor pod used in the Fennek is also used in one of the Polish BRDM-2 upgrades covered later in this section.

This is also fitted to the MOWAG Eagle (4 × 4) artillery observation vehicles ordered by the Swiss Army.

Late in 2001, the Netherlands Ministry of Defence (MoD) awarded the ARGE Fennek consortium a contract worth about €500 million for the supply of 612 Fennek (4 × 4) reconnaissance vehicles for the German and Netherlands armies. The Netherlands MoD is also acting on behalf of the German MoD for this programme.

The ARGE Fennek consortium comprises SP aerospace and vehicle systems of the Netherlands (owned by RDM) and Krauss-Maffei Wegmann (KMW) of Germany.

It is expected that first production Fennek vehicles will be delivered in 2003 with final deliveries due in 2007.

The German Army will take delivery of 202 vehicles of which 178 will be in the reconnaissance version and 24 in the combat engineer version. These will be armed with a 40 mm automatic grenade launcher. Fennek will replace the number of vehicles currently used in the reconnaissance role, including the Rheinmetall Landsysteme Luchs (8 × 8) vehicle.

The Royal Netherlands Army will take delivery of a total of 410 vehicles of which 202 will be in the reconnaissance version, 130 in Medium Range Anti-Tank (MRAT) (with the RAFAEL Armament and Development Authority Gill missile system) and 78 general purpose vehicles.

The Netherlands vehicles will be armed with a 12.7 mm machine gun. These will replace the United Defense LP command and reconnaissance vehicle which has already been phased out of service, as well as specialised Land Rover (4 × 4) unarmoured vehicles.

Status

Ordered by Germany (202) and the Netherlands (410), with production expected to run from 2003 through to 2007.

Manufacturer

SP aerospace and vehicle systems.

UPDATED

Rear view of Fennek of the Royal Netherlands Army with sensor pod almost retracted into hull and left door open (D Caemerlynck) ***2001***/0093539

PAKISTAN

Pakistani 4 × 2 light armoured vehicles

Development/Description

In the late 1980s, Pakistan modified a number of standard commercial 4 × 2 tractors for military applications.

These vehicles are intended to serve as training vehicles to reduce wear on actual combat equipment so saving significant operating costs.

The modifications included fitting a lightly armoured body to the chassis which covers the engine at the front and extends right to the rear.

The driver and commander are seated to the immediate rear of the engine and to their front is a large windscreen. To the rear of the driver and commander is space for a further two or three personnel, with the weapon installation mounted on the roof.

Normal means of entry and exit are believed to be via doors in the rear of the hull. There are minor differences between vehicles with some being provided with small hatches in the sides of the hull either side of the driver's compartment.

Mounted on either side of the hull, just forward of the driver's compartment, is a bank of three electrically operated smoke grenade dischargers that fire to the front of the vehicle. Most vehicles appear to be fitted with a single radio antenna mounted at the rear of the hull. As far as is known, these vehicles are not fitted with an NBC system or any night vision equipment.

Variants

Anti-Tank with TOW

This version is called Cheeta and is fitted with a roof-mounted US Raytheon Systems Company TOW ATGW launcher with additional missiles carried internally for manual reloading. The TOW missile has a maximum range of 3,750 m.

RBS 70 SAM carrier

This is armed with the Swedish SAAB Bofors Dynamics RBS 70 surface-to-air missile system with additional missiles carried internally for manual reloading.

Anza SAM carrier

This has a pedestal-type mount on which are mounted four Anza (Lance) surface-to-air missiles in the ready to launch position. The Anza has been developed in Pakistan by the Dr A Q Khan Research Laboratories and is based on the China National Precision Machinery Import and Export Corporation (CNPMIEC) HN-5 surface-to-air missile system which is of the infrared fire-and-forget type. Full details of the Anza missile are given in *Jane's Land-Based Air Defence 2001-2002* pages 18 and 19.

MG carrier

This has a simple open hatch on which is mounted the Chinese version of the Russian 14.5 mm KPVT heavy machine gun. Due to the limited depression of the weapon the main role of this version is probably air defence.

Status

Production complete. In service with the Pakistani Army. These were not offered on the export market.

Manufacturer

Not known.

UPDATED

Pakistani 4 × 2 vehicle armed with a four-round launcher for the Anza fire-and-forget surface-to-air missile system ***2002***/0080424

Pakistani 4 × 2 vehicle with a 14.5 mm heavy machine gun 0088033

POLAND

Polish BRDM-2 Upgrade

Specifications			
Model	**96i**	**97**	**98**
Crew	5	5	4-5
Weight (combat)	7,400 kg	8,000 kg	8,100 kg
Length	6.29 m	6.29 m	6.29 m
Width	2.425 m	2.425 m	2.425 m
Height	2.31 m	2.43 m	2.43 m
Ground clearance	340 mm	285 mm	280 mm
Max speed			
(road)	95 km/h	100 km/h	100 km/h
(water)	9/10 km/h	9/10 km/h	9/10 km/h
Engine	IVECO diesel developing 165 hp in all versions		
Turret			
(traverse)	360°	360°	360°
(weapon elevation)	+30°/–5°	+32.5°/ –4.5°	+35°/–5°

Development/Description

It is understood that the Polish Army has over 500 Russian-supplied BRDM-2 (4 × 4) amphibious scout cars in service and that at least part of this fleet is being modernised to extend their operational lives.

This upgrade work is being carried out by Military Mechanical Works Siemianowice Slaskie which was established in 1952 and has considerable experience in upgrading and overhauling tracked and wheeled armoured fighting vehicles.

It is believed that there are at least three other Polish BRDM-2 upgrades called the BRDM-2 Model 96, Model 97 Zbik and the Model 98.

The Model 96i features the installation of the new IVECO 165 hp diesel engine, removal of the two belly wheels either side which were lowered to improve the trench crossing capabilities of the vehicle, installation of a door in either side of the hull to allow the occupants to rapidly enter and leave the vehicle, seating for a maximum crew of five compared to three of the standard production BRDM-2, installation of a new heater for the crew compartment, improved internal stowage and installation of a 24 V electrical system. The standard 14.5 mm KPVT and 7.62 mm PKT turret-mounted machine guns are retained.

Optional equipment for this version includes a satellite navigation system, passive night vision, additional feeder current and a digital radio system.

The Model 96i is already in service with the 18 Battalion of the Polish Army, which has seen service in Bosnia.

The Model 97 includes the new IVECO 165 hp diesel engine, removal of belly wheels, installation of a new door in either side of the hull, improved shape to the existing one-person turret with additional external stowage, installation of a roof-mounted AT-4 Fagot anti-tank guided weapon (ATGW) system, replacement of the 14.5 mm KPVT machine gun by a 12.7 mm NSTV weapon and improved mount for the 7.62 mm PKT machine gun, installation of a new day/night sight, installation of laser warning system and new 81 mm smoke grenade launchers, installation of new passive observation devices for the commander and driver seated at the front of the vehicle, installation of a GPS, new fire detection and suppression system, installation of a chemical warning system, new digital radio and intercom, spare wheel carrier, new 24 V generator, new generation engine and crew compartment heater, new internal stowage and eye protection system for nuclear flash.

The Polish Army has ordered 47 BRDM-2 upgraded to this standard with conversion work starting from the second half of 1998.

The Model 98 is the same as the above but has the German STN ATLAS BAA pedestal-mounted day/night sighting and observation system, which is already in volume production for a number of applications.

This version of the Polish BRDM-2 has been developed by Poland's WZM (Wojskowe Zaklady Mechaniczne – Military Mechanical Works) and STN ATLAS Elektronik of Germany.

This features a second-generation Ophelios thermal imager, eye-safe laser range-finder and a day/night sight with correlated fields of view.

A joystick, which can be positioned up to 40 m away from the vehicle, allows the operator to control the system from a safe distance.

After lasing the target, its position is determined automatically. With real-time position updates and interface to command, control, communications and intelligence systems, the new observation and intelligence platform will be capable of providing real-time intelligence.

Optional equipment includes a GPS system, digital map and automatic modes of operation, including scanning and tracking and target lock-on via a video tracker.

Status

Upgrades in service with the Polish Army.

Manufacturer

Military Mechanical Works Siemianowice Slaskie.

UPDATED

Polish BRDM-2 Model 98 (4 × 4) upgraded BRDM-2 amphibious scout car
2002/0114332

Close-up of the STN ATLAS BAA observation and sighting equipment installed on the BRDM-2 Model 98 (4 × 4) upgraded BRDM-2 amphibious scout car
2002/0073902

Polish BRDM Model 96i (4 × 4) upgraded BRDM-2 amphibious scout car
2002/0114337

Polish BRDM-2 Model 97 (4 × 4) upgraded BRDM-2 amphibious scout car
2002/0114343

ROMANIA

ABI armoured car

Development

As well as developing the TABC (4 × 4) reconnaissance vehicle (which is covered in the Armoured personnel carriers (wheeled) section) and the TAB-77 (8 × 8) APC, Romania has also developed the ABI (4 × 4) light armoured car.

This is based on the chassis of the ARO-240 (4 × 4) light vehicle and typical roles of the ABI include the protection of high-value targets such as airfields and counter insurgency roles in rear areas. As far as it is known, the only export customer for the AM 100 ALG is Algeria.

Description

The layout of the ABI is conventional with the power pack at the front, driver and commander in the centre and the troop compartment extending to the very rear.

The hull is of all-welded steel armour that varies in thickness from 4 to 10 mm providing protection from small arms fire and shell splinters.

The driver is seated at the front on the left side with the commander to his right, both being provided with a forward-opening side door. The left door has a flap that opens upwards, and forward of this is a firing port. The right door has only a firing port.

To the front of the commander and driver is a window which, when in action, is covered by an armoured shutter hinged at the top, vision then being via six roof-mounted PF1 day periscopes that give observation to the front and sides. The commander also has a firing port to the immediate right of the windscreen.

The troops normally enter and leave the vehicle via twin doors that open outwards in the rear of the hull and each of these has a firing port. There are an additional two firing ports in each side of the hull and above these is a single vision block. Two fixed PF1 day periscopes are also provided.

Fitted on top of the hull is a one-man manually operated turret which can be traversed through a full 360° and this has a single-piece hatch cover that opens to the rear.

Fitted externally on the turret is a 12.7 mm DSKM machine gun which can be used to engage ground and aerial targets. This weapon can be elevated from –5 to +25° under manual control.

The turret is armed with a 7.62 mm PKMS machine gun which can be elevated from –5 to +25° under manual control.

Turret observation equipment includes two PF1 fixed day observation periscopes, one either side, one mirror reflex sight, one binocular sight and one modified PGO-7 telescope.

Standard equipment includes a light obstacle clearing device which is mounted at the front of the hull, a white light searchlight mounted in front of the commander's position which can be operated from within the vehicle, a hull ventilator and communications equipment. The latter includes an RTM-4MF radio telephone and an RFT-4 internal communications system which links the commander, gunner and driver.

Specifications

Crew: 2 + 4
Configuration: 4 × 4
Combat weight: 4,000 kg (without crew)
Power-to-weight ratio: 17 hp/t
Length: 4.22 m
Width: 2.225 m
Height:
(hull roof) 1.65 m
(turret top) 2.12 m
(top of MG) 2.71 m
Ground clearance:
(front axle) 0.195 m
(rear axle) 0.21 m
Track:
(front) 1.485 m
(rear) 1.495 m
Wheelbase: 2.35 m
Angle of approach/departure: 20°/35°
Fording: 0.30 m
Max speed:
(road) 90 km/h
(cross-country) 30 km/h
(average road) 48.5 km/h
Range: 600 km
Gradient: 20°
Side slope: 15°
Engine: Model D-127 4-stroke diesel engine developing 68 hp at 3,200 rpm
Transmission: manual, 4 forward and 1 reverse gears
Transfer case: G450 mechanical, 2 speed
Steering: hydraulic assisted
Tyres: 7.50 × 16/12PR
Suspension:
(front) independent with oscillation arms, coil springs and double-action hydraulic telescopic dampers
(rear) laminated springs and double-action hydraulic telescopic dampers
Brakes:
(main) drum on front and rear wheels
(parking) mechanical
Electrical system: 12 V DC
Armament:
(turret) 1 × 12.7 mm MG: 1 × 7.62 mm MG
(rifles) 6 × 7.62 mm
Gun control equipment
Turret power control: manual
Gun elevation/depression: +25°/–5°
Turret traverse: 360°
NBC system: yes
Night vision equipment: yes

Variant

AM 100 ALG

This has a very similar hull to the standard production ABI (4 × 4) armoured car but has a different design of turret which can be armed with a 7.62 mm PKMS machine gun. According to the manufacturer this vehicle has been designed mainly for the internal security role.

Specifications are similar to the ABI except that it is powered by a Bulgarian VAMO Model D 3900 A engine developing 82.5 hp at 2,600 rpm which gives a maximum road speed of 100 km/h, has a loaded weight of 3,900 kg and an unloaded weight of 3,320 kg.

More recently, another version called the AM 100 Armoured Off Road Vehicle has been marketed. This appears to have a different weapon station. Mounted on the front left side of the hull is an infra-red searchlight and mounted in the centre of the forward part of the roof is a white light searchlight that can be operated by driver or commander. Finally, there is a controllable infra-red searchlight mounted on the right side of the roof. This is operated by the commander from within the vehicle.

Status

Production as required. In service with Romanian armed forces and paramilitary units.

A quantity of AM 100 ALG vehicles has been supplied to Algeria for use in the internal security role.

Manufacturer

ROMARM SA.

UPDATED

Romanian ABI armoured car showing turret with externally mounted 12.7 mm machine gun

Romanian AM 100 ALG which has a turret which can be armed with a 7.62 mm PKMS machine gun
0045879

RUSSIAN FEDERATION AND ASSOCIATED STATES (CIS)

Kurgan/Rubstovskiy BRM-3K combat reconnaissance vehicle

Development

In 1993, it was revealed that a new version of the Russian BMP-3 Infantry Combat Vehicle (ICV) had been built specifically for the battlefield reconnaissance role under the provisional designation of Obiekt 501.

Early in 1995, it was disclosed that this vehicle is referred to as the BRM combat reconnaissance vehicle in the Russian Army, or the BRM-3K 'Rys' (Lynx) as it is officially designated.

The chassis of the BRM-3K combat reconnaissance vehicle is provided by the Kurgan Machine Construction Plant, which is the prime contractor for the chassis, while the systems integrator is the PO Rubstovskiy Machine Building Plant.

The role of the BRM-3K combat reconnaissance vehicle is to undertake battlefield and artillery reconnaissance for ground troops and coastal units of the navy by day and night and under all weather conditions.

The BRM-3K vehicle can carry out the following missions:

(1) Surveillance of the terrain, search, detection and identification of the ground (surface), fixed, moving and concealed targets
(2) Determination of target rectangular co-ordinates within 360° with an accuracy of 20 m
(3) Continuous computation of navigational data and indication of the directional angle to the assigned point of destination
(4) Round-the-clock radio communications at a range of 100 km whilst moving and up to 350 km when stationary
(5) Establishment of an observation post at up to 500 m from the BRM-3K vehicle by means of wire communication or up to 6 km by means of a portable UHF station
(6) Engaging air and land targets while the BRM-3K is stationary or moving under both day and night conditions.

While the basic BMP-3 ICV has been exported in significant numbers, as of mid-2001, there were no known exports of the BRM-3K combat reconnaissance vehicle.

BRM-3K combat reconnaissance vehicle in travelling configuration with radar antenna lowered

Description

Externally, the BRM-3K reconnaissance vehicle is very similar to the standard BMP-3 ICV which entered service with the Russian Army some years ago and is now also used by Cyprus, Kuwait, South Korea and the United Arab Emirates (Abu Dhabi). Full details of this are given in the BMP-3 entry in the Armoured personnel carriers (tracked) section.

The main external differences include only two firing ports in the rear troop compartment, deletion of the two (one either side) bow-mounted 7.62 mm PKT machine guns and removal of the 100 mm 2A70 rifled gun which fires conventional ammunition as well as the 100 mm 9M117 Bastion laser-guided missile.

The BRM-3K reconnaissance vehicle retains the three crew seats and hatches at the front of the vehicle with the centre one being for the driver. There appear to be additional day observation devices over the frontal arc in the forward part of the hull and a stowage box on the left side of the hull just above the track guard.

The commander and gunner, who also operate the turret-mounted electro-optical devices, are seated in the turret, with the sixth crew member probably being seated to the rear of the turret in the troop compartment to operate the battlefield surveillance radar.

The six-man crew consists of the reconnaissance group commander, vehicle commander, gunner, driver/mechanic, navigator/computer operator and wireless operator.

The fire-control system installed in the BRM-3K includes the BPK2-42 periscopic combined day/night sight, PPB-2 periscopic monocular sight, 2E52-1 armament stabiliser and the BETs-088 control system. As the 100 mm 2A70 gun of the BMP-3 is removed there is no missile laser guidance system for the BRM vehicle.

Specifications

Crew: 6
Combat weight: 19,000 kg
Power-to-weight ratio: 26.31 hp/t
Ground pressure: 0.6 kg/cm²
Length: 7 m
Width: 3.15 m
Height: 2.37 m
Ground clearance: 450 mm
Track: 2.76 m
Track width: 380 mm
Length of track on ground: 4.06 m
Max road speed:
(forwards) 70 km/h
(reverse) 20 km
Max water speed: 10 km/h
Road range: 600 km
Trench: 2.5 m
Vertical obstacle: 0.8 m
Fording: amphibious
Gradient: 60%
Side slope: 30%
Engine: Model UTD-29 10-cylinder 4-stroke diesel developing 500 hp
Transmission: hydromechanical, 4 forward and 2 reverse gears
Steering: differential with hydrostatic drive
Suspension: independent torsion bar with hydraulic shock-absorbers
Electrical system: 27 V
Armament:
(main) 1 × 30 mm 2A72 cannon
(coaxial) 1 × 7.62 mm PKT machine gun
(smoke grenade dischargers) 3 × 81 mm
Ammunition:
(30 mm) 600
(7.62 mm) 2,000
NBC system: yes
Night vision equipment: yes

BRM-3K combat reconnaissance vehicle from the front (Christopher F Foss)
0088034

BRM-3K combat reconnaissance vehicle showing 1RL-133-1 battlefield surveillance radar in elevated position (Christopher F Foss)
0088035

Main armament of the BRM-3K is the dual-feed 30 mm 2A72 cannon which can engage armoured targets at a range of 1,500 to 2,000 m and helicopters out to a range of 4,000 m and is fully stabilised.

Cyclic rate of fire of the 30 mm 2A72 cannon is 300 rds/min and three types of ammunition can be fired, API, HE-I and HE-T.

Mounted coaxially to the left of the 30 mm 2A72 cannon is the standard 7.62 mm PKT machine gun with a practical rate of fire of 250 rds/min and a maximum effective range of 1,500 m.

Like the BMP-3 ICV, the BRM-3K reconnaissance variant has two banks of three 81 mm electrically operated smoke grenade launchers covering the frontal arc, although these are positioned lower down the turret front than on the original BMP-3.

Turret traverse is probably the same as the standard BMP-3 - through a full 360°, with weapon elevation from –5 to +60°. Turret traverse and weapon elevation are all-electric with manual controls for emergency use.

Equipment installed to enable the vehicle to carry out its specialised reconnaissance role includes the mast-mounted 1RL-133-1 battlefield surveillance radar, which can be retracted into the vehicle when not operating; the 1PN71 thermal night observation device; the 1PN61 night observation device; and the ID14 periscopic laser range-finder.

The electro-optical devices are mounted on either side of the turret and, when not being used, the optics are covered by a hinged shutter that opens to the left.

The 1D14 periscopic laser range-finder has a magnification of ×7.3 and ×18 and a maximum range of 25,000 m. This laser is mounted on a turntable and can be used by the vehicle commander for reconnaissance purposes with an arc of 60° regardless of the position of the turret.

The 1PN61 active pulse night vision device with laser illuminator has a magnification of ×7 and a range of 1,500 m in the passive mode and 3,000 m in the active mode. This is mounted externally on the right side of the turret.

The 1PN71 thermal observation device operates in the 8 to 14 μm waveband and has a magnification of ×3.7 and ×11.5 and maximum target identification range is 3,000 m. This is mounted externally on the left side of the turret.

The 1RL-133-1 radar detects vehicles out to a range of 8,000 to 10,000 m and people out to a range of 4,000 m, with the actual ranges depending on the terrain. The radar can be elevated to a height of 1 m and traversed 240° left and right regardless of the turret position.

It is also fitted with the latest TNA-4-6 navigation device, 1T129 co-ordinator and the 1G50 gyrocompass to enable it to determine quickly its own position on the battlefield.

Communications equipment fitted includes R-163-50U, R-163-50K and R-163-10U radios plus a 1V520 electronic computer. Information would normally be collated in the vehicle and then transmitted to the next chain of command.

A 4 kW generator is provided in the rear of the vehicle to allow the electronic equipment to be operated with the main engine shut down.

If required, the vehicle can be fitted with additional reconnaissance equipment to enable an observation post to be established up to 6 km away from the base vehicle.

Standard equipment includes a fire detection and suppression system and an NBC system. The vehicle is also fully amphibious, being propelled in the water by two water-jets mounted one either side at the rear. Before entering the water the trim vane is erected at the front of the vehicle and the bilge pumps activated.

Weapons carried for use by the crew of the BRM-3K include six assault rifles and 15 hand grenades. A 9K113 Konkurs (NATO AT-5 'Spandrel') ATGW system for dismounted use can be carried.

Status

Prototype. May be in low-rate production. There are no known exports of this system.

Manufacturer

Kurgan Machine Construction Plant (prime contractor for chassis).

UPDATED

PRP-4 reconnaissance vehicle

Development

The PRP-4 battlefield reconnaissance vehicle is a further development of the earlier PRP-3 battlefield reconnaissance vehicle which first appeared in 1975. Details of the PRP-3 are given in the entry for the BMP-1 infantry combat vehicle on which the PRP-3 and PRP-4 reconnaissance vehicles are based. The overall prime contractor is the Rubstovskiy Machine Building Plant with the original BMP-1/2 chassis being built at the Kurgan Machine Construction Plant.

The PRP-4 reconnaissance vehicle has been designed to carry out reconnaissance of both stationary and moving targets, including low-flying helicopters, by both day and night, under any weather conditions, at altitudes of up to 3,000 m above sea level and in temperatures ranging from –40 to +40°C.

The system is usually found with self-propelled artillery units, especially those equipped wit 122 mm 2S1 and 152 mm 2S3 self-propelled guns. There has been no recent production of the PRP-4 reconnaissance vehicle. This vehicle is also referred to as the Deyteriy. As far as it is known, there has been no recent production of the PRP-4, although production could commence if significant orders were placed.

Description

The hull of the PRP-4 is virtually identical to the BMP-1 infantry combat vehicle although, as it has a different mission, the layout of the vehicle and roof hatches is very different.

The driver is seated front left with the power pack to his right and another crew member seated to his rear. Both have single-piece hatch covers and associated day observation devices. The driver's hatch lifts and swings to the right while the hatch to the rear of the driver opens forwards.

The two-man power-operated turret is in the centre of the vehicle and is provided with two roof hatches, sensors and a single 7.62 mm PKT machine gun which is aimed and fired under complete armour protection.

Optical devices are mounted on the forward part of the roof and in pods on either side of the turret.

Mounted in the rear of the turret is the IRL-133-1 radar with a flat antenna that folds forwards when not operational. This radar can detect and track MBT-size targets out to a range of 8 to 10 km. When not required, the radar antenna is retracted into the turret under full armour protection.

The IRL-331-1 radar is also referred to as the PSNR-5K (NATO designation Tall Mike) and is also installed in a number of other armoured vehicles such as the older BRM-1K vehicle based on the BMP-1 vehicle and the more recent BRM reconnaissance vehicle based on the BMP-3 ICV.

PRP-4 reconnaissance vehicle from the front with radar system elevated (Gunter Lippert)

The 1PN59 thermal night vision system is mounted on the left side of the turret and has a target detection range of at least 3,000 m or 1,300 m and a target recognition range of at least 2,000 m.

Mounted on the right side of the turret is the 1PN61 infra-red night vision device with a detection range in the active mode of 2,500 to 3,000 m and 1,500 m in the passive mode.

To determine the range to targets, a 1D11M-1 periscopic laser binocular range-finder is also fitted in the right side of the turret. An ID13 laser range-finder is carried inside the vehicle and deployed away from the vehicle when required.

Navigation equipment fitted includes course plotter KP-4, gyro course indicator 1G13 and gyrocompass 1G25-1.

Information processing and communications equipment fitted includes a 1V250 electronic calculator, an R-173 radio system and a 1A30M automatic command receiver.

The fifth crew member is seated in the rear and has a circular roof hatch that opens to the rear. The twin doors in the hull are retained but the vehicle does not have any firing ports.

Specifications

Crew: 5
Combat weight: 13,200 kg
Ground pressure: 0.60 kg/cm²
Length: 6.735 m
Width: 2.94 m
Height: 2.146 m
Ground clearance: 370 mm
Max speed:
(road) 65 km/h
(water) 7 km/h
Range:
(road) 550-600 km
Fording: amphibious
Engine: UTD-20 4-stroke diesel developing 300 hp
Armament: 1 × 7.62 mm PKT MG
NBC system: yes
Night vision equipment: yes

(**Note:** other performance details are identifical to those of the basic BMP-1 which is covered in detail in the following section)

As the PRP-4 has extensive communications equipment and other systems that need additional power, an auxiliary power unit is fitted as standard.

The PRP-4 retains the amphibious characteristics of the BMP-1 and is fitted with an NBC system as standard; it can also lay its own smoke screen by injecting diesel fuel into the exhaust outlet which is located on the right side of the hull.

Variant

Late production models are designated the PRP-4M which is similar to the PRP-4 but has improved 1PN71 thermal imager and a new laser range-finder designated the 1D14. The latest version is designated the PRP-4MU.

Status

Production complete. In service with the Russian Army and some other countries.

Manufacturer

Rubstovskiy Machine Building Plant.

UPDATED

PRP-3 reconnaissance vehicle from rear showing hatch in turret roof for the radar
0088036

BRDM-2 amphibious scout car

Development

The BRDM-2 amphibious scout car was designed by V K Rubtsov of the Dedkov OKB in the early 1960s as the successor to the earlier BRDM-1 and was first seen in public during 1966. Main improvements over the BRDM-1 vehicle are its fully enclosed armament installation, more powerful rear-mounted GAZ-41 petrol engine which gives it a higher road and cross-country performance, and improved amphibious characteristics.

Description

The all-welded steel armour hull of the BRDM-2 provides the crew with protection from small arms fire and shell splinters.

The driver is seated at the front of the hull on the left with the vehicle commander to his right. Both are provided with a bulletproof windscreen to their front which is covered by an armoured shutter, hinged at the top, when the vehicle is in combat areas. When the shutters are in position the driver and commander observe the terrain through day periscopes around the front and sides of their positions, mounted level with the roof of the vehicle. The only means of entry to the BRDM-2 is by two circular roof hatches immediately behind the commander and driver, hinged towards the centre of the vehicle and opening only as far as the vertical.

The turret is mounted in the centre of the vehicle and is the same as that mounted on the Russian BTR-60PB (8 × 8) and Czech OT-64 SKOT-2A (8 × 8) armoured personnel carriers.

There is a single firing port in each side of the hull and immediately behind it are three vision blocks which protrude from the outside of the hull to give some vision to the front and rear of the vehicle.

The engine compartment is at the rear of the hull and there are two air-inlet louvres in the forward part of the engine compartment roof and four smaller air-inlet louvres to the rear. The exhaust pipes are on either side of the hull. The engine compartment is separated from the crew compartment by a bulkhead. The complete drive train of the BRDM-2 comprises the engine, single-plate clutch, manual gearbox, reduction gearbox and transfer case which in turn provides power to the front and rear wheels, each of which have a self-locking differential. The manual transmission is hydraulically assisted with the third and fourth gears being synchronised with lockouts in first, second and reverse gears.

BRDM-2U command vehicle with roof-mounted generator from the rear (Michael Jerchel)
0088039

Polish BRDM-2 R5 command vehicle showing additional radio antenna (Jaroslaw Cislak)
0088040

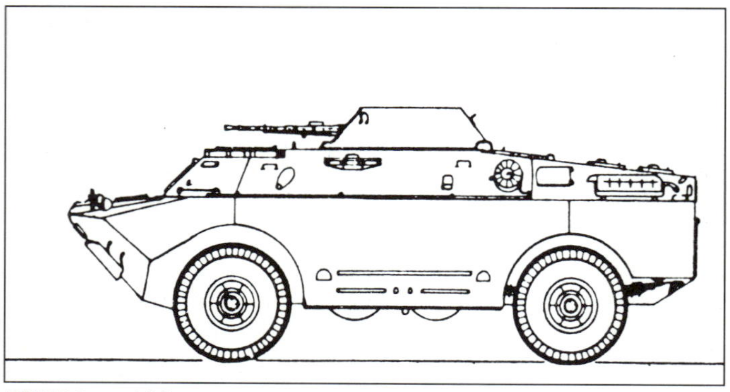

BRDM-2 amphibious scout car

BRDM-2-RKhb (Radiobiological-chemical-biological) reconnaissance vehicle (Michael Jerchel)
2002/0114333

SA-9 'Gaskin' surface-to-air missile system with missiles elevated

Iraqi BRDM-2 (9P148) with two AT-4 (left) and three AT-5 'Spandrel' (right) ATGWs in the ready to launch position (Christopher F Foss) **2000**/0064840

BDRM-2 (4 × 4) with turret removed and additional windows provided for forward observation (T J Gander) **2000**/0064841

Late production BRDM-2 from the rear and on the right a rear ambulance version of the BTR-70 (8 × 8) APC (D Caemerlynck) **2001**/0093540

BRDM-2 reconnaissance vehicle in service with Ukraine (Michael Jerchel) **2002**/0114334

The suspension consists of four semi-elliptical springs with telescopic dual-action shock-absorbers mounted two per axle. Steering is hydraulically assisted on the front two wheels with the sealed brakes having air-assisted hydraulic actuators.

On each side of the vehicle, between the front and rear wheels, are two chain-driven belly wheels, which are lowered by the driver and give the BRDM-2 improved cross-country performance and also allow it to cross ditches. A central tyre-pressure regulation system allows the driver to adjust the tyre pressure to suit the type of ground being crossed. The driver can adjust the pressure on all four tyres or on individual tyres while the vehicle is still moving. If one of the tyres is damaged, the air compressor can maintain the tyre pressure until the vehicle is out of the combat area.

The BRDM-2 is fully amphibious, being propelled in the water by a single water-jet at the rear of the hull. Before entering the water a trim vane, which is stowed under the nose of the hull when travelling, is erected at the front of the hull. When not in use, the water-jet outlet is covered by a triangular plate pivoted at the top. The hydraulic control that activates both the hydrojet drive and the trim vane is located to the front of the driver. When afloat, the driver uses the steering wheel as on land with the water rudders being connected to the steering wheel. At a speed of 6 to 7 km/h, its turning radius is about 10 m.

The vehicle is fitted with an overpressure NBC system and its air inlet is on the top of the hull to the left rear of the turret. Standard equipment includes infra-red driving lights; an infra-red searchlight mounted over the commander's position which can be operated from within the vehicle; a winch mounted internally at the front of the hull; a radio; and a TNA-2 land navigation system. It also has a DP-3B roentgenometer chemical detector, a VPKhR troop reconnaissance device that warns the crew of chemical agents and a heater that provides hot air to the driver's windows so that they do not become frozen in very cold weather. The winch is provided with 30 m of cable and has a maximum load of 4,000 kg.

Armament consists of a 14.5 mm KPVT machine gun with a 7.62 mm PKT machine gun mounted coaxially to the right. A telescopic sight is mounted to the left of the main armament. The day only sight is provided with a wiper. The turret has manual traverse and is provided with an adjustable seat for the gunner.

Variants

BRDM-2-RKhb (radiobiological-chemical-biological reconnaissance vehicle)

This is easily distinguishable from the basic vehicle by two rectangular racks which contain lane-marking poles with pennants on either side of the hull at the rear. The racks are normally covered with a canvas and when required are swung through 90° so that they are vertical over the rear of the vehicle. The poles are used to mark lanes through contaminated areas. There are two versions of this vehicle. The original RKha has the standard 14.5 mm KPVT heavy machine gun while the more usual RKhb has twin 7.62 mm machine guns and improved sensors.

BRDM-2U command vehicles

There are several versions of this specialised command vehicle. The well-known model without a turret is designated the BRDM-2u (also referred to as the BRDM-2U3) and is equipped with a 1 kW generator AB1-P/30 and radios R-105M, R-108M and R-130M.

The BRDM-2uM is an improved version. The BRDM-2U2 is similar but is equipped with a sword-type antenna on the rear decking.

The third version is called the BRDM-2U1 and is distinguished by a large box on the roof and two small boxes/generators at the rear.

BRDM-2K command

There is also a BRDM-2 (command) which retains the turret and is called the BRDM-2K (Kommandnyi). It is equipped with a collapsible telescopic antenna on the right side of the hull and has an additional whip antenna on the left side of the hull.

Polish command vehicles

R-1A is equipped with two R-123Z, a R-107, a generator and fire-control device. The R-1A is used by commanders of anti-tank platoons and batteries.

The R-5 is similar to the R-1A but has different antenna configuration.

Iraqi upgraded BRDM-2

Shown for the first time in Baghdad in 1989 was an upgraded BRDM-2 armoured car. The original 14.5 mm KPVT heavy machine gun was replaced by a 23 mm cannon which has a significant increase in range and armour penetration characteristics over the original weapon. In addition, the elevation of the mount, originally limited to +30°, has been increased to +60° to allow the vehicle to engage slow-moving aircraft and helicopters, with depression of the weapons being -5°. A total of 500 rounds of 23 mm and 2,000 rounds of 7.62 mm ammunition is carried. As far as is known, this upgraded BRDM-2 vehicle did not enter service with the Iraqi Army.

BRDM-2 with 'Sagger' ATGWs

This was first used in combat by the Egyptian and Syrian forces during the 1973 Middle East campaigns.

The base model is called the 9P122 Malyutka and is equipped with a 9S414 sight and 9M14 Malyutka missiles with Manual Command to Line of Sight (MCLOS) guidance.

The 9P122-1 is a slightly improved model while the 9P133 is based on a modified chassis, has a new and bigger sight called the 9S466 and is identified by an additional small window on the front side. This version uses the improved missile 9M14P Malyutka (AT-3c) with Semi-Automatic Command-to-Line Of Sight (SACLOS) guidance.

It is a BRDM-2 with its turret removed and fitted with an arm on top of which are mounted six 'Sagger' AT-3 ATGWs. When travelling, they are within the hull but in action the arm is raised above the top of the hull complete with overhead armour protection. The missiles can be launched from within the vehicle or from up to 80 m away with the aid of a separation sight. When launched from within the vehicle they are guided by the missile operator/vehicle commander who is seated at the front of the vehicle on the right side and is provided with a sight in the forward part of the hull. A total of eight missiles is carried in reserve and the radio antenna is at the rear on the right side rather than at the front. In 1977, it was reported that the 'Sagger' was being converted from its normal wire command guidance system to a semi-automatic infra-red wire guidance system, which would increase the effectiveness of the missile at its maximum range of 3,000 m.

This version is probably no longer in front-line service with the Russian armies.

BRDM-2 with 'Swatter-C' ATGWs

This was first seen in 1973 and is similar in appearance to the BRDM-2 with the 'Sagger' ATGW but has a quadruple launcher for the AT-2 ATGW converted from its original radio command to line of sight guidance to semi-active infra-red/command guidance. This missile weighs 29.48 kg and has a range of 3,500 m compared with the 3,000 m of the original 'Swatter'. A total of eight 'Swatters' is carried including three in the ready to launch position. As far as is known, the improved 'Swatter' has not been fitted to the original BRDM-1 installation. This version is probably no longer in front-line service with the Russian Federation and Associated States (CIS). The official designation for this system is the 9P124.

BRDM-2 with 'Spandrel' ATGWs

The base model is called the 9P137 Konkurs and can only use the 9M113 or 9M113M (AT-5 Spandrel) missile. In addition to the five missiles in the ready to launch position another nine can be carried internally.

Shown for the first time at the military parade held in Moscow in November 1977 was a BRDM-2 fitted with a turret on top of which were mounted five ready to launch missiles which have been given the NATO codename of 'Spandrel', or AT-5. This vehicle is sometimes referred to as the BRDM-3. To the rear of the turret in the top of the hull is a hatch which enables the launcher to be reloaded without the crew leaving the vehicle. It is estimated that in addition to the five visible missiles a further 10 are carried inside the hull. The missile is of the second-generation type (semi-automatic, command to line of sight) and all that the operator has to do is to keep the cross-hairs of his sight on the target to ensure a hit. The sight is in the roof of the vehicle on the right side and can be traversed through 180°.

BRDM-2 with launcher for 'Sagger' ATGWs raised (Jaroslaw Cislak) 0088038

Specifications

Crew	4
Configuration	4 × 4
Combat weight	7,000 kg
Power-to-weight ratio	20 hp/t
Length	5.75 m
Width	2.35 m
Height	
(overall)	2.31 m
Firing height	2.13 m
Ground clearance	0.43 m
Track	1.84 m
Wheelbase	3.1 m
Max speed	
(road)	100 km/h
(water)	10 km/h
Fuel capacity	290 litres
Max road range	750 km
Road fuel consumption	0.35-0.45 litres/km
Fording	amphibious
Gradient	60%
Vertical obstacle	0.4 m
Trench	1.25 m
Engine	GAZ-41 V-8 water-cooled petrol developing 140 hp at 3,400 rpm
Transmission	manual with 4 forward and 1 reverse gears
Suspension	semi-elliptical springs with hydraulic shock-absorbers
Tyres	13.00 × 18
Electrical system	24 V
Turning radius	9 m
Armament	
(main)	1 × 14.5 mm KPVT MG
(coaxial)	1 × 7.62 mm PKT MG
Ammunition	
(main)	500
(coaxial)	2,000
Gun control equipment	
Turret power control	manual
Gun elevation/depression	+30°/-5°
Armour (max)	
Hull front upper	5 mm at 80°
Hull front lower	7 mm at 45°
Hull nose plate	14 mm at 14°
Hull sides upper	7 mm at 18°
Hull sides lower	7 mm at 0°
Hull rear upper	7 mm at 0°
Hull rear lower	7 mm at 0°
Hull top	7 mm at 0°
Belly front	2 mm
Belly rear	3 mm
Turret front	7 mm at 43°
Turret sides	7 mm at 36°
Turret rear	7 mm at 36°
Turret top	7 mm
NBC system	yes
Night vision equipment	yes

The HEAT warhead weighs about 7 kg and will penetrate between 500 and 600 mm of conventional steel armour at an incidence of 0°. The launch tube has a blow-out cap at the front and is flared at the rear. Basic specifications of the AT-5 'Spandrel' ATGW are as follows:
Length: 1.3 m
Diameter: 155 mm
Max range: 4,000 m
Min range: 100 m
Max flight time: approx 20 s

On some BRDM-2 'Spandrel' ATGW vehicles, for example those used by Iraq, the outer launch position has been fitted with the AT-4 'Spigot' ATGW. This arrangement allows the carriage of an increased missile load. This version is designated the 9P148 and can carry and launch the smaller 9M111 or 9M111-2 (AT-4 Spigot). A normal mix would be 10 9M111 and 10 9M113. From 1996 some vehicles were observed with a thermal sight designated the 1PN66.

BRDM-2 with loudspeaker

This is called the BRDM-2ZS (Zvukovetsatelnaya stantsiya) and is used for PSYCOPS and is also referred to as the BRDM-2 M1986/2.

BRDM-2 with loudspeaker

This is a local modification for psychological warfare operations.

SA-9 'Gaskin' System

The SA-9 'Gaskin' mobile surface-to-air missile system is based on a modified BRDM-2 (4 × 4) amphibious chassis with its belly wheels removed. The system is designated the 3P31M Strela 1. The original turret has been replaced by a new one-man turret with an elevating arm on either side of which are mounted two box-type launchers for the SA-9 fire-and-forget missiles.

The SA-9 first entered service in 1968 and has since been widely deployed. In Russia and some other countries it has now replaced by the SA-13 'Gopher' SAM system on a modified MT-LB tracked chassis.

To reduce the overall height of the system for travelling, the missiles are normally lowered into the horizontal position either side of the vehicle. No reserve missiles are carried on the SA-9 system.

Full details of the SA-9 'Gaskin' are given in *Jane's Land-Based Air Defence 2001-2002.*

BRDM-2 ATM-1 Transport Vehicle

With the reduction in size of the RFAS armies, a modified BRDM-2 is now being offered for the civilian market under the designation of the ATM-1 Ingul Universal Emergency Transport Vehicle.

This is essentially the BRDM-2 with its turret removed and on top of the hull is a stowage bed that can take a maximum load of 500 kg. It is also fitted with welding equipment, an acetylene generator and an oxygen tank. It normally has a crew of four and is provided with additional windows and side doors.

Czech Republic BRDM-2 Upgrade

Details of this upgrade carried out by the VOP 026 overhaul facility are given earlier in this section under the Czech Republic. As of early 2001 there were two prototypes of this upgrade for Czech Army trials.

Status

Production complete. In service with the following countries:

Country	Quantity	Comment
Afghanistan	250	includes BRDM-1
Algeria	75	plus SA-9 SAM and ATGW (used by army, gendarmerie and national guard)
Angola	50	plus SA-9 SAM
Benin	14	plus SA-9
Bulgaria	58	includes BRDM-1, plus SA-9 and ATGW
Burundi	30	
Cape Verde	10	
Congo	25	includes BRDM-1
Croatia	7	
Cuba	100	includes BRDM-1, plus SA-9
Czech Republic	150	uses SA-9
Egypt	300	includes BRDM-1, also BRDM-2 ATGW and SA-9
Equatorial Guinea	6	
Eritrea	6	
Estonia	7	
Ethiopia	n/avail	status uncertain
Gaza/Jericho	40	
Guinea	25	includes BRDM-1
Guinea-Bissau	10	plus SA-9 SAM
Hungary	12	uses SA-9
India	110	plus SA-9 SAM
Indonesia	14	marines
Iraq	400	number unconfirmed, also BRDM-2 ATGW and SA-9
Israel	n/avail	internal security role
Kazakhstan	140	
Kyrgyzstan	34	
Latvia	2	
Libya	250	also BRDM-2 ATGW and SA-9
Lithuania	11	
Macedonia	10	
Madagascar	35	plus SA-9 SAM
Mali	20	plus SA-9 SAM
Mauritania	nil	SA-9 SAM only
Mongolia	120	
Mozambique	30	includes BRDM-1, also SA-9 SAM
Namibia	n/avail	
Nicaragua	20	also BRDM-2 ATGW and SA-9
Peru	30	
Poland	510	also BRDM-2 ATGW and SA-9. Some BRDM-2 being upgraded
Romania	121	also BRDM-2 ATGW and SA-9
Russia	2,000	includes BRDM-1, plus SA-9 and ATGW
Seychelles	6	
Slovakia	129	uses BRDM-2 ATGW and SA-9
Slovenia	16	
Somalia	n/avail	status uncertain
Sudan	60	includes BRDM-1
Syria	700	also BRDM-2 and SA-9
Tanzania	40	also SA-9
Turkmenistan	14	
Ukraine	1,000	estimate
Uzbekistan	35	
Vietnam	100	includes BRDM-1
Yemen	80	number unconfirmed
Yugoslavia	86	also BRDM-2 and SA-9
Zambia	88	includes BRDM-1

Polish BRDM-2 upgrade

Details of this are given earlier in this section under Poland.

Manufacturer

Molotov GAZ plant.

UPDATED

BRDM-1 amphibious scout car

Development

The BRDM-1 amphibious scout car first appeared in the Spring of 1959 and was designed by V K Rubtsov of the Dedkov OKB who later designed the more capable BRDM-2 (4 × 4) reconnaissance vehicle covered in the previous entry.

The first prototype was completed in February 1956 and went to the Black Sea area for official government trials. It was accepted for service with the Soviet Army the following year and production began late in 1957.

In Russian units the BRDM-1 has now been replaced by the BRDM-2, which has a higher road speed and is fitted with a one person manually operated turret, armed with a 14.5 mm and a 7.62 mm machine gun rather than weapons on unprotected mounts.

BRDM-1 amphibious scout car with commander's and driver's hatches open
0088042

Specifications
Crew: 5
Configuration: 4 × 4
Unloaded weight: 5,600 kg
Power-to-weight ratio: 16.07 hp/t
Length: 5.7 m
Width: 2.25 m
Height: 1.9 m
Ground clearance: 0.315 m
Track: 1.6 m
Wheelbase: 2.8 m
Max speed:
(road) 80 km/h
(water) 9 km/h
Fuel capacity: 150 litres
Max road cruising range: 500 km
Fuel consumption: 0.3 litres/km
Fording: amphibious
Gradient: 60%
Vertical obstacle: 0.4 m
Trench: 1.22 m
Engine: GAZ-40P 6-cylinder in-line petrol, water-cooled developing 90 hp at 3,400 rpm
Transmission: manual with 4 forward and 1 reverse gears
Transfer case: 2 speed
Clutch: single dry plate for all models
Suspension: semi-elliptical springs and hydraulic shock-absorbers
Tyres: 12.00 × 18
Electrical system: 24 V
Main armament: 1 × 7.62 mm SGMB MG
Ammunition: (7.62 mm) 1,250
Gun control equipment: manual
Armour (max): 10 mm
NBC system: yes
Night vision equipment: yes

The BRDM Model 1957 has an open roof and the BRDM-1 Model 1958 has a closed roof with twin hatches, this being the standard production model.

Total production of the BRDM-1 is believed to have amounted to around 10,000 vehicles. In most countries the BRDM-1 is now held in reserve or used by second-line units. There are no known upgrades of this vehicle and by today's standards it must be considered obsolete.

Description

The hull of the BRDM-1 (4 × 4) amphibious scout car is made of all-welded steel with the engine at the front and the crew compartment at the rear. The driver is seated at the front of the vehicle on the left with the commander to his right. Both are provided with a hatch to their front that is hinged at the top and opens forwards, and a windscreen with a wiper blade. The hatch cover has a vision block for use when closed. A vision slit protected by a vision block is provided to the left of the driver and right of the commander. There are two firing ports in each side of the hull and two large hatches in the forward part of the roof that open to the rear. The rear of the crew compartment slopes at an angle of about 30° and is fitted with a two-piece hatch that opens either side of the superstructure. There is a firing port in each hatch.

On either side of the vehicle, between the front and rear wheels, are two belly wheels which are lowered by the driver and give the BRDM-1 improved cross-country performance and also allow it to cross ditches. The BRDM-1 is fully amphibious, being propelled in the water by a single water-jet at the rear of the hull. Before entering the water a trim board, which is stowed folded under the nose of the vehicle when not in use, is erected at the front of the hull. Standard equipment includes a system that allows the driver to adjust the tyre pressures to suit the type of ground being crossed. The vehicle is not fitted with an NBC system and when introduced was not fitted wit any night vision equipment. A white light searchlight mounted on the left side of the hull is operated by the driver. A communications reel and wire is mounted on the left side of the superstructure towards the rear.

The vehicle is normally armed with a 7.62 mm SGMB machine gun pintle mounted on the forward part of the roof, which has a limited traverse of 45° left and right and can be elevated from -6 to +23.5°. Some vehicles are fitted with a 12.7 mm DShKM heavy machine gun mounted on the forward part of the roof with a 7.62 mm SGMB machine gun mounted at the rear.

Variants

Command vehicle (BRDM-U)

This can be distinguished from the basic vehicle as it has a radio antenna on each side of the hull and two radio antennas at the rear, whereas the basic vehicle only has a single radio antenna on the right side of the hull.

BRDM-1-RKhb (radiobiological-chemical-biological reconnaissance vehicle)

This is easily distinguishable from the basic vehicle as on either side of the hull at the rear of the vehicle are two rectangular racks which contain lane marking poles with pennants. The racks are normally covered with a canvas cover and when required are swung through 90° so that they are vertical over the rear of the vehicle. The poles are used to mark lanes through contaminated areas.

BRDM-1 with ATGWs

As far as is known, the following versions of the BRDM-1 with ATGWs are no longer in service. Full details were given in *Jane's Armour and Artillery 1990-91,* page 252:
BRDM-1 with three 'Snapper' ATGWs
BRDM-1 with four 'Swatter' ATGWs
BRDM-1 with six 'Sagger' ATGWs.

Status

Production complete.

Country	Quantity	Comment
Afghanistan	250	includes BRDM-2
Albania	15	all BRDM-1
Bulgaria	58	includes BRDM-2
Congo	25	includes BRDM-2
Cuba	100	includes BRDM-2
Kazakhstan	140	includes BRDM-2
Guinea	25	includes BRDM-2
Mozambique	30	includes BRDM-2
Russia	2,000	includes BRDM-2
Sudan	60	includes BRDM-2
Vietnam	100	includes BRDM-2
Zambia	88	includes BRDM-2

Note: In most of the above cases the BRDM-1s are in reserve

Manufacturer

Molotov GAZ plant.

UPDATED

GAZ-39344 SIAM reconnaissance vehicle

Development

Late in 1994, it was revealed that the Arzamas Machinery Plant had developed and placed in production a new 4 × 4 special-purpose vehicle called the GAZ-39344, or SIAM as it is also known.

This has been developed to carry out a variety of missions such as border patrol, reconnaissance and the protection of high-risk areas such as airports and fuel installations.

As far as is known the SIAM is not used by the Russian Army. As of mid-2001 there are no known exports of the GAZ-39344 SIAM reconnaissance vehicle.

Previously the Arzamas Machinery Plant was part of the Molotov GAZ Plant which also built the BRDM-2 (4 × 4) and BRDM-1 (4 × 4) amphibious reconnaissance vehicles covered in earlier entries.

Description

The hull of the GAZ-39344 is made of all-welded steel armour construction that provides the occupants with protection from small arms fire and shell splinters. Bulletproof windows are provided in the front and sides of the hull.

The vehicle can carry a total of six personnel, driver seated at the front on the left, vehicle commander to his right rear and four men in the back on individual seats. The last are seated two either side of each side door facing each other.

In either side of the hull is a large forward-opening door which allows the occupants to enter and leave the vehicle rapidly. Each of these doors has a large bulletproof window and is claimed to have a sophisticated locking system to prevent unauthorised entry. The right door also has a firing port.

There are also two firing ports in the left side of the hull, one either side of the door, and one in the right side of the hull, which allow the occupants to fire their small arms from inside the hull with a high degree of safety.

Mounted on the roof of the vehicle is a one-man manually operated turret which is the same as that fitted to the original BTR-80 (8 × 8) APC and armed with one 14.5 mm KPVT and one 7.62 mm PKT machine gun. A white light/infra-red searchlight is mounted over the weapons and moves in elevation with the weapons.

Arzamas Machinery Plant GAZ-39344 (4 × 4) vehicle clearly showing the bulletproof windows which give observation to the front of the vehicle and with side door open (Christopher F Foss) 0004946

Specifications

Crew: 2 + 4
Configuration: 4 × 4
Max weight: 7,000 kg
Empty weight: 5,000 kg
Payload: 2,000 kg
Length: 5.64 m
Width: 2.32 m
Height with turret: 2.65 m
Ground clearance: 0.40 m
Track: 1.74 m
Wheelbase: 2.84 m
Angle of approach/departure: 24.5°/30.5°
Turning radius: 9.5 m
Max road speed: 95 km/h
Range: 700 km
Engine: diesel developing 125 hp
Suspension: independent
Electrical system: 24 V
Armament:
(main) 1 × 14.5 mm KPVT MG
(coaxial) 1 × 7.62 mm PKT MG
Ammunition:
(main): 500
(coaxial) 2000
Gun control equipment
Turret power control: manual
Gun elevation/depression: +30°/–5°
Turret traverse: 360°
Gun stabiliser
(vertical) no
(horizontal) no
NBC system: no
Night vision equipment: yes

Turret traverse is manual through a full 360° while weapon elevation is from –5 to +60°. Like most Russian turrets of this type, no hatch cover is provided.

The engine compartment is located at the rear of the SIAM and separated from the crew compartment by a bulkhead.

Standard equipment includes a central tyre-pressure regulation system that allows the driver to adjust the tyre pressure to suit the type of ground being crossed, air conditioning system, navigation system, vehicle intercom system, communications equipment, loudspeaker and blue flashing lights.

The GAZ-39344 is understood not to be a replacement for the BRDM-2 but is aimed more at the paramilitary and internal security market.

Variants

Other versions of the GAZ-39344 are also available including a bullion transport vehicle.

Status

In low-rate production. In service in the civil/paramilitary sector. As far as is known it is not in service with the Russian Army.

Manufacturer

Arzamas Engineering Plant.

UPDATED

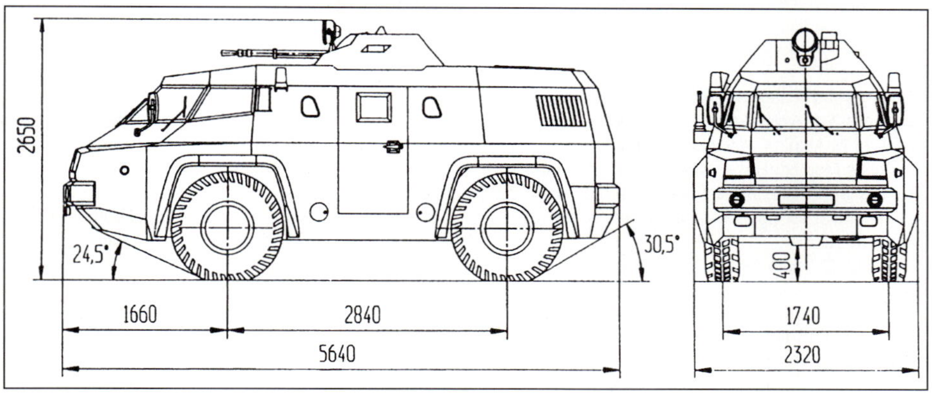

General arrangement drawing of the GAZ-39344 (4 × 4) vehicle

SLOVAKIA

Aligator light armoured reconnaissance vehicle

Development

In 1995, development of the Aligator (4 × 4) light armoured reconnaissance vehicle commenced, with the first of three prototypes being completed in November 1996.

Funding for the development and construction of the prototype vehicles has been provided by Kerametal Company Limited, with the Transmisie Company building the prototype vehicles.

It is envisaged that production of the Aligator vehicle will be undertaken by ZTS Tees of Slovakia which already has considerable experience in the design, development and production of armoured vehicles.

The Aligator has been designed to undertake a wide range of roles on the battlefield including reconnaissance, command and control.

To reduce development and procurement costs, wherever possible the Aligator uses proven commercial components such as engine, transmission and suspension.

The first prototype of the Aligator light armoured reconnaissance vehicle made its first public appearance early in 1997.

Early in 2001 it was revealed that DMD Mobiltec would complete the first of eight production Aligator (4 × 4) scout cars for the Slovakian Army late in 2001, with deliveries to be completed in 2002. The Slovakian Army is the launch customer for the Aligator.

The first Slovakian Army Aligator vehicles will be in the mobile command post (MCP) configuration and have an applique ceramic passive armour package that provides protection from 7.62 mm × 51 armour-piercing attack.

Seating will be provided for four people, commander, driver and two others, and the vehicles will be fully amphibious, being propelled in the water by a propeller mounted under the hull rear.

Description

The hull of the Aligator is of all-welded steel armour, which provides the occupants with protection from 7.62 mm armour-piercing small arms fire and shell splinters.

The floor of the crew compartment is double thickness to provide increased protection against mines.

The engine compartment is at the front with the crew compartment at the rear. The standard commercial Deutz water-cooled and intercooled diesel is coupled to a Renk automatic transmission with manual override.

The commander is seated on the right with the driver on the left. The driver has a single-piece hatch cover above his position with a roof-mounted day periscope for forward observation.

Vehicle commander has a circular low-profile cupola

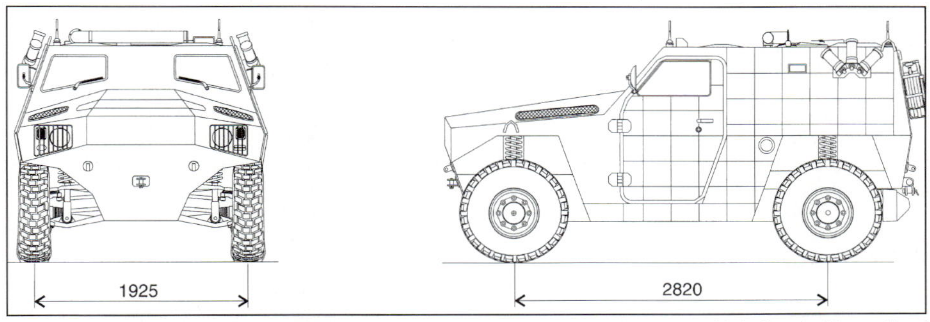

Provisional front and side drawings of Aligator light armoured reconnaissance vehicle fitted with appliqué armour

Specifications

Provisional
Crew: 1 + 5
Configuration: 4 × 4
Power-to-weight ratio: 28.20 hp/t
Weight loaded: 6,700 kg
Weight empty: 6,000 kg
Length: 4.40 m
Width: 2.20 m
Height: 1.95 m
Ground clearance: 385 mm
Track: 1.925 m
Wheelbase: 2.82 m
Max road speed: 120 km/h
Fuel capacity: 160 litres
Maximum range: 600 km
Fording: n/avail (an amphibious version is available)
Gradient: 60%
Side slope: 40%
Vertical obstacle: 0.4 m
Trench: 0.8 m
Turning radius: 6 m
Engine: Deutz BF6 1013 water-cooled turbocharged and intercooled diesel developing 189 hp at 2,300 rpm
Transmission: Renk 606 automatic powershift with 6 forward and 1 reverse gears, primary retarder
Steering: power assisted
Tyres: Michelin, with run-flat core 11.00 R 16 XL
Suspension: independent
Brakes:
(main) disc
(parking) internal shoe
Electrical system: 24 V
Batteries: 2 × 24 V, 180 Ah
Armament: 1 × 7.62 mm or 12.7 mm MG
Smoke grenade launchers: 2 × 3 (optional)
NBC system: optional
Night vision equipment: optional

with a circular hatch cover that opens towards the rear and on the forward part of this can be mounted a 7.62 mm or 12.7 mm machine gun.

The commander and driver are each provided with a large bulletproof window to their front and a door in the side that opens forwards. In the upper part of this is a large bulletproof window. All the bulletproof windows provide the same level of protection as the all-welded steel hull.

Other roof hatch arrangements are possible, including two single forward opening roof hatches for the commander and driver at the front and a single rear opening roof hatch to the rear of the troop compartment.

The troop compartment is at the rear and normal means of entry and exit is via a larger door that opens to the left and is provided with a firing port and vision device.

In either side of the troop compartment is a firing port and vision device, and in the rear part of the roof is a single-piece hatch cover that opens towards the rear. If required, a bank of three electrically operated smoke dischargers can be mounted either side of the hull towards the rear.

Standard equipment on production Aligators includes communications system, run-flat tyres, powered steering, fire detection and suppression system, two handheld fire extinguishers, hull-mounted searchlights that can be controlled from within the hull by the driver or commander and 2 × 3 barrelled electrically operated smoke grenade launchers.

Optional equipment includes an amphibious kit which allows the vehicle to propel itself in the water using a propeller mounted at the rear of the hull; air conditioning system; NBC system; central tyre-pressure regulation system which will allow the driver to adjust the tyre pressure to suit the terrain being crossed; recovery winch; various communications equipment; various types of day and night observation equipment for the commander, and driver; global positioning system, auxiliary generator; and different weapon installations.

Variants

Observation and Reconnaissance Post (ORP)

The vehicle can be fitted with the MOWAG MBK2 one-person observation turret armed with a 7.62 mm machine gun and FORTIS (Forward Observation and Reconnaissance Thermal Imaging System) which operates in the 8-12 μm band.

This version of the Aligator is called the Observation and Reconnaissance Post (ORP). The MOWAG MBK2 turret has been produced in large quantities for Denmark and Switzerland for installation on the MOWAG Eagle (4 × 4) reconnaissance vehicle.

One of the prototypes of the Aligator light armoured reconnaissance vehicle (T J Gander) ***2000***/0064843

Police Modification (PCM)

Early in 2001 negotiations were under way for the supply of 21 Aligator vehicles for the Slovakian police in the Police Modification (PCM) version. These will have a slightly higher roof line for greater internal volume and have seats for six people: commander, driver and four people in the rear.

These PCM Aligator vehicles will have an applique steel armour package that will provide protection from 7.62 mm × 54 armour-piercing attack and be fitted with firing ports and associated vision devices in the hull sides and rear.

Mobile Command Post (MCP)

This has a crew of two, driver and commander, plus two communications operators for the extensive communications equipment installed.

Peace Support Operations (PSO)

This has been developed to meet the requirements of units deployed on police support operations under the auspices of the United Nations or other international security organisations. It can be fitted with various weapons and observation equipment.

Enhanced Aligator

To further enhance the operational capabilities of the Aligator, it was revealed early in 2001 that an international consortium had been established.

In addition to DSD Mobiltec this also includes the German companies of STN ATLAS Elektronik and Rheinmetall Landsysteme. As a first step, one of the Aligator prototypes has been modified to accept the STN ATLAS Elektronik BAA observation and sighting system and a FaKoM command and control system. The BAA is already in production for a number of other vehicles and will be installed on production versions of the German/Netherlands Fennek (4 × 4) reconnaissance platform.

The recently formed Rheinmetall Landsysteme markets a wide range of tracked and wheeled armoured vehicles and the incorporation of the Aligator, or an enhanced version of this vehicle, would broaden its product range even further.

Many of the automotive components of the Aligator, such as the engine (Deutz BF6M 1013 diesel developing 141 kW), transmission (Renk Reco 606) and power steering (ZF) are already supplied by Germany.

Status

Production for Slovakian Army, which ordered a first batch of eight vehicles for delivery between 2001 and 2002.

Manufacturers

Kerametal Ltd (see text).
DMD Mobiltec.

UPDATED

SOUTH AFRICA

Vickers OMC Rooikat 105 mm armoured car

Development

While the Vickers OMC Rooikat armed with the 76 mm gun (fully covered in a separate entry) meets the requirements of the South African National Army, for the export market a Rooikat armed with a 105 mm rifled tank gun has been developed since 1990.

So far two prototypes of this have been built as a private venture with Vickers OMC (which was previously called Reumech OMC and prior to that Reumech Defence OMC) being responsible for the chassis and LIW, part of the Denel Group, for the complete turret and armament.

Vickers OMC is the prime contractor for the system as well as being responsible for international marketing.

Development of the 105 mm Rooikat primary stabilised system was completed in 1997 and production can commence as soon as firm orders are placed. The Rooikat 105 mm has been demonstrated in Abu Dhabi and Malaysia.

The complete turret is now being offered by LIW for installation on other chassis, tracked and wheeled. For trials purposes it has already been fitted to the Austrian Steyr-Daimler-Puch ASCOD tracked chassis and

Head on view of the Rooikat 105 mm turret with commander's and gunner's hatches in open position (Christopher F Foss) 0088043

Specifications
Crew: 4
Configuration: 8 × 8
Combat weight: 28,500 kg
Power-to-weight ratio: 20 hp/t
Length:
(with 105 mm gun) 8.80 m
(hull) 7.1 m
Width: 2.9 m
Height:
(overall) 2.8 m
(turret roof) 2.6 m
(hull top) 1.76 m
Ground clearance: 0.49 m
Track: 2.415 m
Wheelbase: 1.55 + 2.032 + 1.625 m
Angle of approach/departure: 45°/60°
Max road speed: 120 km/h
Cross-country speed: 50 km/h
Acceleration: 0 to 30 km/h in 9 s
Fuel capacity: 540 litres
Max road range: 850 km
Fording: 1.5 m
Gradient: 70% (forward and reverse)
Side slope: 30%
Vertical obstacle: 1 m (soil); 0.6 m (concrete)
Trench:
(crawl speed) 2 m
(60 km/h) 1 m
Engine: V-10 water-cooled turbocharged diesel developing 563 hp
Transmission: automatic, 6 forward and 1 reverse gears
Steering: power assisted, front four wheels
Turning radius:
(tarred surface) 25 m
(tactical, using brake steering) 15 m
Suspension: internally driven trailing arms, coil springs and shock-absorbers
Tyres: 16.00 × R20 radial with run-flat inserts
Brakes: dual-circuit drums on all eight wheels, retarder on transmission, exhaust brake, crawl brake, park brake
Electrical system: 24 V
Batteries: 6 (in banks)
Armament:
(main) 1 × 105 mm GT7 high-pressure rifled gun
(coaxial) 1 × 7.62 mm MG
(anti-aircraft) 1 × 7.62 mm MG
Smoke-laying equipment: 2 × 4 81 mm smoke grenade launchers, exhaust smoke screen generator
Ammunition:
(main) 32 (9 ready use)
(7.62 mm) 3,200
(smoke grenades) 8
Turret power control: electric/manual
Traverse: 360°
Gun elevation/depression: +17°/–8°
Gun stabiliser: yes
Laser rangefinder: yes
Commander's sight magnification: selectable ×3, ×6 and ×9
Gunner's sight magnification:
(day) ×8
(night) ×8
NBC system: optional
Night vision equipment: yes

details of this version are given in the Light tanks section under International.

Description
The all-welded steel hull and turret are virtually identical to those of the 76 mm Rooikat, although the chassis has been fitted with larger tyres. If required, a central tyre-pressure regulation system can be fitted which allows the driver to adjust the tyre pressure to suit the terrain being crossed.

The turret layout is similar to that of the 76 mm Rooikat with the commander being seated on the right side of the turret with the gunner forward and below his position and the loader on the left.

The commander is provided with eight day vision blocks to give observation through a full 360°, and mounted in the roof forward of the commander's station is the C560 day panoramic sight with magnifications of ×3, ×6 and ×9. This enables him to observe the terrain through 360° independently of the gunner.

Once the commander locates a target this is handed over to the gunner so freeing the commander to acquire new targets. During an emergency, the commander can directly engage targets by using his panoramic sight, overriding gun controls and the firing switch.

The gunner has a roof-mounted sight which has ×8 day and night (image intensifying) channels and an integral laser range-finder. He also has an auxiliary day sight as used in the Vickers OMC Ratel 90 (6 × 6) vehicle.

The loader has a single-piece hatch cover which opens to the rear and he is also provided with day periscopes to allow observation to the front and left.

The three-man power-operated turret is fitted with an LIW 105 mm GT7 52 calibre rifled tank gun fitted with a vertical sliding semi-automatic breech block with semi-automatic or manual ejection, concentric glass fibre fume extractor and a thermal sleeve.

The barrel is autofrettaged for enhanced service life. The recoil system is of a unique compact concentric hydropneumatic design. The core of this system is a hydropneumatic recuperator which uses a bladder-type accumulator and a barrel sleeve with different diameters fore and aft of the recoil piston.

The accumulator is precharged to a pressure that keeps the gun in battery at any elevation. On firing, the recoil causes the barrel sleeve to displace oil into the accumulator, compressing the gas in the bladder. That pressurised gas then provides the energy to push the gun back into battery.

The GT7 gun fires standard NATO full-pressure 105 mm ammunition types including APFSDS-T. The gun has a total recoil length of 600 mm.

According to the manufacturer, effective range of the 105 mm GT7 gun firing different natures of ammunition is as follows:

APFSDS-T	4,000 m
HEAT	2,000 m+
HESH	2,000 m+
HE-T	4,000 m

Rooikat 105 mm armoured car **2000**/0064844

A 7.62 mm machine gun is mounted coaxial to the left of the main armament and a similar weapon is mounted at the commander's position for use in the ground and air defence roles. This weapon is skate-mounted so that it can be rapidly moved around the commander's hatch.

In the left side of the turret is an ammunition resupply hatch and mounted on either side of the turret rear is a bank of four forward-firing electrically operated 81 mm smoke grenade dischargers.

The gun drives consist of low-voltage solid-state power electronics. The 105 mm gun is stabilised in two axes and the fire-control system is configured in a network to facilitate up- or downgrading.

The standard fire-control system is configured for the GS60 primary stabilised gunner's sight and the CS60 primary stabilised panoramic commander's sight.

The GS35 slaved gunner's sight is offered as a lower-cost/lower-performance option. A thermal imaging elbow is standard on the GS60 gunner's sight with an image intensifying night capability as a lower-cost option.

The CS60 commander's sight provides the crew with a hunter-killer capability and the two primary stabilised sights extend the system envelope to a fire on the move and rough terrain capability.

The digital fire-control system receives information from a laser range-finder and a number of sensors including cant and windspeed. Ammunition type is set manually. Rooikat can engage targets while it is moving across rough terrain.

Ballistic offsets are automatically computed and implemented when the laser range-finder is triggered. The offsets are implemented both in the sight (electronics sight slaving) and the main weapon (auto lay) and a ready to fire indication appears in the gunner's field. Total reaction time from lasing to firing is typically less than two seconds.

The fire-control system provides automatic fire control by computing and implementing the ballistic offset for the ammunition type selected by taking into account target range (from laser range-finder), target speed (automatically derived by tracking the target), manually entered environmental data (air pressure, ambient temperature, ammunition/temperature, headwind) crosswind speed (automatically sensed), weapon tilt (automatically sensed) and gun jump characteristics of the main weapon.

The integrated fire-control system has built-in test equipment which enables any fault to be immediately traced and rectified through the substitution of line-replaceable units.

Variants
The Rooikat 105 mm is offered with three different types of Fire Direction System (FDS) which are described as low-, medium- and high-work performance systems. The first prototype of the Rooikat 105 mm had the medium-performance system fitted.

The primary stabilised system, comprising the high-performance FDS with thermal imaging night sights included, is fully qualified and ready for production.

The modular design of the system enables the manufacturer to up- or downgrade the system with ease and reliability according to prospective clients' choice.

The three FDS sophistication levels can be summarised as:

(1) Low-work performance comprising electromechanical gun control with stabilised main weapon. In addition, a gunner's sight is fitted with a day/night laser channel which is mechanically slaved to the main weapon.
(2) Medium-work performance comprising all the elements of the low-performance system but with the addition of a fire-control system which will automatically compute and implement the ballistic offsets.
(3) High-work performance comprising all the elements of the medium-performance system but with the addition of a primary stabilised gunner's sight.

Apart from the weapon system changes, other improvements have been made to enhance operations in desert and tropical conditions. These include the installation of a more powerful air conditioning unit and detailed efficiency improvements to the engine cooling system.

Options include:

(1) Commander's stabilised or non-stabilised panoramic sight
(2) Gunner's sight mechanically slaved to the 105 mm GT7 gun with either image intensification or thermal night fighting capability
(3) Choice of communications equipment
(4) Navigation system
(5) A choice in terms of fire directing sensors including crosswind, omnidirectional wind/barometric/ambient temperature and static or dynamic cant sensors
(6) Different coaxial machine gun
(7) BC protection system
(8) Explosion detection and suppression system
(9) Battle management system.

Rooikat 120 mm

It is understood that LIW, prime contractor for the Rooikat three-person turret is studying a 120 mm model.

Status

Development complete. Ready for production.

Manufacturer

Systems supplier – Vickers OMC.

UPDATED

Vickers OMC Rooikat 76 mm armoured car

Development

Development of the Rooikat armoured car commenced in 1976 when three 8 × 8 vehicles were built to verify platform concepts. This was followed by a further three vehicles which were built and evaluated by the South African Defence Force in the early 1980s.

These experimental development models (XDMs) consisted of (6 × 6) 22 tonne 76 mm, (8 × 8) 24 tonne 76 mm and (8 × 8) 32 tonne 105 mm armed concepts. The Rooikat was eventually developed from the (8 × 8) 24 ton 76 mm concept.

Full-scale development commenced in late 1985 with the vehicle first being unveiled in October 1988 as the Rooikat.

Five Advanced Development Models (ADMs) were completed in 1986/87 for extensive operational and technical evaluation trials. In 1987, work started on building four Engineering Development Models (EDMs) for user evaluation trials and also to confirm modifications that resulted from the troop trials with the Advanced Development Models.

First Rooikat vehicles were completed in 1989 with the vehicle becoming operational with the then South African Defence Force in 1990.

Vickers OMC (previously Reumech OMC) is responsible for integration of the weapon platform and the turret, and is also the prime contractor for the system which includes responsibility for the international marketing of the system. The three-person all-welded steel turret is supplied by LIW.

The main role of Rooikat is combat reconnaissance during high-mobility operations using its speed to outmanoeuvre the enemy and strike at his flanks and deep in his rear areas. Its secondary role is that of hunter/killer against a variety of battlefield targets including AFVs, with its third role being interdiction.

It is understood that production of the Rooikat 76 mm is now complete with about 240 vehicles being built for the South African Armoured Corps.

By mid-2001, there were no export sales of the 76 mm Rooikat and marketing was being concentrated on the 105 mm version covered in the previous entry.

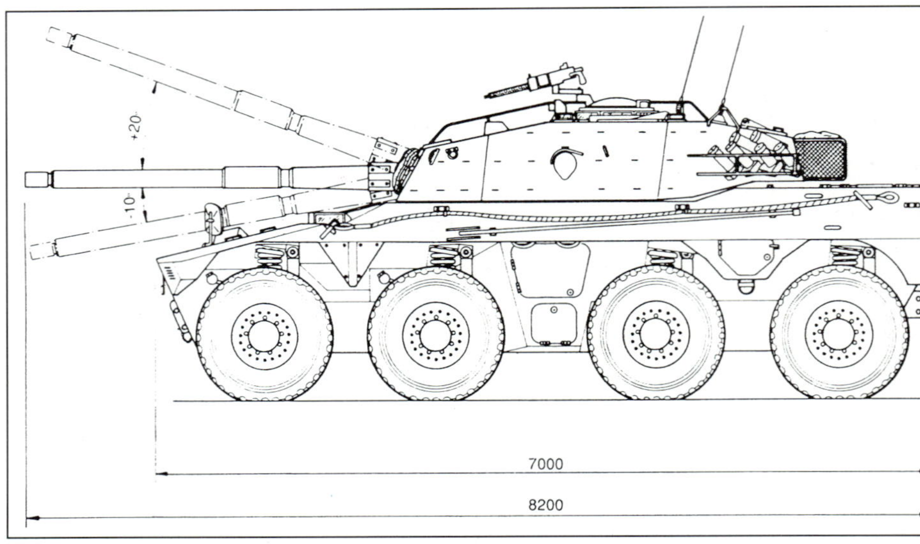

Side drawing of the Rooikat 76 mm (8 × 8) armoured car showing the elevation/depression of the 76 mm gun

Although production of the Rooikat 76 mm armoured car was completed several years ago, Vickers OMC continue to overhaul and upgrade these vehicles for the South African Army.

Description

The hull of the Rooikat is of all-welded steel armour construction with the driver seated at the front, three-man power-operated turret in the centre and the power pack at the rear.

Over the frontal arc protection is provided against penetration from the 23 mm AP (Armour-Piercing) rounds fired from the Russian ZU-23-2 LAAG, widely deployed in both ground and air defence roles in Southern Africa. Trials have demonstrated that the hull design of the Rooikat provides a high degree of protection against anti-tank mines.

The driver, seated on the centreline of the vehicle, can enter through the fighting compartment or through the single-piece hatch cover that opens to the left. The driver's station (including the essential controls such as foot pedals, steering wheel and direction indicators) is fully adjustable and, when driving in the closed-down position, forward observation is via three day periscopes, the centre one of which can be replaced by a passive night periscope. When closed-down the driver's periscopes can be cleaned by a compressed air cleaning system. The headlamps are under armour protection at the front of the vehicle.

A hull escape hatch is provided in each side of the hull, between the second and third axles, thus avoiding the crew being exposed to small arms fire if they were to evacuate.

The commander is seated on the right of the turret with the gunner forward and below his position, and the loader on the left.

The commander is provided with eight day vision blocks to give observation through 360°. Mounted in the roof, forward of the commander's station, is a day panoramic sight with a magnification of ×12. This

Rooikat 76 mm (8 × 8) armoured car showing vertical obstacle climbing ability
(Christopher F Foss) 0088045

Rooikat 76 mm (8 × 8) armoured car showing main armament fitted with thermal sleeve and fume extractor
2000/0064845

enables the commander to observe through a full 360° without moving his/her head. The sight can also be slaved to the 76 mm gun or uncoupled.

The loader has a single-piece hatch cover that opens to the rear, and is provided with two day periscopes. These can be traversed to allow observation of the terrain on the left side of the vehicle.

The gunner has a roof-mounted periscopic sight which has ×8 day and night (image intensification) channels and an integral laser range-finder. The gunner also has an auxiliary day sight as used in the Vickers OMC Ratel (6 × 6) armoured vehicle. If required, the commander of the Rooikat can override the gunner.

Turret traverse and weapon elevation features solid-state electric controls for both gunner and commander, with manual emergency back-up. Traverse is through a full 360° with elevation limits from −10 to +20°. The commander has a single control handle and the gunner twin control handles.

The digital fire-control system receives information from a laser range-finder and a number of sensors including cant and windspeed. Ammunition type is set manually. The Rooikat can engage and hit targets with a high probability while it is moving across rough terrain.

Ballistic offsets are automatically computed and implemented when the laser range-finder is triggered, to compensate for various factors affecting the ammunition ballistics. The offsets are implemented both in the sight (electronics sight slaving) and the main weapon (auto lay) and a ready to fire indication appears in the gunner's field advising him/her to fire. Total reaction time from lasing to firing is typically less than 2 seconds.

The fire-control system provides automatic fire control by computing and implementing the ballistic offsets for the ammunition type selection by taking into account target range (automatic from laser range-finder), target speed (automatically derived by tracking the target), manually entered environmental data (air pressure, ambient temperature, ammunition temperature, headwind), crosswind speed (automatically sensed), weapon tilt (cant automatically sensed), and gun jump characteristics of the main weapon.

The fire-control system has BITE (Built-In Test Equipment) which enables any faults to be immediately traced and rectified through the substitution of line-replaceable units.

Main armament comprises a 76 mm/62 calibre GT4 gun which was developed at LIW and has the same chamber volume as the Italian Otobreda 76 mm Compact naval gun but fires different ammunition in keeping with its role. The ammunition of the naval gun can also be fired by replacing the percussion primer with an electric primer. The 76 mm GT4 gun was chosen for the Rooikat for a number of other reasons including the larger number of rounds capable of being carried when compared to a 105 mm gun, the ease of handling ammunition when moving across country and the optimum selection of recoil forces and firing impulses for maximum crew operational effectiveness.

The 76 mm stabilised gun has a vertical sliding semi-automatic breech block and is fitted with a thermal sleeve and concentric glass fibre fume extractor. The recoil system consists of a concentric hydrospring with an external replenisher with maximum recoil being 370 mm.

Two types of fixed 76 mm ammunition have been developed, HE-T (High Explosive-Tracer) and APFSDS-T (Armour-Piercing Fin-Stabilised Discarding Sabot-Tracer) with a total of 48 rounds of ammunition being carried, nine of which are for ready use, stowed vertically below the turret ring on the left side. For safety reasons no main gun ammunition is stowed above the turret ring.

Rate of fire is stated to be 6 rds/min, with the APFSDS-T projectile having an effective range of 2,000 to 3,000 m. It will penetrate the T-54/T-55 and T-62 from all angles of attack. The HE-T round has a maximum range of 12,000 m in the indirect fire role and 3,000 m in the direct fire role. A range drum is fitted as standard.

Basic details of the 76 mm ammunition are:

Type	**APFSDS-T**	**HE-T**
Weight		
(complete round)	9.2 kg	13.5 kg
Length	873 mm	908 mm
Muzzle velocity	1,610 m/s	915 m/s

Specifications

Crew: 4
Configuration: 8 × 8
Combat weight: 28,000 kg
Power-to-weight ratio: 20.11 hp/t
Length:
(including gun) 8.2 m
(hull) 7.09 m
Width:
(overall) 2.9 m
(hull) 2.8 m
Height:
(overall) 2.8 m
(turret roof) 2.5 m
(hull top) 1.76 m
Ground clearance: 0.4 m
Track: 2.415 m
Wheelbase: 1.55 + 2.032 + 1.625 m
Angle of approach/departure: 45°/60°
Max road speed: 120 km/h
Cross-country speed: 60 km/h
Acceleration: 0 to 30 km/h in 8 s
Fuel capacity: 540 litres
Max road range: 1,000 km
Fording: 1.5 m
Gradient: 70% (forward and reverse)
Side slope: 30%
Vertical obstacle: 1 m
Trench:
(crawl speed) 2 m
(60 km/h) 1 m
Engine: V-10 water-cooled diesel developing 563 hp
Transmission: automatic, 6 forward and 1 reverse gears
Steering: power-assisted, front 4 wheels
Turning radius: 25 m
Suspension: internally driven trailing arms, coil springs and shock-absorbers
Tyres: 16.00 × R20 radial with run-flat inserts
Brakes: dual-circuit drum on all 8 wheels, retarder on transmission, exhaust brake, crawl brake, parking brake
Electrical system: 24 V
Batteries: 6 (2 banks)
Armament:
(main) 1 × 76 mm rifled gun GT4
(coaxial) 1 × 7.62 mm MG
(anti-aircraft) 1 × 7.62 mm MG
Smoke-laying equipment: 8 × 81 mm smoke grenade launchers, exhaust smoke screen generator
Ammunition:
(main) 48
(7.62 mm MG) 3,600
(smoke grenades) 16
Gun control equipment
Turret power control: electric/manual
(by commander) yes
Turret traverse: 360°
Gun elevation/depression: +20°/−10°
Gun stabiliser: yes
Laser range-finder: yes
Commander's sight magnification: ×12
Gunner's sight magnification:
(day) ×8
(night) ×8
NBC system: optional
Night vision equipment: yes

The APFSDS-T round has a tungsten alloy penetrator with the sabot, fins and nose shield made of aluminium and the rotating obturator of Nylon 66.

The HE-T round contains 0.6 kg of RDX/TNT on a 60/40 basis with the tracer being visible for 8 to 11 seconds. The projectile is fitted with a Point Detonating (PD) and selectable delay fuze which provides a super-quick characteristic of 170 μs or a delay action of between 30 and 80 μs, depending on the tactical situation. A canister and practice round are under development.

A 7.62 mm coaxial machine gun is mounted to the left of the main armament, while a similar weapon is mounted at the commander's position for use in the ground and air defence roles. This is skate mounted so that it can be moved around the commander's hatch. An ammunition resupply hatch is provided in the left side of the turret.

Mounted either side of the turret rear is a bank of four forward-firing electrically operated 81 mm smoke grenade dischargers. The Rooikat can also lay its own smoke screen by injecting diesel fuel into the exhaust outlet at the rear of the hull on the left side.

Its 563 hp (420 kW) is provided by a V-10 turbo-charged water-cooled engine through a six-speed fully automatic transmission, drop down gearbox and high/low-range transfer gearbox.

The engine, transmission, drop down gearbox and cooling system are packaged into a power pack unit which has been designed so that field exchange will be possible within 60 minutes.

The engine compartment is fitted with an automatic fire detection and suppression system.

The driver can select full 8 × 8 drive or 8 × 4 drive, depending on the tactical situation. Steering is power assisted on the front four wheels. The wheels, with 16.00 × R20 or 14.00 × R20 (this being the standard version) tyres, can be fitted with run-flat inserts. The Rooikat remains mobile with two same-side wheels blown off by landmine explosion.

Suspension at each wheel station consists of internally driven trailing arms, coil springs and hydraulic shock-absorbers. Total wheel travel is 327 mm and the suspension has been designed for high-speed cross-country operations.

Dual-circuit drum-type brakes are fitted on all eight wheels. In addition there is a retarder on the transmission, an exhaust brake, a crawl brake and a parking brake.

Standard equipment includes a vehicle intercom and radios (2 × VHF and 1 × HF which are mounted in the turret bustle). Optional equipment includes a chemical and biological overpressure collective system which is integrated with a 16 kW air conditioner with air distribution to all crew members.

In 1998, it was revealed the South African company Althane had developed a new puncture proof tyre for the Rooikat (8 × 8) armoured car called Terra Trak.

The company started development work on the Terra Trak puncture proof tyre in response to a South African National Defence Force requirement for a tyre that would be better than standard pneumatic bush tyres which are fitted with reinforced sidewalls and runflat inserts.

These standard tyres survive attack from 7.62 mm small arms fire and the laceration effects of tree stumps or sharp boulders.

Initial tests on the single and two layer Terra Trak tyres installed on a SAMIL cross-country truck commenced in 1996.

Subsequently, further development resulted in a four layer Terra Trak tyre which was tested on a Rooikat armoured car in 1998.

Tests have shown that the mobility of the vehicle is not affected by the Terra Trak tyres with each tyre having a foot print of 405 mm without the requirement to install a central tyre pressure regulation system which is normally fitted to pneumatic tyres to improve their cross country mobility.

According to the manufacturer, the average life of the polyurethane Terra Trak tyre is put at about 16,000 km and during flame trials it has been shown to self-extinguish within 5 seconds.

Variants

Rooikat 105 mm tank destroyer

This version has been developed by Vickers OMC for the export market. Details are given in the previous entry. As of mid-2001, this remained at the prototype stage.

Status

Production complete. In service only with South Africa.

Manufacturer

Systems supplier - Vickers OMC.

UPDATED

Vickers OMC Eland light armoured car

Development/Description

In November 1961, South Africa concluded a contract with the French company Panhard for the supply of 100 AML (4 × 4) armoured cars armed with a 60 mm mortar and AA52 machine guns. An order was also placed for 150 Giat 90 mm DEFA gun turrets and components for a further 500 units.

In January 1962, tenders were requested from 12 local South African companies for the manufacture of 350 Panhard AML armoured cars complete with turrets and 150 without turrets. Only four companies responded to the tender, which was subsequently awarded to Sandock Austral.

The original South African Defence Force order was for 600 AMLs of which 100 came direct from France and the remaining 500 were locally manufactured. Of the latter, 350 were armed with 60 mm mortars, a coaxial 7.62 mm machine gun and a 7.62 mm anti-aircraft machine gun and the remaining 150 were armed with a 90 mm gun.

The first locally manufactured armoured cars, called the Eland Mk 2, were delivered in July 1963 and had a local content of 40 per cent. Between eight and 12 units per month were delivered for the following 16 months. The Eland Mk 1 was the version imported from France. The Mk 2, 3 and 4 versions of the Eland all had an imported Panhard petrol engine which was air cooled. The Mk 2 had an improved steering system, the Mk 3 had an improved brake system and the Mk 4 had an updated fuel system and an electrical clutch. Although the three versions are similar externally they are different internally.

The Mk 5 was the first version with a locally built engine, this was a 2.5 litre water-cooled petrol engine which could be changed in the field in 90 minutes.

Vickers OMC Eland 60 light armoured car with a 60 mm mortar and a 7.62 mm coaxial machine gun. Note the mount on the turret for a 7.62 mm anti-aircraft machine gun (Christopher F Foss)

It also had locally developed hydraulic clutches, communications equipment, springs, shock-absorbers, wheels and run-flat tyres.

During 1972-73, the assembly of 356 Eland Mk 5 armoured cars was completed and delivery was at the rate of 24 vehicles per month. A programme was then completed during 1974-75 for the repair and conversion of earlier Elands, which included a complete rebuild of the vehicle, general repair work and overhauls which converted the vehicle to the Mk 6. This programme involved the conversion of a total of 1,016 vehicles.

The final version of the Eland was the Mk 7, which was introduced in 1979. It has new power brakes, an increased storage capacity resulting from the slightly longer chassis, inclusion of a freshwater tank and a

Specifications
Eland Mk 7 DT
Crew: 3
Configuration: 4 × 4
Combat weight:
(Eland 60) 5,300 kg
(Eland 90) 6,000 kg
Power-to-weight ratio:
(Eland 60) 19.43 hp/t
(Eland 90) 17.16 hp/t
Ground pressure:
(Eland 60) 0.279 kg/cm^2
(Eland 90) 0.284 kg/cm^2
Length:
(Eland 60 gun forward) 4.04 m
(Eland 90 gun forward) 5.12 m
(Eland 60 hull) 4.04 m
(Eland 90 hull) 4.04 m
Height:
(Eland 60 turret top) 1.88 m
(Eland 90 turret top) 2.5 m
Width: 2.015 m
Wheelbase: 2.5 m
Track: 1.62 m
Ground clearance:
(Eland 60) 0.40 m
(Eland 90) 0.38 m
Max road speed: 85 km/h
Safe road speed: 65-70 km/h
Safe cross-country speed: 30 km/h
Acceleration:
(0-40 km/h) 18 s
(0-60 km/h) 42 s
Fuel capacity: 142 litres
Range:
(road) 450 km
(cross-country) 250 km
Fording:
(without preparation) 0.615 m
(with preparation) 0.82 m
Gradient: 51%
Side slope: 27%
Vertical obstacle: 0.3 m
Trench: 0.5 m
Turning circle: 12.6 m
Engine: 2.4 litre 4-cylinder turbocharged water-cooled diesel developing 103 hp at 4,000 rpm
Transmission: manual, 6 forward and 1 reverse gears
Suspension: fully independent active trailing arm with single coil and double-action hydraulic shock-absorber
Tyres: 12.00 × 16 tubeless with run-flat inserts
Batteries: 2 × 12 V, in series
Armament:
(Eland 60 main) 60 mm mortar
(Eland 90 main) 90 mm rifled gun
(Eland 60 coaxial) 7.62 mm MG
(Eland 90 coaxial) 7.62 mm MG
(Eland 60 roof-mounted) 7.62 mm MG
(Eland 90 roof-mounted) 7.62 mm MG
Smoke-laying equipment: 2 × 2 81 mm smoke grenade launchers
Ammunition:
(Eland 60 main) 54
(Eland 90 main) 29
(Eland 60 7.62 mm) 3,800
(Eland 90 7.62 mm) 2,000
(Eland 60 grenades) 12
(Eland 90 grenades) 8
NBC system: no
Night vision equipment: optional

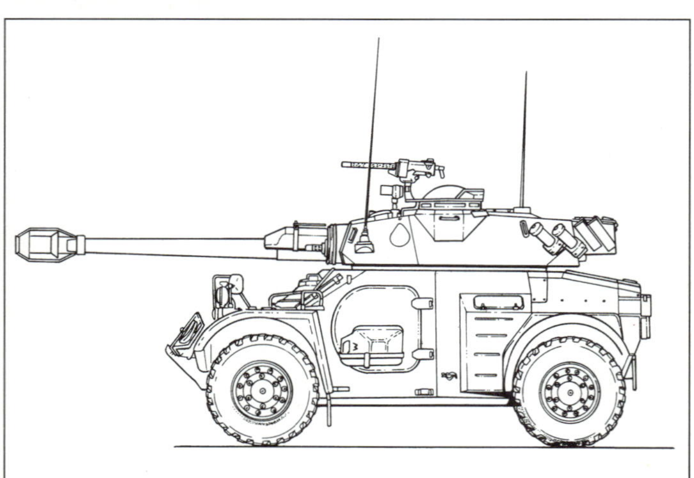

Vickers OMC Eland 90 light armoured car

Vickers OMC Eland 90 Mk 7 DT (4 × 4) armoured car (Christopher F Foss)
0088046

more comfortable driver's compartment. External modifications include a towbar at the front, halogen lights on the mudguards and the commander's cupola with armoured glass windows which allows the commander to see through a full 360°.

South Africa also made the complete 90 mm gun for the vehicle as well as a 60 mm mortar, machine guns, optics and ammunition.

Prime contractor for the weapons was the LIW Division of Denel with other parts of the company providing the ammuntion.

Production models of the Eland were armed with a 90 mm gun/7.62 mm coaxial and 7.62 mm anti-aircraft machine guns or a 60 mm mortar and a 7.62 mm machine gun. A model with a two-man turret armed with a 20 mm cannon and a coaxial 7.62 mm machine gun was also produced. A number of this last version were supplied to Morocco.

Ninety-five per cent of the Eland was produced in South Africa by the time production was completed. Total South African production of the Eland amounted to approximately 1,300 vehicles.

Eland Mk 7 DT Upgrade

In 1994, production commenced of the upgraded Eland Mk 7 DT (4 × 4) armoured car for an undisclosed export customer, believed to be the Congo. The DT in the designation stands for diesel turbocharged.

With the entry into service of the more powerful 76 mm Rooikat (8 × 8) armoured car, most Elands have now phased out of service in South Africa.

The Mk 7 DT is powered by a turbocharged 2.4 litre water-cooled four-cylinder overhead valve diesel engine developing 103 hp at 4,000 rpm, which gives the vehicle a maximum road speed of 85 km/h.

It can stop from 60 km/h to 0 in 45 m on a tar surface. The fuel tank holds 142 litres, including 15 litres in the pick-up tank which gives an operating range of 450 km at 80 km/h on roads and 240 km at 30 km/h cross-country.

The engine and cooling system is mounted on sliding rails to facilitate field maintenance.

Other options include improved vision and surveillance equipment, new communications equipment and run-flat tyres.

Vickers OMC (previously known as Reumech OMC) are no longer offering this upgrade package for the AML/Eland 4 × 4 armoured car.

Status

Production complete. In service with:

Country	Quantity	Comment
Congo, Democratic Republic	20?	Upgraded with diesel engine
Malawi	13	
Morocco	n/avail	20 mm, 60 mm and 90 mm
South Africa	118	
Zimbabwe	20	60 mm and 90 mm

Manufacturer

Vickers OMC.

UPDATED

SPAIN

Santa Barbara Sistemas VEC cavalry scout vehicle

Development

The VEC (*Vehículo de Exploración de Caballeria*), or cavalry scout vehicle, was designed to meet the requirements of the Spanish Army and shares many common components with the Santa Barbara Sistemas (and now owned by General Dynamics of the USA) BMR-600 (6 × 6) infantry fighting vehicle which is fully described in the Armoured personnel carriers (wheeled) section. The VEC is also known as the VEC 3562. The first of five prototypes was completed in 1977-78.

It is understood that total production of the VEC vehicle amounted to about 340 vehicles, all for the Spanish Army as there have been no known exports of the vehicle.

The 32 vehicles of the first production batch were fitted with the Italian Otobreda TC20 turret armed with a 20 mm Rheinmetall MK 20 Rh 202 cannon.

A further 100 vehicles were fitted with the complete H 90 turret of the Panhard AML-90 (4 × 4) armoured car which was phased out of service with the Spanish Army in 1986.

The remaining 208 vehicles were fitted with the Otobreda TC25 turret armed with The Boeing Company 25 mm M242 Chain Gun as covered in the following description.

VEC Cavalry Scout Vehicle fitted with a Hispano-Suiza H 90 turret armed with a 90 mm gun removed from a Panhard AML 90 (4 × 4) armoured car 0088048

Description

The hull of the VEC is of all-welded aluminium armour construction providing protection over its frontal arc against 7.62 mm × 51 armour-piercing ammunition, and standard 7.62 mm × 51 rounds over the remainder of the vehicle. Some areas of the VEC have spaced armour protection.

The driver is seated at the front of the vehicle on the left side and has a single-piece hatch cover that opens to the right, in front of which are three day periscopes for forward observation. The centre day periscope can be replaced by an image intensification periscope for night driving.

The turret is in the centre of the vehicle, between the first and second axles. A variety of turret installations can be fitted including the same turret as fitted to the Italian FIAT - Otobreda Type 6616 (4 × 4) armoured car which is fully described earlier in this section. A variety of other weapon stations ranging from 7.62, 12.7, 20 up to 90 mm in calibre can be installed.

Spanish Army VECs are fitted with an Italian-designed Otobreda TC25 two-man turret manufactured

VEC Cavalry Scout Vehicle fitted with a TC25 turret armed with a 25 mm The Boeing Company M242 Bushmaster cannon, a 7.62 mm coaxial machine gun and electrically operated smoke grenade dischargers This vehicle has been upgraded to the latest VEC 2 standard (Richard Stickland) ***2002***/0114335

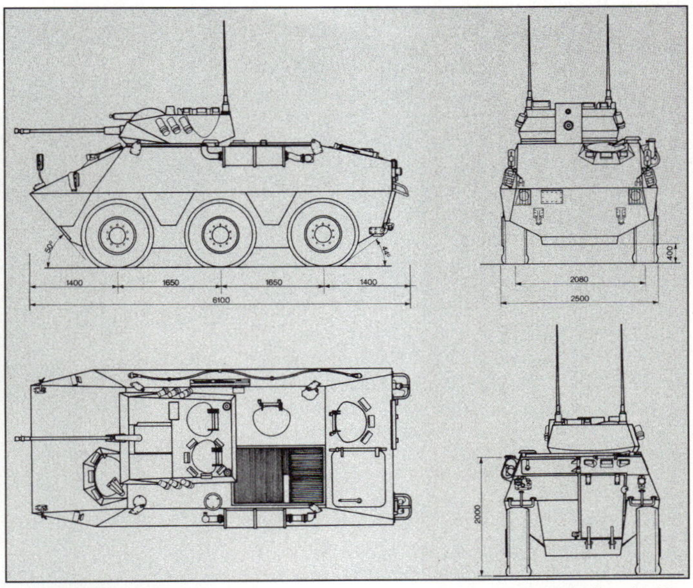

Four-view general arrangement of the VEC Cavalry Scout Vehicle fitted with a Santa Barbara Sistemas Otobreda TC25 turret

Specifications
Crew: 3 + 2
Configuration: 6 × 6
Combat weight: 13,750 kg
Loaded weight:
(front axle) 5,000 kg
(centre axle) 4,450 kg
(rear axle) 4,300 kg
Power-to-weight ratio: 22.25 hp/t
Length: 6.10 m
Width: 2.5 m
Height:
(to hull top) 2 m
Ground clearance: 0.4 m
Track: 2.08 m
Wheelbase: 1.65 + 1.65 m
Angle of approach/departure: 50/45°
Max road speed:
(1st gear position, 1 gear) 17 km/h
(2nd gear position, 1, 2, 3, 4 gear) 56 km/h
(3rd gear position, 1, 2, 3, 4, 5 gear) 69 km/h
(4th gear position, 2, 3, 4, 5, 6 gear) 103 km/h
(reverse) 12 km/h
Max water speed:
(wheels) 3 km/h
(water-jets) 9 km/h
Fuel capacity: 400 litres
Max road range: 800 km
Gradient: 60%
Side slope: 30%
Vertical obstacle: 0.6 m
Trench: 1.5 m
Turning radius: 7.5 m
Engine: Pegaso model 9157/8 6-cylinder in-line turbocharged water-cooled diesel developing 310 hp at 2,200 rpm. Now being replaced by a locally built Scania D9 diesel developing 310 hp at 2,200 rpm which is now being fitted to the BMR-600 vehicle (additional details are given in description)
Transmission: ZF 6 HP 500, fully automatic, 6 forward and 1 reverse gears
Torque converter and hydraulic retarder: ZF
Steering: hydraulic, power assisted on first and third axles
Tyres: 13.00 × 20 (or 14.00 × 20)
Suspension: independent each wheel, MacPherson type, hydropneumatic
Brakes:
(main) disc, hydraulic, dual-circuit, air-assisted on all wheels
(parking) disc, operating on propeller transmission
(emergency) hydraulic, air-operated MGM chambers
Electrical system: 24 V
Batteries: 2 × 12 V, 140 Ah
Armament:
(main) 1 × 25 mm M242 cannon
(coaxial) 1 × 7.62 mm MG
Smoke-laying equipment: 2 × 3 smoke grenade dischargers
Ammunition (ready use):
(main) 170 (135 HE + 35 APDS)
(coaxial) 250
(smoke grenades) 6
Gun control equipment
Turret power control: 360° powered/manual
Elevation/depression: +50/–10°
Gun stabiliser: no
NBC system: optional
Night vision equipment: yes
Note: The maximum road speed of 103 km/h is with 14.00 × 20 tyres. When fitted with 13.00 × 20 tyres maximum road speed is 96 km/h

under licence by the now Santa Barbara Sistemas. It was expected that main armament would be an Oerlikon Contraves 25 mm KBA cannon but, early in 1987, it was announced that Spain had selected The Boeing Company M242 25 mm Bushmaster cannon as installed in the M2 Bradley; 208 weapons were ordered at a total cost of US$7 million and all these have now been delivered to Spain.

The engine compartment is to the rear of the vehicle on the left side with air inlet and outlet louvres in the roof, a silencer on the left side of the hull and a hatch in the rear of the hull on the left side.

Early in 1996, following a competition, Scania Hispania (Scania Spain) was awarded a contract by the now Santa Barbara Sistemas, for the supply of 700 new Scania D9 diesel engines for installation in the VEC cavalry scout vehicles and BMR-600 armoured personnel carriers of the Spanish Army.

These were delivered over a five year period which started in late 1996 with the total value of the contract being SKr100 million. The engines were installed in a total of 642 VEC and BMR-600 series vehicles.

The Scania D9 is a six-cylinder in-line diesel engine which develops 310 hp at 2,200 rpm and is more compact and fuel efficient than the original engine.

In the hull to the rear of the turret on the right side is a seat for the fourth crew member and above this is a single-piece hatch cover opening to the right with a day periscope in the hull roof. The fifth crew member is seated at the very rear on the right side and has a single-piece hatch cover that opens to the rear, and three day periscopes for observation. There is also a downward-opening door in the hull rear on the right side.

All three axles are of Pegaso design and manufacture and are of the double reduction type with self-locking differentials. Suspension is independent for each wheel with variable flexibility and 275 mm suspension travel achieved by an oleopneumatic system with MacPherson arrangement. Steering is power assisted on the front and rear axles.

Standard equipment includes a toolset, a 4,500 kg winch, a semi-automatic fire extinguishing system protecting the engine and transmission and two 5 kg portable fire extinguishers.

Optional equipment includes an NBC system, a water-jet propulsion system (one unit either side of the hull at the rear and a trim vane that folds back onto the glacis plate), a 4,500 kg winch, a communications system, a land navigation system, 14.00 × 20 tyres, run-flat tyres and an air conditioning system.

Variants

The VEC can also be used in the anti-aircraft role (guns or missiles) and the anti-tank role with ATGWs. Apart from the three models in service with the Spanish Army and covered in the development section, no variants of the VEC have ever entered production.

VEC 2

This is the designation given to the upgraded VEC vehicles by the Spanish Army. In addition to the installation of the new Scania diesel engines, other improvements include the installation of additional passive armour to the hull front, sides and rear. Additional external stowage has also been provided.

Status

Production complete, but can be resumed if required. In service only with the Spanish Army.

Manufacturer

Santa Barbara Sistemas (part of the General Dynamics Group of the US)

UPDATED

SWITZERLAND

MOWAG Piranha III (10 × 10) Armoured Combat Vehicle

Development

Development of the Piranha (10 × 10) Armoured Combat Vehicle (ACV) commenced in December 1992 with the decision to build a prototype vehicle taken in April 1993. The vehicle has been developed by MOWAG specifically for the export market.

Development of the MOWAG Piranha (10 × 10) vehicle was completed late in 1994 and early in 1995 went to the Middle East for customer demonstrations.This vehicle was fitted with the French Giat Industries TML 105 turret armed with a 105 mm gun.

During the development programme, some 100 rounds of 105 mm ammunition were fired, including some with the turret traversed at 90° left and right and on an inclined position. Mobility trials covered over 7,500 km on both road and cross-country terrain.

MOWAG Piranha (10 × 10) for the Swedish Defence Administration (FMV) in Armoured Command Vehicle (ACV) configuration and clearly showing the higher roof for this role
2002/0114336

While the MOWAG Piranha (8 × 8) is suitable for most requirements, the new 10 × 10 version gives an increase in payload, a larger internal volume and enhanced ballistic protection. It retains the same high degree of mobility as the earlier 8 × 8 vehicles.

According to MOWAG, the new five-axle configuration offers the following advantages:

(1) Redundancy of drive assemblies in the case of a drive assembly failure due to combat damage
(2) High cross-country mobility with crew comfort
(3) Minimal vehicle width for air-transportability (the 10 × 10 model is air-portable in a C-130)
(4) Enhanced crew protection against mines
(5) Increased protection against direct fire.

Sweden is the first customer for the MOWAG Piranha (10 × 10) ACV with the first order being for three vehicles, two in the Armoured Command Vehicle (ACV) configuration and the third in the Armoured Sensor Vehicle (ASV) radar carrier configuration.

The first of these vehicles was handed over in mid-1997 followed by the second in September 1997. This was followed by the ASV which has an extendable mast, also developed by MOWAG on which the surveillance and target radar is mounted. The vehicles were fitted out with an extensive suite of C^3I equipment in Sweden.

For its specialised mission, the MOWAG Piranha ACV has a number of modifications including a raised rear roof for increased internal volume and a 40 kW auxiliary power unit developed and tested by MOWAG as part of the contract with FMV.

The commander's position, to the rear of the driver on the left side is provided with an externally mounted 7.62 mm machine gun.

With options, the Swedish original requirement was for a total of 44 (27 sensor and 17 command vehicles) in the 10 × 10 configuration and 10 in the 8 × 8 Armoured Escort Vehicle (AEV) role.

Late in 1997, Sweden exercised its second option and placed a contract with MOWAG for a further five ACV (10 × 10) vehicles in ACV configuration. These vehicles were equipped with a 340 hp Scania diesel engine coupled to a ZF seven-speed fully automatic transmission.

In August 2000, the Swiss company of MOWAG, now owned by General Motors Defense, was awarded another contract by the Swedish Defence Material Administration for the supply of a further five MOWAG Piranha IIIC (10 × 10) Armoured Sensor Vehicles (ASV) for use by the Swedish Amphibious Brigade.

These ASV will be delivered to Sweden in 2002 where the new Ericsson Microwave Systems Giraffe AMB (Agile Multi-Beam) radars will be integrated, these have already been ordered under a separate contract. The C in the designation indicates that the vehicle has a conventional independent suspension system.

It was originally expected that Sweden would order a total of 44 MOWAG Piranha IIIC (10 × 10) vehicles, in the ASV and 17 in the ACV configuration. This requirement has now been cut back and it is probable that the latest order received by MOWAG will be the last.

It was also expected that Sweden would order 18 Armoured Escort Vehicles (AEV) in 8 × 8 configuration fitted with a turret to provide cover for the ASC and ACV versions, but this option was never exercised.

Description

The all-welded steel hull of the original MOWAG Piranha (8 × 8) vehicle provides protection against 7.62 mm small arms fire but the new 10 × 10 version has protection against 14.5 mm small arms fire over the frontal arc. Spall liners can be added to the crew compartment to provide additional protection from splinters.

The layout of the vehicle is similar to other members of the MOWAG Piranha family with the driver front left, power pack to his right and turret in the centre. The rear compartment holds 26 rounds of 105 mm ammunition, the self-recovery winch and the fuel tank.

The driver has a single-piece hatch cover that opens to the left rear and in front of this are three day periscopes, the centre one of which can be replaced by a passive periscope for night driving. The driver's seat is fully adjustable. On the prototype there is an additional seat and hatch cover to the rear of the driver.

The prototype is fitted with the Giat Industries 105 TML power-operated turret armed with a Giat 105 mm G2 rifled gun capable of firing standard NATO ammunition natures, including APFSDS (Armour-Piercing Fin-Stabilised Discarding Sabot) with a muzzle velocity of 1,495 m/s.

The commander is seated on the right, with the gunner forward and below his position, and has a single-piece hatch cover that opens to the rear. The loader is seated on the left side of the turret and also has a single-piece hatch cover that opens to the rear.

A 7.62 mm machine gun is mounted coaxial with the main armament and a Giat Industries GALIX grenade launching system is fitted as standard. Turret traverse and weapon elevation are electrohydraulic with manual back-up.

The fire-control system includes a digital gunner's day/thermal sight with a laser range-finder, a ballistic computer and a control panel. The vehicle commander has nine HL 69 day periscopes for all-round observation and a day/night image intensification sight. The gunner also has two M336 day periscopes while the loader has three M223 day periscopes.

The current TML turret is not provided with a stabilisation system for the 105 mm gun but has now been developed and would allow the main armament to be laid and fired with a high first round hit probability when the vehicle is moving. A load assist device is also to be developed to enable rate of fire to be increased to 8 rds/min.

Steering is power assisted on the front four roadwheels with the rear three axles being permanently driven. The two front axles can be engaged when required. The 10 × 10 version has greater mobility as well as providing a more stable firing platform.

The new 10 × 10 vehicle shares many common components with the MOWAG Piranha family of armoured vehicles.

Communications equipment depends on the user but the prototype has two Thales VHF 50 W Jaguar frequency-hopping radios and an intercom system for all four crew members.

A number of options have already been developed and tested for the MOWAG 10 × 10 vehicle including:

(1) Larger tyres size 335/80 × 20 or 375/70 × 20
(2) Central tyre-pressure regulation system that allows the driver to adjust the tyre pressure to suit the ground being crossed from his seat. Three different settings are possible (on road, off road and soft ground) and a warning lamp will alert the driver as soon as the travelling speed becomes excessive for the selected tyre pressure
(3) New power pack consisting of an MTU diesel developing 400 hp (with growth potential to 450 hp) coupled to a fully automatic ZF six-speed automatic transmission
(4) Additional passive armour for the turret and hull for improved ballistic protection
(5) Hydropneumatic suspension for greater cross-country mobility which allows the driver to set one of three different ground clearances.

Various specialised equipment can be fitted including a Rotzler winch with a capacity of 8 tonnes which is installed in the rear of the hull and can be used to the front or rear, NBC system of the overpressure or ventilated face mask type and a 10 kW air conditioning system.

Specifications
(with TML 105 turret)
Crew: 4
Configuration: 10 × 10
Combat weight: 18,000 kg
Power-to-weight ratio: 19.44 hp/t
Length:
(incl armament) 8.75 m
(hull) 7.45 m
Width: 2.60 m
Height:
(with turret) 2.99 m
(hull top) 1.96 m
Ground clearance: 0.57 m
Max road speed: 100 km/h
Road range: 800 km
Fording: 2.00 m
Gradient: 60%
Side slope: 30%
Trench: 2.0 m
Engine: Detroit Diesel developing 350 hp (Swedish vehicles have Scania diesel)
Transmission: Allison automatic with 6 forward and 1 reverse gears (Swedish vehicles have ZF transmission)
Transfer case: MOWAG 2-speed
Suspension: independent with torsion bars and McPherson-type spring struts and A-frame control arms
Tyres: 335/80 R20 R20XZL to 375/70 R20XZL
Steering: power assisted
Armament:
(main) 1 × 105 mm gun
(coaxial) 1 × 7.62 mm MG
Ammunition:
(main) 38 (12 ready use)
(7.62 mm) 2,000 (600 ready use)
Gun control equipment
Turret power control: electrohydraulic/manual
(by commander) yes
(by gunner) yes
Turret traverse: 360°
Gun elevation/depression: +20/–6.5°
Gun stabiliser:
(vertical) no*
(horizontal) no*
NBC system: no*
Night vision equipment: yes*
*optional

Variants

Upgraded TML105 turret

The main drawback of the first generation Giat Industries TML 105 turret is that, as the 105 mm weapon and its associated sights are not stabilised, the vehicle has to come to halt in order to engage moving targets.

Giat Industries has now developed a new version of the 105 TML 105 mm turret which allows the system to engage stationary and moving targets while it is stationary or moving with a high first round hit probability.

This version of the 105 TML 105 mm turret integrates an electrical two-axis stabilised turret drive system, a roof-mounted gunner's sight which is also two-axis stabilised with day and thermal channels. The stabilised sight has already been proven in a number of other applications and has already demonstrated a high level of hit probability when firing on the move.

The commander's station has a removable modular cupola which, in its standard version, is fitted with a panoramic non-stabilised day sight.

As an option this can be replaced by a panoramic two-axis stabilised sight with day/thermal cameras so giving the system a hunter/killer capability similar to that fitted to the latest generation main battle tanks.

In addition, the commander's and gunner's sights can be fitted with an automatic target tracking system. A semi-automatic loader for the 105 mm gun is fitted.

A 7.62 mm machine gun is mounted coaxial with the main armament and, if required, a 7.62 mm machine gun provided with a shield can be installed on the roof and operated by the commander. A total of 14 GALIX grenade launchers is mounted around the turret and these can fire a variety of grenades such as smoke, decoy and fragmentation.

The turret is of all-welded construction and in its basic version provides protection against small arms fire up to 14.5 mm but, if required, a higher level of protection can be provided.

Another option for this turret is the installation of a Battle Management System (BMS) at the commander's station.

Although the prototype of the MOWAG Piranha (10 × 10) is fitted with the Giat Industries TML 105 105 mm turret, a wide range of other variants are possible including:

(1) Thales Blazer air defence turret with 25 mm cannon and Matra BAe Dynamics Mistral SAMs
(2) Bofors Defence 40 mm L/70 turret
(3) 105 mm turret from other manufacturers
(4) Heavy APC with Rheinmetall Landsysteme E4A1 30 mm turret
(5) Logistics carrier with 20 ft container which can be loaded and unloaded using a mechanical handling system

(6) Maintenance and recovery vehicle
(7) Oerlikon Aerospace ADATS missile system
(8) Cadillac Gage 105 mm two-man turret
(9) Krauss-Maffei Wegmann Wildcat 30 mm air defence system
(10) Thales Crotale New-Generation SAM system
(11) Armoured command post vehicle
(12) Radar carrier.

ACV with Bofors 40 mm turret

In 1998, the Armoured Combat Vehicle carried out a series of trials in Sweden fitted with the latest version of the Bofors Defence CV 9040 two person turret armed with a Bofors Defence 40 mm/ 70 calibre cannon.

During the trials, the system engaged stationary and moving targets with target practice – tracer (TP-T) ammunition while the vehicle was stationary and moving.

The latest CV 9040 turret is designated the CV 9040B and has been developed by Bofors Defence under contract to the Swedish Defence Material Administration (FMV). First production turrets were completed in 1999. Additional details are given in the entry for the CV 9040 in the Armoured personnel carrier (tracked) section.

Status

In production for Sweden which has placed three contracts for a total of 13 vehicles, seven in ACV and six in ASV configuration. Final deliveries are due in 2002.

Manufacturer

MOWAG Motorwagenfabrik AG.
In August 1999, MOWAG was taken over by the now General Motors Defense (at the time of the takeover the company was known as the Diesel Division, General Motors of Canada).

UPDATED

MOWAG Piranha (10 × 10) Armoured Combat Vehicle fitted with Giat Industries TML 105 105 mm turret (MOWAG) **2001**/0002329

MOWAG Piranha III (10 × 10) in Armoured Sensor Vehicle configuration with antenna raised and stabilisers lowered
2002/0114338

MOWAG SPY reconnaissance vehicle

Development

The SPY reconnaissance vehicle was developed as a private venture by the MOWAG company specifically for the export market; the first prototypes were completed in 1980. Drive train components, pressure, hydraulic and electrical components of the SPY are also used in the MOWAG Piranha range of 4 × 4, 6 × 6 and 8 × 8 armoured vehicles, which are fully described in the Armoured personnel carriers (wheeled) section.

In 1982, it was announced that an undisclosed Asian country had placed an order for between 20 and 50 SPYs fitted with a MOWAG-designed one-man turret armed with a 12.7 and a 7.62 mm machine gun.

There has been no recent production of the MOWAG SPY vehicle. This vehicle is still being marketed by MOWAG for the export market.

Description

The hull of the MOWAG SPY is made of all-welded steel armour and gives complete protection against small arms fire up to 7.62 mm AP calibre over the front, sides and rear, against anti-personnel mines exploding under the hull and against shell splinters above the hull.

The driver is seated at the front of the hull on the left with the commander immediately behind. The driver has a single-piece hatch cover and three day periscopes for forward observation, the centre one of which can be replaced by an image intensification periscope for night driving. The commander has a cupola mounted on a plinth above the roof to give observation over the driver. The cupola has a single-piece hatch cover and a day periscope. In the side of the hull behind the commander is a small door that opens forwards.

The diesel engine compartment is to the right of the driver with the air inlet and outlets in the roof and the exhaust pipe along the right side of the hull.

A variety of armament installations can be mounted at the rear of the hull including the MOWAG Type V 041

MOWAG SPY (4 × 4) reconnaissance vehicle fitted with a MOWAG one-man cupola armed with externally mounted 7.62 mm machine gun 0045882

MOWAG SPY reconnaissance vehicle fitted with a MOWAG one-man cupola armed with externally mounted 7.62 mm machine gun **2000**/0064849

Specifications

Crew: 3
Configuration: 4 × 4
Combat weight: 7,700 kg
Unloaded weight: 5,900 kg
Power-to-weight ratio: 27.2 hp/t
Length: 4.55 m
Width: 2.60 m
Height:
(hull) 1.66 m
Ground clearance:
(hull) 0.5 m
(axles) 0.39 m
Track:
(front) 2.18 m
(rear) 2.205 m
Wheelbase: 2.5 m
Angle of approach/departure: 50/70°
Max road speed: 110 km/h
Fuel capacity: 200 litres
Road range: 700 km
Fording: 1.3 m
Gradient: 60%
Side slope: 35%
Vertical obstacle: 0.5 m
Turning radius: 6.3 m
Engine: Cummins Diesel model 6 BTA 5.9 developing 210 hp at 2,500 rpm
Transmission: Allison Transmission AT-1545 automatic, 4 forward and 1 reverse gears
Transfer case: 2-speed
Suspension:
(front) coil spring and wishbone
(rear) torsion bars; all wheels have hydraulic shock-absorbers
Tyres: 325/85 R16 or 335/80 R20
Brakes: dual-circuit, hydraulic, air-assisted
Electrical system: 24 V
Armament:
(main) 1 × 12.7 mm M2 MG
(coaxial) 1 × 7.62 mm MG
(smoke grenade dischargers) 2 × 3
Ammunition:
(12.7 mm) 3,000
(7.62 mm) 2,000
(smoke grenades) 20
Gun control equipment
Turret traverse: 360°/manual
Gun elevation/depression: +50/−10°
Gun stabiliser:
(vertical) no
(horizontal) no
NBC system: no
Night vision equipment: optional

and Type V 042 turrets. The V 042 is armed with a remote-controlled 12.7 mm M2 HB machine gun with an elevation of +45°, depression of −8° and turret traverse of 360°. Turret traverse and weapon elevation are manual and 100 rounds of ready use 12.7 mm ammunition are provided. The gunner aims the machine gun with a roof-mounted PERI-Z-12 sight with a magnification of ×1 and either side of this is a day periscope for lateral observation.

The Type V 041 turret is armed with a 12.7 and a 7.62 mm machine gun with an elevation of +50°, depression of −10° and turret traverse of 360°. Turret traverse and weapon elevation are manual and 200 rounds of 12.7 mm and 500 rounds of 7.62 mm ready use ammunition are carried. The gunner aims the weapons via a PERI-Z-11 sight with a magnification of ×2 and ×6 and also has seven day periscopes for all-round vision.

All four wheels have independent suspension and the front axle can be disengaged for road driving. Automatic locking differentials and run-flat core tyres are fitted as standard.

Optional equipment includes various radio installations and an air conditioning system.

Status

Production as required. In service with an undisclosed country.

Manufacturer

MOWAG Motorwagenfabrik AG. In August 1999, MOWAG was taken over by the now General Motors Defense (at that time Diesel Division, General Motors of Canada).

UPDATED

MOWAG Eagle armoured reconnaissance vehicle

Development

The Eagle armoured reconnaissance vehicle was developed by MOWAG to meet the operational requirements of the Swiss Army.

Following extensive trials with three prototype vehicles, the Swiss MoD placed a contract for 156 MOWAG Eagle armoured reconnaissance vehicles at a total cost of SFr105 million.

The first of these were delivered early in 1995 and production ran at the rate of four vehicles a month.

In the Swiss Army the Eagle is used for the reconnaissance role, although other missions suggested by the manufacturer include surveillance, liaison, border patrol and police. It is also suitable for use by units operating as part of United Nations forces.

In addition to the three Swiss Army prototype Eagle vehicles, MOWAG also built a single vehicle as a company demonstrator.

In mid-1995, Denmark placed an order for 10 Eagle vehicles followed by an option on a further 26 vehicles, all of which have now been delivered. These have been used by the Danish Army as part of United Nations forces deployed to Albania and Bosnia.

Late in 1997, the Swiss government's Defence Procurement Agency (GR) awarded a contract worth SFr99 million to MOWAG for the supply of 175 Eagle II (4 × 4) armoured reconnaissance vehicles.

The Swiss Army has already taken delivery of 154 Eagle I vehicles under the designation of Aufklarungsfahrzeug/Vehicule d'Exploration 93/97. These Eagle I vehicles are mainly used as reconnaissance vehicles for the Leopard II MBT equipped armoured brigades.

Production of the latest Eagle II commenced at the MOWAG facilities in Kreuzlingen late in 1998 with first production vehicles being completed in 1999 and production will continue to the year 2001.

Externally, the Eagle II is almost the same as the Eagle I and is fitted with a MOWAG one person turret armed with 7.62 mm machine gun, smoke grenade launchers and day/night observation devices. The only distinguishing feature is that the rear side doors in the latest Eagle II have full depth bullet proof windows as have the front side doors.

The latest batch of 175 Eagle II vehicles are based on the new ECV (Enhanced Capacity Vehicle) HMMWV chassis supplied by AM General of the USA. The version provides a higher engine performance as well as increased payload carrying capacity.

MOWAG Eagle II (4 × 4) armoured reconnaissance vehicle fitted with MOWAG one-man turret armed with 7.62 mm machine gun and day/night sighting system **2001**/0093543

The correct Danish designation for the MOWAG Eagle I is the Pansret Spejdervogn (Spider) M/95 Eagle.

The correct Swiss Army designation for the MOWAG Eagle I is the Auklärungsfahrzeug 93 while the more recent Eagle II is the Aufklärungsfahrzeug 97.

Late in 2000, the Swiss Defense Procurement Agency (GR) awarded the Swiss company of MOWAG, now part of General Motors Defense, a contract covering the supply of 120 MOWAG Eagle III (4 × 4) Artillery Forward Observation Vehicles for use with Swiss Army 155 mm M109 self-propelled artillery units.

Under an earlier contract, the Swiss Army took delivery of two prototype vehicles which underwent extensive user trials. This resulted in the signing of an Option Agreement with MOWAG in March 2000 following the approval of the Defense Procurement Programme from 2001. Deliveries will run from 2002 through to 2004.

Description

The MOWAG Eagle armoured reconnaissance vehicle is based on the US AM General Heavy High Mobility Multipurpose Wheeled Vehicle (4 × 4) chassis which is purchased direct from the United States.

The performance and dimensions of the Eagle are identical to the HMMWV except that the Eagle is 100 mm wider.

The chassis has been fitted with an armoured superstructure designed by MOWAG which provides the occupants with protection from small arms fire and shell splinters. The armour package consists of composite armour plates fitted to the inner face of the aluminium plates.

The protection level of the hull and engine compartment is as follows:
7.62 mm ball fired from a distance of 30 m.
5.56 mm ball fired from a distance of 30 m.
7.62 mm steel core fired from a distance of 100 m.

The protection level around the front and rear wheel area is as follows:
7.62 mm ball fired from a distance of 30 m.
5.56 mm ball fired from a distance of 30 m.

The engine is at the front of the vehicle with the four crew members being seated in the centre, two on either side, each provided with an individual adjustable seat.

The driver is seated front left with the vehicle commander to his right. To their front is a bulletproof windscreen and provision is made in the roof above the

MOWAG Eagle II based forward observation vehicle with day/night sensor pod in raised position **2002**/0064867

MOWAG Eagle I armoured reconnaissance vehicle in service with the Danish Army during peace keeping missions in Albania **2002**/0120448

Specifications

(Eagle I)
Crew: 4
Configuration: 4 × 4
Combat weight: 5,100 kg
Unloaded weight: 3,900 kg
Max weight:
(front axle) 2,400 kg
(rear axle) 2,900 kg
Power-to-weight ratio: 31.3 hp/t
Length: 4.90 m
Width: 2.28 m
Height:
(over hull) 1.75 m
Ground clearance: 400 mm
Track: 1.81 m
Wheelbase: 3.30 m
Angle of approach/departure: 60/50°
Max road speed: 125 km/h
Max reverse speed: 35 km/h
Fuel capacity: 95 litres
Road range: 450 km
Fording: 0.76 m
Gradient: 60%
Side slope: 40%
Turning radius: 7.5 m
Engine: General Motors Type 6.5 1 NA 4-stroke V-8 diesel developing 160 hp at 1,700 rpm
Transmission: Hydra-Matic 4L 80 E automatic with 4 forward and 1 reverse gears
Transfer case: New Venture 2-speed
Steering: power-assisted, front wheels
Suspension: independent with double control arms
Brakes: disc on all wheels
Tyres: 37 × 12.50 R 16.5 LT with run-flat core
Electrical system: 24 V
Battery: 1 × 24 V, 125 Ah
Armament:
(main) 1 × 7.62 mm MG
Smoke-laying equipment: 1 × 6 76 mm smoke grenade launchers
Ammunition: 400 × 7.62 mm (ready use)
Gun control equipment
Turret power control: manual
Elevation/depression: +20/–10°
Turret traverse: 360°
Gun stabiliser:
(vertical) no
(horizontal) no
NBC system: yes
Night vision equipment: yes

The Eagle II is similar to the above but has a combat weight of 5,500 kg and an unloaded weight of 4,100 kg. Maximum weight on the front axle is 2,600 kg while maximum weight on the rear axle is 3,200 kg. Power-to-weight ratio is 34.5 hp/tonne while maximum road speed is 119 km/h. The liquid turbo charged General Motors diesel engine develops 190 hp at 3,400 rpm.

driver's and commander's positions for the mounting of a forward-facing night vision device.

The commander has a circular hatch in the roof that opens to the rear and in each side of the vehicle are two large doors that open forwards. Each of these has a bulletproof window in the upper part. To the rear of the commander and driver are the radio operator and the observer. The latter operates the observation turret when required.

Mounted on the roof of the vehicle is a one-man armoured observation cupola designated the MBK 2 which has also been developed and manufactured by MOWAG.

The cupola can be traversed through a full 360° and is fitted with a day/night thermal imaging device in the forward part of the turret, a roof-mounted day periscope in the right side of the turret roof and a single-piece hatch cover that opens to the rear.

Mounted externally on the right side of the turret is a 7.62 mm machine gun which can be aimed and fired from within the turret. This is provided with 400 rounds of ready use ammunition.

Weapon elevation is from –12 to +20° while the thermal imager has an elevation of +15° and a depression of –12°. If required the thermal imager can be dismounted for use in the ground surveillance role.

A bank of 76 mm six electrically operated smoke grenade dischargers is mounted on the turret rear and fires to the front of the vehicle.

In the rear of the Eagle, provision is made for the stowage of ammunition and personnel equipment and access to this area is via a large door in the hull rear that opens upwards. In addition, there are two emergency seats in the rear.

The Eagle reconnaissance vehicle is air-portable in a Lockheed Martin C-130 transport aircraft and some types of helicopter.

Standard equipment includes an NBC system of the overpressure type, run-flat tyres and communications equipment.

Optional equipment for the Eagle includes a 7 kW heater in the crew compartment, a 7.5 kW preheat unit, an air conditioning system, a central tyre-pressure regulating system and a front self-recovery winch with a pulling capacity of 5,400 kg.

Variant

Artillery observation vehicle

In 1998, under contract to the Swiss Defence Procurement Agency, MOWAG designed, built and delivered two of their latest Eagle II (4 × 4) light armoured vehicles configured specifically for use in the forward artillery observation role.

These two prototypes underwent extensive trials with the Swiss Army, that were completed in 1999.

This contract was placed late in 2000 for the supply of 120 vehicles based on the Eagle III vehicle. The observation and reconnaissance suite is the BAA system supplied by STN ATLAS of Germany. Deliveries are now underway to the Swiss Army.

According to MOWAG, from an observation distance of 7 km, targets can be measured within 30 seconds of the vehicle coming to a halt and within another 15 seconds of the vehicle moving off.

The new forward observation vehicle will be used to accurately locate targets under day and night conditions for the Swiss Army's fleet of 155 mm M109 series self-propelled howitzers, a large part of which is being upgraded by SW Swiss Ordnance Enterprise with a new 155 mm/47 calibre barrel.

The four man crew of the vehicle comprises a forward observer, surveyor, signalman and driver. Electronic equipment installed to carry out its mission includes a land navigation system, video camera, thermal image system and a laser range-finder.

The video camera, thermal imaging system and laser range-finder are installed in a sensor head on top of an electrically activated hinged arm mounted on top of the extended height roof. This can be operated by the forward observer, who is also the vehicle commander or the surveyor/range-finder, from under full armour protection.

The ring laser gyro land navigation system is moveable internally and gives the crew the position of the vehicle in azimuth and terrain angle at all time.

Once the target has been located and its co-ordinates obtained, this information is fed into the Swiss Army Integrated Artillery Fire Command and Control System (INTAFF) and then onto the battalion/battery by online data transmission.

The INTAFF units are mounted above the right-hand rear wheel arch. The appropriate Pentop is movable and mounted over the commander's work table which also allows access by the surveyor/range-finder. The communications system is mounted between the two front seats.

Eagle air defence

It has been proposed that the Eagle could be fitted with the Thales Air Defence Aspic four round surface-to-air missile launcher on the roof of the vehicle with the operator being seated inside.

Status

Production. In service with the following countries:

Country	Quantity	Comment
Denmark	36	Eagle I, all delivered
Switzerland	154	Eagle I, all delivered
Switzerland	175	Eagle II, deliveries under way
Switzerland	120	Eagle III AOV. Order placed late 2000, delivery 2002 to 2004

Manufacturer

MOWAG Motorwagenfabrik AG. In August 1999, MOWAG was taken over by the Diesel Divison, General Motors of Canada.

UPDATED

TURKEY

Otokar Cobra Armoured Reconnaissance/Surveillance Vehicle

Development

The Cobra Armoured Reconnaissance/Surveillance Vehicle has been developed by the Microwave and System Technologies Division of ASELSAN and Otokar to meet the operational requirements of the Turkish Land Forces Command (TLFC). In 1999, it was stated to be in operational service with the TLFC.

The system is essentially the Otokar Cobra (4 × 4) armoured personnel carrier, fully covered in the Armoured Personnel Carriers (Wheeled) section fitted with a new surveillance and sensor package developed by ASELSAN.

Prime role of the vehicle is reconnaissance, surveillance and intelligence, border monitoring and target information transmission to command and control units.

The Turkish company of ASELSAN AS, Microwave & System Technology Division, is now marketing a complete Reconnaissance/Surveillance and Intelligence System.

This comprises the following elements:

(1) Otokar Cobra Modular Armoured Reconnaissance/Surveillance Vehicle (MARS-V) which is essentially the Cobra covered in this entry.

(2) Otokar Armoured Command Vehicle with generator, air conditioning system, command and control software, extensive communications equipment, communication monitoring and audio recording system. There is also a roof-mounted SATCOM antenna which is folded down while travelling.

(3) Mobile Command and Control vehicle which is contained in a shelter carried on the rear of a Mercedes-Benz UNIMOG (4 × 4) truck chassis.

(4) Armoured Tactical DF (Direction Finding) Vehicle based on the Akrep (Scorpion) armoured personnel carrier (covered in the Armoured personnel carriers - wheeled section).

Through BAE Systems, RO Defence, a version of the Otokar Cobra was entered in the British Army's Future Command and Liaison Vehicle Competition, but it was not successful.

Cobra Armoured Reconnaissance/Surveillance Vehicle with mast-mounted sensor pod raised into operating position
***2000**/0064885*

Description

This is generally similar to the Cobra (4 × 4) Armoured Personnel Carrier (APC), covered in detail in the Armoured Personnel Carriers (Wheeled) section, with the hull rear modified to accept the sensor pod. This version retains the two side doors but there are no doors in the hull rear as this is occupied by the electronics and retracted sensor pod.

The Cobra has an all-welded monocoque steel hull which provides the occupants with protection from 7.62 mm armour piercing small arms fire, shell splinters and mines.

The complete running gear and power pack, consisting of a 190 hp diesel engine coupled to a four-speed automatic transmission, is from the US AM General High Mobility Multipurpose Wheeled Vehicle (HMMWV) which has been built in large numbers for the home and export markets.

The hull has been modified at the rear to accommodate the mast-mounted sensor pod. This consists of an ASELSAN ARS 2000 ground-surveillance radar, second-generation thermal imager (option) and a day/TV. As an option, a target co-ordination system can also be installed.

Communications equipment fitted includes a Type 9600 VHF frequency-hopping radio, SK-4000 digital encrypted radio (option), video transmission system (via a microwave link) and a field telephone.

The command and control software includes a geographical information system, data base and a formatted message device. According to ASELSAN, the sensor package has a typical range of up to 38 km on a vehicle convoy and up to 15 km against personnel when using the battlefield surveillance radar.

Although the TLFC uses the Cobra as the platform for this sensor pod, ASELSAN stress that this can be incorporated on other types of platform, tracked and wheeled.

Status

Production. In service with the Turkish Land Forces Command.

Manufacturer

Otokar Otobus Karoseri Sanayi AS.

UPDATED

Otokar Akrep (Scorpion) light reconnaissance vehicle

Development

The Akrep (Scorpion) light reconnaissance vehicle was developed by Otokar Otobus Karoseri Sanayi AS with the first prototypes being completed in May 1993 and the first production vehicles in June 1994.

In addition to being used in the light reconnaissance role other suggested missions include escort, perimeter control, counter-insurgency and light attack.

The vehicle builds on Otokar's extensive experience in the manufacture of light vehicles for both the civilian and military markets. By mid-2001, there were no known exports of the Akrep light reconnaissance vehicle.

Description

The Akrep has an all-welded steel hull with some 70% of the automotive components being from the well-known Land Rover Defender 90/110 (4 × 4) vehicle. This has been manufactured under licence in Turkey for many years and is in service with the Turkish armed forces in large numbers.

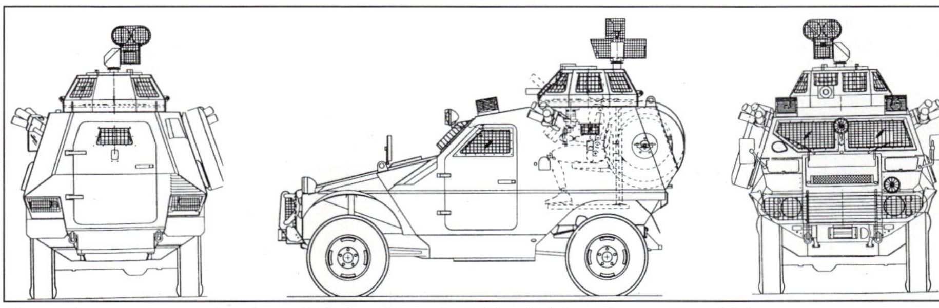

General arrangement drawing of Otokar (Scorpion) (4 × 4) internal security vehicle

The monocoque steel armour hull provides protection from small arms fire up to 7.62 mm ball ammunition fired at 90° incidence and point-blank range, and shell splinters. All windows and vision blocks are of laminated construction armoured glass with a polycarbonate anti-spall liner.

The layout is conventional, with the engine at the front, driver and commander in the centre with the machine gun installation being mounted on the roof towards the rear with the gunner seated below.

To the front of the commander and driver is a bulletproof window and in each side is a door with a bulletproof window in the upper part and a firing port below.

There is also a firing port and an associated vision device in either side of the hull and a large door in the hull rear that opens to the left. This is provided with a firing port and a vision block above. In the roof of the vehicle, towards the rear, is an oblong roof hatch that opens to the rear.

Otokar Akrep (Scorpion) (4 × 4) light reconnaissance vehicle fitted with twin roof-mounted 7.62 mm MGs and FLIR system (Michael Jerchel) ***2002***/0114339

Otokar Akrep (Scorpion) (4 × 4) light reconnaissance vehicle fitted with roof-mounted remote-controlled machine gun ***2000***/0064850

Specifications
Crew: 3
Configuration: 4 × 4
Combat weight: 3,200 kg
Power-to-weight ratio: 37 hp/t
Length:
(with winch) 4.19 m
(without winch) 4.025 m
Width: 1.91 m
Height:
(with MG mount) 2.563 m
(without MG mount) 2.01 m
Ground clearance: 230 mm
Track: 1.486 m
Wheelbase: 2.694 m
Angle of approach/departure: 67/48°
Max road speed: 125 km/h
Fuel capacity: 110 litres
Range:
(petrol engine) 650 km
(diesel engine) 1,000 km
Gradient: 70%
Side slope: 40%
Vertical obstacle: 0.315 m
Fording: 0.65 m
Turning radius: 6.6 m
Engine: Land Rover V-8 3.5 l water-cooled petrol developing 134 hp at 5,000 rpm or Rover 300 TDI 4-cylinder turbocharged diesel developing 111 hp at 4,000 rpm
Transmission: manual, 5-speed forward and 1 reverse or automatic 4-speed
Clutch: single dry plate
Transfer box: 2-speed, high and low ranges
Axles: live beam with single reduction hypoid differentials
Suspension: coil springs and hydraulic dampers Panhard rod and anti-roll bars
Tyres: 8.25 R16 size PR8 tubeless
Brakes:
(main) servo-assisted front and rear disc on all wheels
(parking) drum-type operating on transmission
Steering: power assisted
Electrical system: 12 V
Alternator: 65 A
Armament:
(main) 2 × 7.62 mm MG
Smoke-laying equipment: optional
Ammunition:
(ready use) 230 × 7.62 mm
(reserve) 3,000 × 7.62 mm
Gun control equipment
Turret power control: manual
Elevation/depression: +70°/–20°
Turret traverse: 180° (left and right)
Gun stabiliser:
(vertical) no
(horizontal) no
NBC system: no
Night vision equipment: yes

Note: Above specifications relate to Akrep with twin remote-controlled 7.62 mm machine guns.

The machine gun mount fitted on the roof is normally armed with a 7.62 mm machine gun which can be aimed and fired from within the vehicle. It is understood that the overhead weapon stations used in these Akrep vehicles are supplied by the RAFAEL Armament Development Authority of Israel.

The mount can be traversed 180° left and right and the 7.62 mm machine gun can be elevated from –20 to +70°. In addition to the 230 rounds of ready use ammunition, 3,000 rounds are carried inside the vehicle.

Firing is electrical with safety locking and braking together with a mechanical back-up. A last round indicator is also provided.

The day sight has a wide field of view (22°) with ×1 magnification including a collimated aiming circle, and a narrow field of view (8°) with ×8 magnification and a ballistic graticule.

The night sight has a wide field of view (22°) with ×1 magnification and a collimated aiming circle and a narrow field of view (7°) with ×7 magnification and a 25 mm image intensifier tube and a ballistic graticule.

Standard equipment includes an air conditioning system, a heater and demister, seat belts, two-speed windscreen wipers with wash facility, rifle clips, an infra-red driving headlamp, a blackout lighting system, a smoke extraction fan mounted on the roof, a battery master switch, antenna mounting provision, special insulation for roof and body interior from thick polyurethane for thermal and noise insulation, front and rear recovery hooks, a first aid kit, stowage for extra gun barrel and a weapon cleaning kit, a toolkit, an explosion suppressed fuel tank and Hutchinson run-flat tyres.

Optional equipment includes a front-mounted electrical winch with 25 m of cable with a capacity of 3,600 kg, 66 mm smoke grenade launchers, a pioneer kit including axe, shovel and pickaxe, a communication system, a land navigation system and 24 V suppressed vehicle electrics. The vehicle has been observed with a spare wheel and tyre mounted on the right side of the hull towards the rear.

Variants

As an alternative to a 7.62 mm machine gun turret a similar remote-controlled weapon station with a 12.7 mm M2 heavy machine gun is available.

This turret allows protected firing from inside where all gunner controls, including ammunition feed, are situated. A combined day/night system is fitted.

The latest versions, which were delivered in 1996 have twin 7.62 mm remote-controlled machine guns and a FLIR (Forward Looking Infra-Red) sight and CCD camera as a sight system.

The gunner scans the area of expected approach of the enemy by FLIR and/or CCD (Charge Coupled Device) camera and the image is transferred to two monitors, one in front of the gunner and the other in front of the commander. The sight system is boresighted with the machine guns and incorporates a ballistic reticule. The vehicle is also fitted with VCR (Video Cassette Recorder) equipment to record the incidents.

The vehicle can also be adapted for special missions armed with anti-tank guided weapons or surface-to-air missiles, or it can be fitted out with a ground surveillance radar or as a command and control vehicle.

Armoured personnel carrier

This has a similar hull but extended slightly to the rear to create more internal volume. Full details are given in the Armoured personnel carriers (wheeled) section.

Internal security vehicle

This is the latest version developed for the Turkish Police and has a similar hull to the standard Akrep (Scorpion) light reconnaissance vehicle but has been modified for the internal security role. These modifications include:

(1) Internal security turret which can be traversed through 360° and is fitted with a teargas launcher and mast-mounted video camera which can be raised and lowered as required
(2) Video recording system with two monitors
(3) Wheel arch fire suppression system
(4) Powerful siren and public address system
(5) Wire mesh protection for all windows and observation devices.

Optional equipment includes front-mounted barricade removal equipment, remote-controlled front searchlight, front-mounted electric winch, wire cutter, additional gun ports, blackout lighting system, infra-red driving headlamp system, vehicle-mounted CN/CS gas sprayer and underbelly protection shields.

Status

Production. In service with Turkish Army and security forces. By mid-2002 over 250 vehicles, including variants had been delivered.

Manufacturer

Otokar Otobus Karoseri Sanayi AS.

UPDATED

UNITED KINGDOM

Alvis Vehicles Stormer 30 armoured reconnaissance vehicle

Development

The low-profile Stormer 30 armoured reconnaissance vehicle/light tank has been developed by Alvis Vehicles Limited as a private venture to meet potential user requirements for a high-mobility air-portable armoured reconnaissance vehicle.

It builds on the Alvis Vehicles experience in the design, development and production of some 3,700 Scorpion and Stormer full-tracked light armoured vehicles for the UK and export markets. Production of the Scorpion and Stormer series of vehicles was originally undertaken in Coventry but is now undertaken in Telford.

Development work on the Stormer 30, which uses components already proven and in production wherever possible, commenced in 1995 with the first prototype being completed in 1997.

Stormer 30 can be lifted by a Sikorsky CH-53 helicopter when fully laden, or a Boeing CH-47 Chinook unladen and will fit into a Lockheed Martin C-130 Hercules transport aircraft. It can also be moved by commercial truck, on a rail flat car or in a landing craft.

Stormer 30 has been demonstrated in the UK and Middle East but as of mid-2001, it remained at the prototype stage.

Description

The hull of the Stormer 30 is of all welded aluminium armour with an additional layer of spaced steel armour on the front and sides which provides protection against 14.5 mm armour-piercing rounds through the frontal arc. Internally there is also a spall liner.

The space between the aluminium armour hull and the steel spaced armour is used for the stowage of kit. The hull sides slope slightly inwards towards the bottom to decrease the thermal signature of the vehicle. The armour also provides a high level of protection against mines and artillery splinters.

The driver is seated at the front of the vehicle on the left side with the engine compartment to his right.

The driver has an adjustable seat for driving in the head-out position and a single-piece hatch cover that opens to the rear. This is provided with a single wide-angle day periscope which can be quickly replaced by an image intensification or thermal night vision device. The hatch cover can also be partly opened to allow head-out driving while still retaining some overhead armour protection.

The power pack consists of the Cummins 6B-275 six-cylinder diesel developing 250 hp at 2,600 rpm coupled to a David Brown Defence Equipment T303 crossdrive automatic transmission with integral steering system.

Alvis Vehicles Stormer 30 armoured reconnaissance vehicle from the front (Christian Dumont) ***2000***/0064851

Mounted in the centre of the hull, slightly to the rear, is the Italian Otobreda HITFIST turret with the commander seated on the left and the gunner on the right. Both crew members have a single-piece hatch cover that opens to the rear. The turret is of all welded aluminium armour construction.

The gunner has a stabilised day/night sight while the commander has five day periscopes for all-round observation and a roof-mounted day/image intensification sight mounted in front of his hatch.

Main armament comprises The Boeing Company 30 mm Bushmaster II dual-feed cannon with a 7.62 mm machine gun mounted coaxially to the right.

A bank of four electrically operated smoke grenade launchers is mounted either side of the turret firing forwards, and mounted on either side of the turret is a single launcher for a Raytheon Systems Company TOW anti-tank guided missile with a maximum range of 3,750 m.

Turret traverse and weapon elevation is all electric with manual controls for emergency use. A weapon stabilisation system is fitted as standard and this allows the weapon to be laid onto the target while the vehicle is moving across country.

Although the prototype Stormer 30 has an Otobreda turret, other types of turret could also be fitted according to customer requirements.

For emergency crew exit and ammunition resupply, a large door is located in the hull rear which opens to the right. Access to the hull side stowage areas is via doors located in either side of the hull rear above the tracks.

The suspension is of the torsion bar type with either side having six dual rubber-tyred road wheels with the drive sprocket at the front and idler at the rear. Horstman Defence Systems lever-actuated rotary dampers are provided for the first, second and sixth road wheel stations.

There are two track-return rollers and the upper part of the track is covered by a light steel covering to reduce dust.

Standard equipment includes an NBC system and various other sighting and communication systems can be fitted as required by the user.

Variants

There are no variants of the Stormer 30 although a variety of other turrets could be fitted. The baseline Stormer armoured personnel carrier, covered in detail in the Armoured personnel carriers (tracked) section, can be used for the more specialised roles such as command post vehicle, load carrier, troop carrier and ambulance, for example.

Status

Prototype.

Manufacturer

Alvis Vehicles Limited.

UPDATED

Specifications

Crew: 3
Max combat weight: 13,000 kg
Power-to-weight ratio: 21 hp/tonne
Ground pressure: 0.48 kg/cm²
Length: 5.251 m
Width:
(overall) 2.690 m
(over tracks) 2.475 m
Height:
(hull top) 1.60 m
(turret top) 2.495 m
(top of sight) 2.721 m
Ground clearance: 0.42 m
Track: 2.05 m
Track width: 410 mm
Length of track on ground: 3.115 m
Max road speed:
(forwards) 80 km/h
(reverse) 80 km/h
Min sustainable speed: 4 km/h
Acceleration:
(0/50 km/h) 20 s
(50/65 km/h) 20 s
Fuel capacity: 390 litres
Max range: 400 km (mixed terrain)
Fording: 1 m
Gradient: 60%
Side slope: 40%
Vertical obstacle: 0.60 m
Trench: 2 m
Engine: Cummins 6B-275 6-cylinder diesel developing 250 hp at 2,600 rpm
Transmission: David Brown Defence Equipment T303 crossdrive automatic with integral steering system
Clutch: centrifugal clutch
Final drive: single epicyclic reduction 3.667 : 1
Suspension: transverse torsion bar and lever-actuated rotary Horstman dampers on first, second and sixth road wheel stations
Wheel movement: 275 mm (bump), 96 mm (rebound)
Main brakes: caliper discs, hydraulically operated
Steering brakes: caliper discs, hydraulically operated
Parking brake: hand operated, spring applied, hydraulically released
Electrical system: 28.5 V
Batteries: 4 (located in vehicle hull)
Alternator: 140 Amp at 28 V
Armament:
(main) 1 × 30 mm Bushmaster II cannon
(coaxial) 1 × 7.62 mm MG
(ATGW) 2 × Raytheon Systems Company TOW
Smoke-laying equipment: 2 × 4 electrically operated smoke grenade
Ammunition:
(ready use) 180 × 30 mm
700 × 7.62 mm
2 × Raytheon Systems Company TOW ATGW
Turret power control: electric/manual
(by commander) yes
(by gunner) yes
Max rate power traverse: 60°/s max
Max rate power elevation: 60°/s max
Gun elevation/depression: +60/–10°
Turret traverse: 360°
TOW launcher elevation/depression: +30/–7.5°
Gun stabiliser:
(vertical) yes
(horizontal) yes
NBC system: yes
Night vision equipment: yes, passive
Air transportability:
CH-53 (external, loaded)
CH-47 (external, unladen)
C-130 (internal)

Alvis Vehicles Saladin armoured car

Development

In January 1946, the British Army issued a requirement for a new armoured car as a replacement for the Daimler Mk II (4 × 4) and AEC Mk III (4 × 4) armoured cars developed during the Second World War. Design work on this new vehicle, called the FV601(A), began in the same year at the Fighting Vehicles Design Department (FVDD). This facility then became the Fighting Vehicles Research and Development Establishment (FVRDE) and then the Royal Armament Research and Development Establishment Chertsey. This establishment will close in the near future.

As originally conceived, the FV601(A) was to have had a crew of four and be armed with a 2-pounder gun fitted with a Littlejohn Adaptor to increase the muzzle velocity. But in February 1948 it was decided that the 2-pounder was not powerful enough so a new weapon, the 76 mm gun L5, was designed by the Armament Design Establishment at Fort Halstead but was not ready until 1953. In 1947, a contract was awarded to Alvis Limited of Coventry to build two prototypes of the FV601 which were completed in 1953. The first mock-up, fitted with a 2-pounder gun, was completed in 1948.

As the FV601 prototypes were being built, Alvis was also building the FV603 armoured personnel carrier which eventually became known as the Saracen. As the Saracen was urgently required for operations in Malaya it was decided that Alvis would concentrate on this vehicle and the six preproduction vehicles, the FV601(B), would be built by Crossley Motors at Stockport in Cheshire.

The FV601(B)s were completed in 1955 and had a slightly different shaped turret from the prototype vehicles as well as different vision equipment and different hatches for the commander and gunner.

Production was authorised in 1956 and the vehicle entered production at Coventry in 1958 with first vehicles delivered to the army the following year. Production continued until 1972 by which time 1,177 vehicles had been built. The vehicle was succeeded in the British Army by the Alvis Scorpion CVR(T) which is armed with a new and lighter version of the 76 mm gun of the Saladin called the L23. The Saladin is officially known as the FV601(C) Armoured Car 76 mm Gun (Alvis Saladin Mk 2 6 × 6).

The Saladin has been replaced in all front-line British units by the Alvis Scorpion CVR(T). More recently, the 76 mm armed version of the Combat Vehicle (Reconnaissance) Tracked family of vehicles, Scorpion, has been phased out of service with the British Army. All other members remain in service including the more recent Sabre which had the turret of the Fox armoured car armed with a 30 mm RARDEN cannon.

Kuwait Army Saladin armoured cars were used against Iraqi units during the invasion of Kuwait City in the Summer of 1990.

Production of all of the FV600 series of 6 × 6 armoured vehicles was undertaken at the Alvis Vehicles facility in Coventry which closed down in 1999. Although no longer in service with the British Army, Alvis Vehicles is still the overall design authority.

Specifications

Crew: 3
Configuration: 6 × 6
Combat weight: 11,590 kg
Unloaded weight: 10,500 kg
Power-to-weight ratio: 14.66 hp/t
Ground pressure: 1.21 kg/cm^2
Length:
(gun forward) 5.284 m
(hull) 4.93 m
Width: 2.54 m
Height:
(to top of gunner's periscope) 2.39 m
(to turret roof) 2.19 m
Firing height: 1.854 m
Ground clearance: 0.426 m
Track: 2.038 m
Wheelbase: 1.524 + 1.524 m
Angle of approach/departure: 60/50°
Max road speed: (forwards) 72 km/h
Fuel capacity: 241 litres
Max range: 400 km
Fording: 1.07 m
(with kit) 2.13 m
Gradient: 46%
Vertical obstacle: 0.46 m
Trench: 1.52 m
Turning radius: 7.31 m
Engine: Rolls-Royce B80 Mk 6A 8-cylinder petrol developing 170 bhp at 3,750 rpm
Transmission: preselector 5-speed
Steering: recirculating ball, hydraulic-assisted
Clutch: fluid coupling
Suspension: independent, double-wishbone and torsion bar, hydraulic shock-absorbers
Tyres: 12.00 × 20
Brakes:
(main) hydraulic disc all wheels
(parking) mechanical all wheels
Electrical system: 24 V
Batteries: 2 × 12 V, 60 Ah
Armament:
(main) 1 × 76 mm L5A1 rifled gun
(coaxial) 1 × 7.62 mm L3A3/L3A4 MG
(anti-aircraft) 1 × 7.62 mm L3A3/L3A4 MG
Smoke-laying equipment: 2 × 6 66 mm smoke grenade dischargers
Ammunition:
(76 mm) 42
(7.62 mm) 2,750
Gun control equipment
Turret power control: electric/manual
(by commander) yes
(by gunner) yes
Gun elevation/depression: +20 to –10°
Gun stabiliser:
(vertical) no
(horizontal) no
Range setting device: yes
Traverse indicator: yes
Armour
Hull glacis: 12 mm at 45°
Hull nose: 14 mm at 42°
Hull sides: 16 mm at 20°
Hull top: 10-12 mm
Hull floor: 8-12 mm
Hull rear: 16 mm
Turret front: 32 mm at 15°
Hull sides: 16 mm
Hull rear: 16 mm
Hull top: 10 mm
NBC system: no
Night vision equipment: no

Description

The all-welded steel hull of the Saladin is divided into three compartments: driver's at the front, fighting in the centre and power pack at the rear.

The driver is seated at the front of the vehicle and is provided with a single-piece hatch cover that folds forwards onto the glacis plate for improved visibility. The driver has three No 17 day periscopes, one in the hatch cover and one on either side of his position.

The all-welded turret is in the centre of the hull with the gunner seated on the left and the commander, who also acts as the loader, on the right, both with a single-piece hatch cover that opens to the rear. The commander has four No 17 day periscopes arranged around the forward part of his hatch and a single swivelling day periscope to the rear. The gunner has a day periscope forward of his hatch cover, the lower part with a magnification of x6 for fire control and the upper part with a x1 magnification for observation. On the left side of the turret, just to the rear of the mantlet, is an extractor fan. A small rectangular observation hatch is provided in the rear of the fighting compartment, just below the turret ring. Either side of the turret below the turret ring is a crew escape hatch.

The engine compartment at the rear of the hull is separated from the crew compartment by a fireproof bulkhead and is fitted with a fire warning system and a fire extinguishing system, the latter being operated by the driver.

Air for the engine compartment is drawn in by two fans via six louvred engine covers and then passed through the radiator and expelled through grilles at the very rear of the hull.

The engine is coupled to a Daimler fluid coupling which in turn is coupled to a Daimler epicyclic preselective gearbox which transmits power to the transfer box consisting of a bevel and helical gear, incorporating reverse and a differential and enabling the five gears to be used in both directions. This transfers the drive direct to each centre bevel box and then via transmission shafts to the front and rear axles. Each wheel houses epicyclic reduction gears and is connected to the bevel box by a transmission shaft and two universal joints.

The suspension at each wheel station consists of an upper and lower link, with the upper attached to a longitudinal torsion bar which is splined to a tube and in turn is secured to the hull. Steering is hydraulically power assisted on the front and centre roadwheels.

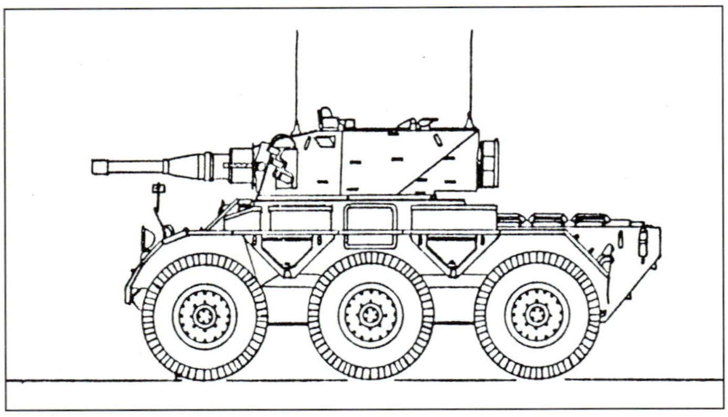

Alvis Saladin armoured car

Alvis Saladin armoured car with driver's, commander's and gunner's hatches open

The Saladin armoured car, as originally built, does not have an NBC system or night vision equipment.

Main armament of the Saladin consists of a now BAE Systems, RO Defence 76 mm L5A1 gun with a vertical sliding breech block and a hydro-spring recoil mechanism. A 7.62 mm (0.30) machine gun model M1919A4(L3A3/L3A4) is mounted coaxially to the left of the main armament and a similar anti-aircraft weapon is mounted on the right of the commander's hatch. Six 66 mm smoke grenade dischargers fitted on either side of the turret are electrically fired from within the vehicle. The 76 mm gun fires the following types of ammunition: canister, HESH, HE, HE/PRAC, SH/P, smoke/BE and illuminating.

The 42 rounds of ammunition are stowed vertically: 11 between the commander and the gunner, 12 on the left and 11 on the right behind the driver and 8 at the left rear of the hull.

The 76 mm ammunition is manufactured by the now BAE Systems, RO Defence.

Saladin repower

In 1991, it was announced that A F Budge Limited and Alvis Vehicles Limited had developed a repower package for the Saladin armoured car. The existing petrol engine is replaced by a Perkins Engines Company 180 MTi diesel developing 180 hp.

Additional improvements have been made to the chassis including a modern cooling pack, a new electrical system, upgraded brakes, a fire suppression system and the replacement of the original transmission with a more modern automatic transmission.

The diesel installation requires only minor changes to the engine compartment and offers the advantages of improved reliability and simplified maintenance, lower fuel consumption, increased operational range and reduced risk of fire.

This conversion package has already been trialled in Asia together with an upgraded Saracen (6 × 6) armoured personnel carrier, while the Alvis Stalwart with the same has been evaluated by the UK.

In May 1994, Alvis Vehicles announced that it had been awarded a contract worth US$10 million for the supply of upgrade kits for Saladin, Saracen and Ferret vehicles from an undisclosed country in Asia, believed to be Indonesia.

Status

Production complete. In service with:

Country	Quantity	Comment
Bahrain	8	
Honduras	50	
Indonesia	69	16 upgraded
Jordan	40	police, quantity unconfirmed
Lebanon	40	
Mauritania	40	
Portugal	29	
Sri Lanka	26	
Sudan	90	
Tunisia	24	
UAE	20	reserve

Manufacturer

Alvis Vehicles Limited.

UPDATED

Alvis Scarab fast attack/scout vehicle

Development

Several years ago, Alvis Vehicles, who have now consolidated all of their research, development and production facilities at Telford, Shropshire, built as a private venture, a 4 × 4 fast armoured scout vehicle called the Scarab.

Scarab was first shown in 1997 in the Middle East fitted with an externally mounted LIW turret armed with a 20 mm cannon.

The first example of Scarab (called P1) was essentially a technology demonstrator and used automotive and running gear from the German Mercedes-Benz UMIMOG 1600 series 4 × 4 chassis cross-country truck.

Following a detailed analysis of potential customers' operational requirements, a decision was taken to build two production-standard Scarab vehicles (D1 and D2) with one additional hull for extensive ballistic tests (D3).

These two production-standard vehicles incorporate a number of improvements in response to potential customer inputs including an increase in internal volume, longer wheelbase and are based on automotive components of the latest production standard Mercedes-Benz 2100 series UNIMOG (4 × 4) truck.

One of the production standard Scarab vehicles was shown for the first time at the Defence Systems and Equipment International Exhibition and Conference held at Chertsey, Surrey, southern England, in September 1999. This was with a full sensor suite to meet the Belgian Army Reconnaissance 2000 requirement.

The first Scarab technology demonstrator had an internal volume of 3.4 m^3 but the latest version has 5.26 m^3 of reconfigurable internal space. This enables the vehicle to meet a wide range of user requirements from C^3I to reconnaissance and liaison.

To reduce overall life-cycle costs, Commercial-Off-The-Shelf (COTS) sub-systems are used wherever possible in the Scarab design with protection and mobility being designed into the vehicle.

The Alvis Vehicles Scarab is one of the remaining contenders for the British Army's Future Command and Liaison (FCLV) requirement (qv). It was one of the three vehicles tested by the Belgian Army in mid-2001 to meet its Reconnaissance 2000 requirement. The latter version was fitted with a special version of the HELIO SWARM turret, covered in the description.

Alvis Vehicles Scarab armed with pintle-mounted 12.7 mm M2 machine gun **2002**/0114340

Alvis Vehicles Scarab fitted with Helio Stabilised Weapon and Reconnaissance Mount (SWARM) armed with a 12.7 mm M2 machine gun (Christopher F Foss) **2002**/0067845

Description

The hull of the Scarab is of all-welded steel armour construction that provides protection against 7.62 mm Armour Piercing (AP) attack through a full 360° in azimuth and from –10 to +30° in elevation.

Over the frontal 60° arc, protection is provided against 12.7 mm AP attack at point-blank range and against 14.5 mm AP attack over a similar arc at all ranges over 550 m.

The armour also provides protection against 155 mm artillery fragments and anti-tank mines containing up to 7 kg of high explosive under the wheels. The panels above the wheels have been designed to blow away in the event of a vehicle running over a mine, so the hull does not contain the mine blast.

If required by the user, an enhanced mine protection kit developed by Mecham Consultants is also available. This provides protection against TMRP-6 and all lesser threat Self-Forging Fragment (SFF) mines exploding anywhere under the vehicle. The increased weight of this kit means there is a reduction in the power-to-weight ratio of the Scarab.

The driver is seated at the front of the vehicle with the other two primary crew members being seated to the rear with observer on left and commander on right.

Specifications

Crew: 3
Configuration: 4 × 4
Combat weight: 10,400 kg*
Weight empty: 9,100 kg
Power-to-weight ratio: 25.4 hp/t
Length overall: 5.17 m
Width overall: 2.45 m
Height overall: 1.95 m (without weapons)
Ground clearance: 0.48 m
Track: 1.963 m
Wheelbase: 2.81 m
Angle of approach/departure: 55/45°
Max road speed: 110 km/h
Fuel capacity: 160 litres
Range:
(road) 600 km (at a constant speed of 70 km/h)
Fording: 1.10 m
Gradient: 100%
Vertical obstacle: 0.35 m
Trench with channels: 0.60 m
Turning radius: 8.6 m
Engine: Mercedes-Benz OM906 6-cylinder turbocharged intercooled diesel developing 231 hp at 2,300 rpm
Transmission: Allison MD 3560 six-speed automatic with torque converter
Steering: hydraulic power assisted
Suspension: coil springs with inner elastomere springs and telescopic shock-absorbers, anti-roll stabilisers for front and rear axles
Tyres: Michelin XZL 365/85 R20 tubeless (run-flats)
Brakes:
(main) disc, double front, single rear
(parking) mechanical
Electrical system: 24 V
Batteries: 4 × 110 Ah Chloride Armasafe
Armament: depends on role
NBC system: optional
Night vision equipment: optional
Armament: see text
* Maximum possible payload is 2,500 kg
(This is for baseline vehicle without additional mine protection)

Close-up of Alvis Vehicles Scarab fitted with Helio Stabilised Weapon and Reconnaissance Mount (SWARM) armed with 12.7 mm M2 machine gun (Christopher F Foss) ***2002***/0067236

The actual seating arrangement and number of crew depends on the user's requirement and up to five people can be carried if required.

To the front and either side of the driver's position is a large bullet-proof window with similar windows in either side. These provide protection against attack from 7.62 mm small arms fire. For driving at night, the driver could use head-mounted night vision goggles.

In either side of the hull is a large door that opens to the rear and there are typically three roof hatches, one for the driver and one for each other crew member, but again this depends on the version.

The power pack is mounted at the rear of the hull and consists of a EURO II (with option of EURO III in the future) standard Mercedes-Benz turbo-charged diesel developing 231 hp coupled to an Allison MD 3560 six-speed fully automatic transmission which, in turn, is coupled to an Alvis Vehicles reversing transfer box.

Mercedes-Benz portal axles are fitted with wheel hub gears and with pneumatically operated four-wheel drive and differential locks in both axles. The suspension consists of coil springs and telescopic shock-absorbers and anti-roll stabilisers for front and rear axles.

Steering is hydraulic power assisted and the tyres are fitted with run-flat inserts as standard. As an option a central type pressure regulation system can be fitted that allows the driver to adjust the tyre pressure to suit the type of terrain being crossed.

Environmental equipment includes an air conditioning system, crew water heater and full collective NBC protection.

Armament depends on battlefield mission but can include 7.62 mm and 12.7 mm machine guns, 40 mm grenade launchers, 20 mm cannon and various anti-armour systems such as the Euromissile MILAN anti-tank guided missile.

There are two baseline versions of the British Army FCLV, scout and utility. It is probable that the former will be fitted with a mast-mounted sensor pod containing day and thermal sights and a laser range-finder. The thermal sight will be the same as that selected for the British Army's Battle Group Thermal Imaging (BGTI) programme.

Both scout and utility will be armed with a 12.7 mm M2 machine gun. The Scarab shown at Chertsey to meet the Belgian Reconnaissance 2000 requirement, features a Helio Stabilised Weapon and Reconnaissance Mount (SWARM) armed with a 12.7 mm M2 machine gun, day/thermal cameras and an eye-safe laser range-finder with electrically operated smoke grenade launchers mounted lower down.

The sensor/weapon pod can be operated by either of the crew members seated to the rear of the driver under complete armour protection, using a Barco flat display screen with joystick type controls. This system, designed by Alvis Vehicles, is called the Turret Control Panel and forms the heart of the system which integrates the SAGEM BMS, Helio SWARM, IRIS thermal imager, laser ranger-finder, day camera and navigation system.

This flat display screen is mounted on a rail, so it can be rapidly slid from one side of the vehicle to the other to be used by either of the two rear crew members. For the Belgian requirement, a SAGEM battlefield management system is fitted.

The Scarab is also suitable for parachute and LAPES (Low-Altitude Parachute Extraction Equipment) operations. Standard equipment includes NATO tow hook at rear, inter-vehicle start socket and 12-pin NATO trailer sockets.

Variants

There are no variants of the Alvis Vehicles Scarab although its design is such, that it can be rapidly adopted for a wide range of battlefield missions.

Status

Prototypes undergoing trials.

Manufacturer

Alvis Vehicles Limited.

UPDATED

Future Command and Liaison Vehicle (FCLV)

Development/Description

In mid-1999, contenders to meet the British Army's new requirement for a wheeled Future Command and Liaison Vehicle (FCLV) submitted their Pre-Qualification Questionnaire (PQQ) to the Defence Procurement Agency (DPA) who are running this competition.

This will be an off-the-shelf purchase or an existing vehicle modified to meet the FCLV mission.

The original requirement was for 500 FCLVs to replace a variety of vehicles currently used for this role such as the Alvis Vehicles Spartan and Alvis Vehicles Saxon (4 × 4) armoured personnel carriers and un-armoured Land Rover (4 × 4) light vehicles. This requirement has now increased to just over 600 vehicles.

These vehicles will be used by the infantry, Royal Armoured Corps, Royal Artillery, Royal Electrical and Mechanical Engineers and Royal Engineers.

There will be two versions, general/utility and scout with about 75 per cent being the former. Typical roles of the FCLV will be reconnaissance, command, liaison and radio rebroadcast.

The scout version is expected to have an all-weather sensor package for day/night observation and target ranging using a laser range-finder.

If the FCLV does move ahead, then it is considered unlikely that the UK DPA will exercise its option for a further 443 diesel upgrade packages for the Spartan armoured personnel carriers. These are currently used for this role following the withdrawal of the Ferret (4 × 4) scout car some years ago.

Armament of the FCLV will consist of a 12.7 mm M2 machine gun, which will be Government Furnished Equipment (GFE) as will be the new Bowman communications system. It is also required to have protection from small arms fire and at least anti-personnel mines.

In early 2000, the UK Defence Procurement Agency (DPA) stated that there were seven remaining contenders to meet the British Army's new requirement for a wheeled (4 × 4) Future Command and Liaison Vehicle (FCLV).

These were:
Alvis Vehicles (UK) with Scarab
BAE Systems and DSD (UK) with Turkish Otocar Cobra
Hunting Engineering (UK) with the French ACMAT TPK 430 BL
IVECO (Italy) with the Puma
Krauss-Maffei Wegmann (Germany) with All Protected Carrier Vehicle-2 (APCV-2)
SP aerospace and vehicle systems with the Fennek
Trading Force (UK) with the Vickers OMC RG-31
Vickers Defence Systems (UK) with the Swiss MOWAG Eagle

In mid-2001, the UK Defence Procurement Agency (DPA) awarded three UK contractors, each a £500,000 one-year Risk Reduction and Trials contract, for the next phase of the British Army competition for the Future Command and Liaison Vehicle (FCLV).

The three companies now shortlisted by the DPA are Alvis Vehicles (with Alvis Scarab Italian MLV 4 × 4 vehicle), Hunting Engineering (ACMAT VLRA) and Vickers Defence Systems (RG-31M and RG-32M).

The Alvis Vehicles Scarab was originally built as a private venture and was also one of the contenders for the Belgian Army Recce 2000 requirement. Hunting Engineering is proposing a vehicle based on the well proven ACMAT VLRA chassis with the support of Marshall Specialist Vehicles. The Vickers Defence Systems RG-31M and RG-32M leverage off the large number of RG-31 and RG-32 vehicles produced in South Africa by Vickers OMC.

All these vehicles are of the 4 × 4 type with varying levels of armour protection.

Some contractors are bidding for more than one vehicle as they believe no single vehicle can meet the FCLV requirement in the area of payload/capacity. FCVL will be used for a wide range of battlefield missions by different combat arms including infantry, artillery, armour, engineer and signals.

Each contractor will be allowed to submit a number of vehicles for the independent vehicle assessment phase due to start in November 2001. During this phase, the vehicles submitted by the contractor will be assessed to confirm their claimed specifications such as weight, speed, gradient, load carrying capacity and range.

The user trials will start early in January 2002 at the Armoured Trials and Development Unit (ATDU) at Bovington, southern England running through to mid-March. For this phase each contractor will be allowed to provide a maximum of two vehicles.

Early next year an ITT (Invitation to Tender) will be issued to all three contractors for full production and support of the FCLV.

A preferred bidder is expected to be announced late in 2002 and the contract placed in 2003. It is expected that between 500 and 600 vehicles will be purchased at a total cost of up to £370 million (including a complete logistic support package) with a projected in-service date of 2006.

The FCLV programme is being run by the DPA's Close Armour Integrated Project Team at Abbey Wood.

Status
British Army requirement.

Manufacturer
Not selected.

UPDATED

Fox light armoured car

Development
In the 1960s, the Fighting Vehicles Research and Development Establishment designed two reconnaissance vehicles, the Combat Vehicle Reconnaissance (Tracked) Scorpion (FV101) and the Combat Vehicle Reconnaissance (Wheeled) Fox (FV721), both of which use the same Jaguar 4.2 litre petrol engine.

Development of the Fox began in 1965 and the following year the Daimler company of Coventry, which was building the Ferret scout car at the time, was awarded a contract to build 15 prototype vehicles. The first was completed in November 1967 and the last in April 1969. User trials began in 1968 and the first official announcement concerning the Fox was made in October 1969.

The following year the Fox was accepted for service with the British Army and a production order was placed with Royal Ordnance Leeds. This facility was subsequently taken over by Vickers Defence Systems but has now closed.

Production began in 1972 and the first vehicle was completed in May 1973. Production of the Fox was undertaken at Royal Ordnance Leeds, which was purchased by Vickers Defence Systems in 1986. It is understood that final deliveries of the Fox were made in 1979.

Daimler completed production of the Ferret in 1971 and then closed down its armoured vehicle production facility.

The Fox has now been phased out of service with the British Army with some of the turrets removed and fitted on to Alvis Scorpion CVR(T)s which have had their turrets removed. These vehicles are then known as the Sabre. Details of this vehicle are given in the entry for the Alvis Scorpion reconnaissance vehicle in the Armoured fighting vehicles families section.

All of the welded aluminium turrets for the Fox light armoured car were manufactured by Alvis Vehicles. These were built at their Alvis facility which has now closed with work being transferred to Telford.

A small number of Fox turrets were fitted on FV432 full tracked armoured personnel carriers. These were deployed with the Berlin Brigade but some are now used for training purposes in the UK.

Vickers Defence Systems were the UK design authority for the Fox armoured car.

Specifications
Crew: 3
Configuration: 4 × 4
Combat weight: 6,120 kg
Unloaded weight: 5,733 kg
Power-to-weight ratio: 30.04 bhp/t
Length:
(gun forward) 5.08 m
(gun rear) 4.242 m
(hull only) 4.166 m
Width: 2.134 m
Height:
(to turret top) 1.981 m
(overall) 2.2 m
Ground clearance: 0.3 m
Track: 1.753 m
Wheelbase: 2.464 m
Max speed:
(road) 104 km/h
(water) 5.23 km/h
Acceleration: (0-48 km/h) 9 s
Fuel capacity: 145.47 litres
Max road range: 434 km
Fording: (amphibious with preparation) 1 m
Gradient: 46%
Vertical obstacle: 0.5 m
Trench with channels: 1.22 m
Turning radius: 6.1 m
Engine: Jaguar XK 4.2 litre 6-cylinder petrol developing 190 bhp at 4,500 rpm
Transmission: Daimler preselective 5-speed unit
Transfer case: Daimler spiral bevel directional control gears and double-helical drop down gear to differential provide, in combination with gearbox, 5 forward and 5 reverse speeds
Steering: power-assisted
Clutch: fluid flywheel
Suspension: independent
Tyres: 11.00 × 20, run-flat
Brakes:
(main) hydraulic disc all wheels
(parking) contracting band on front bevel box input
Electrical system: 24 V
Batteries: 2 × 6TN, 100 Ah
Armament:
(main) 1 × 30 mm RARDEN L21 cannon
(coaxial) 1 × 7.62 mm L37A2 MG
Smoke-laying equipment: 2 × 4 66 mm smoke grenade dischargers
Ammunition:
(main) 99
(coaxial) 2,600
(smoke grenades) 16
Gun control equipment
Turret power control: manual, 2-speed 360°
(by commander) yes
(by gunner) yes
Gun elevation/depression: +40 to –14°
Gun stabiliser:
(vertical) no
(horizontal) no
NBC system: yes
Night vision equipment: yes

Description
The Fox is a further development of the late production Ferret light scout car (covered in detail in a separate entry) and has an all-welded aluminium armour hull and turret which gives protection against medium and heavy machine gunfire and field artillery splinters.

The driver is seated at the front of the vehicle and is provided with a hatch cover that lifts and opens to the right with an integral wide-angle day periscope that can be quickly replaced by a passive night periscope.

The turret is in the centre of the vehicle with the commander/loader on the left and the gunner on the right, both with a single-piece hatch cover that opens to the rear. The commander has a day periscopic binocular surveillance instrument in a rotating mount with a magnification of ×1 and ×10 and seven day observation periscopes. The gunner has a day periscopic binocular daylight sight linked to the main armament, with a magnification of ×1 and ×10, and two day observation periscopes.

Mounted to the right of the main armament is a SPAV L2A1 passive night sight with a high-magnification of ×5.8 (8° field of view) and a low magnification of ×1.6 (28° field of view), the first for gunnery and the second for surveillance. There can be no confusion between the two because when the high magnification is being used a shutter isolates the low magnification, and when the other is being used an iris diaphragm isolates the

CVR(W) Fox with driver's hatch open and muzzle of the 30 mm RARDEN covered up (Michael Jerchel) 0088050

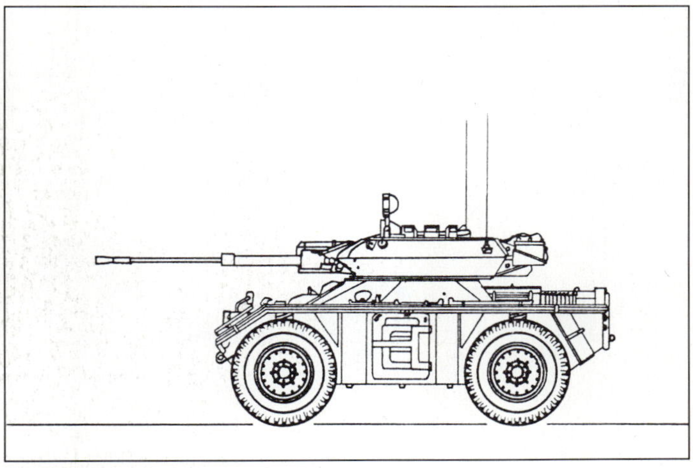

Fox light armoured car

high-magnification objective. The image intensifier tube is protected from the effect of gun muzzle flash by a flash shutter which is operated electrically from the gun firing circuit. When high magnification is selected an illuminated ballistic graticule with brightness control is automatically injected into the optical system. The exposed objective window is cleared by a wiper and washer and the sight is protected by an armoured cowl with a door which is kept closed when the sight is not in use. The L2A1 night sight will also detect infrared devices.

The radios are in the rear of the turret and the batteries are stowed below the false floor in the crew compartment.

The engine compartment is at the rear of the hull and the engine and its auxiliaries, including the main and transfer gearboxes, are mounted as a power pack which can be removed through the rear of the hull. The engine is a militarised 4.2 litre Jaguar XK with a reduced compression ratio (7.75:1) in order to use military petrol. The valve gear is leadproofed and a single two-choke Marcus carburettor is fitted, as is a Ki-gas cold starting system.

The cooling system has twin radiators horizontally disposed across the top and to the rear of the engine and, sealed to the radiator, a pair of ducted centrifugal fans. An oil/water heat exchanger unit serves both the engine and the transmission.

Power is transmitted from the engine to the wheels through a fluid coupling, a five-speed preselecting epicyclic gearbox, and a transfer box providing reversal of all five speeds.

The independent suspension comprises in each case an upper and lower wishbone, a coil spring and a hydraulic telescopic damper. The independent suspension allows wheel movements of 0.279 m and the tyres are of the run-flat type.

The Fox can ford to a depth of 1 m without preparation. A flotation screen carried collapsed around the top of the hull can be erected by the crew in 2 minutes and the vehicle is then propelled and steered across the river or stream by its wheels. The front of the flotation screen contains transparent panels for viewing and a bilge pump with a capacity of 205 litres/min is fitted as standard on all vehicles. The Fox can travel down or up river banks with a gradient of up to 46 per cent.

Fox has been designed to operate in temperatures from -40 to +50°C and is air-transportable. A Lockheed Martin C-130 Hercules can carry three Fox vehicles or two for parachute dropping.

Standard equipment on British Army vehicles included infrared filters for the white light headlamps, a spotlight for the commander, external stowage boxes either side of the hull between the wheels, a drinking water tank containing 13.6 litres and a hull electric distribution box with a cooking vessel socket, inspection socket, intervehicle starting socket and a master switch. Optional equipment includes navigation aids, powered traverse, nuclear and chemical detection equipment and a ZB 298 surveillance radar mounted on the left side of the turret.

Main armament of the Fox is a now BAE Systems, RO Defence 30 mm L21 RARDEN cannon which fires an APDS-T round jointly developed by Royal Ordnance and PATEC of the USA. In addition to the APDS-T round the RARDEN cannon fires standard British APSE-T, HEI-T and PRAC-T rounds as well as standard Oerlikon Contraves HE, AP and Practice rounds.

The gun is fired in rapid single shots but bursts of up to six rounds can be fired. The empty cartridge cases are automatically ejected forward of the turret outside the turret.

A 7.62 mm L37A2 machine gun is mounted coaxially to the left of the main armament and four 66 mm smoke grenade dischargers are mounted either side of the front of the turret. The main armament, coaxial machine gun and smoke grenade dischargers are electrically operated with the main armament and coaxial machine gun having manual overriding control.

Status

Production complete. In service with:

Country	Quantity	Comment
Malawi	20	additional 50 vehicles may have been delivered
Nigeria	55	status uncertain

Manufacturer

Vickers Defence Systems.

UPDATED

Daimler Ferret scout car

Development

Late in 1948, Daimler was awarded a contract to design and build prototypes of a new scout car under the designation FV701. The first prototype was completed in 1949 and delivered to the British Army for trials the following year. After trials it was adopted for service with the British Army and was named the Ferret.

The first production Ferret, the Mk 2, was completed in mid-1952 and the first production Mk 1 was completed late the same year. Throughout its production life the basic hull of the Ferret remained unchanged although slightly more powerful engines were installed as well as different turrets fitted to meet different requirements. Production of the Ferret was finally completed in 1971 by which time 4,409 vehicles of all types had been built.

In 1966, the Daimler company was awarded a contract to build 15 prototypes of a new vehicle called the FV721 (or Fox as it later became known), which were completed between 1967 and 1969. The Fox was accepted for service with the British Army in 1970 but the production order was placed with the Royal Ordnance Factory at Leeds (now owned by Vickers Defence Systems) rather than Daimler at Coventry. When production of the Ferret was completed Daimler closed down its production line for wheeled armoured vehicles. The Fox is based on experience obtained with the rebuilt Ferret Mk 4 and Mk 5. These were sometimes referred to as the Big Wheeled Ferrets.

Alvis is now the design authority for the Ferret and supplies the spare parts for it. Alvis Vehicles built all of the turrets for the Fox light armoured car. These turrets were built at the Alvis Vehicles facility in Coventry which closed down in late 1999. All Alvis Vehicles production is now concentrated at the former GKN Defence site at Telford.

Description

(Mk 2)

The all-welded steel hull of the Ferret is divided into three compartments: driving at the front, fighting in the centre and the power pack at the rear.

The driver has three hatches, one to his front and one to each side, each of which contains an integral No 17 day observation periscope. The front hatch can be folded down onto the glacis plate for increased visibility and can then be fitted with a splinter-proof windscreen with a wiper blade and an electric motor. The two side hatches can be opened upwards on the outside for increased visibility when not in a combat area.

The manually operated turret in the centre of the hull has a single-piece hatch cover in the rear part of the roof which can be locked in three different positions, and the rear part of the turret folds down to the horizontal enabling it to be used as a seat. Mounted in the forward part of the turret roof is a day sight periscope AFV No 3 Mk 1 used for aiming the turret-mounted machine gun.

Mounted either side of the hull below the turret ring is a vision slit protected by a glass block. At the rear of the fighting compartment are two hatches which can be opened for increased observation. On each side of the hull, between the front and rear roadwheels, is a hull escape hatch. The left one is covered by the spare wheel and the right by a stowage box.

The engine is at the rear of the hull and is fully waterproofed and will operate when completely submerged without any preparation other than venting the crankcase breather pipe. Drive is transmitted to all four roadwheels through a fluid coupling, five-speed preselecting epicyclic gearbox and a transfer box, incorporating a forward and reverse mechanism and a differential unit to give five speeds in each direction.

The suspension at each wheel station consists of a single coil spring which encloses a double-acting shock-absorber and is mounted on a stabiliser bracket at the bottom and carried in a bracket at the top attached to the hull plates. The tyres are of the run-flat type.

The Mk 2 is armed with a turret-mounted 7.62 mm (0.30) Browning machine gun which can be elevated manually from −15 to +45°, turret traverse being a full 360°. A total of 2,500 rounds of machine gun ammunition is carried. Mounted either side of the hull at the front of the vehicle are three 66 mm smoke grenade dischargers which are electrically fired from within the vehicle.

The basic Ferret has no NBC system, no night vision equipment and no amphibious capability. In order that the vehicle can cross trenches, Ferret Mks 1, 2 and 3 have provision for carrying steel channels on the front of their hulls.

Daimler Ferret Mk 1 scout car with 0.30 (7.62 mm) MG (Christopher F Foss) **2000**/0064855

Daimler Ferret scout car upgraded by Alvis Vehicles showing additional smoke grenade dischargers either side of the turret

Variants

Ferret Mk 1
This is called Car Scout 4 × 4 Liaison (Ferret Mk 1) FV701(C) and has an open top which can be covered by a canvas. It is armed with a 7.62 mm Bren LMG or a 7.62 mm (0.30) Browning machine gun, with 450 rounds of machine gun ammunition carried. Later production models are known as the Ferret Mk 1/1, FV701(J).

Ferret Mk 1/2 (FV704)
This is identical to the Mk 1 but has a crew of three and was used by infantry units as a light reconnaissance vehicle in forward areas. Its official designation is Car Scout Liaison (Ferret) Mk 1/2. It has an armoured roof with a hatch and is armed with a 7.62 mm pintle-mounted Bren LMG. The single-piece hatch cover opens to the rear and day periscopes and vision blocks are provided for observation by the commander.

Ferret Mk 2
This is basically a Mk 1 fitted with a turret and is officially known as the Car Scout 4 × 4, Reconnaissance (Ferret) Mk 2 FV701(E).

Ferret Mk 2/2
None of these versions remain in service.

Ferret Mk 2/3
This is a later production model of the basic Mk 2 and is officially known as the Scout Car Reconnaissance Mk 2/3 (Daimler Ferret 4 × 4) FV701(H). The description in this entry relates to this model.

Ferret Mk 2/4
This is a Ferret Mk 2/3 with additional armour.

Ferret Mk 2/5
This is a Ferret Mk 2 brought up to Ferret Mk 2/4 standard.

Ferret Mk 2/6 (FV703)
The full designation of this model is the Scout Car Reconnaissance/Guided Weapon Mk 2/6. It is basically a Mk 2/3 with a single Vigilant ATGW in its launcher box mounted on either side of the turret. As far as is known, there are no Ferret Mk 2/6s now operational. In addition to being used by the UK it was sold to Libya and the United Arab Emirates.

Ferret Mk 2/7
This is a Ferret Mk 2/6 with the missile system removed and is therefore basically a Ferret Mk 2/3.

Ferret Mk 3
This is the Ferret Mk 1/1 brought up to the same standards as the Mk 4 but with the machine gun turret.

Ferret Mk 4 (FV711)
This model is basically an early Ferret rebuilt with stronger suspension units, its original 330 mm diameter drum brakes replaced by vacuum-assisted 381 mm diameter disc brakes, and with large wheels and tyres. A flotation screen is carried collapsed around the top of the hull and, when erected, the vehicle has amphibious capabilities. There are watertight stowage containers either side of the hull, and the number of servicing points has been reduced.

Specifications

Model	**Mk 1/1**	**Mk 2/3**
Crew	2-3	2
Configuration	4 × 4	4 × 4
Combat weight	4,210 kg	4,400 kg
Unloaded weight	3,510 kg	3,640 kg
Power-to-weight ratio	30.64 bhp/t	29.35 bhp/t
Length	3.835 m	3.835 m
Width	1.905 m	1.905 m
Height overall	1.448 m	1.879 m
Ground clearance	0.33 m	0.33 m
Track	1.539 m	1.539 m
Wheelbase	2.286 m	2.286 m
Max road speed	93 km/h	93 km/h
Fuel capacity	96 litres	96 litres
Max range		
(road)	306 km	306 km
(cross-country)	160 km	160 km
Fuel consumption		
(road)	0.31 litres/km	0.31 litres/km
(cross-country)	0.6 litres/km	0.6 litres/km
Fording	0.914 m	0.914 m
(with preparation)	1.524 m	1.524 m
Gradient	46%	46%
Vertical obstacle	0.406 m	0.406 m
Trench (with channels)	1.22 m	1.22 m
Turning radius	5.795 m	5.795 m
Engine	Rolls-Royce B60 Mk 6A 6-cylinder in-line water-cooled petrol developing 129 bhp at 3,750 rpm	
Transmission	Daimler preselective epicyclic, 5-speed	
Steering	recirculating ball (not power-assisted)	
Clutch	Daimler fluid coupling	
Suspension	independent, coil spring and double-acting shock-absorber at each wheel station	
Tyres	9.00 × 16	9.00 × 16
Brakes		
(main)	hydraulic on all wheels	
(parking)	hand-operated, mechanical on transmission	
Electrical system	24 V	24 V
Batteries	2 × 12 V, 60 Ah	2 × 12 V, 60 Ah
Armament		
(machine gun)	1 × 7.62 mm	1 × 7.62 mm
Smoke-laying equipment	all Ferret models have 2 × 3 66 mm smoke grenade dischargers	
Armour		
Hull front	12 mm at 50° and 16 mm at 35° all models	
Hull sides	12 mm at 10°, 15°, 40°; 16 mm at 20° all models	
Hull floor	10 mm driver's position, 6 mm remainder of vehicle all models	
Hull rear	12 mm and 6 mm at 45° all models	
Turret front	n/a	16 mm at 15°
Turret sides	n/a	16 mm at 15°
Turret rear	n/a	16 mm at 15°
Turret top	n/a	8 mm
NBC system	no	no
Night vision equipment	optional	optional

Daimler Ferret Mk 2/3 scout car ***2002***/0114341

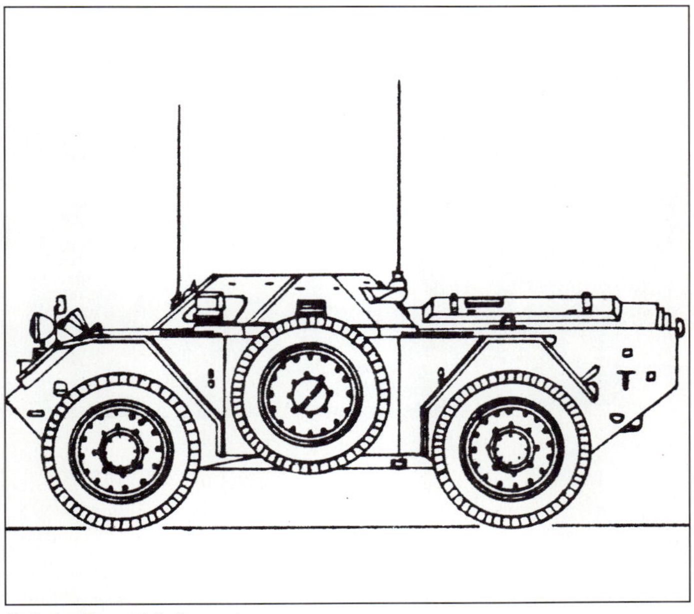

Daimler Ferret Mk 1 scout car

Ferret Mk 5 (FV712)

This model is no longer in service and was used only by the British Army. It was essentially a Mk 4 with a turret mounting four Swingfire ATGWs.

Ferret Upgrade

In March 1988, Alvis announced that it had developed a repower package for the Daimler Ferret scout car. This package involves virtually no hull cutting and comprises a Perkins Engines Company Phaser diesel engine (110MT) developing 109 hp coupled to a Chrysler A727 fully automatic transmission. These are interfaced with the existing transfer box and drivelines while the cooling system has been modified.

Other modifications include the replacement of the existing periscopes by new Helio day periscopes, new smoke/fragmentation grenade launchers, new machine gun mount and a new machine gunsight.

According to Alvis, the weight, road speed and acceleration of the Ferret are unchanged, while the range, reliability and life cycle costs have been much improved. In 1988, Alvis signed a formal joint-venture agreement with the Malaysian Mining Corporation to use the Alvis upgrade package. The Malaysian Army requirement has been met with the Alvis Repower Package. Production contracts were signed during early 1991. This upgrade programme is now complete and it is understood that about 70 vehicles were upgraded.

Status

Production complete. In service with the following:

Country	Quantity	Comment
Bahrain	8	
Burkina Faso	30	
Cameroon	8	
Central African Republic	10	
India	45	police
Indonesia	55	13 upgraded
Jordan	170	quantity unconfirmed
Kenya	12	
Lebanon	5	
Madagascar	10	may not be operational
Malawi	8	
Malaysia	92	70 upgraded
Myanmar	45	
Nepal	40	
South Africa	50+	some with MILAN ATGW and 106 mm RR
Sri Lanka	15	
Sudan	80	

In May 1994, Alvis Vehicles announced that it had been awarded a contract worth US$10 million for the supply of upgrade kits for Ferret, Saracen and Saladin vehicles Indonesia. This upgrade programme has now been completed.

Manufacturers

Daimler Company Limited.
Alvis Vehicles Limited.
Spare parts and post-design work are now the responsibility of Alvis Vehicles Limited.

UPDATED

Alvis Hornet armoured car

Development

The Hornet armoured car, developed as a private venture by Glover Webb Limited, was shown for the first time at the 1986 British Army Equipment Exhibition.

The Glover Hornet has in excess of 90 per cent spare parts commonality with the standard military Land Rover and meets all Land Rover engineering specifications and approval.

There has been no production of the Hornet armoured car in recent years and it is no longer being marketed.

The Glover Webb facility at Hamble was closed down late in 1997 and research, development, production and post delivery work is now carried out at the main GKN Defence facility at Telford in Shropshire. More recently, Alvis Vehicles have merged with GKN Defence to form Alvis Vehicles with headquarters and production being undertaken at Telford. The Alvis Vehicles facility at Coventry closed in 1999.

The Glover Webb facility at Hamble also designed and built the prototype of the Greys Defence Systems Panther multirole combat vehicle which remains at the prototype stage. Full details of this 4 × 4 vehicle were given in *Jane's Armour and Artillery 2000-2001* on pages 241 to 242. In previous editions of *Jane's Armour and Artillery* the Hornet armoured car was listed as a Glover Webb product.

Specifications

Crew: 3
Configuration: 4 × 4
Combat weight: 3,600 kg
Power-to-weight ratio: 31.66 bhp/t
Length: 4.429 m
Width:
(overall) 1.79 m
Height:
(over turret) 2.28 m
(hull roof) 1.85 m
Ground clearance: 0.215 m
Track: 1.5 m
Wheelbase: 2.794 m
Angle of approach/departure: 50/35°
Max road speed: 120 km/h
Fuel capacity: 55 litres
Max road range: 350 km (700 km with long-range tanks)
Gradient: 60%
Side slope: 40%
Vertical obstacle: 0.23 m
Turning radius: 8.75 m
Engine: Rover 4-stroke V8, 3,528 cc petrol developing 114 bhp at 4,000 rpm (or turbocharged diesel)
Transmission: manual with 4 forward and 1 reverse gears
Transfer case: 2-speed
Steering: recirculating ball, worm and nut, power-assisted
Clutch: single dry plate
Suspension: heavy-duty long-travel coil springs, anti-roll bars fitted front and rear
Tyres: 7.50 × 16
Brakes:
(main) dual-line hydraulic with disc front, drum rear, servo-assisted
(hand-brake) mechanical, on transfer box
Electrical system: 12 V
Armament: 1 × 7.62 mm MG

Gun control equipment
Gun elevation/depression: +50 to –10°
Turret traverse: 360°
NBC system: no
Night vision equipment: optional

Hornet armoured car

Hornet armoured car from the rear **2000**/0088051

Description

The Hornet armoured car is based on the chassis of the heavy-duty military Land Rover Defender (4 × 4) vehicle chassis which is already in service with a large number of armed forces around the world and was available powered by a V8 petrol engine or a turbocharged diesel engine.

The ballistically tested steel armour plate hull of the Hornet will defeat NATO 7.62 mm machine gun attack firing at right angles to the target. All welded seams have reinforcing backing strips and all openings have ballistic overlaps. The floor of the vehicle is made of high-yield steel which offers protection against multiple grenade attack. Internally operated fold-down shields with vision blocks are standard equipment for urban situations.

The vehicle configuration is similar to the standard Land Rover with the engine at the front, commander and driver in the centre (left or right-hand drive models are available) and a one-man low-profile machine gun turret on the roof.

The commander and driver each have a forward-opening door. In the upper part of the door is a large vision area and to the immediate front both commander and driver have protected vision areas. Internally operated drop-down flaps with an observation slit offer additional protection.

The low-profile one-man machine gun turret, also designed by Glover Webb, is armed with a 7.62 mm machine gun with manual elevation from –10 to +50°. The turret has a full 360° traverse under manual control with the weapon being aimed by a day periscope sight to the left of the weapon. In addition, the turret is fitted with a single-piece hatch cover that opens to the rear with rectangular vision blocks provided in the sides and rear of the turret for all-round observation. Turret-mounted searchlight and 2 × 4 66 mm or 76 mm electrically operated smoke grenade dischargers are standard options for the turret.

Optional equipment for the Hornet included gun ports, run-flat tyres, sand tyres, an extended range fuel tank, a self-recovery winch, HF/UHF/VHF radios, a barricade remover, passive night vision equipment and urban police equipment.

Status

In service in undisclosed countries in Africa (one of which is believed to be Nigeria) and the Middle East. No longer being actively marketed.

Manufacturer

Alvis Vehicles Limited originally built by Glover Webb which was taken over by GKN Defence and then merged with Alvis Vehicles.

UPDATED

UNITED STATES OF AMERICA

Cadillac Gage (Textron Marine & Land Systems) LAV-600 armoured car

Development

The LAV-600 (6 × 6) armoured car was developed as a private venture by Cadillac Gage and was announced in 1985.

The LAV-600 was previously known as the V-300 A1 but was renamed in mid-1986.

So far one prototype of the LAV-600 has been built and fitted with the private venture Cadillac Gage 105 mm Low-Recoil Force Turret also installed on the Cadillac Gage Stingray light tank, full details of which are given in the Light tanks section.

The LAV-600 armoured car uses many automotive components of the Cadillac Gage LAV-300 Armoured Vehicle Range (6 × 6) which is fully described in the Armoured personnel carriers (wheeled) section. This has been produced in production quantities for a number of export customers.

Late in 1988, the LAV-600 (6 × 6) armoured car carried out extensive firepower and mobility trials in Egypt.

While the vehicle is marketed as the Cadillac Gage LAV-600 it is now actually built by Textron Marine & Land Systems. As of mid-2001, the LAV-600 remained at the prototype stage.

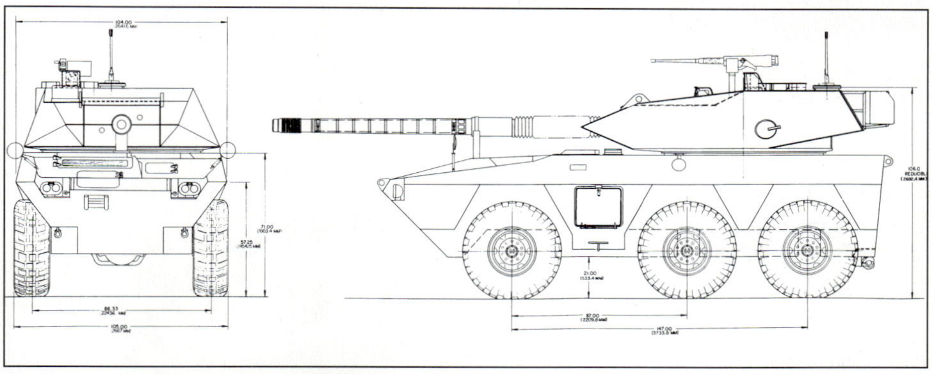

Front and side drawings of the Cadillac Gage LAV-600 (6 × 6) armoured car with 105 mm gun in travel lock

Description

The hull and turret of the LAV-600 are of all-welded Cadloy armoured steel which provides protection from small arms fire and shell splinters.

The driver is seated at the front left of the vehicle with the engine compartment to his right. Above his position is a single-piece hatch cover that opens to the rear and in front of this are three day periscopes that give observation to the front of the vehicle. The centre periscope can be replaced by a passive night vision periscope for driving at night. In the left side of the hull, to the rear of the driver and between the first and second roadwheel stations, is a square hatch that opens downward to provide access to the vehicle with a bulletproof day observation periscope above. Mounted on the glacis plate is a travelling lock for the 105 mm gun.

The turbocharged Cummins V-8 engine is to the right of the driver and develops 270 hp at 3,000 rpm. It is coupled to a fully automatic Allison Transmission transmission. The single-speed transfer case has a spline engagement clutch for the front axle.

All axles are of the single reduction hypoid type with silent locking differentials. Suspension on the front axle is of the coil spring type while the rear axles have torsion bars. Steering is power assisted on the front axle only. The rear two axles of the LAV-600 are fitted with no-spin differentials that practically eliminate wheel slippage.

While the preproduction LAV-600 has the complete turret of the Stingray light tank built for the Royal Thai Army it is envisioned that production vehicles would use the complete turret of the latest Stingray II light tank which has a number of improvements in the areas of fire control and target acquisition capabilities. The computer system used is similar to that installed on the General Dynamics Land Systems M1 seriesof MBTs.

At the rear of the hull are two large outward-opening doors, each of which has a firing port with a vision block above. The prototype has a single door with an integral vision block which is used for ammunition resupply as the turret holds eight rounds, the remaining 28 rounds being carried in the hull.

Mounted internally at the front of the hull is a power-operated winch for self-recovery and recovering other vehicles. This is provided with 46 m of cable and has a maximum capacity of 9,000 kg.

Standard onboard equipment for the LAV-600 includes one 19-litre water can, one 19-litre fuel can, pioneer tool set, tow cable, slave cable, first aid kit, vehicle tool kit, pamplet bag, oddment box and engine-driven air compressor and hose.

Specifications

Crew: 4
Configuration: 6 × 6
Combat weight: 18,500 kg
Power-to-weight ratio: 14.59 bhp/t
Length: (hull) 6.3 m
Width:
(hull) 2.641 m
(tyres) 2.667 m
Height:
(with turret) 2.74 m
(reduced) 2.69 m
(to hull top) 1.8 m
Ground clearance: 533 mm (hull)
Track (front and rear): 2.24 m
Wheelbase:
(1st to 2nd axle) 2.209 m
(1st to 3rd axle) 3.733 m
Max speed: 100 km/h
Range: 600 km
Gradient: 60%
Side slope: 30%
Vertical obstacle: 0.61 m
Engine: Cummins 6CTA 8.3 turbocharged, after-cooled, 6-cylinder in-line diesel developing 275 hp at 2,500 rpm
Transmission: Allison Transmission automatic with 6 forward and 1 reverse gears
Transfer case: 2-speed, spline engagement clutch for front axle
Steering: hydraulically assisted, front wheels only
Suspension: coil springs front, torsion bar rear
Tyres: 14R20-XL tubeless radial (run-flat)
Brakes: dual hydraulic
Electrical system: 24 V
Batteries: 4 × 6TN
Armament:
(main) 1 × 105 mm LRF rifled gun
(coaxial) 1 × 7.62 mm M240 MG
(anti-aircraft) 1 × 7.62 mm M240 MG
Smoke grenade dischargers: 2 × 4 (optional)
Ammunition:
(105 mm) 8 + 26
(7.62 mm coaxial) 400 + 2,000
(7.62 mm anti-aircraft) 200
(smoke grenades) 8 + 8

Gun control equipment
Turret power control: electrohydraulic/manual
(by commander) yes
(by gunner) yes
Max rate of power traverse: 360 at 30°/s
Gun elevation/depression: +20 to –7.5°
Gun stabiliser:
(vertical) optional
(horizontal) optional
NBC system: optional
Night vision equipment: yes

Optional equipment includes NBC protection with M13A1 ventilated face masks for all crew members, chemical-resistant paint, night driving lights and an automatic Halon fire-detection and suppression system as well as two portable Halon bottles. If required, the roof-mounted 7.62 mm machine gun can be replaced by a 12.7 mm M2 HB machine gun with 100 rounds of ready use ammunition.

Other optional equipment for the LAV-600 includes run-flat inserts and a central tyre-pressure regulation system that allows the driver to adjust the tyre pressure to suit the terrain being crossed.

A Vehicular Precision Lightweight GPS receiver can be fitted as can a Third Eye Laser warning system. This is supplied by Moked Engineering (1969) Limited of Israel.

Status
Development complete. Ready for production.

Manufacturer
Textron Marine & Land Systems.

UPDATED

Cadillac Gage LAV-600 (6 × 6) armoured car **2000**/0064857

Cadillac Gage LAV-600 (6 × 6) armoured car firing its 105 mm Low-Recoil Force Gun 0088052

Cadillac Gage (Textron Marine & Land Systems) Light Armored Vehicle — Assault Gun

Development
In December 1988, the US Marine Corps approved a requirement for an LAV-AG and selected the Benet Laboratory 105 mm XM35 gun as the Government Furnished Equipment (GFE) weapon.

Following competitive bidding, in mid-1990 Cadillac Gage was selected to develop the Assault Gun Version of the LAV to meet the requirements of the US Marine Corps.

Under the terms of the original fixed-price Full-Scale Engineering Development (FSED) 40 month contract, Cadillac Gage was to build three vehicles for delivery by May 1992 to the US Marine Corps for Development and Operational Testing.

Late in 1991 however, the US Marine Corps cancelled the research and development phase of the LAV-105 due to lack of production funding between FY93 and FY96 as Pentagon acquisition policy did not allow continued development of a programme unless there was assured production money.

At the time of cancellation, the LAV-105 was on time and within the fixed-price budget with first unmanned firings already completed.

In mid 1993, the US Navy released US$33.8 million to allow Cadillac Gage to complete development and testing of the three prototypes of the LAV-105.

In June 1995, Cadillac Gage (now Textron Marine & Land Systems) completed development of the 105 mm Light Armored Vehicle — Assault Gun (LAV-AG) for the US Marine Corps, although there is currently no funding for production vehicles.

The Cadillac Gage 105 mm LAV-AG 0088053

The company is continuing to market the LAV-AG on the international market, especially the Middle East, and the turret is also being offered for installation on other chassis, wheeled and tracked.

Textron Marine & Land Systems has leased the three prototype LAV-AG and their turrets from the US Marine Corps to enable marketing of the complete system and turret to continue.

Description
The LAV-105 is essentially a new two-man turret designed by Cadillac Gage installed on an upgraded 8 × 8 chassis supplied by the now General Motors Defense (this company was previously called the Diesel Division, General Motors of Canada). The chassis is

Specifications
Crew: 3
Configuration: 8 × 8
Weight:
(combat) 13,864 kg
(reduced) 13,182 kg
Power-to-weight ratio: 19.83 hp/t
Length:
(gun forward) 7.835 m
(hull) 6.553 m
Width:
(over hull) 2.489 m
(over wheels) 2.458 m
Height:
(overall) 2.625 m
(turret roof) 2.501 m
(hull top) 1.893 m
Angle of approach/departure: 30/40°
Max road speed: 100 km/h
Max water speed: 9.656 km/h
Max range: 668 km
Gradient: 60%
Side slope: 30%
Vertical obstacle: 0.5 m
Trench: 2.057 m
Engine: Detroit Diesel 6V-53T, 6-cylinder diesel developing 275 hp at 2,800 rpm
Transmission: Allison Transmission MT-653 DR automatic, 5 forward and 1 reverse gears
Transfer case: Rockwell AG-VST (modified)
Steering: power-assisted on front 2 axles
Suspension:
(front four wheels) independent coil springs and shock-absorbers
(rear four wheels) independent torsion bars and shock-absorbers
Brakes:
(main) 8-wheel dual air hydraulic drum brake
(parking) transmission brake and transfer case lock
Tyres: 11.00 × 16 with Hutchinson run-flat inserts
Armament:
(main) 1 × 105 mm M35 rifled gun
(coaxial) 1 × 7.62 mm M240 MG
(anti-aircraft) 1 × 7.62 mm M240 MG (or 1 × 12.7 mm)
Smoke grenade launchers: 2 × 4
Ammunition:
(105 mm) 8 ready-use
(7.62 mm) 400 rounds (turret)
Gun control equipment: electric/manual
Gun elevation/depression: +15 to –8°
Traverse: 360°
Stabiliser:
(vertical) yes
(horizontal) yes
Range-finder: yes (laser)
NBC system: yes
Night vision equipment: yes

similar to that used for the standard LAV, described in the Armoured personnel carriers (wheeled) section under Canada, but has additional buoyancy aids fitted either side of the turret towards the rear and a different roof arrangement because there will be no hatches in the roof of the vehicle at the rear.

The all-welded steel turret is armed with a 105 mm Watervliet Arsenal M35 (development designation XM35 which is also installed in the United Defense LP full-tracked M8 Armored Gun System) lightweight gun which is fed from a bustle-mounted FHL automatic loader which holds eight rounds of ready use ammunition with additional rounds being carried in the hull.

The 105 mm gun is fitted with a muzzle brake and thermal sleeve and can fire all types of 105 mm ammunition including the latest Armour-Piercing Fin-Stabilised Discarding-Sabot (APFSDS) types. The empty cartridge cases are ejected out of the turret rear through a trap door.

A 7.62 mm M240 machine gun is mounted coaxially with the 105 mm gun and a bank of four electrically operated M257 smoke grenade launchers is mounted either side of the turret firing forward. Provision is made for the installation of a 7.62 mm or 12.7 mm machine gun on the turret roof.

The turret features an HR Textron (previously Cadillac Gage Control Systems) electromechanical gun/drive system in combination with a Computing Devices Canada digital fire-control computer.

Turret traverse is a full 360° with weapon elevation from -8 to +15°. Trials have shown that the LAV-AG can engage stationary and moving targets with a high first round hit probability while the vehicle itself is stationary or moving.

Both the commander and gunner are provided with a single-piece hatch cover that opens to the rear. The vehicle commander is seated on the left and has four day periscopes and a thermal day/night remote monitor. The gunner, seated on the right, has a stabilised Raytheon Day/Night Range Sight (DNRS), four observation sights and a back-up periscope.

The Raytheon DNRS comprises five key subsystems, Raytheon Infra-Red (HIRE) 240-line common thermal image with dual field of view, Raytheon laser range-finder, Raytheon Line of Sight Stabilisation Platform (LSSP), commander's remote display for viewing the day and night sight images and with commander override for all the gunner's controls and the Line of Sight Electronics Unit (LOS-EU). The DNRS has a unity window and a ×10 magnification narrow field of view for the day sight.

Status

Development complete. Ready for production. No funded US Marine Corps requirement. Still being marketed by Textron Marine and Land Systems.

Manufacturer

Textron Marine & Land Systems.

UPDATED

Cadillac Gage (Textron Marine & Land Systems) Scout

Development

The Cadillac Gage Scout (4 × 4) light reconnaissance vehicle was designed as a private venture specifically for the export market and was first announced in October 1977 at the Association of the United States Army meeting in Washington DC.

In 1983, it was confirmed that Indonesia had placed an order for 28 Scout vehicles as well as Cadillac Gage 22 Ranger armoured personnel carriers. Details of this vehicle, which is no longer manufactured or marketed are given in the Armoured personnel carriers (wheeled) section.

In August 1986, it was announced that a foreign country had placed a US$22.8 million contract with Cadillac Gage for an undisclosed number of Scout reconnaissance vehicles with half the vehicles delivered in 1986 and the remainder in 1987. Middle East sources confirm that the country is Egypt and 112 vehicles were purchased.

By mid-2001, a total of 140 Scout (4 × 4) vehicles had been built by Cadillac Gage.

Cadillac Gage is now part of Textron Marine & Land Systems. There has been no recent production of the Scout although marketing continues.

Specifications

Crew:
(depending on role) 1 + 1 or 1 + 2
Configuration: 4 × 4
Combat weight: 7,240 kg
Power-to-weight ratio: 20.58 hp/t
Length: 5.003 m
Width: 2.057 m
Height: 2.159 m
Ground clearance:
(differential) 0.29 m
(bottom of hull) 0.38 m
Track: 1.66 m
Wheelbase: 2.743 m
Max road speed: 88 km/h
Fuel capacity: 378 litres
Max road range: 846 km
Fording: 1.168 m
Gradient: 60%
Side slope: 30%
Vertical obstacle: 0.609 m
Trench crossing: 1.14 m
Turning radius: 7.925 m
Engine: Cummins V-6 diesel developing 155 hp at 3,300 rpm
Transmission: Allison Transmission automatic with 4 forward and 1 reverse speeds
Transfer case: 2-speed with spline engagement clutch for front axle drive
Steering: integral hydraulic
Suspension:
(front) coil springs over solid axle with trailing arms
(rear) independent rear axle with coil springs for each wheel
Tyres: 15.50 × 21 run-flat tubeless
Brakes:
(main) dual-power 4-wheel hydraulic, pump-driven emergency electrical back-up
(parking) drum and shoe, hand-operated
Electrical system: 24 V
Batteries: 2 × 12 V, 100 Ah
NBC system: optional
Night vision equipment: optional

Note: Height and weight are for vehicle with 1 m turret and twin 7.62 mm MGs.

Description

The all-welded hull of the Scout is made of special hardness Cadloy armour plate which will defeat at a minimum 7.62 mm armour-piercing rounds. The front, sides and rear of the hull are well sloped to afford maximum protection within the weight of the vehicle.

The front of the Scout is sloped at an angle of 76° from the top of the driver's hatch to the nose. In addition to providing ballistic protection it allows the vehicle to push its way through underbrush.

The driver is seated on the left side of the hull with the engine to his right and is provided with an adjustable seat and a single-piece hatch cover that slides to the front of the vehicle when he is driving with his head out. Driver vision is provided by three day periscopes mounted forward of the hatch area. One of these can be replaced by a passive night vision periscope.

The fuel tank is at the front of the hull between the wheels. The Cummins engine is coupled to an Allison Transmission four-speed automatic transmission via a Cadillac Gage power transfer unit. The complete power pack, consisting of the engine, transfer unit, transmission and cooling system, can be removed and replaced in two hours. Access to the engine is by a large hatch in the right side of the hull through which all fluid levels (coolant, lubricant, brake fluids and so on) can be checked.

The turret or pod is at the rear of the vehicle and access is by a two-part hatch in the rear of the hull with the bottom part folding downwards and the top part opening to the right.

The front suspension features coil springs over a solid axle with trailing arms. The split rear axle is tied to a heavy-duty coil spring suspension. Cone-shaped passages in the rear of the hull allow for high individual vertical roadwheel travel. Both axles are fitted with

Cadillac Gage Scout armed with twin 7.62 mm machine guns in a 1 m turret
0088055

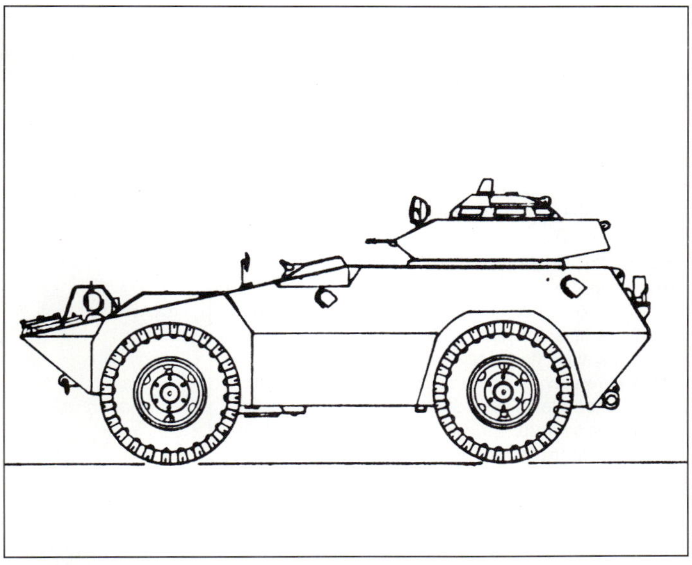

Cadillac Gage Scout with a 1 m turret

positive locking differentials which provide improved traction by preventing one wheel spin-out.

The integral hydraulic power steering is powered by a gear-driven pump working directly off the engine. This approach has eliminated the requirement for belt drives which slip or break and steering cylinders that require maintenance.

The independent front and rear brake systems function through a split master cylinder and a back-up system supports them in the event of an engine failure. The back-up system consists of an electric motor which supplies pressure to the master cylinder. The tyres are of the run-flat type. Standard equipment includes two hand-held fire extinguishers stowed inside the vehicle, a pioneer toolset, a breaker bar and lug wrench, a first aid kit, a vehicle toolkit, and an air compressor with a 15.24 m hose.

The Scout is not fitted with an NBC system, has no night vision equipment and is not amphibious. Optional equipment includes a siren/public address system, radio installations, weapon stowage, water and fuel cans, a slave cable, a 15.24 m auxiliary cable, a camouflage net, a smoke grenade system and a fragmentation grenade system.

Variants

Twin/Combination Machine Gun (1m) turret

This turret is manually/electrically rotated by a Power Assist Traverse (PAT) system and can be armed with twin 7.62 mm or twin 0.50 calibre machine guns, or a combination of 7.62 mm/0.50 calibre machine guns, which can be elevated from –10 to +55°. Manually traversed rate with PAT is a maximum of 45°/s. The amount of ammunition depends on the armament installed but if twin 7.62 mm calibre weapons are fitted each gun will be provided with 200 rounds of ready use ammunition and 2,200 rounds stowed in the vehicle. If twin 0.50 calibre weapons are fitted, each weapon has 100 rounds of ready use ammunition and 1,000 rounds stowed in the hull. It can also mount a 7.62 mm MG and a 40 mm Mark 19 grenade launcher with 100 ready rounds and 200 rounds stowed in the vehicle. The turret is also provided with eight vision blocks, an M28C day sight with a magnification of ×1.5 (×5 magnification optional) and a 500,000 candle-power spotlight mounted coaxially with the weapons.

40 mm/12.7 mm turret

One-man turret armed as on US Marine Corps AAV7A1 armoured amphibious assault vehicles. This is armed with a 40 mm Mk 19 grenade launcher and a 12.7 mm M2 machine gun with manual traverse and elevation.

Twin Machine Gun (1 m) turret

This is armed with twin 7.62 mm machine guns which can be fired singly or together and 400 rounds of ready use ammunition are carried with a further 2,200 rounds stowed in the vehicle. It has manual traverse and elevation and the weapons can be elevated from –10 to +55°. The turret is provided with eight vision blocks, and an M28C sight with a magnification of ×1.5 (×5 magnification optional); a 500,000 candle-power spotlight is mounted coaxially with the weapons.

Command pod

This model has a crew of three, commander, radio operator and driver and has a pod mounted on top of the hull in each side of which is a firing port and a vision block. A circular hatch cover is provided on which a 7.62 mm machine gun and shield can be mounted. There are 2,200 rounds of 7.62 mm ammunition carried. When fitted with radios the antennas are mounted at the rear of the hull.

Cadillac Gage Scout fitted with a command pod and armed with a 12.7 mm M2 machine gun 0088054

Anti-tank armed with TOW

This model has a crew of two and is fitted with a Raytheon Systems Company TOW ATGW launcher which has a traverse of 120° and can be elevated from –10 to +30°. There are two TOW ATGWs carried inside the hull and provision for a further four missiles outside. A standard TOW ground mount is carried inside the vehicle which enables the TOW system to be deployed away from the vehicle. Two positions for mounting one 7.62 mm machine gun are also provided with 2,000 rounds of ammunition being carried for this weapon.

106 mm recoilless rifle

This version incorporates a modified pod with top opening doors with armament consisting of a 106 mm M40 series recoilless rifle on a 150° rotating swing mount.

The weapon backblast cone has been co-ordinated with the slope angles of the vehicle hull to allow 10° of depression and 10° of elevation of the weapon.

The 106 mm M40 recoilless rifle is aimed using a 12.7 mm barrel-mounted spotting rifle for which 100 rounds of ammunition are carried. A total of 15 rounds of 106 mm ammunition are carried for the M40 recoilless rifle.

20 mm Cannon (1 m) turret

This one-man turret is armed with an Oerlikon Contraves 20 mm cannon and a 7.62 mm coaxial machine gun. Turret traverse is powered through 360° with manual back-up, while weapon elevation is from –8 to +55°, also powered with manual back-up.

The turret is also fitted with an internally controlled hydraulic charger, turret ventilating blower, emergency firing triggers and a rate controller for the 20 mm weapon which allows firing of 1, 2, 4 rds/s or full automatic mode.

For night surveillance a 500,000 candlepower searchlight is coaxially mounted to the 20 mm cannon. Sighting equipment consists of a ×8 monocular day sight and unity periscope with projected graticule, eight direct view vision blocks and an external anti-aircraft sight.

The turret is designed to carry 200 rounds of ready use 20 mm ammunition and 220 rounds of ready use 7.62 mm ammunition. Hull stowage is provided for 200 rounds of 20 mm and 2,200 rounds of 7.62 mm ammunition.

Status

Production as required. In service with:

Country	Quantity
Egypt	112
Indonesia	28

Manufacturer

Textron Marine & Land Systems.

UPDATED

General Dynamics Land Systems Reconnaissance, Surveillance, Targeting Vehicle (RST-V) system

Development

Early in 1999, following a competition, the Naval Surface Warfare Centre's Carderock Division awarded a US$22 million contract to General Dynamics Land Systems for Phase Two of the Reconnaissance, Surveillance, Targeting Vehicle (RST-V) system.

Late in 1997, under Phase One, General Dynamics Land Systems and Lockheed Martin Vought Systems (now Lockheed Martin Missiles and Fire Control – Dallas) were each awarded a US$3 million 12 month contract to develop a new reconnaissance vehicle that could be the basis for a family of weapons.

Under the Phase Two contract, which is sponsored by the US Marine Corp and the Defense Advanced Research Projects Agency (DARPA), General Dynamics Land Systems has built four demonstrator vehicles as well as supplying technical data, test support and spare parts.

Specifications (provisional)

Crew:	6 (maximum)	**Wheelbase:**	3.276 m
Configuration:	4 × 4	**Maximum road speed:**	112 km/h
Combat weight:	3,629 kg	**Acceleration:**	
Payload:	1,360 kg		0 to 48 km/h in 4 s
Length:	5.45 m		0 to 96 km/h in 15 s
Width:		**Fuel capacity:**	95 litres
(maximum):	2.057 m	**Range:**	
(minimum):	1.674 m	(main engine)	758 km
Height:		(batteries)	32 km
(maximum):	1.674 m	**Fording:**	0.914 m
(minimum):	1.397 m	**Gradient:**	60%
Ground clearance:	0.457 m (adjustable – see text)	**Side slope:**	30%
		Armament:	1 × 7.62 mm MG (see text)

This contract runs through to April 2002 and includes an option to purchase additional quantities of vehicles up to a maximum of 39.

General Dynamics Land Systems are prime contractor with other members of the team including Magnet Motors of Germany (electric drives and

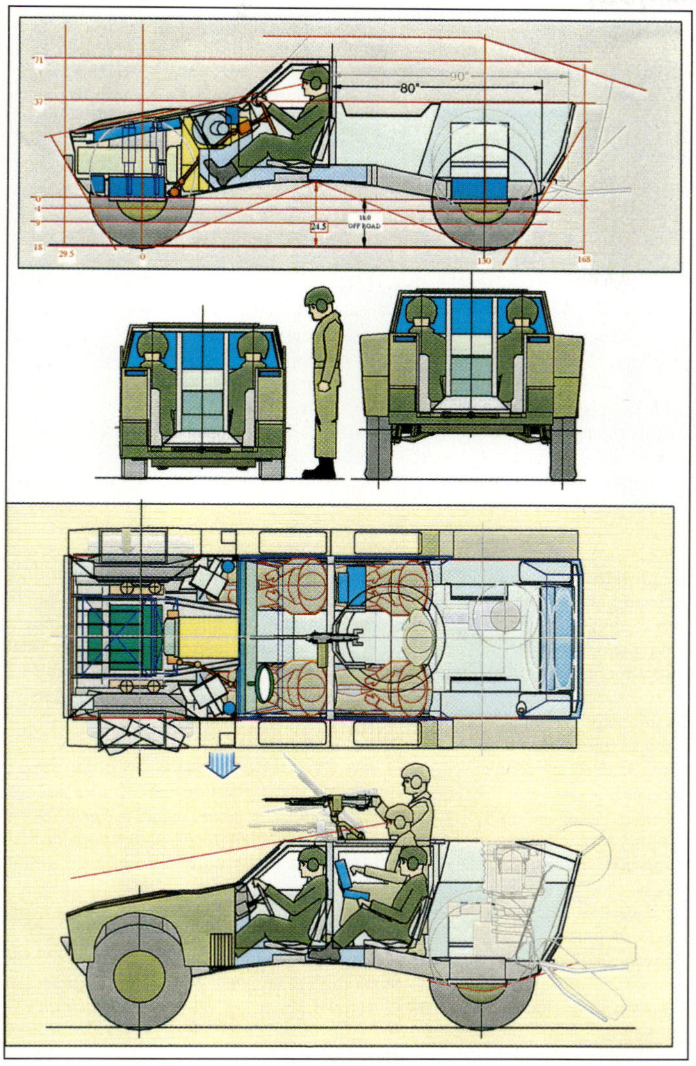

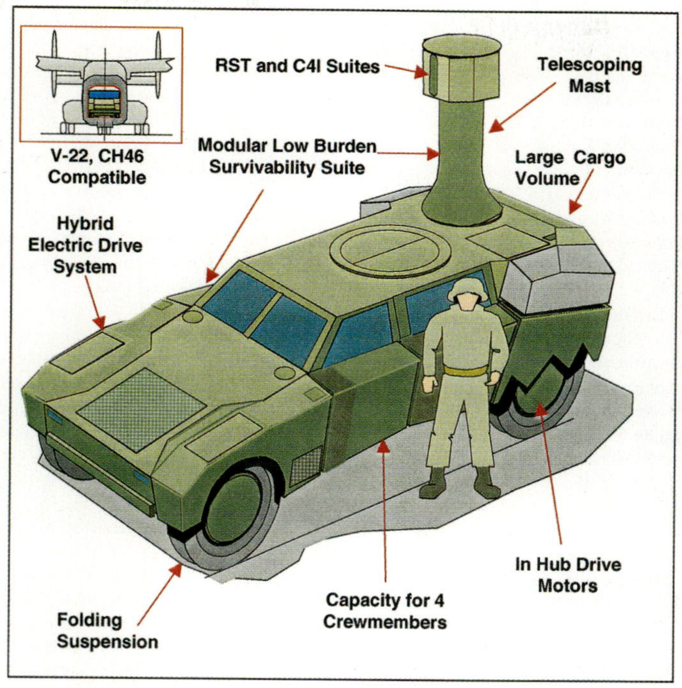

Key components of the General Dynamics Land Systems RST-V system with sensor pod raised
0024986

Drawings of General Dynamics Land Systems RST-V system
0024985

controls), Computing Devices Canada (integration of the reconnaissance, surveillance and targeting suite) and Boeing (signature design and integration).

Actual vehicle design work of the RST-V system was carried out at the General Dynamics Land Systems Technical Center in Muskegon. Communications and power systems engineering support will be provided by the General Dynamics Land Systems facility in Florida.

The first RST-V systems successfully took part in the US Marine Corps' Capable Warrior advanced war fighting experiment scheduled for the first quarter of 2001.

The RST-V is a hybrid electric drive vehicle mounting a reconnaissance, surveillance, targeting suite, command, control, communication and intelligence suite and an integrated survivability package.

When compared with current vehicles the RST-V features improved fuel economy and advanced signature management features.

Description

The RST-V features a hybrid electric drive (4 × 4) vehicle with a mast-mounted sensor pod and will weigh around 3.6 tonnes, have a maximum crew of six and is capable of being carried in a V-22 Osprey tilt-rotor aircraft or a CH-46 helicopter.

To allow it to fit into the V-22 and CH-46 the RST-V has a unique suspension system that can be raised or lowered and extended or retracted as required. The tyres and wheels of the RST-V are from the current in service AM General HMMWV (4 × 4) vehicle used in large numbers by the US Army and Marine Corps.

The crew of four is seated in the centre of the vehicle facing the front with bulletproof windows to the front and sides. They enter via two doors in either side with a roof hatch on which can be mounted 7.62 mm or 12.7 mm machine guns or a 40 mm grenade launcher. A large volume cargo area is at the rear with entry to this via a ramp.

It is expected that the vehicle will provide the occupants with protection from small arms fire and at least anti-personnel mines. Combat survivability will further be enhanced by the fitting of a defensive aids suite. This will include radar, laser and missile warning devices.

Mounted in the rear is an extendable mast on which is mounted the sensor package which is operated by the crew from within the vehicle.

The vehicle has a hybrid electric drive system with in-hub drive motors. It is envisioned that the electric drive will only be used in some tactical situations where a stealth approach is required.

Variants

General Dynamics has suggested that the RST-V could be modified for a wide range of roles and missions including forward observer/forward air control, NBC reconnaissance, C4I/command post with a roof that can be extended, light strike vehicle, ambulance with a roof that can be raised, air defence vehicle, logistics vehicle, personnel carrier and a weapons carrier. The latter include anti-armour and mortar carrier.

Status

Development for US Marine Corps. First prototypes underwent US Marine Corps trials early in 2001.

Manufacturer

General Dynamics Land Systems.

UPDATED

O'Gara-Hess & Eisenhardt Armoring Company M1114 up-armored HMMWV

Development

The AM General High-Mobility MultiPurpose Wheeled Vehicle (HMMWV) is the standard vehicle in its class in the US Army, Navy, Air Force and Marines and is used for a wide range of missions.

By mid-2001, well over 140,000 HMMWV and variants had been built for the home and export markets with production still under way.

The standard AM General HMMWV is unarmoured and in 1993 the US Army Tank Automotive and Armaments Command (TACOM) selected the O'Gara-Hess Eisenhardt Armoring Company as the supplier to manufacture and field the up-armored HMMWV XM1109 vehicle.

Successful live fire tests were carried out on the up-armored HMMWV XM1109 at Aberdeen Proving Ground before the vehicles were deployed to Haiti and Somalia. A total of 159 XM1109 vehicles was delivered to the US Army and some of those have also been redeployed to Bosnia.

In August 1994, the US Army awarded the company a contract to develop an improved version which became the M1114. The first prototype of this was completed in three months and it was in production in six months.

In the Spring of 1998, the company was awarded a US$59.4 million contract for a further 738 up-armoured HMMWVs for the US Army and Air Force which brought the total to 1,800 units.

The M1114 vehicle is based on the latest Enhanced Capacity Vehicle (ECV) version of the HMMWV and has a higher level of armour protection including protection against anti-tank mines and 7.62 mm armour-piercing rounds.

Blast testing on the M1114 included 5.5 kg contact detonated anti-tank mines being detonated under the front wheels and 1.81 kg contact detonated mines in the rear of the vehicle.

AM General delivers an incomplete HMMWV chassis to O'Gara-Hess & Eisenhardt which then integrates the armour system and delivers the completed M1114 vehicle to the customer.

Typical users of the up-armored HMMWV include scouts, military police, commander transport, ordnance disposal and the US Air Force.

In early 1996, Luxembourg placed an order for 24 M1114 vehicles through the FMS programme and these were delivered in September and October 1996 for use in Bosnia. Qatar ordered eight vehicles which have now been delivered.

Specifications

Crew: 1 + 3
Configuration: 4 × 4
Combat weight: 5,489 kg
Unloaded weight: 4,445 kg
Payload: 1,043 kg
Max towed load: 1,906 kg
Power-to-weight ratio: 34.61 hp/t
Length: 4.99 m
Width: 2.3 m
Height:
(overall) 1.9 m
(cargo bed) 0.92 m
Ground clearance: 0.30 m
Track: 1.82 m
Wheelbase: 3.30 m
Angle of approach: 46.5°
Angle of departure: 33.5°
Max road speed: 125 km/h
Acceleration: 0-48 km/h in 6.96 s; 0-80 km/h in 17.84 s
Fuel capacity: 94 litres
Max range: 443 km
Fording: 0.762 m
Gradient: 60%
Side slope: 40%
Vertical obstacle: n/avail
Trench: n/avail
Turning radius: 7.62 m
Engine: V-8 6.5 litre turbocharged, diesel fuel injection developing 190 hp at 3,400 rpm
Transmission: 4-speed full automatic
Transfer case: 2 speed
Steering: power-assisted, variable ratio
Suspension: independent double A-frame with open end coil springs and hydraulic shock-absorbers
Tyres: 37 × 12.5 R 16.5 LT with low-profile run-flat inserts
Brakes:
(main) hydraulic inboard-mounted power disc with dual-reservoir master cylinder
(parking) manual on rear service brakes
Armament: 1 × 12.7 mm M2 HB QCB MG or 1 × 7.62 mm M60 MG or 1 × 40 mm Mk 19 grenade launcher
Electrical system: 12/24 V waterproof, 200 A alternator
NBC system: no
Night vision equipment: optional
Air transportability:
C-130 3 (internal)
C-141E 6 (internal)
C-5A 15 (internal)
CH-53 helicopter (2 underslung)
CH-47 helicopter (2 underslung)

In May 1998, the company completed the 1,000th XM1114 up-armoured HMMWV and at the same time handed over the first of the M1116 vehicles for the US Air Force.

In mid-2000, O'Gara-Hess & Eisenhardt announced that the company had been awarded a contract extension for production of the M1114 through the year 2005.

A rate of 360 vehicles in 2000 is being budgeted and, according to the company, production could exceed 3,000 vehicles. Including the options it could mean as much as US$200 million to O'Gara over its five year life.

By mid-2000, the company had delivered some 1,700 vehicles to the US Army and 449 of the M1114 variant to the US Air Force.

Description

The overall layout of the M1114 up-armored HMMWV is virtually identical to the standard production vehicle, with the engine compartment at the front, driver and three passengers in the centre and a load area at the rear.

The M1114 chassis has been reinforced for this role and other improved features include: modified differentials; improved braking and suspension; 6.5 litre turbo-diesel engine developing 190 hp; upgraded wheels; central tyre-pressure regulation system fitted as standard.

In the middle of the roof is a newly designed turret which provides a more stable weapons platform providing better performance. A brush guard has been added to eliminate problems caused by spent cartridge cases and links becoming trapped in the assembly.

Various types of weapon can be installed including the 7.62 mm M60 or 12.7 mm M2 machine gun or a 40 mm MK19 grenade launcher.

The weapon can be accessed through the bifold door which locks into an upright position so providing the gunner with protection from the rear. The gun pintle is mounted to the turret and rotates with the turret bearing for ease of access.

High light transmission transparent armour ('white glass') within the driver's windshield allows the crew to use night vision equipment. In addition, an electric de-ice system is also installed to enhance visibility in adverse weather conditions.

The roof of the M1116 provides the crew compartment with 155 mm overhead burst protection from 100 m. Armour in the roof assembly is rolled on the edge of the overlap of the perimeter armour, providing protection of the crew compartment from all possible angles.

The entire roof assembly is mounted on shock-absorbing bushings. This counteracts the twisting and torsion of the crew compartment caused by rough terrain and also improves handling of the vehicle.

In each side of the vehicle is a forward-opening door with an operable window in its upper part. All windows provide the same level of protection as the remainder of the vehicle and are counter-balanced to assist the crew member in raising or lowering the window. The doors have reinforced hinges and straps and double locks to prevent the doors opening during a blast.

The armour system, which has passed US Army live fire and blast testing, has the following specifications:

(1) 7.62 mm armour piercing through a full 360° perimeter
(2) 155 mm airburst overhead
(3) 5.5 kg front underbody blast
(4) 1.81 kg rear underbody blast
(5) White glass transparent armour which is fully compatible with Generation III night vision equipment.

Standard equipment on this version includes:

(1) Waterproof dual batteries with a 200 A alternator
(2) Payload is easily accessible through the cargo shell door which opens in two directions
(3) The electric windscreen defrost/de-ice system ensures clear vision in any weather
(4) In the event of blast the seat bases protect occupants from high accelerations and three-point seat belts are fitted as standard equipment. The bucket-type seats have a high back
(5) A specially designed fuel filler door prevents unauthorised access during urban operations.

The heavy-duty air conditioning system features front and rear blowers in the crew compartment and can substantially lower interior temperatures even in desert conditions.

The vehicle is fully air-portable in C-130, C-141, C-5A and C-17 transport aircraft used by the US Air Force.

Variants

The company is also offering HMMWV armour kits to non-US armed forces. In addition to the standard package the company can deliver customised packages that can include overhead burst protection, cargo shells and rear partitions and protected seats.

M1116 up-armoured for USAF

In 1997 the US Air Force placed an order for 71 uparmoured M1116 vehicles from the O'Gara-Hess and Eisenhardt Armoring Company. Details are given in the Armoured personnel carriers (wheeled) section. By mid-2000, the US Air Force had taken delivery of 449 of these with production complete.

M1114 with SWARM

In mid-2000, an M1114 was shown fitted with the UK Helio Stabilised Weapon and Reconnaissance Mount (SWARM) armed with a 12.7 mm M2 machine gun.

Enhanced M1114

In 2000, the company stated that to meet the emerging requirements of the digital battlefield, where premiums have been placed on awareness, agility and crew protection, it was working to integrate into the M1114 land navigation systems, forward looking infra-red, a remote weapon station and advanced communication and surveillance electronics.

Status

Production. In service with following countries:

Country	Quantity	Comment
Luxembourg	24	delivered
Qatar	8	delivered
Slovenia	1	delivered
United States	1,700	additional vehicles on order (army)
	449	delivered (air force)

Manufacturer

O'Gara-Hess & Eisenhardt Armoring Company.

UPDATED

M1114 up-armoured HMMWV (4 × 4) fitted with roof-mounted machine gun (Michael Jerchel) **2002**/0114342

M1114 up-armoured HMMWV (4 × 4) with machine gun armament installed **2001**/0093547

ARMOURED PERSONNEL CARRIERS

Armoured personnel carriers (tracked)
Armoured personnel carriers (wheeled)

ARMOURED PERSONNEL CARRIERS (TRACKED)

ARGENTINA

TAMSE VCTP infantry combat vehicle

Development

In mid-1974, the German company, Thyssen Henschel (which in 2000 became Rheinmetall Landsysteme GmbH) was awarded a contract for the design and development of a new medium tank for the Argentine Army under the designation TAM *(Tanque Argentino Mediano)*, as well as an infantry combat vehicle under the designation VCTP *(Véhiculo de Combate Transporte de Personal)*.

The contract covered the design, development and construction of three prototypes of both the tank and the VCTP. The total requirement was for just over 500 vehicles, 200 TAMs and 300 VCTPs but, for financial reasons, production was completed after 350 had been built. Recent information has indicated that less than 100 VCTP are in service.

It is believed that another batch of around 50 VCTP vehicles may have recently been delivered to the Argentine Army.

Description

The VCTP is similar to the German Rheinmetall Landsysteme Marder 1 ICV but has a more powerful MTU 720 hp engine, a less sophisticated two-man turret armed with a 20 mm cannon and a 7.62 mm roof-mounted machine gun, three gun ports in each side of the troop compartment at the rear, two rectangular roof hatches over the troop compartment that are hinged in the centre, and a remote-controlled 7.62 mm machine gun installation which is fitted over the troop compartment at the rear of the hull. Mounted either side of the hull, slightly forward of the turret, is a bank of four electrically operated 76 mm smoke grenade dischargers firing forwards.

The vehicle is fitted with an NBC system which pressurises the crew compartment for three hours. A crew compartment heating system also preheats the engine for cold starts.

The two-man power-operated turret is mounted in the centre of the hull, with the commander seated on the left and the gunner on the right. Both have a single-piece hatch cover that opens to the rear with the commander being slightly raised over the gunner's position for increased all-round observation.

The commander has a sight and seven day observation periscopes while the gunner has a day sight and three day observation periscopes. Turret traverse and weapon elevation are electrohydraulic. The 20 mm cannon can be elevated from −11° 30' to +60° at a maximum speed of 60°/s with traverse through a full 360° at a maximum speed of 52°/s.

The 20 mm Oerlikon Contraves KAD-B19 cannon has 400 rounds of ready use ammunition in the turret (325 HE and 75 APDS) with a further 1,000 rounds in reserve in the hull.

The remote-controlled turret is called the TPA-1 and is armed with a 7.62 mm machine gun. A similar weapon is mounted externally at the commander's station for anti-aircraft defence.

To increase the operational range of the vehicle, long-range diesel fuel tanks can be mounted either side of the hull at the rear.

Specifications

Crew: 2 + 10
Combat weight: 28,000 kg
Power-to-weight ratio: 25.7 hp/t
Ground pressure: 0.69 kg/cm²
Length: 6.83 m
Width: 3.32 m
Height: (overall) 2.68 m
Ground clearance: 0.45 m
Track: 2.62 m
Max road speed
(forwards) 80 km/h
(reverse) 80 km/h
Fuel capacity
(main tank) 640 litres
(external tanks) 400 litres
Range
(main tank) 590 km
(external tanks) 350 km
Fording: (without preparation) 1.5 m
Gradient: 60%
Side slope: 30%
Vertical obstacle: 1 m
Trench: 2.5 m
Engine: MTU MB 833 Ka-500 V-6 diesel developing 720 hp at 2,400 rpm
Transmission: Renk HSWL 204 with 4 forward and reverse gears
Suspension: torsion bar in all versions
Electrical system: 24 V
Batteries: 6 × 12 V
Armament
(main) 1 × 20 mm Oerlikon Contraves KAD-B19 cannon
(anti-aircraft) 1 × 7.62 mm MG
(remote-controlled) 1 × 7.62 mm MG
Smoke grenade dischargers: 2 × 4 76 mm
Ammunition: (20 mm) 1,400
NBC system: yes
Night vision equipment: optional

Variants

VCTM 120 mm mortar carrier

The mortar version of the VCTP is called the VCTM *(Véhiculo de Combate Transporte de Mortero)* and is fitted with a 120 mm TDA AM-50 mortar for which a total of 61 rounds is carried. Of these, 12 are of the PEPA-LA long-range type with a maximum range of 9,500 m, the remaining 49 being standard mortar bombs with a range of 6,650 m.

The driver is seated front left with the vehicle commander to his rear. The mortar is mounted in the centre of the vehicle firing forwards and, when not in action, a three-part roof cover is positioned over the mortar, the centre part opening forward onto the glacis plate and the other parts opening left and right. There is an additional hatch to the right rear of the vehicle and mounted on the roof at the rear is a remote-controlled 7.62 mm machine gun as fitted to the basic VCTP.

VCPC command post vehicle

This, the command post member of the family, is called the *Véhiculo de Combate Puesto de Comando* (VCPC) and has extensive communications equipment installed. Armament consists of a pintle-mounted 7.62 mm machine gun on the cupola in the centre of the vehicle on the right side. The two oblong hatches over the rear troop compartments are retained, as are the firing hatches and observation periscopes.

VCRT armoured recovery vehicle

Details of this are given in the TAM entry. This version remains at the prototype stage.

VCLC multiple rocket launcher

Details of this are given in the Multiple rocket launchers section. A small number of these vehicles are now in service.

VCA 155 mm self-propelled gun

Details of the *Véhiculo de Combate de Artilleria* based on a modified and stretched TAM chassis are given in the Self-propelled guns and howitzers (tracked) section under Argentina. A small number of these are in service. The turret is the same as the Italian Otobreda Palmaria 155 mm self-propelled artillery system that has been sold to Libya and Nigeria.

Status

Production complete. In service in Argentina.

Manufacturer

Tanque Argentino Mediano Sociedad Del Estado.

UPDATED

VCPC command post vehicle with a cupola-mounted 7.62 mm machine gun, 76 mm smoke grenade dischargers and two radio antennas

VCTP infantry combat vehicle showing a bank of four 76 mm smoke grenade dischargers on the forward part of the hull side

AUSTRALIA

Australian M113A1 fleet

The Australian Army operates a fleet of 771 members of the M113A1 series of full-tracked armoured personnel carriers and has applied the following designations to the vehicle:

M113A1 APC/LRV (Armoured Personnel Carrier/Light Reconnaissance Vehicle)
This is fitted with the US Cadillac Gage T50 one-man turret armed with a 7.62 mm (0.30) and a 12.7 mm (0.50) machine gun. Total ammunition carried is 2,500 rounds of 7.62 mm and 3,000 rounds of 12.7 mm.

M113A1 APC (A) (Ambulance)
Standard M113A1 modified for use in an ambulance role.

M113A1 FSV (Scorpion) (Fire Support Vehicle - Scorpion)
This was renamed the Medium Reconnaissance Vehicle. Following trials with three prototypes of the M113A1 fitted with the turret of the British Scorpion CVR(T) vehicle armed with a 76 mm gun, a further 45 conversions were delivered to the Australian Army. These were finally phased out of service with the Australian Army in 1996. It is understood that a small number of MRV are still retained for Opposition Forces training vehicles. These have extended barrels and other additions to alter the silhouette as well as different camouflage and markings.

M113A1 FSV (Saladin) (Fire Support Vehicle - Saladin turret)
In total 18 were converted; no longer in service.

M577A1 ACV (Armoured Command Vehicle)
Similar to standard M577A1 vehicle.

M125A1 APC (M) (Carrier Mortar)
Similar to standard M125A1 mortar carrier and armed with a turntable-mounted 81 mm mortar.

M806A1 (Armoured Recovery Vehicle - Light)
Although called the M806A1, this was produced only for export. It is fitted with a hydraulically operated winch which leads out through the rear. Additional details are given in the entry for the M113 APC family under USA.

M113A1 APC (F) (Carrier, Repair Fitters)
This is also an export version and is fitted with an HIAB crane for changing components in the field. Additional details are given in the entry for the M113 APC family under USA.

M548A1 TLC (Tracked Load Carrier)
This is identical to the US Army M548A1 and is unarmoured.

Australian Army M113 series APC upgraded by Tenix Defence Systems with a new turret (Australian Army)
2002/0034825

Trials versions

106 mm Recoilless Rifle
In 1995, trials were carried out with an M113A1 APC fitted with the US 106 mm M40A2 recoilless rifle. The weapon is mounted on the right side of the vehicle to the rear of the commander's cupola.

The rear roof hatch has been modified and the hatch covers now open left and right to provide some lateral protection to the two-man crew of the 106 mm M40A2.

To enable them to aim, load and fire the M40A2, a stowage bin for the 84 mm Carl Gustav man-portable anti-tank weapon also carried has been fitted on the right side of the floor of the M113A1 and the two-man crew stands on this to operate the weapon. A total of 16 rounds of 106 mm ammunition is carried for the 106 mm M40A2 recoilless rifle.

MILAN ATGW carrier
This is an M125A1 mortar carrier with the 81 mm mortar removed and a hydraulic ram mounted on the mortar turntable on top of which is a Euromissile MILAN ATGW system. The MILAN ATGW has a maximum range of 2,000 m.

M113A1 Minimum Up-Grade Project
To enable the fleet of M113A1s to remain in service past the year 2010, the M113A1 Product Improvement Project was proposed but, as a result of defence cutbacks, this is now known as the M113A1 Minimum Up-Grade Project.

In September 1996, the Australian Department of Defence selected the now Tenix Defence Systems Pty Ltd, Australia to upgrade 364 of the Australian Army's fleet of M113A1 series of full-tracked Armoured Personnel Carriers (APCs) to extend their operational life through to the year 2010.

The contract was finally signed in the spring of 1997 and is valued at A$50 million and originally included:

(1) New one-man turret from Tenix Defence Systems armed with FN HERSTAL 12.7 mm M2 machine gun
(2) New gunsight from Electro Optic Systems which will display imagery on a flat panel display and provide frame storage capability so images and overlaid data can be retained in the computer buffer

The new one-man turret has been developed by Tenix Defence Systems specifically for the M113 series APC and provides the commander with more space and vision as well as greater reliability.

Turret traverse and weapon elevation is powered and the day/night sight, provided by Electro Optic Systems, enables targets to be engaged under day and night conditions. The turret will be armed with an FN HERSTAL M2 quick change 12.7 mm machine gun.

When upgraded under Phase I they will be equivalent to the US Army's M113A2 series. Night vision equipment will be through the Australian Project Ninox Phase 1B.

These Phase I vehicles were completed in 1998 and it was expected that once in service they would be upgraded to the more enhanced Phase II standard and designated the M113AS3.

In addition to the items previously mentioned this would also have included the following:

(1) Daewoo Heavy Industries for the Daewoo D2848T diesel engine, ZF LSG 1000 fully automatic transmission, upgraded torsion bars and external fuel tanks;

M577A1 command post vehicle of the Australian Army in travelling configuration (Barry Marriott)
2000/0064859

M113A1 Armoured Personnel Carrier/Light Reconnaissance Vehicle of the Australian Army fitted with T50 turret and armed with 12.7 mm and 7.62 mm machine guns (Barry Marriott)
2000/0064858

(2) DEW Enginering of Canada for passive armour kit and spall liners based on the German Army IBD Modular Expandable Armour System;
(3) Santa Barbara Dual Spectrum with the FM200+ automatic fire detection and suppression system;
(4) Aircontrol Technologies for the heating, ventilation and air conditioning system that also includes a cool water drinking system.

Following trials with prototype vehicles a number of problems were encountered, especially in the area of the turret, which was not considered to be suitable.

In April 2001, the Australian Army announced that A$10 million would be spent on Project Definition Phase of the M113 upgrade and that it was expected that 350 vehicles would be upgraded.

The Project Definition Phase will involve the manufacture of a mock-up vehicle and the continuation of turret development. This work is to be carried out by Tenix Defence Systems at its Adelaide based Systems Division and should be completed by the first quarter of 2002.

In the longer term it is expected that the production of the upgraded fleet is expected to occur at the Australian Army's Southern Logistics Group facility in the Albury/Wodonga area.

Within the Australian Army, the M113 upgrade programme is referred to as Project Land 106. It was expected that the in-service date would be 2002/2003, but is now expected to be later, perhaps 2005. It is also possible that some of the subcontractors mentioned above may well change.

Status
Development. Not yet in production or service.

Prime contractor
Tenix Defence Systems Pty Ltd.

UPDATED

AUSTRIA

Steyr-Daimler-Puch 4K 7FA G 127 armoured personnel carrier

Development
The Steyr-Daimler-Puch 4K 7FA G 127 is an up-armoured version of the earlier Saurer 4K 4FA armoured personnel carrier fitted with the Steyr-Daimler-Puch 7FA diesel engine and transmission of the SK 105 tank destroyer which was also manufactured by Steyr-Daimler-Puch. The first prototype of the 4K 7FA G 127 was completed in 1976, with first production vehicles following in 1977. The Steyr-Daimler-Puch 4K 7FA G 127 has been exported to Bolivia (6), Greece (qv) and Nigeria (170).

There have been no recent sales of this vehicle and Steyr-Daimler-Puch is now concentrating marketing on the ASCOD infantry combat vehicle that it has developed in association with SANTA BARBARA of Spain. Details of this are given under International. This is now in volume production for the Spanish Army. In mid-1999, the Austrian Army placed an order for 112 vehicles which will be called the Ulan, with first four production vehicles being completed in mid-2001.

Specifications
4K 7FA-KSpZ
Crew: 2 + 8
Combat weight: 14,800 kg
Power-to-weight ratio: 21.62 hp/t
Ground pressure: 0.55 kg/cm²
Length: (hull) 5.87 m
Width: 2.5 m
Height: (without armament but including MG support) 1.611 m
Ground clearance: 0.42 m
Track: 2.12 m
Track width: 380 mm
Length of track on ground: 3.192 m
Max road speed:
(forward) 70 km/h
(reverse) 7.4 km/h
Fuel capacity: 360 litres
Max road range: 520 km
Fording: 1 m
Gradient: 75%
Side slope: 40%
Vertical obstacle: 0.8 m
Trench: 2.1 m
Engine: Steyr-Daimler-Puch 7FA 6-cylinder liquid-cooled 4-stroke turbocharged diesel developing 320 hp at 2,300 rpm
Transmission: ZF 6-S80 manual with 6 forward and 1 reverse gears (as an option this can be replaced by the ZF 6HP 600 automatic with 6 forward and 1 reverse gears)
Steering: hydrostatic split torque type
Clutch: Fichtel & Sachs single dry plate
Suspension: torsion bar
Electrical system: 24 V
Batteries: 2 × 12 V, 180 Ah
Armament: (main) 1 × 12.7 mm M2 MG
Smoke-laying equipment: 4 smoke grenade dischargers on rear of gunner's cupola
Armour
Hull (max): 25 mm
NBC system: optional
Night vision equipment: optional

Description
The basic model is designated the 4K 7FA G 127 and is armed with a 12.7 mm (0.50) M2 HB Browning machine gun. The hull is made of all-welded steel and is immune to penetration from 20 mm armour-piercing projectiles over its frontal arc.

The driver sits at the front of the hull on the left side with the engine to his right. He has a two-piece hatch cover that opens to the left and right, in front of which are three day periscopes. The centre periscope can be replaced by a passive periscope for driving at night.

The gunner sits behind the driver and has a cupola with a two-piece hatch cover opening left and right which can be locked vertical to give some protection. There are two day observation periscopes in the front part of the cupola. A standard 12.7 mm (0.50) Browning M2 machine gun is mounted on the forward part of the cupola and mounted on the rear of the cupola are four smoke grenade dischargers which point to the rear of the vehicle. When unlocked, the 12.7 mm machine gun has an elevation from –20 to +35° and when locked –10 to +7.5°.

The personnel compartment is at the rear of the hull with the eight infantrymen seated on bench seats, four down each side of the hull.

The section leader, the first man on the right bench, is provided with a day periscope in the forward part of the personnel compartment, on the right side. Two hatches over the roof of the vehicle open outwards. Mounting sockets for the pivot trunnion of an adaptor mount for the MG 74 machine guns are provided on the front, right, left and rear side of the hatch, above the personnel compartment. Normal means of entry and exit for the crew is by the twin doors at the rear of the hull.

The torsion bar suspension has five dual roadwheels with the drive sprocket at the front, idler at the rear and three track-return rollers. The first and last roadwheel stations are provided with a hydraulic shock-absorber. The end-connector tracks are provided with rubber pads and can be fitted with spikes for operation on icy surfaces.

Standard equipment includes a ventilation system for the fighting compartment and a vehicle heater, while optional equipment includes a passive night vision device for the driver, an automatic fire extinguishing system for the engine compartment, an automatic fire suppression system for the fighting compartment, a collective NBC system and an air conditioning system.

Variants

Upgrade package
Steyr-Daimler-Puch has now developed an upgrade package for its 4K 7FA G 127 armoured personnel carrier along similar lines to that offered for its SK 105 tank destroyer.

This includes the installation of a ZF 6HP 600 fully automatic transmission with torque converter and lock-up clutch, improved air cooling system, cyclone-type air filter system and driver's image intensification night viewer.

Other modifications include hydraulic bump stops for improved cross-country mobility, 250 Ah dry batteries and a fire detection and suppression system.

4K 7FA-San armoured ambulance which is unarmed 0018406

4K 7FA armoured personnel carrier fitted with roof-mounted RASIT radar system 0004973

For desert operations the 4K 7FA G 127 can be fitted with sand shields, an improved cooling system and a cyclone-type air cleaner.

4K 7FA-KSPz MICV

This model carries a total of 10 people: section leader/ commander, driver, machine gunner and seven infantrymen. The driver is front left as normal with the section leader seated at the front right corner of the fighting compartment on a hinged single seat. He is provided with a rotatable day periscope that can be traversed through 360°.

The vehicle is fitted with four ball mountings with day periscopes above and the former allow for insertion of a Steyr 5.56 mm calibre AUG 508 assault rifle or a Steyr MP81 sub-machine gun, two either side of the rear fighting compartment. There are also two day periscopes in the hull rear.

Four of the seven infantrymen sit back-to-back down the centre of the vehicle so that they can use the ball-mounted weapons. The remaining members of the infantry section are seated on hinged single seats, one at each remaining corner of the fighting compartment.

Two hatches over the hull roof open outwards. Mounting sockets for the pivot trunnion of an adaptor mount for the 7.62 mm MG of the infantry section are provided on the front, right, left and rear sides of the hatch.

4K 7FA FSCV 90 fire support combat vehicle

This is the basic G 127 with a modified hull that can be fitted with various 90 mm turrets including the French Giat Industries TS 90 or Cockerill LCST 90 turret system already manufactured for Kuwait, Qatar and Saudi Arabia. This model is not yet in production.

4K 7FA MICV 30/1

The prototype was completed in 1984 and is essentially the 4KA 7FA-KSPz fitted with a Steyr-Daimler-Puch SPz/200 one-man turret armed with a BAE Systems, RO Defence 30 mm RARDEN cannon and a 7.62 mm coaxial machine gun. This model has yet to enter production. It can also be fitted with a Steyr-Daimler-Puch two-personnel turret armed with a 30 mm cannon and 7.62 mm coaxial machine gun.

4K 7FA G 127 which is fitted with a cupola armed with a 12.7 mm M2 machine gun with smoke grenade launchers to the rear **2000**/0064860

4K 7FA-Fü command vehicle

This model has a crew of seven: commander, gunner, two radio operators, two messengers and the driver. Communications equipment includes two radios with an automatic tuning unit, a loudspeaker and a crew intercom.

4K 7FA-San armoured ambulance

Its crew of two consists of the driver and a medical orderly. It can carry up to four sitting and two stretcher patients and carries extensive medical equipment.

4K 7FA AMC 81

Mortar carrier with a crew of five (driver, commander, gunner and two mortar men), armed with an 81 mm mortar and carrying 78 mortar bombs.

Status

Production complete in Austria but can recommence if additional orders are placed. In service with:

Country	Quantity	Comment
Bolivia	6	
Cyprus	200	estimate (see Greece)
Greece	400	estimate (see Greece)
Nigeria	170	some sources state 300

Manufacturer

Steyr-Daimler-Puch Spezialfahrzeug AG & Co KG.

The vehicle has been manufactured under licence in Greece by ELBO where it is known as the Leonidas. Photographs and additional details of Greek-built vehicles appear in this section under Greece.

UPDATED

Saurer 4K 4FA armoured personnel carrier

Development

In 1956, Österreichische Saurer-Werke of Vienna (which was taken over by Steyr-Daimler-Puch AG in 1970) started development of a full-tracked armoured personnel carrier. The first prototype, the 3K 3H, was completed in 1958 and the second, the 4K 3H, the following year. They are both powered by a Saurer 3H 200 hp diesel engine.

The next model, the 4K 2P, had a different hull layout, a new torsion bar suspension and was powered by a 250 hp Saurer 2P diesel engine. In appearance it was similar to the first production vehicles which were completed in 1961 and designated 4K 4F. Later production vehicles were the 4K 3FA (230 hp) and the 4K 4FA (250 hp), which differed only in minor details such as their engines.

A total of 445 production vehicles was delivered to the Austrian Army in the following versions: 30 4K 4F, 211 4K 3FA, 38 4K 3FA-FS, 12 4K 4FA-Fü/FIA, 15 4K 4FA-San, 76 4K 4FA-G2, 60 4K 4FA FüA 1 to 4 series, three 4K 3FA Fül and a number of 4K 3FA GrW1 81 mm mortar carriers which were conversions of earlier vehicles. Production was finally completed in 1969. The vehicle was succeeded in production by a new model, described in the previous entry, called the 4K 7FA which uses the same automotive components as the SK 105 light tank also manufactured by Steyr-Daimler-Puch, which is covered in detail in the Light Tanks section under Austria.

These vehicles are now over 30 years old and have never been upgraded. In the future they will be supplemented by the Steyr-Daimler-Puch ASCOD infantry combat vehicle, which is covered in this section under International. In the Austrian Army this will be called the Ulan.

In mid-1999, the Austrian MoD awarded Steyr-Daimler-Puch a contract for the supply of 114 Ulan infantry fighting vehicles for the Austrian Army. These will be delivered from 2002 to 2004 and will replace at least part of the current Saurer 4K 4FA APC fleet of the Austrian Army.

Description

The hull of the Saurer armoured personnel carrier is made of all-welded steel armour and is immune to penetration from 20 mm armour-piercing projectiles over its frontal arc.

The driver sits at the front of the hull on the left side with the engine to his right. He has a two-piece hatch cover that opens to the left and right, in front of which are three day periscopes. The centre periscope can be replaced by a passive periscope for driving at night.

The gunner sits behind the driver and on the basic model has a cupola with a two-piece hatch cover opening left and right and a single day observation periscope. A standard 12.7 mm (0.50) M2 Browning machine gun is mounted on the forward part of the cupola and most vehicles are fitted with a shield to give a measure of protection to the gunner. This model is known under the company designation of the 4K

Saurer 4K 4FA-G1 of the Austrian Army with all roof hatches closed and showing the position of the rear-mounted 7.62 mm MG (Stefan Marx)

Saurer 4K 4FA-San ambulance which is not armed (Stefan Marx)

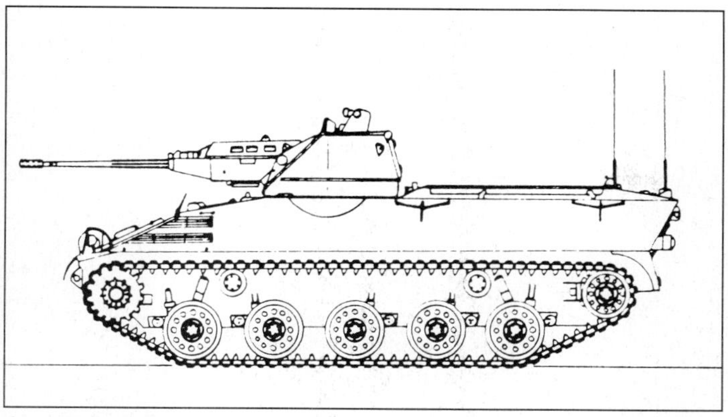

Saurer 4K 4FA-G2 with Oerlikon Contraves turret-mounted 20 mm cannon

4K 4FA-G2 fitted with Oerlikon Contraves 20 mm turret type GAD-AOA

4FA-G1 with its Austrian Army designation being the usMGSPz (Schützenpanzer/Überschweres MG).

The personnel compartment is at the rear of the hull with the eight infantrymen seated on bench seats, four down each side of the hull. The section leader is the first man on the right seat and is provided with a day periscope in the forward part of the personnel compartment roof, on the right side. Access to the personnel compartment is by two doors in the rear of the hull. Two hatches over the roof of the vehicle open to the outside. Mounting sockets for the pivot trunnion of an adaptor mount for the MG 42 machine gun are provided on the front, right, left and rear side of the hatch above the personnel compartment.

The torsion bar suspension consists of five dual roadwheels with the drive sprocket at the front, the idler at the rear and two track-return rollers. The first and last roadwheel stations are provided with a hydraulic shock-absorber.

The 4K 4FA has no NBC system, night vision equipment or amphibious capability.

Specifications

Crew: 2 + 8
Combat weight:
(with 20 mm turret) 15,000 kg
(with 12.7 mm MG) 12,500 kg
Unloaded weight: (without turret or armament) 11,100 kg
Power-to-weight ratio:
(20 mm model) 16.66 hp/t
(12.7 mm model) 20 hp/t
Ground pressure: 0.52 kg/cm²
Length: 5.4 m
Width: 2.5 m
Height:
(including 12.7 mm MG) 2.1 m
(to hull top) 1.65 m
Firing height: 1.95 m
Ground clearance: 0.42 m
Track: 2.12 m
Track width: 375 mm
Length of track on ground: 2.9 m
Max road speed:
(1st gear) 11.3 km/h
(2nd gear) 18 km/h
(3rd gear) 28.3 km/h
(4th gear) 43.8 km/h
(5th gear) 65 km/h
Fuel capacity: 184 litres
Max road range: 370 km
Fording: 1 m
Gradient: 75%
Side slope: 50%
Vertical obstacle: 0.8 m
Trench: 2.2 m
Engine: Saurer model 4FA 4-stroke 6-cylinder diesel developing 250 hp at 2,400 rpm
Transmission: manual with 5 forward and 1 reverse gears
Suspension: torsion bar
Electrical system: 24 V
Batteries: 2 × 12 V, 180 Ah
Armament: 1 × 12.7 mm M2 MG or 1 × 20 mm Oerlikon Contraves Model 204 GK cannon
Armour
Hull nose: 20 mm at 45°
Hull glacis: 20 mm at 30°
Hull sides: 14 mm
Hull top: 8 mm
Hull rear: 12 mm
Turret front: 20-35 mm
Turret, other surfaces: 15-20 mm
NBC system: no
Night vision equipment: no

Variants

4K 4FA-G2 (Grenadier-Schützenpanzer)

This is similar to the basic 4K 4FA-G1 but is fitted with an Oerlikon Contraves GAD-AOA one person turret which is armed with a 20 mm Oerlikon Contraves cannon Model 204 GK. This is belt-fed and can be used in both the ground and anti-aircraft roles. The weapon is provided with 100 rounds of ready use ammunition and the empty cartridge cases and links are ejected outside the turret. The 20 mm cannon has an elevation of +70° and a depression of −12°, turret traverse is 360°. Although its company designation is the 4K 4FA-G2 its Austrian Army designation is the MK66SPz Schützenpanzer/Mashinenkanone 66.

4K 3FA-Fü1 (Command Vehicle) (Führungs-SPz, for Brigade Commander)
4K 3FA-FüA (Artillery Command Vehicle) (Artillerie-Führungs-Schützenpanzer)
4K 3FA-Fü/FIA (Anti-aircraft Command Vehicle) (FIA Führungs-Schützenpanzer)
4K 3FA-FS (Wireless Teleprinter Vehicle) (Funkfernschreiber Panzer)
4K 4FA-San (Ambulance)

Full designation is 4K 4FA-San (Sanitätspanzer). This model is unarmed and has a crew of two, the driver and the medical orderly, and it can carry four sitting and two stretcher patients. The stretchers are carried in the left side of the hull with the four sitting patients on the bench seat down the right side of the hull, strapped in with waist and shoulder straps.

4K 4F GrW1 81 mm Mortar Carrier (Granatwerfer-SPz)

Mounted in the rear of the troop compartment, firing forwards, is an 81 mm mortar. A 120 mm self-propelled mortar was developed to the prototype stage (GrW2) but never entered service. The vehicle is armed with an 8.1 cm mGrW 82 (Mittelschwere Granatwerfer 82). Minimum range of the mortar is 180 m with maximum range being 5,850 m, normal rate of fire is 6 rounds per minute with maximum rate of fire being 18 to 25 rounds per minute.

81 mm Multiple Rocket Launcher

This is the basic APC with an Oerlikon Contraves 81 mm rocket launcher mounted on a turntable over the rear troop compartment. It is understood that these vehicles have now been phased out of service.

Shützenpanzer A1/LWT

This is the section vehicle for the Matra BAe Dynamics Mistral surface-to-air missile systems in the Pz1FAL batteries.

Status

Production complete. In service only with the Austrian Army (443). To be supplemented by the Ulan IFV in the future (see Development).

Manufacturer

Steyr-Daimler-Puch Spezialfahrzeug AG & Co KG.

UPDATED

BELGIUM

CMI, Cockerill Mechanical Industries Defence Division, BMF production of M113A1/AIFV

Development

In April 1980, the Belgian government placed an order for 514 of the then FMC (now United Defense LP) Armored Infantry Fighting Vehicles and 525 M113A1 armoured personnel carriers. These were built under licence by Belgian Mechanical Fabrication (BMF).

This was a consortium between a private owner (the majority owner), a Belgian Bank (BBL) and Cockerill Mechanical Industries (CMI). The Aubange factory of BMF was responsible for the hulls and hull assembly while the subcontractor NV Henschel in Antwerp carried out final assembly.

The first two vehicles were handed over in 1982 and production continued at the rate of 20 to 25 vehicles a month until April 1988. In 1983, the Netherlands ordered 26 M113A1 hulls from BMF which were converted to recovery vehicles by the Dutch company, RDM Technology.

Early in 1985, Cockerill Mechanical Industries of Belgium purchased an M113A1-B APC from BMF and mounted a CM 90 two-man 90 mm turret for use as a Fire Support Vehicle (FSV). Despite intensive marketing this chassis/turret combination was never sold.

Late in 1989, CMI was awarded a contract valued at about US$100 million to build Armored Infantry Fighting Vehicle (AIFV) hulls for Turkey (qv). Cockerill Mechanical Industries built and delivered 285 hulls and then provided the equipment to build the remaining 1,415 vehicles in Turkey. Since then Turkey has placed additional orders with the prime contractor, being FNSS Defense Systems.

Delivery of the hulls was spread over three years with the first being delivered late in 1989. DAF in the Netherlands received a US$50 million subcontract to provide weapon stations and vehicle integration. The first 20 AIFVs were assembled by DAF, after which assembly commenced in Turkey.

Belgian Army AIFV-B (C R Zwart) 0045979

Belgian Army M113A1-B-CP command post vehicle with generator on roof (D Caemerlynck) ***2001***/0093571

Today DAF Special Products is known as SP aerospace and vehicle systems and has developed the Multi Purpose Carrier covered in detail in the Reconnaissance vehicles section.

The CMI facility is still engaged in defence production work and recent projects have included the assembly and delivery of 54 Steyr-Daimler-Puch Pandur (6 × 6) APCs to the Belgian Army and the overhaul and upgrading of M992 FAASVs for the US Army.

M113A1-B

The Belgian variants of the M113A1 are very similar to the standard US Army M113A1 but with the following modifications: a torsion bar suspension system similar to that of the M113A2, a 200 A alternator, an exterior air intake/exhaust personnel heater, an NBC system supplying air to each individual, a Halon fixed fire extinguishing system for the engine compartment and vehicle lights for Belgian and German roads.

There are two basic models produced; the first is the standard two-holed vehicle with a driver's hatch and a commander's cupola, the second is a new three-holed vehicle with a driver's hatch and two transversely adjacent cupolas. The M113A1-B vehicles are all armed with the standard 12.7 mm M2 HB machine gun on the commander's cupola, except for the unarmed ambulance vehicle. The manual machine gun has an elevation of +53°, a depression of −21° and 360° traverse. Additional weapons other than personal equipment are detailed in the variants.

Variants

M113A1-B-ATK

This is a three-holed vehicle with the 12.7 mm M2 machine gun on the right and an unarmed cupola on the left. The vehicle has a crew of 10 and carries a 7.62 mm machine gun, a short-range anti-tank weapon and ammunition.

M113A1-B-MIL MILAN

This is a three-holed vehicle which has the 12.7 mm machine gun on the left of the commander's cupola and a quick-disconnect mount on the right cupola for the 2,000 m range Euromissile MILAN ATGW. The missile is fired from its tripod either attached to the cupola mount or dismounted. The mount allows for the rapid escape of the vehicle over a limited distance with the launcher still attached to the cupola. Two Bofors Lyran mortar launchers are mounted on the top of the vehicle at the left rear. This model has a crew of five.

Belgian Army AIFV-B.50 armed with 12.7 mm M2 machine gun and MILAB ATGW launcher (D Caemerlynck) ***2001***/0093568

M113A1-B-TRG

This is the basic MILAN vehicle with hatches, cupolas, stowage provision and so on. It is equipped with two plexiglas clear canopies, one for the observer and the other for the driver/instructor. Equipment is provided for all-weather manoeuvres and the transport of observers. This vehicle can be converted into the M113A1B-MIL.

Engineer vehicle M113A1-B-ENG

This is a two-holed vehicle for carrying a demolition team and its equipment; it has a self-recovery winch and a crew of eight.

Command post vehicle M113A1-B-CP

This is a two-holed vehicle based on the M113A1, not the M577A1, and is fitted with communications equipment, map boards and other equipment for a mobile command post. It includes an optional side awning or canvas-covered extension. The extension is of lightproof construction and may be connected to extensions of the same design on other M113A1-B-CPs. This has a crew of seven.

Recovery vehicle M113A1-B-REC

This is a two-holed vehicle fitted with a hydraulic winch and a fairlead assembly for recovering disabled vehicles. It also has a manual crane mounted on the top left. This model has a crew of four.

Maintenance vehicle with crane M113A1-B-MTC

This is a two-holed vehicle with an HIAB hydraulic crane on the mid-left top, a self-recovery winch and a portable work bench, special lighting and storage for tools. It has a crew of five.

Ambulance vehicle M113A1-B-AMB

This is a two-holed vehicle, unarmed and with a crew of three. It can carry four stretcher, two stretcher and three sitting, or six sitting patients.

M113A1-B-SVB

This is called the Svagveldbewaking and is fitted with a telescopic arm on top of which is mounted the EL/M-2130A battlefield surveillance radar. This system is only used by the 2e bataljon Jagers te Paard.

M113-B-Mor

This was originally fitted with an M30 4.2 in mortar mounted in the rear for which 28 mortar bombs were carried. These are now being replaced by the French TDA 120 RT rifled mortar which will be towed behind the vehicle rather than mounted internally.

Belgian Army M113A1-B-GN squad vehicle for engineer troops with some of the equipment carried (D Caemerlynck) ***2001***/0093570

M113-B-FAC

Some M113-B-FAC, or ATK have been converted for FAC (Forward Air Control) vehicles for the brigades. Also, some of the M113-B-MIL, without the MILAN equipment but fitted with a support for a portable laser range-finder, serve as forward observer vehicles in the field artillery battalions.

M113A1-B-CP.50

This is command post vehicle with generator and tent. There is also M113A1-B-CP with side awning.

AIFV-B

The Belgian Armored Infantry Fighting Vehicle is very similar to the Dutch AIFV fully described in the AIFV entry later in this section except that the following major differences are incorporated: a suspension similar to the M113A2's (and identical to the M113A1-Bs), an external air intake/exhaust personnel heater, an NBC system supplying air to each individual, and a Halon fixed fire extinguishing system for the engine compartment and lights for Belgian and German roads. There are three versions of this vehicle. The first Belgian production AIFVs were handed over to infantry units in September 1985.

AIFV-B with 25 mm Oerlikon KBA-BO2 cannon, Enclosed Weapon Station (EWS), AIFV-B-25 mm

This is similar to the Dutch AIFV squad vehicle and has a crew of 10. The smoke grenade launchers are mounted on the turret sides while on the YPR-765 PRI they are on the front of the hull.

AIFV-B-TRG

This is a driver training vehicle.

AIFV with 12.7 mm Cupola Weapon Station (CWS), AIFV-B-.50

This vehicle is similar to the AIFV-B-25 mm except that the 12.7 mm CWS is used in place of the EWS.

Belgian Army M113A1-B-SVB with telescopic surveillance mast erected (D Caemerlynck) ***2001***/0093569

This CWS permits operation of the gun in both closed hatch, remote mode and open hatch, manual flex mode. Remote operation permits an elevation range of -10 to +50° and complete 360° traverse. The flex mode has an elevation range from -12 to +53° and 360° traverse.

The left side commander's cupola is fitted with a quick-disconnect mount for the 2,000 m range Euromissile MILAN missile launcher. As with the standard M113A1-B-MIL, the launcher remains fixed to its tripod for use on the quick-disconnect mount or dismounted. With the launcher attached to the mount, limited distance operations are possible. The vehicle also has two Bofors Defence Lyran launchers.

AIFV-B-CP command post

This vehicle is similar to the Dutch AIFV command post with an M113A1 type of cupola in place of a weapon station on the right. A pintle-mounted 12.7 mm M2 HB machine gun is affixed to this cupola. Special communications equipment, map boards and other equipment for operating a mobile command post are provided and a single side awning is included. This version has a crew of seven.

Status

Production of these vehicles was completed in April 1988. In service with the Belgian Army.

Manufacturer

BMF.

BMF was purchased by CMI, one of the minority owners of the BMF consortium, late in the production and no longer exists as an entity. CMI is, however, very active in defence and is prime contractor for the 90 mm two-man turret which has been produced for Kuwait, Oman and the Saudi Arabian National Guard.

More recently, this turret has been selected for the 130 Diesel Division, General Motors of Canada, LAV-AG (Light Armoured Vehicle - Assault Gun) ordered by the Saudi Arabian National Guard.

UPDATED

BULGARIA

BMP-30 infantry combat vehicle

Development

The Bulgarian BMP-23 IFV, covered in detail in the following entry, was followed by the improved BMP-30 which has a similar chassis to the BMP-23 but has a new two-person turret that is almost identical to that fitted to the Russian BMP-2 infantry combat vehicle.

Recent information has indicated that there has been no recent production of the BMP-30 infantry combat vehicle in Bulgaria, although marketing does continue.

The BMP-30 and BMP-23 infantry fighting vehicles are manufactured by the TEREM Joint Stock Company who also built the BRM-23 reconnaissance vehicle. Recent information has indicated that as few as 10 of these BMP-30 vehicles are in service.

Specifications

Crew: 3 + 7
Combat weight: 15,000 kg
Ground pressure: 0.484 kg/cm²
Power-to-weight ratio: 20 hp/t
Length: 7.127 m
Width: 2.85 m
Height: 2.115 m
Ground clearance: 0.40 m
Length of track on ground: 4.445 m
Track: 2.50 m
Road range: 600 km
Fording: amphibious
Max road speed: 61.5 km/h
Max water speed: 4.5 km/h
Gradient: 60%
Side slope: 30%
Vertical obstacle: 0.8 m
Trench: 2.5 m
Engine: YaMZ-238N diesel developing 302 hp at 2,100 rpm
Transmission: manual, 6 forward and 1 reverse gears
Steering: clutch and brake
Suspension: torsion bar
Electrical system: 24 V
Armament:
(main) 1 × 2A42 30 mm cannon
(coaxial) 1 × 7.62 mm PKT machine gun
(anti-tank) 1 × 9P135M launcher
Ammunition:
(30 mm) 1,000
(7.62 mm) 2,000
(ATGW) (US AT-4B) 2 × 9M111-2 Fagot-2
(ATGW) (US AT-5) 2 × 9M113 Konkurs
Turret traverse: 360°
Weapon elevation/depression: +75/–6°
Infantry weapons:
(assault rifles) 5
(9 mm Makarov pistol) 3
(infantry MG) 2
RPG-7V rocket launcher: 1
Rockets for above: 2
RPG-22 rocket launcher: 4
Strela-2M SAM: 1
Flares: 15
NBC system: yes
Night vision equipment: yes

Description

The hull of the BMP-30 is of all-welded steel construction which provides protection against small arms fire of up to 20 mm in calibre as well as shell splinters.

The driver is seated at the front of the BMP-30 on the right with one infantryman to his left, the power pack is behind the driver with the turret in the centre of the hull and the infantry compartment at the rear.

The driver has a single-piece hatch cover that opens to the rear and in front of this are three forward-facing TBPO-170 day periscopes, the centre one of which can be replaced by a TVN-2B infrared night vision device. The infantryman to the right of the driver has a single-piece hatch cover that opens to the rear and three forward-facing periscopes. Both the driver's and commander's hatches can be locked in the vertical position if required.

The turret is virtually identical to that fitted to the Soviet BMP-2 infantry fighting vehicle which was first observed in 1982 and is armed with a 30 mm 2A42 cannon with a 7.62 mm PKT machine gun being mounted coaxial to the left. Turret traverse/weapon elevation is electric, with manual controls for emergency use.

The stabilised 30 mm 2A42 is a much more effective weapon than the 23 mm 2A14 cannon installed in the earlier BMP-23 and has a maximum range against ground targets of 4,000 m and a maximum range against airborne targets of 2,000 m. The gunner can select single shots or one of two automatic rates of fire, low at 200 to 300 rds/min or high at 500 rds/min. The cannon has dual feed with one normally being for HE-T (High Explosive-Tracer) and the other for AP-T (Armour-Piercing-Tracer) rounds, with both having a muzzle velocity of 970 m/s.

The gunner is seated on the left with a single forward-opening hatch cover and has a binocular BPK-1-42 day/night sight and day vision periscopes. He is also provided with a white light searchlight.

The vehicle commander is seated on the right and has a cupola with a single-piece hatch cover that opens to the front and mounted on the forward part of the cupola is an infra-red searchlight.

The BMP-30 commander has a 1PZ-3 monocular designation sight and a TKN-3B day/night sight with a day magnification of ×4.75 and a night magnification of ×4 but night vision range is only 400 m.

Mounted on the roof of the vehicle is a launcher for the AT-4 'Spigot' or AT-5 'Spandrel' ATGW. In its first production version the AT-4 'Spigot' has a maximum range of 2,000 m while the AT-5 has a maximum range of 4,000 m.

These were designed in Russia by the KBP Instrument Design Bureau but have been manufactured under licence in Bulgaria for some years by the Vazov Engineering Plant in Sopot, Bulgaria.

Late production versions of the AT-4 and AT-5 have tandem HEAT (High Explosive Anti-Tank) warheads for increased armour penetration characteristics.

The missile launcher can be elevated from –5 to +15° with traverse being 8° left and 90° right, with this being accomplished manually.

Mounted on each side of the turret is a bank of three 81 mm electrically operated forward-vision smoke grenade dischargers. In addition, the BMP-30 can lay its own smoke screen by injecting diesel fuel into the exhaust. The exhaust output is located on the right side of the hull at the front.

Six infantry personnel are seated three down either side back-to-back and in each side of the troop

Close-up of BMP-30 infantry combat vehicle turret showing roof-mounted anti-tank guided weapon 0018404

BMP-30 showing overall layout of the vehicle with all hatches closed

compartment are three firing ports with associated bulletproof vision devices. The first one on either side is for the machine gun with the other two being for an assault rifle.

The infantry personnel enter and leave the vehicle via two large doors in the hull rear that open outwards, with each of these being provided with a firing port and an associated bulletproof vision block.

Suspension is of the torsion bar type with either side having seven single rubber-tyred roadwheels, drive sprocket at the front and idler at the rear. The first and last roadwheel stations have a hydraulic shock-absorber fitted.

The BMP-30 retains the amphibious characteristics of the MT-LB (details of the Bulgarian-built versions are given later in this section under Bulgaria) being propelled in the water by its tracks. Before it enters the water, the trim vane is erected at the front of the vehicle, oblong slotted devices are removed from the hull sides and fitted either side of the front of the vehicle outside the front drive sprocket and forward suspension system and devices removed and fitted to the rear of the hull either side to improve water flow.

Standard equipment includes a fire detection and suppression system which can be activated manually or automatically and an R-123M VHF radio and R-124 tank intercom system. An NBC system is fitted, as is a GO-27 radiation and chemical detection system.

In addition to the infantry weapons, the BMP-30 also carries a standard RPG-7V anti-tank rocket launcher and disposable RPG-22 rocket launchers. One Strela-2M (US designation SA-7b) man-portable fire-and-forget surface-to-air missile is carried for air defence purposes.

Variants

There are no known variants of the BMP-30.

Status

Production as required. In service with the Bulgarian Army in small numbers. Bulgaria has stated that it has a total of 114 BMP-23 and BMP-30 vehicles in service.

Manufacturer

TEREM Joint Stock Company.

UPDATED

BMP-23 infantry combat vehicle

Development

The BMP-23 infantry combat vehicle was developed by the Bulgarian defence industry to meet the requirements of the Bulgarian Army.

It is based on a redesigned, Russian-developed MT-LB multipurpose tracked vehicle chassis also manufactured in Bulgaria (qv) which in turn shares a number of common components with the Russian 122 mm 2S1 self-propelled artillery system. Both of these have been manufactured under licence in Bulgaria.

The BMP-23 and BMP-30 infantry combat vehicles have been manufactured by the TEREM Joint Stock Company. It is understood that there has been no recent production of the BMP-23 infantry combat vehicle.

Description

The hull and turret of the BMP-23 are of all-welded steel armour construction providing the crew with protection from small arms fire and shell splinters.

The engine compartment is to the immediate rear of the front crew members with access hatches provided in the roof. The exhaust outlet is on the right side of the hull with the air inlet and outlet louvres being in the roof on the left side, in a similar arrangement to the MT-LB.

No details of the engine are available apart from its output of 315 hp. However, both the MT-LB and the 122 mm 2S1 are powered by a YaMZ-238V V-8 water-cooled diesel developing 240 hp at 2,100 rpm. The engine is coupled to a manual transmission with five forward and one reverse gears.

Behind the transmission, seated on the left side of the vehicle, is the driver, who steers using conventional tillers.

Access is provided by a single-piece hatch cover that opens to the rear and in front of this hatch are three forward-facing day periscopes which provide observation over the frontal arc when driving closed down. The central periscope can be replaced by a passive night observation device.

Suspension is of the torsion bar type with seven rubber-tyred roadwheels each side, drive sprocket at the front with the idler at the rear. There are no track-return rollers.

The two-man power-operated turret is on the centre-line of the vehicle roof with the gunner on the left and the commander on the right. Turret traverse is a full 360° with maximum weapon elevation +80°.

The main armament consists of a 23 mm 2A14 cannon fitted with a muzzle brake flash suppressor and is also used in the ZU-23-2 twin 23 mm towed anti-aircraft gun system which is known to have been produced in Bulgaria for the home and export markets. A 7.62 mm PKT machine gun is mounted coaxially to the right. The 23 mm cannon fires HE-IT and AP-IT ammunition with a muzzle velocity of 970 m/s. Maximum vertical range is 1,500 m and maximum horizontal range is 2,000 m. The 7.62 mm PKT machine gun fires 7.62 × 54R cartridges with a maximum muzzle velocity of 827 m/s and a maximum effective range of 1,500 m.

The 23 mm 2A14 cannon day sight has a magnification of ×4.5 and a 40° field of view, while the 7.62 mm machine day gunsight has a magnification of ×5.5 and an 11° field of view.

On the roof of the turret is a launcher for the AT-3 'Sagger' ATGW with a maximum range of 3,000 m. The missile is launched by the gunner from within the safety of the vehicle and an additional three missiles are stowed vertically to the rear of the turret for manual reloading.

The AT-3 'Sagger' was designed by the Russian KBP Instrument Design bureau but has been manufactured under licence in Bulgaria by the Vazov Engineering Plant in Sopot.

At the rear of the turret, opening to the left, is a small rectangular door which is believed to be for ammunition resupply purposes.

The gunner has a single-piece hatch cover that opens to the left and there are day and passive night sights mounted on the forward part of the turret roof. The night sight is believed to be of the image intensification type.

The commander of the BMP-23 is provided with a forward-opening hatch cover with three day-only periscopes and a night searchlight which can be operated from within the turret.

The infantry compartment is at the rear of the hull with the six infantrymen seated in two rows of three, back-to-back. The men enter the vehicle via two doors, opening outwards, in the rear of the hull.

Specifications

Crew: 3 + 7
Combat weight: 15,200 kg
Power-to-weight ratio: 20.72 hp/t
Ground pressure: 0.49 kg/cm²
Length: 7.285 m
Width: 2.85 m
Height:
(overall) 2.53 m
(top of hull) 1.71 m
Ground clearance: 0.40 m
Track width: 350 mm
Max speed:
(road) 61.5 km/h
(water) 4-5 km/h
Range: 550-600 km
Fuel capacity: 500-560 litres
Gradient: 60%
Side slope: 30%
Vertical obstacle: 0.8 m
Trench: 2.5 m
Electrical system: 24 V
Armament:
(main) 1 × 23 mm 2A14 cannon
(coaxial) 1 × 7.62 mm PKT MG
(ATGW) 1 × 'Sagger'
Ammunition:
(23 mm) 600
(7.62 mm) 2,000
(ATGW) 4
Gun stabiliser:
(vertical) no
(horizontal) no
NBC system: yes
Night vision equipment: yes

Over the top of the troop compartment are two roof hatches. The main fuel tank, like that on the BMP-1/BMP-2, is located in the centre of the rear troop compartment, with the seats either side. An SA-7 man-portable SAM is carried above the fuel tank.

One of the seven infantrymen is seated at the front of the vehicle on the right side with a hatch, three roof-mounted day periscopes and a firing port. To his right is a firing port for a 7.62 mm AK-47 assault rifle

BMP-23 ICV with all hatches closed

BMP-23 ICV crew, weapons and ammunition

and rockets for the RPG-7 launcher. There are eight firing ports provided in the rear troop compartment, three on each side and one on each of the rear doors, all ports have associated firing blocks. The first firing port either side is for a 7.62 mm PK machine gun while the other six are for 7.62 mm AK-47 assault rifles.

In addition to the 23 mm 2A14 cannon, 7.62 mm PKT coaxial machine gun and AT-3 ATGW launcher, the following weapons are carried for use by the infantrymen:

1 × SA-7 man-portable SAM
1 × RPG-7 rocket launcher and five rockets
4 × RPG-22 light anti-tank weapons
2 × 7.62 mm PK machine guns, each with 600 rounds
6 × 7.62 mm AK-47 assault rifles each with 240 rounds
20 × hand grenades
18 × signal flares.

The BMP-23 is fully amphibious and is propelled in the water by its tracks. Before entering the water the bilge pumps are switched on and the trim vane erected at the front of the vehicle. To enable the driver to see above the erected trim vane, the standard day periscope is replaced by an extendable one. When the trim vane is not required, it is folded back onto the glacis plate.

Like the 2S1 self-propelled howitzer, a slotted device to improve water flow is attached to either side of the hull at the very front alongside the drive sprocket; when not required the device is stowed on the hull side.

Standard equipment includes a fire detection and suppression system, a heater, a fume extractor for use when the weapons are being fired from within the vehicle, a GPK-59 course corrector, a UHF communications system with a range up to 20 km, internal vehicle lighting and a system that allows the BMP-23 to lay its own smoke screen by injecting diesel fuel into the exhaust.

Variants

BMP-23A infantry combat vehicle

This is similar to the BMP-23 but has a bank of three 81 mm smoke grenade dischargers on either side of the turret firing forwards and new roadwheels. The roof-mounted AT-3 'Sagger' has been replaced by a launcher for the AT-4 'Spigot'. This is only used by Bulgaria.

BRM-23 reconnaissance vehicle

There is a separate entry for this in the Reconnaissance vehicles section. This is only used by Bulgaria.

BMP-30 infantry combat vehicle

This is essentially the chassis of the BMP-23 fitted with the complete two-man turret of the Russian BMP-2 ICV now covered under the Russian Federation and Associated States (CIS). Details of the BMP-30 are given in the previous entry. This is only used by Bulgaria.

Status

Production as required. In service with the Bulgarian Army. Bulgaria has stated that it has a total of 114 BMP-23 and BMP-30 vehicles in service. There are no known exports of the BMP-23 ICV. Recent information has indicated that of these 114 vehicles 10 are BMP-30 vehicles with the remainder being BMP-23s.

Manufacturer

TERM Joint Stock Company.

UPDATED

Bulgarian MT-LB family of multipurpose tracked vehicles

Development

For many years Bulgaria has been manufacturing the Russian MT-LB multipurpose armoured personnel carrier under licence for both the home and export markets.

These MT-LB vehicles are manufactured in Bulgaria by the BETA Joint Stock Company with first production vehicles being completed in 1972. This company also manufactured the 122 mm 2S1 self-propelled artillery system with first production vehicles being completed in 1979. There is a separate entry on the Bulgarian 2S1 in the Self-propelled guns and howitzers (tracked) section under Bulgaria.

The overall layout of the standard MT-LB APC version is identical to the Russian vehicle covered in detail in this section under the Russian Federation and Associated States (CIS).

Bulgaria has, however, developed a number of variants using the basic MT-LB chassis, details of which are given below.

In addition to these, the MT-LB can be used for a wide range of other roles including towing artillery or trailers up to a maximum weight of 1,500 kg and, as an alternative to the normal version which has a turret-mounted 7.62 mm PKT machine gun, a 12.7 mm version is offered with the weapon being mounted externally on the turret. When armed with a 7.62 mm PKT machine gun a total of 1,000 rounds of 7.62 mm machine gun ammunition are carried.

All versions are fitted with an NBC system, a fire detection and suppression system and night vision equipment. They are also fully amphibious, being propelled in the water by their tracks. Before entering the water the bilge pumps are switched on and the trim vane is erected at the front of the vehicle. In addition, an extension is fitted around the air intake/outlet louvres on the left side of the hull roof to the rear of the driver's position.

The total Bulgarian Army fleet of MT-LB vehicles, and variants, is understood to total over 2,350 vehicles.

Variants

MT-LB R-81 command and staff vehicle

This is the basic MT-LB fitted with extensive communications equipment for use in the command post role and retains its turret-mounted machine gun.

It can be recognised by its roof-mounted antenna which is 4 m high when erected as well as its roof-mounted generator and cable reels which hold 1,000 m of cable.

MT-LB recovery vehicle

This is the standard MT-LB used in the recovery role and can tow trailer or vehicles weighing up to 6,500 kg.

MT-LBU

This is essentially an Armoured Command and Reconnaissance Vehicle (ACRV) based on the 122 mm 2S1 family of self-propelled artillery systems and is covered in detail in the Armoured Command Post Vehicles section under RFAS. The MT-LBU is fully amphibious being propelled in the water by its tracks at a maximum speed of 6 km/h.

MT-LB-VB

This is the version armed with an externally mounted 12.7 mm machine gun on the right side of the hull at the front.

MR HR chemical reconnaissance vehicle

This has been developed to undertake the role of chemical reconnaissance and the assessment of radioactive areas as well as the marking of nuclear contamination areas. Standard equipment includes a TNA-3 land navigation system, a flag dispensing system, night vision devices and chemical and radioactive measuring systems.

Maritza automated NBC system

The Maritza is an automated NBC reconnaissance system based on a modified MT-LB chassis. It has been designed for radiation and chemical reconnaissance, data acquisition and data transfer to the computation complex of the Bulgarian Automated Troops Control System (TCS). This is also designated the MT-LB RHR.

82 mm MT-LB mortar carrier

This is the basic MT-LB with the rear modified to carry an 82 mm mortar which fires to the rear through a two-part circular roof hatch. The four-man crew consists of commander, aimer, loader and vehicle driver.

The one-man turret is armed with a 7.62 mm PKT machine gun with a turret traverse of 360° and weapon elevation from –5 to +35°. Maximum range of the 7.62 mm PKT MG is 1,500 m and a total of 1,000 rounds of ammunition is carried. If required, the mortar can be dismounted for ground use. This is also referred to as the MT-LB Podnos and is fitted with the M-82 Podnos 82 mm mortar which, in turn, is a copy of the Russian 82 mm 2B14.

120 mm MT-LB mortar carrier

This has the same layout as the previous vehicle and is fitted with the same 7.62 mm PKT machine gun turret, the major difference is that the 120 mm mortar, which fires to the rear, has a longer oblong two-part roof hatch that opens left and right. If required, the 120 mm mortar can be dismounted for ground use. The first Bulgarian 120 mm MT-LB self-propelled mortar was the SMM B1.10 Tundzha which had the older Russian 120 mm M38/43 mortar.

The more recent model is the SMM 74.B1.10 Tundzha which has the Russian 120 mm 2B11 mortar installed.

MT-LB SE (sanitary evacuation carrier)

This has been designed for the evacuation of wounded from the front line area and can carry four stretcher or eight seated patients. Standard equipment includes oxygen, respirators, splints, a medicine bag, heating vessels, supporting straps, an aspirator and artificial breathing apparatus. Externally this version is recognisable by its raised rear roof.

MT-LB KShM-R-80

Communications vehicle for artillery units. Identified by one large box, or two smaller ones, located on the rear

Bulgarian MT-LB with 82 mm mortar (left) and 120 mm mortar (right) 0018403

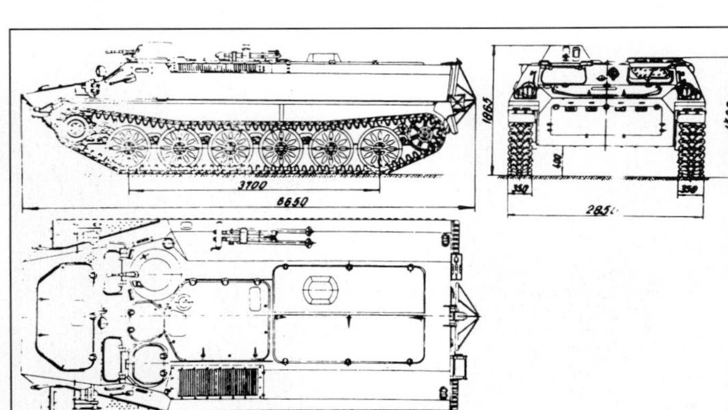

General arrangement drawing of the 120 mm mortar version of the Bulgarian MT-LB

Bulgarian-built MT-LB multipurpose tracked armoured vehicle ***2001***/0093567

MT-LB R-81 command and staff vehicle in travelling configuration 0004979

Specifications
APC
Crew: 2 + 11
Weight: 11,653 kg
Power-to-weight ratio: 20.5 hp/t
Ground pressure: 0.45 kg/cm²
Length: 6.454 m
Width: 3.150 m
Height:
(with turret) 1.865 m
(hull top) 1.56 m
Ground clearance: 0.4 m
Length of track on ground: 3.7 m
Track width: 350 mm
Max road speed: 61.5 km/h
Max water speed: 4-6 km/h
Max range: 500 km
Gradient: 60%
Side slope: 40%
Vertical obstacle: 0.8 m
Engine: YMZ-238B, 4-stroke V-8 direct injection, liquid-cooled diesel developing 240 hp
Suspension: torsion bar and hydraulic shock-absorbers
Armament:
Machine gun: 1 × 7.62 mm PKT
NBC system: yes
Night vision equipment: yes

deck, three whip antennas at the rear and two poles in the centre of the vehicle.

MT-LB KShM-R-81
Communications vehicle, identified by a power generator on the rear deck, a Half Horse (NATO designation) antenna and an additional cupola with infra-red searchlight in the centre of the vehicle. This vehicle is known to be in service with Iraq.

MT-LB BRM Sova
Reconnaissance vehicle, based on MT-LBu.

MT-LB-AT-I
This is an unknown variant.

MT-LB MTP
This is an armoured repair vehicle, similar to the Russian MTP-LB.

Status
Production as required. In service with Bulgaria and other undisclosed countries, including some in the Middle East. Many of the MT-LB multipurpose vehicles used by the Iraqi Army were built in Bulgaria.

Manufacturer
BETA Joint Stock Company.

UPDATED

CANADA

Canadian M113 series

The Canadian Forces have 1,214 M113A1 series armoured personnel carriers including the M577A1 command post vehicle, the M548A1 tracked cargo carrier and the M113A1 engineer vehicle with a front-mounted dozer blade.

M113A1 with 25 mm cannon
Under contract to the Canadian Forces the now General Motors Defense M113A1 with the two-man Delco Defense power-operated turret installed on the LAV-25 (8 × 8) vehicles.

M113A2 engineer vehicle
Under contract to the Canadian Forces the now General Motors Defense has modified an M113A2 engineer vehicle to a new configuration known as the M113A2 Engineering Variant Specially Equipped Vehicle. This retains the front-mounted hydraulic dozer blade but is also fitted with a hydraulic earth auger, hydraulic power tools, a ramp restraining system and external fuel tanks on either side of the hull. A total of 70 vehicles are now in service.

M113A1 TUA
The Canadian Forces have fitted 72 of their M113A1 series APCs with the TOW Under Armour (TUA) weapon system under the Phase I programme. Under Phase II the purchase of an additional 30 systems was planned but this was cancelled in May 1991 as part of the defence review. The TUA weapon systems were manufactured in Canada by Invar Manufacturing Limited under licence from Kvaerner Eureka A/S of Norway.

The Canadian TUA have a 7.62 mm machine gun mounted coaxial to the right and smoke grenade launchers mounted on the turret rear firing forwards. Some of these TUA turrets are to be removed from the M113 series chassis and installed on a wheeled-vehicle platform.

M113 mine clearing vehicles
The Canadian Forces have purchased quantities of the Pearson Engineering Surface Mine Ploughs (SMP) and Mine Rollers (MR) and these are normally fitted to M113 series vehicles.

M113 upgrade programme
This programme is known as the Canadian M113 APCLE (Armoured Personnel Carrier Life Extension).

The first prototype to be upgraded was rolled out at the United States Defense LP facility in San Jose, California, in January 1998.

This prototype is derived from the United Defense LP Mobile Tactical Vehicle Light (MTVL) design which

Canadian Forces M113 series armoured personnel carrier fitted with Pearson Engineering Surface Mine Plough ***2000***/0064861

Canadian Forces M113 series armoured personnel carrier fitted with Pearson Engineering Mine Roller 0079393

Canadian M113A2 Engineering Variant Specially Equipped Vehicle with dozer blade (Michael Jerchel) 0018407

Canadian Forces TOW Under Armour Vehicle (Michael Jerchel) **2000**/0064862

lengthens the chassis and adds an additional road wheel either side. Details of the MTVL are given later in this section under the USA.

The first MTVL will be configured for combat engineer applications similar to that of the M113A2 engineer vehicle covered previously. Three additional MTVL prototypes have been configured as a recovery vehicle, basic personnel carrier and as a non-running hull which will be used for live fire testing.

The MTVL is also the first vehicle to be equipped with the new Detroit Diesel 6V53 series DDEC (Detroit Diesel Electronic Control) III engine. This allows the user to electronically tune the engine to a desired rating up to 400 hp.

The Canadian MTVLs will initially utilise a 350 hp rating while later prototypes will retain the standard hull configuration and will be tested at 300 hp. The lower initial ratings will allow for future horsepower upgrades to compensate for weight add-ons such as additional armour or weapon stations.

Another unique feature of the Canadian MTVL programme involves the placement of vehicle fuel tanks. The current fleet of Canadian M113A2 vehicles slated for upgrade already utilise external 95 gallon (US) fuel tanks mounted one either side of the hull rear. To obtain the required operating range the fuel tanks had to be increased to approximately 130 gallons.

The installation of larger external fuel tanks could have an impact on vehicle stability so it was decided that the MTVL would have four fuel cells located underneath a floor bulkhead.

As well as improving vehicle stability by lowering the centre of gravity, the fuel bladders have already withstood mine blast testing.

Other features include the fitting of an Allison X200-4B transmission, 200 A alternator, thermostically controlled fan drive, improved driver's station, improved suspension/ride and Diehl 513 double pin track.

All of the vehicles will be equipped with mounting points for modular passive add-on armour which will be procured at a later date.

Early in 2000, the Canadian Department of National Defence awarded a C$215 million contract to DEW Engineering and Development of Ottawa for the M113 APCLE programme.

United Defence LP, the original manufacturer, is the main subcontractor and was awarded a C$147 million contract by DEW for the supply of 156 M113A3 upgrade kits and 174 upgrade kits for the stretched MTVL version, engineering data and integrated logistic support.

Canadian Forces M113A2 ADATS with radar erected **2001**/0007562

In total, 341 of the existing fleet of Canadian M113 series will be refurbished, of which 183 will be the MTVL version. There is also an option for another 61 as dedicated air defence units which will bring the total to 402 and DEW's contract to C$280 million.

Of these 61, 28 will be ADATS and their 33 support vehicles which will be to the M113A3 standard. It has yet to be decided as to whether the ADATS vehicles will be upgraded to the M113A3 or MTVL standard.

DEW has also been selected to provide the Protected Weapons Station (PWS) and, in mid-2000, placed a contract with Oerlikon Aerospace for this.

The PWS has been developed jointly by Oerlikon Aerospace and RAFAEL Armament Development Authority of Israel and consists of a day/night sight and a weapon mount for the 7.62 mm C6 machine gun.

This is aimed and fired by the vehicle commander from within the vehicle using a remote-control joystick and flat panel display.

The PWS will be fitted to the Recovery and Fitter variants of the MTVL plus some of the M113A3s with a total of 83 vehicles.

It will also be fitted to a number of 8 × 8 and 6 × 6 wheeled vehicles in service with the Canadian Army. First PWS were delivered in the second quarter of 2001 with deliveries to run over a five year period.

Although DEW Engineering and Development Ltd is the prime contractor, this is a co-operative venture with Canadian Forces 202 Workshop Depot in Montreal which will integrate the upgrade kits into the rebuilt vehicles. Kit deliveries commenced in mid-2000 and are expected to continue through to January 2006.

M113A2 with ADATS

The Canadian Forces operate a fleet of 34 Oerlikon Aerospace ADATS surface-to-air missile systems which are based on a modified M113A2 series APC. Mounted on top of the vehicle is an unmanned power-operated turret with two banks of four surface-to-air missiles either side. Full details of this system, which in its self-propelled version is only used by Canada, are given in *Jane's Land-Based Air Defence 2001-2002*.

UPDATED

CHINA, PEOPLE'S REPUBLIC

NORINCO Type 90 Armoured Personnel Carrier Family

Development

In 1991, China North Industries Corporation (NORINCO) announced that it had developed the Type 90 Armoured Personnel Carrier Family which consists of 10 vehicles. This has also been referred to in some sources as the YW 535 armoured personnel carrier.

The Type 90 family uses some automotive components of the NORINCO Type 531 H APC, which is also referred to as the Type 85, but has a slightly wider and lower profile hull with a distinct chamfer between the hull sides, which slope slightly inwards, and the roof, together with improvements to its automotive subsystems, especially the transmission.

The Iranian Boragh full-tracked APC, covered later in this section, is very similar in some respects to the Chinese Type 90 APC. In recent years China and Iran have signed a number of defence deals and there could well have been one which included the Type 90 APC family, or technology transfer of some aspects of the Type 90 vehicle.

Description

The hull of the Type 90 APC is of all-welded steel armour construction which provides the crew with protection from small arms fire and shell splinters.

The driver is seated at the front of the vehicle and has a single-piece hatch cover and three day periscopes giving observation to the front. The centre periscope can be replaced by a passive low-light level periscope for driving at night. The vehicle is steered using a yoke-type steering wheel rather than tillers.

The engine compartment is to the right of the driver and is provided with a fire detection and suppression system.

Depending on the model, the Type 90 is powered by a KHD BF8L413F air-cooled diesel engine developing 320 hp or a BF8L513C air-cooled diesel engine developing 360 hp. The engine is coupled to a new hydraulic gearbox which includes a hydraulic converter with locking clutch, a hydraulic selector with four forward and one reverse gears and steering having a two-stage planetary steering gear.

The two-stage final drive can be changed to suit the overall weight of the vehicle.

To the rear of the driver is the commander, who has a single-piece hatch cover and a single day periscope.

The troop compartment is at the rear of the hull with the infantrymen entering via a large door hinged on the right.

Over the top of the troop compartment is a total of four hatches, two small ones located towards the front and two oblong ones to the rear. The latter open outwards.

In either side of the troop compartment are three firing ports with an associated roof-mounted day periscope. The latter are angled forwards.

Mounted on the roof of the vehicle is the 12.7 mm Type 54 machine gun which is protected through about 320°. Mounted either side of the hull front is a bank of four forward-firing smoke grenade dischargers.

Suspension either side is of the torsion bar type and consists of five dual rubber-tyred roadwheels with the drive sprocket at the front, idler at the rear and three track-return rollers. The first, second and fifth roadwheel stations are provided with a hydraulic shock-absorber and the upper part of the suspension is covered by a light sheet-steel skirt. The tracks are single pin with an inner and outer pad, the latter being replaceable.

The vehicle is fully amphibious and propelled in the water by its tracks. Before entering the water the trim vane is erected at the front of the vehicle and the bilge pumps are switched on.

For increased battlefield survivability, the diesel fuel tanks are located one either side externally at the rear of the hull.

NORINCO Type 90 APC from above, showing general arrangement of the roof hatches and position of the cupola-mounted 12.7 mm machine gun ***2002***/0110670

Specifications

Standard APC
Crew: 2 + 13
Combat weight: 14,500 kg
Power-to-weight ratio: 22 hp/t
Ground pressure: 0.55 kg/cm²
Length: 6.744 m
Width: (overall) 3.148 m
Height:
(hull top) 1.72 m
(overall) 2.376 m
Ground clearance: 0.42 m
Track: 2.626 m
Length of track on ground: 3.625 m
Max road speed: 67 km/h
Max water speed: 7 km/h
Road range: 500 km
Fording: amphibious
Gradient: 60%
Side slope: 40%
Vertical obstacle: 0.7 m
Trench: 2.2 m
Engine: KHD BF8L413F, turbocharged, 8-cylinder, intercooled diesel developing 320 hp (360 hp engine also available – see text)
Transmission: hydraulic, 4 forward and 1 reverse gears
Steering gear: 2-stage planetary
Steering control: fully hydraulic
Suspension: torsion bar
Armament: 1 × 12.7 mm Type 54 MG
Ammunition: 1,120 × 12.7 mm
Smoke grenade dischargers: 2 × 4
Gun control equipment
Commander's fire control (override): no
Turret power control: manual
Elevation/depression: +85/–5°
Traverse: 360°
Gun stabiliser:
(vertical) no
(horizontal) no
Armour
Hull front: 12 mm
Hull rear: 10 mm
NBC system: yes
Night vision equipment: yes

Armoured personnel carrier

This is covered in the previous description.

Infantry fighting vehicle

This has a longer chassis with six roadwheels and is powered by the 360 hp diesel engine. In appearance it is very similar to the Russian BMP-2 and can be fitted with a two-man power-operated turret armed with a 23, 25 or 30 mm cannon and a 7.62 mm coaxial machine gun.

Anti-tank missile launching vehicle

This has a raised rear roof with a retractable launcher with four NORINCO Red Arrow 8 ATGWs in the ready to launch position with a further 12 missiles being carried inside. The missiles have a maximum range of 3,000 m. The latest version of the Red Arrow 8, called the Red Arrow 8E, has its range extended to 4,000 m and is fitted with a tandem HEAT (High Explosive Anti-Tank) warhead. These missiles are also used with the Red Arrow 8 anti-tank missile launching vehicle based on a 4 × 4 chassis.

Tracked armoured command vehicle
This has additional communications equipment, a 10 m high antenna and a petrol-engined generator to provide additional power for the communications equipment installed.

82 mm self-propelled mortar
This has an 82 mm (or 81 mm) mortar which is turntable-mounted and can be traversed through a full 360°. The mortar has a maximum range of 8,000 m with 112 mortar bombs being carried. A 12.7 mm anti-aircraft machine gun is fitted for air defence purposes.

120 mm self-propelled mortar
This has a 120 mm mortar that fires to the rear with a maximum range of 12,000 m with 50 mortar bombs carried. A 12.7 mm machine gun is fitted on the roof.

122 mm self-propelled howitzer
Mounted on the top of the hull towards the rear is a Chinese-built version of the Russian 122 mm D-30 howitzer which, using an ERFB-BB projectile, has a maximum range of 21,000 m with a total of 40 rounds carried. This version would also have the more powerful 360 hp engine and have a longer chassis with six road-wheels.

130 mm (30 round) self-propelled rocket launcher
Mounted on the roof is a 130 mm 30-round rocket launcher with a maximum range of 15,000 m. Details of the actual 130 mm rocket system, which is marketed on a variety of chassis, tracked and wheeled, are given in the Multiple rocket launchers section later in this volume.

Tracked armoured ambulance
This can carry four stretcher patients or a mixture of seated and stretcher patients, standard equipment includes oxygen supply and an air conditioning system.

Basic NORINCO Type 90 APC armed with roof-mounted 12.7 mm machine gun ***2001***/0093566

Tracked armoured recovery vehicle
This is fitted with a roof-mounted hydraulic crane, a 25 kW generator, welding equipment, a battery charger, a high-pressure air compressor, tools and spare parts.

Mine scattering vehicle
This has six road wheels either side and a modified rear on which are six (three either side) 36-round multiple-tube mine scattering launchers that point to the rear.

Status
Production. In service with the People's Liberation Army.

Manufacturer
Chinese state factories.
Enquiries to China North Industries Corporation.

UPDATED

NORINCO Type WZ 501 infantry fighting vehicle

Development/Description
The NORINCO Type WZ 501 is almost a direct copy of the Russian BMP-1 infantry fighting vehicle, the only major differences being slight reductions in weight and maximum road speed. It is understood that the prototype of the NORINCO (China North Industries Corporation) WZ 501 IFV was reversed engineered from a single BMP-1 supplied by Egypt.

The WZ 501 IFV has also been referred to as the Type 86 or the Type YW 501. It did appear during the large military parade that took place in Beijing late in 1999.

Main armament of the Type WZ 501 IFV is a 73 mm smoothbore gun with 40 rounds of ammunition and a 7.62 mm coaxial machine gun with 2,000 rounds. The original Russian vehicle has a 'Sagger' wire-guided ATGW mounted above the 73 mm gun with an additional four missiles carried internally for manual reloading. The Chinese WZ 501 carries four NORINCO Red Arrow ATGWs which are copies of the 'Sagger' that China is known to have produced for some years. In addition to seven 7.62 mm rifles and two 7.62 mm LMGs, a new 40 mm man-portable rocket launcher is carried, as is a man-portable surface-to-air missile called Red Tassel, a copy of the Russian SA-7 'Grail'.

NORINCO Type WZ 501 Infantry Fighting Vehicle 0004981

The WZ 501 IFV is fully amphibious and is fitted with an NBC system and infra-red night vision equipment for the commander, driver and gunner.

Communications fit on Chinese WZ 501 vehicles consists of one type A-220A radio and one type A-221A intercom set.

The Type WZ 501 infantry fighting vehicle also forms the chassis for the NFV-1. This is essentially the WZ 501 fitted with the former FMC power-operated turret armed with The Boeing Company 25 mm M242 Chain Gun (as fitted to the United Defense LP Bradley). As of 2001, the NFV-1 had not entered production. Details were given

NORINCO Type WZ 503 Infantry Fighting Vehicle with 12.7 mm machine gun covered up

NORINCO Type WZ 504 Missile Launcher Vehicle with launcher in travelling configuration 0018409

Specifications

WZ 501
Crew: 3 + 8
Weight:
(loaded) 13,300 kg
(empty) 12,100 kg
Power-to-weight ratio: 22 hp/t
Ground pressure: 0.621 kg/cm²
Length: (overall) 6.74 m
Width: (overall) 2.97 m
Height:
(top of turret) 1.92 m
(top of searchlight) 2.158 m
Ground clearance: 0.30-0.40 m
Angle of approach:
(on water) 25°
(under water) 15°
Angle of departure:
(on water) 25°
(under water) 15°
Length of track on ground: 3.57 m
Track: 2.547 m
Track width: 300 mm
Max speed:
(road) 65 km/h
(water) 7-8 km/h
Fuel capacity: 460 litres
Max range:
(road) 460-510 km
(water) 100 km
Gradient: 60%
Side slope: 47%
Angle of approach/departure: 15/25°
Vertical obstacle: 0.6-0.8 m
Trench: 2 m
Engine: Type 6V-150 water-cooled diesel developing 293 hp at 2,600 rpm
Transmission: manual
Armament:
(main) 1 × 73 mm low-pressure gun
(coaxial) 1 × 7.62 mm MG
(ATGW launcher) 1 × Red Arrow
Ammunition:
(73 mm) 40
(7.62 mm) 2,000
(ATGW) 4
Gun control equipment
Commander's fire control: (override) no
Turret power control: electric/manual
(by commander) no
(by gunner) yes
Turret traverse: 360°
Gun elevation/depression: +30/–3°
Gun stabiliser:
(vertical) no
(horizontal) no
NBC system: yes
Night vision equipment: yes

Note: One Red Tassel 5 man-portable SAM is normally carried on the scale of issue of one per company and platoon commander vehicles

in *Jane's Armour and Artillery 1991-92*. This turret is also being marketed by United Defense LP for other applications and has recently been installed on a Polish BMP-1 vehicle (see Poland).

Variants

Type WZ 503 infantry fighting vehicle

This was first shown in 1988 and is based on the chassis of the WZ 501 IFV, except that its hull is 100 mm higher. The two doors in the rear troop compartment are replaced by a single door that opens to the right and the troops sit on individual seats down either side of the hull rather than back-to-back down the centre of the vehicle.

Typically the WZ 503 IFV is armed with a manually operated cupola armed with a 12.7 mm Type 54 machine gun with no overhead protection being provided for the gunner.

Standard equipment includes five spherical firing ports, four rotatable day periscopes, an NBC system, a fire extinguishing system and a navigation system for the driver. The vehicle is fully amphibious, being propelled in the water by its tracks.

Type WZ 504 ATGW launcher vehicle

This was also first shown in 1988 and is based on the WZ 501 IFV chassis. Mounted on the roof is a pod that contains four Red Arrow 73B wire guided ATGWs that are raised when required for action, with the sighting system being mounted between the twin launchers. In addition to the four missiles in the ready to launch position, 12 more are carried inside the vehicle for manual reloading. Two ground launchers are also carried enabling missiles to be launched away from the vehicle if required by the tactical situation.

Type WZ 501A infantry fighting vehicle

This is essentially the WZ 501 fitted with a new turret armed with a 25 mm cannon, a 7.62 mm machine gun and eight smoke grenade dischargers. It is the same turret as that fitted to the 25 mm IFV version of the WZ 551 (6 × 6) APC, covered in detail in the Armoured personnel carriers (wheeled) section.

Type WZ 505 armoured ambulance

This is essentially a WZ 501 with a raised rear superstructure which can carry four stretchers, or two stretchers and four seated wounded or eight seated wounded. Specialised equipment includes an air conditioning system, an oxygen system and other medical equipment.

Type WZ 506 command post vehicle

This is based on the WZ 503 and is recognisable externally by its four radio antennas. Communications equipment installed includes two sets of Type 889 or VRC-83 radio sets, one Type 892 or VRC-84 radio set, one Type 70-zB or SR119 receiver. A generator and additional batteries are also provided.

Status

Production complete. In service with the People's Liberation Army.

Manufacturer

Chinese state factories.
Enquiries to China North Industries Corporation.

UPDATED

NORINCO Type 77 armoured personnel carrier

Development

The NORINCO (China North Industries Corporation) Type 77 armoured personnel carrier is very similar to the Russian BTR-50PK APC but has a much more powerful diesel engine giving improved road and water speeds as well as a higher power-to-weight ratio.

Many of the automotive components of the NORINCO Type 77 APC are also used in the Type 63 light amphibious tank. This has the US Army designation of the M1974, indicating the year it was first observed.

Recent information has indicated that total production of the Type 77 APC amounted to about 900 units. The type 77 APC has also been referred to as the WZ 211 and it may have been designed at Factory No 615. As far as it is known there has been no production of the Type 77 APC for many years.

Description

The hull of the Type 77 APC is of all-welded steel construction with the crew compartment at the very front, the troop compartment in the centre and the engine compartment at the rear.

The driver sits at the front on the left side and has a single-piece hatch cover opening to the right and three forward-facing day periscopes. The commander sits in a projecting bay at the front on the right side and has three day periscopes for forward observation and a single circular hatch cover opening to the rear. At the commander's station is the 12.7 mm Type 54 heavy machine gun, a copy of the Russian 12.7 mm DShKM weapon; no protection is provided for the commander while firing this weapon.

The troop compartment has hatches in the roof and a single forward-opening door in the right side. There are three firing ports, two in the left and one in the right side. The troop compartment has a ventilating system and a white light searchlight, which can be operated from within the vehicle, is mounted in the roof at the front left.

NORINCO Type 77-2 armoured personnel carrier armed with 12.7 mm machine gun and with driver's hatch locked open
0004982

The torsion bar-type suspension consists of six rubber-tyred roadwheels with the idler at the front and drive sprocket at the rear; there are no track-return rollers.

The Type 77 APC is fully amphibious and is propelled by two water-jets at the rear of the hull. The only preparation required is to erect the trim vane at the front of the hull and switch on the two electric bilge pumps. There is a manual bilge pump for emergency use.

Communications equipment fitted includes a radio type A-220A with a range of 16 km and a vehicle intercom type A-221A.

Variants

In addition to being used as an APC, the Type 77 APC can also be used as an ambulance, a command post, a

Specifications

Type 77-2
Crew: 2 + 16
Combat weight: 15,500 kg
Power-to-weight ratio: 25.8 hp/t
Length: 7.4 m
Width: 3.2 m
Height:
(overall) 2.436 m
Ground clearance: 0.40 m
Angle of approach: 36°
Angle of departure: 28°
Track: 2.8 m
Track width: 400 mm
Length of track on ground: 4.44 m
Max speed:
(road) 60 km/h
(water) 11-12 km/h
Fuel capacity: 416 litres
Max range:
(road) 370 km
(water) 120 km
Fording: amphibious
Gradient: 60%
Side slope: 40%
Vertical obstacle: 0.87 m
Trench: 2.9 m
Engine: Type 12150L-2A, 4-cycle, compression ignition, direct injection, water-cooled diesel developing 400 hp at 2,000 rpm
Transmission: manual
Steering: clutch and brake
Suspension: torsion bar
Electrical system: 24 V
Armament: 1 × 12.7 mm Type 54 MG
Ammunition: 500
Gun control equipment
Commander's fire control: (override) no
Turret power control: manual
(by commander) no
(by gunner) yes
Gun stabiliser:
(vertical) no
(horizontal) no
NBC system: no
Night vision equipment: optional

fuel resupply vehicle and a load carrier. The Type 77-1 APC has three loading ramps that can be positioned at the rear to enable an 85 mm Type 56 gun (with a crew of eight and 30 rounds of ammunition) or a 122 mm Type 54 howitzer (with a crew of eight and 20 rounds of ammunition) to be carried; the Type 77-2 APC has no loading ramps.

The Type 77-2 APC also has circular firing ports and vision blocks in the troop compartment and a large dome-shaped device on the turret roof which may be a ventilator. The Type 77-1 APC is also fitted with an electrically operated winch with a capacity of 1,500 kg which is operated by the driver. The baseline chassis for the Type 77 is designated the WZ 211.

HQ-2J SAM carrier

A much modified Type 77 APC chassis with seven, rather than six, roadwheels either side has been developed to act as the transporter/launcher platform for the HQ-2J SAM, the Chinese equivalent of the Russian SA-2 'Guideline'.

This system is not amphibious. Full details are given in *Jane's Land-Based Air Defence 2001-2002*. There is a probability that this is the surface-to-surface version of the HQ-2J SAM.

Type 76 KY-Q recovery vehicle

This is based on the Type 77 series APC chassis.

HY-2 anti-ship missile system

A modified Type 77 chassis has been used to transport and launch an HY-2 anti-ship cruise missile.

Type 89 122 mm self-propelled howitzer

This system, first seen in public during a parade held in Beijing late in 1999, uses some parts of the Type 77 armoured personnel carrier. Additional details are given in the Self-propelled guns and howitzers (tracked) section.

Status

Production complete. In service with the People's Liberation Army. There is a possibility that Albania may have a small quantity of these vehicles in service.

Manufacturer

Chinese state factories.
Enquiries to China North Industries Corporation.

UPDATED

NORINCO Type YW 534 armoured personnel carrier

Development

The NORINCO (China North Industries Corporation) Type YW 534 armoured personnel carrier (APC) is believed to be the successor to the older Type YW 531 series of full-tracked vehicles which has been in service with the People's Liberation Army for many years. The Type YW 534 is almost identical to the Type YW 531 H APC (qv) but the latter has a slightly shorter and narrower hull. The Type 534 is also referred to as the Type 89 armoured personnel carrier with its industrial number believed to be the ZSD-89.

Description

The hull of the Type YW 534 APC is of all-welded steel armour construction protecting the crew from small arms fire and shell splinters.

The driver sits at the front of the vehicle with a left-opening single-piece domed hatch cover and three forward-facing day periscopes, one of which can be replaced by an infrared device for night driving. The commander sits to the rear of the driver and also has a single-piece hatch cover with an integral vision device.

The engine compartment is to the right of the driver with the air inlet and outlet louvres in the roof and the exhaust pipe on the right side.

The troop compartment is at the rear of the vehicle with the troops entering via a large door that is hinged on the right; this has a day vision block and a firing port.

Specifications

Crew: 2 + 13
Combat weight: 14,300 kg
Power-to-weight ratio: 22.06 hp/t
Ground pressure: 0.59 kg/cm²
Length: 6.15 m
Width: (overall) 3.134 m
Height:
(hull top) 1.88 m
(over turret) 2.556 m
Ground clearance: 0.48 m
Angle of approach: 25°
Angle of departure: 20°
Track width: 360 mm
Track: 2.626 m
Length of track on ground: 3.425 m
Max speed:
(road) 66 km/h
(water) 6 km/h
Max range: 500 km
Fording: amphibious
Gradient: 60%
Side slope: 40%
Vertical obstacle: 0.7 m
Trench: 2.5 m
Engine: Deutz BF8L413F 4-cycle air-cooled diesel developing 320 hp at 2,500 rpm
Transmission: manual, 4 forward and 1 reverse gears
Steering: clutch and brake
Suspension: torsion bar
Electrical system: 24 V
Armament: 1 × 12.7 mm Type 54 MG
Smoke grenade dischargers: 2 × 4
Ammunition: (12.7 mm) 1,120
Turret power control: manual
(by commander) no
(by gunner) yes
Gun stabiliser:
(vertical) no
(horizontal) no
NBC system: yes
Night vision equipment: yes

In the left side of the vehicle are two firing ports and day periscopes. In the right side there are three or four firing ports and a similar number of periscopes, but other vehicles have been observed with three firing ports and periscopes in the left side and four in the right; these may well be differences between production vehicles. The firing ports are provided with ball and socket-type mountings for rifles. These allow the crew to use their small arms from within the vehicle with protection from small arms fire and shell splinters.

NORINCO Type YW 534 APC with raised rear superstructure fitted with four Red Arrow 8 ATGWs in the ready to launch position and one missile leaving its launcher 0018410

NORINCO Type YW 534 APC

In the forward part of the troop compartment roof is a single circular roof hatch either side. To the rear of this is the 12.7 mm Type 54 machine gun installation, and behind this are two oblong roof hatches.

The 12.7 mm Type 54 machine gun can be used to engage both air and ground targets but has no overhead protection for the gunner. Mounted either side of the forward part of the hull is a bank of four electrically operated smoke grenade dischargers.

Suspension is of the torsion bar type with each side having five dual rubber-tyred roadwheels, drive sprocket at the front, idler at the rear and track-return rollers. Tracks are of the rubber-bushed double pin type with removable rubber pads.

The Type YW 534 is fully amphibious, propelled in the water by its tracks. Before entering the water a trim vane is erected at the front of the vehicle and the bilge pumps are switched on. Standard equipment includes an NBC system. Standard communications fit of the NORINCO Type YW 534 APC is a Type 889 radio and a Type 803 intercom system.

Variants

Type YW 307 IFV

This is essentially the Type YW 534 APC fitted with a one-man power-operated turret armed with an externally mounted 25 mm cannon, a 7.62 mm coaxial machine gun and smoke grenade dischargers. The turret can be traversed through a full 360° and the weapon elevated from -12 to +52°. Totals of 400 rounds of 20 mm and 1,000 rounds of 7.62 mm ammunition are carried. Types of 25 mm ammunition fired include APDS and HE.

Physical characteristics are similar to those of the Type YW 534 except that it has a combat weight of 15,400 kg and carries seven troops plus its crew of three. This has also been referred to as the Type 89-II IFV.

ATGW carrier

Shown for the first time in late 1988 was an anti-tank guided weapons carrier which is believed to be based on the Type YW 534 series of armoured personnel carrier, although it could well be based on the YW 531 H/Type 85 series.

This version has a raised hull rear and within this is a retractable launcher with four NORINCO Red Arrow 8 ATGWs in the ready to launch position with additional missiles being carried in reserve in the hull. The same turret has also been fitted to the YW 531 series of vehicles (qv). The latest version is called the Red Arrow 8E and has a range of 4,000 m and a tandem High Explosive Anti-Tank (HEAT) warhead. This new warhead enables targets to be defeated fitted with Explosive Reactive Armour (ERA).

Status

Production complete. In service with the People's Liberation Army. As far as is known this APC has not been exported.

Manufacturer

Chinese state factories.
Enquiries to China North Industries Corporation.

UPDATED

NORINCO Type 85 armoured personnel carrier

Development

The NORINCO (China North Industries Corporation) YW 531 H armoured personnel carrier was developed at the same time as the Type YW 534 APC and is believed to be the successor to the older Type 531 which has been in service with the People's Liberation Army (PLA) for many years. The Type YW 531 H is almost identical to the YW 534 but is slightly lighter, shorter and narrower and does not have the four electrically operated smoke grenade dischargers either side of the forward part of the hull.

Although referred to as the Type YW 531 H, this vehicle is now normally referred to as the Type 85 within the PLA.

It is understood that total production of the Type 85 APC for the home and export markets amounted to about 2,000 vehicles. The vehicle is no longer being marketed by NORINCO, with emphasis now on the Type 90. The latter is very similar in appearance to the Type 85 but is slightly longer and wider and is powered by a 320 hp air-cooled diesel engine coupled to an improved transmission.

Specifications
Type YW 531 H
Crew: 2 + 13
Combat weight: 13,600 kg
Power-to-weight ratio: 23.5 hp/t
Ground pressure: 0.546 kg/cm^2
Length: 6.125 m
Width: (overall) 3.06 m
Height
(hull top) 1.914 m
(including MG) 2.586 m
Ground clearance: 0.46 m
Angle of approach: 20°
Angle of departure: 25°
Track: 2.526 m
Track width: 360 mm
Length of track on ground: 3.275 m
Max speed
(road) 65 km/h
(water) 6 km/h
Max range: 500 km
Fording: amphibious
Gradient: 60%
Side slope: 40%
Vertical obstacle: 0.6 m
Trench: 2.2 m
Engine: BF8L413F 4-cycle air-cooled diesel developing 320 hp at 2,500 rpm
Transmission: manual, 4 forward and 1 reverse gears
Steering: clutch and brake
Suspension: torsion bar
Electrical system: 24 V
Armament
(main) 1 × 12.7 mm Type 54 MG
(coaxial) nil
(ATGW) nil
Ammunition (main): 1,120
Gun stabiliser
(vertical) no
(horizontal) no
NBC system: yes
Night vision equipment: yes

Description

Almost identical to the Type YW 534 fully described in an earlier entry, but has some minor differences such as three firing ports and vision devices in left hull side.

Standard communications equipment includes a Type 889 radio and a Type 803 vehicle intercom system.

Variants

Type YW 309 infantry combat vehicle

This is essentially the Type YW 531 H fitted with the turret of the WZ 501 IFV which in turn is a copy of the Russian BMP-1. In addition to the single firing port in the rear hull door, there are three firing ports and day periscopes in the left side of the hull and four firing ports and day periscopes in the right side of the hull.

The YW 309 has a crew of three and carries eight fully equipped infantrymen. The turret can be traversed through 360° and the 73 mm gun can be elevated from –3.5 to +30.5°. A 7.62 mm machine gun is mounted coaxially with the main armament and there is a NORINCO Red Arrow 73 ATGW launcher over the main armament. This is also called the Infantry Fighting Vehicle Type 85.

WZ 702 armoured command post

This is similar to the YW 531 H but has a crew of two and carries six staff officers plus extensive communications equipment, for which three radio antennas are provided. The driver sits front left and the commander, who also mans the 12.7 mm anti-aircraft machine gun, is seated to his rear. The command area is at the very rear and has four circular roof hatches, one in each corner. This version weighs 13,800 kg loaded and a four-barrelled smoke grenade discharger is mounted either side of the forward part of the hull.

Type WZ 751 armoured ambulance

This is also based on the YW 531 H but has a new and raised hull rear and a new larger commander's cupola to the rear of the driver which is fitted with an externally mounted 12.7 mm anti-aircraft machine gun. This has a crew of two and also carries two medical personnel plus four stretcher patients, eight seated patients or two stretcher patients and four seated patients. In addition to the stretchers, a medical kit, a splint box and an

NORINCO armoured command vehicle Type 85 0004984

NORINCO Type WZ 751 armoured ambulance 0004985

NORINCO Type 85 maintenance engineering vehicle 0004986

NORINCO Type 85 recovery vehicle showing roof-mounted crane 0004987

oxygen supply are carried and standard equipment includes an air conditioning system. Loaded weight is 13,800 kg. This is now called the Tracked Armoured Ambulance Type 85.

HJ-62C armoured scout vehicle

This has its turret removed and is fitted with an elevated arm on top of which is mounted a battlefield surveillance radar with a detection range of 25 km for vehicles and 10 km for personnel.

Armoured Command Vehicle Type 85

This is the basic vehicle with its turret removed for use in the command role. It has a crew of two plus six command staff. It is externally recognisable by its pintle-mounted 12.7 mm machine gun and three radio antennas. Communications equipment installed includes two Type 889 radio sets or VRC-83 radio sets, one Type 892 radio set or VRC-84 radio set, one 70-2B receiver or one SR119 receiver. An auxiliary generator and ventilation fans are also fitted as standard.

120 mm Self-propelled Mortar Type 85

This has a 120 mm mortar mounted in the rear which fires to the rear with a traverse of 14° left and right with elevation from +45 to +80°. A total of 50 rounds of mortar ammunition is carried, with a maximum range of 7,700 m. This model has a crew of six, including the commander and driver, and weighs 14,100 kg.

82 mm Self-propelled Mortar Type 85

This is similar to the above but with a slightly different hatch arrangement and the 82 mm mortar can be traversed through 360° with elevation being from +45 to +85°. The 82 mm Type 87 mortar has a maximum range of 3,040 m with a total of 120 rounds of ammunition carried. This model has a crew of seven and weighs 13,800 kg. The mortar is also capable of being deployed away from the vehicle if required by the tactical situation.

120 mm turret-mounted mortar

Some years ago it was proposed that the Type 85 APC be fitted with the now BAE Systems, RO Defence 120 mm breech-loaded mortar system. This combination never reached the prototype stage. The first customer for the 120 mm AMS is the Saudi Arabian National Guard with the platform being the General Motors Defense LAV (8 × 8).

122 mm self-propelled howitzer

This is essentially the basic vehicle with its turret removed and the Chinese-built version of the Russian 122 mm D-30 towed howitzer removed from its standard carriage and mounted on the hull rear top. Additional details are given in the Self-propelled guns and howitzers (tracked) section.

Recovery Vehicle Type 85

This is a basic Type 85 retaining its cupola-mounted 12.7 mm MG but fitted with a hydraulic crane on the left side of the hull with a lifting capacity of a tonne, a generator, welding equipment, special mounting and dismounting tools and a tool cabinet; it has a crew of five. Combat weight is 14 tonnes. Details of this vehicle and the Type 85 armoured repair vehicle are given in *Jane's Military Vehicles and Logistics 2001-2002.*

Maintenance Engineering Vehicle Type 85

This has a similar hull to the Type 85 ambulance with a raised roof at the rear but is fitted with a cupola-mounted 12.7 mm MG on the roof rather than the forward part of the hull. Equipment includes a generator, a boom, an inertia dynamometer, an air filter cleaner, an oil filter cleaner and other special tools.

Artillery command post vehicle

A version of the YW 531 H with a raised superstructure to the rear of the hull is used as the command post vehicle for the new PLZ 45 155 mm Chinese self-propelled gun/howitzer which is covered in detail in the Self-propelled guns and howitzers (tracked) section. To the rear of the driver's position is the commander's cupola armed with an externally mounted 12.7 mm MG. This is in service with the Kuwait Land Forces. These are the ZCY45 battalion command vehicle/fire-control centre vehicle, ZCL45 battery command post vehicle and the GCL forward observation/reconnaissance vehicle.

Thai purchase of YW 531 H

In 1987, Thailand took delivery of the first of some 410 Chinese Type YW 531 APCs. These have their Chinese 12.7 mm machine guns removed and replaced by US M2 12.7 mm machine guns which are standard on Thai tanks such as the M48A5 and on the existing M113A1 series of APC in service with Thailand.

Thailand has also taken delivery of a quantity of specialised versions of the Type 85 series including a 130 mm 30-round multiple rocket launcher. It is believed that the launcher is the same as that used on the 130 mm (30-round) Type 82 truck-mounted multiple rocket launcher system covered later in this book.

NVH-1 MICV

This vehicle, a joint development between NORINCO of China and Vickers Defence Systems of the UK, used a chassis almost identical to that of the Type YW 531 H APC and was fully covered in *Jane's Armour and Artillery 1991-92,* pages 357-358. So far it has yet to enter production or service.

Status

Production complete. No longer marketed. In service with:

Country	Quantity	Comment
China	1,330	estimate
Myanmar	150	
Thailand	410	reserve

Manufacturer

Chinese state factories.
Enquiries to China North Industries Corporation.

UPDATED

NORINCO YW 531 armoured personnel carrier

Development

The NORINCO (China North Industries Corporation) YW 531 APC was developed in the late 1960s and at various times has been called the K-63, M1967 and M1970. The official Chinese designation for this vehicle is the Type 63 but within the People's Liberation Army it is referred to as the YW 531. It has seen combat in Vietnam (used by both China and Vietnam), in Angola with Zaïrean forces and with Tanzania against Uganda in the 1978 October/November border war. In 1983, a quantity of these vehicles was delivered to Iraq by China.

The YW 531 has a crew of four which consists of the commander, gunner, loader and driver, and can carry 10 fully equipped infantrymen.

The Type 63 APC was designed and developed at No 618 Factory in co-operation with the Ministry of Ordnance Industry (MOI), No 30 Research Laboratory and other institutions.

Further development resulted in the Type 63-I APC which was also developed by No 618 Factory.

The first improved version of the YW 531 was the YW 531A (Type 63-I) which featured a heater and the replacement of the 12.7 mm machine gun by the Type 56-I 7.62 mm machine gun. The command version of this has raised rear roof and large bedstead type radio antenna and was called the WZ 701.

The YW 531A was followed by the YW 531B which is believed to have appeared in 1981.

Further development resulted in the Type 513C which featured replacement of the Deutz Type 6150L water-cooled diesel by a Deutz BF8L413F air-cooled diesel engine, hydraulically changed transmission, secondary planetary steering mechanism and hydraulic booster system.

Other improvements include an automatic fire extinguisher, a smoke-laying system, an NBC protection system and a cannon or machine gun as main armament. The Type 531C is usually referred to as the YW 531C APC.

It is understood that total production of the NORINCO YW 531 amounted to just under 3,000 vehicles of all types for home and export markets. Production of the YW 531 was completed some time ago and it is no longer being marketed by NORINCO.

Description

The hull of the YW 531 is made of all-welded steel. The driver sits at the front of the hull on the left side and has a single-piece domed hatch cover that opens to the left, in front of which are two day periscopes. The commander sits at the front of the hull on the right side and has a single-piece hatch cover opening to the right with an integral day periscope that can be traversed through 360°. The air-cooled diesel engine is behind the commander, with the air inlet and outlet louvres in the roof and the exhaust outlet in the right side of the roof.

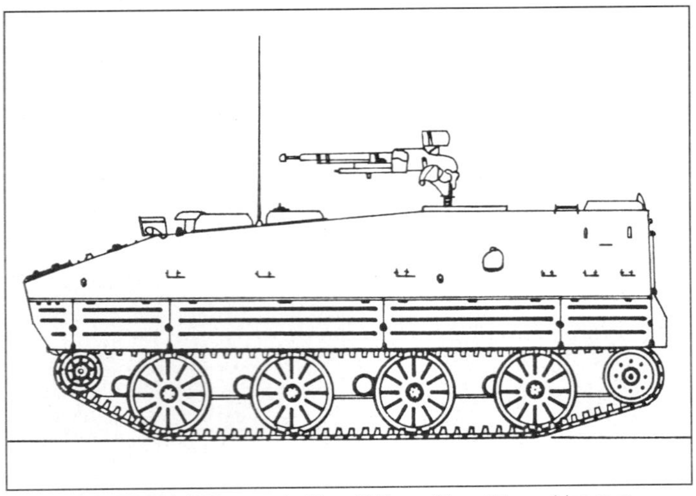

NORINCO YW 531 APC armed with a 12.7 mm Type 54 machine gun

NORINCO YW 750 armoured ambulance showing a roof-mounted 12.7 mm Type 54 machine gun traversed to the rear

NORINCO YW 304 82 mm self-propelled mortar system 0004989

NORINCO YW 381 120 mm self-propelled mortar system 0004988

The third crew member is seated behind the driver and has a single-piece domed hatch cover that opens to the left, with an integral day periscope that can be traversed through 360°.

The gunner's position is to the rear of the vehicle in the roof. Mounted on the forward part of the gunner's hatch is a 12.7 mm heavy machine gun Type 54 (the Russian DShKM weapon manufactured in China). The gunner also has a single-piece hatch cover that opens to the rear.

The infantrymen enter and leave the vehicle by a door in the rear of the hull which is hinged on the right side. A single firing port is provided in each side of the hull. The torsion bar suspension either side consists of four single rubber-tyred roadwheels with the drive sprocket at the front and the idler at the rear. There are no track-return rollers, as the top of the track rests on the top of the roadwheels.

The YW 531 is fully amphibious, being propelled in the water by its tracks. Before entering the water a trim board, which is stowed on the glacis plate when travelling, is erected at the front of the hull. As far as is known the YW 531 has no NBC system or night vision equipment, although the driver can have a TVN-2 night vision device.

There are at least three sub-variants of the YW 531 APC which differ mainly in their communications equipment. These versions are designated the YW 531C, YW 531D and YW 531E.

Specifications

Type YW 531C
Crew: 2 + 13
Weight:
(loaded) 12,600 kg
(empty) 11,100 kg
Power-to-weight ratio: 25.39 hp/t
Ground pressure: 0.57 kg/cm^2
Length: 5.476 m
Width: (overall) 2.978 m
Height:
(overall) 2.85 m
(to hull top) 1.904 m
Ground clearance: 0.45 m
Angle of approach: 25°
Angle of departure: 20°
Track: 2.464 m
Track width: 360 mm
Length of track on ground: 3.095 m
Max speed:
(road) 66 km/h
(water) 6 km/h
Fuel capacity: 450 litres

Max road range: 500 km
Fording: amphibious
Gradient: 60%
Side slope: 40%
Vertical obstacle: 0.6 m
Trench: 2 m
Engine: BF8L413F 4-cycle turbocharged, inter-cooled, V-8 diesel developing 320 hp at 2,500 rpm
Transmission: manual, 5 forward and 1 reverse gears
Suspension: torsion bar
Electrical system: 24 V
Main armament: 1 × 12.7 mm Type 54 MG
(elevation) 0 to +90°
(traverse) 360°
Ammunition: 1,120
Gun stabiliser:
(vertical) no
(horizontal) no
NBC system: no
Night vision equipment: no

Variants

YW 531C (modified)

This has a number of improvements including a shield for the 12.7 mm machine gun, new vision ports and improved ventilation.

The original Type 6150L water-cooled diesel engine was replaced by a licence-produced German BF8L413F 4-cycle turbocharged diesel developing 320 hp.

Additional improvements included an NBC system, automatic fire detection and suppression system and hydraulic booster for the driving controls.

There were also two command versions for use at platoon company level. These were the YW 531D with two Model 889 radios and one firing port in the left side of the hull and the YW 531E with one Model 889 and one Model 892 radio.

YW 531D armoured personnel carrier

This is a further development of the YW 531C and is fitted with one Type 889 radio and one Type 803 intercom system. It also differs from the YW 531C by having only one firing port in the left side compared with two in the YW 531C.

YW 531E armoured personnel carrier

This is also a development of the YW 531C and has one Type 889 radio and one Type 803 intercom. It differs from the YW 531C in having only one firing port in the left side of the vehicle compared with two on the left side of the YW 531C.

Type YW 701 command post vehicle

This has a similar chassis and hull to the YW 750 ambulance and is powered by the 6150L diesel developing 260 hp at 2,000 rpm, giving the vehicle a maximum road speed of 60 km/h. It has a crew of three and can carry five passengers when being used in the command role, and would normally be used by regimental and divisional commanders.

Communications equipment installed includes a vehicle intercom, five radio sets and a remote-controlled intercom for the vehicle commander when he has dismounted from the vehicle. The YW 701 also has an auxiliary power unit to charge vehicle batteries, a radio for receiving meteorological reports and air raid alarms when the vehicle is stationary, a battery commander's telescope and map board, an infra-red night driving device for the commander, 12 observation blocks, an air heater, forced ventilation and fans.

Armament comprises a 7.62 mm Type 56-1 light machine gun with 1,000 rounds of ammunition. Four manually operated CO_2 fire extinguishers are carried.

The Type YW 701A has better communications and mobility than the YW 701. Communications equipment includes two Type 889 FM radios and one Type 892 radio. In addition there are two (Type 889 and Type 892) reserve radio sets. The vehicle is powered by the

NORINCO YW 701A armoured command vehicle with communications antenna raised
0004990

NORINCO Type 531 APC fitted with roof-mounted turret with four Red Arrow ATGW in ready to launch position
0004991

BF8L413F engine developing 320 hp. Armament consists of one 12.7 mm anti-aircraft machine gun with 560 rounds of ammunition. There are three manually operated Type 1211 fire extinguishers provided. Another variant is the YW 701B which has one M889 and one M892 radio set.

YW 750 Ambulance

This has the Chinese designation YW 750; it has a crew of two plus two doctors and can carry four stretchered or eight seated patients. Like the YW 701 command post vehicle, the YW 750 ambulance has a higher roof line.

122 mm SPH

This is called the 122 mm Type 70-1 self-propelled howitzer by the Chinese and is a longer and slightly wider YW 531 fitted with a 122 mm Type 54-1 howitzer in an open-topped troop compartment. The engine is a six-cylinder, four-cycle, water-cooled diesel developing 260 hp at 2,000 rpm. Full details of this version are given in the Self-propelled guns and howitzers (tracked) section.

Rocket launcher

This is called the 130 mm Type 70 19-barrel multiple rocket launcher and has the same engine as the 122 mm self-propelled howitzer. Full details of this version are given in the Multiple rocket launchers section.

Type YW 304 82 mm self-propelled mortar

This is fitted with a turntable-mounted 82 mm mortar which can be traversed through a full 360° and is elevated from +40 to +85°. It has a minimum range of 85 m and a maximum range of 3,040 m. A total of 120 rounds of 82 mm mortar ammunition is carried. A 12.7 mm machine gun is mounted on the left side of the turret roof for anti-aircraft and ground defence and 560 rounds of ammunition are carried for this.

The Type YW 304 has a crew of six or eight and weighs 12,800 kg fully loaded. The physical characteristics are identical to those of the basic YW 531 APC. The mortar can be dismounted for use in the ground role.

Type YW 381 120 mm self-propelled mortar

This has a W86 120 mm mortar with a traverse of 14° left and right with elevation from +45 to +80°. The mortar has a maximum range of 7,500 m and 50 mortar bombs are carried. A baseplate and tripod are carried externally to allow the mortar to be dismounted for use in the ground role. A 12.7 mm machine gun is fitted for anti-aircraft and local defence. The vehicle has a crew of six and weighs 14,100 kg.

Anti-tank vehicle

It is known that a Type 531 APC has been fitted with a turret on the hull top towards the rear with four Red Arrow 8 wire-guided ATGWs ready to launch; additional missiles are carried inside for manual reloading. This missile has a range of 3,000 m. It is believed that this system on the Type 531 APC has yet to enter service. The latest version of the Red Arrow is the 8E with a maximum range of 4,000 m. The latest versions of this missile have a tandem HEAT warhead to defeat MBTs fitted with explosive reactive armour.

Psychological warfare

Fitted with roof-mounted loudspeakers.

Status

Production complete. In service with the following countries:

Country	Quantity	Comment
Albania	103	
China	1,800	estimate
Congo, Democratic Republic	12	
Iraq	150	estimate
North Korea	500	estimate, local version is VTT-323
Sudan	50	status uncertain
Tanzania	30	status/quantity uncertain
Vietnam	132	estimate
Zimbabwe	30	

Manufacturer

Chinese state factories.
Enquiries to China North Industries Corporation.

UPDATED

CZECH REPUBLIC AND SLOVAKIA

OT-90 armoured personnel carrier

Development/Description

The OT-90 is basically the Russian-designed BMP-1 infantry fighting vehicle, which was also built under licence in Czechoslovakia, with its complete two-man turret removed and replaced by the OT-64C (1) (8 × 8) armoured personnel carrier turret armed with a 14.5 mm KPVT and a 7.62 mm PKVT machine gun.

From 1972 to 1988, the plant in Dubnia supplied a total of 5,100 BMP-1 vehicles to the former USSR, this amounted to around 70 per cent of total production in Czechoslovakia during this period.

The reason some BMP-1s were converted to the OT-90 configuration was due to the fact that the BMP-1 was thought to be Tank Category (because of its 73 mm gun barrel) under the Conventional Forces Europe (CFE) treaty. Large scale conversions were subsequently cancelled and they are to be phased out of service. Some have been used by former Czech units operating as part of United Nations forces.

DTP-90M maintenance vehicle in travelling configuration 0018414

OT-R5 command post vehicle with antenna erected (T J Gander) **2000**/0064864

OT-90 armoured personnel carrier (BP) **2001**/0093578

BPzV reconnaissance vehicle from the rear with rear doors open and radar antenna not installed (BP) **2001**/0093579

BVP-1 MA upgraded infantry combat vehicle (T J Gander) **2000**/0064865

Czech BMP-1 variants

BRM-1K is the Russian BRM-1 reconnaissance vehicle.

BVP-1K is the former Czech designation for the commander's version of the Soviet BMP-1K vehicle.

Boure III is the former Czech psychological warfare version of the BMP-1 with a loudspeaker system installed.

DTP-90 is a repair version of the OT-90 APC.

MP-31 is an air defence command version of the Russian BMP-1KShM command post vehicle.

MU-90 is a minelaying version of the OT-90 APC but has no turret, the space being covered by a steel sheet.

SVO is a mineclearing version of the BMP-1 with the turret removed and fitted with a Hedgehog-type launcher in the rear troop compartment that fires to the front of the vehicle.

VPV is a crane-equipped recovery variant of the BMP-1. The turret and troop compartment roof hatches have been removed and a powered crane is mounted on the troop compartment roof.

VP-90 is a reconnaissance version of the OT-90 with an OT-64 one-man turret armed with 14.5 mm and 7.62 mm MGs.

ZT-90 is a recovery version of the OT-90 APC.

BPzV reconnaissance vehicle

This is based on the Russian BMP-1 with modified hatches and a passive sight over the commander's hatch to the immediate rear of the driver's position.

The complete BMP-1 turret, armed with 73 mm gun, 7.62 mm machine gun and six 81 mm smoke grenade launchers at the turret rear, are retained. The full meaning of the BPzV designation is the Bojove pruzkumne Vozidlo. The armament is the same as the original BMP-1 but additional ammunition is carried, for example 76 × 73 mm, 3,500 × 7.62 mm and seven Sagger missiles.

The crew are armed with five Vz.58V automatic rifles, four RPG-75 rocket launchers and 12 grenades. The vehicle has a maximum combat weight of 13.8 tonnes.

Like the Russian BRM-1K, the BPzV is fitted with a battlefield surveillance radar type PSNR-5K which is mounted on a support on the right side of the turret. Mounted on the left side of the turret is an NNP-21 observation device.

On the left and right side of the hull are laser warning receivers. The original side skirts are replaced by the more recent BMP-2 type side skirts.

In addition, the BPzV has fewer firing ports and a telescopic antenna at the left rear which lays horizontally accoss the rear of the hull roof when not required.

Known users of this vehicle are the Czech Republic (129), Slovakia (72) and Poland (38). An illustration of the latter appears in the BMP-1 entry under Russia.

Snezka reconnaissance vehicle

This is based on a stretched BMP chassis with seven roadwheels either side and full details are given in the Reconnaissance vehicles section. This version is designated the BVP-1PPK Snezka. It is now in service with the Czech Republic Army.

Light Observation System (LOS)

This is based on a modified BMP-2 chassis and fitted with a mast-mounted telescopic sensor pod that, when not required, is retracted into the rear part of the turret under full armour protection. Details of this vehicle are given in the Reconnaissance vehicles section under the Czech Republic.

Ambulance

This has a raised superstructure that extends to the rear of the hull and is unarmed. It retains its amphibious capability and is fitted with an NBC system and night vision equipment. The ambulance versions are designated the AMB-S and OT-90 ZDR.

DTP-90M maintenance vehicle

This has been developed by the VOP 026 facility of the Czech Republic and is a BMP-1 series vehicle rebuilt for use in the maintenance role.

To the rear of the commander's and driver's position on the left side a new superstructure has been fitted which extends to the rear of the hull.

On top of the superstructure on the left side is a cupola with an externally mounted 7.62 mm machine gun and on the right side at the rear are two roof hatches that open forwards. The normal means of entry and exit is via two doors in the hull rear which are identical to those of the standard BMP-1.

Mounted on top of the hull, on the left side, is a crane which can lift a maximum load of 1,000 kg and specialised equipment carried includes a portable generator, welding equipment, tow bars, workbenches and tools.

The DTP-90M is fully amphibious, being propelled in the water by its tracks and standard equipment includes an NBC system and a fire suppression system.

Three slightly different models of the DTP-90M are available to support different vehicles, first for the T-72, BMP-1, BMP-2 and OT-90, second for the T-55AM2, BMP-1, BMP-2 and OT-90 and third for the BMP-1 and BMP-2.

OT-R5 command post vehicle

This has been developed by VOP 026 of the Czech Republic and is a BMP-1 series vehicle rebuilt for use in the command post role.

To the rear of the commander's and driver's position on the left side a new raised superstructure has been added, with a cupola on the left side on which is mounted a 12.7 mm anti-aircraft machine gun. The standard two BMP-1 entry doors are retained in the hull rear.

A complete communications system is installed and standard equipment includes an NBC system with air conditioning an option. When deployed in the static role, additional antennas can be quickly erected. The vehicle retains its amphibious capabilities.

120 mm self-propelled mortar

This under-armour 120 mm self-propelled mortar is based on a lengthened BMP-1 chassis and details are given in the Self-propelled mortar systems section under Czech Republic and Slovakia. Only 12 of these have been built so far, six for the Czech Republic and six for Slovakia.

Swedish BMP-1 programme

The VOP 026 facility in the Czech Republic has now completed upgrading a large number of ex-East German Army BMP-1 infantry combat vehicles for the Swedish Army. Additional details of this programme are given later in this section under Sweden. The Swedish designation of the BMP-1 is the Pbv 501.

BVP-1 MA upgraded infantry combat vehicle

This has been developed by VOP 26 as a private venture and was first shown in 1999. It is essentially a modernised BMP-1 chassis fitted with a brand new German Rheinmetall Landsysteme (previously KUKA) E8 one-person turret, armed with The Boeing Company

30 mm Bushmaster II cannon and 7.62 mm coaxial machine gun. Mounted either side of the turret are banks of electrically operated smoke grenade launchers.

The 30 mm cannon is provided with 200 rounds of ready use ammunition and is fully stabilised in both planes. The weapon is laid onto the target using a passive day/night sight with a computerised fire-control system also installed.

Turret traverse and weapon elevation is all electric with manual controls for emergency use. Turret traverse is 360° with weapon elevation from −10 to +45°.

A 60 mm mortar is installed which can, if required, be fired from the vehicle.

For improved crew comfort an independent crew heating system is fitted, sharp edges have been removed or covered, the troop compartment floor has been levelled, thermal insulation provided and anti-skid material provided for the hull.

To enable the ride characteristics to be improved, new shock-absorbers have been fitted and the existing torsion bars replaced.

To allow the vehicle to be driven on public roads, new external lights are fitted as well as indicators. The brakes have been modernised, rear-view mirrors fitted and new tracks with rubber pads installed.

New communications equipment is fitted as is a satellite land navigation system.

For improved battlefield survivability, additional passive armour has been provided to some parts of the hull and improved protection is provided against mines.

In addition, laser detectors are fitted at each corner of the vehicle, an explosion detection and suppression system is fitted and the diesel fuel tanks are filled with a special anti-detonation device. New thermal camouflage paint is also provided.

The crew consists of three: commander, gunner and driver with seats for six infantry in the rear. The normal BMP-1 has four roof hatches over the rear but the BVP-1 MA has only two roof hatches.

The Czech Republic has completed the prototype of a BMP-1 infantry combat vehicle upgraded in a number of areas to meet the possible future requirements of the Czech Army.

New passive night vision devices have been installed as has a laser range-finder, land navigation system and thermal meters.

For improved safety, the tracks have now been fitted with rubber pads and new rear view driving mirrors and anti-slip devices have been installed.

Mounted one either side above the rear troop compartment is a Singapore Technologies Kinetics 40 AGL 40 mm automatic grenade launcher and a Singapore Technologies Kinetics 50MG 12.7 mm machine gun. Both of these weapons have been in volume production for some years in Singapore for vehicle-mounted and infantry applications.

The 73 mm 2A28 low-pressure gun has been retained but the latest types of fin-stabilised ammunition can now be fired, including OG-15VM high explosive, OG-15BGI with improved fragmentation and PG-15VNT with a tandem HEAT (High Explosive Anti-Tank) warhead. It is claimed that the latter will penetrate 700 mm of conventional steel armour protected by explosive reactive armour.

The launcher rail for the Russian 9K11 Malyutka (AT-3 'Sagger') wire-guided anti-tank missile (ATGM) system is retained above the 73 mm gun, but the latest version of the missile is now used, which is fitted with a tandem HEAT warhead to defeat targets fitted with ERA.

The battlefield survivability of the BMP-1 can be further improved by the installation of additional armour protection of the Kevlar and ceramic type.

The electrically operated smoke grenade launcher system has been improved and new special camouflage paint has been applied to the hull and turret.

Slovak artillery reconnaissance models

To work with its new Zuzana 155 mm/45 calibre self-propelled (8 × 8) artillery systems, Slovakia has developed special reconnaissance models based on the BMP chassis, as well as dedicated command post models based on the Tatrapan (6 × 6) chassis.

The BMP-based system is called the DPK *(Delostrelecky Prieskunmny Komplet)* and is fitted with extensive communications equipment and a sensor package which includes day/night devices, a laser range-finder and acoustic sensors.

Status

Production as required. In service with the Czech Republic and Slovakia. Quantities of these vehicles have been exported.

Manufacturer

VOP 026.

UPDATED

DENMARK

Danish M113 armoured personnel carrier fleet

The Danish Army has a fleet of about 600 M113 series APCs in a number of variants including 55 81 mm self-propelled mortars and 56 fitted with the Raytheon Systems Company TOW anti-tank guided missile system.

The latest acquisition is a batch of 50 new M113A2 Mk 1 vehicles fitted with the Italian Otobreda T25 turret armed with an Oerlikon Contraves 25 mm KBA cannon and a 7.62 mm coaxial machine gun. The system has sights from Zeiss of Germany. This version of the M113 is designated the PNMK M/92 Wildcat in Danish Army service. This version has a combat weight of 11.95 tonnes with 135 rounds of AP and 35 rounds of HE ammunition carried. For service in the Balkans additional passive armour is installed.

The M113A2 Mk 1 chassis were purchased from the USA and the Danish company of E Falck Schmidt A/S carried out the actual conversion work.

Early in 1999, following an international competition, the Danish Army Material Command selected FFG (Flensburger Fahrzeugbau GmbH) of Flensburg, Germany, as the prime contractor for the upgrade of 97 M113 series APCs.

The first six vehicles were upgraded in Germany in 2000. In February 2000, the Danish Army Material Command placed an order for another 160 vehicles.

The FFG-developed power pack for the M113 consists of a MTU 6V 183 TC 22 turbocharged diesel developing 220 kW at 2,100 rpm coupled to a ZF LSG 1000 six-speed power shift steering transmission.

The installation of this FFG-developed power pack offers the user a number of significant advantages including higher power-to-weight ratio, greater acceleration and reduced stopping distance. The driver now has a steering wheel rather than tillers and a new independent dual-circuit brake system has also been fitted.

In addition, it has greater reliability and the vehicle now meets the latest German Road Traffic Act as well as meeting the Euro 2 standard regarding exhaust emissions and noise levels.

The first customer for the FFG M113 was the German Army and, following extensive trials with six prototype vehicles, it placed an order for 270 units, the first of which have now been delivered. Additional details of this version are given under Germany.

Danish M113 with ETS

In 1999, the Elevated TOW system (ETS), developed as a private venture by E Falck Schmidt A/S of Denmark and Delco Defense of the US, completed a series of successful firing trials in Denmark.

The ETS has been designed to be installed on light tracked and wheeled armoured fighting vehicles with the first prototype system being fitted onto an M113 series full-tracked armoured personnel carrier of the Danish Army.

During the trials in Denmark, a total of nine Raytheon Systems Company TOW Anti-Tank Guided Missiles (ATGM) were fired, all hit the target.

The ETS is mounted on the roof of the armoured vehicle with the operator being under full armour protection in the hull. It is a combined observation platform and TOW weapon system. A typical ETS system would have a crew of three: driver, commander and gunner.

The ETS can be raised to a maximum height of 6.5 m above the ground and has two launchers for the Raytheon Systems Company 3,750 m range TOW ATGW either side.

The stabilised sensor pod is installed between two banks of missiles and includes day, thermal and colour

M113 of the Danish Army upgraded for operations in Bosnia with additional armour protection and shield for 12.7 mm machine gun (Richard Stickland) 0045974

Elevated TOW System (ETS) installed on an M113 series armoured personnel carrier of the Danish Army **2000**/0064866

M113A1 Mk 1 of the Danish Army fitted with Otobreda T25 turret (Richard Stickland) ***2002***/0110656

Standard M113 series APC of the Danish Army without machine gun installed (Richard Stickland) ***2002***/0110657

optics plus a laser range-finder. The commander and gunner are each provided with displays and the all-electric turret system is controlled by the gunner using a joystick.

As the sighting system is stabilised, the vehicle can carry out reconnaissance missions while on the move. As soon as it comes to a halt, the launcher is ready for immediate action. In addition to the four TOW missiles in the ready to launch position, another 16 TOW missiles are carried internally and loaded from the rear under armour protection. All versions of TOW can be launched.

E Falck Schmidt A/S has been responsible for the mast design and build and M113 modifications, while Delco Defense has been responsible for pod design and electronics integrator.

More recently, for trials purposes, the ETS has been fitted on a MOWAG Piranha wheeled armoured personnel carrier chassis.

UPDATED

EGYPT

Egyptian M113 APC fleet

Since 1980, Egypt has taken delivery of around 2,000 examples of the United Defense M113 series of armoured personnel carriers from the USA, these being a mixture of new and surplus US Army vehicles.

Specialised versions include mortar carriers, M577A2 command post vehicles, M113A2 armoured ambulances, M548A1 ammunition carriers, fitters' vehicles, M806E1 armoured recovery vehicles, M901 Improved TOW Vehicles and M981 Fire Support Team Vehicle which is known locally as the Artillery Target Locating Vehicle. The last were purchased brand new from the USA.

Egypt also has 50 US-supplied Lockheed Martin Chaparral SAM systems on the modified M548A1 chassis which are used in conjunction with the TRACKSTAR radar system, mounted on a modified M577A2 command post vehicle chassis.

Egypt has also developed an appliqué armour kit for the M113A2 with the ballistic steel plates giving protection against 23 mm projectiles fired from a range of 200 m at an angle of 35°. Each panel weighs 25 kg.

Following competitive trials between prototype twin 23 mm self-propelled anti-aircraft gun systems on a modified M113A2 chassis, in 1987 Egypt selected the now Thales turret system. This is fitted with a one-man power-operated turret, armed with twin 23 mm cannon, which is manufactured in Egypt and fitted with an Italian Officine Galileo fire-control system and a French now Thales surveillance radar.

In addition to the two 23 mm cannon, it is also armed with the locally built Sakr Eye SAM, this being an improved version of the Soviet SA-7 'Grail'. Full details of this air defence system are given in *Jane's Land-Based Air Defence 2001-2002.* This system has the local name of Sinai.

The Kader Factory for Developed Industries has produced a kit enabling the M113 APC to be fitted with an Egyptian 120 mm mortar which fires to the rear of the vehicle.

In total, 60 rounds of 120 mm mortar bomb ammunition are carried and, if required, the mortar can also be deployed away from the vehicle, as a baseplate is carried on the ramp externally. A 12.7 mm M2 machine gun and 1,200 rounds of ammunition are also provided.

In 1996, Egypt took delivery from the Netherlands of a total of 611 Armoured Infantry Fighting Vehicles (AIFVs) which were surplus to the requirements of the Royal Netherlands Army.

The 611 vehicles comprised:

(1) 304 YPR AIFV with 25 mm Oerlikon Contraves cannon and 7.62 mm machine guns
(2) six YPR PRCO-B command post vehicles armed with the 25 mm Oerlikon Contraves cannon and 7.62 mm machine gun
(3) 210 YPR 765 PRAT-TOW anti-tank vehicles. This has two Raytheon Systems Company TOW ATGW in the ready to launch position with a maximum range of 3,750 m
(4) 79 YPR 765-C command post vehicles armed with a 12.7 mm machine gun
(5) 12 M577A1 command post vehicles.

Infantry Fighting Vehicle Light

To meet the future requirements of the Egyptian Army, United Defense LP of the USA has developed the Infantry Fighting Vehicle Light, or Egyptian Infantry Fighting Vehicle as it has also been called.

The first prototype of this vehicle, which is essentially the private venture Mobile Tactical Vehicle Light (MTVL) fitted with the complete turret of the Bradley Infantry Fighting Vehicle, was completed in 1996 and has been tested in Egypt.

Additional details of this vehicle are given later in this section under the USA. As of mid-2001, this remained at the prototype stage.

Production of the first batch of 555 General Dynamics Land Systems M1A1 MBT was completed at the Egyptian Tank Plant in December 1998. This is now being used in a co-production programme for the United Defense LP M88A2 armoured recovery vehicle. In the future this facility could undertake production or local assembly of the Egyptian Infantry Fighting Vehicle. This is also referred to as Factory 200.

UPDATED

Thales twin 23 mm self-propelled air defence system which also has Sakr Eye SAMs

M577A2 series command post vehicle fitted with TRACKSTAR radar system 0004994

FRANCE

Giat Industries AMX-10P infantry combat vehicle

Development

The Giat Industries AMX-10P ICV was developed from 1965 by the Atelier de Construction d'Issy-les-Moulineaux to meet a French Army requirement for a vehicle to replace the older AMX VCI based on the chassis of the AMX-13 light tank. The first prototypes of the AMX-10P were completed in 1968.

Production began at the Atelier de Construction Roanne in 1972 and first vehicles were delivered to the 7th Mechanised Brigade at Rheims in 1973.

In 1990, Singapore awarded Giat a contract for 22 AMX-10 PAC 90 Fire Support Vehicles and 22 AMX-10P vehicles with 25 mm Dragar turrets. These have now been delivered and were all in the special Marines configuration which is detailed in the following entry.

Production of the AMX-10P was completed early in 1994 after about 1,810 had been built for the home and export markets.

Giat Industries is no longer marketing the AMX-10P infantry combat vehicle and in French Army service it will eventually be replaced by the new Satory Military Vehicles VBCI (8 × 8) infantry combat vehicle, which was selected late in 2000. The vehicle is fitted with a one-person turret armed with a 25 mm cannon and 7.62 mm coaxial machine gun.

Description

The hull of the AMX-10P is made of all-welded aluminium armour with the driver's compartment at the front of the vehicle on the left, engine compartment to his right and the troop compartment at the rear of the hull. The driver has a single-piece hatch cover that opens to the rear, in front of which are three day periscopes, the centre one of which can be replaced by a passive periscope for driving at night.

The engine compartment is provided with a fire extinguishing system and the complete power pack can be replaced in two hours. The air inlet and outlet louvres are in the top of the hull and the exhaust outlet is in the right side of the hull. The transmission consists of a torque converter with an electromagnetically operated clutch and a steer/drive unit comprising a gearbox, a steering mechanism and a Power Take Off (PTO) for the water-jets.

The two-man Giat Industries Toucan II turret is mounted in the centre of the vehicle, offset slightly to the left of the centreline, with the gunner seated on the left and the commander on the right. Both the commander and the gunner have a single-piece hatch cover that opens either side of their positions. The commander has an M371 day sight with a magnification of ×1 and ×6, a direct sight for anti-aircraft use and an external sight for direct fire. The gunner has an OB-40 day/night periscope, which has a day magnification of ×6 (10° field of view) and a night magnification of ×5 (7° field of view). The vehicle could also be delivered with the gunner's OB-40 sight replaced by an M406 day sight with a magnification of ×2 and ×6, which is replaceable with an OB-37 image intensification sight with a magnification of ×6. The commander and gunner also have seven day periscopes for all-round observation.

Giat Industries AMX-10P ICV (Richard Stickland) **2002**/0110658

The Giat Industries M693 20 mm cannon is mounted externally on the turret and has a dual-feed system allowing the gunner to select either HE or AP projectiles, and a burst selector. The cyclic rate of fire of the 20 mm cannon is 700 rds/min and the maximum effective range is 1,500 m. Of 325 rounds of 20 mm ready ammunition carried in the turret, 260 are HE and 65 AP.

Mounted coaxially above and to the right of the 20 mm M693 cannon is a 7.62 mm machine gun, which has a cyclic rate of fire of 900 rds/min and a maximum effective range of 1,000 m. Of 900 rounds of ammunition for the machine gun carried in the turret, 200 are for ready use. Mounted coaxially with the 20 mm cannon and 7.62 mm machine gun is a PH 9A searchlight and mounted on either side of the forward part of the turret are two electrically operated 80 mm smoke grenade dischargers.

The eight infantrymen are seated on individual bucket seats in the troop compartment at the rear. One man is seated forward and right of the turret facing the rear, one man to the right of the turret facing the turret, four men to the rear of the turret (two each side facing outwards), and one man on each side at the rear facing inwards. They enter and leave by the large electrically operated ramp at the rear, which is hinged at the bottom and has two doors, each with a firing port. Over the top of the troop compartment are two roof hatches which are hinged in the centre and can be locked vertical. Seven periscopes are fitted in the roof of the troop compartment, one at the front on the right, two in either side and two at the rear. The basic AMX-10P is also in service carrying two Euromissile MILAN launchers with 10 missiles carried in the hull. If required, the 2,000 m range Euromissile MILAN launchers can be mounted on the sides of the hull top, one each side.

The torsion bar suspension either side consists of five single rubber-tyred roadwheels with the drive sprocket at the front, idler at the rear and three

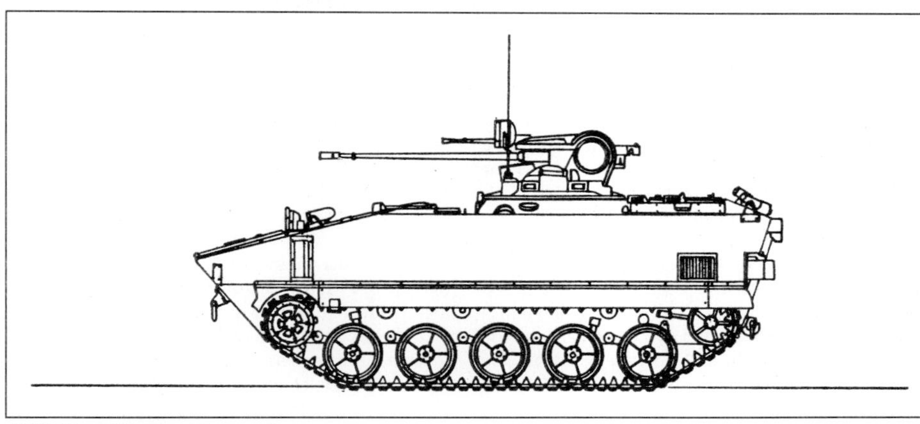

AMX-10P ICV with a Toucan II turret

AMX-10 ECH repair vehicle changing the power pack of an AMX-10P (ECP Armées)

AMX-10P HOT with a turret with four Euromissile HOT ATGWs ready to launch

AMX-10 TM towing a 120 mm TDA mortar

Giat Industries AMX-10 VOA artillery observation vehicle in Bosnia (Richard Stickland) 0045972

track-return rollers. The first and fifth roadwheel stations are fitted with hydraulic shock-absorbers. The tracks have replaceable rubber pads.

The AMX-10P is fully amphibious, being propelled in the water by two water-jets at the rear of the hull, one either side of the ramp. Before entering the water two bilge pumps are switched on (one in the engine compartment and one in the troop compartment) and the trim vane is erected at the front of the hull. When not in use the trim vane is stowed on the glacis plate. The AMX-10P could be delivered without the amphibious capability.

Standard equipment includes an NBC system, in the right side of the hull, and a heater. The vehicle can be delivered without the NBC system or the heater and optional equipment includes an NBC detector and decontamination equipment.

Giat Industries has proposed a retrofit kit for the AMX-10P infantry combat vehicle. This would include the following key parts:

(1) Installation of a more powerful diesel engine developing 300 hp
(2) Installation of a new transmission
(3) Replacement of the existing turret with the more recent Giat Industries Dragar turret armed with a stabilised 25 mm cannon and 7.62 mm coaxial machine gun.

There are no known plans by the French Army or any other army to carry out this upgrade.

Variants

AMX-10P 25 ICV

The prototype of the AMX-10P 25 was shown for the first time in June 1983. It is based on the standard AMX-10P hull without the original two-man turret. The commander now sits to the left of the turret and has a single-piece hatch cover that opens to the front, five day periscopes and a day/night periscope.

Mounted in the centre of the hull roof is the Giat Industries Dragar one-man turret armed with a Giat Industries 25 mm Model 811 dual-feed cannon which has 175 rounds of HE and 45 rounds of armour-piercing ammunition for ready use. A 7.62 mm machine gun is mounted coaxial to the 25 mm cannon on the right side of the turret and has 200 rounds of ready use ammunition. Mounted either side of the turret is a bank of two electrically operated 80 mm smoke grenade dischargers firing forwards.

Turret traverse and weapon elevation are electric at a speed of 60°/s and the weapons can be depressed/elevated from –10 to +45° with turret traverse being a full 360°. The gunner has four M336 day periscopes for forward observation, two fixed rear vision blocks and a day/night sight for aiming the weapons. This has a night channel with a magnification of ×4.5 and a day channel with a magnification of ×1 and ×7. As an option a thermal camera can be fitted and the turret stabilised.

This model weighs 14,300 kg loaded and 11,700 kg empty, it has a crew of three: commander, gunner and driver. It has the capacity to carry eight fully equipped troops.

The AMX-10P 25 ICV has been developed by Giat Industries as a private venture and has so far been ordered by Singapore in the special amphibious version.

The Turkish Army has ordered a quantity of Giat Industries Dragar turrets armed with a stabilised 25 mm cannon for some of its locally built AIFVs. Additional details are given under Turkey in the entry for Turkish Infantry Fighting Vehicle.

AMX-10P Marines

Full details of this version, developed specifically for the use of marines, are given in the following entry. This is only used by Indonesia and Singapore.

AMX-10 Ambulance

This model is unarmed and is fitted with a fixed commander's cupola with three forward-facing vision blocks. It has a crew of three: the driver and two orderlies. The vehicle can carry three stretcher patients or one stretcher and four seated patients. The AMX-10 ambulance has an air conditioning system, a searchlight, washing facilities, an oxygen system and blood transfusion equipment.

AMX-10P Driver Training Vehicle

This is the basic vehicle with its turret removed and replaced by an observation position for the instructor and drivers not under instruction.

AMX-10 ECH Repair Vehicle

This has the Giat Industries Toucan I one-man turret or TOP 7 cupola, and a crew of five consisting of the commander/chief mechanic, three mechanics and the driver. On the right side of the roof at the rear is a crane with a capacity of 6,000 kg. When it is being used, two jacks are positioned under the rear of the hull between the ground and the floor of the vehicle. The AMX-10 ECH is used to carry out power pack changes on other members of the AMX-10P family as well as changing torsion bars on both the AMX-10P and the AMX-30.

Full details of the AMX-10 ECH are given in *Jane's Military Vehicles and Logistics 2001-2002.*

AMX-10 HOT (anti-tank)

This is the basic AMX-10P, with the Toucan II turret replaced by a new two-man Lancelot turret based on the TH 20, with four ready to launch 4,000 m range Euromissile HOT missiles and a further 14 missiles carried in the rear of the hull. The crew of five consists of the commander and gunner in the turret, two missile loaders and the driver. The turret has powered traverse through 360° at a speed of 50°/s and manual traverse at a speed of 10°/s. The missiles can be elevated from –12 to +18° at a speed of 18°/s (powered) and 22°/s (manual). The missiles are launched by the gunner who is seated on the left side of the turret and has an M509 sight with a magnification of ×3 and ×12, while the commander operates the M427 laser range-finder which has a maximum range of 8,000 m and a magnification of ×8 (day), and ×3.5 (night). In addition, the turret has six day periscopes and two vision blocks. The AMX-10 HOT has a combat weight of 14,100 kg and an unloaded weight of 12,400 kg. This model has been produced for export, the only known user being Saudi Arabia. As an option the turret can be fitted with a Castor thermal camera with two magnifications.

AMX-10 PC (command vehicle)

This is the command member of the AMX-10P family and is identical to the basic vehicle except that it has additional communications equipment and a crew of six (two officers, one NCO, two radio operators and the driver). Mounted on the top of the hull at the rear is a portable generator which is positioned on the ground when the vehicle is being used in the static role. When two AMX-10 PCs are back-to-back, a canopy can be erected between them, and in addition an awning can be erected at the side of the hull. This model is in service with the French and several other armies.

Specifications

AMX-10P
Crew: 3 + 8
Combat weight: 14,500 kg
Unloaded weight: 12,700 kg
Power-to-weight ratio: 17.93 hp/t
Ground pressure: 0.53 kg/cm²
Length: 5.90 m
Width: 2.83 m
Height:
(overall) 2.83 m
(to hull top) 1.95 m
Ground clearance: 0.45 m
Track width: 425 mm
Length of track on ground: 2.93 m
Max road speed:
(4th gear) 65 km/h
(3rd gear) 34.5 km/h
(2nd gear) 15.4 km/h
(1st gear) 9.4 km/h
Max water speed: (water-jets) 7 km/h
Fuel capacity: 528 litres
Max road range: 500 km
Fording: amphibious
Gradient: 60%
Side slope: 30%
Vertical obstacle:
(forwards) 0.7 m
(reverse) 0.6 m
Trench: 2.1 m
Engine: Hispano-Suiza (built in France by RVI) HS 115 V-8 water-cooled supercharged diesel developing 260 hp at 3,000 rpm
Transmission: preselective with 4 forward and 1 reverse gears
Suspension: torsion bar
Electrical system: 24 V
Batteries: 4 × 12 V
Armament:
(main) 1 × 20 mm M693 cannon
(coaxial) 1 × 7.62 mm MG
Smoke-laying equipment: 2 × 2 80 mm smoke grenade dischargers
Ammunition:
(main) 760
(coaxial) 2,000

Gun control equipment
Turret power control: electric
(by commander) yes
(by gunner) yes
Max rate power:
(traverse) 50°/s
(elevation) 30°/s
Gun elevation/depression: +50/–8°
Turret traverse: 360°
Commander's override: yes
Gun stabiliser:
(horizontal) no
(vertical) no
NBC system: yes
Night vision equipment: yes

AMX-10P with RATAC radar

This has no turret and is fitted with a roof-mounted RATAC (Radar For Field Artillery Fire) pulse Doppler radar scanner. The vehicle has a crew of five: commander; radar operator; assistant radar operator;

radio operator; driver. The radar has four operating modes: ground surveillance; acquisition and identification; angular deviation measurement; automatic tracking. The radar is in the forward part of the vehicle on the right side, with the radar operating console behind it and the plotting table at the rear on the right. Height including radar is 2.84 m. This model also has a navigation system and an identification system. At least 24 of this version have been sold to Saudi Arabia.

AMX-10RC (reconnaissance vehicle)

This 6 × 6 reconnaissance vehicle armed with a turret-mounted 105 mm gun uses many automotive components of the AMX-10P series. Full details are given in the entry in the Reconnaissance vehicles section.

AMX-10 SAO (artillery observation vehicle)

The AMX-10 SAO is used for battlefield observation and artillery fire control and is an AMX-10P with its Toucan II turret replaced by a new two-man turret armed with an externally mounted 7.62 mm machine gun on the right side. Mounted on the top of the turret are four electrically operated 80 mm smoke grenade dischargers firing forwards. Turret traverse is electric, two-speed, with a maximum speed of 50°/s. The turret has a laser range-finder with a maximum range of 8,000 m coupled to a day sight with a magnification of ×8 and a night sight with a magnification of ×4.5. It also has a binocular telescope with a magnification of ×2.5 and ×10, which has adjustable cross wires, a retractable sun filter and can be elevated from −10 to +45°. The crew of five consists of the commander (who normally operates the binocular telescope), the second in command (usually the laser range-finder operator), two radio operators and a driver. The AMX-10 SAO is fitted with a vehicle attitude corrector which enables it to establish its exact position. The ATILA system message transmitter unit (CMS) can be used in the SAO vehicle.

AMX-10 SAT (artillery survey vehicle)

This is based on the AMX-10 PC command vehicle but has been modified internally to carry a gyrostabilised theodolite, a topographic survey theodolite, distance measuring equipment and a navigation system.

AMX-10 TM (mortar tractor)

This model tows a 120 mm TDA Model MO-120-RT-61 rifled mortar, as well as carrying 60 rounds of ammunition, of which 56 are stored horizontally and four vertically. The AMX-10 TM has a crew of six consisting of commander, aimer, artificer, loader, ammunition carrier and driver, and is fitted with a Toucan I turret armed with a 20 mm cannon and a 7.62 mm machine gun, which have elevations of +50° and depressions of −8°, turret traverse being a full 360°. The turret is manually operated: maximum traverse speed is 12°/s and maximum elevation speed 16°/s. The turret has an optical sight with magnifications of ×1 and ×6, plus periscopes. The AMX-10 TM has been produced in quantity for both the home and export markets.

AMX-10 VOA (artillery observer vehicle)

This is based on the AMX-10PC command vehicle chassis and has been designed for use by artillery observers. Its crew of four consists of a commander, observer, radio operator and driver. The turret, which is manned by one or two men, is provided with equipment for day and night observation and local defence. The equipment fitted in the turret enables the acquisition of target co-ordinates and the manual or automatic transmission of artillery messages. Armament consists of a 7.62 mm machine gun and four smoke grenade dischargers. The onboard laser range-finder has a magnification of ×8 and can be dismounted if required. The turret optical sight has magnifications of ×2.5 and ×10, a thermal camera and periscopes.

AMX-10Ps used in the ATILA control system

Two versions of the AMX-10P have been developed to mount parts of the ATILA automatic artillery fire-control system. The AMX-10 VFA is used at regimental and battery level and the AMX-10 VLA, which does not have a computer, is used as an artillery liaison vehicle.

AMX-10 SAF

This is also known as the AMX-10PC SAF and is used with the ATILA artillery fire-control system. It has a crew of five consisting of the radio operator, driver, two warrant officers and one commanding officer. It is fitted with an auxiliary power unit in the forward part of the vehicle, two air conditioning systems for the electronics and crew, a multipurpose computer, a data processing equipment centre, a man/machine interface (screen/keyboard) and three radio sets.

AMX-10 PAC 90 Fire Support Vehicle

The AMX-10 PAC 90 has been developed primarily for the export market and was first announced in 1978. It basically consists of an AMX-10P ICV hull fitted with the Giat Industries TS 90 turret which is also mounted on the Panhard ERC 90 F4 Sagaie (6 × 6) armoured car as well as the Renault VBC 90 (6 × 6) armoured car. The Indonesian Marine models have automotive characteristics and modifications similar to the AMX-10P Marines described in the following entry. More recently this version has also been adopted by Singapore, which also adopted the Marines version of the AMX-10P with the 25 mm Giat Industries Dragar one-man turret.

The primary role of the AMX-10 PAC 90 is that of anti-tank and fire support on the battlefield at ranges between 1,000 and 2,500 m, but it can also be used as a reconnaissance vehicle or as an APC, as well as transporting 2,000 m range Euromissile MILAN anti-tank or 81 mm mortar teams.

The Giat Industries 90 mm gun fires the following types of ammunition: canister, high-explosive, high-explosive long-range, HEAT, smoke and APFSDS.

There are 20 rounds of ammunition, 12 HEAT and 8 HE, carried in the turret, 16 in the turret bustle to the rear of the commander, who also loads the gun, and four behind the commander's seat. Another 10 rounds are carried in the hull. A 7.62 mm machine gun is mounted coaxially to the left of the main armament, for which 2,000 rounds of ammunition are carried in the turret, with an additional 1,200 rounds carried in the hull. Mounted either side of the turret towards the rear are two 80 mm smoke grenade dischargers.

The four infantrymen sit on individual seats at the rear of the hull and enter and leave by the large ramp in the hull rear, which has two doors, each with a firing port. There are four day periscopes in the rear roof of the troop compartment, one in each side and two at the rear.

Status

Production complete but can be resumed if further orders are placed. In service with:

Country	Quantity	Comment
France	713	
Greece	96	in reserve
Indonesia	34	AMX-10P Marines
Iraq	100	estimate, some used with ATILA artillery fire-control system
Qatar	40	
Saudi Arabia	570	
Singapore	44	AMX-10P Marines
UAE	18	

Note: Giat has stated that 1,810 were built but these figures equal 1,625 so there are either more customers or some of the countries listed have larger holdings than indicated.

Manufacturer

Centre de Roanne, Giat Industries.

UPDATED

Giat Industries AMX-10P Marines

Development

The AMX-10P Marines was developed as a private venture by Giat Industries Roanne facility specifically for the export market for use by marines and other armed forces having an amphibious capability.

The first customer was Indonesia, which placed an order for 34 vehicles in two versions, 24 fitted with the Giat Industries one-man turret armed with a 12.7 mm M2 machine gun and 10 fitted with the Giat Industries TS 90 two-man turret armed with a 90 mm F4 gun and 7.62 mm coaxial machine gun. This turret is also installed on a number of other armoured vehicles including the Panhard ERC Sagaie (6 × 6) and Renault VBC 90 (6 × 6) armoured cars.

More recently, Singapore ordered a total of 44 vehicles in two configurations, 22 with the Giat Industries TS 90 two-man turret (PAC 90) and 22 with the Giat Industries Dragar one-man turret armed with a 25 mm M811 cannon and 7.62 mm coaxial machine gun (P 25 ICV). Delivery of these vehicles to Singapore has now been completed.

The AMX-10P Marines is based on the standard AMX-10P infantry fighting vehicle of which over 1,800 have been built for the home and export markets but has improved amphibious characteristics.

The AMX-10P Marines has been designed to operate in fresh water (river) or salt water (sea) without limitation of depth at a maximum speed of 10 km/h. The vehicle can embark or disembark from LCTs or LSTs without

Giat Industries AMX-10 Marines fitted with Dragar turret armed with 25 mm cannon and 7.62 mm coaxial machine gun from the rear with trim vane erected **2000**/0064808

Giat Industries AMX-10 PAC 90 Marines with trim vane extended 0045969

Giat Industries AMX-10 Marines fitted with turret armed with 12.7 mm M2 machine gun **2001**/0093581

assistance and if required can use its main weapon whilst still afloat.

Although production of the AMX-10P Marines was completed by Giat Industries several years ago, it is still being marketed and if a significant order is placed then production can start again at Roanne.

Description

The overall layout of the AMX-10P Marines is identical to the standard AMX-10P covered in detail in the previous entry. The main differences are its modifications to give it an improved amphibious capability and a wide range of weapon systems can be fitted.

Modifications carried out to the AMX-10P Marines when compared to the standard vehicle can be summarised as follows:

(1) Anti-corrosion treatment of the hull and of different accessories
(2) Installation of a Baudouin 6 F11 SRX V-8 diesel engine developing 300 hp to increase both land and sea mobility
(3) Large diameter (305 mm) water-jets mounted one either side of the hull at the rear to increase the maximum water speed
(4) Installation of four bilge pumps, two in the engine compartment and two in the troop compartment
(5) Large trim vane installed on the glacis plate
(6) Streamlined track links for improved hydrodynamic characteristics
(7) Emergency pneumatic engine starter
(8) Watertight engine air inlets and outlets in order to ensure chassis is completely watertight
(9) Water/water exchanger enabling cooling of the power pack during amphibious operations.

The vehicle is propelled in the water by two large water-jets mounted one either side at the rear. As soon as the driver engages the water-jets the following occurs:

(1) The trim vane, which when travelling is laid back onto the glacis plate, is automatically raised by two hydraulic rotary actuators
(2) The engine air inlets are automatically closed tight
(3) Bilge pumps are automatically switched on
(4) The special power pack cooling circuit is automatically engaged.

The AMX-10P Marines can be fitted with the following Giat Industries turrets:

(1) Turret armed with 12.7 mm M2 machine gun
(2) Dragar turret armed with 25 mm M811 dual-feed cannon and 7.62 mm coaxial machine gun
(3) TS 90 armed with 90 mm F4 gun and 7.62 mm coaxial machine gun.

Specifications

AMX-10 P 25 ICV
Crew: 3 + 8
Combat weight: 15,300 kg
Unloaded weight: 13,600 kg
Power-to-weight ratio: 19.6 hp/t
Ground pressure: 0.56 kg/cm²
Length: 5.90 m
Width: 2.83 m
Height:
(overall) 2.75 m
(hull top) 1.95 m
Ground clearance: 0.45 m
Track width: 425 mm
Length of track on ground: 2.93 m
Max road speed:
(4th gear) 65 km/h
(3rd gear) 34.5 km/h
(2nd gear) 15.04 km/h
(1st gear) 9.4 km/h
Max water speed with water-jets: 10 km/h
Fuel capacity: 518 litres
Max road range: 500 km
Fording: amphibious
Gradient: 60%
Side slope: 30%
Vertical obstacle:
(forwards) 0.7 m
(reverse) 0.6 m
Trench: 2.1 m
Engine: Baudouin 6 F11 SRX V-8 water-cooled supercharged diesel developing 300 hp at 3,000 rpm
Transmission: preselective with 4 forward and 1 reverse gears
Suspension: torsion bar
Electrical system: 24 V
Batteries: 4 × 12 V
Armament:
(main) 1 × 25 mm M811 cannon
(coaxial) 1 × 7.62 mm MG
Smoke-laying equipment: 2 × 2 80 mm smoke grenade dischargers
Ammunition:
(main) 608
(coaxial) 3,000
Turret power control: electric/manual (stabilisation system optional)
Max rate power:
(traverse) 60°/s
(elevation) 60°/s
Gun elevation/depression: +45/–8°
Turret traverse: 360°
Gun stabiliser:
(horizontal) no
(vertical) no
NBC system: optional
Night vision equipment: optional

Status

Production as required. In service with Indonesia (34) and Singapore (44).

Manufacturer

Centre de Roanne, Giat Industries.

UPDATED

Giat Industries AMX VCI infantry combat vehicle

Development

The AMX VCI *(Véhicule de Combat d'Infanterie)* was developed in the early 1950s to meet the requirements of the French Army. The first prototype was completed in 1955 and the first production vehicles in 1957 at the Atelier de Construction Roanne (ARE). When the ARE (which is now part of Giat Industries and is currently prime contractor for Leclerc MBT) started production of the AMX-30 MBT, production of the complete AMX-13 tank family, including the AMX VCI, was transferred to the Creusot-Loire facility at Châlon-sur-Saône.

When originally introduced into service with the French Army the vehicle was called the *Transport de Troupe Chenillé* Model 56 (or TT 12 CH Mle 56 for short). This was later changed to the *Véhicule Transport de Personnel* (or AMX VTP) and to the *Véhicule de Combat d'Infanterie.* It was used in large numbers by the French Army but has been replaced by the Giat Industries AMX-10P amphibious ICV.

It is believed that total production of the vehicle amounted to approximately 3,400 vehicles.

In more recent years production and marketing of the complete range of AMX-13 family of light armoured vehicles, including the VCI infantry combat vehicle, was carried out by Mecanique Creusot-Loire. Giat Industries has now taken over this company and is responsible for after-sales support and provision of upgrade packages. Marketing of new vehicles is no longer being carried out by Giat Industries.

In recent years many countries have phased these out of front line service. Mexico has taken delivery of 401 AMX VCI series vehicles from Belgium. These were overhauled prior to delivery to Mexico by the Belgian company SABIEX International. Deliveries took place as follows:

Year	Quantity
1994	168
1995	97
1996	136

Description

The chassis of the AMX VCI is similar to that of the AMX-13 light tank. The hull of the vehicle is divided into three compartments, with the driver and engine compartments at the front and the troop compartment at the rear.

The driver is seated at the front of the vehicle on the left with the engine compartment to his right. The driver has a single-piece hatch cover that opens to the left in front of which are three day periscopes. The centre periscope can be replaced by an infra-red or image intensification periscope for night driving.

The petrol engine is mounted facing the rear and can be removed through the roof in 40 minutes. The engine transmits power via the clutch to the gearbox at the front of the hull, to the right of which is the Cleveland-type steering differential.

Mounted on the glacis plate is a splashboard to prevent water rushing up the front of the vehicle when fording and a spare wheel is normally carried on the glacis plate.

The troop compartment is behind and above the driver with the gunner seated on the left and the vehicle commander to his right. When the VCI was originally introduced into the French Army, the gunner had a single-piece hatch cover with eight day vision devices which swivelled to open, forward of which was a pintle-mounted 7.5 mm machine gun. This was later replaced by a ring-mounted 12.7 mm machine gun or a Giat Industries CAFL 38 turret armed with a 7.5 mm or a 7.62 mm machine gun.

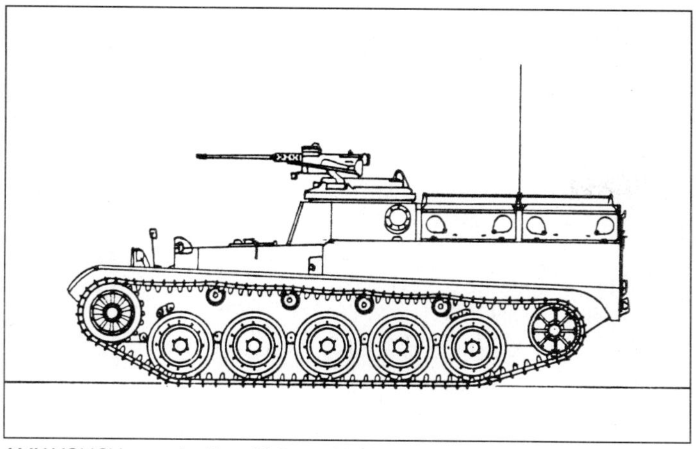

AMX VCI ICV armed with a 12.7 mm (0.50) Browning M2 HB machine gun

AMX VCI vehicle fitted with machine gun turret **2000**/0064809

The vehicle commander is seated to the right of the gunner and has a single-piece hatch cover that opens forwards, to the front and right side of which are three day periscopes.

The troop compartment is at the rear: the 10 infantrymen sit back-to-back down the centre of the hull and enter and leave the vehicle by two doors in the hull rear that open outward. Each door has a single firing port.

In each side of the troop compartment are two two-piece hatch covers. The lower part of each has two firing ports and folds forwards into the horizontal and the upper part folds upwards through 180° to rest on the troop compartment roof.

The torsion bar suspension either side consists of five rubber-tyred roadwheels with the drive sprocket at the front, idler at the rear and four (in some cases three) track-return rollers that support the inside of the track only. The first and last roadwheel stations have hydraulic shock-absorbers. The steel tracks have 85 links each side when new and can be fitted with rubber pads.

When originally introduced into service the VCI was not fitted with an NBC system but one was subsequently fitted to later production vehicles for the French Army. Infra-red driving lights were standard on most vehicles but they do not have any amphibious capability.

Specifications

(1989 production model with diesel engine replacing original SOFAM Model 8Gxb 8-cylinder water-cooled petrol engine developing 250 hp at 3,200 rpm)
Crew: 3 + 10
Combat weight: 15,000 kg
Unloaded weight: 12,500 kg
Power-to-weight ratio: 16.67 hp/t
Ground pressure: 0.7 kg/cm²
Length: 5.7 m
Width: 2.67 m
Height:
(to turret top) 2.41 m
(to hull top) 2.1 m
Ground clearance: 0.48 m
Track: 2.159 m
Track width: 350 mm
Length of track on ground: 3.012 m
Max road speed: 64 km/h
Fuel capacity: 410 litres
Road range: 500-550 km
Fording: 1 m
Gradient: 60%
Vertical obstacle:
(forwards) 0.65 m
(reverse) 0.45 m
Trench: 1.6 m
Turning radius: skid turns
Engine: Detroit Diesel 6V-53T, turbocharged 6-cylinder water-cooled diesel developing 280 hp at 2,800 rpm, or Baudouin 6F 11 SRY turbocharged 6-cylinder water-cooled diesel developing 280 hp at 3,200 rpm
Transmission: manual with 5 forward and 1 reverse* gears
Steering: Cleveland-type differential
Clutch: single disc
Suspension: torsion bar**
Electrical system: 24 V
Batteries: 4 × 12 V, 190 Ah
Armament: 1 × 20 mm cannon, or 1 × 12.7 mm MG, or 1 × 7.62 mm/7.5 mm MG
Turret power control: manual
Gun stabiliser:
(horizontal) no
(vertical) no
Smoke-laying equipment: none
Armour
Hull front: 15 mm at 55° (equivalent to 40 mm)
Hull sides: 20 mm
Troop compartment front: 30 mm
Troop compartment sides: 20 mm
Troop compartment hatches: 15 mm
Hull top: 15 mm
Hull floor front: 20 mm
Hull floor rear: 10 mm
Hull rear: 15 mm
NBC system: optional
Night vision equipment: optional
*or an automatic transmission with 3 forward and 1 reverse gears as an option
**or hydropneumatic suspension as an option (as far as it is known, there have been no sales of this vehicle fitted with the hydropneumatic suspension which will give a significant increase in cross-country mobility as well as an improved ride for the crew and occupants).

AMX VCIs in service with the Mexican Army (Julio Montes) 0018417

Variants

Listed below are the current variants of the AMX VCI family in service, excluding trials vehicles. There are separate entries for the AMX-13 light tank and the AMX 155 mm self-propelled gun Mk F3. As far as is known, none of these is any longer in service with the French Army.

Ambulance (VTT/TB)

The VTT/TB *(Transport de Blessés)* is the ambulance member of the family, is unarmed and has a crew of four: driver, surgeon and two orderlies/stretcher bearers. The vehicle can transport three stretcher and four seated patients at one time.

Command Post (VTT/PC)

The VTT/PC *(Poste de Commandement)* is the basic VCI adapted for use as a command vehicle and has a crew of six to eight depending on its use. Internally the vehicle has two map tables and additional communications equipment.

Cargo (VTT/Cargo)

This version can carry up to 3,000 kg of cargo.

Fire Control Vehicle (VTT/LT)

The VTT/LT *(Véhicule de commandement du tir)* is a development of the basic VTT/PC command vehicle, has a crew of seven and is fitted with additional communications equipment, a plotting table, a fire computing table, a four-channel switchboard for battery communications and an external loudspeaker.

RATAC Radar Vehicle (VTT/RATAC)

This is the basic vehicle fitted with the RATAC *(RAdar de Tir pour Artillerie de Campagne)* radar which detects, identifies, locates and automatically tracks ground targets such as tanks, vehicles and personnel, light aircraft and helicopters or shell bursts. The radar itself is mounted on the roof of the vehicle on the right side. The French Army took delivery of 115 systems on the AMX VCI chassis but have now been transferred to the VAB (4 × 4). Export sales of RATAC on the AMX VCI chassis were made to Ecuador (4) and Morocco (12).

155 mm Support Vehicle (VTT/VCA)

The VTT/VCA *(Véhicule Chenillé d'Accompagnement)* is designed to accompany the 155 mm Mk F3 self-propelled gun and carries the remainder of the gun crew and ammunition. The crew of eight consists of the driver, layer, loader, gunner and four ammunition numbers. In total 25 155 mm projectiles are carried plus 25 charges and 49 fuzes. The vehicle can also tow an ARE 2T F1 ammunition trailer containing 30 projectiles plus charges and fuzes. The vehicle is normally fitted with the CAFL 38 turret armed with a 7.5 mm or a 7.62 mm machine gun or a Giat Industries CB 127 shield gun racer for a 12.7 mm M2 machine gun.

81 mm Mortar Carrier (VTT/PM)

The VTT/PM *(Porte Mortier)* has an 81 mm mortar firing forward through the roof of the personnel compartment. It has an elevation from +43 to +80° and a total traverse of 40°. A total of 88 long mortar bombs can be carried plus 40 short mortar bombs. The crew of six consists of the driver, commander, observer and three mortar men. A baseplate is carried enabling the mortar to be deployed away from the vehicle if required.

120 mm Mortar Carrier (VTT/PM)

The VTT/PM *(Porte Mortier)* has a 120 mm TDA MO 120 LT mortar firing forwards through the roof of the personnel compartment. The mortar has a maximum range with PEPA/LT projectiles of 8,300 m. When fired away from the vehicle with the aid of the baseplate carried the range is increased to 8,950 m. The mortar has an elevation of +45 to +77° and a total traverse of 46°. A total of 60 mortar bombs is carried and the crew of six consists of the driver, commander, layer, gunner and two mortar men.

AMX-13 VTT Version 1987

This incorporates all the automotive improvements of the modernised 105 mm AMX-13 light tank (qv) and has a Detroit Diesel engine Model 6V-53T developing 280 hp or a Baudouin 6F 11 SRY diesel engine which is much more fuel efficient, an automatic transmission and a hydropneumatic suspension.

Status

Production complete. No longer being marketed. In service with the following countries:

Country	Quantity	Comment
Argentina	30	some assembled locally
Cyprus	16	delivered from 1970 in PC and VCA configurations
Ecuador	60	
Indonesia	200	some from the Netherlands
Lebanon	75	
Mexico	401	from Belgium, delivered 1994/1996
Qatar	33	upgraded by RDM Technology of the Netherlands incl PC and VCA
Sudan	5	2 PC and 3 VCA status uncertain
UAE	6	incl VCA
Venezuela	25	plus some 20 LAR-160 rocket versions

Manufacturer

Giat Industries.

UPDATED

GERMANY

Rheinmetall Landsysteme Marder 1 infantry combat vehicle

Development

In January 1960, contracts for the design and construction of prototypes of a new infantry combat vehicle were awarded to two groups of companies: the Rheinstahl group comprising Rheinstahl-Hanomag, Ruhrstahl, Witten-Annen and Büro Warnecke and the partnership between Henschel Werke of Kassel and MOWAG of Kreuzlingen, Switzerland. There were three vehicles built by Hanomag, two by Henschel and two by MOWAG, known as the first series prototypes.

Between 1961 and 1963, the second series of prototypes was built, four by Hanomag, one by Henschel and three by MOWAG. There was then a pause in the development of the vehicle as priority was given to the Jagdpanzer Kanone (which entered production at the works of both Hanomag and Henschel in 1965) and the Jagdpanzer Rakete (which entered production at the works of both Hanomag and Henschel in 1967 when production of the Kanone had been completed).

In 1966, the military requirements were finalised and in 1967 the construction of the third and final series of prototypes began. There were 10 prototypes built, three by Hanomag, four by Henschel and three by MOWAG.

Following the takeover of Henschel Werke in 1964 by the Rheinstahl Group, most of the final development work was completed by Rheinstahl. The 10 preproduction vehicles were completed in 1967-68 and, in October 1968, the first vehicles were delivered to the German Army for troop trials which ran from October 1968 to March 1969. In April 1969, companies were asked to tender for series production of the vehicle, which was officially named the Marder 1 in May 1969. In October of that year Rheinstahl (today Rheinmetall Landsysteme) was nominated as prime contractor with MaK of Kiel as a subcontractor.

A total of 2,136 Marder 1 ICV were built, 976 by the then Atlas MaK at Kiel and 1,160 at Kassel by the then Rheinstahl. Both these companies are now part of the Rheinmetall Landsysteme GmbH group.

The first Marder 1 was handed over on 7 May 1971 and production continued through to 1975. In the same year, the Euromissile MILAN ATGW was adopted to be fired by the commander of the Marder 1 from his open hatch position.

The chassis of the Marder 1 ICV remained in production at Kassel for the Euromissile Roland 2 surface-to-air missile system until 1983.

The Marder 1 series was to have been supplemented by the Marder 2 but this was cancelled after a single prototype had been designed and built by Krauss-Maffei (which in January 1999 became Krauss-Maffei Wegmann).

At present the main emphasis is aimed at providing the vehicle with additional armour protection, especially against anti-tank mines.

Other improvements under consideration are said to include a more powerful engine and a new weapon station with a larger calibre gun, perhaps the Rheinmetall 35 mm/50 mm weapon, which has been under development for some time and was originally earmarked for the Marder 2.

Description

The all-welded steel hull of the Marder 1 provides the crew with protection from small arms fire and shell splinters, with the front of the vehicle giving complete protection from 20 mm armour-piercing discarding sabot (APDS) projectiles.

The driver sits at the front of the hull on the left side and has a single-piece hatch cover that opens to the right, in front of which are three day periscopes; the centre one can be replaced by a passive night driving device. One of the infantrymen sits to the rear of the driver and has a single-piece hatch cover that opens to the right, with a day periscope that can be traversed through a full 360°.

The engine compartment is to the right of the driver with the MTU MB Ea-500 six-cylinder diesel being coupled to a Renk four-speed HSWL 194 planetary gearbox and a stepless hydrostatic steering unit which transmits power to the tracks via two final drive assemblies mounted at the front of the hull. The radiators are mounted at the rear of the hull, one either side of the ramp, with cooling air being sucked in through grills in the top of the hull by an axial cooling fan.

The two-man turret, designed by KUKA (which in 2000 became Rheinmetall Landsysteme), is mounted in the forward part of the roof, with the commander on the right and the gunner on the left. The commander has a single-piece hatch cover that opens to the right while the gunner has a similar hatch cover that opens to the rear.

The 20 mm Rheinmetall MK 20 Rh 202 cannon is mounted externally and has an elevation of +65°, depression of −17° and 360° turret traverse. The empty cartridge cases are automatically ejected outside the turret.

Both the commander and gunner have a PERI-Z11 sight for use in the ground and anti-aircraft role with a magnification of ×2 and ×6 which can be replaced by a night periscope. In addition, the commander has eight day periscopes for all-round observation and the gunner has three day periscopes.

Mounted coaxially above and to the right of the 20 mm cannon is a 7.62 mm MG3 machine gun and mounted to the left of the 20 mm cannon are six electrically operated 76 mm smoke grenade dischargers.

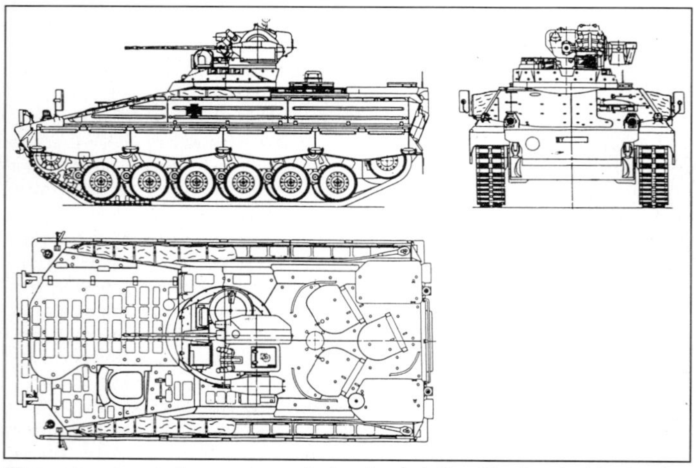

Three-view general arrangement drawing of the Marder 1 A3 showing additional armour and revised roof hatches

Marder 1 A3 infantry combat vehicle (Michael Jerchel) ***2001***/0093572

Marder 1 A3 infantry combat vehicle (Michael Jerchel) **2002**/0110659

M12 Infantry Fighting Vehicle fitted with Rheinmetall Landsysteme E4 turret armed with 30 mm cannon and 7.5 mm coaxial machine gun **2000**/0064810

The six infantrymen are seated in the troop compartment at the rear of the hull, three down each side facing outwards. Their seating unit was designed by MOWAG and can be adjusted so that four men can sleep. Entry and exit is by a power-operated ramp at the rear of the hull that opens downwards. All Marder 1s have an NBC system.

The torsion bar suspension each side consists of six dual rubber-tyred roadwheels with the drive sprocket at the front, the idler at the rear and three track-return rollers. The first, second, fifth and sixth roadwheel stations have a hydraulic shock-absorber. The Diehl tracks have replaceable rubber pads.

The Marder 1 can ford to a depth of 1.5 m without preparation or to 2.5 m with the aid of a kit. An amphibious kit was developed for the Marder 1 but has not been adopted.

Variants

R 397 ICV

The R 397 (previously called the TH 397) is the Marder 1 ICV with the rear external 7.62 mm remote-controlled machine gun removed.

Upgraded Marder 1

From 1982 most Marder 1s were upgraded to one of two new standards, 1 A1 or A1A, at the rate of about 80 vehicles per month.

1 A1(+)

This is the complete upgrade and 674 vehicles have been brought up to this standard. Main improvements are: better firepower due to double belt feed for the 20 mm cannon; improved night capabilities with the installation of a PERI-Z59 (WOE) night sight with sight head PERI-Z16 which uses the image intensification technique, with thermal pointer. The remaining active infra-red searchlight and sight are retained, there is a holder for the commander's NBC mask on the upper mounting for the MILAN ATGW, new racks for water cans inside, stowage for image intensification equipment, flaps for periscopes, storage for helmets and two 5 m tow ropes. This model weighs 30,000 kg, has a crew of four and carries five instead of six infantrymen. This upgrade was carried out between 1979 and 1982.

Specifications

(Key specifications of the upgraded Marder A3 are given in square brackets where applicable)

Crew: 9
Combat weight: 29,207 [33,500*] kg
Unloaded weight: 28,200 [29,900*] kg
Power-to-weight ratio: 20.54 [18] hp/t
Ground pressure: 0.83 [0.94] kg/cm²
Length: 6.79 [6.88] m
Width: 3.24 [3.38] m
Height:
(over turret top) 2.985 [3.015]m
(over hull top) 1.9 m
Firing height:
(20 mm cannon) 2.56 [2.7] m
(7.62 mm coaxial MG) 2.735 m
Ground clearance: 0.44 [0.455] m
Track: 2.62 m
Track width: 450 mm
Length of track on ground: 3.9 m
Max road speed (forward and reverse):
(4th gear) 75 [65] km/h
(3rd gear) 47 km/h
(2nd gear) 31 km/h
(1st gear) 16 km/h
Fuel capacity: 652 litres
Max road range: 520 [500] km
Fording: 1.5 m
(with preparation) 2.5 [2.0] m
Max gradient: 60%
Side slope: 30%
Vertical obstacle: 1 m
Trench: 2.5 m
Turning radius:
(4th gear) 30 m
(3rd gear) 20 m
(2nd gear) 13.5 m
(1st gear) 6.5 m
Engine: MTU MB 833 Ea-500 6-cylinder liquid-cooled diesel developing 600 hp at 2,200 rpm
Transmission: 4-speed HSWL 194 planetary, 4 forward and 2 reverse gears, integral steering and braking system
Suspension: torsion bar
Electrical system: 24 V
Batteries: 6 × 12 V
Armament:
(main) 1 × 20 mm MK 20 Rh 202 cannon
(coaxial) 1 × 7.62 mm MG3 MG
Smoke-laying equipment: six 76 mm smoke grenade dischargers
Ammunition:
(main) 1,250
(7.62 mm) 5,000

Gun control equipment
Commander's fire control:
(override) yes
Turret power control: electrohydraulic/manual
(by commander) yes
(by gunner) yes
Max rate power traverse: 60°/s
Max rate power elevation: 40°/s
Gun elevation/depression: +65/–17°
Turret traverse: 360°
Gun stabiliser:
(vertical) no
(horizontal) no
NBC system: yes
Night vision equipment: yes
* Maximum gross weight of Marder 1 is 35,000 kg, military empty weight is 31,200 kg.

Note: When originally introduced into service with the German Army the Marder 1 had a crew of 4 + 6 and was fitted with a remote-controlled 7.62 mm machine gun over the rear of the troop compartment. Most vehicles also carry one Euromissile ATGW for which six missiles are carried inside the vehicle.

1 A1(–)

In total, 350 vehicles were upgraded to the A1(–) configuration, but unlike the A1(+) standard (see above) they were prepared for, but not fitted with, the thermal pointer. The latter can, however, be transferred from an A1(+) turret that is not operational.

1 A1A

In total 1,112 vehicles were modified to the previous standards but not fitted with passive vision equipment. All Marder 1s of the German Army, except command vehicles, now carry a 2,000 m range Euromissile MILAN ATGW system with five missiles and this has meant that the number of infantrymen carried has been reduced from six to five.

1 A1A2

This is a converted Marder 1 with A1 turret and A2 chassis and fitted out with WOE (Warme-Ortungs-Empfanger) PERI Z 59 sighting system.

1 A1A3

This is Marder 1 A1 with SEM 80/90 radios.

1 A1A4

This is Marder 1 A1A with SEM 80/90 radios.

1 A1A5

This is Marder 1 A1A2 with SEM 80/90 radios.

1 A2

Since 1983, all the German Marder 1s have been upgraded to the A2 standard. The 674 A1(+) were upgraded to A2 standard which excludes thermal imaging sight equipment. The remaining 1,462 ICVs are fitted with thermal imaging sight equipment instead of the image intensifier. At the same time the infra-red searchlight on the left side of the turret is removed and the chassis (fuel tanks, cooling system and water-jet cleaning system) and suspension are modified. This upgrade work was carried out between 1984 and 1991 and included installation of WBG (PERI Z 11A1) sighting system.

1 A2A1

This is Marder 1 A2 with SEM 80/90 radios.

1 A3

Four prototype upgraded Marder 1 infantry combat vehicles were completed, followed by six preproduction vehicles. In 1988, Thyssen Henschel (which in 2000 became Rheinmetall Landsysteme GmbH) was awarded a contract covering the conversion of 2,100 Marder 1 A1/A2 series vehicles to the A3 standard over a 10 year period at the rate of 220 vehicles a year, with the project starting in 1989. The first vehicles were handed over to the German Army on 17 November 1989. The modifications include the following:

Armour package – a new passive armour package weighing 1,600 kg was fitted to provide protection against penetration from the Russian BMP-2 30 mm cannon. This package consists of frontal armour on the hull, add-on armour on the glacis plate, conformal add-on armour on both sides of the turret, three box-type armour components on both sides of the hull (so blocking off the hull firing ports), spaced armour plates on the roof to cause premature detonation of top attack rounds, add-on armour with a stowage compartment at the rear door and improvements in human engineering.

The infantryman's hatch to the immediate rear of the driver's position has been eliminated and over the top of the troop compartment there are now three roof hatches arranged in a triangular pattern, two towards the front and one towards the rear.

As the combat weight of the Marder 1 has increased to 35 tonnes, a number of modifications have been carried out to the suspension including the installation of reinforced torsion bars, a hydraulic duplex brake system, change of gear ratio of the lateral final drives to maintain high mobility, steering agility and acceleration.

The diesel-electric air heating system has been replaced by a warm water heating system coupled and electrically linked to the engine cooling system.

New stowage boxes are provided along the sides of the upper part of the hull and a new radio installation is provided.

In addition to being provided with additional armour protection, the 7.62 mm MG3 machine gun, including the ammunition magazine, has been rearranged on the extreme left in a separate mount.

Internal turret modifications include changed space-saving ammunition feed, newly designed and improved seats, a larger free space in the knee area, an enlarged passage between the commander's and gunner's seats, a redesigned console between the crew seats and optical signal repeating push-buttons in the centre operating console replacing lever functions previously distributed within the turret compartment.

Final deliveries were made to the German Army in 1998.

Future improvement

These could include newly designed and stronger armoured track skirts and an infra-red camouflage for the final drives, a new electronic control for the engine and changing the current maintenance concept to keep the batteries fully charged even during extended parking periods.

Marder 1 with additional mine protection

In 1999, the then Henschel Wehtechnik delivered one prototype of the Marder 1 to the German Army, with an integrated mine-protection system under the crew compartment for enhanced protection against heavy anti-tank mines.

Late in 2000, it was stated that the German Army was to retrofit 60 of its Marder 1 A3 to this enhanced standard by 2003, with Rheinmetall Landsysteme being prime contractor.

It is claimed that this system will ensure safety against 86 per cent of all existing anti-tank mines.

The new mine protection system includes reinforcements to the lower hull at the welding seams by the application of specially formed panels.

Stowage installation and the crew compartment have been redesigned and the benches and seats have been rearranged with roof-mounted fastenings.

Spall liners have been installed and the batteries moved to the track covers. The fuel tank under the two-person turret is now smaller and extra fuel is carried in an additional tank in the former MILAN ATGW compartment at the right rear side near the turret.

When fitted with the additional armour and revised stowage, the combat weight of the Marder is now 37,000 kg, although this could be lowered by reducing the ammunition stowage.

KUKA M12 IFV

The German KUKA (today Rheinmetall Landsysteme M12 infantry fighting vehicle (IFV) was one of the three vehicles evaluated in 1998/1999 to meet the Swiss Army requirement for 310 new Schutzenpanzer 2000 ICV. This competition was won by the Swedish Hagglunds Vehicle CV 9030CH.

The M12 is basically a modified Marder 1A3 chassis fitted with the new advanced two-person turret system designated the E4.

The new E4 turret system is fitted with a Mauser 30 mm MK 30-2 stabilised cannon, all electric gunlaying system and stabilised line of sight fire-control system. Both the commander and gunner have roof hatches and the commander has seven day periscopes for all round observation.

For the Swiss trials a 7.5 mm machine gun was mounted coaxial with the 30 mm cannon and a 7.62 mm machine gun can also be installed, as can other types of cannon with a calibre range of 20 to 30 mm. The empty cartridge cases and links are ejected outside of the turret.

The gunner has a RTWL stabilised ×2 and ×8 magnification day/thermal sighting system with an integrated laser range-finder which gives the M12 a high first round hit probability against stationary or moving targets when the vehicle itself is stationary or moving.

The vehicle commander has a PERI Z 17 sight with a ×2 and ×8 magnification and display of the gunner's thermal sight.

The fire-control/sighting system has three modes of operation. These are the aiming mode, the stabilisation mode (slaving the weapon to the optical line of sight with cross-hairs always on the target) and observation mode (all-round observation through the periscope without slewing the turret).

The turret system also features an automatic munition counting system which provides for the necessary number of rounds depending on the range to the target and the type of target being engaged. Provision has also been made for the integration of an automatic target tracking device.

For ease of use, the main turret control panel is located between the two crew stations so that it can be used by commander or gunner. All functions of the turret are carried out electrically or mechanically. Turret traverse is through a full 360° with weapon elevation from −10 to +45°.

The Mauser 30 mm cannon can fire a variety of ammunition types with a calibre of 30 mm x 173. The 30 mm cannon is provided with 140 ready rounds of high-explosive ammunition and 140 rounds of armour-piercing ammunition while the coaxial machine gun has 450 rounds of ready use ammunition.

The E4 turret has been designed with growth potential, for example the addition of a command and information system or an identification friend or foe system. Turret options include electrically operated smoke grenade launchers, illumination grenade launchers and various radios and intercoms.

The M12 has a crew of three consisting of commander, gunner and driver, with eight infantrymen being seated in the rear. On the standard Marder 1A3 used by the German Army the eight infantrymen are seated back-to-back in the rear, facing outwards. On the latest M12 they are seated four down either side facing inwards and also have a redesigned stowage system. The infantry enter and leave the M12 via a power-operated ramp in the hull rear.

The M12 also has performance-enhanced suspension, a liner to reduce the effectiveness of mines and a number of ergonomic improvements in the fighting compartment.

Marder 1 with radar system

First prototypes of the Radarpanzer TÜR *(Tieffliegerüberwachungs-Radar)* based on the chassis of the Marder 1 ICV were delivered to the German Army late in 1981. This is an extensively modified raised Marder 1 hull with the turret replaced by a hydraulically operated arm, on top of which is mounted a Siemens radar with a range of some 30 km. When raised into the operating position, the radar is approximately 10 m above the ground. The hull of the RadarPz TÜR has been raised to provide increased space for the crew and equipment. The interior of the hull houses the radar unit, the display and control units, the radio and data processing system and the power supply, cooling systems and hydraulic equipment.

Armament consists of two banks of four electrically operated 76 mm smoke grenade dischargers mounted either side of the hull rear and a 7.62 mm machine gun on each cupola. In July 1988 Siemens handed over the matured MPDR 3002-S for the TÜR to the Federal German Office for Military Procurement and Technology for extensive technical trials at the Army Military Technology Department 81.

Marder 1 driver training tank

Krauss-Maffei Wegmann has delivered 36 Driver Training Tank versions of the Marder 1 infantry combat vehicle to the German Army. These have all had their two-man power-operated turrets replaced by a fixed cupola with seats for the instructor and pupils under instruction. These are called the Fahrschul SPz driver training vehicles and are based on the Marder 1 A1A2 chassis.

Roland surface-to-air missile system

In total 144 Euromissile Roland 2 surface-to-air missile systems on the Marder 1 chassis have been built for the German Army, with a further four systems delivered to Brazil. There are two Roland missiles carried in the ready to launch position, with another eight missiles being carried in the hull, four either side of the turret, ready for automatic loading.

Full details of this version are given in *Jane's Land-Based Air Defence 2001-2002*.

TAM medium tank family

Details of this vehicle, developed to meet the requirements of Argentina, are given in the MBTs and Medium Tanks section under Argentina.

Status

Production of the Marder 1 has been completed in both Argentina and Germany. Brazil has four Roland SAM systems based on the Marder 1 chassis in service.

Manufacturers

Production of the Marder was undertaken by the then: Rheinstahl, Kassel and MaK of Kiel. Design authority is the now Rheinmetall Landsysteme GmbH.

UPDATED

German M113 APC fleet

The German Army operates a fleet of over 2,500 M113A1 series vehicles and has modified many versions to meet its own specific operational requirements.

In recent years Germany has transferred surplus M113 series vehicles to a number of countries including:

Country	Quantity
Greece	300 in 1994/1995
Portugal	50 in 1994
Turkey	187 in 1993
United Nations	25 in 1994

The official German Army designations of the M113/M577 versions are given below, although they are also referred to as the M113A1G/M113A2G.

German Army M113A1G in 120 mm self-propelled mortar configuration (Michael Jerchel) **2000**/0064811

All of these feature Diehl tracks, are armed with a roof-mounted 7.62 mm MG3 machine gun and are fitted with a bank of eight 76 mm electrically operated smoke grenade launchers on the front of the vehicle which fire over the forward arc.

The official designators for the M113 series used by the German Army are as follows:
M 113 A 1 is fitted with the 6V-53 Detroit Diesel engine;
M 113 A 2 is fitted with the 6V-53 Detroit Diesel engine and improved US cooling system/radiator;
EFT = external fuel tanks;
GE = fitted with SEM 25/30 radios;
GE AO = fitted with SEM 80/90 radios.

Original overall prime contractor for the M113 series of vehicles in German Army service was Henschel Wehrtechnik based in Kassel which, in 2000, became Rheinmetall Landsysteme GmbH.

This company developed and produced an integrated fire-control system for artillery batteries called the IFAB *(Integrierte Feuerleitmittel fur Artillerie Batterie).*

This consists of three main components, a forward observer vehicle on an M113A1G chassis, a fire direction centre vehicle on an M577A1G chassis and data installation kits for the actual weapons.

The artillery observation vehicle is called the *Beobachtungs-panzer Artilleries* (Beob PzArt) and 320 vehicles were delivered to the German Army between 1982 and 1985.

This vehicle is fitted with a PERI-D11 double-periscope sight which incorporates a laser range-finder, an FNA 615 land navigation system and a data input/output terminal linking it with the battery fire-control centre.

The first IFAB command post vehicle, based on a modified M577A1G chassis, and the first armoured artillery fire-control vehicle, based on the M113A1G chassis, were handed over to the German Army in April 1987.

There are three versions of the M577A1G: command cell, teleprinter cell and information cell; 220 vehicles were converted.

There are also 154 armoured artillery fire-control vehicles fitted with a Falke 2 computer, three data terminals, a matrix printer, a teleprinter, a radio and a field telephone.

437 120 mm self-propelled mortar vehicles are in service. The 120 mm Tampella mortar is mounted in the back and fires to the rear. It has a traverse of 18.5° left and 25° right, with elevation limits being from +45 to +80°. In total, 63 120 mm mortar bombs are carried but when afloat this number has to be reduced to 23.

The now Rheinmetall Landsysteme also offers a low-wear track for the M113 series of vehicle (the Diehl B213), an M113A2 suspension, cooling, rear-mounted external fuel, hydrostatic steering system, more powerful diesel engine, additional armour, NBC protection system and various turret systems with weapons of up to 20 mm in calibre.

The German Army has also taken delivery of 301 M548A1G tracked carrier modified by Krauss-Maffei (which in January 1999 became Krauss-Maffei Wegmann) to take the Dynamit Nobel Skorpion (MiWS) mine launching system, with final deliveries taking place in 1989. The Greek Army has taken delivery of a number of MiWS systems from Germany.

In the future, the M113 series will be supplemented in German Army service by the future GTK wheeled armoured personnel carrier, details of which are given in the Armoured personnel carriers (wheeled) section under International. The artillery observation post version is expected to be replaced by a new vehicle based on a modified Leopard 1 series MBT chassis. Details of this are given in the Leopard 1 MBT entry earlier in this volume. As of mid-2001, no production order for this version had been placed.

FFG M113 upgrade

Under the leadership of FFG Germany, an automotive upgrade has been developed for specialised versions of the M113 used by the German Army.

With the down sizing of the German Army it is expected that the following versions will remain in service:

Quantity	Designation	Mission
612	Fue/FuPz M 113 A2GE AO	command and control vehicle
31	RiFuMuxPz M 113 A2	communications version
100	SchrFuPz VHF M 113 A2GE	communications version
86	TRFz RechVerbArt M 113	artillery data processing version
11	Fue/FuPz M 113 A2GE HF 400 W	communication/control version
16	SchrFuPz HF M113 A2GE	communications version
139	FltPz (Mrs) M 113 A1	fire-control vehicle, mortar
28	TrFz Green Archer M 113 A1GE	fitted with Green Archer location radar
70	SchulPz M 113	driver training vehicle
143	FlgLtPz M 113 A2GE AO	forward air controller
437	PzMrs M 113 A2 (EFT) GE	120 mm mortar carriers
220	GefStdFueFz A 577 A2GE	command/control version
438	KrKw M 113 A2GE	ambulance
154	FltPz (Art) M 113 A1GO	fire-control vehicle artillery
73	TrFz RATAC M 113 A1GE AO	carrier for RATAC radar, also known as ABRA
320	BeobPzArt A1GE AO	forward observer vehicle

Following trials with a prototype upgraded vehicle, another five examples were ordered in 1996 and delivered in 1997. The first production order was for 200 units with the second expected to be for a further 100 units. In early 1999 the Danish Army ordered 97 M113 upgrade packages from FFG (see Denmark), first six were completed in 2000 and the Danish Material Command has since ordered another batch of 160 vehicles.

The actual upgrade is carried out by FFG who have been engaged in upgrading and overhauling a wide range of tracked vehicles for the German Army for many years.

The heart of the upgrade includes the installation of a new power pack which consists of an MTU 6V 183 TC

TrFz Green Archer M113 A1GE (Michael Jerchel) ***2001***/0093577

M113A1GEAO Fire control vehicle (Michael Jerchel) ***2002***/0110660

M113A2GE of German Army in KrKw armoured ambulance configuration and with trim vane removed (Michael Jerchel) ***2002***/0110661

The FFG-developed Waran is essentially a stretched and upgraded M113 series armoured personnel carrier **2001**/0093574

22 diesel engine, based on the commercial Mercedes-Benz OM 441 LA, coupled to a ZF LSG 1000 steering and transmission system.

The 6V 183 TC 22 is a six-cylinder diesel with exhaust turbocharging and charge air cooling which develops 295 hp while meeting the Euro II exhaust emission standards.

This engine has a growth potential of up to 394 hp and is a member of a complete family of MTU 6-, 8- and 12-cylinder diesel engines, all of which share many common components.

The M113A2 pulled fan cooling system has been retained with a few modifications and no alterations have been made to the internal bulkheads.

The ZF LSG 1000 steering and shift transmission consists of a hydrodynamic torque converter with bridge clutch, downstream six-gear power shift transmission and a hydrostatic steering transmission.

FFG, in addition to installing the new power pack, has also developed and integrated the new steering and brake system, upgraded the drive system and redesigned the driver's position. The driver now steers the vehicle using a small steering wheel rather than tillers, which gives more precise steering as well as reducing driver fatigue.

The brake system consists of a dual-circuit power-assisted system with hydraulic boost which is separate from the power pack and is operative even when the power pack has been removed.

With a combat weight of 15 tonnes, the FFG upgraded M113 has a maximum road speed of 61.2 km/h and travelling at this speed a range of over 720 km can be achieved. Acceleration from 0 to 50 km/h is 18 seconds while the vehicle can pivot turn through 360° in 8.4 seconds.

According to FFG, its upgraded M113 has improved braking power, increased steering safety, lower exhaust noise and emissions, reduced fuel consumption and therefore longer range, improved acceleration and greater cross-country speed and mobility, and is easier to handle.

The German Army only intends to install the new power pack and other FFG improvements to part of its fleet of M113 series vehicles.

FFG, teamed with MTU and ZF, Rheinmetall Landsysteme (at that time KUKA), GLS and Diehl, is now offering potential customers a number of further enhancements to the KUKA M113 to extend its capabilities even further.

The potential user can select those parts which meet his current operational, or indeed, financial requirements with the option of further upgrades in the future.

The vehicle commander's cupola of the basic M113 is fitted with an externally mounted 12.7 mm M2 machine gun which is aimed and fired by the commander with his head and shoulders being exposed above the turret ring.

Large and better armed turrets can be fitted. In the case of, for example, a two-man 20 or 25 mm turret, the internal volume taken up by the turret basket means that fewer troops or less equipment can be carried.

More recently, FFG has developed another stretched version of the M113 with an additional road wheeled station either side.

This is called the Waran and also features a higher level of armour protection, especially against anti-tank mines, and the installation of a new power pack.

The latter consists of an MTU 6V 1893 TE 22 diesel engine developing 400 hp coupled to a ZF 1000 six-speed automatic transmission. The final drives have been replaced and the suspension upgraded for improved cross-country mobility.

Cross-country mobility of the upgraded M113 can be improved by upgrading the suspension system which is the responsibility of GLS, a subsidiary of Krauss-Maffei (which in January 1999 became Krauss-Maffei Wegmann), prime contractor for the Leopard 2 MBT.

The M113 suspension upgrade includes the installation of new telescopic shock-absorbers, hydraulic bump stops, new support rollers, new axles for the idler which have been repositioned and improved roadwheel arms.

The existing track can be replaced by the Diehl Remscheid Type 513 track, with replaceable rubber pads, which is claimed to have a much longer life than the existing track.

Other possible options include additional passive armour system from Clouth of Germany and a fire detection and suppression system from Intertechnique of Austria.

UPDATED

Rheinmetall Landsysteme HWK 11 armoured personnel carrier

Development

There were 40 of these vehicles built by the now Rheinmetall Landsysteme (previously Thyssen Henschel) between 1964 and 1965 and these remain in service with the Mexican Army. Full details were given in *Jane's Armour and Artillery 1983-84,* pages 302-303.

The NIMDA company of Israel has modernised a single example of the HWK 11 APC with the original power pack replaced by a new one consisting of a Model 6V-53 Detroit Diesel developing 210 hp coupled to an Allison Transmission TX-100 automatic transmission. This gives the HWK 11 a maximum speed of 60 km/h and a maximum range of 400 km.

Recent information has indicated that the complete fleet of Mexican HWK 11 vehicles has been upgraded. These vehicles now have a distinct exhaust pipe on the right side of the hull.

Some of these vehicles have recently been upgraded and fitted with French one-man turrets, armed with a 7.62 mm machine gun, which were previously fitted to Giat Industries AMX VCI Infantry combat vehicles. These turrets are also installed on many of the ex-Belgian Army AMX VCI full-tracked vehicles also in service with Mexico.

Status

Production complete. In service only with Mexico.

HWK 11 full-tracked APC shown in 1994 and fitted with French turret armed with 7.62 mm machine gun (Julio Montes) 0018425

Manufacturer

Rheinmetall Landsysteme GmbH.

VERIFIED

Rheinmetall Landsysteme Breitkettenfahrzeug R 439 APC

Development

The R 439 full-tracked APC has been developed as a private venture by Rheinmetall Landsysteme (previously known as Thyssen Henschel) to transport men and equipment over marginal terrain especially swamp, sand, mud and snow.

It is basically the German Kassbohrer Flexmobile FM 23.150K all-terrain chassis, already produced for a number of civilian applications. Fitted with a fully armoured body, it provides its occupants with protection from 7.62 mm small arms fire and shell splinters. A higher level of ballistic protection can be provided by using compound armour.

By 2001, volume production of the R 439 had yet to commence. This vehicle was originally referred to as the TH 439 then the H 439, but as it is now part of Rheinmetall Group it is now called the R 439.

Description

The diesel engine is at the front of the vehicle with the personnel compartment to the rear. The commander and driver are seated at the front of the vehicle with a large bulletproof window to their front. When required, these can be covered by a shutter hinged at the top.

Specifications
Crew: 1 + 9
Weight:
(loaded) 6,500 kg
(unloaded) 5,450 kg
Power-to-weight ratio: 23 hp/t
Ground pressure: 0.16 kg/cm^2
Length: (overall) 4.235 m
Width: 2.27 m
Height: 2.36 m
Ground clearance: 0.33 m
Track width:
(summer track) 700 mm
(winter track) 825 mm or 925 mm
Max road speed: 50 km/h
Fuel capacity: 116 litres
Road range: 250 km
Gradient: 80%
Side slope: 40%
Vertical obstacle: 0.45 m
Trench: 1.50 m
Fording: 0.75 m
Engine: Mercedes-Benz OM 352A water-cooled 6-cylinder turbocharged diesel developing 150 hp
Transmission: hydrostatic
Steering: hydrostatic
Turning radius: pivot
Armament: optional
NBC system: optional
Night vision equipment: optional

Rheinmetall Landsysteme Breitkettenfahrzeug R 439 armoured personnel carrier 0018426

A forward-opening door is provided in each side of the hull and in its upper part is a bulletproof window.

In either side of the rear troop compartment are two bulletproof windows and in the hull rear is a door that opens to the right with a window above. The troop compartment is provided with seats, extractor fans and a heater.

The vehicle uses rubber tracks that can be fitted with bolt-on double steel web plates for difficult terrain; one width of summer track and two widths of winter track, all steel, are available. The main roadwheels use a double-track guided system and are self-cleaning. Either side of the vehicle are five dual roadwheels with drive sprocket at the front, idler at the rear and three track-return rollers.

Variants

The interior and interchangeable lateral wall and roof segments enable the R 439 to undertake a wide range of roles including APC, weapons carrier, ambulance, command post vehicle, tractor and specialised reconnaissance vehicle.

Options include a KHD air-cooled diesel engine developing 152 hp in place of the Mercedes-Benz OM 352A diesel engine, hatches for crew in the rear part of the roofing, firing ports with vision blocks within the sidewalls and rear door and a hatch for a 7.62 mm machine gun above the commander's seat.

Status

Development complete. Ready for production.

Manufacturer

Rheinmetall Landsysteme GmbH.

UPDATED

Rheinmetall Landsysteme Wiesel 2 Multi-Purpose Carrier

Development

The prototype of the Wiesel 2 light armoured air-transportable vehicle, developed by the now Rheinmetall Landsysteme as a private venture with the assistance of a number of subcontractors, was completed at Kiel in mid-1994.

Between 1989 and 1992, the now Rheinmetall Landsysteme built a total of 345 Wiesel 1 (covered in the Reconnaissance vehicles section) Armoured Weapon Carriers (AWCs) for the German Army.

These Wiesel vehicles are armed with a Raytheon Systems Company TOW ATGW system (three-man crew) or a KUKA (now Rheinmetall Landsysteme) turret armed with a 20 mm Rheinmetall MK 20 Rh 202 cannon (two-man crew). More recently a small number of Wiesel 1 vehicles have been modified into the reconnaissance role for the German Army by Rheinmetall Landsysteme.

Market studies by the company showed that there is a requirement for a new version of the Wiesel with an increased volume and payload to enable it to carry out a wide range of additional missions on the battlefield, including the support of the existing Wiesel 1 vehicles which are purely weapon carriers.

The result is the Wiesel 2, which has a stretched hull with an additional roadwheel either side and a slightly higher roof line.

The chassis of the Wiesel 2 has been selected as the basis for the German Army's new LeFlaSys Stinger self-propelled air defence system (covered under Variants) as this is the first production application of the Wiesel 2 system. Production of the Wiesel series was originally undertaken in Kiel but has now been transferred to Unterluess.

Description

The layout of the Wiesel 2 is similar to the Wiesel 1, with the engine compartment front left, driver front right and the remainder of the vehicle for the troop compartment.

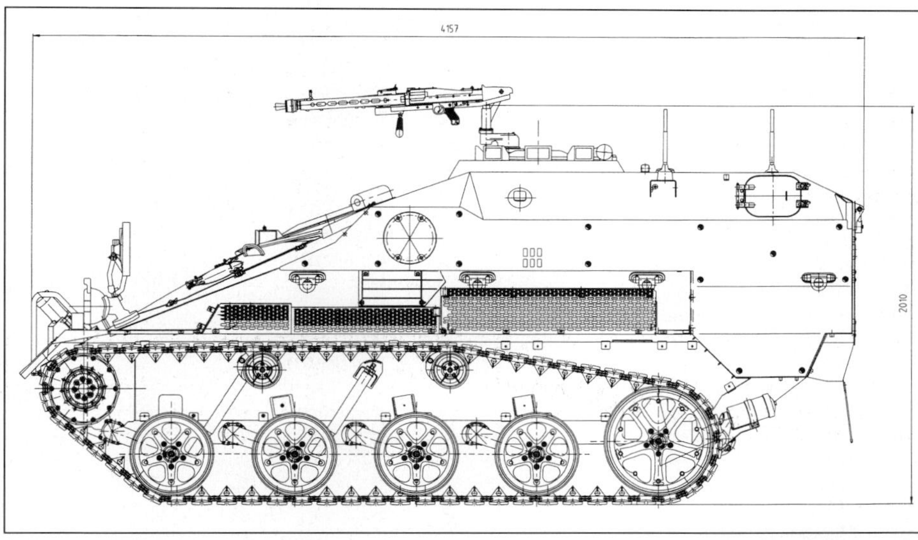

Detailed side drawing of Wiesel 2 in troop carrier configuration and fitted with 7.62 mm MG3 machine gun ***2002***/0110662

The troops enter the vehicle through a large door in the hull rear that opens to the left. The commander's position is to the rear of the driver and on the prototype the commander is provided with a Krauss-Maffei Wegmann cupola armed with an externally mounted 7.62 mm MG3 machine gun.

The Wiesel 2 has twice the internal volume of the earlier Wiesel 1, with the prototype being configured as a half squad carrier.

This version can carry six people including the driver and commander, with the commander manning the Krauss-Maffei Wegmann 7.62 mm MG3 machine gun mount which enables the weapon to be fired from under complete armour protection.

Using the Wiesel 2 as a basis, the company has proposed a wide range of versions including command post, medical evacuation vehicle, radar carrier, reconnaissance vehicle, NBC vehicle and as a weapons platform.

A full-scale mock-up of the ARGUS 2 reconnaissance vehicle has already been completed which features a three-man crew, a retractable mast-mounted sensor package and touch menu displays to allow for extended operations.

A CH-53 helicopter, as used by the German Army, can carry two Wiesel 2 vehicles internally and the Wiesel 2 can also be slung underneath helicopters such as the CH-53 and CH-47.

The new Wiesel 2 shares a number of common components with the earlier Wiesel 1 but has a new Diehl Type 622 track and a new power pack.

The new power pack consists of the new low-emission Audi 1.9 litre four-cylinder turbocharged diesel developing 109 hp, which is already in volume

Wiesel 2 in the ambulance configuration **2000**/0064813

Wiesel 2 command and control vehicle showing raised roof and extended rear **2000**/0064814

production for the Audi 100 car, coupled to a ZF LSG 300/4 automatic transmission. The engine has an electronic control system that automatically tunes the engine and transmission output depending on the vehicle weight.

Variants

Wiesel 2 120 mm mortar

Details of this system, which remains at the prototype stage, are given in the Self-propelled mortars section.

Wiesel 2 ASRAD air defence system

Four German companies were originally awarded fixed-price study contracts for the LeFlaSys *(Leichte Flugabwehrsystem)* and in 1995 STN ATLAS Elektronik was awarded the contract, which included the design, development and manufacture of preproduction systems plus the full production order.

By early 1997, STN ATLAS Elektronik had completed one weapon platform Ozelot, one platoon command post and one reconnaissance vehicle 'low level'. The first two are based on the latest Rheinmetall Landsysteme stretched Wiesel 2 light tracked armoured vehicle chassis while the reconnaissance vehicle is based on the Mercedes-Benz Wolf (4 × 4) vehicle.

The weapon platform Ozelot has an unmanned mount at the rear with four European-built Raytheon Systems Company Stinger fire-and-forget Surface-to-Air Missiles (SAM) in the ready to launch position.

When travelling, the missile launcher system is retracted to allow the complete weapon platform to be driven into the CH-53H transport helicopters of the German Army.

Although the weapon platform Ozelot would normally use the Stinger SAM, one of the requirements of the German Army was that it should also be capable of firing the Russian SA-16 SAMs, as the German Army has large quantities of these available for training purposes.

Each weapon platform Ozelot is also fitted with communications equipment and a Thales Optronics passive Air Defence Alert Device (ADAD) which has been in service with the British Army for some years in both static (for example tripod-mounted) and vehicle-mounted (Starstreak High-Velocity Missile) roles.

The Ozelot platoon command post is fitted with an improved version of the Swedish Ericsson Microwave Systems HARD (Helicopter & Airplane Radio/Radar Detection) 3-D radar system which in its original form is a key part of the Swedish Army's SAAB Bofors Dynamics RBS-90 SAM system.

A third LeFlaSys vehicle, also on a Wiesel 2 chassis, is the battery command post. It is expected that the single LeFlaSys regiment will have three batteries with each of these comprising battery HQ unit and three platoons.

The reconnaissance vehicle 'low level', based on a Mercedes-Benz (4 × 4) Wolf, is foreseen as an interface for allied forces, receiving and transmitting reconnaissance data via the so-called FAST interface.

Each platoon will have one platoon command post with the HARD radar and control five Ozelot weapon platforms.

The total order, including the prototype systems, consists of 50 weapon platforms, 10 platoon command posts (with radar) and seven battery command post vehicles, all on the Wiesel 2 multipurpose carrier chassis.

The first production Wiesel 2 multipurpose carrier chassis was completed in 1999 and then sent to Krauss-Maffei Wegmann for internal work, before being sent to STN ATLAS Elektronik for final integration and delivery to the customer.

Full details are given in *Jane's Land-Based Air 2001-2002*, pages 126-127. The first platoon of the German Army equipped with this system became operational in the first quarter of 2001.

Specifications

Crew: up to 6 (depending on version)
Combat weight: 4,100 kg
Power-to-weight ratio: 26.58 hp/t
Ground pressure: 0.41 kg/cm^2
Length: (overall) 4.20 m (depending on version)
Width: (overall) 1.852 m (depending on version)
Height: 1.7 to 2.11 m (depending on version)
Ground clearance: 0.30 m
Track: 1.62 m
Track width: 200 mm
Length of track on ground: 2.427 m
Gradient: 60%
Side slope: 30%
Trench: 1.5 m
Fording: 0.5 m
Max road speed: 70 km/h
Road range: 550 km
Suspension: torsion bars, hydraulic dampers
Brakes:
(service) fully hydraulic dual-circuit brake system
(parking) mechanical
Engine: Audi 4-cylinder TDI diesel with exhaust turbocharger developing 109 hp
Transmission: ZF LSG 300/4 automatic with hydrostatic cross-drive steering, 4 forward and 2 reverse gears
Electrical system: 24 V
Batteries: 2 × 12 V, 45 Ah
Armament: 1 × 7.62 mm MG3 MG (depending on version)
NBC system: optional
Night vision equipment: optional

Weapon platform Ozelot on Wiesel 2 chassis with launcher elevated and four Stinger SAMs in ready to launch position (left) and platoon command post on Wiesel 2 chassis with Ericsson Microwave Systems HARD 3-D radar in elevated position (right) 0005004

Wiesel 2 ambulance

Late in 1997, Rheinmetall Landsysteme completed the prototype of the Wiesel 2 armoured ambulance and this is now being evaluated by the German Army.

This model is unarmed and is fitted with doors that open at the rear. It has sufficient space for a driver, leader of the medical party, one casualty lying on the stretcher and two seated wounded. The actual ambulance equipment was supplied by the German company of Binz. The German Army requirement is for 18 ambulance vehicles based on the Wiesel 2 multipurpose carrier chassis.

Wiesel 2 command and control vehicle

Rheinmetall Landsysteme has delivered one command and control vehicle to the German Army based on the Wiesel 2 multipurpose carrier chassis and this started trials in 1998. The German Army requirement for this model of the Wiesel 2 is for nine units.

This has been developed to meet the requirements of the German airborne forces and has two workstations HEROS including a 7 inch monitor, latest state-of-the-art VHF and HF radio equipment as well as a GPS system. It would normally have a crew of three: driver, commander and battalion commander.

Wiesel 2 ammunition vehicle

This has also been developed to meet the requirements of the German Army and has a payload of up to 1,000 kg and would typically be used to support the Wiesel 1 armoured weapon carriers. The German Army requirement for this version is 13 units.

Wiesel 2 Syrano robotic vehicle
The Wiesel 2 chassis is used as the basis for the Syrano robotic vehicle developed in France as a technology demonstrator.

Engineer reconnaissance
The German Army has a requirement for an engineer reconnaissance version of the Wiesel 2, the requirement being for six units. The German Army has taken delivery of one vehicle for trials purposes.

PRIMAS
Two so-called digitalised Wiesel 2 vehicles were delivered to the German MoD early in 1998. One is a robot carrier and the second is used as a control station for use in the PRIMUS (Programme for Intelligent Mobile Unmanned Systems) project.

Wiesel 2 ARGUS
Rheinmetall Landsysteme has built a full-scale mock-up of the ARGUS 2 Airtransportable Armoured Surveillance and Reconnaissance Vehicle as a private venture.

Typical roles for this vehicle would include reconnaissance, scout missions, battlefield surveillance, artillery observation, command missions and security missions.

To carry out its missions, the vehicle is fitted with a mast-mounted sensor pod developed by STN ATLAS Elektronik and a command and control system developed by Rheinmetall Landsysteme.

ARGUS also features a hybrid navigation system, intercommunication radio system, NBC system and electric power supply which includes an additional generator.

LLX experimental vehicle
This light air-transportable armoured experimental vehicle has been developed by Rheinmetall Landsysteme and features all-electric drive and power supply.

Digitisation Wiesel 2
This has been developed by Rheinmetall Landsysteme and has proved that it is possible to obtain authorisation for using a safety critical drive-by-wire vehicle for road traffic and also meet the highest safety demands.

The system concept combines a decentral data processing structure with a decentral power distribution structure, thus ensuring the modularity required for future vehicles.

The digitisation concept is based on a CAN (Controller Area Network), ring shaped data bus structure. In addition, a power bus, also ring shaped, has been installed in the vehicle.

Status
In production for the German Army as part of the LeFlaSys air defence system.

Manufacturer
Rheinmetall Landsysteme GmbH.

UPDATED

Schutzenpanzer 3 infantry combat vehicle

Development/Description
In the early part of the 21st century the German Army was expected to field a new generation of tracked combat vehicles as part of its New Generation Platform (NGP) series of full-tracked armoured vehicles.

These would have included a new Main Battle Tank (MBT) to replace the current Krauss-Maffei Wegmann Leopard 2 MBT, a new Infantry Fighting Vehicle (IFV) to replace the current Rheinmetall Landsysteme Marder 1 and a new self-propelled air defence system to replace the current Krauss-Maffei Wegmann Gepard twin 35 mm self-propelled anti-aircraft gun system.

As the NGP has now been cancelled, there are now no immediate plans for the replacement of the Leopard 2 or Gepard. The German Army is now formulating its requirements for a new vehicle to replace the current Rheinmetall Landsysteme Marder 1 IFV, which originally entered service as far back as 1971.

Since then, Marder 1 has been continuously upgraded but still retains its two-person turret armed with a 20 mm cannon, which is now inadequate to combat many potential battlefield targets.

The Marder 1 was to have been replaced by the Marder 2 with a prototype being completed some years ago by the now Krauss-Maffei Wegmann, but this was cancelled in 1992 shortly after the prototype was completed.

It was expected that 699 Marder 2 vehicles would be built. The prototype weighed over 40 tonnes and was fitted with a two-person turret armed with a Rheinmetall MK 35/50 Rh 503 35 mm/50 mm cannon.

Today there are only two companies in Germany capable of developing a vehicle such as the Schutzenpanzer 3, Krauss-Maffei Wegmann and Rheinmetall Landsysteme.

The new Schutzenpanzer 3 may be fitted with the MK 35/50 Rh 503 35 mm/50 mm weapon and be capable of being fitted with different armour packages to meet operational requirements ranging from peacekeeping to full scale mechanised warfare. It may well have an electric drive system and a decoupled running gear that will reduce noise and vibration.

Status
German government studies.

Manufacturer
Not selected.

NEW ENTRY

GREECE

ELBO Kentaurus armoured infantry fighting vehicle

Development
ELBO, Hellenic Vehicle Industry SA, has developed, as a private venture, a new armoured infantry fighting vehicle called the Kentaurus. The first prototype of this was completed late in 1998 and shown in public for the first time in October 1998.

The Kentaurus has been designed as the basis for a complete family of full-tracked armoured fighting vehicles which will be able to undertake a wide range of battlefield roles.

The manufacturer has claimed a number of advantages for the Kentaurus IFV, including a low profile for improved battlefield survivability; smallest width for maximum mobility; air transportability in a Lockheed Martin C-130 Hercules transport aircraft; highest infantry section capacity (option); lethal firepower; high mobility in hot temperature environment; suitability for combat under day and night conditions; high first round hit probability in comparision with other fielded fire-control systems due to the independent line of sight option; digital fire-control system with built-in test equipment as an option; laser warning system with direction indicator option; and finally the option of a two-round Raytheon Systems Company Stinger fire-and-forget surface-to-air missile system mounted on the left side of the turret.

Hellenic Vehicle Industry SA hope that the Kentaurus will meet the future operational requirements of the Hellenic Army and will eventually replace the ELBO built Leonidas armoured personnel carrier covered in a separate entry.

Late in 2000, the Kentaurus underwent a three-week trial with the Greek Army. With the projected introduction into service with the Greek Army of a new MBT it is expected that an initial batch of 140 to 150 Kentaurus AIFV's will be acquired.

First prototype of the ELBO Kentaurus armoured infantry fighting vehicle **2001**/0093582

Description
The hull of the Kentaurus AIFV is of all-welded steel armour construction with the driver seated front left, power pack to his right, turret in the centre and offset to the right and the troop compartment at the rear.

The standard vehicle has ballistic protection against 25 mm APDS projectiles fired from a range of 400 m over the frontal arc and full protection from 7.62 × 51 mm small arms fire through a full 360°.

If the user requires a higher level of protection, additional add-on armour could be mounted on the hull and turret. Against indirect fire, the hull provides protection against 155 mm HE projectiles. In addition, the vehicle provides a high level of protection against non-direction anti-personnel mines. The vehicle also has a low radar and infra-red signature.

As an option, and in order to reduce the behind-armour effects of penetrating projectiles, fragments and hollow charges, the crew compartment and turret can be fitted with spall liners.

The driver enters and leaves his compartment via a hatch cover above his position. For driving closed down there is an STN ATLAS Elektronik ODS uncooled thermal imager combined with a CCD day camera. As a back-up one wide-angle Helio 2100 Mk 29 day periscope with a laser filer is provided.

In addition, the driver has a fully adjustable seat and steers the vehicle using a steering wheel. The commander is seated to the immediate rear of the driver and he leaves the vehicle via the troop compartment. If required, a roof hatch can be provided. As an option, the vehicle commander can be provided with a monitor and a control box with joystick in co-operation with the fire-control system, in order to have the same view as the gunner and to take control and move the turret if required.

The power pack consists of the German MTU 6V 183 TE22 V-90 liquid-cooled turbocharged diesel developing 420 hp at 2,300 rpm coupled to a ZF LSG 1000 fully automatic six-speed transmission with integrated steering system and hydraulic brake.

The final drives are made by Renk. The complete power pack, consisting of engine, transmission and part of the cooling system, can be removed from the Kentaurus in about 15 minutes.

The Type E8 one-man turret has been developed by Rheinmetall Landsysteme (at that time KUKA) and is of a low-profile design. Main armament comprises a Mauser 30 mm MK 30F cannon which has been manufactured in Greece by EBO for some years with 200 rounds (100 + 100) of ready use ammunition being provided, with another 196 rounds in reserve. The gunner can select either single shot or burst modes of fire.

Other weapons, however, can be installed in place of the Mauser cannon, for example, the US The Boeing Company 30 mm Bushmaster II, which has already been built in production quantities.

A 7.62 mm MG3 machine gun, manufactured by EBO, is mounted coaxial with the main armament and is provided with 500 rounds of ready use ammunition with another 500 rounds in reserve.

Other coaxial weapons can be installed, for example, a 12.7 mm machine gun or a 40 mm automatic grenade launcher. Turret traverse and weapon elevation is all electric with manual controls for emergency use.

Mounted either side of the turret is a bank of four 76 mm smoke/fragmentation grenade launchers which are manufactured by Pyrkal and are fitted as standard on all Greek armoured vehicles.

In the basic turret configuration, the Thales Optronics Sabre day/night sighting system is fitted, but as an option the turret can be fitted with a state-of-the-art TOXOTIS digital fire-control system with an independent line of sight which has been manufactured by STN ATLAS Elektronik.

TOXOTIS consists of a two-axis stabilised sensor platform which incorporates a two field-of-view second-generation thermal sight, CCD camera, eyesafe laser range-finder, inertial sensors, weapon stabilisation system and a digital fire-control computer. The gunner is provided with a monitor.

Other turret options include a laser-warning system with a directional indicator; integrated command and control system; launcher rail on left side of turret for two Stinger fire-and-forget surface-to-air missiles; remote control of the turret via a joystick and monitor which enables the infantry section to be increased from eight to 10 men, so that the total crew will consist of 13 men.

The troop compartment is at the rear of the hull with five infantrymen seated on the left side and three on the right side (or six for the increased infantry section). The infantry section commander, seated at the right corner of the vehicle, normally dismounts with the troops.

As an option, single foldable seats for all of the infantrymen can be installed with four-point fastener safety belts for the whole crew.

The infantry section compartment can be entered from the rear by means of two hinged doors which can be opened, closed and secured by one man. Above the doors are two Helio 2100 Mk 29 day periscopes for observation of the terrain, prior to opening the doors.

Suspension is of the torsion bar type with six dual rubber-tyred roadwheels either side, drive sprocket at the front, idler at the rear and three track-return rollers.

First prototype of the ELBO Kentaurus armoured infantry fighting vehicle from the rear with doors open
***2001**/0093583*

Specifications

Crew: 3 + 8 (or 3 + 10)
Combat weight: 19,800 kg
Power-to-weight ratio: 21.12 hp/t
Ground pressure: 0.74 kg/cm²
Length:
(overall) 6.28 m
(hull) 5.98 m
Width:
(with skirts) 2.55 m
(without skirts) 2.50 m
Height:
(hull roof) 1.70 m
(overall) 2.45 m
Firing height: 2.08 m
Ground clearance: 0.42 m
Track: 2.12 m
Track width: 380 mm
Length of track on ground: 3.3 m
Max road speed:
(forward) 75 km/h
(reverse) 42 km/h
Fuel capacity: 390 litres
Cruising range (road): 500 km
Fording: 1.15 m (without preparation)
Gradient: 75%
Side slope: 35%
Vertical obstacle: 0.85 m
Trench: 1.15 m
Turning radius: pivot
Engine: MTU 6V 183 TE22 V-90 diesel developing 420 hp at 2,300 rpm
Transmission: ZF LSG 1000, automatic, 6 forward and 2 reverse gears with integrated steering system and hydraulic brake
Suspension: torsion bar with trailing arm for each roadwheel, rotary dampers for first, second and sixth roadwheels
Electrical system: 24 V
Batteries: 6 × 12 V, 100 Ah each
Armament:
(main) 1 × 30 mm Mauser EBO MK 30F cannon
(coaxial) 1 × 7.62 mm MG3 MG
Ammunition:
(main) 200 × 30 mm ready use (+ 196 reserve)
(coaxial) 400 × 7.62 mm ready use (+ 1,200 reserve)
Smoke grenade launchers: 2 × 4 76 mm
Smoke grenades: 8 + 8
Gun control equipment: electric/manual (by gunner)
Gun elevation/depression: +45/–10°
Turret traverse: 360°
Gun stabiliser:
(vertical) yes
(horizontal) yes
NBC system: yes
Night vision equipment: yes
Air transport: Lockheed Martin C-130 Hercules

Rotary dampers are installed at the first, second and sixth roadwheel stations and an automatic adjustable hydraulic track tensioning system, operated by the driver, is provided. All of this is provided by Horstman Defence Systems.

The vehicle is fitted with 380 mm Diehl Type 224M tracks which are of the double-pin type and have replaceable pads.

Standard equipment includes a fire detection/suppression system for the engine compartment, diesel-operated heater unit for the crew compartment and a ventilation system.

As an option, the vehicle can be fitted with an NBC and air conditioning system developed by Aircontrol Technologies and a fire detection and suppression system for the crew compartment.

Variants

The basic chassis can be used for a wide range of other roles and can be fitted with different turrets such as, light tank, air defence, ambulance, anti-tank missile, armoured engineer, armoured logistics carrier, command post vehicle, mortar carrier and recovery.

Status

First prototype completed late in 1998 and now undergoing company trials.

Manufacturer

ELBO, Hellenic Vehicle Industry SA.

UPDATED

ELBO Leonidas armoured personnel carrier

Development/Description

This is essentially the Austrian Steyr-Daimler-Puch 4K 7FA G127 APC (full details of this vehicle are given earlier in this section under Austria) manufactured under licence in Greece for the Greek Army. Its physical characteristics are identical to the original Austrian version. The first Greek Leonidas 1 armoured personnel carrier was completed in 1982 and by early 1985 100 vehicles had been built for the Greek Army as well as a quantity of vehicles for Cyprus.

Following extensive trials, ELBO selected the German ZF 6 HP 500 automatic transmission as a replacement for the current ZF 6-S80 manual transmission with six forward and one reverse gears installed in the locally built Leonadas APC.

A total of 64 6 HP 500 automatic transmissions have been delivered by ZF to Greece for installation in new build vehicles as well as being retrofitted into older vehicles, with the second order being for 57 transmissions.

It is expected that the whole Greek Army fleet will eventually have their current manual transmissions replaced by the new ZF 6 HP 500 automatic transmission, the installation of which reduces driver fatigue.

Leonidas 2 APC-AIFV

Early in 1987, an order for an additional 344 Leonidas 2 APCs was placed by the Greek Army.

Most of the vehicles in the new order were similar to the existing Leonidas 1 APC but with an additional automatic fire suppression system, commander's rotating periscope and improved smoke grenade dischargers.

By mid-2001 it was estimated that production had reached around 500 vehicles of which 350 were for the Greek Army and 150 for Cyprus. It is believed that the latter figure includes a batch of 84 vehicles, the first of which was handed over in December 1995. According to figures released by the United Nations, Greece exported the following quantities of Leonidas APCs between 1992 and 1998 to Cyprus:

Year	Quantity
1992	49
1993	32
1994	nil
1995	nil
1996	26
1997	39
1998	nil

In 1992, three Leonidas 2 Armoured Infantry Fighting Vehicles (AIFVs) were evaluated by the Greek Army, each with a different turret. These were:

ELBO Leonidas 2 armoured personnel carrier without armament installed ***2000***/0064816

(1) Spanish Santa Barbara TC-13 turret armed with The Boeing Company M242 25 mm Chain Gun and a 7.62 mm MG. This turret is still at the prototype stage.

(2) DAF/FMC Enclosed Weapon Station armed with an Oerlikon Contraves 25 mm KBA-BO2 cannon and a 7.62 mm machine gun. This turret has already been produced in quantity for Belgian and Royal Netherlands AIFVs. More recently this turret has also been adopted by Turkey.

(3) Norwegian NFT T25 turret armed with a 25 mm Mauser Model E cannon and a 7.62 mm machine gun. This turret is also still at the prototype stage.

The chassis of the Leonidas has also been fitted with the Brazilian ENGESA ET-90 turret armed with the EC-90 90 mm gun and the French Giat Industries TS 90 turret armed with a long-barrelled 90 mm gun.

Status

Production complete. In service with Cyprus and Greece.

Manufacturer

ELBO, Hellenic Vehicle Industry SA.

UPDATED

INDIA

Sarath infantry combat vehicle

Development/Description

The Ordnance Factory Medak, owned by the Indian government, is currently manufacturing the Kurgan Machine Construction Plant BMP-2 infantry combat vehicle under licence from the Russian Federation under the local name of the Sarath.

The first vehicle, assembled from components supplied by the Soviet Union, was handed over to the Indian Army in August 1987 and since then the local content of the vehicle has increased year by year and, by 1999, about 90 per cent of the complete vehicle and its associated systems was being built in India.

The Sarath is identical to the BMP-2 covered in detail under the Russian Federation and Associated States (CIS) in this section and is armed with a 30 mm 2A42 dual-feed cannon, a 7.62 mm PKT coaxial machine gun and a roof-mounted AT-5 'Spandrel' ATGW launcher with a maximum range of 4,000 m. This missile is manufactured in India by BDL (Bharat Dynamics Limited).

Sarath ICV of the Indian Army ready for amphibious operations and showing snorkel to rear of turret in the raised position (Indian Army) 0018435

Since production of the BMP-2 commenced in India, a number of improvements have been incorporated including new radios and modified stabiliser, with many smaller improvements expected to be incorporated into future production vehicles.

The Ordnance Factory Medak is responsible for the fabrication of the hull and turret, final assembly and testing as well as manufacturing the suspension system, engine and driver's compartment bulkheads, 30 mm and 7.62 mm ammunition storage and feeding systems, the fuel system and missile launcher and missile control system.

Other companies and facilities involved in the programme include the Ordnance Factory Trichy for the 30 mm 2A42 cannon; MTPF Ambarnath for the turret traversing and elevating mechanism and part of the missile launcher; Gun Carriage Factory Jabalpur for main gun mount and 81 mm smoke grenade launchers; OLF Deharadun, day and night vision devices and the gunsight; BEML KGF for transmission and final drives; BELTEX Madras for gun stabiliser and electrical cables; and BDL Medak for the missile and launcher.

Indian Armoured Dozer based on modified BMP-2 chassis ***2000***/0064817

Indian Engineer Reconnaissance Vehicle based on BMP-2 chassis 0005008

Key Sarath milestones

February 1982	Contract signed for BMP-1
June 1984	Project to make BMP-2 in India preliminarily sanctioned
July 1984	Foundation stone of new factory laid down
February 1985	Contract signed for BMP-2
August 1987	First Sarath rolled out
March 1988	50th vehicle completed
March 1989	200th vehicle completed
March 1990	396th vehicle completed
May 1990	First locally built ATGW launched from locally built BMP-2
March 1991	600th vehicle completed
November 1991	First indigenously built hull and turret rolled out

It is estimated that by early 2001, total production of the BMP-2 in India amounted to about 1,200 vehicles and these serve alongside about 700 (although some sources have quoted a lower figure of 350) BMP-1 infantry fighting vehicles.

Variants

Ambulance

This is already in production and is the basic amphibious BMP-2 modified for the ambulance role with the turret being retained but the armament removed.

The vehicle can be quickly configured to carry four stretcher patients or two stretcher patients and four seated patients or eight seated patients. It normally has a four-man crew consisting of driver, commander and two medical attendants and has a laden weight of 12,200 kg.

Medical equipment fitted includes stretchers, blood/saline bottles, blood transfusion equipment, oxygen equipment, ice container and fresh water container.

Engineer reconnaisance vehicle

The Engineer Reconnaissance Vehicle (ERV) is currently at the prototype stage and has been developed to meet the requirements of the Indian Engineers.

The vehicle retains the hull and turret of the standard production BMP-2, but the armament has been removed apart from the bank of three electrically operated 81 mm smoke grenade launchers either side of the turret. Like the BMP-2, the ERV is fully amphibious, being propelled in the water by its tracks.

The ERV is fitted with all the equipment required to acquire, record and transmit reconnaissance information back to the command post so enabling the engineers to cross both dry and water obstacles.

Equipment installed on the ERV includes a gyro land navigation system, global positioning system, radio navigation and guidance system, hand-held recording conpenetrometer to test soil, electronic distomat and theodolite, water current meter, echo-sounder, laser range-finder and picket driving and trail blazing equipment.

The picket driving and trail blazing equipment is mounted on the left side of the hull towards the rear and enables the ERV to quickly mark the path for the following vehicles. When travelling, this is in the horizontal position and when required for action is swung into the vertical position. The marking rods are fired into the ground using the electropneumatic system, with the magazine holding 50 rods.

All the instruments on the ERV have been integrated through serial interfaces and the data is logged in a 386 ruggedised IBM-compatible PC.

Standard equipment includes a roof-mounted air conditioning system, NBC system, two bilge pumps and a gyro direction indicator. Although originally developed for military applications, consideration is also now being given to promote this for civilian applications. Details of this vehicle are given in *Jane's Military Vehicles and Logistics 2001-2002.*

Armoured dozer

This Armoured Amphibious Dozer (AAD) has also been developed to meet the requirements of the Indian Army Engineers and is a BMP-2 with its turret removed and extensive modifications carried out to its hull to enable it to undertake its new mission.

Namica tank destroyer is based on BMP-2 chassis and shown here with launcher elevated ***2000***/0064818

The vehicle has a two-man crew consisting of the driver and operator who are seated back-to-back and are provided with a dual control system.

Standard equipment includes a hydraulically operated earth bucket at the rear with a capacity of 1.5 m^3, winch with a capacity of 8,000 kg, front-mounted track with mine ploughs and a rocket-propelled earth anchor, which is similar to that fitted to the now BAE Systems, RO Defence Combat Engineer Tractor which has been in service with India for some years.

The rocket-propelled earth anchor is used for self-recovery and has a maximum range of between 50 and 100 m depending on a number of factors.

The vehicle has a maximum road speed of 60 km/h and a water speed of 7 km/h. It is provided with an NBC system and in the future an autonomous version could be developed for use in hazardous areas. Details of this vehicle are given in *Jane's Military Vehicles and Logistics 2001-2002.*

Akash SAM

This uses a modified and stretched BMP-2 chassis with seven roadwheels each side compared to the six of the standard vehicle. Mounted on the roof of the vehicle is a rotatable launcher with three Akash surface-to-air missiles (SAM) in the ready to launch position.

The associated multifunction 3-D phased radar used with the Akash SAM system also has the stretched chassis with seven roadwheels, as does the associated command post vehicle. Both these feature a raised superstructure.

Additional details of this system are given in *Jane's Land-Based Air Defence 2001-2002.*

Trishul SAM

The chassis of the Sarath has been modified to launch the Trishul SAM, with three missiles being carried in the ready to launch position, search and track radars and their associated fire-control system.

In addition, there is a mobile command post with a raised superstructure which is the nerve centre of the missiles command, control, communications and intelligence functions.

Additional details of this system are give in *Jane's Land-Based Air Defence 2001-2002.*

Nag ATGW vehicle

The Indian Army is to deploy a new self-propelled anti-tank missile system called Namica which is based on the locally manufactured Sarath tracked infantry combat vehicle chassis.

The Namica will be armed with the locally designed and manufactured Nag (Cobra) anti-tank guided weapon (ATGW) that has been under development for some years by the Defence Research and Development Organisation (DRDO).

A total of four Nag ATGW will be carried in the ready to launch position, with additional missiles being stowed inside. These can be reloaded from within the vehicle under full armour protection.

Nag is claimed to be a third-generation top attack ATGW and is expected to enter production early in the 21st century. Nag weighs 42 kg at launch, has a range of more than 4,000 m and is of the fire-and-forget type. A tandem High-Explosive Anti-Tank (HEAT) warhead is fitted to counter Main Battle Tanks (MBTs) fitted with explosive reactive armour.

Three different seekers have been developed for Nag: Charge Coupled Device (CCD) based day version, Mercury Cadmium Telluride (MCT) Focal Plane Array sensor-based Imaging Infra Red and Millimetric Wave (MMW) active radar seeker for all-weather target engagement. It uses the Lock On Before Launch (LOBL) concept. The version with the CCD seeker is expected to be the first version of Nag to enter production.

As well as being launched from the Namica, Nag will also be launched from a standard ground firing post and the Advanced Light Helicopter, which will have eight missiles in the ready to launch position.

90 mm light tank

This is the BMP-2 with the turret removed and replaced by the French Giat industries TS-90 two-man turret armed with a 90 mm gun and 7.62 mm coaxial machine gun.

This version was developed to replace the Russian-supplied PT-76 light tank armed with a 76 mm gun. Two prototypes were completed but in the end a decision was taken not to proceed with this model.

105 mm self-propelled gun

The Sarath chassis has been modified to accept the Indian-manufactured 105 mm Field Gun which fires ammunition of the separate loading type that is similar to that used in the BAE Systems, RO Defence 105 mm Light Gun.

This fires a high-explosive projectile to a maximum range of 17,200 m. It is believed that this system is still at the prototype stage. Details of the 105 mm gun are given in the Towed anti-tank guns, guns and howitzers section.

Status

In production. In service with the Indian Army. As far as is known, India has not exported any of its locally built BMP-2 vehicles or variants based on this chassis. As of mid-2001, it is believed that none of these specialised variants has entered volume production.

Manufacturer

Ordnance Factory Medak.

Enquiries to Government of India, Ministry of Defence.

UPDATED

INTERNATIONAL

ASCOD armoured infantry fighting vehicle

Development

In the 1980s, the Austrian company, Steyr-Daimler-Puch (which designed and built the 4K 4FA and 4K 7FA full-tracked armoured personnel carriers covered in this section under Austria) started private venture studies in close co-operation with the Austrian Army for a new Armoured Infantry Combat Vehicle (AICV) which eventually became the ASCOD.

Compared with the earlier vehicles this was required to have significant improvements in the three key areas of armour, mobility and firepower.

The vehicle was also required to be easy to operate and maintain and have high operational readiness rate and the basic chassis had to be capable of being used for a wide range of variants.

The first prototype (PT01) was completed in 1990, with the turret and chassis being built in Austria and systems integration taking place in Spain. It weighed 21 tonnes, was powered by a 500 hp Pegaso diesel and used the track of the Steyr-Daimler-Puch SK-105 light tank. The 30 mm cannon was not stabilised and no fire-control system was fitted.

This was one of the five vehicles tested in Norway and, together with the Hägglunds Vehicle CV 9030 and the United Defense LP M2 Bradley, was shortlisted by the Norwegian Army for a further series of trials.

The second prototype (PT02) was completed in late 1992 and weighed 26 tonnes. The turret was made in Austria, the chassis in Spain, with final systems integration taking place in Spain. Improvements included more powerful 600 hp diesel engine, wider tracks, modified turret and redesigned troop compartment.

The third prototype (PT03) was rebuilt as a result of the original trials with PT01 in Norway and had many further improvements including an additional roadwheel station either side making a total of seven. For improved cross-country mobility Horstman Defence Systems rotary dampers were fitted, as was a track tensioning system. Stowage and internal layout were also improved.

PT04 was completed in 1994 and was built in Spain; this is the baseline for the Spanish Army version and features virtually all the improvements incorporated in the earlier vehicles.

A joint venture company was formed by Steyr-Daimler-Puch and SANTA BARBARA of Spain called ASCOD (Austrian Spanish Co-Operative Development) which carries out marketing of the vehicle on a worldwide basis.

The first customer for the ASCOD AIFV is the Spanish Army, which has placed a contract with SBB Blindados (part of the SANTA BARBARA Group) for an initial 144 Pizarro ICV, worth US$330 million.

Of these 144, 123 will be in the AIFV configuration and 21 in the command post configuration. While SBB Blindados will be the prime contractor in Spain, the Trubia factory of Santa Barbara will supply the hull, chassis, armament and other subsystems.

It is understood that the total Spanish Army requirement, including variants, is for 463 vehicles.

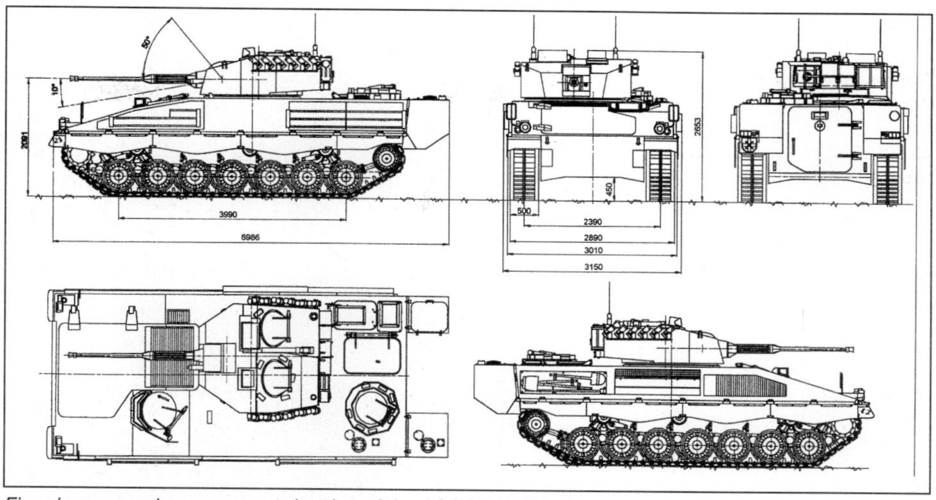

Five-view general arrangement drawing of the ASCOD MICV with seven roadwheel stations either side
0018399

The variants include armoured command post, communications, ambulance, 81 and 120 mm mortar carriers, anti-tank guided missile (with TOW and MACAM), armoured engineer vehicle, armoured recovery vehicle, artillery forward observation post vehicle and anti-tank with 105 mm gun fitted in new turret.

In May 1999, the Austrian MoD finally awarded Steyr-Daimler-Puch Spezialfahrzeug AG & Co KG a contract worth Sch3.2 billion for the delivery of 112 Ulan Armoured Infantry Fighting Vehicles (AIFV) to the Austrian Army.

Production of the Ulan is undertaken at the Vienna facilities of Steyr-Daimler-Puch Spezialfahrzeug AG & Co KG with deliveries to run from 2002 to 2004. The contract awarded to the company also includes a complete training and logistic support package. The first four Ulan vehicles were handed over in mid-2002.

Ulan is the Austrian Army name for ASCOD (Austrian Spanish Co-operative Development) AIFV which was developed to meet the requirements of the Austrian and Spanish armies.

The Ulan will supplement and eventually replace the older Steyr-Daimler-Puch Spezialfahrzeug AG & Co KG 4K 47A family of full-tracked armoured personnel carriers.

The Ulan contract does not at present include any options and in the future the Austrian Army will deploy a mix of Ulan AIFV and Steyr-Daimler-Puch Spezialfahrzeug AG & Co KG Pandur (6 × 6) APC of which 68 have already been delivered with up to 200 additional vehicles required.

Late in 1999, it was revealed that the Royal Thai Marine Corps had selected the ASCOD light tank to meet its future requirements with an order to be placed with Steyr-Daimler-Puch Spezialfahrzeug AG & Co KG in the future. The order is expected to comprise 15 ASCOD 105 light tanks, one command post vehicle and one recovery vehicle. As of mid-2001, this order had not been confirmed.

PT04 ASCOD armoured infantry fighting vehicle showing additional armour protection for hull and turret for increased battlefield survivability
2000/0064819

Specifications
(AIFV)

Crew: 3 + 8
Combat weight: 28,000 kg
Power-to-weight ratio: 25.75 hp/t (721 hp engine)
Ground pressure: 0.62 kg/cm²
Length: 6.986 m
Width: 3.15 m
Height:
(to hull roof) 1.775 m
(overall) 2.653 m
Firing height: 2.091 m
Ground clearance: 0.45 m
Track: 2.39 m
Track width: 500 mm
Length of track on ground: 3.99 m
Max road speed:
(forwards) 70 km/h
(reverse) 35 km/h
Fuel capacity: 860 litres
Cruising range (road): 600 km
Fording:
(without preparation) 1.2 m
(with preparation) 1.5 m
Gradient: 75%
Side slope: 40%
Vertical obstacle: 0.95 m
Trench: 2.5 m
Turning radius: pivot
Engine: V-8 cylinder diesel (MTU 8V 183 TE22 8 V-90) developing 600 hp at 2,300 rpm (Austrian vehicles have MTU 8V 199 TE20 8V-90 diesel engine developing 721 hp at 2,300 rpm)
Transmission: Renk HSWL 106C hydromechanical with integrated reversing and steering gear, automatic 6-speed with lock up clutch
Suspension: torsion bar with trailing arm for each roadwheel, rotary dampers for first and sixth roadwheel
Electrical system: 24 V
Batteries: 6 × 24 V, 300 Ah
Armament:
(main) 1 × 30 mm Mauser Model F cannon
(coaxial) 1 × 7.62 mm MG3 MG or MAG 58 7.62 mm
Ammunition:
(main ready use/stowed) 200/202 × 30 mm
(coaxial ready use/stowed) 700/2,200 × 7.62 mm
Smoke grenade dischargers: 4 × 3 76 mm
Smoke grenades:
(ready use/stowed) 12/24
Gun control equipment: electromechanic/manual
(by commander) yes
(by gunner) yes
Gun elevation/depression: +50/–10°
Turret traverse: 360°
Gun stabiliser:
(vertical) yes
(horizontal) yes
NBC system: yes
Night vision equipment: yes

The ASCOD has been ordered by the Austrian Army under the name Ulan and powered by an MTU 8V 199 diesel engine ***2000***/0064820

Description

The hull and turret of the ASCOD is of all-welded steel armour construction which provides, by a modular system, full protection against attack from 7.62 × 51 mm small arms fire through a full 360°. Over the frontal 60° arc, the vehicle is protected against 14.5 mm API (Armour-Piercing Incendiary) attack.

The vehicle can be fitted, as shown in the photographs, with additional steel armour plates to the turret front and sides, glacis plate and hull sides which provide a higher level of protection.

Option one protects against 30 mm APDS (Armour-Piercing Discarding Sabot) projectiles fired from a range of 1,000 m over a 24° frontal arc.

Option two provides protection against 30 mm APFSDS (Armour-Piercing Fin-Stabilised Discarding Sabot) rounds fired from 1,000 m through the frontal 60° (for example 30° left and right of the vehicle centre-line). Over the remainder of the vehicle protection is provided against 14.5 mm API attack through 360°.

Against indirect fire, the ASCOD provides the crew with protection from a 155 mm HE projectile bursting 10 m from the vehicle with a 90 per cent confidence level.

Special attention was made in the design of the ASCOD to provide a high degree of protection against both anti-tank and anti-personnel mines.

The chassis and turret of the ASCOD are fitted with spall liners to further reduce the behind armour effects of a variety of battlefield weapons. According to the manufacturer, the vehicle also has a low radar and infra-red signature.

The Spanish vehicles will be fitted with additional Explosive Reactive Armour (ERA) packages, developed by Santa Barbara, to the hull and turret.

The driver is seated front left with the power pack to his right. The latter consists of the German MTU 8V-183 TE22 V90 liquid-cooled turbocharged diesel developing 600 hp at 2,300 rpm coupled to a Renk HSWL 106C hydromechanical fully automatic six-speed transmission with integrated steering system. The Austrian Ulan is powered by a more recent MTU 8V 199 TE20 8V-90 diesel engine developing 721 hp at 2,300 rpm.

The complete power pack, consisting of the engine, transmission and cooling system, can be removed from the ASCOD in about 15 minutes.

Mounted in the centre of the hull offset to the right is the two-man power-operated turret with the commander seated on the left and gunner on the right.

Main armament comprises a dual-feed Mauser 30 mm Model F cannon with 200 (120 + 80) rounds of ready use ammunition and an additional 202 rounds in reserve. The gunner can select single shot, 3- 5- and 10-round bursts.

Types of 30 mm ammunition fired include APDS-T, HEI-SD-T and TP-T, with other natures under development including APFSDS-T, MP-T and SAPHEI-T. All of these 30 × 173 mm rounds have a plastic driving band, which is claimed to extend the barrel life by a factor of three.

A 7.62 mm MG3 machine gun is mounted coaxial to the left of the 30 mm cannon and this is provided with 700 rounds of ready use ammunition and a further 2,200 rounds in reserve. Two banks of three Wegmann 76 mm smoke/fragmentation grenade launchers are mounted either side of the turret, firing forwards.

The all-electric traverse and weapon elevation system has been developed by the now Curtiss-Wright Antiebstechnik GmbH, with controls being provided for commander and gunner.

Weapon elevation of +50° enables battlefield helicopters to be engaged and, if required, the commander can override the gunner. A deck clearance system is fitted as standard and the turret will not function while any of the hatches are open.

To engage targets under a wide range of conditions the gunner is provided with Kollsman DNRS (Direct Night Range Sight) fire-control system which includes a digital computer and a day/thermal night sight with integrated laser range-finder. The commander has a day sight with a monitor screen showing a picture of the target seen by the gunner.

The DNRS plus the advanced day/night sighting system and two-axis stabilised 30 mm cannon enable the ASCOD to engage a wide range of battlefield targets under all weather conditions with a high first round hit probability with the vehicle stationary or moving.

The turret is designated the SP-30 and can be fitted to other vehicles, tracked and wheeled, including the reconnaissance version of the Steyr-Daimler-Puch Pandur (6 × 6) vehicle. For installation on lighter vehicles, the armour protection would be to a lower level of protection.

The troop compartment is at the rear with the eight infantry being seated on individual fold-up seats, five down the left side and three down the right side. All the seats are provided with seat belts. Normal means of entry and exit for the troops is via a large power-operated door in the rear that opens to the right which has a single vision block with a spherical firing port below.

Over the left side of the troop compartment roof is a cupola with day periscopes which allows the squad leader all-round vision before he dismounts with the troops. This cupola is also provided with a hatch cover that opens to the rear. It is possible that this cupola will not be fitted on production vehicles. There is an additional rectangular hatch cover in the right side of the roof which opens to the right.

Suspension is of the torsion bar and trailing arm type with seven dual rubber-tyred roadwheels either side, drive sprocket at the front, idler at the rear and three track-return rollers.

Rotary dampers are provided at the first and sixth roadwheel stations and a hydraulic track tensioning system operated by the driver is fitted as standard.

The 500 mm wide Diehl Type 129C4 tracks are of the double-pin type and can be fitted with grousers for use in winter conditions, but when these are fitted maximum road speed is 40 km/h.

Standard equipment includes: fire detection/suppression system for the engine compartment; central climatisation which consists of an NBC/ventilation system with optional air conditioning and water heating system (located in the right side of the hull); passive night vision device for the driver and two bilge pumps.

A wide range of optional equipment can be installed on the vehicle including a fully automatic fire/explosion detection system for the crew compartment; preheating of the coolant; land navigation system, including Global Positioning System (GPS); and an integrated threat countermeasures system can be installed on the vehicle at the user's request. The hull and turret have been designed for the installation of the intercom and communication equipment selected by the user.

A power-operated dozer blade can be mounted at the front of the hull and operated by the driver under full armour protection. This can be quickly replaced by a mineclearing system of the roller or plough type.

According to ASCOD the vehicle has plenty of stretch potential up to a gross vehicle weight of 28 to 30 tonnes with engine upgraded.

Variants

Using the basic chassis, ASCOD has suggested that the chassis of the vehicle can be used for a wide range of other roles and be fitted with different turrets.

ASCOD with LPT

Late in 1998 the ASCOD was tested in the United States fitted with the General Dynamics Land Systems 105 mm armed Low-Profile Turret (LPT). Additional details of this are given in the entry for the ASCOD 105 light tank in the Light tanks section under International.

ASCOD 105 mm light tank

For trials purposes, late in 1997 one of the prototypes of the ASCOD chassis was fitted with the complete LIW turret of the South African Rooikat 105 mm (8 × 8) armoured car. This then underwent a successful series of firing trials. Full details are given in the Light tanks section under International.

ASCOD with Centauro turret

For trials purposes, in early 2000 an ASCOD chassis was fitted with the turret developed by Otobreda which is fitted to the Centauro (8 × 8) tank destroyer in service with Italy (400) and Spain (22). A photograph of this version appears in the Light tanks section under International.

Air defence

Various types of air defence weapons, gun and missile could be mounted on the ASCOD chassis, with the first including the Krauss-Maffei Wegmann twin 30 mm Wildcat and the last including the Oerlikon Aerospace ADATS missile system.

Ambulance

This would have a raised superstructure, have a two-man crew and carry four stretchers or up to eight seated patients or a combination of stretcher and seated patients.

Anti-tank missile

Various types of anti-tank weapons and/or turrets can be fitted onto the roof of the ASCOD.

Armoured engineer

This would be based on the chassis of the ASCOD but be fitted with a 3.3 m wide dozer blade which could be angled left and right.

Pivoted on the left side of the hull would be a hydraulic arm which could be fitted with various engineer attachments including a bucket. Armament would comprise a cupola-mounted 12.7 mm machine gun and, in addition to the two-man crew, the vehicle would also carry seven men.

Armoured logistics carrier

This would be a flatbed-type vehicle and be able to carry a maximum load of 10 tonnes.

Command post

To provide additional space for the command staff this version would have a raised roof at the rear and be fitted with a cupola armed with a 7.62 mm MG.

Mortar carrier

In addition to being fitted with various types of turntable-mounted 81 and 120 mm mortars, the ASCOD could also be fitted with a turret-mounted 120 mm mortar system and have a four-man crew.

Recovery/repair

To carry out the recovery/repair role this model would have a rear-mounted stabiliser blade and winch and a crane which can lift a maximum of 6,500 kg. Like the command post version, this would have a raised superstructure and be fitted with a cupola armed with a 7.62 mm machine gun. Normal crew would consist of four men.

Status

In production. In service with:

Country	Quantity	Comments
Austria	112 ordered in 1999	first four delivered in mid-2001
Spain	144 ordered in 1996	deliveries under way

Manufacturer

ASCOD AIE.

UPDATED

IRAN

Boragh armoured personnel carrier

Development

After establishing its capability in the development and production of unguided surface-to-surface rockets (covered in detail later in this book) and a wide range of infantry weapons and equipment, Iran is now moving into the armoured fighting vehicle area and has developed a full-tracked armoured personnel carrier called the Boragh.

This APC is in production by the Shahid Kolahdooz Industrial Complex of the Iranian Defence Industries Organisation (DIO).

Although no production figures of the Boragh APC have been released, it is estimated that up to 200 vehicles may now have been built.

Description

In many respects the basic hull of the Boragh APC is similar to the Russian BMP-1, or its Chinese equivalent, the China North Industries Corporation (NORINCO) WZ 501 full-tracked infantry combat vehicle, which are described in detail elsewhere in this section.

The Russian BMP-1 has been in service with Iran for many years and it is estimated that up to 300 vehicles are still in use. The BMP-1 has a one-man turret armed with a 73 mm 2A28 gun with a Sagger wire-guided anti-tank missile above and a 7.62 mm PKT coaxial machine gun. First examples of the Boragh APC have a much simpler weapons station.

Borагh APC armed with roof-mounted 12.7 mm machine gun and clearly showing M113 type roadwheels and new track skirts 0045961

According to Iran, the hull of the Boragh APC is of all-welded steel armour which provides protection against 7.62 and 12.7 mm small arms fire and shell splinters.

The driver is seated at the front left and has a single-piece hatch cover that opens to the right and four day periscopes for observation to his front and sides.

The V-8 air-cooled power pack is mounted to the right of the driver with the air inlet and outlets being in the roof and the exhaust outlet on the right side of the hull. The Boragh can lay its own smoke screen by injecting diesel fuel into the exhaust outlet.

The BMP-1 is powered by a Type UTD-20 six-cylinder in-line water-cooled diesel developing 300 hp at 2,000 rpm but the Boragh is powered by a V-8 air-cooled turbocharged diesel developing 330 hp at 2,300 rpm.

The troop compartment is at the rear with the main difference being that the Iranian vehicle is armed with a pintle-mounted Iranian-built 12.7 mm Dooshka

Specifications
Crew: 2 + 8 (or 2 + 12)
Combat weight: 13,000 kg
Power-to-weight ratio: 25.4 hp/t
Ground pressure: 0.485-0.55 kg/cm²
Length: 6.72 m
Width: 3.10 m
Height, hull top: 1.66 m
Ground clearance: 0.40 m
Track: n/avail
Track width: n/avail
Length of track on ground: n/avail
Max speed:
(road) 65 km/h
(cross-country) 40-45 km/h
(water) 5-7 km/h*
Fuel capacity: 500 litres
Max range: 550-625 km
Fuel consumption: 80-90 litres/100 km (road)
Fording: amphibious
Gradient: 60%
Side slope: 40%
Vertical obstacle: 0.70 m
Trench: 2.30 m
Engine: V-8 air-cooled turbocharged diesel developing 330 hp at 2,300 rpm
Transmission: manual, 5 forward and 1 reverse gears
Steering: clutch and brake
Final drive: planetary/single stage
Suspension: torsion bar
Electrical system: 24 V
Armament: 1 × 12.7 mm MG
Ammunition: 1,000 × 12.7 mm
Gun stabiliser:
(vertical) no
(horizontal) no
Smoke-laying equipment: diesel fuel injected into the exhaust
* depends on type of track fitted

The new ATGW turret is installed on the DIO's full-tracked APC (AIO) **2002**/0073606

machine gun (The Russian 12.7 mm DShKM machine gun) and there is no commander's position to the rear of the driver on the left side.

The 12.7 mm machine gun can also be fitted with side and rear armour protection similar to that fitted to the more recent Chinese NORINCO Type 90 series of full tracked APC.

There are only two roof hatches over the rear troop compartment rather than the four on the BMP-1. In each side of the troop compartment are three firing ports with roof-mounted day periscopes above which allow the infantry to fire their weapons from inside. The first firing port on each side would normally be used for a machine gun with the other two for assault rifles.

The infantry would normally enter and leave the vehicle by two doors in the rear of the vehicle which open outwards with the left door having a firing port and associated vision device.

It is probable that like the original Russian vehicle, these rear doors each contain an integral diesel fuel tank which is connected to the main fuel tank.

The suspension either side is of the torsion bar type with the drive sprocket at the front, idler at the rear, track-return rollers and six dual rubber-tyred roadwheels. The tracks have replaceable rubber pads and two different types can be fitted.

The roadwheels are a different design from those on the Russian BMP-1 (which, like the Type 90, only has five roadwheels either side) and appear to be very similar to those of the US United Defense M113 series APC which have been in service with the Iranian Army for many years.

The Iranian DIO has been making replacement M113 series roadwheels and tracks for some years and for logistical reasons it is logical that any locally manufactured APC would use these.

The upper part of the suspension of the Boragh is covered by a four-part rubber skirt which hinges upwards to allow access to the suspension for maintenance purposes. On the BMP-1 a light sheet steel cover was fitted.

The Boragh APC is fully amphibious, being propelled in the water by its tracks. Before entering the water, a trim vane is erected at the front of the vehicle by the driver without leaving his seat and the bilge pumps are switched on.

The basic vehicle is fitted with a ventilation system with air entering the vehicle via a roof-mounted device to the rear of the 12.7 mm machine gun installation.

Standard equipment includes a vehicle intercom and a radio, the antenna for the latter is positioned at the rear of the vehicle on the left side. The vehicle is not fitted with any night vision equipment.

Optional equipment includes NBC system, electrically operated smoke dischargers, fire detection and suppression system and various communications systems.

Variants

Using the same basic chassis it is probable that a complete family of vehicles will be developed such as command post, weapons carriers and better armed versions.

Iran now builds a wide range of rocket launchers, the Russian ZU-23-2 twin 23 mm light anti-aircraft gun system and the Raad anti-tank missile system (essentially the Russian AT-3 Sagger), and it is probable that some of these could well be fitted to the Boragh APC in the future. More recently, Iran has started to manufacture the Russian 9K113 Konkurs ATGW (NATO designation AT-5 'Spandrel') and this may well be installed on the Boragh APC in the future, if not already fitted.

Self-propelled Toophan ATGW

The Aerospace Industries Organisation (AIO) of Iran has revealed that it has developed and placed in production a new one-person turret that can launch the Toophan series of wire-guided Anti-Tank Guided Weapon (ATGW).

The first known application of this turret is on the Borogh full-tracked Armoured Personnel Carrier (APC) which is manufactured by the Shahid Kolahdooz Industrial Complex of the Defence Industries Organisation (DIO).

The first version of the ATGW is called the Toophan and is fitted with a single High Explosive Anti-Tank (HEAT) warhead weighing 3.6 kg that will penetrate 550 mm of conventional steel armour.

The second version, Toophan 2, has a tandem HEAT warhead weighing 4.1 kg that is claimed will penetrate 760 mm of conventional steel armour protected by Explosive Reactive Armour (ERA).

The new turret system can launch both versions of the Toophan ATGW and has the standard Toophan Semi-Automatic Command-to-Line-Of-Sight (SACLOS) guidance system.

In this all the gunner has to do to ensure that the missile hits the target is to keep his sight on the target throughout the target engagement. Once the missile hits the target new missiles are loaded via a hatch in the roof of the Borogh APC.

Maximum range of both missiles is 3,850 m and time of flight to maximum range is 13 seconds. Toophan is very similar to the US Raytheon Systems Company Basic TOW while the Toophan 2 is comparable to the Improved TOW that also has an extendable nose probe.

It is believed that the AOI is also developing a laser beam riding version of Toophan which will have a longer range. This may be designated Toophan 3 or Toophan 4.

Consideration is also being given to the development of a turret with the gunner under full armour protection from small arms fire and shell splinters.

Status

Production. In service with Iranian Army.

Manufacturer

Defence Industries Organisation (DIO), Shahid Kolahdooz Industrial Complex.

UPDATED

Cobra armoured personnel carrier

Development/Description

In addition to the Boragh full-tracked armoured personnel carrier, covered in detail in the earlier entry, Iran has also developed at least one other armoured personnel carrier called the Cobra.

This has a very low-profile hull with the engine compartment at the front and the troop compartment at the rear. Entry to the troop compartment is via a large single door in the hull rear. The upper part of the hull sides of the Cobra slope slightly inwards.

The suspension either side of the Cobra consists of six dual rubber-tyred roadwheels with the drive sprocket at the front and the idler at the rear; there are no track-return rollers.

On the roof of the prototype Cobra is the locally built Russian ZU-23-2 twin 23 mm light anti-aircraft gun.

Variants

In July 1997, it was stated that Iran was also building another type of armoured personnel carrier called the BMT-2, which is armed with a 30 mm gun and can carry seven men. No other details are available. It is possible that the BMT-2 is the Iranian designation for the Russian BMP-2, which is known to be in service with the Iranian Army. According to United Nations figures, the first three BMP-2 vehicles were delivered to Iran in 1998.

Specifications

Not available.

Manufacturer

Defence Industries Organisation (DIO).

VERIFIED

ISRAEL

Puma combat engineer armoured personnel carrier

Development

The Puma heavy armoured personnel carrier is used by armoured engineers rather than the armoured infantry, which uses the Achzarit based on the T-54/T-55 MBT chassis. It is possible that as many as 700 of the Puma vehicles are in service with the Israel Defence Force. It is understood that the Puma entered service with the IDF as far back as 1984.

Description

The Puma can be fitted with various Israeli-developed mineclearing devices of the plough or roller type and carry up to eight men. It can also tow mineclearing equipment or trailers containing specialised engineer equipment.

Puma is basically an upgraded Centurion MBT (see MBTs and medium tanks section) with the turret removed, some building up of the remaining superstructure and the addition of armour protection.

The commander is seated to the rear and above the driver's position and has a single-piece hatch cover that opens to the rear and five periscopes that give observation through the frontal arc. To the left of this is the gunner's hatch that opens to the rear and in front of this is the 7.62 mm OWS. To the rear of the gunner's and commander's hatches is a large oblong hatch that opens to the rear. This is normally used by the troops to enter and leave the vehicle via the rear engine decks which are fitted with steps on either side that fold up when not required. It should be noted that as far as it is known there are no Centurion gun tanks now in front line service with the Israel Defence Force.

Puma combat engineer armoured personnel carrier (IDF) 0045958

The Puma vehicle has a complex appliqué armour package to protect the occupants in its fighting compartment. The driver is seated at the front right side of the vehicle and has three day periscopes for forward observation and a single-piece hatch cover that opens to the rear.

The Toga passive armour system, designed and produced by RAFAEL Armament Development Authority, consists of carbon steel mesh. Toga is applied around the engine air intakes and equipment sponsons on the sides of the vehicle.

The skirts are also of passive armour with those towards the front, where the crew compartment is, providing a higher level of armour protection than those at the rear which protect the engine compartment.

The third type of armour array, the one which offers the most comprehensive protection, consists of Chobham-style passive armour. This material looks very much like that fitted to the Merkava Mk 3 MAGACH-7 (upgraded M60 series MBT). The most prominent place that this armour is visible on the Puma is the glacis plate.

Armament on the Puma consists of three 7.62 mm machine guns, one of these, on the front left of the Puma, is fitted to a RAFAEL Overhead Weapons Station (OWS), allowing the machine gun to be fired remotely from under armour protection. The OWS is fitted to a number of other Israeli AFVs including the Achzarit and some M113A1 series vehicles.

The Puma, as in the case of the Nagmash'ot and Nagmachon, has three Soltam 60 mm mortars fitted to its roof.

In addition, Puma is fitted with two Israel Military Industries CL-3030 instantaneous self-screening smoke grenade systems, one either side at the front. These have been standard equipment on Israeli MBTs for many years and have also been fitted to the M113A1 APC and the Achzarit infantry armoured vehicle.

The Puma has a crew of three and can carry five assault engineers. It is estimated that it has a combat weight of 40 tonnes.

Some Puma vehicles have been observed to have been fitted with the suspension and track of the Merkava Mk1 and Mk2 MBT. It is not known if this was backfitted to early vehicles or only introduced on late production Puma vehicles.

Variants

None apart from earlier models.

Specifications

Not available.

Status

In service with the IDF.

Manufacturer

Israel Defence Force facilities, probably the Israeli Ordnance Corps Workshops at Tel a Shomer, near Tel Aviv, Israel.

UPDATED

Achzarit infantry armoured vehicle

Development

The Achzarit infantry armoured vehicle has been developed to meet the specific operational requirements of the Israel Defence Force for a well-protected vehicle capable of operating with the well-armoured locally designed and built Merkava MBTs.

Development of heavily protected infantry vehicles commenced in Israel some time ago using a variety of chassis including the Merkava and Centurion (qv). The first prototypes of the Achzarit vehicle were completed in 1987.

Following trials with prototype vehicles, the Achzarit infantry armoured vehicle was accepted for service with the Israel Defence Force and entered production in 1988. It is estimated that between 400 and 500 of these vehicles are now in service.

In 1997, Russia unveiled a very similar vehicle based on a much modified T-54/T-55 MBT chassis called the BTR-T. Details of this vehicle, which remains at the prototype stage, are given later in this section.

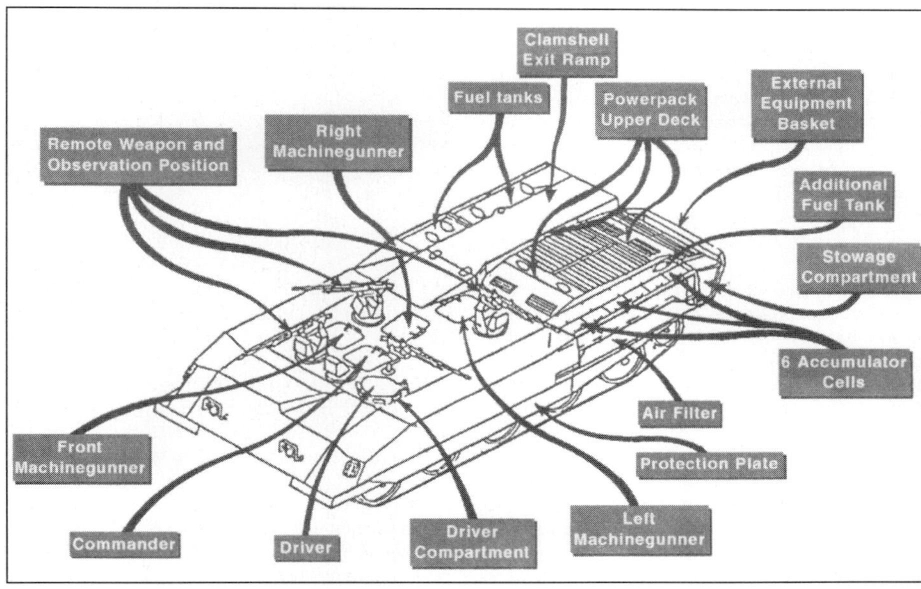

Outline drawing of the Achzarit armoured personnel carrier showing position of the main components
2000/0064821

Description

The chassis of the Achzarit is based on heavily modified Russian T-54/T-55 MBT chassis which have been rebuilt in significant numbers by the Israel Defence Force for some years.

The chassis has been stripped down and completely rebuilt, with the crew and infantry compartment at the front and the new locally developed NIMDA power pack at the rear.

An additional layer of Israeli-developed passive armour is mounted on the chassis, which makes the vehicle the best protected infantry vehicle of its type in the world.

A total of 10 men is carried, with the driver being seated front left with a single-piece hatch cover that opens to the left and four day periscopes. The centre one of these can be replaced by a passive periscope for driving at night. The commander is seated to the right of the driver and has a traversible roof-mounted periscope and a square hatch that opens to the rear. Seated at the front right is a machine gunner who operates the RAFAEL Overhead Weapon Station which is armed with a 7.62 mm M240 machine gun. This can be aimed and fired from within the vehicle in complete safety. Both day and image intensification sights are provided.

There are an additional three 7.62 mm pintle-mounted machine guns, one on the commander's hatch and two over the rear troop compartment.

The seven infantrymen are seated to the rear, three on a bench seat on the left side of the compartment, three on individual folding seats on the right and one in the centre of the rear part of the compartment.

The driver, commander front right machine gunner are each provided with a hatch, with that of the commander capable of being half raised for observation purposes. An additional two hatches are provided to the rear, one over the centre of the crew compartment and the other to its left rear.

The original Russian engine and transmission have been removed and a NIMDA-designed power pack has been installed at the vehicle's rear.

Achzarit armoured personnel carrier showing additional armour and roof hatches (IDF) 0018437

This unusual power pack is extremely compact and has allowed the installation of the passageway at the rear of the vehicle on the right side which allows the infantry to dismount through the rear clamshell door. This has been achieved without modifying the hull.

The power pack integrates in different ways some major components similar to those used on the Samovar (upgraded Russian T-55 MBT) and the US-supplied 155 mm M109 self-propelled artillery system used in large numbers by the Israel Defence Force.

The power pack of the Achzarit Mk 1 consists of the transversely mounted Detroit Diesel 8V-71 TTA water-cooled, two-stroke V-8 diesel developing 650 hp coupled to an Allison XTG-411-4 hydrokinetic automatic transmission.

The Achzarit Mk 2 power pack consists of a Detroit Diesel 8V-92 TA diesel developing 850 hp coupled to an Allison XTG-411-5 automatic transmission. This gives the vehicle an improved power-to-weight ratio and greater acceleration. It is believed that the current production model is the Mk 2.

These engines can draw in air from either the crew compartment or the engine compartment through a precleaner with a blower to eject the dust and a paper filter.

While the five roadwheels, drive sprocket and idler of the T-54/T-55 MBT have been retained, the wheels are now sprung by new torsion bars which allow a greater wheel travel, with the first and last wheel stations being fitted with an hydraulic bump stop produced by Israel Aircraft Industries which is also used on the Merkava MBT.

This suspension upgrade provides the vehicle with better cross-country mobility and allows it to operate with MBTs.

The infantry can enter and leave the Achzarit via the hatches in the roof or via the hydraulically operated clamshell door located in the rear of the vehicle on the right side. The lower part folds down to form a ramp while the upper part opens upwards.

No detailed specifications of the Achzarit infantry armoured vehicle have been revealed, although it has a combat loaded weight of 44 tonnes compared to the 36 tonnes of a T-54/T-55 MBT.

The T-55 turret has been removed from the Achzarit, most of the difference in weight between the present Achzarit 44 tonnes and the 27 tonnes of the T-55 hull is accounted for by the additional armour protection for improved battlefield survivability. Overall height to the top of the hull is about 2 m while overall width is 3.64 m.

Mounted at either side of the front of the hull is a bank of six Israel Military Industries CL-3030 instantaneous self-screening system for combat vehicles which were originally installed on Israeli Centurion, M48/M60 and Merkava MBTs. This can launch various types of grenade over the frontal arc of the vehicle.

Standard equipment for the Achzarit includes a Spectronix fire detection and suppression system, individual NBC system and the vehicle can also lay a smoke screen by injecting diesel fuel into its exhaust outlet on the left side of the hull.

Variants

There is known to be a command post variant of the Achzarit heavy armoured personnel carrier which lacks the roof-mounted machine guns and is fitted with additional communications equipment for its specialised role.

Specifications

Not available.

Status

In service with IDF.

Manufacturer

Israel Defence Force facilities, probably the Israeli Ordnance Corps Workshops at Tel a Shomer, near Tel Aviv, Israel.

UPDATED

Other Israeli heavy armoured personnel carriers

Development/Description

In addition to the Puma (based on the rebuilt Centurion tank chassis) and the more recent Achzarit (based on a rebuilt Russian T-55 tank chassis) covered in detail in the previous two entries, Israel has also developed and fielded a number of other heavy full-tracked Armoured Personnel Carriers (APCs).

These heavy APCs have been designed to provide infantry and engineer units with a higher level of combat survivability when operating with armoured units as well as in high-threat combat zones such as that encountered in southern Lebanon. It is possible that some of these heavy APCs have been fitted with the suspension system of the Merkava MBT for improved mobility.

The first of these is called the Nagmash'ot and is based on the chassis of the Centurion and this appeared in the early 1980s in southern Lebanon. It is understood that this vehicle has now been phased out of front line service with the IDF. This was armed with roof mounted 7.62 mm and some times a single 12.7 mm MG.

The Nagmachon has a new armoured superstructure with vertical sides and pintle-mounted 7.62 mm machine guns. It is also fitted with explosive reactive armour to provide a higher level of protection against HEAT projectiles. This vehicle is believed to be powered by a now General Dynamics Land Systems AVDS series diesel developing 750 hp and has a combat weight of around 55 tonnes. This has four CL-3030 systems with two being mounted either side towards the front of the vehicle.

Nakpadon heavy armoured personnel carrier (IDF) 0045957

The Nakpadon is based on the Centurion chassis and is used in the security zone in northern Israel and also has various missions in a state of emergency. Its armour provides protection from small arms fire, anti-tank ammunition and demolition charges and it is also provided with ballistic protection for its belly.

Up to four 7.62 mm machine guns are mounted on the roof to provide suppressive fire and most vehicles have a 60 mm mortar fitted on the roof. On the upper parts of the troop compartment are shields with bulletproof vision blocks which are very similar to those fitted to the Classical M113 series APC covered later in this section. Armour is understood to be of the passive type. Both the Nagmachon and the Nakpadon have been observed fitted with explosive reactive armour packages on their skirts.

Like virtually all other front-line Israeli combat vehicles, the Nakpadon is also fitted with the Israel Military Industries CL-3030 instantaneous self-screening system for combat vehicles.

Additional details of this system are given in the entry for the Classical armoured personnel carrier later in this section.

It is believed that the Nakpadon also weighs around 55 tonnes and is fitted with a General Dynamics Land Systems AVDS-1790-SA diesel power pack developing 900 hp which is also installed in the Merkava Mk 1 MBT.

To take account of the increased weight of these vehicles and the harsh terrain that they operate over, it is believed that the suspension has been upgraded, possibly the installation of some hydraulic bump stops.

Both of these heavy armoured personnel carriers are understood to have improved protection against mines. It is believed that some vehicles have been fitted with electronic devices to detonate explosive devices ahead of these heavy APCs.

Nagmachon heavy armoured personnel carrier clearly showing explosive reactive armour and special roof mounted observation devices (IDF) ***2002***/0004640

Status

Production complete. In service with the IDF.

Manufacturer

Israel Defence Force facilities, probably the Israeli Ordnance Corps Workshops at Tel a Shomer, near Tel Aviv, Israel.

UPDATED

Israeli M113 APC fleet

It is estimated that the Israel Defence Force (IDF) has a fleet of up to 6,000 United Defense (previously FMC Corporation) M113A1/M113A2 series full-tracked Armoured Personnel Carriers (APC), so making it the largest user outside the USA. Recent M113 deliveries from the United States have included 450 in 1994, 28 in 1995 and one in 1996. It is believed that there were all US Army surplus vehicles.

Most of these M113 series vehicles have their trim vanes removed from the front of the hull and an extension fitted to the exhaust outlet so that the exhaust does not blow back into the crew compartment when the vehicle is being driven with its hatches open.

Israeli M113 series APC fitted with locally developed passive armour package 0018438

The Machbet combines the M163 20 mm Vulcan self-propelled anti-aircraft gun system and a four-round Stinger surface-to-air missile launcher 0036653

Other modifications include mounting up to four 12.7 mm (0.50 in) M2 or 7.62 mm machine guns on top of the hull and racks fitted on either side of the hull enabling personnel kit and other stores to be carried outside the vehicle, thus giving more room inside the troop compartment.

The Israelis call the M113 the Zelda and, in addition to the basic APC, many vehicles have been modified to carry various weapons and other equipment. Variants include:

M113 ambulance
M113 anti-tank with TOW
M113 engineer
M113 radar, including air defence and battlefield surveillance
M163 20 mm Vulcan air defence system; some of these have been fitted with a pod of Stinger fire-and-forget missiles. Full details of this system are given in *Jane's Land-Based Air Defence 2001-2002.*
M577 command post vehicle
M548 ammunition supply vehicle
M548 with Elta EL/M-2310 artillery fire-control system
M48 Chaparral SAM system
M667 Lance surface-to-surface missile system (used in non-nuclear role)
81 and 120 mm mortar carriers.

The RAMTA Improved TOW Vehicle is an M113 fitted with a retractable telescopic mast on top of which is a single Raytheon Systems Company TOW ATGW launcher. When not required, this is retracted into the hull and is raised hydraulically. The gunner stands on a special platform that rises together with the complete system.

The gunner controls the lifting and lowering from the platform as well as the launching and tracking of the missile. The system has a traverse of 180° and takes 60 seconds to be raised. In total, 10 TOW missiles are carried internally. This system can also launch the locally developed MAPATS ATGW.

While the original M113 series were powered by a Chrysler petrol engine, all of the Israeli vehicles are now understood to be of the more recent M113A1/M113A2 standard which has a more fuel efficient Detroit Diesel engine. Not only does this increase the operational range of the vehicle but also reduces the risk of fire.

The single internal fuel tank has been replaced by two new fuel tanks which are mounted one either side of the hull rear. This has the added advantage of making more room in the troop compartment.

In 1987, the RAFAEL Armament and Development Authority announced that it had developed a passive armour system called the Toga, which gives protection against 14.5 mm rounds and this has been successfully used during fighting in southern Lebanon.

The Israeli company, NIMDA, offers a conversion package to enable users of the M113 to convert their vehicles to the M113A1/M113A2 configuration, or from the M113A1/M113A2 to the latest M113A3 standard.

The first conversion package features a Detroit Diesel 6V-53T engine developing 216 or 265 hp coupled to an Allison Transmission TX-100-1A automatic transmission, an improved cooling system, a new transfer case, a steering/brake differential and a new electrical system.

The second conversion package includes a Detroit Diesel 6V-53T developing 275 to 300 hp coupled to an Allison X-200-4 crossdrive transmission with hydrostatic steering.

Other options include improved protection, automatic fire detection and suppression, spall liners, improved suspension, external fuel tanks, NBC system and various remote-controlled weapons ranging from a 7.62 mm machine gun up to a 20 mm cannon.

Details of the Classical version of the Israeli M113 series are given in the following entry.

Status

In service with the IDF.

Manufacturer

United Defense LP of the US, with upgrade work being carried out in Israel.

UPDATED

Classical armoured personnel carrier

Development/Description

In addition to various members of the United Defense built M113 series covered in a separate entry, in 1996 it was revealed that the IDF had taken into service a more advanced version of the M113 series APC which is called Classical.

This is fitted with a new Explosive Reactive Armour (ERA) system developed by the Ordnance Systems Division of the RAFAEL Armament Development Authority. This was first observed fitted to M113s deployed in northern Israel and southern Lebanon in 1996.

This ERA is fitted to the front and sides of the vehicle and, according to the RAFAEL Armament Development Authority, provides complete protection against hand-held anti-tank weapons such as the Russian RPG-7 (which fires a High Explosive Anti-Tank rocket) as well as small arms ammunition up to 14.5 mm in calibre.

Its main advantages have been summarised by the RAFAEL Armament Development Authority as:

(1) High protection level
(2) Unique design ensures the survivability of thin-walled light armoured vehicles
(3) It meets the most stringent safety standards and uses a new generation of very insensitive explosive
(4) Ease of maintenance with easy access to the engine compartment.

The ERA package on the glacis plate and nose of the M113 consists of very thin boxes while that on its sides is of a different design with two arrowhead layers one above the other.

Upgraded M113 series APC of the Israel Defence Force fitted with RAFAEL Armament Development Authority explosive reactive armour package over the frontal arc **2000**/0064822

This side arrangement may have been installed to provide a higher level of protection against light anti-tank weapons fired from above or below the vehicle.

As well as the new ERA package the upgraded M113 has a number of additional new features including an extension to the top-mounted exhaust pipe on the right side of the hull.

The exhaust pipe has now been extended downwards so that the hot exhaust gases are now expelled just above the roadwheels rather than in line with the troop compartment roof.

Mounted either side of the hull front is a pod of electrically operated Israel Military Industries CL-3030 instantaneous self-screening system for combat vehicles. This is standard equipment on most tanks, heavy armoured personnel carriers and M113 series armoured personnel carriers of the IDF.

The CL-3030 consists of two launch pods, one is mounted either side of the hull front, firing forwards. Each pod contains six smoke grenades.

At the vehicle commander's discretion, two smoke grenades are launched simultaneously, one from each pod to create, in less that two seconds, a smoke screen which effectively conceals the vehicle. Under average wind conditions, this screen lasts between one and two minutes, so giving the vehicle time to move its position. The screen is created from a distance of 40 to 50 m from the vehicle and is initially 60 to 70 m wide and 4 to 8 m high.

The commander of the upgraded M113 has vertical steel armour protection either side with thick bulletproof windows. No protection is provided to the immediate front or rear of the commander's position. Either side of the troop compartment at the rear are similar vertical steel panels with thick bulletproof windows. Pintle mounts are also provided for 7.62 mm machine guns on the troop compartment roof.

These Classical armoured personnel carriers normally have the fuel tanks mounted externally at the rear and some vehicles have been observed fitted with an electronic device on the roof which is believed activates roadside munitions.

Status

In service with the IDF. As far as it is known, Israel has not exported any of these Classical armoured personnel carriers.

Manufacturer

United Defense LP of the US, with upgrade work being carried out in Israel.

UPDATED

ITALY

Dardo infantry fighting vehicle

Development

After the completion of feasibility studies into a new infantry fighting vehicle, followed by a project definition stage, a five year development contract was awarded to Otobreda and IVECO by the Italian Army in February 1982. This covered the construction of three prototype vehicles, all of which were completed by 1988 and subsequently evaluated by the Italian Army under the designation of the VCC-80.

This vehicle was fitted with a two-man turret armed with a 25 mm Oerlikon Contraves cannon and a 7.62 mm coaxial machine gun, with the computerised fire-control system being similar to that fitted to the Centauro (8 × 8) tank destroyer and the Ariete MBT which are already in service with the Italian Army. The commander had a roof-mounted stabilised sight.

A decision was then taken to retain the chassis of the VCC-80 but to develop a two-person turret armed with a 25 mm cannon and 7.62 mm machine gun with the addition of a Raytheon Systems Company TOW ATGW system either side of the turret. A different and less expensive fire-control system was also fitted.

Late in 1998, the first production order for the Dardo IFV was placed with the Consorzio Iveco Oto. This covered the supply of a first batch of 200 vehicles including the standard version with a two-person turret armed with a 25 mm cannon and 7.62 mm coaxial machine gun.

In addition, the contract also covered four specialised versions: anti-tank, 120 mm mortar, command post and ambulance. The anti-tank is essentially the Dardo with a launcher for the Raytheon Systems Company TOW ATGW mounted either side of the turret. All Dardo turrets built are fitted to accept the TOW launchers and guidance system.

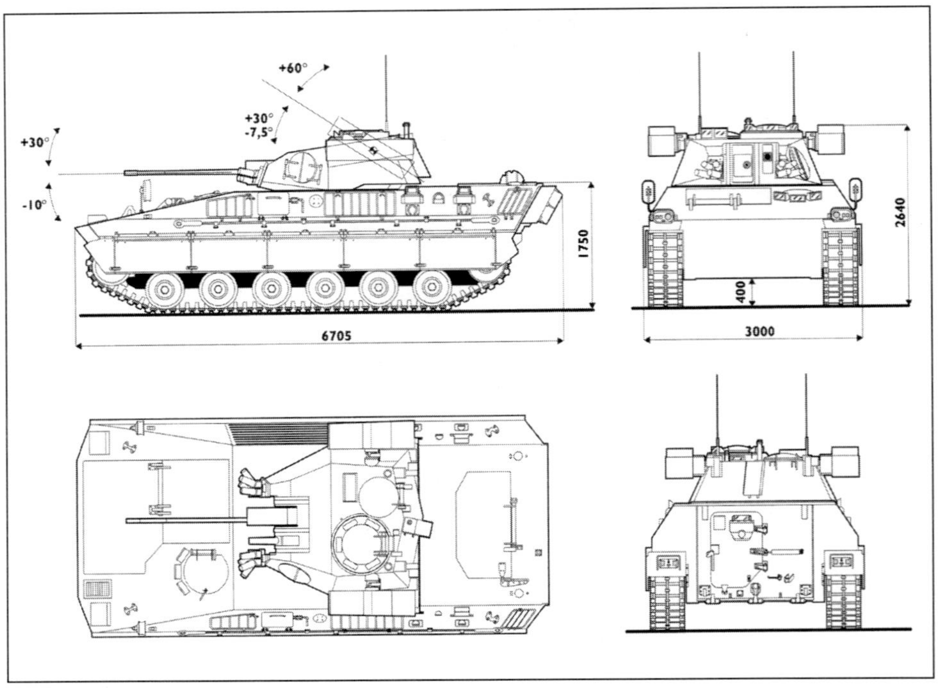

Four-view general arrangement drawing of Dardo infantry fighting vehicle **2002**/0110665

Description

The hull and turret of the Dardo are of all-welded aluminium armour construction, with an additional layer of ballistic steel armour bolted to the hull and turret for increased protection.

The driver sits at the front left side of the vehicle and has a single circular hatch that opens to the right. In front of this are three day periscopes for forward observation, the centre one of which can be replaced by a passive periscope for driving at night.

The engine compartment is to the right of the driver and separated from the remainder of the vehicle by a bulkhead. The air inlet and outlet louvres are in the top of the hull and the exhaust outlet is on the right side of the hull.

The IVECO four-stroke supercharged water-cooled multifuel diesel engine develops 520 bhp, giving the Dardo a power-to-weight ratio of about 22.60 hp/t. This diesel engine is the same as that installed in the Centauro (8 × 8) 105 mm armed tank destroyer of which 400 have been delivered to the Italian Army with a further 22 being delivered to Spain in late 2000.

The engine is coupled to a hydromechanical shift reverse and steering system with a built-in retarder giving a total of four forward and two reverse gears.

The ZF LSG 1500 automatic transmission includes a torque converter, a retarder and three epicycloidal gear trains. The steering system is integral with the transmission, with hydrostatic control of the steering radii and pivot being continuous.

The control of the forward and reverse ranges is electronic and automatic, including the converter lock up clutch. The speed ranges can be preselected manually among four programmes, after which the gears are selected automatically.

The brake discs are of the self-ventilated type and are splined to the shaft of the final drives. The braking system is of the full power dual-circuit type and is fed by a single pump driven directly by the engine.

The cooling system of the vehicle has been designed to enable the Dardo vehicle to operate in an ambient temperature of up to +49°C with a high percentage of dust in the atmosphere.

The air fan is hydraulically operated with speed control depending on the coolant temperature. The air intake grill is located on top of the front plate of the hull.

Exhaust gases expelled from the engine turbocharger enter a manifold, provided with a silencer, from which they are forced outward through a grid on the right side of the vehicle after mixing with the expelled air of the cooling system to reduce the thermal signature.

Dardo infantry fighting vehicle **2002**/0110663

Dardo infantry fighting vehicle fitted with T60/70A turret armed with 60 mm gun **2000**/0064826

Dardo infantry fighting vehicle ***2001***/0093575

Dardo infantry fighting vehicle from the rear ***2002***/0110664

The electrically operated turret is in the centre of the vehicle, with the commander on the left and the gunner on the right, both being provided with a circular single-piece hatch cover that opens to the rear. The turret is called the TC-25 and has been developed by Otobreda. Although the turret has a two-man crew it can also be operated by one man.

The commander has six day periscopes for observation while the gunner has a roof-mounted day/thermal night sight. The fire-control system includes the gunner's day/thermal night sight which includes an eye-safe laser range-finder.

In mid-1999, Otobreda awarded the US company of Kollsman a contract worth US$34 million for the supply of 202 Day/Night Range Sights (DNRSs) for installation in the Dardo IFV.

The first two DNRSs were delivered to Otobreda late in 1999, followed by production ramp up with final deliveries being made in 2004. All production will be undertaken in the US.

These DNRSs will be installed at the gunner's station of the Otobreda TC-25 two-person power-operated turret and will allow the gunner to detect, recognise and engage targets under day and night conditions while the Dardo is stationary or on the move, so increasing first-round hit probability.

For the Dardo application the Kollsman DNRS features a thermal channel, day channel with a magnification of ×12, integrated eye-safe laser range-finder and integrated TOW tracking capability. The system also features a remote display of the gunner's thermal picture at the commander's station.

Main armament comprises a stabilised Oerlikon Contraves 25 mm KBA dual-feed cannon with a 7.62 mm MG 42/59 machine gun being mounted coaxial to the left.

Mounted on either side of the turret is a Raytheon Systems Company TOW ATGW system with a maximum range of 3,750 m. A bank of four 80 mm electrically operated smoke grenade launchers is mounted either side of the turret, firing forwards.

The six infantrymen sit at the rear of the vehicle and enter and leave via the power-operated ramp in the rear. Two are seated on each side facing the front and two in the centre, one of which sits facing the left side of the vehicle and the other facing the rear.

The troop compartment has five firing ports, each with an associated vision block allowing five of the infantrymen to fire their weapons from inside the vehicle. Two of these are in either side of the hull and one is located in the door position in the left side of the power-operated ramp at the rear.

The infantry personnel enter and leave the vehicle via a hydraulically operated ramp at the rear of the hull, which has an integral door should the ramp fail to open. Over the top of the troop compartment is a single roof hatch that opens to the rear. This is also used to reload the TOW missile launcher from inside under armour protection.

The torsion bar suspension each side consists of six dual rubber-tyred roadwheels with the drive sprocket at the front, idler at the rear and three track-return rollers. The upper part of the suspension is covered by a skirt.

Standard equipment on the vehicle includes a full range of passive night vision equipment, an air conditioning system, automatic fire extinguishing system for the crew and engine compartment and equipment to allow the vehicle to ford to a depth of 1.5 m. The vehicle is fitted with an integrated air conditioning system and NBC system.

In place of the current torsion bar suspension system, a hydropneumatic suspension system could also be fitted. Other optional equipment includes a roof-mounted laser warning system and a commander's roof-mounted stabilised sight.

Variants

At present there are no variants of the Dardo in production although the vehicle could be fitted with other types of turret such as the two-man Otobreda T 60/70A turret armed with a 60 mm High-Velocity Gun System, details of which are given below.

The manufacturer states that the Dardo chassis could be the basis for a complete family of vehicles including air defence, engineer vehicle, command post vehicle and specialised liaison.

30 mm turret

As an alternative to the 25 mm version, Otobreda has developed a 30 mm Bushmaster II HITFIST turret. This turret, integrated with TOW missiles, has been fitted and tested on the Alvis Vehicles Stormer 30 chassis covered in the Reconnaissance vehicles section.

T60/70A turret system

Otobreda have developed the OTO T60/70A turret as a private venture to meet user requirements for an advanced weapon system to defeat Light Armoured Vehicles (LAVs) and Infantry Fighting Vehicles (IFVs) well into the 21st century.

The latest OTO T60/70A turret is of all-welded ballistic steel construction which provides protection from 14.5 mm machine gun fire and shell splinters. The basic turret weighs 5,000 kg but can be fitted with an additional layer of passive armour for increased protection against attack from 35 mm armour piercing projectiles through a significant frontal arc.

The gunner is seated on the right and the vehicle commander on the left, with both having a single-piece hatch cover that opens to the rear.

Main armament comprises an Otobreda 60 mm rifled gun designated the OTO 60/70 which has a barrel length of 4,200 mm and is autofrettaged. Maximum recoil length is 270 mm and recoil force is 9,000 kg. The barrel is fitted with a fume extractor but no muzzle brake is required.

Two combat natures of 60 mm ammunition have been developed, High-Explosive - Tracer (HE-T) and Armour-Piercing Fin-Stabilised Discarding Sabot - Tracer (APFSDS-T), both of which have a conventional brass cartridge case. Firing mode is electrical and one round can be fired every 2 seconds.

Mounted vertically in the forward part of the turret basket are the two drum-type magazines, each of which holds eight rounds of ready use ammunition. Typically one would contain APFSDS-T and the other HE-T natures of ammunition.

The APFSDS-T projectile weighs 850 g and has a muzzle velocity of 1,680 m/s and an improved length-to-diameter ratio. No details have been released on its penetration characteristics but these could be expected to be in excess of 120 mm of conventional steel armour at 60° to the vertical at a range of 2,000 m.

The HE-T round weighs 7.4 kg and has a muzzle velocity of 1,000 m/s. The actual projectile weighs 3.07 kg and is fitted with a triple action fuze and has an effective area of about 660 m^2.

Specifications

Crew: 2 + 7
Combat weight: 23,000 kg
Power-to-weight ratio: 22.6 hp/t
Ground pressure: 0.70 kg/cm^2
Length: 6.705 m
Width: 3.00 m
Height:
(overall) 2.64 m
(hull top) 1.75 m
Ground clearance: 0.40 m
Max speed: 70 km/h
Cruising range: 500 km
Fording: (without preparation) 1.5 m
Gradient: 60%
Side slope: 40%
Vertical obstacle: 0.85 m
Trench: 2.5 m
Engine: IVECO 8260 V-6 4-stroke turbocharged, intercooled diesel engine developing 520 bhp at 2,300 rpm
Transmission: ZF LSG 1500 fully automatic with 4 forward and 2 reverse gears. It also has a built-in retarder
Steering: hydrostatic
Suspension: torsion bar with bumper and hydraulic shock-absorber located on each suspension arm
Electrical system: 24 V
Batteries: 6 × 12 V, 180 Ah at 24 V
Armament:
(main) 1 × 25 mm Oerlikon Contraves 25 mm KBA cannon
(coaxial) 1 × 7.62 mm MG 42/59
(anti-tank) 2 × TOW launchers
(smoke grenade dischargers) 2 × 4 80 mm
Ammunition:
(main) 200 (ready use)
(coaxial) 700 (ready use)
(anti-tank) 2 TOW (ready use)
Gun control equipment
Turret traverse: 360° electric/manual
(by commander) yes
(by gunner) yes
Gun elevation/depression: +60/–10°
TOW launcher elevation/depression: +30/–7.5°
Gun stabiliser:
(vertical) yes
(horizontal) yes
NBC system: yes
Night vision equipment: yes

Note: Maximum combat weight with appliqué armour is 26,000 kg.

The triple-action fuze of the HE-T round can be set for impact, time programmable or proximity functions, depending on the actual target to be engaged.

The vertical sliding wedge breech mechanism is first opened manually using a lever, after which operation, for example loading and ramming the projectile, firing and ejection of the spent cartridge cases, is fully automatic. Manual back-up controls are provided in case of power failure.

Either the commander or the gunner can aim and fire the main and secondary armament when the vehicle is stationary or moving. Turret traverse is a full 360° and weapon elevation is from –6 to +40°.

The sighting system of the turret is from the Ariete MBT and Centauro 105 mm tank destroyer which are in service with the Italian Army, while the fire-control system is from the Dardo HITFIST IFV.

The commander has a Galileo two-axis stabilised panoramic day (×2.5 and ×10 magnification)/image intensification sight and in a typical target engagement would first detect the target and then hand over to the gunner who would complete the target engagement using his Galileo stabilised day/thermal sight with eyesafe laser range-finder. As an option, the commander's and gunner's sights can be fitted with second-generation IR thermal cameras. The commander also has six day observation periscopes, while the gunner has two.

The computerised fire-control system includes the commander's sight, commander's monitor screen of gunner's thermal picture, gunner's roof-mounted sight, commander's and gunner's control panels, display panel, automatic loading control electronics, distribution and control box, azimuth sensor and turret and gun gyros.

The heart of the fire-control system is the digital 32 bit parallel processing computer which automatically processes target range (from the laser range-finder), elevation and traverse, angular velocity from gun gyro, longitudinal tilt angle, traverse tilt angle, charge temperature and barrel wear.

A 7.62 mm machine gun is mounted coaxial to the right of the 60 mm gun and a similar weapon can be mounted on the turret roof for local and air defence purposes. A bank of four electrically operated 76 mm grenade launchers is mounted either side of the turret, firing forwards. These can fire smoke or HE/canister grenades. A laser-warning system can be fitted as an option and this can directly control the launching of the 80 mm grenades.

Total turret ammunition supply is 32 rounds of 60 mm (16 + 16), 200 rounds of 7.62 mm ready use ammunition for the coaxial machine gun plus 100 rounds of ready use ammunition for the anti-aircraft machine gun and 1,000 7.62 mm rounds in reserve.

For trials purposes in Italy, the T60/70A turret was installed on one of the prototypes of the Dardo.

If required, a launcher for the 3,750 m range Raytheon Systems TOW anti-tank guided missile could be mounted either side of the turret in a similar manner to that of the 25 mm turret of the Dardo selected by the Italian Army.

In 1997 the OTO T60/70A turret was sent to Poland where it was installed on the locally developed BWP 2000 IFV chassis for an intensive series of firepower and mobility trials with a Polish Army crew.

The OTO T60/70A turret design is such that it can be fitted to a wide range of other chassis, tracked and wheeled, weighing over 10 tonnes. These include the Russian BMP-1, SANTA BARBARA BMR 600 (6 × 6), Swiss MOWAG Piranha (6 × 6 and 8 × 8), Alvis Vehicles Stormer, United Defense M113 and Alvis Vehicles Warrior, to name but a few.

According to Otobreda, in the future the Italian Army could operate a mix of Dardo vehicles and the same chassis fitted with the OTO T60/70A turret system.

Status

In production for the Italian Army which has placed an initial order for 200 units. First production vehicles were completed in 2002.

Manufacturer

Consorzio Iveco Oto.

UPDATED

Otobreda Infantry Armoured Fighting Vehicle (VCC-1)

Development

The Infantry Armoured Fighting Vehicle (IAFV) is the result of a development programme by the Automotive Technical Service (Servizio Tecnico della Motorizzazione) of the Italian Army to increase the combat effectiveness of the M113A1 APC.

The Infantry Armoured Fighting Vehicle is based on the M113A1 APC which Fiat, Lancia and Oto Melara manufactured under licence from the then FMC Corporation (today United Defense, Ground Systems Division) of the USA in the 1960s.

Main improvements over the M113A1 can be summarised as increased firepower, improved armour protection, the infantry's ability to aim and fire their weapons from inside the hull, and improved seating arrangements.

The IAFV is called the VCC-1, or Camillino, by the Italian Army. Final deliveries were made to the Italian Army in 1982 when production of the TOW armed version of the VCC-1 for Saudi Arabia began.

Details of Italian production of the M113 series of full-tracked armoured personnel carriers are given in the following entry.

Production of the VCC-1 was completed some time ago and there are plans to upgrade the current A1 configuration to the A2 plus package comprising a mix of A2 variants (suspension and cooling) together with other derivatives equipped with the most recent A3 RISE (Reliability Improvement for Selected Equipment) package. The standard RISE package includes an upgraded propulsion system (turbocharged engine and new transmission), improved drive controls (new power brakes and conventional steering controls) and 200 A alternator.

From 2001/2002, the VCC-1 will be supplemented in Italian Army service by the Iveco/Oto Dardo, which is covered in detail in a separate entry.

Description

The hull of the IAFV is made of all-welded aluminium armour with an additional layer of steel armour on the front and sides of the hull to improve ballistic protection.

The driver sits at the front of the hull on the left and has a single-piece hatch cover that opens to the right. There are four M17 day periscopes to the front and left of his position and a single M19 day periscope, mounted in the hatch cover, which can be traversed through a full 360°. The M19 can be replaced by a passive periscope for night driving.

The power pack is mounted to the right of the driver with the air inlet and outlet louvres in the roof of the vehicle. The engine compartment is fitted with a fire warning and extinguishing system and the engine is coupled to an Allison Transmission TX-100

Saudi Arabian VCC-1 with a Systems & Electronics Inc Improved TOW System installed on the roof

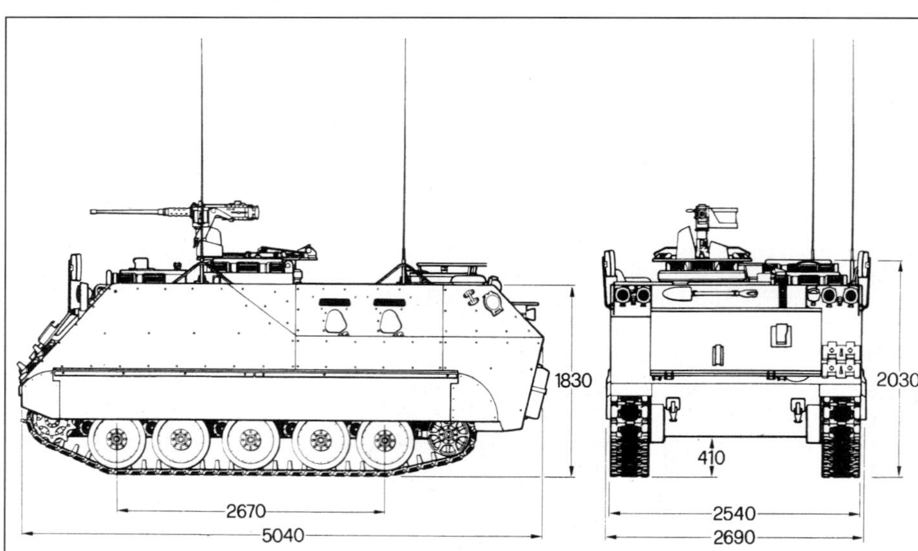

Side and front drawings of the Otobreda IAFV armed with a roof-mounted 12.7 mm MG

Italian Army VCC-1 fitted with appliqué armour deployed to the Balkans (Richard Stickland) **2002**/0110666

Italian Army VCC-2, without 12.7 mm M2 machine gun fitted, deployed to Bosnia (Richard Stickland) **2001**/0093576

transmission which consists of a three-speed gearbox and a two-stage torque converter which gives six forward and two reverse gears. Access to the engine for maintenance is by a panel in the glacis plate and panels inside the crew compartment.

The vehicle commander sits immediately behind the driver and has a single-piece hatch cover that opens to the rear, and five M17 day periscopes. The gunner sits to the right of the commander and has a three-piece hatch cover providing protection to his sides and rear when raised, and five M17 day periscopes. The six infantrymen sit at the rear of the vehicle, three down each side of the hull on individual seats. The first four face outwards and the last two inwards. The upper part of the hull sides of the personnel compartment is sloped to improve ballistic protection and in each side of the hull are two firing ports, each with a rectangular vision block over it. The infantrymen enter and leave the vehicle by a power-operated ramp in the rear of the hull. There is a firing port in the right side of the ramp and an emergency door in the left side should the ramp fail to open. In the roof of the personnel compartment, at the rear, is a rotating hatch which is covered by a three-part triangular hatch cover when not in action. The forward part of the cover folds forwards and is locked vertical and the two side covers open to the outside of the vehicle. These are also locked vertical to provide protection for the machine gunner. A 7.62 mm machine gun can be mounted at this position to provide covering fire to the rear of the vehicle.

The two fans in the top of the hull, one either side of the rear gunner's position, expel fumes from the personnel compartment. There are two aluminium diesel fuel tanks each side of the ramp at the rear.

The torsion bar suspension consists of five dual rubber-tyred roadwheels, with the drive sprocket at the front and the idler at the rear. There are no track-return rollers. The first and last roadwheel stations are provided with a hydraulic shock-absorber.

The basic model of the IAFV is called the Mk 1 and is armed with a pintle-mounted 12.7 mm M2 HB machine gun. The Mk 2 version, which is not in service with the Italian Army, has a remote-controlled 12.7 mm machine gun. The Mk 3 version, still at the prototype stage, has an externally mounted 20 mm cannon.

The IAFV is fully amphibious, being propelled in the water by its tracks. Before entering the water a trim board is erected at the front of the hull and the two electric bilge pumps are switched on. The vehicle is fitted with passive night vision devices but has no NBC system.

VCC-1 with additional armour

The Italian Army deployed some 50 VCC-1 IFVs to Somalia which were subsequently fitted with an Enhanced Appliqué Armour Kit (EAAK) similar to those fitted to the AAV7A1 amphibious assault vehicles of the US Marine Corps.

This armour package provides protection against penetration from small arms fire up to 14.5 mm as well as 155 mm projectile fragments. The total weight of the kit is 750 kg and crew survivability is further enhanced by the installation of spall liners inside the rear crew compartment.

The EAAK armour plates use Special Spaced Armour (SSA) which consists of a laminated structure of special steels with non-metallic layers between them. These steel plates break and deflect kinetic energy projectiles and reduce the residual energy so that the projectile or debris cannot penetrate the hull.

These kits were developed by RAFAEL of Israel and supplied to Italy by the FMS Corporation (now Marvin Land Systems) of Los Angeles under a contract worth US$1.5 million. The contract was awarded in July 1993 and the kits were delivered and fitted to the vehicles in Somalia early in October, ahead of schedule.

VCC-1 with TOW

Saudi Arabia placed an order worth LIt350,000 million with Otobreda for 224 VCC-1 vehicles fitted with the same Systems & Electronics Inc TOW ATGW as the

Specifications

Crew: 2 + 7
Combat weight: 11,600 kg
Power-to-weight ratio: 18.53 bhp/t
Ground pressure: 0.57 kg/cm²
Length: 5.041 m
Width: 2.686 m
(over tracks) 2.54 m
Height:
(over MG) 2.552 m
(over commander's cover) 2.032 m
(to hull top) 1.828 m
Ground clearance: 0.406 m
Track: 2.159 m
Track width: 381 mm
Length of track on ground: 2.667 m
Max speed:
(road) 64.4 km/h
(water) 5 km/h
Fuel capacity: 360 litres
Range: 550 km
Fording: amphibious
Gradient: 60%
Side slope: 30%
Vertical obstacle: 0.61 m
Trench: 1.68 m
Engine: Detroit Diesel Model 6V-53 6-cylinder water-cooled diesel developing 215 bhp at 2,800 rpm
Transmission: Allison Transmission TX-100 with 3 forward and 1 reverse gears coupled to 2-stage torque converter which gives 6 forward and 2 reverse gears
Suspension: torsion bar
Electrical system: 24 V
Armament:
(main) 1 × 12.7 mm M2 MG
(secondary) 1 × 7.62 mm MG 42/49 MG
Smoke-laying equipment: none
Ammunition:
(main) 1,050
(secondary) 1,000
Gun control equipment
Cupola traverse: manual
Gun stabiliser:
(vertical) no
(horizontal) no
NBC system: no
Night vision equipment: optional

IAFV with firing ports open and armour protection for the 12.7 mm machine gun raised

Italian Army VCC-1 fitted with appliqué passive armour (Richard Stickland) **2002**/0110667

M901 Improved TOW Vehicle. The first 10 production vehicles were delivered early in 1983 and final deliveries were made in 1984. Full details of the M901 Improved TOW Vehicle are given in the Tank destroyers section later in this book.

VCC-2

This is the Italian-built M113 with appliqué armour fitted to the front and sides. There are two firing ports with a bulletproof vision block above in either side of the rear troop compartment. The VCC-2 has the standard circular hatch for the driver. To the rear of this is a circular hatch for the commander, while the gunner has an externally mounted 12.7 mm M2 HB machine gun with lateral protection.

Other variants

Otobreda has proposed that the vehicle could be fitted with other turrets including 4 × 25 mm SIDAM anti-aircraft, 20 mm, 25 mm, 60 mm and 90 mm, all of which have been tested on other Otobreda designed and built vehicles.

Status

Production complete. In service with Italy and Saudi Arabia (TOW version only).

Manufacturer

Collaboration between Fiat, Lancia and Oto Breda (later Otobreda) with final assembly being undertaken at the Oto Melara plant in La Spezia.

UPDATED

ARIS Arisgator armoured personnel carrier

Development

The Italian company of ARIS SpA (Applicazioni Rielaborazioni Impianti Speciali) has developed, as a private venture, a special kit to improve the amphibious capability of the United Defense LP M113 series full tracked armoured personnel carrier (APC).

The first prototype of this vehicle, called the Arisgator was completed in 1997 and early in 1999 it was stated that the Italian MoD had awarded a low rate initial production contract to the company. The M113 is used in large numbers by the Italian Army and to a lesser extent the Marines.

Description

The standard United Defense LP designed M113 series APC is fully amphibious being propelled in the water by its tracks at a maximum speed of 5.8 km/h which is adequate for calm water.

The M113 was not designed to operate in fast flowing rivers or to be launched from landing craft off shore.

To improve the amphibious characteristics of the M113 a kit has been developed by ARIS which enables the vehicle to operate safely in ship-to-shore operations as well as rivers.

Specifications
(as M113 except)
Length: 6.87 m
Width: 2.95 m
Height: 2.05 m
Ground clearance: 0.414 m

This kit weighs between 1,350 and 1,700 kg including bow, roof and stern units, all of which are made of welded aluminium as is the basic M113 series hull.

A pointed front bow has been added to improve the buoyancy of the vehicle and two stern units are fitted, one either side of the rear ramp. Each ramp incorporates a hydrostatically operated propeller which can be steered independently allowing the Arisgator to turn upon its axis while afloat.

The modified roof has a new cooling system and there are now two snorkel-type extension tubes in the right side of the hull roof, front and rear. An extension has been fitted to the exhaust pipe on the right side of the hull roof and a special cover system has been fitted over the normal air inlet/outlets in the roof of the M113 to the right of the drivers position.

The normal commander's roof-mounted cupola and oblong hatch over the rear troop compartment have been retained as has the power-operated ramp in the hull rear. The overall sealing of the M113 series vehicle has also been improved.

The Arisgator can carry 11 fully equipped troops and can also be fitted with various external weapon stations.

Variants

Upgraded M548

Based on their experience in the M113, upgraded to enhance its amphibious capabilities, ARIS have carried out a similar upgrade to the M548 full-tracked cargo carrier which is also a member of the M113 family of vehicles.

Status

Low-rate production for the Italian Army. Entering service with Italian Army.

Manufacturer

ARIS SpA.

UPDATED

Arisgator showing its amphibious capabilities, with driver's hatch in open position and air intakes extended 0058575

Arisgator out of the water and clearly showing bow and stern extensions for improved amphibious capability ***2002***/0110655

Italian M113 APC fleet

Otobreda (previously Oto Melara) of La Spezia, built 4,278 M113 series armoured personnel carriers under licence from the then FMC Corporation (now United Defense LP) of the USA, including M577 command post vehicles and M548 tracked cargo carriers.

The production was a collaboration between Oto Melara (now Otobreda), Fiat and Lancia. The first batch of 40 M113 series APCs was delivered to the Italian Army in May 1964.

The first order comprised a total of 3,000 vehicles for the Italian Army, with the second order being for 697 vehicles. A total of 581 vehicles were also built for Turkey.

M577 series command post, built under licence in Italy, showing higher roof (left) with VCC-2 without armament fitted (right) (Richard Stickland) ***2000***/0064829

The initial M113 Chrysler petrol engine versions were subsequently replaced in production by the Detroit Diesel diesel-powered M113A1.

Further development by Otobreda and the Italian Army resulted in the Infantry Armoured Fighting Vehicle described in the previous entry. These are also known as the VCC-1, with some of the modified M113s being called the VCC-2.

One of the latest vehicles, based on the M113 series chassis, to enter service with the Italian Army is the Otobreda SIDAM 25 (4 × 25 mm) self-propelled anti-aircraft gun system. Full details of this are given in *Jane's Land-Based Air Defence 2001-2002.*

This consists of an M113 chassis modified by Astra with a new power-operated turret armed with four 25 mm Oerlikon Contraves KBA cannon and a day-only fire-control system.

Key modifications to the chassis can be summarised as: replacement of the existing Detroit Diesel engine with a more powerful Detroit Diesel 6V-53T developing 265 hp; two new fuel tanks mounted one either side externally at the hull rear, giving a total capacity of 360 litres; modifying the top deck to incorporate an improved reverse-flow cooling system; reinforcement of idler wheel mounts and repositioning the shock-absorber supports and the existing fan; replacement of air cleaner and silencer.

The existing Allison TX-100-1 transmission can be modified or replaced by the TX-100-1A or the new X-200-4. Kits to perform these chassis conversions were purchased by Astra of Italy from Marvin Land Systems of the USA.

New and stronger torsion bars are fitted and the combat engineer vehicle has additional armour protection, a front-mounted dozer blade and a 3 tonne hydraulic crane. Internally there is a 10 tonne winch and stabilisers are provided at the rear.

In Italian Army service the Infantry Armoured Fighting Vehicle is known as the VCC-1 or Camillino. The VCC-2 is the locally built M113 with appliqué armour fitted to the hull and sides. There are two firing ports with a bulletproof vision block above on either side of the rear troop compartment. The M113 upgraded by Astra is known as the VTC-9 in Italian Army Service.

Otobreda SIDAM 4 × 25 mm self-propelled anti-aircraft gun system 0045948

A contract was signed in 1999 with the Italian Army for the upgrading of 100 M577 command post vehicles and M106 mortar carriers. The upgrade will bring these up to the A2 plus configuration.

In 1997, Chile took delivery of 12 M113 series APCs from Italy with up to 128 expected to be supplied. In 1994, Italy supplied Turkey with 97 M113 series APCs.

Albania is also expected to take delivery of 100 vehicles as part of NATO's Partnership For Peace (PFP) programme.

Manufacturer

Collaboration between Fiat, Lancia and Oto Melara (now Otobreda). Final assembly at the Oto Melara plant in La Spezia.

UPDATED

JAPAN

Mitsubishi Type 89 Mechanised Infantry Combat Vehicle

Development

The 1984 Japanese defence budget included a ¥600 million request for the construction of four prototypes of a new mechanised infantry combat vehicle. In FY81 funding was provided for prototypes of the turret and hull. Following trials with the prototype vehicles, it was type-classified as the Type 89 Mechanised Infantry Combat Vehicle and it is now in production for the Japanese Ground Self-Defence Force, with the initial requirement being for 300 vehicles.

Production of the Type 89 has been undertaken at a very low rate and, by early 1997, only 52 vehicles were in service with the Japanese Ground Self-Defence Force. Since then it is understood that production has continued at a low rate and at least 60 vehicles have now been built.

Prime contractor for the Type 89 MICV is Mitsubishi Heavy Industries, although a major subcontractor is Komatsu.

Description

The hull of the Type 89 MICV is of all-welded construction which provides the crew with protection from small arms fire and shell splinters. Mounted under the nose of one of the prototype Type 89 MICVs were attachment points for a mine clearance system.

The power pack is mounted at the front of the vehicle on the left side with the exhaust outlet on the left side of the hull, the driver being seated on the right and one of the seven infantrymen to his immediate rear.

The driver has a single-piece hatch cover that opens to the right. In front of this are three day periscopes. In the driver's roof hatch is a day periscope which can be traversed. One of the driver's day periscopes can be replaced by a passive night vision device. The infantryman to the rear of the driver has a single-piece hatch cover, periscopes that give observation to the front and sides and a spherical firing port in the right side of the hull.

The two-man power-operated turret is in the centre of the hull, with the commander seated on the right and gunner on the left, both having single-piece hatch covers opening to the rear. Both the commander and gunner have a roof-mounted periscopic sight mounted in the forward part of the turret. The gunner has two fixed day periscopes to give observation to his front and left side, and the commander has six day periscopes giving observation to the front, sides and rear. Mounted on the turret roof is a laser warning device, which is also fitted to most recent production Japanese AFVs.

Main armament is an Oerlikon Contraves 35 mm KDE cannon which has a cyclic rate of fire of 200 rds/min. First weapons were supplied direct from Switzerland, with licensed production then being

Type 89 MICV with a launcher for an ATGW either side of the turret (Kensuke Ebata)

Type 89 MICV showing three spherical firing ports in the side of the hull (Paul Beaver)

Close-up of the Type 89 MICV turret showing firing ports and smoke grenade dischargers (Kensuke Ebata)

Type 89 MICV from the rear showing ATGW launcher on turret side (Paul Beaver)

undertaken in Japan. A 7.62 mm Type 74 machine gun is mounted coaxially to the left of the locally manufactured Oerlikon Contraves 35 mm dual-feed cannon and either side of the forward part of the turret is a bank of three electrically operated smoke grenade dischargers firing forwards.

Mounted either side of the turret is a single Jyu-MAT medium-range wire-guided anti-tank missile developed by Kawasaki Heavy Industries. Once the missiles have been launched, new missiles have to be reloaded manually. These have a maximum range of at least 4,000 m.

The raised troop compartment is at the rear of the hull, with the troops entering and leaving via two doors in the rear that open outwards. The right door has a spherical firing port. There are a further three firing ports in the right side of the troop compartment and three in the left side, each of which has a roof-mounted day periscopic sight. Over the top of the troop compartment is a two-part roof hatch that opens left and right of the vehicle's centreline. There are a further two roof-mounted day periscopes which give observation to the rear of the vehicle.

Of the six infantrymen seated in the rear, the three on the left side face outwards, the first two on the right side face outwards and the third sits facing the rear.

Suspension is of the torsion bar type with either side seven dual rubber-tyred roadwheels, drive sprocket at the front, idler at the rear and three track-return rollers.

The vehicle does not have any amphibious capability, but is fitted with an NBC system and a full range of passive night vision equipment.

Specifications

Crew: 3 + 7
Combat weight: 27,000 kg
Power-to-weight ratio: 22.22 hp/t
Ground pressure: 0.73 kg/cm²
Length: 6.7 m
Width: 3.20 m
Height: 2.50 m
(overall) 2.75 m
Ground clearance: 0.45 m
Track: 2.68 m
Length of track on ground: 4.00 m
Track width: 450 mm
Max road speed:
(forwards) 70 km/h
(reverse) 42 km/h
Range: 400 km
Fording: 1 m
Gradient: 60%
Side slope: 30%
Vertical obstacle: 0.8 m
Trench: 2.4 m
Engine: 6 SY 31 WA diesel developing 600 hp at 2,000 rpm
Transmission: automatic
Suspension: torsion bar
Armament:
(main) 1 × 35 mm Oerlikon Contraves KDE cannon
(coaxial) 1 × 7.62 mm Type 74 MG
(ATGW) 2 × Jyu-MAT
Smoke grenade dischargers: 2 × 3
Gun control equipment
Turret power control: powered/manual
Turret traverse: 360°
NBC system: yes
Night vision equipment: yes

Variants

There are no known variants of the Type 89 mechanised infantry combat vehicle although a modified version of its chassis may be used for a new 155 mm self-propelled artillery system.

Type 96 120 mm self-propelled mortar

Available details of this system, which may be related to the Type 89, are given in the Self-propelled mortar systems section.

Status

In production. In service with the Japanese Ground Self-Defence Force. This vehicle has not been offered on the export market.

Manufacturer

Mitsubishi Heavy Industries.

UPDATED

Mitsubishi Type 73 armoured personnel carrier

Development

The development of a new armoured personnel carrier to supplement the older SU 60 began in FY67 under the direction of the Technical Research and Development Institution of the Japanese Defence Agency.

In January 1967, the following requirements were issued for the new APC: maximum speed of over 60 km/h, ability to carry 12 men including the crew, to be fully amphibious, have all-welded aluminium armour, provision for the infantry to use their small arms from inside the vehicle and be armed with a 20 mm cannon, a 12.7 mm machine gun and a 7.62 mm machine gun.

In 1967, the following manufacturers started on the production of specific components: Hitachi Manufacturing Company (transmission), Kobe Iron Works (aluminium alloy), Komatsu Manufacturing Company (tracks), Mitsubishi Heavy Industries (engine, steering and suspension), and Nihon Electric Corporation (night vision equipment).

In 1968, an automotive test rig, called the SUT, was completed and tested. The following year Mitsubishi Heavy Industries and the Komatsu Manufacturing Company each built two prototype vehicles, one steel and one aluminium.

After trials, in December 1972, the second prototype of the Mitsubishi model, with aluminium armour, was

Close-up of the 7.62 mm bow-mounted machine gun position on Type 73 APC (K Nogi)

Type 73 APC with trim vane erected at the front for amphibious operations (Panzer Magazine)

Type 75 self-propelled ground wind measuring system in travelling configuration (Kensuke Ebata)

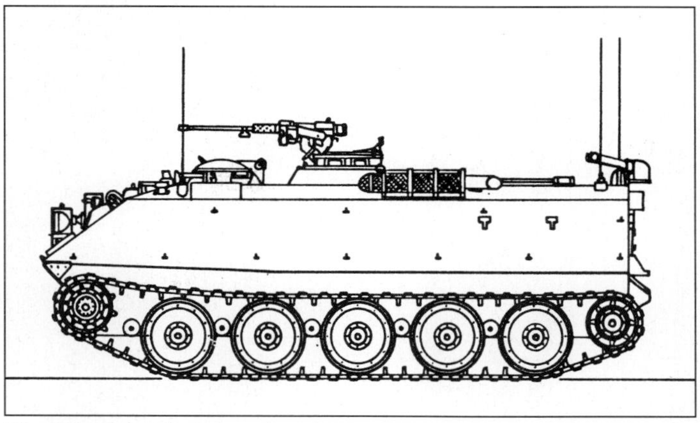

Type 73 armoured personnel carrier

Specifications

Crew: 3 + 9
Combat weight: 13,300 kg
Power-to-weight ratio: 22.6 hp/t
Length: 5.8 m
Width: 2.8 m
Height:
(incl MG) 2.2 m
(to hull top) 1.7 m
Ground clearance: 0.4 m
Track: 2.56 m
Track width: 380 mm
Length of track on ground: 3.67 m
Max speed:
(road) 70 km/h
(water) 7 km/h
Max range: 300 km
Fuel capacity: 450 litres
Fording:
(with kit) amphibious
Gradient: 60%
Vertical obstacle: 0.7 m
Trench: 2 m
Turning radius: 7 m
Engine: Mitsubishi 4ZF air-cooled 2-stroke V4 diesel developing 300 hp at 2,200 rpm
Suspension: torsion bar
Electrical system: 24 V
Armament:
(main) 1 × 12.7 mm M2 MG
(bow) 1 × 7.62 mm MG
Smoke-laying equipment: 2 × 3 smoke grenade dischargers
Gun control equipment
Cupola traverse: 360°, manual
Gun elevation/depression: +60/–10° (manual)
NBC system: yes
Night vision equipment: yes

selected for use by the Japanese Ground Self-Defence Force and after minor modifications and improvements was standardised as the Type 73 APC.

It is believed that total production of the Type 73 amounted to 340 vehicles. The Type 73 and Type SU 60 are now being very slowley supplemented by the new Type 89 MICV covered in the previous entry.

As the Type 89 MICV is being built in very small numbers, the older Type 73 APC will remain in service with the Japanese Ground Self-Defence Force well into the 21st century.

Description

The hull of the Type 73 APC is made of all-welded aluminium armour with the 7.62 mm bow machine gunner seated at the front of the hull on the left side. The 7.62 mm bow machine gun has an elevation, depression and traverse of 30°. The gunner has a single-piece hatch cover that opens to the left and has a single day periscope which can be traversed through 360°.

The driver sits at the front of the hull on the right side and has a single-piece hatch cover that lifts and swings horizontally to the rear to open, and an integral day periscope. To the front of the driver's position are three day periscopes. One of these can be replaced by a passive periscope for night driving.

The commander sits slightly behind the bow machine gunner and driver and has a single-piece hatch cover opening to the rear. Six day vision blocks give the commander all-round observation.

The engine is mounted to the rear of the bow machine gunner on the left side of the hull with the air inlet, outlet and exhaust pipe in the roof. The engine and transmission are mounted as a power pack and can be removed from the vehicle as a complete unit in half an hour. The machine gunner sits behind the driver and has a cupola that can be traversed through a full 360°. The cupola has a single-piece hatch cover that opens to the rear and six day periscopes for all-round observation. The 12.7 mm (0.50) M2 HB machine gun is not provided with a shield, as it can be aimed and fired from inside the cupola. The 12.7 mm machine gun has an elevation of +60°, depression of –10° and can be traversed through 360°.

The Type 73 carries nine infantrymen, one of whom normally mans the 12.7 mm (0.50) M2 machine gun position. The eight infantrymen sit in the troop compartment at the rear of the hull down either side on seats that can be folded up, and enter the vehicle via two doors that open outwards at the rear of the hull. There is a T-shaped firing port in each door and two in each side of the troop compartment. Over the top of the troop compartment are two single-piece hatch covers that open either side of the hull top. Mounted at the very rear of the hull top, one either side, is a three-barrelled smoke grenade discharger that fires forwards over the front of the vehicle.

The torsion bar suspension consists of five dual rubber-tyred roadwheels with the drive sprocket at the front and the idler at the rear. There are no track-return rollers. A hydraulic shock-absorber is provided for the first roadwheel station.

The basic model of the Type 73 APC is not amphibious although an amphibious kit has been developed and fitted to some vehicles. The kit consists of: buoyancy aids attached to the outside of each of the roadwheels; skirts fitted to cover the upper part of the tracks to improve water flow; box-type structures fitted on top of the hull over the air inlet, outlet and exhaust to stop water entering when afloat; and a trim vane fitted on the front of the hull. The trim vane is in two parts and is hinged horizontally. A transparent panel in the right side provides forward vision for the driver. When afloat, the vehicle is propelled in the water by its tracks.

The Type 73 is fitted with passive night vision equipment and an NBC system.

Variants

The Type 75 self-propelled ground wind measuring unit is used in conjunction with the Type 75 self-propelled 130 mm multiple rocket launcher (full details of this are given in the Multiple rocket launchers section) which also uses some components of the Type 73. The measuring unit has a mast which when extended is 12.5 m high and can measure wind velocity in two directions. The wind velocity measuring equipment was developed by Meisei Electric.

The Type 74 105 mm self-propelled howitzer and Type 75 multiple rocket launcher both use the same engine as the Type 73. There is also a command post version of the Type 73 which has a higher roof with hatches to the rear of the commander's cupola.

Mineclearing systems

Components of the Type 73 APC chassis are used in the Type 92 Mine Clearing Vehicle (MCV) which has six roadwheels, compared to five on the normal vehicle, and a larger idler.

Mounted at the front of the vehicle is a dozer blade and mounted on top of the chassis at the rear is a fully enclosed launcher with two rockets which are fired to a maximum range of 800 m.

The Type 92 MCV is related to the High-Speed Ammunition Carrier which also uses components of the Type 73 APC. Prime contractor for the MCV system is Nissan Motors, with the first prototype completed in 1989.

Some Type 73 APCs have also been used to mount the older Type 70 mineclearing system. This is mounted externally on the roof at the rear and can lay a 17 m line charge up to 550 m; it can also clear a path 100 m long and 0.5 m wide through a minefield.

Status

Production complete. In service only with the Japanese Ground Self-Defence Force. This vehicle has not been offered on the export market.

Manufacturer

Mitsubishi Heavy Industries.

UPDATED

Mitsubishi Type SU 60 armoured personnel carrier

Development

In the mid-1950s, the Japanese Ground Self-Defence Force issued a requirement for a full-tracked armoured personnel carrier. Development began in 1956 with overall control of the project managed by the Technical Research and Development Headquarters of the Japanese Self-Defence Agency. In 1957, prototypes of the APC were completed by Komatsu and Mitsubishi. The Komatsu prototype was called the SU-I and the Mitsubishi model the SU-II.

Both prototypes weighed 10.6 tonnes, carried nine men and had a maximum road speed of 45 km/h. Between 1957 and 1958, the SU-I and SU-II were tested alongside a US-supplied M59 APC. Eleven second series prototypes were manufactured by Mitsubishi Heavy Industries and the Komatsu Manufacturing Corporation. Mitsubishi built five SU-II (Kai) Ms, one SV (Kai) M and one SX (Kai) M, M standing for Mitsubishi and Kai for modified. Komatsu built two SU-II (Kai) Ks, one SV (Kai) K and one SX (Kai) K, K standing for Komatsu. The SV was the 81 mm mortar carrier and the SX was the 107 mm (4.2 in) mortar carrier.

In both, the mortar was mounted in the rear compartment but on the SX the top of the hull was sloped to the rear to avoid muzzle blast from the 107 mm mortar. This also decreased the weight of the vehicle.

In July 1959, four prototypes of the SU-II (Kai) were completed and tested through to October 1959. In 1960 the vehicle was standardised as the Type 60 APC,

Type SU 60 APC clearly showing roof-mounted 12.7 mm M2 HB and bow-mounted 7.62 mm machine guns

Type SX 4.2 in (107 mm) mortar carrier in travelling configuration

with the mortar carriers standardised as the Type 60 self-propelled 81 mm mortar and the Type 60 107 mm (4.2 in) mortar.

The first production contract for 24 vehicles was provided for in the 1959 budget and production was undertaken by Komatsu and Mitsubishi. Final orders were placed, with funding from the 1972 Budget.

It is estimated that in 2001 about 300 of these vehicles, including variants, remain in service.

As increasing numbers of the more recent Type 89 MICV's enter service with the Japanese Ground Self-Defence Force, older Type SU 60s will be progressively phased out of service as by today's standards they are obsolete.

Specifications

Crew: 4 + 6
Combat weight: 11,800 kg
Unloaded weight: 10,600 kg
Power-to-weight ratio: 18.64 hp/t
Ground pressure: 0.57 kg/cm²
Length: 4.85 m
Width: 2.4 m
Height:
(incl MG) 2.31 m
(to hull top) 1.7 m
Ground clearance: 0.4 m
Track: 2.05 m
Track width: 350 mm
Length of track on ground: 3.18 m
Max road speed: 45 km/h
Max range: 300 km
Fording: 1 m
Gradient: 60%
Vertical obstacle: 0.6 m
Trench: 1.82 m
Turning radius: 7 m
Engine: Mitsubishi Model 8 HA 21 WT, V-8 air-cooled 4-cycle turbocharged diesel developing 220 hp at 2,400 rpm
Transmission: manual with 3 forward and 1 reverse gears and torque converter
Steering: controlled differential
Suspension: torsion bar
Electrical system: 24 V
Batteries: 2 × 12 V
Armament:
(main) 1 × 12.7 mm M2 MG
(bow) 1 × 7.62 mm MG
Smoke-laying equipment: none
Gun control equipment
Cupola traverse: 360°, manual
Gun stabiliser:
(vertical) no
(horizontal) no
NBC system: no
Night vision equipment: no

Description

The hull of the SU 60 APC is made of all-welded steel with the driver seated at the front of the hull on the right. The driver has a single-piece hatch cover that opens to the right over his position, in front of which are three M17 day periscopes. The bow machine gunner sits at the front of the hull on the left and has a single-piece hatch cover that opens to the left rear, with an integral M6 day periscope for aiming the 7.62 mm (0.30) M1919A4 bow machine gun.

The commander sits immediately behind the driver's and bow machine gunner's position and has a single-piece hatch cover that opens to the left rear. The eight vision blocks provide all-round observation for the commander.

The engine is on the left side of the hull to the rear of the bow machine gunner and power is transmitted to the differential at the front of the hull by a short propeller shaft.

The gunner sits to the rear of the commander on the right side of the hull and has a two-piece hatch cover that opens either side of his position. The 12.7 mm (0.50) M2 HB machine gun can be traversed through a full 360° and is provided with a shield.

The troop compartment is at the rear of the hull with the six infantrymen seated on individual seats, three down each side of the hull. They enter and leave the vehicle by two doors in the rear of the hull that open outwards. The right door has a small observation slit. Over the top of the troop compartment is a single-piece hatch cover that opens forwards. To the rear of this is a two-part hatch cover that opens either side of the hull. Each part of the hatch cover is in two halves, hinged in the centre. There is no provision for the infantry to use their small arms from inside the vehicle.

The torsion bar suspension either side consists of five dual rubber-tyred roadwheels with the drive sprocket at the front, idler at the rear and three track-return rollers. The first, second and fifth roadwheel stations are provided with a hydraulic shock-absorber. The SU 60 APC does not have an NBC system, nor any night vision equipment or amphibious capability.

Variants

Type SV 60 81 mm mortar carrier

This is essentially a Type 60 APC with an 81 mm mortar mounted in the back of the hull firing to the rear with two doors in the rear of the hull. A bipod and baseplate are carried on the glacis plate, enabling the 81 mm mortar to be deployed away from the vehicle. The 81 mm mortar has an elevation of +40 to +85°, a traverse of 40° left and right and a maximum range of 3,000 m. The SV 60 carries 24 mortar bombs. It has a crew of five and weighs 12,100 kg fully loaded. Armament consists of one roof-mounted 12.7 mm M2 HB (0.50) anti-aircraft machine gun. Only 18 of these vehicles were built.

Type SX 60 4.2 in (107 mm) mortar carrier

This has a 4.2 in (107 mm) mortar mounted in the rear of the hull firing to the rear. A baseplate and stand are carried on the glacis plate, enabling the 107 mm mortar to be deployed away from the vehicle. The 107 mm mortar has an elevation of +37 to +65°, a traverse of 40° left and right and a maximum range of 4,000 m. Eight mortar bombs are carried. The SX 60 has a crew of five, weighs 12,900 kg fully loaded but is fitted with the M2 HB 12.7 mm anti-aircraft machine gun only. It can be distinguished from the 81 mm mortar carrier by the lack of a bow machine gun and the distinct chamfer to the rear of the top of the hull of the Type SX 60. The SX 60 has a single-piece ramp at the rear of the hull, hinged at its lower part. Only 18 of these vehicles were built. This has now started to be replaced by the Type 96 120 mm self-propelled full-tracked mortar system.

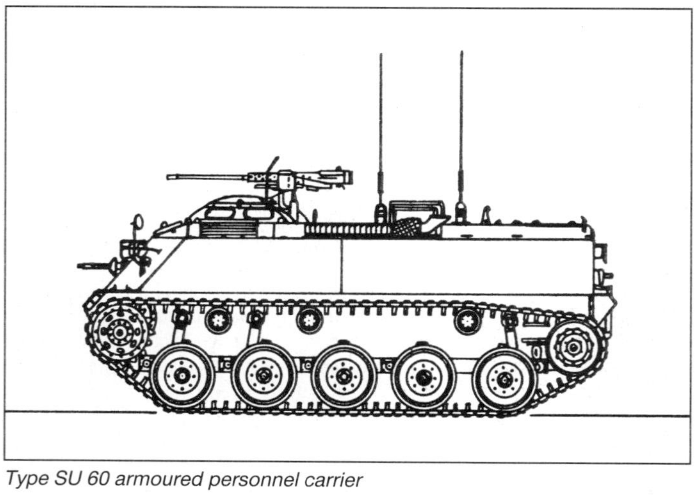

Type SU 60 armoured personnel carrier

Type SU 60 APC with all roof hatches open (K Nogi)

Type SU 60 Anti-tank Vehicle

A small number of Type SU 60 APCs have been fitted with Mitsubishi KAM-3D wire-guided anti-tank missiles in their launcher boxes, one on either side of the hull top rear.

Other variants

There are at least three other variants of the Type SU 60 APC, an NBC detection vehicle, a dozer and a model modified to resemble the Russian BMD-1 airborne combat vehicle used by the Japanese Ground Self-Defence Force for training purposes. Only two dozers are in service, used by the 7th Division for clearing snow. A prototype of a 105 mm self-propelled howitzer was built under the designation SY 60 but was never placed in production.

Status

Production complete. In service only with the Japanese Ground Self-Defence Force.

Manufacturer

Mitsubishi Heavy Industries.

UPDATED

JORDAN

King Abdullah II Design and Development Bureau AB14 Temsah heavy infantry fighting vehicle

Temsah heavy infantry combat vehicle fitted with roof-mounted Helio SWARM system **2002**/0095154

Development

Based on their experience in the design, development and production of a wide range of land systems, the King Abdullah Design and Development Bureau have developed the AB14 Temsah Heavy Infantry Fighting Vehicle (HIFV) which has also been referred to as an 'experimental demonstrator model'.

The first prototype of the Temsah HIFV was completed early in 2001 and was shown for the first time at the IDEX Exhibition held in Abu Dhabi in March 2001.

The King Abdullah Design and Development Bureau has been assisted in the development of this vehicle by the Mechanology Design Bureau located in South Africa.

According to the prime contractor, Temsah is a cost-effective conversion of potentially surplus to requirement Main Battle Tanks (MBTs) to provide a mission capable, multiconfigured heavily protected Infantry Fighting Vehicle (IFV) that can operate with the largest generation MBTs.

Description

The AB14 Temsah (Arabic for crocodile) HIFV is based on a much modified Tariq MBT chassis which is a British-supplied Centurion upgraded some years ago for the Jordanian Army, under the leadership of the now General Dynamics Land Systems.

For this conversion only the lower hull and engine compartment of the original Tariq vehicle have been retained.

With the introduction of ex-British Army Challenger 1 MBTs into the Jordanian Army under the local name of Al Hussein, there are significant numbers of Tariq MBTs that could be upgraded or converted into other applications, such as the AB14 Temsah HIFV.

When compared with other IFV and APCs, the Temsah has a very high level of combat survivability, as the all-welded steel armoured hull provides the occupants with protection from small arms fire and shell splinters. Through the frontal 30° arc it is claimed to provide protection against 120 mm ammunition. Additional passive or explosive reactive armour can be fitted for a higher level of battlefield survivability.

The diesel power pack is mounted at the front of the Temsah with the air inlet louvres in the roof and the outlet louvres either side of the hull. The exhaust outlets are located one either side of the hull.

In the Tariq MBT, the power pack and drive sprockets are at the rear, but in Temsah they are at the front, which meant that the suspension geometry had to be changed.

The commander/weapons operator and the driver are seated to the immediate rear of the power pack with each being provided with a single-piece circular hatch cover that opens to the rear. The commander/weapons operator is on the left and has eight periscopes, while the driver is on the right and has five periscopes.

Mounted on the roof of the prototype of the Temsah is the HELIO Stabilised Weapon And Reconnaissance Mount (SWARM), which for this application is armed with a stabilised Giat Industries 20 mm M621 cannon and 7.62 mm machine gun.

SWARM is fitted with a Thales Optronics STAG thermal viewer and day/low light level thermal imager which transmits the view to a 250 mm colour flat screen display located at the commander/weapon operator's position. The system is operated by remote control and takes up very little space inside of the hull.

The design of SWARM is modular so it can be fitted with other weapons such as a 12.7 mm M2 machine gun or up to four laser-beam riding anti-tank guided weapons.

To the left of SWARM is a rectangular roof hatch that opens to the left while in the roof to the rear of SWARM are two rectangular roof hatches that open either side.

The troops can rapidly enter and leave Temsah via the large power-operated ramp situated in the rear of the hull. The troop compartment is 3.35 m long, 1.77 m wide and 1.445 m high. Sufficient stowage is provided to carry combat supplies to enable a 48-hour

Temsah heavy infantry combat vehicle from the rear with power-operated ramp lowered **2002**/0110654

Specifications

Crew: 2 + 10
Combat weight: 46,900 kg
Payload: 5,400 kg
Power-to-weight ratio: 15.99 hp/tonne
Ground pressure: n/avail
Length: 7.640 m
Width: 3.760 m
Height (hull top): 2.08 m
Ground clearance: 450 mm
Length of track on ground: 4.572 m
Track width: 610 mm
Track: 2.66 m
Max road speed: 47 km/h
Range: 560 km
Fuel capacity: 880 litres
Gradient: 60%
Side slope: 30%
Vertical obstacle: 0.914 m
Trench: 3.352 m
Engine: General Dynamics Land Systems AVDS-1790 series V-12 diesel engine developing 750 bhp
Transmission: Allison CD850 automatic with two forward and one reverse gears
Suspension: hydropneumatic (Model 3870)
Armament:
(main) 1 × 20 mm M621 cannon
(coaxial) 1 × 7.62 mm machine gun
Smoke grenade launchers: 2 × 4

Gun control equipment
Turret power control: electric/manual
Turret traverse: 360°
Gun elevation/depression: +40/-15°
Gun stabiliser:
(vertical) yes
(horizontal) yes
Range-finder: no
NBC system: optional
Night vision equipment: yes

self-supported operation. Compartments located above the track sponsons provide additional external stowage as well as providing additional protection against standoff weapons.

The suspension is of the hydropneumatic type with each side having six dual rubber tyre roadwheels with the drive sprocket at the front, idler at the rear and track return rollers. Total roadwheel travel is +350 to -100 mm.

The upper part of the suspension is covered by an armoured skirt. Tracks are dry steel pin with the option of fitting dry pin tracks with rubber pads.

In addition to having a high level of armour protection, the combat survivability of the Temsah is further enhanced by the installation of an Avimo LWD 2 laser warning system. This can be used in conjunction with the multibarrel (four round) grenade launchers mounted either side of the hull.

Each detector array system has an azimuth field of view from −12 to +47°, with an optional version having a field of view from −12 to +90°. A maximum of eight arrays can be fitted to cover a full 360° with overlap.

A number of options are being offered and, in addition to enhanced protection and different weapon systems, these include an NBC system and an air conditioning system.

Variants

Although the prototype is fitted with the Helio SWARM it can be fitted with various other weapon systems according to customers requirements. In addition, it can be configured as an ambulance or command post.

Status

Prototype.

Manufacturer

King Abdullah Design and Development Bureau.

NEW ENTRY

KOREA, NORTH

VTT-323 armoured personnel carrier

Development

The VTT-323 full-tracked armoured personnel carrier was developed to meet the requirements of the North Korean Army and is understood to be based on the Chinese NORINCO (China North Industries Corporation) Type 63/Type YW 531 vehicle which is fully described under China. The USA has also referred to the VTT-323 as the M1973. This indicates that it was first observed in 1973.

Description

The hull of the VTT-323 is of all-welded steel armour construction, with the driver being seated at the front of the hull on the left side, and has a single-piece domed hatch cover that opens to the left, in front of which are two day periscopes.

The vehicle commander sits at the front of the hull on the right side and has a single-piece hatch cover opening to the right with an integral day periscope that can be traversed through 360°.

The engine compartment is located behind the commander with the air inlet and outlet louvres in the roof and the exhaust outlet in the right side of the roof. The third crew member is seated behind the driver and has a single-piece hatch cover that opens to the left with an integral day periscope that can be traversed through 360°.

Mounted on the roof of the vehicle towards the rear is the one-man turret armed with a 14.5 mm machine gun and mounted externally on the roof is a Russian-designed Igla (NATO SA-16 Gimlet) fire-and-forget surface-to-air missile. It is believed that the turret is manually operated.

The infantrymen enter and leave the vehicle by a door in the rear of the hull, which is hinged on the right side. At least two firing ports are provided in each side of the hull.

The torsion bar suspension either side consists of five single rubber-tyred roadwheels with the drive sprocket at the front, idler at the rear and no track-return rollers.

The VTT-323 is fully amphibious being propelled in the water by its tracks. Before entering the water, a trim board, which is stowed on the glacis plate when not required, is erected at the front of the vehicle by the driver without leaving his seat.

Variants

Anti-aircraft

This has a different superstructure on the rear of which is a mount armed with four 14.5 mm machine guns for use in the air defence role. It is believed that the weapons are the locally built version of the Russian 14.5 mm KPV series ZPU-4 towed anti-aircraft gun system.

Anti-tank

This is fitted with a raised box on the rear which contains a number of Russian designed, North Korean manufactured Sagger AT-3 anti-tank guided missiles. When required for action, the missiles, complete with overhead armour protection, are raised from the box. This version also has a pintle-mounted 14.5 mm machine gun towards the front of the hull. This has the US designation of the M1992 self-propelled anti-tank guided missile system.

107 mm multiple rocket launcher

Mounted on the roof is the 107 mm (12-round) multiple rocket launcher. This system has the US designation of the M1992 Multiple Rocket Launcher (MRL).

120 mm self-propelled gun

This has a different hull and is armed with a turret-mounted 120 mm gun/mortar mounted at the rear. This has the US Army designation of the M1992 as this was the year that it was first observed.

M1985 light tank

This vehicle, covered in the Light tanks section, may incorporate automotive components of the VTT-323 armoured personnel carrier.

Status

Production probably complete. In service with North Korea.

Manufacturer

Democratic People's Republic of Korea.

UPDATED

VTT-323 series full-tracked armoured personnel carriers in service with North Korea ***2000***/0064830

Specifications
(provisional)
Combat weight: 12,700 kg
Power-to-weight ratio: 25.19 hp/t
Length: 6.2 m
Width: 3.06 m
Height: 2.50 m
Engine: 320 hp diesel
Armament: 1 × 14.5 mm machine gun; 1 × SA-16 SAM

KOREA, SOUTH

Daewoo Korean Infantry Fighting Vehicle

Development

In 1981, the Korean Army issued a request for proposals for a new Korean Infantry Fighting Vehicle (KIFV) and Daewoo Heavy Industries was subsequently awarded a contract for development and production of this vehicle.

The KIFV entered production in 1985 and is sometimes referred to as the Type K200 and by 2001 it is estimated that over 2,000 vehicles, including variants, had been built for the home and export markets with production still under way.

Based on the experience in the development and production of the KIFV, in 1994, Daewoo Heavy Industries developed a further model called the KIFV (K200A1). This has a number of improvements over the earlier vehicle including the installation of a 350 hp turbocharged MAN diesel engine and a new fully automatic Allison X200 transmission.

It is expected that earlier K200 vehicles will be brought up to the K200A1 standard as the new power pack gives an improved power-to-weight ratio and greater acceleration. In the earlier K200 the driver used sticks to steer the vehicle but in the K200A1 he uses a steering handle, which reduces driver fatigue.

Late in 1993, Malaysia placed an order and took delivery of 42 KIFVs for use with its units operating as part of United Nations forces in the former Yugoslavia.

K277A1 command post vehicle clearly showing the raised rear roof for increased internal volume **2001**/0093584

K216A1 NBC reconnaissance version of the KIFV has a different hull from the standard KIFV

This order comprised the following versions: 32 KIFVs; 4 mortar carriers; 2 recovery vehicles; 2 command post vehicles; 2 ambulance vehicles.

In mid-1994, Malaysia placed a further order for 22 vehicles in the following versions: 18 KIFVs; 2 recovery vehicles; 1 command post vehicle; 1 ambulance.

In mid-1995, Malaysia placed a further order for 47 KIFV and variants, all of which have now been delivered.

It is not expected that Malaysia will order any additional KIFV from South Korea and late autumn of 2000 Malaysia finalised a contract with FNSS Defence Systems of Turkey to supply 211 Turkish Infantry Fighting Vehicles and variants to Malaysia in a number of configurations.

Description

The hull of the KIFV is of all-welded aluminium armour with an additional layer of spaced laminate steel armour bolted to it. This composite armour provides better protection for less weight.

According to the manufacturer, protection from small arms fire is provided up to 12.7 mm armour-piercing rounds fired from the front and flanks.

The engine compartment is located at the front right of the vehicle and is separated from the remainder of the KIFV by a bulkhead. The engine compartment is fitted with a fire extinguishing system that can be operated by the driver or from outside the vehicle.

The air inlet, air outlet louvres and the exhaust pipe are located on the roof of the vehicle.

The complete power pack consists of the engine, transmission with integrated hydrostatic steering system and hydraulic service and multidisc friction brake, final drives and associated drive shaft and universal joint.

Power is transmitted through two shafts to the final drives which can be disconnected to remove the power pack. The latest KA200A1 is powered by the German MAN D2848T V-8 water-cooled direct injection turbocharged diesel engine which develops 350 hp at 2,300 rpm. This is manufactured under licence by Daewoo Heavy Industries.

The latest KA200A1 has the US Allison X200-5K transmission, which is an automatic shift and steering transmission with four forward and one reverse gears and a hydrostatic steering system.

The driver is seated at the front of the vehicle on the left side and has a single-piece hatch cover that can be opened upwards by means of a locking hook placed on the hatch ring.

Four M17 day periscopes are used for forward observation and a locally produced KAN/VVS-2 passive periscope can be installed in his roof hatch for driving at night.

The gunner is seated behind the engine compartment and the cupola for the gunner can be traversed smoothly through 360° even on a slope. The gunner is provided with five M17 day periscopes and a single-piece hatch cover that opens to the rear and forward mounted on his cupola is a 12.7 mm M2 machine gun. The gunner is provided with front, side and rear protection.

For operations in Bosnia, a number of Malaysian KIFV have had their 12.7 mm M2 machine guns replaced by a 40 mm automatic grenade launcher with the gun shield being retained.

To the rear of the driver is another gunner's position armed with a 7.62 mm machine gun, the cupola is similar to that fitted to the M113 series and has five M17 day periscopes and a single-piece hatch cover that opens to the rear.

The troop compartment is at the rear of the KIFV and the troops enter and leave via a hydraulically operated downward-opening ramp in the rear; this is also provided with a ramp door in the left side.

There is a roof hatch over the troop compartment that opens to the rear and in either side are two firing ports with a vision block above. There are also two firing ports in the rear ramp.

Torsion bar suspension either side consists of five dual rubber-tyred roadwheels with the drive sprocket at the front and idler at the rear, which is connected to the tension adjuster to maintain track tension. There are no track-return rollers.

The first, second and last roadwheel stations are provided with a telescopic shock-absorber and the steel tracks have replaceable pads.

The vehicle is fully amphibious, being propelled in the water by its tracks. Before entering the water a trim vane is erected at the front of the vehicle and the two bilge pumps are switched on.

The two bilge pumps are located in the bottom of the hull, below the floor level, one in the rear of the personnel compartment and the other in the front of the engine compartment. These have a total discharge capacity of 348 litres/min.

Mounted on the forward part of the hull above the trim vane is a bank of six electrically operated smoke grenade launchers that fire forwards.

Variants

K216A1 NBC reconnaissance vehicle

This has been developed to meet the technical requirements of the battlefield environment for NBC warfare. Equipment fitted includes NBC detection equipment such as mobile mass spectrometer (MM-1), collective overpressure system, contamination marking set and a weather observation system.

K242A1 armoured 107 mm mortar carrier

This is similar to the standard K200A1 but in the rear is fitted a turntable-mounted 107 mm (4.2 in) mortar which fires through the roof, which has a three-part hatch. If required, the mortar can be dismounted for ground use. In total, 88 rounds of 107 mm ammunition are carried, with a pintle-mounted 12.7 mm M2 machine gun provided for local protection.

K263A1 20 mm SPAAG

This is basically the K200A1 chassis fitted with one-man electrically powered turret which is armed with a 20 mm M168 Vulcan cannon. Turret traverse is 360° with weapon elevation from –5 to +80° and turret traverse through a full 360°. The gunner can select either 10, 30, 60 or 100 round bursts. This system has a four-man crew and a combat weight of 13,200 kg. Full details are given in *Jane's Land-Based Air Defence 2001-2002*. This model is in service only with South Korea.

K277A1 command post vehicle

This has a similar chassis to the K200A1 except that it has a raised hull rear for increased accommodation. The vehicle has a six-man crew consisting of driver, radio operator and four additional crew members.

The command post vehicle is equipped with map boards, tables, an NBC detection unit and an auxiliary power unit to provide current electrical power. A tent can be erected at the rear to provide more space for the command personnel.

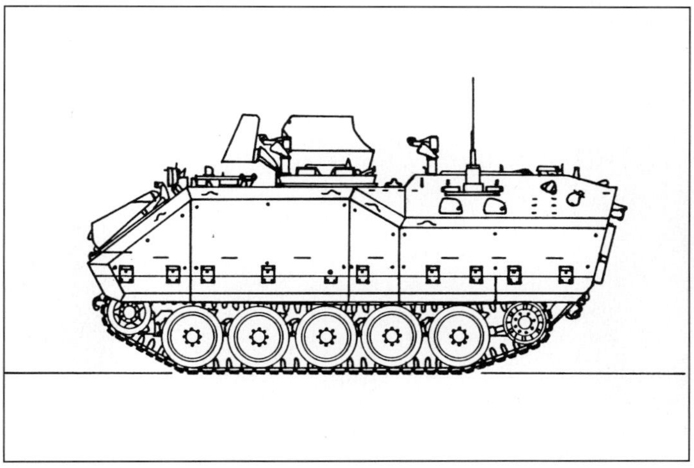

Unarmed K200 Korean Infantry Fighting Vehicle

K263A1 air defence version of the Korean Infantry Fighting Vehicle armed with a 20 mm Vulcan air defence system

Armament consists of a pintle-mounted 12.7 mm M2 machine gun and a bank of six forward-firing electrically operated smoke grenade dischargers.

K281A1 81 mm mortar carrier

This is similar to the standard K200A1 but in the rear is fitted a turntable-mounted 81 mm mortar which fires through the roof, which has a three-part hatch cover. If required, the 81 mm mortar can be deployed away from the vehicle. A total of 114 rounds of 81 mm mortar ammunition are carried with a pintle-mounted 12.7 mm M2 machine gun provided for local protection.

K288A1 recovery vehicle

The K288A1 recovery vehicle is based on the chassis of the K200A1 and is equipped with a hydraulically driven winch for the recovery of damaged and disabled vehicles.

Mounted either side of the hull at the rear are spades which are lowered to the ground during operations. An auxiliary spade unit, carried on the roof when not required, can be used between the two spades in soft soil.

The winch is fitted in the rear crew compartment and has a direct maximum capacity of 10,000 kg, although this can be increased to 20,000 kg by use of a snatch block.

Mounted on the left side of the hull roof is a crane which can lift a load of 6,000 kg through a full 360°.

The four-man crew consists of commander, driver, operator and mechanic. There is provision for mounting a 12.7 mm M2 machine gun at the commander's cupola on top of the superstructure and six smoke grenade dischargers are mounted on the front of the hull.

K200A1 ambulance vehicle

This was developed as one of a series of vehicles for the Malaysian battalion UN peacekeeping forces in Bosnia. The crew compartment of the vehicle is furnished to stow the equipment for medical treatment of wounded troops.

Equipment carried includes four litter sets, two sets of automatic oxygen resuscitators and two splint sets. Exterior configuration is identical to the standard vehicle.

KAFIV 30A, 30 mm turret vehicle

This model has been developed as a private venture for the export market and is fitted with a one-man AV30 turret armed with a 30 mm Boeing M230 Chain Gun with a 7.62 mm M240 coaxial machine gun. The turret also has a bank of four M239 smoke grenade launchers mounted either side of the turret rear, firing forwards.

The turret has a stabilised electric drive system for traverse and weapon elevation and is fitted with a day sight, unity window and seven vision blocks which provide the gunner with a 360° field of vision.

The AV30 turret is designed to provide armour protection for the gunner against small arms fire and artillery bursts at least equal to 7.62 mm at 0° obliquity at 0 m.

KAFIV 30M, 30 mm turret vehicle

This has also been developed as a private venture for the export market and is fitted with the German Mauser MT30K turret armed with an externally mounted 30 mm Mauser MK cannon with a 7.62 mm MG3A1 machine gun being mounted coaxial. A bank of four electrically operated smoke grenade launchers is mounted on the left side of the turret firing forwards.

The turret is electrically operated and fitted with a PERI Z 16 day sight with optional image intensifier and a CAN-BUS-based signal transmission system.

KAIFV, 40 mm grenade turret vehicle

This is fitted with a one person turret designed by Daewoo Heavy Industries and armed with one 40 mm grenade launcher K4 with a 12.7 mm M2 machine gun being mounted coaxial to the right. Turret traverse and weapon elevation is powered with the gunner being provided with a M36E3 sight plus periscopes for all round observation.

KAFIV 90, 90 mm turret vehicle

This has been developed as a private venture for the export market and is fitted with a two-man turret armed with a 90 mm Cockerill gun, 7.62 mm coaxial machine gun and a commander's ring with a 7.62 mm machine gun.

The turret can be traversed through 360° under power at speeds of up to 30°/s with weapon elevation from –8 to +28° at speeds of up to 30°/s.

External vision is by means of three direct-view vision blocks, commander's periscope mounted in the machine gun ring mount that provides 360° traverse, and a gunner's M36E1 day/night sight. For night surveillance a 203 mm 500,000 candle power white light spotlight is coaxially mounted to the 90 mm gun.

Eight rounds of ready use 90 mm ammunition are carried in the turret plus 400 rounds of ready use 7.62 mm ammunition in the turret and a further 200 rounds of 7.62 mm ready use ammunition at the commander's ring mount.

Malaysian KIFV upgrade

Early in 2000, Daewoo Heavy Industries proposed to the Royal Malaysian Army that its current K200 KIFV could be upgraded to the enhanced K200A1 standard in Malaysian Army facilities in a very short period of time.

This offer was accepted and the task carried out by Number 91 Workshop of the Royal Malaysian Army which dismantled the vehicle, carried out modifications to the hull and fitted the new power pack.

As the new power pack is slightly larger, the bulkhead has to be taken out and positioned 75 mm to the rear and there are also a number of small electrical changes. The final drives have also been changed to take into account the increased engine output available.

Work started on 28 February and the vehicle ran for the first time on 30 March and then made its first public appearance at the DSA 2000 exhibition held in Kuala Lumpur in March 2000.

The installation of this new power pack not only improves the power-to-weight ratio of the vehicle from 21.7 hp/t to 26.5 hp/t but also improves acceleration. Maximum road speed is now increased to 70 km/h. Braking is also improved.

It was then demonstrated to the Royal Malaysian Army at its armour school at Port Dickson and Daewoo hopes that a contract will be awarded for two conversions in the future. This could eventually lead to the whole fleet being upgraded to the latest K200A1 production standard.

Specifications

Model	K200	K200A1
Crew	3 + 9	3 + 9
Combat weight	12,900 kg	13,200 kg
Unloaded weight	10,700 kg	11,000 kg
Power-to-weight ratio	21.7 hp/t	26.5 hp/t
Ground pressure	0.63 kg/cm²	0.64 kg/cm²
Length		
(overall)	5.486 m	5.486 m
(hull)	5.345 m	5.345 m
Width		
(overall)	2.846 m	2.846 m
(over tracks)	2.540 m	2.540 m
Height		
(hull top)	1.93 m	1.93 m
(MG shield)	2.518 m	2.518 m
Ground clearance	0.41 m	0.41 m
Track width	381 mm	381 mm
Angle of approach/departure	68°/45°	68°/45°
Max speed		
(road)	74 km/h	70 km/h
(water)	6 km/h	7 km/h
Fuel capacity	400 litres	400 litres
Cruising range	480 km	480 km
Fording	amphibious	amphibious
Gradient	60%	60%
Side slope	30%	30%
Vertical obstacle	0.64 m	0.64 m
Trench	1.68 m	1.68 m
Engine	MAN D 2848T V-8 diesel 280 hp at 2,300 rpm	MAN D 2848T V-8 diesel 350 hp at 2,300 rpm
Transmission	T-300 semi-automatic 7 forward, 7 reverse	X200-5K full-automatic hydrodynamic mechanical, 4 forward, 1 reverse
Suspension	torsion bar in tube	torsion bar in tube
Electrical system	28 V	28 V
Batteries	2 × 6TN, 100 Ah	2 × 6TN, 100 Ah
Armament		
(main)	1 × 12.7 mm M2 MG	1 × 12.7 mm M2 MG
(secondary)	1 × 7.62 mm M60 MG	1 × 7.62 mm M60 MG
Smoke-laying equipment	1 × 6 smoke grenade launchers	1 × 6 smoke grenade launchers
NBC system	optional	optional
Night vision equipment	yes	yes

Daewoo family of AFV

Details of the Daewoo family of armoured fighting vehicles are given in the Armoured fighting vehicles families section later in this book.

KIFV fitted with Daewoo turret armed with 40 mm granade launcher and 12.7 mm coaxial machine gun **2001**/0093585

KIFV in service with Malaysian Army with armament covered up (Richard Stickland) **2000**/0064831

Vehicles based on this larger and more powerful chassis include the Flying Tiger twin 30 mm self-propelled anti-aircraft gun system, Pegasus surface-to-air missile system and the Field Artillery Ammunition Support Vehicle. It is understood that these vehicles entered service with the Republic of Korea Army late in 1999.

Status

In production. In service with South Korea and Malaysia.

Manufacturer

Daewoo Heavy Industries Limited.

UPDATED

NORWAY

Norwegian Army M113A1 vehicles

At present the Royal Norwegian Army operates a fleet of approximately 500 US-supplied United Defense LP M113 series armoured personnel carriers, of which 53 have been upgraded to the Norwegian NM 135 configuration.

More recently, additional surplus M113 series vehicles have been purchased from Germany, the Netherlands and the USA and soon the total fleet, including the many local variants, will amount to 670 vehicles in a total of 16 variants. M113 series deliveries from the USA included 136 vehicles in 1993 and 63 vehicles in 1996.

The NM 135 is an M113A1 fitted with a Swedish Hägglunds Vehicle one-man turret armed with a German Rheinmetall 20 mm MK Rh 202 cannon, with a 7.62 mm machine gun mounted coaxially with the main armament to the right of the turret. Mounted on either side of the turret front is a bank of three electrically operated 76 mm smoke grenade dischargers firing forwards.

The turret is mounted behind the engine compartment and the vehicle commander's hatch is moved to the immediate rear of the driver's position. The rearward-opening hatch in the troop compartment roof is retained.

The Royal Norwegian Army has also fitted 97 M113A1 APCs with the Norwegian Kvaerner-Eureka Armoured Launching Turret (ALT), which has two Raytheon Systems Company TOW ATGWs in the ready to launch position and additional TOW missiles carried inside the hull for manual loading. Mounted on the turret rear are forward-firing smoke grenade dischargers.

The Norwegian M113A1 with the ALT is designated the NM 142 and the first batch comprised 36 vehicles which were conversions of existing M113A1 vehicles. The second batch consisted of 44 vehicles purchased specifically for this purpose.

In addition to being used by Norway, the ALT is used by Canada (on M113A1s), Switzerland (on MOWAG Piranha 6 × 6 light armoured vehicles) and Turkey (on locally built FNSS Defense Systems full-tracked AFVs).

Norway also operates a total of 86 M548A1 tracked cargo carriers which are based on automotive components of the M113A1 armoured personnel carrier. These are used to support self-propelled artillery units equipped with 155 mm M109 series self-propelled howitzers. A locally developed recovery version of the M548A1, called the NM 84, is in service.

In mid-1997, following an international competition, the Norwegian Ministry of Defence awarded a contract to the Canadian Commercial Corporation for the supply of a quantity of upgrade kits for M113 series vehicles used by the Royal Norwegian Army.

These kits were supplied from 1997 through to 1998 and were fitted to the vehicles in Norway and include improvements to the suspension, engine cooling system and final drive.

Following an international competition, late in 1999 the Norwegian Army selected the German company of IBD Deisenroth to upgrade its fleet of M113A2 APCs and M577 series command post vehicles with a passive armour package to provide a higher level of battlefield survivability.

A total of 350 Norwegian M113/M577 series vehicles are to be fitted with the IBD Deisenroth modular MEXAS II+ ceramic faced composite armour system. For the M113 series APC this weighs about 700 kg, while that of the M577 weighs about 800 kg.

This add on passive armour protection kit includes blast mine protection under the belly and sponsons plus ballistic protection of the vehicle front, top, sides and skirts.

The new passive armour package gives all-round protection from 14.5 mm armour-piercing attack at 0 m range and 30 mm armour-piercing attack across the frontal arc.

The Norwegian M113 series vehicles have been supplemented by Swedish Hägglunds Vehicle CV 9030 series vehicles, which were selected in 1994 following an international competition. Additional details are given in the entry for the CV 90 under Sweden. In total 104 vehicles have been ordered as the CV90N; the first four were delivered late in 1995. Final deliveries were made late in 2000. There was also an option on an additional batch of 70 vehicles, but as of mid-2001 this had not been exercised.

UPDATED

Royal Norwegian Army NM 135 0018445

Royal Norwegian Army M113 series APC fitted with an Armoured Launching Turret (ALT) ***2001***/0093565

Norwegian Army M113A2 with appliqué armour and fitted with Pearson Engineering mineclearing rollers (Richard Stickland) ***2002***/0110668

Norwegian Army M577A1 command post vehicle (Richard Stickland) ***2002***/0110669

PAKISTAN

Local production of M113A2

Development/Description

Under an agreement signed in 1989, Pakistan was to build 775 M113A2 Mk1 armoured personnel carriers under licence from United Defence LP of the United States.

Pilot production commenced immediately with hulls provided by the United Defense LP facility at San Jose, California. This was followed by manufacture of hulls locally and assembly of vehicles, with materials supplied by United Defense LP in a new purpose-built facility at Heavy Industries, Taxila, Pakistan.

Shipment of material for the 775 M113A2 Mk 1 APCs was completed in mid-1995. In June 1998, United Defense LP was awarded a follow-on contract for the supply of a substantial number, believed to be 1,200, of additional M113A2 Mk1 vehicles.

The Pakistan Army operates a large fleet of M113 series vehicles including the basic APC, fitted with the Raytheon Systems Company TOW ATGW, Baktar Shikan ATGW, twin 106 mm M40 recoilless rifles and Swedish SAAB Bofors Dynamics RBS-70 surface-to-air missile systems.

The Baktar Shikan (a local version of the Chinese NORINCO Red Arrow 8 system) carrier version employs a launcher which is raised into operational position when required for action and with reserve missiles carried inside the hull.

Since 1990, a new locally built air defence version of the M113 has been in service with the Pakistan Army. Mounted on top of the hull is a pedestal armed with four locally built surface-to-air missiles and two 14.5 mm KPVT machine guns that can be elevated simultaneously.

The locally built M113 is designated M113 in Pakistani service and is identical to the US M113A2 Mk1, except that the 0.50 calibre M2 machine gun has been replaced by a Chinese 12.7 mm Type 54 machine gun manufactured by Pakistan Ordnance Factories.

Early in 1991, Pakistan obtained an Armoured Infantry Fighting Vehicle (AIFV) from United Defense LP for trials. This vehicle has not yet been manufactured in Pakistan, but extensive research and development work to develop an indigenous infantry fighting vehicle is currently being pursued by Heavy Industries Taxila.

M113A2 Mk-1 armoured personnel carrier manufactured in Pakistan and showing roof-mounted 12.7 mm machine gun. Note that the smoke grenade launchers are not fitted **2001**/0093564

Status

US-built M113A1/M113A2 and the indigenously produced M113P are in service with the Pakistan Army. M113P remains in production.

Manufacturer

Heavy Industries Taxila.

UPDATED

M113A1 APC with a twin 106 mm recoilless rifle mount developed by Pakistan Ordnance factories

M113 APC with a Saab Bofors Dynamics RBS 70 surface-to-air missile system in the raised position

POLAND

BWP 2000 infantry fighting vehicle

Development

The origins of the Polish BWP 2000 full-tracked infantry fighting vehicle can be traced back to the days of the Warsaw Pact when Poland was working with Russia on local production of a wide range of armoured vehicles.

This included the Russian OKB-40 design bureau in Mytischi on the MT-S full-tracked transporter. Components of this were used in the development of the BWP 2000 infantry fighting vehicle which is also referred to as the Kalina.

Poland is now moving away from former Russian designs to Western-type systems and in 1997 the BWP 2000 was fitted with a fully developed and tested Italian Otobreda OTO T60/70A turret, full details of which are given under Italy in the entry for the Dardo vehicle which has been developed to meet the requirements of the Italian Army.

The BWP 2000 was designed by OBRUM (Osrodek Badawczo Rozwojowy Urzadzen Mechnicznych) although any volume production is expected to be undertaken by Zaklady Mechaniczne Bumar Labedy where production of the locally developed (from the Russian T-72) PT-91 MBT was undertaken for the Polish Army.

By mid-2001, there were no plans for the Polish BWP 2000 IFV to enter production for the Polish Army. The Polish Army is currently targeting two programmes, first the acquisition of a new wheeled armoured personnel carrier (qv Armoured Personnel Carriers (Wheeled) section) and second the upgrade of at least part of the existing BMP-1 IFV fleet.

Description

The hull of the BWP 2000 is of all welded steel armour construction and over the frontal arc this provides protection against APFSDS-T projectiles of up to 35 mm in calibre. Over the roof and sides protection is provided against 12.7 mm rounds fired from a range of 100 m.

Over the complete vehicle protection is provided against 7.62 mm rounds from any distance and angle. If a higher level of protection is required additional armour can be fitted onto the hull and turret. This can be of the passive or explosive reactive armour type.

The driver is seated at the front left and has a single-piece hatch cover that lifts and swings to the right and in front of this is a single wide-angle periscope which can be replaced by a passive periscope for driving at night.

The power pack is to the right with the air inlet/outlet louvres in the roof and the exhaust outlet on the right side.

Mounted in the centre on the hull is the Italian Otobreda T60/70A turret which is armed with a stabilised 60 mm rifled cannon and a 7.62 mm coaxial machine gun. Mounted either side of the turret is a bank of four electrically operated smoke dischargers.

BWP 2000 fitted with Otobreda T60/70A turret with Raytheon Systems Company TOW missile launcher either side of turret
0008941

BWP 2000 fitted with Otobreda TC25 turret armed with 25 mm Oerlikon Contraves cannon, 7.62 mm coaxial machine gun and single Raytheon Systems Company TOW launcher either side of turret (Stefan Marx)
0018447

Full details of this 60 mm turret which, by mid-2001, remains at the prototype stage, are given in the entry for the Dardo infantry fighting vehicle covered in this section under Italy.

The troop compartment is at the rear of the BWP 2000 with the infantry entering and leaving via a power-operated ramp which is also provided with an emergency door. Over the top of the troop compartment are two circular roof hatches which open to the rear and in each side of the troop compartment are two firing ports.

Suspension is of the torsion bar type with each side having six dual rubber-tyred roadwheels with the drive sprocket at the front, idler at the rear and three track-return rollers. The upper part of the track is covered by a sheet metal skirt like that on the Russian BMP-3.

Standard equipment for the BWP 2000 includes an NBC system and a fire detection and suppression system. An unditching beam is normally carried at the rear of the vehicle.

The vehicle can be fitted with a snorkel which allows it to ford to a depth of 5 m or, when fitted with buoyancy aids, it can be fully amphibious being propelled in the water by its tracks.

Variants

Although the prototype is fitted with the Italian Otobreda T60/70A turret, a wide range of other turrets can be fitted including 40 mm, 35 mm, 25 mm and 23 mm cannon and a larger turret armed with a 105 mm gun.

Specifications

Crew: 3 + 8
Weight: 25-29,000 kg*
Power-to-weight ratio: 28-24.13 hp/t*
Ground pressure: 0.84 kg/cm²
Length:
(overall) 7.344 m
(hull) 6.95 m
Width: 3.25 m
Height:
(incl sights) 3.018 m
(turret roof) 2.513 m
(hull roof) 1.891 m
Max road speed: 70 km/h
Max range: 500 km
Fording: 1.20 m
Gradient: 60%
Side slope: 50%
Vertical obstacle: 0.80 m
Trench: 2.80 m
Engine: Model S 12-K diesel developing 700 hp
Transmission: semi-automatic with hydraulic torque converter, hydrostatic steering mechanism with continuous change of turning radius and pivot turn. Four forward and 1 reverse gear.
Armament:
(main) 1 × 60 mm cannon
(coaxial) 1 × 7.62 mm MG
(anti-tank missiles) 2 × Raytheon Systems Company TOW launchers
Ammunition:
(main) 16 + 16
(coaxial) 2,000
(anti-tank missiles) 2 + 2
Gun control equipment
Turret power control: electric/manual
(by commander) yes
(by gunner) yes
Gun elevation/depression: +40/–6°
Turret traverse: 360°
Gun stabiliser:
(vertical) yes
(horizontal) yes
NBC system: yes
Night vision equipment: yes
* Weight and power-to-weight ratio depends on the turret/weapon system fitted, above specification is for vehicle with Italian Otobreda T60/70 turret

In 1997, the BWP 2000 was also shown fitted with the Italian Otobreda TC25 turret developed for the Dardo HITFIST. This is armed with an Oerlikon Contraves 25 mm KBA cannon, 7.62 mm coaxial machine gun and a single launcher either side of the turret for a 3,750 m range Raytheon Systems Company TOW anti-tank guided missile.

SUM tracked minelayer

A modified version of the BWP 2000 IFV chassis is used as the basis for the SUM tracked minelayer. Full details of this are given in *Jane's Military Vehicles and Logistics 2001-2002* page 204. By early 2000, the SUV remained at the prototype stage.

Status

Prototype.

Manufacturer

See text.

UPDATED

Polish MT-LB multipurpose armoured vehicle

Development/Description

For many years Poland has manufactured the Russian MT-LB multipurpose amphibious armoured vehicle under licence for both the home and export markets.

The Polish-built vehicle is virtually identical to the original vehicle described in detail later in this section.

Variants

Communications

A number of versions are in service including the MT-LB R-137.

Ambulance (Lotos)

This has a raised superstructure at the rear and can carry four stretcher patients, or two stretcher and four seated patients or eight seated patients.

Standard equipment includes radios, an NBC system, electrically heated instruments and windscreen, heater and engine preheater, a searchlight, a bilge pump, fire extinguishers, specialised medical equipment, safety belts, 24 V and 12 V electrical systems and an exhaust fan.

Engineer Vehicle (Mors)

This is similar to the original Russian (CIS) engineer vehicle. Mounted at the rear of the hull is a self-entrenching blade which is stowed vertically when not required. Details of the CIS version are given in *Jane's Military Vehicles and Logistics 2001-2002.*

Armoured engineer reconnaissance vehicle

This vehicle has been developed for the reconnaissance of roads, bridges and their approaches, structures and potential demolition sites, minefields, battlefield obstacles, water resources and other areas.

The vehicle carries eight men (commander/gunner, driver and six engineers) and is covered in detail in *Jane's Military Vehicles and Logistics 2001-2002.* The MT-LB TRI Mors is also an Engineer reconnaissance vehicle. A variant of this is the TRID with the UZR-3 LWD (Snake) or the trailer-mounted P2P-W1WD rocket propelled mine clearing system.

Polish-built MT-LB ambulance (Lotos)

Polish-built MT-LB rescue and service standby vehicle without armament installed (Jaroslaw Cislak) 0018449

Polish-built MT-LB engineer vehicle with entrenching blade lowered at the rear

Polish-built MT-LB TRI 'Mors' armoured engineer reconnaissance (Jaroslaw Cislak) 0018448

Rescue and service standby vehicle

This has been designed for use in the forward battlefield area and has a self-entrenching blade mounted at the rear of the hull and a roof-mounted crane. Typical roles include removing damaged equipment from the immediate battlefield area and providing rapid technical support to damaged and disabled vehicles. This is also referred to as the MT-LB WPT and is similar to TRI but without the 81 mm smoke grenade launchers or Half Horse antenna. Equipped with a hydraulic plough blade at the rear of the hull and A frame.

Air defence vehicle (Sopel)

Poland has completed an air defence vehicle based on a stretched MT-LB chassis which has seven roadwheels either side compared to the normal six. The chassis was manufactured by the Steel Works Stalowa Wola.

This is fitted with a power-operated turret designed and built by the Mechanical Plant Tarnow armed with twin 23 mm water-cooled cannon and a bank of two SA-7 type fire-and-forget surface-to-air missiles.

Details of this system, which is yet to enter production or service, are given in *Jane's Land-Based Air Defence 1999-2000*.

Infantry fighting vehicle

For trials purposes a stretched MT-LB chassis, similar to that used in the previous conversion, has been fitted with the complete Bofors Defence turret of the Swedish CV 9040 infantry fighting vehicle in service with the Swedish Army.

This turret is also fitted to the BMP-40 (Polish BMP-1 modified) and is armed with a Bofors Defence 40 mm L/70 cannon and a 7.62 mm coaxial machine gun. In 2001 this remained at the prototype stage.

The standard MT-LB is propelled in the water by its tracks but this version is propelled in the water by two ducted propellers mounted one either side at the rear, just above the idler station. This is designated the BWO-40.

Minelayer (Kroton)

This is based on the stretched MT-LB chassis and is armed with a roof-mounted 12.7 mm machine gun. Various minelaying systems of the scatter type would be carried on the rear of the vehicle, which to the rear of the fourth roadwheel would be of the flatbed type. As of 2001, this remained at the prototype stage.

This is also referred to as the MT-LB ISM Kroton (Inzynieryniy System Minowania) and has a new cupola with a 12.7 mm NSVT machine gun and 81 mm smoke grenade launchers.

Mineclearing system (Durian)

This is based on the standard six-wheeled MT-LB chassis and would tow a trailer carrying a rocket-propelled mineclearing system. This remains at the prototype stage.

Armoured ammunition carrier (Bor)

This is based on the stretched MT-LB chassis and would be configured for battlefield ammunition resupply.

MT-LB-23M armoured personnel combat carrier (Krak)

This is based on a modified six-wheeled MT-LB carrier chassis with the standard machine gun turret removed. Mounted over the rear of the hull on a raised plinth is a new one-man turret armed with a 23 mm cannon and 7.62 mm machine gun. Turret traverse is a full 360° with weapon elevation from –7 to +70°. In addition to its three-man crew it can also carry eight fully equipped troops.

MT-LB (Opal 2)

This has a stretched chassis with seven road wheels either side, longer nose and 12.7 mm NSVT machine gun.

MT-LB WZT (Durian)

This is a recovery vehicle and is similar to the Russian MT-LB M1980 but with an A frame for lifting AFV components. WZT stands for Woz Zabezpieczenia Technicznego.

MT-LB BETA communications vehicle

This is believed to be a communications vehicle and may be designated the BETA-3M and is a division computer vehicle based on the MT-LBu chassis.

MT-LB R-137 T radio station

Radiostacja Ruchoma UKLF and is basic MT-LB but without the 7.62 mm machine gun turret. Equipped with a telescopic antenna, large box at the right rear and four whip antennas.

MT-LB WD command

This is a command or communications vehicle with rear-mounted generator, three antennas and no machine gun turret.

MT-LB L

This is an unknown version.

MT-LB LA-3 command automation vehicle

This has a raised roof at the rear for greater internal volume and is a mobile anti-aircraft vehicle for air defence units. It is now in service with the Polish Army, the prime contractor being RADWAR.

Status

Production as required. In service with the Polish Army.

Manufacturer

Steel Works Stalowa Wola.

UPDATED

Polish upgraded BMP-1 infantry fighting vehicle

Development/Description

Poland, like a number of former members of the Warsaw Pact, is now looking more to the West to meet its future military requirements.

Several years ago, for trials purposes, Poland installed the Swedish Bofors Defence CV 9040 turret armed with a Bofors Defence 40 mm L/70 gun on a BMP-1 chassis and details of this were given in *Jane's Armour and Artillery 1998-1999*, page 328.

In 1997, a number of Western companies demonstrated their turrets in Poland and brief details of this follow:

(1) Delco Defense demonstrated their LAV-25 turret armed with a 25 mm The Boeing Company M242 cannon and 7.62 mm machine gun. Well over 1,000 examples of this turret have been built. For Poland, Delco Defense is teamed with Steel Works Stalowa Wola (Huta Stalowa Wola SA) which currently manufacture the MT-LB. This turret is available with a wide range of options including the addition of a Raytheon Systems Company TOW anti-tank guided missile either side of the turret.

(2) United Defense demonstrated their private venture one-man powered turret also armed with a 25 mm The Boeing Company M242 cannon and 7.62 mm machine gun. For Poland, United Defense is teamed with Wojskowe Zaklady Motoryzacyjne, Bumar Labedy and OBRUM. When fitted with this turret the vehicle is referred to as the BWP-1P.

(3) The Polish company Tarnow is also proposing the development of a new two-man turret armed with twin 23 mm cannon and two pods of two fire-and-forget surface-to-air missiles.

(4) The Italian company, Otobreda, is offering its latest OTO T60/70A turret armed with a 60 mm cannon and 7.62 mm coaxial machine gun as well as its TC25 turret armed with a 25 mm cannon and 7.62 mm coaxial machine gun. As an option a Raytheon Systems Company TOW launcher can be mounted either side of the turret. These turrets have already been installed for trials purposes on the Polish BWP 2000 which is covered in detail earlier in this section under Poland.

(5) Poland has also completed the prototype of a turret armed with a 23 mm cannon and an externally mounted 73 mm SPG-9 recoilless anti-tank weapon. The turret is also fitted with a 60 mm mortar and a bank of three 81 mm electrically mounted smoke grenade launchers are mounted either side of the turret which is also fitted with locally developed explosive reactive armour. An illustration of this turret on a BMP-1 chassis is in *Jane's Armour and Artillery Upgrades 1998-1999*, page 128. This version has also been referred to as the BMP-95.

Late in 2000, Otobreda of Italy and HSW of Poland signed an agreement to develop a new two person 25 mm turret for installation on the Polish BMP-1 chassis. This would be based on the existing T25 turret but be lighter to enable the BMP-1 to retain its amphibious characteristics.

Status

As of late 2001, the Polish Army has not adopted any of these turrets for retrofit to their BMP-1 infantry combat vehicles. The Polish Army has decided not to procure the locally developed BWP 2000 IFV in the short term. Instead it will concentrate on upgrading at least part of its fleet of BMP-1 vehicles as well as procuring a new 6 × 6 or 8 × 8 armoured personnel carrier.

Manufacturer

Not selected.

UPDATED

Polish BMP-1 chassis fitted with US Delco Defense LAV-25 turret (Stefan Marx) 0018450

Polish BMP-1 chassis fitted with United Defense one-man power-operated turret (Stefan Marx) 0018451

ROMANIA

MLI-84 infantry combat vehicle

Development/Description

In appearance and layout the MLI-84 is virtually identical to the Russian BMP-1 infantry fighting vehicle, but is slightly larger and heavier and has a 12.7 mm anti-aircraft machine gun mounted over the left side of the rear troop compartment.

The MLI-84 is fully amphibious, being propelled in the water by its tracks, and is fitted with an NBC system and night vision equipment for commander, gunner and driver. This vehicle only has three rectangular roof hatches compared with the BMP-1 which has four. In place of the left rear rectangular hatch is a circular hatch cover on which is mounted the 12.7 mm anti-aircraft machine gun.

Variants

MLI-84 weapons upgrade

Following an international competition, the Romanian Army selected the Israeli RAFAEL Overhead Weapon Station 25 mm (OWS-25) to upgrade its MLI-84 vehicles.

Late in 1998 the Rafael Armament Development Authority supplied the first batch of Overhead Weapon Station 25 mm (OWS-25) to the Romanian state owned Industrial Group of the Army (GIA-RA) Arsenalul Armatei for installation on the MLI-84.

Romanian MLI-84 IFV showing the roof-mounted 12.7 mm anti-aircraft machine gun to the rear of the turret

This OWS is armed with an Oerlikon Contraves 25 mm KBA cannon, 7.62 mm PKT machine gun mounted externally on the right side and two Malyutka M2T anti-tank guided missiles, an improved version of the Russian AT-3 Sagger, mounted externally on the left.

A day/night sight is fitted as standard with a thermal sighting system being an option. The day sight has

Specifications
(standard MIL-84)
Crew: 2 + 9
Combat weight: 16,600 kg
Power-to-weight ratio: 21.68 hp/t
Length: 7.32 m
Width: 3.15 m
Height: 1.975 m
Ground clearance: 0.4 m
Max speed:
(road) 65 km/h
(water) 7 km/h
Fuel consumption:
(road) 110/150 litres/100 km
(cross-country) 190/220 litres/100 km
Engine: 8V-1240 DT-S supercharged diesel developing 360 hp
Armament:
(main) 1 × 73 mm 2A28 gun
(coaxial) 1 × 7.62 mm PKT MG
(ATGW launcher) 1 × 'Sagger'
(anti-aircraft) 1 × 12.7 mm DShKM MG
Ammunition:
(main) 40
(coaxial) 2,000
(anti-aircraft) 500
('Sagger') 4

Gun control equipment
Turret power control: electric/manual
Gun elevation/depression: +30/-4°
Turret traverse: 360°
Gun stabiliser:
(vertical) no
(horizontal) no
NBC system: yes
Night vision equipment: yes

Romanian MLI-84 infantry fighting vehicle fitted with RAFAEL OWS-25 armed with 25 mm Oerlikon Contraves cannon, 7.62 mm coaxial machine gun and two anti-tank missiles 0036243

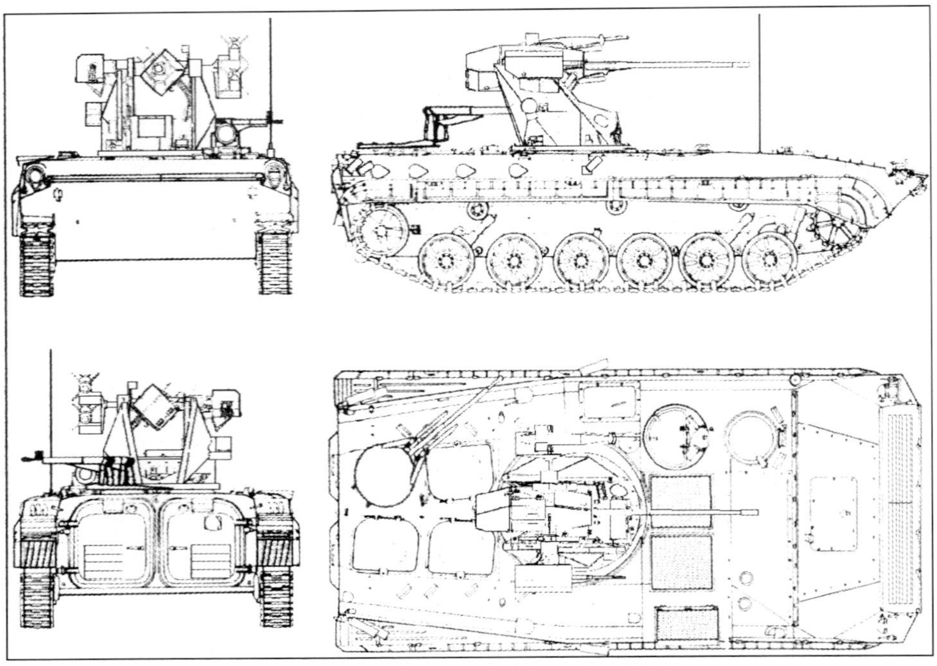

Four view general arrangement drawing of the upgraded Romanian MLI-84 **2002**/0110651

magnifications of ×1 and ×8 while the passive night vision elbow has a magnification of ×7.5, in addition there are also four day observation periscopes. As the weapon is mounted externally it can be aimed and fired with the gunner under complete armour protection.

Turret traverse and weapon elevation is all electric and the main armament is stabilised in both elevation and traverse, manual controls are provided as a back up. Turret traverse is a full 360°, with weapon elevation from −11 to +60°.

A total of 200 (60 armour piercing plus 140 high explosive) rounds of ready use 25 mm ammunition and 200 rounds of ready use 7.62 mm machine gun ammunition are carried.

The 25 mm cannon has four rates of fire, single shot, 100, 200 or 600 rounds per minute with dual belt feed system giving the operator an immediate ammunition selection capability.

Types of 25 mm ammunition fired include armour piercing fin-stabilised discarding sabot tracer, frangible armour-piercing discarding sabot tracer, high explosive incendiary tracer, semi-armour piercing high explosive incendiary tracer and training natures. Reserve ammunition consists of 450 rounds of 25 mm, 2,000 rounds of 7.62 mm machine gun and two guided missiles.

A total of six electrically operated 81 mm smoke grenade launchers are fitted, three on either side of the hull and the MLI-84M can also lay its own smoke screen by injecting diesel fuel into the exhaust outlet on the right side of the hull.

Standard equipment includes NBC system of the over pressure type, automatic fire detection and suppression system and a laser and infra-red detection system.

While first OWS-25 are being supplied direct from Israel the plan is that the Arsenalul Armatei will progressively undertake local production of the complete mount over the coming years. The first Oerlikon Contraves dual feed 25 mm KBA cannon were supplied via Italy with progressive licenced production being undertaken in Romania.

The MLI-84 with the weapons upgrade has a combat weight of 17,080 kg and a maximum height of 2.942 m. Other data is similar to the standard MLI-84.

MLI-84 repower

Perkins Engines Company of the UK has supplied a CV-8 diesel engine rated at 400 hp for installation by RATMIL, the original builder of the MLI-84, as a replacement for the existing 8V-1240 DT-S supercharged diesel developing 360 hp. If and when this goes ahead it is expected that the first batch will consist of 10 units followed by a second batch of 160 vehicles. This will convert all of the operational MLI-84 vehicles in service with the Romanian Army. As of mid-2001 no quantity orders had been placed for this repower package.

Status

Production probably complete. In service with the Romanian Army. In 1999, Romania stated that it had a total of 178 MLI-84 vehicles in service. There are no known exports of the MLI-84 vehicle.

Manufacturers

ROMARM SA.
Regia Autonoma Arsenalul Armatei (prime contractor for upgrade).

UPDATED

MLVM mountaineers combat vehicle

Development/Description

This vehicle has been designed specifically for use in mountainous terrain and can carry an assault group of seven men as well as its two-man crew, which consists of commander and driver. It can also be used to carry stores and ammunition in the forward battlefield area.

The hull and turret are of all-welded steel armour construction providing the crew with protection from small arms fire and shell splinters.

The driver is seated at the front of the vehicle on the left side with three day periscopes for forward observation, one of which can be replaced by a night vision device, and has a single-piece hatch cover. To the rear of the driver is the vehicle commander who has a single-piece hatch cover and three day periscopes, one of which can be raised for forward observation. The commander also has an externally mounted infrared searchlight.

The turret is in the centre of the hull and is armed with one 14.5 mm and one 7.62 mm machine gun, with the sight on the left side of the turret. The turret is the same as that installed on the TAB-71M (8 × 8) APC. This turret is recognisable by the large sight mounted on the left side which is provided with a protective cage.

The troop compartment is at the rear and is provided with two roof hatches, six firing ports (three in each side of the hull) each with an associated day periscope and a door in the hull rear which has two firing ports and two day periscopes as well as containing some diesel fuel. The troops are seated three each side, with the seventh manning the turret.

Suspension is probably of the torsion bar type with six rubber-tyred spoked roadwheels, drive

Romanian MLVR AR 120 mm self-propelled mortar (BP) **2001**/0093563

Artist's impression of upgraded MLVM fitted with OWS armed with 25 mm cannon, 7.62 mm coaxial machine gun and two ATGW **2000**/0034407

sprocket at the front, idler at the rear and track-return rollers.

In addition to the small arms of the infantry an AG-7 anti-tank grenade launcher is also carried.

The MLVM is fully amphibious, being propelled in the water by its tracks.

There has been no production of this vehicle in recent years.

Variants

MLVM upgrade

Building on its experience in the development of a major upgrade package for the MLI-84 (a local version of the Russian BMP-1) infantry fighting vehicle (IFV) and covered in detail in a separate entry, the Romanian company of Regia Autonoma Arsenalul Armatei has developed an extensive upgrade package for the Romanian MLVM mountaineers combat vehicle.

The aim of this upgrade, which has been developed in co-operation with Tractorul UTB, is to improve the overall efficiency of the MLVM vehicle, reduce its operating and maintenance costs, improve its firepower and generally bring the vehicle up to NATO standards.

The current MLVM is fitted with a one-person manually operated turret, armed with one 14.5 mm KPVT and one 7.62 mm PKT machine gun. These old weapons lack the firepower and range to engage many of the targets likely to be encountered on the modern battlefield.

This has been replaced by the Israeli RAFAEL Armament Development Authority OWS 25R Overhead Weapon Station (OWS) which is already used by a number of other countries including Romania, which has adopted it for a major upgrade of its MLI-84 IFVs. Full details of this OWS are given in the previous entry.

To improve the battlefield survivability of the MLVM, a laser illumination warning system is fitted which alerts the crew when the platform has been illuminated by a laser range-finder or laser designator. This informs the crew via a light and acoustic warning system as to the type of radiation, direction of threat and time from detection.

With this information the crew can either try and engage the hostile target, move out of the way or lay a smoke trail or other obscurant allowing the vehicle to escape undetected. The laser illumination warning system has automatic and manual modes of operation.

The six 81 mm smoke grenade launchers are mounted three either side of the hull to cover the frontal arc, with the control panel being mounted at the commander's station.

Specifications
(upgraded MLVM vehicle)
Crew: 2 + 9
Combat weight: 11,300 kg
Power-to-weight ratio: 30.1 to 35.4 hp/t (see text)
Length: 6.75 m
Width: 2.70 m
Height:
(maximum) 2.80 m
(hull roof) 1.60 m
Max speed:
(road) 65 km/h
(water) 15 km/h
Max road range: 800 km
Vertical obstacle: 0.85 m
Trench: 1.6 m
Engine: Mercedes-Benz or Deutz air-cooled diesel developing 340 to 400 hp
Transmission: hydrostatic, automatic with 5 forward and reverse gears
Suspension: torsion bars and hydraulic shock-absorbers
Electrical system: 24 V
Armament:
(main) 1 × 25 mm cannon
(coaxial) 1 × 7.62 mm MG
(ATGW) 2 × AT-3 Sagger

In addition to being provided with a day periscope, the vehicle commander is now provided with an EC2-55 second-generation intensification periscope developed in co-operation with the French SAGEM-ODS company. This has a magnification of ×3 and a 24° field of view. It is a direct replacement for the standard day periscope. The EC2-55 can detect objects at a range of 1,600 m and recognise objects at a range of 600 m.

The standard production MLVM is powered by a Model 798-05M2 4-stroke supercharged diesel engine, developing 154 hp coupled to a manual transmission.

This will be replaced by a German Mercedes-Benz diesel or Deutz air-cooled diesel developing 340 to 400 hp coupled to an automatic transmission. This gives the vehicle not only an increase in road speed from 48 to 65 km/h but as the engine is more fuel efficient, operational range is increased to 800 km. The power-to-weight ratio also shows a significant increase from 17.11 hp/t of the standard vehicle to 30 to 35 hp/t of the upgraded vehicle.

French Matra Defense fire detection and suppression system has been installed. This comprises two independent subsystems which assure a rapid and efficient fire detection and suppression capability in both the crew and engine compartments.

For improved communications, the existing radio system has been replaced by a new system produced in co-operation with Thales of the UK. It is of the VHF frequency hopping type and is said to be fully compatible with NATO systems.

The standard MLVM is fully amphibious, being propelled in the water by its tracks and the upgraded MLVM maintains this amphibious capability and water speed is increased from 6 to 15 km/h.

The basic MVLM has a hull of all-welded steel armour that provides the occupants with protection from small arms fire and shell splinters.

To provide a higher level of protection, an add-on passive armour package can be fitted. Due to the higher power to weight ratio of the vehicle due to the installation of a new diesel engine, the power-to-weight ratio would not be degraded with the installation of additional armour.

120 mm mortar carrier

In order to accommodate the 120 mm muzzle-loaded mortar, the roof of the rear troop compartment has been raised and when travelling this is covered by hatches that open left and right. The 120 mm mortar has the Romanian designation of the M1982 with the complete vehicle being designated the MLVM AR.

A baseplate and bipod are carried externally on the vehicle to enable the 120 mm mortar to be deployed away from the vehicle if required by the tactical situation.

The vehicle has been observed with two types of roadwheel, one spoked and similar to that of the standard MLVM mountaineers combat vehicle and a second of stamped metal similar to that used on the Romanian MLI-84 infantry combat vehicle.

Ambulance

A small number of the ambulance version has been built. These are similar in appearance to the ABAL and are not fitted with a machine gun turret.

Romanian MLVM mountaineers combat vehicle

Specifications
MLVM
Crew: 2 + 7
Combat weight: 9,000 kg
Power-to-weight ratio: 17.11 hp/t
Ground pressure: 0.47 kg/cm²
Length: 5.85 m
Width: 2.714 m
Height:
(turret roof) 1.95 m
(hull top) 1.55 m
Ground clearance: 0.38 m
Max road speed: 48 km/h
Max water speed: 5-6 km/h
Range:
(cross-country) 370-400 km
(road) 680-740 km
Fuel capacity: 480 litres, 360 litres in main fuel tank and 120 litres in door tank
Fording: 0.6 m
Gradient: 60%
Side slope: 30%
Vertical obstacle: 0.6 m
Trench: 1.5 m
Engine: Model 798-05M2 4-stroke supercharged diesel developing 154 hp
Armament:
(main) 1 × 14.5 mm KPVT MG
(coaxial) 1 × 7.62 mm PKT MG
Ammunition:
(main) 600
(coaxial) 2,500
Gun control equipment: manual
Elevation/depression: +82/-3°
Traverse: 360°
Gun stabiliser:
(vertical) no
(horizontal) no
NBC system: yes
Night vision equipment: yes

ABAL armoured vehicle for combat supply

This is very similar to the MLI-84 ICV but has been designed to carry ammunition in the forward area, for example 64 rounds of 100 mm or 104 rounds of 76 mm ammunition, with total payload being 2,100 kg. Some of the ammunition is in ready use lockers in either side of the hull with external access doors provided, while some is stowed inside. This version does not have turret-mounted machine guns although a 7.62 mm machine gun is mounted to the rear of the driver's position with a traverse of 360° and elevation from -5 to +82°. Standard equipment includes an NBC system, firefighting and communications equipment.

Specifications are similar to the MLI-84 except for combat weight of 10,500 kg, unladen weight of 8,400 kg, ground pressure 0.49 kg/cm², width 2.714 m, height 1.90 m, maximum road speed 32 km/h, road range 666 km, cross-country range 383 km and fuel capacity 450 litres.

Status

Production complete. In service with the Romanian Army. In 1999, Romania stated that it had a total of 88 of these vehicles in service.

Manufacturers

ROMARM SA.

Regia Autonoma Arsenalul Armatei (prime contractor for upgrade).

UPDATED

RUSSIAN FEDERATION AND ASSOCIATED STATES (CIS)

Heavy infantry fighting vehicle

Development/Description

Russia has recently released details of a proposed new Heavy Infantry Fighting Vehicle (HIFV) based on a much modified T-72 series MBT chassis.

From the limited amount of information revealed on the HIFV, it appears that the internal volume of the vehicle has been increased by raising the height of the hull above the roadwheels.

The driver is seated at the front of the vehicle in the centre and is provided with a single-piece hatch cover and three forward-facing day periscopes, the middle one of which can be replaced by a passive periscope for driving at night.

Either side and slightly to the front of the driver's position is a seat for one infantryman, who is provided with a roof hatch with two integral day periscopes with an additional day periscope mounted in the hull roof.

Mounted in the centre of the hull is a two-person power-operated turret armed with a 100 mm 2A70 rifled gun, 30 mm 2A72 coaxial cannon and a 7.62 mm PKT coaxial machine gun. Mounted either side of the turret is a bank of six electrically operated 81 mm smoke grenade launchers.

The turret is believed to be a modified version of the turret installed on the BMP-3 infantry combat vehicle, but fitted with additional armour protection over the frontal arc for improved battlefield survivability.

The other five infantrymen are seated in the small troop compartment to the rear of the turet and enter and leave the vehicle via roof hatches. Periscopes are provided for observational purposes.

The hull of the T-72 MBT has been provided with additional armour for the nose, glacis plate, upper hull sides and skirts. It is understood that some of this is of the explosive reactive type. This is understood to provide protection against both Chemical Energy (CE) and Kinetic Energy (KE) attack. Additional armour is also provided to cover the area between the hull roof and lower part of the turret.

Standard equipment includes an NBC system, day/night vision equipment and the vehicle can also lay its own smoke screen by injecting diesel fuel into the exhaust outlet on the left-hand side of the hull above the last roadwheel station.

The HIFV has a crew of three consisting of commander and gunner in the turret and driver seated at the front of the vehicle in the centre. In addition, it can carry seven fully equipped troops.

Artist's impression of Russian Heavy Infantry Fighting Vehicle based on modified T-72 MBT chassis ***2000***/0064823

Specifications

Crew: 3 + 7
Weight: 41,000 kg
Max speed: 60 km/h
Armament:
(main) 1 × 100 mm 2A70
(coaxial) 1 × 30 mm 2A42 cannon
(coaxial) 1 × 7.62 mm machine gun

Status

Concept. By mid-2001, it was understood that a prototype of the heavy infantry fighting vehicle had still to be completed.

Manufacturer

Russian state factories probably located in Nizhnyi Tagil where the T-72 and T-90 MBTs are built.

UPDATED

BTR-T armoured personnel carrier

Development

To provide a higher level of protection for infantry units working with MBTs, Russia has developed to the prototype stage a new heavy armoured personnel carrier called the BTR-T based on the old T-55 MBT chassis.

This was first revealed late in 1997, at a defence exhibition held in Omsk, Russia, by the GUP Design Bureau of Transport Machine Building.

It is probable that the BTR-T heavy APC was developed at the Omsk Machine Production Plant, where the T-80U is still in low-rate production.

Rear view of the BTR-T heavy APC showing side armour arrangement (Steven Zaloga) ***2000***/0064783

The BTR-T heavy APC could have been developed in response to the performance of the poorly protected Russian BMP-2 and wheeled BTR-80 (8 × 8) vehicles in Chechnya where they were ill suited to fighting in an urban environment.

As the Russian Army cannot afford to procure large numbers of the very expensive BMP-3 infantry combat vehicle, covered in detail in a separate entry later in this section, the BTR-T may have been developed to provide a vehicle with a much higher level of battlefield survivability and a similar firepower but at a much lower cost.

The Russian Army has large quantities of T-54/T-55 series MBTs which could be converted into BTR-T APCs.

Description

The new BTR-T heavy APC is based on a heavily modified old T-55 MBT chassis which has had its turret removed and hull raised to provide additional space for the crew and troop compartment at the front with the engine compartment being retained at the rear.

The BTR-T heavy APC has a combat weight of 38.5 tonnes (compared with 36 tonnes of the standard T-55 MBT) and has a crew of two, driver seated front left and and commander/gunner in the turret. Five fully equipped troops can be carried.

Mounted at the front left of the hull is a turret armed with an externally mounted 30 mm 2A42 automatic cannon (which is also fitted to the BMP-2, BMD-2 and BMD-3).

Mounted lower down on the right hand side of the turret is a pintle mount for the Tula KBP Design Bureau for Instrument Building 4,000 m range Konkurs (AT-5 Spandrel) anti-tank guided missile (ATGM). No protection is provided for the AT-5 operator when using the weapon.

It is possible that the turret used on the BTR-T is the same as that developed for the prototypes of the BMP-3 infantry combat vehicle, Obiekt 688, as this was armed with a 30 mm 2A42 cannon, 30 mm grenade launcher and 7.62 mm PKT machine gun.

It is known that weapon fit of the BTR-T is modular so that different weapons can be fitted to meet different operational and user requirements. Other versions being marketed include:

(1) Externally mounted 30 mm 2A42 cannon with a new twin round AT-5 Spandrel launcher on the left with operator under full armour protection
(2) Externally mounted 30 mm 2A42 cannon with 30 mm AGS-17 automatic grenade launcher on the left
(3) Armed with two twin 30 mm 2A38 cannon which are currently installed on the 2S6M Tunguska self-propelled air defence system
(4) Armed with a 12.7 mm machine gun with new twin round AT-5 Spandrel launcher on the left as used in (1)
(5) Armed with 12.7 mm machine gun and 30 mm grenade launcher

In addition to weapons of Russian design and manufacture, the modular design weapon turret can

BTR-T heavy APC fitted with turret armed with externally mounted 30 mm cannon, 7.62 mm coaxial machine gun and AT-5 Spandrel ATGW (Steven Zaloga) **2002**/0110652

The BTR-T heavy APC can be fitted with various one-person turrets armed with externally mounted machine guns, cannon and anti-tank guided weapons, or a combination of these weapons **2000**/0064785

also be fitted with similar weapons from the West. It would also be possible to build a similar heavy APC on redundant Western type MBTs according to Russian sources.

The front and sides of the BTR-T are protected by the latest ERA package which provides a high level of protection against chemical energy and kinetic energy attack. This appears to be similar to that used on recent production T-80U series MBTs.

It has been confirmed that the only means of entry and exit for the troops is through hatches in the roof to the front right and rear of the turret. As the roof line is higher then the rear hull deck two hatches are provided in the very rear of the troop compartment. These open upwards and are provided with vision blocks.

Like many Russian armoured vehicles, the BTR-T can lay its own smoke screen by injecting diesel fuel into the exhaust outlet on the left side of the hull towards the rear. In addition, four banks of three electrically operated smoke dischargers are mounted at the rear of the vehicle firing forwards.

Specifications

No detailed specifications of the BTR-T have been released apart from the following:
Crew: 2 + 5
Combat weight: 38,500 kg
Armament:
1 × 30 mm 2A42 cannon
1 × AT-5 Spandrel ATGW launcher
Max road speed: 50 km/h

The front and sides of the BTR-T are protected by the latest ERA package which provides a high level of protection against chemical energy and kinetic energy attack. This appears to be similar to that used on the latest T-80U series MBTs.

In many respects the Russian BTR-T is similar in design and concept to the Israeli Achzarit infantry armoured vehicle which is also based on a T-55 chassis.

On the Achzarit, the infantry enter and leave the vehicle via roof hatches or a clamshell door in the rear on the right side, on the BTR-T they can only leave through the roof hatches. The troop compartment is provided with observation periscopes to give the troops all-round observation prior to them leaving the vehicle.

Status

Prototype. Not yet in production or service.

Manufacturer

GUP Design Bureau of Transport Machine Building.

UPDATED

BMP-3 infantry combat vehicle

Development

The BMP-3 was developed by the BMP-1/2 design bureau at Kurgan with the original Obiekt 688 prototype being armed with an externally mounted 30 mm 2A42 cannon and two ATGWs. The chassis was a further development of the Obiekt 685 light tank but with a new engine.

This was not accepted for service as its armament was not a significant advance over that of the BMP-2. A new weapons system called the 2K23 was developed at Tula armed with a 100 mm 2A70 rifled gun fed by an automatic loader, a 30 mm 2A72 cannon and a 7.62 mm PKT coaxial machine gun.

This was accepted for service under the designation of the BMP-3 and made its first public appearance in 1990.

The BMP-3 entered service with the Russian Army in 1990 and it is estimated that by late 2001 between 800 and 1000 vehicles had been built, most of which were for the export market. In 1993, the export sale price of the BMP-3 was US$800,000.

The first export customer was Abu Dhabi, part of the United Arab Emirates, which placed an initial order for 330 vehicles which were delivered between 1992 and 1995.

These vehicles have a number of modifications to meet the specific operational requirements of Abu Dhabi, with some of the vehicles being fitted with a Namut thermal sight which includes a thermal camera mounted on the left side of the turret rear and provides a thermal view for the gunner.

According to United Nations sources, between 1992 and 1998 Russia exported a total of 575 BMP-3 series ICVs, including variants:

Country	Quantity	Delivery
Cyprus	43	1995-96
Republic of Korea	23	1996-97
Kuwait	118	1995-96
UAE	391	1992-97

Description

The hull and turret of the BMP-3 are of all-welded aluminium armour construction providing the crew with protection from small arms fire and shell splinters. The glacis plate is well sloped and the hull sides vertical.

The hull floor of the BMP-3 consists of two layers with a space between them and, over the frontal arc of the turret, spaced armour is fitted as standard.

The driver is seated at the front of the vehicle in the centre and has a single-piece hatch cover that opens to the right. He is provided with five day periscopes that give observation to the front and sides of the vehicle.

Like the BMD there is an additional crew member to the left and right of the driver's position, each provided with a roof hatch that opens forwards and a single day periscope in the hull roof in front of the hatch cover.

One 7.62 mm PKT machine gun is mounted either side of the hull, firing forwards and these are operated by one of the front three crew members.

The turret is in the centre of the vehicle with the commander seated on the right and the gunner on the left.

The commander has a single-piece hatch cover that opens forwards with integral day periscopes for observation purposes, day/night sights and a searchlight mounted on the forward part of his cupola.

UAE vehicle with the Athos sight fitted **2000**/0064786

Close-up of BMP-3 suspension (Christopher F Foss) **2000**/0064784

In the roof of the vehicle, forward of the commander's hatch, is a large rectangular hatch that opens to the left which may be used for ammunition/missile resupply purposes.

In addition, there is a rectangular hatch in the turret roof towards the rear that opens forwards, which may be used for ejecting spent cartridge cases.

Over the frontal arc the turret is provided with a layer of spaced armour and mounted on either side of this is a bank of three 81 mm electrically operated smoke grenade launchers.

The complete weapon system of the BMP-3 is known as the 2K23 and consists of a 100 mm 2A70 gun, a 30 mm 2A72 coaxial cannon and a 7.62 mm PKT coaxial machine gun, all in a common mount. The armament is stabilised in two planes via the 2E52 electromechanical system.

The 100 mm 2A70 rifled gun fires 3UOF17 and 3OF32 high-explosive fragmentation rounds at a rate of 8 to 10 rds/min, or the 3UBK-10-3 with the 9M117 laser-guided missiles. The 100 mm high-explosive rounds are fed from an automatic loader while the 100 mm laser-guided projectiles are fed into the breech manually.

The 100 mm missile launched from the 100 mm 2A70 gun has the US designator of the AT-10 and the NATO codename of 'Stabber'. Its official Russian designation is the 9M117 while the complete system for the BMP-3 is designated the 9K116-3. The common Russian name for the missile is the Basnya.

The missile is also fired from the following weapon systems:

(1) MT-12 100 mm towed anti-tank gun, system designation 9K116, US designator AT-10, NATO codename 'Stabber' and with missile called Kastet by the Russians
(2) 100 mm T-55 MBT, system designation 9K116-1, US designator AT-10, NATO codename 'Stabber' and with missile called Bastion by the Russians
(3) 115 mm T-62 MBT, system designation 9K116-2, US designator AT-12, no NATO codename and with missile called Sheksna by the Russians.

The complete round weighs 24.5 kg with the missile itself weighing 17.6 kg for the BMP-3 version, muzzle velocity of 370 m/s and maximum range of 5,500 m, although earlier sources quoted a range of 4,000 m.

The original 9M117 missile was claimed to be able to penetrate between 650 and 700 mm of conventional steel armour but more recently another missile, the 9M117M with a tandem warhead has been developed. This is claimed to be able to penetrate up to 550 mm of steel armour with the original missile now stated to be able to penetrate only 275 mm of steel armour.

There are 30 rounds of high-explosive fragmentation ammunition carried for the 100 mm gun, 22 of which are for ready use in the automatic loader. This round has a maximum range of 4,000 m and a muzzle velocity of 250 m/s.

A total of eight 9M117 anti-tank guided missiles is carried. The 9M117 is a semi-automatic laser beam guided missile and all the gunner has to do to ensure a hit is to keep his sight on the target.

The main armament is fully stabilised as is the gunner's stabilised 1K13-2 sight which has a day channel with a magnification of ×8 and a night channel with a magnification of ×5.5.

The fire-control system includes a 1V539 ballistic computer, a 1D16-2 laser rangefinder which is mounted in a box over the main armament and a met sensor.

The gunner has a secondary PPB-2 (or PPD-1 sight) day sight with a magnification of ×2.6 while the commander has a dual sight 1P3-10 with a magnification of ×1.2 and ×4 and a periscope TKN-3MB for observation purposes with a magnification of ×5 by day and ×3 by night.

The gunner's searchlight is designated the OU-5 while the commander's illuminator is the OU-3. Mounted coaxially to the right of the 100 mm gun is the 30 mm 2A72 cannon which is ballistically identical to the 30 mm 2A42 cannon of the earlier BMP-2. The more recent 30 mm 2A72 only has 349 parts whereas the older 30 mm 2A42 has some 578 parts.

The dual-feed 30 mm 2A72 cannon can engage armoured vehicles at ranges of 1,500 to 2,000 m and helicopters out to 4,000 m. Cyclic rate of fire is 330 rds/min and three types of 30 mm ammunition are fired, AP-I, HE-I and HE-T. The AP-I and the HE-I both have a muzzle velocity of 960 to 980 m/s. The AP-T projectile will penetrate 25 mm of steel at an angle of 60° at a range of 1,500 m. A total of 500 rounds of 30 mm ammunition is normally carried, 305 of which are HE-I and 195 AP-I.

The POZIS State Unitary Enterprise recently stated that it was marketing three 30 × 165 mm rounds for use by the 30 mm 2A38, 2A42 and 2A72 cannon and details of these are given below: -

APT round will penetrate 15 mm of steel armour at an incidence of 60° at a range of 1,500 m.

The 7.62 mm PKT coaxial machine gun has a practical rate of fire of 250 rds/min and an effective range of 1,500 m.

According to Russian sources, the BMP-3 can engage a target at a range of 1,000 m in just under 3 seconds with a 90 per cent accuracy. For a 2,000 m range however, this takes 14 seconds. The second target can be engaged in 4.9 seconds at 1,000 m and

Type	**HEI**	**APT**	**FT**
Length of cartridge	291 mm	291 mm	290 mm
Weight of cartridge	842 g	853 g	826 g
Weight of projectile	389 g	400 g	386 g
Explosive content	123 g	127 g	118 g
Muzzle velocity	960 m/s	970 m/s	970 m/s
Maximum range	4,000 m	n/a	4,000 m

Notes:
HEI = High Explosive Incendiary
APT = Armour Piercing-Tracer
FT = Frangible Tracer

BMP-3 ICV in service with Cyprus (BP) **2001**/0093562

BMP-3 of Abu Dhabi from the rear with both roof hatches and twin rear doors in the open position (Christopher F Foss) 0018454

First photograph released of the KBP Instrument Design Bureau Germes long-range anti-armour system on BMP-3 chassis **2001**/0093561

BMP-3K command vehicle which has additional communications equipment 0018455

19.6 seconds at 2,000 m. Either the commander or gunner can aim the turret weapon systems.

Turret traverse is a full 360° with weapon elevation from –5 to +60°. The latter enables the weapons to engage helicopters. Turret traverse and weapon elevation are powered. The dual-axis stabiliser is designated the 2E52.

The basic crew consists of commander, gunner and driver. A total of seven infantrymen can be carried, one either side of the driver with the remainder being seated to the sides and rear of the turret.

Normally, five infantrymen are carried at the rear, two either side, one behind the other and one in the centre although, if required, a further two infantrymen can be carried bringing the total to seven in the rear.

The troop compartment is at the rear of the hull with entry via two doors in the hull rear opening left and right, with the left door having a firing port. As these doors are opened steps automatically fold down.

Over the top of the troop compartment are large rectangular roof hatches, each of which has an integral hatch that opens to the outside of the vehicle.

In either side of the troop compartment, to the immediate rear of the turret, are two small firing ports of a different design from those fitted to the BMP-1/BMP-2/BMD-1/2/3 vehicles.

The power pack of the BMP-3 is located low down at the very rear of the hull with the hydromechanical transmission, with four forward and one reverse gears, on the left and the UTD-29M 500 hp diesel engine on the right. The air inlet/outlets are located in the right side of the hull roof with the rectangular exhaust outlet located on the right side of the hull just above the track guard.

When the BMP-3 is used amphibiously a short snorkel is extended above the right side of the hull at the rear.

The suspension either side consists of six dual rubber-tyred roadwheels with the idler at the front, drive sprocket at the rear and three track-return rollers.

The roadwheels have individual hydropneumatic suspension which consists of a spring, shock-absorber and an elevation device that allows the ground clearance to be adjusted from 190 mm to 510 mm. An external shock-absorber is attached to the first, second, and sixth roadwheel stations.

The BMP-3 is fitted with an electromechanical track adjustment system allowing the driver to adjust the track tension for both tracks without leaving his seat.

The BMP-3 is fully amphibious being propelled in the water by two water-jets mounted on either side low down at the rear of the hull. These water-jets are run from a power take off to the hydromechanical transmission.

Before entering the water the bilge pumps are switched on, a snorkel is erected at the rear and a trim vane is erected at the front of the vehicle. When the trim vane is not required it folds back under the nose of the BMP-3.

Infantry weapons carried in the BMP-3 include two 7.62 mm PK machine guns, six 5.56 mm AK-74 automatic rifles and a 26 mm signal pistol.

Standard equipment includes the capability to lay a smoke screen by injecting diesel fuel into the exhaust outlet, an NBC system and a dozer/entrenching blade fitted at the front of the vehicle.

The BMP-3 has a fire detection/suppression system in the front and rear crew compartments, turret and engine compartment.

The BMP-3 can be carried by the Mi-26 heavy lift helicopter and can also be dropped by parachute using a specially designed platform.

Optional equipment includes an air conditioning system for use in hot climates and thermal night vision equipment which is covered under variants.

Additional equipment is a directional gyro GPK-59 for all vehicles and the navigation system TNA-3 for the BMP-3K.

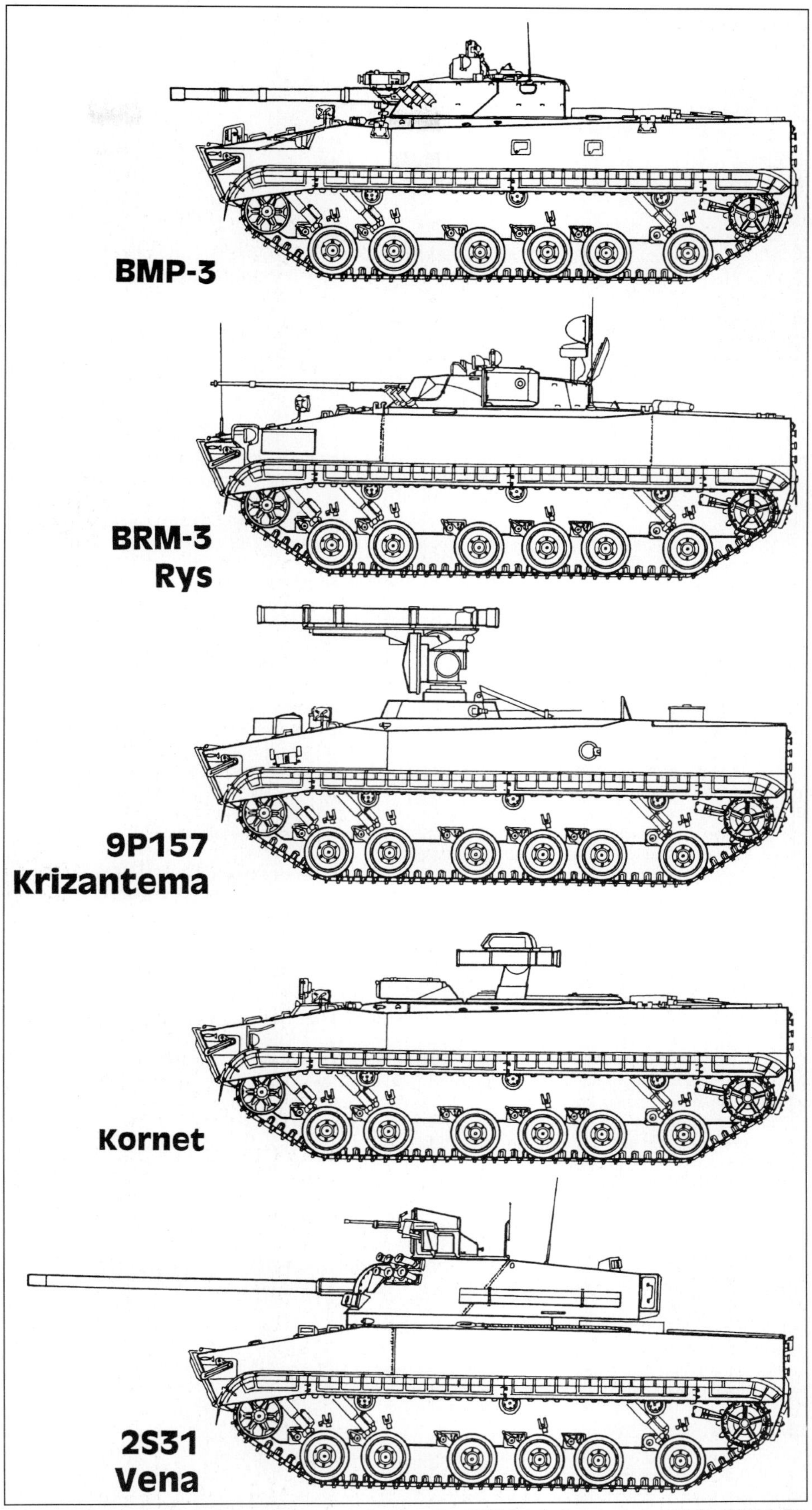

The BMP-3 family now includes, from top to bottom, the BMP-3 infantry combat vehicle, BRM-3 Rys reconnaissance vehicle, 9P157 Krizantema missile armed tank destroyer, Kornet tank destroyer and the 2S31 Vena self-propelled artillery system (Steven Zaloga) 0005032

Variants

BMP-3M

The KBP Instrument Design Bureau has released details of its new upgrade package for the BMP-3 ICV.

This upgrade package is generally referred to as the BMP-3M and for the turret there are three distinct upgrades designated Variant A, B and C.

Variant A is essentially an ammunition upgrade and includes a new 100 mm gun launched laser-guided projectile with enhanced lethality and increased range, new 100 mm High Explosive Fragmentation (HEF) round (3UOF19) with enhanced lethality and increased range, new 30 mm Armour-Piercing Discarding Sabot (APDS) round and a new digital computer. The new APDS round is also referred to by the Russians as the hard core or High-Velocity Armour-Piercing (HVAP) projectile.

The new 100 mm laser-guided projectile is called the Arkan (9M117M1) and has a maximum range of 5,500 m and is fitted with a tandem High-Explosive Anti-Tank (HEAT) warhead. The 100 mm HEF projectile has a maximum range of 7,000 m and is twice as effective as the older round.

Variant B includes the ammunition upgrading (A) plus a fire-control subsystems upgrade which includes a new integrated gunner's sight BZS1 and the installation of a new weapon stabiliser system for increased accuracy.

The integrated BZS1 gunner's sight includes a mirror stabilisation system, day channel with adjustable magnification, laser range-finder, laser-guidance system, SAGEM thermal imager and auto tracker and display.

Variant C is the most advanced and consists of a brand new turret fitted with a high-resolution automatic fire-control system, new ammunition reserve and a single reloading system.

BREM-L recovery vehicle based on a modified BMP-3 chassis (Christopher F Foss) 0058022

BMP-3 Multi-Role Chassis

It includes the integrated gunner's sight as well as a new commander's panoramic stabilised sight with TV channel, laser range-finder and display. This system allows for hunter/killer target engagements to take place.

Variant C also has a single reloading system and new feeding systems for 30 mm cannon and 7.62 mm coaxial machine gun.

Other upgrades being offered include the Arena defensive aid suite, battle management system, KBM-3M air conditioning system, more powerful UTD-32 660 hp diesel engine to name but a few.

BMP-3 with Arena

To further improve the battlefield survivability of the BMP-3 ICV, it can be fitted with the Arena active defence system which was originally developed for installation on MBT. Arena was shown fitted onto a BMP-3 ICV in the Middle East early in 2000. By mid-2001, the Arena system was understood not to have entered production, although, for trials purposes it has been installed on a number of vehicle chassis.

BMP-3 with ERA

In mid-2000, Russia completed delivery to the United Arab Emirates of an advanced explosive reactive armour system for installation on the BMP-3.

In addition to the ERA, the United Arab Emirate BMP-3 vehicles were also fitted with splinterproof shields in the troop and combat compartments and the diesel fuel tanks were given additional protection against inflammation and explosion.

According to Russian sources, some 49 per cent of the frontal projection area and 85 per cent of the lateral projection area is protected by ERA tiles.

The installation of the ERA increased the overall length of the vehicle to 7.1 m and its width to 4 m. Maximum combat weight is now quoted as 23.4 tonnes.

BMP-3 155 mm Hermes multipurpose weapon system

Details of this 155 mm multipurpose weapon system, developed by the Tula Instrument Design Bureau, were given in *Jane's Armour and Artillery 1999-2000.* As far as it is known it never passed the concept phase.

BMP-3 Germes ATGW system

In 1999, it was stated that the KBP Instrument design team was developing a new long-range ATGW system called Germes (or Hermes) based on a BMP-3 chassis. This was the same name previously given to the 155 mm weapon mentioned above.

The first photograph released shows a BMP-3 chassis with new turret. The turret has a radar on top and on either side are six Vikhr-S missiles which can be fitted with various seekers. The supersonic Vikhr-S was originally developed for air-launched applications. In this land application it is said to have a maximum range of 15,000 m under day conditions. It can be fitted with a tandem HEAT warhead or a thermobaric warhead.

Mounted at the front left side of the BMP-3 chassis is a ball-type electro-optical device containing day/thermal sights and a laser rangefinder/designator.

BMP-3F

This has been developed for use by naval infantry and lacks the self-entrenching blade that is normally mounted at the front of the hull. A large snorkel is normally extended at the rear of the hull on the right side when the vehicle is afloat.

According to the prime contractor, the BMP-3F can operate up to sea state 3, while the weapons can be fired with a high level of accuracy up to sea state 2. It can also stay afloat for a maximum of 7 hours.

DZM engineer vehicle

This has a modified hull and is fitted with a front-mounted dozer blade for clearing battlefield obstacles and preparing firing positions. It also has a crane. The DZM is also referred to as the Vostorg.

SOZH sight

The PELENG Joint Stock Company has developed a new sight called the SOZH which can be integrated into existing BMP-3 vehicles or installed in new build vehicles. This sight includes a laser range-finder channel, missile guidance channel, day sighting channel and an image intensification channel.

BMP-3 fire-control system for China

Late in 1999, it was disclosed that Russia had sold to China the complete fire-control system, weapons and associated munitions, including laser-guided projectiles, for installation into a new vehicle of Chinese design.

BMP-3 with AMOS 120 mm mortar

For the fire support role, BMP-3 is being marketed with the Finnish/Swedish Patria Vammas/Hagglunds Vehicle AMOS twin 120 mm mortar system. Details of the AMOS are given in the Self-propelled Mortars section under International.

BMP-3 with Athos sight

The French SAGEM Athos thermal imaging system is now in volume production for the Giat Industries Leclerc MBT and the Russian BMP-3 ICV being supplied to the United Arab Emirates.

The standard BMP-3 is fitted with image intensification night vision equipment which, as shown during the 1991 Middle East conflict, has a number of severe disadvantages.

The two-axis stabilised thermal sight for the BMP-3 is called the Namut and is a joint development between SAGEM of France which is responsible for the thermal imager, Peleng-Belemo of Russia for the stabilisation system and SKBM Kurgan, also of Russia, which is the prime contractor for the BMP-3 ICV.

The new Namut thermal sight is mounted at the rear of the BMP-3 turret on the left side with two armoured shutters opening left and right.

The sight is stabilised in both the horizontal and vertical planes and has two magnifications, ×3 and ×10 and three fields of view; 9 × 6, 3 × 2 and 1.5 × 1. The thermal imager operates in the 8 to 12 µm wavelength with the gunner, seated in the left of the turret, having a monitor screen.

The installation of the Namut thermal sight in the BMP-3 allows the vehicle to use its three turret-mounted weapons at greater effective ranges than the original night vision equipment fitted to the vehicle.

The SAGEM Athos thermal imager features a closed-cycle cooler and built-in test equipment and normally operates from a 28 V power supply. It is based on the French SMT (Système Modulaire Thermique) common modules with the units being interchangeable without adjustment. It has a standard 625-line CCIR TV output.

Specifications

BMP-3
Crew: 3 + 7
Weight: 18,700 kg
Power-to-weight ratio: 26.73 hp/t
Ground pressure: 0.61 kg/cm^2
Length:
(overall) 7.14 m
(hull) 6.715 m
Width:
(over tracks) 3.15 m
(overall) 3.23 m
Height:
(over turret roof) 2.30 m
(overall) 2.65 m
Firing height: 2.02 m
Ground clearance: adjustable between 190-510 mm
Track: 2.76 m
Track width: 380 mm
Length of track on ground: 4.06 m
Max road speed:
(forwards) 70 km/h
(reverse) 20 km/h
Max speed afloat: 10 km/h
Range:
(cruising road) 600 km
Trench: 2.5 m
Vertical obstacle: 0.8 m
Fording: amphibious
Gradient: 60%
Side slope: 30%
Engine: Type UTD-29M 10-cylinder 4-stroke liquid cooled diesel developing 500 hp
Transmission: hydromechanical, 4 forward and 2 reverse gears
Steering: differential with hydrostatic drive
Suspension: Hydropneumatic shock-absorbers
Electrical system: 27 V
Armament:
(main) 1 × 100 mm 2A70 gun
(coaxial) 1 × 30 mm dual-feed 2A72 cannon
(coaxial) 1 × 7.62 mm PKT MG
(bow) 2 × 7.62 mm PKT MG
(smoke grenade dischargers) 2 × 3 81 mm
Ammunition:
(100 mm) 40 (22 in automatic loader)
(ATGW) 8
(30 mm) 500
(7.62 mm) 6,000

Gun control equipment
Turret power control: electric/manual
(by commander) yes
(by gunner) yes
(ballistic computer) yes
(laser rangefinder) yes
Gun elevation/depression: +60/-6°
Turret traverse: 360°
Gun stabiliser:
(vertical) yes
(horizontal) yes
NBC system: yes
Night vision equipment: yes

BMP-3 multirole chassis

The manufacturer is offering the basic BMP-3 chassis without the complete turret, top decking and front decking (for example, either side of the driver's position), as a chassis which can be adopted for a wide range of other roles. The base line chassis, without a turret, weighs 12.1 tonnes and can accept a maximum payload of 6 tonnes.

Roles suggested by the maunfacturer include self-propelled guns, mortars, air defence and anti-armour systems, command and control vehicles, transporters and loaders and engineer vehicles.

BMP-3 reconnaissance vehicle

This is called the Rys (Lynx) (or BRM) and is covered in detail in the Reconnaissance vehicles section. As far as it is known this is only used by the Russian Army.

BMP-3 driver training vehicle

This has had its turret and bow weapons removed to enable it to be used in the training role and mounted on top of the vehicle is a fully enclosed superstructure for the instructor.

BMP-3 recovery vehicle

This has been developed to the prototype stage and is called the BREM-L or Beglianka. To enable it to carry out its mission it is fitted with a wide rangeof equipment including crane with telescopic jib, winch with a capacity of up to 40 tonnes, cutting/welding equipment, front-mounted dozer/stabilising blade and a cargo bed on which can be carried spare components. Armament consists of a turret-mounted 7.62 mm PKT machine gun.

Cyprus has taken delivery of at least one of these vehicles. Additional details are given in *Jane's Military Vehicles and Logistics, 2001-2002.*

BMP-3K command vehicle

This version of the BMP-3 is fitted with additional navigation and communications equipment to enable it to carry out the command role. It normally has a six-man crew and retains the complete weapon system of the standard production BMP-3 vehicle. The BMP-3K also has an auxiliary power unit as standard equipment.

Communications equipment includes two radio sets, receiver, intercom for seven users and a portable power unit. It can also be fitted with a SAGEM thermal sighting system.

BMP-3 with Wildcat turret

GLS, a subsidiary of Krauss-Maffei Wegmann of Germany, has proposed that the chassis of the BMP-3 infantry combat vehicle could be fitted with the complete Wildcat turret armed with two 30 mm cannon, optical tracking system incorporating FLIR and laser rangefinder and roof-mounted Siemens 2D surveillance radar.

The design of Wildcat is modular so that other weapon/sensor suites are also possible. The former could include, for example, one 30 mm Mauser cannon and a pod of four Russian SA-16/SA-18 SAMs or two Mauser 30 mm cannon and a pod of two SA-16/SA-18 SAMs. Other SAMs could also be fitted, for example the Matra BAe Dynamics Mistral.

Sensor options include the Swedish Ericsson HARD 3D radar system which is already in production for the Swedish SAAB Bofors Dynamics RBS 90 SAM system. Details of the turret are given in *Jane's Land-Based Air Defence 2001-2002.*

120 mm 2S31 Vena artillery system

Details of this system, which first appeared in public in 1997, are given in the Self-propelled motar section as it is based on a modified BMP-3 chassis. So far this system has yet to enter production.

BMP-3 Kornet-E anti-tank system

Details of this system on a modified BMP-3 chassis are given in the Tank destroyers section. So far this system has yet to enter production.

BMP-3 Krizantema anti-tank system

Details of this system in a modified BMP-3 chassis are given in the Tank destroyers section. So far this system has yet to enter production.

Status

In production. In service with:

Country	Quantity	Comment
Azerbaijan	4	
Cyprus	43	Including at least one ARV
Republic of Korea	23	debt repayment
Kuwait	118	all delivered
Russia	200	Estimate
UAE	391	all delivered, some have thermal sights
Ukraine	4	

Manufacturer

Kurgan Machine Construction Plant, Russian Federation and Associated States (CIS).

UPDATED

Upgraded BMP-2 infantry fighting vehicle

Development/Description

Early in 1995, the Joint Stock Company Kurganmashzavod announced that it had developed a new version of the BMP-2 infantry combat vehicle for the export market.

This upgraded version of the BMP-2 can be supplied as a new build vehicle or kits can be provided to upgrade existing vehicles to this configuration. T date there are no known export sales of this complete system or upgrade packages. As far as it is known, there has been no recent production of the BMP-2 IFV although marketing of this is still carried out.

The standard BMP-2, which entered service in 1983, is armed with a 30 mm 2A42 cannon and a 7.62 mm PKT coaxial machine gun. Mounted on the turret roof is a launcher for the KBP Instrument Design Bureau Konkurs (NATO AT-5 'Spandrel') ATGW, the launcher can be traversed through 360° with elevation from -5 to +75°. The AT-5 has a maximum range of 4,000 m with the latest versions having a HEAT (High Explosive Anti-Tank) tandem warhead to defeat MBTs fitted with explosive reactive armour.

The new version of the BMP-2 has the following improvements:

(1) Mounted externally on the left side of the turret roof at the rear is a 30 mm AG-17 automatic grenade launcher which is the vehicle-mounted version of the widely deployed AGS-17 automatic grenade launcher. This entered service with the Russian Army in the tripod-mounted configuration many years ago.

(2) The weapon is fed from a magazine mounted externally at the turret rear and moves in elevation with the main armament. Maximum effective range is 1,730 m with the empty cartridge cases being ejected externally. Cyclic rate of fire is 350 to 400 rds/min.

(3) The installation of this grenade launcher enables the vehicle quickly to lay suppressive fire onto a target

Specifications

Crew: 3 + 7
Combat weight: 14,500 kg
Power-to-weight ratio: 20.68 hp/t
Length: 6.735 m
Width: 3.15 m
Height: 2.45 m
Ground clearance: 0.42 m
Max road speed: 65 km/h
Max water speed: 7 km/h
Max road range: 600 km
Gradient: 60%
Side slope: 30%
Vertical obstacle: 0.7 m
Trench: 2.5 m
Engine: UTD-20 4-stroke liquid-cooled 6-cylinder diesel developing 300 hp
Transmission: manual, 5 forward and 1 reverse gears
Armament:
(main) 1 × 30 mm 2A42 cannon
(coaxial) 1 × 7.62 mm PKT machine gun
(grenade launcher) 1 × 30 mm AG-17
(anti-tank) 1 × AT-5 launcher
Smoke laying equipment: 2 × 3 81 mm smoke grenade launchers, diesel fuel injected into the exhaust
Ammunition:
(30 mm cannon) 500
(30 mm grenade) 350
(7.62 mm MG) 2,000
(ATGW) 4

Gun control equipment
Turret power control: electric/manual
(by commander) yes
(by gunner) yes
Gun elevation/depression: +75/–5°
Turret traverse: 360°
Gun stabiliser:
(vertical) yes
(horizontal) yes
NBC system: yes
Night vision equipment: yes

Upgraded BMP-2 infantry fighting vehicle (Christopher F Foss) ***2000***/0064788

Upgraded BMP-2 from the rear showing 30 mm grenade ammunition stowage box to turret rear and unditching beam carried on hull roof at rear (Christopher F Foss) ***2000***/0064789

(4) For use in hot climates an air conditioning system has been installed
(5) The gunner is provided with a thermal night sight to enable targets to be engaged under virtually all weather conditions.
(6) This sight is called the Sanoet-1 and has been developed in co-operation between SAGEM of France (thermal imager), Peleng-Belomo (stabilisation system) and the chassis manufacturer. This consortium also developed the thermal sighting system installed on the BMP-3 vehicles delivered to Kuwait.
(7) The day channel has a ×6 magnification (with 10° field of view) with the thermal channel having a magnification of ×4.6 (9.2° field of view) and ×6.4 (3.2° field of view).

This turret is also being offered for installation on other armoured vehicles, tracked and wheeled and this vehicle has been demonstrated to a number of potential customers in the Middle East.

Kurgan upgrade

In addition to this upgrade the prime contractor of the BMP-2 now offers another upgrade which includes the installation of the following:

(1) 30 mm AG-17 grenade launcher
(2) BPK-3-42 gunner's sight with laser range-finder or BPK-M sight with thermal imaging module
(3) PMW driver's universal night vision device
(4) modernisation of fire-control system
(5) installation of URD-23 engine with output of 350 hp
(6) upgraded suspension and track
(7) additional armour
(8) installation of quick acting firefighting system
(9) installation of KBM-2 air conditioning system

Status

Development complete. Ready for production. The upgraded BMP-2 IFV has been demonstrated in various parts of the world, including the Middle East.

Manufacturer

Kurgan Machine Construction Plant, Russian Federation and Associated States (CIS).

UPDATED

BMP-2 infantry fighting vehicle

Development

The original BMP-1 Infantry Fighting Vehicle (IFV), covered in detail in the following entry, suffered from a number of disadvantages including the position of the commander in the front of the hull to the rear of the driver, the ineffectiveness of the 2A28 73 mm gun and the difficulty of controlling the AT-3 Sagger ATGW mounted over the 73 mm gun.

After evaluating a number of different prototype vehicles, the Obiekt 675 was finally selected as the BMP-2 in the late 1970s with production being undertaken at Kurgan and possibly Rubtsovsk.

The BMP-2 IFV was first seen in public during the parade held in Red Square, Moscow, in November 1982, although by that time it had already been in service for some years.

The BMP-2 has seen combat with the Russian Army in Afghanistan, with the Iraqi Army in the 1990-91 Gulf War and with the Kuwait Army in the recapture of Kuwait in 1991.

In addition to being manufactured in Russia, it is also manufactured under licence in India, where it is known as the Sarath, and in the former Czechoslovakia. In 1993, Russia was marketing the BMP-2 IFV at US$400,000 per unit with the BMP-3 at US$800,000.

It is understood that there has been no recent production of the BMP-2 IFV at the Kurgan Machine Construction Plant, although facilities for production still remain.

According to United Nations sources there were the following exports of BMP-2 IFV vehicles between 1992 and 1998:

From	To	Quantity	Date
Belarus	Uzbekistan	1	1995
Hungary	Angola	7	1993
Poland	Angola	50	1994
Poland	Togo	20	1997
Russia	Finland	84	1992
Russia	Iran	3	1998
Slovakia	Algeria	48	1995
Slovakia	Algeria	6	1996
Slovakia	Indonesia	11	1998
Ukraine	Algeria	32	1998
Ukraine	Slovakia	6	1995
Ukraine	Sri Lanka	4	1994
Ukraine	Sudan	6	1996

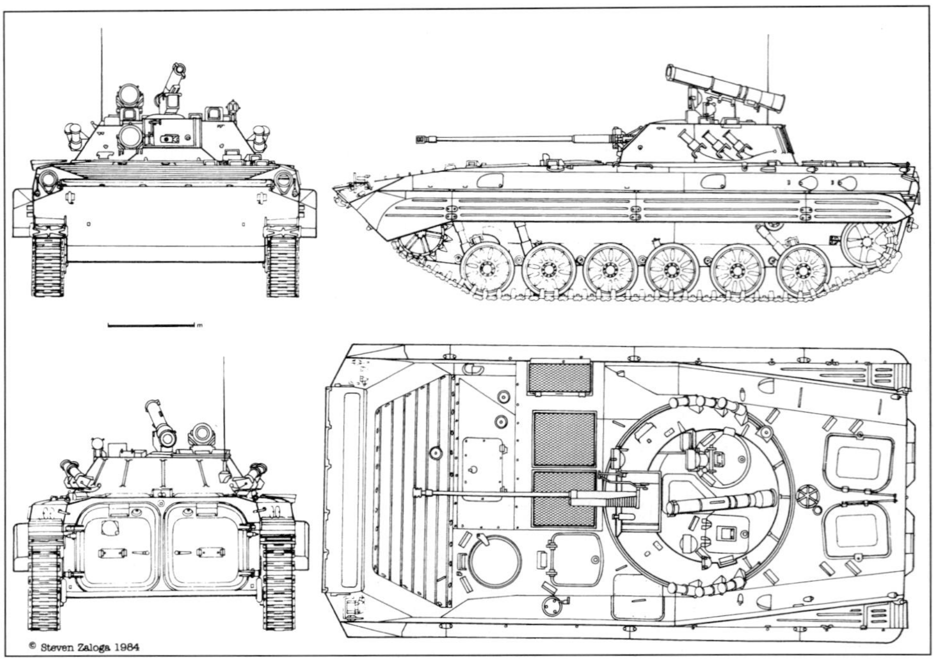

BMP-2 IFV with all hatches closed and with 'Spandrel' AT-5 ATGW mounted on the turret roof (Steven Zaloga)

Description

The chassis of the BMP-2 IFV is almost identical to that of the earlier BMP-1 IFV, but has increased armour protection. The driver sits at the front of the vehicle on the left side with a single-piece hatch cover opening to the right. In front of his position are three day periscopes, the centre one (TNPO-170A) of which can be replaced by a vertically extensible one (TNPO-350B) which allows the driver to see ahead when the trim vane is erected for amphibious operations. For driving at night, the driver's centre TNPO-170A can be replaced by a TVNE-1PA active/passive night vision device.

On the BMP-1 IFV the commander sits to the rear of the driver, but on the BMP-2 IFV is moved into the turret so this position is occupied by an infantryman. He has a single-piece hatch cover that hinges forward to open. In front of this and to the left side is a day periscope facing towards the front of the vehicle. In the left side of the hull, forward of the infantryman's hatch is a single firing port.

The engine and transmission are to the right of the driver's compartment with the air inlet and outlet louvres on top of the hull. The BMP-2 IFV is powered by a four-stroke, six-cylinder Model UTD-20 supercharged diesel engine developing 285/300 hp at 2,600 rpm. This is fitted with a compressed air starting system with an electric starter provided as a back-up. The engine is coupled to a manual transmission with five forward and one reverse gears.

The BMP-2 IFV has a two-man all-welded steel turret with the commander seated on the right and the gunner on the left. The vehicle commander would normally dismount with the infantry.

The gunner has a single rectangular hatch that opens to the front with an integral rear-facing day periscope and three fixed day periscopes, two to the front and one to the left side.

These fixed day periscopes are designated the TNPO-170A and are the same as those used by the driver/mechanic (3), commander (2) and fighting compartment (8), and are electrically heated.

The vehicle commander has a cupola with a single-piece hatch cover that opens forward and in the forward part of this are three fixed day periscopes for observation to the front.

The gunner's binocular sight is designated the BPK-1-42, he also has a TNPT-1 designator.

BMP-2 IFV with Kliver turret. The missiles are not fitted in this photograph (Christian Dumont) **2001**/0099467

BMP-2 IFV from the rear showing appliqué armour. In this photograph neither the 81 mm smoke grenade launchers, nor the skirts are fitted **2001**/0099468

BMP-2 of the Czech Republic with non-standard rubber skirt system fitted (Michael Jerchel) ***2002***/0110653

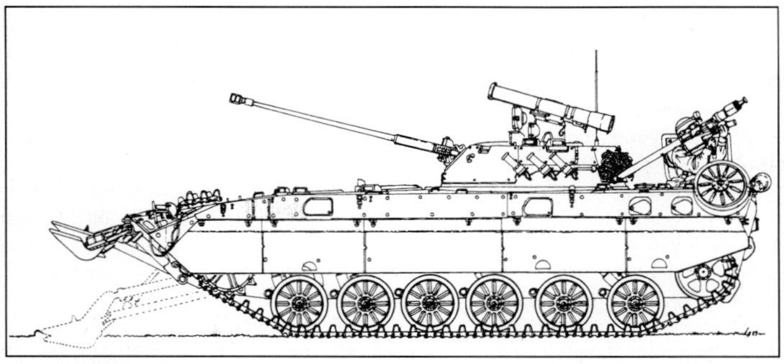

BMP-2 fitted with the KMT-10 mineclearing system in the retracted position with operational position shown by the dotted line. The vehicle also has appliqué armour, 'Spandrel' AT-5 ATGW and an SA-16 SAM being launched from the troop compartment (Miltra)

The commander's day monocular sight Model 1PZ-3 has magnifications of ×1, ×2 or ×4. He also has a TNP-165A designator and a TKN-3B binocular sight with a day magnification of ×4.75 and a night magnification of ×4.

An infra-red searchlight is mounted coaxially to the right of the 30 mm 2A42 cannon and the commander also has a roof-mounted infra-red searchlight Model OU-3GA2.

Main armament comprises a stabilised 30 mm cannon 2A42 with an elevation of +75°. Turret traverse and weapon elevation are powered, with traverse through 360° in 10.28 seconds and elevation through 74° in 12.33 seconds.

The gunner can select single shots or one of two automatic rates of fire, low at 200 to 300 rds/min; high at 500 rds/min. The 30 mm 2A42 cannon has dual feed, one would normally be for HE-T and the other for AP-T, both with a muzzle velocity of 970 m/s. The 30 mm 2A42 cannon, which has a long barrel and a muzzle brake, is fully stabilised.

Russian Federation sources claim that the 30 mm 2A42 cannon has an effective range against ground targets of 1,000 m although it is sighted to 4,000 m. Its high elevation of +75° allows it to be used against aircraft and helicopters.

A total of 500 rounds of 30 mm ammunition is carried of which 160 are AP-T and 340 rounds are HE-T.

A 7.62 mm PKT machine gun is mounted to the left of the main armament and 2,000 rounds are carried for this weapon. Late production BMP-2s have a bank of three electrically operated 81 mm smoke grenade dischargers firing forwards and these have now been backfitted to some earlier vehicles. The smoke grenade launcher system is designated the 902V.

The BMP-2 IFV can also lay its own smoke screen by injecting diesel fuel into the exhaust outlet on the right side of the hull. This system is called the TDA and can only be used when the engine is warm and the vehicle moving.

Mounted on the turret roof between the gunner's and commander's hatches is a 'Spandrel' AT-5 ATGW as on the latest model of the BRDM-2 (4 × 4) amphibious reconnaissance/anti-tank vehicle. The 'Spandrel' is a second-generation type and has a maximum range of 4,000 m. All the gunner has to do to ensure a hit is to keep the cross-hairs of his sight on the target. A ground mount is carried to allow 'Spandrels' to be launched away from the vehicle. Foreign BMP-2 IFV have the 2,500 m 'Spigot' ATGW in place of the 'Spandrel' installed on Russian BMP-2s. The 'Spigot' version can be distinguished from the 'Spandrel' by its shorter launch tube which is parallel whereas the 'Spandrel' is about 200 mm longer and has a distinct taper at the rear end. The roof-mounted ATGW launcher can be traversed through a full 360° with elevation from –5 to +15°.

The BMP-2 IFV normally carries a standard ground launcher so that the ATGW system can be deployed away from the vehicle if required.

In addition to the rifles and machine guns of the infantry carried in the BMP-2 IFV, an anti-tank grenade launcher and one SA-7/SA-14/SA-16/SA-18 surface-to-air gripstock and two missiles in their launcher tubes are carried, although the SAMs are not carried by every BMP-2 IFV.

To the turret rear is the infantry compartment which has only two roof hatches compared to the BMP-1's four. It carries six infantrymen, compared with eight in the BMP-1, who sit down either side of the vehicle back-to-back. In either side of the rear troop compartment are three firing ports, each of which has an associated roof-mounted day periscope angled to the front of the vehicle. The infantry normally enters and leaves the vehicle via the two doors in the hull rear, each of these has an integral fuel tank and day periscope in the upper part of the door. The left door also has a firing port.

The torsion bar suspension either side consists of six roadwheels with the drive sprocket at the front, idler at the rear and track-return rollers. The first, second and sixth roadwheel stations have hydraulic shock-absorbers. The upper part of the track has a sheet metal cover deeper than that of the BMP-1 which is filled with a buoyancy aid.

Like the BMP-1, the BMP-2 IFV is fully amphibious. Before entering the water the trim vane stowed on top of the glacis plate is erected, the bilge pumps are switched on and the driver's centre day periscope is replaced by the TNPO-350B.

Standard equipment on the BMP-2 includes a full range of night vision equipment for commander, gunner and driver, a fire extinguishing system, a GPK-59 gyrocompass system, a PAZ overpressure NBC system, an engine preheater and a turret extractor fan.

Specifications

Crew: 3 + 7
Combat weight: 14,300 kg
Power-to-weight ratio: 20.30 hp/t
Ground pressure: 0.65 kg/cm²
Length:
(gun forward) 6.735 m
(hull) 6.735 m
(gun rear) 7.295 m
Width:
(overall) 3.15 m
(over tracks) 2.85 m
Height:
(to commander's sight) 2.45 m
(for air transport) 2.25 m
Ground clearance: 0.42 m
Track: 2.55 m
Track width: 300 mm
Length of track on ground: 3.60 m
Max road speed:
(5th gear) 65 km/h
(4th gear) 43.3 km/h
(3rd gear) 29.1 km/h
(2nd gear) 19.6 km/h
(1st gear) 10.6 km/h
(reverse) 10.6 km/h
Max water speed: 7 km/h
Fuel capacity: 462 litres
Max road range: 550-600 km
Gradient: 60%
Side slope: 30%
Vertical obstacle: 0.7 m
Trench: 2.5 m
Engine: Model UTD-20 6-cylinder, 4-stroke direct injection diesel developing 300 hp at 2,600 rpm
Transmission: manual, 5 forward and 1 reverse gears
Clutch: multiplate, dry, constant engagement
Suspension: torsion bar, hydraulic shock-absorbers on 1st, 2nd and 6th roadwheels
Electrical system: 22-29 V
Batteries: 2, 140 Ah
Armament:
(main) 1 × 30 mm 2A42 cannon
(coaxial) 1 × 7.62 mm PKT MG
(other) 1 launcher for AT-5 'Spandrel' or AT-4 'Spigot' ATGW
Smoke-laying equipment: 2 × 3 81 mm smoke grenade dischargers, diesel fuel injected into exhaust
Ammunition:
(main) 500
(coaxial) 2,000
(ATGW) 4

Gun control equipment
Turret power control: electric/manual
(by commander) yes
(by gunner) yes
Gun elevation/depression: +75/-5°
Turret traverse: 360°
Gun stabiliser:
(vertical) yes
(horizontal) yes
NBC system: yes
Night vision equipment: yes

BMP-2 IFV complete with roof-mounted Konkurs (NATO AT-5 'Spandrel') ATGW and unditching beam carried on roof at rear of hull ***2000***/0064790

Variants

Improved BMP-2

From the late 1980s, a number of improvements were carried out to production BMP-2s, most of which can be backfitted to the earlier BMP-2s and some are also applicable to the older BMP-1.

These improvements include modifications to the stabilisation system of the 30 mm 2A42 cannon. The BPK-1-42 gunner's sight has the warm-up switch moved to the control panel, improved internal communications and the tracks' rubber bushes have conductive elements installed to ensure that the electricity flows through the track pins and links and into the ground. The tracks also have new end connectors and the track now has a chevroned cleat assembly; the spare track links have been moved from the rear exit doors to the left and right upper sidewalls of the troop compartment and a sensor now indicates the position of the louvre in the exhaust ejector port.

The vehicle also carries a special mat for the transport of seriously wounded personnel and six slings for lightly wounded.

This vehicle has a combat weight of 14,500 kg and has lost its amphibious capabilities.

BMP-2ZS

This is fitted with loudspeakers for PSYOPS operations like the Z-88.

BMP-2D

This is the late production BMP-2 and has a number of improvements including appliqué passive armour on the turret, provision for mounting a mineclearing system under the nose of the vehicle and appliqué armour of the spaced type fitted along either side of the hull.

BMP-2K

This is the command version of the BMP-2 IFV and it has additional communications equipment. Externally it is recognisable by its additional radio antenna and that some of its firing ports are blanked off.

BVP-2

This is the Czech Army designation for the Russian BMP-2 AFV.

BMP-2 with mine plough system

Both the BMP-1 and the more recent BMP-2 IFV can be fitted with the KMT-10 plough-type mineclearing system which is installed on the front of the hull in line with the tracks. The mineclearing ploughs are very narrow with just two digging tines – probably limiting them to clearing smaller surface-laid mines, in particular scatterable anti-tank mines. Tilt-rod fuzes mounted between the right and left-hand ploughs can detonate mines between them. To reduce the travelling width, the outer section of each plough can be folded inwards and held in place by a spring lock.

To enable the vehicle to retain its amphibious capabilities, an extra flotation device has been secured to the lower front of the hull to counteract the weight of the KMT-8 or KMT-10 system when afloat, although this could also be an additional layer of composite armour to improve protection to the forward part of the vehicle.

Upgraded BMP-2

Details of the upgraded BMP-2 IFV are given in a previous entry. As far as it is known, this remains at the prototype stage.

New engine for BMP-2

The BMP-1 and BMP-2 IFV are both powered by a Transmash Model UTD-20 6-cylinder 4-stroke, liquid cooled, direct injection diesel developing 300 hp at 2,600 rpm. In the case of the BMP-2, this gives a power-to-weight ratio of 20.30 hp/tonne.

Further development of the UTD-20 engine by the company has resulted in the improved UTD-23 which has been designed for retrofit applications in the older BMP-1 and BMP-2 IFV.

This upgraded engine develops 360 hp at 2,600 rpm compared to the 300 hp of the earlier UTD-20 and this has been achieved with only minor modifications and turbo-supercharging. The engine is almost a direct replacement for the existing UTD-20.

According to Transmash, or the Barnaultransmash as it is called, the installation of the latest UTD-23, which weighs only 40 kg more than the older UTD-20, improves the power-to-weight ratio of the vehicle to just over 25 hp/t as well as giving an increase in speed and acceleration.

BMP-2 with Kliver turret

The BMP-2 can also be fitted with the new private venture Kliver one-man combat turret armed with a 30 mm cannon, 7.62 mm machine gun and four Kornet ATGWs. Full details of this turret are given in the entry for the BTR-80 (8 × 8) vehicle in the Armoured personnel carriers (wheeled) section. As of mid-2001 this turret is understood not to have entered quantity production.

BMP-2 upgrade

The State Scientific and Technical Centre of Artillery and Rifle Arms (SSTC ARA) of the Ukraine has developed an upgrade package for the BMP-2 IFV.

This includes the replacement of the 30 mm cannon by the locally produced 30 mm KBA-2 cannon and the installation of a 30 mm AG-17 automatic grenade launcher on the left side of the turret towards the rear.

A fire-control system has been installed with the commander being provided with a 1PZ-3 sight and with the gunner having a TKN-45S Agat stabilised day/night sighting system.

Status

Production complete but can be resumed if required. In service with the following countries:

Country	Quantity	Comment
Afghanistan	550	Includes BMP-1 and BMP-2
Algeria	225	
Angola	100	also supplied from Hungary and Poland
Armenia	158	Includes BMP-1 and BMP-2
Azerbaijan	96	
Belarus	1,265	
Czech Republic	186	
Finland	110	
Georgia	12	
India	1,500	local production (qv)
Indonesia	11	from Slovakia
Iran	200	Estimate
Iraq	800	Estimate, includes BMP-1 and BMP-2
Jordan	35	
Kazakhstan	300	Including BMP-1 and BMP-2
Kyrgyzstan	101	
Kuwait	46	
Russia	3,250	
Sierra Leone	2	
Slovakia	93	
Sri Lanka	4	Supplied by Ukraine
Sudan	6	Supplied by Ukraine
Syria	100	
Tajikistan	85	Includes BMP-1 and BMP-2
Togo	20	Supplied by Poland
Turkmenistan	405	from Poland
Ukraine	1,468	
Uzbekistan	172	
Yemen	270	Includes BMP-1 and BMP-2

Manufacturers

Czech – ZTS Dubnica nad Váhom which is now in Slovakia.

Indian – Ordnance Factory Medak (see, India).

Russian Federation – Kurgan Machine Construction Plant.

UPDATED

BMP-1 infantry fighting vehicle

Development

Following extensive trials between various competing designs in the early 1960s, the Obiekt 765 design from Isakov design bureau was accepted to meet the requirements of the Russian Army and was subsequently called the BMP-1 IFV *(Boevaya Mashina Pekhota)*.

A preproduction batch of BMP-1 IFV vehicles was completed for trials and the vehicle made its first public appearance during the November 1967 Moscow Parade. Production of the vehicle was undertaken at Kurgan Machine Construction Plant in the Urals.

The BMP-1 IFV was first used in combat by Egypt and Syria during the 1973 conflict with Israel; subsequently, it has seen combat with the Russian Army in Afghanistan, with the Iraqi Army in numerous Middle East conflicts, with Libyan forces in Chad and in Angola.

The BMP-1 IFV was subsequently replaced in production by the much improved BMP-2 IFV which is a further development of the BMP-1 IFV and is covered in a separate entry.

Production of the BMP-1 IFV was completed in Russia some years ago, but quantities of these continue to be exported from various countries.

According to United Nations sources, the following quantities of BMP-1 IFV vehicles were exported between 1992 and 1998 as laid out in the following table.

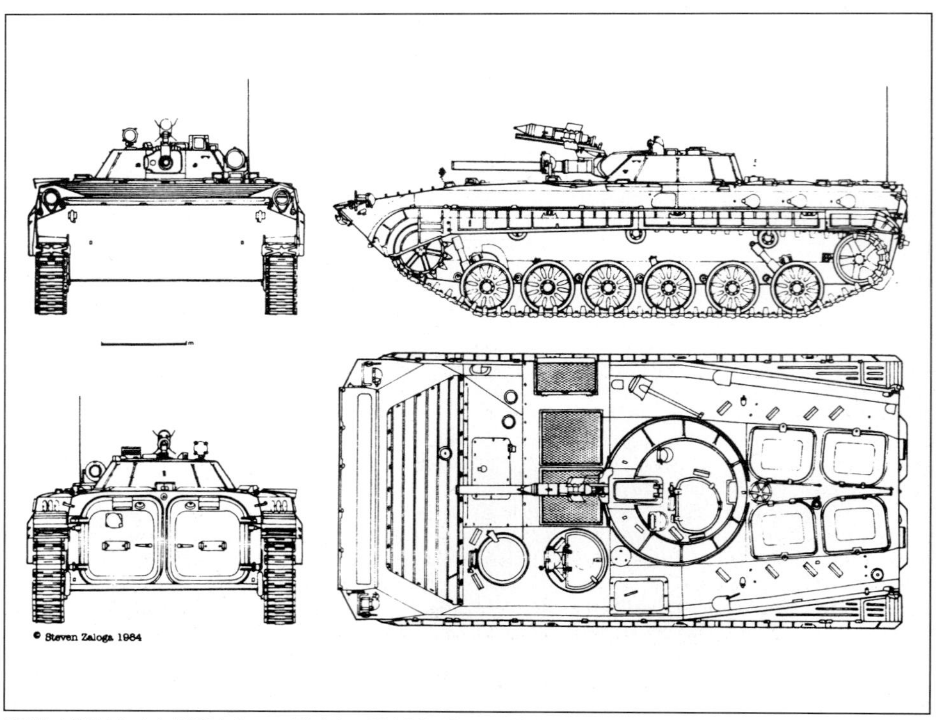

BMP-1 IFV Model 1970 Infantry Fighting Vehicle (Steven Zaloga)

From	To	Quantity	Date
Bulgaria	Angola	50	1993
Czech Republic	Hungary	2	1995
Czech Republic	Uruguay	10	1996
Czech Republic	Uruguay	5	1998
Germany	Finland	110	1993
Germany	Finland	30	1994
Germany	Greece	501	1994
Germany	Sweden	2	1995
Germany	Sweden	12	1996
Germany	Sweden	361	1997
Romania	Bulgaria	100	1996
Ukraine	Sri Lanka	12	1994

Description

The hull of the BMP-1 IFV is made of all-welded steel which provides the crew with protection from small arms fire and shell splinters. According to US reports, the armour provides complete protection from 12.7 mm AP/API rounds over its frontal arc. The well sloped glacis plate is ribbed, with an infra-red driving light at the front on the right and a white light on the left.

The driver sits at the front of the hull on the left side and has a single-piece hatch cover that opens to the right. In front of his position are three day periscopes, the centre one (TNPO-170) can be replaced by a vertically extend one (TNPO-350B) and allows the driver to see ahead when the trim vane is erected for amphibious operations.

The commander is seated behind the driver and has a cupola that can be traversed through a full 360°. In the forward part of the cupola are three day periscopes, the centre one of which can be replaced by a TKH-3 binocular sight or a sight with variable magnification. An infra-red searchlight is mounted on the commander's cupola and controlled from inside the vehicle.

The engine and transmission are to the right of the driver's and commander's positions with the steering system at the front of the hull. The air inlet and outlet louvres are in the top of the hull. A pneumatic system is provided for air starting of the engine, compressed air and liquid clearing of the driver's and commander's periscopes, actuation of the control linkage operating the valve guard, air intake pipe and dust removal valve, the duplicating pneumatic stopping brake linkage and for disengaging the master clutch.

The one-man turret of the BMP-1 IFV is identical to that installed on the BMD-1 Airborne Combat Vehicle. Full details of this vehicle are given in a separate entry. The gunner has a single-piece hatch cover that opens forwards, in front of which on the left side is a dual-mode 1PN22M1 monocular periscopic sight. In the day mode it has a magnification of ×6 and a 15° field of view, and in the image intensification mode it has a magnification of ×6.7 and a 6° field of view. Its maximum range at night is between 400 and 900 m. Graticule lines coincide with the different ranges in the gunner's sight and the gunner also has a stadiametric rangefinder to determine range to the target based on a target height of 2.7 m. The sight is heated and provided with a wiper arm. The gunner has four additional day observation periscopes and a white light or infra-red searchlight is mounted on the right side of the turret.

Main armament of the BMP-1 IFV is a 73 mm Model 2A28 smoothbore, low-pressure, short-recoil gun which weighs 115 kg. This is fed from a 40-round magazine to the right rear of the gunner. After each round is fired the gun is returned to an elevation of +3° 30' for reloading. The weapon fires a fixed fin-stabilised HEAT round with an initial velocity of 440 m/s which increases to 700 m/s once the projectile has left the barrel and a rocket motor cuts in. The projectile is the same as that used in the SPG-9 infantry weapon and has a maximum effective range of 1,300 m. The PG-9 HEAT projectile fired by the 73 mm 2A28 low-pressure gun uses a small PG-15P stub casing to boost the projectile out of the gun tube at which point the PG-9V rocket motor cuts in to provide the main source of propulsion to the target. The HE-FRAG projectile is called the OG-9, but this is not believed to be used with the 73 mm 2A28 low-pressure gun of the BMP-1 IFV. The HEAT projectile will penetrate 300 mm of conventional steel armour at an incidence of 0°. According to US reports, the weapon is not effective in high winds. Maximum rate of fire is 8 rds/min. The gun elevation and turret traverse mechanism are electric with mechanical controls for emergency use.

Ammunition for the 73 mm gun is available from a number of sources including Bulgaria who markets the

BMP-1 IFV with trim vane erected and showing the driver's centre periscope TNPO-350B in the raised position 0005037

BWR-1K of the Polish Army complete with radar in raised position but with no Sagger on launch rail of 73 mm gun (Jaroslaw Cislak) 0018506

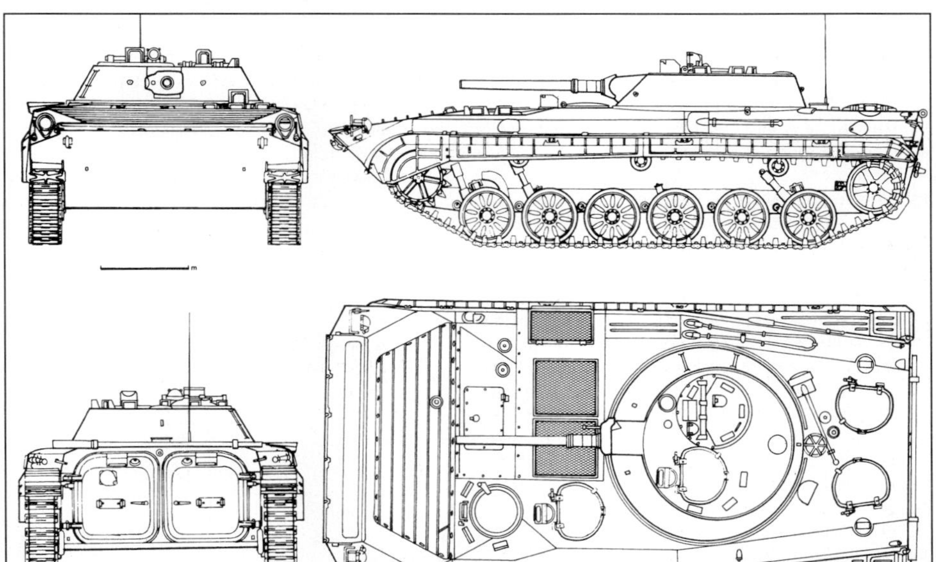

BRM Model 1976 reconnaissance vehicle (Russian designation BRM) (Steven Zaloga)

OG-15V (with HE fragmentation grenade OG-9), OG-15VM (with HE fragmentation grenade OG-9M) and the PG-15V (with HEAT grenade PG-9).

Nammo Liab of Sweden has developed an upgrade for the 73 mm ammunition of the BMP-1 IFV which includes a new propellant and safety and arming mechanism. An improved round is also under development which is expected to have an effective combat range of up to 2,000 m. A new anti-personnel combat round is also under development which will feature a warhead with steel balls.

Mounted coaxially to the right of the 73 mm 2A28 main armament is a 7.62 mm PKT machine gun which is fed from a continuous belt of 2,000 rounds honeycombed in an ammunition box mounted below the weapon. There is a turret ventilator system for extracting fumes.

Mounted over the main armament is a launcher for a Kolomna KBM 9K 11 Malyutka AT-3 'Sagger' wire-guided ATGW. One missile is carried in the ready to launch position with a further two missiles in the turret, these are loaded via a rail through a hatch in the forward

BMP-1KSh regimental and divisional level (command and staff) vehicle in the travelling configuration with antenna lowered

BMP-1P with turret-mounted 81 mm smoke grenade launchers (BP) ***2001***/0099465

part of the turret roof. A further two missiles are carried in the hull. The controls for the 'Sagger' are normally kept under the gunner's seat and when required are released by pulling a handle. The controls are then locked in position between the legs of the gunner who controls the missile by using the joystick in the conventional manner. The 'Sagger' has a minimum range of 500 m and a maximum range of 3,000 m. The 'Sagger' can be used only in daylight and once launched a new missile cannot be loaded until the missile has hit the target. Missile loading time is about 50 seconds.

Some modernised BMP-1 IFV vehicles have had the 'Sagger' launcher removed and a pintle mount fitted onto the turret roof for a 9K111 Fagot (AT-4 Spigot) or 9K113 Konkurs (AT-5 Spandrel) ATGW. In addition, 81 mm smoke grenade dischargers have been fitted either side of the turret towards the rear. These are usually referred to as BMP-1P IFV vehicles.

The troop compartment is at the rear with the eight infantrymen seated back-to-back, four down each side of the hull. They enter and leave by two doors in the rear of the hull which have integral fuel tanks (60 litres in the left door and 70 litres in the right) and are hinged on the outside. Each door has a day vision device and the left door has a firing port. Over the top of the hull are four roof hatches, two each side, which are hinged in the centre and can be locked vertically. There are four firing ports in each side of the hull and four heated day periscopes each side of the roof for aiming. Each weapon station has a vacuum exhaust system attached to the weapon to remove fumes.

The forward firing port on each side is usually used for the 7.62 mm PKM machine gun and the remaining three either side for the AKMS 7.62 mm assault rifle. Each PKM weapon is provided with 950 rounds of ammunition and each AKMS with 120 rounds. In addition, five rounds are carried for the single RPG-7 unguided anti-tank grenade launcher carried in the vehicle and one BMP-1 IFV in each platoon carries an SA-7 'Grail' SAM of the fire-and-forget type, which is launched by one of the infantrymen standing up in the troop compartment with one hatch open.

The torsion bar suspension either side consists of six rubber-tyred roadwheels with the drive sprocket at the front, idler at the rear and three track-return rollers. The first and last roadwheel stations have a hydraulic shock-absorber and the top of the track has a light sheet steel cover, which is normally removed when operating in snow. The track links are of the double-pin type with water scoops between the housings.

The BMP-1 IFV is fully amphibious being propelled in the water by its tracks. Before entering the water the trim vane, which is stowed on the glacis plate of the vehicle when not in use, is erected at the front of the hull, the bilge pumps are switched on, and the driver's centre day periscope is replaced by an extensible day one to enable him to see ahead over the trim vane. When afloat and fully loaded, the vehicle is driven in third gear but when empty second gear is used.

Should a nuclear explosion take place, the protection system ensures automatic shutdown of the engine, closing of the engine louvres, stopping of the ejector and valves of the turret and troop compartment fans, stopping of the fans and supercharger, turning off the electric drive of the turret, and switching on the absorbent filter air delivery system. When the shock-wave has passed the driver/mechanic turns on the supercharger, which provides decontaminated air at overpressure for the inhabited compartment of the BMP-1 IFV.

The overpressure system of the BMP-1 IFV consists of the NBC filter element and the blower/dust separator which is mounted in the roof to the rear of the commander's position. There is also a related scavenger system which removes gases from inside the vehicle when weapons are fired.

The blower/dust separator is similar to that used in the T-62 MBT and the BTR-60PB/BTR-60PK (8 × 8) APCs except that an extra set of fan blades has been added to force the air through the NBC filter. When the hatches are closed the blower/dust separator draws contaminated air into the system through an air intake immediately behind the turret. The contaminated air is then drawn through ducts around the turret ring to the blower/dust separator where fallout is separated from the air by the turning action of the propeller. The air is then forced into a heating chamber (in cold weather) or, if necessary, directed through the NBC filter where the chemical and biological contaminants are removed. The air is then forced through the heating chamber and into the troop and driver's compartments through 11 outlet vents. The forced air creates a build-up of air pressure inside the vehicle and so prevents NBC agents from leaking into the BMP-1 IFV through areas such as the firing ports which do not have airtight seals; if the scavenger system is used when the NBC system is in operation, the effectiveness of the overpressure is reduced.

Variants

Differences between production vehicles

The BMP-1 IFV, first shown in Moscow in 1967, is believed to have been the preproduction vehicle; it was soon followed by the improved version, the BMP Model 1966, initial production batch. This had a curved shock-absorber behind the first roadwheel and the rear fender and the swim vanes were modified by cutting down the number of vanes from seven to four and raising the angle of the fender profile nearer to the horizontal. The fume extraction ports on the hull roof to the rear of the turret were moved outwards and reoriented to expel toward the rear of the BMP-1 IFV. The single torsion bar used to spring the rear roof troop hatches was dropped in favour of a twin torsion bar system.

This was followed by the BMP-1 IFV Model 1966 standard production vehicle on which the two fender tool stowage boxes were removed. The BMP Model 1966 final production type had a modified port for the PKM section machine gun on the centre of the hull side.

In 1970, the BMP-1 IFV Model 1970 was introduced, this was probably the main production variant and has a 200 mm extended nose to improve its amphibious characteristics. The triangular air intake to the turret rear was replaced by a circular telescopic snorkel which is raised when the vehicle is afloat. The roof hatches were also slightly rearranged, the air intake located to the front left of the driver was removed and the NBC sensor cover moved inwards closer to the turret. To the left of the turret is a prominent NBC filter cover and the PKM port has been redesigned.

Some sources have also mentioned an improved BMP-1 IFV which could have a stabilised 73 mm gun, semi-automatic guidance for the 'Sagger' ATGW, an improved NBC system, a more powerful engine, an improved automatic loader and improved night vision devices.

Russian designations

Under the Conventional Forces Europe (CFE) Treaty, the Russians declared the following versions of the BMP-1 IFV and additional details of most of these are given later:

BMP-1 IFV fitted with Delco Defense LAV-25 turret system ***2000***/0064806

Close-up of the BMP-1P IFV turret fitted with an AT-5 'Spandrel' ATGW with a maximum range of 4,000 m (Michael Jerchel) 0018509

PRP-3 (BMP-SON) Model 1975 artillery reconnaissance vehicle 0005181

IRM amphibious engineer reconnaissance vehicle

BMP-1 converted by Iraq into a forward-area armoured ambulance (Christopher F Foss)

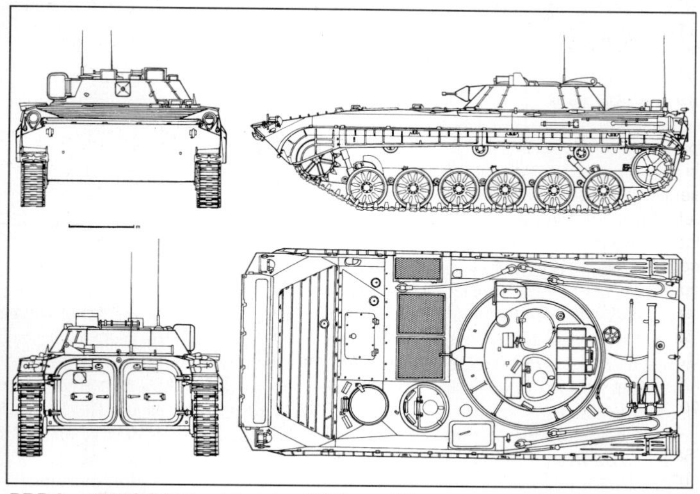

PRP-3 (BMP-SON) Model 1975 artillery reconnaissance vehicle (Steven Zaloga)

BMP, basic BMP-1 Model 1966, now obsolete
BMP-1 IFV, basic BMP-1 Model 1970
BMP-1F, BMP-1 reconnaissance model used by Hungary
BMP-1K, BMP-1 commander's model
BMP-1K3, BMP-1 commander's model
BMP-1P, BMP-1 with AT-4 9M114 ATGW and smoke grenade launchers
BMP-1PK, BMP-1P commander's model
BRM-1K, BRM-1 basic reconnaissance version – armoured cavalry (also known as M1976)
BMP-1KShM, BMP-1 unarmed command variant
BWP, Polish designation for basic BMP-1
IV31, or mashina upravleniya BMP-1 MP-31 is a command and staff vehicle for air defence units. It is equipped with two radios R-111, an R-159, R-123MT and R-130M and has the same HAWK EYE antenna as the similar BMP-1KShM. It is distinguished by a much larger generator. The MP-21-R-1 (an MT-LBu) is the successor for the MP-31. The 9S743, also known as the BMP-1KShM-9S743 (Bulgaria) is similar, but has a large stowage box on the right side of the generator instead of the two smaller ones left and right.
BREM-4 recovery vehicle, this is the Russian designation for the Czech VPV or BREM-Ch.
BREM-2 recovery vehicle.
MLI-84, Romanian-built BMP-1 IFV (qv Romania).

BMP-1M upgrade

In addition to the new engine described below and the BMP-1G upgrade described later, Russia has recently revealed another BMP-1 upgrade called the BMP-1M.

This includes the installation of the KBP Instrument Design Bureau Kliver single-seat combat module which is armed with a 30 mm 2A72 automatic cannon, 7.62 mm PKT coaxial machine gun and a pod of four Kornet laser-guided ATGW on the right side.

The latter has a maximum range of 5,500 m and the turret system is fitted with a computerised fire-control system with stabilised day/thermal sighting system.

Full details of this turret system, which can be fitted to a wide range of tracked and wheeled armoured fighting vehicles, are given in the entry for the BTR-80 (8 × 8) vehicle in the Armoured personnel carriers (wheeled) section. As far as it is known, this turret remains at the prototype stage.

New engine for BMP-1 IFV

The BMP-1 and BMP-2 IFVs are both powered by a Transmash Model UTD-20 6-cylinder 4-stroke, liquid cooled, direct injection diesel developing 300 hp at 2,600 rpm. In the case of the BMP-1 IFV, this gives a power-to-weight ratio of 22.22 hp/tonne.

Further development of the UTD-20 engine by the company has resulted in the improved UTD-23 which has been designed for retrofit applications in the older BMP-1 and BMP-2 vehicles.

This upgraded engine develops 360 hp at 2,600 rpm compared with the 300 hp of the earlier UTD-20 and this has been achieved with only minor modifications and turbo-supercharging. The engine is almost a direct replacement for the existing UTD-20.

According to Transmash, or the Barnaultransmach Joint Stock Company as it called, the installation of the latest UTD-23, which weighs only 40 kg more than the older UTD-20, improves the power-to-weight ratio of the vehicle to over 26 hp/t as well as giving an increase in speed and acceleration.

Czech BMP-1 IFV designations

OT-90, BMP-1 IFV with turret replaced by same turret as fitted to Czech OT-64C (1) (8 × 8) APC armed with 14.5 mm and 7.62 mm machine guns
BVP-1, Czech designation for BMP-1 IFV
BPzV, BVP-1 with modified hatches and massive sight over commander's hatch to rear of driver's position
BRM-1K, BRM-1 reconnaissance vehicle
BVP-1K, Czech designation for commander's version of Russian BMP-1K vehicle
Boure III, Czech psychological warfare version of BMP-1 IFV with loudspeaker system fitted
DP-90, maintenance version of OT-90
DTP-90M, repair version of OT-90 APC
MP-31, air defence command version of BMP-1KShM command post vehicle (also called AKA 9S743)
MU-90, minelaying version of OT-90, no turret with space being covered by steel sheet
SVO, mineclearing version of BMP-1 IFV with turret removed and fitted with Hedgehog type launcher in rear troop compartment. An illustration of this model appears in the entry for this vehicle in the Czech Republic
VPV, crane-equipped recovery version of BMP-1 IFV. Turret and troop compartment roof hatches have been removed and mounted on troop compartment roof and at the rear is a powered crane (Hungarian designation is BMP-1VPV)
VP-90, reconnaissance version of OT-90 with OT-64 turret
ZT-90, recovery version of OT-90
PRAM-S, 120 mm self-propelled mortar system, full details are given in the Self-propelled Mortar Systems section under Czech Republic and Slovakia
AMB-S, ambulance version with turret removed and raised rear superstructure.
OT-R5 command post vehicle
Snezka reconnaissance vehicle, details of this vehicle, based on a stretched BMP-1 IFV chassis with a mast-mounted sensor pod, are given in the Reconnaissance vehicles section. Additional details are given under Czech Republic and Slovakia in this section.

Egyptian BMP-1s

It has been confirmed that from 1979, between 140 and 150 Egyptian Army BMP-1s were retrofitted with a French Poyaud 520 6L CS2 diesel engine developing 310 hp at 2,800 rpm. It is believed that these vehicles are now held in reserve.

Modified BMP-1s

Also for use in Afghanistan some BMP-1s were fitted with a 30 mm AGS-17 automatic grenade launcher on their turrets; as far as is known this is not aimed and loaded from within the turret.

Some BMP-1/BMP-2 vehicles used in Afghanistan were fitted with appliqué 6 mm thick steel armour on the upper side of the hull and covering the upper track and return rollers. It is believed that this modification was incorporated for protection against 12.7 mm rounds, fired by DShKM heavy machine guns used by the guerrillas, which can penetrate the 16 to 18 mm side armour of the BMP-1 IFV. The vehicle has an additional stowage box on the hull roof towards the rear. There is also a spike-type mount on the right side of the roof which is for the AT-4 'Spigot'/AT-5 'Spandrel' ATGW man-portable missile launcher.

BMP-1K (command)

A company level command version of the BMP-1 IFV is known to exist and is called the BMP-1K by the former Soviet Army. It has additional communications equipment, an additional radio aerial on the right rear of the hull, a redesigned troop compartment to incorporate tables and a map board and the firing ports on the right side of the hull have been either eliminated or welded. This model is also known in the US Army as the BMP M1974 Command Vehicle.

The normal BMP-1 ICV has four rectangular roof hatches above the rear troop compartment but the BRM-1K has only two roof hatches. These open outwards and either side is a roof-mounted TNPO-170A device for aiming their weapons.

The 73 mm 2A28 gun is only provided with 20 rounds of ready use ammunition which is loaded manually.

The 7.62 mm PKT machine gun has 2,000 rounds of ready use ammunition. Two man-portable surface-to-air missiles are also carried.

There are several versions of the BMP-1K. The BMP-1K1 and K2 are both equipped with two R-123M and a semi-telescopic antenna. The BMP-1K3 is used on a higher level and is fitted with an R-123M, an R-130M and a 10 m telescopic antenna. Based on the BMP-1P with the AT-4 ATGW system, Russia developed the BMP-1PK1 and PK3, both with two R-123M and a 10 m telescopic antenna. There does not appear to be a PK2 version.

BMP-1KSh command and staff vehicle

This was originally designated the BMP M1978 and mounts a larger telescopic antenna called HAWK EYE and a radio set (R-137, R-140M or R-145BM, each with four or five radios). The original navigation system GPK-59 has been replaced by the TNA-3. For self-defence the crew is provided with a portable 7.62 mm PK machine gun with 2,000 rounds of ammunition. In Polish Army service the BMP-1KSh is designated the BWP-1D. The BMP-1KSh is normally deployed at regimental and divisional level.

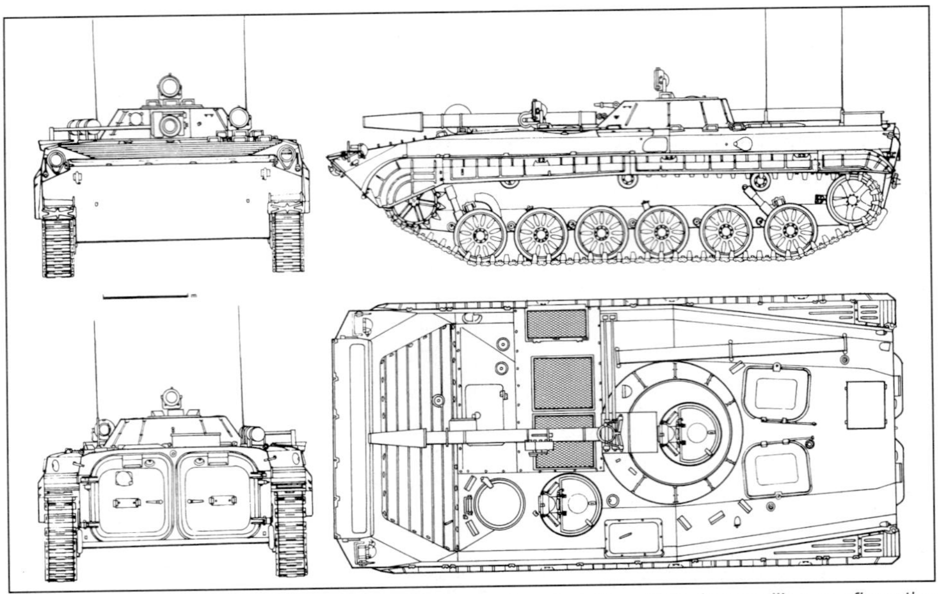

BMP-1KSh regimental and divisional level (command and staff) vehicle in travelling configuration (Steven Zaloga)

BMP-1 IFV with mine plough system

Both the BMP-1 IFV and the more recent BMP-2 can be fitted with the KMT-10 plough-type mineclearing system which is installed on the front of the hull in line with the tracks. The mineclearing ploughs are very narrow with just two digging tines which probably limits them to clearing smaller surface-laid mines, in particular scatterable anti-tank mines. Tilt-rod fuzes mounted between the right- and left-hand ploughs can detonate mines between them. To reduce the travelling width, the outer section of each plough can be folded inwards and held in place by a spring lock.

To enable the vehicle to retain its amphibious capabilities an extra flotation device has been secured to the lower front of the hull to counteract the weight of the KMT-10 system when afloat, although this could also be an additional layer of composite armour to improve protection to the forward part of the vehicle. An illustration of the KMT-10 appears in the previous entry on the BMP-2.

BRM and BRM-1 (reconnaissance)

This is known in NATO as the BMP-R or the BMP M1976 Reconnaissance Vehicle, or BRM by the former Soviet Army. It has the same hull as the BMP-1 IFV but is fitted with a larger two-man turret armed with a 73 mm gun, to the rear of which are two roof hatches instead of four rectangular ones as in the case of the BMP-1 IFV. Some vehicles have been observed fitted with a bank of electrically operated 81 mm smoke grenade dischargers to the turret rear firing forwards. In the designation, BRM stands for *Boevaya Razvedyvatnaya Mashina*.

BRM-1K (reconnaissance)

The NATO designation for this vehicle is the BMP M1976/2 and it is similar to BRM-1 but the 'Sagger' ATGW launcher over the 73 mm gun has been removed and a 'Tall Mike' (NATO designation) surveillance radar is mounted in the rear part of the turret. The radar is retracted into the turret until required when a hatch opens towards the rear and the radar is elevated above the turret roof.

The PSNR-5K radar has a maximum range of 10,000 m and other equipment carried by the vehicle includes a DKRM-1 (1D8) laser rangefinder, an ARRS-1 location device, an IMP mine detector and 1PN33B night binoculars.

Navigation equipment carried includes a TNA-1, an IG11N gyrocompass and a 1T25 survey device.

This version has a six-man crew and armament consists of one 73 mm 2A28 gun, one 7.62 mm PKT coaxial machine gun and six 81 mm smoke grenade dischargers. Because it has a different battlefield mission, it has a smaller number of firing ports and vision devices than the standard BMP-1 IFV. Only 20 rounds of 73 mm ammunition are carried for this version. There is also the BRM-1KSh command and staff version with additional communications equipment.

PRP-3 (radar)

This model appeared in 1975 and has been observed attached to 152 mm 2S3 and 122 mm 2S1 self-propelled units. It is used as an artillery fire adjustment and/or artillery/mortar locating vehicle, as these battalions already use the ACRV-2 Model 2 for the command post/observation role. The front of the vehicle is identical to the BMP-1 IFV but the vehicle has a new two-man turret which has two single-piece hatch covers opening forwards. They both have periscopes for observation plus a large optical device in front of the hatch. Armament consists of a 7.62 mm PKT machine gun. Mounted on the rear of the turret is a radar with a flat antenna that folds forwards when not operating. To the rear of the turret on the left side is a further circular hatch cover and a telescopic antenna. The 'Small Fred' Battlefield Surveillance Radar (1RL-126) operates in the J-band, with a detection range of 20 km and a tracking range of 7 km. This version is believed to have a crew of five. This model is known as the PRP-3 or the BMP M1975 surveillance vehicle by the US Army. It has a five-man crew and is fitted with extensive communications equipment and optical devices.

The PRP-3 is also referred to as the Val and is armed with a 6P7 machine gun with 1,000 rounds and a 90 mm mortar 2P130-1 with 20 rounds. The 90 mm mortar is located on the rear deck and has a range of 100 to 3,040 m and is used for battlefield illumination.

In addition it is equipped with high observation device 1PN29, navigation systems 1V44 (KP-4), 1G13M and 1G25-1, a 1V520 computer, a DS-09 or KTD-1 range-finder, APK data transmission system, TNPO-170A periscope, PAB-2A aiming circle and laser range-finders 1D6M1 (fixed) and 1D13 (portable).

PRP-4 (radar)

This is a further development of the PRP-3 and has an additional housing on the left side of the turret for an electro-optical device similar to that on the right side. Additional details are given in the Reconnaissance vehicles section.

IRM amphibious engineer reconnaissance vehicle

This vehicle was originally believed to be based on the chassis of the 122 mm 2S1 self-propelled howitzer, but in 1986 it was disclosed that it was based on the BMP-1 IFV chassis. It has a new hull and uses the engine and suspension of the BMP-1 vehicle. It has been designed to undertake a variety of specialised engineer reconnaissance roles including mine detection with two devices mounted at the front of the vehicle which can be retracted flush with the front of the hull when not required.

The IRM is fully amphibious, propelled by two shrouded propellers at the rear of the vehicle. When submerged a snorkel is erected on the top of the hull which is kept horizontal when not required.

PRP-3 radar vehicle showing a folded 'Small Fred' radar at the rear of the turret and a 7.62 mm PKT machine gun at the front

BREM-2 armoured recovery vehicle based on BMP-1 chassis (Richard Stickland)

Additional details of this vehicle, which so far has been identified only to be in service with the former Soviet Army, are given in *Jane's Military Vehicles and Logistics 2001-2002*.

RTV repair vehicle

The Rubtsovsk Machine Building Plant has developed the Repair Tracked Vehicle (RTV) to undertake repairs to tanks and BMP-type armoured vehicles in the forward combat area.

The RTV is based on the chassis of the BMP-1 IFV or BMP-2 IFV and has a similar layout with the driver seated front left and the vehicle commander to his rear with both are provided with a single-piece hatch cover and periscopes.

The engine compartment is to the right and the crew compartment at the rear which has a raised welded steel superstructure with two roof hatches with a 7.62 mm PKT machine gun being mounted on the right roof hatch.

Forward of the superstructure on the left side is an electrically operated crane which can lift a maximum load of 3,000 kg. This can be traversed through 220° and has a maximum lifting height of 4 m.

Standard equipment includes an electric welding system, repair set, tools, NBC system, night vision equipment, 81 mm smoke grenade launchers and an engine driven smoke generating system.

The suspension either side is the same as that of the BMP-1 IFV with six single rubber-tyred roadwheels, drive sprocket at the front, idler at the rear and track-return rollers. The upper part of the track is covered by a steel shroud and an unditching beam can be carried on the left side of the hull.

BMP-PPO mobile training centre

The BMP-PPO *(Podvizhniiy Punkt Obucheniya)* mobile training centre is essentially a BMP-1 IFV with its turret removed and fitted with eight roof-mounted cupolas for trainees under instruction, plus seats for the vehicle commander and driver. The commander and driver sit at the front left of the vehicle with the eight trainees at the rear, three down either side and two in the centre. Each trainee has two TNPO-170 and one Type MK-4 observation devices mounted in the forward part of the cupola and an A-2 unit of the R-124 intercom set (used via the two-way radio). The instructor has a console with three cassette recorders, an AGU-10-3 amplifier, three switches, external loudspeakers and a microphone. There is also an R-123M two-way radio, A-1 and A-2 units from the tank intercom R-124 and 10 chest-mounted switches with cables.

For training, each platoon has two standard BMP-1s and one BMP-PPO, with the former being manned as normal practice. The instructor and one trainee take turns to send information while the remainder hear via loudspeakers.

BMP-1 IFV command vehicle

The BMP-1 IFV command vehicle, called the BMP KShM Command and Staff Vehicle by the former Soviet Army, has the same set of communications equipment as the BMD KShM (BMD M1979/3).

BMP-1G IFV

Late in 1993, Russia revealed that it had developed an upgrade package for the BMP-1 IFV for the export market.

The upgrade package includes the removal of the AT-3 'Sagger' ATGW system mounted over the barrel of the 73 mm gun and the installation of a pintle mount on the turret roof for the AT-4 'Spigot' ATGW with a maximum range of 2,500 m or the AT-5 'Spandrel' ATGW with a maximum range of 4,000 m.

Both these ATGWs have a Semi-Automatic Command to Line Of Sight (SACLOS) guidance system for improved accuracy with the latest production models of these ATGWs also having a tandem HEAT (High Explosive Anti-Tank) warhead for increased armour penetration.

The 73 mm 2A28 gun is retained but mounted on the left side of the turret roof is a 30 mm AG-17 grenade launcher that can lay down suppressive fire out to a maximum range of around 1,800 m. The ammunition used for the 30 mm AG-17 grenade launcher is also used for the standard tripod-mounted AGS-17 grenade launcher which is used in large numbers by the Russian Army. A total of 103 rounds of 30 mm ammunition is carried for the AG-17 automatic grenade launcher.

BMP-1 of the Polish Army who calls the vehicle the BWP-1. In this photograph the Sagger ATGW is not mounted on the launcher rail, but an unditching beam is carried at the rear (Michael Jerchel) ***2002***/0116783

The power pack and running gear of the BMP-1 IFV have also been upgraded and modifications to the latter include a new track that is interchangeable with the existing track and is claimed to have twice the life of the older track. The engine has been uprated to 350 hp and to enable the vehicle to operate in high ambient temperatures an air conditioning system has been installed.

The upgraded BMP-1 IFV is known as the BMP-1G, but in 2001 there were no known customers for this upgrade.

BMP-1 IFV with Delco 25 mm turret

This has been developed to meet the requirements of the Polish Army. Details of this and the other Western turrets demonstrated on Polish BMP-1 IFV chassis in 1997 are given earlier in this section under Poland. As of mid-2001, Poland had not upgraded any of its BMP-1 IFV series vehicles with a new turret system.

Swedish BMP-1 IFV

In mid-1994, the Swedish Ministry of Defence agreed to purchase from Germany 350 former East German Army Soviet-designed BMP-1 infantry combat vehicles.

The BMP-1s were manufactured under licence in the former Czechoslovakia between 1974 and 1982 and were upgraded before being taken into service with the Swedish Army. Additional details are given under Sweden.

Finnish BMP-1 IFV

It is understood that Finland has removed the automatic loaders from its BMP-1 IFV as these have proved to be unreliable in service. All of the Finnish BMP-1 IFV and BMP-1P have been converted to the BMP-1PS, with four Wegmann 76 mm smoke grenade launchers and a Bofors Lyran launcher. There is also the BMP-1TJ forward observation vehicle with the original turret replaced by a cupola with observation devices.

Greek BMP-1 IFV

The German Army took delivery of a large number of ex-East German Army BMP-1s, which were subsequently upgraded to the BMP-1A1 OST standard. These were soon phased out of service and in 1993/94 the Hellenic Army received 501 of these vehicles.

The Hellenic vehicles are a mix of older BMP-1 and the later BMP-1P models. Modifications included the removal of the AT-3 Sagger launcher and the automatic loader for the 73 mm 2A28 gun, a mounting lug is provided on the roof for a AT-4 Spigot/AT-5 Spandrel ATGW system, fuel cells in the rear doors have been removed and spare track links have been relocated on the rear-end of the track guards.

New and modern lights have been installed, a boarding ladder has been fitted on the left side of the vehicle, anti-slip coating strips have been fitted on the roof, rubber pads have been fitted to the tracks, a new air inlet for the new crew comparment heater and convoy lights added.

Iraqi BMP-1 IFV with appliqué armour

In 1989, a BMP-1 IFV which had appliqué armour fitted to the sides of the hull to provide protection against 12.7 mm and 14.5 mm armour-piercing projectiles fired from a range of 200 m was first observed in Iraq. The total package weighs 1,250 kg and spaces are cut in the sides of the hull armour package to allow the infantry to use its small arms from within the vehicle. No appliqué armour was observed fitted to the turret or glacis plate. As far as is known this did not enter service with the Iraqi Army in quantity.

Iraqi BMP-1 IFV ambulance

To evacuate wounded personnel from forward battlefield areas, Iraq has developed an armoured ambulance based on the BMP-1 IFV vehicle. The turret has been removed and the rear of the vehicle extended upwards to allow stretchers and walking wounded to be more easily transported. The original doors of the BMP-1 IFV have been removed and replaced by two new higher doors that do not contain the integral fuel tanks of the basic BMP-1 IFV vehicle. As far as is known this did not enter service with the Iraqi Army in quantity.

Civilian BMP-1 vehicles

With the reduction in the size of the former Soviet Army, many of the older BMP-1 vehicles are now surplus to requirements and some of these are being converted for civilian applications, including the BMP-1 IFV.

The Berezina transport vehicle is essentially a BMP-1 with its turret removed and a raised roof over the driver's, commander's and turret areas with windows being provided for all-round observation. Over the top of the hull at the rear is a cargo area with a maximum capacity of 1,000 kg.

It can carry a total of 12 personnel and is fitted with a VG-7500 generator and a welding system.

Status

Production complete. In service with the following countries:

Country	Quantity	Comment
Afghanistan	550	includes BMP-1 and BMP-2 IFVs
Algeria	700	
Angola	50+	from Bulgaria
Armenia	158	includes BMP-1 and BMP-2 IFVs
Azerbaijan	114	
Belarus	388	
Bulgaria	100	
Cuba	400	
Czech Republic	615	plus variants
Egypt	220	held in reserve
Ethiopia	50	status uncertain
Finland	163	
Georgia	67	
Greece	501	from Germany
Hungary	502	

In service with the following countries (con't):

Country	Quantity	Comment
India	700	some sources state 350
Iran	300	Estimate
Iraq	800	
Kazakhstan	300	includes BMP-1 and BMP-2 IFVs
Korea, North	n/avail	not confirmed user
Kyrgyzstan	118	plus variants
Libya	1,000	
Mongolia	400	
Mozambique	40	
Poland	1,367	plus variants
Romania	178	local equivalent, MLI-84 (qv)
Russian Federation	3,000+	
Slovakia	383	plus 207 OT-90
Sri Lanka	12	from Ukraine
Sweden	375	Upgrade, qv Sweden
Syria	2,250	
Tajikistan	85	includes BMP-1 and BMP-2 IFVs
Turkmenistan	156	
Ukraine	1,563	
Uruguay	10	from Czech Republic
Vietnam	300	
Yemen	270	includes BMP-1 and BMP-2 IFVs

Manufacturers
Former Czech and Soviet state factories. The latter included the Kurgan Machine Construction Plant.

Note: The Chinese build a similar version of the BMP-1 IFV called the WZ 501, full details of which are given under China. Romania has built a modified version of the BMP-1 IFV under the designation of the MLI-84. Full details are given under Romania.

UPDATED

Specifications
Crew: 3 + 8
Combat weight: 13,500 kg
Unloaded weight: 12,500 kg
Power-to-weight ratio: 22.22 hp/t
Ground pressure: 0.6 kg/cm^2
Length: 6.74 m
Width: 2.94 m
Height: (over searchlight) 2.15 m
Ground clearance: 0.39 m
Track: 2.75 m
Track width: 300 mm
Length of track on ground: 3.53 m
Max speed:
(road) 65 km/h
(water) 7 km/h
Fuel capacity: 460 litres
Max range: 550-600 km
Fuel consumption: 1 litre/km
Fording: amphibious
Gradient: 60%
Side slope: 30%
Vertical obstacle: 0.8 m
Trench: 2.2 m
Engine: Type UTD-20 6-cylinder in-line water-cooled diesel developing 300 hp at 2,000 rpm
Transmission: manual, 5 forward and 1 reverse gears
Steering: clutch and brake
Suspension: torsion bar
Electrical system: 24 V
Armament:
(main) 1 × 73 mm 2A28 gun
(coaxial) 1 × 7.62 mm PKT MG
(other) 1 launcher rail for 'Sagger' ATGW
Smoke-laying equipment: diesel fuel injected into exhaust
Ammunition:
(main) 40
(coaxial) 2,000
(other) 4 + 1 'Sagger'

Gun control equipment
Turret power control: electric/manual
(by commander) no
(by gunner) yes
Gun elevation/depression: +33 to –4°
Turret traverse: 360°
Gun stabiliser:
(vertical) no
(horizontal) no
Armour
Hull front upper: 7 mm at 80°*
Hull front lower: 19 mm at 57°
Hull side upper: 16 mm at 14°
Hull side lower: 18 mm at 0°
Hull rear upper: 16 mm at 19°!
Hull rear lower: 16 mm at 19°!
Hull top: 6 mm
Hull belly front: 5 mm
Hull belly rear: 7 mm
Turret front: 23 mm at 42°
Turret sides: 19 mm at 36°
Turret rear: 13 mm at 30°
Turret top: 6 mm
Turret mantlet: 26-33 mm
NBC system: yes
Night vision equipment: yes

* Does not include trim vane, does not represent aluminium engine cover
! Does not include fuel or inner armour of door

BMD-3 airborne combat vehicle

Development
First production BMD-3 Airborne Combat Vehicle (ACV) were delivered in 1990 to Russian airborne units followed by naval infantry units.

The latest BMD-3 airborne combat vehicle features a brand new chassis fitted with the complete turret of the BMP-2 infantry fighting vehicle which entered service in the late 1970s and was seen for the first time in Moscow in 1982.

The BMD-3 airborne combat vehicle has been developed by the design bureau headed by Arkadyi Shabalin with production being undertaken at the Volgograd Tractor Plant.

This facility also under took production of the earlier BMD-2 airborne combat vehicle as well as the PT-76 light amphibious tank for the home and export markets. Details of the PT-76 are given in the Light tanks section.

While having similar mobility to the earlier BMD-1 and BMD-2 vehicles, the new BMD-3 has a much better amphibious capability, hence its additional use by the Russian Navy Infantry; more room inside the vehicle; a significant increase in firepower, with the two-man turret offering better overall command and control of the vehicle and its weapons.

The BMD-3 can be dropped from transport aircraft by parachute with the complete crew of seven inside the vehicle. In the past, the crew were dropped separately from the vehicle and it often took considerable time for the crew to locate its vehicle. By dropping the BMD-3 with its crew the vehicle is ready for action much quicker so keeping the element of surprise associated with airborne operations. The Russian Il-76M transport aircraft can carry three BMD-3 airbourne combat vehicles ready for air-drop.

It is estimated that the Russian airborne forces and marines have well over 1,600 BMD-1, BMD-2 and BMD-3 vehicles in service. As far as it is known there have been no exports of the BMD-3 airborne combat vehicle.

It is believed that by mid-2001, less than 100 BMD-3 airborne combat vehicles had been built.

The first unit to receive the BMP-3 was the 104th Guards Airborne Regiment, part of the 70th Guards Airborne Division based at Pskov.

The BMD-3 airborne combat vehicle has a significant increase in firepower over the earlier BMD-1 and BMD-2 vehicles (Steven Zaloga) ***2000***/0064794

Description
Like the earlier BMD-1 and BMD-2 airborne combat vehicles, the new BMD-3 is of all-welded aluminium construction providing the crew with protection from small arms fire and shell splinters. An NBC system of the overpressure type is fitted as standard.

The crew consists of commander, gunner, driver and four infantry with the commander normally dismounting with the infantry. In an emergency, a further three infantry can be carried in the rear.

The driver is seated at the front of the vehicle, and has a single-piece hatch cover and three day periscopes for forward observation, the centre one of which can be replaced by a night vision periscope. There is a man seated either side of the driver and each has a single-piece hatch cover and at least one day periscope. For added crew survivability, if the BMD-3 runs over an anti-tank mine, the driver's and two front gunners' seats are suspended from the roof of the vehicle.

Mounted at the front of the BMD-3 in the left side is a 30 mm AG-17 automatic grenade launcher while in the right bow is a 5.45 mm RPKS MG. The standard infantry version of the AG-17 is called the AGS-17 with the S standing for *Stankovyy* (mount); as the 30 mm AG-17 is vehicle-mounted it does not require the mount.

Each of the bow weapons is operated by one of the infantry seated in the front of the BMD-3, and covers the frontal arc. The 5.45 mm RPKS MG has a high rate of fire for use in the direct role whereas the 30 mm AG-17 automatic grenade has a higher trajectory and can be used to suppress targets at ranges out to 1,200 m using HE fragmentation grenades. If required, these weapons can also be quickly dismounted for use in the ground role.

The two-man power-operated turret is mounted on the hull top and is armed with a stabilised 30 mm 2A42 dual-feed cannon which is fitted with a distinctive muzzle brake and is stabilised in both elevation and traverse. This has two cyclic rates of fire, 200 or 300 rds/min and 550 rds/min. The stabilisation system is designated the 2E36-5.

Turret traverse is through a full 360° with weapon elevation from -5 to +75°. The high elevation allows it to be used against low- and slow-flying aircraft and helicopters. Effective range when firing AP-T (Armour-Piercing – Tracer) ammunition is 2,000 m while when firing HE-I (High-Explosive – Incendiary) rounds the maximum range is 4,000 m.

Specifications

Crew: 7 (3 + 4), 10 (max)
Weight: 13,200 kg
Ground pressure: 0.48 kg/cm^2
Power-to-weight ratio: 34 hp/t
Length: 6.36 m
Width: 3.134 m
Height: 2.170 m
(to top of searchlight) 2.55 m
Ground clearance: 150-530 mm (450 mm normal)
Length of track on ground: 3.2 m
Track: 2.744 m
Track width:
(normal) 320 mm
(wide) 480 mm
Max road speed: 70 km/h
Max water speed: 10 km/h
Fuel capacity: 450 litres
Range:
(road) 500 km
(cross-country) 350 km
(water) 8 hours
Gradient: 60%
Side slope: 30%
Engine: 2V-06-02 water-cooled diesel developing 450 hp
Transmission: hydromechanical with 5 forward and 5 reverse gears
Armament:
(main) 1 × 30 mm 2A42 cannon
(coaxial) 1 × 7.62 mm PKT MG
(ATGW) 1 × AT-4/AT-5
(bow) 1 × 30 mm AG-17 30 mm grenade launcher; 1 × 5.45 mm RPKS MG
Smoke grenade launchers: 2 × 3 81 mm
Ammunition:
(30 mm cannon) 500 ready + 360 reserve
(30 mm grenade) 290 ready + 261 reserve
(7.62 mm) 2,000
(5.45 mm) 2,160
(ATGW) 4 + 2

Gun control equipment
Turret power control: electric/manual
By gunner: yes
Turret traverse: 360°
Gun elevation/depression: +75/–5°
Gun stabiliser:
(vertical) yes
(horizontal) yes
NBC system: yes
Night vision equipment: yes

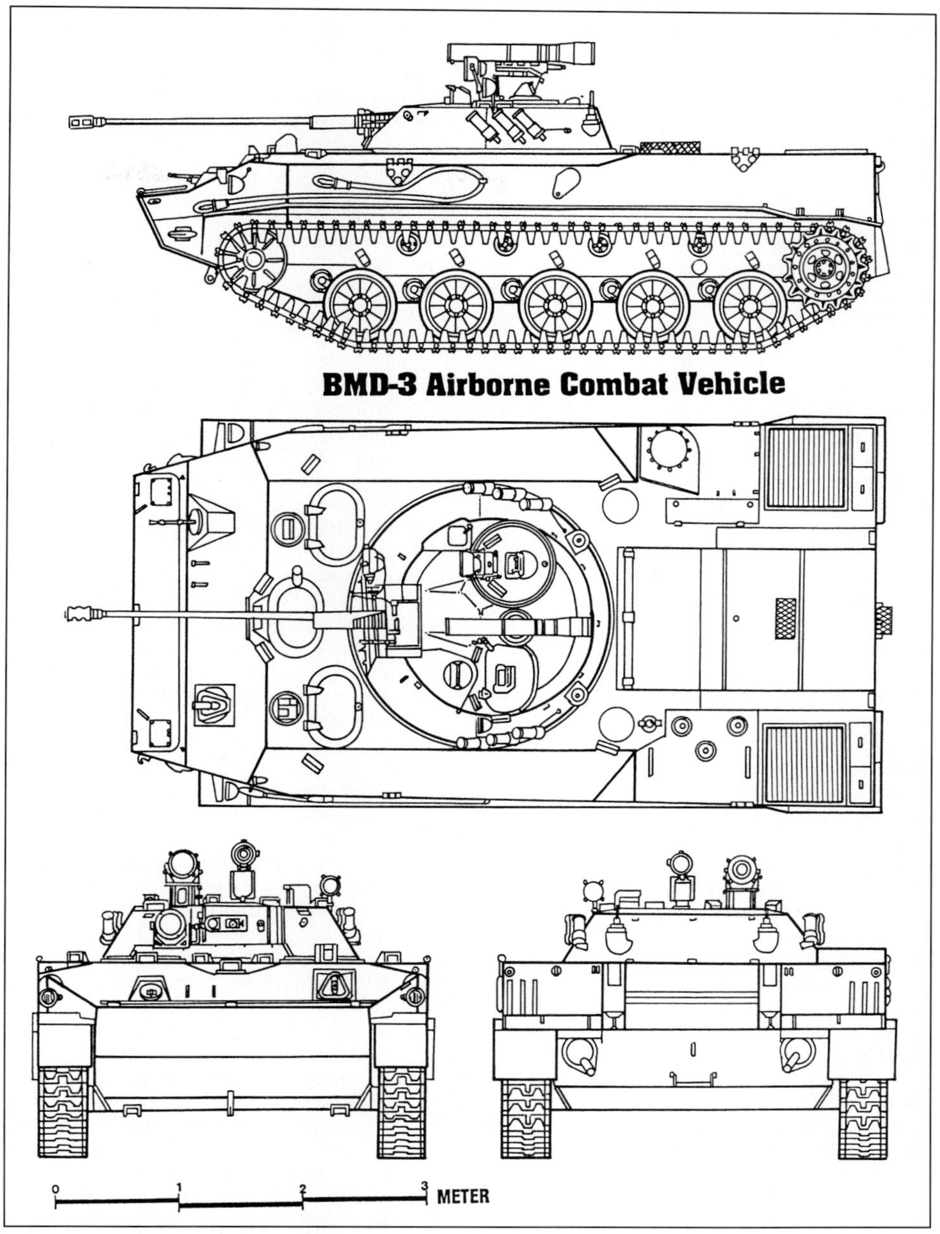

BMD-3 airborne combat vehicle (Steven Zaloga)

The more recent 30 mm 2A72 cannon installed in the BMP-3 infantry combat vehicle and the 30 mm 2A38 cannon installed in the 2S6M Tunguska self-propelled anti-aircraft gun/surface-to-air missile system fire the following types of ammunition:

Type	AP-T	HE-I
Cartridge weight	853 g	837 g
Cartridge length	291 mm	291 mm
Propellant weight	127 g	123 g
Muzzle velocity	960/980 m/s	950/970 m/s
Tracer burn time	not less than 3.5 s	

A 7.62 mm PKT machine gun is mounted coaxially to the right of the cannon and mounted on the roof is an AT-4 'Spigot' or AT-5 'Spandrel' ATGW launcher. The latest versions of the AT-4 'Spigot' have a maximum range of 2,500 m while the AT-5 'Spandrel' has a maximum range of 4,000 m. Current production models of these missiles have a new tandem High Explosive Anti-Tank (HEAT) warhead for increased armour penetration characteristics and can be launched at night using a thermal night sight.

Mounted on either side of the turret is a bank of three 81 mm electrically operated smoke grenade dischargers firing forwards which are now fitted to most Russian armoured vehicles. In addition, the BMD-3 can lay its own smoke screen by injecting diesel fuel into the exhaust.

The commander is seated on the right of the turret and has a cupola with a forward-opening hatch cover, while the gunner is seated on the left and has a single-piece hatch cover that opens to the front. Both the commander and gunner have day/night sights and observation periscopes.

Two of the five infantry are seated at the front, one either side of the driver, with the other three seated in the fighting compartment near the turret, one firing port provided in each side of the hull. Another three infantry can be seated for short distances to the immediate rear of the turret with the roof hatch open.

The power pack is at the rear of the hull and consists of a 2V-06-02 water-cooled diesel developing 450 hp, which gives a very high power-to-weight ratio of 34 bhp/t. The engine is coupled to a hydromechanical transmission with five forward and five reverse gears and a hydraulic steering unit.

Like earlier BMD vehicles, hydropneumatic suspension of the BMD-3 is adjustable giving a ground clearance between 130 and 530 mm, although for normal road travel it is 450 mm. The ground clearance is adjusted by the driver from his seat and it takes 10 seconds to lower or raise the suspension. The hydropneumatic suspension system of the BMD-3 also helps to cushion the landing. The new PBS 950 parachute landing system has been developed for use with the BMD-3.

The running gear either side consists of five rubber-tyred roadwheels which are similar to those of the BMP-2 with a large drive sprocket at the rear, idler at the front and four track-return rollers. The normal tracks are 320 mm wide but these can be replaced by wider 480 mm tracks in snow or other marginal terrain which lower the ground pressure of the BMD-3 as well as improving traction. A hydraulic track tensioning system is fitted as standard to the BMD-3.

The BMD-3 is fully amphibious being propelled in the water by two water-jets mounted either side at the rear of the vehicle. Before entering the water a trim vane is erected at the front of the vehicle and the automatic bilge pump is switched on. A snorkel is also erected from the driver's position.

The BMD-3 can operate at up to Sea State 3 and is very buoyant, hence its use by the Russian Naval Infantry. According to the Russians, it has a buoyancy reserve of more than 30 per cent of its own displacement.

Standard equipment includes an NBC system, fire detection and suppression system and a heating system for the engine to facilitate starting in cold weather conditions.

Variants

BMD-3M

This is fitted with a new turret which is similar to that installed on the BMP-3. Armament consists of a 100 mm gun/launcher 2A70 with 18 plus four round, 30 mm cannon with 300 rounds and a 7.62 mm PKT coaxial machine gun with 2,000 rounds. It also has a new digital fire-control system and day/night sight. It is believed that the BMD-3M remains at the prototype stage.

In 1998, it was stated that the Volgograd Tractor Plant was developing a large number of weapons based on the platform of the BMD-3 chassis.

2S25 SPATG

The new 125 mm 2S25 self propelled anti-tank gun is based on the BMD-3 airborne combat vehicle components. Details of this vehicle are given in the Self-propelled Anti-Tank Guns section. As of mid-2001, this is believed to be at the prototype stage.

Status

Production. In service with the Russian Federation and Associated States (CIS).

Manufacturer

Volgograd Tractor Plant.

UPDATED

BMD-2 airborne combat vehicle

Development

The BMD-2 airborne combat vehicle was designed by Arkadyi Shabalin who later designed the BMD-3 airborne combat vehicle covered in detail in a separate entry.

The first prototype of the BMD-2 was completed in 1985 and first seen in 1988. It was given the designation of the BMD M1988/1 in the West.

Production of vehicles started in 1989 and was originally issued on the scale of nine in each of the three airborne regiments of a Russian air assault division.

It was thought that BMD-2 were rebuilds of older BMD-1 but recent information has indicated that all BMD-2 were in fact new build vehicles.

Production of the BMD-2 was undertaken at the Volgograd Tractor Plant where the original BMD-1 was built. Production of the BMD-2 was completed some time ago. It is now being supplemented by the more advanced BMD-3 but, due to funding problems, this is only being built in small numbers.

Specifications

Crew: 3 + 4
Combat weight: 8,225 kg
Ground pressure: 0.56 kg/cm^2
Power-to-weight ratio: 29.19 hp/t
Length: 5.91 m
Width: 2.63 m
Height: 1.97 m
Ground clearance: 0.10 to 0.45 m
Track width: 230 mm
Length of track on ground: 2.84 m
Max speed:
(road) 60 km/h
(water) 10 km/h
Range: 450-500 km
Fording: amphibious
Gradient: 60%
Side slope: 40%
Vertical obstacle: 0.6 m
Trench: 1.2 m
Engine: Type 5D20 V-6 liquid-cooled diesel developing 240 hp
Transmission: manual, 5 forward and 1 reverse gears
Suspension: hydropneumatic
Electrical system: 24 V
Armament:
(main) 1 × 30 mm 2A42 cannon
(coaxial) 1 × 7.62 mm PKT machine gun
(bow) 1 × 7.62 mm PKT machine gun
(ATGW) 1 × AT-5 Spandrel ATGW
Ammuniton:
(main) 300
(coaxial and bow) 2,540 + 400
(ATGW) 3 AT-5 Spandrel ATGW

Gun control equipment
Turret power control: electric/manual
(by gunner) yes
Turret traverse: 360°
Gun elevation/depression: +75/–5°
Gun stabiliser:
(vertical) yes
(horizontal) yes
Armour:
Hull: (max) 15 mm
Turret: (max) 23 mm
NBC system: yes
Night vision equipment: yes

Description

The chassis of the BMD-2 is almost identical to the BMD-1 covered in detail in a separate entry, but the two-man turret has been replaced by a one-man turret of a new design with the gunner being seated on the left and provided with a single-piece circular hatch cover opening to the front.

Forward of this is the gunner's day/night sight that appears to be identical to that fitted on the BMP-1 infantry combat vehicle. Additional day periscopes give observation to the sides and a white light searchlight is mounted on the forward part of the turret roof.

On the left side of the turret is another sight which moves in elevation with the 30 mm 2A42 cannon and this is believed to be the high angle of fire sight used to aim the weapon when it is being used in the anti-aircraft/anti-helicopter role.

The fire-control/vision equipment fitted to the BMD-2 includes the TKN-3B combined day/night vision device, 1PZ-3 periscopic sight, BPK-2-42 gunner's combined binocular sight and a 2E36-3 electromechanical stabilisation system for the main armament.

Main armament comprises a 30 mm 2A42 dual-feed stabilised cannon with a 7.62 mm PKT machine gun mounted coaxially to the right. The main armament is stabilised in both vertical and horizontal planes.

The 30 mm 2A42 cannon, which is also used in the BMP-2 infantry combat vehicle, has an elevation of +75° and a depression of -5°. There are 300 rounds of 30 mm and 2,980 rounds of 7.62 mm ammunition carried.

Mounted on the right side of the turret roof is a searchlight that moves in elevation with the main armament; this is similar to that fitted to the more recent BMP-3 infantry combat vehicle.

Mounted on the right side of the turret roof is a pintle on which can be mounted a launcher for an AT-4 'Spigot' ATGW with a maximum range of 2,000 m or an AT-5 'Spandrel' ATGW with a maximum range of 4,000 m.

The AT-4 and AT-5 was originally manufactured with a single HEAT warhead, but more recent versions with a tandem warhead have been manufactured.

In addition there is a hatch in the turret roof towards the rear which is believed to be the missile resupply hatch, as there is insufficient room to pass new missiles to the gunner via his small roof hatch.

It is also possible that a ground launcher is carried to allow the ATGW system to be deployed away from the vehicle if required by the tactical situation.

While the BMD-1 has two bow-mounted 7.62 mm PKT machine guns either side, the BMD-2 has only one 7.62 mm PKT bow-mounted machine gun in the right side with the gun port in the left side eliminated.

The front right hatch cover has been eliminated on the BMD-2 indicating that the bow machine gunner is not carried in the BMD-2 and that this space is used for additional stowage.

Other features of the BMD-1 are retained on the BMD-2 including a total of three firing ports for the infantrymen carried, one in either side of the hull and one in the rear troop compartment.

If required, the BMD-2 airborne combat vehicle can also carry and deploy an SA-7/SA-14/SA-16/SA-18 man-portable surface-to-air missile system.

Variants

BMD-2K command

This is the command post version of the BMD-2 airborne combat vehicle and is fitted with an additional R-173 radio, has a crew of six and is also provided with a GPK-58 gyro compass, heater, two tables and an NBC detection system RKhR GO-27. It does not carry a 9P135M ATGW launcher.

BMD-2 turret application

The BMD-2 turret is now being offered for installation on other tracked and wheeled armoured vehicles. It was stated in 1999, that an undisclosed export customer had ordered a batch of these turrets.

Status

Production completed. In service only with Russia (1,600 of all type, BMD-1, BMD-2 and BMD-3) and the Ukraine (78).

Manufacturer

Volgograd Tractor Plant.

UPDATED

BMD-2 airborne combat vehicle with unditching beam on right side of the hull (Michael Jerchel) ***2001***/0099464

BMD-2 airborne combat vehicle from the rear showing ATGW launcher, but not missile, on the roof (Michael Jerchel) ***2001***/0099463

BMD-1 airborne combat vehicle

Development

The BMD-1 (*Boevaya Mashina Desantnaya*, or Airborne Combat Vehicle) entered service with Soviet airborne units in 1969 and was first seen in public during a parade held in Moscow in November 1973. The vehicle was called the M1970 before its Russian designation became known.

It spearheaded the Russian invasion of Afghanistan in December 1979. It is issued to Russian Air Assault Divisions (or Air Landing Divisions) on the scale of 330 per division, three command versions at divisional headquarters and three airborne regiments with 109 BMDs each: 10 command vehicles, nine BMDs without turrets and 90 basic BMDs. It is air-portable and can be dropped by parachute.

Unconfirmed reports have indicated that India has taken delivery of a quantity of BMD-1 airborne combat vehicles and a very small number (believed to be two) of BMD-1 vehicles was captured in Kuwait by US forces early in 1991 during Operation Desert Storm.

The BMD-1 and the BMD-2 airborne combat vehicles have started to be replaced by the more recent and more effective BMD-3 airborne combat vehicle also designed at the Volgograd Tractor Plant. Production of the BMD-3 is however running at a very low rate due to a shortage of funding. As far as it is known there have been no exports of the BMD-3 airborne combat vehicle.

Description

The hull of the BMD-1 is welded aluminium. The driver sits at the front of the vehicle in the centre just forward of the turret and has a single-piece hatch cover that opens to the right. Three day periscopes are mounted forward of this hatch, the centre one of which can be replaced by an infra-red periscope for night driving. The vehicle commander sits to the left of the driver/mechanic and alongside the commander's seat are the radio and gyrocompass. The bow machine gunner sits to the driver's right and aims the two bow-mounted 7.62 mm PKT machine guns using a day TNPP-220 periscopic sight to which the 7.62 mm machine guns are connected by parallelogram drives. The two machine guns are mounted one either side of the vehicle's front firing forwards. A single semicircular hatch cover is positioned either side of the forward part of the turret. There is a single fixed forward-looking day periscope to the left and slightly to the rear of the left turntable-mounted periscope.

The turret is identical to the one fitted to the BMP-1 ICV and has a single-piece forward-opening hatch cover to the left of the turret. The gunner has four periscopes, one mounted each side and two forward of the hatch.

The personnel compartment at the rear of the vehicle has a concertina-type cover that opens towards the front. The only means of entry and exit to this compartment is via this hatch. The senior gunner, grenade launcher and assistant are seated here. There is a roof-mounted day TNPO-170 periscope which can be traversed for observation. In the rear part of this hatch is a normal BMP-1 type firing port with an observation device to enable one of the crew to use a weapon from within the vehicle.

The independent suspension either side consists of five small roadwheels with the idler at the front and the drive sprocket at the rear. There are four track-return rollers. The independent suspension combines a hydraulic system for altering the ground clearance and maintaining track tension with pneumatic springs, enabling the ground clearance to be altered from 100 to 450 mm.

The BMD-1 is fully amphibious, and is propelled in the water by two water-jets at the rear of the hull. Before entering the water a trim vane, which is stowed on the glacis plate when not in use, is erected at the front of the hull. The water entry points (for the water jets) are located one either side under the hull rear. The vehicle has electric and manual bilge pumps, a GPK-59 gyrocompass, an engine preheater, smoke-generating equipment, a PAZ NBC system and a PPO centralised ethylene-bromide fire extinguishing system as fitted to other Russian armoured vehicles.

The gunner is provided with a dual-mode 1PN22M1 monocular periscopic sight mounted on the left side of the turret roof. In the day mode it has a magnification of x6 and a 15° field of view, while in the image intensification night mode it has a magnification of x6.7 and a 6° field of view. Its maximum range at night is between 400 and 900 m. Graticule lines coincide with the different ranges in the gunner's sight and he also has a stadiametric rangefinder to determine range to the target, based on a target height of 2.7 m. The sight is heated and has a wiper arm. A white light searchlight is mounted on the right side of the turret although some vehicles have been observed with an infrared searchlight at this position.

BMD-1 airborne combat vehicle without Sagger ATGW mounted over 73 mm gun. In this photograph the suspension has been retracted (Michael Jerchel) ***2002***/0116784

Main armament of the BMD-1 is a 73 mm 2A28 smoothbore, low-pressure, short-recoil gun. This is fed from an automatic 40-round magazine to the right rear of the gunner. After each round is fired the gun is returned to an elevation of 3° 30' for reloading. The weapon fires a fixed fin-stabilised HEAT round with an initial muzzle velocity of 400 m/s, which increases to 665 m/s once the projectile has left the barrel and a rocket motor cuts in. The projectile is the same as that used in the standard tripod mounted SPG-9 infantry weapon and has a maximum effective range of 1,300 m.

The gun elevation and traverse mechanism is electric with mechanical controls for emergency use.

A 7.62 mm PKT machine gun is mounted coaxially to the right of the main armament and is fed from a continuous belt of 2,000 rounds, honeycombed in an ammunition box mounted below the weapon. A cartridge case and link collector are mounted in the turret basket. A ventilator system is provided in the turret to extract fumes.

Mounted over the 73 mm main armament is a launcher rail for an AT-3 'Sagger' ATGW. There are two missiles carried inside the turret which are loaded via a rail through a hatch in the forward part of the turret roof. The controls for the 'Sagger' are normally kept under the gunner's seat and when required are released by pulling a handle. The controls are then locked in position between the legs of the gunner who controls the missile by using the joystick in the conventional manner. The 'Sagger' has a minimum range of 500 m and a maximum range of 3,000 m.

BMD-1 airborne infantry combat vehicles are reported to be equipped with a radio 'bleeper' device that allows their crews, who are equipped with direction-finding equipment, to locate their own individual BMD-1 when they are parachuted onto the drop zone.

Variants

Differences between BMD-1s

The preproduction models of the BMD-1 first appeared in 1970 and had a squarish NBC filter cover on the right side of the hull roof. The first production model, the BMD-1, appeared in 1973 and had the NBC hatch replaced by a more prominent circular vent cover. To increase the operational range of the vehicle it can be fitted with additional fuel cells externally in the rear well.

Due to poor ventilation in early models of the BMD-1, which could lead to crew asphyxiation when the weapons were fired, a modernised version with new internal ventilation was developed. This was shown in 1980 and is called the BMD-1M. First models had the same wheels as the initial production BMD-1 but were distinguishable by a vent grill on the bow. Late production models have new pattern roadwheels.

BMD-1P airborne combat vehicle

This has a two-man crew consisting of vehicle commander and driver plus six airborne combat soldiers (squad leader plus five soldiers).

The AT-3 'Sagger' ATGW launcher over the 73 mm barrel has been removed and on the roof is a pintle launcher for the AT-4 'Spigot' ATGW which, in its latest version, has a maximum range of 2,500 m. The system can be fitted with a night sight to enable targets to be engaged at night.

The 73 mm 2A28 smoothbore short-recoil gun is normally provided with a total of 50 rounds of ammunition, 26 HE-FRAG with a maximum effective range of 1,600 m and 24 HEAT with a maximum effective range of 1,300 m.

BMD-1K

This is the command-post version of the BMD-1 with additional communications equipment.

BMD-2 airborne combat vehicle

Details of this vehicle, basically a BMD-1 chassis with a new turret, are given in a previous entry.

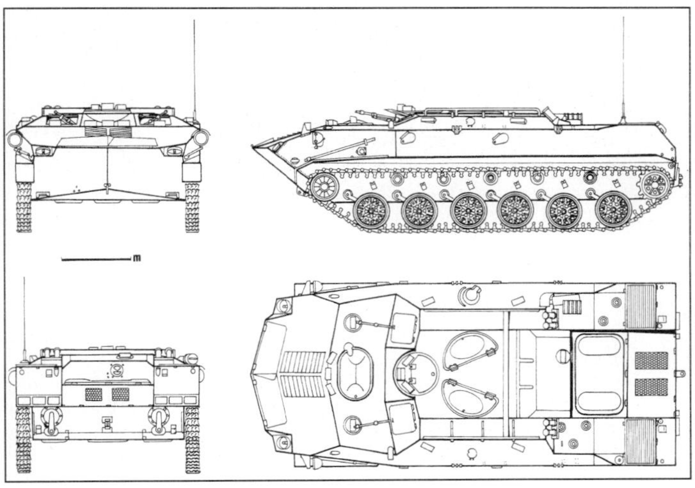

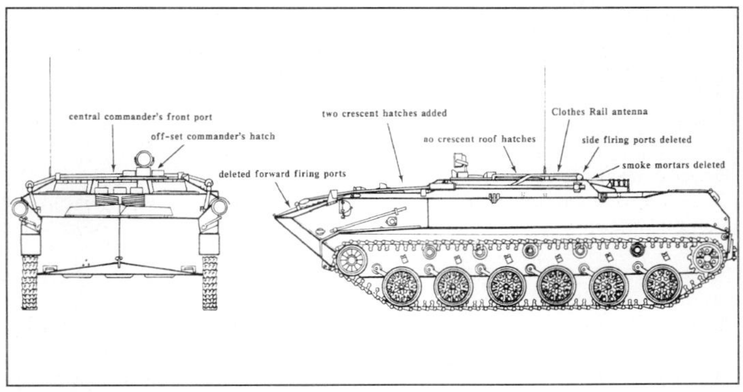

BTR-D Air Assault Command Vehicle (Steven Zaloga)

BTR-D Air Assault Transporter (Steven Zaloga)

82 mm mortar carrier

For use in Afghanistan, some BMD-1s were fitted with a standard 82 mm mortar in the rear troop compartment firing to the rear. It would be more logical to remove the existing turret and mount the 82 mm mortar in this position, rather than in the very small troop compartment at the hull rear.

BTR-D APC

The BTR-D was first seen during the Russian invasion of Afghanistan and is often referred to as the BMD M1979 by NATO. It is distinguishable from the standard BMD-1 by its longer chassis with six rather than five roadwheels, a different hull top and the lack of a turret.

The basic BTR-D has the STANAG designation BMD M1979/1 and is a multipurpose armoured transporter used by the Russian VDV (Air Assault Forces) air assault divisions, for a variety of roles including troop transporter, towing support weapons such as the 23 mm ZU-23-2 light anti-aircraft gun and maintenance support. This version retains the hull front and side firing ports of the BMD-1 and has two forward-firing smoke grenade dischargers on each side of the hull roof in line with the fifth roadwheel station. There are also two crescent-shaped hatches on the roof to the rear of the commander's position which projects forward onto the glacis plate. There are nine firing ports, two in the hull front each with a 7.62 mm machine gun; two in each side of the hull; one in the rear and the two front hatches below and either side of the commander's cupola can be used.

Development of the BTR-D commenced in 1974. It is based on automotive components of the BMD-1 airborne combat vehicle and like this vehicle has a hull of all-welded aluminium construction. The glacis protection of the vehicle has been increased by the use of a dual-slanted angle in the upper plates of the armour at the front.

The BTR-D can carry 13 men, the driver/mechanic at the front, a bow machine gunner either side and 10 infantrymen at the rear. The bow machine gunners normally deploy with the infantry.

There are 2,000 rounds of 7.62 mm ammunition carried for the 7.62 mm PKT machine guns mounted at the front of the vehicle and an additional two 7.62 mm PKT machine guns can be mounted on the roof.

Some early production vehicles had a small one-man turret armed with a 7.62 mm PKT machine gun that could be aimed and fired from within the turret. Some vehicles have also been fitted with a 30 mm AG-17 automatic grenade launcher.

The second version is called the BMD-KShM and is a command post vehicle. This has a folding 'Clothes Rail' antenna around the superstructure but the front firing ports with their 7.62 mm machine guns are faired over and there are no firing ports in the hull sides; the forward-firing 81 mm smoke grenade dischargers have also been removed. The commander's hatch is offset to the left and does not project forward. Known users of the BTR-D armoured personnel carrier include Azerbaijan (22), Belarus (115), Moldova (11), Russia (1,600+), Ukraine (42), and Uzbekistan (70).

BTR-RD Robot

This is the anti-tank version of the BTR-D (which is sometimes referred to as the BMD-1D) and has a five-man crew consisting of vehicle commander, driver and a three-man anti-tank missile team.

A large single-piece hatch cover is provided at the top front part of the superstructure which opens to the rear. To the right side of the opened hatch cover is the retractable AT-5 'Spandrel' ATGW launcher which has a maximum range of 4,000 m.

The basic load is 12 missiles. In addition the missile can also be launched by the operator from a remote-control panel up to 20 m from the vehicle.

The roof-mounted launcher will also fire the shorter range AT-4 'Fagot' ATGW.

An air landing regiment has a battery of six Robot vehicles while the tank destroyer battalion has 18. Some of these systems are now using the latest AT-5 Konkurs-M ATGW which has a tandem HEAT (High Explosive Anti-Tank) warhead.

BMD-1 with Shmel-1 RPV

A BMD-1KShM chassis is used as the launcher and flying control station for the Shmel-1 remotely piloted vehicle. The actual launch vehicle is called the Sterkh. This launch platform is also used as the launch platform for the improved RPV Ptsjela-1T (Bee) of the Stroy-P system.

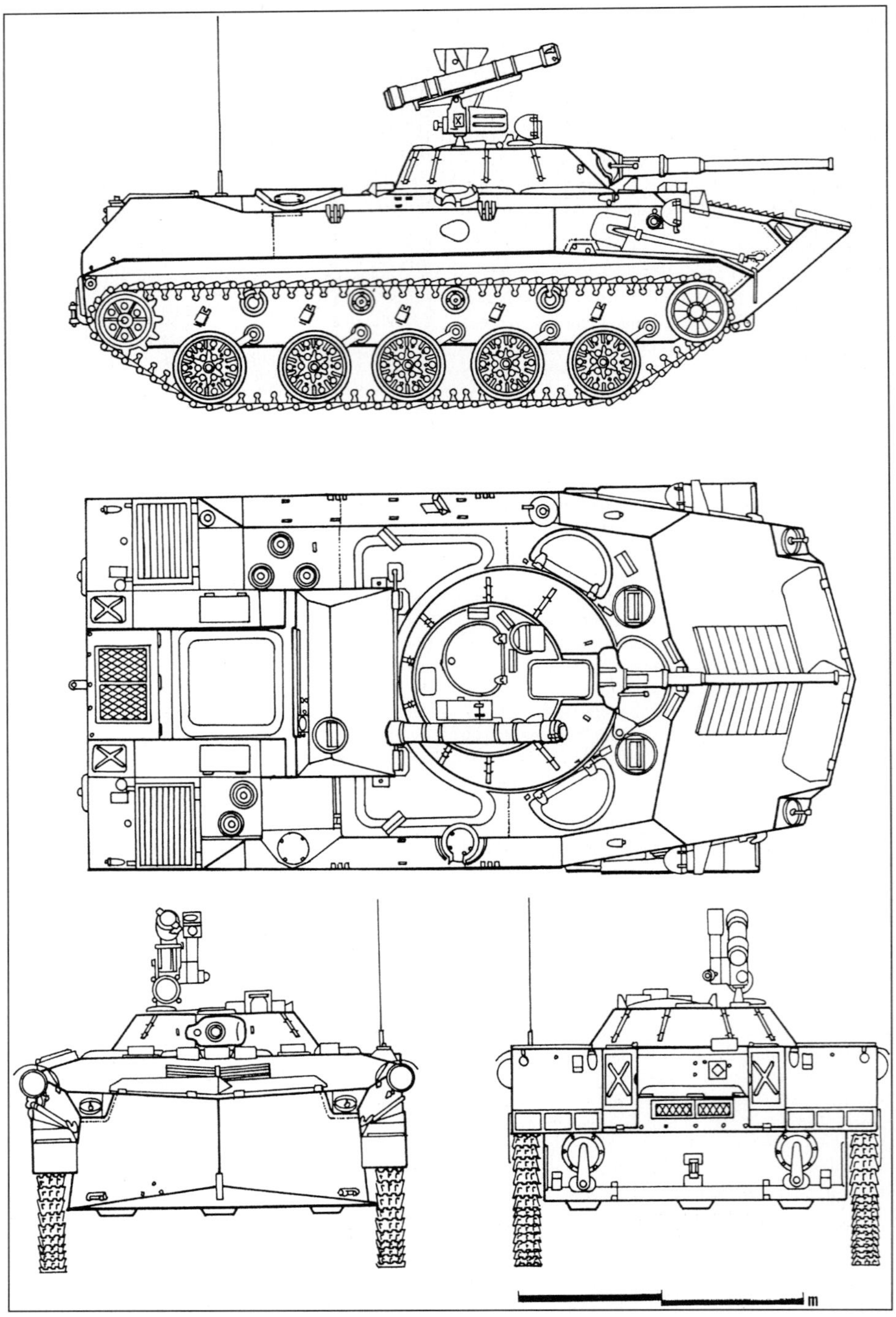

BMD-1 airborne combat vehicle (Steven Zaloga)

BTR-D armoured personnel carrier from the rear (Michael Jerchel) **2001**/0099461

Specifications
(BTR-D)
Crew: 1 + 13
Combat weight: 8,000 kg
Power-to-weight ratio: 30 hp/t
Ground pressure: 0.50 kg/cm^2
Length: 5.885 m
Width: 2.63 m
Height: 1.67 m
Ground clearance: 0.1-0.45 m
Engine: 5D-20 V-6 liquid-cooled diesel developing 240 hp
Max road speed: 61 km/h
Max water speed: 10 km/h
Range: 500 km
NBC system: yes
Night vision equipment: yes

BTR-D armoured personnel carrier which has a longer chassis with six roadwheels

1V119 Spektr used in artillery fire command role 0018695

BRehM-D repair and recovery vehicle

This is based on the chassis of the BTR-D APC and is fitted with specialised equipment for the repair and recovery of BMD-1 type vehicles.

Equipment fitted includes a crane which can be traversed through 180°, a recovery winch, a combination spade and dozer blade, towing equipment, an electric welding system plus tools and ready use spares.

Standard equipment includes a 7.62 mm bow machine gun, a radio, an intercom system and an NBC system.

BTR-ZD

This is the air defence member of the family and is a BTR-D which can tow the twin 23 mm ZU-23-2 anti-aircraft gun system, or carry this on the roof of the vehicle. Ramps are provided to assist in loading and unloading the ZU-23-2 system onto the vehicle. This version is also provided with two roof-mounted stowage racks for man-portable fire and forget surface-to-air missiles, such as the Igla-1.

BMD-1E command post vehicle

This is similar to the KShM but lacks the HORSE RAIL antenna. The telescopic antenna is much bigger and the whip-type antennas are now located at the front side.

1V118 Reostat artillery observation post vehicle with radar antenna erected

BMD-1 R440 ODB communications vehicle

This is a BTR-D fitted with the R-440 satellite communications system and mounted on the roof of the vehicle is a large circular communications dish. This is believed to be a mobile airborne version of the R-440 (NATO code name PARK DRIVE).

1V118 Reostat

This is the BMD-1 chassis with its turret removed. It is used in the artillery observation post vehicle role for the 2S9 120 mm self-propelled howitzer/mortar.

To carry out its mission on the battlefield, this version of the BMD-1 is fitted with a mast-mounted battlefield surveillance radar with an acquisition range of 14 km, laser rangefinder with a range of 8,000 m, day and night observation devices, land navigation system, onboard computer, radios and an intercom system.

Armament consists of a bow-mounted 7.62 mm PKT machine gun, man-portable SAM system with two missiles and three RPG-18 unguided anti-tank systems. A typical self-propelled battery consists of six 2S9 and two Reostat vehicles.

1V119 Spektr

This is the BMD-1 chassis with the turret removed and used in the artillery fire direction centre role for the 2S9 120 mm self-propelled howitzer/mortar.

It is similar to the 1V118 but lacks the PSNR-5 radar system and carries additional communications equipment as well as computers to calculate fire-control information for the 2S9 or other artillery systems such as the 122 mm D-30 and 122 mm BM-21 (40-round) multiple rocket launchers.

SO-120 (Anona) (2S9) 120 mm self-propelled howitzer/mortar

Details of this are given in the Self-propelled guns and howitzers (tracked) section.

Status

Production complete. In service with following countries:

Country	Quantity	Comment
Angola	n/avail	
Armenia	10	
Azerbaijan	41	plus BTR-D
Belarus	142	plus BTR-D
India	n/avail	use not confirmed
Iraq	n/avail	small number
Moldova	54	plus BTR-D
Russia	1,600	all types
Ukraine	61	plus BTR-D
Uzbekistan	131	BMD-1 plus BTR-D

Manufacturer

Former Soviet state factories.

Unconfirmed reports have stated that the BMD-1 was built at Izhevsk, Russian Federation and Associated States (CIS).

UPDATED

Specifications

(BMD-1)
Crew: 3 + 4
Combat weight: 7,500 kg
Power-to-weight ratio: 32 hp/t
Ground pressure: 0.57 kg/cm^2
Length: 5.4 m
Width: 2.63 m
Height: 1.62-1.97 m
Ground clearance: 0.1-0.45 m
Track width: 230 mm
Length of track on ground: 2.84 m
Max speed:
(road) 70 km/h
(water) 10 km/h
Fuel capacity: 300 litres
Range: 320 km
Fording: amphibious
Gradient: 60%
Vertical obstacle: 0.8 m
Trench: 1.6 m
Engine: Type 5D-20 V-6 liquid-cooled diesel developing 240 hp
Transmission: manual, 5 forward and 1 reverse gears
Suspension: hydropneumatic
Electrical system: 24 V
Armament:
(main) 1 × 73 mm 2A28 gun
(coaxial) 1 × 7.62 mm PKT MG
(forward firing) 2 single 7.62 mm PKT MGs
(other) 1 launcher rail for 'Sagger' ATGW or roof-mounted AT-4 'Spigot' on turret
Ammunition:
(main) 40
(coaxial) 2,000
('Sagger') 3

Gun control equipment
Turret power control: electrical and mechanical
(by gunner) yes
Turret traverse: 360°
Gun elevation/depression: +30 to –4°
Gun stabiliser:
(vertical) no
(horizontal) no
Armour
Hull: (max) 15 mm
Turret: (max) 23 mm
NBC system: yes
Night vision equipment: yes

BMPT Tank Support Combat Vehicle

Development

During the Ural 2000 defence exhibition, Russia showed the recently designed Tank Support Combat Vehicle (BMPT). This has been developed by the Urals Transport Machinery Design Bureau Federal State Unitary Enterprise and the Uralvagonzavod Production Association State Unitary Enterprise.

It has been said that the company has already lobbied the Russian Army to adopt the BMPT to start replacing current in service vehicles that are not very well protected against anti-tank weapons.

During the fighting in Chechnya, Russian armoured fighting vehicles, such as the T-72 and T-80 Main Battle Tanks (MBTs), and Infantry Fighting Vehicles (IFVs), such as the BMP-2, were attacked and neutralised at short ranges by man-portable anti-tank weapons.

The role of the new BMPT would be to provide fire support for the MBT with the aim of neutralising infantry, especially those equiped with anti-armour weapons. It has not been designed for use as a heavy infantry fighting vehicle.

Brand new vehicles could be built to the BMPT standard or surplus T-72 MBT chassis could be rebuilt to this configuration.

Scale model of the Russian BMPT well-protected vehicle that is designed to support MBTs (Yihong Zhang) ***2002***/0073248

In Russian the designation BMPT stands for Boyevaya Mashina Podderzhki Tankov – or Tank Combat Support Vehicle.

Description

The prototype of the BMPT is based on the chassis of the well-known T-72 MBT which is used in large numbers by the Russian Army and has been manufactured under licence by many countries.

To the rear of the driver's compartment, at the front of the vehicle, the hull has been raised providing greater internal volume. On top of the hull is mounted a new turret armed with an externally mounted 30 mm 2A42 dual-feed cannon with a cyclic rate of fire of up to 600 rds/min. A total of 500 rounds of ready use ammunition are carried.

This 30 mm 2A42 cannon is already in widespread use with the Russian Army and can fire a wide range of ammunition types including High Explosive – Tracer (HE-T), Armour-Piercing Discarding Sabot (APDS), High-Explosive FRAGmentation (HE-FRAG) and Armour-Piercing – Tracer (AP-T).

Mounted coaxial with the 30 mm cannon is a 30 mm AG-30 or AGS-17A automatic grenade launcher which can be a stabilised in two axis.

Mounted on the left side of the turret is a four-round launcher for the 9M113 Konkors (NATO AT-5 Spandrel) Anti-Tank Guided Weapon (ATGW) which has a maximum range of 4,000 m and can be fitted with various types of warhead.

To enable targets to be engaged under day and night conditions when the BMPT is stationary or moving, a computerised fire-control system is fitted, which uses proven elements from that fitted to the T-90S series MBT.

This includes a PNK range-finding sight, day and thermal sighting systems, Agat-MR course operator's sight, combined day and night sighting system with image converter tube. Either the commander or gunner can lay the weapons onto the target using the PNK sighting system.

There is a 7.62 mm PKT machine gun mounted in either side of the raised superstructure to provide suppressive fire over the frontal arc and aimed via a roof mounted periscope sighting device. As an alternative these can be replaced by a 30 mm automatic grenade launcher with a similar sighting system. Standard 81 mm smoke grenade launchers are positioned on the hull and turret to cover the frontal arc.

The driver, seated in the centre of the hull at the front, has day vision devices and a TBH-5 night driving device with a range of 60 to 180 m.

The BMPT has a combat weight of 47 tonnes and has a crew of five, maximum road speed is 65 km/h and cruising range, with external drum type diesel fuel tanks at the rear, is 550 km.

The power pack, mounted at the rear, consists of a B92C2 four-stroke multifuel diesel with liquid cooling and supercharging, develops 1,000 hp and gives the vehicle a power-to-weight ratio of 21.2 hp/tonne.

For a higher rate of battlefield survivability, the BMPT is fitted with additional explosive reactive armour, on the front and sides. The forward part of the suspension either side is provided with skirts fitted with explosive reactive armour, while the rear part of the suspension is fitted with a grill type passive armour array. This has been designed to detonate the incoming High-Explosive Anti-Tank (HEAT) projectile before it impacts the main armour.

Standard equipment includes NBC system, 5 kW diesel generator which is mounted in an armoured compartment on the right side of the hull and an automatic two-shot fire detection and suppression system. A laser detection system is fitted which can be used in conjunction with the 81 mm smoke grenade launchers.

Mounted at the front of the hull is a standard T-72 type dozer blade that can be used for clearing obstacles or for preparing fire positions. If required, the vehicle can be fitted with various types of mineclearing equipment, such as the KMT-8 plough type device or the EMT system, which has been designed to detonate magnetic mines well ahead of the vehicle.

Specifications

Full technical details have not been released. Available details are given in the description.

Status

Prototype. Not yet in service.

Manufacturer

See development.

NEW ENTRY

MT-LB multipurpose tracked vehicle

Development

The MT-LB multipurpose tracked armoured vehicle was designed at the Kharkov Tractor Plant in the Ukraine under the direction of Anatoliy Belousov in the early 1960s.

Typical roles of the MT-LB include use as an artillery prime mover towing 100 mm T-12/MT-12 anti-tank guns and 122 mm D-30 howitzers, command and radio vehicle (which usually has at least two whip antennas on the roof), artillery fire-control vehicle and cargo carrier. It is also widely used as an armoured personnel carrier, especially in difficult terrain such as snow or swamp.

In all, there have been more than 80 variants of the MT-LB multipurpose tracked vehicle which makes it the most widely deployed armoured vehicle of any type in the Russian Army.

As the Kharkov Tractor Plant is now in the Ukraine, the Russian Army is now seeking a replacement vehicle from within its own borders as well as looking at ways to upgrade the vehicle to extend its life.

It is possible that Russia will establish a modernisation and repair facility for the MT-LB vehicles at the Muromsk Diesel Locomotive Works.

This facility would strip down and overhaul the MT-LB and at the same time the current steering system would be replaced by a new hydrodynamic steering mechanism which has already been tested. This modification would not only improve the ride of the vehicle but also make it easier to handle. A comparison of the standard MT-LB and the upgraded MT-LB is given in Table 1.

Table 1	**MT-LB Standard**	**MT-LB Upgraded**
Max road speed	61.5 km/h	70 km/h
Average speed on dirt road	30 km/h	45 km/h
Steepless band of gear change when in gear	0 km/h	9.8 km/h
Min speed of movement in forward gears	4 km/h	from 0 km/h
Min speed of movement in reverse gear	6.8 km/h	from 0 km/h
Periodic service interval	from 1,000 km	from 4,000 km

Description

The hull of the MT-LB is all-welded steel armour with the crew compartment at the front, engine immediately behind the crew compartment on the left side and the troop compartment at the rear of the hull.

The driver sits at the front of the hull on the left side and has a single-piece hatch cover in front of which are three day periscopes. The commander sits to the right

MT-LB engineer vehicle with plough blade stowed on the right side of the hull top for travelling (Michael Jerchel)

MT-LB multipurpose tracked vehicle showing extension fitted to air intake/outlet on left side of vehicle (Christopher F Foss) ***2000***/0064797

MTP-LB repair vehicle/light armoured recovery vehicle in travelling configuration (Michael Jerchel)

MT-LB with SNAR-10 radar mounted on the hull roof at rear in the operating position (Richard Stickland) **2000**/0064798

of the driver in the centre and has a single-piece rear-opening hatch cover and two day periscopes. In action, the commander also mans the turret. The machine gun turret is mounted to the right of the commander's position and is armed with a 7.62 mm PKT machine gun. Like the turrets fitted to the BRDM-2 (4 × 4) amphibious scout car and BTR-60PB (8 × 8) vehicle, it does not have a hatch cover.

Both the driver and machine gunner have a windscreen in front of their positions which, when in action, is covered by a flap hinged at the top. There is a vision block in each side of the hull, to the left of the driver's and the right of the machine gunner's position.

An aisle provides access from the crew compartment at the front of the vehicle to the personnel compartment at the rear. This has inward-facing folding canvas seats for the 10 infantrymen. The two hatches over the top of the troop compartment open forwards. The infantry enter and leave the vehicle by two doors in the rear of the hull, both of which are provided with a firing port. There is an additional firing port and vision block in each side of the troop compartment. An unditching beam is often carried on the roof or side of the vehicle.

The torsion bar suspension either side consists of six roadwheels with the drive sprocket at the front and the idler at the rear with hydraulic shock-absorbers on swing arms on the first and last roadwheel stations. There are no track-return rollers as the top of the track rests on the top of the roadwheels.

The MT-LB is fully amphibious being propelled in the water by its tracks. Before entering the water a trim vane is erected at the front of the vehicle and the bilge pumps, which have a capacity of 450 litres/min, are switched on. Standard equipment on all MT-LB vehicles includes an NBC system. The MT-LB has air-actuated brakes which can be connected to a trailer. Night vision equipment includes an OU-3GK white/infrared searchlight with a range of 400 m for the commander and a TVN-2 infrared periscope for the driver with a range of 40 m.

It can also tow a trailer or weapon weighing up to 6,500 kg or carry up to 2,000 kg of cargo or stores.

Variants

MT-LB artillery tractor

This vehicle is officially known as the MT-LB T or MT-LB AT (Artilleriskyi Tjagach) and the left side of the crew compartment is fitted with stowage racks for ammunition.

MT-LBM

There are two known versions of this vehicle. The MT-LB6MA is equipped with the complete turret of the BTR-80 (8 × 8) armoured personnel carrier, while the MT-LB6MB is fitted with the complete turret of the BTR-80A (8 × 8) armoured personnel carrier.

Kurgan MT-LB upgrade

As well as the upgrade package for the MT-LB mentioned under the development section, in 1998 it was revealed that the Kurgan Machine Construction Plant, well known for the design, development and production of the BMP-2 and BMP-3 infantry combat vehicles, has developed to the prototype stage an upgrade package for the MT-LB.

This has been offered to the Russian Army as well as being offered on the export market.

For improved cross country mobility, the upgraded MT-LB has been fitted with new stronger torsion bars, new rubber tyred road wheels and new hydraulic shock absorbers.

The diver, seated at the front of the vehicle on the left side has been provided with a new seat and for improved comfort a new climate control system has been installed in the crew and rear troop compartments.

In the standard MT-LB the driver steers the vehicle using tillers but for the upgraded MT-LB these have been replaced by a steering wheel. This feature, coupled with the hydraulic strengthening of the turning controls and with the disc clutches now operating in oil, improved parking brakes, clutch and gear shift controls, all reduce driver fatigue as well as making for more precise vehicle control.

The MT-LB is fully amphibious being propelled in the water by it's tracks at a maximum speed of 4.5 km/hr. The upgraded MT-LB is fitted with two new independent electric bilge pumps so if one is put out of action water that creeps in through door and hatch seals can be quickly pumped out.

Kharkov Tractor Plant Upgraded MTLB

Details of the extensive upgrade package for the MTLB developed by the Kharkov Tractor Plant, based in the Ukraine, are given later in this section.

MT-LB Declared Variants

Under the Conventional Forces Europe (CFE) Treaty signed in 1990, the Russians declared the following variants of the MT-LB multipurpose tracked armoured vehicle, and additional details of some of these are given later in this entry:

Hungarian MT-LB Engineer Vehicle, said to be based on MT-LBU chassis but has six roadwheels

MT-LBO, Bulgarian designation for MT-LB vehicle

1V13, battery fire direction centre vehicle, called 1W13 by Poland, also used by former Czechoslovakia and Hungary

1V14, battery command post vehicle, called 1W14 by Poland, also used by former Czechoslovakia and Hungary

1V15, battalion command vehicle, called 1W15 by Poland, also used by former Czechoslovakia and Poland

1V16, battalion fire direction centre vehicle, called 1W16 by Poland, also used by former Czechoslovakia and Hungary

Note: The 1V12 family consists of the 1V13, 1V14, 1V15 and 1V16. These are the original command and control set with the 1V14, 1V15 and 1V16 each having an APK digital data transmission set and the 1V15 and 1V16 also having the 9V59 artillery fire-control computer. The 1V12M family consists of the 1V13M, 1V14M, 1V15M and 1V16M, they are modified artillery command and control sets with each vehicle having an APPK artillery data computation and digital transmission computer system.

1V21, staff command vehicle, called MP-21 by former Czechoslovakia, also used by Poland. Has new C^3 equipment fitted

1V22, air defence management vehicle, called MP-22 by former Czechoslovakia, also used by Poland. Has new C^3 equipment fitted

1V23, command and control vehicle, called MP-23 by former Czechoslovakia, also used by Poland. Has new C^3 equipment fitted

1V24, artillery C^3 vehicle, called MP-24 by former Czechoslovakia, also used by Poland. Fitted with new C^3 equipment

1V25, air defence management vehicle, called MP-25 by former Czechoslovakia, also used by Poland. Has new C^3 equipment installed

MT-LB R-330B, MT-LB with R-330 VHF jammer, also used by Hungary. This is the MP-31 regimental C^3 system. The Russian Federation uses MP-31 system installed in the BMP-1KShM vehicles

MT-LB 9S743, Bulgarian MT-LB with radio system and generator at hull rear

MT-LB KShM R-80, Bulgarian MT-LB with table and increased headroom

MT-LB KShM R-81, Bulgarian MT-LB with radio equipment

MT-LB TRI, Polish engineer reconnaissance vehicle

MT-LB WPT, Polish recovery and maintenance vehicle

MT-LB R-137, Polish MT-LB with R-137 radio

MT-LB R-137T, Polish MT-LB with R-137 radio

MT-LB Sova, Bulgarian MT-LB with dismountable surveillance radar system

Beta EM, Polish communications vehicle.

MT-LB multipurpose tracked vehicle towing 100 mm T-12 anti-tank gun (Willis A Bullard)

MT-LBV

The MT-LB can also be fitted with 565 mm wide tracks for operation in snow and swampy ground; it is then called the MT-LBV. This version has a ground pressure of 0.28 kg/cm².

MT-LBV M1989

This is fitted with a new one-person manually operated turret armed with a 12.7 mm NSVT machine gun.

ADZM engineer

This engineer variant is fitted with a hydraulic dozer blade and a roof-mounted extendable arm, with a bucket. At the left rear of the hull is a working station for the operator. According to Russian sources, the ADZM is found in airborne brigades.

K-611 and K-612

These are radiological reconnaissance vehicles and equipped with special sensors, three R-173 and one R-171 radios, TNA-3 navigation apparatus, gyrocompass IG17-1, computer with teleprinter P115 and AB-2 power generator. The K-611 would be used for the determination of the power of the explosion while the mission of the K-612 is to find the exact location of ground zero. A complete set consists of a K-617 command post (with four K-618 relay stations and a K-619 information gathering station) and a K-702-I test maintenance vehicle.

MT-LB Artillery Tractor

The MT-LBs used as artillery prime movers have been observed with a fully enclosed box mounted over the troop compartment roof containing the gun section equipment. This is also referred to as the MT-LB-T or MT-LB AT.

MT-LBU (command)

This is the command version of the MT-LB and has additional radios, a generator, a land navigation system and a canvas cover that can be extended to the rear when the vehicle is being used in the static role.

MT-LB with 'Big Fred' radar (SNAR-10)

This has the Russian designation of the SNAR-10, or 'Big Fred' in NATO, and was originally believed to be an artillery radar location system.

It is now known that it is a battlefield surveillance radar with its main role being the location of stationary and moving targets in the Forward Edge of the Battle Area (FEBA).

The radar antenna is mounted on top of a turret positioned at the rear of the hull and when travelling the antenna is folded forwards through 90°. Maximum range against moving targets is 16 km while it can plot artillery projectiles impacting at ranges of 10 km. This mobile radar station is also called the 1RL-232 and is equipped with a 1RL-127 radar, two R-123M radios, a PAB-2A aiming circle, a power generator and a TV-240 observation device for the vehicle commander. In 1982 the modified SNAR-10M or 1RL-232-1 was introduced.

Standard equipment includes an NBC system, a full range of communications equipment, a heater, a land navigation system and an auxiliary power unit.

Unlike other members of the MT-LB family, the SNAR-10 is not amphibious. The turret-mounted 7.62 mm PKT machine gun is retained.

Combat weight of this system is 12.6 tonnes with performance similar to that of the standard MT-LB.

MTP-LB (repair vehicle)

The MTP-LB is designed for field maintenance, repair, and recovery of tanks and other AFVs and is recognisable by its lack of a machine gun turret.

Mounted at the front of the vehicle is an A-frame which can lift a maximum load of 1,500 kg. Standard equipment includes tools, gas welding and cutting equipment, a cable winch with 85 m of cable and a capacity of 6,700 kg, a jacking device, towing attachment hooks and a crane. Full details of the MTP-LB are given in *Jane's Military Vehicles and Logistics 2001-2002.*

MT-LB (ambulance)

This is an MT-LB used as an armoured evacuation vehicle (armoured ambulance) with stretchers fitted in the rear compartment.

MT-LB engineer vehicle

This is similar in appearance to the basic MT-LB but modified to mount a plough blade on the roof. Hydraulic devices at the rear of the vehicle allow manual mounting of the plough blade to the rear only. This has the NATO designation of the MT-LB M1980.

RKhM chemical reconnaissance vehicle

This was originally believed to be based on the chassis of the MT-LB multipurpose vehicle but is now known to be based on the chassis of the 122 mm 2S1 self-propelled howitzer. Details of this version are given in the Self-propelled guns and howitzers (tracked) section.

MT-LB with Vasilek

For use in Afghanistan, a self-propelled version of the towed 2B9 Vasilek 82 mm automatic mortar described and illustrated in the Towed anti-tank guns, guns and howitzers section was developed. The mortar has had its wheels removed and has been propped up on the upper rear deck on steel ammunition boxes.

MT-LB with AT-6 'Spiral' ATGW system

Details of this are given in the Tank destroyers section.

MT-LB 120 mm mortar

The Bulgarian Army has mounted a 120 mm mortar in the rear of the MT-LB multipurpose armoured vehicle. Early reports that the new 2S12 Sani (Sled) self-propelled mortar system used a modified MT-LB chassis are incorrect. The 2S12 in fact uses the GAZ-66 (4 × 4) truck with the 120 mm two-wheeled mortar being carried portée fashion.

Other versions of MT-LB

In addition to those versions of the MT-LB previously mentioned, US Army sources list the following variants, some of which may well refer to the vehicles already mentioned: MT-LB M1979/1, MT-LB M1979/2, ATV M1979/3, ATV M1979/4 and the MT-LB M1981. It is now known that the MT-LB 1979/1, -2 and -3 are three different repair and recovery versions, partly of East German origin and are now simply known as MTP-LB (or MT-LB MTP in Bulgarian service). The M1974/4 was a code name for the RKhM and MT-LB M1981 and is actually the K-611/K-612.

MT-LB with WAT turret

Poland has fitted an MT-LB with the WAT turret, as installed on the OT-64C(2) (SKOT-2AP) and OT-62C APCs, which is armed with a 14.5 mm KPVT and a 7.62 mm PKT machine gun.

SA-13 'Gopher' SAM system

The MT-LB multipurpose armoured vehicle chassis is also used as the basis for the SA-13 'Gopher' SAM system. Mounted in the centre of the hull roof is a turret with an arm, on which is carried a total of four SA-13 missiles in the transport/launch containers, with the operator being seated below and between the missiles.

Specifications

Crew: 2 + 11
Combat weight: 11,900 kg
Unloaded weight: 9,700 kg
Max payload: 2,000 kg
Max towed load: 6,500 kg
Power-to-weight ratio: 20.16 hp/t
Ground pressure: 0.46 kg/cm²*
Length: 6.454 m
Width: 2.86 m
Height: (to turret top) 1.865 m
Ground clearance: 0.4 m
Track width: 350 mm*
Length of track on ground: 3.7 m
Max speed:
(road) 61.5 km/h
(water) 4.5 km/h
Fuel capacity: 450 litres
Max range: 500 km
Fuel consumption: 0.9-1.2 litres/km
Fording: amphibious
Gradient: 60%
Side slope: 30%
Vertical obstacle: 0.61 m
Trench: 2.41 m
Engine: YaMZ 238 V, V-8 cylinder diesel developing 240 hp at 2,100 rpm
Transmission: manual with 6 forward and 1 reverse gears
Steering: clutch and brake
Suspension: torsion bar
Electrical system: 24 V
Armament: 1 × 7.62 mm PKT MG
Ammunition: (7.62 mm) 2,500

Gun control equipment
Turret traverse: manual, 360°
Gun elevation/depression: +30 to –5°
Gun stabiliser:
(vertical) no
(horizontal) no
Armour: 3.19-9.87 mm
NBC system: yes
Night vision equipment: yes
*The MT-LB can also be fitted with 565 mm wide tracks which reduce ground pressure to 0.27 or 0.28 kg/cm²

Full details of the SA-13 are given in *Jane's Land-Based Air Defence 2001-2002.*

MT-LB with ZU-23-2

During a number of recent conflicts, for example during the Georgian civil war and the more recent fighting in Chechnya, a number of MT-LB multipurpose tracked armoured vehicles has been observed fitted with the 23 mm ZU-23-2 towed anti-aircraft gun system mounted on the rear decking. This appears to be a local modification with the weapons being used both in the air defence and ground roles.

Iraqi 120 mm self-propelled mortar

Early in 1989, an MT-LB was shown in Iraq for the first time with wide tracks modified locally to carry a 120 mm mortar in the rear of the vehicle. The roof hatches have been modified and now open to either side of the hull to allow the 120 mm mortar to fire to the rear. A baseplate is carried on the right side of the hull

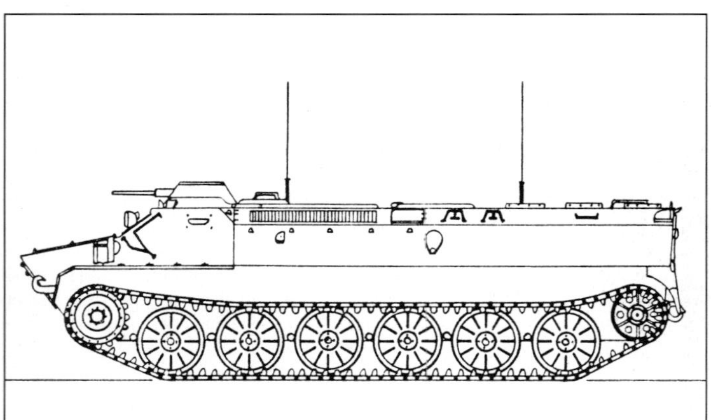

MT-LB multipurpose tracked vehicle

SA-13 'Gopher' SAM system which uses the MT-LB chassis (Christopher F Foss)

to allow the mortar to be deployed away from the vehicle if required by the tactical situation. This is not believed to have entered service with the Iraqi Army.

Iraqi 4 × 120 mm self-propelled mortar

Details of this system, shown for the first time in 1989 in Baghdad, were given in *Jane's Armour and Artillery 1992-93,* page 573. As far as is known it never entered service with the Iraqi Army.

Bulgarian MT-LB

This country has developed a number of versions of the MT-LB vehicle and details are given under Bulgaria earlier in this section. These are manufactured by the BETA Joint Stock Company who has also built the 122 mm 2S1 self-propelled artillery system.

Polish MT-LB

This country has developed a number of versions of the MT-LB vehicle, including one with an additional pair of roadwheels, and details are given under Poland. These are manufactured by Steel Works Stalowa Wola.

Swedish MT-LB

Late in 1993, it was announced that Sweden was to purchase 800 MT-LB multipurpose armoured vehicles from Germany under a deal worth US$1.7 million.

Details of the Swedish MT-LB and variants are given later in this section under Sweden.

Status

Production complete. In service with:

Country	Quantity	Comment
Armenia	75	
Azerbaijan	196	
Bangladesh	n/avail	unconfirmed user
Belarus	71	
Bulgaria	2,350	local production (qv)
Finland	220	including variants
Hungary	30	plus 300 variants
Iraq	500	estimate
Kazakhstan	200	estimate
Moldova	62	
Russia	2,300	including variants
Sweden	460	qv
Ukraine	2,300	plus variants
Yugoslavia	50	estimate

Note: The above figures are for those used in the APC role; many other countries use them for the specialised roles such as radar system and SA-13 SAM.

Manufacturer

State facilities of Bulgaria, Poland and the Russian Federation and Associated States (CIS).

UPDATED

SINGAPORE

Singapore Technologies Kinetics Bionix Infantry Fighting Vehicle

Development

The origins of the Bionix Infantry Fighting Vehicle (IFV) can be traced back to 1988 when the Singapore Armed Forces (SAF) decided to supplement its existing large fleet of United Defense LP M113 full-tracked Armoured Personnel Carriers (APC).

At that time, Singapore Technologies Automotive (today called Singapore Technologies Kinetics) were already working on the upgrading of the French AMX-13 light tank to the enhanced AMX-13 SM1 configuration and were looking for a means to continue their work on armoured vehicles once this was completed.

Using its own funding, the company built two XV series full-tracked test beds, one based on the AMX-13 light tank and one on the M113A3 APC. The XV 1 was completed in January 1990 and was followed by the XV 2 in December 1990. These were put through six months of extensive tests, including some by the Singapore Ministry of Defence (MINDEF).

These XV series trials proved to MINDEF that the company had the technical capability to design, develop and build a new IFV to meet the operational requirements of the SAF.

A tripartite team was formed to run the whole new IFV programme consisting of SAF Armour Formation (who would use the vehicle), the Defence Material Organisation (DMO) of MINDEF and Singapore Technologies Kinetics.

This unique arrangement allowed for a very rapid development time scale, as well as making for the maximum utilisation of the available existing resources in Singapore.

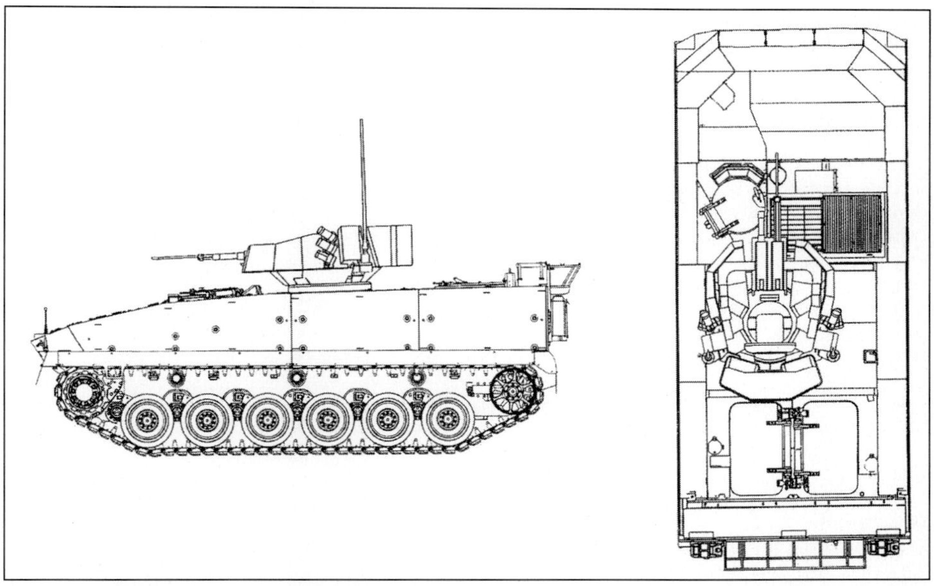

Side and top elevation of the Bionix 40/50 showing cupola weapon station offset to the left and the twin hatch covers over the rear troop compartment ***2000***/0064800

To reduce development time and cost, three Testbeds (T1, T2 and T3) were built to test the main subsystems of the new IFV, one was fitted with a locally developed two-person 25 mm turret, one with a foreign 25 mm one-person turret and one with the locally designed 40/50 cupola which was developed to upgrade the SAF M113 fleet. The first testbed was completed in January 1993 followed by the second and third in June 1993.

The Bionix 25 infantry fighting vehicle is now fully operational with the Singapore Armed Forces (T J Gander) ***2002***/0064799

Not only did these testbeds test the three different weapon stations but also automotive components, suspension, man-machine interface and the modular passive armour protection system.

These three testbeds were followed by three Final Prototype (FP) vehicles which undertook more than 30,000 km of Reliability Growth Development Testing. These were fitted with a power pack consisting of a Detroit Diesel engine coupled to the then General Electric HMPT 500 series automatic transmission. Turrets were either the locally developed 25 mm or the 40/50 cupola.

Finally a preproduction unit was built which incorporated all of the improvements requested by the user, following trials with the three testbeds and the three prototype vehicles. This was completed in March 1996 although by that time a decision had been taken to place the vehicle in production.

Singapore Technologies Kinetics was awarded a production contract for the new IFV officially called the Bionix. While under development the overall programme was called Project Funby.

First Bionix 25 production vehicles were completed in September 1997 and under the Phase 1 production contract awarded to Singapore Technologies Kinetics, production will continue until 2001.

The 42nd Battalion of the Singapore Armoured Regiment (42 SAR) is the first unit to be fully equipped with the Bionix 25 and Bionix 40/50 and was declared fully operational in mid-1999.

The 42 SAR is the Singapore Army's first armoured battle group and has a strength of 700 officers and men, 40 Bionix IFV, 13 upgraded M113 APCs, 12 AMX-13 SM1 light tanks plus supporting vehicles.

From the time a decision was taken to design develop and manufacture a local IFV to the time the first

Bionix 40/50 infantry fighting vehicle **2001**/0099459

Bionix armoured recovery vehicle in with crane extended **2002**/0073072

Bionix vehicle rolled off the production line was just seven years.

When compared to the current in-service upgraded M113 series (called Ultra) APC, the Bionix 25 IFV has a significant increase in armour, mobility and firepower.

It is capable of further development, which is a very important factor when a major weapon system such as an IFV has to stay in service for anything from 20 to 30 years.

Two main variants of the Bionix IFV are currently in production and service with the SAF, Bionix 25 and Bionix 40/50.

The former is fitted with a two-person turret armed with the combat proven Boeing 25 mm M242 Chain Gun, while the 40/50, has the latest version of the Chartered Industries of Singapore 40/50 cupola station which has been built in large numbers for the M113 upgrade, which was completed several years ago.

The Bionix 25 has been demonstrated in the UK with Vickers Defence Systems having a teaming arrangement with Singapore Technologies Kinetics for Europe.

Both versions of the Bionix were demonstrated in the US late in 1999 and early 2000. As of October 2001 no export contracts for the Bionix or its variants had been announced.

Description

Export marketing of the Bionix is being concentrated on the Bionix 25 so the description here relates to this version as this and the Bionix 40/50 have the same hull, the only difference being in their weapon stations, roof hatch arrangement and seating arrangements for the troops.

The hull and turret of the Bionix IFV are of all-welded steel to which an additional layer of modular passive armour is attached.

The layout of the Bionix is conventional, with the driver front left, power pack situated front right, turret in the centre and troop compartment at the rear.

The driver enters via a roof hatch above his position that opens to the left rear and can be locked in the vertical position when driving in the head out position. The driver has a seat that can be rapidly adjusted front/rear and up/down.

When driving, closed-down observation is via three day periscopes which give a good view over the frontal arc, the centre periscope can be rapidly replaced by an image intensification periscope for driving at night.

The driver steers the vehicle using a small steering wheel rather than tillers and as an automatic transmission is fitted, he has just two pedals, accelerator and brake. The instrument panel is mounted on the left with the transmission selection box on the right.

The integrated power pack is mounted to the right of the vehicle and can be removed from the Bionix as a complete unit in less than 15 minutes. The engine compartment is fitted with an automatic fire detection and suppression system that uses dry powder rather than Halon, as this is banned in Singapore. A manual back-up system is also fitted.

The power pack consists of the well-known Detroit Diesel 6V-92TA turbocharged diesel engine with electronic control diesel injection developing 475 hp at 2,400 rpm with a stretch potential of 550 hp. Detroit Diesel engines are also installed in the AMX-13 and M113 series of vehicles of the SAF.

The engine is coupled to the General Dynamics Land Systems HMPT-500EC fully hydromechanical transmission, with the final drives being provided by General Dynamics Land Systems. HMPT-500EC is the latest version of the HMPT-500 installed in the Bradley fighting vehicle.

The cooling system has been designed by Galley, with the radiator being at the very rear of the power pack compartment. The air intake and outlet louvres are in the roof with the exhaust pipe running through the hull on the right side and with the outlet on the right rear of the vehicle. This not only helps to reduce the signature of the Bionix but avoids diesel fumes blowing into the faces of the crew when they are in the head out position.

The Bionix has two fuel tanks which are situated one either side in the upper part of the hull side near the roof. These hold a total of 562 litres of fuel.

Suspension is of the hydrostrut type and is a further development of that installed on the AMX-13 SM1 light tank. This was developed by Horstman Defence Systems of the UK to meet the requirements of Singapore Technologies Kinetics. It provides a good ride for the crew and troops, on both roads and cross-country and also provides a more stable firing platform.

Specifications

Bionix 25	
Crew:	3 + 7
Combat weight:	23,000 kg
Power-to-weight ratio:	20 hp/t
Ground pressure:	0.76 kg/cm²
Length:	5.92 m
Width:	2.70 m
Height:	
(hull top)	1.80 m
(turret roof)	2.57 m
Track width:	430 mm
Max speed:	
(road)	70 km/h
(cross-country)	25-40 km/h
Acceleration:	10 s (0 to 32 km/h)
Fuel capacity:	
(total)	562 litres
(usable)	527 litres
Range:	415 km at 45 km/h
Gradient:	60%
Side slope:	30%
Vertical obstacle:	0.60 m
Fording:	1 m
Turning radius:	9.90 m
Pivot turn:	6.50 m
Trench:	2.00 m
Angle of approach/departure:	40/30°
Engine:	Detroit Diesel Model 6V-92TA diesel developing 475 hp at 2,400 rpm
Transmission:	General Dynamics Land Systems HMPT-500-3C fully automatic, hydro-mechanical
Steering:	Hydrostatic
Brakes:	multidisc oil-cooled
Suspension:	Hydropneumatic
Electrical system:	28 V
Batteries:	4 × 12 V, 200 Ah
Alternator:	300 A
Armament:	
(main)	1 × 25 mm M242 cannon
(coaxial)	1 × 7.62 mm GPMG
(anti-aircraft)	1 × 7.62 mm GPMG
(smoke grenade launchers)	6 × 76 mm
(rear)	1 × 7.62 mm GPMG
	electric/manual
Gun control equipment:	
Turret traverse:	360°
Gun stabiliser:	
(vertical)	Yes
(horizontal)	Yes
NBC System:	No
Night vision equipment:	Yes

The running gear either side consists of six dual rubber tyre road wheels with drive sprocket at the front, idler at the rear and three track return rollers. The first and last track return rollers support the inside of the track only, with the middle one supporting the entire track. The idler at the rear can be adjusted manually.

The tracks are of the single-pin cast type with replaceable rubber pads and when new, each has 77 links. A lighter weight track is also being studied.

The turret is of all-welded steel armour construction with an additional layer of spaced passive armour added. The gunner is seated on the right and the vehicle commander is seated on the left, with both turret crew members being provided with a single-piece hatch that opens to the rear and can be locked into the vertical position.

The main armament comprises the combat proven stabilised Boeing M242 25 mm dual-feed cannon with a 7.62 mm coaxial General Purpose Machine Gun. It is mounted coaxial to the left, with a similar weapon being mounted externally on the turret roof. This can be used by either the commander or gunner. Mounted either side of the turret is a bank of three 76 mm electrically operated smoke grenade launchers firing forwards.

The 25 mm M242 cannon would normally fire HEI and APFSDS ammunition which is manufactured under licence from Oerlikon Contraves by Chartered Industries of Singapore. The cartridge cases are ejected externally from the turret with the links being collected internally. A total of 600 rounds of 25 mm ammunition is carried, of which 180 is for ready use.

Turret traverse and weapon elevation is digital all electric with either the gunner or commander being able to lay and fire the weapons. The gunner has twin control handles and a day/thermal sight with a magnification of ×8 and two fields of view, high and low, with stadia for the 25 m cannon and 7.62 mm coaxial machine gun. The commander has an optical relay from the gunner's sight and a single control handle. If required, a laser range-finder module could be fitted to the day/night sight for improved first round hit probability.

The vehicle commander has an additional five day periscopes to give observation to the turret front, left side and rear while the gunner has three day periscopes to give observation to the right and rear.

The gunner has the turret control box mounted to the right of his position and both turret crew members are provided with an adjustable seat with the communications equipment being installed in the turret bustle. Either crew member has an emergency turret stop button and a deck limiter is fitted to stop the 25 mm canon hitting any part of the vehicle.

There is also a small external stowage basket on the turret rear and two replacement 7.62 mm machine gun barrels are carried externally in boxes on the left side of the turret. In addition to entering the turret via the roof hatches, the commander and gunner can also enter via the hull rear and sliding covers in the lower part of the turret.

The troop compartment is situated at the rear of the hull and the troops enter the vehicle via a power-operated ramp at the rear which also has an emergency door. There are no periscopes or observation devices for the infantry, as this was not a requirement for the SAF.

Seven fully equipped infantry are carried in the rear. Of these, three are seated either side facing the centre of the vehicle on bench seats with the seventh to the rear of the turret facing back.

Over the top of the troop compartment is a single hatch that opens to the rear and can be locked in the vertical postion. A 7.62 mm machine gun is mounted on the right side of the roof providing suppressive fire through the rear arc of the Bionix 25.

The Bionix IFV can ford to a depth of 1 m with preparation, but with the aid of a specially developed amphibious kit is fully amphibious.

The Bionix 25 does not have an air conditioning system fitted as this was not required by the user, nor was an NBC system. This could be fitted on export vehicles if required by the user.

Variants

Bionix 40/50

The only variant of the Bionix in service so far is the Bionix 40/50 which is fitted with the locally developed 40/50 twin weapon cupola, an earlier model of which is fitted to some of the M113 series APCs upgraded by Singapore Technologies Kinetics for the SAF. This is mounted slightly to the left of the vehicle's centre line.

The Bionix 40/50 has a combat weight of 21,500 kg and in addition to the commander, gunner and driver, can carry nine fully equipped infantry. Of these four are seated down either side facing the centre of the vehicle with the ninth man being seated to the right side of the turret facing the rear.

The 40/50 twin weapon cupola is armed with the locally developed 12.7 mm 50MG and the 40 mm 40AGL with a bank of SDS-93 three-round 76 mm smoke grenade launchers being mounted on either side of the turret and firing forwards.

Both weapons are mounted on a common cradle and firing of both weapons takes place with the hatch cover open, but there is no provision for firing the weapons with the gunner under armour protection. Observation is provided when closed down using three M17 day and two L794D periscopes.

The 40 mm AGL is provided with 32 rounds of ready use ammunition (for example HE or HEDP) while the 12.7 mm 50MG is provided with 100 rounds of ready use ammunition, which can include the locally developed SLAP (Saboted Light Armour Penetrator). The all-welded steel armour provides the gunner with protection from 7.62 mm AP attack.

Turret traverse and weapon elevation is powered, with manual controls being provided for emergency use. Turret traverse is 110° left and right with weapons elevation being from -8 to +45°.

The roof hatch arrangement of the Bionix 40/50 is also different and has two roof hatches in the centre which open left and right and can be locked in the vertical position.

Bionix infantry carrier vehicle

This has been developed as a private venture for the export market and was unsuccessfully offered to the US Army. The Bionix ICV is much lighter than the Bionix 25 and the overall height of the vehicle has been reduced by the installation of an external weapon station. it is armed with a 12.7 mm machine gun that can be aimed and fired from within the hull.

This version of Bionix weighs 16,964 kg and is fully air portable in a Lockheed Martin C-130 Hercules transport aircraft. With the special armour package developed by IBD of Germany the combat weight of the vehicle is 23,965 kg. This version has a crew of two and can carry nine fully equipped troops.

Although the infantry carrier vehicle was not selected by the US Army it is still being marketed alongside the Bionix 25 and 40/50 used by the Singapore Armed Forces.

Projected variants of the Bionix for the US Army requirement included the following versions:

ATGW vehicle
Commander's vehicle
Engineer squad vehicle
Fire support vehicle
Medical evacuation vehicle
Mortar carrier
NBC reconnaissance vehicle
Reconnaissance vehicle

Bionix AVLB

The Bionix armoured vehicle launched bridge (AVLB) was first revealed in 2000 and is currently undergoing trials with the Singapore Armed Forces.

It is essentially a modified Bionix chassis with turret removed and fitted with a bridge laying mechanism and bridge developed by MAN Technologies of Germany.

Mounted on top of the hull is the Military Load Class 30 (MLC 30) two part bridge which, when open, is 22 m long and can be used to span wet and dry gaps up to 20 m. Two MLC 30 bridges can be used side by side to form a MLC Class 60 bridge. It has a crew of two and weighs 28,000 kg. Details of this system are given in *Jane's Military Vehicles and Logistics 2001-2002.*

Bionix armoured recovery vehicle

The Bionix ARV is currently being tested by the Singapore Armed Forces and is based on a modified Bionix chassis. It is fitted with a boom type crane, winches and other specialised equipment to enable it to recover damaged and disabled tracked and wheeled armoured vehicles. It has a crew of three and a combat weight of 24,000 kg. Details of this system are given in *Jane's Military Vehicles and Logistics 2001-2002.*

Stretched Bionix

Studies have been undertaken on a stretched and wider version of the Bionix which would have an additional roadwheel either side. This would allow for greater internal volume and the ability to fit a larger turret system.

Status

Production. In service with Singapore Armed Forces.

Manufacturer

Singapore Technologies Kinetics.

UPDATED

Bionix AVLB in travelling configuration **2001**/0099458

Bionix infantry carrier vehicle, developed from the Bionix infantry fighting vehicle in service with the Singapore Armed Forces. It is fully air portable in a Lockheed Martin C-130 Hercules transport aircraft and is armed with an externally mounted 12.7 mm M2 machine gun. It is shown here without its appliqué side armour installed **2002**/0073309

Singapore Technologies Kinetics all-terrain tracked carrier (Bronco)

Development

Early in 2000, after almost five years of extensive research and development work, Singapore Technologies Kinetics, a part of Singapore Technologies Engineering, revealed their new articulated armoured All-Terrain Tracked Carrier (ATTC).

Although similar in concept to the Swedish Hägglunds Vehicle Bv 206S and more recent Hägglunds Vehicle BVs 10 armoured articulated vehicles, Singapore Technologies Kinetics believe that the ATTC is a new generation fully armoured articulated vehicle.

When compared with other vehicles of this type, the ATTC has greater internal volume, greater load carrying capacity and can be modified for a wide range of battlefield missions.

Following extensive trials with an automotive test bed which ran for the first time in 1997, two preproduction ATTC were completed in 1999, one in the troop carrier configuration and the other in a cargo carrying role.

Early in May 2001, Singapore Technologies Kinetics delivered the first of an undisclosed quantity of their new articulated armoured All Terrain Tracked Carrier (ATTC) vehicles to the Singapore Armed Forces (SAF).

It is understood that the ATTC will operate with the Singapore Technologies Kinetics Bionix infantry fighting vehicle. For export marketing purposes the ATTC is referred to as the Bronco by Singapore Technologies Kinetics. Although developed for military applications, it is also being marketed for a number of possible civil applications including forestry and oil exploration operations.

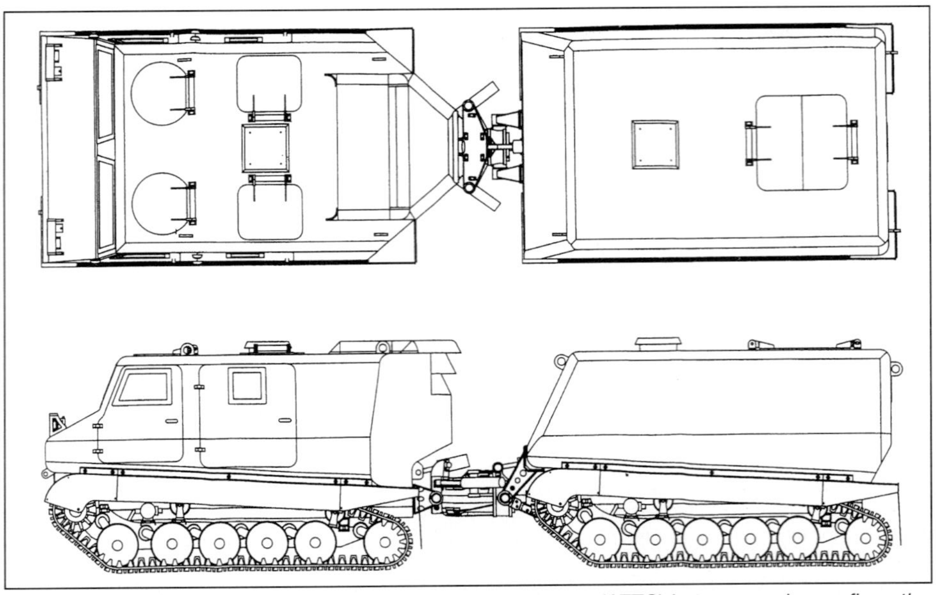

General arrangement drawing of the All Terrain Tracked Carrier (ATTC) in troop carrier configuration ***2002***/0116782

Specifications
(troop carrier)
Crew: 6 + 10
Combat weight: 16,000 kg
Curb weight: 11,400 kg
Payload: 4,600 kg
Power-to-weight ratio: 21.87 hp/t
Length: 8.6 m
Width: 2.3 m
Height: 2.2 m
Track width: 600 mm
Max speed:
(road) 60 km/h
(water) 4.5 km/h
Gradient:60%
Side slope: 30%
Vertical obstacle: 0.6 m
Trench: 1.5 m.
Engine: Caterpillar 3126B four-stroke turbocharged and charged air-cooled diesel developing 350 bhp at 2,400 rpm
Transmission: Allison MD3560P fully automatic
Steering: hydraulic, articulated
Brakes:
(service) hydraulic disc
(parking) hydro mechanical disc brake
Suspension: rubber bush
Armament: optional
NBC system: optional
Night vision equipment: optional

Description

The ATTC consists of two fully armoured units, front and rear, connected by a specially designed hydraulic articulated joint that features two steering cylinders and two damping cylinders.

The all-welded steel armoured hull provides the occupants with protection from 7.62 mm ball small arms fire and shell splinters but an enhanced modular passive protection package has been developed for the ATTC by IBD of Germany.

The driver is seated at the front on the right (with a left hand model also being offered) with the vehicle commander to his left with large bulletproof windows to their front.

There are two entry doors in either side of the hull with a bulletproof window in the upper part. The bulletproof windows provide the same level of protection as the hull.

The front unit has four roof hatches, one each for the commander and driver, that open to the rear and another two in the centre of the roof that open inwards.

Entry to the rear unit is via a large door that opens to the right with a two part roof hatch also provided, although there is considerable flexibility in the design of the vehicle as regards positioning of the hatches and doors. Both front and rear units are fitted with an air conditioning system as standard.

The ATTC is powered by a Caterpillar model 3126B four stroke turbocharged and charge air-cooled diesel engine developing 350 bhp at 2,400 rpm coupled to an Allison MD3560P fully automatic transmission.

The engine is located in the rear part of the forward unit with the air inlet and outlet in the roof towards the rear. Power is taken from the transmission to the drop box and then to the drive sprockets at the front and rear units.

Suspension is of the rubber bush type with the front and rear unit having six lightweight dual road wheels either side with the drive sprocket at the front and two track return rollers. A mechanical track tensioner is fitted as standard. To save weight the road wheels and drive sprockets are of reinforced plastic.

The endless moulded rubber band tracks have been developed specifically for the ATTC by Soucy of Canada and are 600 mm wide. These are lighter than conventional steel tracks and are also much quieter. Nominal ground pressure of 60 kpa gives the vehicle exceptional mobility in marginal terrain such as snow and swamp.

The service braking system is of the hydraulic disc type operating on each track with the parking brake being of the hydro-mechanical type. Steering is hydraulic and very positive even at low speeds. The fully articulated steering makes the vehicle easy to handle in confined spaces and all four tracks are in constant contact with the ground. As an option, a differential lock can also be provided for the ATTC.

The ATTC is fully amphibious being propelled in the water by its tracks at a maximum speed of 4.5 km/h without any preparation apart from activating the two electrically operated bilge pumps in the front and rear units.

The troop carrier version can carry six people in the front unit, three either side, and ten in the rear unit, five either side facing inwards, with plenty of space for their equipment. Maximum payload is 1,200 kg for the front unit and 3,000 kg for the rear unit. When fully loaded the vehicle is well balanced as the front and rear units weigh the same.

Typical basic weapon fit would be a roof-mounted 7.62 mm machine gun and banks of three forward firing electrically operated smoke grenade launchers.

The cargo carrier can carry 1,200 kg in the front unit and 3,700 kg in the rear unit which is fitted with drop sides and a drop tailgate. This version retains the amphibious characteristics of the troop carrying version.

The design of the ATTC is flexible and the system has further growth potential as well as different unit configurations with various engine/transmission combinations.

All Terrain Tracked Carrier resupply variant showing pallet of cargo being unloaded ***2002***/0095438

All Terrain Tracked Carrier in basic troop carrier configuration role ***2002***/0095440

The ATTC is fully air portable in a Lockheed Martin C-130 Hercules transport aircraft.

Variants

In addition to the basic general troop carrier, three specialised versions have already been revealed: armoured engineer, resupply and ambulance.

Armoured engineer

The armoured engineer version has sufficient space to carry explosives and obstacle clearing equipment and can also be configured to carry and launch mine clearing systems.

Ambulance

The rear cabin of the ambulance version can carry four stretcher or nine seated patients. It is provided with a swing out stretcher system for ease of loading/unloading and an integrated tentage system for additional workspace. The rear cabin is air conditioned for improved comfort.

Resupply

The resupply variant has the rear unit fitted with a mechanical handing system and can carry pods of fuel or cargo.

Other variants

These include a weapons carrier; for example armed with anti-tank guided missiles, reconnaissance, command and control and internal security vehicle. There is also a possibility of a non-armoured version for the civil market.

Status

Production. In service with Singapore Armed Forces.

Manufacturer

Singapore Technologies Kinetics

UPDATED

Singapore M113 fleet

Development/Description

In 1995, it was revealed that the Singapore Armed Forces armour battalions were putting some of its M113A1 APCs through an upgrade programme developed by engineers from the Ministry of Defence (MINDEF) Defence Technology Group and the now Singapore Technologies Kinetics.

Tailored to extend the service life of the M113A1 and enhance its survivability and firepower, this project mainly involved a total rebuild and reconditioning of the hull, refurbishment of the vehicle's automotive components and installation of a new locally developed CIS 40/50 cupola for some variants.

The company won the contract, believed to be worth over S$300 million, in 1993 and began delivering upgraded M113s to the Singapore Armed Forces in 1994 for trials. The upgrade programme is now complete. Singapore Technologies Kinetics is now marketing this M113 upgrade programme to other countries.

A related MINDEF contract, worth some S$53 million was awarded to Chartered Industries of Singapore (CIS) in 1993. Under this contract CIS supplied appliqué armour kits and 40/50 weapon stations for installation on some variants of the upgraded M113, with delivery spread over three years from 1995-98.

The Singapore Armed Forces have more than 1,000 M113A1s in service with its armour, artillery and combat engineer battalions, making Singapore's M113 fleet one of the largest in Southeast Asia.

Apart from the standard APC version, modified Singapore Armed Forces M113A1s serve as 120 mm self-propelled mortars, artillery command posts, fitters' vehicles, armoured ambulances, recovery, scout/reconnaissance and armoured engineer carriers. The 120 mm self-propelled mortar carrier carries a baseplate on the left side of the hull that allows the mortar to be deployed away from the vehicle if required.

The Singapore Armed Forces upgraded M113s are known to its operators as the A2 model because they feature improvements which bring the APC close to the US M113A2 marque in terms of combat performance.

For improved mobility, the existing diesel engine has been replaced by a Detroit Diesel Model 6V-53T developing 265 hp, coupled to an upgraded Allison TX110-1A automatic transmission.

Changes in the gear ratios have reduced the top speed of the vehicle, thus increasing acceleration by 23 per cent. An enlarged radiator capacity has improved the cooling system and extended vehicle life.

Upgraded M113 of the Singapore Armed Forces fitted with RAFAEL Armament Development Authority Overhead Weapons Station armed with a 25 mm M242 cannon **2000**/0064801

The installation of new torsion bars have resulted in larger wheel travel and the installation of improved shock-absorbers and stronger rear-wheel idler assembly has improved the ride and therefore helps to minimise troop fatigue.

The existing electrical system has been upgraded with a new 200 A generator as well as a new wiring harness. The double amperage improves the reliability of the M113 electrical system as well as the growth potential to meet future demands.

The company is able to handle upgrades of mobility kits from 210 to 300 hp diesel engines. The power train can be further customised and upgraded, depending on the vehicle payload and demands made on the vehicle.

Enhancements include the addition of slabs of bolt-on composite armour panels on the glacis plate and hull sides, armoured track skirts, spall liners in the interior and external diesel fuel tanks positioned one either side at the rear.

The rear-mounted bolt-on armoured fuel tanks replace the single fuel tank in the interior of the M113A1, creating additional internal space for stowage of crew equipment, ammunition or weapons. They also contribute substantially to reducing the risk of internal fires should the vehicle get hit.

The upgraded M113 is the first Singapore Armed Forces armoured vehicle to receive passive armour protection. These Singapore-developed appliqué armour slabs, understood to be made of ceramic, are also designed to negate the penetrating effects of shaped charges from HEAT and HESH rounds. An enlarged trim vane has been added for improved stability during amphibious operations due to the increase in combat weight of the up-armoured M113A2.

Improvements in firepower take the form of the CIS 40/50 cupola which replaces the single 12.7 mm M2 heavy machine gun fitted with a steel gun shield on the commander's station of older M113s. The Singapore Armed Forces is the first customer for the 40/50 weapon station.

The CIS 40/50 cupola incorporates the CIS-developed 40 mm Automatic Grenade Launcher (CIS

Singapore Armed Forces Upgraded M113 series vehicle from the front (David Boey)

Singapore Armed Forces Upgraded M113 series vehicle from the rear (David Boey)

40AGL) as main armament (32 ready rounds) and a 12.7 mm heavy machine gun (CIS 50MG) as secondary armament (100 ready rounds) on a common weapons cradle in a one-man weapons station. An armoured gun shield provides the firer with ballistic protection against 7.62 mm armour-piercing rounds.

Upgraded M113s also carry two pintle-mounted 7.62 mm GPMGs on either side of the rear hatch to provide maximum suppressive fire when dismounting armoured infantry.

The 40/50 weapon station is also fitted with a CIS SDS-92 smoke defence system with a bank of three electrically fired 76 mm smoke grenade dischargers on either side of the gun shield. This system allows the M113 to lay down a bank of obscurants up to 150 m from the vehicle using 76 mm smoke grenades.

The electrically powered weapon station, which can be traversed 340°, allows elevation of the weapon station from -8 to +45°. Weapons are sighted and fired from the open hatch position using the weapons' optical sights, with one weapon being fired at any one time.

It is believed that the Singapore Armed Forces has now taken delivery of 300 MUGS sights from Electro Optic Systems Pty Ltd of Australia for installation on the 40/50 cupola.

It has been disclosed that several years ago Singapore placed an order with the RAFAEL Armament Development Authority of Israel for an undisclosed quantity of Overhead Weapon Stations (OWS) armed with a 25 mm The Boeing Company M242 Chain Gun and 7.62 mm coaxial machine gun.

It has been confirmed that these 25 mm armed OWSs have now been installed on some of the M113 series APCs used by Singapore. This 25 mm M242 cannon is also installed in one of the two versions of the new Singapore Bionix IFV covered in a previous entry.

Status

Upgrade programme now complete. In service with Singapore Armed Forces. The complete upgrade is now being offered on the export market as well as some of the locally built subsystems.

Manufacturer

Singapore Technologies Kinetics Ltd.

UPDATED

SPAIN

Santa Barbara Sistemas Pizarro infantry combat vehicle

Development/Description

To develop a new Infantry Combat Vehicle (ICV) to meet the requirements of the 1990s a joint venture company called ASCOD (Austrian Spanish Co-Operative Development) was formed.

The ASCOD company consists of the Austrian Company, Steyr-Daimler-Puch Spezialfahrzeug AG & Co AG and the Spanish company, Santa Barbara Sistemas.

Following extensive trials with prototype vehicles, the Spanish government placed the first contract for the ASCOD early in 1996.

This contract is for the acquisition of 144 ASCOD ICV under the Spanish name of the Pizarro, with the actual contract being placed with SBB Blindados, a member of the now Santa Barbara Sistemas.

The initial contract covers the delivery of 123 Pizarro fighting vehicles and 21 command vehicles at a total value of US$330 million. The first two pre-production Pizarro vehicles were completed early in 1998. The first production batch are being issued to X Brigade which will be the first Spanish contingent assigned to the Eurocorps.

As of 2001 the total Spanish Army requirement was for 463 vehicles.

Santa Barbara Sistemas will be supported by the Trubia factory of its parent company Santa Barbara, which will supply the hull, turret body, armament and other subassemblies.

Pizarro infantry combat vehicle of the Spanish Army, fitted with explosive reactive armour to hull and turret, developed by Santa Barbara Sistemas **2001**/0099456

Full details of the ASCOD ICV are given in this section under International. The Spanish vehicles are virtually identical to the detailed description but will be fitted with Santa Barbara-developed explosive reactive armour.

In mid-1999 the Austrian Army finally awarded a contract to Steyr-Daimler-Puch Spezialfahrzeug AG & Co AG for the supply of 112 ASCOD vehicles under the local name of Ulan. These will be built in Austria and delivered from 2002 through to 2004.

Status

Production. In service with the Spanish Army.

Manufacturer

Santa Barbara Sistemas (part of General Dynamics).

UPDATED

SWEDEN

Hägglunds Vehicle CV 9030 infantry fighting vehicle

Development

The CV 9030 infantry fighting vehicle has been developed by Hägglunds Vehicle (now owned by Alvis plc of the UK) of Sweden specifically for the export market.

The basic CV 9030 chassis is similar to the CV 9040 currently in production for the Swedish Army and covered in detail in a separate entry.

To meet the specific operational requirements of the Royal Norwegian Army the CV 9030 has a number of improvements including a more powerful Scania diesel engine and the installation of a new Hägglunds Vehicle two-man power-operated turret armed with a stabilised 30 mm The Boeing Company Bushmaster II Chain Gun.

Five vehicles, the Austrian/Spanish ASCOD, Alvis Vehicles Warrior (UK), Hägglunds Vehicle CV 9030 (Sweden), Krauss-Maffei (today Krauss-Maffei Wegmann) Puma (Germany) and United Defense LP M2A2 Bradley (USA), undertook an extensive series of trials in Norway and early in 1994, Norway selected the Hägglunds Vehicle CV 9030 to meet its future infantry fighting vehicle requirement.

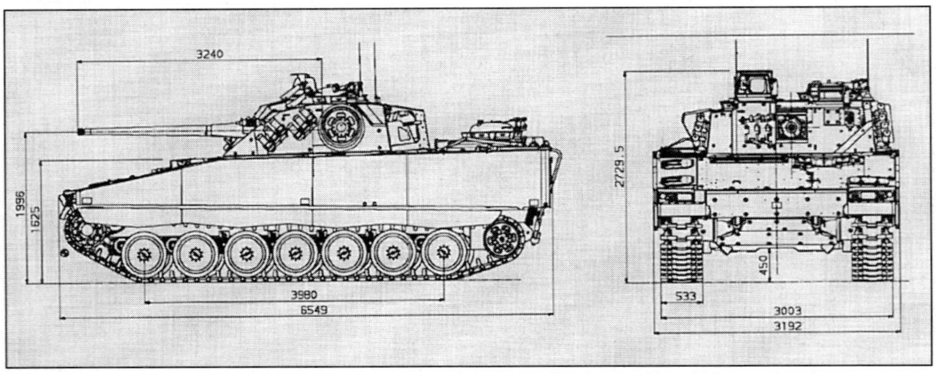

Side and front drawing of the Hägglunds Vehicle CV 9030N IFV for the Royal Norwegian Army

In April 1994, the Royal Norwegian Army Matériel Command placed a contract with Hägglunds Vehicle of Norway worth almost SKr2 billion covering the supply of 104 CV 9030 infantry fighting vehicles, or CV 9030N as it is more often known.

The first four preproduction CV 9030N vehicles were delivered early in 1996 to the Royal Norwegian Army for extensive user trials.

Within the Royal Norwegian Army the CV 9030N is called Project 5033 and the following technical modifications were carried out:

(1) Alteration of main armament to 30 mm Bushmaster II
(2) Fastener for water can on track shelf
(3) Tools in tool cabinet
(4) Adapter for funnel, fuel (for tank up jerrycan)
(5) Towing rope as on Leopard 1A5 NO MBT
(6) Shackle as on Leopard 1A5 NO MBT
(7) Infantry headset
(8) Eryx anti-tank guided missile installation
(9) Devibration of thermal sight/launching unit for Eryx and ballistic armour plating.
(10) Change NBC filter to NATO standard
(11) MRR installation
(12) Changes to due to configuration freeze (37 items)
(13) Dual component paint
(14) Practice cartridge 30 mm 173 TP
(15) Minimum distance fire-control system of 50 m
(16) Prepared for separate maintenance of each battery group

Fielded by the
Maximum fighting power at minimum risk
Ease of operation
Austrian Army and
Strong sustainable force
Advanced mobility
High reliability under all conditions
by Nato Member Spain.
Superior survivability
State of the art equipment
Family concept ranging from Infantry Combat Vehicle to Light Tank with NATO standard 105 mm gun.
The secure choice.
ASCOD
SPAIN: Head Office
ASCOD A.I.E.
c/o Santa Barbara
Manuel Cortina, 2, E-28010 Madrid, Spain
Phone: +34-91-585 02 36, Fax: +34-91-585 02 44
AUSTRIA:
ASCOD A.I.E.
c/o Steyr-Daimler-Puch Spezialfahrzeug AG & CO KG
2. Haidequerstraße 3, A-1111 Vienna, Austria,
Phone: +43-1-76 0 64, Fax: +43-1-769 81 49

(17) Prepared for GPS
(18) Prepared for laser warner
(19) Fitted with ROVIS (Royal Ordnance Vehicle Information System)
(20) Fitted with Swedish SaabTech Vetronics UTAAS (Universal Tank and Anti-aircraft System) sighting system
(21) Improved protection
(22) Prepared engine for 666 hp power output
(23) Change of IR camera
(24) Electrical heated batteries

Of the 100 production vehicles ordered, 24 were delivered in 1998, 36 in 1999 and 40 in 2000. The four preproduction vehicles have been modified to full production standard and the last of these were delivered in 1999. In addition, the Royal Norwegian Army has an option on another 70 CV 90N vehicles and variants which can be exercised in the future. Recent information has indicated that this option will not be exercised.

The turret of the CV 9030N is produced by Kvaerner Eureka in Norway, while final integration of the turret and the Swedish-built chassis was carried out by Hägglunds Vehicle's Norwegian subsidiary Hägglunds Moelv.

Following an international competition, in mid-1999 the Swiss Army selected a further development of the CV 9030 to meet its future operational requirements. Additional details are given under variants.

Late in 2000, the Finnish Defence Forces selected the CV 9030 to meet its TA2000 requirement. The first contract will be for 57 vehicles for delivery from 2002 to 2005. Additional details are given later in this entry.

Description

The hull of the CV 9030N is of all-welded steel construction with the driver seated front left, power pack to his right, turret in the centre and offset to the left and troop compartment at the rear of the hull.

No details of the protection level of the CV 9030N are available, but over the frontal arc the armour probably provides protection against penetration from 23 mm armour-piercing projectiles at least.

The well-sloped glacis plate and the vertical hull sides allow additional armour to be fitted, if required by the user, to improve battlefield survivability of the CV 9030N.

The driver has a well laid out compartment and normally enters and leaves via a single-piece hatch cover above his position which is hinged at the left. There are three integral day periscopes for driving when closed down, the centre one of which can be replaced by a passive periscope for driving at night.

The driver has a fully adjustable seat and steers the vehicle using a yoke-type steering wheel rather than tillers. As the transmission is fully automatic, driver fatigue and training is much reduced.

The power pack consists of Scania DI14 V-8 diesel developing 605 hp, coupled to a Perkins Engine Company X-300-5N fully automatic transmission with torque converter and lock up clutch giving a total of four forward and two reverse gears.

The well-sloped glacis plate can be quickly opened to allow access to the engine compartment. To keep the profile of the CV 9030N as low as possible the radiators are located in the right side of the hull at the rear. The engine compartment is fitted with a fire detection and suppression system as standard.

The complete power pack of the CV 9030N can be removed from the vehicle in 15 minutes. The exhaust outlet is on the right side of the hull with the air inlet/outlet at the rear.

The upper part of the hull sides are used to stow fuel and subsystems of the CV 9030N, for example, in the left side of the hull is the NBC pack and one of the battery packs, while in the right side of the hull is the cooling system.

Suspension system is of the torsion bar type with each side having seven dual rubber-tyred roadwheels,

Specifications

Crew: 3 + 8
Combat weight: 26,000 kg
Power-to-weight ratio: 23 hp/t
Ground pressure: 0.586 kg/cm²
Length: (hull) 6.549 m
Width:
(over tracks) 3.003 m
(overall) 3.192 m
Height:
(hull top front) 1.625 m
(axis of fire)1.996 m
(overall) 2.779 m
Ground clearance: 0.45 m
Track width: 533 mm
Length of track on ground: 3.98 m
Max road speed:
(forwards) 70 km/h
(reverse) 43 km/h
Acceleration 0 to 30 km/h: 7 s
Fuel capacity: 840 litres
Range: 600 km
Fording: 1.40 m
Gradient: 60%
Side slope: 30%
Vertical obstacle: 1.20 m
Trench: 2.40 m
Engine: Scania DI 16 V-8, 4-cycle diesel developing 680 hp at 2,200 rpm
Transmission: Perkins Engines Company X-300-O automatic with 4 forward and 2 reverse gears
Steering: hydrostatic
Brakes: multiple wet
Suspension: torsion bar
Electrical system: 24 V
Battery groups: 3
Armament:
(main) 1 × 30 mm The Boeing Company Bushmaster II cannon (Finland and Switzerland have selected the more recent Boeing Bushmaster MK 44 30 mm/40 mm cannon)
(coaxial) 1 × 7.62 mm MG3 machine gun
Smoke-laying equipment: 2 × 6 grenade launchers
Ammunition:
(30 mm) 160 ready + 240 stowed
(7.62 mm) 600 ready + 3,200 stowed

Gun control equipment
Turret power control: electric/manual
Gun elevation/depression: +45 to –10°
Turret traverse: 360°
Max elevation/traverse rate: 60°/s
Gun stabiliser:
(vertical) yes
(horizontal) yes
NBC system: yes
Night vision equipment: yes

Note: The combat weight of the CV 9030 series depends on a number of factors including the armour package required and is variable between 24,000 and 28,000 kg.

Hägglunds Vehicle CV 9030CH for the Swiss Army which is very similar to the CV 9030N in production for Norway
2002/0055601

Main armament of the Hägglunds Vehicle CV 9030N consists of a fully stabilised 30 mm The Boeing Company Chain Gun with a 7.62 mm coaxial machine gun and computerised fire-control system and with driver's hatch locked in open position
2001/0099455

drive sprocket at the front and idler at the rear. There are no track-return rollers. The upper part of the suspension is covered by a skirt.

The CV 9030N is fitted with a semi-active track tensioning system as standard and this allows the driver to adjust both tracks at once without leaving his seat. Horstman rotary dampers are fitted at the first, second, sixth and seventh roadwheel stations and give the CV 9030N vehicle an improved ride across country.

The United Defense LP tracks are 533 mm wide and this, together with the length of track in contact with the ground, gives a nominal ground pressure of 0.586 kg/cm^2.

The two-man Hägglunds Vehicle power-operated turret is of all-welded steel construction with the commander seated on the left and the gunner on the right, each of whom has an adjustable seat.

Both turret members have a single-piece hatch cover. The commander's hatch cover opens to the rear, the gunner's hatch cover opens vertically in two positions to allow observation from under the hatch cover. Both turret members can also leave the turret via the turret basket and troop compartment if required.

Turret traverse and weapon elevation is all-electric with manual controls for emergency use. Unlike the Bofors 40 mm cannon fitted to the CV 9040 of the Swedish Army, the 30 mm cannon installed in the CV 9030N is fully stabilised and can be aimed and fired whilst the vehicle is moving across country.

The commander has a SOPELEM sight with a x1, x16 magnification day and a thermal remote display from the gunner's sight and six day periscopes for all-round observation.

The gunner is provided with a SaabTech Vetronics UTAAS which has x1, x8 with integrated laser rangefinder and thermal channel, plus four day periscopes.

Main armament of the CV 9030N selected by Norway is a 30 mm The Boeing Company Bushmaster II Chain Gun with a 7.62 mm MG3 machine gun mounted coaxial to the left.

Other weapons can be fitted such as the German 30 mm Mauser MK 30 cannon or the 35 mm Bushmaster III cannon, if this were required by the customer. Finland and Switzerland have selected the Boeing 30 mm/40 mm MK 44 Bushmaster cannon which will also be installed in the US Marine Corps AAAV (Advanced Amphibious Assault Vehicle).

Two banks of three electrically operated grenade launchers are mounted either side of the turret covering the frontal arc and one Bofors Lyran illuminating grenade launchers can be mounted on the turret rear.

The infantry enters and leaves via a large door in the rear that opens to the right, which is provided with a single vision block with associated firing port in the centre.

The eight infantrymen are seated four each side facing each other, with each man having a bucket-type seat which is also provided with a seat belt.

The CV 9030N has a different troop compartment arrangement from that of the Swedish Army with a single-piece hatch on the left side and a cupola on the right side which has six day periscopes for all-round observation and a single-piece hatch cover that opens to the rear.

Variants

A wide range of variants of the CV 9030N is available along similar lines to those of the CV 9040 covered in a following entry.

CV 9030CH IFV

Late in 2000, the Swiss Parliament approved the acquisition of 186 Swedish Hägglunds Vehicle CV 9030CH Infantry Fighting Vehicles (IFV).

Following extensive trials between the Alvis Vehicles Warrior 2000, the then KUKA Wehrtechnik (now Rheinmetall Landsysteme GmbH) Marder M12 and the Hägglunds CV 9030CH, the latter was selected in August 1999.

The Swiss Procurement Agency is expected to complete negotiations for the acquisition of 186 CV 9030CH IFVs and variants by the end of 2000. Total value will be around £300 million (SEK 4 billion) with deliveries to run from 2002 through to 2005.

Of the order for 186 vehicles about 85 per cent will be in the CV 9030CH IFV variant, with the remaining 15 per cent in the CV 9030CH-COM command post vehicle for use at squadron level. There is also an option for an additional 124 CV 9030CH which will include command post vehicles for use at battalion and brigade level.

Final integration and assembly of the CV 9030CH will be undertaken at the SW Swiss Ordnance Enterprise Corp facility at Thun who will also build and integrate the turret assembly. The chassis will come from Hägglunds Vehicle in Sweden.

In mid-August 2000, Hägglunds Vehicle delivered the CV 9030CH verification vehicle to Switzerland which included all of the modifications and new systems requested as a result of the user trials several years ago.

The CV 9030CH will be partly based on the CV 9030N currently in service with Norway who ordered 104 units with final deliveries taking place later in 2000.

It will however have many improvements, including built-in test equipment, battlefield management system and video links for commander, gunner, driver and squad leader. It will be armed with The Boeing Company 300 mm/40 mm MK 44 cannon which will be manufactured under licence by Oerlikon Contraves of Switzerland. There will be separate competition for the ammunition.

The CH 9030CH IFV will replace the upgraded United Defence M113 series (called the Spz 63/89) fitted with a 20 mm turret. Like many armies, the Swiss Army is to undergo major changes under the Army Structure 21 that is expected to be finalised in the future.

CV 9030FIN

The Finnish Defence Force selected the CV 9030FIN to meet their operational requirements for their TA2000 Infantry Fighting Vehicle programme.They have placed an order for a first batch of 57 vehicles with deliveries to take place from 2002 to 2005. Total value of this contract is about €210 million (about FMk1.3 billion).

Main contractor for the CV 9030FIN is Patria Hägglunds located in Tampere, Finland. This company is jointly owned by Patria Industries of Finland and Hägglunds Vehicle of Sweden.

Under the terms of this contract, Hägglunds Vehicle will be responsible for the system integration and manufacture of the chassis for the CV 9030FIN, while Patria Vehicles will build the turrets and then integrate this with the chassis.

E30 turret for Warrior

The two-person electronically-operated turret installed on the CV 9030 ordered by Finland, Norway and Switzerland is referred to as the E30. This turret is also suitable for installation on other vehicles and in September 2001, was shown on the private venture Alvis Vehicles Warrior 2000 chassis.

Status

Production. In service with:

Country	Quantity	Comment
Finland	57	delivery 2002 to 2005
Norway	104	final deliveries 2000
Switzerland	186	selected 1999, deliveries 2002 onwards

Manufacturer

Hägglunds Vehicle AB (now owned by Alvis plc of the UK).

UPDATED

Bofors/Hägglunds Vehicle Combat Vehicle 90 (Stridsfordon 90)

Development

The Combat Vehicle 90 (CV 90) was designed to meet the specific operational requirements of the Swedish Army for a vehicle with improved armour, mobility and firepower for use by the armed forces' mechanised/ armoured units.

The Combat Vehicle 90 has been jointly developed by Hägglunds Vehicle (now owned by Alvis plc of the UK) and the now Bofors Defence under contract to the Swedish Defence Matériel Administration (the FMV).

Development of the CV 90 started in 1984 and, following trials with an unarmed automotive test chassis, five prototype vehicles and turrets were built for extensive trials under a wide range of operational and environmental conditions.

Following trials, the Swedish Army placed its first contract for the Strf 9040, now called the CV 9040, in March 1991, with the first production vehicle, a CV 9040, handed over to the Swedish Army on 1 November 1993.

The CV 9040 is armed with a Bofors Defence 40 mm L/70 gun and a 7.62 mm coaxial machine gun
0018522

In mid-1993, the Swedish Defence Matériel Administration (FMV) placed its second production contract, worth SKr1,300 million, for a further four versions of the CV 90 family of vehicles for the Swedish Army.

This second contract covered the 40 mm air defence version (TriAD), the CV 90 Forward Observation Vehicle (FOV), the CV 90 Forward Command Vehicle (FCV) and the CV 90 Armoured Recovery Vehicle (ARV), with deliveries taking place from 1995-96 onwards.

In April 1994, the Swedish Defence Matériel Administration placed its third order for a further 150 CV 9040 vehicles.

Late in 1997 the Swedish Defence Matériel Administration (FMV) placed an order worth SKr800 million for an additional 87 Combat Vehicle 90 vehicles in the improved CV 9040 version armed with the Bofors Defence 40 mm L/70 gun.

This was the fourth Swedish Army order for the CV 90 family of vehicles. The most significant improvements to the fourth production batch is the installation of a stabilisation system for the 40 mm L/70 gun to enable moving targets to be engaged with a high first round hit probability. A new suspension system is fitted for improved cross-country mobility.

Under earlier contracts, a total of 600 members of the CV 90 family of vehicles are being built for the Swedish Army and the latest order will extend the production line to the first quarter of 2002.

Prime contractor for the CV 90 family of vehicles for the Swedish Army is HB Development AB.

The Swedish Defence Matériel Administration (FMV) awards contracts to HB Development AB, which in turn

CV 90 Forward Command Vehicle clearly showing two roof-mounted Lyran launchers 0058035

awards contracts to Bofors Defence and Hägglunds Vehicle.

Bofors Defence builds the 40 mm turrets and installs these on the chassis supplied by Hägglunds Vehicle, with the complete system then being delivered to the Swedish Army. Hägglunds Vehicle builds all the Combat Vehicle 90 chassis as well as having complete responsibility for the armoured recovery member of the family.

In the Swedish Army, the CV 90 is issued to new mechanised brigades that have been formed for operations in northern Sweden and in infantry and armoured brigades. Each CV 90 battalion has three mechanised companies with CV 90s, one combined staff and mortar company and one logistics company.

Following extensive trials between the Hägglunds Vehicle CV 9030, the Austrian/Spanish ASCOD and the US United Defense LP M2A2 Bradley infantry fighting vehicle, in April 1995 Norway selected the Hägglunds Vehicle CV 9030 to meet its future requirements.

Details of the CV 9030, which has been developed by Hägglunds Vehicle for the export market and is used by Norway, are given in a separate entry. Norway has now taken delivery of 104 vehicles. Following an international competition, in 1999 the Swiss Army selected the CV9030CH to meet its future requirements with 186 vehicles ordered.

The latest customer for the CV 9030 is Finland who has ordered 57 vehicles under the local designation of the CV 9030FIN with deliveries to run from 2002 through to 2005. Additional details are given in the previous entry.

Description (CV 9040 Infantry Fighting Vehicle)

The hull of the CV 90 is of all-welded steel construction with the driver seated front left, the power pack to his right, the turret in the centre and offset 200 mm to the left, and the troop compartment at the rear of the hull.

No details of the protection level of the CV 90 are available, but armour over the frontal arc probably provides protection against penetration from 23 mm projectiles at least. The well-sloped glacis plate and the vertical hull sides allow additional armour to be fitted if required by the user.

According to the manufacturers, battlefield survivability of the CV 90 is also enhanced by its low profile and compact size, minimal radar and infrared signature, low noise level and reduced vulnerability to electromagnetic pulses. The smooth features of the CV 90 make the vehicle easier to decontaminate.

The driver has a well laid out compartment and normally enters and leaves via a single-piece hatch cover above his position which is hinged at the left. There are three integral day periscopes for driving when closed down, the centre one of which can be replaced by a passive periscope for driving at night.

The driver has a fully adjustable seat and steers the vehicle using a yoke-type steering wheel rather than tillers. As the transmission is fully automatic, driver fatigue and training is much reduced.

The power pack consists of a Swedish-built Scania DI 14 V-8 diesel developing 550 hp, coupled to a Perkins Engines Company X-300-5N fully automatic transmission with torque converter and lock up clutch and a total of four forward and two reverse gears. The engine has further growth potential up to approximately 600 hp. The latest CV 9030 has a Scania D116 V-8 four-cycle diesel developing 680 hp coupled to a Perkins Engines Company X-300-8 fully automatic transmission.

The glacis plate can be quickly opened to allow access to the engine compartment. To keep the profile of the CV 90 as low as possible, the radiators are located in the right side of the hull at the rear. The engine compartment is fitted with a fire detection and suppression system as standard.

The complete power pack can be removed from the CV 90 in 15 minutes. The exhaust outlet is on the right side of the hull with the air inlet/outlet at the rear.

The upper part of the hull sides are used to stow fuel and subsystems of the CV 90; for example, in the left side of the hull is the NBC pack and one of the battery packs while in the right side of the hull is the cooling system.

Suspension is of the torsion bar type with either side having seven dual rubber-tyred roadwheels, drive sprocket at the front, and idler at the rear with no track-return rollers. A track tension adjusting system is fitted as standard on the CV 90 allowing the driver to adjust both tracks at once without leaving his seat. Friction dampers are fitted at the first, second, sixth and seventh roadwheel stations.

The United Defense LP tracks are 533 mm wide and this, together with the length of track in contact with the ground, gives a nominal ground pressure of 0.53 kg/cm^2.

The two-man power-operated turret is of all-welded steel construction, with the commander seated on the left and the gunner on the right, with both crew members having a single-piece hatch cover that opens to the rear.

Turret traverse and weapon elevation are all-electric with manual controls provided for emergency use.

The commander's cupola has six day periscopes to give all-round observation. The front periscope is provided with a wiper and an armoured shutter and, if required, this can be replaced by an enhanced sighting system.

The gunner has a SaabTech Vetronics Universal Tank and Anti-Aircraft System (UTAAS) sight with an armoured shutter. The CV 90 version has both day and night channels and a laser rangefinder. The thermal imager is provided by Kollsman together with the fire-control computer.

Main armament of the CV 9040 comprises a 40 mm Bofors Defence 40/70B cannon with the ammunition being fed from underneath and the empty cartridge cases ejected out of the roof. The magazine holds a total of 24 rounds in three sections of eight rounds with a carousel of 48 rounds under the turret floor. The ammunition mix depends on the tactical situation, but if required each of the three magazines could hold a different type of ammunition with the change from one magazine to another accomplished by a hydraulic device. The 40 mm gun can be aimed and fired by the commander or gunner selecting single shots at 60 rds/min, or automatic with 300 rds/min cyclic rate of fire. Maximum effective ranges are 2,000 m for ground targets and 4,000 m for air targets.

Types of Bofors Defence 40 mm ammunition that can be fired by the 40/70B gun include the well-known Bofors Defence PFHE (PreFragmented High Explosive) Mk 2 for use against aircraft and helicopters, HE-T (High Explosive – Tracer) for engaging ground targets and TP-T (Target Practice – Tracer). It can also fire the MultiPurpose Tracer (MP-T) and the new Prefragmented Programmable Proximity-Fuzed (3P) round.

For engaging combat vehicles at ranges of 2,000 m, Bofors Defence developed the APFSDS-T (Armour-Piercing Fin-Stabilised Discarding Sabot – Tracer) round. This has a tungsten rod penetrator with a length-to-diameter ratio of at least 15:1 and will penetrate over 120 mm of conventional steel armour at combat ranges. Basic details of the 40 mm ammunition are given in Table 1.

Table 1

Type	**APFSDS-T**	**3P**	**PFHE Mk 2**	**MP-T**
Muzzle velocity	1,470 m/s	1,012 m/s	1,025 m/s	1,025 m/s
Weight				
(projectile)	0.500 kg	0.975 kg	0.880 kg	0.940 kg
(explosive)	nil	0.12 kg	0.12 kg	0.105 kg
(round)	2.3 kg	2.5 kg	2.4 kg	2.5 kg
Fuze type	nil	Prog GP	Prox/Impact	Pyrotech
Length				
(projectile)	520 mm	534 mm	534 mm	534 mm
Time to self-destruct	nil	Prog	8.5 s	8.5 s

CV 9040B with commander's stabilised Lemur sighting system **2000**/0080316

First production CV 9040 vehicles had the Bofors Defence 40 mm 40/70 Ba gun, but latest production CV 9040B vehicles have the 40 mm 40/70Bc weapon. The difference between the two versions is that the latest Bc provides higher hit probabilities in the stabilised mode due to its controlled strike-pin release.

A 7.62 mm M/39 machine gun is mounted coaxially on the left side of the turret and this can be fired by the commander or gunner. Mounted either side of the turret is a bank of three Giat Industries GALIX grenade launchers which can fire a wide range of grenades. Fitted on the turret rear are two Bofors Lyran 71 mm mortar tubes which can fire illuminating bombs out to a maximum range of 1,600 m. In addition, the turret is fitted with a GPS system and a KVH compass.

The infantry enters and leaves via a large door in the rear of the hull that opens to the right. The eight infantrymen are seated four each side facing each other with each man having a bucket-type seat. These seats can be quickly folded up to allow the CV 9040 to carry cargo.

Over the top of the troop compartment is a spring-loaded five-part roof hatch which opens to the left (three parts) and to the right (two parts).

CV 90 TriAD air defence version with radar system in elevated position 0018525

Variants

Fire control options

First production CV 9040 IFV for the Swedish Army were fitted with the SaabTech Vetronics stabilised UTAAS (Universal Tank and Anti-Aiircarft System) gunner's sight with day channel, integrated laser range-finder and ballistic reticule. One in three vehicles were also fitted with a thermal imaging channel. They also have 7.62 mm coaxial machine gun and Giat Industries Galix grenade launchers.

Under a second add on contract, a fire-control computer was added for increased first round hit probability against stationary and moving targets and preparation to fire the advanced 3P GP round which is currently only fired from the CV 9040 TriAD.

A third add-on contract gives the CV 9040 a fire on the move capability, this version is called the CV 9040A and entered production in 1997.

The forth add-on contract also provides for an increased hit capability and includes a new elevation drive, coincidence window (closed breech firing) and improved ergonomics for the commander and gunner (especially the gunner's ability to use the full functions of the fire-control systems.

Additional improvements under this contract include a back-up sight, prepared for fitting of the Bofors Defence Lemur sight, prepared for fitting of the VCCS (Vehicle Command and Control System) (FUM/SLB) and improvements to the suspension system. These vehicles are designated the CV 9040B and first production vehicles were completed in the autumn of 1999.

In the future it is expected that all CV 9040A IFV will be updated to a common CV 9040B build standard with the obvious training and logistic advantages.

Following demonstrations of a prototype turret developed by Bofors Defence as a private venture, the Swedish Defence Material Administration (FMV) awarded Bofors Defence a contract for three commander's roof mounted stabilised sights called Lemur.

The Lemur stabilised sight can be configured as required by the customer with the Swedish Army version having a day/thermal capability with integrated laser range-finder. This will enable the CV 9040B to carry out hunter killer target engagements under day/night conditions.

In a typical target engagement the commander would observe and track the target which will then be handed over to the gunner who would carry out the actual target engagement. The commander would then start to acquire a second target while the first target is being engaged. The commander also has a coincidence window back up sight.

The CV 9040B will also incorporate a number of other improvements including the SaabTech Vetronics developed Vehicle Command and Control System (VCCS).

The VCCS is a computerised command and control system which has been designed to assist the commander to make faster, better and more informed decisions based on all relevant data from a combat radio network, fire control system and where applicable, integrated warning sensors.

It provides a single display unit for tactical information and images from available sensors presented as overlays to a background digital map. The VCCS (FUM/SLB) is a tactical command and control system for the battalion level and is not the same system installed on the Leopard 2's (Strv 122) of the Swedish Army.

Up-armoured CV 9040 IFV

Early in 2001 it was revealed that the Swedish Defence Material Agency (FMV) had awarded a contract valued at about SKr200 million (about £14 million) for the supply of passive armour kits. These were to be installed on 55 Bofors Defence/Hägglunds Vehicle CV 9040 infantry fighting vehicles (IFV), currently in service with the Swedish Army.

These CV 9040 IFV were returned to the Hägglunds Vehicle facility at Ornsköldsvik, Northern Sweden, where they were fitted with the new passive armour package supplied by IBD of Germany.

This consists of additional passive armour on the hull front, sides and floor as well as the turret, with internal spall liners also being fitted to minimise possible damage from fragments and projectiles that penetrate the hull or turret armour.

These up-armoured CV 9040s will then be redelivered to the Swedish Army through 2002 for use by the Swedish international rapid reaction forces. These are intended to participate in the European Union (EU) rapid reaction forces from 2003.

CV 9025 IFV

A prototype of this has been built and tested by the Swedish Army but is not being procured.

TriAD (Autonomous Armoured Air Defence) System

This was originally designated the Lvkv A2 but is now known as the TriAD, with the first prototype being handed over for trials with the Swedish Army early in 1992.

The TriAD has not replaced an existing vehicle within the Swedish Army and will be deployed at platoon/battery level.

This has the same hull as the CV 9040 but is modified for air defence, for example anti-helicopter and anti-aircraft roles. A total of 234 rounds of 40 mm ammunition is carried, 24 rounds of which are for ready use with 48 carried in the carousel magazine.

Turret traverse is a full 360° with weapon elevation from –8 to +50°. The commander is seated on the left and the gunner on the right, with the latter having the UTAAS sight as in the first version of the CV 9040. The gunner's UTAAS sight has magnifications of ×1 and ×7, a laser rangefinder plus a monitor for the thermal image camera mounted on the right side of the turret. In addition, the gunner has three observation periscopes.

This version has a seven-man crew consisting of the driver in the hull front, the vehicle commander and gunner in the turret and the other four men in the rear of the vehicle, which, unlike the CV 9040, does not have overhead roof hatches. The other four crew members are the radar operator, combat controller, external co-ordinator and observer/loader. Three of the latter are seated on the right side of the hull, these are the radar operator, combat controller and external co-ordinator.

Mounted on the rear of the turret is the radar, which is the French Thales Air Defence Gerfaut system. The Gerfaut radar has a very high detection rate against aircraft and helicopters with the operator selecting the aircraft or helicopter detection mode.

An IFF system is fitted as standard on this version of the CV 90 vehicle and a maximum of six targets can be prioritised. If required by the tactical situation, the radar can be switched off and target information received from another source.

Specifications

CV 9040
Crew: 3 + 8
Combat weight: 22,800 kg
Power-to-weight ratio: 24.12 hp/t
Ground pressure: 0.53 kg/cm²
Length: (hull) 6.471 m
Width:
(over tracks) 3.003 m
(overall) 3.192 m
Height:
(hull top rear) 1.73 m
(hull top front) 1.640 m
(turret roof) 2.5 m
Ground clearance: 0.45 m
Track width: 533 mm
Length of track on ground: 3.98 m
Max road speed:
(forwards) 70 km/h
(reverse) 43 km/h
Fuel capacity: 525 litres
Gradient: 60%
Side slope: 30%
Vertical obstacle: 1.00 m
Trench: 2.4 m
Engine: Scania DI 14 diesel developing 550 hp
Transmission: Perkins Engines Company X-300-5N fully automatic with torque converter with 4 forward and 2 reverse gears
Steering: cross-drive
Suspension: torsion bar
Electrical system: 24 V
Batteries: 2 battery groups, main group 204 Ah, start group 105 Ah
Armament:
(main) 1 × 40 mm Bofors L/70
(coaxial) 1 × 7.62 mm machine gun
(grenade dischargers) 2 × 3
(Lyran launchers) 2
Ammunition:
(40 mm) 238 rounds
(7.62 mm) 3,000 rounds

Gun control equipment
Turret power control: electric/manual
Gun elevation/depression: +35 to –8°
Turret traverse: 360°
Gun stabiliser:
(vertical) no
(horizontal) no
NBC system: yes
Night vision equipment: yes

The PPI can be set for a range of 7 or 15 km and displays speed, range and course for each target tracked. Automatic tracking is performed after marking the targets by means of a joystick. Automatic threat evaluation is performed by the fire-control computer and displayed on the PPI.

The combat controller analyses the aerial threat situation based on external sources and radar information. In combat he confirms the priority and permits engagement. The Fire-Control Computer (FCC) slews the weapon and sight to the target bearing and performs search in elevation.

The gunner selects ammunition type and function mode for the 3P rounds and performs electro-optic tracking of the target. The FCC predicts lead angles and provides target data for programming of the 40 mm 3P rounds.

When within effective range of the system, the gunner opens fire. Typically a burst of four rounds would be fired to engage the aircraft or helicopter.

Full details of the TriAD are given in *Jane's Land-Based Air Defence 2001-2002.*

CV 90 Forward Command Vehicle (FCV)

This is the command post version of the CV 90 and has extensive communications equipment, seats and equipment for the command staff. This has a two-man turret armed with a 7.62 mm MG. The development contract for this version was awarded in April 1990 and production commenced in mid-1995.

CV 90 Forward Observation Vehicle (FOV)

This FOV is also referred to as the Epbv. It has an extensive array of sensors and communications equipment and is fitted with a two-man 7.62 mm MG turret.

The development contract for this version was awarded in April 1990. Production deliveries commenced in mid-1995. This version is very similar to the forward command vehicle.

CV 90 Armoured Recovery Vehicle (ARV)

The Armoured Recovery Vehicle (ARV) is designated the Bgbv and was developed by Hägglunds Vehicle which previously developed the Bgbv 82 ARV.

Assisting Hägglunds Vehicle is Hagglunds Moelv of Norway which has considerable experience in the design, development and production of recovery vehicles and equipment.

The CV 90 ARV is fitted with two ITAG hydraulic winches for recovery operations, a hydraulic crane and a front-mounted dozer blade to stabilise the vehicle during recovery or lifting operations. Armament comprises a cupola-mounted 7.62 mm MG.

Trials with the prototype, based on one of the original CV 90 prototype vehicles, were completed in 1993. In June 1993 FMV placed a production order and first deliveries were made in mid-1995. Details of this version are given in *Jane's Military Vehicles and Logistics 2001-2002.*

CV 90105 TML

The first prototype of this was completed in mid-1994. It is a modified CV 90 chassis with the new Giat Industries 105 mm TML turret. So far this remains at the prototype stage. Full details were given in *Jane's Armour and Artillery 2000-2001.*

CV 90120 T

Full details of this are given in the Tanks section under Sweden. By mid-2001 this remained at the prototype stage.

AMOS 120 mm mortar system

Details of this twin 120 mm self-propelled mortar system, which could be fitted to the CV 90 chassis, are given in the Self-propelled mortar systems section under International. The first prototype system, based on an Ikv-91 chassis, was completed in mid-1997. The second, on a CV 90 chassis was completed in 1998. The Swedish Defence Material Administration has placed an additional contract for CV 90 series chassis which are earmarked for the AMOS system. Additional details of AMOS are given in the Self-propelled mortar systems section.

Other 40 mm turret applications

For trials purposes it has been proposed that this turret could be fitted on the South African Rooikat (8 × 8) chassis. It has already been installed on the Swiss MOWAG ACV (10 × 10) chassis.

Status

In production. In service with:

Country	Quantity	Comment
Finland	57	CV 9030FIN, delivery 2002-2005
Norway	104	CV 9030N, all delivered by late 2000
Sweden	687	CV 9040 and variants, still being delivered
Switzerland	186	CV 9030CH, ordered late 2000

Manufacturers

Bofors Defence AB.

Hägglunds Vehicle AB (now owned by Alvis plc of the UK).

UPDATED

Hägglunds Vehicle Pansarbandvagn 302 armoured personnel carrier

Development

Late in 1961, Hägglunds and Söner was awarded a contract to develop a new full-tracked amphibious armoured personnel carrier (APC).

The first of two prototypes was completed in December 1962 and in October the following year the company was awarded an initial production contract for 700 Pansarbandvagn 302 APCs (or Pbv 302 for short). The first production Pbv 302s were completed in February 1966 and production continued until December 1971.

The Pbv 302 was never exported and is no longer being marketed.

It is believed that the Swedish Army currently employs 540 Pbv 302 APCs and 203 specialised versions.

Hägglunds and Soner subsequently became Hägglunds Vehicle and in 1997 this was taken over by Alvis plc of the UK.

Description

The hull of the Pbv 302 armoured personnel carrier is made of all-welded rolled steel with the front of the hull providing complete protection against projectiles up to 20 mm in calibre. Above the tracks the hull sides are double-skinned, which gives both increased buoyancy and a measure of protection against attack from HEAT projectiles.

The driver sits at the front of the vehicle in the centre and has a single-piece hatch cover that opens to the rear, in front of which are three day periscopes. The driver is also provided with a collapsible windscreen that is stowed in a box on the right rear door when not required.

The Hägglunds-designed one person turret to the left and slightly to the rear of the driver is armed with a 20 mm Hispano-Suiza cannon with an elevation of +50° and a depression of −10°, turret traverse being a full 360°. The 20 mm cannon has a cyclic rate of fire of 500 rds/min and fires both HE and AP projectiles. Three belts each holding 135 rounds of HE ammunition are carried, plus 10 magazines of AP ammunition, each magazine holding 10 rounds. The gunner has a monocular sight with a magnification of ×8 for use against ground targets. For aerial targets the turret hatch is opened to the rear and the gunner uses the open sights mounted on the weapon itself. There are three day periscopes for forward and one for rear observation.

The commander sits to the right and slightly to the rear of the driver and has a single-piece hatch cover that opens to the rear, plus five day periscopes for all-round observation.

The power pack, which consists of the engine, clutch and transmission, is mounted under the floor of the vehicle. From the transmission power is taken forward to a cross shaft with clutches and disc brakes at either end which provide a clutch and brake steering system. The air inlet, air outlet and exhaust pipe are immediately behind the driver. The Pbv 302 vehicles deployed to the Balkans have been fitted with a large extension to their exhaust pipe so that fumes do not enter the crew compartment.

The nine infantrymen sit at the rear of the vehicle, three on each side facing inwards, one at the front facing forwards and two at the rear back-to-back. The personnel enter and leave the Pbv 302 by two large doors in the rear of the hull.

Over the top of each side of the troop compartment is a hydraulically operated hatch controlled by the commander which, when opened, allows the infantry to use their small arms from inside the vehicle. The hatches are operated by the commander of the vehicle and can be opened halfway if required. A pressure-sensitive bar is fitted along the top of the hull under the

Hägglunds Vehicle Pbv 302C armoured personnel carrier with appliqué armour (Richard Stickland) **2002**/0116789

Bplpbv 3023 armoured fire direction post vehicle in travelling configuration showing access panels in glacis

hatch and if a weapon or hand is in contact with the strip the hatch remains open.

The torsion bar suspension either side consists of five dual rubber-tyred roadwheels with the drive sprocket at the front and the idler at the rear. Hydraulic shock-absorbers are fitted at the first and last roadwheel stations. The tracks are the Hägglunds Vehicle M70 type which have a life of over 10,000 km with one recondition, and have 20 per cent better traction on soft soil than earlier tracks.

The Pbv 302 is fully amphibious being propelled in the water by its tracks. Before entering the water a two trim vane is erected at the front of the hull and the bilge pumps switched on. The Pbv 302 has no NBC system, although one could be fitted.

The basic Pbv 302 can be used, with minor modifications, as a cargo carrier (carrying 2,000 kg of cargo) and ambulance (carrying four stretchers in its basic form and up to six with a special kit).

Variants

Pbv 302 upgrade

In mid-1997, it was stated that the Pbv 302 vehicles were to be upgraded from 2001 through to 2007 so that they can remain in service to at least 2015. By late 2001 no orders had been placed for this upgrade work.

Pbv 302 with appliqué armour

A number of Hägglunds Vehicle Pbv 302 APCs have been deployed to the former Yugoslavia for UN operations. These have been fitted with additional armour of the passive type to their turrets and hulls for additional battlefield survivability. They have also been fitted with spall liners, additional external stowage boxes and a modified exhaust outlet. These are sometimes referred to as the Pbv 302C.

Pbv 302 turrets for 6 × 6 APC

Some of the Finnish built Patria Vehicles XA-203 (6 × 6) armoured personnel carriers delivered to the Swedish Army from October 2001, are fitted with surplus Pbv 302 turrets.

Armoured command vehicle

This is designated the Stripbv 3021 and can be distinguished from the Pbv 302 by its four rather than two radio antennas. The radios are mounted on the right side of the troop compartment. Four radio operators are normally carried plus command staff who also have map boards.

Armoured observation post vehicle

This is designated the Epbv 3022 and is issued on the scale of 10 per armoured brigade. The major difference is that the commander's hatch has been replaced by a new cupola manned by the fire-control officer. The cupola has a combined binocular and optical rangefinder which can be changed between ×4 binocular and ×10 coincidence rangefinder using the same eyepieces. The cupola is hand-cranked in traverse and the line of sight is manually elevated. An azimuth sensor driven by the turret ring gives the accurate azimuth to the target. In the personnel compartment are three radios, two wire links and a navigation system which displays information at one of the operator's positions. Externally the Epbv 3022 is recognisable by its three radio antennas, additional cupola and additional buoyancy aids on the two trim vanes at the front of the hull.

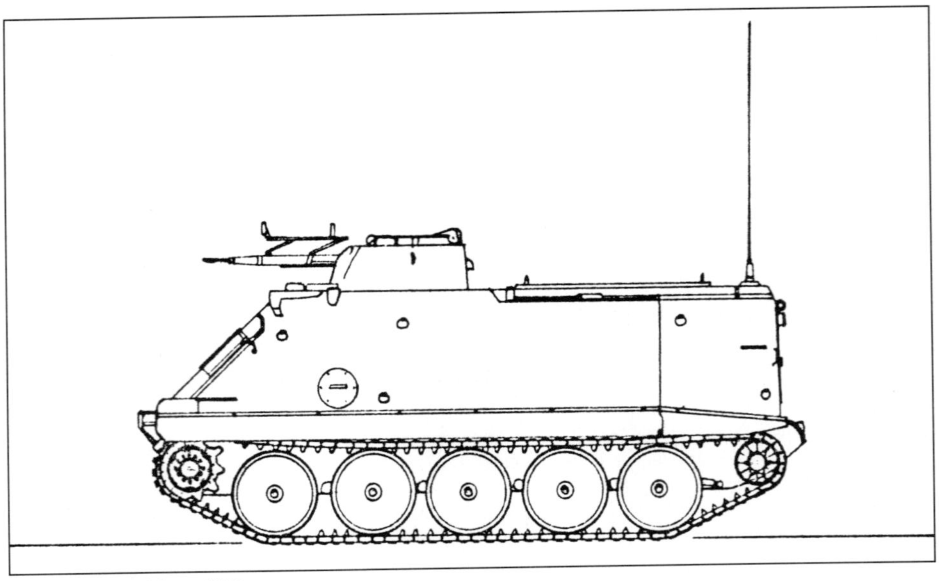

Pbv 302 amphibious APC

Armoured fire direction post vehicle

This is designated the Bplpbv 3023 and has a crew of three (commander, gunner and driver). It also carries a ranging section of seven. Externally it is recognised by its four radio antennas and additional equipment carried on the roof of the troop compartment. Internally it has four radios and a fire direction computer.

Status

Production complete. In service only with the Swedish Army. Upgrade expected to take place in the future.

Manufacturer

Hägglunds Vehicle AB (taken over by Alvis plc in 1997).

UPDATED

Specifications

Crew: 2 + 10
Combat weight: 13,500 kg
Power-to-weight ratio: 20.74 hp/t
Ground pressure: 0.6 kg/cm^2
Length: 5.35 m
Width: 2.86 m
Height:
(to turret top) 2.5 m
(to hull top) 1.9 m
Ground clearance: 0.4 m
Track: 2.42 m
Track width: 380 mm
Length of track on ground: 2.98 m
Max speed:
(road) 66 km/h
(water) 8 km/h
Fuel capacity: 285 litres
Max road range: 300 km
Fording: amphibious
Gradient: 60%
Vertical obstacle: 0.61 m
Trench: 1.8 m
Engine: Volvo-Penta Model THD 100B horizontal 4-stroke turbocharged 6-cylinder in-line diesel developing 280 hp at 2,200 rpm
Transmission: Volvo-Penta T60 manual with 8 forward and 2 reverse gears
Steering: clutch and brake
Clutch: double-disc clutch plate
Suspension: torsion bar
Electrical system: 24 V
Batteries: 2 × 12 V
Armament:
(main) 1 × 20 mm Hispano-Suiza cannon
Smoke-laying equipment: smoke grenade dischargers
Ammunition:
(main) 405 HE plus 100 AP

Gun control equipment
Turret power control: manual
Gun elevation/depression: +50 to –10°
Turret traverse: 360°
Gun stabiliser:
(vertical) no
(horizontal) no
NBC system: no
Night vision equipment: no

Hägglunds Vehicle Pbv 302C armoured personnel carrier with appliqué armour (Richard Stickland) ***2002***/0116790

Epbv 3022 armoured observation post vehicle

Pbv 401 armoured personnel carrier

Development/Description

In August 1991, the Swedish Army decided to purchase from Germany over 1,000 of the Russian designed and built MT-LB family of vehicles which had been used by the former East German Army.

A further five MT-LB vehicles, which had been upgraded to meet a number of Swedish requirements, were purchased from Germany. Three of these were in the armoured personnel carrier configuration, one was an armoured recovery vehicle configuration and the fifth was a further improved ARV with a turbocharged diesel engine.

In 1993, four MT-LBU vehicles were tested, one of these being a mineclearing vehicle.

In November 1993, the final contract was signed with Germany which covered the supply of over 1,000 vehicles, of which 450 were to be renovated and improved in 110 areas to meet the requirements of the Swedish Army. This work was carried out in Germany by Neubrandenburg Fahrzeug Werke.

The first five vehicles were delivered to Sweden in June 1994 and delivery of the main batch soon started at the rate of 20 vehicles a month. According to United Nations sources, deliveries of the MT-LB to Sweden were as follows:

Year	Quantity
1994	9
1995	137
1996	230
1997	152
1998	57
1999	34

Of the 660 MT-LBs purchased, 450 have been renovated as the Pbv 401 with the remaining 210 being delivered complete and scrapped as spare parts.

Modifications to meet the requirements of the Swedish Army include new lighting system, new roof hatch and large box on the roof rear for equipment. The turret-mounted 7.62 mm PKT machine gun has been retained, this having the Swedish designation of the MG 95. All the vehicles have provision for the installation of a 7.62 mm MG 58 machine gun on the roof.

The Pbv 401 can quickly be converted for use as an armoured personnel carrier, command post vehicle or missile carrier, for example the now SAAB Bofors Dynamics RBS 56 Bill.

In total 90 MT-LB armoured recovery vehicles have been purchased and these have the local designation when refurbished of the Bgbv 4102.

Of the 1,059 vehicles purchased, 81 are in the MT-LBU command post version, or ACRV as it is referred to in the west. It is expected that 35 of these will be rebuilt as command post vehicles and 25 as ambulance vehicles.

Finally, 228 122 mm 2S1 self-propelled howitzers were acquired, all of which have been scrapped, with their engines, transmissions and tracks being used to support the MT-LB.

MT-LB with RBS 56 Bill ATGW system

Early in 1996, the now SAAB Bofors Dynamics started firing trials of the prototype Bill Under Armour (BUA) turret installed on the MT-LB full-tracked Armoured Personnel Carrier (APC). This programme has now been cancelled.

Pbv 501 APC

Details of this vehicle, essentially a modified Russian BMP-1 infantry combat vehicle, are given in the previous entry.

Status

Production complete. In service with Swedish Army. Upgrade to chassis is still under way.

Manufacturer

Most of the MT-LBs used by Sweden are believed to have been built in Bulgaria.

UPDATED

Pbv 501 armoured personnel carrier

Development

In May 1997, the Swedish Defence Matériel Administration awarded a contract worth around US$25 million to the Czech Republic VOP 026 repair facility for the modification and renovation of 350 ex-East German Army BMP-1 infantry combat vehicles for the Swedish Army.

These upgraded BMP-1 vehicles, designated the Pbv 501 in Swedish Army service, were delivered to Sweden from 1998 through to 2001 to mechanise the Swedish Army infantry brigades.

Description

A detailed description of the Russian designed BMP-1 is given under the Russian Federation and Associated States earlier in this section. The main differences between the Pbv 502 and the BMP-1 are as follows:

(1) Removal of the launcher rail for the Sagger ATGW over the 73 mm gun 2A28 and its associated control box inside of the turret
(2) Weapon monitor circuits include devices inhibiting use of weapons when the hatches are open
(3) Removal of asbestos elements in certain parts of the vehicle and their replacement by harmless material, while retaining the original functions. This includes brake linings, main clutch lining as well as sealing and insulation
(4) Installation of a new fire detection and suppression system that does not contain Freon, this can be operated manually or automatically.
(5) External lights, including indicators, now conform to NATO standards
(6) Sockets for outside starting, preventive charging of batteries and connection of devices are unified in accordance with NATO regulations
(7) Camouflage system corresponds to NATO regulations
(8) The accumulator batteries are now stored in a hermetic box ventilated outside of the vehicle space
(9) Observation devices are fitted with protective covers
(10) New footsteps, handles and non-slip coverings are installed to ensure safe movement of the crew on the hull
(11) Holders for Swedish weapons including sub-machine guns, machine guns and ammunition
(12) Installation of an independent crew heater
(13) Improved exhaust outlet
(14) Additional radio installation

Specifications

Similar to Russian BMP-1.

Status

Production complete. In service with Swedish Army.

Manufacturer

VOP 026.

UPDATED

Pbv 501 armoured personnel carrier of the Swedish Army **2001**/0099453

Pbv 501 armoured personnel carrier of the Swedish Army **2001**/0099452

Hägglunds Vehicle BvS 10 armoured personnel carrier

Development

In mid-1998, Hägglunds Vehicle, which was purchased by Alvis PLC of the UK late in 1997, launched a new and improved all-terrain armoured vehicle called BvS 10.

The BvS 10 has been developed by Hägglunds Vehicle as a private venture and is being marketed alongside the existing unarmoured Bv 206 (of which over 11,000 have been built for the home and export markets) and the more recently armoured Bv 206 S which is covered in detail in the following entry.

According to Hägglunds Vehicle, the BvS 10 is a new-generation all-terrain armoured vehicle with increased payload, a higher level of armour protection (which can be up to 7.62 mm armour piercing) and a road speed of 65 km/h.

As the BvS 10 has increased load capacity and volume it can be used for a much wider range of battlefield missions than the earlier Bv 206/Bv 206 S.

So far the company has built three examples of the BvS 10, two prototypes and one to production standard which was shown for the first time in mid-1998 at the Eurosatory defence equipment exhibition held in Paris, France.

In early 1999 the BvS 10 was selected to meet the Royal Marines' requirement for a new All-Terrain Vehicle (Protected) for use by 3 Commando Brigade, Royal Marines.

In March 2000, the UK Defence Procurement Agency (DPA) awarded a combined development and production contract worth MSEK 700 to Hägglunds Vehicle for the BvS 10 vehicles to meet the Royal Marines requirement for an All-Terrain Vehicle (Protected) for use by 3 Commando Brigade.

In June 2001, just over 15 months after the UK Defence Procurement Agency (DPA) awarded Hägglunds Vehicle the contract, the first two

BvS 10 in service with Royal Marines and armed with roof-mounted 12.7 mm M2 machine gun (MoD CCR) ***2002***/0116803

BvS 10 in service with Royal Marines and armed with roof-mounted 7.2 mm machine gun (MoD CCR) ***2002***/0116804

preproduction units were delivered to the Royal Marines.

A total of 108 vehicles are being supplied, two preproduction units plus 106 production vehicles which will be delivered from 2003 through to 2005.

The first two vehicles were delivered to the Royal Marines base at Chivenor, North Devon and will undergo 12 months of extensive trials, not only in the UK, but also overseas in areas that the Royal Marines are expected to operate.

The Royal Marines will deploy three basic versions of the ATV(P): Troop Carrying Vehicle (TVC), Command Vehicle (CV) and Repair and Recovery Vehicle (RRV).

The latter will be developed by Hägglunds Moelv AS of Norway who have extensive experience in the design, development and production of specialist vehicles such as recovery vehicles.

The Royal Marines already operate a large fleet of the unarmoured Bv 206 series vehicles and the new and more capable ATV(P), which is called the Viking by the Royal Marines, will now start to supplement these. The front unit can be fitted with roof-mounted 7.62 or 12.7 mm machine guns.

When compared to the current Bv 206, the new vehicle has protection against small arms fire up to 7.62 mm armour-piercing as well as shell splinters and has an increased load carrying ability. It can carry 12 fully equipped troops and has a maximum payload of 3,160 kg and a maximum gross vehicle weight of 10,600 kg.

Alvis Vehicles will supply the armoured bodies to Hägglunds Vehicle while Hägglunds Moelv AS of Norway will develop and manufacture the Recovery Vehicle variant.

Description

The layout of the BvS 10 is similar to the earlier Bv 206 S and consists of two units coupled together by a special articulation mechanism. The front unit has 2.5 m² of cargo space while the rear unit has 10 m² of cargo space.

Both units are of all-welded steel armour construction, which provides protection from small arms fire and shell splinters.

With the standard appliqué armour kit the vehicle is protected against attack from 7.62 mm × 54 armour-piercing rounds (B32).

The BvS 10 consists of two tracked units linked by an ingenious steering mechanism. The front and rear units are articulated in relation to each other, horizontally for steering and vertically for following the contours of the terrain. The automatic transmission, which is also available commercially, ensures smooth gear changes and requires a minimum of driver training.

The power pack is located in the front unit, with four individual seats to the rear, two either side. The driver is seated on the left with the commander on the right and both crew members have a bulletproof window to their immediate front. To their side is a front-opening door with a bulletproof window in the upper part.

To the rear of the driver's and commander's position is another forward-opening door which can be fitted with a bulletproof window in the upper part if required.

The rear unit has a single large door in the rear although there are various options for vision devices. The ten occupants are seated five down either side facing each other on individual seats. Stowage racks are provided on the roof of the BvS 10.

Each unit has a molded rubber band track with an internal cord for additional strength. Either side of the front unit are six rubber-tyred road wheels with the last acting as the idler, drive sprocket at front and return device. Either side of the rear unit are six roadwheels with the last of these acting as the idler, drive sprocket at the front and track-return device.

The payload of the BvS 10 depends on the level of armour protection but typically would be 850 kg for the front unit and 2,250 kg for the rear unit. If cargo is not being carried then four passengers can be carried in the front and 10 in the rear.

Like all other Bv 206/Bv 206 S all-terrain vehicles, the BvS 10 is fully amphibious, being propelled in the water by its rubber band tracks at a maximum speed of 5 km/h.

Optional equipment includes a higher level of armour protection, winch, NBC system, air conditioning system, ring mount for various weapons, power take-off and cyclone preheater. A more powerful diesel engine can also be fitted.

Specifications

Crew: 4 + 10
Max combat weight: 10,500 kg to 11,500 kg*
Unloaded weight: 760 kg
(front unit) 4,670 kg
(rear unit) 3,090 kg
Payload: 2,840 kg*
(front unit) 730 kg
(rear unit) 2,110 kg*
Power-to-weight ratio: 23.80 hp/t
Length: 7.55 m
Width: 2.10 m
Height:
(front) 2.19 m
(rear) 2.08 m
Ground clearance: 0.35 m
Track: 1.48 m
Track width: 620 mm
Length of track on ground: n/av
Max road speed: 65 km/h
Max water speed: 5 km/h
Gradient: 100% (limited by friction)
Range (road): 300 km
Engine: Cummins 5.9 litre 6-cylinder in-line heavy-duty diesel developing 250 hp at 2,500 rpm with 300 hp engine available in future
Transmission: Allison MD 3560 automatic with 6 forward and 1 reverse gears
Transfer box: Hägglunds single speed
Steering: hydrostatic, articulated, damped
Turning radius: 6 m
Electrical system: 24 V
Alternator: 2 × 100 A

* depends on level of protection

Variants

Like the earlier Bv 206/Bv 206 S, the latest BvS 10 is of modular design and various modular subsystems such as add-on armour, weapon mount, load changer and cargo platforms can be provided. In addition to the Troop carrying Vehicle (TCV) the Royal Marines will have at least two other variants: Command Vehicle (CV) and Repair and Recovery Vehicle (RRV).

Status

Prototypes undergoing trials. Selected by UK to meet the Royal Marines requirement for an All-Terrain Vehicle (Protected). Development and production contract awarded in March 2000. First two vehicles were delivered to the UK in mid-2001 with main production run to be delivered from 2003 through to 2005.

Manufacturer

Hägglunds Vehicle AB (owned by Alvis plc of the UK).

UPDATED

Hägglunds Vehicle Bv 206 S armoured personnel carrier

Development

The Bv 206 S was developed by Hägglunds Vehicle (purchased by Alvis plc of the UK late in 1997) and the Swedish Army Matériel Command (FMV) to meet the increased need for all-terrain mobility under armour protection.

The Bv 206 S is essentially an armoured version of the Bv 206 all-terrain vehicle, originally developed to meet the requirements of the Swedish Army and fully described in *Jane's Military Vehicles and Logistics 2001-2002.* Total sales of the Bv 206 had reached some 11,000 units and it was in service with more than 40 countries.

Four prototype vehicles were completed from 1989 and the FMV subsequently ordered two vehicles which were delivered late in 1989.

Following trials with pre-production vehicles configured as a troop carrier, armoured command post vehicle and ambulance, in early 1999 the German Army approved the Bv 206 S for service.

The German Army is expected to place a significant production contract for the Bv 206 S for use by its Rapid Reaction Forces in the near future.

In addition to the troop carrier, armoured command post vehicle and ambulance (fitted out by Binz) a 120 mm mortar carrier will also be procured.

The first significant order for the Bv 206 S was placed by France. In September 1999 it was announced that

Hägglunds Vehicle Bv 206 S all-terrain armoured personnel carrier **2002**/0073380

Hägglunds Vehicle Bv 206 S all-terrain armoured vehicle in ambulance configuration for the German Army (Michael Jerchel) **2002**/0116781

the French DGA had awarded Hägglunds Vehicles a contract worth SKr50 million (£3.8 million) for 12 Bv 206 S APCs for use by French Army units operating in Kosovo. These were delivered in February 2000.

Some of these French Bv 206 Ss are fitted with a roof-mounted 12.7 mm M2 machine gun.

In June 2000, Hägglunds Vehicle was awarded a contract by the Italian Ministry of Defence (MoD) for the supply of two Bv 206 S articulated all-terrain armoured personnel carriers for the Italian Army.

These were delivered late in 2000 and will be used for extensive winter trials prior to the production option being exercised for the main production run, which is expected to be for a total of 58 vehicles.

The Italian Army already operates 90 of the earlier Bv 206 unarmoured vehicles which share many common components with the more recent armoured Bv 206 S.

In 1999 the German Army approved the Bv 206 S for service and has a requirement for over 100 vehicles. The first German order, which is expected to be placed in the near future, will be for around 30 vehicles.

In September 2000, the Spanish Army placed its first order for 10 Bv 206 S vehicles with the total requirement being 50 units for operating in the Pyrennees.

In March 2001, the Swedish Defence Material Agency awarded a contract worth SKr 70 million (£5 million) for the supply of 15 Bv 206 S armour protected all terrain tracked carriers for the Swedish Army.

These Bv 206 S are earmarked for use by Sweden's rapid reaction force for international service which will form part of the European Union's joint force. The order for 15 Bv 206 S comprises three versions, personnel carrier, cargo carrier and ambulance. This is the first major order for Bv 206 S from Sweden.

In September 2001, Spain ordered its second batch of 10 Bv 206 S vehicles bringing to 20 the number of vehicles ordered, out of a total requirement for 50 units.

Specifications

Crew: 4 + 8
Combat weight: 7,000 kg
Empty weight: 5,330 kg
Payload: 1,670 kg
Power-to-weight ratio: 19.42 hp/t
Length: 6.92 m
Width: 2 m
Height overall:
(rear body) 2.0 m
(front body) 2.10 m
Ground clearance: 0.35 m
Track width: 620 mm
Max road speed: 52 km/h
Max water speed: 4.7 km/h
Range: (road) 300 km
Gradient:
(hard surface) 100%
(snow) 31%
Engine: Steyr M16 direct-injected 6-cylinder in-line diesel developing 186 hp at 4,300 rpm
Transmission: Mercedes-Benz W4A 040 automatic with 4 forward and 1 reverse gears
Steering: hydrostatic, articulated
Swept diameter: 8 m
Electrical system: 24 V
Batteries: 2 × 12 V 800 CCA, RC 120 min, 56 Ah
NBC system: no
Night vision equipment: no

Description

The chassis, drive train and steering system on the Swedish Army configuration and are identical to the Bv 206, but the all-welded steel bodies and the cooling system are new. For other customers, new enhancement has been accomplished on sub-systems such as steering, brakes and track assemblies.

The basic Bv 206 S consists of two tracked units which are linked together with a steering unit. The front unit has seats for the driver and three passengers and the rear unit has eight seats.

The track assemblies are attached to the central beam by two transversal leaf springs. Each track assembly is built up around a tubular bar which carries the sprocket assembly, roadwheels and idler.

In each track assembly there are four pairs of roadwheels on trailing arms sprung by rubber tension springs. The idler at the rear with a tensioning device is also supported by a rubber spring. All four track assemblies of the vehicle are identical and interchangeable. The tracks are rubber with longitudinal textile cord and integral steel profile reinforcements.

Steering is accomplished by changing the direction between the front and rear unit by two hydraulic cylinders, servo-controlled from a conventional steering wheel. The hydraulic system is built with commercially available components and the steering system is designed to permit a large freedom of movement between the two bodies.

The engine and transmission are mounted in the front unit. A shaft connects the gearbox with the drop-down gearbox. A disc brake is mounted in front of the drop-down gearbox inside the body. Cardan shafts transmit power to the final drives on the front end of both chassis.

The all-welded steel armour of the Bv 206 S provides the crew with protection against 7.62 mm ball ammunition at point blank range as well as shrapnel from artillery fire. All windows have the same protection level as the body. The front unit also has a circular roof hatch that opens to the rear. The rear unit has a large door in the rear that opens to the right and a roof hatch that opens to the right. This door has a single window in its upper part.

The Bv 206 S is fully amphibious, being propelled in the water by its tracks and has been designed to operate in temperatures from –46 to +49°C.

The Bv 206 S can be equipped with a ring mount for a 7.62 or 12.7 mm machine gun or a 40 mm grenade launcher.

The Bv 206 S is fully air-portable and can be transported inside a CH-47 or CH-53 helicopter. It can also be air-dropped from fixed-wing aircraft.

Variants

In addition to the basic armoured personnel carrier, other versions suggested by the manufacturer include mortar carrier, command post vehicle and recovery vehicle. A Bv 206 S front unit with a modified standard Bv 206 rear unit has been used for trials with a 140 mm twin recoilless rifle system, the latter being mounted on top of the rear unit.

Status

Production. In service with:

Country	Quantity	Comment
France	12	delivered 2000
Germany	100	order for first batch of 30 due soon, 100 is total requirement
Italy	60	2 + 58, first deliveries late 2000
Spain	50	10 ordered September 2000, total requirement is for 50 units
Sweden	2+15	first customer (2), another 15 ordered in 2001

Manufacturer

Hägglunds Vehicle AB (taken over by Alvis plc late in 1997).

UPDATED

SWITZERLAND

Swiss Army M113 series armoured personnel carriers

Development/Description

The Swiss Army operates a fleet of about 1,475 United Defense LP, Ground Systems Division, M113 series Armoured Personnel Carriers.

The basic APC is called the Schützenpanzer 63 by the Swiss Army but many have been fitted with the Swedish Hägglunds Vehicle turret armed with a 20 mm cannon. This turret is also fitted to the Swedish Hägglunds Vehicle (now owned by Alvis plc of the UK) Pbv 302 full-tracked armoured personnel carrier. When fitted with this one-man turret, the vehicle are called the Schützenpanzer 63/73. It is understood that a total of 192 of these vehicles is in service.

Other variants in service include:

(1) Engineer fitted with dozer blade called the Geniepanzer 63

Swiss Army 120 mm self-propelled mortar which is designated the Mw Pz 64 ***2000***/0080319

Upgraded Spz 63/89 armoured personnel carrier showing appliqué armour and smoke grenade launchers 0018528

Command post vehicle is designated the Kdo Pz 63/73 0018529

Upgraded Spz 63/89 from the rear showing rear-mounted fuel tanks 0058039

(2) 120 mm self-propelled mortar called Minenwerferpanzer 64 (Mw Pz 64) with 132 in service
(3) Command vehicle designated the Kommandopanzer Schützenpanzer 63 (Kdo Spz 63)
(4) Command vehicle with turret-mounted 20 mm cannon called the Kommandopanzer 63/73 (Kdo Pz 63/73)
(5) Repair vehicle called the Kranpanzer 63
(6) Wireless vehicle called the Ubermittlungspanzer 63
(7) Artillery command vehicle called the Feuerleitpanzer 63

In 1986, it was decided that 316 of the Schützenpanzer 63/67 vehicles and 66 of the command post equivalent would be upgraded to enable them to operate with the Swiss Army Leopard 2 MBTs.

A total of SFr124 million was allocated for this programme. The modifications to the vehicles were extensive and included additional passive armour protection to the hull, installation of a bank of three electrically operated smoke grenade launchers either side of the hull front, more powerful Detroit Diesel engine and external fuel tanks either side of the rear hull ramp.

These kits were provided by United Defense LP and when upgraded became the Schützenpanzer 63/89. This programme commenced in 1990 and was completed by 1994 with the work carried out at the Federal Armament Factory in Thun which is today known as SW Swiss Ordnance Enterprise Corporation.

From 2002 these upgraded M113s will be supplemented by the Swedish Hägglunds Vehicle CV 9030CH with 186 to be delivered from 2002 onwards. Hägglunds Vehicle will supply the chassis with facilities in Switzerland at Thun undertaking production of the turret and integration of this with the CV 90 chassis.

M113 in mineclearing role

In 2001 the French state owned Giat Industries was awarded a contract by Switzerland for the supply of 33 of its Demeter 2 vehicle signature duplication system, which is used to detect and neutralise anti-tank mines.

The Demeter 2 is the latest version of the system which has been in service with the French Army for several years. It is installed on special AMX-30B2 DT mineclearing tanks which are also fitted with Pearson Engineering Track Width Mine Ploughs (TWMP).

The system radiates a false magnetic field in front and to the side of the carrier vehicle, causing magnetically fuzed anti-tank mines to detonate before the actual vehicle reaches them. According to Giat Industries, Demeter 2 coverage extends a distance of 5 m in front of the vehicle and 2.5 m either side.

The system is controlled from within the carrier vehicle, which typically has two Demeters mounted one either side at the front of the vehicle. It is normally used in conjunction with mineclearing devices of the plough or roller type to allow for maximum level of mine clearance.

The Swiss Army will install these Demeter 2 mine neutralisation systems onto modified United Defense M113 series full-tracked armoured personnel carriers.

Delivery is expected to be made to Switzerland between 2001 and 2002. This is the first export order for the Demeter 2 mine clearance system.

UPDATED

TAIWAN

Armoured Infantry Fighting Vehicle

Development

The Armoured Infantry Fighting Vehicle (AIFV) was developed by the Republic of China Armoured Fighting Vehicle Development Centre in the 1970s and the first prototype was completed in 1979. The Armoured Infantry Fighting Vehicle was manufactured by the Republic of China Fighting Machines Command which is responsible for final assembly and testing. There were approximately 40 companies involved in supplying major components for the AIFV, including the Taiwan Machinery Manufacturing Corporation (suspension and power plant) and the Taiwan Aluminium Corporation (hull and tracks). This is understood to have the local designation CM-21 while the variant with the Raytheon Systems Company TOW ATGW is referred to as CM-25. This is the standard TOW ATGW system, not the M901 originally developed to meet the operational requirements of the US Army.

Description

The AIFV is based on the US United Defense LP M113 APC which has been in service in Taiwan for many years but incorporates features of the M113A2, M113A3 and the private venture United Defense LP AIFV.

The layout of the AIFV is similar to the M113 series with the driver at the front left, engine compartment to his right and the troop compartment at the hull rear.

The hull is of all-welded aluminium armour construction with an additional layer of laminate steel armour bolted to the hull front, sides and rear. Closed cell polyurethane foam is used to fill the space between the two layers of armour and this gives increased buoyancy for amphibious operations.

The driver has a single-piece hatch cover that opens to the rear and day periscopes for observation. The engine compartment has a firefighting system. The vehicle is powered by a British Perkins TV8.640 diesel developing about 215 bhp, coupled to a transmission developed by the Institute of Industrial Technology in Taiwan, and is similar to the M113A1. The air inlet, air outlet louvres and exhaust pipe outlet are in the roof of the vehicle.

Armoured Infantry Fighting Vehicle (DTM)

Armoured Infantry Fighting Vehicle (L J Lamb)

The commander sits to the rear of the driver and has a cupola which can be traversed through a full 360°. This is provided with a single-piece hatch cover, day periscopes for all-round observation, a 12.7 mm M2 HB machine gun and a shield that gives front and side protection against small arms fire. Either side of the shield is a bank of three electrically operated grenade dischargers firing forwards.

The troop compartment is at the rear of the vehicle and the infantry enter and leave via a power-operated ramp in the hull rear. This has an emergency door in the left side. There are five firing ports, each with a day periscope or a vision block provided, two in each side of the hull and one at the rear. These allow the infantry to use its 5.56 mm weapons from within the vehicle in safety. It is believed that between six and eight infantrymen are carried plus the crew of two (commander/gunner and driver). In addition, there is a hatch in the troop compartment roof.

The suspension is of the torsion bar type and consists either side of five rubber-tyred roadwheels with the drive sprocket at the front and idler at the rear. There are no track-return rollers. The upper part of the suspension of the vehicle is covered by a rubber skirt.

The electrical system is similar to that of the M113. Standard equipment includes bilge pumps and night vision equipment. The AIFV is fully amphibious being propelled in the water by its tracks. Before entering the water, bilge pumps are switched on and a trim vane erected at the front of the hull. An NBC system is not installed at present.

An armoured fuel tank, similar to that offered by United Defense LP for the M113A2, is fitted either side of the hull ramp at the rear.

Physical and performance characteristics of the AIFV are similar to those of the M113A2.

More recently another version of the basic CM 31 has been revealed called the CM 21A1. This is believed to have a number of automotive improvements. The latest CM 21A2 has a new and enhanced armour package.

Variants

The following variants are believed to be in service or under development:

Mortar carrier (81 mm, 107 mm and 120 mm)

Command post (command, communications and fire direction)

AIFV with 20 mm or 30 mm cannon

AIFV with Kung Feng IV 126 mm multiple rocket system; details of this version are given in the Multiple rocket launchers section.

AIFV with Raytheon System Company TOW ATGW system. Maximum range of the TOW anti-tank missile is 3,750 m

AIFV with Kuen Wu ATGW. In many respects this is very similar to the Russian AT-3 Sagger ATGW system.

Others could include fire support vehicle, reconnaissance vehicle, flamethrower, cargo carrier and ambulance.

Status

Production probably complete. In service only with the Taiwanese Army. It is believed that total production of this vehicle amounted to between 200 and 300 vehicles. To this must be added over 650 US-supplied M113 series vehicles and numerous specialised models.

Manufacturer

Fighting Machines Command, ROC Army.

UPDATED

TURKEY

FNSS Defense Systems Armoured Combat Vehicle – New Generation

Development

Early in 2000, FNSS Defense Systems revealed that they had developed, as a private venture, the Armoured Combat Vehicle - New Generation (ACV-NG) to meet potential future home and export requirements for a vehicle with increased armour, mobility and greater internal volume.

The baseline for the ACV-NG is the FNSS Defense Systems Armoured Infantry Fighting Vehicle (AIFV) of which 1,698 have been delivered to the Turkish Land Forces Command (TLFC) in four basic versions. Full details of this vehicle, which has also been selected or is in service with Malaysia and the United Arab Emirates, are given in the following entry.

FNSS Defense Systems is a joint venture company formed between United Defense LP of the US and the Turkish Company of Nurol Holding and was founded in 1989 specifically for the Turkish ACV programme, which was the largest armour programme undertaken in Turkey.

The first prototype of the ACV-NG was completed in late 1998 and tested by the US Army with the second prototype completed in the year 2000.

The ACV-NG is essentially an enhanced version of the AIFV with an additional roadwheel either side, slightly larger hull, upgraded suspension and a new power pack.

It draws upon the considerable experience obtained by United Defense LP in the enhanced version of the Mobile Tactical Vehicle Light (MTLV). This in turn is a further development of the M113, the most widely deployed vehicle of its type in the world.

Like the M113, the ACV-NG is fully air portable in a Lockheed Martin C-130 Hercules transport aircraft and retains its amphibious characteristics with the baseline vehicle, capable of carrying a section of infantry together with sufficient supplies for 72 hours.

A second model of the Armoured Combat Vehicle is now being built by FNSS Defense Systems and this should be completed in 2002.

Description

In appearance it is almost identical to the AIFV with the driver seated front left with power pack to the right. The driver has a one-piece hatch cover that opens to the right and four day periscopes, the forward one of which can be replaced by a passive periscope for driving at night.

There is considerable flexibility in the roof-hatch arrangement depending on the turret fitted. If fitted with the Boeing 25 mm Sharpshooter turret, this is installed on the right with commander being seated to the immediate rear of the driver with a single-piece hatch cover and periscopes for all round observation.

In the roof compartment, to the immediate rear of the turret, is a one-piece hatch that opens to the rear. The troops sit on a bench-type seat positioned in the centre

Armoured Combat Vehicle - New Generation fitted with a modified two-person Bradley armed with a 30 mm cannon and 7.62 mm co-axial machine gun ***2001***/0067712

Armoured Combat Vehicle - New Generation fitted with turret with four Hellfire ATGW in ready to launch position (Christopher F Foss) ***2001***/0099451

of the troop compartment facing outwards with firing ports and vision devices being provided in each side of the hull. The troops enter and leave by a power-operated ramp in the rear that also has a door.

The current ACV-NG is powered by a Detroit Diesel series 6V-53T turbo-charged diesel developing 350 hp coupled to an Allison X-200 series fully automatic transmission. This gives a maximum road speed of between 65 to 75 km/h and an acceleration of 0 to 30 km/h in 7.5 sec.

An alternative engine is the electronically tuned version, the 6V-53TIA which develops 400 hp and also features built-in prognostics and diagnostics. Other power packs are however possible, for example an MTU diesel coupled to a ZF automatic transmission.

The suspension upgrade includes improved final drives, more aggressive suspension with more than 380 mm of roadwheel travel, which is said to give improved cross-country mobility. Suspension either side consists of six roadwheels with the drive sprocket at the front and idler at the rear. There are no track return rollers. The upper part of the suspension is covered by a skirt.

The first example of the ACV-NG is fitted with a modified version of the United Defense LP two-person Bradley IFV turret. This has the twin Raytheon Systems Company TOW Anti-Tank Guided Weapon (ATGW) launcher on the left side of the turret removed and the Boeing 25 mm M242 cannon replaced by the Boeing 30 mm Bushmaster II cannon with the 7.62 mm co-axial machine gun being retained.

A day/thermal sighting system is provided and as the main armament is stabilised, the gunner can aim and fire the main armament while moving across country.

Other turret options are available including the FNSS Defense Systems and United Defense LP jointly developed one-person Sharpshooter turret armed with a stabilised 25 mm M242 cannon and 7.62 mm machine gun and other small turrets and cupolas armed with 7.62 mm or 12.7 mm machine guns.

Compared with an M113/AIFV, the ACV-NG has 30 per cent more under armour volume and 70 per cent more load carrying capability. The internal seating depends on the weapon system fitted, but with a one-person turret, it would carry eight to ten troops plus driver, commander and gunner with firing ports being optional. Maximum combat weight depends on weapon fit but would be between 18 and 20 tonnes.

The baseline ACV-NG has a passive armour package that provides protection against attack from small arms fire up to 14.5 mm Armour Piercing (AP) through a full 360° and artillery projectiles up to 155 mm in calibre bursting overhead.

If required, internal spall liners can be provided as can additional armour to protect the occupants from cannon fire up to 30 mm in calibre and anti-tank weapons such as the RPG-7 firing a rocket propelled grenade with a single HEAT (High-Explosive Anti-Tank) warhead.

Special protection has been provided against mines. This includes a 18 mm thick rolled homogenous armour plate under the belly and floor spall liners. The sponsons have also been upgraded with armour with additional protection for the driver, while collapsible seats help to absorb some of the shock.

Variants

In addition to the IFV variant, a number of other variants have been proposed. These include the 120 mm self-propelled mortar carrier fitted with either the 120 mm Armoured Mortar System from BAE Systems, RO Defence/Delco Defense or the 120 mm TDA 2RM mortar system.

Other specialised versions of the ACV-NG could include a reconnaissance vehicle with a one-person turret armed with 25 mm cannon and 7.62 mm machine gun and a mast-mounted sensor pod, artillery forward observation vehicle and command post vehicle.

It can also be fitted with ATGW systems such as the 3,750 m range TOW or the longer range Hellfire, surface-to-air missiles such as Stinger and even a Cockerill two-person turret armed with a 90 mm gun.

FNSS Defense Systems have stressed that there is plenty of stretch in the ACV-NG design and that new technology will be inserted as it becomes available, including significant survivability enhancements.

These could include signature management techniques to reduce the radar signature of the vehicle. This will involve reshaping the vehicle exterior with appliqué panels and/or new passive armour tiles.

To save weight and reduce the internal and external noise of the vehicle, serious consideration is being given to the fitting of composite tracks of the rubber band type that have been tested on M113 series vehicles with great success.

Standard equipment includes fire detection and suppression system for the crew and engine compartment and NBC system of the face mask or overpressure type.

To enable the ACV-NG to be integrated into the digitised battlefield, a data bus can be installed in the vehicle together with a commander's display unit and BCIS. A defensive aids suite including a laser warning system could be fitted.

In September 2001, the prototype of the ACV-NG was shown fitted with the Helio SWARM turret armed with a 12.7 mm Boeing Chain Gun.

Specifications

Not available.

Manufacturer

FNSS Defense Systems.

UPDATED

FNSS Defense Systems Turkish infantry fighting vehicle

Description/Development

To replace its current fleet of armoured vehicles, many of which are over 30 years old, the Turkish Army asked 11 companies worldwide to submit detailed proposals for a new infantry fighting vehicle family. Seven companies responded. A key part of this proposal was technology transfer to Turkish industry with 50 to 70 per cent being built in Turkey.

In 1986, three vehicles were shortlisted, the United Defense LP Armored Infantry Fighting Vehicle (25 mm cannon), the GKN Defence (now Alvis Vehicles) Warrior (30 mm RARDEN cannon) and the Krauss-Maffei/Diehl Puma (20 mm cannon).

These underwent extensive mobility, high-altitude, cold weather, road, cross-country and various obstacle trials in Turkey and the United Defense LP Armored Infantry Fighting Vehicle was subsequently selected.

In June 1988, FMC-NUROL Savunma Sanayii AS (FNSS), a joint venture company between United Defense LP (sold to Carlyle Group by FMC Corporation in 1997) of the USA and the Turkish-based NUROL Group, signed a contract to produce the Armored Infantry Fighting Vehicle (AIFV) to meet the requirements of the Turkish Army.

However, with the purchase of United Defense LP by the Carlyle Group from the FMC Corporation, it was deemed inappropriate for FNSS to maintain the original name of FMC-Nurol. The Board of Directors issued a resolution to change the name of the company. As from 15 August 1998 the company name has been changed to FNSS Defense Systems Inc (FNSS Savunma Sistemleri AS in Turkish).

The original contract called for the delivery of the vehicles over an eight year period but, in 1995, this was extended to 10 years. The contract had a total value of US$1.076 billion to manufacture 1,698 armoured combat vehicles to meet the requirements of the Turkish Army.

The project is being supervised by the Undersecretariat for Defence Industries of Turkey with the new purpose-built factory built at Golbasi, near Ankara.

Assembly started in August 1990 and production late in 1991. Local content increased to 76 per cent by the end of the last programme year from 24 per cent, which was achieved at the end of the first programme year.

Late in 1989, Cockerill Mechanical Industries was awarded a contract to build 285 hulls for the Turkish programme and subsequently provide the equipment required to build the remaining 1,413 hulls in Turkey. Delivery of the hulls was spread over a three year period with the first hulls completed in Belgium late in 1990.

The first 135 AIFVs were fitted with a DAF Special Products (today called SP aerospace and vehicle systems) one-man turret armed with an Oerlikon Contraves 25 mm cannon and a 7.62 mm coaxial machine gun.

In May 1992, it was stated that future vehicles would be armed with a French Giat Industries one-person Dragar turret armed with a 25 mm Giat Industries M811 cannon and a 7.62 mm coaxial machine gun. The installation of Giat turrets on to the AIFVs began in 1997.

The total order is for 515 turrets of which the first 67 came direct from France with the remainder being manufactured in Turkey, NUROL building the turret and MKEK the 25 mm cannon.

The Turkish order comprised a total of 1,698 vehicles in four different versions:

Model	**Total**
AAPC	830
AIFV	650
ATV	48
AMV	170

Armoured TOW Vehicle (ATV)

Armoured Mortar Vehicle (AMV)

0018531

AAPC fitted with one-man turret armed with 12.7 mm and 7.62 mm machine guns

AAPC fitted with pintle-mounted 12.7 mm MG

The order for the 1,698 vehicles being built in Turkey comprised the following basic variants:

Armoured Infantry Fighting Vehicle (AIFV)
A total of 650 vehicles were built and fitted with a one-man turret armed with a 25 mm cannon and a 7.62 mm coaxial machine gun. Total of 13 personnel carried with a maximum combat weight of 13,687 kg.

Advanced Armoured Personnel Carrier (AAPC)
A total of 830 vehicles built and fitted with a one-man turret armed with a 12.7 mm and a 7.62 mm machine gun. Total of 13 personnel carried, with a maximum combat weight of 12,948 kg.

Armoured Mortar Vehicle (AMV)
A total of 170 vehicles built and fitted with a turntable-mounted 81 mm mortar and also armed with a 7.62 mm machine gun. Total of six personnel carried and a maximum combat weight of 12,965 kg.

Armoured TOW Vehicle (ATV)
A total of 48 vehicles built and fitted with the Norwegian ALT one-man turret with two TOW ATGWs in the ready to launch position. Total of four personnel carried, with a maximum combat weight of 13,715 kg.

Specifications
AIFV
Crew: 13
Combat weight: 13,687 kg
Power-to-weight ratio: 21.92 hp/t
Ground pressure: 0.67 kg/cm²
Length: 5.26 m
Width:
(overall) 2.82 m
(over tracks) 2.54 m
Height:
(hull roof) 2.01 m
(with turret) 2.62 m
Ground clearance: 0.43 m
Track width: 381 mm
Length of track on ground: 2.68 m
Fuel capacity: 416 litres
Max road speed:
(forwards) 65 km/h
(reverse) 11 km/h
Max water speed: 6.3 km/h
Max road range: 490 km
Fording: amphibious
Gradient: 60%
Side slope: 30%
Vertical obstacle: 0.74 m
Trench: 1.83 m
Engine: Detroit Diesel Model 6V-53T developing 300 hp
Transmission: Allison X-200-4 crossdrive with 4 forward and 1 reverse gears
Steering: hydrostatic
Suspension: torsion bar
Electrical system: 24 V
Armament:
(main) 1 × 25 mm cannon
(coaxial) 1 × 7.62 mm MG
Smoke-laying equipment: 6 smoke grenade launchers

Gun control equipment
Turret power control: powered/manual
Gun elevation/depression: +50 to –10°
Turret traverse: 360°
Gun stabiliser:
(vertical) optional
(horizontal) optional
NBC system: optional
Night vision equipment: optional

Additional Turkish order
Deliveries of the original 1,698 vehicles for the Turkish Land Forces Command have now been delivered. An additional order has now been placed for a further batch of 551 vehicles which will be delivered from late 2001. All of these will be in the AAPC configuration. These will be delivered to the Turkish Land Forces Command between December 2001 and December 2004 and be armed with a 12.7 mm M2 machine gun.

Private Venture Variants

120 mm Armoured Mortar Vehicle (AMV)
The chassis has been modified to incorporate a TDA 120 mm Recoiling Rifled Mounted Mortar 2R 2M in the rear of the vehicle. Additional details of this system are given in the Self-propelled mortar systems section.

The 120 mm mortar, which is provided with a semi-automatic loading system, has a traverse of 220° left and right and an elevation from +42 to +85°. The mortar fires through a two-part roof hatch that opens left and right.

A prototype of this system was completed several years ago and it has been demonstrated in the Middle East. For this programme, FNSS developed a vehicle which incorporated a new hatch and base plate.

SPM-120 Self-Propelled Mortar Vehicle (SPMV)
In addition to the above 120 mm AMS, FNSS Defence Systems have also trialled the Swiss 120 mm Bighorn smoothbore mortar system. This is fitted with a semi-automatic loading system and the mortar has a traverse of 45° left and right with elevation from +40 to +85°. As of October 2001, the 120 mm Bighorn mortar system remained at the prototype stage.

AIFV with stabilised 25 mm cannon
FNSS Defense Systems has developed a stabilised turret (called Sharpshooter) for the AIFV which is electrically operated and controlled and is armed with a 25 mm The Boeing Company M242 Chain Gun and 7.62 mm coaxial machine gun. This has been developed as a private venture, with the first export customer being Malaysia.

ACV + Hellfire ATGW
To meet the requirements of potential customers in the Middle East for a vehicle mounted long range anti-tank guided weapon (ATGW) system, FNSS Defense Systems of Turkey and Lockheed Martin Missiles and Fire Control of the US have teamed to market the Hellfire ATGW on the Turkish Armoured Combat Vehicle (ACV) chassis.

The Manned Hellfire Turret System (MHTS) has been developed as a private venture by Lockheed Martin Missiles and Fire Control and is a self-contained weapon system with four Hellfire long-range ATGW in the ready-to-launch position.

For trials purposes the MHTS has already been installed on an AV Technology (now owned by General Dynamics Land Systems) Pandur (6 × 6) light armoured vehicle and successfully demonstrated in Kuwait late in 1997.

AIFV with Giat Industries 25 mm Dragar turret **2000**/0080320

ACV-ENG for UAE with air conditioning and NBC system, this is also referred to as the ACV 350 **2000**/0080321

Export ACV in IFV configuration and fitted with Sharpshooter turret, armed with Boeing 25 mm cannon and 7.62 mm coaxial machine gun **2002**/0116805

Export ACV in Armoured Recovery Vehicle configuration with hydraulic crane installed on the left side of roof **2002**/0116806

When being launched from the MHTS in the autonomous mode the Hellfire missiles have a maximum range of 5,000 m but when used in the co-operative mode with a forward observer this is increased to over 8,000 m.

Targets can be engaged under day and night conditions and the AGM-114 missiles are the same as that launched from the AH-64, AH-1W and S-70B helicopters.

The turret has powered traverse through a full 360° with elevation from −10 to +20°. In addition to the four missiles in the ready-to-launch position another four missiles are carried internally for manual reloading.

Although the MHTS was recently shown to potential customers installed on the current production ACV chassis which has five road wheel stations either side, it is envisioned that production systems would be fitted to the latest ACV-NG (New Generation) chassis.

Armoured Combat Vehicle – New Generation

This was completed late in 1999 and has a stretched chassis with six road wheels either side and not only increased internal volume but also has increased mobility and a modular armour package. Available details of this vehicle, which remains at the prototype stage, are given in the previous entry.

UAE vehicles

In 1997 the United Arab Emirates awarded FNSS Defense Systems a contract for 136 specialised variants with first deliveries being made in 1999.

This covered the supply of 53 ACV Engineering Squad Vehicle (ACV-ENG), eight repair/recovery vehicles fitted with a crane and winch and 75 artillery support vehicles.

The UAE ACV were the latest production standard and have the company designation of the ACV 350 with the latter figure indicating the rating of the engine in hp.

The UAE vehicles are a new enhanced tropicalised version of the AAPC and are fitted with a Detroit Diesel 6V-53TA engine developing 350 bhp coupled to an Allison X-200-4B automatic transmission.

A new high temperature cooling system is fitted that allows the vehicle to operate in ambient temperatures of up to +52°C.

The ACV-ENG has a crew of three, commander, gunner and driver and has been designed to transport a squad of combat engineers and their specialised equipment. Variants of the ACV-ENG include command post and reconnaissance.

The vehicles are fully amphibious, being propelled in the water by their tracks, and also feature a new FNSS Defense Systems developed weapon station which allows the gunner to aim and fire the 12.7 mm M2 machine gun under full armour protection. Mounted at the front of the hull is a bank of six electrically operated smoke grenade launchers.

Standard equipment includes an NBC system of the overpressure type, full air conditioning and external stowage on the sides of the vehicle for specialised engineer equipment.

The artillery support vehicles includes a specialised Artillery Forward Observation Vehicle (ACV-FOV) which will be used to acquire targets for self-propelled artillery systems. This has a mast-mounted sensor pod that includes day/thermal cameras, laser range-finder and surveillance radar. Final deliveries took place in late 2000.

Malaysian vehicles

Late in 2000, Malaysia placed an order for 211 vehicles in 10 variants based on the ACV 300 version with a 300 hp Detriot Diesel.

The 11 variants are:

- Infantry fighting vehicle with one-person Sharpshooter turret armed with Boeing 25 mm M242 Chain Gun and 7.62 mm coaxial machine gun
- Anti-tank with undisclosed anti-tank guided weapon system
- Armoured personnel carrier with Helio turret armed with 40 mm grenade launcher and 7.62 mm machine gun
- Armoured personnel carrier with Helio turret armed with 12.7 and 7.62 mm machine guns
- Ambulance
- Command post vehicle
- Mortar carrier (81 mm)
- Recovery vehicle
- Fitters vehicle
- Signals vehicle
- Cargo carrier

M113 upgrade programme

FNSS Defense Systems is also offering to upgrade the M113 family of vehicles into the latest A300 configuration which includes a 300 hp 6V-53T Detroit Diesel engine coupled to an Allison X200-4 automatic transmission.

According to FNSS Defense Systems, the A300 configuration provides higher mobility, higher internal volume, improved cooling performance in extreme conditions and better survivability. A full NBC and air conditioning system is also available in this configuration.

AIFV equivalent suspension, steering, fuel and electrical systems are available in M113 A300 configuration with various other optional equipment to meet different customer requirements.

Status

Production. In service with:

Country	Quantity	Comment
Malaysia	211	ordered late 2000 in 10 versions, first deliveries late 2001
Turkey	1,698	first order, all delivered
	551	second order placed 2000 with deliveries from December 2001 through to December 2004
UAE	136	delivery will take place between 1999 and 2002

Manufacturer

FNSS Defense Systems.

UPDATED

Turkish Army M113 fleet

It is estimated that the Turkish Army operates a fleet of over 3,000 United Defense, Ground Systems Division M113 series full-tracked armoured personnel carriers. Most of these were supplied by the US although 581 were supplied brand new from the Italian M113 production line in the 1960s.

More recently Turkey has received significant quantities of surplus M113 series vehicles from US Army stocks:

Year	Quantity
1992	220
1993	61
1994	74
1995	nil
1996	25
1997	179
1998	nil

The Turkish Army's 109th Ordnance Factory at Kayserie has developed an upgrade package for the M113 series APC under the local designation of the M113A2T2. It is understood that at least 1,000 M113 series vehicles have been upgraded to this standard.

The original Chrysler petrol engine has been replaced by a more fuel-efficient Detroit Diesel Model 6V-53T turbocharged diesel engine developing 275 hp coupled to an Allison TX-100-1 fully automatic transmission.

The installation of the diesel engine increases its operational range to 483 km with maximum road speed being 67 km/h. As well as increasing the operational range there is also a reduced risk of fire.

The upgraded M113 is still fully amphibious, being propelled in the water by its tracks at a speed of 5.8 km/h.

Other improvements carried out to the M113A2T2 vehicle include an upgraded braking system, new troop compartment heater, new and stronger trim vane mounted on the glacis plate and the installation of an NBC system.

The new trim vane is of welded Aluminium Type 5052 with an interior filling of polyurethane.

To provide more interior room in the M113, the large fuel tank has been removed from inside and replaced by two new fuel tanks mounted externally either side at the rear of the hull.

These fuel tanks are made of welded Aluminium Type 5083 and provide protection against small arms fire up to 12.7 mm in calibre. Each fuel tank holds 180 litres of fuel.

The normal armament of a roof mounted 12.7 mm M2 Browning machine gun is retained and this has been provided with additional armour protection. Mounted on the front of the hull is a bank of eight electrically operated 76 mm smoke grenade launchers that cover the front of the vehicle.

Turkish M113A2T2 upgraded armoured personnel carrier (Christopher F Foss) ***2000***/0080322

Atilgan Pedestal Mounted Air Defence Missile system (PMADS) is based on a modified M113A2 APC chassis ***2000***/0080323

The Kayserie facility has also completed the prototype of a self-propelled air defence version of the M113. Mounted on top of the modified hull roof is the complete upper part of the old US-supplied M55 4 × 12.7 mm towed anti-aircraft gun system fitted with an optical day sight.

A modified M113A2T2 APC is also used as the basis for the Microwave and System Technologies Division of Aselsan Atilgan Pedestal Mounted Air Defence Missile System (PMADS).

This has a roof-mounted power-operated pedestal with eight Stinger surface-to-air missiles in the ready-to-launch position. Mounted between the two banks of four missiles is the day/night sensor suite and a 12.7 mm machine gun for local defence. Full details of this system which is soon expected to enter production for the Turkish Land Forces Command are given in *Jane's Land-Based Air Defence* 2001-2002. The first production contract for this system was placed in late September 2001, with first deliveries to be made within 23 months.

Status

The upgraded M113A2T is in service with the Turkish Army.

Manufacturer

Turkish Land Forces Command facilities.

UPDATED

UKRAINE

BTMP-84 Heavy Infantry Fighting Vehicle

Development

Late in 2001, it was revealed that the Kharkov Morozov Machine Building Design Bureau in the Ukraine had built the prototype of a new Heavy Infantry Fighting Vehicle (HIFV) called the BTMP-84.

This is based on a much modified chassis of the T-84 series MBT that is already in service in small mumbers with the Ukrainian Army and covered in detail in the MBTs and Medium Tanks section.

The T-84 MBT is a further development of the T-80UD of which 320 have been delivered to Pakistan. While the Kharkov Morozov Machine Building Design Bureau is the overall design authority, actual production of the BTMP-84 would be undertaken by the local Malyshev Plant.

The role of the BTMP-84 is to work in close co-operation with other MBTs and provides the five infantrymen carried with the same level of protection and battlefield survivability as the T-84 MBT.

Description

The BTMP-84 HIFV has a crew of three, driver, commander and gunner. The driver is seated at the front of the vehicle in the normal central position with the turret to the rear.

The driver has a single-piece hatch cover above his position and three day periscopes for forward observation. One of these can be replaced by a passive periscope for driving at night.

The two-person turret installed on the BTMP-84 is the same as that fitted to the T-84 MBT. It is armed with a stabilised 125 mm KBA-3 smoothbore gun with a 7.62 mm machine gun mounted coaxial and a 12.7 mm machine gun mounted on the right side of the roof. The commander and gunner are each provided with a roof hatch, with the commander being seated on the right and the gunner on the left.

The 125 mm KBA-3 smoothbore gun was developed in the Ukraine and is similar to the Russian 125 mm 2A46 series weapon. A total of 450 rounds of 12.7 mm and 1,250 rounds of 7.62 mm machine gun ammunition are carried.

A computerised day/night fire-control system is fitted to engage stationary and moving targets with a high first-round-hit probability.

The normal T-84 MBT carries a total 43 rounds of separate loading 125 mm ammunition of which 28 is in the automatic loader located below the turret.

First photograph released of the Ukrainian BTMP-84 heavy infantry fighting vehicle based on a T-84 series MBT chassis ***2002***/0096258

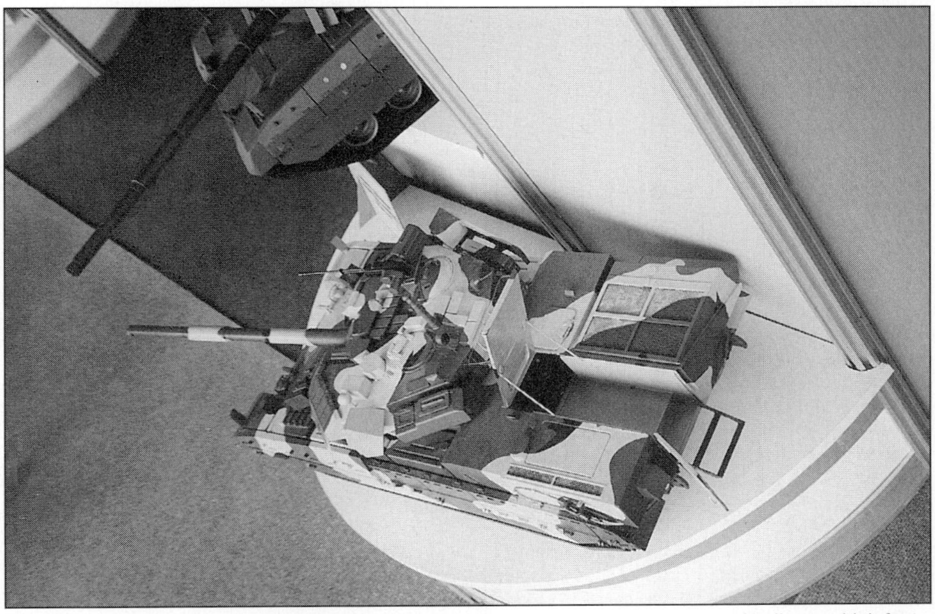

BTMP-84 heavy infantry fighting vehicle from above, clearly showing the redesigned hull rear with infantry compartment hatches open (Christopher F Foss) ***2002***/0096259

To provide additional space for the troops, the 125 mm ammunition load of the BTMP-84 has been reduced to a total 30 rounds and the engine compartment redesigned to provide additional internal volume for the five troops.

The compact troop compartment is to the immediate rear of the turret with a passageway now provided in the left side at the rear. The upper part of the roof hinges upwards, with the door in the rear opening to the right and a step being provided at the rear to allow the troops to rapidly dismount the vehicle. Additional space is provided to the rear of the turret on the right side and above this is a roof hatch and day vision periscope.

In many respects this arrangement is similar to the Israeli Achzarit infantry armoured vehicle based on a rebuilt T-55 tank chassis. Details of this are given earlier in this section under Israel.

The redesigned powerpack installed in the BTMP-84 consists of the very compact 6TD-2 diesel engine developing 1,200 hp coupled to manual transmission.

The exhaust outlet appears to be located in the roof of the engine compartment on the left side, with the air entry and air outlet louvres on the right side of the engine compartment roof.

The BTMP-84 has a combat weight of 50 tonnes, power-to-weight ratio of 24 hp/tonne, maximum road speed of 70 km/h and an operating range of 450 km.

The BTMP-84 has the same level of protection as the T-84 MBT which consists of advanced passive armour and advanced explosive reactive armour package.

This gives protection over the frontal arc against both High Explosive Anti-Tank (HEAT) and kinetic energy projectiles. The vehicle can also be fitted with a defensive aids suite as well as a satellite navigation system.

Russia has also developed two similar vehicles to support MBTs, the BMPT Tank Support Vehicle based on a much modified T-72 series MBT chassis and the BTR-T heavy armoured personnel carrier, based on a much modified T-55 tank chassis. Both of these however have their normal large calibre turrets removed and replaced with medium calibre weapons in remote controlled mounts.

Status
Prototype. Not yet in production or service.

Manufacturer
Malyshev Plant

NEW ENTRY

Upgraded MT-LB

Development
The MT-LB multirole full tracked armoured vehicle was designed and manufactured in the Ukraine at the famous Kharkov Tractor Plant Joint Stock Company (KhTP JSC), with large numbers being built for the home and export markets. Additional details of the basic vehicle and its many variants are given elsewhere in this section under Russian Federation and Associated States.

It was also manufactured under licence in Bulgaria and Poland and these countries also developed many of their own specialised versions which are covered in detail in this section under their respective countries.

The basic MT-LB has a crew of two, commander/gunner and driver and can carry 11 fully equipped troops. It is fully amphibious, being propelled in the water by its tracks. The vehicle has excellent cross-country mobility and is normally fitted with 350 mm wide tracks with the option of fitting 560 mm tracks for improved traction as well as lowering the ground pressure still further.

The MT-LB is probably used for more roles than any other armoured vehicle in eastern Europe, for example towing 122 mm D-30 howitzers, ambulance, command post, radar carrier, engineer vehicle, anti-tank missile platform and for the SA-13 'Gopher' surface-to-air missile system, to name but a few.

The KhTP JSC have recently developed an upgraded MT-LB to meet the requirements of the 21st century in the key areas of mobility and firepower, although protection has also been improved.

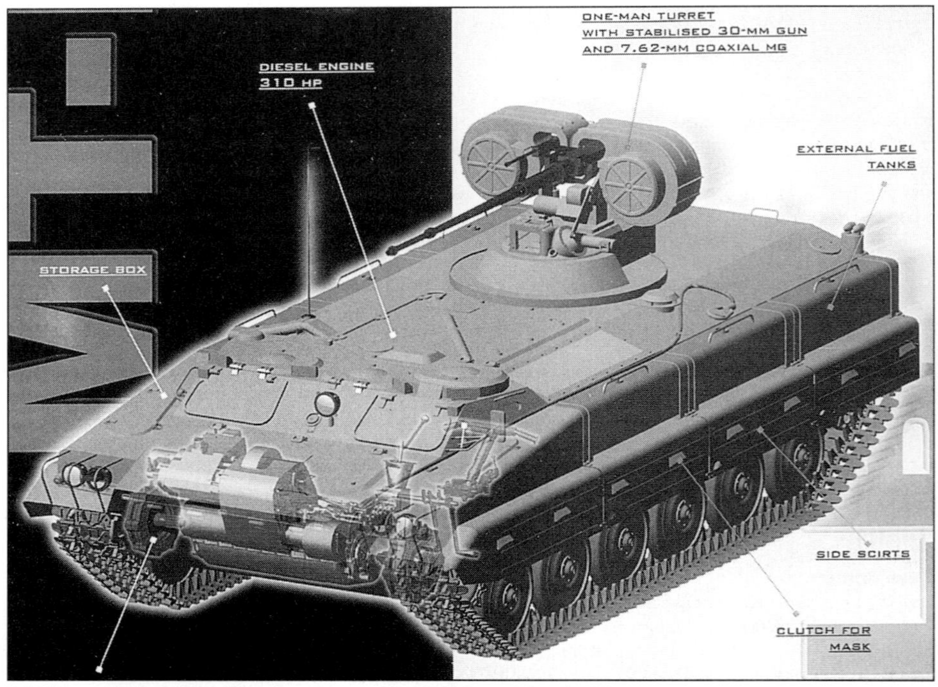

MT-LB upgrade by the Kharkov Tractor Plant JSC and fitted with new one-person turret armed with 30 mm cannon and 7.62 mm c-axial machine gun and showing position of main improvements
2002/0034819

Description
The overall layout of the Upgraded MT-LB is almost identical to the original version. The standard MT-LB is powered by a V-8 diesel developing 240 hp coupled to a manual transmission. The Upgraded MT-LB is powered by a V-8 diesel developing 310 hp coupled to a manual transmission but with hydrostatic steering, rather than the more conventional clutch and brake steering.

The driver's tillers have been replaced by a motorcycle type T-bar with the shoe brakes replaced by disc brakes. This upgrade not only gives a higher road and cross-country speed but also improved power-to-weight ratio, as well as making the vehicle easier to drive and improving safety. Maximum road speed of the upgraded vehicle is 70 km/h compared with 60 km/h of the original vehicle.

The standard MT-LB is fitted with a one-person manually operated turret mounted on the forward part of the hull on the right side and armed with a 7.62 mm PKT series machine gun.

The existing armament can be retained, but for increased fire power, the Upgraded MT-LB can be fitted with a new one-person turret armed with an externally mounted, fully stabilised, 30 mm dual-feed cannon with a 7.62 mm PKT machining gun mounted coaxial above and to the right. This allows the gunner to be under complete armour and NBC protection inside of the vehicle.

The 30 mm cannon can either be the Russian 30 mm 2A72 or the locally developed 30 mm KBA-2 with elevation being from –10 to +60° and with traverse through a full 360°.

The high angle of elevation gives the system the capability to engage slow flying aircraft and helicopters. Mounted either side of the 30 mm cannon is a drum type magazine holding 130 rounds of ready use ammunition. One of these would typically be filled with armour piercing – tracer (AP-T) and the other with high explosive (HE) ammunition.

Total weight of the turret is 1,250 kg and the sighting system consists of a TKN-4S day/night sight with the PZU-7 day sight being fitted for the engagement of aerial targets. This turret has also been fitted to the Ukrainian upgraded BTR-70 (8 × 8) armoured personnel carrier.

The standard MT-LB suspension consists of torsion bars with hydraulic shock-absorbers. For improved cross-country mobility the original spring-type bump stops have been replaced by hydraulic bump stocks and three track return rollers have been added either side.

Standard MT-LB equipment includes NBC system and night vision equipment. The diesel fuel tanks are now mounted externally along either side of the hull and additional stowage boxes have been integrated into the hull. New side skirts have been fitted and additional protection provided against mines.

Specifications
Upgraded MT-LB
Crew: 2 + 11
Combat weight: 10,700 kg
Payload:
(land) 2,500 kg
(water) 2,000 kg
Towed load: 6,500 kg
Power-to-weight ratio: 28.97 hp/t
Ground pressure:
(350 mm wide tracks) 0.47 kg/cm^2
(560 mm wide tracks) 0.29 kg/cm^2
Length: 6.325 m
Width:
(overall) 3.35 m
(tracks) 3.15 m
Height:
(hull top) 1.68 m
(overall 30 mm turret) 2.695 m
Ground clearance: 0.51 m
Length of track on ground: 3.70 m
Track: 2.50 m
Track width: 350 mm or 560 mm
Max speed:
(road) 70 km/h
(water) 6 to 7 km/h
Fuel capacity: 450 litres
Max range: 500-620 km
Fording: amphibious
Gradient: 70%
Side slope: 46%
Vertical obstacle: 0.80 m
Trench: 2.50 m
Engine: V-8 diesel developing 310 hp
Transmission: manual
Steering: hydrostatic
Suspension: torsion bar with hydraulic shock-absorbers
Electrical system: 24 V
Armament: see text
Armour: 3.19 - 9.87 mm
NBC system: yes
Night vision equipment: yes

Status
Development complete. Ready for production.

Manufacturer
Kharkov Tractor Plant Joint Stock Company.

NEW ENTRY

UNITED KINGDOM

Alvis Vehicles Warrior 2000 infantry fighting vehicle

Development

In June 1998 Alvis Vehicles, at that time GKN Defence, completed trials of the new private venture Warrior 2000 Infantry Fighting Vehicle (IFV).

Warrior 2000 was developed to meet the requirements of the Swiss Army who eventually selected the Hägglunds Vehicle CV 9030CH to meet its future requirements. The Warrior 2000 continues to be marketed by Alvis Vehicles which is part of Alvis plc which now owns Hägglunds Vehicle of Sweden.

The Warrior 2000 was designed and built by Alvis Vehicles in nine months on the Warrior production line and is based on experience in the design, development and production of 789 Warriors for the British Army and 254 Desert Warriors for the Kuwait Land Force.

Description

The hull of the Warrior 2000 is of all-welded aluminium armour and is slightly longer (230 mm) than previous Warriors. It incorporates stealth technology. Wherever possible, the exterior surfaces of the Warrior 2000 are smooth and the upper hull sides slope slightly inwards.

Over the frontal arc, the armour of the Warrior 2000 provides protection against attack from 30 mm Armour Piercing Discarding Sabot (APDS) attack. The layout of Warrior 2000 is similar to earlier Warriors with driver front left, power pack to the right, turret in centre and troop compartment at the rear.

The driver has a single-piece hatch cover that opens to the rear with three integral day periscopes. This hatch cover is being backfitted to some British Army Warrior vehicles as a direct replacement for the current Warrior hatch cover which has a single periscope. The central day periscope in the driver's hatch can be replaced by an image intensification or thermal viewer from within the vehicle.

The Perkins power pack has been uprated to 650 hp and Warrior 2000 has a maximum road speed of 75 km/h, range of 500 km and a combat weight of 31.5 tonnes.

The engine is provided with a twin fan cooling group and this, together with the rear exhaust (right hand side) and air-cooled silencer has led to a reduction in both the thermal and acoustic signatures of the vehicle. The engine decks are lifted by an integral hydraulic ram for ease of maintenance.

The hull incorporates spaced armour which provides protection against 30 mm armour piercing fin stabilised discarding sabot (APFSDS) projectiles but a higher level of protection is offered similar to that fitted to the British Army and export Warrior vehicles.

A new double pin track has been fitted which is claimed to provide high durability, ease of maintenance and reduced running noise. Quick release track pads for replacement, or installation of ice grousers, are fitted as specified by Switzerland.

Warrior 2000 is fitted with the new Delco Defense all electric two-person turret armed with a fully stabilised The Boeing Company 30 mm Bushmaster II cannon with a 7.5 mm machine gun mounted coaxial to the left. Either side of the turret is a bank of four electrically operated 76 mm smoke grenade launchers.

This turret is a further development of the combat proven LAV-25 turret but has a larger turret ring and greater internal volume. It has the same level of armour protection as the hull.

The 30 mm Bushmaster II, which is already in production, has three rates of fire, single shot, 200 or 400 rds/min.

The digital fire control system, stabilised gun and stabilised sights ensure a high first round hit probability against stationary and moving targets while Warrior 2000 is stationary or moving. The gunner has twin control handles with the commander having a single control handle, manual controls are provided for emergency use.

The gunner, seated on the left, has a Raytheon stabilised day/thermal sighting system with a laser range-finder and one periscope. The commander, seated on the right, has eight day periscopes for all round observation plus a flat screen display which receives a colour picture from a camera mounted over the 30 mm cannon.

This camera has an autotrack facility as one of the key Swiss requirements was that it should engage helicopters. The commander also has a monitor of the gunner's sight. The commander and gunner each have a single piece hatch cover that opens to the rear.

In addition to the crew of three, Warrior 2000 also carries seven fully equipped troops in the rear who enter and leave via a power operated ramp with an emergency door. There is a large stowage box either side of the hull rear.

Four are seated on the right hand side and three on the left with one of the latter being the dismounted section commander who also has a roof mounted cupola with seven day periscopes and a single piece hatch cover and an electronic display screen. There is also a single piece roof hatch over the right side of the troop compartment that opens outwards.

The video distribution system of the Warrior 2000 provides images to the infantry section in the rear of either the driver's thermal viewer, reversing camera, turret mantlet camera or the day and thermal sights in the turret.

Standard equipment includes spall liners, air conditioning/NBC system, fire detection and suppression system and rear mounted TV camera for the driver.

Specifications
Crew: 3 + 7
Combat weight: 31,500 kg
Power-to-weight ratio: 20.6 hp/t
Ground pressure: 0.74 kg/cm^2
Length: 6.968 m
Width: 3.375 m
Height:
(hull roof) 1.990 m
(overall) 2.952 m
Vertical obstacle: 0.8 m
Trench: 2.5 m
Max road speed: 75 km/h
Range: 500 km
Engine: Perkins Engine Company V-8 Condor diesel developing 650 hp
Transmission: Perkins Engine Company X-300-7 fully automatic with torque converter and lock up clutch, 4 forward and 1 reverse gears
Armament:
(main) 1 × 30 mm Bushmaster II cannon
(coaxial) 1 × 7.5 mm machine gun
(smoke launchers) 2 × 4 76 mm
Stabilisation system:
(vertical) yes
(horizontal) yes
NBC system: yes
Night vision equipment: yes

Variants

The Warrior 2000 chassis can be used to mount other weapon systems such as the South African LIW 105 mm turret which has a stretch potential to 120 mm.

Warrior 2000 with E30 turret

Late in 2001, the Alvis Vehicles Warrior 2000 IFV was shown fitted with the latest version of the Hägglunds Vehicles E30 two-person electrically operated turret, which is fitted to the CV 9030 IFV in service with Norway and selected by Finland and Switzerland.

Key features of this turret include a fully digitised open electronic architecture, CAN-based Hägglunds Vehicle Control System and armed with a Boeing Bushmaster II MK 44 30/40 mm cannon with a 12.7 mm M2 machine gun mounted coaxially.

All electric turret drives are fitted and the fire-control computer includes a digital processor and a dual-axis stabilised independent line of sight stabilisation system is also fitted.

Status

Development complete. Ready for production.

Manufacturer

Alvis Vehicles Limited.

UPDATED

Alvis Vehicles Warrior 2000 ICV during trials in Switzerland. This is fitted with a Delco Defense turret armed with a 30 mm cannon ***2002***/0116797

Alvis Vehicles Warrior armoured combat vehicle

Development

The initial 1967 proposal to consider future armoured personnel carrier requirements for the British Army was followed by in-house Ministry of Defence feasibility studies from 1968 and 1971. Initial industrial work (Project Definition 1) and Ministry of Defence option studies to define an affordable programme took place between 1972 and 1976. Full project definition (Project Definition 2) was undertaken in 1977-78 by GKN Sankey (which became GKN Defence which in turn was then merged with Alvis Vehicles late in 1998) after its competitive selection as prime contractor for the MCV-80.

In 1978, a parallel evaluation of the American XM2 (now known as the M2 Bradley Infantry Fighting Vehicle) took place. The following year full development of the MCV-80 started in parallel with continued study of the XM2.

In June 1980, the following parliamentary statement was made concerning the British Army's APC replacement:

'The army's present armoured personnel carrier, the FV432 series of vehicles, has been in service since the 1960s and will need to be replaced from the mid-1980s.

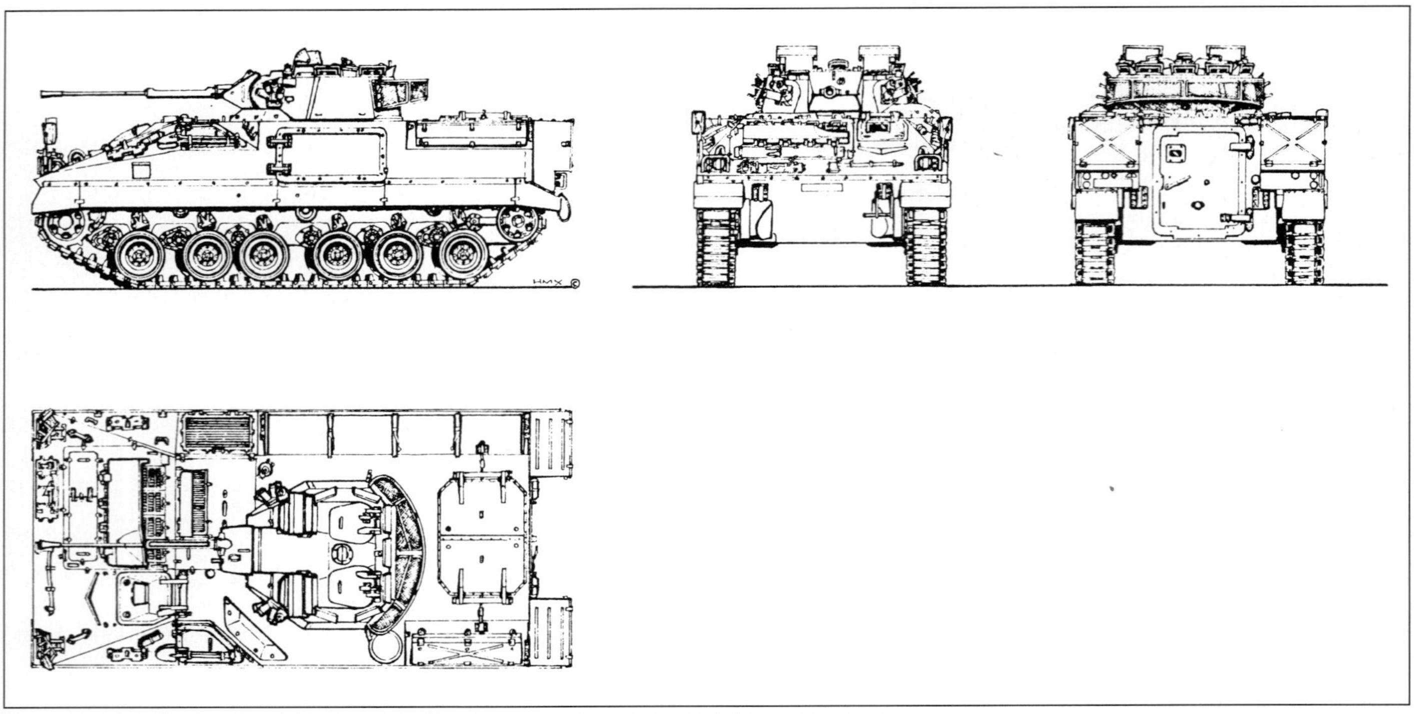

Four-view general arrangement drawing of the Warrior armoured combat vehicle (Henry Morshead)

There have been two vehicles considered for this requirement, the Mechanised Combat Vehicle (MCV-80), designed by the British company GKN Sankey, and the American Infantry Fighting Vehicle, which would be manufactured under licence in this country.

After careful assessment of the relevant operational, financial and industrial factors I have decided to select MCV-80 to meet this requirement. The total estimated cost of the replacement programme is about £1,000 million and full development will be launched shortly'.

Actual design work commenced in 1977 and, by 1980, three prototypes of the MCV-80 were running and the company was also using test rigs.

In January 1984, the British Ministry of Defence announced that negotiations with GKN Sankey for an initial production order were well advanced, but that future orders would be open to competitive tendering by both GKN Defence and other qualified manufacturers. At the same time it was announced that the MoD would be contracting with GKN Defence for full development of variants and derivative vehicles, subject to satisfactory completion of preliminary studies. MCV-80 (now known as Warrior) was accepted for service with the British Army in November 1984 with a reliability rating well in excess of the trials targets.

By 1984, 10 prototypes had been built; one of these had a highly successful demonstration in the Middle East in 1983 and again in another Gulf country in Summer 1984, when a Warrior Platoon Vehicle covered 850 km, 650 km of which was completed in two days of running. This has led to the development of a Desert Fighting Vehicle configuration with 25 mm, 30 mm or 90 mm cannon and a crew of 10.

Production of Warrior began at Telford in January 1986, with first production vehicles completed in December. The first production batch comprised 290 vehicles, 170 of which were section vehicles with two-man 30 mm RARDEN turrets and the remaining 120 vehicles specialised variants. The first production Warrior was officially handed over to the British Army in May 1987. The first Warrior battalion in the British Army of the Rhine was fully operational in mid-1988.

The second and third batches were to have totalled 763 vehicles to bring the total British Army order to 1,053 vehicles, including variants, which would be sufficient to equip 13 armoured infantry battalions.

As a result of 'Options for Change' the size of the British Army has been reduced and only eight armoured infantry battalions are now equipped with the Warrior, six in Germany and two in the UK.

The total requirement of 1,053 (or 1,048 as a result of some additional changes) vehicles was subsequently reduced to 789 vehicles, with final deliveries to the British Army made in the spring of 1995.

Variant	**Original quantity**	**Final quantity**
Section vehicle	595	387
MR TRIGAT	nil	105*
Infantry command vehicle	146	84
MCRV	110	103
MRV(R)	67	9
Battery command vehicle	35	19
Observation post vehicle	95	52
Total	1,048	789

Note: * currently used with the Euromissile MILAN ATGW which was to have been replaced by the EMDG TRIGAT-MR. The UK pulled out of this programme late in 2000 and is now studying alternative ATGW solutions such as the Israeli Gill/Spike and the US Javelin.

Following extensive trials in Kuwait, the Desert Warrior was selected by Kuwait late in 1992, with an order for 254 being placed in August 1993. Additional details are given later in this entry.

Final deliveries were made to Kuwait on schedule, late in 1997. The production line for Warrior is still being retained for the building of demonstrators, such as the Swiss version of Warrior with a 30 mm cannon, repair of damaged hulls and other work. The production line can open when other orders are received.

Warrior in Operation Desert Storm

During Operation Desert Storm, the recapture of Kuwait early in 1991, all six variants were deployed including the Artillery Observation and Battery Command vehicles which entered service for the first time.

A number of Warriors were converted for use in the MILAN ATGW role. The order was placed on 22 October 1990 and the first kits were designed, produced and delivered in one week. This allowed MILAN to be launched from the vehicle with provision being made for internal missile stowage. This version was subsequently adopted by the British Army and additional details are given under Variants.

As an interim measure before the issue of Artillery Observation Post vehicles to units (and afterwards), some Infantry Command Vehicle Warriors were converted for the Forward Observation Role.

Warrior Command Vehicles were also deployed in the forward observation role for armoured regiments, a new task for the vehicle. An 81 mm mortar carrier was

Alvis Vehicles Warrior armoured combat vehicle with commander's, gunner's and driver's hatches open **2001**/0099491

Alvis Vehicles Warrior armoured combat vehicle with appliqué armour, deployed to the Balkans (Shaun C Connors) **2002**/0116792

Warrior Mechanised Recovery Vehicle (Repair) with earth anchor raised (Shaun C Connors) ***2001***/0099489

Warrior Mechanised Combat Repair Vehicle in travelling configuration (Shaun C Connors) ***2002***/0116793

developed to a stage to allow early production when required in the future.

To improve battlefield survivability, the Warriors deployed to Saudi Arabia were fitted with passive armour developed and built by Vickers Defence Systems and the then GKN Defence. This was fitted in-country. This armour is now fitted on vehicles in Bosnia. It is also now a standard feature in wartime/high-threat areas.

Many minor modifications were also carried out, such as installation of maintenance-free batteries, installation of Mini Pipe Fascines and satellite navigation aids for section, infantry command and REME variants.

An order for air conditioning for deployed Warriors was placed in January 1991, and design and development commenced in conjunction with Galley and NGL. Full habitability trials were completed and 1,500 km of trundling trials were undertaken but no deliveries were made due to the speedy end of the conflict.

Description

The hull of Warrior is of all-welded aluminium construction which provides a significant increase in protection over that of the current FV432 APC. The driver sits at the front of the vehicle on the left side and has a single-piece hatch cover hinged at the rear, with a single wide-angle day periscope that can be replaced by an image intensification periscope for driving at night.

British Army Warriors are now being fitted with a new driver's hatch which is also fitted to the Kuwait vehicles, this has three day periscopes for forward observation and therefore gives the driver a better view of the terrain.

The power pack is to the right of the driver and drives the front sprockets through spur final drive units. The engine is a Perkins Engines Company Condor CV-8 TCA diesel developing 550 hp which is coupled to an X-300-4B transmission manufactured by Perkins Engines Company. The automatic transmission has four forward and two reverse gears, and steering is through a hydrostatically controlled differential steer assembly.

The power-assisted vehicle brakes are integral with the transmission. The cooling system covers the engine, generator, transmission and vehicle brakes. The Lucas oilspray-cooled 300 A generator is driven from the gearbox by a Power Take Off (PTO) which also drives the hydraulic pump for the fan.

The two-man steel turret is in the centre of the vehicle, offset slightly to the left of the vehicle's centreline. The turret design, which was subcontracted to Vickers Defence Systems of Newcastle-upon-Tyne, has seats for the commander on the right and the gunner on the left, with both crew members having individual hatch covers that open to the rear. The turret has power traverse with manual controls for emergency use. Both the commander and gunner have a Thales Optronics Raven day/night sight. In the day mode this has magnifications of ×1 and ×8, while in the image intensification night mode it has magnifications of x2 and x6. The commander's Raven sight has additional traverse. Additional day periscopes give vision to the sides and rear of the turret.

After evaluating the British RARDEN 30 mm cannon and the American 25 mm M242 Chain Gun, the former was selected for the Warrior. Following competitive bidding by Royal Ordnance Enfield, which had previously built all production RARDEN weapons, and the British Manufacture And Research Company (BMARC), the latter was awarded the Warrior RARDEN production contract in 1985. In 1992, BMARC was taken over by Royal Ordnance, with production of the RARDEN cannon transferred to the now BAE Systems, RO Defence facility at Nottingham. This facility is now closed and all UK gun barrel production will be concentrated at RO Defence facility at Barrow-in-Furness.

Mounted coaxially with the main armament is a 7.62 mm EX-34 Chain Gun (British designation L94A1) with the expended cartridge cases being ejected outside the turret.

Mounted either side of the turret are four electrically operated 66 mm grenade dischargers.

Desert Warrior with TOW missile launchers on side of turret in retracted position 0018540

Specifications
(British Army Warrior with appliqué armour)
Crew: 3 + 7
Combat weight: 28,000 kg
Unloaded weight: 24,500 kg
Power-to-weight ratio: 19.6 bhp/t
Length: 6.34 m
Width:
(overall) 3.034 m
(over tracks) 3.00 m
Height:
(hull roof) 1.93 m
(turret top) 2.735 m
(overall) 2.791 m
Ground clearance: 0.49 m
Track width: 460 mm
Length of track on ground: 3.816 m
Max road speed:
(forwards) 75 km/h
(reverse) 48 km/h
Acceleration: (0 to 48 km/h) 13.5 s
Fuel capacity: 770 litres
Max road range: 660 km (at 60 km/h)
Fording: 1.3 m
Gradient: 60%
Side slope: 40%
Vertical obstacle: 0.75 m
Trench: 2.5 m
Engine: Perkins Engines Company CV-8 TCA V-8 diesel developing 550 hp at 2,300 rpm
Transmission: Perkins Engines Company X-300-4B fully automatic with torque converter and lock up clutch, 4 forward and 2 reverse gears
Steering system: differential steering system based on a variable hydrostatic drive
Suspension: torsion bar
Electrical system: 24 V
Batteries: 6 × 12 V
Armament:
(main) 1 × 30 mm L21 RARDEN cannon
(coaxial) 1 × 7.62 mm L94A1 Chain Gun
Smoke-laying equipment: 2 × 4 66 mm grenade dischargers
Ammunition:
(30 mm) 250
(7.62 mm) 2,000

Gun control equipment
Turret power control:
(traverse) electric/manual
(elevation) manual
Gun elevation/depression: +45 to –10°
Turret traverse: 360°
Gun stabiliser:
(vertical) no
(horizontal) no
NBC system: yes
Night vision equipment: yes

The Alvis Vehicles Warrior ICV vehicle fitted with a Thales Optronics BGTI suite **2002**/0116788

Gunner's position on Alvis Vehicles Warrior ICV fitted with BGTI showing flat panel display with monitor of gunner's thermal sight on left **2002**/0116794

The troop compartment is at the rear of the Warrior with the troops entering and leaving via a single power-operated door in the rear of the hull which has a vision block. In the roof of the troop compartment are two day rotating periscopes, one either side, for use by the infantry. The sides of the troop compartment are sloped inward but there is no provision for the infantry to fire its small arms from within the vehicle. Firing ports/vision devices could be provided on export vehicles if required. Over the top of the troop compartment are double roof hatches. The Warrior carries eight infantry, one of which also commands the vehicle and therefore dismounts with the infantry. The remaining seven sit in the troop compartment, four on the right and three on the left, in separate seats with seat belts. Personal kit can be stowed under the seats or in the sponsons which have interior grills to hold the kit in position. Warrior carries all the equipment required to fight on the battlefield for a period of 48 hours.

The suspension is of the torsion bar type and consists either side of six rubber-tyred aluminium roadwheels with the drive sprocket at the front, idler at the rear and three track-return rollers. Dampers, built into the suspension pivot housing, are fitted on the first, second and sixth roadwheel stations. The tracks are conventional rubber-bushed cast steel with rubber pads. Following a competition, William Cook Defence was in 1999 awarded a contract for a new generation of single pin track designated the TR30.

Standard equipment includes an NBC system which is mounted in the side of the hull to the left rear of the driver and a full range of passive night vision equipment. Either side of the rear troop doors is a stowage box. The Warrior has no amphibious capability as this was not required by the British Army.

In 1986, Chubb Fire Security was awarded a contract for the supply of engine fire protection systems for Warrior. This system uses Halon gas as the extinguisher. Situated in the front of the vehicle, the system is carried on two lightweight, rechargeable Halon canisters which vent into sparge pipes running around the engine compartment. In the event of an engine compartment fire, the crew manually activates the system releasing a 4 second burst of Halon. A second Halon canister provides a back-up, with further manual triggers on the outside of the vehicle's hull providing a remote extinguishing facility. In addition, Warrior carries five hand-held fire extinguishers, three inside and two outside.

BGTI

In August 2001, the UK Defence Procurement Agency (DPA) awarded Thales Optronics a £230 million contract for British Army's major Battle Group Thermal Imaging (BGTI) programme.

Three UK contractors, Avimo, Hunting Engineering and Thales Optronics submitted their BGTI bids late in 2000. Of the £230 million contract, it has been estimated that about two-thirds will be for procurement and one third for support.

When fielded, BGTI will replace the current image intensification sights installed on most members of the Alvis Vehicles Warrior Infantry Combat Vehicle (ICV) and the Alvis Vehicles Scimitar, used in the close reconnaissance role,

The contract covers the supply of two systems, Group 1 and Group 2, with the first vehicles to be equipped in 2003. The installation of the BGTI system will allow the Warrior and Scimitar vehicles to engage targets at longer ranges and under almost all-weather conditions.

In addition, BGTI will enable the crew to rapidly establish its own position and the six figure co-ordinates of other targets on the battlefield. It can also establish the range of targets using a laser range-finder with the driving route map controlled by the commander and displayed to the driver.

Group 1 vehicles are the Warrior ICV and the Scimitar reconnaissance vehicles, armed with a 30 mm RARDEN cannon. These will have a new gunner's sight with day/thermal imager and eye-safe laser ranger-finder, with the commander having a monitor and a flat pannel display. This will provide access to Bowman and battle management systems being procured under separate contracts. The driver has a navigation capability.

For the Group 2 vehicles, which include the Warrior repair and recovery vehicles armed with a cupola mounted 7.62 mm machine gun, a less advanced system will be procured.

The first Group 1 contract covers 450 Warrior and 150 Scimitar vehicles with the Group 2 contract covering 100 Warrior repair and recovery vehicles. There is also also an option on another 200 Group 1 and 50 Group 2 systems.

Warrior Upgrades

The UK MoD is currently considering a number of further upgrades to the Alvis Vehicles Warrior vehicles used by the British Army. These include enhancements in the key areas of armour, mobility and firepower.

The latter is expected to be the first upgrade and could include anything from stabilising the existing 30 mm RARDEN cannon to the installation of a new turret armed with a larger calibre weapon, such as the Bushmaster 30/40 mm MK 44 or the CTA International 40 mm Case Telescoped Weapon System.

British Army Variants

Section Vehicle

This is the basic Warrior MCV fitted with a two-man turret armed with a 30 mm RARDEN cannon and a 7.62 mm L94A1 Chain Gun as detailed in the Description.

Command Vehicles

The company has produced three command versions of Warrior: platoon, company and battalion. These all left Telford to the same configuration, with the army fitting the communications systems and other equipment. These all have a two-man turret armed with a 30 mm RARDEN cannon and a 7.62 mm L94A1 Chain Gun as previously described.

Warrior Mechanised Recovery Vehicle (Repair) (MRV(R)) (FV513)

This has been designed under contract to the British Army and two prototypes were completed by mid-1986.

The MRV(R) has a crew of five consisting of the commander, driver, gunner and two fitters, and mounted on the left side of the hull is a hydraulically operated crane which can lift a maximum of 6,500 kg. A hydraulically operated stabiliser leg is lowered to the ground before the crane is used. The crane can be traversed through a full 360° and can lift a complete Challenger 1/2 MBT power pack.

Mounted in the rear is a hydraulically operated winch with a capacity of 20,000 kg which can be increased to 38,000 kg when double reeved.

The vehicle is provided with suspension lockout and, mounted at the rear of the hull, to facilitate winching is an earth anchor is provided with a pilot winch and 200 m of cable giving a capacity of 1.25 tonnes.

An air compressor is provided for the use of air tools and to assist work in the field an electrohydraulic pump allows the vehicle to change its own power pack.

The manually operated one-person turret is armed with a 7.62 mm L94A1 Chain Gun and 76 mm smoke grenade dischargers.

Warrior Mechanised Combat Repair Vehicle (MCRV) (FV512)

This is virtually identical to the Warrior MRV(R) previously described, except it does not have the rear-mounted earth anchor or recovery winch. It retains the 7.62 mm machine gun turret.

The MCRV can tow the now Alvis Vehicles T4 high-mobility trailer which weighs 10,500 kg fully loaded and carries a complete Challenger 1 or 2 MBT power pack.

Warrior Mechanised Artillery Observation Vehicle (MAOV) (FV514)

The MAOV is used by the forward observation officers and battery commanders in Royal Artillery field regiments and the first prototype was completed in 1988.

Externally the MAOV is similar in appearance to the basic Warrior vehicle but has a dummy 30 mm RARDEN cannon. The observer has a Thales Optronics Osprey combined day/thermal/laser system. This has magnifications of ×1 and ×8 in the day mode and built-in laser protection. The thermal imaging system is based on BAE Systems Class II Thermal Imaging Common Modules already in service with the British Army. The assistant observer has a Thales Optronics Raven day/night sight as fitted to the standard Warrior vehicle.

Between the observer and the assistant's position is a remote display for the thermal imaging system. The vehicle is also to be fitted for the Thales Manportable Surveillance and Target Acquisition Radar (MSTAR).

The vehicle is fitted with an inertial land navigation and attitude reference system, BATES (Battlefield Artillery Target Engagement System) and extensive communications equipment.

MAOV has a crew of six consisting of the driver, two observers in the turret, one of whom commands the vehicle, and three in the rear to operate communications equipment and Thales MSTAR.

First production vehicles were delivered to the Royal Artillery early in 1990.

Battery command vehicle

This has been designed to meet the requirements of the Royal Artillery with first production vehicles completed in 1990.

Warrior with MILAN

This conversion was originally developed for Operation Desert Storm with a number of kits supplied to convert existing vehicles to carry out this role. Basically, the Euromissile MILAN ATGW launcher is pintle mounted on the right side of the turret roof, with racks provided internally for additional missiles.

Late in 1991, the UK MoD ordered another batch of Warriors to carry the MILAN ATGW system, a role previously undertaken by FV432s which lack the cross-country mobility to work with Warrior vehicles and Challenger 1 and 2 MBTs.

Warrior with mineclearing attachments

The Warrior can be fitted with the front-mounted Pearson Engineering Surface Mine Plough or the Combat Dozer Blade UDK-2.

Warrior export vehicles

Alvis Vehicles is offering a wide range of variants of the Warrior for the export market in addition to those already manufactured for the British Army.

Reconnaissance vehicle

This was developed as a private venture and is no longer being marketed. Additional details were given in *Jane's Armour and Artillery 2000-2001*.

Warrior LMT105 light tank

This is a modified Warrior chassis fitted with the latest version of the South African LMT 105 turret system. Additional details were given in *Jane's Armour and Artillery 2000-2001*. This so far remains a concept only.

Warrior 2000

This is covered in detail in a previous entry.

Warrior utility vehicle

As a private venture, Alvis Vehicles has completed the prototype of a full-tracked Warrior utility vehicle in a command post configuration.

The baseline hull is similar to that used for the Warrior infantry fighting vehicle.

For the command post role the Alvis Vehicles Warrior utility vehicle is equipped with the Aerosystems Apache planning station, Giat Industries FINDERS (Fast Information, Navigation, DEcision and Reporting System) battle management system, Thales Bowman PD1 radio fit, Siemens GP3 software elements, European Data Systems IARRCIS liaison computer and Lynwood Genesis rugged computer.

This model of the Warrior utility vehicle is fitted with an Alvis Vehicles turret armed with a 7.62 mm Chain Gun and two banks of four electrically operated smoke grenade launchers. Stowage bins are provided on the roof as are roof hatches and in place of the standard door, a power-operated ramp could be fitted.

The Warrior utility vehicle has also been demonstrated, fitted with the BAE Systems, RO Defence 81 mm mortar.

Arctic Warrior

A vehicle of similar configuration to Desert Warrior, but with a high-performance integrated diesel air-to-air heater replacing the air conditioning system.

Desert Warrior

Shown for the first time at the 1986 British Army Equipment Exhibition was Desert Warrior.

Desert Warrior is aimed at the export market, mainly the Middle East, and is fitted with a stabilised Delco Defense two-man power-operated turret armed with a 25 mm The Boeing Company 7.62 mm Chain Gun and a bank of four electrically operated grenade dischargers either side of the turret.

In 1990, Desert Warrior was shown with the Delco Defense turret with a Raytheon Systems Company TOW ATGW launcher mounted either side of the turret. The optional TOW system has an integrated Raytheon Systems Company sight providing thermal imaging and TOW 2 ATGW tracking. All vehicles will be provided fitted for TOW to allow a TOW retrofit by field engineers if the system is not fitted to all turrets initially.

Desert Warrior seats seven infantry facing each other, with up to seven firing ports, three in each side of the hull and one in the redesigned door. These are an option and not fitted to Kuwaiti vehicles.

The Desert Warrior is also fitted with a high-performance combined air conditioning and NBC filtration system.

Desert Warrior for Kuwait

In August 1993, the now Alvis Vehicles was awarded a contract for the supply of 254 Desert Warrior infantry fighting vehicles and variants to Kuwait.

The bulk of the order was for the Desert Warrior which is fitted with the US Delco Defense LAV-25 two-person power-operated turret armed with a 25 mm The Boeing Company M242 Chain Gun, a 7.62 mm coaxial machine gun and a single launcher either side of the turret for a Raytheon Systems Company TOW ATGW system.

In addition to the Desert Warrior infantry section vehicle, other variants were ordered including repair, repair and recovery and command post vehicles.

In November 1994, the company handed over the first of eight Desert Warrior vehicles to the Kuwait Land Forces; these were used for training instructors in the UK. Desert Warrior entered service with the Kuwait Land Forces in mid-1995.

In addition to the installation of the Delco Defense turret with TOW ATGWs, 25 mm Chain Gun, day/thermal sights and additional turret armour, the Desert Warrior in service with Kuwait has a number of other modifications.

The Warrior chassis has been modified and is fitted with additional passive armour to its front and sides. There are no firing ports in the rear troop compartment.

The driver has a new hatch cover with three day periscopes to give improved vision to the front and sides and the central periscope can be replaced by a passive day periscope for night driving.

The Perkins Engines Company power pack has been improved to operate in the high ambient temperatures encountered in the Middle East. A full engine-driven air conditioning system has been fitted, as has an NBC system of the overpressure type.

The other variants in service or proposed for Warrior are available to the Desert Warrior standard.

Other Variants

Warrior Engineer Vehicle, mechanised engineer combat vehicle with cupola-mounted 7.62 mm machine gun and minelaying system on hull roof at the rear.

MCV/ATGW, with various ATGW systems, for example TOW, HOT or MILAN.

MCV/AA, with two-man turret armed with twin 30 mm cannon.

MCV/MLRS, carrying MLRS at rear (longer chassis).

MCV/MLC, modular load carrier (longer chassis).

Warrior 81 mm or 120 mm mortar carrier.

120 mm AMS, Warrior can also be fitted with the Delco Systems/RO Defence 120 mm Armoured Mortar System.

Warrior with CM 90 Turret

In mid-1984, Warrior fitted with the Belgian Cockerill CM 90 two-man power-operated turret with a Cockerill 90 mm Mk III gun, a 7.62 mm coaxial machine gun, an OIP fire-control system and a two-axis power drive system was tested in Belgium. This combination remains at the prototype stage.

Vehicle Electronics Research Defence Initiative — 2

A modified Warrior chassis has been used as the basis for the Vehicle Electronics Research Defence Initiative — 2 (VERDI-2) programme. Details of this were given in *Jane's Armour and Artillery 1998-1999*, pages 579 to 580.

Status

Production complete, but can be resumed. In service with Kuwait (Desert Warrior) and UK.

Manufacturer

Alvis Vehicles Ltd.

UPDATED

Alvis Vehicles FV432 armoured personnel carrier

Development

In the late 1950s, the then GKN Fighting Vehicle Development Division (FVDD) was awarded a contract for the construction of four prototype and 10 trials vehicles of the FV420 unarmoured light tracked vehicle, all of which were delivered by 1958.

Subsequently, GKN's FVDD was awarded the contract for the design and development of the FV432 family of APCs with the first contract covering four prototypes and 13 vehicles for troop trials. In addition, Royal Ordnance (today BAE Systems, RO Defence) was awarded a contract to build a further seven troop trials vehicles under GKN's design parentage. All of these vehicles were delivered by 1961.

In 1962, GKN Sankey, which is now Alvis Vehicles, was awarded a production contract for the FV432 APC and first production vehicles were completed in 1963. In addition, the company was awarded design and development contracts for the FV431 Light Tracked Load Carrier, which did not enter service, and the FV434 Armoured Fitters' Vehicle. Production continued until 1971, by which time around 3,000 FV432s had been built. The first production models were the Mk 1, followed by the Mk 2 and finally the Mk 2/1. Late production vehicles are distinguishable by the NBC pack in the right side, which is almost flush against the hull rather than protruding as on earlier vehicles.

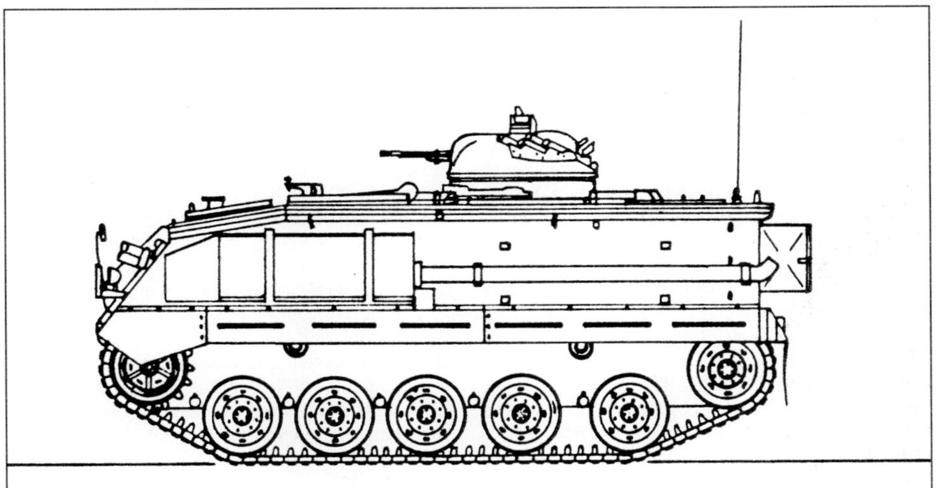

FV432 APC fitted with a Peak Engineering lightweight turret armed with a 7.62 mm GPMG

The FV432 replaced the Alvis Saracen (6 × 6) APC in British Army service years ago and was itself supplemented from 1988 by the Warrior mechanised combat vehicle.

FV432s were deployed to Saudi Arabia and took part in Operation Desert Storm in early 1991. A simple air conditioning system for the FV432 was developed, but this did not enter service due to the rapid end to the conflict. More recently the FV432 APC and the Alvis Vehicles Warrior have been deployed to the former Yugoslavia.

The FV432, in its many specialised roles, such as mortar locating radar, engineer, ambulance and 81 mm mortar, is expected to remain in service for some years.

FV432 with a Thales Cymbeline mortar-locating radar in the operating position (Richard Stickland)

FV432 with Fox turret modified to resemble Russian BMP-2 infantry combat vehicle (Richard Stickland) **2000**/0080329

Late in 2001 there were unconfirmed reports that the UK was considering extending the life of the FV 432 by carrying out a number of mobility enhancements.

Description

The all-welded steel hull of the FV432 provides complete protection against small arms fire and shell splinters. The driver sits at the front of the vehicle on the right side and has a single-piece hatch cover that opens to the left. This has an AFV No 33 Mk 1 wide-angle day periscope which can be replaced for night driving by an L5A1 passive periscope.

The commander sits behind the driver and has a cupola that can be traversed manually through 360°. This cupola has a single-piece hatch cover and three AFV No 32 Mk 1 day periscopes. Mounted on the forward part of the commander's cupola is a 7.62 mm GPMG. Many FV432s have been fitted with a Peak Engineering lightweight turret, armed with a 7.62 mm GPMG which has an elevation of +50°, depression of –15° and 360° traverse. The turret has a single-piece hatch cover that opens to the rear, three day periscopes and eight 66 mm grenade dischargers mounted four either side of the turret. The turret is mounted over the forward part of the circular troop compartment hatch. The original four-part hatch has been removed and replaced by a circular steel piece which contains the turret, with a hatch to the immediate rear of the turret. Some of these turrets have been removed from FV432s and fitted to Alvis Vehicles Saxon (4 × 4) vehicles deployed to Bosnia.

The engine compartment is to the left of the driver with the air inlet (forward), air outlet (rear) louvres in the roof and the exhaust pipe running along the left side of the hull. The complete engine with its oil tanks and filters is mounted on a common subframe which can be removed from the vehicle as a complete unit and reconnected on the ground by extended cables and fuel lines for testing. The engine compartment is fitted with a FIREWIRE detection system. The engine is coupled to the General Motors Allison Division TX-200-4A semi-automatic transmission, which was built under licence in the UK by Rolls-Royce. This in turn supplies power to the steering unit at the front of the hull via a universally jointed propeller shaft. Access to the steering system for maintenance is through a forward-opening hatch in the glacis plate.

The troop compartment is at the rear of the FV432 with 10 (five a side) infantrymen seated on bench seats that run down either side of the hull. The seats are hinged to the lower side plates on either side of the compartment and fold upwards, enabling the vehicle to carry up to 3,670 kg of cargo. The infantrymen enter and leave the vehicle via a large door in the rear of the hull which opens to the right and is provided with a vision block. Over the top of the troop compartment is a circular hatch cover that opens to the left and right of the vehicle, each part being hinged in the middle, 'concertina' fashion.

The torsion bar suspension either side consists of five dual rubber-tyred roadwheels with the drive sprocket at the front, idler at the rear and two track-return rollers. The first and last roadwheel stations have a friction shock-absorber and the upper part of the track is covered by a light sheet-metal covering. The steel tracks are rubber-bushed and fitted with removable rubber pads.

There is an NBC system mounted in the right side of the hull which provides fresh air via ducts to both the driver's and troop compartments.

Variants

(excluding trials variants)

Ambulance

This model is unarmed and carries four stretcher patients or two stretcher and five seated patients, plus a crew of two. The stretchers are easily loaded and unloaded from the vehicle by means of sliding swivel racks.

Command

This model is fitted internally with two map boards and additional communications equipment. To increase the available working area a penthouse measuring 3.66 × 2.74 × 1.98 m can be quickly erected at the rear of the hull. The command model has a crew of seven and weighs 15,500 kg.

81 mm mortar carrier

This is the basic vehicle fitted with a now BAE Systems, RO Defence 81 mm L16 mortar in the rear of the hull on a turntable which can be traversed through a full 360°. The mortar has a maximum range of 5,660 m and 160 mortar bombs are carried. The vehicle has a crew of six, weighs 16,400 kg and is issued on the scale of six per armoured infantry battalion.

Maintenance carrier

This model is designated the FV434 and is operated by the Royal Electrical and Mechanical Engineers (REME). On the right side of the vehicle is an HIAB crane which has a lifting capacity of 1,250 kg at 3.96 m radius to 3,050 kg at 2.26 m radius. The suspension on the FV434 can be locked when the crane is being used. The FV434 has a crew of four and weighs 17,750 kg fully loaded.

Minelayer

The FV432 is used by the Royal Engineers to tow the Barmine layer which can lay anti-tank mines at a rate of 600/h.

Radar

The FV432 is used to mount the Thales Cymbeline mortar locating radar (range 20,000 m).

Recovery

This is the basic FV432 with a winch installed in the rear troop compartment. The conversion kits were supplied by the British Rail Derby Locomotive Works from 1972. The winch is driven from the Power Take Off (PTO) on the engine transfer case and has both high and low-speed gears. The winch has a maximum pull, using a three part tackle, of 18,299 kg and maximum winching speeds are 18.34 m/min at 6,608 kg and 122.4 m/min at 3,050 kg.

Royal Artillery vehicles

In addition to using the FV432 with the Thales Cymbeline mortar locating radar, the Royal Artillery uses the FV432 modified for a number of other roles including fitted with the Battlefield Artillery Target Engagement System (BATES). This version is recognisable by its external roof stowage and pole-type antenna mounted externally on the side of the hull.

Specifications

Crew: 2 + 10
Combat weight: 15,280 kg
Unloaded weight: 13,740 kg
Power-to-weight ratio: 15.7 bhp/t
Ground pressure: 0.78 kg/cm²
Length: 5.251 m
Width: 2.8 m
(over tracks) 2.527 m
Height:
(including machine gun) 2.286 m
(to hull top) 1.879 m
Ground clearance: 0.406 m
Track: 2.184 m
Track width: 343 mm
Length of track on ground: 2.819 m
Max road speed:
(6th gear) 52.2 km/h
(5th gear) 37.6 km/h
(4th gear) 27 km/h
(3rd gear) 19.5 km/h
(2nd gear) 13.7 km/h
(1st gear) 9.8 km/h
(reverse) 8.65 km/h
Fuel capacity: 454 litres
Max road range:
(diesel) 480 km
(petrol) 424 km
Fording: 1.066 m
Gradient: 60%
Vertical obstacle: 0.609 m
Trench: 2.05 m
Engine: Rolls-Royce K60 No 4 Mk 4F 2-stroke, 6-cylinder multifuel developing 240 bhp at 3,750 rpm
Transmission: TX-200-4A semi-automatic with 6 forward and 1 reverse gears
Suspension: torsion bar
Electrical system: 24 V
Batteries: 6 × 12 V
Armament: (main) 1 × 7.62 mm MG
Smoke-laying equipment: 2 × 3 66 mm grenade dischargers
Ammunition: 1,600 rounds of 7.62 mm (belted when fitted with GPMG)
Armour: 6-12 mm
Gun stabiliser:
(vertical) no
(horizontal) no
NBC system: yes
Night vision equipment: yes

Royal Engineers vehicles

In addition to using the FV432 for towing the Barmine layer, the Royal Engineers use it to tow the Giant Viper mine clearance trailer. The Bar mine system is now being supplemented by the US Volcano system installed on the Alvis Vehicles Shielder chassis, which is a further development of the Stormer.

Royal Signals vehicles

The Royal Signals use a number of specialised communication versions of the FV432 under the

FV434 maintenance carrier (Shaun C Connors) **2001**/0099486

FV432 used in the ambulance role with additional roof stowage box (Richard Stickland) **2002**/0116780

designation FV439, most of which have extensive antenna arrays on the roof. It is also used to house parts of the Wavell and Ptarmigan systems.

FV438 Wavell

The FV438 was originally fitted with a twin launcher for the Swingfire ATGW; these have now been withdrawn from service and some have been fitted with the Wavell Automatic Data Processing (ADP) system, designed to assist formation headquarters' staff to carry out their command and control functions. The Wavell vehicle is usually referred to as the FV435 and is recognisable by the large box mounted on the roof.

Opposition force vehicles

The British Army has modified a number of members of the FV432 family as opposition force vehicles. Examples of these include the FV432 with the 30 mm RARDEN turret which has been fitted with a mockup of the Russian Sagger ATGW over the 30 mm cannon so that the vehicle now resembles a Russian BMP-2 infantry combat vehicle. Another model has been modified to resemble the Russian ZSU-23-4 self-propelled air defence system.

Status

Production complete. In service with the British Army.

Manufacturer

Alvis Vehicles Limited.

UPDATED

Alvis Stormer combat vehicle

Development

In 1980, Alvis purchased the manufacturing and marketing rights of the FV4333 from the British Ministry of Defence.

The vehicle was originally designed by the then Military Vehicles and Engineering Establishment and the first prototype made a brief appearance at the mobility demonstration at Bovington during the 1978 British Army Equipment Exhibition. In June 1981, Alvis announced that its development of the FV4333 had been designated Stormer.

The Stormer has taken advantage of the existing technology of the Alvis Scorpion CVR(T) family and many of the proven components have been used. The transmission is the same as on the CVR(T) but with a modified input gear train (with the designation changed from TN15 to T300) and the suspension uses CVR(T) components with wheel stations increased to six per side.

The hull of the Stormer is wider than the CVR(T) and its original petrol engine has been replaced by a Cummins or Perkins Engines Company turbocharged diesel engine.

Late in 1981, Malaysia placed an order with Alvis for 26 Scorpion 90 vehicles and 25 Stormer APCs, production of which commenced in 1982 with first deliveries made in 1983. Of the Stormer APCs, 12 vehicles are fitted with Helio FVT 900 one-person turrets armed with a 20 mm Oerlikon Contraves cannon and a 7.62 mm machine gun and the remainder with a now Rheinmetall Landsysteme TH-1 turret armed with twin 7.62 mm machine guns.

The first-generation Stormer was sold to the US government (three for the Light Armored Vehicle competition) and to Malaysia, which took delivery of 25 vehicles between 1983 and 1985.

Further development, especially in the area of Reliability, Availability, Maintainability and Durability (RAM-D), resulted in the second-generation Stormer. This vehicle, following a competition, was selected as the platform for the Thales Air Defence UK Starstreak High-Velocity Missile (HVM) system ordered by the British Army.

In 1995, Alvis Vehicles completed the last of 145 Stormers for this contract of these, 135 were for the Starstreak High-Velocity Missile (HVM) system and the remaining 10 for the Troop Reconnaissance Vehicle (TRV) requirement.

Alvis Stormer vehicle with the Starstreak HVM installation in service with the Royal Artillery (Michael Jerchel) **2001**/0099485

For operation Desert Storm, Alvis Vehicles designed and built six specialised platform versions of the vehicle called the Stormer Vehicle Launched Scatterable Mine System (VLSMS). This was fitted with the Giat Industries Minotaur scatterable anti-tank minelaying system to meet an urgent operational requirement.

Although it had taken delivery of the Giat Industries Minotaur system, the British Army subsequently had a competition to meet its main requirement and two systems were then trialled, the Giat Industries Minotaur and the US Alliant Techsystems Volcano. The latter system was selected with the chassis again being the Stormer flatbed.

Alvis Vehicles has also delivered to Oman four Stormer vehicles in the command post configuration. These form part of the £140 million contract placed by Oman in 1993 which included Vickers Defence Systems Challenger 2 MBTs and variants as well as UNIPOWER (6 × 6) tank transporters.

For all the specialised roles previously undertaken by members of the Scorpion family, for example armoured personnel carrier (Spartan), ambulance (Samaritan), command post vehicle (Sultan) and recovery vehicle (Samson), the Stormer chassis is now the preferred choice as not only is it a later development but also has a significant increase in internal volume over that of the older versions.

Early in 1995, Indonesia placed a major order for Scorpion 90 vehicles and a complete family of Stormer variants. All these feature the latest upgraded suspension, new Perkins Engines Company Phaser diesel engine and a complete air conditioning system.

The Phaser in the Scorpion 90 is rated at 180 hp and coupled to the David Brown Engineering Limited TN15D transmission, while the Stormer diesel is rated at 250 hp and coupled to a David Brown Engineering Limited T300 transmission.

All the Indonesian Scorpion 90s and Stormer vehicles (with the exception of the recovery and flatbed versions) are fully amphibious and are propelled in the water by their tracks, or using a sprocket-mounted propulsion system.

In addition to the basic armoured personnel carrier version of the Stormer, Indonesia ordered a complete family of variants including command post vehicle, ambulance, recovery, logistic carrier and bridgelayer. The last two vehicles are flatbeds with the bridgelayer carrying a launcher and a class 30 scissors-type bridge which when opened out is 15 m long.

The first Stormer command post vehicles were shipped to Indonesia in the summer of 1995, within six months of the production order being placed. Final deliveries were made in 1997.

Late in 1996, Alvis won a second contract from Indonesia for the supply of approximately 50 members of the Scorpion family of light tracked armoured vehicles.

Indonesian Alvis Stormer in armoured recovery vehicle configuration for Indonesia **2002**/0116787

Indonesian Alvis Stormer with scissors bridge in travelling configuration **2002**/0116798

Total value of this contract was £80 million with first deliveries in 1998. According to United Nations sources deliveries of the Scorpion/Stormer family of vehicles to Indonesia was as follows:

Year	**Quantity**
1995	27
1996	16
1997	11
1998	39

The Indonesian Scorpion/Stormer order is in three phases of which the first two have now been let.

The Indonesian vehicles are powered by a Perkins Engines Company Phaser diesel, have upgraded suspension system and air conditioning system.

In addition to the latest Scorpion 90 vehicles, the second contract also includes Stormer command post vehicles, armoured personnel carriers, armoured recovery vehicles, ambulances, flatbed cargo carriers and an armoured vehicle-launched bridge carrying a Class 30 scissors bridge.

Production of all members of the Stormer was undertaken at Alvis Vehicles facility in Coventry. This was closed in mid-1999 and today production is undertaken, as required, at the Alvis Vehicles factory in Telford, which was previously operated by GKN Defence.

Description

The hull of the Stormer is made of all-welded aluminium armour with the driver at the left front, engine compartment to the right and the troop compartment at the rear. The driver has an adjustable seat and a single-piece hatch cover that opens forwards. In front there is a single ×1 day periscope which can be replaced by a passive night driving periscope.

The Perkins Engines (Shrewsbury) T6.3544 diesel is mounted to the right of the driver with air induction through armoured louvres and Farr air cleaners. Engine cooling is by a single mixed-flow fan drawing air through the radiator, over the engine and through the outlet louvres. The exhaust is also cooled by some of this air. The engine is coupled to a David Brown Engineering Limited T300 hot shift, foot-operated, seven-speed gearbox with centrifugal clutch. An automatic shift option for the transmission is available. The transmission contains a controlled differential steering system giving turning circle radii varying with the gear ratio, and a neutral or pivot turn when needed.

From the gearbox the drive goes to the front sprockets and then the tracks. Track adjustment is by hydraulic ram and hand pump.

Alvis Vehicles is now offering the Stormer with a Cummins diesel engine and the David Brown Engineering Limited T303 automatic transmission. The latter is installed in the Shielder vehicles delivered to the British Army.

The vehicle commander sits behind the driver and has a No 16 cupola with eight ×1 day periscopes and a periscope with a magnification of ×1 and ×10 for aiming the 7.62 mm GPMG mounted on the right side of the cupola. This can be aimed and fired from within. The cupola has a single-piece hatch cover that opens to the left.

The section leader sits to the right of the vehicle commander and has four day periscopes and a hatch cover.

The number of troops carried depends on the armament installation but it would normally be eight, seated four down either side of the troop compartment on bench seats that can be folded up to enable stores to be carried. The infantry enters and leaves the vehicle through a large single door that opens to the right and incorporates a single vision block.

Over the top of the troop compartment there is a large two-piece hatch that opens either side of the hull. On the prototype there was no provision for the infantry to aim and fire its small arms from inside the vehicle, but this capability has been provided on one of the trials Stormer vehicles.

The torsion bar suspension either side consists of six dual rubber-tyred aluminium roadwheels with the drive sprocket at the front and the idler at the rear. Tracks are single pin, steel links, with rubber bushes and integrally mounted rubber pads. The first, second and sixth roadwheel stations have hydraulic shock-absorbers. Early Stormers did not have track-return rollers but the latest versions have two per side.

The Stormer has an NBC and/or air conditioning system in the left side of the hull, below the commander's station, and can be fitted with passive night vision equipment. There are two four-barrelled electrically operated grenade dischargers mounted on the front of the hull. Appliqué armour can be added to the vehicle for increased protection.

The basic model can ford to a depth of 1 m without preparation but with a front flotation screen it is fully amphibious, propelled in the water by its tracks at a speed of 6.5 km/h. When fitted with an appliqué propeller kit its water speed is 9.5 km/h.

Optional equipment includes a No 27 cupola, a 7.62 mm machine gun, an NBC system, an air conditioning system, firing ports and vision blocks, a driver's night periscope, night devices for the commander, fire warning and protection equipment, an automatic shift transmission, a flotation screen and fittings, a propeller kit and washboard, communications equipment and a ground navigation system.

Further evaluation of the Stormer design is evident as in early 1991 the Stormer VLSMS (Vehicle-Launched Scatterable Mine System) vehicles were shown. These have a distinctive sharper front end and revised lighting layout. The vehicle is also longer and incorporates radar Stealth technology in the outer structure of the vehicle. Further details of this version are given later.

Variants

The prototype of the Stormer was fitted with a No 16 cupola armed with a 7.62 mm GPMG but the basic hull can be adapted to mount either a gun or missile turret. This is achieved by modification of the roof section and installation of compatible turret rings. Typical variants include:

- Two-person turret armed with a 76 or 90 mm gun
- Two-person turret armed with 20, 25 or 30 mm cannon
- ATGW vehicle with HOT, TOW or MILAN systems
- Anti-aircraft vehicle, gun or missile
- Engineer vehicle
- Load carrier
- Minelayer
- 120 or 81 mm mortar carrier
- Repair/recovery vehicle
- Ambulance with higher roof
- Command/control vehicle.

Stormer Infantry Fighting Vehicle

The Stormer Infantry Fighting Vehicle can carry eight fully equipped infantry plus the gunner and driver. The example shown in 1984 was fitted with a Helio FVT 900 one-person turret armed with a 20 mm Oerlikon Contraves KAA cannon (with 100 ready rounds) and a 7.62 mm coaxial machine gun (200 ready rounds). It has firing ports and vision devices in the rear troop compartment.

Stormer 30

Details of the Stormer 30, unveiled in 1997, are given in the previous Reconnaissance vehicles section. As of late 2001, this remained at the prototype stage.

Stormer VLSMS

Under Operation Granby, a quantity (believed to be six) of Stormer Vehicle-Launched Scatterable Mine Systems (VLSMS) was ordered from Alvis late in 1990

Alvis Shielder anti-tank minelayer now in service with British Army **2000**/0080335

Alvis Stormer ambulance **2002**/0116779

Specifications
APC
Crew: 3 + 8
Max combat weight: 12,700 kg
Power-to-weight ratio: 19.68 bhp/t
Ground pressure: 0.46 kg/cm²
Length: 5.27 m
Width: 2.4 m
(over stowage boxes) 2.764 m
(over tracks) 2.31 m
Height: (top of MG) 2.27 m
Ground clearance: 0.40 m
Track: 1.892 m
Track width: 410 mm
Length of track on ground: 3.112 m
Max road speed: 80 km/h (forward and reverse)
Max water speed:
(tracks) 5 km/h
(propeller) 10 km/h
Acceleration: (0-50 km/h) 24 s
Fuel capacity: 405 litres
Max road range: 650 km
Fording: 1.1 m
(with preparation) amphibious
Gradient: 60%
Vertical obstacle: 0.6 m
Trench: 1.75 m
Theoretical turning circle radii:
(1st gear) 2.04 m
(2nd gear) 4.64 m
(3rd gear) 6.58 m
(4th gear) 11.01 m
(5th gear) 20.09 m
(6th gear) 26.2 m
(7th gear) 40.89 m
Engine: Perkins Engines Company T6.3544 water-cooled 6-cylinder turbocharged diesel developing 250 bhp at 2,600 rpm (or Cummins diesel)
Transmission: David Brown Engineering Limited T300 crossdrive, semi-automatic hot-shift type, providing 7 speeds in each direction and pivot turns (or T303 fully automatic transmission)
Steering: regenerative triple-differential system controlled by external disc brakes
Suspension: torsion bar with trailing arm geometry and lever-type hydraulic dampers
Wheel movement:
(static to bump) 274 mm
(static to rebound) 97 mm
Handbrake: hand lever connected by cable to contracting bands operating on periphery of main brake discs
Foot brake pedal: hydraulically connected to main brakes
Electrical system: 28.5 V
Batteries: 4 × 6 TN, 125 Ah (2 for automotive, 2 for radio) or low maintenance
Armament: depends on role
Gun stabiliser:
(vertical) optional*
(horizontal) optional*
NBC system: optional
Night vision equipment: optional

*Depends on turret model. Latest production Stormer vehicles have Perkins Phaser diesel engine

and integrated in Coventry with the Minotaur anti-tank mine scattering system from Giat Industries of France. The UK MoD awarded Alvis the first of two contracts early in October 1990 and the first vehicle was built and delivered in just 14 weeks.

These Stormer vehicles were the first to enter service with the British Army in February 1991.

While automotive components of the new Stormer are identical to those of future British Army production Stormers, the hull is of a new design and is both longer and wider.

The driver is seated at the front left of the vehicle with the engine to his right. He has a single-piece hatch cover that opens forward and a single wide-angle day periscope that can be replaced by a passive night periscope if required.

The power pack is to the right of the driver with the vehicle commander seated to the rear of the driver. He operates the minelaying controls and a pintle-mounted 7.62 mm machine gun. There is space to his right for a third crew member.

The driver's and commander's compartment has both air conditioning and NBC protection allowing mines to be laid without the crew being exposed to contamination. A bank of four electrically operated, forward-firing grenade dischargers is mounted on either side of the vehicle.

The vehicle hull is all-welded aluminium armour and the hull sides have been sloped inwards at the bottom to reduce the radar signature.

Suspension is similar to the standard Stormer except that this version has longer torsion bars and a greater length of track in contact with the ground.

Mounted on the flatbed is the Minotaur scatterable anti-tank mine system.

Although the Minotaur system was adopted for Operation Granby, the British Army had another competition and, early in 1996, the US Alliant Techsystems Volcano system was selected with the Alvis Stormer flatbed being the platform. A total of 29 Stormer flatbed systems have been ordered for this application under the name Shielder.

The first three vehicles were built at Coventry with the main production run being undertaken at Telford with first vehicles being delivered ahead of schedule late in 1999.

The Stormer flatbed can be adopted for a wide range of other roles on the battlefield including repair/recovery and for towing minelaying systems.

Stormer SPAAG

The Stormer chassis can be fitted with a number of different power-operated air defence turrets, both gun and gun/missile.

Stormer High-Velocity Missile carrier

In 1986, the Stormer chassis was selected by the British Army as the platform for its Thales Air Defence Starstreak High-Velocity Missile (HVM) system. Series production of this version commenced at the Alvis manufacturing facility in Coventry in late 1991. In this application the Stormer is armed with eight Starstreak missiles in the ready to launch position with additional missiles being carried in reserve. Mounted on the turret roof to the right of the commander's cupola is the Avimo acquisition sight and mounted between the two banks of four Starstreak missiles is the Thales Optronics Air Defence Alert Device (ADAD) which is passive.

Production of the chassis for the Thales Air Defence Starstreak HVM has been completed, with a total of 135 being built, plus 10 standard Stormer for use in the Troop Reconnaissance Vehicle (TRV) role.

The Thales Air Defence Starstreak HVM system is now operational with the Royal Artillery. Full details of this system, which is only used by the British Army, are given in *Jane's Land-Based Air Defence 2001-2002*.

Status

In production. In service with following countries:

Country	Quantity	Comment
Indonesia	n/avail	many variants, delivery from 1995
Malaysia	25	APC, delivered from 1983
Oman	4	command post vehicle, delivered 1995
UK	6	flatbed, delivered 1991, plus another 29 with Volcano mine scattering system
	135	Starstreak HVM, delivered
	10	Stormer, Troop Reconnaissance Vehicle, delivered

Manufacturer

Alvis Vehicles Limited.

UPDATED

Armoured Battlegroup Support Vehicle

Development

The Armoured Battlegroup Support Vehicle (ABSV) is required to operate with the Vickers Defence Systems Challenger 2 Main Battle Tank (MBT) and have similar protection to the current Alvis Vehicles Warrior Infantry Combat Vehicle (ICV). It will work in the direct fire zone and is required to have a high level of protection.

The ABSV was previously called the M1P1 armoured personnel carrier with the M2P2 being the Multi-Role Armoured Vehicle (MRAV). The latter is an 8 × 8 vehicle under development by ARTEC to meet the requirements of Germany, Netherlands and the UK. Details of this are given in the MRAV/GTK entry under International in the Armoured Personnel Carriers (Wheeled) section.

In mid-2000 the following companies registered an interest with the UK Defence Procurement Agency for the ABSV competition:

- Alvis Vehicles (Warrior and Warrior 2000)
- Hägglunds Vehicle (CV 90)
- Henschel Wehrtechnik (now Rheinmetall Landsystems) (Marder 1)
- Krauss-Maffei Wegmann (technology test bed)
- MOWAG (Piranha 8 × 8)
- Multi-Drive (an advanced wheeled vehicle)
- Otobreda (Dardo)
- Singapore Technologies Kinetics (Bionix)
- Steyr-Daimler-Puch (ASCOD/ULAN)
- United Defense (Mobile Tactical Vehicle Light)

While most of the contenders are currently fitted with a two-person turret armed with a 25 mm or larger cannon, the ABSV requirement calls for a much lighter weapon, for example a 12.7 mm machine gun mounted in a one-person turret or overhead weapon station.

The Armoured Personnel Carrier (APC) version of the ABSV will carry eight troops and their equipment plus the vehicle crew of two for a 48-hour battlefield mission.

It must also be air portable in larger transport aircraft such as the C-5, C-17 and C-141 and two vehicles should be carried on the Heavy Equipment Transporter (HET).

The ABSV will replace those members of the Alvis Vehicles FV432 series of APC and Alvis Vehicles Combat Vehicle Reconnaissance (Tracked) that are not being replaced by the MRAV (8 × 8) for which the development contract was awarded late in 1999.

As well as the basic APC, other possible versions of the ABSV will include command post, armoured treatment and evacuation vehicle, mortar carrier, fire control and repair and recovery.

Total value of the ABSV programme is estimated to be around £1.5 billion with a possible in-service date of 2009. The figures below come to 1,105 vehicles, but it is expected that only between 800 and 1,000 ABSVs will be procured.

Possible quantities of ABSV are:

(1) Armoured personnel carrier (400)
(2) Command post vehicle (270)
(3) Armoured treatment and evacuation vehicle (150)
(4) Armoured mortar vehicle (90)
(5) Mortar fire-control vehicle (45)
(6) Armoured repair and recovery vehicle (150)

It is expected that the DPA will eventually down select to about three vehicles which will then be tested, and following bids, one will be selected for production.

Under current plans, production of ABSV could start in 2007 with first deliveries in 2008 and an in-service date of 2009.

In 2001, it was reported that some consideration was being given to integrate the ABSV requirement into the Future Rapid Effects System (FRES) requirement. Another alternation was to upgrade some of the FV432 series vehicles.

Status

British Army operational requirement.

Manufacturer

Not selected.

UPDATED

UNITED STATES OF AMERICA

United Defense LP M2 Infantry Fighting Vehicle/M3 Cavalry Fighting Vehicle

Development

In April 1972, the US Army issued a Request For Proposals (RFP) for a new MICV. A number of companies submitted proposals to meet this requirement and three companies were shortlisted: Chrysler Corporation, (today General Dynamics Land Systems) the then FMC Corporation (now United Defense LP) and Pacific Car and Foundry.

In November 1972, an Engineering Development and Advanced Production Engineering contract was awarded to the Ordnance Division of FMC Corporation (now United Defense LP). The total value of this contract was US$29.3 million which covered the cost of the design, development and fabrication of three prototype vehicles, a ballistic vehicle, 12 pilot vehicles and associated systems engineering, product assurance and test support.

The prototypes, called the XM723, had a crew of three (commander, gunner and driver) and carried eight fully equipped infantrymen. The driver sat at the front on the left, with the commander to his rear and the gunner in the turret on the right side of the hull. The turret was armed with a 20 mm cannon and a coaxial 7.62 mm machine gun. The troop compartment was at the rear of the hull and the infantrymen could aim and fire their weapons from inside the vehicle. All prototypes were completed by the Summer of 1975.

In August 1976, an MICV task force was formed by the US Army to make an independent examination of the whole XM723 programme to determine whether the vehicle would meet the future requirements of the army.

The task force made a number of recommendations, of which the following were accepted by the army in October 1976:

A common vehicle would be developed for both the infantry and scout roles as the Armored Reconnaissance Scout Vehicle (prototypes of which were built by the FMC Corporation and Lockheed Missile and Space) had been cancelled.

The vehicle would be fitted with the Raytheon Systems Company TOW ATGW system and 25 mm cannon in a two-man TOW/Bushmaster Armored Turret (TBAT-II).

A two-tube TOW ATGW launcher would be mounted on the left side of the turret to give the vehicle an anti-tank capability.

The firing ports would be retained.

The vehicle would be amphibious.

The vehicle would have the same level of armour protection as the XM723.

The vehicle would be issued on the scale of four per platoon, 13 per company and 41 per mechanised infantry battalion. (The M2 would not replace the M113 one-for-one, but would replace it for specific roles such as the APC in the mechanised infantry battalions.)

The complete programme was then renamed the Fighting Vehicle System (FVS) which consisted of two vehicles, the XM2 Infantry Fighting Vehicle and the XM3 Cavalry Fighting Vehicle and was expanded in June 1977 to include a carrier for the General Support Rocket System (now known as the Multiple Launch Rocket System).

Bradley M2 infantry fighting vehicle fitted with explosive reactive armour system developed by General Dynamics Ordnance Systems and RAFAEL Armament Development Authority of Israel

The IFV and CFV are almost identical in appearance except the CFV carries five men (the crew of three and two scouts), has the firing ports blanked off and has greater ammunition capacity.

The first two prototypes of the XM2 Infantry Fighting Vehicle were handed over to the US Army at San Jose, California in December 1978 and the remaining six were completed by March 1979. In December 1979, the XM2 was type-classified as the M2 and the XM3 as the M3. In February 1980, the US Army announced that it planned to purchase 100 M2/M3s with FY80 funds. The first production contract deliveries of 100 vehicles started in May 1981 and were completed on schedule in July 1982. Second year deliveries of 400 vehicles were completed in May 1983.

In October 1981, during the first production contract, the M2 IFV/M3 CFV was named the Bradley Fighting Vehicle after the late General of the Army, Omar N Bradley.

The first unit equipped with the M2 was a battalion at Fort Hood, Texas, in March 1983, with the US Army Training Centre in Vilseck, Germany, receiving deliveries in September 1983.

Funding for the M2/M3 series for the US Army has been as follows:

Year	Quantity	Year	Quantity
FY80	100	FY88	554
FY81	400	FY89	644
FY82	600	FY90	400
FY83	600	FY91	300
FY84	600	FY92	300
FY85	655	FY93	200
FY86	716	FY94	54
FY87	662		

In the spring of 1991, the FMC Corporation was awarded a US$1.126 billion contract to supply the US Army with a further 1,600 Bradley Fighting Vehicles and 250 carriers for the Multiple Launch Rocket System. The contract included approximately 1,200 Bradleys for the US Army and 400 for sale to Saudi Arabia, with deliveries running from June 1991 through to October 1994.

In March 1992, the US Army again revised its requirement for the Bradley to 8,709 (in the end this figure was not achieved) after approving 560 Bradleys for the role of Fire Support Vehicle replacing the current M981 vehicle which uses the M113 series chassis. A total of 267 Bradleys was also authorised to perform the MANPAD Under Armor role when the US Army terminated the LOS-F-H (ADATS) programme. Company and battalion officers in mechanised infantry and cavalry squadrons were also authorised Bradleys. The M981 ITVs in the Echo company of the Mechanised Infantry Battalions have now been replaced by Bradleys.

In February 1995, the last Bradley Fighting Vehicle was delivered to the US Army. A total of 6,785 M2/M3s were built of which 400 were for the Saudi Arabian Army with the remainder for the US Army.

Marketing of the Bradley continues and the vehicle can be placed back in production if additional significant orders are received.

On the same day as the last production Bradley Fighting Vehicle was handed over, the first Bradley A0 to A2 remanufactured vehicle was also handed over.

International Acquisition

Early in 1988, Congress was advised of a Letter of Offer and Acceptance (LOA) to Saudi Arabia for 200 Bradley IFVs which was accepted. There were two vehicles delivered in FY89, 98 in FY90 and the remaining 100 in FY91.

Bradley A3 fighting vehicle launches a Raytheon TOW missile during trials (US Army) 0058052

M2A2 Bradley showing appliqué passive armour fitted to hull and turret and with TOW launcher retraced **2000**/0080336

In December 1990, a second LOA for 200 additional Bradley IFVs was approved and accepted by Saudi Arabia. The additional 200 Bradley IFV procurement by Saudi Arabia was included in the Bradley multiyear contract over and above the 1,200 vehicle US Army procurement. Deliveries for the second buy were 88 in FY92 and 112 in FY93.

Bradley in the Middle East

In March 1991, the US Army issued an initial report on Army Weapons Systems Performance in Southwest Asia with the following comment on the M2/M3: "crews reported that the sights were very effective, even during sand storms. Other crews reported that the 25 mm Bushmaster cannon was more lethal than they expected. There were no reports of transmission failure during offensive operations. Of the 2,200 Bradleys in theatre, three were disabled. To date, we have no information on the number damaged. Overall Bradley operational readiness rates remained at 90 per cent or above during combat".

Description

The hull of the original M2 Infantry Fighting Vehicle was made of all-welded aluminium armour with spaced laminate armour fitted to the hull, sides and rear. Current M2A2/M3A2 Bradleys have replaced the spaced laminate armour with a layer of appliqué steel armour plus provisions for explosive reactive or passive armour for increased battlefield survivability.

The driver sits at the front of the vehicle on the left and has a single-piece hatch cover that opens to the rear and four day periscopes, three to the front and one to the left side. The centre front day periscope can be replaced by an AN/VVS-2 passive night periscope.

The engine compartment is to the right of the driver. The Cummins VTA-903T engine is coupled to a General Dynamics Land Systems HMPT-500 hydromechanical transmission. There are three speed ranges with overall transmission ratios infinitely variable in all ranges.

The vehicle is equipped with one fixed fire extinguisher in the engine compartment, two automatic fire extinguishers in the personnel compartment and two 1.2 kg portable fire extinguishers.

The turret of the Bradley is made of welded aluminium armour with an additional layer of steel appliqué armour and is mounted in the centre of the vehicle offset to the right with the gunner seated on the left and the commander on the right. Each crew member is provided with a single-piece hatch cover that opens to the rear. The gunner has a combined day/thermal sight with an optical relay for the commander with magnifications of ×4 and ×12, as well as periscopes for front and side observation. In addition, production vehicles are fitted with a fixed power, daylight back-up sight which will allow the gunner or commander secondary sighting capability in the event of primary sight failure.

Main armament consists of The Boeing Company M242 25 mm Bushmaster Chain Gun with a 7.62 mm M240C machine gun mounted coaxially to the right of the main armament. The 25 mm cannon has dual-feed rate capability and the gunner can select single shots, 100 or 200 rds/min rates of fire. The cannon will fire both Oerlikon Contraves 25 mm and US M790 series ammunition including M791 APDS-T, M792 HEI-T and M793 TP-T, and the M919 APFSDS-T round which is now in production. The older M791 will defeat the BMP-1 at a range of 2,500 m. The M919 has even greater penetration characteristics as compared to the current M791 APDS-T round. The empty cartridge cases for both the 25 mm cannon and 7.62 mm machine guns are automatically ejected outside the turret.

The turret has 360° electric traverse and the weapons can be elevated from −10 to +60°. The General Dynamics Land Systems turret drive and stabilisation system allows the armament to be laid and fired while moving across rough country.

The TOW weapon subsystem has been developed by the now Raytheon Systems Company. When travelling, the twin tube TOW launcher is lowered and lies along the left side of the turret. The TOW system enables the M2/M3 vehicles to engage enemy armour out to a maximum range of 3,750 m. The TOW missile launcher has an elevation of +30° and a depression of −20°.

There are two M257 electrically operated smoke grenade launchers, with four smoke grenades in each, mounted on the forward part of the turret, one on either

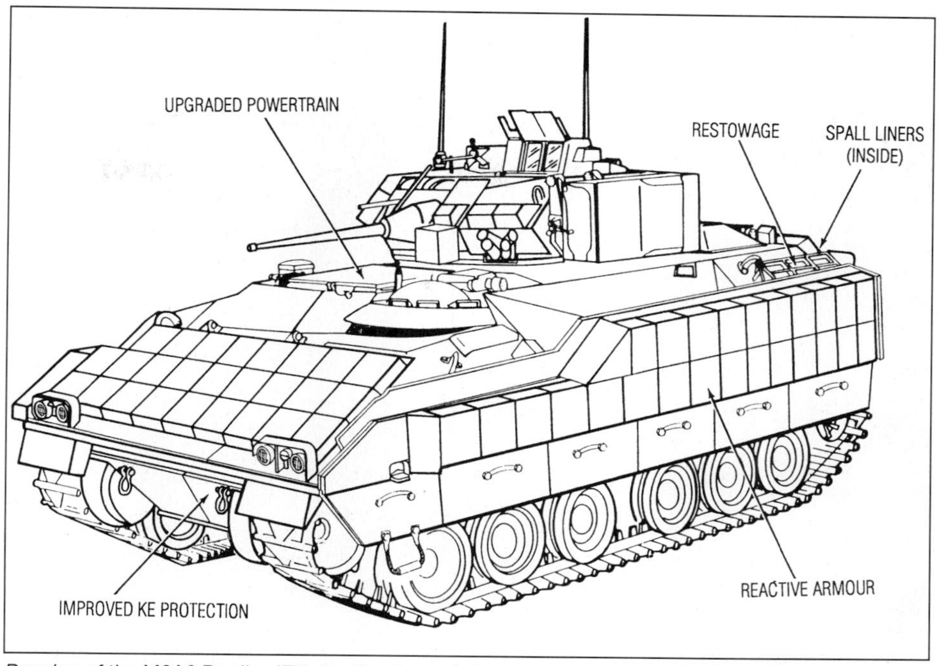

Drawing of the M2A2 Bradley IFV showing the main improvements over the earlier M2A1 vehicle

United Defense LP M7 Bradley Fire Support Team Vehicle clearly showing targeting enclosure on left side of turret which enables it to locate targets under day and night conditions when the vehicle is stationary or moving
0005210

United Defense Bradley M2A3 with Commander's Independent Viewer mounted on the right rear (behind commander's hatch) of the turret
2002/0116799

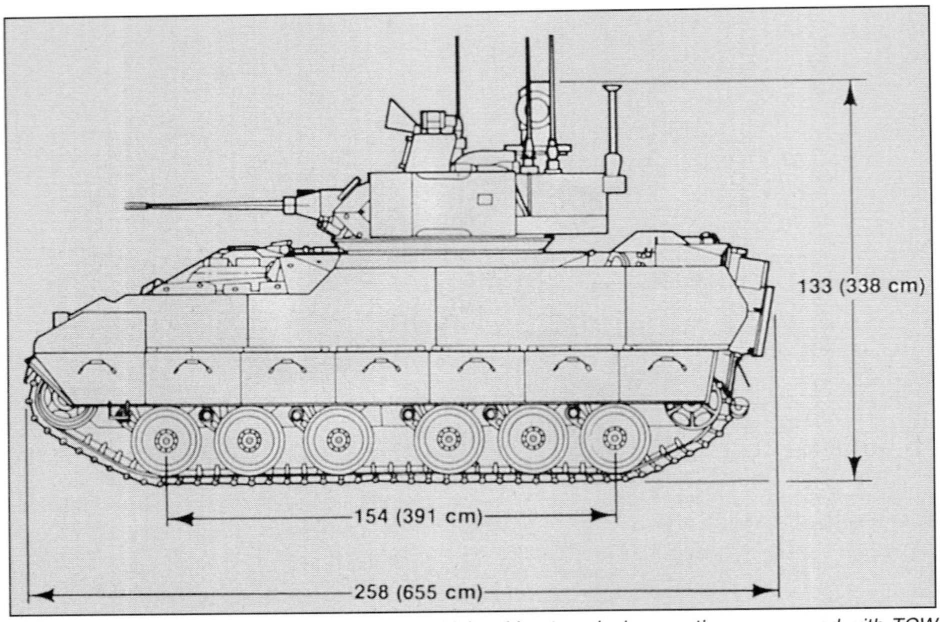

Side view of Bradley M2A3 infantry fighting vehicle without explosive reactive armour and with TOW launcher retracted **2001**/0099483

side of the main armament. New production vehicles were fitted with an engine smoke-generating system.

The M2 carries seven infantry. One sits forward of the turret on the left side facing the rear, one to the left of the turret facing the front, one at the left rear of the vehicle facing inwards, two sit at the right rear facing the back and two sit to the back of the turret facing the front. The M2A2 only carries a six-person squad as the seat to the rear of the driver's position has been eliminated. In both vehicles, the commander also dismounts with the infantry. The infantry enter and leave the vehicle through a large hydraulically operated ramp at the rear of the hull, which has an integral door in the left side. A single-piece hatch cover opening to the rear is over the top of the troop compartment. Six firing ports, two in each side of the hull and two at the rear, each with a day periscope over it, enable the crew to fire XM231 5.56 mm weapons from inside the vehicle. The M2A2 has no side firing ports but those in the ramp are retained.

The suspension system includes torsion bars, and on each side there are six dual rubber-tyred roadwheels with the drive sprocket at the front and the idler at the rear. There are two track-return rollers that support the inside of the track only and one double roller. Hydraulic shock-absorbers are fitted to the first, second, third and sixth roadwheel stations. The tracks are the single pin type with replaceable rubber pads. In 1991, the Bradley was upgraded with the bigfoot track which has a larger rubber pad area for increased track life.

In M2A1/M3A1 production vehicles a central NBC M13A1 gas particulate filter system is provided for the commander, gunner and driver but the dismounted infantry must wear individual masks when in an NBC environment.

All M2/M3 vehicles are fully amphibious and propelled in the water by their tracks. Before entering the water a special water barrier must be erected. This takes approximately 30 minutes with a trained crew.

The M2A2/M3A2 Bradley has a redesigned water barrier, but the A3 version is fitted with an inflatable pontoon system.

This system will be attached to the front and sides of the vehicle and be deployable in less than 15 minutes using an onboard inflation system that will operate continuously during water operations.

As an additional protection against loss of the vehicle due to the pontoons being punctured by small arms fire, each pontoon will be compartmentalised.

The pontoon swim system will allow for increased mobility, both on land and water, over the current curtain system, as well as allowing the firing of the weapon systems during all phases of swim operations.

The M2/M3 series of vehicle is transportable by truck, rail, cargo ship or aircraft. The Bradley can be airlifted by C-141, C-5 and C-17 aircraft. When being airlifted in the C-141 Starlifter, the head of the gunner's integral sight and the skirt plates are removed and all roadwheel arm positions, six each side, are snubbed in the up position using the steel cables provided.

Variants

M2A1 and M3A1 Bradleys

Further development of the M2 and M3 has resulted in the M2A1 (development designation M2E1) and M3A1 (development designation M3E1), which entered production in 1985, with first production deliveries in May 1986.

The M2A1 and M3A1 can fire all versions of the Raytheon Systems Company TOW ATGW including the full diameter TOW 2 with a maximum range of 3,750 m.

The NBC system is the M13A1 gas particulate filter system that has been added in both models. On the M2A1/M3A1 the commander, gunner and driver are all connected to the central NBC system, while the dismounted infantrymen in the rear have their own NBC suits and masks which allow them to leave the vehicle quickly. These are generally known as Block I improvements.

Other improvements in the M2A1/M3A1 include a turret power indicator, padding for the driver's station, redesigned weapons interlock and top deck clearance arrangements, simplified stowage strapping, new onboard rations, two camouflage nets carried, and replacement of the single water tank by two smaller ones with the same total capacity.

On the M2A1, the Dragon AN/TAS-5 night sight is now carried and up to five Dragons can be carried in place of TOWs. Mine and flare stowage has been rearranged so that more mines and fewer flares are carried. The M3/M3A1 is not equipped with the firing ports or the associated vision blocks of the M2/M2A1. The two scouts carried now have individual seats that can be raised and lowered and four day periscopes in the missile loading hatch for improved observation. For increased survivability the M2A1/M3A1 also incorporates a revised fuel system and a revised fire suppression system.

A total of 1,371 A1 vehicles was built of which 20 have been lost. These have all been upgraded to the A2 standard at Mainz (Germany) or Red River (USA).

M2A2 and M3A2 Bradleys

In late 1984, the US Army initiated a Bradley Block II study that evaluated many enhancements to improve the survivability of the Bradley vehicle and its crew. The Block II study evolved into the High Survivability Bradley programme in 1986 which was the development period for the M2A2 and M3A2 production vehicles.

The improvements include the following: enhanced vehicle survivability by restowing ammunition; internal armour protection of key components; spall liners and mounting provisions for improved vehicle smoke screening; additional armour protection including additional armour for the hull and turret; a ballistic shroud provided for the commander's back-up sight; and improved drive train and suspension.

The additional armour protection has increased weight to around 30,000 kg. It is powered by the Cummins VTA-903 engine developing 600 hp compared with 500 hp of the original production vehicle.

The General Dynamics Land Systems HMPT-500 transmission has been upgraded and the intake air cleaner expanded to accommodate the additional airflow requirements of the more powerful engine. A larger exhaust pipe has been installed and modifications have also been made to the existing air muffler.

The final drives are similar to those fitted on the MLRS system but have been improved. The suspension has been upgraded. All these improvements, except the ballistic shield of the commander's back-up sight, were included in the M2A2/M3A2 series vehicles.

The 600 hp engine was not included in the initial M2A2/M3A2 production of 662 vehicles from May

Bradley-Linebacker M6 with pod of four Stinger fire-and-forget missiles in the elevated position **2002**/0116800

Fighting Vehicle Systems Carrier fitted with Thales Air Defence Crotale New Generation air defence system 0005213

Prototype of the United Defense Armoured Medical Evacuation Vehicle

0058053

Fighting Vehicle Systems Carrier

1988 but was included in all follow-up M2A2/M3A2 production from May 1989. The US Army has now completed their retrofit with the 600 hp diesel engine.

The cut-in dates for the various survivability modifications to the M2A2/M3A2 Bradley have been as follows:

Modification	Production cut-in date
Addition of steel appliqué armour, spall liners, relocation of ammunition and addition of attachment points for armour tiles	May 1988
Addition of armour tiles (passive or explosive reactive)	backfitted
Addition of 600 hp engine	May 1989
Modification of transmission	May 1989
Changes to fuel system	May 1986
Rerouteing of fire extinguishing system cables	May 1986

The US Army plans to retrofit 2,300 of its Bradley AOs to the high-survivability A2 configuration.

Of the 6,785 Bradley vehicles produced, 5,212 are currently of the A2 configuration.

In 1991, the US Army released its initial contract for conversion kits for the high-survivability modification to the FMS Corporation (now Marvin Land Systems), Los Angeles, California, for both the M2A1 to M2A2 and the M3A1 to the M3A2 vehicles. With the exercise of an option, FMS Corporation (now Marvin Land Systems) supplied 705 kits to the US Army through 1993. The high-survivability kit consists of three key elements, turret kit, hull kit, new final drives and a 300 A alternator, with a total weight of just over 7,257 kg.

The turret kit is identical to both the M2 and M3 and consists of 150 items including passive armour blocks and their associated mountings. There are minor differences between the hull kit for the M2 and M3, although each kit consists of some 800 items including passive armour blocks and their associated mountings.

The kits are supplied to the US Army which then installs it. The first batch of 286 was supplied by the then FMC Corporation, Ground Systems Division.

In September 1995, Marvin Land Systems Incorporated (previously the FMS Corporation) was awarded a US$19.045 million fixed price contract from the US Army Tank Automotive and Armaments Command, for the supply of 177 turret kits and 177 hull kits for the Bradley.

Bradley model	FY90	FY91	FY92	FY93	FY94
A1 to A2 high survivability	54	239	395	443	240
A2, plus upgraded power train	249	292	121	0	0

Specifications (M2A3)

Crew: 3 + 7
Combat weight: 30,391 kg
Power-to-weight ratio: 17.9 hp/t
Ground pressure: 0.79 kg/cm²
Length: 6.55 m
Width:
(over hull) 3.28 m
(over tracks) 2.97 m
Height:
(to commander's sight) 3.38 m
(turret roof) 2.565 m
Ground clearance: 0.46 m
Track width: 533 mm
Length of track on ground: 3.91 m
Max speed:
(road) 61 km/h
(water) 6.4 km/h
Fuel capacity: 662 litres
Cruising range: 400 km
Fording: amphibious (with preparation)
Gradient: 60 %
Side slope: 40 %
Vertical obstacle: 0.91 m
Trench: 2.54 m
Engine: Cummins VTA-903T turbocharged 8-cylinder diesel developing 600 hp at 2,600 rpm
Transmission: General Dynamics Land Systems HMPT-500-3EC TEC hydromechanical
Steering: hydrostatic
Suspension: torsion bar
Electrical system: 28 V
Batteries: 4 × 12 V 6TN
Armament:
(main) 1 × 25 mm M242 cannon
(co-axial) 1 × 7.62 mm M240C MG
(ATGW) 2-tube TOW launcher
Smoke laying equipment: 2 × 4 smoke grenade launchers plus exhaust generator
Ammunition:
(25 mm) 300 ready, 600 reserve
(7.62 mm) 800 ready, 3,600 reserve
TOW missiles: 2 in launcher, 5 in reserve

Gun control equipment
Turret power control: electric/manual
(by commander) yes
(by gunner) yes
Max rate power traverse: 60°/s
Max rate power elevation: 60°/s
Gun elevation/depression: +60/–10°
Turret traverse: 360°
TOW launcher elevation/depression: +30/–20°
Gun stabiliser:
(vertical) yes
(horizontal) yes
NBC system: yes
Night vision equipment: yes
Armour:
(top and front slopes) 5083 aluminium*
(vertical sides and rear) spaced laminate*
Bottom: 5083 aluminium with anti-mine appliqué
*plus additional steel armour and ERA

These kits were supplied to the US Army to upgrade existing Bradley Fighting Vehicle Systems from the older M2A1 configuration to the upgraded M2A2 standard.

Operation Desert Storm modifications

As a result of Operation Desert Storm, a number of improvements to part of the Bradley fleet have been proposed. This includes a laser range-finder, a Global Positioning System with digital compass, a simple IFF capability, Driver's Viewer Enhancer, a missile countermeasure device and improvements to the stowage of internal equipment.

The resultant vehicle is referred to as the M2A2ODS/M3A2ODS. Current plans call for 1,433 A2 Bradley conversions.

The first United Defense LP, York assembled Bradley fighting vehicle was accepted in September 1995, with first production vehicles following in 1996.

Bradley Modernisation Plan M2A3

After Operation Desert Storm, the US Army initiated a Bradley Modernisation Plan (BMP) that included:

(1) Operation Desert Storm Improvements (see above)
(2) Development of a Bradley A3.
(3) Remanufacturing some old A0 model Bradleys to the A2 configuration
(4) Moving some M113-based mission packages to the Bradley chassis.

The research and development contract for US$280 million was awarded to United Defense LP in May 1994. This was for the development of the Bradley M2A3/M3A3 with the following six improvements:

Core electronic architecture — 1553 databus, central processors and memory, digital information displays for commander, driver and squad leader.

Lethality — improved target acquisition, full-solution ballistic fire control, auto-track and dual target tracking, automatic boresighting and a hunter/killer capability.

Survivability — combat identification system, roof fragmentation protection and squad ventilated facepiece.

Command and Control — command and control software, navigation software, digital communications, commander's independent thermal viewer and squad leader's FLIR monitor.

Sustainability — digital electronics, electronic training manuals, built-in test equipment and diagnostics.

Mobility — improved driver's vision.

In July 1997, the US Army signed a Low Rate Initial Production (LRIP) contract with United Defense for the supply of 35 Bradley A3 vehicles worth US$68 million with an option on another 18 vehicles.

This contract followed trials with the eight prototype vehicles, the last of which was completed early in 1997.

The US Army is expected to upgrade around 1,113 vehicles to the latest A3 standard over the next eight to 10 years. The conversions are carried out at the United Defense facility at York, Pennsylvania.

In December 1998 the US Army awarded United Defense LP a contract for a further 70 vehicles for delivery between March 2000 and March 2001.

Due to problems with the Embedded Battle Command (EBC) system, LRIP of the M2A3 will continue through FY00. Through the FY99 contract award, 126 Bradley A3s have been ordered.

The M2A3 Bradley will in the future has a new internal exhaust system with the outlet at the right rear rather than near the driver's position.

In mid-2001, the US Army M2A3 Bradley was approved for full-rate production and under a multiyear contract with United Defense LP, Ground Systems Division, 926 existing Bradleys will be fitted with the Force XXI Battle Command Brigade and Below system and the Improved Bradley Acquistion System.

Improved Bradley Acquisition System (IBAS)

In the spring of 1994, Texas Instruments Defense Systems & Electronics Group, now Raytheon Systems Company, was awarded a US$57.9 million letter contract from the US Army Aviation and Missile Command to develop a thermal imaging and fire-control system for the Bradley Fighting Vehicle A3 configuration.

Under the 41-month engineering and manufacturing development programme, the company developed an IBAS as part of the future planned upgrades to the Bradley Fighting Vehicle.

The contract included both hardware and software development, the delivery of 16 prototype systems and development of a pilot production line.

IBAS includes an integrated day and night, second-generation thermal sight, a missile control subsystem and associated software. The installation of IBAS enables the gunner to acquire targets at longer ranges and with a greater probability of a hit by using an automated target tracking device.

Bradley with ERA

Late in 1995, General Dynamics Ordnance Systems was awarded a contract valued at US$9.021 million for production of Explosive Reactive Armour (ERA) tiles for the M2 Bradley.

This contract was a follow-on to a previous US$6 million contract placed in 1993 with the company for the qualification of the ERA tiles, which followed extensive trials of three different armour (passive and ERA) packages fitted to the Bradley to improve its battlefield survivability.

During the extensive trials programme over 1,500 tiles were tested, with the actual tests including shaped charge protection, sensitivity to small arms fire, sequential environmental, transportation, artillery fragment sensitivity, sympathetic detonation, collateral damage, tile mapping and junction tests and special safety tests.

Teamed with General Dynamics Ordnance Systems, is Rafael Armament Development Authority of Israel.

Each Bradley will be provided with a total of 105 ERA tiles, which will provide a high degree of protection against weapons fitted with a HEAT (High-Explosive Anti-Tank) warhead such as the widely deployed Russian RPG-7 and most tube-launched anti-tank weapons.

The set of 105 tiles consists of 26 M3 frontal, nine M4 frontal wedge, 55 M5 side, seven M6 turret front and eight M7 turret wedge. It is not envisioned that these ERA tiles will be fitted for normal peacetime use but would be fitted in times of tension or war.

Battle Command Vehicle

Details of the Battle Command Vehicle based on a modified M1A1 Abrams MBT and a Bradley infantry fighting vehicle are given in the Armoured Command Post Vehicles section.

Bradley with 35 mm Bushmaster III

Early in 1997, The Boeing Company completed initial manned firing trials of their private venture 35 mm Bushmaster III Chain Gun™ Weapon System installed in an M2 Bradley Infantry Fighting Vehicle.

The Boeing Company is also working with Mauser of Germany on future ammunition developments. The Bushmaster III will have ammunition interoperability with the German Rheinmetall 35 mm/50 cannon which has been under development for some years to meet the requirements of the German Army.

The firing trials to date have shown that the 35 mm Bushmaster III has excellent functional reliability and dispersion characteristics.

In production weapons the gunner would be able to select single shots, controlled burst or full automatic at 200 or 300 rds/min. Muzzle velocity firing 35 mm Armour-Piercing Fin Stabilised Discarding Sabot (APFSDS) ammunition is 1,400 m/s while firing 50 mm APFSDS ammunition it is 1,600 m/s. The empty cartridge cases are ejected forwards out of the turret.

The linked feed 35 mm Bushmaster III has completed its development testing and is now ready for system integration.

While the weapon is currently chambered for standard 35 mm natures of ammunition, it can be quickly reconfigured for 50 mm ammunition if this were required. This requires only a barrel change and minor part changes.

Bradley with 40 mm CTWS

Late in 1999 the Bradley was fitted, for trials purposes, with the CTA International 40 mm CTWS (Case Telescoped Weapon Station).

Multiple Launch Rocket System

Details of this are given in the Multiple rocket launchers section.

The complete MLRS system is designated the M270 which comprises the MLRS carrier (M993) and the MLRS rocket itself (M269). The FVSC by itself (generic) is the XM987. It has not yet been approved as a stand-alone type-classified M987 chassis. When referred to as the 'FVSC on which the MLRS (M270) is carried', the carrier has to carry the M993 designation. The difference between the M993 and the XM987 is the removal of the mounting provisions in the bed for the MLRS rocket system.

This chassis is also being offered as a platform for surface-to-air missile systems such as the Euromissile Roland and Thales Air Defence Crotale New Generation.

XM5 Electronic Fighting Vehicle Systems carrier

The Electronic Fighting Vehicle Systems (EFVS) carrier has been type redesignated by the US Army as the XM5 EFVS. The prototype of the EFVS carrier was completed in 1986 and purchased by the US Army in 1989.

A further two vehicles were delivered in 1991 and 1992 and an additional four vehicles were ordered of which one was delivered. These were originally earmarked for the J-STARS Ground Station Module — Heavy but, with the termination of the J-STARS Ground Station Module — Heavy, the remaining three were transferred to the C2V programme.

By early 2000, the EFVS effort has been placed on hold while the Ground Based Common Sensor — Heavy programme is being restructured. The programme has been reconfigured to provide only a light version and redesignated as the Prophet Programme.

M4 Command and Control Vehicle

Details of this vehicle are given in the Armoured Command Post Vehicles section. United Defense LP, Ground Systems Division, has proposed that the M4 Command and Control Vehicle could be converted into Armoured Medical Treatment Vehicles.

Armored Medical Treatment Vehicle

Early in 1996, United Defense LP was awarded a US Army contract to convert one of the prototypes of the XM4 Command and Control Vehicle into an armored medical treatment vehicle and this was delivered in mid-1996.

The Armored Medical Treatment Vehicle (AMTV) has been designed to carry up to 9 litter patients or up to 10 ambulatory patients.

This prototype has been highly successful during Task Force XXI trials, both at Fort Hood and at the National Training Center, Fort Irwin, California.

A US Army requirement was identified for a total of 675 systems. However, the programme is currently unfunded. If and when the AMEV moves ahead it will be based on rebuild Bradleys rather than new build Bradleys. With the cancellation of the M4 Command and Control Vehicle programme, United Defense LP has suggested that those vehicles built could be converted into Armored Medical Treatment Vehicles.

Bradley FIST M7

Following a competition, in which they were the only bidder, United Defense LP was selected by the US Army to develop the XM7 Bradley FIST (Fire Support Team) vehicle to meet the requirements of the US Army field artillery.

The US Army previously used the M981 Fire Support Team Vehicle (FISTV), which is a modified version of the M901 series Improved TOW Vehicle (ITV) which entered service with the US Army over 20 years ago.

The FISTV and ITV are both based on the chassis of the M113 series Armoured Personnel Carrier (APC) which entered service with the US Army 40 years ago.

Both have a one-man power-operated hammerhead-type device mounted on the hull top, which in the case of the FISTV contains the targeting equipment. When travelling, this folds down onto the top of the hull to reduce the overall height of the vehicle.

In order to use the targeting equipment the M981 has to come to a halt and raise the hammerhead device and, once it has finished its mission, this has to be lowered again. The vehicle cannot be moved with the hammerhead device in the raised position.

This takes too long to accomplish and, in addition, the M981 lacks the armour, mobility and firepower to operate with mechanised units for which it is to provide artillery fire support. Apart from a 7.62 mm machine gun and smoke grenade launchers it has no armament with which to defend itself.

Before being awarded the US Army contract, as a private venture, United Defense LP converted an M2 Bradley to the XM7 configuration and demonstrated this to the US Army at Fort Sill, home of the US Army field artillery, and at other locations in the US. As a result of this the vehicle was further refined as the user suggested a number of additional improvements.

Under the US Army Engineering and Manufacturing Development (EMD) phase contract, United Defense LP has converted four standard M2 Bradley IFV into the new XM7 Bradley FIST configuration, with the first of these vehicles completed late in 1996.

The XM7 has now been type classified as the M7 Bradley FIST and late in 1997 United Defense was

CTA International 40 mm CTWS installed in a United Defense, Ground Systems Division, Bradley (US Army) **2002**/0067923

The Boeing Company Bushmaster III 35 mm/50 mm automatic cannon installed in a United Defense, Ground Systems Division, Bradley **2002**/0116801

awarded a low rate initial production contract for the delivery of 22 vehicles commencing in January 1999.

An option for additional 27 M7s for delivery in the year 2000 was also included. The total US Army requirement is for more than 300 vehicles which are based on the M2A2 Operation Desert Storm (ODS) vehicle.

It is expected that these will be followed by a further 200 vehicles based on the more recent M2A3 Bradley IFV.

These vehicles have the development designation of the XM7E1 and will be eventually type-classified as the M7A1. United Defence LP will carry out this upgrade.

The new M7 FIST is based on the current in-service Bradley M2 chassis which has the same level of armour, mobility and firepower, with the exception that the launcher on the left side of the turret which normally carries two TOW anti-tank guided weapons now carries the targeting equipment.

The latter is taken from the M981 and consists of the AN/TVQ-2 Ground Vehicle Laser Locator Designator and the AN/TAS-4B TOW night sight. In addition, the vehicle will have an inertial navigation system together with a new Ring Laser Gyro (RLG) with accuracy expected to be 80 m CEP at a range of 10 km.

Externally the M7 is virtually identical to the Bradley which makes it hard for the potential enemy to detect which vehicles are M7s and which are M2 Bradley IFVs.

The M7 Bradley FIST retains the two-man power-operated turret of the M2 Bradley which is armed with a stabilised 25 mm The Boeing Company M242 Chain Gun, 7.62 mm coaxial M240C machine gun and two banks of four forward-firing electrically operated smoke grenades. The weapons are fully stabilised, allowing the vehicle to engage targets whilst on the move across country.

As well as the standard day/night sight the gunner will also have the Integrated Sight Unit (ISU) with the eye-safe laser range-finder, with the commander having an optical relay from the gunner's sight. The eye-safe laser rangefinder is to be fitted to current Bradley vehicles as part of the ODS (Operation Desert Storm) improvement package.

Communications, command and control equipment fitted is expected to include FOCC, IFSAS, FOS, AFATDS (Advanced Field Artillery Tactical Data System) plus extensive hull and turret communications equipment.

Standard equipment for the M7 will include fire detection and suppression system, NBC system and day/night vision equipment for commander, gunner and driver. With preparation it will be fully amphibious, being propelled in the water by its tracks.

Bradley-Linebacker M6

Following a competition, in March 1995 Boeing Defense & Space Group, Missiles & Space Division, was awarded a contract by the US Army Aviation and Missile Command for the Bradley-Linebacker, with the first prototype units being completed in March 1996. Full details of the Bradley-Linebacker M6 are given in *Jane's Land-Based Air Defence 2001-2002*.

The Bradley-Linebacker Short-Range Air Defense (SHORAD) system is essentially a United Defense Bradley infantry fighting vehicle with the twin Raytheon Systems Company TOW anti-tank guided missile system replaced by a pod containing four Stinger fire-and-forget Surface-to-Air Missiles (SAM).

The stabilised 25 mm M242 Chain Gun is retained, as is the 7.62 mm coaxial machine gun.

According to the manufacturer, the Bradley-Linebacker is capable of engaging all threat targets including cruise missiles and unmanned aerial vehicles, performing reconnaissance, intelligence and target acquisition within the boundaries of the Stinger SAM.

The system is designed to perform these functions using hardware and software (common or derived) from the Avenger self-propelled SAM system on the AM General HMMWV chassis, of which over 1,000 have been built for the US Army and Marine Corps.

The main advantage of the Bradley - Linebacker is that the crew are under complete armour protection and the vehicle has the capability to keep up with mechanised units.

The four-round Stinger pod is a Standard Vehicle-Mounted Launcher (SVML). By early 1997, Boeing had completed one prototype system and eight initial production units which were used for extensive trials.

The first production Bradley-Linebacker was handed over to the US Army in November 1997, by which time two production orders had been placed, the first for 60 units followed by another for 25 units. A further 14 were built in the first half of 1999. There is a possibility of a follow on order for an additional 100 plus vehicles in 2000-2003.

Boeing provides the kits which are then installed on the M2A2 Bradley by United Defense at its York facility.

Bradley with Stinger SAM

Following the withdrawal of the M163 20 mm self-propelled Vulcan air defence systems, the US Army has been using existing Bradley Fighting Vehicles to transport Stinger man-portable surface-to-air missile teams.

The main armament of the Bradley is retained but the ammunition load is reduced. A total of five TOW missiles is carried, two of which are ready use and three inside for manual reloading.

This version has a five-man crew consisting of squad leader, senior gunner, driver and two Stinger gunners. The Stinger gunners are seated in the rear facing the rear. Six Stingers are carried, two for ready use and four in reserve.

The US Army has fielded 267 of these vehicles, mostly converted from the M2A2 configuration. The US Army has designated the Bradley carrying the MANPADS team as the Bradley Stinger Fighting Vehicle.

These were interim vehicles pending the full scale introduction of the Bradley-Linebacker M6.

Stingray

Stingray is a protection system that increases the survivability of the host vehicle and adjacent forces against Optical and Electro-Optical (OEO) equipped threats by disrupting threat fire-control systems long enough for the user to engage the hostile system by fire or to take other appropriate action.

Stingray will detect, track and CounterMeasure (CM) systems by neutralising the target acquisition system on the threat force vehicle and other ground and airborne systems beyond the threat's effective ranges.

The system consists of four basic components: (1) sensor assembly with optics, stabilisation and high-resolution detector; (2) laser transmitter; (3) commander's controls and displays; (4) system electronics which contain signal processing.

Stingray has been in advanced development since 1982 and two systems had been built by 1991; both were deployed to the Middle East for Operation Desert Storm. Six additional vehicles ordered were delivered in 1995.

Electric Drive Bradley

Details of this experimental vehicle developed under the leadership of United Defense LP were given in *Jane's Armour and Artillery 2000-2001*.

Bradley turret applications

The turret of the Bradley has been offered for installation on other chassis and is fitted to the prototype of the private venture United Defense Infantry Fighting Vehicle Light (IFVL) covered in a separate entry.

Status

Production complete but can be resumed if additional orders are received. Chassis remains in production for other applications. In service with the US Army (6,720) and Saudi Arabia (400).

Manufacturer

United Defense LP, Ground Systems Division.

UPDATED

United Defense LP Armored Infantry Fighting Vehicle

Development

In 1967, under contract to the US Army, the then Ordnance Division of the FMC Corporation built two MICVs under the designation XM765. These were based on the proven M113 APC modified to incorporate firing ports and a fully enclosed weapon station.

Further development of the XM765 by FMC as a private venture resulted in the Product Improved M113A1, the first prototype of which was completed in 1970. This had a fully enclosed weapon station in the centre of the hull, behind the driver and engine, and the commander's cupola to the rear of the turret.

This arrangement gave the commander very little forward vision so the vehicle was redesigned with the commander sitting behind the driver on the left and the turret to the rear of the engine on the right. This vehicle was subsequently called the Armored Infantry Fighting Vehicle.

The first country to order the AIFV was the Netherlands which eventually placed orders for a total of 2,079 vehicles in two batches, one of 1,264 which was co-production with the FMC Corporation and the second for 815 vehicles which was local production with a number of companies in the Netherlands.

The second order of 815 vehicles included 119 in the Improved TOW Vehicle configuration. Daf Special Products was involved in the production and final assembly of these vehicles. Daf Special Products (which recently became SP aerospace and vehicle systems bv) purchased all the materials to support its production from the now Marvin Land Systems Inc (previously FMS Corporation).

YPR 765 PRMR 120 mm mortar tractor of the Royal Netherlands Army (Richard Stickland) ***2001***/0073004

The Philippines received 45 AIFVs. In July 1979, Belgium placed an order for 514 AIFVs and 525 M113A2s to be built under licence in Belgium by Belgian Mechanical Fabrication to replace M75 and AMX VCIs in service with the Belgian Army.

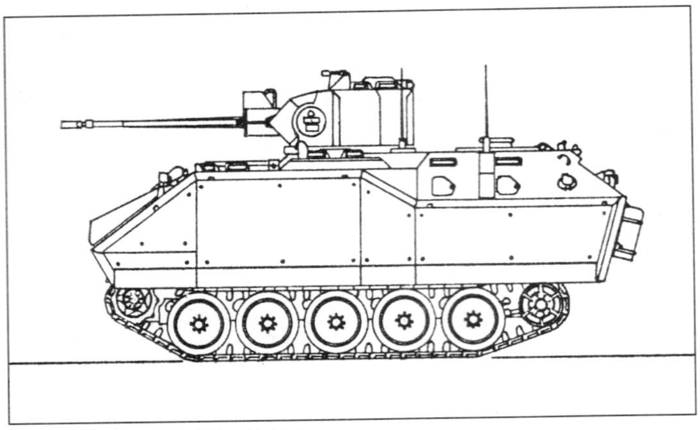

Armored Infantry Fighting Vehicle

The YPR 765 KMAR is a modified YPR PRI/I for use by the Royal Netherlands Army Military Police (D Caemerlynck) **2001**/0099482

First production vehicles were completed in 1982 and production ran at the rate of 20 to 25 vehicles (AIFV and M113A2) a month until May 1988. Additional details of the Belgian Armoured Infantry Fighting Vehicles are given in this section under Belgium.

In 1998 the AIFV was selected by Turkey to meet its future requirements and additional details of this programme are given under Turkey. In total, 1,698 vehicles are covered, with the first 285 hulls coming from Cockerill Mechanical Industries which then shipped the equipment to enable the remainder of the order to be built in Turkey by FNSS Defense Systems (previously called FMC-Nurol). More recently the Turkish Land Forces Command has placed another order for an additional 665 vehicles.

The AIFV has also been tested in Pakistan but no production has yet been undertaken in this country.

In January 1993, the Royal Netherlands Army announced substantial cuts including the phasing out of 336 of the AIFV (YPR 765) infantry fighting vehicles and 201 YPR 765 PRATs with the TOW ATGW system. In addition, all M113 series vehicles would be phased out of service.

In mid-1994, the Netherlands stated that Egypt was to purchase a total of 611 AIFVs and variants under a contract worth US$150 million. All of these were delivered by the end of 1996 in the following versions:

304 YPR 765 PRI AIFVs
6 YPR 765 PRCO-B command post vehicles
210 YPR 765 PRAT-TOW anti-tank vehicles
79 YPR 765 PRCO-C command post vehicles
12 M577A1 command post vehicles (these are not actually based on the AIFV chassis being a member of the M113 family of armoured personnel carriers).

By mid-2001, it was estimated that the Royal Netherlands Army had about 1,200 vehicles in service.

Description

The hull of the AIFV is made of welded aluminium armour with an additional layer of spaced laminate steel armour bolted onto the hull. Closed cell polyurethane foam within the armour system gives increased buoyancy for amphibious operations.

The driver sits at the front of the hull on the left and has a single-piece hatch cover that opens to the right. To the front and left side of his position are four M27 day periscopes, the centre one of which can be replaced by a passive periscope for driving at night.

The commander sits to the rear of the driver under a cupola which can be rotated through a full 360°. It has a single-piece hatch cover that opens to the rear and five day periscopes, four standard M17 day periscopes and a day periscope with a magnification of ×1 and ×6.

In the Dutch, Belgian, Filipino and first 200 Turkish versions, the engine is to the right of the driver and coupled to the three-speed automatic transmission by a transfer gear case.

The complete power pack can be removed through a front hatch, which is fitted with a door for daily maintenance tasks.

Normal steering and braking are by oil-cooled band-type brakes in the controlled differential. There is a set of air-cooled disc brakes on the differential output shafts for low-speed tight turns on land and for more efficient manoeuvring in the water. A large radiator and belt-driven high-capacity fan mounted in the top of the engine compartment provide cooling to normal high-ambient temperature limits.

The power train is similar to the M113A1's except for improved features such as the engine turbocharger, larger capacity radiator, transmission modified with heavy-duty components, heavy-duty universal joints and final drives as fitted to the United Defense M548 tracked cargo carrier.

On the first 200 Turkish vehicles the cooling system differs from the Dutch, Belgian and Filipino AIFVs in that it is the same as that of the M113A2. In the Turkish versions of the AIFV, after the first 200 vehicles, the Detroit Diesel 6V-53T 300 hp turbocharged diesel engine sits to the right of the driver and is coupled to an Allison X-200-4 four-speed automatic transmission. This is similar to the M113A3 power pack but at a higher power level. The transmission provides four forward speeds, reverse, and hydrostatic steering. The cooling system has been modified with a larger radiator and improved fan to provide improved cooling as compared to the older AIFVs and M113A3s. These vehicles are now designated the ACV-300 family.

The fully enclosed weapon station is on the right side of the hull behind the engine. It is all-welded with an additional layer of spaced armour as fitted to the hull. The gunner has a single-piece hatch cover that opens to the rear and four M27 day periscopes (two either side). In the original Belgian and Dutch vehicles a Philips sight with a magnification of ×2 or ×6 for day use and ×6 for night use is installed. In the Netherlands the Philips sights have been replaced with thermal sights produced by Oldelft. For Turkey, thermal sights have also been installed.

Main armament consists of an Oerlikon Contraves 25 mm KBA-B02 cannon with dual feed, which has 180 rounds of ready use ammunition and another 144 rounds in reserve. Mounted coaxially to the left of the main armament is an FN MAG-58 (US M240C) 7.62 mm machine gun with 230 rounds of ready use ammunition and a further 1,610 rounds in reserve in the hull (Turkish vehicles have a 7.62 mm MG3 machine gun built under licence by MKEK). Turret traverse and gun elevation are electrohydraulic, elevation is from −10 to +50° at a speed of 60°/s and turret traverse is through a full 360° at 60°/s. Dutch vehicles also have six smoke grenade launchers mounted on the front of the hull.

The troop compartment at the rear of the hull accommodates seven seated infantry. One man sits between the commander and the turret facing the rear and the remaining six sit three each side, back-to-back. In the Turkish AFV and AAPC versions, the troop compartment accommodates 10 seated infantry. One man sits between the commander and turret facing the rear, seven are seated on bench seats at the side of the vehicle facing inwards and two are seated on jump seats facing the rear. Over the top of the troop compartment is a single-piece hatch cover that opens to the rear and contains the troop compartment ventilator. There are five firing ports, two in each side of the hull and one in the ramp at the rear. Over each of the side firing ports is an M17 day periscope and over the rear firing port an M27 day periscope. To reduce accidental damage inside the vehicle, holders support the individual weapons during firing. Spent brass catching bags prevent ejected cartridge cases from hitting adjacent men.

The infantrymen enter and leave the vehicle by a power-operated ramp which opens downwards at the rear of the hull. An emergency door is provided in the left side of the ramp.

The two armoured fuel tanks at the rear of the vehicle, one either side of the ramp, are separated from the vehicle interior by armour plate.

The torsion bar-in-tube spring arrangement, which almost doubles the effective length of the torsion spring, gives a much improved ride compared with the M113A1 APC. The track, roadwheels and most of the other suspension elements are identical to the M113A1's. The suspension either side consists of five dual rubber-tyred roadwheels with the drive sprocket at the front and the idler at the rear. There are no track-return rollers. The first, second and last roadwheel stations either side have a hydraulic shock-absorber. The track is a T130E1 steel block, rubber-bushed pin type with removable rubber pads.

Due to advances in metallurgy, the Belgian and Turkish AIFVs use torsion bar only suspension, similar

Armored Infantry Fighting Vehicle of the Royal Netherlands Army fitted with ITV turret shown in elevated position (Directorate of Matériel, Royal Netherlands Army, Communications Office)

YPR 765 PRI Armoured Infantry Fighting Vehicle of the Royal Netherlands Army in Bosnia with Dragon ATGW in foreground (Richard Stickland)
2002/0116791

Dutch YPR 806 PRBRG recovery vehicle in travelling configuration with vehicle painted in United Nations markings for operations in Bosnia (Directorate of Matériel, Royal Netherlands Army, Communications Office)

to the M113A2, to achieve the same results as the older AIFV bar-in-tube springs.

The Belgian and Turkish vehicles are fitted with NBC systems with clean air provided individually to each person. All AIFVs are fitted with a personnel heater.

The AIFV is fully amphibious and is propelled in the water by its tracks. Before entering the water a trim vane, which is stowed flat on the glacis plate when not in use, is erected at the front of the hull.

Optional equipment includes a 25 mm The Boeing Company M242 cannon mounted in the electrical drive stabilised United Defense Sharpshooter turret, or a 12.7 mm (0.50) M2 HB machine gun in place of the 25 mm cannon, various day/night and thermal sights and a heater for the engine coolant and battery.

Variants

Belgian vehicles

Details of these are given in this section under Belgium.

Netherlands models

The basic AIFV is designated the YPR 765 PRI by the Royal Netherlands Army and variants in service are:

YPR 765 PRCO-B command vehicle, crew nine, combat weight 13,700 kg.

YPR 765 PRCO-C1 to C5, crew nine, combat weight 12,400 kg, armed with a 12.7 mm (0.50) M2 machine gun. The C1 is a battalion commander's vehicle, C2 a battalion gunnery centre, C3 a mortar fire-control vehicle, C4 an anti-aircraft command vehicle and C5 an observation vehicle with a crew of four. All these versions have an M113-type cupola with a 12.7 mm M2 HB pintle-mounted machine gun. The YPR-765 PRCO-3 is no longer used by the Royal Netherlands Army. The C3 designator is now used by the combat engineer command vehicle YPR-765 PRCGN-C3. This stands for Pantser Rups Commando, Genie.

YPR 765 PRRDR radar vehicle fitted with the UK ZB 298 battlefield surveillance radar.

YPR 765 PRRDR-C radar/command vehicle.

YPR 765 PRGWT ambulance, crew four, weight 11,600 kg, unarmed.

YPR 765 PRI/I squad vehicle fitted with M113-type cupola with a 12.7 mm (0.50) M2 machine gun, carries nine men plus driver.

YPR 765 PRMR mortar tractor (tows 120 mm TDA mortar), crew seven, combat weight 13,000 kg, armament one 12.7 mm (0.50) M2 machine gun pintle-mounted on a standard M113-type cupola.

YPR 765 PRVR-A and PRVR-B cargo vehicles, crew two, combat weight 13,400 kg, armament one 12.7 mm (0.50) M2 machine gun pintle-mounted on a standard M113-type cupola.

YPR 765 PRAT fitted with the Systems & Electronics Inc TOW launcher system as fitted to the M901 Improved TOW Vehicle. The PRAT (*Pantser Rups* Anti-Tank) entered service with the Royal Netherlands Army in 1982 and has a crew of four. Full details of the launcher, which has two Raytheon Systems Company TOW ATGW in the ready to launch position, are given in the Tank destroyers section under the M981 Improved TOW Vehicle.

YPR 806 PRBRG armoured recovery vehicle, crew four, weight 12,200 kg, armament one 12.7 mm (0.50) M2 machine gun. Details of this model are given later in this entry.

The first prototype of a mid-life upgrade vehicle for the Netherlands Army was completed in 1993.

With the addition of a variety of kits, the basic AIFV Universal Vehicle Chassis can be converted to the following roles:

Command post

This has revised seating arrangements for a command staff of six plus its crew of three (commander, gunner and driver), and is fitted with additional communications equipment, a table and a stowage rack for 12.7 mm (0.50) ammunition. It is normally armed with a 12.7 mm (0.50) M2 HB machine gun on a pintle mount.

Mortar prime mover

This tows a 120 mm TDA or similar mortar and has racks at the rear of the troop compartment on the right side for 51 mortar bombs. A crew of seven is normally carried, three vehicle crew plus four mortar crew. It is normally armed with a 12.7 mm (0.50) M2 HB machine gun on a pintle mount, but this can be replaced by an M26 weapon station fitted with a 12.7 mm (0.50) M2 HB machine gun.

TOW ATGW vehicle

In addition to being fitted with the Systems & Electronics Inc ITV kit, other TOW installations are possible, and the vehicle can also be fitted with other types of ATGW system such as the HOT. All users with an anti-tank requirement have so far selected the ITV launching 3,750 m range Raytheon Systems Company TOW ATGW.

Cargo carrier

This model can carry up to 2,040 kg of cargo and has a crew of two, driver and gunner. Armament consists of a 12.7 mm (0.50) M2 HB machine gun. The cargo kit includes protective grating behind the driver and gunner and tie-down devices to secure the load while travelling across country.

Ambulance

Four stretchers suspended from hanger harnesses are fitted in the troop compartment at the rear and two seats are provided for medical assistants. This model is unarmed.

Recovery vehicle

This is similar to the XM806 which is a member of the M113A1 family but the Dutch vehicles have grenade dischargers mounted on the front of the hull, an HIAB crane mounted on the roof and additional buoyancy pods attached to the hull sides. This vehicle features the same suspension and power train as the AIFV. The trim vane at the front of the hull also has an additional buoyancy attachment.

Specifications

Belgian and Dutch vehicles
Crew: 3 + 7
Combat weight: 13,687 kg
Unloaded weight: 11,405 kg
Power-to-weight ratio: 19.29 hp/t
Ground pressure: 0.67 kg/cm^2
Length: 5.258 m
Width: 2.819 m
(over tracks) 2.54 m
Height:
(to top of turret sight) 2.794 m
(turret roof) 2.619 m
(to hull top, front) 1.854 m
(to hull top, rear) 2.007 m
Firing height: 2.33 m
Ground clearance: 0.432 m
Track width: 381 mm
Length of track on ground: 2.667 m
Max speed:
(road) 61.2 km/h
(on 10% gradient) 28.2 km/h
(water) 6.3 km/h
Acceleration:
(0 to 32 km/h) 10 s
(0 to 48 km/h) 23.1 s
Fuel capacity: 416 litres
Max road range: 490 km
Fording: amphibious
Gradient: 60%
Side slope: 30%
Vertical obstacle: 0.635 m
Trench: 1.625 m
Turning radius: 8.53 m
Engine: Detroit Diesel 6V-53T V-6 liquid-cooled diesel developing 264 hp at 2,800 rpm
Transmission: Allison Transmission Division of General Motors TX-100-1A automatic with 3 forward and 1 reverse gears, torque converter and lock up clutch
Steering: FMC DS 200 mechanically controlled differential and pivot steer
Suspension: torsion bar-in-tube for Dutch and bar only for Belgian vehicles
Electrical system: 28 V
Batteries: 2 × 12 V 6TN
Armament:
(main) 1 × 25 mm Oerlikon Contraves KBA-B02 cannon
(coaxial) 1 × 7.62 mm MAG-58 MG
Smoke-laying equipment: 6 smoke grenade launchers vehicle-mounted for Dutch vehicles and turret-mounted for Belgian vehicles
Ammunition:
(25 mm) 180 ready + 144 reserve
(7.62 mm) 230 ready + 1,610 reserve

Gun control equipment
Turret power control: electrohydraulic/manual
Max rate power traverse: 60°/s
Max rate power elevation: 60°/s
Gun elevation/depression: +50/–10°
Turret traverse: 360°
Gun stabiliser:
(vertical) no*
(horizontal) no*
NBC system: no* (standard on Belgian vehicles)
Night vision equipment: no* (I2 sight on early Belgian and Dutch vehicles and thermal on later Dutch vehicles as a retrofit)
* optional

Military police vehicle

This is designated the YPR-765 KMAR (Koninklijke Marechaussee) and is a YPR-765 PRI/I with a number of modifications for use by the military police for interventional purposes.

YPR-2000 Programme

Netherlands YPR-765 fleet is currently being modified with additional stowage boxes and 3 tone NATO camouflage.

Engineer support vehicle

This is called the YPR-765GN (Genie) and is a support vehicle for engineer squads.

Philippines models

In 1979, the Philippines government procured 45 AIFVs which have the same basic configuration as the Netherlands YPR 765 PRI vehicles. Although the enclosed weapon station was delivered ready to accept the Oerlikon Contraves 25 mm KBA-B02 cannon, a kit was developed to convert the station to accept a 12.7 mm (0.50) M2 HB machine gun in its place. Six armoured recovery vehicles were also procured (see previous variant entry). The designation of the version with the 12.7 mm M2 machine gun is the PIFV.

Turkish variants

The Turkish AIFVs have been designated the Armoured Combat Vehicle (family) consisting of four basic variants:

Advanced Armoured Personnel Carrier (AAPC)
Armoured Fighting Vehicle (AFV)
Armoured TOW Vehicle (ATV)
Armoured Mortar Vehicle (AMV).

The first 200 vehicles produced utilised the same basic power train as the Dutch and Belgian vehicles with the exception of the engine, which was the improved Detroit Diesel Silver diesel 53T producing 265 hp.

Starting with vehicle 201 the power train was changed to a 300 hp Detroit Diesel Silver diesel 53T coupled to the Allison X-200-4 four-speed transmission. This is similar to the M113A3 but at a higher power level. Other changes included improved cooling for the increased power and an improved air intake and exhaust system. The vehicles with the new power trains are designated the AFV-300, AAPC-300, ATV-300 and AMV-300.

Additional details and photographs of the Turkish vehicles are given earlier in this section under Turkey.

Status

Production completed but can be restarted if required.
In service with:

Country	Quantity	Comment
Bahrain	25	from Netherlands and delivered in 1996
Belgium	514	local production, qv Belgium
Egypt	611	from Netherlands and delivered in 1996
Netherlands	1,200	approx number from US
Philippines	45	
Turkey	1,698	built in Turkey (qv) who is also building modified vehicles for Malaysia and the United Arab Emirates

Manufacturer

United Defense LP, Ground Systems Division.

UPDATED

United Defense LP Infantry Fighting Vehicle Light

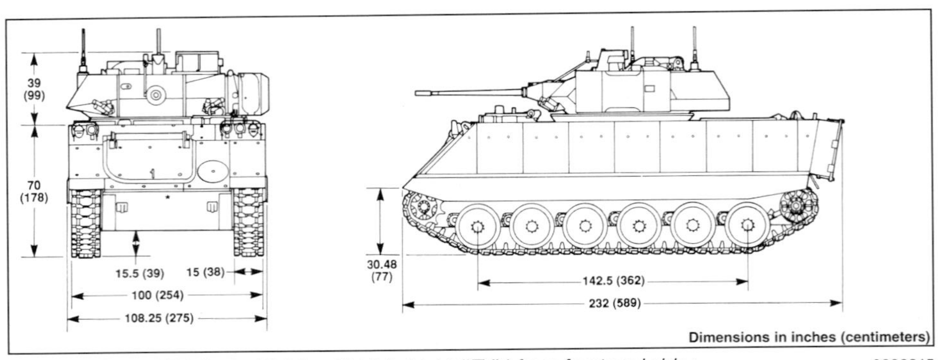

United Defense LP Infantry Fighting Vehicle Light (IFVL) from front and side 0002215

Development

The Infantry Fighting Vehicle Light (IFVL) has been developed by United Defense LP under contract to the Egyptian MoD. This vehicle is also referred to as the Egyptian Infantry Fighting Vehicle.

It essentially consists of the chassis of the United Defense LP Mobile Tactical Vehicle Light (covered in the following entry), fitted with the complete two-man power-operated turret of the United Defense LP M2A2 Bradley Infantry Fighting Vehicle (covered in an earlier entry) which is currently in service with Saudi Arabia and the USA.

Design activities commenced in 1995 as a conversion of an existing M113 series chassis, but United Defense stresses that new vehicles could also be built to the IFVL configuration.

The first prototype was assembled in the Egyptian Tank Plant where assembly of the General Dynamics Land Systems M1A1 MBT is carried out for the Egyptian Army. The first prototype ran for the first time in 1996 and performance validation trials were completed in mid-1997.

As of late 2001, there was no funded production programme for the Egyptian Infantry Fighting Vehicle.

Description

The IFVL chassis is based on the proven M113 which has been lengthened to provide additional weight carrying capability with a torsion bar suspension system that has been reconfigured to provide the vehicle with a cross-country mobility similar to that of the M1A1 MBT.

The chassis shares many features with the recently completed Mobile Tactical Vehicle Light and is powered by a Detroit Diesel 6V-53 TIA turbocharged intercooled and after-cooled diesel engine which develops 400 hp. This is coupled to an Allison X200 cross-drive, hydrostatic four-speed transmission and redesigned severe duty rated final drives. As an option a 350 hp Detroit Diesel engine can be installed.

According to United Defense LP, the IFVL is capable of sustained cross-country travel over rough terrain as well as high-speed operation on improved roads and desert hard pan. Fuel capacity is sufficient to provide for a range of 482 km.

The hull of the Infantry Fighting Vehicle Light is of all-welded 5083 aluminium armour which provides basic protection from small arms fire and shell splinters although optional armour packages are available to defeat armour-piercing projectiles up to 14.5 mm in calibre over a 360° arc of attack.

The driver is seated at the left front of the hull with the engine compartment to his right. The driver has a single-piece pop-up hatch cover which rotates over the engine cooling air intake grill when open, four day periscopes are provided for observation to the front and side and the centre one can be replaced by a passive periscope for driving at night.

Mounted on top of the hull in the centre is the complete turret of the Bradley M2A2 which is armed with The Boeing Company M242 Chain Gun, 7.62 mm M240C machine gun mounted coaxial to the right and a twin launcher for the 3,750 m range Raytheon Systems Company TOW anti-tank guided missile on the left side of the turret.

The commander is seated on the right of the turret with the gunner on the left. Turret traverse and weapon elevation/depression is all electric and the main armament is fully stabilised. A bank of four electrically operated grenade launchers is mounted either side of the turret.

In the roof of the IFVL, to the rear of the turret, is a standard M113A3 cargo hatch. A total of six infantrymen is seated in the rear, three down either side facing each other. Normal means of entry and exit is via the power-operated ramp in the rear which is also provided with an emergency door.

Provision for ammunition stowage is also made in the crew compartment including four TOW missiles, 600 rounds of 25 mm ammunition and 3,600 rounds of 7.62 mm ammunition.

The suspension either side consists of six roadwheels with the drive sprocket at the front and idler at the rear. The upper part of the suspension is covered by a skirt.

Various equipment upgrades for the chassis are available and include an NBC system which utilises an M8A3 or M13A1 gas particulate filter unit to provide clean, safe filtered air for the vehicle crew and mounted infantrymen and the vehicular intercommunication system (VIS/VIC-3) manufactured by Northrop Grumman which is a time division multiple access system that routes communications signals between crew stations and onboard radios.

An automatic fire suppression system, which employs thermal or optical sensors, can be fitted for the

United Defense LP Infantry Fighting Vehicle Light (IFVL) with TOW launcher retracted **2001**/0099480

Specifications
Crew: 3 + 6
Combat weight: 18,140 kg
Power-to-weight ratio:
(400 hp diesel) 22.05 hp/t
(350 hp diesel) 19.29 hp/t
Ground pressure: 0.67 kg/cm²
Length: 5.89 m
Width:
(overall) 2.75 m
(over tracks) 2.54 m
Height:
(gunner's sight) 2.78 m
(hull roof) 1.78 m
Ground clearance: 0.39 m
Track: 2.159 m
Track width: 380 mm
Length of track on ground: 3.62 m
Max speed: 66 km/h
Acceleration:
(0 to 32 km/h) 7.1 s
(0 to 48 km/h) 16.7 s
Fuel capacity: 492 litres
Cruising range: 482 km
Turning: pivot to infinite
Fording: n/avail
Gradient: 60%
Side slope: 40%
Vertical obstacle: 0.71 m
Trench: 2.18 m
Engine: Detroit Diesel 6V-53TIA 6-cylinder diesel developing 400 hp at 2,800 rpm (or 350 hp)
Transmission: Allison X200-4 TIB series hydrokinetic
Steering: hydrostatic
Brake type: multiple wet type
Suspension: torsion bar
Electrical system: 24 V
Generator: 200 A (300 A optional)
Batteries: 4 × 6TL, 120 Ah, 12 V
Armament:
(main) 1 × 25 mm M242 cannon
(coaxial) 1 × 7.62 mm M240C MG
(ATGW) 2-tube TOW launcher
Smoke-laying equipment: 2 × 4 grenade launchers
Ammunition:
(main) 300 ready + 600 reserve
(coaxial) 800 ready + 3,600 reserve
TOW missiles: 2 in launcher + 4 reserve

Gun control equipment
Turret power control: electric/manual
(by commander) yes
(by gunner) yes
Max rate power traverse: 60°/s
Max rate power elevation: 60°/s
Gun elevation/depression: +60/–10°
Turret traverse: 360°
TOW launcher elevation/depression: +30/–20°
Gun stabiliser:
(vertical) yes
(horizontal) yes
NBC system: optional
Night vision equipment: yes

crew compartment and/or the engine compartment to detect both explosions and fires and trigger single or multiple fire extinguisher bottles to minimise personnel injury and equipment loss.

The driver can be provided with a passive night viewer AN/VVS-2 which uses a second-generation image intensifier tube system which relies on objects being illuminated by ambient low light but provides protection from bright lights as well as shell bursts and flares.

Other options include a swim kit, firing ports, air conditioning and various communication/navigation packages.

Variants
None announced so far.

Status
Development complete. Ready for production.

Manufacturer
United Defense LP, Ground Systems Division. (If production of this vehicle is under taken in Italy it will be carried out at the Egyptian Tank Plant, or Factory 200 as it is also known).

UPDATED

United Defense LP Mobile Tactical Vehicle Light

Development
The Mobile Tactical Vehicle Light (MTVL) has been developed by the Ground Systems Division of United Defense LP as a private venture with the first prototype being completed at its San Jose facility in 1994.

United Defense is offering the MTVL as a new build vehicle or a conversion of existing M113 series vehicles.

The prototype is based on an existing M113 series vehicle to demonstrate that such a conversion is possible.

According to United Defense, the MTVL has 70 per cent more payload capacity, 30 per cent more volume under armour and 50 per cent greater cross-country mobility than the latest production M113A3.

The additional payload and volume under armour creates essentially a heavy APC variant based on the M113 chassis.

When mounting a one- or two-man weapon station and fitted with additional armour, the chassis is the basis for the United Defense LP Infantry Fighting Vehicle Light, covered in detail in the previous entry.

The MTVL was designed by United Defense to provide continued growth capability for the M113 series of vehicles beyond the capabilities of the world-renowned M113A3.

By early 1998, a total of six MTVL vehicles had been built, all of which were conversions of existing vehicles.

The first was the prototype MTVL and also used for the Combat Engineer Squad Vehicle, the second was for the Egyptian Infantry Fighting Vehicle Light programme and numbers three, four, five and six are for the Canadian M113APCLE (Armoured Personnel Life Extension).

The first Canadian M113APCLE was rolled out at the United Defence San Jose facility in January 1998. This was completed in 106 working days. Additional details of the M113APCLE are given in this section under Canada.

The Canadian version of the MTVL is the first vehicle to be equipped with the new Detroit Diesel 6V-53T series DDEC (Detroit Diesel Electronic Control) III electronic engine which allows the user to electronically tune the engine to a desired rating up to 400 hp.

Late in 1998, United Defense proposed a modified version of the MTVL fitted with a Sharpshooter turret armed with a 25 mm cannon and 7.62 mm coaxial machine gun to the Royal New Zealand Army. This is called the Infantry Fighting Vehicle Lightweight Sharpshooter. The New Zealand Army however selected the General Motors Defense Light Armoured Vehicle (8 × 8) III to meet its future operational requirements.

United Defense is offering to convert current M113A2 hulls to the MTVL/IFVL configuration to meet this requirement.

Description
The MTVL is essentially an M113 APC with a new lower hull configured for the six roadwheel high-mobility suspension and an uprated power pack that features the NATO-approved electronically controlled Detroit Diesel 6V-53 TIA, turbocharged, intercooled and after-cooled engine that develops 400 hp.

The hull is of all-welded 5083 aluminium armour which provides protection from small arms fire and shell splinters. An add-on optional armour package is available that is understood to provide protection against attack from 30 mm armour-piercing projectiles over the frontal arc. Details are as follows:

Upper and lower front glacis and sides – titanium/steel appliqué

Bottom – spaced laminate steel

Upper front glacis and sides – armour tiles.

The driver is seated at the front left and has a single-piece hatch cover that opens to the rear and to his front and left side are four M17 day periscopes with provision for an AN/VVS-2 passive night vision device of the image intensification type or an AN/VAS-5 thermal type.

The engine compartment is to the right of the driver's position and is fitted with a fire extinguishing system that can be operated by the driver or from outside the vehicle. The air inlet and outlet louvres and the exhaust pipe outlet are in the roof and there is an engine compartment access door that opens upwards in the front of the sloping glacis plate.

United Defense Mobile Tactical Vehicle Light armed with pintle-mounted 12.7 mm M2 machine gun during US Army trials ***2001***/0099479

United Defense Mobile Tactical Vehicle Light armed with pintle-mounted 12.7 mm M2 machine gun and showing passive 30 mm armour package fitted ***2001***/0099478

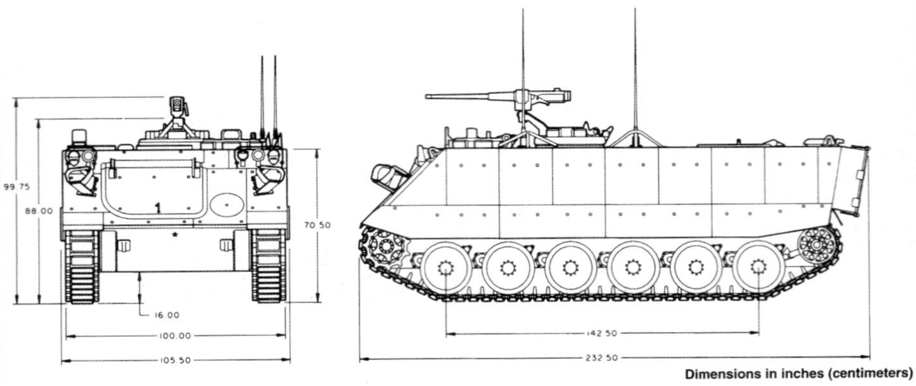

Detailed front and side drawings of the Mobile Tactical Vehicle Light armed with a cupola mounted 12.7 mm M2 machine gun ***2002***/0116795

The commander is seated to the rear of the engine compartment and has a cupola that can be traversed through a full 360°, five M27 day periscopes and a single-piece hatch cover.

The armament is pintle-mounted on the forward part of the commander's cupola and can be a 12.7 mm M2 HB machine gun, 40 mm MK 19 Mod 3 grenade launcher or 7.62 mm M240 machine gun.

If the 12.7 mm M2 HB machine gun is fitted then elevation limits are from -21 to +53° with traverse through a full 360°. There are 100 rounds of ready use ammunition carried for this weapon. Mounted either side of the hull, firing towards the front, is a bank of four electrically operated grenade dischargers.

The infantrymen sit on bench seats that run down either side of the hull facing the centre and enter and leave via a large power-operated ramp in the hull rear. There is also a door in the left side of the ramp.

Over the top of the troop compartment is a large roof hatch that opens to the rear and to the rear of this is a domed adjustable ventilator.

The torsion bar suspension either side consists of six dual rubber-tyred roadwheels with the drive sprocket at the front and the idler at the rear. There are no track-return rollers. The vehicle can be supplied with the T130 or T150 series track.

Compared to earlier versions of the M113 series, the idler and drive sprocket of the MTVL have been repositioned and the reconfigured suspension geometry gives more than 389 mm of roadwheel travel.

Although the basic MTVL can be configured as an armoured personnel carrier, it can be configured for a wide range of other battlefield support roles. It is roll-on/roll-off air-transportable in a Lockheed Martin C-130 Hercules, but is not amphibious. It can be fitted with a swim kit.

The diesel fuel tanks of the MTVL can either be in external fuel tanks one either side at the rear as normal or under the floor of the vehicle in special fuel cells designed not to rupture in an anti-tank mine blast. The Canadian vehicles have the fuel cells under the floor of the vehicle. As with all members of the M113 family of tracked vehicles, a wide range of optional equipment is available including spall liners, mine protection, NBC system (ventilated face piece, overpressure or regenerative), swim kit, firing ports and air conditioning system.

120 mm AMS on MTVL

Shown for the first time in September 1995 was the second example of the 120 mm Armored Mortar System (AMS) installed on the chassis of the Mobile Tactical Vehicle Light (MTVL) full-tracked armoured vehicle.

There are three members in the 120 mm Armored Mortar System consortium: Delco Defense (electronics and systems integration), BAE Systems, RO Defence (turret and 120 mm mortar) and United Defense LP (MTVL chassis).

The first complete AMS was fitted to the chassis of the General Motors Defense LAV 8 × 8 chassis. A total of 73 AMS or an LAV (8 × 8) have been built for the Saudi Arabian National Guard by General Motors Defense.

The second AMS is aimed at those customers in the Middle East who require a 120 mm mortar system on a tracked rather than a wheeled chassis.

The actual 120 mm mortar system in the second AMS is identical to that in the first system but the turret has been reconfigured to make the system more user friendly and a new keypad has been fitted. As the MTVL has more interior volume than the LAV, this version can carry around 90 120 mm mortar bombs. Full details of the 120 mm Armoured Mortar System are given in the self-propelled Mortars section later in this volume.

Combat Engineer Squad Vehicle

The Combat Engineer Squad Vehicle (CESV) has been proposed to the US Army Engineers. It is based on the current in-service M113A3 series APC and would be rebuilt to expand its internal volume. In many respects the CESV is identical to the MTVL.

MTVL with M1 cupola

For trials purposes an MTVL has been fitted with the complete commander's cupola from the M1 MBT. This is referred to as the M113 Protected Weapon Station (PWS) and is armed with an externally mounted 12.7 mm M2 machine gun which can be aimed and fired from under armour.

MTVL for Interim Armoured Vehicle requirement

To meet the US Army requirement for an Interim Armoured Vehicle, United Defense offered a number of variants based on the MTVL chassis. These include the following variants:

- Anti-tank guided missile vehicle with twin round TOW launcher
- Commander's vehicle with remote controlled 12.7 mm M2 machine gun and extensive communications equipment
- Engineer squad vehicle with remote controlled 12.7 mm M2 machine gun

Specifications

Max combat weight: 18,144 kg
Max payload: 7,756 kg
Power-to-weight ratio: 22.05 hp/t
Ground pressure: 0.67 kg/cm²
Length: 5.84 m
Width:
(overall) 2.69 m
(over tracks) 2.54 m
Height:
(overall) 2.50 m
(hull top) 1.78 m
Ground clearance: 0.39 m
Track: 2.159 m
Track width: 380 mm
Length of track on ground: 3.66 m
Max road speed: 66 km/h
Acceleration:
(0 to 32.187 km/h) 7.1 s
(0 to 48.28 km/h) 16.7 s
Fuel capacity: 492 litres
Max cruising range: 483 km
Turning radius: pivot to infinite
Vertical obstacle: 0.91 m
Gradient: 60%
Side slope: 40%
Trench: 2.18 m
Engine: Detroit Diesel 6V-53 TIA 6-cylinder diesel developing 400 hp at 2,800 rpm
Transmission: Allison X200-4B hydrokinetic
Steering: Hydrostatic
Brake type: multiple wet type
Suspension: torsion bar with 4 shock-absorbers either side
Electrical system: 28 V
Generator: 200 A (300 or 400 A optional)
Batteries: 4 × 6TL, 120 Ah, 12 V
Armament: 1 × 12.7 mm M2 MG (see text)
Ammunition: (12.7 mm) 1,000
Armour: 12-44 mm (see text)
NBC system: optional
Night vision equipment: optional
Turrets: optional

- Fire support vehicle with Striker mission equipment package
- Medical evacuation vehicle with raised rear roof
- Mortar carrier with 120 mm M121 mortar
- NBC reconnaissance vehicle
- Reconnaissance vehicle

All of the above versions would have been powered by the Detroit Diesel 6V-53TIA diesel developing 400 hp at 2,800 rpm, coupled to an Allison X200-4B hydrokinetic transmission.

Standard equipment would have included NBC system, fire detection and suppression system and be fully airportable in C-130, C-17 and C-5 aircraft.

More specialised versions included two Mobile Gun Systems, one with the complete 105 mm turret of the M8 Armoured Gun System and the other with the Belgium Cockerill LCTS Mk 8 system. There was also a anti-tank version which consisted of the MTVL chassis fitted with the complete turret of the Bradley infantry fighting vehicle covered in the earlier entry.

Status

Pre-production vehicles completed for Egypt and Canada (see development). The first production customer is Canada, who placed a production contract in 2000 to upgrade part of its existing fleet to the MTLV standard (qv Canada).

Manufacturer

United Defense LP, Ground Systems Division.

UPDATED

United Defense LP M113 armoured personnel carrier family

Development

In January 1956, development began of an air-transportable, armoured, multipurpose vehicle family, 'to provide a lightweight, armored personnel carrier for armour and infantry units capable of amphibious and air-drop operation, superior cross-country mobility and adaptation to multiple functions through applications of kits and/or modifications of its superstructure'.

Prototypes of both aluminium (T113) and steel (T117) were built and tested, but development of the T117 was cancelled in favour of further development of the T113. This became the T113E1, which was standardised as the M113 in 1960 and entered production at FMC's San Jose facility in early 1960. Trials of a diesel-engined model called the T113E2 showed a substantial improvement in operating range (from 321 to 483 km) as well as a reduction in fire risk. This model was standardised as the Carrier, Personnel, Full-Tracked, Armored, M113A1, in May 1963. After standardisation, 10 preproduction vehicles were built and the M113A1 replaced the M113 in production from September 1964.

Recent production figures for the M113 family of tracked vehicles are:

Spanish M113A1 VCZ Combat Sappers' Vehicle (F A Marin)

M113 ambulance of US Army (Richard Stickland) **2002**/0116785

Year	US Army	Foreign Military Sales	Direct Sales
1977	707	2,166	307
1978	1,167	1,734	292
1979	1,302	777	213
1980	1,229	905	629
1981	317	1,211	518
1982	119	1,186	524
1983	479	566	112
1984	587	603	23
1985	287	311	666
1986	464	130	426
1987	567	202	65
1988	706	93	17
1989	571	110	58
1990	424	63	72
1991	346	31	104
1992	228	109	24

In addition, Otobreda of Italy built 4,500 members of the M113 series of vehicle under licence for home and export markets. Details of these are given for the entry on the Italian M113 APC fleet earlier in this section.

Production of the M113 series of APC was completed at San Jose in December 1992 when the last vehicle, an M577 series command post vehicle, rolled off the production line.

Late in 1993, Kuwait placed an order for 30 M577A3 command post vehicles and 16 M113A3 armoured ambulances worth US$32 million.

The M113 production line was reopened at San Jose with deliveries running November 1994 to March 1995.

In mid-1995, the Royal Thai Army placed an order with United Defense LP for the supply of 82 M113 series armoured personnel carriers and variants were completed by August 1997.

This order comprised the following vehicles:

12 M577A3 command post vehicles
12 120 mm M1064A3 mortar carriers
21 81 mm M125A3 mortar carriers
18 M901A3 Improved TOW Vehicles
10 M113A3 maintenance/recovery vehicles
9 M113A3 ambulances.

This was the first time that A3 versions of the M1064, M125, M901 and M113 maintenance/recovery had been placed in production and in addition United Defense integrated the TOW missile system with the M113A3 chassis to make the M901A3 Improved TOW Vehicle.

Production was undertaken at the United Defense San Jose facility in California but assistance was also provided by Aiken, Steel Products Division and FNSS Defense Systems (then FMC-Nurol) in Turkey. Deliveries commenced in November 1996 and continued through to August 1997. In April 1997 the Royal Thai Army placed a follow-on order for 12 M113A3 maintenance/recovery vehicles which were delivered by June 1998.

M113 Designations

Late in 1992, the US Army changed a number of its designations in an effort to standardise vehicle nomenclatures within the M113 family of vehicles. The new designations apply to all M113 family of vehicle variants containing the RISE power pack upgrade utilised in the latest M113A3 APC. The M113A3 was type-classified in September 1979 and fielding started in 1997.

Under the new designator system, all variants (except the M730A2 Chaparral) utilising the RISE power package will be identified as A3 models. The US Army is in the process of upgrading seven different M113 variants to the RISE configurations and the new designations will be applied to the following vehicles:

(1) M981A3 Fire Support Team Vehicle (FISTV)
(2) M1064A3 120 mm self-propelled mortar system
(3) M1068A3 Standard Integrated Command Post
(4) M548A3 cargo carrier for the Volcano mine scattering system
(5) M577A3 command post
(6) M1059A3 smoke generator vehicle
(7) M901A3 Improved TOW Vehicle (ITV).

Description

The all-welded aluminium armour hull of the M113 protects the crew from small arms fire and shell splinters.

The driver sits at the front of the hull on the left side and has a single-piece hatch cover that opens to the rear. To his front and left side are four M17 day periscopes and there is also an M19 infra-red periscope in his roof hatch.

The M19 infra-red periscope system has now been replaced by the Litton AN/VVS-2 night viewer system in all M113A3 variants of the US Army.

The engine compartment to the right of the driver's position is fitted with a fire extinguishing system that can be operated by the driver or from outside the vehicle. The air inlet and outlet louvres and the exhaust pipe outlet are in the roof and there is an engine access door in the front of the hull that hinges forward.

The power train consists of the power plant (engine, transfer gear case and transmission), steering control differential, pivot steer, final drive and associated drive shafts and universal joints.

Grand Total 1977 to 31 December 1992: 23,747
Procurement history of M113 family as of 1 March 1992:

Vehicle type	US	FMS	Direct Sales	Total
M113A1 ITV	139			139
M113A1 TOW	257	211	12	480
M113A2 ITV	1,403	196		1,599
M113A2 TOW	5	367		372
Ambulance		13	8	21
Bulldozer		4		4
C&R			266	266
C&R2			173	173
Fitters		517	110	629
M106	841	20	1	862
M106A1	990	326	93	1,409
M106A2	53	286	11	350
M113	10,260	4,553		14,813
M113A1	10,193	13,036	2,094	25,323
M113A1R			1	1
M113A2	932	3,478	1,964	6,374
M113A2DR			8	8
M113A2H			20	20
M113A2IT			6	6
M113A2MK			50	50
M113A2R		29	87	116
M113A3	1,740	55	1	1,796
M125A1	854	1,398	133	2,385
M125A2	69	492	36	597
M132		201		201
M132A1	150			150
M548	2,362	195	2,868	5,425
M548G			290	290
M548A1	256	133	80	469
M577	179	973		1,152
M577A1	2,201	1,440	222	3,863
M577A2	1,817	454	146	2,417
M577A2R2			7	7
M667		26		26
M730	52	64	1	117
M730A1	119	44	29	162
M741		220		220
M741A1		10		10
M741A2		13		13
M901	145	56		201
M901 ITV			14	14
RTL		42		42
XM474	57			57
XM474E2	240			240
XM667E1	302	3		305
XM727	124			124
XM730	424	37		461
XM741	381	46		427
XM806E1			54	54
XM806E2		27		27
Totals	36,546	28,965	8,785	74,296

M113 of Greek Army (Richard Stickland) **2002**/0116786

M58 Wolf smoke generating vehicle from the rear with ramp lowered **2001**/0099477

Spanish minelayer based on a stretched and modified M548 carrier chassis (F A Marin)

M113A1 of the Taiwanese Army fitted with a TOW ATGW launcher and a 12.7 mm MG (L J Lamb)

The commander sits to the rear of the engine compartment and has a cupola that can be traversed through a full 360°, five M17 day periscopes and a single-piece hatch cover. Pintle mounted on the forward part of the commander's cupola is a 12.7 mm (0.50) Browning M2 HB machine gun with an elevation of +53°, a depression of –21° and a total traverse of 360°. There are 100 rounds of ready use ammunition carried for this weapon.

The infantrymen enter and leave the M113 via a power-operated ramp in the rear of the hull that opens downwards and has a door in the left side. Behind the commander's cupola is rectangular cargo hatch that opens to the rear, behind which is a domed ventilator. The infantrymen travel on seats down each side of the hull, which can be folded up to enable the vehicle to be used as an ambulance or to carry cargo.

The torsion bar suspension either side consists of five dual rubber-tyred roadwheels with the drive sprocket at the front and the idler at the rear. There are no track-return rollers and the first and last roadwheel stations are fitted with a hydraulic shock-absorber. A rubber track shroud on each side of the hull controls the flow of water over the tracks when the vehicle is afloat. The M113 is fully amphibious, being propelled in the water by its tracks. Steering when afloat is the same as on land. Before entering the water the two bilge pumps are switched on and the trim vane, which folds back onto the glacis plate when not in use, is extended at the front of the hull.

The following kits are available for members of the M113 family: anti-mine armour bolted on front half of vehicle bottom (includes buoyancy aids), anchor kit (set of two for use with capstans for self-recovery), buoyant side pods, combination bulldozer/snow plough, NBC detector and automatic alarm, full width buoyant trim vane, gun shields, heater for personnel and cargo areas, heater for engine coolant and battery, stretcher kit which provides support for four stretchers when the vehicle is being used as an ambulance, M8A3 gas-particulate unit (includes M2A2 air purifier with flexible hoses to fit M14A1 tank gas masks of driver and commander and up to two others), non-skid ramp plate kit and windscreen for driving with hatch open.

M113A2 (formerly Product Improved M113A1)

Following successful trials with prototype vehicles, in 1978, the US Army decided to carry out a Product Improvement Programme on its fleet of M113 series of APC's. The army hopes to have 21,323 Product Improved M113A1s in service. The US Army also ordered an initial batch of 2,660 M113A2s at a cost of US$154 million, the first of which were completed at the San Jose facility in July 1979. These improvements can be summarised as follows:

Improved engine cooling design that reverses the position of the fan and radiator and incorporates a new radiator, surge tank and a new cooling fan. This new cooling system draws in ambient air through the radiator, providing increased cooling efficiency for both the engine and transmission; reduces oil film and dust build-up on the core of the radiator; provides a negative pressure engine compartment which reduces the possibility of exhaust fumes leaking into the crew compartment; and finally provides a longer engine life and greater tractive effort at reduced temperatures.

The improved suspension system incorporates high-strength steel torsion bars that provide 228 mm of roadwheel travel, improved shock-absorbers on the first, second and fifth roadwheel stations, stronger rear idler assembly which has also been raised to reduce the chance of ground impact and finally increased ground clearance to minimise sprocket ground contact on rough terrain.

Other M113A2 variants such as the M577A2 and the M125A2 are currently available. The M548 cargo carrier with A2 improvements has been redesignated the M548A1 and first production vehicles were completed in 1982.

Rear-mounted armoured fuel cells provide more interior space and reduce the danger of internal fire.

M113A3

M113A3 production commenced in early 1987 and US Army units started receiving these vehicles in the summer of 1987. In the autumn of 1987, the US Army released its initial contract for M113A3 kits to the FMS Corporation, Los Angeles, California (now Marvin Land Systems Inc), to support the conversion of M113A2 vehicles to the M113A3 configuration.

With exercise options and subsequent contracts, in total 1,901 kits have been purchased from the FMS Corporation.

This M113A3 kit contains the Detroit Diesel 6V-53T turbocharged engine developing 275 bhp and the Allison cross-drive X-200-4 hydrostatic four-speed transmission (purchased separately), external fuel tanks, a variable-speed cooling fan drive and hardware for mounting the spall liners and the external add-on armour.

The M113A3 incorporates the cooling and suspension improvements of the M113A2 but also has better performance and reliability. Major improvements include the replacement of the 6V-53 (212 hp) diesel engine by the turbocharged Detroit Diesel 6V-53T (275 hp) and the replacement of the present TX-100-1 transmission, transfer gearcase, steering differential and pivot steer with the X-200-4 Allison transmission which provides four forward speeds instead of the present three. Hydrostatic steering provides smoother turning with less effort and reduced shock loading on the suspension system and greater power efficiency, which results in more usable horsepower and fuel savings.

The driver's controls have also been changed and the conventional sticks have been replaced by a steering wheel and brake pedal.

The US Army refers to this improvement as the RISE power train (Reliability Improvement of Selected Equipment). The M113A3 has a 200 A generating system, four batteries and improved electrical diagnostic capabilities to replace the 100 A system with two batteries in the M113A2. It also has armoured external fuel tanks and internal spall protection liners. The two improvements (RISE and AEFT) were applied to the M901A1 and M981 vehicles which became the M901A3 and M981A3. The first M113 family variant to be converted to the RISE power train was the M730 Chaparral SAM carrier. The entire US fleet of 495 vehicles was converted to the M730A2 between 1987 and 1989. The M730 Chaparral SAM system has now been phased out of US Army service.

Possible additional improvements

Several other improvements have been investigated for the M113 family. Each member of the M113 family is now being studied by the US Army for both active and passive NBC protection. Beginning in 1987, all M113A2/M113A3 production vehicles were fitted with grenade launchers.

In mid-1986, it was announced that the US Army was to modify its M113 series vehicles in a two-phase programme. For Block I, the RISE power train installed as well as internally mounted and movable spall suppression liners, external fuel tanks and mounting provisions for external armour kits.

Block II includes bolt-on external kits of armour; the exact requirement is to be determined.

The US Army still maintains a fleet of around 17,500 members of the M113 family of vehicles and, as these will remain in front-line service well into the 21st century, they will continue to be upgraded.

In February 1997, NAPCO International was awarded a contract by the US Army Tank-automotive and Armaments Command for the supply of 401 conversion kits for the M113 family of vehicles with work completed by March 1998.

The following month United Defense LP, Steel Products Division, was awarded a contract worth a total of US$21.43 million for conversion/upgrade work on the 332 members of the M113 series of vehicles. The vehicles involved are 254 M1064A3, 51 M113A3 and 27 M577A3 vehicles with work completed by March 1998.

M113 with Band Track

The Canadian company of Souncy has developed a Band Track for installation on the M113 series of full tracked vehicles. The rubber band track offers the user a number of significant advantages including a reduction in vibration, infra-red signature, reduction in noise and weight. During trials this Band Track also gave an improvement in traction as well as decreasing rolling resistance.

Mobile Tactical Vehicle Light

Details of this private venture vehicle based on the M113A3 are given in the previous entry. First customer for this variant is Canada.

XM1108 Universal Carrier

This has been designed as a private venture replacement for the M548 cargo carrier member of the M113 family, the most recent variant for which a prototype has been built as the XM1108 Universal Carrier.

The XM1108 consists of a stretched (six roadwheels either side) lower hull mounting the hardened cab from the standard production M993 Multiple Launch Rocket System.

The vehicle provides an armoured cab and a maximum payload capacity of 6,804 kg at a gross vehicle weight of 16,329 kg. The power train and suspension have been upgraded with the 350 hp 6V-53TA Detroit Diesel turbocharged/after-cooled diesel engine, the X-200-4A Allison cross-drive hydrostatic four-speed transmission, and the high-mobility suspension configuration of the MTVL.

According to United Defense LP, potential applications of the XM1108 include the Chaparral missile system, the MICLIC unguided mineclearing system HIMARS (High Mobility Artillery Rocket System) or a medical evacuation vehicle to name but a few.

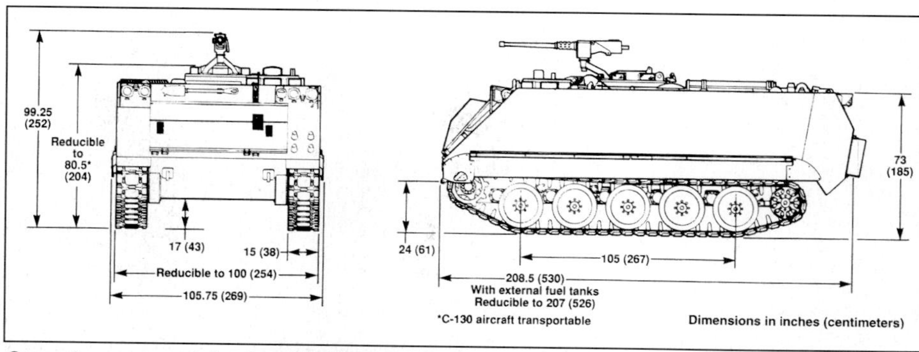

General arrangement drawing of M113A3 APC without appliqué armour fitted

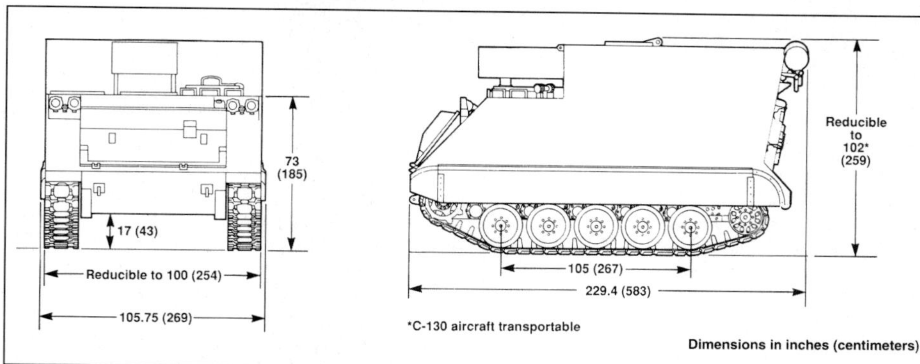

General arrangement drawing of the M1068A3 Standard Integrated Command Post

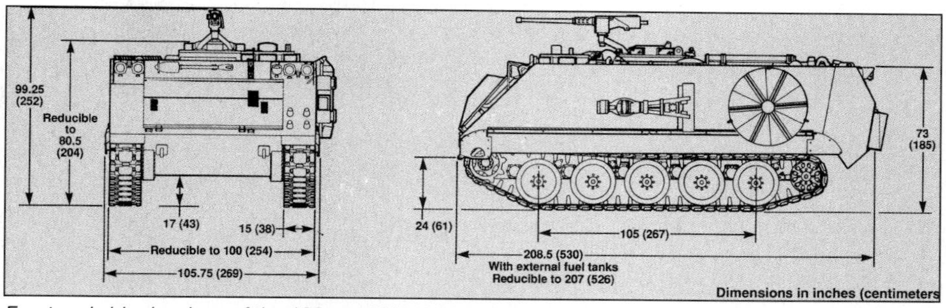

Front and side drawings of the 120 mm M1064A3 self-propelled mortar carrier

Details of this system are given in *Jane's Military Vehicles and Logistics 2001-2002*.

DAREOD

Under contract to the Canadian Forces, a prototype of the United Defense LP Damaged Airfield Reconnaissance Explosive Ordnance Disposal (DAREOD) vehicle has been completed.

This is based on an M113A2 Mk 1 hull with a 265 hp power pack, a hydraulically operated V-type dozer blade at the front of the vehicle, additional spotlights to the front and sides of the hull and the one-man Helio turret and provision for night vision equipment.

Variants

M113/M113A1/M113A2 with dozer kit

The bulldozer kit adapts the M113A1 APC to carry out general bulldozing work, improvement of water entrances and grading. The installation of the kit does not change the basic vehicle or its load-carrying ability. It is fully amphibious as the front-mounted buoyant blade acts as a trim vane when raised during amphibious operations, after removal of the standard trim vane.

M106/M106A1/M106A2 107 mm mortar carrier

Full designation is Carrier, Mortar, 107 mm, Self-propelled. This is the basic APC with a 4.2 in (107 mm) M30 mortar mounted on a turntable in the rear of the hull that fires to the rear through an opening in the roof covered by a three-part hatch. The mortar has a traverse of 43° right and 46° left and 88 (M106A1/M106A2) or 93 (M106) rounds of mortar ammunition are carried. Mounted externally on the left side of the hull is a mortar baseplate and bridge that allows the mortar to be fired away from the vehicle if required. The 12.7 mm (0.50) M2 HB machine gun is retained.

Following extensive trials with three competing 120 mm mortar systems, the Israeli Soltam system was selected in early 1988 and this has now replaced the old 107 mm mortar system.

M106A1 107 mm (4.2 in) self-propelled mortar system (C R Zwart)
2000/0080339

M163A1 20 mm Vulcan self-propelled anti-aircraft gun system (Pierre Touzin)
2000/0080338

M548 tracked cargo carrier with bows and a tarpaulin cover over the rear compartment (Israel Defence Force) ***2000***/0080340

M730 series Chaparral surface-to-air missile launcher, based on the M548 tracked cargo carrier with four Chaparral surface-to-air missiles in the ready to launch position. These are late production systems used by Taiwan (L J Lamb)

M1064A3 120 mm mortar carrier

This is the existing 107 mm mortar carrier previously covered, upgraded to take the new 120 mm mortar. It also has external fuel tanks, a modified turntable and modified stowage racks. This version is fitted with the RISE power pack and designated the M1064A3. It has a four-man crew, with the 120 mm mortar being mounted on a turntable in the rear of the hull and firing to the rear. In total 69 mortar bombs are carried. The US Army expects to have a total of 1,081 of these systems in service.

M125/M125A1/M125A2 81 mm mortar carrier

Full designation is Carrier, Mortar, 81 mm, Self-propelled. This is the basic APC with an 81 mm M29 mortar mounted on a turntable with a 360° traverse in the rear of the hull. It fires through an opening in the roof, covered by a three-part hatch. The mortar can also be deployed away from the vehicle and 114 mortar bombs are carried. The 12.7 mm (0.50) M2 HB machine gun is retained. Current US Army plans call for the conversion of all M125 series to the M1064 or other vehicle configurations.

M163 Vulcan self-propelled anti-aircraft gun

This is essentially an M113A1 chassis fitted with a one-man electrically driven turret which is armed with a 20 mm M168 Vulcan gun, Navy Mk 20 Mod. A gyro lead-computing sight and a range-only radar are mounted on the right side of the turret. The chassis is designated the M741. Full details of this system are given in *Jane's Land-Based Air Defence 2001-2002*.

The M741A1 is a product-improved version of the original M741 and has the improved cooling system common to the M113A2. This system has now been phased out of service with the US Army but is still used by Israel, Jordan, Morocco, Portugal, the Sudan, Thailand and the Yemen. Israel has some M163 systems upgraded with a pod of Stinger fire-and-forget SAMs in the ready to launch position. An illustration of the latter version, called the Machbet in Israel Defence Force service, is provided in the entry under Israel for the M113 series.

M548 cargo carrier

This unarmoured cargo carrier was designed from 1960 under the designation XM548 and uses many automotive components of the M113A1 APC. The XM548 was followed by a diesel-engined version called the XM548E1 which was subsequently standardised as the M548. First production vehicles were completed early in 1966 by the FMC Corporation at its Charleston, West Virginia, facility. M548A1 vehicles were produced with M113A2 improvements, such as suspension and cooling system. The M730A1 Chaparral surface-to-air missile carrier includes suspension and cooling improvements. The M730A2 incorporates the 6V-53T diesel engine and the X-200-4 transmission. A full description of the vehicle is given in *Jane's Military Vehicles and Logistics 2001-2002*.

One of the more recent US Army roles for the M548A1 is carrying the Volcano minescattering system (*Jane's Military Vehicles and Logistics 2001-2002*).

In US Army service, some M548A1s are being upgraded to the M548A3 standard. M548A3 upgrade programmes were established at both the Red River Army Depot (26 vehicles) and in Korea (14 vehicles) during FY93. All conversions were accomplished utilising residual conversion kits from the M730A2 Chaparral SAM programme. These residual kits allowed additional conversions of 34 in FY94 and 47 in FY95.

In addition to the M548A3 RISE power pack upgrade programme, planners at the US Army TACOM envisage a number of other potential improvements that could be applied to the M548 series in the future including an armoured cab, a materials handling crane, drop side enhancements, decontamination apparatus, a global positioning system, ammunition racks, a winch and trailer brakes.

MSM-Fz: This is a modified M548 with an anti-tank minelaying system for the German Army. Details are given in *Jane's Military Vehicles and Logistics 2001-2002*. Greece is understood to have taken delivery of a quantity of these systems from Germany.

The Lance surface-to-surface tactical missile system uses two variants of the M548 chassis, the **M752 launch vehicle** (chassis designation M667) and the **M688 loader transporter** which carries additional missiles. The Lance missile system has been phased out of NATO and US service but is still used by Israel fitted with a conventional warhead.

M1015 electronics carrier: This is used to carry tactical electronic reconnaissance and countermeasures systems but is now being phased out of service.

The basic M548 can carry 5,443 kg of cargo. The manufacturer has built and tested, as a private venture,

M1068A3 Standard Integrated Command Post System which is a further development of the M577 series ***2001***/0099476

M1064 series mortar carrier which is being upgraded to the M1064A3 configuration (US Army) 0018555

a stretched M548 which has a payload of 7,258 kg. This has an additional roadwheel each side and all the improvements mentioned in the stretched M113A2 described previously.

M730 Chaparral SAM launcher

This is based on a modified M548 series tracked cargo carrier, and mounted over the rear area is a one-man power-operated turret with four Chaparral SAMs in the ready to launch position; a further eight missiles are carried in reserve. Following its withdrawal from US Army service Chaparrel remains in service with Egypt, Israel, Morocco, Portugal, Taiwan and Tunisia.

Details of this system are given in *Jane's Land-Based Air Defence 2001-2002.*

M577/M577A1/M577A2/M577A3 command post

Full designation is Carrier, Command Post: Light Tracked, M577 or M577A1. The first four prototypes were built at Detroit Arsenal under the designation XM577 and were conversions of existing M113 APCs. The XM577 was standardised as the M577 in March 1963, after the vehicle was already in production.

The M577/M577A1 is essentially an M113/M113A1 with a higher roof to the rear of the driver's position. In the static role a tent can be erected at the rear of the hull to increase the work area. A generator is mounted externally to provide power for the additional communications equipment installed. As well as being used as a command post the vehicle can also be used as a fire direction centre, a communications vehicle and a mobile medical treatment facility.

The M577A2 is a product-improved version of the M577A1 utilising the improved cooling, suspension system and personnel heating system. When fitted with the RISE power pack the vehicle is known as the M577A3.

Stretched M577A2E2 command post

Late in 1988, the prototype of the Stretched M577A2E2 command post vehicle was completed. This has six, rather than five, roadwheels either side as well as external fuel tanks similar to those fitted to recent production M113 series vehicles. Other improvements include the complete M113A3 power pack, the option of additional armour on the roof for increased protection against top attack weapons and the installation of an environmental control unit. This remains at the prototype stage.

M1068A3 command post vehicle

In 1990, prototype modification kits for a new command post vehicle, then designated the XM1068 Carrier, Standard Integrated Command Post System (SICPS Carrier) were field tested by the US Army and prototype SICPSs were used during Operation Desert Storm in 1991.

The conversion kit, mounted on the basic M577A2/M577A3 chassis, consists of the following items:

(1) A new 5 kW Tactically Quiet Diesel APU mounted on the front of the vehicle
(2) A new tent extension carried on the top of the vehicle and attached around the vehicle's rear ramp
(3) An AC/DC power supply and distribution system (wires mounted on the inner walls of the vehicle and external covered ports)
(4) An electronic equipment data distribution system (wires mounted on the inner walls of the vehicle and external covered ports)
(5) Mounting/stowage provisions for the Army Tactical Command and Control Systems (ATCCS) hardware.

In July 1992, United Defense LP was awarded a contract by the US Army for 162 SICPS kits to be used in depot conversion of existing M577 vehicles to the M1068 configuration. In June 1993, the US Army increased the kit quantity to a total of 330 units.

It is expected that up to 1,534 M577A2/M577A3s will be upgraded to the M1068A3 SICPS configuration, funding permitting.

M113A2 recovery vehicle

Full designation is Recovery Vehicle: Full Tracked, Light Armored M113A2 (previously called the XM806) and is essentially an M113A2 fitted with a hydraulically driven winch and a fairlead assembly for recovering disabled vehicles. Mounted either side of the ramp at the rear of the hull are spades which are lowered to the ground during recovery operations. An auxiliary spade unit, carried on the roof when not required, can be used between the two spades in soft soil. The winch has a capacity of 5,103 kg on a full drum and 9,070 kg on a bare drum. Mounted on the top of the hull on the left side is an auxiliary crane which can lift a maximum load of 1,361 kg at a radius of 1.53 m. There is also a fitters' model (United Defense calls this the M113A2 maintenance recovery vehicle) of the M113/M113A1/M113A2 which has an HIAB crane mounted on the left side of the roof which can lift a maximum load of 3,085 kg. Full details of the ARVs and Repair Vehicles based on the M113 chassis are given in *Jane's Military Vehicles and Logistics 2001-2002.* Neither of these vehicles was ever adopted by the US Army.

M901/M901A1 Improved TOW vehicle

Details of this version are given in the Tank destroyers section. This is no longer used by the US Army. The M901 series is still used by many other countries and the actual launcher is fitted onto a LAV (8 × 8) for the Saudi Arabian National Guard and on AIFV's in service with Egypt and the Netherlands.

M981 Fire Support Team Vehicle

The Fire Support Team Vehicle (FISTV) M981 has been developed under contract to the US Army by Electronics and Space Corporation (now Systems & Electronics Inc) and is essentially the M901 Improved TOW vehicle fitted with the AN/TVQ-2 GLLD with a north-finding module and line of sight subsystem, an AN/TAS-4 night sight, a land navigation system and extensive communications equipment.

Three preproduction units were used in the engineering development tests with an additional three vehicles completed late in 1982.

In August 1983, the company was awarded a production contract for the first 70 systems with a further 459 systems being ordered in March and December 1984. The first batch of 37 production units was delivered in May 1985 with first fielding completed at Fort Stewart in August 1985 and delivery of first units to Europe in August 1986.

Major subcontractors were United Defense LP for the M113A2 chassis, the now Raytheon Systems Company for the AN/TVQ-2 (GLLD), Bendix for the north-seeking gyroscope and Systems & Electronics Inc integrated the whole system as well as providing the targeting station, the intercommunications system and the image transfer assembly.

In early 1988, Egypt took delivery of 20 Artillery Target Locating Vehicles (ATLV), which are almost identical to the FISTV used by the US Army, with another five following.

Vehicles fitted with the RISE power package are now designated the M981A3.

This unmanned turret is also installed on a number of other platforms. Additional details of the M901 are given in the Tank destroyers section. In US Army service the M981 is now being supplemented by the M7 Bradley Fire Support Team Vehicle which is an earlier vehicle refitted for its specialised role.

M113/BMP-2 Opposing Forces Surrogate Vehicle (OSV) which is now in service with the US Army
2001/0099475

M1059 smoke generating system

This entered service with the US Army in 1988 and is an M113A2 with two M54 smoke generators mounted on top of the hull, which can be operated either singly or in tandem.

An electrical air compressor mounted inside the M113A2 provides air to the pulse jet engine and purges it before restarting. The air pressure is monitored by the operator at the control panel allowing hot restarts by the press of a button.

The fog oil system includes a 120 (US) gallon fog oil tank mounted inside the vehicle with two fog oil pumps. Fog oil flow can be controlled by adjusting the rheostat on the control panel.

Each smoke generator has its own remote 5 gallon jerrican external fuel supply which is submersible. The integral equipment box contains a fuel pump, a fuel solenoid and fuel pressure switches. Associated controls allow remote control from inside the vehicle. In total 275 M1059 vehicles were fielded between 1988 and 1990.

M58 smoke generating vehicle

This is also referred to as the M58 Wolf and is based on an M113A3 series armoured personnel carrier. Major components include a turbine powered smoke generator system which is capable of continuously producing 90 minutes of visual and 30 minutes of infra-red screens. According to United Defense, the M58 Wolf has the ability to encounter the threat arising from the wide proliferation of advanced visual and infra-red sensors.

The US Army uses the M113A3 chassis but United Defense could also build this system on an MTVL chassis which has greater internal volume.

Opposing Forces Surrogate Vehicles

For many years the US Army has been using over 300 old M551 Sheridan light tanks as Opposing Forces Surrogate Vehicles(OSVs).

These were modified externally to resemble vehicles of Eastern origin such as the BMP-1 (73 mm) and BMP-2 (30 mm) infantry combat vehicle, T-72 tank, ZSU-23-4 self-propelled air defence gun system and 122 mm 2S1 self-propelled artillery system.

They are used in large numbers at the US Army's Combat Training Centres (CTC) including the National Training Center at Fort Irwin, California.

To meet the requirements of the US Army Simulation and Training and Instrumentation Command (STRICOM), the US Army's Red River Army Depot modified an M113 series armoured personnel carrier (APC) as a replacement for the M551.

This has been designed to visually and tactically simulate the widely deployed Russian BMP-2 infantry combat vehicle and is usually referred to as the OSV M113/BMP-2.

This is a hybrid vehicle consisting of an M113A3 chassis, itself converted from surplus M90l Improved TOW Vehicle (ITV), fitted with a Bradley Fighting Vehicle (BFV) surrogate turret. This comprises a new turret mounted on a Bradley basket and all electric drive system.

Specifications

Model	M106A2	M113A1	M113A2	M113A3	M125A2	M577A2
Crew	6	2 + 11	2 + 11	2 + 11	6	5
Combat weight	12,094 kg	11,070 kg	11,253 kg	12,150 kg	11,365 kg	11,616 kg
Unloaded weight	11,270 kg	9,630 kg	9,957 kg	10,746 kg	10,541 kg	10,864 kg
Power-to-weight ratio	17.63 hp/t	19.27 hp/t	18.51 hp/t	22.29 hp/t	18.76 hp/t	18.36 hp/t
Ground pressure	0.59 kg/cm²	0.54 kg/cm²	0.55 kg/cm²	0.59 kg/cm²	0.56 kg/cm²	0.57 kg/cm²
Length	4.926 m	4.863 m	4.863 m	5.30 m	4.863 m	4.863 m
Width	3.117 m	2.686 m	2.686 m	2.686 m	2.686 m	2.686 m
(over tracks)	2.54 m	2.54 m	2.54 m	2.54 m	2.54 m	2.54 m
Height						
(overall)	2.52 m	2.52 m	2.52 m	2.52 m	2.52 m	2.70 m
(to hull top)	1.85 m	1.85 m	1.85 m	1.85 m	1.85 m	2.489 m
Ground clearance	0.43 m	0.41 m	0.43 m	0.43 m	0.43 m	0.43 m
Track	2.159 m	2.159 m	2.159 m	2.159 m	2.159 m	2.159 m
Track width	381 mm	381 mm	381 mm	381 mm	381 mm	381 mm
Length of track on ground	2.667 m	2.667 m	2.667 m	2.667 m	2.667 m	2.667 m
Max speed						
(road)	58 km/h	60.7 km/h	60.7 km/h	65.7 km/h	60.4 km/h	58.4 km/h
(water)	5.8 km/h	5.8 km/h	5.8 km/h	5.8 km/h	5.8 km/h	5.8 km/h
Fuel capacity	360 litres	360 litres	360 litres	360 litres	360 litres	454 litres
Max cruising range	480 km	480 km	480 km	480 km	480 km	480 km
Fording	amphibious	amphibious	amphibious	amphibious	amphibious	amphibious
Vertical obstacle	0.61 m	0.61 m	0.61 m	0.61 m	0.61 m	0.61 m
Gradient	60%	60%	60%	60%	60%	60%
Trench	1.68 m	1.68 m	1.68 m	1.68 m	1.68 m	1.68 m
Engine	Detroit Diesel model 6V-53 6-cylinder water-cooled diesel developing 212 bhp at 2,800 rpm	Detroit Diesel model 6V-53 6-cylinder water-cooled diesel developing 212 bhp at 2,800 rpm	Detroit Diesel model 6V-53 6-cylinder water-cooled diesel developing 212 bhp at 2,800 rpm	Detroit Diesel model 6V-53T 6-cylinder water-cooled diesel developing 275 bhp at 2,800 rpm	Detroit Diesel Model 6V-53 6-cylinder Water-cooled Diesel developing 212 bhp at 2,800 rpm	Detroit Diesel model 6V-53 6-cylinder water-cooled diesel developing 212 bhp at 2,800 rpm
Transmission	GMC, Allison TX-100-1 with 3 forward and 1 reverse ranges	GMC, Allison TX-100-1 with 3 forward and 1 reverse ranges	GMC, Allison TX-100-1 with 3 forward and 1 reverse ranges	GMC, Allison X-200-4 with 4 forward and 1 reverse ranges	GMC, Allison TX-100-1 with 3 forward and 1 reverse ranges	GMC, Allison TX-100-1 with 3 forward and 1 reverse ranges
Suspension	torsion bar	torsion bar	torsion bar	torsion bar	Torsion bar	torsion bar
Electrical system	24 V	24 V	24 V	24 V	24 V	24 V
Batteries	2 × 12 V 6TL	2 × 12 V 6TL	2 × 12 V 6TL	4 × 12 V 6TL	2 × 12 V 6TL	2 × 12 V 6TL
Armament	1 × 12.7 mm M2 MG	1 × 12.7 mm M2 MG	1 × 12.7 mm M2 MG	1 × 12.7 mm M2 MG	1 × 12.7 mm M2 MG	1 × 7.62 mm MG
Ammunition	1,000	2,000	2,000	2,000	1,000	1,000
Armour	12-44 mm	12-44 mm	12-44 mm	12-44 mm	12-44 mm	12-44 mm
NBC system	yes	yes	yes	yes	Yes	yes
Night vision equipment	yes	yes	yes	yes	Yes	yes

The vehicle is also equipped with SAWES/MILES II (Small Arms Weapons Effect Simulator/Multiple Integrated Laser Engagement System) to simulate indirect threats during force-on-force training exercises. The latter replicates the 30 mm cannon, 7.62 mm coaxial machine gun and roof mounted AT-5 Spandrel of the original BMP-2 vehicle.

The hull of the M113A3 has been modified to resemble the BMP-2 in a number of key areas. This includes extending the nose of the vehicle using heavy aluminium plate and extending the rear hull for greater internal volume.

Following trials with three prototype vehicles, 15 Low-Rate Initial Production (LRIP) vehicles were built at the Anniston Army Depot (ANAD).

Supporting ANAD is Rock Island Arsenal, Benet Labs, Tank Automotive and Armaments Command and other facilities as well as United Defense LP, the original manufacturer of the M113 series.

The first batch of 160 production vehicles will be completed by FY2002 and be used by the National Training Center. A further 36 will be built for the Joint Readiness Training Center at Fort Polk. More recently the prototype of an OSV built to resemble a T-72 MBT has been built.

Civilian M113 vehicles

United Defense LP has recently proposed that some members of the M113 family of vehicles could undertake a number of specialised civilian roles.

These include the M113 Firefighting Vehicle (M113 FFV) and the M113 Hazardous Material Response Vehicle (M113 HAZMAT) which are based on the M577 series command post vehicle using automotive components of the M113A3 but with six roadwheels either side.

Armored Infantry Fighting Vehicle

The AIFV was developed from the M113A1 as a private venture by the then FMC Corporation and is fully described in its own entry earlier in this section.

Egyptian Infantry Fighting Vehicle (EIFV)

Based on an MTVL (stretched M113) chassis and fitted with a Bradley 25 mm two-man turret. The first prototype was completed in 1997 in Egypt. Additional details are given earlier in this section.

Sharpshooter infantry fighting vehicle

Based on an M113 series chassis with a United Defense one-man Sharpshooter turret armed with a 25 mm cannon and 7.62 mm coaxial machine gun. Developed to the prototype stage.

Automatic cannon vehicle

This is an M113A2 fitted with the then FMC two-man 25 mm weapon station that is a development of the weapon station installed on the M2 Bradley IFV. The ACV is a private venture and has yet to enter production.

Lynx command and reconnaissance vehicle

This uses many automotive components of the M113 APC and was fully described in *Jane's Armour and Artillery 1999-2000*. Sales were made to Canada (174 units) and the Netherlands (250 units). Both of these countries have phased this vehicle out of service. The Netherlands has sold 35 of these vehicles to Bahrain and these were delivered between 1997 and 1998.

M113 with RBS 70 SAM

The now Saab Bofors Dynamics of Sweden has developed a kit which enables a standard M113 series APC to be fitted with its RBS 70 SAM system on a pedestal in the rear troop compartment. This was announced in 1988 and is in service with the Pakistan Army. Conversions were undertaken locally. An illustration of this version appears under Pakistan in this section.

M113 with 120 mm mortar

In 1987, an M113A2 was fitted with a turret-mounted 120 mm breech-loaded mortar and demonstrated in the UK. The mortar is aimed, loaded and fired under complete armour protection.

The turret and 120 mm breech-loaded mortar have been developed as a private venture by BAE Systems, RO Defence and can be installed on a number of vehicles. The first customer for this 120 mm AMS is the Saudi Arabian National Guard which has 73 systems based on a General Motors Defense, 8 × 8 LAV chassis.

More recently, United Defense, BAE Systems, RO Defence and Delco Systems have fitted a stretched M113A3 chassis (the Mobile Tactical Vehicle Light) with the 120 mm turret mortar system and conducted firing trials. This vehicle will be used for demonstrations in a number of countries.

M113 with ADATS

The Oerlikon Aerospace ADATS (Air Defence Anti-Tank System) can be installed on a wide range of tracked and wheeled armoured vehicles. The Canadian Forces selected a modified M113 series APC and a total of 36 systems has now been delivered; final deliveries took place in 1992. Mounted on top of the modified M113 vehicle is a turret with eight ADATS missiles, four either side, in the ready to launch position. Mounted between the missiles is the electro-optical module and mounted on the turret rear is the surveillance radar. Full details of ADATS are given in *Jane's Land-Based Air Defence*

2001-2002. A photograph of this version appears earlier in this section under the entry for Canadian M113 series vehicles.

Infantry Armoured Fighting Vehicle (VCC-1)

This is a further development of the M113A1 by Otobreda of Italy and is fully described in this section under Italy.

M113 fire support vehicle

This is essentially an M113A2 APC fitted with the Cadillac Gage two-man turret armed with a 90 mm Cockerill Mk III gun, a coaxial 7.62 mm machine gun, a 7.62 mm anti-aircraft machine gun and, if required, smoke grenade dischargers. This model has yet to enter production. For trials purposes the Belgian company Cockerill has fitted an M113A1-B APC with a Cockerill CSE 90 mm two-man turret for use in the fire support role. More recently an M113A2 has been fitted with a Delco Defense two-man LAV-25 turret armed with 25 mm cannon, 7.62 mm MG and TOW launchers.

M113A2/TS 90 light tank

This is a joint development between United Defense LP and Giat Industries and consists essentially of an M113A2 APC with the roof modified to accept the Giat Industries TS 90 two-man turret armed with a 90 mm gun. In 2001, the M113A2/TS 90 system was still at the prototype stage.

Other variants of the M113 family

Almost every user of the M113 has developed variants of the M113/M113A1/M113A2/M113A3 chassis. There are separate entries on the following countries earlier in this section: Australia, Belgium, Canada, Denmark, Egypt, Germany, Israel, Italy, Norway, Pakistan, Singapore, Switzerland and Turkey.

Other countries that have carried out modifications to the M113 series include:

Argentina: A total of 319 in service in M113 and M577 series versions. Original 12.7 mm M2 HB MG replaced by either 7.62 mm MAG or 12.7 mm Browning M1915.

Belgium: Details are given earlier in this section under Belgium.

Brazil: the Brazilian Army has about 600 M113 series vehicles and Moto Pecas SA of Sorocaba, São Paulo state, has overhauled these and replaced the original petrol engine with a Brazilian-built Mercedes-Benz OM 352A diesel.

Denmark: Details of these are given in this section under Denmark.

Lebanon: this country operates a fleet of over 500 vehicles, many of which have been modified to suit local operational requirements. Types used include M113A1, M113A2, ACAV, M577A1, fitted with MILAN and TOW ATGWs, fitted with a 106 mm recoilless rifle, fitted with 2 × 23 mm, 2 × 14.5 mm and 4 × 14.5 mm anti-aircraft machine guns, fitted with an 82 mm B-10 recoilless rifle and 81 mm M125A1 self-propelled mortar carrier.

New Zealand: New Zealand purchased 72 members of the M113 family including the M113A1 armoured personnel carrier (some of which have the T50 turret, as on Australian vehicles), M125A1 mortar carrier, M577A1 command vehicle, M579 fitters' vehicle and the M806A1 armoured recovery vehicle. The Ministry of Works and Development (MWD) has rebuilt 64 of these over a six-year period beginning in 1980, at a cost of US$6 million.

For operations in Bosnia, part of the New Zealand fleet of M113A1 vehicles has been upgraded.

Specific vehicles modified included 20 M113A1 APCs with the T-50 turret, one M113 ambulance vehicle, two M125 81 mm mortar carriers, one M806 recovery vehicle and two M579 fitters' vehicles.

Some of the initial vehicle modifications were carried out by the New Zealand Army Support Command and included upgrades to the T-50 weapon turrets to increase the gun's stability as well as the installation of a new day/night sighting system and belly armour plates to provide additional mine protection.

Other modifications were carried out by the FMS Corporation (now Marvin Land Systems) and included external armour packages, interior spall liners, gun shield kits and personnel heater kits.

Saudi Arabia: Under a US$413 million contract awarded early in 1997 to FMC Arabia, a joint venture company between United Defense LP and the local Saudi Arabian company of Al Hejailan, formed to overhaul 523 M113 series armoured personnel carriers (APCs) for the Royal Saudi Land Forces (RSLF), a new facility has been built at Al Kharj, 80 km south of Riyadh.

There was some delay in starting the RSLF M113 Modernisation Programme due to funding problems but these are now being overcome and the major programme is now moving ahead.

Under terms of the contract, a first batch of 523 M113A1/M113A2 series full tracked APCs are to be upgraded to the latest and improved M113A3 standard using parts supplied from the US.

It is estimated that Saudi Arabia has between 1,500 and 1,700 M113 series vehicles and variants and that over the coming years many of these will be upgraded to the latest M113A3 standard which includes a more powerful 6V-53T Detroit Diesel engine, Allison X-200-4 automatic transmission, external fuel tanks and variable speed cooling fan.

Spain: Spain uses the M113, M113A1, M113A2, M548 tracked cargo carrier, M577 command post vehicle, M113 series recovery vehicle and the M125 81 mm mortar carrier. Talbot carries out the complete overhaul of M113s in Spain and can also bring M113 petrol-engined vehicles up to the M113A1/M113A2 standard. In place of the M106 107 mm mortar carrier the Spanish Army uses a similar vehicle with a Spanish-built 120 mm mortar which fires to the rear of the hull. A baseplate is carried externally on the left side of the hull to allow the mortar to be dismounted for use away from the vehicle.

Two new versions of the M113A1 series have recently entered service with the Spanish Army, the M548 minelayer and the M113A1 VCZ combat sappers' vehicle.

The M548 minelayer has been developed by EXPAL under the designation of the SEM-11 and is based on a stretched M548 vehicle that has six roadwheels and has been modified to carry a mechanical minelaying system. Details are given in *Jane's Military Vehicles and Logistics 2001-2002*.

The M113A1 VCZ combat sappers' vehicle entered service in 1991 and is fitted with a front-mounted dozer blade and a winch. In the future it is expected to tow a trailer-mounted rocket-propelled mineclearing system.

Thailand: the Royal Thai Army has a large fleet of M113 series armoured personnel carriers and, in 1992, it was announced that 100 of these were to be upgraded over a two year period with kits provided by the FMS Corporation (now Marvin Land Systems). The kits upgrade the existing M113A1 vehicles to the M113A2 configuration. In 1996 United Defense LP was awarded

Status

Production as required. In service with the following countries:

Country	Quantity	Comment
Argentina	319	
Australia	771	many variants, upgrade under way, see under Australia
Bahrain	220	M113A2 inc 212 from USA in 1992/1996
Belgium	434	plus variants, see under Belgium
Bolivia	50	
Bosnia	80	from USA in 1996
Brazil	584	includes M577, plus 28 with marines
Cambodia	50	Estimate
Canada	1,214	228 plus variants, including M577, many variants, some to be upgraded, see under Canada
Chile	228	plus 128 from Italy
Colombia	80	
Congo, Democratic Republic	12	including 54 from USA in 1992
Denmark	600	includes 25 mm turret, 81 mm mortar and TOW ATGW, see under Denmark
Ecuador	20	
Egypt	2,000	includes M113A2, 82 mm and 120 mm mortar, M577 command post, M901 TOW and air defence, see under Egypt
Ethiopia	?	
Germany	2,700	many variants, see under Germany
Greece	1,700	includes 107 mm mortar and M901 TOW, including 996 from USA between 1994 and 1997
Guatemala	10	
Iran	300	Estimate
Iraq	100	Estimate
Israel	6,000	many variants, see under Israel including 300 from Germany, 175 from Netherlands and 361 from USA
Italy	3,019	includes 738 M113, 1,792 VCC1/VCC2, 270 TOW and 219 SIDAM 25 mm SPAAG (marines have 40 VCC2 and 91 M113 series), see under Italy
Jordan	1,300+	includes M113, 81 mm mortar, M901 TOW and 20 mm SPAAG
South Korea	560	includes 420 M113 series and 140 M577
Kuwait	108	includes 60 M113, 40 M577 and 8 ITV
Lebanon	867	many variants
Libya	40	
Macedonia	30	
Morocco	440	variants include TOW
New Zealand	78	including mortar and M548, to be replaced by LAV (8 × 8) vehicles
Norway	670	including mortar carriers, 25 mm, basic APC and TOW, see under Norway
Pakistan	1,600+	many variants including M901 TOW, local production, see under Pakistan
Peru	130	
Philippines	100	
Portugal	534	including TOW and mortar, including 50 from Germany and 104 from Netherlands
Saudi Arabia	1,700	variants include VCC-1 with TOW
Singapore	1,000+	many variants, see under Singapore
Somalia	50	believed non-operational
Spain	1,700	variants include 81 mm and 120 mm mortar and M577
Sudan	36	
Switzerland	1,475	including 120 mm mortar carriers and APC with 20 mm cannon (qv)
Taiwan	800	estimate, includes local model, 81 mm and 120 mm mortars, M113A2 ambulance, M901 TOWs and M577A2 command post
Thailand	430	many variants, including M577A3, M1064A3, M125A3, M901A3, M113A3 ambulance and fitters
Tunisia	140	including M577A1, M125A1 and TOW
Turkey	3,000	approx number, includes 81 mm and 107 mm mortars, plus TOW including 187 from Germany (delivered 1993) and 97 from Italy (delivered 1994)
USA	25,000	includes many variants and 4,650 M577 command post
Uruguay	15	
Vietnam	100	Estimate
Yemen	60	

a follow-on contract for a second batch of 100 vehicles.
Turkey: Details of these are given under Turkey.
Vietnam: in 1985, the Vietnamese People's Army was observed to be still using significant numbers of US M113 series armoured personnel carriers including a version similar to the ACAV but with the 12.7 mm M2 HB machine gun replaced by a 12.7 mm DShkM machine gun.

Manufacturer

United Defense LP, Ground Systems Division.

UPDATED

General Dynamics Land Systems Advanced Amphibious Assault Vehicle (AAAV)

Development

Following a competition between United Defense LP, which built all of the previous armoured amphibious assault vehicles for the US Marines Corps and General Dynamics Land Systems, the US Marine Corps awarded General Dynamics Land Systems the Program Definition and Risk Reduction (PDRR) contract for the Advanced Amphibious Assault Vehicle (AAAV) in June 1996.

Following the award of this contract, General Dynamics Land Systems opened a new division and facility in Woodbridge, Virginia, called General Dynamics Amphibious Systems. This is dedicated to amphibious systems research and development, as well as prototype integration and assembly.

The total value of the cost-plus-fixed feed contract was US$216.9 million and this included design, virtual prototyping, systems integration, fabrication and testing of three AAAV(P) prototype vehicles.

In mid-1998, the AAAV completed a successful critical design review, clearing the way for the fabrication of the first prototype vehicles. Assembly of the first vehicle commenced at the Woodbridge facility in December 1998, with the first prototype being completed in June 1999.

Early in 2001, General Dynamics Land Systems was awarded a US$712 million contract for the Systems Development and Demonstration (SSD) phase of the AAAV.

Under the terms of this cost-reimbursable contract, the company will provide all required materials, services, personnel and facilities to complete the design and development of the AAAV.

The company will also manufacture and test nine new prototypes, refurbish three early development prototypes and support The US Marine Corps initial operational test and evaluation and prepare for the production phase of the programme.

These vehicles will undergo extensive reliability and live fire testing. Of these nine vehicles, eight will be in the troop carrier version and one in the command and control configuration.

If all goes to plan, then the US Marine Corps plans to acquire 1,013 AAAV through FY2015 of which 935 will be personnel and 78 command and control. Initial operational capability is expected to be 2006 with one platoon of vehicles. This will consist of 14 AAAV (personnel) and one AAAV (command).

Low-rate initial production is sheduled to commence in 2004, followed by full-rate production of the AAAV in 2006.

When compared to the current United Defense AAV7A1, the new AAAV has three times the water speed, additional armour protection, more powerful armament and can also operate ashore with the US Marine Corps M1A1 Main Battle Tanks.

The AAAV will form one part of the US Marine Corps 'Amphibious Triad' with the other two elements being the MV-22 Osprey tilt rotor aircraft and the current Landing Craft Air Cushion (LCAC).

Until the AAAV enters service, a major part of the current fleet of AAV7A1 are being upgraded under a joint United Defense/US Marine Corps programme called the RAM/RS (Reliability, Availability and Maintainability/Rebuild to Standard). A total of 680 vehicles are being rebuilt with the first of these delivered in 1999.

The Lima, Ohio Army Tank Plant will handle initial structure fabrication for the SSD phase of the programme, with the final production/assembly location to be determined during the SSD phase.

Description

The AAAV(P) will have a three-man crew and carry 17 or 18 fully equipped marines. The crew consists of the driver, who is seated in front left and the vehicle commander and gunner who are seated in the turret.

The hull of the AAAV is made of 2519-T87 welded aluminium and built in a space frame configuration, which allows a low-risk modular ceramic armour to be added.

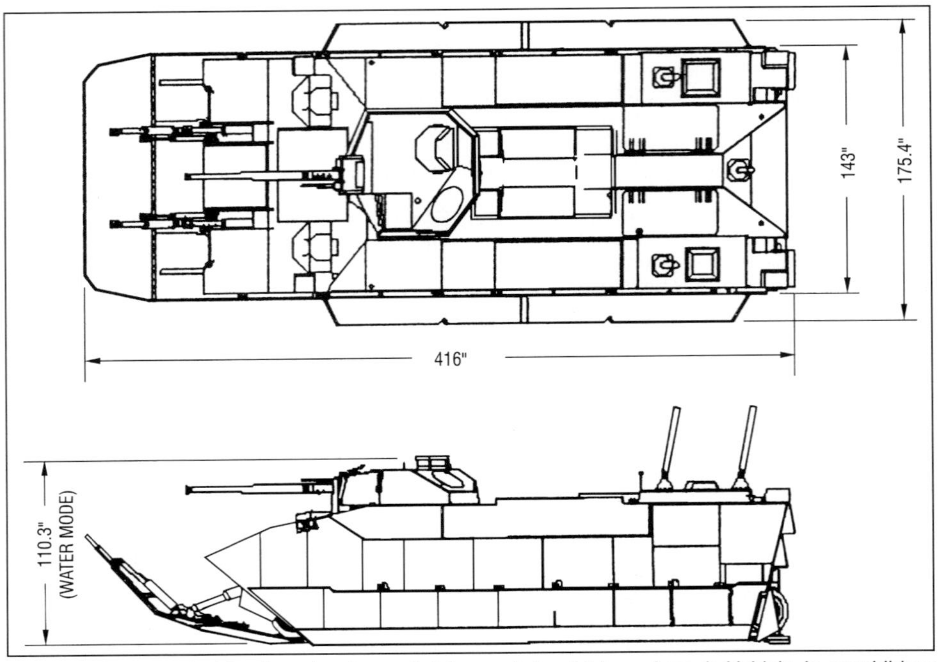

Provisional top and side view drawings of Advanced Amphibious Assault Vehicle in amphibious configuration **2001**/0099474

The armour provides excellent multi-hit protection and can easily be upgraded and repaired when damaged.

The armour package will provide protection against attack from 14.5 mm armour piercing rounds fired at 300 m and against 155 mm artillery fragments.

To enable the AAAV to achieve the high water speeds required by the US Marine Corps, the AAAV uses a planing hull design that was initially tested on the General Dynamics Land Systems Hydrodynamic Test Rig (HTR) during the Concept Exploration Phase of the programme. Many of the design features of the HTR were incorporated into the AAAV(P).

The weapons station of the AAAV, which is designated the MARK 46 (MK 46), is a two-man electronically power-operated turret. This is mounted towards the front of the vehicle and has seats for the commander on the right and the gunner on the left.

The vehicle commander is provided with five forward and one rear looking periscopes to give him observation through a full 360°. As the 30 mm cannon is stabilised, it will be able to engage moving targets with a high first-round hit probability.

The MK 46 turret uses a General Dynamics Land Systems Compact Modular Sight (CMS) that incorporates second-generation Forward Looking Infra-Red (FLIR), day optics and an eye-safe laser range-finder and a Kearfott Dual Axis Head Assembly (DAHA).

These components allow the system to engage targets under day, night and adverse weather conditions. The fire-control system installed is a derivative of that installed in the General Dynamics Land Systems M1A2 MBT.

Late in 1997, the US Marine Corps awarded The Boeing Company a contract for the supply of two 30 mm Bushmaster II cannons for delivery as Government Furnished Equipment (GFE) for installation in the first two prototypes of the AAAV.

These were delivered in 1998. The design of these cannon was later modified to permit easy upgrade to a 'Super 40 mm' configuration, navalised by the addition of stainless steel components and to accommodate a titanium integral gun mount rather than a separate barrel support and turret trunnion block. The modified configuration is designated the MARK 44 in the US Navy/Marine Corps.

The first customer for the Bushmaster II cannon was Norway who selected the weapon for the 104 CV 9030N infantry fighting vehicles which have now been delivered by Hägglunds Vehicle of Sweden.

Later deliveries to the company will be in the MK 44 configuration. Switzerland has also recently selected the MK 44 for its CV 9030 CH infantry fighting vehicles. Finland has also selected this weapon for its recently ordered CV 9030FIN vehicles recently ordered. The complete MK 44 turret is also being marketed for other applications. Smoke grenade launchers are controlled from the turret, but the actual launchers are installed on the hull of the AAAV.

Maximum land speed of the AAAV will be over 72 km/h while the maximum water speed will be over 37 km/h.

Prototype of the Advanced Amphibious Assault Vehicle (AAAV) in land configuration with turret traversed to front **2001**/0099473

Specifications

Crew: 3 + 18
Weight loaded: 34,476 kg
Weight empty: 28,576 kg
Ground pressure: 0.625 kg/cm^2
Length:
(overall, land) 9.271 m
(overall, water) 10.566 m
Width across skirts: 3.632 m
Height to turret roof: 3.20 m
Ground clearance: 0.406 m
Track width: 533 mm
Max road speed: 72.41 km/h
Max water speed: 37-46.61 km/h
Transition mode speed: 16.6 km/h
Range:
(land) 643 km
(water) 121 km
Fuel capacity: 1,506 litres
Engine: MTU MT 883 Ka-523 diesel developing 2,700 hp in water mode and 850 hp in land mode
Transmission: Allison Transmission X4560
Suspension: hydropneumatic (retractable)
Armament:
(main) 1 × 30 mm MK 44 cannon
(coaxial) 1 × 7.62 mm M240 MG
Ammunition:
(main) 200 (ready) + 400 (stowed)
(coaxial) 800 (ready) + 1,600 (stowed)
Gun stabiliser:
(vertical) yes
(horizontal) yes
NBC system: yes
Night vision equipment: yes

The diesel engine is mounted in the centre of the AAAV and consists of an MTU 883 Ka-523 12-cylinder diesel engine that develops 2,700 hp for water operations and 850 hp for land operations.

The engine is coupled to an Allison Transmission Division Power Transfer Module that distributes power for land and sea modes of operation. During land mode the PTM sends power through its torque converter to the X4560 fully automatic six-speed transmission.

When in the sea mode, the PTM will transfer power to two Allied Signal 584 mm water jets mounted on both sides on the vehicle. These water jets will give over a 400 per cent increase in water speed when compared to the current AAV7A1 as well as the stability, manoeuvrability and control required for movement in surf.

In the first quarter of 1997, General Dynamics Amphibious Systems awarded Textron Marine & Land Systems, an initial US$4 million contract for the design and development of the suspension system for the AAAV.

Under the terms of this contract, they developed and built a retractable, actively damped, hydropneumatic suspension system for the AAAV. A total of 42 units were built for installation on prototypes of the AAAV, plus spares.

The new suspension system will retract during high-speed water operation to minimise hydrodynamic drag. It will also employ a computer-controlled active damping system which reduces terrain disturbance to offer a smoother ride. The track used in the AAAV is a double-pin aluminium track system developed by United Defense.

When afloat, the hydropneumatic suspension system is retracted and protected by plating, a prow plate opens outwards to extend the planing surface and at the rear two flaps open to expose the water jets.

The troop compartment begins on the sides of the engine compartment and continues through the rear of the AAAV. The troops enter the vehicle via a hydraulically power-operated ramp that also contains a troop door. In addition, there are two sliding roof hatches over the troop compartment.

The signature of the AAAV is minimised through the incorporation of signature treatments for the exterior surface and other features.

Two electric and three hydraulic bilge pumps are strategically located throughout the hull with reserve buoyancy assuring stability and safety. The diesel fuel cells are isolated from the crew and troop compartments to minimise potential hazards.

Standard equipment will include an automatic fire detection and suppression system and an environmental control system for climate control and an overpressure collective system against NBC contamination.

Variants

Command and control vehicle

The only variant currently being developed is the command and control vehicle. The vehicle will have seven work stations operating on a tactical data network. Each will be capable of simultaneously running Tactical Combat Operations (TCO), Advanced Field Artillery Tactical Data System (AFATDS), Intelligence Analysis System (IAS) and Command and Control Personnel Computer (C2PC) applications.

It will have two Single Channel Ground and Airborne Radio Systems (SINCGARS), two Ultra-High Frequency (UHF) Enhanced Position Location Reporting System (EPLRS) and two UHF Have Quick II radios.

Status

Under development for US Marine Corps. Production is expected to commence in 2006 with deliveries to run from 2006 through to 2013.

Manufacturer

General Dynamics Amphibious Systems.

UPDATED

United Defense LP AAV7A1 (LVTP7A1) Amphibious Assault Vehicle

Note: This vehicle was originally called the LVTP7 and is still referred to by this designation by some users. In 1985, the US Marine Corps changed this to the Amphibious Assault Vehicle 7 (AAV7). All new production vehicles were built to the AAV7A1 configuration and existing US Marine vehicles were upgraded to the new production standard.

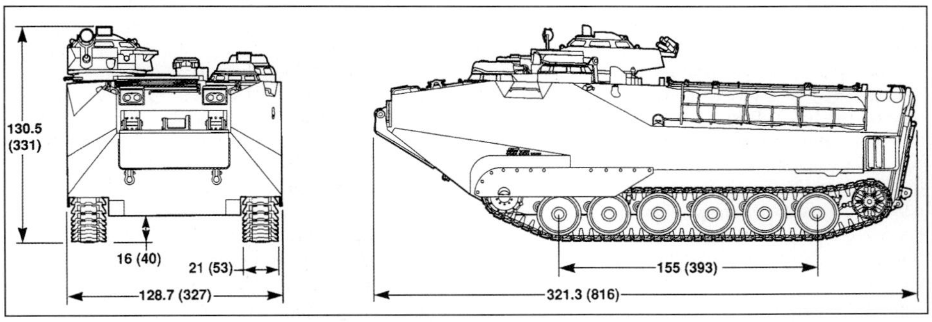

General arrangement drawings of AAV7PA1 amphibious assault vehicle ***2002***/0116796

Development

In March 1964, the United States Marine Corps issued a requirement for a new LVTP and, after evaluating a number of proposals, a contract for the development of a new LVTP was awarded to the then Ordnance Division of the FMC Corporation. In January 1994, FMC's Defense Systems Group (which included the Ground Systems Division) and BMY Combat Systems Division formed a new joint venture company called United Defense LP. United Defense became a private company in 1997 when the Carlyle Group purchased the defence interest of the FMC Corporation and the Harsco Corporation for US$850 million.

Engineering development began in February 1966 and the first of 15 prototypes was completed in September 1967 and delivered to the Marine Corps the following month. Trials with the prototypes, designated the LVTPX12, were completed in September 1969.

In June 1970, FMC was awarded a contract worth US$78.5 million for the production of 942 vehicles without weapon stations. When the government-supplied weapon stations were included, the unit cost was around US$129,000. The first production vehicles,

AAV7A1 fitted with the RAFAEL Armament Development Authority Enhanced Appliqué Armor Kit (EAAK). This vehicle also has the Up Gunned Weapons Station and a trim vane mounted under the nose of the vehicle (Michael Jerchel) ***2000***/0080342

AAVP7A1 clearly showing new turret armed with 40 mm grenade launcher and 12.7 mm machine gun and bow plate retracted under nose of vehicle ***2002***/0116802

	LVT7 Personnel	Command	Recovery	Total	AAV7A1 Personnel	Command	Recovery	Total
Argentina	19	1	1	21	0	0	0	0
Brazil	0	0	0	0	22	2	2	26
Italy	24	1	0	25	0	0	0	0
Korea, South	53	5	3	61	41	1	0	42
Spain	16	2	1	19	0	0	0	0
Thailand	12	0	0	12	18	2	1	21
USA	0	0	0	0	1,153	106	64	1,323
Venezuela	9	1	1	11	0	0	0	0
Total	130	10	6	149	1,234	111	67	1,412

Note: These figures exclude the order placed by South Korea in 1995, these are given at the end of the Development section. Spain has recently completed an upgrade to their vehicles, while Italy is now starting an upgrade.

which were designated the LVTP7 (Landing Vehicle, Tracked, Personnel, Model 7) were handed over to the United States Marine Corps in August 1971 and the first unit was equipped with the vehicle by March 1972. Final deliveries were made in September 1974 after which the LVTP5A1 and its variants were phased out of service.

The US Marine Corps has upgraded all its LVTP7s to the new AAV7A1 configuration and total production/conversion figures for both vehicles are as above.

The AAV7A1 is expected to be replaced in US Marine Corps service by the Advanced Amphibious Assault Vehicle (AAAV) covered in an earlier entry. To enable the current AAV7A1 to be viable through to the year 2010 now that the RAM/RS programme, covered in detail later in this entry, is now underway.

The AAV7A1 was used by the US Marine Corps during Operation Desert Storm.

In April 1995, through the Marine Corps Systems Command, United Defense LP was awarded a firm fixed price contract for a total of 14 AAV7A1 Assault Amphibious Vehicles for the Brazilian Marines under the Foreign Military Sales Programme (FMS).

The order for 14 vehicles comprised 12 AAVP7A1 troop carriers, one AAVR7A1 recovery vehicle and one AAVC7A1 command post vehicle. The Brazilian Marines already operate 12 AAV7A1 (10 AAVP7A1 personnel carriers, one AAVR7A1 recovery vehicle and one AAVC7A1 command post vehicle).

Engineering, management, purchasing and hull fabrication was carried out at San Jose, where production of all previous AAV7A1 vehicles was carried out, with assembly and testing being carried out at Anniston, Alabama. Production was completed by late 1996 and all vehicles have now been delivered to Brazil.

It has been disclosed that, late in 1995, United Defense LP and Samsung Aerospace of South Korea signed a major contract to co-produce a total of 57 AAV7A1 vehicles in three versions between 1996 and the year 2000.

United Defense and Samsung Aerospace rolled out the first two AAV7A1 vehicles in July 1997 at Samsung's research and development centre in Changwon, South Korea; this was the first time that the AAV7 series had been produced outside of the USA.

United Defense are manufacturing the hull and procuring most of the major sub-components in the knockdown kits. Assembly of the vehicles, manufacture of local components and vehicle testing is being carried out in South Korea in this co-operation agreement.

Actual production of the AAV7A1 commenced in South Korea in November 1997 and will run through to 2001.

In May 1999, Samsung Aerospace, Special Products Division signed a follow-on contract with United Defense for the supply of Korean Armoured Amphibious Vehicle (KAAV) Manufacturing Licence Agreement (MLA) for an additional batch of kits.

United Defense supplies knockdown kits and hulls while Samsung Aerospace provides some components from companies in the ROK and is also responsible for final assembly, testing and delivering completed vehicles to the customer.

Under earlier contracts, Samsung Aerospace is building 57 AAV7A1. First vehicles were completed at the Samsung Aerospace facility at Changwon in 1998 with final deliveries due in 2001.

This latest US$120 million contract for an undisclosed quantity will extend deliveries to 2006. Prior to the local production/assembly of the AAV7A1 in the ROK, a total of 61 AAV7 and 42 of the improved AAV7A1 vehicles and variants were purchased direct from the United Defense production line in San Jose, California.

In September 1997, United Defense was awarded a US$40 million contract with Spain to rebuild AAV7 vehicles of the Spanish Marines to the AAV7A1 standard in Spain.

The contract covered the modernisation and upgrade of 16 AAV7, two AAV7 command vehicles and one AAVR7 recovery vehicle, this being the whole Spanish fleet. In addition, the contract covered technology transfer, tools and spare parts.

The actual rebuilds were carried out by a Spanish sub-contractor CAF who have previously undertaken work on these vehicles. First six upgraded vehicles were handed over to Spain in July 1999 with final deliveries in July 2000.

For many years Italy has operated a fleet of 24 United Defense LP, Ground Systems AAVP7 personnel carriers and one AAVC7 command vehicles that have never been upgraded.

Of the 24 Italian vehicles, 14 AAVP7 are operated by the Italian Army and are based at Brindisi with the remaining 10 AAVP7 and one AAVC7 command vehicles being operated by the Marines and based in Venice.

Under a contract worth US$90 million awarded late in 1999 to United Defense LP, all of these Italian vehicles are to be upgraded to the latest US Marine Corps Reliability, Availability, Maintainability/Rebuild to Standard (RAM/RS).

Italy will be the first export customer for this upgrade package which is now being marketed to other potential customers by United Defense LP.

The RAM/RS work will be carried out in Italy by Goriziane with United Defense LP providing the expertise and kits.

In addition, Italy will take delivery of a further nine vehicles. These will be ex-US Marine Corps hulls which will be sent to Italy where they will be fitted out by Goriziane to the latest standard.

Description

The all-welded aluminium hull of the LVTP7 protects the crew from small arms fire, shell splinters and flash-burns.

The driver sits at the front of the hull on the left side and has a single-piece hatch cover that opens to the right rear and seven day vision blocks for all-round observation. A night vision periscope can be fitted in the forward part of the driver's hatch cover for driving at night.

The commander sits behind the driver and has a single-piece hatch cover that opens to the rear and seven day vision blocks for all-round observation. Forward of the commander's station is an M17C periscope which can be extended enabling the commander to see forward over the driver's hatch cover.

The engine is in the forward part of the hull on the centreline and power is transmitted to the FMC HS-400 transmission via a torque converter with lock-up. The transmission combines shifting, steering and braking functions in one unit and gives four gears forward and two in reverse. The engine can be removed from the vehicle in 45 minutes and can be run outside for test purposes if required. Power is transmitted to the drive sprockets through hull-mounted final drives. A Power Take Off (PTO) mounted on the converter housing supplies power for the water-jets and the cooling fan through electrohydraulically controlled clutches.

Mounted over the power pack are a fan and radiator. Air is drawn in and discharged through ballistic grills in the roof. When afloat, hydraulically actuated doors below the grills seal off the air openings. Air for the engine and troop compartment then enters the vehicle through a hydrodynamically actuated air valve mounted on the top of the hull to the right of the driver. Cooling takes place in the contact cooler which is an integral part of the hull floor.

The armament installation is beside the engine on the right side of the hull. The turret has a single-piece hatch cover that opens to the rear, eight vision blocks, a sight with a magnification of ×1 and ×8 and one direct ring sight. The turret is armed with a 12.7 mm (0.50) M85 machine gun with two rates of fire, 1,050 rpm (high) and 450 rpm (low). Of 1,000 rounds of 12.7 mm ammunition carried, 400 are for ready use. The machine gun has an elevation of +60° and a depression of −15°. Turret traverse, which is electrohydraulic, is a full 360°.

The 25 fully equipped personnel sit on three bench seats in the troop compartment at the rear of the hull, one in the centre and one on either side. The centre seat can be removed and the two side ones folded up to allow up to 4,536 kg of cargo to be carried. The removable centre seat is stowed on the left bulkhead when not required. The marines enter and leave the vehicle through a power-operated ramp at the rear of the hull, which has a door in the left side of the ramp in case of ramp failure and a single vision block. Over the

AAV7 amphibious assault vehicles showing amphibious capabilities with commander's, gunner's and driver's hatches in open position (US Navy)

AAVR7 recovery vehicle with the crane retracted into the travelling position

top of the troop compartment are three torsion spring-balanced roof hatches.

The tube-over-bar suspension consists of six dual rubber-tyred roadwheels with the drive sprocket at the front and the idler at the rear. There are no track-return rollers. The first and last roadwheel stations have a hydraulic shock-absorber. The single-pin, rubber-bushed tracks have a moulded rubber top surface and replaceable rubber pads.

The LVTP7 is fully amphibious without preparation, being propelled in the water by two water-jets which are mounted one either side of the hull at the rear. The water-jets are driven through right-angled gearboxes on top of the sponsons and drive shafts that extend to the rear of the hull. The water-jets, which are aluminium, are mixed flow pumps with special exhaust nozzles to maximise thrust and each pumps 52,990 litres of water per minute. At the rear of each pump is a hinged steering deflector which reverses the direction of the jet when closed. Maximum speed forwards is 13.51 km/h and maximum speed in reverse is 7.24 km/h. The tracks can also be used to propel the vehicle when afloat.

Four bilge pumps are fitted: two electric (each with a capacity of 379 litres/min) and two hydraulic (each with a capacity of 435 litres/min).

The LVTP7 has no NBC system; it is fitted with night driving lights. A number of kits has been developed for use with the LVTP7, including one for converting the vehicle to an ambulance that can carry six stretcher patients, a navigation kit, a visor kit for the driver and a winterisation kit which keeps the crew and personnel compartment heated when the outside temperature is down to –54°C, the normal heater being effective down to –32°C.

Modernisation

AAV7A1

In March 1977, the FMC Corporation was awarded a contract to convert 14 LVTP7s into prototypes of a new configuration known as the LVTP7A1. The entire Marine Corps' LVTP7 fleet was converted between 1982 and 1986. The objectives of the conversion were to increase the vehicle's reliability, availability, maintainability and durability while reducing life cycle costs. New items added to the LVTP7 to bring it up to the LVTP7A1, subsequently redesignated AAV7A1, standard are:

A power pack using the Cummins VT400 diesel engine

Offboard and onboard diagnostic instrumentation (fault isolation)

Smoke-generating capability

Passive night driving and night firing devices

Improved fire suppression system

Secure voice radio installation

Provision to mount Position Location And Reporting System (PLARS)

Improved troop/crew compartment ventilation

Improved hydraulic and electrical system

Improved electric weapon station

Improved suspension system.

In addition to previously announced improvements, the AAV7A1 has a non-integral fuel tank, watertightness improved by reworking seals which surround air intakes, exhaust plenums and cargo hatches, a main wiring harness segmented for rapid repair without requiring an entirely new harness, removable headlights, a troop commander's hatch and a vision block raised for better visibility over the more forward driver's hatch.

The programme was known as the LVTP7 Service Life Extension Programme (SLEP) with US$10.5 million requested in FY80, US$47.3 million in FY81 (25 conversions plus US$7.2 million for RDT & E), US$171.3 million in FY82 (149 conversions), US$116 million in FY83 (307 conversions), US$150.5 million in FY84 (263 conversions), and US$115.9 million in FY85 (240 conversions).

In addition to the conversion of LVT7 to AAV7A1 configuration (this conversion programme also applied to the other two members of the family, AAVC7A1 and AAVR7A1), funding was provided to build new vehicles to AAV7A1 configuration: US$12 million in FY81, US$61.6 million (30 vehicles) in FY82, US$116.0 million (146 vehicles) in FY83 and US$115.3 million (153 vehicles) in FY84. With the later funding the additional 329 AAV7A1s for the Maritime Prepositioning Ships (MPS) were completed. In FY85, a further four vehicles were ordered at a cost of US$3.5 million.

The first production AAV7A1 was handed over to the US Marine Corps in August 1983. By mid-1984, production was running at about 50 vehicles a month with final deliveries made in 1986.

In March 1988, following two years of extensive tests, the US Marine Corps selected the KVH Industries MV103 militarised digital compass for use on its AAV7A1 family of vehicles.

Further Improvements

The AAV7A1 Product Improvement Programme commenced with the initial fielding of the Up Gunned Weapons Station during 1987. Up Gunned Weapons Station, bow planes and magnetic heading systems were fully fielded by 1990 as was the automatic fire sensing and suppression system. The Enhanced Appliqué Armour Kit and improved transmissions were fully fielded by 1992.

US Marine Corps AAV7A1 series vehicles have also been fitted with a driver-operated extendable/retractable bow flap for increased water speed, improved visibility during amphibious operations, and to correct negative trim problem when the vehicle is afloat.

RAM/RS programme

In 1997, United Defense LP, Ground Systems Division, was awarded a contract for the AAV7A1 Reliability, Availability and Maintainability/Rebuild to standard (RAM/RS) programme. It is expected that 680 vehicles will be upgraded over a four year period at the rate of up to 180 per year.

Over the years, the addition of a heavier weapons turret, additional passive armour protection, bow plane and other improvements has increased the overall weight of the vehicle to the point of reducing some of the vehicles operational capabilities.

Key parts of the RAM/RS includes installation of a Cummins diesel developing 525 hp which is a derivative of that installed in the United Defense Bradley and the installation of the suspension system from the Bradley infantry fighting vehicle used in large numbers by the US Army.

These changes will restore the performance capabilities of the original design requirements of the AAV7A1, especially the power-to-weight ratio.

Specifications

(data in square brackets relates to the AAV7A1 where different from the basic vehicle)

Crew: 3 + 25
Combat weight: 22,838 [25,261] kg
Unloaded weight: 17,441 [19,100] kg
Power-to-weight ratio: 17.51 [15.83 hp/t
Ground pressure: 0.57 [0.59] kg/cm^2
Length: 8.16 m
Width: 3.27 m
Height: (overall) 3.31 m
Ground clearance: 0.406 m
Track: 2.609 m
Track width: 533 mm
Length of track on ground: 3.94 m
Max speed:
(road, forwards) 64 [72.42] km/h
(water, water-jets) 13.5 [13.2] km/h
(water, tracks) 7.2 km/h
Fuel capacity: 681 [647 litres
Max range: (land at 40 km/h) 482 km
Max endurance (water): 7 h at 2,600 rpm
Fording: amphibious
Gradient: 60%
Side slope: 40%
Vertical obstacle: 0.914 m
Trench: 2.438 m
Turning radius: pivot
Engine: Detroit Diesel model 8V-53T, 8-cylinder, water-cooled, turbocharged diesel developing 400 hp at 2,800 rpm [Cummins VT400, 8-cylinder, water-cooled, turbocharged diesel developing 400 hp at 2,800 rpm]
Transmission: HS-400 [HS-400-3A1 with hydraulic torque converter, 4 forward and 2 reverse ratios
Suspension: tube-over-bar
Electrical system: 24 V
Batteries: 4 × 12 V 6TN
Armament: (main) 1 × 12.7 mm M85 MG (1 × 40 mm MK 19 grenade launcher and one 12.7 mm machine gun)
Smoke-laying equipment: none [yes (smoke grenade launchers)
Ammunition: (12.7 mm) 1,000
Gun control equipment
Turret power control: electrohydraulic/manual [electric/manual]
(by commander) no
(by gunner) yes
Gun elevation/depression: +60/–15° (+45/–8°)
Turret traverse: 360°
Gun stabiliser:
(vertical) no
(horizontal) no
NBC system: no
Night vision equipment: yes
Armour
Hull sides: 31-45 mm
Hull top: 30 mm
Hull floor: 30 mm
Hull rear: 35 mm
Ramp outer: 25.4 mm
Ramp inner: 12.7 mm

United Defense LP, Ground Systems Division, as the design authority for the AAV7A1 will conduct hull modifications to the AAV7A1, provide teams (technical engineering and management support) and be the sole source for FMS (Foreign Military Sales) international opportunities. The assembly and test of the upgraded AAV7A1 vehicles is carried out at Marine Corps facilities at Albany and Barstow.

The first rebuilt vehicle was handed over to the US Marine Corps in the spring of 1999 and under current plans a total of 680 vehicles are to be upgraded.

Late in 1998, the US company of Twin Disc, confirmed that they had received a contract from United Defense LP, Ground Systems Division, for 170 parts sets to rebuild transmissions installed in the current AAV7A1 series vehicles. This was the first of a four year option.

Late in 1999, Italy awarded United Defense LP a contract to upgrade its fleet of LVTP7 series vehicles, additional details are given in the development part of this entry.

Appliqué Armour

Early in 1986, the US Marine Corps Systems Acquisition Review Council considered the armour programme and recommended adopting as Phase I, an armour kit for forward deployed AAV7A1s, as well as those with the maritime prepositioning assets.

The first contract was for 189 kits of P900 (this is also fitted to US Army M113s), which consisted of a set of punched and perforated armour plates attached to the sides and sloping parts of the vehicle.

The remaining 1,137 vehicles were fitted with an enhanced appliqué armour kit which protects the sides, slopes, top surfaces and all three hatches. The enhanced armour package has two layers, the first of which is about 150 mm away from the hull with an outer layer a further 25 mm away. The enhanced appliqué armour package is known as the Phase II Programme, which is also known as the Enhanced Appliqué Armour Kit (EAAK). This is required to meet the following specifications:

14.5 mm Armour-Piercing (AP) at 300 m, 95 per cent probability of no penetration

12.7 mm AP, 95 per cent probability of no penetration

155 mm AP at 15 m overhead, 99 per cent probability of no penetration.

The RAFAEL Armament Development Authority was selected as the sole winner of the open competition to design and fabricate three preproduction EAAKs for ballistic and field testing.

Development of the EAAK was completed in 1989 and subsequently three complete preproduction prototype systems were delivered to the US Marine Corps for extensive tests. Following the successful completion of the performance and operational tests, the production option of the contract was exercised. Production of the EAAKs commenced in 1991.

Production of the EAAK peaked at 80 per month with final deliveries made in early 1993.

Variants

AAVC7A1 command vehicle

This had the development designation LVTCX2 and is similar to the basic AAV7 except that its machine gun turret has been replaced by a simple hatch cover similar to that fitted at the driver's and commander's position. Armament consists of a pintle-mounted 7.62 mm M240G machine gun, with 1,000 rounds of ammunition. The AAVC7A1 also has a navigation system, a loaded weight of 21,537 kg and an unloaded weight of 19,785 kg.

AAVR7A1 recovery vehicle

This had the development designation of the LVTRX2 and like the AAVC7A1 has had the machine gun turret replaced by a 7.62 mm M240G machine gun. Equipment carried includes benches, tools, an air compressor, a battery charger, a generator, a 14 to 35 V DC MIG welder, and a portable welding kit. Mounted on the roof of the AAVR7A1 is a hydraulic crane with an extending boom which can lift 2,722 kg. The boom can be elevated from 0 to +65°. The vehicle also has a recovery winch with 84.73 m of 19 mm diameter cable with a breaking strength of 22,044 kg. The two-speed (high and low) winch has a maximum capacity of 13,605 kg on a bare drum at a low speed and 1,878 kg on a full drum at high speed. The AAVR7A1 has a crew of five, a loaded weight of 23,601 kg and an unloaded weight of 22,563 kg. See *Jane's Military Vehicles and Logistics 2001-2002*.

AAV7A1 with mine plough

United Defense LP has a demonstration model of a track width mine system which can be mounted on the front of an AAV7A1 which will proof a lane previously cleared by an M58/M59 linear demolition charge. Its secondary role is to provide a mineclearing capability from the surf zone exit point, across the beach and inland, in support of amphibious operations. A contract for full-scale engineering development was awarded late in 1988. Caterpillar Incorporated has built the prototype of a track width mine plough in response to a US Marine Corps requirement.

AAV7A1 with VEMASID

For trials purposes the AAV7A1 has been fitted with the Vehicle Magnetic Signature Duplicator (VEMASID) under development for the US Army.

AAV7A1 with mine clearance system kit

With the cancellation of the LVTE7 many years ago, the US Marine Corps had no vehicles with a mineclearing capability.

A modular line charge system can be bolted into the cargo compartment of the AAV7A1. This is a modified M125 mine clearance system which uses a Mk 22 rocket motor fired from an elevated launcher rail. When launched, the rocket pulls a linear charge to a range of about 550 m. Once the charge is on the ground it is detonated and this should detonate about 80 per cent of the mines for a distance of 7 m either side of the line. Production AAV7A1 vehicles have been adapted to mount this mine clearance system kit. This system is called the Mineclearing Line Charge M58/59 (MICLIC), details are given in *Jane's Military Vehicles and Logistics 2001-2002*.

AAV7A1 with Up Gunned Weapons Station

After studying bids from Cadillac Gage (now Textron Marine & Land Systems) and the then Ordnance Division of FMC Corporation, the US Navy Sea Systems Command, acting for the US Marine Corps, placed an order with Cadillac Gage early in 1986 for the supply of 240 Up Gunned Weapons Stations for installation on AAV7A1s.

The total value of this contract was US$17.236 million and the first turrets were delivered in November 1986.

The Up Gunned Weapons Station is armed with a 40 mm grenade launcher and a 12.7 mm M2 HB machine gun, has electric traverse and manual elevation, and replaces the turret armed with a single 12.7 mm M85 machine gun.

The original contract for 240 Up Gunned Weapons Stations included an option for a further 100 turrets which was exercised in January 1987 at a cost of US$5.823 million.

In September 1988, AV Technology Corporation was awarded a competitive rebuy contract for the Up Gunned Weapons Station for the US Marine Corps.

Electric Drive AAV7A1

Details of this testbed, which was completed by United Defense in early 1994, were given in *Jane's Armour and Artillery 1998-99*.

Status

Production complete but can be resumed if required. In service with Argentina, Brazil, Italy, South Korea, Spain, Thailand, the USA and Venezuela. In late 1995, United Defense LP and Samsung Aerospace of South Korea signed a major contract to co-produce a total of 57 AAV7A1 vehicles in three versions, AAVP7A1, AAVC7A1 and AAVR7A1, between 1996 and 2000. A follow on contract has been placed which will extend deliveries to 2006. There was also co-production programme with CAF of Spain to upgrade the Spanish fleet of AAV7 to AAV7A1 standard with deliveries running from 1999 through to mid-2000. Italy is also upgrading its fleet under a co-production programme signed late in 1999.

Manufacturer

United Defense LP, Ground Systems Division.

UPDATED

YUGOSLAVIA, FEDERAL REPUBLIC

BVP M80A mechanised infantry combat vehicle

Development

The BVP M80A mechanised infantry combat vehicle is a further development of the M-980 (although in early 1991 this was redesignated the M-80), and was shown in public for the first time at the Cairo defence equipment exhibition held in late 1984.

It is believed that the BVP M80A entered production in 1980 as the replacement for the M-980 vehicle and by the time production was completed it is estimated that up to 1,000 vehicles of all types had been completed.

Description

The overall layout of the BVP M80A is identical to the M-80 but it is slightly heavier (14 tonnes compared to 13 tonnes), slightly wider and higher, has a greater ground clearance (0.4 m compared to 0.3 m) and has a slightly higher road speed (64 km/h compared to 60 km/h). It has a more powerful engine which, in addition to giving the higher road speed, gives an improved power-to-weight ratio.

Standard equipment includes an NBC system, fire extinguishing equipment, a system for eliminating gases in the vehicle when the weapons are fired, a heater and bilge pumps; and with preparation the vehicle is fully amphibious, being propelled in the water by its tracks. It can also lay its own smoke screen by injecting diesel fuel into the exhaust outlet on the right side of the hull.

BVP M80AK IFV showing climbing ability during trials

BVP M80A MICV vehicle is used by the Croatian Army (Roderick de Normann)
0005220

BVP M80A MICV from the rear with 'Sagger' ATGWs covered and showing the firing ports and vision blocks in the rear doors

Specifications
Crew: 3 + 7
Combat weight: 14,000 kg
Power-to-weight ratio: 22.5 hp/t
Ground pressure: 0.67 kg/cm²
Length: 6.42 m
Width: 2.995 m
Height overall:
(turret roof) 2.20 m
(with ATGW) 2.67 m
Ground clearance: 0.4 m
Track: 2.526 m
Length of track on ground: 3.30 m
Max speed:
(road) 64 km/h
(water) 7.8 km/h
Fuel capacity: 510 litres
Max range:
(road) 500 km
(cross-country) 300-350 km
Fording: amphibious
Gradient: 66%
Side slope: 55%
Vertical obstacle: 0.8 m
Trench: 2.4 m
Engine: FAMOS 10V003 4-stroke 10-cylinder direct-injection diesel developing 315 hp at 2,500 rpm
Transmission: manual, 5 forward and 1 reverse gears
Clutch: mechanical, dry plate
Suspension: torsion bar
Electrical system: 24 V
Armament:
(main) 1 × 20 mm cannon
(coaxial) 1 × 7.62 mm MG
(other) twin 'Sagger' ATGW launcher
Ammunition:
(20 mm) 1,400
(7.62 mm) 2,000
('Sagger') 2 + 2
Turret traverse: (powered/manual) 360°
Gun elevation/depression: +75/–5°
Gun stabiliser:
(vertical) no
(horizontal) no
NBC system: yes
Night vision equipment: yes

Variants

BVP M80AK infantry fighting vehicle

This is essentially the BVP M80A previously described but fitted with a new one-man turret armed with a 30 mm M86 cannon, a 7.62 mm coaxial machine gun, a twin launcher for the Yugoslav-built 'Sagger' ATGW and two two-round smoke grenade dischargers that fire to the front of the vehicle.

Turret traverse is a full 360° with weapon elevation from –10 to +65°. Infantry weapons carried are four 64 mm rocket launchers, two 7.62 mm machine guns, six 7.62 mm automatic rifles and a 26 mm signal pistol.

Specifications are almost identical to the BVP M80A except for a combat weight of 14,400 kg, height to turret roof 2.398 m, height over anti-tank missiles 2.87 m, track 2.526 m, length of track on ground 3.30 m, maximum road speed 65 km/h and maximum water speed 6.8 km/h.

This is also referred to as the BVP M-80 'Vidra'.

Model 30/2 anti-aircraft system

This has a new two-man power-operated turret armed with twin 30 mm automatic anti-aircraft cannon and a four man crew consisting of the commander and gunner in the turret and the driver in the hull. This vehicle was developed to the prototype stage but never entered production. This may have the designation of the M80A1.

SAVA low-altitude SAM system

This is the Yugoslav equivalent of the Russian SA-13 SAM system based on an MT-LB chassis. The Yugoslav system is based on a modified BVP M80 chassis but it is understood that this has yet to enter production. Details are given in *Jane's Land-Based Air Defence 2001-2002*. Croatia has developed a similar system on a 6 × 6 armoured chassis. Details of this are given in *Jane's Land-Based Air Defence 2001-2002*.

BVP M80A KC company commander's vehicle

This is the standard vehicle modified for use by the company commander with additional communications equipment installed.

BVP M80A KB battalion commander's vehicle

This is the standard vehicle modified for use by the battalion commander with additional communications equipment installed.

BVP M80A LT anti-tank vehicle

Standard M80A with the turret removed and fitted with a new turret armed with six Yugoslav-built 'Sagger' ATGWs in the ready to launch position which have a maximum range of 3,000 m. Mounted between the two banks of missiles is a 7.62 mm machine gun. This pod is probably identical to that fitted to the BOV-1 (4 × 4) ATGW carrier (qv). Details of the Vozila BOV (4 × 4) are given under Slovenia in the Armoured personnel carriers (wheeled) section. This is also referred to as the LT-M-92 (Lovac Tenova) and is also known to be used by the Bosnian Serbs.

BVP M80A Sn ambulance

This is the standard vehicle with its turret removed and plated over and modified for use in the ambulance role. There is a single oblong hatch in the roof that opens left and a single door in the hull rear. Its four-man crew consists of a driver and three medical orderlies and it can transport four stretcher patients, four seated and two stretcher patients or eight seated patients.

BVP M-80 VK command post vehicle

This command post does not feature the turret installed on the standard BVP M-80A vehicle and has a single hatch at the rear, a telescopic antenna on the right side and six whip antennas. This is known to be used by the Bosnian Serbs.

Status

Production complete. In service with:

Country	Quantity	Comment
Bosnia	30	estimate
Croatia	109	estimate
Macedonia	2	estimate
Slovenia	52	estimate
Yugoslavia	568	estimate

Note: These figures include all models of the M-80 series.

Manufacturer

Yugoimport - SDPR.

UPDATED

M-80 mechanised infantry combat vehicle

Development

The M-80 mechanised infantry combat vehicle was seen for the first time in public during a parade in Belgrade in May 1975. It was designed in the former Yugoslavia but uses a number of foreign components; for example the engine is the same as in the French Giat Industries AMX-10P armoured infantry carrier, the ATGW is the Russian AT-3 'Sagger' manufactured in Yugoslavia and the 20 mm HS 804 cannon was manufactured under licence from Switzerland.

The vehicle is no longer being marketed by the Yugoimport - SDPR. It is estimated that total production of all versions of the M-80 amounted to about 1,000 vehicles.

Description

The hull of the M-80 is made of all-welded steel. Over the front arc of the M-80 the armour provides complete protection against 20 mm armour-piercing projectiles at a range of 100 m, while the remainder of the vehicle's armour provides protection against 7.62 mm armour-piercing projectiles fired from a range of 100 m. The M-80 carries 10 men: driver and turret gunner who remain with the vehicle at all times, vehicle commander, and seven fully equipped infantrymen. The driver sits at the front of the hull on the left and has a single-piece hatch cover that lifts and swings to the right to open. In front of his hatch are three day periscopes, the centre one of which can be replaced by an infra-red or image intensification periscope for driving at night.

The squad commander sits behind the driver and has a cupola that can be traversed through a full 360°, in the forward part of which is a single day periscope. The cupola has a single-piece hatch cover that opens forwards and can be locked vertical if required. An infra-red searchlight mounted externally on the forward part

M-80 MICV from the front with twin 'Sagger' ATGWs

M-80 MICV from the rear showing two ready to launch 'Sagger' ATGWs mounted above the turret rear

of the commander's cupola can be operated from inside the vehicle.

The engine is mounted to the right of the driver with the air inlet and outlet louvres in the roof and the exhaust outlet in the right side of the hull.

The one-man turret is to the rear of the engine compartment and offset to the right of the vehicle's centreline. The turret is armed with an HS 804 20 mm cannon which has a cyclic rate of fire of 800 rds/min and is fed from a 50-round drum magazine. The cannon fires an HE projectile weighing 0.122 kg, with a muzzle velocity of 880 m/s or an AP projectile weighing 0.142 kg with a muzzle velocity of 880 m/s, which will penetrate 18 mm of conventional steel armour at an incidence of 0° at 500 m. Maximum anti-aircraft range is 1,500 m and maximum range in the ground role is 2,000 m.

Mounted coaxially to the right of the 20 mm M55 cannon is a 7.62 mm machine gun with a cyclic rate of fire of 1,200 rds/min, which is fed from a 50- or 300-round belt. Both weapons have a maximum elevation of +75° and a depression of –5°.

Mounted to the rear of the turret on the right side are two launcher rails for the 'Sagger' ATGW. This missile weighs 11.29 kg and has a maximum range of 3,000 m. It is estimated that a further two to four missiles are carried in reserve in the hull, although the gunner has to expose himself in order to load another missile.

The turret is provided with a day sight for engaging ground targets (magnification of ×1.1 and a 40° field of view) and a night sight (range of 800 m, magnification of ×7.3 and 7° field of view). There are three day vision blocks in the left side and rear of the turret. In the turret roof on the left side are two forward-facing periscopes and a single-piece hatch cover that opens to the rear.

The troop compartment is at the rear of the vehicle and the infantrymen enter and leave by two doors in the rear of the hull, each of which has a firing port and a vision block. In each side of the hull are three firing ports with associated day periscopes in the roof of the vehicle for weapon aiming. Over the top of the troop compartment are two oval roof hatches hinged in the centre, which can be locked vertical.

The torsion bar suspension either side consists of five single rubber-tyred roadwheels similar to those used on the French AMX-10P, with the drive sprocket at the front and idler at the rear. There are two track-return rollers and the top of the track is covered by a skirt. Hydraulic shock-absorbers are fitted at the first and last roadwheel stations.

The M-80 is fully amphibious, being propelled in the water by its tracks. Before entering the water a trim vane, which is stowed on the glacis plate when not in use, is erected at the front of the hull. Standard equipment includes an NBC system which allows the vehicle to stay in a contaminated area for 10 hours, a smoke-laying system and a central fire extinguishing system with automatic and manual operation.

Specifications

Crew: 3 + 7
Combat weight: 13,700 kg
Unloaded weight: 11,700 kg
Power-to-weight ratio: 18.97 hp/t
Ground pressure: 0.64 kg/cm²
Length: 6.4 m
Width: 2.59 m
Height:
(over missile launcher) 2.5 m
(to turret roof) 2.3 m
(to hull roof) 1.8 m
Ground clearance: 0.4 m
Track: 2.47 m
Track width: 300 mm
Length of track on ground: 3.484 m
Max speed:
(road) 60 km/h
(water) 7.5 km/h
Max range:
(road) 500 km
(cross-country) 300-350 km
Fording: amphibious
Gradient: 60%
Vertical obstacle: 0.8 m
Trench: 2.2 m
Engine: HS 115-2 V-8 water-cooled turbocharged diesel developing 260 hp at 3,000 rpm
Transmission: preselector gearbox with hydraulic torque converter, 4 forward and 1 reverse gears
Suspension: torsion bar
Electrical system: 24 V
Armament:
(main) 1 × 20 mm M55 cannon
(coaxial) 1 × 7.62 mm MG
(other) twin AT-3 'Sagger' ATGW launcher
Ammunition:
(20 mm) 400
(7.62 mm) 2,250
('Sagger') 4

Gun control equipment
Turret power control: n/avail
(by commander) no
(by gunner) yes
Turret traverse: 360°
Gun elevation/depression: +75/–5°
Gun stabiliser:
(vertical) no
(horizontal) no
Armour
Hull front: 25-30 mm
Hull sides: 15 mm
Hull top: 8 mm
Hull floor: 12 mm
Hull rear: 12 mm
Turret front: 20 mm
NBC system: yes
Night vision equipment: yes

Status

Production complete. No longer marketed. In service with:

Country	Quantity	Comment
Bosnia	30	estimate
Croatia	109	estimate
Macedonia	2	estimate
Slovenia	52	estimate
Yugoslavia	568	estimate

Note: These figures include all models of the M-80 series.

Manufacturer

Yugoimport - SDPR.

UPDATED

M-60P armoured personnel carrier

Development

The M-60P armoured personnel carrier was first seen in public during a parade held in Yugoslavia in 1965, and for a short period was called the M-590.

Its engine is an Austrian Steyr diesel which also powers the Yugoslav GJ-800 artillery tractor (which in turn is based on the Hungarian K-800 light tracked artillery tractor). Its suspension is based on the Soviet SU-76 self-propelled gun which was supplied to Yugoslavia in the immediate post-Second World War period.

This vehicle is used by a number of factions in the former Yugoslavia, although production was completed many years ago.

The vehicle is no longer being marketed by the Yugoimport - SDPR. The M-60P was supplemented by the M-80 series which are covered in the previous two entries.

Description

The hull is made of all-welded steel providing the crew with protection from small arms fire and shell splinters.

The driver sits at the front of the hull on the left side and has a single-piece rear-opening hatch cover with an integral day periscope, which can be replaced by an infra-red (or more recently image intensification) periscope for driving at night. The co-driver is seated on the right side of the hull and also has a hatch cover with an integral day periscope that opens to the rear. The co-driver also operates the bow-mounted 7.92 mm machine gun which is based on the German MG42 weapon.

The commander sits to the rear of the driver and has a single-piece hatch cover. This has an integral day periscope that can be traversed through a full 360°. The machine gunner sits to the rear of the co-driver and has a cupola which can be traversed through 360°. The cupola has a two-piece hatch cover that opens either side of the machine gunner, giving him some protection when he is using the 12.7 mm (0.50) M2 HB machine gun which is mounted on the forward part of his cupola. A tripod carried on the right side of the hull enables the 12.7 M2 mm machine gun to be removed from the vehicle and used in the ground role.

The infantry sits on bench seats down either side of the hull and enters and leaves the vehicle by twin doors in the hull rear. There are also hatches in the roof of the troop compartment. There are three firing/observation ports provided in each side of the hull and one in each of the rear doors. A replacement roadwheel is carried on the left side of the hull rear and a jerrican is carried on the right side of the hull rear.

The torsion bar suspension either side consists of five single rubber-tyred roadwheels with the drive

M-60PB armoured personnel carrier with twin recoilless 82 mm M60 anti-tank guns of the Serbian Baranja Division in Eastern Slovenia (D Caemberlynck)
2001/0099471

M-60P APC of the Baranja Division (D Caemberlynck) ***2001***/0099470

sprocket at the front and the idler at the rear, and three track-return rollers. The running gear is outside the hull armour and its top part is normally covered by a light sheet metal guard.

The M-60P APC is not amphibious and is not fitted with an NBC system.

Variants

M-60PB anti-tank

Mounted on top of the hull on the left side are two 82 mm M60A recoilless rifles on a rotating mount with a mechanical traverse and elevating mechanism. The gunner has a suspended seat which moves with the rifles and a single-piece hatch cover that opens forwards. To the right of the mount is a hatch for the loader. The M-60PB has also been observed with the recoilless rifles mounted on the right side of the hull top.

The 82 mm M60A recoilless rifles have a maximum range of 4,500 m and a maximum rate of fire of 4 to 5 rds/min. They have an elevation of +6°, depression of –4° 30′ with a prohibited zone of action of 76° in the plane of the anti-aircraft machine gun and 135° to the rear of the vehicle.

Effective range when engaging stationary targets is 1,500 m, and 1,000 m when engaging moving targets. The HEAT projectile will penetrate 250 mm of conventional steel armour at an angle of 90° with dispersion being 2 m at 500 m and 4 m at 1,000 m. The HEAT projectile itself weighs 4.2 kg, and the complete round weighs 7.2 kg; 10 rounds are carried. The M-60PB carries one less crew member than the basic M-60P APC.

It is also probable that there are other versions of the M-60 such as ambulance, command and radio vehicles. Recent information has also indicated that there is a 82 mm self-propelled mortar version of this vehicle.

Status

Production complete. No longer marketed. In service with Croatia (40+) and Yugoslavia (165+).

Manufacturer

Yugoimport - SDPR.

UPDATED

Specifications

Crew: 3 + 10
Combat weight: 11,000 kg
Power-to-weight ratio: 12.73 hp/t
Ground pressure: 0.7 kg/cm²
Length: 5.02 m
Width: 2.77 m
Height:
(without armament) 1.86 m
(including MG) 2.385 m
(MG at max elevation) 3.51 m
Ground clearance: 0.4 m
Track: 2.385 m
Track width: 400 mm
Length of track on ground: 2.94 m
Max road speed: 45 km/h
Fuel capacity: 150 litres
Max range:
(road) 400 km
(cross-country) 250 km
Fuel consumption: (road) 0.38 litres/km
Fording:
(still water) 1.35 m
(running water) 1.25 m
Gradient: 60%
Side slope: 40%
Vertical obstacle: 0.6 m
Trench: 2 m
Engine: FAMOS 6-cylinder in-line water-cooled diesel developing 140 hp
Transmission: manual, 5 forward and 1 reverse gears
Suspension: torsion bar
Electrical system: 24 V
Batteries: 2 × 12 V
Armament:
(anti-aircraft) 1 × 12.7 mm M2 MG
(bow) 1 × 7.92 mm M53 MG
Gun stabiliser:
(vertical) no
(horizontal) no
NBC system: no
Night vision equipment: yes
Armour:
(max) 25 mm
(min) 10 mm

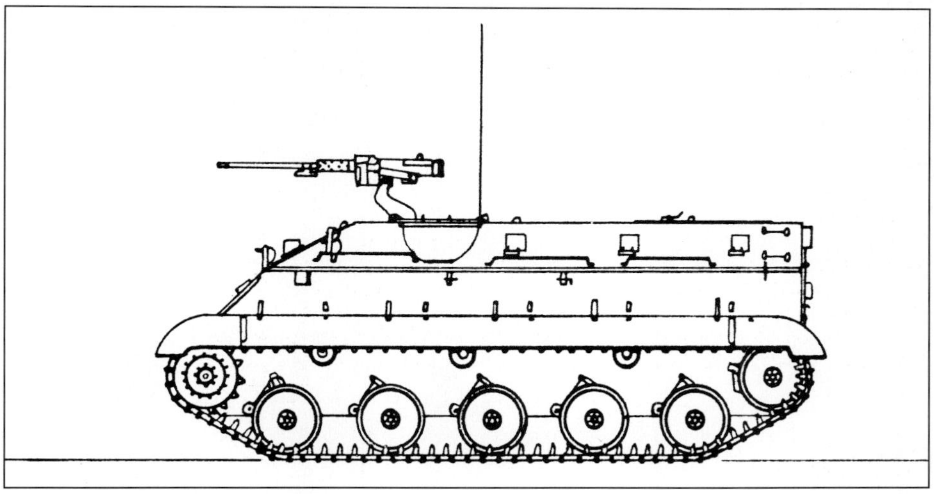

M-60P APC with a 12.7 mm machine gun

ARMOURED PERSONNEL CARRIERS (WHEELED)

AUSTRALIA

Australian LAV (ASLAV) programme

Development/Description

Under Phase 1 of the Australian Light Armoured Vehicle (ASLAV) programme, the Australian Army purchased 15 LAV (8 × 8) vehicles from the US Marines.

These vehicles were built by the Diesel Division, General Motors of Canada (which in 2000 became GM Defense) and comprised 14 LAV-25 and one maintenance/recovery vehicle.

Australia also evaluated four Bison (8 × 8) APCs which are currently in service with the Canadian Army.

The original 15 vehicles were used by the 2nd Cavalry Regiment to develop operating and support procedures for wheeled armoured vehicles.

Under Phase 2, late in 1992, Australia signed an A$120 million (US$85 million) contract with the Canadian Commercial Corporation for 97 LAV (8 × 8) vehicles to be built in Canada. The first of these was completed in 1994. Final fitting was carried out in Australia by British Aerospace Australia Limited and all vehicles have now been delivered.

In June 1995, Australia contracted for an additional 14 ASLAV-25 vehicles bringing the total to 47.

The order for 111 LAV (8 × 8) vehicles comprises:

Variant	Quantity	Designation
ASLAV-25	47	ASLAV-25
Bison APC	33	ASLAV-PC
Command	9	ASLAV-C
Fitters	10	ASLAV-F
Surveillance	10	ASLAV-S
Ambulance	2	ASLAV-A

ASLAV-25 (8 × 8) in operational service with the Australian Army (Barry Marriott) **2002**/0121125

The six variants use three different hull types and by the use of non-permanent, Mission Role Installation Kits (MRIK), four of the variants are created from a single hull design.

Although similar to the US Marine Corps' vehicles/Canadian Forces' vehicles, the Australian models do differ in a number of details.

The ASLAV-25 is generally similar to the Canadian LAV-25 late build standard and armed with 25 mm The Boeing Company M242 Chain Gun, 7.62 mm MAG 58 coaxial and 7.62 mm MAG 58 anti-aircraft machine gun.

ASLAV-PC is generally similar to the Canadian Bison and has a crew of two, commander and driver with the former manning the 12.7 mm M2 machine gun which can be aimed and fired under complete armour protection and is provided with a day/night sight. A total of seven troops can be carried.

The ASLAV-C uses the same hull as the ASLAV-PC and is converted to the command role with an MRIK. This has an enhanced radio installation and radio masts, mapboards, stowage compartments, appropriate seating and an annex. It has a crew of five.

The ASLAV-S is based on the same hull as the ASLAV-PC and is converted by an MRIK. The vehicle carries the RASIT ground surveillance radar and other observation equipment and has a crew of four.

The ASLAV-A ambulance is based on the same hull as the ASLAV-PC using an MRIK which comprises medical equipment and litter stations. It has a crew of three and can carry up to six patients. It is capable of fitting a Mobile Intensive Resuscitation Facility (MIRF).

The ASLAV-F and ASLAV-R function as either fitters or recovery vehicles, depending on which of the two non-permanent installation kits is installed, with hull type being similar to ASLAV-PC with some structural modifications to mount the winch or crane and to provide anchor points and winch cable entry.

The ASLAV-F will have two non-permanent installation kits, one for a winch for recovery and the other with a crane and stowage for use as a fitters' vehicle.

All versions are fitted with a climate control system to bring internal temperatures down to ambient levels and all vehicles are fitted with Michelin XML tyres which are wider. For improved battlefield survival, the ASLAV fleet can be fitted with a ceramic appliqué armour system.

These LAV (8 × 8) vehicles are being used by the Australian Army 2nd Cavalry Regiment which has moved from New South Wales to new facilities near Darwin in the Northern Territory.

The only operational user is the 2nd Cavalry Regiment with the remainder being allocated to the School of Armour, Royal Australian Electrical and Mechanical Engineers Training Centre and logistics.

Additional details of the LAV (8 × 8) family of vehicles is given in this section under Canada.

British Aerospace Australia fitted out all of the ASLAV vehicles but early in 1999 Tenix Defence Systems acquired all of the vehicle business of British Aerospace Australia.

This includes current work for the ASLAV as well as the Shorland series of 4 × 4 light armoured vehicles covered later in this section.

In 1998, the now GM Defense LAV fitted with the 120 mm Armoured Mortar System and the LAV-25 turret with TOW anti-tank guided missiles were demonstrated in Australia.

ADI has developed a new lightweight weapon station called the Commander's Weapon Station (CWS) for installation on light armoured vehicles.

CWS has powered elevation and traverse and is armed with a 12.7 mm M2 machine gun which can be aimed and fired from under armour. A day night sighting system is fitted as standard. The Australian Army is expected to purchase 69 CWS for installation on ASLAV-PC and ASLAV-S vehicles.

In mid-2000, a contract was placed valued at A$350 million for a further 150 ALAV vehicles under the

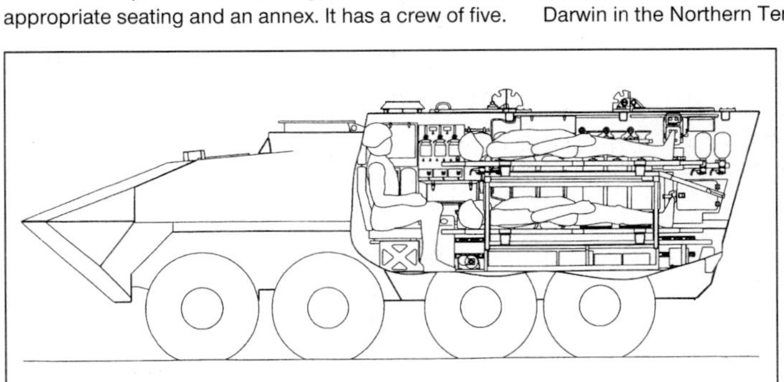

Cutaway drawing of ASLAV (8 × 8) in ambulance configuration

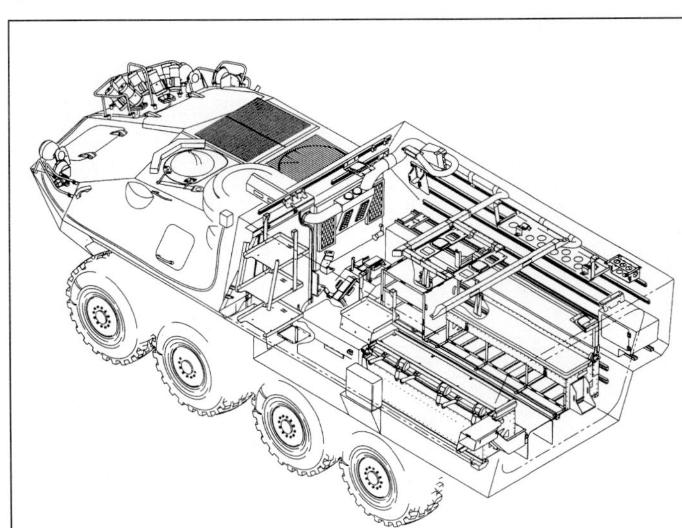

Cutaway drawing of ASLAV (8 × 8) in armoured personnel carrier configuration

Australian Light Armoured Vehicle (ASLAV) project Land 112.

Under the terms of this contract a total of 150 vehicles will be supplied to equip four reconnaissance squadrons in the following configurations:

81 ASLAV-25 (two person 25 mm turret)
15 ASLAV-S surveillance
14 ASLAV-C command
11 ASLAV-PC personnel carrier
11 ASLAV-F fitters
11 ASLAV-R recovery
7 ASLAV-A ambulance

Under the terms of this contract, Tenix Defence Systems of Australia will build the 25 mm turret fabrication and then send this to Canada for fitting out. Australia will become the only 25 mm turret fabrication facility for GM Defense.

Under a contract awarded to Racal Australia (now Thales Australia) in August 2000, the Australian Defence Forces will receive 51 Ground Surveillance Radars from 2002. This is called the Australian Manportable Surveillance and Target Acquisition Radar (AMSTAR) and is an improved version of the MSTAR used by the UK. Some of these will be installed on the ASLAV-R version.

In the longer term, Australia has a requirement for 20 120 mm ASLAV mortar carriers. The front runner for this is considered to be the BAE Systems, RO Defence/Delco Defense 120 mm AMS. This has already been shown in Australia.

Status

In service with the Australian Army.

Manufacturers

GM Defense.
Tenix Defence Systems Pty Ltd (main Australian subcontractor).

UPDATED

ADI Bushmaster Infantry Mobility Vehicle

Development

In the mid-1990s, Perry Engineering developed the Bushmaster (4 × 4) armoured vehicle to meet the Australian Army requirement for an infantry mobility vehicle. The Bushmaster is an infantry mobility vehicle, not a combat vehicle and is required for tactical and strategic movement of infantry troops to and within the area of operations.

It is not required to have an integral weapon, but can mount, on a weapon ring, up to 12.7 mm machine guns. It is required to carry three days sustenance for 10 personnel including ammunition, rations and 270 litres of water and sufficient fuel for 1,000 km.

Other key requirements include air conditioning and cold water drinking system, smoke grenade launchers, full protection against 5.56 and 7.62 mm small arms fire and capable of being upgraded against 5.56 and 7.62 mm armour piercing. Also required is full mine protection against 9.5 kg TNT blast under any wheel station or under the hull.

It must be spacious with large glassed areas for crew situational awareness and comfort. Finally, although currently a 4 × 4 vehicle, it should be capable of being expanded to 6 × 6 configuration.

Early in 1997, ADI purchased the design and prototype of the Bushmaster vehicle from Boral Limited which owned Perry Engineering.

Stewart & Stevenson of the USA and Timoney Technologies of Ireland continue to be involved in the programme.

By early 1998, the prototype Bushmaster vehicle had undergone an exhaustive test programme, during which it covered more than 30,000 km

In February 1998, the Australian Army took delivery of three Bushmaster vehicles for an extensive 44 week trial in competition with the Taipan (4 × 4) vehicle.

These trials were completed early in 1999 and in March 1999 the ADI Bushmaster was selected to meet the requirements of the Australian Army.

In June 1999 the Australian Department of Defence awarded ADI a contract worth A$180 million for the supply of 370 Bushmaster Infantry Mobility Vehicles.

ADI Bushmaster Infantry Mobility Vehicle ***2000***/0080345

The first of six pre-production vehicles was delivered late in 1999 with another five following in 2000. Full production started late in 2000.

In August 2001 it was revealed that delivery of the Bushmaster IMV to the Australian Army would be delayed by two or three years because of reliability problems with the suspension and driveline.

Under the original plans, production of the Bushmaster should have started in mid-2000. The Australian Army will receive a minimum of 323 vehicles with the Air Force taking an additional 18 vehicles for airfield defence.

Description

The monocoque hull of the Bushmaster is of all-welded steel armour construction which provides protection from both anti-vehicular and anti-personnel mines, together with protection from 5.56 and 7.62 mm small arms fire.

ADI Bushmaster Infantry Mobility Vehicle ***2000***/0080344

The vehicle can be armoured to provide protection from 5.56 and 7.62 mm armour piercing rounds using an ADI-developed composite appliqué armour kit.

It is possible that production versions of the Bushmaster will now provide all-round armour protection against 12.7 mm ball projectiles. The previously 'desirable' criteria of protection against 7.62 and 5.56 mm armour piercing rounds is now expected to be standard.

Other crew survivability features incorporated into production vehicles are expected to include the location of the fuel and hydraulic oil tanks externally to shield the occupants from fire. In addition, the fuel tank will be fitted with fire retardant elements to prevent explosion caused by ballistic overmatch or rupture.

The wheel arches are designed to give way in the event of the vehicle running over a mine and the extensive external storage lockers are also expendable. The crew compartment will also be fitted with an automatic fire detection and suppression system.

To deflect mine blast, one of the key requirements of the Australian Army, the hull has a specially profiled V-design underbody. The drinking water tank is mounted internally, below floor level, for added protection and to lower the centre of gravity of the Bushmaster. The fuel and oil tanks are mounted externally for safety reasons.

The Bushmaster carries a section of 10 fully equipped infantry with sufficient room for personal equipment and stores.

The driver and commander are seated at the front with large bulletproof windows to their front and sides.

A further eight troops are seated towards the rear facing inwards, four on the left and four on the right. The troop compartment is provided with bulletproof windows in the sides and rear.

As the vehicle has a range of over 1,000 km, considerable attention has been paid to crew comfort with the seats being designed to provide protection during a mineblast, frontal impact, roll over and side impact.

The crew normally enter and leave through a door in the rear of the Bushmaster, but there are also five roof hatches, the most forward of which can be fitted with a 5.56, 7.62 or 12.7 mm machine gun. Mounted either side of the hull is a bank of four electrically operated smoke grenade dischargers.

Specifications

Crew: 2 + 8
Configuration: 4 × 4
Combat weight: 15,000 kg
Unloaded weight: 12,500 kg
Payload: 2,500 kg
Power-to-weight ratio: 22 hp/t
Length:
(hull) 6.60 m
(overall) 7.02 m
Width: 2.50 m
Height: 2.65 m
Ground clearance:
(hull) 430 mm
(axle housings) 400 mm
Track: 2.10 m
Wheelbase: 3.90 m
Max road speed: 120 km/h
Fuel capacity: 340 + 15 litres (usable)
Max road range: 1,000 km
Fording: 1.20 m
Gradient: 60%
Side slope: 40%
Trench: n/avail
Vertical obstacle: 0.50 m
Turning radius: 8.3 m
Engine: Caterpillar 3126 ATAAC 6-cylinder, turbocharged and air-to-air after-cooled diesel developing 330 hp at 2,400 rpm with 25% torque rise
Transmission: ZF 7HP502 full automatic with front end retarder providing controlled descent of 60% slopes.
Steering: power assisted
Tyres: 395/85 R20 with beadlocks or optional run-flat, central tyre pressure regulation system fitted as standard
Brakes: Rockwell air-operated disc, dual circuit
Suspension: Coil spring and short-long-arm wishbone with double-acting hydraulic shock-absorbers and bump-stop rubbers
Electrical system: 24 V

The Caterpillar 3126 6-cylinder turbocharged and air-to-air after-cooled diesel engine develops 330 hp and is coupled to a ZF 7HP500 fully automatic transmission. This has seven forward and one reverse gears and converter lock-up which provides constant four-wheel drive.

The power pack installation has been designed for ease of replacement and all engine and transmission service points have been positioned to simplify daily maintenance checks.

Bushmaster is equipped with Meritor independent axle systems with each wheel station provided with coil spring and short/long-arm wishbone independent suspension with a double-acting hydraulic shock-absorber. This system is claimed to provide excellent ride and handing on both roads and cross-country, with exceptional cross-country speed at very low energy impact to the occupants.

Steering is power assisted and standard equipment includes a central tyre inflation system operated by the driver, which provides four settings, road, sand, mud/snow and emergency. Tyres have bead locks with run-flat being an option. One spare wheel and tyre is carried.

As the Bushmaster will be operating in high ambient temperatures, a split system air conditioner is fitted as standard which provides air to the front and rear sections. A 270 litre water storage tank is located under the vehicle at the rear and is dispensed through a cooled drinking water supply system.

A hydraulically operated 10 tonne capacity self-recovery winch is mounted on the left side of the hull and this can be deployed to the front or rear of the hull.

With a combat weight of 14 tonnes, the Bushmaster has a maximum road speed in excess of 110 km/h and a power-to-weight ratio of 21.42 hp/t. Bushmaster is fully airportable in a Lockheed Martin Hercules C-130 transport aircraft.

Variants

There are at least six variants of the Bushmaster at the design phase:

(1) Ambulance
(2) Command
(3) Mortar carrier with internally mounted 81 or 120 mm mortar
(4) Direct fire weapons with a combination of 12.7 mm machine gun and 40 mm grenade launcher
(5) Troop carrier
(6) Engineer/Assault pioneer

For the civil market a bushfire fighting vehicle has been developed called the FireKing.

Status

On order for Australian Army and Air Force. Not yet in production.

Manufacturer

ADI Limited.

UPDATED

Tenix Defence Systems Australia S600 Armoured Personnel Carrier

Specifications

Crew: 1 + 11
Configuration: 4 × 4
Combat weight: 12,500 kg
Power-to-weight ratio: 17.1 hp/t
Length: 5.76 m
Width: 2.95 m
Height: 2.70 m
Ground clearance: 0.44 m
Track: 1.92 m
Wheelbase: 3.25 m
Angle of approach/departure: 40/40°
Max speed: 110 km/h
Fuel capacity: 320 litres
Max range: 1,000 km
Fording: 1.2 m
Gradient: 75%
Side slope: 58%
Vertical obstacle: 0.5 m
Turning radius: 7.5 m
Engine: Mercedes-Benz OM-366LA 6-cylinder in-line turbocharged and intercooled diesel developing 214 hp at 2,600 rpm
Transmission: UG3/65 manual with 8 forward and 4 reverse gears
Steering: hydraulic, power assisted
Suspension: coil springs and hydraulic shock-absorbers
Tyres: 365/85 R 20 XZL
Brakes: pneumatic servo-assisted, hydraulic braking system
Electrical system: 24 V
Batteries: twin
NBC system: no
Night vision equipment: no

Development

Building on its experience in the design, development and production of the S52 (4 × 4) armoured patrol car and the S55 (4 × 4) armoured personnel carrier, both based on the Land Rover Defender 110 chassis, Shorts developed the S600 armoured personnel carrier as a private venture, with the vehicle being shown for the first time in September 1995.

While the earlier S 55 series meet the operational requirements of many army and police forces, some requirements now call for a vehicle with increased carrying capacity, more internal volume and improved cross-country mobility.

In 1992, Shorts started looking at existing chassis which could be used as the basis for a new armoured personnel carrier and finally selected the proven German Mercedes-Benz Unimog U1550L/U2150L (4 × 4) chassis.

Some 80 per cent of the components used in the S600 are standard commercial components which have already demonstrated high reliability and a low maintenance requirement. In order to fit the armoured body, some parts of the Unimog chassis have been repositioned.

Detailed design work on the S600 started in 1993 and was completed in 1994, with the prototype being rolled out early in 1995.

The S600 is available with a choice of two different engines, the Mercedes-Benz 366A in-line water-cooled turbocharged diesel developing 156 hp (U1550L chassis) or the 366LA in-line turbocharged and intercooled diesel version which develops 214 hp (U2150L).

Total weight depends on the version and application but gross vehicle weight can be between 8,000 and 9,500 kg, or even up to 12,500 kg with the U2150L chassis. With a gross vehicle weight of 12,500 kg its maximum payload is 4,500 kg.

The description is for the baseline S600 as there is considerable flexibility in the design of the S600, allowing it to be tailored to meet individual operational requirements.

When being used in the internal security role, for example, 12 men can be carried while for the military role a section of eight men and three days' supplies can be carried.

In 1996, Short Brothers plc of Northern Ireland sold the complete design, development and production rights of the S600 and the Shorland range of 4 × 4 vehicles to British Aerospace Australia Limited.

In future, production of these vehicles will be undertaken in Australia or Northern Ireland depending on the quantity and end user of the vehicles.

Early in 1997, the Kuwait National Guard placed an order with British Aerospace Australia for the supply of 22 S600 Shorland armoured personnel carriers and variants.

Production of these vehicles was undertaken in Australia with first deliveries made early in 1998.

This is the first order for the S600 Shorland and so far two preproduction S600 Shorland vehicles have been built, one in Australia and one in Northern Ireland.

Early in 1999, Tenix Defence Systems acquired all of the vehicle business of British Aerospace Australia including the S600 and other members of the Shorland family of light armoured vehicles. As of late 2001 it is understood that the S600 is the only member of the Shorland vehicle to be built in Australia to date.

In 2001 Tenix Defence Systems restarted production of its S600 Shorland (4 × 4) Armoured Personnel Carrier (APC) for the Belgian Gendarmerie who placed an order valued at A$5 million for an undisclosed quantity of vehicles.

The S600 Shorland (4 × 4) APC in the Internal Security Vehicle (ISV) configuration was selected by the Belgian Gendarmerie following extensive competitive trials alongside a number of other vehicles, including the Alvis Vehicles Tactica and the Vickers OMC RG-12.

According to Tenix Defence Systems its S600 Shorland (4 × 4) was selected by the Belgian Gendarmerie as it met a number of key operational criteria including mobility, protection, obstacle clearance, through-life cost and supportability.

The ISV configuration of the S600 provides seating for up to 12 fully equipped personnel who can rapidly enter and leave the vehicle via the rear and/or sides of the vehicle.

Belgian vehicles are fitted with an air conditioning system, run-flat tyres, powered steering, powered sliding door, spotlights and wire-mesh protection for the windscreen and a rapidly removable front-mounted power-operated light obstacle removal device.

Description

The hull of the S600 is of all-welded steel armour construction that provides all-round protection from

Tenix Defence Systems S600 (4 × 4) armoured personnel carrier **2001**/0103746

Shorland S600 in ISV configuration with front-mounted light-barricade removal device and sliding door **2002**/0121124

5.56 and 7.62 mm ball small arms fire. If required, appliqué armour can be added to provide a higher level of protection against 5.56 and 7.62 mm armour-piercing attack. Underneath protection is provided against grenade and small mine threats.

The engine and transmission are at the front of the vehicle with access to the engine for maintenance purposes being through hatches. If required, the complete armoured body can be removed from the chassis to allow full access to the power pack.

The diesel engine is coupled to a manual transmission which gives a total of eight forward and four reverse gears. The portal axles have hub drive and torque tubes. Both axles have pneumatically operated differential locks which can be operated while the vehicle is moving.

Suspension at each wheel station consists of coil springs and hydraulic shock-absorbers with steering being power assisted to reduce driver fatigue, especially when moving across rough country.

Both left- and right-hand drive models are available. In the case of the former, the driver is seated front left with the vehicle/detachment commander to his right. To their front and sides is a bulletproof windscreen which gives excellent visibility through 180°. The bulletproof windows provide the same level of protection as the hull.

The troops are seated on bench seats that run down either side of the hull facing each other with each person being provided with a seat belt. Space is provided under the seats for the stowage of equipment and stores.

To allow for the rapid entry and exit of the troops, which is considered essential for many of the roles that the S600 is expected to undertake, the S600 has three means of entry. In each side of the hull is a two-part hatch, the upper part with an integral vision block opens upwards while the lower part folds down to form a step.

In the rear of the hull is a similar but wider hatch with the upper part, which can contain a firing port and vision device, opening upwards and the lower part folding downwards to form a step. If required, the upper part can be locked in the open position while travelling. Other door configurations are available if required.

If the S600 is operating away from its base for extended periods, without extensive logistic support, a spare wheel can be carried on either side of the hull towards the rear.

The prototype S600 has three bulletproof vision blocks in either side of the troop compartment, but again there is flexibility in the design and, if required, firing ports can also be fitted.

The prototype has a single-piece circular hatch cover in the roof and a wide range of armament systems up to and including a 12.7 mm M2 machine gun or a 40 mm Mk 19 Mod 3 grenade launcher can be fitted.

Other roof hatch arrangements are possible on the S600, for example a circular roof hatch above commander's and driver's positions, standard circular roof hatch for machine gun (as on the prototype) and twin hatch covers at the rear.

Standard equipment on production vehicles would include an air conditioning system.

Optional equipment for the S600 includes appliqué armour, front-mounted self-recovery winch, wire cutters, smoke grenade launchers, land navigation system (such as a global positioning system), Mercedes-Benz central tyre-inflation system that allows the driver to adjust the tyre pressure from his position to suit the type of ground being crossed, Hutchinson run-flat inserts, an automatic transmission, Euro 2 diesel engine, heater, fire detection and suppression system, night vision equipment and various communications systems.

Variants

Tenix Defence Systems has suggested that the S600 could form the basis for a wide range of variants including:

Barricade removal vehicle

This would be fitted with a front-mounted heavy-duty self-clearing barricade removal blade.

Ambulance

Accommodating a crew of three including driver and two medical staff; it can carry three stretcher patients or two stretcher patients plus four seated patients.

Mortar carrier

Carrying a standard BAE Systems, RO Defence 81 mm mortar with a crew of three including driver, mortar detachment commander and mortar crew member. This would fire through an opening roof hatch.

Command post vehicle

Fitted with up to five radios and folding work bench, fitted mapboard and an enclosed annex for use in the vehicles stationary role as a command post. The vehicle will have a crew of six including commander, driver and four radio operators.

Internal security

This could be fitted with various systems including a high pressure hose fed by an onboard water tank for crowd control.

Surveillance vehicle

This would have a stabilised mast-mounted sensor package comprising thermal imager, radar, TV camera, laser range-finder and operators console inside of the vehicle. This model would have a crew of four.

Airport security vehicle

This is fitted to allow the crew to remain closed up for long periods of time in comfort and provides airfield communication equipment and provision for a concealed weapon. It would normally have a crew of four.

Status

In production. On order for Belgian Gendarmerie (A$5 million contract placed in 2001) and in service with the Kuwait National Guard (22 units).

Manufacturer

Tenix Defence Systems Pty Ltd.

UPDATED

Tenix Defence Systems Australia Shorland S 55 Armoured Personnel Carrier

Development

The first Shorland armoured personnel carrier, the SB 301, was developed as a private venture by Shorts with the prototype completed in 1973 and the first production vehicle the following year. It has been designed specifically for internal security operations and is based on operational experience obtained with the Shorland armoured patrol car (4 × 4), with which it shares many common components. The Shorland SB 401, with strengthened armour and the more powerful V-8 petrol engine was introduced in 1980.

The latest version, the S 55, was designed to take advantage of the Land Rover Defender 110 (4 × 4) chassis with its improved coil spring suspension and wider wheel track. Other improvements introduced with the new chassis were improved performance, a tighter turning circle and power steering to reduce driver fatigue.

Late production Shorland vehicles have a redesigned bonnet for easier engine access, bullet-resistant glass windscreen and windows and eight firing ports, each with a vision block above. Of these, there are three in either side of the hull and one in each of the rear doors.

The complete Shorland Series 5 family comprises the S 52 armoured car, the S 53 air defence vehicle, the S 54 anti-hijack vehicle and the S 55 armoured personnel carrier. Details of the first three versions are given in the entry for the Shorland armoured patrol car in the Reconnaissance vehicles section.

Short Brothers sold the complete design, development and production rights of its extensive range of 4 × 4 armoured vehicles to British Aerospace Australia Limited.

Future production of the Shorland S 55 range of armoured vehicles will take place in Australia or the United Kingdom depending on the quantity of vehicles required and the eventual end customer.

As of late 2001, it is understood that no production of the S 55 Shorland or other members of the Shorland Series 5 had been undertaken in Australia.

Early in 1999, Tenix Defence Systems acquired all of the vehicle business of British Aerospace Australia including the S 55 and other members of the Shorland family of light armoured vehicles. This also included the new S600 covered in the previous entry. The S600 is in service with the Kuwait National Guard and in 2001 was ordered by Belgium.

Description

The S 55 is based on the 110 in heavy-duty Land Rover Defender chassis which has permanent four-wheel drive and a lockable central differential. Normal power pack is the 3.5 litre Rover V-8 petrol engine with the Rover 2.5 litre turbo-diesel available as an alternative.

The chassis has been strengthened to take the armoured body and is fitted with stronger axles and suspension and different final drive ratios.

Specifications
Crew: 2 + 6
Configuration: 4 × 4
Combat weight: 3,600 kg
Power-to-weight ratio:
(petrol) 37.2 bhp/t
(diesel) 29.7 bhp/t
Length: 4.25 m
Width: 1.8 m
Height: 2.28 m
Ground clearance: 0.324 m
Track: 1.5 m
Wheelbase: 2.795 m
Max road speed: 120 km/h
Fuel capacity: 136 litres
Max range: over 630 km
Fording: 0.50 m
Vertical obstacle: 0.23 m
Turning radius: 6.6 m
Engine: Land Rover 3.5 litre V-8 water-cooled 4-stroke petrol developing 134 bhp at 5,000 rpm or Land Rover Tdi 2.5 litre water-cooled 4-stroke diesel developing 107 bhp at 3,800 rpm
Transmission: manual with 5 forward and 1 reverse gears
Transfer case: 2 speed
Steering: power-assisted worm and roller
Clutch: single dry plate
Suspension: coil springs and telescopic hydraulic shock-absorbers, anti-roll bars fitted front and rear
Tyres: 9.00 × 16
Brakes:
(main) hydraulic, dual-circuit front and rear, disc front, drum rear
(parking) mechanical drum operating on transmission
Electrical system: 12 V (24 V optional)
Battery: 1 × 12 V, 58 Ah
NBC system: no
Night vision equipment: no

Nevertheless, some 85 per cent of the automotive and mechanical components are stock Land Rover parts, so reducing spares and maintenance problems.

The vehicle body of the S 55 provides protection against right-angle fire by NATO 7.62 × 51 mm ball ammunition down to 25 m range while the bullet-resistant glass provides the same level of protection as the steel and meets the test requirements of BS 5051 Part 1 Class R2.

The engine and radiator are fully armour protected while the floor is made from reinforced glass fibre to give protection against mines and improvised explosive devices. The fuel fillers are under armoured hatches which can only be released from inside the vehicle.

The roof of the S 55 is sloped so that hand-thrown missiles roll off even when the vehicle is stationary, and the bonnet and windscreen assemblies are designed to ensure that burning fuel from petrol bombs thrown against the windscreen is directed to the ground without entering the engine compartment.

A searchlight, controlled from within the vehicle, is mounted in the forward part of the roof between the driver's and commander's seats.

The vehicle is designed to carry eight personnel including the driver and commander. The two side and two rear doors permit easy entry and exit and a grab rail suspended from the roof assists rapid movement in operational emergencies.

The large bullet-resistant windscreen and side windows provide excellent visibility for the driver and commander and eight gun ports with associated vision blocks are fitted, three on each side and one in each of the rear doors.

The vehicle is trimmed internally with fire-retardant material which provides thermal insulation. An air conditioning system is fitted as standard to allow for vehicle operation under all climatic conditions.

Latest Shorland S 55 with new low-profile front, bulletproof windscreen, spare tyre and wheel on roof and armed with roof-mounted 7.62 mm MG 0021131

Variants

Optional equipment for the S 55 includes run-flat tyres, drop-down visors with bullet-resistant vision blocks over the bullet-resistant windscreen, police/emergency lights either as individual lamps or as part of a multifunction unit, a public address system, automatic fire protection and a roof hatch either with or without a machine gun mounting. Fixed grenade launchers can be fitted which can fire 66 or 76 mm electrically initiated smoke/illumination/CS gas grenades.

A special barricade remover can be installed at the front of the vehicle if required. Unlike other conventional designs this does not reduce ground clearance when not in use.

Status

Production as required. The S 55 is in service in over 20 countries including Malaysia. In January 1988, Shorts Brothers announced that it had been awarded a contract to supply 40 S 55 Shorland APCs to the Turkish Ministry of the Interior, to be used by the gendarmerie. In June 1990, Shorts Brothers announced that it had been awarded a contract to supply 24 Shorland S 55 APCs to the Sindh Police of Pakistan. Late in 1993, Shorts Brothers announced that vehicles had been ordered by the Papua New Guinea Defence Force and by television companies for the protection of film crews operating in the former Yugoslavia.

Manufacturer

Tenix Defence Systems Pty Ltd.

UPDATED

AUSTRIA

Steyr-Daimler-Puch Pandur II (8 × 8 and 6 × 6) Armoured Personnel Carrier

Development

Based on its experience in the design, development and production of the Pandur (6 × 6) family of light armoured vehicles Steyr-Daimler-Puch Spezialfahrzeug has now developed the enhanced Pandur II family as a private venture.

This will be offered in both 6 × 6 and 8 × 8 configurations with these two vehicles having some 90 per cent common components to reduce total life cycle costs.

The first pre-series Pandur II (6 × 6) prototype commenced its company trials in September 2001 and was followed by the Pandur II (8 × 8).

According to Steyr-Daimler-Puch, the Pandur II is an evolutionary development of the Pandur I and features an extended wheelbase and modified hull while still retaining the Lockheed Martin C-130 Hercules roll on/roll off capability.

It is envisioned that existing customers will probably elect to stay with the current Pandur (6 × 6) which is in service with Austria, Belgium, Kuwait, Slovenia and the United States, but new customers would require the enhanced Pandur II 6 × 6 and 8 × 8 series of vehicles.

Description

The hull of the Pandur II is of all-welded steel armour construction that provides the occupants with full protection against 7.62 mm armour piercing (AP) attack through a full 360°.

This can be upgraded to a higher level of protection by the installation of additional passive appliqué armour including against 14.5 mm AP at 100 m. Spall liners are fitted as well as additional protection being provided against mines.

The overall layout of the Pandur II family is similar to that of the original Pandur with the driver seated at the front left with the powerpack to his immediate right.

The driver has a single-piece hatch cover that opens to the right, and in front of this are three day periscopes, the centre one of which can be replaced by a passive periscope for driving at night.

The current Pandur, now sometimes referred to as the Pandur I, is powered by a Steyr 260 or 285 hp diesel engine, but the Pandur II is powered by a Cummins ISC 350 diesel developing 355 hp (with option of 400 hp model) at 2,000 rpm coupled to a ZF 6HP 602C fully automatic transmission.

This gives Pandur II (6 × 6) a maximum road speed of 100 km/h with a fully armour protected cooling system now installed. Both models share a common powerpack that can be removed in 15 minutes and replaced in a similar time.

All wheels have an independent suspension system and the driveline and steering linkages are within the armoured hull. The hull is provided with special

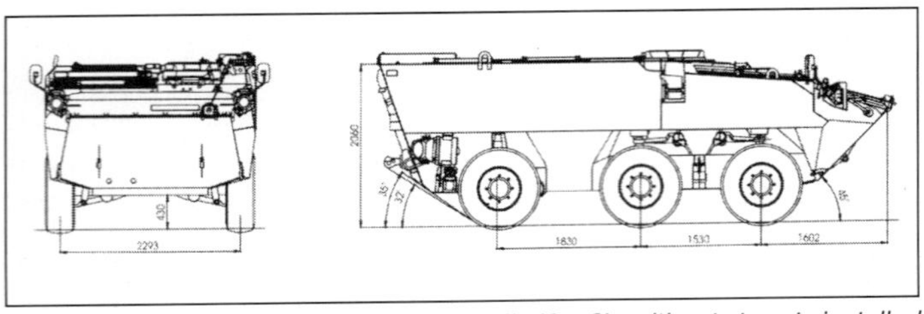

Front and side drawings of Pandur II (6 × 6) without turret installed
2002/0121123

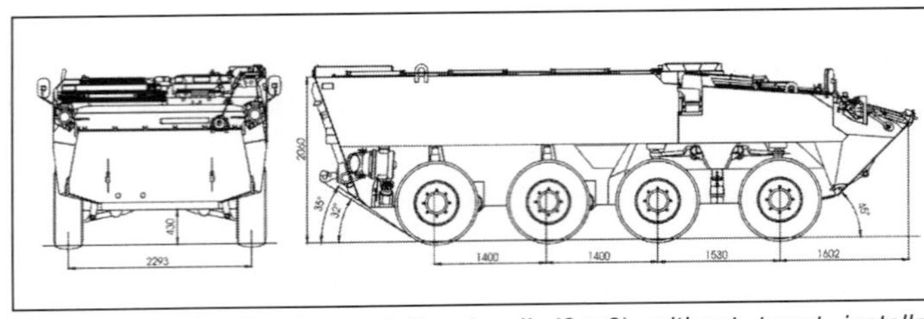

Front and side drawings of Pandur II (8 × 8) without turret installed
2002/0121122

Specifications

Model	**6 × 6**	**8 × 8**
Crew:	2 + 10	2 + 12
Weight:		
(combat)	15,000 kg	18,300 kg
(unloaded)	12,000 kg	13,600 kg
Payload:	3,000 kg	4,700 kg
Power-to-weight ratio:	23.66 hp/t	19.39 hp/t
Length	6.573 m	7.543 m
Width:	2.679 m	2.679 m
Height (hull top):	2.06 m	2.06 m
Ground clearance:	0.43 m	0.43 m
Track:	2.293 m	2.293 m
Wheelbase:	1.53 + 1.83 m	1.53 + 1.40 + 1.40 m
Angle of approach:	0 0	0 0
Angle of departure:	38/45	38/45
Max speed:		
(road)	100 km/h	100 km/h
(water)	11 km/h	10 km/h

Acceleration:		
(0 to 30 km/h)	6 s	6.5 s
(0 to 50 km/h)	13 s	14 s
Fuel capacity:	n/avail	n/avail
Range:	600 km	550 km
Fording:	1.5 m	1.5 m
Gradient:	70%	70%
Side slope:	40%	40%
Trench:	1.2 - 1.4 m	2.00 - 2.2 m
Vertical obstacle:	0.5 m	0.5 m
Turning radius:	9 m	10 m
Engine:	Cummins ISC 350 6-cylinder in-line, turbocharged and intercooled diesel developing 355 hp at 2,000 rpm (with option for 400 hp diesel)	
Transmission:	ZF 6HP602 C automatic with torque converter, lockup clutch and retarder, 6 forward and 1 reverse gears	
Transfer case:	2-speed	2-speed
Steering:	power assisted recirculating ball steering, first and second axle steered	

Tyres:	12.5R20 or 365/80 R20 plus CTIS	
Brakes:		
(main)	hydraulic, dual circuit with disc brakes on each wheel	
(parking)	sping-loaded hydro-pneumatically operated disc type parking brake shaft in transfer case	
Suspension:	independent, control arms with telescopic shock absorbers on each wheel. Coil springs and hollow rubber springs on 1st and 2nd axle. Torsion bars and hollow rubber springs on 3rd and 4th axle	
Electrical system:	24 V	24 V
Armament:	depends on mission	depends on mission
NBC system:	optional	optional
Night vision equipment:	optional	optional

Prototype of the Pandur II (6 × 6) armoured personnel carrier fitted with two-person turret armed with 30 mm cannon
2002/0096107

protection against mines and the troop seats hang from the hull sides rather than being attached to the floor.

The roof hatch arrangement depends on the type of weapon station installed. If a two-person turret armed with a 25 mm cannon was fitted then the rear troop compartment would probably have two roof hatches. Normal means of entry and exit for the troops is via two doors in the hull rear, although a power operated ramp could also be provided.

Standard equipment for the Pandur II series includes steering on front two axles, independent suspension, central tyre inflation sytstem, run-flat elements, disc brakes, anti-skid braking system, Automatic Drivetrain Management (ADM) system, electronic engine management and a self-recovery winch. A fire detection and suppression system is fitted in the engine compartment.

According to Steyr-Daimler-Puch the ADM offers a number of advantages including:

- Sensor-controlled automatic engagement of all wheel drive and differential locks depending on surface and driving conditions
- Reduced wear and tear on driveline components
- Reduced training effort
- Superior off road mobility
- Mechanical override by driver

A water heating unit is provided for engine pre-heating (cold start capability) and crew compartment heating. The ventilation system has a cyclone and paper filter.

Optional equipment includes an air circulating system, NBC protection system and fire detection and suppression system for the crew compartment.

Pandur II (6 × 6) can have a maximum crew of up to 12 people while the Pandur II (8 × 8) will be able to carry up to 14 people. The latter vehicle will also be able to be fitted with wider range of weapon stations up to 105 mm. The latter could include the two-person General Dynamics Land Systems 105 mm Low Profile Turret System.

Weapon fits will include a version of the two-person turret fitted to the Austrian Army's new Ulan infantry fighting vehicle, armed with a stabilised 30 mm Mauser cannon, 7.62 mm co-axial machine gun, advanced day/night fire control system and an auto tracker.

Variants

Amphibious model

The Pandur II will be able to ford to a depth of 1.5 m but a fully amphibious model will also be offered, propelled in the water by two water jets at a maximum speed of 10 km/h. The two water jets are mechanically driven and mounted at the front of the vehicle is a hydraulically operated trim vane which is rapidly erected by the driver without leaving his seat.

Other variants

As well as the basic armoured personnel carrier these would include a command post vehicle, infantry combat vehicle, mobile gun system, armoured recovery vehicle, ambulance, anti-tank vehicle with guided missiles and a mortar carrier.

Polish APC requirement

Steyr-Daimler-Puch has confirmed that it will offer Pandur II (6 × 6 and 8 × 8) to meet the Polish Army requirement for a new wheeled armoured personnel carrier. For this competition the company is teamed with OBRUM. The Pandur (6 × 6) has already been evaluated in Poland twice, the basic version in 1991 and the new amphibious version in 1997.

Status

Prototype vehicles undergoing trials. Not yet in production or service.

Manufacturer

Steyr-Daimler-Puch Spezialfahrzeug AG.

NEW ENTRY

Steyr-Daimler-Puch Pandur armoured personnel carrier

Development

The Pandur armoured personnel carrier was developed from 1979 as a private venture by Steyr-Daimler-Puch and was unveiled late in 1985, by which time two prototypes had been built.

The first of six preproduction vehicles was completed in June 1986 and the remaining five were completed by December 1986.

Wherever possible, for example with the diesel engine and transmission, standard and in-production commercial components have been incorporated into the vehicle.

The basic Pandur 6 × 6 vehicle has the so-called Type 330 hull and has an overall length of 5.697 m, but there is also a longer hull called the Type 332 with an overall length of 6.172 m (+500 mm). The latter version has a longer wheelbase between the second and third axles.

During the Iraqi invasion of Kuwait in August 1990, two prototypes of the Steyr-Daimler-Puch Pandur (6 × 6) APC, one of which was fitted with the MECAR 90 mm Kinetic Energy Weapons Station, were captured by the Iraqi Army. Since the loss of these two prototype vehicles, Steyr-Daimler-Puch has built two replacement vehicles.

In 1994, the Austrian Army placed an order for 68 Pandur vehicles and the first of these were completed in December 1995 and production was completed late in 1996 but was resumed again for export.

These 68 Pandurs, of which 17 are painted white ready for United Nations operations with the remaining 51 being in standard army green, are issued to two battalions of the Austrian Army. One battalion is in southern Austria and the other is in eastern Austria.

Standard equipment on the Austrian Army Pandur includes a cupola-mounted 12.7 mm machine gun, two banks of three electrically operated smoke grenade launchers mounted on the right side of the hull, spherical firing ports and associated day vision devices in either side of the rear troop compartment and an air conditioning system which is mounted on the left side of the vehicle.

The Austrian Army Pandurs also have a fire detection and suppression system installed in both the engine and troop compartments, special dust filters for engine and crew compartments and Semperit tyres. They are designed for, but not fitted with an NBC system as this is not required for UN operations.

Some vehicles are fitted with a capstan-type recovery winch which is installed at the rear of the vehicle on the right side. This has 60 m of cable and can be used to the front and rear of the Pandur.

Austrian Army Pandur vehicles are fitted with spall liners and anti-mine mats as standard equipment.

In October 1996, the Kuwait National Guard placed an order with AV Technology International of the USA for the supply of 70 Pandur (6 × 6) light armoured vehicles in six variants with first production vehicles being completed in November 1997. Additional details of these vehicles and their variants are given in the USA section.

The six variants are: armoured personnel carrier fitted with lightweight two-man MultiGun Turret System (MGTS), ambulance, 81 mm mortar carrier, recovery vehicle, command vehicle and armoured reconnaissance fire support vehicle fitted with a Cockerill two-man turret armed with a Cockerill 90 mm gun and computerised fire-control system with day/night sights.

The production of these vehicles and turrets was undertaken at the AV Technology LLC (an affiliate of General Dynamics Land Systems) facility, at Chesterfield, Michigan.

In January 1997, it was announced that following intensive trials between the Austrian Steyr-Daimler-Puch Pandur (6 × 6) and Swiss MOWAG Piranha Generation III (6 × 6) armoured personnel carriers, the Belgian Army had selected the Pandur.

In November 1998, the Belgian Army took delivery of the first of 54 Pandur APCs for use by the Para Commando Brigade.

The order for 54 Pandur vehicles comprised 41 APCs, five command post vehicles, four maintenance and recovery vehicles and four ambulance vehicles.

The final four vehicles were built in the ambulance model as these have a longer wheel base to give greater internal volume for their specialised mission. Since then an additional six vehicles have been ordered by Belgium.

As part of the 100 per cent off-set required by Belgium, production of the complete all welded steel hull and final assembly of the Pandur took place in the facilities of EMI (Espace Mobile International sa) in Aubange, Belgium.

The Pandurs built for the Belgium Army are the first production vehicles to be fitted with the Steyr-Daimler-Puch developed Automatic Drivetrain Management (ADM) system and also have Belgium BMAS radios, Global Positioning System and air conditioning system.

The Belgian Army Para Commando Brigade has three battalions 1st (Para), 2nd (Commando) and 3rd (Para), each of which will have one company equipped with Pandurs with the APC being armed with a roof mounted 7.62 or 12.7 mm machine gun.

The Austrian Army has also recently stated that it intends to place a contract for a further 270 Pandur Vehicles in the future.

Following extensive trials, Slovenia selected the Pandur to meet its future requirements for a new wheeled APC.

The Pandur APC is now manufactured under licence in Slovenia by the local company of STO RAVNE doo who have been awarded the production contract by the Slovenian Ministry of Defence.

In Slovenian Army service the Pandur is be called the Valuk, with the first batch consisting of 10 vehicles which have now been completed. A further batch of 10 vehicles has been ordered. The total requirement is for 70 vehicles in the first phase.

Key automotive components such as the drivetrain will be supplied by Steyr-Daimler-Puch while STO RAVNE will fabricate the all welded steel armour hull, fit all of the sub-systems, including weapons and then deliver the complete Valuk to the Slovenian Army.

First deliveries were made to the Slovenian Army in 1999. These are similar to the Austrian Army vehicles but have a ramp at the rear.

Late in 1999 the US Army awarded AV Technology a contract with a total value of US$51 million for up to 50 Pandur vehicles to meet the US Army requirement for an Armored Ground Mobility System (AGMS). Additional details are given under the USA.

Description

The hull of the Pandur is of all-welded steel armour construction that provides protection against penetration from 7.62 mm × 51 armour-piercing projectiles fired from a distance of 30 m all round (0 to +7° vertical) and over the frontal arc (–39 to +17° vertical and –30 to +30° horizontal).

A higher degree of armour protection is also available which provides protection against 12.7 mm × 99 armour-piercing projectiles fired from a distance of 100 m.

Other options include the installation of spall liners inside the hull which significantly reduces the behind armour effects of penetrating projectiles, fragments and hollow charges in the area of the driver's and combat crew compartment. The Pandur can also be fitted with mine protection mats.

The underside of the hull is smooth so that it does not catch on any rough objects. In addition, this also improves its water speed.

The driver is seated at the front left side with the power pack to his right. The driver has a left-opening single-piece hatch cover in front of which are three day periscopes, the centre one of which can be replaced by a passive night vision device.

The Steyr diesel engine is coupled to a fully automatic Allison transmission with torque converter and lock-up clutch. The complete power pack of the Pandur, which consists of the engine, transmission and cooling/ventilation system, can be removed as a complete unit in about 20 minutes. The air intake/outlets are in the roof of the vehicle with the exhaust outlet on the right side.

Future vehicles will be powered by a Cummins ISC 350 six-cylinder diesel developing 355 hp, coupled to a ZF 6HP602 C fully automatic transmission with torque converter giving six forward and one reverse gears.

From the transmission, a drive shaft leads to a central transfer box interlocked with the axial drive of the second axle. Two further shafts lead to the differential gears of the first and third axles and from these power is transmitted to the planetary gears in the wheels. All parts of the power train are either situated within the hull or protected by tunnels or depressions of the hull.

When being driven on roads the vehicle is normally in (6 × 4) configuration, with the first and third axles driven via a central differential. For improved traction in rough terrain the second axle is engaged. The axle differentials as well as the central differential can be locked.

The hull to the rear of the driver's position has been raised to provide more room for the troops and height to the hull top is between 2.06 and 2.16 m compared with the 1.82 m of the basic Pandur vehicle.

To the rear of the driver can be installed a 12.7 mm machine gun mount with the gunner being provided with day periscopes and a single-piece hatch cover that opens to the rear.

The troop compartment has two outward-opening roof hatches on the left side while on the right side is the commander's cupola with a circular hatch cover that opens to the right and five day periscopes. To the rear of this is a single rectangular roof hatch that opens to the right.

Pandur (6 × 6) armoured personnel carrier as delivered to the Austrian Army armed with 12.7 mm M2 HB machine gun **2001**/0103747

Belgian Army Pandur shown in APC configuration and armed with 12.7 mm machine gun 0058016

Pandur (6 × 6) Armoured Reconnaissance Fire Support Vehicle fitted with Cockerill 90 mm gun

Pandur fitted with a Euromissile UTM 800 turret launching HOT missile
2000/0058475

The troops enter and leave via two doors in the hull rear that open left and right with the right door having a vision block. In either side of the hull are two firing ports with associated vision devices below.

Mounted on the right side of the hull, firing forwards, are two banks of three electrically operated smoke grenade launchers. Mounted on the left side of the hull rear is the air conditioning system while on the right side of the hull is the recovery winch.

All three axles have independent suspension with power steering on the first and second axles. The suspension of the first and second wheel stations consists of an upper longitudinal control arm with a lower transverse control arm and coil springs. The rear wheel station consists of a trailing arm and torsion bar. The identical spacing of the axles gives an even load distribution. The maximum vertical wheel travel is 280 mm. All wheels have disc brakes with the parking brake being of the drum type and manually operated.

The Pandur has a central tyre-pressure regulation system which allows the driver to adjust the tyre pressure to the type of ground being crossed at any given time. The tyre pressure can be changed while the vehicle is moving. All wheels of the Pandur are fitted with an emergency run-flat system allowing for a 50 km ride even with all tyres punctured.

A wide range of armament installations can be mounted on the Pandur.

Standard equipment on production vehicles would include an automatic fire extinguishing system for the engine compartment and a heater.

Optional equipment would include, depending on variant, an NBC protection/NBC detection system, a fire suppression system for the crew compartment, an air conditioning system, a recovery winch and a land navigation system. Various types of passive night vision equipment can be provided, including thermal sights for the commander and gunner when the Pandur is fitted with a two-man turret.

Pandur with ADM

Steyr-Daimler-Puch has developed, as a private venture, the Automatic Drivetrain Management (ADM) system to improve the performance of its Pandur (6 × 6) Armoured Personnel Carrier (APC) and variants, especially in off-road conditions.

ADM is essentially a computer-controlled system for the automatic control of all shifting functions in the vehicle drivetrain with the capability of carrying the full load.

ADM continuously checks, via sensors, the actual surface conditions, traction on the ground, driving conditions and then automatically engages or disengages all-wheel drive, the longitudinal lock or the specific differential locks when required.

The Pandur driver still retains the option to override the ADM so that all functions remain under his direct control.

The ADM consists of four major components, the Electronic Control Unit (ECU), sensors, pneumatic actuators and dog clutch.

The ECU processes all of the data coming from the sensors for speed of the roadwheels, steering angle, position of throttle and braking pedals as well as parameter tables designed for the configuration of the Pandur vehicle.

Depending on the actual speed of the vehicle, this parameter table provides limits of percentile slip for each separate lock. The computer is integrating the difference from the slip curve to the slipsum curve.

As soon as the limits of the slipsum curves are exceeded the specified locks are automatically actuated. At certain levels of the slipsum curves, additional locks can be actuated to the specified ones, for example all-wheel drive.

All of the data provided by standard ABS sensors are continuously corrected by the steering angle. Throttle performance and braking give additional information for different characteristics to the system.

If required, the ECU gives the order to the onboard computer system to engage the lock. A standard pneumatic system engages the dog clutch whether for all-wheel drive, for longitudinal lock or for differential lock of a separate axle.

As soon as there is no more torque difference between the single wheels and the axles, the lock disengages automatically.

ADM has been developed exclusively for the Pandur vehicle and is integrated into the vehicle when it is built. It is envisioned that ADM will be offered for any future sales of the Pandur family. It is also possible that new build Pandurs could be fitted for but not with ADM to enable the vehicle to be fitted with the system at a later time.

The main advantages of ADM have been summarised by Steyr-Daimler-Puch as increased safety, as all the driver now has to concentrate on is driving the actual vehicle and using the ground to the best tactical advantage, reduced driver training and a higher level of comfort for the crew and passengers as all drivetrain shifting is now fully automatic.

In addition, maximum possible traction is available at all times due to 100 per cent locking of the drive train system whenever required. Cross-country performance is improved as the system automatically adapts the drive train functions to the actual conditions.

It is also claimed that the vehicle is more stable when fitted with ADM and the life of the automotive components is extended as shifting errors by the driver are eliminated and there is reduced stress on the drive train components. With ADM, the locks are only actuated when required.

For trials purposes the ADM was first installed on one of the older prototypes of the Pandur APC and this commenced extensive trials at various locations in Austria early in 1996.

Steyr-Daimler-Puch has now installed ADM in its new company demonstrator Pandur vehicle, which is to the

Amphibious version of Steyr-Daimler-Puch Pandur (6 × 6) armoured personnel carrier showing its amphibious characteristics **2001**/0103750

Pandur with Euromissile HOT/ATM turret system with sensor package lowered
2002/0121149

Pandur fitted with 120 mm Armoured Mortar System (AMS) **2002**/0121150

Pandur 6 × 6 Armoured Personnel Carrier prepared to be equipped with calibre 40 mm automatic grenade launcher MK 19 **2002**/0121148

latest build standard and also has a raised roof in the rear troop compartment.

Amphibious Pandur

The amphibious Pandur is available in all variants and features water-to-water cooling, closed air intake and electric bilge pumps.

Mounted on either side at the rear of the hull is an integrated waterjet which is run from a power take-off, these give a maximum water speed of 11 km/h and the Pandur can turn through a full 360° in 8 seconds.

Before entering the water the driver only has to push one button which opens the hydraulically activated front trim vane, bilge pumps, water-to-water cooling, air-intake shut down and change of the exhaust system.

Ambulance Pandur

The ambulance model has a higher raised rear compartment and its rear wheelbase (second to third axles), has been increased from 1.53 to 1.83 m for greater internal volume.

This vehicle has a maximum gross weight of 13 tonnes and a three-man crew consisting of commander, driver and medical attendant.

Casualties can be carried in three different configurations, these are four stretcher patients, two stretcher and three sitting or six seated patients. A full range of medical equipment is carried.

Variants

Pandur II (6 × 6 and 8 × 8)

Late in 2001 Steyr-Daimler-Puch announced the enhanced Pandur II 6 × 6 and 8 × 8 series of armoured personnel carriers and variants, details of these are given in a separate entry.

US Market

For the USA and some other markets, such as the Middle East, Steyr-Daimler-Puch is working with the US company AV Technology which has extensive experience in the design and production of wheeled armoured vehicles (including the Dragoon 4 × 4) as well as turrets. Additional details are given in this section under the USA.

Austrian Army APC

The detailed description relates to this model.

The troop compartment roof has two outward-opening roof hatches on the left side while on the right side is a fixed cupola with a circular hatch cover that opens to the right and five day periscopes. To the rear of this is a single rectangular hatch cover that opens to the right.

The troops enter and leave via the two doors in the hull rear that open left and right with the right door having a vision block. In either side of the hull are two firing ports with associated vision devices.

Mounted on either side of the hull, firing forwards, is a bank of three electrically operated smoke grenade launchers. Mounted on the left side of the hull rear is the air conditioning system.

Mechanised Infantry Combat Vehicle (MICV 127)/Armoured Personnel Carrier (APC 127)

This version of the Pandur would be fitted with the AV Technology MultiGun Turret System (MGTS) armed with: 40 mm grenade launcher and 7.62 mm MG; 30 mm M230 Chain Gun and 7.62 mm MG; 12.7 mm M2 MG and 7.62 mm MG; or a cupola weapons station armed with a 12.7 mm M2 MG and a 7.62 mm MG.

The commander would be seated to the rear of the driver and have a raised cupola with periscopes to give observation over the frontal arc of the Pandur. Firing ports can be provided on request.

Armoured reconnaissance fire support vehicle

This would be fitted with a two-man turret armed with a 90 mm gun, for example the Cockerill 90 mm LCTS turret system which is armed with the Mk 8 gun system.

Other 90 mm turret options are the Giat Industries TS-90 turret armed with a 90 mm CN90F4 gun or the Cockerill CSE90 turret armed with a Cockerill 90 mm Mk III gun.

Anti-tank guided missile carrier

A Pandur (6 × 6) in the version ATGMC/HOT and fitted with the Euromissile UTM 800 launcher and, after re-equipment, in the version ATGMC/TOW, fitted with the Norwegian Kvaerner-Eureka Armoured Launching Turret (ALT) has been evaluated by the Austrian Army to meet its ATGM 4000 requirement.

This version of the Pandur has a crew of four. In addition to the missiles in the ready to launch position, four in the case of the UTM 800 and two in the case of the Armoured Launching Turret, a further eight missiles are carried in the hull for manual loading.

AV Technology have tested the Hellfire missile on the Pandur chassis in Kuwait in 1997.

More recently Pandur has been fitted, for trials purposes, with the private venture HOT/ATM system which has four 4,000 m range HOT ATGW and a mast mounted sensor pod.

Ambulance (AAMB)

This model would be unarmed.

Armoured Reconnaissance Scout Vehicle (ARSV 25 and 30)

This would be fitted with the two-man Steyr-Daimler-Puch SP3/300 turret armed with a 30 mm Mauser MK 30 cannon. This turret is similar to the full-tracked ASCOD IFV turret (covered under International) but with reduced armour protection, similar to the hull protection level of the Pandur on which it is mounted. The ASCOD is now in service with Austria under the name Ulan, with first four vehicles delivered in mid-2001.

Specifications

(in brackets the amphibious version, where different)

Basic vehicle
Crew: 2 + 8 [2 + 10]
Configuration: 6 × 6
Max combat weight: 13,500 kg [14,000 kg]
Unloaded weight: 10,300 [10,800 kg]
Max permissible payload: 3,200 kg
Power-to-weight ratio:
(max weight) 19.25 hp/t [20.3 hp/t]
Length: 5.697 m [6.297 m]
Width: 2.5 m [2.6 m]
Height:
(to hull top) 1.82 m [2.060 m]
Ground clearance: 0.43 m
Track:
(first axle) 2.148 m [2.248 m]
(second axle) 2.148 m [2.248 m]
(third axle) 2.155 m [2.255 m]
Wheelbase: 1.53 + 1.53 m [1.53 + 1.83 m]
Angle of approach/departure: 45/49° [48/38°]
Max road speed: 100 km/h
Acceleration: (13,000 kg)
(0 to 50 km/h) approx 15 s
(0 to 80 km/h) approx 35 s
Fuel capacity: 275 litres [300 litres]
Range cruising: 700+ km
Fording: 1.5 m (amphibious without preparation)
Gradient:
(at 13,500 kg) 70% (stop and start)
Side slope:
(at 13,500 kg) 40% (stop and start)
Trench: 1.1 m, 1.6 m (diagonal to ditch edge)
Vertical obstacle: 0.5 m
Turning radius: 8.5 ± 0.25 m
Engine: Steyr WD 612.95 (WD 612.35) 6-cylinder turbocharged water-cooled diesel with air-intake cooling, developing 260 hp (285 hp at 2,500 rpm) at 2,400 rpm
Transmission: Allison Transmission MT-653 DR 5-speed with torque converter and lock up clutch
Steering: ZF power-assisted on first and second axle
Tyres: 12.5 R 20 or 365/80 R20 XZL with central tyre pressure regulation
Brakes:
(main) pneumatically controlled ventilated disc, dual circuit, all wheels
(parking) drum brake acting on transmission output shaft
Suspension: independent
Vertical wheel travel: +150 mm/–135 mm [+150 mm/–130 mm]
Electrical system: 24 V
Armament: see text, ranges from 7.62 mm MG up to 90 mm gun, 81 mm or 120 mm mortars
NBC system: optional
Night vision equipment: optional

Mortar carrier

This could be fitted with an 81 mm turntable-mounted mortar (AMC81) or a 120 mm mortar. The latter, AMC120, would feature a recoil device, hydraulic pointing system, GPS firing aids, laser gyro bearing indicator, altitude sensors and ballistic computer.

Anti-aircraft

Fitted with a power-operated turret armed with twin 20 mm.

Pandur 40 mm

For trials purposes a Pandur has been fitted with a new one-person turret armed with a 40 mm automatic grenade launcher.

Other variants

A wide range of command and communication carriers, provided for installation of various C[3]I equipment, armoured personnel carriers and a recovery vehicle are currently under development by Steyr-Daimler-Puch.

Status

Production. In service with the following countries:

Country	Quantity	Status
Austria	68	built by Steyr-Daimler-Puch, delivered + another six ordered
Belgium	54	assembled in Belgium
Kuwait	70	built by AV Technology International
Slovenia	10	first deliveries 1999 + another 10 ordered
USA	50	first deliveries 2000, includes option

Manufacturer

Steyr-Daimler-Puch Spezialfahrzeug AG and Co KG.

UPDATED

Steyr-Daimler-Puch Survivor Armoured Vehicle

Development

The Survivor has been developed as a private venture by the Austrian company of Steyr-Daimler-Puch Fahrzeugtechnik AG & Co KG.

It is being marketed as a high mobility multi-purpose wheeled armoured vehicle which is capable of being modified to undertake a wide range of battlefield roles and missions.

It is currently available in two wheelbase lengths and can be fitted with various diesel or petrol engine/ automatic transmission combinations to suit specific users' operational requirements.

Description

The Survivor is based on a commercial 4 × 4 chassis that has been fitted with an all-welded monocoque steel armour body that provides the occupants with protection from ball type small arms fire up to level B7. In addition it provides protection against hand grenades and 155 mm overhead burst artillery projectile fragments.

The engine is at the front with the commander and driver to the immediate rear. Both left and right hand drive models of the Survivor are available with commander and driver each being provided with a forward opening door with a bullet proof window in its upper part. To the immediate front of the driver is a large bullet proof window that is provided with wipers and washers.

The troop compartment is at the rear of the vehicle and the occupants enter and leave via doors in the hull rear. Bullet proof windows with firing ports below can be provided if required by the customer in the rear troop compartment.

Various petrol or diesel engine options are available for the Survivor with all of these coupled to a four or five speed fully automatic transmission.

Steyr-Daimler-Puch Fahrzeugtechnik Survivor (4 × 4) high-mobility multipurpose wheeled armoured vehicle **2002**/0121147

Standard equipment on the Survivor includes power steering on the front two wheels, permament 4 × 4 drive with the driver selecting high or low range, 100 per cent differential locks, heating and heavy duty air conditioning system, tyres with run flat inserts, 12 V or 24 V electric power supply and fitted for radio.

Optional equipment includes various roof-mounted fixed or retractable weapon stations, smoke grenade launcher system, intercom, global positioning system, various external radio systems, fire detection and suppression system, camouflage system, various types of day and night observation equipment.

Specifications

Crew: 2 + 6/8
Configuration: 4 × 4
Max combat weight: n/avail
Unloaded weight: n/avail
Max permissable payload: n/avail
Power-to-weight ratio: 71 hp/tonne (depending on weight/engine)
Length:
(Type A) 4.50 m
(Type B) 4.70 m
Width: 1.90 m
Height:
(Type A) 1.88 m
(Type B) 2.06 m
Ground clearance: n/avail
Track: n/avail
Wheelbase: n/avail
Angle of approach/departure: n/avail
Max road speed: 138 to 170 km/h (depending on engine installed)
Accelaration: n/avail
Fuel capacity: 96 litres
Range: 600 km (cruising)
Fording: 0.70 m
Gradient: 100%
Side slope: 80%
Trench: n/avail
Vertical obstacle: n/avail
Turning radius: 6.5 m
Engine: intercooled turbocharged diesel engine developing up to 205 hp OR fuel injected 6-cylinder petrol engine developing up to 240 hp OR 8-cyclinder fuel injected petrol engine developing up to 300 hp
Transmission: 4- or 5-speed automatic
Steering: power assisted
Tyres: have run flat inserts as standard
Brakes: heavy duty
Suspension: heavy duty coil spring with hydraulic shock absorbers
Electrical system: 12 V or 24 V
Armament: depends on mission
NBC system: no
Night vision equipment: optional

Variants

The manufacturer has suggested that the Survivor could be used for a wide range of missions including the following:-

- Ambulance
- Anti-tank fitted with Euromissile HOT ATM system with 4,000 m range HOT ATGW and/or 20 mm cannon
- Armoured personnel carrier carrying 10 personnel including driver and commander
- Border patrol vehicle with mast-mounted day/night sensor pod
- Command and control vehicle
- Internal security vehicle with front-mounted obstacle clearing blade and various surveillance devices
- Reconnaissance vehicle with various roof-mounted weapon stations and sensors
- Fitted with various turret and cupola-mounted machine guns and cannon up to a maximum calibre of 25 mm

Status

Development complete. Ready for production.

Manufacturer

Steyr-Daimler-Puch Fahrzeugtechnik AG & Co KG.

NEW ENTRY

BELGIUM

SIBMAS armoured personnel carrier

Development

The SIBMAS (6 × 6) wheeled armoured vehicle range was originally designed as a private venture from 1975 by B N Constructions Ferroviaires et Métalliques. A prototype was completed in 1976 and was extensively tested both in Belgium and Malaysia. In many respects the SIBMAS is very similar to the South African (now Vickers OMC) Ratel (6 × 6) infantry fighting vehicle.

A demonstration vehicle, with improved vision for the driver and a more powerful engine, was completed in mid-1979.

Wherever possible, standard MAN automotive components such as the engine, transmission and axles were used in the design of the SIBMAS.

Late in 1981, Malaysia placed a £50 million order with the company for 186 vehicles with first deliveries made in June 1983. Two versions have been delivered:

AFSV-90: Armoured Fire Support Vehicle with a Cockerill CM 90 two-person turret and a 90 mm Mk III gun and an OIP LRS-5 fire-control system; 162 have been supplied and production was completed in late 1984.

ARV: Armoured Recovery Vehicle with front and rear spades, a weight of 20,000 kg, a winch with a capacity of 20,000 kg and a crane with a capacity of 10,500 kg; 24 have been supplied and production was completed in mid-1985.

Although CMI took over this vehicle in the mid-1980s, the company is no longer marketing the SIBMAS range of 6 × 6 vehicles and is concentrating on its range of AFV turrets.

Early in 1998, it was revealed that three Malaysian contractors had each overhauled a single example of the SIBMAS (6 × 6) AFV. In the future, one of these is expected to be awarded a contract to overhaul part or all of the Malaysian fleet of SIBMAS vehicles. As of late 2001, it is understood that no orders had been placed with any of these contractors.

Description

The hull of the SIBMAS is made of all-welded steel and provides the crew with complete protection against 7.62 and 5.56 mm armour-piercing rounds.

The driver sits at the very front of the vehicle and has a single-piece hatch cover over his position that opens to his right, an adjustable seat and an adjustable steering wheel. In front and at both sides of the driver are large bulletproof windscreens providing excellent vision at all times. If required, these can be quickly covered by armoured shutters which are hinged at the lower part.

The turret is mounted immediately behind the driver and in each side of the hull is a large door that opens forwards.

The troop compartment is to the rear of the turret and the engine at the very rear on the left side. Six of the infantrymen sit back-to-back down the centre of the vehicle with the remaining three in the aisle between the troop compartment and the rear door. One small and three large hatches are provided over the top of the troop compartment and, if required, a remote-controlled machine gun can be pintle-mounted over the aisle. There are spherical firing ports and vision blocks in the troop compartment. These allow the infantry to fire their weapons from within the vehicle with a high level of safety.

The engine can be removed through the roof of the vehicle in 30 minutes by two men. Quick disconnect couplings are provided for both the power and hydraulic circuits. The cooling system is behind and separated from the engine compartment by a waterproof bulkhead. Power is taken from the engine to the transmission, which is under the floor between the first and second axles, by a propeller shaft. The differentials of the front and rear two axles are lockable and a lockable differential between the two rear axles allows them to be longitudinally locked.

The SIBMAS is a fully amphibious vehicle propelled in the water by its wheels at a speed of 4 km/h. There was also a version with two propellers at the rear of the hull which can be traversed through 360° and give the vehicle a maximum water speed of 11 km/h. The propellers are driven from the Power Take Off (PTO) on the engine. One advantage of the SIBMAS is that when approaching a river bank the driver can engage first gear and as soon as the front wheels touch the bank maximum tractive effort is available as the propellers are still pushing the vehicle forward. Before entering the water a trim vane is automatically erected at the front of the hull. There are two bilge pumps, each with a maximum capacity of 180 litres per minute, fitted as standard.

The tyres are of the run-flat type. Optional equipment includes passive night vision equipment, a hydraulic winch mounted at the front of the hull with 60 m of cable and a maximum capacity of 8,000 kg (which can be increased to 16,000 kg with snatch blocks), a heater, an air conditioning system and an NBC system.

The SIBMAS could be supplied fitted with a wide variety of armament installations but the CM 90 turret armed with the Cockerill 90 mm gun, 7.62 mm coaxial machine gun and 7.62 anti-aircraft machine gun is the only combat model to have been produced in production quantities.

Specifications

Crew: 3 + 11
Configuration: 6 × 6
Combat weight: 14,500-18,500 kg depending on role
Unloaded weight: 13,200 kg
Max axle loads:
(first axle) 7,500 kg
(second axle) 8,500 kg
(third axle) 8,500 kg
Power-to-weight ratio: 20 hp/t*
Ground pressure: 2.4 kg/cm²
Length:
(hull) 7.32 m
Width: 2.5 m
Height:
(to turret top) 2.77 m
(to hull top) 2.24 m
Ground clearance: 0.4 m
Track: 2.07 m
Wheelbase: 2.8 m + 1.4 m
Angle of approach/departure: 50/46°
Max road speed: approx 100 km/h
Max water speed:
(propelled by wheels) up to 4 km/h
(propelled by propellers) up to 11 km/h
Acceleration: 0-60 km/h in 22 s
Fuel capacity: 400 litres
Max road range: approx 1,000 km
Fuel consumption: 0.36 litres/km
Fording: amphibious
Gradient: 70%
Side slope: 40%
Vertical obstacle: 0.7 m
Trench: 1.5 m
Turning radius: 8 m
Engine: MAN D 2566 MK 6-cylinder in-line water-cooled turbocharged diesel with intercooler developing 320 hp at 1,900 rpm
Max torque: 1,270 Nm (130 kgm) between 1,300 and 1,600 rpm
Transmission: fully automatic ZF 6 HP 500, 6 forward and 1 reverse gears, with hydrodynamic torque converter
Steering: power assisted
Suspension: longitudinal suspension arms (lower) and wishbone above axle, helical springs and hydraulic shock-absorbers
Tyres: 14.00 × 20 or 16.00 × 20 with run-flat elements
Brakes:
(main) hydraulic air-assisted
(parking) pneumatic on gearbox
Transmission: hydraulic retarder
Electrical system: 24 V
Batteries: 4 × 12 V, 100 Ah
Armament: see text
NBC system: optional
Night vision equipment: optional
* based on a combat weight of 16,000 kg

SIBMAS (6 × 6) vehicle of the Royal Malaysian Army fitted with CM 90 turret armed with 90 mm gun, 7.62 mm coaxial and 7.62 mm anti-aircraft machine gun

Variants

Many versions were proposed but apart from the model with a 90 mm CM 90 turret the only model built in production quantities was the recovery vehicle.

This has a front and rear stabiliser blade, 20,000 kg winch and a 10,500 kg crane.

Status

Production complete. In service with Malaysia (162 with 90 mm turret and 24 in armoured recovery vehicle configuration). No longer being marketed by CMI. The company is, however, ready to transfer the technology and expertise to another country which could undertake local production.

Manufacturer

CMI, Cockerill Mechanical Industries. This company is no longer involved in the design, development or production of armoured vehicles.

UPDATED

BDX armoured personnel carrier

Development

In 1976, the Engineering Division of Beherman Demoen negotiated successfully with the Irish company Technology Investments Limited for a licence to produce an improved version of the Timoney (4 × 4) armoured personnel carrier.

The first production order for the BDX was awarded by the Belgian government in 1977 and called for the delivery of 43 vehicles to the Belgian Air Force for airfield protection duties and 80 vehicles to the Belgian Gendarmerie.

The vehicles for the air force were armed with a ring-mounted 7.62 mm light machine gun. The gendarmerie order consisted of 41 APCs without a turret, 26 vehicles with a dozer blade for clearing obstacles and 13 vehicles fitted with an 81 mm mortar in the rear of the hull. First production vehicles were completed late in 1978 and production was completed for the Belgian government in February 1981.

BDX delivered five vehicles to Argentina and the vehicle has also been demonstrated to a number of other countries including Indonesia and Malaysia. Further development of the BDX APC by Vickers Defence Systems of the UK has resulted in the Valkyr.

Despite intensive marketing, only two of these were sold to Kuwait. Details were given in *Jane's Armour and Artillery 1995-96,* pages 441-442.

Beherman Demoen Engineering (BDX) is no longer involved in the design, development or production of armoured vehicles.

According to United Nations arms transfer list, Mexico took delivery of 95 vehicles in 1994. This left a total of 28 vehicles remaining in Belgium.

Early in 2001, following an international competition, the Australian Tenix Defence Systems S600 (4 × 4) internal security vehicle was selected to meet the requirements of the Belgian Gendarmerie.

Recent information has indicated that a total of 152 BDX vehicles were built. The above figures quote a total of 128 vehicles thus indicating additional export sales to other undisclosed countries.

Description

The hull of the BDX is made of all-welded steel which varies in thickness from 9.5 to 12.7 mm and provides complete protection against 5.56 and 7.62 mm armour-piercing projectiles fired at point-blank range and splinters from 105 mm artillery projectiles exploding on the ground 10 m from the vehicle. The BDX will withstand the blast from a 9 kg mine exploding 10 m from the vehicle.

The driver sits at the front of the hull in a central position and has three windscreens, one in front and one on either side, made of multiplex glass 81 mm thick which provides the same level of protection as the rest of the hull. The large front windscreen has a wiper and can be hinged forward for improved ventilation in a non-combat area.

The crew enters and leaves the BDX by three doors, one in each side of the hull and one at the rear. All the doors open through a full 180° and can be locked half open if required. Ball mounts with vision blocks are provided in each side of the hull to the rear of the side entrance door and in the rear door.

The 10 infantrymen have individual seats with waist seatbelts. The interior of the hull is padded to reduce both noise and the internal temperature of the vehicle.

The engine is mounted to the immediate rear of the driver and is coupled to an automatic transmission which in turn provides power to the transfer box at the back of the vehicle. This transmits power to the front and rear axles by a propeller shaft. Both front and rear axles are equipped with a self-locking differential of the 'no-spin' type. The air inlet and air outlet louvres are in the roof over the driver's position with the air outlet louvres located either side of the engine compartment in the roof. Protection against Molotov cocktails can be fitted if required. The exhaust pipes run along the outside of the hull.

The machine gun turret is fitted on the roof of the BDX to the rear of the engine compartment and a further circular hatch cover is provided in the roof at the rear. The vehicle, when fitted with the 81 mm mortar, has two rectangular hatches in the roof that open either side of the hull.

The BDX is fully amphibious and propelled in the water by its wheels. Optionally, two water-jets can be fitted to increase its water speed.

Optional equipment includes an air conditioning system, a front-mounted dozer blade for clearing obstacles, a heater, an NBC system and hull-mounted smoke grenade dischargers.

Variants

In addition to being used as an armoured personnel carrier, the BDX could be used for a wide range of other roles such as an ambulance, command vehicle and reconnaissance vehicle. The BDX could be fitted with a turret-mounted 20 mm cannon, a turret-mounted 90 mm Cockerill gun, a turret with single or twin 7.62 mm machine guns, a Euromissile MCT turret with two MILAN ATGWs in ready to launch position, a turret with an 81 mm breech-loaded mortar and a coaxial 7.62 mm machine gun or an 81 mm mortar mounted in the hull, firing forwards. Apart from the version with an 81 mm mortar, none of these specialised versions of the BDX were ever placed in quantity production.

Argentine BDX

A total of five vehicles was purchased in 1980, four in APC configuration and one in command post role. These are used by Federal police and are fitted with a locally developed 38.1 Venturini multiple rail grenade launcher on vehicle roof at front right side.

Status

Production complete. In service with Argentina, Belgium (to be replaced by S600) and Mexico (see also development).

Specifications

Crew: 2 + 10
Configuration: 4 × 4
Combat weight: 10,700 kg
Unloaded weight: 9,750 kg
Power-to-weight ratio: 16.82 bhp/t
Length: (overall) 5.05 m
Width: 2.5 m
Height:
(overall with twin 7.62 mm MG turret) 2.84 m
(to hull top) 2.06 m
Ground clearance: 0.4 m
Track: 1.93 m
Wheelbase: 3.003 m
Max road speed:
(forwards) 100 km/h
(reverse) 18 km/h
Fuel capacity: 248 litres
Max road range: 500-900 km
Fording: amphibious
Gradient: 60%
Side slope: 40%
Vertical obstacle: 0.4 m
Turning radius: <7 m
Engine: Chrysler V-8 water-cooled petrol developing 180 bhp at 4,000 rpm*
Transmission: Allison Transmission Division of General Motors AT-540 automatic with 4 forward and 1 reverse gears
Transfer case: 2 speed
Steering: hydraulically assisted
Clutch: hydraulic, torque converter
Suspension: independent wishbone with helical springs and telescopic dampers
Tyres: 12.00 × 20
Brakes: disc (dual circuit)
Electrical system: 24 V
Batteries: 4 × 12 V
Armament: see text

Armour
Hull front: 12.7 mm
Hull sides: 12.7 mm
Hull top: 9.5 mm
Hull floor: 9.5 mm
Hull rear: 12.7 mm
NBC system: optional
Night vision equipment: optional
* Diesel engine available if required

Manufacturer

Beherman Demoen Engineering (BDX). This company is no longer involved in the design, development or production of armoured vehicles.

UPDATED

BDX fitted with an 81 mm mortar showing the open two-part roof hatch

BDX APC fitted with a retracted nose-mounted dozer blade, shown here in the travelling position

BRAZIL

ENGESA EE-11 Urutu armoured personnel carrier

Development

The EE-11 Urutu armoured personnel carrier was designed by ENGESA from January 1970 with the first prototype completed in July the same year. After trials, a production order was placed for the EE-11 Urutu by the Brazilian Army which calls the vehicle the Carro de Transporte Sobre Rodas Anfibio, or CTRA for short, and production began at São José dos Campos in 1974.

The EE-11 shares many common components with the ENGESA EE-9 (6 × 6) armoured car which was developed from July 1970 and is also in service with the Brazilian and other armies.

The ENGESA company is no longer trading, although large numbers of EE-11 Cascavel armoured personnel carriers and EE-9 armoured cars remain in service in many countries. Final deliveries of the EE-9 were made to Colombia in 1992 together with EE-11 Urutu armoured personnel carriers.

While some elements of the suspension and drive line were built specifically for the EE-9 and EE-11 series of 6 × 6 armoured vehicles, the engines and transmissions were proven commercial subsystems for which spare parts are still available.

Description (Mk V)

The hull is of all-welded steel construction and consists of an outer layer of hard steel and an inner layer of softer steel roll-bonded and heat treated to give the maximum ballistic protection. The armour gives complete protection against small arms fire at point-blank range.

The driver sits at the front of the hull on the left side and is provided with a single-piece hatch cover, with three day periscopes fitted in the glacis plate forward of the hatch. If required, a windscreen can be erected over the driver's position when he is driving head out. The centre periscope can be replaced by a passive night periscope. Steering is power assisted and the steering wheel is adjustable according to the driver's position.

The engine compartment is at the front of the vehicle on the right side. The Detroit Diesel 6V-53T engine is coupled to an Allison Transmission MT-643 automatic transmission. The cooling system comprises a water-to-air radiator for the engine water, oil coolers for the engine and transmission and transfer case oils. Air circulation is forced through the engine compartment by a propeller shaft and driven fan. While swimming, the cooling is performed through a keel cooler with the air inlet and outlet louvres closed. The air inlet louvres are on the hull top to the right of the driver while the exhaust outlets are on the right side of the hull.

The troop compartment is at the rear of the hull and bench seats are provided down each side of the vehicle with an additional bench seat behind the driver's seat across the troop compartment.

There are two doors in the hull, one on the left side immediately behind the front axle and one at the back which is hinged on the left side. The rear door is automatically operated by the driver.

Table 1

Model	I	II	III	IV	V	VI	VII
Engine manufacturer	MB	MB	DD	MB	MB	MB	DD
Engine model	OM-352A	OM-352A	6V-53	OM-352A	OM-352A	OM-352A	6V-53T
Output	170 hp	170 hp	212 hp	170 hp	190 hp	190 hp	260 hp
Transmission	manual	manual MB 63/40	automatic Allison MT-643	automatic Allison AT-540	automatic Allison AT-545	automatic Allison MT-643	automatic Allison MT-643
Tyre-pressure regulation system	no	yes	yes	yes	yes	yes	yes

Note: All are diesel engines, MB being Mercedes-Benz and DD being Detroit Diesel; Mk VII also has epicycloidal reduction on wheels.

Two arrangements of firing ports and vision blocks are possible: five ball mount firing ports, two in each side of the hull and one in the rear door, and two glass bullet- and splinter-proof vision blocks in the side door; 11 firing ports, five on each side of the hull and one in the rear door, and five bullet- and splinter-proof vision blocks, one in the left door, one on the left side, two on the right side of the hull and one in the rear door. Four roof hatches over the troop compartment open towards the outside of the vehicle.

The main armament station is behind the driver, slightly offset to the left of the vehicle's centreline. The basic Model EE-11 Urutu APC is armed with a ring-mounted 12.7 mm (0.50) M2 HB machine gun, but the vehicle could be fitted with many other types of armament installation including a ring-mounted 7.62 mm machine gun, a turret-mounted 60 mm TDA mortar and an ENGESA ET-90 turret armed with an ENGESA EC-90 90 mm gun. ENGESA developed three new turrets that could be fitted to the EE-11 Urutu armoured personnel carrier or the EE-3 Jararaca scout car. Brief details of these are:

ENGESA ET-MD one-man turret armed with a 7.62 mm machine gun and a 12.7 mm M2 HB machine gun, both internally operated

ENGESA ET-20 one-man turret armed with a 20 mm Oerlikon Contraves cannon and a 7.62 mm machine gun

ENGESA ET-25 two-man turret armed with a 25 mm Oerlikon Contraves cannon and a 7.62 mm coaxial machine gun.

The front suspension is independent and consists of double-wishbone arms, coil springs and double-action hydraulic shock-absorbers. The front wheels are driven by hypoid angular transmission and bevel differential gears. For amphibious operations, the front and rear axles can be pressurised. The rear suspension consists of semi-elliptical springs and double-action telescopic shock-absorbers. The articulated rear axle is the ENGESA 'Boomerang' type with walking beams and a differential locking system. The 'Boomerang' axles allow the wheels to travel 0.87 m upwards and downwards and ensure that all wheels are in constant contact with the ground. All tyres are of the run-flat type with an automatic inflation system. Disc brakes and planetary reduction gears on the wheel hubs are provided for all wheels. The two fuel tanks are situated either side of the hull rear and are filled from the top. The electrical system has a double 24 V system, a battery set is used for vehicle operations while a second set is just for engine start to avoid lack of power after long radio watch operations.

The EE-11 Urutu was originally developed for the Brazilian Marines, who required a vehicle capable of operating in heavy surf. This vehicle is propelled in the water by two propellers with kort-nozzles and is steered by two rudders at the rear of the hull which are connected to the steering wheel. Before entering the water, a trim vane is erected pneumatically at the front of the hull by the driver. An air compressor integrated to the engine feeds three air reservoirs and all operating systems and accessories; brakes, transfer case, gear shifting, tyre inflation, rear door and amphibious elements.

It has a maximum water speed of 8 km/h and, in addition to the electric bilge pumps, a hand-operated emergency bilge pump is provided. The Urutu is also equipped with disc brakes and an automatic tyre inflation system.

Optional equipment originally marketed for the EE-11 Urutu included night vision equipment, various radio installations, a winch with a capacity of 5,000 kg and a Mercedes-Benz OM-352A (190 hp SAE) engine with an Allison Transmission MT-643 in place of the standard Detroit Diesel 6V-53N diesel developing 212 hp. The EE-11 Urutu could be supplied fitted with an NBC system.

Variants

ENGESA produced seven distinct models of the EE-11 Urutu, which can be summarised in Table 1, above.

81 mm mortar carrier

Mounted in the hull rear is an 81 mm mortar firing through the roof of the vehicle. A 7.62 mm machine gun is fitted for local protection. The mortar can be elevated from +40 to +80° and if required can be dismounted from the vehicle and used in the ground role.

Ambulance

This model is unarmed, has a higher roof and can carry between six and eight walking wounded, or a smaller number of stretcher cases. Standard equipment includes an air conditioning system, breathing

ENGESA EE-11 Urutu armoured personnel carrier as used by Brazil and armed with 12.7 mm M2 machine gun

Recovery model of ENGESA EE-11 Urutu APC with stabilisers lowered and jib crane extended. The opening in the front of the vehicle is for the hydraulically operated winch which leads to the front of the vehicle

ENGESA EE-11 APC (6 × 6) entering the water (Scott R Gourley)

ENGESA EE-11 Urutu Armoured Fire Support Vehicle with turret-mounted 90 mm gun

Specifications

Mk VII
Crew: 13 (12 + 1)
Configuration: 6 × 6
Combat weight: 14,000 kg
Unloaded weight: 11,000 kg
Power-to-weight ratio: 18.6 hp/t
Length: 6.1 m
Width: 2.65 m
Height:
(to top of 12.7 mm MG mount) 2.9 m
(to hull top) 2.125 m
Ground clearance: 0.38 m
Track: 2.2 m
Wheelbase: 3.05 m
Angle of approach/departure: 60/58°
Max road speed: 105 km/h
Max water speed: 8 km/h
Fuel capacity: 380 litres
Max range:
(road) 850 km
(water) 60 km
Fording: amphibious
Gradient: 60%
Side slope: 30%
Vertical obstacle: 0.6 m
Turning radius: 9.8 m
Engine: Detroit Diesel 6V-53T 6-cylinder water-cooled diesel developing 260 hp at 2,800 rpm
Transmission: Allison Transmission MT-643 automatic with 4 forward and 1 reverse gears
Transfer box: 2 speed
Steering: ZF integral hydraulic
Suspension:
(front) independent with helical springs and double-action telescopic shock-absorbers
(rear) ENGESA 'Boomerang' with walking beams, leaf springs and double-action shock-absorbers
Tyres: 14.5 × 20
Brakes:
(main) air over hydraulic, dual circuit
(parking) mechanical on transfer case output
Electrical system: 24 V
Batteries: 4 × 12 V
Armament: depends on role of vehicle
Smoke-laying equipment: 2 or 3 grenade dischargers either side of turret (optional)
NBC system: optional
Night vision equipment: optional

equipment and a refrigerator for storing medicines. It can also be equipped with four stretchers.

Cargo

When being used for cargo, the EE-11 can carry 2,000 kg.

AFSV

The ENGESA EE-11 Urutu AFSV is an amphibious Armoured Fire Support Vehicle fitted with the ENGESA ET-90 turret armed with the ENGESA EC-90 90 mm gun. This model has a crew of eight: driver, commander, gunner and five infantrymen who are seated at the very rear of the hull. A total of 48 rounds of 90 mm ammunition is carried, 12 rounds of ready use in the turret and 36 rounds in the hull. The turret could be fitted with a power traverse system and a gunner's sight with an integral laser range-finder. The first customer for this was Tunisia. In mid-1988, Venezuela placed an order for the supply of 100 EE-11 Urutu APCs including some in the Armoured Fire Support Vehicle configuration. This turret is the same as that fitted to the ENGESA EE-9 (6 × 6) armoured car.

Command vehicle

Basic vehicle fitted with radios and all the modern requirements for a command post.

Recovery

This model has different roof hatches to permit personnel to work standing upright and is fitted with a front-mounted winch, a hydraulic crane, an oxy-acetylene welding machine, a portable generator, an electric welding transformer, an electric grinder, a complete range of tools and specialised equipment and a tent.

Riot control

The ENGESA EE-11 Urutu in the riot control and internal security operation version has the same basic features as the EE-11 Urutu armoured personnel carrier and retains the same mobility and ballistic protection. Mounted at the front of the vehicle is a blade for clearing obstacles.

Status

Production complete, no longer being marketed. In service with the following countries:

Country	Quantity	Comment
Angola	n/avail	
Bolivia	24	
Brazil	219	army
	5	marines
Chile	30	
Colombia	76	final deliveries 1992
Cyprus	10	
Ecuador	20	
Gabon	12	most have 20 mm turret
Guyana	n/avail	unconfirmed user
Iraq	150	status uncertain
Jordan	30	police role
Libya	100	
Morocco	50	
Nigeria	n/avail	unconfirmed user
Paraguay	10	
Surinam	15	
Tunisia	18	most 90 mm
UAE	120	
Uruguay	10	some 60 mm mortar
Venezuela	60	some 90 mm

Manufacturer

ENGESA Engenheiros Especializados SA. This company is no longer involved in the design, development or production of armoured vehicles.

UPDATED

ENGESA EE-11 Urutu APC fitted with a Swedish Hägglunds Vehicle turret with a 20 mm cannon and a pintle-mounted 7.62 mm machine gun

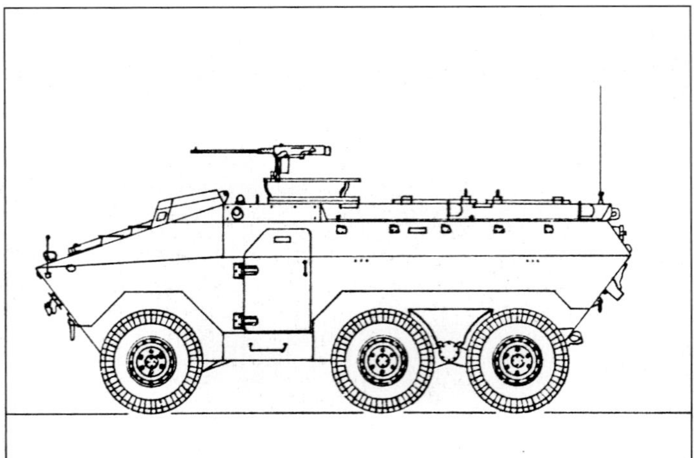

ENGESA EE-11 APC with a 12.7 mm M2 MG

Bernardini AM-IV armoured security vehicle

Development

The Bernardini Model AM-IV armoured security vehicle has been built as a private venture and has been designed mainly for use in the internal security role although its 4 × 4 drive gives it good cross-country mobility. The first prototype was completed in 1988 and, by late 2001, about 60 vehicles had been built for the home and export markets. Bernardini has also been involved in the upgrading of Brazilian Army M113 series APCs and M41 light tanks. Details of the latter upgrade are given in the Light tanks section under the United States. This is no longer being marketed by Bernardini. It is understood that Bernardini is no longer involved in the design, development and production of tracked armoured vehicles.

Description

The layout of the Bernardini AM-IV armoured security vehicle is conventional, with the engine at the front, commander and driver in the centre and four men in the rear of the vehicle. The commander and driver are both provided with a forward-opening side door while the rear door is electrically operated by the driver. The side and rear windows are provided with wire mesh protection while the driver's and commander's windows can be covered by a shutter, hinged at the top. There is also a roof hatch enabling heavier weapons to be used from within the vehicle.

The all-welded steel armoured hull provides protection against 7.62 mm and 9 mm small arms fire with the windows providing the same level of protection as the hull. The interior of the vehicle is insulated.

Bernardini AM-IV armoured security vehicle with all doors closed

There are nine firing ports provided in the front, side doors, hull sides and rear hull door allowing the occupants to use their small arms in safety from within the vehicle.

Early production versions of this vehicle were fitted with a chemical filter for clean air.

Standard equipment includes an air conditioning system, a fixed Halon fire suppression system in the engine compartment and wheel arches, three-point internal seat belts for all six crew members, flashing lights on the roof and a searchlight mounted well above the roof that can be controlled from inside the vehicle by the commander or driver.

Some vehicles have recently been observed with a bank of electrically operated grenade dischargers mounted on the roof of the vehicle either side at the rear.

Status

Production as required. In service with Brazil (police) and Chile (police).

Manufacturer

Bernardini S/A Indústria e Comércio.

UPDATED

Specifications

Crew: 2 + 4
Configuration: 4 × 4
Weight: (loaded) 3,370 kg
Payload: 600 kg
Power-to-weight ratio: 27.89 hp/t
Length: 3.8 m
Width: 1.8 m
Height: 2.4 m
Wheelbase: 2.285 m
Max road speed: 100 km/h
Fuel capacity: 102 litres (2 tanks)
Max range: 860 km
Gradient: 60%
Turning radius: 5.3 m
Engine: Mercedes-Benz OM-364 4-cylinder diesel developing 94 hp at 2,800 rpm
Transmission: manual, 5 forward and 1 reverse gears
Transfer case: 2 speed (high and low)
Clutch: hydraulically assisted, single dry disc
Axles: floating type front and rear
Steering: power assisted with recirculating balls
Brakes:
(main) front, ventilated disc; rear, drum
(parking) drum
Suspension: front and rear, leaf springs, rolling bars and hydraulic shock-absorbers
Tyres: 7.50 × 16 (bulletproof radial)
Electrical system: 12 V
Batteries: 2 × 12 V, 60 Ah
Alternator: 55 A
NBC system no
Night vision equipment: no

CANADA

General Motors Defense, LAV-III armoured personnel carrier

Development

In January 1997, the Diesel Division, General Motors of Canada (now General Motors Defense), was awarded a contract worth C$552 million for the supply of 240 new LAV-III 8 × 8 armoured personnel carriers for the Canadian Forces.

Production of the new LAV-III is undertaken in London, Ontario, with first vehicles rolled out in September 1998.

Since then a further three options have been exercised which brings the grand total up to 651 vehicles worth a total of C$2.043 billion. These were fielded from March 2000.

A major subcontractor to General Motors Defense is Delco Defense of the USA which will supply LAV-25 two-man power-operated turrets for these vehicles.

The Delco Defense LAV-25 two-man turret will be similar to previous turrets of this type but feature a new all-electric turret drive system in place of the previous electrohydraulic system used on earlier US Marine Corps' vehicles.

In addition, the turret has a Raytheon Systems Company thermal imaging system, eye-safe laser range-finder, laser warning receiver and Improved M36 and related vision devices.

The turret is armed with a US The Boeing Company 25 mm M242 Chain Gun with a 7.62 mm C6 coaxial machine gun and a 5.56 mm C9 machine gun on the turret roof for air and local defence. Mounted on either

Canadian Armed Forces LAV-III armoured personnel carrier **2001**/0103751

side of the turret is a bank of four 76 mm electrically operated grenade launchers.

The baseline for the new Canadian APC is the new Swiss MOWAG Generation III Piranha (8 × 8) vehicle which was developed in Switzerland from 1994-95 and unveiled for the first time early in 1996.

Following trials in Switzerland, the prototype Generation III Piranha was sent to Canada in 1996 where it was evaluated by the Canadian Forces.

Compared to earlier MOWAG Piranha (8 × 8) vehicles, the new Generation III has a higher level of mobility and armour protection with the model selected

Specifications
Crew: 3 + 8
Configuration: 8 × 8
Combat weight: 16,329 kg
Unloaded weight: 14,243 kg
Power-to-weight ratio: 21.43 hp/t
Length: 6.934 m
Width: 2.667 m
Height: (hull top) 1.955 m
Ground clearance: 0.595 m
Max road speed: 99.77 km/h
Range: 500 km
Fuel capacity: 300 litres
Gradient: 60%
Side slope: 30%
Vertical obstacle: 0.6 m
Trench: 1.981 m
Engine: Caterpillar 3126B diesel developing 350 hp
Transmission: Allison MD 3066 P automatic, 6 forward and 1 reverse gears
Transfer case: 2 speed
Steering: power assisted on front 2 axles
Turning radius: 8 m
Suspension: independent hydropneumatic with height control
Brakes: power brakes
Tyres: 1200 R20 with central tyre inflation system
Electrical system: 24 V
Batteries: 4
Armament:
(main) 1 × 25 mm M242 cannon
(coaxial) 1 × 7.62 mm C6 MG
(anti-aircraft) 1 × 5.56 mm C9 MG
(smoke grenade launchers) 2 × 4 76 mm
Ammunition:
(main) 210 rounds ready + 210 stowed
(coaxial) 440 rounds ready + 1,300 stowed
(anti-aircraft) 3,200 rounds (total)
(smoke grenade) 8 + 8
Gun control equipment
Turret power control: electric/manual
Elevation/depression: +60/–10°
Turret traverse: 360°
Gun stabiliser:
(vertical) yes
(horizontal) yes
NBC system: yes
Night vision equipment: yes

by Canada being powered by a Caterpillar 3126B diesel coupled to an Allison MD 3066 P fully automatic six-speed transmission. All previous Canadian vehicles have been powered by a Detroit Diesel 6V-53T diesel developing 275 hp.

The new APC is similar to the LAV-25 (Reconnaissance), or Coyote, of which 203 have been delivered to the Canadian Forces. This uses the earlier Generation II Piranha (8 × 8) chassis with the all-electric LAV-25 turret.

The LAV-III will be used for the troop transport role with a total of eight troops being carried in the rear troop compartment, seated four down each side facing each other.

Other equipment fitted as standard includes a 500 A slave receptacle, automatic fire detection and suppression systems in the engine and crew compartments, a climate control system consisting of an air conditioning/heating system, ATGW wire cutter for the driver and the turret stations and a self-recovery winch.

The LAV-III is fully air-transportable in the C-130, C-141, C-5A and C-17 transport aircraft.

Canadian Armed Forces LAV-III armoured personnel carrier **2000**/0080350

Description
As for Piranha Generation III covered later in this section.

Variants

Specialised variants
The final contract for 171 vehicles includes 71 TOW Under Armour, 39 pioneer, 47 forward observation officer/mortar fire controller vehicles and 14 infantry carriers. The TOW Under Armour turrets are being removed from M113s, refurbished and then installed on the LAV-III vehicles.

105 mm LPT
To replace its Cougar (6 × 6) vehicles armed with a 76 mm gun and eventually its 105 mm armed Leopard 1 MBTs, the Canadian Forces have a requirement for a new vehicle called the Armoured Combat Vehicle (ACV).

Details of this programme are given in the Reconnaissance Vehicles section under Canada.

One of the contenders for this requirement is the latest LAV III (8 × 8) chassis fitted with the General Dynamics Land Systems 105 mm armed Low-Profile Turret system. There is a separate entry for this vehicle in the Reconnaissance vehicles section under Canada.

US Army LAV-III
Late in 2000, the US Army selected the LAV-III (8 × 8) to meet its Brigade Combat Team (BCT) requirement. Additional details are given in this section under the United States.

New Zealand
New Zealand is purchasing 105 LAV-III from General Motors Defense to replace its current M113 series vehicles. Total projected cost is $611.764 million. These will be issued to two battalions of the New Zealand Army.

Status
In service or on order:

Country	Quantity	Comment
Canada	651	delivery from 1998
New Zealand	105	selected 2000
USA	366	selected late 2000, this is the first order
USA	32	on loan from Canada

Manufacturer
General Motors Defense.

UPDATED

General Motors Defense, Bison armoured personnel carrier (8 × 8)

Development
In 1988, the Diesel Division, General Motors of Canada (which in 2000 became General Motors Defense) designed and built within seven days, as a private venture, a new 8 × 8 armoured personnel carrier based on the chassis of the LAV (8 × 8) vehicle described in the previous entry.

The turret of the LAV was removed and a raised commander's cupola with five M17 day periscopes and an externally mounted 7.62 mm machine gun was fitted to the immediate rear of the driver's station at the front left side of the hull.

The troop compartment had a raised roof that ran to the rear of the vehicle with roof hatches and a power-operated ramp rather than doors as on the US Marine Corps vehicles. No firing ports were provided although two vision blocks were mounted in the rear ramp.

In addition to the commander and driver, the vehicle carried eight fully equipped troops, seated four down each side facing each other.

Standard equipment included a front-mounted winch, while optional equipment included a fire detection and suppression system, a wire cutter, an NBC defence system and night vision devices.

Following trials with the prototype vehicle in July 1989, the Canadian Department of National Defence awarded the now General Motors Defense, a contract worth C$100 million to build a total of 199 vehicles to a slightly modified design for the Canadian Militia.

The full Canadian designation of the Bison is Militia Light Armoured Vehicle (MILLAV).

First production vehicles were delivered in the Summer of 1989 with deliveries running through to 1993. The Canadian Forces call the vehicle the Bison with the 199 vehicles consisting of the following versions:

149 APCs
18 command post vehicles
16 81 mm mortar carriers
16 maintenance repair vehicles with the same crane as the Husky 6 × 6 vehicle already in service with the Canadian Forces. This can carry a complete power pack internally.

The Bison can be quickly adapted for specialised roles as the baseline vehicle has a rail system fitted as standard.

In addition, four vehicles for trials were loaned to the Royal Australian Army, which has since placed an order for a total of 111 vehicles in six different configurations. All of these have now been delivered to Australia where they were fitted out by British Aerospace (Australia). The armoured fighting vehicle business of British Aerospace (Australia) has now been taken over by the Australian company Tenix Defence Systems.

Canada loaned 12 vehicles to the US National Guard to support the US Drug Enforcement Agency. Since the start of this operation, the vehicles have taken part in several counter-narcotic operations. The evaluation was concluded in December 1992.

Late in 1994, an order was placed for 12 vehicles which were delivered in 1995 to the US National Guard.

The Canadian Army has deployed its Bison (8 × 8) vehicles and Armoured Vehicle General Purpose (6 × 6) vehicles to the former Yugoslavia for use in peacekeeping operations.

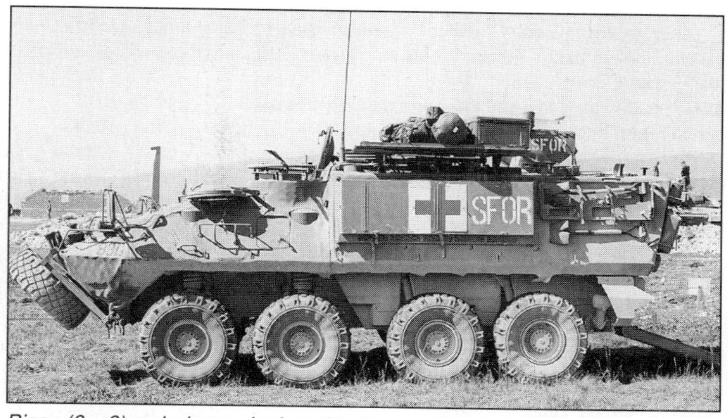

Bison (8 × 8) ambulance deployed to Bosnia (Richard Stickland) ***2000***/0080352

Bison 81 mm mortar carrier with roof hatches open, as deployed to Bosnia (Richard Stickland)
2001/0103752

Description

The basic automotive layout of the Bison APC is similar to the LAV (8 × 8) with a hull of all-welded steel armour construction providing protection from small arms fire and shell splinters.

The driver is seated at the front left with the engine compartment to his right. The driver has a single-piece hatch cover that opens to the left and three M17 day periscopes, the centre one of which can be replaced by an AN/VVS-2(V)4 passive night vision device.

The commander is seated to the rear of the driver and has a single-piece hatch cover that opens to the rear and five M17 day periscopes, the centre one of which can be replaced by an AN/VVS-2(V)4 passive night periscope. A 7.62 mm C6 MG is mounted on a skate-type mount at the commander's station.

The troop compartment is at the rear of the hull with the troops entering and leaving via a large power-operated ramp that also has a door and two vision blocks.

The infantrymen sit on bench-type seats down either side of the hull with stowage provided under their seats and to their immediate rear. There is an additional seat to the immediate rear of the engine compartment bulkhead.

Over the forward part of the roof compartment is a three-part roof hatch which opens left and right of the vehicle centreline. On the roof at the rear is a large stowage box with stowage racks being provided externally on either side of the vehicle.

Mounted on the top either side of the hull rear is a bank of M257 four-barrelled smoke grenade dischargers that cover the frontal arc of the vehicle.

Standard equipment includes a Halon 1301 fire suppression system in the engine and crew compartments, a front-mounted hydraulic winch with a capacity of 6,800 kg, an M8A1 ventilated face mask NBC system, a wire cutter forward of the driver's position and complete communications installation.

Specifications

Crew: 2 + 8
Configuration: 8 × 8
Combat weight: 12,936 kg
Curb weight: 11,072 kg
Payload: 1,864 kg
Power-to-weight ratio: 21.25 hp/t
Length: 6.452 m
Width: 2.5 m
Height: 2.21 m
Max road speed: 100 km/h
Max water speed: 9.7 km/h
Max range: 665 km
Gradient: 60%
Side slope: 30%
Vertical obstacle: 0.381-0.508 m
Trench: 2.06 m
Engine: Detroit Diesel, 6V-53T, 6-cylinder diesel developing 275 hp at 2,800 rpm
Transmission: Allison Transmission MT-653 DR automatic, 5 forward and 1 reverse gears
Transfer case: Rockwell AG-VST (modified)
Steering: power assisted on front 2 axles
Turning radius: 7.75 m
Suspension:
(front four wheels) independent coil springs and shock-absorbers
(rear four wheels) independent torsion bars and shock-absorbers
Brakes:
(main) 8-wheel dual air hydraulic drum brake
(parking) transmission brake and transfer case lock
Tyres: 11.00 × 16 with Hutchinson run-flat inserts
Electrical system: 24 V
Batteries: 4
Armament: 1 × 7.62 mm C6 MG
Smoke grenade dischargers: 2 × 4
Gun stabiliser:
(vertical) no
(horizontal) no
NBC system: yes
Night vision equipment: yes

The Bison is fully amphibious with two minutes' preparation, being propelled in the water by two propellers mounted under the rear of the hull; to the rear of the propellers are two rudders. As the Bison vehicles go through their upgrade programme the amphibious capability is being removed.

A number of Bison vehicles will be fitted with a new wider tyre developed by Michelin (the 325/85 R15 XML). Trials have shown a significant improvement in mobility.

Some of the vehicles deployed to the former Yugoslavia have been fitted with additional passive armour protection.

Bison (8 × 8) armoured personnel carrier deployed to Balkans showing additional passive armour and extensive external stowage (Michael Jerchel) ***2002***/0121145

81 mm mortar fire support vehicle

This is armed with a turntable-mounted 81 mm mortar for which a total of 90 mortar bombs is carried. It has a four-person crew consisting of driver, vehicle commander and two mortar crew. The vehicle is equipped with an 81 mm C5 mortar. In 1995, these vehicles received new roof hatches and the 81 mm mortar can now be fired from within the vehicle.

Command post

This has a five-person crew consisting of driver, vehicle commander and three radio operators and is fitted with extensive communications equipment.

Mobile repair team vehicle

Two-person crew consisting of driver and vehicle commander/rigger. Standard equipment includes a boom crane with a traverse of 265° and a maximum payload of 4,982 kg which can be controlled remotely and is also fitted with an overload protection system, two outriggers, two stabilisers, floodlights, towbars, a spare wheel and other specialised equipment.

Future conversions

With the introduction of the new LAV-III (8 × 8) armoured personnel carrier, the Canadian Forces are now to re-role their existing fleet of 199 Bison (8 × 8) vehicles. In the future the Bison fleet will consist of:

- 81 mm mortar carrier (60)
- Command post vehicle (74)
- Mobile Repair Team Vehicle (16)
- Electronic Warfare (14)
- Maintenance and recovery (35).

The current ambulance version of Bison will cease to exist. Improvements to be carried out to the Bison fleet include an increase in engine power to 300 hp, radiator stack from the General Motors Defense Coyote Reconnaissance vehicle fitted, command post vehicles and electronic warfare variants will be fitted with air conditioning system as standard, remainder of the fleet will be fitted for but not with air conditioning, stowage boxes will be added and the marine drives will be removed.

Other improvements will include the installation of heavier torsion bars, ventilated respirator system, IBD passive armour kit, ballistic crew seats and a Protected Weapon Station (PWS).The first upgraded vehicles will be delivered back to the Canadian Army in 2003.

Following a competition, Oerlikon Aerospace was selected, in mid-2000, to provide the PWS, which it has developed in association with the RAFAEL Armament Development Authority of Israel.

The PWS is armed with a 7.62 mm C6 machine gun which can also be aimed and fired by remote control and will go on the following vehicles:

Mobile Tactical Vehicle Light and some of the M113A3 series vehicles (83)
LAV-III (8 × 8) pioneer (39)
Bison (8 ×8) 60
AVGP (6 × 6) 40.

Prior to the above conversion programme, a number of the 149 Bison APCs were converted for a number of specialised roles. These include the following:
five to engineer vehicles
one or two to control vehicles for the Jingoss mine searching system
one to TROLIP with Raytheon TOW ATGW
23 to ambulance
two to electronic warfare.

Status
Production as required. In service with Australia (33 under designation ASLAV-PC), Canada (199 in service) and US Army National Guard (12 delivered 1995).

Manufacturer
General Motors Defense.

UPDATED

General Motors Defense, Armoured Vehicle General Purpose (6 × 6)

Development
In June 1974, the Canadian Forces issued a requirement for an Armoured Vehicle General Purpose (AVGP) for use in Canada. After an initial study of 14 possible vehicles three were selected for field and engineering tests, as a result of which the Swiss MOWAG Piranha (6 × 6) vehicle was chosen.

In February 1977, the Canadian Minister of Defence signed a contract worth C$171 million with the Diesel Division, General Motors of Canada Limited (which in 2000 became General Motors Defense), for the supply of 350 vehicles manufactured under licence from MOWAG. The first was handed over in January 1979. The original order for 350 vehicles was subsequently increased to 441 units owing to lower production costs and, early in 1981, was again extended by 35 units bringing the total to 476 units. A further extension of 15 units brought the total order to 491 vehicles. Delivery of this contract was completed by September 1982.

The Canadian vehicles are used by both regular and militia units; they have heaters for the engine and crew compartment and are fitted with cold-starting aids. All vehicles have an AN/VVS-501 passive night driving periscope for the driver. The Canadians have three versions of the Piranha: Cougar; Grizzly; Husky.

For operations in Bosnia some of these vehicles have been fitted with the US LAST appliqué armour system which has been fitted to some US Marine Corps LAV-25 vehicles and is covered in detail in the earlier LAV-25 entry.

Additional details of the MOWAG Piranha family of 4 × 4, 6 × 6 and 8 × 8 vehicles are given in this section under Switzerland. The MOWAG company has now been taken over by the General Motors Defense.

Specifications
Grizzly WAPC
Crew: 3 + 6
Combat weight: 10,500 kg
Power-to-weight ratio: 20.46 hp/t
Length: (hull) 5.968 m
Width: 2.53 m
Height:
(overall) 2.53 m
(hull roof) 1.85 m
Ground clearance: 0.392 m
Max road speed:
(forward) 101.5 km/h
(reverse) 17.9 km/h
Max water speed: 7 km/h
Fuel capacity: 204 litres
Max road range: 603 km
Gradient: 60%
Side slope: 30%
Vertical obstacle: 0.381-0.508 m
Trench: 0.406 m
Turning radius: 6.55 m
Engine: Detroit Diesel 6V-53T, 6-cylinder diesel developing 215 hp
Transmission: Allison Transmission MT-650 automatic, 5 forward and 1 reverse gears
Steering: power assisted, front wheels
Suspension: independent, all wheels
Tyres: 11.00 × 16 with Hutchinson run-flat inserts
Armament:
(main) 1 × 12.7 mm M2 HB MG
(coaxial) 1 × 7.62 mm C5A1 GPMG
(smoke grenade dischargers) 2 × 4 No 12 Mk 1
Ammunition:
(main) 1,000
(coaxial) 4,400
Gun control equipment
Turret power control: electric/manual
(by commander) no
(by gunner) yes
Gun elevation/depression: +55/–8°
Turret traverse: 360°
Gun stabiliser:
(vertical) no
(horizontal) no
Turret vision devices: 1 × M28C sight, 8 vision blocks
Armour
Hull front upper: 8 mm at 23°
Hull front lower: 8 mm at 45°
Hull sides: 10 mm at 55°
Turret: 9.5 mm
NBC system: no
Night vision equipment: yes

Cougar 76 mm gun Wheeled Fire Support Vehicle (WFSV)
The Cougar, 195 of which were built, is the basic vehicle fitted with the turret mounted on the British Alvis Vehicles CVR (T) Scorpion. The turret is armed with a 76 mm L23A1 gun with a 7.62 mm machine gun mounted coaxially and four smoke grenade dischargers mounted on either side of the turret. The weapons have an elevation of +35° and depression of –8°, with 360° powered turret traverse. There are 10 rounds of 76 mm and 220 rounds of 7.62 mm machine gun ammunition carried in the turret, with a further 30 rounds of 76 mm and 15 boxes of 7.62 mm ammunition in reserve in the hull. The Cougar has a crew of three: commander, gunner and driver. It carries an AN/GVS-5 hand-held laser range-finder.

The UK has withdrawn the 76 mm armed Scorpion and the Cougars are being converted to more specialised roles.

Grizzly Wheeled Armoured Personnel Carrier (WAPC)
The Grizzly APC, of which 269 were built, is the basic vehicle fitted with a Cadillac Gage (now Textron Marine & Land Systems) 1 m turret armed with a 12.7 mm machine gun and a 7.62 mm machine gun, with four smoke grenade dischargers mounted on either side of the turret. The weapons have an elevation of +55° and a depression of –8°, turret traverse being a full 360°. Ten boxes of 12.7 mm and 18 boxes of 7.62 mm machine gun ammunition are carried. The Grizzly WAPC has a crew of three and carries eight fully equipped infantrymen. The vehicle has firing ports and vision blocks in the hull sides and rear.

In 1991, the Canadian Forces awarded a contract to the now General Motors Defense, to upgrade the suspension of their current 6 × 6 fleet to the heavier 8 × 8 suspension. This upgrade was completed by 1994.

Husky Wheeled Maintenance and Recovery Vehicle (WMRV)
There were 27 of these built for field repair and recovery work. The Husky has a crew of three and is fitted with an HIAB 650A crane with a capacity of 4,536 kg. Full details of this version are given in *Jane's Military Vehicles and Logistics 2001-2002*.

Life extension of AVGP
The Canadian Armed Forces are to re-role some of their fleet of Grizzly AVGP into the following more specialised versions:
Ambulance (81)

Wheeled Armoured Personnel Carrier Grizzly deployed to Bosnia (Michael Jerchel) ***2001***/0103753

Wheeled Fire Support Vehicle Cougar deployed to Bosnia with appliqué armour (Michael Jerchel) ***2001***/0103754

Javelin surface-to-air missile carrier (23)
Radio Rebroadcast Vehicle (10)
Surface Plough Mine Disruption System (6)
Mobile Repair Team Vehicle (23)
Personnel carrier (91)

Following a competition, Oerlikon Aerospace was selected in mid-2000 to provide the Protected Weapon System (PWS) to meet the requirements of the Canadian Armed Forces. This has been developed by the company in association with the RAFAEL Armament Development Authority of Israel.

The PWS is armed with a 7.62 mm C6 machine gun which can be aimed and fired by remote control, and will go on the following vehicles:

Mobile Tactical Vehicle Light and some of the M113A3 series vehicles (83)
LAV-III (8 ×8) pioneer (39)
Bison (8 × 8) 60
AVGP (6 × 6) 40.

In early 2001 it was stated that of the 269 Grizzly vehicles built a total of 247 would be upgraded, although only 234 were, according to the figures above. First upgraded vehicles are expected to be returned to service in 2003.

Each of the above vehicle types will receive role specific modifications. In some cases this will include extending the Grizzly chassis to increase internal volume.

In addition, the upgrade will include the same wheel, steering and torsion bar suspension system of the latest Canadian Army General Motors Defense LAV-III (8 × 8) series vehicles which are now in service.

These upgraded vehicles will be able to be fitted with the modular appliqué armour kits developed by DEW Defense Technologies in Canada using licensed technology from IBD of Germany. In addition, improved smoke grenade launchers will be installed.

Status
Production complete. In service only with Canadian Forces.

Manufacturer
General Motors Defense.

UPDATED

General Motors Defense, light armoured vehicle (8 × 8)

Development
In September 1981, the Diesel Division, General Motors of Canada (which in 2000 became General Motors Defense), was awarded a US$3.1 million contract by the United States Marine Corps to supply four 8 × 8 models of the Swiss MOWAG Piranha vehicle, and supporting services, for use in competitive evaluation in the United States Marines Light Armored Vehicle (LAV) programme.

By October 1981, the company had supplied two Light Assault Vehicles fitted with a two-man Arrowpointe turret armed with a 25 mm The Boeing Company 25 mm M242 Chain Gun, a 7.62 mm M240 coaxial machine gun and eight smoke grenade launchers, and one Assault Gun Vehicle fitted with an Arrowpointe two-man turret armed with a 90 mm Cockerill Mk III gun, a 7.62 mm M240 coaxial machine gun and eight smoke grenade launchers. The second Assault Gun Vehicle was delivered early in 1982.

In September 1982, it was announced that the then Diesel Division, General Motors of Canada, had won the LAV competition. The original five year contract was to have been for 969 LAVs, 680 for the US Army and 289 for the US Marine Corps with a total value of US$477.8 million, and an option for a further 598 vehicles. The first buy, with FY82 funding, was for 60 vehicles, all of which were for the US Marine Corps. The first of these was handed over in October 1983. The second year buy, with FY83 funding, was for 170 vehicles, 134 of which were for the US Marine Corps and the remaining 36 for the army. Early in 1984, the army withdrew from the programme and its vehicles were taken over by the US Marine Corps.

At the same time as the original order was placed, the company was awarded a US$21.6 millon contract for the development and testing of five specialised versions: 81 mm mortar carrier, logistic support, command and control, maintenance/recovery and anti-tank fitted with an Electronics and Space Corporation (now Systems & Electronics Inc (SEI) twin launcher as on the M901 Improved TOW Vehicle (ITV). All these vehicles had been delivered by early 1984.

The quantities of LAV-25s procured by the US Marine Corps between FY82 and FY85 were as follows:

	FY82	FY83	FY84	FY85	Total
LAV-25	60	170	123	69	422
LAV(R)	-	-	33	13	46
LAV(L)	-	-	80	14	94
LAV(M)	-	-	-	50	50
LAV(C)	-	-	-	50	50
LAV(AT)	-	-	-	96	96
Total	60	170	236	292	758

Final deliveries of the LAV were made to the US Marine Corps in 1988.

Late in 1988, it was announced that LAV battalions would in future be called Light Armored Infantry Battalions with each battalion having five companies. Three companies operate 14 LAV-25 and two LAV command vehicles. The weapons company has 16 LAV anti-tank, eight LAV 81 mm mortars and eight LAV-25s for reconnaissance. The HQ and service company have one LAV recovery vehicle, three LAV logistic vehicles and two LAV command vehicles.

US Marine Corps LAVs were used during the USA's invasion of Panama in 1989 by the 2nd Light Armored Infantry Battalion. The US Marines also used them during Operation Desert Storm, the recapture of Kuwait, in early 1991 and in Somalia in 1992 during Operation 'Restore Hope'. The US Marine Corps has loaned 15 LAV-25s and a recovery vehicle to the US Army's 82nd Airborne Division. Two of these vehicles were modified for paradrop certification.

The LAV is air-deployable in current US Air Force cargo transports. The C-5A can carry eight, the C-17 six, the C-141 two and the C-130 one. The US Marine Corps CH-53E transport helicopter can transport one in the sling position.

In February 1987, an LAV-25 was successfully Low Altitude Parachute Extraction System (LAPES) deployed from a C-130 transport aircraft and was operationally ready within 12½ minutes of landing. The vehicle was also air-dropped from 1,500 ft and was again operationally ready within 12½ minutes.

Description
The LAV (8 × 8) is based on the Swiss MOWAG Piranha family of vehicles described in detail under Switzerland and the overall layout and description is virtually the same. All US Marine Corps vehicles are fully amphibious, being propelled in the water by two propellers mounted one either side at the rear.

Variants

LAV-25(MC)
This is for the US Marine Corps as are all the following variants. The US Army designation for this vehicle was the M1047: this was very similar to the Marine Corps version but had increased ammunition stowage as it was not intended to carry troops.

The LAV-25(MC) has a Delco Defense two-man turret armed with a 25 mm The Boeing Company M242 Chain Gun, a 7.62 mm M240 coaxial machine gun and a pintle mount for an M60 7.62 mm machine gun. Mounted either side of the turret is a bank of M257 smoke grenade launchers. A stabilisation system allows the 25 mm M242 cannon to be aimed while the vehicle is moving across country.

Up-armoured LAV for Operation Desert Storm
In April 1991, the US Defense Advanced Research Projects Agency (DARPA) announced that it had successfully completed a project to take a technology from proof-of-principle concept to fieldable prototype system within three months.

The objective of this US$2 million effort was to provide 75 prototype kits of Light Appliqué System Technique (LAST) armour to the Marine Corps for application to LAVs deployed to Saudi Arabia by the US Army and Marine Corps.

The LAST is an attachment system for appliqué armour developed by Foster-Miller Inc under contract to DARPA. It consists of a hook-and-loop attachment system (from Velcro USA), armour modules (from Lanxide Armor Products and Coors Ceramics) and a protective cover (Bell Avon).

In the three-month effort, in which an extensive list of technical and operational issues were addressed and overcome, the most challenging was the creation of an industrial base that produced over 60,000 ceramic armour tiles in less than 90 days. In addition, the armour tiles also had to undergo ballistic validation testing to verify tile performance. The LAST system underwent development testing and an LAV equipped with a LAST armour kit underwent live fire testing and was tested against chemical agents, all during the 90 day effort.

While the timing of the ground war in the Gulf made it impossible to field prototype LAST systems in support of Operation Desert Storm as planned, the Marine Corps decided to equip marine units on the US east and west coasts with LAST kits. It is expected that this armour will now be replaced by the new RAFAEL armour system described below.

RAFAEL armour kit for LAV
Late in 1997, following a competition between Foster-Miller and Simula of the USA and the Ordnance Systems Division of RAFAEL, Israel, the latter company was awarded a contract worth US$7.1 million to provide appliqué passive armour kits for the LAV-25 (8 × 8) light armoured vehicles used by the US Marine Corps.

This RAFAEL passive armour system is known as the Composite Ceramic Armour (CCA) and is fitted to the hull and turret of the LAV-25 to provide a higher level of protection against Kinetic Energy (KE) attack.

The lightweight ceramic tiles can be quickly fitted or removed from the LAV-25 and protrude less than 40 mm from the hull of the vehicle. Any damaged tile can be quickly replaced. They can defeat threats of up to 20 mm armour piercing at 15 m and 12.7 mm armour piercing at 100 m.

According to RAFAEL, its CCA system was the only one of the three systems tested that passed the 100 per cent environmental and functional tests conducted in the USA. First CCA systems were delivered to the USA in 1998.

Additional LAVs
The USMC has a requirement for a further 400 LAVs, 200 of which could be APCs (similar to the Canadian Bison) with the remainder being specialised variants similar to existing vehicles.

Future possible LAV improvements
As the LAV series will remain in service well into the 21st century, a number of improvements to the family are already under way including the Mobility Block Improvement Programme which converts the current 'green' Detroit Diesel 6V-53T engine to the latest 'silver' 6V-53T standard.

The conversion is carried out through a remanufacture process with only those parts necessary to change the configuration from the green to the silver being replaced. First conversions have already been made and full operational capability was reached in 1999.

The first order was for a total of 463 remanufactured engines with the second buy expected to be for 462 engines which would allow the entire US Marine Corps LAV fleet to be refitted.

Upgrades to the engine include the oil cooler core, blower, governor, oil pump, crank, pulleys, rocker cover and hardware as well as an improved extraction case for the engine.

The LAV-25 currently has two M36E1 day/image intensification sights, one for the commander and one for the gunner. The latter is now being replaced by a thermal sight which will allow targets to be acquired and hit under poor weather conditions.

LAV SLEP
In mid-2000, following a competition, the US Marine Corps selected Metric Systems of Fort Walton Beach, Florida, to perform its SLEP (Service Life Extension System) for its LAV vehicles.

The first phase, covering research and development, was for US$4.4 million and covered the delivery of five prototypes SLEP LAV by April 2001.

LAV 81 mm mortar carrier vehicle with roof hatches closed

US Marine Corps Light Armored Recovery Vehicle in travelling configuration

It is expected that production will commence at the end of FY2002 with the research and development phase, which also covers production options, worth up to US$60 million.

This SLEP is part of the LAV Product Improvement Programme which is expected to cost US$287 million through FY2005 and enable the force of some 772 vehicles to remain in service through 2015. Other elements include upgrading its command and control equipment.

LAV-25 gun upgrade

In mid-2001 it was revealed that US Marine Corps planners were investigating possible upgrade packages that would increase the lethality of the current LAV-25 armed with The Boeing Company 25 mm M242 cannon.

Although a brand new turret could be installed on the existing 8 × 8 chassis, consideration is also being given to replacing the current 25 mm M242 cannon with a new 30 mm cannon for which there are a number of candidates. Contractors were required to submit their Rough Order of Magnitude (ROM) cost estimates for the upgrade of 408 vehicles in July 2001.

LAV HYDRA rocket launcher

In 1986, an LAV-25 was fitted with a 19-round HYDRA 70 70 mm (2.75 in) rocket launcher on the right side of the turret with powered elevation and depression (-200 to +600 mils). Approximately 100 unguided rockets were successfully fired during trials in July 1986. The rockets are identical to those fired from aircraft and helicopters and would be used in both the surface and anti-helicopter roles. This was a trials installation and there are no plans to fit this to production vehicles.

Logistics vehicle

This has a higher roof with twin hatches for the rapid loading and unloading of cargo, a crew of three consisting of the commander, driver and loadmaster, and a crane for loading and unloading cargo at the left hull rear. Armament consists of a pintle mount with a 7.62 mm M60 machine gun and two four-barrelled M257 smoke grenade launchers. Production of this version began in November 1985 and was completed in August 1986.

Specifications

LAV-25
Crew: 3 + 6
Configuration: 8 × 8
Combat weight: 12,792 kg
Unloaded weight: 10,932 kg
Power-to-weight ratio: 21.49 hp/t
Length: 6.393 m
Width: 2.499 m
Height:
(overall) 2.692 m
(reduced) 2.565 m
Max road speed: 100 km/h
Max water speed: 9.656 km/h
Max range: 668 km
Gradient: 60%
Side slope: 30%
Vertical obstacle: 0.5 m
Trench: 2.057 m
Engine: Detroit Diesel 6V-53T, 6-cylinder diesel developing 275 hp at 2,800 rpm
Transmission: Allison Transmission MT-653 automatic, 5 forward and 1 reverse gears
Transfer case: Rockwell AG-VST (modified)
Steering: power assisted on front 2 axles
Suspension:
(front four wheels) independent coil springs and shock-absorbers
(rear four wheels) independent torsion bars and shock-absorbers
Brakes:
(main) 8-wheel dual air hydraulic drum brake
(parking) transmission brake and transfer case lock
Tyres: 11.00 × 16 with Hutchinson run-flat inserts
Armament:
(main) 1 × 25 mm M242 cannon
(coaxial) 1 × 7.62 mm M240 MG
(anti-aircraft, optional) 1 × 7.62 mm M60 or 1 × 12.7 mm M2 HB MG
(smoke grenade launchers) 2 × 4 (M257)
Ammunition:
(25 mm) 210 + 420 (stowed)
(7.62 mm) 420 + 1,200 (stowed)
Gun control equipment
Turret power control: hydraulic/manual
Elevation/depression: +60/-10°
Turret traverse: 360°
Gun stabiliser:
(vertical) yes
(horizontal) yes
Periscopes: 7 × M27
Sights: 2 × M36E1
Gun stabiliser:
(vertical) yes
(horizontal) yes
NBC system: yes
Night vision equipment: yes

81 mm mortar carrier

This has a crew of five consisting of commander, driver and three mortar men and is fitted with an 81 mm mortar (94 rounds carried). The mortar is mounted in the centre of the vehicle and fires through the three-part roof hatch. It is also armed with a pintle-mounted 7.62 mm M60 machine gun and two four-barrelled M257 smoke grenade launchers. Production of this version began in September 1985 and was completed in December 1986.

Maintenance/recovery

This has a crew of four consisting of commander, driver and two fitters and is armed with a 7.62 mm M60 pintle-mounted machine gun and two M257 four-barrelled smoke grenade launchers. Equipment fitted includes a portable auxiliary power unit, a crane with a capacity of 1,814 kg and a rear-mounted winch with a capacity of 13,608 kg. There are four stabilisers fitted giving a more stable platform when the crane is being used. Production of this version began in May 1986 and was completed in July 1987.

Anti-tank

This has a twin TOW launcher mounted on top of the hull above the third wheel station carrying two ready to launch 3,750 m range Raytheon Systems Company

US Marine Corps LAV anti-tank vehicle with TOW launcher

General Dynamics Armament Systems Light Armored Vehicle – Air Defense
0058065

MODA LAV (8 × 8) assault gun vehicle with Giat Industrie's 90 mm TS-90 turret

MODA LAV-25 (8 × 8) with Delco Defense 25 mm turret

TOW anti-tank guided missiles and a further 14 missiles in the hull. Secondary armament consists of a 7.62 mm or 12.7 mm machine gun on a pintle and two four-barrelled M257 smoke grenade launchers. A TOW ground mount is also carried as is a hand-held laser range-finder. This model has a crew of four: driver, commander, gunner and loader. Production of this version started in January 1987 and finished in June 1987.

Command and control

This has a similar hull to the logistics vehicle and has extensive communications equipment installed. The crew consists of the vehicle commander and driver and there is space for a unit commander and four staff members or radio operators. Armament consists of one pintle-mounted 7.62 mm machine gun and two four-barrelled smoke grenade dischargers. Production started in January 1987 and finished in May 1987.

Anti-aircraft

Following extensive trials between two competing designs, in June 1992 the now General Dynamics Armament Systems was selected to meet the US Marine Corps' requirement for the LAV - Air Defense vehicle.

Although designed for installation on the LAV-25 (8 × 8) chassis, the turret can be installed on other chassis types including the M113 series APC and the M2/M3 Bradley Fighting Vehicle.

Details of the LAV - Air Defense are given in *Jane's Land-Based Air Defence 2001-2002.* Late in 1995, General Dynamics Armament Systems was awarded a production contract for a total of 17 LAV-AD systems valued at US$70 million with first production systems delivered in 1997 and final deliveries taking place in 1998.

Further development of this turret with the now Thales Air Defence of France has resulted in the Blazer turret system which retains the 25 mm gun system of the LAV-AD but adds a Thales Air Defence radar and replaces the Stinger SAMs with a bank of four or eight Matra BAe Dynamics Mistral SAMs. Details of this turret, developed as a private venture, are given in *Jane's Land-Based Air Defence 2001-2002.* This system has been developed for the export market. By late 2001 no orders had been placed for this version.

Mobile Electronic Warfare Support System (MEWSS)

In August 1985, the US Marine Corps ordered from the now General Motors Defense, two prototypes of the Mobile Electronic Warfare Support System (MEWSS), which were delivered in 1986. After extensive testing, a production order for 12 vehicles was placed in September 1987. Delivery was made in early 1989.

The MEWSS has a crew of four with EW equipment including the WJ-8618B radio receiver, AN/PRD-10 radio direction-finding systems, an AN/VLQ-19 communications jammer and radios. It also has a radio communications system, provision for the Position Location Reporting System, a Land Navigation System, vehicle and shore power capabilities and two air conditioning systems.

This has been type-classified as the AN/MLQ-36 by the US Marine Corps.

Assault gun vehicle

Under contract to Tank Automotive Command, Cadillac Gage was awarded a contract to build three prototypes of a 105 mm armed Light Armored Vehicle - Assault Gun (LAV-AG) based on an 8 × 8 chassis supplied by the now General Motors Defense.

This contract was cancelled late in 1991 as there was no assured funding for production systems. By that time, the first prototype had already started its unmanned firing trials. In mid-1993, funding was released to enable the three prototypes of the LAV-AG to be completed. Details of the LAV-AG are given in the Reconnaissance vehicles section under the USA. Development of this model was completed in June 1995 but no contracts have been placed for production vehicles.

120 mm Mortar Carrier

Within the 1,117 LAV (8 × 8) vehicles for the Saudi Arabian National Guard (SANG), there was a requirement for 73 120 mm mortar systems. Following extensive trials, in the Spring of 1995 the 120 mm Armoured Mortar System (AMS) was selected.

The 120 mm AMS is a joint development between General Motors Defense, Delco Defense and BAE Systems, RO Defence.

Details of the 120 mm AMS and the French TDA 120 mm 120 2R 2M mortar system are given in the Self-propelled mortar systems section later in this book.

Australian LAV

Details of the Australian programme for the LAV-25 vehicle and variants as well as the Bison 8 × 8 APC are given under Australia.

US Army LAV

Late in 1989, the US Marine Corps loaned the 82nd Airborne Division of the US Army 16 LAV-25s and in return the army will loan the Marine Corps 11 Multiple Launch Rocket Systems. Two of these vehicles have been modified by General Motors Defense for certification trials for heavy paradrop. If successful, the US Army intends to purchase these vehicles from the US Marine Corps and modify all vehicles for paradrop.

Bison (8 × 8) APC

There is a separate entry in this section for the Bison 8 × 8 APC family, 199 of which have been built for the Canadian Forces. More recently, Australia has trialled four of these vehicles and placed an order for 33 vehicles which have now been delivered.

Canadian LAV-III APC

Details of this new Canadian 8 × 8 APC based on the latest MOWAG Generation III Piranha vehicle are given in the previous entry. A total of 651 has been ordered by the Canadian Forces. This has also been ordered by a number of other countries including New Zealand and the United States.

Canadian LAV-25 (RECCE)

To replace the tracked Lynx Command and Reconnaissance Vehicles in service with the Canadian Forces, in April 1992 a modified version of the LAV-25 was selected and the following year a contract was placed for 203 vehicles. Details of this version, first examples of which were completed in 1996, are given in the Reconnaissance Vehicles section.

LAV - Saudi Arabian National Guard (SANG)

In June 1990, a request was made to the US Congress for the sale of 1,117 LAVs (8 × 8) to Saudi Arabia through the Foreign Military Sales (FMS) programme. These vehicles are operated by the Saudi Arabian National Guard and replace Cadillac Gage (now Textron Marine & Land Systems) LAV-150 (4 × 4) vehicles currently in service.

US Marine Corps LAV-25 light armoured vehicle fitted with a Delco Defense turret armed with a 25 mm M242 cannon, 7.62 mm coaxial machine gun and roof-mounted 7.62 mm machine gun (Michael Jerchel) **2002**/0121146

The Saudi Arabian National Guard order for 1,117 vehicles comprises the following:

Table 1

Variant	Quantity	Comment
LAV-25	384	as US Marine Corps
Anti-tank	111	as US Marine Corps
Personnel carrier	47	
Recovery	67	
Command and control	182	
Ambulance	71	
Engineer	34	
LAV-Assault Gun	130	will have two-man 90 mm gun
120 mm mortar system	73	120 mm AMS selected 1995
Ammunition carrier	18	

Delivery of vehicles for the Saudi Arabian National Guard is under way and by early 2001 well over 900 had been delivered.

Late in 1999 the Diesel Division, General Motors of Canada, was finally awarded the Foreign Military Sales (FMS) contract for 130 Light Armoured Vehicle - Assault Gun (LAV-AG) for SANG.

This will be the last variant of the 1,117 members of the LAV (8 × 8) family in 10 variants to be delivered to SANG. These new LAV (8 × 8) are replacing older Cadillac Gage LAV-150 (4 × 4) vehicles.

The 130 chassis for the LAV-AG have already been built and these will now be fitted with a Belgian CMI Defence LCTS two-person turret armed with a 90 mm Cockerill Mk 8 gun system. This turret has already been built in production quantities for sale to Kuwait and Qatar (36) for installation on other wheeled chassis.

The Belgian company of MECAR will supply a complete family of ammunition for the 90 mm Cockerill Mk 8 gun including armour-piercing fin-stabilised discarding sabot - tracer (APFSDS-T), high explosive plastic (HEP), smoke and target practice - tracer (TP-T). The first ammunition order will be used for US certification of the turret and weapon system. More recently another 132 vehicles will be supplied for the SANG for its 9th battalion.

LAV - Ministry of Defence and Aviation (MODA)

Before the order for 1,117 LAV (8 × 8) vehicles for the Saudi Arabian National Guard, the Saudi Arabian Ministry of Defence and Aviation placed a significant order for members of this family which have now been delivered. Unlike the US Marine Corps vehicles they have no amphibious capability.

This order was placed with MOWAG of Switzerland with some vehicles being built by the new General Motors Defense and Alvis Vehicles of the UK (qv). Known variants are shown in Table 1.

General Motors Defense

(1) Anti-tank with Giat Industries Lancelot turret with four Euromissile HOT ATGWs in the ready to launch position
(2) Assault with Giat Industries TS-90 turret armed with 90 mm gun and 7.62 mm coaxial machine gun
(3) LAV-25 with same turret as US Marine Corps and armed with a 25 mm M242 cannon and 7.62 mm coaxial machine gun
(4) Command post vehicle with raised roof.

Alvis Vehicles

(1) Ambulance with raised roof at rear
(2) APC in VIP configuration with raised roof and armed with externally mounted 12.7 mm MG
(3) Battlefield surveillance with roof-mounted radar
(4) Maintenance and recovery vehicle, raised roof at rear and crane
(5) Mortar carrier, 120 mm, with TDA mortar carried externally on the roof
(6) Mortar carrier, BAE Systems, RO Defence 81 mm, with mortar carried internally.

It is believed that Alvis Vehicles built over 110 vehicles for MODA.

LAV Roadrunner

This has been developed as a private venture by General Motors Defense, and is essentially an LAV with the area to the rear of the engine compartment cut away and lowered to form a flatbed on to which a variety of weapons can be installed, for example, a power-operated Multiple Purpose Launch System armed with four Chaparral surface-to-air missiles, seven Hydra 70 unguided rockets and two Hellfire anti-tank missiles.

LAV - National Guard

In 1992, the US Army National Guard borrowed 12 Bison vehicles from Canada to conduct an evaluation of these vehicles in support of the US drug enforcement agencies.

Since the start of this operation, the vehicles have taken part in several counter-narcotic operations with the evaluation being concluded in December 1992 and the final report being issued in March 1993.

In November 1994, a contract was received to provide the US National Guard with 12 LAVs which were delivered during 1995. These LAVs will assist the National Guard with its various state missions.

US Army LAV-III

Late in 2000, the US Army selected the LAV-III (8 × 8) to meet its Brigade Combat Team (BCT) requirement. Additional details are given in this section under the United States. Production and assembly will take place in Canada and the USA.

New Zealand

New Zealand is purchasing 105 LAV-III from General Motors Defense to replace its current M113 series vehicles. These will be issued to two battalions of the New Zealand Army. The contract was signed in January 2001 and is worth NZ$483.74 million. The vehicles will be delivered from January 2003 and May 2004. These will replace M113 series vehicles which will then be phased out of service.

Status

Production. In service or on order for the following countries from the Canadian production line:

Country	Quantity	Comment
Australia	15	LAV-II from US Marine Corps stocks, all delivered
	111	LAV-II, brand new from Canada, all delivered
	150	LAV-II, brand new, ordered from Canada in mid-2000
Canada	199	Bison, LAV-II, all delivered
	203	Coyote, LAV-II, all delivered
	651	LAV-III, delivery from 1998
	491	6 × 6, all delivered
New Zealand	105	LAV-III, selected 2000, ordered January 2001
Saudi Arabia	1,117	National Guard, deliveries under way, plus another 132 being ordered
	n/avail	Ministry of Defence and Aviation, all delivered
USA	772	US Marine Corps, all delivered
	15	US Army, loaned by US Marines
	12	US Army National Guard, delivered 1995
	366	US Army, LAV-III selected late 2000, this is the first order
	32	LAV-III, on loan from Canada

Manufacturer

General Motors Defense.

UPDATED

CHILE

FAMAE Piranha 8 × 8D armoured personnel carrier

Development/Description

FAMAE (Factories & Metalworks of the Army) today holds the licence for the manufacture of the Swiss MOWAG (now owned by General Motors Defense) Piranha range of armoured vehicles in Chile; previous production of this vehicle was undertaken by Cardoen and is covered in the following entry.

In early 1993, FAMAE commenced production of the 8 × 8D (with the D standing for diesel) APC for the Chilean Army. The first production lot was delivered in December 1994. The vehicle became operational with the Chilean Army in 1995.

This is similar in appearance to the original MOWAG version but, as the Chilean Army did not require an amphibious capability, the propellers at the rear of the hull have been removed and their space taken by stowage boxes. If required, the Piranha 8 × 8D can be supplied in the amphibious configuration fitted with propellers, trim vane and bilge pumps.

The infantry sits back-to-back down the centre of the vehicle and enters and leaves via the large power-operated ramp in the hull rear that is provided with an emergency door in the left side.

FAMAE Piranha 8 × 8D fitted with Cockerill turret armed with 90 mm gun 0005240

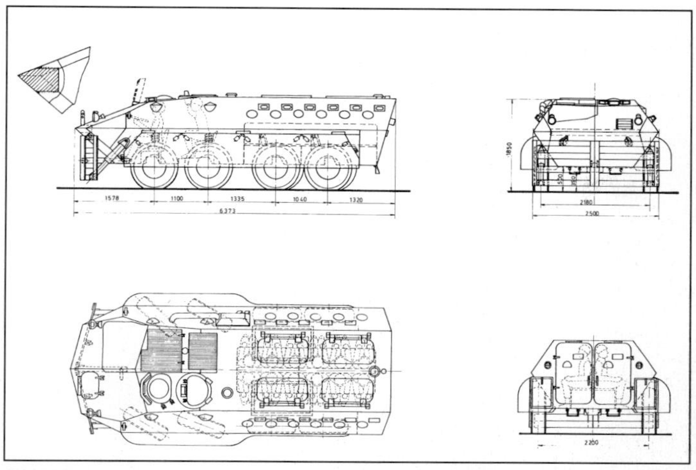

FAMAE Piranha (8 × 8) for use in the internal security role and fitted with a front-mounted obstacle-clearing blade

FAMAE Piranha 8 × 8D APC from the left side, armed with a 12.7 mm MG

The troop compartment is provided with four firing ports and day vision devices in either side and roof hatches above. When fitted with a machine gun installation, the troop compartment is normally fitted with four roof hatches but, when fitted with a 90 mm turret system, only two roof hatches are normally fitted.

Mounted above the commander's hatch to the rear of the driver's position is a cupola with an externally mounted 12.7 mm M2 machine gun which can be used for local and anti-aircraft defence.

In 1992, examples were shown fitted with a turret armed with a 90 mm gun (on 8 × 8 chassis), an Israel Military Industries 60 mm Hypervelocity Medium Support Weapon (on 6 × 6 chassis) and a RAFAEL Overhead Weapon Station (on 6 × 6 chassis).

Other versions that were marketed by FAMAE include a standard 8 × 8 APC with a 12.7 mm machine gun, an 8 × 8 APC with a turret-mounted 25 mm cannon, a 6 × 6 anti-aircraft with twin 20 mm cannon, a 6 × 6 command post vehicle with additional communications equipment and a 6 × 6 81 mm mortar carrier. A 120 mm model mortar is also marketed by FAMAE.

More recently an 8 × 8 armoured logistics carrier and an 8 × 8 internal security vehicle have been developed. The latter is fitted with a barricade and wall remover, fire suppression devices, gun ports for individual use of personnel weapons, tear gas dischargers, a machine gun assembly and a protection device for operations in areas contaminated by toxic gases.

For trials purposes, at least one example of the FAMAE Piranha 8 × 8D has been fitted with the Belgian Cockerill Mechanical Industries CSE two-person 90 mm turret.

As far as is known, all production 8 × 8 vehicles built in Chile are fitted with a 12.7 mm M2 machine gun.

Specifications

Crew: 14
Configuration: 8 × 8
Combat weight: 13,000 kg
Empty weight: 9,700 kg
Power-to-weight ratio: 22 hp/t
Length: (overall) 6.373 m
Width: 2.50 m
Height: (top of hatch) 1.9 m
Ground clearance:
(under differential) 0.46 m
(under hull) 0.57 m
Track: 2.18 m
Wheelbase: 1.1 + 1.335 + 1.04 m
Angle of approach/departure: 43°
Max road speed: 100 km/h
Max water speed: 10 km/h
Fuel capacity: 300 litres
Cruising range: (road) 850 km
Fording: 1.10 m
Gradient: 70%
Side slope: 37%
Trench: 2.10 m
Vertical obstacle: 0.57 m
Turning radius: 15 m
Gun stabiliser:
(vertical) no
(horizontal) no
NBC system: optional
Night vision equipment: optional

Note: Remaining specifications are identical for Piranha 8 × 8 described under Switzerland.

Status

Production as required. In service with the Chilean Army. As far as it is known FAMAE has not exported any of these vehicles.

Manufacturer

FAMAE.

UPDATED

Cardoen/MOWAG armoured personnel carriers

Development/Description

Cardoen Industries is believed to have built over 225 MOWAG (now owned by General Motors Defense of Canada) Piranha 6 × 6 vehicles under licence for the Chilean Army and a single prototype of the basic 4 × 4 model.

The Chilean-built Piranha is identical to the original Swiss model except that its diesel fuel capacity has been increased to 400 litres compared with 250 litres. This has increased its operational range from 600 km to about 1,000 to 1,200 km.

The Chilean Army has taken delivery of 150 vehicles in the APC role, 50 120 mm mortar carriers, five command post vehicles and 20 with two-man turrets armed with French 90 mm F1 guns. The command post vehicle has a power-operated ramp in the rear in place of the normal two doors.

Variants built to the prototype stage or converted from production vehicles for trials purposes include an aerial surveillance radar carrier, anti-aircraft with the Israel Aircraft Industries, RAMTA Structure and Systems, TCM-20 twin 20 mm light anti-aircraft gun system, variants fitted with an Oerlikon Contraves one-man 20 mm GAD-AOA turret, a French Hispano-Suiza 90 mm turret, a Brazilian ENGESA ET-90 90 mm turret and a Cardoen-designed and built 90 mm turret armed with a Cockerill 90 mm Mk III gun. Prototypes have also been built of the vehicle fitted with a 70 and 102 mm multiple rocket launcher. Ambulance, anti-tank, workshop and repair models were marketed, but as far as it is known, none of these entered production or service.

In 1988, a Cardoen-built MOWAG Piranha (6 × 6) vehicle was fitted, for trials purposes, with a Chinese NORINCO one-man turret and four Red Arrow 8 ATGWs in the ready to launch position. This turret has been built by NORINCO for installation on its full-tracked vehicles.

Specifications of the Chilean-built vehicles are identical to those 6 × 6 built by MOWAG in Switzerland and are fully covered in that entry.

Cardoen/MOWAG (6 × 6) command post vehicle (T J Gander)

Status

Production complete. In service with the Chilean Army in 6 × 6 configuration only. As far as it is known, Chile has not exported any Piranha series 6 × 6 vehicles.

Manufacturer

Industrias Cardoen Chile Ltda. (No longer involved in the design, development or production of armoured vehicles)

Note: Industrias Cardoen Chile Ltda is now known as Metalnor Industria Metalurgica Del Norte Ltda. The licence to manufacture the MOWAG Piranha family of light armoured vehicles is now held by FAMAE, for which there is a separate entry.

UPDATED

CHINA, PEOPLE'S REPUBLIC

NORINCO WZ 551 armoured personnel carrier

Development

The existence of a new Chinese 6 × 6 armoured personnel carrier was first revealed in early 1986 and one of the 16 prototype vehicles made its first public appearance during an exhibition that took place in Beijing in November 1986.

At that time it was confirmed that the vehicle had been designed and built by the NORINCO (China North Industries Corporation) Vehicle Research Institute outside Beijing which had also built the prototype vehicles.

Using automotive components of the 6 × 6 WZ 551, NORINCO has also developed 4 × 4 and 8 × 8 versions with the former having been built in production quantities for the NORINCO Red Arrow 8 ATGW systems covered later in this entry.

This family of vehicles will be capable of being used for a wide range of roles, including 82 mm and 120 mm mortar carriers, and be armed with a wide range of weapons including 12.7 mm machine gun, 25 mm cannon, 90 mm or 105 mm guns.

NORINCO WZ 551B (6 × 6) APC fitted with a one-man machine gun turret 0005242

Description

There are minor differences between prototype WZ 551 vehicles and the following description should be taken as provisional.

Although the 6 × 6 version of the WZ 551 is very similar in appearance to the French VAB (6 × 6) vehicle, it has a number of important differences as well as being longer, wider and heavier.

The hull of the WZ 551 is of all-welded steel construction, protecting the crew from small arms fire and shell splinters.

The driver sits at the front on the left with the vehicle commander to his right and the 12.7 mm machine gunner to his rear. Their compartment is separated from the troop compartment by an airtight door.

The commander and driver each have a windscreen to the front which is covered, when in a combat area, by an armoured shutter hinged at the top. When these are lowered, forward observation for each is via three roof-mounted day periscopes, the centre one of which can be quickly replaced by a passive night vision device.

Both the commander and driver have a rear-opening circular roof hatch.

On at least one of the prototypes there is a door in the lower half of the hull between the first and second roadwheels. This is for maintenance purposes.

The engine compartment is to the rear of the driver's compartment and separated from the remainder of the vehicle by airtight bulkheads. The air inlet and outlet louvres are in the roof of the engine compartment with the exhaust pipe running along the upper left side of the hull.

The air-cooled diesel engine is coupled to a fully automatic transmission with five forward and one reverse gears with torque converter. Axles have double reduction and differential locking.

The troop compartment is at the rear of the hull and the troops enter via a large door that is hinged on the right and has a firing port and a vision block. The infantry sits on bench-type seats down either side of the hull facing each other. However, on the 25 mm version the troop seating layout is different.

Over the top of the troop compartment are four oblong roof hatches, two either side, that are hinged on the outside and can be locked vertical if required.

In each side of the troop compartment are four firing ports with a day periscope mounted above in the roof of the vehicle. The first firing port in each side is for a 7.62 mm machine gun, with the remaining three being for assault rifles. Each firing port has a gas extractor device. The diesel fuel tanks are located one either side of the troop compartment.

The WZ 551 is fully amphibious, propelled in the water by two shrouded propellers mounted one either side under the rear of the vehicle. These are swung backwards through 180° for improved propulsion when in the water. Before entering the water, a trim vane is erected at the front of the vehicle and the bilge pumps, located in the floor at the rear, are switched on.

Standard equipment includes an NBC system, night vision equipment for the driver, an automatic fire extinguishing system and run-flat tyres. If the tyres are punctured by bullets or splinters it can still travel a distance of 100 km at a speed of 40 km/h.

Variants

WZ 551B

In 1996, it was revealed that NORINCO had developed a new version of the WZ 551 called the WZ 551B.

The new WZ 551B is fitted with a one-man manually operated turret armed with a 12.7 mm or 14.5 mm machine gun with a 7.62 mm machine gun being mounted coaxially. Mounted on the rear of the turret is a bank of electrically operated grendade dischargers that fire to the front of the turret. In many respects this turret is very similar to that fitted to the Russian BTR-70/BTR-80 series of 8 × 8 APCs.

The power pack consists of the Deutz BF8L413FC air-cooled turbocharged diesel developing 320 hp coupled to a 5S111GPA manual transmission with nine forward and one reverse gears. Maximum road speed is 95 km/h and maximum road range is over 800 km.

More recent production vehicles have the manual transmission replaced by a ZF 6HP500 automatic transmission with torque converter providing five forward and one reverse gears.

12.7 mm APC (6 × 6)

The basic vehicle carrying 11 fully equipped troops plus a crew of two. It has a circular roof hatch and is armed with an externally mounted 12.7 mm machine gun for which 500 rounds are carried. Its empty weight is 12,500 kg, maximum payload on land 7,000 kg and maximum payload on water 3,500 kg.

Ambulance (6 × 6)

This has a raised rear troop compartment, a crew of three and can carry four stretchers. Its empty weight is 12,600 kg. Mounted on the forward part of the roof is a cupola with an externally mounted 12.7 mm machine gun for which 500 rounds are carried.

73 mm IFV (6 × 6)

This has the same turret as fitted to the full-tracked WZ 501 IFV described and illustrated in the previous Armoured personnel carrier's (tracked) section which in turn is a copy of the turret installed on the Russian BMP-1 vehicle. This is armed with a 73 mm gun (40 rounds carried), a coaxial 7.62 mm machine gun (2,000 rounds carried) and a NORINCO Red Arrow ATGW mounted over the main armament, for which four missiles are carried. Combat weight is 14.2 tonnes.

25 mm IFV (6 × 6)

This has a crew of two and is fitted with a one-man power-operated turret on the roof to the immediate rear of the engine compartment. To the rear of the turret are two rectangular roof hatches. The turret is also installed on the full-tracked YW 307 IFV.

This has also been referred to, at various times, as the Type 82, Type 90 and Type 92. It is now known to be in service with the PLA and took part in the parade held in Beijing late in 1999.

The turret can traverse through a full 360° and the 25 mm externally mounted cannon has an elevation of −8 to +55°. A 7.62 mm machine gun is mounted coaxially to the left of the 25 mm cannon and either side of the lower part of the turret are electrically operated smoke grenade dischargers. The gunner has a single-piece hatch cover that opens to the right, two day periscopes and a day/low-light level night sight which is mounted in the forward part of the turret roof.

The 25 mm cannon fires HE and API rounds and the gunner can select single shots or rates of fire of 100, 300 or 500 rds/min. Two hundred rounds of ready use 20 mm ammunition are carried plus a further 200 rounds in reserve. One thousand rounds of 7.62 mm ammunition are carried.

WZ 554 self-propelled anti-aircraft gun

This is an 8 × 8 chassis fitted with a turret, armed with twin 23 mm cannon, it is understood that this never entered production or service.

Type NGV-1 infantry fighting vehicle (6 × 6)

This was shown for the first time in late 1988 and is essentially the NORINCO WZ 551 (6 × 6) vehicle fitted with the French Giat Industries Dragar one-man power-operated turret. The Dragar turret weighs 1,250 kg and, in addition to being armed with the 25 mm Giat Industries M811 cannon, is armed with a 7.62 mm machine gun and two banks of three electrically operated smoke grenade dischargers. As far as is known, this version of the WZ 551 has not entered production.

Anti-tank vehicle (4 × 4) Red Arrow 8

This is fitted with a one-man turret with four NORINCO Red Arrow 8 ATGWs ready to launch; an additional eight missiles are carried in reserve. The missiles have a range of 100 to 3,000 m. The turret can be traversed through 360° and the launcher elevated from −7 to +13°. This version has a crew of four.

This version is known to have been produced in production quantities and in addition to being in service with the People's Liberation Army it is understood that a quantity has been obtained by the army of Bosnia-Herzegovina in the former Yugoslavia. The latter also has a quantity of the standard tripod-mounted infantry version of the Red Arrow 8.

As an alternative, the 4 × 4 version can be fitted with the one-man turret armed with the 73 mm gun, a 7.62 mm coaxial machine gun and the Red Arrow 73 ATGW.

Anti-tank vehicle (4 × 4) Red Arrow 9

A similar 4 × 4 chassis is used for the latest NORINCO Red Arrow 9 heavy long-range anti-tank guided weapon

NORINCO WZ 551 (6 × 6) APC with 25 mm cannon 0058071

NORINCO WZ 551 with one person cupola armed with 12.7 mm machine gun 0058072

system, which was shown for the first time in late 1999. This chassis is also referred to as the Type 92B and is covered in detail in the tank destroyers section.

120 mm self-propelled mortar

Available details of this system, which is fitted with what appears to be a modified Russian 2S23 NONA-SVK turret system, are given in the Self-propelled mortars section under China.

Armoured personnel carrier (4 ×4)

This system is fitted with a one-person turret armed with machine guns for use in the internal security role. Also referred to as the WZ 901 or Type 92B.

NORINCO PL-9 SAM system

The chassis of the WZ 551 is used as the basis for the NORINCO PL-9 low-altitude surface-to-air missile system which remains at the proposal stage. This system is also referred to as the WZ 551D air defence vehicle and additional details are given in *Jane's Land Based Air Defence 2001-2002*.

122 mm SPH (8 × 8)

This is based on a lengthened 6 × 6 chassis and mounted on the hull top at the rear is a turret armed with a 122 mm howitzer which has a maximum range of 15,000 to 21,000 m, depending on the type of ammunition used. The turret has a traverse of 25° left and right and the ordnance can be elevated from –3 to +70°. Forty rounds of ammunition are carried and, before firing commences, two spades can be lowered at the rear of the hull to provide a more stable firing platform. Available details of this are given in the Self-propelled guns and howitzers (wheeled) section. This remains at the prototype stage.

NORINCO is also suggesting that this vehicle could be fitted with the 73 mm turret or one with four 25 mm cannon for air defence use.

Specifications

(fitted with one-man 25 mm turret)
Crew: 2 + 11
Configuration: 6 × 6
Combat weight: 15,000 kg
Power-to-weight ratio: 21.33 hp/t
Length: 6.73 m
Width: 2.86 m
Height:
(hull top) 2.10 m
(over 25 mm turret) 2.89 m
Ground clearance: 0.41 m
Track: 2.44 m
Wheelbase: 1.9 + 1.9 m
Angle of approach/departure: 45°/41.5°
Max speed:
(road) 90 km/h
(water) 8.5 km/h
Fuel capacity: 400 litres
Max road range: 800 km
Fording: amphibious
Gradient: 70%
Side slope: 30%
Trench: 1.2 m
Vertical obstacle: 0.5 m
Turning radius: 9 m
Engine: Deutz BF8L413FC, V-8 4-cycle air-cooled diesel developing 320 hp at 2,500 rpm
Transmission: ZF 6HP500 automatic with torque converter providing 5 forward and 1 reverse gears
Steering: hydraulic, first and second axles
Suspension: coil springs and hydraulic shock-absorbers
Tyres: 14.00 × 20 (run-flat)
Brakes: dual-circuit, air/hydraulic
Electrical system: 24 V
Batteries: 2 × 24 V/180 Ah
Armament:
(main) 1 × 25 mm cannon
(coaxial) 1 × 7.62 mm MG
Ammunition:
(25 mm) 400
(7.62 mm) 1,000
Gun control equipment: electric/manual
Turret power control:
(by commander) no
(by gunner) yes
Gun elevation/depression: +55/–8°
Turret traverse: 360°
Gun stabiliser:
(vertical) no
(horizontal) no
NBC system: yes
Night vision equipment: yes

Other variants

NORINCO has stated that other variants could include 82 mm and 120 mm self-propelled mortar systems. Ambulance and armoured command post vehicles have also been proposed by NORINCO.

Status

Production. In service with the People's Liberation Army and the army of Bosnia-Herzegovina. The latter uses the 4 × 4 version with NORINCO Red Arrow ATGW. Pakistan is believed to have evaluated the vehicle.

Manufacturer

Chinese state factories. Enquiries to China North Industries Corporation.

UPDATED

NORINCO Type WZ 523 armoured personnel carrier

Development/Description

During a parade held in Beijing in October 1984, a new 6 × 6 armoured personnel carrier was seen in public for the first time. In appearance it is very similar to the Belgian SIBMAS and South African Vickers OMC Ratel 6 × 6 APCs developed many years ago. The US Army refers to this as the M1984 armoured personnel carrier. The Chinese designation of this vehicle is the NORINCO (China North Industries Corporation) Type WZ 523.

It is understood that the WZ 523 uses components from the existing HY472 and LT665 trucks. The WZ 523 can be distinguished from the WZ 553 as the rear two axles on the former vehicle are much closer together.

The WZ 523 APC was deployed to Hong Kong by the People's Liberation Army (PLA) in mid-1997.

The hull of the WZ 523 APC is of all-welded steel armour construction providing protection from small arms fire and shell splinters. The driver sits at the front of the hull with the commander to his side, probably on the right. Forward vision is through two large windows; in a combat area these would be covered by an armoured shutter hinged at the top. Above the commander's and driver's positions is a single-piece hatch cover, hinged in the centre, which can be locked vertically. There is a door in the right side of the hull, to the rear of the first roadwheel. This door opens to the front.

The engine compartment is in the centre with drive to all six wheels and transmission is probably manual. The exhaust pipe is on the right side of the hull and the engine is probably offset to the left (as on some Western vehicles such as the French Giat Industries VAB) to allow individual crew members to move from the front of the vehicle to the rear. Steering is on the front wheels only and probably power assisted to reduce driver fatigue.

The troop compartment is at the rear of the hull and the troops enter and leave via a large door in the hull rear hinged on the right with a single circular firing port. There are rectangular roof hatches above the troop compartment, opening outwards, and two observation/firing hatches in either side that open upwards.

On the hull roof, slightly forward of the troop compartment, is a 12.7 mm Type 54 machine gun with lateral protection from small arms fire.

Specifications

Crew: 2 + 10
Configuration: 6 × 6
Combat weight: 11,200 kg
Power-to-weight ratio: 14.73 hp/t
Length: 6.02 m
Width: 2.55 m
Height: (including MG) 2.73 m
Max road speed: 80 km/h
Max water speed: 7 km/h
Fuel capacity: 255 litres
Max range: 600 km
Gradient: 60%
Side slope: 30%
Engine: EQ 6105 water-cooled petrol, developing 165 hp
Armament: 1 × 12.7 mm MG
Ammunition: 600 × 12.7 mm
Gun stabiliser:
(vertical) no
(horizontal) no
NBC: yes
Night vision equipment: yes

NORINCO Type WZ 523 (6 × 6) armoured personnel carrier fitted with a roof-mounted 12.7 mm machine gun

NORINCO Type WZ 523 (6 × 6) APC showing the roof-mounted 12.7 mm machine gun and the roof hatches locked open (via G Jacobs)

More recent vehicles have a fully enclosed one-person turret armed with what appears to be a 7.62 mm machine gun on the right and a larger calibre weapon on the left. Mounted either side of the turret is a bank of electrically operated smoke grenade launchers while mounted on the turret roof is a white light searchlight and flashing beacons. Some sources have stated that this version is called the WZ 531 or ZFB 91.

The vehicle is fully amphibious and propelled by two water-jets at the rear of the hull; before entering the water a trim vane is erected at the front of the hull.

The driver has an infra-red periscope (or image intensification) for night driving and an A220 radio is fitted as standard. The engine for the WZ 523 is produced by the Second Automotive Plant which may also produce the vehicle.

Variants

Other variants of this 6 × 6 APC could have been developed including command post, ambulance, mortar carrier and anti-tank vehicle with anti-tank missiles.

China has also built a 4 × 4 internal security vehicle which was covered in *Jane's Armour and Artillery 1997-98*. Mention has also been made of a vehicle called the ZFB91.

Status

Production is probably complete. In service with the PLA. There are no known exports of this vehicle.

Manufacturer

Chinese state factories. Enquiries to China North Industries Corporation.

UPDATED

CROATIA

RH ALAN LOV armoured personnel carrier family

Development

As early as 1992, the Torpedo company of Rijeka, Croatia, was awarded a contract to design, develop and manufacture a 4 × 4 light armoured personnel carrier (APC) for the Croatian Army (HV) whose basic design could be used for a wide range of other roles with the minimum of modification. First production vehicles were completed by RH ALAN in 1993.

These vehicles are called the LOV with the APC model being the LOV – OP (*Opce Potpore – Transporter APC*) whose existence was first revealed during a military parade held in Croatia in mid-1995. This has a crew of two and can carry 10 fully equipped troops.

The LOV is based on the chassis of the locally manufactured Torpedo TK – 130 T-7 (4 × 4) vehicle. As well as being used as a cargo truck it can also be used to tow artillery and other weapons up to a maximum weight of 2,500 kg.

By using standard commercial automotive components in the design of the LOV series, research, development, production and total life cycle costs have been kept to a minimum.

By late 2001 there were no exports of the RH ALAN LOV family of 4 × 4 light armoured vehicles. The basic armoured personnel carrier is called the LOV-OP in Croatia.

Description

The LOV series of vehicles has a hull of all-welded steel armour which provides protection from 5.56 mm and 7.62 mm API (Armour Piercing Incendiary) small arms fire from a range of 30 m, shell splinters from a range of 40 m and anti-personnel mines. Some models have additional hull protection against more powerful mines.

RH ALAN LOV-AP (4 × 4) armoured personnel carrier armed with 12.7 mm M2 machine gun **2000**/0080356

The driver is seated at the front left with a bulletproof window to his front which is provided with a windscreen wiper and an additional bulletproof window either side.

Above the driver's position, which is raised slightly above the hull top for improved visibility, is a single-piece hatch cover that opens to the front.

The commander is seated to the right and has a rectangular roof hatch that opens to the rear and to his immediate front is a day periscope which can be traversed left and right.

In each side of the hull, to the side of the driver and commander, is a single forward-opening door which is not provided with any observation or firing devices.

The power pack, which consists of a Deutz model BT6L 912S turbocharged air-cooled diesel engine coupled to a ZF model Z5-35S manual transmission and two-speed transfer case, is mounted towards the front of the vehicle below and between the commander's and driver's positions.

Suspension consists of leaf springs and hydraulic shock-absorbers which are attached to the hull by rubber cushions.

A winch with 38 m of 10 mm diameter cable and a capacity of 50 kN is fitted and this leads to the front of the vehicle and can be used for self-recovery or for recovering other damaged and disabled vehicles.

The troop compartment is at the rear with the troops entering and leaving via two doors in the rear that open outwards with the right door having a spare wheel and tyre and the left door a firing port.

In each side of the hull is a single firing port and a small rectangular window which can be covered by a shutter. Over the top of the troop compartment are two roof hatches that open left and right

Armament consists of a 12.7 mm M2 machine gun with the gunner being provided with lateral and rear protection. Eight 82 mm electrically operated grenade launchers are also fitted.

Standard equipment includes a central tyre pressure regulation system that allows the driver to adjust the tyre pressure between 0.45 and 0.70 bar to suit the type of terrain being crossed. The tyres also have a run-flat device which enables the vehicle to travel a distance of up to 50 km when the tyres are damaged by small arms fire.

It is also fitted with an engine independent heating and ventilation system and adjustable steering column.

RH ALAN LOV-Z (4 × 4) command post vehicle 0021147

Strijela 10 CROA1 (6 × 6) surface-to-air missile system with SAM launcher elevated (Roderick de Normann)

Specifications

LOV-OP APC
Crew: 2 + 10
Configuration: 4 × 4
Combat weight: 8,200 kg
Unloaded weight: 7,200 kg
Payload: 2,000 kg
Power-to-weight ratio: 15.85 hp/t
Length: 5.89 m
Width: 2.39 m
Height:
(hull top) 1.98 m
(overall) 2.10 m
Ground clearance: 0.315 m
Track: 1.86 m
Wheelbase: 2.85 m
Angle of approach/departure: 40°/40°
Maximum road speed: 100-120 km/h
Fuel capacity: 170 litres
Maximum road range: 500-700 km
Fording: 1.0 m
Gradient: 65%
Side slope: 35%
Vertical obstacle: 0.50 m
Trench: n/a
Turning radius: 13 m
Steering: power assisted
Engine: Deutz BT6L 912S turbocharged diesel developing 130 hp at 2,650 rpm
Transmission: ZF manual type Z5-35S with 5 forward and 1 reverse gears
Transfer case: 2 speed
Suspension: leaf springs and hydraulic shock-absorbers
Tyres: 14.5 R20 with run-flat inserts, central tyre pressure regulation system is fitted as standard
Brakes: hydraulic servo unit/drum
Electrical system: 24 V
Batteries: 2 × 100 Ah
Armament: 1 × 12.7 mm M2 MG
NBC system: optional
Night vision equipment: optional

A wide range of optional equipment can be fitted including an air conditioning system, intercom, various communications systems, electric winch, NBC ventilation system, additional sound- and waterproofing equipment.

An auxiliary power unit can be fitted if required, especially when the vehicle has additional communications equipment or electrical equipment requiring additional power.

Variants

LOV T-1 OP

This is an armoured personnel carrier and can carry five troops plus its crew.

LOV-IZV reconnaissance

The reconnaissance member of the family is called the LOV-IZV and is also fitted with a roof-mounted 12.7 mm M2 machine gun and grenade launchers. It has additional communications equipment as well as an eight-barrel 60 mm rocket launcher, 20 mm RT-20 anti-matériel rifle and 40 mm grenade launcher.

LOV-Z command post

This has a six-man crew and additional communications equipment and field telephone switchboard with 10 channels. It also has soundproof insulation, an air conditioning system and a 2.5 kW generator.

LOV-ABK NBC reconnaissance

For the NBC reconnaissance role the LOV-ABK version has been developed and this is fitted with equipment to detect radioactive radiation and chemical agents. It is fitted with a collective protection filter and ventilation equipment.

LOV-RAK 24/128 mm rocket launcher

This is the LOV APC with its hull rear cut down and fitted with a 128 mm (24-round) multiple rocket launcher. Full details are given in the Multiple rocket launchers section.

LOV-ED electronic warfare

This is fitted with an extensive range of jamming equipment with a large externally roof-mounted antenna.

Future variants

At least two other versions of the LOV are under development, the UP1 and UP2. The UP1 is an artillery fire-control vehicle which sends information to the fire-control centre.

Strijela 10 CROA1 SAM system

The Strijela 10 CROA1 air defence vehicle (6 × 6) is not a member of the LOV (4 × 4) family of light armoured vehicles and may be based on the TAM (6 × 6) chassis fitted with an all-welded steel hull.

Mounted on the hull top, above the twin rear road-wheels, is the updated upper part of the Russian KB Tochmash 9K35 Strela 10 (SA-13 'Gopher') surface-to-air missile system which is normally mounted on the MT-LB full-tracked chassis.

Some sources have indicated that locally developed enhanced computer software has been incorporated into the launch, target tracking and missile guidance subsystems. Details of this are given in *Jane's Land Based Air Defence 2001-2002*.

Status

Production as required. In service only with Croatia.

Manufacturer

RH ALAN.

UPDATED

CZECH REPUBLIC AND SLOVAKIA

OT-64 armoured personnel carrier

Development

The OT-64 (*Obrneny* Transporter) was jointly developed from 1959 by Czechoslovakia and Poland and is used by them in place of the similar Russian BTR-60 (8 × 8) APC.

The vehicle, which uses many automotive components of the Tatra 813 series of 8 × 8 high-mobility trucks, entered service in 1964 and has since been widely exported.

The Polish designation for the vehicle is the *Sredni Kolowy Opancerzny* Transporter. Tatra of Czechoslovakia provided the chassis and automotive components and FSC/Lubin in Poland the armoured body and weapon system. Production of the OT-64 family was completed in 1990 after about 10,300 had been built for the home and export markets.

Although production of the OT-64 was completed many years ago the Czech Republic still has quantities of these vehicles available for export, see later under OT-93. These are currently being offered on the export market by Calibre Prague Limited.

Description

(OT-64C(1))

The hull of the OT-64 (8 × 8) APC is made of all-welded steel, with the crew compartment at the front, engine behind and the troop compartment at the very rear. This model has recently been called the OT-64A by the Czech Republic.

The driver is seated at the front of the vehicle on the left side with the commander to his right, both with a single-piece rear-opening door beside them. The driver has a single-piece hatch cover that opens to the rear and three day periscopes for vision to the front and sides. The commander has a single-piece hatch cover that opens forwards with a day periscope in its forward part which can be traversed through 360°. On the roof, between the driver's and commander's positions, is a searchlight which can be operated from inside the vehicle.

The Tatra 928-18 V-8 diesel engine is behind the commander and driver with the air inlet and outlets on the top of the hull and the exhaust pipes on either side of the hull.

OT-93 (8 × 8) armoured personnel carrier fitted with one-man turret armed with 7.62 mm machine gun 0058074

OT-64C(2) which has the same armament as the OT-64C(1) but mounted in a new turret

The turret is in the centre of the vehicle and is mounted on an eight-sided plinth to enable the weapon to have some depression.

The troop compartment is at the rear of the vehicle and the troops enter and leave by two doors in the rear of the hull, each of which has a firing port. Over the top of the troop compartment are four roof hatches, each with a firing port, hinged on the outside. If required, the roof hatches can be locked vertical. There are two firing ports in each side of the troop compartment. The infantrymen are seated down each side of the vehicle on seats which can be folded up to allow the vehicle to be used for cargo.

Steering is powered on the front four wheels and all eight tyres are provided with a tyre pressure regulation system which allows the driver to adjust the tyre pressure to suit the type of ground being crossed. The vehicle is fully amphibious being propelled in the water by two propellers mounted at the rear of the hull. Before entering the water, a trim vane is erected at the front of the vehicle; when travelling, it is stowed on the glacis plate. Standard equipment on all vehicles includes an overpressure NBC system, a winch mounted at the front of the hull and bilge pumps.

A tow hook is provided at the front of the hull with a trailer coupling being provided at the rear. Maximum trailer weight is 3,000 kg, including payload.

The one-man manually operated turret is similar to those fitted to the Russian BRDM-2 (4 × 4) reconnaissance vehicle and BTR-60PB (8 × 8) APC and is armed with a 14.5 mm KPVT machine gun with a 7.62 mm PKT machine gun mounted to the right of the main armament and the sight to the left. The weapons have a maximum elevation of +29°, depression of –4° and 360° turret traverse. Some models have been observed with a single AT-3 'Sagger' ATGW mounted on each side of the turret with lateral armour protection.

OT-64C(1) deployed with the Czech Republic Army in Bosnia and with weapons elevated (Richard Stickland) **2001**/0103755

Variants

Note

In both former Czech and Polish service these vehicles are officially designated SKOT, SKOT-2, SKOT-2A and SKOT-2AP. The designations OT-64A, OT-64B, OT-64C (1) and OT-64C(2) are not always used. The Czech Republic has declared its vehicles under the CFE Treaty as being the OT-64, not SKOT, with the basic vehicle being the OT-64 and the model with the Russian type 7.62/12.7 mm machine gun turret being the OT-64A.

OT-64A (SKOT)

This is the basic member of the series. Some are unarmed while those used by Poland are fitted with a pintle-mounted 7.62 mm machine gun. There are five hatches over the troop compartment: two open to the left and right of the vehicle, the centre one opens forwards and the last two open to the left and right. All of them can be locked vertical. Some OT-64As have been observed with two AT-3 'Sagger' ATGWs mounted over the top of the troop compartment. Four 'Sagger' ATGWs are carried.

OT-64B (SKOT-2)

This model has only been used by Poland and has a square plinth with a pintle-mounted 7.62 mm or 12.7 mm machine gun. It has a shield which gives front and side protection to the gunner, but has an open top and is open at the rear. The OT-64B has three hatches over the troop compartment, the first one opens forwards and the other two open either side. This was probably a temporary expedient pending the delivery of turrets.

OT-64C(1) (SKOT-2A)

This is the model described in detail earlier.

OT-64C(2) (SKOT-2AP)

This is used only by Poland and is similar to the OT-64C (1) except that it has a new turret with a curved top. It is armed with a 14.5 mm KPVT machine gun with a 7.62 mm PKT machine gun mounted to the right of the main armament and the sight to the left. The weapons have a depression of –4° and an elevation of +89.5° which allows them to be used against helicopters and aircraft. Some vehicles have been observed with a single AT-3 'Sagger' ATGW mounted either side of the turret. This turret is also installed on the OT-62C tracked armoured personnel carrier.

OT-64C(1) with OT-62B turret

Morocco is known to have some OT-64C(1) (SKOT-2A) APCs with the original turret replaced by the same turret installed on the OT-62B.

DPT-64 repair

This is an OT-64 modified for use in the repair role.

OT-64 R-2M command

Former Czech command variant of the OT-64A. Polish Army designation for this model is the SKOT R-2AM with another variant being the SKOT R-2M. These are

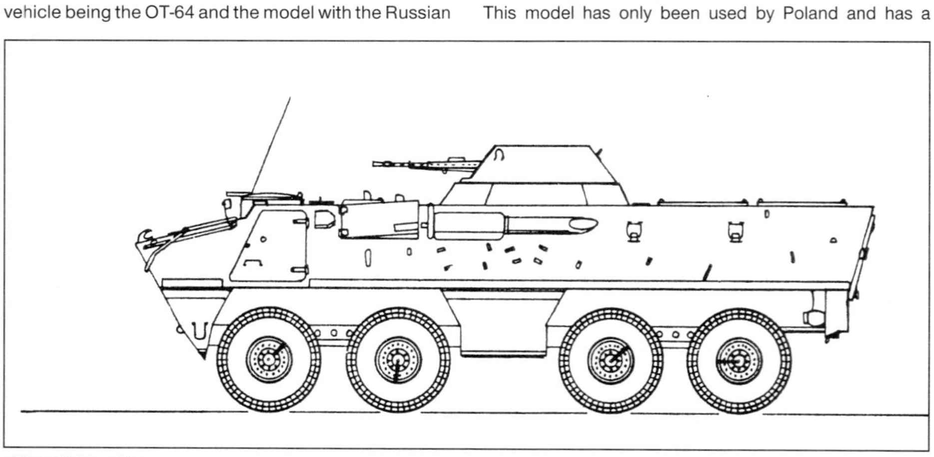

OT-64C(1) APC

Specifications
SKOT-2A
Crew: 2 + 10
Configuration: 8 × 8
Combat weight: 14,500 kg
Unloaded weight: 12,800 kg
Power-to-weight ratio: 12.41 hp/t
Length: 7.44 m
Width: 2.55 m
Height:
(to turret top) 2.71 m
(to hull top) 2.06 m
Firing height: 2.5 m
Ground clearance: 0.46 m
Track: 1.86 m
Wheelbase: 1.3 + 2.15 + 1.3 m
Angle of approach/departure: 42°/43.5°
Max speed:
(road) 94.4 km/h
(water) 9 km/h
Fuel capacity: 330 litres
Max road range: 710 km
Fuel consumption:
(road) 0.45 litres/km
Fording: amphibious
Gradient: 60%
Side slope: 30%
Vertical obstacle: 0.5 m
Trench: 2 m
Engine:
(model) Tatra 928-18
(type) V-8 air-cooled diesel developing 180 hp at 2,000 rpm
Transmission: semi-automatic dual-range 5-speed plus overdrive providing 20 forward and 4 reverse gears
Steering: power assisted
Suspension: telescopic springs and hydraulic shock-absorbers
Tyres: 13.00 × 18
Electrical system: 24 V
Batteries: 4 × 12 V
Armament:
(main) 1 × 14.5 mm KPVT MG
(coaxial) 1 × 7.62 mm PKT MG
Ammunition:
(main) 500
(coaxial) 2,000
Gun control equipment
Turret power control: manual
Gun elevation/depression: +29/–4°
Weapon traverse: 360° (turret)
Gun stabiliser:
(vertical) no
(horizontal) no
Armour
Hull: 10 mm (max)
Turret: 14 mm
NBC system: yes
Night vision equipment: yes

the OT-64A/SKOT-2A with periscopes and the eight-sided plinth, generator and antenna. All other R versions also have this plinth but without the armed turret.

OT-64/SKOT R-2AM command
This is an artillery command vehicle equipped with an R-130, two R-123Z and R-809 radios. This model can be identified by a collapsible Half Horse antenna on the right side of the roof, a generator on the rear deck and a telescopic antenna on the right rear.

OT-64 R-3MT radio
Former Czech version of the OT-64 with radio equipment. Polish Army designation for this model is the SKOT R-3M. These are also command vehicles and are similar to the R-2AM, but with the telescopic antenna on the left side of the hull.

Communications equipment includes an R-130, two R-123Z, an R-111, an R-405PT-1, an R-126, a K-1 and a K-2.

SKOT R-3Z command
This command vehicle is similar to the R-2AM but with a telescopic antenna on the left and right rear.

OT-64 R-4MT radio
Former Czech version of the OT-64 with radio equipment. It is identified by a sword antenna on the front side of the plinth and has an additional door in the left side of the hull, behind the exhaust pipe. No frame type antenna is fitted.

OT-64 R-4T command
This is an unknown Czech command version.

SKOT-R4 command
This command vehicle is similar to the R-3Z but can be identified by a sword antenna and a generator to the immediate rear of the plinth.

SKOT R-6 command
This is an air defence artillery command vehicle with the same mission as the Russian BTR-60Pu-12. It has no frame antenna but two telescopic antennas.

SKOT S-260 engineer
This is an engineer vehicle based on the basic SKOT and can be identified by grab handles over the rear engine deck and another two behind the exhaust pipes. These bars are believed to be fitted to avoid mine clearance hoses being ignited by the heat of the engine.

DTP-64 repair
This is a repair version with two gas bottles on the rear of the turret roof.

SKOT propaganda
This version has been declared under the CFE Treaty and is believed to be fitted with loudspeakers.

OT-93 APC
More recently a new version of the SKOT (8 × 8) APC, called the OT-93, has been developed and this is now being offered for the export market.

SKOT-2A minus turret and with roof-mounted machine gun covered up (Richard Stickland) ***2002***/0121144

The OT-63 is basically the SKOT-2A with the original one-person turret armed with a 14.5 mm KPVT and 7.62 mm PKT machine gun removed and replaced by a turret armed with a single 7.62 mm UK-59T machine gun.

Turret traverse is 360° with weapon elevation from –5 to +80°. Turret traverse and weapon elevation is manual.

This turret is the same as that fitted to the OT-65A (4 × 4) reconnaissance vehicle and the OT-62B full-tracked APC.

With the downsizing of the Czech Republic Army, quantities of OT-64 vehicles are now surplus to requirement and are being offered on the export market in a number of configurations.

As well as the OT-93, other versions of the OT-64 vehicle are being offered including police, firefighting, recovery/repair and bullion carrier.

PCO upgrade package for SKOT APC
The Polish-based company of PCO (Przemyslowe Centrum Optyki) is now offering a complete electro-optical upgrade suite for the SKOT (8 × 8) Armoured Personnel Carrier (APC). These PCO systems are of the Image Intensification (II) type.

The driver is seated at the front of the vehicle on the left side. PCO is offering two alternatives, the PNS-1 driver's night-vision goggles and the PNK-55 night-vision periscope.

The PNS-1 night-vision goggles are typically attached to the driver's head using a harness and have a magnification of ×1 and a minimum field of view of 38°.They are powered by an integral standard AA size lithium battery.

The PNK-55 night-vision periscope is a direct replacement for the day periscope and has a magnification of ×1 and a 30° field of a view and requires an 18-30 V DC power supply. It has an effective range of about 300 m compared to the 100 m of the original equipment.

The commander is seated to the right of the driver and PCO is offering two solutions. First is the TKN-1Z passive night-vision periscope with a magnification of ×3 and a 10° field of view.

Second is the more sophisticated modernised TKN-3Z periscope that has two channels, day and night. The day channel magnification of ×4.75 and a 9° 30′ field of view while the night channel has magnification of ×4.2 and a 11° 30′ field of view. Both of these are powered from an 18-30 V DC power supply. In the night mode these have a maximum range of 1,100 m compared to 400 m of the first generation system.

Some versions of the SKOT are fitted with a one-person turret armed with a 14.5 mm KPVT and one 7.62 mm PKT machine gun. This turret is identical to that installed on a number of other AFVs, such as the Russian BRDM-2 (4 × 4) scout car which is still widely deployed around the world.

For this version of the SKOT, PCO is offering its CDN-1 periscopic day/night sight. The day and night channels have a magnification of ×3.7 and a 10.5° field of view, with an 18-30 V DC power supply required.

SKOT-2A minus turret
In Bosnia, a number of SKOT-2A vehicles were observed. These had their turrets removed and plated over and a simple cupola with a rear-opening hatch cover fitted. On the forward part of this cupola is a 12.7 mm machine gun in an unprotected mount.

WPT armoured repair and maintenance vehicle
Repair and maintenance version of the OT-64 which is fitted with a light crane for changing components in

the field. Armament consists of a 7.62 mm PKT machine gun. It is fitted with an A-frame and is also referred to as the SKOT WPT.

Other Polish versions of the OT-64

SKOT R-3Z is Polish OT-64 R-3Z
SKOT R-4 is Polish OT-64 R-4
SKOT R-6 is Polish OT-64 R-6
SKOT S-260 is Polish engineer vehicle for laying mines
Unidentified Polish OT-64 is an artillery resupply vehicle
SKOT WPT is Polish repair and maintenance vehicle and is equivalent of Czech WPT.

OT-62 APC

This full-tracked armoured personnel carrier is the Czech equivalent of the Russian BTR-50 with some enhancements. It is still used by some countries and details are given in *Jane's Armour and Artillery 1997-98*, pages 277 to 279.

Tatrapan armoured all-terrain vehicle

Details of this 6 × 6 vehicle based on a Tatra truck chassis are given in the **Armoured Command Post Vehicles** section later in this volume.

Status

Production complete. In service with the following countries:

Country	Quantity	Comment
Algeria	150	from Slovakia in 1994
Angola	9	from Slovakia in 1994
Cambodia	26	from Czech Republic in 1994
Czech Republic	100	estimate
India	100	estimate, for disposal (also has OT-62)
Iraq	500	estimate (also has (OT-62)
Libya	50	estimate (also has OT-62)
Morocco	50	status uncertain (also has OT-62)
Poland	734	
Sierra Leone	10	from Slovakia in 1994
Slovakia	33	
Sudan	40	estimate (also has OT-62)
Syria	100	estimate
Uganda	4	status uncertain
Uruguay	60	from Czech Republic in 1995

Manufacturer

Former Czech state factories.
Marketing is now carried out by Calibre Prague Limited.

UPDATED

EGYPT

Kader Fahd armoured personnel carrier

Development

The Fahd (4 × 4) APC was developed to meet the requirements of the Egyptian Army, based on its experiences in Middle East conflicts. The Fahd was designed by the then Thyssen Henschel (today Rheinmetall Landsysteme and part of Rheinmetall DeTech) of Germany under the designation TH 390; Thyssen Henschel also built in Germany the first prototypes of the vehicle for Egypt.

Production of the Fahd is undertaken by the Kader Factory for Developed Industries, part of the Arab Organisation for Industrialisation. In the Egyptian Army it has now replaced old armoured personnel carriers such as the BTR-40 as well as the Egyptian-built Walid (or Waleed as it is called in Egypt) (qv) armoured personnel carrier.

The Fahd essentially consists of a Mercedes-Benz LAP 1117/32 truck (4 × 4) chassis fitted with an armoured body. Wherever possible, standard and proven commercial components are used in the construction of the vehicle. Production of the Fahd commenced in 1985, with the first production vehicles being completed in 1986 for the Egyptian Army, which is believed to have placed an initial order for 300 vehicles with an option on a further 110.

By late 2001, it was estimated that total production of the Fahd APC and its variants, for the home and export markets, had reached over 1,000 vehicles. The basic production model is the Fahd 240, details of which are given later in this entry.

Description

The hull of the Fahd is of all-welded steel armour construction giving complete protection from attack by 7.62 × 54 mm AP rounds and shell splinters.

The driver sits at the front of the vehicle on the left side with the commander to the right. Both have a large bulletproof windscreen to the front which can be covered by a shutter hinged at the top, and a side door that opens to the front. This has a window in its upper part which can be covered by a shutter if required. Above the commander's position is a single-piece, rear-opening hatch cover. The driver has a forward-facing, roof-mounted day periscope which can be replaced by a passive periscope for night driving.

The troop compartment is at the rear of the hull with the infantry entering via a door in the rear, the upper part of which folds upwards and the lower part downwards to form a step. Over the top of the troop compartment are two rectangular roof hatches hinged in the centre which can be locked vertical. The infantry sits on individual bucket seats down the centre of the vehicle facing outwards. In either side of the troop compartment are four firing ports with a vision block above so that the troops can fire their weapons from

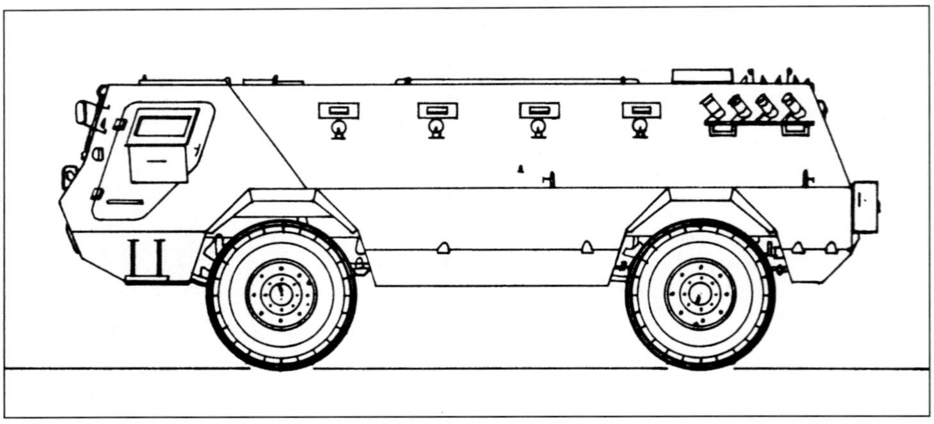

Unarmed Fahd (4 × 4) armoured personnel carrier

within the vehicle. Either side of the rear door is another vision block with a firing port underneath.

The engine is mounted towards the front of the vehicle with ventilation for the engine and cooling system provided through an opening with armoured shutters at the front of the vehicle. The air inlet for the engine is in the roof of the vehicle.

The Fahd is fitted with a central tyre pressure regulation system allowing the driver to adjust the tyre pressure to suit the terrain being crossed. The driver can select front or rear tyres, or both together. The central tyre pressure system stops automatically after reaching the maximum or minimum tyre pressure selected by the driver. Low-pressure, run-flat tyres are fitted as standard and allow the vehicle to travel a distance of 50 km at 30 km/h in the event of a puncture. Power-assisted steering is fitted as standard, and there is a differential lock between the front and rear axles and between the rear wheels.

Up to three machine guns can be mounted externally on the roof of the Fahd for suppressive fire.

Optional equipment includes a front-mounted hydraulic winch with a capacity of 5,000 kg, an NBC and ventilating system, passive night vision equipment and grenade dischargers mounted either side of the hull.

In March 1989, it was announced that the Kader Factory for Industrialisation, a member of the Arab Organisation for Industrialisation, had placed an order with a British consortium for the supply of 110 air cooling systems for the Fahd (4 × 4) APC being built for export. Galley Limited supplied the air cooling systems to Mantech Limited which shipped them to Egypt, with first deliveries made in early 1989.

The Galley-designed air cooling system is a totally self-contained unit mounted at the rear of the vehicle and driven by its own three-cylinder diesel engine. The unit takes warm air from within the crew compartment, cools it in a compact vapour-cycle refrigeration system and then returns it through ducts directing the air around the vehicle.

Cooling air for the radiator and condenser is drawn into the system from the surrounding atmosphere and is discharged, together with the engine exhaust cases, through louvred panels in the hull of the vehicle. The unit is sealed for use under NBC conditions.

The air cooling unit is capable of delivering more than 7 kW of power and weighs 260 kg. It has been designed to operate in ambient temperatures of up to +52°C and relative humidities up to 90 per cent, delivering dehumidified air to the interior of the vehicle at approximately +20°C.

Variants

The design of the Fahd is such that a wide range of armament installations can be mounted on the hull top, for example, a one-man turret with single or twin 7.62 mm machine guns, a 12.7 mm machine gun or a 20 mm cannon. Other variants suggested by the manufacturer include ambulance, command post vehicle, missile carrier, multiple rocket launcher, minelaying, internal security and recovery.

Specialised versions of the Fahd already announced include:

Combat vehicle with 20 mm turret, which weighs 11 tonnes fully loaded or 9,800 kg empty, has a three-man crew and can carry 10 fully equipped troops.

Police security vehicle with turret armed with 7.62 mm and/or 12.7 mm MG, same crew as above, weight 11 tonnes fully loaded.

Police security vehicle with turret armed with 12.7 mm and/or 7.62 mm MG and smoke bomb launchers, combat weight 11.3 tonnes, same crew as before. Options include an air conditioning system, a night periscope for driver and commander, a wireless and intercom system, a navigation compass, a hydraulic towing winch with a capacity of 5 tonnes, an obstacle removal blade at the front of vehicle, a loudspeaker, a

Fahd (4 × 4) APC from the front showing the winch in the lower part of the hull (Christopher F Foss) 0021152

Fahd (4 × 4) APC as sold to Kuwait with a SAMM one-man turret and with the two-part rear door open

siren, a searchlight and a flashing light.

Minelaying vehicle The prototype of an anti-tank minelaying version of the Fahd has been completed.

Fahd 240 APC

This was first announced in 1996 and differs from the earlier production Fahd (4 × 4) APC in that it is based on the latest Mercedes-Benz LAP 1424/32 (4 × 4) chassis and is powered by a Mercedes-Benz model OM 366 LA four-stroke turbocharged diesel developing 240 hp at 2,600 rpm.

In addition to the standard APC version, other Fahd 240 variants being proposed include:

(1) APC with SAMM BTM 208K one-man turret armed with 12.7 mm and 7.62 mm machine guns
(2) APC with complete turret of BMP-2 as described above and designated Fahd 240-30
(3) Command post vehicle with 7.62 mm machine gun, positioning and navigation system and a complete communications system
(4) Minelaying vehicle capable of carrying 250 anti-tank mines and laying these via a chute at the hull rear
(5) Ambulance for four stretcher patients or four seated patients with medical staff
(6) Recovery and workshop fitted with 2.5 tonne hydraulic crane and repair equipment. Before the crane is used, two stabilisers are lowered at the rear of the hull
(7) Riot control vehicle with two-gun turret fitted with water cannon and tear gas grenade launcher

All versions can be fitted with air-cooling system dependent or independent of the vehicle, 5 tonne winch, various optical devices and communications equipment.

AFV Fahd 240/30

This was announced for the first time late in 1990. It consists essentially of the standard Fahd (4 × 4) vehicle with the roof modified. On this has been fitted a plinth, on which is mounted the complete turret of the Russian BMP-2 Infantry Fighting Vehicle (IFV). At least 58 of these vehicles have been built. Production is now understood to be completed. This version was never exported.

Some modifications have been carried out to the hull. In the right side of the hull to the rear of the driver's position is an additional side entrance. The lower part folds down to form a step while the upper part with an integral firing port and vision block opens to the left. At the very rear of the hull on the right side is a single vision block with a firing port underneath.

In the left side of the hull rear are two firing ports with associated vision blocks, one in front and one to the rear of the turret.

The commander is seated on the right of the turret with the gunner on the left. The commander's binocular sight has a day system with a magnification of ×5 and a 10° field of view, and a night system with a magnification of ×4.2 and an 8° field of view. In addition he has a day monocular periscope with a magnification of ×1.2 or ×4 and two day periscopes, one to the front and one to the rear of the vehicle.

Specifications
(Fahd 240/30)
Crew: 10
Combat weight: 13,500 kg
Unloaded weight: 11,900 kg
Height: 2.85 m
Max road speed: 80 km/h
Road range: 700 km
Cross-country range: 400 km
Trench: 0.8 m
Armament:
(main) 1 × 30 mm 2A42 cannon
(coaxial) 1 × 7.62 mm PKT MG
(ATGW) 1 × AT-5 'Spandrel'
Smoke grenade dischargers:
(number) 2 × 3 turret
(calibre) 81 mm turret
(range) 200-300 m turret
(smoke screen) 80 × 5 m turret
Ammunition (ready):
(main) 500 (160 AP + 340 HE-I)
(coaxial) 2,000
(ATGW) 1 + 4
Gun control equipment
Turret power control: powered/manual
Max traverse speed:
(rapid) 36°/s
(slow) 9°/s
Max elevation speed: 30°/s
Gun elevation/depression: +80/–5°
Turret traverse: 360°
Stabilisation system: yes
Commander's override: yes

The BMP-2 turrets were purchased from the Soviet Union before the break-up with first production Fahd 30 vehicles delivered to the Egyptian Army in mid-1992.

Specifications
Standard Fahd APC
Crew: 2 + 10
Configuration: 4 × 4
Combat weight: 10,900 kg
Unloaded weight: 9,100 kg
Payload: 1,800 kg
Power-to-weight ratio: 15.4 hp/t
Length: 6 m
Width: 2.45 m
Height: (to hull top) 2.1 m
Ground clearance: 0.31-0.37 m
Angle of approach/departure: 35°/35°
Wheelbase: 3.2 m
Max road speed: 90 km/h
Max road range: 800 km
Fording: 0.7 m
Gradient: 70%
Side slope: 30%
Trench: 0.9 m
Vertical obstacle: 0.5 m
Engine: Mercedes-Benz OM 352A 6-cylinder direct-injection, turbocharged, water-cooled diesel developing 168 hp at 2,800 rpm
Transmission: manual with 5 forward and 1 reverse gears
Transfer case: 2 speed
Steering: power assisted
Turning radius: 7 m
Brakes: hydropneumatic
Suspension: leaf springs and shock-absorbers
Electrical system: 24 V
Armament: see text
NBC system: optional
Night vision equipment: optional

Note: Current production vehicles are powered by the Mercedes-Benz OM 366 LA 4 turbocharged intercooled diesel developing 240 hp at 2,600 rpm. This gives a maximum road speed of 81 km/h, a road range of 700 km and an off-road range of 450 km. Empty weight is quoted as 9,900 kg with payload being 3,600 kg. This version is known as the Fahd 240.

Fahd 240/30 (4 × 4) armoured personnel carrier deployed with the Egyptian Army in Sarajevo (Michael Jerchel) ***2001***/0103756

The gunner has a day/night binocular periscopic sight with a day magnification of ×6 and a 10° field of view, and a night magnification of ×5.5 and a 7° field of view, and four day periscopes, three to the front and one to the rear of the vehicle.

The two-man power-operated turret is armed with a 30 mm 2A42 dual-feed cannon, a 7.62 mm PKT machine gun mounted coaxially with the main armament and a roof-mounted AT-5 'Spandrel' ATGW launcher.

The AT-5 can also be deployed away from the vehicle if required by the tactical situation and has a maximum range of 4,000 m (with time of flight to this range being 20 seconds). The launcher has a traverse of 360°, elevation +15° and depression −5°. The AT-5 has a single HEAT warhead and will penetrate 550 mm of RHA.

In addition to the one missile in the ready to launch position, a further four missiles are carried internally, two either side of the hull.

Mounted either side of the turret is a bank of three 81 mm electrically operated grenade dischargers that fire to the front of the vehicle. Basic specifications where different from the standard vehicle:

Status
In production. In service with Algeria (at least 200 delivered for use in IS role), Egypt, Kuwait (IS role), Oman, the Sudan and Democratic Republic of Congo (IS role with BTM 105 turret). The exact status of the 110 Kuwaiti vehicles is uncertain, although it is believed most were captured during the Iraqi invasion of 1990.

Manufacturer
Kader Factory for Developed Industries.

UPDATED

Kader Walid armoured personnel carrier

Development
The Walid 1 armoured personnel carrier was designed and developed by the Kader Factory for Developed Industries near Cairo, with first production vehicles completed in 1960. It is also referred to as the Waleed and was used in combat for the first time in the 1967 Middle East campaign.

It basically consists of a German Magirus Deutz (4 × 4) chassis, which was manufactured under licence in Egypt, fitted with an armoured body. The vehicle is very similar in appearance and performs a similar role to the Russian BTR-40 (4 × 4) armoured personnel carrier which was also used by the Egyptian Army. The BTR-40 can be distinguished from the Walid by the vertical rather than sloped sides to the troop compartment. The Walid 2 was introduced in 1981 and is based on Mercedes-Benz automotive components.

In Egyptian Army service the Walid was replaced by the Fahd (4 × 4) armoured personnel carrier covered in detail in the previous entry.

Description
The hull of the Walid is made of all-welded steel with the diesel engine at the front, driver and commander in the centre and the troop compartment at the rear. The driver sits on the left with the commander to his right, both with a door in the side of the hull, the upper part of which folds down on the outside for increased visibility. Both the commander and driver have an individual windscreen to the front which, when in action, can be covered by an armoured shutter hinged at the top.

The open-topped troop compartment is at the rear of the hull and entry is by a door on which the spare wheel and tyre are carried. Three observation/firing ports are provided in each side of the hull with a further two at the rear of the hull, one either side of the spare wheel.

The Walid has no NBC system, no night vision equipment and no amphibious capability. A central tyre pressure regulation system is fitted as standard.

Walid (4 × 4) armoured personnel carrier

Specifications

Crew: 2 + 8-10	**Height:** 2.3 m
Configuration: 4 × 4	**Max road speed:** 86 km/h
Combat weight: 12,000 kg	**Max range:** 800 km
Unloaded weight: 9,200 kg	**Engine:** 168 hp diesel
Max payload: 2,800 kg	**NBC system:** no
Length: 6.12 m	**Night vision equipment:** no
Width: 2.57 m	

The basic model is armed with a 7.62 mm pintle-mounted machine gun but additional weapons can be mounted around the top of the hull.

Variant

Minelayer
This is a basic Walid fitted with a minelaying chute attached to the rear door. Anti-tank mines slide down the ramp and are laid on the ground where they can subsequently be buried by hand.

Status
Production complete. In service with Burundi, Israel (border guard), Sudan (120) and Yemen. Egypt may have up to 650 of these vehicles in reserve. No longer being marketed.

Manufacturer
Kader Factory for Developed Industries.

UPDATED

Kader G 320 armoured personnel carrier

Development
The G 320 (4 × 4) armoured personnel carrier has been designed as a private venture by the Kader Factory for Developed Industries and is based on the chassis of the commercial German Mercedes-Benz MB 320G (4 × 4) light vehicle.

In many respects, the G 320 is very similar to the German former Thyssen Henschel (today Rheinmetall Landsysteme) TH 444 (4 × 4) fully described in this section under Germany but the description relates to the Kader Weapons Carrier.

The standard armoured personnel carrier has two doors in the side and one in the rear, with large bulletproof windows on the front and sides.

This vehicle is based on the long-wheelbase version of the German Mercedes-Benz G 320 with the chassis frame being modified by a heavy-duty suspension which has been designed to withstand the additional weight imposed by the armour package.

The vehicle has been designed to be capable of moving on and off the road as well as over rocky and dusty terrain in temperatures of up to 50°C and 90 per cent humidity.

Description
The Mercedes-Benz MB 320G (4 × 4) light chassis has been fitted with a fully armoured steel body that

Kader G 320 armoured personnel carrier

provides full ballistic protection against attack from 7.62 mm × 39 rounds through 360° and against 7.62 mm × 39 ball over the roof. The windscreen, side and rear windows provide the same level of protection as the roof. Up to five firing ports can be provided if required by the user. The fuel tank is also protected by armoured shields.

The chassis is of the off-set ladder type with high bending and torsional resistance with the radiator and engine being provided with bottoming protection.

The independent suspension includes coil springs and twin-tube telescopic shock-absorbers, anti-roll bar (stabiliser) on front and rear axles.

Standard equipment includes run-flat tyres and a fully automatic transmission with the off-road ratio capable of being engaged even when the vehicle is on the move.

All the differential locks (transfer case, rear axle and front axle) can be engaged by push buttons, providing higher traction progressively and they are only engaged in the logical sequence.

The ventilated disc brakes and ABS system have been adapted to this off-road vehicle. The ABS system is switched off automatically if one of the differential locks is selected. The vehicle also has an electronic cruise control and locking system with infra-red remote control and driver's airbag.

Specifications

Crew: 1 + 4
Configuration: 4 × 4
Combat weight: 3,820 kg
Unloaded weight: 3,000 kg
Payload: 820 kg
Power-to-weight ratio: 56.28 hp/t
Length: 4.679 m
Width: 1.76 m
Height: 1.931 m
Ground clearance: 0.21 m
Track: 1.475 m
Wheelbase: 2.850 m
Angle of approach: 36°
Angle of departure: 37°
Max road speed: 150 km/h
Acceleration: 0-100 km/h in 11 s
Fuel capacity: 96 litres
Range: 700 km
Gradient: 80%
Side slope: 30%
Vertical obstacle: 0.3 m
Trench: 0.6 m
Fording: 0.6 m
Engine: Mercedes-Benz M112E32 6-cylinder water-cooled petrol developing 215 hp at 5,600 rpm
Transmission: automatic, 5 speeds forward, 1 speed reverse, torque converter
Transfer box: for mechanical front axle engagement and synchronised cross-country gear
Steering: recirculating ball, power assisted
Turning radius: 6.3 m
Suspension: front and rear, coil springs and telescopic shock-absorbers. Both front and rear axles are located by 1 transverse and 2 longitudinal links
Tyres: 265/70 R/6 in 7.5 J × 16
Number of tyres: 4 + 1 spare
Electrical system: 24 V
Batteries: 1 × 12 V
NBC system: no
Night vision equipment: no

G320 armament carrier

The engine is at the front, commander and driver in the centre and troop compartment at the rear.

To the front of the commander and driver is a bulletproof windscreen which, when in action, can be covered by an armoured shutter hinged at the top. The commander and driver are each provided with a forward-opening door with a bulletproof window in its upper part.

The troop compartment is at the rear of the hull with entry via a large door that opens to the left. This is provided with a bulletproof vision block and underneath this is a firing port.

Mounted on the rear door is a spare wheel and tyre and in either side of the troop compartment is a single bulletproof vision block with a firing port underneath.

Mounted on top of the roof is a manually operated turret armed with a 7.62 mm machine gun. This is provided with a single-piece roof hatch that opens to the rear with vision blocks being fitted for all-round observation.

Variants

Kader 594 armament carrier
Kader 596 ambulance
Kader 597 APC
Kader command and communications vehicle
Kader reconnaissance vehicle
Kader riot control vehicle

Status

Production as required. Not believed to be in service with Egyptian Army but may be used by paramilitary units.

Manufacturer

Kader Factory for Developed Industries.

UPDATED

EL SALVADOR

VAL (light assault vehicle) Cashuat

Development

To meet its requirements for an armoured personnel carrier, El Salvador built a mild-steel prototype based on the chassis of the old US unarmoured M37B1 (4 × 4) ¾ ton truck. Trials with this prototype showed that it lacked ballistic protection. The government of El Salvador then turned to the USA for assistance and the US Army Tank Automotive Command's RDE Center's Design and Manufacturing Technology Directorate became involved.

After studying the first prototype in El Salvador in February 1985 it was apparent that the country lacked the facilities to build the required numbers of vehicles.

Tank Automotive Command then offered a three-part plan to El Salvador. First, to build the prototype of the APC in its own facilities in Detroit and then take it to El Salvador to be used as a reference model. Second, to precut dual-hardness armour and carry out critical welding on the side panels, side rails and windshield before shipment to El Salvador, with a total of 66 kits to be provided. Finally, to train a team of El Salvadoreans in the USA in welding techniques.

This package was accepted by El Salvador and after carrying out some redesign work, including changing the driver's compartment doors and rear crew compartment doors from a flush to a surface mount design, modifying the rear fenders and cutting the hood configuration straight across in front of the windscreen, the first prototype was completed in Detroit in March 1985.

Trials in the USA showed that the armour could not be penetrated by 5.56 mm ball ammunition fired from a range of 25 m. This was followed by two days of mobility trials during which the prototype was driven over 500 km of varying terrain.

In April 1985, the prototype was shipped from Tank Automotive Command for its flight to El Salvador, 74 days after receiving the go-ahead. In May 1986, TACOM shipped the last of 66 kits to El Salvador, which assembled them at the rate of two a week.

By early 2001 it is estimated that 14 weapon carriers and 30 armoured personnel carriers remained in front-line service.

Description

This vehicle is essentially an M37B1 (4 × 4) truck chassis fitted with an armoured body to protect the occupants from small arms fire. Dual-hardness armour protects the crew compartment, high-hardness armour protecting less crucial areas. Kevlar is fitted inside the troop compartment and on the outside of the 12.7 mm turret. At present, belly armour is not fitted.

The layout is conventional with the armour-plated engine compartment at the front, driver and vehicle commander to its immediate rear, and the troop compartment at the very rear of the hull. Both commander and driver have a forward-opening side

Cashuat weapons carrier in latest configuration and fitted with twin rear wheels (Julio A Montes) ***2001***/0103757

El Salvadorean Cashuat weapons carrier with a turret-mounted 12.7 mm machine gun (Julio A Montes) ***2001***/0103758

door with a bulletproof window in the upper part and a firing port underneath. To their front is a bulletproof window with a firing port underneath and armour protection is provided above their position. Provision has been made for installing a radio.

The open-topped troop compartment is at the rear and the troops normally enter and leave via two outward-opening doors in the rear that have a firing port on either side. An additional three firing ports are provided in the upper part of each side of the troop compartment.

Before being fitted with the new armoured body, the chassis was stripped down, cleaned and reinforced, the suspension was uprated, and new brakes, shock-absorbers and springs were fitted.

The tyres are filled with a Du Pont component to allow the vehicle to travel up to 90 km with a perforated tyre.

Variants

There are two versions of this vehicle in service with the El Salvadorean Army. One is the armoured personnel carrier which has a crew of two and carries eight fully equipped troops. The second version is a gun carrier and is fitted with an armoured turret, to the immediate rear of the commander and gunner, which is armed with a 12.7 mm M2 HB machine gun and a 7.62 mm M60 machine gun and shield, mounted either side towards the rear.

Two prototypes of the Cashuat have been upgraded by adding the front and rear axles from the US-supplied Reo 2.5 ton (US) (6 × 6) trucks. This has resulted in increased off-road performance and a more stable vehicle.

Some vehicles have also been fitted with wire mesh screens to detonate RPG-7-type rocket propelled anti-tank projectiles before they contact the main armour of the vehicle.

More recently, some of these vehicles have been upgraded further by replacing the existing front and rear axles with new axles taken from 2.5 t (US) (6 × 6) Reo trucks. Two 12.7 mm M2 machine guns have been installed on the roof of the vehicle.

Specifications

Crew: 2 + 8
Configuration: 4 × 4
Weight:
(combat gun carrier) 4,223 kg
(loaded APC) 4,454 kg
(unloaded gun carrier) 3,769 kg
(unloaded APC) 3,547 kg
Length: 5.38 m
Width: 1.83 m
Height: 2.63 m
Ground clearance: 0.35 m
Max road speed: 80 km/h
Max range: 480 km
Fuel capacity: 90 litres
Engine: Detroit Diesel Model 353 developing 110 hp
Transmission: Spicer 3053, 5 forward and 1 reverse gears
Electrical system: 24 V
NBC system: no
Night vision equipment: no

Other El Salvadorean APCs

A wide range of other armoured personnel carriers, tracked and wheeled, had been designed and built in El Salvador. Most of these have now been phased out of front-line service. Details were given in *Jane's Armour and Artillery 1997-98*, pages 422 to 423. Details of the modified now Rheinmetall Landsysteme UR-416 (4 × 4) armoured personnel carrier are given in the entry for this vehicle under Germany.

Status

Conversions complete. In service with the El Salvadorean Army which is the only user of this vehicle.

Manufacturer

Chassis and armour package, USA. Final assembly and testing, National Armoury.

UPDATED

FINLAND

Patria Vehicles Armoured Modular Vehicle (AMV)

Development

The Armoured Modular Vehicle (AMV) has been developed by Patria Vehicles in close co-operation with the Finnish Defence Force (FDF) and leverages off the design, development and production of its highly successful XA series of 6 × 6 Armoured Personnel Carrier (APC).

These have been in continuous production since 1984, with the total order book now standing at around 1,000 units. Since it was first introduced, the XA series has been expanded to undertake a much wider range of battlefield missions.

User requirements are now demanding vehicles with not only greater mobility, but better protection, increased payload and lower overall life cycle costs.

This has led to the development of the AMV (8 × 8). A non-running engineering development model of the vehicle was shown in mid-2000 and fitted with the twin 120 mm AMOS (Advanced MOrtar System) which has been developed by Patria Hägglunds (a company comprising Patria Vammas of Finland and Hägglunds Vehicle of Sweden).

The first prototype of the AMV was completed in late 2001 and was in the basic hull configuration, which can be fitted with a wide range of 25 and 30 mm turrets for use in the Infantry Fighting Vehicle (IFV) role. Alternative kits are being provided for APCs, command posts and ambulances.

The second model will be in the system platform configuration that will have a higher roof line to the rear of the front of the vehicle where the power pack (right) and driver and vehicle commander (left) are positioned. This will have greater internal volume for more specialised roles such as command, communications, ambulance, workshop and the installation of a power-operated mast.

In the near future, Patria Vehicles hopes to get an FDF contract for two additional AMVs for field testing and serial production could start as early as 2004.

To reduce life cycle costs the AMV uses proven Commercial-Off-The-Shelf (COTS) components wherever possible. The prototype will be powered by a Scania 360 kW diesel of the same generation to that installed in the recently ordered Finnish Defence Force CV 9030 tracked vehicles, coupled to a German ZF 900 series fully automatic transmission with seven forward and one reverse gears. Other engine/transmission options will also be offered for the AMV.

Typical weapons fit would include a two-person turret armed with a 25 or 30 mm cannon and an AMOS (Advanced Mortar System) which has twin 120 mm breech-loaded mortars and is being developed to meet the operational requirements of Finland and Sweden. Full details of AMOS are given in the Self-propelled mortar section later in this volume.

The AMV has a high area of modularity in two dimensions. The first dimension is at the systems level that allows the customer to specify the power pack (engine/transmission), heating, air conditioning system, NBC system and wheel stations, for example. Diesel engines with outputs of 335 to 536 hp can be installed.

The second dimension is the mission level which means different versions and/or role kits. Patria Vehicles intends to offer the customer the following versions and/or modular role kits for the AMV:

AMV 8 × 8 basic model

Infantry Fighting Vehicle
Armoured personnel carrier kit
Command kit
Ambulance kit

AMV 8 × 8 system platform

Command kit
Communication kit
Mast version
Large ambulance kit
Workshop version

AMV 8 × 8 module carrier

Container carrier version
Missile module carrier

Patria Vehicles has stressed that the new AMV will be marketed alongside its existing XA series of 6 × 6 vehicles that entered production in the early 1980s, with the latest XA-200 series being referred to as the third generation. The XA-200 series has been constantly developed in recent years with the latest models having significant improvements in the areas of armour protection and mobility.

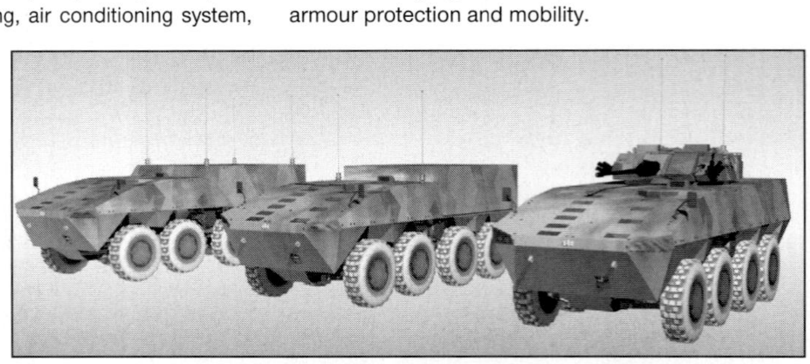

AMV family comprising from left to right systems platform configuration, modular carrier and basic platform fitted with 25 mm two-person turret
2002/0095615

Artist's impression of the new Patria Vehicles Armoured Modular Vehicle (AMV) (8 × 8), fitted with AMOS turret armed with twin 120 mm mortars
2001/0073063

Description

The hull of the AMV is of all-welded steel construction that provides the occupants with protection from small arms fire and shell splinters. To this, an additional layer of passive armour can be added to provide a higher level of protection.

Patria Vehicles has stated the overall survivability of the AMV is enhanced by not only its modular protection system but also signature management and a defensive aids suite, if fitted.

The layout of the new AMV is similar to the earlier XA 6 × 6 series with power pack front right, driver front left and the whole of the remainder of the hull rear being left free for the troop compartment.

The driver is provided with a single-piece hatch cover that opens to the left and to the immediate front of this are three day periscopes, the centre one of which can be replaced by a passive night device.

To the rear of the commander is another seated crew position, for example the crew commander. This position would typically have a cupola or a single-piece hatch cover that opens to the rear. The exact arrangement would depend on the mission requirement.

A number of different hull heights will be available in the AMV design for different user requirements, for example the higher roof one would be well suited for use as a command post or forward ambulance.

The weapon station would be mounted in the centre of the vehicle and the roof hatch arrangement would depend on the weapon fit. The normal means of entry and exit would be via the large door in the hull rear.

Suspension is of the hydropneumatic type with height adjustment being an option. Steering is power-assisted on the front two wheels. The AMV will feature full 8 × 8 drive with power steering on the front four wheels. Tyres will be 14.00 R 20 with run-flat devices fitted as standard. Options will include a central tyre pressure regulation system and anti-skid braking system.

The vehicle will also feature signature management and a variety of passive armour packages will be available up to 30 mm Armour Piercing Fin Stabilised Discarding Sabot (APFSDS) over the frontal arc. With standard armour and without weapons or crew, the AMV will weigh 14,000 kg with maximum combat weight being 24,000 kg.

With a combat weight of 18,000 kg the AMV is fully amphibious being propelled in the water by its wheels; with a higher combat weight the vehicle is not amphibious.

Specifications

Crew: 1+9
Configuration: 8 × 8
Combat weight:
(max) 24,000 kg
(min) 14,000 kg
Power-to-weight ratio: 22.33 hp/tonne (with a combat weight of 24 tonnes and fitted with a 536 hp diesel engine)
Length: 7.3 m
Width: 2.9 m
Height: 2.3 m (hull top)
Ground clearance: n/avail
Track: 2.3 m
Wheelbase: n/avail
Angle of approach/departure: n/avail
Max speed:
(road) 100/120 km/h
(water) 8 km/h (see text)
Fuel capacity: n/avail
Max range: 800 km
Fording: 1.50 m
Gradient: 60%
Side slope: 30%
Vertical obstacle: 0.70 m
Trench: 2 m
Turning radius: n/avail
Engine: diesel with an output of between 335 and 536 hp
Transmission: manual or automatic with 6/7 forward and 1 reverse gears
Suspension: hydropneumatic with optional height adjustment
Steering: power-assisted
Tyres: 14.00 R 20 with run-flat inserts and central tyre-pressure inflation system
Brakes:
(main) n/avail
(parking) n/avail
Electrical system: 24 V
Batteries: n/avail
Generator: n/avail
Armament: depends on mission
Ammunition: depends on armament
NBC system: optional
Night vision equipment: optional

Note: Specification depends on customer requirements and mission to be undertaken

Variants

See development.

Status

Development.

Manufacturer

Patria Vehicles Oy.

UPDATED

Patria Vehicles Oy XA-series armoured wheeled vehicles

Development

To meet the requirements of the Finnish Defence Forces for a new 6 × 6 armoured personnel carrier, SISU (now Patria Vehicles Oy) built the XA-180 (6 × 6) and Valmet the Model 1912-6 (6 × 6), with both prototypes completed in 1982.

Following trials with these vehicles, in December 1983 the Finnish government placed an initial order for 59 SISU XA-180 APCs at a cost of FMK42 million, with deliveries taking place over a three year period.

Of the initial order, 50 were for the Finnish Defence Forces and the remaining nine for the Finnish United Nations Battalion.

The first production XA-180 series vehicles were followed by the improved XA-185. By late 2001 total production of the XA series 6 × 6 APCs and variants amounted to 1,000 units, with production scheduled to continue to at least 2004.

Main improvements of the XA-185 over the original production XA-180 can be summarised as a more powerful Valmet 612 DWI six-cylinder diesel engine and new stronger axles.

Further development of the XA-185 resulted in the XA-186 which has a number of detailed improvements for UN-type missions.

These include additional passive armour protection, wire cutting device on front of the hull, weapon mount in hull front, firing devices in the hull sides with associated roof-mounted periscopes and a UK Helio one-man turret armed with a 12.7 mm M2 machine gun with a bank of five electrically operated grenade dischargers mounted either side of the turret.

Unlike the standard XA-185, which is propelled in the water by two water-jets mounted one either side of the hull at the rear, the XA-186 is not amphibious.

To further extend the operational capabilities of the XA-186, the vehicle has been fitted with the one-man Norwegian Kvaerner-Eureka Armoured Launching Turret (ALT) with four, two either side, 3,750 m range Raytheon Systems Company TOW anti-tank guided weapons in the ready to launch position.

This is the latest version of the ALT which, in its standard two-round version, has been built in large numbers for Canada (M113A1), Norway (M113A1), Switzerland (MOWAG Piranha 6 × 6) and Turkey (Turkish Infantry Fighting Vehicle).

As of late 2001 there were no customers for the XA-186/ALT combination.

In mid-1999, the Finnish Defence Force (FDF) placed a contract for 48 XA-203 series vehicles for use by the Pori Readiness Brigade with deliveries running from late 1999 through to early 2001.

Late in 2000, Patria Vehicles was awarded a contract worth about FMk400 million from the FDF for the supply of about 100 XA-202 (6 × 6) armoured personnel carriers (ACP). In addition there is also an option of another 70 units.

Under the terms of this contract, the first small batch of XA-202 vehicles were delivered late in 2001 with the main production run being delivered between 2002 through to 2004.

As of late 2001, the current production models are the XA-202 and the XA-203. The former is the baseline APC version while the XA-203 has a raised rear roof for greater internal volume for use in more specialised roles such as command post and artillery fire control. Both vehicles also feature a higher level of protection against small arms fire, anti-tank and anti-personnel mines.

The FDF has already taken delivery of 48 XA-203 vehicles ordered in 1999 for use in specialised missions. One of these was fitted with the Patria

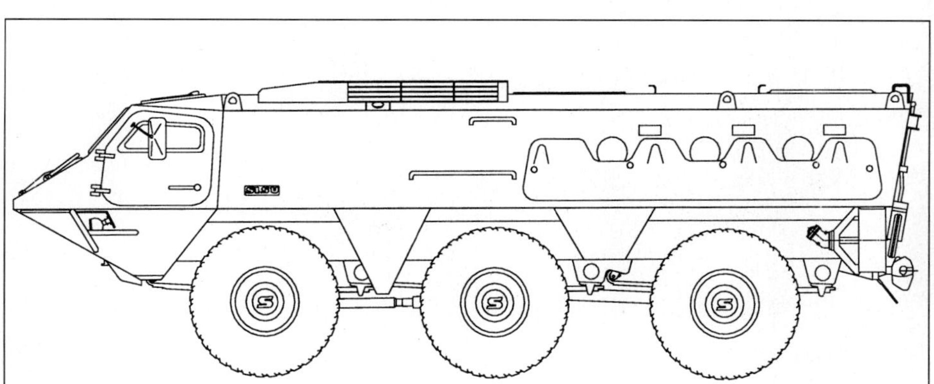

XA-185 (6 × 6) APC without armament installed

XA-180 (6 × 6) APC fitted with the Swedish Ericsson Giraffe radar system which was fitted in Finland by Jantronic Oy. When travelling, the radar folds down on top of the hull

Finnish Crotale New Generation on XA-181 (6 × 6) vehicle

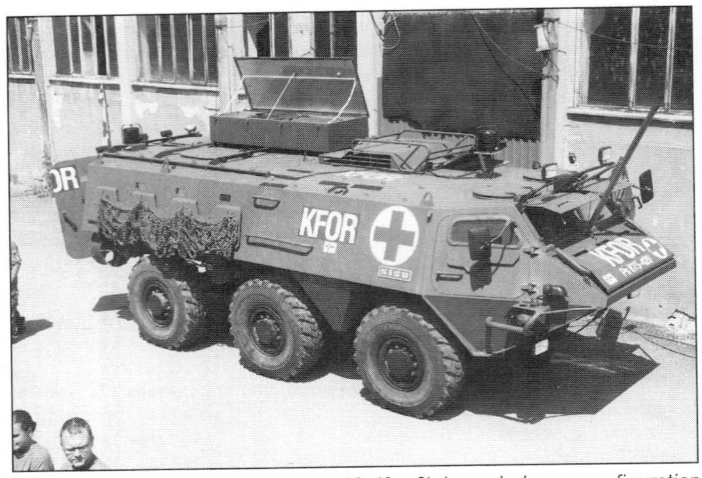

Finnish Army Patria Vehicles XA-180 (6 × 6) in ambulance configuration (Richard Stickland) **2002**/0121142

Vammas twin 120 mm AMOS (Advanced MOrtar System) and this is now undergoing trials with the FDF.

Sweden has ordered a total of 104 vehicles designated the XA-202S and XA-203S and some of these will be fitted with a one-person turret armed with a 20 mm cannon. The S in the designation stands for Sweden.

In September 2001, the first eight XA-203S vehicles were handed over to the Swedish Army in Stockholm. Of the 104 vehicles, 86 are XA-203S and the remaining are XA-202S command post vehicles. Some of the XA-203S vehicles will be fitted with turrets removed from Swedish Army Pbv 302 series APCs.

Patria vehicles have already completed 92 XA-200 series vehicles for the Netherlands, of which 72 were for the Army and 20 for the Marines. These were ordered late in 1996 and deliveries took place from 1998 to 1999 and are designated the XA-188. These were up-armoured to provide the occupants with protection from 14.5 mm armour-piercing ammunition and this increased the weight of the vehicle from 13.2 to 17.4 tonnes.

Norway has now taken delivery of its 32 latest vehicles (22 plus the option on 10) that are designated the XA-203N. These have been designed so that the user can quickly change roles according to mission requirements such as command, ambulance and workshop. It should be noted that the first Norwegian models were XA-186 versions with specialised models including the XA-186 CPV command post vehicle and the XA-186 MV repair and recovery vehicle which is fitted with a crane.

Description

The hull of the XA-185 is of all-welded steel armour construction providing the crew with protection from small arms fire and shell splinters. If required by the customer, a higher level of armour protection can be provided as can different levels of EMP and NBC protection. The commander sits at the front of the hull on the right with the driver on the left. Both enter the vehicle via forward-opening doors at the front of the vehicle, one in each side. In the upper part of the door is a bulletproof window which can be partly covered by a shutter. To the front of both commander and driver is a large bulletproof windscreen that can be covered by a shutter hinged at the top when in a combat area. Above the driver's position is provision for installing a night observation device and rear-opening single-piece hatch covers.

The engine compartment is to the rear of the driver on the left side with the air inlet and outlets in the roof. An aisle on the right side of the hull allows passage from the front crew compartment to the troop compartment at the rear of the hull.

The infantry sits on bench seats in the rear troop compartment five down either side, and enters via two doors in the rear which both have firing ports and vision blocks. On either side of the troop compartment are three vision blocks with a firing port underneath allowing the troops to fire their weapons from inside the vehicle. Over the top of the troop compartment are two roof hatches hinged in the centre which can be locked vertical if required, and a mount for a 12.7 mm machine gun.

The suspension of the XA-185 consists of leaf springs and hydraulic shock-absorbers. Steering is power assisted on the front two axles. The vehicle is fully amphibious, being propelled in the water by two propellers mounted under the hull rear, one either side. Before entering the water a trim vane is erected at the front of the vehicle. Standard equipment includes a winch with a capacity of 9 tonnes and 50 m of 16 mm cable, and an automatic fire suppression system for the engine compartment. If required, the XA-185 can be fitted with an NBC system and its tyres fitted with snow chains.

When originally delivered to the Finnish Defence Forces the XA-180 (6 × 6) vehicles were unarmed, but many vehicles have now been fitted with a cupola-mounted 12.7 mm NSV machine gun which is also installed on the more recent Russian AFVs such as the T-72 which is also used by the Finnish Defence Forces. The Finnish United Nations vehicles have an all-welded turret armed with a Russian 12.7 mm DShKM machine gun with a bank of five smoke grenade dischargers mounted either side of the turret. Some Finnish Defence Forces XA-180 vehicles are also fitted with a roof-mounted AT-4 'Spigot' ATGW system.

Variants

XA-180 with 12.7 mm MG

In 1992, Manroy Engineering of the UK fulfilled a contract to convert the machine gun mounts on 12 XA-180 (6 × 6) vehicles for the United Nations forces in the Lebanon.

This contract covered the supply of new .50 cal (12.7 mm) M2 HB (Heavy Barrel) Quick Change Barrel (QCB) weapons and the modification of the existing Russian mount (which was designed for the 12.7 mm NSV heavy machine gun) to accept 0.50 cal M2 HB standard or QCB machine guns.

Initial work carried out by Manroy led to the production of conversion kits to allow local fittings of the mounts. The bolt-on kits were designed to allow quick (20 minutes) fitting of the mounts. Acceptance tests and firing trials were then successfully completed in Ireland.

Specifications

Crew: 3 + 8
Configuration: 6 × 6
Combat weight: 22,000 kg
Unloaded weight: 18,000 kg
Power-to-weight ratio: 13.40 hp/t
Payload: 4,000 kg
Length: 7.45 m
Width: 2.95 m
Height: 2.6 m
Ground clearance: 0.4 m
Max speed: (road) 90 km/h
Fuel capacity: 290 litres
Max road range: 800 km
Gradient: 60%
Side slope: 60%
Engine: Valmet 612DWIBIC (EURO 2), 6-cylinder in line turbocharged and intercooled diesel developing 295 hp at 2,000 rpm
Transmission: Allison MD 3560PR electronically controlled automatic with torque converter, power take-off and with 6 forward and 1 reverse gears
Steering: hydraulic power assisted, front two axles
Tyres: 14.00 × 20 or 14.00 × R20 with run-flat device
Suspension: leaf springs, hydraulic shock-absorbers
Electrical system: 24 V
Armament: 1 × 12.7 mm MG
Smoke-laying equipment: 2 × 5 smoke grenade dischargers
NBC system: optional
Night vision equipment: optional

Note: Above specification relates to new X-200 announced in early 1999

XA-185 with LAV-25 and LAV-30 turrets

For trials purposes, an XA-185 has been fitted with the US Delco Defense LAV-25 turret armed with a stabilised 25 mm The Boeing Company M242 Chain Gun and 7.62 mm coaxial machine gun. This turret is standard equipment on all members of the General Motors Defense LAV-25, although the latest version of this

Norwegian Army Patria Vehicles XA-186 (6 × 6) APC fitted with Helio FVT 800 series turret (Richard Stickland) **2002**/0121143

turret has many improvements including all-electric gun control equipment. It has also been fitted with the LAV-30 turret armed with a 30 mm cannon.

Air defence radar

The Finnish Defence Forces have taken delivery of a number of XA-180 (6 × 6) vehicles fitted with the Swedish Ericsson Giraffe air surveillance radar system on a hydraulic arm. This version of the vehicle has been developed by the Finnish company, Jantronic Oy (which is now part of Patria Vehicles Oy), which has manufactured the hydraulic mast, electrics, heating, raising and armouring system. To provide a more stable platform, stabilisers are lowered to the ground before the roof-mounted scissors-type mast is raised into the operating position.

The system is used to provide key target information to anti-aircraft guns and surface-to-air missile systems in the forward battlefield area. This model is designated the Jantronic J-1000.

Crotale New Generation on XA-181 chassis

In mid-1988, Finland placed a contract with Thales Air Defence for the Crotale New Generation (Crotale NG) air defence system mounted on the XA-181 (6 × 6) chassis. This was the first order for Crotale New Generation which has been developed as a private venture by Thales Air Defence, with the VT-1 missiles being developed by the now Lockheed Martin (then LTV) of the USA.

The system is mounted on the roof of the XA-181 and has a total of eight missiles in the ready to fire position plus tracking and surveillance radars.

Finland sent the XA-181 chassis to Thales Air Defence complete with an NBC system, an air conditioning system, an auxiliary power unit and stabilisers. Thales Air Defence then integrated the Crotale NG SAM system, completed trials and delivered the complete system to Finland.

It is believed that the Finnish Defence Forces ordered a total of 20 of these systems based on the XA-181 chassis, first production systems being delivered to Finland in September 1992 with final deliveries in late 1993.

Full details of the Thales Air Defence Crotale New Generation air defence system are given in *Jane's Land-Based Air Defence 2001-2002.*

Swedish Army Patria Vehicles XA-180 (6 × 6) with cupola mounted 12.7 mm M2 machine gun (Richard Stickland)
2002/0121141

AMOS 120 mm Advanced Mortar System

Details of this twin 120 mm turret mortar system, developed as a private venture between Patria Vammas and Hägglunds Vehicle are given in the Self-propelled mortar systems section. In 2000 an XA-203 fitted with AMOS was delivered to the Finnish Defence Force for trials.

Other variants

The Finnish Defence Forces also have an ambulance and command post vehicle based on the XA-180 (6 × 6) APC.

The manufacturer has also proposed other versions of the XA-series of wheeled armoured vehicles including ambulance, command, IS, NBC reconnaissance, mortar carrier, anti-aircraft gun, fire support/anti-tank with two-man turret armed with 90 mm gun, anti-tank with the Euromissile Mephisto ATGW system which launches 4,000 m range HOT missiles, minelayer and repair and recovery.

Status

Production of the XA-185 has been completed. Production of the improved XA-188/XA-200 is now under way. In service with the following countries:

Country	Quantity	Comment
Finland	600+	additional vehicles on order
Ireland	2	for United Nations roles
Norway	80	some for United Nations roles
Sweden	32	all delivered, plus 104, first deliveries September 2001
Netherlands	92	72 Army and 30 for Marines

Manufacturer

Patria Vehicles Oy.

UPDATED

FRANCE

Satory Military Vehicle VBCI (8 × 8) infantry combat vehicle

Development

Late in 2000, following a competition, the French Delegaton Generale Pour l'Armement (DGA) awarded a contract to the Giat Industries/Renault Vehicules Industriels (RVI) consortium for the design, development and production of 700 VBCI (Vehicule Blinde de Combat d'Infanterie) for the French Army. To handle this programme the two well established companies formed a new company called Satory Military Vehicles.

Artist's impression of the VBCI (8 × 8) vehicle being used in an urban situation with infantry dismounting at the rear
2002/0095253

The headquarters of Satory Military Vehicles is at Satory, outside of Paris which is also the headquarters of the state owned Giat Industries. The DGA awards contracts to Satory Military Vehicles which in turn awards contracts to Giat Industries and RVI.

As well as handing the VBCI programme, Satory Military Vehicles will also be responsible for the current VAB family of 4 × 4 and 6 × 6 armoured personnel carriers, AMX-10P full tracked vehicles and variants and the AMX-10RC 6 × 6 armoured cars for the home and export markets as well as upgrades for these vehicles.

Satory Military Vehicles in effect merges the strategic outlines of both companies in the field of light and medium armoured vehicles.

According to the terms of the contract concluded between the two groups, RVI designs and controls the mobility and onboard network functions and the driver's station. For prototypes and production vehicles it will produce all of the mobility components including engine, transmission, suspension, axles, electrical circuits and driver's station.

Giat Industries will design and integrate the protection, fire, observation and command functions and ensure the complete integration for all vehicles. They will be responsible for the armoured hull, interior fitting out, assembling of the IFVs with the 25 mm turret and the complete integration of all vehicles.

Of the 700 vehicles, 550 will be in the VBCI configuration with the remaining 150 in the command post role. The first contract is for 65 vehicles, 54 in the infantry fighting vehicle configuration and 11 in the command post role. In the future it is expected that other variants of the VBCI will be developed.

At this stage no decision has been taken as to where final production and integration of the VBCI will

Specifications

Crew: 3 + 8
Configuration: 8 × 8
Combat weight: 28,000 kg
Unloaded weight: not available
Power-to-weight ratio: 19.82 hp/tonne
Length: 7.60 m
Width: 2.98 m
Height: 2.20 m
Ground clearance: 0.50 m
Track: not available
Wheelbase: not available
Angle of approach/departure: 45°/45°
Max road speed: 100 km/h
Fuel capacity: not available
Max road range: 750 km
Fording: 1.2 m
Gradient: 60%
Side slope: 30%
Trench: 2 m
Vertical obstacle: 0.70 m
Turning radius:
(normal) 11 m
(skid) 8.6 m
Engine: diesel developing 555 hp
Transmission: automatic with four central differentials
Steering : power-assisted
Tyres: Michelin 395/90 R22 XML
Brakes: not available
Suspension: independent hydropneumatic at all wheel stations
Electrical system: 24 V
Batteries: not available
Armament:
(main) 1 × 25 mm cannon
(co-axial) 1 × 7.62 mm machine gun
(grenade launchers) 2 × 4 (turret); 2 × 3 (hull)
NBC system: yes
Night vision equipment: yes
Internal volume: 13 m^3
Air transportability:
(C-130) no
(A-400M) yes

The VBCI will be fitted with the latest version of the Giat Industries Dragar one-person turret armed with a 25 mm cannon and 7.62 mm coaxial machine gun **2002**/0095254

be undertaken. This could be the former RVI facility at Saint Chammond where production of the VAB series of 4 × 4 and 6 × 6 APCs is undertaken or at the Giat Industries facilities at Roanne, where final integration of the Leclerc MBT is currently undertaken.

It is expected that first prototype of the VBCI will be completed late in 2002 or early 2003 with production starting in 2004. Export marketing of the VBCI has already commenced, especially in the Middle East.

In French Army service the VBCI will first replace the current Giat Industries tracked AMX-10P series vehicles and also the VAB (4 × 4) in the longer term.

To reduce operating and procurement costs, proven Commercial-Off-The-Shelf-components will be used wherever possible in the design of the vehicles.

Description

The hull of the VBCI will be of all-welded aluminium with an additional layer of modular passive armour and internal spall liners. This is claimed to provide protection over the frontal arc against medium-calibre attack. An anti-tank guided missile decoy system will be mounted on the turret front and a Galix grenade system fitted as standard to further improve battlefield survivability.

The Galix system is already in service with the French Army installed on the Giat Industries Leclerc MBT. On the VBCI Galix will be installed on the 25 mm Dragar turret (four launchers either side) and at the rear of the hull (three launchers either side).

It will have a crew of three (commander, gunner and driver) and carry eight fully equipped infantry, who can rapidly enter and leave the vehicle via the power operated ramp in the hull rear.

The driver is seated at the front of the vehicle on the left side and has a single-piece circular hatch cover that opens to the left, in front of this are three day periscopes that give observation over the frontal arc. The centre day periscope can be rapidly replaced by a passive periscope for driving at night.

The power pack compartment is front right and separated from the remainder of the hull. The power pack is expected to consist of RVI diesel developing 555 hp coupled to fully automatic transmission.

A large access hatch is provided in the glacis plate that opens to the right. Two wire cutters will be mounted on the glacis plate, on either side. These will fold back down onto the glacis plate when not required.

The vehicle commander, who dismounts with the troops, is seated to the immediate rear of the driver and has a roof-mounted hatch and three periscopes that cover the frontal arc. The vehicle commander can also lay and fire the armament of the Dragar turret using a panoramic sight.

The VBCI will be fitted with the latest version of the Giat Industries one-person Dragar turret armed with a 25 mm dual-feed cannon and a 7.62 mm machine gun mounted coaxial to the right. The gunner has a roof mounted day/thermal sight with day periscopes giving observation around the turret. The Dragar turret is offset slightly to the right.

The troop compartment is at the rear, with the troops being seated down either side facing inwards. They enter and leave the VBCI by a large power-operated ramp in the hull rear that opens downwards and which is also provided with an emergency door in the left side.

Over the top of the troop compartment are two rectangular roof hatches that open outwards. Mounted in the roof of the troop compartment at the rear are four periscopes situated two either side that give the troops observation through an arc to the rear of about 215°.

Mounted on either side of the hull, towards the rear, are two external stowage boxes for the troops personal equipment.

Standard equipment will include a central tyre inflation system, run flat device, hydropneumatic suspension, air conditioning, NBC system and powered steering on the front four wheels. In addition, there is skid steering for very sharp turning. The engine and crew compartments are fitted with a fire detection and suppression system as standard. It will also have a battle management system and battlefield IFF.

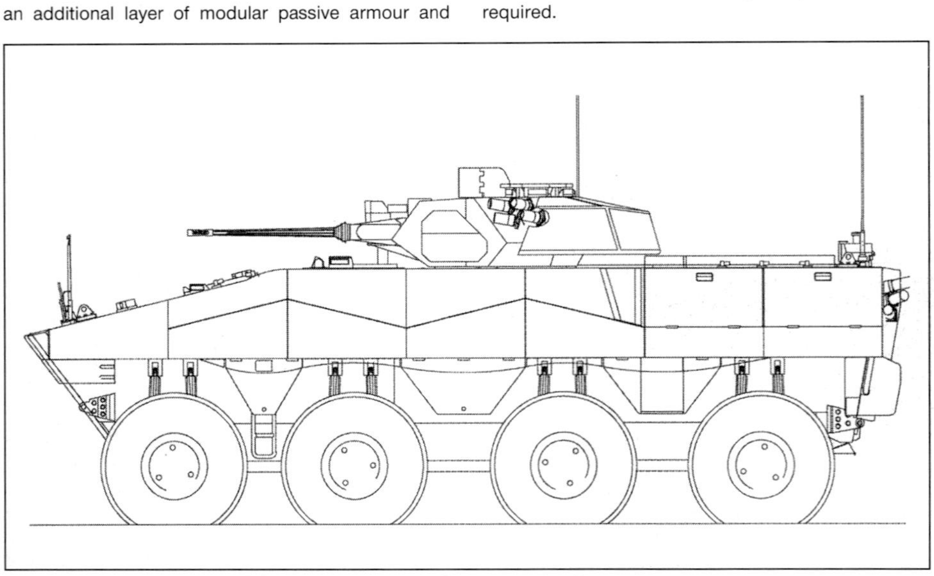

Detailed side drawing of VBCI (8 × 8) fitted with Giat Industries Dragar turret **2002**/0114260

Variants

VBCI/VPC command post vehicle

This is the only variant of the VBCI announced to date and will be fitted with a one-person turret armed with a 12.7 mm machine gun and have extensive communications equipment for its specialist role. It will weigh 24 tonnes and have a total crew of seven of which five will be command staff.

Status

Development. French Army requirement is 700 units (550 in infantry fighting vehicle and 150 in command post configurations).

Manufacturer

Satory Military Vehicles.

NEW ENTRY

Giat Industries Improved VAB armoured personnel carrier

Development

Since 1996, Giat Industries has been offering an improved version of the VAB armoured personnel carrier in 4 × 4 and 6 × 6 configurations.

The Improved VAB was developed and funded by Renault VI and Giat Industries in co-operation with the French Army and is the updated successor of the original VAB covered in detail in the following entry.

By late 2001, no firm production orders had been placed for the Improved VAB although marketing is continuing, especially in Asia.

The two main customers in this area were Indonesia and Thailand but current economic problems may well delay any procurement by these two countries.

While Giat Industries is now responsible for overseas marketing of the VAB family of 4 × 4 and 6 × 6 vehicles, within France, Renault VI is responsible for any upgrade work carried out on the vehicles of the French Army.

The French Army is considering upgrading part of its fleet of VAB vehicles to extend their operational life pending the introduction of the new Satory Military Vehicles VBCI (8 × 8) infantry fighting vehicle.

Description

The Improved VAB is identical in overall layout to the original production VAB covered in detail in the following entry.

The main features of the Improved VAB cover the areas of armour protection, mobility and ergonomics with the full list of improvements being as follows:

(1) Driver and commander have a high-protection bulletproof windscreen
(2) As an option, the vehicle can have armour protection against medium calibre armour-piercing ammunition
(3) Direction indicators at the front of the hull are now recessed low in the hull
(4) Hydraulically powered front-mounted winch
(5) Car-type driver's dashboard and steering wheel
(6) Option of integrated air conditioning system with controls on driver's dashboard
(7) Reinforced down transmission
(8) Six-speed automatic transmission with lock up device
(9) Integrated NBC system (option)
(10) RVI 300 hp turbocharged six-cylinder diesel engine
(11) Reinforced suspension
(12) Remote-controlled tyre inflation/deflation system as an option
(13) Option of 2 × 3 GALIX grenade launchers mounted either side of turret firing forwards
(14) Troop compartment floor is flat and washable
(15) As an option the troops can have individual fold-up seats
(16) New engine cooling system
(17) Optional add-on armour to sides of troop compartment
(18) Splinter-absorbent liner for crew compartment as an option
(19) As an option, the engine compartment can be fitted with a Molotov cocktail protection system
(20) More room in commander's and driver's compartment
(21) As an option, the commander's and driver's shutters can be supplied made of composite armour
(22) Wide range of weapon installations including Giat Industries Dragar turret armed with 25 mm cannon and 7.62 mm coaxial machine gun.

Giat Industries Improved VAB (6 × 6) fitted with Giat Industries cupola armed with 12.7 mm M2 machine gun
0058076

Specifications
(4 × 4 version: data in brackets relate to the 6 × 6 model where different)
Crew: 2 + 10
Configuration: 4 × 4 [6 × 6]
Combat weight: 13,600 [14,800] kg*
Power-to-weight ratio: 22.0 [20.3] hp/t
Length: 6.102 m
Width: 2.5 m
Height: (to hull top) 2.1 m
Ground clearance:
(axles) 0.4 m
(hull) 0.5 m
Wheelbase: 3 m [1.5 + 1.5]
Angle of approach/departure: 45°/45°
Max road speed: 110 km/h
Min road speed: 4 km/h
Fuel capacity: 300 litres (+ four 20-litre feeders)
Range: (at 60 km/h) 1,000 km
Gradient: 60%
Side slope: 40%
Trench: n/avail [1.5 m]
Vertical obstacle: 0.6 m
Engine: Renault MIDR 06.20.45 EURO 2 in-line water-cooled turbocharged diesel developing 300 hp at 2,100 rpm
Transmission: ZF automatic with 6 forward and 1 reverse gears, torque converter with lock up clutch
Steering: hydraulic assisted
Tyres: 14.00 × R20 run-flat
Brakes: disc on all wheels, dual circuit
Suspension: independent with torsion bar and telescopic shock-absorbers
Electrical system: 24 V
Batteries: 4 × 12 V, 6TN
NBC system: yes
Night vision equipment: optional

Status

Development complete. Ready for production on receipt of orders.

Manufacturer

Giat Industries.

UPDATED

Giat Industries VAB armoured personnel carrier

Development

In the late 1960s, the French Army decided to equip its infantry units with both tracked and wheeled troop carriers. The mechanised units would be issued with the tracked Giat Industries AMX-10P ICV, then already under development, and the remaining units with wheeled APCs as the tracked AMX-10P was considered too expensive and sophisticated for many of the roles it was expected to undertake.

In 1970, the French Army issued a requirement for a Forward Area Armoured Vehicle (VAB or *Véhicule de l'Avant Blindé*) which would meet the requirements of the remaining infantry units and would be capable of undertaking a wide range of roles, including use as an APC, a cargo carrier and a mortar towing vehicle.

To meet this requirement, Panhard and Saviem/Renault Group built prototypes of both 4 × 4 and 6 × 6 vehicles between 1972 and 1973. These were tested by the *Section Technique de l'Armée de Terre* between May 1973 and early 1974 and, in May 1974, the 4 × 4 version of the Saviem/Renault Group entry was selected for service.

There were no preproduction vehicles built and first production vehicles were delivered to the French Army in 1976.

The largest export order so far is for 394 units from Morocco, including APCs and command vehicles. All of these were in the 6 × 6 version.

The VAB was used by France and Qatar during Operation Desert Storm, the recapture of Kuwait, early in 1991. The VAB with HOT ATGWs successfully engaged Iraqi AFVs during this conflict.

By late 2001, total production of the VAB family amounted to over 5,000 vehicles in both 4 × 4 and 6 × 6 configurations for the home and export markets of which 3,975 were for the French Army.

Prime contractor was SMS jointly formed by Mécanique Creusot-Loire and Renault Véhicules Industriels, but export marketing and manufacture are now carried out by Giat Industries. Production of the VAB was completed in 1993 but production can resume again if further orders are placed.

Further development has resulted in the Improved VAB which is covered in the previous entry. In early 2001 this vehicle also remained at the prototype stage.

In the future, production and marketing of this vehicle will be carried out by Satory Military Vehicles, a joint venture company between Giat Industries and RVI.

Description

The basic model used by the French Army is the 4 × 4 VAB VTT *(Véhicule Transport de Troupe)* which has a crew of two (commander/machine gunner and driver) and carries 10 fully equipped infantrymen. The following description relates to this vehicle.

The all-welded steel hull of the VAB provides the crew with protection from small arms fire and shell splinters. The driver sits at the front of the vehicle on the left with the commander/machine gunner to his right. Both crew members have a side door that opens to the front, with a bulletproof window in its upper part hinged at the top and opening outwards that can be covered by a shutter. In front of the crew is a heated bulletproof windscreen that can be covered by a flap hinged at the

VAB (6 × 6) armed with 12.7 mm machine gun in service with Morocco (Michael Jerchel) **2001**/0103761

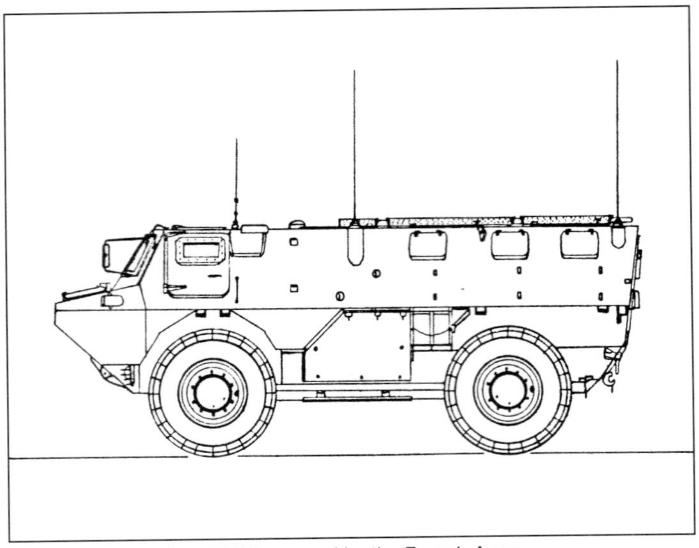

Basic (4 × 4) version of VAB as used by the French Army

top. Over the driver's position is a single-piece hatch cover that opens to the front. A similar hatch is mounted over the commander's position but vehicles issued to the French Army have a Giat Industries rotating gun mount type CB 127 armed with a 12.7 mm M2 machine gun with an elevation of –15 to +45° with the shield in the normal position and from –20 to +80° in the anti-aircraft position; in both cases traverse is a full 360°.

The engine compartment, which is fitted with a fire extinguishing system, is immediately behind the driver with the air inlet and outlet louvres in the roof and the exhaust pipe running along the top of the hull on the right side. Power is transmitted from the engine to the wheels by a hydraulic torque converter and a gearbox with five forward and one reverse gears. Gears are shifted using a small electrically operated lever which also operates the clutch. The axles have differential reduction gears with double reduction and differential locking. The wheels are independently suspended by torsion bars and hydraulic shock-absorbers. Steering is hydraulically assisted on the front wheels, or, in the case of the 6 × 6 model, on the front four wheels. The tyres are Michelin run-flat radials, the pressure of which can be adjusted to suit the type of ground being crossed. A remotely controlled tyre inflation/deflation device is available as an option on the VAB family.

There is a passageway on the right side of the hull which connects the crew compartment at the front with the personnel compartment at the rear with a flat floor. The infantrymen enter and leave the VAB via a double door without a central pillar in the rear of the hull which opens outwards. Each door has a window that can be opened to the outside and is covered by an armoured shutter.

The infantrymen sit on bench seats that run along either side of the hull. In each side of the hull are three windows which can be opened to the outside and can be covered by an armoured shutter. They can be locked open on the outside to allow the infantrymen to use their small arms from inside the vehicle.

On the basic VAB there are three hatches over the troop compartment, one in the centre and two at the rear, all of which open towards the front. The troop compartment is 2.46 m long, 1.375 m wide and 1.35 m high and has a total floor area of 3.38 m^2. The seats can be quickly folded up to enable cargo to be carried. The lack of a central pillar at the rear enables cargo, up to 2,000 kg, to be quickly loaded with the aid of a fork-lift truck.

The VAB is fully amphibious, being propelled in the water by two water-jets at the rear of the hull, each fitted with a deflector for steering and reverse thrust, which is hydraulically controlled by a joystick mounted on the dashboard. Before entering the water, the bilge pumps are switched on and the trim vane, which is folded back onto the glacis plate when not required, is erected at the front of the hull. Not all export customers have water-jets fitted.

In the French Army, the VAB is also used to transport MILAN anti-tank teams. Optional equipment for the VAB includes an air conditioning system, an armoured grenade thrower hood, a firing gun port with incorporated observation port, a gas dispenser, an NBC system (standard for French Army vehicles), passive night vision equipment, a public address system and a front-mounted winch with a capacity of 7,000 kg (which can be doubled with the aid of a pulley) and 60 m of cable.

VAB upgrade

Since the VAB series of 4 × 4 and 6 × 6 vehicles entered service with the French Army in 1976, some 5,000 vehicles have been built for the home and export markets. Over the years over 1,100 detailed improvements have been carried out on the vehicle.

Giat Industries and RVI are now offering a number of upgrades for the VAB series to extend their operational lives into the 21st century.

These include:

(1) Conversion of 4 × 4 models into 6× 6 version for improved cross-country mobility. The 4 × 4 VAB comprises a steering and driving front axle assembly and a driving rear axle assembly. The 6 × 6 VAB comprises a middle axle assembly identical to the front one. Each axle assembly comprises roadwheels with independent suspension. Modular components are used to obtain the steering 6 × 6 from the 4 × 4 configuration and the braking system is adjusted for the 6 × 6 configuration.
(2) Fitting of more powerful armament installations up to the Giat Industries Dragar one person turret armed with 25 mm cannon and 7.62 mm coaxial machine gun. In this configuration the vehicle has a crew of three, commander, gunner and driver and can carry four troops.
(3) Fully automatic transmission to replace current manual transmission for vehicles powered by a MAN or MIDS engine.
(4) Upgraded suspension including the installation of suspension rod pins to provide constant lubrication to articulation bearings. For improved road and cross-country drive, enhanced shock absorbers are provided at each wheel station.
(5) Improved braking system to comply with the latest standards. This involves the replacement of most of the existing braking system with a brand new system and the installation of an air dryer.
(6) Installation of central tyre pressure regulation system that allows the driver to adjust tyre pressure to suit the ground being crossed. The air pressure is transmitted to the roadwheels by a rotary joint incorporated in the gear unit fitted to each wheel hub.
(7) Installation of air conditioning system with cooling being accomplished by a compression refrigeration system using coolant. The compressor is mechanically driven by the engine with the condenser being fitted outside the vehicle.
(8) Special equipment for use in the internal security role including front-mounted obstacle clearing blade, crowd pusher and special anti-Molotov cocktail equipment.

Variants

VCI (infantry combat vehicle)
VAB VCI T20

VAB VCI T20 has a crew of 9 to 11 and is fitted with a Giat Industries T20 one-person electrically operated turret armed with a 20 mm cannon and a 7.62 mm

VAB (4 × 4) of the French Army without machine gun fitted (Pierre Touzin) **2000**/0080380

VAB (4 × 4) of the French Army fitted with SAGEM DANAOS day/night targeting system (Pierre Touzin) **2001**/0103763

VTM 120 (4 × 4) 120 mm mortar towing vehicle of the French Army (Pierre Touzin) 0058081

French Army NBC reconnaissance vehicle from the rear 0058082

coaxial machine gun. Totals of 720 rounds of 20 mm and 2,200 rounds of 7.62 mm ammunition are carried.

VAB VCI Dragar

VAB VCI Dragar has a crew of eight and is fitted with the Giat Industries Dragar one-person turret armed with one 25 mm cannon and one 7.62 mm coaxial machine gun. A total of 620 rounds of 25 mm and 1,400 rounds of 7.62 mm ammunition is carried.

VAB VCI Toucan

VAB VCI Toucan with a crew of 11 and fitted with a Toucan I turret armed with a 20 mm cannon and a 7.62 mm coaxial machine gun. Totals of 700 rounds of 20 mm and 200 rounds of 7.62 mm ammunition are carried. This vehicle is in service with the French Air Force as the VIB *(Véhicule d'Intervention sur Base)*. Production complete.

VAB for the French Gendarmerie

In late 1988, it was stated that the French Gendarmerie was to purchase an initial batch of 33 VABs, which would be ordered at the rate of 11 per year from 1989 to 1991.

VAB Echelon (repair vehicle)

This is fitted with welding equipment, a work bench and tools for carrying out vehicle repairs in the field and has a crew of three. Armament consists of a Giat Industries TLi 52 A turret with a machine gun over the commander's position. This model has been adopted by the French and other armies.

VCAC HOT (anti-tank) Mephisto

This model is now in service with the French Army in its 4 × 4 configuration and is basically the VAB with the Euromissile Mephisto system with four ready to launch HOT missiles. Details of this version are given in the Tank destroyers section.

VCAC HOT (anti-tank) UTM 800

This has been developed for the export market and is a VAB fitted with the Euromissile UTM 800 turret with four ready to launch HOT missiles and a further 16 missiles in reserve in the hull. Details of this version are given in the Tank destroyers section.

VAB RATAC and VAB ATILA

Developed to meet the requirements of the French Army, these two variants of the VAB belong to the inventory of the '155 AUF 1' artillery regiments of the armoured divisions.

The VAB RATAC is a mobile firing radar system for field artillery and is fitted with a roof-mounted 20 km range Doppler radar with the generating set installed on the right side of the hull. It has a five-man crew of whom two are radar operators and one a radio operator.

The VAB ATILA is a mobile automatic data transmission system in use at various command post levels in artillery regiments and is fitted out as a mobile data teleprocessing centre.

The French Army also uses the VAB RASIT ground surveillance radar system in its reconnaissance squadrons.

VAB PC (command vehicle)

This has been developed to meet the requirements of the French Army and is the basic VAB fitted with map boards and additional communications equipment, with a crew of six. The VAB is also used as a Fire Direction Centre (VAB FDC) and Forward Observation Officer vehicle (VAB FOO) acting as part of the ATILA system and to mount parts of the RITA communications system.

VAB transmission

This version is fitted with communications equipment and has seats for the operators. It is also used for the RITA Automatic Integrated Transmission Network and normally has a crew of four.

Engineer vehicle

The French Army has taken delivery of a version of the VAB modified for use by engineers. This is called the VAB Genie and carries engineer equipment and a rubber dinghy on the roof. This model has a crew of nine.

Electronic warfare vehicle

The French Army has taken delivery of an electronic warfare version of the VAB called the Bromure. This has the electronic warfare equipment at the very rear and the operators in the centre. This model has a crew of three.

VAB with SAGEM DANAOS

In 1996, the French Army took delivery of the first of about 80 VAB (4 × 4) APCs rebuilt for use in the forward artillery observation role. Mounted on the roof is a one-person turret designated LOA 20 with day and thermal channels and an integrated laser range-finder. Two remote displays are provided, one at the forward observer's station and one in the rear of the vehicle. Inside vehicle is an NSM 20 land navigation system.

VAB Sanitaire (ambulance)

This version can carry either 9 seated wounded or 3 stretcher and 1 seated wounded or 2 stretcher and 5 seated wounded or 1 stretcher (intensive care) and 5 seated wounded. This version, which has been adopted by the French Army, is fitted with an air conditioning system and carries first aid equipment. This model has a different roof hatch arrangement. Over the rear of the vehicle is a circular observation cupola and to the rear of this is the air conditioning system.

VMO (internal security vehicle)

This is called the *Véhicule de Sécurité* and has been developed for export. It can be fitted with a variety of light armament installations over the commander's position including the Giat Industries TLi 52 A armed with a 7.62 mm machine gun or the Giat Industries TLi 52 G turret. Mounted at the front of the hull is a hydraulically operated obstacle-clearing blade, and special protection is fitted against petrol bombs. Mounted in each side of the hull and at the rear is a special firing port with a vision block over it, which allows three of the crew members to use their small arms from inside the vehicle in complete safety. Equipment installed includes a loudspeaker system while a wide range of optional equipment is available including a front-mounted winch, a heater, an air conditioning system, radios and various communications systems. Abu Dhabi has a quantity of these in service in the 6 × 6 configuration with a 7.62 mm GPMG in a CB 80 mount.

VMO (internal security vehicle) with TOI

In 1990, the prototype of a new VMO internal security vehicle was revealed, fitted with a Giat Industries TOI

Specifications

(4 × 4 version; data in square brackets relates to the 6 × 6 model where different)
Crew: 2 + 10
Configuration: 4 × 4 [6 × 6]
Combat weight: 13,000 [14,200] kg
13,600 [14,800] kg with limited amphibious capabilities
Unloaded weight: 10,200 [11,400] kg
Power-to-weight ratio: 16.92 [15.49] hp/t
23 [21.13] hp/t (300 hp engine)
Length: 5.98 m
Width: 2.49 m
Height:
(to hull top) 2.06 m
Ground clearance:
(axles) 0.4 m
(hull) 0.5 m
Track: 2.035 m
Wheelbase: 3 m [1.5 + 1.5 m]
Angle of approach/departure: 45°/45°
Max road speed:
(5th gear) 92 km/h
(4th gear) 58 km/h
(3rd gear) 34 km/h
(2nd gear) 20 km/h
(1st gear) 12 km/h
(reverse) 13 km/h
Max water speed: 8 km/h
Fuel capacity: 300 litres
Max road range: 1,000 km
Fording: amphibious
Gradient (starting from static): 60%
Side slope: 35%
Trench: n/avail [1 m]
Vertical obstacle: 0.6 m
Turning radius: 9 m
Engine: Renault MIDS 06-20-45 in-line water-cooled turbocharged 6-cylinder diesel developing 220 bhp at 2,200 rpm. Original engine was a MAN D.2356 HM 72 in-line water-cooled 6-cylinder diesel developing 220 hp at 2,200 rpm (standard on VAB until 1983) (option of 300 hp diesel)
Transmission: Transfluide with 5 forward and reverse gears
Steering: recirculating ball, hydraulically assisted
Tyres: 14.00 × 20 (run-flat)
Brakes: disc
Suspension: independent (torsion bar and telescopic shock-absorbers)
Electrical system: 24 V
Batteries: 4 × 12 V 6TN
Armament: see text
NBC system: no*
Night vision equipment: no*
* standard on French Army vehicles

VAB (4 × 4) with Toucan turret armed with 20 mm cannon and 7.62 mm machine gun (Richard Stickland) **2002**/0121140

VAB Sanitaire (ambulance) of the French Army which is unarmed and has a different roof arrangement (Richard Stickland) **2002**/0121139

observation and interventional cupola. Thirteen production vehicles have been built for an export customer.

The TOI cupola has manual traverse through a full 360° and can be fitted with a wide range of optional equipment according to mission requirements, including a TV camera with power zoom, a white or active spotlight, washers, a reclining seat, an electrical console, monitors and various grenade dischargers, a 7.62 mm machine gun and GALIX decoy launchers.

VTM 120 (mortar towing vehicle)

This is called the *Véhicule Tracteur de Mortier*, is already in service with the French Army and tows a TDA 120 mm mortar. The VTM 120 has a crew of six, a Giat Industries TLi 52 A turret at the front of the hull and carries 70 120 mm mortar bombs.

VPM 81 (mortar carrier)

This model is fitted with an 81 mm mortar that is fired through a two-part opening in the roof. Qatar uses this version in a 6 × 6 configuration armed with a BAE Systems, RO Defence 81 mm mortar and a 7.62 mm machine gun turret installed at the front.

VAB NBC reconnaissance vehicle

Early in 1997, the French Army took delivery of the first of 40 NBC reconnaissance vehicles which have been designed to identify and characterise all threats related to the use of NBC (Nuclear, Biological and Chemical) warfare.

Prime contractor the now Thales (previously Thomson-CSF) which is also responsible for developing the application software, building the central processing units, system hardware and software integration studies, system validation through a joint qualification programme, logistic back-up and design and supply of training facilities including a system simulator. Giat Industries carried out the actual conversion work.

The vehicle is commonly referred to as the VAB Reco NBC and is normally operated by a crew of four consisting of commander, driver and two operators (one analyser and one sampler).

The three main missions of the vehicle are identification and qualification of the threat, evaluation of the threat and zone marking and information transmission to the rear command centre in real time via the French Army communications networks.

VAB recovery

This has been developed as a private venture and is available in both 4 × 4 and 6 × 6 configurations and has a crew of four. Equipment fitted includes a hydraulic crane with a maximum lifting capacity of 6 tonnes, four hydraulic stabilisers which are lowered to the ground to provide a more stable platform, a front-mounted winch with 60 m of cable, a flashing light, a warning triangle, block and tackle, tow hooks and a ladder. The rear compartment is fitted with a work bench, tools, welding equipment and a generator. Armament consists of a 7.62 mm or a 12.7 mm machine gun.

VAB TOW ALT

This is the VAB fitted with the Norwegian Kvaerner-Eureka armoured launching turret for Raytheon Systems Company TOW ATGWs. A total of two missiles would be carried in the ready to launch position with a further 16 missiles carried internally, plus a standard ground mount allowing for dismounted operations. This model of the VAB has yet to enter production.

VAB TOW PIVOT

This is the basic VAB with a pivot-mounted Raytheon Systems Company TOW launcher. It has one missile in the ready to launch position and a further 16 carried internally. This model of the VAB has yet to enter production.

VAB VDAA TA 20 twin 20 mm SPAAG

The Omani Royal Guard has taken delivery of nine VAB (6 × 6) vehicles fitted with the Thales one-man TA 20 turret armed with twin 20 mm cannon for use in the air defence role. Three of the vehicles are also fitted with radar. Full details of this are given in *Jane's Land-Based Air Defence 2001-2002*. This system is no longer being marketed.

Status

Production was completed for the French Army early in 1993 but the production line is still being maintained for future orders and to upgrade VAB.

Country	Quantity	Comment
Argentina	n/avail	according to Giat
Brunei	26	IS role, delivered 1996
Central African Republic	10	
Côte d'Ivoire (Ivory Coast)	13	6 × 6
Cyprus	145	incl 27 (20 mm), 100 (12.7 mm) and 18 with UTM-800 HOT ATGW turret
France	3,975	army incl variants
	33	Gendarmerie (VAB)
	28	Gendarmerie (VBC-90)
Indonesia	18	from French Army stocks
Kuwait	n/avail	according to Giat
Lebanon	85	20 are VTT and 65 are VBCI
Mauritius	11	6 × 6
Morocco	394*	all 6 × 6 incl 12 120 mm mortar towing, 68 PC, 20 ECH repair, 49 VCI
Norway	n/avail	according to Giat
Oman	14**	VAB (6 × 6)
	9**	VAB (6 × 6) SPAAG
	9**	VBC-90 (6 × 6)
Qatar	162	6 × 6, incl 24 UTM-800 ATGW
UAE	20	6 × 6, VMO, some sources say total is 80

* Original number supplied, some lost in fighting
** Royal Guard

VAB VDAA ALBI SAM System

This has been developed as a private venture and consists of a 6 × 6 or 4 × 4 VAB with a manually operated turret on the roof with two Matra BAe Dynamics Mistral surface-to-air missiles in the ready to launch position, with an additional six missiles being carried internally for manual reloading. Turret traverse is 360° with elevation from −10 to +45°. A standard SATCP tripod is carried inside the vehicle to allow missiles to be launched away from the vehicle if required by the tactical situation.

The first prototype of the VAB VDAA ALBI SAM system was completed in 1988. As of 2001 this remained at the prototype stage. The first customer for the ALBI system is Oman who ordered the system late in 2000 for installation on the Panhard VBL (4 × 4) scout vehicle chassis.

VBC 90

Full details of this armoured car which uses components of the 6 × 6 VAB are given in the Reconnaissance vehicles section. This vehicle is no longer being marketed and was only sold to France (Gendarmerie) and Oman.

Manufacturer

Giat Industries.

UPDATED

Berliet VXB-170 armoured personnel carrier

Development

In the mid-1960s, as a private venture, Berliet started the development of a 4 × 4 armoured personnel carrier which would use automotive components from its wide range of tactical military vehicles then in production. The first prototype of this vehicle, called the BL-12, was completed in March 1968 and shown for the first time at the 1969 Satory Exhibition of Military Equipment. A second vehicle was built for the French Army, which did not place a production order for the BL-12.

The French Gendarmerie placed an order for five slightly modified versions called the VXB-170 which were delivered from 1971. Following successful trials

Berliet VXB-170 APC with a SAMM one-man turret (Pierre Touzin) ***2001***/0103766

Berliet VXB-170 APC from the rear (Pierre Touzin) ***2001***/0103765

with these vehicles the French Gendarmerie placed an initial order for 50 vehicles for internal security use, the first of which were completed at Berliet's Bourge facility in 1973. The total Gendarmerie requirement was for about 400 vehicles but only 155 were ordered and delivered.

The VXB-170s are called the VBRG-170 by the French Gendarmerie. They have been supplemented by the VAB. Large numbers of these vehicles remain in service with the French Gendarmerie.

Description

The hull of the VXB (4 × 4) APC is made of all-welded steel. The driver is seated at the front of the hull on the left side with the commander to his right. In front of the driver and commander is a large windscreen which, when in action, is covered by a two-piece armoured screen hinged at the top. To the left of the driver and right of the commander is a smaller window which, in action, is covered by a flap hinged at the bottom. Over the top of the driver's position are three day observation periscopes for use when the windscreen is covered and over the commander's position is a single-piece hatch cover that opens to the right with a swivel-mounted day periscope for all-round observation.

The main armament, if fitted, is installed in the centre of the vehicle behind the driver's and commander's positions. There is an entry door in each side of the hull and a third door at the rear of the hull on the right side. In addition to the roof hatch, on which the main armament is installed, there are three other roof hatches: two in line with the doors in the sides of the hull, one of which opens forwards and the other to the rear, and a third in the aisle which connects the personnel compartment with the door in the rear of the hull.

In the original model there were seven firing ports, two in the left side (one of which was in the door), four in the right side (one of which was in the door) and one in the door in the rear of the vehicle. Vehicles delivered to the French Gendarmerie have five bulletproof windows, two in each side of the hull (one of which is in the door) and one in the door in the rear of the vehicle. These can be opened to allow the troops to fire their weapons from inside the vehicle. The Gendarmerie vehicles also have a single-piece hatch cover over the driver's rather than the commander's position and there is no hatch cover over the aisle.

Specifications

Crew: 1 + 11
Configuration: 4 × 4
Combat weight: 12,700 kg
Unloaded weight: 9,800 kg
Power-to-weight ratio: 13.38 hp/t
Length: 5.99 m
Width: 2.5 m
Height:
(without armament) 2.05 m
Ground clearance:
(transfer case) 0.45 m
Track: 2.04 m
Wheelbase: 3 m
Angle of approach/departure: 45°/45°
Max speed:
(road) 85 km/h
(water) 4 km/h
Fuel capacity: 220 litres
Max road range:
(at 50 km/h) 750 km
Fording: amphibious
Gradient: 60%
Side slope: 30%
Vertical obstacle: 0.3 m
Turning radius: 7.8 m
Engine: Berliet Model V 800 M V-8 diesel developing 170 hp at 3,000 rpm
Transmission: preselective with 6 forward and 1 reverse gears
Transfer box: 2-speed
Steering: hydraulic power assisted
Suspension: helical springs and hydraulic shock-absorbers
Tyres: 14.00 × 20
Electrical system: 24 V
Batteries: 4 × 12 V
Armament: 1 × 7.62 mm MG
Smoke-laying equipment: 2 smoke grenade dischargers either side of hull (optional)
Armour: 7 mm
Gun stabiliser:
(vertical) no
(horizontal) no
NBC system: optional
Night vision equipment: optional

The power pack is at the rear of the hull on the left side and access panels are provided in the top and rear of the hull for maintenance.

Power is transmitted from the engine to the manual gearbox and then by a propeller shaft to the two-speed transfer box in the centre of the vehicle under the floor. Power is then taken from the transfer box to the front and rear axles via propeller shafts. The front and rear wheels have planetary reduction gears and the rear axle has a pneumatically controlled interwheel differential lock. The suspension at each wheel station consists of a concentric helical spring and a hydraulic shock-absorber.

The basic vehicle is fully amphibious, being propelled in the water by its wheels. Optional equipment included a heater, a hydraulic obstacle-clearing blade at the front of the hull (as fitted to French Gendarmerie vehicles), passive night vision equipment, an NBC system, a winch with a capacity of 3,500 to 4,500 kg (and 60 m of cable) and bulletproof tyres.

The vehicles delivered to the French Gendarmerie were originally fitted with a simple pintle-mounted machine gun but some were subsequently fitted with the SAMM S365-3 (now designated the BTM 103) light turret which is armed with a 7.62 mm machine gun and a 40 mm grenade launcher. Elevation limits of the weapons are from −15 to +60°.

It is believed that some consideration is being given to upgrading these vehicles to extend their operational life with the French Gendarmerie. This could include the installation of a new engine or complete power pack.

Status

Production complete. In service with France (Gendarmerie, 155), Gabon (15) and Senegal (Gendarmerie, 12).

Manufacturer

Société des Automobiles Berliet. (This company no longer exists as a separate entity.)

UPDATED

Panhard VCR-2 armoured personnel carrier

Development

Further development of the French Panhard VCR TT 2 (6 × 6) APC, covered in detail in the following entry, has resulted in the Panhard VCR-2 (8 × 8) armoured personnel carrier which has been developed for the export market.

This was one of the original contenders for the Polish wheeled APC competition. Recent information has indicated that this is no longer one of the contenders.

Description

In general terms the VCR-2 (8 × 8) is a stretched version of the VCR TT 2 (6 × 6) armoured personnel carrier and would use independent driving wheels with trailing arms and with the lateral transmission under armour.

Specifications

Provisional
Crew: 8-12
Configuration: 8 × 8
Combat weight: 14,000-15,000 kg
Power-to-weight ratio: 20-30 hp/t
Length: 6 m
Width: 2.7 m
Height: 1.8 m (hull top)
Max road speed: 100 km/h
Max water speed: 10 km/h
Range: 800 km
Gradient: 60%
Side slope: 40%
Trench: 0.80 m
Fording: amphibious

The first prototypes could be completed within 24 months of the go ahead, according to Panhard.

Weapon fit would depend on mission but various armament options would be offered including turret-mounted 20 mm/25 mm cannon up to two-man turret armed with 90 mm gun.

Steering would be power assisted on the front four wheels and the standard version could ford to a depth of 1.8 m, although an amphibious version would also be offered.

Wherever possible, proven automotive components from existing vehicles already in production would be used to reduce development time and cost as well as reducing total life cycle costs.

Turbo diesel engines and automatic or preselective gearboxes locally manufactured are typical examples of the type of automotive components which could be incorporated into the VCR-2 (8 × 8) family of vehicles.

Variants

Panhard has suggested that a complete family of 4 × 4, 6 × 6 and 8 × 8 light armoured vehicles could be developed using the same automotive components

Status

Proposal. Developed specifically for the export market.

Manufacturer

Société de Constructions Mécaniques Panhard et Levassor.

UPDATED

Arrtist's impression of the Panhard VCR-2 (8 × 8) armoured personnel carrier
0005357

Panhard VCR TT 2 armoured personnel carrier

Development

The VCR *(Véhicule de Combat à Roues)* TT *(Transport de Troupes)* 2 was originally developed by Panhard to meet the requirements of the Algerian Army.

One of the prototypes undertook some 3,000 km of trials in Algeria. It was intended that production of this vehicle would be undertaken in Algeria, but in the end the project was discontinued by Algeria owing to a shortage of funds.

The VCR TT 2 shares many automotive components, for example the power pack and suspension, with the Panhard ERC 90 F4 Sagaie 2 (6 × 6) armoured car, which was unveiled together with the VCR TT 2 for the first time in June 1985. At that time three prototypes of the VCR TT 2 had been built but, by late 2001, production of this vehicle had yet to commence.

Specifications

Crew: 2 + 10 to 12
Configuration: 6 × 6
Combat weight: 9,600 kg
Power-to-weight ratio: 30.2 hp/t (petrol engines)
Length: 6.14 m
Width: 2.7 m
Height: 2.27 m
Ground clearance: 0.37 m
Track: 2.35 m
Wheelbase: 2.6 + 1.09 m
Max speed:
(road) 110 km/h
(water) 7.2 km/h
Road range:
(petrol engines) 600 km
(diesel engines) 1,000 km
Gradient: over 50%
Side slope: 30%
Vertical obstacle: 0.8 m
Trench: 0.8 m
Engines: 2 × Peugeot PRV V-6 petrol engines developing a total of 290 hp or two Peugeot XD 3T diesels developing a total of 196 hp
Transmissions: 2 × ZF 4 HP 22 automatic, 4 forward and 1 reverse gears
Transfer box: Panhard, 2 speed
Brakes:
(main) hydraulic, disc
(parking) mechanical
Tyres: 13.00 × 16
Electrical system: 24 V
NBC system: optional
Night vision equipment: optional

Description

The hull of the Panhard VCR TT 2 is of all-welded steel armour construction providing protection from small arms fire and shell splinters. The driver sits at the front of the vehicle in the centre and has three day periscopes for forward observation, the centre one of which is mounted in a forward-opening hatch for driving in the head-out position.

The engine compartment is to the immediate rear of the driver. The VCR TT 2 is available powered by two Peugeot XD 3T diesel engines developing a total of 196 hp at 4,150 rpm or two Peugeot PRV V-6 petrol engines developing a total of 290 hp at 5,500 rpm. These are each coupled to a ZF 4 HP 22 fully automatic transmission with four forward and one reverse gears which are in turn connected to a Panhard two-speed, road and cross-country transfer box. Power is transmitted to all roadwheels via drive shafts in the armoured hull.

Steering is power assisted on the front two roadwheels and, when driving on roads, the centre two roadwheels are normally raised clear of the ground.

Each roadwheel is independent and mounted on trailing arms with helicoidal spring suspension. The front and rear wheels have hydropneumatic suspension units while the centre ones have hydraulic jacks.

There is a single door in each side of the hull, to the rear of the first roadwheel station, and a large door in the hull rear that opens to the left. This has two Giat Industries firing ports each with an observation block above. These allow the troops to fire their weapons from within the vehicle.

Six troops sit down either side of the hull facing each other, with two upward-opening hatches in the upper part of each side of the troop compartment. Over the top of the troop compartment are four roof hatches, two either side, that are hinged on the outside.

The main armament installation is mounted on the roof to the rear of the engine compartment. If, for example, a SAMM BTE 105 cupola is fitted, armed with a 12.7 mm M2 HB machine gun, then 12 troops can be carried, but when heavier two-man turrets are mounted, then the number of troops is reduced.

Optional equipment includes an air conditioning system, NBC equipment, a front-mounted winch and various passive night vision devices.

The VCR TT 2 is fully amphibious, propelled in the water by two steerable propellers mounted under the hull rear.

Variants

In addition to being fitted with a wide range of armament installations, Panhard has suggested a number of specialised versions of the VCR TT 2, including command post, ambulance and workshop.

Panhard VCR-2 APC (8 × 8)

Additional details of this 8 × 8, which still remains at the proposal stage, are given in the previous entry.

Status

Development complete. Ready for production.

Manufacturer

Société de Constructions Mécaniques Panhard et Levassor.

UPDATED

Panhard VCR TT 2 APC without armament installed 0018680

Panhard VCR TT 2 without armament and showing propellers under the hull rear 0018681

Panhard VCR armoured personnel carrier

Development

The VCR *(Véhicule de Combat à Roues)* range of 6 × 6 armoured personnel carriers was developed by Panhard as a private venture from 1975 and is aimed specifically at the export market. The range was first shown in 1977 at the Satory Exhibition of Military Equipment and entered production at Panhard's Marolles factory the following year.

Although the VCR range was developed as a private venture, it has recently been disclosed that it was originally designed specifically for the Iraqi programme to find a platform for the 100 Euromissile UTM-800 four-round HOT anti-tank guided missile turrets ordered in September 1974.

Early in 1975, an international competition was held by Iraq to find a chassis suitable for the UTM-800 HOT turret following which the Panhard VCR was selected. The suspension system was taken from the existing Panhard M3 (4 × 4) APC fitted to a new chassis, the front part of which was very similar to the existing Panhard M3. From the VCR family evolved the Panhard ERC (6 × 6) armoured car. This has been built in large numbers for the French Army and export customers and additional details are given in the Reconnaissance vehicles section. There has been no recent production of the ERC armoured car although it is still being marketed by Panhard.

First production Panhard VCR vehicles, equipped with the Euromissile UTM 800 turret, were completed in 1979. The VCR range of armoured personnel carriers uses 95 per cent of the automotive components of the Panhard ERC range of 6 × 6 armoured cars which was also shown for the first time in 1977. Late production vehicles have a redesigned hull, a troop compartment roof 400 mm above the commander's position, external stowage boxes, more internal hull space, twin doors at the hull rear and petrol fuel tanks repositioned from inside the troop compartment to under the floor.

In June 1983, Panhard issued new specifications for the VCR range of 6 × 6 APCs. These showed that there had been slight changes to the vehicle with the wheelbase increased from 1.425 + 1.425 m to 1.66 + 1.425 m. This has made the hull slightly longer and increased the overall weight of the vehicle.

The specifications for the VCR/TH anti-tank version remain unchanged except for an increase in loaded and empty weights.

By late 2001, production of the Panhard VCR range of 6 × 6 APCs amounted to 262 vehicles, all for export.

As far as is known, there has been no recent production of the Panhard VCR series of 6 × 6 vehicle.

Recent infomation has indicated that Gabon has taken delivery of Panhard VCR/TT (6 × 6) armoured personnel carriers.

Description

The all-welded steel hull of the VCR/TT *(Transport de Troupes)* protects the crew against small arms fire and shell splinters. The floor of the hull consists of two plates welded together to form a flat V which helps the vehicle to slide off obstacles and also offers resistance to damage from mines.

The driver sits at the front of the hull in the centre and has a single-piece hatch cover that opens to his right with three integral day periscopes for forward observation, the centre one replaceable by an image intensification periscope for driving at night.

The commander sits to the rear of the driver on the left side at a higher level and has a single-piece hatch cover that opens forwards and seven day periscopes around his position.

The engine is to the rear of the driver on the right side and power is transmitted through the single clutch plate to a transversely mounted Panhard gearbox (as fitted to the AML armoured car and M3 APC) with six forward and one reverse gears and incorporating a limited slip differential. Drive from the bevel gears to the roadwheels is by a train of gears housed within trailing arms which independently locate the wheels.

The front and rear wheels are sprung by single coil springs with telescopic hydraulic dampers. The centre wheels, which are raised off the ground when on roads, are sprung by hydropneumatic units. The centre wheels are powered even when raised clear of the road surface. All wheels have disc brakes. Only the front wheels are steered, and steering is hydraulically assisted. The low-pressure tyres are fitted with Hutchinson cellular inner tubes which enable the vehicle to travel for at least 100 km at 30 km/h when the tyres have been punctured.

The 10 infantrymen are seated on bench seats: five along each side of the hull facing each other. They enter and leave the vehicle by a large door in the rear of the hull which is hinged on the right side and has a single firing port. Two observation/firing hatches in each side of the troop compartment open upwards and can be locked open. The VCR/TTs built for the United Arab Emirates have a wider hull with double doors at the rear to give increased leg room.

The armament is mounted to the rear of the commander's position in the roof. When armed with a 7.62 mm machine gun the VCR/TT can carry nine men plus its crew of three (commander, driver and gunner, who sits on one of the bench seats when not manning the machine gun) but when heavy armament is installed, such as a 20 mm cannon, only six infantrymen can be carried plus a crew of three. The VCR/TT can be fitted with a variety of armament installations including: a SAMM BTM 263 turret with a 60 mm mortar and a 7.62 mm machine gun; a Giat Industries STB ring mount with a shield (which forms the hatch cover when not in use) and a 7.62 mm machine gun; a Giat Industries CB rotating gun ring with a 7.62 mm machine gun; a Giat Industries turret with twin 7.62 mm machine guns; a Giat Industries gun ring Model CB 20 with a 20 mm cannon; a Giat Industries gun ring with a TDA 60 mm breech-loaded mortar; a Giat Industries one-man turret armed with a 20 mm cannon; and a 7.62 mm machine gun. There is also a single-piece hatch cover in the roof of the troop compartment at the rear which opens towards the front. A 7.62 mm machine gun on a Giat Industries STR rail mount can be mounted behind it. Two grenade dischargers can be mounted on either side of the hull if required.

The VCR is fully amphibious being propelled in the water by its wheels and steered by turning the front roadwheels as for land operations. Optional equipment includes a front-mounted winch with 60 m of cable and a capacity of 3,000 kg, NBC system, air conditioning system and passive night vision equipment.

Variants

VCR/AT repair vehicle

The VCR/AT *(Atelier Véhicule)* is the basic vehicle modified to undertake repairs in the field. It has no recovery capability although it can tow other vehicles.

Specifications
VCR/TT APC
Crew: 3 + 9
Configuration: 6 × 6
Combat weight: 7,900 kg
Unloaded weight: 6,800 kg
Power-to-weight ratio: 18.35 hp/t
Length: 4.875 m
Width: 2.50 m
Height:
(including armament) 2.56 m (7.62 mm MG)
(to hull top) 2.13 m
Ground clearance:
(4 wheels) 0.315 m
(6 wheels) 0.375 m
Track: 2.115 m
Wheelbase: 1.66 + 1.425 m
Angle of approach/departure: 53°/45°
Max speed:
(road) 90 km/h
(water) 4 km/h
Max road range: 700 km
Fuel capacity: 242 litres
Fording: amphibious
Gradient: 60%
Side slope: 30%
Vertical obstacle: 0.8 m
Trench: 1.1 m
Engine: Peugeot PRV V-6 petrol developing 145 hp at 5,500 rpm
Transmission: Panhard with 6 forward and 1 reverse gears
Steering: power assisted
Clutch: hydraulic control
Tyres: 11.00 × 16
Electrical system: 24 V
Armament: 1 × 7.62 mm MG, or 1 × 20 mm cannon (see text)
Smoke grenade dischargers: optional
Armour (est): 8-12 mm
NBC system: optional
Night vision equipment: optional

Panhard VCR/TT (6 × 6) armed with 7.62 mm machine gun and front-mounted winch

Panhard VCR/IS (6 × 6) ambulance which has a higher roof and is unarmed

Panhard VCR/TT (4 × 4) model, as used by Argentina, being propelled in the water by its waterjets 0058083

Panhard VCR/TH with a Euromissile UTM 800 turret with four ready-to-launch HOT ATGWs and a Mascot 7.62 mm remote-controlled machine gun mount at the rear. Turret is traversed to rear and HOT missile is being launched

Equipment installed includes a generator, tools, work benches, an inspection lamp, towbars and tow cables. A penthouse can be erected at the rear to enable repairs to be carried out in bad weather. A block and tackle can be erected at the rear allowing the VCR/AT to carry out engine and other component changes. This model is used by the United Arab Emirates.

VCR/AA anti-aircraft vehicle

This was unveiled in 1984 and is fitted with the now SAAB Bofors Dynamics RBS 70 surface-to-air missile system. The VCR/AA weighs 7,500 kg and has a crew of three. Mounted in the rear troop compartment is the RBS 70 missile launcher with one missile ready to launch and a further eight in reserve. For travelling, the launcher is retracted within the hull making the vehicle almost indistinguishable from other members of the VCR family. This model has yet to enter production.

VCR/TH (Tourelle HOT) anti-tank vehicle

This is the basic VCR/TT fitted with the Euromissile UTM 800 HOT turret mounted on a raised plinth to the rear of the commander's position. The turret has four ready to launch 4,000 m range HOT missiles and another 10 missiles carried in reserve in the hull are loaded via a hatch in the right rear of the roof, which opens to the right. The turret has an optical sight with a magnification of ×3 (18° field of view) and ×12 (5° field of view), elevation being +22°, depression −10° and turret traverse 360°.

Mounted on the roof at the rear on the left is a Giat Industries Mascot remote-controlled 7.62 mm machine gun which can be elevated from −10 to +50° and traversed through a full 360°. There are 200 rounds of ready use ammunition provided for this weapon. Two grenade dischargers are mounted on the front of the hull at either side. Panhard has delivered 100 of these vehicles to Iraq. The total number of these vehicles now operational is uncertain as many were lost during the 1991 conflict.

VCR/IS (Intervention Sanitaire) ambulance

The VCR/IS *(Véhicule Blindé d'Intervention Sanitaire)* is the ambulance member of the family and has a much higher hull behind the commander than the basic VCR. The vehicle has a crew of three (commander, driver and orderly) and can carry four stretcher patients or six sitting and two stretcher patients.

The casualties enter the vehicle through a door at the rear of the hull that opens to the left and has a single vision port. Unlike the basic APC, there are no observation flaps in the sides of the hull.

Equipment carried includes folding stretchers, cupboards and instrument storage lockers, a water tank, a refrigerator and breathing and blood transfusion equipment. If required, a tent, which is stowed in boxes on either side of the hull, can be erected at the rear. This model is used by the United Arab Emirates.

VCR/PC (Poste de Commandement) Command Vehicle

This is basically the VCR/TT adapted for use as a command vehicle with map boards, tables and additional communications equipment. This model is used by the United Arab Emirates.

VCR/TT (4 × 4)

This was developed to meet the requirements of the Argentine Marines and is essentially the 6 × 6 VCR/TT with the central roadwheel either side removed and a water-jet installed in its place.

This gives the VCR/TT a maximum water speed of 7.2 km/h. Before entering the water, the bilge pumps are switched on and a trim vane with transparent upper part for forward observation is erected at the front of the vehicle.

Status

Production as required. In service with the following countries:

Country	Quantity	Comment
Argentina	24	VCR/TT (4 × 4) amphibious
Gabon	n/av	VCR/TT (6 × 6)
Iraq	100	VCR/TH HOT ATGW, some lost in fighting
Mexico	2	VCR/PC command post role
United Arab Emirates	82	VCR/TT, VCR/AT, VCR/IS and VCR/PC

Manufacturer

Société de Constructions Mécaniques Panhard et Levassor.

UPDATED

Panhard Buffalo armoured personnel carrier

Development

The Buffalo (4 × 4) APC has been developed as a private venture by Société de Constructions Mécaniques Panhard et Levassor and was shown for the first time in mid-1985.

Early in 1986, it was announced that the Buffalo would replace the older Panhard M3 (4 × 4) APC in production. The Buffalo is essentially a much improved version of the original Panhard M3 described in the following entry but with additional external stowage, a slightly longer wheelbase and a choice of a Peugeot V-6 petrol engine developing 145 hp or a Peugeot diesel engine developing 95 hp. Both these engines are also used in other members of the Panhard range of 4 × 4 and 6 × 6 armoured vehicles. The XD 3T is installed in the Panhard VBL.

By 2001, total production of the Buffalo and the M3 APCs amounted to 1,222 vehicles, all of which were for export. There has been no recent production of the Panhard Buffalo APC or the AML armoured car.

Description

The hull of the Buffalo is of all-welded steel armour construction with the driver seated at the front of the

Panhard Buffalo armoured personnel carrier with a front-mounted winch and a Puma 7.62 mm machine gun cupola

hull, the engine to his immediate rear and the troop compartment at the rear of the hull.

The troop compartment has a single large door in each side of the hull and twin doors in the hull rear. Each of the latter has a single circular firing port. The spare wheel and tyre are carried on the right door.

Extensive external stowage is provided above the wheels and in the event of the vehicle running over a mine this will blow off, maintaining hull integrity.

A variety of light weapon stations can be mounted on the roof of the Buffalo to the rear of the engine, for example a cupola with an externally mounted 7.62 mm machine gun. There is an additional hatch in the roof of the vehicle at the rear and three upward-opening roof hatches along each side of the troop compartment which allow some of the troops to fire their weapons from within the vehicle.

Optional equipment includes a front-mounted winch and an air conditioning system. The basic vehicle is fully amphibious, propelled in the water by its wheels.

Variants

In addition to being used as an APC, the Buffalo can be used as an internal security vehicle with a hydraulically operated obstacle-clearing blade at the front of the hull, a command vehicle with extensive communications equipment installed, an 81 mm mortar carrier, a workshop and an ambulance.

Specifications

Crew: 2 + 10
Configuration: 4 × 4
Combat weight: 6,600 kg
Power-to-weight ratio: 21.96 hp/t (PRV petrol engine)
Length: 4.585 m
Width: 2.4 m
Height: (without armament) 2 m
Ground clearance: 0.33 m
Track:
(front) 2.06 m
(rear) 2.08 m
Wheelbase: 2.835 m
Max speed:
(road) 90 km/h
(water) 3.6 km/h
Road range: 600 km
Fording: amphibious
Gradient: 60%
Side slope: 30%
Vertical obstacle: 0.4 m
Engine: Peugeot PRV V-6 water-cooled petrol developing 145 hp at 5,500 rpm or Peugeot XD 3T turbocharged water-cooled diesel developing 95 hp at 4,150 rpm
Transmission: manual with 6 forward and 1 reverse gears
Suspension: independent, coil and spring with hydropneumatic shock-absorbers acting on suspension trailing arm of wheel mechanism
Tyres: 11.00 × 16
Brakes:
(main) hydraulic, dual circuit
(parking) handbrake operating on gearbox output shaft
Electrical system: 24 V
NBC system: no
Night vision equipment: optional

Status

Production as required. In service with Benin (one in command post version), Columbia (Police) and Rwanda (18). In view of the recent conflict in this country the exact status of these last vehicles is uncertain.

Manufacturer

Société de Constructions Mécaniques Panhard et Levassor.

UPDATED

Panhard M3 armoured personnel carrier

Development

The M3 armoured personnel carrier was designed as a private venture by Panhard with the first prototype completed in 1969. This had square-cut sides with a single door in each side of the hull and twin doors in the hull rear. It was intended that each door would have a single circular firing port. Mounted on the top of the hull was a Giat Industries one person turret armed with a 7.5 mm machine gun.

The first production vehicle, with a redesigned hull incorporating three hatches either side of the troop compartment, was completed in 1971. The automotive components of the Panhard M3 are 95 per cent identical to those of the Panhard AML series of light armoured vehicles of which over 4,800 have been built. The Panhard M3 APC has now been replaced in production by the Buffalo fully described in the previous entry.

By late 2001, total production of the Buffalo and M3 APC amounted to 1,222 vehicles, all of which were for export. There has been no recent production of this vehicle.

Bosnia Herzegovina has taken delivery of a quantity of Panhard VAT, VTS and VPC vehicles from an existing user rather than direct from Panhard. Recent information indicated that these vehicles were supplied from a country in the Middle East.

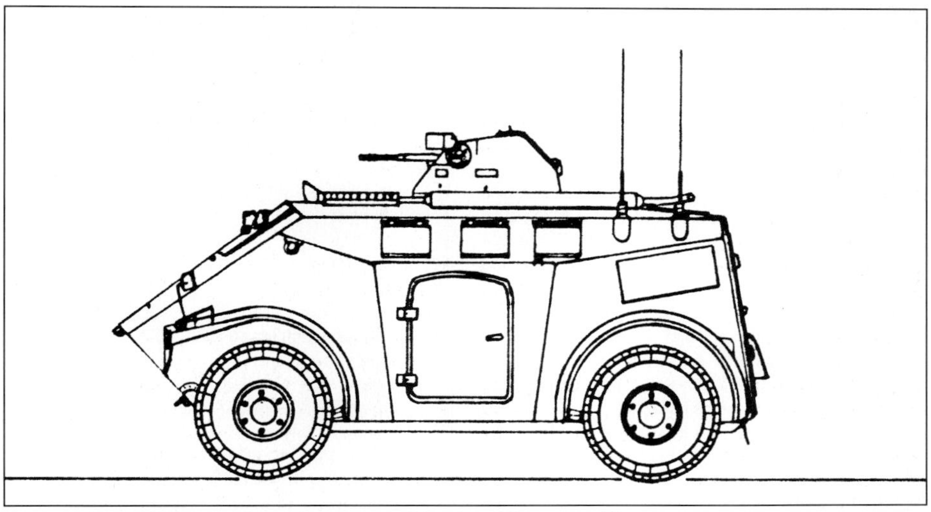

Panhard M3 APC fitted with a turret armed with twin 7.62 mm machine guns

Description

The hull of the Panhard M3 is made of all-welded steel with the driver seated at the front of the vehicle. He has a single-piece hatch cover that opens to the right in which there are three day periscopes, the centre one replaceable by an image intensification periscope for night driving.

The engine is immediately behind the driver with the air inlet louvres above and to the rear of the driver, air outlet louvres on either side of the roof and the two exhaust pipes running along the top of the hull, one on each side.

The gear casing is crosswise and consists of two gearboxes in one coupled on both sides of the bevel pinion. The low-range gearbox comprises two low gears, top and one reverse for use in rough country and the high-range gearbox has three low ranges and one overdrive. When the low-range gearbox is in direct drive, the four ratios of the high-range box command the four upper gears of the range (6th, 5th, 4th and 3rd). Drive is transmitted from the main gearbox to the two

Panhard M3 internal security vehicle with front-mounted obstacle removal blade 0058084

Panhard M3/VDA twin 20 mm self-propelled anti-aircraft gun system with stabilisers retracted for travelling 0018683

lateral boxes, which transmit power to the front wheels by pinions and to the rear wheels by drive shafts that run along the inside of the hull.

Ball-type differentials in the gearbox and in each of the rear lateral transfer boxes prevent wheel slip. The suspension at each wheel station consists of coil springs and hydropneumatic shock-absorbers acting on the suspension arms of the wheel mechanism. The tyres have puncture-proof Hutchinson inner tubes.

There are four doors in the M3, one in each side of the hull and two in the rear. The rear doors both have a firing port. Along the upper part of each side of the hull are three hatches hinged at the top that can be locked open. When being used as an APC, a maximum of 10 men can be carried in addition to the crew of two, or it can carry 1,360 kg of cargo.

There are two circular hatches in the roof, one behind the engine compartment and a second one at the rear with a single-piece hatch cover that opens forwards and is normally fitted with a Giat Industries STR rail armed with a 7.62 mm machine gun. A wide range of armament installations could be mounted on the forward position; typically these are a 7.62 mm or a 12.7 mm machine gun, but some users have fitted a turret-mounted 20 mm cannon.

The basic Panhard M3 is fully amphibious without preparation. It is propelled in the water by its wheels. Steering, when afloat, is by turning the front wheels as on land. Optional equipment includes an air conditioning system and smoke grenade dischargers.

Variants

M3/VDA anti-aircraft vehicle

The M3/VDA *(Véhicule de Défense Antiaérienne)* is a joint development for the export market by Panhard (responsible for the chassis) and the now Thales. It consists essentially of a modified Panhard M3 (4 × 4) chassis fitted with the now Thales turret armed with twin 20 mm cannon and a Galileo sight, with the option of an RA 20 series radar on the turret rear. One M3/VDA with the radar can control two other systems without the radar.

The first prototype of the Panhard M3/VDA was completed in December 1973 and, following trials, production commenced in April 1975. It is known to be in service with the Côte d'Ivoire (6), Niger (10) and the United Arab Emirates (Abu Dhabi) (48).

Full details of this system are given in *Jane's Land-Based Air Defence 2001-2002.*

M3/VAT repair vehicle

This model has a crew of five (commander, driver and three mechanics) and equipment fitted includes a pulley block with struts and tackle, cutting equipment, a generator, benches, a vice, towbars, tow cables and a complete range of tools.

M3/VPC command vehicle

This model has additional communications equipment, two additional batteries and map tables. Its basic crew consists of commander, assistant commander, driver and two radio operators plus command staff. Armament normally comprises a 7.62 mm machine gun on a Giat Industries STB mount.

M3/VLA engineer vehicle

This model is essentially the M3 APC fitted with a removable hydraulically operated dozer blade at the front of the hull and has a crew of six consisting of commander/gunner, driver, pioneer section commander and three pioneers.

M3 internal security vehicle

This is fitted with a front-mounted obstacle clearing blade and other equipment for use in the internal security role.

M3/VTS ambulance

This model has a crew of three (driver and two orderlies) and can carry four stretcher patients, six sitting wounded or two stretcher patients and three sitting wounded. It is unarmed and has a large single-piece door in the rear of the hull that is hinged on the left. Medical equipment carried includes oxygen, blood transfusion and a complete first aid outfit.

M3/VSB radar

The Panhard M3 can also be used to carry electronic warfare equipment. This model is called the M3/VSB,

Panhard M3 APC fitted with a one-man Giat Industries STB rotary support shield armed with 7.62 mm MG and mount at rear not fitted with 7.62 mm MG (Terry J Gander)

Specifications

Crew: 2 + 10
Configuration: 4 × 4
Combat weight: 6,100 kg
Unloaded weight: 5,300 kg
Power-to-weight ratio: 14.75 hp/t
Length: 4.45 m
Width: 2.4 m
Height:
(with turret armed with twin 7.62 mm machine guns) 2.48 m
(to hull top) 2 m
Ground clearance: 0.35 m
Track: 2.05 m
Wheelbase: 2.7 m
Angle of approach/departure: 68°/50°
Max road speed:
(6th gear, high range) 90 km/h
(5th gear, high range) 61 km/h
(4th gear, high range) 35 km/h
(3rd gear, high range) 18.8 km/h
(3rd gear, low range) 18.8 km/h
(2nd gear, low range) 9.3 km/h
(1st gear, low range) 4.5 km/h
(reverse) 5.5 km/h
Max water speed: 4 km/h
Fuel capacity: 165 litres
Max road range: 600 km
Fuel consumption: 0.26 litres/km
Fording: amphibious
Gradient: 60%
Side slope: 30%
Vertical obstacle: 0.3 m
Trench:
(1 channel) 0.8 m
(5 channels) 3.1 m
Turning radius: 6.55 m
Engine: Panhard Model 4 HD 4-cylinder air-cooled petrol developing 90 hp at 4,700 rpm
Transmission: manual with 6 forward and 1 reverse gears
Clutch: centrifugal with electromagnetic automatic control
Suspension: independent, coil spring and hydropneumatic shock-absorbers acting on suspension trailing arms of wheel mechanism
Tyres: 11.00 × 16
Brakes:
(main) hydraulic, dual circuit
(parking) handbrake operating on gearbox output shaft
Electrical system: 24 V
Batteries: 2 × 12 V
Armament: see text
Armour: 8-12 mm
NBC system: no
Night vision equipment: optional

Status

Production complete. Replaced in production by the Panhard Buffalo APC. In service with the following countries:

Country	Quantity	Comment
Algeria	44	gendarmerie, variants including APC with SAMM MG turret, five VPC, two VAT, two VLA and two VTS
Bahrain	110	
Bosnia Herzegovina	n/avail	VAT, VTS and VPC
Burkina Faso	13	
Burundi	9	
Chad	10	approx number, some have 20 mm turret (may be non-operational)
Congo, Democratic Republic	60	
Côte d'Ivoire (Ivory Coast)	16	
Gabon	10	approx number
Iraq	200	estimate, status uncertain
Ireland	47	
Kenya	10	believed now in reserve
Lebanon	15	
Malaysia	37	
Morocco	50	estimate
Niger	22	
Nigeria	18	four with RASIT radars, four VTS, four VAT and six VPC
Saudi Arabia	150	
Senegal	16	
Sudan	n/avail	including some with RASIT surveillance radar, one VAT, two RA-20 air surveillance radars and 13 with Toucan I turret armed with 20 mm cannon
Togo	6	estimate
UAE	240	some sources state 370 vehicles

being fitted with a variety of radar systems such as the RASIT battlefield surveillance radar or the now Thales RA-20S air surveillance radar. The latter is used with the Panhard M3/VDA twin 20 mm self-propelled anti-aircraft gun system when this is not equipped with its own individual radar.

Manufacturer

Société de Constructions Mécaniques Panhard et Levassor.

UPDATED

Panhard P4 Protected armoured personnel carrier

Development

From March 1995, the French Army took delivery of 50 Peugeot P4 (4 × 4) 750 kg light vehicles fitted with a passive armour kit originally developed by Panhard as a private venture.

The French Army calls this version the P4 Protected. The first batch of 10 vehicles was delivered in March with eight of these going to the French Army operating in Bosnia. The remaining 40 were then delivered at the rate of 10 per month.

Since then, an additional 30 kits have been ordered to bring the total supplied to the French Army to 80 kits. Similar kits are available for other light wheeled vehicles, for example the Land Rover. As far as it is known there have been no exports to other customers by Panhard.

In French Army service the Panhard P4 will be supplemented by the projected Petit Véhicule Protégé (PVP) currently at the requirements stage and covered in detail in the previous entry. As of October 2001, no contracts had been awarded for the PVP.

Description

The layout of the P4 Protected is similar to the P4 (4 × 4) light vehicle which is the standard vehicle in its class in the French Army.

The engine is at the front, with driver and commander in the centre and a bench seat to the rear. At the very rear of the vehicle is a small load area which is covered over by a steel sheet. Access to this is via an oblong door at the rear which opens to the right and on this is carried the spare wheel. To the rear of the crew compartment is a sliding door that opens left.

The installation of the passive protection kit on the P4 allows the occupants to carry out their peacekeeping operations with full protection from 7.62 mm small arms fire and from the secondary effects of anti-personnel mines.

The passive armour kit is made of high-hardness armour plate with bulletproof windows in the front, sides and rear. Four firing ports (one in front, one each side and one in the rear) are also provided and are covered by circular hatch covers when not required.

The upper part of each door with the integral bulletproof window, hinges down on the outside and, if required, can be locked open at an angle of 15°. At the rear is a sliding door for emergency access or for when carrying a stretcher.

The passive armour kit can be mounted on the standard P4 vehicle at battalion level, without welding or drilling, using standard tools. It can also be quickly removed from the vehicle and stored for future use.

Standard equipment includes a defrosting system for the front window, folding front seat and a bench seat to the rear for an additional two passengers. All four occupants are provided with seat belts.

In addition to offering the passive armour kit for installation on existing P4 vehicles, Panhard could also install this kit on new build Peugeot P4 vehicles.

This could also be fitted with the Peugeot XD 3TE turbocharged diesel engine developing 105 hp and other options include power steering, power winch installed at the front of the vehicle and run-flat devices for the tyres.

The Peugeot P4 (4 × 4) 750 kg light vehicle is essentially the Mercedes-Benz 750 kg light vehicle manufactured under licence in France. Panhard delivered a total of 13,500 vehicles, with final deliveries being made in 1993. The Panhard-built Peugeot P4 is, however, powered by a Peugeot four-cylinder diesel engine.

Status

Production as required. In service with the French Army (80).

Manufacturer

Société de Constructions Mécaniques Panhard et Levassor.

UPDATED

Specifications

Crew: 1 + 3
Configuration: 4 × 4
Combat weight: 2,900 kg
Payload: 500 kg
Power-to-weight ratio: 24.31 hp/t
Length: 4.20 m
Width: 1.83 m
Height: 1.85 m
Ground clearance: 0.24 m
Track: 1.40 m
Wheelbase: 2.40 m
Angle of approach/departure: 42°/37°
Fording: 0.50 m
Gradient: 55%
Side slope: 30%
Max road speed: 108 km/h
Fuel capacity: 75 litres
Range: 600 km
Engine: Peugeot 2.498 litre 4-cylinder diesel developing 70.5 hp at 4,500 rpm
Gearbox: manual, 4 forward and 1 reverse gears
Clutch: single dry disc with ball-bearing thrust mechanism
Transfer case: 2-speed
Steering: rack and pinion with damper
Turning radius: 5.5 m
Suspension:
(front) coil springs, anti-sway bar and double-acting telescopic shock-absorbers
(rear) coil springs and double-acting telescopic shock-absorbers
Tyres: 700 RC 16 × C Type L
Brakes: hydraulic dual-circuit, disc front and drum rear
Electrical system: 24 V
Generator: 1,200 W
NBC system: no
Night vision equipment: no

Peugeot P4 (4 × 4) light vehicle, fitted with passive armour kit by Panhard, from the rear 0058085

Peugeot P4 (4 × 4) light vehicle, fitted with passive armour kit by Panhard (Christopher F Foss) **2002**/0121138

ACMAT armoured personnel carrier (TPK 420 BL)

Development

The TPK 420 BL armoured personnel carrier was developed as a private venture by Ateliers de Construction Mécanique de l'Atlantique (ACMAT) and is essentially the well-known VLRA (4 × 4) 2,500 kg long-range reconnaissance vehicle fitted with an armoured body.

The VLRA is used by 42 countries in Europe, Africa, the Middle East and Asia and is fully described in *Jane's Military Vehicles and Logistics 2001-2002*.

The first prototype of the ACMAT TPK 420 BL was completed in 1980 and production began the same year. The company is now offering the TPK 650 BL armoured personnel carrier which has a 6 × 6 configuration. In 2001, the 6 × 6 version was still at the prototype stage.

The current production model is the ACMAT TPK 420 BL which has been designed to carry 10 people, the commander and driver in the front and eight troops, seated four down either side, at the rear.

This version has a combat weight of 7,800 kg with the armoured body providing protection against 7.62 × 51 mm ball small arms fire.

Description

The hull of the TPK 420 BL is made of all-welded nickel chrome molybdenum steel, 5.8 mm thick. The vehicle has a similar layout to the basic unarmoured VLRA, with the engine at the front, behind which are seats for the commander and driver, and the troop compartment at the rear.

Immediately in front of the driver and commander is a bulletproof windscreen which is covered, when in a combat area, by a hatch hinged at the top. The commander and driver enter and leave the vehicle by a side door that opens to the rear. This has a bulletproof window which is covered by a flap that opens upwards, below this is a circular firing port if required. Above the commander's and driver's positions is a single-piece hatch cover opening to the rear.

The troops are positioned on bench seats down either side of the hull. The sides of the rear troop compartment are carried upright to give the maximum possible protection to the occupants. If fire is to be returned or if rapid dismounting is required, these sides

ACMAT 420 BL anti-riot armoured vehicle fitted with a 7.62 mm Wagschal turret and Helio grenade launcher as delivered to the Kingdom of Saudi Arabia, Ministry of Interior, Public Security Department 0058086

ACMAT VBL with sides and rear of body folded down and fitted with an 81 mm mortar firing to the rear

can be hinged outward and downwards at their mid-point as shown in the accompanying photograph. The vehicle can also be delivered with a fully enclosed, removable rear compartment to give improved protection from roof-mounted snipers.

If required, an 81 mm TDA mortar can be mounted in the rear. Various 12.7 mm and 7.62 mm machine guns can be pintle mounted in the rear troop compartment.

TPK 420 BL internal security (4 × 4)

This is fitted with a one-person manually operated turret with hinged shields and armed with a 7.62 mm machine gun. The system comprises one fixed-ring basis, one rotating ring on a 360° bearing, three hinged protection shields with automatic locks forming a cover in the closed position, five day periscopes, azimuth lock every 12° and a weapon cradle with support and balancing device.

If required, the turret can be fitted with eight grenade launchers on either side which are controlled from within the vehicle.

Other equipment on this version includes a loudspeaker system and a full air conditioning system.

Specifications

(TPK 420 BL)
Crew: 2 + 8
Configuration: 4 × 4
Combat weight: 7,800 kg
Unloaded weight: 6,000 kg
Length: 5.98 m
Width: 2.1 m
Height:
(hull top) 2.205 m
Ground clearance:
(axle) 0.273 m
(hull) 0.5 m
Track: 1.66 m
Wheelbase: 3.6 m
Angle of approach/departure: 45°/35°
Max speed: 95 km/h
Fuel capacity: 310 litres
Max range: 1,600 km
Fording: 0.8 m
Gradient: 65%
Turning radius: 8.5 m
Engine: Perkins Engines Company 6-cylinder diesel developing 138 hp (or 145 hp) at 2,800 rpm
Gearbox: Type 4.42 manual with 4 forward and 1 reverse gears
Clutch: single dry plate
Transfer box: 2 speed ACMAT
Steering: worm and nut
Suspension: leaf springs and double-action hydraulic shock-absorbers
Tyres: 12.50 × 20 run-flat
Number of tyres: 4 + 1 spare (run-flat)
Brakes:
(main) air/hydraulic
(parking) mechanical
Electrical system: 24 V
Batteries: 2 × 12 V, 100 Ah
Armament: see text
Armour: 5.8 mm
NBC system: no
Night vision equipment: no

BL light armoured car (4 × 4)

This has the same chassis and hull front as the BL armoured personnel carrier but has a fully enclosed rear troop compartment. The troops are seated on bench seats down either side of the hull. In each side of the hull are four bulletproof sliding vision blocks.

In the rear of the BL are two outward-opening doors, in each of which is a bulletproof vision block covered by an armoured flap.

Mounted in the centre of the roof is the one-man turret armed with a 7.62 mm machine gun which is aimed by using a roof-mounted periscope or a Giat Industries one-man turret Model TLi armed with a 12.7 mm (0.50) M2 HB machine gun. A 2,000 m range Euromissile MILAN ATGW system can also be installed.

The BL can be fitted with optional equipment including different radios and an air conditioning system, as well as being modified for more specific roles such as ambulance or command post.

Reconnaissance ACMAT

Giat Industries teamed with Thales Optroniques and ACMAT to develop a new 4 × 4 reconnaissance vehicle based on a much modified ACMAT vehicle, but with a new body. This was for the Belgian Army Reconnaissance 2001 requirement. INSYS is also offering a similar vehicle for the British Army Future Command and Liaison Vehicle (FCLV) requirement. INSYS (previously Hunting Engineering) was one of the three companies shortlisted by the UK DPA for the next phase of the competition.

Armoured ambulance (4 × 4)

In 1996, the prototype of the Type TPK 420 BL-SAM (4 × 4) armoured ambulance was completed. This has been developed as a private venture and is similar to the APC but configured for the ambulance role.

In addition to its three-man crew, it can carry four stretcher patients in the rear or a mixture of seated and stretcher patients. The wounded are loaded into the vehicle via the rear which is provided with a two-part door, the lower part opens downwards to form a step while the upper part opens upwards. To provide more room for the occupants, this model has a higher roof line. In late 2001, this remained at the prototype stage.

Logistics Cargo Transport Vehicle (6 × 6)

The prototype of this vehicle, designated the Type TPK 650 BL-CTL was shown in 1996 and is based on the 6 × 6 chassis with a fully armoured front cab and the load area at the rear. Maximum payload is quoted as 4,500 kg. In late 2001, this model remained at the prototype stage.

Status

Production as required. In service with a number of countries including Central African Republic (25), Côte d'Ivoire, Gabon, Zimbabwe (delivered in 1999) and the Kingdom of Saudi Arabia (Public Security Department).

Manufacturer

ACMAT, Ateliers de Construction Mécanique de l'Atlantique.

UPDATED

ACMAT TPK 420 BL (4 × 4) armoured personnel carriers delivered to Zimbabwe in 1999 **2000**/0080382

ACMAT Logistics Cargo Transport Vehicle TPK 650 BL-CTL (6 × 6) showing armoured cab 0058088

Auverland A3 SL armoured personnel carrier

Development

The Auverland A3 SL armoured personnel carrier (APC) was developed as a private venture by Auverland SA and is essentially their standard Auverland A3 (4 × 4) light cross-country vehicle fitted with an all-welded steel armour body.

Well over 5,000 of the standard unarmoured Auverland A3 (4 × 4) light vehicle have been manufactured for the home and export markets. Development of the Auverland A3 SL APC is complete and production can commence as soon as orders are placed.

Details of the base line Auverland Type A3 (4 × 4) light vehicle are given in *Jane's Military Vehicles and Logistics 2001-2002*.

Description

The layout of the Auverland A3 SL armoured personnel carrier is conventional, with the engine compartment at the front and the crew compartment at the rear. The all-welded steel hull provides the crew with protection against attack from 7.62 mm ball small arms fire.

The commander and driver are seated at the front of the vehicle with the driver on the left and the commander on the right. Both have a bulletproof window to their front and a door in the side that opens forwards. This is provided with a bulletproof window in its upper part.

The troop compartment is at the rear and entry to this is via a large door in the hull rear which is provided with a bulletproof window with a firing port in its upper part. Additional firing ports and vision devices can be provided in the sides and rear of the rear compartment as required.

Standard equipment includes blackout lighting and towing hooks front and rear. It can also be fitted with a roof-mounted one-person central mount onto which a variety of weapons can be fitted including a 5.56, 7.62 or 12.7 mm machine gun, 20 mm cannon or even a 40 mm grenade launcher.

Variants

The design of the Auverland A3 SL armoured personnel carrier is such that it can be adapted for a number of other roles such as command post vehicle.

Status

Development complete. Ready for production.

Manufacturer

Auverland SA.

UPDATED

Auverland A3 SL armoured personnel carrier 0058579

Specifications

Crew: 2 + 1
Configuration: 4 × 4
Weight:
(loaded) 2,500 kg
(unloaded) 1,880 kg
(payload) 620 kg
Power-to-weight ratio: 36.8 hp/t
Length: 3.70 m
Width: 1.65 m
Height: 1.80 m
Ground clearance: 0.25 m
Track: 1.342 m
Wheelbase: 2.65 m
Angle of approach/departure: 50°/50°
Max road speed: 130 km/h
Fuel capacity: 80 litres
Range: 800 km
Fording: 0.60 m
Gradient: 100%
Vertical obstacle: n/avail
Trench: n/avail
Engine: Peugeot XUD 9 TF 4-cylinder turbocharged diesel developing 92 hp at 4,000 rpm
Transmission: Peugeot 5 speed manual or ZF automatic
Transfer case: Auverland A80, 2 ratio
Clutch: dry plate
Differential: limited rear slip
Steering: power assisted
Turning radius: 5.5 m
Suspension: rigid axles, coil springs and hydraulic dampers
Tyres: 7.00 × 16
Brakes: servo-assisted, front and rear
Electrical system: 24 V
Armament: optional
Armour: steel
NBC system: no
Night vision equipment: optional

Petit Véhicule Protégé (PVP) Programme

Development/Description

The French Army has established an operational requirement for between 500 and 1,000 new 4 × 4 light protected armoured vehicles called the PVP (Petit Véhicule Protégé).

This new vehicle will bridge the existing gap between the Panhard P4 (4 × 4) light protected vehicle and the Panhard VBL (4 × 4) scout vehicle used by the French Army. Details of the Panhard P4 are given later in this section, while details of the Panhard VBL are given in the Reconnaissance vehicles section.

The Panhard P4, which entered service with the French Army in 1995, is essentially a Peugeot P4 (4 × 4) 750 kg light vehicle fitted with a passive-armour package originally developed by Panhard as a private venture. It can carry four people, including the driver and a small amount of cargo and is unarmed.

The French Army has taken delivery of over 1,000 Panhard VBL (4 × 4) scout vehicles with deliveries to continue for some years. These are used for a wide range of battlefield missions including command and control, reconnaissance and control, and anti-tank fitted with the Euromissile MILAN ATGW (anti-tank guided weapon) system.

To reduce procurement and support costs the Délégation Générale pour l'Armement (DGA) is looking for a Commercial Off-The-Shelf (COTS) solution to meet the PVP requirement.

It is envisioned that the PVP will be an in production 4 × 4 chassis fitted with a passive armour package that will provide the occupants with protection from small arms fire and shell splinters, and possibly with greater internal volume.

The current Panhard VBL is fully amphibious and fitted with an NBC system, but, to lower costs, these two items have been omitted from the new PVP requirement.

The Invitation to Tender (ITT) has already been issued and a short list will be drawn up with a request for proposals and/or quotations being issued in the near future.

In addition to being an off-the-shelf solution with procurement costs at least half that of the VBL, low operating and maintenance requirements are also required for the PVP. Experience has shown that the vehicle selected would clock up high rates during peacekeeping operations such as that in the former Yugoslavia.

Status

French Army programme.

Manufacturer

Not selected.

UPDATED

GERMANY

Rheinmetall Landsysteme New Armoured Wheeled Vehicle

Development

To meet emerging requirements for an air portable family of wheeled armoured fighting vehicles (AFV), the recently formed Rheinmetall Landsysteme is now developing, as a private venture, the New Armoured Wheeled Vehicle (NAWV).

The NAWV leverages off the former Thyssen Henschel TH 400 (6 x 6) vehicle developed some time ago that demonstrated a high level of cross-country mobility due to its advanced suspension system. This is now called the R 400 and is covered in detail in the following entry.

Thyssen Henschel subsequently became Henschel Wehrtechnik and in 2000 this company and a number of other companies owned by Rheinmetall were

Rheinmetall Landsysteme New Armoured Wheeled Vehicle (NAWV) (8 × 8) currently under development and shown here fitted with a three-person turret armed with a Rheinmetall 105 mm smoothbore gun
2001/0073694

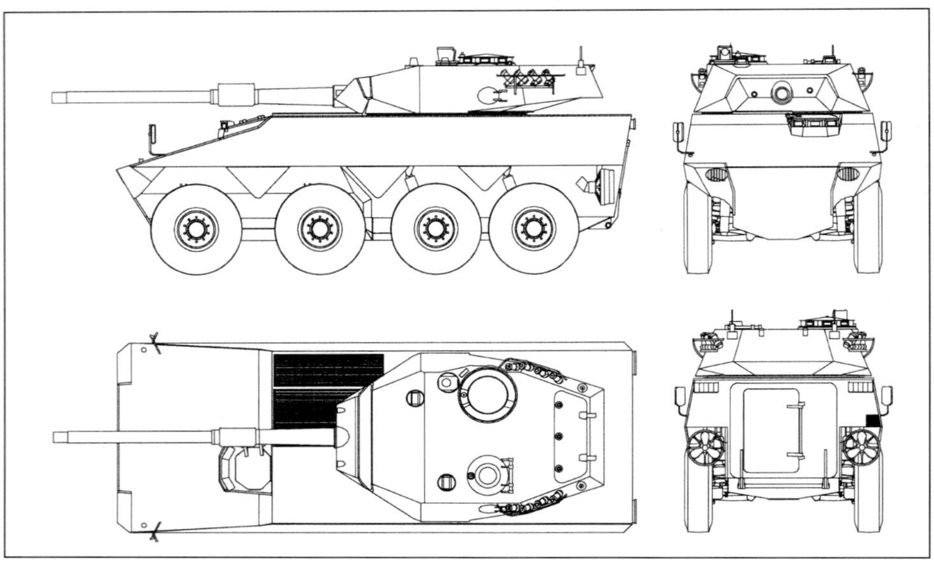

Four view general arrangement drawing of Rheinmetall Landsysteme New Armoured Wheeled Vehicle (NAWV) (8 × 8) currently under development and shown here fitted with a three-person turret armed with a Rheinmetall 105 mm smoothbore gun
2001/0105855

Specifications

Crew: 4
Configuration: 8 × 8
Combat weight: 20,000 kg
Unloaded weight: n/avail
Power-to-weight ratio: 21.75 hp/tonne
Length: 7.10 m (chassis)
Width: 2.720 m
Height: 2.838 m
Ground clearance: 0.43 m
Max road speed: 100 km/h
Max water speed: 10 km/h
Fuel capacity: n/avail
Max range: 800 km
Gradient: 60%
Side slope: 40%
Vertical obstacle: 0.60 m
Trench: 2.50 m
Turning radius: 13 m
Engine: Caterpillar, MTU or RVI diesel developing 435 hp
Transmission: automatic
Steering: power assisted: front four wheels
Suspension: hydropneumatic
Brakes: n/avail
Tyres: n/avail
Electrical system: 24 V
Batteries: n/avail
Armament
(main) 1 × 105 mm smoothbore gun
(coaxial) 1 × 7.62 mm MG
(air-defence) 1 × 7.62 mm MG
(smoke grenade launchers) 2 × 4 76 mm
Ammunition: 40 x 105 mm
Gun control equipment
Turret power control: electric/manual
Elevation/depression: n/avail
Turret traverse: 360°
Gun stabiliser
(vertical) yes
(horizontal) yes
NBC system: yes
Night vision equipment: yes
Airtransportability: Lockheed Martin C-130 Hercules

merged to form Rheinmetall Landsysteme which has built more wheeled vehicles than any other company in Germany.

As of late 2001, no decision had been taken by Rheinmetall Landsysteme to build a prototype of the NAWV. According to Rheinmetall Landsysteme, a prototype of the New Armoured Wheeled Vehicle could be running within 24 months of being given the go-ahead.

Description

The hull and turret of the NAWV is of all welded steel armour construction to which an additional layer of armour can be attached for a higher level of battlefield survivability. Enhanced protection is provided against anti-tank mines.

The driver is seated at the front of the NAWV on the left side and has a single piece hatch cover that opens to the right. In front of this are three day periscopes, the centre one of which can be replaced by a passive periscope for driving at night.

The power pack, which consists of a diesel engine coupled to an automatic transmission, is mounted to the right of the driver with the air inlet and outlet louvres in the roof. Various types of engine options will be available including Caterpillar, MTU and Renault as well as different automatic transmission options.

A wide range of roles is envisaged for the NAWV including a troop carrier and reconnaissance vehicle/tank destroyer. The latter would be fitted with a three-person turret armed with a Rheinmetall 105 mm low recoil smoothbore gun, with a 7.62 mm machine gun mounted coaxial and a similar weapon for air defence purposes. A bank of four 76 mm forward firing smoke grenade launchers are mounted either side of the turret.

Various types of STN ATLAS computerised fire-control systems and night vision equipment would be fitted to the turret such as the MOLF of EMES series. The 105 mm main armament would be fully stabilised.

The commander is seated on the right side of the turret and has a roof mounted stabilised day/night periscope and a single piece hatch cover with day periscopes for all round observation.

The gunner is seated forward and below the commander and also has a roof mounted stabilised day/night sighting system. The loader is seated on the left side of the turret and has a single piece hatch cover that opens to the rear. Forward of this is a rotatable day observation device.

The crew can enter via the roof hatches while the turret crew can also enter via a large ramp in the hull rear that opens downwards. This is also provided with a door in the left side. If the NAWV is configured as a troop carrier then there would also be roof hatches in the rear troop compartment.

The combat weight of the NAWV depends on the turret/armament installation and the level of armour protection required. A typical combat weight would be around 20 tonnes with a maximum road speed of 100 km/h and an operational range of up to 800 km.

Depending on its armour, the NAWV would be fully air portable in a Lockheed Martin C-130 Hercules transport aircraft and depending on the weight would be fully amphibious, being propelled by two propellers mounted one either side at the rear.

Other features of the NAWV, of which a full scale mockup has already been completed at the Rheinmetall Landsysteme Kassel facility, include built-in test equipment, can-bus digital onboard system, hydropneumatic suspension, powered steering on the front two axles and the option of a central tyre pressure regulation system.

Variants

As stated above, it is envisioned that the NAWV will be capable of being adapted for a wide range of battlefield missions.

Status

Private venture development.

Manufacturer

Rheinmetall Landsysteme GmbH.

UPDATED

Rheinmetall Landsysteme Fuchs 2 armoured personnel carrier

Development

Late in 2001, Rheinmetall Landsysteme revealed that it had designed and built as a private venture a new version of its well-established Fuchs Transportpanzer 1 (6 × 6) armoured personnel carrier called the Fuchs 2.

The first example of this ran for the first time early in 2001 and has already started extensive company automotive trials. It was shown in public for the first time at the MSPO International Defence Industry Exhibition in Poland and will be offered to Poland to meet its requirement for a new 6 × 6 or 8 × 8 armoured personnel carrier (APC).

Rheinmetall Landysteme built 1,236 Fuchs 1 vehicles, most for the German Army. They did have a number of significant export sales including Netherlands, Saudi Arabia, UK, USA and Venezuela.

Fuchs 2 (6 × 6) in armoured personnel carrier configuration fitted with remote controlled 12.7 mm M2 machine gun

2002/0096114

Description

The overall layout of the Fuchs 2 is almost identical to that of the original Fuchs covered in detail in a separate entry. The all-welded steel armour hull provides the occupants with protection from 7.62 mm armour piercing attack.

There is also an option to upgrade the level of protection further to provide protection against 12.7 and 14.5 mm armour pirercing attack, artillery fragments and mines.

When compared to the current Fuchs the new Fuchs 2 has a number of significant improvements. The troop compartment roof has been raised 145 mm for greater internal volume and the roof-mounted air inlet/outlet louvres have been redesigned.

Stonger DaimlerChrysler axles have been installed on the Fuchs 2 which has enabled the payload to be increased by 4,000 kg to a maximum of 9,400 kg. The braking system has been upgraded to take into account the increased weight of the vehicle.

To reduce overall life cycle costs, proven automotive parts are used wherever possible in the design of the Fuchs 2.

Fuchs 2 is powered by a MTU 6V 199 TE20 turbocharged low-mission diesel engine which develops 425 hp and meets EURO III emission standards. This is coupled to a ZF 6 HP 602 fully automatic six-speed transmission with integrated retarder.

A completely new digital electrical system has been fitted which meets CAN Standard SAE J 1939 and this provides for the automatic control of the vehicle status and easy component failure analysis. The driver of the Fuchs 2 is provided with a Driver Information System (DIS) which provides for online system status monitoring and failure identification.

At the request of the customer, the digital bus system of the Fuchs 2 may be expanded and data exchanged between the vehicle CAN BUS and mission kit bus system by means of a gateway.

Power steering is on the front four wheels and the tyres are 14.00 – 20 all round. An optional central tyre pressure regulation system is fitted which allows the driver to adjust the tyre pressure to suit the terrain being crossed. The tyres are run flat which enables the vehicle to remain mobile with one tyre damaged.

With a maximum combat weight of 20,000 kg the Fuchs 2 is fully amphibious, being propelled in the water by two Schottel propellers mounted one either side of the hull at the rear. These propellers can be traversed through 360° and with the wheels and propellers being used together Rheinmetall Landsysteme has stated that the Fuchs has excellent water ingress and egress capabilities.

The weapon fit of the Fuchs 2 depends on the operational requirement with a typical example being fitted with a new RLS 609 external weapon system armed with a 12.7 mm M2 machine gun or a 40 mm automatic grenade launcher. If required, this can be fitted with a Charge Coupled Device (CCD) camera, Liquid Crystal Display (LCD) and joystick control.

The Fuchs 2 vehicle can be fitted with a wide range of weapon systems up to and including their own E8 turret armed with a 30 mm cannon and 7.62 mm machine gun as well as various missile installations such as anti-tank.

A wide range of optional equipment is available for the Fuchs 2 including collective NBC system, air conditioning system, various weapons, grenade launchers, amphibious capability, run-flat tyres, central tyre pressure regulation system, automatic fire detection and suppression system, intercom and radio, driver's and commander's night vision devices, winch, vehicle navigation system with integrated global positioning system, 5 kW auxiliary power unit and a variety of mission kits.

The baseline Fuchs 2 armoured hull provides protection against 7.62 mm Armour Piercing (AP) attack at zero degrees obliquity through a full 360°. The vision blocks and windows provide the same level of protection as the hull.

To meet different threat levels the Fuchs 2 has been designed to be fitted with an enhanced armour package developed by IBD including 12.7 and 14.5 mm AP, artillery fragments and mines. There is also a passive armour package that provides protection against attack from 30 mm Armour Piercing Discarding Sabot (APDS) attack over the frontal arc.

The first example of the Fuchs 2 has the standard twin doors at the rear of the hull but for overseas trials, expected to be carried out in the Middle East in 2002, a power operated ramp will be fitted.

Standard equipment includes an NBC system and an air conditioning system and there are a host of options available including automatic fire detection and suppression system, winch, global positioning system and an auxiliary power unit, to name but a few.

Specifications

Crew: 2 + 10 (max)
Configuration: 6 × 6
Combat weight: 22,000 kg
Unloaded weight: 14,600 kg
Maximum payload: 7,400 kg
Power to weight ratio: 19.31 hp/t
Length: 6.83 m
Width: 2.98 m
Height: 2.42 m
Ground clearance: 0.41 m
Track: n/avail
Wheelbase: n/avail
Angle of approach/departure: n/avail
Max speed:
(road) 96 km/h
(water) 10 km/h
Fuel capacity: n/avail
Max road range: 700 km
Fording: amphibious (see text)
Gradient: 60%
Side slope: 30%
Vertical obstacle: 0.70 m
Trench: 1.2 m
Turning radius: 8.5 m
Engine: MTU 6V 199 TE20 turbocharged diesel developing 315 kW at 1,800 rpm (EURO III standard)
Transmission: ZF 6HP 602 6-speed automatic with integrated retarder and STU 13000 secondary gearbox
Steering: LS7FS DaimlerChrysler on front four wheels
Tyres: 14.00 × 20 radial with 11-20 rims (with run-flat as an option)
Brakes:
(main) air assisted, dual circuit
(parking) spring loaded acting on gearbox output shaft
Electrical system: 24 V
Batteries: 4 × 200 Ah
Generator: 6.5 kW
Armament: 1 × 12.7 mm M2 machine gun (typical)
Smoke laying equipment: 6 × 76 mm grenade dischargers
NBC system: yes
Night vision equipment: yes

Variants

As well as being used as an APC and fitted with various weapon systems, Rheinmetall Landsysteme can offer a complete family of vehicles on the same Fuchs 2 platform including NBC, C^3I/C^4I, electronic warfare, reconnaissance, explosive ordnance disposal and mortar carrier, to name but a few.

Status

Prototype. Not yet in production or service.

Manufacturer

Rheinmetall Landsysteme GmbH.

NEW ENTRY

Rheinmetall Landsysteme Transportpanzer 1 (Fuchs) armoured personnel carrier

Development

In 1964, the German Ministry of Defence examined its requirements for a new generation of wheeled military vehicles for the 1970s. This generation was to have included an 8 × 8 armoured amphibious reconnaissance vehicle, 4 × 4 and 6 × 6 armoured amphibious load carriers and a complete range of 4 × 4, 6 × 6 and 8 × 8 tactical cargo trucks, some of which were to be amphibious.

In 1964, a Joint Project Office was formed to undertake development of the complete range of vehicles. Companies in the Joint Project Office were Büssing, Klöckner-Humbolt-Deutz, Friedrich Krupp, MAN and Rheinstahl Henschel. Daimler-Benz did not join the Joint Project Office but went ahead on its own

Transportpanzer 1A4 of the German Army with additional armour for hull and turret (Michael Jerchel) ***2000***/0080383

M93A1 Fox Nuclear Biological Chemical Reconnaissance System with sensor pod in raised position 0018673

to develop a similar range of vehicles which were finally chosen.

After severe and extensive trials the 8 × 8 armoured amphibious reconnaissance vehicle became the Spähpanzer Luchs and 408 vehicles were built by Rheinstahl Wehrtechnik between 1975 and 1978. The 4 × 4, 6 × 6 and 8 × 8 tactical trucks were built by MAN and all were delivered to the German Army by 1980. The 4 × 4 armoured amphibious load carrier became the Transportpanzer 2 and the 6 × 6 armoured amphibious load carrier became the Transportpanzer 1. The Transportpanzer 2 was not placed in production but further development was undertaken by EWK and the eventual result was the APE Amphibious Engineer Reconnaissance Vehicle but this never entered production or service.

In 1977, Rheinstahl Wehrtechnik (which in 1996 became Henschel Wehrtechnik (today Rheinmetall Landsysteme)), under licence from Daimler-Benz, was awarded a production contract for 996 Transportpanzer 1 vehicles for delivery from 1979 at the rate of 160 vehicles per year. The first production vehicle was handed over to the German Army, which also calls the vehicle the Fuchs, in December 1979. Deliveries of the Transportpanzer 1 were completed in late 1986.

The Netherlands has taken delivery of a small quantity of Transportpanzer 1 vehicles for the electronic warfare role. Late in 2001, Rheinmetall Landsysteme stated that total production of the Fuchs amounted to 1,236 units.

Recent information has stated that some years ago Saudi Arabia took delivery of 36 Transportpanzer 1 series vehicles direct from Germany. Of these, 14 were in the armoured personnel carrier configuration, eight in ambulance role, four command post vehicles and 10 NBC vehicles.

Under contract to the German BWB the Fuchs KRK was developed but following trials with two prototype vehicles, this was not ordered. Further developments by Rheinmetall Landsysteme has resulted in the new Fuchs 2 first shown in 2001.

Description

The all-welded steel hull of the Transportpanzer 1 protects the crew from small arms fire and shell splinters. The hull has a rhomboid cross-section and incorporates spaced armour in certain areas.

The driver sits at the front of the hull on the left with the vehicle commander to his right. A large bulletproof windscreen to the front can be covered by an armoured shutter hinged at the top. There are four day periscopes fitted in the roof of the vehicle to the front of the driver's hatch for use when the windscreen is covered by the armoured shutter, the centre one of which can be replaced by a passive periscope for driving at night. The commander has a circular roof hatch, which opens to the left rear. Both the driver and commander have a door in the side of the hull with an integral window that can be covered by a shutter in a combat area.

The engine compartment behind the driver and commander is provided with an automatic fire extinguishing system. There is a small passage on the right side between the front of the vehicle and the passenger/cargo area at the rear. The air inlet and outlet louvres are in the roof of the vehicle with the exhaust pipe running along the left side of the hull.

The power pack consists of an eight-cylinder Mercedes-Benz OM 402A exhaust turbocharged diesel, developing 320 hp and a six-speed planetary gear torque converter transmission on the side of the engine and connected to it by spur gearing. Above the engine is the air cleaner system and the cooling system, with a hydraulically driven fan above the transmission. Attached to the power pack are the vehicle parking brake, 5 kW alternator, brake supply system and hydraulics, including the oil supply reservoir. The power pack can be removed in about 10 minutes as all disconnecting points for mountings, functionally connected shafts for automotive and amphibious devices, pneumatic, hydraulic and electric supply or control lines are designed as quick-disconnect points. The power pack of the Fuchs can be run outside the vehicle for test purposes.

The troop/cargo compartment is at the rear of the vehicle and is 3.2 m long, 1.25 m high and 1.5 m wide at its widest point. In each side of the hull is a small vision port protected by an armoured shutter. There are three hatches in the roof of the troop compartment: a large 1 m circular one in the forward part on which the main armament is normally mounted, a single-piece hatch cover to the rear of the first one on the right side which opens to the left, and to the rear and left of the second hatch is the third one which opens to the right.

The 10 infantrymen are seated on individual bucket seats, five down each side of the hull. The seats can be folded up, enabling cargo to be carried. Normal means of entry for the infantrymen is through two large doors in the rear of the hull. The right door has a vision port covered by an armoured shutter when not in use.

Nominal amphibious payload is 4,000 kg but, in the non-amphibious mode, it can carry up to 5,000 kg of cargo.

Steering is hydraulically assisted on the front four wheels. An electrically driven back-up pump automatically maintains the supply of hydraulic fluid to the steering box and propellers in case of hydraulic failure. The rigid axles incorporate hub-mounted planetary gears and are positioned by guide rods supported by progressively acting coil springs. All axles have differential lock-ups and hydraulic shock-absorbers. The tyres are of the run-flat type.

The Transportpanzer 1 is fully amphibious, being propelled in the water by two Schottel four-bladed propellers beneath the floor level of the vehicle at the rear of the hull. For steering when afloat the propellers can be traversed through 360°. Before entering the water a trim vane, which is stowed on the glacis plate when travelling, is hydraulically erected at the front of the hull. The three bilge pumps each have a capacity of 180 litres/min.

An NBC system, fitted as standard, can also be used to ventilate the crew and personnel compartments. Armament varies according to mission requirements but can consist of a 7.62 mm MG3 machine gun mounted over the commander's position or a 20 mm

HELAS electronic warfare system with mast-mounted sensor pack in the raised position

TPz 1A7 A2 ambulance uparmoured for use in Bosnia (Michael Jerchel) ***2001***/0103767

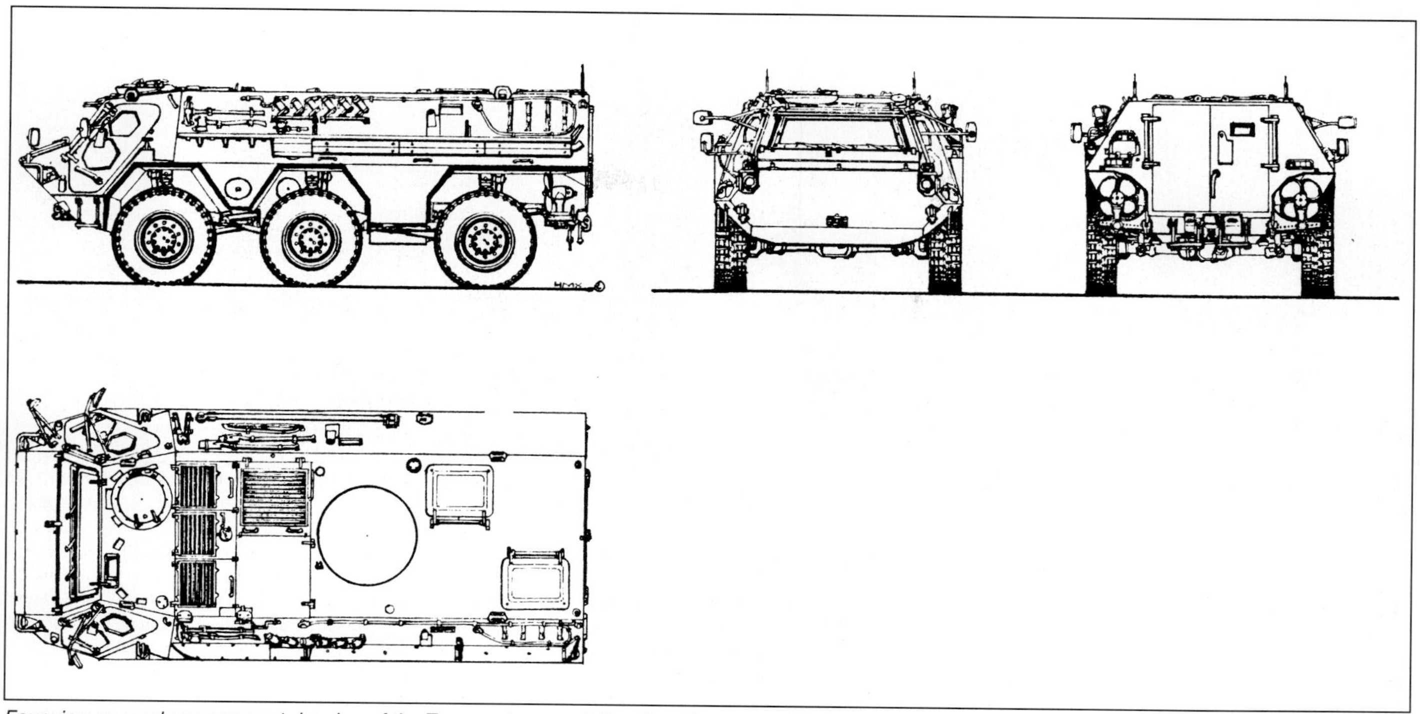

Four-view general arrangement drawing of the Transportpanzer 1 (6 × 6) vehicle without armament fitted (Henry Morshead)

Rheinmetall cannon mounted on a now Rheinmetall Landsysteme (previously KUKA) E-6 ring mount over the first circular roof hatch. The 7.62 mm machine gun has an elevation of +40° and a depression of −15°. The 20 mm cannon has an elevation of +75° with depression being determined by a contour limiter; 150 rounds of ready use ammunition are provided. All vehicles have six 76 mm grenade dischargers mounted on the left side of the hull firing forwards.

The Transportpanzer 1 is used by the German Army in the following roles:

Transportpanzer 1 designations

Original German Army designations are as follows:

TPz 1/Standard, 504 delivered (APC, ambulance and engineer support vehicle)

TPz 1A1/EloKa, 87 delivered (not amphibious, subvariants are Hummel jammer and HELAS direction finding vehicle)

TPz 1A2/Funk, 265 delivered (subversions are TPz 1A2/FuFu command and communications vehicles and TPz 1A2/PARA with RASIT battlefield radar)

TPz 1A3/ABC or Spurpanzer Fuch, 140 delivered (NBC reconnaissance vehicle).

The TPz 1 Standard could be fitted with various installation kits for different battlefield missions, these included:

Ambulance (SAN)
MILAN ATGW for use by recce units (MILAN)
Engineer vehicle (Pioniergruppe)
Command/control for engineer units (PiFue)
Command/ control for recce units (PzAufklFue).

The TPz 1A2 was fitted with a Rustsatz (installation kit) for the following roles:

RASIT radar (PARA)
Command/communication for brigade tactical operation centre (Fue/Fu)
Command air defence vehicle (FlaFue)

The TPz 1A1 and 1A3 are specialised versions so do not have installation kits.

With the installation of the new SEM 80/90 radios, the designations for some of these changed to:

TPz 1 Standard with SEM 80/90 radios became the TPz 1A4

TPz 1A2 with SEM 80/90 radios became the TPz 1A5

The TPz 1A5 A1 Flugabwehr is a modified air defence control vehicle with the new SEM 90/93 radios.

TPz 1A3 ABC with SEM 80/90 radios is the TPz 1A6 ABC

TPz 1A6 A1ABC with GPS and met probe is the TPz 1 A6 A1 NBC.

For service in the former Yugoslavia, in 1997 a further 55 TPz 1 Standard/TPz 1A4 were upgraded with additional armour and a number of other modifications and this led to new designators:

TPz 1A7 A1 - data processing/command/support UN, mobile command post
TPz 1A7 A2 - ambulance
TPz 1A7 A3 - MILAN ATGW
TPz 1A7 A4 - engineer (modified)
TPz 1A7 A5 - engineer command/control
TPz 1A7 A6 - forward air controller.

RASIT radar

This model is fitted with the RASIT battlefield surveillance radar (radar unit, target tracking and locating unit and radio) manufactured by LMT of France. The radar is mounted in the forward part of the hull and, when operating, is raised hydraulically to a maximum height of 1.8 m above the vehicle's roof. A 30 m cable allows remote operation of the system. This model is known in the German Army as the Panzeraufklärungsradargerät, 110 were built.

Command and communications vehicle

A total of 134 of these was built under the designation of FüFu. This model has a 5 kW generator in the left wing of the rear door, fuel being supplied from the main tank, communications equipment, a map board, two folding tables with personnel lamps, three folding seats and a box with a padded seat. Up to four antennas can be fitted, including a very high one at the right rear of the hull.

NBC reconnaissance vehicle

Some 140 of these were built. The NBC detection equipment includes radiation and chemical agent detection units, sampling instruments and markers. All operations are carried out inside the vehicle which gives protection against NBC agents. The German Army designation for this vehicle is the ABC Spürfuchs. The first vehicles were handed over to the German Army in February 1988.

This version carries radiation and chemical agent detectors, including the MM-1 mobile mass spectrometer, sampling instruments and markers.

In early 1990, a modified version of this vehicle, called the Nuclear, Biological and Chemical M93 Reconnaissance System, was selected by the US Army. Additional details of the M93 are given later in this entry.

In March 2000, the German Army still had 114 NBC reconnaissance vehicles in service.

Engineer vehicle

There are 220 of these used by combat engineers. They carry mines, demolition and other specialised equipment.

Electronic warfare

This model, known as the TPz 1 Eloka, is fitted with the EK 33 jammer kit. Unlike other models of the Transportpanzer 1 it has no amphibious capability, only one roof hatch and a 15 kW generator to provide power for the extensive electronic warfare system installed.

Supply carrier

There have been 220 of these vehicles issued to supply units for use in the forward area. It can also be used as an ambulance, carrying four stretcher patients or two stretcher and four sitting patients.

HELAS EW system

The radio and systems division of Siemens has developed the HELAS passive electronic radar reconnaissance system.

HELAS consists of two reconnaissance sets and a central processing set which are mounted in a Transportpanzer 1 (6 × 6) vehicle. The sensor package is carried on a mast pivoted at the rear and when required for action is raised into the vertical position.

HELAS can carry out a number of roles on the battlefield including searching for a reception of all electromagnetic radiation from radar and fire-control systems in the surveillance area and localising and signal processing. Signal processing includes identification, classification (especially assignment to weapon platforms), threat detection, contributing towards definition of the situation, detection of new radars, operating modes and strategies and information for central evaluation.

Extra armour for Transportpanzer 1

By early 1998, the now Rheinmetall Landsysteme had upgraded 55 Transportpanzer 1 vehicles of the German Army.

Standard Transportpanzer 1 vehicles have an all-welded steel armour hull that provides protection against penetration from small arms fire up to 7.62 mm Armour Piercing (AP).

For service with the United Nations Implementation Force (IFOR) operating in Bosnia, a higher level of protection was required.

The original vehicles are returned to the Rheinmetall Landsysteme facility in Kassel where the hull was stripped down and the new MEXAS passive fibre compound add-on armour package developed by IBD-Deisenroth is fitted.

This armour package includes an add-on armour mounted externally on the sides of the hull; internal spall liners fitted to the hull sides, rear and roof to reduce the secondary effects of particles having penetrated the main armour; additional elements in the floor area of the front and rear compartments to provide increased protection against mines; new windows with increased protection and modification of the window protective shields.

When fitted with this armour package the Transportpanzer 1 is provided with protection against attack from small arms fire up to 14.5 mm AP, 155 mm artillery and mortar projectiles bursting near the vehicle and protection against 7 kg anti-tank anti-personnel mines detonating under a wheel.

As the doors in the sides and rear have been fitted with additional armour they are now heavier and the higher actuating forces required to open and close the doors have been reduced to acceptable levels by integrating new torsion bar springs.

TPz 1A6 ABC/NBC from the rear (Michael Jerchel) ***2001***/0103768

TPz 1A1 Hummel ECM vehicle of the British Army (Andreas Kirchoff via Michael Jerchel) ***2001***/0103769

Other improvements incorporated include relocation of the vent valves for brake hydraulics, wheel hub cover of steel and slave socket for electrical power.

These upgraded Transportpanzer 1 vehicles, known as the Improved Fuchs, are now operational in Bosnia and have a combat weight of 19 tonnes and a payload of 5 tonnes.

A wide range of weapon fits is available, for example the Rheinmetall Landsysteme (previously KUKA) one-person turret Model 606A2 armed with a 12.7 mm M2 machine gun and 40 mm Mk 19 grenade launcher.

For operations in Bosnia the 7.62 mm MG3 machine gun has been provided with front and side armour protection. Rear protection is provided by the roof hatch.

Fuchs KRK

Late in 1997, the now Rheinmetall Landsysteme delivered the two preproduction examples of the new Fuchs KRK (Crisis Reaction Forces) Armoured Personnel Carrier (APC) to the German Army.

The Fuchs KRK is a further development of the Transportpanzer 1 Fuchs and has a number of significant improvements including a new and more powerful Mercedes-Benz 428 hp diesel engine, improved fully automatic ZF transmission and stronger axles.

This upgraded suspension allows for a gross vehicle weight of 24 tonnes of which 6.5 tonnes is the payload.

The all-welded steel hull has appliqué armour which provides protection against 14.5 mm armour piercing attack and 155 mm artillery fragments, as well as mortar shrapnel. Mine protection is provided against both 7 kg anti-tank mines and below wheel detonating anti-personnel mines.

Two Rheinmetall Landsysteme 605 one-person turrets were currently trialled on the preproduction vehicles, one fitted with a Rheinmetall 7.62 mm MG3A1 machine gun (6057.62M) and the other with a Rheinmetall 20 mm MK 20 cannon (60520M) and a 12.7 mm machine gun. In addition, a 7.62 mm MG3A1 machine gun can be installed over the commander's hatch at the front of the vehicle.

Standard equipment includes an air conditioning system and a central tyre pressure inflation system that allows the driver to adjust the tyre pressure to suit the type of terrain being crossed. Unlike earlier Transportpanzer 1 vehicles, the Fuchs KRK is not amphibious.

The model with the 20 mm gun turret was called the TVM I while the model with the machine gun turret was called the TVM II. These will not now enter service with the German Army.

Fuchs 2

This was previously called the TPz 2000 and builds on work carried out on the Fuchs RKR, originally developed to meet the requirements of the German Army. Details of the Fuchs 2 are given in a separate entry.

Export versions

In 1983, Venezuela placed an order for 10 Transportpanzer 1s which were delivered in September that year. These vehicles did not have the standard NBC system installed but had 76 mm smoke grenade dischargers, an 8,000 kg capacity winch, an air conditioning system and two roof-mounted weapon systems. To the rear of the engine compartment is a Rheinmetall Landsysteme one-person turret with an externally mounted 12.7 mm machine gun while behind this and facing the rear is a Krauss-Maffei Wegmann 11/1 mount with a shield and a 7.62 mm machine gun. These vehicles have full amphibious capability.

The Rheinmetall 105 mm Super Low Recoil gun has been successfully tested on the MC 601 chassis, a derivative of the Transportpanzer 1 (6 × 6) vehicle.

Export vehicles will not have the extensive range of equipment fitted as standard to vehicles of the German Army and Rheinmetall Landsysteme is therefore offering the following equipment as options: 12,000 Kcal/h heater, induction warming device, NBC system with an output of 3 m^3/min, vehicle navigation system, six electrically operated launcher cups for smoke or fragmentation grenades on both sides of the hull, 5,000 Kcal/h air conditioning system with two evaporators, automatic fire extinguishing system, intercom system with external and internal telephones, passive night vision equipment for the driver, trailer coupling, non-skid chains, self-recovery winch and a maximum of six ball mounts to enable the crew to use its small arms from inside the vehicle.

US Army M93 Nuclear, Biological and Chemical Reconnaissance System

The US Army purchased five Rheinmetall Landsysteme Fox (6 × 6) NBC reconnaissance vehicles from Germany which were tested alongside two 8 × 8 vehicles from the then Diesel Division, General Motors of Canada, with TRW being the prime contractor.

In the Spring of 1990, the US Army selected the Fox to meet its requirement for the M93 Nuclear, Biological and Chemical Reconnaissance System.

In March 1990, General Dynamics Land Systems was awarded a US$13.878 million contract to improve the M93 system with 66 per cent of the work being carried out in the US and the remaining 34 per cent in Germany with Rheinmetall Landsysteme as the main subcontractor.

Phase I was a three month competitive vehicle evaluation that culminated with the Fox being selected in July 1989.

The US Army subsequently awarded two contracts to the General Dynamics/Rheinmetall Landsysteme team to build Fox vehicles for the US Army.

Phase II, for Interim System Production (ISP) and fielding of 48 slightly modified vehicles in two batches, one of eight vehicles and the other of 40, was completed by 1993.

The first vehicles were handed over by Rheinmetall Landsysteme at its Kassel facility in October 1991 and have been slightly modified to include air conditioning.

Following the Iraqi invasion of Kuwait in the Summer of 1990, the German government offered a number of standard production German Army NBC reconnaissance vehicles to the Allied Coalition Forces who had no vehicles of this type. Israel (8), Turkey (4), the UK (11) and the USA (60) were supplied with these vehicles.

The upgrading of the vehicles included integrating US communications equipment, weapons, smoke grenade launchers, air conditioning and other engineering changes.

There were expected to be two further phases to the US M93 system, these being Phase III and IV, with the latter including licensed production in the US of an additional 210 vehicles.

In mid-1995, it was announced that as the US Army had already received a batch of 60 vehicles for

Specifications

standard production vehicle
Crew: 2 + 10*
Configuration: 6 × 6
Combat weight: 19,000 kg
Unloaded weight: 14,000 kg
Power-to-weight ratio: 16.84 hp/t
Length:
(external) 6.83 m
(internal) 3.2 m
Width:
(external) 2.98 m
(internal) 1.42 m
Height:
(to hull top) 2.30 m
(internal) 1.25 m
Ground clearance:
(axles) 0.406 m
(hull) 0.506 m
Track:
(front) 2.54 m
(rear) 2.56 m
Wheelbase: 1.75 + 2.05 m
Angle of approach/departure: 45°/45°
Max speed:
(road) 105 km/h
(water) 10.5 km/h
Fuel capacity: 390 litres
Max road range: 800 km
Fording: amphibious
Gradient: 70%
Vertical obstacle: 0.6 m
Trench: 1.10 m
Turning radius: 8.5 m
Engine: Mercedes-Benz Model OM 402A V-8 liquid-cooled diesel developing 320 hp at 2,500 rpm
Transmission: ZF Model 6 HP 500 6-speed automatic with torque converter
Steering: hydraulic, recirculating ball, operating on 1st and 2nd axles
Tyres: 14.00 × R20
Brakes:
(main) air-assisted, dual circuit
(parking) spring-loaded acting on gearbox output shaft
Electrical system: 24 V, 250 Ah
Batteries: 4 × 12 V, 125 Ah
Generator: Bosch T3, 28 V, 180 A
Armament: 1 × 20 mm cannon or 1 × 7.62 mm MG
Smoke-laying equipment: 6 × 76 mm grenade dischargers
NBC system: yes
Night vision equipment: yes

* Maximum is 14 for export vehicles
All versions are fully amphibious using propellers, except EW version which is not amphibious.

Operation Desert Storm and as the US Army had been downsized, there would be no production in the US.

The US Army Project Manager for Nuclear, Biological and Chemical Defense Systems awarded the prime contract for the Fox Nuclear Biological Chemical Reconnaissance System Block I Modification Program to General Dynamics Land Systems.

General Dynamics Land Systems has two major subcontractors, Anniston Army Depot and Rheinmetall Landsysteme of Germany which originally built the Fuchs vehicle.

For the M93A1 programme, General Dynamics Land Systems is responsible for vehicle disassembly and reassembly, systems integration and final test.

Anniston Army Depot is responsible for hull preparation, NBC trail modifications, asbestos removal and final paint. Rheinmetall Landsysteme is responsible for the overall design, pilot verification, engineering assistance and technical assistance.

All US Army vehicles are now expected to be brought up to the Block I standard with the first three vehicles upgraded in Germany and the remainder in the United States at the Anniston Army Depot in association with General Dynamics Land Systems.

The main improvements of the M93A1 can be summarised as:

(1) Reduction of the crew from four to three
(2) A mount that allows 180° rotation and levelling in all three axes of the roof-mounted M21 Remote Sensing Chemical Agent Alarm
(3) Improved sensor to measure windspeed, direction, air and ground temperature while the vehicle is stationary
(4) Improved reporting capability
(5) Capability of digital transmission via the SINgle Channel Ground/Air Radio System (SINCGARS)
(6) Global Positioning System for driver and commander
(7) Warning lights for the M21 aiming camera and the sample wheels added to the dashboard
(8) Monitor provided for the commander to display either mass spectrometer information or pictures from the M21.

Integrated/automated sensors on the M93A1 include:

(1) Nuclear radiation detection
(2) Chemical detection (blood, blister, nerve and choking)
(3) Standoff chemical agent detection up to 5 km (M21)
(4) Position/Location System (VOS 25)
(5) Central Processing Unit (CPU)
(6) Fox Commander's Display (FCD)
(7) Printer/keyboard
(8) Met sensors for ground and air sampling
(9) SINCGARS digital communications.

The M93A1 automatically integrates contamination information from detectors with input from onboard navigation and meteorological systems and transmits digital NBC warning messages through the current Maneuver Control System.

The M21 Multipurpose Integrated Chemical Agent Detector (MICAD) is the basis of the Fox system's digital architecture and is the resident processor that generates the digital NBC reconnaissance products and warning messages.

Uparmoured TPz 1A4 MILAN also fitted with armour protection for roof-mounted 7.62 mm MG3 machine gun (Michael Jerchel)
2002/0121137

The M93A1 was type classified in June 1995 and approved for production and fielding in Fiscal Year 1995.

The first upgraded M93A1 to be upgraded in the US was delivered to the US Army by General Dynamics Land Systems and the Anniston Army Depot in March 1998.

The US Army has taken delivery of 123 M93 systems of which 120 will be modified, but so far only 86 of these have been funded. The US Marine Corps has 10 vehicles but in the future may field a similar system based on the LAV (8 × 8) or modified AM General HMMWV chassis.

Red Fox

This is a private venture by Rheinmetall Landsysteme and has been designed for civilian applications such as disaster relief with variants including fire extinguishing, NBC reconnaissance, recovery/rescue and transport.

TPz 1 Fox EOD vehicle

Rheinmetall Landsysteme has completed the prototype of the Fox EOD (Explosive Ordnance Disposal) vehicle based on the Transportpanzer 1 (6 × 6) chassis to meet the German Air Force requirement for an airfield EOD vehicle (EOD – EinsFz).

This is essentially a modified Transportpanzer 1 chassis fitted with a one-man Rheinmetall TS-15 turret and carries specialised EOD equipment.

The modifications to the chassis include:

(1) Partially reinforced armour plating, mainly in the front part, to increase protection for the crew
(2) Modified roof to allow for the installation of the turret
(3) Omission of the vehicle's smoke launcher unit and lateral vision blocks in the crew compartment
(4) Omission of the water propulsion unit
(5) Internal modifications to accept the EOD operational kit.

The one-man Rheinmetall TS-15 turret system is armed with a 20 mm cannon and a 7.62 mm machine gun mounted coaxially to the right with three smoke grenade dischargers mounted either side of the turret firing to the front.

The targets to be engaged are selected by the commander seated in the front of the vehicle next to the driver. EODs are engaged using the 20 mm cannon or 7.62 mm machine gun with the turret being provided with a Steinheil LZL 2000 sight with an integrated laser range-finder. A coaxial Xenon searchlight enables targets to be engaged at night.

A total of 220 rounds of 20 mm and 500 rounds of 7.62 mm ammunition is carried for ready use.

Export variants

Armoured mortar carrier with hull-mounted 81 mm mortar or towed 120 mm mortar.

Armoured missile launcher with turret-mounted TOW, HOT or MILAN ATGWs.

Armoured support vehicle for use in a variety of roles such as ambulance, cargo, recovery and maintenance.

Armoured infantry fighting vehicle with various armament options including 20 or 25 mm cannon with the option of a 7.62 mm remote-controlled machine gun over the rear part of the troop compartment.

Status

Production complete. Known users of the Transportpanzer 1 include:

Country	Quantity	Comment
Germany	1,125	including 114 NBC and 87 EW
Netherlands	18	all electronic warfare
Saudi Arabia	36	in four versions
UK	11	all NBC plus some electronic warfare
USA	130	120 Army and 10 Marines
Venezuela	10	all APC

Manufacturer

Rheinmetall Landsysteme GmbH.

UPDATED

Rheinmetall Landsysteme Condor armoured personnel carrier

Development

The Condor has been developed as a private venture by Rheinmetall Landsysteme a successor to its UR-416 (4 × 4) armoured personnel carrier. Main improvements over the earlier vehicle are that it is fully amphibious, has an increased road speed and load-carrying capacity, and improved ballistic protection. Although designed primarily as an armoured personnel carrier, the vehicle can also be adopted for a wide range of other roles including anti-tank, cargo carrier, command vehicle, ambulance, fitters' vehicle and reconnaissance vehicle.

The first prototype was completed in 1978. Wherever possible, standard automotive components have been used in the design of the Condor, both to reduce costs and to enable spare parts to be obtained from a commercial source.

Late in 1981, Rheinmetall Landsysteme was awarded a contract by the Malaysian government for 459 Condor armoured personnel carriers. The last of these was completed in March 1984. Some of the Malaysian Condors were fitted with a British Helio FVT 900 one-man turret armed with an Oerlikon Contraves 20 mm cannon and variants included an ambulance, one-person turret armed with twin 7.62 mm machine guns, fitters' vehicle with a crane and a command post vehicle.

Production is undertaken as required and, in late 2001, it is estimated that over 600 vehicles have been built, all for the export market.

Description

The hull of the Condor is made of all-welded armoured steel protecting the crew from 7.62 mm armour-piercing rounds and 5.56 mm ball-type rounds, shell splinters and anti-personnel mines.

Bulletproof windows are provided to the driver's front and sides. When in combat, the windows are covered by armoured flaps. A periscope and slit between the armoured flaps provide sufficient vision for the driver in combat situations.

The two rigid axles are supported by Panhard-type bars and torque tubes with axle motion damped by progressively acting coil springs. Each axle has double-action hydraulic shock-absorbers and a differential lock.

The engine, which is the same as that used in the well-known Mercedes-Benz Unimog range of cross-country vehicles, is mounted to the right of the driver with the air inlet and outlet louvres in the top of the hull to the right and rear of the driver and the exhaust pipe on the right side of the hull. The engine is mounted on three shock-absorbing suspension points and power is transmitted to the gearbox by a short drive shaft. The gearbox is a Daimler-Benz Model UG3/40-8/13.85 GPA and has eight gears which can be used in both directions. Power is transmitted to the front and rear axles by Cardan shafts mounted inside torque tubes.

The tyres allow a maximum road speed of 50 km/h to be maintained, even after they have been punctured and deflated, to a range of 30 km. If the vehicle is driven at a lower speed the range is longer.

The commander sits behind the driver facing the centre of the vehicle, with a single-piece hatch

Specifications

Standard version
Crew: 2 + 12
Configuration: 4 × 4
Combat weight: 12,400 kg
Unloaded weight: 9,200 kg
Power-to-weight ratio: 13.54 hp/t
Length:
(hull) 6.13 m
(including winch) 6.47 m
Width: 2.47 m
Height:
(to turret top) 2.79 m
(to hull top) 2.18 m
Ground clearance: 0.475 m (axles)
Track: 1.84 m
Wheelbase: 3.275 m
Angle of approach/departure: 45°/52°
Max speed:
(road) 100 km/h
(water) 10 km/h
Fuel capacity: 280 litres
Max road range: 900 km
Fording: amphibious (with propeller)
Gradient: 60%
Side slope: 30%
Vertical obstacle: 0.55 m
Engine: Mercedes-Benz OM 352A 6-cylinder supercharged water-cooled diesel developing 168 hp at 2,800 rpm (or Daimler-Benz OM 366 LA diesel with exhaust gas turbocharger and intercooler developing 204 hp)
Transmission: manual giving 8 speeds in both directions
Clutch: single dry plate
Suspension: coil springs with double-acting hydraulic shock-absorbers
Tyres: 13.00 × 20 with run-flat inserts
Brakes: dual-circuit hydraulic disc brakes
Electrical system: 24 V
Batteries: 4 with total capacity of 250 Ah
Armament:
(main) 1 × 20 mm cannon
(coaxial) 1 × 7.62 mm MG
Smoke-laying equipment: 2 × 3 76 mm smoke grenade dischargers
Ammunition (in turret):
(20 mm) 220 rounds
(7.62 mm) 500 rounds
Gun control equipment
Turret power control: manual
(by commander) no
(by gunner) yes
Gun elevation/depression: +60°/–7°
Turret traverse: 360°
Gun stabiliser:
(vertical) no
(horizontal) no
NBC system: optional
Night vision equipment: optional

Condor (4 × 4) APC in service with Portugal and used in command post role (Richard Stickland) ***2002***/0121136

cover. The Condor has doors in both sides and the rear allowing rapid deployment of the crew. Firing ports in the hull sides and rear door allow some of the troops to fire their weapons from within the vehicle.

The Condor is fully amphibious, propelled in the water by a propeller mounted under the rear of the hull. This can be fully traversed through 360° by the driver who is provided with an indicator to show the position of the propeller.

Before entering the water, a trim vane is erected at the front of the vehicle. Optional equipment includes air conditioning equipment, radios, an intercom, night vision equipment, an NBC system, a propeller kit and a winch with 50 m of cable which can be operated to the front or rear of the Condor by remote control.

Variants
Other armament installations can be fitted including a turret armed with twin 7.62 mm machine guns and a turret armed with ATGWs such as HOT, and 12.7 mm and 7.62 mm machine guns.

Condor with SAMs
For trials purposes, a Condor vehicle has been fitted with the German STN Atlas Elektronik ASRAD air defence system with four Stinger fire-and-forget missiles in the ready to launch position.

This system, based on a modified Wiesel 2 full-tracked armoured chassis, is now in service with the German Army.

Malaysian upgrade
Between 1991 and 1993, MMC Engineering Services Sdn Bhd upgraded 100 Condor vehicles and, in mid-1997, another contract was awarded to the company for the upgrade of a further batch of 150 vehicles. This contract is worth US $7.7 million with the vehicles redelivered to the Malaysian Army between 1997 and 2000 at the rate of 50 units a year.

Status
Production as required. In service with:

Country	Quantity	Comment
Argentina	20	estimate
Malaysia	459	incl fitters, command post and APC
Portugal	12	
Thailand	18	delivered 1993
Turkey	25	used by police
Uruguay	50	

Manufacturer
Rheinmetall Landsysteme GmbH.

UPDATED

Condor (4 × 4) APC fitted with Helio FVT 900 turret in service with the Malaysian Army (Richard Stickland) ***2002***/0121135

Condor (4 × 4) repair and recovery vehicle used by the Malaysian Army, shown here in the travelling configuration (Richard Stickland) ***2002***/0121134

Rheinmetall Landsysteme TM 170 armoured personnel carrier

Development

The now Rheinmetall Landsystems (originally Thyssen Henschel) TM 170 is based on the chassis of a Mercedes-Benz Unimog (4 × 4) cross-country vehicle and spare parts are therefore easily obtainable worldwide.

The TM 170 was announced for the first time in 1978 and entered production in 1979. It has been designed primarily for use as an armoured personnel carrier or an internal security vehicle, but can be adapted for a wide range of other roles such as an ambulance, anti-tank vehicle, command post or reconnaissance vehicle.

The TM 170 was selected by the German Border Guard and state police to replace the old MOWAG MR 8 series of APCs, designated the SW1 and SW2. The TM 170 has been designated the SW4, the SW3 being the armoured version of the Mercedes-Benz 4 × 4 light vehicle. The first order for 87 vehicles was placed in July 1982 with deliveries running from 1983 onwards. By 2001, 380 had been built for home and export markets. The only known recent sale of the TM 170s is to Kuwait which took delivery of 15 vehicles in 1994.

The German Border Guard had a total of 121 TM 170, but it is understood that these were handed over to the German Army some years ago. It is understood that Macedonia has acquired a quantity of these TM 170 vehicles.

Description

The hull of the TM 170 is of all-welded steel providing protection from small arms fire and shell splinters. The engine is at the front of the hull, with the commander and driver immediately behind the engine and the troop compartment at the rear. The body is attached to the chassis at five points by gimbal mounts and can be removed quickly and lifted for repair or maintenance and overhaul such as changing the engine. The engine is coupled to a manual gearbox and the driver can select either 4 × 4 drive for cross-country use or 4 × 2 drive for roads, with the differential lock. All vehicles have run-flat tyres as standard. The exhaust outlet is on the left side of the hull.

The driver sits on the left with the commander to his right. There is a large windscreen which can be covered in action by two individually operated armoured shutters hinged at the bottom which fold forwards onto the front when not in use. The commander and driver each have three roof-mounted day periscopes for observation to the front and sides of the vehicle when the armoured windscreen shutters are raised. On each side is a window which is covered in action by a shutter hinged at the bottom, while over each of their positions is a single-piece hatch cover that opens to the rear of the vehicle.

Apart from the driver and commander a maximum of 10 fully equipped troops can be carried seated on individual seats down each side of the hull. There is a door in each side of the hull which opens parallel to the vehicle's body, immediately behind the commander and driver, with a single firing port in each. There is also an entrance hatch at the rear of the hull folding downwards on the outside. In the centre of the hatch is a vision block with a spherical firing port underneath and a small firing port either side. In each side of the hull, to the rear of the side entrance door, is one firing port and one vision block with a spherical firing port underneath. The spherical firing ports enable the crew to use its small arms from within the vehicle in complete safety. A rifle or SMG fitted in one of the spherical firing ports has a traverse of 16° left and right, an elevation of +23° and a depression of –8°.

In the centre of the roof is a large circular single-piece hatch where a variety of armament installations can be mounted.

All vehicle types have a fording depth of 1.3 m. There is also a fully amphibious version with a trim vane which moves upwards at the front of the bonnet where it is stowed when travelling on land. When afloat, the TM 170 has a maximum speed of 10 km/h, being propelled and steered by two hydraulically operated propellers which can be turned through a full 360°.

The vehicle can be fitted with the following optional equipment: auxiliary heater, fire warning and extinguishing system, hydraulically operated 5,000 kg capacity winch with 40 m of 13 mm diameter cable, public address system, police flashing lights, radio, NBC filtering system and run-flat tyres. A single roof-mounted tear gas/smoke discharger with aiming device can be traversed through a full 360°. This can be reloaded from within the vehicle and triggered by remote control. As an alternative, banks of four 76 mm smoke dischargers, each with a fixed direction, can be mounted on the roof. Another option is a hydraulically operated obstacle-removal blade at the front of the hull. When not in use this can be carried at the rear of the vehicle.

The following armament installations are available for the TM 170: pintle- or ring-mounted 7.62 mm machine gun, turret armed with twin 7.62 mm machine guns, Rheinmetall Landsysteme (previously KUKA) turret armed with 20 mm cannon which can be aimed and fired from within the vehicle, turret armed with four Euromissile HOT or MILAN ATGWs in the ready to launch position. A surveillance cupola is also available.

Variants

A low-priced version of the TM 170 is known as the Armoured Special Vehicle TM 170 Hardliner which has a number of features comparable to those of the original TM 170 BGS version.

The two front and two side windows for the driver and commander are fitted with bulletproof glass. When in a combat area, the front windows are covered by armoured shutters hinged at the top which are manually lowered. The windows either side of the commander and driver are in the upper parts of low doors which open forwards, and can be covered by swivel-type armoured plates. When the windows are covered by bulletproof shutters, a set of three day periscopes each for the commander and driver provide forward observation. Above their seats are two rear-opening circular roof hatch covers.

Either side of the vehicle are four firing ports and the crew enters via a large ramp in the hull rear which is provided with one firing port.

A turret armed with 7.62 mm or 12.7 mm machine guns or a 20 mm or 25 mm cannon can be mounted on the roof of the troop compartment. To the rear of this is a single-piece circular hatch cover that opens towards the front. The TM 170 Hardliner has a fording depth of 1.3 m.

The TM 170 Hardliner Patrol is a pick-up-type vehicle but with a fully enclosed cab for the commander and driver and an open troop compartment at the rear with bench seats down either side.

For riot control the TM 170 can be fitted with a 1,000 litre water tank plus a CN/CS gas tank. The single CN/water cannon (monitor) ejects intermittent water/gas mixture shots up to 50 m through 360°.

TM 170 A8 airport surveillance

This version of the TM 170 has been developed for the surveillance of high-value strategic areas such as airports, public and government buildings.

Special equipment fitted includes video recording units for day and night observation, radio sets and an information decoder. The day/night camera is mounted on a telescopic arm which can be elevated to a height of 6 m above the ground.

Specifications

Crew: 2 + 10
Configuration: 4 × 4
Combat weight: 11,650 kg
Unloaded weight: 8,800 kg
Power-to-weight ratio: 18 hp/t
Length: 6.140 m
Width: 2.47 m
Height:
(without armament) 2.32 m
Ground clearance: 0.48 m
Track: 1.84 m
Wheelbase: 3.25 m
Angle of approach/departure: 45°/48°
Max speed:
(road) 100 km/h
(water) 9 km/h
Fuel capacity: 200 litres
Max road range: 870 km
Fording: 1.2 m
Vertical obstacle: 0.6 m
Gradient: 80%
Engine: Mercedes-Benz OM 366 turbocharged diesel developing 240 hp at 1,400/1,700 rpm
Steering: hydraulic, power assisted
Electrical system: 24 V
Batteries: 2 × 12 V, 125 Ah
Armament: optional
Gun stabiliser:
(vertical) no
(horizontal) no
NBC system: no
Night vision equipment: optional

TM 170 M3

This is the latest version of the TM 170 which was announced in 1993 and has a number of improvements including the installation of a Mercedes-Benz OM 366 LA EURO 1 turbocharged 6-cylinder in-line diesel with direct injection developing 214 hp which gives the vehicle a maximum road speed of 100 km/h. The fuel tank holds 170 litres of fuel which gives an operational range of about 680 km. Overall specifications are very similar to the earlier TM 170 vehicle.

Optional equipment includes:

(1) Wheels with run-flat tyres
(2) Air conditioning system
(3) Additional engine independent heater for the crew compartment
(4) Automatic fire warning and extinguishing system
(5) Public address system with siren and blue flashing light
(6) NBC ventilation system
(7) Hydraulic system for winch and dozer blade
(8) Winch
(9) Dozer blade
(10) Protection grids
(11) Searchlight
(12) Spherical gun mount with vision block
(13) Tear gas/smoke grenade launchers
(14) Special device for lifting the superstructure.

Status

Production as required. In service with Germany, Kuwait (15), Macedonia and in undisclosed Near and Middle East countries.

Manufacturer

Rheinmetall Landsysteme GmbH.

UPDATED

Amphibious version of TM 170 with trim vane erected at front of vehicle 0018664

Standard production TM 170 armoured personnel carrier 0018665

Rheinmetall Landsysteme UR-416 M armoured personnel carrier

Development

The UR-416 was developed by Rheinstahl Maschinenbau (which has changed its name on numerous occasions and in 2000 became Rheinmetall Landsysteme) as a private venture, the first prototype being completed in 1965. Production commenced in 1969 and, by the time production had been completed, a total of 1,030 vehicles had been built, most for the export market.

The UR-416 series is no longer being marketed and has been replaced by the now Rheinmetall Landsysteme TM-170 (4 × 4) vehicle covered in detail in the previous entry.

The UR-416 is essentially the chassis of a Mercedes-Benz Unimog (4 × 4) cross-country vehicle fitted with an armoured body. Spare automotive parts are identical to those used in the truck and are therefore available from commercial sources.

Description

The last production model of the UR-416 was called the UR-416 M. The modifications and improvements have been included in the following detailed description.

The hull of the UR-416 M is of all-welded steel armour construction which protects the crew against small arms fire, shell splinters and anti-personnel mines.

The driver is seated at the front of the vehicle on the left, just behind the engine, with the vehicle commander to his right. Both have a large bulletproof front and side window for improved visibility. Armoured flaps covering the commander's and driver's front windows are lowered by gas pressure. Side windows are protected by swivelling armoured flaps. Forward observation is maintained by two single day periscopes in the forward part of the roof.

The diesel engine is coupled to a manual gearbox which has six forward and two reverse gears. For normal road use only the rear axle is engaged but for cross-country travel the front axle is also engaged and when travelling across very rough country the front and rear axle differential locks are engaged.

The eight fully equipped troops are seated to the rear of the commander and driver, three down each side of the hull facing outwards and two at the back facing the rear.

Specifications

Crew: 2 + 8
Configuration: 4 × 4
Combat weight: 7,600 kg
Unloaded weight: 5,400 kg
Power-to-weight ratio: 16.5 hp/t
Length: 5.1 m
Width: 2.25 m
Height: 2.25 m
Ground clearance: 0.44 m
Track: 1.78 m
Wheelbase: 2.9 m
Angle of approach/departure: 47°/51°
Max road speed: 81 km/h
Fuel capacity: 150 litres
Max road range: 600-700 km
Fuel consumption:
(road) 0.25 litres/km
Fording: 1.3 m
Gradient: 70%
Vertical obstacle: 0.55 m
Turning radius: 6.45 m
Engine: Mercedes-Benz OM 352 6-cylinder water-cooled diesel developing 120 hp at 2,800 rpm
Transmission: manual with 6 forward and 2 reverse gears
Steering: hydraulic, power assisted
Clutch: single dry plate
Tyres: 12.50 × 20
Brakes:
(main) hydraulic, air assisted
(parking) mechanical on rear wheels
Electrical system: 24 V
Batteries: 2 × 12 V, 110 Ah
Armament:
(basic model) 1 × 7.62 mm MG
Smoke-laying equipment: optional
Gun stabiliser:
(vertical) no
(horizontal) no
NBC system: no
Night vision equipment: optional

UR-416 (4 × 4) APC in service with El Salvador and armed with roof-mounted 12.7 mm M2 machine gun (Julio Montes) ***2001***/0103773

Each man has an individual seat which can be folded upwards. There are three doors, one in each side of the hull and one at the rear. The side doors are in two parts: the lower part folds downwards to form a step and the upper part is hinged on one side and folds flat onto the side of the hull. The rear door is a wide gas spring supported ramp allowing fast mounting and dismounting of the crew with the spare wheel and tyre mounted externally.

There are six firing ports altogether, two in the rear door, one in each of the side doors and a further two ports in the hull side. In addition, each sidewall has an observation block fitted with a spherical ball mount underneath.

In the roof of the UR-416 M, behind the commander and driver, is a circular single-piece hatch that opens to the rear and is secured on the roof. The basic model is normally armed with a 7.62 mm machine gun with a shield, which has an elevation of +75°, a depression of -10° and a total traverse of 360°. It could also be delivered with a horseshoe-shaped rail on which a 7.62 mm machine gun and shield with a traverse of 235° can be mounted. To the rear of this armament installation is a forward-opening single-piece hatch cover, normally with a single firing port which can be used when the hatch is locked vertical.

Forward of the commander's and driver's position are two side-mounted ventilators which draw in fresh air from the outside and pipe it to two channels which run down either side of the crew compartment. Each channel has six controlled air outlets which can be adjusted by the crew. The crew compartment also has handrails, lights and mountings for personnel equipment.

A feature of the UR-416 M is that the armoured body can be removed and lifted from the chassis for major repairs and overhaul such as changing the engine. The lifting equipment consists of three jacks, one of which is fitted under each side of the hull and the third at the rear which acts as a pivot. The chassis can be driven away once the superstructure has been lifted.

Optional equipment includes an automatic fire warning and extinguishing system, an independent heater, night vision periscopes, run-flat tyres, smoke dischargers (either turret- or hull-mounted) and an air conditioning system. A 5,000 kg capacity winch with 40 m of 12 mm diameter cable can also be fitted. This can either be driven by the PTO of the engine or an independent unit mounted on one half of the exit door.

Variants

Ambulance

Carrying eight sitting, or four sitting and two stretcher patients plus a crew of two.

Command and communications

Fitted with additional communications equipment and map boards.

Internal security

The internal security model can be fitted with an obstacle-clearing blade at the front of the hull.

The vehicle can be fitted with the same turrets as the reconnaissance model but one cupola has been specifically developed for the internal security role. This cupola can be rotated through 360° and is infinitely lockable in any required position.

Reconnaissance

This could be fitted with most types of turret on the market.

Workshop

This has a full range of tools, work benches, a vice and cutting equipment, and an A-frame can be erected at the front of the hull to enable the vehicle to change engines and other components. When the A-frame is in use, two stabilisers are lowered at the front of the hull.

El Salvadorean UR-416 vehicles

A total of 10 of these vehicles was purchased from Germany many years ago and of these it is understood that eight remain in service. For additional protection against RPG-7-type rocket propelled unguided anti-tank grenades, some have been fitted with wire mesh screens. These detonate the HEAT warhead of the projectile before it comes into contact with the main armour of the vehicle. They are normally armed with a roof-mounted 12.7 mm M2 machine gun. It has been observed that some vehicles have been fitted with a shield for the 12.7 mm M2 machine gun and some of these also have a 7.62 mm M60D machine gun and shield mounted on the roof at the rear. Some UR-416 vehicles in service with El Salvador have now been fitted with appliqué armour on their hull sides.

Status

Production complete. No longer being marketed. In service with the following countries:

Country	Quantity	Comment
Argentina	40	gendarmerie
Ecuador	n/avail	police
El Salvador	8	
Germany	20*	police
Greece	15	gendarmerie
Kenya	52	army
Morocco	20*	police
Netherlands	10*	police
Pakistan	45	national guard, some are fitted with one-person turret armed with 20 mm cannon
Peru	130	army
Qatar	10*	police
Saudi Arabia	100	Special Security Service
Spain	20*	police
Togo	30	army
Turkey	34	national guard
Venezuala	20	national guard
Zimbabwe	n/avail	

Note: * Estimated figure. Given that a total of 1,030 UR-416 vehicles were built, there are probably additional undisclosed customers and some of the above figures are probably higher.

Manufacturer

Rheinmetall Landsysteme GmbH.

UPDATED

Krauss-Maffei Wegmann All-Protected Carrier Vehicle (Dingo)

Development

The All-Protected Carrier Vehicle was developed by Krauss-Maffei Wegmann under contract to the BWB, but with additional funding from Krauss-Maffei Wegmann. The first demonstrator for mine testing was completed in 1995.

As a result of trials with this demonstrator, a preproduction vehicle called the APCV 1, incorporating a number of improvements, was completed in mid-1996 and subsequently evaluated by the German Army and the BWB.

To meet user requirements for a vehicle with increased payload and volume, a third vehicle was completed in 1997 under the designation APCV 2. This is based on a larger and more powerful UNIMOG (4 × 4) chassis. A high level of protection against mines is required.

The All-Protected Carrier Vehicle has been designed and built using proven and in-production automotive components to reduce both procurement and total life cycle costs.

It has been designed to have a high degree of road and cross-country mobility as well as providing the crew with modular protection against a variety of threats, especially those found in peacekeeping operations.

As the All-Protected Carrier Vehicle is based on a proven Mercedes-Benz UNIMOG (4 × 4) chassis, spares and technical support are available in most parts of the world.

According to Krauss-Maffei Wegmann, typical roles for this vehicle include:

(1) Armoured personnel carrier
(2) Light reconnaissance vehicle
(3) Command and control vehicle
(4) Material transport vehicle
(5) Weapon carrier for various calibres
(6) Ambulance vehicle
(7) Light air defence system
(8) Missile launching vehicle
(9) Forward observer vehicle.

Late in 1999, the German BWB awarded Krauss-Maffei Wegmann a contract worth DM42 million for the supply of 56 vehicles.

In August 2000, just nine months after being awarded the contract, Krauss-Maffei Wegmann handed over to the German Army two All Protected Carrier Vehicles (APCV-2) and all 56 were delivered by the end of 2000.

The official German Army designation for the APCV is the Allschutz Transport Fahrzeug (ATF), or Dingo.

APCV-2, selected by the German Army, is based on the Mercedes-Benz UNIMOG U-1550L (4 × 4) chassis which is supplied direct to Krauss-Maffei Wegmann who then fabricate and install the armoured body with the armour system being provided by IBD.

This version of the vehicle has a maximum combat weight of 8,000 kg with maximum payload being 1,400 kg. It has individual seats for the commander and driver at the front with another three individual seats to the rear. The seats have been designed to absorb shock if the vehicle should run over a mine and special protection is provided under the vehicle hull against anti-tank mines.

Fitted on the roof is a Krauss-Maffei Wegmann mount armed with an externally mounted 7.62 mm MG3 machine gun that can be aimed and fired under complete armour protection. If required a heavier weapon can be installed such as a 12.7 mm machine gun or a 40 mm grenade launcher.

Standard equipment on German Army vehicles includes an air conditioning system, parking heater, anti-skid breaking system, GPS navigation system, run-flat tyres, communications systems and a reversing camera system.

Description

The Krauss-Maffei Wegmann All-Protected Carrier Vehicles are based on the chassis of the Mercedes-Benz UNIMOG U100 L or U1550L (4 × 4) cross-country truck.

The engine compartment is at the front, the heavily armoured safety crew compartment is in the centre and the cargo/load area is at the rear. The engine compartment can be easily accessed for routine checks by simply folding the hinged bonnet forwards.

If the vehicle runs over an anti-personnel/anti-tank mine, the blast will be deflected away from the armoured crew compartment by means of a specially designed mine deflector.

The crew compartment armour provides protection against any hand-held rifles, anti-tank and anti-personnel mines, artillery fragments from 155 mm shells and NBC agents.

The crew compartment is of thin plate steel armour to which composite armour can be added if increased protection is required.

The driver is seated front left with the vehicle commander to his right and a further two to four personnel in the rear. To the front of the commander and driver is a large bulletproof window which provides the same level of protection as the remainder of the crew module.

In each side of the vehicle is a large forward-opening door with a bulletproof window in its upper part. The occupants are provided with individual seats which have been designed to absorb shock if the vehicle should run over a mine.

The cargo area is at the rear and maximum payload is between 1,000 and 1,400 kg depending on the vehicle type.

Specifications

Model	**Dingo**	**Dingo-S**
Crew	1 + 3/5	1 + 5/9
Configuration	4 × 4	4 × 4
Combat weight	8,800 kg	10,800 kg
Payload	1,400 kg	3,200 kg
Length	5.40 m	6.05 m
Width	2.30 m	2.34 m
Ground clearance	0.42 m	0.48 m
Track	1.56 m	1.56 m
Wheelbase	3.25 m	3.70 m
Volume		
(safety compartment)	6.5 m²	8.0 m²
(stowage compartment)	2.0 m²	2.0 m²
Max road speed	90 km/h	120 km/h
Range	700 km	700 km
Engine type	diesel	diesel
Engine output	240 hp	240 hp
Transmission	manual or automatic	
Steering	powered	powered
Electrical system	24 V	24 V
NBC system	optional	optional

Note: Height is without armament

The vehicle is air-transportable in a Lockheed Martin C-130 (three APCV 1 or two APCV 2) or C-160 transport aircraft and the APCV 1 can be lifted by a CH-53 helicopter.

Standard equipment includes a central tyre pressure regulation system that allows the driver to adjust the tyre pressure to suit the terrain being crossed.

Variants

The prototype of the All Protected Carrier Vehicle 1 is built on the chassis of the Mercedes-Benz U100L (4 × 4) vehicle, but the APCV 2 prototype is based on the U1550L with a wheelbase of 3.25 m. A version with a wheelbase of 3.7 m would also be possible.

The latter is the largest member of the family and would be able to carry 10 soldiers, including the commander and driver.

Various weapon mounts can be fitted in the centre of the roof including 40 mm automatic grenade launcher, 7.62 mm or 12.7 mm machine gun and a 20 mm low-recoil cannon.

Status

Production as required. The first batch of 56 vehicles was delivered to the German Army from August to December 2000.

Manufacturer

Krauss-Maffei Wegmann GmbH & Co KG.

UPDATED

In August 2000, the German Army took delivery from Krauss-Maffei Wegmann of the first of 56 All Protected Carrier Vehicles based on a UNIMOG (4 × 4) chassis and armed with a roof-mounted 7.62 mm MG3 machine gun ***2001***/0103774

The first of 56 All Protected Carrier Vehicles have just been handed over to the German Army at Krauss-Maffei Wegmann's facility in Munich, Germany (Christopher F Foss) ***2002***/0073406

Rheinmetall Landsysteme TH 444 APC

Development

The TH 444 APC has been developed as a private venture by the now Rheinmetall Landsysteme, with the first prototype being completed in 1987. The vehicle has been demonstrated in Germany, Turkey and the USA.

It is based on the long-wheelbase version (2.85 m) of the Mercedes-Benz 250 GD (4 × 4) cross-country chassis which is already in service with some 20 armed forces, so making training and logistics easier. To improve its power-to-weight ratio to 30.3 hp/t it has been fitted with a more powerful 112 hp diesel engine.

By late 2001, the TH 444 APC had not entered volume production in Germany. A further development of this vehicle has entered production in Egypt and is made by

Specifications
Crew: 5-8
Configuration: 4 × 4
Weight:
(loaded) 3,700 kg
(empty) 2,900 kg
Power-to-weight ratio: 30.3 hp/t
Length: 4.35 m
Width: 1.73 m
Height: 1.85 m
Ground clearance: 0.24 m
Max road speed: 140 km/h
Road range: 600 km
Gradient: 60%
Vertical obstacle: 0.3 m
Trench: 0.6 m
Fording: 0.6 m
Engine: 112 bhp diesel
Transmission: manual with 4 forward and 1 reverse gears
Clutch: single dry plate
Transfer box: for mechanical front axle engagement and synchronised cross-country gear
Steering: recirculating ball, power assisted
Turning radius: 6.3 m
Suspension: front and rear, coil springs and telescopic shock-absorbers. Both front and rear axles are located by 1 transverse and 2 longitudinal control links
Tyres: 7.50 × 16
Number of tyres: 4 + 1 spare
Brakes (main): hydraulic, dual circuit with vacuum booster, disc front, drum rear
Electrical system: 24 V
Batteries: 2 × 12 V, 45 Ah
NBC system: optional
Night vision equipment: optional

the Kader Factory for Developed Industries under the designation of the G320. Details of this are given in the earlier section under Egypt. The chassis for the Egyptian vehicles are supplied as required from Germany.

Description
The layout of the TH 444 is conventional, with the engine at the front, commander and driver in the centre and the remainder of the crew at the rear.

The welded steel armour provides protection against 7.62 mm attack and its frame body design is such that the armour package (sides, rear and roof) can be manufactured to meet specific user requirements.

The prototype has a single-piece bulletproof windscreen and two forward-opening doors in each side. Each of the latter has a bulletproof window in its upper part. The rear area also has a bulletproof window in its upper part on either side, and at the back a rear door with a bulletproof window in its upper part is provided with the spare wheel being mounted below. Various roof hatch arrangements can be provided, including a machine gun mount above the commander's position and forward-opening roof hatch.

Status
Development complete. Ready for production.

Manufacturer
Rheinmetall Landsysteme GmbH.

UPDATED

Rheinmetall Landsysteme TH 444 (4 × 4) APC with cupola above commander's position 0018667

GUATEMALA

Danto (Tapir) armoured personnel carrier

Development
The Danto (Tapir) (6 × 6) armoured personnel carrier was developed to meet the requirements of the Guatemalan Army for a support vehicle armed with a 20 mm cannon to bridge the gap between the 12.7 mm M2 heavy machine gun and the 106 mm M40 recoilless rifle.

The decision to build a prototype vehicle was taken in 1992 and this was subsequently designed and built by the Guatemalan War Matériel Service with the first example being completed in 1994.

By late 2001, it is understood that production had yet to commence but, if placed in production, 10 vehicles would be built over a six month period.

Description
The Danto essentially consists of a US-supplied M54A2 5 ton (US) (6 × 6) truck chassis fitted with an armoured steel body which is 8 mm thick on the front, sides and rear with the floor being 15 mm thick.

The armoured engine compartment is at the front and to the rear of this are the commander and driver who are provided with bulletproof windows 50 mm thick to their front and sides.

The troop compartment extends right to the rear and in each side is a two-part door, the lower part opening downwards to form a step with the upper part opening to the rear. There is also a door in the rear of the hull and a single roof hatch.

Internally, the troop area is divided into three compartments which can be separated by armoured doors to limit damage if the vehicle is hit.

Mounted on top of the hull is a Swiss Oerlikon Contraves GAD-AOA one-man manually operated turret

Danto (Tapir) (6 × 6) armoured personnel carrier fitted with Oerlikon Contraves GAD-AOA turret (Julio A Montes) 0058095

Specifications
Crew: 3 + 17
Configuration: 6 × 6
Length: 6.25 m
Width: 2.35 m
Height: 3.45 m
Track:
(front) 1.88 m
(rear) 1.829 m
Wheelbase: 3.55 + 1.371 m
Max road speed: 84 km/h
Range: 1,000 km
Engine: Cummins 6-cylinder diesel developing 290 hp at 2,100 rpm
Transmission: manual, 5 forward and 1 reverse gears

armed with an Oerlikon Contraves Model 204 GK 20 mm cannon. This turret can be traversed through a full 360°, with weapon elevation from –12 to +70°.

It is believed that production vehicles would have a new hull design with the chassis being reduced by 1 m between the front and rear axles while the height of the vehicle would be reduced by about 0.5 m. Production vehicles would also be powered by a Detroit Diesel engine.

Status

Prototype. Not yet in production or service.

Manufacturer

Not selected.

UPDATED

Armadillo armoured personnel carrier

Development

Since new armoured vehicles could not be obtained from the USA, in the late 1970s, Guatemala decided to develop its own armoured personnel carrier/ reconnaissance vehicle meeting specific local operational requirements.

Design and subsequent production of this vehicle, called the Armadillo, was undertaken by the Servicio de Material de Guerra (SMG) in Guatemala City.

The first prototype of the Armadillo was completed in June 1981 using, wherever possible, standard commercial components. The first five production vehicles were completed in 1983 and since then at least five batches, each of three vehicles, have been completed.

Vehicles from number 10 onwards have a more powerful diesel engine and many other detailed improvements including firing ports. It is reported that all vehicles have recently been fitted with transfer boxes from United States built Reo (6 × 6) 2.5 t trucks.

It is estimated that by late 2001, a total of 35 vehicles were in service and these serve alongside four US-supplied Textron Marine & Land Systems LAV-100 (originally designated the V-100) light armoured vehicles.

Description

In many respects, the Armadillo is very similar in layout to the US Cadillac Gage (now Textron Marine & Land Systems) LAV-100 (4 × 4) APC, which was built in large numbers for the home and export markets.

The hull of the Armadillo is of all-welded steel armour construction with the commander and driver seated at the front of the vehicle. There are windows to their front and sides and twin hatch covers above their position.

The troop compartment is in the centre with the engine compartment at the left rear. The exhaust is located on the left side of the hull.

In each side of the hull, between the front and rear wheels, is a two-part entry door, the upper part opening upwards and the lower part opening downwards to form a step. There is a similar door in the hull rear on the right side.

Each door has a vision block with a firing port below and additional firing ports and associated vision blocks are provided in the sides of the vehicle, two on the left and three on the right.

Mounted on the top of the hull is a cupola with roof hatches that open left and right and mounted on this is a 12.7 mm M2 HB machine gun. Some vehicles have a one-man manually operated turret armed with 12.7 mm and 7.62 mm machine guns.

Some Armadillos have also been fitted with one or two pods of six unguided rockets which were previously fitted to aircraft or helicopters for use in the air-to-ground role.

More recently, some vehicles have been fitted with an externally mounted pod containing two 7.62 mm machine guns which can be aimed and fired from within the hull.

Status

Production as required. In service with Guatemala (35 estimated). There are no known exports of this vehicle outside Guatemala.

Manufacturer

Servicio de Material de Guerra, Guatemala City, Guatemala.

UPDATED

Specifications
Crew: 16 (3 + 13)
Configuration: 4 × 4
Weight: 10,000 kg
Power-to-weight ratio:
(212 hp diesel) 21.2 hp/t
(300 hp diesel) 30 hp/t
Length: 6.15 m
Width: 2.8 m
Height: 2.5 m (without weapons)
Ground clearance: 0.37 m
Max road speed: 100 km/h
Max range: 1,200 km
Fuel capacity: 454 litres
Engine: Detroit Diesel Model 6V-53 water-cooled diesel developing 212 bhp at 2,800 rpm (first 10 vehicles), Detroit Diesel 6V-53T water-cooled diesel, turbocharged, developing 300 bhp at 1,800 rpm (last 15+ vehicles)
Transmission: Allison Transmission AT-475 automatic with 4 forward and 1 reverse gears
Brakes: hydraulic
Electrical system: 24 V (first 10); 12 V (next 15+)
Armament:
(front) 1 × 12.7 mm MG;
(rear) 1 × 7.62 mm MG
Armour:
(floor) max 15 mm
(body) max 8 mm
Gun stabiliser:
(horizontal) no
(vertical) no
NBC system: no
Night vision equipment: no

Note: Armament depends on version as some have turret-mounted weapons while others have pintle-mounted weapons.

Armadillo (4 × 4) APC with a 12.7 mm MG at the front and a 7.62 mm MG at the rear (Julio A Montes)

Armadillo (4 × 4) APC from the rear (Julio A Montes)

HUNGARY

PSZH-IV armoured personnel carrier

Development

After developing and producing the FUG (4 × 4) amphibious scout car the Hungarian defence industry went on to produce a wheeled APC. This is used in the Hungarian Army in place of the Russian or Czech/ Polish models, in spite of the fact that it carries a smaller landing party and has a reduced cross-country capability.

A certain amount of confusion has arisen over this vehicle and its predecessors. The original Hungarian scout car was officially known as the FUG and, because of the small size and 4 × 4 configuration, observers in the West promptly dubbed the 1966 prototype and the PSZH-IV as FUG-66 and FUG-70. The so-called FUG-66, which mounted an egg-shaped turret and dummy automatic cannon, appeared only on a single manoeuvre parade in Bratislava.

PSZH in Hungarian is *Páncélos Szálltó Harcjármü* (or armoured personnel carrier). It is also referred to as the FUG-2, although its correct Hungarian designation, under the Conventional Forces Europe (CFE) Treaty, is the PSZH D-944.

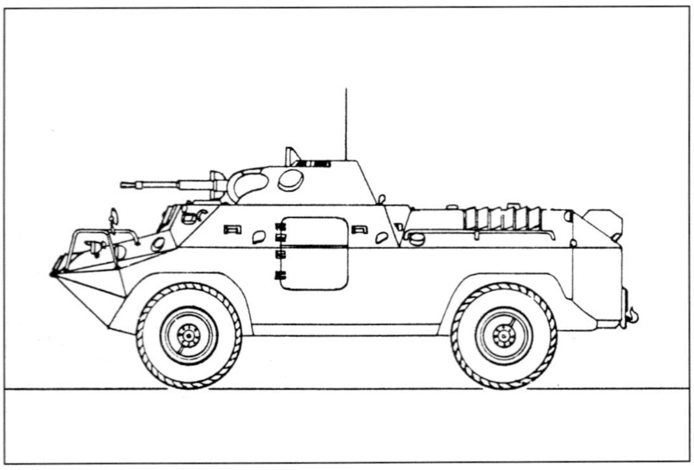

PSZH-IV armoured personnel carrier

PSZH-IV APC showing its amphibious capabilities

Description

The all-welded steel armour hull of the PSZH-IV has the driver and commander seated at the front of the vehicle. Both have a hatch cover in front of them hinged at the top and opening forwards, and a windscreen with a wiper blade. The hatch cover is equipped with an integral vision block for observation when the hatch is closed.

A door, four vision blocks and two firing ports are provided in each side of the hull. The ports are positioned one either side of the hull door, next to a vision block. The six infantrymen are seated around the sides and rear of the fighting compartment in what must be very cramped positions. The side door is in two parts, top and bottom, with both parts opening towards the front of the vehicle.

The turret is in the centre of the vehicle and is a different shape from that of the Russian BRDM-2 (4 × 4) amphibious scout car, with a stepped top and forward-facing day periscopes in the top part. Like the turret fitted to the BRDM-2 it has no roof hatch. The sight for aiming the armament is in the roof of the turret, just in front of the periscope. A ventilator fan is provided in the left side of the turret.

The water-cooled diesel engine is behind the turret and is much more powerful than that installed in the earlier FUG. The PSZH-IV is fully amphibious and propelled in the water by two water-jets at the rear of the hull. Before entering the water a trim board, which is stowed folded on the glacis plate when not in use, is erected at the front of the hull. Standard equipment includes infra-red driving lights, an NBC system, a power-operated winch and a central tyre-pressure system that allows the driver to adjust the tyre pressure to suit the type of ground being crossed.

Main armament comprises a 14.5 mm KPVT machine gun and there is a 7.62 mm PKT machine gun mounted coaxially to its left. An infra-red searchlight is mounted coaxially to the right of the main armament.

Specifications

Crew: 3 + 6
Configuration: 4 × 4
Combat weight: 7,600 kg
Power-to-weight ratio: 13.15 hp/t
Length: 5.695 m
Width: 2.5 m
Height: 2.308 m
Firing height: 2.05 m
Ground clearance:
(hull) 0.42 m
(rear differential) 0.3 m
Track: 1.9 m
Wheelbase: 3.3 m
Max speed:
(road) 80 km/h
(water) 9 km/h
Fuel capacity: 200 litres
Max cruising range: (road) 500 km
Fording: amphibious
Gradient: 60%
Vertical obstacle: 0.4 m
Trench (with channels): 0.6 m
Engine: Csepel D.414.44 4-cylinder in-line water-cooled diesel developing 100 hp at 2,300 rpm
Transmission: manual, 5 forward and 1 reverse gears
Clutch: single dry plate
Steering: hour-glass worm with pneumatic assistance
Brakes: dual-circuit, air/hydraulic, parking on rear axle
Electrical system: 24 V
Tyres: 30.00 × 18
Armament:
(main) 1 × 14.5 mm KPVT MG
(coaxial) 1 × 7.62 mm PKT MG
Ammunition:
(main) 500
(coaxial) 2,000
Gun control equipment
Turret traverse: 360°, manual
Gun elevation/depression: +30/–5°
Armour: max 14 mm
Gun stabiliser:
(vertical) no
(horizontal) no
NBC system: yes
Night vision equipment: yes

Variants

A turretless version of the PSZH-IV is used as a command vehicle and other variants of the vehicle exist such as an ambulance and a radiological-chemical reconnaissance vehicle. There is also a turreted command vehicle which does not carry the landing party of six but is armed with a 14.5 mm KPVT machine gun.

There is also known to be an NBC reconnaissance version of the PSZH-IV (4 × 4) in service which has similar equipment to that fitted to the NBC reconnaissance version of the Russian BRDM-2 (4 × 4) vehicle (see Reconnaissance vehicles section).

It has recently been disclosed that some of these vehicles have designators; these include:

PSzH-IV-10

This is the turretless version and was probably only used by the former East German Army.

D-944Z.PK and D-944Z.TOF

These are command vehicles.

PSzH-Ch/D-944VS

These are NBC reconnaissance vehicles.

PSzH M1992

This is the ambulance version.

Status

Production complete. In service with:

Country	Quantity	Comment
Czech Republic	182	incl BRDM-2
Hungary	400	called PSZH D-944
Iraq	100	estimate
Slovakia	90	called OT-66

Manufacturer

Hungarian state factories.

UPDATED

PSZH-IV (4 × 4) APC with commander's and driver's hatches in the open position

NBC reconnaissance version of the PSZH-IV APC

INDIA

Rakshak Gypsy armoured patrol car

Development

The Rakshak Gypsy armoured patrol car was developed by the Indian Defence Research and Development Organisation (DRDO) mainly for internal security operations and was shown for the first time in 1994. By late 2001 there were no known exports of this vehicle.

Description

The Rakshak Gypsy vehicle is based on the locally built Maruti Gypsy (4 × 4) light cross-country vehicle, although other types of light armoured vehicle (wheeled) can be modified to a similar standard.

The vehicle has been fitted with panels of 'Kackal' armoured steel and according to the manufacturer provides complete protection against penetration from 7.62 mm SLR and 7.62 mm AK-47 ball burst fire.

The power pack is at the front, the driver and commander are in the centre and the troop compartment is at the rear. The windscreen is of bulletproof glass as is the upper part of the forward-opening side doors.

The troop compartment is at the rear and the four men are carried two down either side on bench seats facing each other. In each side of the vehicle are two firing ports which, when not required, are covered by sliding shutters.

The troops enter and leave via two doors in the hull rear, each of which is provided with a circular firing port.

As the basic vehicle has been up-armoured the suspension has been upgraded, with one extra leaf being added at the front and three extra leafs at the rear.

Rakshak Gypsy armoured patrol car (Christopher F Foss)

Specifications

Crew: 2 + 4
Configuration: 4 × 4
Length: 4.01 m
Width: 1.46 m
Height: 1.86 m
Armament: nil

Status

Production as required. In service with paramilitary units in India.

Manufacturer

Ordnance Factory Medak.

UPDATED

INTERNATIONAL

GTK/MRAV/PWV Wheeled Armoured Vehicle Programme

Development

Late in April 1998, it was announced that ARTEC had been selected to design, develop and manufacture the new 8 × 8 wheeled armoured personnel carrier to meet the requirements of France (Véhicule Blindé de Combat d'Infanterie - VBCI), Germany (Gepanzertes Transport Kraftfahrzeug - GTK) and the United Kingdom (Multi-Role Armoured Vehicle – MRAV). France subsequently withdrew from the programme.

This programme is managed by the new Organisation for Joint Armament Co-Operation (OCCAR) and a Memorandum of Understanding (MoU) was signed late in 1999 between Germany and the UK covering their involvement in this programme, the first of its type in Europe.

Following negotiations, a contract was awarded in November 1999 for eight periscope vehicles, four for Germany and four for the UK. Total value of this contract was £70 million sterling.

Of these eight prototypes, four were to be assembled in Germany and four in the UK. The first of these will be rolled out in 2001 and of these eight vehicles, four were to be in the Command Post Variant (CPV) and four in the Armoured Personnel Carrier (APC) configuration.

Trials with the prototypes will be undertaken between 2002 and 2004. Germany will be responsible for ballistic testing while the UK will be responsible for reliability testing.

It also included an option for 300 production vehicles per nation, all in the 8 × 8 configuration. Follow on production will be subject to later national approvals. According to ARTEC, the total requirement could be for as many as 5,000 vehicles. In addition, significant export contracts are expected.

Under the terms of the bilateral contract awarded to ARTEC late in 1999, 300 vehicles were to be built for the two armies with deliveries to take place from June 2006 through to March 2009. These will mostly comprise CPV and APC vehicles.

In February 2001, the Netherlands joined the MRAV/GTK programme and a total of 12 prototypes will now be built, four for each of the three countries. The first production run will now be for 600 vehicles, 200 for each country. The value of the development contract has been increased to £110 million while the UK's share of the costs has dropped by £5 million.

According to the UK MoD, the Netherlands and the UK have common requirements for ambulance and mortar variants and share similar views on integrated logistics support.

ARTEC International (ARmoured TECnology) comprises Krauss-Maffei Wegmann and MaK (now Rheinmetall Landsysteme) of Germany and Alvis Vehicles of the UK. With the Netherlands joining

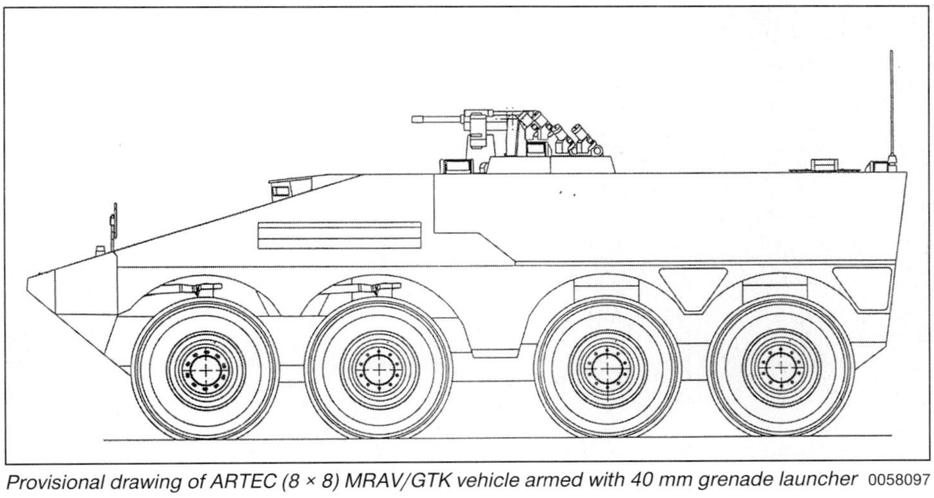

Provisional drawing of ARTEC (8 × 8) MRAV/GTK vehicle armed with 40 mm grenade launcher 0058097

Specifications

Configuration: 8 × 8
Crew: 3 + 8 (depending on national variant)
Max combat weight: 33,000 kg
Unloaded weight: 25,000 kg
Payload: 8,000 kg
Length: 7.880 m
Width: 2.99 m
Height (roof): 2.376 m
Ground clearance: 504 mm
Wheelbase: 1.55 m + 1.90 + 1.55 m
Track: 2.582 m
Engine (diesel): MTU diesel developing 710 hp coupled to an Allison fully automatic transmission
Max road speed: 103 km/h
Minimum speed: 3.2 km/h
Range: 1,050 km
Trench: 2.00 m
Vertical obstacle: 0.8 m
Fording: 1.5 m
Turning circle:
(between curbs) 20 m
(with steering brake) 15 m
NBC system: yes
Gradient: 60%
Sideslope: 30%
Night vision equipment: yes

Artist's impression of ARTEC (8 × 8) MRAV/GTK PWV vehicle with 40 mm grenade launcher **2002**/0121133

the programme it now also includes Stork NV of the Netherlands. The Netherlands has a stated requirement for 384 vehicles.

In 1998, Krauss-Maffei Wegmann completed a company-funded 6 × 6 demonstrator.

The Netherlands designation for the vehicle is PWV (*Pantser Wiel Voertuig*).

The MRAV design will provide for APC and command vehicle variants but will also allow for the development of other variants using the same basic vehicle.

The exact details of the workshare have not been disclosed but it is understood that Germany will provide the engine, transmission and complete running gear, UK the baseline armoured hulls (drive modules) and the mission module.

Each country will have its own production line and be responsible for final assembly and delivery of completed vehicles and their mission modules for its own country.

Development work commenced in 1999 and it is expected that the first vehicles will be completed in 2001 for the trials programme and that production will commence in 2004 with first production vehicles being completed in 2005.

The first lot of 8 × 8 vehicles to be built will share common and production commercial components, wherever possible, to reduce total life cycle costs.

According to ARTEC, the main advantages of the 8 × 8 model are increased mobility, increased load capacity (8 tonnes) and an internal volume of 14 m³. For further variants of vehicles a 6 × 6 configuration could be easily developed.

Description

The hull of the MRAV vehicle is of thin wall bending technology steel armour construction with an applique layer of passive armour to provide a higher level of battlefield survivability.

According to ARTEC the vehicle will have a high level of protection against mines, artillery fragments and projectiles. The overall radar and infra-red signature is minimised and full NBC protection is integrated.

The overall vehicle is configured in two major parts, a driving module which accommodates the driver and the automotive component and a separate mission module which mounts the national specific weapons, communications and stowage items.

The mission module is at the rear of the vehicle and can be quickly detached to allow a different module to be fitted. In the APC variant, four troops will be seated either side on individual seats and integral seat belts. They enter and leave via a power operated ramp in the rear which is also provided with an emergency door.

Specially designed stowage compartments allow individual crew members sufficient space for mission and personal kit. Outside bins between the wheel arches provide space for maintenance spares and tools.

In the Autumn of 2000, Alvis Vehicles selected the Swedish company of Hägglunds Vehicles, which is part of the Alvis plc group, to build the weapon stations for the British Army version of the MultiRole Armoured Vehicle (MRAV) (8 × 8).

This contract, valued at SKr20 million, covers development work and the construction of a small number of prototype weapon stations. According to Hägglands Vehicle, the long-term value of the contract could be in the range of SKr200 to 300 million.

The weapon station will be armed with a 12.7 mm or a 7.62 mm EX-34 Chain Gun (British Army designation L8A2) and be laid onto the target with the gunner remaining under armour in the hull.

The rear-mounted mission module weighs around 8 tonnes, depending on version and is interchangeable within one hour.

The power pack is in front and to the left of the driver and consists of a commercially available diesel engine that meets EURO III norms and is coupled to an automatic transmission.

The vehicle features permanent all-wheel drive, independent suspension with large spring travel and a Central Tyre Inflation System (CITS) that allows the driver to adjust the tyre pressure to suit the terrain being crossed, and run flat tyres.

Variants

The first two models will be armoured personnel carrier and command post vehicles with each country developing special two-role versions. The UK intends to acquire six versions of the MRAV:

- Armoured Personnel Carrier (APC)
- Command Vehicle (CV)
- Communications Variant (COMV)
- Mortar Variant (AMV)
- Anti-Tank Platoon Variant (ATVP)
- Treatment and Evaluation Variant

Variants required by Germany and the Netherlands also include:

- Cargo vehicle
- Battle damage repair vehicle
- Pioneer vehicle

At the present time there is no German or UK requirement for a 6 × 6 version.

Status

Under development to meet the operational requirements of Germany, the Netherlands and the UK.

Manufacturer

ARTEC GmbH.

UPDATED

Terrier Light Armoured Unit (LAU)

Development

The Terrier (4 × 4) Light Armoured Unit (LAU) has been jointly developed by Krauss-Maffei Wegmann of Germany and the Defence Vehicles Division of IVECO of Italy.

It is a further development of the Italian Boneschi MAV 5 (4 × 4) light armoured vehicle of which 200 units were supplied to the Italian Army with an additional 12 being supplied to the Italian Carabinieri. Full details of this vehicle, which is no longer marketed, are given in this section under Italy. Large numbers of MAV 5 vehicles remain in service in Italy.

The first customer for the Terrier LAU was the United Arab Emirates which took delivery of 10 vehicles in 2000. Single examples of the Terrier LAU have also been supplied to Egypt and Poland.

Terrier (4 × 4) Light Armoured Unit in unarmed armoured personnel carrier configuration **2002**/0103775

According to the Krauss-Maffei Wegmann facility at Munich, where the Terrier is assembled, the German Army has a possible future requirement for 250 vehicles of this type for use in peacekeeping operations.

The first batch would consist of 174 units with the second batch consisting of 76 units. It is expected that the Terrier will be qualified for use by the German Army in the future with a possible order being placed in 2002.

Description

The Terrier LAU essentially consists of a standard production IVECO 40.12WM (4 × 4) truck chassis that uses a proven commercial diesel power pack fitted with a new armoured body. Over 15,000 of the standard unarmoured trucks are used by a number of different armies in various parts of the world.

The all-welded steel armoured body provides the occupants with protection from 7.62 mm × 51 NATO small arms fire, with the floor providing protection from anti-personnel mines. When compared to the earlier MAV 5, the Terrier LAU has much improved protection under the hull and around the doors.

The power pack is at the front of the vehicle with the driver being seated on the left and the commander on the right. There is a large one-piece bulletproof window to their front and a smaller bulletproof window in either side of the hull. Under this window is a circular firing port. To the immediate front of the vehicle commander is a circular firing port. Mounted on the forward part of

Specifications

Crew: 2 + 6
Configuration: 4 × 4
Weight combat: 5,200 kg
Weight empty: 4,200 kg
Power-to-weight ratio: 23.40 hp/t
Length: 5.242 m
Width: 2.094 m
Height: 2.148 m
Ground clearance: 0.248 m
Track: 1.67 m
Wheelbase: 2.80 m
Angle of approach/departure: 40°/34°
Max road speed: 110 km/h
Fuel capacity: 75 litres
Range: 600 litres
Fording: 0.70 m
Gradient: 60%
Side slope: 30%
Turning radius: 6.0 m
Engine: IVECO turbocharged 4-cylinder diesel developing 122 hp
Gearbox: manual with 5 forward and 1 reverse gears (with option of automatic transmission)
Clutch: single plate, mechanically operated
Transfer box: rack and pinion
Steering: power-assisted
Suspension: torsion bar front, leaf springs with hydraulic and telescopic dampers rear
Axles: single reduction front and rear, self-locking differential
Wheels: disc type, 6.50H 15 rim size
Tyres: 9.00.16 14 PR
Brakes: disc type front, drum type at rear, split circuits
Electrical system: 24 V
Batteries: 2 × 12 V, 110 Ah
Generator: 1,200 W
Armament: 1 × 7.62 mm MG (optional)
Gun stabiliser:
(horizontal) no
(vertical) no
Armour: 6 mm
NBC system: no
Night vision equipment: no

the roof is a searchlight that can be operated from within the vehicle.

In either side of the hull between the front and rear axles is a forward opening side door with a bulletproof window in its upper part. There is another door in the rear side of the hull slightly offset to the right. This has a bulletproof window in its upper part with a firing port below. A spare wheel is carried at the rear of the hull on the left side. There are an additional two bulletproof windows in either side of the rear roof compartment.

The standard Terrier LAU has a circular roof hatch on which a 7.62 mm machine gun or similar weapon can be installed. Other weapon options are also possible including a one-person turret armed with 7.62 mm machine gun.

The seating arrangement is flexible but a typical Terrier LAU would have the commander and driver at the front and another six seated at the rear. Four would be to the immediate rear on individual seats facing the front and enter and leave the vehicle via the side doors. The remaining two would be seated at the very rear, one either side facing each other, and enter and leave via the rear door.

For added protection the diesel fuel tanks and batteries are under armour protection, a fire suppression system is provided for the tyres and engine and run-flat tyres are fitted as standard.

If required a 7.62 mm machine gun can be mounted on the roof of the vehicle. The troops are seated on individual sets and the six seats in the rear can be rapidly folded up. The Terrier has a total internal volume 9.2 m^2.

To reduce hull vibration, special elastatic pads have been fitted between the hull and chassis. Power steering and an air conditioning system is fitted as standard. Various optional equipment is available including roof-mounted blue flashing lights.

Variants

As well as being used as a troop carrier other roles suggested for the Terrier LAU include command post, explosive ordnance disposal unit, ambulance and transport vehicle for high-value goods.

Status

Production as required. In service with United Arab Emirates who took delivery of 10 vehicles in 2000.

Manufacturers

Krauss-Maffaei Wegmann GmbH (Munich facility).
IVECO SpA, Defence Vehicles Division.

NEW ENTRY

RN-94 armoured personnel carrier

Development

The RN-94 (6 × 6) armoured personnel carrier has been jointly developed Nurol Machinery and Industry Co Inc based in Ankara and S N Romarm SA Filiala S.C Moreni of Bucharest, Romania.

Development commenced in 1994 with the prototype making its first appearance in late 1995. Between 1995 and 1996 three prototypes were built with the last one being produced in Turkey. It was subsequently presented to the Turkish Land Forces Command for test and qualification purposes.

After the qualification was complete in 1997, the Turkish MoD placed an order for a preproduction batch of five RN-94 vehicles for an extensive series of trials all over Turkey.

Based on the same basic vehicle chassis, three different types of turret have been adapted to the five vehicles:

(1) One of the five vehicles was fitted with the Kollmorgen Compact Lightweight Armoured Weapon Station (CLAWS) equipped with 40 mm grenade launcher
(2) Two of the vehicles were fitted with a locally produced standard Gunner's Cupola equipped with a 12.7 mm M2 machine gun.
(3) The last two vehicles were fitted with the General Dynamics Dragon turret armed with a 12.7 mm M2 machine gun.

The five RN-94 vehicles that were delivered to the Turkish Land Forces Command early in 1998 each covered about 18,000 km by the end of 1998 which included road and cross-country running.

For trials purposes the RN-94 has also been fitted with a Giat Industries Dragar one-person power-operated turret armed with a 25 mm M811 cannon and 7.62 mm coaxial machine gun.

This turret has been manufactured under licence by Nurol in Turkey for installation on some of the locally built FNSS Defence Systems Armoured Infantry Fighting Vehicles (AIFV). The 25 mm M811 cannon has been produced under licence in Turkey by MKEK.

The Nurol Machinery and Industrial Co Inc is hopeful that the Turkish Land Forces Command will place a significant production order for the RN-94 in the future. A number of other contractors are also competing for the Turkish order for wheeled armoured personnel carriers.

The RN-94 has been certified by the Turkish Land Forces Command but, as of October 2001, no production order had been placed.

Description

The hull of the RN-94 is all-welded steel armour structure which provides maximum protection against 7.62 mm armour-piercing ammunition and shell splinters all round. If required, a higher level of armour protection could be provided but this would mean the vehicle losing its amphibious capability.

The driver is seated at the front left with the vehicle commander to the right, with each being provided with a roof hatch that opens to the rear and four day periscopes for forward observation.

The engine compartment is to the rear of the driver with an aisle on the right side connecting the troop compartment at the rear. The exhaust outlet is on the left side of the hull with air inlet and outlet louvres in the roof. There is a forward-opening door in the left side of the hull that allows access to the engine for maintenance purposes.

In addition to the driver and commander, a further 11 fully equipped troops can be carried, with the capability of dismounting quickly through the hull rear, which has

RN-94 (6 × 6) armoured personnel carrier fitted with a Giat Industries Dragar turret armed with a 25 mm cannon and 7.62 mm coaxial machine gun (Christopher F Foss) ***2001***/0103778

RN-94 (6 × 6) armoured personnel carrier fitted with 120 mm mortar system (Christopher F Foss) ***2000***/0081980

Specifications

Crew: 2 + 11
Configuration: 6 × 6
Combat weight: 13,000 kg
Power-to-weight ratio: 18.46 hp/t
Length: 6.69 m
Width: 2.80 m
Height:
(hull top) 2.05 m
(turret) 2.74 m
Ground clearance: 0.43 m
Wheelbase: 1.50 + 1.90 m
Max road speed: 110 km/h
Max water speed: 8 km/h
Range: 600 km
Gradient: 60%
Side slope: 30%
Vertical obstacle: 0.48 m
Trench: 1.0 m
Fording: amphibious
Turning radius: 9.5 m
Engine: Cummins 6 CTA 8.3-10 diesel developing 240 hp at 2,400 rpm or Cummins 6 CTAA 8.3-20 diesel developing 260 bhp at 2,400 rpm
Transmission: Allison fully automatic
Suspension: independent with torsion bars and shock-absorbers
Tyres: Michelin 12.00 R 20 XML with Hutchinson VFI run-flat system and automatic inflation and deflation system or Michelin 12.00 R20 XZL with Hutchinson VFI run-flat or 365/85/R20 XZL
Brakes: two-stage hydraulic
Steering: hydraulic, first and second axles
Electrical system: 24 V
Armament: Dragar turret armed with 25 mm cannon and 7.62 mm coaxial machine gun or General Dynamics Dragon turret armed with 12.7 mm machine gun or Gunner's Cupola armed with 12.7 mm machine gun
NBC system: optional
Night vision equipment: optional
Air conditioning system: optional

RN-94 (6 × 6) armoured personnel carrier fitted with General Dynamics Dragon Turret armed with 12.7 mm machine gun
0058100

two doors. The upper part opens left and right with the lower part opening downwards to form a step. The upper part also has a circular firing port with integral sighting system with a vision block above.

In the left side of the troop compartment are three firing ports with integral day sighting systems and two periscopes while in the right of the hull are three firing ports with integral day sighting systems and three periscopes.

Over the top of the troop compartment are two roof hatches which open upwards with each of these being provided with a circular firing port. These can be locked in the vertical position if required.

Eight of the infantrymen are seated in the rear of the troop compartment, four either side facing outwards on individual seats which can quickly be folded up. The remaining three men are seated with individual seats.

Steering is power assisted on the front four wheels and the RN-94 is fitted with the Hutchinson VFI run-flat system and an automatic inflation/deflation system.

The RN-94 is fully amphibious being propelled in the water by two Ultra Hydraulic water-jets mounted one either side at the rear. Before entering the water the trim vane is erected at the front of the hull and the bilge pumps are switched on. Maximum water speed is 8 km/h.

The RN-94 can be fitted with a wide range of optional equipment as well as night vision aids and additional armour protection.

Variants

The RN-94 can be adopted for a wide range of battlefield missions such as command post, ambulance and specialised weapons carrier.

120 mm mortar carrier

This was shown for the first time in 1999 and is armed with a 120 mm mortar system firing through a two-part opening roof hatch. The rifled mortar barrel has been supplied by MKEK while the recoiling mortar mount is from Soltam Systems of Israel. The Turkish company of Deha Insaat is also involved.

The mortar can be traversed through a full 360°, with elevation and traverse being powered with manual back-up. Firing a standard mortar bomb it has a maximum range of 8,300 m, although longer range mortar bombs are under development. Rate of fire of the system is claimed to be 14 rounds per minute rapid fire and 10 rounds per minute sustained.

A key feature of the 120 mm mortar system is that it has a BGT inertial navigation system for incorporating a Global Positioning System (GPS) and north pointing gyro.

This, used in conjunction with a Windows-based ballistic computer with built-in test functions, enables the mortar to be rapidly and accurately laid onto the target with an accuracy of ±1 mil. A standard M3 optical sight is mounted on the left side of the mortar as a back-up.

Mounted towards the front of the hull on the right sight is a Kollmorgen Compact Lightweight Armoured Weapon Station (CLAWS) armed with an externally mounted 12.7 mm M2 machine gun (MG).

Battlefield surveillance vehicle

The Nurol Machinery and Industry Co has also completed a battlefield surveillance model of the RN-94 fitted with a hydraulically operated arm on which is mounted a locally produced ASELSAN battlefield surveillance radar. Prior to raising the hydraulic arm, two stabilisers are lowered to the ground.

Status

Five preproduction vehicles built for Turkish Land Forces Command trials.

Manufacturers

Nurol Machinery and Industry Co Inc (Turkey).
S N Romarm.
SA Filiala.
S C Moreni.

UPDATED

TRACE Bravia armoured personnel carriers and reconnaissance vehicles

Development

In the 1960s, a range of 4 × 4 armoured amphibious vehicles was developed by Joao Donas-Botto to meet the requirements of the Portuguese armed forces.

The first prototype was called the Chaimite (4 × 4) completed in 1963 and, by late 2001, total production of this vehicle and its many variants amounted to about 600 vehicles.

The original Chaimite is now known as the Bravia Mk I (4 × 4) with production being undertaken by TRACE (Trading Company & Consulting Engineers) which is the holding company of the Bravia Group SA and Ordnance Engineering.

In addition to the Bravia Mk I (4 × 4) vehicles, 6 × 6 and 8 × 8 models have been developed, but these have not yet entered production. All these vehicles are fully amphibious, being propelled in the water by their wheels.

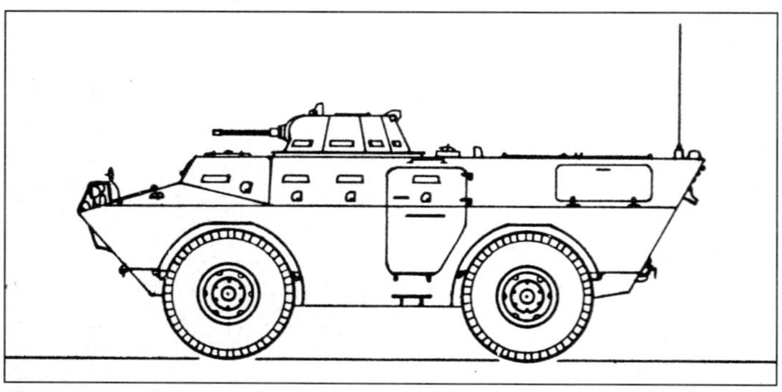

Bravia Mk I with a one-man turret armed with twin 7.62 mm machine guns

Bravia Mk I V-200 with turret-mounted 7.62 mm and 12.7 mm machine guns

Specifications

Bravia Mk I (V-200)
Crew: 11
Configuration: 4 × 4
Combat weight: 7,300 kg
Unladen weight: 6,300 kg
Power-to-weight ratio: 32 hp/t
Length:
(hull) 5.606 m
Width: 2.26 m
Height:
(incl turret) 2.26 m
(hull top) 1.84 m
Ground clearance:
(hull) 0.61 m
(differential) 0.41 m
Track: 1.87 m
Wheelbase: 2.667 m
Angle of approach/departure: 52°/46°
Max road speed: 110 km/h
Max water speed: 9 km/h
Fuel capacity: 300 litres
Max road range: 1,050 km
Max cross-country range:
(petrol engine) 563-724 km
(diesel engine) 1,046-1,207 km
Fording: amphibious
Gradient: 65%
Side slope: 40%
Vertical obstacle: 0.9 m
Engine: V-8 turbocharged water-cooled developing 200 hp at 3,600 rpm
Transmission: Allison Transmission automatic with 6 forward and 1 reverse or manual with 5 forward and 1 reverse gears
Transfer case: 1 or 2 speed
Steering: power assisted
Clutch: single dry plate, hydraulically operated
Suspension: semi-elliptical springs and hydraulic double-action shock-absorbers
Tyres: 14.00 × 20
Brakes:
(main) air/hydraulic, power assisted
(parking) on output shaft of transmission
Electrical system: 24 V
Armament: see text
Armour
Hull front upper: 6.35 mm at 75°
Hull front lower: 7.94 mm at 40°
Hull sides upper: 6.35 mm at 30°
Hull sides lower: 7.94 mm at 10°
Hull top: 6.35 mm
Hull floor: 9.35 mm
Hull rear: 6.35 mm at 30°
Turret front: 7.94 mm
Turret sides: 6.35 mm
Turret top: 6.35 mm
NBC system: no
Night vision equipment: no

The basic APC is called the V-200. As far as is known, there has been no recent production of the Bravia series of armoured personnel carriers and reconnaissance vehicles in Portugal.

Description

(V-200)

The hull of the Bravia (4 × 4) APC is made of all laser-welded ballistic steel. The driver sits towards the front of the vehicle with the second crew member, normally the commander, to his right. Over their positions is a two-piece hatch cover that opens either side of the vehicle. The driver and commander each have a vision block to their front, between which is a single firing port. To the left of the driver and the right of the commander is a vision block and firing port.

There is a two-part door in each side of the hull, the upper part opening to the rear and the lower part folding downwards to form a step. There are two vision blocks and two firing ports in each side of the hull between the front of the vehicle and the side door.

The main armament is installed immediately behind the driver's and commander's position.

The power pack is at the rear of the vehicle on the left side and is separated from the personnel compartment by a fireproof bulkhead. Access to the power pack for maintenance is through two hatches in the roof and a single hatch in the left side of the hull. The power pack compartment is equipped with a fire extinguisher operated by the driver.

A door at the rear of the hull on the left side opens to the left and has a single vision block and a single firing port. There is an additional forward-opening hatch in the roof of the vehicle to the right of the power pack compartment and a firing port and vision block below it in the right side of the hull.

Power is transmitted from the engine to the transmission, then to the transfer case and finally to the front and rear axles. The axles are of the fully floating double-reduction type with locking differentials. The tyres are run-flat combat or bulletproof type.

Mounted in the front of the vehicle is a hydraulic winch with a maximum capacity of 4,530 kg, provided with

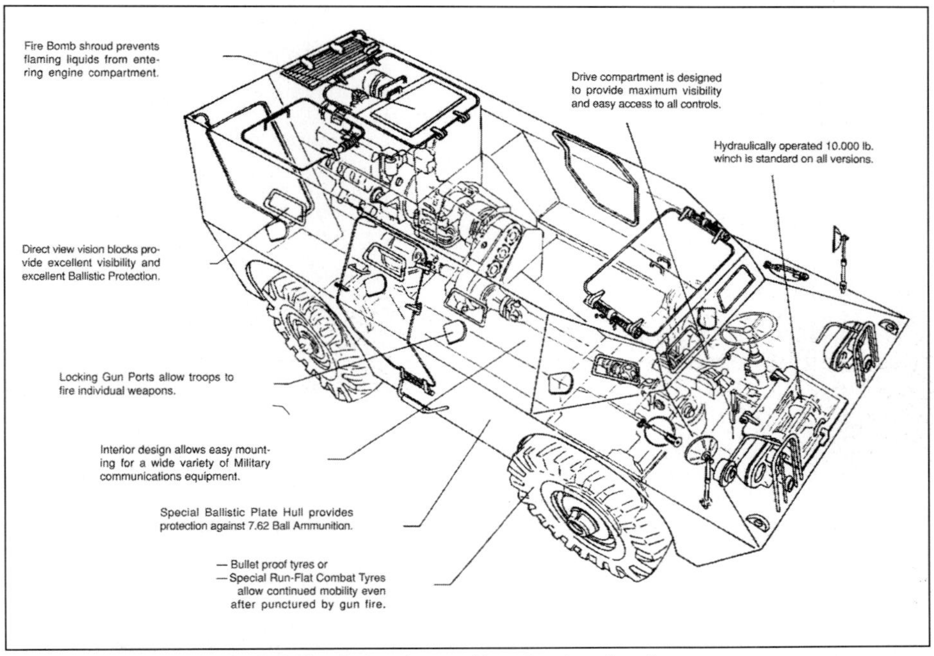

Key features of the Bravia Mk 1 (4 × 4) showing position of main subsystems **2001**/0103776

38.1 m of 11 mm diameter cable with a maximum breaking strength of 8,090 kg. Optional equipment includes passive night vision equipment and various weapon fits.

The Bravia is fully amphibious and propelled in the water by its wheels.

The V-200 is fitted with a Bravia-designed one-person turret armed with twin 7.62 mm (0.30) machine guns, twin 5.56 mm machine guns or one 7.62 mm (0.30) and one 12.7 mm (0.50) machine gun. The weapons can be elevated manually from −15 to +50° and traverse is manual through a full 360°. The turret with twin 7.62 mm (0.30) machine guns is provided with 500 rounds of ready use ammunition (250 rounds per gun) plus a further 9,000 rounds carried in the vehicle in reserve. The turret has a single-piece hatch cover that opens to the rear, vision blocks and an M28C periscopic day sight mounted in the left side of the turret for aiming the machine guns. If required, the turret can be fitted with a device for launching five 60 mm anti-personnel, anti-tank, smoke, illuminating or incendiary grenades. The turret can also be fitted with four 3.5 in rocket launchers either side of the turret, which can be fired from inside the vehicle but are reloaded from outside.

A wide range of optional equipment can be fitted, including an air conditioning system, a fire extinguishing system and an explosion detection/suppression system.

Bravia Mk I V-200 without armament fitted deployed to Bosnia (Richard Stickland) **2002**/0121131

Bravia Mk I V-200 used in mortar carrier role with roof hatches open (Richard Stickland) **2002**/0121130

Variants

V-300 reconnaissance vehicle

This model has a crew of four or five and can be fitted with a one- or two-man turret. The former is the Oerlikon Contraves GAD-AOA one-person turret armed with a 20 mm KAA-001 cannon.

The following two-man turrets can be fitted to the V-300: Bravia turret armed with a 20 mm cannon (with 100 ready use rounds and 300 rounds in reserve), 7.62 mm coaxial machine gun (500 ready rounds), a 7.62 mm anti-aircraft machine gun (500 ready rounds) and 3,000 rounds of 7.62 mm ammunition in reserve. Various other single and twin 20 mm turrets were also offered for this vehicle, including twin 20 mm air defence turrets.

V-400 reconnaissance vehicle

This can be fitted with a variety of two-man turrets currently manufactured and armed with a 90 mm gun, 7.62 mm coaxial machine gun and 7.62 mm anti-aircraft machine gun.

V-500 armoured car

This is a command and communications vehicle.

V-600 armoured car

This is a mortar carrier with the 4 × 4 model being armed with an 81 mm/82 mm mortar, the 6 × 6 model with a 120 mm mortar and the 8 × 8 with a 160 mm mortar.

V-700 armoured car

ATGW vehicle armed with HOT, TOW or MILAN systems.

V-800 armoured car

This is the ambulance version and is unarmed.

V-900 armoured car

Crash rescue vehicle. The 4 × 4 model has its own winch while the 6 × 6 and 8 × 8 models can also be fitted with a crane with a maximum lifting capacity of 5,000 kg.

V-1000 armoured car

This is a riot control vehicle and can be fitted with various armament installations including a water cannon.

Bravia Mk I with SS-11 ATGW

Portugal has fitted a number of its Bravia Mk I (4 × 4) vehicles with a single SS-11 wire-guided ATGW either side of the turret.

Bravia Mk II (6 × 6)

The all laser-welded ballistic steel hull of the Bravia Mk II (6 × 6) protects the crew from small arms fire and shell splinters.

The driver is seated at the front of the hull on the left side and has a single-piece hatch cover that opens to the rear. In front of this hatch cover are three day periscopes, the centre one of which can be replaced by a passive night vision periscope.

The V-8 turbocharged diesel engine is to the right of the driver with the air inlet and air outlet on the right side of the hull.

The main armament for the APC version, the V-200, is installed in the centre of the hull. The troop compartment is at the rear and the troops enter and leave the vehicle via two doors in the rear of the hull that open outwards.

There are two hatches over the roof compartment which open outwards and in each side of the hull are four vision blocks each with an associated gun port. Each of the rear doors also has a vision block and associated gun port.

Each axle is provided with locking differentials and hydraulic shock-absorbers. Tyres are of the run-flat combat or bulletproof type and are 11.00 × 20.

Bravia Mk III (8 × 8)

This is generally similar to the Bravia Mk II (6 × 6) model described previously except that while this model has front axles with coil springs and rear two axles have torsion bars, in the case of the Mk III (8 × 8) the first two axles have coil springs and last two have torsion bars.

Status

Production as required. The Chaimite and Bravia Mk I 4 × 4 APC are known to be in service with a number of countries including:

Country	Quantity	Comment
Lebanon	15	internal security role
Libya	n/avail	status uncertain
Peru	15	marines
Philippines	20	
Portugal	96	army
	4	navy

Manufacturer

Bravia Sarl.

UPDATED

TRACE Bravia Commando Mk III armoured personnel carrier and scout car

Development

The Commando Mk III APC was developed by Bravia in 1977 to meet the requirements of the Portuguese National Guard and has since been exported to at least two countries for internal security use. The first prototype of the Commando Mk III APC was completed in 1977 with production vehicles following the same year. The vehicle is basically a short wheelbase version of the Bravia Gazela (4 × 4) cross-country 1 tonne truck chassis fitted with an armoured body.

By 2001, total production of the Bravia Commando Mk III armoured personnel carrier amounted to approximately 200 vehicles. As far as is known there has been no recent production of the Bravia Commando Mk III vehicle for the home or export market.

Production is undertaken as and when required by TRACE (Trading Company & Consulting Engineers) which is the holding company of the Bravia Group SA and Ordnance Engineering.

Description

The V-8 turbocharged diesel engine is at the front of the vehicle, driver and commander in the centre and the personnel compartment at the rear. The driver sits on the left of the vehicle with the commander to his right. They each have a windscreen to the front which can be covered quickly by an internally controlled drop-down armoured visor containing a laminated glass observation block. The commander and the driver have a side door, the top part of which folds down on the outside for improved visibility and a laminated glass observation block.

The five troops sit to the rear of the driver's and commander's position and are provided with four firing ports with a laminated observation window over each, one in each side of the hull and two in the rear. The interior of the vehicle is lined with non-inflammable PVC

Specifications

Crew: 3 + 5
Configuration: 4 × 4
Combat weight: 4,855 kg
Unloaded weight: 4,330 kg
Loaded weight:
(on front axle) 2,236 kg
(on rear axle) 2,619 kg
Length: 4.975 m
Width: 1.93 m
Height:
(to turret top) 2.42 m
(to hull top) 2.05 m
Ground clearance: 0.21 m
Track:
(front) 1.72 m
(rear) 1.665 m
Wheelbase: 3.03 m
Angle of approach/departure: 49°/47°
Max road speed:
(diesel engine) 90 km/h
(petrol engine) 110 km/h
Fuel capacity: 160 litres
Max road range:
(diesel engine) 800 km
(petrol engine) 600 km
Gradient: 70%
Turning radius: 7.65 m
Engine: V-8 turbocharged diesel developing 155 bhp at 3,600 rpm
Transmission: manual with 4 forward and 1 reverse gears or automatic with 3 forward and 1 reverse gears
Transfer case: 2 speed
Steering system: integral power with pump assistance
Clutch: single dry plate
Suspension: heavy-duty semi-elliptical springs and telescopic shock-absorbers at each wheel station
Tyres: 9.00 × 16 or 10.50 × 16
Brakes:
(main) hydraulic on all wheels
(parking) mechanical operating on rear wheels
Electrical system: 24 V
Batteries: 2 × 12 V, 105 Ah
Armament: see text
Armour
Hull: 6.35-7.94 mm
Turret front: 7.94 mm
Turret sides: 6.35 mm
Turret top: 6.35 mm
NBC system: no
Night vision equipment: no

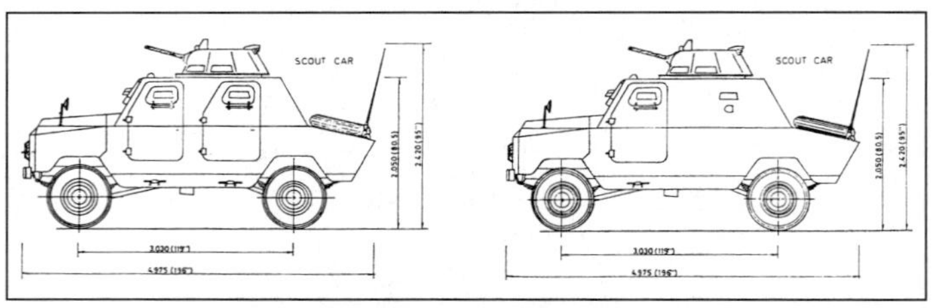

Bravia Commando Mk III fitted with turret-mounted machine guns and showing different hull door arrangement

Bravia Commando Mk III APC and scout car with driver's and commander's hatches open
2001/0103777

foam and ventilation is provided by a fresh air system with electric fan boost.

The Bravia-designed one-person manually operated turret is mounted in the centre of the roof and is armed with twin 7.62 mm machine guns or one 7.62 mm and one 12.7 mm machine gun. The weapons can be elevated from −15 to +60° manually and traverse is manual through a full 360°. The turret has a single-piece hatch cover that opens to the rear, vision blocks and a periscopic day sight in its forward part for aiming the machine guns. The turret can be fitted with a device for launching five 60 mm anti-personnel, anti-tank, smoke, illuminating or incendiary grenades.

Optional equipment includes sand tyres, run-flat tyres, an air conditioning system and a fire extinguishing and explosion suppression system.

Variants

The Commando Mk III is available in a 4 × 2 configuration when there is a limited off-road requirement. The vehicle is also available with an additional door to the immediate rear of the commander's and driver's doors. This is similar to the driver's door.

If required, the Commando Mk III can also be supplied without the manually operated machine gun turret for use in the armoured personnel carrier role.

For use in the dedicated armoured personnel carrier role, a version of the Commando Mk III is available with the hull roof extended to the rear. The troop compartment has two doors in the rear, each of which has a vision block with a firing port underneath. There are also two vision blocks and two firing ports in either side of the rear troop compartment. This version can be delivered with or without the 7.62 mm/12.7 mm machine gun turret.

Status

Production as required. In service with the Portuguese National Guard and at least two other countries. As far as is known there has been no recent production of this vehicle.

Manufacturer

Bravia Sarl.

VERIFIED

IRAN

Defence Industries Organisation Iranian wheeled (4 × 4) armoured personnel carrier

Development

In 1996, it was revealed that the Shahid Kolah Dooz Industrial Complex of the Iranian Defence Industries Organisation had developed, to the prototype stage, a new 4 × 4 armoured personnel carrier.

Description

This APC is based on a modified ladder-type truck chassis which has been fitted with an all-welded steel hull that provides the occupants with protection from 7.62 mm small arms fire and shell splinters.

To provide a higher level of protection, an appliqué passive armour kit can be fitted to the hull.

The vehicle can carry 13 troops, plus the driver and has a combat weight of 11,000 kg. The engine is at the front and the commander and driver are seated at the rear. Each of these is provided with a roof hatch that opens to the rear while to the front is a windscreen which can be covered by an armoured shutter in combat areas.

The troop compartment extends to the rear and entry is via doors in either side of the hull with a single door in the rear. Circular firing ports are provided in the sides and rear of the hull and above each of these is a bulletproof window.

Mounted in the centre of the roof is a one-man manually operated turret armed with a 7.62 mm MG3 and a 14.5 mm KPVT machine gun with a day searchlight mounted on the left. The turret has a single rear-opening roof hatch. There is an additional hatch in the troop compartment roof to the rear. An extractor fan is mounted in the turret roof on the right side.

The vehicle is powered by a Deutz BF 6 L 913 six-cylinder air-cooled diesel developing 168 hp at 2,500 rpm coupled to an S 6-36 manual transmission with five forward and one reverse gears and a two-speed transfer case.

To reduce driver fatigue, ZF hydraulic steering is fitted as standard. The front and rear axles are of the banjo type with the rear axle having a differential lock. Each wheel station is provided with leaf springs and a telescopic hydraulic double-acting shock-absorber.

The wheels are fitted with 13.00 × R20 tyres with a run-flat system fitted as standard. These allow the vehicle to cover a distance of about 60 km, when the tyres have been penetrated by small arms fire. Front and rear brakes are hydraulic while the parking brake is spring actuated.

As far as is known, the APC is not fitted with an NBC system and the prototype does not appear to be fitted with any night vision equipment. It is understood that the vehicle has no amphibious capability, although it can ford. A radio system is fitted as standard with antennas being mounted, one either side of the roof at the rear. It is possible that a power operated winch is fitted to this vehicle.

Recent information has indicated that this may have entered low-rate production.

Iranian 4 × 4 armoured personnel carrier fitted with a one-man turret armed with a 14.5 mm and 7.62 mm machine gun 0018668

Status

Prototype. May be in low rate production. There are no known exports of this vehicle.

Manufacturer

Defence Industries Organisation Shahid Kolah Dooz Industrial Complex

UPDATED

Specifications

Crew: 1 + 13
Configuration: 4 × 4
Combat Weight: 11,000 kg
Power-to-weight ratio: 15.27 hp/t
Max road speed: 87.8 km/h
Gradient: 78.4%
Engine: Deutz BF 6 L 913 6-cylinder air-cooled diesel developing 168 hp at 2,500 rpm
Transmission: S 6-36 manual with 5 forward and 1 reverse speeds
Transfer case: 2 speed
Steering: ZF power-assisted (front wheels)
Suspension: leaf springs and hydraulic double-acting shock-absorbers
Electrical system: 28 V
Batteries: 2 × 12 V, 110 Ah
Armament:
(main) 1 × 14.5 mm KPVT MG
(coaxial) 1 × 7.62 mm PKT MG

ITALY

Centauro VBC armoured personnel carrier

Development

Late in 1996, Consortium IVECO-Otobreda delivered the last of 400 Centauro (8 × 8) 105 mm armed tank destroyers to the Italian Army. Details of this vehicle are given in the Reconnaissance vehicles section.

The Centauro was selected by Spain to meet its operational requirement and 22 vehicles were delivered late in 2000.

Early in 1996, the prototype of the Centauro (8 × 8) VBC Armoured Personnel Carrier (APC) was completed and this was shown in public for the first time at the Eurosatory 1996 defence equipment exhibition held in Paris in June 1996.

This vehicle was developed as a private venture but is aimed at the Italian Army requirement for a wheeled armoured personnel carrier to operate alongside the Centauro tank destroyer.

The prototype Centauro VBC armoured personnel carrier is fitted with an Otobreda TC-25 two-man power-operated turret armed with a 25 mm Oerlikon Contraves KBA cannon and a 7.62 mm coaxial machine gun.

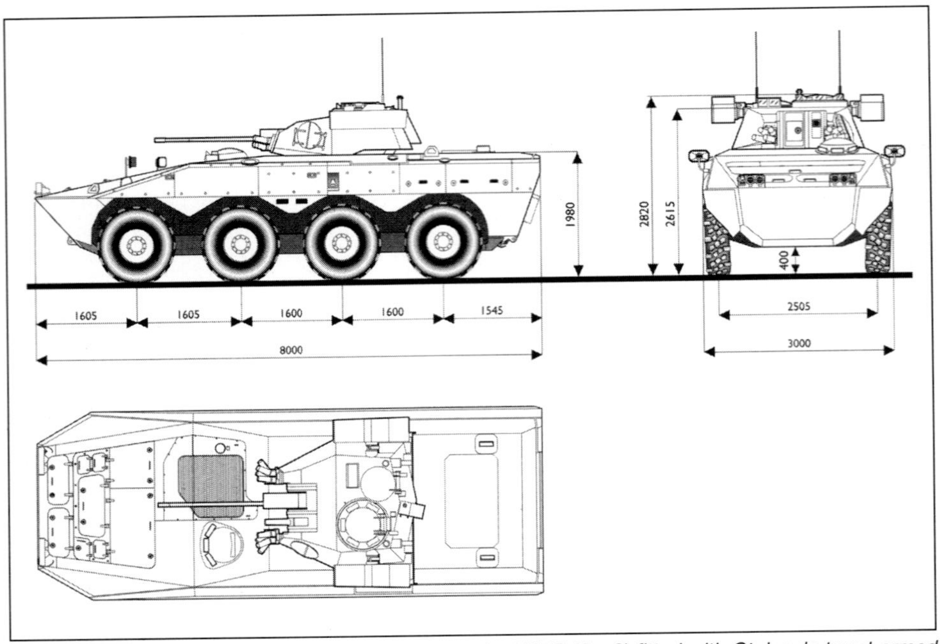

General arrangement drawing of a production standard VBC (8 × 8) fitted with Otobreda turret armed with 25 mm cannon, 7.62 mm coaxial machine gun and TOW ATGW **2001**/0103808

Centauro VBC with two-person turret armed with 25 mm cannon and mock-up launchers for TOW ATGW **2002**/0121129

Specifications
(with Dardo HITFIST turret)
Crew: 3 + 6 or 3 + 8 (depending on turret)
Configuration: 8 × 8
Combat weight: 24,000 kg
Power-to-weight ratio: 21.66 hp/t
Ground pressure: n/avail
Length: 8.0 m
Width: 3.0 m
Height:
(hull top) 1.98 m
(overall) 2.82 m
Ground clearance: 0.40 m
Track: 2.505 m
Wheelbase: 1.605 + 1.6 + 1.6 m
Max speed: 105 km/h
Fuel capacity: n/avail
Range: 800 km
Fording: 1.5 m
Gradient: 60%
Side slope: 30%
Vertical obstacle: n/avail
Trench: n/avail
Engine: IVECO MTCA V-6 turbocharged after-cooled diesel developing 520 hp at 2,300 rpm
Transmission: ZF 5 HP 1500 automatic with 5 forward and 2 reverse gears
Steering: power-assisted, front four and rear two roadwheels
Turning radius: 9 m
Suspension: independent hydropneumatic
Tyres: 14.00 × 20 (run-flat)
Electrical system: 24 V
Batteries: n/avail
Armament:
(main) 1 × 25 mm KBA cannon
(secondary) 1 × 7.62 mm MG
(ATGW launchers) 2 × TOW
Smoke-laying equipment: 2 × 4 80 mm smoke grenade launchers
Ammunition:
(main) 200 (ready use)
(secondary) 700 (ready use)
Gun control equipment
Turret power control: electric/manual
Turret traverse: 360°
Gun elevation/depression: +60/–10°
Gun stabiliser:
(vertical) yes
(horizontal) yes
NBC system: yes
Night vision equipment: yes

Early in 1999 the Italian Army awarded IVECO-Otobreda a contract for the design, development and construction of three prototypes of the Centauro VBC in the APC/IFV configuration, fitted with an Otobreda turret armed with a 25 mm cannon and 7.62 mm coaxial machine gun. This turret is the same as that fitted to the tracked Dardo HITFIST now in quantity production for the Italian Army (see Armoured personnel carriers – tracked). It is under stood that the total Italian Army requirement is for 240 basic VBC vehicles.

In addition, four other specialist versions of the VBC are being developed for the Italian Army:

- Mortar carrier
- Command post
- Recovery vehicle
- Ambulance

Description

The hull of the VBC (8 × 8) is virtually identical to that of the Centauro tank destroyer, with the driver being seated at the front of the hull on the left side with the power pack to the right.

The driver has a single-piece hatch cover that opens to the rear and three day periscopes for forward observation, the centre one of which can be replaced by a passive periscope for driving at night.

Production Centauro VBC vehicles would have a wheelbase of 1.6 + 1.6 + 1.6 m compared to the current Centauro tank destroyer which has a wheelbase of 1.6 + 1.45 + 1.45 m.

The two-person turret is in the centre of the vehicle. A wide range of turrets can be fitted to the VBC including the new Otobreda 25 mm/TOW HITFIST turret fitted to the Italian Army's new Dardo infantry fighting vehicle now in production for the Italian Army.

The HITFIST turret is armed with an Oerlikon Contraves 25 mm KBA cannon with a 7.62 mm machine gun being mounted coaxial to the left and on either side of the turret is a launcher for a Raytheon Systems Company TOW ATGW with a maximum range of 3,750 m.

The commander is seated on the left and the gunner on the right. The commander has a single-piece hatch cover that opens to the rear and five day periscopes.

The gunner has a single-piece hatch cover that opens to the rear and a stabilised day/thermal sight that also incorporates a laser range-finder. This feeds

Centauro VBC with two-person turret armed with 25 mm cannon and mock-up launchers for TOW ATGW **2002**/0121128

information to the fire-control computer so that the 25 mm cannon can be laid on to the target.

The troop compartment is at the rear with three men being seated either side facing each other. They enter and leave the vehicle via a large power-operated ramp in the rear which is also provided with a door.

The prototype vehicle has two firing ports in either side of the troop compartment with associated roof-mounted day periscopes, but the mix and type of firing ports/periscopes would depend on the user and specific operational requirements.

Standard equipment includes an appliqué passive armour package for the hull which provides a higher level of protection against small arms fire, power steering on the front four and rear two wheels, central tyre pressure inflation system, run-flat tyres, NBC system, integrated air conditioner, laser warning system and fire detection and suppression system.

If the vehicle was fitted with a turret armed with a 7.62 mm or 12.7 mm machine gun, then a typical crew would consist of a commander/gunner, driver and 10 infantry.

As previously stated, the first prototype of the Centauro VBC (8 × 8) APC was based on the standard production Centauro (8 × 8) 105 mm tank destroyer chassis, but the three examples now being built for the Italian Army will be on a purpose-built chassis.

Not only will this have a different wheelbase but also a slightly narrower hull which will be 3.00 m wide, compared to the standard Centauro which has a width of 3.28 m. The hull will also be slightly longer and higher for its more specialised troop-carrying role. The breaking system of the VBC will also be improved.

According to the Consortium IVECO-Otobreda, more powerful diesel engines can be installed while the gross weight of the vehicle could be increased to a maximum weight of 28 tonnes.

The command post vehicle is expected to have a higher roof line for its specialised role. The vehicles are expected to have the same diesel engine as installed in the Dardo IFV now in production for the Italian Army.

The baseline APC version is expected to be fitted with a standard cupola armed with an externally mounted 12.7 mm machine gun or possibly the Otobreda remote-controlled overhead weapon system already tested on the Puma (6 × 6) covered in a separate entry.

Variants

Armoured personnal carrier with a one-person turret in the centre of the vehicle, commander sitting to the left of the turret and eight men sitting in the rear troop compartment.

In addition to the versions for the Italian Army mentioned at the end of the development section, a number of other versions of the VBC have been proposed:

- Centauro armoured combat vehicle with 120 mm low recoil force gun turret
- Centauro VBC with OTO 60/70A turret armed with 60 mm gun (photographed in Centauro entry in Reconnaissance vehicles section)
- Centauro amphibious
- Centauro 155 mm self-propelled gun
- Centauro NBC reconnaissance vehicle
- Centauro air defence (gun or missile)
- Centauro armoured vehicle launched bridge

Status

Under development to meet requirements of the Italian Army.

Manufacturer

Consorzio Iveco Oto.

UPDATED

Puma armoured personnel carrier

Development

In the early 1980s, IVECO built a 4 × 4 armoured vehicle called the 6634 Armoured Vehicle Light (AVL), the first of a family of projected 4 × 4 and 6 × 6 armoured vehicles.

The Italian Army, however, wanted a larger vehicle that could undertake a wider range of roles on the battlefield and operate with its IVECO Centauro (8 × 8) tank destroyers (see Reconnaissance vehicles section) already under development.

The end result was the Puma, the first prototype of which was completed in 1988, with the second prototype being completed in 1989. These two were followed by a further three prototypes of the definitive armoured personnel carrier version in mid-1990.

The Italian Army also awarded the company a contract for the design of six specialised versions of the Puma; three are weapon platforms, one each for the Raytheon TOW and Euromissile MILAN ATGW systems and one for the Matra BAe Dynamics Mistral SAM (all these have three or four crew), the fourth is an 81 mm mortar carrier (four crew including the driver), the fifth is an ambulance (carrying two stretcher cases plus crew) and the sixth is a command post vehicle (crew of four or five).

Development of the Puma family of 4 × 4 and 6 × 6 series of armoured personnel carriers were completed and late in 1999 Consorzio Iveco Oto was awarded a production contract worth Lit585 billion for the supply of 250 Puma (6 × 6) and 330 Puma (4 × 4) APCs for the Italian Army.

Production will be undertaken at the IVECO facilities at Bolzano and the first production vehicles are due to be delivered in 2001.

Production will peak at 16 Puma vehicles per month with final deliveries to take place in 2004. The Italian Army will call the Puma the VBL (Veicolo Blindato Leggero), or light armoured vehicle.

The Italian Army will deploy the four-wheeled version of the Puma in the five reconnaissance regiments which will in future operate a mixture of Centauro (8 × 8) and Puma (4 × 4). Within the three reconnaissance squadrons each one of the three sections will deploy two Centauro and four Puma vehicles. The latter will carry a five-person team, plus the driver, allowing it to deploy two two-man scout teams.

The Puma 6 × 6 will be deployed within light infantry regiments, alpini, paratroop and the army amphibious regiment. This will carry a six-person combat team plus the driver as well as its equipment.

In addition to the supply of 580 Puma vehicles in 4 × 4 and 6 × 6 configurations the contract also includes the supply of two prototype vehicles, one will be fitted with the Otobreda turret armed with a 12.7 mm machine gun which can be aimed and fired under armour protection and the other with additional add on passive armour protection.

If trials with these two prototype vehicles are successful then it is expected that all production vehicles will be ready to accept these upgrades which would be the subject of additional contracts.

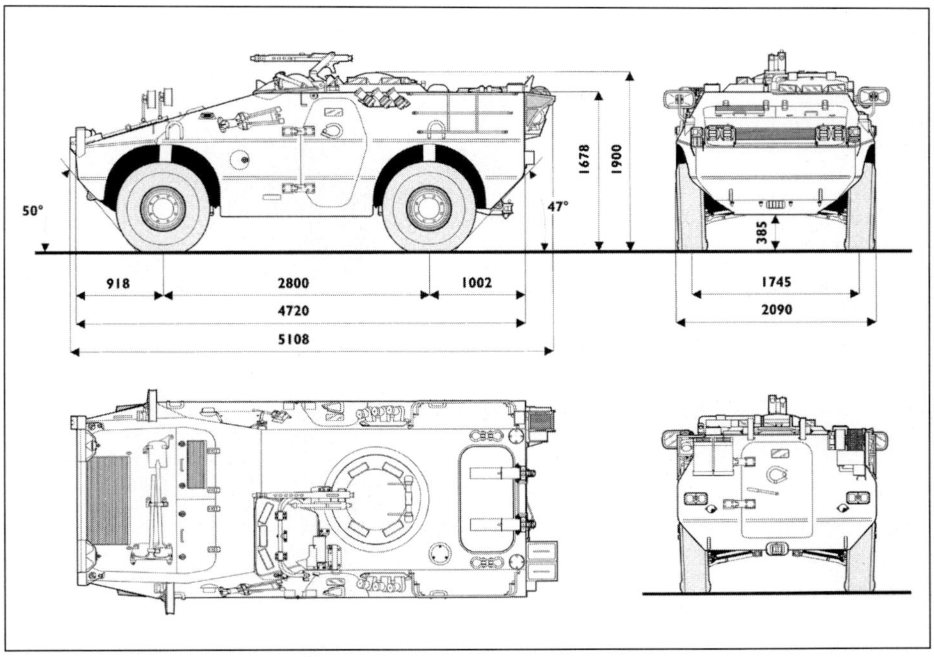

General arrangement drawing of the Puma (4 × 4) armoured personnel carrier **2001**/0103807

One of the prototypes of the Puma (4 × 4) armoured personnel carrier with doors and hatches closed and without armament installed 0005373

Specifications

Configuration	4 × 4	6 × 6
Crew (typical)	6 + 1	8 + 1
Combat weight	5,700 kg	7,500 kg
Power-to-weight ratio	31.57 hp/t	24 hp/t
Length	5.108 m	5.526 m
Width	2.09 m	2.284 m
Height		
(to hull top)	1.678 m	1.678 m
Ground clearance	0.385 m	0.385 m
Track	1.745 m	1.945 m
Wheelbase	2.8 m	1.45 + 1.95 m

Angle of approach/departure	50/47°	51/49°
Max speed	105 km/h	100 km/h
Fuel capacity	150 litres	270 litres
Range	800 km	700 km
Fording	1 m	1 m
Gradient	60%	60%
Side slope	30%	30%
Engine	IVECO Type 8042 TCA 4-cylinder diesel developing 180 hp at 3,000 rpm	
Transmission	Renk Reco automatic with 5 forward and 1 reverse gears	

Suspension	independent, hydropneumatic	independent, hydropneumatic
Brakes	disc, all wheels	disc, all wheels
Steering	power assisted	power assisted
Turning radius	6 m	7.5 m
Tyres	11.00 × 16 (run-flat)	11.00 × 16 (run-flat)
Electrical system	24 V	24 V
Batteries	2 × 12 V, 100 Ah	2 × 12 V, 100 Ah
NBC system	yes	yes
Night vision equipment	yes	yes

The specialised versions of the Puma will also be subject to additional future contracts.

The Puma in its 4 × 4 and 6 × 6 versions was proposed to the UK MoD to meet the British Army requirement for a Future Command and Liasion Vehicle (FCLV).

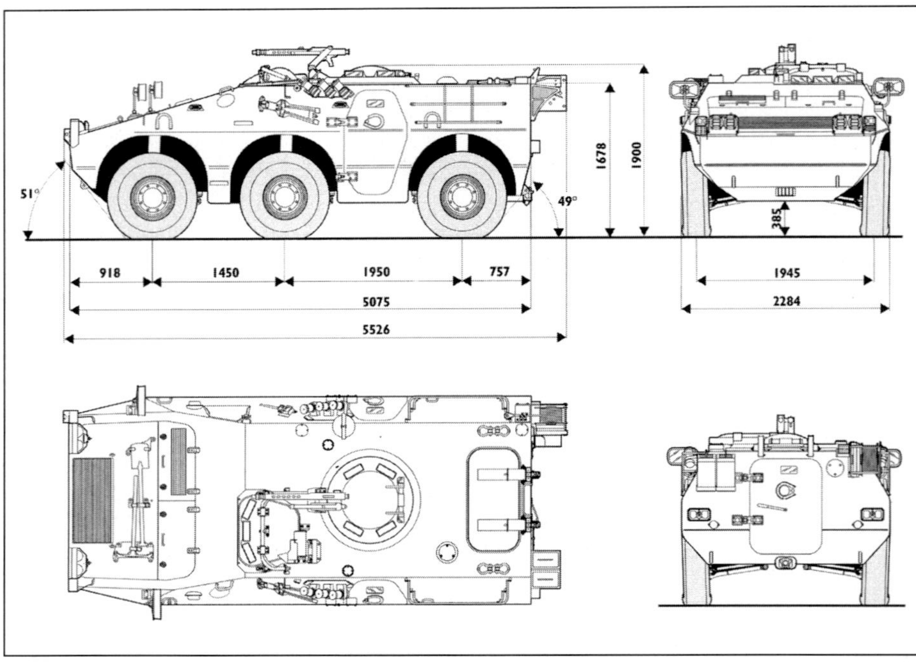

General arrangement drawing of the Puma (6 × 6) armoured personnel carrier **2001**/0103809

Description

The hull of the Puma is of all-welded steel armour construction that provides protection from small arms fire and shell splinters with the upper part of the hull sides sloping inwards. The hull design is unusual in that the sides are in one piece rather than two pieces welded together.

The driver sits towards the front of the vehicle on the left side and has a single-piece domed hatch cover that opens to the right. To his front are three day periscopes, the centre one of which can be replaced by a passive night periscope.

The engine compartment is to the front of the driver and is separated from the remainder of the vehicle by a bulkhead. Access panels are provided in the glacis plate for maintenance purposes, and the infantry is provided with individual seats that fold up.

The prototypes were powered by an IVECO 8141.47 diesel engine developing 125 hp coupled to a ZF transmission but later vehicles are powered by an IVECO 8042.45 four-cylinder turbocharged diesel developing 180 hp at 3,000 rpm coupled to a Renk Reco automatic transmission with five forward and one reverse gears.

From the automatic transmission, which incorporates a torque converter, power is taken to the transfer box and then by propeller shafts to the front and rear differentials. It then goes via short universally or constant-velocity jointed shafts to the front and rear wheels.

Each of the four roadwheels is independently located by means of an upper telescopic strut and a lower arrangement of transverse links, which is made up of anti-roll torsion bars and radial links. The hydropneumatic struts are made by Messier.

Steering on the front wheels is power assisted and the Michelin 11.00 × 16 tyres are of the run-flat type.

The engine air inlets are under the forward part of the hull and the exhaust outlet is on the right side of the hull.

In either side of the hull is a door that opens forwards and in its upper part is a bulletproof vision block with a firing port underneath. In the rear of the hull is a large door that opens to the left, with a vision block in its upper part and a firing port underneath. On the right side of the hull at the rear is a winch.

The commander's cupola is mounted in the centre of the roof and has a single-piece hatch cover that opens to the rear and five day periscopes that give all-round observation. A 7.62 mm MG 42/59 or 12.7 mm M2 machine gun can be mounted on the cupola. To the immediate rear of the commander's cupola is a single-piece hatch that opens to the rear.

Prototype of the 6 × 6 version of the Puma (6 × 6) armoured personnel carrier fitted with Otobreda turret armed with externally mounted 12.7 mm MG 0058101

Mounted on either side of the hull towards the rear is a bank of three electrically operated smoke grenade dischargers firing forwards.

Standard equipment includes an NBC system, a fire detection and suppression system and an integrated air conditioning system.

Variants

As well as being used as an armoured personnel carrier carrying a total of seven persons, the Puma is suitable for a wide range of other roles including reconnaissance, anti-tank with ATGWs such as MILAN, air defence with SAMs, 81 mm mortar carrier, ambulance, NBC reconnaissance and internal security.

Late in 1990, the first prototype of the 6 × 6 version of the Puma armoured personnel carrier was built, sharing a number of common components with the 4 × 4 version including the same engine, transmission and suspension. A number of specialised versions of the 6 × 6 Puma has been proposed including an 81 mm mortar carrier and a command post vehicle.

The prototype of the 6 × 6 version of the Puma is armed with an externally mounted 12.7 mm M2 HB machine gun with a bank of three electrically operated smoke grenade dischargers mounted either side of the turret firing forwards, although a wide range of other armament systems is possible including ATGWs and surface-to-air missiles such as the Mistral.

For trials purposes, one of the prototypes of the Puma (6 × 6) has been fitted with a new Otobreda 12.7 mm power-operated turret armed with a 12.7 mm M2 machine gun which can be aimed and fired under complete armour protection.

Status

Development complete. Production order placed for 250 (6 × 6) and 330 (4 × 4) late in 1999 with deliveries to run from 2001 to 2004.

Manufacturer

Consorzio Iveco Oto.

UPDATED

FIAT-Otobreda Type 6614 armoured personnel carrier

Development

The Type 6614 (4 × 4) APC was a joint development between FIAT and Otobreda, the former responsible for the hull and the automotive components and the latter for the armament installation. The vehicle shares many common components with the FIAT-Otobreda Type 6616 (4 × 4) armoured car for which there is a separate entry in the Reconnaissance vehicles section. Production of the Type 6614 and Type 6616 was completed some years ago and they are no longer marketed.

The Type 6614 Armoured Personnel Carrier (APC) has been manufactured under licence in South Korea by the Asia Motors Company Incorporated. Production has been completed in South Korea and it is understood that between 200 and 275 vehicles were built for the South Korean Army. There were no vehicles exported from South Korea.

The FIAT Type 6614 was also used by the Italian contingent serving in the Lebanon and was tested by the Italian Army in Denmark and Norway while operating as part of the Allied Command Europe Mobile Force.

The Italian Air Force has 110 vehicles for patrolling airfields while 60 were delivered to the Italian Police. More recently the Italian Army has taken delivery of 14 vehicles for use by its Alpine units. Total production of the Type 6614 APC for home and export markets amounted to 1,160 vehicles. This vehicle is no longer being marketed.

Description

The hull of the Type 6614 APC is of all-welded steel which provides the crew with protection from small arms fire and shell splinters.

The driver sits at the front of the hull on the left side and has a single-piece domed hatch cover that opens to the right. To the front and side of the driver are five vision blocks which cover an arc of 200°, the centre three with wiper blades.

The power pack is mounted at the front of the vehicle to the right of the driver and is separated from the personnel compartment by a fireproof bulkhead. The IVECO engine is coupled to a manual five-speed transmission from which power is transmitted to the two-speed transfer case and then to the front and rear axles by propeller shafts. Both front and rear differentials are mechanically lockable. The planetary final drives, front and rear, consist of epicyclic gear trains in the wheel hubs.

Entry doors in each side of the hull both open forward and have a vision block and a firing port. There are another three vision blocks and firing ports in each side of the hull. At the rear of the hull is a large ramp hinged at the bottom, either side of which is a single vision block and firing port. The 10 infantrymen, one of whom acts as gunner, have individual seats which can be folded up. In the centre of the roof is mounted the main armament installation. On the basic vehicle this consists of an M113-type cupola which has five day periscopes and a single-piece hatch cover that opens to the rear. Mounted on the forward part of this is a 12.7 mm (0.50) M2 HB machine gun. Other types of armament can be mounted if required, for example a fully enclosed turret armed with single or twin 7.62 mm machine guns. The two roof hatches over the top of the

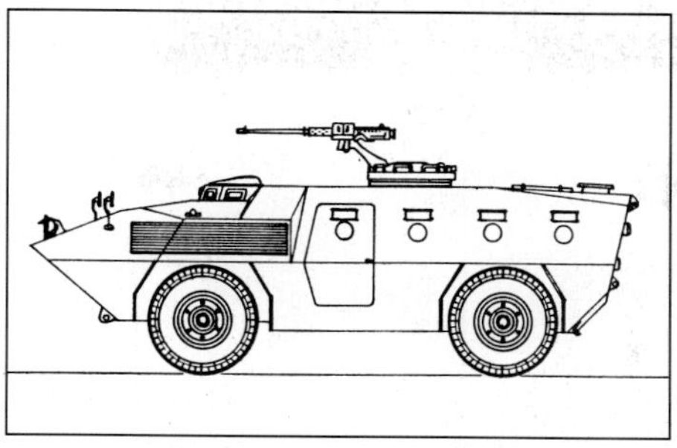

FIAT-Otobreda Type 6614 APC with a 12.7 mm M2 machine gun

troop compartment open either side of the vehicle. A ventilator is mounted in the roof at the rear.

The suspension, front and rear, is of the independent strut and link type with a helical spring and coaxial rubber bump stop. Each wheel station has two hydraulic shock-absorbers and anti-roll bars are fitted as standard. The tyres are of the run-flat type.

The vehicle is fully amphibious, being propelled in the water by its wheels. Standard equipment includes four electrically operated bilge pumps with a capacity of 180 litres/min each. Optional equipment includes passive or active night vision equipment, an air conditioning system, a fire extinguishing system in the wheel arches, a spare wheel and holder, two three-barrelled 76 mm smoke grenade dischargers and a front-mounted 4,500 kg capacity winch with 40 m of 12 mm diameter cable.

Variants

Although designed primarily for use as an APC, the Type 6614 can also be used for other roles such as reconnaissance and internal security. Projected variants included an ambulance, cargo, command and mortar carrier.

The Italian Army has deployed a number of Type 6614 (4 × 4) armoured personnel carriers to Bosnia. These have been fitted with a bank of three electrically operated 76 mm grenade launchers either side towards the rear, passive armour for a higher protection level and a spare wheel is carried on the roof towards the rear. The 12.7 mm machine gun installation is also provided with additional armour protection for the gunner.

Status

Production complete. In service with the following:

Country	Quantity	Comment
Argentina	20	air force
Italy	110	air force
	40	police
	154	army
South Korea	275	called KM900 series
Peru	15	some have 81 mm mortars
Somalia	270	non-operational
Tunisia	110	
Venezuela	24	national guard

Manufacturer

Collaboration between FIAT and Oto Melara (later Otobreda) with final assembly taking place at the Fiat (later Iveco Defence Vehicles Division) plant in Bolzano.

UPDATED

Specifications

Crew: 1 + 10
Configuration: 4 × 4
Combat weight: 8,500 kg
Loaded weight:
(front axle) 4,500 kg
(rear axle) 4,000 kg
Power-to-weight ratio: 18.82 hp/t
Length:
(hull) 5.86 m
Width: 2.5 m
Height:
(to hull top) 1.78 m
(incl MG mount) 2.18 m
Ground clearance:
(axles) 0.37 m
(hull front) 0.485 m
(hull centre) 0.385 m
(hull rear) 0.385 m
Track: 1.96 m
Wheelbase: 2.9 m
Angle of approach/departure: 40°/45°
Max speed:
(road) 100 km/h
(water) 4.5 km/h
Fuel capacity: 142 litres
Max road range: 700 km
Fuel consumption:
(road) 0.2 litres/km
Fording: amphibious
Gradient: 60%
Side slope: 30%
Vertical obstacle: 0.4 m
Turning radius: 7.9 m
Engine: IVECO Model 8062.24 supercharged liquid-cooled in-line diesel developing 160 hp at 3,200 rpm
Transmission: manual with 5 forward and 1 reverse gears
Transfer case: 2 speed
Steering: hydraulic
Clutch: single dry plate, hydraulically operated
Suspension: independent
Tyres: 14.50 × 20 (run-flat)
Brakes:
(main) drum, air over hydraulic, dual circuit
(parking) drum, hand-operated, on transfer front output shaft
Electrical system: 24 V
Batteries: 2 × 12 V, 100 Ah
Armament: 1 × 12.7 mm M2 MG

Gun control equipment
Cupola traverse: 360° manual
Armour: 6-8 mm
Gun stabiliser:
(vertical) no
(horizontal) no
NBC system: no
Passive night vision equipment: optional

FIAT-Otobreda Type 6614 APC, with appliqué armour, deployed in Bosnia (Michael Jerchel) ***2000***/0081981

FIAT-Otobreda Type 6614 APC, with appliqué armour, deployed in Bosnia (Richard Stickland) 0005375

Boneschi MAV 5 light armoured vehicle

Development

The MAV 5 light armoured vehicle was developed as a joint venture by Carrozzeria Boneschi Srl and IVECO, and has been designed to carry six fully equipped persons (commander, driver and four crew) and their equipment both on roads and across country.

The Italian Army took delivery of over 200 units which have been used for peacekeeping duties in Albania, Bosnia and Somalia. Some of these have been fitted with a ring on the roof on which a 7.62 mm MG 42/59 machine gun has been mounted for local protection. Twelve vehicles have been supplied to the Italian Carabinieri. Within the Italian Army these vehicles are known as the VM/90P.

Recently, IVECO has teamed with the German company Krauss-Maffei Wegmann to develop a similar vehicle called Terrier. Details of this vehicle are given in this section under International.

The first customer for the Terrier is the United Arab Emirates which has taken delivery of 10 vehicles, while Egypt and Poland have each taken delivery of one vehicle.

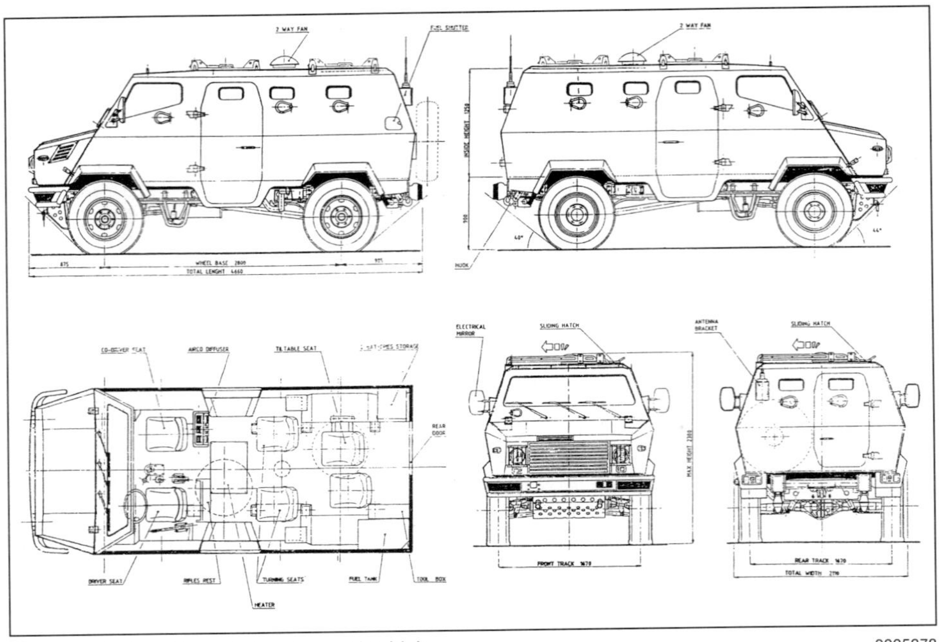

Boneschi MAV 5 (4 × 4) light armoured vehicle 0005378

Description

The MAV 5 light armoured vehicle is essentially an IVECO 40.10 (with 100 hp engine) 4 × 4 truck chassis, fitted with a new steel armoured body 6 mm thick providing the occupants with complete protection from small arms fire. The floor and roof are also of ballistic steel armour with a thickness of 2 to 3 mm.

All windows and vision blocks of the MAV 5 are of bulletproof glass with a total thickness of 38 mm which provides the same level of protection as the armoured hull.

The driver is seated at the front of the vehicle on the left side with the vehicle commander to his right, both being provided with a bulletproof window to the front and sides. The vehicle commander also has a circular firing port below the front window on the right side.

The four troops are seated on individual turntable seats facing the centre or the side of the vehicle and they can enter the vehicle via three doors, one in each side and one at the rear. The rear door opens to the right and has an integral vision block with a circular firing port underneath. The spare wheel is normally carried on the left side of the hull rear and above this is a firing port and vision block. All doors are provided with locks and anti-intruder devices.

In the left and right sides of the troop compartment are two vision blocks each with a firing port underneath. There are two circular roof hatches, one towards the front on the left and one towards the rear on the right.

The engine is located at the front of the vehicle and is protected by ballistic steel.

An extension pipe for the diesel engine is on the forward right side of the hull and extends just above the roof line.

Standard equipment includes provision for carrying six rifles, ceiling lights for internal lighting and special protection for electrical cables. A spare wheel and tyre is carried at the rear of the vehicle on the left side.

Optional equipment includes an air conditioning system, an anti-tear gas filtering system and a firefighting system that protects the engine.

The MAV 5 can also be provided with different tyres, including run-flat type, an exhaust brake and an air/electrical connection for a trailer.

Variants

More recently the company has developed a new version with increased mine protection called the MAV 5 (MP).

MAV 4

Details of this large 4 × 4 APC, a number of which were purchased by the Italian Army for use in the command post role, were given in *Jane's Armour and Artillery 1997-98*, pages 462 to 463.

Status

Production complete. In service with the Italian Army (200 units) and with the Italian Carabinieri (12 units). Details of the similar Terrier Light Armoured Unit (LAU) developed by Krauss-Wegmann of Germany and IVECO of Italy are given in this section under International.

Manufacturer

Carrozzeria Boneschi Srl.

UPDATED

Specifications

Crew: 2 + 4
Configuration: 4 × 4
Weight:
(max) 4,450 kg
(unloaded) 3,850 kg
(towed load) 1,500 kg
Power-to-weight ratio: 23.1 hp/t
Length: 4.66 m
Width: 2.00 m
(internal) 1.78 m
Height:
(roof) 2.15 m
(fan) 2.268 m
Track: 1.67 m
Wheelbase: 2.8 m
Angle of approach/departure: 45/40°
Max road speed: >100 km/h
Fuel capacity: 75 litres
Range: 600 km
Fording: 0.7 m
Gradient: 60+ %
Side slope: 30%
Turning radius: 6 m
Engine: IVECO turbocharged 4-cylinder diesel developing 103 hp
Gearbox: manual with 5 forward and 1 reverse gears
Clutch: single plate, mechanically operated
Transfer box: rack and pinion
Steering: power assisted, LHD
Suspension: torsion bar front, leaf springs with hydraulic and telescopic dampers rear
Axles: single reduction front and rear, self-locking differential
Wheels: disc type, 6.50 H 16 rim size
Tyres: 9.00.16 14 PR
Brakes: disc type at front, drum type at rear, split circuits
Electrical system: 24 V
Batteries: 2 × 12 V, 110 Ah
Generator: 1,200 W
Armament: 1 × 7.62 mm MG 42/59 (optional)
Gun stabiliser:
(vertical) no
(horizontal) no
Armour: 6 mm
NBC system: no
Night vision equipment: no

Boneschi MAV 5 (4 × 4) light armoured vehicle deployed to Kosovo (Richard Stickland) **2002**/0121127

Boneschi MAV 5 (4 × 4) light armoured vehicle deployed to Kosovo (Richard Stickland) **2002**/0121126

JAPAN

Type 96 armoured personnel carrier (8 × 8)

Development/Description

The Japanese Ground Self-Defence Force has type-classified a new 8 × 8 armoured personnel carrier called the Type 96. This is now in production and in service with the Japanese Ground Self-Defence Force. As of late 1999, a total of 46 Type 96 (8 × 8) APCs had been delivered to the Japanese Ground Self-Defence Force. This included 28 manufactured in late 1999 with another 28 ordered in FY2000. Since then production has continued.

The prime contractor is probably Komatsu or Mitsubishi. The former company designed and built the Type 87 (6 × 6) reconnaissance and patrol vehicle while Mitsubishi Heavy Industries designed and built the Type 82 (6 × 6) command and communications vehicle. Full details of both these vehicles are given in the Reconnaissance vehicles section under Japan.

The hull of the Type 96 APC is of all-welded construction with the engine compartment being located at the front of the vehicle on the left side. The exhaust outlet is on the left side of the hull roof.

One of the prototypes of the Japanese 8 × 8 armoured personnel carrier Type 96

The driver is seated at the front of the vehicle on the right side and is provided with a single-piece hatch cover that opens to the rear and three day periscopes for forward observation.

The commander/gunner is seated to the rear of the driver and has a raised cupola which is provided with day periscopes for all-round observation and a single-piece hatch cover that opens to the rear. Mounted on the forward part of the cupola is a 12.7 mm M2 machine gun which can probably be aimed and fired from within the cupola. More recently, some vehicles have been fitted with a locally developed 40 mm grenade launcher.

The troop compartment is at the rear of the vehicle and normal means of entry and exit for the troops is via a power-operated ramp located at the rear of the vehicle. Over the top of the troop compartment are five roof hatches, three on the left and two on the right, all of which open outwards. At least two firing ports and associated vision devices are provided in each side of the rear troop compartment.

Mounted on top of each side of the troop compartment, towards the rear, is a bank of four forward-firing smoke grenade launchers.

Standard equipment includes powered steering on the front four roadwheels, central tyre pressure regulation system which allows the driver to adjust the tyre pressure to suit the terrain being crossed, NBC system and a laser warning device which is mounted at the front of the vehicle on the left side of the roof. This is of the same type as fitted to a number of other Japanese armoured fighting vehicles such as the Type 90 MBT.

Variants

None known.

Status

Production. In service with the Japanese Gound Self-Defence Force.

Manufacturer

Not known.

UPDATED

JORDAN

M35 Mine Protected Carrier

Development

The M35 Mine Protected Carrier (MPC) is a joint development between AM General of the United States (who originally built the M35 6 × 6 truck), the King Hussien Main Workshops of Jordan and Mechem Consultants of South Africa.

Marketing of the M35 MPC is carried out by the National Resources Development Company of Jordan with the assistance of the King Abdullah Design and Development Bureau (KADDB).

The first two production standard M35 MPCs were completed early in 1998 although tests of the new mine protected body on a 6 × 6 chassis had been completed well before this date, in South Africa.

It has been designed as a low cost mine protected carrier for the transport of troops in areas where there is a significant risk from anti-tank mines.

The King Hussien Main Workshops have been responsible for the chassis while Mechem has been responsible for the new mine protected all-welded steel armour cab and troop compartment.

AM General supplied the automotive parts of the upgrade as most of these are the same as those used in its Extended Service Plan for the M35 truck used by the US Army National Guard and Reserve.

Description

The M35 Mine Protected Carrier (MPC) essentially consists of a modified US supplied M35 2.5 ton (US) (6 × 6) truck chassis fitted with a mine protected cab and mine protected rear troop area.

Both the cab and rear troop area have a V-shaped bottom to provide the maximum possible protection from the blast of anti-tank mines. A special kit can be supplied to provide enhanced protection against the Russian TMRP-6 Self-Forming Fragmentation (SFF) anti-tank mine.

The standard level provides protection against 7.62 mm × 51 ball ammunition all round and against a conventional TM-57 anti-tank mine under any wheel station.

The chassis has been modified by the installation of a new power pack consisting of a Caterpillar 3116 170 hp diesel coupled to a ZF fully automatic transmission.

The standard M35 (6 × 6) truck has dual rear wheels but on the M35 MPC these have been replaced by single

Specifications
Crew: 2 + 10
Configuration: 6 × 6
Weight: 9,000 kg
Length: 7.00 m
Width: 2.4 m
Height: 2.65 m
Range: 500 km
Engine: Caterpillar 3116 170 hp diesel
Transmission: ZF 5-speed automatic

M35 Mine Protected Carrier **2001**/0104372

M35 Mine Protected Carrier **2002**/0121121

wheels and tyres which have also been fitted to the front axle.

A central tyre-pressure regulation system has been installed which allows the driver to adjust the tyre pressure to suit the terrain being crossed. A spare wheel and tyre are under the left side of the chassis.

The commander and driver are seated in the fully enclosed all-welded steel cab to the rear of the engine compartment. This is provided with bulletproof windows in the front, sides and rear. The crew enter via a square door in each side of the upper part of the cab which opens forwards.

The rear troop compartment has a large door in the rear with opens to the left and this is supplied with a large bulletproof window. The are also large bulletproof windows in either side of the troop compartment and one at the front. Hatches are also provided in the roof. Ten troops are carried in the rear and these are seated five down each side facing each other. The troops enter and leave via the rear door and additional step provided at the rear of the vehicle.

Status
Prototypes. Not yet in production or service.

Manufacturer
National Resources Development Company (but see text).

UPDATED

KOREA, NORTH

M1992/1 Armoured Personnel Carrier

Development
The M1992/1 4 × 4 Armoured Personnel Carrier (APC) was developed in the late 1980s to meet the operational requirements of the North Korean Army. It was first observed in 1992 and was subsequently allocated the US Army designation of the M1992/1. This designation indicates the year that the vehicle was first observed by US sources.

Description
It is believed that the M1992/1 (4 × 4) APC uses commercial automotive components that may well have been procured from overseas. The hull of the vehicle is of all-welded steel armour to protect the occupants from small arms fire and shell splinters. The air intake for the engine is located over the top of the engine compartment with the air outlet louvres either side.

The layout of the vehicle is similar to a truck with the engine compartment at the front, commander and driver to the immediate rear and the whole of the remainder of the vehicle being taken up by the crew compartment that extends to the rear of the vehicle. The commander and driver have windows to their immediate front and sides. The upper part of the hull either side slopes slightly inwards.

It is believed that the troops enter the M1992/1 APC by doors in the hull rear and/or roof hatches. In either side of the hull are five firing ports to allow some of the troops to fire their small arms from inside the hull. There do not appear to be any vision blocks associated with these firing ports.

Specifications
Configuration: 4 × 4
Weight: 12,000 kg
Power-to-weight ratio: 20.83 hp/t
Length: 6.2 m
Width: 2.8 m
Height: 2.5 m
Engine: diesel believed to develop 250 hp
Armament:
1 × 30 mm AGS-17 automatic grenade launcher
1 × AT-4 'Spigot' ATGW
NBC system: possible
Night vision equipment: possible
Note: above specifications should be taken as provisional

Mounted over the forward part of the crew compartment is a Russian 30 mm AGS-17 grenade launcher, while mounted on the roof towards the rear is a launcher for the Russian Tula KBP 9K111 Fagot (AT-4 Spigot) ATGW which has a maximum range of 2,000 m or 2,500 m depending on the version.

A trim board or splash board is believed to be mounted under the front of the hull and extends forward, thus indicating that the vehicle has an amphibious capability. There do not appear to be any water jets so it is understood that it is propelled in the water by its wheels.

Variants
M1992/2 multiple rocket launcher
This is similar to the M1992/1 APC except the hull roof line extends just to the rear of the commander's and driver's position and then drops vertically in line with the lower hull. There is just one firing port in either side.

Mounted at the rear of the hull is a 107 mm (12-round) multiple rocket launcher which is understood to have similar physical characteristics to the NORINCO (China North Industries Corporation) Type 63 towed system. Additional 107 mm rockets are probably carried for manual reloading.

North Korean M1992/1 (4 × 4) wheeled armoured personnel carrier **2001**/0104373

Status
Production probably complete. In service with North Korea.

Manufacturer
North Korea state factories.

UPDATED

MEXICO

Mexican armoured personnel carriers

Development/Description
Since the 1980s Mexico has been engaged in the design, development and production of wheeled light armoured vehicles for the home market.

It was originally believed that these were based on foreign designs, especially from the USA, but recent information has indicated that these have all been designed, developed and manufactured in Mexico.

The confusion has been caused largely by the use of the designation DN, plus a numerical suffix, for the members of a widely varied family of light wheeled AFVs.

These vehicles have been developed since 1967, under the leadership of the Mexican Department of National Industry (DINA). In the designation the DN stands for Defense Nacional.

The first two armoured vehicles developed were designated the DN-I and DN-II, but it is understood that these never passed the prototype stage.

Mexican DN-IV Caballo armoured personnel carrier (Mexican Army)
0058318

Mexican DN-III Sedena 1000 armoured personnel carrier (Mexican Army)
0058317

These were followed in 1979 by the DN-III, which is similar in appearance to the then Cadillac Gage (now Textron Marine & Land Systems) V-100 (recently renamed the LAV-100) series of 4 × 4 light armoured vehicles.

A total of 24 of these were built in the early 1980s under the designation DN-III or the SEDENA 1000. It is believed that these were based on a US designed Dodge 4 × 4 truck chassis with the diesel engine being mounted at the rear.

Fitted on the roof was a turret with an externally mounted 7.62 mm machine gun which is very similar to that fitted to the Swiss MOWAG Roland (4 × 4) armoured personnel carrier. Details of this vehicle, production of which was completed some time ago, are given later in this section under Switzerland.

This was followed by the DN-IV, which is based on a 6.5 tonne chassis with the diesel engine mounted at the rear. These are also referred to as the Caballo.

At least two versions of these were developed, one fitted with a turret armed with a 20 mm cannon and the other with a turret similar to that fitted on the DN-III, armed with an externally mounted 7.62 mm machine gun. A 7.62 mm machine gun could also be installed at the rear of the hull. The troop compartment is provided with firing ports and associated vision devices.

In the early 1960s these Caballo vehicles were overhauled and modernised and some were fitted with a French SAMM one-man manually operated turret armed with 7.62 and 12.7 mm machine guns.

The Caballo is powered by a DINA Cummins V8-504 eight-cylinder diesel engine. Basic specifications are:

Specifications
Crew: 2 + 7
Configuration: 4 × 4
Weight: 9,500 kg
Length: 5.78 m
Width: 2.50 m
Height: 1.90 m
Ground clearance: 0.67 m
Wheelbase: 2.65 m
Max road speed: 80 km/h
Range: 750 km
Engine: Cummins diesel model V8-504
Transmission: automatic

The DN-IV was followed by the DN-V, which appears in at least two versions: the Toro reconnaissance and the Bufalo fire support vehicle. Both are similar in some respects to the LAV-150 (which in the past has been called the Commando) and basic specifications are:

Specifications
Configuration: 4 × 4
Weight: 7,500 kg
Length: 5.86 m
Width: 2.50 m
Height: 2.365 m
Ground clearance: 0.64 m
Wheelbase: 2.66 m
Max road speed: 80 km/h
Range: 750 km
Engine: Cummins diesel model V8-504
Transmission: automatic

The standard Toro is fitted with a British Helio FVT 900 series one person turret armed with a 20 mm Oerlikon Contraves cannon and 7.62 mm coaxial machine gun with a bank of four electrically operated smoke grenade launchers being mounted either side of the turret.

The Bufalo version appears to mount the complete turret of the old US-supplied full-tracked 75 mm M8 self-propelled howitzer, which dates back to the Second World War, with a 12.7 mm M2 machine gun mounted on the open-topped turret for air defence purposes.

It is believed that five of these vehicles are in service. These are the only self-propelled artillery systems in service with Mexico.

There are believed to be at least two mortar versions, one mounting an 81 mm mortar and the other towing a French 120 mm TDA mortar. Recent information has indicated that all of these vehicles have the ability to tow a 120 mm TDA mortar.

The DN-VI is a 4 × 4 APC which in some respects is similar to the French Giat Industries VAB series of vehicles that have been built in large numbers for the home and export markets.

This has also been referred to as the Ocelote and is understood to be based on a US designed Dodge truck chassis. Later vehicles, however, are based on a Mercedes-Benz UNIMOG (4 × 4) U1300L or U1600L (4 × 4) truck chassis.

This vehicle is powered by a Mercedes-Benz six-cylinder diesel which gives a maximum road speed of 100 km/h and a maximum range of 1,100 km on roads. Armament comprises two externally mounted HK21A1 7.62 mm machine guns.

There may be a number of other variants including reconnaissance vehicle, command post vehicle, armoured cargo carrier and an armoured ambulance. A fire support vehicle with a 90 mm turret has been projected.

Specifications
Configuration: 4 × 4
Crew: 3 + 10
Weight: 8,000 kg
Length: 6.15 m
Width: 2.50 m
Height: 2.32 m
Ground clearance: 0.58 m
Wheelbase: 3.25 m
Max road speed: 100 km/h
Range: 1,000 km
Engine: Mercedes-Benz OM-352 6-cylinder diesel developing 120 bhp

The latest vehicle to enter service with the Mexican Army is the DN-VII Lince scout car. This is very similar to the Renault VBL (4 × 4) light scout car, which was developed to the prototype stage for the French Army but not placed in production. The three-man crew consists of commander, gunner and driver and the vehicle is armed with a single HK21A1 7.62 mm machine gun. The DIM-I is a reconnaissance vehicle while the DIM-II is a VIP transport vehicle. Basic specifications are:

Specifications
Crew: 3
Configuration: 4 × 4
Weight: 4,000 kg
Length: 4.10 m
Width: 1.80 m
Height: 2.10 m
Ground clearance: 0.21 m
Max road speed: 90 km/h
Max range: 280 km
Engine: Dodge diesel

Mexico has also recently acquired from Belgium a very large number of BDX (4 × 4) and AMX VCI full-tracked armoured personnel carriers.

United Nations' sources state that a total of 95 BDX (delivered in 1994) and no less than 401 AMX VCI tracked vehicles which were overhauled by the Belgian company, SABIEX, prior to Mexico. The AMX VCI were delivered between 1994 and 1996.

Some of the US-supplied half-tracks have also been upgraded using kits supplied by NAPCO of the USA and it is probable that some of the German-supplied Henschel Wehrtechnik (which in 2000 became Rheinmetall Landsysteme) HWK10 series full-tracked APCs, for which Mexico was the sole customer, have also been upgraded. This work is believed to have been carried out by NIMDA of Israel.

Status
Production complete. In service with Mexican Army.

Manufacturer
Not known.

UPDATED

MYANMAR (BURMA)

Myanmar light armoured vehicles

Development/Description
Since the early 1990s the State Law and Order Restoration Council (SLORC) has made major efforts to enable Myanmar to be virtually self-sufficient in the area of small arms ammunition, small arms (sub-machine guns, rifles and light machine guns), light mortars, mortar ammunition, grenades and mines.

More recently a number of light armoured vehicles have been developed and placed in service. These are based on modified imported commercial-type chassis fitted with an armoured body. All of these are believed to use a Japanese commercial chassis in 4 × 4, 4 × 2 or 6 × 6 configuration, which has been modified in some cases to take into account the increased weight of the armoured body.

Reports have mentioned a scout car based on a Japanese Mazda (4 × 4) chassis with about 30 vehicles in service and an armoured personnel carrier based on a Japanese Hino (6 × 6) chassis with up to 10 vehicles in service.

There are at least four types of light armoured vehicle built or assembled in Myanmar and these have the designations of the BAAC-83, BAAC-84, BAAC-85 and BAAC-86. Some reports have also mentioned the BAAC-87. Use of Japanese Nissan and Toyota chassis has also been reported.

Mention has also been made of an 'Armoured Fighting Car' which was first observed late in 1996 patrolling the streets of Rangoon.

This is believed to be based on a Ford (4 × 4 or 4 × 2) light commercial truck chassis powered by a Nissan diesel engine. This is normally armed with a Chinese 12.7 mm Type 85 machine gun and a German 7.62 mm MG3 machine gun.

The 'Armoured Fighting Car' is of conventional appearance with the engine compartment at the front and troop compartment at the rear. The latter has sloping sides and a sloping rear.

The commander and driver have a large oblong windscreen which can be covered by an armoured shutter hinged at the bottom.

All of these vehicles are understood to have been developed under the leadership of the state-owned Electrical and Mechanical Engineering Corps (EMEC).

Specifications

Not available.

Status

Production as required. In service with Myanmar Armed Forces. There are no known exports.

Manufacturer

State factories.

UPDATED

PAKISTAN

HIT Mohafiz Armoured Personnel Carrier/Internal Security Vehicle

Development

The Heavy Industries Taxila (HIT) facility is the largest facility of its type in Pakistan and is known mainly for its extensive experience in the overhaul and upgrade of tracked armoured fighting vehicles for the Pakistani Army.

More recently the HIT has played a key role in the development of the Al Khalid Main Battle Tank (MBT) which has been developed in association with NORINCO (China North Industries Corporation) with an initial batch of 15 vehicles now being built for the Pakistani Army.

Early in 2000, the HIT started design work on a new 4 × 4 Internal Security (IS) vehicle called the Mohafiz. This was first shown in mid-November 2000 and in December 2000 started extensive trials.

Possible customers for this vehicle are the Pakistani Rangers and Police. Some interest has already been shown by banks that would use the Mohafiz to transport bullion and cash.

So far one prototype of the Mohafiz IS vehicle has been completed in the armoured personnel carrier configuration and based on the chassis of a Toyota Land Cruiser (4 × 4) chassis. This chassis is already in service for a number of roles in the Pakistani Army and paramilitary units.

Description

The Mohafiz APC has an all-welded aluminium armour body that provides the occupants with protection from small arms fire and shell splinters. The fully enclosed engine compartment is at the front with the crew compartment extending to the rear.

The commander and driver are seated at the front and provided with bullet proof windows to their front and a forward opening door with a bullet proof window in the upper part either side. These bullet proof windows are covered by wire mesh screens.

The troop compartment is at the rear with the troops being seated either side on bench seats and enter the vehicle via a large door in the rear opening to the right.

Bullet proof windows are provided in the sides and rear of the troop compartment. A total of 10 firing ports are provided to allow rifles and other small arms to be used from within the vehicle. The upper parts of the hull slope slightly inwards as does the hull rear.

Mounted on the forward part of the roof is a manually operated one person turret armed with a 7.62 mm machine gun and/or a tear gas grenade launcher and a two part roof hatch which opens left and right.

To the immediate rear of the turret are two forward opening roof hatches although the design of the vehicle is flexible so that various types of roof hatch arrangement are possible.

Standard equipment for the Mohafiz includes run flat tyres, a public address system, powerful siren, powered steering and a searchlight operated from within the vehicle. Optional equipment includes an air conditioning system.

Specifications
Configuration: 4 × 4
Crew: 3 + 6
Gross vehicle weight: 3,500 kg
Length: 4.876 m
Width: 1.625 m
Height
(hull roof) 2.032 m
(turret) 2.463 m
Engine: 4.2 litre diesel
Transmission: manual

The prototype is powered by a 4.2 litre diesel engine coupled to a manual transmission with five forward and one reverse gears and a two speed transfer case. If required, a more powerful 4.5 litre diesel engine could be installed to provide a higher power-to-weight ratio.

Although the first example of the Mohafiz is based on a Toyota Land Cruiser chassis, it could also be built on other 4 × 4 chassis. This includes the Land Rover Defender 110 (4 × 4) which is already in service with the Pakistani Army in large numbers.

Variants

None announced although it can also be used, with some modifications in some cases, as an ambulance or command post vehicle.

Status

Prototype undergoing trials.

Manufacturer

Heavy Industries Taxila.

UPDATED

The Mohafiz internal security vehicle developed by HIT (Christopher F Foss) **2002**/0121207

The Mohafiz internal security vehicle developed by HIT (Christopher F Foss) **2002**/0121208

POLAND

Polish Army wheeled armoured personnel carrier requirement

The Polish Army has a requirement for up to 720 8 × 8 Armoured Personnel Carriers (APCs) which would be a foreign design manufactured under licence in Poland.

At present, the Polish Army uses mainly tracked APCs such as the Russian BMP-1 and a few of the more recent Russian BMP-2 and the Czechoslovakian OT-64 (8 × 8) APCs, which were developed and built in the 1960s.

Late in August 2001, the Polish Ministry of Defence issued a formal Invitation To Tender (ITT) to five Polish defence contractors for a new 8 × 8 APC. Additional background to the competition was given in *Jane's Armour and Artillery 2001-2002.*

It is possible that the vehicles selected will be tested in Poland between March and June 2002 with the first vehicles to be delivered by 2003, but it remains to be seen if this objective can be achieved.

The first batch includes 120 vehicles equipped with a two-person turret armed with a 25 /30 mm cannon and 7.62 mm coaxial machine gun and 100 specialised variants such as command vehicles, mortar carriers and ambulances.

The five Polish companies receiving this ITT were:

- Bumar Labedy
- OBRUM
- Huta Stalowa Wola
- Wojskowe Zaklady Motoryzacyne (WZM)
- WZM-5

By the end of 2001, each contractor is required to submit a detailed bid together with details of a foreign partner.

Possible contenders for the Polish APC requirement include:

- GAZ (BTR-90) (Russia)
- Kharkov Morozov Machine Building Design Bureau (BTR-94) (Ukraine)
- MOWAG (Piranha III or IV) (Switzerland)
- Rheinmetall Landsysteme Fuchs 2 (Germany)
- Satory Military Vehicles (VBCI) (France)
- Singapore Technologies Kinetics (AV 81)
- Steyr-Daimler-Puch (Pandur II) (Austria)

Status

As of early 2001, the Polish Army had not made a final selection on its future APC.

UPDATED

Austrian Steyr-Daimler-Puch Pandur (6 × 6) APC in its amphibious version (Steven Zaloga) ***2000***/0081984

Swiss MOWAG Piranha (8 × 8) APC is being marketed in Poland (Steven Zaloga) ***2000***/0081985

ROMANIA

ROMARM B33 armoured personnel carrier

Development

For some years Romania built a slightly modified version of the Russian BTR-70 (8 × 8) amphibious Armoured Personnel Carrier (APC) called the TAB-77 and details of this are given in the following entry.

The TAB-77 was followed in production by the B33 that is essentially the Romanian version of the Russian BTR-80 (8 × 8) APC with some minor modifications to meet Romanian requirements, including the installation of a different diesel engine.

Description

The hull and turret of the B33 (8 × 8) APC is of all-welded steel armour construction that provides the occupants with protection from 7.62 mm small arms fire and shell splinters.

Over the frontal arc this provides protection against 12.7 mm × 107 B43 rounds fired from a range of 100 m. Over the remainder of the vehicle protection is provided against 7.62 mm × 39 rounds fired from a range of 20 m. The turret is provided with protection against 7.62 mm × 39 rounds fired from 20 m.

The driver is seated at the front of the vehicle on the left side and is provided with a single-piece hatch cover above his position that opens to the right. Arranged around the forward part of his hatch are four day peiscopes.

The commander is seated to the right of the driver and has a single-piece hatch cover that opens forward and can be locked in the vertical position if required. He is also provided with four periscopes and a roof-mounted searchlight that can be operated from within the vehicle.

The commander and driver are each provided with a bullet-proof window fitted with a wiper to their immediate front, which can be covered when in a combat area by a shutter hinged at the top. To the right of the commander's windscreen is a firing port.

ROMARM B33 (8 × 8) armoured personnel carrier with snorkel erected to rear of turret for amphibious operations ***2002***/0081986

The one-person turret is mounted on the roof to the rear of the comander and driver and is armed with a 14.5 mm KPVT machine gun with a 7.62 mm PKT machine gun mounted coaxial.

Mounted coaxially above the two machine guns is an infra-red searchlight and mounted on the turret rear is a bank of six electrically operated 81 mm smoke grenade launchers that fire over the frontal arc.

The troop compartment is in the centre of the vehicle with the troops entering and leaving via a door in the side of the hull between the second and third roadwheels.

The lower part of the door folds down to form a step, while the upper part opens to the front. The troop compartment is provided with roof hatches and firing ports with associated vision devices.

Each of the roof hatches has a single circular firing port. The six to eight infantrymen are seated three or four down each side facing outwards on bench seats that are over the transmission and gear divided box.

The engine compartment is at the rear of the hull with the air inlet and outlet louvres in the roof and the exhaust outlets on either side.

The B33 is fully amphibious with the minimum of preparation which includes erecting the trim vane which is folded back onto the glacis plate when not in use, switching on the bilge pumps and extending the snorkel to the turret rear. The vehicle is propelled in the water by a single water jet mounted at the rear of the hull.

Standard equipment on the B33 includes powered steering on the front four wheels with all tyres being fitted with a central tyre pressure regulation system. This allows the driver to adjust the tyre pressure to suit the type of terrain being crossed.

Mounted at the front of the B33 is a mechanically-operated winch driven from the transfer box with 60 m of cable and a force of 4,500/6,000 daN. A VHF frequency hopping radio with a range of 20 to 50 km is fitted plus a vehicle intercom system. An automatic fire suppression system is also fitted.

Specifications

Crew: 3 + 6/8
Configuration: 8 × 8
Combat weight: 17.85 hp/tonne
(with crew) 14,000 kg
(without crew) 13,000 kg
Power-to-weight ratio: 17.85 hp/t
Length: 7.65 m
Width: 2.90 m
Height: 2.35 m
Ground clearance: 0.475 m
Track: 2.46 m
Wheelbase: 4.4 m (first to last axle)
Angle of approach/departure: n/avail
Max road speed: 80 km/h
Max water speed: 9 km/h
Fuel capacity: 300 litres (estimated)
Road range: 300 km (estimated)
Fording: amphibious
Gradient: 60%
Side slope: 40%
Vertical obstacle: 0.50 m
Trench: 2 m
Turning radius: 13.2 m
Engine: Type 1240-V8, V-8 diesel developing 250 hp at 2,600 rpm
Transmission: manual
Transfer box: 2-speed
Steering: hydraulic, on first two axles
Tyres: 13.00 × 18 with central tyre inflation system
Suspension: independent at each wheel station with torsion bars and hydraulic double-action shock-absorbers
Brakes:
(main) dual, drum and floating shoe type on each wheel, power-assisted
(parking) drum, outer shaft of gear box, mechanically actuating
Electrical system: 24 V
Batteries: 1 × 12 V @ 143 Ah
Armament:
(main) 1 × 14.5 mm KPVT machine gun
(coaxial) 1 × 7.62 mm PKT machine gun
(smoke grenade launchers) 6 × 81 mm
Ammunition:
(14.5 mm) 500
(7.62 mm) 2,000

Gun-control equipment: manual
Turret traverse: 360°
NBC system: yes
Night vision equipment: yes

Variants

According to the manufacturer, the B33 can be fitted with alternative weapon stations from France (Giat), Germany (Rheinmetall Landsysteme) and Italy (Otobreda) armed with various 25 to 30 mm stabilised cannnon.

Status

Production as required. In service with Romania.

Manufacturer

ROMARM SA.

NEW ENTRY

ROMARM TAB-77 armoured personnel carrier

Development/Description

For some years the ROMARM company, previously known at RATMIL, has been building a modified version of the Russian BTR-70 (8 × 8) armoured personnel carrier under licence. This has the Romanian designation of the TAB-77 and differs from the BTR-70 in a number of areas.

The most significant difference is that the Romanian TAB-77 is powered by two diesel engines whereas the Russian BTR-70 has two petrol engines.

The all-welded steel hull of the TAB-77 provides the crew with protection from 7.62 mm small arms fire fired from a range of 100 m.

The layout of the TAB-77 is similar to the BTR-70, with the driver seated at the front left with the vehicle commander to his right. Both have forward-opening hatch covers that can be locked in the vertical position.

Both these crew members are provided with a windscreen to their front which, when the vehicle is in a combat area, are covered by a hinged shutter, forward observation then being by roof-mounted day periscopes with additional observation devices being provided to give observation to the sides. The commander also has a roof-mounted searchlight operated from within the vehicle.

TAB-77 (8 × 8) armoured personnel carrier

The one-man manually operated turret is mounted in the centre of the hull roof and, like that fitted to the BTR-60/BTR-70 vehicles, is not fitted with a hatch cover. Armament comprises one 14.5 mm KPVT and one PKT 7.62 mm machine gun with the day sight being on the left side of the turret. Turret traverse and weapon elevation is manual.

The troop compartment is in the centre and has two roof hatches with the troops normally entering and leaving the vehicle via a small forward-opening door situated between the second and third roadwheel stations.

The engine compartment is located at the rear of the vehicle with a single exhaust pipe being located either side of the hull at the rear.

The TAB-77 is fully amphibious, being propelled in the water by a single water-jet situated in the rear of the hull. Before entering the water the bilge pumps are switched on and the trim vane is hydraulically erected by the driver from his seat. When travelling on land this is stowed retracted on the glacis plate.

Standard equipment includes an automatic fire detection and suppression system, front-mounted winch with a capacity of 5,500 kg, an NBC system, infra-red night vision equipment, power steering on the front four wheels, a central tyre-pressure regulation system, that allows the driver to adjust the tyre pressure from his seat while the vehicle is in motion, and an engine preheater.

Variants

TAB-77 PCOMA

This is an armoured artillery command post vehicle, with the turret removed and a larger cupola with a hatch cover fitted that opens to the rear. The cupola contains an optical range-finder. Mounted above the commander's hatch at the front of the vehicle is a 7.62 mm PKMS machine gun. This version has additional but unspecified communications equipment.

Specifications

TAB-77
Crew: 2 + 9
Configuration: 8 × 8
Combat weight:
(with crew) 13,350 kg
(without crew) 12,250 kg
Power-to-weight ratio: 19.77 hp/t
Length: 7.42 m
Width: 2.95 m
Height:
(without turret) 1.92 m
Ground clearance: 0.525 m
Track: 2.39 m
Wheelbase: 4.392 m
Angle of approach/departure: 41/30°
Max road speed: 83 km/h
Max water speed: 10 km/h
Fuel capacity: 290 litres
Road range: 550 km
Fording: amphibious
Gradient: 60%
Side slope: 30%
Vertical obstacle: 0.50 m
Trench: 2 m
Turning radius: 12 m
Engines: 2 × Model 797-05M1 diesel developing 132 hp at 3,000 rpm each
Transmission: manual
Armament:
(main) 1 × 14.5 mm KPVT MG
(coaxial) 1 × 7.62 mm PKT MG
Ammunition:
(14.5 mm) 600
(7.62 mm) 2,500

Gun control equipment: manual
Turret traverse: 360°
NBC system: yes
Night vision equipment: yes

TAB-77A 145M

This is a command post vehicle which has a new low-profile turret armed with a 14.5 mm KPVT machine gun and additional communications equipment. There are eight antennas including a pole-type mast on the right side of the roof. Mounted externally at the rear of the vehicle are two large stowage boxes. There are, in fact, at least two versions of the TAB-77A command post vehicle. The TAB-77R-1451/M (or R-1450M) is similar to the basic vehicle but is equipped with additional antennas and one or two power generators at the rear of the vehicle. The TAB-77R-1452 is equipped with the mentioned low profile turret but only has a dummy machine gun. At the rear are the same generators and on the roof are mounted seven whip antennas and a single telescopic antenna.

TERA-77L

This is a dedicated maintenance and recovery vehicle and is fitted with a hydraulic crane which is mounted on the roof and traversed to the rear when not required, a front-mounted dozer/stabiliser blade and a winch.

B33 (8 × 8) APC

Although originally believed to be an improved version of the TAB-77, which is based on the Russian BTR-70 APC, it is now known that the B33 is, in fact, based on the more recent Russian BTR-80 APC and is in service with Romania under the name TAB Sumbra. Full details of the B33 armoured personnel carriers are given in the previous entry.

RN-94 (6 × 6) APC

This 6 × 6 APC is a joint development between Romania and Turkey and details are given in this section under International. This has been evaluated by the Turkish Land Forces Command but, as of late 2001, no production orders had been placed.

Status

Production probably complete. In service with the Romanian Army. In 1999, Romania stated that it had a total of 167 of these vehicles in service. As far as it is known there has been no production of this vehicle in recent years.

TERA-77I (8 × 8) recovery vehicle in travelling configuration (Paul Beaver) 0018651

Manufacturer

ROMARM SA.

UPDATED

ROMARM TAB-71 armoured personnel carrier

Development/Description

In the late 1960s, the Romanian Socialist Army (ASR) approached the Military Industrial Committee of COMECON with the request to licence produce the Russian BTR-60PB (8 × 8) APC. This request was granted and the first model, called TAB-71, entered production in 1970 and was first seen in public during a military parade in Bucharest held in August 1972. Production was originally undertaken by RATMIL but today this company is known as ROMARM.

The TAB-71 is very similar to the Russian BTR-60PB (8 × 8) armoured personnel carrier, but the two 90 hp petrol engines have been replaced in the TAB-71 by two 140 hp petrol engines.

The TAB-71 retains the central tyre-pressure regulation system, amphibious capability, front-mounted winch and NBC system of the Russian vehicle.

In addition to the hatches above the commander's and driver's position there is a door on either side of the hull above the second/third roadwheels and two rectangular roof hatches.

Production of the TAB-71 (8 × 8) APC and variants was completed some time ago. It was followed in production by the TAB-77 APC which is covered in detail in the previous entry.

TAB-71M (8 × 8) APC clearly shown sight on left side of turret (PB) **2002**/0104376

Specifications

Crew: 3 + 8
Configuration: 8 × 8
Combat weight: 11,000 kg
Power-to-weight ratio: 25.4 hp/t
Length: 7.22 m
Width: 2.83 m
Height:
(overall) 2.7 m
Ground clearance: 0.47 m
Wheelbase: 4.21 m
Max speed:
(road) 95 km/h
(water) 10 km/h
Fuel capacity: 290 litres
Max road range: 500 km
Fording: amphibious
Gradient: 60%
Vertical obstacle: 0.4 m
Trench: 2 m
Engines: 2 × V-6 liquid-cooled petrol developing 140 hp each
Transmission: manual, 5 forward and 1 reverse gears
Tyres: 13.00 × 18
Armament:
(main) 1 × 14.5 mm KPVT MG
(coaxial) 1 × 7.62 mm PKT MG
Ammunition:
(main) 500 rounds
(coaxial) 2,000 rounds

Gun control equipment
Turret power control: manual
Gun elevation/depression: +85/–5°
Turret traverse: 360°
Gun stabiliser:
(vertical) no
(horizontal) no
Armour
Hull (max): 9 mm
Turret (max): 7 mm
NBC system: yes
Night vision equipment: yes

Variants

TAB-71M

TAB-71 fitted with a locally designed one-man manually operated turret armed with one 14.5 mm KPVT and one 7.62 mm PKT machine gun with a distinctive day gunsight mounted on the left side of the turret. This turret is also fitted to the TAB-77 (8 × 8) and TABC-79 (4 × 4) APCs, and the full tracked MLVM Mountaineers Combat Vehicle. Details of this vehicle, which are only used by Romania, are given in the Armoured personnel carriers – tracked section. Production was completed some time ago.

TAB-71AR mortar carrier

This version has its turret removed and instead is fitted with an 82 mm mortar (the Russian-designed M-1937) that fires through the roof of the vehicle after opening the circular roof hatches. To the rear of the mortar hatches are two rectangular roof hatches, one on the right and one on the left, to the rear of which the 7.62 mm PKMS machine gun is mounted.

TERA-71 maintenance and recovery

This vehicle has been modified for the maintenance and recovery role with its turret removed, and mounted on the hull above the first and second wheel stations is a hydraulic crane that is traversed to the rear when not required.

This model has full overhead protection and mounted at the front of the vehicle, under the nose, is a hydraulic stabiliser which is lowered when the crane is being used. It is not known whether or not this version has a winch.

TERA-71L maintenance and recovery

This vehicle has been modified for the maintenance and recovery role, and has a different roof arrangement from the previous vehicle. The enclosed troop compartment is at the front, with an open compartment in the centre and the engine compartment at the rear.

The light hydraulic crane is normally traversed to the front when not in use. Armament comprises a 7.62 mm PKMS machine gun.

TAB-71A R 450 command vehicle

This has been modified for the armoured command role with additional communications equipment. In addition to the normal antenna at the front right side, it has a second antenna on the left side of the hull roof.

TAB-71A R 451 command vehicle

This is modified for the armoured command role and has a different turret armed with a 7.62 mm PKT machine gun, a box (which may be for a generator) on the right side of the roof and four antennas (one of these being a pole type) on the right side of the hull roof.

TAB-71A R1 452 command vehicle

This is modified for the armoured command role with additional communications equipment and externally is similar to the previous model.

Status

Production complete. In service with:

Country	Quantity	Comment
Moldovia	110	delivered 1994/1995
Romania	965	Army
	172	Naval Infantry
	33	Border Guard
Yugoslavia	n/a	small number

Manufacturer

ROMARM.

UPDATED

ROMARM TABC-79 armoured personnel carrier

Development/Description

The TABC-79 (4 × 4) armoured personnel carrier has been developed in Romania and uses some automotive components of the TAB-77 (8 × 8) armoured personnel carrier. The TABC-79 was originally designed and built by RATMIL but this company is now called ROMARM.

Although classed as an armoured personnel carrier, its correct Romanian designation is the TABC-79 reconnaissance armoured personnel carrier.

It has a hull of all-welded steel construction that provides protection from small arms fire of up to 20 mm in calibre as well as shell splinters.

The commander and driver are seated at the front of the vehicle, with the driver on the left and the vehicle commander on the right, both having a single-piece hatch cover that opens forward and four PF2 day periscopes for observation to the front and sides, while the commander has an AON1 infra-red searchlight that is mounted on the vehicle's roof and can be operated from inside. To the immediate front of the commander and driver is a windscreen that can quickly be covered by an armoured shutter when in a combat area.

The one-man manually operated turret is in the centre of the roof and is the same as that installed on the TAB-71M (8 × 8) and TAB-77 (8 × 8) armoured personnel carriers as well as the full tracked MLVM Mountaineers Combat Vehicle. This turret is armed with one 14.5 mm KPVT machine gun and one 7.62 mm PKT machine gun. Turret traverse and weapon elevation is manual.

The engine compartment is located at the rear of the TABC-79 in the centre, with the air inlet and outlet louvres in the roof and the exhaust pipe at the left rear. There is a single door in either side at the rear that opens outwards, with the right door having firing port.

There is also a hull entry door in either side of the hull, as well as a single roof hatch to the right of the engine compartment.

The TABC-79 is provided with firing ports and periscopes with the latter including turntable-mounted roof periscopes.

The TABC-79 is fully amphibious, being propelled in the water by a single water-jet situated in the rear of the hull. Before entering the water the bilge pumps are switched on and the trim vane is hydraulically erected by the driver from his seat. When travelling on land this is stowed retracted on the glacis plate.

Standard equipment includes an automatic fire detection and suppression system, a front-mounted winch with 50 m of cable and a capacity of 5,500 kg, an NBC system, infra-red night vision equipment, a central tyre-pressure regulation system that allows the driver to adjust the tyre pressure from his seat while the vehicle is in motion, and an engine preheater.

More recently the vehicle has been referred to by Romanian sources as the ABC-79M 4 × 4 armoured reconnaissance vehicle. It is understood that there has been no recent production of this family of vehicles.

Specifications

Crew: 3 + 4
Configuration: 4 × 4
Combat weight:
(with crew) 9,275 kg
(without crew) 8,575 kg
Power-to-weight ratio: 16.66 hp/t
Length: 5.65 m
Width: 2.805 m
Height:
(with turret) 2.335 m
Ground clearance: 0.485 m
Track: 2.25 m
Wheelbase: 2.9 m
Angle of approach: 48°
Max road speed: 80 km/h
Max water speed: 8 km/h
Range: 700 km
Fording: amphibious
Gradient: 60%
Side slope: 30%
Vertical obstacle: 0.30 m
Trench: 0.70 m
Turning radius: 10 m
Engine: Saviem 798.05N2 6-cylinder in-line 4-stroke turbocharged diesel developing 154 hp at 2,800 rpm
Transmission: manual with 4 forward and 1 reverse gears (with power take off for water jet system)
Transfer box: 2 speed
Steering: mechanical, hydraulic assisted on first axle
Tyres: 14.00 × 20
Suspension: independent with torsion rods and hydraulic damper and hydraulic telescopic shock absorbers
Brakes:
(main) hydraulic
(parking) mechanical
Electrical system: 24 V
Batteries: 2 × 12 V × 88 Ah
Armament:
(main) 1 × 14.5 mm KPVT MG
(coaxial) 1 × 7.62 mm PKT MG
Ammunition:
(14.5 mm) 500
(7.62 mm) 2,000
Gun control equipment: manual
Turret traverse: 360°
Gun elevation/depression: +80/–4°
Gun stabiliser:
(vertical) no
(horizontal) no
NBC system: yes
Night vision equipment: yes

Variants

TAB-C APC

In 1997, Romania started to market the TAB-C (4 × 4) vehicle but all information released to date is identical to that of the TABC-79 and the vehicles are therefore probably identical. Another version of this vehicle is called the TAB-CG.

AM 425 APC

Romania has recently revealed another variant called the AM 425 although this may well relate to one of the versions below or a more recent model of the TAB-79. Combat weight is 9,000 kg.

TAB-79A POMA artillery observation post carrier

This has a similar cupola to that installed on the TAB-77 PCOMA (8 × 8) vehicle which is mounted on the roof to the rear of the commander's and driver's hatches, with the 7.62 mm PKMS machine gun being mounted externally over the rectangular roof hatch to the right of the engine compartment.

TAB-79AR mortar carrier

This is a self-propelled 82 mm mortar system. To the rear of the commander's and driver's roof hatches is a two-part circular hatch cover that opens left and right enabling the 82 mm Model 1937 mortar to be used.

A 7.62 mm PKMS machine gun is installed on the circular roof hatch at the rear. A baseplate is carried at the rear of the hull on the left side enabling the mortar to be deployed away from the vehicle if required by the tactical situation.

TAB RCH-84

This is an armoured chemical and radiological reconnaissance vehicle and is fitted with chemical and

Romanian TABC-79 (4 × 4) armoured personnel carrier (PB) **2001**/0104377

Romanian ABC-79M (4 × 4) armoured vehicle with roof-mounted 7.62 mm machine gun **2000**/0081987

radiological detection systems with racks of pennants mounted either side at the hull rear to mark contaminated areas.

On the roof, to the rear of the commander's and driver's roof hatches, is another circular roof hatch with an externally mounted 7.62 mm PKMS machine gun. For the export market this has also been referred to as the TAB-CG (CA-SG).

TCG-80

This is a turretless specialist vehicle based on the TABC-79 chassis and is armed with a 7.62 mm PK series machine gun. It is probably used for engineer reconnaissance.

Self-propelled SAM system

A modified version of the TABC-79 (4 × 4) armoured personnel carrier chassis is used as the basis for the Romanian A95 short-range air defence system.

The one-man turret is mounted on top of the hull and mounted either side is a bank of two passive infra-red fire and forget surface-to-air missiles in the ready to launch position. Details of this system are given in *Jane's Land-Based Air Defence 2001-2002.* This system is also referred to as the CA-95. As far as it is known, it has not been exported.

Status

Production probably complete. In service with the Romanian Army. In 1999, Romania stated that it had a total of 394 (army) and 36 (naval infantry) TABC-79 vehicles in service. In addition, there are many variants in service.

Manufacturer

ROMARM.

UPDATED

RUSSIAN FEDERATION AND ASSOCIATED STATES (CIS)

BTR-90 (GAZ-5923) armoured personnel carrier

Development

During the defence equipment exhibition held in Nizhny Novgorod, Russia, late in 1994, the GAZ (Gorky Automobile Plant) briefly showed the prototype of a new 8 × 8 armoured personnel carrier called the Obiekt 51, or GAZ-5923.

The BTR-90 was designed from 1993, at the Arzamas Machinery Construction Factory with the all-welded steel armour hull being built at the Chelyabinsk Factory for Heavy Machinery Construction. This provides a higher level of protection, when compared to the earlier BTR-80 (8 × 8) APC.

The Chelyabinsk Tractor Factory supplied the diesel engine while the Ural Truck Factory supplied the tyres. The first prototype was completed in early 1994.

The BTR-90, is virtually a scaled-up version of the BTR-80 and is fitted with the complete turret of the tracked Kurgan Machine Construction Plant BMP-2 infantry combat vehicle.

At the present time it is not clear whether the BTR-90 has been developed to meet the requirements of the Russian Army or aimed mainly at the export market. Recent information has indicated that a small quantity of these vehicles has been supplied for Ministry of Interior troops. The BTR-90 has the nickname of the Rostock.

Description

The hull of the BTR-90 is of all-welded steel armour construction that provides protection from small arms fire and shell splinters. It is believed that over the frontal arc protection is provided against penetration from 14.5 mm projectiles. Recent information has indicated that production versions of the BTR-90 have a higher level of protection than the prototypes.

Whereas the BTR-80 hull has a flat bottom, that of the BTR-90 has a flatter V shape to provide a higher level of protection against mines.

The layout is very similar to the BTR-80, with the driver and one person at the front, turret in the centre and engine compartment at the rear.

The driver seated on the left has a total of five day periscopes which give observation through about 180°, while the infantryman seated to his right has two day periscopes. The driver's seat can be raised when he is driving in the head-out position.

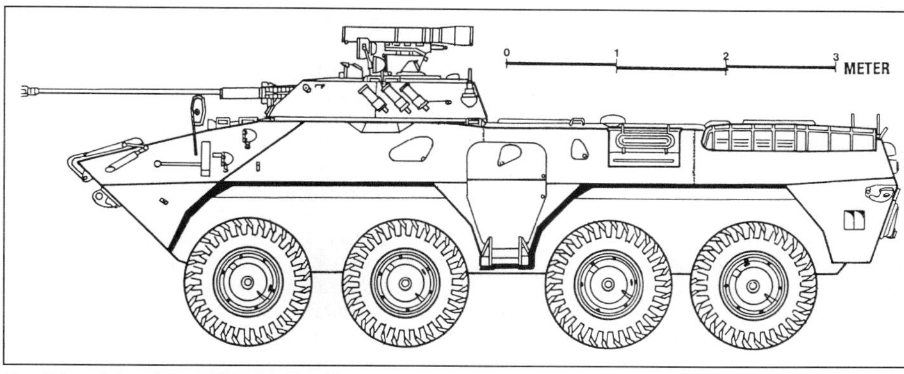

Provisional drawing of BTR-90 (8 × 8) armoured personnel carrier (Steven Zaloga)

The driver has a single-piece hatch cover that lifts and swings to the left while the infantryman to the right has a single-piece hatch cover that opens to the rear.

As with the BTR-80, the only means of entry and exit is via the roof hatches or the two-part door between the second and third roadwheels, the lower part of which folds downwards to form a ramp while the upper part, with an integral firing port, hinges upwards. Additional firing ports are provided to the front and rear of the side door.

Mounted on top of the hull is the complete turret of the BMP-2 infantry combat vehicle, which is described in detail in the entry for this in the Armoured personnel carriers — tracked section.

The fire-control system includes the BPKZ-42 gunner's combined day/night sight and the 1P-13 commander's optical sight. If required, the BPK-M gunner's sight with thermal imaging module supplied by SAT of France can be fitted.

The turret is armed with a stabilised 30 mm 2A42 cannon, a 7.62 mm PKT machine gun and a roof-mounted AT-5 'Spandrel' ATGW system which has a maximum range of 4,000 m. Early versions had a single HEAT warhead but the latest versions have a tandem HEAT warhead. The latest version of the BTR-90 has a 30 mm AGS-17 grenade launcher mounted externally on the left side of the turret. Internally, two man-portable SAM are carried. A standard tripod is carried to enable the AT-5 "Spandrel" ATGW system to be deployed away from the vehicle, if required by the tactical situation.

The troop compartment is at the rear with two small rectangular roof hatches being located to the rear of the turret, both of which are hinged on the inside and can be locked in the vertical position. The rear troop compartment is provided with four roof-mounted day periscopes, two either side.

The roof hatches allow infantry to fire other weapons, for example, surface-to-air missiles and grenade launchers, with some level of protection.

Between eight and 10 infantrymen can be seated in the rear troop compartment.

Standard equipment includes power-assisted steering on the front four wheels, a central tyre-pressure regulation system that allows the driver to adjust the tyre pressure to suit the terrain being crossed, an NBC system and passive night vision equipment for the commander, gunner and driver.

The BTR-90 is fully amphibious, being propelled in the water by two water-jets mounted at the rear of the hull. Before entering the water a trim vane is erected at the front of the vehicle (when this is not required it lies back on the glacis plate) and the bilge pumps are switched on.

Standard communications equipment on Russian BTR-90 vehicles includes the R-163-50U radio, R-163UP receiver and the R-174 vehicle intercom system.

Optional equipment for the BTR-90 APC includes an air conditioning and cooling unit, to allow the vehicle and its crew to remain effective in a high ambient temperature. An Onboard Information System (OICS) is fitted, to allow the driver to monitor the engine and transmission.

To provide a higher level of battlefield survivability, the BTR-90 can be fitted with additional armour, including explosive reactive armour, as well as an active protection system.

BTR-90 (8 × 8) armoured personnel carrier in recent configuration with 30 mm grenade launcher on left side of turret **2000**/0081988

BTR-90 (8 × 8) armoured personnel carrier from the rear

Specifications
(Provisional)
Crew: 3 + 7
Configuration: 8 × 8
Combat weight: 20,920 kg
Power-to-weight ratio: 24.37 hp/t
Length: 7.64 m
Width: 3.20 m
Height: 2.975 m
Ground clearance: 0.510 m
Max road speed: 100 km/h
Max water speed: 9 km/h
Range: 800 km
Fording: amphibious
Gradient: 60%
Side slope: 30%
Vertical obstacle: 0.8 m
Trench: 2 m
Turning radius:
(front wheels) 14 m
(all wheels) 6 m
Engine: turbocharged multifuel developing 510 hp
Transmission: automatic reversible hydraulic
Suspension: torsion bar suspended from A-frame mounts, dual hydraulic shock-absorbers on first and last roadwheel stations and single hydraulic shock-absorbers on second and third wheel stations
Steering: power assisted, front four wheels
Tyres: 14.00 × 20
Armament:
(main) 1 × 30 mm 2A42 cannon
(coaxial) 1 × 7.62 mm PKT machine gun
(ATGW) 1 × AT-5 Spandrel
1 × 30 mm AGS-17 grenade launcher
Ammunition:
(main) 500
(coaxial) 2,000
(ATGW) 4
(grenade launcher) 400
Grenade launchers: 2 × 3 81 mm

Gun control equipment
Turret power control: electric/manual
(by commander) yes
(by gunner) yes
Gun elevation/depression: +75/–5°
Turret traverse: 360°
Gun stabiliser:
(vertical) yes
(horizontal) yes
NBC system: yes
Night vision equipment: yes

Russian BTR-90 (8 × 8) APC fitted with advanced turret armed with 100 mm gun, 30 mm cannon and 7.62 mm coaxial machine gun and advanced day/night fire-control system (Christopher F Foss)
2002/0095242

Variants

BTR-90 with 100 mm turret

In 2001 the Russian KBP Instrument Design Bureau launched their latest turret system installed on a modified BTR-90 (8 × 8) series (APC) chassis.

This turret retains the same armament as the turret fitted to the BMP-3 Infantry Combat Vehicle (ICV) in service with a number of export customers including Cyprus, Korea (South), Kuwait and the United Arab Emirates. This consists of a full stabilised 100 mm 2A70 gun, 30 mm 2A72 cannon and 7.62 mm PKT machine gun.

The actual turret is of a new design with additional armour protection, enhanced ammunition as well as having a much more advanced computerised day/night fire-control system. Due to the depth of the turret basket, the hull of the BTR-90 has been raised to the immediate rear of the driver's compartment at the front of the vehicle.

This version of the BTR-90 APC can carry 10 people, driver and one person at the front, commander and gunner in turret and six troops in the rear. It retains the amphibious characteristics of the BTR-90 being propelled in the water by two water jets at the rear.

While the BMP-3 armament is retained, the 100 mm rifled gun can fire a new 100 mm laser-guided projectile with a tandem High Explosive Anti-Tank (HEAT) warhead, which will penetrate explosive reactive armour out to a maximum range of 5,500 m.

The original missile has a maximum range of only 4,000 m. A 100 mm new high-explosive fragment (HE-FRAG) round is also available. Four guided projectiles are now carried plus four in reserve, with 34 rounds of ready use 100 mm HE-FRAG ammunition and an additional 18 rounds in reserve. A new Armour Piercing Discarding Sabot (APDS) 30 mm round is also provided with enhanced penetration characteristics.

The advanced computerised fire-control system includes a ballistic computer with the gunner having a roof-mounted stabilised sight with day/thermal channels, laser range-finder, laser guidance system and an auto tracker. The gunner has a panoramic sight with a TV channel, laser range-finder and a display of the gunner's sight. This sighting system enables hunter/killer target engagements to take place.

As well as providing brand new turrets, the KBP Instrument Design Bureau can also offer three upgrade packages to enable existing users to upgrade their turrets. The user can select just the new ammunition, new fire-control system or a complete turret upgrade.

Other variants

The manufacturer has also proposed that the chassis could be used as the basis for an anti-tank guided missile complex.

Status

Production as required. Small batch built for specialised use.

Manufacturer

Arzamas Machinery Construction Plant.

UPDATED

BTR-80A armoured personnel carrier

Development

The BTR-80 (8 × 8) APC, covered in detail in the following entry, entered service with the Soviet Army in the mid-1980s and is a further development of the BTR-70 (8 × 8) which in turn is a development of the even older BTR-60 (8 × 8).

The main armament of the BTR-80 is a one-man manually operated turret armed with a 14.5 mm KPVT and a 7.62 mm PKTM coaxial machine gun, which in some operational situations is no longer adequate.

Further development of the BTR-80 has resulted in the latest BTR-80A which was first seen in Nizhny Novgorod in September 1994. Prime contractor for the BTR-80A is the Arzamas Machinery Construction Plant.

The first five BTR-80A vehicles were delivered to the Main Directorate for Presidential Security based in Moscow and these were seen in public for the first time in May 1995, when they defended key government buildings. The BTR-80A has also undertaken firepower and mobility demonstrations in Abu Dhabi. By early 2000 the only known export customer is Hungary and, between 1996 and 1998, Russia supplied Hungary with 439 BTR-80 series APCs. As of late 2001, known export customers for the BTR-80A included Hungary, Korea (North and South) and the Sudan.

Description

The BTR-80A is virtually the same as the BTR-80 but has a new one-person turret armed with a 30 mm 2A72 cannon (also installed in the tracked BMP-3 infantry fighting vehicle) with a 7.62 mm PKTM machine gun being mounted coaxially to the right. Either side of the 30 mm cannon is a bank of three forward-firing electrically operated 81 mm smoke grenade launchers.

According to Russian sources, the 30 mm cannon has an effective range of 2,000 m when firing armour-piercing tracer projectiles and 4,000 m when firing high-explosive incendiary projectiles. Maximum range of aimed 7.62 mm machine gun fire is quoted as 1,500 m.

The power-operated turret is mounted on the roof of the vehicle above the front two axles with the armament being mounted on the upper part of the turret which pivots for elevation/depression.

The sighting system is mounted coaxially with the weapons and a searchlight is mounted coaxially above the armament.

The day sight is designated the 1PZ-9 and has magnifications of ×1.2 and ×4 while the night sight is designated the TPN-3 and has a magnification of ×5.5. It can also be fitted with the TPN-3-42 Kristall night sight which enables targets to be engaged out to a range of 900 m.

The commander and driver are seated at the front, as in the BTR-80, with the troop compartment in the centre. The troops are seated back-to-back and enter and leave the vehicle via a door in either side of the hull between the second and third axles.

The lower part of the door folds down to form a step while the upper part, which contains a firing port and associated vision device, opens forwards.

There are an additional two firing ports/vision devices either side of the troop compartment plus roof-mounted day periscopes. There are also two roof hatches that hinge upwards and each of these has a firing port.

The engine compartment is at the rear of the vehicle. Until 1993 the vehicle was powered by a V-8 KamAZ-7403 turbocharged diesel developing 260 hp but the vehicle is now offered with a YaMZ-238M2 diesel which develops 240 hp.

Standard equipment on the BTR-80A includes power steering on the front four roadwheels, central tyre-pressure regulation system that allows the driver to adjust the tyre pressure to suit the type of ground being crossed and an NBC system. A self-recovery winch is provided, with 50 m of cable. This has a rated pull of 6 tonnes which can be increased to 12 tonnes with a pulley block.

It is fully amphibious, being propelled in the water by a single water-jet mounted at the rear of the hull. Before entering the water a trim vane is erected at the front of the hull and the bilge pumps are switched on.

Russian BTR-80A vehicles are fitted with a R-4163-50U UHF radio set, R-163UP UHF receiver and R-174 intercom.

Improvements to the hull include increased rigidity, removal of one firing port, different seating arrangements and revised hull stowage.

For trials purposes, a BTR-80 APC has been fitted with a Western Cummins model 6CTA 8.3-C250 diesel engine coupled to a ZF 9S-75 automatic transmission

BTR-80A (8 × 8) APC well camouflaged but clearly showing crew entry/exit door in side of hull (PB) **2001**/0104380

BTR-80A (8 × 8) APC of the Hungarian Army (Michael Jerchel) **2001**/0104383

Specifications

Crew: 2 + 8
Configuration: 8 × 8
Combat weight: 14,550 kg
Power-to-weight ratio: 17.9 hp/t
Length: 7.65 m
Width: 2.90 m
Height: (overall) 2.80 m
Ground clearance: 0.475 m
Track: 2.41 m
Wheelbase: 4.40 m (1st to 4th axle)
Max road speed: 90 km/h
Max water speed: 10 km/h
Range: 600-800 km
Fuel capacity: 300 litres
Vertical obstacle: 0.5 m
Trench: 2.0 m
Gradient: 60%
Side slope: 42%
Fording: amphibious
Engine: V-8 KamAZ-7403 V-8 turbocharged diesel developing 260 hp or, from 1993, YaMZ-238M2 V-8 turbocharged diesel developing 240 hp
Transmission: manual with 5 forward and 1 reverse gears
Transfer case: 2 speed
Steering: power assisted
Tyres: 11.50 × 400-457 (tubeless)
Suspension: torsion bar with hydraulic shock-absorbers
Brakes: hydraulic
Electrical system: 24 V
Armament:
(main) 1 × 30 mm 2A72 cannon
(coaxial) 1 × 7.62 mm PKTM machine gun
(smoke grenade launchers) 6 × 81 mm
Ammunition:
(30 mm) 300
(7.62 mm) 2,000
Weapon elevation/depression: +70/–5°
Turret traverse: 360°
Gun stabiliser:
(vertical) no
(horizontal) no
NBC system: yes
Night vision equipment: yes

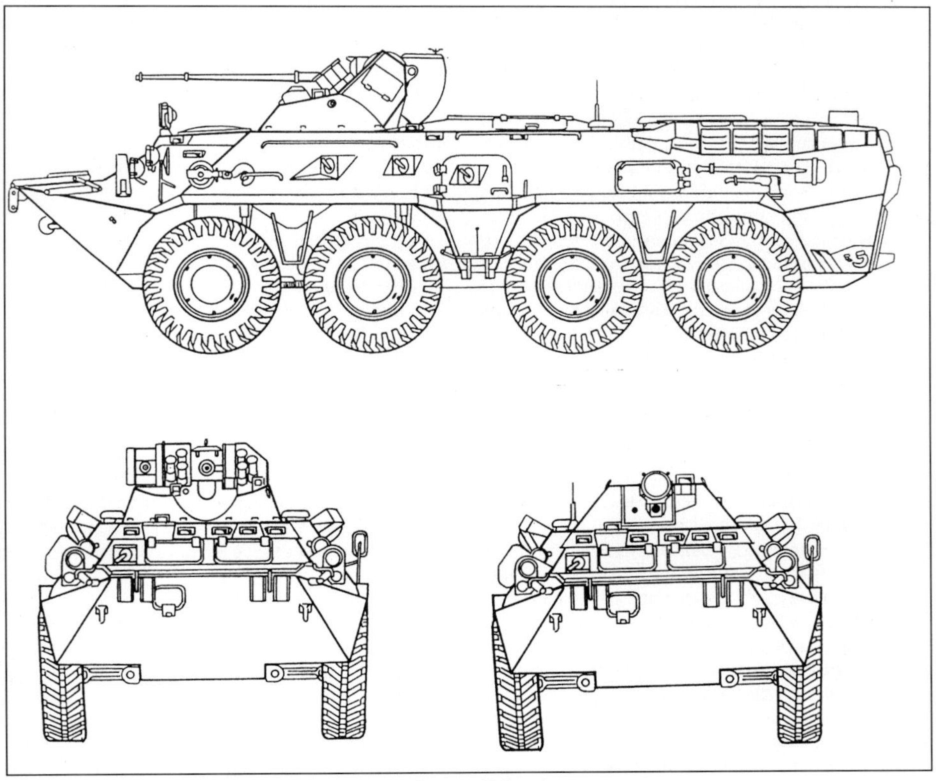

BTR-80A (8 × 8) armoured personnel carrier (Steven Zaloga)

and this is also being offered for the more recent BTR-80A. Additional details are given in the following entry.

Variants

The only known variant is the BTR-80S which has the 30 mm 2A72 cannon replaced by a 14.5 mm KPVT heavy machine gun. This version has been designed for use by MVD Internal Troops.

The turret of the BTR-80A, which was designed by the Arzamas Machinery Construction Factory, has been designed as a direct replacement for the turrets currently fitted to the BRDM-2 (4 × 4) reconnaissance vehicle and the BTR-60PB, BTR-70 and BTR-80 (8 × 8) armoured personnel carriers, without any modification to the hull.

Status

Production as required. In service with the following countries:

Country	Quantity	Comment
Hungry	586	Delivered 1996/1999, including BTR-80A version
Korea, North	10	delivered 2000
Korea, South	10	delivered 2000
Russia	n/avail	
Sudan	60	delivered 2000

Manufacturer

Arzamas Machinery Construction Factory.

UPDATED

BTR-80 armoured personnel carrier

Development

While the BTR-70 (8 × 8) did have a number of improvements over the BTR-60 (8 × 8) introduced in 1960-61, it still suffered from a number of significant disadvantages including its poor means of entry and exit for troops and the fact that it was powered by two petrol engines which were not fuel efficient and could catch fire.

In 1984, the Soviet Army took delivery of its first production BTR-80 (8 × 8) armoured personnel carrier which features many improvements over the earlier vehicle.

From 1988, the BTR-80 APC started to replace the older BTR-60PB APCs used by the naval infantry.

The BTR-80 was designed and built at the Arzamas Machinery Construction Factory which was previously the Gorky Automobile Plant.

Description

Like the BTR-70 and BTR-60, the BTR-80 has a hull of all-welded steel armour construction. Its layout is similar to the earlier vehicles except that the two petrol engines have been replaced by a single V-8 diesel developing 260 hp, which gives a significant increase in the power-to-weight ratio.

This gives a slightly increased road speed and is more fuel efficient, giving a greater operating range as well as reducing the risk of fire.

The engine compartment has an automatic fire extinguishing system, a preheater, a fuel primer, heat exchangers for the radiator and crank case, diesel fuel tanks and a bilge pump.

Over the frontal arc the welded steel armour provides protection against attack from 12.7 mm AP rounds at a distance of 100 m, the upper part of the hull provides protection against 7.62 mm AP rounds at a range of 100 m while the lower part of the hull provides protection against 7.62 mm AP rounds fired from a range of 750 m.

Turkish BTR-80 (8 × 8) APC (Michael Jerchel) **2002**/0121206

BREM-K (8 × 8) ARV which is based on the BTR-80 (8 × 8) APC chassis (Christopher F Foss) 0018656

The one-man manually operated BPU-1 turret is similar to that fitted to the BTR-70 and BTR-60PB but the 14.5 mm KPV heavy machine gun can be elevated to +60° compared to the +30° of earlier vehicles, increasing its effectiveness against low-flying aircraft and helicopters. The standard 7.62 mm PKVT coaxial machine gun is retained. The turret also has an improved day sighting system (magnification of ×2 and ×4) and has been retrofitted to older vehicles such as the BRDM-2 (4 × 4) amphibious scout car and the BTR-70 APC.

Specifications
Crew: 3 + 7
Configuration: 8 × 8
Combat weight: 13,600 kg
Power-to-weight ratio: 19.1 hp/t
Length: 7.65 m
Width: 2.90 m
Height:
(loaded) 2.35 m
(unloaded) 2.46 m
Ground clearance: 0.475 m
Track: 2.41 m
Wheelbase: 4.4 m
Max road speed: 90 km/h
Max water speed: 9.5 km/h
Fuel capacity: 300 litres
Max road range: 600 km
Cross-country range: 223-480 km
Fording: amphibious
Gradient: 60%
Side slope: 42%
Vertical obstacle: 0.5 m
Trench: 2 m
Engine: V-8 KamAZ-7403 V-8 turbocharged diesel developing 260 hp or from 1993, YaMZ-238M2 V-8 turbocharged diesel developing 240 hp
Transmission: manual, 5 forward and 1 reverse gears
Transfer case: 2 speed
Steering: power assisted
Turning radius: 13.2 m
Tyres: 13.00 × 18 (tubeless)
Suspension: torsion bar with hydraulic shock-absorbers
Brakes:
(main) hydraulic on all wheels
(parking) mechanical
Electrical system: 24 V
Armament:
(main) 1 × 14.5 mm KPVT MG
(coaxial) 1 × 7.62 mm PKVT MG
(smoke grenade launchers) 6 × 81 mm
Ammunition:
(main) 500
(coaxial) 2,000

Gun control equipment
Turret traverse: manual, 360°
Gun elevation: +60/–4°
Gun stabiliser:
(vertical) no
(horizontal) no
NBC system: yes
Night vision equipment: yes

The turret for the BTR-80 vehicle was designed and built by the Arzamas Machinery Construction Factory.

Mounted on the rear of the turret is a bank of six 81 mm Model 902V forward-firing electrically operated smoke grenade dischargers. This is operated from within the turret and each smoke grenade can create a smoke screen up to 30 m wide and 10 m high depending on the strength and direction of the wind.

Late production BTR-80 vehicles have the OU-ZGA2M spotlight on the 14.5 mm machine gun frame and a new periscope on the left side of the turret.

A new hatch arrangement has been provided between the second and third axles, the upper part of which opens to the front while the lower part folds down to form a step. This allows the infantry to dismount from the BTR-80 much more quickly and with less exposure to enemy fire than with earlier vehicles. The BTR-80 has a crew of three, commander, gunner and driver and carries seven fully equipped infantry.

The BTR-70 has three firing ports in either side of the troop compartment, whereas the BTR-80 has three firing ports in the side of the hull angled to fire obliquely forward, so helping to cover the previously dead ground towards the front of the vehicle. There is also a single firing port to the right of the commander's bow position, which also appears on the BTR-70, and an additional firing port in each of the two roof hatches.

The two forward-firing ports are designed for the 7.62 mm PK general purpose machine guns while the three firing ports either side allow the AKMS/AK-74 weapons to be fired from under cover. Small arms carried by the crew are two 7.62 mm PK general purpose machine guns (with 3,000 rounds), eight 7.62 mm AKMS or 5.56 mm AK-74 assault rifles and nine Type F1 hand grenades.

In addition, the BTR-80 normally carries two SA-14/SA-16/SA-18 man-portable fire-and-forget surface-to-air missiles and a grenade launcher.

The commander and driver each have a total of four day vision devices while the gunner has three. The BTR-80 is fully amphibious, and an overpressure NBC system, night vision equipment and a central tyre-pressure regulation system are fitted.

Before entering the water the trim vane is erected at the front of the vehicle, the drain valves are closed and locked, the tyre pressure is lowered and the cover over the water-jet system is opened. If the waves are over 0.5 m, air intake extension pipes are provided and stored low in the rear of the engine compartment. Low gear is then selected, the transfer case locked in all-wheel drive and the dust cover over the air intake filter is closed. Also shut and closed are the louvres over the air intake and exhaust for the radiators. The BTR-80 drives into the water and once afloat, wheel drive is disengaged and water-jet drive engaged. Steering is accomplished by means of a steering wheel controlling the rudder, which in turn steers the vehicle.

Standard equipment on all BTR-80 vehicles includes a DP-3B radiation reconnaissance device, a VPX-R chemical radiation reconnaissance device and the P-123M short-range radio system.

The BTR-80 is also fitted with a crew compartment heater, engine preheater and a front-mounted power-operated winch which is provided with 50 m of cable and has a capacity of six tonnes which can be increased with the aid of a pulley to 12 tonnes. A fire detection and suppression system is also fitted as standard on the BTR-80.

The BTR-80 is fully airportable in transport aircraft such as the An-22 and IL-76.

Variants

BTR-80 with ERA

To improve the combat survivability of the BTR-80, especially in urban conflict, a small number of vehicles based in the Moscow area have been fitted with Explosive Reactive Armour (ERA) blocks removed from the widely deployed T-72 MBT.

The turret has eight blocks of armour while the forward part of the hull has a total of 34 blocks of armour. This ERA provides protection against anti-tank munitions, guided and unguided, fitted with a High Explosive Anti-Tank (HEAT) warhead.

The running gear and lower sides of the hull are fitted with skirts removed from T-72 MBTs. These panels are attached by hinges. There is a gap of about 1 m between the front and rear pair of road wheels either side to allow access to the troop compartment.

BTR-80 with Cummins diesel

For trials purposes, a BTR-80 (8 × 8) APC has been fitted with a Western Cummins model 6CTA 8.3-C250 diesel engine developing 250 hp coupled to a German ZF 9S-75 automatic eight-speed transmission. Maximum road speed is 100 km/h and operational range is between 600 and 800 km.

Other improvements carried out include adjustable steering wheel, new bulletproof windshields, Sanden air conditioning system, thermal insulation wall

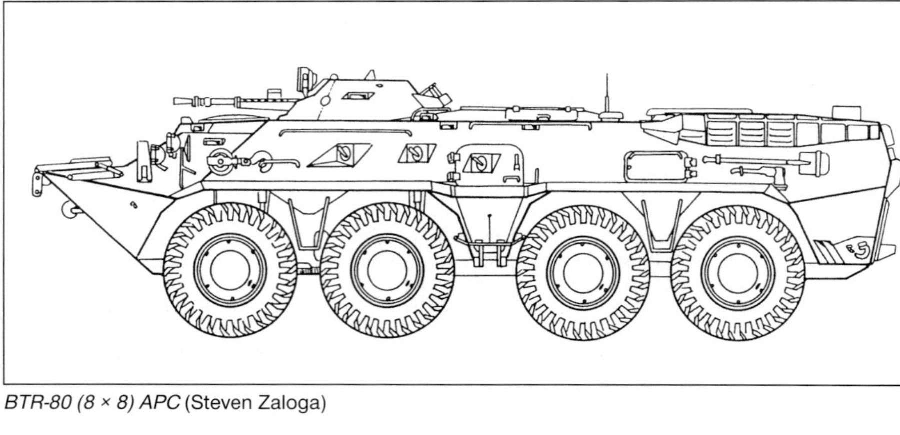

BTR-80 (8 × 8) APC (Steven Zaloga)

between the engine compartment and the crew compartment, sunscreens for all crew members, increased vision for rear view mirrors and improved headlamps.

BTR-80 M1989/1
A command version of the BTR-80 is in service under the designation BTR-80 M1989/1 (Command Variant), this has additional communications equipment and antennas.

The Russian designation for this vehicle is the Kushetka-B command/staff armoured vehicle, although it is also referred to as R-149BMR.

Externally, it is recognisable by its redesigned and much higher roof line, the five radio antennas (one on the left side of the turret, one either side of the hull at the front and one either side at the very rear) and the turret is not fitted with a machine gun. In the forward part of the turret is an observation window with an armoured shutter hinged at the top.

Communications equipment fitted includes Berkut-M, R-171M, R-163-50V and R-163-10 transceivers, an intercom switching system and a secure communications capability.

BTR-80K command post vehicle
This has a six-man crew and communications equipment fitted includes two VHF R-173 radio sets, antenna device set for UHF antennas, two VHF R-159 portable radio sets, IFF automatic interrogator, TNA-4 navigational aids, PAB-2M artillery periscopic aiming ring, TNP-165A vision device and TSh-4 interphone headsets with GVSh-T-13 head and chest set. The turret with a 14.5 mm KPVT and 7.62 mm PKT machine gun is retained and externally it is recognisable by its additional radio antenna.

Other BTR-80 command post vehicles
These include:

BTR-80Pu-16M6 and M7 which are air defence command posts and are similar to the Russian BTR-60Pu-12 vehicles.

P-240BT3 Zenit-B command vehicle which is a successor to the BTR-60 P-240BT command vehicle.

BTR-80K1Sh1 Dedyutant-B (NATO designation M1989/6).

1V153 Kapustnik-B, command system for MRL battalions equipped with the Splav 220 9P140 (16-round) Urgan System.

BTR-MBP is an artillery command vehicle.

BTR-80 M1988 command vehicle.

BTR-80 M1989/3 command vehicle. Could be command post vehicle with Cloths Rail frame antenna.

BTR-80 M1989/5 command vehicle. As BTR-80 but with generator and whip antennas.

BTR-80 M1991 command vehicle.

ZS-88 is BTR-80 M1989/4 command vehicle.

BTR-80M APC
This version has the new YaMZ-238 diesel engine and is also fitted with improved KI-128 tyres.

BTR-80S
Based on standard BTR-80 series APC chassis.

BTR-80 Pioner
Fitted with 'Darth Vader' turret of IV18/19 system, understood to have been built in prototype form only.

BTR-80 R-975
Understood to be forward air-control vehicle.

BTR-80 R-145
Command-post vehicle fitted with Hawk Eye antenna, turret is unarmed.

BTR-MBP
Based on BTR-80 series chassis.

BTR-80 with Kliver turret
For trials purposes the BTR-80 has been fitted with the KBP Instrument Design Bureau Kliver one-person turret armed with four Kornet anti-tank guided missiles, one 30 mm 2A72 cannon and a 7.62 mm PKTM machine gun. The company design designation of this turret is the TKB-799.

During a defence equipment exhibition held in Turkey late in 1997, the Kliver turret was shown on a modified BTR-80 (8 × 8) Armoured Personnel Carrier (APC) which had its firing ports plated over.

In this particular application the Kliver turret was mounted on the centre of the hull roof with the turret basket projecting into the hull in such a manner that some internal parts of the floor covering had to be removed, so exposing some parts of the drive train.

The four Kornet ATGWs, mounted on the right side of the turret, can be removed and replaced by a pod of Igla (SA-18 Grouse) fire-and-forget surface-to-air missiles.

Although shown installed on the BTR-80, Kliver has been designed to be installed on the Russian BMP-2 Infantry Combat Vehicle (ICV) and in order to be fitted with the Kliver turret the BTR-80 must be fitted with a new turret ring.

The standard laser beam guided Kornet ATGW, also developed by the KBP Instrument Design Bureau, is designated the 9M133 and is fitted with a tandem High Explosive Anti-Tank (HEAT) warhead which will penetrate 980 mm of steel armour protected by Explosive Reactive Armour (ERA).

The second version of the Kornet is designated the 9M133F and is fitted with a thermobaric high-explosive warhead which is highly effective against light armoured vehicles, bunkers and other field fortifications.

It has also been confirmed that the sighting system, designated the IK13-2, is a further development of the IK13 sight that is fitted to the BMP-2 ICV. The main improvement of the IK13-2 day/thermal sight is that it has two-axis stabilisation.

As of late 2001, the Kliver turet remained at the prototype stage.

BTR-80A APC
This is covered in detail in a separate entry. This is used by Hungary, Korea (North), Korea (South), Russia and the Sudan.

BTR-80K
This is another command version of the BTR-80A.

SPR-2 wide band jamming system
This system is based on a modified BTR-80 (8 × 8) chassis with the turret removed and replaced by a large telescopic mast with an antenna array on top. This system was designed by the Bryansk Electromechanical Plant and has been developed to neutralise artillery projectiles fitted with proximity fuzes.

RKhM-4-01 NBC reconnaissance
This has been developed to undertake chemical and radiological reconnaissance in all terrain conditions by both day and night.

The equipment carried enables the vehicle to automatically measure the strength of gamma radiation dosage and the level of active radiological contamination in specific areas. This is also referred to by NATO as the BTR-80 1989/2.

BREM-K ARV
This is the recovery member of the BTR-80 (8 × 8) APC variants and is also referred to as the GAZ-59033. Standard equipment includes an A-frame which is fitted to the front of the turret when required, towbars, small stowage platform to the turret rear and stabilisers under the nose of the vehicle.

120 mm 2S23 self-propelled gun
Available details of this are given in the Self-propelled guns and howitzers (wheeled) section. This is also referred to by NATO as the BTR-80 M1986. Details of this are given in the Self-propelled mortar systems section.

BTR-80 with loudspeaker
A small number of BTR-80 vehicles has been fitted with a loudspeaker, probably for the psychological/riot control role. This is called the ZS-88 or BTR-80ZS (Zvukovetsatelnaya Stantsiya) with the NATO designation being the M1989/4.

BMM armoured medical vehicle
There are at least three versions of the BMM armoured medical vehicle based on the chassis of the BTR-80 (8 × 8) armoured personnel carrier:

BMM-1, battlefield evacuation vehicle to search, collect and evacuate wounded

BMM-2, battalion medical station

BMM-3, mobile dressing station with doctors and appropriate equipment.

This is also referred to as the GAZ-59039.

UNSh standardised chassis
Common base vehicle for command and staff vehicle, fire control, radar stations, mobile radio stations and so on. All that is required is to bolt on required equipment and select auxiliary generator unit. Combat weight is 13,600 kg of which 1,600 kg is payload.

Civilian versions
As part of the defence diversification programme, a number of vehicles based on the BTR-80 (8 × 8) APC chassis has been developed including the GAZ-59032, which has a higher roof line and can carry 10 people, and the GAZ-59037 universal cross-country armoured car, which has a fully enclosed heated cabin and a load platform with folding sides. Both versions retain the full amphibious capability of the standard BTR-80 and also retain the central tyre-pressure regulation system.

Other variants include the GAZ-59039 with a scissors-type elevated arm on top of which is mounted a cage for the operators and another version with hydraulic lifting equipment.

Vetluga fire-fighting vehicle
This is a civilian version of the BTR-80 with the turret removed and replaced by 22 launcher tubes for launching fire suppressants.

BTR-80 (8 × 8) APC fitted with Kliver turret showing Kornet ATGW in ready to launch position and 30 mm 2A72 cannon (Christopher F Foss) ***2000***/0081990

Romanian B33 (8 × 8) APC

This is a modified version of the BTR-80 (8 × 8) APC manufactured in Romania under the local designation of the B33 and is also referred to as the TAB Zimbru. Details are given earlier in this section under Romania.

Status

Production. In service with following countries:

Country	Quantity	Comment
Afghanistan	n/avail	
Armenia	50	incl BTR-60/BTR-70 and BTR-80
Azerbaijan	11	
Bangladesh	58	UN role
Belarus	195	
Estonia	32	incl BTR-60 and BTR-70
Finland	2	
Georgia	3	
Hungary	586	Including BTR-80A
Kazakhstan	500	
Korea, North	10	BTR-80A
Korea, South	10	BTR-80A
Kyrgyzstan	45	
Macedonia	60	from Germany 1998
Moldova	11	
Russia	2,000	Estimate
Sudan	60	BTR-80A, delivered 2000
Sri Lanka	n/avail	first observed 1999
Tajikistan	40	incl BTR-60, BTR-70 and BTR-80
Turkey	239	Gendarmerie. Delivered 1993/1995
Turkmenistan	728	incl BTR-60 and BTR-80
Ukraine	457	
Uzbekistan	290	

Manufacturer

Arzamas Machinery Construction Plant.

UPDATED

BTR-70 armoured personnel carrier

Development

The BTR-70 (8 × 8) amphibious armoured personnel carrier was shown in public for the first time during the November 1980 parade in Moscow. It was subsequently replaced in production by the BTR-80 (8 × 8) series of armoured personnel carrier.

For a short period, the US Army referred to the BTR-70 as the M1978 as this was the first time that it was observed; the East German Army referred to the vehicle as the SPW-70.

The BTR-80 was designed by the Dedkov design bureau in Gorkiy-Arzamas under the development designation of the Obiekt 60. This entered production as the BTR-70 in 1972. Although the manufacturer is normally given as the Gorky Automobile Plant, production of this and the later BTR-80 was undertaken in the neighbouring Arzamas Machinery Construction Factory.

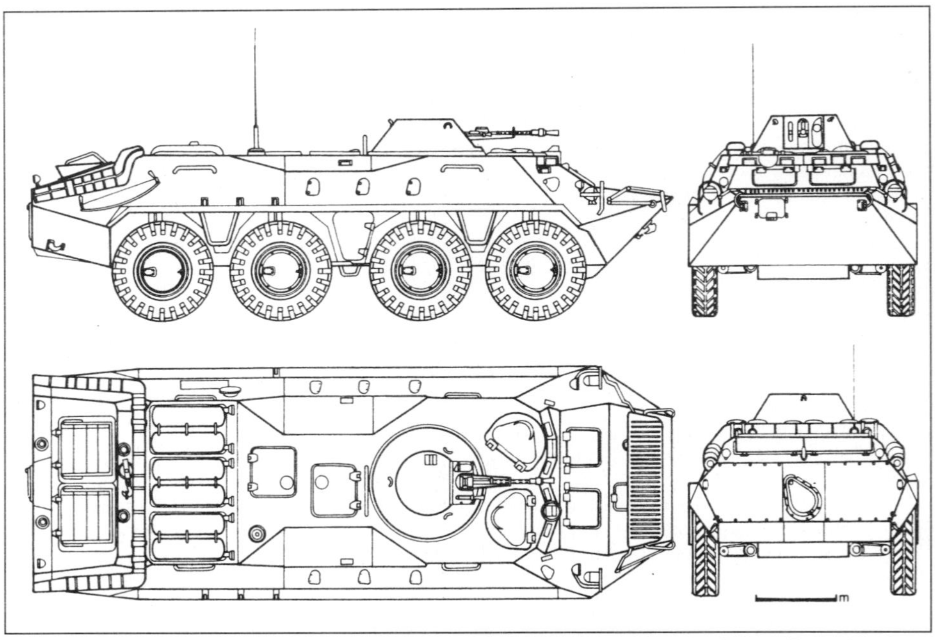

Standard production version of the BTR-70 (8 × 8) armoured personnel carrier (Steven Zaloga)

Description

The hull of the BTR-70 is of all-welded steel armour construction with improved protection over its frontal arc compared to the original BTR-60 series. The nose is also wider and the front of the vehicle provides added protection to the front wheels. The bow section of the BTR-70 may incorporate special layered armour for increased protection.

The commander sits at the front of the hull on the right with the driver to his left. The commander has a single-piece hatch cover that opens to the front. The driver also has a single-piece hatch cover that opens at a 45° angle towards the centre of the vehicle. Both the commander and driver have a window to their front which is covered in combat by an armoured hatch cover hinged at the top. Each also has three forward- and one side-facing day periscopes, with a single firing port under the commander's last periscope only. To the rear of the commander and driver are two infantrymen facing the front.

Over the second axle is the one-person turret, armed with a 14.5 and a 7.62 mm machine gun. This turret is identical to that fitted to the BTR-60PB (8 × 8) armoured personnel carrier and the BRDM-2 (4 × 4) armoured car. Some BTR-70 (8 × 8) APCs have been observed fitted with the complete turret of the more recent BTR-80 (8 × 8) APC described in the previous entry.

The troop compartment is to the rear of the turret with three firing ports and one vision block in each side of the hull. Over the top of the troop compartment are two square roof hatches each with a circular firing port, opening to the front and rear respectively. Between the second and third axles is a small door that opens to the front. The six infantrymen sit on a bench seat down the centre, back-to-back, enabling them to use the firing ports.

The engine compartment is at the rear of the hull and is fitted with a fire extinguishing system. The air inlets are in the top of the engine compartment and the air outlets at the rear. The exhaust pipes run from the top of the engine compartment down either side towards the rear. The original two GAZ-49B six-cylinder in-line water-cooled petrol engines have been replaced by two ZMZ-4905 8-cylinder petrol engines developing 120 hp each compared to the 90 hp each of the BTR-60.

Each engine has its own transmission with the right engine providing power for the first and third axles, the left for the second and fourth axles. If one engine is damaged or disabled then the vehicle can still travel on the other, although at a reduced speed.

Steering is power assisted on the front two axles. Early models of the BTR-70 had the same wheels and tyres as the BTR-60 series. Standard equipment includes an NBC system, a central tyre-pressure regulation system and a front-mounted winch which is provided with 50 m of cable and has a maximum pull capacity of 6,000 kg. It also has a fire detection/suppression system, a DP-3B roentgenometer, a service device for chemical reconnaissance (VPKhR), special toolsets, towing mountings and lifejackets.

The wheels have split rims with pneumatic directional tread tyres fitted as standard. The driver can adjust the tyre pressure of all eight wheels from 2.8 to 0.5 kg/cm^2 without leaving his seat. If a tyre is damaged the onboard compressor compensates for the leakage of air.

BTR-70 (8 × 8) APC in service with the Ukraine (Michael Jerchel) ***2001***/0104386

BTR-70 (8 × 8) APC with commander's and driver's windscreen shutters locked in the closed position (Michael Jerchel) ***2002***/0121205

BTR-70 (8 × 8) APC in service with the Ukraine and fitted with surveillance radar at the rear turret (Michael Jerchel) **2000**/0081993

BTR-70 (8 × 8) APC from above with the commander's and driver's hatches locked in the vertical position and armoured windscreen shutters lowered

Specifications

Crew: 2 + 9
Configuration: 8 × 8
Combat weight: 11,500 kg
Power-to-weight ratio: 20.86 hp/t
Length: 7.535 m
Width: 2.8 m
Height:
(to top of sight) 2.32 m
(to top of turret) 2.235 m
Ground clearance: 0.475 m
Track: 2.38 m
Wheelbase: 4.4 m
Max speed:
(road) 80 km/h
(water) 10 km/h
Fuel capacity: 350 litres (est)
Max range: 600 km (with external fuel tanks)
Fording: amphibious
Gradient: 60%
Side slope: 40%
Vertical obstacle: 0.5 m
Trench: 2 m
Engines: 2 × ZMZ-4905 8-cylinder petrol developing 120 hp each
Transmissions: manual with 4 forward and 1 reverse gears
Transfer case: 2 speed
Steering: power assisted
Turning radius: 6.3 m
Suspension: torsion bar with hydraulic shock-absorbers
Electrical system: 12 V
Armament:
(main) 1 × 14.5 mm KPVT MG
(coaxial) 1 × 7.62 mm PKVT MG
Ammunition:
(main) 500
(coaxial) 2,000

Gun control equipment
Turret power control: manual
Gun elevation/depression: +30/–5°
Turret traverse: 360°
Gun stabiliser:
(vertical) no
(horizontal) no
NBC system: yes
Night vision equipment: yes

The BTR-70 is fully amphibious and propelled in the water by a single water-jet at the rear of the hull. On the BTR-60 this had a two-part cover but on the BTR-70 a one-part cover is fitted, hinged at the top left. Before entering the water the trim vane is erected at the front of the hull. When travelling, this is stowed on top of the hull front whereas on the BTR-60 series it is under the nose of the vehicle.

Internal stowage space is provided for a light anti-tank launcher such as the RPG-7 and two 30 mm AGS-17 automatic grenade launchers plus nine hand grenades.

Variants

BTR-70 with BTR-80 turret

A number of BTR-70 (8 × 8) APCs have been observed fitted with the turret of the more recent BTR-80. Details of this turret are given in the entry for the BTR-80. This version has the NATO/US designation of BTR-70 M1986/1. In addition to being fitted with the complete BTR-80 (8 × 8) turret, it has a modified wave deflector, rear bumpers, headlight brushguards, additional armour brackets on the vehicle sides and two additional firing ports in the very top of each side of the troop compartment between the hull sides and the roof.

BTR-70Kh

Chemical reconnaissance vehicle.

SPR-2 wide band jamming station

This system is based on a modified BTR-70 or the more recent BTR-80 (8 × 8) chassis with the turret removed and replaced by a large telescopic mast with an antenna array on top. This system was designed by the Bryansk Electromechanical Plant and has been developed to neutralise artillery projectiles fitted with proximity fuzes.

BTR-70 MBP

This is an artillery command vehicle.

BTR-70Kh APC

This was developed by Russia for the then East German Army and was also referred to as the SPW-70Ch. Mounted on the turret rear is a bank of forward firing 81 mm smoke grenade launchers similar to that fitted to the more recent BTR-80.

BTR-70 with AGS-17

A number of BTR-70s were seen in Afghanistan fitted with the 30 mm AGS-17 automatic grenade launcher on the roof to the rear of the driver and commander. These vehicles retain the machine gun turret.

BTR-70MS

This is a turretless communications vehicle with the MS standing for *Mashina Svyazi* (signals vehicle).

BTR-70 KShM

This is a command/staff vehicle with the KShM standing for *Komandno-Shtabnaya Mashina*. This is fitted with the R-145BM communications system.

BTR-70 ZHALO-S

This is understood to be a tank destroyer based on the BTR-70 (8 × 8) chassis and is fitted with a new turret armed with a long barrelled gun with a calibre of between 85 and 100 mm which is fitted with a fume extractor and an unusual pepperpot muzzle brake. It is possible that the main armament is fed by a bustle mounted automatic loader and there is a cartridge case ejection port in the turret rear. The gunners sight is believed to have come from the BMP-1 infantry fighting vehicle. It is not known whether this system entered production.

BREM repair and recovery vehicle

This is a turretless BTR-70 fitted with a bow-mounted jib crane and other equipment; BREM stands for *Bronirovannaya Remontno-Evakuatsionannaya Mashina* (armoured repair and recovery vehicle).

TAB-77 APC

This is the Romanian version of the Russian BTR-70 and details are given earlier in this section under Romania. It was subsequently replaced in production by the B33 which is based on the more recent BTR-80 chassis.

BTR-70 upgrade

The Ukraine has developed a major upgrade for the Russian designed BTR-70 APC. Details are given under the Ukraine.

Status

Production complete. In service with the following countries:

Country	Quantity	Comment
Afghanistan	n/avail	
Armenia	32	BTR-80
Azerbaijan	28	
Bangladesh	58	
Belarus	446	
Bosina Herzegovina	n/avail	
Estonia	32	incl BTR-60, BTR-70 and BTR-80
Georgia	19	
Hungary	309	
Kazakhstan	45	
Kyrgyzstan	45	
Lithuania	29	delivered in 1992/1993
Macedonia	60	from Germany 1998
Moldovia	20	
Nepal	135	
Pakistan	169	UN role
Romania	167	local model is TAB-77 (qv)
Russia	2,000	estimate
Spain	100	
Tajikistan	40	incl BTR-60, BTR-70 and BTR-80
Turkmenistan	728	incl BTR-60, BTR-70 and BTR-80
United Nations	227	from Germany 1994
Ukraine	1,105	
Uzbekistan	36	

Manufacturer

Arzamas Machinery Construction Plant.

UPDATED

BTR-60P armoured personnel carrier

Development

In the late 1950s, a Russian Army requirement was issued for a new wheeled APC to replace the non-amphibious BTR-152 (6 × 6) APC. Development of an 8 × 8 APC began in two competing design bureaus in 1957-58, this competition was called Project 49. The Dedkov OKB team won with its vehicle, the Type 62B. This was accepted for service with the Russian Army in 1960 under the designation BTR-60P.

The vehicle was seen in public for the first time during the November 1961 parade in Moscow. The BTR-60P and its variants were normally used by the motorised rifle divisions, whereas the tank divisions have the tracked BMP-1/BMP-2 ICV. The original BTR-60P and BTR-60PA have been relegated to reserve, second-line or training use.

Production of the BTR-60 series was completed in 1976 and it is believed that total production amounted to around 25,000 vehicles, including the many variants.

The BTR-60P APC is not used by former Czech or Polish motor rifle units. Regular units use the former Czech OT-64 (8 × 8) amphibious APC which has a number of advantages over the former Soviet vehicle. The Polish internal security forces have also used the BTR-60 series vehicle.

Production of the BTR-60 series of 8 × 8 APC was undertaken at the Gorky Automobile Plant which is today known as the Arzamas Machinery Construction Factory where production of the latest BTR-80 series is also undertaken.

A further development of the BTR-60PB has been manufactured in Romania under the designation of the TAB-71 (qv Romania). More recently, Romania has produced another version called the TAB-77 which is now known to be based on the latest BTR-80 series APC.

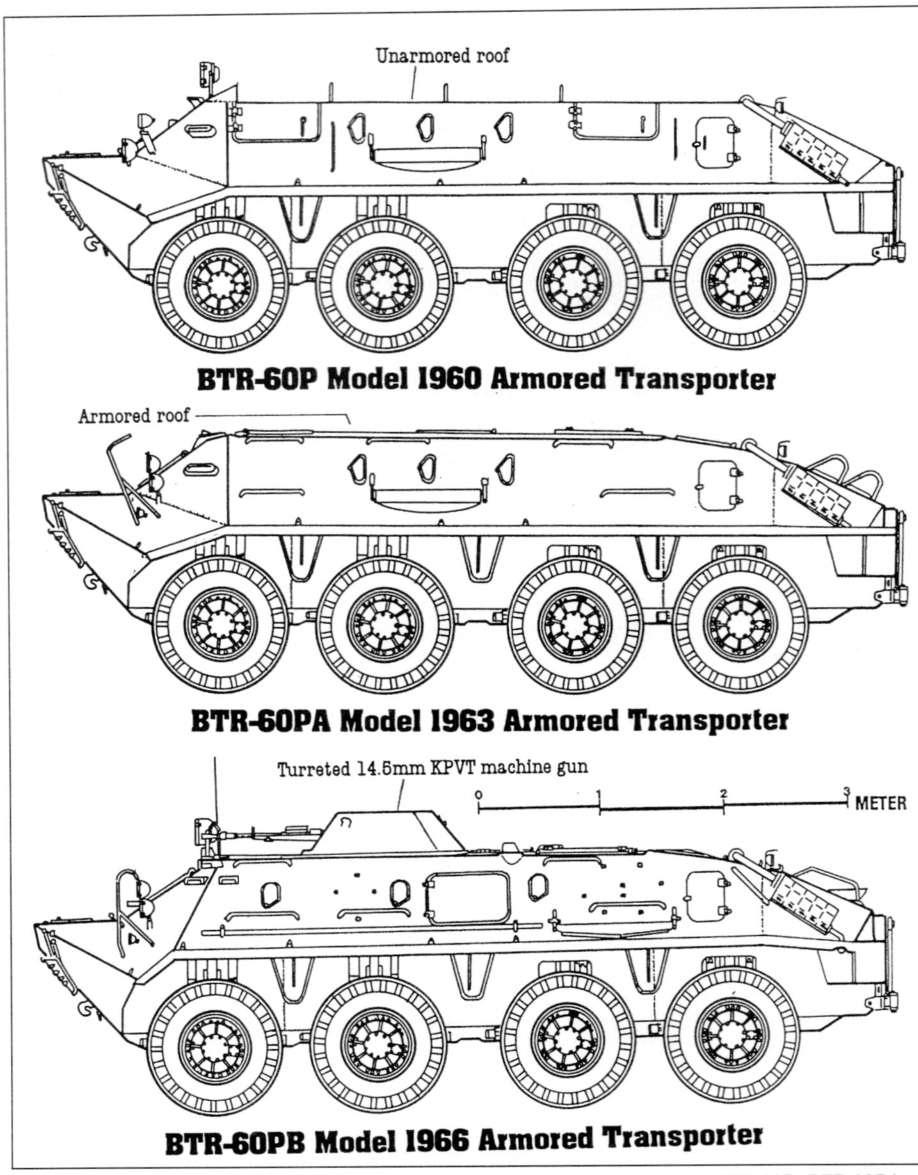

The three key versions of the BTR-60 (8 × 8) APC family, from top to bottom BTR-60P, BTR-60PA and BTR-60PB (Steven Zaloga)

Description

The hull of the BTR-60P is of all-welded steel armour with the driver and commander seated at the front of the hull, the open-topped personnel compartment behind them and the engine compartment at the very rear of the hull. The interior of the BTR-60P is open from the driver's and commander's position to the engine compartment at the rear.

The driver sits on the left with the vehicle commander to his right. Both have a windscreen in front of them which, when in action, is covered by a flap hinged at the top. This has an integral day vision device which can be replaced by an infra-red periscope for night use. To the left of the driver and the right of the commander is a vision block. Mounted above the flap in front of the commander is an infra-red searchlight which can be operated from inside the vehicle.

The infantrymen are seated on bench seats that run across the full width of the hull. In each side of the hull are two half doors and three firing ports. If required, bows and a tarpaulin cover can be fitted over the top of the troop compartment.

Main armament comprises a 7.62 mm SGMB or PKB machine gun which is pintle-mounted on the forward part of the hull. Some vehicles have been seen with the heavier 12.7 mm DShKM machine gun at this position. Mounts are also provided for 7.62 mm machine guns on either side of the hull.

The two 8-cylinder water-cooled petrol engines are mounted at the rear of the hull; the first and third axles are powered through the transmission of the right engine and the second and fourth axles through the transmission of the left engine. Torque is transmitted from each engine to the wheels through a single-plate clutch with hydraulic control, a four-speed gearbox with synchromesh in third and fourth gears, a two-speed transfer case, the final drives of the two driving axles with gearless differentials and four wheel reducers. Torque from the final drives to the wheel reducers is transmitted through half-axles and gimbal drive gears.

All eight wheels are powered and the first four, which are used for steering, are power assisted. The vehicle can be driven with one wheel missing from the second axle. The suspension is of the torsion bar type with the first and second roadwheels on each side provided with two hydraulic shock-absorbers and the third and fourth roadwheels with a single hydraulic shock-absorber. A central tyre-pressure regulation system, fitted as standard on all BTR-60 series APCs, enables the driver to adjust the tyre pressure to suit the ground being crossed.

The BTR-60P is fully amphibious and is propelled in the water by a single water-jet mounted at the rear of the hull. When not in use the exit at the rear of the hull is covered by a two-part circular plate. Before entering the water a trim vane is erected at the front of the hull and the bilge pump is switched on. When not in use the trim vane is stowed flat under the nose of the vehicle. Power is transmitted to the water-jet via a Power Take-Off (PTO) and water-jet reducer. Direction is changed when afloat by using a hydrodynamic rudder in the water-jet propeller and by turning the front four roadwheels. To reverse direction, the shutters of the water-jets are closed and water is ejected through side ducts in the rear part of the vehicle that face forwards.

The basic BTR-60P does not have an NBC system as it is open-topped. All vehicles are fitted with infra-red night vision equipment and a front-mounted winch with a maximum capacity of 4,500 kg.

BTR-60 (8 × 8) IV 18 artillery observation post vehicle **2000**/0081995

BTR-60PB (8 × 8) APC with commander's and driver's windscreen shutters in the open position and with BRDM-2 (4 × 4) armoured car in background (Richard Stickland) **2000**/0081996

BTR-60PA (8 × 8) APC showing the arrangement of roof hatches and machine guns

BTR-60Pu-12 (8 × 8) command post vehicle with antenna retracted
2002/0121204

Variants

BTR-60PA

This model is also referred to as the BTR-60PK, and entered service in 1963. It has complete overhead armour protection for the troop compartment and is fitted with an NBC system. The commander's and driver's positions are identical to those on the original BTR-60P except that the driver has a roof-mounted day periscope and both have a single-piece hatch cover that opens to the rear.

Behind the commander's and driver's hatches is a single rear-opening rectangular hatch, in front of which is a single pintle-mounted 7.62 mm SGMB or PKB machine gun. Many vehicles also have a 7.62 mm machine gun mounted either side of this hatch cover. There is another hatch cover in the roof of the personnel compartment towards the rear on the right side, which opens to the right.

Three firing ports are provided in each side of the hull but no entry doors. The infantrymen are seated on bench seats along the side and rear of the troop compartment.

BTR-60PB

The BTR-60PB is essentially the BTR-60PA fitted with a machine gun turret and other modifications. In place of the vision blocks in the hatches immediately in front of the commander and driver, there are a number of roof-mounted day periscopes for observation to the front and sides of the vehicle.

The turret, which is identical to that fitted to the Russian BRDM-2 (4 × 4) amphibious scout car and the Czech OT-64 (8 × 8) APC, is armed with a 14.5 mm KPV machine gun and a 7.62 mm PKT machine gun mounted coaxially to the right, with the telescopic day sight mounted coaxially to the left. The weapons have an elevation of +30°, a depression of –5° and the turret can be traversed through a full 360°. Gun elevation and depression and turret traverse are all manual.

To the rear of the turret on the right side is a single-piece hatch cover that opens to the left and to the left rear of this is a single-piece hatch cover that opens to the right.

There are single doors that open forwards either side of the hull and two or three firing ports and vision blocks.

The BTR-60PB was preceded by the BTR-60PAI which also had a turret. Late production BTR-60PBs have the same sighting improvement as the BTR-70: a small additional periscopic sight on the turret roof.

BTR-60PBK

This is a company commander's model of the BTR-60PB with two R-123 and one R-148 radios.

BTR-60 1V18

This is an artillery observation post vehicle and is the wheeled equivalent of the tracked 1V14M with the same communications fit and the APPK computer/ communications system.

The 1V18 Klen-1 has the US Army designation of the ACRV M1979/2a with the 1V18-1 being an improved version.

BTR-60 1V19

This is a fire direction centre vehicle and the wheeled equivalent of the 1V15M with the same communications fit and the APPK computer system. Wheeled variants of the other vehicles are the 1V110 on the GAZ-66 (1V13M) and the 1V111 on the ZIL-131 (6 × 6) (1V16M) chassis.

The 1V19 Klen-2 is also known as the ACRV M1979/2b and is equipped with a telescopic antenna for the R-130 on the left side of the hull. The IV19-1 is an improved model. Known users are Bulgaria, Finland, Romania, Russia and the former Yugoslavia.

BTR-60 R-145BM

This is fitted with two R-111/171, one R-123/173, one R-130M and one R-8xx ground-to-air radio sets.

BTR-60 R-975

This is a forward air control vehicle with an unknown fit of R-8xx radio systems. This model has a turret. BTR-60 R-975MI is very similar.

BTR-60-Z-351BR

Fitted with an armoured generator system for high-power (for example 15 kW and larger) requirements.

BTR-60-P-238BT

Armoured field switchboard with one R-123M radio set. BTR-60-P-239BT is very similar.

BTR-60-P-240BT

Radio/wire integration switchboard with one R-405, one R-123M and switching equipment.

BTR-60-P-241BT

BTR-60 with a P-241 field telephone system. Fitted with a secure telephone terminal set.

BTR-60-R-137B

BTR-60 with an armoured communication system with one R-137 VHF set, one R-405 and one R-123M radio. This is fitted with a turret but does not have a 14.5 mm machine gun.

BTR-60-R-140BM

BTR-60 with an armoured command and control communications system with one R-140 HF set, one R-405 and one R-123M radio. This is fitted with a turret but does not have a 14.5 mm machine gun.

BTR-60-R-145

BTR-60 with the R-145 radio system. This does not have a turret.

BTR-60-R-156BTR

Armoured command and control communications system vehicle with one R-156 HF set, one R-405 and one R-123M radio. This is not fitted with a turret.

Specifications

(BTR-60PA; data in square brackets relates to BTR-60PB where different)
Crew: 2 + 16* [2 + 14]**
Configuration: 8 × 8
Combat weight: 9,980 [10,300] kg
Power-to-weight ratio: 18.03 [17.47 hp/t]
Length: 7.56 m
Width: 2.825 m
Height:
(to turret top) n/a [2.31 m]
(to hull top) 2.055 m
Firing height: n/a [2.13 m]
Ground clearance: 0.475 m
Track: 2.37 m
Wheelbase: 1.35 + 1.525 + 1.35 m
Max speed:
(road) 80 km/h
(water) 10 km/h
Fuel capacity: 290 litres
Max road range: 500 km
Fuel consumption: 0.58 litre/km
Fording: amphibious
Gradient: 60%
Side slope: 40%
Vertical obstacle: 0.4 m
Trench: 2 m
Engines: 2 GAZ-49B 6-cylinder in-line water-cooled petrol developing 90 hp at 3,400 rpm each
Transmission: manual with 4 forward and 1 reverse gears
Transfer case: 2 speed
Steering: power assisted
Tyres: 13.00 × 18 (with central tyre pressure regulation system)
Suspension: torsion bar with hydraulic shock-absorbers
Electrical system: 12 V
Armament:
(main) 1 × 7.62 mm PKT [1 × 14.5 mm KVP MG
(coaxial) n/a [1 × 7.62 mm PKT MG]
Smoke-laying equipment: none
Ammunition:
(main) 2,000 [500]
(coaxial) n/a [2,000]
Gun control equipment
Turret power control: n/a [manual]
Gun elevation/depression: +23.5/–6° [+30/–5°]
Gun traverse: 90° [+360° (turret)]
Armour
Hull front upper: 7 mm at 86°
Hull front lower: 9 mm at 47°
Hull side upper: 7 mm at 35°
Hull side lower: 7 mm at 0°
Hull rear upper: 5 mm at 65°
Hull rear lower: 7 mm at 0°
Hull top: 7 mm at 86°
Hull belly front: 5 mm
Hull belly rear: 5 mm
Turret front: n/a [7 mm at 43°]
Turret sides: n/a [7 mm at 36°]
Turret rear: n/a [7 mm at 36°]
Turret top: n/a [7 mm]
NBC system: yes [yes]
Night vision equipment: yes [yes]

* Max is 16, 12 are normally carried
** Max is 14, 8 are normally carried

BTR-60-R-409BM

BTR-60 used as an armoured radio relay station with one R-409 and one R-123M radio. This is not fitted with a turret.

MEP

Infantry command post vehicle.

BTR-60PAu

Bulgarian BTR-60 with unidentified communications equipment for use by artillery units.

BTR-60Pu command vehicle

This is essentially the BTR-60P fitted with bows and a tarpaulin cover and internally with seats for the command staff, map boards and additional communications equipment. A command version of the BTR-60PA is also in service with a roof-mounted generator, a 10 m high radio antenna, which is used in the static role and a rail antenna running along the front, left side and rear of the hull. Variants of the BTR-60Pu include the BTR-60PuM and BTR-60PuM1.

BTR-60PuM command vehicle

This is a Finnish variant of the Pu (R-145BM) without the Hawk Eye or Clothes Rail antenna and is fitted with the 12.7 mm NSVT machine gun and modern communications equipment. This entered service with Finland in 1996.

BTR-60PB-MD Bulgaria

This is an upgraded Bulgarian BTR-60PB with 81 mm smoke grenade launchers, VAMO DT 3900 or Rover TD-200 diesel engine, Melopa night vision equipment, new NBC system and modern radios.

BTR-60 MBP command

This is an artillery command vehicle with modified turret and a box on the right side of the hull.

BTR-60Pu-12 command vehicle

The BTR-60Pu-12 command and control vehicle has been in service with the former Soviet and now Russian Army since the late 1970s and has also been identified in service with Bulgaria, the former Czechoslovakia, Hungary, Iraq and the former Yugoslavia.

The vehicle is essentially a BTR-60PA with a large stowage box on the right side of the hull to the rear of which is a water or fuel can. On the roof is a box which may house a generator to power the additional communications equipment installed. To the immediate rear of the commander's roof hatch is a telescopic mast which is raised into the operating position by turning a handle located on the right side of the mast.

The BTR-60Pu-12 operates with air defence units equipped with systems such as the ZSU-23-4 self-propelled anti-aircraft gun, and 'Gecko' SA-8 and 'Gopher' SA-13 surface-to-air missile systems.

The BTR-60Pu-12 is also known as the 9S482 with the BTR-60Pu-12M being the 9S482M.

BTR-60Pu-12M

BTR-60 air defence command post vehicle.

BTR-60P maintenance assistance

A number of older BTR-60P (8 × 8) APCs have been converted for use in the maintenance assistance role and have a raised tarpaulin cover over the troop compartment that runs almost to the rear. This may have the Russian designation of the MTR-2.

BTR-60MS radio vehicle

This is the BTR-60P series used in the radio role and fitted with a 'High Ball' (NATO designation) telescopic antenna.

BTR-60 twin 30 mm SPAAG

The Cuban Army has used a modified BTR-60P with the former Czech twin 30 mm towed anti-aircraft gun system mounted in an open mount on top of the vehicle.

BTR-60 VVS

This is essentially the BTR-60PB with the armament removed from the turret and the resulting port covered by a plexiglas observation window. To provide power for the additional communications equipment carried, a generator is mounted externally on the top of the hull at the rear.

BTR-60 ACRV M1979(2)

The BTR-60 Artillery Command and Reconnaissance Vehicle (ACRV) M1979(2) is used by towed artillery batteries and battalions as a Command Observation Post (COP) vehicle. The designation ACRV M1979(2) was allocated by the US Army and may well relate to one of the variants already described.

The ACRV M-1979 set turned into the 1V18, 1V19, 1V110 and 1V111. The entire set is the 1V17 (à la 1V12).

Other BTR-60 command versions

These include P-219, P-238BT, P-239BT and P-241BT, P-240, P-240BR, P-240BT and P-240TBR, P241TBR, P251, P-286, P-537, R118AM3, R-137, R-137B, R-137BR, R-140, R-145BM, R-145K and MB R-145, R-156, R-156B, R-156BTR, R-396, R-409, R-409BM and R-409BR, R-419B, R-419BR, R-419BTR, R-440B, R-975, R975M and R-975M1, Z-351BR, BMU, EPS, KOAM, BTR-60KShM, BTR-60MS and BTR-60 Salon.

BTR-60 90 mm Ethiopia

Ethiopia has fitted at least one of its BTR-60PB (8 × 8) APCs with the complete turret from the French Panhard AML-90 (4 × 4) armoured car (covered in detail in the Reconnaissance vehicles section) which is armed with a 90 mm gun and a 7.62 mm coaxial machine gun.

Civilian BTR-60 versions

Following the reduction in the size of the Russian Army, many BTR-60 vehicles are now surplus to requirements and some are being converted into civilian vehicles. These include the technical assistance vehicle MTP-2B and the disaster assistance transporter ATM-2, both of which have been developed by the Kaunas Vehicle Repair Factory.

Status

Production complete. In service with the following countries:

Country	Quantity	Comment
Afghanistan	300	estimate
Algeria	530	incl BTR-50/BTR-60
Angola	100	incl BTR-60 and BTR-152
Armenia	14	incl BTR-60/BTR-70/ BTR-80
Azerbaijan	25	
Belarus	188	
Botswana	30	most not operational
Bulgaria	781	
Cambodia	210	incl BTR-60 and BTR-152
Congo	30	
Cuba	400	estimate
Djibouti	12	status uncertain
Eritrea	30	
Estonia	32	incl BTR-60/BTR-70/ BTR-80
Ethiopia	80	estimate
Finland	120	
Guinea	8	
Guinea-Bissau	35	incl BTR-40/BTR-60/ BTR-152
Iran	300	incl BTR-50/BTR-60
Iraq	500	estimate
Korea, North	1,200	500 BTR-60PA and 700 BTR-60PB
Laos	70	incl BTR-40/BTR-60/ BTR-152
Libya	750	incl BTR-50/BTR-60
Lithuania	29	from Russia in 1992/ 1993
Mali	10	
Moldova	2	
Mongolia	300	incl BTR-40, BTR-60 and BTR-152
Mozambique	150	
Nicaragua	64	
Peru	12	
Romania	965	called TAB-71, locally built (qv)
Russia	2,000	estimate
Somalia	n/avail	status uncertain
Syria	600	estimate
Tajikistan	40	incl BTR-50/BTR-60/ BTR-70 and BTR-80
Turkey	302	estimate including 105 from Germany in 1992
Turkmenistan	728	incl BTR-60 and BTR-80
Uganda	20	
Ukraine	203	
Vietnam	400	estimate
Yemen	500	Incl BTR-40/BTR-60/ BTR-152
Zambia	13	

Manufacturer

Arzamas Machinery Construction Plant.

UPDATED

BTR-60-145BM command vehicle with the telescopic mast to the rear of the commander's station, the roof-mounted generator and the rail-type antenna mounted around the front, left and rear sides of the hull roof shown in raised position (Michael Jerchel) 0058326

BTR-60-R137B armoured communications vehicle with antenna erected (Michael Jerchel) **2000**/0081997

BTR-152 armoured personnel carrier

Development

In November 1946, the BM Fitterman design bureau started development work on a 6 × 6 armoured personnel carrier called the BTR-140. This was based on the chassis of the ZiS-123 truck, itself a derivative of the ZiS-151 truck but with an improved engine. The first prototypes of the BTR-140 were completed in May 1947 and tests were completed in 1949. The BTR-140 was accepted for service on 24 March 1950 under the designation BTR-152 and first production vehicles were completed in 1950.

The first production models were based on the chassis of the ZIL-151 (6 × 6) 2,500 kg truck but later vehicles (from and including the BTR-152V1) were based on the chassis of the improved ZIL-157 (6 × 6) 2,500 kg truck.

By today's standards the BTR-152 (6 × 6) is obsolete as it lacks cross-country mobility and has poor protection for its occupants from small arms fire, shell splinters and mines.

Description

The hull of the BTR-152 is made of all-welded armoured steel with the engine at the front, commander and driver immediately behind the engine and the troop compartment at the rear.

Armoured shutters, controlled from inside the driver's compartment, protect the radiator from damage by small arms fire. Power is transmitted from the engine to the transmission and then by a propeller shaft to the transfer box in the centre of the vehicle. Power is then transmitted by a propeller shaft to the front axle and each of the rear axles.

The driver sits on the left of the vehicle with the commander to his right, both have a windscreen which can be covered by an armoured shutter hinged at the top. This shutter has a vision block for observation when the shutter is closed. Both the commander and driver have a door in the side of the hull and the upper part, which has a small vision slit, folds down on the outside for increased visibility. The roof over the commander's and driver's position is armoured.

The 17 infantrymen are seated in the open-topped troop compartment at the back of the vehicle, the top of which can be covered by a tarpaulin which is normally carried in the vehicle. The seating arrangements depend on the model: in some there are bench seats across the vehicle and others have them down either side. The infantrymen enter and leave the vehicle by twin doors in the rear of the hull which open outwards, with the left door carrying the spare wheel. There are three firing ports in each side of the hull and another two in the rear, one each side of the doors.

The BTR-152 has no NBC system, no night vision equipment and no amphibious capability. A saw is carried externally on the left side of the hull. The winch, if fitted, is mounted at the very front of the vehicle and has a maximum capacity of 5,000 kg and 70 m of cable. The winch is also equipped with a special block which can be used as a pulley block to increase pulling capacity, or as a conventional block to change the direction of pull. The winch incorporates a three-speed Power Take-Off (PTO) unit (low, high and reverse) controlled by a lever in the driver's compartment.

Some models (see variants) have a central tyre pressure regulation system which allows the driver to adjust the tyre pressures to suit the type of ground being crossed.

There are three sockets for mounting machine guns, one on the top of the armour cover over the commander's and driver's position and the other two either side of the troop compartment. The forward mounting normally has a 7.62 mm SGMB machine gun, which can be elevated from –6 to +23.5° and has a traverse of 45° left and right, or a 12.7 mm DShKM heavy machine gun. The side mountings normally have 7.62 mm SGMB machine guns.

Specifications

BTR-152V1
Crew: 2 + 17
Configuration: 6 × 6
Combat weight: 8,950 kg
Power-to-weight ratio: 12.29 hp/t
Length: 6.55 m
Width: 2.32 m
Height:
(hull top) 2.36 m
Ground clearance: 0.295 m
Track:
(front) 1.742 m
(rear) 1.72 m
Wheelbase: 3.3 + 1.13 m
Max road speed: 75 km/h
Fuel capacity: 300 litres
Max road range: 600 km
Fuel consumption:
(road) 0.46 litres/km
Fording: 0.8 m
Gradient: 55%
Vertical obstacle: 0.6 m
Trench: 0.69 m
Engine: ZIL-123 6-cylinder in-line water-cooled petrol developing 110 hp at 3,000 rpm
Transmission: manual with 5 forward and 1 reverse gears (5th is overdrive)
Transfer box: 2 speed
Clutch: dry, 2-plate
Suspension:
(front) 2 leaf springs and 2 dual-action hydraulic shock-absorbers
(rear) equalising type with 2 leaf springs working in conjunction with torsion bars
Tyres: 12.00 × 18 (with central tyre pressure regulation system)
Electrical system: 12 V
Batteries: 1 × 12 V
Armament: 1 × 7.62 mm SGMB MG
Ammunition:
(7.62 mm) 1,250 rds
Gun elevation/depression: +23.5/–6°
Gun traverse: 90°
Armour
Hull front: 13.5 mm at 35°
Hull sides: 9 mm at 7°
Hull top: 6 mm
Hull rear: 9 mm at 0°
Belly: 4 mm
NBC system: no
Night vision equipment: no

BTR-152U command vehicle clearly showing the higher superstructure of this model (Michael Jerchel)

BTR-152K (6 × 6) APC showing roof armour and used in the ambulance role (Stefan Marx)

BTR-152V1 (6 × 6) APC with tarpaulin cover over roof and external air line system removed (Michael Jerchel)

BTR-152A anti-aircraft vehicle with twin 14.5 mm KPV machine guns

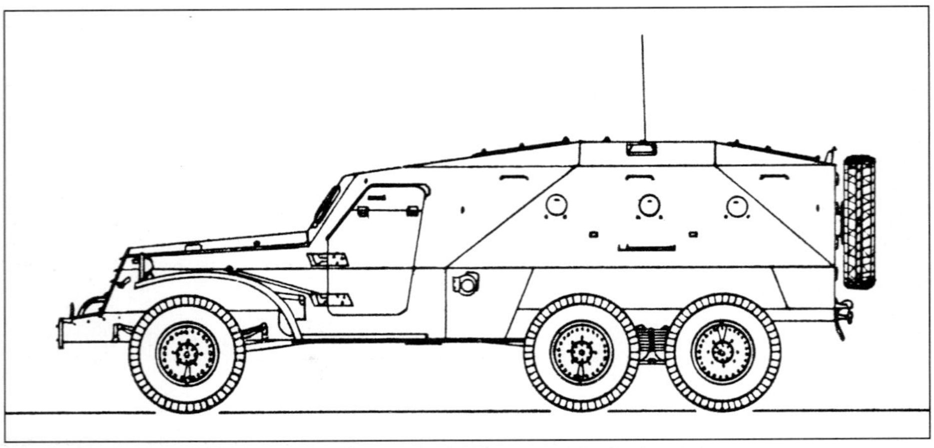

BTR-152K APC with overhead protection

Variants

BTR-152
This was the first model to enter service and is also known in the West as the Model A. It has an open top, no winch and no central tyre-pressure regulation system.

BTR-152V
This was first seen in 1954 and was accepted for service in October 1955. This model has a central tyre pressure regulation system for improved cross-country performance. The artillery command vehicle is called the BTR-152I.

BTR-152V1
This has been referred to as the Model B in the West and has an open top, front-mounted winch and a central tyre-pressure regulation system with external air lines. The BTR-152V1 was developed in 1959 and entered service in 1962.

BTR-152V2
Basic BTR-152 fitted with a central tyre-pressure regulation system but no winch.

BTR-152V3
This is also known as the Model C in the West and has an open roof, a front-mounted winch, a central tyre pressure regulation system with internal air lines and infra-red driving lights.

BTR-152K
This model, which is also known as the Model D (or 1961) in the West, has all the modifications of the BTR-152V3 plus full overhead armour protection. Over the top of the troop compartment are two roof hatches that open to the right, one with a single firing port and one with two.

BTR-152U command vehicle
The command vehicle is usually based on the BTR-152V1 or BTR-152V3 APC but has a much higher superstructure to allow the command staff to work upright. There are four windows in the upper part of the superstructure, two on the left, one on the right and one at the back. Access to stowage racks on the roof is by a ladder forward of the roof. The BTR-152U normally tows a trailer carrying additional equipment such as a generator. The first of these PUA (artillery command post) vehicles with a higher roof entered service in 1952 as the BTR-152B but did not have the central tyre pressure regulation system, unlike the BTR-152S which does.

BTR-152A anti-aircraft vehicle
This is armed with twin 14.5 mm KPV heavy machine guns in a manually operated turret (ZTPU-2) with a traverse of 360°. The guns have an elevation of +80° and a depression of –5°. The weapons have a maximum horizontal range of 8,000 m and a maximum vertical range of 5,000 m. Effective anti-aircraft range however is 1,400 m and effective range when being used in the ground role is 2,000 m. These weapons and mount are also used on the BTR-40A (4 × 4) anti-aircraft vehicle (which has the same turret as the BTR-152A); towed models of the KPV are called the ZPU-1 (single barrel), ZPU-2 (twin barrel) and ZPU-4 (quad barrel). KPV machine guns are also mounted in armoured vehicles. Details of these 14.5 mm ZPU machine guns are given in *Jane's Land-Based Air Defence 2001-2002*.

The BTR-152A was designed at the same time as the BTR-152 APC and entered service in 1952. The BTR-152D also had four KPV machine guns for anti-aircraft use, plus 2,000 rounds of ammunition and a crew of five. In 1955, the BTR-152E was introduced with a twin 14.5 mm KPV machine gun system, but based on the BTR-152V chassis.

BTR-152 with Quad 12.7 mm M53 MGs
The Egyptians fitted a number of their vehicles with the Czech Quad 12.7 mm M53 anti-aircraft system. The M53 consists of four Russian-designed 12.7 mm DShKM machine guns on a Czech-designed two-wheeled mount. It is believed that these are no longer in front-line service with the Egyptian Army. Some of these have, however, been encountered in Afghanistan, operated by the Afghan Army.

BTR-152 with 23 mm ZU-23-2 AAG
During the fighting in Lebanon in the summer of 1982, the Israel Defence Force captured from the PLO, several BTR-152 APCs which were fitted with the towed twin 23 mm automatic anti-aircraft gun ZU-23-2 in the rear of the troop compartment. The 23 mm ZU-23-2 has a maximum vertical effective anti-aircraft range of 2,500 m. There was also another with the ZU-23-2 mounted lower down at the hull rear which was cut away at an angle of about 45°. Details of the twin 23 mm ZU-23-2 light anti-aircraft gun are given in *Jane's Land-Based Air Defence 2001-2002*.

Status
Production complete. In service with the following countries:

Country	Quantity	Comment
Afghanistan	n/avail	
Angola	100	incl BTR-60
Cambodia	210	incl BTR-60
Central African Republic	4	
Congo	20	
Cuba	700	estimate, incl BTR-40, BTR-50, BTR-60 and BTR-152
Equatorial Guinea	10	
Ethiopia	n/avail	
Guinea	6	
Guinea Bissau	35	incl BTR-40/ BTR-60/ BTR-152
Iraq	n/avail	
Korea, North	n/avail	
Laos	70	incl BTR-40, BTR-60, BTR-152
Mali	10	
Mongolia	300	incl BTR-40, BTR-60 and BTR-152
Mozambique	100	
Namibia	n/avail	
Nicaragua	102	
Sri Lanka	10	
Sudan	90	incl BTR-50
Syria	n/avail	
Tanzania	66	incl BTR-40
Vietnam	200	estimate
Yemen	500	incl BTR-40/ BTR-60/ BTR-152

Note: In many of these countries the BTR-152 is in non-front-line units or held in reserve.

Manufacturer
Russian state factories.

UPDATED

BTR-40 armoured personnel carrier

Development
Development work on a new 4 × 4 armoured personnel carrier (under the designation BTR-141) began in 1944 at the VA Dedkov OKB design bureau under the chief designer VK Rubtsov. This was based on a lengthened GAZ-63 (4 × 4) 2,000 kg light truck chassis with a slightly more powerful engine (increased by 10 hp). The first prototype was completed in 1947 and had angled sides that tended to restrict its carrying capacity. The BTR-141 was then redesigned with vertical sides and rear. Trials with this model were carried out from 1949 and in 1950 it was accepted for service under the designation BTR-40. It entered service in 1950. The vehicle was originally used both as an armoured personnel carrier and as a command and reconnaissance vehicle.

BTR-40 series vehicles produced from 1956 had a central tyre-pressure regulation system and were known as the BTR-40Vs. BTR-40 series vehicles are often called the Sorokovke in Russia.

By today's standard the BTR-40 APC is obsolete because it has limited cross-country mobility and provides a low level of protection against small arms fire, shrapnel and mines.

Description
The hull of the BTR-40 is made of all-welded steel with the petrol engine at the front, commander and driver in the centre and the personnel compartment at the rear.

The driver sits on the left of the vehicle with the vehicle commander to his right. The windscreen in front of them can be covered by an armoured shutter hinged at its upper part. The shutter has a vision block for observation when it is lowered. Both the commander and driver have a door in the side of the hull, in the upper part of which is a vision flap that hinges open downwards on the outside and has a vision slit.

The eight infantrymen sit in the back of the vehicle, three on a bench seat at the very rear of the hull and the other five on individual seats. The troops enter and leave the vehicle by twin doors in the rear of the hull. The troop compartment has an open top which can be covered by a tarpaulin which is normally carried in the vehicle.

There are three pintle mounts for machine guns, one on the rail that runs across the top of the troop compartment immediately behind the commander's and driver's position, and the other two on either side of the troop compartment. The basic vehicle is armed with a single 7.62 mm SGMB machine gun which is normally mounted on the front pintle. This weapon has an elevation of +23.5°, a depression of –6° and a traverse of 45° left and right of the vehicle's centreline.

Early production vehicles had no firing ports but later production vehicles have two or three in each side of the hull and two at the rear of the hull, one either side of the twin doors.

A spare wheel is carried on the rear of the hull in the centre and a saw is carried on the left side of the hull. Some vehicles have a winch with a capacity of 4,500 kg mounted at the front of the hull. The basic BTR-40 does not have an NBC system, night vision equipment, an amphibious capability or a central tyre-pressure regulation system.

Variants

BTR-40A anti-aircraft vehicle
This is armed with twin 14.5 mm KPV heavy machine guns in a turret (ZTPU-2) with a traverse of 360°.

Specifications

Crew: 2 + 8
Configuration: 4 × 4
Combat weight: 5,300 kg
Power-to-weight ratio: 15 hp/t
Length: 5 m
Width: 1.9 m
Height:
(without armament) 1.75 m
(BTR-40A) 2.5 m
Firing height:
(BTR-40A) 2.05 m
Ground clearance: 0.275 m
Track:
(front) 1.588 m
(rear) 1.6 m
Wheelbase: 2.7 m
Max road speed: 80 km/h
Fuel capacity: 120 litres
Max road range: 285 km
Fuel consumption: 0.42 litre/km
Fording: 0.8 m
Gradient: 60%
Side slope: 30%
Vertical obstacle: 0.47 m
Trench (with channels): 0.7 m
Turning radius: 7 m
Engine: GAZ-40 6-cylinder water-cooled in-line petrol developing 80 hp at 3,400 rpm
Transmission: manual with 4 forward and 1 reverse gears
Transfer box: 2 speed
Steering: cone worm and dual-ridge roller
Clutch: single dry plate
Tyres: 9.75 × 18
Brakes: hydraulic drum on all wheels
Suspension:
(front) longitudinal semi-elliptical springs with hydraulic shock-absorbers
(rear) longitudinal semi-elliptical springs with check springs
Electrical system: 12 V
Main armament: 1 × 7.62 mm SGMB MG
Ammunition: (7.62 mm) 1,250
Armour: 8 mm max
NBC system: no
Night vision equipment: no

The guns have an elevation of +80° and a depression of –5°, both elevation and traverse being manual. The BTR-40A has a crew of five and carries 2,400 rounds of 14.5 mm ammunition. The guns have a maximum horizontal range of 8,000 m and a maximum vertical range of 5,000 m, but effective anti-aircraft range is 1,400 m and effective ground range is 2,000 m. These weapons are also mounted on the BTR-152A (6 × 6) vehicle and towed models of the KPV are called the ZPU-1 (single barrel), ZPU-2 (twin barrel) and the ZPU-4 (quad barrel).

This vehicle was developed at the same time as the BTR-40 armoured personnel carrier. Although normally referred to as an anti-aircraft gun system, it was also employed as a long-range support weapon against ground targets.

BTR-40B armoured personnel carrier

This is similar to the basic model but has overhead armour protection and four roof hatches, two per side, that open to the outside of the vehicle, with a circular firing port. This entered service in 1957 and can carry six troops plus its crew of two.

BTR-40Kh chemical reconnaissance vehicle

This is fitted with equipment at the rear of the hull which enables the crew to set marking pennants into the ground while the vehicle is crossing contaminated ground.

BTR-40 zhd

This rail scout vehicle was introduced in 1969 and is a BTR-40 with a metal wheel system for running on railway lines. It was derived from the earlier BA-64G rail scout vehicle based on the BA-64 (4 × 4) light armoured car.

Walid APC

The Egyptian Walid (4 × 4) APC is very similar to the BTR-40 and was built in Egypt by the Kader Factory for Developed Industries. Full details of this vehicle are given under Egypt. This vehicle has now been phased out of service with Egypt although it still remains in service with some other countries.

Status

Production complete. In service with the following countries:

Country	Quantity	Comment
Afghanistan	n/avail	
Burundi	40	
Cuba	700	estimate, incl BTR-40, BTR-50, BTR-60 and BTR-152
Guinea	10	
Guinea Bissau	35	incl BTR-40/BTR-60/ BTR-152
Indonesia	85	
Korea, North	n/avail	
Laos	70	incl BTR-40, BTR-60, BTR-152
Mali	30	
Mongolia	300	incl BTR-40, BTR-60 and BTR-152
Syria	n/avail	
Tanzania	66	incl BTR-152
Vietnam	200	estimate
Yemen	500	incl BTR-40/BTR-60/ BTR-152

Manufacturer

Russian state factories.

VERIFIED

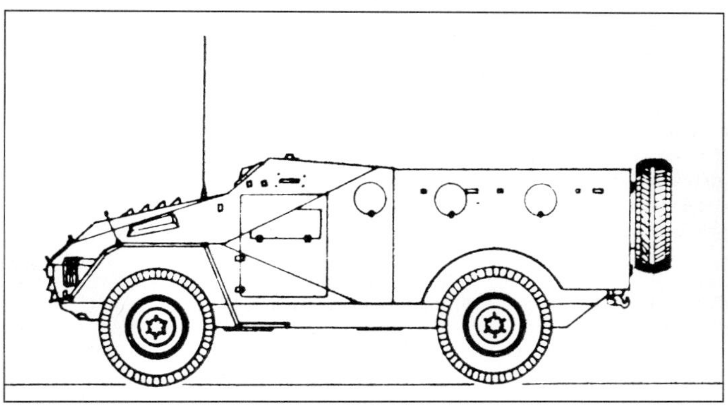

Unarmed BTR-40 (4 × 4) armoured personnel carrier

BTR-40 (4 × 4) APC with side door open and front observation hatches locked in open position (Michael Jerchel)

GAZ-3937 Vodnik light armoured vehicle

Development

The GAZ-3937 series of 4 ×4 light armoured vehicles, also referred to as the Vodnik, has been developed by the Arzamas Machinery Plant as the basic member of a complete family of light armoured vehicles which can be used for a wide range of para-military and military roles.

Trials with the first prototype GAZ-3937 vehicles commenced around 1999 and it has been produced in small pre-production batches for trials purposes.

The original vehicle was called the GAZ-47 which was produced in wheeled (GAZ-4701) and tracked (GAZ-4707) versions but it was decided to concentrate on the wheeled version. As far as it is known, no variants of the Vodnik have entered service with the Russian Army so far.

Description

The construction of the GAZ-3937 is of modular construction so that it can be rapidly adapted for a wide range of missions. The wheeled chassis, which uses parts from the Russian BRDM-2 (4 × 4) amphibious scout car and BTR-80 (8 × 8) Armoured Personnel Carrier (APC), has two removable modules, front and rear. These are of all welded steel armour construction that provides protection from small arms fire and shell splinters.

The front module consists of the engine compartment and drivers compartment divided by the hermetically sealed partition.

The engine is at the front with driver seated on left and vehicle commander on right. Each has a forward opening side door and a circular roof hatch that opens to the rear.

The rear module is of functional volume, which can be used for the transportation of people or cargo as well as the installation of specialised equipment.

According to Russian sources, the design of the GAZ-3937 allows for the use of no less than 26 replaceable modules for a wide range of missions.

The APC version has a slightly higher roofline with roof hatches, side hatches and two doors in the rear, firing ports are also provided. Another version has the projected drivers/commanders position on the left side.

The main advantage claimed for the vehicle is the quick exchange of different modules on the same vehicle chassis even under field conditions.

It has been designed to be operated with outdoor temperatures ranging from –40 to +50°C, in a wind speed of up to 30 m/s and in altitudes of up to 4,500 m.

It is fully amphibious being propelled in the water by its wheels and standard equipment includes powered steering on the front axles and a central tyre-pressure regulation system that allows the driver to adjust the tyre-pressure to suite the terrain being crossed.

It can tow trailers and weapons, for example a 122 mm D-30 howitzer if required.

Although the Vodnik is a 4 × 4 vehicle, the front wheels can be disengaged if required. Standard communications equipment on Russian vehicles includes an R-174 intercom system with provision for installation of an R-163-50U radio set.

Variants

GAZ-3937-10 armoured car

This was developed in 1995 and first seen in 1996 and is essentially the GAZ-3937 fitted with the complete one person turret of the BTR-80 (8 × 8) APC armed with one 14.5 KPVT and one 7.62 mm PKT machine guns and a bank of six 81 mm smoke grenade launchers at the rear. In addition to its crew of three it can carry seven infantrymen.

GAZ-3937 APC

In this version the driver is seated on the left in his own cab with the fighting compartment at the rear with side and roof hatches and doors in the hull rear. The roof has a distinct chamfer to its sides.

GAZ-39371

This has seats for the commander, driver and radio operator and also features full size doors at the front for the driver and commander that open to the front. This can be fitted with various turrets including a one person turret armed with a 14.5 mm KPVT and 7.62 mm PKT machine gun or the same turret as fitted to the BTR-80A armed with a 30 mm cannon and 7.62 mm co-axial machine gun. Basic specifications of the GAZ-39371 are:

Crew: 3 + 8
Configuration: 4 × 4
Combat weight: 7,500 kg
Payload: 2,000 kg
Length: 5.50 m
Width: 2.60 m
Height: 2.15 m (rear module)
Ground clearance: 0.475 m
Track: 2.20 m
Wheelbase: 3.0 m
Max road speed: 115-130 km/h
Max water speed: 4-5 km/h
Max range: 1,000 km

GAZ-3937 Vodnik (4 × 4) APC with fully enclosed rear body 0018632

Specifications

(GAZ-3937)
Crew: 1 + 10
Configuration: 4 × 4
Combat weight: 6,600-7,500 kg
Payload: 1,500-2,000 kg
Power-to-weight ratio: 23/33 hp/t (depends on vehicle weight)
Length: 5.10 m
Width: 2.60 m
Height: 2.30 m
Ground clearance: 0.475 m
Track: 2.20 m
Wheelbase: 3.0 m
Max road speed: 120-140 km/h
Max water speed: 4.5 km/h
Max range: 1,000 km
Engine: GAZ-5423 5-cylinder turbo-charged diesel developing 175 hp

Vodnik SAM system

In 2001 it was revealed that the Russian Defence System consortium was marketing a new long-range Infra-Red Search and Track (IRST) system called Phoenix. This can be installed on a variety of land and sea platforms to detect targets without revealing its presence.

Phoenix operates in the 8 to 12 μm spectral range and is claimed to be able to detect low-altitude missiles at a range of 5 to 7 km, helicopters at 8 to 9 km, fighter type aircraft at 15 to 18 km and cargo aircraft at over 20 km. More than 50 targets can be tracked simultaneously. It can also track a variety of land and sea-based targets.

The IRST includes the sensor pod, display and control console with the signal processing and servo control unit mounted below and the power supply. It has full 360° coverage in azimuth and from −10 to plus 40° in elevation.

The first application of Phoenix is a self-propelled Surface-to-Air Missile (SAM) system based on the GAZ-3937 (4 × 4). Mounted on the roof of the GAZ-3937 is the Phoenix IRST and mounted either side are four fire-and-forget SAMs on the ready to launch position and one 7.62 mm PKT machine gun. The missile sighting system is mounted between the pods of weapons and moves in elevation with these weapons.

In a typical target engagement the target would first be detected by the IRST and if confirmed as hostile would be engaged by the fire-and-forget SAMs. The two 7.62 mm PKT machine guns are fitted for close in targets as well as providing a self-defence capability for the platform.

GAZ-3937 120 mm mortar

This has a new flat platform at the rear of which is mounted a 120 mm muzzle loaded mortar, with a shorter crew compartment at the front. If required the flatbed can be fitted with bows and a tarpaulin cover.

GAZ-39344 Siam

Details of this are given in the reconnaissance vehicles section. This has been developed mainly for internal security applications.

GAZ-3937 IS vehicle

This is GAZ-3934 adopted for the internal security role and also has the turret from the BTR-80 (8 × 8) APC installed.

Status

It is understood that low-rate production of at least one version of the GAZ-3937 is now under way.

Manufacturer

Arzamas Machinery Plant.

UPDATED

SINGAPORE

Singapore Technologies Kinetics Terrex AV81 armoured infantry fighting vehicle

Development

The Terrex AV81 (8 × 8) Armoured Infantry Fighting Vehicle (AIFV) has been developed as a private venture by Singapore Technologies Kinetics in close association with Timoney High Mobility Vehicles of Ireland who have considerable experience in the design, development and manufacture of wheeled armoured fighting vehicles.

The first example of the Terrex AV81 was completed in Ireland in July 2001 and, after initial running trials, went to the UK and was prepared for its first public appearance in September 2001, by ABRO at Warminister.

Following its first public appearance at the Defence Systems and Equipment International Exhibition and Conference held in London, the Terrex 81 underwent extensive series of reliability trials in the UK and then returned to Singapore. Information from these trials will be incorporated into the next vehicle which is expected to be completed in 2003.

Artist's impression of Singapore Technologies Kinetics Terrex AV81 (8 × 8) AIFV fitted with two person turret armed with 25 mm cannon and 7.62 mm coaxial machine gun **2002**/0095379

Specifications

Crew:
(APC variant) 2 + 12
(25 mm variants) 3 + 8
Configuration: 8 × 8
Maximum weight: 24,000 kg
Combat weight: 17,500 kg
Unloaded weight: 13,000 kg (without turret)
Power-to-weight ratio: 22.85 hp/tonne at a combat weight of 17.5 tonnes
Length: 6.8 m
Width: 2.7 m
Height:
(hull top) 2.1 m
(turret top) depends on turret
Ground clearance: 0.35 m
Track: n/available
Wheelbase: 1.45 + 1.50 + 1.45 m
Angle of approach/departure: not available
Max speed:
(road) 120 km/h
(water) 10 km/h
Acceleration:
(0 to 30 km/h) 4 s
Fuel capacity: 400 litres
Max range: not available
Fording: 1.5 m (or amphibious)
Gradient: 70%
Side slope: 40%
Vertical obstacle: 1 m
Trench: 2.5 m
Turning radius:
(normal) 7.5 m
(electronic control system) 4.5 m
Engine: Caterpillar C-9 four-stroke turbocharged air-to-air aftercooled diesel developing 400 bhp at 2,200 rpm
Transmission: Allison MD4560P fully automatic
Steering: power-assisted, front four wheels
Suspension: Timoney independent double wishbone with coil springs and shock-absorbers (interchangeable with hydrostruts)
Tyres: 395/85 R20 XML with central tyre inflation system with option for 475/80 R20
Brakes:
(main) hydraulic, assisted by compressed air disc brake
(parking) self-adjusting. Hydraulic actuated disc brake
Electrical system: 24 V
Alternator: not available
Batteries: not available
Armament: depends on role
Ammunition: depends on role

Gun control equipment: depends on weapon installation
Turret power control: depends on weapon installation
Gun elevation/depression: depends on weapon installation
Turret traverse: depends on weapon installation
NBC system: yes
Night vision equipment: yes

The Terrex 81 has a nominal payload of 11 tonnes which includes the crew, weapon station and ammunition. Maximum gross vehicle weight is 24,000 kg, which includes weapons, ammunition, crew and fuel. This version is not however amphibious.

Although the Terrex AV81 has been developed as a private venture it is already being marketed on a worldwide basis.

Today the Singapore Armed Forces only uses tracked Armoured Personnel Carriers (APCs). These include the Singapore Technologies Kinetics Bionix and United Defense M113 series vehicles upgraded by Singapore Technologies Kinetics.

As the Singapore Armed Forces is now placing increased emphasis on peacekeeping operations, there may well be a local requirement for a vehicle such as the Terrex AV81.

Description

The hull of the Terrex AV81 is of all welded steel armour construction to which an additional layer of passive armour can be added for a higher level of battlefield survivability.

The Singapore Technologies Kinetics Terrex AV81 (8 × 8) AIFV fitted with a Remote Weapon Station armed with 40 mm grenade launcher and 7.62 mm machine gun (Christopher F Foss) ***2002***/0096171

The floor of the Terrex AV81 is of a V shape to help deflect the blast of anti-tank mines. It has spaced armour in the floor and other parts of the hull for increased battlefield survivablity.

The driver is seated at the front of the vehicle on the left side with the power pack to his right. The driver has a single-piece hatch cover to the immediate front of which are three day periscopes, the centre one of which can be replaced by a passive periscope for driving at night. The driver can also enter the vehicle via the troop compartment to the immediate rear of his position.

Power pack of the Terrex AV81 consists of a Caterpillar C9 four-stroke turbocharged diesel engine developing 400 bhp, coupled to an Allison MD4560P fully automatic transmission. The air intake and outlet louvres are in the roof with the exhaust outlet on the right side of the hull.

This power pack gives a maximum road speed of 120 km/h. A Timoney axle box is fitted and provides automatic traction control.

The weapon station is mounted in the centre of the hull with the first example of the Terrex AV81 being fitted with a a Remote Weapon Station that is armed with a 40 mm Automatic Grenade Launcher (AGL) and 7.62 mm machine gun. These can be aimed and fired with the gunner seated within the hull under full armour protection. On the lower part of the Remote Weapon Station is a bank of five electrically operated smoke grenade launchers firing forwards.

Additional 7.62 mm machine guns can be fitted over the rear troop compartment to provide suppressive fire over the rear of the vehicle.

When being used as a troop carrier it has an internal volume of 12.2 m³. In its baseline troop carrier version it has a crew of two and can carry up to 12 fully equipped troops.

The normal means of entry and exit for the troops is via the large power-operated ramp in the hull rear which is also provided with an integral door. In addition there are two hatches over the troop compartment which open to the left and right.

The Terrex AV81 uses a patented double wishbone suspension that is claimed to greatly improve ground mobility and ride comfort over the roughest terrain.

Its mobility is further enhanced by the built-in central tyre inflation system. This allows the tyres to be inflated and deflated automatically to suit different terrain while the vehicle is on the move.

Steering is power-assisted on the front four wheels. It is also fitted with an all-wheel Anti-Lock Braking System (ABS) which is claimed to improve the safety of the crew and also allow better control of the vehicle.

The Terrex AV81 is fully amphibious, being propelled in the water by two hydraulically driven and steered water jets mounted one either side of the hull at the rear. An NBC system is fitted as standard. The vehicle is also fully airportable in a Lockheed Martin C-130 Hercules transport aircraft.

The Terrex is of modular design with a flexible top-deck configuration for rapid role changes. It has also been designed with technology insertions in mind, for example, different power packs, hydraulic or electric hybrid drive systems, appliqué armour, hydro pneumatic suspension and a power management system.

Variants

25 mm two-person turret

A wide range of other turrets can be fitted to the Terrex AV81 including the complete two-person turret installed on the Bionix 25 infantry fighting vehicle, already in service with the Singapore Armed Forces (SAF). This is armed with a stabilised Boeing 25 mm M242 cannon and 7.62 mm coaxial machine gun. Details of this turret are given in the entry for the Bionix 25 in the Armoured personnel carriers (tracked) section.

105 mm turret

Singapore Technologies Kinetics is also considering the installation of a turret armed with a low recoil 105 mm gun and a 7.62 mm coaxial machine gun.

155 mm self-propelled artillery system

This would have a fully enclosed turret mounted on the hull rear fitted with a 155 mm/52 calibre ordnance.

120 mm mortar carrier

A self-propelled mortar carrier armed with the new Singapore Technologies Kinetics 120 mm Super Rapid Advanced Mortar System (SRAMS) has also been proposed.

Command post

The Terrex AV81 can be adopted for use in the command post role.

Sensor platform

The Terrex AV81 can be adopted as a platform for a wide range of battlefield sensors including EO devices.

Rocket launcher

It can also be fitted with various guided missile systems such as air defence or anti-tank or surface-to-surface rocket launchers.

Status

Prototype undergoing trials. Not yet in production or service.

Manufacturer

Singapore Technologies Kinetics

NEW ENTRY

SLOVENIA

Vozila BOV-VP armoured personnel carrier

Development

In the early 1980s, Yugoslavia developed a new family of 4 × 4 wheeled armoured fighting vehicles which shares the same basic 4 × 4 chassis, engine, transmission and suspension. This family is known as the BOV and is used both by the army and police.

The BOV family of vehicles was originally manufactured in Yugoslavia but as a result of the break up of Yugoslavia into a number of separate elements, the production line is now in Slovenia at the facilities of MMP Vozila doo.

Recent information is that production has now ceased and that the BOV-VP series of 4 × 4 armoured personnel carriers is no longer being marketed.

In the future Slovenia is expected to acquire additional quantities of the Austrian Steyr-Daimler-Puch Pandur (6 × 6) series vehicles that are manufactured under licence by Ravne na Koroskem. These Pandur vehicles are known locally as the Valuk.

Description

The basic APC is called the BOV-VP and carries a total of eight fully armed men plus commander and driver.

The hull of the vehicle is of all-welded steel construction that provides complete protection from small arms fire and shell splinters. It is believed that maximum armour thickness of the BOV is 8 mm. The driver sits at the front left with the commander to the right, both with a rear-opening single-piece hatch cover. The driver has a single forward-facing day periscope in the roof which can be replaced by a passive periscope. To the immediate front and either side of the commander and driver is a bulletproof window covered by a wire mesh screen.

The eight infantry are seated four either side down the centre of the vehicle and enter and leave via a single door in each side that opens towards the front. Over the top of the troop compartment is a cupola with a two-part roof hatch that opens left and right, an externally mounted 7.62 mm M86 machine gun and a searchlight. The 7.62 mm machine gun is manned by one of the eight infantry. In either side of the troop compartment are three firing ports each with an associated vision block and there is an additional firing port to the right of the commander's windscreen. Windows are also provided in the rear of the troop compartment.

The engine compartment is at the rear of the hull with air inlet and outlet louvres on the top and an engine access door at the rear.

The vehicle is powered by a German Deutz F 6L 413 F six-cylinder diesel developing 150 hp at 2,650 rpm. It is coupled to a manual transmission with five forward and one reverse gears and a two-speed transfer case.

Steering is power-assisted on the front wheels to reduce driver fatigue and a central tyre pressure regulation system is fitted as standard, allowing pressure to be adjusted from 0.7 to 0.35 bar to suit the type of ground being crossed. Main brakes are air-hydraulic with a hand-operated parking brake. Suspension front and rear consists of leaf-type springs with telescopic shock-absorbers with 1300 - 18 PR10 cross-country tyres fitted as standard. The differential locks are controlled electropneumatically.

Standard equipment includes a Jugo-Webasto 7.5 kW heater, day and night vision equipment, intercom and radios.

Mounted either side of the hull is a bank of three forward-firing electrically operated smoke grenade dischargers and either side of the forward part of the hull is a wire mesh screen that lies alongside the hull when travelling and when required, for example in a riot situation, is swung out through 90°.

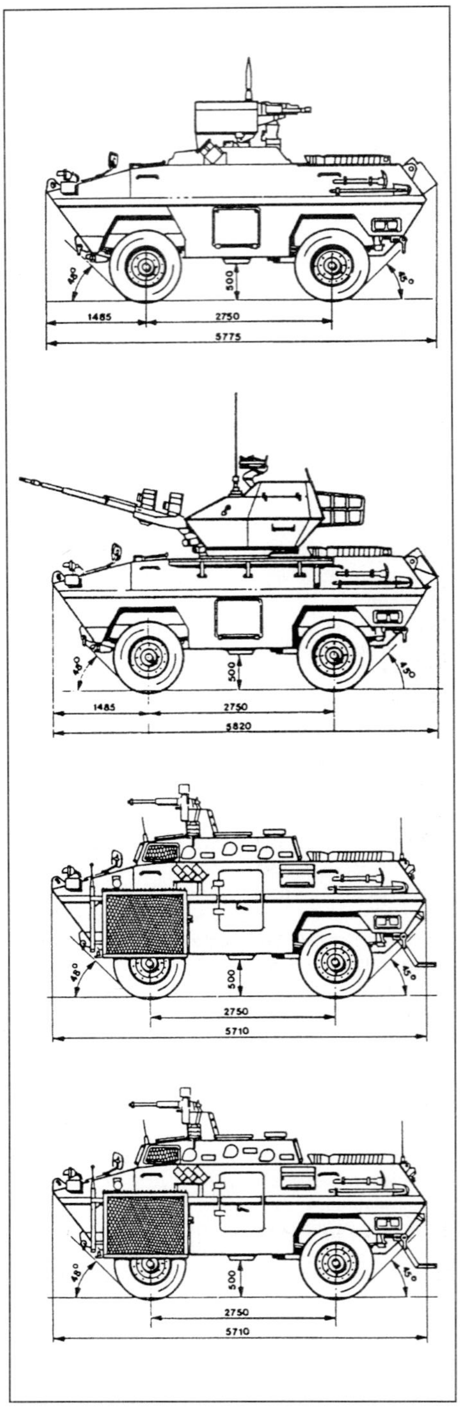

BOV family of 4 × 4 armoured vehicles from top to bottom includes the BOV-1 anti-tank, BOV-3 SPAAG, BOV-AP police and BOV-SN ambulance **2000**/0082000

Specifications
BOV-VP
Crew: 2 + 8
Configuration: 4 × 4
Weight: 9,400 kg
Power-to-weight ratio: 15.94 hp/t
Length: 5.70 m
Width: 2.534 m
(with screens) 2.92 m
(with screens fully extended) 7.6 m
Height
(hull roof) 2.335 m
(hull top) 1.99 m
Ground clearance: 0.325 m
Track: 1.90 m
Wheelbase: 2.75 m
Angle of approach/departure: 48°/45°
Max speed: 95 km/h
Range: 500 to 800 km
Fuel capacity: 220 litres
Fording: 1.1 m
Gradient: 55%
Side slope: 30%
Vertical obstacle: 0.54 m
Trench: 0.64 m
Turning radius: 10 m
Engine: Deutz type F 6L 413 F 6-cylinder diesel developing 150 hp at 2,650 rpm
Transmission: manual, 5 forward and 1 reverse gears
Suspension: leaf springs and telescopic shock-absorbers
Tyres: 1300 - 18 PR10
Electrical system: 24 V
Armament: 1 × 7.62 mm M86 MG
Smoke grenade dischargers: 2 × 3
NBC system: no
Night vision equipment: yes

BOV-1 ATGW system from rear showing six missiles in the ready to launch position **2000**/0081998

BOV-3 triple 20 mm self-propelled anti-aircraft gun system showing magazines mounted above barrels **2000**/0081999

Variants

BOV-1 ATGW carrier

Mounted on the roof of this vehicle is a pod containing six ATGWs based on the Russian designed and locally manufactued AT-3 'Sagger' but fitted with a semi-automatic guidance system. Additional missiles are carried under armour protection within the hull. Maximum rate of fire is two missiles per minute and maximum range of the missiles is 3,000 m. Between the two pods of missiles is mounted a 7.62 mm machine gun. The BOV-1 ATGW and the two SPAAG chassis are slightly different from the APC described previously as they do not have the raised troop compartment roof.

The official designation for this version is the POLO M-83/BOV-1. In addition to the six missiles in the ready to launch position, there are two banks of two electrically operated smoke grenade launchers located below the missile launchers.

BOV-3 SPAAG

This is fitted with a power-operated turret armed with three 20 mm cannon, each of which has a drum-type magazine holding 60 rounds of ammunition with a total of 1,500 rounds of ammunition being carried. Maximum anti-aircraft engagement altitude is quoted as 2,000 m with effective anti-aircraft range being between 1,000 and 1,500 m. The turret is fitted with a J-171 optical sight. BOV-3 is a clear weather system only. Full details of the BOV-3 SPAAG are given in *Jane's Land-Based Air Defence 2001-2002.*

BOV-30 SPAAG

This is the latest SPAAG version and is fitted with a larger enclosed turret armed with twin 30 mm cannon with a cyclic rate of fire of 600 rds/gun/min. There are two three-barrelled smoke grenade dischargers mounted either side of the turret. This model was first seen in 1985. Details of this are given in *Jane's Land-Based Air Defence 2001-2002.* It is not believed that this was manufactured in very large numbers for the former Yugoslav Army.

BOV-SN ambulance

This has a similar hull to the BOV-SN but has been modified for use in the ambulance role.

Status

Production as required. In service with Croatia, Slovenia and Yugoslavia.

Manufacturer

MPP Vozila doo.

UPDATED

SOUTH AFRICA

Vickers OMC Ratel infantry fighting vehicle

Development

In 1968, Sandock-Austral (which in the spring of 1994 became Reumech Sandock, which became Reumech OMC only to be purchased by Vickers Defence Systems, of the UK, late in 1999 and then renamed Vickers OMC), then building an improved version of the French Panhard AML (4 × 4) armoured car called the Eland, started design work on a new infantry fighting vehicle, the Ratel, to meet the requirements of the South African Infantry Corps and, at a later date, the South African Armoured Corps.

The first prototype was completed in July 1974, followed by four preproduction vehicles, the first completed in 1976. The first production models were designated the Mk 1 and were followed in 1979 by the Mk 2, final deliveries of which were made in early 1987. The differences are mainly the result of operational experience in Namibia and long-range penetration raids into Angola.

Most of the modifications were subsequently retrofitted to earlier vehicles when they returned for overhaul.

The Ratel Mk 3 upgrading programme commenced in 1985 and mainly included the following modifications to the vehicle:

(1) Improved cooling system
(2) Improved bush protection
(3) Automatic cocking system for the 20 mm G12 cannon built by LIW division of Denel
(4) Smaller functional modifications to the vehicle.

Although production of the Ratel infantry fighting vehicle was completed some time ago, the vehicle is still being marketed by Vickers OMC.

The company is also upgrading earlier vehicles to the latest Ratel Mk 3 standard at its Benoni facility.

It is understood that the South African National Defence Force has a total of 1,243 Ratel vehicles in service, plus an additional 53 in the Swift anti-tank vehicle configuration. Details of this are given in the Tank Destroyers section. This model is only in service in South Africa.

Description

(FSV 90)

The hull of the Ratel is made of all-welded steel construction providing complete protection from 7.62 mm small arms fire and shell splinters, with protection over the frontal arc against 12.7 mm armour-piercing rounds. Emphasis has also been placed on providing protection against mines.

The driver sits at the front of the vehicle with excellent forward visibility through three large bulletproof windows to the front and sides. In combat areas these can be quickly covered by armoured shutters hinged at the bottom and activated by the driver with a single handle from within the vehicle. With the shutters in position the driver observes the terrain through three day periscopes, one to the front and one either side. Over the driver's position is a single-piece hatch cover that opens to the left. The driver's seat and steering wheel are adjustable and he can also enter his compartment from the rear so enabling drivers to be changed without either of them leaving the vehicle.

Ratel FSV Mk 3 complete with stowage and with 90 mm gun and roof-mounted 7.62 mm machine gun covered up 0005389

Mounted on the roof of the vehicle immediately behind the driver's position is the two-man all-welded steel turret, identical to that of the Vickers OMC Eland (4 × 4) armoured car. The commander sits on the left and the gunner on the right. The commander's cupola has vision blocks for all-round observation and a single-piece hatch cover opening to the rear, plus an orientation vane sight. The gunner has four day periscopes for observation plus an M494 optical day sight for aiming the main and secondary armament, and a single-piece hatch cover that opens to the rear.

The main armament is an LIW (a Division of Denel) 90 mm GT2 semi-auto rifled quick-firing gun which is a further development of the weapon fitted in the 90 mm version of the Eland (4 × 4) armoured car. This has an elevation of +15° and a depression of –8° with turret traverse being a full 360°. Depression can be obtained only through 285° because of the troop compartment to the turret rear. Gun elevation and turret traverse are manual, operated by the gunner. When travelling, the turret is often traversed to the rear. The 90 mm gun fires HEAT (effective range 1,200 m), HE (effective range 2,200 m) and practice (effective range 1,200 m) projectiles. There are 29 rounds of 90 mm ammunition carried in the turret and 42 rounds in the hull. A 12.7 mm M2 HB machine gun can be mounted over the 90 mm gun for training.

Mounted to the left of the 90 mm gun is a 7.62 mm coaxial machine gun, for which 2,000 rounds of 7.62 mm ammunition are carried in the turret with a further 4,000 rounds in the hull. Both the main and coaxial weapons are fired electrically and above the coaxial machine gun is an extractor fan to remove fumes from the turret.

Mounted either side of the turret towards the rear is a bank of two 81 mm electrically operated smoke grenade dischargers which can be operated by the commander or gunner. A 7.62 mm anti-aircraft machine gun is mounted externally at the gunner's station and on top of the turret is a hand-operated searchlight. In the left side of the turret there is an ammunition resupply hatch.

The crew consists of the vehicle commander, driver, main gunner (90 mm), rear anti-aircraft gunner, section commander and five infantrymen.

In either side of the hull behind the driver's position is a large door opened pneumatically by the driver. It opens forwards and has a vision block with a firing port underneath.

The section commander sits near the left door and the five infantrymen on bench seats down the middle of the vehicle behind the turret, two on the left and three on the right. There are three bulletproof vision blocks, with firing ports underneath, and over the top in both sides of the troop compartment are four roof hatches that are hinged on the outside and can be locked vertical. There is a passageway from the troop compartment to the rear of the vehicle on the right side, above which are two roof hatches that open outwards and a circular mount with a single-piece hatch cover on which the 7.62 mm machine gun can be mounted.

In the rear of the hull on the right side is a door. Its lower part folds down to form a step and the upper part opens to the left.

The engine compartment is at the rear of the Ratel on the left side with access panels in the roof for maintenance.

The Bussing D 3256 BTXF six-cylinder direct injection turbocharged diesel is coupled to a fully automatic power shift gearbox with a hydrodynamic torque converter. The gearbox can also be operated manually and has a mechanical emergency gearshift. Power is transmitted to the three axles in two stages with final reduction by means of planetary gearing housed in the wheel hubs. The three axles are provided with their own lockable differential and longitudinal differential locks. Suspension consists of progressively acting coil springs and large hydraulic shock-absorbers. The hydropneumatic braking system can be assisted by an exhaust brake provided through the engine.

The power pack can be removed from the vehicle by two men with a crane in less than 30 minutes. Quick disconnect couplings and connectors are provided in all power, hydraulic, electric and pneumatic circuits.

Standard equipment includes a complete set of tools, spades, picks, shovels and axes, a towbar and towing cables, a first aid kit, two petrol stoves, fire extinguishers, emergency spares, two 50 litre drinking water tanks, two external stowage boxes, radios, an intercom and a remote handset with 1,000 m of cable.

Optional equipment includes night vision equipment, an air conditioning and ventilation system, and the replacement of the standard D3256 BTXF six-cylinder diesel developing 282 hp with an ADE 407 TI diesel developing 315 hp.

Ratel IFV fitted with LIW LCT-35 turret armed with 35 mm cannon and 7.62 mm machine gun **2001**/0080938

Specifications

Model Ratel 90
Crew: 10
Configuration: 6 × 6
Combat weight: 19,000 kg
Unloaded weight: 17,000 kg
Power-to-weight ratio: 14.84 hp/t
Length: (hull) 7.212 m
Width: (hull) 2.516 m
Height
(overall) 2.915 m
(to hull top) 2.105 m
Ground clearance: 0.34 m
Track: 2.08 m
Wheelbase: 2.809 + 1.4 m
Angle of approach/departure: 44/45°
Max road speed: 105 km/h
Fuel capacity: 430 litres
Max range
(road) 860 km
(cross-country) 430 km
Fuel consumption
(road) 0.4 litres/km
(cross-country) 1 litre/km
Fording: 1.2 m
Gradient: 60%
Side slope: 30%
Vertical obstacle: 0.6 m
Trench: 1.15 m
Turning radius: 8.5 m
Engine: Bussing D 3256 BTXF 6-cylinder in-line turbocharged diesel developing 282 hp at 2,200 rpm
Transmission: Renk HSU 106 automatic, 6 forward and 2 reverse gears with hydrodynamic torque converter
Steering: mechanical, recirculating ball with hydraulic assistance
Tyres: 14.00 × 20*
Brakes
(main) hydropneumatic on front wheels and air/mechanical on rear 4 wheels
(crawl) pneumatic/mechanical on rear 4 wheels
(parking) mechanical on rear 4 wheels
Axles
(front) steering axle with steering damper and intermediate steering arm, lockable differential
(middle)non-steering with transfer train, lockable differential (transfer and longitudinal)
(rear) non-steering and lockable differential
Suspension: solid axle with single-coil springs and double acting hydraulic shock absorbers supported by wishbones and longitudinal arms
Electrical system: 24 V
Batteries: 2 × 12 V, 100 Ah
Armament
(main) 1 × 90 mm gun
(coaxial) 1 × 7.62 mm MG
(anti-aircraft, turret) 1 × 7.62 mm MG
(anti-aircraft, rear) 1 × 7.62 mm MG
Ammunition
(90 mm) 69
(7.62 mm) 6,000

Gun control equipment
Turret power control: manual, 360°
(by commander) yes
Gun elevation/depression: (main) +15/−8°
Armour
Hull front: 20 mm at 60°
Hull sides upper: 8 mm at 65°
Hull sides lower: 10 mm at 90°
Hull top: 6 mm
Hull floor: 8 mm and 10 mm
Hull rear: 10 mm
NBC system: no
Night vision equipment: optional

* with run-flat inserts

Variants

Ratel 60 IFV

This model has a crew of 11 consisting of a vehicle commander, driver, main gunner, anti-aircraft gunner, section commander and six infantrymen. It is fitted with a two-man turret armed with a 60 mm M2 mortar and 7.62 mm coaxial and 7.62 mm anti-aircraft machine guns. An additional 7.62 mm anti-aircraft machine gun is mounted at the rear of the vehicle on the right side. The 60 mm mortar is breech loaded and is capable of firing HE, smoke and illuminating bombs to a maximum range of 1,500 m. A typical ammunition load would consist of 38 to 45 HE bombs, three smoke and three illuminating bombs.

Ratel 20 IFV

This model has the same crew as the Ratel 60 IFV, but is fitted with a two-man turret armed with a 20 mm F2 cannon (M693), and a 7.62 mm machine gun mounted coaxially to the left, and a similar weapon can be fitted on the turret roof for anti-aircraft defence. There is also a 7.62 mm anti-aircraft machine gun at the rear on the right side. Turret traverse and weapon elevation are manual. The weapons have an elevation of +38° and a depression of −8°. The 20 mm cannon fires armour-piercing (effective range 1,000 m), high-explosive (effective range 2,000 m) and practice (effective range 1,000 m) projectiles.

Ratel 12.7 mm command

This command vehicle has a crew of nine: the vehicle commander, driver, main gunner and six command post personnel. The two-man turret is armed with a 12.7 mm L4 machine gun and a 7.62 mm machine gun, mounted at the hull rear as on other members of the Ratel family. This command vehicle is fitted with three radios, a tape recorder with time injection, a civilian-type combined receiver and cassette recorder in the hull, an intercom for all key personnel, internal loudspeakers, a public address system, a pneumatic mast and map boards.

Ratel armoured repair vehicle

This is simply a Ratel 20 with a fixed lifting jib made up of four tubular steel posts mounted on the rear hull with a block and tackle arrangement to raise and carry damaged vehicles such as the Ratel. A wide range of spare parts and repair equipment is carried inside the vehicle and in a wire mesh stowage box on the front of the hull. An air compressor is carried, as are spare wheels.

Ratel 81 mm mortar carrier

Production of the Ratel was to have been completed late in 1985 but this was extended to January 1987 as the South African Army wanted an 81 mm mortar version of this vehicle.

The 81 mm mortar is mounted on a turntable which can be quickly traversed through a full 360°. It fires through a two-part circular roof hatch which opens front and rear.

The 81 mm mortar carrier has a crew of six including the driver. There are 150 rounds of 81 mm mortar ammunition carried. The commander sits on a drop-down seat to the rear of the driver and has a raised cupola armed with a 7.62 mm machine gun. This version has only two firing ports and associated vision blocks, one in either side.

More recently the South African company Teklogic has developed a fire-control computer for the Ratel 81 mm self-propelled mortar system. This computer performs all the ballistic calculations required to enable the mortar to be fired quickly and accurately. The Global Positioning System is used to ensure vehicle position accuracy.

The first 81 mm mortar bomb can be fired within 10 seconds of the vehicle coming to a halt with first shot accuracy claimed to be better than 100 m. Full ballistic calculations are performed in real time by the fire-control computer compensating for the attitude (roll and pitch) of the vehicle.

Ratel 120 mm mortar carrier

Shown for the first time in 1994 was the prototype of a Ratel fitted with a 120 mm mortar system.

Mounted in the rear of the Ratel is an LIW 120 mm muzzle-loaded mortar which is used in conjunction with a computer weapon aiming system.

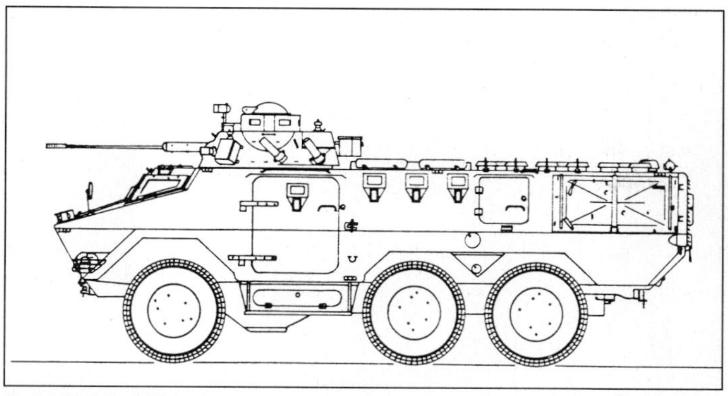

Ratel 20 Mk 2 IFV with a two-man turret armed with a 20 mm cannon and a 7.62 mm machine gun

Ratel 60 Mk 3 infantry fighting vehicle with all doors closed 0005391

This uses data (type of mortar bomb, charge, environmental conditions and target location) in conjunction with a land navigation system to enable targets to be rapidly engaged.

The 120 mm mortar is mounted on a turntable that can be traversed through 360° and is laid onto the target automatically.

The Ratel 120 mm self-propelled mortar carrier can carry a total of 46 mortar bombs with the normal load consisting of 28 standard and 18 rocket-assisted 120 mm mortar bombs.

Firing with charge 9, a maximum range of over 9,000 m can be achieved, while with the rocket-assisted projectile using charge 10 a range of over 12,000 m is possible.

As of late 2001, there were no plans for this 120 mm mortar carrier to enter production or service.

Ratel with Swift ATGW system

This was first used in Angola in 1988 and is a Ratel (6 × 6) hull fitted with a new two-man turret on top of which is mounted a bank of three Kentron Swift (ZT-3) laser command-guided ATGWs with an additional 12 missiles being carried inside for manual reloading. The turret also has a 7.62 mm machine gun and two banks of smoke grenade launchers, one on either side, that fire to the front of the vehicle.

Full details of this vehicle, which is only used by the South African National Defence Force, are given in the Tank Destroyers section.

Ratel with LCT-35 turret

For trials purposes a Ratel IFV has been fitted with the LIW Division of Denel LCT-35 turret system.

This is based on the envelope of the Vickers OMC Eland 90 turret with enhanced fire control, fire detecting and fire power. Armament comprises a 35 mm EMAK cannon and 7.62 mm co-axial machine gun with various gunner and commander sight options as well as fire control options.

Enhanced artillery observation system

This forms a key part of the AS-2000 artillery engagement system.

Other elements of the system include a portable target acquisition system for dismounted artillery observers, a target acquisition RPV, datalinks between these systems and regimental and battery fire-control posts, ballistic software that runs on IBM-compatible PCs, a gun management system, a muzzle velocity analyser, a meteorological system and a rugged computer and data communication system.

This is a rebuild of an existing Ratel with its turret removed and a raised superstructure added to the rear of the driver's position. Pivoted either side of the hull is a hydraulically operated mast on top of which is a multisensor head which contains a day/night observation payload consisting of a three-in-one video camera for daytime acquisition, a long-range zoom and FLIR TV cueing, an Eloptro-developed FLIR and an Eloptro LH-31 laser range-finder with a video channel. This system provides accurate daytime observation to 20 km and night observation to 3.5 km.

Onboard systems include an integrated vehicle navigation system and accurate pointing control. A computer automatically adds new targets to the database and maintains the target lists, fire plans and the observation post log. Data communications to the battery fire-control centre are provided.

The vehicle has a five-man crew consisting of the forward observation officer, technical assistant, two signallers and a driver. As of late 2001, this remained at the prototype stage.

Ratel 8 × 8 Logistic Support Vehicle

A prototype of the Ratel 8 × 8 logistic support vehicle was built and tested but it was not placed in production.

Ratel replacement programme

Since the Ratel first entered service with the South African Army in the late 1970s, it has been constantly upgraded, mainly in the automotive area to improve the overall reliability of the vehicle.

Under the South African ICV2000 programme it was expected that some 700 new infantry combat vehicles would have been purchased to replace the current Ratel fleet.

Recent information has indicated that due to funding constraints, between 240 and 300 vehicles will now be procured to equip two or three of the South African regular mechanised infantry battalions. The fourth battalion and the reserve mechanised battalions would retain the Ratel until around 2015.

The new South African ICV will be an 8 × 8 vehicle. It is required to have sufficient armour protection to work with the Rooikat (8 × 8) tank destroyer/armoured car that is currently armed with a 76 mm gun.

To meet this requirement, Vickers OMC will be offering a locally built variant of the Swiss MOWAG Piranha IV (8 × 8) vehicle which was shown in public for the first time in September 2001. Armament could be the new LIW 35 mm EMAK which, in a two-person turret, has already been shown on a Ratel (6 × 6) chassis.

There is also a requirement to replace the current Ratel (6 × 6) tank destroyer with a new system, perhaps armed with the new South African Kentron Ingwe laser beam-riding missile. A possible in service date of 2015 has been mentioned with up to 75 systems being required, funding permitting.

Status

Production complete but can be resumed if further orders are placed. In service with Morocco (60 to 80) and South Africa (1,243 plus 53 Swift anti-tank).

Manufacturer

Vickers OMC.

UPDATED

Vickers OMC Casspir Mk III armoured personnel carrier

Development

The original South African Mine-Protected Vehicle (MPV) was based on a Bedford (4 × 4) truck chassis. This was followed by the Casspir Mk I vehicle, about 200 of which were built between 1979 and 1980.

In 1981, production of the Casspir was taken over by TFM Limited, which developed the improved Mk II vehicle. Externally there is not a lot of difference between the Mk I and Mk II vehicles. The Mk I had a left-side escape hatch which was ineffective, a longer rear overhang and early models had a 120 hp engine.

TFM overhauled a large number of the original Casspir Mk I vehicles from 1981.

The actual name Casspir refers both to the complete family of vehicles and to the armoured personnel carrier model, some 700 of which, in both Mk I and Mk II versions, had been built by late 1986.

The main difference between the Mk II and Mk III vehicles is a change in axle type and ratio, as well as the equipment it is required to carry, hence the overall increase in weight of the vehicle.

Using the same unitary hull, dimensions and driveline, TFM developed a range of supporting vehicles which have also been produced in quantity.

In designing the Casspir range of vehicles, special attention has been paid to protecting the vehicle against land mines. According to the manufacturer, field repair times after a single anti-tank mine, the Russian TM-57 or equivalent, are between 1 and 2 hours and after a multiple detonation of mines, 8 to 12 hours. The practical limit to this mine protection is a mine or mines equivalent to 18 to 20 kg of TNT/RDX.

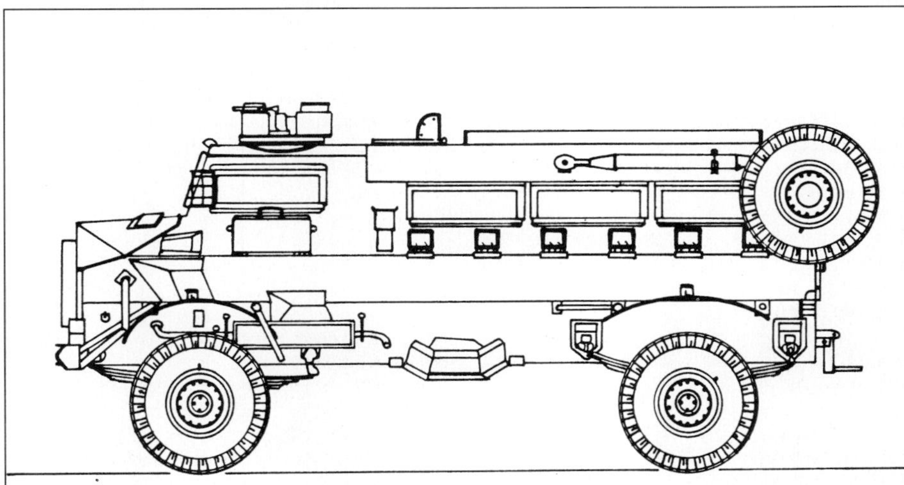

Casspir Mk II APC

By early 1996, total production of the Casspir family of armoured personnel carriers amounted to around 2,500 vehicles, about 460 of which had been refurbished or upgraded by TFM.

Casspir Mk III APC (Christopher F Foss)
2002/0121203

Casspir APC without armament (Christopher F Foss) **2002**/0121202

Production of the Casspir family of 4 × 4 vehicles has not taken place since 1993/94, however, an extensive rebuilding programme of old series Casspirs is currently taking place.

Early in 1997, the TFM Defence and Security Division was taken over by and its product range integrated with Reumech OMC. This company, in turn, was taken over by Vickers Defence Systems of the UK late in 1999 and renamed Vickers OMC.

Within South Africa, most Casspir vehicles are used by the police and internal security forces but in 1996 the South African National Defence Force stated that it had a total of 372 Casspir weapon platforms in service.

Vickers OMC are now offering reconditioned Casspir vehicles for the export market with recent customers including India and Uganda. Additional details of the Indian Casspir vehicles are given later in this entry.

Description

The monocoque hull of the Casspir is of all-welded steel and will withstand attack by 5.56 or 7.62 mm NATO ball ammunition as well as grenade and mortar splinters and shrapnel.

The running components of the Casspir are derived from a commercial 15 tonne truck and the vehicle is powered by a locally produced ADE six-cylinder diesel developing 170 hp coupled to a five-speed manual transmission and a two-speed transfer case. Limited slip differential axles are fitted as standard.

The engine, transmission and fuel tank are all contained within the armoured hull of the Casspir for maximum protection.

The protected engine compartment is at the front with the armoured crew compartment extending right to the rear.

The commander and driver are seated to the rear of the engine compartment with the commander on the left and the driver on the right. The commander also operates the weapons if installed. To the front and sides of the commander and driver are bulletproof windows 52 mm thick which provide the same level of protection as the hull.

A double machine gun mounting has been designed for the optional installation of either twin 7.62 mm Browning machine guns or twin LIW 5.56 mm SS-77 machine guns. A single 7.62 mm Browning machine gun mounting is provided in the front left window.

The troop compartment extends right to the rear of the vehicle and the 10 troops sit five each side facing each other on individual seats with four-point safety harnesses.

In each side of the hull are six rectangular firing ports and three oblong bulletproof windows.

The troops enter and leave via two doors in the rear, both of which have a bulletproof vision block and firing port underneath. The rear doors are power assisted and remote controlled. There are six hatches in the roof and the crew compartment is force-ventilated by two impeller fans.

The original versions of the Casspir had an open roof, but steel roofs are fitted to vehicles used in the IS role for protection against petrol bombs and grenades.

All versions of the Casspir armoured personnel carriers have a 200 litre drinking water tank, two spare wheels and tyres and are designed for an operational range between 600 and 800 km lasting for five to seven days.

Standard equipment also includes two fire extinguishers, a tow hook and D-eyes at the rear and a towing pin at the front.

Optional equipment includes floodlights, searchlights and a pneumatically controlled drop-down front bumper for clearing obstacles.

More recently, the Casspir has been marketed with a number of additional options including run flat inserts, armoured roof and protection against armour piercing ammunition.

Variants

Indian Casspir APC's

In September 1998, the Indian Army awarded the then Reumech OMC a contract worth US$11.5 million to supply 90 re-conditioned Casspir Mk II (4 × 4) mine protected vehicles.

All 90 vehicles were overhauled and delivered to the Indian Army by September 1999 where they are used in counter-insurgency operations. India subsequently placed additional orders for the Casspir, which brought the total number of vehicles supplied by Vickers OMC to 165 units.

Two Casspir vehicles were supplied for an extensive series of tests in India. Four different mine and blast detonations were carried out on one of the vehicles and it was still able to be driven back the 1,657 km to Bombay.

Police version

This has larger windows to increase visibility for use in urban areas and the windows have grills. The grill for the commander's and driver's windscreen can be pneumatically raised for improved visibility. A wire cutter can be mounted on the roof and the front buffer can be lowered from inside the vehicle to bulldoze barricades and other obstructions.

Blesbok freighter

This has a fully armoured two-man cab at the front and a dropside rear cargo area that can carry 5 tons. One hundred and sixty units have been built. All three variants (Blesbok, Duiker and Gemsbok) have individual side doors for both the commander and driver.

As an option, a 1,000 litre tank can be installed at the rear of the vehicle. Armament options include the provision for mounting a single light machine gun on the roof as well as a light machine gun mounting in the front left window.

Duiker tanker

This has a fully armoured front cab, to the rear of which is a 5,000 litre diesel fuel tank. Thirty units have been built. The tank is fitted with a gravity feed system for dispensing fuel with an electrical pump as an option. Provision is made for the mounting of a single light machine gun on the roof or in the front left windscreen.

Gemsbok 15 tonne recovery vehicle

This has a fully armoured cab which extends to the centre of the vehicle with seats for five people, including the crew of two, and is fitted with recovery equipment. An additional side door is fitted on the left-hand side. There have been 30 units built.

Casspir (4 × 4) APC of the Indian Army **2000**/0080989

VAMIDS mine sensor vehicle system which is based on the chassis of the Casspir (4 × 4) vehicle **2000**/0080990

Casspir (4 × 4) vehicle fitted with 106 mm recoilless rifle 0005395

Casspir (4 × 4) vehicle fitted with LIW 81 mm mortar 0005396

Artillery fire control

The South African Army has a number of artillery fire-control vehicles based on the Casspir. Externally, these are recognisable by their additional radio antenna and a larger telescopic mast.

Casspir ambulance

This is an armoured mine-protected ambulance version of the standard Casspir APC and uses a hull and power-train identical to that of the standard Casspir. The cabin interior has been adapted to carry two stretcher cases and three seats in the rear compartment with a driver and co-driver in the front. Ample storage is provided for standard medical equipment including a storage rack for drips. Standard equipment includes a toolkit stored inside the cabin compartment plus externally mounted pick and shovel and the cabin is also fitted with blackout curtains.

Plofadder mineclearing system

The rear compartment of the Casspir has been modified to accept the containerised Plofadder 160AT rocket-propelled mineclearing system (*Jane's Military Vehicles and Logistics 2001-2002*).

This system is slid into the Casspir on rails and launched through the open roof. The rails for loading the container are carried on the side while on the right side is the cable drum for remote control of the system when the rocket is fired.

Specifications

Crew: 2 + 10
Configuration: 4 × 4
Power-to-weight ratio: 13.51 hp/t
Weight:
(combat) 12,580 kg
(empty) 11,040 kg
(permissible gross) 14,000 kg
(permissible gross combination) 22,000 kg
Length: 6.87 m
Width: 2.45 m
Height: 3.125 m
Ground clearance:
(axles) 0.41 m
Track:
(front) 2.07 m
(rear) 2.07 m
Wheelbase: 4.2 m
Overhang:
(front) 1.234 m
(rear) 1.436 m
Angle of approach/departure: 46/40°
Max road speed: 90 km/h
Range:
(road) 850 km
(cross-country) 564 km
Fuel capacity: 220 litres
Gradient:
(forward and reverse) 65%
Side slope: 40%
Vertical obstacle: 0.5 m
Trench: 1.06 m
Fording: 1.00 m
Turning radius: 10.5 m
Engine: ADE-352T, 5.7 litres, 6-cylinder diesel developing 170 hp at 2,800 rpm
Transmission: MB G3 manual, 5 forward and 1 reverse gears
Clutch: single dry plate
Transfer case: MB VG-500-3W 2 speed
Steering: power-assisted, ball and nut
Suspension: leaf spring plus telescopic shock-absorbers
Tyres: 14.00 × 20 tubeless
Number of tyres: 4 + 2
Brakes:
(main) dual-circuit, air over hydraulic
(parking) spring brakes on rear axle
Electrical system: 24 V
Battery: 2 × 12 V, 115 Ah
Armament: 1 to 3 × 7.62 mm MG
NBC system: no
Night vision equipment: no

Mechem mineclearing vehicle

This version of the Casspir has its four standard rubber-tyred roadwheels removed and replaced by four steel wheels which are used to detonate anti-personnel mines by the vehicle running over them. This version has been successfully used in South Africa and Mozambique. Another South African mine detection system, developed by Denel, is the Vehicle Array Mine Detection System (VAMIDS) which is also based on the Casspir.

MEDDS mine sensor vehicle

The Casspir is used as the sensor element of the MEDDS mineclearing system. This takes samples by remote contol for later analysis and subsequent location and clearance. Details of this system are given in *Jane's Military Vehicles and Logistics 2001-2002*

Weapon carriers

More recently, two more variants of the Casspir Mk III have been developed for use as weapons platforms, one for the LIW 81 mm mortar and one for the US-developed 106 mm M40 recoilless anti-tank rifle.

Seven examples of each have been delivered for extensive user trials and a production order for 32 vehicles has been placed. These are used by South African motorised infantry units.

The South African mechanised infantry units use special versions of the Ratel (6 × 6) vehicle for these roles, for example turntable-mounted 81 mm mortar and Kentron 127 mm ZT3 Swift anti-tank guided missile carrier.

The new weapon platform versions are conversions of existing vehicles and not new build vehicles.

In the case of the 81 mm mortar version, the fully enclosed crew compartment is at the front and provided with all-round protection.

The 81 mm mortar is mounted in the rear and is provided with armour protection and bulletproof windows to the sides and rear. A total of 192 81 mm mortar bombs is carried in ready use racks plus their associated fuzes and propelling charges.

By installing the 81 mm mortar in the Casspir, time into and out of action is reduced and this leads to quicker target engagement and reduced possibility of the mortars being located and neutralised.

The 106 mm M40 series recoilless rifle weapons carrier also has the fully enclosed troop compartment at the front but the sides and rear can be quickly folded down into the horizontal position to allow the weapon to be laid onto the target. A total of 12 rounds of 106 mm High Explosive Anti-Tank (HEAT) ammunition is carried for ready use.

In both cases the weapons can also be removed from the vehicle and used in the ground role if required by the tactical situation or if the vehicle becomes damaged or disabled. Weapon tactics and drills are similar to those of the weapons mounted on other South African Army vehicles. A roof-mounted 7.62 mm machine gun is provided for local defence.

Status

Production as required. In service with a number of countries including:

Country	Quantity	Comment
Angola	3	Mk 2B
India	165	delivered 1998 to 2001
Namiba	n/avail	
Nigeria	5	delivered 2001
Peru	20	
South Africa	390	plus many more police
Uganda	70	delivered 1996/97

Manufacturer

Vickers OMC.

UPDATED

Vickers OMC RG-12 armoured patrol vehicle

Development

Based on its experience in the development and production of the Casspir range of 4 × 4 vehicles, TFM, (now Vickers OMC) developed the RG-12 armoured patrol vehicle.

The first prototype vehicles were completed early in 1990, with the first preproduction RG-12 armoured patrol vehicle completed in June 1990, and series production commenced in the September of that year.

The RG-12 armoured patrol vehicle has been developed for the urban riot control situation and according to Vickers OMC its main function is the patrol and surveillance of riot areas, while giving full ballistic protection to occupants.

In mid-1997, the TFM Defence and Security Division was taken over and its product range integrated into that of Reumech OMC. Late in 1999 Vickers Defence Systems, of the UK, purchased Reumech OMC and the company ws renamed Vickers OMC.

RG-12 armoured patrol vehicle showing the wire mesh protection fitted to the windows and obstacle clearing blade at the front of the hull 0005398

RG-12 armoured patrol vehicle from the rear

In 2001, Vickers OMC stated that they had won their first order for the RG-12 internal security vehicle as two undisclosed countries in the Middle East had placed contracts worth US$5 million for an undisclosed quantity of vehicles. Deliveries were completed by the end of 2001 and two versions were supplied, in armoured personnel carrier and patrol vehicle configurations.

These two export sales bring the total number of RG-12 internal security vehicles built to well over 700 units and total number of customers to eight, with most sales being for the police/internal security roles.

The RG-12 internal security vehicle is also referred to as the Nyala. Late production vehicles are powered by a Mercedes Benz 6-cylinder diesel engine coupled to an Allison automatic transmission.

Description

The RG-12 armoured patrol vehicle is designed to allow, without modification to the basic vehicle, a 4 × 2 and 4 × 4 fleet mix within budget constraints.

The standard vehicle carries eight personnel, including driver and commander, but a variant is available to carry 12 personnel with a rearranged internal layout.

Ballistic protection levels of the RG-12 are to 5.56 × 45 mm NATO ball.

Entrance to the vehicle is via two front side doors and two pneumatically operated plug-type side doors. The rear of the vehicle is used for storage or as a dog cage, which can be entered from inside the vehicle or via a rear door. A hatch is provided in the roof above the commander's seat.

The RG-12 armoured patrol vehicle is powered by a locally manufactured ADE 366T six-cylinder diesel engine delivering 170 hp at 2,800 rpm and power is transmitted via a five-speed gearbox and two-speed transfer box (4 × 4 version) to the axles.

The power pack, including engine, gearbox, air cleaner and cooling system, is mounted on a slide frame which can be removed quickly for maintenance purposes. To assist in this process, engine-driven pneumatic and hydraulic systems are all connected with quick-release couplers.

Steering is power assisted on the front wheels which, taken together with the short wheelbase and narrow width, gives high manoeuvrability in confined areas such as those encountered in an internal security operation.

A tropical roof is fitted to reduce heat load to the interior of the vehicle, with the additional benefit of improved hand grenade protection from above. The vehicle is fitted with a roof-mounted air conditioning system to maximise crew comfort.

The suspension system consists of long travel leaf springs and double-acting telescopic shock-absorbers, tyres are 12.50R × 20 radial.

The standard electrical system works at 12 V and is supplied by a 140 Ah battery and 90 A alternator.

The dashboard and driver controls have been ergonomically laid out for comfort with a full range of instrumentation. The vehicle has five floodlights mounted on the roof as well as two rotating beacons. In addition a hand-held spotlight is fitted. A 250 W public address system is installed in the vehicle with klaxon and wail facilities. All lights of the RG-12 are protected with mesh screens and wire cutters are fitted to protect the roof fittings and vehicle commander.

The spare wheel is housed under the rear storage compartment and is accessed through a rear door. The wheel can be removed and replaced by one person with the aid of a roller and strap arrangement.

The braking system is a fully pneumatic system which also meets ISO specifications.

Standard equipment of the RG-12 fitted to the vehicle includes fire extinguishers and a 25 litre drinking water tank.

Specifications

Version	4 × 2	4 × 4
Crew	2 + 6	2 + 6
Weight		
(combat)	9,200 kg	9,200 kg
(empty)	7,050 kg	7,420 kg
Length	5.2 m	5.2 m
Width	2.45 m	2.45 m
Height	2.64 m	2.64 m
Ground clearance	0.322 m	0.322 m
Track		
(front)	1.80 m	1.80 m
(rear)	1.80 m	1.80 m
Wheelbase	3.4 m	3.4 m
Overhang		
(front)	1.36 m	1.36 m
(rear)	0.84 m	0.84 m
Angle of approach	30°	30°
Angle of departure	41°	41°
Max road speed	104 km/h	100 km/h
Acceleration (0-60 km/h)	29.5 s	30.5 s
Range	1,000 km	1,000 km
Fuel capacity	250 litres	250 litres
Gradient	60%	50%
Side slope	40%	40%
Engine	ADE 366T diesel water-cooled developing 170 hp at 2,800 rpm	
Transmission	manual ASTAS S5-35 with 5 forward and 1 reverse gears (as an option an Allison AT-545 automatic transmission with 4 forward and 1 reverse gears can be fitted)	
Transfer box	none	2 speed
Steering	power-assisted	power-assisted
Tyres	12.5R x 20 16-ply rating	12.5R x 20 16-ply rating
Electrical system	12 V	12 V
Battery	12 V, 140 Ah	12 V, 140 Ah
Alternator	12 V, 90 A	12 V, 90 A
NBC system	no	no
Night vision equipment	no	no

Status

Production. By 2001, a total of eight customers had purchased the RG-12 armoured patrol vehicle including Cote d'Ivoire (Ivory Coast) (3 units), Swaziland (police) and South Africa (police).

Manufacturer

Vickers OMC.

UPDATED

Vickers OMC RCV 9 light armoured vehicle

Development

The RCV 9 light armoured vehicle was developed as a private venture by Sandock-Austral (subsequently incorporated into Reumech OMC which, late in 1999, was taken over by Vickers Defence Systems of the UK and renamed Vickers OMC)), mainly for use in the internal security role and for protecting high-risk areas such as airports and power stations. The first prototype of the RCV 9 light armoured vehicle was completed in 1986, with first production vehicles completed in August 1987. The RCV 9 is also referred to as the Nongqai, Uklebe or Falcon.

The RCV 9 light armoured vehicle is based on proven commercial components to reduce procurement, operating and life cycle costs, but a purpose-designed suspension of the double-wishbone type with coil springs is utilised to meet specific operational requirements.

Production of this vehicle has been completed and it is no longer being marketed. The RCV 9 light armoured vehicle was never used by the South African Defence Force/South African National Defence Force.

RCV 9 light armoured vehicle fitted with twin side doors and wire mesh protection for all windows (Christopher F Foss) 0005397

Description

The hull of the RCV 9 light armoured vehicle is of all-welded steel armour construction which provides the crew with complete protection from 39 × 7.62 mm ball rounds fired from a range of 30 m.

The driver sits at the front in the middle, and has a large forward-opening door with a bulletproof window in the upper part. To the front is a large bulletproof windscreen.

The troops sit on bench seats down the centre of the hull facing outwards and can rapidly enter and leave the vehicle via two pneumatically operated doors in the sides that open outwards, each of which is provided with a bulletproof window in the upper part with an integral firing port. In each side of the troop compartment, to the immediate rear of the twin side doors, is another bulletproof window with a firing port in its upper part.

The engine compartment is at the rear of the vehicle with the spare wheel and tyre carried on the roof. In the roof of the vehicle are three roof hatches, one above the commander's and driver's position which opens to the rear and one above either side of the troop compartment which opens to the outside.

Optional equipment includes 4 × 4 drive, communications equipment, floodlights and spotlights, weapons storage fittings, a roof-mounted wire cutter, wire mesh protection for all windows and lights, a ball- or pin-type towbar, drinking water containers, a barricade bumper with pneumatic control, fire extinguishers, air conditioning, a front-mounted electric winch and run-flat tyres.

Specifications

Crew: 1 + 8
Configuration: 4 × 2
Power-to-weight ratio: 17.39 hp/t
Weight:
(combat) 6,900 kg
(empty) 6,020 kg
Length:
(overall) 5.38 m
(hull) 5.15 m
Width:
(over hull) 2.10 m
(doors open) 3.16 m
Height:
(over spare wheel) 2.28 m
(hull top) 2.08 m
Ground clearance: 0.29 m
Track: 1.8 m
Wheelbase: 2.995 m
Angle of approach/departure: 37.5/42°
Max road speed: 100 km/h
Range (road): 675 km
Gradient: 40%
Side slope: 45%
Vertical obstacle: 0.25 m
Trench: 0.54 m
Engine: turbocharged diesel developing 120 hp at 2,800 rpm
Transmission: ZF S5-35 manual, 5 forward and 1 reverse gears
Steering: hydraulic
Turning radius: 13 m
Suspension: independent on all wheels with double wishbones and coil springs, hydraulic double-acting shock-absorbers
Tyres: 11.00 × 16 tubeless
Brakes:
(main) dual-line, air over hydraulic disc front and rear
(handbrake) mechanical on transfer case
Electrical system: 12 V
NBC system: no
Night vision equipment: no

Variants

In addition to being used as an armoured personnel carrier, the RCV 9 light armoured vehicle could be used as a bullion transport vehicle, an ambulance, a reconnaissance vehicle, a command post vehicle and a weapons carrier. The RCV 9 was also available with a single forward-opening door in each side of the hull rather than the twin doors.

As far as it is known, none of these variants ever entered production or service.

Status

Production complete. No longer being marketed. In service in an internal security role in Colombia (60) and South Africa.

Manufacturer

Vickers OMC.

VERIFIED

Other South African armoured personnel carriers

Development/Description

In addition to the South African Vickers OMC Ratel (6 × 6) series and the Casspir (4 × 4) series, the South African National Defence Force and police use a wide range of mine-protected vehicles. Brief details of some of these are given below.

In addition, South West Africa/Namibia developed two ranges of mine-protected vehicles called the Wolf and the Albatross series; brief details of these are also provided.

Bulldog APC

The Bulldog APC has not been adopted by the South African National Defence Force although a quantity of vehicles has been supplied to the South African Air Force for airbase defence. It is based on the chassis and automotive components of the SAMIL 20 range of (4 × 4) 2,000 kg trucks which are the standard vehicles in their class in the South African National Defence Force (*Jane's Military Vehicles and Logistics 2001-2002*. This differs from the current in-service Buffel in that it incorporates the driver's cab into the main troop compartment which has back-to-back seating for 10 men, five down either side. The only means of entry and exit is by climbing over the sides of the vehicle.

If required, a 7.62 mm machine gun can be fitted for local protection.

Specifications
Bulldog APC
Crew: 1 + 10
Configuration: 4 × 4
Weight: 7,700 kg
Length: 5.34 m
Width: 2.3 m
Height: 3.1 m
Ground clearance: 0.46 m
Max road speed: 90 km/h
Cruising speed: 80 km/h
Road range: 900 km
Gradient: 60%
Vertical obstacle: 0.60 m
Trench: 0.865 m
Fording: 1.2 m
Turning radius: 6.5 m
Engine: 4-cylinder air-cooled diesel
Transmission: manual, 6 forward and 1 reverse gears
Suspension: hydraulic telescopic shock-absorbers
Steering: power-assisted
Brakes: air/hydraulic, dual circuit
Tyres: 14.50 × 20 PR 12

Buffel (4 × 4) series

This has been the standard mine-protected APC since 1978 and it is estimated that around 2,400 were built before production was completed. The engine is at the front of the vehicle, with the driver seated above on the left side in a fully armoured cab which is provided with bulletproof windows to the front and sides. A spare

Bulldog (4 × 4) mine-protected APC 0005402

Buffel (4 × 4) mine protected vehicle under suspended tow and clearly showing V-shaped hull which provides a high level of protection against mines (Christopher F Foss) 0005399

wheel and tyre is carried on the right side of the engine compartment.

The hull floor is V-shaped to provide maximum possible protection against mines. The rear troop compartment has an open top, with 10 infantrymen seated five down either side back-to-back. Each seat is provided with a full seat belt harness and the only means of entry and exit is by climbing over the sides of the vehicle. The latter can be hinged downwards if required. Two machine guns can be pintle mounted on top of the hull. These are normally positioned front right and rear left of the hull. More recently the vehicle has been marketed with twin machine guns at each station and the gunner being provided with a shield. Specialised variants include a 81 mm mortar carrier. Prime contractor for the Buffel family of vehicles was Armscor, with a Mercedes-Benz Model 416/162 (4 × 4) Unimog chassis being used as the basis.

A quantity of these vehicles, estimated to be 30 to 40 units, has been supplied to Sri Lanka.

There is also the Buffel Mk I tractor, which has a fully armoured cab at the front but is fitted with a cargo area at the rear with a drop tailgate. This can carry 2,500 kg and has also been used as a prime mover, for example, towing a 20 mm light anti-aircraft gun.

Specifications

Buffel APC
Crew: 1 + 10
Configuration: 4 × 4
Weight:
(loaded) 6,140 kg
Length: 5.1 m
Width: 2.05 m
Height: 2.995 m
Ground clearance: 0.42 m
Max road speed: 96 km/h
Road range: 1,000 km
Fuel capacity: 200 litres
Engine: Mercedes-Benz OM-352 6-cylinder water-cooled diesel developing 125 hp at 2,800 rpm
Transmission: manual, 8 forward and 4 reverse gears (driver can select 4 × 4 or 4 × 2 drive)
Tyres: 12.50 × 20

Buffels for export

During the International Countermine Conference held in Copenhagen, Denmark, in 1995, a request was made to conference delegates to make a lower-cost Mine Protected Vehicle (MPV) available.

Armscor, the South African government procurement agency, responded to this request by offering the combat proven locally designed and built Buffel (4 × 4) MPV.

On behalf of the South African National Defence Force, a number of remanufactured Buffels are now being made available to the international demining world at a cost of only US$59,660 per vehicle.

Before delivery, the Buffels would be stripped down and completely rebuilt so that they are virtually a new vehicle. They come with a complete 12 month or 20,000 km guarantee. Driver and maintenance training can also be provided by Armscor as can a complete spare parts package.

As well as offering the crew protection from small arms fire, the Buffel provides the following levels of mine protection:

(1) Single Russian TM57 anti-tank mine or equivalent, detonated anywhere under the vehicle
(2) Double Russian TM57 anti-tank mine or equivalent, detonated under any wheel.

Mk II Buffel

There are two basic models, Mk IIA with drum brakes and Mk IIB with disc brakes. This is a rebuild of the earlier Buffel vehicles by 61 Base Workshop of the South African National Defence Force. The vehicle has a new fully enclosed rear troop compartment with a large access door at the rear below which is a series of steps. To the front, sides and rear are bulletproof windows, each of which is provided with one or two circular firing ports.

Anti-aircraft command and control vehicle

This was unveiled for the first time in 1990, by which time it was already in production and service with the South African National Defence Force air defence units.

It is a mobile air defence command and control vehicle for mobile and towed anti-aircraft defence systems, providing information on the position of both friendly and hostile aircraft.

It is based on an armoured SAMIL 20 (4 × 4) truck chassis with a Bulldog braking system and a Rhino nine-man troop carrier hull.

Prime contractors for the system were ESD and Armscor. A pneumatic lifting system elevates the antenna to a height of 6 m to provide for an improved communications capability.

Communications equipment installed includes one HF radio, one VHF ground-to-air radio and a 1.4 kVA diesel generator is fitted to power the onboard electrical equipment independently from the vehicle's battery system, allowing silent running for a period of up to 3 hours.

Armament comprises a 7.62 mm machine gun which is pintle mounted on the roof.

Standard equipment includes a 50 litre drinking water tank, camouflage nets and a compressor system.

Ystervark 20 mm self-propelled anti-aircraft gun system

This is based on a SAMIL 2,000 kg (4 × 4) truck chassis with a fully enclosed armoured cab. Mounted on the rear is an Oerlikon Contraves 20 mm GAI-CO1 manually operated anti-aircraft cannon. Additional details of this system are given in *Jane's Land-Based Air Defence 2001-2002.*

Zumlac self-propelled anti-aircraft gun

This is in service with the South African National Defence Force and consists of the SAMIL 100 (6 × 6) mine protection truck fitted with the larger four door cab. On the flatbed to the rear is a Russian designed and built twin 23 mm ZU-23-2 light anti-aircraft gun system. If required, the system can be removed from the truck and used in the ground emplaced configuration. The Zumlac has a five-man crew. Full details of this system are given in *Jane's Land-Based Air Defence 2001-2002.* A total of 36 systems are in service.

Rinkhals Mk I

This is the standard mine-protected ambulance of the South African Medical Service's Medical Battalion

Rhino protected (4 × 4) troop carrier (Terry J Gander)

Withings MK 1A (6 × 6) recovery vehicle on Samil 100 chassis with armoured cab (Christopher F Foss) 0005400

Zumlac twin 23 mm self-propelled anti-aircraft gun system on a mine-protected SAMIL 100 (6 × 6) truck chassis deployed in the firing position (Christopher F Foss) 0018648

Ystervark 20 mm self-propelled anti-aircraft gun system (Christopher F Foss)

Corps. It was the first monocoque mine-protected vehicle to be taken into service with the South African National Defence Force, first vehicles entering service in the early 1980s. The engine is at the front and the crew compartment to its immediate rear. The vehicle can carry both sitting and walking wounded entering the vehicle via the rear. Bulletproof windows are provided all-round and both commander and driver are provided with a side door. Provision is made for a roof-mounted 7.62 mm machine gun. This vehicle was built by TFM (since taken over by Reumech OMC which late in 1999 was taken over by Vickers Defence Systems of the UK and renamed Vickers OMC) on the SAMIL 50 (4 × 4) truck chassis.

SAMIL 100 Kwevoel Mk II A

This the SAMIL 100 (6 × 6) 10 tonne truck (*Jane's Military Vehicles and Logistics 2001-2002*) fitted with a mine-protected two-man cab. Other versions include the Mk I horse transporter, the Mk I tanker and there is also a mine-protected recovery vehicle. This version also forms the basis for the 127 mm (40 round) Valkiri multiple artillery rocket system described later in this book.

SAMIL 20

This is the SAMIL 20 (4 × 4) 2 tonne truck (*Jane's Military Vehicles and Logistics 2001-2002*) fitted with a mine-protected two-man cab. There is also another version called the Rhino protected troop carrier. This has a fully armoured rear compartment for 10 men which is also provided with bulletproof windows, firing ports, side and rear doors and roof hatches.

SAMIL 50

As far as is known, the only mine-protected version of the SAMIL 50 (4 × 4) 4,800 kg truck (*Jane's Military Vehicles and Logistics 2001-2002*) is the SAMIL 50 recovery vehicle (*ibid* page 86).

Wolf (4 × 4) series

These were developed to meet the requirements of the South West Africa CO-IN unit in 1983 with first production vehicles completed the following year. The Wolf II entered production in 1985.

Wolf S500 series 4 × 4 mine protected vehicle being used for EOD teams in the Balkans (Caemerlynck Dirk) ***2001***/0104413

Specifications

Rinkhals Mk 1
Configuration: 4 × 4
Weight: 11,970 kg
Length: 7.21 m
Width: 2.55 m
Height: 3.15 m
Ground clearance: 0.355 m
Fuel capacity: 360 litres
Max road range: 900 km
Fording: 1 m
Engine: V-6 fuel-injected diesel developing 146 hp at 1,600 rpm
Transmission: manual, 6 forward and 1 reverse gears with high and low ratios
Steering: power-assisted
Brakes: air

The engine is at the front and fully enclosed within the armour envelope of the vehicle. The whole floor is in the usual V-configuration providing maximum protection against mines.

The hull is of all-welded steel armour construction with the driver and co-driver at the front and the troop compartment extending right to the rear of the vehicle. Bulletproof windows are provided in the front, sides and rear as are firing ports. The Wolf I had spherical firing ports while the Wolf II had simple firing ports.

Armament comprises one or two 7.62 mm MGs mounted above the co-driver's position in a remote

Buffel Mk IIB from the front (Desmond Dempsey)

SAMIL 100 (6 × 6) mine-protected truck (T J Gander)

Specifications

Wolf APC
Crew: 1 + 10
Configuration: 4 × 4
Weight:
(loaded) 14,200 kg
Max road speed: 120 km/h
Cruising speed: 90 km/h
Range:
(road) 2,000 km
(off-road) 1,600 km
Fuel capacity: 610 litres
Engine: diesel developing 260 hp
Transmission: manual, 16 speed
Transfer case: 2 speed
Suspension: leaf springs and hollow spring shock-absorbers
Tyres: 16.00 × 20 18-ply

controlled mounting with a traverse of 180° left and right and elevation from –20 to +30°. There is also a roof-mounted 1 × 12.7 mm MG or twin 7.62 mm MGs, or 1 × 7.62 mm MG to the rear of the driver's and co-driver's position. In a spherical mounting in the armour glass in front of the co-driver is a 7.62 mm MG.

In 1998, it was stated that Windhoeker Maschinenfabrik (WMF) would refurbish 30 Wolf Mk 4 mine protected vehicles plus building 30 Wer'wulf Mk 2 vehicles. These will be powered by a MAN 220 bhp intercooled turbocharged diesel engine.

The supply truck version of the Wolf is called the Strandwolf. This retains the fully enclosed and armoured cab at the front but has a cargo area at the rear. Armament comprises a single 7.62 mm machine gun on the roof which is remote controlled. There is also a 6 × 6 recovery vehicle called the Sterk Hans which has a mine-protected cab capable of carrying four passengers.

Albatross series

These were designed and built by TFM (taken over by Reumech OMC and late in 1999 was taken over by Vickers Defence Systems of the UK and renamed Vickers OMC) for South West Africa. The basic vehicle is a 10 tonne (6 × 6) cargo carrier with a fully armour protected two-man cab at the front and the cargo area at the rear. The latter is provided with drop down sides and a rear tailgate. Variants include a recovery vehicle and a tractor truck for towing semi-trailers. All versions have a 7.62 mm remote-controlled machine gun above the co-driver's position.

Status

Production of these mine-protected vehicles was completed some time ago. In service with the South African National Defence Force and other countries including Sri Lanka.

Manufacturer

These were made by various companies in South Africa but the current design authority is Armscor. Quantities of these are available for sale by Armscor.

UPDATED

Vickers OMC Kobra armoured personnel carrier

Development

The Kobra armoured personnel carrier has been developed as a private venture for the export market by Vickers OMC and builds on experience in the design, development and production of the Mamba Mk 2 (4 × 4) armoured personnel carrier which is covered in detail in the following entry.

Vickers OMC stresses that the Kobra is being marketed alongside the earlier Mamba, which was based on the German Mercedes-Benz UNIMOG U416 (4 × 4) chassis, and is being aimed at those customers who require a more capable 4 × 4 armoured personnel carrier.

A further development of the Kobra called the Taipan was developed to meet the requirements of the Australian Army. Following extensive trials between the Taipan and the Australian ADI Bushmaster, the latter was selected early in 1999.

The Cobra was originally designed and built by Reumech OMC, the sole manufacturer of armoured fighting vehicles in South Africa. Late in 1999 this company was taken over by Vickers Defence Systems of the UK and renamed Vickers OMC. As of late 2001, there were no known sales of the Kobra.

Vickers OMC Kobra (4 × 4) armoured personnel carrier (Christopher F Foss) 0018731

Description

The Kobra is based on the Mercedes-Benz UNIMOG U2150 4 × 4 cross-country chassis fitted with an all-welded monocoque steel body that provides the occupants with protection from 5.56 and 7.62 mm ball small arms fire and shell splinters.

As an option, an add-on armour package is available which increases the protection level up to 7.62 mm armour piercing attack.

The standard Kobra provides protection against a 14 kg TNT detonation anywhere under the vehicle, for example under the belly or under a wheel.

The engine compartment is at the front with the forward part opening left and right to gain access to the power pack and cooling system. Over the engine is an upward opening cover.

The diesel engine drives through a manual gearbox with eight forward and eight reverse gears. The final drives incorporate portal hub reduction with pneumatically operated differential locks.

For the Australian programme a new fully automatic transmission was developed. This consisted of a Reco 606 transmission from Germany and transfer and drop down gearbox designed by Ermetek of South Africa.

The crew compartment is to the immediate rear of the engine compartment and extends to the very rear of the vehicle. The driver is seated on the right (although a left-hand version is also available) with the commander to the left with large bulletproof windows to their immediate front and sides.

A total of six roof hatches is provided in the Kobra, three either side which open outwards and can be locked in the vertical position. If required, a circular roof hatch can be installed on which can be mounted a 5.56 or 7.62 mm machine gun.

Specifications

Crew: 2 + 9
Configuration: 4 × 4
Combat weight: 12,500 kg
Unloaded weight: 10,481 kg
Power-to-weight ratio: 16.8 hp/t
Length: 5.950 m
Width: 2.465 m
Height:
(overall) 2.675 m
(reduced) 2.554 m
Ground clearance: 0.482 m
Wheelbase: 3.250 m
Track: 1.872 m
Angle of approach/departure: 40/40°
Max road speed: 100 km/h
Vertical obstacle: height of loaded tyre radius
Trench: 1.00 m
Gradient:
(forwards) 70%
Side slope: 71%
Turning radius:
(between curbs) 15 m
(between walls) 16 m
Range:
(road) 1,000 km at 90 km/h
Fuel capacity: 270 litres (an additional armour-protected fuel tank of 18 litres is also provided)
Fording: 1.2 m
Engine: Mercedes-Benz OM 366LA 6-cylinder in-line turbocharged and intercooled diesel developing 210 hp at 2,600 rpm
Transmission: Mercedes-Benz UG 3/65-8/13.01 GPA BM18.815 8-speed manual (6-speed automatic is an option)
Transfer box: none
Clutch: single dry plate
Steering: power-assisted, recirculating ball
Suspension:
(front) single coil springs (primary only) with double-acting hydraulic telescopic shock-absorbers, Panhard rod and torque tube with strut bar
(rear) double coil springs (primary and secondary) with double-acting hydraulic telescopic shock-absorbers, Panhard rod, torque tube and strut bar
Tyres: Michelin XZL 365/85 R20
Brakes:
(foot, front) disc
(foot, rear) disc
(parking) air on rear disc brakes
Electrical system: 24 V
Batteries: 4
NBC system: no
Night vision equipment: no
Air transportability: Lockheed Martin C-130 Hercules

A single wire cutter is fitted at the centre front end of the roof of the vehicle and a single hose with air pressure gauge is provided as standard.

The troops sit facing inwards and enter the vehicle via a large door in the rear. The troop compartment is fitted with large bulletproof windows which can be fitted with special circular firing ports if required by the user.

In its basic APC configuration, the Kobra has a crew of two, commander and driver and carries nine passengers. Mounted either side at the rear of the Kobra is a spare wheel and tyre.

The suspension uses single coil springs at the front and double coil springs at the rear with double-acting hydraulic telescopic shock-absorbers, Panhard rod, torque tube and strut bar to absorb cross-country energy and loadings.

The braking system comprises a primary dual-circuit air over hydraulic system, activating disc brakes. The secondary brake is the parking brake which is an air pressure-actuated cylinder on the rear disc brakes with a spring on, air off activation. The tertiary supplementary brake system is an engine exhaust brake, pneumatically activated by a floor-mounted valve.

Standard equipment includes an SME 11 environmental control system with a total of six vents being provided. On either side of the vehicle are three stowage bins and, if required, a 5.56 or 7.62 mm machine gun can be mounted on the roof of the vehicle. A water tank with a capacity of 270 litres is fitted.

A wide range of optional equipment can be fitted to the Kobra including add-on passive armour for a higher level of armour protection, 55 mm glass all round which provides protection from armour-piercing small arms fire, fire and explosion suppression system for the diesel engine and crew compartments and a central tyre inflation system that allows the driver to adjust the tyre pressure to suit the terrain being crossed.

The diesel fuel tanks of the Kobra can also be filled with a fire retardant insert if required by the user.

The Kobra is air-transportable in a Lockheed Martin C-130 Hercules transport aircraft. As the vehicle is based on a UNIMOG (4 × 4) design, spare parts are available on a worldwide basis.

Variants

The Kobra can be used to carry out a wide range of roles on the battlefield including mortar carrier, command post vehicle, ambulance, assault pioneer vehicle, maintenance and repair vehicle, supporting arms vehicle, flatbed/container carrier vehicle and a direct fire weapons platform.

Concept design work by the company has also shown that a 6 × 6 version is possible and can be developed in response to user requirements.

Status

Development complete. Ready for production.

Manufacturer

Vickers OMC.

UPDATED

Vickers OMC Mamba Mk 2 armoured personnel carrier

Development

The original Mamba 4 × 2 vehicle was developed by Mechem Consultants, part of the Denel Group, and a number of production vehicles was built.

Further development by Reumech OMC resulted in the Mamba Mk 2 (4 × 4) which is some 70 mm higher than the earlier two-wheel drive version.

Between 1993 and 1997 a total of 653 Mamba vehicles was manufactured by Vickers OMC for the home and export markets with production ranging from three vehicles a week up to four vehicles a day.

The first contract was awarded early in 1993 and covered the supply of three prototypes and 151 production vehicles, with all vehicles being completed by March 1994.

By the time production was completed late in 1997 a total of 67 Mambas had been exported to seven different countries. Nine Mamba variants had been developed for specific clients requirements and marketing.

According to Vickers OMC, production of the Mamba will only commence again for substantial and worth while export orders.

Further development by Alvis Vehicles of the UK has resulted in the Alvis 4 and Alvis 8 series of Mine Protected Vehicles which is covered under the UK. A small quantity of Alvis 4 vehicles have been sold to the British Army for use by Royal Engineer units operating in the Balkans. Production of this was undertaken by Reumech OMC which was taken over by Vickers Defence Systems of the UK late in 1999 and renamed Vickers OMC. Alvis vehicles are no longer marketing the Alvis 4 and 8 family of armoured vehicles.

Specifications

Crew: 2 + 9
Configuration: 4 × 4
Combat weight: 6,800 kg
Unloaded weight: 5,710 kg
Power-to-weight ratio: 18.47 hp/t
Length: 5.46 m
Width: 2.205 m
Height: 2.495 m
Ground clearance: 410 mm (skid plate)
Wheelbase: 2.90 m
Track: 1.79 m
Angle of approach/departure: 40/38°
Max road speed: 102 km/h
Vertical obstacle:
(forwards) 400 mm
(reverse) 300 mm
Trench: 0.9 m
Gradient:
(forwards) 70%
(reverse) 60%
Side slope:
(static) 60%
(moving) 40% at 10 km/h
Turning radius: 6.25 m
Range:
(road) 900 km
(off-road) 500 km
Fuel capacity: 200 litres
Fording: 1.0 m
Engine: Mercedes-Benz OM 352 6-cylinder water-cooled diesel developing 123 hp at 2,800 rpm
Transmission: Mercedes-Benz UG2/30 717.801 4-speed manual, 2 ranges, high and low
Transfer box: integral with gearbox
Clutch: single dry plate, foot-operated
Steering: recirculating ball, power-assisted
Suspension:
(front) single coil springs with double-acting hydraulic telescopic shock-absorbers, Panhard rod, torque tube and strut bar
(rear) double coil springs with double-acting hydraulic telescopic shock-absorber, Panhard rod, torque tube and strut bar
Tyres: 12.5 × 20 radial
Brakes:
(foot, front) disc
(foot, rear) disc
(parking) pneumatic
Electrical system: 24 V
Batteries: 2 × 12 V, 55 Ah in series
NBC system: no
Night vision equipment: no

Description

The monocoque V-shaped hull of the Mamba Mk 2 is of all-welded steel armour construction that provides a very high degree of protection against anti-tank mines as well as small arms fire and shell splinters.

The standard Mamba Mk 2 steel hull and glass provides protection against 5.56 × 34 mm and 7.62 × 51 mm NATO hull ammunition with an optional protection against 7.62 × 51 mm NATO armour piercing ammunition.

The driveline and suspension of the Mamba Mk 2 is from the well-known German Mercedes-Benz UNIMOG 416 (4 × 4) range of cross-country vehicles which is in service with civil and military users all over the world. This gives the Mamba exceptional cross-country mobility.

According to the manufacturer, the Mamba protection level includes a double anti-tank land mine under any wheel station, a centre anti-tank land mine, small arms fire up to 7.62 mm rifle, shrapnel or incendiary and limpet mine.

The layout is conventional, with the engine at the front, commander (left) and driver (right) and troop compartment at the rear. Normal means of entry and exit is through a large door in the rear that is hinged on the left.

Nine troops are carried, five seated on the left side and four on the right side facing each other on individual seats with a full harness seat belt. Above the commander's and driver's position is a roof hatch that opens to the rear and can be locked in the vertical position, with additional roof hatches being provided in the rear troop compartment. These are located three either side, are hinged on the outside and can be locked in the vertical position.

Vickers OMC Mamba Mk 2 (4 × 4) APC (Richard Stickland) ***2001***/0104414

Vickers OMC Mamba in service with Swedish Army (Richard Stickland) ***2002***/0121210

Vickers OMC Mamba in service with Swedish Army (Richard Stickland) ***2002***/0121211

Stowage lockers are located on the exterior of the hull and mounted at the rear is a tow hitch with seven-pin electrical connector. A replacement wheel and tyre is located on the left-hand side of the hull just to the rear of the front roadwheels.

Bulletproof windows are provided for the front, sides and rear of the Mamba Mk 2 and these provide the same degree of protection as the steel hull. If required, these windows can be provided with integral firing ports.

Drive is taken from the diesel engine to the four-speed full synchromesh gearbox with the high and low ranges giving a total of eight forward and four reverse gears.

The pneumatically controlled differential locks operate front and rear separately. Two- or four-wheel drive may be selected while the vehicle is in motion.

The front and rear axles are fitted with coil springs and hydraulic dampers. The air-assisted foot-operated duplex braking system acts on discs all-round. A cable-operated parking brake and an exhaust brake are also provided.

Standard equipment includes a 100 litre water tank and optional equipment includes air conditioning system, a roof-mounted light machine gun mount, firing ports, run flat inserts, higher level of armour protection, various communications equipment, customised internal layout, electric 5-tonne winch mounted at the front of the vehicle and various seating arrangements.

Variants

The design of the Mamba Mk 2 (4 × 4) vehicle is such that it can undertake a wide range of roles in addition to its basic role as an armoured personnel carrier.

Projected roles include ambulance, command post, utility vehicle, recovery vehicle, VIP transport and logistic support. If required, lower levels of protection against small arms fire and land mines can be provided.

A short wheelbase version has been developed called the Komanche, which can carry up to seven people. There is also a platform version called the Sabre, which has a four-man fully enclosed cab at the front and a platform at the rear on which can be mounted a variety of weapons and other systems, or it can be used as a cargo carrier. Left-hand drive models of the Mamba Mk 2 are also available.

Status

Production complete. In service with eight countries including:

Country	Quantity	Comment
Congo	18	delivered 1997
South Africa	544	army
South Africa	33	air force
Sweden	n/avail	Have been used in Kosovo

Manufacturer

Vickers OMC.

UPDATED

Vickers OMC RG-31 Charger armoured personnel carrier

Development

The RG-31 Charger armoured personnel carrier has been developed by the now Vickers OMC to undertake a wide range of roles such as border patrol and to secure perimeters in times of conventional warfare and land mine threat.

Further development of the RG-31 Charger has led to the RG-31 Quick Intervention Vehicle, which has been designed to ferry key personnel in maximum security into situations of high risk.

Commonality of body and drivetrain components allows for a range of design alternatives for accommodation of weapons platforms and modular ammunition stowage.

The RG-31 was originally developed and built by TFM who was subsequently taken over by Vickers Defence Systems of the UK and renamed Vickers OMC.

Late in 1996, an international consortium consisting of TFM (South Africa), the Trading Force (UK) and LNY Inc (USA) were awarded a contract by the US Army for three RG-31 Nyala mine protected vehicles.

In July 1998, the London based Trading Force Limited was awarded a contract from the United Nations (UN) for a large quantity of RG-31 mine protected and RG-32 mine hardened wheeled vehicles.

The initial contract was for a period of three years with the option of two 12 month extensions to make a total of five years with a total value of up to US$300 million.

Under the terms of this contract, Trading Force Limited supplied the latest generation upgraded RG-31 Nyala (4 × 4) mine protected armoured personnel carriers and RG-32 (4 × 4) Scout mine-hardened vehicles for use by UN forces operating in various parts of the world.

The first RG-31 and RG-32 vehicles were completed in South Africa late in 1998. In addition, Trading Force Limited supplied a wide range of in-service support activities, back-up and driver training for these vehicles.

The UN currently operates a wide range of armoured vehicles to support its peacekeeping and humanitarian missions which have come from many different sources. These have proved difficult to support and maintain and has led to high operating costs and reduced vehicle availability.

By standardising on the RG-31 and RG-32, the UN anticipates that there will be a significant reduction in life cycle costs and higher vehicle availability.

Vickers OMC RG-31 (4 × 4) APC in service with the United Nations in Bosnia 0018650

Description

The V-shaped hull is of all-welded steel armour construction that provides the occupants with protection against 5.56 × 45 mm small arms fire, although also available is protection against 7.62 × 51 mm armour-piercing small arms fire.

Land mine protection is provided against one Russian TM-57 anti-tank mine under the body and two TM-57 anti-tank mines under a wheel.

The RG-31 is built upon a monocoque omni-protection capsular concept. The vehicle does not rely on a conventional chassis ladder frame to mount the cabin or automotive systems but rather the cabin assembly is built from high-alloy armour-resisting steel.

According to Vickers OMC the main benefits of this construction methodology can be summarised as:

(1) Maximum safety of the crew from ballistic and other forms of attack by virtue of the entire envelope being of an armoured capsule
(2) High degree of safety in the event of collision or vehicle roll-over
(3) The effect of extending the life cycle of the vehicle almost indefinitely, as the body construction does not suffer any material degree of fatigue degeneration
(4) A minimum period of downtime following the detonation of land mines under the vehicle
(5) The installation of robust beam axles and leaf spring suspension systems, together with the power-assisted steering system which is conveniently accessible, due to this configuration.

The engine compartment is at the front. The driver is seated on the right side and has an adjustable seat with the commander to his left. Both have armoured steel side doors which hinge upwards and are provided with laminated glass windows to their front and sides.

The personnel compartment is at the rear and entry to this is via a large pneumatically assisted rear entry door of fabricated steel armour plate. Large obstruction-free windows made of laminated plate glass are provided in the side and rear. If required, these can be fitted with gun ports.

There are also eight hatches incorporated in the roof of the vehicle and these can be locked in the vertical

Vickers OMC RG-31 in quick intervention vehicle configuration and fitted with two 12.7 mm M2 machine guns

position if required. A machine gun can be fitted above the commander's position if required.

The troops are seated on bench seats five down each side facing inwards and there is a fold-up seat to the immediate rear of the commander and driver for an eleventh man.

A differential is incorporated in the transfer box to adjust for the differences in the drive of the front and rear axles. A differential lock can be engaged to prevent any slippage between the front and rear axles. The transfer case incorporates a high and low range which enables the vehicle to serve in a variety of urban and rural driving conditions.

The axle hubs are fitted with planetary reduction gears and pneumatically operated 'wedge'-type brakes. A pneumatically actuated differential lock can be engaged to prevent slippage between the rear wheels.

The front axle assemblies are designed so that major components can be accessed and if necessary removed from the axles without removing the axles from the vehicle.

The front and rear axles are attached to the body with semi-elliptic multileaf springs fitted on shackles attached longitudinally on the underside of the body. Hydraulic double-acting telescopic shock-absorbers are fitted to the front and rear axles.

Steering is powered to reduce driver fatigue and the front and rear wheel brakes are operated by a pneumatic brake system. The rear brakes are of the fail-to-safety spring type which are released by applying air pressure from the pneumatic system.

Several storage compartments are located in the vehicle's exterior body section to facilitate the storage and carrying of various types of logistical equipment and ammunition. An air conditioning system is fitted and, as an option, a winterisation package can also be fitted to the RG-31.

A variety of communications equipment and various types of surveillance equipment can be fitted to the RG-31.

Optional equipment for the RG-31 includes left or right hand drive, run flat inserts, armour piercing protection, various communications equipment, various weapon mountings, customised internal layout and a front-mounted electric winch with a capacity of 5 tonnes.

Variants

Quick intervention vehicle

Alternative configurations of the RG-31 include the deployment of the unit in a quick intervention vehicle configuration. This version is sometimes referred to as the Scorpion.

In this configuration, a fully protected crew cab for four personnel is constructed as a monocoque hull with an open reinforced weapons platform fitted to the rear which is capable of carrying a wide range of palletised weapon systems such as a multiple rocket launcher system, cannon, mortar and anti-aircraft weapons. A 7.62 or 12.7 mm machine gun can also be mounted on the platform.

Other variants

In addition to the standard 12-man armoured personnel carrier, the following additional versions have been proposed by the manufacturer:

(1) Anti-tank vehicle with recoilless rifle or ATGW
(2) Bomb disposal vehicle
(3) Anti-riot/internal security vehicle
(4) Scout/reconnaissance vehicle
(5) Anti-tank/air defence vehicle
(6) General weapons carrier
(7) Command post vehicle with additional communications equipment installed.

Specifications

Model APC
Crew: 2 + 10
Configuration: 4 × 4
Combat weight: 8,400 kg
Unloaded weight: 7,368 kg
Payload: 1,032 kg
Length: 6.40 m
Width: 2.47 m
Height: 2.72 m
Ground clearance: 0.40 m
Wheelbase: 3.4 m
Max road speed: 105 km/h
Cruising speed: 95 km/h
Gradient:
(low range) 60%
(high range) 35%
Engine: Mercedes-Benz OM 366A 6-cylinder in-line water-cooled turbocharged diesel with direct injection developing 168 hp at 2,800 rpm
Transmission: Allison AT-545 4-speed automatic
Transfer box: Rockwell BT232 with 2 speed permanent 4 wheel drive
Suspension: semi-elliptical leaf springs and hydraulic telescopic double-acting telescopic shock-absorbers
Steering: full power
Turning radius: 8 m
Brakes:
(main) pneumatic, front and rear
(parking) pneumatic release/spring applied
Tyres: 12.5 × 20 radial, 16-ply
Electrical system: 24 V
NBC system: no
Night vision equipment: no
Air transportability: Lockheed Martin C-130 Hercules

RG-31M for FCLV

Vickers Defence Systems of the UK are offering a further development of the RG-31 called RG-31M for the British Army's FCLV programme, together with the RG-32M covered in more detail in the following entry.

Status

Production. In service with a number of countries including:

Country	Quantity	Comment
Colombia	4	
Cote d'Ivorie	6	
NATO	2	
Rwanda	5	
South Africa	n/avail	police
Swaziland	n/k	police
United Nations	30	ordered in 1995 and delivered
United Nations	n/avail	additional order placed in 1998
United States	3	ordered in 1996 (trials)

Manufacturer

Vickers OMC.

UPDATED

Vickers OMC RG-32 Scout armoured patrol vehicle

Development

The RG-32 Scout armoured patrol vehicle was developed as a private venture to fill the gap between ineffective partially armoured commercial vehicles and a heavy-duty security vehicle. Typical roles include police, security, VIP and bullion transport.

In mid-1997, the TFM Defence and Security Division was taken over by Reunert Group and the former's range of products was integrated with Reumech OMC. Late in 1999 Vickers Defence Systems took over Reumech OMC and the company was renamed Vickers OMC.

By late 2001, well over 500 Scout vehicles had been built for the home and export markets with first production models being in the 4 × 2 configuration with the 4 × 4 model entering production in 1995. Within South Africa, the RG-32 Scout is only found in service in the private sector. It is not used by the South African National Defence Force or the South Africa Police.

The UN High Commission for Refugees procured 27 RG-32 Scouts for use in Georgia and Burundi.

In July 1998, the London based Trading Force Limited was awarded a contract from the United Nations (UN) for a large quantity of RG-31 mine protected and RG-32 mine hardened wheeled vehicles.

The initial contract was for a period of three years with the option of two 12 month extensions to make a total of five years with a total value of up to US$300 million.

Under the terms of this contract, Trading Force Limited supplied the latest generation upgraded RG-31 Nyala (4 × 4) mine protected armoured personnel carriers and RG-32 (4 × 4) Scout mine-hardened vehicles for use by UN forces operating in various parts of the world.

The first RG-31 and RG-32 vehicles were completed in South Africa late in 1998. In addition, Trading Force Limited has supplied a wide range of in service support activities, back-up and driver training for these vehicles.

Description

The hull of the RG-32 Scout is of all-welded steel armour construction that provides the occupants with protection against 5.56 × 45 mm ball small arms fire, although protection is also available against 7.62 × 51 mm armour-piercing small arms fire.

Versions of the RG-32 Scout are available in 4 × 4 or 4 × 2 configurations with either a petrol or diesel engine. The vehicle can be equipped with a lockable

First photograph released of the Vickers Defence Systems/Vickers OMC RG-32M (4 × 4) vehicle for the UK Army's Future Command and Liaison Vehicle requirement **2002**/0121218

In 2002, the restyled and mine hardened Scout was shown for the first time **2002**/0121219

rear steel body with screen window or glazed load body canopy, or have adaptations to fit individual customer needs.

The vehicle is built upon a monocoque omni-protection capsular concept. The vehicle does not rely on a conventional chassis ladder frame to mount the cabin or automotive systems but rather the cabin assembly is built from high-alloy armour-resisting steel. The front and rear subframes accommodate the automotive driveline, suspension and steering systems.

According to the manufacturer the main benefits of this construction methodology can be summarised as:

(1) Maximum safety of the crew from ballistic and other forms of attack by virtue of the entire envelope being an armoured capsule
(2) High degree of safety in the event of collision or vehicle roll-over
(3) The effect of extending the life cycle of the vehicle almost indefinitely as the body construction does not suffer any material degree of fatigue degeneration
(4) The installation of robust beam axles and leaf sprung suspension, together with the power-assisted steering system which is conveniently accessible, due to this configuration.

Vickers OMC RG-32 Scout in four-door configuration deployed by the United Nations in Kosovo **2002**/0121212

The crew cab consists of an all-welded, armour-protected steel shell which comprises the floor, roof, sides, front firewall and rear panel. Apertures for the doors, windows, harness, pipes and linkages are located at strategic locations with regard to structural integrity and the anti-ballistic design of the capsule.

The engine compartment is at the front, the driver is seated on the right side and has an adjustable seat with the commander to his left, with a three-man bench seat to the rear.

The doors are manufactured from heavy armour-protected steel plates and are hinged to the body by heavy-duty hinges. Each door is fitted with a heavy-duty, positive locking slam lock.

The windscreen and all door windows are manufactured of high impact-resistant laminated plate glass sections, each of which is supplied with a thick rubber surround. The glass sections are bolted into position in steel frames.

The windows of the crew cab are statically mounted and the crew cab is provided with a power air conditioning system. The roof of the cab is fitted with a tropical roof.

The subframes are bolted directly to the crew cab. They consist of welded frames and make provision for the installation of the axles and suspension, fuel tanks, load body and engine.

The front axle is a solid, reverse Elliot I-beam type with a total safe carrying capacity of 1,700 kg on the 4 × 2 version. A torsion bar assembly aids in improving road holding and steering response. On the 4 × 4 version a drive steer axle with a carrying capacity of 1,550 kg is used. The rear axle is a rigid, single reduction driven type with a gear ratio of 4.8:1 and a total safe carrying capacity of 2,200 kg.

The RG-32 Scout is fitted with a conventional heavy-duty suspension system with both front and rear axles mounted on semi-elliptic leaf springs. Each wheel station is fitted with a single double-acting hydraulic shock-absorber.

To reduce driver fatigue, the RG-32 Scout is equipped with a conventional power-assisted steering system. The vehicle is also equipped with conventional vacuum-assisted hydraulic service brakes. Each front wheel is fitted with a double hydraulic calliper acting on a disc of 280 mm diameter. The rear wheels are each fitted with drums with a diameter of 260 mm.

The crew compartment is also fitted with a tropical roof and air conditioning as an aid to crew comfort.

Specifications

Crew: 1 + 4 (or 1 + 6)
Configuration: 4 × 4 or 4 × 2 (LH or RH drive)
Combat weight:
(4 × 2 petrol) 3,500 kg
(4 × 2 diesel) 3,500 kg
(4 × 4 petrol) 3,660 kg
(4 × 4 diesel) 3,660 kg
Unloaded weight:
(4 × 2 petrol) 2,630 kg
(4 × 2 diesel) 2,670 kg
(4 × 4 petrol) 2,598 kg
(4 × 4 diesel) 2,998 kg
Payload:
(4 × 2 petrol) 870 kg
(4 × 2 diesel) 830 kg
(4 × 4 petrol) 702 kg
(4 × 4 diesel) 662 kg
Length: 5.00 m
Width: 1.96 m
Height:
(4 × 4) 1.91 m
(4 × 2) 1.81 m
Wheelbase: 3.05 m
Ground clearance: 0.20 m
Angle of approach/departure: 25/28°
Acceleration:
(petrol) 0-80 km/h in 23 s
(diesel) 0-80 km/h in 25 s
Max speed:
(petrol) 120-140 km/h
(diesel) 100-120 km/h
Range:
(petrol) 400 km
(diesel) 600 km
Gradient:
(4 × 2) 25%
(4 × 4) 60%
Engine: V-6 petrol developing 147 hp at 5,700 rpm or 6-cylinder turbocharged Detroit Diesel developing 134 hp at 2,000 rpm
Transmission: 3-speed automatic
Transfer box: 2-speed chain drive
Suspension: semi-elliptic leaf springs mounted on rubber bushes in fixed shackle boxes on chassis, double-action telescopic hydraulic shock-absorbers
Steering: recirculating ball, power-assisted
Turning radius: 6 m
Brakes:
(main) dual-line, boost-assisted, disc front, drum rear
(parking) cable-operated on rear drums
Tyres:
radial 205R14C (optional 225R15)
Electrical system: 12 V
Alternator: 14 V, 55 A
NBC system: no
Night vision equipment: no

Options include left or right hand drive and a higher level of ballistic protection.

Variants

The RG-32 Scout can be fitted with a lockable rear steel body or a glazed load body canopy. If required, gun ports can also be fitted into the body area.

RG-32 Scout weapons carrier

This is a further development of the Scout armoured patrol vehicle and was announced in late 1996. This is essentially a standard five-man Scout patrol vehicle with the following modifications:

(1) Larger wheels
(2) Modified rear body which can carry a variety of weapons
(3) A roof hatch with a pintle mount for 20 mm cannon, machine guns or twin grenade launchers
(4) Additional protection to assure survival of the crew in a single mine blast under any wheel
(5) Ceramic armour tiles affording protection against 7.62 mm armour-piercing ammunition.

The prototype vehicle is fitted with the South African Mechem RO-107 107 mm (12-round) multiple rocket launcher covered later in this volume.

Other armament installations include a 60 mm standard or long-range mortar, 106 mm M40 recoilless rifle, anti-tank guided missiles and a 20 mm cannon.

RG-32 Military Scout

In 1996, the prototype of the RG-32 Military Scout was shown for the first time and this has a number of improvements to suit it for military use. These include:

(1) Larger 406 mm diameter wheels
(2) More robust axles
(3) Differential locks on both axles
(4) Adaptable military-type rear body with canopy
(5) More powerful six-cylinder diesel engine
(6) Roof-mounted hatch with gun ring
(7) Adaptability as a weapons carrier
(8) Left- or right-hand drive
(9) Additional mine protection including capping on weld seams and special door locks
(10) Manual or automatic transmission.

RGM-32M for FCLV

To meet the British Army requirement for between 500 and 600 Future Command and Liaison Vehicle's (FCLV), Vickers Defence Systems are offering two vehicles from its South African subsidiary Vickers OMC for trials purposes.

These are the RG-31M and RG-32M, both of which are enhanced versions of proven vehicles that have been built in large quantities for the home and export markets.

RG-32M consists of an all-welded armoured monocoque citadel to which the front and rear subframes are attached. The baseline vehicle provides protection from 7.62 mm ball attack and a 6 kg anti-tank mine exploding under any wheel station. Higher levels of armour protection are available including protection against 7.62 mm armour piercing attack.

It has a maximum gross vehicle weight of 4,650 kg (including crew and weapons) with a payload of 1 tonne. A version with seating capacity of seven people is also being marketed and its design is such, that it can be adopted for a number of other missions.

Pull out compartments and bins for the crew's equipment are provided for a 48-hour mission. A Lockheed Martin C-130 Hercules transport aircraft can carry three RG-32M vehicles.

The baseline RG-32M has space for five people with a load area at the rear. Standard equipment includes full 4 × 4 drive, diesel engine, four speed automatic transmission, coil-spring suspension and an electrically operated 4,000 kg winch. Maximum road speed is 122 km/h and road range is around 750 km.

To reduce overall life cycle costs, the RG-32M uses standard commercial components wherever possible, with the complete drive line being provided by Mercedes-Benz in Germany, with the axles being uprated for their new role.

Scout 2

In late 2001, Vickers OMC unveiled a more streamlined version of the RG-32 scout called the Scout 2, which is being marketed in two versions. First a five-seat Mine Hardened Patrol Vehicle (MHPV) and second a seven-seat estate vehicle. These share a common driveline and controls and are powered by a 6-cylinder water-cooled turbocharged diesel engine coupled to an automatic transmission.

The basic Scout 2 is provided with protection from 5.56 × 45 mm attack and a variety of grenades, but a higher level of protection against armour-piercing attack is also available.

Status

Production. In service with various users in South Africa, Swaziland (Police) and with various elements of the United Nations.

Manufacturer

Vickers OMC.

UPDATED

SPAIN

Santa Barbara Sistemas BMR-600 infantry fighting vehicle

Development

In 1972, the Comisión de Desarrollo de Vehículos Blindados of the Dirección de Investigación of the Spanish Army Ministry, and the Dirección de Proyectos of ENASA started development work on a wheeled IFV to meet the requirements of the Spanish Army. ENASA was responsible for the automotive side of the vehicle while the army concentrated on the armour and the armament installation. Between 1975 and 1976 the prototype, called the Pegaso 3.500, was tested under a wide range of operational conditions. The engine was behind the driver on the left side of the hull and the armament installation to the right of the engine compartment. Redesigned prototypes had the engine on the right side of the hull and the armament installation to the rear of the driver.

The company designation for this vehicle is the BMR 3560. The specific variants offered by the now Santa Barbara Sistemas (and owned by General Dynamics of the United States) are designated:

3560.50	armoured personnel carrier
3560.51	command vehicle
3560.56	radio communications vehicle
3562.03	VEC cavalry scout vehicle (see entry in Reconnaissance vehicles section)
3560.57	missile launcher vehicle (prototype only)
3560.53	81 mm mortar is 3560/53E while 120 mm mortar model is 3560/59E
3564	combat vehicle with 90 mm cannon
3560.54	ambulance
3560.55	recovery and maintenance vehicle.

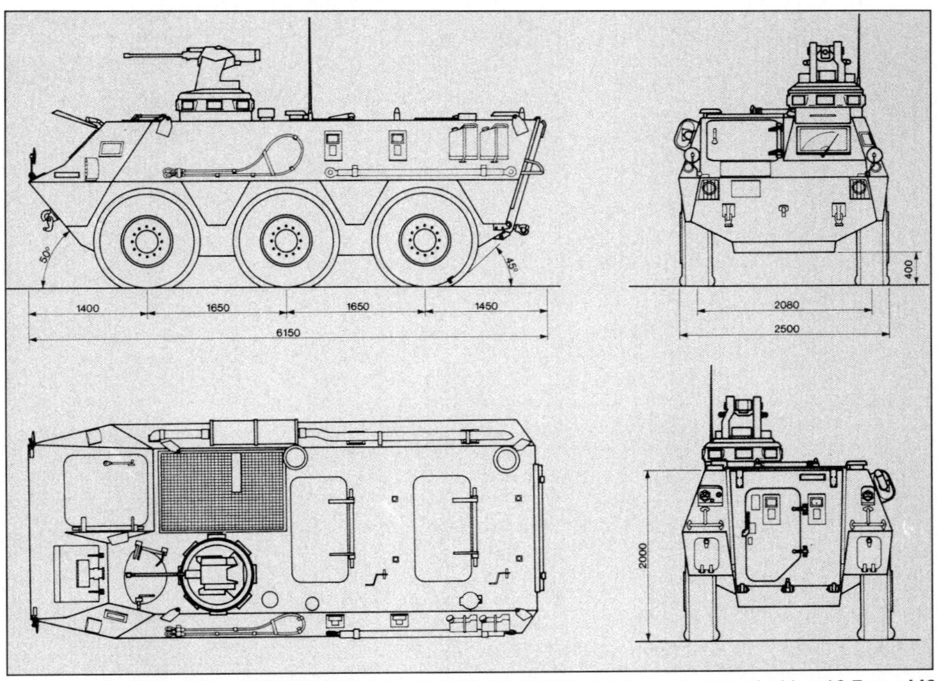

Four-view drawing of the BMR-600 IFV with a SANTA BARBARA TC-3 cupola armed with a 12.7 mm M2 HB MG

After comparative trials between the BMR-600 (Blindado Medio de Ruedas), the French Renault VAB and the Swiss MOWAG Piranha vehicles, the BMR-600 was selected for the Spanish Army and an initial production order for 15 vehicles was placed in 1979. First production vehicles were handed over to the Spanish Army in 1979.

In 1982, Egypt signed a contract for 250 BMR-600s, the first of which were delivered by late 1984. This comprised 217 APCs, 13 command post vehicles, 10 recovery and maintenance vehicles and 10 ambulances. These are used by the Egyptian Marines.

In 1983, Saudi Arabia ordered 140 BMR-600s for delivery from 1985. These are used by the marines.

In 1986, Peru purchased a total of 25 BMR-600 vehicles which comprised 20 APCs, one command post vehicle and four fire support vehicles with a two-man turret armed with a 90 mm gun. These are used by the marines.

Although the BMR-600 family of vehicles was originally designed and built by ENASA, production and marketing of all of the ENASA wheeled armoured vehicles is now carried out by Santa Barbara Sistemas. Production of the BMR-600 was undertaken at Seville.

Total production of the BMR-600 series of vehicles amounted to about 1,500 of which about 1,000 were for the Spanish Army and the remaining 500 for the export market.

Early in 1999 it was reported that negotiations were under way between Saudi Arabia and Spain for the supply of a new batch of 200 BMR-600 vehicles for the Saudi Arabian Marines plus the upgrade of the existing fleet of vehicles. It is possible that some of this work would be carried out in Saudi Arabia. As of late 2001, Saudi Arabia had not placed any additional orders for this vehicle.

Although production was completed some years ago, production could restart if further significant orders were placed.

An upgrade programme for Spanish Army BMR-600 and VEC vehicles is now under way by Santa Barbara and additional details are given later in this section.

Description

The hull of the BMR-600 is made of all-welded aluminium armour and incorporates spaced armour at the front.

BMR-600 fitted with Euromissile MILAN anti-tank guided tank weapon (PB)
2001/0104374

Upgraded BMR-2 of the Spanish Army showing passive armour and additional stowage cages
2000/0080993

Specifications

Crew: 2 + 10
Configuration: 6 × 6
Combat weight: 14,000 kg
Power-to-weight ratio: 22 hp/t
Length: 6.15 m
Width: 2.5 m
Height:
(including armament) 2.36 m
(to hull top) 2 m
Ground clearance:
(adjustable) 0.4 m
Track: 2.08 m
Wheelbase: 1.65 + 1.65 m
Angle of approach/departure: 50/45°
Max speed:
(1st gear) 17 km/h
(2nd gear) 56 km/h
(3rd gear) 69 km/h
(4th gear) 103 km/h
(reverse) 12 km/h
(water-jets) 9 km/h
(wheels) 3 km/h
Fuel capacity: 400 litres
Max range: 1,000 km
Fording: amphibious
Gradient: 60%
Side slope: 30%
Vertical obstacle: 0.6 m
Trench: 1.35 m
Turning radius: 7.5 m
Engine: Pegaso 9157/8 6-cylinder in-line diesel developing 310 hp at 2,200 rpm (now being replaced by a Scania DS9 diesel, see text)
Transmission: ZF 6 HP 500 automatic with 6 forward and 1 reverse gears. Torque converter and hydraulic retarder
Steering: recirculating ball, power-assisted on first and last axles
Suspension: Messier hydropneumatic independent each wheel, McPherson type
Tyres: 13.00 × 20 (run flat) (14.00 × 20 are options)
Brakes:
(main) dual-circuit hydraulic disc, air-assisted on all wheels
(parking) disc operating on propeller transmission, hydraulic
(emergency) air-operated
Electrical system: 24 V
Alternator: 28 V/120 A
Batteries: 2 × 12 V, 150 Ah
Armament:
(main) 1 × 12.7 mm M2 MG
Ammunition: 2,500 rounds

Gun control equipment
Turret power control: manual
Gun elevation/depression: +60/–15°
Turret traverse: 360°
NBC system: optional
Night vision equipment: optional

Note: With 13.00 × 20 tyres the vehicle has a maximum speed of 96 km/h and with 14.00 × 20 tyres a speed of 103 km/h.

Over the frontal arc of 60° (30° left and 30° right), protection is provided against small arms fire up 12.7 mm calibre.

Through a full 360° protection is provided against 7.62 mm small arms fire and shell splinters up to 155 mm in calibre. The underneath of the hull provides protection against anti-personnel mines containing up to 3 kg of high explosive.

The BMR-600 has a crew of two, consisting of the machine gunner/radio operator and driver, and can carry nine fully equipped infantrymen and a commander who dismounts with them. The driver sits at the front of the vehicle on the left side and has a bulletproof windscreen in front of him and a smaller bulletproof windscreen either side (or a periscope). In action, the front windscreen can be covered by an armoured flap and forward vision is then through a periscope mounted in the roof. There is a single-piece hatch cover over the driver's position that lifts and swings to the right. The engine is to the right of the driver with the air inlet and outlet louvres in the roof and the exhaust pipe on the right side of the hull. There is an engine access plate in the front of the hull. The engine compartment has a semi-automatic fire detection and suppression system.

The machine gunner/radio operator sits behind the driver. He has a SANTA BARBARA TC-3 cupola that can be traversed through a full 360° and is provided with a single-piece hatch cover that opens to the rear and eight day periscopes. A 12.7 mm M2 machine gun is mounted externally.

Over the top of the troop compartment are two rectangular roof hatches that open to the rear. The infantrymen enter and leave the vehicle by a ramp in the rear of the hull that opens downwards. There is also a door in the left rear side in case the ramp fails to open. Depending on the model, six firing ports with vision blocks can be fitted to the sides and rear of the vehicle.

The BMR-600 has an independent suspension system for each wheel station, which is a combination of the hydropneumatic and McPherson type. This allows each wheel station to be raised or lowered through 275 mm, according to the terrain being crossed. Steering is power-assisted on both the front and rear axles. All three Pegaso designed and built axles are of the double-reduction type with self-locking differentials. The tyres are of the run-flat type.

The vehicle is fully amphibious, being propelled in the water by water-jets either side of the hull at the rear, which have deflectors for steering when afloat. When propelled by the two water-jets the BMR-600 has a maximum speed of 10 km/h. It is also available without the water-jets and is then powered by its wheels at a speed of 4.5 km/h. Before entering the water a trim vane, which is folded back onto the glacis plate when not in use, is erected at the front of the hull and the bilge pumps are switched on. Spanish Army vehicles are not fitted with the water-jets for amphibious operations.

Mounted at the rear of the BMR-600 is a winch with a capacity of 4,500 kg. Optional equipment includes various radio systems, 14.00 × 20 tyres, run-flat tyres, a 300 litre fuel tank, night vision equipment, an NBC system, communications equipment, a land navigation system, smoke grenade dischargers and an air conditioning system. It can also be delivered without the amphibious capability.

BMR-600 3560.54 (6 × 6) armoured personnel carrier in ambulance configuration with roof-mounted generator (Richard Stickland) **2002**/0121213

BMR-600 3560.55 (6 × 6) armoured recovery and repair vehicle in travelling configuration (Richard Stickland)
2002/0121214

Variants

The vehicle has already been fitted with other armament installations, such as the French Giat Industries Toucan I turret and the MOWAG 7.62 mm remote-controlled machine gun mounting. Other turrets can also be fitted, for example the Santa Barbara TC-7 turret armed with two 106 mm recoilless rifles and a 12.7 mm ranging machine gun.

The BMR-600 is suitable for a wide range of other roles including an ambulance carrying four stretcher patients and two attendants; battalion, brigade or division command vehicle; fire support vehicle (for trial purposes it has already been fitted with the French Giat Industries TS 90 turret armed with a 90 mm gun); section or company command vehicle; radio vehicle; anti-tank vehicle (it has already been successfully tested fitted with the Euromissile HCT turret, with four HOT ATGWs ready to launch); and anti-aircraft vehicle with 20 mm cannon or fire-and-forget surface-to-air missile systems.

Mortar carriers

The BMR-600 is offered in two mortar carrier configurations, one armed with a turntable-mounted 81 mm mortar, the Model 3560/53E, and the other with a turntable-mounted 120 mm mortar, the Model 3560/59E. Both versions have a total crew of five men, including commander and driver, with the mortar firing through a two-part roof hatch that opens left and right. A baseplate and mount are also carried, allowing the mortar to be deployed away from the vehicle if required by the tactical situation. The 3560/59E 120 mm mortar carrier has four hydraulic stabilisers which are lowered before firing commences to provide a more stable firing platform. The Spanish Army uses both of these versions.

BMR 3560.55 armoured recovery and repair vehicle

This model is in service with Egypt, Saudi Arabia and Spain. Equipment installed includes a roof-mounted hydraulic crane with a 360° traverse and a lifting capacity of 18,000 kg, four hydraulic stabilisers that are lowered to the ground (two either side of the hull) when the crane is in use, a winch in the hull rear with a capacity of 7,000 kg and 100 m of cable, tools, lighting, welding equipment and a penthouse. Full details of this model are given in *Jane's Military Vehicles and Logistics 2001-2002.*

BMR 3560.54 ambulance

This model is in service with the Egyptian Army (and Spain which has 16). It has a crew of four (driver, commander and two orderlies) and can carry four stretcher patients or two stretcher and four seated patients.

BMR 600 with 40 mm grenade launcher

The Spanish Army has deployed a number of BMR-600 series vehicles with the 12.7 mm M2 machine gun removed and replaced by a new cupola with a single-piece hatch cover that opens to the rear. On the forward part of the cupola is a Santa Barbara 40 mm SB40 LAG automatic grenade launcher.

BMR VMA

This model has improved amphibious characteristics and remains at prototype stage.

BMR-600 3560.53E 81 mm mortar carrier with roof hatches open (Richard Stickland) ***2002***/0121215

Anti-tank vehicle

This is the BMR-600 with the TC-3 cupola removed and fitted with a roof-mounted Euromissile MILAN ATGW launcher, with the missile having a maximum range of 2,000 m. Spain has also fitted some BMR-600 with the Raytheon Systems Company TOW ATGW system.

Armoured command vehicle

This has additional communications equipment, map board and a folding table with personnel maps. To increase the work area a tent can be erected at the rear of the vehicle. This is used by the Spanish Army and has been exported.

Radio communications vehicle

This is fitted with a teletype, AM and FM radio equipment and is used by the Spanish Army.

Engineer vehicle

This is used by the Spanish Army under the designation of the BMR-600VZ and is fitted with a winch with a capacity of 10,000 kg and a front-mounted dozer blade.

BMR upgrade (BMR-2)

Early in 1996, following a competition, Scania Hispania (Scania Spain) was awarded a contract by Santa Barbara Sistemas, for the supply of 700 new Scania DS9 diesel engines for installation in the VEC Cavalry Scout Vehicles and BMR-600 armoured personnel carriers of the Spanish Army.

These are being delivered over a five year period starting in late 1996 with the total value of the contract being SKr100 million. The engines are being installed in a total of 642 VEC and BMR-600 series vehicles.

The Scania DS9 is a six-cylinder in-line diesel engine which develops 310 hp at 2,200 rpm and is more compact and fuel efficient than the original Pegaso engine.

The weight saved by the installation of a lighter engine will enable additional passive armour to be fitted to the vehicle for improved battlefield survivability. This armour could be of the aluminium or ceramic type and spall liners could also be fitted in the troop compartment.

The crew compartment is fitted with a fire detection and suppression system and the layout of the troop compartment will be improved. A new heating and ventilating system is installed.

The maintenance of the vehicle will be improved, with new indicators in the driver's panel in order to prevent failures before they damage the vehicle.

At present, the BMR-600 is fitted with a TC-3 cupola with an externally mounted 12.7 mm M2 machine gun but in the future, this could be replaced by a TC-13 turret armed with a 25 mm cannon or a light mount for the SANTA BARBARA 40 mm grenade launcher.

An additional stowage cage has been mounted on the right side of the hull roof with a larger stowage cage mounted on the hull roof at the rear, this extends over the rear of the hull. Additional vertical armour has been provided above the rear troop compartment. This allows some of the infantry when standing to have some degree of protection when using their rifles.

When upgraded the BMR-600 is called the BMR-2 by the Spanish Army.

Status

Production complete. In service with the following countries:

Country	Quantity	Comment
Egypt	260	used by marines
Saudi Arabia	140	used by marines
Peru	20	used by marines
Spain	682	used by army, and now being upgraded

Manufacturer

Santa Barbara Sistemas (owned by General Dynamics of the united States).

UPDATED

Santa Barbara Sistemas BLR armoured personnel carrier

Development

The BLR (*Blindado Ligero de Ruedas*) APC was designed by Empresa Nacional de Autocamiones to meet the requirements of the Spanish Army and police for an APC with good cross-country mobility and provision for the crew to observe the terrain. It is capable of being used for a wide range of operational roles such as airport security, border patrol and IS operations.

Its company designation is BLR 3545. Production and marketing of this and the BMR-600 IFV is now carried out by Santa Barbara Sistemas (now owned by General Dynamics of the United States).

As far as is known, there has been no recent production of the BLR armoured personnel carrier. An upgrade programme for the BMR-600 (6 × 6) armoured personnel carrier, covered in the previous entry, has been started, but there are no known plans for the BLR (4 × 4) APC to be upgraded.

Although production of the BLR was completed some time ago, export marketing of this vehicle is still being carried out.

Description

The all-welded steel hull of the BLR gives the crew complete protection against small arms fire (up to 7.62 mm × 51 armour-piercing ammunition) over its frontal arc, and against 7.62 mm × 51 ball ammunition over the remainder of the vehicle. The glass vision blocks give the same degree of protection as the vehicle's frontal arc. The driver sits at the front of the hull on the left with the vehicle commander to his right. In front of each is a large bulletproof windscreen which, in action, is covered by a steel shutter hinged at the top with an integral vision slit. The windscreens have a double-wiper facility, one is the conventional type and the other uses a solvent to deal with paints and spray. To the left of the driver and to the commander's right is a smaller bulletproof window, and above each of their positions is a single hatch cover that opens to the rear. In front of the commander and below the windscreen are two firing ports to give covering fire over the frontal arc.

There are four doors in the BLR, one in each side of the vehicle opening forwards and two at the rear, one either side of the engine compartment. The rear two open outwards and both have a vision block. In each side of the hull are five bulletproof windows. In the centre of the hull roof, behind the commander's and driver's positions, is a cupola with eight vision blocks and a single-piece hatch cover. This cupola can be traversed through a full 360° and has a 7.62 mm machine gun and shield for the gunner. In the roof, over each of the side doors, is a single-piece hatch cover

Specifications

Crew: 1 + 12
Configuration: 4 × 4
Combat weight: 12,000 kg
Unloaded weight: 9,600 kg
Power-to-weight ratio: 17.5 hp/t
Length: 5.65 m
Width: 2.5 m
Height:
(hull top) 2 m
Ground clearance:
(axles) 0.32 m
(hull sides) 0.47 m
Track:
(front) 1.96 m
(rear) 2.135 m
Wheelbase: 3.15 m
Angle of approach/departure: 42/35°
Max speed: 93 km/h
Fuel capacity: 200 litres
Road range: 570 km
Fording: 1.1 m
Gradient: 60%
Side slope: 30%
Turning radius: 9 m
Engine: Pegaso 4-stroke, turbocharged, direct-injection 6-cylinder in-line diesel developing 210 hp at 2,100 rpm
Transmission: ZF manual, 6 forward and 1 reverse gears
Transfer case: ZF, single speed
Clutch: single dry plate
Steering: hydraulic, power-assisted
Turning radius: 9 m
Transfer case: ZF, 2 speed
Suspension: semi-elliptical springs and double-action hydraulic shock-absorbers at each wheel station
Tyres: 13.00 × 20
Brakes:
(main) dual-circuit, air
(parking) mechanical
Electrical system: 24 V
Batteries: 2 × 12 V, 150 Ah
Alternator: 55 A
Armament: see text
NBC system: optional
Night vision equipment: optional

BLR (4 × 4) APC with a turret-mounted 7.62 mm machine gun and the cupola hatch open

that opens to the centre of the vehicle and similar hatches are provided in the roof either side of the engine compartment.

The personnel compartment has a volume of 14 m^3 and all personnel can enter the vehicle in 16 seconds and leave in 10 seconds. The BLR is smoke- and gas-proof, ventilated by two rear suction units and an interior intake equipped with a filter.

The engine compartment is at the rear of the vehicle in the centre, with the air inlet and outlet in the top and rear of the vehicle. It could be delivered with a 170 hp diesel engine coupled to a manual transmission or a 224 hp diesel engine coupled to an automatic transmission.

An unusual feature of the BLR is that in addition to the engine compartment being fitted with a semi-automatic fire extinguishing system, each of four roadwheels has outlets for extinguishing fires. The tyres of the BLR are fitted with puncture-proof Hutchinson O-rings.

Optional equipment includes smoke or CS gas dischargers which can be activated by the vehicle commander. It can also be fitted with various types of communications equipment, a siren and loudspeakers. Optional equipment includes a Power Take-Off (PTO), run-flat tyres, a front-mounted winch with a capacity of 4,500 kg, a front-mounted obstacle-removing blade, night vision devices, radios and specialised anti-riot equipment. According to the manufacturer the BLR can be fitted with a wide range of other armament systems including 12.7 mm M2 machine guns, 20 and 25 mm cannon and guns of up to 90 mm. As far a it is known, all production BLR vehicles have been armed with 7.62 mm or 12.7 mm machine guns.

Variants

There are no variants of the BLR but it can be adopted for other roles such as a command/communications vehicle or an ambulance.

Status

Production as required. In service with Ecuador, the Spanish Marines (28) and the Guardia Civil (rural paramilitary police) (22). There has been no recent production of this vehicle.

Manufacturer

Santa Barbara Sistemas (now owned by General Dynamics of the United States).

UPDATED

SWITZERLAND

MOWAG Piranha IV multipurpose armoured vehicle

Development/Description

In September 1999, the Swiss company of MOWAG, now owned by General Motors Defense (previously known as the Diesel Division, General Motors of Canada), released outline information on its new Piranha IV (8 × 8) armoured vehicle.

Under a recently signed agreement Vickers Defence Systems has a licence to market this vehicle for the UK and other undisclosed markets. Alvis Vehicles, also of the UK, holds the licence to build the current Piranha II and III (8 × 8) vehicles.

The first prototype of the Piranha IV (8 × 8) was built in Switzerland and was completed in 2001. This was shown for the first time in September 2001. A second prototype of the Piranha IV is now being built in Switzerland.

The most significant improvements over the current Piranha II and III vehicles are its higher level of protection due to a modular armour package, greater carrying capacity, increased internal volume (12 m^3) and larger tyres.

The hull also includes stealth characteristics to make the vehicle more difficult to detect on the battlefield.

The Piranha IV can be powered by a Caterpillar diesel engine developing 522 hp or a German MTU diesel developing 544 hp coupled to a ZF automatic transmission. These give a maximum road speed of 100 km/h and an operating road range of 750 km.

A central tyre pressure regulation system is fitted as standard as an anti-lock braking system. For improved

MOWAG Piranha IV APC (8 × 8) fitted with Rheinmetall Landsysteme 12.7 mm weapon station **2002**/0120455

MOWAG Piranha IV APC (8 × 8) fitted with Rheinmetall Landsysteme 12.7 mm weapon station **2002**/0120454

MOWAG Piranha IV APC (8 × 8) fitted with Rheinmetall Landsysteme 12.7 mm weapon station **2002**/0120460

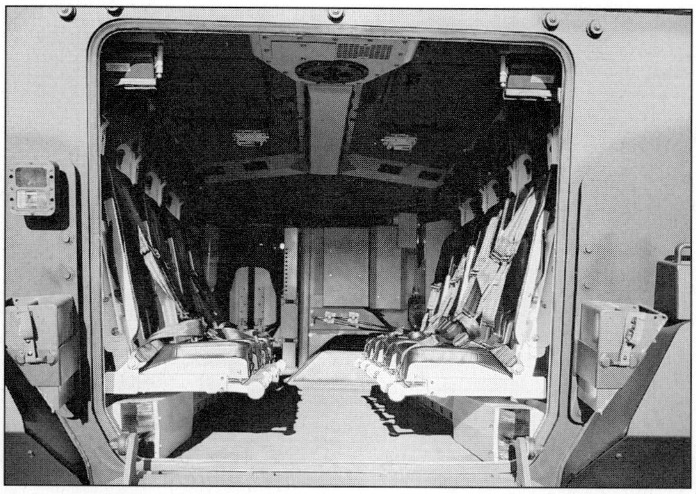

Troop compartment of Piranha IV APC (8 × 8) from the rear with seats folded down **2002**/0120462

cross-country mobility a height adjustable hydropneumatic suspension system is provided at each wheel station. A traction control system is fitted as standard.

The baseline Piranha IV chassis weighs 15 tonnes and has a maximum possible combat weight of 25 tonnes.

The difference between these figures is made up of the payload, including the crew, weapons and the specific armour package.

Weapon installation would depend on mission requirements but could typically be a Delco Defense two person turret armed with a stabilised 30 mm Bushmaster II cannon and 7.62 mm coaxial machine. According to the manufacturer weapons up to 105 mm in calibre can be fitted to the chassis of the Piranha IV vehicle.

To meet specific users requirements, the Piranha IV can be fitted with an air conditioning system, heater, fire/explosion detection/suppression system, winch and a NBC system.

Variants

As with the earlier generation MOWAG Piranha family of vehicles, the new Piranha IV will be able to undertake a wide range of roles such as troop carrier, command post vehicle and weapons carrier.

Specifications

Crew: 3 + 8 (depends on mission)
Configuration: 8 × 8
Combat weight: 25,000 kg*
Unloaded weight: 15,000 kg*
Power-to-weight ratio: 22 hp/tonne (with 544 hp engine @ 25 tonnes)
Ground Pressure: n/avail
Length: 7.24 m
Width: 2.80 m
Height: 2.20 m (hull top)
Ground Clearance: 0.58 m
Track: n/avail
Wheelbase: n/avail
Angle of approach/departure: 42°/38°
Max road speed: 100 km/h
Fuel capacity: n/avail
Max range: 750 km
Fording: n/avail
Gradient: 60%
Side slope: 30%
Vertical obstacle: 0.7 m
Trench: 2.00 m
Engine: Caterpillar diesel or MTU 6V199 TE20 diesel developing 544 hp at 2,300 rpm or Scania diesel
Transmission: ZF automatic model 7HP 902 electronically controlled, 7 speed, automatic
Steering: assisted on front four wheels
Suspension: semi-active with height management system
Tyres: n/avail (but CTIS fitted as standard)
Brakes: n/avail
Electrical system: 24 V
Batteries: n/avail
Armament: depends on version
Ammunition: depends on version
Armour: depends on version
NBC system: optional
Night vision equipment: optional
* Depends on weapon and armour. 15,000 kg is the baseline chassis without weapons or armour

Status

Development. First prototype completed in 2001.

Manufacturer

MOWAG motorwagenfabrik AG.

UPDATED

MOWAG Piranha Category III multipurpose armoured vehicle

Development

In response to market requirements for a vehicle with increased mobility and armour protection, MOWAG developed as a private venture the Piranha III series of 6 × 6, 8 × 8 and 10 × 10 armoured vehicles.

By early 1996, two preproduction MOWAG Piranha III vehicles had been built: the first in 6 × 6 configuration and fitted with a roof-mounted 7.62 mm machine gun, with the second in 8 × 8 configuration and fitted with a Delco Defense two-man LAV-25 power-operated turret armed with The Boeing Company 25 mm M242 Chain Gun and 7.62 mm coaxial machine gun.

Since then, two further MOWAG Piranha Category III (8 × 8) vehicles have been built, one by the Diesel Division, General Motors of Canada (who in August 1999 took over the Swiss MOWAG company) and one by MOWAG in the amphibious configuration. Late in 2000, the Diesel Division, General Motors of Canada, became General Motors Defense.

The Canadian-built MOWAG Piranha Category III (8 × 8) vehicle is the baseline for the new Canadian Forces LAV-III (8 × 8) APC which is now in production in Canada following completion of production of the Coyote (8 × 8) reconnaissance vehicle, which is fully covered in the Reconnaissance vehicles section earlier in this volume.

The major difference in the new LAV-III is that it is powered by a US-built Caterpillar 3126 diesel engine whereas all previous MOWAG Piranha vehicles have been powered by a US-supplied Detroit Diesel engine.

Early in 1998, following extensive competitive trials between the Finnish Patria Vehicles XA-186 (6 × 6), German Rheinmetall Landsysteme Fuchs (6 × 6) and the Swiss MOWAG Piranha III (8 × 8) Armoured Personnel Carriers (APCs), the latter was selected by the Danish Army.

Denmark placed an initial order with MOWAG for the supply of two Piranha III (8 × 8) APCs, with an option on an additional 20 vehicles. The Danish vehicles are Piranha IIIH with the letter standing for hydropneumatic

MOWAG Piranha III in 8 × 8 configuration as built for Denmark without applique armour fitted **2000**/0080992

MOWAG Piranha III in 6 × 6 configuration armed with 40 mm grenade launcher. This version has firing ports and additional armour protection 0058344

MOWAG Piranha III (8 × 8) in amphibious configuration and fitted with Delco Defense turret armed with 25 mm cannon **2002**/0120449

MOWAG Piranha III in 8 × 8 configuration as built for Denmark without appliqué armour fitted **2001**/0104379

suspension and are also fitted with a central tyre pressure regulation system.

The first two Danish vehicles were manufactured by MOWAG in Switzerland with the remainder being manufactured in Denmark by E Falck Schmidt A/S with key subsystems being provided by MOWAG.

The Danish vehicles also have a power-operated ramp in the hull rear rather than two doors and are fitted with a roof-mounted Krauss-Maffei Wegmann external weapon system armed with a 12.7 mm M2 machine gun. A special passive armour package can be installed for a higher level of battlefield survivability.

Of the 22 vehicles, two will be used in the command post role, two maintenance vehicles and the remaining 18 in the APC configuration.

The Danish Piranha III vehicles are powered by an MTU 6V183 TE 22 diesel developing 400 hp coupled to an ZF Ecomat 6HP 600 six-speed automatic transmission.

In addition to having improved mobility and armour protection, MOWAG III vehicles also have higher payload and gross vehicle weights, so allowing them to undertake an increased range of battlefield missions.

The increase in payload has been brought about by the redesigned hull, which is slightly longer and wider and has more internal volume. The 6 × 6 version has a payload of 3,000 kg and an internal volume of 9.5 m^3, the 8 × 8 a payload of 6,000 kg and an internal volume of 11 m^3 while the 10 × 10 has a payload of 7,500 kg and an internal volume of 14 m^3. The payload is the amount available for weapons, extra armour, men and other equipment.

Late in 1999 the Piranha Category III was selected by the Irish Defence Forces and a contract awarded to MOWAG for the supply of 40 Piranha IIIH (8 × 8) vehicles in four variants.

These are APC (34), command (4), recovery (1) and ambulance (1). The first two vehicles were delivered late in 2000 with all vehicles delivered by early 2002.

The APCs are fitted with a Helio FVT800 turret armed with a 12.7 mm and 7.62 mm MG. The turret will also be fitted to the command post vehicle.

Ireland selected the Piranha IIIH version with standard equipment including a fire detection and suppression system, NBC system, air conditioning system and a central tyre pressure regulation system. The Irish vehicles are powered by an MTU 6V183 TE22 diesel developing 400 hp, coupled to a ZF Ecomat 6HP 600 automatic transmission.

In the first quarter of 2001, it was stated that Ireland was expected to exercise its option for an additional batch of 20 Piranha IIIH vehicles in the near future, with the longer term requirement being for 40 vehicles to bring the fleet up to 80 vehicles.

Late in 2000, the US Army selected a version of the LAV-III to meet its Brigade Combat Team (BGT) requirement and additional details are given in this section under the United States of America.

The Swiss Army has placed an initial contract with MOWAG for 12 Piranha III (8 × 8) vehicles for use in the armoured command post role.

New Zealand has ordered 105 LAV-III (8 × 8) vehicles and variants which will be manufactured by General Motors Defense in Canada. These will replace United Defense M113 series full tracked vehicles currently used by the New Zealand Army.

Description

The hull of the MOWAG Piranha III series is of all-welded high-hardness steel armour construction that provides the occupants with protection from small arms fire and shell splinters.

To this can be added a replaceable add-on armour package, for example steel, composite or Kevlar, for an increased level of protection. Spall liners can also be fitted to the inside of the vehicle. As the armour system is of modular design, it can be tailored to meet the specific user's operational requirement.

The top armour package weighs 980 kg and provides the Piranha III vehicle with protection against 30 mm APDS (Armour-Piercing Discarding Sabot) attack through a 30° frontal arc and with protection against 14.5 mm AP attack from a range of 500 m through 360°.

Many countries are placing increased emphasis on protecting their vehicles against mines. Consequently, a mine protection kit (which can be quickly fitted) has also been developed for the vehicle.

To minimise radar reflection and heat radiation, special attention has been paid to designing the hull plate angles, add-on armour spacings and surface coating.

The layout of Piranha III vehicles is virtually identical to the earlier Category I and II vehicles. The driver is seated at the front left and has an adjustable seat and a single-piece hatch cover that opens to the left. In front of this are three day periscopes that give observation over the frontal arc. The centre day periscope can be replaced by a passive night vision periscope.

Another person is seated to the rear of the driver and this position is provided with a cupola with a single-piece hatch cover that opens to the rear and three day periscopes for forward observation.

To the right of the driver is the power pack. The complete power pack, consisting of the engine, transmission and cooling system, can now be changed much more quickly under field conditions and there are now no less than three different high-performance diesel power pack options:

(1) MTU 6V183 TE 22 developing 400 hp, with a potential of 450 hp, coupled to a ZF Ecomat 6HP 600 automatic six-speed transmission (Denmark is the first customer for this version)
(2) Caterpillar 3126 developing 350 hp, with a potential of 400 hp, coupled to Allison MD 3066 P fully automatic six-speed transmission. This is the engine selected by Canada to power its new LAV-III 8 × 8 vehicles
(3) Scania DS19-48A diesel developing 340 hp (with growth potential of 420 hp) coupled to ZF Ecomat 7HP 600 automatic 7-speed transmission.

A special combustion air filter has been fitted for desert operations and the cooling system has also been optimised for tropical conditions.

The turret is mounted in the centre of the hull and varies according to the vehicle type. The prototype of the 6 × 6 model, for example, has an externally mounted 7.62 mm machine gun which can be aimed and fired from within the vehicle in complete safety.

The 8 × 8 model can be fitted with a heavier turret, for example the two-man LAV-25, while the 10 × 10 can be fitted with a wide range of turrets up to a calibre of 105 mm.

The troop compartment is at the rear of the hull with the troops sitting on a bench-type seat down either side facing each other, with each person being provided with a seat belt. The 6 × 6 version has two roof hatches that can be opened out either side and a single day periscope in each side at the rear facing the rear to give observation to the rear of the vehicle.

The troops enter and leave via the power-operated ramp in the hull rear which is provided with a door in the left side. Firing ports and associated vision devices can

MOWAG Piranha IIIH (8 × 8) as selected by Ireland and fitted with Helio one-person turret armed with 12.7 mm machine gun **2001**/0104381

be fitted in the sides and rear of the troop compartment as required by the user.

The driveline has been enhanced and for increased cross-country mobility the Piranha III vehicles have been fitted with a new hydropneumatic suspension system developed by MOWAG which provides a much better ride across country. The front suspension has a total vertical travel of 320 mm while the rear has a total vertical travel of 340 mm.

According to MOWAG, cross-country mobility of the Piranha III vehicle is now similar to that of a tracked vehicle.

In addition, the vehicle can be delivered with a hydropneumatic suspension system with height adjustment which enables the overall height of the vehicle to be positioned to suit its operating environment or tilted front or rear.

Wider and larger diameter tyres with increased footprint area for improved mobility have been fitted. Tyres are of the run-flat type.

The two-speed transfer case allows for a travelling speed of up to 65 km/h in low off-road range and 100 km/h in the road range. All axles are provided with 100 per cent manually lockable differentials.

The vehicle is fitted with a French Syegon Central Tyre Inflation System (CTIS) that allows the driver to adjust the tyre pressure to one of four settings to suit the terrain being crossed.

This can be accomplished by the driver while the vehicle is travelling. A warning lamp will alert the driver as soon as the travelling speed becomes excessive for the selected tyre pressure.

One of the many options for the Piranha III vehicles is the addition of a Bosch Anti-locking Brake System (ABS) for added safety when travelling on roads.

The ABS has already been adopted by the Swiss Army for its new 8 × 8 armoured personnel carriers, deliveries of which started in 1995, with the initial order being for 205 vehicles out of a total requirement of 615. For a number of reasons, this was subsequently reduced to 515 units with final deliveries to be made by the end of 2002.

MOWAG emphasises that the latest Piranha III vehicles are an extension to the existing range and do not replace the earlier vehicles which have now been designated Category I and II and are covered in detail in the following entry.

These are available in 4 × 4, 6 × 6, 8 × 8 and 10 × 10 versions, whereas Piranha III vehicles are in 6 × 6, 8 × 8 and 10 × 10 configurations.

With a gross combat weight of 18.5 tonnes, the Piranha III vehicle in the 8 × 8 configuration is still air-transportable by a C-130 Hercules transport aircraft.

The cost of the Piranha III vehicle depends on many variables but if the Category I vehicle has a cost factor of 1, then the Category II would be 1.2 and III would be 1.4.

All the Piranha family of vehicles have a number of common features including independent suspension on all wheels, central driveline system, engine front right, driver front left, crew compartment in the rear; they can be adopted for a wide range of missions and are fully amphibious.

The latter versions are fitted with two high-performance propellers mounted one under either side of the hull at the rear, with each of these being provided with two rudders. Special provisions for seawater operations are available.

An 8,000 kg pulling capacity (drum or capstan) type self-recovery winch can be front or rear mounted. Other optional equipment includes air conditioning system, arctic heater kit, NBC system (collective or overpressure), fire and explosion detection and suppression system and various communications systems.

Specifications

Configuration	**6 × 6**	**8 × 8**	**10 × 10**
Max crew	14	16	18
Weight			
(combat loaded)	12,500 kg	18,500 kg	20,000 kg
(unladen)	9,500 kg	12,500 kg	12,500 kg
(payload)*	3,000 kg	6,000 kg	7,500 kg
Power-to-weight ratio	28 hp/t	21.6 hp/t	22.5 hp/t
Length overall	6.250 m	6.96 m	7.45 m
Width overall	2.66 m	2.66 m	2.66 m
Height overall			
(hull front)	1.985 m	1.985 m	1.985 m
(hull rear)	2.170 m	2.170 m	2.170 m
Ground clearance	0.595 m	0.595 m	0.595 m
Wheelbase	1.22 + 2.16 + 2.268 m	1.22 + 1.40 + 1.22 + 2.268 m	1.22 + 1.40 + 1.22 + 1.22 + 2.268 m
Track	2.268 m	2.268 m	2.268 m
Angle of approach	42°	42°	42°
Angle of departure	37°	37°	37°
Max speed			
(road)	100 km/h	100 km/h	100 km/h
(road creep speed)	3 km/h	3 km/h	3 km/h
(water)	10 km/h	10 km/h	10 km/h
Fuel capacity	200 litres	300 litres	400 litres
Max range			
(road)	500/800 km	500/800 km	500/800 km
Fording	amphibious	amphibious	yes
Gradient	60%	60%	60%
Side slope	30%	30%	30%
Vertical obstacle	0.6 m	0.6 m	0.6 m
Trench	2.0 m	2.0 m	2.0 m
Turning radius	7.5 m	8 m	9.5 m
Engine	350 hp diesel	400 hp diesel	450 hp diesel
Transmission	automatic	automatic	automatic
Transfer case	1-speed	1-speed	1-speed
Suspension	independent	independent	independent
Tyres	12.00 R20 365/80 R20	12.00 R20 365/80 R20	12.00 R20 365/80 R20
Brakes			
(main)	dual circuit, hydraulic-assisted by compressed air		
(parking)	parking lock in transfer case		
Electrical system	24 V	24 V	24 V
Batteries	2 × 12 V	2 × 12 V	2 × 12 V
NBC system	optional	optional	optional
Night vision equipment	optional	optional	optional

* This is the total available amount, for example turrets, weapons and ammunition

Variants

The Piranha 6 × 6, 8 × 8 and 10 × 10 all share the same hull front and driveline system and can be adopted for a wide range of battlefield roles including reconnaissance, anti-tank, armoured personnel carrier, infantry fighting vehicle, command post vehicle and so on.

Swiss Army command post vehicle

In 2000, the Swiss Army placed a contract with MOWAG for the supply of 12 Piranha III (8 × 8) in the command post role. These will be delivered from mid-2001 through to early 2002.

This version will feature a higher roof line giving greater internal volume, with provision for the erection of a tent at the rear for more space when being used in the static role. As the command post vehicle will have extensive communications equipment installed, an auxiliary power unit will be fitted.

MOWAG ACV 10 × 10

Additional details of this vehicle, for which Sweden is the first and so far only customer, are given in the Reconnaissance vehicles section.

MOWAG Piranha IIIH (8 × 8) as selected by Ireland and fitted with Helio one-person turret armed with 12.7 mm machine gun **2002**/0120452

MOWAG Piranha III (8 × 8) from the rear showing ramp 0058346

Status

Production. In service with:

Country	Quantity	Comment
Canada	651	total requirement, first production vehicles completed late 1998, all in 8 × 8 configuration, built in Canada
Denmark	22	18 APC, two command post, two recovery, all in 8 × 8, built in Switzerland
Ireland	40	delivery 2000 to 2002 in 4 versions, built in Switzerland
New Zealand	105	from Canadian production line
Sweden	13	specialised roles, all in 10 × 10 configuration in Switzerland
Switzerland	12	delivery 2001/2002 from Swiss production line
USA	366	from Canada/US (qv USA)

Manufacturer

MOWAG Motorwagenfabrik AG.

In August 1999 MOWAG was taken over by Diesel Division, General Motors of Canada and today known as General Motors Defence.

UPDATED

MOWAG Piranha armoured personnel carrier

Development

The Piranha range of 4 × 4, 6 × 6 and 8 × 8 armoured vehicles was developed by MOWAG as a private venture in the early 1970s and is designed for the home and export markets. The first prototype was completed in 1972 with first production vehicles following in 1976. In February 1977, Canada placed an initial order for 350 6 × 6 Piranhas (subsequently increased to 491), which were built under licence in Canada by the Diesel Division of General Motors Canada (late in 2000 this was renamed General Motors Defense). Additional information on the Canadian vehicles will be found in this section under Canada. The company has built 758 vehicles in the 8 × 8 configuration for the US Marine Corps under the designation Light Armored Vehicle. Final deliveries were made in mid-1988.

More recently the Canadian Forces have ordered an 8 × 8 APC version called the Bison, for which there is a separate entry under Canada, as there is for the Light Armoured Vehicle (8 × 8) and the Armoured Vehicle General Purpose (6 × 6). Details of the LAV-25 (Reconnaissance), known as the Coyote by the Canadian Forces, are given in the Reconnaissance Vehicles section under Canada. A total of 203 vehicles has been built with final deliveries taking place late in 1997. In August 1999, MOWAG was taken over by the now General Motors Defense.

Cardoen Industries manufactured the 6 × 6 MOWAG Piranha under a licence which has now expired, while FAMAE has built the 8 × 8 version of the MOWAG Piranha under the designation 8 × 8D. Full details of these are given under Chile.

The Piranha range has been designed to undertake a wide range of roles such as ambulance, anti-tank (armed with ATGWs), cargo, command, internal security, mortar carrier, recovery and reconnaissance. Many components of the vehicle are common to all members of the family, such as the front and rear hull sections, doors, hatches, wheel drives, wheels, differentials, suspension, steering and propellers.

In 2001, MOWAG stated that over 4,800 Piranha vehicles were in service in 4 × 4, 6 × 6 and 8 × 8 configurations, with a further 1,150 Piranha vehicles on order or in production. This includes previous production in Canada, Chile and the UK.

Specifications

Piranha Category I vehicles

Model	**4 × 4**	**6 × 6**	**8 × 8**
Crew (max)	10	14	15
Combat weight*	7,800 kg	10,500 kg	12,300 kg
Unloaded weight	6,700 kg	8,000 kg	8,800 kg
Power-to-weight ratio			
(diesel engine)	25 hp/t	28.5 hp/t	24.4 hp/t
Length	5.32 m	5.97 m	6.365 m
Width	2.5 m	2.5 m	2.5 m
Height			
(without armament)	1.85 m	1.85 m	1.85 m
Ground clearance	0.5 m	0.5 m	0.5 m
Track			
(front)	2.18 m	2.18 m	2.18 m
(rear)	2.2 m	2.2 m	2.2 m
Wheelbase	2.42 m	2.04 + 1.04 m	1.1 + 1.135 +1.04 m
Angle of approach/ departure	40/45°	40/45°	40/45°
Max speed			
(road)	100 km/h	100 km/h	100 km/h
(water)	9.5 km/h	10.5 km/h	10.5 km/h
Fuel capacity	200 litres	200 litres	300 litres
Max road range	700 km	600 km	780 km
Fording	amphibious	amphibious	amphibious
Gradient	70%	70%	70%
Side slope	35%	35%	35%
Vertical obstacle	0.5 m	0.5 m	0.5 m
Turning radius	6.3 m	7.3 m	7.7 m
Engine	Cummins 6 BTA 5.9 diesel, developing 195 hp at 2,800 rpm	Detroit Diesel 6V-53T developing 300 hp at 2,800 rpm	Detroit Diesel 6V-53T developing 300 hp at 2,800 rpm
Transmission	Allison Transmission AT-545 automatic, 4 forward/1 reverse gears	Allison Transmission MT-653 automatic, 5 forward/1 reverse gears	Allison Transmission MT-653 automatic, 5 forward/1 reverse gears
Suspension	independent	independent	independent
Tyres	11.00 × 16	11.00 × 16	11.00 × 16
Brakes (main)	air hydraulic (dual circuit) on all vehicles		
Electrical system	24 V	24 V	24 V
Armament	depends on role	depends on role	depends on role
NBC system	optional	optional	optional
Night vision equipment	optional	optional	optional

* Dependent on role

Alvis Vehicles-built Piranha (8 × 8) AFVs

Details of the UK-built MOWAG Piranha range of vehicles is given under Alvis Vehicles Piranha under the UK later in this section. This company has supplied 8 × 8 Piranha vehicles, including many variants, to Oman, Saudi Arabia and Qatar. In late 2000, Oman placed another order for a further 80 vehicles in 7 configurations. In September 2001, less than a year after the order was placed, Alvis Vehicles handed over the first of 80 Piranha II (8 × 8) Armoured Personnel Carriers (APCs) to the Royal Army of Oman (ROA), ahead of schedule.

Under the terms of this contract, a total of seven variants will be supplied with deliveries to be completed from the UK in 2002. These variants are understood to include APC (with and without armament), 81 mm mortar carrier, command and control, repair and recovery, ambulance and artillery observation.

The first batch of vehicles was shipped to Oman late in 2001 with final work, such as the installation of locally designed and built turrets, being carried out in the country. It is expected that final vehicles will be handed over to Oman in 2003.

Description

The all-welded steel hull of the Piranha protects the crew from small arms fire and shell splinters. The driver sits at the front of the vehicle on the left side and has a single-piece hatch cover that opens to the rear. In front of his hatch cover are three day periscopes, the centre one of which can be replaced by a passive night vision periscope. The engine is to the right of the driver with the air inlet and outlet louvres in the top of the hull and the exhaust outlet on the right side of the hull.

The main armament is normally installed in the centre of the hull. The troop compartment is at the rear and the troops enter and leave the vehicle by two doors in the rear of the hull that open outwards. There are two hatch covers over the roof of the troop compartment, which open outwards. MOWAG-designed spherical firing ports, with a vision block over each, can be fitted into each of the rear doors and two in either side of the hull, enabling the crew to fire their weapons from inside the vehicle.

The automatic locking differentials are flexibly mounted within the hull. The suspension is of the independent type with coil springs on the front axle and torsion bars on the rear. All wheel stations have hydraulic shock-absorbers. On the 6 × 6 vehicle the first axle has coil springs and the rear two axles have torsion bars, and on the 8 × 8 vehicle the front two axles have coil springs and the rear two axles have torsion bars. The suspension allows the roadwheels a maximum travel of 0.32 m. The tyres are of the run-flat type.

All members of the family are fully amphibious, propelled in the water by two propellers mounted at the rear of the hull. Before entering the water a trim vane, which is stowed under the nose of the vehicle when not in use, is erected at the front of the hull. This is erected by the driver without leaving his seat. Optional equipment includes NBC system, air conditioning equipment and various types of night vision devices.

Current production Piranha vehicles are known as the Category II vehicle. Details of the new Category III vehicles are given in the previous entry. These were first shown in 1996 and are available in 6 × 6, 8 × 8 and 10 × 10 models.

Variants

Piranha (4 × 4)

A variety of armament installations can be mounted, up to and including a 20 mm turret-mounted cannon, for example one or two 7.62 mm remote-controlled machine guns, a turret armed with an externally mounted 12.7 mm (0.50) M2 HB machine gun or a 20 mm turret.

Piranha (6 × 6)

A variety of armament installations can be fitted including a 12.7 mm MG turret, a 20 mm turret, a

MOWAG Piranha (6 × 6) in ambulance configuration and fitted with two propellers for amphibious operations 0018638

MOWAG Piranha II (8 × 8) fitted with Giat Industries Dragar turret armed with 25 mm cannon and 7.62 mm machine gun ***2000***/0080996

25 mm turret, a 30 mm turret, a turret armed with a 90 mm gun, or an 81 mm mortar mounted in the rear. More specialised versions include an unarmed ambulance fitted with a roof-mounted air conditioning system and a radar carrier. This has a RASIT battlefield surveillance radar mounted on the roof which can be lowered inside the vehicle under armour protection.

Piranha (6 × 6) anti-tank vehicle

Full details of this vehicle, designed to meet the requirements of the Swiss Army, are given in the Tank destroyers section later in this volume. The Swiss Army is the only user of this variant.

Piranha (8 × 8) anti-tank vehicle

MOWAG has completed the prototype of an anti-tank version of the Piranha (8 × 8) vehicle fitted with the Euromissile Mephisto system, consisting of a retractable launcher with four Euromissile 4,000 m range HOT ATGWs and additional missiles carried in reserve. This system is also used by the French Army on a VAB (4 × 4) APC.

This version has a combat weight of 12,600 kg and is 6.4 m long, 2.5 m wide and 2.8 m high with the Mephisto sight raised. There are four HOT missiles ready to launch and a further eight rounds in reserve.

Piranha (8 × 8)

A variety of armament installations can be fitted, including all those of the 6 × 6 model plus twin 20 or 30 mm anti-aircraft guns, a multiple rocket launcher with two banks of 15 81 mm launcher tubes, and a mortar tractor towing a 120 mm mortar.

Swiss Army Piranha (8 × 8) APC

Following extensive trials, the Swiss Army placed an initial order with MOWAG for the supply of 205 Piranha (8 × 8) APCs with the first of these delivered in January 1995 against a total requirement for 515 vehicles. The second order for 205 vehicles was placed in December 1996. The third order for 105 units was placed in December 1999.

The vehicles are delivered in three configurations, all of which are externally identical. These versions include armoured personnel carrier, command and control and a version to carry Dragon anti-tank guided weapon teams.

All vehicles are fitted with a Rheinmetall Landsysteme (previously KUKA) 606 A1 one-man turret armed with an externally mounted 12.7 mm M2 machine gun and forward-firing 76 mm smoke grenade dischargers.

The new 8 × 8 Piranhas have wider tyres for improved cross-country mobility, a slightly longer hull, external fuel tanks, additional external stowage and are fitted for, but not with, additional armour protection over the frontal arc.

The Swiss Army designation for this vehicle is the Füsilier-Radschutzerpanzer 93 and Kommando-Radschutzerpanzer 93.

Second-generation Piranha

New production Piranha vehicles are known as the Category II versions and have many improvements including larger size 12.5 R20 tyres, a more efficient braking system, a 100 per cent manually locking differential and a two-speed main transfer gearbox for improved cross-country mobility.

Specifications

Piranha Category II vehicles

Model	**4 × 4**	**6 × 6**	**8 × 8**
Max crew	12	14	16
Combat weight	8,000 kg	11,500 kg	14,000 kg
Unloaded weight	6,500 kg	8,300 kg	10,000 kg
Payload	1,500 kg	3,200 kg	4,000 kg
Power-to-weight ratio	26.2 hp/t	23.9 hp/t	19.6 hp/t
Length	5.40 m	6.00 m	6.98 m
Width	2.60 m	2.60 m	2.63 m
Height			
(without armament)	1.85 m*	1.85 m*	1.85 m*
Ground clearance	0.50 m*	0.50 m*	0.50 m*
Angle of approach/ departure	40/45°	40/45°	40/45°
Max speed			
(road)	100 km/h	100 km/h	100 km/h
(water)	9.5 km/h	10.5 km/h	10.5 km/h
Fuel capacity	200 litres	200 litres	300 litres
Max road range	700 km	600 km	780 km
Fording	1.4 m	1.4 m	1.4 m
Gradient	70%	70%	70%
Side slope	40%	40%	40%
Vertical obstacle	0.60 m*	0.60 m*	0.60 m*
Turning radius	6.3 m	7.3 m	8.25 m
Engine	Cummins 6 BTA 5.9 developing 210 hp at 2,500 rpm	Detroit Diesel 6V-53T developing 275 hp at 2,800 rpm	Detroit Diesel 6V-53T developing 275 hp at 2,800 rpm
Transmission	Allison Transmission AT-1545 automatic, 4 forward/1 reverse gears	Allison Transmission MT-653 DR automatic, 5 forward/1 reverse gears	Allison Transmission MT-653 DR automatic, 5 forward/1 reverse gears
Transfer case	all have single- or two-speed type		
Axles	front axle drive(s) disengageable, automatic locking differentials or 100% lock up		
Suspension	all have independent for each wheel		
(front)	coil springs and control arms		
(rear)	torsion bar		
(shock-absorbers)	hydraulic on all wheels		
Tyres	325/85 R16 or 335/80 R20* with run-flat cores		
Brakes	dual-circuit, hydraulic, assisted by compressed air		
Electrical system	24 V	24 V	24 V
Armament	depends on role and mission		
NBC system	optional	optional	optional
Night vision equipment	optional	optional	optional

* When fitted with the larger 335/80 R20 tyres, height over hull is 1.92 m, ground clearance is 0.57 m, vertical obstacle is 0.60 m and fording depth is 1.50 m. All models are fully amphibious, being propelled in the water by their propellers.

The driveline has a lower noise emission and the engine has a lower environmental impact. The engine compartment has been fitted with a new fire detection and suppression system. A new environmental control system is also fitted and the electrical system is of the latest Multiplex type.

Seawater cooling and special seawater equipment increase the amphibious capability mainly for operations in the sea. The hull has been stretched by an additional 290 mm for more interior space. For operation in deserts a special combustion air filter has been installed.

The driver has been provided with a new dashboard for easier and better operation and there are now three engine options, Detroit Diesel 6V-53T Silver developing 275 hp, Detroit Diesel 6V-53TA Silver developing 350 hp or Cummins 6 CTA 8.3 developing 300 hp.

For increased battlefield survivability, the vehicle can be fitted with various types of armour systems. An externally mounted spill winch is also provided.

90 mm assault gun vehicle

Late in 1986, MOWAG demonstrated the AGV-90 in Switzerland and France. This is essentially an improved MOWAG Piranha 8 × 8 fitted with the Giat Industries TS 90 Weapons Station. This is armed with the 90 mm gun with a coaxial 7.62 mm machine gun.

The 90 mm gun fires APFSDS-T, HEAT-T, HE, smoke and canister rounds with 18 ready rounds carried in the turret. A further 25 rounds are in the hull.

MOWAG Piranha (8 × 8) APC of the Swiss Army showing the mounting points on the front of the hull for additional armour **2000**/0080997

Optional equipment includes a land navigation system. Other improvements include a fuel tank with increased fuel capacity which increases operational range to 1,000 km, a small access hatch in the left side of the hull and a winch with a direct pulling capacity of 6.8 tonnes.

Combat weight of the AFV-90 is 13,000 kg, length with gun forwards 7.28 m, width 2.5 m and height 2.7 m.

Significant quantities of the Canadian-built MOWAG Piranha (8 × 8) vehicles fitted with the Giat Industries TS 90 turret were supplied to the Saudi Arabian Ministry of Defence and Aviation (MODA) by the now General Motors Defense. Additional details of this contract are given under Canada.

More recently, the MOWAG Piranha has been fitted with the latest Cockerill LCTS Mk 8 90 mm turret armed with a Cockerill 90 mm gun and this version has been ordered by Qatar (built by Alvis Vehicles of the UK) and the Saudi Arabian National Guard (SANG), additional details are given under Canada as this forms part of a complete order for 1,117 vehicles.

Piranha SPAAG

For trials purposes, a MOWAG Piranha (8 × 8) chassis has been fitted with a now Thales Blazer air defence turret armed with a 25 mm cannon, Matra BAe Dynamics Mistral fire-and-forget missiles and a Thales TRS 2630 radar system.

In late 2001, this remained at the prototype stage. In addition to being installed on the MOWAG Piranha (8 × 8) chassis it can also be installed on a wide range of other chassis, both tracked and wheeled. Details of this system are given in *Jane's Land-Based Air Defence 2001-2002*. The US Marine Corps has taken delievery of a different version which retains 25 mm cannon but has Stinger fire-and-forget SAMs. Details of this are given in *Jane's Land-Based Air Defence 2001-2002.*

Piranha 120 mm SPM

For trials purposes, the MOWAG Piranha chassis has been fitted with the French TDA 120 mm MO 120 2R 2M mortar system. Details of this are given in the Self-propelled motor systems section. It can also be fitted with the BAE Systems, RO Defence 120 mm mortar system. Detail of the latter, for which the first customer is Saudi Arabian National Guard (SANG), are given in the Self-propelled motor systems section.

Piranha Jernas

In 1995, the Dynamics Division of British Aerospace Defence (now Matra BAe Dynamics) used a MOWAG Piranha (8 × 8) to tow the Jernas launcher with eight Rapier Mk 2 surface-to-air missiles in the ready to launch position.

The MOWAG Piranha was fitted out as the Tactical Operations Cabin (TOC) for the Jernas SAM system which is the export version of Rapier Field Standard C (FSC) in service with the British Army and Royal Air Force Regiment. Additional details are given in *Jane's Land-Based Air Defence 2001-2002.* As of late 2001, there had been no orders placed for the Jernas surface-to-air missile system.

Manufacturer

MOWAG Motorwagenfabrik AG.

In August 1999, MOWAG was taken over by the Diesel Division, General Motors of Canada and now renamed General Motors Defense. Also made under licence in Chile (qv) and UK (qv).

UPDATED

Status

In production. In service with numerous countries including:

Country	Quantity	Comment
Australia	126	all delivered, 8 × 8 (see Australia), plus additional order
Canada	199	8 × 8 Bison, all delivered (see Canada)
	203	8 × 8 Recce, all delivered (see Canada)
	651	8 × 8 APC, under delivery (see Canada)
	491	6 × 6, all delivered (see Canada)
Chile	250+	6 × 6 and 8 × 8 (see Chile)
Denmark	2 + 20	8 × 8, ordered 1998 (Generation III)
Ghana	63	4 × 4, 6 × 6 and 8 × 8
Ireland	40	8 × 8, ordered 1999 (Generation III) (delivery 2000/2001)
Liberia	n/avail	4 × 4
New Zealand	105	from Canadian production line (Generation III)
Nigeria	140	delivered
Oman	80	all delivered, 8 × 8 (qv UK), plus another 80 to be delivered from 2001/2003
Qatar	40	8 × 8, all delivered
Saudi Arabia	1,117	8 × 8 (SANG) (see Canada) deliveries still under way
	400+	all delivered, 8 × 8 (MODA) (see Canada)
Sierra Leone	n/avail	believed 6 × 6
Switzerland	310	6 × 6 anti-tank, all delivered
	10	delivery 2001/2002 from Swiss production line (Generation III CPU)
	515	8 × 8 APC, deliveries under way
USA	758	US Marines, 8 × 8, delivered (see Canada)
	12	US Army National Guard, 8 × 8
	366	from Canada/US (qv USA) (Generation III)

MOWAG Piranha II (8 × 8) in command post configuration with front-mounted APU and tent erected at rear **2001**/0104382

MOWAG Piranha II (8 × 8) fitted with Giat Industries Dragar turret armed with 25 mm cannon and 7.62 mm machine gun **2001**/0104384

MOWAG Roland armoured personnel carrier

Development

The Roland 4 × 4 armoured vehicle was developed as a private venture by MOWAG in the early 1960s and is suitable for a wide range of roles including use as an ambulance, armoured personnel carrier, command and radio vehicle, internal security vehicle and as a reconnaissance vehicle.

The prototypes of the Roland were completed in 1963, with series production following in 1964. It is most widely used as an internal security and reconnaissance vehicle, especially in Africa and South America, and its crew normally consists of commander, gunner, driver and three or four infantry.

Production of the MOWAG Roland was completed in 1980 and it is no longer being marketed by the company.

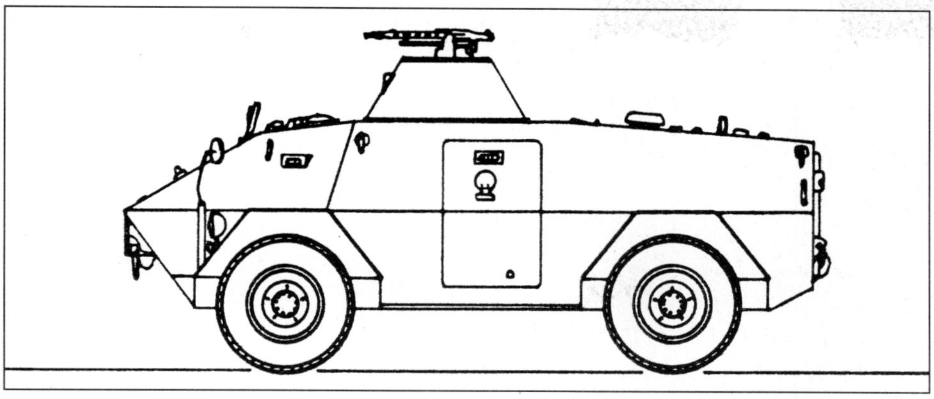

MOWAG Roland APC armed with a MOWAG remote-controlled 7.62 mm GPMG on the turret roof

MOWAG still continues to supply spare parts and after sales support for this vehicle as well as other vehicles that are no longer in quantity production by the company.

Description

The hull of the Roland is all-welded steel and protects the crew from 7.62 mm small arms fire. The driver sits at the front of the hull on the left side and has a single-piece hatch cover that opens to the right, in front of which are three day periscopes. A windscreen and wiper can be fitted in front of the driver's position in wet weather. To the left of the driver is a vision block protected by a grill.

In either side of the hull, to the rear of the driver's position, is an entry hatch hinged at the top. The left hatch has a vision block protected by a grill beneath which is a firing port. A vision block and firing port are provided to the immediate rear of the right side hatch. The gunner sits on an adjustable seat in the centre of the vehicle with one man to his left facing the front and another to his right facing the rear. Over the gunner's position is a turret which can be manually traversed through a full 360°. There is a single-piece hatch cover that opens to the rear and a vision block, to the right of which is a small circular hatch that opens to the right. To the rear of the turret on the right side is a circular hatch cover that opens to the rear, to the left of which is the ventilator.

Another door at the rear of the vehicle on the right side opens to the right and has a vision block and a firing port. Two men are seated in the passageway that connects the main crew compartment and the rear door, one facing forward and one facing the rear.

The engine is at the rear of the hull on the left side and is separated from the crew compartment by a fireproof bulkhead. The suspension system, front and rear, consists of semi-elliptical springs and hydraulic shock-absorbers. The rigid axles have hypoid gears and self-locking differentials.

Late production models of the MOWAG Roland had a slightly longer wheelbase and were fitted with an automatic instead of a manual transmission.

Standard equipment for the internal security version includes blue flashing lights, an electric roof fan, a heater, a two-tone intermittent siren and wire mesh protection for the headlamps.

The basic vehicle could be fitted with a number of different types of light armament installation, including a MOWAG-designed remote-controlled 7.62 mm machine gun mount on top of the low turret. The gunner is provided with a day sight with a magnification of ×4 for aiming the weapon. Smoke grenade dischargers could be fitted on either side of the turret if required.

Optional equipment included an air conditioning system; passive night vision equipment; an obstacle-clearing blade mounted at the front of the hull; searchlights; MOWAG-designed firing ports which allow the infantrymen to aim and fire their weapons from within the vehicle in safety; and MOWAG bulletproof cross-country wheels consisting of metal discs either side of the tyre, which support the tyre when it has been punctured and also provide additional traction when crossing rough country.

Variants

There are no MOWAG built variants of the Roland apart from the various roles that the vehicle has been designed to undertake.

Specifications

(Vehicle armed with remote-controlled 7.62 mm MG. Specifications in brackets relate to the model with automatic transmission where it differs from the manual version)
Crew: 3 + 3
Configuration: 4 × 4
Combat weight: 4,700 [4,900] kg
Unloaded weight: 3,900 [4,000] kg
Power-to-weight ratio: 42.9 [41
hp/t
Length: 4.44 [4.73 m]
Width: 2.01 [2.05 m]
Height:
(to turret top, without armament) 2.03 [2.1 m]
(to hull top) 1.62 [1.67 m]
Ground clearance:
(hull) 0.4 [0.42 m]
Track:
(front) 1.71 [1.66 m]
(rear) 1.655 [1.66 m]
Wheelbase: 2.5 [2.65 m]
Angle of approach/departure: 40/36° [46/40°]
Max road speed: 110 km/h
(with bulletproof wheels) 80 km/h
Fuel capacity: 154 [170
litres
Max road range: 550 [570] km
Fuel consumption: 0.28 [0.3
litres/km
Fording: 1 m
Gradient: 60%
Side slope: 30%
Vertical obstacle: 0.4 m
Turning radius: 6.45 [6.85 m]
Engine: V-8 4-stroke water-cooled petrol developing 202 hp at 3,900 rpm
Transmission: manual with 4 forward and 1 reverse gears [automatic with 3 forward and 1 reverse gears]
Transfer case: 2 speed
Clutch: single dry plate
Suspension: semi-elliptical springs and hydraulic shock-absorbers
Tyres: 9.00 × 16
Brakes:
(main) hydraulic on all wheels
(parking) mechanical on rear wheels
Electrical system: 12 V
Battery: 1 × 12 V, 125 Ah
Armament: 1 × 7.62 mm MG
Gun stabiliser:
(vertical) no
(horizontal) no
NBC system: no
Night vision equipment: optional

Status

Production complete. In service with at least the following countries:

Country	Quantity	Comment
Bolivia	24	army
Chile	50	carabineros
Ghana	40	army
Greece	20	gendarmerie
Liberia	8	
Mexico	25	some have been fitted with surface-to-surface unguided rockets
Sierra Leone	24	

Note: Argentina also uses the MOWAG Grenadier with about 60 being used by the Army and six by the Marines. These were delivered in 1967/68. In the 1980s one vehicle was upgraded by TENSA in Argentina with a new diesel engine, transmission, electrical system and 90 mm turret.

Manufacturer

MOWAG Motorwagenfabrik AG.
In August 1999, MOWAG was taken over by the Diesel Division, General Motors of Canada (in late 2000, this company was renamed General Motors Defense).

UPDATED

MOWAG Roland armed with a pintle-mounted 12.7 mm M2 HB MG

MOWAG Roland APC with side and rear doors/hatches closed

Armoured Duro armoured personnel carrier

Development

The Armoured Duro armoured personnel carrier is a joint development between Bucher-Guyer AG of Switzerland and O'Gara-Hess & Eisenhardt of the US.

Bucher-Guyer is responsible for the chassis and final integration while O-Gara-Hess & Eisenhardt is responsible for the supply of the passive armour package. It is available in both 4 × 4 and 6 × 6 configurations.

According to the manufacturers, typical roles of the Armoured Duro include reconnaissance, patrol and humanitarian missions during peacekeeping operations.

The first prototype complete early in 1999 was in the troop carrying role and was followed by ambulance and command post versions.

Late in 1999 the Swiss Army took delivery of a small quantity of Armoured Duro (6 × 6) chassis with a forward control armoured cab and seven rear armoured units of which four were in the armoured personnel carrier configuration and three in the ambulance configuration. These are being procured for possible humanitarian duties outside Switzerland.

Three versions of the Armoured Duro are being marketed, SP (4 × 4), LP (4 × 4) and P (6 × 6) which can carry 10, 10 or 16 troops respectively in the rear compartment. All of these have a two- three-person forward control cab.

The basic unarmoured Duro family of 4 × 4 and 6 × 6 cross-country chassis can be adopted for a wide range of roles. The Swiss Army has taken delivery of 3,000 in the 4 × 4 configuration. Early in 2001 Venezuela placed an order with Bucher-Guyer for the supply of 150 units, this being the first export order for the basic Duro vehicle.

O'Gara-Hess & Eisenhardt are well known for up-armouring civil and military vehicles and already have contracts for almost 2,000 HMMWV (4 × 4) vehicles.

Description

The Armoured Duro essentially consists of a Duro L (4 × 4) cross-country truck chassis fitted with a new fully armoured cab of the forward control type and a rear body which is capable of providing protection against 5.56 mm and 7.62 mm ball small arms fire up to NATO level B6, M67 grenades and anti-personnel mines.

The armour package consists of transparent armour on the front cabin windscreen and doors combined with a high hardness steel armour.

The torsion-free chassis is provided with an integral roll stabiliser with the frame being of the tubular ladder type.

The engine for all three models is a 4.2 litre six-cylinder diesel with direct injection, electronically controlled fuel injection and exhaust gas turbocharger with inter cooler.

This is coupled to an automatic transmission with torque converter and a two-speed transfer case. There is permanent all-wheel drive with Torson self-locking differentials. Steering is power assisted on the front two wheels.

The rigid De Dion axles with long suspension travel and the patented stabilisers are claimed to provide maximum traction over rough terrain and stability when loaded.

The Armoured Duro family from left to right, ambulance (6 × 6), armoured personnel carrier (4 × 4) and armoured personnel carrier (6 × 6) **2001**/0104412

Specifications

Configuration	**LP**	**SP**	**P**
Configuration	4 × 4	4 × 4	6 × 6
Crew	2 + 10	2 + 10	2 + 16
Weight (chassis and cab)	6,500 kg	6,500 kg	8,000 kg
Max load			
(chassis and cab)	2,700 kg	2,700 kg	3,700 kg
(crew compartment)	1,200 kg	1,000 kg	2,000 kg
Length	5.94	5.175 m	6,325 m
Width	1.96 m	1.96 m	1.96 m
Height	2.60 m	2.60 m	2.60 m
Ground clearance	0.36 m	0.36 m	0.36 m
Track	1.68 m	1.68 m	1.68 m
Wheelbase	3.23 m	3.88 m	3.23 m + 1.15 m
Max speed	110 km/h	110 km/h	110 km/h
Fuel capacity	120 litres	120 litres	120 litres
Max gradient	80%	80%	80%
Fording	0.8 m	0.8 m	0.8 m
Engine	VM 4.162 litre 6-cylinder electronically controlled direct injection exhaust gas turbocharged with inter cooler developing 160 hp at 3,000 rpm		
Transmission	aurtomatic with torque converter with 4 forward and 1 reverse gears		
Transfer box	2 speed	2 speed	2 speed
Steering	ZF power	ZF power	ZF power
Turning radius	6.25 m	7.25 m	7 m
Suspension	coil spring with gas pressure shock absorber and roll stabiliser		
Tyres	275/80 R 20	275/80 R 20	275/80 R 20
Brakes	hydraulic disc	hydraulic disc	hydraulic disc
Electrical system	24 V	24 V	24 V

The front cabin has seats for the driver and commander and due to the added weight of the armour each side door has been equipped with an automatic door assist. Once the door lever has been disengaged, the door opens automatically.

Since the door windows are fixed, crew members can electronically adjust the rear view mirrors in all directions. These mirrors come equipped with a defroster as standard.

The rear troop compartment, which can seat 10 (in 4 × 4 model) passengers, six down either side facing inwards, provides a similar level of protection to the cab and has interior lighting.

The troops enter the Armoured Duro through a single door in the rear which is provided with a bulletproof window with an additional two windows in either side of the troop compartment.

Additional features of the Armoured Duro include climate-controlled front and rear compartments and run flat tyres which allow the vehicle to travel for some 45 km.

Status

Production as required. First deliveries made to Swiss government late in 1999.

Manufacturer

Bucher-Guyer AG.

UPDATED

Armoured Duro (6 × 6) in ambulance configuration **2001**/0104387

Armoured Duro (6 × 6) in armoured personnel carrier configuration with side door **2001**/0104411

TURKEY

Otokar Cobra armoured personnel carrier

Development

The Otokar Cobra (4 × 4) armoured personnel carrier has been developed by Otokar as a private venture, with the first prototype being completed in late 1995.

The Cobra has a one-piece all-welded steel hull with the automotive components and suspension coming from the US-built AM General Expanded Capacity Vehicle (ECV) (4 × 4). The ECV in turn builds on the HMMWV, of which over 150,000 examples have been built for the home and export markets.

A total of eight prototype and preproduction Cobra vehicles had been built by 1997 in two versions, armoured personnel carrier and scout vehicle.

The Turkish Turkish Land Forces Command placed its first production contract for the Cobra and first production vehicles were completed late in 1997.

It is understood that the first export customer for the Otokar is the Maldives which took delivery of three amphibious versions in 1997.

Specifications

Crew: 2 + 11
Configuration: 4 × 4
Combat weight: 6,300 kg
Unloaded weight: 4,800 kg
Power-to-weight ratio: 30 hp/t
Length: 5.229 m
Width: 2.2 m
Height: 2.1 m (without turret)
Ground clearance: 0.35 m
Track:
(front) 1.819 m
(rear) 1.819 m
Wheelbase: 3.302 m
Angle of approach/departure: 68/68°
Max road speed: 115 km/h
Fuel capacity: 120 litres
Range: 500 km
Gradient: 70%
Side slope: 40%
Vertical obstacle: 0.32 m
Fording: 1 m
Turning radius: 7.6 m
Engine: General Motors V-8 water-cooled turbocharged diesel developing 190 hp at 3,400 rpm
Transmission: 4-speed automatic
Transfer box: permament 4-wheel drive, 2 speed
Axles: independent, double A-frame with hypoid differentials
Suspension: helical springs, hydraulic shock-absorbers
Tyres: 37 × 12.5 × 16.5 radial with run-flat system
Brakes: power-assisted, front and rear disc type
Steering: power-assisted
Electrical system: 24 V, suppressed
Alternator: 200 A
Batteries: 2 × 12 V, 100 Ah
Armament (optional): 1 × 7.62 mm or 12.7 mm machine gun
Stabiliser:
(vertical) no
(horizontal) no
NBC system: optional
Night vision equipment: optional

Description

The Cobra armoured personnel carrier has an all-welded monocoque steel hull for improved protection against both 5.56 mm and 7.62 mm small arms fire as well as mines. If required, the Cobra can be delivered with modular bolt-on type add-on passive armour providing a higher level of protection.

The power pack is at the front, with the driver being seated at the front left and the vehicle commander to his right. To the front and side of the commander and driver are bulletproof windows, with those to the front being provided with a washer and wiper. An electric de-icing system is provided for the front windscreen.

In either side of the hull is a forward-opening door. If required, these single-piece doors can be configured as two-part hatches.

To the rear of the commander and driver are the seats, the centre one being occupied by the person who mans the roof-mounted machine gun. Three men are seated down each side facing each other on individual seats which fold up when not required.

A large door is provided in the rear of the hull which opens to the left and is provided with a bulletproof window and firing port in the upper part.

In either side of the rear troop compartment are two bulletproof vision blocks with a firing port below, which allow some of the troops to fire their weapons from within the vehicle.

The weapon station is normally mounted on the roof to the rear of the commander and driver with a single-piece hatch to the rear.

The suspension and complete power pack (engine and transmission) is from the recent US AM General Expanded Capacity Vehicle (ECV). The General Motors water-cooled turbocharged diesel engine develops 190 hp and is coupled to a four-speed automatic transmission with permanent 4 × 4 drive and a two-speed transfer case. The exhaust pipe runs from the engine compartment up to the right side of the roof.

Single-speed Torsen differentials are also standard, with the suspension at each wheel station being of the double A-frame type with coil springs which provide good cross-country mobility. Hypoid single reduction differentials are mounted under the hull and are coupled with independent half shafts.

Power steering is fitted as standard as is an air conditioning system, blackout system, start socket, main cut-off switch, gun clips, combat-type seat belts, front and rear recovery hooks, thermal insulation inside hull, pioneer tools kit and run-flat tyres.

The standard vehicle is configured as an armoured personnel carrier with seats for 11 people, including the driver and commander.

Firing ports with associated observation blocks are provided in the sides and rear. The weapon station is suitable for mounting a 12.7 mm machine gun or 40 mm automatic grenade launcher. The pedestal is a 360° manually rotating gun ring incorporating ballistic protection shields all round.

The gun mounting is spring-assisted and can be locked by the gunner in any required angle for better control of the gun.

The turret version of the production Cobra vehicles have a one-man manually-operated turret armed with a 12.7 mm machine gun. The sight system is of the day and night type with the gun being electrically fired. The turret has banks of smoke grenade launchers, automatic ventilation and an intercom.

Optional equipment includes a Central Tyre Inflation System (CTIS), electric winch, NBC system, smoke grenade launchers, roof periscopes, night vision equipment and various types of communications equipment.

The 7.62 mm and 12.7 mm Overhead Weapon Stations (OWS) fitted to some versions of this vehicle are believed to be supplied by RAFAEL Armament Development Authority of Israel.

Variants

Amphibious Cobra

The amphibious version of the Cobra has a combat weight of 6.3 tonnes and is propelled in the water at a maximum speed of 8 km/h by two propellers mounted one either side under the hull rear. Before entering the water a trim vane is erected at the front of the vehicle manually. The vehicle takes three minutes to prepare for amphibious operations.

To the right of the drivers position at the front side of the vehicle is the amphibious configuration control panel. The propellers are powered by a hydraulic drive to the main engine. When afloat, engine cooling is via a seawater heat exhanger. The driver has a simple joystick control for amphibious operations and the propellers have reverse thrust.

Otokar Cobra (4 × 4) armoured personnel carrier fitted with cupola-mounted 12.7 mm M2 machine gun **2001**/0104417

The latest amphibious version of the Otokar Cobra (4 × 4) amoured personnel carrier in the water **2001**/0104418

Otokar Cobra with pintle mounted 40 mm automatic grenade launcher (Scott R Gourley) **2002**/0121201

Otokar Cobra with one-person turret but without machine gun installed (T J Gander) **2002**/0121200

It is believed that the Maldives is the first customer for this version and these are fitted with a one person turret armed with a 12.7 mm M2 machine gun.

Cobra Armoured Reconnaissance/ Surveillance Vehicle

Details of this vehicle, which is now in service with the Turkish Land Forces Command, are given in the Reconnaissance vehicles section.

APC

As in description, 8 to 12 personnel and armed with 12.7 mm machine gun.

Armoured reconnaissance vehicle

Crew of four and armed with 7.62 mm machine gun and fitted with telescopic mast with various day/night sensors.

Armoured reconnaissance vehicle

Armed with externally mounted 25 mm cannon with FLIR, crew of four.

Armoured command and control vehicle

Crew of six and armed with 7.62 mm machine gun and fitted with various types of communications equipment.

Armoured ambulance

Crew of three and can carry four stretchers or six seated patients plus medical equipment.

TOW missile carrier

Crew of five and fitted with roof-mounted launcher for Raytheon Systems Company TOW ATGW system. The TOW missile has a maximum range of 3,750 m and the launcher can have a day or night sighting system.

NBC reconnaissance vehicle

This would be fitted with special sensors to carry out its specialised NBC reconnaissance role on the battlefield.

Armoured command post

This is fitted with specialised communications equipment.

EOD vehicle

This has been developed to carry out Explosive Ordnance Disposal Base Recovery After Attack (EOD BRAAT) missions.

Armoured Tactical DF Vehicle

This is based on the Cobra and is fitted with extensive equipment to intercept, direction-find, monitor and record target communications under battlefield conditions that require armour protection and high mobility.

The system can be rapidly set up and is fitted with the ASELSAN DFINT-3T communications direction-finding equipment which has been designed to intercept, direction-find and analyse communications in the VHF/ UHF band.

If required, the DFINT-3T system and its associated communication equipment can be detached and operated outside of the vehicle. A mobile generator is fitted as standard to provide additional power.

This vehicle forms parts of the ASELSAN Reconnaissance/Surveillance and Intelligence System (RSIS).

Armoured Command Control Vehicle

This, together with other elements including the Armoured Tactical DF Vehicle (see previous) and the Armoured Reconnaissance/Surveillance Vehicle (covered in the Reconnaissance vehicles section), also forms part of the RSIS system developed by ASELSAN.

The Armoured Command Control Vehicle collects and evaluates operations and intelligence data and tracks the situation in the area of interest of the level that it is deployed.

This data is collected using various communications subsystems including VSAT satellite terminal and is evaluated in a totally computerised environment, using command and control software. An external generator is provided as standard.

Cobra for FCLV

With BAE Systems acting as prime contractor, Cobra was unsuccessfully entered for the British Army's Future Command and Liaison Vehicle requirement.

Status

The first batch of vehicles in three different variants has entered service with the Turkish Army. The Maldives took delivery of three amphibious vehicles in 1997.

Manufacturer

Otokar Otobus Karoseri Sanayi AS.

UPDATED

Otokar Akrep (Scorpion) armoured personnel carrier

Development

The Otokar Akrep (Scorpion) (4 × 4) armoured personnel carrier is a further development of the Otokar Akrep (Scorpion) (4 × 4) light reconnaissance vehicle, which is covered in detail in the Reconnaissance vehicles section.

The main difference between the two vehicles is that the hull rear of the armoured personnel carrier version has been extended and modified as a troop compartment.

Otokar Akrep (4 × 4) armoured personnel carrier **2000**/0081000

Description

The Akrep armoured personnel carrier has an all-welded steel hull with about 70 per cent of the automotive components being from the well-known Land Rover Defender 90/100 (4 × 4) vehicle, which is manufactured under licence in Turkey by Otokar.

The interior of the vehicle is insulated with fire-resistant type semi-rigid polyurethane material.

The monocoque all-welded steel hull provides protection from small arms fire up to 7.62 mm ball ammunition at 90° incidence and point-blank range, and shell splinters. All windows and vision blocks are of laminate construction armoured glass with a polycarbonate anti-spall liner. The lower part of the armoured hull slopes inwards.

To meet different users' requirements the Akrep is marketed with a petrol or diesel engine coupled to a manual or fully automatic transmission.

The layout is conventional, with the engine at the front, driver and commander in the centre and the troop compartment at the rear.

To the front of the commander and driver is a bulletproof window and in each side is a door with a bulletproof window in the upper part and a firing port below. Mounted in the roof of the vehicle, between the commander's and driver's positions, is a white light searchlight which can be operated by either crew member from within the safety of the vehicle.

The troops are seated three down each side facing the centre of the vehicle and enter and leave via a large door in the rear that opens to the left.

In total, seven firing ports (each with a day vision device above) are provided in the rear troop compartment, three in the left side, two in the right side, one in the rear door and one in the right corner of the vehicle.

There is a circular hatch cover in the roof on which is mounted a 7.62 mm machine gun with front, side and rear protection being provided for the gunner, who also has a single-piece hatch cover that opens to the rear. Elevation and traverse is manual. A vision block is provided in the upper part of either side of the gunner's position.

Mounted on the roof to the rear of the machine gunner's position is a spare wheel and tyre.

Standard equipment includes power-assisted steering, air conditioning and ventilating system, roof-mounted fan for smoke extraction, blackout lighting system, seat belts, battery master switch, fire extinguisher, antenna mounting provision, front and rear recovery hooks and a first aid kit.

Optional equipment includes run-flat tyres, a front-mounted winch with a capacity of 3,600 kg, smoke grenade dischargers mounted on either side of the vehicle, roof-mounted rotating beacon, searchlight mounted on the forward part of the roof which can be operated from inside by the commander, rifle clips and barricade remover, to name but a few.

Specifications

Crew: 2 + 6
Configuration: 4 × 4
Combat weight: 3,600 kg
Unloaded weight: 3,000 kg
Power-to-weight ratio:
(petrol engine) 37.22 hp/t
(diesel engine) 37.00 hp/t
Length: 4.40 m
Width: 1.95 m
Height:
(with MG mount) 2.41 m
(without MG mount) 2.20 m
Ground clearance: 200 mm
Track: 1.486 m
Wheelbase: 2.794 m
Angle of approach/departure: 41/49°
Max road speed: 125 km/h
Fuel capacity: 85 litres
Range:
(petrol engine) 650 km
(diesel engine) 1,000 km
Gradient: 70%
Side slope: 40%
Vertical obstacle: 0.315 m
Fording: 0.60 m
Turning radius: 6.7 m
Engine: V-8 3.5 litre water-cooled petrol developing 134 hp at 5,000 rpm or 4-cylinder turbo diesel with intercooler developing 111 hp at 4,000 rpm
Transmission: manual, 5 speed forward and 1 reverse or automatic 4 speed
Clutch: single dry plate
Transfer box: 2-speed, high and low ranges
Axles: live beam with single reduction hypoid differentials
Suspension: coil springs and hydraulic dampers with Panhard rod and anti-roll bars
Tyres: 7.50 R16 with run-flat inserts
Brakes:
(main) servo-assisted front and rear disc on all wheels
(parking) drum-type operating on transmission
Steering: power-assisted
Electrical system: 24 V
Alternator: 65 A
NBC system: no
Night vision equipment: no

Status

Production. In service with Turkish security forces. There are no known exports of this vehicle.

Manufacturer

Otokar Otobus Karoseri Sanayi AS.

VERIFIED

Otokar armoured personnel carrier

Development

This light armoured personnel carrier (or Armoured Patrol Vehicle as it is referred to by Otokar) was developed by the Turkish company, Otokar Otobus Karoseri Sanayi AS, to meet the operational requirements of the Turkish security forces.

The first prototypes were completed in 1991 and, following extensive trials, the vehicle was accepted for service, with the first production vehicles being completed in 1992.

The vehicle builds on Otokar's extensive experience in the manufacture of light vehicles for both the civilian and military markets. Since 1987 the company has built the Land Rover range of 4 × 4 vehicles under licence.

Description

The Otokar armoured personnel carrier has an all-welded steel hull with some 70 per cent of the automotive components being from the well-known Land Rover Defender 90/110 (4 × 4) vehicle for which spare parts are available worldwide.

It also shares many parts with the Otokar Akrep (Scorpion) (4 × 4) light reconnaissance vehicle covered in the Reconnaissance vehicles section.

The Otokar armoured personnel carrier is offered with a diesel or petrol engine coupled to a manual transmission.

The monocoque all-welded steel armour hull provides protection from small arms fire up to 7.62 mm ball ammunition fired at 90° incidence and point-blank range, and shell splinters. All windows and vision blocks are of laminated armoured glass construction with a polycarbonate anti-spall liner.

Machine-pressed GRP plates fitted to the floor area provide protection against grenade fragmentation, anti-personnel mines and blast.

The layout is conventional, with the engine at the front, driver and commander in the centre and the troop compartment at the rear.

The machine gun installation is on the roof to the rear of the commander and driver and on the roof behind this is the spare wheel and tyre.

To the front of the commander and driver is a bulletproof window and in each side is a door with a bulletproof window in the upper part. There is one horizontally sliding type gun port in the right door and one hinged type gun port to the front of the commander who is seated on the right.

Mounted in the roof of the vehicle, between the commander's and driver's positions, is a white light searchlight which can be operated by either crew member under full armour protection.

Otokar (4 × 4) armoured personnel carrier armed with a cupola-mounted 7.62 mm MG

Specifications

Crew: 2 + 6
Configuration: 4 × 4
Combat weight: 3,600 kg
Unloaded weight: 3,000 kg
Power-to-weight ratio: 37 hp/t
Length: 4.155 m
Width: 1.81 m
Height:
(with MG mount) 2.75 m
(without MG mount) 2.26 m
Ground clearance: 200 mm
Track: 1.486 m
Wheelbase: 2.794 m
Angle of approach/departure: 48/32°
Max road speed: 125 km/h
Fuel capacity: 85 litres
Range:
(petrol engine) 500 km
Gradient: 70%
Side slope: 40%
Vertical obstacle: 0.315 m
Fording: 0.60 m
Turning radius: 6.7 m
Engine: V-8 3.5 litre water-cooled petrol developing 134 hp at 5,000 rpm or Land Rover 300 TDI 4-cylinder turbocharged diesel developing 111 hp at 4,000 rpm
Transmission: manual, 5 forward and 1 reverse speeds
Clutch: single dry plate
Transfer box: 2-speed, high and low ranges
Axles: live beam with single reduction hypoid differentials
Suspension: coil springs and hydraulic dampers, Panhard rod and anti-roll bars
Tyres: 7.50 R16 size PR8 run-flat tyres
Brakes:
(main) servo-assisted front and rear disc on all wheels
(parking) drum type operating on transmission
Steering: power-assisted
Electrical system: 12 V
Alternator: 65 A
NBC system: no
Night vision equipment: no

The six troops are seated three either side down the rear facing inwards, with each man being provided with a seat belt. The troops enter and leave via two doors in the rear that open outwards, with each of these being provided with a small bulletproof window with a firing port underneath.

The 7.62 mm machine gun mount has front, rear and side protection, with the gunner being provided with a single-piece circular hatch cover that opens to the rear and that can be locked in the vertical position. A 12.7 mm machine gun can be fitted as an alternative.

The 7.62 mm machine gun mount can be traversed through a full 360° with weapon elevation from −12 to +65°. An adjustable platform for the gunner's height is mounted on the floor of the vehicle and can be adjusted to three heights.

Standard equipment includes power-assisted steering, a roof-mounted extractor fan, an air conditioning system, a heater and demister, seat belts, three-speed windscreen wipers with wash facility, run-flat tyres, an armour-protected explosion-suppressed fuel tank, a tool set, a fire extinguisher, small arms clips, a blackout system, a battery master switch, antenna provision, recovery hooks and a first aid box.

Optional equipment includes a front-mounted barricade remover, a public address system, a siren, rotating roof beacons, a front-mounted electrical winch with 25 m of cable and a capacity of 3,600 kg, internally controlled searchlights, communications equipment, 66 mm smoke grenade launchers, a roof hatch, fog lights and a 24 V electrical system.

Variants
In addition to being used as an armoured personnel carrier, the manufacturer has suggested that the vehicle can be used for a wide range of other missions including as an armoured police vehicle, an armoured security vehicle, an armoured ambulance, an armoured repair vehicle and an armoured command post vehicle. As far as it is known there has been no volume production of these specialised variants.

Status
Production as required. In service with Turkish armed and security forces and two other undisclosed countries.

Manufacturer
Otokar Otobus Karoseri Sanayi AS.

UPDATED

UKRAINE

Guardian Armoured Personnel Carrier

Development
In the late 1980s, the Kharkov Morozov Machine Building Design Bureau built an 8 × 8 APC which was originally called the BTR-94 or BTR-94K, full details of this are given in the following entry.

The BTR-94 has also been referred to as the BTR-DA and for the export market the improved Hunter or Guardian has been developed, which made its first appearance outside the Ukraine in 2001.

Some sources have indicated that the United Arab Emirates has already placed an order for 80 Guardian units for use by its Marine forces.

Although the overall design authority is the Kharkov Morozov Machine Building Design Bureau, actual production of the BTR-94 is carried out by the Malyshev Plant.

Description
The hull and turret of the Guardian APC are of all-welded steel armour, providing the occupants with protection from small arms fire and shell splinters. A Kevlar liner can be provided for increased protection.

The driver is seated at the front of the hull on the left side with the vehicle commander to his right. Both crew members are provided with individual roof hatches and periscopes.

To their immediate front is a bullet-proof window which is covered when in action by an armoured shutter hinged at the top. The commander also has a roof-mounted searchlight that can be operated from within the vehicle as well as a firing port in the front right side of the vehicle.

Although similar in appearance to the Russian BTR-80, a close examination of the BTR-94 shows that it has a higher roof line to the immediate rear of the commander and driver who are seated at the front of the vehicle.

This not only gives greater internal volume for the six troops carried but also allows it to be fitted with a wider and more powerful range of weapon systems.

This is possible as the raised roof line allows turrets to be fitted with a deeper basket than normal. The Guardian is however heavier than the original BTR-94 which weighed 13.6 tonnes, compared with 16.4 tonnes of the more recent Guardian.

The original BTR-94 was fitted with a locally designed and built 3TD multifuel engine developing 280 to 300 hp and coupled to a manual transmission.

For the export market a new Western power pack has been fitted to the Guardian which consists of a Deutz BF6M1015 diesel developing 326 hp coupled to an Allison MD3066 fully automatic transmission. The new power pack has been fitted by the ADCOM Group International of Abu Dhabi.

The BTR-94 is fully amphibious, being propelled in the water at a speed of 8 km/h by a single water jet mounted at the rear of the hull. Before entering the water a trip vane is erected by the driver from inside the vehicle and the bilge pumps are switched on.

The six troops can enter and leave the vehicle by means of a door in either side of the hull between the second and third roadwheel station. The lower part of the door folds downwards to form a step, with the upper part opening forwards. In addition there are also hatches in the roof to the rear of the turret and firing ports, with associated vision devices provided in the sides and front of the vehicle.

Weapon fit depends on the specific role that the Guardian is required to undertake. In Abu Dhabi the Guardian was shown fitted with the one-person Shakval unified battlefield module which can be fitted with various weapon systems. This is already in service with the Ukraine, installed on a BMP-1 chassis.

These include a stabilised 30 mm KBA-3 cannon with a 7.62 mm PKT or 12.7 mm machine gun, 30 mm AGS-17 automatic grenade launcher (or a Western 40 mm automatic grenade launcher) and Anti-Tank Guided Weapons (ATGW).

The latter include the Chinese NORINCO (China North Industries Corporation) Red Arrow or the Russian Konkurs or Kornet. The version shown in Abu Dhabi was fitted with two Russian Konkurs (9M113)(NATO AT-5 Spandrel) ATGW in the ready to launch position, these have a maximum range of 4,000 m.

Six 81 mm electrically operated smoke grenade launchers are also fitted, these are mounted three either side of the turret rear firing forwards.

Guardian (8 × 8) armoured personnel carrier fitted with Shakval unified battlefield module armed with 30 mm cannon, 7.62 mm machine gun, 30 mm grenade launcher and two Spandrel ATGW ***2002***/0121220

Specifications:
Crew: 3 + 6
Configuration: 8 × 8
Power-to-weight ratio: 19.87 hp/tonne
Combat weight: 16,400 kg
Length: 7.65 m
Width: 2.90 m
Height: 2.86 m
Ground clearance: 0.47 m
Track: 2.41 m
Wheelbase: 4.4 m
Max road speed: 100 km/h
Max water speed: 8 km/h
Fuel capacity: 300 litres
Max road range: 600 km
Fording: amphibious
Gradient: 60%
Side slope: 40%
Vertical obstacle: 0.50 m
Trench: 2 m
Engine: Deutz BF6M1015 air-cooled diesel developing 326 hp
Transmission: Allison MD3066 fully automatic
Transfer case: 2-speed
Steering: power-assisted on front four wheels
Turning radius: 13.2 m
Tyres: 13.00 × 18 (tubeless)
Suspension: torsion bar with hydraulic shock-absorbers
Brakes:
(main) hydraulic on all wheels
(parking) mechanical
Electrical system: 24 V
Armament:
(main) 1 × 30 mm KBA-3 cannon
(secondary) 1 × 30 mm AGS-17 automatic grenade launcher
(secondary) 1 × 7.62 mm or 12.7 mm machine gun
(ATGW) see description
(smoke grenade launchers) 6 × 81 mm
Ammunition:
(30 mm cannon) 300 (ready use)
(30 mm grenade launcher) 29 (ready use)
(7.62 mm) 2,000 (total)

Gun control equipment: ?
Turret traverse: powered/manual
Gun stabiliser:
(vertical) no
(horizontal) no
NBC system: yes

The 30 mm dual-feed KBA-3 cannon has 300 rounds of ready use ammunition with the 30 mm AGS-17 grenade launcher, mounted on the left side of the turret, having 29 rounds of ready use ammunition, with additional ammunition being carried in reserve (3 × 29). A total of 2,000 rounds of 7.62 mm ammunition is carried for the co-axial PKT machine gun.

Sighting system typically includes a commander's 1PZ-3 and a TKN-4S Agat stabilised sight which can be integrated the missile fire-control system plus observation periscopes. A laser range-finder is fitted with a maximum range of 4,000 m.

The unified battlefield module was developed by the State Scientific Technical Centre of Artillery & Rifle Arms (SSTC ARA) based in the Ukraine. It has been designed for installation on not only wheeled vehicles such as the Guardian, but also on full tracked armoured vehicles such as the BMP-1 infantry combat vehicle and the BRDM-2 (4 × 4) scout car.

Standard equipment on the Guardian includes powered steering on the front four roadwheels and a central tyre pressure regulation system that allows the driver to adjust the tyre pressure to suit the terrain being crossed. The Guardian can also be fitted with French Michelin tyres.

The engine compartment at the rear of the hull is fitted with an automatic double-action fire extinguisher and a Webasto C6 air conditioning system is provided as standard equipment.

Jordan took delivery of 50 BTR-94 (8 × 8) APCs each fitted with a one-person turret armed with two 23 mm externally mounted cannon and a 7.62 mm PKT coaxial machine gun. These were first seen in public in 2000, although this did not have the raised roof line of the example recently shown in Abu Dhabi.

Status
Production as required.

Manufacturer
Maylshev Plant (see development)

NEW ENTRY

BTR-94 Armoured Personnel Carrier

Development
The BTR-94 (8 × 8) armoured personnel carrier is essentially the Russian BTR-80 (8 × 8) APC, fully described under Russian Federation and Associated States, fitted with a new engine and weapon station designed and built in the Ukraine.

The vehicle has been developed by the Malyshev Plant and the Morozov Design Bureau, both of whom are known mainly for the development and production of the T-80UD and T-84 series Main Battle Tanks (MBTs).

It is understood that the first customer for the BTR-94 was Jordan who, in 2000, took delivery of 50 vehicles fitted with the turret armed with twin 23 mm cannon and a 7.62 mm coaxial machine gun. These were fitted with standard US HF (PRC-104) and VHF radios.

Description
The overall layout of the BTR-94 is identical to the BTR-80 covered in detail earlier in this volume but fitted with a new weapon station and new engine. For a higher level of battlefield survivability, internal spall liners of Kevlar have been fitted.

The standard production BTR-80 is fitted with a one person turret armed with a 14.5 KPVT and 7.62 mm PKT machine gun with a bank of electrically operated 81 mm smoke grenade launchers to the rear.

BTR-94 (8 x 8) armoured personnel carrier in service with Jordan (Rupert Pengelly) **2001**/0067812

On the BTR-94, this has been replaced by a new locally developed turret with the weapons mounted externally so that the gunner is seated within the protection of the chassis.

Main armament comprises two 23 mm AZP-23-2 cannon or one 30 mm cannon with a 7.62 mm PKT machine gun being mounted co-axial and below these weapons. Turret traverse and weapon elevation is powered with manual controls being provided as a back up.

Mounted on either side of the forward part of the turret is a bank of three electrically operated 81 mm smoke grenade launchers firing forwards and an infra-red searchlight moves in elevation with the main armament.

If required by the user, the turret can also be fitted one fire-and-forget SAM, for example the Russian Strela-2M or Igla, mounted externally either side. Another alternative is to have an anti-tank guided missile mounted externally either side of the turret.

The standard Russian BTR-80 is powered by a KamAZ-7403 V-8 diesel developing 260 hp coupled to a manual transmission. The upgraded BTR-94 is powered by a 3TD (UTD-20) multi-fuel engine developing 280 to 300 hp.

The BTR-94 has been designed to operate in temperatures of –40 to +55° with a humidity of 95 to 98 at 35°C.

Specifications
Crew: 3 + 10
Configuration: 8 × 8
Power to weight ratio: 22.05 hp/t
Combat weight: 13,600 kg
Length: 7.65 m
Width: 2.90 m
Height: 2.95 m
Ground clearance: 0.475 m
Track: 2.41 m
Wheelbase: 4.4 m
Max road speed: 90 - 100 km/h
Max water speed: 9 - 10 km/h
Fuel capacity: 300 litres
Max road range: 800 km
Fording: amphibious
Gradient: 60%
Side slope: 42%
Vertical obstacle: 0.5 m
Trench: 2 m
Engine: 3TD (UTD-20) multi-fuel developing 280 to 300 hp
Transmission: manual, 5 forward and 1 reverse gears
Transfer case: 2-speed
Steering: power assisted on front four wheels
Turning radius: 13.2 m
Tyres: 13.00 × 18 (tubeless)
Suspension: torsion bar with hydraulic shock absorbers
Brakes
(main) hydraulic on all wheels
(parking) mechanical
Electrical system: 24 V
Armament
(main) 2 × 23 mm cannon (see text)
(secondary) 1 × 7.62 mm PKT machine gun
(smoke grenade launchers) 2 × 3 81 mm
Ammunition: n/avail

Gun control equipment
Turret traverse: powered/manual
Gun stabiliser:
(vertical) no
(horizontal) no
NBC system: yes
Night vision equipment: yes

Variants
None known apart from the different armament installations. Further development of the BTR-94 resulted in the Guardian, details of which are given in the previous entry.

Status
Production as required. In service with Jordan (50).

Manufacturer
Malyshev Plant (see development).

UPDATED

BTR-70 APC Upgrade

Development/Description
The Kharkov Morozov Design Bureau in the Ukraine, known in the past as the design authority for the T-84 and T-64 Main Battle Tanks (MBTs), as well as upgraded versions of the T-72 and T-80 MBTs, has developed an upgrade package for the Russian designed BTR-70 (8 × 8) APC.

The two most significant parts of the Ukrainian BTR-70 upgrade concern the armament and the power pack.

The standard BTR-70 APC is fitted with a one-person manually-operated turret weighing 640 kg armed with a 14.5 mm KPVT and a 7.62 mm PKT machine gun with elevation from –5 to +30° and turret traverse through a full 360°. The gunner aims these weapons using a PP-61AM day sight.

The Ukrainian upgraded BTR-70 has a new one-person weapon station armed with a 30 mm 2A72 cannon (or the locally developed 30 mm KBA-2 cannon) with a 7.62 mm PKT coaxial machine gun.

The new weapon station weighs a total of 1,250 kg and has powered elevation from –10 to +60° with powered traverse through a full 360°. The higher level of elevation allows the weapons to engage slow flying aircraft and helicopters as well as being very useful in hilly and mountainous terrain.

The main 30 mm armament is stabilised in the vertical plane with the weapons mounted externally and the gunner under full armour protection in the hull.

The 30 mm 2A72 is a dual feed cannon and when fitted to the BTR-70 is fed by two drum type magazines, each holding 130 rounds of 30 mm ammunition, mounted externally one either side of the weapon. One of these would typically be filled with 30 mm Armour

BTR-70 upgraded by the Ukraine with new 30 mm/7.62 mm weapon station and more powerful diesel power pack **2000**/0034327

Piercing - Tracer (AP-T) ammunition and the other High Explosive (HE) ammunition.

Compared with the standard BTR-70 APC, the Ukrainian upgraded BTR-70 has a much more powerful weapons station and is capable of engaging a wider range of battlefield targets at longer ranges.

The alternative weapon, the 30 mm KBA-2 cannon, developed by the local State Scientific and Technical Center of Artillery and Rifle Arms (SSTC ARA) has a cyclic rate of fire of 300 rds/min and fires the same family of 30 mm ammunition.

The new weapon station is fitted with a TKN-4S day/night weapon sight which is stated to have a maximum effective day range of 5,000 m and a maximum effective night range of 700 m. For anti-aircraft use, a PZU-7 day sight is fitted, which has a maximum viewing distance of 2,000 m.

The standard BTR-70 APC is powered by two ZMZ-4905 four-stroke petrol engines developing 120 hp each, giving the vehicle a maximum road speed of 80 km/h and a road range, excluding additional diesel fuel tanks, of 400 km, or 600 km with additional diesel fuel tanks. The average road speed of the standard Ukrainian BTR-70 is said to be 50 km/h while the upgraded BTR-70 has an average road speed of 55 km/h.

The upgraded BTR-70 is fitted with two SMD-21 four-stroke diesel engines which develop 145 hp each, which give the vehicle a similar road speed, but most important, are more fuel efficient so road range is increased to 560 km using standard fuel tanks or 840 km with additional fuel tanks.

In addition, the diesel engine has a service life of 7,000 hours compared with 4,000 hours of the petrol engine and there is a reduced risk of fire with a diesel engine when compared to a petrol engine. The installation of the new diesel engines has increased the overall length of the vehicle from 7.355 m to 7.535 m.

The standard BTR-70 APC has a combat weight of 11,500 kg but the upgraded BTR-70 has a combat weight of 13,500 kg. Power-to-weight ratio of the standard BTR-70 is 20.86 hp/tonne while the upgraded BTR-80 is 21.48 hp/tonne.

The upgraded BTR-70 is fully amphibious, being propelled in the water by a single waterjet mounted at the rear of the hull, maximum water speed is 9 to 10 km/h. Standard equipment includes an NBC system, powered steering and winch.

The BTR-70, like the earlier BTR-60 and more recent BTR-80 series of 8 × 8 APCs, were designed and built in Russia by the now Arzamas Machinery Construction Plant who is also offering upgrades to the vehicle.

Guardian APC

Details of the 8 × 8 APC are given in an earlier entry.

BTR-94 APC

Details of this new 8 × 8 APC are given in an earlier entry.

Status

Development complete. Ready for production.

Manufacturer

Kharkov Morozov Design Bureau.

UPDATED

UNITED KINGDOM

Alvis Vehicles Piranha armoured vehicle family

Development/Description

In January 1991, GKN Defence (now merged with Alvis Vehicles) announced that MOWAG of Switzerland (who late in 1999 was taken over by now General Motors Defense (previously Diesel Division, General Motors of Canada)) had contracted it to build a significant quantity of Piranha 8 × 8 armoured fighting vehicles for one of MOWAG's customers in the Middle East.

The customer is believed to be the Saudi Arabian Ministry of Defence and Aviation (MODA), with six variants being supplied by Alvis Vehicles between 1992 and 1994. It is believed that between 100 and 120 vehicles were built by Alvis Vehicles for MODA.

These were ambulance, recovery, surveillance with roof-mounted battlefield surveillance radar, special VIP armoured personnel carrier and BAE Systems, RO Defence 81 mm and TDA 120 mm mortar.

MOWAG and Alvis Vehicles signed an agreement which licensed Alvis Vehicles in its own right to be System Prime Contractor for the design, integration, production and marketing of Alvis Vehicles Piranha 8 × 8 vehicles for agreed customers.

All Piranha (8 × 8) built to date by Alvis Vehicles are Category II models but Alvis Vehicles also holds the UK rights to the Category III models. Vickers Defence Systems of the UK holds the UK rights to the Category IV vehicles.

In addition to producing a number of variants for MOWAG's Middle East customer, Alvis Vehicles produced a company demonstrator vehicle which was evaluated by the Royal Army of Oman in early 1991.

Subsequently, in May 1994, Alvis Vehicles announced that the Sultanate of Oman had ordered 80 Piranha 8 × 8 vehicles, with an agreed option for a further 46 vehicles. The latter was not exercised.

A 25 year support contract was signed which covers the provision of spares, technical support, enhancements, and other in-service support activities during the operational life of the vehicles.

The Alvis Vehicles Piranha variants for Oman include armoured personnel carrier with 12.7 mm M2 MG, recovery, ambulance, command, mortar carrier (81 mm) and artillery observation. First vehicles were completed in 1995, with final deliveries of the first batch taking place in 1997.

Late in 2000, as part of its next five year defence plan, Oman placed a contract with Alvis Vehicles for 80 new Piranha (8 × 8) light armoured vehicles and the second contract for the upgrading of 60 Scorpion Combat Vehicle Reconnaissance (Tracked) vehicles.

In September 2001, less than a year after the order was placed, Alvis Vehicles handed over the first of 80 Piranha II (8 × 8) Armoured Personnel Carriers (APCs) to the Royal Army of Oman (ROA), ahead of schedule.

Under the terms of this contract, a total of seven variants will be supplied with deliveries to be completed from the UK in 2002. These variants are understood to include APC (with and without armament), 81 mm mortar carrier, command and control, repair and recovery, ambulance and artillery observation.

The first batch of vehicles was shipped to Oman late in 2001 with final work, such as the installation of locally designed and built turrets, being carried out in Oman.

Alvis Vehicles built Piranha (8 × 8) in 81 mm mortar configuration **2000**/0080268

Alvis Vehicles built Piranha (8 × 8) fitted with Cockerill 90 mm LCTS turret system as supplied to Qatar **2000**/0080269

Alvis Vehicles built Piranha II (8 × 8) armoured personnel carrier in service in Oman showing extensive external stowage **2002**/0121221

It is expected that final vehicles will be handed over to Oman in 2003.

The Piranha (8 × 8) vehicles supplied to Oman are to the Category II build standard with many improvements including a higher payload, larger tyres, a more efficient braking system, a quieter driveline, a 100 per cent locking differential, a two-speed main transfer box, a longer hull with more space and a new driver's dashboard.

In 1993, Alvis Vehicles showed a Piranha 8 × 8 vehicle fitted with the US Delco Defense two-man power-operated turret armed with The Boeing Company M242 25 mm Chain Gun, a 7.62 mm coaxial machine gun and a launcher for a 3,750 m Raytheon Systems Company TOW ATGW either side.

This is the same turret as that fitted to the Alvis Vehicles Desert Warrior vehicles ordered by Kuwait in 1993. The chassis was the Alvis Vehicles demonstrator with a number of improvements including larger tyres.

Part of the £500 million contract placed by the State of Qatar with British Aerospace late in 1996 for a wide range of defence equipment includes 40 Alvis Vehicles licence-built Piranha (8 × 8) light armoured vehicles.

First Piranha deliveries were made to Qatar in 1997 with final deliveries in 1999. The order for 40 Piranha vehicles comprised 36 fitted with the Belgian Cockerill LCTS turret armed with Cockerill KEnerga Mk 8 90 mm gun and 7.62 mm coaxial machine gun, two repair and recovery vehicles and two command post vehicles plus spares and training.

The Qatar vehicles are Category II models and are powered by a Detroit Diesel Model 6V-53T turbocharged diesel developing 300 hp coupled to an Allison Transmission MT-653 fully automatic transmission. MOWAG of Switzerland supplied the driveline system for these vehicles.

In anticipation of any future Turkish Land Forces Command requirement for 8 × 8 armoured personnel carriers and variants, Alvis Vehicles has an agreement with the Turkish FNSS Defence Systems company covering the Piranha 8 × 8. This vehicle has already been trialled in Turkey. In September 2001, Alvis Vehicles and FNSS Defense Systems signed an agreement forming a partnership for the production of Piranha (8 × 8) vehicles in Turkey to meet the future Turkish wheeled armoured vehicle requirement.

Late 2001 a Piranha II (8 × 8) vehicle was shown in Turkey fitted with an FNSS Defense Systems Sharpshooter turret armed with a 25 mm M242 cannon and a 7.62 mm coaxial machine gun.

Desert Piranha

Alvis Vehicles is now marketing Desert Piranha specifically for the Middle East market. This features an Anti-lock Braking System (ABS), central tyre inflation system and a digital engine mangement system. These are now incorporated as standard rather than being optional items.

NBC Piranha

In association with the Intellitec Division of Advanced Technical Products of the US, Alvis Vehicles has developed to the prototype stage a new integrated NBC reconnaissance vehicle. This has a full suite of sensors to detect and analyse a wide range of NBC threats and send this information to the next chain of command.

Status

Production as required. In service with the following countries:

Country	Quantity	Comment
Oman	80	six versions, all delivered, additional order for 80 placed late in 2000 with deliveries to run from 2001 through to 2002 in seven different versions
Qatar	40	three versions, all delivered
Saudi Arabia	100 to 120	six versions, all delivered

Manufacturer

Alvis Vehicles.

UPDATED

Alvis Vehicles built Piranha (8 × 8) in NBC reconnaissance role **2000**/0067446

Alvis Vehicles built Piranha (8 × 8) fitted with Delco turret armed with 25 mm cannon, 7.62 mm coaxial machine gun and TOW missile launchers **2002**/0002309

Alvis Saracen armoured personnel carrier

Development

Shortly after the end of the Second World War, the Fighting Vehicles Research and Development Establishment at Chertsey, which no longer exists as a separate entity, began the design of the FV600 series of 6 × 6 armoured fighting vehicles, to meet the operational requirements of the British Army. The first three models in this series were the FV601 Saladin armoured car, the FV602 command vehicle (which was cancelled in 1949 although the FV604 later took over its original role) and the FV603 Saracen APC. Owing to the situation in Malaya in the late 1940s, development of the FV603 was given precedence over the FV601 Saladin armoured car.

Design parent for the FV600 series was Alvis Limited of Coventry which completed the first prototype of the Saracen in 1952 and first production vehicles in December the same year. The first Mk 1s were powered by a Rolls-Royce B80 Mk 3A petrol engine which had BSF threads but later production Mk 1s were powered by a B80 Mk 6A petrol engine and had UNF threads. Early production vehicles had a slightly different turret; the rear of the turret opened either side rather than folding down horizontally as on later turrets.

Throughout the 1950s and early 1960s the Saracen was the standard APC of the British Army and was replaced from 1963 by the FV432 tracked APC. Production continued for export until 1972. Production of the Saracen APC amounted to 1,838 vehicles. Saracen APCs were widely used by the British Army in Northern Ireland in the IS role but these have now been withdrawn.

The Saracen was last used in the British Army by the Saracen Troop of 28 Squadron, the Queen's Own Gurkha Transport Regiment based in Hong Kong. This unit was disbanded early in 1993 leaving no Saracens in operational use with the British Army. All British Army Saracen vehicles have now been disposed of.

Production of the Saracen was undertaken at the Alvis facility in Coventry. With the merger of Alvis Vehicles and GKN Defence in late 1998, the Coventry facility was closed in mid-1999 and all vehicle production is now undertaken in the former GKN Defence facility at Telford.

Description

The hull of the Saracen armoured personnel carrier is made of all-welded steel with the engine at the front and the personnel compartment at the rear.

Air enters the engine compartment at the front of the vehicle through armoured louvres and is expelled through the four hatch covers over the engine compartment. Power is transmitted from the engine to the roadwheels via a fluid coupling, five-speed

Specifications

FV603 (C) (with reverse-flow cooling)
Crew: 2 + 10
Configuration: 6 × 6
Combat weight: 10,170 kg
Unloaded weight: 8,640 kg
Power-to-weight ratio: 15.73 hp/t
Ground pressure: 0.98 kg/cm²
Length:
(hull) 5.233 m
Width: 2.539 m
Height:
(to turret top) 2.463 m
(to hull top) 2 m
Ground clearance: 0.432 m
Track: 2.083 m
Wheelbase: 1.524 + 1.524 m
Angle of approach/departure: 53/53°
Max road speed: 72 km/h
Fuel capacity: 200 litres
Max range: 400 km
Fording: 1.07 m
(with preparation) 1.98 m
Gradient: 42%
Vertical obstacle: 0.46 m
Trench: 1.52 m
Turning radius: 7 m
Engine: Rolls-Royce B80 Mk 6A 8-cylinder petrol developing 160 hp at 3,750 rpm
Transmission: preselector 5 speed
Steering: recirculating ball, hydraulically assisted
Clutch: fluid coupling
Suspension: independent double-wishbone and torsion bar, hydraulic shock-absorbers
Tyres: 12.00 × 20
Brakes:
(main) hydraulic disc on all wheels
(parking) mechanical on all wheels
Electrical system: 24 V
Batteries: 2 × 12 V
Armament:
(turret) 1 × 7.62 mm MG
(ring-mounted) 1 × 7.62 mm MG
Smoke-laying equipment: 2 × 3 66 mm smoke grenade dischargers
Ammunition: (7.62 mm) 3,000
Armour
Hull visor plate: 12 mm at 40°
Hull skid plate: 10 mm at 40°
Hull sides: 12 mm at 20°
Hull top: 8 mm
Hull floor: 12 mm
Turret front: 16 mm at 15°
Turret sides: 16 mm at 15°
Turret rear: 16 mm at 15°
Turret top: 8 mm
NBC system: no
Night vision equipment: no

preselective gearbox, transfer box with forward and reverse change, and differential transverse drive direct to each centre wheel hub through universally jointed shafts. Drive to the front and rear wheel hubs is by transmission shafts and bevel boxes. Each wheel hub houses an epicyclic train which provides the final gear reduction to the wheels. Steering of the front two and centre wheels is by positive mechanical linkage, hydraulically assisted between the steering wheel and the roadwheels. Hydraulic pressure for both steering and the foot brake is provided by an engine-driven hydraulic pump.

The suspension is of the double-wishbone type with torsion bars and sleeves. There are two telescopic double-acting hydraulic shock-absorbers fitted at each front and rear wheel station. The centre wheel stations each have a single telescopic double-acting hydraulic shock-absorber. The gearbox and the transfer box are under the floor of the crew compartment in the centre and the fuel tank is under the floor of the crew compartment at the rear.

The driver sits at the front of the vehicle behind the engine compartment and has three No 17 day periscopes, one in front and one either side of his position. The periscopes are mounted in hatch covers. The hatch cover to the driver's front is folded forward onto the engine compartment for increased visibility, and those either side fold upwards.

The section commander sits behind the driver on the left, with the radio operator in a similar position to his right. The eight infantrymen sit on individual seats, four down each side of the vehicle facing each other. They enter and leave the vehicle by two doors in the rear of the hull which open outwards and have a rectangular firing port that folds downwards on the outside. There are three similar firing ports in each side of the hull and a hull escape hatch over the second axle on each side.

The turret, which is similar to that fitted to the Ferret scout car Mk 2/3, has a basket-type seat and is mounted in the forward part of the roof. The rear part of the turret roof folds forwards and can be locked horizontal if required, while the turret rear folds downwards on the outside to form a seat for the gunner. The turret is armed with a 7.62 mm (0.30) Browning machine gun which has an elevation of +45° and a depression of −12°, except over the rear of the vehicle and is fitted with a No 3 Mk 1 day periscopic sight, which is mounted in the forward part of the turret roof.

At the rear of the roof is a ring mount which is normally fitted with a 7.62 mm Bren LMG or sometimes a 7.62 mm (0.30) Browning machine gun. Three electrically operated 66 mm smoke grenade dischargers mounted on each mudguard at the front of the vehicle launch smoke grenades to the front of the vehicle.

The Saracen APC is not fitted with an NBC system or any night vision equipment, although the latter could be fitted if required. A fresh air ventilation system is standard on all vehicles and pipes air from the outside (via two dome-shaped ventilators on the right side of the hull) to ducts running down each side of the personnel compartment. The Saracen has no amphibious capability, although a deep fording kit was developed for the vehicle.

Variants

Saracen with reverse-flow cooling (FV603 (C))

This was designed by Alvis specifically for operations in the Middle East. On the basic model air is drawn in through the louvres at the front of the engine compartment and expelled through the four hatch covers over the top. On the reverse-flow cooling model air is drawn in through raised louvres at the back of the engine compartment, passed forward over the engine by the fans, the blades of which have been reversed, and then expelled through the radiators and front louvres. To prevent air rushing into the radiator and impeding passage of air through it in the opposite direction, a cowl has been mounted over the front of the vehicle.

Alvis Saracen (6 × 6) APC with driver's and commander's hatches in open position (Christopher F Foss)
2000/0075715

Saracen with open roof

Some of the Saracens delivered to Kuwait in the 1960s were supplied with an open roof and a reverse-flow cooling system. None of these remain in service.

FV604 command vehicle

The FV604 is essentially the FV603 APC modified for use as a command vehicle.

The inside has been fitted with mapboards and modified to accommodate a crew of six: three staff officers, two radio operators and a driver. Externally, stowage baskets have been fitted as well as an auxiliary charging plant and additional batteries to power the communications equipment installed. To increase the working area for the command staff, a tent can be erected at the rear of the FV604.

FV610 command vehicle

The full designation of the FV610 is Armoured Carrier 6 × 6 Command GPO/CPO FV610(A). It has a crew of six which normally consists of three staff officers, two radio operators and a driver. Its main difference from the FV604 is a wider and higher hull as it is anticipated that it will be used in the static role for longer periods than the FV604. Equipment fitted to the FV610 includes additional mapboards, communications equipment, an auxiliary charging plant and batteries.

FV611 ambulance

This has the same hull as the FV610 and can carry 10 seated patients, three stretcher and two seated patients, or two stretcher and six seated patients, in addition to the crew of two, which consists of the driver and the medical orderly.

Saracen repower

In 1991, it was announced that A F Budge Limited and Alvis had developed a conversion package for the Saracen 6 × 6 APC with the backing and assistance of Perkins Engines Company. The conversion involves the replacement of the existing petrol engine with a Perkins Engines Company Phaser 180 MTi diesel developing 180 hp. Further modifications include enhanced driveline components, an improved braking system, a new cooling pack and a modern electrical system.

This conversion package offers the advantages of improved fuel efficiency, improved reliability, increased operational range and a reduced fire risk.

In early 1994, Alvis Vehicles was awarded an initial contract worth US$10 million for the supply of AFV upgrade kits from an undisclosed country in Asia, believed to be Indonesia.

The vehicles concerned are the 76 mm Alvis Saladin (6 × 6) armoured car, the Alvis Saracen (6 × 6) armoured personnel carrier and the Daimler Ferret (4 × 4) scout car.

Key parts of the upgrade are understood to include the replacement of the original petrol engine by a more fuel efficient diesel engine.

This upgrade programme for Indonesia was completed several years ago.

South African Saracens

The South African National Defence Force has placed its entire fleet of 250 Saracen APCs up for sale. All of these vehicles were refurbished during 1985 to 1988 and during this time underwent modifications to improve their RAM-D characteristics. Since then they have been placed in War Reserve. Full details of this extensive upgrade were given in *Jane's Armour and Artillery 1997-98*, page 511.

Status

Production complete. In service with the following countries:

Country	Quantity	Comment
Indonesia	55	some upgraded
Jordan	30	police
Lebanon	20	
Mauritania	5	
Nigeria	10	
South Africa	250	for sale
Sri Lanka	30	

Manufacturer

Alvis Vehicles Limited.

UPDATED

Alvis Vehicles Simba light combat vehicle

Development

The Simba (4 × 4) family of light combat vehicles was designed as a private venture by GKN Defence (which in late 1998 was merged with Alvis Vehicles) specifically for the export market, to meet the requirement for a family of wheeled vehicles capable of local manufacture and of being fitted with a variety of armament installations to meet different operational requirements.

In mid-1992, GKN Defence announced that it had received confirmation of the award of a £56.273 million contract to supply the Armed Forces of the Philippines (AFP) with 150 Simba wheeled armoured vehicles, and associated spares and training. The cost increased to £63.385 million following requirements for 17 vehicles of the order to be configured to accommodate ambulance and command fits.

The Philippine's Simbas were customised to suit specific AFP requirements and have larger capacity diesel fuel tanks for increased operating range.

In 1989, a joint venture company was established in the Philippines called Asian Armoured Vehicle Technologies Corporation (AAVTC), with GKN Defence having a 20 per cent interest. The other 80 per cent is held by Multi-Parts Motors International Incorporated (MMIIC) and Philippines Veterans Investment and Development Corporation (PHIVIDEC).

Of the 150 Simba vehicles ordered by the AFP eight vehicles were delivered complete, two in 'part knocked down' form, two in kit form and the remaining 138 were assembled in a facility in Subic Bay operated by the Asian Armoured Vehicle Technologies Corporation. All vehicles have now been delivered to the AFP.

The basic Simba used by the AFP is fitted with an Alvis Vehicles one-man turret armed with a 12.7 mm M2 machine gun and two banks of four electrically operated smoke grenade dischargers but a number of other variants have been built including ambulance, command post and a more heavily armed version fitted with a Helio FVT 925 one-man turret armed with a 25 mm Oerlikon Contraves cannon.

Production of the Simba was completed some years ago and it is no longer being marketed by Alvis Vehicles.

Description

(Typical)

The hull of the Simba is of all-welded steel armour construction that provides the occupants with protection from 7.62 mm ball small arms fire and shell splinters. Thermal insulation on the roof and forced air circulation are fitted as standard. A heater and/or full air conditioning are available as optional extras, as are various communications systems.

The driver sits at the front of the vehicle on the left and has bulletproof windows to his front and either side and a single-piece hatch cover that opens to the rear.

The Perkins Engines Company diesel engine is to the right of the driver, with the air inlet louvres on the lower front glacis plate and the outlet on the upper side of the glacis plate to the right rear of the driver's position. The Perkins Engines Company diesel engine is coupled to a Clark combined gearbox/transfer box which transmits power to the front and rear axles.

The troop compartment is at the rear of the Simba. Armoured vision blocks are provided in each side of the troop compartment with firing ports positioned between the windows.

The troops enter and leave the Simba via a single large rear door or a door in the left-hand side. Both have a bulletproof vision block in their upper part with a firing port underneath.

A wide range of armament installations can be fitted to the Simba. When used as an APC, the vehicle normally transports up to 12 crew members including the commander and driver. When a one-man turret with single or twin 7.62 mm GPMG, or a 12.7 mm M2 machine gun, is fitted, a crew of 10 is normally carried.

Specifications

Typical
Crew: 2 + 8/10
Configuration: 4 × 4
Combat weight: 9,900-11,200 kg
Unloaded weight: 9,500 kg
Length: 5.35 m
Width: 2.50 m
Height:
(to top of low-profile cupola) 2.19 m
(to top of 12.7 mm turret) 2.59 m
Ground clearance:
(hull) 0.45 m
(axle) 0.33 m
Wheelbase: 2.972 m
Track:
(front) 2.06 m
(rear) 2.12 m
Angle of approach/departure: 45/30°
Fuel capacity: 296 litres
Max road speed: 100 km/h
Road range: 660 km
Fording: 1 m
Gradient: 60%
Side slope: 40%
Vertical obstacle: 0.45 m
Turning radius: 8.23 m
Engine: Perkins Engines Company 210Ti turbocharged intercooled 6 litre diesel developing 210 bhp at 2,500 rpm
Transmission: Clark 13.7 LHR 28422 semi-automatic with power change, 4 forward and 1 reverse gears. Integral transfer box with 2- or 4-wheel drive selection
Suspension: multileaf semi-elliptical springs and hydraulic telescopic shock-absorbers
Steering: power-assisted
Tyres: 13.00R x 20, Hutchinson run-flat
Brakes:
(main) drum, dual circuit/air hydraulic, power-assisted
(parking) hand-operated brake on transfer box
Electrical system: 24 V
Battery: 90 Ah capacity
NBC system: no
Night vision equipment: no

Variants

Simba low-profile APC

The basic APC version of Simba carries 10 to 12 men depending on the operational role, for example the driver front left, one man to his rear facing the rear, five men down the left side facing inwards, four men down the right side facing inwards and one man under the cupola facing the front.

Simba 90 mm fire support vehicle

For trials purposes, a Simba was fitted with the Belgian Cockerill 90 mm CM90 turret armed with a 90 mm gun and a 7.62 mm coaxial machine gun.

Simba anti-tank

This would be fitted with various types of anti-tank guided missiles, either turret- or pintle-mounted.

Simba mortar

Fitted with a turntable-mounted 81 mm mortar in the rear of the hull.

Alvis Vehicles Simba Light Combat Vehicle fitted with turret armed with 12.7 mm MG and lower part of side door open **2002**/0121222

Alvis Vehicles Simba Light Combat Vehicle in service with the Philippines and fitted with one-person turret armed with 12.7 mm M2 MG **2002**/0121223

Simba APC

With an Alvis Vehicles turret armed with a 7.62 mm or 12.7 mm machine gun, this version has a crew of 10 including the commander, gunner and driver. Most of the Philippine's Simba vehicles are in this configuration.

Simba AIFV

This version has a removable roof plinth that can accept a wide range of 20 mm and 25 mm turrets. The Philippine's have a batch of vehicles fitted with the Helio FVT 925 turret armed with a 25 mm cannon and 7.62 mm machine gun.

Simba IS vehicle

This version can be fitted with a wide range of equipment including side riot screens, a barricade remover, a public address system, a siren or wailer, a turret firing 38 mm CS and smoke rounds from under armour, a collective CS gas filtration system and a winch.

Status

Production complete in the Philippines with 150 vehicles, including variants, delivered to the AFP.

Manufacturer

Alvis Vehicles Limited.

UPDATED

Alvis Vehicles Saxon armoured personnel carrier

Development

The AT105 was developed by GKN Defence (which was merged with Alvis Vehicles late in 1998) as the successor to its earlier AT104. The first prototype of the AT105 was completed in 1975 and the first production vehicles in 1976.

Main improvements over the original AT104 can be summarised as: the armoured hull has been designed to give the same level of protection to the radiator, engine and transmission as to the crew; redesigned hull floor to give increased protection against damage from mines; more powerful engine and a shorter wheelbase for improved mobility and turning.

In mid-1982, the AT105 was named the Saxon. Early in 1983, the British Ministry of Defence placed an order for an initial 47 Saxon APCs for the British Army, with first deliveries made in early 1984.

The second order for the British Army comprised 247 vehicles. Late in 1985, the British Army exercised its option for a further 200 Saxons which were all completed by early 1989. The Royal Artillery ordered 30 Saxons for use as command post vehicles with Rapier surface-to-air missile units and final deliveries of this version were made in early 1987.

Each British Army Saxon carries a fully equipped rifle section of 10. The British Army Saxons have a fixed observation cupola for the commander, with a socket in each corner of the cupola for a 7.62 mm GPMG DISA mount. British Army vehicles have no door in the left side of the hull, a stowage rack on the roof and external fully enclosed bins. Originally there were two variants supplied to the British Army: command and recovery. The latter has a side-mounted 5 tonne hydraulic winch that can be used to the front or rear and is driven from a Power Take-Off (PTO) from the gearbox. There are interior seats for a crew of four in addition to stowage for recovery equipment.

Recently, some Saxons deployed to Bosnia have been fitted with a Peak Engineering one-man turret armed with a 7.62 mm MG. These turrets were taken from FV432 APCs.

In 1988, the Hong Kong Police Force took delivery of seven Saxon vehicles for use by the Hong Kong Police Tactical Unit and these replaced older Saracen vehicles. Of the seven vehicles, two are command post vehicles and the remaining five are personnel carriers carrying 10 men plus commander and driver.

By early 1989, the British Army had taken delivery of 524 production Saxon (4 × 4) APCs and variants.

At the 1990 British Army Equipment Exhibition, Saxon was relaunched with a Cummins 6BT 5.91 litre turbocharged six-cylinder diesel engine developing 160 bhp coupled to a fully automatic transmission from Allison Transmission.

Late in 1990, the UK Ministry of Defence placed an order for an additional batch of Saxon (4 × 4) APCs worth approximately £20 million. It also included an order for the Saxon ambulance. This was the first production order for the Saxon ambulance which was originally developed as a private venture. It is understood that around 100 Saxon vehicles were built under this contract, with first deliveries taking place in 1991. Deliveries are now complete.

These vehicles had a number of improvements over current vehicles including a Cummins 6BT 5.91 litre turbocharged six-cylinder diesel engine developing 160 bhp. These are officially known as the Saxon Patrol vehicle by the British Army. Additional details are given under Variants later in this section.

Although production of the Saxon was completed several years ago, it is still being marketed by Alvis Vehicles.

In British Army service the Saxon without the Peak Engineering turret is sometimes referred to as the Mk 1, while the vehicle with the turret is referred to as the Mk 2.

At least six battalions of the British Army infanty currently use the Saxon and in the longer term this is expected to be replaced by a new tracked or wheeled vehicle with improved cross-country mobility.

Description

(Right-hand drive model; left-hand drive also available) The basic model has a hull of all-welded steel, with the driver seated at the very front of the hull on the right side and the personnel compartment at the rear. The hull provides the crew with complete protection against small arms fire up to and including 7.62 mm armour-piercing rounds, and HE shell fragments up to 155 mm bursting at 10 m from the vehicle. The floor is V-shaped to give maximum protection against mine blast, except for the axles, which are outside the armoured section. The engine, transmission, fuel, batteries, vacuum servos and hydraulics are all contained within the armoured compartment. The engine, transmission and other automotive components are standard

Specifications

Crew: 2 + 8 (or 10)
Configuration: 4 × 4
Combat weight: 11,660 kg
Unloaded weight: 9,940 kg
Power-to-weight ratio: (164 bhp engine) 14.06 bhp/t
Length: 5.169 m
Width: 2.489 m
Height:
(to commander's cupola) 2.628 m
(top of turret, if fitted) 2.86 m
Ground clearance:
(hull) 0.41 m
(axles) 0.29 m
Track:
(front) 2.08 m
(rear) 1.99 m
Wheelbase: 3.073 m
Angle of approach/departure: 48/40°
Max road speed: 96 km/h
Fuel capacity: 153 litres
Max road range: 480 km
Fording: 1.12 m
Gradient: 60%
Vertical obstacle: 0.41 m
Turning radius: 7.57 m
Engine: Bedford 500 6-cylinder diesel developing 164 bhp at 2,800 rpm*
Transmission: Allison Transmission AT-545 automatic with 4 forward and 1 reverse gears
Transfer case: 2 speed
Steering: Burman, power-assisted
Suspension: semi-elliptical springs and hydraulic shock-absorbers
Tyres: 14.00/74 × 20 or 14.00 × 20. British Army vehicles are fitted with French Hutchinson run-flat cores
Brakes:
(main) drum, air/hydraulic, dual circuit
(parking) transmission brake
Electrical system: 24 V
Batteries: 2 × 12 V, 100 Ah
Armour: proof against 7.62 mm/0.30 AP rounds at point-blank range
NBC system: no
Night vision equipment: no

* **Note:** Saxon could be fitted with a Perkins Engines Company T6.3544 diesel developing 195 bhp at 2,500 rpm. From mid-1990 the Saxon was fitted with the Cummins 6BT 5.91 litre turbocharged six-cylinder diesel developing 160 bhp coupled to a fully automatic transmission.

Alvis Vehicles Saxon command post of the British Army with tent erected at rear (Richard Stickland) ***2001***/0104420

Alvis Vehicles Saxon armoured personnel carrier with 7.62 mm machine gun on DISA mount (Michael Jerchel) ***2001***/0104421

Alvis Vehicles Saxon armoured personnel carrier of the British Army with Peak Engineering turret (Richard Stickland) **2002**/0121224

Alvis Vehicles Saxon patrol vehicle of the British Army showing barricade removal equipment at front of vehicle (Richard Stickland) 0005295

commercial items. British Army vehicles have a high degree of commonality in engine and axle components with other in-service vehicles.

The driver can reach his seat either from inside the vehicle or by a forward-opening hatch cover over his position. There are bulletproof windscreens in front and on both sides.

The personnel sit four down each side of the hull on padded bench seats which are usually equipped with seat belts, and enter and leave the vehicle by two doors in the rear of the hull and a single door in each side of the hull. The British Army specification has minor differences from other in-service vehicles, these include exterior stowage bins and no left-hand door. Each of the doors has a firing port and a vision block. There are additional firing ports in each side of the hull and a single firing port in the forward part of the hull on the left side. Specially designed ball mounts have also been developed which can be installed in the firing ports to allow the crew to fire their weapons from inside the vehicle in safety. The interior of the hull is lined with 26 mm thick thermal insulation and a forced air circulation system is fitted as standard on all vehicles.

The commander's cupola has four sides each with a vision block, and a single-piece hatch cover that opens forwards. A 7.62 mm machine gun can be pintle mounted at this position if required. The commander's cupola is welded to an oblong sheet of armour which in turn is bolted to the roof of the vehicle. This allows the command module to be removed and replaced by another module with a different armament installation. When the engine needs replacement the roof module is removed first and the complete engine and transmission can be lifted through the roof. Run-flat cross-country tyres are fitted as standard and in a 'shot through' condition the vehicle can travel 96 km at 48 km/h.

A total of 515 Saxon vehicles of the British Army have been fitted with two banks of L8 66 mm smoke grenade launchers.

Optional equipment for the Saxon included air conditioning systems, an auxiliary power unit, a barricade remover, a door safety mechanism, grenade launchers, a hand-operated searchlight, a heater, rotating/flashing beacons and a siren, loudspeakers and a front-mounted winch.

Variants

Saxon LHD armoured personnel carrier

This is fitted with a turret armed with a single 7.62 mm GPMG, with an elevation of +50° and a depression of −15°, or a turret armed with twin 7.62 mm GPMGs with similar elevation and depression. The Saxon can also be used to deploy and operate a remote-controlled EOD vehicle for bomb disposal duties.

Saxon command vehicle

This has the commander's cupola as fitted to the Saxon but can be fitted with an Alvis Vehicles turret armed with single or twin 7.62 mm GPMGs or a 12.7 mm HMG. The interior is modified for use as a command vehicle: on the right side of the hull are staff positions with individual communication control modules and, on the left, is a mapboard with strip lighting and storage space behind. A teleprinter operator's position with seat and table behind are also provided. A similar configuration has been adopted for the British Army command vehicles.

Saxon recovery vehicle

This model is in service with the British Army and is the standard Saxon modified to recover other vehicles. Mounted on the left side of the hull is a Hudson Wharton capstan 5,000 kg hydraulic winch that can handle vehicles weighing up to 16,000 kg with the aid of a block and tackle.

The winch has 40 m of cable, a slow speed of 9.1 m/min and a fast speed of 30.84 m/min. It can pull vertically at angles from 10 to 40° and horizontally from +15 to −15°. The winch can be operated remotely.

The vehicle is normally operated in the British Army by four members of the Royal Electrical and Mechanical Engineers (REME). If required, a lightweight tent can be erected at the rear so that repair work can be carried out in all weathers.

Standard equipment includes tools, a spotlight, Scotch lorry skid pads and a front and rear differential lock.

Ambulance

The Saxon ambulance is the standard vehicle with internal fittings to enable four general service stretchers to be carried, two either side, or two stretcher patients plus five walking wounded.

The crew consists of the commander, the driver and one medical orderly. There is ample space inside the vehicle for medical equipment and, in addition, there is external stowage on the roof and sides. This model was included in the last batch of about 100 Saxon vehicles ordered by the British Army for delivery from 1992.

It is believed that the British Army was the first customer for this version.

Incident control vehicle

This specialised vehicle is fitted with a front-mounted obstacle-clearing blade, wire mesh protection for the driver's windows, a turret armed with a machine gun and smoke grenade dischargers, wire mesh screens and a low light level TV surveillance system on a telescopic mount at the rear of the vehicle with a monitor/record/playback facility inside the vehicle.

Saxon EOD

The Saxon is available in an Explosive Ordnance Disposal (EOD) configuration version.

Alvis Vehicles light turret

Alvis Vehicles has designed a light turret which can be fitted to a variety of APCs including its own Saxon vehicle.

This is armed with a five-shot 37 mm riot gun with 20 rounds of ready use ammunition and a further 100 rounds carried in the turret. It can also be armed with a 7.62 or 12.7 mm machine gun. The turret shell is the same as that of other Alvis Vehicles light turrets and can be configured for alternative fits such as a light machine gun or water cannon. In addition to the main option, 66 mm smoke grenade dischargers can be fitted to give 180° forward spread or directional employment.

Other features include a ×1 day sight or ×1.4/×6 day/night sight, a 150 W coaxial spotlight, 150 cfm fume extraction, a 100 A and 12 channel slip ring, a 70 W flashing beacon and various communications fittings.

Saxon Patrol

This is the latest version of the Saxon to enter service with the British Army and has now replaced the Humber one ton (4 × 4) APC in Northern Ireland, with first production vehicles being completed in 1992.

The Saxon Patrol has a number of improvements over the original Saxon (4 × 4) APCs currently in service with the British Army including the replacement of the Bedford 6-cylinder diesel with a Cummins 6BT 5.91 litre 6-cylinder turbocharged diesel engine developing 160 bhp at 2,800 rpm.

The engine is coupled to a fully automatic transmission and the driver can select either 4 × 4 or 4 × 2 drive depending on the tactical situation.

Other improvements include the installation of barricade removal devices at the front of the hull, roof-mounted searchlights, an anti-wire device, wire netting for the crew's bulletproof windows, a revised cowl for the armoured radiator cover on the front of the hull, improved armour and an upgraded braking system with run-flat tyres fitted as standard.

Status

Production complete. In service with the following countries:

Country	Quantity	Comment
Bahrain	10	
Hong Kong	7	police
Malaysia	44	police, some sources state 140
Nigeria	75	police
Oman	22	police
UAE	10	
UK	664	

Manufacturer

Alvis Vehicles Limited.

UPDATED

Alvis Vehicles Tactica armoured personnel carrier

Development

The Tactica armoured personnel carrier was originally developed as a private venture by Glover Webb, to provide the occupants with a high level of protection and mobility while maintaining clean lines with a non-aggressive appearance.

The standard armoured personnel carrier is the basic model of a wide range of variants for a number of different missions. The Tactica was developed in 1988 using Glover Webb's own semi-monocoque frame, and the first vehicle entered service with an Asian customer late in 1989. Early in 1993, the vehicle entered service with the British Army. These are used for special duties in Northern Ireland, including EOD. The vehicle is also used for a number of civilian applications.

In late 1994, the Glover Webb company was taken over by GKN Defence but still traded from Hamble in Hampshire and retained the name Glover Webb.

Late in 1997, GKN Defence closed down its Glover Webb subsidiary at Hamble and transferred development, production, sales and marketing to its main facility located at Telford, in Shropshire, Northern England. In 1998, GKN Defence merged with Alvis Vehicles with all research, development and production being concentrated at Telford.

In mid-1997, it was disclosed that Alvis Vehicles was building nine unarmed Tactica (4 × 4) armoured personnel carriers for Argentina. These were built at the Hamble facility and were delivered late in 1997 for use with United Nations units operating in Cyprus. These are used by the United Nations island Mobile Force Reserve (MFR). This is a multinational company of five officers and 100 soldiers. It is formed along a traditional infantry line with an added MT platoon.

In addition, Alvis Vehicles has recently built a further 14 Tactica vehicles for Indonesia, with the order including water cannon, 14-seater APCs, explosive ordnance disposal and command post vehicles.

In early 2000, it was announced that Alvis Vehicles had been awarded a contract from an undisclosed African country for the supply of Tactica vehicles worth £6 million.

It is understood that the order was for Ghana with a total of 20 vehicles being supplied in three variants, ambulance, command and control and patrol.

The Glover Webb name has been dropped and these vehicles are now marketed under the Alvis Vehicles brand name.

Alvis Vehicles Tactica 4 × 4 light armoured vehicles in bonneted (front) and forward control (rear) vehicles
2002/0121226

Specifications

Typical	**Gradient:** 60%
Crew: 2 + 12	**Side slope:** 30%
Configuration: 4 × 4	**Engine:** Mercedes or Renault (choice of two) turbocharged diesel
Weight: (combat) 10,000 kg; (unladen) 6,500 kg	**Transmission:** ZF manual with 5 forward and 1 reverse gears
Length: 5.6 m	**Clutch:** single dry plate
Width: 2.2 m	**Steering:** power-assisted
Height: 2.35 m	**Suspension:** parabolic leaf with adjustable shock-absorbers
Ground clearance: 0.30 m (axles)	**Tyres:** 10R x 22.5 on/off road or 12.5R x 20 sand
Track: 1.96 m	**Brakes:** full air, twin system
Angle of approach/departure: 35/35°	**Electrical system:** 12 V or 24 V
Max road speed: 120 km/h	**NBC system:** no
Fuel capacity: 167 litres	**Night vision equipment:** yes
Max road range: 650 km	

Description

The basic Tactica vehicle is of the forward control configuration. The engine is mounted centrally between the driver and commander, with a half bulkhead separating the driver's area from the troop compartment which extends to the rear of the vehicle. Both left- and right-hand drive versions of the Tactica are available.

The semi-monocoque hull is made from high-hardness certified armour plate, providing protection from 7.62 mm armour-piercing attack. All vision blocks and driver's vision blocks provide the same level of protection, and the hull will also provide protection against fire bombs and other improvised devices. All six doors and vision areas have ballistic overlaps and the AL1-DCSR high-speed 'shoot-through' tyre system is standard on all wheels.

The Tactica has been designed to accept four different turbocharged diesel engines, giving the end-user the option to select the most appropriate for that individual requirement. All four engine types are now in service, each engine being matched and type approved to the power train providing a maximum road speed of up to 120 km/h.

The advanced permanent 4 × 4 drive system incorporates selectable ratios, three lockable differentials with power-assisted steering and brakes fully air-powered.

The driver and commander are seated at the front of the vehicle, each with large protected vision areas to their immediate front and also in the side doors.

The rear troop area has five individually shaped seats down either side, with a further two seats facing rearwards from the half bulkhead. This provides a total troop capacity of 14, including the commander and driver.

In addition to the two forward doors, the rear troop area has four full-sized doors and a flat floor with no obstructions, which allows rapid troop deployment.

The vehicle is internally clad in ABS, with a foam lining to reduce noise levels and to act as a thermal barrier. Twin independent under-armour air conditioning units and powered ballistic smoke extractor fans are standard. There are 10 adjustable firing ports, with individual vision blocks provided for the troops to fire from within the vehicle.

Day/night lighting systems and an internally operated roadwheel and engine fire extinguisher system are also standard equipment on all variants of the Tactica.

Optional armament installations include a one-man turret armed with a 7.62 mm machine gun, a one-man riot control turret and 7.62 mm or 12.7 mm machine gun rings.

Alvis Vehicles 7-man patrol version of Tactica fitted with Alvis Vehicles turret armed with 12.7 mm MG
0018744

Alvis Vehicles Tactica in armoured personnel configuration used by Saudi Arabia in Police role
2002/0121225

Variants

In 1993, the patrol variant of the Tactica went into service with a Middle Eastern customer, and the armoured water cannon version was selected by an African country shortly afterwards. The British Army also selected two variants and deliveries of these started in 1993.

Since development of the Tactica started, the range of variants has been extended and now covers the following versions:

APC in 10-, 14-, 16, and 18-man configurations
Armoured command with a high roof
Armoured ambulance, three-man crew and capable of carrying four stretcher cases
Armoured EOD
Armoured water cannon, three-man crew, complete protection
Armoured patrol, high-speed, 10-man crew
Armoured assault, 5- or 10-man crew
Armoured patrol, 10-man APC.

Status

Production as required. In service with numerous countries in both military, paramilitary and police roles. Known users of the Tactica include:

Country	Quantity	Comment
Argentina	9	for UN duties, delivered in 1997
Ghana	20	3 versions delivered in 2000
Indonesia	134	IS role
Kuwait	20	police role
Mauritius	7	police role
Saudi Arabia	n/avail	police role
Singapore	n/avail	police role
UK	n/avail	special army roles (forward control)
	n/avail	MoD police (bonnet version)

Manufacturer

Alvis Vehicles Limited.

UPDATED

Alvis 4 armoured personnel carrier

Development

In the 1980s, Mechem Consultants of South Africa designed and built a 4 × 2 armour-protected vehicle called the Mamba, based on proven automotive components of the German Mercedes-Benz UNIMOG 416 series of 4 × 4 cross-country vehicles.

Further development of this by Reumech OMC (which late in 1999 was taken over by Vickers Defence Systems of the UK and renamed Vickers OMC) resulted in the Mamba series of mine-protected vehicles of which 653 were built by the company for the South African National Defence Force (including air force and army) and for export. Production of the Mamba series has now been completed in South Africa. Details of the Mamba are given earlier in this section under the entry for the Vickers OMC Mamba under South Africa.

Further development of the Mamba Mk II resulted in the Alvis 4 and Alvis 8 series of mine-protected armoured personnel carriers which the company developed for the home and export markets in association with Mechem and the now Vickers OMC.

Although Alvis Vehicles originally marketed the Alvis 4 and 8, they subsequently dropped the Alvis 8 from their product range.

According to Alvis Vehicles, their Alvis 4 has improved protection when compared with the original South African vehicle.

In mid-1996, the British Army placed an order with Alvis Vehicles for the supply of six Alvis (4 × 4) mine-protected vehicles for use by the British Army Contingent of the Implementation Force (IFOR) currently operating in Bosnia.

As an interim solution, Reumech OMC, through Alvis Vehicles, supplied six vehicles from the South African production line, three in the Alvis 4 and three in the Alvis 8 configurations.

These interim vehicles have now been replaced by new vehicles from the Alvis Vehicles Coventry production and are now operational with the British Army.

These vehicles are being manned by Explosive Ordnance Disposal (EOD) teams operating in conjunction with mine clearance detachments and are fitted with additional armour to provide protection against the TRMP-6 SFF (Self-Forming Fragment) anti-tank mine which has been laid in the former Yugoslavia in significant numbers.

The Alvis 4 has a shorter wheelbase than the Mamba and was developed by Alvis to provide a smaller, light vehicle which could be used for liaison, reconnaissance and command/control tasks. It has six seats, three either side facing the front, although the back four can be altered to suit the requirements of the user.

In 1996, Alvis Vehicles supplied one of these vehicles to the US Army for trials purposes.

By late 1999, the UK had placed orders for a total of 18 vehicles (6 + 3 + 9) all for use by Royal Engineer teams operating in the former Yugoslavia.

All production of the Alvis 4 was undertaken at the Alvis Vehicles facility at Coventry, which has now

Specifications

Alvis 4
Crew: 2 + 4
Configuration: 4 × 4
Combat weight: 6,120 kg
Unloaded weight: 5,360 kg
Power-to-weight ratio: 20 hp/t
Length: 4.900 m
Width: 2.160 m
Height: 2.595 m
Ground clearance: 0.38 m
Track:
(front) 1.89 m
(rear) 1.96 m
Wheelbase: 2.380 m
Angle of approach/departure: 45/37°
Max road speed: 110 km/h
Fuel capacity: 200 litres
Max range:
(road), 900 km
(cross-country) 500 km
Fording: 1.00 m
Gradient:
(forwards) 70%
(reverse) 50%
Side slope:
(static) 60%
(moving) 40%
Trench: n/avail
Vertical obstacle:
(forwards) 0.40 m
(reverse) 0.30 m
Turning radius: 6.10 m
Engine: Mercedes-Benz 6-cylinder diesel, water-cooled, with exhaust brake, developing 123 hp at 2,800 rpm
Transmission: manual, 4 speed with 4 high-range, 4 low-range and 4 reverse
Transfer box: integral with gearbox, 4 × 2 and 4 × 4 pneumatically controlled
Clutch: single dry plate, foot-operated
Steering: recirculating ball, power-assisted
Suspension:
(front) single coil spring with double-acting hydraulic telescopic shock-absorbers, Panhard rod, torque tube and strut bars
(rear) double coil spring with double-acting Panhard rod, torque tube and strut bar
Tyres: 12.5 R20 tubeless, 18 ply with Hutchinson VFI run-flat system
Brakes:
(front) disc
(rear) disc
(parking) pneumatic
Electrical system: 24 V
Batteries: 2 × 12 V, 55 Ah in series
NBC system: no
Night vision equipment: optional

Alvis 4 (4 × 4) in service with British Army (Richard Stickland) ***2002***/0048514

Alvis 8 (4 × 4) in service with British Army (Richard Stickland) ***2002***/0121199

closed down. Production is now complete. Alvis Vehicles are no longer marketing this Alvis 4 vehicle.

Description

The monocoque hull of the Alvis 4 is of all-welded steel armour construction that provides a very high level of protection against mines as well as small arms fire and shell splinters.

The driveline and suspension of the Alvis 4 is from the well-known Mercedes-Benz UNIMOG (4 × 4) range of cross-country vehicles, which are used by many countries around the world for civil and military purposes.

According to Alvis Vehicles, the Alvis 4 protection level includes a double 7 kg TNT land mine under any single wheel station, up to a 7 kg TNT land mine under the centre of the hull, small arms fire up to 7.62 mm rifle through a full 360°, shrapnel, incendiary and limpet mine.

The baseline vehicle has undergone extensive trials in South Africa to prove that the vehicle and its crew will survive these mine blasts.

If required by the user, lower or higher levels of protection against small arms fire can be provided. Already type classified and in production is an appliqué armour kit which provides protection from the latest Russian TRMP-6 SFF mine, which has been encountered in Bosnia.

The specially shaped hull is integrated into the driveline and suspension system to help lower the overall profile of the vehicle.

The layout of the Alvis 4 is conventional, with the engine compartment at the front, commander (left) and driver (right) and troop compartment at the rear.

Normal means of entry and exit for the crew and troops is through a large door in the rear that is hinged on the left.

In addition to the commander and driver, four troops are carried, two on the left side and two on the right, facing each other on individual seats with a full harness seat belt.

Above the commander's and driver's position is a square roof hatch that opens to the rear and can be locked in the vertical position.

Stowage lockers are located on the exterior of the hull and mounted at the rear is a tow hitch with a seven-pin electrical connector. A spare wheel and tyre is located on the left-hand side of the hull just to the rear of the front roadwheels.

Bulletproof windows are provided for the front, sides and rear of the Alvis 4 and these provide the same level of protection as the steel hull.

If required by the user, these windows can be provided with integral firing ports and with wire mesh protection. The vehicle could also be supplied without armoured glazing in the rear.

Drive is taken from the diesel engine to the four-speed full synchromesh gearbox with the high and low range giving a total of eight forward and four reverse gears.

The pneumatically controlled differential locks operate front and rear separately. Two- or four-wheel drive may be selected while the vehicle is in motion.

The front and rear axles are fitted with coil springs and hydraulic dampers. The air-assisted foot-operated duplex braking system acts on discs all round. A cable-operated parking and an exhaust brake are also provided.

Standard equipment includes a 100 litre water tank and optional equipment includes a roof-mounted light machine gun mount, firing ports, smoke grenade launchers and alternative seating arrangements.

Other optional equipment included air conditioning system, luxury seating, winch, roof rack/roof-mounted equipment, ambulance fittings, run-flat tyre system, command vehicle fittings, full radio suppression, intercom and various radio communications equipment.

The Alvis 4 is fully air-transportable in a Lockheed Martin C-130 Hercules transport aircraft and variants may also be delivered by Low-Altitude Parachute Extraction System (LAPES).

Status

Production complete. In service with the British Army. One supplied to the US Army for trials purposes in 1996. This vehicle is no longer being marketed by Alvis Vehicles.

Manufacturer

Alvis Vehicles Limited.

UPDATED

Hotspur Hussar armoured personnel carrier

Development

The Hussar APC and Hussar internal security vehicle were originally developed as a private venture by Hotspur Armoured Products of Wales and were shown for the first time at the 1984 British Army Equipment Exhibition.

Following trials with two prototypes, the first export order was placed and, by early 1987, four countries had placed firm orders for the vehicle.

In August 1986, Hotspur Armoured Products became a division of Penman Engineering Limited in Scotland, where the design and manufacturing facility is now located.

Hotspur Armoured Products continues to sell and market wheeled light armoured vehicles but no recent sales figures have been released.

Penman Engineering also designed and built the Tavern 4 × 4 armoured personnel carrier to meet the British Army Staff Requirement (Land) 4037 for an Armoured Patrol Vehicle Mk II for use in Northern Ireland.

Description

The Hussar APC is based on the chassis of the Land Rover One Ten (4 × 4) with an additional powered axle, modified and strengthened to accommodate the weight of the armoured hull.

The Hotspur steel armoured hull, which fully encloses the engine and troop compartment, has been designed to provide ballistic immunity at 25 m at a 90° attack angle from ball ammunition fired from 7.62 mm type weapons and from point-blank attack by lighter weapons such as the 9 mm Sterling SMG, M1 carbine and .44 Magnum. The floor armour protects the crew from mine fragments and grenades.

The engine and transmission are at the front of the vehicle with the troop compartment at the rear. The engine is coupled to a manual transmission and the driver can select either 6 × 6 or 6 × 4 (first and second axles) depending on the terrain encountered.

The commander and driver sit to the rear of the engine and each has a laminated armoured glass windscreen to his front and side with anti-splinter visors of polycarbonate. The visors can be activated from inside the vehicle either manually or electrically. In each side of the hull is a single door that opens to the front. In the upper part of these is a bulletproof vision block with a firing port underneath.

The troop compartment is at the rear of the hull, and the 12 troops sit on individual flip-up seats either side of the hull. They enter via the side doors or the two doors in the hull rear. In each side of the troop compartment are two vision blocks with a firing port underneath and each of the rear doors also has a vision block with a firing port underneath. The troop compartment is 3.15 m long, 1.441 m wide and 1.194 m high.

If required, a one person light armour turret or a one person fully rotating machine gun hatch mounting a 7.62 mm machine gun can be fitted to the roof. The basic turret has manual traverse but powered traverse is an option.

Standard equipment for the Hussar includes twin air conditioning system, twin extractor fans, explosion protection system for fuel tanks and armour-protected fuel cap.

Optional equipment for the Hussar includes a more powerful 3.9 litre diesel engine, fully rotating one-man armoured turret with gun mount, electrically operated grenade launchers, fully rotating machine gun hatch, fire suppression system, run-flat tyres, searchlights,

Hotspur Hussar (6 × 6) APC with one-man turret but without armament fitted 0005297

Specifications

Crew: 1 + 13
Configuration: 6 × 6
Combat weight: 5,350 kg
Power-to-weight ratio: 25 bhp/t
Length: 5.74 m
Width: 1.85 m
Height:
(with turret) 2.62 m
(to hull roof) 2.10 m
Ground clearance: 0.21 m
Track: 1.486 m
Wheelbase: 2.764 + 1.016 m
Angle of approach/departure: 40/30°
Max road speed: 120 km/h
Fuel capacity: 98 litres
Max range: 300 km
Turning radius: 8.65 m
Engine: 3.528-l V-8, water-cooled petrol developing 134 hp at 4,000 rpm
Transmission: manual with 5 forward and 1 reverse gears
Steering: power, worm and roller
Clutch: single dry plate
Transfer box: 2 speed
Suspension:
(front) dual-rate coil springs, live beam axle, double-acting hydraulic dampers
(middle and rear) A-frame, live rear axle, single-rate coil springs, double-acting hydraulic dampers
Brakes:
(front) hydraulic, servo-assisted disc
(rear and middle) hydraulic, servo-assisted drum
Tyres: 7.50 × 16
Electrical system: 12 V
Battery: 2 × 12 V, 95 Ah
NBC system: no
Night vision equipment: yes

siren and public address system, armoured louvres over ballistic glass, barricade removal bar on front of hull, 24 V electrical system and various night vision systems. Additional non-metallic passive armour protection to the crew compartment is also available.

Variants

Several variants of the Hussar are available, including a field ambulance, an internal security vehicle and a communications/command vehicle.

Hotspur Skirmisher 4585

Details of the Hotspur Skirmisher 4585 (4 × 4) armoured patrol vehicle, based on the Land Rover One Ten chassis, were given in *Jane's Armour and Artillery 1996-97*, page 503. This was designed and built for the export market and has been exported to a number of undisclosed countries.

Status

Production as required. In service with Bahrain (police), Egypt (110), Sri Lanka (9) and two other undisclosed countries.

Manufacturer

Hotspur Armoured Products Limited, a Division of Penman Engineering Limited.

UPDATED

Alvis Vehicles Armoured Patrol Vehicle

Development

The Armoured Patrol Vehicle was developed by Glover Webb to meet the requirements of the British Army and, following extensive competitive trials against other vehicles, was selected by the British Army in 1985 for use in Northern Ireland. Over 100 vehicles were ordered and first deliveries were made in 1986.

The vehicle is based on the chassis of the heavy-duty military Land Rover Defender (4 × 4) vehicle which is already used by a number of armed forces. This also forms the basis of the Glover Hornet armoured car which is fully described in the Reconnaissance vehicles section. The Glover Webb Armoured Patrol Vehicle could be supplied powered by a Rover V-8 petrol engine or a Rover turbocharged diesel engine.

The manufacturer suggested a wide range of military and paramilitary roles including border patrol, escort duties, anti-terrorism activities, internal security, riot control and airfield perimeter patrol.

In late 1994, Glover Webb was taken over and run by GKN Defence.

Late in 1997, GKN Defence closed down its Glover Webb subsidiary at Hamble and transferred development, production, sales and marketing to its main facility at Telford, Shropshire.

Late in 1998, GKN Defence was merged with Alvis Vehicles and aftersales support for Glover Webb vehicles is now carried out by Alvis Vehicles from their Telford facility.

There has been no recent production of the Glover Webb Armoured Patrol Vehicle and it is no longer being marketed. Alvis Vehicles markets the Tactica (4 × 4) for the internal security role and details of this are given in an earlier entry in this section.

This vehicle has now been withdrawn from service by the British Army.

Description

The hull of the Glover Webb Armoured Patrol Vehicle is of all-welded steel armour construction which provides protection from 7.62 mm ball and 5.56 mm small arms fire. The front windscreen and cab windows are of bulletproof multilaminate glass with a polycarbonate anti-spall liner giving a similar level of protection to the steel armour.

The layout of the vehicle is conventional, with the engine compartment at the front, driver and commander in the centre and the troop compartment at the rear. The commander and driver both have a forward-opening side door and these, like the rear doors, are ballistically sealed with 20 mm overlaps.

The crew compartment floor is constructed of tough glass fibre-reinforced plastic providing protection for the crew against blast and grenade attack. The fuel tank is protected by Explofoil explosive suppression and is also located behind the belly plate.

The troops enter and leave via two doors in the hull rear which each have a bulletproof vision block. There are no firing ports or vision devices in the sides of the troop compartment, although these could be fitted if specified by the user. The troops sit on bench seats that run down either side of the hull and each man has a seat belt.

In the forward part of the roof is a searchlight protected by a grill which can be traversed from within the vehicle, and over the rear troop compartment is a forward-opening hatch cover.

An obstacle removal device is fitted at the front and a wire mesh grill protects the headlamps and radiator. Wire mesh protection is provided for the front windscreen and, when this is not required, it can be folded forward onto the bonnet. A vertical wire cutter is provided to prevent injury to or decapitation of personnel using the roof hatch.

Steel well filler bands are fitted to all wheels giving a run-flat capability and a minimal loss of steering control in the event of penetration of the tyre by small arms fire.

Options included increased armour protection, an integrated air conditioning system, gun ports and additional vision blocks, electrically operated smoke grenade launchers, hi-lo sirens, a machine gun hatch with a 7.62 mm GPMG pintle, flashing lights, riot screens, a front self-recovery winch, internal and external fire extinguishers, radios and alternative internal layouts.

Variants

A number of variants is available for police roles. Vehicles are in service with the London Metropolitan Police and a number of other tactical UK police forces, as well as overseas.

Specifications

Crew: 2 + 6
Configuration: 4 × 4
Length: 4.55 m
Width: 1.79 m
Height: 2.08 m
Ground clearance: 0.32 m
Track: 1.468 m
Wheelbase: 2.794 m
Max road speed: 120 km/h
Vertical obstacle: 0.23 m
Turning radius: 8.75 m
Engine: Rover 4-stroke V-8 3,528 cc petrol developing 114 bhp at 4,000 rpm (or turbocharged diesel)
Transmission: manual with 5 forward and 1 reverse gears
Transfer case: 2 speed
Steering: recirculating ball, worm and nut, power-assisted
Clutch: single dry plate
Suspension: heavy-duty long-travel coil springs, anti-roll bars fitted front and rear
Tyres: 9.00 × 16
Brakes:
(main) dual-line hydraulic disc front, drum rear, servo-assisted
(handbrake) mechanical, on transfer case
Electrical system: 12 V
NBC system: no
Night vision equipment: no

Status

Production complete. No longer being marketed. In service with a number of police forces in the UK and overseas.

Manufacturer

Alvis Vehicles Limited.

UPDATED

Land Rover Armoured Patrol Vehicle as delivered to the British Army for use in Northern Ireland

Armoured Patrol Vehicle from the rear with doors open

NP Aerospace CAV 100 armoured personnel carrier

Development

Following a competition in the spring of 1991, the British MoD placed a contract valued at about £50 million with the now NP Aerospace, for the supply of about 1,000 lightly protected 4 × 4 vehicles for the British Army.

The original contract covered the development of prototype vehicles for validation of the light armoured concept and the engineering systems of the vehicle assembly.

The new light armoured vehicle is based on the Land Rover Defender (4 × 4) vehicle (*Janes Military Vehicles and Logistics 2001-2002* with NP Aerospace fitting the special ballistic protection system, which is an integral structure manufactured from advanced fibre-reinforced composite materials.

Typical roles of this vehicle include internal security, ambulance, explosive ordnance disposal, logistics, peacekeeping, communications and use by relief agencies in high-threat areas.

Early in 1993, 16 CAV 100 vehicles were purchased by the United Nations High Commissioner for Refugees (UNHCR) to protect relief workers in Bosnia. These were built and delivered in less than four weeks. Three CAV 100s have been supplied to the BBC for use in Yugoslavia.

Early in 1994, it was announced that the French Ministere des Affaires had selected the CAV 100D light armoured vehicle to protect diplomatic personnel, with the initial order being for three vehicles.

Two versions were delivered, VIP and escort, with both versions based on the proven Land Rover Defender 110 (4 × 4) chassis.

These CAV 100Ds are used by French embassy personnel operating in various trouble spots around the world.

In 1994, the Canadian Department of National Defence ordered three CAV 100D armoured vehicles to be used in Bosnia. These were delivered in left-hand drive configuration, powered by the Land Rover 200 Tdi 2.5 litre diesel engine and fitted with the optional steel mesh riot protection window system. This includes a lowerable mesh cover for the windscreen that is operated from inside the vehicle.

It is estimated that about 1,100 light armoured vehicles on the Land Rover Defender 110 (4 × 4) chassis, with the largest customer being the British Army. The British Army vehicles have a different build standard and specification from the export vehicles.

Finland has purchased two CAV 100 series vehicles for use in the former Yugoslavia.

Description

(Typical)

The CAV 100 vehicle protection system is an armour concept based on S2 Glass composite bonded assembly which can be fitted to a wide range of wheeled vehicles.

The layout of the vehicle is conventional, with the engine at the front, driver and commander in the centre (with both left- and right-hand drive versions available) and twin doors at the rear. There is accommodation for four or six men in the rear on seats down either side.

Transparent armour is fitted to the windscreen, side and rear windows and this comprises laminated glass/'Hardcote' Polycarbonate/PVB construction. SHU (Self-Healing Urethane) coating is also available as an option.

The manufacturing programme for the armour is based on NP Aerospace's own extensive experience in composites and also incorporates specific technology licensed from Owens Corning Fibreglas in the US.

The S2 Glass* has been found to be most effective in achieving the design solutions for the projects that have been addressed so far, but the capability and database background to use other reinforcements, such as E glass, aramid, hybrids and combinations with ceramics, is available.

The ballistic protection level can be tailored to defeat the specific threat or range of threats covering fragmentation, 5.56 mm or 7.62 mm small arms fire and blast protection.

By virtually eliminating metal from the armour protection system, the behind-armour effects are claimed to be substantially reduced and the risk of spall damage is almost nil. According to the manufacturer, the armour is good against multiple hits and effective repair schemes have been validated. It is also up to 20 per cent lighter than metal armour systems.

The standard Land Rover Defender (4 × 4) chassis is fitted with a modular monocoque body to which the doors, windscreen and other subcomponents are added. A monocoque structure was selected to provide the maximum possible protection and structural integrity.

The armour concept is based on the use of high-performance glass-reinforced composites made by high-pressure compression moulding processes.

Armour components are moulded as contoured shapes and flat panels for combination into assemblies and subassemblies using advanced adhesive bonding techniques.

Standard equipment includes a PROMEL explosion protection system for the fuel tank, an in-tank electric fuel pump via a full flow filter, a water-to-air vehicle heater with a two-speed fan in the front cabin, and an extractor/ventilator fitted in the rear cabin, which is capable of moving air at the rate of 300 m^3/h or more. Electrical suppression is to ECE 10 standard but as an option it can be upgraded to DEF-STAN 5941.

The standard version is powered by a 3,500 cc V-8 petrol engine, while the CAV 100D is powered by a 2.5 litre Tdi diesel engine, which gives a maximum road range of 700 km and an average road speed of 120 km/h.

All CAV 100 vehicles are now fitted with an air conditioning system as standard.

Optional equipment includes a surveillance roof hatch, document ports, additional windows for rear cab, riot/public disorder protection, seating options for up to eight crew, enhanced comfort front seats, paintwork colour options (for example NATO green/sand/and so on) and signwriting to end-user requirements.

* S2 Glass is the registered trademark of Owens Corning Fibreglas Corporation.

Variants

While the CAV 100 series relates to the Land Rover Defender (4 × 4) chassis, the CAV 200 series covers armour system assemblies based on other vehicle chassis configurations. The CAV 300 series covers armour components and panels.

CAV 100 VIP

This has the rear configured for the VIP role with seats for two senior personnel plus seats for two bodyguards.

CAV 100 ambulance

This has a rear configured for two stretcher cases, one either side, and attendant medical personnel, plus overhead storage lockers and resuscitation equipment.

CAV 100 armed response vehicle

This is similar to the standard vehicle but has specialised police equipment such as blue flashing lights, rotating spotlights and additional communications equipment.

CAV 201 APC

In 1995, NP Aerospace completed the first prototype of the CAV 201 (4 × 4) armoured personnel carrier, which is based on the chassis of the well-known Mercedes-Benz G-Wagon 461 (4 × 4) vehicle.

This model has a similar layout to the CAV 100 with gross vehicle weight being 3,800 kg and unladen weight being 3,000 kg. It is offered with a 2,298 cc petrol engine (maximum road speed 135 km/h) or a 2,874 cc diesel engine which gives a maximum speed of 120 km/h.

Both engine options are coupled to a five-speed manual transmission and a two-speed transfer box and the driver can select 4 × 4 or 4 × 2 drive depending on the tactical situation. Air conditioning is standard, as is a 96 litre fuel tank.

For operations in the Balkans some British Army vehicles have been observed fitted with a spare wheel and tyre mounted on the front of the hull and an anti-wire cutting device installed on the left side of the hull.

Specifications

(Typical on Land Rover Defender chassis)

Crew: 2 + 6
Configuration: 4 × 4
Combat weight: 3,500 kg
Unloaded weight: 2,900 kg
Payload: 600 kg
Power-to-weight ratio: 38.28 hp/t
Length: 4.45 m
Width: 1.79 m
Height: 2.10 m
Ground clearance: 0.216 m
Track: 1.486 m
Wheelbase: 2.794 m
Max road speed: 135 km/h
Max cruising speed: 105 km/h
Road range: 350 km
Fuel capacity: 80 litre tank filled with PROMEL explosion protection system
Engine: 3.528 1 V-8 water-cooled petrol, low compression, developing 134 bhp at 5,000 rpm
Transmission: manual with 5 forward and 1 reverse gears
Clutch: diaphragm spring
Transfer box: 2 speed
Steering: manual, recirculating ball
Turning radius: 6.4 m
Suspension: beam axles front and rear with coil springs controlled by hydraulic telescopic dampers
Tyres: Michelin XZY 7.50 × 16 with inner tubes, TYLOCK safety bands for run-flat capability
Brakes:
(main) dual servo-assisted, discs front, drums rear
(parking) drum, mechanical on transmission
Electrical system: 12 V
Alternator: 45 A
NBC system: no
Night vision equipment: no

Status

Production as required. In service with the British Army and a number of civilian agencies (see Development).

Manufacturer

NP Aerospace Limited.

UPDATED

NP Aerospace CAV 100 APC on Land Rover Defender chassis in service with British Army in the Balkans (Richard Stickland) ***2002***/0121216

NP Aerospace CAV 100 APC on Land Rover Defender chassis in service with British Army (Richard Stickland) ***2002***/0121217

UNITED STATES OF AMERICA

General Motors Defense LAV-III for Brigade Combat Team

Development

In November 2000, following an international competition, the US Army Tank Automotive and Armament Command selected the General Motors Defense LAV-III (8 x 8) (Light Armoured Vehicle – III) to meet the US Army requirement for a Medium Interim Armoured Vehicle (MIAV) to equip the new US Army Brigade Combat Teams (BGT). The actual contract was awarded to a joint venture company called GM GDLS Defense Group.

This comprises General Motors Defense of Canada and General Dynamics Land Systems, prime contractor for the M1 Abrams family of Main Battle Tank. It is estimated that the total value of the contract, if all options are exercised could be as much as US$4 billion through to 2008.

Under current plans the US Army expects to equip up to six BCT with a total of 2,131 LAV-III vehicles. The initial contract was worth US$61.7 million in research, development, test and evaluation with a second contract worth US$578 million for an initial batch of 366 LAV-III production vehicles.

Production will be undertaken by General Motors Defense in London, Ontario (Canada) and by General Dynamics Land Systems Lima, Ohio (USA). In addition, final assembly of the Infantry Carrier Vehicle (ICV) will take place at the Anniston Army Depot in Anniston, Alabama.

Late in 2000 however, United Defense LP, who were offering a tracked solution, protested the US Army decision and a stop work order was issued. In April 2001, the General Accounting Office rejected the protest of United Defense which allowed work to start on the LAV-III programme for the US Army again.

In May 2001 the US Army stated that plans to field the first Interim Armored Vehicles (IAV) in early 2002 were on track and have an Interim Brigade Combat Team equipped with the vehicles and ready for operation by the third quarter of Fiscal Year 2003.

The first brigade will include about 290 LAV as well as substitute vehicles that will be used temporarily while some variants are still in development. The Mobile Gun System, for example, is not expected to be fielded to at least 2004.

A C-5A will be able to carry four LAV-III, a C-17A two and a C-130 one.

Description

Full details of the General Motors Defense LAV-III are given earlier in this section under Canada where the vehicle is already in production for the Canadian Army in a different configuration with the armoured personnel carrier being fitted with a two-person Delco Defense turret armed with a 25 mm cannon.

The LAV-III's for the US Army will be fully air portable in a Lockheed Martin C-130 Hercules transport aircraft as well as the larger C-17 and C-5 aircraft but will not be amphibious. Standard equipment will include a fire/explosion detection and suppression system, NBC system and all vehicles will be powered by a Caterpillar diesel engine coupled to an Allison fully automatic transmission.

The overall layout of the LAV-III for the US Army is similar to the Canadian LAV-III vehicles but with a slightly higher roof line to the rear of the drivers station for greater internal volume.

In the case of the Infantry Carrier Vehicle the remote weapon station is on the right side of the hull roof with a rear opening hatch to the left. To the immediate rear of this is a stowage box with additional roof hatches at the rear.

The passive armour will provide protection against attack from 14.5 mm armour-piercing all round with applique armour providing protection against attack from RPG-7 anti-tank rockets. Overhead protection is against 152 mm air burst artillery munitions. The passive armour package has been developed by IBD of Germany.

The US Army will procure a number of versions of the LAV-III including the base line Infantry Carrier Vehicle (ICV) and a Mobile Gun System (MGS) all of which will share a common 8 x 8 chassis to reduce procurement and operating costs.

The baseline ICV will have a crew of two, commander/gunner and driver and carry a nine strong infantry squad. A total of 118 ICVs plus one for trials are included in the first batch of 366 units.

The basic ICV will be fitted with a remote weapon station on the right side of the roof armed with a 12.7 mm M2 machine gun or 40 mm MK 19 automatic grenade launcher. This will be able to be laid and fired with the gunner under complete armour protection inside of the hull.

When armed with a 12.7 mm M2 machine gun a total of 2,000 rounds would be carried and when armed with a 40 mm MK 19 automatic grenade launcher, 448 rounds of 40 mm ammunition will be carried. Each vehicle will also carry two Javelin fire-and-forget anti-tank guided missiles.

Two Javelin fire-and-forget anti-tank guided weapons will also be carried in each ICV.

In addition, there will be eight versions of the ICV:

Variant	Quantity in 1st batch
Mortar carrier (120 mm)	51
Reconnaissance vehicle	66
Anti-tank guided missile vehicle (TOW)	40 plus 1 for test
Fire support vehicle	nil
Engineer support vehicle	22
Command and control vehicle	40
Medical evacuation vehicle	27
NBC reconnaissance vehicle	nil

The MGS will be an LAV-III fitted with the General Dynamics Land Systems Low Profile Turret armed with a 105 mm rifled tank gun and a 7.62 mm coaxial machine gun. Details of this system, currently at the prototype stage, are given in the Reconnaissance vehicles section under Canada.

The TOW ATGW vehicle will have an elevated launcher and be fitted with Raytheon Improved Target Acquisition System (ITAS). The reconnaissance vehicle will have the LRAS-3 (Long Range Advanced Scout Surveillance System) fitted.

Infantry Carrier Vehicle based on the LAV-III (8 × 8) is fitted with a Remote Weapon Station armed with a 12.7 mm MG or a 40 mm automatic grenade launcher ***2002***/0120451

Specifications

Similar to Canadian built LAV-III, except that this is heavier as it is fitted with a Delco Systems turret armed with a 25 mm M242 cannon and 7.62 mm coaxial machine gun. The US Army vehicle will be lighter.

Status

Selected by the US Army.

Manufacturers

General Motors Defense.
General Dynamics Land Systems.

UPDATED

Cadillac Gage LAV-300 armoured vehicle range

Development

The LAV-300 armoured vehicle range was developed by Cadillac Gage Textron (which in 1994 became part of Textron Marine & Land Systems) as a private venture specifically for the export market in the late 1970s.

The vehicle was originally known as the V-300 and builds on experience in the development and production of the LAV-150 range of 4 × 4 light armoured vehicles covered in detail in a later entry.

The first two prototypes of the LAV-300 were completed in 1979 and the first customer was Panama which, in 1982, placed an order for 12 LAV-300 vehicles in four different configurations including fire support vehicle with a two-person turret armed with a 90 mm Cockerill Mk III gun, APC with twin 7.62 mm machine guns and recovery. All of these were delivered by the end of 1983. In 1984, the Kuwait National Guard ordered 62 LAV-300 Commandos. These were lost during the Iraqi invasion of Kuwait in 1990.

Production of the LAV-300 resumed in 1993 as, late in March 1993, the US Army Tank Automotive Command awarded Cadillac Gage a contract for the supply of 24 LAV-300s (6 × 6) for the Philippines.

Of the 24 vehicles, 12 have a two-man Cadillac Gage turret armed with a 90 mm gun and a 7.62 mm machine gun, and the remaining 12 a Cadillac Gage one-person 1 m turret armed with one 12.7 mm and one 7.62 mm machine gun.

Total value of the contract was US$18.243 million, with production undertaken at the Textron Marine & Land Systems facility at New Orleans, Louisiana. First deliveries were made in September 1994.

There has been no recent production of the LAV-300 series of 6 × 6 armoured vehicles, although marketing continues.

The latest model of the LAV-300 series is designated the Mk II and details of this are given at the end of this entry.

LAV-300 Mk I (6 × 6) light armoured vehicle fitted with Textron Marine & Land Systems one-person turret armed with 40 mm grenade launcher and 12.7 mm machine gun **2000**/0075723

LAV-300 Mk I (6 × 6) light armoured vehicle fitted with two roof-mounted 7.62 mm machine guns each provided with a shield **2000**/0075724

Description

The hull of the LAV-300 is of all-welded unitised construction of special high-hardness steel ballistic plate that protects the crew from small arms fire and shell splinters. If required, a higher level of passive armour protection is available for the LAV-300 series.

The driver sits at the front of the hull on the left side and has a single-piece square hatch cover that opens to the rear. To his front are three day periscopes for forward observation and one of these can be replaced by a passive periscope for driving at night. In the side of the hull to the left of the driver is a bulletproof vision block with a firing port underneath. To the rear of the driver is seated another member of the crew who is provided with a single-piece hatch cover, although he can also enter his position via the rear troop compartment. To his left is a small half door that opens to the rear which is provided with a vision block and a firing port.

The engine compartment is to the right of the driver with the air inlet and outlet in the roof and the exhaust pipe in the right side of the hull. Power from the Cummins diesel engine is transmitted to the final drives via an Allison Transmission MT-643 six-speed automatic gearbox. The transfer box contains a two-speed system with a spline engagement clutch for the axle drive.

On the glacis plate is a holder for a shovel and pick axe and under the nose is an internally mounted hydraulic winch which has a maximum capacity of 9,000 kg. This is provided with 46 m of 12.7 mm diameter steel core cable.

The LAV-300 can be fitted with a number of armament installations mounted in the centre of the hull over the second axle and these include:

Cadillac Gage two-man turret with a 90 mm Cockerill Mk III gun (described in the LAV-150 entry)
One-man turret armed with 40 mm Mk 19 grenade launcher and 12.7 mm machine gun (described in the LAV-150 entry)
Two-man turret armed with 30 mm cannon, 7.62 mm coaxial machine gun and 7.62 mm anti-aircraft machine gun (described in the LAV-150 entry)
Turret armed with a 25 mm M242 The Boeing Company M242 Chain Gun, 7.62 mm coaxial and 7.62 mm anti-aircraft machine guns and smoke grenade dischargers.
Cadillac Gage two-man turret with a 20 mm Oerlikon Contraves cannon (described in the LAV-150 entry)
Cadillac Gage one-man 1 m machine gun turret (described in the LAV-150 entry)
Cadillac Gage one-man machine gun turret (described in the LAV-150 entry)
Ring mount with a 7.62 mm or 12.7 mm M2 machine gun.

The troop compartment is at the rear of the LAV-300 and in each side of the hull are three bulletproof vision blocks each with a firing port underneath. Over the top of the troop compartment are two rectangular roof hatches hinged on the outside. The infantry enters and leaves the vehicle through two doors in the rear of the hull, each of which has a bulletproof vision block with a firing port beneath. More recent production vehicles have a power-operated ramp in the rear which is provided with a single door in the left side with a vision block/firing port. There is another vision block/firing port to the right. As an alternative to the two-part hatch cover, a circular hatch cover with provision for mounting a 7.62 mm pintle-mounted GPMG can be fitted.

The front suspension is a swing-mounted solid axle, with rear suspension independent with coil springs at each wheel station. The solid front axle is restrained longitudinally by trailing arms and laterally by stabilising rods attached to the axle and hull. The two rear axles consist of short drive shafts extending from the hull-mounted differentials to the independently sprung trailing arm-mounted wheel stations. The roadwheels have a combined travel of 30 cm, a jounce of 17.5 cm and a rebound of 12.5 cm. All the swing-mounted axles are of the double-reduction top-mounted type. The front and rear axles have positive no-spin differentials to maintain traction in off-road operations. The central axles contain the inter-axle differential. All wheels have hydraulic brakes and steering is power-assisted on the front wheels. All wheels have run-flat tubeless tyres with a self-cleaning tread design.

The LAV-300 is fully amphibious without preparation, being propelled in the water by its wheels at a speed of 3 km/h. As an option, the vehicle can be fitted with water-jets which increase the water speed to 8 km/h. The amphibious version has a number of modifications to the hull including the installation of a trim vane on the front of the hull which is erected when the vehicle is afloat.

The LAV-300 can be carried slung under helicopters and a Lockheed Martin C-130 Hercules transport aircraft can carry one internally.

Standard equipment includes a 19 litre water can, 19 litre diesel fuel can, fire suppression system, pioneer tool kit, tow cable, slave cable, first aid kit, vehicle tool kit and engine-driven air compressor with hose. All vehicles are fitted with a front-mounted winch with a capacity of 9,000 kg with 46 m of cable and a snatch block.

The basic LAV-300 has a payload of 2,720 kg but with an uprated suspension this can be increased to 4,080 kg.

A wide range of optional equipment can be fitted to the LAV-300 series including various types of passive night vision devices, central tyre-inflation system that allows the driver to adjust the tyre pressure regulation system to suit the type of terrain being crossed, air conditioning system, heater, wiper kit, various smoke/ grenade launchers, NBC system, automatic fire suppression system, to name but a few.

Variants

TOW anti-tank

Fitted with an external top-mounted swing-up launcher with an elevation of +30°, depression of –10° and traverse of 360°. There are nine Raytheon Systems Company TOW ATGW missiles carried plus a 7.62 mm machine gun and 2,000 rounds of ammunition. This version has yet to enter production or service.

81 mm mortar

Fitted with a turntable-mounted 81 mm mortar with elevation from +43 to +85°, total traverse of 360°, range of 150 to 4,400 m and a total of 62 mortar bombs carried. Also fitted with a 7.62 mm machine gun for which 2,000 rounds of ammunition are carried. The 81 mm mortar can also be dismounted for ground use.

120 mm mortar

This proposed version would be fitted with the French TDA 120 mm 120R 2M mortar system which is fully

Specifications
(Mk I vehicle)
Crew: 3 + 9 (max)
Configuration: 6 × 6
Combat weight: 14,969 kg
Power-to-weight ratio: 17.36 hp/t
Length: 6.40 m
Width: 2.54 m
Height:
(MG turret) 2.59 m
(20 mm turret) 2.692 m
(hull) 1.98 m
(ITV) 2.756 m
(hull top) 1.981 m
Ground clearance:
(hull) 0.533 m
(axle) 0.381 m
Track: 2.243 m
Wheelbase: 2.209 + 1.524 m
Angle of approach/departure: 45/40°
Max speed:
(road) 105 km/h
(water) 3 km/h
Acceleration:
(0 to 32 km/h) 10 s
Fuel capacity: 435 litres
Max range: 925 km
Fording: amphibious
Gradient: 60%
Side slope: 30%
Vertical obstacle: 0.609 m
Turning radius: 10 m
Engine: Cummins 6 TCA 8.30-C260 turbocharged, aftercooled 6-cylinder in line diesel developing 260 hp at 2,200 rpm
Transmission: Allison MD3560 automatic 6-speed with electronic control and self-diagnostics
Transfer case: single speed with spline engagement clutch for front axle drive (when travelling cross-country drive is 6 × 6 and when travelling on roads drive is 6 × 4)
Steering: variable ratio power
Suspension: solid front axle, middle and rear independent with coil springs, direct acting telescopic shock-absorber acting at each wheel station. Centre and rear axle system is independently mounted using trailing arms, shock-absorbers and coil springs
Tyres: 14.5 × R20 (run-flat)
Brakes:
(main) hydraulic (all wheels) power booster with electric back-up
(parking) hand-operated disc
Electrical system: 24 V
Batteries: 2 × 12 V, 100 Ah
NBC system: optional
Night vision equipment: optional

Note: Combat weight and power-to-weight ratio depend on turret installed.

covered in the Self-propelled mortars section. A baseplate would also be carried to allow the 120 mm mortar to be used in the dismounted ground role.

Command post vehicle

This version would have a raised roof at the rear for use in the command post role and would typically have a total crew of 10 and a combat weight of 13,154 kg.

Maintenance vehicle

This version would have a raised roof at the rear and be fitted with a hydraulically operated crane capable of lifting a load of 3 tonnes at a distance of 3 m. A wide range of optional equipment can be fitted such as generator, work bench, searchlight, cutting and welding equipment.

Recovery vehicle

This version would have a raised roof and be fitted with an A frame which when erected can lift a maximum load of 4.536 tonnes with stabilisers deployed. It would also have a winch with a maximum payload of 11.34 tonnes.

Ambulance

This model would be unarmed and fitted with a raised roof and ramp in place of twin doors at the rear to facilitate the loading of stretchers and walking wounded.

Others

These include cargo carrier, and anti-aircraft with a missile system such as ADATS, or a gun system such as the 20 mm Vulcan as installed on a number of LAV-150 4 × 4 vehicles supplied to the Saudi Arabian National Guard (SANG).

LAV-300 Mk II

Late in 1999, Textron Marine and Land Systems started marketing an enhanced version of the LAV-300 (6 × 6) series of light armoured vehicles called the Mk II.

In the Mk II design, the overall layout of the Mk I has been retained with the driver front left, power pack to the right and crew/weapon compartment at the rear.

According to Textron Marine & Land Systems, the overall general shape of the LAV-300 has been retained by choice to retain ballistic protection obliquity angles, keep a low silhouette/profile for passive defence and ease of air transportability.

The all-welded steel armour hull has been retained, but the steel ballistic plate chemistry and hardness levels have been improved to provide a better protection level. The baseline hull provides protection from 7.62 mm ball attack through 360° and against 7.62 mm armour-piercing over the frontal 60° arc. A higher level of protection is available if required with the addition of a German IBD developed passive armour package.

New brush guards have been added to the hull front as have new larger rear-view mirrors for improved vision. Additional interior/exterior equipment racks have been added to allow for increased stowage.

For improved safety and comfort, an improved fully adjustable driver's seat has been fitted as has a new design of instrument panel. The driver also has a new adjustable steering wheel that can be quickly altered to suit the stature of the driver.

A new power pack has already been installed which is similar to that recently installed in the Swiss MOWAG Piranha Generation III vehicles. A new engine air intake grill/door has been added to improve air flow and delete any special pre-preparation prior to swimming operations.

The power pack consists of a Cummins QSC 8.3 turbocharged and after-cooled diesel engine developing 208 kW at 2,000 rpm and with a peak torque of 1,183 Nm at 1,400 rpm.

This engine meets the latest environmental emission standards and is fitted with electronic controls to provide an engine management system and automotive diagnostics for maintenance and repair.

The engine is coupled to a new Allison MD3560 six-speed automatic transmission with electronic control and self-diagnostics and a new single-speed transfer case. The driver has push-button controls for full 6 × 6 or 6 × 4 drive and an engine-driven exhaust retarder has been fitted to provide additional braking capacity.

The transfer case also has a neutral position so allowing the vehicle to be towed without any preparation. In some other vehicle designs the drive shaft has to be disconnected to allow the vehicle to be towed without damaging the transmission.

The vehicle has a maximum road speed of 105 km/h and is propelled in the water by its wheels at a maximum speed of 3 km/h.

The single diesel fuel tank of the Mk I has been replaced by twin diesel fuel tanks which has increased fuel capacity to 435 litres so increasing the operational range of the LAV-300 vehicle to 925 km.

The front of the hull has been redesigned and widened, with the winch compartment moved from the lower glacis to the front top glacis. The winch has a capacity of 9,000 kg and is provided with 46 m of cable leading out through the front of the hull.

Early vehicles had twin doors in the hull rear but the Mk II has a redesigned hull rear fitted with a power operated ramp with an integral door. This enables easier entry and exit by the troops as well as enabling the vehicle to be more easily used in specialised roles, such as ambulance and cargo.

Space for integral water jets has also been designed into the hull, this being an optional extra and gives the LAV-300 Mk II a maximum water speed of 8 km/h. Before entering the water, the bilge pumps are switched on and the trim vane is hydraulically erected at the front of the vehicle.

The top of the hull of the Mk II has been redesigned and a commander's hatch and two top-rear troop hatches added. Larger driver's and commander's periscopes/vision blocks has also been provided.

On the Mk I the troops were provided with individual seats, but the Mk II has bench-type seats either side which are said to allow quicker entry and exit from the vehicle. For improved comfort a new high-capacity air conditioning/heater system has been installed and a new ration heater.

For increased load carrying capacity the LAV-300 Mk II has been fitted with new springs and shock-absorbers and maximum vehicle weight (vehicle plus payload) is 16,330 kg.

The original bias ply hard sidewall run-flat tyres have been replaced by radial ply tyres that have run-flat inserts. According to Textron Marine & Land Systems these offer the user improved mobility, ride and lower costs.

The new tyres have been fitted onto new wheels to improve tyre fitting and removal. A new driver operated central tyre inflation system has been fitted which allows the driver to automatically adjust the tyre pressure to suit the type of terrain being traversed at any time. Steering is power-assisted on front wheels.

In the future it is expected that the present trailing arm independent suspension and wishbone type of the LAV-300 Mk II, will be replaced with the new fully independent front and rear suspension from the Textron Marine & Land Systems M1117 Armored Security Vehicle (ASV), which is in production for the US Army Military Police. This will allow the gross weight of the LAV-300 to increase to over 21,000 kg.

Status

Production as required. In service with the following countries:

Country	Quantity	Comment
Kuwait	62	national guard, ordered 1984, no longer in service
Panama	12	delivered 1983, believed no longer in service
Philippines	24	ordered 1993, delivered 1994 to 1995

Manufacturer

Textron Marine & Land Systems, Combat Vehicle Marketing.

UPDATED

LAV-300 Mk II (6 × 6) light armoured vehicle fitted with Textron Marine & Land Systems two-person turret armed with 90 mm gun and 7.62 mm coaxial machine gun **2000**/0075725

LAV-300 Mk II (6 ×6) light armoured vehicle with 90 mm turret from the rear, showing the new power operated ramp in the lowered position and water jets under hull rear for improved amphibious characteristics **2000**/0075726

Cadillac Gage ASV 150 (M1117)

Development

Following an international competition for an Armored Security Vehicle (ASV) in which the two finalists were the Textron Marine & Land Systems, Cadillac Gage ASV 150 (4 × 4) and the AV Technology, Limited Liability Company, with a modified version of its Dragoon (4 × 4) light armoured vehicle, the ASV 150 was selected by the US Army in December 1995.

Under Phase One, the US Army Tank Automotive Command awarded the company a contract worth US$3.2 million to build four prototypes of the ASV 150 (4 × 4) vehicle, which were completed early in 1997.

In early 1999 Textron Marine & Land Systems was awarded a contract by the US Army Tank-automotive & Armaments Command valued at US$50 million for the

ASV 150 fitted with turret armed with one 12.7 mm M2 machine gun and one 40 mm Mk 19 grenade launcher **2002**/0121227

Specifications

Crew: 3 + 1
Configuration: 4 × 4
Combat weight: 13,408 kg
Unloaded weight: 11,884 kg
Power-to-weight ratio: 19.39 hp/t
Length: 6.07 m
Width:
(overall) 2.56 m
(hull) 2.46 m
Height:
(overall) 2.59 m
(hull roof) 2.05 m
Wheelbase: 3.02 m
Track: 2.14 m
Ground clearance: 0.46 m
Angle of approach/departure: 40°/40°
Max road speed: 100 km/h
Acceleration:
4.8 s (0 to 32.8 km/h)
28.5 s (0 to 80.46 km/h)
Fuel capacity: 264 litres
Range: 708 km at 64 km/h
Turning diameter: 8.35 m
Gradient: 60%
Side slope: 30%
Vertical obstacle: 0.60 m
Fording: 1.50 m
Engine: Cummins 6CTA 8.3 turbocharged and after-cooled diesel developing 260 hp at 2,200 rpm
Transmission: Allison model MD 3560 six-speed automatic
Transfer case: single speed with split clutch for front axle drive
Differentials: Rockwell Model 611
Steering: TRW Model TAS-65 hydraulically assisted
Suspension: fully independent front and rear, upper and lower control arms, coil springs with telescopic double-acting shock-absorbers at each wheel station
Tyres: Michelin 14.00R20 with Hutchinson run-flat inserts
Brakes:
(main) dual hydraulic with electric power back-up
(parking) hand-operated, transmission-mounted
Electrical system: 24 V
Batteries: 4 × 6TL at 100 Ah
Armament:
(main) 1 × 40 mm Mk 19 Mod 1 grenade launcher
(coaxial) 1 × 12.7 mm M2 HB machine gun
Smoke grenade launchers: 2 × 4 M257
Ammunition:
(40 mm) 100 ready use + 500 stowed
(12.7 mm) 200 ready use + 600 stowed

Gun control equipment
Turret power control: powered/manual
Gun elevation/depression: +45°/–8°
Turret traverse: 360°
Gun stabiliser:
(vertical) no
(horizontal) no
NBC system: yes
Night vision equipment: yes
Airtransportability: Lockheed Martin C-130 Hercules; one role on/off C-17: 6

supply of 94 ASV 150 (4 × 4) vehicles to the US Army over a five year period.

First vehicles were delivered in the second quarter of the year 2000 and the contract also includes additional production options which, if exercised, could increase the total buy to 250 vehicles.

The US Air Force has an option on two ASV 150s worth US$1.5 million but by late 2001 the option had yet to be exercised.

The US Army ASV 150 vehicles that are used by the military police with standard equipment including protection against 14.5 mm small arms fire, independent suspension and the installation of a one-man turret armed with a 12.7 mm machine gun and a 40 mm automatic grenade launcher. A bank of four electrically operated smoke grenade launchers is mounted either side of the turret, firing forwards.

This turret is the latest version of that originally designed for installation on US Marine Corps United Defense, Ground Systems Division AAV7A1 amphibious assault vehicles. This turret is called the Up Gunned Weapon Station (UGWS) and is also referred to as the 40/50 turret.

Turret improvements over the standard UGWS are also claimed to provide significant benefits to the gunner and include superior survivability, reduced silhouette and greater compliance with human factors design. The weapons are aimed and fired using an M36E2 day/night sight.

The US Air Force version of the ASV 150 if fielded would be fitted with a high-energy laser for destroying unexploded munitions on runways and other parts of an airfield. This version is called the Mobile Ordnance Disrupter System (MODS).

The ASV 150 builds on the current production LAV-150ST (4 × 4) LAV with the S in the designation meaning stretched and the T turbocharged. The major change is the replacement of the old solid axles, semi-elliptical leaf springs and double-acting hydraulic shock-absorbers by a new fully independent coil spring suspension which gives an improved ride on road and cross-country.

The first production vehicles were completed on schedule in April 2000, with the first six M1117 ASVs being sent to the US Army's 18th MP Brigade in Germany for training in two phases, first vehicle familiarisation and operation and second tactical employment and firing.

Three vehicles were issued to the 615th MP Company with the other three to the 527th MP Company. Eventually each MP Company will have 12 ASVs.

Of the 13 vehicles delivered by early 2001, six were deployed to Kosovo with another six in Germany, with one vehicle still undergoing trials at the Cold Regions Test Center at Fort Greenley, Alaska.

The second batch of six vehicles were delivered in April 2001. According to Textron Marine and Land Systems, production is on schedule and one vehicle will be completed every 21 working days. The US Army also refers to the M1117 as the Guardian.

The M1117 ASV is replacing the current up-armoured HMMWV (High Mobility Multipurpose Wheeled Vehicle) and provides a significant increase in armour protection, firepower and mobility for the Military Police mission.

This includes law enforcement, area security, battlefield circulation and enemy prisoner of war operations over the entire continuum of war and operations other than war.

Textron Marine & Land Systems has also studied a stretched vehicle with a longer wheelbase and wider doors to be used as a medical evacuation vehicle or a command and control vehicle.

Description

The overall layout of the ASV 150 is similar to that of the earlier LAV-150ST described in detail in the following entry, with the commander and driver at the front, crew compartment and turret in the centre and power pack at the left rear.

The commander is seated on the right with the driver to the left and bulletproof windows are provided to their front and sides; above each of their positions is a single-piece hatch cover.

In either side of the hull is a two-part door with the lower part folding downwards to form a step and the upper part opening to the rear. The latter has a bulletproof vision block with a circular firing port below. In addition, there is a door in the right side at the rear with the lower part folding downwards to form a step and the upper part opening upwards.

The hull of the ASV 150 is of all-welded steel armour similar to the LAV-150ST but has been fitted with advanced ceramic armour to provide the occupants with a higher level of protection against heavy machine gun armour-piercing rounds, conventional artillery splinters, land mines and the effects of rocket-propelled grenades.

According to Textron Marine & Land Systems, the armour package fitted to the ASV 150 provides protection against 12.7 mm M2 armour piercing ammunition at a range of 250 m, 155 mm artillery projectile fragments overhead at 15 m, 5.44 kg of TNT under any wheel station and 1.81 kg of TNT at any location under the vehicle.

The advanced armour system consists of an IBD Modular EXpandable Armour System (MEXAS) that utilises a ceramic composite appliqué-mounted on the exterior of the welded steel hull of the ASV 150, with a spall liner on the interior surfaces.

Although the US Army ASV 150s are fitted with the one-man UGWS, the vehicle can be fitted with a wide range of other weapon systems to meet different operational requirements.

The turret is in the centre of the vehicle, with the engine compartment left side at the rear. Firing ports and vision devices allow some of the crew to fire their weapons from within the vehicle.

The 40 mm grenade launcher has a cyclic rate of fire of 325 to 375 rounds per minute while the 12.7 mm machine gun has a cyclic rate of fire of 450 to 600 rounds per minute. There is also provision for mounting a 7.62 mm M60 or 5.56 mm M249 machine gun externally.

The gunner is provided with an M36E2 day/night sight with ballistic graticule and seven unity periscopes to given vision through a full 360°.

Turret traverse is powered and manual through a full 360° while weapon elevation is from –8 to +45° under manual control.

The ASV 150 has a new independent axle coil spring suspension system which gives a significant increase in cross-country mobility as well as providing a better ride for the crew. This gives a jounce front/rear of 17.7/15.2 cm and rebound front/rear of 15.2 cm/12.7 cm.

Standard equipment includes a central tyre-pressure regulation system that allows the driver to adjust the tyre pressure to suit the type of ground being crossed,

Rear view of ASV 150 fitted with turret armed with one 12.7 mm M2 machine gun and one 40 mm Mk 19 grenade launcher
0005300

NBC system and measures to reduce noise and minimise the thermal signature of the vehicle. The central tyre pressure system can be adjusted for sand/snow, off-road and highway conditions.

Other equipment on US Army vehicles includes a fire detection and suppression system, passive AN/VVS 501 driver's night viewer and a front-mounted Braden winch with a capacity of 9,072 kg on a bare drum.

To reduce both procurement and operating costs, wherever possible proven and in-production subsystems have been used in the design of the ASV 150.

The ASV 150 can be transported in cargo aircraft, aboard ships or landing craft, or by road or rail and be ready to roll off in a mission-ready condition. With the appliqué armour kit it is fully transportable in a C-130 Hercules transport aircraft. The C-130 can carry one ASV 150, the C-141 two, the C-17 four and the C-5 eight.

Variants

None announced apart from the different versions for the US Army and Air Force.

Textron Marine and Land Systems has proposed that the baseline ASV 150 chassis could be used for a wide range of specialised variants including NBC reconnaissance.

Status

Production. In service with the United States Army.

Manufacturer

Textron Marine & Land Systems, Combat Vehicle Marketing.

UPDATED

Cadillac Gage LAV-150 armoured vehicle range

Development

In 1962 the then Cadillac Gage Company developed, as a private venture, a 4 × 4 light armoured vehicle which could be used for a wide range of battlefield missions such as armoured personnel carrier, weapons platform and reconnaissance vehicle.

The first prototype of this vehicle, called the V-100 Commando, was completed in 1963, with first production vehicles being completed the following year.

The V-100 was widely used in Vietnam for both convoy escort work and for patrolling airbases and other high-value targets. The V-100 was powered by a Chrysler 361 V-8 petrol engine which developed 200 hp and gave the vehicle a maximum road speed of 100 km/h. Gross vehicle weight was 7,370 kg and a maximum of 12 men including the crew could be carried.

The V-100 was supplemented in production by the larger V-200, which was powered by a Chrysler 440 CID petrol engine which developed 275 hp and gave the vehicle a maximum road speed of 96 km/h. Gross vehicle weight of the V-200 was 12,730 kg. The V-200 was sold only to Singapore and is no longer being produced. Versions of the V-200 used by Singapore include: anti-aircraft, with the Swedish SAAB Bofors Dynamics RBS 70 surface-to-air missile system; recovery, fitted with an A-frame at the front of the hull; fire support, fitted with a two-man turret armed with a 90 mm MECAR gun with 7.62 mm coaxial and 7.62 mm anti-aircraft machine guns; and APC, fitted with a two-man turret armed with a 20 mm Oerlikon Contraves cannon, 7.62 mm coaxial and 7.62 mm anti-aircraft machine guns. Cadillac Gage submitted their final proposal for the larger V-200 in April 1968, with the first prototype being completed and ready for trials in April 1969. First batch of V-200 vehicles were delivered to Singapore in July 1970.

The US Air Force still uses a number of original V-100 Commando vehicles at the Tactical Fighter Weapons Centre at Nellis Air Force Base in Nevada. Typical roles include simulating the SA-9 SAM system and command and control systems associated with Russian air defence systems.

There is a separate entry for the more recent LAV-300 (6 × 6) armoured vehicle range. This was developed specifically for the export market and is offered with a similar range of turrets as the latest LAV-150 ST family of vehicles.

There are two other vehicles on the market today which are very similar in appearance to the Commando, the US Dragoon series and the Portuguese Chaimite, which is known to have been sold to Lebanon, Libya, Peru, the Philippines and Portugal. Both these vehicles are described in this section.

In 1984, Kuwait placed an order for 20 V-150 S Commandos and 60 V-300 Commandos. It is believed that all of these were lost during the Iraqi invasion of Kuwait in August 1990.

In 1985, Cadillac Gage announced that production of the V-150 was to cease and that it would be replaced in production by the V-150 S.

In August 1988, US Army Tank Automotive Command placed an order with Cadillac Gage Textron for 14 V-150 S vehicles, at a cost of US$7.014 million, with another 10 ordered in December 1988 at a cost of US$3.113 million.

In 1992, Turkey ordered a total of 124 V-150 S 4 × 4 APCs with first deliveries being made late in 1992.

The first order was for 50 vehicles valued at US$14 million and fitted with turrets, and the second order for 74 vehicles valued at US$25.6 million, 20 of which were supplied in water cannon configuration, 10 are APCs with turrets and 44 standard APCs.

In 1994, Cadillac Gage Textron became part of Textron Marine & Land Systems and the V-150 series was renamed the LAV-150 family. In the following description and variants, the current LAV-150 designation is used.

Production of the LAV-150 was undertaken at the Cadillac Gage Textron facility in Coco, Florida, but production is now undertaken at the Textron Marine & Land Systems facility in New Orleans, Louisiana.

In mid-1995, it was announced that the US Army Tank-automotive and Armaments Command had awarded the company a contract worth US$2 million for the supply of five LAV-150 ST light armoured vehicles for an undisclosed country in Asia.

By late 2001, total production of the LAV-100/LAV-150/LAV-200 series and variants, amounted to over 3,000 vehicles, with production being undertaken as and when orders are placed.

With the type classification of the Cadillac Gage ASV 150 by the US Army in March 1999 and the placing of a five year production contract, the LAV-150 will now be phased out of production. Details of the ASV are given in an earlier entry in this section. This is designated M1117 by the US Army and is now in service with Military Police units.

Description

The all-welded steel hull of the LAV-150 armoured personnel carrier protects the crew from small arms fire up to 7.62 mm in calibre, overhead blast, shell splinters and Molotov cocktails.

The driver sits at the front of the hull on the left side and has two vision blocks in front of him and one on his left. The vehicle commander sits on the driver's right and has one vision block to his front and one on his right. Both crew members have a single-piece hatch cover over their position that opens to the outside. Early production vehicles had a firing port just behind the driver's and commander's side vision block.

The troop compartment is behind the commander's and driver's position, with three infantrymen seated down each side of the hull and one to the rear of the

LAV-150 fitted with a turret armed with a 20 mm Oerlikon Contraves cannon, a 7.62 mm coaxial machine gun, a 7.62 mm anti-aircraft machine gun and six smoke grenade dischargers fitted to the turret side

LAV-150 armed with a turret-mounted Belgian Cockerill Mk III 90 mm gun, 7.62 mm coaxial and 7.62 mm anti-aircraft machine guns and smoke grenade dischargers
0018750

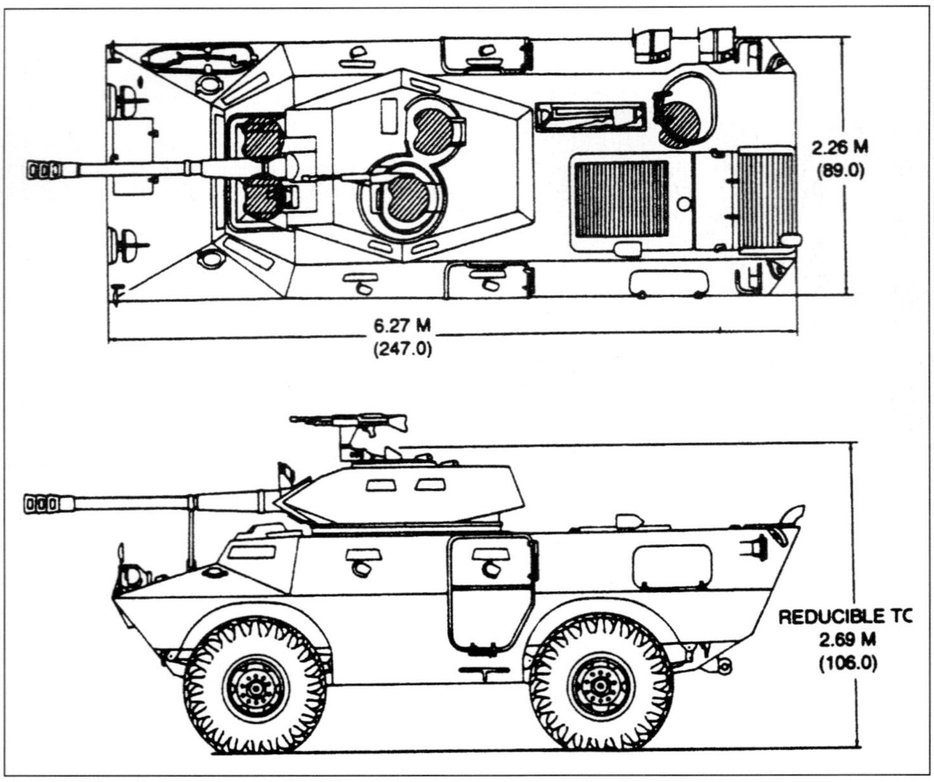

Top and side drawings of V-150 ST fitted with turret armed with 90 mm gun, 7.62 mm coaxial machine gun and roof-mounted 7.62 mm anti-aircraft machine gun 0005302

machine gunner's position. They enter and leave the vehicle by two doors in the side of the hull, one on each side. The lower part of the door folds downwards and the upper part, which contains the vision block and firing port, opens to the rear. In each side of the hull, forward of the side door, is a firing port with a vision block over the top. All vision blocks in the LAV-150 have crash pads and spall shields for additional protection.

In the centre of the troop compartment roof is a cupola which can be traversed manually through 360° and has a single-piece hatch cover that opens to the rear. On the forward part of the cupola, which has no observation devices, is a pintle-mounted 7.62 mm machine gun with 200 rounds of ready use ammunition and another 3,000 rounds in reserve in the hull.

There is also a door at the rear of the hull on the right side, with the lower part folding downwards and the upper part, which contains the firing port and vision block, opening upwards. Over the top of the passageway that connects the troop compartment with the rear door is a single-piece circular hatch that opens forwards. A further two seats are provided in this passageway.

The engine compartment is at the rear of the hull on the left side with access panels in the top and side of the hull. The engine cooling system is a conventional automotive water-cooled system utilising a fin and tubed radiator with an additional surge tank for reserve capacity. The engine fan and venturi are designed for maximum airflow efficiency and minimum horsepower draw. The cooling system is capable of operating in ambient temperatures of 130°F. The engine compartment has a fire extinguishing system that is operated manually by the driver.

Wherever possible, standard military automotive components are used in the LAV-150, such as the engine, transmission and axles. When originally introduced, the Commando LAV-150 was powered by a Chrysler 361 V-8 petrol engine developing 200 hp at 4,000 rpm coupled to a manual transmission with five forward and one reverse gears and a single-speed transfer case. Late production vehicles have a Cummins V-8 diesel engine that develops 202 hp at 3,300 rpm coupled to an Allison automatic transmission and a two-speed transfer case.

The Allison MD3560 automatic transmission fitted to the latest LAV-150 ST is electronically controlled and has a built in computer that tells the transmission when to shift. All gears are helical to ensure quiet operation and the transmission also has a built in diagnostic system.

The Allison MD 3560 transmission is push button controlled and a small box with the push buttons is mounted in the driver's compartment and connected to the transmission by an electrical cable.

The solid axles have silent locking differentials and the suspension consists of semi-elliptical springs and hydraulic double-acting shock-absorbers with rebound control at each wheel station. Jounce control is by hollow rubber springs at each wheel station. The run-flat tyres, which are capable of operation for a distance of 40 km, have self-cleaning treads. Mounted internally at the front of the hull is a winch with a capacity of 4,536 kg, which has 48.76 m of 9.5 mm diameter cable. All members of the LAV-150 family are fully amphibious, being propelled in the water by their wheels.

Standard equipment on all vehicles includes fuel and water cans, two hand-held fire extinguishers, a pioneer toolset, a tow cable, a breaker bar and lug wrench, a first aid kit, a vehicle toolkit (including an 8,000 kg hydraulic jack stowed inside the vehicle), a pamphlet bag, an oddment box, a slave cable, spare vision blocks and an air compressor with hose.

Optional equipment includes smoke or smoke/fragmentation launchers, a fixed pintle socket for the rear, a rear ring mount with a 7.62 mm pintle and cradle, a ring mount and cradle for a 0.50 calibre machine gun or a 40 mm automatic grenade launcher, a gun shield for a fixed pintle socket or a ring mount weapon, a spotlight for a fixed pintle socket or a ring mount weapon with or without a shield, night vision equipment, a periscope for the ring mount (standard on 20 mm and 90 mm turret ring mounts), a spare periscope and stowage, weapons (where applicable), extra spare barrel stowage, hand-held weapon stowage, M26 hand grenade stowage, smoke hand grenade stowage, binocular stowage, radios and associated mountings, a canopy with poles (pod-type vehicles only), vision blocks in pod (standard on police ERV), gun ports in pod (standard on police ERV), sight M28C ×5.6 power (machine gun turret vehicles only), a land navigation system and/or vehicle heading reference system, an NBC system, a public address system (standard on police ERV), a heater kit, an air conditioning system, a wiper kit for the driver, blackout covers, a handbrake warning light, a cooker kit and a lifting boom.

The final production V-150 ST has a larger range of optional equipment including a central tyre-pressure inflation system that allows the driver to adjust the tyre pressure to suit terrain being crossed and the option of a higher level of armour protection.

With standard suspension payload is 2,720 kg but with uprated suspension this can be increased to 4,530 kg. The V-150 ST has a front-mounted winch with a capacity of 5,454 kg and 46 m of cable.

Variants

MG turret

Basic vehicle fitted with a Cadillac Gage one-man turret which can mount twin 7.62 mm machine guns or a combination of 7.62 mm/0.50 calibre machine guns. Laying and control of the weapons are mechanical, with a hand-operated gearbox that allows 360° continuous traverse, and a gunner's handle at the rear of the balanced gun cradle which allows elevation from –13 to +55°. The turret has a single-piece hatch cover that opens to the rear, eight vision blocks and an M28C day sight with a magnification of ×1.5. Mounted coaxially with the weapon is a 500,000 candlepower spotlight. When fitted with twin 7.62 mm machine guns 3,800 rounds of ammunition are carried, 800 of which are in the turret for ready use, and 3,000 in the hull. This model has a crew of three and can carry seven infantrymen. Height to top of turret is 2.54 m and gross vehicle weight is 8,981 kg.

1 m MG turret

Basic vehicle fitted with a Cadillac Gage 1 m turret that can be armed with twin 7.62 mm, or twin 0.50 calibre machine guns or a combination of these weapons. The turret has a manual traverse through 360° and the guns have an elevation of +45° and a depression of –8°. The turret has a single-piece hatch cover that opens to the rear, an M28C day sight with a magnification of ×1.5, eight vision blocks and a 500,000 candlepower spotlight. Depending on the armament installed, 900 rounds of 0.50 calibre ammunition are carried, of which 200 are for ready use in the turret, or 2,400 rounds of 7.62 mm ammunition, 400 of which are for ready use in the turret. This model has a crew of three and can carry seven infantrymen. Height to the top of the turret is 2.59 m and gross vehicle weight is 8,754 kg.

V-200 fitted with a two-man Cadillac Gage turret armed with a 20 mm Oerlikon Contraves cannon 0018751

LAV-150 S used by Taiwan in the APC role and armed with a 12.7 mm MG (L J Lamb)

V-200 used by the Singapore Air Force and armed with SAAB Bofors Dynamics RBS 70 surface-to-air missile system (David Boey) 0018752

LAV-150 ST fitted with Cadillac Gage 40 mm/12.7 mm Up Gunned Weapons Station (UGWS)

1 m 20 mm turret

Basic vehicle with a one-man turret armed with a 20 mm Oerlikon Contraves 204 GK cannon with provision for a coaxial 7.62 mm machine gun. The main armament can be elevated from –8 to +60° at 40°/s and turret traversed through 360° at 60°/s. Traverse is electrohydraulic with manual back-up. The turret has eight vision blocks and a day sight with a magnification of ×8 and ×1, a 500,000 candlepower searchlight that moves in elevation with the armament and a blower for removing turret fumes.

There are 400 rounds of 20 mm ammunition carried, 200 in the turret ready to use and 200 in the hull, and 3,200 rounds of 7.62 mm ammunition, 220 for ready use and the remainder in the hull.

This model has a crew of three consisting of commander, gunner and driver and can carry five men; it has a loaded weight of 9,072 kg and a height of 2.895 m.

20 mm turret

Basic vehicle fitted with a two-man turret armed with a 20 mm Oerlikon Contraves 204 GK cannon, a 7.62 mm machine gun mounted coaxially to the left of the main armament and a 7.62 mm anti-aircraft machine gun mounted on the turret roof. The cannon has an internally controlled hydraulic charger, an emergency firing trigger for the commander and a rate controller which allows firing of 1, 2 or 4 rds/s or fully automatic. Turret vision equipment consists of four vision blocks (two in each side of the turret), a commander's periscope sight that can be traversed through a full 360°, a gunner's day sight with a magnification of ×8 and a day periscope with a projected graticule. In addition to the main sights, the turret has an external anti-aircraft sight. It is also provided with a 500,000 candlepower spotlight and a turret blower. Of 400 rounds of 20 mm ammunition carried, 200 are for ready use; of 3,200 rounds of 7.62 mm ammunition, 400 are in the turret for ready use and 200 in the anti-aircraft gun for ready use. This model has a crew of three and can carry two infantrymen. The gross vehicle weight is 9,888 kg and the height to the top of the turret is 2.54 m.

25 mm turret

This turret is armed with a 25 mm M242 The Boeing Company M242 Chain Gun, a 7.62 mm M240 coaxial MG and a similar weapon on the turret roof for anti-aircraft defence. The turret carries 230 25 mm ready use rounds (170 HE and 60 AP) with a further 400 in the vehicle. There are 400 rounds of 7.62 mm ready use ammunition in the turret, 200 at the anti-aircraft station and a further 1,000 rounds in the vehicle.

The turret has powered traverse (360° at 30°/s) and elevation (from –8 to +60° at 30°/s). Vision equipment includes an M36E1 day/night sight for the gunner and eight day periscopes for the commander. The commander can also have an M36E1 day/night sight but only six day periscopes are then fitted.

40 mm/12.7 mm turret

This is fitted with a one-man turret armed with a 40 mm Mk 19 Mod 1 grenade launcher on the left, a 12.7 mm M2 HB machine gun on the right and four smoke grenade dischargers mounted either side of the turret at the rear. The weapon can be elevated from –8 to +45° and turret traverse is a full 360°. In total 600 rounds of 40 mm and 700 rounds of 7.62 mm machine gun ammunition are carried.

LAV-150 S

This was announced in 1981 with two vehicles delivered for evaluation in the LAV competition late in the same year. Both vehicles had a two-man turret, one armed with a 90 mm Cockerill Mk III gun and the second with a 25 mm M242 The Boeing Company Chain Gun, both having a 7.62 mm coaxial and a 7.62 mm anti-aircraft machine gun. The LAV-150 S is 46 cm longer than the LAV-150, which gives additional internal space, increasing the payload by 726 kg to give a total payload of 3,357 kg. The wheelbase has been lengthened and the suspension strengthened, giving an improved ride across country.

Additional vehicle improvements include a new transfer case giving better grade-climbing performance, a new hydraulic boost brake system with an electrohydraulic back-up and an improved cooling system with increased fan efficiency and increased cooling capacity. Further development of the LAV-150 S has resulted in the LAV-150 ST (with the ST standing for stretched and turbocharged), which is the current production model.

Air defence vehicle

This is the basic vehicle fitted with the same turret as mounted on the M167 towed system armed with a 20 mm six-barrelled General Dynamics Armament Systems cannon and fitted with an optical sight. To provide a more stable firing platform, three outriggers are lowered to the ground before firing commences: one of these is located at the front of the vehicle and one either side of the hull towards the rear. This version was developed to meet the requirements of the Saudi Arabian National Guard. Full details of this version are given in *Jane's Land-Based Air Defence 2001-2002.* This system is no longer being marketed.

90 mm turret-mounted gun

This was originally armed with the Belgian MECAR gun but is now offered with the Belgian Cockerill Mk III gun. Mounted coaxially with the main armament is a 7.62 mm machine gun and a similar weapon is fitted on the turret roof for anti-aircraft defence. Turret traverse is a full 360° at 30°/s, and elevation is from –8 to +28° at 30°/s. Vision equipment consists of three vision blocks (two in the left and one in the right side of the turret), a commander's day periscope that can be traversed through 360°, a gunner's day periscope with a magnification of ×8 and a unity day periscope with a projected graticule. A 500,000 candlepower spotlight is mounted coaxially with the main armament. Of 42 rounds of 90 mm ammunition carried, eight are for ready use in the turret. Of 3,200 rounds of 7.62 mm machine gun ammunition, 400 are for ready use in the turret and 200 for the anti-aircraft machine gun. This model has a crew of three and can carry four infantrymen; gross vehicle weight is 9,888 kg.

81 mm mortar

This is armed with an 81 mm M29 turntable-mounted mortar that fires through the roof of the vehicle and can be traversed through a full 360°. Elevation is from +43 to +85°. Its minimum range is 150 m and maximum range 4,400 m. When not in action the roof is covered by concertina doors that open either side of the vehicle. Provision is made for mounting a 7.62 mm machine gun at any one of four positions around the top of the hull and 2,000 rounds of ammunition are carried for it. In total, 62 81 mm mortar bombs are carried. If required, the mortar can be removed from the vehicle and fired from the ground. This model has a crew of five and a gross vehicle weight of 8,845 kg.

TOW ATGW system

This is almost identical to the above model but is fitted with a Raytheon Missile Systems TOW ATGW system

V-100 (4 × 4) APC used by Guatemala and fitted with a one-man turret armed with 12.7 mm and 7.62 mm machine guns (via Julio Montes)

Specifications

(LAV-150 with a turret-mounted 20 mm cannon. Specifications in square brackets relate to the LAV-150 ST)
Crew: 3 + 2
Configuration: 4 × 4
Combat weight: 9,888 [10,909] kg
Power-to-weight ratio: 20.42 [22.91 bhp/t]
Length: 5.689 [6.274 m]
Width: 2.26 [2.39 m]
Height:
(to turret roof) 2.54 [2.692 m]
(to hull top) 1.981 m
Ground clearance:
(axles) 0.381 m
(hull) 0.648 m
Track:
(front) 1.914 [1.98 m]
(rear) 1.943 [1.98 m]
Wheelbase: 2.667 [3.12 m]
Max speed:
(road) 88 [112] km/h
(water) 5 [5] km/h
Acceleration:
(0-32.2 km/h) 10 s
Fuel capacity: 303 litres
Max range: 643 [800] km
Fording: amphibious
Gradient: 60%
Side slope: 30%
Vertical obstacle: 0.609 m
Turning radius: 8.382 [8.5 m]
Engine: V-504 V–8 diesel developing 202 bhp at 3,300 rpm [Cummins 6 CTA 8.3 turbocharged, after-cooled, 6-cylinder in-line diesel developing 250 hp at 2,500 rpm]
Transmission: automatic, 4 forward and 1 reverse gears [Allison MD 3560 automatic, 6 speed]
Transfer case: 2 speed (with spline engagement for front axle drive) [single speed with spline engagement clutch for front axle drive]
Steering: power-assisted
Suspension: solid axles with semi-elliptical springs and double-acting hydraulic shock-absorbers
Tyres: 14.50 × 21 run-flat [14 × R20]
Brakes:
(main) hydraulic, dual circuit
(parking) drum acting on output shaft of transmission
Electrical system: 24 V
Batteries: 2 × 12 V, 100 Ah
Armament:
(main) 1 × 20 mm cannon
(coaxial) 1 × 7.62 mm MG
(anti-aircraft) 1 × 7.62 mm MG
Smoke-laying equipment: optional
Ammunition:
(main) 400
(7.62 mm) 3,200

Gun control equipment
Turret power control: electrohydraulic/manual
(by commander) yes
(by gunner) yes
Max rate power traverse: 60°/s
Max rate power elevation: 40°/s
Gun elevation/depression: +60/–8°
Turret traverse: 360°
Gun stabiliser:
(horizontal) no
(vertical) no
NBC system: optional
Night vision equipment: optional
Airtransportability: Lockheed Martin C-130 Hercules (2 vehicles)

Note: In 1982, Cadillac Gage announced that it had developed a Cummins V-6 diesel conversion kit as a direct replacement for the V-8 diesel installed in the LAV-150. When fitted with this engine the LAV-150 has an operational range of 800 km, compared to the standard V-8 model with 644 km, with an acceleration of 0 to 32 km/h in 9 seconds. This kit can be installed with no special tools or other equipment at base workshop level.

with a maximum range of 3,750 m, which has an elevation of +30°, depression of –10° and 360° traverse. There are seven TOW missiles carried. When not in action the top of the vehicle is covered by hatches that open to the front and rear. Provision is made for mounting a 7.62 mm machine gun at any one of four positions around the top of the hull and 2,000 rounds of ammunition are carried for this weapon. This model has a crew of four and can also carry two infantrymen. Gross vehicle weight is 8,958 kg.

Command vehicle

This is the standard vehicle fitted with a fixed armoured pod mounted at the roof opening. In each of the four sides of the pod is a firing port and in the roof is a single-piece hatch cover that opens to the rear, forward of which is a 7.62 mm pintle-mounted machine gun. Of 2,000 rounds of 7.62 mm ammunition carried, 200 are for ready use. Internally the vehicle has a mapboard, a table and additional communications equipment. In addition to the crew of three, seven staff members are carried. Overall height without the machine gun is 2.311 m.

Ambulance

This version would be unarmed and have the raised roof. In addition to the two-man crew it would also carry two stretcher patients and four seated patients. Optional equipment includes a public address system, siren and flashing light system.

Base security vehicle

This was originally developed for the US Air Force and, like the mortar carrier, has concertina hatch covers that open either side of the vehicle. Provision is made for mounting 7.62 mm machine guns or 40 mm grenade launchers in four positions at the front, sides and rear. A total of 3,000 rounds of 7.62 mm ammunition is carried. This model has a crew of three and can carry eight fully equipped infantrymen; it weighs 9,072 kg and is 2.311 m high.

Malaysian upgraded LAV-150

In 1987, a 90 mm LAV-150 of the Malaysian Army was modernised by MMC Engineering Services Sdn Bhd and subsequently evaluated by the Malaysian Army over 10,000 km, during which trials were carried out for road endurance and cross-country, vertical obstacle and amphibious capabilities. The original petrol engine has been replaced by a more fuel-efficient diesel engine while the manual transmission has been replaced by a fully automatic transmission. This upgrade has not yet entered production for the Malaysian Army. Details of these upgrades are given in *Jane's Armour and Artillery Upgrades 2001-2002.*

Police Emergency Rescue Vehicle (ERV)

This model has a fixed pod with six vision blocks and eight vertical gun ports. In the roof is a single-piece hatch cover that opens to the rear, in front of which is a 7.62 mm or 12.7 mm machine gun. This model has a crew of three, can carry nine additional men and weighs 9,162 kg. It is used by a number of police authorities and a quantity has been purchased by Turkey.

Status

Production as required. In service with the following countries:

Country	Quantity	Comment
Bolivia	10	V-100
Botswana	36	LAV-150, most 90 mm but also some TOW
Cameroon	43	includes 8 20 mm, 14 90 mm and 21 APC
Chad	9	LAV-150 S, some 90 mm
Dominican Republic	8	LAV-150
Gabon	9	LAV-150
Guatemala	7	V-100
Haiti	6	LAV-150, believed to be non-operational
Indonesia	200	LAV-150
Jamaica	13	LAV-150
Malaysia	184	V-100 and LAV-150, 20 mm and 90 mm
Mexico	28	26 LAV-150 ST water cannon and 2 LAV-150 ST recovery, delivered 1994
Philippines	165	LAV-150 and LAV-150 S, including 10 in 1994
Qatar	8	LAV-150
Saudi Arabia	1,100	LAV-150, national guard, includes ambulance, APC, ARV, 20 mm anti-aircraft, 81 mm mortar, 90 mm gun and TOW ATGW. With the introduction of the LAV-25 (8 × 8) series these are rapidly being withdrawn from front-line service
Singapore	280	V-100, LAV-150 and V-200, includes 20 mm, 90 mm and ARV. V-200 used by the Singapore Air Force and armed with SAAB Bofors Dynamics RBS 70 SAM
Somalia	n/avail	LAV-150, non-operational
Sudan	100	V-100, LAV-150, LAV-150 S, non-operational
Taiwan	300	LAV-150 S, includes APC mortar and TOW anti-tank
Thailand	162	LAV-150, including 12 in 1995
Turkey	124+	LAV-150 S, delivered 1992 to 1994
USA	n/avail	V-100, training
Venezuela	130	V-100 (30), LAV-150 (100)

Note: The current production model is designated the LAV-150 ST. The original and first production model was the V-100 which had the US Army designation of the M706. The vehicle is no longer used by Ethiopia, Oman, Panama, Tunisia or Vietnam.

Recovery vehicle

This is fitted with a heavy-duty winch with a maximum capacity of 11,348 kg which is provided with 60.9 m of 19 mm diameter cable. The boom is pivoted at the front of the vehicle and, when required, hinges forward to the front of the vehicle. When it is being used, two stabilisers are lowered at the front. The boom has a maximum capacity of 4,536 kg with the stabilisers in position. Armament consists of a 7.62 mm or a 12.7 mm machine gun with 2,200 rounds of ammunition.

Taiwan

This country operates a fleet of 300 LAV-150 S (4 × 4) vehicles including some used in the armoured personnel carrier role, with a 12.7 mm machine gun mounted on the forward part of the roof to the rear of the commander's and driver's position, and another in the mortar role. This has a roof hatch arrangement that opens left and right with the 4.2 in (107 mm) mortar firing to the front. A 7.62 mm machine gun is fitted for local protection.

LAV-150 ST

The LAV-150 ST is a stretched version with a turbocharged diesel which can also be fitted with a larger side hatch that is in two parts. The upper part opens upwards while the lower part folds down to form a step.

Manufacturer

Textron Marine & Land Systems, Combat Vehicle Marketing.

UPDATED

AV Technology Pandur armoured personnel carrier

Development/Description

Late in 1996, the Kuwait National Guard awarded AV Technology International (owned by Steyr-Daimler-Puch and AV Technology Limited Liability Company) of the US, a contract worth US$170 million for the supply of 70 Pandur (6 × 6) vehicles in six versions.

These were manufactured in the USA by AV Technology with deliveries taking place over a two year period. In addition, the contract includes two options which could bring the total number of vehicles up to 200 units, which would be sufficient for three battalions. As of late 2001, the two options for follow on orders for the Kuwait National Guard had not been exercised.

The first Pandur vehicle for Kuwait was completed four months ahead of schedule, early in November 1997, and was used to train Kuwait National Guard crews at Aberdeen Proving Ground.

The vehicles were built in the USA with such items as power pack and drivelines being supplied by Steyr-Daimler-Puch of Austria.

The Kuwait National Guard took delivery of six versions:

(1) Armoured personnel carrier fitted with AV Technology two-man Multi Turret Gun System (MTGS) armed with The Boeing Company 25 mm M242 Chain Gun and 7.62 mm coaxial machine gun and two banks of four Galix electrically operated smoke grenade launchers

The turret is also fitted with a General Dynamics Land Systems all-electric digital gun control and stabilisation system, allowing the weapons to be laid onto the target while the vehicle is moving

To allow targets to be engaged under day and night conditions, the gunner has the Kollsman Day/Night Range Sight (DNRS) which incorporates day/thermal channels and a laser range-finder which provides target information to the computer

The commander has an M36 day/image intensification sight and a monitor screen that allows him to see the gunner's thermal view of the target

This version has a crew of three and can carry four troops in the rear, two each side facing outwards. A total of four firing ports and associated vision devices are provided, one in each side of the hull and one in each of the two rear doors

(2) Assault gun vehicle fitted with two-man Cockerill turret armed with a Cockerill 90 mm Kenerga gun and 7.62 mm machine gun. This also has the Galix grenade system installed

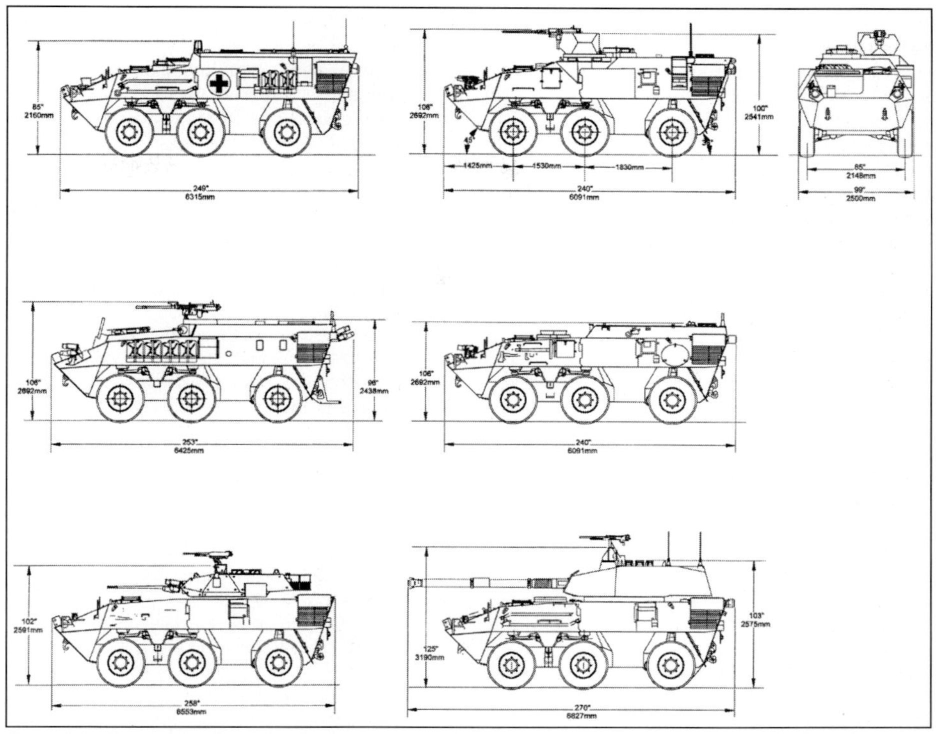

Part of the Pandur family of 6 × 6 armoured vehicles marketed by AV Technology, a General Dynamics Company **2001**/0104427

Pandur (6 × 6) for the Kuwait National Guard fitted with a BAE Systems, RO Defence, 81 mm mortar and with roof hatches open **2001**/0104428

Armoured ground mobility system based on the Austrian Pandur (6 × 6) armoured personnel carrier, armed with 12.7 mm M2 MG **2001**/0048089

Pandur (6 × 6) for the Kuwait National Guard with AV Technology two-person turret armed with 25 mm M242 cannon and 7.62 mm coaxial MG **2000**/0075728

Pandur (6 × 6) for the Kuwait National Guard fitted with CMI two-person turret armed with 90 mm gun and 7.62 mm coaxial MG **2000**/0075729

(3) Ambulance
(4) Command post
(5) Mortar carrier with BAE Systems, RO Defence 81 mm mortar
(6) Repair and recovery.

A detailed description of the Pandur is given earlier in this section under Austria. The vehicles built by AV Technology LLC for the Kuwait National Guard have some differences from those built for Austria.

For the Kuwait National Guard, the hull of the Pandur has been extended by 500 mm with the wheelbase of the 2nd/3rd axle extended by 300 mm to provide more internal space and improved vehicle stability.

All versions of the Pandur are protected against 7.62 mm armour-piercing attack through a full 360° with added protection over the frontal arc of the hull and turret.

All of the Kuwaiti vehicles feature a full air conditioning system; dual-mode (overpressure and face mask) nuclear, biological, chemical warfare defensive system; central tyre-pressure regulation system to allow adjustment to suit terrain; run-flat tyre inserts and a rear-mounted Rotzler hydraulic winch that can be used to the front or rear of the vehicle.

Communication equipment fitted includes UHF and VHF radios and 25 of the 70 vehicles use the global positioning system.

In addition to the six variants built for the Kuwait National Guard, AV Technology is also proposing a number of other variants including an anti-tank missile vehicle.

Late in 1997, a Pandur fitted with an AV Technology one-man turret with four Hellfire Anti-Tank Guided Missiles (ATGM) underwent a series of trials and demonstrations in Kuwait.

During these trials, conducted under day and night conditions, six Hellfire ATGM were launched at target ranges of between 500 and 8,000 m, with all missiles hitting the target.

Hellfire is a Semi-Active Laser (SAL) guided ATGM and can be used in the Lock-On-Before-Launch (LOBL) or Lock-On-After-Launch (LOAL) modes, both of which were demonstrated in Kuwait. Although normally launched from helicopters such as the AH-64, the Hellfire has been demonstrated installed on a number of tracked and wheeled platforms.

The one-man turret used on the Pandur has four Hellfire ATGM in the ready to launch position, two either side, with a standard 12.7 mm M2 machine gun being fitted for close protection.

The all-electric turret drive system was developed by General Dynamics Land Systems who also build the system for the United Defense LP Bradley infantry fighting vehicle.

Armored ground mobility system

In March 1999, following an international competition, AV Technology was awarded a contract by the US Army Tank Automotive and Armaments Command (TACOM) for the Armored Ground Mobility System (AGMS).

This is a Firm Fixed Price Indefinite Quantities contract and covers the supply of up to 50 vehicles, including the first batch for the Production Qualification Test (PQT), for delivery over a five-year period, with a value of US$51 million.

The AGMS is based on the Austrian Steyr-Daimler-Puch Spezialfahrzeug AG & Co KG Pandur (6 × 6) Armoured Personnel Carrier (APC) which has been manufactured by the company for the Kuwait National Guard

Production of these AGMS vehicles is undertaken at the new AV Technology facility at Shelby Township, Michigan, with the assistance of General Dynamics Land Systems and Steyr-Daimler-Puch Spezialfahrzeug AG & Co KG. First vehicles were delivered to the US Army in March 2000 with all vehicles expected to be completed by 2003.

The Austrian company supply key subsystems, for example complete power pack and drive line, for integration into the hull, which is built in the United States.

The AGMS will have a number of additional features including an advanced passive applique armour system, roof-mounted 12.7 mm M2 machine gun, automatic drive management system and a central tyre pressure regulation system that allows the driver to adjust the tyre pressure to suit the terrain being crossed.

Status

Production. In service with Kuwait National Guard (70 units) and the US Army (50 units).

Manufacturer

AV Technology LLC.

UPDATED

AV Technology Dragoon family of armoured vehicles

Development

This expanded family of wheeled armoured vehicles has been developed as a result of field experience gained with the earlier Dragoon 300 4 × 4 carrier design. The Dragoon 300 was originally developed by Verne Corporation and marketed by Arrowpointe Corporation.

In 1984, the two companies merged to form the privately held AV Technology Corporation, a company then owned by the two founders of Verne and Arrowpointe Corporations. Details of the original Dragoon 300 armoured vehicle family were given in *Jane's Armour and Artillery 1988-89,* pages 418-420.

The latest design of Dragoon has been purchased by the US military and several foreign governments and is available in 10 different variants. One- and two-man turrets are available that have been easily adapted to a variety of other armoured vehicles.

The current Dragoon, like its predecessors, makes extensive use of proven US military, NATO and world class commercial components. There is more than 70 per cent logistics commonality with in-service tactical vehicle fleets such as the M113A2 series full-tracked APC and the M809 and M939 series of 6 × 6 5-ton (US) tactical trucks used by many countries.

The first production vehicles were completed for the US Army and Navy in 1983. Since then, Venezuela, Thailand and the US Navy have placed orders for Dragoon armoured vehicles, which have subsequently been delivered.

In 1988, AV Technology Corporation introduced the Dragoon Armored Security Vehicles (ASVs) and the Patroller (ASV-XC) to police departments and other law enforcement agencies around the world. AV Technology Corporation has already delivered the standard ASV to some police departments, and the newly designed Patroller ASV-XC (extended cab) to the New York and New Jersey Port Authority, for use by security police at John F Kennedy Airport. Similar vehicles have also been sold to the Macomb County Sheriff Department, Houston Police Department, Detroit Police Department and Sureté de Quebec.

Late in 1997, AV Technology completed, ahead of schedule, the first of 70 Pandur vehicles for the Kuwait National Guard. Additional details of the US-built Pandur vehicles are given in the previous entry.

Late in 1998, AV Technology, now owned by General Dynamics Land Systems, prime contractor for the M1 Abrams Main Battle Tank (MBT), secured loan guarantees for the 140 Dragoon (4 × 4) armoured personnel carriers (APCs) that it is building for the Turkish National Police.

The company had sought a US$38 million loan guarantee from the US Export-Import Bank for the deal. The Export-Import Bank then asked the US State Department to review the deal to determine as to whether it violated the law because the Turkish National Police Units have been accused of torture and various other human rights abuses.

In the end the US State Department decided that the US loan guarantee could be used to finance 101 Dragoon vehicles going to the police in 32 provinces but not the remaining 39 vehicles for 11 provinces where there have been credible reports of abuse. General Dynamics Land Systems has used its funds to guarantee the remaining 39 vehicles.

Description

The hull of the current Dragoon is constructed of high-hardness steel ballistic plate, which meets MIL-A-12560 and is of monocoque construction with welded seams. The standard ballistic resistance of the Dragoon is 7.62 mm NATO ball (147 g M80 2,750 ft/s) with a 0.95 probability of no penetration at 0 m and 0° obliquity on all surfaces. Additional armour protection is available that upgrades the protection to 7.62 mm NATO AP (147 g M80 2,800 ft/s) with 0.95 probability of no penetration on the front, rear, sides and top of the vehicle.

The hull has an angular surface on the front, right and left sides which projects outward from the top and bottom, meeting at the mid-section around the perimeter. The rear surface of the hull projects outward from the bottom at an angle of 25° vertical. The basic design of the hull provides for easy changes of configuration. The Dragoon is coated with a two-part epoxy primer and can be top coated with CARC. The interior of the vehicle is finished with a heavy-duty epoxy coating.

The two double-width side doors are 1.066 m wide and 0.889 m high. The rear door is 0.609 m wide and 0.916 m high. There are three hatches on the top of the vehicle. All doors and hatches allow for swift entry and exit while providing security and watertightness. Power up and power down adjustment of the driver and

Dragoon fitted with AV-30 turret armed with 30 mm cannon and 7.62 mm machine gun and with hull fitted with appliqué armour (Christopher F Foss)
0018755

Dragoon Light Forces Vehicle with a 90 mm two-man turret

Dragoon Patroller showing front-mounted obstacle blade in extended position at front of hull (Christopher F Foss) 0018756

AV Technology Dragoon (4 × 4) with raised roof and appliqué armour (Scott R Gourley) ***2002***/0073368

co-driver seats allows the vehicle to be driven with the hatches open. When the seats are lowered, visibility is through the vision blocks. The driver has four vision blocks, which provide him with a 180° field of view, while the commander has two vision blocks, giving him a view of the front and right side of the vehicle. An optional weather cover kit with a windshield and wiper is available for the driver and co-driver, to be used with the hatches open in adverse weather or temperatures. The vehicle is equipped with two outside rear-view mirrors. The standard vehicle comes fitted with 12 ballistic vision blocks which are located for easy viewing from all sides of the vehicle by seated personnel.

There is seating for 12 occupants including the driver and all seats are equipped with seat belts. Other optional seating arrangements are available to meet specific requirements. There are five gun ports, two each side and one in the rear that will allow the use of hand-held weapons. All have ballistic covers that can be positively secured from the inside of the vehicle.

Radio and internal communications can be easily mounted in the hull. Mounting pads and power for the more common AN/VRC-12 family of radios are standard in the Dragoon, but other radio communication systems have been installed and special configurations can be tailored to meet user specifications.

The Dragoon comes with a 50,000 BTU heating system and a 30,000 BTU air conditioning system as standard. There are seven adjustable direct-flow air ducts mounted for easy access and for even distribution throughout the crew compartment. There is an additional heater in the front area of the vehicle that provides an extra level of comfort.

The 24 V water-resistant electrical system is equipped with a master power switch, two 100 Ah heavy-duty batteries and a 200 A heavy-duty radio-suppressed military-type alternator. The Dragoon is equipped with blackout headlights with high and low beam selection, stop and tail lamps, instrument lights, interior lights, a horn and turn signals, all mounted in protected areas. An optional one million candlepower spotlight is available and it can be mounted to the top hatch ring mount for 360° rotation.

A fixed Halon fire extinguisher system is installed with direct discharge into the engine compartment, and there are two hand-held Halon fire extinguishers installed in the crew compartment.

The power pack system allows for the fast and easy removal of the complete engine, transmission, hydraulic, cooling and electrical system. This is removed as a single unit in less than 30 minutes without any loss of operating fluids or Freon from the air conditioning system. Power packs can be quickly exchanged between vehicles without any modifications.

The Dragoon has a patented cooling system including a thermostatically controlled two-speed hydraulically driven fan. This allows all vehicles to operate in a temperature range of –25 to +125°F. All routine operator fluid checks are easily accessed at the engine hood opening on the top of the vehicle. Engine oil and hydraulic filters are the spin-on type for quick service and preventative maintenance.

The Dragoon is equipped with two 170 litre armoured fuel tanks and includes an emergency fuel pump prime system. This feature is used in the event of the main fuel pump failing or to prime the engine if it should run out of fuel.

The patented Dragoon system driveline provides the most durable and rugged system with the least number of components. The Allison Transmission MT-653 heavy-duty five-speed forward and single-speed reverse transmission has a torque converter that locks up in the fourth and fifth gears providing better fuel economy. The single-speed transfer case has an easy quick-disconnect four-bolt universal joint to the drive shaft. This allows for rapid removal of the power pack. The Dragoon's single-speed in/out transfer box allows for selection of two- or four-wheel drive at any safe speed without stopping. The hydraulic system has an engine-mounted three-section hydraulic pump that supplies hydraulic fluid to the winch, power steering, cooling fan, brakes, power seat and other optional accessory features such as the 10 kW generator.

Front and rear axles are standard off-the-shelf heavy-duty Rockwell military type (the same axles used on the 6 × 6 M939). The axles may be equipped with the optional Central Tire Inflation System (CTIS) which increases flotation and traction by the driver selecting and automatically adjusting tyre pressure for different terrains. All Dragoons are equipped with power brakes and a disc-type parking brake.

Power steering is equipped with a manual override in the event of power loss. The front and rear leaf springs have a capacity of 12,700 kg and 304 mm of suspension travel. The heavy-duty shock-absorbers are interchangeable. Military standard all-terrain tyres may be equipped with a run-flat device that permits continuous operation.

The Dragoon is fully amphibious, propelled by all wheels. Turning is accomplished by steering the front wheels. There are three electric bilge pumps with a capacity of 190 litres/min. All Dragoons come equipped with a front-mounted hydraulic self-recovery winch that has 53.34 m of cable.

Variants

New turrets for Dragoon

In addition to the variants listed below, AV Technology has recently developed four new turrets which can be installed not only on the Dragoon family of 4 × 4 light armoured vehicles, but also on a wide range of other vehicles, tracked and wheeled. These turrets are:

(1) Two-man 90 mm turret armed with 90 mm gun, 7.62 or 12.7 mm coaxial machine gun and optional 7.62 mm anti-aircraft machine gun. This turret was designed and built by AV Technology
(2) Two-man MultiGun Turret System (MGTS) which can be armed with a 35, 30 or 25 mm cannon and a 7.62 mm coaxial machine gun. This turret was designed and built by AV Technology) The two-person turret was installed on some of the AV technology Pandur (6 × 6) armoured personnel carriers, recently supplied to the Kuwait National Guard.
(3) 12.7 mm/40 mm Up Gunned Weapons Station (UGWS) fitted with a 40 mm Mk 19 Mod 3 automatic grenade launcher and 12.7 mm M2 HB coaxial machine gun
(4) One-man MultiGun Turret System which can be armed with various combinations of weapons, for example one 7.62 mm and one 12.7 mm machine gun
(5) AV-30 turret which is armed with The Boeing Company 30 mm M230 cannon (as installed on the Apache attack helicopter) and a 7.62 mm coaxial machine gun. All-electric gun control and stabilisation system is fitted, as is a day/thermal sighting and fire-control system.

Armoured personnel carrier

The Dragoon armoured personnel carrier (also referred to as an armoured security vehicle) is the base vehicle configuration and is a fully equipped vehicle capable of carrying 12 occupants with equipment. It is outfitted for use as a reconnaissance, security and escort vehicle as well as a troop carrier. Each vehicle is equipped with a ring- or pintle-mounted machine gun up to 12.7 mm in calibre and a variety of communications systems. This variant has a payload capacity of 3,402 kg and a gross vehicle weight of 13,068 kg with a top speed of 116 km/h.

Dragoon Patroller armoured security vehicle (ASV-XC Extended Cab)

The Patroller has all the performance features of the standard Dragoon APC, but the design allows for two 406 × 660 mm windshields and angular 304 × 685 mm side windows that have 'no-spall' 7.62 mm NATO AP protection. The Patroller is equipped with windshield wiper and washers. Extra large vision blocks and gun ports are also installed. The ASV-XC has 1.995 m of interior headroom allowing the occupants sufficient area to stand upright in the vehicle if necessary for surveillance activities. This variant can also be equipped for amphibious operations. The Patroller will accommodate a crew of 12 with weapons, gear and supplies.

Dragoon Armoured Command Vehicle (ACV)

The Dragoon command control and communications vehicle is another variant of the APC which can be configured to the user's preferred interior layout and will accommodate a wide variety of specialised equipment to satisfy the needs of field commanders, tactical communication and electronic warfare systems. A 360° rotating machine gun ring mount with a removable ballistic shield, a work table, a 3 kW generator, a camouflage net, a communication harness and extra brackets for equipment and radios are all standard equipment. The base model will accommodate a crew of five with gear, weapons, ammunition and rations.

Dragoon Armoured Mortar Carrier (AMC)

The Dragoon Armoured Mortar variant mounts an 81 mm mortar on a 360° rotating turntable. The mortar may be traversed and locked at a number of azimuth positions. Fine adjustment in azimuth and elevation is provided as part of the mortar-mounting system. Full width armoured bi-fold doors are provided over the weapons/crew compartment. The mortar may be fired from within the vehicle. The crew is protected from small arms fire and artillery fragments by the vehicle hull. The 60 rounds of various types of mortar ammunition may be stowed in the weapons compartment and are easily accessible to the loader. Additional stowage is provided for the equipment necessary for firing off-vehicle when required. The Dragoon AMC variant will accommodate a crew of five with standard field gear, weapons, ammunition and rations.

Dragoon Light Forces Vehicle 90 mm and two-man turret (LFV-90 mm) (LFV-2)

The Dragoon Light Forces Vehicle comes equipped with the two-man turret, a 90 mm cannon and a coaxial 7.62 or 12.7 mm (0.50) calibre MG. The turret shell is constructed from the same high-hardness steel armour plate as the vehicle hull and provides the same level of ballistic protection. The fire-control power system is

Specifications
(Fitted with 90 mm turret)
Crew: 1 + 4
Configuration: 4 × 4
Power-to-weight ratio: 23.62 bhp/t
Combat weight: 12,700 kg
Unloaded weight: 11,204 kg
Load: 1,497 kg
Towed load: 4,536 kg
Length: 5.89 m
Width: 2.49 m
Height:
(overall) 2.819 m
(hull top) 2.082 m
Ground clearance:
(axles) 0.381 m
(hull) 0.658 m
Track: 1.981 m
Wheelbase: 3.10 m
Angle of approach/departure: 60/45°
Max road speed: 105 km/h
Max water speed:
(wheels) 5.5 km/h
(propeller kit) 8.8 km/h
Fuel capacity: 350 litres
Range cruising: 869 km
Fording: amphibious
Gradient: 60%
Side slope: 30%
Vertical obstacle: 0.609 m
Turning radius: 8.77 m
Engine: Detroit Diesel 6V-53T 6-cylinder, liquid-cooled, turbocharged diesel developing 300 bhp at 1,800 rpm
Transmission: Allison Transmission MT-653 DR automatic, 5 forward and 1 reverse gears
Transfer case: single-speed, gearless
Steering: hydraulic, power-assisted, variable ratio
Suspension: semi-elliptical springs and hydraulic shock-absorbers
Tyres: 14.50 × 21, 12-ply radial
Brakes:
(main) hydraulic, power-boosted
(parking) dual grip
Electrical system: 24 V
Batteries: 2 × 12 V, 100 Ah
Armament:
(main) 1 × 90 mm gun
(coaxial) 1 × 7.62 mm MG
(anti-aircraft) 1 × 7.62 mm MG (optional)
Ammunition:
(90 mm) 46
(7.62 mm) 2,600

Gun control equipment
Turret power control: electrohydraulic/manual
(by commander) yes
(by gunner) yes
Gun elevation/depression: +45/–8°
Turret traverse: 360°
Gun stabiliser:
(vertical) no
(horizontal) no
NBC system: optional
Night vision equipment: optional
Airtransportability: Lockheed Martin C-130 Hercules (1)

electrohydraulic and is operated by a single handgrip controller in azimuth and elevation. Secondary or back-up manual controls are provided for both azimuth and elevation. The fire direction system is the standard M36E1 passive day/night sight for the gunner (thermal imaging and other sight options are available). Seven M27 day periscopes are provided for the commander for 360° visibility. Space is available in the turret bustle for mounting a variety of radio communications systems. Stowage is provided in the turret for 500 ready rounds of 7.62 mm ammunition and 10 ready rounds of 90 mm cannon ammunition. Additional stowage is provided for up to 50 rounds of 90 mm and 2,000 rounds of 7.62 mm ammunition in the vehicle.

The turret weighs less than 1,905 kg at full load and has solid-state electronics and auto-deck interrupt features. NBC features and commander's control and override are optional. The LFV-90 mm will accommodate a crew of four with standard field gear, weapons, ammunition and rations.

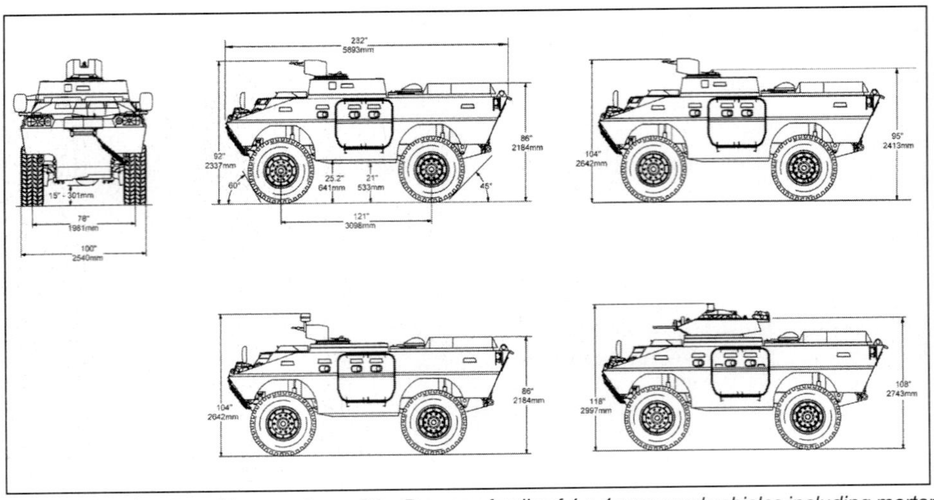

General side drawing of four versions of the Dragoon family of 4 × 4 armoured vehicles including mortar carrier, command vehicle, armoured personnel carrier and reconnaissance vehicle **2001**/0104429

Dragoon Light Forces Vehicle 40/.50 one-man turret (LFV-40 mm) (LFV-1)

The Dragoon LFV-40 mm (one-man turret) has mounts for a Mk 19 Mod 3 40 mm grenade machine gun and a coaxially mounted M2 HB 12.7 mm (0.50) calibre MG. The Mk 19 Mod 3, firing the M430 High Explosive Dual-Purpose (HEDP) round, can penetrate up to 50 mm of conventional steel armour plate at over 2,000 m. Like the two-man weapon system, this turret is equipped with the standard M36E1 sight. Stowage is provided in the turret for 100 rounds of 40 mm and 200 rounds of 12.7 mm ammunition, plus eight ready smoke grenades. Additional ammunition stowage is provided in the vehicle. The LFV-40 mm vehicle will accommodate a crew of eight with standard field gear, weapons, ammunition and rations.

Dragoon Armored Maintenance Vehicle (AMV)

The Dragoon Armored Maintenance Vehicle and Engineer variant is equipped to provide field recovery, maintenance and logistics support for all Dragoons and many other tactical vehicles. The 2,268 kg hydraulic boom crane is capable of performing a wide variety of maintenance and engineering tasks including installation and removal of vehicle power packs and turrets, tyre changing, cargo loading and unloading and general construction.

Bi-fold top doors are provided for cargo loading and equipment access. Auxiliary power is provided for operating both air hydraulic accessories and tools. The vehicle comes standard with hydraulic levelling jacks, a fuel transfer pump, a portable pump, a towbar, a toolset, an air compressor, a power cable, a snatch block and a spare tyre mount. The interior is equipped with built-in tool and accessory storage and a work bench with a vice. The Dragoon AMV will accommodate a crew of five with standard field gear, weapons, ammunition and rations.

Mobile Electronic Warfare System (MEWS)

The MEWS carries out advanced battlefield direction-finding, high-speed communications jamming and is fitted with an antenna that can be quickly raised well above the vehicle. The video optical surveillance vehicle has a modified AV Technology two-man turret which has been fitted with a long-range day/night surveillance system that can be retracted quickly under armour protection. This provides the field commander with a highly mobile, armour-protected observation capability, which, when connected to an onboard data link, gives clear front-line view of the battlefield for rear commanders. The Teampack electronic detection system is normally carried in the horizontal position on the left side of the vehicle. When required for action it can be erected quickly into a vertical position. A 7.62 mm machine gun is ring-mounted on top of the hatch.

Armored Logistics Support Vehicle (ALSV)

The Dragoon ALSV is configured to facilitate the transport of standard pallets and containers. It is equipped with a hydraulic boom crane capable of self-loading and unloading cargo up to one tonne. The ALSV may be fitted with a variety of self-defence weapons and will accommodate a crew of three with standard field gear, weapons, ammunition and rations.

Dragoon TOW carrier

The Dragoon TOW system mounts the TOW pedestal on a 360° rotating turntable. It is protected from small arms fire and artillery fragments by the vehicle's armoured wall and top bi-fold doors. Provision is made for stowage of six TOW missiles in the vehicle. The vehicle will accommodate a crew of three with field gear, weapons, ammunition and rations.

Dragoon 2

In 1997, a further improvement of the Dragoon family of light armoured vehicles was introduced under the name Dragoon 2.

Improvements include larger self-cleaning tyres, upgraded suspension, improved exterior lighting, brush guards and mirrors, adapters for mission kit installation, water sprayer system, central tyre-inflation system, uprated heating and air conditioning system, relocated steering gear, new instrument panels, improved electrical system, relocated alternator, new four-battery starting and charging system, new hydraulic pump, improvements to the rear-mounted power pack, increased frontal armour and improved armour under hull.

Dragoon 2, also has a reduced infra-red signature and improved toxic gas protection. If required, the crew member seated to the right of the driver can take over control of the Dragoon 2 for a short period of time.

It can also be provided with an emergency fuel pump and slam locking doors. A wide range of up-armour kits is also available, ranging from the baseline 7.62 mm armour piercing up to 30 mm APDS at 1,000 m.

The three mission kit installations are a bodyguard push bumper for obstacle removal, obstacle ram and a multipurpose blade. These mission kits can be installed or removed in less than 10 minutes.

Dragoon 3

This is the designation of a projected 6 × 6 version of the Dragoon 2, and would use the same power pack, axles, suspension, tyres, central tyre-inflation system, heating and air conditioning system, front-mounted winch and cable, electrical system, lighting, instrument panels, seals and service parts.

For the Dragoon 3 application, the Dragoon 2 hull has been stretched with an additional axle to give a 6 × 6 configuration.

On the left side of the hull, between the second and third axles on the right side, between the first and second axles is a two-part door. The upper part contains a vision block with a firing port below while the lower part folds down to form a step.

Dragoon 3 can be fitted with a wide range of weapons stations up to a two-man turret armed with a 90 mm gun, 7.62 mm coaxial machine gun and roof-mounted 7.62 mm machine gun.

Status

Production as required. In service with the US forces, Thailand, Turkey and Venezuela (100) plus civil authorities.

Manufacturer

AV Technology LLC.

UPDATED

O'Gara-Hess & Eisenhardt Armoring Company M1116 armoured personnel carrier

Development

Late in 1997, following extensive trials, the US Air Force placed a contract with O'Gara-Hess & Eisenhardt Armoring Company for the supply of 71 M1116 (4 × 4) up-armoured AM General HMMWV (High Mobility Multipurpose Wheeled Vehicles) to replace the current Textron Marine & Land Systems Ranger (or Peacekeeper as called by the US Air Force) covered in the following entry.

These 71 M1116 vehicles were delivered between February and May 1998 and by late 2000 a total of 449 had been delivered to the US Air Force.

Typical roles of the M1116 include Security Police (SP), Civil Engineer (CE) and Explosive Ordnance Disposal Base Recovery After Attack (EOD BRAAT).

Description

The M1116 vehicle is based on the Enhanced Capacity Vehicle (ECV) member of the AM General HMMWV family with the armour package being designed, produced and fitted by O'Gara-Hess & Eisenhardt Armoring Company.

It is very similar to that of the O'Gara-Hess & Eisenhardt Armoring Company M1114 up-armoured HMMWV covered in full detail in the Reconnaissance vehicles section.

The armour package provides protection against 7.62 mm Armour-Piercing attack (AP) all round, 155 mm airburst overhead, underbody blast and M67 grenade attack.

No less than five basic kits are available. The first (kit A) provides the gunner with protection through 360° from 7.62 mm AP attack and the turret can be fitted with a 7.62 mm M60 or 12.7 mm M2 machine gun or a 40 mm Mk 19 grenade launcher with Mk 64 mount.

When fitted with kit B, the gunner has overhead protection from 155 mm shell bursts at 100 m and also allows the use of a 12.7 mm M82A1 Barrett Rifle in the Barrett soft mount.

With the gunner protection kit C, the gunner and crew are provided with underbody protection with the kit installed inside the vehicle. An adjustable gunner platform permits height adjustments. Double layers of ballistic steel and aramid blanks reinforce the floor and prevent grenade fragmentation entry into the crew compartment.

M1116 up-armoured HMMWV built by O'Hara-Hess & Eisenhardt Armoring Company specifically to meet the requirements of the US Air Force and fitted with kit A and 40 mm grenade launcher 0018757

Kit D provides added protection against underbody blasts from M67 grenades to provide crew and vehicle survivability. In addition, it protects the engine from off-road hazards.

Finally, kit E provides bolt-on protection for the engine and radiator from 7.62 mm armour-piercing rounds fired at 100 m or more. The design allows for easy installation, using four existing bolt holes to attach the grill assembly to the front bumper of the vehicle.

The new design traversing mechanism provides a continuous 360° traversing of the entire turret assembly using kits A and B. The new environmental cover was designed to meet the specific requirements of the US Air Force. The rear storage compartment has been enlarged to provide greater payload flexibility. A padlock handle latches the tailgate and hatch.

Additional features include air conditioning system, heating system, additional sound insulation and storage for 7.62 mm M60 machine gun.

The M1116 is fully air transportable with the Lockheed Martin C-130 able to carry 2 or 3 vehicles and the Lockheed Martin C-141 3 or 4 vehicles.

Specifications

These are almost the same as those of the M1114 covered in the earlier section.

Status

Production complete.In service with the US Air Force.

Manufacturer

O'Gara-Hess & Eisenhardt Armoring Company.

UPDATED

Cadillac Gage Ranger armoured personnel carrier

Development

The Ranger (4 × 4) armoured personnel carrier was developed by Cadillac Gage as a private venture to meet a requirement for a low-cost vehicle capable of transporting personnel at a high road speed while still retaining an off-road capability.

Late in 1977, the US Air Force issued a requirement for a Security Police Armored Response/Convoy Truck for patrolling airbases. The Air Force decided to initiate a two-step procurement, technical and price. In April 1978, a number of US manufacturers submitted detailed technical proposals in response to the requirement and in October the same year the US Air Force selected three manufacturers' proposals as meeting the technical requirements: Cadillac Gage, Oshkosh Truck Corporation and the Vehicle Systems Development Corporation.

In March 1979, the US Air Force chose the Cadillac Gage Ranger and in June 1979 the company was awarded a contract worth US$4.4 million for the supply of a number of vehicles plus technical and administrative data.

The first Cadillac Gage Ranger, called the Peacekeeper by the US Air Force, was handed over in April 1980. By early 1981, 560 vehicles had been ordered by the US Air Force and Navy. By 1994, 708 Cadillac Gage Ranger armoured personnel carriers had been built for the US forces and overseas sales. Since then there have been no sales of the Ranger and it is no longer being marketed.

Small quantities of these vehicles that are surplus to the requirements of the US military have been made available to some US police departments.

In 1996, some of these Ranger vehicles were deployed to Bosnia and operated by the Peace Implementation Force (IFOR).

In US Air Force service the Cadillac Gage Ranger has now been replaced by the O'Gara-Hess & Eisenhardt Armoring Company M1116 up-armoured HMMWV which is covered in detail in an earlier entry.

Description

The hull of the Ranger is made of all-welded 'CADLOY' high-hardness armour, which provides the crew with complete protection from small arms fire. The crew and vital components are protected from the effects of grenades exploding under the vehicle by bolt-on armour shields.

The Ranger armoured personnel carrier is based on a standard Chrysler 4 × 4 truck chassis but with a shorter wheelbase.

The diesel engine is at the front of the hull and is coupled to an automatic transmission with three forward and one reverse gears, which is equipped with an auxiliary oil cooler. Coupled to the transmission is a

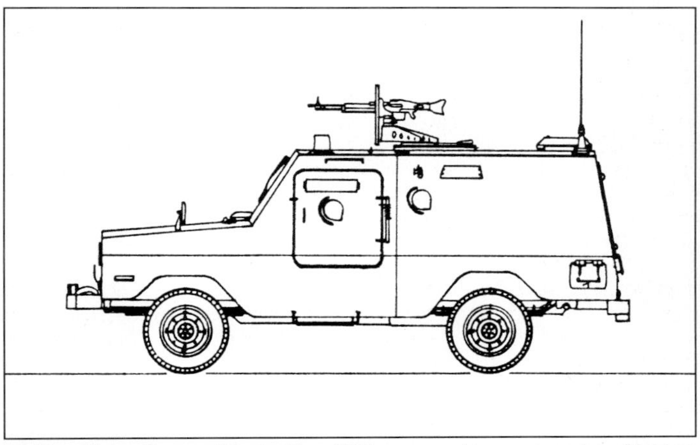

Cadillac Gage Ranger fitted with a 7.62 mm machine gun and a shield to protect the gunner

Cadillac Gage Ranger deployed to Bosnia as part of IFOR (Richard Stickland) 0018760

two-speed transfer case. The manually selected four-wheel drive case with an interaxle differential delivers proportionalised torque to both front and rear axles. The interaxle differential enables the wheels on one axle to travel at a different speed from those on the other. Both axles are of the full-floating type.

The vehicle has a dual-hydraulic service brake system and a foot lever-actuated parking brake. The service brake system consists of two separate brake fluid lines and a split, front and rear, master cylinder. Power assist is supplied by means of a tandem-diaphragm brake booster that utilises engine vacuum for power. The front brakes are disc with self-adjusting calipers and the rear are drum with self-adjusting shoes.

The suspension, front and rear, consists of leaf springs with double-acting shock-absorbers and jounce stops at each wheel station. The combat tyres are filled with foam and when penetrated by small arms fire allow the vehicle to travel up to 80 km at 56 km/h.

The driver sits at the front of the vehicle on the left with the commander to his right. The steering wheel can be tilted to assist the driver in entering and leaving the vehicle. Both crew members have in front of them a bulletproof windscreen with a wiper. There is a firing port between the two windscreens. Each also has a rear-opening side door with a bulletproof vision block in its upper half under which is a firing port.

The troop compartment is at the rear of the hull with the personnel seated three down each side facing each other. They enter the vehicle through two doors in the rear of the hull which open outwards. Both doors have a firing port and the left one has a vision block. In each side of the troop compartment is a vision block with a firing port underneath. The top of the hull is angled to allow hand grenades to roll off before they explode. If required, the vehicle can be delivered with a roof hatch, on which can be fitted a variety of light armament installations such as a pintle-mounted 7.62 mm M60 machine gun with a shield and 200 rounds of ready use ammunition; a further 800 rounds carried in the hull.

The interior of the Ranger is insulated, which reduces noise, retains cool air as the vehicle is fitted with an air conditioning system and also aids crew safety. An air conditioning system is fitted as standard and there is also provision for expelling air. Standard equipment includes two-speed wipers with washers, a heater, a windscreen defogger and an internal dome light.

Optional equipment includes flashing lights, a public address system, radios, a spotlight, Wegmann 76 mm grenade launchers, a front-mounted winch and a 24 V electric system in place of the standard 12 V system.

Specifications

Crew: 2 + 6
Configuration: 4 × 4
Combat weight: 4,536 kg
Power-to-weight ratio: 40 hp/t
Length: 4.699 m
Width: 2.019 m
Height: 1.981 m
Ground clearance:
(front axle) 0.203 m
(rear axle) 0.203 m
Track: 1.689 m
Wheelbase: 2.641 m
Angle of approach/departure: 25/24°
Max road speed: 112.65 km/h
Fuel capacity: 121 litres
Max range: (at 72 km/h) 482 km
Trench: 0.71 m
Fording: 0.60 m
Gradient: 60%
Side slope: 30%
Vertical obstacle: 0.254 m
Engine: Dodge 360 CID V-8 liquid-cooled petrol developing 180 hp at 3,600 rpm
Transmission: Dodge model A727 automatic with 3 forward and 1 reverse gears
Transfer case: 2 speed
Steering: integral power with pump assistance
Suspension: leaf springs and double-acting hydraulic shock-absorbers
Brakes:
(main) hydraulic, dual circuit
(parking) foot lever-actuated
Electrical system: 12 V
Battery: 1 × 12 V, 70 Ah
Armament: see text
NBC system: no
Night vision equipment: no

Variants

In addition to being used as an APC, the Ranger could be used as a command vehicle (crew of two plus two command staff), an ambulance (crew of two and carrying two stretcher plus two seated patients) and as a light reconnaissance vehicle, which could be fitted with a Cadillac Gage one-man turret armed with twin 7.62 mm (or 0.30), or one 7.62 mm (0.30) and one 12.7 mm (0.50) machine guns. These have an elevation of +55°, depression of −14° and 360° turret traverse, which is manual. The turret has an M28C roof-mounted day sight with a magnification of ×1.5, eight vision blocks and a single-piece hatch cover that opens to the rear.

Status

Production as required. In service with Indonesia (22, of which 7 are in VIP configuration), Luxembourg (5 with one-man turret armed with twin 7.62 mm MGs), the US Air Force and Navy. In US Air Force service this has started to be replaced by the M1116 which is covered in detail in the previous entry.

Manufacturer

Textron Marine & Land Systems, Combat Vehicle Marketing.

UPDATED

ARMOURED FIGHTING VEHICLE FAMILIES

ARMOURED FIGHTING VEHICLE FAMILIES

GERMANY

Rheinmetall Landsysteme R 495 family

Development

In September 1992, Thyssen Henschel (which recently became known as Rheinmetall Landsysteme) completed the first prototype of a new family of tracked armoured fighting vehicles which has been given the company designation R 495.

This family of AFVs has been developed specifically as a private venture and is already being marketed to a number of countries.

The R 495 has a typical combat weight of 26 tonnes. To reduce total life cycle costs, proven and production components have been used in the design of the R 495 wherever possible.

Although some design work was carried out in 1989-90, the decision to go ahead and build a complete prototype was only taken in October 1991. The first prototype was in the Infantry Fighting Vehicle (IFV) configuration, was built in just 11 months and, following the official roll-out at Kassel, started automotive trials.

In addition to the IFV, Rheinmetall Landsysteme is proposing that the chassis could be used for a wide range of roles including anti-aircraft, anti-tank, ambulance, command post vehicle, reconnaissance, recovery, light tank with 105 mm turret and armoured personnel carrier.

Although development and construction of the prototypes of the R 495 has been undertaken in Germany, Rheinmetall Landsysteme is offering this for the export market on technology transfer/co-production basis.

As of late 2001 there were no known plans for the Rheinmetall Landsysteme R 495 to enter production.

Description

The hull of the R 495 is of all-welded steel armour construction to which additional layers of spaced steel have been added for increased battlefield survivability. The additional passive armour is in modular sections that two men can change in the field very quickly using simple tools. Spaced armour is fitted to the top of the hull to provide protection from top attack weapons.

The driver is seated at the front left and has a single-piece hatch cover that opens to the left rear with three integral day periscopes, one of which can be replaced by a passive periscope for driving at night.

The power pack is mounted to the right of the driver and consists of an MTU 183 TE22 diesel coupled to a modified ZF LSG 1500 fully automatic transmission.

The IFV has a three-man crew consisting of driver, commander and gunner and seven infantrymen seated either side of the hull facing each other, four on the left and three on the right.

The infantry can rapidly dismount via two doors in the rear of the hull that open outwards with each of these having a new spherical observation device combined with a firing port. Over the rear part of the troop compartment are two circular roof hatches.

Each infantryman has an individual seat complete with a seat belt and the seats can be adjusted so that a total of four men, two each side, can sleep at any one time.

The prototype of the R 495 was fitted with an Italian Otobreda power-operated turret armed with a 25 mm Oerlikon Contraves cannon and a 7.62 mm coaxial machine gun. The 25 mm cannon is fully stabilised and day and night sights are fitted as standard. The turret has the commander on the left and the gunner on the right. The commander is provided with a roof-mounted stabilised sight that can be traversed through a full 360°. Mounted either side of the turret is a bank of four 76 mm electrically operated smoke grenade launchers firing forwards.

More recently the R 495 has been fitted with the new private venture German Rheinmetall Landsysteme (previously known as KUKA) 30 mm E4 two-man power-operated turret armed with a Mauser MK 30 mm cannon and 7.62 mm coaxial machine gun.

The suspension of the R 495 is the torsion bar type. side has six dual rubber-tyred roadwheels with the drive sprocket at the front, idler at the rear and three track-return rollers. The first, second, fifth and sixth roadwheels have hydraulic shock-absorbers and the double-pin tracks are from Diehl.

Standard equipment includes digital electronics, a video monitoring system for the infantry, consisting of a TV camera mounted in either side of the hull with a single TV monitor inside, explosion-proof fuel tanks, an NBC system and spall liners.

One of the key requirements of the R 495 is that it should be fully air-transportable in a Lockheed Martin C-130 Hercules transport aircraft with just the armour skirts removed.

The second prototype of the R 495 is slightly shorter and fitted with only five dual rubber-tyred roadwheels. This has an overall length of 5.97 m and, when fitted with a cupola armed with a 12.7 mm M2 machine gun, has a combat weight of 21.6 tonnes.

Variants

Using the same basic chassis and automotive components, a wide range of variants can be developed, some of which are mentioned under Development.

Second prototype

This was completed late in 1993 and made its first appearance at the Eurosatory 94 exhibition held in June 1994.

The second prototype has five roadwheels either side and is fitted with a Krauss-Maffei Wegmann one-man cupola armed with a 12.7 mm M2 machine gun that can be aimed and fired under complete armour protection. Mounted either side of the cupola is a bank of four electrically operated 76 mm smoke grenade dischargers firing forwards.

In addition to the crew of two (commander/gunner and driver) it can also carry eight fully equipped infantrymen in the rear troop compartment.

The power pack consists of the MTU 183 TE22 diesel developing 600 hp coupled to a ZF LSG 1500 fully automatic transmission. Other features are similar to the first prototype of the R 495.

The second prototype of the TH 495 has been tested in Malaysia. This version of the TH 495 has a combat weight of 21.6 tonnes and is 5.97 m long, 2.84 m wide and has an overall height of 2.39 m. As with other members of the R 495 the actual weight and height, and therefore power-to-weight ratio does depend on the armament system installed. When armed with the 12.7 mm M2 machine gun 800 rounds of ammunition are carried.

Specifications

(Provisional-version with 6 roadwheels)
Crew: 3 + 7
Combat weight: 25,900 kg
Power-to-weight ratio: 23 hp/t
Ground pressure: 0.72 kg/cm²
Length:
(overall) 6.75 m
Width:
(overall) 2.84 m
(over tracks) 2.72 m
Height:
(overall) 2.83 m
(hull top) 1.865 m
Ground clearance: 0.40 m
Track width: 450 mm
Length of track on ground: 3.90 m
Max road speed: 73 km/h
Range: 500 km
Fuel capacity: 700 litres
Vertical obstacle: 0.8 m
Trench: 2.70 m
Engine: MTU 183 TE22 diesel developing 600 to 680 hp (max)
Transmission: ZF LSG 1500 automatic with 4 forward and 2 reverse gears
Suspension: torsion bar
Gun stabiliser:
(vertical) no
(horizontal) no
NBC system: yes
Night vision equipment: yes

Rheinmetall Landsysteme R 495 in six roadwheel configuration and fitted with a two-person power-operated turret armed with a Mauser MK 30 mm cannon and 7.62 mm machine gun has been offered to the UK for its FRES requirement **2002**/0096347

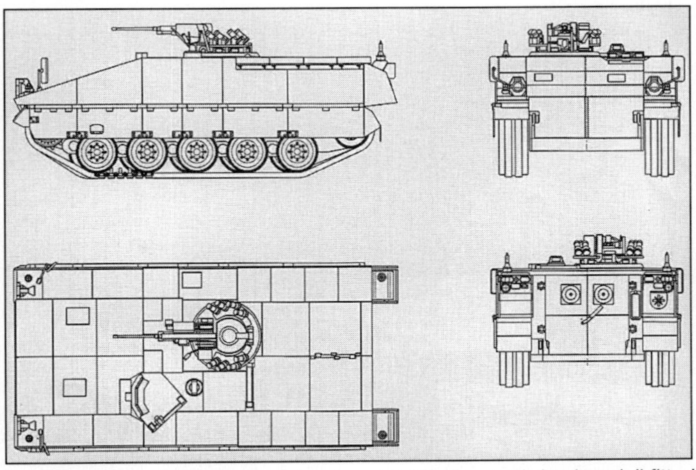

Four-view general arrangement drawing of R 495 (five roadwheel model) fitted with a Krauss-Maffei Wegmann cupola armed with a 12.7 mm M2 machine gun

R 495 for FRES

Late in 2001, it was revealed that Rheinmetall Landsysteme GmbH had proposed its private venture R 495 full-tracked armoured fighting vehicle family to meet the UK Army requirement for its Future Rapid Effects System (FRES) and Armoured Battlegroup Support Vehicle (ABSV) requirements.

According to Rheinmetall Landsysteme, one of the key design factors was that the R 495 should be fully airportable in a Lockheed Martin C-130 Hercules transport aircraft with its armoured skirts removed to save weight.

The weight of the R 495 depends on the armour package and the weapon station fitted, but a six-wheeled version fitted with a Rheinmetall two-person turret and armed with a stabilised Mauser 30 mm MK 30-2 cannon, 7.62 mm coaxial machine gun, advanced fire-control system with hunter/killer capability, would weigh 26.6 tonnes complete with add on armour. A five wheeled version, armed with an externally mounted 7.62 mm machine gun, would weigh 21.2 tonnes with add-on armour.

For the UK requirement, the R 495 is being referred to as the Future MultiRole Vehicle (FMRV) and if the UK did adopt this to meet FRES, then Rheinmetall Landsysteme would team with a UK contractor.

A full range of variants of the FMRV would be offered for FRES, including 120 mm Armoured Mortar System (Delco/RO Defence turret), command post, ambulance, infantry fighting vehicle and direct-fire platform.

The latter could be fitted with a 105 mm rifled gun or the more recent 105 mm Rheinmetall smoothbore gun which leverages off the 120 mm smoothbore gun already in service with an increasing number of countries.

Status

Development complete. Ready for production.

Manufacturer

Rheinmetall Landsysteme GmbH.

UPDATED

KOREA, SOUTH

Daewoo family of armoured fighting vehicles

Development

In addition to developing the Korean Infantry Fighting Vehicle (KIFV) and its many variants, full details of which are given in the Armoured personnel carriers (tracked) section, Daewoo Heavy Industries has also developed a new and larger chassis on which another family of armoured fighting vehicles has been developed to meet the requirements of the South Korean Army.

The Flying Tiger and Pegasus air defence systems used the same chassis while the armoured military logistic support vehicle has a chassis which is similar in concept to that used by the US Multiple Launch Rocket System. All three vehicles are powered by a Daewoo D2840L 10-cylinder diesel developing 520 hp coupled to a General Dynamics Land Systems HMPT-500-3 automatic transmission.

Flying Tiger twin 30 mm self-propelled anti-aircraft gun system

This has been designed to engage aircraft and helicopters and consists of the new all-welded aluminium chassis on which is installed a new two-man powered turret armed with twin 30 mm Oerlikon Contraves KCB automatic cannon which have a cyclic rate of fire of 600 rds/min and an effective range of 3,000 m.

Mounted on the turret roof is the surveillance radar and mounted on the forward part of the roof is the day/thermal electro-optical tracker with laser range-finder which automatically tracks the target.

Following extensive trials with two prototype systems the first production order was placed late in 1996. Full details of this system are given in *Jane's Land-Based Air Defence 2001-2002*.

Pegasus surface-to-air missile system

This has a similar chassis to that used for the Flying Tiger twin 30 mm self-propelled anti-aircraft gun system but has an unmanned turret with eight (four either side) surface-to-air missiles in the ready to launch position.

The system has tracking and surveillance radars and an electro-optical package which is the same as that used for the French Thales Air Defence Crotale (New Generation) surface-to-air missile system.

This is now in service with the Republic of Korea Army. Full details are given in *Jane's Land-Based Air Defence 2001-2002*.

Military logistics support vehicle

This has been designed to support 155 mm self-propelled artillery systems and can carry 139 155 mm projectiles, 146 propellant charges and 167 fuzes.

The crew is seated in the fully armoured protected cab at the front of the vehicle with the armour protected ammunition area extending to the rear.

Suspension either side is of the torsion bar type and consists of six dual rubber-tyred roadwheels with the drive sprocket at the front, idler at the rear and four track-return rollers. Hydraulic shock-absorbers are provided for the first two and last two roadwheel stations.

In the forward part of the roof is a hatch for loading ammunition while mounted over the forward part of the

Daewoo Heavy Industries Military Logistic Support Vehicle without 12.7 mm M2 anti-aircraft gun fitted
0005312

Daewoo Heavy Industries Flying Tiger twin 30 mm self-propelled anti-aircraft gun system 0005311

Specifications

Military logistics support vehicle
Crew: 4
Combat weight: 26,700 kg
Empty weight: 18,400 kg
Power-to-weight ratio: 19.5 hp/t
Ground pressure: 0.70 kg/cm^2
Length: 6.47 m
Width: 3.15 m
Height: 3.30 m
Track width: 392 mm
Max road speed: 60 km/h
Max cruising range: 360 km
Fuel capacity: 500 litres
Vertical obstacle: 0.7 m
Trench: 2.0 m
Gradient: 60%
Side slope: 40%
Engine: Daewoo D2840L 10-cylinder inter-cooled diesel developing 520 hp
Transmission: General Dynamics Land Systems HMPT-500-3 hydromechanical
Steering: hydrostatic
Brakes: multidisc, oil cooled
Suspension: torsion bar
Electrical system: 28 V
Batteries: 4 × 6 TN, 100 Ah
Generator: 100 Ah
NBC system: yes
Night vision equipment: yes

roof, where the crew is seated, is a cupola which can be armed with a 12.7 mm M2 HB machine gun.

In the rear of the hull is a two-part hatch with the lower part opening to the right while the upper part opens upwards. When deployed with the 155 mm self-propelled artillery system, ammunition passes from the military logistic support vehicle to the self-propelled artillery system via a conveyor belt at the rate of six projectiles per minute.

Status

Production of the Flying Tiger twin 30 mm SPAAG and the Pegasus SAM is now underway. In service with the Republic of Korea Army.

Manufacturer

Daewoo Heavy Industries Limited.

UPDATED

SAUDI ARABIA

Al-Faris AF-40-8-2 armoured car and AF-40-8-1 armoured personnel carrier

Development

The Saudi Arabian company, Abdallah Al-Faris & Company for Heavy Industries (one of the Military Industries Corporation companies), has been developing a 6 × 6 armoured fighting vehicle, suitable for a wide range of operational roles on the battlefield, since 1977.

The first vehicle, an armoured personnel carrier, was completed in 1981, while the second, which incorporated many improvements as a result of trials with the prototype, was completed in 1988 (*Jane's Armour and Artillery 1989-90*, page 776). This was fitted with a Belgian Cockerill LCTS two-man turret armed with a Cockerill 90 mm Mk 7 gun and a 7.62 mm coaxial machine gun.

Although the first prototypes of the Al Fahd were in 6 × 6 configuration a decision was taken to concentrate future development on an 8 × 8 version as this would have greater internal volume and increased cross-country mobility.

Two versions of the 8 × 8 Al Fahd have been developed, both of which share many common automotive components to reduce procurement and total life cycle costs.

These are the AD-40-8-1 armoured personnel carrier (APC) and the AD-40-8-2 reconnaissance vehicle.

The AD-40-8-1 is the first of these to enter production it can carry 12 troops and be fitted with a wide range of weapon stations up to 40 mm in calibre.

First production vehicles were completed in 1998 and the first production batch of 100 vehicles is expected to be completed in the near future. As of late 2001, there were no exports of the AF-40-8-1 or AF-40-8-2 8 × 8 armoured vehicles, although the latter has been demonstrated in a number of countries in Europe and the Middle East.

Description

The hull of the AF-40-8-1 APC is of all welded steel armour which provides protection against attack from 14 mm projectiles over the frontal arc and against 7.62 mm attack through a full 360°.

The armour also provides protection against shell splinters and anti-personnel mines. The interior of the hull is lined with Kevlar.

For a higher level of battlefield survivability the vehicle can be fitted with a German IBD passive armour system.

The vehicle has the commander and driver at the front, powerpack in the centre and troop compartment at the rear. There is no direct access between the two compartments.

The commander and driver each have a single piece hatch cover which opens upwards and to the rear which can be locked in the vertical position. This has a single day periscope which can be replaced by a passive periscope for driving at night.

The powerpack consists of a Deutz air-cooled 400 hp diesel coupled to a ZF fully automatic transmission, hydraulic oil pumps, oil radiators and the main hydraulic reservoir. The complete power pack can be removed in 10 min. The exhaust pipe is on the right side of the hull.

Entry to the troop compartment at the rear is via a power operated ramp but there are also two square roof hatches in the rear part of the troop compartment roof.

A wide variety of weapon systems can be fitted, for example:

(1) Pintle or turret mounted 7.62 mm and 12.7 mm machine guns

Saudi Arabian Al Fahd AF-40-8-1 (8 × 8) in armoured personnel carrier configuration without armament installed (Terry J Gander) **2001**/0103424

Saudi Arabian Al Fahd AF-40-8-1 (8 × 8) fitted with Delco Defense LAV-25 turret with TOW missile launcher either side (Christopher F Foss) **2001**/0103426

(2) Delco Defense turret armed with a 25 mm cannon and 7.62 mm coaxial machine gun with the option of a Raytheon Systems Company TOW ATGW either side of turret
(3) Delco Defense/BAE Systems, RO Defence 120 mm Armoured Mortar System with breech loaded mortar
(4) Anti-tank fitted with 106 mm M40 recoilless rifle

It has full 8 × 8 drive for cross-country use with the front two and last two wheels powered for road use. Powered steering is on all wheels.

Standard equipment includes NBC system, air conditioning system and adjustable suspension which allows the driver to adjust the ground clearance from 150 to 600 mm.

Options include water cooled diesel, Deutz central tyre inflation system, firing ports, winch with a capacity of 9,060 kg and provided with 46 m of 14.3 mm diameter cable, passive night vision equipment, automatic fire detection and suppression system and auxiliary power unit.

For amphibious operations, a water-jet can be located under either side of the hull at the rear and maximum water speed is between 10 and 12 km/h. Before entering the water a trim vane is erected at the front of the vehicle and bilge pumps switched on.

Variants

In addition to the wide range of variants proposed for the AF-40-8-1 design work on the AF-40-8-2 has been completed. This has the powerpack at the rear and can be fitted with a wide range of armament installations up to a turret armed with a 105 mm low recoil gun.

AF-40-8-35 fighting and air defence

The Abdallah Al Faris company and Oerlikon Contraves of Switzerland have proposed another version of the AF-40-8-1 called the AF-40-8-35 fighting and air defence vehicle.

This is fitted with a turret armed with the Oerlikon Contraves 35 mm Revolver Gun which fires the Advanced Hit Efficiency And Destruction (AHEAD) ammunition which is already in quantity production for a number of applications, including upgraded Oerlikon Contraves 35 mm GDF series twin 35 mm towed anti-aircraft guns.

According to the two companies, this weapon system would have a number of roles. These would include ground-to-air such as the protection of mechanised troops, protection of high-value assets and self-protection. Ground-to-ground applications include escort, support and covering fire.

The two-person fully enclosed power-operated turret would be of all-welded steel armour and in addition to being armed with the 35 mm Revolver Cannon, would also be fitted with a 7.62 mm coaxial machine gun and two banks of electrically operated smoke grenade launchers.

Specifications

Model	**AF-40-8-1**	**AF-40-8-2**
Type	**APC**	**armoured car**
Crew	1 + 11	1 + 3
Configuration	8 × 8	8 × 8
Power-to-weight ratio	20.51 hp/t	25.22 hp/t
Weight		
(combat)	19,500 kg	21,800 kg
(unloaded)	16,000 kg	16,300 kg
(payload in amphibious mode)	2,500 kg	2,200 kg
(payload land)	3,500 kg	5,500 kg
Length	7.90 m	7.90 m
Width	2.94 m	2.94 m
Height		
(max)	2.36 m	2.36 m
(medium)	2.165 m	2.165 m
(min)	1.91 m	1.91 m
Ground clearance		
(max)	600 mm	600 mm
(medium)	405 mm	405 mm
(min)	150 mm	150 mm
Wheelbase	1.42+1.8+1.42 m	1.42+1.8+1.42 m
Angle of approach/departure	50°/50°	50°/50°
Max road speed	90 km/h	90 km/h
Fuel capacity	550 1	550 1
Range (cruising)	800 km	600 km
Fording	amphibious	amphibious
Gradient	80%	80%
Side slope	55%	55%
Vertical obstacle	1.25 m	1.25 m
Trench	2.50 m	2.50 m
Turning radius	7.5 m	7.5 m
Engine		
(type)	Deutz air-cooled* 10-cylinder	Deutz air-cooled* 12-cylinder
(output)	400 hp	550 hp
Transmission	ZF automatic 6 forward and 1 reverse gears	ZF automatic 6 forward and 1 reverse gears
Steering	hydraulic	hydraulic
Suspension	hydropneumatic	hydropneumatic
Tyres	1,600 - 20 radial	1,600 - 20 radial
Brakes		
(main)	dual circuit	dual circuit
(hand brake)	mechanical	mechanical
Electrical system	12 - 24 V	12 - 24 V
Batteries	4 × 12 V 125 Ah	3 × 12 V 125 Ah

* option of water cooled engine

Various fire-control systems are being proposed, with the most advanced version including a tracking device incorporating a laser range-finder.

Status

AF-40-8-1 in production for Saudi Arabian Ministry of Defence. There are no known exports of this vehicle.

Manufacturer

Abdallah Al-Faris & Company for Heavy Industries.

UPDATED

Saudi Arabian Al Fahd AF-40-8-1 (8 × 8) fitted with Delco Defense LAV-25 turret with TOW missile launcher either side (Christopher F Foss) ***2000***/0075733

Saudi Arabian Al Fahd AF-40-8-1 (8 × 8) fitted with 106 mm recoilless rifle (Christian Dumont) ***2001***/0103425

UNITED KINGDOM

Alvis Scorpion reconnaissance vehicle

Development

In the late 1950s, the British Army issued a requirement for an Armoured Vehicle, Reconnaissance (AVR) to undertake the roles of reconnaissance, fire support and anti-tank. After studying a variety of proposals the then Fighting Vehicles Research and Development Establishment (which no longer exists as a separate entity) came to the conclusion that two vehicles would be required to carry out these roles. These became the Combat Vehicle Reconnaissance (Tracked) Scorpion range and the Combat Vehicle Reconnaissance (Wheeled) Fox vehicle, both of which use the same Jaguar petrol engine.

In September 1967, Alvis Limited of Coventry, which was at that time building the FV600 range of 6 × 6 armoured vehicles, was awarded a contract to build 17 prototypes. The first prototype was completed in January 1969 and the first official announcement of the vehicle was made in September the same year.

All 17 prototypes were completed by the middle of 1970 and in May 1970 the Scorpion was accepted for service with the British Army. In October 1970, the Belgian Army placed an order for 701 Scorpions and variants, which were assembled at a British Leyland facility at Malines in Belgium.

The first production Scorpions were delivered to the British Army in January 1972 with first deliveries being made to the Belgian Army in February 1973.

It was announced early in 1979 that the Royal Air Force Regiment had ordered 150 members of the Scorpion family (including Scorpion, Spartan, Sultan and Samson) for the defence of RAF airfields in Germany. In November 1981, the Royal Air Force Regiment at Catterick formally took delivery of its first Alvis CVR(T) Scorpions for airfield defence. This order was followed by another in 1982 for Scimitar vehicles. In 1992 however, the Royal Air Force announced that it was to phase out its fleet of Scorpion light armoured vehicles. Some of these vehicles were subsequently taken over by the British Army.

Scorpion has been designed to operate in temperatures from -30 to +50°C and a Lockheed Martin C-130 transport can carry two of them.

Many countries have ordered Scorpions modified to meet their own particular operational requirements. The Malaysian vehicles have the Perkins Engines Company diesel engine, 90 mm gun and German Wegmann 76 mm grenade launchers.

In June 1988, Alvis won an US$85 million contract from Venezuela for the supply of an undisclosed number (believed to be 84) of Scorpion 90 light tanks, support vehicles, ammunition, communications equipment and both turret and desktop training equipment. First vehicles were completed late in 1989 and deliveries were completed in 1992.

The Venezuelan Scorpion 90s are armed with the Cockerill 90 mm Mk III gun which is capable of firing APFSDS ammunition, an Avimo fire-control system incorporating a laser range-finder, Helio periscopes and an external machine gun mount. The vehicles are powered by a Perkins Engines Company diesel engine coupled to the David Brown Engineering TN15D transmission.

Support vehicles included in the Venezuelan order are the Samson recovery vehicle and the Sultan command post vehicle, this being the first time that diesel-powered versions of these vehicles had been placed in production. By 1 January 1989, Alvis had completed the following quantities of Scorpion family vehicles:

Scorpion 76	1,241
Scorpion 90	32
Striker	132
Spartan	967
Samaritan	101
Sultan	372
Samson	136
Scimitar	486

Production has continued at a low rate for the export market as and when required although no recent updated production figures have been released. There has been no recent production of the 76 mm armed Scorpion and all recent production has been with a Perkins Engines Company diesel engine.

Early in 1995, Indonesia placed a contract with Alvis worth around £90 million for the supply of Scorpion 90s and Stormer APCs with deliveries running from 1995 through to 1997.

Late in 1996, Alvis Vehicles won a second contract from Indonesia for the supply of about 50 members of the Scorpion family of light tracked armoured vehicles. Total value of this contract is around £80 million with deliveries starting in 1997.

The Indonesian vehicles are powered by Perkins Engines Company Phaser diesel, have an upgraded suspension system and an air conditioning system. In addition to the latest Scorpion 90 vehicles the first contract also included Stormer command post vehicles, armoured personnel carriers, armoured recovery vehicles, ambulances, flatbed cargo carriers and an armoured vehicle-launched bridge carrying a Class 30 scissors bridge.

It is believed that there are three batches of Scorpion 90 vehicles involved: 36 + 45 + 40 making a total of 121 vehicles.

The Chilean Marines are understood to have taken delivery of a total of 30 second-hand Scorpion light armoured vehicles in two batches of 15. These are believed to retain their 76 mm guns.

Current production vehicles are referred to by Alvis as the Scorpion 2 and these incorporate many improvements, some of which were incorporated in earlier production vehicles for the export market. These include the proven Perkins Engines Company diesel engine which is much more fuel efficient, uprated suspension system, track and running gear, and further enhanced RAM-D characteristics with options including an NBC system, an air conditioning system, a fire detection and suppression system, a fully automatic transmission and power steering. The two-man turret systems have been refined and upgraded and options now include 30 mm, 76 mm or 90 mm main armaments, combined day/night sights for commander with laser range-finder options for the gunner, an advanced electromechanical gun and turret drive system, a refined electrical system, enhanced capacity for communications equipment and improved ergonomics.

Many of these improvements are now available as retrofit options that can be undertaken by the user, or the vehicles can be returned to Alvis Vehicles for overhaul and modernisation.

In 31 October 1992, the British Army and the Royal Air Force had the following quantities of CVR(T)s in service or in reserve:

Type	Army	RAF
Scimitar	313	11
Scorpion	283	25
Striker	89	nil
Samaritan	50	nil
Sultan	287	4
Samson	90	4
Spartan	533	81
Spartan MCT	67	nil
Total	1,712	125

Scorpions are no longer in operational service with the British Army or Royal Air Force Regiment. The British Army retains a number of 76 mm armed Scorpions in reserve and the vehicle is also used for training purposes as an opposing force vehicle. Production of the CVR(T) was undertaken at the Alvis Vehicles facility in Coventry. This was closed in mid-1999 and all future production will be undertaken in the Alvis Vehicles facility at Telford, Shropshire.

Description

The hull of the Scorpion is made of all-welded aluminium armour and provides the crew with protection against attack over its frontal area from 14.5 mm projectiles and against 7.62 mm armour-piercing rounds over the remainder of the vehicle. The aluminium armour is also particularly effective against shell splinters. The hull is divided into three compartments: driver's at the front on the left, power pack at the front on the right and the fighting compartment at the rear.

The driver has a single-piece hatch cover that swings to the left, in front of which is a single wide-angle day periscope which can be replaced by a Thales Optronics passive periscope for night driving.

The original version of the Scorpion family was powered by a derated (from 265 to 190 bhp) and militarised Jaguar 4.2 litre engine with a Marcus carburettor and a compression ratio reduced from 9:1 to 7.75:1, which permits the use of military fuels. The David Brown Engineering TN15 transmission, which is at the front of the vehicle, was developed on the same principles as that used for the Chieftain MBT. It is a hot-shift, foot-operated, seven-speed gearbox with a controlled differential steering system. For engine cooling a single mixed-flow fan draws in air through the

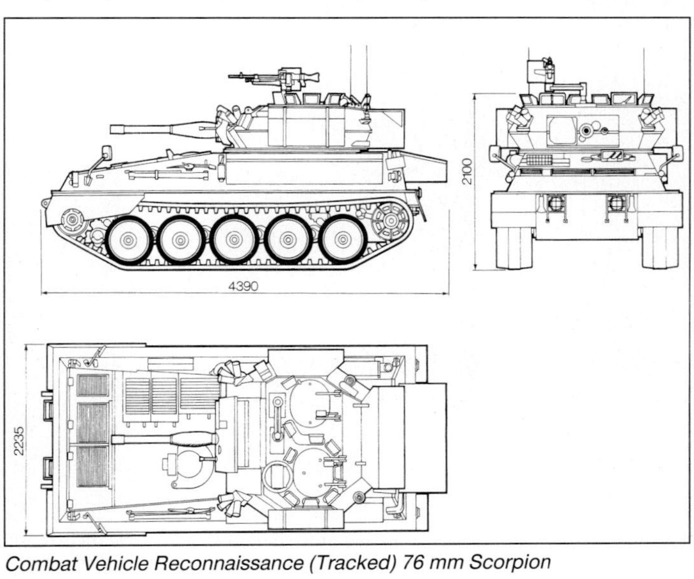

Combat Vehicle Reconnaissance (Tracked) 76 mm Scorpion

Combat Vehicle Reconnaissance (Tracked) 76 mm Scorpion fitted with flotation screen
2002/0121605

radiator over the gearbox, over the engine and out through the louvres.

The other two crew members are seated in the all-welded aluminium armour turret with the commander on the left and the gunner on the right, both with a one-piece hatch cover that opens to the rear. The commander has seven day periscopes and a roof-mounted day sight in front of his hatch cover with magnifications of ×1 and ×10. The roof-mounted sight is capable of limited rotation allowing approximately 85° of horizontal field of view. The gunner has two day periscopes and a roof-mounted day sight with magnifications of ×1 and ×10. Mounted to the right of the main armament is a now BAE Systems passive night sight with a magnification of ×5.8 (8° field of view) and a low magnification of ×1.6 (28° field of view).

New production Scorpion vehicles have a more advanced sighting system which can include day/thermal sights and a laser range-finder for increased first round hit probability.

The radios are in the turret bustle and a light metal stowage box is mounted externally at the rear of the hull. British Army Scimitar, now have a stowage box on the left side of the hull.

The torsion bar suspension either side consists of five rubber-tyred aluminium roadwheels with the drive sprocket at the front and the idler at the rear. There are no track-return rollers. Hydraulic lever-type shock-absorbers are provided for the first and last roadwheel stations either side. The tracks are of light steel with rubber bushes and pads, and have a life of over 5,000 km of mixed road and cross-country running.

In mid-1999, following a competition, William Cook Defence of the UK were selected to provide a new generation of single pin track for installation on the CVR (T) family of vehicles used by the British Army. This track is called the TR10 with first deliveries being made in 2000.

British Army Scorpions have an NBC system at the rear of the hull, but if it is not installed an additional five rounds of 76 mm ammunition can be carried. Optional equipment includes an NBC detector kit, a vehicle navigation system and a Galley air conditioning system which has already been fitted on the vehicles supplied to the United Arab Emirates.

The Scorpion can ford to a depth of 1 m without preparation. A flotation screen carried collapsed around the top of the hull can be erected by the crew in 5 minutes and the vehicle is then propelled and steered across the stream or river by its tracks at a speed of 6.5 km/h. A propeller kit has also been developed and when fitted the Scorpion has a maximum water speed of 9.5 km/h. The British Army has removed the flotation screens from all members of its Alvis Scorpion CVR(T) tracked series of vehicles.

The Scorpion is armed with a BAE Systems RO Defence 76 mm L23 gun which is a lighter version of the L5 used in the Saladin armoured car. This weapon has a vertical sliding breech and is loaded with fixed ammunition. It has a recoil of approximately 280 mm and is returned to the firing position by a hydropneumatic recuperator. During runout, the breech is opened by a semi-automatic cam, the empty cartridge case is ejected and the breech then remains open, ready for reloading. The 76 mm gun has an elevation of +35° and a depression of -10°, turret traverse being 360°. Elevation and traverse are both manual but if required the vehicle can be delivered with powered traverse. The following types of ammunition can be fired: canister, HESH, HE, HE/PRAC, SH/P, smoke/BE and illuminating.

Alvis Sabre reconnaissance vehicle which is now in service with the British Army (Richard Stickland)
2001/0103431

A 7.62 mm machine gun mounted coaxially to the left of the main armament can be used as a ranging machine gun. Mounted each side of the turret is a four-barrelled electrically operated 66 mm (British Army, 76 mm export vehicles) smoke grenade discharger.

Belgian Scorpion and Scimitar

Between 1981 and 1984, the Belgian Army's 286 Scorpion and Scimitar vehicles underwent two major repair and maintenance programmes conducted by the 65e Section of the Arsenal du Matériel Mécanique et de l'Armement at Rocourt. The first programme, called SCOREPOLE, included modifications to rectify cracks in the armour including needle peening, shot peening, adding of reinforcement ARPs (Armoured Repair Plates), screw-up of the hairline cracks, sewing of the plates (metalock) and welding. The BARGEPOLE programme, which was combined with a mid-life general overhaul, was a vehicle improvement programme aimed at improving safety for the crew, better operational capability, improving reliability and easing maintenance.

Following the withdrawal of the 73 mm armed Scorpion after their return from UNOSOM in Somalia, other members of the CVR(T) were modified.

From 1992, vehicles were fitted with Krauss-Maffei Wegmann 76 mm smoke grenade launchers. The Scimitar vehicles, used in the command role by platoon commanders, were fitted with a BLU radio which is located in a new type of stowage box at the rear of the turret.

For use in the anti-tank role, some Spartan APC's have been fitted with a pintle mounted Euromissile MILAN ATGW installation with stowage racks being provided inside for additional MILAN ATGW

The Striker ATGW system has now been phased out of service with the Belgian Army.

The Belgian vehicles are expected to be replaced by a new, wheeled (4 × 4) reconnaissance vehicle. This programme is called Reconnaissance 2000/2001 for which there are a number of competitors.

Irish Scorpions

Ireland ordered 14 Scorpion vehicles, the first of these being taken into service in March 1980 and operated by 1 Tank Squadron. The Irish Scorpions have a 12.7 mm M2 HB machine gun on an M63 anti-aircraft mount installed on the turret roof.

Jordanian Scorpions

Early in 1999 the Jordanian King Hussein Main Workshops upgraded a Alvis Vehicles Scorpion member of the Combat Vehicle Reconnaissance (Tracked) (CVR(T)) family of light armoured vehicles to meet the possible requirements of the Royal Jordanian Army and other Scorpion operators in the Middle East.

Alvis Vehicles supplied a large number of Scorpion vehicles to Iran prior to the overthrow of the Shah of Iran. Some of these were subsequently captured in the 1980s by Iraq who in turn passed them onto Jordon.

A total of 19 Scorpions was supplied to Jordan together with 60 M47 tanks, 90 Chieftain Main Battle Tanks (MBTs) and 35 armoured personnel carriers.

The first example of the upgraded Scorpion was completed in February 1999 under project AB4. It has been developed by the King Hussein Workshops in partnership with a UK consortium headed by CLS Systems which includes AEME Services, the now Thales Optronics and Sabre Engineering.

The consortium regard the first example of the upgraded Scorpion as a concept demonstrator and if response from the market place is positive then a further two vehicles would be converted in Jordan.

The original Jaguar 4.2 litre petrol engine has been replaced by a Perkins Engines Company Phaser 200

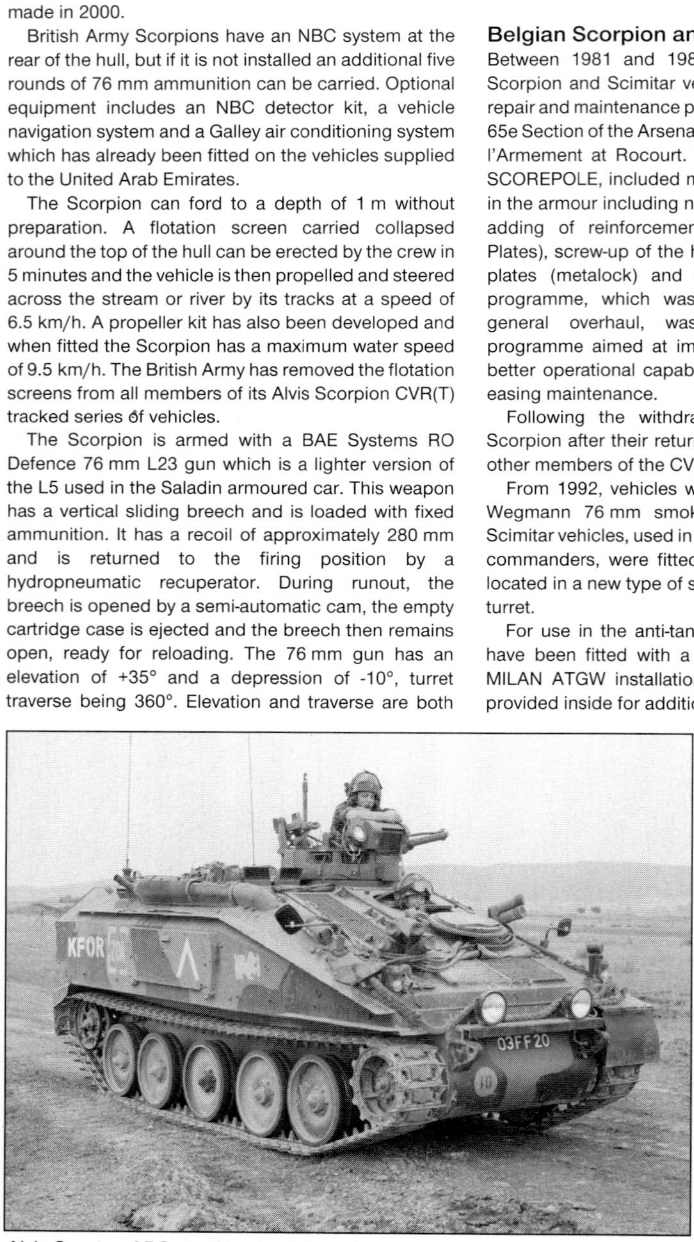

Alvis Spartan APC used by the British Army (Richard Stickland) **2002**/0121604

Alvis Samaritan armoured ambulance in travelling order (Richard Stickland)
2002/0121603

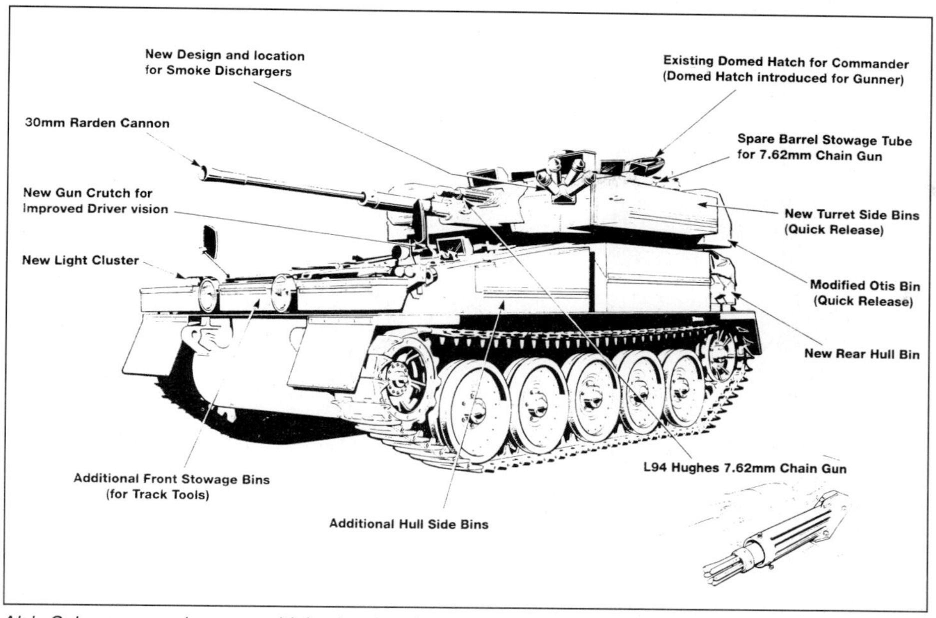

Alvis Sabre reconnaissance vehicle showing the key areas of improvement to the hull and turret

series turbocharged diesel developing 200 hp which for this application has been supplied by Sabre Engineering Limited. This is coupled to a TN15D seven speed transmission that has been upgraded to include an electro-hydraulic fully automatic control system.

The complete electrical system of the Scorpion has been redesigned for ease of operation and repair and new electronic drivers control panel and centralised turret electrical panel have been fitted.

The upgraded Scorpion has been fitted with the Thales Optronics Mithras thermal fire-control system which is based on the companies High Definition Thermal Imager (HDTI) product range.

The thermal camera is mounted on the right side of the mantlet so that it is coaxial with the 76 mm gun. This can be used for reconnaissance and fire control purposes under day and night conditions. Roof mounted Sabre day/night sights can be fitted.

The driver can have a Thales Optronics night driving periscope or a new system which comprises an un-cooled thermal device for forward movement and a day CCD (Charge Coupled Device) reversing camera with the picture being displayed on a flat panel display.

The BAE Systems, RO Defence 76 mm L23 gun is retained as is the 7.62 mm coaxial machine gun. As an option the 76 mm L23 gun could be replaced by a more recent and effective Cockerill 90 mm gun.

New Zealand Scorpions

New Zealand purchased 26 Scorpions as reconnaissance vehicles and tank training expedient and these entered service in August 1982. These have been withdrawn from service and been sold.

Omani Scorpions

Scorpions replaced the Alvis Saladin armoured car in Oman and between 30 and 50 vehicles were delivered between late 1982 and late 1983, plus three Samson armoured recovery vehicles. A second order for up to 30 vehicles was delivered in 1985; this included Scorpion, Sultan, Spartan and Samson vehicles. The Omani Scorpions had many modifications, including installation of a central warning system; fitting of a coolant temperature gauge connected to a fire warning horn; larger rubber mud flaps at rear; glass fibre track shrouds replaced by rubber shrouds; driver's mine plate increased to 20 mm; larger rear hull stowage bin; left tool bin replaced by larger bin with facility to stow two extra jerricans; additional stowage bin on right side of hull; additional insulation for engine/driver bulkhead; no NBC pack so five rounds of 76 mm ammunition and three boxes of machine gun ammunition in its place; No 25 reel and bracket installed; turret fume extractor fitted; Helio MG pintle fitted adjacent the commander's hatch aperture; gun selector box lights duplicated at commander's station; stowage bins either side of turret; time switch connected to fire warning circuit to isolate horn after 5 to 10 minutes operation in event of false alarm; removable sealed plates fitted to engine/ driver bulkhead to permit access to distributor and oil filter; latest type fuel filler and gauze fitted; heater removed and spare driver's sight carried; Fox-type silencer fitted; latest jacking strut fitted; electrical feed to elevating handle to be slit-ring type; rear turret bin catches strengthened; warning notices in Arabic; speedometer in km/h in lieu of mph and fitted with trip distance recorder as well as total distance recorder (also in km); 7.62 mm ammunition box brackets to be modified to suit pattern used in Oman; two-colour camouflage paint; air-circulation system fitted using fan normally incorporated in NBC system; mounting points to enable Clark-type radio mast to be erected at front of vehicle; AV No 54 sight mounting and turnbuckle supplied and L20 laser sights to complement the L22 Tank Laser Sight fitted in Omani Chieftain MBTs. The latter have now been withdrawn from service and placed in reserve.

Late in 2000, Oman placed a contract with Alvis Vehicles for the upgrade of 60 CVR(T) vehicles. This upgrade is similar to that carried out to British Army vehicles under its LEP programme and included the replacement of the current Jaguar petrol engine by a more fuel efficient Cummins diesel engine. The work was carried out in Oman with Alvis Vehicles supplying the actual conversion kits.

Scorpion 90

For the export market Alvis has fitted a Scorpion with a 90 mm Cockerill Mk III gun. The turret of the Scorpion 90 is designated the AC 90 and is almost identical to that fitted to the basic vehicle.

The 90 mm gun has an elevation of +30° and a depression of -8° which is almost the same as that of the basic Scorpion. The first production orders were fitted with the American Cadillac Gage (now HR Textron) power traverse and elevation system which has now been superseded by the now BAE Systems, RO Defence equipment.

The basic Scorpion carries 40 rounds of 76 mm ammunition and the Scorpion 90 carries 34 rounds. Late in 1981, Malaysia ordered 26 Scorpion 90s fitted with the Perkins Engines Company diesel engine. Production commenced in 1982 and deliveries are now complete.

The Malaysian vehicles also have German 76 mm Wegmann smoke/fragmentation grenade dischargers fitted to give 360° coverage. Early in 1983, it was confirmed that 33 Nigerian Scorpions were fitted with the Belgian OIP-5 fire-control system and a Cockerill 90 mm Mk III weapon.

In mid-1988, Venezuela placed an order with Alvis for the supply of an undisclosed number (believed to be 84) of Scorpion 90 mm vehicles (see Development).

Scorpion improvement programme

In 1980, Alvis announced that a Scorpion with its Jaguar XK petrol engine replaced by a Perkins Engines Company turbocharged six-cylinder diesel model T6.3544 would be offered for export. This model entered production for Malaysia in 1982 with first deliveries made in 1983. The engine, for this application, develops 200 bhp at 2,700 rpm. Performance comparison between the diesel engine and the basic petrol engined versions is given below:

Model	**Diesel**	**Petrol**
Weight	8,260 kg	8,073 kg
Ground pressure	0.384 kg/cm²	0.376 kg/cm²
Road range	over 840 km	Over 640 km
Acceleration to 48.3 km/h	18.5 s	16 s

In October 1986, Ferranti announced that the now BAE Systems had ordered an additional batch of Ferranti Type 520 laser range-finders for incorporation in the Vickers L20 gunner's sight. A total of 160 L20 sights was provided to an Asian customer over a four year period. It is believed that the customer was Thailand which is the largest user of Scorpion in Asia.

Some of the Scorpions used by the United Arab Emirates have been fitted with a Barr & Stroud (now Thales Optronics) Tank Laser Sight similar to the one fitted to the Chieftain MBT but with a Nd:YAG laser in place of the original ruby laser and a graticule for a 76 mm instead of 120 mm gun.

The 76 and 90 mm Scorpion series of light armoured vehicle could be fitted with a wide range of fire-contol systems, although none of these was adopted by the British Army.

Alvis Sultan command vehicle in travelling configuration (Richard Stickland) **2001**/0103429

Alvis Scimitar fitted with SPIRE (Richard Stickland) **2001**/0103430

A complete package of improvements is now offered on export vehicles. This included:

Avimo NV 3001 system
Avimo NV53 day/night commander's sight
Avimo NVL53 day/night gunner's sight with integral laser range-finder
Power traverse and elevation drives
External 7.62 mm GPMG mount which can be used by commander or gunner
Revised final drive units
Revised sprocket carrier and rear idler unit
Longer life tracks
New fume extraction system
New electrics with low-maintenance batteries
Improved transmission
Improved suspension and enhanced cross-country mobility
Fire detection and suppression system.

A range of thermal imaging sights is now offered to future and existing Scorpion customers.

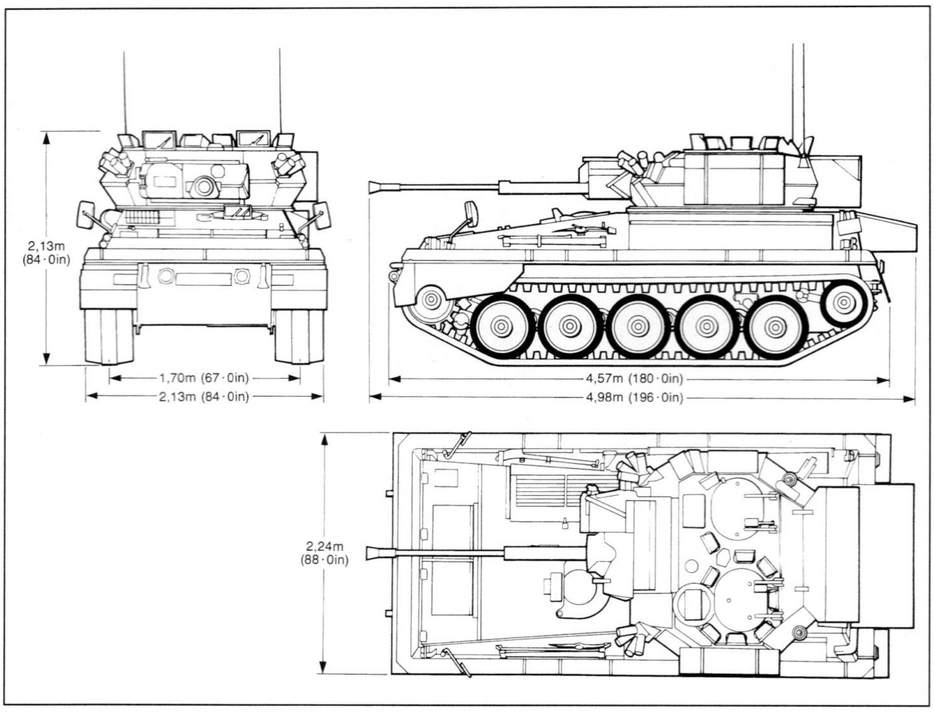

Three-view drawing of the Alvis Scimitar reconnaissance vehicle armed with a 30 mm RARDEN cannon

Latest production Alvis Scorpion armed with 90 mm gun and revised external stowage **2002**/0121602

Sabre reconnaissance vehicle

Following extensive trials with two prototype Sabre vehicles at the Armoured Trials and Development Unit at Bovington, this vehicle was accepted for service with the British Army in August 1993.

The two prototype Sabre vehicles were prepared by Alvis Vehicles Limited, the design authority for the Combat Vehicle Reconnaissance (Tracked) family of vehicles. Sabre has now replaced the Alvis Scorpion vehicle armed with a 76 mm gun as it has been withdrawn from service with the British Army.

Sabre essentially consists of the Scorpion chassis fitted with a turret previously installed on the Combat Vehicle Reconnaissance (Wheeled) Fox armoured car which has been withdrawn from service with the British Army. The Fox turret is armed with a BAE Systems, RO Defence 30 mm L21 RARDEN cannon and a 7.62 mm L37A2 coaxial machine gun.

In addition to the installation of the two-man manually operated Fox turret, the following modifications have been carried out to the hull and turret: new design and location of 66 mm smoke grenade dischargers; new domed hatch for gunner with commander retaining existing domed hatch; spare barrel stowage tube for 7.62 mm Chain Gun; replacement of 7.62 mm L37A2 coaxial machine gun with 7.62 mm The Boeing Company Chain Gun (made under licence in the UK) also fitted to the Warrior mechanised combat vehicle and Challenger 2 MBT; new turret side bins with quick-release mechanism; new gun crutch for improved driver vision; new light cluster on front of hull and new front stowage bins for track tools; modified OTIS bin on turret rear with quick-release mechanism; new rear hull bin and additional side bins.

Alvis Vehicles supplied the kits to the British Army which carried out the work at ABRO (Army Base Repair Organisation) facility at Donnington, with first vehicles completed in 1994. It is understood that 138 Scorpion vehicles were converted to the Sabre configuration.

Striker ATGW vehicle (FV102)

Details of this vehicle, which is now only used by the British Army, are given in the Tank destroyers section.

Spartan armoured personnel carrier (FV103)

This entered service in 1978 and is used for a variety of roles such as carrying Royal Artillery Javelin/Starburst man-portable SAM teams, as a missile resupply carrier for the Striker and for carrying Royal Engineer assault teams. It was not the replacement for the FV432 APC.

The Spartan has a similar hull to the Striker and can carry four infantrymen in addition to the driver, vehicle commander/gunner and section commander/radio operator.

The driver is seated at the front of the hull on the left and has a one-piece hatch cover that opens forwards. He is provided with a single wide-angle day periscope which can be replaced with a passive periscope for night driving.

The vehicle commander/gunner is seated behind the driver and has a No 16 cupola with eight day periscopes and a monocular sight with a magnification of ×1 or ×10, the line of sight elevating with the machine gun. The ×1 mirror assembly can be replaced by a ×1.8 image intensifier. Mounted on the right side of the cupola is a 7.62 mm machine gun which can be aimed and fired from inside the vehicle. The cupola has a single-piece hatch cover that opens to the left.

To the right of the commander/gunner is the section commander/radio operator who is provided with three day observation periscopes and a single-piece hatch cover that opens to the right.

Entry to the personnel compartment at the rear of the vehicle is by a single door in the rear of the hull hinged on the right side, which has an integral vision block. Over the top of the troop compartment are two roof hatches that open either side of the vehicle. There are two day periscopes provided in the left side and one in the right side of the troop compartment but there is no provision for the crew to fire their weapons from inside the vehicle. The three infantrymen are seated on the left side of the vehicle with a fourth to the rear of the vehicle commander/gunner and section commander/radio operator, facing the rear.

Stormer armoured personnel carrier

Details of this vehicle, previously known as the FV4333, are given in the Armoured personnel carriers (tracked) section, as are details of a new family of light tracked combat vehicles.

Samaritan Armoured Ambulance (FV104)

This model entered production early in 1978, and has the same hull as the Sultan command vehicle and has a crew of two: commander/medical orderly and driver. It is unarmed and can carry four stretcher cases, or five sitting cases, or two stretcher and three sitting cases. There is a large door at the rear of the hull, hinged on the right, with an integral vision block. The commander has five day periscopes for all-round observation, plus a single wide-angle day periscope with a forward-sloping window with an interchangeable passive night sight. The single-piece hatch cover opens to the left. Medical supplies can be carried on the top of the hull and at the rear of the vehicle.

Sultan Armoured Command Vehicle (FV105)

First production Sultans were delivered to the British Army in April 1977. The vehicle has a similar hull to the Samaritan and has a crew of five or six: commander/radio operator, radio operator and driver plus two or three additional crew members. The command area is at the rear of the vehicle and to increase the amount of working space a tent can be quickly erected at the rear of the vehicle. The vehicle is normally equipped with at least two radios (one forwards and one rearwards), and mapboards. A radio antenna can be erected at the front of the vehicle when it is being used in the static role. The commander has a No 27 cupola with five day periscopes for all-round observation, plus a single wide-angle day periscope with a forward-sloping window and an interchangeable passive night sight. The single-piece hatch cover opens to the left. Armament consists of a pintle-mounted 7.62 mm machine gun and four 66 mm smoke grenade dischargers mounted on either side of the vehicle towards the front.

Alvis Samson armoured recovery vehicle with A frame erected and spades lowered (Christopher F Foss) 0058337

Alvis Vehicles 76 mm Scorpion upgraded in Jordan by the King Hussein Workshops and showing Thales Optronics thermal camera to right of 76 mm gun (Christopher F Foss) 0058335

Samson Armoured Recovery Vehicle (FV106)

This entered service with the British Army in 1978 and has a hull similar to that of the Spartan APC. The driver is seated at the front of the vehicle on the left and has a single-piece hatch cover that opens forwards and a single wide-angle day periscope which can be replaced by a passive night periscope. The commander's No 27 cupola has a single wide-angle day periscope which can be replaced with a passive night periscope, five day periscopes for all-round observation and a single-piece hatch cover that opens to the left.

The heavy-duty winch fitted inside the hull is driven from the main engine and has a variable speed of up to 122 m/min on the 229 m of wire rope. Maximum pull, with a 4:1 snatch block, is 12,000 kg. The winch leads to the rear of the vehicle and when being used for recovery operations two spades are lowered manually at the rear of the vehicle. An A-frame can be erected at the rear of the hull. Tools, a bench, towbars and tow cables are fitted as standard. Armament comprises a pintle-mounted 7.62 mm machine gun and four 76 mm smoke grenade dischargers mounted on either side of the vehicle towards the front. Full details of the Samson ARV are given in *Jane's Military Vehicles and Logistics 2001-2002.*

Specifications

Model	**Scorpion**	**Scorpion 90**	**Spartan**	**Samaritan**	**Sultan**	**Samson**	**Scimitar**
Designation	FV101	n/a	FV103	FV104	FV105	FV106	FV107
Crew	3	3	3+4	2	5-6	3	3
Combat weight	8,073 kg	8,723 kg	8,172 kg	8,664 kg	8,664 kg	8,738 kg	7,800 kg
Power-to-weight ratio	23.54 bhp/t	22.92 bhp/t	23.25 bhp/t	21.93 bhp/t	21.93 bhp/t	21.74 bhp/t	24.5 bhp/t
Ground pressure	0.36 kg/cm²	0.37 kg/cm²	0.338 kg/cm²	0.358 kg/cm²	0.358 kg/cm²	0.358 kg/cm²	0.338 kg/cm²
Length							
(gun forward)	n/a	5.288 m	n/a	n/a	n/a	n/a	4.985 m
(hull (incl stowage boxes where applicable))	4.794 m	4.794 m	5.125 m	5.067 m	4.8 m	4.788 m	4.794 m
Width	2.235 m	2.226 m	2.242 m	2.242 m	2.254 m	2.43 m	2.242 m
(over tracks)	2.134 m	2.134 m	2.134 m	2.134 m	2.134 m	2.134 m	2.134 m
Height							
(overall)	2.102 m	2.102 m	2.26 m	2.416 m	2.559 m	2.254 m	2.096 m
Ground clearance	0.356 m	0.419 m	0.356 m	0.356 m	0.356 m	0.356 m	0.42 m
Track	1.708 m	1.708 m	1.708 m	1.708 m	1.708 m	1.708 m	1.708 m
Track width	432 mm	432 mm	432 mm	432 mm	432 mm	432 mm	432 mm
Number of links per track	79	79	84	84	82/84	84	79
Max road speed	80.5 km/h	72.5 km/h	80.5 km/h	72.5 km/h	72.5 km/h	72.5 km/h	80.5 km/h
Fuel capacity	423 litres	391 litres	386 litres	395 litres	395 litres	404 litres	423 litres
Max road range	644 km	756 km	483 km	483 km	483 km	483 km	644 km
Fording	1.067 m	1.067 m	1.067 m	1.067 m	1.067 m	1.067 m	1.067 m
	all members of the Scorpion family are amphibious with preparation*						
Freeboard							
(forward with screen erected)	0.965 m	0.965 m	0.914 m	0.832 m	0.883 m	0.965 m	0.965 m
(aft with screen erected)	0.711 m	0.711 m	1.168 m	0.559 m	0.457 m	0.815 m	0.711 m
Gradient	60%	60%	60%	60%	60%	60%	60%
Vertical obstacle	0.5 m	0.5 m	0.5 m	0.5 m	0.5 m	0.5 m	0.5 m
Trench	2.057 m	2.057 m	2.057 m	2.057 m	2.057 m	2.057 m	2.057 m
Turning radius	all members can pivot turn in neutral						
(in 1st gear)	1.71 m	1.71 m	1.84 m	1.84 m	1.84 m	1.84 m	1.71 m
(in 7th gear)	33.22 m	33.22 m	35.97 m	35.97 m	35.97 m	35.97 m	33.22 m
Engine	Jaguar J60 No 1 Mk 100B 4.2 litre 6-cylinder petrol developing 190 hp at 4,750 rpm**						
Transmission	TN15 crossdrive, semi-automatic hot-shift type, providing 7 speeds in each direction and pivot turns						
Steering	Merritt system incorporated in transmission						
Suspension	torsion bar on all members of Scorpion family						
Electrical system	28.5 V	28.5 V	28.5 V	28.5 V	28.5 V	28.5 V	28.5 V
Batteries	4	4	4	4	8	4	4
Main armament	76 mm	90 mm	n/a	none	n/a	n/a	30 mm
Machine gun	7.62 mm	7.62 mm	7.62 mm	none	7.62 mm	7.62 mm	7.62 mm
Smoke grenade dischargers	2 × 4	2 × 4	2 × 4	2 × 4	2 × 4	2 × 4	2 × 4
Ammunition							
(main)	40	33	n/a	none	n/a	n/a	165
(machine gun)	3,000	3,000	3,000	none	2,000	2,000	3,000
Gun control equipment	manual	powered	n/a	none	n/a	n/a	manual
Gun elevation/depression	+35/-10°	+30/-8°	n/a	none	n/a	n/a	+35/-10°
NBC system	yes	yes	yes	yes	yes	yes	yes
Night vision equipment	yes	yes	yes	yes	yes	yes	yes

* This capability has been removed from British vehicles.
** All members of the Scorpion are now offered with a Perkins Engines Company diesel engine developing 200 bhp at 2,700 rpm and details of the Scorpion 90 relate to a vehicle with this engine. British Army CVR(T) vehicles have been refitted with a new Cummins diesel engine.

Spartan 2000 has been developed by Repaircraft for the export market
2000/0075740

Scorpion 2000 has been developed by Repaircraft for the export market
2000/0075739

Scimitar Reconnaissance Vehicle (FV107)

The first prototype of the Scimitar was completed in July 1971 and accepted for service in June 1973. First deliveries were made to the British Army in March 1974 and first deliveries to the Belgian Army in April 1974. It has the same hull and turret as the Scorpion but is armed with a 30 mm L21 RARDEN cannon instead of the 76 mm gun.

The 30 mm cannon fires a number of British rounds and an APDS-T round developed jointly by BAE Systems, RO Defence and PATEC of the USA. In addition to this APDS-T round the 30 mm L21 RARDEN fires British APSE-T, HEI-T and PRAC-T ammunition and Oerlikon Contraves HE, AP and PRAC rounds.

Following an international competition, in 1996, the UK MoD awarded BAE Systems, RO Defence a contract to design, develop and manufacture an Armour-Piercing Fin-Stabilised Discarding Sabot - Tracer (APFSDS-T) round to meet its requirement for a 30 mm Armour-Piercing Enhanced Performance (APEP) round. Due to development problems, this contract was cancelled by the MoD late in 1998 and in 2000 BAE Systems, RO Defence, was again selected to be overall prime contractor for APEP although no in service date is currently available.

The gun is fired in rapid single shots but bursts of up to six rounds can be fired if required. The empty cartridge cases are automatically ejected forwards outside the turret.

A 7.62 mm machine gun is mounted coaxially to the left of the main armament and four 66 mm smoke grenade dischargers are mounted either side of the front of the turret. The main armament, coaxial machine gun and smoke grenade dischargers are electrically operated, with the main armament and coaxial machine gun having manual overriding control. The Royal Air Force uses the Scimitar in the EOD role on airfields.

CVR(T) LEP

In addition to the introduction of the Sabre vehicle covered in a separate entry, there is also the Life Extension Programme (LEP) for part of the remainder of the British Army Combat Vehicle Reconnaissance (Tracked) fleet of vehicles which will remain in service well into the next century.

LEP is a three-phase programme and includes the replacement of the current petrol engine by a more fuel efficient diesel engine, installation of thermal night observation equipment and installation of a secure communications system.

The CVR(T) was introduced over 25 years ago and, apart from some improvements in the suspension, has never been upgraded or improved. The night vision equipment on the Scimitar is still of the first-generation image intensification type.

Experience in the Gulf demonstrated the superiority of thermal imaging equipment, not only for acquiring and engaging targets under both day and night conditions but also for essential battlefield surveillance.

In mid-1997 the UK MoD confirmed that it was to install a thermal observation and sighting system into its current fleet of 170 Scimitar vehicles.

In mid 1998, following an international competition, the UK MoD awarded Avimo a contract for the supply of an undisclosed quantity, believed to be 104, SPIRE (Sight Periscopic Infra-Red Equipment), thermal imaging systems, plus spares.

Deliveries commenced within nine months and the new batch of SPIRE systems were fitted with an enhanced linear cooling system which is quieter and more reliable. This was backfitted to the existing SPIRE systems already deployed by the British Army.

Under previous contracts, Avimo has already supplied and installed 66 SPIRE systems under an Urgent Operational Requirement (UOR) programme for Scimitar vehicles deployed with the British contingent of SFOR (Stabilisation FORce) operating in Bosnia.

SPIRE replaced the current image intensification sight and allows the commander or gunner of the Scimitar to lay the 30 mm RADEN cannon onto the target under almost all weather conditions.

SPIRE was developed by Avimo as a private venture and includes an integral laser range-finder and fire-control computer together with the HIRE thermal imaging module manufactured by Raytheon Systems Corporation of the US. HIRE consists of the sensor and gunners display, commanders remote display and a power/electronics unit.

The British Army version of SPIRE is called E-SPIRE but other versions have been developed by Avimo including second generation (M-SPIRE) and stabilised second generation (U-SPIRE).

In August 1997, following extensive trials between Alvis Vehicles Combat Vehicle Reconnaissance (Tracked) Scorpion vehicles fitted with diesel engines supplied by the Cummins Engine Company (UK), Perkins Engines Company and Steyr-Daimler-Puch (Austria), the UK MoD selected the Cummins 6BTA 5.9 litre six-cylinder diesel developing 235 hp.

The UK MoD awarded Alvis Vehicles a contract worth £32 million for the supply of 1,107 diesel conversion kits for the British Army's Combat Vehicle Reconnaissance Vehicle (Tracked) (CVR(T)) fleet of vehicles and to covert the first 200 of these vehicles.

Following the placing of this contract, Alvis Vehicles was awarded a contract to convert seven CVR(T) vehicles for a further series of extensive trials.

Four converted vehicles, two Scimitar and two Spartan, underwent a 40,000 km trial at the Armoured Trials and Development Unit (ATDU) and another three variants were used for supporting trials.

The existing David Brown Engineering (previously Self Changing Gears) semi-automatic TN15 transmission is retained with some modifications.

Work on the upgrade of the 200 CVR(T) started in the first quarter of 1999 and were completed early in 2000.

The current CVR(T) Jaguar 4.2 litre petrol engine has been removed and replaced by a more fuel efficient Cummins 6 BTA 5.9 diesel engine which is also more reliable, durable as well as meeting the latest European emission regulations. The conversion will allow the vehicle to run on either diesel fuel or AVTUR aviation fuel without adjustment.

The installation of the new Cummins diesel engine will extend the life of the CVR(T) to 2015. Vehicles to be converted are 30 mm Scimitar .and Sabre and specialised variants such as Samaritan ambulance, Spartan armoured personnel carrier, Sultan command vehicle, Samson recovery vehicle and Striker anti-tank vehicle.

Alvis Vehicles and ABRO then each converted a further 35 vehicles and ABRO was awarded the contract for the remaining vehicles. In the Spring of 2000, ABRO was selected to undertake the upgrading of the remainder of the British Army's CVR(T) fleet of vehicles at its Donnington facility. The kits were provided by Alvis Vehicles.

The CVR(T) dieselisation project is one of ten pilot programmes chosen by the UK MoD to test their Integrated Project Team (IPT) concept, which is a key part of the new Smart Procurement Initiative.

Battle Group Thermal Imaging programme

In August 2001, the UK Defence Procurement Agency (DPA) awarded Thales Optronics a £230 million contract for British Army's major Battle Group Thermal Imaging (BGTI) programme.

Three UK contractors, Avimo, Hunting Engineering and Thales Optronics submitted their BGTI bids late in 2000. Of the £230 million contract, it has been estimated that about two-thirds will be for procurement and one-third for support.

When fielded, BGTI will replace the current image intensification sights installed on most members of the Alvis Vehicles Warrior Infantry Combat Vehicle (ICV) and the Alvis Vehicles Scimitar used in the close reconnaissance role.

The contract covers the supply of two systems, Group 1 and Group 2, with first vehicles to be equipped in 2003. The installation of the BGTI system will allow the Warrior and Scimitar vehicles to engage targets at longer ranges and under almost all weather conditions.

In addition, BGTI will enable the crew to rapidly establish its own position and the six figure co-ordinates of other targets on the battlefield. It can also establish the range of targets using a laser range-finder with the driving route map controlled by the commander and displayed to the driver.

Group 1 vehicles are the Warrior ICV and the Scimitar reconnaissance vehicles armed with a 30 mm RARDEN Cannon. These will have a new gunner's sight with day/thermal imager and eye-safe laser range-finder. The commander has a monitor and flat panel display which will provide access to Bowman and battle management systems being procured under separate contracts. The driver has a navigation capability.

For the Group 2 vehicles, which include the Warrior repair and recovery vehicles armed with a cupola mounted 7.62 mm machine gun, a less advanced system will be procured.

The first Group 1 contract covers 450 Warrior and 150 Scimitar vehicles with the Group 2 contract covering 100 Warrior repair and recovery vehicles. There is also an option on another 200 Group 1 and 50 Group 2 systems.

Cummins engine

While all production CVR(T) for the Army were powered by a Jaguar 4.2 litre petrol engine, export vehicles for some years have been powered by a Perkins Engines Company diesel engine developing 200 hp.

Alvis Vehicles are now offering the more recent Cummins B series engine which has been installed in most of the British Army's operational CVR(T) vehicles. The first export customer for this is Oman.

Suspension upgrade

The Cummins B series engine is also installed in the British Army new Shielder minelaying vehicle, the first three examples of which were completed in 1998. Prime contractor for this system is Alliant Techsystems of the US.

Part of the CVR(T) fleet has already been fitted with Messier hydraulic dampers giving a 25 per cent improvement in cross-country mobility over the existing vehicle.

Sturgeon and Salamander

These are vehicles converted by the British Army for use in the Opposing Forces (OPFOR) role at the British Army Training Unit Suffield (BATUS) in Canada. The Sturgeon is based on the Spartan while the Salamander is based on the 76 mm Scorpion which is no longer used in the reconnaissance role.

Scimitar with SPIRE

Following extensive trials with competing systems, late in 1995 the UK MoD placed an order with Avimo for the supply of the SPIRE sighting system for installation on the Alvis Scimitar vehicles deployed with the British Army in Bosnia.

SPIRE includes an integral laser range-finder which is a standard production system, a fire-contol computer developed by Avimo and a HIRE thermal imaging module manufactured by the Raytheon Systems Company.

SPIRE consists of three main assemblies, sensor and gunner's display, commander's remote display and power supply/electronics unit.

Avimo delivered the systems to the British Army who carried out the actual installation work in Bosnia. No actual modifications are required to the vehicle itself.

A total of 66 SPIRE systems was delivered to the British Army to meet an Urgent Operational Requirement in Bosnia. Since then an additional 104 units have been ordered to an improved design. These have now been delivered.

Streaker HMLC

Late in 1982, Alvis completed the prototype of the private venture Streaker high-mobility load carrier which is based on the chassis of the Spartan APC. This is powered by the standard 4.2 litre Jaguar petrol engine which could be replaced by the Perkins Engines Company T6.3544 200 bhp turbocharged diesel coupled to an automatic transmission. Maximum road speed is 84 km/h, road range is 480 km, the unloaded weight is 5,465 kg and it has a payload of 3,000 kg.

The baseline vehicle has a flat deck and offers two front configurations, one with a lightweight cab structure which is unarmoured, the other with the driver and engine compartment having full armour protection.

By 1984, two prototypes of the Streaker high-mobility load carrier had been built. Typical roles suggested for Streaker include towing the BAE Systems, RO Defence Barmine layer, being fitted with the Belgian SA LAU 97 40-round 70 mm multiple rocket launcher (details in the Multiple rocket launchers section), excavator, refueller, trench digger, missile carrier, towing the 105 mm Light Gun, carrying anti-aircraft systems, general cargo carrier anti-tank minelayer. Despite extensive marketing the Streaker was never placed in production. Further development by Alvis on the Stormer chassis resulted in the Stormer HMLC of which a quantity was supplied to the British Army in the minelaying role in early 1991. A modified version is used by Indonesia, also based on the Stormer chassis.

Repaircraft Scorpion 2000 upgrade

In 1999 Repaircraft launched its Scorpion 2000 upgraded reconnaissance vehicle/light tank which has been developed specifically for the export market.

The overall aim of Repaircraft is to improve the fighting capability of the Scorpion vehicle as well as increasing its reliability and reducing its operating costs.

Standard equipment on Scorpion 2000 includes auxiliary power unit secondary generator for standby battery charging, new commander's and gunner's night vision sights, turret fume extractor, improved engine cooling, brushless heavy-duty generator/electronic regulator, twin automatic fire suppression systems and ballistic fibre fuel cell with integral anti-slosh and explosion suppression foam.

The original Jaguar 4.2 litre petrol engine is retained in the Scorpion 2000 vehicle, but the installation of the auxiliary power unit secondary generator, allows the vehicle to use all of its onboard electrical equipment with the main engine switched off, so saving engine life. The 76 mm gun of the Scorpion is also retained in the Scorpion 2000.

Optional equipment on the Scorpion 2000 includes laser range-finder, driver's passive night periscope, externally mounted 7.62 mm or 12.7 mm machine guns, larger additional hull stowage bins, upgraded suspension, spall liners, applique passive armour, power turret traverse and weapon elevation and various navigation, communications and fire-control options.

A number of companies are subcontractors to Repaircraft including Avimo for the Day Night Gunnery Sight (DNGS) which is available in a range of modular upgradable options including image intensification, thermal image (uncooled 8 to 12 μm), thermal image (cooled 3 to 5 μm) and integral eye-safe laser range-finder.

The Firetrace automatic fire suppression system is a fully automatic system which attacks fires at their sources. Trace tubers are routed through risk areas and if the temperature exceeds 150°C it erupts where heated, at the source of the fire, automatically discharging halon extinguishant. Simultaneously an alarm is sounded and the fuel cuff-off valve is actuated. The system is self-monitoring and requires only periodic servicing and uses environmentally friendly recycled halon gas.

Galley Limited has provided the auxiliary power unit secondary generator and heavy-duty air conditioning system (APU SG/AC) that allows battery charging for the vehicle radios and vetronics which enables the main engine to be shut down when the vehicle is stationary. According to Repaircraft this can increase the main engine life by up to 200 per cent so reducing maintenance and most importantly, minimising vehicle operating costs.

Premier Fuel Systems have provided the fuel cell for Scorpion 2000 which consists of a flexible ballistic fibre reinforced container filled with military specification safety foam.

The C E Niehoff brushless generator/regulator reduces radio interference, is more efficient and has a greatly increased mean time between failure. In addition, the built-in electronic regulator, with reverse polarity and split charging facility, is now not only more simple but also more reliable.

The company is also offering to replace the current 76 mm gun with a Cockerill 90 mm gun and replace the Jaguar petrol engine by a Steyr-Daimler-Puch M16 diesel engine.

As well as providing kits to enable the user to upgrade its existing Scorpion vehicles to the latest standard, Repaircraft can also provide Scorpion 2000 vehicles upgraded and ready for use, as it has a stock of surplus Scorpion vehicles and variants.

Repaircraft can also supply kits to upgrade other members of the Scorpion family such as the Spartan to the same automotive standard as the Scorpion 2000, this is called the Spartan 2000.

Scorpion with Helio turrets

For trials purposes, an Alvis Scorpion chassis has been fitted with the Helio FVT925 and Helio FVT940 one-person turrets. The former is armed with a 25 mm M242 cannon and 7.62 mm coaxial machine gun, while the latter is armed with a 40 mm grenade launcher and 12.7 mm machine gun.

Digitised British Army Scimitars

About 40 British Army Scimitars have been upgraded with a Tactical Navigation and Target Location System. This integrates the laser range-finder part of SPIRE (qv) with a KVH Industries TacNav fluxgate compass-based position and navigation system, a north pointer and a Rockwell PLGR GPS receiver.

Manufacturer

Alvis Vehicles Limited.

UPDATED

Status

Production. In service with:

Country	Quantity	Comment
Belgium	701	all members of family, 76 mm Scorpion and Striker phased out of service*
Botswana	20	delivered 1995/96, 76 mm, diesel, ex-Belgian Army
Brunei	19	16 Scorpion, 2 Sultan, 1 Samson
Chile	30	ex-British Army, 76 mm version, delivered 1995 (12 + 18)
Honduras	15	12 Scorpion, 3 Scimitar, 1 Sultan
Iran	80	Scorpion 76 mm, approx number
Indonesia	121	Scorpion 90 diesel, plus Stormer APCs with deliveries from 1995 through to 1997
Ireland	14	14 Scorpion 76 mm, delivered 1980
Jordan	19	Scorpion, from Iraq (captured from Iran), non-operational
Malaysia	26	26 Scorpion 90 diesel, delivered 1983
Nigeria	100	Scorpion 76 mm, approx number, some sources state 140
Oman	60	Scorpion 76 mm, Spartan, Samson, Sultan, 4 Stromer command post vehicles (upgraded with Cummins diesel)
Philippines	41	Scorpion 76 mm, approx number
Spain	17	Scorpion 76 mm diesel (marines), delivered 1985
Tanzania	40	Scorpion 76 mm
Thailand	154	Scorpion 76 mm, fitted with laser range-finder
Togo	12	Scorpion 90 diesel, delivered 1986
UAE	76	Scorpion 76 mm
UK	1,500	76 mm no longer used
Venezuela	80	Scorpion 90, Samson, Sultan (diesel), delivered 1989 to 1992

* Some of these vehicles have been sold and by 1998 the only combat version in service was the Scimitar with 141 in the fleet of which 29 were in reserve.

TANK DESTROYERS

TANK DESTROYERS

INTERNATIONAL

Other tank destroyers

The vehicles described in this section have mainly been developed specifically for the tank destroyer role. There are, however, many other tank destroyers (tracked and wheeled) in service or developed to the prototype stage that are based on other chassis, normally an armoured personnel carrier's or a reconnaissance vehicle's. Listed here is a *résumé* of the tracked and wheeled tank destroyers that are covered elsewhere. It should be noted that almost any vehicle can be fitted with a standard infantry man-portable MILAN or TOW system so the listing of these missiles is not exhaustive. Recoilless rifles are excluded as these are in service in declining numbers.

Missile	Platform	Comment
HOT	AMX-10P	Has Lancelot turret with four missiles ready to launch, in service with Saudi Arabia.
	Panhard VCR (6 × 6)	Has UTM 800 turret, called VCR/TH, in service only with Iraq, photograph in the Armoured personnel carriers (wheeled) section
	Piranha (8 × 8)	Fitted with HOT Mephisto system. This remains at prototype stage.
	VAB (6 × 6)	Fitted with Mephisto system, in service with the French Army (4 × 4)
	VAB (6 × 6)	Has UTM 800 system, used by Qatar and Cyprus
MILAN	EE-3 Jararaca (4 × 4)	Pintle-mounted, in service with Cyprus
	Marder 1	Most German Army Marder 1s have pintle-mounted MILAN on turret
	M113	Belgium uses this vehicle with pintle-mounted MILAN
	Panhard VBL (4 × 4)	Pintle-mounted, photograph in the Reconnaissance vehicles section (adopted by France and Mexico)
	VAB (6 × 6)	Oman uses pintle-mounted MILAN on this vehicle
	Warrior MCV	Pintle-mounted, UK
	Spartan	Pintle-mounted Belgium
'Sagger'	BMP-1	Launcher rail over 73 mm gun, photograph in the Armoured personnel carriers (tracked) section, standard on all vehicles
	BMD-1	Launcher rail over 73 mm gun, photograph in the Armoured personnel carriers (tracked) section, standard on all vehicles
	BOV-1 (4 × 4)	Slovenian chassis with six 'Sagger'-type ATGWs, photograph in the Armoured personnel carriers (wheeled) section, standard vehicle
	BRDM-1	Sextuple launcher, photograph in the Reconnaissance vehicles section, standard vehicle
	BRDM-2	Five missiles ready to launch, photograph in the Reconnaissance vehicles section, standard vehicle
	M-80	Two missiles ready to launch on turret, photograph in the Armoured personnel carriers (tracked) section, standard vehicle
	OT-64A (8 × 8)	Two missiles ready to launch above troop compartment (probably no longer in service)
	OT-64C (1) (8 × 8)	One missile either side of turret (probably no longer in service)
	BMP-23/BRM-23	One missile on turret roof
'Spandrel'	BMP-2	One missile ready to launch, photograph in the Armoured personnel carriers (tracked) section, standard vehicle
	BRDM-2 (4 × 4)	Five missiles ready to launch, photograph in the Reconnaissance vehicles section, standard vehicle
	BMD-3	One missile ready to launch, photograph in Armoured personnel carriers (tracked) section, standard on all vehicles
'Spigot'	BMD-1	One missile pintle-mounted on turret, photograph in the Armoured personnel carriers (tracked) section
	BMP-1	Some vehicles have pintle-mounted 'Spigot' in place of 'Sagger'
	BMP-2	Some vehicles have 'Spigot' in place of 'Spandrel'
	BRDM-2 (4 × 4)	Some vehicles have three 'Spandrel' and two 'Spigot' ATGWs
	BMP-30	One missile pintle-mounted on vehicle turret
Swingfire	Striker	Five missiles ready to launch, standard vehicle, photograph in this section. Only used by UK.
TOW	AIFV	Fitted with ITV launcher, used by the Royal Netherlands Army and Egypt
	LAV (8 × 8)	Fitted with ITV launcher for US Marine Corps and Saudi Arabian National Guard, photograph in the Armoured personnel carriers (wheeled) section
	M8 (6 × 6)	Fitted with normal infantry TOW, used by Colombia
	Piranha (6 × 6)	Swiss Army has version with Kvaerner-Eureka turret
	RAM V-1 (4 × 4)	Fitted with normal infantry TOW
	VCC-1	Fitted with ITV launcher, 224 supplied to Saudi Arabia, photograph in the Armoured personnel carriers (tracked) section
	M113	M113 with ITV launcher is known as M901 Improved TOW Vehicle (ITV), standard in many countries
	M113	Fitted with basic infantry TOW
	M113	Norwegian Army has Kvaerner-Eureka turret for its M113 vehicles, also used by Canada (M113A1) and Turkey (AIFV)
	Warrior	Single TOW launcher either side of turret, used by Kuwait

UPDATED

CHINA, PEOPLE'S REPUBLIC

120 mm PTZ89/Type 89 SPATG

Development

Late in 1999 the People's Liberation Army (PLA) finally showed in public for the first time their Type 89 120 mm self-propelled anti-tank gun (SPATG), which is also referred to as the PTZ89. Recent information has stated that trials with the first prototype were carried out early in 1988.

The Type 89 120 mm SPATG is believed to have been introduced into service with the PLA over 10 years ago and is based on the chassis of the Type 83 152 mm self-propelled artillery system which is fully covered in the Self-propelled guns and howitzers (tracked section).

As far as it its known, neither the 120 mm PTZ89/Type 89 120 mm SPATG or the Type 83 152 mm self-propelled gun have been exported by NORINCO (China North Industries Corporation.

Description

The layout of the PTZ89 is similar to the Type 83 152 mm self-propelled artillery system with the driver's compartment front left, engine compartment to the right and fighting compartment at the rear. In the rear of the hull is a door that opens to the right which is not fitted with a vision block or firing port.

This chassis has been fitted with a new all-welded steel turret positioned further forward with an extended bustle which is believed to contain an automatic loader for the 120 mm smoothbore gun and fitted with a thermal sleeve and a fume extractor.

The 120 mm smoothbore gun fires various types of Chinese-developed (one piece) ammunition with a semi-combustible cartridge case. All that remains after firing is the stub cartridge case.

The latest 120 mm Armour Piercing Fin Stabilised Discarding Sabot-Tracer (APFSDS-T) projectile is called the 120-II which has a higher muzzle velocity and greater armour penetration characteristics.

According to NORINCO (China North Industries Corporation), the complete 120-II round weighs 22.5 kg and has an overall length of 1.15m. The projectile is 655 mm long, weighs 7.4 kg and has a muzzle velocity of 1,725 m/s. This will penetrate 600 mm of Rolled Homogenous Armour (RHA) at a range of 2,000 m.

The NORINCO 120 mm smoothbore gun will also fire western types of 120 mm ammunition, such as that manufactured by Rheinmetall of Germany.

The gunner is seated on the left side of the turret with the commander on the right side. The former has a roof-mounted sight and a direct fire sight.

The Type 89 SPATG is also fitted with a roof-mounted 12.7 mm anti-aircraft machine gun for air defence and local protection and mounted on either side of the turret is a bank of four forward firing electrically operated smoke grenade launchers.There is also a stowage basket mounted on either side of the turret.

This 120 mm smoothbore gun is also being offered for installation in Russian T-54/T-55 and Chinese Type 59 Main Battle Tanks currently armed with a 100 mm rifled tank gun.

Suspension is of the torsion bar type with either side having six dual rubber tyred road wheels with the drive sprocket at the front and the idler at the rear. Track return rollers are provided.

Standard equipment includes an NBC system and night vision equipment. The system can ford but has no amphibious capability.

Details of the 120 mm upgrade package for the Type 59 are given in the entry for this vehicle in the MBTs and medium tanks section. As far as it is known, this system remains at the prototype stage.

The 120 mm armed Type 89 self-propelled anti-tank gun **2000**/0085189

The Type 89 has a combat weight of about 30 tonnes, other details are not currently available.

Status
Production complete. In service with China in small numbers.

Manufacturer
Chinese state factories.
NORINCO, China North Industries Corporation.

UPDATED

NORINCO Type 92B Red Arrow tank destroyer

Development
In 2000, NORINCO (China North Industries Corporation) finally released details of their new Red Arrow 9 (Hong Jian 9) anti-tank guided missile (ATGM) system which made its first public appearance during a major parade held in Beijing late in 1999.

The name Red Arrow 9 is for export purposes only, within the People's Liberation Army (PLA) it is referred to as the SD-1 Lightning and it is understood that it first entered service in 1992.

The first application for the Red Arrow 9 ATGM is self-propelled, installed on a modified version of the NORINCO WZ 551 (6 × 6) armoured personnel carrier, but in a 4 × 4 configuration.

This 4 × 4 chassis is also used as the basis for the older Red Arrow 8 ATGM system that has been in service with the PLA for some years.

It is believed that the chassis for the latest Red Arrow 9 application is designated Type 92B with the industrial number being WXLZ602B.

The new Red Arrow 9 anti-tank guided missile system is based on a modified WZ 551 (4 × 4) chassis and has four 5,000 m range missiles in the ready to launch position **2002**/0121596

Description
The hull of the Type 92B 4 × 4 vehicle is of all welded steel armour construction which provides the occupants with protection from small arms fire and shell splinters.

The driver is seated at the front on the left side and has a single-piece hatch cover above his position that opens to the left. To his front are three periscopes, the centre one of which can be replaced by a passive periscope for driving at night. There is another seat at the front right with a single-piece hatch cover that opens to the rear. It is believed that this seat is occupied by the vehicle commander.

There is an additional roof hatch on the right side of the hull just in front of the launcher. This has a single-piece hatch cover that opens to the rear and in front of this is a single rotatable periscope.

To the immediate front of both of these positions is a bullet proof windscreen with a hinged shutter above, which can be rapidly lowered to cover this windscreen. Forward vision is then obtained via roof-mounted periscopes. In addition is a roof-mounted periscope at the front right side of the hull roof. To the front are periscopes.

The power pack is to the immediate rear of the driver's position with the air intake and outlet louvres being located in the roof. The exhaust pipe runs along the left side of the upper part of the hull to the rear. In the rear of the hull is a single larger door that is used for missile resupply purposes. There are also firing ports in either side of the upper part of the hull.

There is a door in the lower part of the hull on either side between the front and rear roadwheels. Another person is seated on the right side of the hull, alongside the engine compartment.

Mounted at the rear is the Red Arrow 9 missile installation with the day/night sighting system in the centre and two missiles in the ready to launch position either side.

The Red Arrow 9 ATGM is of a different design to the older Red Arrow 8 and apart from its warhead, is very similar in appearance to the US Raytheon Systems Company 3,750 m range TOW ATGM, which is the most widely deployed missile of its type in the world.

In some respects Red Arrow 9 is also similar to the Israel Military Industries MAPATS ATGM system, which was first revealed in1984.

The launcher can be traversed 200° left and right with elevation and depression being 10°. In addition to the four Red Arrow 9 missiles in the ready to launch position another eight rounds are carried in the hull. These can be loaded onto the retracted launcher automatically or manually.

The gunner is seated under full armour and NBC protection. The method of guidance is Semi-Automatic Command-to-Line Of Sight (SACLOS) and all the operator has to do to accomplish a hit on the target, is to keep his optical sight on the target.

In addition to the standard two field of view day sighting system, a thermal sight is also fitted to enable targets to be engaged under conditions of smoke and fog. This operates in the 8 to 12 um range and can detect a target at a range of 4,000 m and recognise a target at a range of 2,500 m.

The major difference between Red Arrow 9 and the US Raytheon TOW is that the former is laser guided, while TOW is wire guided. The laser has a range of more than 5.5 km and operates in the 0.9 um waveband.

The Red Arrow 9 missile is ejected out of its launcher after which the main motor cuts in. The missile is 152 mm in diameter and with its associated launch tube weighs 37 kg.

It has four fins at the rear that unfold after launch with the four control surfaces almost two thirds down the missile towards the rear. Maximum rate of fire is quoted as two missiles per minute when engaging targets at maximum range.

The missile has a minimum range of 100 m and a maximum range of 5,000 m. The main HEAT (High Explosive Anti-Tank) warhead will penetrate 320 mm of rolled homogenous armour at an angle of 68° protected by Explosive Reactive Armour (ERA). The nose-mounted HEAT precursor charge activates the ERA, allowing the main charge to penetrate the armour.

There are at least two versions of Red Arrow 9, the Red Arrow 9A millimeter wave command guidance version and the Red Arrow 9B laser beam riding version.

Some sources have stated that there are three versions of the Red Arrow 9 marketed, the SACLOS, millimeter wave command guidance and laser.

Although the first known application of the Red Arrow 9 is the WZ 551 (4 × 4) wheeled armoured vehicle, it can also be installed on a number of other platforms, including full-tracked armoured vehicles, trucks, helicopters and coastal craft.

Mounted either side of the hull firing forwards is a bank of four electrically operated smoke grenade launchers.

No detailed specifications of the chassis/platform have been revealed, although it is estimated to have a combat weight of about 13.75 tonnes and powered by a Deutz air-cooled diesel Model BF8L41. This gives the system a maximum road speed of 95 km/h and an operational road range of up to 800 km. The vehicle has an NBC system but is not amphibious.

Status

Production as required. In service with the People's Liberation Army.

Manufacturer

NORINCO, China North Industries Corporation.

UPDATED

FRANCE

Giat Industries VAB tank destroyers

Development

Using the basic chassis of the Giat Industries VAB (4 × 4 and 6 × 6) armoured personnel carrier, a number of specialised anti-tank versions has been developed and placed in production.

While Giat Industries is the prime contractor for the chassis, Euromissile is responsible for the actual HOT launcher, missile and systems integration.

In addition, a number of countries fit the standard pintal mounted Euromissile MILAN ATGW system on the roof of the VAB.

Production of the VAB was completed for the home and export markets several years ago but if sufficient orders are placed then production can start again.

Description

The actual chassis is identical to the standard production VAB (4 × 4 and 6 × 6) armoured personnel carrier and is fully described in the Armoured personnel carriers (wheeled) section. Details are also given in the latter on a number of automotive upgrades that are now being marketed for the VAB 4 × 4 and 6 × 6 chassis.

VCAC HOT Mephisto

This has been in service with the French Army for some years in its 4 × 4 configuration and is basically the VAB with the Euromissile Mephisto system armed with four HOT ATGWs. The missile has a maximum range of 4,000 m and a minimum range of 75 m.

The launcher is powered and can be traversed through a full 360° with elevation from −10 to +10°.

When travelling, the launcher is retracted flush with the roof of the VAB and raised when required.

In addition to the four HOT ATGWs in the ready to launch position, a further eight HOT missiles are carried in reserve in two four-round magazines. These are loaded automatically when the launcher is retracted.

The gunner is provided with a Giat APX 590 gyrostabilised mirror-type sight with day magnifications of ×3 and ×12 and if required this can be raised independently of the launcher module. A thermal camera can be fitted as an option to engage targets under night and poor weather conditons. The thermal camera is now standard on all French Army VAB VCAC HOT Mephisto systems.

Specifications

Model	**Mephisto**	**UTM 800**
Configuration	4 × 4	6 × 6
Crew	4	4
Combat weight	13,400 kg	14,150 kg
Power-to-weight ratio	16.41 hp/t	15.54 hp/t
Length	5.98 m	5.98 m
Width	2.49 m	2.49 m
Height		
(hull top)	2.10 m	2.10 m
(module retracted)	2.81 m	n/a
(module raised)	3.25 m	n/a
(over turret)	n/a	2.885 m
Ground clearance		
(axles)	0.40 m	0.40 m
(hull)	0.50 m	0.50 m
Track	2.035 m	2.035 m
Wheelbase	3 m	1.5 + 1.5 m
Angle of approach/ departure	45/45°	45/45°
Max speed	92 km/h	92 km/h
Max water speed	7 km/h	7 km/h
Fuel capacity	300 litres	300 litres
Max road range	1,000 km	1,000 km
Fording	amphibious	amphibious
Gradient	50%	50%
Side slope	35%	35%
Trench	n/a	1 m
Vertical obstacle	0.5 m	0.5 m
Turning radius	9 m	9 m
Engine	Renault MIDS 06-20-45 in-line water-cooled turbocharged 6-cylinder diesel developing 220 bhp at 2,200 rpm	
Transmission	Transfluide with 5 forward and reverse gears	
Steering	recirculating ball, hydraulically assisted	
Tyres	14.00 × 20 (run-flat)	14.00 × 20 (run-flat)
Brakes	disc	disc
Suspension	independent (torsion bar and telescopic shock-absorbers)	
Electrical system	24 V	24 V
Batteries	4 × 12 V 6TN	4 × 12 V 6TN
Armament		
(ATGW)	4 + 8	4 + 12
(MG)	nil	1 × 7.62 mm
NBC system	optional	optional
Night vision equipment	optional	optional

Giat Industries VAB (6 × 6) fitted with Euromissile UTM 800 turret with four HOT missiles in ready to launch position

Giat Industries VAB (4 × 4) with Euromissile Mephisto launcher in raised position and trim vane erected for amphibious operations (C R Zwart)

VCAC UTM 800

This has been developed specifically for the export market and is a VAB fitted with a roof-mounted Euromissile UTM 800 turret with four HOT ATGWs in the ready to launch position and a further 16 HOT missiles in reserve in the hull. New missiles are loaded manually through the roof hatches to the rear of the turret.

The one-man power-operated turret can be traversed through a full 360° and the missiles, two either side of the turret, can be elevated from –6 to +22°. Turret traverse and weapon elevation are electric with manual controls for emergency use.

The turret is fitted with an APX M509 day sight with magnifications of ×3 and ×12 and the option of a thermal camera for engaging targets at night.

A 7.62 mm machine gun on a Giat Industries STBV gun ring shield is located on the roof at the front right side and 2,000 rounds of machine gun ammunition are carried for this weapon.

This version is not used by the French Army but is known to be in service with Cyprus (18) and Qatar (24).

Status

Production as required. In service with Cyprus (18), France (135) and Qatar (24). There has been no recent production of the VAB although production of the Euromissile HOT missile continues.

Manufacturers

Giat Industries (chassis).
Euromissile (missile and launcher).

UPDATED

GERMANY

Rheinmetall Landsysteme Jagdpanzer Jaguar 1 self-propelled ATGW vehicle

Development

The first prototype of the Jagdpanzer Rakete was completed by Hanomag under the designation RU 234. This was followed by a further six prototypes which were completed from 1963, three of which were built by Hanomag (the RU 341, RU 342 and RU 343) and three by Henschel (RU 344, RU 345 and RU 346). The Jagdpanzer Rakete succeeded the Jagdpanzer Kanone in production in 1967 and 370 were built by the following year, 185 by Hanomag and 185 by Henschel.

These were fitted with two launchers for the Aerospatiale SS-11 wire-guided ATGW with a maximum range of 3,000 m. Following successful trials with prototype vehicles, 316 of the 370 SS-11 armed vehicles were rebuilt with the Euromissile K3S HOT ATGW system. The first vehicles were handed over by Thyssen Henschel (which subsequently became Henschel Wehrtechnik) in 1978 with final deliveries in 1983.

As previously stated, the German Army took delivery of 316 vehicles fitted with the Euromissile K3S HOT installation under the name of Jaguar 1.

More recently, 90 of these have been transferred to the Austrian Army so the German Army now has 226 of these in service serving alongside the 136 Jaguar 2 vehicles which have a pedestal mounted Raytheon Systems Company TOW launcher fitted. Of the 90 Jaguar 1 systems for the Austrian Army, three were delivered in 1995 and the remaining 87 in 1997.

As of early 2001, the prime contractor is Rheinmetall Landsysteme, previously called Henschel Wehrtechnik.

Description

(Jaguar 1)

The all-welded steel hull of the Jagdpanzer Jaguar 1 has the fighting compartment at the front and the engine compartment at the rear. The driver sits at the front of the hull on the left side and has a single-piece hatch cover that opens to his left, in front of which are three day periscopes, the centre one can be replaced by a passive night vision periscope. The 7.62 mm MG3 bow machine gunner sits on the right side of the hull and has a single-piece hatch cover that opens to the right, forward of which are three day periscopes.

The engine compartment is at the rear of the hull with the engine on the left side, cooling system on the right side and the Renk HSWL 123 transmission at the rear. The engine compartment is equipped with an automatic fire extinguishing system, which is activated automatically in the event of fire.

The torsion bar suspension system either side consists of five dual rubber-tyred roadwheels with the idler at the front, drive sprocket at the rear and three track-return rollers. The first, second, fourth and fifth roadwheel stations have a hydraulic shock-absorber. The Jagdpanzer Jaguar 1 is equipped with passive night vision equipment and an NBC system. The basic vehicle can ford to a depth of 1.2 m and with a kit to 1.8 m.

The missile fire control system operates as follows. The missile operator tracks the target through his periscopic day sight which is mounted in the forward part of the roof and has a two-stage magnification, the first stage being ×4 (with a 13° field of view) and the second ×12 (with a 4.3° field of view). Traverse is 140° and elevation from –20 to +20°. The missile is launched and when it has travelled between 30 and 50 m is armed. The booster stage accelerates the missile to 240 m/s in one second and this speed is then kept constant throughout the flight by the sustainer stage. Once the missile hits the target the launcher arm is depressed vertically into the magazine, which holds eight missiles, and a new missile is loaded. This takes several seconds and once accomplished the launch cycle can begin again. Average firing rate is 3 rds/min. Hit probability against a fixed target (2.3 × 2.3 m) or moving target (2.3 × 4.6 m) is 80 per cent at ranges of up to 500 m and almost 100 per cent at ranges between 500 and 4,000 m. A thermal night sight has been developed to enable the vehicle to engage enemy armour at night and in bad weather.

The RZ 1201 thermal imaging sight cost about DM10 million to develop and uses Raytheon Systems Company thermal imaging modules. First company trials of the HOT with this sight, also called the NZBG (Night Target and Observation Equipment), took place in November 1984 at the Meldorf range in Germany. During these trials stationary and moving targets were

Specifications

Jaguar 1
Crew: 4
Combat weight: 25,500 kg
Power-to-weight ratio: 19.6 hp/t
Ground pressure: 0.7 kg/cm²
Length
(hull) 6.61 m
Width: 3.12 m
Height:
(incl missiles) 2.54 m
(to hull top) 1.98 m
Ground clearance:
(front) 0.45 m
(rear) 0.44 m
Track: 2.54 m
Track width: 450 mm
Length of track on ground: 3.8 m
Max road speed:
(forward and reverse) 70 km/h
Fuel capacity: 470 litres
Max road range: 400 km
Fording: 1.2 m
(with preparation) 1.8 m
Gradient: 58%
Side slope: 30%
Vertical obstacle: 0.75 m
Trench: 2 m
Engine: Daimler-Benz MB 837 8-cylinder water-cooled diesel developing 500 hp at 2,000 rpm
Transmission: Renk HSWL 123
Suspension: torsion bar
Electrical system: 24 V
Armament:
(main) 1 HOT ATGW launcher
(bow) 1 × 7.62 mm MG3 MG
(anti-aircraft) 1 × 7.62 mm MG3 MG
Smoke-laying equipment: 8 smoke 76 mm grenade dischargers on roof
Ammunition:
(missiles) 20
(machine gun) 3,200
Armour: 12-50 mm (est)
NBC system: yes
Night vision equipment: yes

Jagdpanzer Jaguar 1 showing the bank of electrically operated 76 mm smoke grenade dischargers on the hull rear (Michael Jerchel)

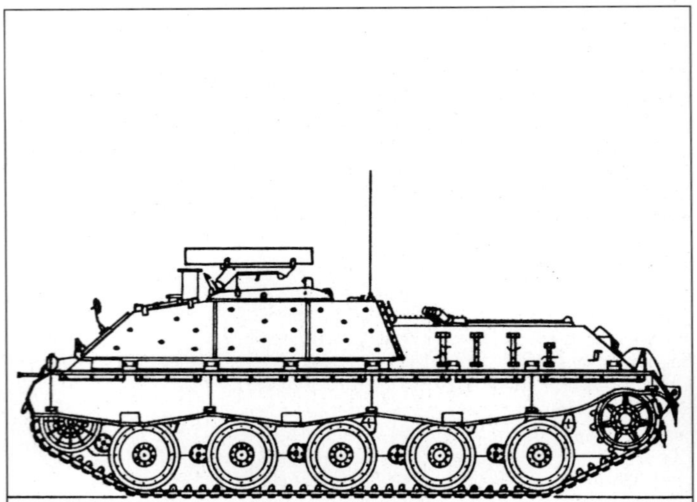

Jagdpanzer Jaguar 1 armed with the HOT ATGW system

Jaguar 2 tank destroyer with TOW launcher in elevated position (Michael Jerchel)

Jaguar 1 from the front showing the new sighting system for the 4,000 m range HOT anti-tank guided missile with cover in open position (C R Zwart)

0018770

engaged and hit at ranges of over 2,000 m by both day and night.

The HOT missile is carried in its launch tube and is treated like a normal round of ammunition. The total weight of missile and launch tube is 32 kg; it is 1.3 m long and 175 mm in diameter. The missile itself weighs 23 kg at launch with the HEAT warhead weighing 6 kg including fuze. The warhead is of the HEAT type, contains 3 kg of explosive and its diameter is 136 mm. Flight times are 9 seconds to 2,000 m, 13 seconds to 3,000 m and 17.3 seconds to 4,000 m.

In 1985, the HOT 2 entered production with delivery of first production missiles in 1986 after which time production of the HOT 1 ceased. HOT 2 has its warhead diameter increased from 136 mm to 150 mm and the explosive charge increased from 3 kg to 4.1 kg with no actual increase in missile weight. The HOT 2 warhead will penetrate 1,300 mm of conventional steel armour.

Other modifications to the vehicle have included the addition of spaced armour to the front and sides of the hull, consisting of a layer of steel bolted to the vehicle with rubber mountings providing the space and acting as a buffer between the armour and the hull. The commander sits behind the bow machine gunner and has a single-piece lift-and-swing hatch cover. He is provided with day periscopes for all-round observation and mounted to the rear of his hatch is a periscope which can be traversed through a full 360°. A 7.62 mm MG3 machine gun is mounted at the commander's position.

Mounted in the bow of the vehicle on the right side is a 7.62 mm MG3 machine gun with an elevation of +15°, a depression of –8° and a traverse of 15° left and right. Some vehicles have had their 7.62 mm bow machine guns removed. A second 7.62 mm MG3 machine gun can be mounted on the roof of the vehicle at the bow gunner's position for anti-aircraft use. There are eight 76 mm smoke grenade dischargers mounted on the rear decking firing forwards over the front of the vehicle.

Austrian Jaguar 1 vehicle

Austria took delivery of 90 vehicles from the German Army which are designated the PAL 4000. These use HOT 3 missiles and 76 of the 90 will be upgraded in the near future with a thermal imager.

Jagdpanzer Rakete with TOW (Jaguar 2)

In 1978, Thyssen Henschel completed two prototypes of the Jagdpanzer Rakete armed with the Raytheon Systems Company TOW system, known as the Jaguar 2. Mounted on the top of the vehicle is a Raytheon Systems Company TOW system which can be elevated from -10 to +15° and has a traverse of 30° left and right. The main advantage of this system is that the missiles are much cheaper than the HOT and the system already has the Raytheon Systems Company AN/TAS-4 night sight. The TOW missile is faster than the HOT and has an almost identical range (3,750 m compared to 4,000 m with the HOT).

Early in 1983, the German BWB placed an order worth DM25 million with the then Thyssen Henschel of Kassel for the conversion of 162 Jagdpanzer 90 mm self-propelled anti-tank guns to TOW ATGW carriers. The vehicles were converted between 1983 and 1985.

The bow-mounted 7.62 mm machine guns on the Jaguar 2 anti-tank systems have now been removed.

Although a total of 162 vehicles were converted to the Jaguar 2 configuration, only 136 vehicles remain in service.

Jagdpanzer Kanone

A total of 750 Jagdpanzer Kanone (JPZ 4-5) self-propelled anti-tank guns was built for the German Army between 1965 and 1967. In addition, 80 of an improved model were built for the Belgian Army. None of these remains in front-line service. Details were given in *Jane's Armour and Artillery 1994-95*, pages 500-501. Some of these German vehicles have been converted into observation vehicles with the 90 mm guns removed.

Status

Production complete. No longer marketed. In service with:

Country	Quantity	Comment
Austria	90	Jaguar 1 (HOT)
Germany	226	Jaguar 1 (HOT)
	136	Jaguar 2 (TOW)

Manufacturer

Original manufacturers were Henschel and Hanomag. Current manufacturer would be Rheinmetall Landsysteme GmbH (Kassel facility).

UPDATED

JAPAN

Komatsu Type 60 twin 120 mm self-propelled recoilless gun

Development

Two prototypes of the self-propelled recoilless anti-tank system were built under the direction of the Technical Research and Development Headquarters of the Japanese Ground Self-Defence Force, one by Komatsu (called the SS1) and the other by Mitsubishi Heavy Industries (called the SS2), both powered by a six-cylinder diesel which developed 110 hp. The prototypes were completed in December 1956 and were tested until December 1958.

In 1957, it was proposed to arm the vehicle with four recoilless rifles, instead of the two 105 mm weapons fitted to the prototypes, to increase the hit probability, but with the adoption of the 106 mm recoilless rifle this reverted to two.

As a result of trials with the second series of prototypes it was decided to build a third and final series of prototypes with weight increased to 7,500 kg, a modified diesel engine, a new transmission with four forward and one reverse gears, a two-speed auxiliary transmission and a wet multidisc hydraulically operated clutch. Three of the third series prototypes were built by Komatsu under the designation SS4 and were tested

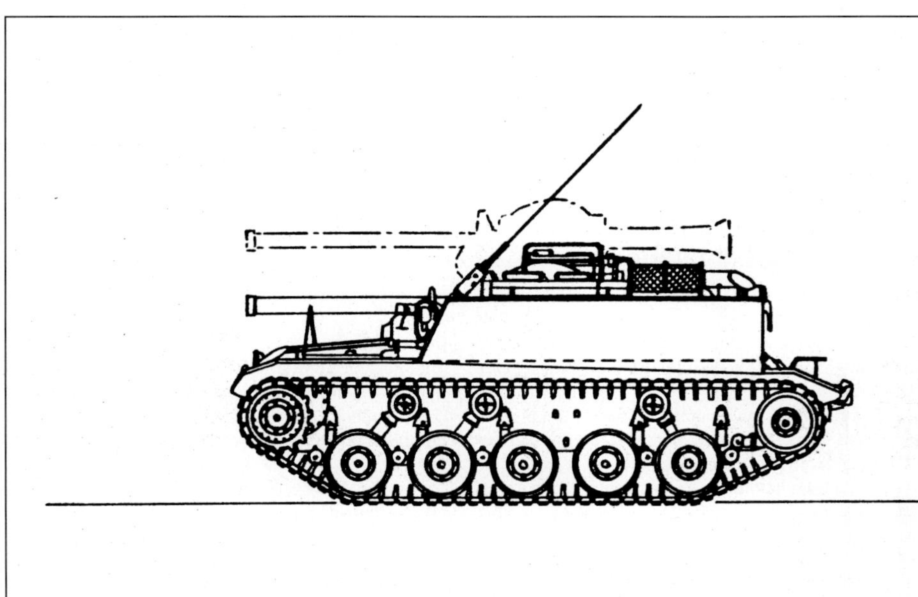

Type 60 self-propelled recoilless gun with dotted line showing position of weapons when raised

from May 1959 to March 1960. As a result, the clutch was changed to a wet multidisc mechanical type. In September 1960, the vehicle was accepted for service as the Type 60 Self-propelled 106 mm Recoilless Gun.

Production of the vehicle was undertaken by Komatsu. The first production model, the Model A, was replaced in 1967 by the Model B which was basically a strengthened Model A. In 1974 the Model C powered by a commercial model Komatsu SA4D105 four-cylinder water-cooled diesel engine developing 150 hp at 2,800 rpm, entered production. The Type 60 twin 120 mm self-propelled recoilless gun was produced at the rate of 10 to 20 vehicles a year and by December 1977, 252 vehicles had been built. Production was completed in FY79.

There are no known plans for these systems to be upgraded. Japan is the only country in the world still to deploy a system of this type.

Each year small numbers of Type 60 twin 120 mm self-propelled anti-tank guns have been withdrawn from service and by 1999 it was stated that 160 remained in front line service.

Type 60 self-propelled recoilless gun with twin 106 mm recoilless rifles in lowered position (K Nogi)

Description

The hull of the Type 60 is of welded and riveted steel armour construction. The driver is seated on the left side of the vehicle towards the front and has a single-piece hatch cover that opens to the rear in front of which are three day periscopes. The commander is seated in the middle of the hull to the left of the two recoilless rifles and his position is attached to the recoilless rifle mounting so when they are raised into the firing position he is at the same height as the weapons. The commander has a single domed hatch cover that opens to the rear, in front of which is a day periscopic sight. There is also a vision block to his rear. The loader is seated to the left of the commander and has a single-piece hatch cover that opens to the left and a retractable day periscope for observation to the rear.

The diesel engine is at the rear of the hull and power is transmitted to the manual transmission at the front by a propeller shaft. The torsion bar suspension either side consists of five dual rubber-tyred roadwheels with the drive sprocket at the front, idler at the rear and three track-return rollers. The first, second and fifth roadwheel stations have a hydraulic shock-absorber. The tracks can be fitted with rubber pads when operating on roads and wider tracks can be fitted for operating in snow. The Type 60 has no NBC system and no amphibious capability.

Main armament consists of two 106 mm Type 60 recoilless rifles which were manufactured by Nihon Seikojyo and also fitted singly to Japanese Ground Self-Defence Force jeeps. They have a maximum range of 7,000 m and an effective range of 1,100 m against tanks. A 12.7 mm spotting rifle manufactured by Howa Kogyo is mounted on the right recoilless rifle. The two recoilless rifles are mounted parallel to each other on the right side of the hull. When lowered they have an elevation of +10°, a depression of –5° and a traverse of 10° left and right. When raised they have an elevation of +15°, a depression of –20° and a traverse of 30° left and right. The weapons are raised into the firing position by a hydraulic/electric power system, but if this fails they can be raised by a hand pump. Eight rounds of HEAT or HE ammunition are carried. A 75 cm stereo range-finder is also carried on the vehicle and recently a night sight has been mounted on the left 106 mm recoilless rifle.

The effectiveness of the 106 mm HEAT rounds against modern conventional steel armours is now highly questionable.

Specifications

Crew: 3
Combat weight: 8,000 kg
Unloaded weight: 7,600 kg
Power-to-weight ratio:
(120 hp diesel engine) 15 hp/t
(150 hp diesel engine) 18 hp/t
Ground pressure: 0.67 kg/cm²
Length: 4.3 m
Width: 2.23 m
Height: 1.59 m
Ground clearance: 0.35 m
Track: 1.97 m
Track width:
(normal tracks) 260 mm
(snow tracks) 680 mm
Length of track on ground: 2.3 m
Max road speed:
(forwards, Models A and B) 45 km/h
(forwards, Model C) 55 km/h
Fuel capacity: 140 litres
Max road range: 250+ km
Fording: 0.7 m
Gradient: 60%
Side slope: 30%
Vertical obstacle: 0.6 m
Trench: 1.8 m
Engine:
(Models A and B) Komatsu model 6T 120-2 air-cooled, 4-cycle, 6-cylinder diesel developing 120 hp at 2,400 rpm
(Model C) Komatsu SA4D105 air-cooled, 4-cycle, 6-cylinder diesel developing 150 hp at 2,800 rpm
Transmission: manual with 4 forward and 1 reverse gears, auxiliary, 2 speed
Clutch:
(Models A and B) dry, multidisc
(Model C) wet, multidisc
Suspension: torsion bar
Armament:
(main) 2 × 106 mm recoilless rifle
(ranging machine gun) 1 × 12.7 mm MG
Ammunition: (main) 8
Gun elevation/depression: see text
Gun traverse: see text
Armour: 12 mm max (est)
NBC system: no
Night vision equipment: no

Status

Production complete. In service only with the Japanese Ground Self-Defence Force.

Manufacturer

Komatsu Ltd.

UPDATED

RUSSIAN FEDERATION AND ASSOCIATED STATES (CIS)

Volgograd Tractor Plant 2S25 self-propelled anti-tank gun

Development

The 125 mm 2S25 self-propelled anti-tank gun was developed by the Volgograd Tractor Plant Joint Stock Company to meet the requirements of the Russian Air Assault Divisions.

It is understood that prototypes of the 2S25 were completed some time ago but as of 2001 volume production had yet to commence.

It is already being offered on the export market as the Russian Army has insufficient funds to purchase this vehicle at the present time.

In mid-2001, it was stated that the Volgograd Tractor Plant had been working on the 2S25 for about seven years.

It was also stated that the 2S25, which is also referred to as the SPRUT-SD, could operate in the water and still be able to fire its 125 mm gun while afloat. It can also operate in altitudes of up to 3,000 m.

Description

The hull and turret of the 2S25 self-propelled anti-tank gun is of all-welded construction with the crew compartment at the front, turret and fighting compartment in the centre and engine compartment at the rear.

Through the frontal arc of 40° left and right the armour provides protection against attack from small arms fire up to 12.7 mm and through the remainder of the vehicle against attack from 7.62 mm small arms fire.

Many of the automotive parts of the 2S25 are the same as that installed in the BMD-3 airborne combat vehicle that was also designed and built at the Volgograd Tractor Plant. As of early 2001 the BMD-3 was only in service with the Russian Army.

The driver is seated at the front of the hull in the centre and has a single-piece hatch cover and three day periscopes, the centre one of which can be replaced by a night vision periscope.

There is an additional seat either side of the driver for the commander and driver as they are normally seated here rather than in the cramped turret when the vehicle is travelling in a non operational area. There is a single hatch cover above their position to enable them to quickly leave the vehicle.

Main armament consists of a 125 mm 2A75 smoothbore tank gun developed at Artillery Plant No 9 at Ekaterinburg and is based on the 125 mm 2A46 smoothbore tank gun installed in the T-72, T-80 and

125 mm 2S25 self-propelled anti-tank gun with suspension retracted to lower the overall height of the system ***2001***/0048275

T-90 MBTs. For this lighter chassis application it has been fitted with a new low recoil system.

The 125 mm 2A75 smoothbore tank gun is fitted with a fume extractor and a thermal sleeve but is not fitted with a muzzle brake. It is fully stabilised in elevation and traverse.

This fires the same family of separate loading ammunition as the 125 mm 2A46 smoothbore gun installed in the T-72, T-80 and T-90 series MBTs. It can also fire a laser guided projectile out to a range of 4,000 m. Maximum rate of fire is quoted as 7 rds/min.

The 125 mm 2A75 smoothbore gun is fed by a horizontal automatic loader that is located below the turret. This holds a total of 22 projectiles and charges ready for immediate use. The projectile is loaded first, followed by the charge with its semi-combustible cartridge case. A manual back up loading system is also provided.

A 7.62 mm PKT machine gun is mounted coaxial with the main armament and this is fed by a belt of 2,000 rounds of ammunition. Additional 7.62 mm ammunition is stowed externally in boxes on the turret rear.

The commander is seated on the right and the gunner on the left with both crew members being provided with roof mounted day/night observation devices and a single piece hatch cover. The gunner also has a sight in the forward part of the turret roof.

The commander's sight is stabilised in elevation and traverse and has a laser sighting device to guide the 125 mm laser guided projectiles onto the target. The gunner's sight is stabilised in elevation and incorporates a laser range-finder which feeds target information into the ballistic computer.

There is an infra-red searchlight mounted externally on the right side of the turret and a smaller infra-red light on the right side of the hull front. There is a standard white light searchlight on the turret roof and a smaller one on the left side of the hull front for driving at night.

The suspension is of the hydropneumatic type and allows the ground clearance to be adjusted by the driver from his seat between 100 and 500 mm in 6 to 7 seconds.

The running gear either side consists of seven single rubber tyred road wheels, idler at the front, drive sprocket at the rear and track return rollers. Two types of track are fitted, standard and the wider snow/mud track which reduces the ground pressure of the vehicle.

The 2S25 is fully amphibious being propelled in the water by two water jets situated one either side low down at the rear. Before entering the water a trim vane is erected at the front of the vehicle and the bilge pumps are switched on. It can operate in up to Sea State 3 and can also fire the 125 mm gun while afloat over a 70° frontal arc.

Standard equipment includes night vision equipment for the commander, gunner and driver and an NBC system.

According to the manufacturer, the 2S25 can be air dropped from transport aircraft with the crew inside the vehicle.

Optional equipment includes a thermal channel for the gunner's sight, improved amphibious capability and additional armour protection. A defensive aids suite can also be installed, for example the Russian Arena.

Variants

There are no known variants of the 2S25 self-propelled anti-tank gun. Some Russian sources have indicated that the 2S25 chassis could be the basis for a complete family of airborne light armoured vehicles to replace systems based on the older BMD-1/BMD-2 series airborne combat vehicle chassis.

Specifications

Crew: 3
Combat weight: 18,000 kg
Power-to-weight ratio: 28.33 hp/t
Ground pressure:
(standard track) 0.53 kg/cm^2
(wide track) 0.36 kg/cm
Length:
(hull) n/avail
(gun forwards) n/avail
Width: n/avail
Height: n/avail
Ground clearance: 100 to 500 mm
Track width: n/avail
Length of track on ground: n/avail
Max speed:
(road) 71 km/h
(water) 10 km/h
Range (road): 500 km
Gradient: 60%
Side slope: 40%
Vertical obstacle: n/avail
Trench: n/avail
Engine: Model 2B06-2C diesel developing 510 hp
Transmission: hydromechanical/hydrostatic drive
Suspension: hydropneumatic
Electrical system: n/avail
Armament:
(main) 1 × 125 mm 2A75 smoothbore gun
(coaxial) 1 × 7.62 mm PKT machine gun
Ammunition:
(main) 40 × 125 mm (of which 22 are ready for use)
(coaxial) 2000 × 7.62 mm

Gun control equipment
Turret power control: powered/manual
(by commander) yes
(by gunner) yes
Turret traverse: 360°
Gun elevation/depression: +15/–5° (bow); +17/–3°(aft)
Gun stabiliser:
(vertical) yes
(horizontal) yes
NBC system: yes
Night vision equipment: yes

Status

Prototypes. Not yet in production or service.

Manufacturer

Volgograd Tractor Plant.

UPDATED

KBM Khrizantema BMP-3 tank destroyer

Development

The Khrizantema (Chrysanthemum) BMP-3 based tank destroyer has been developed under the leadership of the KBM Engineering Design Bureau in Moscow and was announced for the first time in 1996.

It is believed that the first prototype was completed around 1990 with the BMP-3 launch platform being designated the 9P157-2 and the original missile, with a single HEAT (High Explosive Anti-Tank) warhead being designated 9M123.

Although the overall prime contractor for the Khrizantema BMP-3 tank destroyer is the KBM Engineering Design Bureau, the manufacturer of the chassis is the Kurgan Machine Construction Plant where production of the BMP-3 Infantry Combat Vehicle (ICV) and its other variants is currently undertaken.

The KBM Khrizantema BMP-3 tank destroyer was first seen in public during an air show in Moscow in late 1999 and at that time it remained at the prototype stage. Russia refers to the system as the Khrizantema-S.

Description

For its new role, the BMP-3 (ICV) chassis has been modified with the complete turret being removed as well as the two 7.62 mm PKT bow-mounted forward-firing machine guns, firing ports and associated vision devices, troop compartment roof hatches and rear doors.

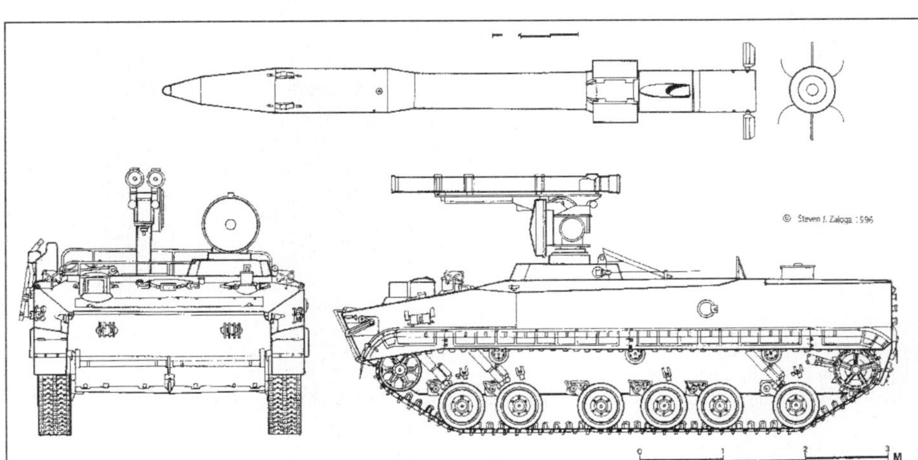

BMP-3 Chrysanthemum ATGW system with radar and twin missile launcher in raised position (lower) and with missile (above) (Steven Zaloga) 0008271

The hull roof has been modified and, mounted on the hull roof, between the second and third roadwheels, is an elevated arm which carries two Chrysanthemum (US designation AT-15) ATGW in the ready to launch position.

The unique feature of the missile is that it has two modes of guidance. There is no known comparable anti-tank missile with a similar guidance system in the West under development or in service.

First mode of guidance is automatic under both day and night conditions where it is guided by a millimetre-waveband radar system mounted on the left side of

BMP-3 Chrysanthemum ATGW system ready for action with two-round launcher elevated 0008272

Close-up of the Khrizantema long-range ATGW system installed on a BMP-3 chassis showing two missiles in the ready to launch position (Peter Felstead) ***2000***/0085190

the roof; the second is an automatic laser beam rider using a sight mounted on the front right side of the glacis plate.

In the first mode the gunner does not actually have to see the target but for the second semi-automatic laser-guided mode he must keep his sight on the target until the missile impacts the target. Russian sources state that the millimetre-waveband radar system would be the prefered method of target emgagement.

It is possible to engage two targets at the same time using a different mode of guidance. It is also possible to fire two missiles in the automatic and semi-automatic modes respectively. When not required, the radar system can be retracted into the hull under full armour protection.

The Chrysanthemum tank destroyer has a two-man crew consisting of commander/gunner and driver, both under full armour protection. They are seated at the front of the vehicle with the driver in the centre.

Once the missiles have been launched, the launcher arm is retracted into the hull and loaded automatically with a total of 15 missiles carried in their disposable launch tubes. The launcher has powered traverse through 170° and powered elevation/depression.

Once the 15 missiles have been expended new missiles are loaded from the right side of the vehicle using an onboard load assist device.

Two models of the supersonic 9M123 missile have been developed, one with a tandem High Explosive Anti-Tank (HEAT) (9M123) warhead and the second with a fuel-air explosive warhead (9M123F). The missiles are selected by the gunner prior to engaging the target.

The missile is 2.057 m long and 150 mm in diameter with the warhead a slightly larger diameter. This would give a penetration of over 1,000 mm of steel armour protected by Explosive Reactive Armour (ERA).

The missile has a maximum range of 6,000 m and maximum speed is 400 m/s. The two movable control surfaces are at the rear with four wraparound wings about three quarters of the way down the missile towards the rear. Minimum range of the missile is being quoted as 400 m.

As the datalink is in the rear of the missile, the rocket motor exhausts are above and below the missile at right angles to the control surfaces.

Key role of the Chrysanthemum missile would be to destroy MBTs and other armoured targets at long range, although the manufacturer also states that it can be used to destroy fortified structures, pillboxes and small ships. Given its high speed it can also engage slow and low-flying helicopters.

The chassis, designated the 9P157-2, retains the full amphibious capability of the standard BMP-3, being propelled in the water at a speed of 10 km/h by two water-jets mounted at the rear of the hull. Combat weight of the system is 19.4 tonnes. Standard equipment for the vehicle includes an NBC system and a front-mounted dozer blade.

Although the prototype system is based on the BMP-3 full-tracked chassis, of which an estimated 1,000 have been built for the home and export markets, especially the Middle East, it can be fitted to other types of chassis, tracked and wheeled. An air-launched version is currently being studied.

Specifications

Similar to BMP-3 covered in Armoured personnel carriers (tracked) section apart from armament and ammunition load.

Status

Prototype. Not yet in production or service.

Manufacturer

KBM Engineering Design Bureau (overall systems design authority).

UPDATED

Volsk Kornet-E BMP-3 tank destroyer

Development

The tripod-mounted infantry version of the Russian Kornet (Western designation AT-14) Anti-Tank Guided Weapon (ATGW) made its first public appearance late in 1994 and since then is understood to have entered low-rate production for the Russian Army.

Recent information has indicated that at least one country in the Middle East (believed to be Syria) has acquired the standard tripod-mounted version of the Kornet.

The main advantages claimed for the Kornet ATGW are its longer range of 5,500 m (against the 4,000 m of Euromissile HOT and 3,750 m of Raytheon Systems Company TOW) and its laser guidance system which is difficult to jam.

The Kornet tripod launcher is provided with a 1P45-1 day sighting and guidance unit with the option of a clip-on thermal device to identify and engage targets at night or under conditions of poor visibility.

The Kornet-TP thermal sight has been developed by the SPA State Institute of Applied Optics and operates in the 8 to 13 μm spectral band.

According to the manufacturer, it has a detection range of no less than 4,000 m, a recognition range of no less than 3,500 m and a firing range of no less than 4,000 m.

It has two fields of view, 5.4 and 7.8° in the target search mode and 1.8 × 2.6° in the target recognition mode. Power consumption is quoted as 750 W.

More recently, two further versions of the Kornet have been shown in public for the first time as complete systems, the self-propelled Kornet-E on the full-tracked BMP-3 Infantry Combat Vehicle chassis (ICV) and the new Kliver gun/missile turret system which can be fitted to Eastern (for example Russian BTR-80) or Western armoured vehicle chassis.

Kornet-E tank destroyer on BMP-3 chassis from the side with twin missile launcher in elevated position (Christopher F Foss) 0008276

Late in 1997 the Kliver turret was shown installed on a BTR-80 (8 × 8) chassis and an illustration of this appears in the entry for this vehicle in the Armoured personnel carriers (wheeled) section.

More recently the Kliver turret has been offered for installation on Western type chassis, tracked and wheeled. In 1999 it was shown installed on a BMP-2 chassis.

Both of these systems have been developed by the KBP Instrument Design Bureau which is also the prime contractor for the basic Kornet ATGW system.

It is probable that at least the Kornet-E self-propelled system was developed to meet the operational requirements of the Russian Army but due to a shortage of funding no production systems have so far been purchased.

While the prime contractor for the complete system is the Volsk Mechanical Plant the prime contractor for the chassis is the Kurgan Machine Construction Plant where production of the standard BMP-3 ICV is undertaken for the home and export markets.

Description

For its new role, the BMP-3 chassis has been modified by the Volsk Mechanical Plant with the complete turret being removed as well as the two 7.62 mm PKT bow-mounted forward-firing machine guns. The complete Kornet-E system is designated the 9P162.

Firing ports have been removed as have the troop compartment roof hatches and rear doors and there is now a single small door in the right side at the rear.

The Volsk Mechanical plant was also responsible for the design and development of the Sturm-S ATGW system based on the MT-LB multipurpose full-tracked armoured vehicle chassis which is covered in detail in a separate entry.

Mounted on top of the BMP-3 hull in the centre is an unmanned turret with two external arms each of which has a single Kornet ATGW in the ready to launch position. These retract into the hull for automatic loading with 12 missiles carried internally in the two loading drums, each of which holds six missiles. An additional four missiles are stowed in the lower part of the hull and are manually loaded into the drums as required.

The vehicle also carries a standard infantry tripod launcher for Kornet so that if the vehicle is disabled then the two-man crew can still deploy Kornet away from the vehicle.

Although the main role of the system is the engagement of enemy MBTs at long range, the system can also engage hovering helicopters.

In its standard version Kornet can engage targets in daylight from 100 m out to a maximum range of 5,500 m and with the aid of the thermal sight, which can be quickly fitted, targets can be engaged under poor weather conditions out to a range of 3,500 m.

The standard 9M133 missile has a tandem High Explosive Anti-Tank (HEAT) warhead which will penetrate 980 mm of conventional steel armour protected by Explosive Reactive Armour (ERA) but a version fitted with a fuel-air explosive (or thermobaric) warhead has been developed which is highly effective against light armoured vehicles and personnel.

Typically, one of the reloading drums could be loaded with HEAT missiles and the other with fuel-air explosive missiles.

More recently a new HEAT warhead has been developed of the tandem type to penetrate vehicles fitted with explosive reactive armour.

Most Russian anti-tank weapons (guided and unguided) are now produced with the option of a fuel-air explosive warhead which was developed as a result of combat experience in Afghanistan.

The driver is seated in the centre of the vehicle at the front, as in the normal BMP-3 with the operator's station to the rear on the left with the actual firing unit with two arms to the right. When travelling, the commander would normally be seated to the right of the driver and this position is provided with a single-piece hatch cover that opens to the rear and two day forward-facing periscopes.

To the right of the operator's sighting system is a single roof hatch which opens to the rear and is probably used for missile resupply purposes.

Although Russian sources have quoted the Kornet-E as having a two-man crew it has ample space and seating for a three-man crew which would consist of commander, gunner and driver.

Method of guidance is Semi-Automatic Command to Line Of Sight (SACLOS) and all the operator has to do is to keep the sight on the target to ensure a hit.

The BMP-3 Kornet-E anti-tank system has a maximum road speed of 70 km/h and a road range of 600 km. It can quickly be made amphibious by erecting the trim vane at the front of the hull and when afloat is propelled in the water at a speed of 10 km/h. Standard equipment includes an NBC system of the overpressure type and a land navigation system.

Although the Kornet-E system is based on the BMP-3 chassis, the missile installation could be adopted for a number of other platform types, tracked and wheeled.

Specifications

Similar to BMP-3 covered in Armoured personnel carriers (tracked) section.

Status

Prototype. Not yet in production or service.

Manufacturer

Volsk Mechanical Plant (VMP).

UPDATED

Volsk AT-6 'Spiral'/MT-LB tank destroyer (Shturm-C)

Development

In 1990, the then Soviet Union revealed that it had deployed a version of the MT-LB multipurpose armoured vehicle, fully described in the Armoured personnel carriers (tracked) section, armed with the AT-6 'Spiral' (US/NATO designation) ATGW under the designation of the Shturm-C (industrial index number 9P149).

The radio command-guided tube-launched AT-6 'Spiral' (industrial index number of the basic missile is the 9M114 or Kokon) Semi-Automatic Command to Line Of Sight (SACLOS) anti-tank missile first entered service as a helicopter-launched anti-tank weapon for the Mil Mi-24 'Hind'. It was subsequently fitted to other helicopters such as the Mil Mi-28 'Havoc' and the Kamov Ka-29 'Helix-B'.

As far as is known, this system has not entered service outside the former Soviet Union, although it has been demonstrated in the Middle East.

The chassis for the MT-LB multipurpose tracked armoured vehicle on which the system is based was built at Kharkov in the Ukraine. Production of this was completed some years ago.

Both Russia and the Ukraine are offering a number of automative upgrades for the AT-6 'Spiral'/MT-LB tank destroyer. Available details of these are given in the Armoured personnel carriers (tracked) section under the entries for the MT-LB 'RFAS and Ukraine (Upgraded MT-LB).

The system designation for the missile system as distinct from the MT-LB based version is the 9K114.

Description

Externally, the MT-LB with the AT-6 'Spiral' is virtually identical to the standard MT-LB, but has a large gunner's sight mounted at the front of the hull front on the right side and a single rail launcher mounted at the rear. This retracts under armour protection when not required and for reloading purposes.

This sight is called the Pyl (Dust) and according to the manufacturer, allows the gunner to acquire and track targets under limited vision conditions such as those encountered in the desert.

All the gunner has to do is to keep the day/night sight cross-hairs on the target until the missile impacts. Airborne targets with crossing/approach speeds of 60 m/s and altitudes from ground level up to 3,000 m can be engaged. Rate of fire of the system aboard the 9P149 is 3 to 4 rds/min with a total of 12 missile container/launchers carried. Time of flight of missile to maximum range of 5,000 m is quoted as 17.6 seconds.

Crossing targets travelling at a speed of up to 60 km/h can be engaged by the system as can head-on targets travelling at a speed of up to 80 km/h. The target engagement arc is 85° left and right with elevation arc from –5 to +15°.

In addition to the basic 9M114 missile with a single HEAT warhead, at least two other versions have been developed, the 9M114M1 with a tandem HEAT warhead to defeat targets fitted with explosive reactive armour and the 9M114M2 which is believed to have a fuel air or thermobaric type warhead.

Like the standard MT-LB, the AT-6 'Spiral' MT-LB is fully amphibious, being propelled in the water by its tracks. It has a two-man crew and a combat weight of 12.3 tonnes.

Although the system was originally marketed with the AT-6 'Spiral' ATGW, more recently it has been marketed with the 9M120 Ataka ATGW, or 'AT-9' as it is referred to in the West.

This missile weighs 33 kg and can be supplied with four different types of warhead, HEAT, tandem HEAT, fuel-air explosive and the so called 'rod'.

The Ataka ATGW weighs 33 kg, has a maximum range of 6,000 m and is supersonic.

Specifications

(Chassis details are almost identical to MT-MB given in Armoured personnel carriers (tracked) section)

AT-6 Missile
Type: SACLOS
Length: 1.83 m
Diameter: 0.13 m
Wing span: 0.25 m
Launch weight: 40 kg
Warhead: 7-8 kg HEAT
Range: 400-5,000 m
Armour penetration: 800 mm +
Speed:
(at launch) 55 m/s
(in flight) 350-400 m/s

Standard equipment on this tank destroyer includes night vision equipment and an NBC system of the overpressure type.

Status

Production as required. In service with the RFAS armies. The AT-6 'Spiral'/MT-LB tank destroyer was used during the fighting in Georgia in 1991. As the actual chassis on which this system is based is no longer manufactured it is considered that there will be no brand new systems built in the future.

Manufacturer

Volsk Mechanical Plant (VMP), (overall design authority).

UPDATED

MT-LB tracked vehicle with AT-6 'Spiral' launcher in elevated position

MT-LB tracked vehicle from the rear with AT-6 'Spiral' launcher in elevated position and missile ready to launch (Christopher F Foss) 0005323

SOUTH AFRICA

Vickers OMC/Kentron Ratel ZT-3 tank destroyer

Development

The Ratel ZT-3 tank destroyer was developed under the leadership of the Kentron Division of Denel to provide the South African Armoured Corps with a long-range anti-tank capability.

The system is essentially a Vickers OMC Ratel (6 × 6) infantry fighting vehicle fitted with a new turret weapon system designed by the Kentron Division of Denel which has three ZT-3 Swift anti-tank guided missiles on its upper part in the ready to launch position.

Following extensive trials with prototype systems, some of which were successfully used in combat in Angola in 1987 and 1988, a total of 53 production Ratel ZT-3 Swift tank destroyers was delivered to the South African Army.

Production of the platform and its associated missile, was completed some years ago but production could commence again if a significant order was placed. Further development of the missile is currently underway by Kentron and details of this are given under Variants.

Ratel (6 × 6) IFV fitted with Kentron ZT-3 Swift ATGW system with three missiles in ready to launch position (Christopher F Foss) **2002**/0008277

Description

The hull of the Ratel ZT-3 tank destroyer is all-welded steel construction providing complete protection from 7.62 mm small arms fire and shell splinters, with protection over the frontal arc against 12.7 mm armour-piercing rounds. Emphasis has also been placed on providing a high level of protection against anti-tank mines.

The driver sits at the front of the vehicle with excellent forward visibility through three bulletproof windows to the front and sides. In combat areas these can be quickly covered by armoured shutters hinged at the bottom and activated by the driver with a single handle from within the vehicle. With the shutters in position the driver observes the terrain through three day periscopes, one to the front and one either side. The centre one can be replaced by a passive periscope for driving at night.

Over the driver's position is a single-piece hatch cover that opens to the left. The driver's seat and steering wheel are adjustable and can be accessed from the rear enabling drivers to be changed without either of them leaving the vehicle.

In either side of the hull behind the driver's position is a large door opened pneumatically by the driver. It opens forwards and has a vision block with a firing port underneath.

Mounted on the roof to the rear of the driver is the two-man power-operated turret on the top of which is a pod with three ZT-3 Swift anti-tank guided missiles in the ready to launch position, a 7.62 mm coaxial machine gun and a bank of two electrically operated smoke grenade launchers either side of the turret.

The gunner is seated on the right and the commander on the left with the latter having a single-piece hatch cover that opens to the rear and day vision devices for all-round observation.

The ZT-3 Swift is a laser beam command-guided Semi-Automatic Command to Line Of Sight (SACLOS) weapon and all the gunner has to do is to keep the cross-hairs of his sight aligned on the target until the missile impacts.

After launch, the deviation between the missile and the line of sight is measured by the system with either a goniometer or laser beam decoder.

The laser received in the tail of the missile detects and decodes the pulsed laser commands and feeds the information to the onboard flight control unit which transforms the information into servo commands.

The servos are connected to the control fins at the rear of the missile. Commands from the flight control unit cause the control fins to manoeuvre the missile onto the line of sight. The missile then follows the line of sight until it strikes the target. In a typical target engagement two missiles can be launched every minute with a total of 12 missiles being carried.

The ZT-3 missile is 127 mm in diameter and has a maximum range of at least 5,000 m with its HEAT warhead, activated by a contact fuze, capable of penetrating over 650 mm conventional of steel armour.

The system has a crew of four consisting of driver, commander, missile operator and reloader. Missile reloading takes place via hatches in the roof to the rear of the turret.

There is a passageway from the troop compartment to the rear of the vehicle on the right side, above which are two roof hatches that open outwards. At the rear on the right side is a cupola on which is mounted a 7.62 mm machine gun for use in the air defence role.

In the rear of the hull on the right side is a door. Its lower part folds down to form a step and the upper part opens to the left. The engine compartment is at the rear of the Ratel on the left side with access panels in the roof for maintenance.

The six-cylinder direct injection turbocharged diesel is coupled to a fully automatic power shift gearbox with a hydrodynamic torque converter. The gearbox can also be operated manually and has a mechanical emergency gearshift. Power is transmitted to the three axles in two stages with final reduction by means of planetary gearing housed in the wheel hubs.

The three axles are provided with their own lockable differential and longitudinal differential locks. Suspension consists of progressively acting coil springs and large hydraulic shock-absorbers. The hydropneumatic braking system can be assisted by an exhaust brake provided through the engine.

The power pack can be removed from the vehicle by two men with a crane in less than 30 minutes. Quick disconnect couplings and connectors are provided in all power, hydraulic, electric and pneumatic circuits.

Standard equipment includes a complete set of tools, spades, picks, shovels and axes, a towbar and towing cables, a first aid kit, two petrol stoves, fire

Specifications

Crew: 4
Configuration: 6 × 6
Combat weight: 19,230 kg
Unloaded weight: 17,740 kg
Power-to-weight ratio: 14.66 hp/t
Length: (hull) 7.212 m
Width: (hull) 2.520 m
Height:
(overall) 3.335 m
(to hull top) 2.235 m
Ground clearance: 0.34 m
Track: 2.08 m
Wheelbase: 2.809 + 1.4 m
Angle of approach/departure: 44/45°
Max road speed: 105 km/h
Fuel capacity: 430 litres
Max range:
(road) 1,000 km
(cross-country) 14 h
Fuel consumption:
(road) 0.4 litres/km
(cross-country) 1 litre/km
Fording: 1.2 m
Gradient: 60%
Side slope: 30%
Vertical obstacle: 0.6 m
Trench: 1.15 m
Turning radius: 8.5 m
Engine: D 3256 BTXF 6-cylinder in-line turbocharged diesel developing 282 hp at 2,200 rpm (or 315 hp as an option)
Transmission: automatic, 6 forward and 2 reverse gears with hydrodynamic torque converter
Steering: mechanical, recirculating ball with hydraulic assistance
Tyres: 14.00 × 20
Brakes:
(main) hydropneumatic on front wheels and air/mechanical on rear 4 wheels
(crawl) pneumatic/mechanical on rear 4 wheels
(parking) mechanical on rear 4 wheels
Axles:
(front) steering axle with steering damper and intermediate steering arm, lockable differential
(middle) non-steering with transfer arm, lockable differential (transfer and longitudinal)
(rear) non-steering and lockable differential
Suspension: solid axle and single coil springs and double-acting hydraulic shock-absorbers supported by wishbones and longitudinal arms
Electrical system: 24 V
Batteries: 2 × 12 V, 100 Ah
Armament:
(main) 1 × ZT-3 Swift missile launcher
(coaxial) 1 × 7.62 mm MG
(anti-aircraft, rear) 1 × 7.62 mm MG
Smoke grenade launchers: 2 × 2
Ammunition:
(missiles) 12
(7.62 mm) 800

Gun control equipment
Turret power control: electric/manual
(commander) yes
(gunner) yes
Weapon elevation/depression: +20/−10°
Turret traverse: 360°
Armour:
(hull front) 20 mm at 60°
(hull sides upper) 8 mm at 65°
(hull sides lower) 10 mm at 90°
(hull top) 6 mm
(hull floor) 8 mm and 10 mm
(hull rear) 10 mm

extinguishers, emergency spares, two 50 litre drinking water tanks, two external stowage boxes, radios, an intercom and a remote handset with 1,000 m of cable.

Optional equipment includes day/night vision equipment, an air conditioning and ventilation system and the replacement of the standard D3256 BTXF six-cylinder diesel developing 282 hp with an ADE 407 TI diesel developing 315 hp.

Variants

As previously stated, the current in service model of the ZT-3 is fitted with a single HEAT warhead activated by a contact fuze that will penetrate over 650 mm of RHA.

Currently under development by Kentron is the ZT-3B missile which is fitted with the same HEAT warhead and a nose-mounted precursor charge and it is estimated that this warhead will penetrate around 1,000 mm of RHA protected by ERA.

The ZT-3B missile will also feature a new laserbeam guidance system and a more compact sighting system.

ZT-3 can also be fitted with a fully automatic command to line of sight guidance system through the use of a stabilised sight and automatic tracker.

This is the guidance system used on the ZT-3 installation for helicopters. The Rooivalk development helicopter was equipped with this system as was the Polish W-3 Sokol helicopter that was modified by Kentron and called the Huzar.

Ratel replacement programme

Since the Ratel first entered service with the South African Army in the 1970s it has been constantly upgraded, mainly in the automotive area to improve the overall reliability of the vehicle.

Under the South African ICV2000 programme it was expected that some 700 new ICVs would have been purchased to replace the current Ratel fleet.

Recent information has indicated that due to funding constraints between 240 and 300 vehicles will now be procured to equip two or three of the South African regular mechanised infantry battalions. The four battalions and the reserve mechanised battalions would retain the Ratel until around 2015.

The new South African ICV will be an 8 × 8 vehicle and is required to have sufficient protection to work with the Rooikat (8 × 8) tank destroyer/armoured car that is currently armed with a 76 m gun.

To meet this requirement, Vickers OMC will be offering a locally built variant of the Swiss MOWAG Piranha IV (8 × 8) vehicle which was shown in public for the first time in September 2001. Armament could be the new LIW 35 mm EMAK which, in a two-person turret, has already been shown on a Ratel (6 × 6) chassis.

There is also a requirement to replace the current Ratel (6 × 6) tank destroyer with a new system, perhaps armed with the new Ingwe laser beam-riding missile. A possible in service date of 2015 has been mentioned with up to 75 systems being required.

Status

Production complete. In service with South African National Defence Force (53).

Manufacturers

Vickers OMC (chassis).
Kentron Division of Denel (missile system).

UPDATED

SWEDEN

Hägglunds Vehicle Pvrbv 551 anti -tank missile carrier

Development

In February 1983, following successful trials with a prototype vehicle, the Swedish Defence Matériel Administration awarded a contract to Hägglund and Söner (now Hägglunds Vehicle) for the conversion of a number of obsolete Ikv-102 and Ikv-103 self-propelled infantry cannons, which were withdrawn from front-line use some years previously, into Pvrbv 551 anti-tank missile carriers. At the same time the company was awarded a contract to convert the Ikv-102 and Ikv-103 into the now SAAB Bofors Dynamics RBS 70 SAM vehicles. First vehicles were delivered to the Swedish Army in 1984 and production continued until mid-1986.

The modifications to the chassis and suspension were extensive and include replacement of the engine and transmission, extending the crew compartment to the rear of the hull, improving protection and fitting new communications equipment and observation devices.

The vehicles are organised into companies and supplied to both armoured and mechanised brigades.

It is believed that a total of 55 of these Pvrbv 551 anti-tank missile carriers armed with Raytheon Systems Company TOW missiles is in service with the Swedish Army.

Specifications

Crew: 4
Combat weight: 9,700 kg
Power-to-weight ratio: 14 hp/t
Ground pressure: 0.4 kg/cm²
Length: 4.81 m
Width: 2.54 m
(over tracks) 2.23 m
Height:
(hull top) 1.97 m
(missile and launcher raised) 2.46 m
Ground clearance: 0.33 m
Track: 1.83 m
Track width: 400 mm
Length of track on ground: 3 m
Max road speed:
(high range forwards) 41 km/h
(low range forwards) 21 km/h
(reverse) 7 km/h
Fuel capacity: 240 litres
Max range: 300 km
Fording: 0.9 m
Gradient: 60%
Trench: 1.5 m
Engine: Ford Model 2658E V-6 petrol developing 136 hp at 5,200 rpm
Transmission: Mercedes W 4A-018, 4 forward, 1 reverse gears
Armament: 1 × TOW ATGW launcher
Electrical system: 24 V
Batteries: 4 × 12 V, 57 Ah
NBC system: no
Night vision equipment: no

Description

The gun has been removed from the glacis plate and the hole plated over. The driver sits at the front of the hull on the left side and has a single-piece hatch cover and fixed day periscopes for observation to the front and sides. The commander sits to the right of the driver and has a single-piece hatch cover that opens to the right rear, and day observation periscopes. The commander's position is raised above the driver for improved observation.

In the centre of the hull roof is a larger circular opening covered by a two-piece hatch cover opening either side. The Raytheon Systems Company TOW ATGW system is carried inside the hull and raised up when required for action. Reserve missiles are stowed in the hull rear above the engine and transmission compartment and access to these is via hatches in the roof that open either side.

The suspension, which is identical to that of the Ikv-102 and Ikv-103, consists of six rubber-tyred roadwheels either side with the drive sprocket at the rear, idler at the front and two track-return rollers. The original engine has been replaced by a Ford V-6 petrol engine, also installed in the Hägglunds Vehicle AB Bv 206 all-terrain vehicle in service with the Swedish Army. The crew of four consists of the commander, driver, gunner and loader.

Variant

Lvrbv 701 RBS 70

This has a similar hull to the Pvrbv 551 anti-tank missile carrier but is armed with the SAAB Bofors Dynamics RBS 70 surface-to-air missile system. When travelling, the launcher is retracted under armour protection and raised when required. Reserve missiles are stowed in

Pvrbv 551 anti-tank vehicle with the Raytheon Systems Company TOW ATGW (FMV)

Lvrbv 701 with the SAAB Bofors Dynamics RBS 70 SAM launcher in the elevated position ready to engage targets

the hull rear above the engine and transmission compartment with access via the front edge of the magazine which folds upwards. This version has a four-man crew. Full details of this version are given in *Jane's Land-Based Air Defence 2001-2002*. A thermal sight can be fitted to engage targets under night and poor weather conditions.

Status
Production complete. In service only with the Swedish Army. There are no known plans for this system to be upgraded and this will probably be replaced by a new system , perhaps based on the SEP chassis currently at the prototype stage, later this decade.

Manufacturer
Hägglunds Vehicle AB (now owned by Alvis plc of the UK).

UPDATED

SWITZERLAND

MOWAG Piranha TOW anti-tank vehicle

Specifications
Crew: 5
Configuration: 6 × 6
Combat weight: 11,000 kg
Unloaded weight: 10,200 kg
Power-to-weight ratio: 27.27 hp/t
Length: 6.30 m
Width: 2.50 m
Height: 1.85 m
Ground clearance: 0.5 m
Track:
(front) 2.18 m
(rear) 2.205 m
Wheelbase: 2.04 + 1.04 m
Angle of approach/departure: 40/45°
Max speed: (road) 100 km/h
Fuel capacity: 180 litres
Max road range: 500 km
Fording: 1.40 m
Gradient: 70%
Side slope: 35%
Vertical obstacle: 0.5 m
Turning radius: 7.3 m
Engine: Detroit Diesel Model 6V-53T developing 300 hp at 2,800 rpm
Transmission: Allison Transmission MT-653 automatic, 5 forward/1 reverse gears
Suspension: independent
Tyres: 11.00 × 16 XL
Brakes:
(main) air hydraulic
(parking) mechanical
Electrical system: 24 V
Armament: twin TOW ATGW launcher
Smoke grenade launchers: 2 × 4 76 mm
Ammunition: 2 + 8 ATGW
NBC system: no
Night vision equipment: yes

Development
After extensive trials of three MOWAG Piranha (6 × 6) AFVs fitted with the Norwegian Kvaerner Eureka A/S Armoured Launching Turret (ALT), and three vehicles fitted with a pedestal-mounted TOW launcher which retracted under full armour protection when not required, the former was selected by the Swiss Army. The system has the official Swiss Army designation of the Panzerjager 90 (PzJg 90).

The Swiss 1986 Armament Programme proposed SFr911 million for the procurement of 310 Swiss MOWAG Piranha (6 × 6) anti-tank vehicles fitted with the Armoured Launching Turret (ALT), to be issued to 31 anti-tank companies at that time equipped with the 106 mm 58 (US designation M40A1) recoilless anti-tank gun.

These 310 vehicles were delivered between 1989 and 1992 with MOWAG building the Piranha (6 × 6) vehicle and installing the ALT which was manufactured under partial licence by the Ateliers Fédéraux with the total Swiss share being 70 per cent.

The Federal Aircraft Factory at Emmen built about 49 per cent of the aiming system, including day and night sights, and almost 55 per cent of the Raytheon Systems Company TOW 2 guided missile and simulators.

This was the first MOWAG Piranha to enter service with the Swiss Army and more recently the improved MOWAG Piranha 8 × 8 has entered service. The original requirement was for a total of 615 vehicles but this was subsequently reduced to 515 with deliveries still under way.

Description
The MOWAG Piranha (6 × 6) vehicle used for this mission is virtually identical to that covered in the Armoured personnel carriers (wheeled) section earlier in this volume. The five-man crew consists of commander, gunner, loader, driver and combat orderly.

The hull is of all-welded steel armour construction which provides the occupants with protection from small arms fire and shell splinters, with the driver seated at the front left with the engine compartment to his right. The driver has a single-piece hatch cover that opens to the left and three forward-facing day periscopes, the centre one of which can be replaced by a passive night driving device.

The commander is seated to the rear of the driver with a raised cupola for improved observation. This is provided with a single-piece hatch cover, three day observation periscopes and a sighting periscope.

The ALT is mounted in the centre of the hull roof with two Raytheon Systems Company TOW 2 missiles with a maximum range of 3,750 m being mounted one either side. An additional eight TOW missiles are carried in racks in the right side of the rear compartment and are loaded manually via the rear roof hatch. It takes about 45 seconds to reload two new TOW 2 missiles. The ALT has day/thermal sights which allow targets to be engaged under all weather conditions.

Turret traverse is powered through 360° with launcher elevation from −12 to +15°. The turret is also provided with eight 76 mm smoke grenade launchers which are identical to those fitted to other Swiss armoured fighting vehicles. These are mounted on the upper part of the turret rear and fire forwards.

The rear crew compartment has a bench seat down the left side and in the roof, one either side, is a fixed day periscope to give observation to the side of the vehicle. Normal means of entry and exit is via the ramp in the hull rear which is provided with an emergency door.

The MOWAG Piranha anti-tank vehicle is not amphibious as this was not a requirement of the Swiss Army. Standard equipment includes a front-mounted winch with a capacity of 7,000 kg which is provided with 40 m of cable. A Raytheon Systems Company TOW ground mount is carried to enable missiles to be launched away from the vehicle if required by the tactical situation.

Variants
None.

Status
Production complete. In service with the Swiss Army. In August 1999, MOWAG was taken over by the Diesel Division, General Motors of Canada. The company is now known as General Motors Defense Limited.

Manufacturer
MOWAG Motorwagenfabrik AG.

UPDATED

MOWAG Piranha (6 × 6) anti-tank vehicle of the Swiss Army with Armoured Launching Turret and crew of five **2000**/0085197

MOWAG Piranha (6 × 6) anti-tank vehicle of the Swiss Army with Armoured Launching Turret 0018777

UNITED KINGDOM

Alvis Striker anti-tank guided weapon vehicle (FV102)

Development

The Alvis Striker anti-tank guided weapon vehicle is a member of the Alvis Scorpion Combat Vehicle Reconnaissance (Tracked) family of vehicles which is covered in detail in the earlier Armoured fighting vehicle families section.

Alvis was responsible for the chassis with what is now Matra BAe Dynamics responsible for the missile system and the Swingfire ATGW.

Following trials and acceptance for service by the British Army, first production vehicles were completed in June 1975 and the Striker entered service in 1976.

Within the British Army it is issued to the anti-tank units of the Royal Armoured Corps deployed with the reconnaissance regiments.

First deliveries to the Belgian Army were made in June 1976 and it took delivery of a total of 43 Striker vehicles, which have now been withdrawn from service.

In British Army service, the Alvis Striker was to have been replaced by the Formation Reconnaissance Guided Weapon member of the UK/US TRACER/FSCS. This will no longer be funded by the UK after mid-2002.

Production of all members of the Alvis Vehicle Scorpion Combat Vehicle Reconnaissance (Tracked) was undertaken at Coventry. This facility closed down in 1999 and all future production and upgrade work on the chassis will be carried out at the Alvis Vehicles facility in Telford, Shropshire, previously the home of GKN Defence.

Alvis Striker vehicle launching Matra BAe Dynamics Swingfire ATGW ***2000***/0085203

Specifications

Crew: 3
Combat weight: 8,346 kg
Power-to-weight ratio: 22.77 bhp/t
Ground pressure: 0.345 kg/cm^2
Length:
(hull) 4.826 m
Width:
(overall) 2.28 m
(over tracks) 2.134 m
Height:
(overall) 2.28 m
Ground clearance: 0.356 m
Track: 1.708 m
Track width: 432 mm
Max road speed: 80.5 km/h
Fuel capacity: 350 litres
Max road range: 650 km
Fording: 1.067 m
Gradient: 60%
Vertical obstacle: 0.5 m
Trench: 2.057 m
Turning radius: pivot turn in neutral
(in 1st gear) 1.84 m
(in 7th gear) 35.9 m
Engine: Cummins diesel developing 200 bhp
Transmission: TN15 crossdrive, semi-automatic hot-shift type, providing 7 speeds in each direction and pivot turns
Steering: Merritt system incorporated in transmission
Suspension: torsion bar
Electrical system: 28.5 V
Batteries: 4
Main armament: Swingfire ATGWs
Machine gun: 1 × 7.62 mm
Smoke grenade dischargers: 2 × 4 66 mm
Ammunition:
(main) 10 × ATGW
(machine gun) 3,000
NBC system: yes
Night vision equipment: yes

Description

The Striker is based on the hull of the Spartan APC with the driver at the front on the left with the power pack to his right. The commander is to the rear of the driver with the missile controller to his right. Automotive aspects are identical to the standard Alvis Vehicles Scorpion already described in detail earlier in this book.

The commander has a No 26 cupola which has eight day periscopes and a monocular day sight with a magnification of ×1 or ×10, the line of sight elevating with the 7.62 mm machine gun. The ×1 mirror assembly can be replaced by a ×1.8 image intensifier.

Mounted on the right side of the cupola is a 7.62 mm machine gun which can be aimed and fired from within the vehicle. The cupola has a single-piece hatch cover that opens to the left.

The missile controller has the Barr & Stroud (now Thales Optronics) Combined Sight (CS) which has both thermal and visual optics with magnifications of ×2 or ×6 (thermal) or ×2 and ×10 (day) respectively. The sight can be traversed through 55° left and right, and a single-piece hatch cover that opens to the right is provided to the rear of this sight.

On the top at the rear of the vehicle is a launcher box with five of the now Matra BAe Dynamics Swingfire ATGWs. A further five missiles carried inside the vehicle are loaded manually and one of the crew has to leave the vehicle to accomplish this. The launcher box is pivoted at the rear and is elevated to an angle of about 35° before the missiles are launched.

The Swingfire missile weighs 28 kg and has a HEAT warhead. Minimum range is between 150 and 300 m depending on the distance of the vehicle and the controller when being used in the separated mode.

Maximum range of the Swingfire is 4,000 m. An advantage of the Swingfire ATGW system over other systems is that the missiles can be launched from behind a crest or cover so that the launcher position cannot be seen by the enemy.

The missiles can be launched by the controller in the vehicle, or with the controller at a vantage point (up to 100 m from the launcher) separated from the launcher giving him protection from counterfire.

Mounted at the front of the vehicle are two four-barrel 66 mm electrically operated smoke grenade dischargers which are operated from within the vehicle.

Standard equipment for the Striker tank destroyer includes an overpressure NBC system.

Upgraded Striker

Following on from 'proof of principle' research programmes by the then Royal Armament Research and Development Establishment (RARDE) and the then British Aerospace, in mid-1990 the UK MoD awarded a £35 million five year contract to the now Matra BAe Dynamics, for the implementation of the Swingfire Improved Guidance (SWIG) programme into the Alvis Striker.

First production standard modified vehicles were delivered in July 1995 and the last of 60 upgraded vehicles were handed over to the British Army late in 1996.

The SWIG modification involved virtually a complete replacement of the existing analogue electronics ground control equipment by new digital technology electronic units.

The system retains the existing dual-mode, visual/thermal Combined Sight which entered service in 1980. The updated ground control equipment incorporates a comprehensive Built In Test (BIT) facility which, combined with the much improved reliability, substantially increases system availability.

The improved system offers a simple to use ACLOS (Automatic Command to Line Of Sight) guidance mode so reducing operator training needs. The existing MANCLOS (Manual Command to Line Of Sight) mode is retained, for direct fire reversionary capability, as is the versatile separated fire mode, where the operator can be positioned remotely, up to 100 m from the launching vehicle.

The ACLOS guidance system uses thermal image data from the combined sight to automatically track both the target and the missile and effect command to line of sight guidance, independently of the operator.

To engage in the ACLOS mode the operator positions a movable electronic cross-wire into the region of the target and initiates the target tracker. Target tracking is now autonomous, while the operator launches and gathers the missile.

Once the missile enters the field of view, guidance is fully automatic until target impact. Sophisticated tracking logic allows the system to cope with a wide variety of potentially decoying effects.

The SWIG modification also incorporates a mini-compressor for thermal sight cooling, freeing the vehicle from the logistic constraints of high-pressure bottled air. A comprehensive workstation-based, second-line test facility for the new SWIG line-replaceable units has been incorporated into an existing support vehicle.

Repower

In August 1997, following extensive trials between Alvis Vehicles Scorpion light armoured vehicles fitted with diesel engines supplied by the Cummins Engine Company (UK), Perkins Engines Company (UK) and Steyr-Daimler-Puch (Austria), the UK MoD selected the Cummins 6BTA 5.9 litre six-cylinder diesel developing 235 hp.

All operational British Army CVR(T) family of tracked vehicles will have had their current Jaguar 4.2 litre petrol engines replaced by the more fuel efficient Cummins diesel engine.

The 73 mm Scorpion member of the CVR(T) has been withdrawn from service with the British Army but all other versions remain in service including the Scimitar and Sabre (30 mm), Samson recovery vehicle, Sultan command post vehicle, Striker anti-tank vehicle, Spartan troop carrier and Samaritan ambulance.

The installation of the Cummins diesel engine also requires some modifications to other subsystems of the CVR(T) such as the transmission.

Alvis Vehicles has converted the first 200 Scorpion's with the Cummins diesel engine at Telford and the first of these was handed over to the British Army in April 1999. There was a competition for the next batch of 600 vehicles and this was won by the Army Base Repair Organisation (ABRO) facility at Donnington. The upgrade is now complete.

Status

Production complete. In service only with the British Army. Upgrade missile system programme is now complete. Chassis upgrade programme has been completed and was carried out by Alvis Vehicles and ABRO.

Manufacturers

Alvis Vehicles Limited (chassis).
Matra BAe Defence Limited (missiles).

UPDATED

UNITED STATES OF AMERICA

Lockheed Martin Missiles and Fire Control – Dallas Line Of Sight Anti-Tank (LOSAT) weapon system

Development/Description

The Line Of Sight Anti-Tank (LOSAT) weapon system consists of the Kinetic Energy Missile (KEM) and its Improved Bradley Acquisition System (IBAS) fire-control system mounted on a modified Expanded Capacity AM General HMMWV (4 × 4) wheeled chassis.

The system is being developed by Lockheed Martin Missiles and Fire Control – Dallas (previously Loral Vought Systems Corporation) under contract to the US Army Aviation and Missile Command, to provide a highly lethal, long-range deployable anti-tank system which can engage and defeat all present and future armoured threats.

The KEM missile developed out of the old Hypervelocity Missile (HVM) programme and is designed to destroy a target using its high velocity of approximately 1,500 m/s, combined with a long rod penetrator that can penetrate protective armour. According to Lockheed Martin, the KEM has 'decisive lethality against all known or projected armour threats through the year 2020 including active protection systems'.

Missile update on the onboard guidance package is accomplished through an aft looking, cryogenically cooled, infra-red receiver in the rear of the missile on the rocket motor nozzle.

A second generation Forward Looking Infra-Red (FLIR) sensor is used for target acquisition and target and missile tracking. A CO_2 laser is used for target ranging as well as to transmit position information to the in-flight missile.

An all boost rocket motor accelerates the missile to hypervelocity speed and attitude control motors are placed radially around the missile to provide a full three-dimensional manoeuvre capability.

Late in October 1994 the now Lockheed Martin Missiles and Fire Control – Dallas was awarded a contract worth over US$42 million to integrate the LOSAT system into the United Defense Armored Gun System (AGS) chassis to form the LOSAT/AGS.

This had a new two-man turret with a total of 12 Kinetic Energy Missiles (KEM) in the ready to launch position, these being mounted six either side.

Although the AGS was type classified as the M8 by the US Army the whole programme was cancelled so leaving the US Army with no LOSAT chassis.

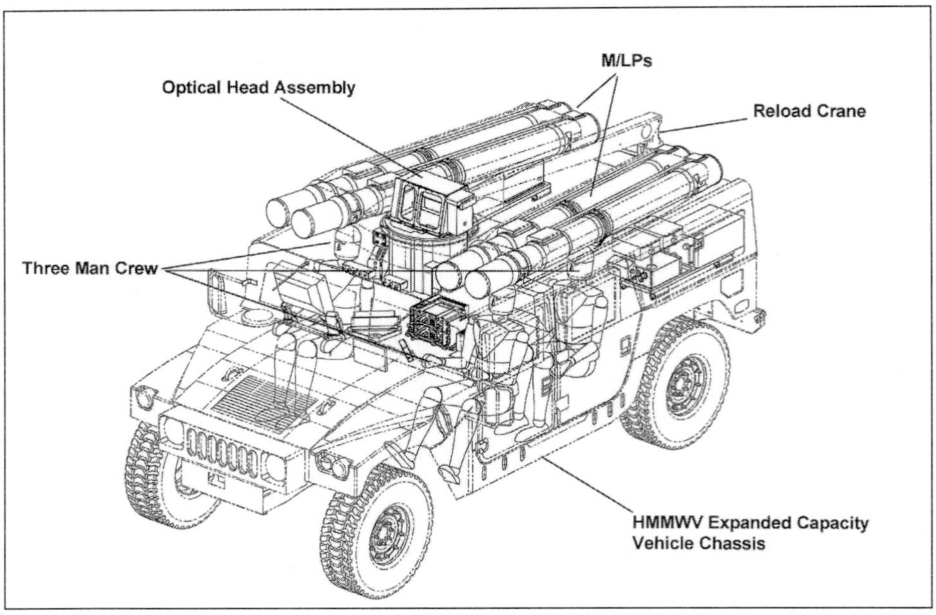

Detailed drawing of the latest configuration of the Enhanced Capacity HMMWV fitted with LOSAT, which differs from the prototype system and showing position of key subsystems ***2002***/0121607

Lockheed Martin Missiles and Fire Control – Dallas have now integrated the LOSAT system on a modified AM General Expanded Capacity High Mobility Multi-Purpose Wheeled Vehicle (4 × 4) chassis. The platform is also fitted with Position Navigation (POSNAV) and Global Positioning System (GPS).

The system has a crew of three with four missiles being carried above their position. The roof-mounted sight is in the centre of the roof with replacement missiles being loaded in pairs using an onboard crane. This tows a resupply trailer carrying a further eight missiles.

In June 1996, the LOSAT/KEM was successfully test fired from an expanded capacity HMMWV chassis at White Sands Missile Range, New Mexico. According to Lockheed Martin Missile and Fire Control – Dallas, the successful firing proved that the HMMWV was a suitable platform.

Instrumentation in and around the vehicle measured launch effects on a simulated crew and the vehicle itself, with all post launch readings falling well within the US Army's acceptable ranges.

Missile Specifications
Missile launch weight: 80 kg
Missile length: 2.845 m
Missile diameter: 0.162 m
Missile range: over 4,000 m

According to Lockheed Martin Missile and Fire Control – Dallas, LOSAT/HMMWV (4 × 4) transportability is as follows: C-5 aircraft, 5; C-17 aircraft, 7; C-141 aircraft, 3; C-130 aircraft, 2; CH-47D helicopter, 1 external load; UH-60L helicopter, 1 external load.

Major subcontractors to Lockheed Martin Missiles and Fire Control – Dallas are Raytheon Systems Company for the electro-optical subsystem of the fire-control system, Alliant Techsystems for the solid propellant rocket motor, ARC for the attitude control motors and BAE Systems for the laser.

In April 1998, Lockheed Martin Missiles and Fire Control – Dallas was awarded a US$180 million

LOSAT on HMMWV (4 × 4) chassis launching a LOSAT at White Sands Missile Range 0018779

Two LOSAT missiles being reloaded onto HMMWV (4 × 4) using onboard crane 0018780

Advanced Concept Technology Demonstration contract under which it will build 12 fire units and 44 missiles with the work to be completed in 2003.

Although the HMMWV LOSAT only has four missiles in the ready to launch position the vehicle would also tow a trailer carrying additional missiles. These would be loaded on to the launcher with the aid of an onboard crane.

According to Lockheed Martin, LOSAT can also be integrated into other platforms such as the LAV-III (8 × 8) selected by the US Army for service with its Brigade Combat Teams (BGT).

Status

Technology demonstration phase for US Army.

Manufacturer

Lockheed Martin Missiles and Fire Control - Dallas.

UPDATED

Systems & Electronics Inc (SEI) M901 Improved TOW Vehicle

Development

To meet a US Army requirement for a turret-mounted and elevated TOW ATGW system, three companies built prototype systems and, after trials in December 1976, Emerson (now Systems & Electronics Inc (SEI)) was awarded a low-rate initial production contract for 10 vehicles.

In June 1978, the M901 Improved TOW Vehicle (ITV) was approved for full-scale production and by 1998 well over 5,500 systems had been completed for the home and export markets. This figure includes 3,315 M901 Improved TOW Vehicles and 1,306 M981 Fire Support Team Vehicles for the US Army. Production for the US Army was completed some time ago but it remained in production for the export market. Production is now complete with last customer being Thailand.

The latest production model, the M901A1, is capable of firing any of the TOW missile variants (Basic TOW, Improved TOW (I-TOW) or TOW-2 series) produced by the now Raytheon Systems Company.

The US Army mounts the system on a modified United Defense LP M113 series chassis but it can be installed on a wide range of other chassis. For example, the US Marine Corps uses the Light Armored Vehicle (8 × 8), the Netherlands uses the Armored Infantry Fighting Vehicle (with 210 of these recently sold to Egypt) while Saudi Arabia uses the Italian Otobreda VCC-1 full-tracked chassis.

Recent export orders include 18 for Thailand (the first customer for the latest M901A3 with deliveries in 1997) and the Saudi Arabian National Guard (111 systems on a now General Motors Defense (at that time Diesel Division, General Motors of Canada) 8 × 8 Light Armored Vehicle chassis). Production of the M901 series has now been completed.

When fitted with the latest RISE power pack, the M901A1 is designated the M901A3 ITV. The Fire Support Team Vehicle (FISTV) is related to the M901 ITV.

It is understood that significant numbers of M901 Improved TOW Vehicles are now surplus to US Army requirements and these are being made available to a number of other countries with Egypt (130), Morocco (80) and Norway (126) being possible future recipients.

Description

The M901/M901A1 Improved TOW Vehicle consists of a standard United Defense LP M113A1 APC with an M27 cupola mounted on the roof. On the forward part of this is an arm containing the image transfer assembly and on top of the arm is the armoured launcher assembly which contains two TOW launcher tubes, the TOW sight assembly and the acquisition sight.

The turret has full powered 360° traverse at 35°/s with the launcher head having an elevation of +34° and a depression of −30°. This enables the ITV to operate on gradients and side slopes with only the launcher head showing above the crest.

When travelling, the launcher head is retracted level onto the hull top so making the ITV difficult to distinguish from the standard M113A1 series APC.

Once the ITV has come to a halt, it takes only 20 seconds for the launcher to be elevated and the target to be engaged. Time from first round impact to second round triggering with a 12.5° target separation is 4.25 seconds.

The gunner identifies and tracks the target through the image transfer assembly, which contains the coupling optics and allows the gunner to select the standard day sight or night tracking by remote control.

In addition, the coupling optics have a wide field of view so allowing the gunner to scan for targets. The guidance and sights of the ITV are identical to the normal TOW ground-based systems and if required, the components can be removed from the vehicle and fitted to a standard TOW series ground mount; this can be carried inside the hull.

Once the two TOW missiles have been launched, the launcher assembly is depressed at the rear and two new missiles are loaded, this takes only 40 seconds. The loader is provided with both side and overhead armour protection during the manual loading operation.

A further 10 TOW missiles are carried internally and a 7.62 mm machine gun with a total of 2,000 rounds is mounted on the cupola for local defence. Mounted on either side of the hull front is a bank of four electrically operated smoke grenade dischargers.

Specifications

(Similar to M113A1 APC except for following)

Crew:
(deployed with mechanised infantry) 4
(deployed with cavalry) 5
Combat weight: 11,794 kg
Ground pressure: 0.58 kg/cm²
Height:
(launcher stowed) 2.91 m
(launcher erected) 3.35 m
NBC system: yes
Night vision equipment: yes

Status

Production complete. Known to be in service with the following countries:

Country	Quantity	Comment
Egypt	52	M113A1 chassis
	210	YPR 765 PRAT-TOW (from Netherlands)
Greece	24	M113A1 chassis, some sources state 156 systems in service
Jordan	50	M113A1 chassis
Kuwait	58	M113A1 chassis, only 8 may be operational
Netherlands	304	YPR 765 PRAT-TOW
Pakistan	24	M113A1
Thailand	18	M113A3 chassis
Saudi Arabia	224	VCC-1 chassis (army)
	111	LAV 8 × 8 chassis (SANG)
USA	3,315	M113A1/M113A2 chassis (being phased out)
	96	LAV 8 × 8 chassis (Marine Corps)

Manufacturer

Systems & Electronics Inc (SEI).

UPDATED

M901 Improved TOW vehicle in travelling configuration with launcher retracted (Julio Montes) 0058360

Improved TOW vehicle with launcher deployed in the firing position (Michael Jerchel) 0058361

ARMOURED COMMAND VEHICLES

ARMOURED COMMAND VEHICLES

RUSSIAN FEDERATION AND ASSOCIATED STATES (CIS)

Artillery Command and Reconnaissance Vehicle (ACRV)

Development

To operate with its 122 mm 2S1 and 152 mm 2S3 full tracked self-propelled artillery systems, Russia developed the Artillery Command and Reconnaissance Vehicle (ACRV) family of full tracked armoured vehicles in the early 1970s.

Within the Russian Army these are normally referred to as the MT-LBus as they are related to the MT-LB series of vehicle.

These were subsequently adopted for a wide range of specialised support roles for other arms such as air defence command and control, electronic warfare, NBC vehicle and radar carrier to name but a few.

Many of the automotive components of the ACRV are also used for the MT-LB full tracked armoured personnel carrier that was built in the Ukraine and the 2S1 self-propelled artillery system. Details of the MT-LB given in the Armoured personnel carriers (tracked) section under RFAS.

The ACRV has also been manufactured in Bulgaria by the Beta Joint Stock Company who also built the 122 mm 2S1 and MT-LB variants. Additional details of their versions of the ACRV are given in the entry for the 2S1 122 mm self-propelled artillery system.

Description

The hull of the ACRV is of all-welded steel armour construction that provides the occupants with protection from small arms fire and shell splinters.

The commander and driver are seated at the front of the vehicle and in front of the hull are two large windows which can be rapidly covered by an armoured shutter hinged at the top.

Vision is then obtained by three roof-mounted day periscopes in the case of the driver and a roof-mounted swivelling day periscope in the case of the commander.

The driver is seated at the front of the vehicle on the left and above his position is a single-piece hatch cover that opens to the rear.

The commander is seated on the right and above his position is a single-piece hatch cover that opens forwards. Both can be locked in the vertical position.

The engine compartment is located to the rear of the commander and driver with the air inlet and outlets in the roof. The exhaust pipe is also located on the roof and extends to the right side of the hull.

This arrangement of the driver, commander and engine compartment leaves the whole of the rear of the hull clear for the specialised equipment carried.

Suspension is of the torsion bar type with each side having seven rubber tyred road wheels with the drive sprocket at the front and idler at the rear, there are no return rollers.

All variants of the ACRV normally have a single door in the hull rear that opens to the right which is also provided with a vision device and an associated firing port. The number and type of roof hatch depends on the role of the variant.

The basic ACRV has seating for up to seven people and typically has a five-man crew consisting of driver, commander, radio operator, navigator and range-finding operator.

All versions normally have an auxiliary power unit, NBC system of the overpressure type and are fully amphibious, being propelled in the water by their tracks.

Before entering the water the bilge pump is switched on, the trim vane erected at the front of the hull, shrouds are fitted to the hull above the drive sprocket and the front road wheels are covered.

Any water that enters the hull during amphibious operations is removed via the exhaust outlet using the bilge pump.

Variants

The Russian Army is now known to refer to these vehicles as the 1V12 system, which applies to those command and control vehicles assigned to 122 mm 2S1 and 152 mm 2S3 self-propelled artillery systems.

Each battery has two vehicles, with the 1V13 deputy battery commander's vehicle serving as the battery fire direction centre. The 1V14 serves as the observation post for the battery commander. In addition, the battalion command platoon also includes one 1V15 for use by the battalion commander and one 1V16 for the battalion fire direction under the control of the deputy battalion commander.

Designation	Role	NATO designation
1V13	deputy battery commander	M-1974-1
1V14	battery commander	M-1974-2A
1V15	battalion commander	M-1974-2B
1V16	deputy battalion commander	M-1974-3

The ACRV chassis is also used as the basis for the Taran 1 electronic warfare system

1V12 series ACRV showing 12.7 mm anti-aircraft machine gun (Willis A Bullard)

The ACRV chassis is also used as the basis for the Taran 3 electronic warfare system **2002**/0121601

ACRV 1V14/1V15 in travelling configuration (Michael Jerchel) **2001**/0103435

Crew/special equipment of 1V12 family

Type		1V13	1V14	1V15	1V16
Crew	battalion commander			1	
	chief of staff				1
	battery commander		1		
	battery officer	1			
	vehicle commander	1	1	1	1
	driver	1	1	1	1
	radio operator	2	2	3	3
	spotter		1	1	
	fire controller	1			1
	total crew	6	6	7	7
Radios	R-107M		1	1	
	R-123M	3	3	2	2
	R-111			1	1
	R-130M			1	1
	R-326				1
Vehicle equipment	survey equipment 1T121-M	1	1	1	
	gyrocompass 1G25-1	1	1	1	
	directional gyro PAB-2AM	1	1	1	
	range-finder DSP-30	1	1	1	
	periscope PV-1	1			
	laser range-finder 1D11M-1		1	1	
	day/night sight 1PN-44		1	1	
	range-finder DS-1		1	1	
	co-ordinate processor 1T804	1	1	1	
	all-round periscope VOP-7A		1	1	
	meteorology box PMK				1
	computer 9V59				1
	teleprinter STA-M67B				1
	auto commander receiver 9Sch34	1			

Zoopark-1 radar system with antenna erected (Christopher F Foss) **2002**/0121597

Specifications
(depends on variant)
Crew: 6
Combat weight: 15,500 kg
Power-to-weight ratio: 19.35 hp/t
Length: 7.475 m
Width: 2.85 m
Height: 2.425 m
Ground clearance: 0.40 m
Track width: 400 mm
Max speed:
(road) 61.5 km/h
(water) 4.5 km/h
Fuel capacity: 550 litres
Road range: 500 km
Fording: amphibious
Gradient: 77%
Side slope: 55%
Vertical obstacle: 0.7 m
Engine: YaMZ-238 V-8 water-cooled 4-stroke diesel developing 300 hp
Transmission: manual, 4 forward and 1 reverse gears
Steering: clutch and brake
Armament: 1 × 12.7 mm machine gun
Armour:
(hull) 15 mm (estimate)
(turret) 20 mm (estimate)
NBC system: yes
Night vision equipment: yes

ACRV M-1974-1 (1V13)
(Battery fire direction centre)

This model is issued on the scale of three per self-propelled battalion. It has a crew of six, with the driver seated at the front on the left with a single-piece hatch cover that opens to the rear and the vehicle commander to his right with a single-piece hatch cover that opens forwards. Both have a windscreen in front with a shutter which is lowered in a combat area. The remainder of the crew is seated at the back. This vehicle is often fitted with an externally mounted 12.7 mm DShKM anti-aircraft machine gun which is mounted on a one-person rotating cupola. The circular cupola is provided with vision equipment and a single-piece hatch cover.

ACRV M-1974-2 (1V14/1V15)
(Battery and battalion command vehicle)

The battery commander's vehicle is designated the ACRV M-1974-2A and has whip antennas, while the battalion commander's vehicle, the ACRV M-1974-2B, has a 10 m telescopic antenna which is stowed on the upper left hull when not required.

The ACRV M-1974-2A, used both as a command and observation vehicle, is fitted with a laser range-finder in the turret and also carries a dismountable range-finder such as the DAK-1 (NATO codename 'Sage Gloss').

ACRV M-1974-3 (1V16)
(Battalion fire direction centre)

Main identification features of this version are the lack of a box on the right side of the hull, clean turret top and sides and an additional antenna on the hull roof.

This version contains the battalion's electronic field artillery computer and is fitted with a 10 m telescopic antenna which is stowed on the upper left hull when not deployed.

Late production ACRVs

These are designated as follows:

Original designation	New designation
1V12	1V12M-2
1V13	1V13M-1
1V14	1V14M-1
1V15	1V15M-1
1V16	1V16M-2

Note: The modified 1V16M-2 lacks the original turret.

1V21 series air defence management vehicles

Russia also deployed a whole family of ACRV-based vehicles for use in the air defence command and control roles and these are designated the 1V21, 1V22,

1V14M-1 ACRV fitted with 7.62 mm roof-mounted PKT machine gun (Gunter Lippert) **2002**/0121600

MT-LBus jamming vehicle in travelling position with telescopic mast in retracted position **2002**/0121599

1V23, 1V24 and 1V25. They were exported to Czechoslovakia and Poland with the former designating them the MP21, MP22, MP23, MP24 and MP25.

These are also referred to as the Rangir air defence command vehicle. Additional details are given in *Jane's Land-Based Air Defence 2001-2002*.

Finnish MT-LB

These are designated the MT-LBu-TP, P and PS and all have a combat weight of 15.5 tonnes and a crew of eight. They entered service in 1995 and are equipped with Krauss-Maffei Wegmann 76 mm smoke grenade launchers, 12.7 mm NSVT machine gun, generator, additional stowage boxes and modern communications equipment.

The MT-Lbu-PS *(Tulipatterin komentopaikkapssarivaunu)* is a battery command vehicle based on the 1V13. The MT-LBu-P (*paikantamipanssarivaunu*) is a navigation vehicle based on the 1V14 or 1V 15. The MT-LBu-PS (*patteriston komentopaikkapanssarivaunu*) is a battalion command post and based on the 1V16

Other Russian versions

These include the R-378B tactical HF jamming vehicle, R-412B tropospheric scatter communications vehicle, PPRI-5 communications and computer vehicle, MT-LB KShM 9V514 and 9S714-N-18 command vehicles, R-161B communications vehicle, BETA-3 and -4 computer vehicles, 1RL243 and R-330K jammer and MZS-83 PSYOPS vehicle.

Hungarian variants

These include the MT-LB MFAJ engineer vehicle used for reconnaissance and obstacle clearence and easily identified by the 81 mm smoke grenade launchers.

MT-LBus jamming vehicle

This was first identified in Afghanistan and is an ACRV series chassis with an auxiliary power unit mounted at the rear of the hull. Mounted on the roof is an antenna for the Type R-330P VHF jamming set. When in use the 11-element fan-type antenna is in the vertical position, but it can be lowered into the horizontal position if required.

The ACRV chassis is also used as the basis for a number of other electronic warfare systems such as the Taran 1, 2 and 3.

Dal NBC reconnaissance vehicle

The ACRV chassis has been used as the platform for the Dal NBC reconnaissance vehicle, or to give the vehicle its correct designation, the KDKhR-1N complex for standoff chemical reconnaissance. Trials with prototype systems took place in 1987 and series production commenced in 1990.

Mounted on top of the chassis is a large turret with a laser system which rotates through a full 360° to scan the surrounding atmosphere for the presence of chemical agents, especially nerve agents.

The data received by the laser is processed by an onboard computer which also controls an automated data transmission system that sends data to a central command station.

In addition to the laser detection system, the KDKhR-1N also carries other radiological and chemical agent sensors to allow the system to conduct a full range of NBC reconnaissance missions.

Zoopark-1 radar

An ACRV chassis is used as the basis for the 3-D multifunction phased-array battlefield locating radar designated the Zoopark-1, or IL219.

This advanced radar has been designed to pinpoint enemy mortars, artillery and multiple rocket systems as well as correcting artillery fire from friendly systems.

The system has an instrumented detection range of 45 km and between four and six trajectories can be tracked at any one time. The system is normally operated by a crew of three.

The multifunctional roof-mounted phased-array radar can be traversed through a full 360° and, when travelling, the antenna is lowered into the horizontal position on top of the hull.

The Zoopark-1 is fitted with an auxiliary power unit and an NBC system and retains the amphibious characteristics of the standard ACRV.

This system entered series production in 1992 and is now in service with the Russian Army. There are no known exports of the system so far.

'Dog Ear' radar vehicle

A variant of the basic M-1974 chassis carrying surveillance radar known by the NATO reporting name of 'Dog Ear' has been observed. This radar uses a curved grid rotating aerial mounted over the rear of the hull and is used in conjunction with air defence units equipped with ZSU-23-4 self-propelled anti-aircraft guns, SA-9 Gaskin or SA-13 Gopher SAM systems and 2S6M gun/missile systems, indicating that the 'Dog Ear' is a target surveillance acquisition radar. The official designation of the 'Dog Ear' is the PPRU (*Podvizhnyj Punkt Razvedkii Uprauleniya* – Mobile Reconnaissance and Control Post).

Status

Production complete. In service with many countries.

Manufacturers

Russian state factories and Beta Joint Stock Company (Bulgaria).

UPDATED

SLOVAKIA

Tatrapan armoured all-terrain vehicle

Development

The Tatrapan armoured all-terrain vehicle has been developed as a private venture by Podpolianske Strojarne Detva and is based on the chassis of the Tatra T-815 VP 21 265 6 × 6 1R all-terrain military truck.

This has been fitted with a fully enclosed armoured driver's cab to the rear of which has been mounted a fully enclosed box-type armoured superstructure which can be used for a wide range of military roles.

A small production batch of Tatrapan all-terrain vehicles have been built as command post/artillery fire-control vehicles for use with the new Zuzana 155 mm/45 calibre self-propelled artillery systems now in service with Slovakia.

Description

The all-welded steel armoured forward control cab has a single forward-opening door in either side provided with a window which can be covered by armoured shutters if required. In each side of the hull to the immediate rear of the side door is a circular firing port.

The all-welded steel armour hull of the Tatrapan provides the occupants with protection from 7.62 mm armour-piercing small arms fire and shell splinters. Special protection is provided against attack from mines.

To the front of the driver and passenger is a windscreen which, when in a combat area, can be covered by horizontal armoured shutters.

The cab is also provided with a roof hatch and to the rear of the driver and commander is a folding couch. A 7.62 mm PKT machine gun can be mounted on the roof of the cab for local and anti-aircraft defence.

The front cab is also provided with communications equipment, heating, ventilating and filtration equipment and an air conditioning system. For access to the power pack the front cab is tiltable under hydraulic control and can be mechanically locked.

Specifications

Crew: 1 + 1 (cab only)
Configuration: 6 × 6
Combat weight: 21,300 kg
Unloaded weight 17,900 kg
Power-to-weight ratio: 16.66 hp/t
Length: 8.46 m
Width: 2.50 m
Height:
(cab roof) 2.75 m
(without weapons) 2.94 m
(with weapons) 3.36 m
Ground clearance: 0.39 m
Track:
(front) 2.044 m
(rear) 1.988 m
Wheelbase:
(1st to 2nd axle) 1.65 m
(2nd to 3rd axle) 3.20 m
Angle of approach: 32°
Angle of departure: 32°
Max road speed: 120 km/h
Fuel capacity: 460 litres
Fuel consumption:
(road) 64 litres/100 km
(cross-country) 84 litres/100 km
Range: 850 km
Fording: 1.4 m
Gradient: 60%
Side slope: 40%
Trench: 1.10 m
Vertical obstacle: 0.6 m
Turning radius: 12.5 m
Engine: Tatra T3 - 930 - 51 V-12 air-cooled supercharged diesel developing 355 hp at 2,200 rpm
Transmission: manual, 10 forward and 2 reverse gears
Clutch: friction, single disc with Belleville springs, hydraulically operated with pneumatic booster
Transfer case: two-stage downwards, wheel in permanent gear, changing with claw clutch electropneumatic
Differential: 1 interaxle differential, 3 axle-mounted differentials and torque divider
Differential locks: with claw clutches, operated electropneumatically
Steering: mechanical with hydraulic booster
Tyres: 15.00 - 21 to 4 (tubeless)
Brakes:
(main) dual-circuit, foot-operated, overpressure
(engine) on exhaust, active on all wheels and operated by means of electropneumatic switch
(parking) spring, active on second axle
Suspension:
(front two) swing axles in permanent gear, independently spring loaded, leaf spring with telescopic shock-absorbers
(rear) swing axle in permanent gear, independently spring loaded, air suspension with telescopic shock-absorbers
Electrical system: 24 V
Batteries: 2 × 12 V, 180 Ah
Armament:
(main) 1 x12.7 mm NSV MG
(secondary) 1 × 7.62 mm PKT
NBC system: yes
Night vision equipment: yes

Note: Recent information released shows that the Tatrapan is now offered with the German KHD BF 6M 1015C water-cooled 6-cylinder diesel developing 450 hp at 2,100 rpm. This is coupled to a Twin Disc 8 FLW-1754-1 fully automatic transmission with six forward and one reverse gears.
Michelin 16.00 R 20 XZL tyres are now fitted with a central tyre pressure regulation system that allows the driver to adjust the tyre pressure from 100 to 525 kPa.

The rear unit is of all-welded armour construction which provides protection from small arms fire and shell splinters. It is fitted with an NBC system, heaters and is lined with thermo-insulating materials, as is the cab.

In the rear is a large door that opens to the left which is provided with a firing port and there is an additional side door in the lower part of the hull between the second and third axles. It is believed that the complete rear door is from the Russian BMP-1 full-tracked infantry combat vehicle.

Three hatches are provided in the roof and on the centre one can be mounted a 12.7 mm NSV machine gun.

Standard equipment includes power-assisted steering on the front axles and a central tyre pressure regulation system that allows the driver to adjust the tyre pressure to suit the type of ground being crossed. A fire detection and suppression system is fitted as standard for the power pack tank areas.

Variants

Two models have been produced for the Zuzana 155 mm/45 calibre self-propelled artillery systems in service with the Slovak Republic.

The Delsys operates at the brigade, battalion and battery level while the second vehicle, the Bazus, provides the C^3 element for the Zuzana batteries.

The manufacturer has suggested the following possible roles for the Tatrapan (6 × 6) armoured vehicle:

(1) Personnel carrier
(2) Command post vehicle
(3) Patrol vehicle
(4) Supply vehicle (with or without an hydraulic crane for loading purposes)
(5) Ammunition vehicle
(6) Recovery vehicle
(7) Mobile workshop
(8) Ambulance
(9) Police vehicle
(10) United Nations missions
(11) Mortar vehicle
(12) Transport for pilots and other aircrew

Tatrapan (6 × 6) armoured all-terrain vehicle being used in the command post role with 155 mm Zuzana (8 × 8) self-propelled artillery systems (PB) 0121598

ARSYS artillery fire-control system

This is being offered for export and is for use at battery, regiment and brigade command levels.

Air defence command post

For use with the 2K12 Kub (SA-6 'Gainful') surface-to-air missile systems used by Slovakia, ALES and the VTU Military Technical Institute have developed a new air defence mobile command post, which is now fully operational.

This is known as Mobile Velitelske Stanovisko – mobile command post (MOVES) and is part of the LETVIS Radar Data Processor (RDP) programme. This was previously called the Tatrapan - PVO.

Status

Production as required. In service with Slovakia.

Manufacturer

Podolianske strojarne as (manufacturer).
Karametal Company Limited (marketing).

UPDATED

SOUTH AFRICA

Vickers OMC Okapi command post vehicle

Development

The Okapi command post vehicle has been designed by the now Vickers OMC (previously Reumech OMC) to provide a high level of mobility to keep up with other South African wheeled armoured systems such as the Vickers OMC Ratel (6 × 6) infrantry combat vehicle and the 155 mm LIW Division of Denel G6 (6 × 6) self-propelled artillery system.

A small number of production Okapi vehicles have been built with some of these being fitted with electronic warfare equipment for use in the forward battlefield area.

The Okapi command post vehicle is now being offered by Denel as part of a complete artillery system which also includes the G6 155 mm/45 calibre 6 × 6 self-propelled gun, ammunition support vehicles and fire-control system as well as long-range reconnaissance systems.

In mid-2001 it was stated that the South African Army had taken delivery of nine mobile electronic warfare systems as part of an extensive upgrade of its electronic warfare capability.

This new system was developed by Grintek Systems Technologies and installed in Vickers OMC Okapi mine-protected vehicles.

Vickers OMC Okapi (6 × 6) armoured command post vehicle 0021349

Description

The Okapi (6 × 6) armoured command post vehicle has a monocoque all-welded steel hull which provides the occupants with protection from 5.56 mm × 46 NATO ball ammunition fired at a range of 30 m and 7.62 mm × 51 NATO ball fired from 30 m.

The armour also provides protection against air burst shrapnel. If required, a higher level of protection is possible.

As with most South African wheeled armoured vehicles, the Okapi has a very high level of protection against anti-tank land mines.

According to the manufacturer, two Russian TM-57 anti-tank landmines can be placed anywhere under the hull or three TM-57 anti-tank landmines can be placed under any wheel without any injury to the occupants.

The power pack is installed at the front of the vehicle with the commander and driver being seated above and to the rear. To their front is a large bulletproof windscreen which is provided with a washer and wiper.

The commander and driver each have a forward opening door with a bulletproof window in the upper part. Above each position is a hatch cover and a 7.62 mm machine gun can be mounted above the commanders position on the left side for local and air defence purposes.

In the right side of the hull is the APU which is separated from the crew compartment by a bulkhead. There are bulletproof windows in the sides of the hull and there is also a single door in the hull rear on the left side which is provided with a bulletproof window in its upper part. Roof hatch arrangement depends on the role of the vehicle. The axles incorporate hub reduction with pneumatically operated lateral differential locks.

A spare wheel and tyre is mounted at the rear of the hull on the right side and standard equipment includes 200 litre drinking water tank and power steering on the front two axles. An air conditioning system is also fitted as standard.

Variants

In addition to being used as a command post vehicle it can also be adopted for a number of other specialised roles including fire-control post or specialised anti-mine equipment carrier.

Mfezi ambulance

This 4 × 4 vehicle is very similar to the Okapi and has twin doors at the rear.

Status

Production as required. A small number of vehicles are in service with the South African National Defence Force.

Manufacturer

Vickers OMC.

UPDATED

Specifications
Crew: 2 + 4
Configuration: 6 × 6
Combat weight: 22,000 kg
Unloaded weight: 18,000 kg
Power-to-weight ratio: 17.90 hp/t
Length: 8.047 m
Width: 2.48 m
Height: 3.30 m
Ground clearance: 0.48 m
Track: 2.03 m
Wheelbase: 2.10 + 2.645 m
Angle of approach/departure: 40/49°
Max speed: 100 km/h
Fuel capacity: 420 litres
Range
(road) 900 km (at 80 km/h)
(cross-country) 400 km (at 25 km/h)
Fording: 1.2 m at 5 km/h
Gradient: 60%
Side slope: 45%
Vertical obstacle: 0.60 m (at crawling speed)
Trench: 1.20 m (at crawling speed)
Engine: ADE 447 TI 6-cylinder 4-stroke turbocharged diesel developing 394 hp at 2,200 rpm
Transmission: Allison HT 740 4-speed automatic with input retarder
Torque converter: three element with lock-up clutch
Steering: power-assisted
Tyres: n/avail
Suspension: semi-elliptical leaf spring packs and double-acting telescopic shock-absorbers with all axles stabilised by torsion bars
Electrical system: 24 V
Batteries: n/avail

UNITED STATES OF AMERCIA

United Defense M4 Command and Control Vehicle

Development

The United Defense M4 Command and Control Vehicle (C2V) uses a design similar to that of the XM5 Electronic Fighting Vehicle Systems Carrier, but mission equipment differences dictate areas of change.

The M4 C2V uses a 10 m telescoping mast and has a primary power pack, environmental control unit and NBC overpressure system integrated into the mission equipment enclosures.

It is based on the standard United Defense MLRS carrier chassis which is used in large numbers by the US Army. The main modifications include the integration of the Bradley A2 600 hp Cummins diesel engine, General Dynamics Land Systems automatic transmission and the upgraded suspension.

Engineering development commenced in the second quarter of 1992, with the six XM4 prototypes delivered in 1995 and subsequently sent to Fort Hood for extensive trials with the 4th Infantry Division.

Following trials the US Army type classified the system as the M4 Command and Control vehicle.

The US Army Tank-automotive and Armaments Command awarded United Defense LP a contract worth US$12.5 million for Low-Rate Initial Production (LRIP) of the M4 in the fourth quarter of 1996.

The first contract was to build five M4 vehicles with warranty during FY97 but the contract had three options years worth US$82 million for a total of 46 vehicles with deliveries running through to February 2002.

Production of the M4 C2V commenced at the United Defense LP York facility late in 1997 and the first four vehicles were delivered to the US Army by late 1998. The first vehicles were built in San Jose but all subsequent production was undertaken in York.

In December 1998 a US$18.9 million contract was awarded for another 10 vehicles for delivery between August 2000 and May 2001. This brought the total order to 20 vehicles (5 + 5 + 10).

The total US Army requirement was for up to 439 M4 Command and Control Vehicles but the programme was cancelled in late 1999 to fund other programmes.

United Defense LP has proposed that the 20 M4 vehicles already fully funded could be converted into Armored Medical Treatment Vehicles (AMTV). An illustration of the AMTV appears in the United Defense Bradley IFV in the Armoured personnel carriers (tracked) section.

In addition to tactical command and control vehicle applications, the M4 is reconfigurable to an artillery or an air defence fire direction center as well as an armoured medical evacuation vehicle.

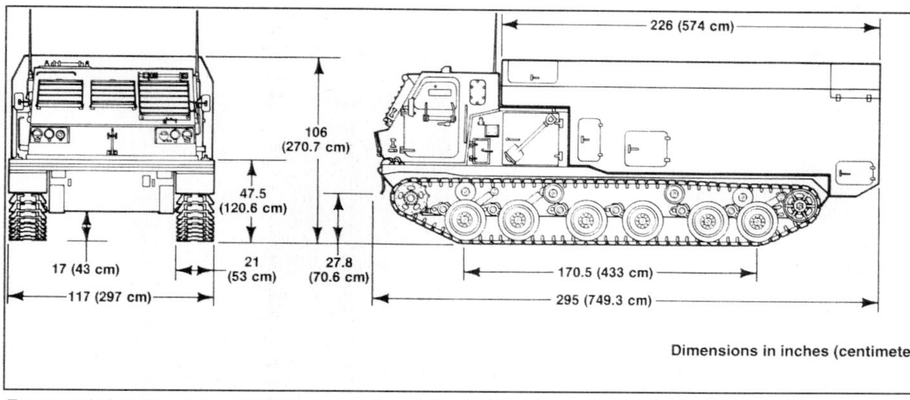

Front and side drawings of M4 Command and Control Vehicle

M4 Command and Control Vehicle based on a similar chassis to that used for the MLRS ***2002***/0098884

Specifications
Crew: 9 (depends on role)
Combat weight: 25,401-29,937 kg (depends on role)
Power-to-weight ratio: 23.6/20.04 hp/tonne (depends on role)
Length: 7.493 m
Width: 2.97 m
Height: 2.707 m
Ground clearance: 0.43 m
Track width: 533 mm
Length of track on ground: 4.33 m
Max speed: 64 km/h
Acceleration: 20 s (0-48 km/h)
Stopping: 10.668 m (0-32 km/h)
Turning radius: pivot to infinite
Fuel capacity: 617 litres
Cruising range: 483 km
Fording: 1.01 m
Gradient: 60%
Side slope: 40%
Vertical obstacle: 2 m
Trench: 2.29 m
Engine: Cummins VTA-903T turbocharged 8-cylinder diesel developing 600 hp at 2,600 rpm
Transmission: General Dynamics Land Systems HMPT-500-3EC 'TEC' hydromechanical
Steering: hydrostatic
Brakes: multidisc, oil cooled
Suspension: torsion bar in tube
Electrical system: 28 V
Batteries: 4 × 12 V 6TN

Description

The cab and mission module is of all-welded aluminium construction that provides the occupants with protection from small arms fire and shell splinters. A modular armour kit is available for more advanced threats.

The three-person cab is at the front and is provided with forward-opening side entrance doors and roof hatches. To the immediate front of the crew are three larger windows which can be rapidly covered by armoured shutters from within the cab.

The fully enclosed mission module is at the rear with entrance via a large door in the rear, there is also an emergency hatch in the roof. The interior communications and workstation fit depends on the mission requirement and arm of service.

Suspension is of the torsion bar type with each side having six dual rubber-tyred road wheels with the drive sprocket at the front and the idler at the rear. There are four track return rollers with hydraulic shock-absorbers being provided for the first, second and last road wheel stations.

Specialised equipment includes a Cummins diesel-powered APU, 10 m hydraulically operated mast which can be raised or lowered in 30 seconds (maximum).

The M4 is completely self-contained without need for trailers or support vehicles for extended mission performance. Onboard subsystems provide adequate power for mission operations plus NBC protection and environmental control.

Status

Production as required. 20 built for US Army (see Development).

Manufacturer

United Defense LP.

UPDATED

United Defense LP Battle Command Vehicles (BCV)

Development/Description

The US Army has tested two Battle Command Vehicles (BCV) which have been designed to enhance battlefield operations. One of these is based on the General Dynamics Land Systems M1A1 Abrams MBT (full details of which are given in the MBTs and medium tanks section) while the second is based on the United Defense LP Bradley M2 infantry fighting vehicle (full details given in Armoured personnel carriers (tracked) section.

The US Army's BCV programme has been designed to give manoeuvre commanders the ability to move forward on the battlefield and to evaluate the situation and provide real-time battle command.

The BCV concept has emerged since the mid-1990s including an early prototype system built by United Defense LP with independent research and development funding that was demonstrated in the Advanced Warfighting Experiment (AWE) 'Focused Dispatch' at Fort Knox, in August 1995.

The concept proved valuable enough that the Mounted Space Battle Lab at Fort Knox funded United Defense LP to build additional prototypes and make it an initiative in the AWE Task Force XXI Experiment. These prototypes have been built under a purely IPT (Initial Prototype Trials) concept.

By 1997, United Defense had converted two Bradley Infantry Fighting Vehicles (IFV) into the BCV configuration. Both of the Bradley BCVs maintain the Raytheon Systems Company TOW launcher and the Boeing 25 mm M242 Chain Gun. However, in order to fit the required communications equipment and workstations, the ammunition load has been decreased.

The M1A1 Abrams BCV modified by United Defense is visually identical to the other MBTs but features significant internal modifications. Remaining defensive weapons are the roof-mounted 12.7 mm (.50 calibre) M2 machine gun in the commander's cupola and the 7.62 mm machine gun in the loader's hatch area.

The 120 mm M256 main armament on the Abrams BCV has been replaced with a full articulating mockup of the gun barrel. In addition, the dummy gun barrel has been covered by the actual shrouds removed from the original barrel so that only about 200 mm of the dummy barrel is exposed.

The 120 mm main armament was removed at the direction of the Battle Lab to provide interior space to create a BCV that was functionally identical to the Bradley version.

Each BCV carries a unit commander and two staff officers. The M1A1 MBT BCV has an additional driver and vehicle commander while the Bradley also retains a driver and vehicle commander as well as a gunner for the 25 mm M242 cannon.

In 1997, the three BCVs were issued to units at Fort Hood and during the Task Force XXI AWE they serve as the brigade commander's vehicle, mechanised infantry battalion commander's vehicle and the tank battalion commander's vehicle.

In order to carry out its mission, the BCV has the following systems installed:

V3 appliqué computer

This is a militarised computer unit that consists of the following components: six slot VME card chassis, active matrix liquid crystal display, SCSI – II external mass memory unit and a keyboard.

Bradley infantry fighting vehicle, converted for use as a command vehicle at brigade level, and below showing position of main subsystems ***2001***/0103434

V2 appliqué computer

This V2 computer system is a ruggedised computer unit that consists of the following components: computer unit with intel processor, VGA liquid crystal display, keyboard and external power module.

High-capacity Computer Unit (HCU)

This is a US Army Common Hardware Software II computer consisting of the following components: V2 HCU, SVGA high-resolution monitor and a keyboard.

Frequency-Hopping Multiplexer (FHMUX)

The FHMUX couples four frequency-hopping Single Channel Ground/Air Radio Systems (SINCGARSs) to one antenna to reduce antenna visual signature.

Surrogate Digital Radio (SDR)

This is a secure packet radio that provides moderate rate data communications for the HCU computer system. It has a 20 MHz bandwidth and a 300 Kbps transmission rate.

Mobile Subscriber Radio Telephone (MSRT)

This provides the command and battle staff with secure point-to-point telephone communications.

Enhanced Position Location Reporting System (EPLRS) User Unit (EPUU)

This is a digital radio that, when connected to the SINCGARS SIP/INC, functions as a tactical internal gateway.

SINCGARS SPI/INC radios

The US Army's current secure combat net radio configured to transmit and receive voice and digital data and function in the 'tactical internet.'

21 in flat panel display

This display is a ruggedised colour plasma display that provides wide viewing without image distortion. It is mounted in such a way as to allow the commander to use it as a briefing tool.

Meshnet intercom system

This is a state-of-the-art intercom that provides the following capabilities: dual-intercom configuration with override, point-to-point intercom, all devices addressable, interfaces with SINCGARS and MSRT and headset binaural or monaural.

Other US Army command post vehicles

The United Defense M577 command post vehicle is the most widely used vehicle of its type in the US Army. This is a member of the M113 series of APC and is covered in the Armoured personnel carriers (tracked) section.

The US Army has type classified the M4 Command and Control Vehicle which is based on the chassis of the MLRS. Details of this are given in a separate entry in this section. Funding for the M4 has now stopped.

Status

Prototype systems being tested by the US Army. These have not been exported. A battle command vehicle remains a valid requirement for the US Army, but this programme has not progressed for a lack of funds.

Manufacturer

United Defense LP, Ground Systems Division.

UPDATED

SELF-PROPELLED GUNS AND HOWITZERS

Self-propelled guns and howitzers (tracked)
Self-propelled guns and howitzers (wheeled)

SELF-PROPELLED GUNS AND HOWITZERS (TRACKED)

ARGENTINA

TAMSE VCA 155 155 mm self-propelled artillery system

Development

The VCA 155 *(Vehiculo de Combate de Artilleria)* 155 mm full tracked self-propelled artillery system was developed by TAMSE to meet the operational requirements of the Argentine Army.

Until the introduction of the TAMSE VCA 155 155 mm self-propelled artillery system the only self-propelled guns in service with Argentina were 24 French-supplied Giat Industries (previously Creusot-Loire) 155 mm Mk F3 systems on a modified AMX-13 full-tracked chassis.

The VCA 155 is essentially a lengthened TAM MBT chassis with seven roadwheels either side, fitted with the same turret as installed on the Italian Otobreda 155 mm Palmaria self-propelled howitzer described in full later in this section.

Otobreda supplied a total of 20 turrets to TAMSE and it is believed that by 2001 10 systems had been completed. Previous information was that the TAMSE production line for the TAM had closed.

Description

The layout of the VCA 155 155 mm self-propelled artillery system is virtually identical to the TAM medium tank fully described in the MBTs and medium tanks section, with the driver front left, power pack to his right and turret in the centre of the hull slightly to the rear. However, an auxiliary generator is mounted in the left side of the hull and, to increase the operational range of the system, two long-range diesel fuel tanks are mounted at the rear of the hull.

TAMSE VCA 155 155 mm self-propelled artillery system with ordnance in travel lock

The all-welded aluminium turret is identical to that of the Palmaria covered in this section under Italy. Firing standard ammunition, a maximum range of 24,000 m can be achieved while a range of 30,000 m can be achieved with special ammunition. An automatic projectile loading system is installed, with charges being loaded manually.

Status

Production complete. In service with Argentinian Army.

Manufacturer

TAMSE.

UPDATED

TAMSE VCA 155 155 mm self-propelled artillery system with ordnance at maximum elevation (T J Gander)

Specifications

Crew: 5
Combat weight: 40,000 kg
Power-to-weight ratio: 18 hp/t
Ground pressure: 0.82 kg/cm²
Length:
(overall) 10.30 m
(chassis) 7.69 m
Width:
(overall) 3.30 m
Max height: 3.20 m
Ground clearance: 0.45 m
Track: 2.62 m
Max road speed: 55 km/h
Fuel capacity: 873 + 400 litres
Max road range: 719 km (with external fuel tanks)
Fording: 1.4 m
Gradient: 60%
Side slope: 30%
Vertical obstacle: 1 m
Trench: 2.5 m
Engine: MTU MB 833 Ka-500 6-cylinder diesel developing 720 hp at 2,400 rpm
Transmission: Renk HSWL 204 planetary with 4 speeds forward and reverse
Steering: double differential with hydrostatic continuous steering
Suspension: torsion bar
Electrical system: 24 V
Armament:
(main) 1 × 155 mm 41 calibre howitzer
(anti-aircraft) 1 × 7.62 mm AA MG
Smoke grenade dischargers: 2 × 4 76 mm
Main armament elevation/depression: +70/–5°
Turret traverse: 360°
Ammunition carried: 28 × 155 mm (23 turret, 5 hull)
NBC system: yes
Night vision equipment: optional

BULGARIA

122 mm 2S1 self-propelled howitzer

Development/Description

For some years Bulgaria has manufactured the Russian 122 mm 2S1 self-propelled howitzer for both the home and export markets. As far as is known this is identical to the Russian system although there may well be local modifications. The 2S1 has also been built in Poland, while Romania has designed and built a similar system called the Model 1989 which uses a 2S1 turret on a local chassis. Details of this are given later in this section under Romania.

It has been confirmed that production of the 2S1, ACRV and the various models of the MT-LB full-tracked vehicles was undertaken in Bulgaria by BETA plc of Cherven Briag. Production of the 2S1 commenced in Bulgaria in 1979.

Like the original Russian 2S1, the Bulgarian version is fully amphibious being propelled in the water by its tracks. The direct sight has a magnification of ×5.5 while the indirect sight has a magnification of ×3.7.

Bulgarian-built 122 mm 2S1 self-propelled artillery system in travelling configuration
2000/0085211

MTP-1 technical support vehicle in travelling configuration

Bulgarian-built ACRV without weapons or equipment installed **2002**/0126550

Specifications
Crew: 4
Weight: (combat) 15,400 kg
Max range: 500 km
Max speed:
(road) 61 km/h
(water) 4.5 km/h
Length: 7.31 m
Width: 2.88 m
Height: 2.775 m
Engine: YMZ-238H diesel
Armament: 1 × 122 mm howitzer
Ammunition:
(HE) 35
(HEAT) 5
Main armament elevation/depression: +70/—3°
Turret traverse: 360°
Max range: 15,200 m

Specifications
ACRV
Crew: 2 + 8
Weight:
(loaded) 15,000 kg
(empty) 11,085 kg
Ground pressure: 0.485 kg/cm²
Length: 7.21 m
Width: 2.85 m
Height: 2.035 m
Max water speed: 6 km/h
Gradient: 60%
Side slope: 30%
Engine type: YaMZ-238h diesel
NBC system: yes
Night vision equipment: yes

Standard equipment includes an NBC system, intercom system and a radio system.

Variants

MTP-1 Engineer Vehicle

This is based on the Armoured Command and Reconnaissance Vehicle (ACRV) chassis.

The MTP-1 has been designed to carry out a wide range of roles on the battlefield including the recovery of damaged and disabled vehicles, changing vehicle components, preparing fire positions and vehicle hides and general lifting duties.

The MTP-1 is virtually identical to the ACRV (covered in the Armoured command vehicles section Russian Federation and Associated States) but mounted on the front right side is a turret that is identical to that fitted to the MT-LB multipurpose tracked vehicle which is also manufactured in Bulgaria. Details of the MT-LB are given in the Armoured personnel carriers (tracked) section under Russian Federation and Associated States (CIS).

This manually operated turret is armed with a 7.62 mm machine gun, although some Bulgarian turrets have been observed upgunned to 12.7 mm.

Mounted on top of the roof is a crane with a telescopic jib and, when travelling, this is traversed to the front and the hook restrained by an eye at the front of the hull.

The crane is mounted in a small turret-like cupola with the operator being provided with full armour protection. The crane can lift a weight of 3,000 kg with the jib extended out to 3.4 m or 2,000 kg with the jib extended out to 5 m.

Mounted at the rear of the vehicle is an entrenching blade similar to that fitted to the specialist engineer variant of the MT-LB vehicle. This is mounted on hydraulic arms and can prepare a vehicle scrape in about 110 minutes.

The MTP-1 has a combat weight of 14,000 kg and is provided with an onboard winch with a maximum capacity of 30 tonnes with the entrenching device lowered or 10 tonnes with the entrenching device raised.

Like the ACRV (this being its Western designation, its correct designation being the MT-LBus), it is fully amphibious, being propelled in the water by its tracks. An NBC system is fitted as standard.

ACRV

Bulgaria manufactures the ACRV (MT-LBus) which can be used for a wide range of roles in addition to the version described previously. Bulgarian sources provided the information given under Specifications on the standard ACRV.

KshTMS

This is a Bulgarian ACRV vehicle modified to carry out the role of battlefield command, control and communications and is fitted with an automated equipment complex, life support complex and a power supply complex.

There is an onboard computer which stores, processes and distributes information and controls the input/output devices, to which it is linked by a Bulgarian standard parallel interface, Type IZ.

A shelter is provided which allows the crew to work up to 100 m from the vehicle, to which they are linked by a fibre optic cable.

Communications equipment installed on the vehicle includes the R-17M, R-173M and R-134 and the data transfer is capable of feeding four radio or wire networks. There are two modes of online encryption.

R55 Radio System

This is an ACRV vehicle fitted with extensive communications equipment by the Electron company with equipment including R55P radio system, automatic antenna matching device and an operator's control panel.

Status

Production as required. In service with the Bulgarian Army. It is understood that there has been no production of the 2S1 or ACRV in Bulgaria in recent years.

Manufacturer

BETA plc.

UPDATED

CANADA

Medium Indirect Fire System (MIFS)

The Canadian Armed Forces (CAF) have taken delivery of a total of 76 United Defense 155 mm M109 series self-propelled artillery systems which have been upgraded over the years to the M109A4 configuration.

Late in 2001, it was stated the CAF wanted a new medium, indirect fire capability with an automated, interoperable and integrated fire-control system to replace its current fleet of M109A4s.

This is called the Medium Indirect Fire System (MIFS) and the CAF is currently considering a number of options that includes:

(1) Retaining the current M109A4 with its manual voice-based fire-contol system
(2) Upgrading the M109A4 and incorporating a more modern fire-control system
(3) Purchasing a new 155 mm self-propelled artillery system, wheeled or tracked with a modern fire-control system
(4) Purchasing a new 155 mm towed artillery system with a modern fire-control system
(5) Acquiring a multiple launch rocket system with a modern fire-control system
(6) Acquiring a mix of rockets and howitzers with modern fire-control systems

It is believed a total of 56 systems will be procured wiyth a budget of C$350 million being allocated. It is possible a request for proposals could be issued in September 2004, with contract award in 2005 and an initial operational capability of 2007. This programme could, however, be effected by changes in the force structure of the CAF or budget changes.

UPDATED

CHINA, PEOPLE'S REPUBLIC

NORINCO long range artillery system

Development

Late in 1999, reports from Hong Kong stated China North Industries Corporation (NORINCO) had developed a new long-range artillery system, a so called super-range rocket gun, with a maximum range of 360 km and a calibre of 406 mm.

In the mid-1990s, it was revealed China had built a prototype long-range 'Super Gun' similar to that designed by the late Dr Gerald Bull for Iraq. This weapon is understood to have been a technology testbed rather than an actual weapon system.

Dr Gerald Bull was heavily involved in developing towed and self-propelled artillery systems in China in the early 1980s with calibres of 155 mm and 203 mm.

Available details of the latter system, based on the chassis of the US M107/M110 self-propelled artillery system, are given in the following entry. As far as is known, this system never entered production or service with the People's Liberation Army (PLA).

It is believed Dr Gerald Bull was also involved in some work on a 'Super Gun' along similar lines to that developed for Iraq. The prototype of the latter had a calibre of 350 mm with the eventual aim of having one with a calibre of 1,000 m.

An earlier Chinese super-range rocket gun had a maximum range of 180 km. No firm details about the design of the actual rocket or its propellant system have yet been released. There is no hard information this system is in service with the People's Liberation Army.

Description

The new Chinese super-range rocket gun is said to have a maximum range of 360 km, much further than any known artillery system. This gives it sufficient range to strike at most of Taiwan from mainland China.

The new system is mounted on a modified version of the same 8 × 8 chassis as the M-11 ballistic missile carrier that has good cross-country mobility.

According to China, each platform can carry up to six launch tubes with an automatic loading system to increase its rate of fire. At this time it is not clear as to whether the rocket tubes are reloaded or complete new rocket tubes are loaded.

A typical super-range rocket gun company would consist of six launchers, command vehicle, communications vehicle and ballistic correction radar system.

The guided rockets can be fitted with various types of warhead including high explosive, incendiary, ground penetration, cluster and anti-armour submunition. Effective radius depends on the warhead fitted, but is typically said to be 450 m.

The rockets are said to be highly accurate and it is claimed that each rocket is fitted with a global positioning system, television, infra-red and laser sensors for ballistic correction.

While Taiwan has an adequate air defence capability against aircraft and, to a lesser extent, missiles, no current air defence system could provide an effective air defence against a system such as the Chinese super-long-range rocket gun with its long range, accuracy and high rate of fire.

Status

Development. A small batch may have been built for trials purposes.

Manufacturer

China North Industries Corporation (NORINCO). (Unconfirmed).

UPDATED

NORINCO 203 mm (8 in) self-propelled gun

Development/Description

The prototype of the 203 mm (8 in) self-propelled gun was built in the mid-1990s by NORINCO (China North Industries Corporation) but by late 2001 the system is not known to have entered volume production.

It is probable that the M107/M110 chassis was used only to prove the basic concept of the system and that any production systems would use a Chinese-developed full-tracked chassis. Full details of the M107/M110 chassis are given later in this section under the USA.

Production of this chassis was completed in the United States many years ago and it has now been phased out of service with the US Army and replaced by the Lockheed Martin Missiles and Fire-Control – Dallas MLRS. The only reason that the M110A2 was retained in service for so long was that it could fire a tactical nuclear projectile.

The exact origins of the M107/M110 chassis used in this upgrade are not clear. It could have come from Pakistan, which has very close ties with China in armoured fighting vehicle research, development and production.

Another alternative would be to back fit this system for customers who are already using the 175 mm M107 or 203 mm (8 in) M110 systems which, in many countries, have been withdrawn from service.

In the case of the M107, only one nature of ammunition was used, while in the case of the M110 its range was too short. With the new NORINCO 203 mm (8 in) ordnance fitted, both of these problems would be overcome.

It essentially consists of the US-developed United Defense LP, Ground Systems Division 175 mm M107/M110 self-propelled gun/203 mm (8 in) self-propelled howitzer chassis with the elevating mechanism and recoil system modified to accept the ordnance of Chinese 203 mm (8 in) towed gun, which is covered in detail in the Towed anti-tank guns, guns and howitzers section.

NORINCO 203 mm (8 in) self-propelled gun based on a US M107/M110 self-propelled artillery chassis

At least two types of projectile can be fired by the new system: Extended-Range Full-Bore Hollow Base High Explosive (ERFB HB HE) out to a maximum range of 40 km and an Extended-Range Full-Bore Base Bleed High Explosive (ERFB BB HE) out to a maximum range of 50 km.

Status

Prototype. Not yet in production or service.

Manufacturer

NORINCO, China North Industries Corporation.

UPDATED

NORINCO 155 mm 45 calibre self-propelled gun-howitzer PLZ45

Development

Late in 1988, the prototype of a new Chinese 155 mm self-propelled gun-howitzer was shown for the first time. In many respects this is very similar to the US M109 series, but it is heavier and is fitted with a 45 calibre barrel based on that used in the Chinese NORINCO towed 155 mm gun-howitzer Type WAC 21 system which is already in production and is covered in the Towed anti-tank guns, guns and howitzers section.

Late in 1997, Kuwait placed a contract with NORINCO for a complete self-propelled artillery package which included 27 155 mm PLZ45 155 mm 45 calibre self-propelled artillery systems. The first batch of 18 PLZ45 self-propelled guns and their associated support vehicles were delivered to Kuwait in March 2000.

All of these systems have now been delivered and are in service with Kuwait.

Late in 2001 it was stated Kuwait had placed a contract for another batch of PLZ45 self-propelled gun-howitzers, sufficient to equip an additional battalion. This second batch may also consist of 27 systems, plus a complete support package.

Description

The hull and turret of the PLZ45 system are of all-welded steel armour construction that provides protection from small arms fire and shell splinters. The layout is conventional, with the driver's and engine compartments at the front and the fighting compartment and turret at the rear. The PLZ45 has a crew of five consisting of commander, layer, two loaders and a driver.

The driver is seated at the front of the hull on the left side with the engine compartment to his right. The driver has a single-piece hatch cover that opens to the rear and in front of this are two day periscopes for use when the hatch is closed. One of these can be replaced by a passive night driving periscope.

The turret is at the rear of the hull with the commander's cupola on the right side. This has five day periscopes for observation, a single-piece hatch cover that opens to the rear and an externally mounted 12.7 mm machine gun for use in the ground defence and anti-aircraft roles.

Chinese 155 mm self-propelled gun-howitzer PLZ45 with ordnance partly elevated ***2001***/0106217

China North Industries Corporation Ammunition Support Vehicle PCZ45 which has the same chassis as the PLZ45 self-propelled artillery system 0014399

On either side of the turret is a large forward-opening door with an observation block in its lower part and there is also a large hatch in the left side of the turret towards the rear.

A small door in the turret rear and a larger door in the hull rear both open to the right for ammunition resupply.

The direct fire telescope is mounted in the front of the turret to the left of the 155 mm gun-howitzer while the indirect fire sight is mounted in an armoured cupola on the left side of the turret roof. The panoramic telescope has a magnification of ×4 and a 9° 30′ field of view, while the direct fire telescope has a magnification of ×5.5 and a 1.1° field of view.

The 155 mm 45 calibre barrel is provided with a slotted muzzle brake and a fume extractor and, when in the travelling position, is held in place by a manually released travelling lock mounted on the glacis plate.

The 155 mm/45 calibre ordnance has a semi-automatic horizontal sliding-wedge breech block, a semi-automatic loader and an electrically controlled and hydraulically operated rammer that enables projectile loading to take place at any angle of elevation with the charge being loaded manually.

The actual ammunition handling system consists of the loader, rammer and shell trolley.

The loader and the rammer are driven hydraulically with electric control. The projectile handling process starts with the projectile moving automatically sideways to the take-off position.

When the loader presses the unloading handle the projectile drops into the shell tray. Then the shell loader draws out the shell trolley together with the projectile and moves it to the loading tray.

The rammer automatically swivels the loading tray into the axis of the barrel and rams the projectile into the chamber. The bagged propellant charges are then loaded by hand.

Ammunition is stored in the rear compartment of the turret. A total of 30 155 mm projectiles is carried in two positions, 24 rounds in the loader and six rounds on the right side below the loader.

A total of 30 propellant charges is carried in two positions with eight bagged charges placed on the left side of the loader and 22 bagged charges longitudinally placed on the left and right sides of the gun gear.

The 155 mm self-propelled gun-howitzer is fitted with an inertial direction-finder, a gun point display system and a gun display unit, with fire information coming from the battery or battalion command system. Chinese sources state that the system can have either automatic or manual operating and laying. It is understood that the PLZ45 systems sold to Kuwait have a new fire-control system which includes an automatic laying system, optical sighting system, gun orientation and navigation system and a GPS receiver. The automatic laying system can automatically carry out azimuth and elevation and tilt correction. The hydraulic equilibrator for elevation/depression is mounted on the roof of the turret to make more space for the crew. Although the turret has full 360° turret traverse, when firing it is normally 30° left and right.

When in the firing position, two spades are lowered at the rear of the hull, one either side and in line with the track.

Using ERFB-BB ammunition a range of 39,000 m can be achieved with maximum rate of fire being 4 to 5 rds/min.

Specifications

Crew: 5
Combat weight: 33,000 kg
Weight unloaded: 29,000 kg
Power-to-weight ratio: 16.40 hp/t
Length:
(gun forward) 10.15 m
(chassis) 6.1 m
(gun barrel, without muzzle brake) 7.046 m
Width: 3.23 m
Height:
(with AA MG) 3.417 m
(to turret roof) 2.597 m
Axis of fire: 2.16 m
Ground clearance: 450 mm
Turret diameter: 2.60 m
Max road speed: 56 km/h
Max range: 450 km
Fording: 1.2 m
Gradient: 58%
Side slope: 47%
Vertical obstacle: 0.7 m
Trench: 2.7 m
Engine: Deutz BF12 L413FC air-cooled supercharged diesel developing 525 hp
Gearbox: manual with dual power take-off with 6 forward and 1 reverse gears
Steering: hydraulic
Suspension: torsion bars and hydraulic shock-absorbers
Armament:
(main) 1 × 155 mm gun-howitzer
(anti-aircraft) 1 × 12.7 mm MG
Ammunition:
(155 mm) 30
(12.7 mm) 400
Rate of fire: 4-5 rds/min

Gun control equipment
(elevation) hydraulic/manual
(traverse) electric/manual
Gun elevation/depression: +72/–3°
Turret traverse: 360°
Normal firing arc: 30° left and right
NBC system: yes
Night vision equipment: yes

	ERFB	ERFB-BB	ERFB-WP	ERFB III	ERFB BE
Length	938 mm	938 mm	938 mm	940 mm	940 mm
Weight	45.54 kg	47.6 kg	47.9 kg	45.8 kg	45.7 kg.
Muzzle velocity	897 m/s	903 m/s	795.5 m/s*	789.2 m/s*	974.5 m/s*
Max range	30 km	39 km	25 km	24 km	26 km

(The system can also fire NATO 155 mm M107 projectiles)
* with M2 charge 9 system
Under development are three new carrier projectiles, ERFB-BB illuminating, ERFB-ICM and ERFB-ICM-BB.

The torsion bar suspension either side consists of six dual rubber-tyred roadwheels with a distinct gap between the first, second and third roadwheels and the drive sprocket at the front, idler at the rear and three track-return rollers. The first and last roadwheel stations are fitted with a hydraulic shock-absorber.

Standard equipment includes an auxiliary power unit, NBC system, an explosion detection and suppression system and a fire suppression system. A muzzle velocity measuring device is fitted over the rear of the ordnance.

The types of 155 mm ERFB ammunition that can be fired are shown in the table.

PLZ45 artillery system

In addition to marketing the 155 mm 45 calibre PLZ45 self-propelled artillery system, for which the Kuwait Army is the first export customer, NORINCO is now offering this as part of a complete artillery package.
The complete PLZ45 self-propelled artillery system includes:

(1) PLZ45 155 mm/45 calibre self-propelled gun
(2) ZCY45 battalion command post/fire-control vehicle that is based on Type 85 (WZ 751) armoured ambulance chassis
(3) ZCL45 battery command post vehicle based on the Type 85 (WZ 751) armoured ambulance
(4) GCL Type (WZ 751) tracked chassis or a Chinese-built lightweight 4 × 4 vehicle used for forward observation/reconnaissance vehicles
(5) PCZ45 ammunition resupply vehicle that uses the same chassis as the PLZ45 155 mm self-propelled gun
(6) Type 84 (modified Type 653A) engineering/recovery vehicle
(7) optional Type 704-1 fire direction radar mounted on a 6 × 6 truck chassis
(8) optional LLX-05 meteorological radar system mounted on a 6 × 6 chassis

The ammunition resupply vehicle carries 90 155 mm projectiles and associated charges which are fed to the PLZ45 via a conveyor belt at the rate of eight rounds a minute. Standard equipment includes an auxiliary power unit and a crane for loading pallets of ammunition through the roof.

It is believed that the Kuwait Army ordered the complete PLZ45 artillery system.

Status

Production as required. In service with China (at least one regiment) and Kuwait (27 systems). A second batch of systems was ordered by Kuwait late in 2001.

Manufacturer

NORINCO, China North Industries Corporation.

UPDATED

NORINCO Type 83 152 mm self-propelled gun-howitzer

Description

The NORINCO (China North Industries Corporation) Type 83 152 mm self-propelled gun-howitzer was first seen in public in October 1984, although it is thought it entered service some years before. Recent information has stated that the first prototype was completed in 1981 with the system being type classified in 1983. The Type 83 uses an entirely new full-tracked chassis that is shared by the 122 mm Type 89 (40-round) tracked self-propelled rocket launcher (see entry under Multiple rocket launchers).

The 155 mm gun-howitzer carried by the Type 83 is a derivative of the towed 152 mm gun-howitzer Type 66 fitted with a fume extractor. The howitzer is mounted in a traversing enclosed turret at the rear of the hull. The vehicle driver sits at the front left with the engine compartment to the right. The turret has a full 360° traverse and there are doors for the four-person turret crew at the hull rear and on both sides of the turret.

A 12.7 mm Type 54 anti-aircraft machine gun is mounted on the roof and there is also a 7.62 mm tank machine gun fitted next to the main barrel. It is assumed that the Type 83 152 mm self-propelled gun-howitzer fires the same 152 mm ammunition as the towed Type 66, HE and smoke. The maximum range is 17,230 m and the muzzle velocity at full charge is 655 m/s. An RAP known as the MP-152 has been developed and tested. Four rounds can be fired in 1 minute. Some sources have stated the type 83 152 mm self-propelled gun-howitzer has a maximum rate of fire of six to eight rounds per minute.

The fire-control system is operated from a panel and all turret movements are power controlled. Time in and out of action is 1 minute. Panoramic and direct fire sights are provided along with night vision equipment. Communications equipment includes a Type 889D radio set and a Type 803 intercom system.

The suspension system either side uses six roadwheels with the central four wheels arranged in two pairs. There are three track-return rollers, idler at the rear and drive sprocket at the front.

Specifications

Crew: 5
Combat weight: 30,000 kg
Power-to-weight ratio: 17.33 hp/t
Ground pressure: 0.68 kg/cm²
Length:
(gun forward) 7.005 m
(gun to rear) 9.805 m
(hull) 6.405 m
Width:
(overall) 3.236 m
Height:
(overall) 3.502 m
(centre of AA MG) 3.19 m
(top of turret) 2.682 m
(bore of gun) 2.08 m
(top of hull) 1.47 m
Ground clearance: 0.45 m
Track: 2.62 m
Track width: 480 mm
Length of track on ground: 4.601 m
Max speed:
(road) 55 km/h
(average on road) 40-42 km/h
(country roads) 30-32 km/h
Fuel capacity: 885 litres
Max range: 450 km
Fording: 1.3 m
Gradient: 60%
Side slope: 40%
Vertical obstacle: 0.7 m
Trench: 2.5 m
Engine: Deutz Type 12150L diesel developing 520 hp
Electrical system: 24 V
Batteries: 4 × Type 65
Armament:
(main) 1 × 152 mm howitzer
(anti-aircraft) 1 × 12.7 mm Type 54 AA MG
(local defence) 1 × 7.62 mm MG
1 × Type 40 rocket launcher
Main armament elevation/depression: +63/–5°
Turret traverse: 360°
Ammunition:
(152 mm) 30
(12.7 mm and 7.62 mm) 650
(rocket launcher) 4
Gun control equipment: electric/manual
NBC system: no
Night vision equipment: yes

Variants

130 mm SPG

Details of the 130 mm self-propelled gun, which uses the same turret and chassis as the Type 83, are given in the following entry. As far as is known, this system remains at the prototype stage. It is probable this system was developed for the export market.

Trench digger

The chassis of the Type 83 152 mm self-propelled gun-howitzer is used as the basis for a trench-digging machine capable of digging 250 to 300 m³ of earth per hour. The digging machine head is carried over the rear of the hull and the crew of three is in a raised cab at the front. A dozer blade is also carried. Combat weight is 30,000 kg.

425 mm mineclearing rocket launcher

The Type 83 is also used as the basis of the carrier/launcher for the Type 762 425 mm mineclearing rocket. Details of this system are given in the Multiple rocket launchers section.

120 mm PZT89 self-propelled anti-tank gun

The 120 mm self-propelled anti-tank gun, covered in the earlier Tank destroyers section, is based on the chassis of the NORINCO Type 83 152 mm self-propelled gun-howitzer. This system has a new turret armed with a 120 mm smoothbore gun which fires fixed ammunition with a semi-combustible cartridge case. This system is only used by China and has not been offered on the export market by NORINCO. This is also referred to as the Type 1989 or the PZT89. This was first seen in public during a parade held in China late in 1999.

Status

Production complete. In service with the People's Liberation Army. There are no known export sales of this system.

Manufacturer

NORINCO, China North Industries Corporation.

UPDATED

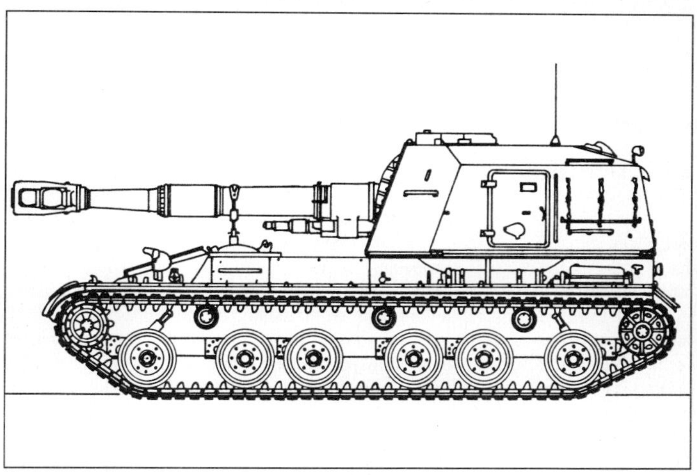

NORINCO Type 83 152 mm self-propelled gun-howitzer

NORINCO Type 83 152 mm self-propelled gun-howitzer showing ordnance with large double-baffle muzzle brake
0003976

NORINCO 130 mm self-propelled gun

Development/Description

The 130 mm self-propelled gun is essentially the NORINCO (China North Industries Corporation) Type 83 152 mm self-propelled gun-howitzer, fully described in a separate entry, with the 152 mm weapon replaced by a 130 mm ordnance taken from the towed NORINCO 130 mm gun Type 59. Full details are given

Specifications

Identical to the Type 83 152 mm except for slightly longer overall length due to the longer barrel of the 130 mm gun Type 59. The latter has an elevation from –2.5 to +45° whereas the self-propelled 130 mm on the Type 83 chassis and turret has an elevation from –5 to +65°. In total, 38 rounds of 130 mm ammunition and associated charges are carried.

This 130 mm self-propelled artillery system is also fitted with a roof-mounted 12.7 mm anti-aircraft machine gun and internally there are two 7.62 mm rifles. The crew consists of five people, including the driver. Communications equipment includes a radio set Type 889 and intercom.

in the Towed anti-tank guns, guns and howitzers section. Recent information has indicated this system was developed for the export market as it has never been identified to be in service with the People's Liberation Army. The system has not been marketed in recent years.

Status
Prototype. As far as is known, this system has yet to enter series production.

Manufacturer
NORINCO, China North Industries Corporation.

UPDATED

NORINCO 130 mm self-propelled gun showing long ordnance with double-baffle muzzle brake
0003977

NORINCO 122 mm Type 89 self-propelled Howitzer

Development
The NORINCO (China North Industries Corporation) 122 mm Type 89 self-propelled howitzer was one of the many new weapon systems that were shown for the first time during a large display of land, sea and air systems that took place in China in late 1999.

It is understood that the Type 89 entered service with the People's Liberation Army some time ago. It is believed to use some automotive components of the Type 77 full-tracked armoured personnel carrier which in turn is an improved version of the Russian BTR-50 that was phased out of service with the Russian Army many years ago.

As of late 2001 there were no known exports of the 122 mm Type 89 self-propelled howitzer.

Description
The hull and turret of the 122 mm Type 89 self-propelled howitzer is of all-welded aluminium construction that provides the occupants with protection from small arms fire and shell splinters.

The driver is seated at the front of the vehicle on the left side with the vehicle commander to his rear and the engine compartment to his right.

The driver has a single-piece circular hatch cover that opens to the right, in the front of which are three day periscopes, the centre one of which can be replaced by a passive periscope for night driving. The commander's position to his rear has a single-piece hatch cover that opens to the rear and can be locked in the vertical position.

The engine compartment has air inlet and outlet louvres in the roof with the large rectangular exhaust outlet on the right side.

Suspension is of the torsion bar type with either side having six large single rubber-tyred roadwheels with the drive sprocket at the front and idler at the rear, there are no track-return rollers. It is believed that side skirts are normally fitted to the Type 89 self-propelled howitzer.

The power-operated turret is positioned to the rear and mounted in the forward part is the 122 mm howitzer which is understood to be the Chinese equivalent of the Russian 122 mm D-30 towed howitzer.

When travelling, the 122 mm ordnance is held in position by a travelling lock mounted on top of the hull.

First seen in public late in 1999 was the NORINCO 122 mm Type 89 self-propelled howitzer **2001**/0093548

The 122 mm ordnance is fitted with a multislotted muzzle brake and a fume extractor.

Assuming that the ordnance and ammunition are the same as the 122 mm D-30, the following ranges can be achieved:

- 15,300 m with standard ammunition
- 18,000 m with special charge
- 18,000 m with extended range full bore – hollow base
- 21,000 m with extended range full bore – base bleed

It is believed that some form of semi-automatic ammunition handling system is fitted in the turret bustle and that maximum rate of fire is between six and eight rounds per minute.

Additional details of the 122 mm ammunition are given for the entry on the NORINCO 122 mm howitzer D-30 in the Towed anti-tank guns, guns and howitzers section.

Mounted on either side of the turret is a bank of four electrically operated smoke grenade launchers while mounted on the turret roof on the right side is a 12.7 mm machine gun which is mainly used for anti-aircraft defence. This is mounted on the forward part of a cupola that is provided with a single-piece hatch cover that opens to the side. There are additional hatches in the roof.

Observation devices are provided in either side of the turret and the indirect fire sight is mounted in the roof on the left side.

It is believed standard equipment on the Type 89 self-propelled howitzer includes NBC, night vision equipment, fire detection and suppression system and an onboard, computerised fire-control system.

Some sources have indicated the 122 mm Type 89 self-propelled howitzer has a combat weight of 20.5 tonnes and a maximum road speed of 56 km/h.

Specifications
Not available.

Status
Production probably complete. In service with the PLA. There are no known exports of this system.

Manufacturer
China North Industries Corporation (NORINCO).

UPDATED

NORINCO 122 mm self-propelled howitzer Type 85

Development
The NORINCO (China North Industries Corporation) 122 mm self-propelled howitzer Type 85 is the self-propelled artillery component of the range of vehicles based on the Type 85 armoured personnel carrier, formerly known as the Type YW 531 H; for details of this vehicle refer to the entry in the Armoured personnel carriers (tracked) section. This system has also been referred to as the Type 85/Type 54-II self-propelled 122 mm howitzer.

Recent information has indicated that the 122 mm self-propelled howitzer Type 85 was not considered to be a success and its design left a lot to be desired.

It is now being supplemented by the more recent and effective 122 mm Type 89 which features an improved weapon system mounted in a fully enclosed turret.

Description
The Type 85 self-propelled howitzer uses the basic all-welded steel chassis of the Type 85 APC with a Chinese-produced copy of the Russian 122 mm D-30

122 mm self-propelled howitzer Type 85 from the front with ordnance covered and in travelling lock

NORINCO 122 mm self-propelled howitzer Type 85 with ordnance elevated and showing hoops on hull side for pontoons to give system an amphibious capability
0003978

towed howitzer mounted in a semi-open superstructure at the rear.

The driver is seated at the front of the hull on the left side and has day periscopes to the front and a single-piece hatch cover that opens to the rear. Another hatch cover is positioned to the rear of the driver.

The engine compartment is to the right of the driver with the crew and gun compartment extending to the rear. The semi-open superstructure is equipped with a canvas cover for weather protection. Although the crew compartment has an open top, sides and rear, a shield provides the gun crew with some protection from small arms fire and shell splinters from the front.

The 122 mm D-30 howitzer is fitted with a multibaffle muzzle brake and has a maximum elevation of +70°. Traverse is limited to 22.5° either side. The D-30 fires a 122 mm HE projectile weighing 21.76 kg to a maximum range of 15,300 m, but an ERFB projectile has been developed which has a maximum range of 21,000 m. The maximum rate of fire is 6 to 8 rds/min.

When travelling, the 122 mm ordnance is held in a travel lock which folds down on to the hull front when the weapon is deployed in action.

The vehicle is amphibious but requires additional pontoons attached to both sides of the hull to achieve full flotation. When afloat, the vehicle is propelled in the water by its tracks. No secondary armament is carried other than the crew's weapons and the vehicle is equipped with a Type 889 or VRC-83 radio set.

This vehicle is no longer being marketed as a self-propelled version as the D-30 is now offered on the export market based on the NORINCO Type 90 APC chassis. Details of this vehicle, which first appeared in a military parade held in China late in 1999 are given in a separate entry.

Specifications

Crew: 6
Combat weight: 16,500 kg
Power-to-weight ratio: 19.4 hp/t
Ground pressure: 0.67 kg/cm²
Length:
(overall) 6.663 m
Width: 3.068 m
Height:
(overall) 2.85 m
(roof) 1.914 m
Ground clearance: 0.46 m
Max road speed: 60 km/h
Speed in water:
(with pontoons) 6 km/h
Fuel capacity: 450 litres
Max range: 500 km
Fording: amphibious
Gradient: 53.1%
Vertical obstacle: 0.6 m
Trench: 2.2 m
Engine: Deutz Type BF8L413F air-cooled diesel developing 320 hp
Transmission: manual with 5 forward and 1 reverse gears
Armament:
(main) 1 × 122 mm D-30 howitzer
Ammunition:
(122 mm) 40 rounds
Gun control equipment:
(elevation/depression) manual
Gun elevation/depression: +70/–5°
Gun traverse: 22.5° left, 22.5° right
NBC system: no
Night vision equipment: yes

Status

Production complete. In service with the People's Liberation Army. There are no known exports of this system.

Manufacturer

NORINCO, China North Industries Corporation.

UPDATED

NORINCO 122 mm self-propelled howitzer Type 54-1/Type 70-1

Description

There are two types of this 122 mm self-propelled howitzer. The first version had the 122 mm howitzer mounted directly on to the chassis of the NORINCO (China North Industries Corporation) Type 531 (YW 531B) APC, but the second version uses a lengthened version of this chassis with an extra roadwheel (five instead of four). It is now known that the latter version is called the Type 70-1 system.

The first version, the Type 54-1 is understood to have entered service around 1967 and was probably used to provide mobile fire support for Chinese armoured divisions.

The 122 mm self-propelled howitzer Type 54-1 carries a 122 mm Type 54-1 howitzer (copy of the Soviet M1938 (M-30)) in an open position on top of the chassis rear. The original howitzer shield is retained and the sides and top have a canvas cover that is

122 mm self-propelled howitzer Type 70-1 with cover in position over the main armament (Christopher F Foss)

122 mm self-propelled howitzer Type 70-1 showing the shield for the 122 mm howitzer

Specifications

Crew: up to 7
Combat weight: 15,400 kg
Power-to-weight ratio: 16.88 hp/t
Ground pressure: 0.592 kg/cm²
Length: 5.654 m
Width: 3.06 m
Height:
(top of gun) 2.729 m
(top of hull) 1.91 m
Ground clearance: 0.456 m
Track: 2.526 m
Length of track on ground: 3.425 m
Max road speed: 56 km/h
Fuel capacity: 450 litres
Max range: 500 km
Fording: 1.5 m
Gradient: 25°
Vertical obstacle: 0.6 m
Trench: 2 m
Engine: Deutz Model 6150L 6-cylinder in-line water-cooled diesel developing 260 hp at 2,000 rpm
Armament:
(main) 1 × 122 mm howitzer
(auxiliary) 1 × 7.62 mm MG
Smoke-laying equipment: none
Ammunition:
(main) 40
(MG) 1,000

Gun control equipment:
(elevation/depression) manual
Gun elevation/depression: +63/–2.5°
Gun traverse: 22.5° left, 22.5° right
Armour:
(front) 12 mm
NBC system: no
Night vision equipment: yes

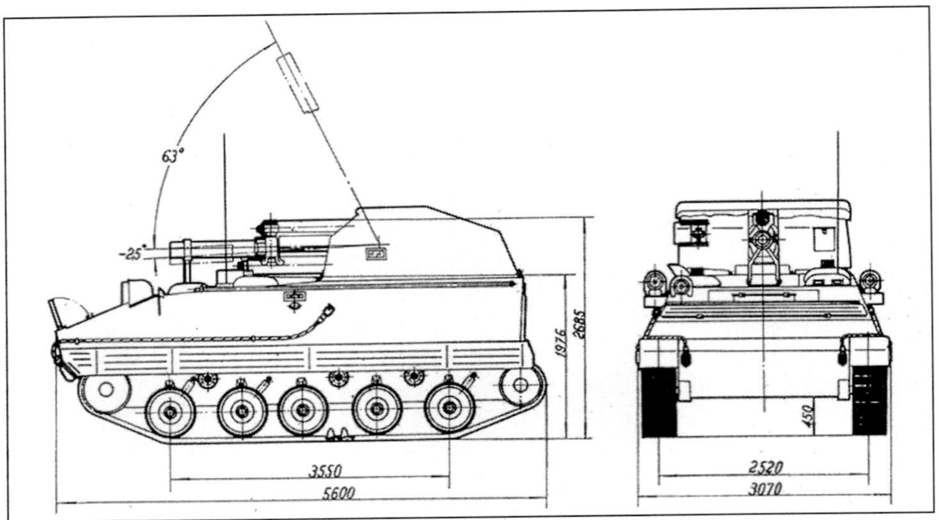

Side and front views of the 122 mm self-propelled howitzer Type 70-1 in travelling configuration

Ammunition type	HE	Smoke	Illuminating*
Designation	Type 54	Type 54	Type 54
Weight:			
(projectile)	21.76 kg	22.55 kg	21.9 kg
(filling)	3.5 kg	3.65 kg	1.515 kg
Type of filling	TNT	WP	n/a
Muzzle velocity	515 m/s	509 m/s	496 m/s
Max range	11,800 m	11,930 m	11,000 m
Type of fuze	Liu-4	Yan-2	Shi-1

* The illuminant burns for 25 seconds with 450,000 candela at a burst height of 450m

stowed around the periphery of the fighting compartment when not in use. The driver sits forward to the left of the engine compartment and has an overhead hatch and two forward-facing day periscopes. There are two white light headlamps, one each side, and on the right is a night driving infra-red lamp with a night vision device for use with one of the driver's periscopes. The suspension on either side consists of five dual rubber-tyred roadwheels and there are three track-return rollers. The first, second and fifth roadwheel arms either side are fitted with shock-absorber cylinders. The track has a cover along its upper length, probably for use during wading when an anti-splash board on the front hull is raised. Rubber-covered track links are used.

The Type 54-1 122 mm howitzer has a maximum range of 11,800 m firing an HE projectile. It fires the variable-charge, case-type, separate loading ammunition detailed in the table.

The actual calibre of the weapon is 121.92 mm with normal recoil being 960 to 1,065 mm and maximum recoil being 1,100 mm. Height of the trunnions is 2,304 mm.

The main drawback of this system by today's standards is the short range of the 122 mm howitzer and the very small amount of protection provided for the gun crew from small arms fire and shell splinters.

Also carried on the Type 54-1 self-propelled howitzer is a 7.62 mm Type 67 machine gun (Chinese version of the Russian RP-46) with 1,000 rounds of ammunition. Each vehicle has a Model A-220A radio set with a range of 16,000 m. This has a total of 288 channels. The intercom is a Type A-221A. Two manually operated CO_2 fire extinguishers are also carried.

Variants

There are no known variants of the Type 54-1/Type 70-1.

Status

Production complete. In service with the People's Liberation Army. It is understood this system is now being supplemented by the more recent 122 mm Type 89 self-propelled howitzer.

Manufacturer

NORINCO, China North Industries Corporation.

UPDATED

EGYPT

SP 122 (122 mm) self-propelled howitzer

Development

The SP 122 122 mm is based on the chassis of the United Defense LP 155 mm M109A2 self-propelled howitzer/M992 Field Artillery Ammunition Support Vehicle covered later in this section, both of which are used by the Egyptian Army.

Originally there were two competitors for the Egyptian Army requirement for a 122 mm self-propelled artillery system, the now BAE Systems, RO Defence of the UK with the SP 122 based on the RO2000 full-tracked chassis (*Jane's Armour and Artillery 1992-93*) and the then BMY Combat Systems SP 122, based on the chassis of the 155 mm M109/M992 systems.

Initial Egyptian Army trials with these two competing prototypes were completed in 1985, with the final series of trials being completed in 1987.

Following these trials, the BMY Combat Systems (now part of United Defense LP) system was selected and a total of 76 was built in the USA, with the 122 mm weapon and associated elevating mechanism being built in Egypt. Final deliveries were made from USA to Egypt in early 1995.

In March 1996, the Arab Republic of Egypt signed a follow-on contract with United Defense for the supply of an additional batch of 24 SP 122 howitzers with a value of US$27.2 million.

Prototype of the United Defense LP SP 122 for the Arab Republic of Egypt with 122 mm D-30 ordnance in travel lock (Christopher F Foss)

Specifications
Crew: 5
Combat weight: 23,182 kg
Unloaded weight: 20,909 kg
Power-to-weight ratio: 17.47 hp/t
Length:
(with armament) 6.957 m
Width: 3.15 m
Height:
(without AA MG) 2.821 m
Ground clearance: 0.46 m
Track: 2.788 m
Track width: 381 mm
Length of track on ground: 3.962 m
Max road speed: 56.3 km/h
Fuel capacity: 511 litres
Max range:
(cruising) 349 km
Fording: 1.07 m
Gradient: 60%
Side slope: 40%
Vertical obstacle: 0.53 m
Trench: 1.83 m
Engine: Detroit Diesel Model 8V-71T, turbocharged, 2-stroke, liquid-cooled 8-cylinder diesel developing 405 hp at 2,350 rpm
Transmission: Allison XTG-411-2A cross-drive with 4 forward and 2 reverse gears
Suspension: torsion bar
Electrical system: 24 V
Armament:
(main) 1 × 122 mm D-30 howitzer
(anti-aircraft) 1 × 12.7 mm M2 MG
Smoke-laying equipment: none
Ammunition:
(main) 77 HE, 4 HEAT, 4 illuminating
(MG) 500
Gun control equipment: manual
(gun elevation limits) –5/+70°
(gun traverse limits) 30° left and 30° right

The first of these vehicles were delivered to Egypt late in 1997 and final deliveries were made late in 1998. In mid-1999, United Defense was awarded a contract by Egypt for the supply of another batch of 24 SP122 systems for delivery between August and December 2000. All of these were built at the United Defense facility at York, Pennsylvania and have now been delivered to Egypt.

Description
The United Defense LP SP 122 uses the same basic aluminium armour construction as other vehicles in the M109 range, which provides protection from small arms fire and shell splinters.

The driver sits in the driver's compartment at front left and the engine compartment is at front right with the raised fighting compartment extending to the rear. The driver has three periscopes for day driving and the centre one of these can be replaced by a passive periscope for driving at night.

The 122 mm D-30 howitzer is mounted in the forward part of the fighting compartment and has manual traverse of 30° left and 30° right with elevation from –5 to +70°. Firing an HE projectile, a maximum range of 15,300 m can be achieved with conventional ammunition.

A dual-purpose sighting system is fitted which comprises an indirect fire sighting device and a direct fire telescopic sight. The former incorporates a range drum which directly computes the required angle of elevation corresponding to the target without additional calculation. The latter is used for engagements of both stationary and moving targets.

The standard towed 122 mm D-30 howitzer was manufactured in Egypt for many years by the Abu Zaabal Engineering Industries Company (qv) which is also known as Factory 100.

The howitzer breech mechanism is a mechanical-type semi-automatic system with automatic case extraction after firing. The breech block is prevented from interfering with the extraction of the fired case and the breech remains open for reloading of the next round.

Hatches are located in the superstructure sides and rear and there is a roof hatch over the gunner's position with accommodation for a 12.7 mm Browning M2 HB machine gun. The driver's position, engine, transmission and suspension of the M109 vehicle family are retained.

The main fighting compartment houses four of the crew of five and there is space for the stowage of 77 HE (High-Explosive), four HEAT (High-Explosive Anti-Tank) and four illuminating rounds. There is also space for the crew's personal 7.62 mm rifles.

D-30 on T-55 MBT chassis
Egypt has started a programme to fit surplus T-54/T-55 tank chassis with a new locally designed turret armed with the locally built 122 mm D-30 howitzer. Factory 100 is responsible for the turret and ordnance and Factory 200 for the chassis.

Status
Production as required. A total of 124 SP122 systems has been built for Egypt with the last of these being completed in late 2000.

Manufacturer
United Defense LP (Abu Zaabal Engineering Industries Company of Egypt is also involved in the programme).

UPDATED

FRANCE

Giat Industries 155 mm GCT self-propelled gun

Development
The 155 mm GCT *(Grande Cadence de Tir)* was developed by the now Giat Industries from 1969 to meet a French Army requirement for a self-propelled gun to replace the 105 mm and 155 mm self-propelled weapons, both based on the chassis of the AMX-13 light tank, then in service.

The first prototype was completed in 1972 and was shown for the first time at the 1973 Satory Exhibition of Military Equipment.

Between 1974 and 1975 six preproduction vehicles were completed and the vehicle entered production in 1977. First production GCTs were delivered in 1978 to Saudi Arabia.

The 155 mm GCT self-propelled howitzer was officially selected by the French Army in July 1979 and is now deployed in five-gun batteries, with each regiment having four batteries.

By late 2001, total production of the GCT amounted to just over 400 systems with the following sales made:

Country	Quantity
France	273
Iraq	86
Kuwait	18
Saudi Arabia	63

The French Army designation for the GCT is the 155 AUF1 of which 179 are in service. From 1988, production switched to the improved 155 AUF1 T with a total of 74 delivered to the French Army. This has a number of improvements including a more powerful APU, an improved and more reliable automatic loading system and an inertial goniometer.

Production of the 155 mm GCT self-propelled gun for the French Army commenced again in 1995 with first production systems being completed in 1996. This batch consisted of 20 vehicles.

The GCT has seen action with the French Army in Bosnia.

Although production of the Giat Industries 155 mm GCT is now complete, marketing of the complete system and the turret still continues.

Description
The 155 mm GCT basically consists of a modified AMX-30 MBT chassis fitted with a new turret armed with a 155 mm gun and an automatic loading system. The vehicle has a crew of four which consists of the commander, gunner, loader and driver.

The hull is almost identical to that of the AMX-30 MBT apart from the following modifications which result in the chassis being 2,000 kg lighter than the MBT's: the 155 mm ammunition racks in the hull have been removed, and a 5kVA 28 V generator and a ventilator system to supply the turret with cold air have been installed.

The driver is seated at the front of the vehicle on the left side and has a single-piece hatch cover that opens to the left, in front of which are three day periscopes. The centre day periscope can be replaced by an image intensification periscope for night driving or a combined SOPELEM OB-16A day/night periscope.

The turret is in the centre of the vehicle. The power pack is immediately behind the bulkhead that separates the engine compartment from the turret compartment. The engine is water cooled and air is

Giat Industries 155 mm GCT self-propelled gun of the French Army (Pierre Touzin) ***2001***/0106218

Giat Industries 155 mm GCT self-propelled guns of the French Army in the travelling configuration (Pierre Touzin) ***2001***/0106219

GCT 155 mm turret installed on Indian T-72 series MBT for trials purposes 0058373

Artist's impression of upgraded 155 mm GCT with 155 mm 52 calibre ordnance **2002**/0126551

drawn in above the chassis in the rear part of the vehicle and blown through the radiator by a fan driven by the engine. The fan's speed is governed by the temperature of the water. The transmission consists of an automatic clutch, a combined gearbox and steering unit, brakes and two final drives. The centrifugal-type clutch is activated electrically by a gear shift lever. The combined gearbox and steering mechanism contains a mechanically operated gearbox giving five speeds both forward and reverse, and a triple-differential steering system. The brakes are hydraulically operated and are used both as service and parking brakes. Each final drive comprises a spur-type right-angle and an epicyclic gear train.

The torsion bar suspension either side consists of five, dual, rubber-tyred roadwheels with the idler at the front, drive sprocket at the rear and five track-return rollers that support the inside of the track only. The first, second, fourth and fifth roadwheels are mounted on bogies and the first and last roadwheel stations have a hydraulic shock-absorber.

The turret is all-welded steel with the commander and gunner seated on the right and the loader on the left. The commander's cupola is equipped with day periscopes for all-round observation and has a single-piece hatch cover that opens to the rear. The loader, who also operates the anti-aircraft machine gun, has a single-piece rear-opening hatch cover. The crew enters the turret via a door in each side.

The 155 mm 40 calibre barrel has a vertical sliding wedge breech block hermetically gas-sealed by a metal blanking plate. The breech is hydraulically opened with manual controls for emergency use. The barrel has a double-baffle muzzle brake. The 155 mm gun has an elevation of +66°, depression of –4° and the turret can be traversed through 360°. Turret traverse and gun elevation are hydraulic with manual controls for emergency use.

The weapon takes 1 to 2 minutes to bring into action and 1 minute to bring out of action. Average rate of fire is 8 rds/min with automatic loading and 2 to 3 rds/min with manual loading. For burst fire, six rounds can be fired in 45 seconds. During French Army tests, 108 projectiles were fired in one hour.

Forty-two projectiles and 42 cartridge cases are carried in the turret rear arranged in seven racks of six identical shells and seven racks of six cartridge cases; in addition, 22 propelling charges for short range are housed near and under the loader's seat. A typical ammunition load would consist of 36 (6 × 6) HE shells plus six smoke (1 × 6) or 30 HE (5 × 6), six smoke (1 × 6) and six illuminating (1 × 6); but any combination of HE, illuminating or smoke shells is possible in racks of six similar shells. The various types of shell can be selected and fired with automatic loading without any preparation of the loading system.

The gunner can select single shots or bursts of six rounds. The turret is resupplied with ammunition by two large doors in the turret rear which fold down into the horizontal to provide a platform for the crew. The turret can be reloaded by a crew of four in 15 minutes or by a crew of two in 20 minutes. The gun can continue to fire while being reloaded.

The 155 mm gun can fire the following types of ammunition:

155 mm HE 56/69 with a maximum range of 23,000 m;
155 mm HE (hollow base) with a maximum range of 23,000 m;
155 mm HE (base bleed) with a maximum range of 29,000 m;
155 mm HE US M107 with a maximum range of 18,000 m; as well as other types of projectiles including illuminating, smoke and carrier.

It will also fire the new BONUS top attack anti-tank munition developed jointly by Giat Industries of France and Bofors Defence of Sweden.

The fire-control system of the current GCT consists of the following:

(1) Gimbal suspension providing for the mounting of the goniometer in the turret and permitting it to move freely by 10° relative to the axis in all directions
(2) One hermetically sealed conventional optical goniometer. Traverse angles are read on engraved drums and elevation angles appear on an elevation scale
(3) One azimuth plate that can be manually adjusted and is provided with a spirit level to set the goniometer vertically
(4) A contrarotation device making it possible to preserve the vertical position of the goniometer during turret rotation
(5) A direct fire sight for anti-tank firing.

A 7.62 mm or a 12.7 mm machine gun can be mounted at the loader's station for anti-aircraft defence. The 7.62 mm machine gun has an elevation of +45°, a depression of –20°, total traverse of 360° and is provided with 2,050 rounds of ammunition, of which 550 rounds are for ready use. The 12.7 mm machine gun has an elevation of +50°, a depression of –20°, total traverse of 360° and is provided with 800 rounds of ammunition of which 100 are for ready use. Two electrically operated smoke grenade dischargers are mounted on the forward part of the turret, below the main armament.

The standard optical control system can be replaced by a CITA 20 system which consists of a land navigator coupled with a gyroscope. When the 155 mm GCT comes to a halt the crew knows immediately the co-ordinates of its position and the bearing of its ordnance.

Variants

T-72 chassis with GCT turret

During a defence equipment exhibition held at Le Bourget in mid-1992, Giat Industries showed a Russian T-72 MBT chassis fitted with the complete turret of the GCT self-propelled artillery system. This was aimed at the Indian market which has a requirement for a 155 mm system on a T-72M1 MBT chassis which is already manufactured in India. This system was demonstrated in India together with the South African LIW T6, the Slovak Republic Zuzana and the now BAE Systems, RO Defence AS90 turret, all of which were installed on a locally supplied T-72M1 MBT chassis. Additional details of this competition and the Indian 155 mm artillery programme are given in this section under India. As of late 2001, India had not selected any of these systems.

GCT with 52 calibre ordnance

In September 1999, the French DGA (Délégation Générale pour l'Armement) awarded Giat Industries a contract worth FFr325 million for the development of the 155 mm AUF 1 self-propelled artillery system to the enhanced AUF 2 configuration, plus the upgrading of the first batch of 10 systems with the Thales ATLAS computerised fire-control system.

Under current plans, the French Army is expected to upgrade a total of 174 155 mm AUF 1 weapons. Of these, 70 will be fitted with the 155 mm/52 calibre ordnance supplied by Giat Industries at Bourges and the complete Thales ATLAS computerised fire-control system. This version is designated the 155AUF1 TA.

Specifications

Crew: 4
Combat weight: 42,000 kg
Unloaded weight: 38,000 kg
Power-to-weight ratio: 17.14 hp/t
Length:
(gun forward) 10.25 m
(gun rear) 9.51 m
(hull) 6.7 m
Width: 3.15 m
(over tracks) 3.115 m
Height:
(to turret top) 3.25 m
Firing height: 2.16 m
Ground clearance: 0.42 m
Track: 2.53 m
Track width: 570 mm
Length of track on ground: 3.9 m
Max road speed: 60 km/h
Max road range: (at 40 km/h) 450 km
Fording: 2.1 m
Gradient: 60%
Side slope: 30%
Vertical obstacle:
(forwards) 0.93 m
(reverse) 0.48 m
Trench: 1.9 m
Engine: Hispano-Suiza HS 110 12-cylinder, water-cooled supercharged multifuel developing 720 hp at 2,000 rpm (to be replaced by Mack E9 diesel)
Transmission: mechanical with 5 gears in both directions
Steering: triple differential
Clutch: centrifugal
Suspension: torsion bar
Electrical system: 28 V
Batteries: 8 × 12 V, 100 Ah
Auxiliary power unit: Microturbo gas turbine
Armament:
(main) 1 × 155 mm gun
(anti-aircraft) 1 × 7.62 mm or 12.7 mm MG
Smoke-laying equipment: 2 × 2 smoke grenade dischargers
Ammunition:
(main) 42
(anti-aircraft) 2,050 rounds of 7.62 mm or 800 rounds of 12.7 mm

Gun control equipment
Turret power control: hydraulic/manual
Max rate of power traverse: 10°/s
Max rate of power elevation: 5°/s
Gun elevation/depression: +66/–4°
Turret traverse: 360°
Max range: 29,000 m
Rate of fire: 6 rounds in 45 s
NBC system: yes
Night vision equipment: yes

The remaining 104 will retain the 155 mm/40 calibre barrel and be fitted with the Thales ATLAS fire-control system.

The substantial upgrade to the GCT will include the installation of a new 155 mm/52 calibre ordnance fitted with a new cradle, new screw-type breech mechanism, new double-baffle muzzle brake and a modified recoil system. A muzzle velocity radar system will also be installed over the main armament.

A new modular charge system is being developed by Giat Industries and SNPE, firing an extended range full-bore - base bleed projectile, a maximum range of over 40 km will be achieved.

A total of 42 155 mm projectiles, 180 top charge modules and 20 bottom charge modules will be carried.

According to Giat Industries, the upgraded GCT with the 155 mm/52 calibre ordnance and new modular charge system will enable a rate of fire of up to 10 rounds a minute to be achieved.

The installation of an automatic loading system relays the ordnance after each round has been fired and also gives the upgraded GCT a Multiple Rounds Simultaneous Impact (MRSI) capability.

To come into action, fire a 10 round salvo and come out of action, takes 2 minutes. It is also fitted with a chamber monitoring device and an automatic feeding system for the primers.

Range depends on projectile and charge combination, but the BONUS anti-tank projectile can be fired to a maximum range of 35 km, the LU 211 base bleed high-explosive shell to a maximum range of 39 km and firing an extended range, full-bore projectile, a range of 42 km can be achieved.

A total of 198 modular charges will normally be carried, 180 top charges and 18 bottom charges. It takes 20 minutes to replenish the complete system with projectiles and charges.

The upgraded AUF2 turret weighs 19 tonnes and, when installed on the upgraded AMX-30 chassis, the complete system weighs 45 tonnes.

Armament will consist of the roof-mounted 12.7 mm M2 machine gun with the option of the GALIX armoured vehicle protection system. Add-on armour is also possible and a collective NBC system is fitted as standard.

The chassis will also be refurbished and the current power pack, consisting of a Hispano-Suiza HS 110 12-cylinder engine coupled to a mechanical transmission will be replaced by a new power pack consisting of a Renault Mack E9 series diesel coupled to a new ENC 200 transmission. The suspension will also be upgraded for improved cross-country mobility.

Giat Industries has already completed prototypes of two other 155 mm/52 calibre artillery systems, a TR towed weapon and the CAESAR (6 × 6) based on a Mercedes-Benz UNIMOG chassis. Both these now have a 155 mm/52 calibre gun. The French Army has placed an order for five CAESAR systems.

Status

Production as required. In service with:

Country	Quantity	Comment
France	273	174 to be upgraded to AUF2 standard
Iraq	86	some were lost in fighting
Kuwait	18	believed to be in reserve
Saudi Arabia	63	

Manufacturer

Giat Industries. (Production is undertaken at its Roanne facility for chassis and Bourges for weapon and turret.)

UPDATED

Giat Industries 155 mm self-propelled gun Mk F3

Development

The 155 mm self-propelled Gun Mk F3, or the Canon de 155 mm Mle F3 Automoteur (Cn-155-F3-Am), was developed in the early 1950s and is essentially a modified AMX-13 tank chassis with the rear idler removed and the hull modified to accept a 155 mm ordnance and its associated recoil, elevating and traversing mechanism.

The 155 mm gun was designed by the Atelier de Construction de Tarbes (ATS), the chassis by the Atelier de Construction Roanne (ARE) while integration of the gun with the chassis and firing trials were undertaken by the Etablissement d'Etudes et de Fabrications d'Armement de Bourges (EFAB).

As the ARE was tooling up for production of the AMX-30 MBT, production of the whole AMX-13 tank family, including the 155 mm self-propelled gun, was transferred to Mécanique Creusot-Loire.

Total production of the Mk F3 amounted to over 600 guns and, in 1993, Mécanique Creusot-Loire became part of Giat Industries. This system is no longer being marketed by Giat Industries although spare parts and upgraded systems can still be supplied. In 1997 France supplied 10 155 mm Mk F3 systems to Morocco.

Description

The hull of the 155 mm self-propelled gun Mk F3 is of all-welded steel armour with the driver's compartment at the front on the left, engine compartment to the right and the main 155 mm armament at the rear. A splashboard mounted at the front of the hull stops water rushing up the glacis plate when the vehicle is fording. One spare roadwheel is often carried on the glacis plate.

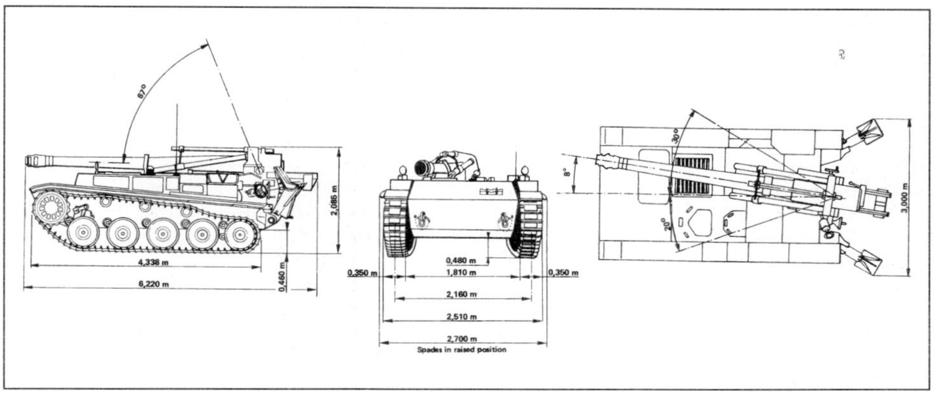

155 mm self-propelled gun Mk F3

The driver is provided with a single-piece hatch cover that opens to the left, in front of which are three day periscopes, the centre one of which is replaceable by an image intensification periscope for night driving. The commander is seated behind the driver and has a two-piece hatch cover that opens either side of his position and three day periscopes in front of him.

The torsion bar suspension either side consists of five, single, rubber-tyred roadwheels with the drive sprocket at the front and the fifth roadwheel acting as the idler, and three track-return rollers. The first and last roadwheel stations have hydraulic shock-absorbers. The steel tracks can be fitted with rubber pads if required. There are stowage containers along each side of the upper part of the hull and standard equipment includes a loudspeaker and a cable reel with 400 m of cable. Each side of the hull at the rear is a large spade which, on arrival at the gun position, is released manually; the vehicle then reverses slightly to bed them in. The vehicle does not have an NBC system and has no amphibious capability.

The other eight members of the gun crew are usually carried in a 6 × 6 truck or a now Giat Industries AMX VCA tracked vehicle which also carries 25 projectiles, 25 charges plus fuzes. For short distances, for example when moving from one gun position to another, four of the gun crew can be carried on the self-propelled gun itself.

When travelling, the 155 mm gun is horizontal and locked 8° to the right of the vehicle's centreline. The gun has a 155 mm 33 calibre barrel, screw breech, double-baffle muzzle brake and a sustained rate of fire of 1 rd/min which can be increased to 3 rds/min for short periods. The weapon is provided with an APX L877G direction-finder with a magnification of ×4 and 175° field of view and an APX M321 sighting telescope for direct fire with a magnification of ×4 and 180° field of view.

155 mm self-propelled gun Mk F3 deployed in firing position 0018868

155 mm self-propelled gun Mk F3 0018869

Specifications

Crew:
(on weapon) 2
Combat weight: 17,400 kg
Power-to-weight ratio:
(petrol engine) 14 hp/t
(diesel engine) 16 hp/t
Ground pressure: 0.84 kg/cm²
Length:
(including gun) 6.22 m
Width: 2.7 m
(over tracks) 2.51 m
Height:
(travelling) 2.085 m
Ground clearance: 0.48 m
Track: 2.16 m
Track width: 350 mm
Length of track on ground: 2.997 m
Max road speed:
(petrol engine) 60 km/h
(diesel engine) 64 km/h
Fuel capacity: 450 litres
Max road range:
(petrol engine) 300 km
(diesel engine) 450 km
Fuel consumption: 1.4 litres/km for petrol engine version, 1 litre/km for diesel engine
Fording: 1 m
Gradient: 40%
Vertical obstacle:
(forwards) 0.6 m
(reverse) 0.4 m
Trench: 1.5 m
Turning radius: skid turns
Engine: SOFAM Model 8Gxb 8-cylinder water-cooled petrol developing 250 hp at 3,200 rpm or Detroit Diesel 6V-53T turbocharged 6-cylinder water-cooled diesel developing 280 hp at 2,800 rpm or Baudouin 6F 11 SRY turbocharged 6-cylinder water-cooled diesel developing 280 hp at 3,200 rpm
Transmission: manual with 5 forward and 1 reverse gears (2nd, 3rd, 4th and 5th are synchronised) or automatic transmission with 3 forward and 1 reverse gears
Steering: Cleveland-type differential
Clutch: single dry plate
Suspension: torsion bar
Electrical system: 24 V
Batteries: 4 × 12 V, 190 Ah
Armament: 1 × 155 mm howitzer
Gun elevation/depression: +67°/none
Gun traverse:
(at 0 to +50° elevation) 20° left and 30° right
(at +51 to +67° elevation) 16° left and 30° right
Armour
Hull front: 15 mm at 55°
Hull sides: 20 mm
Hull top: 10 mm
Hull floor front: 20 mm
Hull floor rear: 10 mm
Hull rear: 15 mm
NBC system: no
Night vision equipment: yes

The gun has an elevation of 0 to +67°. When elevated between 0 and +50° it has a traverse of 20° left and 30° right, and when elevated from +51 to +67° it has a traverse of 16° left and 30° right. Elevation and traverse are both manual.

The 155 mm self-propelled gun can fire the following types of separate loading ammunition:
155 mm HE projectile Mk 56 weighing 43.75 kg to a maximum range, with charge 9A, of 20,047 m, maximum muzzle velocity 725 m/s
155 mm illuminating shell weighing 44 kg to a maximum range of 17,750 m, illumination time of two minutes and intensity of 800,000 candlepower
155 mm smoke shell weighing 44.25 kg to a maximum range of 17,750 m
(With base bleed projectiles ranges are enhanced, for example to 23,800 m for OE ammunition weighing 42.5 kg to a maximum range of 25,300 m)
155 mm HE projectile (type M107 USA) weighing 43.1 kg to a maximum range of 14,700 m (maximum muzzle velocity of 570 m/s) with US charge and 18,000 m (maximum muzzle velocity 690 m/s) with French charge.

Variants

Giat Industries did offer the Mk F3 with a Detroit Diesel 6V-53T diesel engine and a kit is available to convert existing petrol-engined versions to this configuration. More recently the Baudouin 6F 11 SRY diesel developing 280 hp at 3,200 rpm and an automatic transmission have been offered by Giat Industries for this and other members of the AMX-13 family.

RDM Technology upgrade for Qatar

Under a contract awarded late in 1991, RDM Technology of the Netherlands upgraded Qatar's fleet of 155 mm Mk F3 self-propelled artillery systems.

RDM Technology carried out a complete overhaul of these vehicles as well as fitting a new power pack consisting of a Detroit Diesel Model 6V-53T turbocharged diesel developing 280 hp coupled to a Rockford automatic transmission.

In total, 57 vehicles were upgraded plus 22 155 mm self-propelled guns, 33 members of the VCI full-tracked family including ammunition support vehicles and command post vehicles and two armoured recovery vehicles.

155 mm Mk F3 with 155 mm/39 calibre ordnance

This was first shown in 1990 but has not so far been purchased. Details were given in *Jane's Armour and Artillery 1993-94,* pages 469-470.

105 mm self-propelled gun

There was also a 105 mm self-propelled gun on the AMX-13 series chassis and this was used by France, Netherlands, Israel and Morocco (which is believed to have five in service).

Status

Production complete. No longer being marketed. In service with the following countries:

Country	Quantity	Comment
Argentina	24	
Chile	12	
Cyprus	12	
Ecuador	12	some sources state 10
Kuwait	18	reserve
Morocco	98	10 delivered in 1997
Peru	12	
Qatar	28	upgraded from 1991
Sudan	6	
UAE	18	
Venezuela	20	some sources state 10

Manufacturer

Giat Industries.

UPDATED

GERMANY

Krauss-Maffei Wegmann Panzerhaubitze 2000 (PzH 2000)

Development

Following the cancellation of the international 155 mm SP-70 programme in July 1986, the German government awarded contracts totalling DM183 million to two consortia for the development and construction of prototypes of a new 155 mm self-propelled howitzer for the German Army called the Panzerhaubitze 2000 (PzH 2000).

The German Army requirements include:
Ability to carry 60 projectiles and corresponding charges
Installation of automatic loader to realise a high rate of fire
Maximum range of 30 km with unassisted projectiles
NBC system
Good mobility and reliability
Ability to operate as an independent unit
Increased protection against top attack weapons.

PzH 2000 in travelling configuration with 155 mm/52 calibre ordnance in travelling lock (PB) ***2001***/0106220

The first consortium consisted of Krauss-Maffei, KUKA and Rheinmetall and the second of Wegmann and MaK.

Under Phase I, each of the two consortia built a prototype system which was tested by the German Army and in late 1990 the Wegmann and MaK consortium version was selected.

Under Phase II a development contract was subsequently awarded for an additional four vehicles for further trials. In March 1996, Wegmann and Co GmbH of Kassel was awarded a contract from the German Ministry of Defence for the supply of 185 new 155 mm/52 calibre PzH 2000 self-propelled artillery systems.

On 1 January 1999, Krauss-Maffei Wehrtechnik GmbH and Wegmann and Co GmbH merged to form Krauss-Maffei Wegmann GmbH and Co KG. This company is now prime contractor and is responsible for systems integration, with the main subcontractors being:

Company	Subsystem
Rheinmetall Landsysteme	chassis
Rheinmetall	elevating mass
MTU	engine
Renk	transmission
Honeywell	inertial reference unit
ESW	gun laying system
Moog	electric drives
Hensoldt	optical sights
Blohm + Voss	chassis housing

PzH 2000
KMW
KRAUSS-MAFFEI WEGMANN
Tomorrow´s Technology Today
155 mm L 52 gun with ranges of 30 km with NATO standard ammo, 40 km with extended range ammo
High protection level against AP, artillery fragments, bomblets and NBC attacks
Automatic shell loading system with 60 round magazine and automatically selected projectiles
Resupply, by two crew members, of 60 complete rounds (shells and charges) in less than 11 minutes
10 rounds in 56.2 sec, 20 rounds in 2 min 10 sec
With improved autoloader: 12 rounds in 59.7 sec, 20 rounds in 1 min 47 sec
Autonomous navigation and fire control
Speed on roads >61 kmh, cross-country >45 kmh range >420 km
Combat crew of 3 + 2, only 3 required for full operability
RHEINMETALL LANDSYSTEME
THYSSEN
B+V INDUSTRIETECHNIK GMBH
RHEINMETALL W&M
EADS
ZEISS
Hensoldt Systemtechnik GmbH
Honeywell
MOOG
Luftfahrt und Sondertechnik
OTOBREDA
SCHLEIFRING
We are proud to have been selected by the Royal Netherlands Army and the Italian Army
Kassel Facilities
Krauss-Maffei Wegmann GmbH & Co KG
August-Bode-Strasse 1
D-34127 Kassel
Telephone: +49/561/105 2466
Telefax: +49/561/105 2145
info@kmweg.de
www.kmweg.com

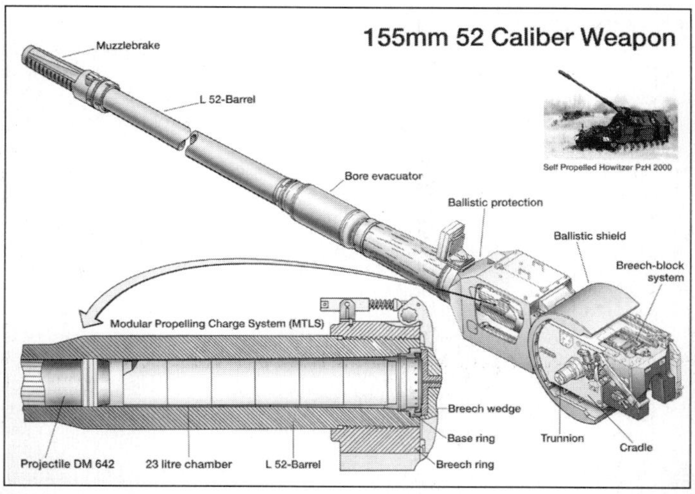

Rheinmetall 155 mm/52 calibre ordnance with chamber below complete with propelling charge (MLTS) and DM642 projectile **2000**/0085209

155 mm PzH 2000 self-propelled artillery system **2002**/0126552

Production of the 155 mm/52 calibre PzH 2000 self-propelled artillery system for the German Army commenced at the Krauss-Maffei Wegmann facility in Kassel in the second half of 1997.

The first production system was completed early in 1998 and, following company trials, was handed over to the German Army in July 1998. Under Phase I of 40 units a year, final deliveries under the current contract should take place in 2002.

The German Army has placed an initial order for 185 PzH 2000 systems to replace the 155 mm M109A3G self-propelled howitzers currently in service. Each PzH 2000 battalion will have three batteries each with six guns. In the future each PzH 2000 battery may have eight guns.

The German Army is the first army in the world to deploy a 155 mm/52 calibre self-propelled artillery system.

Italy is expected to manufacture 68 systems under a licence/co-production programme with IVECO/ Otobreda being prime contractor in Italy.

A government to government MoU was signed in October 2000 and delivery will start in 2004 with two additional systems out of the German series production to be delivered in 2001. These will be used for trials and training purposes.

Otobreda/IVECO is already under contract to Krauss-Maffei Wegmann, Rheinmetall Landsysteme and Rheinmetall to build some sub-components for the 185 PzH 2000 being built for the German Army.

In September 2000, the Royal Netherlands Army selected the PzH 2000 to meet its future artillery requirements and a total of 57 systems is expected to be ordered with deliveries to start in 2004.

The PzH 2000 is one of the systems taking part in the Swedish self-propelled artillery trials to take place from October 2001 through to September 2003 (qv this section under Sweden).

In mid-2001, the Hellenic Army signed a contract worth around €190 million with Krauss-Maffei Wegmann for the supply of 24 PzH 2000 155 mm/52 calibre Self-Propelled Howitzers (SPH), with delivery to take place from early 2003 to 2004. These will have a number of improvements to meet the requirements of the Hellenic Army, including an improved cooling system.

The contract was signed following trials in Germany that demonstrated the PzH 2000 could achieve a maximum range of over 40 km.

These trials used the South African Naschem M2000 BB (Base Bleed) Assegai high-explosive projectile and the Rheinmetall DM72 Modular Charge System (MCS). The Hellenic Army contract also includes Rheinmetall 155 mm ammunition and the DM72 MCS which has been in production for the German Army for some years.

Greece will become the second NATO country to operate the PzH 2000 and, as part of the deal, a number of Greek defence contractors will be involved in coproduction work on the PzH 2000.

Under current contracts, Krauss-Maffei Wegmann is supplying 185 PzH 2000s to the German Army, with final deliveries due late in 2002. These will be followed by two systems for Italy and then the 24 Hellenic Army systems.

Other potential customers according to Krauss-Maffei Wegmann include:

Country	Quantity	Delivery
Denmark	12 + 36	2007
Finland	20 - 30	was 2003 but late in 2000 it was decided to postpone buy of SPG and purchase additional towed guns
Norway	54	2005
Spain	90	2005
Sweden	48	2006/2007

In May 1996, two of the four preproduction PzH 2000 systems were demonstrated in Sweden in the coastal defence role. Additional details of this are given later in this entry.

According to the prime contractor, the main features of the PzH 2000 can be summarised as:

(1) High level of protection - the crew as well as the ammunition are protected by armoured steel in the turret and hull areas as well as by add-on roof protection elements against top attack bomblets and fragments; the hull and turret housings are covered by an additional internal liner.

After intensive tests, the add-on roof protection as well as the internal spall liner have passed the qualification programme. The current series production PzH 2000 are prepared for the adaptation of the roof protection and the internal liner. The retrofit of all series PzH 2000 started in 2001. This protection also allows for the quick adaptation of future armour technologies

(2) High armament performance - the high rate of fire of three rounds in 10 seconds, up to 10 rounds in less than 60 seconds, is achieved with the assistance of an automatic shell-loading device. Ammunition replenishment of the PzH 2000 with both projectiles and charges (with a total weight of about 3,400 kg) can be carried out in 11 minutes by two crew members using the automatic shell-loading device. Graduated semi-automatic and manual modes are provided in the event of failure of individual parts or the complete automatic system

(3) Increased effective range - the required range of 30 km with standard projectiles or almost 40 km with assisted projectiles is achieved with the new Rheinmetall 155/52 calibre ordnance as well as by the new Rheinmetall Modular Charge System (MTLS), which is already in service with the German Army and has been ordered by Malaysia and Norway. The 155 mm/52 ordnance as well as the MTLS are completely compliant with the NATO Joint Ballistic Memorandum of Understanding (MoU). Continued use of the in-service bag charge system is also possible with the minimum range for indirect firing only 2.7 km. The fire-control system with the hybrid navigation system and newly developed and tested flick rammer, with elevation controlled pressure guarantees high accuracy and lowest dispersion at all firing ranges

(4) Increased ammunition load - in its shell magazine located at the centre of the hull, the PzH 2000 can carry a total complement of 60 155 mm projectiles. 288 MTLS modules or the corresponding number of bag charges are stowed in the turret bustle. The distribution of weight is said to be well balanced resulting in a stable weapon platform with only very short movements after firing high rates of fire even with high charges. If required, automatic relaying is performed by the system. Loading and firing is possible within the entire gun and turret control range without any restriction

(5) Autonomy - an onboard hybrid navigation system permits independent determination of the northern direction, vehicle co-ordinates, including elevation above sea level, as well as spatial positioning of the barrel. This capability and the onboard ballistic computer make each PzH 2000 a stand-alone weapon system and independently combat ready at all times. In addition, the PzH 2000 is linked to a higher-level command and fire-control system via radio data transmission. The system is operated and monitored via menu-controlled displays. The 155 mm/52 calibre main

Specifications

Crew: 5
Combat weight: <55,330 kg
Power-to-weight ratio: 18 hp/t
Ground pressure: 0.98 kg/cm^2
Length:
(overall) 11.669 m
(hull) 7.30 m
Width: 3.58 m
(over tracks) 3.37 m
Height:
(to top of periscopic sight) 3.46 m
(to turret roof) 3.06 m
Ground clearance: 0.44 m
Track width: 550 mm
Length of track on ground: 4.91 m
Road speed: <60 km/h
Cruising road range: 420 km
Gradient: 50%
Side slope: 25%
Vertical obstacle: 1 m
Trench: 3 m
Engine: MTU 881 diesel developing 1,000 hp
Transmission: Renk HSWL 284 automatic 4 speed forward, 2 speed reverse, integrated steering and clutch
Suspension: torsion bar
Electrical system: 24 V
Armament:
(main) 1 × 155 mm L52 howitzer
(anti-aircraft) 1 × 7.62 mm MG3 MG
Smoke-laying equipment: 2 × 4 76 mm smoke grenade dischargers
Ammunition:
(155 mm) 60 projectiles and 288 modular charge modules (or equivalent bag charges)

Gun control equipment: electric/manual
Elevation/depression: +65/–2.50°
Turret traverse: 360°
NBC system: yes
Night vision equipment: yes

armament of the PzH 2000 howitzer is automatically laid to the required angle and is automatically relaid during firing. In addition, the gunlaying equipment permits semi-automatic or manual back-up operation. The PzH 2000 is equipped with a commander's sight and gunner's direct fire sight, both of which have day and night capabilities

(6) Improved mobility - due to the position of the turret at the rear of the chassis, minimum barrel overhang is achieved which results in best cross-country mobility and minimum overall vehicle length for good transportability. The PzH 2000 fulfills STANAG 2832B transportation requirements with regard to external dimensions. Maximum interior space by maximum width limited to 3,370 mm is achieved by using 550 mm wide tracks. The new chassis, equipped with an MTU 881 1,000 hp diesel engine coupled to a Renk HSWL 881 transmission, has a high degree of mobility

(7) High endurance and survivability under combat conditions - the onboard fire-control system MICMOS in combination with the hybrid navigation system, the high-mobility performance and the high rate of automation results in a superior shoot-and-scoot capability. After receiving a fire mission on the march it takes the PzH 2000 less than 2 minutes after stopping to carry out the order, for example fire a 10-round volley and leave its firing position again. All-electric and fully digitised turret, automatic loader and chassis with no hydraulic system in the turret, together with the Built-In Test Equipment (BITE) for the crew and maintenance personnel allow for the immediate fault localisation down to circuit board level as line-replaceable units. Due to decentralised vehicle electronics and many back-up systems the crew is able to operate the PzH 2000 in a variety of semi-automatic or manual modes in order to achieve the best performance even in the case of breakdown.

(8) 3 (+2) person crew. The PzH 2000 can be operated in the current configuration by three crew members. This crew is capable of operating the gun, loading the gun and carrying out preventative maintenance. The crew enters the PzH 2000 through a large door in the hull rear independent of the position of the turret. Easy exiting and changing of crew stations even in an emergency situation is possible. The PzH 2000 offers sufficient space for a complete crew of up to five people including storage space for additional equipment and components. In addition to the commander, gunner and driver an additional two loaders can complete the crew. The German Army vehicles have a crew of five.

(9) Logistic support - In order to minimise time, cost and risk during the fielding phase, for initial training, initial spare parts, special tools and test equipment, as well as for the development of technical documentation, a new logistics concept was developed for the German Army. This concept provides for logistic support to be conducted by industry personnel who train the military and gradually transfer the logistic support activity to the military personnel.

(10) Force multiplier potential - rationalisation potential - one PzH 2000 is capable of performing the mission of three existing 155 mm M109A3G howitzers.

During trials that took place at the WTD 91 proving ground in Meppen, Germany, in 1997, one of the four prototypes (PT01) of the 155 mm/52 calibre PzH 2000 self-propelled artillery system, using 48 V technology for some of the autoloader's electric drives attained a rate of fire of 12 rounds in just under 60 seconds.

Other achievements during this trial included: 3 rounds in 8.4 seconds; 8 rounds in 37.1 seconds; 10 rounds in 48.1 seconds; 20 rounds in 1 minute 47 seconds.

In May 1999, the PzH 2000 demonstrated its MRSI (multiple round simultaneous impact) capability. A production standard system fired five rounds at a range of 17,000 m with 1.2 seconds between the impact of the first and fifth rounds.

Description

The hull and turret of the PzH 2000 are of all-welded steel armour construction, with the driver seated at front right, the power pack to the left and the turret towards the rear.

The driver has a single-piece hatch cover that opens to the rear for driving in the head-out position and mounted forward in this hatch are two day periscopes that give observation to the front of the vehicle, one of which can be replaced by a passive night vision periscope.

The driver, who can also exit to his or her immediate rear, controls the vehicle using a steering wheel with essential information being provided by means of digital data processing with multifunction colour display and colour displays on the steering wheel.

The engine and transmission can be removed as a complete unit, with the air intakes being located on the top of the hull and the exhaust outlet at the front of the hull on the left side.

The suspension is of the torsion bar type with drive sprocket at the front, idler at the rear, seven double-tyred roadwheels plus track-return rollers. Linear shock-absorbers and hydraulic bump stops using proven components from the Leopard MBT are installed. The upper part of the track is covered by a skirt.

The turret is at the rear of the vehicle and is provided with two roof hatches, one forward on the left side and one on the right side towards the rear. Mounted on the left hatch is a 7.62 mm MG3 machine gun for defence purposes.

The turret has a total of eight 76 mm smoke grenade dischargers, four either side of the 155 mm ordnance, firing forwards.

To achieve the long ranges required, Rheinmetall developed the 155 mm 23 litre L52 calibre ordnance for the PzH 2000 and this, together with the matching Rheinmetall Modular Charge System (MTLS), gives the required range and rate of fire.

The L52 ordnance system coupled with the six-zone MTLS has a maximum range of 30 km with the standard L15A2 projectile and up to 40 km with extended-range ammunition.

The weapon system includes the cannon assembly with an enhanced slotted muzzle brake, a wedge breech block assembly, recoil and recuperator system, ballistic protection, elevating mechanism interface and a temperature sensor for thermal management.

The 155 mm 52 calibre ordnance is 8.06 m long and is connected to the breech block by an interrupted thread allowing for the quick removal of the gun from the outside of the howitzer without dismantling the weapon system.

Intensive computer simulations of the dynamic firing and recoil process resulted in an optimised barrel having enhanced accuracy with reduced weight. A glass fibre-reinforced plastic bore evacuator ensures an almost gas free crew compartment and also contributes to weight reduction.

The slot-type muzzle brake contributes to an increase in muzzle velocity and also reduces the muzzle flash. Integrated into the breech wedge is an exchangeable primer magazine with an endless conveyor for automatic primer transport, loading and unloading, controlled by the wedge movement. The firing mechanism allows automatic solenoid-actuated firing as well as manual firing. A long wear life is achieved as the barrel is partly chrome-plated.

The Krauss-Maffei Wegmann-developed automatic electrically driven projectile loading system enables loading and firing to take place within the full traverse and elevation range. The 60-round projectile magazine as well as the complete shell handling system is controlled by an automatic data management system.

The PzH 2000 has three ammunition flow modes:

(1) Automatic/manual ammunition loading and unloading
(2) Automatic/semi-automatic/manual loading from the magazine to the gun
(3) Automatic/manual through-loading from the rear of the vehicle directly to the gun.

A burst rate of fire of three rounds in 9.2 seconds has been achieved with the sustained rate of fire being up to 10 rds/min.

Fresh ammunition is loaded into the PzH 2000 from the rear of the vehicle in the lower part of the hull.

The integrated fire-control system and the all-electric gun-control equipment installed in the PzH 2000 enable firing with indirect laying from unprepared firing positions as well as firing with direct laying with a high first round hit probability and short reaction times.

The onboard fire-control equipment, integrated with the gun control system for indirect target laying, consists of an integrated hybrid navigation system with the appropriate digital computers for automatic operation. This is the primary mode and there is optical/mechanical equipment for the back-up mode.

Geographical north, the vehicle's own position and altitude above sea level are determined automatically with great accuracy. The onboard ballistic computer, in conjunction with the remaining equipment and its datalink to an external fire-control command post enables the gun to conduct a fire mission quickly and independently from an unprepared firing position after receiving target information and data.

The PzH 2000 is also able to lay the main armament in accordance with laying and ammunition data radio transmitted by a command post.

The 155 mm armament is automatically laid at high speed and precision, its position checked after every round fired and, if required, relaid automatically.

The commander has direct control as well as monitoring the gunner and the two loaders during system operation and target engagement.

The PzH 2000 is operated from a GUI (Graphic User Interface) terminal using a mission-oriented menu-driven operating system which completely automates the system operation and communication with the fire-control command via data digital radio link, via wire link or via voice radio communications.

A roof-mounted panoramic day/night sight with laser range-finder and a day/night direct fire sight are fitted to the PzH 2000 system.

For improved survivability, especially against top attack weapons, additional armour protection can be fitted to the roof of the vehicle. This can be quickly attached by the crew using onboard equipment. Survivability is also enhanced by compartmentalised charge storage separate from the crew compartment.

Standard equipment includes a fire warning and extinguishing system for the engine compartment and an NBC protection and ventilation system for the crew compartment.

Coastal defence role

Although originally developed as a self-propelled field artillery system, in May 1996 two 155 mm/52 calibre PzH 2000 systems carried out a major demonstration in Sweden at the invitation of the Royal Swedish Coastal Artillery.

In addition to the artillery inspectors of the 'Nordic Group' (Denmark, Finland, Norway and Sweden), officials from Germany, Italy and the Netherlands also attended.

The key objective of the firing trials carried out at Ravlunda in Sweden was to engage moving sea targets (4 × 4 m) from different firing positions, with the target moving at any speed up to a maximum of 50 kt at ranges of between 3 and 20 km.

Another requirement was the capability of PzH 2000 to operate under the control of the Swedish ARTE 734 fire-control system.

The fire mission was with two PzH 2000 firing from three different fire positions with the target moving at a speed of 10 kt, at an oblique angle at a range of 8,100 to 6,700 m.

The mean point of impact relative to the sea target was 0 m in azimuth and 15 m behind the target, this being rated as a centre hit.

After coming into action, the PzH 2000 took 45 seconds to track the target, fire a five-round burst and come out of action.

Although not designed to engage moving targets, Krauss-Maffei Wegmann states that these trials clearly demonstrated that the very low ballistic dispersion of the PzH 2000 makes it suitable for the engagement of moving sea targets.

Typically, sea targets would be engaged firing a three-round burst, with the PzH 2000 having left the firing position before the first rounds had landed on the target.

In addition to the firing trials, two further trials were conducted, automotive and tactical, which were also considered to be a success. During the automotive trials in Karlskrona, a distance of 190 km was covered at an average sustained speed of 59 km/h.

Variants

The German Army has a small number of PzH 2000 driver training vehicles. This is essentially a PzH 2000 with its turret removed and replaced by a cabin with seats for the instructor and other pupils. A dummy 155 mm/52 calibre gun is also fitted.

Status

Production. In service or selected by the following countries:

Country	Quantity	Comment
Germany	185	final deliveries late 2002
Greece	24	delivery 2003 to 2004
Italy	70	2 from Germany rest manufactured in Italy under co-production scheme
Netherlands	57	order expected 2002

Prime contractor

Krauss-Maffei Wegmann GmbH and Co KG (production and systems integration is undertaken at its facilities in Kassel).

UPDATED

Rheinmetall 155 mm M109A3G product improvement programme

Description

In order to update the existing fleet of US-supplied United Defense 155 mm M109G self-propelled howitzers of the German Bundeswehr, the now Rheinmetall DeTec AG has undertaken a programme to modify the vehicles to a new standard. The main modification is the replacement of the existing 155 mm barrel with a new 39 calibre barrel based on the ordnance of the FH-70 towed howitzer, but other modifications have been introduced at the same time.

The installation of the new 39 calibre barrel allows the full family of improved 155 mm ammunition to be fired. This family includes high-explosive, illuminating and smoke, all of which can be fired to a maximum range of 24,700 m. Enhanced range projectiles, such as base bleed, can be fired to a range of 30,000 m. Various 155 mm carrier projectiles such as the Rheinmetall RB 63 and Rh 49 can also be fired.

The 155 mm/39 calibre new ordnance is mounted in the cradle and recoil system of the M109G and has a revised obturation system. An external rotation key is located in a cradle extension under the barrel and the barrel dust cover of the original M109G has been modified. Other innovations to the barrel include a new flexible bellows and a fume extractor, located halfway along the tube. The original M109G muzzle brake is retained but is held in place by new fasteners. Other modifications to the ordnance involve machining alterations to the breech ring and buffer.

Modifications to the turret include a new 155 mm projectile magazine at the rear of the turret bustle. This can be reloaded through two outward-opening doors and holds 22 projectiles. The vehicle can carry a total of 34 155 mm projectiles. New provisions are made for propellant charge stowage. Other changes include door locks, a blast cover for the panoramic sight, and new sealing for the shields. A hand-drive mechanism is provided for the traversing gear and new elevation balancing cylinders are provided.

Changes to the rest of the vehicle include a new barrel clamp, revised ventilators, a new air filter system, a new instrument layout for the driver, the location of a driver's control panel in the main compartment, a turbo-supercharger system, reinforced torsion bars on the suspension and a driver's hatch lock. Alterations have been made to the power pack to allow it to be replaced more readily.

Other modifications are that the turret hydraulics have been altered to suit the metric system and the traversing gear hydraulic motor is now the same as that used on the Krauss-Maffei Wegmann Leopard 1 MBT. An electrical firing circuit has been introduced and various new fire-control data panels have been incorporated.

Specifications	
Calibre:	155 mm
Number of grooves:	48
Twist, clockwise, constant:	8° 55 35″
Length:	
(barrel)	6,874 mm
(rifled part)	5,057.7 mm
Chamber volume:	18.845 litres
Weight:	
(complete barrel)	2,380 kg
(liner)	1,420 kg
Max range:	
(standard round)	24,700 m (increased range up to 30,000 m)
Laying range:	
(elevation)	–44 to +1,245 mils
(traverse)	360°
Effect value of muzzle brake:	40%
Max recoil over 910 mils:	465 mm
Max gas pressure with charge 8 DM 52:	2,300 bar
Max gas pressure with charge FH 155-1:	3,628 bar
Muzzle velocity with charge 8 DM 52:	685 m/s
Muzzle velocity with charge FH 155-1:	827 m/s

Rheinmetall Landsysteme upgrade kit for German M109A3G

Although the German Army has only recently taken delivery of the first of 186 brand new Krauss-Maffei Wegmann 155 mm/52 calibre PzH 2000 self-propelled artillery systems, its older 155 mm M109A3G systems are scheduled to remain in service for some time. This is designated the M109A3GE A2 by the German Army.

To improve ammunition stowage and flow inside the 155 mm M109A3G turret, the German BWB (Federal Office for Defence Technology and Procurement) invited four companies to submit detailed proposals.

The now Rheinmetall Landsysteme (then KUKA) won this competition and was subsequently awarded a contract from the German BWB for two prototypes of the Ammunition Handling Kit (AHK) which the company had originally developed as a private venture in 1994/95.

These were installed in two German Army 155 mm M109A3G self-propelled artillery systems for an extensive series of user trials which led to the type classification of the AHK system early in 1997.

It was expected that a production order would soon follow but funding shortages meant that the company was not finally awarded the contract by the BWB for the AHK system until late in 1998.

The contract was for the supply of 262 AHK to the German Army who carried out the actual installation in its own facilities. These were delivered between July 1999 and June 2000. In addition, the contract contains an option for a further batch of 262 AHKs.

The AHK consists of seven major components. A drawer-type magazine for 24 fuzed 155 mm artillery projectiles is positioned in the turret bustle. Holders for an additional 10 fuzed 155 mm projectiles are mounted one either side of the rear hull entrance door. An electrically driven and electronically controlled projectile hoist is installed in the turret.

A projectile chute is also installed, as is an electrically operated rammer which speeds up the loading of 155 mm projectiles at all angles of elevation as well as reducing crew fatigue. Four magazines for propellant charges are provided in the turret and hull. Four large stowage boxes are provided externally at the turret front and rear.

According to the company, the installation of an electrically operated projectile hoist provides for almost weightless projectile transport and handling within the tight confines of the M109A3G turret.

The design of the different magazines for 155 mm fuzed projectiles and their associated propellant charges allows for the entire supply of onboard ammunition, 34 standard NATO 155 mm artillery projectiles and 34 NATO standard artillery charges, to be easily stowed in the M109A3G and be quickly accessible for fire missions.

The M109A3G upgraded with the AHK can also carry 170 of the Rheinmetall Modular Propelling Charge System (MPCS) which is already in volume production for the German Army under the designation of the DM72. More recently it has been selected by Norway.

Overall, the installation of the AHK increases the rate of fire of the M109, first fielded over 35 years ago, as well as allowing the system to be operated with no standing personnel behind the weapon during firing missions.

As there is a significant reduction in crew fatigue, the upgraded M109A3G with the AHK installed is more suited to carrying out several 24 hour battlefield days than earlier systems.

The German Army is the launch customer for the AHK, which is already being marketed to other M109 customers. In addition, the company has also developed the Automatic Ammunition Flow (AMF) system, which can be fitted to the M109 and other self-propelled systems.

The AMF system, which has already been successfully tested on an M109A2, allows for 155 mm projectile transfer, ramming and primer loading to be carried out automatically and allows the crew of an M109 to be reduced by one person.

Status

This upgrade programme is complete with the work being carried out by the German Army.

Manufacturer

Rheinmetall DeTec AG
(responsible for upgrade package)

UPDATED

Rheinmetall M109A3G in travelling configuration (C R Zwart) 0018874

M109A3G of the German Army fitted with Rheinmetall Landsysteme-developed Ammunition Handling Kit 0058369

M44T 155 mm self-propelled howitzer

M44T showing the new 155 mm ordnance with its double-baffle muzzle brake

Development

Late in 1986, a German consortium consisting of MTU, GLS and Rheinmetall completed the prototype of an upgraded 155 mm M44 self-propelled howitzer subsequently designated the M44T.

This was built to meet the requirements of the Turkish Land Forces Command and the prototype commenced mobility and fire-power trials at the Turkish School of Artillery in 1987. Compared to the original M44 whose original design can be traced back to shortly after the end of the Second World War, the more recent M44T has increased mobility, fire power and reliability.

In total, 168 M44s of the Turkish Army were upgraded in Turkey with some of the component parts being made under licence in Turkey.

The M44T conversion is of modular design so that the user can select only those parts that meet his operational requirements and financial situation.

By 2001, Turkey was the only customer for the M44T upgrade programme.

The 155 mm M44 was originally developed to meet the requirements of the US Army but from 1962 was replaced by the 155 mm M109 series.

Description

The M44T has its original petrol engine replaced by an MTU MB 833 Aa-501 V-6 water-cooled diesel developing 450 hp at 2,300 rpm coupled to the original Allison CD-500-3 transmission via a ZF gearbox.

Increased fuel tankage of 780 litres, together with the diesel engine, provides a range of approximately 620 km compared to the 122 km of the original petrol-engined M44.

On the prototype M44T upgrade, the driver remained in the gun compartment but in production vehicles he is seated in the chassis.

Various other components and subsystems have been modernised and a Halon fire warning and extinguishing system has been installed. A new driver's instrument panel has also been fitted.

Replacement of the existing electrical system has also been carried out using new cables, fuse boxes and batteries. New actuators for steering, braking and acceleration have been fitted, and for cold starting of the engine at temperatures below −18°C a preheating system has been fitted. The top engine deck and exhaust system have also been modified.

Modifications to the suspension include new hydraulic shock-absorbers and torsion bars and some adaptation of the hydraulic bump stops. A new track (with replaceable rubber pads) has also been installed and snow grips can be fitted if required. New rubber dust skirts have been fitted to help keep dust down.

The original ordnance has been replaced by a 155 mm 39 calibre barrel and breech similar to that used on the Rheinmetall M109A3G (qv). This provides a range, firing standard projectiles, of 24,700 m and enhanced range projectiles can now reach 30,000 m. Elevation is from −5 to +65°, with ammunition stowage for 30 155 mm projectiles and charges being provided.

The ordnance has automatic primer feeding as well as automatic opening and closing of the breech mechanism; the recoil system, balancers and ammunition racks have also been modified. The ordnance has a life of 10,000 standard rounds.

The 155 mm/39 calibre ordnance used by the M44T and M52T is now built in Turkey by MKEK.

Status

Production was completed in 1992. As a follow-on programme, the modification of the M52 (which has a similar chassis to the M44 but with a turret-mounted 105 mm gun) with main components and weapon of the M44 has recently been completed. This is a more extensive modernisation programme as it includes the removal of the complete turret and rebuild of the hull. It is thought that 425 M52s were upgraded by 1998. Additional details of the M52 upgrade programme are given under Turkey. Both the upgraded M52 and M44 have seen operational use with the Turkish Army during operations in Northern Iraq.

Manufacturers

MTU GmbH.
Rheinmetall DeTec AG.
GLS GmbH.

UPDATED

INDIA

New Indian 155 mm self-propelled artillery system

Arjun MBT chassis fitted with South African LIW 155 mm/52 calibre T-6 turret system during trials
0014888

Apart from the locally produced 130 mm Catapult self-propelled gun covered in a separate entry the Indian Army does not have any modern self-propelled artillery systems in service.

The Indian Army has a requirement for up to 600 155 mm/52 calibre self-propelled artillery systems which would be based on the locally produced T-72M1 MBT chassis or, in the future, the new Arjun MBT chassis.

Early in 1995, four contenders were shortlisted for an extensive series of firepower and mobility trials in India, which were completed by July 1995.

The four systems tested on a locally manufactured T-72M1 MBT chassis were:

(1) French Giat Industries 155 mm/40 calibre GCT turret
(2) UK BAE Systems, RO Defence 155 mm/39 calibre AS90 turret system (with a 155 mm/52 system tested in mid-1995)
(3) South African LIW (a division of Denel) 155 mm/52 calibre T-6 turret system
(4) Slovakian ZTS 155 mm/45 calibre Zuzana turret system.

All four manufacturers submitted their bids before the extensive trials commenced, these being for 60, 120 and 200 complete 155 mm turret systems with eventual production to be undertaken in India, with up to 600 systems being required.

For the local production phase, BAE Systems, RO Defence was teamed with Bharat Heavy Electricals Ltd (BHEL), ZTS with the Ordnance Board Factory at Avadi (where the T-72 MBT was built), with Giat Industries and LIW both teamed with Bharat Earth Movers Limited (BEML).

By late 2001, none of the above proposals had been accepted and the Indian Army was starting a new competition to acquire a new towed and self-propelled 155 mm/52 calibre artillery system with seven companies being shortlisted:

Bofors Defence (owned by United Defence) (Sweden)
BAE Systems, RO Defence (UK)
Giat Industries (France)
LIW Division of Denel (South Africa)
Singapore Technologies Kinetics
Patria Vammas (Finland)
Santa Barbara (Spain)

While some of these can provide 155 mm/52 calibre weapons to meet both of the Indian Army requirements, others, for example Singapore Technologies Kinetics, Patria Vammas and Santa Barbara can only meet the towed artillery requirement.

The front runners for the self-propelled requirement are considered to be the South African LIW T-6 on the T-72M1 or Arjun MBT chassis (this is also called the Bhim) Bofors Defence FH-77BD on a VME A25C (6 × 6) chassis.

There have also been consistent reports that Russia has offered a 155 mm version of its 152 mm 2S19 full tracked self-propelled artillery system which is based on a fully tracked chassis that uses T-72 and T-80 MBT components.

UPDATED

130 mm self-propelled gun (Catapult)

Development/Description

As a result of the effectiveness of the Russian-supplied towed 130 mm Field Gun M-46 in the Indo-Pakistan conflicts, the Indian Army has mounted some on Vijayanta MBT chassis. The conversion is now in Indian Army service which has an inventory of about 2,200 Vijayanta MBTs and about 750 130 mm M-46 guns.

The gun has a range of 27,150 m and fires an HE projectile weighing 33.4 kg with a muzzle velocity of 930 m/s. It can also fire an APHE projectile weighing 33.6 kg, also at a muzzle velocity of 930 m/s, which can penetrate 230 mm of conventional steel armour at an incidence of 0° at a range of 1,000 m.

It would appear that the Vijayanta/130 mm M-46 combination can carry 30 rounds of ammunition. It is known to the Indian Army as 'Catapult'.

The latest information available indicates that a total of 170 Catapult systems were built of which 100 are operational and the remainder are in reserve. Some have recently been observed fitted with a horizontal shield over the gun and crew compartments to provide protection against top attack weapons.

Status

Production complete. In service with the Indian Army. This is expected to be replaced by a new 155 mm/52 calibre self-propelled gun which has still to be selected. It could well be a mixture of tracked and wheeled systems.

UPDATED

The Indian 130 mm Catapult self-propelled gun consists of a modified Vijayanta MBT chassis with the Russian 130 mm M-46 field gun firing to the rear

IRAN

155 mm Thunder-2 self-propelled artillery system

Development/Description

The first Iranian-developed self-propelled artillery system was called the 122 mm Thunder-1 and details of this are given in the following entry.

The 155 mm Thunder-2 is a new design and was first shown in public carried on a low loader in late 1997, together with the 122 mm Thunder-1.

A detailed analysis of photographs of Thunder-2 shows it has a similar layout to the United Defense M109A1 155 mm/39 calibre Self-Propelled Howitzer (SPH) that has been in service with the Iranian Army for some years.

Prior to the fall of the Shah of Iran, it is understood, 440 M109/M109A1s were supplied by the United States.

The 155 mm/39 calibre ordnance of the Thunder-2 is manufactured by the Hadid facility of the Iranian Defence Industries Organization (DIO) which also builds a wide range of other tank and artillery barrels as well as towed artillery systems, rocket launchers and ground-based mortar systems.

In appearance, the 155 mm/39 calibre ordnance of the Thunder 2, which has the local designation of the 155 mm Cannon HM44, appears almost identical to that of the 155 mm/39 calibre M185 ordnance used in the M09A1. It is fitted with a double baffle muzzle break, fume extractor, screw breech mechanism, hydropneumatic recuperator and a hydraulic recoil brake.

Firing standard 155 mm M107 high-explosive projectiles, a maximum range of 18.1 km can be achieved but, with the enhanced projectiles, longer ranges can be achieved. Maximum stated rate of fire is 4 rds/min. Weapon evaluation is from −3 to +75°. When deployed in the firing position two spades are lowered either side of the hull rear of Thunder-2 to provide a more stable firing platform.

Although the general layout of the Thunder-2 is similar to the United Defense 155 mm M109A1 it does feature a new turret and chassis. The former has hatches in the sides and two roof hatches with a 12.7 mm M2 machine gun mounted externally on the cupola on the right side of the turret roof. The indirect fire sight is located under an armoured hood on the left side of the turret roof, just in front of another roof hatch.

Iranian 155 mm Thunder-2 self-propelled artillery system deployed in the firing system clearly showing the two spades deployed at the rear **2002**/0095234

The Thunder-2 chassis is of a new design with the driver seated front left and the power pack to the right. The driver has a single-piece hatch cover that opens to the right in front of which are three periscopes for forward observation.

The air inlet/outlet louvers are in the forward part of the hull with the exhaust output on the right side, provided with a cowl. There also appear to be stowage boxes integrated into the upper part of the hull towards the front of the vehicle.

Suspension is probably of the torsion bar type with each side having six dual rubber tyred road wheels with the drive sprocket at the front and the idler at the rear. There are also return rollers and the upper part track is covered by a rubber skirt to help keep dust down.

The road wheels of the Thunder-2 appear to be from the Russian T-72 Main Battle Tank (MBT) which has been manufactured under licence in Iran for some years. In the lower part of the hull rear is a large door for ammunition re-supply purposes.

The Hadid facility also manufactures a wide range of other barrels for armoured fighting vehicles and artillery applications including the 122 mm barrels for armoured fighting vehicle and artillery applications including the 122 mm barrel for the Thunder 1 (HM51 and the 125 mm barrel for the T-72 series MBT (HM50).

Thunder-2 was tested early in 1997 when Iran claimed that it had a range of more than 30 km, a maximum rate of fire of 5 rds/min and a maximum road speed of up to 70 km/h.

Iranian sources state that both the 122 mm Thunder-1 and 152 mm Thunder-2 are fitted with a laser range-finder, semi-automatic loading system and a state-of-the-art fire-control unit. The laser range-finder could be for direct fire target engagements.

For some years Iran has been self-sufficient in many natures of artillery ammunition including 122 mm and 155 mm. This is made by the Ammunition Industries Division of the Defence Industries Organisation (DIO).

Natures of 155 mm ammunition manufactured include standard 155 mm HE, smoke and illuminating, with a maximum range of 15 km, and HE base bleed, with a maximum range of 34 km, depending on the weapon/calibre length used.

The actual 122 mm and 155 mm ordnance of the two new weapons is probably made by the Armament Industries Division of the DIO which has for some years been making the standard 122 mm D-30 towed artillery system. The chassis of the two weapons are probably made by the Vehicle and Equipment Industries Division of the DIO and may be powered by an air-cooled diesel.

The development of the Thunder-1 and Thunder-2 self-propelled artillery systems is yet another important step by Iran towards becoming self-sufficient in many key weapon systems.

Status

Low rate production. In service in Iran.

Manufacturer

Defence Industries Organisation (DIO) of Iran.

UPDATED

122 mm Thunder-1 self-propelled artillery system

Development/Description

In May 1996, it was revealed that the Defence Industries Organisation (DIO) of Iran had developed and successfully tested a 122 mm self-propelled artillery system called Thunder-1.

Thunder-1 is based on a new high-profile full-tracked chassis with the engine compartment and driver being at the front and the fully enclosed 122 mm turret mounted at the rear.

The turret appears to be identical to that fitted to the Russian 122 mm 2S1 self-propelled howitzer which was first fielded in 1971, with more than 10,000 systems being built.

As well as being built in Russia, the 2S1 was also built in Bulgaria and Poland for the home and export markets. Russia also sold 122 mm 2S1 turrets to Romania which then integrated them onto a locally designed and built chassis to make the Model 89 122 mm self-propelled howitzer, but this was never exported and only 42 were built.

At the present time it is not clear if the 122 mm turret of the Thunder-1 is coming from abroad, which is possible given the surplus of these systems in a number of Eastern European countries, or if it is being manufactured in Iran. As the 2S1 turret is of welded steel construction, it could easily be made in Iran.

Iran is believed to have taken delivery of at least 60 122 mm 2S1 systems in recent years and it is possible that the turret used on Thunder-1 could have been taken from one of these as an expedient before production of the turret gets under way in Iran.

The 122 mm 2S1 is armed with a 122 mm 2A31 ordnance fitted with a fume extractor and a muzzle brake which is itself a modified version of the 122 mm ordnance used in the 122 mm D-30 towed artillery system which has been manufactured in Iran by the Defence Industries Organisation (DIO) for some years. When travelling, the 122 mm ordnance of the Thunder-1 is held in a travel lock fitted at the front of the hull.

It is now known the 122 mm ordnance of the Thunder-1 self-propelled artillery system is manufactured by the Hadid facility of the Iranian Defence Industries Organisation and is designated the HM51.

The ordnance for the 155 mm Thunder-2 self-propelled artillery system, covered in detail in a separate entry, is designated the HM44. This could indicate that development and/or production of the Thunder-2 self-propelled artillery system started first.

Iranian sources state the HM51 has a calibre length of 32.7 and weighs 1,440 kg. It is fitted with a fume extractor and a double-baffle muzzle brake. Turret traverse is 360° with weapon elevation from –3 to +70°. Maximum rate of fire is stated to be 4 to 5 rounds per minute.

To enable the 122 mm Thunder-1 to have a higher rate of fire it is understood it has been fitted with an automatic loading device of some type. This could well include a flick rammer which would also enable the weapon to be loaded at high angles of elevation as well as increasing its rate of fire. The flick rammer would probably flick the projectile with the charge being loaded manually.

Maximum range of Thunder-1 firing a standard 122 mm high-explosive projectile is 15,200 m. The Russians did develop a Rocket-Assisted Projectile (RAP) for the 2S1/D-30 with a maximum range of 21,900 m, but as far as it is known this has not so far been made in Iran.

No technical details of the Thunder-1 are available although it has a maximum road speed of 65 km/h and is fully amphibious, being propelled in the water by its tracks. The large size of the hull could be influenced by the requirement to be amphibious without significant preparation.

The suspension is of the torsion bar type with the upper part being covered by a rubber shroud. The roadwheels are similar to those of the M107/M110 self-propelled artillery systems supplied to Iran many years ago. Iran is understood to have been making roadwheels for the M107/M110 for some years.

The 122 mm D-30 is the most widely used weapon of its type and has also been made under licence or copied by such countries as China, Egypt, Iran, Iraq and Yugoslavia.

The Iranian DIO has been making the 122 mm separate loading ammunition for some years, including high-explosive, smoke and illuminating natures.

Iranian sources claim that the Thunder-1 features 'rapid firing ability, excellent accuracy, first class mobility and computer-controlled fire'. It is also claimed that Thunder-1 has a laser range-finder and onboard computer. This could also be fitted to the 155 mm Thunder-2 self-propelled artillery system covered in a separate entry.

Status

Low rate production. In service with Iranian Army.

Manufacturer

Defence Industries Organisation (DIO) of Iran.

UPDATED

Iranian Thunder-1 122 mm self-propelled artillery system 0058365

Iranian Thunder-1 122 mm self-propelled artillery system 0058364

ISRAEL

Romach 175 mm self-propelled gun

Development/Description

The Romach 175 mm self-propelled gun is a modified version of the US-supplied United Defense LP. Ground Systems Division 175 mm M107 self-propelled artillery system fully described later in this section.

In addition to firing the standard US-developed 175 mm M437 series HE projectile to a maximum range of 32.7 km, it will also fire the special Extended Range Sub Calibre Mk 7 MOD 7 projectile, developed for Israel by the now defunct Space Research Corporation, which has a maximum range of 40 km.

As the weight of the Romach is almost the same as that of the US M107 system, it is probable that any

Specifications

Crew: 6
Combat weight: 28,200 kg
Length: 11.3 m
Width: 3.15 m
Height: 3.47 m

modifications are limited to overall improvements in the areas of RAM-D.

Some 175 mm M107 systems have been observed to be fitted with additional stowage boxes and a pintle-mounted 7.62 mm machine gun for local and anti-aircraft defence, which is mounted on the left side of the hull to the rear.

Status

Production complete. In service with the Israel Defence Force. SIBAT is now offering ex-Israel Defence Force M107 for export either in 'as is condition' or refurbished. These are also referred to as the 'Lance'.

Manufacturer

United Defense LP.
Now United Defense LP of the US but modifications are understood to have been carried out by the Israeli Ordnance Corps facility at Tel a Shomer, near Tel Aviv, Israel.
Enquiries to SIBAT - Foreign Defence Assistance and Defence Export.

UPDATED

Romach 175 mm self-propelled gun from the front (IDF) **2000**/0085206

Doher 155 mm self-propelled howitzer

Development/Description

The Israel Defence Force has taken delivery of at least 530 United Defense LP 155 mm M109 series self-propelled howitzers from the USA.

It is believed that Israel placed its first production order for the M109 in 1969 and this consisted of 24 units. Many additional systems were ordered from the USA after the 1973 conflict.

Most of these have now been upgraded in Israel to a new standard under the local name of the Doher. According to the Israel Defence Force, the upgraded Doher has a number of significant operational advantages including:

(1) Shorter response time
(2) High mission availability
(3) Capability to provide a wide deployment
(4) Maximum flexibility in operating the artillery battery among others, by leap frogging
(5) High firing accuracy that results from technological improvements in the systems
(6) Fighting under NBC conditions and completing the mission under battery fire
(7) Improving the survivability of the crew by installing a number of new and more modern subsystems including automatic fire detection and suppression system, central NBC filtration system, ability to lay a smoke screen by injecting diesel fuel into the engine exhaust on the right side of the forward part of the hull, installation of another machine gun on the roof and passive night vision equipment for the driver and commander.

The crew of seven consists of the commander, navigator, gunner, person in charge of parking and refuelling, driver and two soldiers who join the crew from the Alfa ammunition vehicle.

It is understood that all M109s in service with the Israel Defence Force are fitted with a 155 mm/39 calibre ordnance.

Minor modifications carried out to the Doher in IDF service are understood to include additional track links, improved lighting and external bars on the hull sides for additional equipment stowage.

While the Israeli defence industry does not offer completed 155 mm M109s for the export market, it can provide a number of upgrades for this system, especially in the areas of fire control and turret modernisation.

Doher 155 mm self-propelled artillery system which is an M109 with many improvements to meet local operational requirements and shown here laying a smoke screen by injecting diesel fuel into the exhaust outlet (IDF) 0005998

Status

Production complete. In service with the Israel Defence Force.

Manufacturer

United Defense LP.
Now United Defense LP of the US but modifications are understood to have been carried out by the Israel Ordnance Corps facility at Tel a Shomer, near Tel Aviv, Israel.
Enquiries to SIBAT - Foreign Defence Assistance and Defence Export.

UPDATED

Specifications

Crew: 7
Weight: 28,200 kg
Length: 9.014 m
Width: 3.14 m
Height: 3.06 m
Max road speed: 50 km/h
Range: 350 km

Soltam Systems Slammer 155 mm self-propelled howitzer

Development

The Slammer 155 mm self-propelled howitzer was originally developed by Soltam Systems to meet the requirements of the Israeli Artillery Corps and was first revealed in early 1990. Assisting Soltam Systems was Israel Aircraft Industries Limited, Electronic Division/ MBT Weapons Systems.

The first prototype of the Soltam Systems Slammer (or Sholef as it is normally referred to in Israel) was completed in June 1983 and was delivered to the Israel Defence Force for testing in 1984. The second prototype was completed in 1986.

By late 2001, no production orders for the Soltam Systems 155 mm Slammer self-propelled howitzer had been placed and it remained at the prototype stage. Some effort has been made to market the turret of the Slammer for installation on other tank MBT chassis.

The Merkava MBT chassis is manufactured at the Israeli Ordnance facility at Tel a Shomer near Tel Aviv. Full details of the Merkava MBT chassis, of which over 1,000 have been built for the Israeli Defence Force, are given in the MBTs and Medium tanks section.

The Merkava has been offered on the export market, but as of late 2001, no overseas sales have been made.

Description

The Slammer 155 mm self-propelled howitzer is based on a modified Merkava MBT chassis fitted with a new welded steel turret, designed by Soltam Systems, that has a 155 mm 52 calibre ordnance with a total of 75 projectiles and charges being carried.

The ordnance is fitted with a fume extractor and a double-baffle muzzle brake. When travelling, the ordnance is held in position by a travel lock that is mounted on the forward part of the glacis plate and this is remotely operated from the crew compartment.

Firing an ERFB-BB projectile, the 155 mm 52 calibre ordnance has a maximum range of 40,000+ m.

The 155 mm 52 calibre ordnance and recoil system is of the companies well-proven type already used in its towed weapons. The breech block assembly is of the semi-automatic wedge type that contains an automatic primer feeding system that enables manual reloading of the primer without opening the breech. Turret traverse and weapon elevation is hydraulic, with manual controls for emergency use.

A maximum rate of fire of 9 rds/min can be achieved due to the automatic computerised loading system, and a burst rate of fire of three rounds in 15 seconds.

The high rate of fire can be achieved using the onboard ammunition supply or from ground-piled ammunition.

The loading cycle is operated by two turret crewmen only, with the commander operating the computer and charge loader.

The automatic loader has five main subsystems: projectile storage system; projectile transfer system; loading tray with flick rammer; charge loading tray and elevator for external charge supply; and projectile elevator for reloading the external storage or directly loading the gun.

The internal projectile storage contains 60 projectiles ready for automatic loading with the remaining 15 stored in other locations. The system enables the handling of all kinds of projectiles in use without any adaptation.

Charge loading is accomplished manually using a loading tray with the ignition primer being inserted automatically.

All systems have a manual back-up so that, in the case of failure, the loading system may be operated partly or completely manually by only three crewmen, so allowing a continuous firing rate of 4 rds/min.

The computer also controls the functioning of the gun. The Loader Control System (LCS) consists of five main units:

(1) The commander's panel provides the means for the commander to control the automatic loader and has a dedicated keyboard and supporting electronic circuits
(2) The Central Control Unit (CCU) is based on the 286 CPU-8086 and produces all of the system's logic equations. The unit transfers commands through the serial communications (RS-422) to the computerised units and controls the display on the commander's panel

Slammer 155 mm self-propelled howitzer with ordnance elevated 0058376

(3) The Terminal Units (TUs) are based on the 8031 controller for purposes of independent control of the drive elements according to a functionally determined division. With the assistance of the terminal unit, a local mode can also be used in working with selected elements
(4) For guiding operators and making round identification and fuzing, the Operator's Panel (OP) includes an LCD display with fixed instructions and one dot matrix line
(5) The Loader Keyboard Panel (LKP) includes breech block closing switch, fire and local activation of the trays.

The main operational roles are: firing from internal storage; firing for elevator - ground-piled ammunition; loading from elevator - external pile; synthesising fire programs; unloading; manual firing; identification; and fuzing and checks.

Standard equipment includes an NBC system of the overpressure type and an inertial navigation and aiming system designed for autonomous operations.

According to Soltam Systems, the ordnance and automatic loader, or parts of the system, could be installed in other self-propelled artillery systems and used to upgrade other self-propelled systems such as the US-designed and built 155 mm M109 and M44.

Outline specifications
Crew: 4
Length:
(L/52 ordnance) 11 m
Width: 3.7 m
Height: 3.4 m
Armament: 1 × 155 mm/52 calibre howitzer
Ammunition: 60 + 15 × 155 mm

Gun control equipment: powered/manual
NBC system: yes
Night vision equipment: yes

Status
Development complete. Ready for production.

Manufacturer
Soltam Systems Limited.

UPDATED

Soltam Systems L33 155 mm self-propelled gun-howitzer

Development
The L33 (the length of the barrel in calibres) 155 mm self-propelled gun-howitzer was developed in the late 1960s by Soltam Systems Limited. It is basically a Sherman M4A3E8 tank chassis fitted with a Cummins diesel engine, a new all-welded steel superstructure and a Soltam Systems 155 mm M-68 gun-howitzer taken from the towed system.

Production of the L33 (which has also been referred to as the 155 mm M68 self-propelled gun commenced just after the 1967 war and it was used operationally for the first time during the 1973 Middle East conflict.

Some sources have stated 200 of these systems were built and held in reserve, together with 120 155 mm M-50 and 34 105 mm M7 Priest systems.

SIBAT, Foreign Defence Assistance and Defence Export, Land Sales Division, at the Israel Ministry of Defence, is now offering quantities of the L33 systems for sale with the standard Soltam Systems 155 mm/33 calibre barrel or fitted with a new 155 mm/39 calibre barrel.

Description
The driver is seated at the front of the hull on the left side and has a bulletproof windscreen to the front and sides and a single-piece hatch cover that opens to the rear over the driver. The commander is seated to the rear and above the driver's position and has a windscreen to the front and sides and a single-piece hatch cover. The anti-aircraft gunner's position is at the front of the hull on the right side. A 7.62 mm (0.30) machine gun is pintle-mounted on the forward part of

Soltam Systems L33 155 mm self-propelled gun-howitzer with ordnance in travelling lock and roof-mounted 7.62 mm machine gun in position (Israel Defence Force)

the cupola which can be traversed through a full 360° and has a two-piece hatch cover that opens either side. The crew enter and leave the vehicle by a single door in each side of the hull that opens to the rear. At the rear of the hull are two large doors for ammunition resupply.

The Detroit Diesel diesel engine is under the gun compartment at the rear of the hull. Power is transferred to the manual transmission in the nose of the vehicle by a two-part propeller shaft. The controlled differential transmits engine power to the final drive unit and also contains a brake system for steering and stopping the vehicle. The suspension each side consists of three two-wheeled bogies, three track-return rollers, the idler at the rear and the drive sprocket at the front.

The 155 mm M-68 gun-howitzer is mounted in the forward part of the superstructure and has an elevation of +52°, a depression of –3° and a total traverse of 60°. Both elevation and traverse are manual. The weapon has a monobloc barrel and incorporates a horizontal sliding semi-automatic breech block. Mounted on the cradle are the hydropneumatic recuperator, a variable hydraulic damper which varies as a function of elevation (for example 800 mm at 52° elevation and 1,100 mm at 0° elevation) and a pneumatic rammer which permits loading of the gun at all angles of elevation. The barrel has a single-baffle muzzle brake, a bore evacuator and a travelling lock.

The L33 has a panoramic telescope for indirect fire engagements and a telescope for direct fire.

Ammunition is separate loading with obturation achieved by self-sealing obturator discs. Firing is accomplished by a cartridge primer inserted into the breech block and fired by a trip action mechanism. The HE M107 projectile weighs 43.7 kg, has a maximum muzzle velocity of 725 m/s and a maximum range of 20,000 m. The following propelling charges are used: M1A2 zone 3 (muzzle velocity 284 m/s), M1A2 zone 4 (muzzle velocity 326 m/s), M1A2 zone 5 (muzzle velocity 390 m/s), M9A2 zone 6 (muzzle velocity 468 m/s), M9A2 zone 7 (muzzle velocity 562 m/s), M9A2 zone 8 (muzzle velocity 647 m/s) and M9A2 zone 9 (muzzle velocity 725 m/s). The M-68 fires all NATO 155 mm projectiles including illuminating and smoke. Of the 60 rounds of 155 mm ammunition carried, 16 are for ready use.

This self-propelled howitzer can be supplied with a 39 calibre barrel that increases the range to 24,000 m using conventional ammunition or 30,000 m with a base bleed projectile. This is called the L39 but has not entered production or service.

Specifications

L33
Crew: 8
Combat weight: 41,500 kg
Power-to-weight ratio: 10.08 hp/t
Length:
(gun forward) 8.47 m
(with 39 calibre barrel) 9.36 m
(hull) 6.47 m
Width: 3.5 m
(over tracks) 2.86 m
Height: 3.45 m
Ground clearance: 0.43 m
Track: 2.28 m
Track width: 580 mm
Length of track on ground: 4.6 m
Max road speed: 38 km/h
Fuel capacity: 820 litres
Max road range: 260 km
Fording: 0.9 m
Gradient: 60%
Side slope: 30%
Vertical obstacle: 0.91 m
Trench: 2.3 m
Turning radius: 9.5 m
Engine: Cummins VT 8-460-Bi diesel developing 460 hp at 2,600 rpm
Transmission: manual with 5 forward and 1 reverse gears
Electrical system: 24 V
Armament:
(main) 1 × 155 mm L33 gun-howitzer
(anti-aircraft) 1 × 7.62 mm MG
Ammunition:
(main) 60 × 155 mm

Gun control equipment
Gun elevation/depression: +52/–4°
Gun traverse:
(total) 60°
Armour: 12-64 mm
NBC system: no
Night vision equipment: optional

Variants

155 mm self-propelled howitzer M-50

An earlier 155 mm self-propelled howitzer was the M-50 which was essentially a modified Sherman tank chassis with the French 155 mm M-50 howitzer mounted in an open compartment at the rear. This is no longer in front-line service with the Israel Defence Force but it is believed that up to 120 systems may be held in reserve. Full details were given in *Jane's Armour and Artillery 1994-95*, page 546.

155 mm Rascal

Details of the Soltam Systems Rascal lightweight tracked self-propelled artillery system were given in *Jane's Armour and Artillery 2000-2001* pages 599 to 600. This is no longer being marketed.

160 mm self-propelled mortar

Details of this system, only used by the IDF, are given in the Self-propelled mortar systems section. This is based on a much modified Sherman tank chassis.

Status

Production complete. In service with the Israel Defence Force reserves.

Manufacturer

Soltam Systems Limited.

UPDATED

ITALY

Otobreda Palmaria 155 mm self-propelled howitzer

Development

The Palmaria 155 mm self-propelled howitzer was developed from 1977 by Otobreda (then called Otomelara) specifically for export and first production vehicles followed in 1982. The first customer to order the Palmaria was Libya, with an order for 210. Nigeria ordered 25 during 1982 and a further 20 Palmaria turrets were ordered by Argentina to be fitted on locally built TAM medium tank chassis. These were completed late in 1986 by which time final deliveries had been made to Libya and Nigeria.

Argentina has built lengthened TAM medium tank chassis which have been fitted with the turret of the Palmaria. The Argentine system is called the VCA 155 and was described in detail earlier in this section. It is believed that a small batch of VCA 155 systems has now entered service with Argentina.

In 1990, Otobreda completed a further batch of 25 155 mm Palmaria systems for an undisclosed customer. Some sources indicate that this was a repeat order from Nigeria.

United Nations sources state that in 1992 Otobreda sent nine Palmaria systems to Nigeria followed by two in 1993 and a further two in 1994. No exports were declared for the years 1995 through to 1999.

There has been no recent production of the Palmaria but export marketing still continues.

Description

The chassis is similar to the Otobreda OF-40 MBTs, apart from the MTU 750 hp diesel engine, and is fully described in the MBTs and medium tanks section. The driver is seated at the front of the hull on the right side and is provided with a single-piece hatch cover and three day periscopes, the centre one of which can be replaced by a passive periscope for night driving. The other four crew members, commander, gunner, charge handler and magazine operator, are seated in the aluminium turret in the centre of the hull.

The commander is seated in the right forward part of the turret which has eight day periscopes for all-round observation and a single-piece hatch cover that opens to the rear. There are two external stowage racks at the rear of the turret. In either side of the turret there is a large rectangular hatch, the left one opening to the rear and the right one opening forwards. A 7.62 or 12.7 mm anti-aircraft machine gun can be mounted on the roof for anti-aircraft defence and four 76 mm smoke grenade dischargers are mounted either side of the turret firing towards the front.

The vehicle has an auxiliary power unit to supply the hydraulic and electric power required to operate the turret, consisting of a diesel engine driving through a gearbox, a generator and a hydraulic pump. Standard equipment includes an NBC system, an automatic fire-extinguishing system, bilge pumps and an emergency escape hatch in the floor.

Production version of the Otobreda Palmaria 155 mm self-propelled howitzer 0058377

Otobreda Palmaria 155 mm self-propelled howitzer with turret traversed to rear 0018883

Specifications

Crew: 5
Combat weight: 46,000 kg
Unloaded weight: 43,000 kg
Power-to-weight ratio: 16.3 hp/t
Length:
(gun forward) 11.474 m
(gun rear) 9.6 m
(hull) 7.265 m
Width: 3.35 m
Height:
(without machine gun) 2.874 m
(axis of bore) 2.195 m
Ground clearance: 0.4 m
Track width: 584 mm
Max speed:
(forwards) 60 km/h
Fuel capacity: 800 litres
Max cruising range: 500 km
Gradient: 60%
Side slope: 30%
Vertical obstacle: 1 m
Trench: 3 m
Fording: 1.2 m
(with preparation) 4 m
Engine: MTU MB 837 Ea-500 (or Ka-500 developing 1,000 hp) 4-stroke, precombustion chamber, turbocharged, 8-cylinder multifuel developing 750 hp
Transmission: Renk RK 304 power shifting 4-speed (4 forward and 2 reverse) planetary type with hydraulic torque converter and lock up clutch
Suspension: torsion bar
Electrical system: 24 V
Batteries: 6 × 12 V, 100 Ah
Generator: 9 kW, 3-phase with rectifier
Armament:
(main) 1 × 155 mm howitzer
(anti-aircraft) 1 × 7.62 mm MG
Smoke-laying equipment: 2 × 4 76 mm smoke grenade dischargers
Ammunition:
(main) 30 × 155 mm

Gun control equipment
Turret power control: hydraulic/manual
(by commander) no
(by gunner) yes
Gun elevation/depression: +70/–5°
Turret traverse: 360°
NBC system: yes
Night vision equipment: yes

The Otobreda-designed 155 mm barrel is 41 calibres long, is fitted with a fume extractor and a muzzle brake and is loaded automatically. Using the automatic loading system, the howitzer can fire one round every 15 seconds, one round every minute for 1 hour or one round every 3 minutes continuously. The automatic loader is at the rear of the turret and can be loaded via a hatch in the turret side. The barrel is loaded at an angle of +2° and the propellant charge is loaded into the chamber manually. Using automatic loading the turret can accommodate 23 rounds ready for use, with a further seven ready for use elsewhere in the hull. The projectiles are stowed in a linked row of tubes at the rear of the turret ready for use. The firing percussion mechanism is actuated by an electrohydraulic valve. Firing is by a push-button and using this system an initial burst rate of three rounds in 20 seconds can be achieved. The barrel elevating and balancing mechanism consists of a hydraulically actuated cylinder linked to the cradle and pivoting on the turret structure. The limits of elevation are from –4 to +70°. There are two modes of operation for the mechanism, hand and power. Using the hand mode the barrel can be moved using a wheel to operate a hand pump linked to the elevation cylinder. In the power mode, with the hydraulic power on, the gunlayer can lay the weapon directly, using a control handle. After firing, the barrel is returned to the loading position of +2° automatically. Traversing is by a hydraulically operated and controlled gearbox; traverse is a full 360°.

The recoil system consists of two buffers and one recuperator. Firing charge 8 at an elevation of +70° the recoil force is approximately 45,000 kg.

The Italian company Simmel developed a new range of 155 mm ammunition for the Palmaria self-propelled howitzer, brief details of which are given in the table.

Type	P3 HE	P3 HE LT	P3 HE RAP	P4 Illuminating	P5 Smoke
Weight					
(projectile)	43.5 kg	43.5 kg	43.5 kg	43.5 kg	43.5 kg
(explosive)	11.7 kg	11.7 kg	8 kg	n/a	n/a
(propellant)	n/a	1 kg	2.8 kg	n/a	n/a
Length	933 mm	933 mm	933 mm	933 mm	933 mm
Max range	24,700 m	27,500 m	30,000 m	24,700 m	24,700 m

It can also fire other types of ammunition including that used by the towed FH-70 and Bofors Defence FH-77B, as well as US 155 mm ammunition types including the M107, M549A1 and M483A1.

The Palmaria has three sighting systems installed: a P170 direct firing periscope, a P186 panoramic sight for indirect sighting and a dial sight.

The P170 can be used for both observation and sighting and in the latter mode has a magnification of ×8. The optical unit of the sighting system can be replaced by a light intensification system that transforms the day periscope into a low-light device.

The P186 panoramic sight is used for indirect firing and is provided with an illuminated reticle for day or low-light level use. The reticle is fixed and prealigned with the optical axis.

The Palmaria can also be provided with an electronic plane converter for rapid and accurate aiming both in direct and indirect fire.

The primary fire-control system of the Palmaria is the Alenia P186 telescope which is mounted on a Bofors telescope mount. This is used for laying the weapon in azimuth and for indirect fire control. The sight has a magnification of ×4 with a 10° field of view and weighs 8 kg. Internal communications are via the normal intercommunication system and an extra external telephone point can be provided. The turret has adequate space for normal radio installations.

Variants

Otobreda has proposed that the chassis of the Palmaria should be used for other applications, for example fitted with a similar turret to the OTOMATIC 76 mm self-propelled anti-aircraft gun system. Full details of the Otobreda OTOMATIC are given in *Jane's Land-Based Air Defence 2001-2002*. This system has been demonstrated to a number of potential customers but as of 2001 remained at the prototype stage.

Status

Production as required. In service with the following countries:

Country	Quantity	Comment
Argentina	20	total number of turrets supplied
Libya	210	delivered from 1982
Nigeria	25	believed that a second batch of 25 systems has been delivered

Manufacturer

Otobreda SpA.

UPDATED

Otobreda 155 mm M109L self-propelled howitzer

Development

Italy purchased 221 United Defense LP, Ground Systems Division M109s without their 155 mm armament from the USA. The armament was manufactured and installed in Italy by Otobreda (today called Otomelara) and subsequently issued to the Italian Army.

In 1970-71 Otobreda produced a long-barrelled version of the M109 capable of firing FH-70 projectiles to a maximum range of 24,000 m. At that time there was no Italian Army requirement for this increase in range but, early in 1984, the Italian Ministry of Defence ordered three examples for extensive trials. The trials were successful and a decision was taken in 1986 to upgrade a total of 280 Italian Army M109 series self-propelled weapons to the M109L configuration.

In addition to taking delivery of 221 M109s without armament, the Italian Army also took delivery of 62 M109A1B systems. The first batch of 32 M109Ls was completed in 1986 and final deliveries were made in 1992.

Otobreda 155 mm M109L deployed to Bosnia (Richard Stickland) 0003987

Otobreda 155 mm M109L deployed to Bosnia (Richard Stickland) 0003988

In the future this system will be supplemented in Italian Army service by the German Krauss-Maffei Wegmann PzH 2000 155 mm/52 calibre self-propelled artillery system. Total Italian Army requirement is 70 systems. Additional details of Italian involvement in PzH 2000 are given in the entry for this system under Germany.

The first two will be supplied from the German Krauss-Maffei Wegmann production line in Kassel, Germany with the remainder manufactured under a licence/co-production agreement in Italy at the Otobreda facilities at La Spezia.

Description

The hull of the M109 has been modified so that it is the equivalent of the United Defense M109A3 which is compatible with the new elevating mass and the aiming system of the M109G used by the German Army (qv).

The main change on the M109L is the replacement of the existing 155 mm/23 calibre barrel with a 155 mm/39 calibre barrel. Externally this is configured to the M109 mounting, but internally it is ballistically similar to the towed FH-70 configuration and can thus fire existing 155 mm M107 ammunition or FH-70 ammunition. Firing L15 high-explosive projectiles, the M109L has a maximum range of 24,000 m or 30,000 m with Rocket-Assisted Projectiles (RAPs).

In addition to the installation of a new ordnance, the following have been modified:

Recoil buffers and run-out buffer
Cradle to fit the new ordnance
Replacement of the equilibrator and elevating cylinder
Barrel clamp to fit new barrel.

The new 155 mm/39 calibre barrel is fitted with a revised muzzle brake, with the maximum recoil increase on the M109L being only 100 mm. The existing breech and breech ring of the original M109 have been retained.

The elevating mass, as modified, is interchangeable with the one on the basic M109; furthermore, the mechanical interface between the elevating mass and saddle, as well as the mountings of the optical device, do not require any modification.

Specifications

Crew: 6
Combat weight: 24,800 kg
Unloaded weight: 20,000 kg
Power-to-weight ratio: 16.33 hp/t
Ground pressure: 0.68 kg/cm^2
Length:
(overall) 8.133 m
(hull) 6.256 m
Width:
(overall) 3.295 m
(over tracks) 3.149 m
Height:
(incl AA MG) 3.28 m
Ground clearance: 0.467 m
Track: 2.768 m
Track width: 381 mm
Length of track on ground: 3.962 m
Max road speed: 56 km/h
Fuel capacity: 511 litres
Max road range: 349 km
Vertical obstacle: 0.533 m
Trench: 1.828 m
Engine: Detroit Diesel Model 8V-71T, turbocharged, 2-stroke, liquid-cooled 8-cylinder diesel developing 405 bhp at 2,300 rpm
Transmission: Allison Transmission XTG-411-2A cross-drive with 4 forward and 2 reverse gears
Suspension: independent torsion bar
Electrical system: 24 V
Batteries: 4 × 12 V, 6TN
Armament:
(main) 1 × 155 mm 39 calibre howitzer
(anti-aircraft) 1 × 12.7 mm M2 MG
Smoke-laying equipment: 2 × 3 76 mm smoke grenade dischargers
Ammunition:
(main) 28
(12.7 mm) 500

Gun control equipment
Turret power control: hydraulic/manual
(by commander) no
(by gunner) yes
Gun elevation/depression: +75/−3°
Turret traverse: 360°
NBC system: no
Night vision equipment: yes

Specifications of the Otobreda 155 mm/39 calibre barrel are as follows:

Calibre: 155 mm
Length: 39 calibres
Powder chamber volume: 18.84 dm^3
Number of rifling grooves: 48
Twist uniform right hand: 1 turn in 20 calibres
Weight:
(gun barrel) 1,670 kg
Elevating mass: 4,200 kg
Max recoil length: 915 mm

Status

Conversions complete. In service with the Italian Army. Production can be resumed if further orders are placed. In the future this will be supplemented by the German 155 mm/52 calibre PzH 2000 self-propelled artillery system. Under subcontract to the German companies, Krauss-Maffei Wegmann, Rheinmetall Landsysteme and Rheinmetall, Otobreda is already carrying out some subcontract work for the 185 155 mm PzH 2000 currently on order for the German Army, with first production systems completed early in 1998.

Manufacturer

Otobreda SpA.

UPDATED

JAPAN

Type 99 155 mm self-propelled howitzer

Development

In the FY00 defence budget, the Japanese Defence Agency revealed that it was to procure seven Type 99 155 mm self-propelled howitzers for the Japanese Ground Self-Defence Force. It is known that four systems were delivered to the Japanese Ground Self-Defence Force in 1999.

It is understood that the Type 99 had been under development for some years under the leadership of Mitsubishi Heavy Industries (hull and final assembly) and the Japan Steel Works (gun and turret system).

The Type 99 155 mm self-propelled howitzer is to supplement the older 155 mm Type 75 self-propelled howitzer. It is known that four Type 99 155 mm self-propelled howitzers were procured in 1999.

Description

The overall layout of the Type 99 155 mm self-propelled gun is similar to other recent weapons of this type.

The driver is seated at the front right with a single-piece hatch cover above his position and forward of these are three day periscopes, the centre one of which can be replaced by a passive periscope for driving at night.

The power pack, consisting of diesel engine, transmission and cooling system is mounted to the left of the driver, with the air inlet and outlet louvres in the roof and the exhaust outlet in the left side of the hull.

The fighting compartment is at the rear with normal means of entry and exit via a door in the lower part of the hull rear.

The turret is at the rear with the commander seated on the right side and provided with a roof-mounted 12.7 mm M2 machine gun mount that is also fitted with a shield.

In addition to the roof hatches there is also a single door in either side of the turret. The one on the left side opens to the rear while the one on the right side opens to the front.

The 155 mm ordnance, which is believed to be a 39 calibre weapon, is provided with a muzzle brake but no fume extractor. A travelling lock is provided at the front of the hull and when not in use, this folds back onto the glacis plate. The ordnance is believed to be based on that used in the towed FH-70 artillery system that has been manufactured under licence in Japan for some years.

It is considered probable that some type of automatic loading system is provided and mounted over the rear part of the ordnance in what appears to be a muzzle velocity measuring device.

Suspension either side consists of seven dual rubber-tyred roadwheels with the drive sprocket at the front, idler at the rear and track return rollers.

No details of the Type 99 155 mm self-propelled artillery system have been released. Standard equipment includes a NBC system and passive night driving aids.

Specifications

Not available.

Status

Production. In service with Japanese Group Self-Defence Force.

Manufacturers

Mitsubishi Heavy Industries
(hull and final assembly) (unconfirmed)
Japan Steel Works/Nihon Seiko Jyo
(gun and turret) (unconfirmed)

UPDATED

Type 75 155 mm self-propelled howitzer

Development

In 1969, the Technical Research and Development Headquarters of the Japanese Ground Self-Defence Force started design work on a new 155 mm self-propelled howitzer. Mitsubishi Heavy Industries built the hull while Nihon Seiko Jyo/Japan Steel Works built the 155 mm gun and turret. The two prototypes, completed in 1971-72, differed only in their ammunition loading systems. Trials with the two prototypes were undertaken between 1973 and 1974 and in October 1975 the vehicle was standardised as the Type 75 155 mm self-propelled howitzer. By late 1978, the first 20 vehicles had been delivered to the Japanese Ground Self-Defence Force. By the end of 1988, 201 Type 75 self-propelled howitzers were in service with the Japanese Ground Self-Defence Force and production was complete. This system was never offered on the export market. There are no known plans for this system to be upgraded.

The 155 mm Type 75 serves alongside the locally built 105 mm Type 74 and larger numbers of locally manufactured US-designed 203 mm M110A2 systems. This has now started to be supplemented by the new 155 mm Type 99 full tracked self-propelled artillery system, details of which are given in a separate entry.

Type 75 155 mm self-propelled howitzer in travelling configuration (Kensuke Ebata) 0003989

Type 75 155 mm self-propelled howitzer fitted with a 12.7 mm anti-aircraft machine gun on the turret roof **2002**/0126554

Description

The hull and turret of the Type 75 155 mm self-propelled howitzer are made of all-welded aluminium. The crew of six consists of the commander, layer, two ammunition loaders, radio operator and driver.

The hull is divided into three compartments: driver at the front on the right, power pack to his left and the turret at the rear. The driver has a single-piece hatch cover that opens to the right, in front of which are three day periscopes. The centre one can be replaced by a passive night vision device. The driver can also enter and leave his compartment via the turret.

The turret is fully enclosed, with the commander and gunner seated on the right and the ammunition loaders on the left. The commander has a single-piece hatch cover that opens to the rear and a single day periscope for forward observation. One of the two ammunition loaders is also provided with a single-piece hatch cover that opens to the rear, in front of which is a 12.7 mm (0.50) M2 HB anti-aircraft machine gun with a small shield. On each side of the turret is a square hatch that opens to the rear.

At the rear of the hull are two doors which open either side of the vehicle. The torsion bar suspension either side consists of six dual rubber-tyred aluminium roadwheels, with the drive sprocket at the front and the sixth roadwheel acting as the idler. The first and fifth roadwheel stations are provided with a hydraulic shock-absorber. There are no track-return rollers.

The Type 75 is provided with an NBC system which is in the left side of the fighting compartment, a fire extinguishing system, a heater, and night vision equipment. An amphibious kit was developed for the vehicle but was not adopted.

Main armament consists of a 155 mm 30 calibre weapon with a double-baffle muzzle brake, a fume extractor and a stepped thread interrupted screw-type breech block which opens to the right. A gun travelling lock is fitted on the glacis plate.

The weapon fires a Japanese HE projectile (maximum charge 9) to a maximum range of 19,000 m or an American M107 HE projectile (maximum charge 7) to a maximum range of 15,000 m. A 155 mm Rocket-Assisted Projectile (RAP) has been developed with a range of around 24,000 m.

The gun has an elevation of +65°, a depression of –5° and the turret can be traversed through a full 360°. Gun elevation and turret traverse are hydraulic by means of a single lever. Movement in the horizontal plane controls the turret position in azimuth and movement in the vertical plane controls the gun in elevation. The traverse of the turret and the elevation of the gun are stopped whenever the layer halts the movement of the lever. Manual controls are provided for emergency use and the turret can also be traversed by the commander.

The gun has a variable recoil mechanism which consists of twin hydraulic cylinders, one on the left above the cradle and the other on the right side under the cradle. The variable recoil system is fitted to prevent the breech hitting the floor when the gun is being fired at a high angle. The recuperator is pneumatic and the recoil damper, which is on the centreline under the cradle, is hydraulic.

When the gun is fired, the 155 mm weapon automatically moves to an elevation of +6° for reloading and returns to its firing elevation once reloaded.

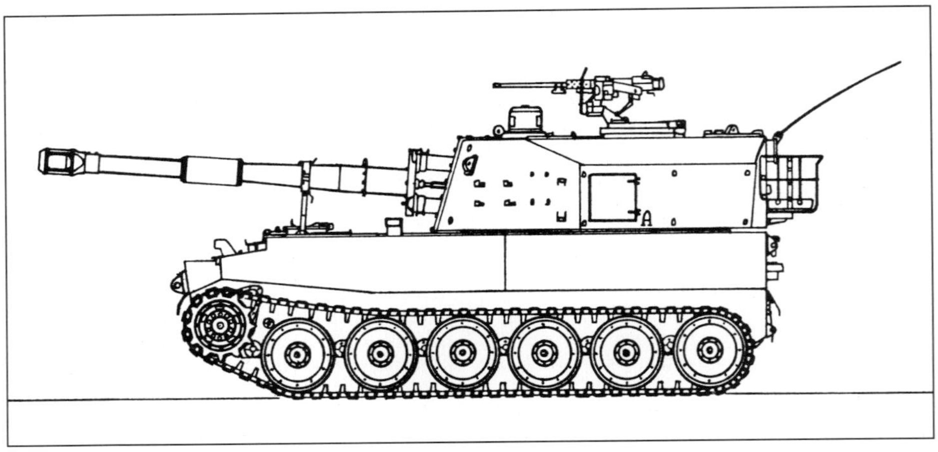

Type 75 155 mm self-propelled howitzer

Specifications

Crew: 6
Combat weight: 25,300 kg
Power-to-weight ratio: 17.78 hp/t
Ground pressure: 0.64 kg/cm²
Length:
(gun forward) 7.79 m
(hull) 6.64 m
Width: 3.09 m
Height:
(to turret roof) 2.545 m
Ground clearance: 0.4 m
Track: 2.5 m
Track width: 500 mm
Length of track on ground: 5.2 m
Max road speed: 47 km/h
Fuel capacity: 650 litres
Max range: 300 km
Fuel consumption: 2.1 litres/km
Fording: 1.3 m
Gradient: 60%
Side slope: 30%
Vertical obstacle: 0.7 m
Trench: 2.5 m

Engine: Mitsubishi 6ZF 2-cycle 6-cylinder air-cooled diesel developing 450 hp at 2,200 rpm
Transmission: manual with 4 forward and 1 reverse gears
Suspension: torsion bar
Electrical system: 24 V
Batteries: 4 × 12 V
Armament:
(main) 1 × 155 mm howitzer
(anti-aircraft) 1 × 12.7 mm M2 MG
Smoke-laying equipment: none
Ammunition:
(main) 28
(anti-aircraft) 1,000

Gun control equipment
Turret power control: hydraulic/manual
(by commander) yes
(by gunner) yes
Max rate power traverse: 360° in 30 s
Gun elevation/depression: +65/–5°
Commander's override: yes
NBC system: yes
Night vision equipment: yes

Mounted in the rear part of the turret are two drum-type rotating magazines, an extensible loading tray and a hydraulic rammer. Each drum holds nine 155 mm projectiles, is rotated electrically or manually and can be reloaded from inside the vehicle or from outside via two small circular doors in the turret rear. The axis of the drums, two-stage extensible loading tray and the rammer are on the same axis as the gun when it is returned from its firing position for loading. Each of the two magazines holds nine 155 mm projectiles which, coupled with the use of the extensible loading tray and the hydraulic rammer, enable 18 rounds to be fired in 3 minutes, that is 6 rds/min. A further 10 projectiles are carried inside the gun compartment plus 56 fuzes and 28 bagged charges.

Fire-control equipment includes a J2 day panoramic sight, with a magnification of ×4 and a 10° field of view, a J3 day direct telescopic sight with a magnification of ×4 and a 10° field of view, an electric elevation sight which includes a cant angle sensor, an angle indicator and an inverter, and finally a collimator. All fire-control equipment is mounted in front of the commander's position and is operated by the layer.

Status

Production complete. In service with the Japanese Ground Self-Defence Force. There have been no exports of this system.

Manufacturers

Mitsubishi Heavy Industries (hull and final assembly).
Japan Steel Works/Nihon Seiko Jyo (gun and turret).

UPDATED

Type 74 105 mm self-propelled howitzer

Development

The first two prototypes of this 105 mm self-propelled howitzer were completed in 1969-70, Komatsu being responsible for the hull and Japan Steel Works for the turret and main armament. In 1974, the vehicle was standardised as the Type 74 105 mm self-propelled howitzer and production began in 1975.

Only 20 were built, the last being completed in 1978, as it was decided to concentrate on the more efficient 155 mm self-propelled howitzer which was eventually standardised as the Type 75. The 17 remaining Type 74s are attached to the 2nd Artillery Regiment at Asahikawa, Hokkaido, the northern part of the Japanese islands.

There are no known plans for this system to be upgraded in the future and, as its range is so short by today's standards, it is expected that it will soon be phased out of service.

Description

In appearance the Type 74 105 mm self-propelled howitzer is very similar to the British Vickers Defence Systems Abbot 105 mm self-propelled gun. (This has been phased out of service with the UK Army and details were given in *Jane's Armour and Artillery 1995-96.*)

The all-welded armoured hull is divided into three compartments: driver at the front on the right, engine to the left and gun compartment at the rear.

The driver has a single-piece hatch cover that opens to the right, in front of which are three day observation periscopes. In the driver's hatch is a rotatable periscope. The engine air intake and outlets are in the roof, with the exhaust outlet being on the left side.

The fully enclosed turret is at the rear and normal means of entry and exit for the gun crew is via a rectangular door in the chassis rear that opens to the right.

Type 74 105 mm self-propelled howitzer with a 12.7 mm M2 HB anti-aircraft machine gun mount with a shield for the gunner installed on the turret roof

There is a single forward-opening door in each side of the turret and two circular opening roof hatches towards the rear, with the 12.7 mm M2 anti-aircraft machine gun mounted on the right side of the turret roof and provided with a shield.

The turret of the Type 74 is not provided with a basket. The torsion bar suspension either side consists of five, dual, rubber-tyred roadwheels with the drive sprocket at the front and idler at the rear. There are no track-return rollers. The first and fifth roadwheel stations either side have a hydraulic shock-absorber.

A flotation screen is carried collapsed around the top of the hull and when it is erected the vehicle can propel itself across rivers using its tracks.

The 105 mm gun has a muzzle brake and a fume extractor. No details of the range of the 105 mm weapon have been released but it can be assumed that it has a similar range to that of the US 105 mm M2A1 towed howitzer which fires an HE projectile weighing 19.06 kg to a maximum range of 11,270 m.

Recent Japanese information has stated that the 105 mm weapon has a maximum range of 14,500 m with maximum muzzle velocity being 645 m/s. Ammunition is of the separate loading type, for example, projectile and charge in a brass cartridge case.

Specifications

Crew: 4
Combat weight: 16,300 kg
Unloaded weight: 13,300 kg
Power-to-weight ratio: 18.40 hp/t
Ground pressure: 0.60 kg/m^2
Length: 5.78 m
Width: 2.87 m
Height: 3.20 m
Ground clearance: 0.40 m
Max speed:
(road) 50 km/h
(water) 6 km/h
Fuel capacity: 410 litres
Range: 300 km
Fording: 1.3 m
(with screen erected) amphibious
Gradient: 60%
Side slope: 30%
Vertical obstacle: 0.70 m
Trench: 2.1 m
Engine: Type 4ZF 2-stroke air-cooled V-4 diesel developing 300 hp at 2,200 rpm
Transmission: manual
Suspension: torsion bar
Electrical system: 24 V
Armament:
(main) 1 × 105 mm howitzer
(anti-aircraft) 1 × 12.7 mm M2 MG
Ammunition:
(main) 30 × 105 mm
(12.7 mm) 500 × 12.7 mm (estimate)
NBC system: yes
Night vision equipment: yes

Status

Production complete. In service only with the Japanese Ground Self-Defence Force. This system was never offered on the export market.

Manufacturers

Komatsu (hull and final assembly).
Japan Steel Works (turret and main armament).

UPDATED

KOREA, NORTH

North Korean self-propelled artillery

Development/Description

North Korea appears to have followed in the footsteps of countries such as China which have mated Russian-developed artillery barrels and mounts onto locally designed and built full-tracked chassis. The US Defense Intelligence Agency has reported the existence of the weapons in the table.

A typical North Korean mechanised infantry brigade has two artillery battalions equipped with 18/20 M1977 self-propelled gun-howitzers. Some of these brigades have substituted 12 M1981/M1985s for the M1977s.

North Korean armoured brigades have two artillery battalions equipped with 12/14 M1981/M1985 self-propelled guns. Within armoured and mechanised infantry divisions, the divisional self-propelled artillery regiments consist of two battalions of M1977 self-propelled howitzers, one battalion of M1974

North Korean self-propelled artillery system 170 mm M-1989 which appears to use the same ordnance as the 170 mm M-1978

North Korean 122 mm self-propelled artillery system M1977 **2001**/0106222

North Korean 122 mm self-propelled artillery system M1981 **2001**/0106223

Designation	Calibre	Weight	Max range	Comment
M1992 SPM	120 mm	16 t	8,700 m	turret-mounted mortar on VT-323 chassis
M1997 SPH	122 mm	18 t	15,400 m	D-30 on modified AT-S chassis
M1985 SPG	122 mm	19 t	20,800 m	A-19 on modified chassis ATS chassis
M1981 SPG	122 mm	19 t	24,000 m	D-74 on modified AT-S chassis
M1991 SPG	122 mm	22 t	24,000 m	D-74 on modified AT-S chassis
M1975 SPG	130 mm	20 t	27,150 m	M-46 on modified ATS chassis
M1991 SPG	130 mm	22 t	27,150 m	M-46 on modified ATS chassis
M1992 SPG	130 mm	20 t	29,500 m	130 mm SM-4-1 coastal gun on modified ATS chassis
M1974 SPH	152 mm	17 t	17,410 m	152 mm D-20 on modified VT-323 APC chassis
M1977 SPH	152 mm	15 t	12,400 m	152 mm D-1 on K-63 chassis
M1991 SPH	152 mm	19 t	17,265 m	152 mm ML-20 on modified AT-S chassis
M1978 SPG	170 mm	40 t	n/known	170 mm gun on T-55 MBT chassis
M1989 SPG	170 mm	40 t	n/known	170 mm gun on modified ATS chassis

Notes: 122 mm Corps gun M1931/37 (A-19) (see *Jane's Armour and Artillery 1990-91*), 152 mm howitzer M1943 (D-1) (see *Jane's Armour and Artillery 1995-96*), 152 mm gun-howitzer D-20 (see Towed anti-tank guns and howitzers), 122 mm field gun D-74 (see Towed anti-tank guns and howitzers), 122 mm howitzer D-30 (see Towed anti-tank guns and howitzers), 130 mm field gun M-46 (see Towed anti-tank guns and howitzers), 130 mm SM-4-1 (see Coastal artillery guns and missiles).

self-propelled guns and one battalion of M1981/M1985 self-propelled guns.

Corps and army level self-propelled artillery brigades are equipped with M1975 or M1978 self-propelled guns and it is believed that by the early 1980s North Korea had deployed one army level artillery brigade with 24/36 M1978 self-propelled guns and at least three army level artillery brigades with 36/60 M1978 self-propelled guns each.

The infantry divisions and regiments are usually still equipped with 122 mm and 152 mm towed artillery.

In recent years, the Korean People's Army has fielded a number of new self-propelled artillery systems including one similar to the Russian 203 mm 2S7.

Status

In service with Korean People's Army.

UPDATED

170 mm M1978 self-propelled gun Koksan

Development/Description

This 170 mm self-propelled gun was first observed by US sources in the city of Koksan in 1978 and is therefore referred to as the M1978 self-propelled gun Koksan.

A number of these guns were supplied to Iran by North Korea and some were subsequently captured by Iraq in 1987-88.

In North Korean service, it is believed that the 170 mm Koksan is employed in regiments with a total of 36 weapons.

The system is based on a modified Chinese Type 59 MBT chassis with the existing turret removed and plated over. A 170 mm gun of a previously unknown type in an open mount has been mounted on top of the chassis. Only the driver of the Koksan is under full armour protection and it is probable that the remainder of the gun crew are carried in other vehicles which also carry the separate loading (for example projectile and charge) type ammunition.

At the very rear of the chassis are two large spades that are lowered to the ground manually before firing commences; when travelling these are vertical. A large ordnance travelling lock is pivot-mounted on the glacis plate and this folds forward onto the ground when deployed in the firing position.

The long 170 mm ordnance is fitted with a multislotted muzzle brake but has no fume extractor.

170 mm M1978 Koksan self-propelled gun from the rear showing the two large spades

170 mm M1978 Koksan self-propelled gun with ordnance in travelling lock

The screw-type breech mechanism opens to the right with a loading tray mounted on the left side of the weapon. In addition, there is probably a power rammer.

Elevation and traverse appear to be powered, although traverse is very limited to avoid overstressing the chassis.

It is possible that the 170 mm weapon used may be a Russian naval or coastal weapon supplied to North Korea in the 1950s for use in the coastal defence role. When these were replaced by surface-to-surface missiles, the coastal guns were surplus to requirements and could then have become available for mounting on the tracked chassis.

As there is insufficient space for the gun crew and ammunition on the Koksan, another vehicle is required to support the system in action and this would carry the remainder of the crew and some ready use ammunition.

It is probable that standard equipment includes an NBC system and infra-red night vision equipment for the driver.

Like the Type 59 MBT on which it is based, the 170 mm M1978 self-propelled gun can lay its smoke screen by injecting diesel fuel into the exhaust outlet on the left side of the hull.

Variant

170 mm M1989 SPG

This is understood to use the same weapon and mount as the M1978 but is mounted on a modified AT-S chassis.

Status

Production complete. In service with Iran and North Korea. Recent information has indicated that these systems may no longer be deployed in a fully operational role. They have not been seen in any of the recent North Korean military parades.

Manufacturer

North Korean state factories.

Specifications
(M1978)
Weight: 40 t
Length: 14.9 m
Width: 3.27 m
Height: 3.1 m
Engine: 520 hp diesel

UPDATED

KOREA, SOUTH

Samsung Techwin 155 mm/52 calibre K9 Thunder self-propelled artillery system

Development

For many years the mainstay of the Republic of Korea (ROK) self-propelled field artillery units has been the United Defense LP 155 mm/39 calibre M109A2 self-propelled howitzer.

Samsung Techwin Defense Program Division, was the ROK prime contractor for the co-production of 1,040 M109A2 SPH, deliveries of which were completed several years ago.

In the late 1980s, the ROK Army drew up its requirements for a new 155 mm/52 calibre tracked SPH to meet the requirements of the 21st century which are based on the following missions.

Deep supporting fire of the corps, qualitatively superior to overcome the numerical inferiority and effective firing support in mountain areas with its longer range. Agency for Defense Development (ADD) started development of a new SPH in 1989.

Key operational requirements included a higher rate of fire, longer range, more accurate fire, faster into and out of action times and greater mobility, all of which lead to a significant increase in battlefield survivability of the weapon system.

Following a competition, the now Samsung Techwin was selected to be the prime contractor for the K9 Thunder 155 mm/52 calibre SPH, although for a short time it was also referred to as the XK9.

First prototypes were completed in 1994 under the designation XK9 and put through an extensive series of Development Test/Operational Test (DT/OT) by the ROK Army and the ADD.

These were followed by three preproduction vehicles manufactured during the full-scale development phase (Phase II), the last of which was completed in 1998.

The mobility and firing reliability of the K9 Thunder had been verified through the trial of the prototype vehicles in Korea and by the end of 1998 these had travelled 18,000 km and fired 12,000 rounds.

The XK9 Thunder was type classified by the ROK Army in 1998 as the K9, with the first production vehicles being completed earlier in 1999. The first ROK field artillery battalion is now operational and has three batteries each of six K9s. Each battery is normally controlled by one battery fire direction centre.

The first production batch of K9 consisted of six units which started to be fielded for the fire support of a corps artillery without replacing the existing towed and self-propelled 155 mm artillery systems.

The ROK Army is the first army in Asia to deploy a 155 mm/52 calibre SPH and the second in the world. The German Army was the first to deploy a 155 mm/52 calibre system with the introduction into service of the Krauss-Maffei Wegmann PzH 2000 in 1998.

Late in 2001, it was revealed the Turkish Land Forces Command (TLFC) had signed a contract with the Defense Program Division of Samsung Techwin for the 155 mm K9 self-propelled artillery system.

Under the terms of this deal, South Korea will supply a first batch of K9 systems with progressive licensed production then being undertaken in Turkey.

Total value of this contract could be as high as US$1 billion with up to 300 systems being provided. Germany was also competing for this contract with the PzH 2000 self-propelled artillery system from Krauss-Maffei Wegmann but the German government blocked this sale.

Description

The hull and turret of the K9 are of all-welded steel armour construction with a maximum thickness of 19 mm. This provides the occupants with protection from medium calibre small arms fire and 155 mm high-explosive shell fragments.

Specifications
Crew: 5
Combat weight: 46,300 kg
Power-to-weight ratio: 21.6 hp/t
Length gun forwards: 12 m
Length hull: 7.44 m
Width: 3.40 m
Height
(turret roof): 2.73 m
(incl 12.7 mm M2 MG): 3.5 m
Ground clearance: 0.41 m
Track: double pin with replaceable rubber pads (GS-2)
Track width: 548 mm
Length of track on ground: 4.72 m
Max road speed: 67 km/h
Max towing speed: 27 km/h
Acceleration: 12 s (0-32 km/h)
Braking distance: 12 m (from 32 km/h)
Fuel capacity: 850 litres
Range (cruising): 360 km
Fording: 1.5 m
Gradient: 60%
Side slope: 30%
Vertical obstacle: 0.75 m
Trench: 2.8 m
Engine: MTU MT 881 Ka-500 water-cooled diesel developing 1,000 hp at 2,700 rpm
Transmission: Allison Transmission X1100-5A3 automatic with 4 forward and 2 reverse gears
Suspension: hydropneumatic
Electrical system:
(generator voltage) 28 V DC
(battery voltage) 24 V DC
Armament:
(main): 1 × 155 mm/52 calibre
(anti-aircraft): 1 × 12.7 mm M2 machine gun
Ammunition :
(155 mm) 48
(12.7 mm MG) 500

Gun control equipment:
Turret power control: hydraulic/manual
Turret traverse: 360°
Gun elevation/depression: +70°/–2.5°
Armour: 19 mm max
Armour type: steel
NBC system: yes
Night vision equipment: yes

155 mm/52 calibre K9 Thunder self-propelled artillery system with ordnance at maximum elevation and showing projectile leaving barrel **2000**/0085201

152 mm/52 calibre K9 Thunder self-propelled artillery system in travelling configuration **2000**/0085205

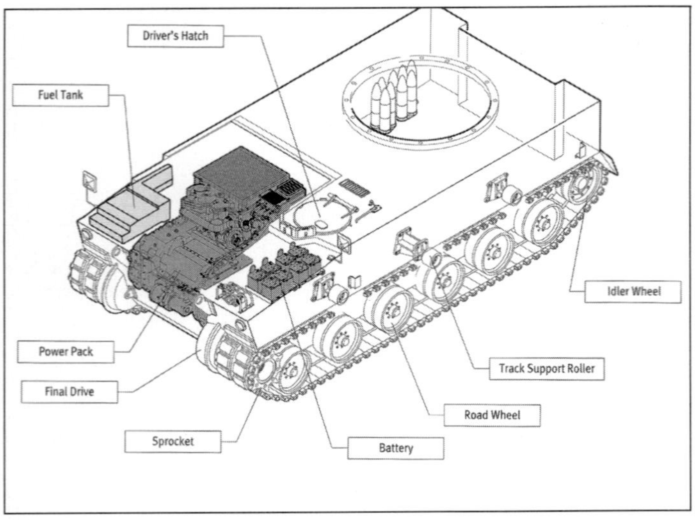

Chassis detail of the 155 mm K9 Thunder self-propelled artillery system showing position of main components ***2002***/0126555

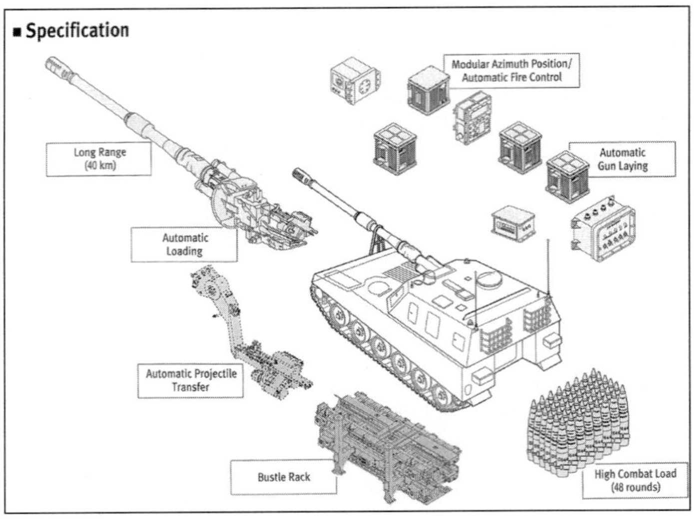

Main components of the 155 mm K9 Thunder self-propelled artillery weapon system showing main subsystems ***2002***/0126556

The K9 has a crew of five, driver and four in the crew compartment, with the latter consisting of commander, gunner, assistant gunner, and ammunition loader.

In appearance and overall layout the K9 is similar to other systems of this type, with the driver being seated front left with the power pack to his right and with the remainder of the vehicle being taken up by the turret and crew compartment.

The driver enters his compartment by a rear opening roof hatch that can be locked in the horizontal position. The driver has three forward-facing day periscopes, the centre one of which can be replaced by a passive night vision device.

Wherever possible, to reduce development time and costs, proven subsystems were used in the development of the K9. The power pack for example consists of the German-designed MTU MT 881 Ka 500 V8 water-cooled diesel developing 1,000 hp, coupled to an Allison Transmission Division X1100-5A3 fully automatic transmission with four forward and two reverse gears.

The MTU MT 881 Ka 500 is a proven engine and is also installed in the German PzH 2000 SPH, while the X1100-5A3 is a diesel version of the X1100-3B transmission developed for the turbine powered US M1/M1A1/M1A2 Abrams MBT.

The fuel tank is located at the front right side of the hull with the battery compartment being front left. The air intake and outlet louvres are in the roof, with the exhaust outlet on the right side.

The indirect fire sight is mounted on the left side of the turret roof and is provided with a ballistic cover.

The commander and gunner are seated on the right side of the turret. Mounted on the forward part of the commander's cupola is a 12.7 mm M2 machine gun for air defence/local defence purposes and a single-piece hatch cover is provided which opens to the rear. A door is provided in the left side of the turret and there is also a large door in the chassis rear for crew entry and ammunition resupply purposes. The 155 mm/52 calibre main armament has a chamber volume of 23 litres and is fitted with a muzzle brake and fume extractor.

The large multiple slotted muzzle brake reduces the firing impulse applied to the recoil system and also provides a superior capability of attenuating the blast overpressure and flash. The vertically sliding breech mechanism automatically opens upward after firing and obturates the propellant gases with the high-strength obturator ring in the barrel assembly.

The primer magazine automatically feeds, inserts and extracts the primers and movement of the breech block allows for the continuous firing capability of 21 rounds.

The gun mount, which consists of two hydraulic recoil brakes with an internal buffer and a pneumatic recuperator, reduces the firing impulse to the chassis. The high-rigidity cradle is fitted to the autoloader to the rear which has enhanced the firing accuracy by minimising the lateral motion of the barrel when firing.

When travelling the 155 m/52 calibre barrel is held in position with a travel lock installed on the forward part of the hull, this is operated by the driver by remote control, so ensuring the NBC integrity of the vehicle.

Mounted over the rear part of the 155 mm/52 calibre ordnance is a muzzle velocity measuring system that feeds information to the onboard computer. This Doppler-type system has a muzzle velocity recording range of from 20 to 2,000 m/s.

A burst rate of fire of three rounds can be fired in 15 seconds, with a maximum rate of fire of six to eight rounds a minute for a period of 3 minutes. Sustained rate of fire is two to three rounds a minute for 1 hour. A thermal warning device is fitted as standard and this feeds the temperature of the barrel to the Automatic Fire-Control System (AFCS).

Firing a standard M107 high-explosive (HE) projectile a maximum range of 18 km can be achieved. Maximum range firing an HE Rocket Assisted Projectile (RAP) with uni-charge of five zones is 30 km. Firing the new K307 projectile with a six zone charge, a maximum range of 40 km can be achieved with a chamber pressure of 52,000 psi and a muzzle velocity of 924 m/s.

Using its onboard fire-control system it can carry out Time On Target (TOT) procedures and put three rounds onto the target at once.

The bustle rack of the turret has been designed to load four types of 155 mm projectile and is fitted with four independently operated electrical drives which can be operated manually in an emergency. A total of 48 projectiles and their associated charges are carried for ready use.

Internally an automatic loading system takes projectiles from the storage position and places them onto the ammunition tray ready for ramming. The charges are loaded manually. Although it can use a conventional bag type charge system (M3A1, M4A2, M119A1 and M203) the K9 has been optimised for use with a locally developed modular charge system.

The new modular charge system comprises two charges. One is a uni-charge, used for up to five zones (five increments) and another is a module-type charge for zone six only (two increments).

The new K307 projectile is a high-explosive extended-range full-bore base bleed (HE ER FB BB) type which features an aerodynamic body and base bleed to achieve its enhanced range.

Standard equipment includes a Honeywell Azimuth Position System, Automatic Fire-Control System, powered gun elevation/depression and turret traverse system. The K9 Thunder can open fire within 30 seconds if it is already stationary or within 60 seconds if it is moving.

Information to lay the weapon onto the target can be transmitted from the battery command post to the K9 Thunder via a data digital radio link or voice communications and the system can also calculate laying data using its onboard fire-control equipment.

Suspension system is of the hydropneumatic type, with each side having six dual rubberised roadwheels and with the drive sprocket at the front, idler at the rear and three track-return rollers. Track is of the double-pin type with replaceable rubber pads.

Subcontractor for the suspension is Tong Myung Heavy Industries who signed a contract with Air Log of the UK in 1995 for the supply of an undisclosed quantity of 'Hydrogas' suspension units and licence production of these in South Korea. This is similar to that used on the now BAE Systems, RO Defence, 155 mm AS90 self-propelled artillery in service with the Royal Artillery. Air Log is no longer in existence and production is now undertaken by Horstman Defence Limited.

The K9 Thunder is also fitted with an NBC system, heater, internal and external communications system and a manual fire suppression system.

Using the same basic chassis as the K9 Thunder an ammunition resupply version is under development and funding permitting, this could enter service around 2005.

Status

Production. In service with the Republic of Korea Army. On order for Turkish Land Forces Command.

Manufacturer

Samsung Techwin, Defense Program Division.

UPDATED

NETHERLANDS

RDM Technology SPGs

Development/Description

Previous RDM Technology production has included the manufacture of parts for Royal Netherlands Army 155 mm M109A2 (supplied by the now United Defense LP, Ground Systems Division) self-propelled howitzers; major assemblies for the Leopard 2 MBT for both Dutch and German armies; fitting FMC (now United Defense LP) command and reconnaissance carriers of the Royal Netherlands Army with Oerlikon Contraves 25 mm, one-person GBD-A0A turrets; modifying Royal Netherlands Army FMC (now United Defense LP) AIFVs to carry the Systems and Electronics Inc Improved TOW mount; and production of a large series of AIFVs (together with DAF, Eindhoven) for the Royal Netherlands Army.

In 1988, RDM Technology delivered the first upgraded M114/39 calibre towed howitzers; details of this programme are given in the Towed anti-tank guns, guns and howitzers section under Netherlands.

For the M109A2 the company manufactured the gun and mount assembly, installing the turret and completing system integration before delivering the complete vehicle.

Canadian M109 upgrade

Late in 1992, RDM Technology received a contract to overhaul 30 M109 series self-propelled howitzers of the Canadian Forces, as well as carrying out a small number of modifications including the installation of an auxiliary power unit. The 155 mm ordnance was retained. This upgrade programme is now complete.

Royal Netherlands Army M109 upgrade

In early 1991, the Royal Netherlands Army artillery order of battle included 76 M110A2 203 mm self-propelled howitzers, 222 M109 155 mm series self-propelled howitzers, 123 M114 155 mm towed howitzers (of which 82 had been upgraded with the 39 calibre ordnance) and 22 Multiple Launch Rocket Systems.

However, under subsequent plans the artillery strength has now been cut to 137 M109 155 mm self-propelled howitzers, 51 155 mm M114/39 calibre towed howitzers and 22 Multiple Launch Rocket Systems.

It was intended to refurbish all 137 M109 155 mm self-propelled howitzers but late in 1993 it was stated that only 126 would be upgraded, which would be issued to five M109 regiments, each having 24 weapons, with the remaining six to be used as driver training vehicles.

The upgrade work was carried out by RDM Technology under a contract worth US$62.6 million. Of the 126 vehicles upgraded, 88 were M109A2s and the remaining 38 M109A3s.

The upgrade was combined with a programme of basic maintenance and included a better dynamo and voltage regulator, an improved starter motor, improved turret rotation installation and anti-corrosion measures. Modifications were carried out to the aluminium hull to prevent cracks and the 155 mm ammunition stowage racks were replaced. No improvements have been carried out to the actual 155 mm weapon. The upgrade has now been completed.

UAE M109 upgrade

In August 1995, Abu Dhabi, one of the states of the United Arab Emirates, purchased 85 155 mm M109A3 self-propelled howitzers that were surplus to the requirements of the Royal Netherlands Army.

RDM Technology M109L47 self-propelled artillery systems for the United Arab Emirates 0003990

These M109A3s were then put through a major upgrade programme by RDM Technology, with the first upgraded systems being delivered to Abu Dhabi in 1997.

Early in 1999, the UAE took delivery of the last of the 85 upgraded 155 mm M109L47 self-propelled guns ordered from RDM Technology.

These have been upgraded in many areas including the installation of a 155 mm/47 calibre barrel supplied by SW Swiss Ordnance Enterprise which is the same as that being fitted to the upgraded Panzerhaubitze 88/95 of the Swiss Army.

All of the 85 systems delivered to the UAE have the 155 mm/47 calibre barrel with 60 up to full standard which includes auxiliary power unit (APU), bustle-mounted air conditioning system, new propellant charge racks (NATO and South African), semi-automatic projectile rammer, remote-control travel lock for 155 mm ordnance and revised hydraulic unit.

A total of three UAE artillery regiments are now operational with the M109L7 and each of these regiments has three batteries, each with eight guns.

The M109L47 has a maximum range, firing an ERFB-BB projectile, of 38 km.

The design of the RDM Technology M109L47 is modular so that the user can select those subsystems that meet current operational requirements.

M109 155 mm/52 calibre upgrade

The first example of this system, now called the RDM Technology M109L52 started its trials late in 2000. Details of this system are given in a separate entry. This remains at the prototype stage.

105 mm MOBAT SPG

Details of the RDM Technology 105 mm MOBAT 9 (4 × 4) self-propelled gun are given in the Self-propelled guns and howitzers (wheeled) section. This remains at the prototype stage.

Manufacturer

RDM Technology bv.

UPDATED

RDM Technology 155 mm/52 calibre M109L52 self-propelled gun

Development

Late in 2000, RDM Technology started firing trials of its private venture M109L52 self-propelled artillery system fitted with 155 mm/52 calibre Extended Range Ordnance (ERO).

The new RDM Technology 155 mm/52 calibre M109L52 is aimed at existing users of the US United Defense M109 whose original development can be traced back well over 40 years.

RDM Technology has already entered into a co-operation agreement with the Hellenic Arms Industry (EBO) to offer not only the M109L52 but also the 105 mm M101 and 155 mm M114 towed artillery system upgrades to the Hellenic Army (qv Towed anti-tank guns, guns and howitzer section).

The new M109L52 is a logical development from work that RDM Technology has carried out on a variety of towed and self-propelled artillery systems over the last few years for both the home, and increasingly, for the international export market. Details of the M109 upgrades are given in the previous entry.

RDM Technology upgraded M109L52 fitted with a 155 mm/52 calibre ordnance in travel lock ***2001***/0106226

Description

The latest RDM Technology M109L52 upgrade has many improvements over the earlier M109L47 developed for the United Arab Emirates and is fitted with a new 155 mm/52 calibre ERO (Extended Range Ordnance).

For the new application the ERO has been fitted with a new breech ring, RDM Technology designed multi-slotted muzzle brake and a new breech mechanism. Also fitted is an automatic 10 round primer magazine with laser ignition being an alternative.

The existing M109 cradle and recoil system has been modified and a remote travel lock, operated by the driver, fitted as standard. This enables the system to come into action without one of the crew leaving the vehicle so reducing time and allowing the driver to remain under complete armour cover.

The maximum range of the M109L52 depends on the type of projectile and charge combination but is typically 40 km with an Extended Range Full Bore – Base Bleed (ERFB-BB) projectile.

To obtain a higher rate of fire and to reduce crew fatigue, an Austrian Intertechnik flick rammer and German Rheinmetall Landsysteme (previously KUKA) Ammunition Handing Kit (AHK) has been installed. The German Army has already taken delivery of 262 of the latter for its M109A3G systems. Burst rate of fire is claimed to be three rounds in 10 seconds. New propellant racks have also been installed.

The upgraded M109L52 has a number of other improvements including installation of the BAE Systems Laser Inertial Artillery Pointing System (LINAPS) of which 137 have been supplied for installation on 105 mm Light Guns of the British Army.

For trials purposes LINAPS has also been fitted to a number of other artillery systems including the BAE Systems, RO Defence 155 mm XM777 and the RDM Technology MOBAT 105 mm (4 × 4) system. Details of the latter are covered in the Self-propelled guns and howitzers (wheeled) section.

The installation of LINAPS provides the M109L52 with its precise location as well as a route navigation capability. It provides the crew with orientation and location data without the requirement for much of the existing sighting equipment. Time into action is reduced and accuracy increased. LINAPS also automatically compensates for gun cant and provides a continuous determination and display of gun barrel direction and orientation.

The turret traverse and weapon elevation system has been upgraded and a muzzle velocity radar mounted over the 155 mm/52 calibre ordnance. A laser range-finder for the direct fire role is also fitted.

Mounted on the bustle is an auxiliary power unit and the NBC/air conditioning system. The current 450 hp 8V-71T Detroit Diesel engine coupled to an Allison Transmission XTG-411-4A automatic transmission is retained but a more powerful diesel engine can be fitted if required to give the system a higher power-to-weight ratio.

The normal M109 has a crew of six people but for the upgraded M109L52 this has been reduced to four, commander (who also lays the weapon), assistant gunner, loader and driver.

The new 155 mm/52 calibre ordnance gives the M109L52 an enhanced range capability as well as faster response time which in turn improves its battlefield survivability against counter battery fire as it can come into action, complete its fire mission and redeploy before counter battery fire arrives.

RDM Technology have stressed their M109L52 is a modular upgrade with the customer only taking those parts that meet their operational requirements, or more likely, budget considerations.

Status
Development. Not yet in production or service.

Manufacturer
RDM Technology bv.

UPDATED

RDM Technology M109 with 155 mm/39 calibre ordnance

Development/Description
In association with Bofors Defence of Sweden (now owned by United Defense of the USA), RDM Technology has proposed a major upgrading of the M109 including the installation of a new 155 mm/39 calibre ordnance, a new gun mount, a semi-automatic loading and ramming mechanism, improved protection (mainly to the turret) and the installation of an automatic fire-control system.

Other options could include an ammunition supply trailer, installation of an air conditioning and NBC protection system, an automatic fire detection and suppression system and a remotely actuated 155 mm/39 calibre ordnance travel lock.

When fitted with the 155 mm/39 calibre ordnance, the upgraded M109 would fire a standard M107 high-explosive projectile to a maximum range of 18.1 km, an extended-range full-bore high-explosive projectile to a maximum range of 25.5 km and an extended-range full-bore base bleed projectile to a maximum range of 32.4 km.

The actual 155 mm/39 calibre ordnance used in this upgrade would be supplied by Bofors Defence which has already supplied 155 mm/39 calibre tubes for the M114/39 towed weapons. Details of this are given in the Towed guns and howitzers section.

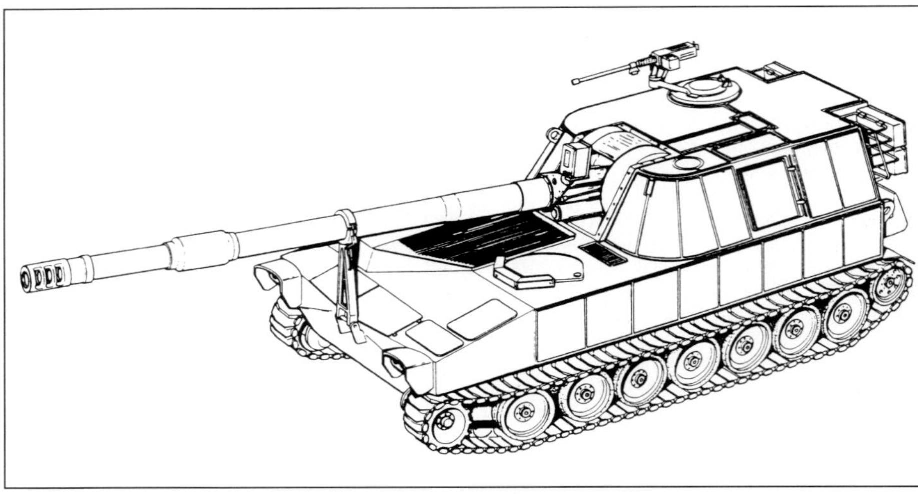

Drawing of M109 series self-propelled artillery system upgraded by RDM Technology with 155 mm/39 calibre barrel and additional armour

Status
By late 2001, no prototype of the M109 with 155 mm/39 calibre ordnance had been completed. It is possible that elements of this upgrade have been included in the M109 series upgraded for the United Arab Emirates. RDM Technology has more recently developed a 155 mm/52 calibre M109 and details of this are given in a separate entry.

Manufacturer
RDM Technology bv.

UPDATED

POLAND

HSW 155 mm/52 calibre Krab self-propelled howitzer

Development
Following an international competition, Poland selected a modified version of the UK's BAE Systems, RO Defence, 155 mm/52 calibre Braveheart self-propelled artillery turret system to meet its future self-propelled artillery requirement.

This 155 mm/52 calibre turret has been installed on a local chassis designed and built by OBRUM with HSW being the overall prime contractor for the complete system which is called the Krab.

By late 2001, two prototypes of the Krab had been designed and built for an extensive series of trials by the Polish Army. During these initial trials, 50 155 mm rounds were fired of which 32 were with the maximum number of modular charges.

South African modular charges will be used during the development phase, which has also been selected by the British Army for its upgraded AS90 155 mm/52 calibre self-propelled artillery systems. It is expected that user tests of the Krab will be completed by June 2002. Although the first two turrets were supplied by the UK, RO Defence has transferred the technical data pack to Poland so that production turrets will be manufactured in Poland.

If all goes to plan the Polish Army expects to field a single battalion of 155 mm/52 calibre Krab systems by 2007. This will be used with the Azalia artillery command, control, communications and intelligence system developed by WB Electronics.

Description
The chassis of the 155 mm/52 calibre Krab self-propelled howitzer is a new design that uses proven parts wherever possible and is of all-welded steel construction that provides the occupants with protection from small arms fire and shell splinters.

The driver is seated at the front of the vehicle on the left side and has a single-piece hatch cover in front of which is a single, wide-angle day periscope that can be replaced by a passive periscope for driving at night. Mounted on the forward part of the hull is a gun travel lock for the 155 mm/52 calibre ordnance, which is operated by the driver from within the vehicle.

Late in 2001 the HSW 155 mm/52 calibre Krab self-propelled howitzer commenced its first test firings
2002/0096185

The powerpack is located to the right of the driver with the air inlet and out louvres in the roof and the exhaust outlet on the right side of the hull so the exhaust gases blow downwards. The diesel fuel tanks are located under the floor of the vehicle.

The suspension is of the torsion bar type with either side consisting of seven, dual, rubber tyred roadwheels with the drive sprocket at the front and idler at the rear. There are also track return rollers and the upper part of the suspension is covered by a skirt to help reduce dirt.

The turret is positioned at the rear of the vehicle with the crew entering and leaving the vehicle via a large door in the hull rear.

The turret is almost identical to that installed on the AS90 Braveheart covered later in the section under the United Kingdom. Main difference in that it has a locally developed fire-control system, local 12.7 mm machine gun mounted on the left side of the turret roof and a bank of four 81 mm electrically operated smoke grenades either side of the turret, firing forwards.

According to HSW, the Krab has a burst rate of fire of three rounds in 10 seconds and a maximum rate of fire of 18 rounds in three minutes. A flick rammer is fitted as standard. Minimum range is 4.7 km and maximum range is 40 km. Aiming accuracy in range is 0.50 percent and in azimith 0.1 per-cent.

The gun commander is provided with a DD-9620 terminal which receives target information from the battery command post. Mounted over the 155 mm/52 calibre ordance is a muzzle velocity measuring radar that feeds information to the fire-control system. A navigation system is also fitted as standard.

In addition to the internal and external communication system, Krab has a fire-control system that includes a computer to control the turret mechanism, gun layer computer and a ballistic calculation and gun commander computer.

There is also a direct fire sight and a laser rangefinder mounted co-axial with the 155 mm armament. An auxiliary power unit is provided for the turret which allows all of the Krab systems to be run with the main diesel engine shut down.

Standard equipment on the Krab includes an NBC system, front-mounted dozer blade, fire/explosion detection and suppression system and a laser warning system that is coupled to the two banks of 81 mm electrically operated smoke grenade launchers mounted either side of the turret.

Specifications

Crew: 5:
Combat weight: 49, 680 kg
Power-to-weight ratio: 16.86 hp/tonne
Ground pressure: 0.55 kg/cm²
Length: 11.717 m
Width: 3.48 m
Height: 3.412 m
Ground clearance: 0.44 m
Max speed (road): 60 km/h
Range: 650 km
Fording: 1.0 m
Gradient: 60%
Slide slope: 40%
Vertical obstacle: 0.8 m
Trench: 2.8 m
Engine: diesel developing 838 hp
Transmission: manual
Armament:
(main) 1 × 155 mm/52 calibre
(air-defence) 1 × 12.7 mm machine gun
Ammunition:
(main) 60
Gun control equipment: electric/manual:
Turret traverse:360°
Gun elevation: +70/–3.5°
NBC system: yes
Night vision equipment: yes

Variants

There are no variants of this system although its chassis could be developed for a wide range of other applications.

Status

Two prototypes undergoing extensive trials.

Manufacturer

Huta Stalowa Wola SA.

UPDATED

ROMANIA

Model 89 122 mm self-propelled howitzer

Development/Description

In the late 1980s, Romania developed a 122 mm full-tracked self-propelled artillery system called the Model 89.

Since then it has been disclosed that the Romanian Army took delivery of about 42 of these systems which are in service alongside six of the Russian equivalent system, the 122 mm 2S1 self-propelled howitzer.

The turrets of the two systems are identical and it has been confirmed that the turrets for the Romanian vehicles were supplied by BETA plc of Bulgaria (which also manufactured the Russian MT-LB multirole tracked armoured vehicle) which has undertaken licence production of the 2S1 for the home and export markets for some time.

Additional details of Bulgarian production of the 122 mm 2S1 and the associated ACRV vehicle are given earlier in this section under Bulgaria. Prime contractor was the BETA.

The chassis of the Romanian Model 89 122 mm self-propelled howitzer is of a new design that incorporates parts of the Romanian MLI-84 Infantry Combat Vehicle (ICV), itself a Romanian equivalent of the older Russian BMP-1. Full details of this are given in the Armoured personnel carriers (tracked) section earlier in this volume. Like the Model 89 122 mm self-propelled howitzer, the MLI-84 has not been exported outside Romania.

Specifications

Crew: 5
Combat weight: 17,500 kg
Power-to-weight ratio: 20.5 hp/t
Length: 7.305 m
Width: 3.15 m
Height: 2.665 m
Max speed:
(road) 65 km/h
(water) 7 km/h
Fuel capacity: 600 litres
Max range: 500 km
Fording: amphibious
Gradient: 60%
Side slope: 40%
Vertical obstacle: 0.80 m
Trench: 2.50 m
Engine: 1240 V8DTS supercharged V-8 diesel developing 360 hp
Transmission: manual
Steering: clutch and brake
Suspension: torsion bar
Armament:
(main) 1 × 122 mm 2A31 howitzer
Turret power control: electric/manual
Gun elevation/depression: +70/–3°
Turret traverse: 360°
NBC system: yes
Night vision equipment: yes

The turret and hull of the Model 89 are of all-welded steel armour construction which provides protection from small arms fire and shell splinters. The driver is seated at the front of the vehicle with the power pack to the right and the turret mounted at the very rear. The air inlet and air outlet louvres are in the roof.

There is a single door in the rear of the hull with a firing port and associated vision device.

The turret is armed with a 122 mm 2A31 howitzer which can fire a wide range of separate loading ammunition types, with the standard OF-462 high-explosive projectile having a maximum range of 15,300 m; firing an HE rocket-assisted projectile a range of 21,900 m can be achieved. The ordnance has a semi-automatic vertical sliding wedge breech block, a double-baffle muzzle brake and a fume extractor.

Like the 122 mm 2S1, there is a long stowage box on the left side of the turret and carried on the turret rear are slotted covers that cover the forward part of the suspension when afloat to improve water flow.

The torsion bar suspension either side consists of seven single rubber-tyred roadwheels with the drive sprocket at the front, idler at the rear and three track-return rollers.

Romanian 122 mm Model 89 self-propelled howitzer showing the water-jets at the rear

Romanian 122 mm Model 89 self-propelled howitzer

The major difference between the two systems is the improved amphibious capabilities of the Romanian vehicle, as it is propelled in the water by two water-jets mounted one either side of the entry door in the hull rear. Steering is accomplished by opening and closing the water-jets. As the Romanian system has a more powerful diesel engine its power-to-weight ratio is also superior to the Russian system.

The Romanian system is distinguishable from the Russian 2S1 system as the former has seven roadwheels and three track-return rollers whereas the 2S1 has seven roadwheels and no track-return rollers.

Standard equipment includes an NBC protection system of the overpressure type, infra-red night vision equipment for commander and driver, an R-123 radio and an R-124 intercom system.

Status

Production complete. In service with Romania. There are no known exports of this system.

Manufacturers

Romanian state factories (chassis and vehicle integration).

Bulgarian state factories (turret).

UPDATED

RUSSIAN FEDERATION AND ASSOCIATED STATES (CIS)

203 mm self-propelled gun M-1975 (SO-203) (2S7)

Development

In December 1967, the requirement for a new large calibre longer range self-propelled artillery weapon was issued although at this time neither the type of mounting nor the calibre of the weapon was specified. The key requirement was for a maximum range of 25 km.

Various calibres were studied including 180 mm and 210 mm but, in 1969, 203 mm was selected. Late in 1969, the Leningrad Kirov Factory proposed the chassis of the T-64 MBT with a 203 mm gun in an open mount while the Barrikady Factory proposed a model using the Obiekt 429 chassis with the 203 mm gun also in an open mount.

The Pion (Peony) chassis was developed by the Kirov Factory under the designation of the Obiekt 216, with the gun and mount being developed by the Barrikady plant. The latter company has more recently developed the 180 mm Barrikady 130 mm Bereg self-propelled (8 × 8) coastal artillery system covered in detail later in this book. As of late 2001, it is understood that this system has yet to enter production.

Following the usual trials and modifications with prototype systems, it was accepted for service as the 203 mm 2S7 Pion (Peony) although it is also referred to as the Malka or the SO-203. In the West it has been referred to as the M-1975 as this is the year when the system was first observed.

The former Soviet ground forces began receiving the 203 mm Model 1975 (M-1975) self-propelled gun during 1975. It is estimated that well over 1,000 of these systems have been completed, although the Russian Federation has not released any production figures.

It is believed that a 2S7 artillery regiment has a total of 24 weapons in three batteries, each with eight weapons.

Production was completed some time ago and it is no longer being marketed. In 1999 Russia was offering an upgrade to extend the range of the 203 mm projectile.

203 mm M-1975 (SO-203) (2S7) self-propelled gun in travelling configuration, from the front (Ted Dyke)
0058379

Description

The chassis of the 2S7 is of all-welded steel armour construction which provides the occupants with protection from small arms fire and shell splinters. It is divided into four compartments: driver's, engine, crew and rear compartment.

When travelling, the commander, gunner and driver/mechanic are seated in the driver's compartment at the front of the vehicle, which is also provided with space to carry a Russian Strela-2 man-portable SAM system.

The commander and driver are each provided with a circular roof hatch and in front of these are day periscopes for forward observation. To their immediate front is a windscreen that, when in a combat area, is covered by an armoured shutter hinged at the top. In addition, there is a single forward-facing day periscope in the forward part of the roof between the commander and driver.

In the left side of the cab towards the rear is an upward-opening hatch cover while in the right side there is a rearward-opening hatch. It is probable that both of these give access to subsystems of the 2S7 rather than the crew compartment.

Mounted under the nose of the vehicle is a dozer blade that can be used to clear obstacles or to provide a firing pit for the weapon.

The Model V-46-I V-12 liquid-cooled diesel engine is located to the rear of the crew compartment cab and is coupled to a manual transmission with eight speeds. The exhaust outlets are either side of the hull top.

203 mm M-1975 (SO-203) (2S7) self-propelled gun in travelling configuration, from the rear (Richard Stickland)
0003993

To the rear of the engine is another crew compartment for the remaining four crew members who enter the vehicle via two circular roof hatches.

On top of the roof compartment is the antenna base for the R-123M radio system, with the later 2S7M probably having the more recent R-173. A 1V116 crew communication system is fitted as standard.

Along either side of the hull is a series of large stowage boxes with the 203 mm gun, which is not fitted with a fume extractor or a muzzle brake, being mounted at the rear.

The 203 mm ordnance is designated the 2A44 and consists of a composite tube, casing, mount, breech ring and sleeve. The breech mechanism is of the two-cycle screw type with the breech block being opened to the right automatically with the aid of a mechanical drive. In an emergency it can be operated by hand.

The 203 mm barrel has an overall length of 56.2 calibres and weighs a total of 7,800 kg and has a barrel life of about 450 rounds. It is not fitted with a fume extractor or muzzle brake.

Recoil braking is hydraulic with recuperation being pneumatic. Total recoil length is a maximum of 1.4 m and, to provide a more stable firing platform, the rear idler wheels are locked, as are the hydraulic shock-absorbers in the suspension.

Mounted at the very rear of the 2S7 is a large recoil spade which, when lowered to the ground, provides stability at all angles and charges. If required, the 2S7 can fire at reduced charges and at low angles of elevation while still in the travel mode and without the spade lowered to the ground.

When travelling, the 203 mm ordnance is held in position by a manually operated travel lock mounted on top of the cab. Elevation is via a sector-type drive, with traverse being accomplished by a screw gear drive. Gun elevation, traverse, loading and operation of the spade are all hydraulic with manual controls for emergency use.

The gun operator is seated at the rear of the vehicle on the left side and for the engagement of targets has a standard PG-1M panoramic day telescope that is used in conjunction with the K-1 collimator. For engaging targets in the direct fire mode he uses the day OP4M-87 telescope.

Ammunition is of the separate loading type, projectile and charge, with a maximum muzzle velocity of 960 m/s. Maximum range, using unassisted ammunition, is 37.5 km.

The standard HE round is designated the ZOF 43 and weighs 110 kg, with a total of four projectiles and charges being carried on the 2S7 for immediate use. The remainder of the ammunition load is carried by another vehicle, usually a truck. In addition, there is a rocket-assisted high-explosive projectile that weighs 103 kg and has a maximum range of 47,500 m.

Full details of the ammunition are shown in the table.

It is understood that other natures of ammunition were developed including concrete-piercing, tactical nuclear and chemical but the last two are understood to be no longer in front-line service.

Ammunition is brought up to the 2S7 using a two-wheeled hand cart which consists of a frame with two wheels and a carrier. The carrier separates from the frame when loading the projectile from the ground and carries the latter up to the rammer. If the hand cart is not used then an additional six men are required to move the ammunition.

Typical rates of fire can be summarised as: 8 rounds in 5 minutes; 15 rounds in 10 minutes; 24 rounds in 20 minutes; 30 rounds in 30 minutes; 40 rounds in 60 minutes.

Completely combustible bag charges are used in the 2S7 and, as in the case of other Russian weapons, the choice of charges is regulated and these range from the largest fixed charge, weighing almost 45 kg, to the increments which weigh from 11.8 to 24.9 kg.

The 2S7 is provided with an ammunition handling system that enables a rate of fire of 2 rds/min to be achieved, this being operated by the loader.

The torsion bar suspension has seven dual rubber-tyred roadwheels either side, with the drive sprocket at the front and the idler at the rear, and six track-return rollers that support the inside of the track only. Shock-absorbers are provided for the first, second, third, sixth and seventh roadwheel stations. When deployed in the firing position, the chassis is lowered at the rear to help provide a more stable firing platform.

The 2S7 takes 5 to 6 minutes to come into action and 3 to 5 minutes to come out of action.

Standard equipment includes an NBC system of the overpressure type, heater and night vision equipment. A 24 hp 9R4-6U2 auxiliary power unit is provided for the 2S7 so the main turbo charged engine does not have to be kept running as a power supply during firing.

Specifications

Crew: 7
Combat weight: 46,500 kg
Power-to-weight ratio: 18.06 hp/t
Ground pressure: 0.80 kg/cm^2
Length:
(overall) 13.12 m
Width: 3.38 m
Height: 3.0 m
Ground clearance: 0.4 m
Engine: Model V-46-I V-12 liquid-cooled 4-stroke turbocharged diesel developing 840 hp
Max road speed: 50 km/h
Max range: 650 km
Fording: 1.2 m
Gradient: 40%
Side slope: 20%
Vertical obstacle: 0.7 m
Trench: 2.5 m (approx)
Armament:
(main) 1 × 203.2 mm 2A44 gun
Ammunition: 4 × 203.22 mm
Gun control equipment: powered/manual
Gun elevation/depression: 0 to +60°
Gun traverse: 15°, left and right
Armour: 10 mm (max, estimate)
NBC system: yes
Night vision equipment: yes

Type	Projectile weight	Filler weight	Charge weight	Muzzle velocity	Max range
HE-FRAG	110 kg	17.8 kg	44 kg	960 m/s	37,500 m
RAP	103 kg	13.8 kg	44 kg	n/avail	47,500 m

Variants

The final production model of the 2S7 version is designated the 2S7M which has more recent R-173 communications equipment and can carry a total of eight 203 mm projectiles and charges. In addition, rate of fire is increased from 1.5 to 2.5 rds/min, durability of the system has been improved and firing data can be transmitted directly to the gun.

Status

Production complete. In service with:

Country	Quantity	Comment
Belarus	48	
Poland	8	
Russia	80	may be higher number
Ukraine	100	
Uzbekistan	48	

Manufacturer

RFAS state factories.

UPDATED

152 mm self-propelled artillery system 2S19 (MSTA-S)

Development

The 152 mm 2S19 self-propelled artillery system was developed under the codename MSTA-S from the 1980s at the Uraltransmash facility, as the replacement for the older 152 mm 2S3 and 152 mm 2S5 self-propelled artillery systems.

Following extensive trials, the 2S19 was accepted for service with the former Soviet Army in 1989. Initial production was undertaken by the Uraltransmash facility before being transferred to the Sterlitamak Machine Construction Factory in Bashkiriya. First production systems were completed in 1989. In the Russian Army the 2S19 is normally deployed in batteries of six guns.

The 2S19 is also referred to as the MSTA-S with the S standing for *Samokhodnyj*, or self-propelled. The 2S19 uses the same ordnance as the 152 mm 2A65 towed gun-howitzer which is referred to as the MSTA-B with the B standing for *Buksiruemyj*, or towed.

It is estimated that by late 2001 total production of the 2S19 amounted to well over 500 systems.

The 2S19 has been demonstrated in the Middle East but as of late 2001 there were no known exports outside the Russian Federation.

In the Russian Army service, this system is also sometimes called the Ferma (Farm) because of the large side of the turret.

Description

The chassis of the 2S19 is based on passive MBT assemblies, with the suspension and running gear from the T-80 MBT and the power pack from the T-72 MBT.

The 2S19 has the driver's compartment at the front, turret in the centre and power pack at the rear.

The driver has three day periscopes for forward observation, the centre one of which can be replaced by a passive night vision periscope, and a single-piece hatch cover that opens to the right. On the glacis plate is a V-shaped splashboard to stop water rushing up the glacis plate and into the driver's compartment when the vehicle is fording.

The hull and turret is of all-welded steel armour construction which provides protection from small arms fire, shell splinters and mines.

Mounted under the nose of the vehicle, similar to the T-72/T-80 MBTs, is a dozer blade that can be used to prepare fire positions or clear obstacles without special engineer support.

152 mm 2S19 from the rear with projectile transfer arm deployed (Christopher F Foss)

152 mm 2S19 self-propelled gun deployed in firing position with weapon elevated and loading chute at rear deployed
0058380

The suspension is of the torsion bar type with either side having six roadwheels, idler at the front, drive sprocket at the rear and five track-return rollers, with the upper part of the track being covered by a skirt. The roadwheels, idler wheels and drive sprocket are the same as those of the T-80 MBT. The torsion bar suspension is from the T-80, the first, second and sixth roadwheels have a telescopic shock-absorber and these are locked when the weapon is fired. The track is the same as that used on the T-80 and 2S7.

The gunner is seated in the left side of the turret and operates the roof-mounted panoramic sight and the direct aiming sight, which is mounted in the front left side of the turret.

The vehicle commander is seated on the right and operates the roof-mounted 12.7 mm NSVT machine gun (which can be remotely controlled) and the searchlight (which can be remotely controlled from under full armour protection). The 12.7 mm NSVT machine gun can be used to engage both ground and air targets. Mounted externally on the right side of the turret are boxes of 12.7 mm NSVT machine gun ammunition for ready use. This weapon can be elevated to +70° with a traverse of 9° left and 255° right.

There are three roof hatches, one for each member of the turret crew.

Main armament comprises a long-barrelled 152 mm gun, the 2A64, fitted with a fume extractor and a muzzle brake. When the 2S19 is travelling, the ordnance is held in position by a travelling lock mounted on the glacis plate.

The 152 mm 2A64 main armament of the 2S19 is manufactured by the Barrikady State Production Association

The 152 mm 2A64 gun fires an HE-FRAG (high-explosive fragmentation) round designated OF-45 to a maximum range of 24.7 km. Other types of projectile include OF-61 base bleed high-explosive with a maximum range of 28.90 km, 30-23 cargo containing 42 HEAT bomblets, 3NS30 jammer projectile with a maximum range of 20 km, smoke, and the Krasnopol laser-guided projectile covered later. Empty brass plastic cartridge cases are automatically ejected forwards out of the turret just below the 152 mm ordnance.

Turret traverse is powered through a full 360° with weapon elevation from −4° to +68°. The electric/mechanical turret control system is designated the 2Eh46. Manual elevation and traverse controls are also provided.

When using onboard ammunition, the 2S19 has a rate of fire of 7 to 8 rds/min. The fuzed projectile is loaded automatically and the charge semi-automatically.

When using ammunition from an external source this is reduced to 6 to 7 rds/min. When using ammunition on the ground, an extra two crew members are required to load the ammunition onto the conveyor, one for the projectile and one for the charge.

Additional details of the ammunition are given in the table.

The 2S19 can also fire all of the ammunition natures used by the older 2S3 (self-propelled) and D-20 (towed) artillery systems.

In addition to selecting the projectile from the magazine, the automatic loading mechanism controls the number of rounds to be fired and has built-in test equipment. Laying in elevation is also automatic with laying in traverse being semi-automatic.

In total, 50 155 mm projectiles and charges are carried. Spades are not required by the 2S19 as it is very stable in the firing position.

While the system is firing, the 2S19 can be reloaded by transporters through the turret rear with the help of special mechanisms of feeding from the ground. This enables the system to move off to its next fire position with a full load of ammunition on board.

The 152 mm 2S19 is fitted with a 1P22 indirect sight with a magnification of ×3.7 which is stabilised in the horizontal plane and a 1P23 direct fire sight.

Three Type 902 81 mm smoke grenade dischargers are mounted either side of the turret.

A 16 kW autonomous AP-18D gas-turbine Auxiliary Power Unit (APU) is provided in the turret and this provides power within 30 to 60 seconds of being switched on. This allows the system to be fully operational at temperatures ranging from −50 to +50°C with the main diesel engine being switched off to conserve fuel supplies.

152 mm 2S19 self-propelled gun in travelling configuration **2002**/0126557

Specifications
Crew: 5
Weight: 42,000 kg
Power-to-weight ratio:
(with 780 hp engine) 18.57 hp/t
(with 840 hp engine) 20 hp/t
Length:
(overall) 11.917 m
Width:
(over tracks) 3.38 m
(over skirts) 3.584 m
Height:
(turret roof) 2.985 m
Ground clearance: 0.435 m
Track width: 580 mm
Track: 2.80 m
Length of track on ground: 4.704 m
Max road speed: 60 km/h
Range: 500 km
Fording:
(without preparation) 1.2 m
(with preparation) 5 m
Gradient: 47%
Side slope: 36%
Vertical obstacle: 0.5 m
Trench: 2.8 m
Engine: V-12 model V-84A multifuel developing 780 hp (or 840 hp diesel)
Transmission: mechanical, 7 forward and 1 reverse gears
Suspension: torsion bar
Armament:
(main) 1 × 152 mm 2A64 gun
(anti-aircraft) 1 × 12.7 mm NSVT MG
Smoke grenade dischargers: 2 × 3 81 mm, can also lay smoke screen by injecting diesel fuel into exhaust outlet
Ammunition:
(152 mm) 50
(12.7 mm) 300

Gun control equipment
Turret power control: electric/mechanical
Gun elevation/depression: +68/−4°
Turret traverse: 360°
Electrical system: 27 V
Batteries: 4
NBC system: yes
Night vision equipment: yes

Projectile designation	Projectile weight	Filler weight	Muzzle velocity	Max range
OF-45, HE	43.56 kg	7.65 kg	810 m/s	24,700 m
OF-61, HE-RAP	42.86 kg	7.80 kg	828 m/s	28,900 m
30-23, cargo	42.80 kg	-	-	29,000 m
3NS30, jammer	43.56 kg	-	-	20,000 m
3OF39, LGP Krasnopol	50.00 kg	-	-	20,000 m

Standard equipment for the 2S19 includes the 1V116 intercom system, R-173 radio set and PPO three-way automatic fire extinguishing system, passive night vision equipment for the driver, two filter and ventilation systems and the ability to lay a smoke screen by injecting diesel fuel into the exhaust outlet on the left side of the hull.

155 mm 2S19M

In mid-2000 it was confirmed that two new versions of the 152 mm 2S19 (MSTA-S) self-propelled artillery system had been developed.

The first version, the 152 mm 2S19M, has been developed to meet the requirements of the Russian Army and has the same physical characteristics and is armed with the same 152 mm/47 calibre ordnance as the original 2S19.

The most significant feature of the 2S19M is the installation of the ASUNO Automatic Gun Fire-Control System (AGFCS). This features an onboard ballistic computer which, used in conjunction with the onboard satellite navigation system and automatic gun laying systems gives a faster reaction time and increase in target accuracy.

Data can be stored for up to 10 fire missions. The system is ready for action within 3 minutes of coming to a halt and can open fire 30 seconds after receipt of the target co-ordinates.

155 mm/52 calibre 2S19M1

As an increasing number of eastern countries are now moving away from 152 mm to 155 mm artillery weapons, since 1995 Uraltransmash has been working on 155 mm artillery systems, towed and self-propelled.

The 2S19M1, development of which was expected to be completed in 2001, also has the same ASUNO AGFCS as the 2S19M but is fitted with a NATO standard 155 mm/52 calibre ordnance for which a total of 46 projectiles and charges are carried.

Maximum quoted range, firing a NATO 155 mm L15A1 high-explosive projectile is quoted as 30 km while firing an Extended Range Full Bore - Base Bleed (ERFB-BB) projectile range is 41 km. Firing a 155 mm M107 high-explosive projectile, a maximum range of 22.6 km can be achieved, while firing Russian 155 mm high-explosive projectiles, a range of 30 km can be achieved. More advanced, smart projectiles such as the French/Swedish 155 mm BONUS top attack projectile can also be fired, according to the Russians.

According to Russian sources, their future 155 mm artillery systems, towed and self-propelled, could have chromed barrels for longer life and advanced cooling systems for enhanced rates of fire.

In addition to ASUNO AGFCS already mentioned, the system can also be used in conjunction with the Falstet-M automated battery/battalion fire-control system.

This includes a number of sub-systems including battery executive officer control vehicle, Kredo-1E portable ground reconnaissance radar, Zoopark-1 reconnaissance and fire-control complex, battalion commander control vehicle, battery commander control vehicle and an observation helicopter.

In the past this system has been referred to as the 155 mm/52 calibre 2S19M1 but, more recently, it has been referred to as the MSTRA-S-155 self-propelled howitzer with 52 calibre barrel.

Krasnopol guided projectile

Designated the 9K25, the 152 mm Krasnopol laser-guided projectile can be fired by the 152 mm 2S19 and 2S3M self-propelled artillery systems as well as the towed 152 mm 2A36, 2A65 and D-20 systems.

During extensive firepower and mobility demonstrations in the Middle East in 1993, the 2S19 fired a total of 40 Krasnopol laser-guided projectiles at targets up to 15 km away. Of these, 38 hit the target.

When the Krasnopol is in flight, four fixed fins unfold to the rear with four movable control surfaces unfolding to the front. It uses inertial guidance for the middle part of the trajectory with semi-active laser homing being used for the terminal phase when the target is illuminated by the tripod-mounted laser designator.

The tripod-mounted laser designator system is called the 1D22 and can be used to designate all of the laser-guided artillery weapons in the Russian inventory and under development.

Late in 1999 it was reported that China had started licensed production of the 152 and 155 mm versions of Krasnopol. It was also stated India has taken delivery of 1,000 155 mm Krasnopol-M projectiles and 10 laser guidance systems from the KBP Instrument Design Bureau.

Status

In production. In service with:

Country	Quantity
Belarus	13
Russia	600
Ukraine	40

Manufacturers

Uraltransmash, Ekaterinburg.
Sterlitamak Machine Construction Factory, Bashkiriya.
Design authority is the Uraltransmash at Ekaterinburg (former Sverdlovsk) with production being undertaken at both the Uraltransmash and Sterlitamak Construction Factory in Bashkiriya in parallel.

UPDATED

152 mm self-propelled gun Giatsint (2S5)

Development

Development of the 152 mm 2S5 Giatsint self-propelled artillery system commenced at the Special Design Bureau of the Sverdlovsk (now Ekaterinburg) Machinery Construction Factory late in 1968.

The original intention was to develop two weapons, towed (Giatsint-B) and self-propelled (Giatsint-S), both of which would have the same 152 mm ordnance and fire a new family of ammunition.

While the Perm Machinery Construction Plant concentrated on the actual weapon, the Sverdlovsk Transport Machinery Plant concentrated on the chassis (which is similar to that used for the 2S3 covered in the following entry), and NIMI the ammunition.

By late 1969, the user was presented with various designs and the version with an open mount was finally selected. Full-scale design work commenced in mid-1970, and in early 1972 prototypes of the self-propelled and towed versions were presented.

There were at least two gun designs, the 2A37 ordnance used in the Giatsint-S used separate loading ammunition while the 2A43 Giatsint-BK used a bagged charge loading system. In the end it was decided to use a weapon with separate loading ammunition (projectile and charge) and the 2A37 ordnance.

Production of the 152 mm 2S5 self-propelled system, which became simply the Giatsint-S, commenced in 1976 and was followed by the 152 mm towed model, the 2A36 which is covered in detail in the Towed anti-tank guns, guns and howitzers section.

Specifications

Crew: 5
Combat weight: 28,200 kg
Power-to-weight ratio: 18.50 hp/t
Length: 8.33 m
Width: 3.25 m
Height: 2.76 m
Ground clearance: 0.45 m (approx)
Max road speed: 63 km/h
Max road range: 500 km
Fording: 1.05 m
Gradient: 58%
Side slope: 47%
Vertical obstacle: 0.7 m
Trench: 2.5 m
Engine: V-59 V-12 4-cycle, liquid-cooled, direct-injection, multi-fuel, diesel supercharged developing 520 hp
Armament:
(main) 1 × 152 mm 2A37 gun
(extra) 1 × 7.62 mm PKT MG
Ammunition:
(main) 30 × 152 mm

Gun control equipment (electric/manual)
(traverse) 15° left and right, manual
(elevation/depression) +57/–2°, manual
NBC system: yes
Night vision equipment: yes

It is normally deployed in batteries of six weapons, with each battalion having three batteries. The 2S5 is commonly known as the Giatsint (Hyacinth) by the Russian Army.

Production of the 2S5 was completed some time ago and the facility is now building the more recent 152 mm 2S19 which is covered in detail earlier in this section.

Some sources have stated that total production of the 152 mm 2S5 could have amounted to 2,000 units. It will eventually be replaced by the 2S19. The development designation of the 2S5 was the Obiekt 307.

Description

The chassis of the 2S5 is of all-welded steel armour construction that is believed to have a maximum thickness of 13 mm, providing the crew with protection from small arms fire and shell splinters.

As with a number of former Russian Army vehicles, including the 152 mm 2S3 self-propelled artillery system and the T-72 MBT, a dozer blade is mounted under the nose of the 2S5. This is used for clearing obstacles and preparing firing positions without specialised engineer support.

The driver is seated at the front of the vehicle on the left and has a single-piece hatch cover that opens to the rear. In front of the hatch cover are two day periscopes, one of which can be replaced by a passive periscope for night driving.

The vehicle commander is seated in a raised superstructure to the rear of the driver and has a cupola that can be traversed through 360°. Mounted externally

152 mm 2S5 self-propelled gun deployed in firing position 0058387

Close-up of 152 mm 2S5 from the rear showing gunlayer on the left and ordnance in elevated position 0018889

152 mm 2S5 self-propelled guns deployed in the firing position **2000**/0085204

152 mm self-propelled gun 2S5 with ordnance in travel lock and base plate at rear in raised position (Richard Stickland) **2000**/0085195

on the forward part of the cupola is a 7.62 mm PKT machine gun that can be operated by remote control, with an IR white light searchlight mounted on the left side. The commander is provided with day periscopes and other viewing devices.

The power pack is to the right of the driver with the air inlet and air outlet louvres in the roof. The engine will run on diesel or aviation fuel. The engine is coupled to a mechanical stepped transmission with constant-mesh gears, two-power flow, four-gearshift, six forward and two reverse gears.

The other three crew members are seated in the crew compartment at the rear of the hull and enter and leave via a ribbed ramp in the rear. The rear crew compartment is provided with roof hatches and roof-mounted day periscopes to give observation to the sides of the vehicle.

Suspension is the torsion bar type with each side having six dual rubber-tyred roadwheels, the drive sprocket at the front, idler at the rear and four track-return rollers. The first two and last two road stations either side are provided with shock-absorbers.

The 152 mm 2A37 ordnance consists of a monobloc tube, breech ring and the five-part multislotted muzzle brake which is screwed to the ordnance and has an efficiency of 55 per cent. When travelling, the ordnance is held in position by a travelling lock. The breech mechanism is the semi-automatic horizontal loading type.

Recoil braking is hydraulic while recuperation is pneumatic. The recoil component cylinders are attached to the barrel and recoil with it; length of recoil is 730 to 950 mm. Elevation and traverse of the weapon is by sector drives while the balancing system is of the pneumatic rod type with an extra spring mechanism.

The gunlayer is seated to the left of the gun and is provided with a shield to his immediate front only. The weapon is aimed using the same sight as installed on the Russian 122 mm BM-21 (40-round) artillery rocket system. This mechanical sight is used with the PG-1M panoramic head sight; direct fire sight is the OP4M-91A. The gun layer operates the loading system and fires the gun.

When deployed in the firing position, a large spade is lowered to the ground to provide a more stable firing platform. Time into and out of action is not more than two minutes.

Ammunition is of the separate loading type, projectile and charge, and crew fatigue is reduced by means of a semi-automatic loading system. This is operated by remote control with the operator normally being positioned on the left side of the 2S5 on the ground.

The system consists of an electrically driven chain rammer located to the left of the breech which, when not required, folds back through 90° so that it is parallel to the breech, and the projectile and charge loading system.

The charge loading system is pivoted on the right side and has a projectile tray and a charge tray. The projectile and charge are loaded into the trays and then swung upwards through almost 90° where the rammer first rams the projectile and then the charge.

The 2S5 can be supplied with ammunition either from on board via the rear of the vehicle or from ammunition on the ground.

In total, 30 projectiles and charges are carried, with the projectiles being stowed vertically in a carousel device in the left side of the rear compartment with the 30 charges to the right in three rows of 10. Each row of 10 charges is also stowed vertically and they are on a horizontal conveyor belt that returns under the floor of the vehicle.

The charge consists of a conventional cartridge case containing the actual charge and, once fired, the breech automatically opens and the spent cartridge case is ejected.

Maximum range, firing conventional ammunition with a muzzle velocity of 942 m/s, is 28,400 m, while a range of 37,000 m may be achieved using HE projectiles with an advanced aerodynamic form.

The normal high-explosive fragmentation projectile weighs 46 kg, with propellant and cartridge weighing a maximum of 34 kg. Other types include concrete-piercing and improved conventional munition. A unit of fire is 30 rounds.

The mechanical ammunition-handling system enables a maximum rate of fire of 5 to 6 rds/min to be achieved and, according to Russian sources, a battery of 2S5s can have 40 projectiles in the air before the first projectile lands on the enemy position.

Standard equipment for the 2S5 includes an NBC system and infra-red night equipment for the driver and commander.

Status

Series production complete. In service with:

Country	Quantity	Comment
Belarus	120	
Finland	18	called Telak 91 (delivered in 1994)
Georgia	1	
Russia	1,000	estimate
Ukraine	24	
Uzbekistan	n/av	unconfirmed user

Manufacturer

Uraltransmash Federal State Unitary Enterprises.

UPDATED

152 mm self-propelled gun-howitzer 2S3 Akatsiya (M1973)

Development

The requirement for the 152 mm 2S3 Akatsiya (Acacia) self-propelled artillery system was drawn up in July 1967 with Uralmash being responsible for the 152 mm weapon and UZTM in Sverdlovsk responsible for the chassis.

The 152 mm ordnance of the 2S3 was ballistically identical to that used for the 152 mm D-20 towed artillery system and used the same ammunition. The new 152 mm gun had the factory designation of the D-22 and the industrial designation of the 2A33.

The chassis was developed under the designation of the Obiekt 303 and was a further development of the Obiekt 123 used for the Krug (SA-4 Ganef) surface-to-air missile system. Details of this system are given in *Jane's Land-Based Air Defence 2001-2002*. The chassis was upgraded in many areas including the installation of a more powerful 520 hp diesel engine. The track and suspension were also upgraded.

The first two prototypes of the 2S3 were completed by late 1968 and testing continued through to 1969 but problems were encountered with contamination in the turret so an additional four improved systems were built. The system was accepted for service in December 1971 although production had commenced the previous year. Until its correct designation became known it was referred to in the West as the 152 mm M1973. The M1973 indicates the year that it was first observed.

It is issued on the scale of 18 per tank division and 18 per motorised rifle division and has now replaced the 130 mm Field Gun M-46 in the former Russian Army artillery brigades and artillery divisions. The 2S3 is normally deployed in batteries of six weapons, with three batteries making up a regiment.

It is estimated that well over 10,000 of these systems were manufactured, although no production figures have been released.

Production of the 2S3 was completed some years ago and its replacement is the more recent 152 mm self-propelled artillery system 2S19. Due to the reduction in the overall size of the Russian Army, numbers of weapons are becoming available for sale including quantities of artillery systems such as the 2S3. United Nations sources state that Germany transferred four 2S3 systems to the USA in 1993.

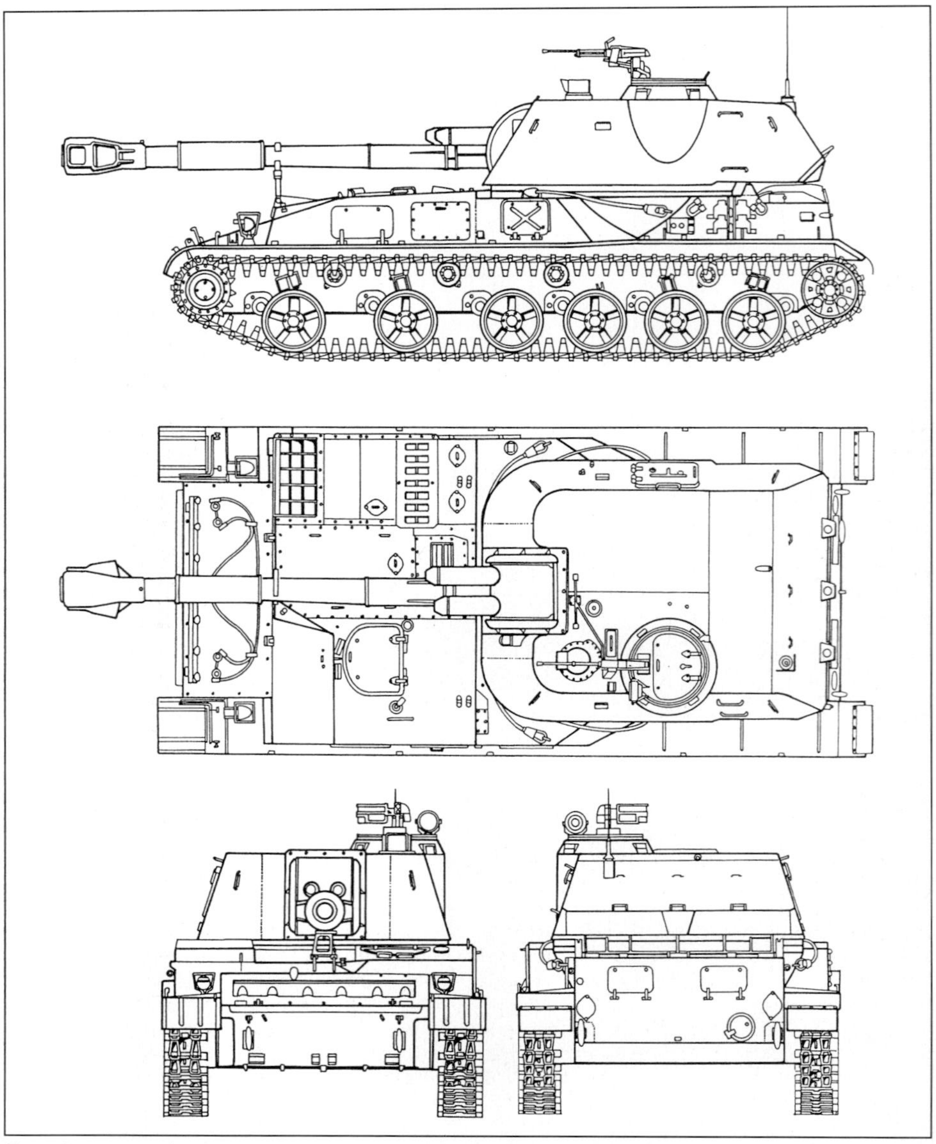

Four-view general arrangement drawing of the 2S3M Akatsiya 152 mm self-propelled howitzer (Steven Zaloga)

152 mm 2S3M1 self-propelled gun-howitzer with ordnance in travelling lock 0058385

155 mm 2S3M3 with a 155 mm/39 calibre ordnance fitted with a fume extractor and a large muzzle brake to reduce recoil forces **2001**/0034535

The chassis of the 152 mm 2S5 self-propelled gun shares many common components with the 2S5 chassis and is also referred to as the Izdeliye 303.

According to the United Nations Arms Transfer register, only two countries took delivery of the 2S3 between 1992 and 2000. Bulgaria supplied 4 for Angola in 1999 while Germany supplied 4 to the US in 1993.

Description

The all-welded steel hull of the 2S3 is divided into three compartments: driver's at the front on the left, engine to the right of the driver and the turret at the rear. The driver has a single-piece hatch cover that opens to the rear. In front of this are two day periscopes, the left one of which can be replaced by an infra-red device (or, more recently, a passive device) for driving at night. The driver has an adjustable seat and under this is an emergency hatch. On the forward part of the glacis is a splashboard to stop water rushing up the glacis plate when the vehicle is fording a stream. The air-inlet and air-outlet louvres are on the top of the hull, with the exhaust outlet on the right side of the hull, just above the track guard.

The large all-welded turret has a sloped front and well-sloped sides with a vision block in each side. The commander is seated on the left of the turret with the gunner forward and below him and the loader on the right. The commander has a cupola which can be traversed through a full 360° and a single-piece hatch cover opening to the rear. Vision devices are provided around the lower part of the cupola. Including a TKN-3A device. Mounted on the forward part of the hatch is a 7.62 mm PKT machine gun which can be aimed and fired from inside the turret. There is an OU-3K infra-red/white light searchlight mounted to the left of the 7.62 mm machine gun.

Gunlaying, including firing, is accomplished by the gunner. In addition to the PG-4 sight there is an OP5-38 gunner's telescope. Cants of up to 5° in pitch and roll axes can be offset.

The panoramic telescope has a 10.5° field of view and ×3.7 magnification. The OP5-38 gunner's telescope has an 11° field of view, a magnification of ×5.5 and is normally used for direct fire engagements.

Turret traverse and weapon elevation/depression are powered, with manual controls for emergency use.

The loader feeds the projectiles and charges to the ordnance via a loading tray with a chain-driven rammer used to seat the projectile and then the charge in the chamber.

The loading mechanism is lowered when the gun is fired. The breech opens automatically.

In the rear half of the hull are 33 projectiles, complete with fuzes, which are stowed in three horizontal layers.

On the 2S3M/2S3M1 is a projectile carousel which holds an additional 12 projectiles in the vertical position.

As each round is fired the carousel automatically revolves 30° so positioning the next projectile.

Stowed below the projectiles are 16 charges in two layers of eight with an additional eight being stowed near the loader's position and more being stowed in brackets in the vehicle. An ammunition resupply hatch is provided in the right side of the turret and there is an oval hatch at the rear of the hull which opens downwards.

Wherever possible, ammunition would be used from outside the 2S3 with the onboard ammunition supply being retained ready for immediate use.

The two ammunition members are in a 6 × 6 truck and, on arriving at the fire position, one of these plugs into the external connector of the 2S3's intercom system at the rear and listens into the firing commands.

Specifications

Crew: 4 + 2 in ammunition carrier
Combat weight: 27,500 kg
Power-to-weight ratio: 17.33 hp/t
Ground pressure: 0.59 kg/cm²
Length:
(gun forwards) 8.4 m
(hull) 7.765 m
Width: 3.250 m
Height: 3.05 m
Ground clearance: 0.45 m
Track width: 480 mm
Length of track on ground: 4.94 m
Max road speed: 60 km/h
Fuel capacity: 830 litres
Max range:
(road) 500 km
(cross-country) 270 km
Fording: 1.0 m
Gradient: 60%
Side slope: 30%
Vertical obstacle: 0.7 m
Trench: 3.0 m
Engine: V-59 V-12 water-cooled fuel-injected diesel developing 520 hp
Transmission: manual, 6 forward and 2 reverse gears
Suspension: torsion bar
Armament:
(main) 1 × 152.4 mm 2A33 gun
(anti-aircraft) 1 × 7.62 mm PKT MG
Rate of fire:
(max) 4 rds/min
(sustained) 60 rds/min (first hour)
Unit of fire: 60 rounds
Ammunition:
(main) 46 (42 HE, 4 HEAT)
(7.62 mm) 1,500

Gun control equipment
Turret power control: powered/manual
Gun elevation/depression: +60/–4°
Turret traverse: 360°
Armour:
(turret) 15 mm (max)
(hull) 20 mm (max)
NBC system: yes
Night vision equipment: yes

The prepared 152 mm projectiles (complete with set fuze) and charge are then passed through the circular openings in the rear of the vehicle. The later 2S3M/2S3M1 has a single opening.

Main armament comprises a 152 mm weapon, designated the 2A33, which is based on the towed 152 mm D-20 gun-howitzer but with a bore evacuator added behind the muzzle brake. The barrel has a double-baffle muzzle brake, fume extractor and a gun barrel travelling lock. This is operated by the driver, from the driving position, without leaving the vehicle. The normal projectile fired by the 2S3 is the HE-FRAG OF-540, which is fitted with an RGM-2 fuze, weighs 43.5 kg, contains 5.76 kg of TNT, has a maximum muzzle velocity of 655 m/s and a maximum range of 18,500 m. Other types of projectile fired include BP-540 HEAT-FS (also referred to as HEAT-SS, for spin-stabilised), HE/RAP (High-Explosive Rocket-Assisted Projectile) with a range of 24,000 m, AP-T, illuminating, smoke, incendiary, flechette, scatterable mines (anti-tank and anti-personnel) and the Krasnopol laser-designated projectile covered in the entry for the 152 mm 2S19. The gun has an elevation of +60° and a depression of –4°. Turret traverse is a full 360°.

The torsion bar suspension consists of six dual rubber-tyred roadwheels each side with the drive sprocket at the front, idler at the rear and four track-return rollers. The first and last return rollers support the inside of the track only. The first, second and sixth roadwheel stations have a hydraulic shock-absorber and there is a distinct gap between the first and second and second and third roadwheels.

The 2S3 has infra-red night vision equipment and an NBC system. According to Russian sources, the NBC system also keeps the turret clear of fumes when the gun is fired. An OV-65G heating unit is also fitted. It has no amphibious capability and normally carries an unditching beam at the rear of the hull.

Most 2S3 systems have an entrenching blade mounted at the front of the hull to enable the system to prepare its own fire position without engineer support. This enables the 2S3 to prepare its own firing position in between 20 and 40 minutes depending on the type of terrain.

Command and control of the 2S3 series is provided by the 1V12 series of vehicles based on the MT-LBu vehicle covered in the Armoured command vehicles section.

Variants

Late production versions of the 2S3 are designated the 2S3M and 2S3M1. The 2S3M and 2S3M1 have an ammunition load of 40 projectiles and charges and also a loading carousel which facilitates loading and therefore increases the rate of fire.

These versions can be distinguished from the rear as they have two small stowage compartment covers positioned to the right and left of the hull rear as opposed to a larger oval cover in the centre.

The 2S3M1 is, in addition, equipped with a data terminal which automatically displays the fire command transmitted to the battery officer's upgraded 1V13M command vehicle. It also has an IP-5 sight.

More recently, Russia has offered a modernised version of the 2S3M1 with a much longer barrel. This fires a standard HE projectile to a maximum range of 20,550 m or a HE base bleed projectile to a maximum range of 23,900 m.

Specifications are similar to the standard system but combat weight is 27,800 kg while overall length with ordnance in travelling lock is 9.48 m.

152 mm 2S3M2

For several years Uraltransmash Federal State Unitary Enterprise in Ekaterinburg, Russia, has been working on a number of enhancements to the 152 mm 2S3 which originally entered service in the early 1970s.

The standard 2S3M has a 152 mm/28 calibre barrel designated the 2A33 but the company is now offering two further enhancements to bring the system up to the 2S3M2 or 2S3M3 standard.

As production of the complete 2S3 series systems was completed some time ago, these upgrades are being offered to existing customers. The upgrades could be carried out by Uraltransmash but it would be most cost effective for the company to provide kits to enable the user to upgrade existing systems in their own facilities.

In the 2S3M2 version the 152 mm/28 calibre barrel is replaced by a new 152 mm/39 calibre barrel which, firing a 3OF61 high-explosive base-bleed projectile enables a maximum range of 25.1 km. It can also fire the Krasnopol laser-guided artillery projectile which is also manufactured in 155 mm and which has been sold to China and India.

155 mm 2S3M3

The 2S3M3 features a western type 155 mm/39 calibre barrel which fires standard NATO types of ammunition. Firing the old 155 mm United States M107 high-explosive projectile a maximum range of 18.2 km is achieved.

Firing the German/Italian/UK L15A1 a range of 24.7 km is quoted while firing the US M549A1 rocket-assisted projectile a maximum range of 30 km is quoted. It can also fire the 155 mm Krasnopol-M laser-guided projectile out to 17 km.

Status

Production complete. In service with the following countries:

Country	Quantity	Comment
Algeria	35	
Angola	4	from Bulgaria in 1999
Armenia	28	
Belarus	166	
Cuba	20	
Georgia	1	
Iraq	50	estimate
Kazakhstan	50	estimate
Libya	60	
Russia	2,000	estimate
Syria	50	
Turkmenistan	16	
Ukraine	501	
Uzbekistan	17	
Vietnam	30	

Manufacturer

RFAS state factories.

UPDATED

122 mm self-propelled howitzer 2S1 Gvozdika (M1974)

Development

The requirement for the 122 mm 2S1 Gvozdika (Carnation) self-propelled artillery system was drawn up in July 1967 at the same time as the 152 mm 2S3 self-propelled artillery system covered in the previous entry.

Uralmash was responsible for the weapon, based on that used in the towed 122 mm D-30 howitzer, and the Kharkov Tractor Plant (which is now in the Ukraine) for the full-tracked chassis which is based on the automotive and running gear of the MT-LB multipurpose tracked armoured chassis.

The first four prototypes of the 2S1 were completed in the second half of 1969, it was accepted for service in September 1970 and entered volume production the following year.

The 122 mm 2S1 entered service with Poland and Russia in 1972 and was first seen in public in 1974, hence its western designation of M-1974. It is estimated that over 10,000 vehicles were built with production completed around 1991.

122 mm 2S1 has been issued on the scale of 36 per artillery division, 36 per motorised rifle division and 72 per tank division.

Each 2S1 battalion originally had three batteries each with six guns.

In the past the Russian Army main artillery calibres were 122 mm and 152 mm but in the future will standardise on the latter calibre. There is no direct replacement for the 2S1 apart from the 152 mm 2S19.

The 2S1 Gvozdika had the development designation of the Izdeliye 26.

The 2S1 has also been manufactured under licence in Bulgaria (qv) and Poland, while Romania has developed a different version called the Model 89 (qv).

With the end of the Cold War a number of European countries have considerable quantities of weapons and equipment that are no longer required. According to United Nations sources, the following quantities of 122 mm 2S1 self-propelled howitzers have recently be exported:

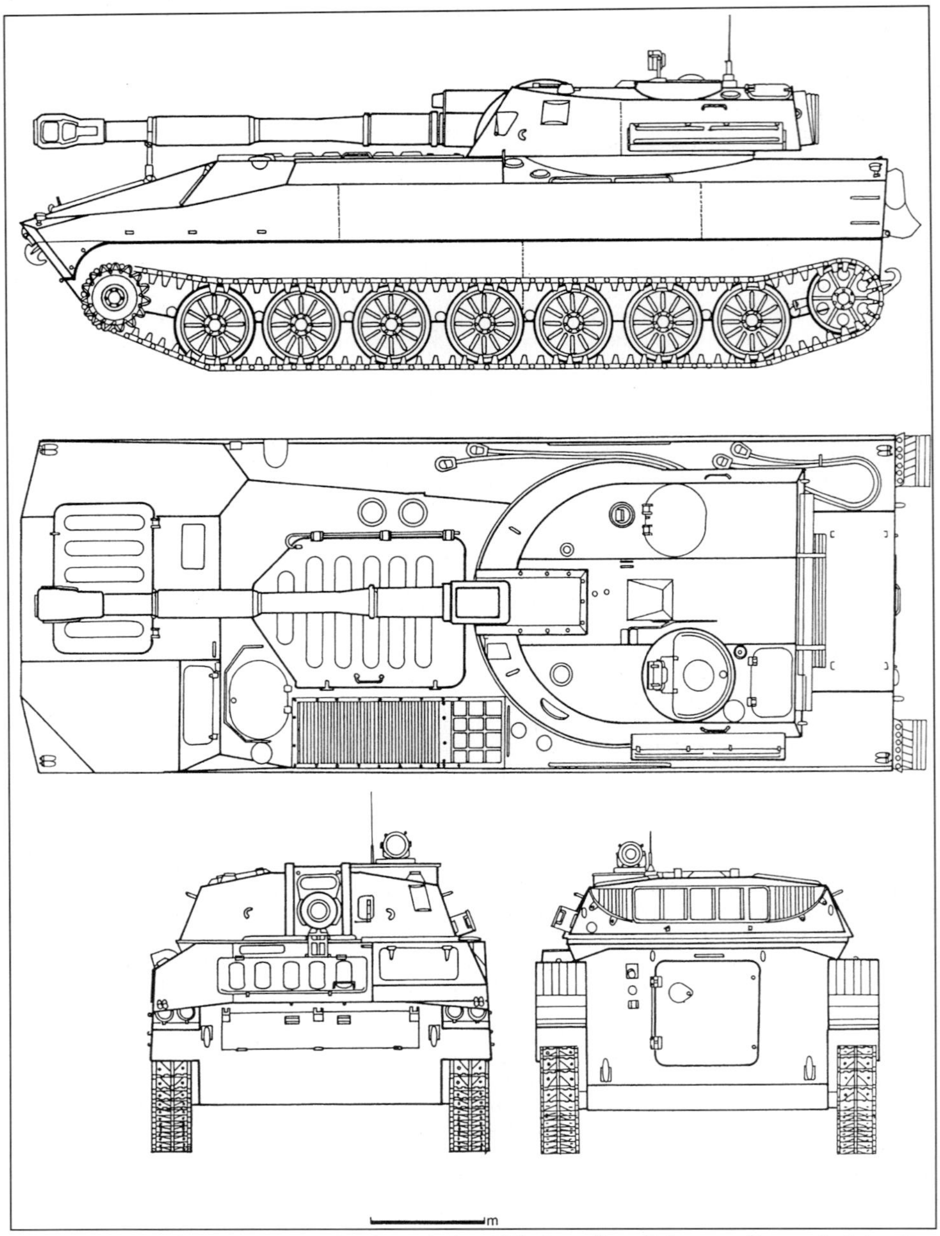

Four-view general arrangement drawing of the 2S1 Gvozdika 122 mm self-propelled howitzer (Steven Zaloga)

From	To	Quantity	Delivery Dates	Comments
Bulgaria	Syria	210	1992	believed to be 2S1
Bulgaria	Togo	6	1997	
Czech Republic	Uruguay	6	1997	
Germany	Finland	75	1992/1994	48 in 1992 and 27 in 1994
Germany	USA	6	1993	

Description

The all-welded steel hull of the 2S1 is divided into three compartments: driver's at the front on the left, engine behind the driver and the turret at the rear.

The driver has a single-piece hatch cover that opens to the rear and, in front of his position, a windscreen covered by a hatch hinged at the top. He has three day periscopes, the middle one of which can be replaced by a TVN-2B infra-red night periscope. In the top of the hull, to the right of the driver's position, is a large engine access hatch that opens to the right. There is a smaller hatch on the right side of the glacis plate. The exhaust pipe is on the right side of the hull towards the front.

The welded turret, which does not overhang the sides of the hull, has a well-sloped front and sides with the commander seated on the left, the gunner in front of and below the commander and the loader on the right. The commander has a cupola which can be traversed through a full 360° and a single-piece hatch cover that opens to the front. This cupola has three day periscopes for forward observation and an infra-red searchlight. Next to the commander's position is fitted the R-123M radio. The indirect sight (PG-2 with a field of view of 10.5° and a magnification of ×5.5) is mounted in the roof of the turret, forward of the commander's cupola with the direct sight (OP5-37 with a field of view of 10.5° and a magnification of ×5.5°) mounted to the left of the main armament. There is also the 1OP40 periscopic sight. The loader's hatch opens forward, and in front of it is a swivel-type day periscope. A long stowage box is often mounted on the left side of the turret and there is a large door in the rear of the hull, hinged on the left side. The rear door has a single firing port in the centre.

The 2S1's main armament, designated the 2A31, is a modified version of that fitted to the widely deployed 122 mm D-30 towed howitzer. This has a semi-automatic vertical sliding wedge breech block and a firing pin that can be recocked in the event of a misfire. The 122 mm ordnance is fitted with a fume extractor and muzzle brake and is held in position, when travelling, by a lock on the hull glacis plate which is operated by remote control by the driver. A power rammer and extractor are fitted to the folding guard rail to enable a higher rate of fire to be achieved and to permit loading at any angle of elevation. A maximum sustained rate of fire of 5 to 8 rds/min can be obtained with a sustained rate of fire of 70 rounds for the first hour. Of the 40 122 mm projectiles normally carried, 32 are HE, six smoke and two HEAT-FS.

Of these, 16 are in standby stowage on the left and right of the sidewalls. When in action, ammunition would normally be used from outside the 2S1 system

122 mm self-propelled howitzer M-1974 (2S1) from the rear (Richard Stickland)
0058383

RKhM chemical reconnaissance vehicle showing marking pennants at rear of vehicle

and fed to the crew inside via a ramp. One of the two ammunition members outside the vehicle is connected to the R-124 vehicle intercom and would fuze the projectiles before they are passed into the vehicle.

The remaining 24 projectiles and charges are stowed near the loader's position with the empty cartridge cases being ejected outside the turret.

The HE projectile has a maximum range of 15,300 m and, in addition to the ammunition listed in the table, leaflet, HE/RAP (range of 21,900 m), armour-piercing high-explosive, flechette, and chemical (no longer used) projectiles are also available.

The 122 mm 2S1 can also fire the Kitolov-2 laser-guided artillery projectile which operates in a similar manner to that of the 152 mm Krasnopol laser-guided artillery projectile covered in the entry for the 152 mm 2S19 self-propelled artillery system. The Kitolov-2 has a maximum range of 12,000 m, with the complete projectile weighing 25 kg.

The suspension system is similar to the MT-LB multipurpose tracked vehicle's and either side consists of seven roadwheels with the drive sprocket at the front, idler at the rear, and no track-return rollers. An unusual feature of the M-1974 is that the suspension can be adjusted to give different heights, which is of particular use when the vehicle is being transported by tactical transport aircraft. The 2S1 is normally fitted with 400 mm wide tracks but when operating in snow or swampy ground 670 mm wide tracks are fitted to give improved traction and a lower ground pressure.

Standard equipment includes infra-red night vision lights and an NBC system. The NBC system for the driver's compartment is designated the FVA-100 while that for the turret compartment is designated the FVA-200. The 2S1 is fully amphibious, being propelled in the water by its tracks.

Before entering the water the bilge pump is switched on, the trim vane is erected at the front of the hull, shrouds are fitted to the hull above the drive sprocket and front roadwheels and water deflectors on the rear track covers are lowered. Any water that enters the hull during amphibious operations is removed via the exhaust outlet using the bilge pump. while afloat only 30 rounds (projectiles and charges) are carried. Covers are also fitted around the engine air intakes to prevent water ingestion into the engine compartment.

Specifications

(2S1)
Crew: 4 + 2 in ammunition carrier vehicle
Combat weight: 15,700 kg
Power-to-weight ratio: 19.1 hp/t
Ground pressure: 0.49 kg/cm^2
Length: 7.26 m
Width: 2.85 m
Height:
(turret roof) 2.287 m
(overall) 2.732 m
Ground clearance: 0.40 m (+15/–5 mm)
Track width: 400 or 670 mm
Max speed:
(road) 61.5 km/h
(water) 4.5 km/h (some sources state 6 km/h)
Fuel capacity: 550 litres
Max road range: 500 km
Fording: amphibious
Gradient: 77%
Side slope: 55%
Vertical obstacle: 0.7 m
Engine: YaMZ-238N, V-8 water-cooled 4-stroke diesel developing 300 hp
Transmission: manual, 5 forward and 1 reverse gears
Steering: clutch and brake
Suspension: adjustable
Armament:
(main) 1 × 122 mm (actual calibre is 121.92 mm) 2A31 howitzer
Ammunition:
(main) 40 (normally 35 high explosive and 5 HEAT)
Unit of fire: 80 rounds

Gun control equipment
Turret power control: electric/manual
Gun elevation/depression: +70/–3°
Turret traverse: 360°
Armour:
(hull) 15 mm
(turret) 20 mm
NBC system: yes
Night vision equipment: yes

Type	**HE**	**HEAT-FS***	**Illumination**	**Smoke**
Projectile designation	OF-462	BK-6M	S-463	D4
Weight of projectile	21.72 kg	21.63 kg	22 kg	21.76 kg
Max muzzle velocity	690 m/s	740 m/s	690 m/s	690 m/s

* will penetrate 460 mm of conventional steel armour at 1,000 m

For improved amphibious capability some of the Polish vehicles have been observed fitted with additional buoyancy aids and are powered in the water by two small propellers situated one either side of the hull at the rear.

Variants

Artillery Command and Reconnaissance Vehicle (ACRV)

Full details of this family of vehicles are given in the Armoured command vehicles section under RFAS.

Mineclearing vehicle M-1979 (MTK-2)

This has a turret-like superstructure that contains three rockets on launch ramps. These, together with the upper part of the superstructure, are hydraulically elevated for firing. It is estimated that the range of the rockets is between 200 and 400 m, with each rocket connected, via a towing line, to about 150 m of mine clearance hose that is stowed folded in the uncovered base of the turret on the roof of the vehicle.

The hose is connected by a cable to the vehicle, which allows the crew to position the hose in the optimum breaching position after launching. The hose is then command detonated.

Scale model of MTK-2 mineclearing system with rocket launcher raised
0018895

122 mm 2S1 self-propelled howitzer with 122 mm ordnance partly elevated (Jaroslaw Cislak)
2001/0106215

RKhM chemical reconnaissance vehicle

It is now known that the RKhM chemical reconnaissance vehicle is based on the chassis of the M-1974 (2S1) self-propelled howitzer and not the MT-LB multipurpose vehicle. It has a raised superstructure on which a 7.62 mm machine gun cupola is mounted and on the rear decking are boxes for the marking pennants. The vehicle is equipped with a single KZO-2 lane marking system with 19/20 poles on the right side, automatic gas detection apparatus GSP-1M or GSA-12, semi-automatic detection systems PPKhR and VPKhR and radios R-107M, R-123M and R-130M. The RKhM-K is a command version with additional communications systems, large antenna, navigation system but is not fitted with the KZO-2.

Status

Production complete. In service with:

Country	Quantity	Comment
Algeria	150	
Angola	n/avail	unconfirmed user
Armenia	10	
Azerbaijan	14	
Belarus	239	
Bosnia Herzegovina	n/avail	
Bosnia Serbs		
Bulgaria	256	see also under Bulgaria
Croatia	n/avail	
Cuba	20	
Czech Republic	91	
Ethiopia	n/avail	status uncertain
Finland	75	called PsH 74
Hungary	151	
Iran	60	
Iraq	50	estimate
Kazakhstan	100	estimate
Libya	130	
Poland	533	
Romania	6	plus 42 of local model designed and called Model 89
Russia	3,000	estimate
Slovakia	49	
Slovenia	8	
Syria	400	include some from Bulgaria
Toga	6	from Bulgaria in 1997
Ukraine	621	
Uruguay	6	from Czech Republic in 1997
Uzbekistan	18	
Yemen	50	estimate
Yugoslavia	83	

Manufacturers

Bulgarian, Polish and RFAS state factories. (The Kharkov facility was overall prime contractor and built the complete 2S1, while the 122 mm weapon and its associated aiming system was supplied by the Uralmash facility and the diesel engine by the Yaroslavl Avtodizal facility).

Note: Polish 2S1 vehicles are fitted with a locally designed and built SW680T diesel engine, new design road wheels and changes to the hydronamics shields for amphibious operations. Bulgarian built vehicles were also used by the Russian Army.

UPDATED

SLOVAKIA

ZTS 155 mm self-propelled gun-howitzer Zuzana T-72M1 A 40

Development/Description

This system has been developed for the export market and essentially consists of the complete turret of the ZTS 155 mm self-propelled gun-howitzer Zuzana, normally mounted on an specially modified and armoured Tatra 8 × 8 wheeled chassis, fitted onto the chassis of the locally manufactured T-72M1 MBT. This system is also referred to as the 155 mm self-propelled gun howitzer Model 2000.

This turret, installed on an Indian-manufactured T-72M1 MBT chassis, was one of the four systems evaluated in India in the summer of 1995 (see entry in this section under India) to meet an Indian Army requirement for a new 155 mm self-propelled artillery system.

As of early 2001 India had not selected a self-propelled artillery system to meet its future requirements.

The 155 mm/45 calibre ordnance fires an Extended Range Full Bore – Base Bleed (ERFB-BB) projectile to a maximum range of 39.6 km with a total of 40 projectiles and associated charges being carried.

Additional details of the turret and the T-72M1 chassis are given in their respective entries.

Due to the installation of an automatic loader, a maximum rate of fire of 6 rds/min can be achieved or 30 rounds in 6 minutes. Using manual loading, a maximum rate of fire of one round every 2 minutes can be obtained.

When travelling, the 155 mm/45 calibre ordnance is held in a travel lock and when not required, folds forwards onto the glacis plate of the T-72M1. Mounted at the rear of the hull on the left side is an auxiliary power unit which allows the complete turret system to be operated with the main engine shut down. An unditching beam is carried under the rear of the hull.

Mounted over the 155 mm/45 calibre ordnance is a muzzle velocity measuring radar, also a meteorological sensor has been added. These feed information to the onboard fire-control computer.

As an option a new T100A power pack consisting of an S-1000 V-12 diesel developing 1,000 hp at 2,000 rpm coupled to a new seven speed electronically controlled speed transmission can be fitted.

A new cooling system has also been fitted with improved air cleaner and radiators

Another engine alternative is the S-850 diesel which develops 850 hp.

Variant

Based on the requirements from a number of potential export customers, ZTS is now studying the installation of this turret onto the chassis of the widely deployed Russian-designed T-55 MBT.

Status

Development complete. Ready for production.

Manufacturer

ZTS Dubnica nad Váhom (Slovakia).

UPDATED

Specifications

Crew: 4
Combat weight: 43,000 kg
Power-to-weight ratio: 18.13 hp/t
Ground pressure: 0.72 kg/cm²
Length:
(overall) 11.10 m
(hull) 6.86 m
Width: 3.59 m
Height:
(with AA MG) 3.45 m
Ground clearance: 450 mm
Track width: 580 mm
Trench: 2.6-2.8 m
Vertical obstacle: 850 mm
Fording: 1.20 m
Max road speed: 60 km/h
Range: road 650 km
Engine: Model W 46.6 V-12 diesel developing 780 hp at 2,000 rpm (or S-1000 diesel developing 1,000 hp at 2,000 rpm)
Transmission: manual with 7 forward and 1 reverse gears
Steering: clutch and brake
Suspension: torsion bar
Electrical system: 24 V
Armament:
(main) 1 × 155 mm gun
(AA) 1 × 12.7 mm MG
Ammunition:
(main) 40 × 155 mm (12.7 mm) 300
Smoke grenade launchers: 2 × 3 81 mm

Gun control equipment
Turret power control: hydraulic/manual
Traverse: 360°
Gun elevation/depression: +68.5/–5°

T-72M1 chassis fitted with Zuzana 155 mm/45 calibre artillery turret

T-72M1 chassis fitted with Zuzana 155 mm/45 calibre artillery turret

SOUTH AFRICA

LIW 155 mm/52 calibre T6 artillery turret

Development/Description

The LIW, a division of Denel, T6 155 mm/52 calibre turret system has been developed from the turret used in the combat proven G6 (6 × 6) 155 mm/45 calibre self-propelled gun system, which in turn builds on experience gained with the combat-proven G5 155 mm/45 calibre towed artillery system.

The T6 155 mm/52 calibre turret has been developed from the G6 turret and is a completely autonomous turret system that differs from the G6 in that provision has been made for the storage of 155 mm projectiles and charges in the turret. This required a repackaging of the turret power unit.

The T6 turret system sent to India was fitted with a 155 mm/52 calibre ordnance which, firing an Extended-Range Full-Bore - Base Bleed (ERFB-BB) projectile, gives a maximum range of 41,000 m.

The chamber volume is 23 litres, similar to the chamber volume of the standard production 155 mm/45 calibre system. The recoil system has been upgraded and now has two diametrically opposed buffers similar in design to the single buffer used on the G6 but with integral oil replenishers. Two diagrammatically opposed gas-operated recuperators are fitted.

The design of the barrel and cradle has been modified slightly to make it possible to remove and replace the 155 mm barrel from the front. The barrel is fitted with a double-baffle muzzle brake and a steel fume extractor.

The breech mechanism used in the T6 is a newly patented design and is of the slide and swing type with double split rings and an obturator pad. An automatic primer-loader with a 20-round magazine is integrated with the breech.

A barrel cooling fan is fitted on the cradle to reduce recovery time in the event of unsafe barrel temperatures being reached. A barrel temperature measurement system has also been installed, which gives a warning to the commander in case of high barrel temperatures being reached.

The turret is fitted with a fully automated charge and projectile replenishment, handling and loading system.

The T6 turret has been designed to fire all types of NATO ammunition. Using the Denel-manufactured ERFB (Extended-Range Full-Bore) boat tail and Extended-Range Full-Bore - Base Bleed (ERFB-BB) projectiles and modular combustible case charge system, muzzle velocity and maximum range (under standard conditions) are as follows:

ERFB projectile	Muzzle velocity	Max range
Charge M51	356 m/s	9,400 m
Charge M52	497 m/s	13,800 m
Charge M51 + 1 increment	660 m/s	19,800 m
Charge M52 + 2 increments	815 m/s	26,000 m
Charge M53	925 m/s	31,900 m

ERFB-BB projectile	Muzzle velocity	Max range
Charge M52 + 2 increments	808 m/s	31,500 m
Charge M53	922 m/s	42,000 m

The ammunition system (projectile and charge) is identical to that fired by the G5 and G6.

The charge system is the modular combustible case cartridge system developed by SOMCHEM, a division of Denel, which consists of an M51 charge, an M52 charge which can be fitted with one or two charge increments and a stand-alone charge M53 that gives a total of five zones. These charges have special barrel erosion reduction and flash suppression properties.

According to LIW, specifications for dispersion are similar to those of the current production 155 mm/45 calibre G5 and G6 systems. Probable errors of 0.48 per cent of range and 1 mil in deflection at 75 per cent of maximum range are quoted.

In firing trials, probable errors of 0.3 to 0.4 per cent of range and 0.35 to 0.4 mil in deflection have been obtained consistently with both boat tail and base bleed ammunition.

Using the M53 A2 top charge, barrel life is more than 3,500 rounds.

The 155 mm/52 calibre system has already been qualified with high-explosive, smoke, illuminating and red phosphorus projectiles, of boat tail and base bleed types, to the full range potential.

The T6 turret has full power traverse through 360° with the 155 mm/52 calibre ordnance capable of being elevated from –5 to +75°. When travelling, the ordnance is held in position by an A frame which is pivoted at the front of the hull.

The ammunition handling system of the T6 is extremely flexible and can be changed depending on the tactical situation and the customer's exact operational requirements.

The T6 turret carries a total of 40 complete 155 mm rounds and, depending on the chassis type selected, an additional 10 complete rounds can be carried.

In the ammunition handling system the projectile and charges are transferred from the carousels to the same position into the charge and projectile loading arms. This facilitates extremely rapid loading of all onboard ammunition. Replenishment of the onboard ammunition can be accomplished either through two doors in the turret rear, or via the turret compartment.

The T6 can also fire ammunition directly from a ground pile, so bypassing the onboard ammunition storage system. For this purpose two openings and trays have been provided in the turret to chassis interface, one on the left side and one on the right side.

Loading is by means of two automatic rammer systems, one for the projectiles and one for the charges.

The maximum rate of fire is 6 rds/min for all onboard ammunition. The sustained rate of fire is 2 rds/min for as long as safe barrel temperatures are indicated on the temperature warning system.

The T6 uses the same laying and navigation system as used in the G6 and full details of this are given in the entry for the G6 in the Self-propelled guns and howitzers (wheeled) section.

An autolaying system is coupled to the laying and navigation system as in the G6 and enables rapid relaying between rounds.

As an independent 36 kW turret power unit is fitted in the turret bustle, the main engine of the chassis can be switched off during firing.

The interface ring between the T6 turret and the T-72 chassis allows for the removal of the T-72 power pack with the turret traversed 90° to the side.

The turret is of all-welded steel armour that provides the crew with protection from small arms fire and shell splinters. In total, eight 81 mm smoke grenade launchers are mounted on the front of the turret. An LIW 7.62 mm SS77 machine gun is fitted on the left turret cupola for local and anti-aircraft defence.

LIW 155 mm/52 calibre T6 turret system on T-72 chassis with turret traversed to the right 0058391

Specifications

(with T-72 chassis)

Crew: 4 (commander, gunlayer, ammunition handler and the driver who also acts as an ammunition handler). In the fully automatic mode, only the commander is required to operate the gun (when firing onboard ammunition).

Rates of fire:
(burst) 3 rounds in 15 s
(intense) 6 rds/min
(sustained) 2 rds/min

Max range: 41,200 m with M53 charge and ERFB-BB projectile
42,100 m with new M90 bimodular charge (see section on G6) and ERFB-BB projectile
52,500 m with VLAP projectile and M53 charge

Min range: 4,700 m

Length:
(barrel forward) 12.49 m
(barrel retracted) 11.40 m

Width: 4.75 m

Height: 3.10 m

Ground clearance: 0.47 m

Weight:
(turret) 16,000 kg
(turret + T-72 chassis) 46,000 kg

Armament:
(main) 1 × 155 mm/52 calibre
(air defence) 1 × 7.62 mm S77 machine gun
(smoke grenade launchers) 2 × 4 81 mm

Elevation/depression: +75/–5°

Traverse: 360°

Into action: 30 s

Out of action: 30 s

Mobility: 60 km/h

Road range: 550-650 km

Standard G6 systems fitted to the T6 are the crew communication system (intercom and radio) and the automatic fire detection and suppression system. Mounted under the cradle is a muzzle velocity measurement radar.

VLAP projectile

In early 1997, the SOMCHEM Division of Denel revealed that it had developed a new 155 mm Velocity-enhanced Long-range Artillery Projectile (VLAP) which has a range of more than 50 km.

In addition to being fired from the current LIW division of Denel 155 mm/45 calibre G5 (towed) and G6/T6 (self-propelled) artillery systems, VLAP can also be fired from other 155 mm/39 and 155 mm/52 calibre systems using standard artillery charges.

No modifications are required to the actual 155 mm weapon in order to fire the VLAP and there would be no changes to drills or logistics.

Extensive firing trials have shown that the VLAP has excellent reliability, dispersion and range.

The new VLAP has the standard elongated shape with welded nubs which is also used for the ERFB and ERFB-BB projectiles.

In addition to the base bleed unit it also has a rocket motor assist which typically extends the range by 10 km depending on the weapon and charge used.

As the rocket motor extends into the VLAP body there is some loss of HE content compared to standard ERFB projectiles.

SOMCHEM stresses that the VLAP is complementary to Denel's existing 155 mm ERFB and ERFB-BB projectiles as it would enable friendly artillery systems to engage threat artillery systems while remaining safe from counter battery fire.

In addition, it would enable targets well to the rear to be engaged, such as command centres, air defence systems and logistics, which are now beyond the range of conventional artillery systems.

The design of the VLAP has been completed, first production units were delivered to an undisclosed export customer in 2001.

Status

Development complete. Ready for production. The 155 mm/52 calibre T6 was one of the four turret systems evaluated in India in 1995 installed on the locally built T6 chassis. The T6 was the only system tested in India in 1998 but as of early 2001 India had not placed a production order for the T6 or any other turret system.

Manufacturer

LIW, division of Denel.

UPDATED

SWEDEN

Future Swedish Self-propelled Artillery Requirement

Late in 2000, the Swedish Defence Material Administration (FMV) signed leasing arrangements for 155 mm/52 calibre self-propelled artillery systems with Krauss-Maffei Wegmann of Germany (PzH 2000) and BAE Systems, RO Defence of the UK (AS90).

The lease covers the period 1 October 2001 through to 30 September 2003 and these two systems are being evaluated in Sweden by Swedish artillery crews together with the existing Bofors Defence 155 mm Bandkanone 1A, Bofors Defence 155 mm FH-77BD (6 × 6) and Bofors Defence 120 mm Karin (6 × 6) self-propelled artillery systems.

The latter two systems in their 6 × 6 configuration are at the prototype system although their actual weapons are well proven.

Sweden has a requirement for 60 new 155 mm/52 calibre self-propelled artillery systems for use by its field artillery and coastal artillery units.

Today all Swedish field artillery is all towed with the exception of the 22 Bofors Defence 155 mm Bandkanone 1A first fielded in 1966 and which is covered in the following entry.

These extensive trials are expected to will lead to one system being selected in 2003. This will be modified to meet specific Swedish operational requirements and tested again before the final order is placed for up to 60 systems.

Production would commence in 2006 with first two vehicles being completed in 2007 building up to six in 2008 and with production extending until around 2011.

In addition to testing these existing self-propelled artillery systems, Sweden is also considering alternative systems. For some years Bofors Defence and Giat Industries of France have been studying a 6 × 6 155 mm/52 calibre wheeled system with a high rate of fire called the SPA. Work on this was cancelled in late 2000 but available details were given in *Jane's Armour and Artillery 2001-2002.*

It should also be noted that Bofors Defence is now owned by United Defense of the US who is also prime contractor for the US Army 155 mm Advanced Field Artillery System Crusader. Technology from this is also applicable to lighter tracked and wheeled artillery systems.

The Krauss-Maffei Wegmann PzH 2000 is already in production for the German Army (185 units) and has been selected by Italy (2 from German production line and remaining 68 being co-produced in Italy by the consortium IVECO/Otobreda) and late in 2000 by the Royal Netherlands Army (57 units). Greece has placed a firm order for 24 units.

According to Krauss-Maffei Wegmann, the fact that the PzH 2000 has now been selected by four countries in Europe as a replacement for the old US-developed M109 series means that PzH 2000 is now well on the way to becoming the future European standard in self-propelled artillery.

The BAE Systems, RO Defence, 155 mm AS90/39 calibre is in service with the UK's Royal Artillery (179 units) with a large part of the fleet being upgraded to the 155 mm/52 calibre standard which is the version currently being offered for export.

So far, the AS90/52 calibre, or Braveheart as it is now referred to, has been selected by Poland which, late in 2000, took delivery of the first two turret systems for installation on its own locally designed full tracked chassis (qv Poland).

AS90 and PzH 2000 were considered the two front runners for the Finnish Defence Force requirement for a 155 mm self-propelled artillery system which has recently been postponed for lack of funding.

UPDATED

Bofors Defence 155 mm Bandkanon 1A self-propelled gun

Development

The 155 mm Bandkanon 1A was developed by the now Bofors Defence (now owned by United Defense of the USA) in the late 1950s, with the first prototype completed in 1960. In 1965, the company was awarded a production contract for the vehicle worth SKr60 million. The first production vehicle was completed the following year and production continued until 1968.

The 155 mm Bandkanon 1A uses many components of the Bofors S-tank including the suspension, power pack and transmission. The S-tank has now been phased out of service with the Swedish Army. The single prototype of the Bandkanon 1A was followed by 26 production vehicles which are deployed in northern Sweden.

In August 1988, the Defence Materiel Administration (FMV) awarded a contract to Bofors Defence for the modernisation of the Swedish Army's 155 mm Bandkanon 1As. Under the terms of the contract, Bofors Defence replaced the existing power pack with a new unit consisting of a Detroit Diesel 6V-53T developing 290 hp coupled to a new automatic transmission developed and produced by Bofors Defence. (This power pack was the same as that installed in Strv 103B S-tanks.) The contract, the first part of which is worth SKr75 million and included training and maintenance documentation, followed Swedish Artillery trials held over a 12 month period using a prototype fitted with the new power pack.

In April 1993, the FMV awarded Bofors Defence a contract to develop and build two prototypes of the 155 mm Bandkanon 1A fitted with the Honeywell MAPS (Modular Azimuth Positioning System) and the gun interfaced with a new fire-control system, the SKER. Following successful trials of the two prototype vehicles, Bofors Defence subsequently upgraded the whole fleet to this standard.

In the future the Bofors Defence 155 mm Bandkanon 1A is expected to be replaced by a new 155 mm/52 calibre self-propelled artillery system with trials of a number of systems expected to take place in 2001 to 2003 (see previous entry).

Upgraded Bofors Defence-designed 155 mm Bandkanon 1A self-propelled gun in travelling order 0018902

Description

The hull of the 155 mm Bandkanon 1A is made of all-welded steel with the engine and transmission at the front. The driver is seated immediately in front of the turret and the all-welded steel turret is at the rear.

The power pack consists of a diesel and a gas turbine geared together to a common output. The diesel is normally used all the time and the gas turbine is engaged when driving across country. The transmission consists of a Volvo torque converter coupled to a gearbox with two forward and two reverse gears (one gear is for road use and the other for cross-country). The steering system is regenerative double-differential with a hydrostatic system and steering is by handlebars on the tiller column. The power pack, complete with transmission, can be removed from the vehicle as a complete unit to facilitate replacement in the field. The cooling system for the diesel is on the right side and the cooler for the steering system on the left.

The driver, who sits slightly to the left of the vehicle's centreline, has a single-piece hatch cover in front of which is a day periscope. This can be replaced by a passive periscope for night driving.

The all-welded turret is in two halves, left and right, both of which have a door in the front and three vision blocks, two for observation to the front and one to the side of the vehicle. In the left half of the turret are the commander, gunlayer and radio operator and in the right half are the loader and 7.62 mm anti-aircraft machine gunner. Both turret halves have a single-piece hatch cover.

The hydropneumatic suspension can be locked in position when the gun is being fired to provide a more stable firing platform. The suspension either side consists of six dual rubber-tyred roadwheels with the drive sprocket at the front and the last roadwheel acting as the idler. There are no track-return rollers.

The L/50 155 mm gun weighs 4,800 kg complete with muzzle brake and replaceable liner and is 7.75 m long. The liner itself weighs 1,340 kg. The gun barrel has a pepperpot muzzle brake and a semi-automatic wedge-shaped breech block that opens downwards. When travelling, the barrel is held by a lock at the front of the hull.

The elevating mechanism is counterbalanced by a spring balancing gear mounted in the pivot of the cradle. The elevating device consists of a handwheel mounted in the left cabin, which operates a worm gear on the elevating pinion shaft. The elevating device also has an auxiliary electric motor for rapid laying between the loading and firing position. The control lever for the auxiliary motor is in the left cabin and the motor itself is in the right cabin.

The gun can be elevated by the electric motor from +2 to +38° at 5°/s. Manual elevation is from –3 to +40°. Traverse is manual, 15° left and 15° right at an elevation of above 0° and a traverse of 15° left and 4° right when the barrel is below 0°. The gun is fed from a 14-round clip. The loading device consists of a loading tray which is suspended on rollers beneath the magazine. The loading tray transports the ammunition from the magazine to the ramming position, it is then fed by a rammer built into the loading tray. The loading tray, as well as the rammer, is operated by springs which are cocked by the run-out of the gun and the movement of the loading tray. The rounds are fed to the loading tray by alternating spring-operated feeding trays, one each side of the loading tray. Maximum rate of fire is 14 rounds in 48 seconds.

The magazine itself consists of seven drop compartments, holding two rounds each, from where the rounds drop down onto the feeding trays. Both the magazine and the feeding devices are protected by armour and the two armour-plated doors on the top of the magazine, which open vertically for reloading, are operated by a single electric motor. Reloading with 14 rounds takes 2 minutes. When the magazine has been expended, the gun is elevated to an angle of +38° and the two covers on top of the magazine are opened vertically. A new clip of ammunition is fed into the magazine from an articulated ammunition hauler. The doors are closed and the gun is then ready to fire again. The complete loading sequence takes only two minutes. The first round is manually loaded, after which loading is automatic. The gunner can select either single shots or fully automatic. The gun fires a complete (for example, projectile and charge system) HE projectile weighing 48 kg to a maximum range of 25,600 m. Maximum muzzle velocity with charge 3 is 865 m/s and with charge 1 is 600 m/s. The sight has a magnification of ×4 and a 10° field of view.

Specifications

Crew: 5
Combat weight: 53,000 kg
Power-to-weight ratio: 10.18 hp/t
Ground pressure: 0.9 kg/cm^2
Length:
(with gun) 11 m
(without gun or loading arm) 6.55 m
Width: 3.37 m
(over tracks) 3.26 m
Height:
(with AA MG) 3.85 m
(to turret top) 3.25 m
Ground clearance:
(hull centre) 0.37 m
Track: 2.59 m
Track width: 670 mm
Length of track on ground: 4.3 m
Max road speed:
(direct gear) 28 km/h
(cross-country gear) 9 km/h
Fuel capacity: 1,445 litres
Max road range: 230 km
Fording: 1 m
Gradient: 60%
Side slope: 30%
Vertical obstacle: 0.95 m
Trench: 2 m
Turning radius:
(direct gear) 40 m
(cross-country gear) 13 m
Engines: Detroit Diesel 6V-53T developing 290 hp at 2,800 rpm. Boeing Model 502/10MA gas turbine developing 300 shp at 38,000 rpm
Transmission: Bofors automatic
Steering: regenerative double-differential with hydrostatic steering drive
Suspension: hydropneumatic
Electrical system: 24 V
Batteries: 4 × 12 V
Armament:
(main) 1 × 155 mm gun
(anti-aircraft) 1 × 7.62 mm MG
Smoke-laying equipment: none
Ammunition:
(main) 14

Gun control equipment
Turret power control: manual
(by commander) no
(by gunner) yes
Max rate of traverse: 0.25°/turn of wheel
Max rate of elevation:
(electric) 5°/s
(manual) –3 to +40°
Gun traverse:
(0° elevation and above) 15° left and 15° right
(below 0° elevation) 15° left and 4° right
Armour:
Hull: 10-15 mm
Turret: 20 mm
Magazine sides: 20 mm
Magazine covers: 10 mm
Magazine bottom: 15 mm
NBC system: no
Night vision equipment: no

Autonomous gun

In place of the normal sight a Honeywell MAPS land navigation system is installed. This system gives the gun position and the attitude of the barrel (that is, traversing, elevation and trunnion angle). This negates the requirement for surveying and alignment normally associated with bringing a gun into action and permits 'shoot and scoot' tactics.

Status

Production complete. In service only with the Swedish Army. An upgrade programme for this system has been completed by Bofors Defence.

Manufacturer

Bofors Defence AB (A United Defense Company).

UPDATED

SWITZERLAND

Swiss Ordnance Enterprise Corp 155 mm M109 upgrade

Development/Description

Under contract to the Swiss Defence Procurement Agency, Swiss Ordnance Enterprise Corp (previously k + w Thun) has developed an upgrade programme for the US-designed 155 mm M109 series of self-propelled howitzers which is used in large numbers by the Swiss Army. Additional details of earlier Swiss M109 systems are given under the entry for the M109 under the United States.

Following trials with three prototype systems, Swiss Ordnance Enterprise Corp is now upgrading 348 M109 series self-propelled artillery systems to this standard, with first units delivered in 1998.

The upgraded systems have been designated the Panzerhaubitze 88/95 by the Swiss Army.

Increased firing range

The firing range is increased by some 60 per cent (with charge 10) with the installation of a new 47 calibre barrel which is similar to that used in the 15.5 cm Fortress Gun L52 Bison. This is covered in detail in the Coastal artillery guns and missiles section later in this book.

Swiss Ordnance Enterprise Corp believes that the retrofitted 155 mm/47 calibre gun is an optimal solution in terms of performance and cost-effectiveness. The recoil force of the new gun is

Panzerhaubitze 88/95 of the Swiss Army upgraded by Swiss Ordnance Enterprise Corp showing new 155 mm/47 calibre barrel and new extended bustle 0058393

practically unchanged and the mass which must be elevated is only 100 kg more than the original M109 barrel.

The new 155 mm/47 calibre ordnance is fully chrome-plated with 60 groove rifling and is fitted with a double-baffle muzzle brake and a new fume extractor.

Using charge 9, a maximum range of 23,000 m can be achieved or 30,100 m with a cargo round with base bleed. Using charge 10, a maximum range of 26,100 m can be achieved or 36,400 m with a base bleed cargo round (with an air density of 1,200 g/m³). The upgraded M109 carries a total of 40 projectiles and associated charges.

A barrel temperature sensor warns of an imminent cook-off condition.

This 155 mm/47 calibre barrel is also used in the Dutch RDM Technology 155 mm M109L47 self-propelled artillery upgrade which has been adopted by the United Arab Emirates. Additional details are given under the Netherlands. A total of 85 systems has been delivered to the United Arab Emirates.

Higher rate of fire

The effective engagement of fast-moving targets requires a rapid rate of fire. A high firing rate is achieved with the upgraded M109 through the installation of a semi-automatic loader. This allows a burst rate of fire of three rounds in 15 seconds. The semi-automatic loader can be used with any 155 mm projectile.

The telescopic rammer enables a maximum rate of fire of up to 8 rds/min.

Attached to the right side of the 155 mm weapon is a charge standby magazine for the temporary stowage of four preprepared charges.

Survivability

Increased sophistication in battlefield surveillance technology has shown the need for an artillery weapon which not only provides for increased range and higher firing rates but also higher mobility and the ability to 'shoot and scoot'.

Navigation and positioning system

The Swiss M109 navigation and gun positioning system is based on a ring laser gyro unit and supplies a continuous flow of accurate data to aid navigation and gunlaying. Computer-based command and display units for the commander, gunner and driver allow optimal fire mission management.

Night vision system

The vehicle is fitted with passive night vision equipment for the commander and driver to allow for 24-hour mobility.

Increased ammunition stowage

Swiss Ordnance Enterprise Corp has increased the ammunition stowage capacity of the M109 to take advantage of the firing rate of the semi-automatic loading system. The original load of 28 155 mm projectiles and associated charges was considered to be insufficient to accommodate the variety of ammunition used on today's battlefield.

The additional 24 charges are stowed in a propellant magazine at the rear of the turret, separated from the crew compartment by a blast-proof sliding door. Doors are provided in both sides of the propellant magazine for ammunition resupply purposes. The base and two resupply doors of the propellant magazine open easily to allow gas expulsion in the case of a deflagration or explosion.

Travel lock

When travelling, the 155 mm ordnance is held in position by a travel lock which is remotely controlled by the driver under complete armour protection. This feature reduces the time taken to come into and be taken out of action.

Fire suppression system

The upgraded M109 has an optimised halon fire suppression system in the crew and driver's compartment and there is also a fire extinguishing system in the engine compartment.

Electrical improvements

As the standard electrical system of the M109 has insufficient power to satisfy the upgrade requirements, Swiss Ordnance Enterprise Corp has developed and tested a new electrical system in the M109.

The upgraded electrical system provides main power with the new hull batteries and a fine power with newly added batteries with a capacity of 60 A.

The system is designed for ease of maintenance and repair and is oversized to provide for possible further requirements in the future. The new alternator, which is fitted into the standard alternator mounts, is driven by the main diesel engine and produces 80 per cent more power than the standard unit.

Specifications

Upgraded M109
Crew: 6
Combat weight: 27,000 kg
Length:
(overall) 11.33 m
(hull) 7.28 m
Width: 3.18 m
Height: 3.20 m
Ground clearance: 0.38 m
Max road speed: 60 km/h
Fuel capacity: 500 litres
Range: 350 km
(remainder similar to M109A1/A2 covered under USA)
NBC system: no
Night vision equipment: yes

The power distribution system is designed with state-of-the-art electronic components to ensure a high degree of reliability. A new 16-way slip-ring developed in Switzerland transfers power and other electrical signals from hull to the turret. The complete system is designed with the best in EMC safety.

The cabling in the vehicle has also been completely redesigned and now has rugged wiring and distribution boxes. The driver's instrument panel is to a new design and the auxiliary driver display is fixed and well protected. The new instrument box contains all-electronic command, control and test circuits for the entire vehicle.

All critical units of the electrical system are equipped with the appropriate test connectors for fault isolation to LRU level by computer-supported test equipment.

Status

Development complete. In production for the Swiss Army with first upgraded systems completed in 1998. Delivery of modules of this system have also been made for M109 systems used by Austria and the United Arab Emirates (qv RDM Technology entry).

Manufacturer

Swiss Ordnance Enterprise Corp.

UPDATED

TAIWAN

XT-69 155 mm self-propelled extended range gun

Development/Description

The XT-68 155 mm self-propelled extended range gun was developed by the Combined Service Forces and is identical in layout and concept to the basic XT-69 described in the following entry.

It is fitted with a 155 mm/45 calibre weapon which may have been developed from the towed South African LIW 155 mm G5 or the Belgian/Canadian SRC International GC 45, although the elevating mass is more like the Israeli Soltam Systems 155 mm M-71 gun-howitzer than these two weapons.

The 155 mm/45 calibre ordnance has a multi-baffle muzzle brake, a fume extractor and a travelling lock that folds back onto the glacis plate when not in use. Maximum range is said to be 30,000 m with the Extended-Range Full-Bore projectile, and over 35,000 m with a base bleed projectile. A maximum rate of fire of 6 rds/min can be attained due to a pneumatically assisted loading device which enables the weapon to be uniformly loaded at any angle of elevation. Elevation is estimated to be from 0 to +70° and traverse 30° left and right. The weapon is used in conjunction with a battery fire-control system developed by the Sun Yat-sen Scientific Research Institute in Taiwan.

This system has a crew of five and is fitted with night vision equipment for the driver and a 12.7 mm M2 HB anti-aircraft machine gun. No NBC system is fitted.

The chassis is based on the US United Defense M108/M109 self-propelled artillery system covered under the United States.

XT-69 155 mm self-propelled extended range gun in travelling order (L J Lamb)

Status

In service with Taiwan. Production complete. Some sources have indicated that a total of 45 of these systems is in service under the production designation of the T-69. This system was never offered on the export market. Taiwanese artillery capability has been further strengthened by the recent arrival of 28 United Defense LP M109A5 systems from the United States.

Manufacturer

Combined Service Forces, Taiwan.

UPDATED

XT-69 155 mm self-propelled howitzer

Development/Description

The XT-69 155 mm self-propelled howitzer was developed by the Combined Service Forces (CSF) with the assistance of a number of civilian companies including the Taiwan Aluminium Corporation (hull and tracks) and the Taiwan Machinery Manufacturing Corporation (chassis and parts of the 155 mm weapon, possibly the diesel engine and suspension). The Hsing-Ho Arsenal was responsible for the hydraulics and the Chi-Ning Arsenal for the 155 mm ordnance. Both the arsenals come under the control of the Combined Service Forces.

The XT-69 is basically the US United Defense LP 155 mm M109/105 mm M108 self-propelled howitzer chassis, with the fully enclosed turret removed and the 155 mm howitzer mounted in the open-topped compartment at the rear of the hull. This has no overhead armour protection and therefore no NBC system. When deployed in the firing position a platform is lowered to the immediate rear to provide the crew with space to operate the weapon.

The 155 mm howitzer is a modified version of the US M114 towed system, but there is the possibility that the weapon has a new and longer ordnance fitted with a multibaffle muzzle brake. A travelling lock is provided for the 155 mm ordnance on the glacis plate, which folds down when not required. When originally introduced the 155 mm ordnance had no muzzle brake.

A ramp is fitted at the rear of the vehicle and is used both as a means of exit and for ease of ammunition loading. Spades are hydraulically lowered to the ground before firing.

It is assumed that the 155 mm weapon has the same traverse of 24° left and 25° right and elevation from –2 to +63° as the towed M114 weapon; there is a possibility that traverse and elevation are hydraulic, with manual controls for emergency use.

XT-69 155 mm self-propelled howitzer with muzzle brake (L J Lamb)

Maximum range of the M114 is 14,600 m, although some Taiwanese sources have given the XT-69 a range of 15,000 m. It is estimated that 25 155 mm projectiles and charges are carried and that there is a crew of five.

Physical and automotive characteristics of the XT-69 are similar to the US M109 described later in this section except for a lower height and weight. Standard equipment includes driving lights and a 12.7 mm M2 HB anti-aircraft machine gun mounted on the right side to the rear of the howitzer.

Status

Production complete. Recent information has indicated that these weapons are no longer in service, although the version with the 155 mm/45 calibre ordnance, covered in detail in the previous entry and now called the T-69, remains in service.

Manufacturer

Combined Service Forces, Taiwan.

UPDATED

TURKEY

155 mm M52T self-propelled artillery system

Development

The Turkish Land Forces Command (TLFC) had a large fleet of US-built 155 mm M44 and 105 mm M52 self-propelled artillery systems which were manufactured in the 1950s.

By today's standards these are obsolete as not only do their weapons have a very short firing range but, as they are powered by a petrol engine, their operational range is very short, around 160 km.

The original 105 mm M52 fired a US-developed HE M1 projectile to a maximum range of 11,270 m while the 155 mm M44 fired a US-developed HE M107 projectile to a maximum range of 14,600 m.

In addition to the calibre of their weapons, the other main difference between the two systems was that the M44 had an open-top gun compartment with the weapon mounted in the forward part while the M52 had the 105 mm gun in a fully enclosed turret. An unusual feature of the M52 was that the driver was seated in the turret, which could be traversed 60° left and right.

In total, 222 M44 were upgraded to the M44T standard (additional details are given under Germany which designed and developed the upgrade and supplied most of the original components).

In total, 365 M52 have been upgraded to the M52T standard. Trials of the M52T were completed in late 1994 and series production commenced in 1995. The system is also referred to as the M52T K/M Obus L/39 with the 39 being the calibre length (39 × 155 mm).

Turkish 155 mm M52T self-propelled gun with ordnance elevated (Michael Jerchel) **2000**/0085191

Description

The modifications to the chassis have been extensive and in the case of the M52 these have included the replacement of the short-barrelled 105 mm howitzer by a 155 mm 39 calibre barrel fitted with a muzzle brake, the same as that fitted to the German-developed 155 mm M109A3, which in turn is based on the NATO standard 155 mm FH-70 towed howitzer.

The 155 mm/39 calibre ordnance has been manufactured under licence in Turkey at the facilities of MKEK which manufactures a wide range of other weapons and ammunition for the home and export markets.

As the 155 mm ammunition is considerably larger than the 105 mm natures fired by the M52, the ammunition racks have also been modified to accommodate the larger calibre ammunition.

A similar modification was carried out to the M44T and both weapons can fire a standard HE M107 projectile to a maximum range of 18,000 m, a rocket assisted projectile to a maximum range of 24,000 m or a base bleed projectile to a maximum range of 30,000 m.

The main difference between the M44T and the M52T is that the former's 155 mm ordnance has limited traverse while the M52T 155 mm ordnance is mounted in a turret which can be traversed left and right. When deployed in the firing position a spade is lowered at the rear to absorb some of the recoil.

The Turkish company TT - Izmur has developed the computer and command system for the M52T which allows automatic gunlaying. The system consists of the keyboard and display unit, control unit and sensors including cant and ambient air pressure.

The front-mounted power pack has been upgraded, with the original petrol engine having been replaced by a more fuel-efficient German MTU MB 833Aa 501 diesel which develops 450 hp coupled to the Allison CD-500-4 automatic transmission. Operational range has been increased to 400 km with fuel consumption being quoted as 1.28 litres/km.

Other improvements include fire detection and suppression systems, new electrical system, new driver's instrument panel as well as improvements to the steering and a cold starting system for the engine.

In addition the top engine deck and exhaust system have been modified and the suspension upgraded to take into account the increased weight of the vehicle. The original US-type tracks have been replaced by a Turkish T317 track.

A roof-mounted 12.7 mm M2 machine gun is provided for local and anti-aircraft defence and a bank of four 76 mm electrically operated smoke grenade launchers is mounted either side of the turret, firing forwards.

The original 105 mm M52 did not have a spade due to the low recoil force of the 105 mm main armament. The upgraded M52T has a spade which is lowered to the ground before firing commences.

Status

Production complete. In service with the Turkish Army.

Manufacturer

Upgrade work was carried out in Turkish Army facilities.

UPDATED

Specifications

Crew: 5
Combat weight: 29,500 kg
Power-to-weight ratio: 15.25 hp/t
Length:
(overall) 8.80 m
Width: 3.20 m
Height:
(with AA MG) 3.30 m
Ground clearance: 0.49 m
Track: 2.602 m
Track width: 533 mm
Length of track on ground: 3.793 m
Max road speed: 60 km/h
Fuel capacity: 530 litres
Max range: 420 km
Fording: 1.219 m
Gradient: 60%
Vertical obstacle: 0.914 m
Trench: 1.828 m
Engine: MTU MB 833Aa 501 V-6 water-cooled diesel developing 450 hp at 2,300 hp
Transmission: Allison, CD-500-4 with 4 forward and 2 reverse gears
Suspension: torsion bar
Electrical system: 24 V
Armament:
(main) 1 × 155 mm 39 calibre
(anti-aircraft) 1 × 12.7 mm MG
Smoke-laying equipment: 2 × 4 76 mm smoke grenade launchers
Ammunition:
(main) 30 × 155 mm
Gun control equipment
Turret power control: power/manual
Gun elevation/depression: +65/–5°
Turret traverse: 60° (left and right)
Max range:
(standard ammunition) 18,000 m
(extended range ammunition) 24,000 m
(extended range base bleed) 30,000 m
NBC system: no
Night vision equipment: optional

UNITED KINGDOM

BAE Systems, RO Defence AS90 (Braveheart) 155 mm self-propelled gun

Development

During production of the FH-70 155 mm towed howitzer the now BAE Systems, RO Defence (at that time Vickers Shipbuilding and Engineering Limited) identified a world need for improved self-propelled artillery. Subsequent concept studies offered a cost-effective solution using the companies private venture GBT 155, a turret based on the 155 mm 39 calibre ordnance of the FH-70 and designed specifically for export.

Conceptual studies commenced in 1982 for the design of a modern artillery chassis using a variant of the GBT 155, incorporating all and more of its features but having a larger turret ring to maximise crew space and a frontally positioned power pack.

The three principal factors in the AS90 concept, (if it were to survive the increased firepower, improved armour, electronic and NBC warfare of the future battlefield) were judged to be performance, crew comfort and survivability and a systems approach to the vehicle's equipment features. As a result, in addition to the stretch potential mentioned earlier, key features in the design of AS90 include: ease of operation and maintenance, reliability and modular construction to suit different user requirements. Rapid rates of fire and greater accuracy than FH-70 were stipulated, with rapid into and out of action time to avoid counter-battery fire, and total manual back-up in the event of battle damage or power failure.

AS90 has been designed to be easily manufactured and wherever possible standard and proven components have been used in the design of the vehicle.

The project definition for AS90 was completed in March 1985, and the first prototype was shown at the British Army Equipment Exhibition in June 1986. Following the exhibition, a series of firing trials augmented by mobility trials were undertaken, leading to the award of Live Crew Clearance from the British Ordnance Board in March 1987. A series of endurance and automotive trials was then initiated, with AS90 travelling some 3,000 km and firing in excess of 1,500 rounds without a mission-relevant failure. These trials, completed in June 1987, covered all aspects of army use.

Following the withdrawal of the United Kingdom from the International 155 mm SP70 project late in 1986, the search for a new self-propelled howitzer started.

This eventually resulted in the selection in June 1989 of the 155 mm AS90 self-propelled howitzer to meet future requirements for the British Army.

The British Army offcial designation for the AS90 is Howitzer, Self-propelled 155 mm, L131 AS 90 155 mm.

The fixed-price contract covered the completion of AS90 development, production of 179 systems and the provision of initial spares. The AS90 artillery system

Armoured Ammunition Supply Vehicle (AASV) (left) feeding projectiles and charges to 155 mm AS90 self-propelled artillery system (right) 0018906

was accepted for service with the British Army in May 1992.

Each regiment of AS90s has 32 weapons in four batteries of eight guns each. First production AS90s were completed in late 1992, with the first regiment becoming operational in 1993.

AS90s are operational with the Royal Artillery in Germany and the UK and also deployed to Canada for training purposes. They have also served with the Peace Implementation Force (IFOR) operating in the former Yugoslavia.

While each AS90 regiment has four batteries each with eight weapons, these are not all manned in peacetime although they would be in time of war. The British Army has five AS90 regiments but it was announced in 1998 that another regiment will be formed. As a result it is expected there will be a reduction in the number of weapons allocated to each regiment.

Throughout 1989 and 1990, AS90 underwent a full series of reliability growth trials for the British Army as part of the trials programme. By the Spring of 1991, the two prototype vehicles had travelled 28,630 km and fired 6,170 rounds (3,380 EFCs).

In 1994, a five-week In Service Reliability Demonstration (ISRD) trial took place in the UK involving a complete battery of eight AS90s. These trials covered mobility and firing reliability equivalent to 80 battlefield days. A total of 6,000 km was travelled and 11,200 rounds fired.

A BAE Systems, RO Defence 155 mm 52 calibre barrel was fitted to one of the guns, proving its performance reliability for future enhancements to the gun.

Final AS90s were delivered to the British Army in the Spring of 1995, with four systems retained by the company for trials.

In mid-1992 the AS90 (Braveheart) turret system was selected by Poland. Additional details are given under Poland. For the export market the latest 155 mm/52 calibre AS90 is known as Braveheart. The first two AS90 155 mm/52 turrets were delivered to Poland late in 2000 for installation on a locally designed full tracked chassis.

Description

The chassis of AS90 is of all-welded steel armour construction with a maximum thickness of 17 mm. Steel rather than aluminium was chosen as it is not only simpler to fabricate but also easier to repair in the field.

The driver sits at the front left with the power pack to his right. He normally enters via a single-piece hatch cover over his position that hinges to the left side. This can be locked open for driving in the head-out position. In front of the driver is a single day periscope with a blackout blind and laser filter. The periscope can be changed quickly for a passive night vision device by the driver while under armour and without breaking the NBC seal. The driver has a fully adjustable seat and can also enter and leave his position via the fighting compartment at the rear. The driver steers the vehicle using a steering wheel rather than sticks.

The power pack, consisting of the Cummins engine, Renk transmission and cooling system, can be removed and replaced as a complete unit in well under an hour. The engine compartment is provided with a fire detection and suppression system which can be operated automatically or manually. Hinged louvres and access covers are provided above the engine compartment for ease of access.

The barrel clamp is located on the front of the vehicle to provide a travelling support for the gun. The barrel clamp is operated from within the vehicle from the driver's position.

AS90 is powered by the Cummins VTA-903T-660 14.8 litre V-8 diesel developing 660 bhp at 2,800 rpm. This is a more powerful version of the engine installed in the US Army's M2/M3 Bradley Infantry Fighting

Vehicle/Cavalry Fighting Vehicle and the Multiple Launch Rocket System. The VTA-903T-660 has plenty of stretch potential, having already been run at 750 hp for trial purposes.

The engine is coupled to a fully automatic German Renk Model LSG 2000 automatic transmission, with four forward and two reverse gears.

A diesel-powered auxiliary power unit is installed in the forward part of the fighting compartment so the main engine does not need to be constantly run to operate the turret systems.

The area to the rear of the engine compartment is the fighting compartment, which extends to the rear of the vehicle. There is a large door opening to the rear of the chassis for access and ammunition resupply, and mounted either side of this on the outside are two integral armoured containers which can be used to store supplies and other equipment. Externally mounted on the turret shell are five stowage bins, one for each crew member's use.

The vehicle runs on 12 double roadwheels, each having hydropneumatic suspension, with a drive sprocket at the front and an idler at the rear. The use of hydropneumatic suspension contributes to the increased headroom in the fighting compartment by eliminating the false floor needed with torsion bar suspension, in addition to giving AS90 a lower profile. The system has been designed to give the vehicle, with its large wheel travel, progressive rates of springing and powerful damping on every wheel station and a very smooth ride, enabling high average speeds to be maintained across rolling irregular terrain without impairing the safety and comfort of the crew. The suspension was developed by Air Log with the current design authority being Horstman Defence Systems Limited.

The turret is of all-welded steel construction, with a maximum thickness of 17 mm and it has a large turret ring diameter of 2.7 m. The commander has a cupola on the right side of the turret roof. An air sentry's hatch with pintle mounting for secondary armament is provided on the left-hand side of the turret. There is also a large hatch on the left-hand side of the turret.

The commander is positioned behind the layer on the right-hand side of the turret. Both are provided with seats which have adjustments for height and rotation. The shell and charge loaders stand on the left-hand side of the ordnance and are provided with seats which fold down from the wall of the chassis. Turret traverse is through a full continuous 360° and weapon elevation is from –5 to +70°. Elevation and traverse drives are both electric and operate at a rate of 10°/s. Full manual controls are provided as a back-up.

An electric traverse and elevation system, with silent operation from the integral power supply was chosen rather than hydraulic, since it was considered to be safer, more reliable and would require less maintenance. The only hydraulics in the turret are those applicable to the balancing gear, loading tray and shell rammer.

155 mm 39 calibre AS90 in service with Royal Artillery (Richard Stickland) ***2002**/0126559*

AS90 is fitted with the BAE Systems, RO Defence 155 mm 39 calibre barrel as standard. The recoil system has two diametrically opposed buffers and one recuperator, each with its own integral reservoir, and a maximum recoil length of 800 mm. The barrel can be drawn out through the front of the turret for replacement in less than 1 hour.

The removal of the complete elevating mass is simple as the saddle is constructed in two parts and can easily be detached from the trunnions.

The 155 mm ordnance has a double-baffle muzzle brake, fume extractor and a split-block breech. The breech combines the speed of action associated with the sliding block principle and the robustness of the Crossley pad obturating system and has an integral 12-round primer magazine.

AS90 can fire a burst of three rounds in less than 10 seconds, an intense rate of 6 rds/min for 3 minutes and a sustained rate of fire of 2 rds/min. The ordnance incorporates an electrically initiated percussion firing mechanism using standard DM191A1 or US M282 igniter tubes. There are 48 projectiles and charges carried in the turret and the ordnance will fire all standard NATO projectiles including the HE M107 and extended range types.

Of the 48 projectiles carried, 31 are stowed in the turret bustle in four magazine modules, each of which has a motor that moves the required projectile to the correct position. This enables it to be pulled forward onto a Shell Transfer Arm (STA). The STA pivots about the left trunnion and aligns with either magazine tray, irrespective of gun elevation. The STA realigns to lock with the elevating mass and the projectile is moved sideways by a motor-driven shell clamp on the loading tray. The loading tray is brought in line with the breech and then rammed by the flick rammer. The charge is then loaded, the breech closed and the ordnance fired by the commander or layer. The loading system has full manual back-up and safety interlocks. The remaining 17 projectiles are statically stowed in the chassis.

Using standard ammunition and charges, the 39 calibre barrel enables a range of 24,700 m to be achieved. Using an assisted projectile, a range of over 30,000 m can be achieved. AS90's area coverage is enhanced even more when the 52 calibre barrel is fitted, with a maximum range of 30,000 m with conventional projectiles and 40,000 m with assisted projectiles.

AS90 is fitted with a full navigation and auto-lay capability using the MAP's standard DRU based on ring laser gyro technology giving the gun complete autonomy. This enables the gun to be laid automatically in bearing as well as elevation.

Muzzle velocity measuring, air conditioning and NBC equipment are provided as standard, the last two being mounted on the rear of the turret bustle.

The then Air Log supplied the complete running gear for AS90. This includes an advanced hydropneumatic suspension system, a Diehl double pin track and sprockets, GLS top rollers and roadwheels and the then Air Log designed and manufactured idler gear. The idler gear incorporates a hydraulic track tensioner with an overload protection device, allowing the idler wheel to retract if excessive track tension is built up.

Avimo designed and developed the day/night direct fire telescope for the AS90 which is designated the DFS90.

Specifications

Crew: 5
Combat weight: 45,000 kg
Power-to-weight ratio:
(combat loaded) 14.66 hp/t
Ground pressure: 0.9 kg/cm^2
Length:
(gun forward) 9.90 m
(hull) 7.20 m
Width: 3.40 m
Height:
(overall) 3.00 m
Ground clearance: 0.41 m
Track: double pin
Track width: 550 mm
Length of track on ground: 4.59 m
Max road speed: 55 km/h
Fuel capacity: (approx) 750 litres
Road range: (approx) 370 km at 45 km/h
Fording: 1.50 m
Gradient: 60%
Side slope: 25%
Vertical obstacle: 0.88 m
Trench: 2.8 m
Engine: Cummins VTA-903T-660 V-8 diesel developing 660 bhp at 2,800 rpm
Transmission: Renk LSG 2000 fully automatic with 4 forward and 2 reverse gears
Auxiliary generator: 4-stroke diesel
Suspension: hydropneumatic
Electrical system: 24 and 48 V
Batteries:
(turret) 4 × 12 V, 100 Ah
(hull) 4 × 12 V, 100 Ah
Armament:
(main) 1 × 155 mm 39 calibre (some British Army AS90, to have a 155 mm/52 calibre barrel)
(anti-aircraft) 1 × 12.7 mm or 7.62 mm MG
Smoke-laying equipment: 2 × 5 smoke grenade dischargers
Ammunition:
(155 mm) 48 projectiles and charges
(anti-aircraft) 1,000 rounds
Gun control equipment: electrical/manual
Turret traverse: 360°
Gun elevation/depression: +70/–5°
Armour: 17 mm
NBC system: yes
Night vision equipment: yes

Variants

AS90 with 155 mm/52 calibre ordnance

Late in 1998, the UK MoD announced that it was to award the now BAE Systems, RO Defence, Phase Two of a major programme to equip the current in service 155 mm AS90 self-propelled artillery system with a 155 mm/52 calibre Extended Range Ordnance (ERO) and a new Modular Charge System (MCS).

Phase 1, which ran from May 1997 through to October 1998, covered design and development of the ERO and selection of the MCS.

The ERO is a 155 mm 52 calibre barrel manufactured by BAE Systems, RO Defence which will replace the current 155 mm 39 calibre barrel. The new barrel, which is not required to be internally chrome plated, will be fitted at unit level.

The new 52 calibre barrel will use the existing muzzle brake, fume extractor and breech mechanism. The ERO conforms to the Joint Ballistic Memorandum of Understanding signed by France, Germany, Italy, UK and USA.

The current 155 mm 39 calibre barrel achieves a maximum range, firing assisted ammunition, of 30 km, but when fitted with a 155 mm 52 calibre barrel this is raised to 40 km.

A total of 179 AS90s were built by the now BAE Systems, RO Defence, but at this stage only three regiments (96 systems) will be fitted with the new ERO, although others may follow in the future.

Eight AS90 systems arrived at Barrow-in-Furness in March 2001 from 26 Regiment, Royal Artillery. These will be the first to be fitted with the new 155 mm/52 calibre barrel and will be used for system qualification tests and firing trials before rejoining the regiment.

Three MCS suppliers were short listed for the MCS requirement, the now BAE Systems, RO Defence, Somchem Division of Denel (South Africa) and team Modular Artillery Charge System (MACS) of the US.

Somchem is the winner of the MCS competition and the contract could be worth £100 million over a period of 10 years if all options are exercised. The introduction of MCS will offer the user a number of advantages especially in ammunition handling.

The Somchem MCS for the AS90 is designated the M90 Bi-Modular Charge System (BMCS) and comprises M91A1 low-zone (Z1 and 2) and M92A1 high-zone (Z3 to 6).

In order to fire the existing 155 mm L15 projectile from the 155 mm/52 calibre ordnance, an enlarged nylon obturator band will have to be fitted to each L15 projectile.

The MoD awarded the now BAE Systems, RO Defence the Phase 2 contract in 1999, which is expected to run through to May 2004 to cover full development, trials and production of the ERO and MCS.

The In Service Date (ISD) with first regiment of the Royal Artillery fully converted and trained with ERO/MCS is November 2003.

Technical Insertion Packages

It was expected the AS90 systems deployed by the Royal Artillery would have a Mid-Life Improvement (MLI) but this proposal has been dropped in favour of a series of Technology Insertion Packages (TIPs).

Two TIPs are being developed, TIP1 and TIP2. TIP1 is to be installed at the same time as the Bowman communications system while TIP2, which is concerned mainly with survivability improvements, has a possible in service date of around 2010.

TIP1 will cover electronic enhancements to AS90, including an automatic fuze setter, muzzle velocity prediction, measurement of barrel bend, thermal warning/planning, onboard ballistic computation, data highway improvements, onboard data collection, low cost course corrected munitions compatibility, centralised Control and Display Unit (CDU), integration of GPS and embedded training.

The installation of a CDU could see a possible reduction in the crew of the AS90 system.

AS90 turret on T-72 chassis

Details of this combination, first shown in 1994, are given in the following entry.

AS90 in the USA

Early in 1993, a production AS90 completed trials with the US Army Research Laboratory at Aberdeen Proving Ground. For these trials the AS90 was fitted with a Magnavox Howitzer Fire-Control Computer (HFCC), integrated into the AS90's existing fire-control, navigation and pointing system. This made the AS90 fully autonomous. The system was also able to receive directly from the forward observers, digital information which was computed on board, shortening response time in fire missions.

The AS90 was fitted with the now BAE Systems, RO Defence 155 mm 39 calibre ordnance; the ability to fit a 155 mm 52 calibre ordnance in 75 minutes was shown.

The AS90 demonstrated a burst rate of fire of three rounds in under 10 seconds, with a six-round burst rate of fire in 50 to 55 seconds. Also demonstrated was a MultiRound Simultaneous Impact mission (MRSI) in which four rounds hit the target. Moving target missions were performed in which the target was straddled.

Desert AS90

As a private venture, the now BAE Systems, RO Defence, has developed Desert AS90 which has a number of improvements for operations in the high ambient temperatures encountered in the Middle East.

Major improvements incorporated in Desert AS90 include:

(1) improved engine cooling system
(2) improved transmission oil cooling system
(3) enhanced transmission gear range
(4) upgraded auxiliary power unit cooling system
(5) solar-reflective paint and thermal cover for turret roof
(6) enhanced air conditioning
(7) improved resistance to sand ingress in magazine
(8) German Diehl Type 940 double pin track.

Desert AS90 has undergone extensive climatic trials in the UK, at Yuma Proving Ground in Arizona and two series of trials in the Middle East during which it operated in temperatures of up to +60°C.

The first series of trials took place in 1995 in Dubai, Saudi Arabia and Kuwait during which over 560 rounds of 155 mm ammunition were fired with key achievements including:

(1) firing 100 rounds in 52 minutes at temperatures of +42°C
(2) firing 18 rounds in under 3 minutes using onboard ammunition
(3) carrying out a shoot and scoot trial which included stopping AS90, firing three rounds and moving off again, all in just 30 seconds
(4) clocking up over 500 km of mobility trials.

At the request of Kuwait, Desert AS90 was flown back to Kuwait in 1996 for another series of trials as the ones conducted in 1995 were not during the hottest summer period.

As 204 rounds were fired by Desert AS90 in Kuwait in 1995 the latest series of trials were essentially confirmation trials in higher temperatures and were carried out by a Kuwait Land Forces crew.

During two days of firing trials over 120 rounds were fired from a 155 mm 39 calibre barrel including M107 HE, M483 cargo, L15 HE, L19 practice, smoke, illuminating and Extended Range Full Bore Base Bleed (ERFB-BB).

According to the company the Kuwait firing and mobility trials meet all of the objectives laid down and during the former 30 rounds were fired in under five minutes with ranges of 30 km being achieved with the 155 mm 39 calibre barrel.

Using the onboard auto-lay system with modified software, AS90 also demonstrated its MRSI (Multiple Round Simultaneous Impact) capability with a night fire mission involving four rounds illumination in a lateral spread pattern range and followed by three rounds HE burst in 10 seconds.

On the third day of the trials, the 155 mm 52 calibre ordnance was fitted and a total of 10 rounds was fired. The L15 projectile achieved a range of 30 km while an extended range full bore - base bleed projectile achieved a range of 41 km.

Repair and recovery vehicle

Using the AS90 chassis as a basis, the company is proposing a family of vehicles including an armoured ammunition resupply vehicle and a repair and recovery vehicle.

The repair and recovery vehicle will have a raised superstructure at the rear which would also contain a winch for recovery operations. Mounted at the front of the hull would be a hydraulically operated blade which could be used for clearing obstacles or as a stabiliser blade when the winch is in use. Mounted on the right side of the hull is a hydraulically operated crane with a telescopic jib which can lift complete powerpacks. As of late 2001 this remained at the proposal stage.

Ammunition resupply vehicle

Two versions of the ammunition resupply vehicle are being proposed, one of which will back up to the AS90 and feed ammunition (projectiles and charges) similar to that of the US M992 Field Artillery Ammunition Support Vehicle (FAASV) while the other will drive up behind the AS90 and feed projectiles and charges using a conveyor. As of late 2001 this remained at the proposal stage.

Status

Production complete. In service with the British Army (179). First two turrets delivered to Poland late in 2000 for installation on locally developed chassis. There is a separate entry for the Polish system, called Krab, in this section under Poland. As of 2001, two prototypes of this had been completed for extensive trials with the Polish Army.

Manufacturer

BAE Systems, RO Defence.

UPDATED

BAE Systems, RO Defence AS90 155 mm turret system

Development/Description

The AS90 turret has been developed as a private venture by the now BAE Systems, RO Defence (at that time Vickers Shipbuilding and Engineering Limited) from the turret system used for the AS90 self-propelled artillery system which is now in service with the British Royal Artillery as its only self-propelled gun system. Full details of the AS90 and further improvements to this system were covered in the previous entry.

The BAE Systems, RO Defence, autonomous AS90 turret was first shown in 1993 and has been successfully trialled on a Russian T-72 MBT chassis. Further firing, mobility and reliability trials took place in 1994 and a small number of improvements have been incorporated as a result.

In mid-1999, Poland selected a version of the AS90 turret system with a 155 mm/52 calibre barrel for installation on a locally designed chassis with a contract being placed for six turrets which was subsequently trimmed back to two that were delivered late in 2001. There is a follow-on order for 72 expected in the future. Additional details of this system, called Krab, are given in this section under Poland. As of 2001, two prototypes of this system had been built for extensive trials with the Polish Army.

155 mm AS90 Turret on T-72 MBT chassis with ordnance in travel lock, during trials in India 0006012

The AS90 turret is of all-welded steel armour construction that provides protection against both small arms fire and shell splinters.

Access to the turret is via a hatch in the left side, the commander's cupola on the right side of the turret roof and the air sentry hatch on the left side of the turret roof. A 7.62 or 12.7 mm machine gun is mounted on the air sentry hatch.

The magazine is in the rear of the turret and holds 31 projectiles. It incorporates simple indexing

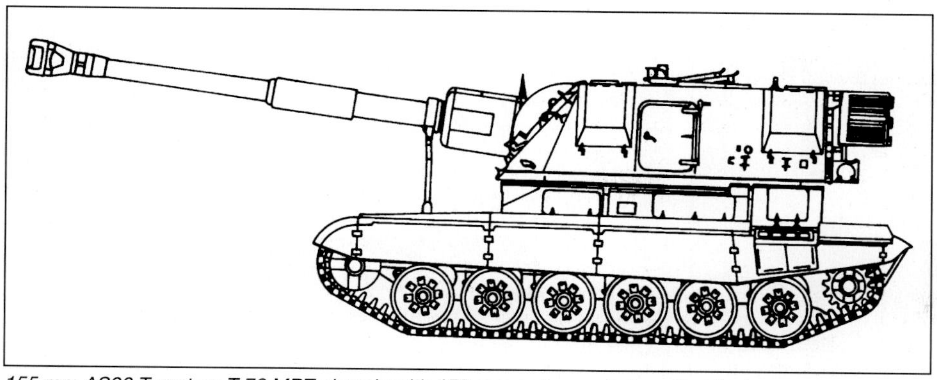

155 mm AS90 Turret on T-72 MBT chassis with 155 mm ordnance in travelling lock

Specifications
With T-72 chassis
Rates of fire:
(burst) 3 rounds in under 10 s
(intense) 18 rounds in under 2 min
(sustained) 2 rds/min
Range: 24,700 m (30,000 m with extended-range projectile) using 155 mm/39 calibre barrel, 30,000 m (40,000 m with extended-range projectile) using 155 mm/52 calibre barrel
Crew: 4 plus driver
Overall length: 10.18 m
Width:
(turret) 3.52 m
Height: 2.98 m
Weight: 45,000 kg (with chassis)
Armament:
(main) 1 × 155 mm/39 calibre (or 155 mm/52 calibre)
(anti-aircraft) 1 × 7.62 mm or 1 × 12.7 mm MG
Ammunition stowage: over 40 rounds of 155 mm stowed internally
External replenishment: manually assisted
Shell loading system: automatic
Mobility: 60 km/h

mechanisms which transfer the projectiles to a central point from where they are transferred to the loading tray.

The sealed containers, mounted above the shell magazine, house 21 propellant charges with a further 12 at the front left and centre right of the turret. Additional ammunition can be stowed in the lower part of the turret compartment taking the stowage capacity to over 40 complete rounds.

A shell replenishment aid was fitted for the Indian trials and this allows projectiles to be transferred from outside straight into the turret. The projectiles are loaded onto the shell replenishment aid manually with the charges also being loaded manually through an opening in the turret rear on the left.

The turret has a standard crew of four with the gunlayer at the forward right, commander behind and the two loaders on the left. If required, the turret can be operated with a crew of just two.

The design of the ordnance, cradle and saddle is such that the 155 mm 39 calibre barrel can be easily withdrawn from the front of the turret, leaving the breech in place.

In addition, the whole of the elevating mass, complete with saddle, can be quickly removed from the turret leaving all other turret-mounted equipment intact.

The recoil system has two diametrically opposed buffers and one recuperator, each with its own integral reservoir to reduce external piping and space requirements. The system is protected by an easily removed armoured cover.

The ordnance is a BAE Systems, RO Defence 155 mm 39 calibre barrel fitted with a vertical sliding split-block breech mechanism ensuring positive obturation. This is fitted with a double-baffle muzzle brake and a fume extractor. When travelling, the ordnance is held in an A-type travelling lock that is pivoted on the front of the hull.

The 155 mm 39 calibre ordnance conforms to the MoU governing the standardisation of 155 mm ordnance and ammunition between France, Germany, Italy, UK and USA.

The unique design of self-obturation breech mechanism combines the rapid action of a sliding breech block with a Crossley pad obturator to provide a positive seal.

The breech can operate in both manual and semi-automatic modes and in the latter case the run-out energy is used to open the breech. Both German DM191 and US M82 percussion igniter tubes can be used.

The breech is connected to the barrel by a conventional interrupted screw thread. The barrel is machined from an Electro-Slag Refined (ESR) forging of monobloc construction and is autofrettaged.

The system can fire the full range of ammunition used for the towed FH-70 155 mm field howitzer and the towed M198 155 mm howitzer to a range of 24,700 m with standard projectiles and to 30,000 m with base bleed projectiles.

Turret traverse is powered through a full 360° with weapon elevation from –5 to +70°. Over 150 projectiles have been fired during trials in the UK and these have shown that the Russian T-72 series chassis is a very stable firing platform and no spades are required.

To enable a rapid burst rate of fire to be achieved and a high rate of continuous fire to be sustained, an autoloader has been incorporated at the rear of the elevating mass.

This system provides a burst rate of fire of three rounds in under 10 seconds and an intense rate of 18 rounds in less than 2 minutes. The equipment comprises a loading tray and power rammer for loading the projectiles with the propellant charges being loaded manually.

Ammunition replenishment is through the rear of the turret. Projectiles are loaded directly into the magazine and tray where they can be either stowed in the magazine or loaded through to the gun for firing. Power for the autoloader is provided by a small hydraulic power pack installed in the front left corner of the turret.

A now BAE Systems, RO Defence, designed Automatic Gun Laying System (AGLS) provides the facilities automatically to lay the gun and also track the vehicle's position. These two functions are achieved by an inertial navigation system which is fully integrated with the turret electrical system.

The inertial navigation system provides the accurate position and attitude of the gun and also the position of the vehicle, thus alleviating the prerequisite of surveying the firing position.

The required gun aiming orders are entered on the gunlayer's display unit and fed to the turret control computer where a demand signal is generated to drive the elevation and traverse power amplifiers to lay the gun.

The layer's display unit shows gun azimuth and quadrant elevation and the target bearing and elevation together with the difference between the two. Target data can be entered directly from the layer's keypad or received from a fire-control system.

Trials have shown that the onboard navigation system combined with the automatic gunlaying and burst rates of fire enable the AS90/T-72 to come into action, fire 18 rounds and move off again in less than 5 minutes.

For direct fire against targets up to 2,000 m away a day/night sight is mounted on the elevating mass adjacent to the layer. The graticule is calibrated for range and crossing target velocity and, since the gunlayer has a two-axis joystick, simple aiming of the armament can be achieved quickly.

The power supply for all the turret services is provided by a turret-mounted battery pack which is charged by an Auxiliary Power Unit (APU) mounted on the rear of the turret. The APU is based on a Honda diesel which is already manufactured in India.

This APU obviates the need to run the vehicle's main engine during fire missions and so helps to conserve fuel. A ventilation unit is fitted as standard to provide adequate air circulation for the crew.

The T-72 chassis can lay its own smoke screeen by injecting diesel fuel into the exhaust outlet on the left side of the hull. In addition, there is a bank of five electrically operated smoke dischargers either side of the turret. Extensive external stowage for personnel kit is provided on the turret sides and between the top of the hull and base of the turret.

A number of readily available growth options can be provided for the AS90 Turret. These include:

(1) a 155 mm 52 calibre barrel which increases the maximum range of conventional ammunition to 30,000 m and extended-range ammunition to 40,000 m. The 155 mm 52 calibre ordnance has already been tested in a British Army AS90 and will be fitted as part of that vehicle's mid-Life upgrade. The 155 mm 52 calibre weapon was also demonstrated in India
(2) onboard ballistic computation
(3) shell Transfer Arm (STA)
(4) barrel cooling.

Status

Development complete. Ready for production. The AS90 155 mm turret system was one of the four systems evaluated in India in 1995. As of December 2001, India had not placed any production orders for any of these turret systems.

A version of this turret with a 155 mm/52 calibre ordnance was selected by Poland in mid-1999.

Manufacturer

BAE Systems, RO Defence.

UPDATED

UNITED STATES OF AMERICA

United Defense M110 series of 203 mm (8 in) self-propelled howitzers

Development

Future US Army requirements for new heavy artillery emphasised air transportability, time into and out of action, common parts and interchangeability area of fire. A feasibility study encompassing these requirements was carried out, presented and approved at a meeting in January 1956.

The Pacific Car and Foundry Company (subsequently PCF Defense Industries) submitted a concept study for a new family of self-propelled weapons and was subsequently awarded a contract for the design, development and construction of six prototype vehicles: two 175 mm self-propelled guns designated the T235, three 203 mm (8 in) self-propelled howitzers designated the T236 and one 155 mm self-propelled gun designated the T245.

Major design features included the interchangeability of the 175 mm gun, 8 in howitzer and 155 mm gun in a common mount, on a common chassis; use of the 8 in howitzer and 155 mm gun field pieces and portions of the M17 standard mount; and drastic reductions in size and weight over conventional equipment which was made possible through a new hydraulic lockout system.

The chassis was also considered suitable for use as a light recovery vehicle and, in 1957, the programme was expanded to include both armoured (T120) and unarmoured (T119 and T121) recovery vehicles.

M110A2 203 mm (8 in) self-propelled howitzer of the Pakistani Army with muzzle brake covered up (Christopher F Foss) ***2001***/0106211

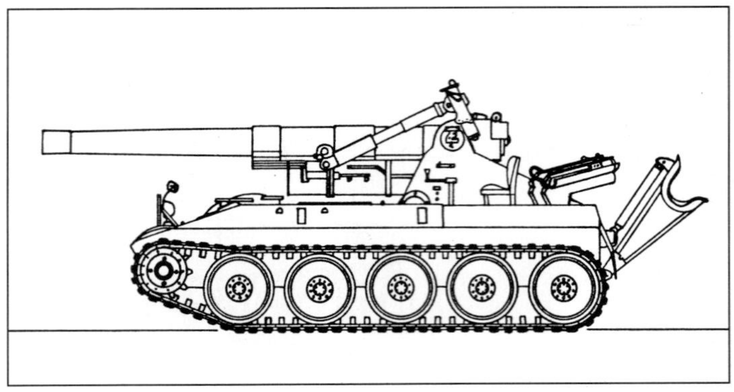

M110 203 mm self-propelled howitzer in travelling order

Ordnance service tests with the first prototype chassis began late in 1958. However, in 1959, a policy was established that diesel rather than petrol engines would be used for future vehicles and three of the prototypes, the T235, T236 and the T120, were retrofitted with Detroit Diesel Model 8V-71T diesel engines and designated the T235E1, T236E1 and the T120E1. Trials with the T235E1 and the T236E1 were completed early in 1961 and in March both vehicles were standardised, the T235E1 as the M107 and the T236E1 as the M110. The T120E1 was subsequently standardised as the M578 ARV (but the T245, T119 and the T121 were not developed further).

In June 1961, the Pacific Car and Foundry Company (which is no longer involved in defence) was awarded an initial production contract for both the M107 and the M110. First production vehicles were completed in 1962 and the first M110 battalion was formed at Fort Sill early in 1963. Production of the M110 was also undertaken at a later date by the FMC Corporation of San José and Bowen-McLaughlin-York (BMY) of York, Pennsylvania. Both these companies merged to become United Defense LP. Original production of the M110 was completed in the late 1960s, by which time about 750 vehicles had been produced. In FY78, 209 M110A2s were ordered at a cost of US$109.3 million for delivery from 1979.

The first M110A1 was produced by Bowen-McLaughlin-York in May 1980 (it produced its first M110 during 1965). The first M110A2 was produced in February 1980.

The M110A2 self-propelled howitzer was used by the UK and United States ground forces in the 1991 Middle East conflict.

In many countries, for example the Netherlands and the UK, the M110A2 has been replaced by the Lockheed Martin Missiles and Fire Control, 227 mm Multiple Launch Rocket System. The main reason the M110 series remained in service for so long was its tactical nuclear capability.

According to United Nations sources, the following quantities of surplus M110 self-propelled howitzers have been exported by Germany, Netherlands and the USA between 1992 and 1998:

From	To	Quantity	Delivery dates
Germany	Greece	72	1994
Germany	Turkey	131	1994
Netherlands	Bahrain	13	1994
USA	Bahrain	49	1997
USA	Greece	84	1992/1993
USA	Jordan	18	1997
USA	Morocco	60	1997
USA	Spain	52	1992/1994
USA	Turkey	72	1992/1993

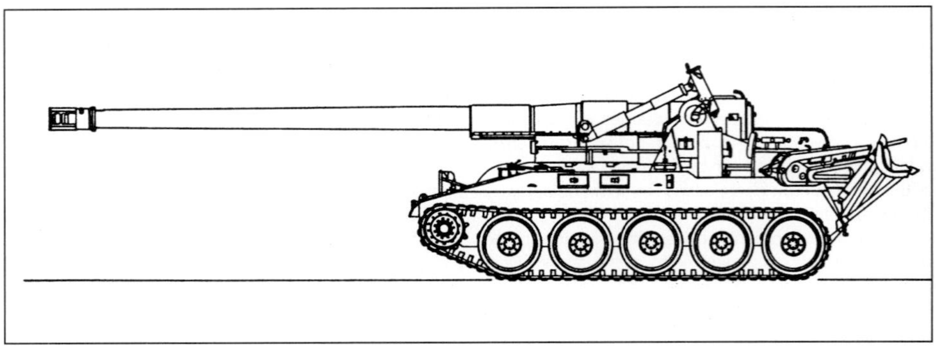

M110A2 203 mm self-propelled howitzer in travelling order

Specifications

M110
(data in square brackets relates to M110A2)
Crew: 5
Combat weight: 26,534 [28,350] kg
Unloaded weight: 24,312 [25,492] kg
Power-to-weight ratio: 15.26 [14.28] bhp/t
Ground pressure: 0.76 kg/cm^2
Length:
(gun forward, spade up) 7.467 [10.731] m
(hull, without gun or spade) 5.72 m
Width: 3.149 m
Height:
(to top of barrel, travelling) 2.93 [3.143] m
(to top of mount) 2.809 [2.743] m
(to top of hull) 1.475 [1.47] m
Ground clearance: 0.44 [0.393] m
Track: 2.692 m
Track width: 457 mm
Length of track on ground: 3.936 m
Max road speed: 56 [54.7] km/h
Fuel capacity: 1,137 [984] litres
Max road range: 725 [523] km
Fuel consumption: 1.568 [1.88] litres/km
Fording: 1.066 m
Gradient: 60%
Side slope: 30%
Vertical obstacle: 1.016 m
Trench: 2.362 [1.905m
Engine: Detroit Diesel Model 8V-71T, turbocharged 2-stroke, liquid-cooled 8-cylinder diesel developing 405 bhp at 2,300 rpm
Transmission: Allison Transmission XTG-411-2A cross-drive with 4 forward and 2 reverse gears
Suspension: torsion bar
Electrical system: 24 V
Batteries: 4 × 12 V Model 6TN
Armament: 1 × 203 mm M2A2 [M201 Howitzer
Smoke-laying equipment: none
Ammunition: 2 rounds (additional ammunition carried in support vehicle)
Gun control equipment
Turret power control: hydraulic/manual
Gun elevation/depression: +65/–2°
Gun traverse: 30° left and right
NBC system: no
Night vision equipment: yes

Description

The M110 is operated by a team of 13, five of whom (commander, driver and three gunners) are carried on the gun, with the rest in the M548 tracked cargo carrier which also carries the ammunition. Full details of the M548 are given in *Jane's Military Vehicles and Logistics2001-2002*. Production of the M548 is now complete.

The hull of the M110 is identical to that of the M107 and is made of all-welded armour and high-tensile alloy steel with the driver at the front of the hull on the left, the power pack to his right and the main armament at the rear.

The driver, who is the only member of the crew to be seated under armour, has a single-piece hatch cover in front of which are three M17 day periscopes. This can be replaced by a passive night vision device. The engine is coupled to the Allison Transmission XTG-411-2A cross-drive transmission at the front of the hull.

The torsion bar suspension either side consists of five dual rubber-tyred roadwheels with the drive sprocket at the front and the fifth roadwheel acting as the idler. There are no track-return rollers. Attached to each roadwheel is a hydraulic cylinder which serves as a shock-absorber, a hydraulic bump stop and a suspension lockout which transmits recoil shock directly to the ground. The tracks are of the single-pin type with removable rubber pads.

The M110 has night vision equipment but no NBC system or amphibious capability.

The M110 is armed with a 203 mm (8 inch) M2A2 howitzer in an M158 mount and has an elevation of +65°, a depression of –2° and a traverse of 30° left and right. Elevation, depression and traverse are hydraulic with manual controls for emergency use. The weapon has a hydropneumatic recoil system, an interrupted screw breech block and a percussion firing mechanism. Mounted at the rear of the chassis on the left side is a rammer and loader assembly which lifts a projectile from the rear or left side of the vehicle, positions it and rams it into the chamber. Hydraulic power for operating the system is obtained from the vehicle's system but it can also be operated by handcranks. Mounted at the rear of the chassis is a large hydraulically operated spade which is lowered into position before firing begins.

The 8 in (203 mm) weapon fires the following types of conventional ammunition:
HE (M106) with the projectile weighing 92.53 kg, a maximum muzzle velocity of 587 m/s and a maximum range (charge 7) of 16,800 m;
HE (M404) (carries 104 M43A1 grenades) with a projectile weighing 90.72 kg, a maximum muzzle velocity of 587 m/s and a maximum range (charge 7) of 16,800 m.

The ammunition is separate loading. Two rounds are carried on the vehicle and the rest in the supporting vehicle. The normal rate of fire is one round every two minutes but two rounds a minute can be fired for short periods.

Fire-control equipment consists of a panoramic sight M115 (magnification of ×4 and 10° field of view) for indirect fire, telescope M116C (magnification of ×3 and 13° field of view) for direct fire, elevation quadrant M15 and gunner's quadrant M1A1.

M110A1 and M110A2

In 1969, the United States Armament Command began the development of a new version of the M110 which would have a longer range and fire new types of improved ammunition. This was standardised as the M110A1 in March 1976 and entered service in January 1977. The M110A1 replaced both the M110 and M107, which were phased out of service with the US Army in Europe by 1980. It then cost less than US$100,000 to convert the M107/M110 to the M110A1/M110A2 standard.

The M110A1 has a new and much longer 203 mm barrel called the M201, a direct fire elbow telescope, the M139, and a chassis identical to the M110's. The M110A2, which was standardised in 1978, is the M110A1 fitted with a double-baffle muzzle brake and can fire charge 9 of the M118A1 propelling charge whereas the earlier M110A1 can fire only up to charge 8. The weapon can fire the following projectiles:

HE (M106) with the projectile weighing 92.53 kg, a maximum muzzle velocity of 711 m/s and a maximum range (charge 8) of 22,900 m;

HE (M404) with the projectile weighing 90.72 kg, a maximum muzzle velocity of 711 m/s and a maximum range (charge 8) of 17,200 m. This is also referred to as an Improved Conventional Munition and carries 104 anti-personnel grenades;

HE (M509A1) carrying 180 anti-personnel/anti-materiel grenades with a maximum range (charge M188A1) of 22,900 m. This is also referred to as an Improved Conventional Munition and carries dual-purpose grenades;

HERA (M650) with a maximum range (charge 9) of 30,000 m.

User	M110	M110A1	M110A2
Bahrain	-	-	62 (13 from Netherlands, 49 from USA)
Greece	-	-	181
Iran	38**	-	about 30 left
Israel	36	-	-
Italy	-	-	26 (reserve)
Japan	-	-	201 (locally built)
Jordan	-	-	120 (29 new, rest upgraded)
Korea, South	99	-	estimate
Morocco	-	-	60 (from USA)
Pakistan	-	-	40
Spain	-	-	64
Taiwan	-	-	60
Turkey	-	-	219
USA	-	-	250 (reserve)

** Total delivered, some destroyed or captured in Iran-Iraq War
In some of the above countries these weapons are held in reserve.

Status

Production of the M110 has been completed. No longer marketed.

Manufacturers

Pacific Car and Foundry Company, (now PCF Defense Industries), Renton, Washington;
FMC Corporation, San José, California;
Bowen-McLaughlin-York (BMY), York, Pennsylvania.
The last two companies have now merged to become United Defense LP.

UPDATED

United Defense M107 175 mm self-propelled gun

Development

Future US Army requirements for new heavy artillery emphasised air transportability, time into and out of action, common parts and interchangeability area of fire. A feasibility study encompassing these requirements was carried out, presented and approved at a meeting held in January 1956. The Pacific Car and Foundry Company (which subsequently became PCF Defense Industries) submitted a concept study for a new family of self-propelled weapons and was subsequently awarded a contract for the design, development and construction of six prototype vehicles: two 175 mm self-propelled guns designated the T235, three 203 mm (8 in) self-propelled howitzers designated the T236 and one 155 mm self-propelled gun designated the T245.

Major design features included the interchangeability of the 175 mm gun, 8 in howitzer and 155 mm gun in a common mount, on a common chassis; use of 8 in howitzer and 155 mm gun field pieces and portions of the M17 standard mount; and drastic reductions in the weight over conventional equipment which was made possible through a new hydraulic lockout system.

The chassis was also considered suitable for use as a light recovery vehicle and, in 1957, the programme was expanded to include both armoured (T120) and unarmoured (T119 and T121) recovery vehicles.

Ordnance service tests with the first prototype chassis began late in 1958. In 1959, a policy was established that diesel rather than petrol engines would be used for future vehicles and three of the prototypes, the T235, T236 and T120, were retrofitted with diesel engines and designated the T235E1, T236E1 and the T120E1. Trials of the T235E1 and T236E1 were completed early in 1961 and, in March 1961, both vehicles were standardised, the T235E1 as the M107 and the T236E1 as the M110. The T120E1 was subsequently standardised as the M578 ARV but the T245, T119 and T120 were not developed further. Details of the M578 were given in *Jane's Military Vehicles and Logistics 2001-2002*.

In June 1961, the Pacific Car and Foundry Company was awarded an initial production contract for both the M107 and M110. First production vehicles were completed in 1962 and the first M107 battalion was formed at Fort Sill in January 1963. Production of the M107 was also undertaken at a later date by the FMC

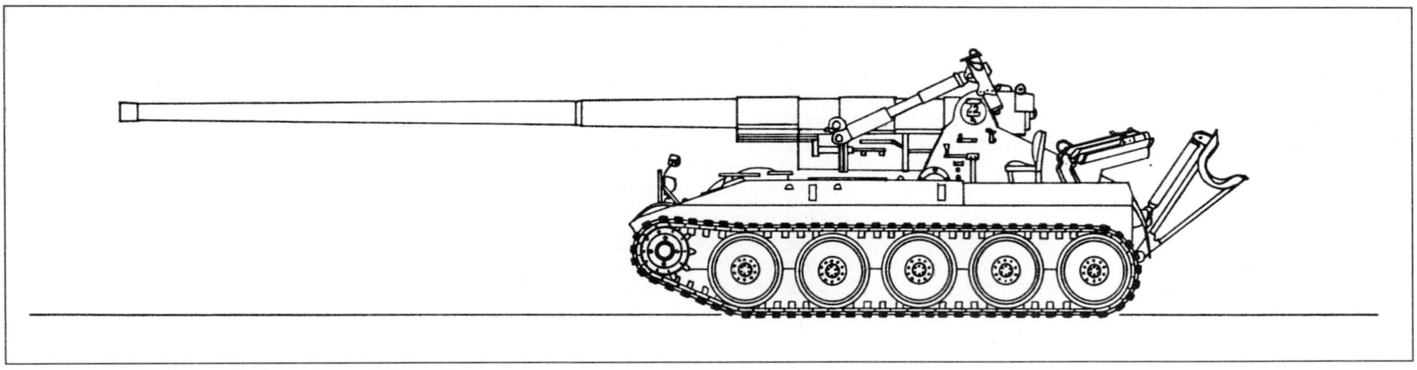

M107 175 mm self-propelled gun in travelling order

M107 175 mm self-propelled gun in travelling configuration 0018910

M107 175 mm self-propelled gun deployed in firing position (US Army)

Corporation of San José and Bowen-McLaughlin-York (BMY) of York, Pennsylvania, (both companies have now merged to become United Defense LP) with the first production model coming off the York line in 1965. Production was completed in May 1980, by which time 524 units had been completed by BMY.

In the US Army M107s were deployed in battalions of 12 guns, held at corps level. All M107s in the US Army and Marine Corps were converted to M110A2s by 1981 and other countries are, or already have, converted their M107s to the M110A2 configuration. Most NATO countries have now replaced their M107 and M110 system with the Lockheed Martin Missiles and Fire Control, Multiple Launch Rocket System.

A list of arms transfers issued by the United Nations for the period 1992 to 1998 show that no surplus M107s were exported by any country between these years.

Description

The M107 is operated by a team of 13, of which the (commander, driver and three gunners) are carried on the gun, with the rest in the M548 tracked cargo carrier, which also carries the ammunition. Details of this are given in *Jane's Military Vehicles and Logistics 2001-2002*.

The hull of the M107 is identical to the M110's and is made of all-welded cast armour and high-tensile alloy steel, with the driver at the front of the hull on the left, the engine to his right and the main armament at the rear.

The driver, who is the only member of the crew seated under armour, has a single-piece hatch cover in front of which are three M17 day periscopes. The engine is coupled to the Allison Transmission XTG-411-2A cross-drive transmission which is at the front of the hull.

The torsion bar suspension either side consists of five dual rubber-tyred roadwheels with the drive sprocket at the front and the fifth roadwheel acting as the idler. There are no track-return rollers. Attached to each roadwheel arm is a hydraulic cylinder which serves as a shock-absorber, a hydraulic bump stop and a suspension lockout which transmits recoil shock directly to the ground. The tracks are of the single-pin type with removable rubber pads.

The M107 has infra-red night vision equipment but does not have an NBC system or any amphibious capability. In most countries passive night vision equipment has replaced the infra-red equipment.

The M107 is armed with a 175 mm M113 gun (development designation T256E3) in mount M158, and has an elevation of +65°, a depression of −2° and a traverse of 30° left and right. Elevation, depression and traverse are all hydraulic, with manual controls available for emergency use. The gun has a hydropneumatic recoil system, minimum recoil being 0.711 m and maximum recoil 1.778 m. The breech block is of the Welin-step thread type with a percussion firing mechanism. Mounted at the rear of the vehicle on the left side is a rammer and loader assembly which lifts a 175 mm projectile from the rear or left side of the vehicle, positions and rams it into the chamber. Power for operating this system is obtained from the vehicle's hydraulic system but it can also be operated by handcranks.

The 175 mm gun only fires an HE projectile M437A2 or M437A1. The projectile weighs 66.78 kg, with the M437A1 containing 13.6 kg of TNT and the M437A2 containing 14.96 kg of Composition B. Range of the 175 mm gun with the standard M437 series projectile is as follows:

	Muzzle velocity	Max range
Charge 1	509 m/s	15,100 m
Charge 2	702 m/s	22,100 m
Charge 3	912 m/s	32,700 m

The M107 can fire the former Space Research Corporation ERSC Mk 7 MOD 7 projectile to a maximum range of 40,000 m. This was not, however, adopted by the US Army, although it has been used by Israel. Production of this round was completed by the Space Research Corporation many years ago and the company no longer exists.

The 175 mm ammunition is the separate loading type; two rounds are carried on the vehicle and the rest in the supporting United Defense M548 tracked cargo vehicle. The normal rate of fire is one round every 2 minutes but 2 rds/min can be fired for short periods. When travelling, the barrel is retracted slightly to the rear to reduce the overall length of the complete vehicle. Mounted at the rear of the chassis is a large hydraulically operated spade which is positioned before firing begins.

Fire-control equipment includes a panoramic sight M115 on mount M137 (magnification of ×4 and a 10° field of view), elevation quadrant M15, gunner's quadrant M1A1 and direct fire telescope M116C on mount M138 (magnification of ×3 and 13° field of view).

Specifications

Crew:
(on gun) 5
Combat weight: 28,168 kg
Unloaded weight: 25,915 kg
Power-to-weight ratio: 14.37 hp/t
Ground pressure: 0.81 kg/cm²
Length:
(gun forward, spade up) 11.256 m
(hull, without gun or spade) 5.72 m
Width: 3.149 m
Height:
(to top of barrel, travelling) 3.679 m
(to top of mount) 2.809 m
(to top of hull) 1.475 m
Ground clearance: 0.466 m
Track: 2.692 m
Track width: 457 mm
Length of track on ground: 3.936 m
Max road speed: 56 km/h
Fuel capacity: 1,137 litres
Max road range: 725 km
Fuel consumption: 1.568 litres/km
Fording: 1.066 m
Gradient: 60%
Side slope: 30%
Vertical obstacle: 1.016 m
Trench: 2.362 m
Engine: Detroit Diesel Model 8V-71T, turbocharged, 2-stroke, liquid-cooled 8-cylinder diesel developing 405 bhp at 2,300 rpm
Transmission: Allison Transmission XTG-411-2A cross-drive with 4 forward and 2 reverse gears
Suspension: torsion bar
Electrical system: 24 V
Batteries: 4 × 12 V Model 6TN
Armament: 1 × 175 mm M113 gun
Smoke-laying equipment: none
Ammunition:
(main) 2 rounds
Gun control equipment
Turret power control: hydraulic/manual
Max rate power traverse: 100 mils/s
Max rate power elevation: 100 mils/s
Gun elevation/depression: +65/−2°
Gun traverse: 30° left and right
NBC system: no
Night vision equipment: yes

Variants

There are no variants of the M107 although the M110 8 in (203 mm) self-propelled howitzer and the M578 ARV use the same chassis as the M107. The Israel Defence Force calls its 175 mm M107s the Romach and additional details of this as well as an illustration appear under Israel earlier in this section.

Status

Production complete. No longer marketed. In service with the following countries:

Country	Quantity	Remarks
Greece	12	
Iran	30	
Israel	140	some sources state 210 systems (some of which are surplus to requirements)
Korea, South	100	estimate
Turkey	36	

Manufacturers

Pacific Car and Foundry Company, (now PCF Defense Industries), Renton, Washington;
FMC Corporation, San José, California;
Bowen-McLaughlin-York (BMY), York, Pennsylvania.
The last two companies have now merged to become United Defense LP.

UPDATED

United Defense LP 155 mm Crusader advanced field artillery system

Development/Description

The Crusader advanced field artillery system consists of the 155 mm self-propelled howitzer XM2001 and the companion resupply vehicle XM2002.

Crusader is expected to the US Army's current M109-series 155 mm self-propelled howitzer, whose basic design is over 40 years old, and the M992/M992A2 Field Artillery Ammunition Support Vehicle (FAASV).

The Crusader programme definition and risk reduction (formerly demonstration/validation) contract started in December 1994. The US Army's total contract value for this phase of the development is US$1 billion.

The originally covered performance period until April 2003 and calls for the design, development, testing, and delivery of two prototype Crusader self-propelled howitzers and one prototype Crusader resupply vehicle.

Crusader prime contractor and systems integrator is United Defense LP with Team Crusader members General Dynamics Land Systems responsible for

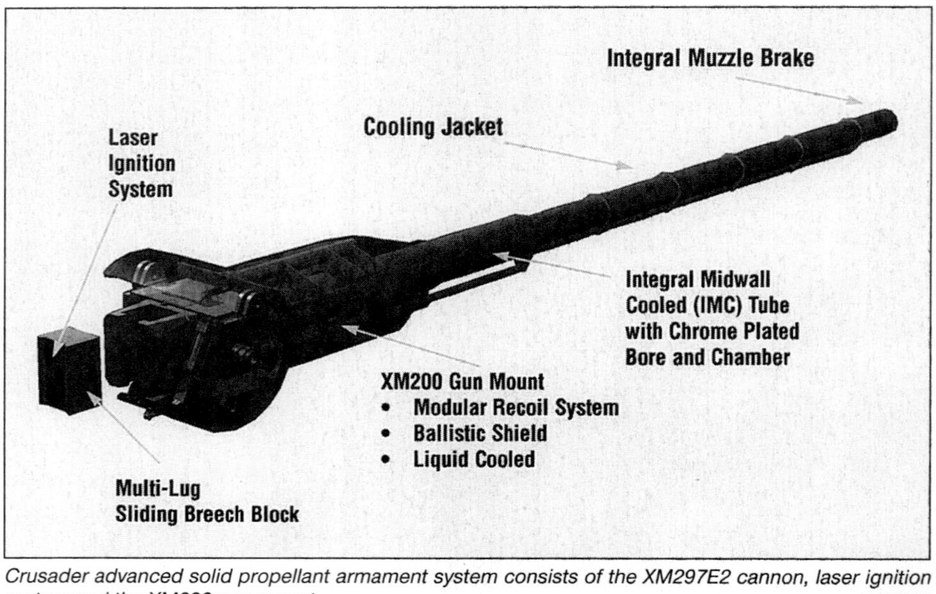

Crusader advanced solid propellant armament system consists of the XM297E2 cannon, laser ignition system and the XM200 gun mount
0006016

Crusader resupply vehicle wheeled based on the Oshkosh PLS (10 × 10) chassis (Artist's Impression) **2001**/0073706

mobility subsystem integration and Honeywell responsible for software development.

Key Team Crusader subcontractors include Alliant Techsystems, Applied Dynamics, Architecture Technologies, Ascent Logic, The Boeing Corporation, Digital Pilot Corporation, Kollmorgen, Lynx, MPC Products, Rational Software, Raytheon Systems, Saunders, Simplex Technologies, Technovative Applications, Upper Mohawk and Vector Research Incorporated.

In June 1999, United Defense LP announced that it would carry out final assembly and test of the Crusader system in a new facility to be established in Eglin, Oklahoma, adjacent to the Fort Sill artillery firing ranges.

At Eglin, United Defense LP intends to use subassemblies and major components provided by other manufacturing locations, including current United Defense and General Dynamics Land Systems manufacturing facilities.

It was expected that the first production XM2001 and XM2002 systems would be completed early in 2004, but as indicated later in this entry this has now slipped for a variety of reasons, including problems with software development.

Funding has been cut to help fund the US Army's new medium force. In February 2000 it was stated that a major redesign effort was underway to reduce the vehicles transport weight to 38 from 42 tonnes. To accomplish this goal the system will be reduced in size, carry modular armour and carry two smaller 25 round magazines instead of twin 30 round magazines as originally envisioned.

Crusader is being developed as a totally automated system for transfer, inventory, selection and loading of ammunition, fuze setting, fire control and secondary armament.

The fully automated features of the howitzer and resupply vehicles require only three crew members each for operation and maintenance. Automated ammunition handling enables crews to remain effective at extremely high operational tempos for extended periods of time.

Crusader will be equipped with integrated command and control providing onboard tactical and technical fire control as well as embedded training and decision

The Crusader will be supported by the resupply vehicle tracked (shown here) or wheeled 10 × 10 (Artists impression) **2001**/0106210

aids. Co-located crews in each vehicle will operate with full situational awareness, and will be fully linked with the combined arms force.

The Crusader self-propelled howitzer will carry 50 155 mm rounds on board (in two magazines each holding 25 155 mm projectiles, two identical propellant charge increment magazines and two identical increment magazines), have a maximum rate of fire of 10/12 min, with a minimum range of 5 km and a maximum range of 40+ km and will be able to fire within 25 seconds of coming to a halt. The howitzer will have the capability to fire multiple rounds to achieve simultaneous impact on target.

The Crusader resupply vehicle will carry 100 155 mm rounds on board and will rearm the self-propelled howitzer at a rate of 50 rounds in 12 minutes as well as resupplying the vehicle with fuel.

Automated rearmament of the Crusader self-propelled howitzer will be accomplished with all systems and crew under armour to ensure continual high availability for missions, reduced resupply time, and improved crew survivability. Automated rearmament will include projectiles, charges, fuel, lubricant and water.

The Crusader main armament will use a 155 mm solid propellant cannon system. The armament system consists of an XM297E2 cannon, a laser ignition system, and a new XM200 gun mount based on the existing XM183E1. The cannon tube is Integral Midwall Cooled (IMC) and employs external cooling jackets to reduce thermal signatures and maintain tube temperatures in a range that is safe for sustained firing.

The multilug, slide-block breech assembly provides fast, automated actuation to support high rates of fire with the laser ignition system being integrated for compatibility with breech automation to support crew safety when firing at high rates for long durations.

The XM200 gun mount utilises a Schneider-type recoil mechanism consisting of two modular recoil cylinders with built in internal replenisher, recoil buffer and two modular recuperator cylinders.

Crusader SPH1 prototype with the original heavier chassis design firing a zone six at Yuma proving ground **2001**/0073707

Specifications (provisional)

	XM2001	XM2002
Crew	3	3
Weight		
(combat)	43,636 kg	44,634 kg
(unloaded)	35,834 kg	33,929 kg
Power-to-weight ratio	n/avail	n/avail
Ground pressure	n/avail	n/avail
Length		
(overall)	12.9 m	7.5 m
(hull)	7.5 m	7.5 m
Width		
(overall)	3.5 m	3.5 m
(hull)	3.5 m	3.5 m
(turret)	3.2 m	3.2 m
Height		
(top of structure)	2.9 m	2.9 m
Ground clearance	0.437 m	0.437 m
Track width	558 mm	558 mm
Length of track on ground	4.3 m	4.3 m
Max speed		
(road)	67 km/h	67 km/h
(cross-country)	39 km/h	39 km/h
Fuel capacity	1,068 litres	1,068 litres
Range	405 km	405 km
Fording	1.22 m	1.22 m
Gradient	60%	60%
Side slope	40%	40%
Vertical obstacle	0.9 m	0.9 m
Trench	2.5 m	2.5 m
Engine	Honeywell LV-100-5 turbine	
Transmission	Allison 5060-3 automatic	
Suspension	hydropneumatic	
Electrical systems	270 V DC	270 V DC
Armament		
(main)	1 × 155 mm/56 calibres	nil
(secondary)	1 × 12.7 mm M2 MG	1 × 12.7 mm M2 MG
Ammunition		
(main)	50	100
Gun control equipment	electric	n/a
Gun elevation/ depression	+75°/–3°	n/a
Turret traverse	20° left/right	n/a
NBS system	yes	yes
Night vision equipment	yes	yes

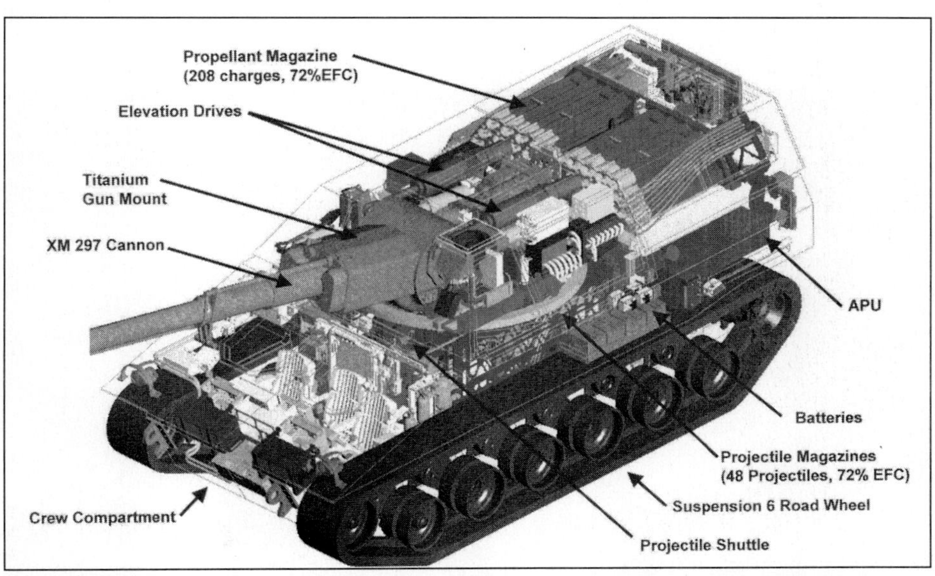

Cutaway drawing of the latest redesign of the Crusader 155 mm self-propelled automated howitzer showing position of main components **2001**/0073708

Crusader 155 mm self-propelled artillery system in latest configuration showing 155 mm ordnance at maximum elevation ***2001***/0106209

The actual cannon tube is chrome plated to minimise wear and erosion and an integral pepper pot muzzle brake is incorporated as standard.

The Crusader armament system chamber volume, shape, and details are tailored for compatibility with the Modular Artillery Charge System (MACS) solid propellant propulsion system.

The MACS has been jointly developed by the US Army and an industrial team comprising ARMTEC, General Dynamics Ordnance and Tactical Systems and Alliant Techsystems. MACS is a modular, increment-based propelling charge system consisting of the XM231 and XM232 configurations. MACS charges are compatible with Crusader and all existing 155 mm cannon designs. MACS eliminates the need to dispose of unused propelling charges, and enables automated ammunition handling due to its sturdy case composition.

Crusader will take advantage of the pending fielding of the Multi-Option Fuze for Artillery (MOFA). The fuze is compatible with all 105 and 155 mm cannon artillery systems. It integrates four function settings for bursting projectiles: proximity, time, point detonating and delay. MOFA has manual and inductive (man-out-of-the-loop) setting compatibility for high rates of fire. MOFA also supports pre-fuzing of projectiles, allowing the Crusader system to inductively set and verify time settings before ramming projectiles.

Both Crusader vehicles will be capable of 48 km/h cross-country speed and 67 km/h highway speed. To avoid hostile counterfire, Crusader vehicles will be able to move 750 m in less than 90 seconds under combat conditions.

Crusader's advanced mobility suite will also include external hydropneumatic suspension from Textron Marine & Land Systems, and advanced track and drive by wire controls.

System survivability will be enhanced by mobility and additional features such as designed-in signature management, integrated defence systems and a macron NBC system. It will also have position navigation/movement planning decision aids and embedded diagnostics/prognostics.

Crusader low-rate production may begin in 2006 and full-rate production is scheduled for 2008. The first fielded unit is expected to be equipped in the year 2007.

For the extensive trials programme two XM2001 and one XM2002 systems were completed in 2001 but, in the Summer, the ammuntion supply vehicle Minus was completed. This will not have the ammunition resupply system installed but will be used for reliability trials of the powerpack and the hydropneumatic suspension system.

Providing that there are no additional problems, the US Army is scheduled to make a Milestone II decision in April 2003 to enable it to enter the Engineering Manufacturing and Development (EMD) phase.

During this phase it is expected that seven Crusader self-propelled artillery systems, five tracked resupply vehicles and four wheeled resupply vehicles will be completed. The latter will be based on an Oshkosh M1075 (10 × 10) truck chassis which is already used in large numbers by the US Army.

Currently the original and much heavier Crusader platforms are being used to maintain the development programme.

In October 2000, it was revealed that the Crusader had recently fired three-round bursts at a rate of fire of six rounds per minute. These tests were carried out with the weapon being aimed and fired by remote control with the automatic loader first loading the projectile and the charge system.

The original intention of the US Army was to procure as many as 1,138 systems but this has now been cut back to 480 systems that will be issued to 13 active and six National Guard battalions.

In addition to software problems, a major redesign of the Crusader and its associated tracked resupply vehicle has been carried out to reduce weight and enable a C-5B aircraft to carry two systems.

By reducing the length and width of the vehicle, reducing the ammunition load (projectiles and charges), designing a modular armour kit and changing the power pack, it is estimated that the weight of the system has been reduced from 60 tons (US) to between 48 to 50 tons (US) complete with fuel, ammunition and armour. Its air portable weight is expected to be around 40 tonnes, for example without ammunition, most of the fuel and applique armour package.

Crusader was originally expected to be powered by a British Perkins Engines Company CV-12 diesel but late in 2000 it was decided to use the new Abrams/Crusader Common Engine (ACCE) which will consist of a Honeywell/General Electric LV100-5 turbine coupled to an Allison automatic transmission.

The hydropneumatic suspension either side consists of six dual rubber-tyred roadwheels with the idler at the front and the drive sprocket at the rear and track-return rollers.

The 155 mm liquid-cooled ordnance, and firing characteristics such as range and rate of fire are not affected by the weight reduction programme and it still has a crew of three who will load, aim and fire the 155 mm ordnance by remote control.

Crusader will have a 155 mm/56 calibre ordnance (including the muzzle brake) and during recent trials 15 rounds (projectile and charge) were fired in 1 minute 43 seconds. For comparison, the current 155 mm M109A6 Paladin took 9 minutes 52 seconds to fire the same number of rounds.

In March 1998, the US Army approved the design of the Crusader and granted Team Crusader authority to commence the detailed design and fabrication of the four Program Definition/Risk Reduction (PD/RR) phase prototypes.

Each Crusader battalion will have three batteries each with six XM2001 155 mm self-propelled howitzers. It was originally intended that each XM2001 would be supported by a dedicated XM2002 but it is possible that these may well be pooled and deployed to the XM2001s as required.

According to United Defense, The Crusader increases overall force effectiveness by more than 50 per cent and force survivability by 30 per cent or more.

Two Crusader systems provide the same firepower as a six gun battery of Paladins, yet ownership costs are reduced by 5 per cent. Crusader uses 32 per cent less ammunition per target and requires 38 per cent less support than the current system.

Maximum road speed will be 67 km/h with cross country speed being 39 km/h. It has a dash speed of 750 m in 90 seconds.

Late in 2001 it was stated that the US Army was looking at ways to accelerate developing and fielding the Crusader system which is fully funded in the Fiscal Year 2002 budget.

Improved Crusader

Prior to the anticipated cutback in Crusader procurement and the anticipated restructuring of the whole programme, it was stated that a number of improvements were already under consideration to enable the system to become the M2001A1/M2002A1.

These included a new advanced power pack, active armour protection, possible electrothermal chemical cannon, use of more advanced materials in the hull and turret, enhanced command and control, communications and intelligence and other improved survivability measures.

A notional M2001A2/M2002A2 upgrade package could have included a new crew cockpit, real-time cannon operating system, new composite armour, simulation-based acquisition processes, real-time electronics, robotics, hit/detection avoidance, prognostics/diagnostics and a new suspension system.

Status

Programme definition/risk reduction (but see text).

Manufacturer

United Defense LP (prime).

UPDATED

United Defense 155 mm/52 calibre International Howitzer

Development

Based on its experience in the design, development and production of 950 155 mm/39 calibre M109A6 Paladin Self-Propelled Howitzers (SPH) (covered in detail in the following entry) for the US Army, United Defense, Ground Systems Division, have developed a new for export only version called the International Howitzer.

Based on the proven M109A6 Paladin system, the new International Howitzer improves upon lethality, survivability and tactical mobility, having been designed to meet emerging international artillery requirements.

United Defense 155 mm/52 calibre International Howitzer which is now being offered for export ***2001***/0073766

When compared to the US Army's latest 155 mm M109A6 Paladin, the most significant improvement of the new International Howitzer is the installation of a 155 mm/52 calibre ordnance.

Description

In overall layout the 155 mm International Howitzer is identical to the M109A6 Paladin but has major differences at the ordnance and sub-system level.

The basic M109A6 includes an integrated automatic fire-control system with onboard ballistic computation, digital communications, inertial and GPS navigation, automatic weapon pointing and an integrated muzzle velocity radar system.

It also includes embedded prognostics/diagnostic capability, improved suspension and drive train as well as a micro climatic conditioning system that provides either heated or cooled NBC filtered air to the standard crew of four.

Unlike the earlier versions of the M109, the M109A6 Paladin system also includes a new cab structure that has been specifically designed to meet the higher firing loads of extended range ordnance.

Some M109 upgrades use the existing cab/turret but United Defense believe that this requires to be modified for the 155 mm/52 calibre application.

The International Howitzer system features a 155 mm/52 calibre ordnance built by Watervliet Arsenal that meets the Joint Ballistic Memorandum of Understanding (JMBOU) and is installed in a modified M182 mount. The ordnance is fitted with a fume extractor and a double baffle muzzle brake. This enables targets to be engaged at longer ranges than current M109s.

To increase its rate of fire and reduce crew fatigue, the International Howitzer is fitted with a semi-automatic rammer, with user selectable options for a semi-automatic primer feed or an integrated laser ignition system. Maximum rate of fire is about six rounds a minute. A total of 39 155 mm projectiles and associated charges are carried.

According to United Defense, the 155 mm International Howitzer receives fire missions whilst on the move, computes firing data, selects and takes up firing position, unlocks and points its 155 mm/52 calibre ordnance, shoots and then moves. It can fire the first round within 45 seconds.

Specifications

Crew: 4
Combat weight: 30,000 kg
Power-to-weight ratio: 14.66 hp/t
Ground pressure: 0.98 kg/cm²
Length:
(overall) 11.68 m
(hull plus bustle) 6.807 m
Width:
(overall, excluding stowage baskets) 3.149 m
(including stowage baskets) 3.922 m
Height:
(to top of MG mount) 3.236 m
(turret roof) 2.794 m
Ground clearance: 0.449 m
Track width: 381 mm
Track: 2.778 m
Length of track on ground: 3.692 m
Max road speed: 64 km/h
Fuel capacity: 504 litres
Max road range: 343 km
Fording: 1.05 m
Gradient: 60%
Side slope: 40%
Vertical obstacle: 0.53 m
Trench: 1.83 m
Engine: Detroit Diesel Model 8V-71T LHR, turbocharged, 2-stroke, liquid-cooled 8-cylinder diesel developing 440 bhp at 2,300 rpm
Transmission: Allison Transmission XTG-411-4A cross-drive with 4 forward and 2 reverse gears
Suspension: independent torsion bar with high-capacity shock absorbers
Electrical system: 24 V
Batteries: 4 × 12 V
Armament:
(main) 1 × 155 mm/52 calibre
(anti-aircraft) 1 × 12.7 mm M2 MG
Ammunition:
(main) 39 × 155 mm
(anti-aircraft) 500
Gun control equipment: automatic fire-control system
Turret power control: hydraulic/manual
Gun elevation/depression: +75°/–3°
Turret traverse: 360°
NBC system: yes, individual crew member
Night vision equipment: optional

The suspension has been upgraded from the standard M109A6 Paladin to include redesigned suspension with hydro-pneumatic struts in lieu of standard shock absorbers.

The power pack of the M109A6 has been retained, this consists of a Detroit Diesel 8V-71T engine developing 440 hp coupled to an Allison XTG-411-4A automatic transmission with quick disconnect.

The all-welded aluminium hull and turret is provided with internal spall liners and supplemental appliqué passive armour for higher battlefield survivability.

Standard equipment on the International Howitzer includes an Automatic Fire Control System (AFCS) with modified tactical software to support the new main armament and a unique International Howitzer mission capability, with integrated fibre-optic inertial and Global Positioning System (GPS) navigation capability.

Options for the International Howitzer include an auxiliary power unit as well as an integrated direct fire sight with laser range-finder. Taken overall, the International Howitzer maintains a full 95 per cent component commonality with the proven M109A6 Paladin.

Customers for the International Howitzer can purchase either complete brand new systems or new turrets for installation on upgraded chassis. With some 6,000 M109s in various configurations still in service, United Defense believe that the export market for International Howitzer is considerable.

Status

Prototype. Not yet in production or service.

Manufacturer

United Defense LP, Ground Systems Division.

UPDATED

United Defense LP M109A6 155 mm Paladin self-propelled howitzer

Development/Description

Following the demise of the Divisional Artillery Support Weapon System (DASWS), several companies were involved with the M109 related project known as the Howitzer Extended Life Programme (HELP). This programme eventually merged with the M109 Howitzer Improvement Programme (HIP) to bring the M109 up to a new M109A6 standard. Approval for the programme was given in November 1984 and invitations to tender were released in February 1985.

In October 1985, BMY was awarded a development contract to carry out the HIP programme. This involved the conversion of eight prototypes (six US and two for the Israeli Defence Forces) over a three-year period. A Special In-Process review of the programme was carried out in April 1986, with a further In-Process Review carried out during the second quarter of FY87.

The first of the US and Israeli HIP prototypes were rolled out on 30 March 1988 at BMY's York facility and subsequently undertook extensive prototype qualification tests at Aberdeen and Yuma proving grounds.

From September 1988 to February 1989, the US prototypes (the M109A3E2) underwent a technical testing programme, and operational tests were carried out at Fort Sill from May 1989 onwards.

The US Army made a production decision for the M109A6 on 5 September 1990, with the first Low-Rate Initial Production (LRIP) options being awarded in September 1990 for 44 vehicles. First production M109A6 systems were completed in April 1992. This was followed by two additional LRIP options for 60 vehicles each, which brought the total up to 164 vehicles. The LRIP portion of the M109A6 programme was completed in November 1994.

M109A6 Paladin self-propelled howitzer with ordnance in travelling lock (Richard Stickland) ***2002***/0126560

For the remainder of the M109A6 Paladin requirement there was a competition between BMY Combat Systems, FMC Corporation, Ground Systems Division and General Dynamics, Land Systems. In April 1993, FMC Corporation, Ground Systems Division was selected.

An initial contract worth US$30.5 million was awarded to FMC on 9 April 1993, to upgrade 60 M109s to the M109A6 Paladin standard, with the whole batch of 630 systems to be completed by October 1998. An additional 83 systems were ordered under options on the basic contract, bringing the total value to US$376 million.

In April 1997, the US Army placed an order with United Defense for the supply of 37 additional M109A6 systems. In November 1997 the US Army exercised an option for a further 36 systems which brought the grand total up to 950 units with final deliveries taking place in June 1999.

The 37 systems ordered in early 1997, marked the beginning of production for the National Guard. It is expected that the National Guard will field a total of 16 battalions with M109A6 Paladins the first three fully operational by mid-1999.

United Defense built the new M109A6 turrets with the upgraded chassis supplied by Letterkenny Army Depot. United Defense then integrates the chassis and turret and delivers the complete system to the US Army.

The US Army did deploy the M109A6 in batteries of eight weapons but these have now reverted to six gun batteries.

In mid-2000 the US Army placed an order for an additional seven M109A6 Paladin systems for the

National Guard at a total value of US$8.3 million. Deliveries took place from November 2001 through to January 2002.

Work will be carried out at the United Defense facilities at Aiken and York. Chassis work will be carried out at the Anniston Army Depot where they will strip down the chassis and then prepare it for shipment to United Defense.

United Defense LP offers the option of SW Swiss Ordnance Enterprise, positive ram semi-automatic projectile loader and automatic primer for the M109A6, although this was not adopted by the US Army. This is offered for the export market.

The M284 cannon is a modified M185 cannon with many improvements including a reinforced muzzle brake, advanced bore evacuator, a redesigned chamber and forcing cone, two shallow grooves for torque key replacing one deep groove, a new breech housing assembly, wiring of fasteners, added M49 firing mechanism and improved firing block, improved crank lead springs, strengthened breech handle springs and modified plunger, modification of the breech ring and a longer ordnance.

The M182A1 gun mount is a modified M182 with an added temperature sensor, a redesigned recuperator cylinder, a redesigned torque key, improved seals in the recoil cylinders and improved seals and new bearing installed in the elevation/equilibration cylinders.

With M203/M203A1 charge a range of 30,000 m can be achieved.

The new Automatic Fire-Control System (AFCS XXI), is in full rate production and is being retrofitted into all M109A6 Paladin fielded systems.

This new fire-control system is based on commercial technology with reduced system costs and parts obsolescence issues while providing significant improvements in computation speed, storage capacity and growth capability.

Features of the M109A6 include:

New turret with improved armour and Kevlar ballistic lining
Full-width turret bustle
Automatic Fire-Control System (AFCS)
Onboard ballistic fire-control computer and navigation system providing automatic gun pointing
New gun drive servos
Muzzle velocity management system
Modular azimuth positioning System (MAPS)
Global positioning system
VIS intercom
SINCGARS radio (Single Channel Ground/Air Radio System)
Built in training software
M182A1 gun mount
M284 cannon which allows the M203 charge to be used
New traversing mechanism
Upgraded suspension system
Driver's passive night vision system
Electrical and hydraulic system improvements, including a new 650 A generator
Increased ammunition stowage
Microclimate cooling system
Remotely operated barrel travel lock
Segregated hydraulic compartment and hydraulic line fuzing
Engine starter protective circuit
Air cleaner blower relay
Engine sensor and connector covers
Crossover tube protective cover
New crew compartment water and oil drains
Slave start circuit
New sensors to monitor power pack parameters
Ventilated face piece system for NBC conditions
Halon fire extinguishing system
Improved hatch and night vision provisions for the commander
Mounted water ration heater
External stowage baskets
Final drive quick disconnect
Propellant segregation
Digital communications link
Four-man crew.

A flick rammer loader assist was installed on the Israeli version of HIP which yielded burst rates of fire of three rounds in less than 13 seconds.

Description

The M109A6 has a similar layout to the M109, covered in the following entry, apart from the improvements mentioned in the development section.

155 mm M109A6 Paladin self-propelled howitzer from the rear clearly showing extended turret bustle and extensive external stowage (Richard Stickland) ***2002***/0126561

M109A6 HIP

A number of additional potential improvements have been identified by the US Army for the United Defense LP 155 mm M109A6 Paladin self-propelled howitzer and these have been prioritised in a System Improvement Plan (SIP).

The 15 identified subsystem upgrades fall into six major categories: computer growth, improved accuracy, increased rate of fire, survivability, reliability, availability and maintainability and crew safety/comfort.

Additional AFCS upgrades under development include an advanced digital display allowing for future expansion to include items such as Interactive electronic technical manuals and situational awareness screens.

The current microclimate conditioning system which provides NBC protection as well as conditioned air to reduce crew stress, is now being redesigned to replace the current R-12 refrigerant with a more environmentally friendly alternative. This is expected to be retrofitted to all current US Army M109A6 Paladin systems.

Improvements to the rate of fire are being studied including addition of a laser ignition system, a semi-automatic loader and an automatic fuze setter. Prototypes of these improvements have already been developed, tested and demonstrated to the US Army for consideration.

Other individual upgrades proposed by the SIP include an upgrade to the digital communications speed, laser range-finder, installation of a driver's thermal viewer to replace the current image intensification device, hull vulnerability reduction and an upgrade to the Prognostic/Diagnostic Interface Unit (PDIU).

Enhanced Paladin Demonstrator

As a private venture, the company has also built a single Enhanced Paladin Demonstrator which successfully underwent US Army trials in 1994.

This is very much a testbed and has a 155 mm 52 calibre barrel, additional armour, semi-automatic ammunition handling, upgraded suspension and additional ammunition capacity. There is no current work being carried out on this vehicle.

M109 export upgrades

Details of this are given in the following entry.

International Howitzer

This has been developed from the M109A6 Paladin for the export market and additional details are given in the previous entry. As of late 2001 this remains at the prototype stage.

Status

Production. In service with US Army (950 with another seven on order).

Manufacturer

United Defense LP.

UPDATED

Specifications

Crew: 4
Combat weight: 28,849 kg
Max payload: 3,130 kg
Power-to-weight ratio: 15.25 hp/t
Ground pressure: 0.951 kg/cm²
Length:
(overall) 9.677 m
(hull plus bustle) 6.807 m
Width:
(overall, excluding stowage baskets) 3.149 m
(including stowage baskets) 3.922 m
Height:
(to top of MG mount) 3.236 m
(overall) 3.62 m
Ground clearance: 0.457 m
Track width: 381 mm
Track: 2.778 m
Length of track on ground: 3.962 m
Max road speed: 64.4 km/h
Fuel capacity: 504 litres
Max road range: 344 km
Fording: 1.05 m
Gradient: 60%
Side slope: 40%
Vertical obstacle: 0.53 m
Trench: 1.83 m
Engine: Detroit Diesel Model 8V-71T LHR, turbocharged, 2-stroke, liquid-cooled 8-cylinder diesel developing 440 bhp at 2,300 rpm
Transmission: Allison Transmission XTG-411-4A cross-drive with 4 forward and 2 reverse gears (with quick disconnect)
Suspension: independent torsion bar with high-capacity shock absorbers
Electrical system: 24 V
Batteries: 4 × 12 V
Armament:
(main) 1 × 155 mm M284 howitzer
(anti-aircraft) 1 × 12.7 mm M2 MG
Ammunition:
(main) 39 (37 + 2 CLGP)
(anti-aircraft) 500
Gun control equipment: automatic fire-control system
Turret power control: hydraulic/manual
Gun elevation/depression: +75/–3°
Turret traverse: 360°

United Defense LP M109 Series of 155 mm self-propelled howitzers

Development

In January 1952, a conference was held in Washington DC on the subject of self-propelled artillery which indicated an urgent need for improved self-propelled artillery. Shortly after this conference, preliminary concept studies began for a self-propelled howitzer to replace the 155 mm M44. The first studies for the design of the new vehicle, called the Howitzer 156 mm Self-Propelled T196, were presented to CONARC in August 1952, but were rejected, as were additional studies presented in September 1953. At a conference in May 1954, a concept was finally approved for presentation to CONARC.

In June 1954, a conference was held at Fort Monroe to review the military characteristics and to determine the whole self-propelled programme. It was then decided that future concepts of the T196 would be prepared along the design proposed for the Howitzer 110 mm Self-Propelled T195. Concept studies continued along these lines and, in June 1956, it was finally decided to use the basic hull and turret of the T195 but with a 155 mm howitzer instead of the original 156 mm howitzer.

In October 1956, the mockup of the T196 was reviewed and verbal authority was given to proceed with development of the first prototype.

The first prototype of the T196 was completed in 1959, about six months later than the 105 mm T195. During preliminary user evaluation at Fort Knox a number of failures occurred in the suspension. The prototype differed from later vehicles in that it had a different shaped hull and turret, the seventh roadwheel acted as the idler and it was powered by a Continental petrol engine. In 1959, a policy was established that diesel rather than petrol engines would be used for future combat vehicles and the prototype of the T196 was then fitted with a Detroit Diesel diesel engine and redesignated the T196E1.

In February 1961, an order was placed for two T196E1 preproduction vehicles which were delivered within six months. After further trials the T196E1 was classified as a Limited Production Type in December 1961. Two months before this decision, in October, a letter order was given to the Cadillac Motor Car Division for one year's production of the T196E1 at the Cleveland Army Tank Plant. The first production vehicles were completed in October 1962. In January 1963, an extension was authorised to continue the classification of Limited Production and, in July the same year, the T196E1 was classified as standard A and designated the Howitzer, Medium, Self-Propelled: 155 mm, M109. Early in 1963, a contract was awarded to Cadillac for the second year of production. The contract for the third year of production, awarded in December 1963, went to the Chrysler Corporation, although production remained at the Cleveland Army Tank Plant. First M109s were issued to the US Army in June 1963.

Further development of the M109 resulted in the M109A1 and the M109A2 and, by early 1979, production had exceeded 4,000 units. M109 production for the US armed forces by the Cleveland Army Tank Plant was as follows:

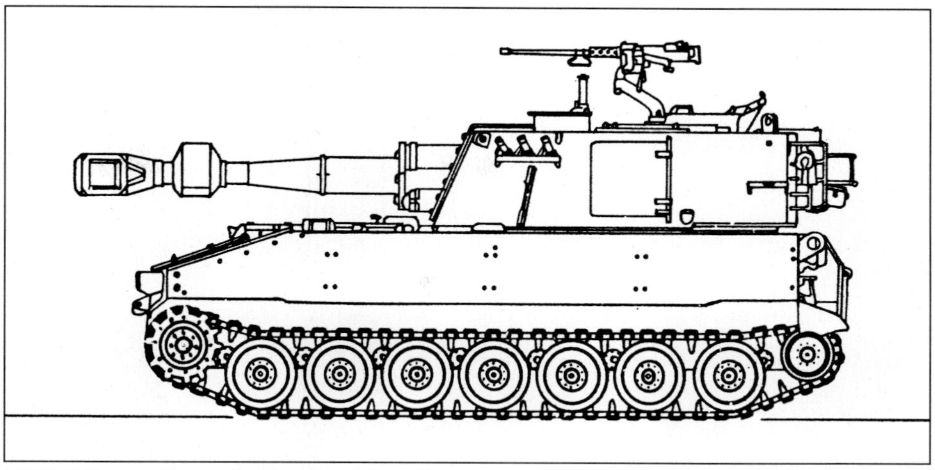

M109 155 mm self-propelled howitzer in travelling order

FY	US Army	US Marine Corps
1962	217	30
1963	150	58
1964	360	-
1965	360	-
1966	454	-
1967	420	36
1968	-	-
1969	-	26

BMY, now United Defense LP, began production of the M109 in 1974 and in 1994 released the following production figures on the 155 mm M109 series of self-propelled howitzer:

Year	M109A1B	M109A2	M109A5	Total
1974	0	99	0	99
1975	0	222	0	222
1976	0	169	0	169
1977	0	183	0	183
1978	0	178	0	178
1979	71	45	0	116
1980	231	70	0	301
1981	335	53	0	388
1982	276	260	0	536
1983	199	120	0	319
1984	194	27	0	221
1985	231	75	0	306
1986	209	0	0	209
1987	256	0	0	256
1988	106	0	0	106
1989	119	0	0	119
1990	117	0	0	117
1991	182	0	0	182
1992	105	0	0	105
1993	110	0	0	110
1994	0	0	10	10

There has been no recent new build production for the US Army but the M109A5 remains in production for the export market as and when required. Recent exports sales include Austria, Greece, Taiwan and Thailand.

M109 HIP (M109A6)

Further development has resulted in the new M109A6 Paladin for which there is a separate entry (qv).

Description

(M109)

The M109 has a crew of six, consisting of commander, gunner, three ammunition members and the driver. The hull is made of all-welded aluminium armour. The driver is at the front of the hull on the left, the power pack is to his right and the turret is at the rear.

The driver has a single-piece hatch cover that opens to the left, with three M45 day periscopes in front that can be covered by small metal flaps to prevent damage. The driver can replace one of his M45 day periscopes with a passive night vision periscope from within the vehicle.

The Detroit Diesel Model 8V-71T engine is coupled to an Allison Transmission XTG-411-4A cross-drive transmission which is at the front of the hull.

The all-welded aluminium armour turret at the rear of the hull has a square hatch in each side that opens to the rear, and twin doors in the turret rear. The commander is seated on the right side of the turret and has a cupola that can be traversed through 360°, a single-piece hatch cover that opens to the rear and an M27 day periscope. Pintle mounted on the forward part of the commander's cupola is a 12.7 mm (0.50) M2 HB local defence machine gun. The gunner is seated on the left side of the turret and has a square single-piece hatch cover that opens to the right. Twin doors are provided at the rear of the turret for ammunition resupply. Mounted at the rear of the hull, each side of the hull door, is a large spade which is lowered manually to the ground before firing.

The torsion bar suspension either side consists of seven dual rubber-tyred roadwheels with the drive sprocket at the front and the idler at the rear. There are no track-return rollers. The tracks are of the single-pin, centre guide type with replaceable rubber pads.

The M109 is fitted with night vision equipment but does not have an NBC system. The basic vehicle can ford to a depth of 1.828 m without preparation. It can be fitted with an amphibious kit consisting of nine air bags, four each side of the hull and one at the front. The bags, which are not carried on the vehicle as part of its normal equipment, are inflated from the vehicle and it can then propel itself across rivers by its tracks at 6.43 km/h.

Main armament is an M126 155 mm howitzer in an M127 mount, with a fume extractor and a large muzzle

M109A2 155 mm self-propelled howitzer in service with Belgian Army (Pierre Touzin) **2000**/0085181

M109A5 155 mm self-propelled artillery system with ordnance in travelling lock and muzzle brake covered up (Michael Jerchel) **2001**/0106070

M992 Field Artillery Ammunition Support Vehicle (FAASV) 0021017

Upgraded M109A3GN of the Norwegian Army showing different external stowage (Svein Wilger) ***2002***/0126558

brake. The recoil system is hydropneumatic and the breech block is of the Welin-step thread type. The weapon has an elevation of +75°, a depression of –5° and the turret can be traversed through a full 360°. Gun elevation and depression and turret traverse are hydraulic with manual controls for emergency use. Normal rate of fire is 1 rd/min but 3 rds/min can be fired for a short period. Fire-control equipment includes an elbow telescope M118C for direct fire with a magnification of ×4 and a 10° field of view, panoramic telescope M117 for indirect fire with a magnification of ×4 and a 10° field of view, quadrant fire control M15 and gunner's quadrant M1A1. The weapon can fire the following types of ammunition:

Copperhead cannon-launched guided projectile M712 with a range of 16,400 m;

HE (M107) with the projectile weighing 42.91 kg, a maximum muzzle velocity of 562.4 m/s and a maximum range (charge 7) of 14,600 m;

HE (M795) which is the replacement for the M107; the projectile weighs 46.90 kg;

HE (M449A1) (carries 60 anti-personnel grenades) with the projectile weighing 43.09 kg, a maximum muzzle velocity of 563 m/s and a maximum range of 14,600 m;

HE (M483A1 series) (carries 64 M42 and 24 M46 dual-purpose grenades) with the projectile weighing 46.53 kg, a maximum muzzle velocity of 535.2 m/s and a maximum range (charge 7) of 14,320 m;

HE (M692) (carries 36 anti-tank mines) with the projectile weighing 46.49 kg, a maximum muzzle velocity of 535.2 m/s and a maximum range (charge 7) of 14,320 m;

M718/M741 (containing nine anti-tank mines) with a maximum range of 14,320 m;

HERA (M549) with the projectile weighing 43.54 kg, with a maximum range (charge 7) of 19,300 m;

Illuminating (M485) with the projectile weighing 41.73 kg, a maximum muzzle velocity of 576.9 m/s and a maximum range (charge 7) of 13,586 m;

Illuminating (M818) with the projectile weighing 46.26 kg, a maximum muzzle velocity of 536 m/s and a maximum range of 11,600 m;

Smoke (M825) with the projectile weighing 46.53 kg, a muzzle velocity of 535.2 m/s and a maximum range of 14,320 m;

Smoke BE (M116 series) with the projectile weighing 42.22 kg, a maximum muzzle velocity of 562.4 m/s and a maximum range (charge 7) of 14,600 m;

Smoke WP (M110 series) with the projectile weighing 44.4 kg, a maximum muzzle velocity of 563.9 m/s and a maximum range (charge 7) of 14,600 m;

M804 Practice.

Variants

M109A1

The M109A1 is basically the M109 fitted with a new and much longer barrel, designated the M185, an improved elevation and traverse system and a strengthened suspension system. The M109A1 fires an HE projectile to a range of 18,100 m compared with the 14,600 m of the M109, while a rocket-assisted projectile extends the range to 24,000 m. The M109A1 was type-classified as Standard A in October 1970 with first conversion kits made available early in 1972 and the first vehicles, which were converted from standard M109s, becoming operational in 1973. The M109A1 weighs 24,070 kg fully loaded and is 9.042 m long including the barrel. The weapon can fire the following projectiles:

HE (M795) which is the replacement for the M107, the projectile weighs 46.9 kg;

HE (M107) with the projectile weighing 42.91 kg, a maximum muzzle velocity of 684.3 m/s and a maximum range (charge 8) of 18,100 m;

HE (M483A1) (carries 64 M42 and 24 M46 dual-purpose grenades) with the projectile weighing 46.53 kg, a maximum muzzle velocity of 650 m/s and a maximum range (charge 8) of 17,500 m;

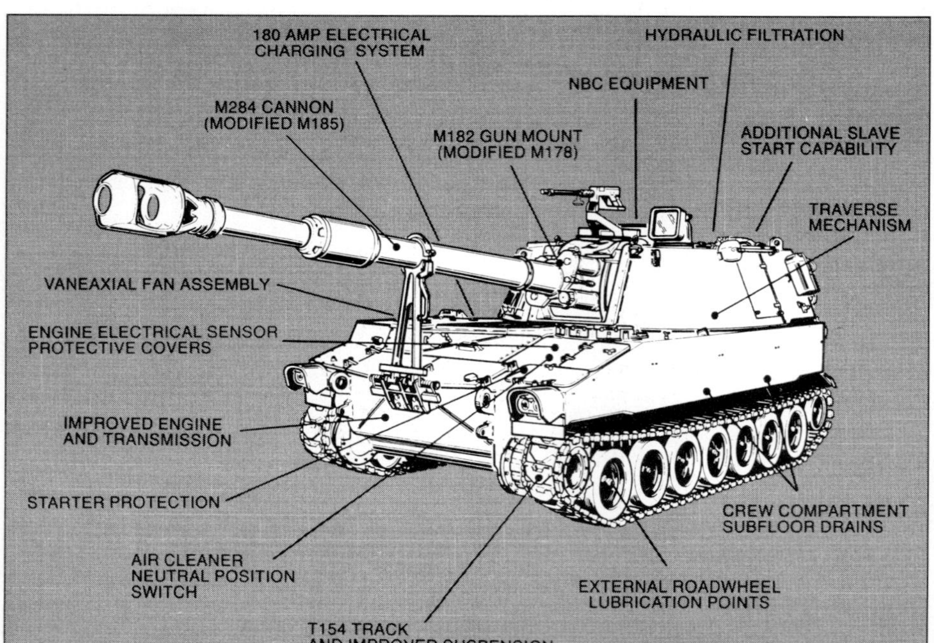

Main features of the United Defense LP 155 mm M109A5 self-propelled howitzer

HE (M692) (carries 36 anti-tank mines) with the projectile weighing 46.49 kg, a maximum muzzle velocity of 650 m/s and a maximum range (charge 8) of 17,740 m;

HERA (M549A1) with the projectile weighing 43.54 kg and a maximum range of 23,500 m;

M864 Improved Conventional Munition, base bleed, type-classified in December 1987;

Illuminating (M485) with the projectile weighing 41.73 kg, a maximum muzzle velocity of 684.5 m/s and a maximum range (charge 8) of 17,500 m;

Illuminating (M818) with the projectile weighing 46.26 kg, a maximum muzzle velocity of 684.3 m/s and a maximum range of 18,100 m;

Smoke BE (M116) with the projectile weighing 42.22 kg, a maximum muzzle velocity of 684 m/s and a maximum range (charge 8) of 18,100 m;

Smoke WP (M110 series) with the projectile weighing 44.4 kg, maximum muzzle velocity of 684.3 m/s and a maximum range of 18,100 m;

Smoke (M825) with the projectile weighing 43.53 kg, a muzzle velocity of 660 m/s and a maximum range of 17,500 m;

HE ICM M449A1 carrying 60 anti-personnel grenades with a maximum range of 18,100 m; weight approximately 43.1 kg and length 899 mm;

ADAM M692/M731 Area Denial Artillery Munition containing 36 self-destruct anti-personnel mines that are ejected in flight. Maximum range 17,740 m, weight 46.72 kg and length 899 mm;

M712 Copperhead, maximum glide range 16,400 m;

M823 Trainer is used for crew training and is not fired;

AT RAAMS M718/M741 containing nine anti-tank mines that are ejected during flight. Weight is 46.72 kg and length 899 mm, maximum range 17,740 m.

M109A2

The M109A2 entered production at BMY Combat Systems in 1978, with first deliveries made early in 1979. The first production quantity of 103 vehicles was with FY77 funding. Major changes from the M109A1 include a redesigned rammer and improved recoil mechanism, engine operation warning devices, a redesigned hatch and door latches, an improved hydraulic system and a bustle designed to carry an additional 22 rounds of ammunition.

M109A2 production carried out by BMY for the US Army was as follows:

FY	US Army total
1976	12
1977	103
1978	250
1979	136
1980	96
1981	-
1983	-
1984	123
1985	70

In addition to the above, 36 M109A2s were produced for the National Guard using FY82 funds. There has been no recent procurement of the M109A2/M109A3 by the US Army or Marine Corps.

M109A3

The M109A3 is a US Army depot converted M109A1. The M109A3 conversion includes:

boresight alignment driver;

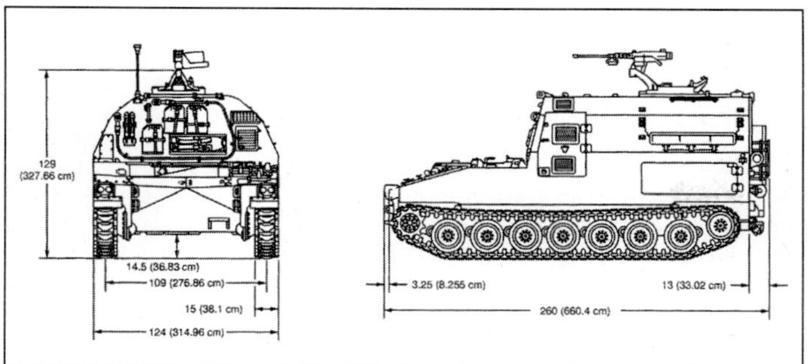

Front and side drawings of United Defense M992A2 Field Artillery Ammunition Support Vehicle (FAASV) ***2002***/0126562

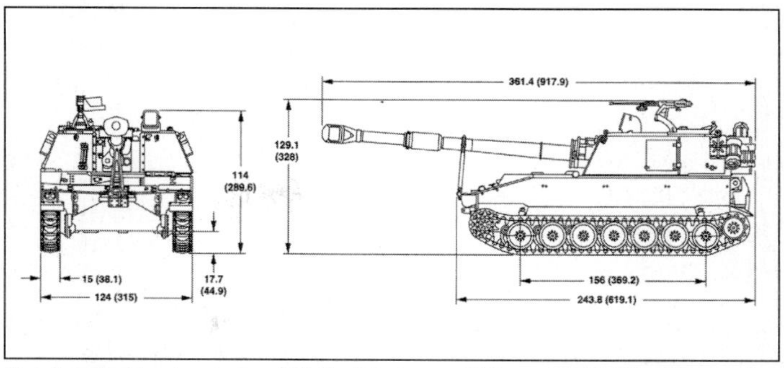

Front and side drawings of United Defense 155 mm M109A5 self-propelled howitzer in travelling configuration ***2002***/0126563

M127 to M178 gun mount kit;
selected RAM and safety kits, which included fuel system air purge, driver's instrument panel, bustle/rack, propellant stowage, torsion bar, counter recoil buffer and upper recoil cylinder.

M109A4

The M109A4 is a converted M109A2 or M109A3. The major changes are the incorporation of Nuclear, Biological and Chemical (NBC) requirements and Reliability And Maintainability (RAM) improvements. Selected improvements are found in the following kits: traverse mechanism, NBC equipment, hydraulic power pack filter, starter protection, external slave start, floor drains.

M109A5

The M109A5 is an upgraded M109A4, with the incorporation of the Reserve Component/Modified Armament System (RC/MAS). The RC/MAS is a range capability improvement kit and, firing the M203A1 charge, ranges of 30 km can be achieved with rocket-assisted projectiles. The RC/MAS kit includes:
M284 cannon assembly;
M182 gun mount.

The M109A5 is available as a new production vehicle from the manufacturer or an M109A2 upgrade kit.

The M109A5 readily accommodates such individual customer requirements as the Global Positioning System (GPS) or the Automatic Fire-Control System (AFCS).

The new higher horsepower Low Heat Rejection (LHR) engine and uprated transmission deliver increased power in all climates due to higher engine torque and more efficient cooling/heat rejection systems.

The engine starter protection device prevents starter burnout. The 180 A alternator extends battery life, even with the increased electrical load from the NBC equipment.

The elevation/equilibration cylinder and the turret traverse clutch assembly are redesigned to prevent failures. Protective covers for engine electrical sensors and conductors rule out accidental damage.

Relocated and easily accessible filters improve hydraulic power pack filtration and easy to replace track pads and external roadwheel lubrication points simplify maintenance in the track area.

An external NATO power receptacle permits the M109A5 to be electrically powered by the M992A1 Field Artillery Ammunition Support Vehicle. The NATO slave start receptacle has been relocated to the driver's compartment for ease of access. The crew compartment sub-floor drains allow easy washdown of NBC contaminant and of other fluids during maintenance.

A ventilated facepiece system gives crew members filtered, conditioned air during NBC operations and as needed.

In addition to the Austrian order for modified M109A5s (covered below), other customers have ordered the M109A5 including Greece (12 delivered from 1998), Taiwan (28 delivered in 1998) and Thailand (20 delivered).

US and Swiss M109 upgrades

In 1998, United Defense LP and the SW Swiss Ordnance Enterprise, signed an agreement to market on a worldwide basis a series of upgrades for the United Defense LP M109 family of 155 mm self-propelled artillery systems.

The main part of the upgrade is the replacement of the existing 155 mm 39 calibre barrel by a new Swiss designed and built chrome-plated 155 mm 47 calibre barrel fitted with a new double baffle muzzle brake and fume extractor.

Using charge 9, a max range of 23,000 m can be achieved or 30,100 m with a base bleed cargo round. Using charge 10 a maximum range of 26,100 m can be achieved or 36,400 m with a base bleed cargo round.

The 155 mm 47 calibre ordinance is already in volume production for the Swiss Army, which is upgrading 348 of its M109s to the latest Pz Hb 88/95 standard (qv Switzerland). It is also fitted to the 85 M109A3s upgraded by RDM Technology of the Netherlands for the United Arab Emirates under a major contract awarded in August 1995.

In addition, the two companies will also market other elements of the Swiss upgrade which also includes a barrel temperature warning device, semi-automatic loader, navigation and positioning system based on ring laser gyro unit, increased ammunition stowage, gun travel lock, fire suppression system and new electrical system. An auxiliary power unit (APU) can be fitted to allow the main engine to be switched off.

United Defense LP are leading marketing of the upgrade package with SW Swiss Ordnance Enterprise providing back up. The design of the upgrade package is modular so that the user can procure only those parts required. Other sub-systems can be selected, for example a different navigation and positioning system, if required by the user.

Depending on the level of upgrading required, the upgrade could be accomplished in the users own facility, Switzerland or the US.

The 155 mm 47 calibre barrels would be provided from the Swiss production line or from the US governments Watervliet Arsenal where all US artillery and tank barrels are built.

The installation of a 155 mm 47 calibre ordnance, together with new ammunition gives a significant increase in range and rate of fire over current M109s

Early in 1999, it was announced that United Defense LP and Barnes & Reinecke had signed an agreement to market M109 self-propelled howitzers on an international basis.

The two companies will combine capabilities to provide the best solution to customer needs. United Defense has manufactured and sold the M109 for more than 15 years while Barne's & Reinecke has provided systems technical support for nearly as long.

Together the two companies will offer international customers the capabilities and options provided by both firms in a cost effective manner.

Customers for the M109 can now upgrade existing howitzers to an A5 model, product improve A5 models by integrating communications, information and automated features, or purchase new build howitzers through a single US source.

M109 (Austria)

Austria purchased 38 M109 and 18 M109A2 self-propelled howitzers from the United States.

In mid-1994, Austria purchased 27 M109 series self-propelled artillery systems which were surplus to the requirements of the British Army and these have now been delivered.

M109A50e (Austria)

Early in 1995, through the US Army Armament and Chemical Acquisition and Logistics Activity at Rock Island, Austria awarded United Defense LP a contract worth US$48.6 million for the supply of 54 155 mm M109A50e self-propelled howitzers for the Austrian Army.

The M109A50e has a number of unique features that differ from the standard M109A5 including an Austrian semi-automatic flick rammer, inertial positive navigation system, Swiss electrical system and the same 155 mm ordnance as used on the M109A6 Paladin currently in production for the US Army.

The M109A50e is armed with a 155 mm M284 cannon in an M182 mount. Using the US M203A1 charge, ranges of 30 km can be achieved firing a rocket-assisted projectile.

The first production M109A50e for the Austrian Army was rolled out on 6 May 1997 and six vehicles were delivered each month until delivery was complete.

M109A2 (Belgium)

In August 1983, the Belgian Minister of Defence signed a contract with BMY for the purchase of 127 M109A2 self-propelled howitzers. The first vehicles were delivered in August 1984 and all were delivered by the end of 1985. These vehicles are used by the six field artillery battalions, the Brasschaat Artillery School and the maintenance reserve. The older M109s were upgraded to M109A3 standard by the Arsenal du Matériel Mécanique et de l'Armement, at Rocourt, near Liège.

From 1998 the operational M109A2's were fitted with a North Finding Module and between 2002 and 2006 a total of 108 M109A2's will be upgraded to the M109A4 standard.

M109A3 (Brazil)

Brazil has taken delivery of 40 M109A3 SPH from Sabiex International with all delivered by mid-1999 and issued to three artillery regiments of the Brazilian Army.

M109A3 (Denmark)

Early in 1989, the Danish Army stated that the modernisation of its 76 M109 155 mm self-propelled howitzers to the M109A3 standard had been completed. The modernisation was carried out in Denmark with kits supplied by FMS Corporation (now Marvin Land Systems) of Los Angeles, California.

Egyptian surplus US Army M109A2/A3

In November 2001, US Defense Security Co-operation Agency notified the US Congress of a possible Foreign Military Sale (FMS) to Egypt of 201 surplus US Army 155 mm M109A2/A3 self-propelled howitzers (SPH) at an estimated cost of US$77 million.

The deal, which has to be approved by Congress, also includes communications equipment, spare and repair parts, publications and technical data, US government and contractor support and logistics personnel services.

In the Spring of 2001, a similar deal for 179 surplus M109A2/A3 SPH at a cost of US$48 million was proposed to Congress but, in the end, Egypt did not take up this offer.

If Egypt does accept the offer for the latest batch of 201 M109A2/A3 SPHs then these would be overhauled and refurbished in the US prior to delivery by United Defense LP, Ground Systems Division.

These have to be overhauled before delivery as they are not currently in an operational condition having been replaced in US Army service by the more modern 155 mm M109A6 Paladin.

According to the US Defense Security and Co-operation Agency, Egypt will use these M109A2/A3 SPH to replace ageing Russian-supplied towed artillery equipment (for example the 122 and 152 mm) and to compliment the modernisation of General Dynamics Land Systems M1A1 Abrams and M60A3 Main Battle Tank (MBT) equipped divisions.

Currently the Egyptian Army operates 164 United Defense 155 mm M109A2 self-propelled howitzers delivered some years ago. These are supported by 51 M992 Field Artillery Ammunition Support Vehicles

(FAASV) and 72 Fire Direction Centre (FDC) vehicles also supplied new by United Defense form their York facility. These are also based on a M109 series chassis.

In 2001 the Egyptian Army took delivery from United Defense of another batch of 24 SP122 122 mm self-propelled howitzers and these joined the 100 SP122 122 mm systems supplied under earlier contracts.

M109 (Germany)

The original German Army vehicles were designated the M109G, with a sliding breech developed by Rheinmetall extending the effective range of the weapon to 18,000 m.

In 1983, the German Army purchased from the FMS Corporation (now Marvin Land Systems), Los Angeles, California, 586 kits to convert its M109Gs to the M109A3G. The newly modified M109A3G has an extended range of 24,000 m and uses a recently developed gun tube made by Rheinmetall Weapons & Munitions but still uses the original M109G sliding breech mechanism. The M109A3G also has a different fire-control system and three forward-firing 76 mm smoke grenade dischargers either side of the turret. There is a separate entry under Germany for the modernised Rheinmetall M109A3G.

The German fleet has now undergone the AURORA modification at the now Rheinmetall Landsysteme, with the modified vehicles designated the M109A3GEA1. Additional details of the German M109A3G are given in this section under Germany.

M109A5 (Greece)

In May 1998, Greece placed an order with United Defense for the supply of 12 M109A5 systems for delivery from late 1998. The contract also included an option for a further 12 M109A5 systems. Total value of the contract, with options, is about US$50 million.

M109 (Israel)

The M109A1s of the Israel Defence Force are designated the M109 AL. They have additional stowage racks on the hull and turret sides, allowing them to carry more ammunition inside. The IDF is planning to upgrade some of its M109s to incorporate selected features of the M109A6. Additional details of these systems, which have the local name of Doher, are given in this section under Israel.

M109L (Italy)

There is a separate entry under Italy for the M109L developed by Otobreda to meet the requirements of the Italian Army.

Korea, South

A total of 1,040 have been built in South Korea with final deliveries in 1997.

M109 (Netherlands)

The Netherlands has M109s with their turrets removed for training drivers. In place of the turret is a fully enclosed cabin with observation windows. The Royal Netherlands Army was the first in Europe to upgrade the M109 to the M109A3. In 1980, the Royal Netherlands Army released its contract to RDM Technology for the conversion of 136 of its M109s to the M109A3 configuration. RDM Technology purchased kits from the now Marvin Land Systems of the USA, which was also its contractor for an earlier parts contract for 86 M109A2s co-produced in the Netherlands for the Royal Netherlands Army.

Canada and the Netherlands awarded RDM Technology contracts to refurbish parts of their M109 fleets of self-propelled howitzers, details of these are given under the Netherlands. This upgrade programme is now complete.

M109 (Norway)

These have all been upgraded to the M109A3GN standard which is almost the same as the German M109A3G configuration covered earlier in this section under Germany.

M109 (Spain)

It is believed that Spain now has a total of 96 M109A1B and six M109A2 series 155 mm self-propelled howitzers in service and 48 M108 105 mm self-propelled howitzers. The Spanish Army plans to modernise its fleet to the M109A5 standard under a United States government FMS contract.

M109 (Switzerland)

In 1968, the Swiss Parliament authorised an order for 140 M109s from the US and these entered service in 1974 under the designation of the Panzerhaubitze 66.

A further 120 M109A1Bs were subsequently purchased and the original M109s were upgraded to the M109A1B standard.

The new M109A1Bs are known as the Panzerhaubitze 74 and the refitted vehicles as the Panzerhaubitze 66/74.

The 1979 armament programme called for the purchase of approximately 207 additional M109A1Bs which entered service as the Panzerhaubitze 79.

The last fleet of 110 M109A1Bs, called Panzerhaubitze 88, entered service as part of the 1988 armament programme.

The Swiss Army M109s have a bank of three forward-firing electrically operated 76 mm smoke grenade launchers mounted either side of the turret front. For improved traction, Swiss M109s are fitted with German Diehl tracks.

Of the current Swiss Army fleet of 577 M109s (some sources have stated 581) at least 348 are now being upgraded in a number of key areas including the installation of a new Swiss designed and built 155 mm/47 calibre barrel. Additional details are given in this section under Switzerland.

M109 (Taiwan)

Details of the Taiwanese 155 mm self-propelled howitzers with the ordnance in an open mount are given in this section under Taiwan. More recently Taiwan has taken delivery of a further 28 M109A5 self-propelled howitzers.

Taiwan has also developed its own ammunition support vehicle for the M109A5, called the CM24A1, based on a full-tracked chassis. This has an all-welded hull with the crew compartment at the front and the ammunition compartment at the rear.

The driver and commander enter the vehicle via a large door either side that opens to the front and, in the upper part of this, there is a bullet proof window. In the front of the vehicle are three bullet proof windows and a roof-mounted machine gun can be fitted.

The suspension system is of the torsion bar type and consists of seven, dual, rubber-tyred roadwheels with the drive sprocket at the front, idler at the rear and three tracked return rollers

The CM21A1 backs up to the rear of the M109A5 and feeds ammunition using a conveyor belt. There is a similar vehicle for the 203 mm (8 in) self-propelled artillery system called the CM27.

M109A5 (Thailand)

Thailand ordered 20 M109A5 self-propelled artillery systems and 20 Field Artillery Ammunition Support Vehicles through the FMS programme for delivery in 1994. These have been formed into a new self-propelled artillery regiment with three batteries each of six weapons, the remaining two being used for training.

Egyptian SPH122

The chassis of the M109/M992 is used as the basis for the Egyptian Army SP122 122 mm self-propelled howitzer, final deliveries of which were made in 1995. Details are given in this section under Egypt. By early 2001 Egypt had taken delivery of 124 SP122 systems.

LIW M109 upgrade

In late 2000 it was revealed that the United Arab Emirates (UAE) was upgrading the major part of its 155 mm self-propelled (SP) artillery fleet with a new gun laying and navigation/integral gun fire-control system supplied by the LIW Division of Denel, South Africa.

This is being installed in the 78 155 mm/45 calibre G6 (6 × 6) SP guns supplied by the LIW Division of Denel from 1991 and the 85 M109A3's SP guns upgraded by RDM Technology of the Netherlands to the enhanced M109L47 configuration. The latter were only recently delivered and are fitted with a 155 mm/47 calibre ordnance supplied by SW Ordnance Enterprise Corp.

According to LIW, the new system navigates accurately to within a distance of about 16 m using its two integrated sensors. The first of these is inertial and used where the system starts from a known point and then keeps track of the exact movement of the gun and continually calculates the current position. Second is using GPS (Global Positioning System) navigation that calculates its position with signals received from navigation satellites.

The BAE Systems ring laser gyro makes it possible to lay the 155 mm gun accurately to within 1 mil or 0.05° which is considered necessary for long-range artillery fire.

This upgrade enables the weapon to come into action, fire and move away before any counter battery fire is returned so enhancing its speed of response as well as improving its survivability.

In addition, the land navigation computer also receives firing data from the Battery Command Post (BCP) and displays this information on the same screen as that used for navigation. This allows the gun commander to carry out the gun laying roles so saving one crew position.

Finally the system is linked to the projectile muzzle velocity measurement radar mounted over the gun. This measures and stores accurate muzzle velocity information as well as telling the BCP that a round has been fired and updates ammunition information.

This is the first time that LIW has supplied an upgrade package for the US-designed M109 self-propelled artillery system. In addition to the navigation and laying system/integral gun fire-control system, the M109 was also fitted with a semi-automatic hydraulic push rammer which is used to ram projectile so increasing rate of fire and reducing crew fatigue.

Other M109 improvements include a remote-operated barrel clamp, auxiliary power unit installed on the turret rear and fitting an air conditioning system to allow the M109 to be operated in high ambient temperatures.

LIW see a considerable export potential for their M109 upgrade kit. Other divisions of Denel, for example NASCHEM and SOMCHEM can also provide enhanced projectiles and charges for the M109.

For both the G6 and M109, LIW provided kits which are then used to upgrade these systems in the United Arab Emirates using local resources.

M992 Field Artillery Ammunition Support Vehicle (FAASV)

Originally developed as a private venture by Bowen-McLaughlin-York, the XM992 was type-classified M992 in 1983.

The M992 FAASV is basically the chassis of a standard M109 155 mm self-propelled howitzer with the turret replaced by a fully enclosed superstructure. Inside this superstructure, which has the same protection as the rest of the vehicle, can be stacked 93 (90 conventional and 3 Copperhead) 155 mm projectiles, 99 propellant charges and 104 fuzes. The ammunition can be loaded into the superstructure by a front-mounted crane and is fed from the vehicle along a conveyor to the recipient self-propelled howitzer; this crane was not fitted to US Army M992 vehicles but is offered for export. All ammunition handling within the M992 is mechanical. In use, the M992 backs up to the user M109 and the conveyor delivers the 155 mm ammunition at the rate of up to 8 rds/min to the M109 bustle or lower hatches. The M992 can have a crew of two plus six passengers and the weight fully loaded is 25,900 kg. Intended primarily for use with M109 field units, the M992 can also be used by units with 175, 203, 120 and 105mm self-propelled weapons. With 203 mm units the M992 can carry 48 203 mm projectiles, 53 charges and 56 fuzes.

It has been suggested that the M992 could form the basis for a family of battlefield vehicles including a Fire Direction Centre Vehicle (see below), Command Post Vehicle (CPV), Medical Evacuation Vehicle (MEV), Maintenance Assistance Vehicle (MAV) and Armoured Forward Area Rearm Vehicle (AFARV).

US Army testing and decisions relating to the M109A6 Paladin reaffirmed the need for an FAASV in direct, organic support of each self-propelled gun. The US Army has improved the basic M992 through a series of upgrades to M992A1 and M992A2 configurations. The US Army recently type-classified the M992A1. The M992A1 contains a series of improvements based on lessons learned during the extensive operational testing of the M109A6 Paladin self-propelled howitzer.

By 2001, total production of the M992 FAASV for the home and export markets amounted to:

United States	921
Egypt	51 (delivered)
Saudi Arabia	60 (delivered)
Spain	6 (delivered)
Thailand	20 (delivered 1993-94)

In addition 119 Fire Direction Centre vehicles have been built:

Egypt	72
Greece	41
Taiwan	6

Full details of the M992 FAASV and its variants are given in *Jane's Military Vehicles and Logistics 2001-2002*.

Since then the US Army has ordered an additional 132 M992A2 FAASV for the National Guard and the first of these was completed in June 1998. Most of these were new build vehicles but the last 36 are rebuilds of M109A2 self-propelled howitzers and these were delivered between May and September 1999. Letterkenny Army depot will also be involved in this programme. All 132 vehicles have now been delivered and fielded to the US Army National Guard. In addition, the savings achieved through the conversion programme, saved enough money for the US Army to purchase an additional six more vehicles which were delivered by the end of 2000.

Under a separate programme, some 166 M992 FAASV have been upgraded to the M992A2 standard by Espace Mobile International in Belgium with final deliveries taking place in mid-1999.

Total production/conversion of FAASV for the US Army is now:

M992A0 to A2 conversions	664
M992A1 to A2 conversions	125
New production M999A2	96
M109A2/M109A3 conversions	36
Total	**921**

(Note: to the above figures should be added the six additional vehicles delivered in 2000.)

Field Artillery Operation Centre Vehicle (FAOCV)

During TF-XXI, United Defense converted three FAASV into the FAOCV configuration for use at the National Training Centre (NTC).

Modifications included four compartmentalised workstations, 10 m high telescoping antenna mast, environmental control unit and 400 A auxiliary power unit.

According to United Defense, results of the exercise demonstrated improved mobility for M109A6 units, FAOCV improved survivability and maintainability for units. Also demonstrated was on-the-move computer operations.

Fire Direction Centre Vehicle (FDCV)

The FDCV, based on the chassis and hull of the M992 FAASV, has been tested at the US Army School of Artillery at Fort Sill to prove the feasibility and function of the concept under field and NBC conditions. The equipment used in the FDCV is a mix of 'off-the-shelf' or prototype products assembled for the concept, for which the internal headroom and armour of the M992 hull is eminently suitable. If required, bolt-on armour can be added to protect the specialist personnel and equipment carried.

The following equipment was carried in the prototype FDCV: mast system with OE254 antenna which is 11.28 m high when elevated; mast elevation control box; mast elevation control power unit; packet digital radio; RT524/VRC receiver/transmitter radios (4); printer; liquid cooling vest connect points (9 stations); tactical graphics terminal; tactical display; ANS 2000 land navigation system display; ANS 2000 land navigation system heading reference unit; M13A1 filtered air system for NBC ventilated facepieces; M43 NBC detector; M42 alarm; M49 NBC collective protective system; APU; air-conditioner (36,000 BTU cooling/heating); ventilation filter/blower; HEL computer (PDP1134); plotter; CRT terminal; digitiser; liquid cooling vest control box; M10 protective entrance; battery boxes and batteries; liquid cooling vest chilling unit; four extra personnel seats and additional stowage compartments.

Other M109 chassis applications

The chassis of the M109, or versions of the M109, have been used for a wide range of other applications. These include the XM-11S threat radar simulator from General Dynamics. This was developed for the US Army Missile and Space Intelligence Centre and is used to simulate the Russian SA-11 ('Gadfly') surface-to-air missile system.

Specifications

M109A2
Crew: 6
Combat weight: 24,948 kg
Unloaded weight: 21,110 kg
Length:
(gun forward) 9.12 m
(hull) 6.19 m
Width: 3.15 m
Height:
(incl anti-aircraft MG) 3.28 m
(reduced) 2.8 m
Ground clearance: 0.46 m
Track: 2.788 m
Track width: 381 mm
Length of track on ground: 3.962 m
Max road speed: 56.3 km/h
Fuel capacity: 511 litres
Max road range: 349 km
Fording: 1.07 m
Gradient: 60%
Side slope: 40%
Vertical obstacle: 0.53 m
Trench: 1.83 m
Engine: Detroit Diesel Model 8V-71T, turbocharged, 2-stroke, liquid-cooled 8-cylinder diesel developing 405 bhp at 2,300 rpm
Transmission: Allison Transmission XTG-411-4A cross-drive with 4 forward and 2 reverse gears
Suspension: independent torsion bar
Electrical system: 24 V
Batteries: 4 × 12 V Model 6TN
Armament:
(main) 1 × 155 mm M185 howitzer
(anti-aircraft) 1 × 12.7 mm or 7.62 mm MG
Smoke-laying equipment: none
Ammunition:
(main) 34 plus 2 CLGP
(12.7 mm anti-aircraft) 500
Gun control equipment
Turret power control: hydraulic/manual
(by commander) no
(by gunner) yes
Gun elevation/depression: +75/–3°
Turret traverse: 360°
NBC system: no
Night vision equipment: yes

The XM975 Roland chassis, based on the M109, was used as the basis for the Robotic Command Center for use in various robotic vehicle programmes.

Status

The M109 and its derivatives are in service with the US Army and National Guard. Other nations using M109 series weapons are listed below.

During the first production phase, carried out initially by the Cleveland Army Tank Plant, US-configured M109s were delivered to the following nations: Austria, Belgium, Canada, Denmark, Ethiopia, Iran, Israel, Jordan, Netherlands, Spain, Switzerland and the UK. Italy, the Netherlands and Norway formed co-production agreements. Canada, Iran, Israel, the Netherlands, Spain, Switzerland and the UK converted their M109s to the M109A1 configuration. M109Gs were delivered to Germany, Italy and Norway.

M109A1B FMS production since 1972 has resulted in systems for Greece, Iran, Israel, Italy, Jordan, Kuwait, Morocco, Netherlands, Pakistan, Portugal, Spain, Taiwan and the UK. M109A2 production since 1980 has resulted in deliveries to Austria, Jordan, Kuwait, Netherlands, Pakistan, Portugal, Spain, Taiwan and the UK.

Known sales of the M109 series of self-propelled howitzers are in the table.

Country	M109	M109A1 kit	PIP kit	M109A1B	M109A2	M109A5
Austria****	38	-	-	-	18	54***
Belgium	-	-	41	-	127D	-
Canada	50	-	50	-	26D	-
Denmark	76	-	-	-	-	-
Egypt	-	-	-	-	164D	-
Ethiopia	12	-	-	-	-	-
Germany	609*	-	2	-	-	-
Greece	-	-	-	51	84	12*****
Iran	50	50	-	390	-	-
Israel	60	60	-	294 + 75D	-	-
Italy	221*	-	-	62	-	-
Jordan	126	-	126	-	108	-
Korea, South	-	-	-	-	1,040	-
Kuwait	-	-	-	6**	31**	-
Morocco	-	-	-	36	3	-
Netherlands	135	-	135	3	91	-
Norway	127*	-	-	-	-	-
Pakistan	-	-	-	-	152	-
Peru	-	-	-	12	-	-
Portugal	-	-	-	-	6	-
Saudi Arabia	-	-	-	87	24	-
Spain	18	18	-	60	6	-
Switzerland	-	146	-	435	-	-
Taiwan	-	-	-	-	197	28
Thailand	-	-	-	-	-	20

D - direct sales
* M109G
** Status uncertain following Iraqi invasion and Operation Desert Storm
*** M109A5 plus
**** 67 M109A2 and 48 M109A3 from UK in 1994/1995 and five from Netherlands in 1997
***** Ordered 1997 with option on additional 12 (see text)

In addition to the above, there are the following additional M109 series users:

Country	Quantity	Comment
Brazil	40	M109A3 delivered by 1999
Iraq	n/avail	may have some in service captured from Iran
Libya	20	believed to be in service
Portugal	14	possible sale of M109A5 announced late 1997, see development
UAE	85	supplied by RDM Technology of the Netherlands (qv) and upgraded with 47 calibre barrel
USA	2,568	includes all models
Venezuela	5	

Manufacturers

United Defense LP, Ground Systems Division.
Original manufacturer was Cadillac Motor Car Division at Cleveland Army Tank Plant, later Chrysler Corporation at the same plant. Current manufacturer is United Defense LP, Ground Systems Division.

UPDATED

M108 105 mm self-propelled howitzer

Development

In 1958, the first prototype of the T195 self-propelled howitzer was completed but a number of failures occurred in the suspension. The prototype differed from later vehicles in that it had a different shaped hull and turret, the seventh roadwheel acted as the idler and the 105 mm howitzer was fitted with a muzzle brake.

In 1959, a policy was established that diesel rather than petrol engines would be used for future vehicles and the prototype of the T195, which was powered by a Continental petrol engine, was refitted with a Detroit diesel engine and called the T195E1. Trials with the latter were carried out in 1960 but there were failures of critical suspension and final drive components. These deficiencies were subsequently corrected and, in December 1961, the T195E1 was classified for limited production as the howitzer, light, self-propelled, 105 mm, M108.

Production of the M108 was undertaken by the Cadillac Motor Car Division of the General Motors Corporation: the first production vehicle was completed in October 1962 and production continued until 1963. The vehicle was in production for only a short period, as the US Army decided to concentrate on the self-propelled 155 mm M109 rather than the 105 mm M108. The M108 is almost identical to the M109, and differs only in its armament, loading system, fire-control system, ammunition stowage racks and in that it is fitted with spades at the rear of the hull.

Description

The 105 mm M108 self-propelled howitzer has a crew of five: the commander, gunner, two ammunition numbers and the driver. The hull is made of all-welded aluminium armour with the driver at the front of the hull on the left, the power pack to the right and the fighting compartment at the rear.

The driver has a single-piece hatch cover that opens to the left, in front of which are three M45 day periscopes that can be covered by small metal flaps to prevent damage. The centre day periscope can be replaced by a passive periscope for driving at night.

The engine is coupled to the Allison Transmission XTG-411-2A cross-drive transmission which is at the front of the hull.

The all-welded aluminium turret is at the rear of the hull. It has a square hatch in each side which opens to the rear and twin doors in the rear. The commander is seated on the right and has a cupola which can be traversed manually through 360°, a single-piece hatch cover that opens to the rear and an M27 day periscope. Pintle mounted on the forward part of the commander's cupola is a 12.7 mm (0.50 in.) M2 HB machine gun. The gunner is seated on the left side of the turret and has a square single-piece hatch cover that opens to the right. Twin doors are provided at the rear of the hull for ammunition resupply.

The torsion bar suspension either side consists of seven dual rubber-tyred roadwheels with the drive sprocket at the front and the idler at the rear. There are no track-return rollers. The tracks are of the single-pin, centre guide type with replaceable rubber pads.

The M108 is fitted with night vision equipment but no NBC equipment. The basic vehicle can ford to a depth of 1.828 m without preparation. It can also be fitted with an amphibious kit which consists of nine air bags, four each side of the hull and one at the front of the hull. The bags, which are not carried on the vehicle as part of its normal equipment, are inflated from the vehicle and when they are inflated the vehicle can propel itself across rivers by its tracks at 6.43 km/h.

Main armament is an M103 105 mm howitzer in an M139 mount. The recoil system is of the hydrospring, constant retarding force type while the breech block is of the sliding wedge drop block type. The weapon has an elevation of +74°, depression of −4° and 360° turret traverse. Gun elevation/depression and turret traverse are manual. Normal rate of fire is 1 rd/min, but 3 rds/min can be fired for short periods. Fire-control equipment consists of an M15 elevation quadrant, an M117 panoramic telescope with a magnification of ×4 and a 10° field of view, and an M118 elbow telescope with a magnification of ×4 and a 10° field of view.

The following types of ammunition can be fired:

HE (M1) with the complete round weighing 18.107 kg, a maximum muzzle velocity of 494 m/s and a maximum range (charge 7) of 11,500 m;

HE (M444) (carries 18 M39 grenades) with the complete round weighing 19.05 kg, a maximum muzzle velocity of 494 m/s and a maximum range (charge 7) of 11,500 m;

HE (M413) (carries 18 M35 grenades) with the complete round weighing 19.05 kg, a maximum muzzle velocity of 494 m/s and a maximum range of 11,500 m;

HEP-T (M327) with the complete round weighing 15.17 kg, a maximum muzzle velocity of 559 m/s and a maximum range of 8,685 m;

HERA (M548) with the complete round weighing 17.46 kg, a maximum muzzle velocity of 548 m/s and a maximum range of 15,000 m;

Illuminating (M314 series) with the complete round weighing 21.06 kg, a maximum muzzle velocity of 494 m/s and a maximum range (charge 7) of 11,500 m;

Illuminating (M314A3) with the complete round weighing 21.06 kg, a maximum muzzle velocity of 494 m/s and a maximum range of 11,500 m;

Leaflet (M84B1) with the complete round weighing 18 kg, a maximum muzzle velocity of 433 m/s and a maximum range (charge 7) of 9,091 m;

Smoke (M60 series) with the complete round weighing 19.46 kg, a maximum muzzle velocity of 494 m/s and a maximum range (charge 7) of 11,500 m;

M108 105 mm self-propelled howitzer with ordnance in travelling lock (C R Zwart)

Specifications

Crew: 5
Combat weight: 22,452 kg
Unloaded weight: 18,436 kg
Power-to-weight ratio: 18.038 bhp/t
Ground pressure: 0.71 kg/cm²
Length: 6.114 m
Width: 3.295 m
(reduced) 3.149 m
Height:
(incl AA MG) 3.155 m
(reduced) 2.794 m
Ground clearance: 0.451 m
Track: 2.768 m
Track width: 381 mm
Length of track on ground: 3.962 m
Max speed:
(road) 56 km/h
(water) 6.43 km/h
Fuel capacity: 511 litres
Max road range: 390 km
Fording: 1.828 m (amphibious with kit)
Gradient: 60%
Vertical obstacle: 0.533 m
Trench: 1.828 m
Engine: Detroit Diesel Model 8V-71T, turbocharged, 2-stroke, liquid-cooled 8-cylinder diesel developing 405 bhp at 2,300 rpm
Transmission: Allison Transmission XTG-411-2A cross-drive with 4 forward and 2 reverse gears
Suspension: torsion bar
Electrical system: 24 V
Batteries: 4 × 12 V 6TN
Armament:
(main) 1 × 105 mm M103 howitzer
(anti-aircraft) 1 × 12.7 mm M2 MG
Smoke-laying equipment: none
Ammunition:
(main) 87
(anti-aircraft) 500
Gun control equipment
Turret power control: manual
(by commander) no
(by gunner) yes
Gun elevation/depression: +74/−4°
Turret traverse: 360°

Belgian Army M108 converted to VBCL configuration (PB)

2001/0106069

Smoke (M84) with the complete round weighing 19.03 kg, a maximum muzzle velocity of 494 m/s and a maximum range (charge 7) of 11,500 m.

Variants

Belgian Army M108A2B training vehicles

The Belgian Army has modified at least 14 M108 105 mm self-propelled howitzers by mounting these turrets on surplus M109A2 hulls. This is designated the M108A2B and is used for training purposes as 105 mm ammunition is cheaper than 155 mm ammunition.

Belgian Army VBCL fire direction and command post

The Belgian Arsenal du Materiel Mecanique et de l'Armement at Rocourt has converted a total of 45 surplus M108 chassis into the VBCL (Vehicule Blinde de Commandement et de Liaison) with final deliveries being made in 1994.

The VBCL is a fire direction centre and command post for Belgian Army field artillery battalions and batteries. It is fitted with four BAMS radios, two telescopic antennas PU-8, a 2.7 kW generator DE-CA, tent to increase working area, ABM-1045 field telephone system and a roof mounted 12.7 mm M2 machine gun.

The Fire Direction Centres are fitted with ruggedised portable computers PCI but are otherwise similar to the CP versions. It is expected that all of these will be modernised between 2002 and 2006.

Status

Production complete. In service with:

Country	Quantity	Comment
Brazil	50	estimate (believed to be in reserve)
Spain	48	to be modified to M109A5 standard with 155 mm ordnance
Taiwan	100	estimate
Turkey	26	

Manufacturer

Cadillac Motor Car Division of General Motors Corporation. No longer involved in armoured fighting vehicle development or production).

UPDATED

SELF-PROPELLED GUNS AND HOWITZERS (WHEELED)

CHINA, PEOPLE'S REPUBLIC

NORINCO 155 mm self-propelled gun (6 × 6)

Development

Early in 2002, it was revealed that China had developed a new 6 × 6 self-propelled artillery system that uses the upper part of the standard China North Industries Corporation (NORINCO) 155 mm/45 calibre WA021 towed artillery system. Full details of this are given in the later Towed anti-tank guns, guns and howitzer section.

The WA021 has been in service with the People's Liberation Army (PLA) for some years and has also been offered on the export market with or without its Auxiliary Propulsion System (APU).

The 155 mm/45 calibre WA021 is ballistically identical to the NORINCO PLZ45 full-tracked armoured self-propelled artillery system, for which the first export customer was Kuwait who took delivery of 27 systems in 2000.

In many respects this new Chinese 155 mm (6 × 6) self-propelled gun is very similar to the French Giat Industries CAESAR 155 mm/52 calibre self-propelled artillery system, based on a modified Mercedes-Benz (6 × 6) truck chassis. The French Army has placed an order for five CAESAR systems for delivery from 2002. CAESAR is fitted with an advanced onboard fire control and navigation system as well as a flick rammer.

This Chinese 155 mm/45 calibre (6 × 6) self-propelled gun is the latest in a number of wheeled Chinese self-propelled systems to be revealed. The others include:

- 122 mm (8 × 8) self-propelled gun
- 120 mm (6 × 6) self-propelled mortar
- 120 mm (8 × 8) tank destroyer/armoured car
- 105 mm (6 × 6) tank destroyer/armoured car

As far as it is known, none of the above vehicles have entered quantity production or service for the PLA. These and the development and fielding of earlier Chinese 6 × 6 armoured personnel carriers, such as the WZ 551, reflects the PLA's greater emphasis on wheeled armoured vehicles with their greater strategic mobility.

Description

This new Chinese 6 × 6 self-propelled artillery system, which is understood to be at the prototype stage, is based on a 6 × 6 forward control unarmoured cross-country truck chassis.

Mounted at the very rear of the chassis above the rear axle, is the complete mount, elevating mass and ordnance of the standard 155 mm/45 calibre WA021 towed artillery system. When travelling the ordnance is held in a travelling lock, mounted on the forward part of the chassis.

When deployed in the firing position, a large spade is lowered to the rear to provide a more stable firing platform. It is probable that a central tyre pressure regulation system is fitted allowing the driver to adjust the tyre pressure from within the cab to suit the type of ground being crossed.

While some ready use ammunition (projectiles, charges and fuzes) are carried on the rear decking. It is probable that the new 6 × 6 self-propelled artillery system would be accompanied by another vehicle with the remainder of the crew and additional ammunition.

The 155 mm/45 calibre WA021 was based on the former Space Research Corporation 155 mm/45 calibre weapon developed by the late Gerald Bull.

Firing a 155 mm Extended Range Full-Bore (ERFB) High Explosive (HE) projectile, a maximum range of 30 km can be achieved. This can be increased to 39 km using ERFB - base bleed ammunition. Maximum rate of fire is 4 to 5 rds/min.

Specifications

Not available.

Status

Prototype undergoing trials.

Manufacturer

Chinese state factories. NORINCO would be responsible for export sales of this system.

NEW ENTRY

NORINCO 122 mm (8 × 8) self-propelled artillery system

Development

It has been revealed that China has developed, to the prototype stage, a new 122 mm 8 × 8 self-propelled artillery system, that is believed to be currently undergoing trials.

At this stage it is not certain as to whether this has been developed for use by the People's Liberation Army (PLA) or for the export market.

Given that the PLA is now placing increased emphasis on expanding its fleet of wheeled Armoured Fighting Vehicles (AFV), it is probable that this system has been initially developed for the home market.

The long-term aim would be to also offer the system on the export market through China North Industries Corporation (NORINCO) who market virtually all of Chinese ground weapon systems.

This is the latest in an increasing number of new light wheeled AFVs that have been developed in China, the others include:

- 120 mm (8 × 8) tank destroyer.armoured car
- 120 mm (6 × 6) self-propelled mortar system
- 105 mm (6 × 6) tank destroyer/armoured car

The increasing emphasis being placed by China on the development and production of wheeled light armoured vehicles, clearly shows that the PLA requires its units to have greater strategic mobility than in the past, when most of its AFV fleet was tracked.

Description

The chassis used for this as yet unnamed 122 mm self-propelled artillery system appears to be very similar to that recently revealed for the new Chinese 120 mm self-propelled anti-tank gun, which is also understood to be still at the prototype stage. Details of this vehicle are given in the Reconnaissance vehicles section.

The hull of the new 122 mm self-propelled artillery system is of all-welded steel armour which provides the occupants with protection from small arms fire and shell splinters.

The driver and one person are seated at the front of the vehicle, each are provided with a window to their immediate front with another to the side. It is probable that there is a hatch cover above each of their positions to enable some of the crew to rapidly enter and leave the vehicle.

In the centre of the vehicle is the all-welded turret with almost vertical sides. It has hatches in the roof and a bank of four electrically operated smoke grenade launchers either side. Pintle mounted on the roof is a 12.7 mm anti-aircraft machine gun that can also be used for self-defence purposes.

The 122 mm ordnance is provided with a slotted muzzle brake and a fume extractor and when travelling is held in position by an small A frame which is pivoted on top of the hull. When not required this folds forwards.

For many years NORINCO has been manufacturing a reverse engineered copy of the Russian 122 mm D-30 towed artillery system. This has a maximum range, firing a standard high-explosive projetectile, of 15.30 km.

In addition, there is also a special charge with a maximum range of 18 km and an extended range full bore - base bleed (ERFB-BB) projectile with a maximum range of 21 km. This 122 mm ordnance is also fitted to the NORINCO 122 mm full-tracked self-propelled howitzer Type 85, as well as the more recent Type 89 system.

The diesel power pack is at the rear of the vehicle and steering is understood to be power-assisted on the front four wheels. A central tyre pressure regulation system is probably fitted, allowing the driver to adjust the tyre pressure to suite the terrain being crossed.

There also appears to be an emergency door/hatch in either side of the lower part of the hull, between the second and third roadwheel stations.

There is a possibility that the vehicle is fully amphibious, being propelled in the water by a single propeller mounted one either side at the rear.

Status

Prototype undergoing trials.

Manufacturer

Chinese state factories

NEW ENTRY

CZECH REPUBLIC AND SLOVAKIA

ZTS 155 mm self-propelled gun-howitzer Zuzana (8 × 8)

Development

In December 1992, ZTS completed the first of two prototypes of a 155 mm 45 calibre version of the Dana 152 mm (8 × 8) self-propelled gun-howitzer, which is fully described in the following entry.

According to the manufacturer, an initial production batch of 10 155 mm Zuzana self-propelled guns has been completed. This turret has also been installed on a modified T-72M1 MBT chassis for the Indian competition which took place in 1995. As of 2002, India had not placed any production orders for a 155 mm turret system.

For the T-72/Zuzana there is a separate entry in the Self-propelled guns and howitzers (tracked) section.

Early in 1997, the Slovak Defence Minister confirmed that the Slovak Army had formally accepted the 155 mm 45 calibre Zuzana (8 × 8) self-propelled gun/howitzer under the designation Howitzer Model 2000.

The Slovak Army is the first member of the former Warsaw Pact to adopt the NATO 155 mm calibre for its self-propelled artillery systems.

The first Slovak order was for eight Zuzana systems but further systems are expected to be purchased in the future as additional funding becomes available.

Unit cost of the 155 mm 45 calibre Zuzana is between US$3.4 and US$3.6 million. Compared with tracked self-propelled artillery system, Zuzana has greater strategic mobility (maximum road speed 80 km/h) and lower operating and maintenance costs.

It has been reported that the Slovakian Army requirement is for a total of 78 systems, funding permitting.

Slovakian 155 mm/45 calibre Zuzana (8 × 8) self-propelled artillery system deployed in the firing position **2002**/0095532

Slovakian 155 mm/45 calibre Zuzana (8 × 8) self-propelled artillery system with ordnance at maximum elevation **2002**/0130602

In May 1998 a further order was placed for 40 systems plus a quantity of Tatrapan (6 × 6) command vehicles.

Late in 2000 it was disclosed that the Greek Army had placed a contract with the Slovakian company of ZTS Defence (part of the DMD Group) covering the supply of 12 Zuzana 155 mm/45 calibre (8 × 8) self-propelled guns (SPGs).

It is understood that these were delivered sometime in 2001, but later that year during a parade held in Cyprus, it was revealed that this country had also taken delivery of the Zuzana. At this stage it is not certain as to whether these systems were supplied direct to Cyprus or came via Greece.

Description

The Zuzana is essentially the earlier 152 mm Ondava self-propelled gun-howitzer fitted with a new 155 mm 45 calibre ordnance capable of firing all Western types of 155 mm ammunition.

Specifications

Crew: 4
Configuration: 8 × 8
Combat weight: 28,600 kg
Length:
(overall) 12.97 m
(hull) 8.87 m
Width:
(overall) 3.015 m
Height:
(overall) 3.525 m
Ground clearance: 0.41 m
Wheelbase: 1.65 + 3.07 +1.45 m
Track:
(front) 2.044 m
(rear) 1.988 m
Max road speed: 80 km/h
Fuel capacity: 460 litres
Max road range: 750 km
Fording: 1.4 m
Gradient: 58%
Side slope: 30%
Vertical obstacle: 0.6 m
Trench: 2 m
Turning radius: 12.8 m
Engine: Tatra 3-930-52 V-12 air-cooled diesel developing 355 hp
Transmission: dual range, 5-speed, plus overdrive providing a total of 20 forward and 4 reverse speeds.
Steering: power assisted, front four wheels
Tyres: 15.00 - 21 T03
Electrical system: 26 V
Armament:
(main) 1 × 155 mm/45 calibre howitzer
(anti-aircraft) 1 × 12.7 mm NSV MG
Smoke grenade launchers: 2 × 3 81 mm
Ammunition:
(main) 40 × 155 mm
Gun control equipment
Turret power control: hydraulic/manual
Gun elevation/depression: +70/–3.5°
Traverse: 60° (left and right)
Max rate of fire:
(automatic loading) 6 rds/min
(sustained) 30 rds in 6 min
(90 rounds) 90 min
(manual) 2 rds/min
NBC system: yes
Night vision equipment: yes

Firing a 155 mm ERFB-BB projectile, a maximum range of 39.6 km can be achieved. Minimum range is quoted as 5.9 km.

The 155 mm 45 calibre ordnance has a new breech mechanism and is fitted with a double-baffle muzzle brake but no fume extractor. The automatic loading system enables a maximum of 30 rounds to be fired in 6 minutes with a total of 40 projectiles and charges being carried. Of the 40 155 mm projectiles carried, 36 are in the automatic magazine for ready use with the remaining four in a fixed magazine. Of the 40 charges carried, 30 are in the automatic magazine and 10 in the fixed magazine. Ammunition handling is fully automatic with ramming and loading of projectiles and charges. The primer magazine holds a total of 30 primers.

Turret traverse is 60° left and right with weapon elevation from –3.5 to +70°. When in the firing position, the system is stabilised by three hydraulic jacks. These are positioned one at the rear and one either side between the second and third axles.

The Zuzana has a crew of four and standard equipment includes an NBC system, air conditioning system, a computerised fire-control system, a land navigation system and a sight with automatic levelling.

The chassis is the Tatra 815 VP31 29265 8 × 8.1R. This features full 8 × 8 drive, powered steering and a central tyre pressure system which allows the driver to adjust the tyre pressure to suit the terrain being crossed.

Variants

Improved Zuzana

The Slovakian company of ZTS - Dubnica nad Vahom is now developing a new version of the Zuzana (8 × 8) 155 mm self-propelled artillery system that will feature a NATO type 155 mm/52 calibre ordnance.

Other improvements will include a gun turret with increased room, a higher level of protection against

Slovakian 155 mm/45 Zuzana (8 × 8) self-propelled artillery system from the rear **2002**/0130600

Ammunition resupply vehicle on Tatra (8 ×8) chassis for use with 155 mm/45 calibre Zuzana showing fully armoured cab (PB) **2002**/0130601

small arms fire and shell fragments and a direct fire sight that will enable targets to be engaged under day and night conditions.

So far all production Zuzana systems have been powered by a Tatra 3-930.52 12-cylinder diesel developing 355 hp coupled to a manual transmission. This will be replaced by a German MAN diesel that meets Euro II emission standards, coupled to an Allison fully automatic transmission.

Zuzana 2000G

The Greek Army has taken delivery of 12 systems under the local designation of the Zuzana 2000G. These feature Greek type grenade launchers, no 12.7 mm anti-aircraft machine gun, new radios and a new intercom system. Greece also has an option on another six systems, making a total of 18. Although originally developed for the field artillery role, Zuzana is also now being marketed for the mobile coastal defence role.

Ammunition resupply vehicle

ZTS have also completed the prototype of an ammunition resupply vehicle based on a Tatra (8 × 8) chassis that has a fully armoured cab at the front. This carries a total of 120 155 mm projectiles and charges and has a maximum weight of 30,562 kg. This also features a power pack consisting of a MAN diesel coupled to an Allison automatic transmission. ZTS hope to get a Slovakian Army contract for the ammunition resupply vehicle in 2002, funding permitting.

Skoda 35 mm SPAAG

The Skoda company of the Czech Repbulic is leading an international team developing a new 35 mm Self-Propelled Anti-Aircraft Gun (SPAAG) system.

This is being developed as a private venture to meet the possible requirements of the Czech Army which currently uses the locally developed twin 30 mm M53/59 (6 × 6), which is now obsolete.

The new system is based on the armoured Tatra (8 × 8) truck chassis which is already used for a number of other military applications including the Dana (152 mm) and Zuzana (155 mm) self-propelled artillery systems, 122 mm RM-70 (40-round) multiple rocket system and 122 mm minelaying system.

This new SPAAG will have the fully armoured crew and engine compartment at the front and Swiss Oerlikon Contraves 35 mm gun mount at the rear. Details of this system are given in *Jane's Land-Based Air Defence 2002-2003.*

Status

In production. In service with Cyprus, Greece and Slovakia.

Manufacturer

ZTS Dubnica nad Váhom (Slovakia).

UPDATED

ZTS 152 mm self-propelled gun-howitzer Dana

Development

The Dana 152 mm self-propelled gun-howitzer was developed in the late 1970s by ZTS to meet the operational requirements of the Czech Army for a highly mobile self-propelled artillery system.

Dana is based on automotive components of the well-known Czech Tatra 815 VP 31 29265 8 × 8.1R truck which has a good demonstrated cross-country capability.

Following trials with prototype systems, the Dana was accepted for service with the Czech Army and first production models were issued to the Kiev Artillery Battalion Jan Zizka of Trocnov based at Tabor, Southern Bohemia, in 1981.

The Czech Army designation for the system is the vzor 77 self-propelled howitzer Dana.

By early 1994, total production of this system, for the home and export markets, had reached over 750 units. As far as is known there has been no recent production of the 152 mm Dana system.

This system, and further developments of this system, are now manufactured and marketed by ZTS Dubnica nad Váhom of Slovakia.

All export marketing is now being concentrated on the ZTS 155 mm self-propelled gun howitzer Zusana, which is covered in a separate entry. This is now in service with Cyprus, Greece and Slovakia.

Description

The Dana is based on automotive components of the Tatra 815 (8 × 8) truck with the driver's compartment at the front, fully enclosed turret in the centre and the engine compartment at the rear. All three compartments are armoured and provide protection against small arms fire and shell splinters.

A central tyre-pressure regulation system is fitted which allows the driver to change the tyre pressure to suit the type of ground being crossed. The tyre pressure can be changed when the vehicle is on the move. Power steering is provided for the front four wheels.

The driver is seated at the front of the vehicle on the left with the vehicle commander to his right. Both are provided with a windscreen to the front and when in a combat area this is protected with two shutters, vision to the front then being maintained through two vision blocks in each shutter. The commander and driver are each provided with a forward-opening roof hatch and a vision block to give observation to the sides of the vehicle. Four circular firing ports are provided, one in each side of the compartment and one either side of the shutters.

In addition to operating the communications system, the commander also has access to the night sight. The driver operates the turret unlocking system as well as lowering or raising the stabilisers for firing.

The 152 mm ordnance is fitted with a muzzle brake and fires the same ammunition of the separate loading type, projectile and charge, as the Russian 152 mm 2S3 tracked self-propelled artillery system, as well as locally produced ammunition.

The breech mechanism is of horizontal wedge type with the empty cartridge cases being ejected as the breech opens. The cartridge cases are then removed by means of a moving belt.

The Dana is fitted with a fully automatic loading system which loads the projectile and then the charge using an automatic rammer. The weapon can be loaded at all angles of elevation.

The automatic loading system is hydraulic and is controlled by electrical elements and the gunner can select either single shots or fully automatic fire.

The system is ready to fire within 2 minutes of coming to a halt and can move off again 1 minute after the last round is fired. To provide a more stable firing platform, three hydraulic jacks are lowered to the ground, one at the rear and one either side between the second and third roadwheels.

Optical systems fitted include a ZZ-73 rocking bar sight collimated with a PG1 M-D telescope for indirect fire and an OP 5-38-D sight for direct fire. All the sights

Specifications

Crew: 5
Configuration: 8 × 8
Weight:
(with fuel) 25,100 kg
(with fuel and 40 rounds) 28,100 kg
(with fuel and 60 rounds) 29,250 kg
Power-to-weight ratio: 13.74 hp/t
Length:
(overall) 11.156 m
Width:
(overall) 3.00 m
Height:
(to turret roof) 2.85 m
Firing height: 2.41 m
Ground clearance: 0.41 m
Wheelbase: 1.65 + 3.07 + 1.45 m
Track:
(front) 2.044 m
(rear) 1.988 m
Max road speed: 80 km/h
Max road range: 740 km
Max dirt-road range: 650 km
Fording: 1.4 m
Gradient: 60%
Side slope: 30%
Vertical obstacle: 0.6 m
Trench: 2 m
Turning radius: 12.8 m
Engine: Tatra 2-939-34 diesel, air-cooled, developing 345 hp
Transmission: dual range, 5 speed, plus overdrive providing 20 forward and 4 reverse speeds
Steering: power assisted, front four wheels
Tyres: 15.00 - 21 T03
Armament:
(main) 1 × 152.4 mm
(length of ordnance) 5,580 mm
(max recoil) 1,300 mm (long) 750 mm (short)
(anti-aircraft) 1 × 12.7 mm NSV MG
Ammunition (main):
60 (limits max road speed to 70 km/h)
40 (limits max road speed to 80 km/h)
Gun control equipment
Turret power control: hydraulic/manual
Gun elevation/depression: +70/-4°
Traverse: ±225°
Indirect fire traverse: 45°
Max rate of fire:
(automatic loading) 5 rds/min
(sustained) 30 rds in 7 min
(90 rounds) 90 min
(manual) 2 rds/min
NBC system: yes
Night vision equipment: yes

152 mm Dana self-propelled gun-howitzer deployed in the travelling system with ordnance in travel lock (Michael Jerchel) ***2001***/0106068

are in the left side of the turret with two openings being provided in the turret front as well as a roof-mounted periscope.

Turret traverse is limited to 225° to each side due to cables, so no full turn through 360° is possible.

The turret is in two parts with the 152 mm ordnance being mounted in the aisle between them. Each half of the turret is fully enclosed and is provided with access doors, roof hatches and vision devices. As the weapon is outside the turret, no fumes can enter the cabin.

The left cabin houses the gunner and loader operator while the ammunition handler is in the right cabin and sets the fuzes of the projectiles manually. Before the weapon is fired for the first time, the breech has to be opened manually. This is achieved by one of the crew members in the left cabin using his hand which is encased in a rubber glove attached to the side of the turret.

On the right side of the turret roof is mounted the 12.7 mm NSV machine gun which can be used in the direct or air defence roles.

The automatic loading system enables 30 rounds of ammunition to be fired in 30 minutes or 90 rounds in 1 hour with the projectiles and charges stowed in magazines. The projectiles and charges are rammed using a chain-driven rammer which is located above the ordnance. Firing is accomplished using a foot pedal.

A total of 60 projectiles and charges is carried in the chassis and inside the turret. This limits the maximum road speed to 70 km/h. Approximately half the ammunition is stowed in the turret vertically, with the charges in the left cabin and the projectiles in the right cabin. The standard HE projectile (EOF D-20) weighs 43.56 kg and has a maximum velocity of 693 m/s. This has a maximum range of 18,700 m. Using a HEAT projectile maximum muzzle velocity is 717 m/s. There is also an enhanced projectile, the Czech EOFd, with a maximum range of 20,080 m.

More recently the EKK projectile has been developed which is of the base bleed type and contains 42 fragmentation bomblets, each of which will penetrate 100 mm of armour. This projectile has a maximum range of 28,230 m.

Minimum range, at an elevation angle of +70° and using charge 6, is 4,600 m.

Standard equipment includes an NBC system which is provided with filter ventilation and an air conditioning system which distributes fresh air to both the cab and the turret.

Pressure for the hydraulic system of the self-loading system as well as for weapon elevation and turret traverse is supplied by a pump driven from the main engine, although manual back-up systems are provided for all powered elements.

N-22 low-level radar system based on armoured Tatra 815 (8 × 8) chassis in the travelling configuration (Michael Jerchel) ***2001***/0106067

Variants

152 mm ShKH Ondava

This is a further development of the Dana and has a 152 mm 47 calibre ordnance which fires a new high-explosive projectile with a base bleed unit which enables a maximum range of 32,000 m to be achieved.

Onboard ammunition supply is only 40 projectiles and charges. As far as is known, this system remains at the prototype stage.

155 mm Zuzana

Details of this system, which is fitted with a 155 mm 45 calibre ordnance, are given in the previous entry. Main marketing emphasis is now on this system which has been built for Cyprus, Greece and Slovakia.

Tatrapan ACV

In 1993, the prototype of an armoured command post based on the Tatra T 815 (6 × 6) chassis was completed. Full details of this model, a small quantity of which have been built to support the new ZTS 155 mm self-propelled gun-howitzer Zuzana (8 × 8) systems now in service with Slovakia, are given in the Armoured personnel carriers (wheeled) section.

MUR-20 EW system

The MUR-20 is a mobile system for the identification of radar systems in the forward battlefield area, for example ELINT/ESM (electronic intelligence/ electronic support/surveillance measures). Typically three or four MUR-20 systems would be used together. It is based on an armoured Tatra 8 × 8 chassis with automatic levelling and an antenna mast.

RADWAR N-22 low-level surveillance radar

This is mounted on a similar Tatra 815 8 × 8 chassis and has a fully enclosed armoured cab in the centre and a elevatable mast at the rear. It is used to provide target information to low-level air defence systems such as the SA-8B Gecko SAM. Prime contractor for the N-22 is the RADWAR company of Poland.

Status

Production complete in service with:

Country	Quantity
Czech Republic	273
Libya	80
Poland	111
Slovakia	134

Manufacturer

ZTS Dubnica nad Váhom (Slovakia).

UPDATED

FRANCE

Giat Industries CAESAR 155 mm self-propelled gun

Development

Giat Industries has developed a 155 mm self-propelled technology demonstrator called CAESAR (*CAmion Equipe d'un Système d'ARtillerie*, or truck-mounted artillery system) and this was shown in public for the first time in June 1994.

Following trials with this system a decision was taken by Giat Industries to build a pre-production CAESAR and following company trials this was evaluated by the French Army late in 1998. During these trials over 120 rounds of 155 mm ammunition were fired.

Giat Industries believe that there is a significant untouched market for a weapon of this type to bridge the gap between 155 mm heavy self-propelled artillery systems, tracked and wheeled, and 155 mm towed artillery systems.

The CAESAR has been developed by Giat Industries, Weapons and Ammunition Systems Division facility at Bourges to fill this gap in the market for a 155 mm 52 calibre self-propelled artillery system for deployment by rapid reaction type forces.

An early example of CAESAR was demonstrated in the Middle East several years ago, but since then much further development work has taken place.

In September 2000, the French Délégation Générale pour l'Armement (DGA) awarded Giat Industries a contract for the supply of a batch of five 155 mm/52 calibre CAESAR (6 × 6) self-propelled artillery systems.

CAESAR 155 mm/52 calibre self-propelled artillery system deployed in the firing position 0058412

These will be integrated at Giat Industries facility at Bourges with the cab being supplied by SOFRAME (previously LOHR) which will be installed on the modified Mercedes-Benz (6 × 6) truck chassis.

By this time Giat Industries had built two CAESAR systems. The first CAESAR was essentially a technology demonstrator to prove that it was possible to mount a 155 mm/52 calibre artillery system on a modified Mercedes-Benz UNIMOG U 2450 (6 × 6) truck chassis.

This was followed by a second model which has been evaluated by the French Army and late in 1999

Giat Industries CAESAR 155 mm/52 calibre (6 × 6) self-propelled artillery system during trials in Malaysia **2001**/0106066

CAESAR 155 mm/52 calibre self-propelled artillery system in travelling configuration during trials in Malaysia **2002**/0130603

went to Asia. It was then evaluated by the Royal Malaysian Army during extensive trials during which the system travelled over 2,800 km and fired 57 rounds of 155 mm ammunition. It was then taken to Brunei where it was demonstrated but not fired.

The third CAESAR has now been built and features all of the lessons learned with the French and Malaysian trials and will be essentially a preproduction vehicle.

It will feature a revised fully enclosed crew cab with improved air conditioning and upgraded computer software. This CAESAR will be the baseline for the French Army version which will have the ATLAS computerised fire-control system and a land navigation system which will also be fitted to the French Army AUF1 and future AUF2 full tracked self-propelled artillery systems.

The first CAESAR systems will be delivered to the French Army late in 2002 with final deliveries taking place the following year. These will be used by the French Army to form one complete artillery troop of four CAESARs with the 5th being used for training and reserve. The DGA contract also covers the maintenance and support of the five systems for a three-year period.

This CAESAR troop will be used to evaluate operational procedures for the deployment of the system by the French Army. Giat Industries hopes that the French Army will eventually replace some of its current fleet of TR 155 mm/39 calibre towed systems with the CAESAR.

Specifications

Crew: 6
Configuration: 6 × 6
Weight:
(combat) 17,680 kg
(unloaded) 15,800 kg
Power-to-weight ratio: 13.63 hp/t
Length: 10 m
Width: 2.55 m
Height: 3.65 m
Track: 1.87 m
Ground clearance: 0.40 m
Wheelbase: 3.9 + 1.4 m
Max road speed: 100+ km/h
Range: 600 km
Fording: 1.20 m
Gradient: 60%
Side slope: 40%
Fuel capacity: 230 litres
Engine: Mercedes-Benz LA 6 cylinder turbo charged diesel developing 240 hp at 2,600 rpm
Transmission: manual, 8 forward and 1 reverse gear
Clutch: single dry plate
Axles/suspension: portal axles with hub drive, torque tubes, all-wheel drive and differential locks in each axle, coil springs and telescopic shock absorbers
Brakes:
(main) pneumatic
(parking) spring loaded on rear axles
Steering: hydraulic
Tyres: 14.5R 20 (with central tyre pressure inflation system)
Electrical system: 24 V
Batteries: 2 × 12 V, 125 Ah
Armament: 1 × 155 mm 52 calibre
Ammunition: 18 projectiles and charges
Elevation/Depression: +66°/–3°
Traverse: 17° (left and right) (total 34°)
NBC system: no
Night vision equipment: no

Description

CAESAR is based on a modified Mercedes-Benz UNIMOG U 2450 L (6 × 6) truck chassis that has a demonstrated high level of cross-country mobility. UNIMOG trucks are used by many armed forces and spare parts and logistic support are available in most parts of the world.

Mobility of the CAESAR is further improved by the installation of a central tyre pressure regulation system. This allows the driver to adjust the tyre pressure to suit the terrain being crossed.

All tyres are of the run-flat type and power-assisted steering is fitted as standard to reduce driver fatigue. CAESAR is powered by a Mercedes-Benz six-cylinder diesel engine developing 240 hp which gives a power-to-weight ratio of 13.63 hp/tonne. Maximum road speed is over 100 km/h while maximum road range is 600 km.

As CAESAR is mounted on a wheeled chassis its overall life cycle costs are much lower than a comparable tracked vehicle. This is of increasing importance as many users are now looking at the total life cycle costs of weapon systems rather than only the initial procurement cost.

A new fully enclosed cab has been mounted at the front of the vehicle and has individual seats for the crew of six. The cab is of welded steel armour and provides protection from small arms fire up to 7.62 mm in calibre and shell splinters. The cab windows are 26 mm thick and the cab is fitted with an air conditioning system.

Mounted at the rear of the CAESAR is the complete upper part of the 155 mm 52 calibre upgraded version of the TRF1 towed artillery system. The 155 mm 52 calibre ordnance is fitted with a double baffle muzzle brake and when travelling the ordnance is held in position by a clamp, located to the immediate rear of the cab, which is operated by remote control.

When the system is deployed in the firing position a large spade is hydraulically lowered at the rear to provide a more stable firing platform. The rear four wheels are raised clear of the ground so that the large spade absorbs all the firing stresses. CAESAR can come into action in less than 1 minute and come out of action in a similar period.

The 155 mm 52 calibre ordnance, which was developed some years ago for the 155 mm 52 calibre upgraded version of the TRF1, is fitted with a screw breech mechanism that opens upwards automatically with a revolving automatic primer feed mechanism holding 14 primers. It conforms to the latest NATO Joint Ballistic Memorandum of Understanding. An electrical firing device is fitted as standard.

CAESAR is a self-sufficient self-propelled artillery system as no survey team is required due to the installation of a SAGEM SIGMA 30 onboard reference package/position data system (RP/PDS) which includes a Global Positioning System and which is mounted on the actual gun.

An Intertechnique RDB4 muzzle velocity radar is mounted over the ordnance and this feeds information to the Matra Défense Equipements & Systèmes CS 2002-G onboard fire-control computer which is located in the cab together with its printer.

This receives target information from the battery command post with ballistic computation taking place on the system. Communications is via Thales PR4G radios.

The CS 2002-G computer also carries out a number of other functions including 3-D display of friend or foe local situation. With the aid of integral sensors it can also carry out ammunition status management, gun status management and ammunition resupply management.

There is also a temperature control device that advises the crew if the ordnance is becoming too hot and there is a danger of a cook off.

To reduce crew fatigue and increase the rate of fire an automatic projectile loader is mounted to the right side of the breech. The propelling charges, which can be of the conventional type or the more recent modular charge type, are loaded manually.

Maximum rate of fire is 6 rds/min and according to Giat Industries a burst rate of fire of three rounds in 18 seconds is normally attained. According to Giat Industries, CAESAR can come into action, fire a burst of six rounds and move off again in 3 minutes.

Elevation and traverse is hydraulic with manual controls being provided as a back-up. For indirect usual firing elevation ranges from +17 to +66° with traverse being 17° left and right. For direct firing elevation ranges from –3 to +10° right, with traverse between 21° left and 27° right. Aiming is carried out automatically via the onboard computer but optical sights are provided as a back-up.

CAESAR 155 mm/52 calibre self-propelled artillery system deployed in firing position with spade lowered to the ground **2002**/0073443

Maximum range depends of the nature of ammunition and charge system but firing an ERFB-BB (Extended Range Full Bore-Base Bleed) a maximum range of 42 km can be achieved. Full range details are given in the ammunition specification in Table 1.

Table 1

Projectile designation	Maximum range	Type
155 OGRE	35 km	Dual Purpose ICM
155 OMI	21.5 km	Cargo, anti-tank mines
155 mm BONUS	35 km	anti-tank, smart
155 mm LU 211 BB	39 km	HE, base bleed
155 mm LU 211 HB	30 km	HE, hollow base
155 mm LU 214 BB	39 km	smoke, base bleed
155 mm LU 214 HB	30 km	smoke, hollow base
155 mm PRAC SR F1	9 km	short range practice
155 LU 112	29 km	practice
M107	18.3 km	HE
M485A2	18.3 km	illuminating
ERFB NR 269 BB	42 km	Dual Purpose ICM, base bleed
ERFB NR 265 BB	42 km	HE, base bleed
ERFB NR 173 BT	33 km	HE, boat tail

The main fire-control computer is located in the cab of the CAESAR but at the rear on the left is the gun display unit for the crew when the system is deployed in the firing position. This provides such information as elevation, traverse, projectile, charge and fuze type.

The main advantages of the CAESAR are that it has similar firepower to existing towed and self-propelled artillery systems with greater strategic mobility and quicker in/out of action times. The last feature also enhances the combat survivability of the system to counter battery fire. The standard 155 mm TRF1 has a crew of eight while CAESAR has crew of only six.

The design of CAESAR is modular so that the customer can select which radio, fire-control system or muzzle velocity measuring system and type of cab required.

As well as being transported by air, for example Lockheed Martin Hercules C-130 and Transall C160, CAESAR can also be carried by landing craft, surface ships and railway flatcars.

Status

Prototypes. Five systems ordered by French Army for delivery in 2002.

Manufacturer

Giat Industries.

UPDATED

INTERNATIONAL

BAE Systems, RO Defence 155 mm lightweight self-propelled artillery system

Development

To meet the potential requirements of the US Army 'medium force' Brigade Combat Team (BGT) BAE Systems, RO Defence, has teamed with the US-based United Defense LP, to develop concepts for a new 8 × 8 155 mm/39 calibre lightweight self-propelled (LW SP 155) artillery system.

This system leverages off work done on the BAE Systems, RO Defence 155 mm AS90 Braveheart 39 calibre full tracked SP artillery system, in service with the British Army and selected by Poland, and the 155 mm/39 calibre XM777 UFH towed artillery weapon, under development for the US Army/Marines and the Royal Artillery.

United Defense LP, is prime contractor for the US Army's current 155 mm M109A6 Paladin and future 155 mm Crusader tracked SP artillery systems.

Description

LW SP 155 will be based on the chassis of the now General Motors Defense Light Armored Vehicle (LAV), or Piranha as it is also known. Well over 5,000 of these have been ordered in 6 × 6 and 8 × 8 configuration by an increasing number of countries around the world.

The US Marine Corps already operates a fleet of almost 800 8 × 8 LAVs in seven configurations. The latest LAV-III (8 × 8) is now in quantity production for the US Army with first production vehicles being handed over early in 2002.

Mounted on the rear of the LAV 8 × 8 chassis is a 155 mm/39 calibre ordnance and associated elevating mechanism which is the same as that used in the towed XM777. When firing an extended range projectile such as the XM982, currently under development, the system will achieve a range of 40 km.

Specifications
Crew: 3
Weight (combat): 22,000 kg
Weight (transport):
(52 calibre ordnance) 18,500 kg
(39 calibre ordnance) 18,000 kg
Length: 11.5 m
Width: 2.667 m
Height: 2.64 m
Max range:
(52 calibre ordnance) 40 km
(39 calibre ordnance) 30 km
Rate of fire:
(burst) 3 rounds in 15 s
(intense) 8 rds/min
Armament: 1 × 155 mm 39 or 45 calibre
Ammunition:
(projectiles) 42
(charges) 36

Projected BAE Systems/United Defense LW SP 155 (8 × 8) in travelling configuration **2000**/0085188

It may be fed by an automatic ammunition handling system that will load the projectile and charge and allow the gun crew to remain under full armour and NBC protection. An onboard computerised gunlaying system will be fitted to reduce into action time and increase weapon effectiveness.

BAE Systems, RO Defence, stresses that a wide range of designs is being studied, including a manually loaded system with the crew deployed outside the weapons.

The crew of three will be seated at the front down the left side with the diesel power pack on the right and the whole rear occupied by ammunition. About 42 155 mm projectiles and associated modular artillery charges will be carried.

For improved cross-country mobility the LW SP 155 will be fitted with a central tyre pressure regulation system that will allow the driver to adjust the tyre pressure to suit the terrain being crossed. Standard equipment will include a winch and powered steering.

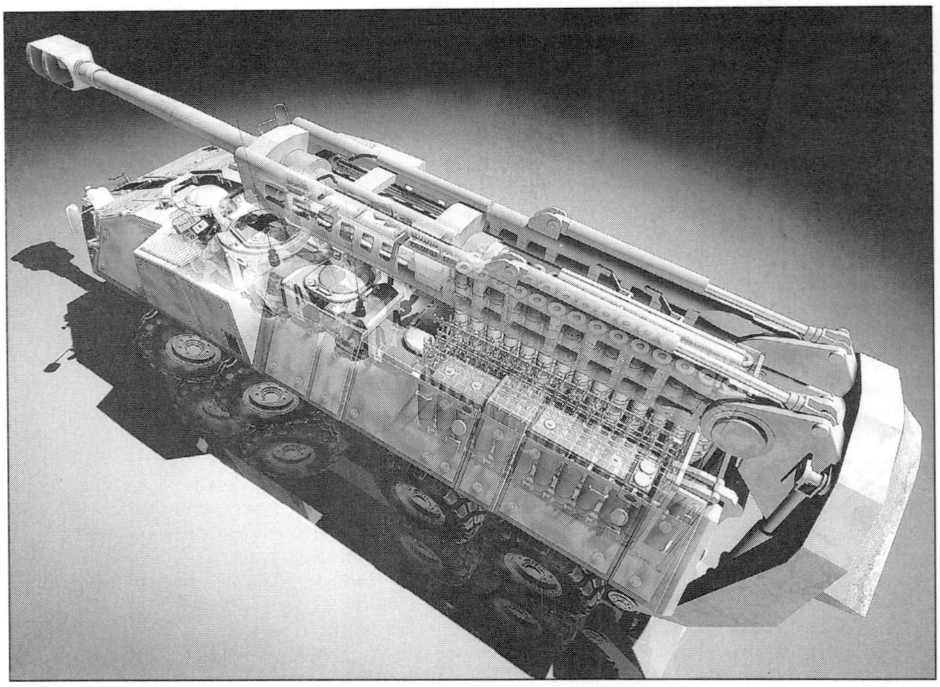

Provisional cutaway drawing of the BAE Systems/United Defense LWSP 155 showing position of crew and ammunition and with ordnance in travelling configuration **2002**/0130604

It will also be fully air portable in a Lockheed Martin C-130 Hercules transport aircraft which is one of the key requirements of the medium force BCT.

The system will also be available in a 155 mm/52 calibre configuration and can be mounted on a variety of other light armoured vehicle chassis of the appropriate size and load-bearing capability.

The LWSP is configured so that the principle firing loads are transmitted to the ground, rather than being absorbed by the vehicle chassis. A large spade is provided at the rear and the 155 mm ordnance fires to the rear rather than over the front of the chassis.

Status

Concept studies.

Manufacturers

BAE Systems, RO Defence.
United Defense LP

UPDATED

RO Defence/United Defense Tilt Bed Carrier 155 mm self-propelled howitzer

Development

Shown for the first time in 2001 at Fort Sill, Oklahoma, home of the US Army's Field Artillery, was a new Tilt Bed Carrier (TBC) concept demonstrator. It allows the 155 mm/39 calibre XM777 towed artillery system to be carried inside a Lockheed Martin C-130 Hercules transport aircraft complete with its prime mover.

The TBC concept was evolved over a five month period with extensive analysis and trade-offs involving 23 different concepts. These included mounting the XM777 155 mm system on a United Defense LP, Mobile Tactical Vehicle Light (MTVL), LAV-III (8 × 8) as well as mounting the system on the rear of a 6 × 6 truck.

The TBC has been developed as a private venture by BAE Systems, RO Defence, prime contractor for the 155 mm/39 calibre XM777 towed artillery system, and United Defense LP, prime contractor for all of the US Army's self-propelled artillery programmes.

The US Army's new Brigade Combat Teams (BCT) based around the General Motors Defense LAV-III (8 × 8) chassis, will be in the short term equipped with the heavy 155 mm M198 howitzer towed by a 5 ton (US) (6 × 6) truck. This will require two Lockheed Martin C-130 Hercules transport aircraft to carry them.

According to the contractors, the TBC 155 mm system is intended to combine the lightweight and operational advantages of the new XM777 towed howitzer with the mobility of self-propelled artillery systems.

This combination provides tactical versatility, mobility equivalent with manoeuvre forces, ease of operations, combat effectiveness and C-130 transportability.

Wherever possible, Non-Developmental Items or near Non-Developmental Items are used in the TBC wherever possible.

Computer generated model of the M1086A1 (6 × 6) truck in Tilt Bed Carrier configuration carrying an XM777 ***2002***/0095348

Description

The TBC concept consists of a standard Stewart & Stevenson M1086A1 (6 × 6) 5 ton (US) truck fitted with a standard commercial tilt bed, on which the complete 155 mm XM777 weapon is carried.

Prior to being loaded onto the TBC the XM777 is slightly elevated and the spades and stabilisers are stowed. The suspension is raised until the base is about 8° relative to the ground. A draw bar assembly is then attached to the gun at the cradle front and travel lock using a tall upstand.

The TBC has been modified with eight tie-down points as well as pockets for stabilisers and base grouser. To move the XM777 onto the TBC the tilt bed is deployed and a powered guide device is lowered to interface with the draw bar with suspension height adjusted to allow hook up.

The sled is then moved forward and stops, and the suspension of the XM777 is lowered until its base contacts the carrier bed. The XM777 is then tied to the bed and the tilt bed is tilted into the level position and rolled forward, where it is locked in position. The complete system is then ready to move off.

To enable it to fit into a C-130 aircraft the tilt bed is rolled back and tilted downwards. The XM777 is slightly depressed and a short upstand is attached to the cradle legs with the tall upstand being folded forwards. Once inside the C-130 aircraft a carrier stabiliser is lowered to the floor of the aircraft to distribute the load.

Once offloaded from the C-130 the weapon can be transported as before or towed behind the truck.

In its standard tactical manoeuvre configuration the XM777 would be towed by the TBC with the FMTV pintle hitch attached to the tilt frame. The pintle can be moved vertically or extended with hydraulic power.

It is described as a low-cost and low-risk solution to meet the possible future requirements of the US Army. The next stage would be to integrate ammunition stowage and space for the crew and their equipment. To meet the C-130 weight constraints it has been calculated that at least 16 155 mm artillery projectiles, associated charges and fuzes would be carried.

Although aimed at the US Army the TBC could also reduce the artillery footprint when deployed on US amphibious ships. The prototype of the TBC is on the M1086A1 (6 × 6) chassis it could be fitted to other chassis.

The 155 mm XM777/39 calibre towed artillery system is currently in the Engineering Manufacturing Development (EMD) phase, with eight weapons built in the UK using an ordnance supplied by Watervliet Arsenal in the US. The first XM777 was shipped in 2000 and all were supplied by the end of 2001. Full details of this are given in the Towed anti-tank guns, guns and howitzer section

Status

Demonstrator

Manufacturers

BAE Systems, RO Defence.
United Defense LP.

NEW ENTRY

ISRAEL

Soltam Systems 155 mm ATMOS 2000 SPG

Development

Late in 2001, Soltam Systems released details of the latest version of its ATMOS 2000 155 mm/52 calibre self-propelled artillery system whose existance was first revealed late in 1999. At that time it was also referred to as the 155 mm Self-Propelled Wheeled Gun (SPWG).

ATMOS 2000 has been developed by Soltam Systems as a private venture. It is aimed mainly at the export market, although it has already been demonstrated to the Israel Defence Force.

There are nine subcontractors to Soltam Systems on the ATMOS 2000 programme, including SAGEM of France.

According to Soltam Systems, wheeled self-propelled guns are not only cheaper to procure than their more common tracked counterparts, but have lower life cycle costs and are easier to operate and maintain.

In addition they also have greater strategic mobility as they do not have to rely on Heavy Equipment Transportors (HET) to be redeployed over extended distances as do their full-tracked counterparts.

By late 2001 the prototype ATMOS 2000 has now fired over 1,000 rounds of various natures of 155 mm ammunition, using bagged charges during extensive company trials in Israel.

Description

The Soltam Systems ATMOS, 155 mm/52 calibre self-propelled artillery system essentially consists of a 6 × 6 cross-country truck chassis, on the rear of which is mounted the upper part of a Soltam Systems 155 mm/52 calibre towed TIG 2000 system.

The prototype is fitted with a 155 mm/52 calibre ordnance that conforms to the NATO Joint Ballistic Memorandum of Understanding (JBMoU), although Soltam Systems 155 mm/39 and 155 mm/45 calibre barrels can also be fitted.

When fitted with a 155 mm/52 calibre barrel a maximum range of 41 km can be obtained firing an Extended Range Full-Bore – Base Bleed (ERFB-BB) projectile, 30 km firing the NATO L15 High Explosive (HE) projectile and 24.5 km firing the older M107 HE projectile.

ATMOS 2000 carries a total of 32 155 mm projectiles and associated charges and is normally operated by a crew of four, of whom two are loaders positioned one either side at the rear.

A rate of fire of between 4 and 9 rds/min are claimed and to reduce crew fatigue a flick rammer is fitted as standard. To provide a more stable firing platform, two large spades are lowered to the ground one either side at the rear of the system by remote control.

It also has an onboard Advanced Fire-Control System (AFCS). This includes a main computer and a navigation and aiming device and enables ATMOS 2000 to rapidly engage the target after receiving the position of the target from the Forward Observation Officer (FOO).

The AFCS processes the target information, presents this on a display to the gunner and then lays the weapon onto the target in traverse and elevation. Manual back-up is also provided for in case of power failure.

Although the prototype system is installed on a modified Tatra (6 × 6) truck chassis (which has been built under licence in India), Soltam Systems have emphasised that it could be fitted to a wider range of other wheeled chassis.

The forward-control armour-protected cab provides the occupants with protection from small arms fire and shell splinters. It is envisioned that production versions of the ATMOS 2000 would be fully air-portable in a Lockheed Martin C-130 Hercules transport aircraft.

Variant

ATMOS 2000 155 mm M46S

Soltam Systems have stated that the ATMOS 2000 could be fitted with the saddle and elevating mass of the Russian 130 mm M-46 field gun, with the ordnance replaced by a new 155 mm/45 calibre system.

Soltam Systems has already won a major contract from India to upgrade some 180 Russian supplied 130 mm M-46 field guns with a 155 mm/39 calibre and 155 mm/45 calibre ordnance in 2000.

India has a requirement for a new self-propelled artillery system. So far all of the systems that have been tested are full tracked but recent information has indicated that a mix of tracked, and wheeled systems are now being considered to give greater operational flexability.

Specifications

Not available.

Status

Prototype undergoing trials. Not yet in production or service.

Manufacturer

Soltam Systems Limited

NEW ENTRY

Soltam Systems ATMOS 2000 155 mm/52 calibre (6 × 6) self-propelled artillery system deployed in the firing position **2002**/0096537

Soltam Systems ATMOS 2000 155 mm/45 calibre (M46S) self-propelled artillery system deployed in the firing position from the rear **2002**/0096538

NETHERLANDS

RDM Technology MOBAT 105 mm (4 × 4) self-propelled gun

Development

The MOBAT 105 mm (4 × 4) self-propelled gun has been developed as a private venture by RDM Technology and builds on experience gained in the development and production of a wide range of 105 and 155 mm upgrades for towed and self-propelled artillery systems for the home and export markets.

According to RDM Technology, the MOBAT 105 mm (4 × 4) self-propelled gun is suitable for use by rapid deployment forces. It has been designed to use shoot-and-scoot techniques for higher battlefield survivability of the platform.

Following trials in the Netherlands in 1998, MOBAT was tested in the United Arab Emirates and was shown for the first time in public during a major defence exhibition held in Abu Dhabi in early 1999.

Late in 2000, a preproduction example of the RDM Technology MOBAT (Mobile Artillery) 105 mm self-propelled gun (SPG) went to Thailand for extensive firepower and mobility trials.

During this time, some 300 rounds of 105 mm ammunition were fired with a maximum rate of fire of 12 rounds per minute being obtained. Types of 105 mm ammunition fired included ERM1 base bleed.

Specifications

Configuration: 4 × 4
Combat weight: 10,000 kg
Unloaded weight: 8,600 kg
Length:
(travelling) 7.20 m
(air transport) 5.40 m
Width:
(travelling) 2.50 m
(air transport) 2.70 m
Height:
(travelling) 3.30 m
(air transport) 2.70 m
Armament: 1 × 105 mm
Ammunition: 40 × 105 mm
Max range: 19,200 m (ERM1 projectile)
Rate of fire:
(burst) 8 rds/m
(sustained) 5 rds/m

Projectile	Charge	Zone	Muzzle velocity	Range (min)	Range (max)
M1	M67	7	507 m/s	7.3 km	11.9 km
ERM1	RO633A2	2	710 m/s	13.5 km	19.2 km
M760	M200	8	648 m/s	9.3 km	14.4 km
M548 RAP	M176	7	557 m/s	9.9 km	15.5 km
XM913 RAP	M200	8	648 m/s	12.6 km	19.6 km

During one demonstration, it took the MOBAT system 3 minutes, after a call while driving, to stop, go into a firing position, carry out ballistic calculations, lay the gun, fire 12 rounds in 1 minute and start driving again.

Description

The prototype of the MOBAT (Mobile Artillery) consists of a DAF YA 4440 (4 × 4) forward control cross-country truck with a flatbed rear body. On this is mounted a new saddle and elevating mass fitted with the latest RDM

RDM Technology MOBAT 105 mm (4 × 4) self-propelled gun deployed in the firing position with stabilisers lowered to the ground **2001**/0106065

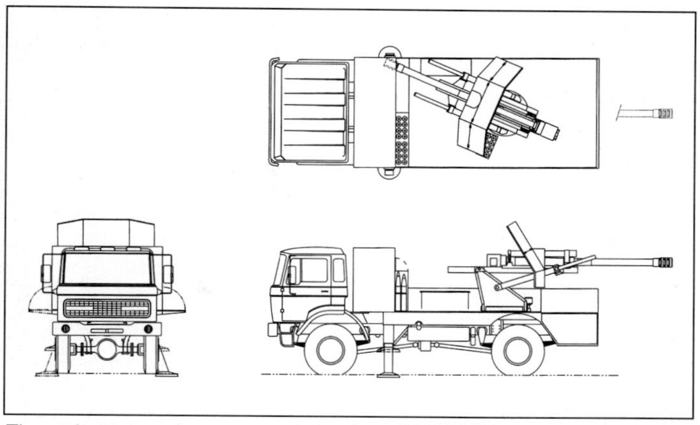

Three view general arrangement drawing of MOBAT in travelling and C-130 Hercules transport configurations **2001**/0106064

A total of 40 105 mm projectiles and associated charges are carried with these being stowed to the immediate rear of the cab under cover.

The baseline MOBAT has a standard gun sight but options include the integration of an automatic gunlaying and navigation system coupled with an onboard computerised fire-control system.

For the Far East trials, MOBAT was fitted with the BAE Systems Laser Inertial Pointing System (LINAPS) of which 137 are now being supplied for installation on the 105 mm Light Guns of the Royal Artillery.

LINAPS provides the precise location of the MOBAT with a route navigation capability and also provides the gun crew with orientation and location data without the requirement for the standard sighting system.

The complete MOBAT system is of modular construction and can be removed from the vehicle in one hour so that the vehicle can be converted rapidly back into a truck.

When travelling, the complete weapon would normally be traversed to the rear and when being transported by air, for example in a Lockheed Martin C-130 transport aircraft, it would normally be traversed to the front and offset slightly to the right.

Although the prototype MOBAT is based on a DAF YA4440 4 tonne (4 × 4) truck, this was only selected by RDM Technology as quantities of these were available in the Netherlands. The weapon system can be mounted on a wide range of other 4 × 4 chassis.

Status

Development complete. Ready for production.

Manufacturer

RDM Technology bv.

UPDATED

SINGAPORE

Singapore Technologies Kinetics 155 mm/39 calibre Light Weight Self Propelled Howitzer (LWSPH) (4 × 4)

Development

Early in 2001, Singapore Technologies Kinetics, which includes the former Singapore Technologies Automotives, Ordnance Development and Engineering (ODE) and Chartered Industries of Singapore (CIS), announced that it was developing a new 155 mm/39 calibre Light Weight Self Propelled Howitzer (LWSPH) (4 × 4).

This system is being developed as a private venture for the export market, with first firing trials of the complete ordnance and possibly the upper carriage taking place in South Africa in 2000.

It is understood that the first complete prototype of the 155 mm/39 calibre LWSPH will be completed some time in 2002 and it will be ready for production in 2003-2004.

The 155 mm/39 calibre ordnance used in the prototype of the LWSPH was supplied by BAE Systems, RO Defence, Nottingham facility.

The 155 mm/39 calibre LWSPH is being developed for use by rapid deployment forces with key design factors stated by Singapore Technologies Kinetics to include:-

(1) Excellent firepower
(2) Superior strategic and tactical air or land mobility
(3) Good survivability
(4) Minimal logistics support

As the weight of the complete LWSPH system is only about 7,000 kg, the system can be transported inside of a Lockheed Martin C-130 Hercules transport aircraft and carried slung under a CH-47 or CH-53 helicopter.

First artist's impression released of the Singapore Technologies Kinetics 155 mm/39 calibre Light Weight Self-Propelled Howitzer deployed in firing position **2002**/0114261

Singapore Technologies Kinetics has considerable experience in the design, development and production of 155 mm artillery systems. Earlier towed artillery systems include the 155 mm/39 calibre FH-88 and the 155 mm/52 calibre FH-2000.

These two 155 mm artillery systems have been developed to meet the operational requirements of the armed forces of Singapore which was the first country in the world to field a 155 mm/52 calibre towed artillery system. The FH-88 has been exported to Indonesia who took delivery of five systems.

Description

The Singapore Technologies Kinetics 155 mm/39 calibre LWSPH is built around a very strong space frame type chassis with the power pack and seats for at least four crew members at the front. When travelling the crew are protected by a steel frame which is swung out either side when deployed in the firing position.

The driver, seated at the front right side can select either 4 × 4 or 4 × 2 drive. The 170 hp turbocharged intercooled diesel engine supplies power to the hydromechanical transmission and then to the hub-radial piston motor at each wheel station.

The front suspension is of the independent multilink type with hydropneumatic struts, while the rear is an independent trailing arm type with hydropneumatic struts. All wheels are fitted with disc brakes and an ABS system is fitted as standard.

The 155 mm/39 calibre ordnance is mounted at the rear and when deployed in the firing system, the complete system is stabilised by two spades that are lowered one either side at the rear of the chassis.

The overall battlefield survivability of the system is enhanced as it can come into and be taken out of action in less than 1 minute, so allowing it to use shoot and scoot tactics. According to Singapore Technologies Kinetics, a quick dash across 500 m takes less than 1 minute.

It can also be integrated with an inertial navigation system and various types of computerised fire-control systems for autonomous operations.

Variants

In March 2001, a MoU was signed between Singapore Technologies Kinetics and Otobreda of Italy to jointly develop a variant of the 155 mm/39 calibre LWSPH.

Technology version of the upgraded 105 mm/33 calibre M101 towed artillery system.

The main improvements for the MOBAT version are the replacement of the existing 105 mm barrel by a new and longer barrel with a multislotted muzzle brake, installation of larger calibre equilibrators, new metering and recoil rod in the recoil system, replacement of the original arc, modified breech and new breech ring.

By mounting the system on a truck chassis, it is claimed that it can come into action and be taken out of action much quicker than a conventional towed 105 mm artillery system. It is therefore less likely to come under immediate counterbattery fire.

MOBAT also has a higher level of strategic mobility than a conventional towed artillery system and can therefore be quickly deployed from one part of the country to another as required without logistic support. If required, the system can also be made to resemble a standard army truck.

To provide a more stable firing platform, a hydraulically operated stabiliser is lowered to the ground on either side.

Weapon elevation is all-electric using joystick control with traverse limits being 45° left and right. An auxiliary power unit is fitted as standard to keep the batteries charged. Elevation range is from –90 to +1,260 mils with traverse being 800 mils left and right. A burst rate of fire of 8 rds/min can be obtained while a sustained rate of fire of 5 rds/min can be achieved.

The range of the 105 mm/33 calibre weapon depends on the projectile/charge combination:

This MoU provides a framework for joint collaboration on design, manufacturing and agreement on marketing rights with regards to the LWSPH variants. In addition, the MoU sets out the agreement for Otobreda to provide their 155 mm ordnance for integration with the chassis of the LWSPH variant.

Specifications (provisional)
Crew: 4 (on platform)
Weight: 7,000 kg (excluding ammunition)
Length: 10.00 m
Width: 2.5 m
Height: 2.6 m
Max road speed: 80 km/h
Max road range: 600 km (cruising)
Gradient: 60%
Side slope: 30%
Steering: power-assisted
Turning radius:
(kerb to kerb) 8.0 m
(wall to wall) 10.6 m
Engine: turbo-charged inter-cooled diesel developing 170 hp
Armament: (main) 1 × 155 mm/39 calibre
Max range:
(M107 HE projectile) 19,000 m
(ERFB BB) 30,000 m
Accuracy:
(unassisted) 0.30%
(assisted) 0.5%
Rate of fire:
(burst) 3 rds in 15 s
(max) 6 rds/min
(sustained) 2 rds/min for 30 min

Status
Development. Not yet in production or service.

Manufacturer
Singapore Technologies Kinetics.

NEW AND UPDATED 11/03/2002

SOUTH AFRICA

LIW G6 155 mm self-propelled gun-howitzer

Development
In the 1960s, the South African Artillery deployed three main artillery weapons, the UK 25-pounder Field Gun (G1), the 5.5-inch Medium Gun (G2) and the Canadian-built Sexton 25-pounder self-propelled gun.

During early South African operations in Angola the poor range of South African artillery compared to opposing Russian-supplied 122 mm D-30 and 130 mm M-46 towed artillery weapons and BM-21 122 mm multiple rocket launchers led to the fielding of the LIW 155 mm G5 gun-howitzer and the SOMCHEM 127 mm (24-round) Valkiri multiple rocket system.

While the towed G5 is perfect for many South African Artillery requirements, it does not have the cross-country mobility of the Vickers OMC Ratel (6 × 6) infantry vehicle which has been in service since 1976.

The South Africans weighed the options of wheels or tracks for their new self-propelled artillery system and came down firmly in favour of wheels. The possible areas of operation of the G6 are likely to be some distance from their main bases, perhaps up to 1,000 km, and tracked vehicles have distinct disadvantages over long distances. Fuel consumption of a wheeled vehicle is a lot less than a tracked vehicle so less logistic support is required. The decision was therefore taken to choose a wheeled artillery system.

The G6 prototype was completed in 1981 and shown for the first time in public in September 1982. By late 1986, four complete G6s had been built plus one additional turret. Of these four, two can be considered as prototypes and two advanced development vehicles, for they incorporated many improvements as the result of user trials and changes in operational requirements.

The first advanced development vehicle was completed early in 1984 and the second during 1986.

There were four engineering development models completed during 1987 for final troop trials and confirmation of the modifications carried out on earlier vehicles.

The first production G6s were completed during 1988, with LIW being the prime contractor. Vickers OMC delivers the chassis to LIW which integrates the turret and delivers the complete system to the user.

The production G6s were the first self-propelled howitzers with a full autonomous laying and navigation capability.

As with the earlier G5, the G6 is a complete system and includes not only the vehicle but also the ammunition system, meteorological station, muzzle velocity analyser, fire-control system and a special artillery helmet radio communication system. Some of these systems have been in production for the G5 for

Latest production LIW G6 with bush guard protectors for the smoke grenade dischargers and with 155 mm/45 calibre ordnance elevated **2000**/0085182

Specifications
Crew: 6
Configuration: 6 × 6
Combat weight: 47,000 kg
Unloaded weight: 42,500 kg
Power-to-weight ratio: 11.17 hp/t
Length:
(gun forward) 10.335 m
(hull) 9.2 m
Width: 3.4 m
Height:
(incl MG) 3.8 m
(turret top) 3.3 m
(fighting compartment top) 1.9 m
Ground clearance: 0.45 m
Track: 2.8 m
Max speed:
(road) 90 km/h
(cruising) 70 km/h
Fuel capacity: 700 litres
Road range: 700 km
Fording: 1 m
Gradient: 40% (hard surface)
Side slope: 30%
Vertical obstacle: 0.50 m
Trench: 1 m
Turning radius: 12.5 m
Engine: air-cooled diesel developing 518 hp
Transmission: automatic with torque converter providing 6 forward and 1 reverse speeds
Suspension: independent with torsion bars and hydropneumatic damping system
Tyres: 21.00 × 25 (run-flat) and central tyre inflation system
Electrical system: 24 V
Batteries:
(hull) 24 V, 175 Ah
(turret) 24 V, 370 Ah
Armament: (main) 1 × 155 mm howitzer/45 calibre
(anti-aircraft) 1 × 12.7 mm M2 HB or 1 × 7.62 mm AA MG
(smoke-laying equipment) 2 × 4 81 mm grenade launchers
Ammunition: 45 × 155 mm projectiles
(50 × 155 mm charges)
Turret traverse: 80° (40° either side)
Weapon elevation: –5 to +75°
NBC system: integrated into turret power unit
Night vision equipment: optional

LIW G6 155 mm (6 × 6) self-propelled gun-howitzer deployed in the firing position with stabilisers lowered **2001**/0106063

Using the new V-LAP projectile and the M64 bimodular charge system, the LIW G6 155 mm 45 calibre G6 (6 × 6) self-propelled artillery system has a maximum range of 53.6 km **2002**/0096018

some years. This support equipment is continuously upgraded or replaced by improved products.

The G6 is deployed in batteries of six with the South African Artillery, each regiment having three batteries.

Three of the engineering development G6 155 mm self-propelled gun-howitzers were successfully used during the final South African campaign in Angola.

In 1990, Abu Dhabi, part of the United Arab Emirates, placed an order for the supply of 78 G6 155 mm self-propelled howitzers with the first of these being delivered early in 1991. All have now been delivered and are organised into three regiments each of 24 weapons, with the remaining vehicles being used for training and war reserve. A comprehensive long-term product support package was also successfully delivered and implemented.

Early in 1994, Oman placed an order for 24 G6 155 mm self-propelled howitzers, spares, logistic support, ammunition and a training package. All have now been delivered to Oman.

Late in 1996, it was disclosed that the South African National Defence Force had a total of 43 G6 systems in service under the designation GV6.

The 155 mm G6 is one of the contenders for a significant order from Saudi Arabia. Due to financial considerations this acquisition, which comprised a complete artillery system including weapons, ammunition, fire-control and target acquisition systems, has been cancelled for the time being.

Description

The hull of the G6 is all-welded steel armour, with protection against a variety of threats including artillery shrapnel. The G6 has a double armoured floor for increased protection against mines.

The chassis is essentially divided into three compartments, driver at the front, engine in the centre and fighting compartment at the rear.

The driver sits at the front between the front two roadwheels and, as there is no hull armour above these, any blast resulting from the vehicle running over a mine will be vented upwards and away from the driver.

To the front and sides of the driver are large bullet-proof windows which afford excellent visibility through the front in excess of 180°. If required, the central front window can be covered by an armoured shutter, hinged at the bottom, simply by activating a lever. When this is raised the driver observes through a periscope. He enters and leaves the vehicle via a roof hatch that opens to the left. Communication between the driver and the gun crew is via the gun's intercom system.

Steering is power assisted and the G6 now has permanent 6 × 6 drive (prototypes had the option of 6 × 4 with the front axle being disengaged). The driver can select either fully automatic or semi-automatic drive.

Forward of the driver's compartment is a lateral wedge-shaped box that doubles as a container for projectiles and as a bush-clearing device capable of cutting through trees and shrubs.

The engine compartment is to the rear of the driver's position with air inlet louvres in the front of the engine venting through the roof and the air outlet louvres through the sides, all of which can be opened to allow access to the engine for maintenance. The engine compartment is fitted with a fire detection and suppression system.

The G6 is powered by an air-cooled diesel developing 518 hp, it is coupled to an automatic transmission with a torque converter giving a total of six speeds forward and one in reverse.

Drive shafts take the power from the transmission via differentials to each roadwheel station.

The crew compartment, with the turret above, has seats for the commander, gun layer/navigator, breech operator, loader and an ammunition handler. There is an escape hatch in the floor. The turret crew enter via the main door in the rear right-hand side of the turret or via two roof hatches in the turret, one either side.

The turret has firing ports either side, with vision blocks above for close-in defence. Standard rifles and grenades are carried for close defence.

Mounted either side of the forward part of the turret is a bank of four 81 mm electrically operated grenade launchers which fire smoke grenades.

The gun commander sits at the right front of the crew compartment. A Gun Management System (GMS) is provided to assist the commander with command, control, communication and information functions in and around the G6. The GMS contains a Gun Display Unit (GDU) situated at the commander station which is used for sorting and displaying navigation data, gunlaying data and fire orders received via radio data communication link to each G6 from a Battery Command Post (BCP). Data transmission is also performed when the gun is mobile, ensuring an in action time of 1 minute and out of action time of 30 seconds.

Navigation and gunlaying data is transmitted via serial communication to the Laying and Navigation System (LNS). Navigation data contains UTM map datum points, routes containing waypoints and danger points with selectable warning radii. The GMS interfaces with a Muzzle Velocity Radar System (MVRS). The MVRS calculates a predicted mean normalised muzzle velocity. This calculated value is obtained utilising a moving average of muzzle velocities of an already fired group of ammunition. This value is used during target laying data calculations.

The GMS also interfaces with a Thermal Warning Device (TWD) warning the crew of unsafe barrel and recoil oil temperatures. Digital and analogue values are displayed on a separate TWD display by the commander. The TWD is connected to the intercom system providing an audible warning. A barrel-cooling fan is provided to reduce recovery time in the event of unsafe temperatures being reached.

The GMS contains an Ammunition Display Unit (ADU) mounted on the left rear of the vehicle providing the ammunition replenishment crew outside the gun with ammunition preparation and gunlaying data ensuring silent operation on the deployment site.

Data transmission makes use of the UHF/VHF frequency-hopping system of up to 50 km in range or back-up voice UHF radio link of up to 10 km in range. The commander in turn is in constant communication with the crew of the gun by using an internal intercom system or external local VHF band system of up to 100 m in range with the crew carrying belt-mounted transceivers.

Safety during firing is ensured with the aid of firing safety interlocks. The gun can only be fired once the breech operator, gun leader and gun layer have depressed their safety/ready switches and the gun has been layed within 1 mil onto target.

The gun layer/navigator sits at the left front of the crew compartment.

A Laying and Navigation System (LNS) assists the gun layer with laying and navigation functions. The LNS, using an RLG (Ring Laser Gyro) inertial unit, provides an autonomous all-weather day and night fighting capability from unsurveyed sites with all time

LIW G6 155 mm self-propelled gun-howitzer **2000**/0085183

ready state navigation and gun orientation ability. Navigation and gun orientation is performed without the use of external reference points or signals.

An additional Fibre Optic Gyro (FOG) inertial unit can be provided as an option giving a full laying and navigation back-up capability ensuring continuous availability of the gun.

Both the RLG and FOG inertial units receive inputs from a Distance Transmitting Unit (DTU) mounted on the gearbox of the vehicle. An additional embedded Global Positioning System (GPS) input complements the navigation accuracy. Advanced filtering techniques are used to optimise gunlaying and navigation performance.

Navigation information and centre of arc is displayed for the driver ensuring orientation of the gun during deployment.

The LNS requires no scheduled service or recalibration during operation and no scheduled maintenance during storage of the gun. Cold start is reduced to 12 minutes for both inertial systems. The Mean Time Between Failures (MTBF) of the RLG inertial sensor is 7,000 hours compared to 1,000 hours of the previous mechanical gyro system used in previous models.

A fully automatic gunlaying capability in both traverse and elevation directions is provided. This automatic gunlaying system enables rapid and accurate relaying of the 155 mm ordnance between rounds at high rates of fire. Accurate gunlaying is activated by a single touch button on the laying and navigation touchscreen. Target data is downloaded with the data communication link from the BCP via radio to the GMS in the gun. The GMS in turn downloads the target data to the LNS ensuring ease of operation. Target data can also be keyed into the LNS via the touchscreen display.

Direct firing is done by the gun layer under the commander's command. A telescopic direct sight is provided for initial aiming onto a direct target by the gun layer. Automatic onboard calculations are done by the LNS utilising direct firing range tables. A laser rangefinder is integrated with the LNS to provide instant target range. The ordnance is aimed automatically on target utilising a single touch button on the laying and navigation touchscreen.

The 155 mm 45 calibre autofrettaged ordnance of the G6 is a modified version of that used in the towed G5. It has a single-baffle muzzle brake, a semi-automatic screw and swing breech mechanism and an electrical firing system. A steel fume extractor is fitted. The recoil system consists of a single buffer and single recuperator.

The turret can be traversed through 180° (90° left and right) under electrohydraulic control with manual controls provided for maintenance, although for practical purposes the firing arc of the turret is 40° left and right. The 155 mm ordnance can be elevated from −5 to +75° under electrohydraulic servo control.

Although a single-touch auto layer provides the main laying function, a digital hand controller is provided for manual elevating and traversing the ordnance. A Hydraulic Control Panel (HCP) provides hydraulic system status. The HCP also has a joystick providing additional back-up for traversing and elevating the ordnance during emergency conditions or during maintenance.

A semi-automatic hydraulic flick rammer is mounted on the left of the breech with the projectile being placed on the rammer tray by hand. This is then aligned with the breech and rammed. The rammer can be used at all elevation angles and is operated by the ammunition loader. Propelling charges are loaded directly into the breech by hand by the breech operator.

A rate of 4 rds/min is achieved when firing with the maximum charge and can be continued for a period of 15 minutes, after which chamber temperatures have to be monitored. A rate of 5 rds/min is achieved when firing charge 2. A burst capability of 3 rounds in 25 seconds can be achieved.

The ammunition used with the G5 and G6 is of the ERFB type. Rounds available include HE, Smoke, Illuminating, RP, Submunition and Leaflet. The HE projectiles are forged from A151 9260 high-fragmentation steel with welded nubs. Filling may be TNH at the choice of the customer. The cargo rounds mentioned above are of the base ejection type. All rounds can be field-fitted with base-bleed. Full details of the projectiles are contained in table 1.

Table 1

Type	**HE**	**HE BB**	**Smoke**	**Illum**	**RP**	**Leaflet**	**Submunition**
Length (fuzed)	938 mm	958 mm	938 mm	938 mm	938 mm	938 mm	938 mm
Weight (nominal)	45.3 kg	47.7 kg	44.3 kg	44.3 kg	44.3 kg	44.3 kg	44.3 kg
Payload	8.7 kg	8.7 kg	10.9 kg	11.8 kg	±120 pellets	±3,500 message discs	13.72 kg

The Illuminating projectile has 1,000,000 candela for 120 seconds while the Smoke (SCR SMK) produces a smoke screen for 120 seconds from four canisters. The submunition round contains 42 bomblets and has an anti-personnel as well as a 120 mm armour-piercing capability. This round is sometimes referred to as the Cluster.

The charge system comprises five zones made up of three charges (combustible cartridge cases) and is optimised for use with 155 mm 45 calibre ballistic systems. Details are in table 2.

Table 2

Charge	**Muzzle velocity**	**Range (sea level)**
1 M51A1	350 m/s	4.5 to 9.1 km
2 M52A1	483 m/s	8.7 to 13.4 km
2 + 1 increment	645 m/s	12.6 to 19.0 km
2 + 2 increment	795 m/s	17.7 to 25.4 km
2 + 2 increment (BB)	789 m/s	22.0 to 31.0 km
3 M53A2	897 m/s	21.6 to 30.0 km
3 M53A2 (BB)	895 m/s	27.5 to 39.3 km
3 M53A2 VLAP	895 m/s	50 km

Area coverage: ERFB BB 1,062 km², VLAP 1,700 km²

The fuzes used with the projectiles include the PD M841, the M9220 electronic time fuze and the M8513 proximity. The PD M841 and M8513 are used with HE. The M9220 is used with Smoke, Illuminating, RP, Submunition and Leaflet. The PD M841 fuze incorporates a rain/cloud desensitising device to replace the US PD M572/PD M739 fuze and is compatible with base bleed projectiles.

At 75 per cent of maximum range, the 50 per cent zone for dispersion (2PE) is 0.96 per cent of range in length and less than 2 mils in deflection.

All 155 mm projectiles and charges are stowed below the turret bearing ring. It takes three men 15 minutes to reload the G6 with 45 projectiles and 50 charges. Fuzed projectiles and charges are transferred from an ammunition support vehicle through the rear of the G6 on conveyor chutes.

The G6 can be fired from the wheels, but is more stable when the four outriggers are lowered hydraulically to the ground by the driver or commander, two at the rear and one between the first and second wheels either side.

The crew compartment is provided with an explosion detection and suppression system with four outlets for the suppressant around the turret. Blowout panels are provided in the hull rear and the stowage of all ammunition is below the turret to improve the survivability.

The driver's compartment is provided with forced ventilation with the option of air conditioning, while the crew compartment has full air conditioning. Temperatures of 25°C lower than ambient can be maintained at ambient temperatures of up to 55°C.

A 38 kW Turret Power Unit (TPU) is mounted in the turret bustle to charge batteries and drive the air conditioning system. The latter is fitted with an NBC filtration system functioning on an overpressure principle.

The production G6 has permanent 6 × 6 drive and diff-locks on all differentials for improved traction. It is fitted with a central tyre-pressure regulation system that enables the driver to adjust the tyre pressure to suit the type of terrain while on the move.

The G6 has an independent suspension with torsion bars, hydraulic shock-absorbers and bump stops all-round. The 21.00 × 25 tyres are fitted with run-flat elements.

Vickers Ermetek and Horstman Defence Systems of the UK have developed a new switchable rotary damper for the G6 for improved cross-country mobility.

The G6 is supported in action by an ammunition supply vehicle carrying 156 rounds in four ammunition units and support equipment.

The turret and driver air conditioning systems of the G6 have been upgraded to accommodate ozone friendly R134a refrigerant gas. Halon gas in the fire extinguishing system has been replaced by NAF SIII gas, these being ozone friendly products.

155 mm T6 turret

This was developed for the Indian Army competition for a 155 mm turret system that could be fitted onto the chassis of the locally built T-72M1 chassis. It underwent extensive trials in India in 1995 and is covered in the Self-propelled guns and howitzers (tracked) section. As of early 2002 the T6 turret system remained at the prototype stage. It has been stated that the T6 turret could be offered as a retrofit for the current G6 turret.

The T6 features a 155 mm/52 calibre barrel that meets the Joint Ballistic Memorandum of Understanding and has a 23-litre chamber. A fully automatic fire-control system is fitted.

Forty rounds of 155 mm ready use ammunition are provided in the left rear of the T6 turret with 40 charge assemblies in the right side of the turret rear.

There are three loading modes, fully automatic, operator controlled and manual using hand ramming. A maximum rate of fire of 10 rds/min has been obtained during trials.

More range for G6

In 2001, Denel announced that their 155 mm G6/45 calibre self-propelled artillery system could now fire accurately out to a range of 53.6 km.

According to the company, this makes the G6 the first fully qualified and in-service gun and ammunition system breaching the 40 km range barrier by a considerable margin.

The 53.6 km range was achieved using the new Velocity enhanced Long range Artillery Projectile VLAP which combines base bleed and rocket motor technology and the new M64 bimodular charge system. The latter has also been fully qualified and is in quantity production for export. The new charge gives a maximum muzzle velocity of 910 m/s.

During extensive qualification firings in South Africa from a G6, probable errors in ranges were constantly better than 0.38 per cent. To make space for the base bleed and rocket assist elements, the V-LAP carries less high explosive. Its lethality is however still considerably higher than the older 155 mm M107 high-explosive projectile.

The V-LAP can be used with the 155 mm G5 towed and G6 self-propelled artillery systems and has already entered production for use with an undisclosed export customer equipped with the G6 system. It is probable that the country concerned is the United Arab Emirates (UAE).

PRO-RAM projectile

In 2001, it was disclosed that South Africa was working on yet another new 155 mm long-range artillery projectile called the PROjectile RAM (PRO-RAM). This has an integrated ramjet propulsion system and will allow ranges of at least 70 km to be achieved. It will also be very accurate when fitted with a low-cost course correction module.

Extended-range G5 and G6

Late in 1992, it was revealed that LIW had built the prototype of a 155 mm ordnance with a calibre length of 52. It is being used in development work which could be used to upgrade the G5 and G6 in the future, enabling their range advantage to be retained.

A number of developments is also under way on 155 mm ammunition in South Africa to give the user greater operational flexibility. Now available is the new M90 bi-modular charge system comprising identical modules of around 2.2 kg which can be used in 39, 45 and 52 calibre systems. A new 155 mm high-explosive Extended-Range Full-Bore - Base Bleed (ERFB-BB) projectile is also under development, containing 8.4 kg of high explosive which gives a range of 41 km. When fired from a 155 mm 52 calibre ordnance a range of 43.5 km is achieved.

The base bleed unit is of the screw-on type and can be removed under field conditions if not required.

More recently, the Velocity-enhanced Long-range Artillery Projectile (VLAP) has been introduced which is a rocket motor-assisted 155 mm projectile combined with the base bleed unit. This extends the range of the 45 calibre G5, G6 and T6 systems to 50 km and the 52 calibre gun to 52.5 km. Additional details are given in the entry for the T6 turret system in the previous section.

UAE G6/M109L47 upgrade

The United Arab Emirates has awarded the LIW Division of Denel a contract to upgrade their G6 and RDM Technology M109L47 to a common gunlaying and navigation integral fire-control system. Additional details are given in the entry for the United Defense M109 self-propelled howitzer in the previous section.

Marksman on G6

Late in 1994, it was announced that the now BAE Systems, RO Defence was marketing its Marksman twin 35 mm air-defence turret installed on the 155 mm G6 self-propelled chassis. Details of the Marksman turret system are given in *Jane's Land-Based Air Defence 2002-2003.*

The Marksman twin 35 mm air defence turret was developed as a private venture and is already in service with the Finnish Army installed on the T-55 chassis. In early 2002, there are no known plans for trials of a complete Marksman G6 and/or demonstration programme.

155 mm GV6

In 1996, it was stated that the first upgraded and retrofitted GV6 had been delivered to the South African Army. It is not known at present what this upgrade includes.

G6 M1A3

This is the latest model of the G6 which is being marketed in the Middle East. This retains the 155 mm/45 calibre barrel with a 1:20 twist rifling and 48 grooves 1.27 mm deep, which are common to 155 mm 39 and 52 calibre tubes that meet the NATO Joint Ballistic Memorandum of Understanding, as well as a 23 litre chamber.

The original glass fibre fume extractor has been replaced by a new steel fume extractor. Additional safety interlocks have been installed, as has an improved rammer, to increase rate of fire from 3 to 5 rds/min.

The Somchem M64 bi-modular charge system and the ERFB-BB give a maximum range of 40.5 km. Firing the latest VLAP, a range of 53.6 km can be achieved.

It also fitted with the latest AS2000 fire-control system to give a four round MRSI capability. The commander has an African Defence Systems launcher management computer with a Windows based touch screen, while the gunner has a Denel WM205 gun data display and a new digital hand controller. The BAE Systems FIN3110 RLG and integrated GPS sensor is retained.

The chassis has been fitted with a three-level central tyre pressure regulation system. Other enhancements include a more powerful turret APU, improved insulation and a cool water dispenser.

As well as featuring improved reliability, these upgraded systems were also fitted with barrel temperature measurement, a barrel cooling fan on the cradle, improved air conditioning system and a backup laying system.

Status

Production as required. In service with the following countries:

Country	Quantity	Comment
Oman	24	delivery from 1996
South Africa	43	deliveries complete, upgrade programme under way
United Arab Emirates	78	delivery from 1991

Manufacturer

LIW, a Division of Denel.

UPDATED

SWEDEN

Bofors Defence 155 mm FH-77 (6 × 6) self-propelled artillery system

Development/Description

Under contract to the Swedish Defence Materiel Administration (FMV), the now Bofors Defence (part of United Defense of the USA) has fitted a 155 mm FH-77 artillery system on the rear of a modified VME A25C (6 × 6) all-terrain chassis.

The first driving and firing rig was completed late in 1995 and, following company trials, was delivered to the Swedish Army for six months of troop trials.

This chassis is widely used in the construction industry and has proved to be very reliable and to have good cross-country mobility. It has been designed to carry a maximum load of 25 tonnes.

In addition to the artillery application, this chassis is also being used as the basis for the TRIDON 40 mm L/70 air defence system unveiled in 1994 (*Jane's Land-Based Air Defence 2002-2003).* This system has yet to pass the prototype stage.

For its new mission the chassis has been modified with a fully armoured cab fitted at the front which would also cover the engine compartment. The cab has bullet-proof windows and can accommodate a six-man crew. A machine gun is mounted on the roof for anti-aircraft/local defence. In addition to the circular hatch in the roof there is also a single door on either side at the rear.

The actual 155 mm 38 calibre ordnance is the same as that used in the standard towed 155 mm FH-77 covered in detail in the Towed anti-tank guns, guns and howitzers section, but some modifications have been carried out to the loading and recoil systems.

Mounted on the rear of the chassis is the Bofors Defence 155 mm FH-77 series weapon which, when travelling, would point to the front. When deployed in the firing position, two stabilisers would be lowered to the ground to provide added stability. The 6 × 6 system would carry a total of 24 155 mm projectiles in a magazine and 36 charges.

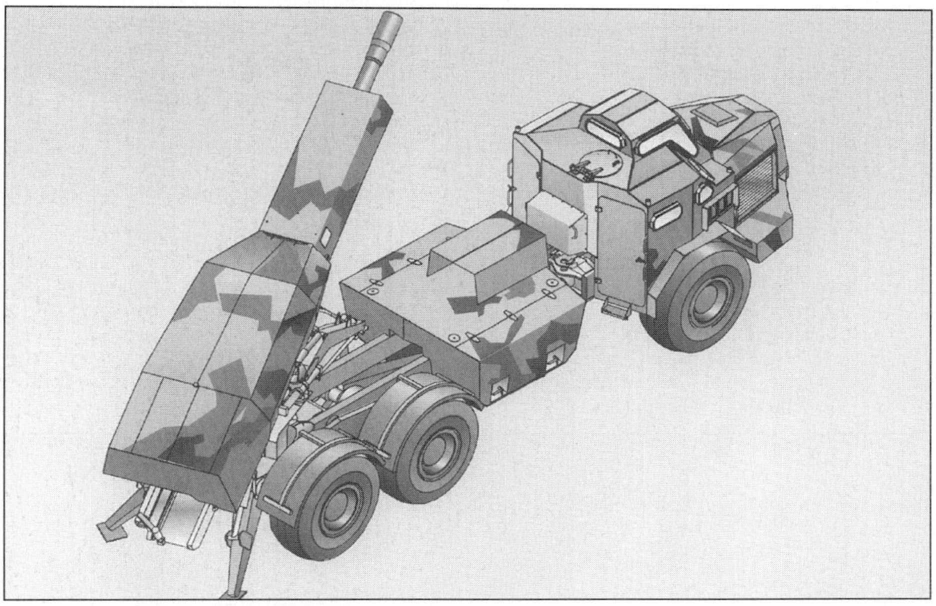

Artist's impression of Bofors Defence 155 mm Haub 77AD deployed in the firing position and with stabilisers lowered at the rear 0021004

At present, the 155 mm FH-77A/FH-77Bs used by the Swedish Army are towed by Saab-Scania 6 × 6 trucks which also carry the crew and some ammunition.

Between 1978 and 1984, the Swedish Army took delivery of 200 Bofors Defence FH-77A towed artillery systems. Of these 200 systems, 100 are in reserve and 100 are for sale.

More recently, the Swedish Army has taken delivery of 50 of the FH-77B (which has also been sold to India and Nigeria). This has a number of improvements including a new muzzle brake and is deployed in northern Sweden with four battalions each having 12 guns.

By mounting the system on a 6 × 6 chassis this combination will not only have greater mobility but, when travelling, the crew will be under armour protection and the system will also be able to come into and out of action faster and therefore be less vulnerable to counter battery fire.

The prototype system uses the FH-77A weapon with the complete system being designated the FH-77AD (or Haub 77AD) by the Swedish Defence Materiel Administration.

Following extensive Swedish Army trials with the driving and firing test rig, the user suggested a number of improvements that were carried out by Bofors Defence.

These included the requirement for the crew to stay in the armoured cabin when carrying out the fire mission with a magazine for 24 155 mm projectiles being mounted above the weapon and protected by armour.

The first prototype driving and firing rig was then rebuilt and tested in 1999 by the Trials Wing of the Swedish Artillery School. The main features of this enhanced model can be summarised as:

Specifications
Crew: 6
Configuration: 6 × 6
Weight: 30,000 kg
Length: 13 m
Width: 3.1 m
Height: 3.3 m
Max road speed: 65 km/h
Range: 500 km
Armament: 1 × 155 mm/45 calibre
Ammunition: 24 × 155 mm
Max range: 40,000 m
Traverse: 30° left and right
Elevation: 0 to +70°
Rate of fire: 3 rounds in 13 s

(1) Possibility of deploying and taking the gun out of action from inside the cabin
(2) Facilities that make it possible to lay and fire the gun from inside the cabin
(3) The onboard gun computer is also capable of taking care of the meteorological information as well as acting as a ballistic computer
(4) The same computer will in future be capable of storing information about the enemy, friends and also different target co-ordinates
(5) A magazine for 24 155 mm projectiles
(6) The charge box has been removed to the right side of the vehicle
(7) The crew has been reduced to five but seating is still provided for six people
(8) When filling up the magazine the computer will be told which type of projectile is where in the magazine

As the weapon has plastic charge cases (in Sweden) and bag charges elsewhere, it means that at least one person has to leave the cabin to feed the mechanism (the rammer) with the charge when firing.

When preparing the gun for a fire mission you also prepare which charge you are going to fire. Six and most probably nine charges will be stored in an alert rack close to the rammer.

Communication between the crew members is performed with a short-range radio set. Time to go into and come out of action is about 30 seconds.

The mean speed when driving on normal roads during the tests was about 60 km/h and, by 2000, the total distance travelled was about 27,000 km under Summer and Winter conditions.

Bofors Defence has offered to send the FH-77BD for field trials in India on a no cost, no commitment basis.

The latest FH-77BD is a further development of the Bofors Defence prototype system based on a commercial VME A25C (6 × 6) articulated all-terrain chassis evaluated by the Swedish Army.

For the Indian trials the FH-77BD has been fitted with a Bofors Defence-designed and built 155 mm/45 calibre barrel with a 25 litre chamber. This, firing an HE ER FB BB (high explosive extended range full bore base bleed) projectile, will achieve a range of around 40 km. It is envisioned that production systems would have a 155 mm/52 calibre barrel.

The FH-77BD has a crew of six of whom the gun commander, layer and driver remain in the fully armoured cabin, while the other three leave the cabin and load the weapon with projectiles and charges.

An ammunition assist system loads the projectiles onto a tray which are then rammed, with the charges being loaded manually.

First photograph released of the Bofors Defence 155 mm FH-77BD L/45 on a Volvo 6 × 6 chassis shown here in the ready-to-fire position **2002**/0095232

As an option, a magazine could also be fitted which would allow the crew to be under complete armour protection when a fire mission was being carried out. A total of 24 155 mm projectiles and associated charges is carried ready for immediate use.

To reduce time into and out of action the FH-77BD is fitted with an integrated position and navigation system and, according to Bofors Defence, it takes only 30 seconds to come into or be taken out of action. This compares to an average of 300 seconds for the normal towed FH-77B used by India.

In addition, the FH-77BD has an onboard ballistic computation capability, a muzzle velocity radar which feeds information to the computer and a short-range communications set. The FH-77BD has been designed to carry out shoot-and-scoot fire missions to reduce its exposure to counter battery fire.

When compared with conventional full-tracked self-propelled artillery systems, the FH-77BD is not only cheaper to procure, operate and maintain but also has greater strategic mobility.

In 1999 India started to place contracts with Bofors Defence for spare parts for the FH-77B artillery systems. Following allegations of corruption, India froze the following two stages of the contract which would have included local production in India.

From 2001, the Swedish Army tested the Bofors Defence 155 mm FH-77BD L/45 self-propelled artillery system based on a Volvo VME A25C (6 × 6) all terrain chassis.

It was expected to have gone to India for an extensive series of firepower and mobility trials in 2000 but, as of January 2002, it was still in Sweden.

The Bofors Defence 155 mm FH-77BD L/45 was tested alongside a number of other Swedish artillery systems, including the towed Bofors Defence 155 mm FH-77B and the Bofors Defence 155 mm Bandkanon 1A self-propelled system.

These trials are the first part of the Swedish Army programme to acquire a new self-propelled artillery system for the 21st century that will probably have field and coastal defence roles.

The Swedish Army has leased a German Krauss-Maffei Wegmann 155 mm/52 calibre PzH 2000 and, from the UK a BAE Systems, RO Defence 155 mm/52 calibre AS90 Braveheart which will also be tested through to 2003.

Status

Development. Not yet in production or service.

Manufacturer

Bofors Defence AB (A United Defence company).

UPDATED

SELF-PROPELLED MORTAR SYSTEMS

CHINA

NORINCO 120 mm SP mortar

Development

In mid-2001, it was revealed that China had developed a 120 mm self-propelled mortar system. It is based on a modified chassis of the WZ 551 (6 × 6) Armoured Personnel Carrier (APC), that has been in service with the People's Liberation Army (PLA) for many years.

Full details of the WZ 551 (6 × 6) APC are given in the Armoured personnel carriers (wheeled) section.

Mounted on the roof in the centre of the WZ 551 hull is the fully enclosed turret. This is very similar to that used by the Russian 120 mm 2S23 NONA-SVK self-propelled mortar system mounted on a modified BTR-80 (8 × 8) APC chassis. Details of this system are given later in this section.

In 1997, it was reported that Russia was to supply China with 100 120 mm 2S23 NONA-SVK self-propelled mortar systems, whose existence was first revealed in 1990.

Russian military sources however never confirmed this deal or the transfer of technology to China allowing local production of the 2S23 system, ever took place.

While the three-person turret resembles the Russian 2S23 NONA-SVK armed with a 120 mm 2A60 rifled ordnance, the actual ordnance used for the Chinese system is rather longer. It is similar to the Russian 120 mm 2A80 installed in the 2S31 Vena self-propelled gun/mortar system. This is based on a considerably modified BMP-3 chassis and has yet to enter volume production.

The 120 mm 2A60 rifled mortar fires an HE-FRAG bomb to a maximum range of 8.85 km and a rocket-assisted bomb to a maximum range of 12.85 m.

There is also a High Explosive Anti-Tank (HEAT) bomb for direct-fire engagements and a laser-guided anti-armour projectile. Turret traverse is 70° left and right with elevation from -4 to +80°.

The more recent 2S31 can fire the same family of ammunition as the 2S23, but a new family of ammunition has been developed with increased range and lethality.

It is understood that China has developed this 120 mm (6 × 6) self-propelled mortar system to strengthen the PLA's ability to engage targets in mountainous terrain.

First photograph released of a modified WZ 551 (6 × 6) chassis fitted with a turret armed with a 120 mm breech-loaded mortar (BQZS via Yihong Zhang) ***2002***/0095563

The WZ 551 (6 × 6) APC is powered by a locally produced German Deutz BF8L413FC air-cooled diesel engine developing 320 hp (235 kW) coupled to a German ZF 6HP 500 fully automatic transmission with six forward and one reverse gears. The vehicle is fully amphibious, being propelled in the water by two water jets mounted one either side at the rear of the hull.

China has also developed a new 105 mm (6 × 6) armoured car that uses components of the WZ 551 (6 × 6) APC. It is considered that the PLA, like a number of other armies around the world, are placing increased emphasis on wheeled armoured vehicles that have greater strategic mobility than their tracked counterparts.

Specifications

Not available.

Status

Prototype systems undergoing trials.

Manufacturer

Chinese state factories.
Enquiries to China North Industries Corporation.

NEW ENTRY

CZECH REPUBLIC AND SLOVAKIA

ZTS PRAM-S 120 mm self-propelled mortar system

Development

In 1990, ZTS completed the prototype of a 120 mm self-propelled mortar system called the PRAM-S.

The Russian-designed BMP-1 and the more recent BMP-2 were manufactured under licence in Czechoslovakia for some years and additional details of the original vehicles are given under the Russian Federation and Associated States (CIS) in the Armoured personnel carriers (tracked) section.

In early 1992, ZTS stated that the PRAM-S was in production and would be deployed in batteries of six systems. With the break-up of the former Czechoslovakia, production and marketing of this system is now carried out by ZTS in Slovakia.

It is understood that of the 12 systems built, six are in service with the Czech Republic and the other six with Slovakia.

Although production of the PRAM-S 120 mm self-propelled mortar system was completed some years ago, it is still being offered on the export market by ZTS of Slovakia and production can resume if further significant orders are placed.

ZTS PRAM-S 120 mm self-propelled mortar

ZTS PRAM-S 120 mm self-propelled mortar with trim vane extended at the front of the vehicle for amphibious operations

Specifications

Crew: 4
Weight: 16,970 kg
Power-to-weight ratio: 17.67 hp/t
Length: 7.47 m
Width: 2.94 m
Height:
(hull top) 2.25 m
(to top of 120 mm mortar (mortar at +40°)) 2.50 m
(to top of 120 mm mortar (mortar at +80°)) 3.38 m
Ground clearance: 0.42 m
Track: 2.55 m
Track width: 300 mm
Length of track on ground: 3.60 m
Max road speed: 63 km/h
Max water speed: 7.4 km/h
Fuel capacity: 462 litres
Max road range: 550 km
Gradient: 60%
Side slope: 30%
Vertical obstacle: 0.7 m
Trench: 2.5 m
Engine: Model UTD-20, 6-cylinder, 4-stroke injection diesel developing 300 hp at 2,600 rpm
Transmission: manual, 5 forward and 1 reverse gears
Clutch: multiple, dry, constant engagement
Suspension: torsion bar, hydraulic shock-absorbers on 1st, 2nd and 6th roadwheels
Electrical system: 22-20 V
Batteries: 2, 120 Ah
Armament:
(main) 1 × 120 mm mortar
(anti-aircraft) 1 × 12.7 mm MG
Gun control equipment:
(elevation/depression) manual
(traverse) manual
Elevation/depression: +40/+80°
Traverse: 15° (left and right)
NBC system: yes
Night vision equipment: yes

Description

The PRAM-S 120 mm self-propelled mortar system is based on a modified BMP-2 chassis which has had its turret removed and roof raised to accept a new 120 mm mortar system that can be aimed, loaded and fired under complete armour protection.

The chassis of the original BMP-2 has been lengthened and raised at the rear and now has a total of seven roadwheels compared to the standard vehicle which has six. The drive sprocket is at the front, idler at the rear and there are three track-return rollers. Hydraulic shock-absorbers are provided for the first and last roadwheel stations.

The driver is seated at the front left with the engine compartment to his right, with the area to the rear of the vehicle being raised to provide sufficient space for the 120 mm mortar, ammunition and crew.

Mounted in the forward part of the roof is the 120 mm breech-loaded mortar which has a traverse of 15° left and right and can be elevated from +40 to +80°. Elevation and traverse of the 120 mm mortar is manual.

Maximum range of the 120 mm mortar is 8,036 m with minimum range being 504 m, which in both cases depends on the type of 120 mm mortar ammunition being used.

Maximum rate of fire using the Czech D charge is 18 to 20 rds/min, 40 rounds in 5 minutes and 70 rounds in 10 minutes. The mortar can be fired on longitudinal slopes of up to 5° and transverse slopes of up to 5°.

On coming to a halt, the first rounds can be fired in 1 minute and the system can come out of action 1 minute after the last round has been fired.

When travelling, the 120 mm mortar is at the minimum elevation angle and when required for action armoured covers unfold either side of the mortar to allow for greater elevation.

The commander is seated to the right of the 120 mm mortar and is provided with a roof hatch, three day periscopes and an infra-red searchlight.

The gunner is seated opposite the commander, has a roof-mounted day laying system and can enter the vehicle via a forward-opening door in the left side of the hull that is also provided with a vision block.

The vehicle commander operates the R-123 UHF/VHF radio with the antenna for this being located forward and to the left of the driver's hatch.

In addition, there is a hatch in the roof of the vehicle to the rear of the commander on which an externally mounted 12.7 mm NVS machine gun is fitted for anti-aircraft defence.

The loader is seated towards the rear of the mortar and is responsible for preparing the mortar bombs and filling the automatic loader from the ammunition containers to the left and right rear of the combat compartment.

The 120 mm mortar bombs are stowed vertically, the loading system being chain driven with a rammer provided to ram the mortar bombs into the mortar.

In the rear of the vehicle is a door for ammunition resupply and 80 rounds of 120 mm mortar ammunition are carried, of which 21 rounds are in the conveyor and 59 rounds in the magazines.

The PRAM-S is fully amphibious, being propelled in the water by its tracks, and is fitted with an NBC system and a full range of night vision devices. The vehicle can lay its own smoke screen by injecting diesel fuel into the exhaust outlet located on the right side of the hull which, when compared to Russian BMP-1/BMP-2 type vehicles, has a distinctive cowl-type shroud.

Variants

In late 2001, it was reported that a number of enhancements were under active consideration for this system, including integrating on a wheeled platform.

Status

Production complete but can be resumed if further orders are placed. In service with the Czech Republic and Slovakia.

Manufacturer

ZTS Dubnica nad Váhom (Slovakia).

UPDATED

FRANCE

TDA 120 mm 120R 2M self-propelled mortar system

Development/Description

The 120 mm 120R 2M recoiling rifled mortar system was developed as a private venture for the export market by TDA (previously Thomson Brandt Armements) and can be fitted to a wide range of vehicles, tracked and wheeled, in the 10 to 15 tonnes class.

The 120 mm 120R 2M recoiling rifled mortar was developed from 1992 and was first revealed in 1993 with the first complete prototype being shown in 1994.

By early 2002 production of the complete system had yet to commence. The US Marines have purchased the actual weapon system and a quantity of ammunition for trials purposes.

The system has already been demonstrated in France and Saudi Arabia installed in the MOWAG Piranha (8 × 8) APC and has also been installed in a Turkish FNSS Defense Systems (previously called FMC-NUROL) Armoured Infantry Fighting Vehicle (AIFV) for trials in Turkey.

The 120 mm mortar and its associated hydraulic recoiling system is mounted on a 1 m circular turntable with ball-bearing enabling it to be quickly traversed under power through 360°. Elevation and traverse is controlled using a joystick with elevation from +42 to +85° in elevation.

When travelling, the 120 mm mortar barrel is lowered into the horizontal position and the two roof hatches are closed making the vehicle difficult to distinguish as a mortar carrier.

The complete 120 mm mortar system weighs around 1,000 kg and comprises the rifled mortar barrel, two hydraulic recoil brakes and a bearing-mounted base plate. Recoil force on the platform is about 35 tonnes and maximum recoil travel is about 300 mm.

The mortar is ballistically identical to the well-known TDA 120 mm MO-120-RT rifled towed mortar which is used by more than 20 countries.

Maximum range using rocket-assisted 120 mm ammunition is 13 km and maximum rate of fire is 6 to 10

MOWAG Piranha 8 × 8 APC fitted with TDA 120 mm 120R 2M mortar system
0058409

Close-up of TDA 120 mm 120R 2M mortar system installed in the Turkish FNSS Defense Systems AIFV chassis

rds/min for 3 minutes and sustained rate of fire is 2 to 3 rds/min. When fitted with a semi-automatic loading system a rate of fire of 18 rds/min can be achieved.

The system fires the same family of 120 mm mortar bombs as the standard 120 mm towed mortar including high explosive (with impact/delay fuze or with proximity fuze), smoke (white phosphorus), anti-armour personnel carrier (with pre-fragmented body), illumination, cargo (with 20 grenades) and practice (black powder). It will also fire 120 mm smart mortar projectiles such as the Swedish Saab Bofors Dynamics Strix. Accuracy of standard mortar bombs is 22 × 5.5 m.

These have a maximum range of 8,135 m. The rocket-assisted projectile can be fitted with an impact/delay fuze or a proximity fuze. The rocket-assisted projectile has a maximum range of 13,000 m and an accuracy of 26 × 18 m.

About 40 mortar bombs, charges and associated fuzes are carried with the actual number depending on the chassis used.

Specifications

Calibre: 120 mm
Total weight: 1,000 kg
Barrel weight: 100 kg
Recoil brake: 350 kg
Baseplate: 400 kg
Recoil brake: strain reduced from 150 t to less than 35 t
Firing from horizontal platform:
(elevation) +42/+85°
(traverse) 360°

A semi-automatic loading device driven by a programmable automatic controller is fitted as standard as is a weapon aiming and pointing system developed by Allied Signal. This comprises a north finder (ring laser gyro), chief or section control display and gunners display unit. A manual back-up mode system is fitted for traverse, elevation and loading.

As an option, a modern location and positioning system (GPS and land navigation system) and fire-control system (ballistic computer and automatic data transmission system) can be fitted.

The 120 mm 120R 2M mortar system does not have a direct fire capability as TDA believes that there is little requirement for this.

Variants

A smoothbore version of this system is available under the designation of the 120 RS2M system.

Status

Development complete. Ready for production on receipt of orders.

Manufacturer

TDA.

UPDATED

GERMANY

Rheinmetall Landsysteme Wiesel 2 120 mm mortar system

Development

Like some other countries, the German Army is placing increased emphasis on its rapid reaction forces. This has lead to the development and fielding of much lighter vehicles and weapon platforms that can more readily be transported by aircraft and helicopter. The latter includes the Sikorsky CH-53G that is used in large numbers by the German Army.

Some years ago, as a private venture, MaK System GmbH of Kiel (today known as Rheinmetall Landsysteme), started work as a private venture on an enhanced version of the Wiesel 1 light armoured vehicle.

Between 1989 and 1992 a total of 345 Wiesel 1s were built for the German Army in two configurations, 20 mm cannon and TOW anti-tank guided weapon.

This enhanced Wiesel finally emerged as the Wiesel 2 and has a significant increase in internal volume and load carrying capacity, enabling it to undertake a much wider range of battlefield missions. Full details of the Wiesel 2 are given in the Armoured personnel carrier's (tracked) section.

Specifications

Crew: 3
Weight: 4,100 kg
Power-to-weight ratio: 26.58 hp/t
Ground pressure: 0.41 kg/cm²
Length: 4.517 m
Width: 1.820 m
Height: 1.807m
Ground clearance: 0.30 m
Track: 1.62 m
Track width: 200 mm
Length of track on ground: 2.427 m
Max road speed: 70 km/h
Cruising range: 280 km
Fuel capacity: 95 litres
Gradient: 60%
Side slope: 30%
Vertical obstacle: n/av
Trench: 1.5 m
Fording: 0.30 m
Engine: Audi 4-cylinder TDI diesel developing 109 hp
Transmission: ZF LSG 300/4 automatic with hydrostatic cross-drive steering
Electrical system: 24 V
Armament: 1 × 120 mm mortar
Ammunition: 30 × 120 mm
Traverse: 30° left and right
Elevation: +35 to +85°
Rate of fire:
(rapid) 3 rds in 20 seconds
(sustained) 6 rds/minute for 3 min
NBC system: yes
Night vision equipment: Yes

Wiesel 2 is now in production as the basis for the STN ATLAS Elektronik Atlas Short-Range Air Defence (ASRAD) system for the German Army. A total of 50 firing platforms (Ozelot), 10 platoon command post vehicles (with HARD radar) and seven battery command post vehicles have been ordered and the first of these was delivered to the German Army in mid-2001.

The Wiesel 2 has since been developed by the company as the baseline for a complete family of other vehicles including the 120 mm self-propelled mortar system.

This has been developed as a private venture under the leadership of Rheinmetall Landsysteme with the assistance of a number of other leading German defence contractors.

Rheinmetall Landsysteme is responsible for the complete system including the vehicle, ammunition handing, fire-control concept, computer system and the electric system.

Rheinmetall Weapons & Munitions is responsible for the actual 120 mm mortar system and its associated family of 120 mm ammunition. Other members of the Wiesel 2 120 mm mortar team are Honeywell who are responsible for the navigation and laying system and rear support drive, ESG for the fire-control software and Wittenstein Motion Control for the actual gear drives.

The first prototype of the Wiesel 2 120 mm mortar system was completed in 1997 and based on a surplus Wiesel 1 chassis. This proved that the concept was viable and during company trials, over 200 rounds of 120 mm ammunition were fired. Live firing from a manned vehicle has been demonstrated to the German Army.

In the future it is expected that the German BWB will award a contract to enable Rheinmetall Landsysteme to complete development of the vehicle and build two preproduction vehicles.

The German Army has already identified an operational requirement for 65 of these 120 mm self-propelled mortar systems plus additional ammunition support vehicles.

Description

The chassis is almost identical to that used for the Wiesel 2 vehicle. The driver is seated at the front right with the power pack, consisting of the diesel engine, transmission and cooling system, front left. The commander and ammunition loader are seated on the left and face inwards while the commander is also provided with a hatch and periscopes for all-round observation.

The actual 120 mm mortar is pivoted at the rear and when being loaded, is lowered into the horizontal position with the mortar bombs being fed into the muzzle manually.

The 120 mm mortar then returns to its designated firing position, fired and then returns to the loading position again. Loading the 120 mm mortar is carried out under complete NBC protection.

The 120 mm muzzle loaded smoothbore mortar has been fitted with a special recoil system and firing a standard 120 mm mortar bomb, a maximum range of 6,000 m is obtained.

The 120 mm mortar mounted on this modified Wiesel 2 chassis can be loaded, aimed and fired with the crew under complete protection **2002**/0018429

Firing a recently developed Rheinmetall 120 mm HE bomb, which has yet to enter production, a maximum range of 8,000 m can be achieved. It can also fire smart 120 mm rounds, for example the Swedish terminally guided Strix up to a maximum length of 1,000 m. A total of 30 projectiles are carried in a magazine in the right side of the vehicle.

Before firing commences a stabiliser mounted to the immediate rear of each of the last roadwheel station is lowered to the ground under electric power from within the vehicle. The mortar can be traversed 30° left and right and elevated from +35 up to +85° without moving the vehicle.

Using its onboard computer and hybrid navigation and laying system, the platform can automatically determine, not only its heading angle, but also its exact position and height. The 120 mm mortar can be brought into action much quicker than a conventional mortar system and can also quickly redeploy to a new fire position before counter battery fire arrives.

The main computer sends and receives information from the laying system, rear support system, safety electronics and navigation systems. Information is displayed on the commander's MRT 86 Central Operating and Display Unit (CODU) that is also linked to the communications system. The 120 mm mortar is laid onto the target and fired from the CODU.

According to Rheinmetall Landsysteme it can open fire in under 60 seconds of coming to a halt. After each 120 mm mortar bomb is fired, the computer automatically adjusts the mortar so that firing accuracy is maintained. A rapid rate of fire of three rounds in 20 seconds can be achieved with a sustained rate of fire of six rounds a minute for three minutes.

The 120 mm mortar can be aimed, loaded and fired by the crew of three, with the commander, loader and driver being under complete protection from small arms fire and NBC attack.

The Wiesel 2 120 mm self-propelled mortar system can be carried internally in a CH-53G helicopter or, as an alternative it can be slung underneath of the helicopter.

Wiesel 2 fitted with 120 mm mortar shown here from the rear with stabilisers lowered to the ground
2002/0067927

Standard equipment includes an NBC system but as an option an air conditioning system can be fitted. Other options include onboard ballistic computation and automatic fuze selection.

The small size of the Wiesel 2 not only makes it easier to deploy and conceal but its low-ground pressure enables it to be deployed to areas which are inaccessible to other tracked vehicles.

A typical Wiesel 2 120 mm mortar platoon would consist of five systems which would come under the control of a fire direction centre and be supported by two ammunition transport vehicles, all based on a common Wiesel 2 chassis.

Target information would come from the battery/battalion command post or direct from the forward observer vehicle, both of which would also use the Wiesel 2 chassis.

Status
Prototype. Not yet in production or service.

Manufacturer
Rheinmetall Landsysteme GmbH.

NEW ENTRY

INTERNATIONAL

AMOS 120 mm self-propelled mortar system

Development/Description
In June 1996, Hägglunds Vehicle of Sweden and Patria Vammas of Finland signed an agreement to develop a new turret-mounted twin 120 mm smoothbore mortar system called AMOS (Advanced Mortar System).

Hägglunds Vehicle (now owned by Alvis plc of the UK) has considerable experience in the design, development and production of tracked vehicles and turrets while Patria Vammas has developed and produced a complete range of mortar systems and their associated ammunition.

Under the terms of the agreement, Hägglunds Vehicle is responsible for the turret with Patria Vammas responsible for the twin 120 mm smoothbore mortar system.

Two versions of the AMOS 120 mm self-propelled mortar system were originally developed: Model A muzzle loaded and Model B, breech loaded. A decision was subsequently taken to concentrate all work on the breech-loaded version.

AMOS has two breech-loaded mortars, each of which is 3 m long. The barrels are mounted in a common cradle but in this case both of them have a hydraulic recoil brake and pneumatic recuperator of their own, thus enabling each barrel to recoil independently.

Each barrel is fitted with a fume extractor to keep toxic gases inside the vehicle within acceptable levels.

The breech system is a vertical downward-opening wedge type which is operated by a toggle link-type semi-automatic mechanism. Breech obturation and positive shell positioning is achieved by means of a special reusable stub case adapted to each projectile.

Each barrel is provided with a revolver-type projectile feeder, although for the prototype stage loading is manual by means of a loading tray.

The first breech-loaded example was completed in 1997 and was shown for the first time in June of that year. Since then, the system has been test fired in Finland and Sweden and has also been demonstrated to various other potential customers.

The turret is of all-welded steel armour construction which provides protection from small arms fire and shell splinters. It also provides muzzle blast and NBC protection for the turret crew of two. Turret traverse and weapon elevation is all electric, with manual controls being provided as a back-up for emergency use.

A wide range of fire-control options is available, depending on user requirements.

The prototype system had an optical laying system but growth potential could include an automatic loading system and an electronic fire-control system.

Ammunition load would depend on the chassis but typically would be 40 conventional mortar bombs (for example HE, smoke and illuminating) and six guided mortar bombs (for example the Saab Bofors Dynamics Strix). Ammunition resupply would take 10 minutes.

As well as Hägglunds Vehicle CV90 full-tracked chassis, the AMOS 120 mm mortar system could be fitted to a wide range of other chassis including the

AMOS 120 mm self-propelled mortar testbed on Hägglunds Vehicle CV90 series chassis 0058368

Specifications (Turret)
Crew: 2
Turret weight: 5,800 kg
Armament: 2 × 120 mm mortar
Barrel length: 3,000 mm
Loading: semi-automatic
Secondary armament: 1 × 7.62 mm MG

Gun control equipment
Turret power: electric/manual
Turret traverse: 360°
Elevation/depression: +85/–3°
Ammunition: 40 HE + 6 guided
Time to fire: 30 s
Time to shoot: 15 s
Max rate of fire: 26 rds/min
First six rounds: 10 s
Max range: 13 km
Simultaneous impact on target: 16 rounds

AMOS 120 mm self-propelled mortar system installed on Patria Vehicles XA-185 (6 × 6) chassis for trials
2000/0085194

Russian MT-LB multipurpose tracked armoured vehicle (used in some numbers by Finland and Sweden), the United Defense LP M113 series tracked vehicle and the Swiss MOWAG Piranha (8 × 8) light armoured vehicle.

In mid-1999 the Finnish Defence Force awarded Patria Hägglunds a contract worth €4.5 million for the twin 120 mm AMOS.

This contract, the first to be awarded to the new company formed by Hägglunds Vehicle and Patria, covered the delivery to the Finnish Army of a study of some aspects of AMOS, including the loading system.

It also covered the supply of one complete AMOS installed on a Patria Vehicles XA-203 (6 × 6) chassis.

In the Spring of 2000, the Swedish Defence Materiel Administration (FMV) and its Finnish counterpart signed a project agreement for the development of the vehicle based twin 120 mm AMOS (Advanced Mortar System).

This agreement is based on a 1994 agreement signed between Denmark, Finland, Norway and Sweden for the joint development of material to meet the operational requirement of all four countries. For the AMOS project Denmark and Norway have observer status.

The new agreement covers a study and prototype phase that runs through to December 2002 with a total value of SKr100 million. Prime contractor is the Patria Hägglunds Oy which is jointly owned by Patria of Finland and Hägglunds Vehicle of Sweden.

The twin 120 mm AMOS has been under development as a private venture for several years and under the terms of this contract the system will be adapted to meet Nordic operational requirements and enable it to be integrated in new and existing chassis, tracked and wheeled.

In mid-2001, the Finnish Defence Forces completed extensive firepower and mobility trials of the AMOS installed on a Patria Vehicles XA-203 (6 × 6) Armoured Personnel Carrier (APC) chassis.

These trials started in November 2000 and were carried out by a FDF crew under a wide range of operational conditions all over Finland. Over 550 120 mm mortar bombs were fired and 3,000 km of road and cross-country running were successfully carried out.

This was the second AMOS turret and was the one ordered in June 1999 and installed on the latest production Patria Vehicles XA-203 (6 × 6) chassis. This was delivered late in 2000 and is referred to as the PT1 and incorporates all of the improvements suggested with the earlier AMOS.

This AMOS turret is of all-welded steel construction providing the occupants with protection from 7.62 mm ball small arms fire and shell splinters.

To this can be added a layer of additional passive armour to provide a higher level of protection. A 7.62 mm or 12.7 mm machine gun can be fitted on the roof for local protection and electrically operated smoke grenade launchers can be mounted either side of the turret. In addition, PT1 has a new and more streamlined shape and incorporates stealth characteristics in its design.

The current PT1 turret weighs 5,800 kg but this does depend on armour protection and if the bustle-mounted ammunition handling system is fitted.

Without the bustle-mounted ammunition handling system, turret weight is reduced to 4,400 kg. The turret crew consists of three, commander (left), gunner (right) and loader.

These can fire all types of 120 mm smoothbore mortar bombs (with the addition of a reusable brass or steel stub cartridge case) including the Patria Vammas 120 mm HE mortar bomb 120VAM15.00. This weighs 15 kg and contains 3 kg of TNT having a maximum muzzle velocity of 480 m/s and a maximum range, using charge six, of over 10,000 m. It also fires smoke and illuminating mortar bombs.

AMOS has also successfully fired the Saab Bofors Dynamics Strix 120 mm terminally guided mortar bomb and the Swiss Ammunition Enterprise Corp 120 mm Mortar Cargo Bomb, which contains 32 grenades fitted with a HEAT warhead that will penetrate 70 mm of steel armour. The grenades are fitted with a self-destruct mechanism.

When installed on the XA-203 (6 × 6) chassis, a total of 90 120 mm mortar bombs can be carried, the type depending on the mission, typically this could be 84 120 mm HE mortar bombs and six Strix guided mortar bombs.

Of these rounds, 30 are in the ammunition handling system in the turret bustle that has two magazines each holding 15 rounds. The remainder are stowed in horizontal racks in the right rear of the vehicle.

Turret traverse is through a full 360° with weapon elevation from -3 to +85°. Elevation and traverse is electric with manual back-up in case of power failure.

AMOS is fitted with a computerised fire-control system developed by Hägglunds Vehicles which enables the system to use shoot-and-scoot techniques. This is called the Hägglunds Vehicle Control System (HVCS) and also includes training, built in test and battle management functions.

According to Patria Hägglunds, AMOS can open fire within 30 seconds of coming to a halt and move off again within 10 seconds of completing its fire mission. A total of four rounds can be fired in about 8 seconds and with a well-trained crew, a maximum rate of fire is stated to be 26 rds/min.

A key feature of AMOS is its ability to employ Multiple Rounds Simultaneous Impact (MRSI) techniques where all rounds hit the target at once for maximum effect.

Following extensive trials with the FDF the AMOS turret was removed from the XA-203 (6 × 6) chassis and sent to Sweden where it was integrated into a Hägglunds Vehicle CV 90 full-tracked chassis.

This will undergo extensive trials with the Swedish Army in 2002-2003. Sweden has a requirement for between 50 and 70 AMOS which will be installed on CV90 series chassis provided by Hägglunds Vehicle.

The first customer is expected to be the FDF who, funding permitting, are expected to order four production standard (or O series) turrets in mid-2002 for delivery in 2004.

These will be installed on the Patria Vehicles XA-203 (6 × 6) APC or more probably in the longer term on the new Patria Vehicles Armoured Modular Vehicle (8 × 8) which ran for the first time late in 2001.

The FDF will deploy AMOS on a scale of four per regiment/battalion with production probably being undertaken from 2005 through to 2007.

Variant
Although originally developed for installation on tracked and wheeled AFVs, AMOS can also be installed on coastal craft and has already been tested on a Swedish Combatboat 90H raiding craft of the Swedish Navy.

Typical AMOS chassis applications

Type	**Combat weight**
BMP-3 tracked	18.7 tonnes
CV90 tracked	24 tonnes
LAV-III (8 × 8)	16.5 tonnes
MRAV (8 × 8)	31 tonnes
MTVL tracked	17.5 tonnes
XA-203 (6 × 6)	22.5 tonnes

Status
Prototypes. Not yet in production.

Manufacturers
Hägglunds Vehicle AB.
Patria Vammas.

UPDATED

Delco Defense/RO Defence 120 mm Armoured Mortar System (AMS)

Development

The 120 mm Armoured Mortar System (AMS) comprises a 120 mm smoothbore breech loading, recoiling mortar integrated with a full solution fire-control system and mounted in a lightweight steel turret that can be fitted to a wide variety of light armoured vehicle chassis.

The now BAE Systems, RO Defence first developed the 120 mm breech-loading mortar as a private venture in 1985. The first prototype of the mortar, complete with its elevating mass, was completed and test fired in mid-1986. During 1987, the turret system was integrated onto a United Defense M113A2 chassis and test fired.

In the autumn of 1991, the now General Motors Defense, Light Armoured Vehicle (LAV) (8 × 8) was fitted with the system and successfully completed initial trials at a UK weapons range. During these trials around 150 rounds of 120 mm mortar bomb were fired at high and low angles in the indirect fire mode as well as in the direct fire mode. For the trials the turret was also fitted with elevation and traverse drives and a mockup of a fire-control system.

BAE Systems, RO Defence is responsible for the turret and 120 mm ordnance, Delco Defense of the USA is responsible for the electronics and sighting system and General Motors Defense is responsible for the LAV (8 × 8) chassis and turret integration.

During 1992, General Motors/BAE Systems, RO Defence undertook the design and construction of a complete turret system with integrated fire control that they then demonstrated. The system comprised a new all-welded steel turret, the 120 mm ordnance and a complete fire-control/sighting system integrated on an 8 × 8 LAV chassis.

In late 1995, another 120 mm AMS turret was produced and then fitted to the private venture United Defense LP Mobile Tactical Vehicle Light chassis, a further development of the widely deployed M113 series. This system was then subjected to extensive trials and a customer demonstration in a country in the Middle East.

Early in 1996, the now Delco Defense awarded the now BAE Systems, RO Defence a contract worth £37 million for the supply of 73 120 mm AMS and associated ammunition.

Under the terms of the contract, BAE Systems, RO Defence supplied the turret, including the 120 mm mortar, Delco Defense supplied the computerised fire-control system and sights.

The complete turret is then fitted to the General Motors Defense, Light Armoured Vehicle (LAV) (8 × 8) chassis.

The BAE Systems, RO Defence order included a substantial quantity of 120 mm ammunition which has been developed by MECAR of Belgium and includes high-explosive (Composition B), illuminating and smoke (white phosphorus) natures.

The AMS has also successfully carried out a series of firing trials at a UK range with the Swiss Ruag Munitions 120 mm mortar cargo round which is designated the 120 mm Ka G 98 in Swiss Army service.

120 mm Armoured Mortar System on Light Armoured Vehicle (LAV) chassis (Christopher F Foss) 0058408

The Swiss 120 mm mortar cargo bomb is already in quantity production for the Swiss Army and contains 32 grenades each of which has a High Explosive Anti-Tank (HEAT) warhead that will penetrate 70 mm of steel armour. It also has a good fragmentation effect. A key feature of this grenade is that it has a mechanical impact fuze with a self-destruct mode and self-neutralisation.

Under a Foreign Military Sales (FMS) agreement, General Motors Defense is supplying a total of 1,117 LAVs in 10 versions to the Saudi Arabian National Guard (SANG) with well over 900 vehicles already delivered. The 73 120 mm AMS variants are within this figure of 1,117 vehicles.

The 120 mm Armoured Mortar System can, however, be fitted on a wide range of other chassis both wheeled and tracked.

Rheinmetall Landsysteme of Germany has proposed that the 120 mm AMS could be installed on the R 495 full-tracked chassis. This would have a combat weight of 23,800 kg and carry 100 rounds of 120 mm ammunition.

Description

The 120 mm mortar has a semi-automatic conical screw, swinging breech mechanism with obturation being achieved with a Crossley pad. Firing is percussion mechanical activated by a solenoid. The recoil mechanism consists of two buffers and a pneumatic recuperator.

Sustained rate of fire is 4 rds/min, rapid rate is 8 rds/min (sustained for 3 minutes maximum). The ability to fire in direct mode at targets in excess of 1,000 m provides the ground force commander with enhanced operational capabilities while also improving the speed of response to conventional indirect fire roles.

The mortar can fire all standard 120 mm conventional smoothbore ammunition including the new generation smart bombs currently under development. With the 120 mm high-performance mortar ammunition, ranges in excess of 9,000 m have been achieved. Also, as the weapon can be fired in the indirect mode of fire at much lower angles of elevation than a conventional mortar, it is much more difficult to detect with mortar locating radar.

The turret is fabricated from armour steel and has stations for two crew. The commander is to the right and the loader to the left. Each crewman has a hatch for access and egress, and an array of vision periscopes. Mounted either side of the turret is a bank of four electrically operated smoke grenade launchers which fire over the frontal arc and mounted on the turret roof is a 7.62 mm machine gun for the commander.

The complete system is all-weather capable and can revert to manual back-ups should a power failure occur. The turret traverse and elevation drives are hydraulically operated.

The full solution fire-control system employs a differential Global Positioning System (GPS) aided Turret Attitude Sensor System (TASS). The latter

United Defense LP Mobile Tactical Vehicle Light fitted with 120 mm Armoured Mortar System

Specifications

Turret

Crew: 2 (commander and loader)

Length:
(without barrel) 2.59 m
(with barrel) 4.46 m
(barrel) 3 m

Width: 2 m

Height: 0.7 m

Armament:
(main) 1 × 120 mm BAE Systems, RO Defence smoothbore breech-loading mortar
(secondary) 1 × 7.62 mm MG
(smoke grenade dischargers) 2 × 4

Control:
(traverse) 360° electrohydraulic with manual back-up
(elevation) –5 to +80° electrohydraulic with manual back-up

Fire-control system: multitarget input capability with automatic weapon positioning, gun angle calculations and weapon compensation for vehicle attitude and meteorological conditions

Navigation: integrated GPS system for vehicle position, heading and attitude

Optics: integrated all-weather day/night system with thermal imager and integrated laser range-finder for automatic range FCS input

Weight: (typical) 2,630 kg

continuously and automatically updates location, turret bearing and tilt and cant parameters to the fire-control system. This allows the AMS quickly to engage the enemy with into action times of under a minute being possible. With the information and target location data stored in the computer, re-engagement of any target from any location within range limitations requires only the selection of that target from the computer menu. A number of targets can be entered into the system to allow rapid switching between them.

For the direct fire role a thermal sight assembly with integral laser range-finder is fitted.

Status

Production as required. A total of 73 systems are being supplied to the Saudi Arabian National Guard on LAV 8 × 8 chassis and Australia has a requirement for 20 systems on a LAV 8 × 8 chassis.

This programme is called the 120 mm Light Armoured Mortar System (LAMS). It is expected that 20 LAMS will be purchased to enter service in 2006 under Project Land 135. The 120 mm AMS is one of the contenders for the market.

Manufacturers

General Motors Defense (LAV).
Delco Defense (turret integration).
BAE Systems, RO Defence (turret and mortar).
United Defense LP (MTVL chassis).

UPDATED

ISRAEL

Soltam Systems 160 mm self-propelled mortar system

Development

The Soltam Systems 160 mm self-propelled mortar system was developed to meet the operational requirements of the Israel Defence Force for a system that was capable of carrying out close-fire support missions and for the battlefield illumination role for night armour actions.

The conversion work was undertaken by the Israeli company of Keren-Or based in Haifa in the late 1960s and is understood to have entered service with the Israel Defence Force around 1969. The complete 160 mm mortar system, which was originally developed as a towed mortar system and its associated family of ammunition, is supplied by Soltam Systems.

Soltam Systems also manufacture a wide range of other mortar systems. Their 120 mm mortar was selected by the US Army following a competition. This is in service in two models, self-propelled (on a United Defence, modified M113 series APC, designated the M1064A3 and trailer mounted.

Soltam Systems 160 mm self-propelled mortar system with front of superstructure lowered for firing and equipment carried displayed in front of vehicle
2000/0085192

Description

It consists of a Sherman tank, or 105 mm M7 Priest self-propelled artillery system chassis with the turret and superstructure removed and replaced by a new all-welded steel superstructure with vertical sides, front and rear.

The front of the superstructure can be lowered forwards into the horizontal position to provide more space for the crew to operate the 160 mm Soltam Systems mortar system.

The Soltam Systems 160 mm breech loaded mortar is mounted in the centre of the vehicle and fires forwards, this has limited traverse and can be elevated from +43 to +70°. Elevation and traverse are manual.

The 160 mm mortar fires a high-explosive bomb weighing 40 kg up to a maximum range of 9,600 m. The high-explosive content is 5 kg. In addition there are other types of mortar bomb, including at least two types of smoke bomb and an illuminating bomb.

A total of 56 120 mm mortar bombs is stowed along the sides of the vehicle and below the floor on which the mortar is mounted. Access to the below floor bombs is via a hatch to the right of the driver's position.

A 7.62 mm or 12.7 mm machine gun is mounted on each side of the vehicle for use in the anti-aircraft or self-defence role. The vehicle has a crew of eight, with a loaded weight of 36,654 kg and is powered by a Cummins diesel engine developing 430 hp. This gives the vehicle a maximum speed of 43 km/h and an operational speed of about 300 km.

Status

Production complete. In service only with the Israel Defence Force. Recent information has indicated that these systems are now held in reserve.

Manufacturer

Soltam Systems Limited.

UPDATED

JAPAN

Type 96 120 mm self-propelled mortar

Development/Description

The Type 96 120 mm self-propelled mortar system was developed in the early 1990s to meet the operational requirements of the Japanese Ground Self-Defence Force for a mobile and protected mortar system to provide fire support for its mechanised infantry units.

Following trials with prototype systems it was type classified as the Type 96 120 mm self-propelled mortar and placed in production. It is believed that production has been at a very low rate with just three being procured in 1999.

The system is based on a full-tracked armoured chassis that provides the occupants with protection from small arms fire and shell splinters. It is probably fitted with an NBC system and passive night vision devices for some of the crew.

The front of the vehicle is well sloped with the hull sides vertical and with a slight chamfer to their upper part. The hull roof slopes slightly down at the rear with the very rear of the vehicle being vertical.

The diesel power pack is situated front left of the hull with the air inlet louvres in the roof and the air outlet louvres and exhaust outlet in the left side of the hull. The driver is seated at the front of the hull on the right side and is provided with a single-piece hatch cover that opens to the left. In front of his position are day periscopes, the centre one of which can be replaced by a passive periscope for driving at night.

Above and to the rear of the driver's position, in the front of the vehicle, is another crew station, probably that of the vehicle commander. This is provided with roof hatches and day vision devices below.

Mounted in the forward part of the roof is a third crew station which is provided with twin opening roof hatch covers. Mounted in front of this is a 12.7 mm M2 machine gun that is provided with a shield.

Mounted in the rear of the hull firing to the rear is the locally built French TDA 120 mm RT rifled mortar which has been manufactured under licence in Japan for many years. According to the Japanese Ground Self-Defence Force a total of 287 of the standard towed 120 mm RT rifled mortars were in service at the end of 1999.

This 120 mm mortar has a maximum rate of fire of 15 to 20 rds/min with maximum range using standard ammunition of 8,100 m. Using a rocket assisted projectile a maximum range of 13,000 m can be obtained. Various types of 120 mm ammunition can be fired including high explosive, smoke and illuminating.

The suspension either side is understood to be of the torsion bar type with each side having the drive sprocket at the front, six dual rubber-tyred roadwheels and the idler at the rear. There are no track-return rollers.

Specifications

Crew: 5
Combat weight: 23,500 kg
Length: 6.70 m
Width: 2.99 m
Height: 2.95 m
Max road speed: 50 km/h
Armament:
(main) 1 × 120 mm rifled mortar
(self-defence) 1 × 12.7 mm M2 MG

Other Japanese self-propelled mortars

The Japanese Ground Self-Defence Force currently deploys at least two other self-propelled mortar systems, both based on the old Type SU 60 full tracked APC. These are the 81 mm Type 60 and the 107 mm Type 60. In both cases the mortars fire to the rear of the vehicle.

Status

Production. In service with the Japanese Ground Self-Defence Force.

Manufacturer

Not known.

NEW ENTRY

RUSSIAN FEDERATION AND ASSOCIATED STATES (CIS)

240 mm self-propelled mortar M-1975 (SM-240) (2S4)

Development

The 240 mm self-propelled mortar, known to the West as the M-1975, is known as the SM-240 (2S4) by the Russian Army, although its more common name is the Tyul'pan, or Tulip Tree. This system reached full operational capability with the former Soviet Army in 1975.

It is estimated that well over 400 of these systems have been completed, although the Russian Federation has not released any production figures.

Recent information has indicated that very few, if any, of these systems now remain in front line service as the weapon has such a short range and low rate of fire.

Recent information has indicated that the system was used during the fighting in Chechnya.

The 2S4 was developed in the early 1970s by the Uraltransmash Works, with the 240 mm 2B8 mortar system being developed by the Perm Machine Construction Works. The development designation of the system was Obiekt 305.

The chassis used for this system is a modified version of that used in the 2K11 Krug (SA-4 Ganef) surface-to-air missile system. Full details of this system are given in *Jane's Land-Based Air Defence*.

The 2S4 has also been used by the Czech Republic and Slovakia, Iraq and Lebanon but as far as it is known it is no longer in front line service with these countries.

M-1975 (SM-240) (2S4) self-propelled 240 mm mortar

Description

The hull of the 240 mm self-propelled mortar M-1975 (SM-240) 2S4 is of all-welded steel armour construction that provides the crew with protection from small arms fire and shell splinters.

The 2S4 has a crew of four with the other five crew members being carried in the support vehicle.

The driver and vehicle commander are seated front left with the power pack to the right and the crew/ammunition compartment extending to the rear.

The driver has a single-piece hatch cover that opens to the rear with day periscopes for forward observation. The centre one can be replaced by a passive night vision device. Behind this is the raised commander's position. The commander's cupola can be rotated through a full 360° and is provided with an externally mounted 7.62 mm PKT machine gun and an infra-red searchlight, with additional hatches being provided in the roof.

The 240 mm smoothbore mortar, designated the 2B8, is carried complete with its baseplate in the horizontal position on top of the hull and is pivoted at the rear.

When in the firing position the mortar is hydraulically swung to the rear under remote control so that the mortar barrel faces rearwards.

The 240 mm mortar has powered elevation from +50 to +80° with powered traverse of 10° left and right.

Inside the hull of the 2S4 are two drum-type magazines, each of which holds 20 240 mm mortar bombs. When the drums rotate, they line up with a hatch in the roof of the vehicle and a specialised telescoping mechanism flips the muzzle down and loads the barrel first with the projectile and then with the bagged charge.

The mortar is fired by either electrical or mechanical means and two basic types of 240 mm mortar bomb are fired. A conventional high-explosive fragmentation bomb (3OF-864), which weighs 130 kg and has a maximum range of 9,650 m, and a high-explosive fragmentation rocket-assisted projectile weighing 228 kg with a maximum range of 18,000 m. Minimum range is 800 m.

The mortar system is aimed with a hydro-electric drive system by remote control. At an elevation of +60° rate of fire is one round in 62 seconds and with an elevation of +80° a maximum rate of fire of one round every 77 seconds can be achieved. Other types of 240 mm projectile are understood to include chemical, concrete-piercing and tactical nuclear. Russian sources have stated that the tactical nuclear and chemical projectiles have now been phased out of service.

Specifications

Crew: 4 + 5 (2S4 plus ammunition vehicle)
Combat weight: 27,500 kg
Power-to-weight ratio: 18.9 hp/t
Ground pressure: 0.6 kg/cm^2
Length:
(travelling) 7.94 m
Width:
(travelling) 3.25 m
Height:
(travelling) 3.225 m
Ground clearance: 0.4 m
Engine: V-59 V-12 diesel developing 520 hp
Max road speed: 60 km/h
Max range: 500 km
Fuel capacity: 850 litres
Fording: 1 m
Gradient: 60%
Side slope: 30%
Trench: 2.79 m
Vertical obstacle: 1.10 m
Armament:
(main) 1 × 240 mm 2B8 mortar
(anti-aircraft) 1 × 7.62 mm PKT MG
Ammunition:
(240 mm) 40
(7.62 mm) 1,500
Gun control equipment: powered/manual
Elevation/depression: +50/+80°
Traverse: 10° left and right
NBC system: yes
Night vision equipment: yes

240 mm self-propelled mortar M-1975 (SM-240) (2S4) in travelling configuration with ammunition cart to rear (Richard Stickland)
2000/0085193

Suspension is of the torsion bar type with each side having six roadwheels and the drive sprocket at the front, idler at the rear and four track-return rollers.

Smel'chak guided projectile

In late 1993, it was revealed that the 240 mm 2S4 self-propelled mortar system and the 240 mm M-240 towed mortar could both fire the Smel'chak (Daredevil) laser-guided mortar projectile. Details of the 240 mm M-240 towed mortar were given in *Jane's Armour and Artillery 1998-99,* pages 713-714. This is no longer in service with the Russian Army.

The Smel'chak was developed in the 1980s and was successfully trialled in Afghanistan, being fired from the 240 mm M-240 towed mortar.

In addition to the actual projectile and mortar, other elements of the system include the 1D15 laser target indicator/laser range-finder, the 1A35K/1A35J synchronisation system and communications equipment.

Typically, the forward observer first locates the target and this information is relayed back to the command post which calculates the relevant target-laying information. The 240 mm guided projectile is then loaded, aimed at the target and fired.

As the projectile gets to within 400 to 800 m of the target the self-guidance system is activated. This reacts to the illumination of the target by the laser designator and the projectile then homes onto the target. If the projectile is off course, a rocket motor is engaged to correct its flight.

Compared to the US Copperhead 155 mm Cannon-Launched Guided Projectile, the Smel'chak is very difficult to jam as the target is illuminated for such a short period of time.

The 1D15 laser target indicator/designator has a ×10 monocular sight and can designate targets from 200 to 10,000 m to an accuracy of 10 m.

The Smel'chak has a minimum range of 3.6 km and a maximum range of 9.2 km with laser target designation range being 0.2 to 5.0 km. The actual projectile, the 3F5, weighs 134.2 kg of which 32 kg is the warhead.

Status

Production complete. It is understood that this vehicle is no longer in front line service with any countries although quantities may be held in reserve.

Manufacturer

RFAS state factories.

UPDATED

120 mm 2S31 Vena self-propelled gun/mortar system

Development

In early 1997 Russia finally revealed the 120 mm 2S31 Vena self-propelled gun/mortar system which had been under development for some years.

Early development examples of the 2S31 were based on the Obiekt 934 light tank chassis developed by the Volgograd Tractor Plant (*Jane's Armour and Artillery 1996-97*, pages 603-604) but the first complete system, shown for the first time in the Middle East, is based on a much modified BMP-3 Infantry Fighting Vehicle (IFV) chassis.

The BMP-3 IFV entered service with the Russian Army in 1990 and is also being used for an increasing number of more specialised supporting roles of which the 2S31 is the latest.

While the prime contractor for the BMP-3 chassis is the Kurgan Machine Construction Plant, the prime contractor for the 2S31 Vena system is the Motovilikha Plants Corporation based in Perm which is involved in a number of other Russian artillery systems as well as multiple launch rocket systems.

Description

For the 2S31 application, the chassis of the BMP-3 IFV has been modified in a number of areas. The crew position either side of the driver has been removed as have the two bow-mounted forward-firing 7.62 mm PKT machine guns.

The existing two-person turret has been removed and replaced by a much larger turret with vertical sides and rear and a curved front. The hull rear of the vehicle has been modified and all roof hatches and rear doors have been removed.

A new and much smaller downward-opening hatch is provided in the rear of the hull on the left side. There are two ammunition resupply hatches in the right side, one in the hull and one in the turret.

The 2S31 Vena has a four-man crew consisting of driver seated at the front in the centre, commander in the turret on the right, gunner in the turret on the left and the ammunition loader who sits to the rear of the commander when travelling.

Turret traverse is powered through a full 360° with weapon elevation from –4 to +80°, manual controls are provided for emergency use.

Each vehicle is also fitted with an onboard automatic survey and orientation system. The commander's

Russian 2S31 Vena 120 mm self-propelled artillery system with ordnance at maximum elevation (Christopher F Foss)
0006003

cupola is mounted on the right side of the turret roof and has a large electro-optical sensor package on the right side which is covered by a door when not required. This cupola can be traversed through a 90° arc.

This sensor package includes 1P51 day/image intensification sights and a 1D22S laser range-finder/designator with some components believed to be used in the BMP-3K reconnaissance vehicle. Full details of this vehicle, which has yet to be exported, are given in the Reconnaissance Vehicles section. A 7.62 mm PKT machine gun is mounted on top of the cupola which can be laid and fired under complete armour protection and can be used to engage ground and air targets. The gunner has both direct and indirect fire sights.

The 2S31 Vena is fitted with a bank of six 81 mm electrically operated 81 mm smoke/decoy grenade dischargers either side of the turret.

A laser detector is mounted on the low part of the turret front, either side of the 120 mm 2A80 ordnance and on either side of the turret roof at the rear which is used in conjunction with the smoke/decoy grenade system.

The long 120 mm 2A80 barrel is rifled and like other Russian 120 mm systems is not fitted with a fume extractor or a muzzle brake as it is of the low pressure type.

This weapon has a direct and indirect fire capability and fires a new family of 120 mm ammunition. It can also fire the same family of ammunition as the current 120 mm 2S9 (tracked) and 2S23 (8 × 8) self-propelled gun mortar systems. According to Russian sources the high-explosive projectile has the equivalent destructive power of a 122 mm/152 mm artillery projectile.

The Russian BAZALT State Research and Production Enterprise has developed two new 120 mm projectiles for the 2S31 Vena self-propelled gun/mortar system, both of which weigh 26 kg.

The high-explosive fragmentation projectile has a minimum range of 2,000 m and a maximum range of 18,000 m with each projectile having a damage area of 2,200 m^2.

The second is a cargo round which contains 35 high explosive/fragmentation sub-munitions each of which will penetrate about 100 mm of conventional steel armour. This projectile has a minimum range of 1,500 m and a maximum range of 11,000 m. Each projectile has a damage area of about 10,000 m^2 and the 2S31 Vena can also fire western natures of 120 mm rifled mortar ammunition.

More recently the Motovilikha Plants Corporation has stated that the maximum range when firing normal HE projectiles is 13,000 m and when firing 'mines' projectiles is 7,200 m.

A total of 70 120 mm projectiles is carried of which 22 (11 + 11) are in ready use magazines for rapid loading using a ramming system. Maximum rate of fire is quoted as 8 to 10 rds/min.

Also carried are 10 Kitolov-2M laser-guided projectiles which have a maximum range of 13 to 14 km with a 0.8 to 0.9 hit probability. Targets can be designated from the 2S31 Vena or from a normal tripod-mounted designator. The Kitolov-2M laser guided projectiles are loaded manually.

The Kitolov was first revealed some time ago in 122 mm calibre for the 2S1 self-propelled artillery system.

Maximum quoted range for HE projectile is 13 km while for a standard mortar projectile it is 7.2 km.

The 2S31 has a combat weight of 19,500 kg and a power-to-weight ratio of 26.31 hp/tonne, maximum road speed is 70 km/h and cruising range on roads is 600 km. Other chassis details and specifications are almost identical to those of the BMP-3 covered earlier in this volume.

Like other members of the BMP-3 family, Vena is fully amphibious being propelled in the water by two water-jets mounted low down one either side of the hull at the rear. Before entering the water the trim vane is erected at the front of the vehicle by the driver without leaving his seat and the bilge pumps are switched on.

Standard equipment for the 2S31 Vena includes an NBC system of the overpressure type, land navigation system, night vision equipment, vehicle washdown capability and a toilet.

Although the prototype of the 2S31 Vena is based on the chassis of the BMP-3 infantry combat vehicle, the turret system can be installed on a wide range of other chassis, tracked and wheeled.

Status

Prototype. Not yet in production or service.

Manufacturers

Motovilikha Plants Corporation(turret and prime contractor).

Kurgan Machine Construction Plant(chassis).

UPDATED

120 mm Anona 2S9 self-propelled gun/mortar

Development

The 120 mm SO-120 (2S9) Anona (Anemone) self-propelled gun/mortar was first seen in public in May 1985 and is an airborne artillery assault vehicle that was developed to carry out two tactical functions.

One role is as a conventional artillery equipment to replace existing mortars and howitzers in the Russian Army's air assault divisions and the second role could be as a direct fire anti-tank weapon system firing HEAT projectiles.

The SO-120 is used by the Russian Army *Vozdushno-desantnyye Voyska* (or VDV) air assault divisions. The 2S9 entered service in 1981 and was deployed in Afghanistan.

Although it was originally developed to meet the requirements of former Soviet air assault divisions, more recently it has also been identified in service with regular ground forces and naval infantry. A quantity was left behind in Afghanistan when the Soviets withdrew. During the fighting in Afghanistan in 2001 to 2002 none of these systems were observed in action.

The chassis used in the S-120 (2S9) system was developed originally by the I V Gavalov Design Bureau located at the Volgograd Tractor Plant.

It is estimated that well over 1,000 of these systems have been completed, although Russia has not released any production figures.

More recently, Russia has developed the 120 mm 2S23 self-propelled gun/mortar system. This is essentially the BTR-80 (8 × 8) armoured personnel carrier, fully described in the Armoured personnel carriers (wheeled) section, fitted with a modified version of the turret fitted to the 2S9. Details of the 2S23 (8 × 8) are given in the following entry.

The 2S9 has also been observed being transported by the Pomornik class air cushion vehicle used by the Russian Navy.

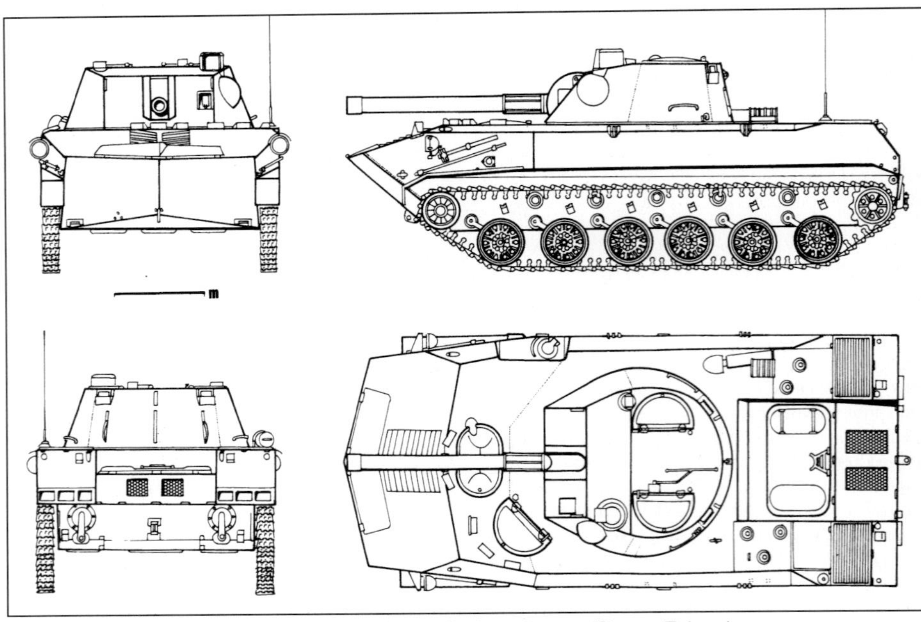

Four-view drawing of the 120 mm 2S9 self-propelled gun/mortar (Steven Zaloga)

The 120 mm 2S9 Anona is also referred to as the Nona-S with the S standing for Samokhodnaya, or self-propelled.

The 2S9 has seen action in Chechnya.

Description

The SO-120 (2S9) has a crew of four: commander, driver/mechanic, gunner and loader. The driver/mechanic is seated in the front centre of the hull and is provided with three day periscopes with the centre block replaceable by an infra-red night driving device. The driver has a single-piece hatch cover that opens to the right.

The commander is located at the front left of the hull. He is provided with three day vision devices and the communication sets, a gyrocompass and land navigation devices. The commander has a single-piece hatch cover that opens towards the rear.

The gunner is on the left of the turret. His position has a panoramic day sight fitted to the turret roof and a telescopic sight for direct fire. The loader is to the right of the gun. He is provided with a single vision block and a seat that folds away while he is serving the gun.

The SO-120 (2S9)'s aluminium hull is a version of the BTR-D airborne APC, which is covered in the entry for the BMD-1 airborne combat vehicle in the Armoured

120 mm 2S9 self-propelled gun/mortar (Michael Jerchel) ***2000***/0085198

120 mm 2S9 self-propelled gun/mortar from the rear (Michael Jerchel) ***2000***/0085199

120 mm 2S9 self-propelled gun/mortar with trim vane on glacis plate in stowed position (PB)
2001/0105772

120 mm 2S9 self-propelled gun/mortar (left) with 1V119 artillery fire direction vehicle (right)

personnel carriers (tracked) section. It is divided into three compartments: the command compartment, the fighting compartment and the power pack compartment. The command compartment is in the front of the hull, forward of the turret mount. It accommodates the commander and the driver/mechanic. The fighting compartment has stowage for 25 rounds of ammunition behind the turret and is in the middle of the hull.

The turret is located above the fighting compartment. At the rear of the hull is the power pack compartment. The engine is a 5D20 diesel developing 240 hp, coupled to a five-speed manual transmission. Fuel is stored in the engine compartment.

The two-man turret is of welded aluminium construction with 16 mm thick frontal armour. The turret roof has two hatches, one for the gunner and the other for the loader. Turret traverse is limited to 35° either side. Mounted in either side of the turret rear is a TNPO-170A day observation device.

The track is the same as that used on the BMD-1 airborne combat vehicle and the suspension is hydropneumatic with an adjustable ground clearance of between 100 and 450 mm. The ground clearance is adjustable by the driver/mechanic from within the vehicle. There are six roadwheels each side and five track-return rollers.

The SO-120 (2S9) is amphibious and is powered in the water by two water-jets at the rear. Power can be delivered to the water-jets and the tracks at the same time. When in the water, splash vanes can be erected in front of the driver's position. Electrically operated bilge pumps are activated to expel any water that enters the vehicle during amphibious operations.

The SO-120 (2S9) is armed with a 120 mm breech-loaded gun 2A51 with a barrel approximately 1.8 m long. The gun is probably provided with an interrupted-screw breech mechanism and a chamber detent to retain a round in place when the barrel is elevated. Ammunition is fixed and loading is manual, although ramming is automatic. After the loader has selected a round from a ready rack it is placed in a feed tray. After an electrical button has been pressed a rammer automatically seats the round in the chamber and closes the breech. The rammer is a pneumatic device operating at a pressure of 150 kg/cm². As the breech opens after firing, the rammer bleeds compressed air into the chamber to force firing fumes from the muzzle.

The 120 mm mortar has a rate of fire of 6 to 8 rds/min. Muzzle velocity is 560 m/s for the HEAT round and 367 m/s for the artillery rounds. The direct fire HEAT projectile is estimated to defeat up to 650 mm of conventional steel armour. In the indirect fire role the mortar fires high-explosive, white phosphorus and smoke rounds. The standard HE mortar bomb has a maximum range of 8,855 m while the 120 mm rocket-assisted mortar bomb has a maximum range of 13,000 m. Additional details of the ammunition are given in the entry for the 120 mm 2B16 in the Towed anti-tank guns, guns and howitzers section later in this book. The family of 120 mm ammunition for the 2S9 is manufactured by the Bazalt State Research and Production Enterprise.

Mounted below the turret rear is an ammunition loading hatch and mounted on top of the hull is a device for loading ammunition from the ground directly into the fighting compartment. This allows sustained fire missions to be carried out while preserving onboard ammunition supply.

The SO-120 (2S9) takes 30 seconds to come into action and a similar time to come out of action. When in the firing position the suspension is raised to provide a more stable firing platform.

The SO-120 (2S9) has no coaxial machine gun in the turret and there are no bow or turret-mounted machine guns. Standard equipment includes an NBC system, night vision equipment and an electric bilge pump.

For air transport the SO-120 (2S9) can be carried by any Russian Military Air Transport (VTA) medium or heavy cargo aircraft. It can be paradropped from the An-32 'Cline', Il-76 'Candid' or An-22 'Cock' aircraft using the PRSM-915 heavy drop system, from heights of between 300 and 1,500 m.

Specifications

Crew: 4
Combat weight: 8,700 kg
Power-to-weight ratio: 27.58 hp/tonne
Ground pressure: 0.5 kg/cm²
Length: 6.02 m
Width: 2.63 m
Height:
(at max ground clearance) 2.3 m
(in firing configuration) 1.9 m
Ground clearance:
(variable) 100 to 450 mm
Track: 2.38 m
Track width: 250 mm
Length of track on ground: 3.23 m
Max speed:
(road) 60 km/h
(water) 9 km/h
Range:
(road) 500 km
(water) 75 to 90 km
Fording: amphibious
Gradient: 60%
Side slope: 33%
Engine: 5D20 diesel developing 240 hp
Transmission: manual, 5 forward and 1 reverse gears
Suspension: hydropneumatic, adjustable
Armament: 1 × 120 mm 2A60 gun
Ammunition: 25 × 120 mm
Gun control equipment
Gun elevation/depression: +80/–4°
Turret traverse: 35° left/35° right
NBC system: yes
Night vision equipment: yes

Variants

There are no known variants of the 2S9 although also used with these units are the 1V118 artillery observation post vehicle and the 1V119 artillery fire direction vehicle, both of which are based on the chassis of the BMD-1 airborne combat vehicle.

Status

Production complete. In service with:

Country	Quantity	Comment
Afghanistan	n/avail	
Azerbaijan	28	
Belarus	54	
Kyrgystan	12	
Moldova	9	
Russia	600	estimate, Army and Marines
Turkmenistan	12	
Ukraine	62	
Uzbekistan	69	
Vietnam	n/avail	unconfirmed user

Manufacturer

RFAS state factories.

UPDATED

120 mm NONA-SVK 2S23 self-propelled gun-mortar system

Development

The 120 mm 2S23 self-propelled gun-mortar system, which is also referred to as the NONA-SVK, is now in service with the Russian Army in small numbers. For the export market it has been offered some years ago at around US$810,000 per system.

Compared with conventional towed mortar systems, self-propelled systems such as the 2S9 (tracked) and the 2S23 (wheeled) have many advantages. These include the ability to keep up with the units for which they are to provide fire support; rapid deployment times; the crew is provided with protection from small arms fire, shell splinters and NBC attack and they also have a direct fire capability, which can be very useful in some tactical situations.

The actual 120 mm weapon is also referred to as the NONA with the 2S9 being the NONA-S, the 2S23 the NONA-SVK and the towed 120 mm 2B16 combination gun (covered in the Towed anti-tank guns, guns and howitzers section) the NONA-K, all of which use the same family of 120 mm ammunition.

The existence of the 2S23 was first revealed at a defence exhibition in Asia in late 1990 when a complete system was displayed for the first time. According to Russian sources the 2S23 entered service with its ground forces in 1990.

It is believed that chassis integration of the 2S9 and more recent 2S23 was undertaken at the I.V.Gavalov Design Bureau which is located at the Volgograd Tractor Plant where the BMD-2 and BMD-3 airborne assault vehicles has been undertaken. Production of the old PT-76 light amphibious tank was also undertaken at this facility.

In 1997, it was reported that China had purchased 100 2S23 systems plus a quantity of 120 mm laser guided projectiles.

More recently a new 120 mm self-propelled gun-mortar system based on a modified BMP-3 chassis has been developed called the 2S31. This has a longer barrel than the 2S23 but fires the same family of ammunition. Details of this system, which remains at the prototype stage, are given in the previous entry.

Description

The 2S23 is based on a modified BTR-80 (8 × 8) Armoured Personnel Carrier (APC) chassis with the turret-mounted mortar over the forward part of the hull.

The BTR-80 is widely used by Russian motorised rifle divisions while the tank divisions use the BMP-1/BMP-2 and, in small numbers, the latest BMP-3. Compared to the BMPs and the 2S9, the BTR-80 and 2S23 have

120 mm 2S23 self-propelled gun-mortar system with thermal shielding (Christopher F Foss) 0018998

2S23 showing the two sights, one mounted in the turret front and one on the turret roof (Christopher F Foss) 0018999

much greater strategic mobility and also lower life cycle costs as they are wheeled rather than tracked. The 2S23 may also be used by Russian naval infantry units.

The hull of the 2S23 is of all-welded steel armour construction that provides protection over the frontal arc from 12.7 mm armour-piercing rounds at a distance of 1,000 m with the remainder of the vehicle providing protection against 7.62 mm armour-piercing rounds.

The driver is seated at the front left with the commander to his right, turret and ammunition compartment in the centre and the power pack compartment at the rear. The latter is provided with an automatic fire detection and extinguishing system, a preheater, fuel primer, heat exchangers for the radiator and crank case, diesel fuel tanks and a bilge pump.

Unlike the BTR-80 APC, the 2S23 does not have any firing ports and apart from the emergency hatch between the second and third roadwheels the only means of entry are the hatches in the turret roof and above the commander's and driver's station.

The turret of the 2S23 is of a slightly different design from that of the 2S9 but has a later 120 mm 2A60 rifled gun-mortar and fires the same family of new ammunition. The direct fire sight is to the left of the 120 mm ordnance and the indirect sight is in the turret roof on the left side.

The commander has a raised cupola which can be traversed through a full 360° and also has vision devices, a 7.62 mm PKT externally mounted machine gun and a TKN-3A infra-red searchlight which is mounted to the left of the machine gun. The machine gun be aimed and fired from within the turret by remote control as it is connected to the TKN-3A sighting device.

The 120 mm ordnance, which has a calibre length of 24.2, has an elevation from –4 to +80° with turret traverse being 35° left and right.

A total of 30 120 mm mortar projectiles and their associated charges is carried and the automatic loading system enables a rate of fire of 8 to 10 rds/min to be achieved. New ammunition is loaded into the 2S23 via a small door in the lower part of the hull between the second and third roadwheels.

Ammunition for the 120 mm 2S23 is manufactured by the Bazalt State Research and Production Enterprise and brief details are as follows:

TYPE	**HE-FRAG**	**HE-RAP**	**HEAT**
Calibre	120 mm	120 mm	120 mm
Round weight	19.8 kg	19.8 kg	13.17 kg
HE content	4.9 kg	3.25 kg	n/a
Round length	820 mm	825 mm	960 mm
Muzzle velocity	367 m/s	367 m/s	560 m/s
Max pressure	100 MPa	100 MPa	110 MPa
Max range	8,850 m	12,850 m	800 m
Min range	1,000 m	1,000 m	40 m

The High-Explosive Fragmentation (HE-FRAG) round is designated the OF-54, with the actual projectile being called the OF-49, and has a maximum range of 8,850 m while the OF-50 High-Explosive Rocket-Assisted Projectile (HE-RAP) has a maximum range of 13,000 m and has less explosive content. Both of these mortar bombs have a body of C60 steel and a nose-mounted point-detonated B35 fuze.

The fin-stabilised High-Explosive Anti-Tank (HEAT) round has a point-blank range of 500 m and will penetrate 650 mm of rolled homogeneous conventional armour. When in flight the fins unfold to provide flight stabilisation. This bomb has a point-initiated base-detonated fuze.

While this 120 mm HEAT round will not penetrate the latest MBTs over their frontal arc, it will inflict damage to their sides and rear as well as being capable of destroying most light armoured fighting vehicles and self-propelled artillery systems likely to be encountered on the battlefield.

All of these three 120 mm mortar bombs feature a pre-engraved copper driving band to the rear with the propellant charge being secured to the base.

In addition to the Russian types of 120 mm ammunition, the weapon can fire ammunition designed for the French TDA 120 RT-61 towed rifled mortar system which is used by many countries around the world.

Russia has disclosed that it has developed a 120 mm laser-guided mortar projectile called the Gran which has been designed to be launched from muzzle-loaded 120 mm mortar systems, including vehicle-carried systems. Basic specifications of the Gran are:

Specifications

Crew: 4
Configuration: 8 × 8
Combat weight: 14,500 kg
Power-to-weight ratio: 17.93 hp/t
Length: 7.40 m
Width: 2.90 m
Height: 2.495 m
Ground clearance: 0.475 m
Track: 2.41 m
Wheelbase: 4.4 m
Max road speed: 80 km/h
Max water speed: 10 km/h
Fuel capacity: 290 litres
Range: 600 km
Fording: amphibious
Gradient: 60%
Side slope: 30%
Vertical obstacle: 0.5 m
Trench: 2 m
Engine: V-8 water-cooled diesel developing 260 hp
Transmission: manual, 5 forward and 1 reverse speeds
Transfer case: 2 speed
Steering: power-assisted
Tyres: 13.00 × 18
Brakes: hydraulic on all wheels
Electrical system: 24 V
Turning radius: 7 m
Armament:
(main) 1 × 120 mm gun-mortar 2A60
(anti-aircraft) 1 × 7.62 mm PKT MG
(smoke grenade launchers) 2 × 3
Ammunition:
(main) 30 × 120 mm (some Russian sources have stated 35)
(anti-aircraft) 500
Gun control equipment
Gun elevation/depression: +80/–4°
Turret traverse: 35° left/35° right
NBC system: yes
Night vision equipment: yes

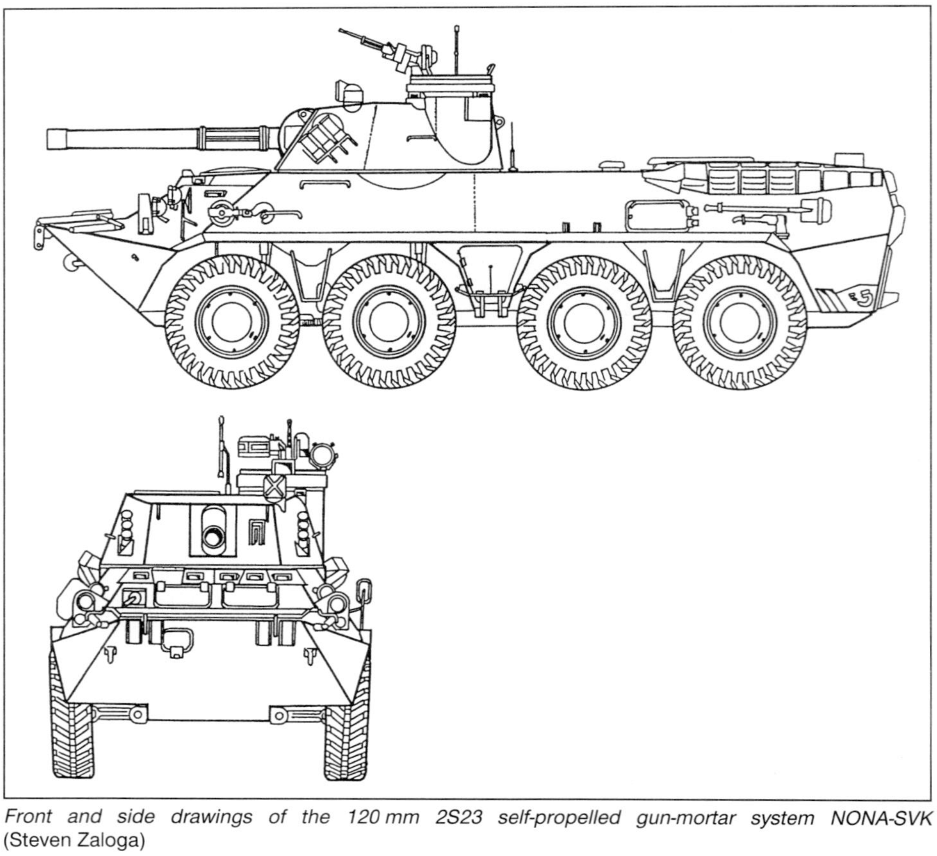

Front and side drawings of the 120 mm 2S23 self-propelled gun-mortar system NONA-SVK (Steven Zaloga)

Calibre: 120 mm
Warhead type: HE-FRAG
Weight:
(projectile) 25 kg
(warhead) 11 kg
(HE) 5.1 kg
Length:
(projectile) 1,225 mm
Max range: 7.5 km

It is probable that the Gran laser-guided mortar projectile has been designed to be fired only from 120 mm smoothbore mortar tubes. Russia has, however, developed to the prototype stage the 122 mm Kitolov-2 laser-guided artillery projectile which can be fired from the 122 mm 2S1 self-propelled artillery system. Late in 1993, it was revealed that this could also be produced in 120 mm version and an artist's impression showed this being fired from the 120 mm 2S23 system. For the 2S23 application a maximum range of 13 to 14 km would be achieved with a 0.8 to 0.9 hit probability. The latest Kitolov-2M has a tandem HEAT (High-Explosive Anti-Tank) warhead which will defeat armoured vehicles fitted with explosive reactive armour. A bank of three standard 81 mm electrically operated smoke grenade dischargers is mounted either side of the turret firing forwards.

Like the BTR-80 APC, the 2S23 is fully amphibious, being propelled in the water by a single water-jet which is located in the lower part of the hull rear. Before entering the water a trim board is erected at the front of the hull by the driver without leaving his seat and the bilge pumps are activated.

The 2S23 is transportable by transport aircraft, can be dropped by parachute and has been designed to operate in temperatures ranging from –50 to +50°C.

Standard equipment includes an NBC system of the overpressure type, a central tyre-pressure regulation system that allows the driver to adjust the tyre pressure to suit the ground being crossed without leaving his seat, and night vision equipment for the driver.

During a defence equipment exhibition held in Nizhny Novgorod in September 1995, a 2S23 was observed with what appeared to be a thermal shield on its turret and chassis.

Status

Production as required. In service with China (100 reported ordered) and Russia.

Manufacturer

Motovilikha Plants Corporation (prime contractor).

UPDATED

TOWED ANTI-TANK GUNS, GUNS AND HOWITZERS

TOWED ANTI-TANK GUNS, GUNS AND HOWITZERS

ARGENTINA

155 mm L 45 CALA 30/2 Cannon

Development

The 155 mm L 45 CALA 30/2 Cannon towed artillery system was developed by the Argentine Armed Forces Research and Technology Institute (CITEFA) with the prototypes being built by the local military manufacturing establishment (DGFM) which is also responsible for production of the weapon.

Final trials of the second prototype, which incorporated a number of improvements over the first prototype, were carried out in November 1996 during which the L 45 CALA 30/2 Cannon achieved a maximum range of 39,000 m firing an Extended Range Full Bore - Base Bleed (ERFB-BB) projectile.

The Argentine armed forces placed an initial production order for this system which comprises not only the 155 mm L 45 CALA 30/2 Cannon but also a new family of 155 mm artillery projectiles including the PACU (HE) with hollow base and the PALA 37 (Extended Range Full Bore) in two versions, hollow base and base bleed. In addition, it can fire all standard types of 155 mm HE projectiles such as the US-designed high-explosive M107 projectile.

Funding permitting, it is probable that in the longer term, the 155 mm L 45 CALA 30/2 Cannon will supplement and perhaps eventually replace the older L 33 X1415 CITEFA Models 77 and 88 which are covered in detail in the following entry.

Description

The 155 mm L 45 CALA 30/2 Cannon artillery system is mounted on a conventional split trail carriage and when deployed in the firing position rests on the two spades and a hydraulic firing jack under the carriage.

The 155 mm 45 calibre ordnance is fitted with a double-baffle muzzle brake, screw breech with automatic opening mechanism and has a fixed recoil system - length does not vary with elevation.

Elevation limits are from –3 to +67° while traverse is 30° left and right. The equilibrators are gas-operated. The aiming unit consists of a panoramic optical periscope. The sighting system is mounted on the left side.

When being towed, the upper carriage and ordnance is traversed through 180° and locked in position over the closed trails. The trails are provided with wheels to assist in bringing the weapon into action.

The system is fitted with an auxiliary power unit with a Deutz air-cooled diesel engine that provides power to the four large roadwheels. When in the self-propelled mode the ordnance is traversed to the rear with the driver being seated at the front left of the carriage.

The suspension of the carriage is of the walking-beam type with air brakes and anti-unhitch safety device.

As previously stated, the L 45 CALA 30/2 Cannon can fire all standard NATO projectiles as well as locally developed projectiles with ranges of the latter being as follows:

Projectile	Muzzle velocity	Max range
M56 or CX4 (HE)	820 m/s	23,000 m
PACU (HE)	900 m/s	27,000 m
PALA 37/CH (HE)	897 m/s	31,000 m
PALA 37/BB	897 m/s	39,000 m

155 mm L 45 CALA 30/2 Cannon in travelling position with ordnance traversed over closed trails
0006031

Specifications

Calibre: 155 mm
Barrel length: 6.975 m
Rifling: 48 right turn, variable pitch grooves, 20/30 cal
Maximum normal service pressure: 3,100 bar
Muzzle brake: double baffle
Carriage: split trail
Shield: no
Elevation/depression: +67/–3°
Traverse: 30° (left and right)
Rate of fire:
(2 minutes, full charge) 6 rds
(short bursts) 2-4 rds/min
(sustained - 60 min) 60-120 rds/h
(prolonged) 1 rd/min
Range:
(M56 or CX4 projectile) 23,000 m
(PACU HE projectile) 27,000 m
(PALA 37/CH HE projectile) 31,000 m
(PALA 37/BB HE projectile) 39,000 m
Towing speed:
(road) 70 km/h
(dirt road) 40 km/h
(cross-country) 20 km/h

Status

Production as required. Believed to be in service with Argentina in small numbers. United Nations sources stated that there were no exports of 155 mm artillery systems from Argentina between 1992 and 2000.

Manufacturers

Instituto de Investigaciones Cientificas y Tecnicas de la Fuerzas Armadas, (CITEFA) (Design authority).
Direccion General de Fabrications Militares (DGFM), (Production).

UPDATED

155 mm L 45 CALA 30/2 Cannon deployed in the firing position
0006030

155 mm howitzer L33 X1415 CITEFA Models 77 and 81

Development

These weapons are basically the top carriage (barrel, cradle, recoil system and equilibrators) of the French Giat Industries (previously Creusot-Loire) 155 mm self-propelled gun Mk F3 (of which 24 are in service) mounted on a new bottom carriage designed by the Scientific and Technical Research Institute of the armed forces (CITEFA). A small number of these 155 mm weapons believed to be 12, has been observed in service in Croatia.

Total production of this system is believed to have amounted to around 120 systems. Recent information has indicated that Argentina has a total of 109 of these systems in service.

In the future, these systems will be supplemented by the more recent 155 mm L 45 CALA 30/2 Cannon also developed in Argentina. This features a 155 mm 45 calibre ordnance and an auxiliary propulsion unit. Details of this weapon are given in the previous entry.

Specifications

Calibre: 155 mm
Barrel length: 5.115 m
Muzzle brake: double baffle
Carriage: split trail
Shield: no
Weight:
(travelling order) 8,000 kg
Length: 10.15 m
Width: 2.67 m
Height: 2.2 m
Ground clearance: 0.3 m
Track: 2.31 m
Elevation/depression: +67/0°
Traverse:
(total) 70°
Rate of fire:
(max) 4 rds/min
(sustained) 1 rd/min
Max range:
(normal round) 22,000 m
RAP: 25,300 m

155 mm howitzer L33 X1415 CITEFA Model 77 in firing position with barrel at maximum elevation

Description

The top carriage is a welded steel structure which contains the elevating brackets and traverse mechanism, and forms the connection with the cradle trunnion attachment and the bottom carriage. The traversing mechanism is mounted inside the front and bottom of the top carriage and, to obtain acceptable handwheel loads, two pneumatic equilibrators are connected to the top carriage and cradle.

The bottom carriage is a welded and machined steel unit. For firing, the wheels are raised clear of the ground and the weapon rests on a firing base at the front and on the trails at the rear. The firing base is a circular steel structure mounted on a ball socket, suspended underneath the bottom carriage, which compensates for uneven ground contour. The firing base is normally carried on the bottom carriage when travelling, as the ground clearance of 0.3 m makes removal unnecessary during movement. The suspension is of the spring type.

The L33 X1415 fires an M107 HE projectile weighing 43 kg with a maximum muzzle velocity of 765 m/s, and illuminating and smoke ammunition.

The Model 81 is a slightly revised version of the Model 77, fitted with a locally produced copy of the 155 mm Mk F3 barrel. Some changes have been made to the trail legs and to the recuperator and equilibrators. The Model 81 has exactly the same specifications as the Model 77 and fires the same 155 mm ammunition.

105 mm exports

According to United Nations sources, in 1995 Argentina exported 18 105 mm guns of an undisclosed type to Venezuela.

Status

Production complete. In service with Argentina. A quantity of these weapons has been observed in Croatia. United Nations sources stated that there were no exports of 155 mm artillery systems from Argentina between 1992 and 2000.

Manufacturer

Instituto de Investigaciones Cientificas y Tecnicas de la Fuerzas Armadas (CITEFA), (Design authority).
Direccion General de Fabrications Militares (DGFM), (Production).

UPDATED

BELGIUM

SRC International GC 45 155 mm gun-howitzer

Development

The GC 45 155 mm gun-howitzer was developed from 1975 as a private venture by SRC International of Belgium. This was a company established jointly by the now defunct Space Research Corporation of Canada and PRB of Belgium (which subsequently went bankrupt).

The first prototype of the GC 45 was completed in 1977 and 12 weapons were subsequently purchased by the Royal Thai Marines. Of these, two were produced and assembled in Canada, the remaining 10 being produced in Canada and assembled in Austria by Voest-Alpine.

Licence for the production of a weapon based on this design was transferred to Voest-Alpine which manufactured this, with a number of improvements, as the GH N-45.

More recently, the Swiss company T & T Technology Trading purchased the complete manufacturing rights of this weapon as well as all of the remaining weapons and spare parts. Additional details of this weapon, now marketed by the NORICUM Division of T & T Technology Trading, are given under International. T & T Technology Trading is now back in production with the GH N-45 for a number of export customers including Thailand.

According to United Nations sources, Thailand took delivery of 18 weapons in 1992, 18 in 1996 and 36 in 1998.

GC 45 stands for Gun Calibre 45 with the figure indicating the length of the 155 mm gun in calibres.

This weapon was the baseline for many other 155 mm 45 calibre weapon systems built in other countries.

The origins of the South African LIW 155 mm G5 (towed) and G6 (6 × 6 self-propelled) artillery systems can be traced back to the SRC International GC 45.

The Space Research Corporation also developed an upgrade kit for the US 155 mm M114 towed howitzer and a further development of this is now marketed by RDM Technology of the Netherlands and covered in detail later in this section. Known sales include Denmark, Netherlands and Norway.

Specifications
Calibre: 155 mm
Barrel length: 6.975 m
Muzzle brake: multibaffle
Carriage: split trail
Shield: none
Weight:
(travelling order) 8,222 kg
(firing position) 8,222 kg
(at lunette, towed mode, muzzle aft) 1,815 kg
(at lunette, towed mode, muzzle forward) 1,134 kg
Length:
(travelling, muzzle aft) 13.614 m
(travelling, muzzle forward over trails) 9.144 m
(firing) 10.82 m
Width:
(travelling) 2.692 m
(firing) 10.364 m
Height:
(travelling) 3.28 m
Ground clearance: 0.355 m
Elevation/depression: +69/–5°
Traverse:
(total) 80°
Rate of fire:
(sustained) 2 rds/min
(max for 15 min) 4 rds/min
Range:
(normal) 30,000 m
(base bleed) 39,000 m
Crew: 8
Towing vehicle: 5 ton (6 × 6) truck

SRC International GC 45 gun-howitzer in travelling mode with ordnance over closed trails

Description

The GC 45 is of the split trail carriage configuration. It has a 155 mm 45 calibre barrel designed to fire all standard and new 155 mm NATO ammunition and charges, with the same muzzle velocity and range as would be achieved with a 39 calibre barrel of the International FH-70 or US M198 type. This performance is achieved at a lower breech pressure owing to a larger chamber, matched to the 45 calibre length. A significant feature of the GC 45 is that it meets the NATO requirement for a range of 30,000 m with a completely ballistic solution, without rocket booster or other external devices. When fired with the SRC 155 mm ERFB projectile and the top zone M11 charge, the maximum range of 30,000 m is achieved. At high angle and low charge (M3A1 zone 2) the GC 45 achieves a minimum range of 2,800 m with the same projectile. Using an ERFB/Base Bleed projectile with the M11 charge, the maximum range is 39,000 m.

All structural components are built of a high-strength alloy steel, the barrel is made of high-yield steel and autofrettaged throughout. The breech assembly incorporates an automatic breech-opening cam system: the breech opens automatically when the barrel assembly returns to battery and closes automatically when the cam is swung out of engagement.

The elevating system incorporates pneumatic equilibrators and results in small handwheel loads during elevation. Adjustment capability in the equilibrator system geometry, together with different nitrogen pressures, allows for ease of elevation at all gun system operating ambient temperatures.

A low-friction traverse bearing eases traversing. A friction device system is also incorporated to minimise feedback loads into the traversing system from any gun barrel whip.

A swing-aside loading tray mounted on the rear cradle of the recoil system moves in elevation with the barrel assembly. A pneumatically operated telescoping ramming cylinder is rotated into place behind the projectile for loading at high-elevation angles, obviating the need to lower the weapon between shots for ramming. The pneumatic rammer also provides a high degree of uniformity from shot to shot. A pneumatic system consisting of a bottle, mount, valving, regulator, accumulator, hoses and fittings is installed to power the ramming cylinder. An added feature of this system is that the bottle can be recharged in the field.

The suspension of the GC 45 incorporates a walking beam (two wheels each side) enabling the gun to be towed at speeds of up to 90 km/h. The freely pivoting walking beams also allow greater mobility during towing in rough terrain.

SRC International GC 45 155 mm gun-howitzer in firing position. Note that the trails are not bedded in

The design of the screw jack and float system, which is mounted under the forward part of the carriage, incorporates a telescoping system and ball screw jack. One crew member can easily jack the gun in 90 seconds. Disengagement of the ball jack will result in the immediate settling of the gun on its undercarriage wheels.

The trails and spades, with the built-in towbar system, allow for efficient handling during emplacement and disemplacement. The spades are pivoted, allowing four angular positions, so that handling of these heavy components is minimised. It is possible to spread the trails to very large angles permitting large traverses of the barrel assembly at all angles of elevation.

The trail lifting assemblies, at the outboard rear of each trail, are designed to assist the gun crew during the emplacement and disemplacement operations. They are fully swivelling and fitted with rubber tyres, and are raised and lowered by handwheels at the forward end of each assembly. The force at the handwheel is transmitted to the wheel through a ball screw actuator connected to the wheel forks by a ball and crank arrangement. The ball screw is reversible and the trails can be simply dropped to the ground during emplacement. An additional feature of these assemblies is that they enable the gun crew to manhandle the gun through 360° in traverse.

The GC 45 fires the following separate loading ammunition: an SRC ERFB Mk 10 Mod 2 projectile weighing 45.4 kg and containing 8.8 kg of HE, illuminating projectile weighing 45.2 kg, smoke projectile weighing 45.7 kg and WP projectile weighing 47.7 kg. An ERFB/Base Bleed projectile weighing 47.5 kg fired with an M11 charge achieves a maximum range of 39,000 m. Firing the ERFB Mk 10 Mod 2 projectile, the weapon has the following range capabilities:

Charge	Muzzle velocity	Max range
M3A1 zone 3	266 m/s	5,900 m
M3A1 zone 4	302.1 m/s	7,300 m
M3A1 zone 5	359.9 m/s	9,300 m
M4A2 zone 6	458.1 m/s	12,400 m
M4A2 zone 7	541.9 m/s	15,100 m
M119 zone 8	658 m/s	19,200 m
M2 zone 9	795.5 m/s	24,700 m
M11	897 m/s	30,000 m

In addition to the above natures of ammunition it will also fire NATO/US natures of ammunition. The SRC family of ammunition for the GC 45 was originally manufactured in Canada. Production ceased there many years ago. Many other countries now manufacture ERFB and ERFB-BB for the GC 45 and other 155 mm towed and self-propelled artillery systems.

Status

Production complete. In service with the Royal Thai Marines (12). No longer marketed. Details of the current NORICM 155 mm GH N-45 155 mm gun-howitzer are given in this section under International. The latter remains in production with Austria supplying 72 to Thailand between 1992 through to 2000 according to United Nations sources.

Manufacturer

SRC International SA.
(No longer trading).

UPDATED

CHINA, PEOPLE'S REPUBLIC

NORINCO 203 mm (8 in) howitzer

Development

In 1991, it was revealed that China North Industries Corporation (NORINCO) had developed, possibly as a private venture, a new 203 mm (8 in) self-propelled howitzer with the prototype based on the chassis of the US M110 203 mm (8 in) self-propelled howitzer (or the 175 mm M107). Available details of this are given in the Self-propelled guns and howitzers (tracked) section. Until the introduction of the 203 mm (8 in) weapon, the largest calibre weapon produced in China was of 155 mm calibre.

At the same time, it was revealed that China had also developed to the prototype stage a new 203 mm (8 in) towed howitzer that used a similar ordnance and a new carriage.

As of early 2002, no volume production of either the towed or self-propelled model of the NORINCO 203 mm (8 in) howitzer is known to have taken place and it is possible that this was developed in response to an export requirement, possibly from a Middle East customer, rather than a People's Liberation Army (PLA) requirement. As far as it is known there are no plans for quantity production. Recent marketing information released by NORINCO has not mentioned either the NORINCO 203 mm (8 in) towed or self-propelled howitzers.

NORINCO 203 mm (8 in) howitzer deployed in the firing position from the rear with spades yet to be embedded

Specifications

Calibre: 203.2 mm
Length of barrel: 9.235 m (45.5 calibres)
Recoil:
(long) 1,524 mm
(short) 1,041 mm
Muzzle brake: double baffle
Shield: no
Carriage: split trail
Weight: 16,396 kg
Length:
(travelling, barrel forwards) 17.70 m
(travelling, barrel rear) 12.087 m
(firing) 14.831 m
Width:
(travelling, barrel forwards) 2.786 m
(travelling, barrel rear) 2.789 m
(firing) 9.097 m
Height:
(travelling, barrel forwards) 3.80 m
(travelling, barrel rear) 2.527 m
Ground clearance: 320 mm
Elevation/depression: +55/-5°
Traverse: 25° left and right
Range:
(ERFB projectile) 40,000 m
(ERFB-BB projectile) 50,000 m
Towing speed:
(road) 90 km/h
(cross-country) 50 km/h
Towing vehicle: 10 ton (6 × 6) truck

Description

A single prototype of a towed version of the new-generation 203 mm (8 in) howitzer has been completed. This has a conventional split trail carriage with a new NORINCO 203 mm (8 in) 45.5 calibre ordnance fitted with a muzzle brake.

The new ordnance is of Electro-Slag Refined (ESR) steel and has 64 grooves. This, together with a low flame temperature propellant charge and low gas pressure, is intended to give a barrel life of more than 1,000 rounds. The breech mechanism is of the vertical sliding type and is opened and closed hydraulically.

There is a possibility that some parts of the weapon, including perhaps the recoil system, are based on the Chinese 152 mm Type 66 gun-howitzer which in turn is based on the Russian 152 mm D-20 system. The carriage of the 203 mm system appears to be based on the NORINCO 155 mm Type WA 021 towed artillery system which was first revealed in 1986 and is in service with the PLA. Some have also been supplied to Iran.

That system was based on a design of Dr Gerald Bull, father of the 155 mm 45 GC 45 calibre system that, with further development, became the South African G5 and the Austrian GH N-45 towed artillery systems.

When in the travelling configuration, the 203 mm (8 in) ordnance is retracted 1.5 m to the rear and the complete upper part of the weapon swung through 180° and locked in position over the closed trails. The limber is then attached to the closed trails.

When deployed in the firing position, it rests on three points, a central baseplate under the carriage and the two spades. When targets are encountered outside its traversing arc, the trail spades can be quickly raised and the gun traversed.

When the weapon is deployed, it has an upper and lower hydraulic system with miniaturised assembly for lightness. The loading tray and rammer elevates with the ordnance and enables a rate of fire of 2 rds/min to be achieved.

According to China North Industries Corporation, the integration of the elevating system and the equilibrator effectively lowers the total height and muzzle height of the weapon. The sighting system and elevation and traverse handwheels are mounted on the left side of the carriage.

The weapon can fire standard NATO types of 203 mm (8 in) ammunition as well as a new NORINCO developed Extended-Range Full-Bore (ERFB) projectile, with a muzzle velocity of 933 m/s and a maximum range of 40 km. The Extended-Range Full-Bore – Base Bleed (ERFB-BB) projectile has a muzzle velocity of 936 m/s and a maximum range of 50 km. Both use a new combustible cartridge case.

According to NORINCO, these new 203 mm projectiles can also be fired from Western 203 mm (8 in) towed and self-propelled artillery systems but, since most of these have a shorter ordnance, the range will be somewhat less than that achieved with the Chinese systems.

Basic specifications of the ammunition are as follows:

Type	ERFB	ERFB-BB
Calibre	203.2 mm	203.2 mm
Weight	95.9 kg	100.1 kg
Muzzle velocity	933 m/s	936 m/s
Max range	40 km	50 km

In addition to firing HE projectiles, Chinese sources also state that it can fire cargo (dispensing anti-tank mines), chemical, terminally guided and tactical nuclear projectiles.

Status

Prototype. Not yet in production or service.

Manufacturer

China North Industries Corporation.

UPDATED

NORINCO 155 mm gun-howitzer Type WA 021

Development

The NORINCO (China North Industries Corporation) 155 mm gun-howitzer Type WA 021 (also referred to as the WAC 21) is very similar in many respects to the Austrian NORICUM GH N-45 gun-howitzer, to the extent that the carriage of the Type WA 021 can accommodate the ordnance of the NORICUM weapon.

The Type WA 021 is produced by the Heping Machinery Factory of the China North Industries Corporation (NORINCO). It would appear that an interim 155 mm weapon known as the MF 45 was produced in China, probably as part of the development programme that led to the Type WA 021.

By the end of 1986, 10 Type WA 021s had been produced, two as prototypes and the remainder as preproduction examples.

The WA 021 was first seen in public during a defence equipment exhibition held in Beijing in 1986. The WA 021 system is based on a design supplied by the now defunct SRC International of Belgium headed by Dr Gerald Bull who also designed the GC 45 (qv).

The Type WA 021 was observed during the major military parade held in Beijing late in 1999 towed by a 6 × 6 truck.

In 1991, officials of NORINCO stated that the WA 021 was in service with the People's Liberation Army (PLA).

NORINCO has also developed and placed in production the PLZ45 155 mm 45 calibre self-propelled artillery system. This uses the same 155 mm/45 calibre ordnance as the WA 021 and details are given in the Self-propelled guns and howitzers (tracked) section.

The first customer for the 155 mm 45 calibre PLZ45 is Kuwait which placed an order for 27 systems and all

Specifications

Calibre: 155 mm
Length:
(barrel assembly) 7.582 m
(barrel) 7.045 m
(rifling) 5.83 m
(travelling, barrel forward) 13.512 m
(travelling, barrel rear) 9.068 m
(firing) 11.4 m
Muzzle brake: multibaffle
Carriage type: split trail
Shield: no
Weight:
(travelling) 9,500 kg
Width:
(travelling) 2.67 m
(firing) 9.931 m
Height:
(travelling, barrel forward) 3.048 m
(travelling, barrel rear) 2.23 m
(firing overall, 0° elevation) 2.187 m
(to bore of gun) 1.659 m
Ground clearance: 0.28 m
Recoil length:
(long) 1.524 m
(short) 1.041 m
Elevation/depression: +72/-5°
Traverse:
(left/right) 30/40°
Range:
(ERFB M11 charge) 30,000 m
(ERFB BB M11 charge) 39,000 m
(min, ERFB) 3,500 m
Rate of fire:
(sustained) 2 rds/min
(max) 4-5 rds/min
EFC: 1,000
Crew: 9
Towing vehicle: 6 × 6 7-10 ton truck
Max towing speed:
(road) 90 km/h
(cross-country) 15-50 km/h

NORINCO 155 mm gun-howitzer Type WA 021 deployed in firing position 0006032

NORINCO 155 mm gun-howitzer Type WA 021 fitted with auxiliary propulsion unit and in travelling configuration 0006033

of these have now been delivered to Kuwait together with associated fire control and support equipment.

Description

The NORINCO Type WA 021 has an auto-frettaged barrel 45 calibres long, fitted with a multibaffle muzzle brake which has an efficiency of 30 per cent. A screw-type breech mechanism opening to the right is employed and the chamber volume is 22.95 litres. The rifling employs 48 grooves with a twist of 1 in 20 calibres, and the grooves are understood to be three times deeper than the rifling depths found on comparable Western 155 mm designs.

The recoil system consists of the recuperator cylinder (top) and the recoil cylinder (below). The recoil mechanism can change the recoil distance with the variation of elevation angle so there is no need to dig a firing pit when the weapon fires at a high elevation.

The upper carriage includes the elevating mechanism, traversing mechanism, equilibrator and gear for traverse. The equilibrator is a pneumatic-hydraulic balancer. When the temperature changes and the unbalanced torque increases or decreases, the torque of the handwheel in the elevating mechanism will change. In this case the hydraulic pressure can be adjusted to obtain suitable torque of the handwheel in the elevating mechanism.

The carriage is of the Type WA 021 and employs a screw jack firing platform under the carriage pivot. For travelling, the barrel is swung through 180° and held in a clamp on the left-hand trail leg. A walking beam suspension is used and pivoting dolly wheels are provided on each trail leg for ease of handling. A pneumatic flick rammer is located on the left-hand trail leg. Elevating and traversing systems are to th left of the barrel. Sighting equipment includes the M115 panoramic sight and the M137 telescopic sight.

Batteries of up to six WA 021s can be controlled by a fire-control system mounted on a command vehicle. The system includes a 16-bit fire-control computer, a command control console, a tape driver/printer, a company command relay display unit, gun display units, automatic observation equipment and a laser range-finder.

The Type WA 021 fires a range of NORINCO developed Extended-Range Full-Bore (ERFB) projectiles. These include ERFB HE, ERFB BB HE, ERFB BB WP, ERFB illuminating, ERFB smoke, ERFB BB illuminating, ERFB improved conventional munition and ERFB improved conventional munition BB. The improved carries 72 bomblets. These are of the HEAT type and are designed to attack the vulnerable upper surfaces of armoured vehicles and are also highly effective against trucks and troops. The ERFB projectiles have welded nubs. Range with ERFB is 30,000 m and 39,000 m with ERFB BB. Dispersion at maximum range is stated to be 0.45 per cent longitudinal and 0.7 mil in traverse. Barrel service life is 1,000 EFCs.

The normal sustained rate of fire is 2 rds/min. This can be increased to between 4 and 5 rds/min for short periods.

According to NORINCO, maximum muzzle velocity when firing an ERFB projectile is 897 m/s and when firing an ERFB-BB projectile is 903 m/s.

155 mm gun-howitzer with APU

The 155 mm gun-howitzer Type WA 021 is also available fitted with an APU. This enables the gun to move in to or out of the firing position as well as travel in the self-propelled mode for a short distance.

The APU can also be mounted on other towed artillery systems such as the 130 and 152 mm types.

The APU consists of a four-cylinder diesel developing 77 hp at 3,600 rpm. It is mounted on the forward part of the carriage giving the system a maximum speed of 18 km/h. When fitted with the APU the complete system has a turning radius of 7 m and a maximum range without refuelling of 80 km. It can also be towed on roads up to a maximum speed of 90 km/h.

The four main wheels are hydraulically powered with the front wheels being used for differential steering.

As far as is known, the APU version has yet to enter volume production.

This version has a total weight of around 12,000 kg and in the travelling position is 9.8 m long, 2.74 m wide and 2.25 m high.

As far as it is known, the WA 021 with an APU is not currently in service with the PLA.

155 mm coastal defence system

Details of this, which is based on the WA 021, are given in the Coastal artillery guns and missiles section. As far as is known, this remains at the prototype stage.

Status

In production and service with the PLA. Recent information has indicated that Iran has taken delivery of at least 15 of these systems from China.

Manufacturers

Heping Machinery Factory.
China North Industries Corporation.

UPDATED

NORINCO 155 mm gun XP52

Development/Description

Early in 1993, China North Industries Corporation (NORINCO) announced that it had completed the prototype of a new 155 mm towed gun with a 52 calibre ordnance called the XP52.

This builds on its experience with the NORINCO 155 mm 45 calibre towed (WA 021 covered in the previous entry) and PLZ45 (now in service with Kuwait) full-tracked self-propelled artillery systems which use a 155 mm/45 calibre ordnance based on a design by the late Dr Gerald Bull.

The prototype of the 155 mm XP52 is based on an existing carriage and recoil system, believed to be the Chinese NORINCO 152 mm gun Type 83 covered later in this section.

The 155 mm 52 calibre rifled ordnance has 40 grooves and, firing a NORINCO developed Extended-Range Full-Bore (ERFB) projectile weighing 45.54 kg, a maximum range of 34,000 m can be achieved. When firing an Extended-Range Full-Bore – Base Bleed projectile weighing 48 kg, a maximum range of 44,000 m is achieved and maximum muzzle velocity is 980 m/s.

Specifications

Calibre: 154.94 mm
Barrel length: 8.850 m
Muzzle brake: yes
Carriage: split trail
Shield: yes
Weight:
(travelling) 10,500 kg
(firing) 9,700 kg
Length:
(travelling) 9.40 m
Width:
(travelling) 2.65 m
Height:
(travelling) 2.20 m
(axis of bore) 1.40 m
Recoil: 910 mm
Ground clearance: 300 mm
Elevation/depression: +45/–2° 30′
Traverse: 50° (24° left and 26° right)
Range:
(ERFB) 34,000 m
(ERFB-BB) 44,000 m
Rate of fire: 3-4 rds/min
Towing speed: 15-60 km/h

Status

Development complete. Ready for production. It is probable that this is a developmental weapon only. There are no known export sales of this weapon.

Manufacturer

China North Industries Corporation.

VERIFIED

NORINCO 155 mm gun-howitzer Type GM-45

Development/Description

The 155 mm Type GM-45 is a conversion of the existing NORINCO 130 mm Gun Type 59-1 (Russian M-46), or the 152 mm gun-howitzer Type 66 (Russian D-20) to accommodate the ordnance of the NORINCO 155 mm gun-howitzer WA-021.

The resultant performance is the same as the WA 021 and it is stated that the barrel and chamber structure used for the ordnance is the same as the Austrian GH N-45.

The GM-45 is fired with its travelling wheels on the ground. The firing rate is 2 to 4 rds/min. Recoil is variable and loading and breech operation are manual.

To increase the rate of fire a flick rammer has been installed on the left side and hydraulically operated wheels have been added to each trail assembly to ease the opening and closing of the trails. Both systems also have a two wheeled dolly which is attached under the rear of the closed trails.

There are some differences between the Russian M-46 and its Chinese equivalent in that the latter has a two wheel carriage while the Russian system has four wheels and a walking beam suspension system.

According to NORINCO, when firing an ERFB 155 mm projectile it has a maximum muzzle velocity of 897 m/s.

Specifications

Details for the GM-45 are the same as the WA 021 except for the following:
Weight:
(travelling) 8,800 kg
(firing) 8,500 kg
Length:
(travelling) 12 m
Width:
(travelling) 2.5 m
Height:
(travelling) 2.2 m
Axis of bore: 1.38 m
Elevation/depression: +55/–2° 30′
Traverse:
(at elevation of –2° 30′ to +15°) 25° left and right
(at elevation of +16 to +55°) 14° left and 11° right
Recoil length:
(max) 775 mm
(min) 125 mm
Rate of fire: 4 rds/min
Range:
(ERFB M11 charge) 30,300 m
(ERFB-BB M11 charge) 39,000 m
(min, ERFB) 3,500 m
Towing speed:
(by truck) 15-60 km/h
(by tracked vehicle) 35 km/h

Status

Development complete. Ready for production upon receipt of orders. As far as it is known this system has been developed for the export market. United Nations sources have stated that China did not declare the exports of any of these systems between 1992 and 2000.

Manufacturer

China North Industries Corporation.

UPDATED

NORINCO 155 mm gun-howitzer Type GM-45 in travelling configuration with limber attached **2002**/0130010

NORINCO 155 mm gun-howitzer Type GM-45 deployed in firing position **2001**/0105756

NORINCO 152 mm gun Type 83

Development/Description

The NORINCO (China North Industries Corporation) 152 mm Type 83 gun is believed to now equip reserve artillery units. It uses a long barrel, believed to be about 45 calibres long, on a much revised carriage and recoil system based on the NORINCO 152 mm gun-howitzer Type 66 (see following entry). Quantities of these have been supplied by China to Iraq and some of these were captured by Allied forces in the 1991 campaign to retake Kuwait.

As far as it is known, the NORINCO 152 mm gun Type 83 is no longer being built or marketed.

The 152 mm barrel is fitted with a single-baffle muzzle brake and the right-hand pitch rifling uses a uniform twist. A shield is fitted and the split trail legs are equipped with hydraulically operated wheels to assist handling. The carriage rests on a firing platform when emplaced, with the main wheels raised off the ground. When deployed in the firing position, a large spade is fitted to each trail. When not required these are carried one on each trail. When travelling, the barrel is swung through 180° and clamped over the closed trail legs. The trails are then attached to a steerable limber. Short moves can be made without having to stow the barrel in the reverse position.

The Type 83 gun fires an HE projectile also designated the Type 83. This projectile weighs 48 kg, is 830 mm long and uses a propellant case 960 mm long weighing 37 kg with a full charge and maximum muzzle velocity of 955 m/s. It can also fire a NORINCO ERFB-BB projectile to a maximum range of 38,000 m.

The fire-control system installed on the Type 83 includes a Type 58 panoramic sight weighing 1.942 kg with a magnification of ×3.7, Type 58 collimator sight weighing 1.3 kg and a Type CS-43 direct fire sight weighing 4.6 kg with a magnification of ×5.5.

A number of these weapons was supplied to Iraq and captured during Desert Storm in 1991. Some of these were provided with vertical attachments on their long barrels so that camouflage netting could be quickly installed.

Status

Production complete. In service with China, Iraq and possibly other countries in the Middle East. As far as it is known the NORINCO 152 mm gun Type 83 is no longer marketed. United Nations sources state that China did not declare any exports of these systems between 1992 and 2000.

Specifications

Calibre: 152 mm
Muzzle brake: single baffle
Carriage type: split trail
Shield: yes
Weight:
(travelling) 10,500 kg
(firing) 9,700 kg
Length:
(travelling) 9.4 m
Width:
(travelling) 2.67 m
Height:
(travelling) 1.865 m
(to bore of gun) 1.4 m
Recoil:
(normal) 910 mm
(max) 950 mm
Elevation/depression: +45/–2.5°
Traverse: 24° left, 26° right
Range:
(max) 30,000 m
(min) 10,000 m
(direct fire) 1,170 m
Rate of fire: 3-4 rds/min
Travelling speed: 15-60 km/h

Manufacturer

China North Industries Corporation.

UPDATED

NORINCO 152 mm Gun Type 83 in travelling configuration with ordnance over closed trails

NORINCO 152 mm Gun Type 83 in travelling configuration and showing limber attached (Steven Zaloga)

NORINCO 152 mm gun-howitzer Type 66 (D-20)

Development/Description

The NORINCO (China North Industries Corporation) 152 mm gun-howitzer Type 66 is the Chinese version of the Russian 152 mm gun-howitzer D-20. It differs in appearance from the original in few details.

According to United Nations sources, between 1992 and 2000 there was only one export sale of the NORINCO 152 mm gun-howitzer Type 66 and that was to Angola. This country took delivery of four systems from China in 1999. It was revealed that Bolivia took delivery of 18 152 mm weapons in 1992. It is possible that these were in fact Type 66s.

The Type 66 may or may not be fitted with a shield. It is normally towed by a 6 × 6 truck and can be towed at speeds of up to 60 km/h. The carriage is equipped with a Type 58 panoramic sight weighing 1.942 kg with a magnification of ×3.7 and a 10° field of view, a Type 58 collimator weighing 1.3 kg and a Type 66 telescopic sight weighing 4.6 kg with a magnification of ×5.5 and an 11° field of view.

The weapon is mounted on a split trail carriage with each trail being provided with a castor wheel to assist in bringing the weapon into action. When deployed in the firing position these castor wheels rest on the top of each trail.

When deployed in the firing position, the weapon is supported on a circular firing jack and the spade on each trail leg.

The ammunition fired by this system is also fired by the 152 mm Type 83 self-propelled gun-howitzer and both systems may use a similar ordnance. Full details of the 152 mm Type 83, which is understood to be only in service with the People's Liberation Army, are given in the earlier Self-propelled guns and howitzer (tracked) section.

Specifications

Calibre: 152.4 mm
Barrel length:
(overall) 5.195 m
Muzzle brake: double baffle
Breech mechanism: semi-automatic, vertical sliding block
Carriage: split trail with castor wheel on each
Shield: optional
Weight: 5,720 kg
Length:
(travelling) 8.69 m
Width:
(travelling) 2.42 m
Height:
(travelling) 2.52 m
(to axis of bore) 1.22 m
Elevation/depression: +45/–5°
Traverse: 58°
Rate of fire: 6-8 rds/min
Range:
(max) 17,230 m
(min) 4,300 m
(direct fire) 800 m
Crew: 10-12
Towing vehicle: 6 × 6 truck
Max towing speed: 60 km/h
More recently three additional 152 mm projectiles have been developed by NORINCO:

NORINCO 152 mm gun-howitzer Type 66 in firing position

The Type 66 is used to fire at least three types of projectile using a variable six-charge bag propellant system. Two types are HE and smoke. Details are as follows:

Type	HE	Smoke
Designation	Type 66	Type 66
Weight		
(projectile)	43.56 kg	43.56 kg
(filling)	5.86 kg	6.626 kg
Length:		
(projectile)	702 mm	689.7 mm
Type of filling	TNT	WP
Muzzle velocity	655 m/s	655 m/s
Max range	17,230 m	17,230 m
Fuze	Liu-4	Yan-2

Type	Incendiary	Smoke (red)	Illuminating
Weight (projectile)	43.56 kg	43.5 kg	40.6 kg
Length (projectile)	704.48/707.3 mm	699.81/707.14 mm	699.25/701 mm
Max muzzle velocity	650 m/s	650 m/s	651 m/s
Max range	17,300 m	17,200 m	16,828 m

The third type is an RAP projectile known as the Type MP-152. This weighs 43.56 kg and provides a range increase of 27 per cent to 21,880 m. Accuracy is claimed to be the same as that for a conventional projectile.

Status

Production complete. In service with the Chinese armed forces. It is believed that Albania has 90 of these systems in service. Angola took delivery of four systems from China in 1999.

Manufacturer

China North Industries Corporation.

UPDATED

NORINCO 130 mm field gun Type 59-1

Development/Description

The first 130 mm field gun produced by China was the Type 59, a virtually direct copy of the Russian 130 mm M-46. At around the same time, China was also producing its 122 mm field gun Type 60, a virtual copy of the Russian D-74. Some of the features of the Type 60 were carried over to the Type 59 to produce the Type 59-1.

Thus, the NORINCO (China North Industries Corporation) Type 59-1 uses a scaled-up version of the Type 60's muzzle brake, recoil mechanism and breech system, all mounted on a carriage featuring some components from the Type 60.

This amalgamation of Type 59 and Type 60 features has produced a weapon that is more manoeuvrable than the original Type 59 (M-46). The carriage no longer requires a limber for towing and weight has been saved by using a smaller shield.

According to United Nations sources, between 1992 and 2000 export sales of the NORINCO 130 mm field gun Type 59-1 were made to Iran (106 delivered in 1992) and Pakistan (27 in 1998 and 60 in 2000).

When the 130 mm field gun Type 59-1 is deployed in the firing position, it rests on the jack under the carriage and the two spades at the rear. To assist bringing the weapon into action each spade is provided with a castor wheel. When travelling or deployed in the firing position this rests on top of the trail.

Much of the development work on the Type 59-1 weapon was carried out by the No.127 Factory of the Fifth Ministry of Machine Building.

The Type 59-1 fires two NORINCO types of projectile, HE and illuminating, although at one stage of the gun's development an AP-T projectile was proposed, capable of penetrating 250 mm of conventional steel armour set at 0° at a range of 1,500 m. Details of the two original projectiles are as follows:

Specifications
Calibre: 130 mm
Muzzle brake: double baffle
Breech mechanism: semi-automatic, vertical sliding breech block
Carriage: split trail with castor wheel on each
Shield: yes
Weight: 6,300 kg
Length:
(travelling) 10.8 m
Width:
(travelling) 2.42 m
Height:
(travelling) 2.75 m
(to axis of bore) 1.214 m
Elevation/depression: +45/–2.5°
Traverse: 58°
Rate of fire: 8-10 rds/min
Range:
(max) 27,150 m
(min) 7,800 m
(direct fire) 1,170 m
Crew: 8-10
Towing vehicle: 6 × 6 truck
Max towing speed:
(truck) 15-60 km/h
(tracked vehicle) 35 km/h

NORINCO 130 mm field gun Type 59-1 in firing position and showing castor wheel on top of each carriage trail 0006035

NORINCO 130 mm field gun Type 59-1 fitted with auxiliary propulsion unit and shown in travelling configuration 0006036

Type	HE	Illum*
Designation	Type 59	Type 59
Weight:		
(projectile)	33.4 kg	29.1 kg
(filling)	3.6 kg	1.96 kg
Length:		
(projectile)	668 mm	663 mm
Muzzle velocity	930 m/s	950 m/s
Max range	27,490 m	24,500 m†
Fuze	Liu-5	Shi-1

* burns for 40 seconds producing 500,000 candela
† air burst

Both projectiles are fired using a two-charge full-variable charge or a three-charge reduced variable charge.

There have been two enhanced-range projectiles developed for the Type 59-1 gun. One is an Rocket Assisted Projectile (RAP) known as the MP-130. It weighs 33.4 kg and provides a maximum range of 34,360 m, an increase of 25 per cent. A RAP known as the MP-132 has been developed and tested. The second projectile is an ERFB HE which is fired using a multisection charge system employing four bags, an increment and 2.5 kg of loose propellant. Details of this ERFB round are as follows:

Weight:
(complete round, with full charge) 58 kg
(complete round, with reduced charge) 50.7 kg
(complete round, packed) 79 kg
(projectile) 32.41 kg
(charge) 3.3 kg
Length:
(projectile) 799 mm
Type of charge: TNT
Muzzle velocity: 944 m/s
Max chamber pressure: 3,150 kg/cm^2
Max range: 32,000 m
Fuze: Liu-5

Early in 1993, NORINCO announced that a new family of ammunition had been developed for the 130 mm field gun Type 59-1 and the Russian 130 mm M-46, brief details of these are as follows:

HE, weight of projectile 33.4 kg with maximum range of 27.49 km.

HE incendiary, weight of projectile 33.4 kg with maximum range of 27.49 km

ERFB-BB, weight of projectile 33.4 kg with maximum range of 38 km

ERFB (boat tail), weight of projectile 32.7 kg with maximum range of 30 km

ERFB-A (Nubb), weight of projectile 32.41 kg with maximum range of 32 km

ERFB-B, weight of projectile 32.7 kg with maximum range of 30.7 km

Illuminating, weight of projectile 29.1 kg with maximum range of 25 km

Shrapnel (10,000 flechettes), weight of projectile 33 kg with maximum range of 25 km

Cargo (35 HEAT bomblets), weight of projectile 33 kg with maximum range of 25 km

Smoke, weight of projectile 32.47 kg with maximum range of 27 km

The 130 mm field gun Type 59-1 is fitted with a panoramic sight Type 58 with a magnification of ×3.7 and a 10° field of view which weighs 1.942 kg, collimator sight Type 58 with a 10° 40′ field of view and weighing 1.3 kg and a direct fire sight with a magnification of ×5.5° and an 11° field of view and weighing 4.6 kg.

Variant

NORINCO has developed an Auxiliary Propulsion Unit (APU) which can be fitted to 155, 152 and 130 mm towed artillery weapons such as the Type 59-1 field gun to increase their battlefield mobility.

This APU consists of one diesel-driven engine which is mounted on the front of the carriage, with the driver being seated on the left, and the hydraulic transmission systems.

The total weight of the APU is 1,800 kg with the engine having an output of 75 hp and a maximum road speed of 15 km/h.

Sufficient fuel is carried for 3 hours of continuous operations and time taken to go from travelling mode into firing mode is 2 minutes. Steering is accomplished by using the two smaller roadwheels at the rear of the carriage.

Status

Production as required, but still being marketed. In service with the following countries:

Country	Quantity	Comment
Albania	100	
Bangladesh	40+	
Cambodia	n/avail	
Cameroon	12	
China	n/avail	
Congo, Democratic Republic	8	
Egypt	n/avail	local production (qv)
Iran	1,000	estimate
Korea, North	n/avail	local production, also three self-propelled models
Lebanon	n/avail	may be non-operational
Oman	12	understood to be held in reserve
Pakistan	250	including 27 in 1998 and 60 in 2000
Somalia	n/avail	may be non-operational
Sri Lanka	12	
Sudan	n/avail	includes Russian M-46
Tanzania	40	
Thailand	15	gift from China
UAE	20	reserve
Vietnam	n/avail	includes Russian M-46

Manufacturer

China North Industries Corporation.

UPDATED

NORINCO 122 mm howitzer D-30

Development/Description

The 122 mm howitzer D-30, produced by China North Industries Corporation (NORINCO) is the same weapon as the Russian D-30 and differs from the original only in detail. United Nations sources state that China exported no 122 mm howitzer D-30 artillery systems between 1992 and 2000.

The Chinese D-30 fires an HE projectile weighing 21.76 kg that is identical to the Russian original. However, the Chinese D-30 also fires a locally developed long-range HE projectile using a special charge that produces a maximum range of 18,000 m (standard range is 15,300 m).

The long-range HE projectile is longer and uses a more aerodynamic body. It has a wide knurl at the bottom of the body that serves to increase friction between the driving band and the body surface to prevent sliding and shedding of the drive band under high chamber pressures. The knurl on the driving band edge also improves the chamber gas sealing. A 122 mm cargo round with six layers of HEAT (High Explosive Anti-Tank) bomblets has been developed.

In 1990, NORINCO stated that it offered the following ammunition for its versions of the Russian D-30 as well as for the original weapon:

ERFB BB with a maximum range of 21,000 m
ERFB HB with a maximum range of 18,000 m
HE/HEAT cargo round with a maximum range of 15,000 m
HE/incendiary with a maximum range of 15,000 m
Smoke with a maximum range of 15,000 m.
Illuminating with a maximum range of 15,000 m

On-carriage fire-control instruments mounted on the Chinese D-30 include a Type PG-1M panoramic sight weighing 1.9 kg (with a magnification of ×4 and a 10° field of view), a Type OP4M-45 telescopic sight weighing 5 kg (with a magnification of ×5.5 and an 11° field of view) weighing 1.3 kg and a Type K-1 indicator. The latter weighs 1.3 kg and has a 10° field of view.

The D-30 is used as the main armament of the Type 85 self-propelled howitzer. For details refer to the Self-propelled guns and howitzers (tracked) section.

Late in 1999, during a major parade held in China, a new 122 mm self-propelled artillery system called the Type 89 was shown for the first time. This is understood to use the ordnance of the NORINCO 122 mm howitzer D-30. Details of the 122 mm Type 89 are given in the Self-propelled guns and howitzers (tracked) section under China.

Status

Production as required. In service with the Chinese armed forces. Quantities of this weapon may have been exported to the Middle East.

Manufacturer

China North Industries Corporation.

UPDATED

Specifications

Calibre: 122 mm
Muzzle brake: multibaffle
Weight: (firing) 3,200 kg
Length:
(travelling) 5.4 m
Width:
(travelling) 1.95 m
Height:
(travelling) 1.66 m
Elevation/depression: +70/–7°
Traverse:
(at elevation from –7 to +22°) 360°
(at elevation from +22 to +70°) 66° between movable trails 124° between fixed mount and left and right trails
Muzzle velocity:
(HE projectile) 690 m/s
Rate of fire: 6-8 rds/min
Range:
(max with standard ammunition) 15,300 m
(min with standard ammunition) 4,070 m
(special charge) 18,000 m
(ERFB – HB) 18,000 m
(ERFB – BB) 21,000 m
(direct fire) 1,000 m
Max towed speed:
(road) 60 km/h

NORINCO copy of the Russian 122 mm howitzer D-30 in firing position

NORINCO 122 mm howitzer Type 54-1

Development/Description

The NORINCO (China North Industries Corporation) 122 mm howitzer Type 54-1 is the Chinese version of the Russian 122 mm howitzer M1938 (M-30). It is virtually identical to the original apart from some small manufacturing expedients.

This was the first large calibre artillery weapon to be manufactured in China and may have been produced as early as 1953.

According to United Nations sources, between 1992 and 1998 the only export sales of the Type 54-1 were to Bolivia, which took delivery of 36 systems in 1992.

The 122 mm Type 54-1 howitzer is fitted with a Type 58 panoramic sight weighing 1.3 kg (with a magnification of ×3.7 and a 10° field of view) and a collimator sight Type 58 (with a 10° 40′ field of view).

The Type 54-1 uses a variable nine-charge propellant system and fires four types of projectile: HE, smoke, illuminating and incendiary. No details are available regarding the incendiary projectile and it is thought to be little used. Details of the other three 122 mm projectiles are as follows:

Type	HE	Smoke	Illum*
Weight:			
(projectile)	21.76 kg	22.55 kg	21.9 kg
(filling)	3.5 kg	3.65 kg	1.515 kg
Length:			
(projectile)	559 mm	551 mm	531 mm
Muzzle velocity	515 m/s	509 m/s	496 m/s
Max range	11,800 m	11,930 m	11,000 m†
Fuze	Liu-4	Yan-2	Shi-1

* burns for up to 30 seconds producing 450,000 candela
† air burst

A leaflet shell containing 1.1 kg of leaflets is also used. A 122 mm cargo round with six layers of HEAT bomblets has been developed to the prototype stage.

This howitzer is mounted on the NORINCO 122 mm self-propelled howitzer Type 54-1. For details refer to the entry in the Self-propelled guns and howitzers (tracked) section.

Specifications

Calibre: 121.92 mm
Barrel length:
(overall) 2.8 m
Muzzle brake: no
Recoil system: hydraulic buffer and hydropneumatic recuperator
Breech mechanism: hinged screw
Carriage: split trail
Shield: yes
Weight:
(total) 2,500 kg
Length:
(travelling) 5.9 m
Width:
(travelling) 1.975 m
Height:
(travelling) 1.82 m
(to axis of bore) 1.2 m
Elevation/depression: +60.5/–3°
Traverse: 49°
Rate of fire: 5-6 rds/min
Range:
(max range) 11,800 m
(min range) 5,350 m
Crew: 8
Towing vehicle: 6 × 6 truck
Max towing speed: 20/25 km/h

NORINCO 122 mm howitzer Type 54-1 with trails closed for travelling

Status

Production complete. The NORINCO 122 mm howitzer Type 54-1 is known to be in service with the following countries. As the weapon is virtually identical to the Russian 122 mm M1938 (M-30) it is possible that there are additional users.

Country	Quantity	Comment
Bangladesh	20	
Bolivia	36	delivered in 1992
China	n/avail	
Iran	100	estimate
Pakistan	400	
Sudan	n/avail	
Tanzania	100	
Zimbabwe	4	

Manufacturer

China North Industries Corporation.

UPDATED

NORINCO 105 mm light howitzer

Development/Description

Early in 1997, it was revealed that China North Industries Corporation (NORINCO) was marketing a 105 mm portable light howitzer which appears to be identical to the Italian Otobreda 105 mm Pack Howitzer covered in detail later in this section.

The Otobreda Division of Alenia Defence has confirmed that it supplied two examples of the 105 mm Pack Howitzer to China some time ago, although no follow-up orders were placed by the People's Liberation Army (PLA). According to United Nations sources there were no declared exports of the NORINCO 105 mm light howitzer between 1992 and 2000.

Like the Italian Otobreda 105 mm Pack Howitzer, the NORINCO 105 mm weapon can be used as a howitzer or a low-profile anti-tank weapon firing HEAT (High Explosive Anti-Tank) ammunition.

NORINCO-built 105 mm portable light howitzer deployed in anti-tank firing position 0021005

Specifications

Calibre: 105 mm
Barrel length: 1.478 m
Muzzle brake: multibaffle
Recoil: hydraulic buffer and helical recuperator
Breech mechanism: vertical sliding wedge
Carriage: split trail
Shield: yes (can be removed)
Weight: 1,310 kg
Length:
(travelling) 1.90 m
(howitzer role) 4.20 m (firing)
(anti-tank role) 4.70 m (firing)
Width:
(travelling) 1.50 m
(howitzer role) 2.53 m (firing)
(anti-tank role) 4.00 m (firing)
Height:
(travelling) 3.65 m (vehicle)
(howitzer role) 1.93 m (firing)
(anti-tank role) 1.550 m (firing)
Axis of fire:
(howitzer role) 1.07 m
(anti-tank role) 690 mm
Elevation/depression:
(howitzer role) +65/–5°
(anti-tank role) +25/–5°
Traverse:
(howitzer role) 18° (left and right)
(anti-tank role) 28° (left and right)
Rate of fire:
(howitzer role) 3-4 rds/min
(anti-tank role) 8 rds/min
Max range: 10,222 m
Min range: 2,200 m
Crew: 7

This has adequate performance against conventional steel but is considered ineffective against the latest advanced armour systems.

Firing a standard US High Explosive (HE) M1 projectile, a maximum range of 10,222 m can be achieved. NORINCO now has the capability to manufacture 105 mm ammunition for this and other Western weapons of this calibre including the HE M1 round.

The NORINCO 105 mm light howitzer can be quickly disassembled in 11 key parts which can be carried by men or horses and a crew of seven can disassemble the weapon in 3 or 4 minutes and assemble the weapon again in 4 to 5 minutes.

The NORINCO 105 mm light howitzer can be used in two configurations, high configuration as a normal howitzer and in the low-profile position for use as an anti-tank weapon.

Status

Production as required. There are no known export sales of this system. As far as it is known this system is not used by the PLA. United Nations sources list no export sales of this weapon from 1992 through to 1998.

Manufacturer

China North Industries Corporation.

UPDATED

NORINCO 100 mm anti-tank gun Type 86

Development/Description

The NORINCO (China North Industries Corporation) 100 mm Type 86 anti-tank gun is very similar to the 85 mm field gun Type 56 (see separate entry). It incorporates a new 100 mm ordnance and recoil system, but it may be a development of the NORINCO Type 73 anti-tank gun, which is believed to be a copy of the Russian 100 mm T-12 towed anti-tank gun which has a smooth barrel.

There are no known export sales of the Type 73 and no mention of the weapon has been made in any recent NORINCO sales literature.

The Type 86 ordnance is smoothbored and features a distinctive muzzle brake. The breech mechanism has a vertical sliding block and the sighting mechanism is mounted to the left of the gun. The carriage uses split trails with a castor wheel on the left trail leg to assist in bringing the gun into action. A shield with sloping sides is fitted with an opening for the sight on the left side.

Although the Type 86 is primarily an anti-tank weapon it can be used for the indirect fire role, even though the potential maximum range is limited by the maximum gun elevation of +38°. Maximum rate of fire is quoted as 8 to 10 rds/min.

A night vision device for sighting purposes is available.

Specifications
Calibre: 100 mm
Muzzle brake: slotted
Trails: split
Shield: yes
Weight:
(travelling) 3,660 kg
Length:
(travelling) 9.52 m
Width:
(travelling) 2.12 m
Height:
(travelling) 1.838 m
(centre of bore) 1.088 m
Axis of fire: 1.08 m
Recoil:
(normal) 1.05 m
(max) 1.18 m
Elevation/depression: +38/–4°
Traverse: 50° left and right
Max range:
(firing HE projectile) 13,654 m
Max towing speed: 50 km/h
Rate of fire: 8-10 rds/min

The Type 86 fires the following types of 100 mm fixed ammunition:
APFSDS Type 86 with a tungsten penetrator at a muzzle velocity of 1,610 m/s with an effective range of 1,800 m; when firing this round the maximum rate of fire is 4 to 5 rds/min
APDS Type 73 with a tungsten penetrator at a muzzle velocity of 1,500 m/s with an effective range of 1,700 m
HEAT Type 73 fin-stabilised projectile at a muzzle velocity of 1,010 m/s.
APFSDS Type 73 with tungsten alloy penetrator with a maximum direct fire range of 1,700 m
APFSDS Type 73 with the projectile having a length-to-diameter ratio of 13.3:1 and a maximum range in the direct fire role of 1,730 m
HE Type 73 fin-stabilised projectile at a muzzle velocity of 900 m/s with a maximum range of 13,700 m.

When firing these projectiles, the usual recoil length is between 1 and 1.14 m; maximum recoil length is 1.18 m.

Status

Production complete. In service with the People's Liberation Army. It is understood that Albania has 50 of these weapons in service.

Manufacturer

China North Industries Corporation.

VERIFIED

NORINCO 100 mm anti-tank gun Type 86 in firing position

NORINCO 100 mm anti-tank gun Type 86 in travelling mode

NORINCO 85 mm field gun Type 56

Development/Description

The Russian 85 mm Divisional gun D-44 was first supplied to China during the Korean War. China began production of this gun during the early 1960s under the designation Type 56. Some slight changes were made to the D-44 design to suit Chinese production methods, one of them being the location of a push-rod firing device in the centre of the elevating handwheel.

The NORINCO (China North Industries Corporation) Type 56 does not appear to use the Russian infra-red searchlight and sight system. It has been observed that some NORINCO Type 85 field guns have a castor wheel under the left trail leg to assist in bringing the weapon into action.

Ammunition currently available for the Type 56 includes HE, HEAT-FS and HESH. In the past, AP-T and

Specifications
Calibre: 85 mm
Barrel length:
(overall) 4.693 m
Muzzle brake: double baffle
Recoil system: hydraulic recoil buffer and hydropneumatic recuperator
Breech mechanism: semi-automatic vertical sliding block
Carriage: split tubular trail
Shield: yes
Weight:
(travelling) 1,750 kg
Length:
(travelling) 8.34 m
Width:
(travelling) 1.73 m
Height:
(travelling) 1.42 m
Elevation/depression: +35°/-7°
Traverse: 54°
Rate of fire: 15-20 rds/min
Range:
(max) 15,650 m
(HEAT) 970 m
Crew: 6-8
Max towing speed:
(road) 60 km/h
(cross-country) 15 km/h
Towing vehicle: 4 × 4 truck

HE-FRAG have been used. The only current projectile for which definite information is available is the Type 56 HEAT-FS. This one-piece round weighs 12.5 kg and is 990 mm long. The projectile weighs 7 kg and the TNT/RDX 50:50 shaped charge weighs 0.647 kg. The direct fire maximum range is 970 m at which it will penetrate 100 mm of conventional steel armour set at an angle of 65°. A Dian-1A fuze is used and the muzzle velocity is 845 m/s.

Optical equipment fitted includes a Type 58 panoramic sight with a magnification of ×3.7, Type 85 direct fire sight with a magnification of ×5.5 and a collimator Type 58.

Variants

Russia developed and fielded an auxiliary propelled version of the D-44 called the SD-44 but as far as is known this was never built in China for the People's Liberation Army or offered on the export market by NORINCO. Full details of the Russian D-44 are given separately in this section.

Details of the SD-44 were given in *Jane's Armour and Artillery 1994-95* pages 602 to 603.

Status

Production complete. The NORINCO 85 mm field gun Type 56 is used by the following countries:

Country	Quantity	Comment
Afghanistan	n/avail	unconfirmed user
Albania	61	
Cameroon	n/avail	
China	n/avail	
Congo, Democratic Republic	20	
Korea, North	n/avail	
Mozambique	12	
Pakistan	180	
Sri Lanka	12	
Tanzania	75-80	
Vietnam	n/avail	

Manufacturer

China North Industries Corporation.

VERIFIED

NORINCO 85 mm field gun Type 56 in firing position

NORINCO 82 mm Type W99 mortar

Development

China North Industries Corporation (NORINCO) is now offering on the export market an 82 mm towed rapid-firing mortar system called the Type W99. This was first revealed early in 2001, but it is not known as to whether this system is currently deployed by the People's Liberation Army.

In appearance the NORINCO Type W99 appears to be almost identical to the Russian 82 mm AM 2B9 Vasilek automatic mortar system. This entered service with the Russian Army around 1971 and has also been manufactured under licence in Hungary.

Description

The 82 mm Type W99 mortar weighs 650 kg and is normally operated by a crew of four or five. It has a minimum range of 800 m and a maximum range of 4,270 m. When deployed in the firing position it rests on a screw jack lowered under the forward part of the split trail carriage and the two trails, each of which are provided with a spade.

Compared with conventional muzzle-loaded 81 mm and 82 mm mortars the 82 mm Type W99 offers a number of operational advantages including a higher rate of fire and the gunner can select single shots or burst firing. It can be also used in direct fire role as well as the normal indirect fire role. A conventional mortar can only be used in the direct fire role.

According to NORINCO, their 82 mm Type W99 mortar is fitted with a standard sighting system and a firing calculator, although no details of the latter have been released.

It can fire a W99 82 mm high-explosive mortar bomb or modified versions of existing 82 mm mortar bombs with maximum muzzle velocity being 272 m/s. Clips of four 82 mm mortar bombs are loaded from the right and four rounds can be fired in 1.5 seconds.

The 82 mm ordnance has a traverse of 30° left and right with elevation from −1 to +85°. Breech loading is normally used at the lower angles of elevation but it can also be muzzle loaded.

The complete Russian 2K21 Vasilek mortar system comprises the actual 82 mm AM 2B9 Vasilek mortar, the 2F54 transport vehicle based on the GAZ-66 (4 × 4) 2,000 kg truck and 226 82 mm mortar bombs (of which 96 are fitted with fuzes ready for use in 24 clips)

Rear view of Chinese NORINCO 82 mm Type W99 mortar deployed in the firing position **2002**/0095233

Specifications

Calibre: 82 mm
Weight: 650 kg
Length: (travelling) 4.115 m
Width: (travelling) 1.576 m
Height: 1.18 m
Ground clearance: 0.26 m
Elevation/depression: +85/-1°
Traverse: 30° left and right
Max range: 4,270 m
Min range: 800 m
Rate of fire:
(maximum) 170 rds/min
(practical) up to 120 rds/min
Max towed speed:
(earth roads) 20 km/h
(asphalt) 60 km/h
Crew: 4-5

The 2B9 Vasilek is normally carried in the rear of the GAZ-66 and unloaded when required for action, using onboard handling gear. NORINCO has stated that it is also offering a complete system built around the 82 mm Type W99 mortar.

According to NORINCO, the Type W99 has a mean time between failure of around 360 rounds and can be deployed in an operational ambient temperature range from −40 to +50°C.

Status

Development complete. Production as required. It is not known as to whether this system is currently in service with the People's Liberation Army.

Manufacturer

NORINCO, China North Industries Corporation.

NEW ENTRY

Other Chinese towed artillery systems

In addition to the NORINCO (China North Industries Corporation) towed anti-tank guns, guns and howitzers previously covered, a number of other weapons have also been designed and built by NORINCO and brief details of these are given below. It is understood that none of the systems listed below is currently marketed by NORINCO.

NORINCO 152 mm Type 86 cannon

Chinese sources have stated that there is a Type 86 152 mm cannon system which was successfully developed in 1986 and has a maximum range of 30,000 m. Development work on this system started in the late 1950s but was suspended several times.

Final development work was carried out by No.123 and No.127 Factories. This weapon is of the towed type with a welded steel carriage and a 152 mm ordnance with a highly efficient muzzle brake. Types of ammunition fired include a hollow base extended range.

NORINCO 152 mm Type 54 howitzer

This is the Chinese-built version of the Russian D-1. It has a combat weight of 3,600 kg, rate of fire of 4 rds/min and a maximum range of 12,400 m.

NORINCO 122 mm Type 83 howitzer

This new light howitzer was first seen in public in 1984 during the National Day Parade and the following outline specifications are available:

Calibre: 121.92 mm
Length:
(travelling) 7.862 m
Width: 2.157 m
Weight:
(travelling) 2,668 kg
Elevation/depression: +65/–3°
Traverse:
(total) 65°
Rate of fire: 7-8 rds/min
Max range: 18,000 m with extended-range ammunition
Min range: 3,700 m

Recent Chinese sources have indicated that this system had a protracted development period and started again, in 1978, for the fourth time.

Development was carried out by the No 247 and 724 Factories and features subsequently incorporated include a movable equilibrator, integrated counter recoil system, semi-automatic breech block plus ammunition including a new single-base propellant system and new fuze. Also involved in the ammunition development was No 202 Research Institute.

There is a possibility that the NORINCO 122 mm Type 83 howitzer system is the updated Chinese version of the Russian D-30 covered earlier in this section. The 122 mm Type 83 howitzer is in service with the People's Liberation Army as its standard divisional towed artillery system.

NORINCO 122 mm Type 60 field gun

This is the Chinese-built version of the Russian D-74, it has a combat weight of 5,500 kg and a maximum range of 24,000 m.

Known export customers for the 122 mm Type 83 include:

Country	Quantity
Iran	100
Pakistan	200
Sudan	n/k
Zimbabwe	12

NORINCO 100 mm Type 73 anti-tank gun

This is possibly a copy of the Russian 100 mm T-12 anti-tank gun and has a maximum range of 1,700 m.

Status

In service with the People's Liberation Army. It is understood that no marketing of these weapons is now being carried out and that none of these weapons is currently in production.

Manufacturer

China North Industries Corporation.

VERIFIED

CROATIA

RH ALAN 122 mm Howitzer D-30 HR M94

Development/Description

This 122 mm howitzer D-30 HR M94 is almost identical to the Russian 122 mm D-30 towed weapon and is used for general fire support.

According to RH ALAN, its 122 mm howitzer D-30 HR M94 has the following modifications when compared to the original Russian D-30 towed weapon:

(a) The original slotted muzle brake has been replaced by a new double-baffle muzzle brake
(b) The fixed trail leg has been redesigned and parts of the braking system and electrical installations are fitted onto it
(c) On the prototype a sighting device with a scale division 360 (= 6,000 mils) was retained, whereas on the serial production weapons a sighting device with a scale division 360 (= 6,400 mils) has been fitted together with the mechanisms that enable simple use of firing tables according to NATO standards
(d) For handling of the weapon in the firing position, a manual hydraulic firing jack is fitted which enables the turning of the weapon with deployed trail legs
(e) The hydraulic recoil brake is of a single-chamber design with a hydraulic compensator placed inside the hydraulic brake cylinder. The standard D-30 had a compensator outside the hydraulic brake which RH ALAN claims to be unreliable.

The 122 mm ordnance consists of a monobloc barrel,with a high-efficiency double-baffle muzzle brake and collars are used for suspension on slides on the cradle. A semi-automatic vertical wedge-type breech block enables a high rate of fire to be achieved.

The counter-recoil system, which consists of a hydraulic brake, compensator and hydropneumatic recuperator, are accommodated in the cradle of the weapon.

The transition from travelling into the firing position and vice versa is readily and quickly made by means of a hydraulic jack mounted under the carriage. For manoeuvring in the firing position, an auxiliary wheel is attached to the towing lunette mounted under the muzzle brake.

A well trained crew can bring the weapon into action in 1.5 to 2.5 minutes. The 122 mm Howitzer D-30 HR M94 is normally manned by a six-man crew plus the section commander.

For towing, a vehicle fitted with a standard hook is used. This vehicle must have a payload of at least 3 tonnes and tow a trailer with a loaded weight of at least 3.30 tonnes.

122 mm howitzer D-30 HR M94 deployed in the firing position with stakes ready to be driven into ground
0021006

Specifications

Calibre: 122 mm
Barrel length: (with muzzle brake) 4.785 m (36 grooves with progressive angle of twist and chamber length of 594 mm)
Muzzle brake: double baffle
Breech mechanism: semi-automatic vertical wedge
Shield: yes
Weight:
(firing) 3,350 kg
(travelling) 3,440 kg
Length:
(travelling) 5.330 m
Width:
(travelling) 1.95 m
Height:
(travelling) 1.66 m
Axis of fire: 0.90 m
Elevation/depression: +70/–7°
Traverse:
(elevation of –5/+18°) 360°
(elevation of +18/+70°) 66°
(between fixed and mobile trails) 29° left and right
Rate of fire: 6 to 8 rpm
Muzzle velocity: HE projectile M76: 735 m/s (full charge); HE projectile TF462: 690 m/s (full charge); HE projectile TF462: 276-600 m/s (reduced charge, variable)
Max range:
(TF462 HE) 15,300 m
(M76 HE) 17,133 m
(ER 122 full charge) 16,530 m
(ER 122 super charge) 17,630 m
(ER BB 122 full charge) 18,850 m
(ER BB 122 super charge) 20,050 m

When the weapon is elevated from –5 to +18° it can be traversed through a full 360°, but when elevated from +18 to +70° traverse is limited to between two of the trails.

The 122 mm D-30 HR M94 fires ammunition of the separate loading type, for example projectile and charge with reduced, variable and full-type propellant charges. Maximum operating pressure of the ammunition must not exceed 2,600 bars and stated rate of fire is between 6 and 8 rds/min.

As well as the normal natures of 122 mm ammunition, a locally developed family of separate loading (projectile and charge) ammunition has been developed for this weapon and its foreign equivalent weapons.

Two standard natures of 122 mm projectile are manufactured in Croatia, the TF462 and the M76. The former has a maximum range of 15,300 m and the latter a maximum range of 17,133 m. In addition there is the M95 HE round that has a maximum range of 16,000 m with a full charge and 17,133 m with super charge.

In addition, there are two extended-range 122 mm projectiles, the ER 122 and the ER BB 122. The former is of the Extended Range Full Bore (ERFB) type with nubs and has a maximum range of 16, 530 m with full charge and 17,630 m with super charge.

The ERFB ER BB 122 is similar and has a base bleed unit attached, this has a maximum range of 18,850 m with full charge and 20,050 m with super charge.

Variants

According to RH ALAN, the carriage can also be fitted with 100 or 105 mm barrels and the former is in service in Croatia in the coastal defence role. Details of the latter are given in the Coastal artillery guns and missiles under Yugoslavia.

Status

Production as required. In service with Croatia.

Manufacturer

RH ALAN.

UPDATED

CZECH REPUBLIC AND SLOVAKIA

100 mm field gun M53

Development/Description

The 100 mm field gun M53 developed in Czechoslovakia in the early 1950s fulfils a similar role to the Russian 100 mm M1944 (BS-3) field gun and fires the same ammunition. Its ammunition is not interchangeable with the more recent Russian smoothbore 100 mm T-12 anti-tank gun. However, the M53 can be used in both the field and anti-tank roles and has a much longer range than the 100 mm M1944, 15,400 m, as its barrel can be elevated to +42° compared with the +20° of the M1944.

The 100 mm field gun M53 can be distinguished from the 100 mm M1944 (BS-3) by single roadwheels, a shield with a straight top and sides that slope to the rear rather than the M1944's dual roadwheels and a shield with a wave shaped top. The M53 can be fitted with the large diameter APN-3-5 infra-red night device over the rear part of the 100 mm barrel.

There are no known plans to upgrade these weapons to extend their operational lives. The 100 mm field gun M53 is also referred to as the 100 mm field gun K53.

Specifications

Calibre: 100 mm
Barrel length:
(overall) 6.735 m
Muzzle brake: double baffle
Recoil system: hydraulic buffer and hydropneumatic recuperator
Breech mechanism: horizontal sliding wedge
Carriage: split trail with castor wheel on each
Shield: yes
Weight:
(travelling order) 4,280 kg
(firing position) 4,210 kg
Length:
(travelling) 9.1 m
Width:
(travelling) 2.36 m
Height:
(travelling) 2.606 m
Axis of bore:
(at 0°) 1.25 m
Ground clearance: 0.35 m
Track: 1.98 m
Tyres: 11.00 × 22
Elevation/depression: +42/–6°
Traverse:
(total) 60°
Rate of fire: 8-10 rds/min
Max range: 21,000 m
(effective anti-tank) 1,000 m
Muzzle velocity:
(HE) 900 m/s
(HVAPDS-T) 1,415 m/s
Crew: 6
Towing vehicle: Tatra-138 (6 × 6) truck

Status

Production complete. In service only with the Czech Republic (75) and Slovakia (52). According to United Nations sources the Czech Republic supplied 26 of these weapons to Latvia in 1995 under the designation K53.

Manufacturer

Former Czechoslovak state factories.

UPDATED

100 mm field gun M53 in firing position
(via Franz Kosar)

85 mm field gun M52

The 85 mm field gun M52 is the Czech equivalent of the Russian 85 mm divisional gun D-44 (covered later in this section) and fires the same ammunition. A slightly modified version of the M52 is designated the M52/55, and both can be fitted with the infra-red night device.

The 85 mm field gun M52 is very similar in appearance to the Czech 100 mm field gun M53 (covered in the previous entry), which also has a double-baffle muzzle brake but on which the sides of the shield slope to the rear and the top of the shield is straight rather than wave shaped in the case of the M52. The trails of the M53 also have castor wheels to assist in bringing the weapon into and out of action.

The 85 mm field gun M52 is based on a conventional split trail carriage which is not fitted with a castor wheel to assist in bringing the weapon into action.

The M52 fires the same family of ammunition as the Russian 85 mm D-44 weapon which is covered in detail later in this section.

Specifications

(M52; data in square brackets relates to M52/55 where different)
Calibre: 85 mm
Barrel length:
(overall) 5.07 m
Muzzle brake: double-baffle
Recoil system: hydraulic recoil buffer and hydropneumatic recuperator
Breech mechanism: semi-automatic vertical sliding wedge
Carriage: split trail
Shield: yes
Weight:
(travelling order) 2,130 [2,168] kg
(firing position) 2,095 [2,111] kg
Length:
(travelling) 7.52 m
Width:
(travelling) 1.98 m
Height:
(travelling) 1.515 m
Axis of bore: (at 0°) 1.135 m
Ground clearance: 0.35 m
Track: 1.63 m
Tyres: 8.25 × 20
Elevation/depression: +38/–6°
Traverse:
(total) 60°
Rate of fire: 20 rds/min
Range:
(max) 16,160 m
(effective, FRAG) 925 m
(effective, HVAP-T) 1,150 m
Crew: 7
Towing vehicle: Praga V3S (6 × 6) truck
Max towed speed:
(road) 50 km/h
(cross-country) 10 km/h

One user of the M52, Austria, uses a Belgian-designed HEAT-T round called the Leuchtspur-Hohlgranate 72 which has a muzzle velocity of 820 m/s and will penetrate 350 mm of conventional steel armour.

The effectiveness of the M52 weapon against the more recent MBTs over their frontal arc is very doubtful.

Status

Production complete. In service with Austria (250), Czech Republic and Slovakia. Recent information has indicated that these weapons may have been phased out of front-line service with all three countries and placed in reserve. This weapon has not been built for many years and is no longer being marketed. These weapons have also been referred to as the 85 mm field gun M52/M55.

Manufacturer

Former Czech state factories.

UPDATED

85 mm field gun M52 deployed in firing position (Austrian Army)

EGYPT

Factory 100 artillery systems

Adu Zaabal Engineering Industries Company, also known as Factory 100, has reverse-engineered the Russian 122 mm D-30 towed howitzer, Chinese 130 mm field gun Type 59 and the Russian twin 23 mm ZU-23-2 towed anti-aircraft gun.

The first 122 mm D-30 was completed in Egypt early in 1984 with Royal Ordnance Nottingham providing some part-machined components which were completed in Egypt.

Factory 100 has also been closely involved with Royal Ordnance Nottingham in the upgunning of the Russian-supplied T-55 MBT with the Royal Ordnance 105 mm L7A3 rifled gun.

Additional details of this programme are given under Egypt in the MBTs and medium tanks section.

Egyptian 130 mm M59-1M guns in firing position (Christopher F Foss)

Egyptian 122 mm D-30-M towed howitzer in travelling position (Christopher F Foss)

Specifications

Original designation	**D-30**	**Type 59**
Egyptian designation	D-30-M	M59-1M
Calibre	122 mm	130 mm
Barrel length	4.27 m	7.15 m
Max pressure	2,500 kg/cm²	3,150 kg/cm²
Muzzle brake	multibaffle	double baffle
Breech mechanism	semi-automatic vertical sliding wedge	horizontal sliding wedge
Shield	yes	yes
Weight		
(travelling)	3,210 kg	6,300 kg
(firing)	3,150 kg	6,100 kg
Length		
(travelling)	5.4 m	10.8 m
Width		
(travelling)	1.95 m	2.42 m
Height		
(travelling)	1.66 m	2.75 m
Ground clearance	0.325 m	0.3 m
Elevation/depression	+70/−7°	+45/−2.5°
Traverse		
(total)	360°	58°
Rate of fire	6-8 rds/min	6-8 rds/min
Max range	15,300 m	27,150 m
Min range	1,170 m	1,170 m
Muzzle velocity	690 m/s	930 m/s
Time into action	90-210 s	180 s
Time out of action	90-210 s	180 s
Crew	7	9
Max towing speed	60 km/h	60 km/h

The Egyptian defence industry also manufactures ammunition for these weapons. For the 122 mm system natures include HE, HEAT, illumination and smoke while for the 130 mm natures include HE, armour-piercing, smoke and illuminating.

SP122

The complete ordnance and elevating system of the Egyptian-built D-30 is fitted to the SP122 self-propelled artillery system. This was developed by United Defense LP of the US under contract to the Egyptian Army. Details are given in the Self-propelled guns and howitzers (tracked) section under Egypt. A total of 124 systems have been purchased with final deliveries taking place late in 2000.

T-54/T-55 with D-30

Early in 1999, it was revealed that a new 122 mm self-propelled artillery system was entering production in Egypt. This consisted of a modified T-54/T-55 tank chassis fitted with a new all welded steel turret armed with the locally produced 122 mm D-30. Prime contractor for the chassis was to be Factory 200 (Egyptian Tank Plant) while prime contractor for the turret and D-30 was to be Factory 100.

Egyptian production of 155 mm 155 GH 52

In August 1999, Patria Vammas Oy of Finland signed an agreement with the Arab Republic of Egypt under which Finland would transfer technology for the local production of the 155 mm 155 GH 52 weapon system. This model is not fitted with an APU and has the designation of the 155 mm EH52. Finland delivered the first complete system to Egypt in 2000.

Status

Production as required. In service with Egyptian Army. Egypt has recently supplied Bosnia-Herzegovina with 12 D-30-M and 12 M59-1M weapons. According to United Nations sources Egypt delivered six D-30 weapons to Rwanda in 1992.

Manufacturer

Abu Zaabal Engineering Industries Company.

UPDATED

FINLAND

Patria Vammas 155 mm 155 GH 52 APU

Development

In early 1994, it was revealed that the Finnish company VAMMAS, previously known as Tampella and more recently Patria Vammas, had completed the prototype of a new 155 mm 52 calibre towed artillery system fitted with an Auxiliary Propulsion Unit (APU).

This weapon is known as the 155 GH 52 APU and has been developed as a private venture together with the Finnish defence forces. It is a further development of the earlier 155 mm M-83 towed howitzer which has been produced in two versions, neither of which has an APU, which are covered in a separate entry.

Development of the new 155 GH 52 APU commenced in 1990 with the first prototype being completed the following year. Firing trials were successfully completed in 1993.

It would be possible to upgrade earlier weapons to the new 155 mm 52 calibre version.

Following trials with prototype weapons under the designation of the KX, the Finnish Army placed an order for a quantity of weapons under the designation of the 155K98 with first deliveries taking place in 1998.

Late in 2000, another batch of 155 mm 155 GH 52 APU systems was ordered for the Finnish Defence Force, which brought the total number of systems up to 36 units.

This figure of 36 consists of 27 + 9 with final deliveries under current contracts due late in 2002. The first regiment was fully operational by 2001 and has a total of 18 guns in three batteries each of six guns, which can be further split into two troops each of three guns.

In the FDF the 155K98 is deployed as the Corps level artillery system with a total of three regiments expected to be eventually formed with a total of 54 guns.

Current production models of the 155K98 for the FDF are being fitted with a TALIN (Tactical Advanced Land Inertial Navigation) system, a modern version of the Honeywell Modular Azimuth Position System (MAPS).

MAPS was originally developed for the US Army M109A6 with over 2,000 built for the home and export markets. The installation of TALIN on the 155K98 reduces the time taken to come into action time as well as increasing the accuracy of the weapon.

In August 1999, Patria Vammas Oy signed an agreement with the Arab Republic of Egypt under which Finland would transfer technology for the local production of the 155 mm 155 GH 52 which has the local designation of the 155 mm EH52 towed howitzer.

Close-up of Patria Vammas 155K98 field gun of the Finnish Defence Force fitted with the TALIN (Christopher F Foss) **2002**/0130007

Patria Vammas 155K98 field gun of the Finnish Defence Force in travelling configuration with ordnance locked in position over closed trails (Christopher F Foss) **2002**/0130009

Description

The 155 GH 52 APU weapon consists of five key components, barrel assembly, top carriage, bottom carriage, sighting devices and the APU.

The barrel assembly consists of the barrel, muzzle brake, fume extractor, breech piece, primer housing lock, firing safety lever, breech block safety lever and flick rammer safety lever.

The top carriage consists of the saddle, elevating mechanism, traversing mechanism gearbox, equilibrators, semi-automatic mechanism cocking device, layers pressure trigger, cradle, buffer, buffer replenisher, recuperator and the flick rammer.

The bottom carriage consists of the gun jack, firing platform, central structure, trail, trail support cylinder, walking beam, wheel, dolly wheel, trail connecting lock, trail support plate, towing eye, cabin, APU, mechanical control of hydraulic units, electric control of hydraulic units, external output for accessories and manual hydraulic pump.

Sighting devices covers indirect and direct firing. The sighting devices for indirect firing include elevation setting mechanism, pitch levelling mechanism, cross-levelling mechanism, telescope mount holder and panoramic telescope illuminating equipment.

The panoramic telescope can also be used for direct firing but a special telescope, an elbow telescope sight, is normally used for direct firing. There is a mount for the elbow telescope on the gun layer's side of the gun.

The APU includes the control devices, steering panel, gauge panel, electric control of hydraulic units, mechanical control of hydraulic units, steering joystick, external output for accessories and manual hydraulic pump.

The 155 mm 52 calibre ordnance has a 23 litre chamber, semi-automatic horizontal sliding wedge breech mechanism, pneumatic rammer located on the left side to ram projectiles and charges at all angles of elevation and a single-baffle muzzle brake. A fume extractor is fitted as standard. The 23 litre chamber conforms to the Quad Ballistic MoU (France, Germany, Italy, UK and USA).

The recoil brake is hydraulic while the recuperator is pneumatic. Recoil length depends on the charge and elevation but maximum recoil stroke is 1,500 mm.

The manual elevating and traverse mechanism is the same as that fitted to the earlier Patria Vammas M-83 system and is fitted on the left side of the carriage together with the direct and indirect sights. The split trail carriage is of a new design.

When deployed in the firing position, the forward part of the carriage rests on a hydraulically operated firing platform under the carriage, with each of the trails being provided with a spade, each of which is fitted with pegs to avoid the need to dig them in.

The weapon takes 2 minutes to come into action and is normally operated by a nine-man crew although two

Specifications
Calibre: 155 mm
Length of ordnance: 8.06 m/52 calibre
Chamber volume: 23 litres
Pressure: 480 MPa
Max working pressure: 350 MPa
Weight: 13,500 kg
(at towing eye) 1,000 kg
Muzzle brake: single baffle
Recoil brake: hydraulic (automatic adjustment)
Recuperator: pneumatic
Breech mechanism: automatic horizontal wedge
Carriage: split trail
Shield: no
Length:
(travelling) 10.90 m
(self-propelled mode) 11.00 m
Width:
(travelling) 2.82 m
Height:
(travelling) 2.25 m
Track: 2.40 m
Wheelbase: 1.45 m
Ground clearance: 0.40 m
Elevation/depression: +70/–5°
Traverse: 35° left and right
Rate of fire:
(max) 10 rds/min
(normal) 8 rds/min
(burst) 3 rds in 12 s
Range:
(standard ammunition) 27,000 m
(base bleed unit) 41,300 m
Crew:
(normal) 1 + 8
(min) 2
Max driving speed:
(two-wheel drive) 15 km/h
(four-wheel drive) 7.5 km/h
Max towing speed: 80 km/h

Patria Vammas 155K98 field gun of the Finnish Defence Force deployed in the firing position (Christopher F Foss) ***2002***/0130008

men can operate the system for a short period of time.When travelling, the 155 mm 52 calibre ordnance is traversed through a full 180° and locked in position over the closed trails. The travelling lock for the 155 mm ordnance is located on the right trail.

The carriage has four rubber-tyred roadwheels, two either side, for high-speed towing with the suspension being of the walking beam type for improved cross-country mobility. In addition, there is a single dolly wheel on each trail leg to assist in bringing the weapon into action.

Mounted on the forward part of the carriage is the APU, a Deutz air-cooled turbocharged diesel developing 104.6 hp, coupled to a hydrostatic transmission which in turn powers the four motors located one at each wheel station.

The APU gives the system a maximum road speed of 15 km/h with two of the four wheels engaged or 7.5 km/h with all four wheels engaged. The driver is seated on the forward part of the carriage on the left side and steers the system using a small joystick. The fuel tank holds 5 litres of diesel fuel.

The installation of the APU enables the 155 GH 52 APU to be taken quickly out of action and deployed into positions that are not normally accessible by the prime mover.

A burst rate of fire of three rounds can be fired in 12 seconds with a normal rate of fire being 10 rds/min.

The FDF uses a local version of the French Giat Industries LU111 high-explosive projectile under the designation 155 tkr88 and this has a maximum range of 23.8 km without a base bleed unit.

The FDF have also adopted and are taking into service the German Rheinmetall 155 mm DM 642 cargo projectile that has 63 bomblets each with a self-destruct mechanism.

Patria also market two of their own 155 mm projectiles, the 155 ERFB-N (Extended Range Full Bore – Nubbs) which has a maximum range with base bleed unit fitted of 42.5 km and the 155 mm ERFB-S (streamlined) which has a maximum range of 40.5 km with base bleed.

Variants
There is also a 155 mm 45 calibre version, without an APU, which is called the 155 GH 45. This remains at the prototype stage.

Status
Production. In service with the Finnish Army. Now being manufactured under licence in Egypt.

Manufacturer
Patria Vammas Oy.

UPDATED

Patria Vammas 155 mm M-83 howitzer

Development/Description
This weapon was originally developed and produced by Tampella which then became known as VAMMAS Ordnance and more recently Patria Vammas.

The Finnish Army designation for this system is the 155 K 83. United Nations sources state that Finland did not export any 155 mm artillery systems between 1992 and 1998. Between 1999 and 2000 Finland exported only one 155 mm artillery system. This was an 155 mm/52 calibre system to Egypt based on the Patria Vammas 155 mm 155 GH 52 APU. This is called the EH52 as it is not fitted with an APU.

Description
The 155 mm M-83 howitzer has a split trail carriage and can be fitted with a 39 or 45 calibre ordnance which has a distinctive 'T' type muzzle brake. As an alternative to the normal 155 mm ordnance, the gun can be supplied with a 122, 130 or 152 mm barrel allowing Chinese/Russian types of ammunition to be fired rather than 155 mm NATO-type ammunition.

The semi-automatic horizontal sliding breech mechanism and a special loading mechanism enable a high rate of fire to be achieved with obturation being obtained by a steel ring.

The special loading system is operated by compressed air, with the cylinder for the latter located on the right side of the mount. The sights, elevation and traverse wheels are located on the left side of the mount.

Patria Vammas 155 mm M-83 howitzer deployed in firing position

Specifications

Calibre	**155 mm**	**155 mm**
Calibre length	39	45
Barrel length	5,990 mm	7,100 mm
Muzzle brake	single baffle	single baffle
Breech mechanism	semi-automatic horizontal wedge	semi-automatic horizontal wedge
Obturation	steel ring	steel ring
Carriage	split trail	split trail
Shield	no	no
Weight in firing position	9,500 kg	9,500 kg
Elevation/depression	+70/–5°	+70/–5°
Traverse	45° (left/right)	45° (left/right)
Rate of fire		
(burst)	3 rds/12 s	3 rds/12 s
(normal)	10 rds/min	10 rds/min
Max range	30,000 m	39,600 m
Max towing speed	80 km/h	80 km/h

Patria Vammas 155 mm M-83 howitzer in travelling configuration

According to Patria Vammas, the weapon is very easy to handle with only a few men required to bring the weapon into firing or travelling configuration. When in the travelling configuration the ordnance is traversed through 180° and locked in position over the closed trail legs. When in the firing position the weapon is supported on a firing jack under the forward part of the carriage and on the two trails, each of which is fitted with a spade. Steel stakes are provided for the spades and these are carried in racks on the inside of each trail.

The 155 mm M-83 fires the standard NATO M107 HE projectile as well as hollow base and two types of base bleed projectile, one type with nubs and the other without.

The design of the charge system is optimised so that range areas cover each other in low- and high-elevation angles. Compared to the standard NATO charge system (at least four different charges), the Patria Vammas system consists of three different charges, incremental charge and two types of full charges.

The make-up of the incremental charge is similar to NATO charges M4A2 and M3A1, including four increments, zones 5, 6, 7 and 8. Full and top charges are similar to NATO M2 and M11, zones 9 and 10. The latter is only used in the 45 calibre gun.

Some 155 mm M-83 systems have been observed fitted with a muzzle velocity radar measuring system over the ordnance.

Variants

Details of the latest Patria Vammas version with a 155 mm 52 calibre ordnance and an APU are given in the previous entry. This system is designated the Patria Vammas 155 mm GH 52 APU and in late 1997 was ordered by the Finnish Army with first deliveries taking place early in 1998.

Status

Production as required. In service with the Finnish Army. There has been no recent production of this weapon.

Manufacturer

Patria Vammas Oy.

UPDATED

Other Finnish towed artillery systems

In addition to the towed artillery systems of local design and manufacture previously covered, the Finnish Army also uses a number of other systems, some of which date back to the Second World War.

Specifications

Designation	**152 H 88-31**	**152 H 88-40**	**152 H 37A**
Calibre	152.4 mm	152.4 mm	152.4 mm
Barrel length	n/avail	n/avail	4.295 m
Muzzle brake	single baffle	single baffle	multibaffle
Breech mechanism	hinged screw	horizontal sliding wedge	hinged screw
Carriage	split trail with limber	split trail	split trail with limber
Shield	yes	yes	yes
Weight			
(travelling)	7,760 kg	6,790 kg	8,330 kg
(firing)	6,940 kg	6,770 kg	7,510 kg
Elevation/depression	+42/–6°	+45/0°	+57/–3°
Traverse			
(total)	50°	58°	53°
Range	using 152 H 37 projectiles (weight 43.7 kg) and charges, maximum muzzle velocity is 655 m/s, maximum range is 16.4 km and rate of fire is 5 rds/min		
Max towing speed	80 km/h	80 km/h	80 km/h

152 mm M1937 (ML-20) gun-howitzer

This was developed in Russia with Finland capturing 37 weapons during the conflict with Russia. A further 27 were purchased from Germany during the Second World War. During 1986-89 these were fitted with a new limber and driving gear, pneumatic brakes, driving lights, fitting for muzzle velocity radar above the 155 mm ordnance plus trail spades and means for sinking these into the ground. The modernised weapon is called the 152 H 37 A whereas the unmodernised weapon was known as the 152 H 37. All the conversion work on this and other artillery systems in Finland has been carried out by the Patria Vammas Oy Refurbishing and Modernisation Division which rebuilds and modernises armoured vehicles and artillery systems for the Finnish Army.

122 mm Corps Gun M1932/37 (A-19)

Similar modifications have been carried out to these Russian-designed weapons and a 152.4 mm ordnance has also been fitted. This has a larger chamber than the 152 H 37 and is also longer. It enables the gun to achieve a longer range if new projectiles and charges are introduced. The new designation of this weapon is the 152 H 88-31 while the old one was the 122 K 31 in Finnish service. The modernised weapons have been transferred to the field artillery; they had previously been used as mobile coastal artillery. A total of 21 122 K 31 guns survived the Second World War, having been captured from the Russian Army.

105 mm sFH18 howitzer

These were supplied by Germany and designated the 105 H 40 in Finnish Army service. A total of 47 were modernised to the 152 H 88-40 configuration. These changes included a new barrel, removal of the limber, new trail spades and earth pins modelled after the 155 K 83, a shield with brush-holster and driving lights. More recently a number of small changes has been made to the cradle and recoil system. The carriage equipment and protective covers have been changed and the breech mechanism modified.

152 mm 2A36 gun

This is the Russian 152 mm 2A36 (NATO designation M1976) weapon and has the Finnish designation of 152 K 89.

105 mm H 61-37 gun

A total of 54 of these were in service but late in 1997 it was stated that 19 were to be transferred to Estonia free of charge to form their first artillery unit. According to the United Nations, Estonia received 6 weapons in 1997 and 13 weapons in 1998.

Self-propelled artillery

The Finnish Army has boosted its self-propelled capability by the acquisition of 72 ex-East German Army Russian 122 mm 2S1 self-propelled artillery systems (local designation PsH 74) and 18 Russian 152 mm 2S5 (local designation Telak 91). In the Spring of 2000 the Finnish Defence Force evaluated three competing 155 mm self-propelled artillery systems to meet future requirements. Late in 2000, the Finnish Defence Force decided not to purchase a new self-propelled artillery system and instead purchased additional 155 mm 155 GH 52 APU towed artillery systems from Patria Vammas. A total of 36 155 mm GH 52 APU have been ordered by Finland.

UPDATED

FRANCE

Giat Industries 155 mm towed gun TR

Development

This 155 mm Towed Gun *(Le Canon de 155 mm Tracte)* has been developed to meet the requirements of the French motorised infantry divisions and rapid action forces as the replacement for the older 155 mm howitzer Model 50. The weapon was shown for the first time at the 1979 Satory Exhibition of Military Equipment. Following trials with prototype systems, a preproduction batch of six weapons was delivered by Giat Industries facility at Bourges for troop trials from December 1987. The main production run commenced in 1989 and a total of 105 weapons have been delivered to the French Army.

Export sales of the 155 mm towed gun have been made to at least two countries, Cyprus (12) and Saudi Arabia (28).

The Giat Industries 155 mm towed gun TR was deployed to Saudi Arabia and took part in Operation Desert Storm, the liberation of Kuwait, in the spring of 1991.

Although production of this weapon was completed several years ago it can be placed back in production if required.

Production of components for the 155 mm towed gun TR was undertaken by various Giat Industries facilities but system's integration was undertaken at Bourges where all French artillery and tank barrels are produced.

Giat Industries 155 mm TR gun with 52 calibre barrel deployed in firing position

Specifications
(with 39 calibre barrel)
Calibre: 155 mm
Barrel length: 6.2 m
Muzzle brake: double baffle
Recoil: hydropneumatic
Breech mechanism: horizontal wedge
Carriage: split trail
Shield: no
Weight:
(travelling order) approx 10,750 kg
Length:
(travelling, barrel over trails) 8.75 m
(firing) 10 m
Width:
(travelling) 3.09 m
(firing) 8.4 m
Height:
(firing) 1.65 m
Ground clearance:
(travelling) 0.4 m
Elevation/depression: +66/–6°
Traverse:
(total) 65° (27° left, 38° right)
Rate of fire: 3 rounds in first 18 s
6 rds/min for 2 min
Range:
(standard projectile) 24,000 m
(base bleed projectile) 32,000 m
Tube life: 3,000 rounds
Crew: 7
Towing vehicle: Renault TRM 10 000 (6 × 6) truck (French Army service)
Autonomous speed: 9 km/h
Gradient: 60%
Fording:
(without preparation) 1.00 m
(with preparation) 1.50 m
Towing speed (TRM 10 000):
(hard roads) 80 km/h
(dirt roads) 30 km/h

Description

A crew of eight is required for the tractor and gun: commander; layer/gunner who also drives the gun in the self-propelled mode; loader who operates the loading device during firing, controls the raising and lowering of the gun and operation of the trails when moving into and out of action; two men who are responsible for the preparation of the propelling charges and supplying the gun; two loaders responsible for the preparation of the projectiles and supplying the loading mechanism; and finally the tractor driver. If the tasks are divided differently, the crew can be reduced to seven, six of whom serve the gun. If required, the gun can be brought into and out of action by a crew of three.

The 155 mm barrel has a double-baffle muzzle brake which when travelling is swung through 180° and secured in position over the closed trails.

Mounted on the front of the carriage is a 39 hp engine which drives three hydraulic pumps, one for each of the two roadwheels and one to provide power for functions such as elevation, traverse, raising the suspension, trail wheel jacks and the loading mechanism. Travelling under its own power the gun has a maximum road speed of 9 km/h, can climb a gradient of 60 per cent and ford to a depth of 1 m without preparation. With preparation it can ford to a depth of 1.2 m.

In French Army service the 155 mm TR is towed by a Renault TRM 10 000 (6 × 6) truck which carries the crew of eight and 48 rounds of ammunition. This combination has a maximum road speed of 80 km/h and a road range of 600 km, but it can be towed by various other types of prime mover.

On arriving at the battery position, the weapon is first uncoupled from the prime mover and then the barrel is swung from its travelling position through 180° into its firing position. The wheels are raised off the ground so that the weapon rests on its firing jack under the carriage. The trails are then swung apart, which is a simple operation as castor wheels are fitted on each trail. When opened out, the small wheels are raised clear of the ground and the trails staked into position. The maximum time taken to bring the weapon into battery is 90 seconds and it can be brought out of battery in the same time due to the hydraulic system.

The layer is seated on the left of the mount and has hydraulic aiming controls. The sights, which can be used by day and night, are the GA 81 goniometer, indirect sight and the direct fire telescopic sight. After indicating the elevation angle on a graduated drum, the barrel position is checked using the level shown on the elevating mechanism's elevation meter. The ranging (GR) and firing (GT) angles can be read directly through the goniometric sight. The direct fire telescope can be used to engage targets up to a maximum range of 2,000 m. In an emergency the gun can be fired without the trails being deployed.

The automated hydraulic system for ramming the projectile into the ordnance operates at all angles of elevation, reduces crew fatigue and ensures all projectiles are rammed at the same pressure, giving a stable muzzle velocity. The ramming system permits a high rate of fire of three rounds in the first 18 seconds and 6 rds/min thereafter.

Due to its total interoperability, the Giat Industries 155 mm towed gun TR can fire projectiles with the French or NATO propelling system to the following ranges:

HE hollow base to 24,000 m
HE base bleed to 30,000 m
HE 56/59 to 23,000 m
HE M107 to 18,500 m
HE ERFB BB to 32,900 m

155 mm HE hollow base shell Type F1 (HE 155 F1) or Type F2 (LU 111 HB) with the projectile weighing 43.25 kg (containing 8.83 kg of HT 50/50), a muzzle velocity of 830 m/s and a maximum range of 24,000 m

155 mm phosphorus smoke shell HB Type F2A (OFUM 155 F2A) with the projectile weighing 43.15 kg (containing 8.7 kg of white phosphorus), a muzzle velocity of 830 m/s and a maximum range of 24,000 m

155 mm training shell Type F1 (OX 155 F1) with the projectile weighing 43.5 kg (containing 1.2 kg of black powder), a muzzle velocity of 830 m/s and a maximum range of 24,000 m

155 mm illuminating shell Type F1 (OECL 155 F1) with the projectile weighing 43.5 kg, a muzzle velocity of 760 m/s and a maximum range of 19,250 m

155 mm anti-tank mine shell Type H1 (OMI 155 H1) with the projectile weighing 46 kg. This shell contains six anti-tank mines each weighing 0.55 kg and although fired at a maximum of charge 5, no maximum range has yet been announced

155 mm HE BB shell Type H2 (BBE 155 H2) with the projectile weighing 43.5 kg (containing 10 kg of HT 50/50), a muzzle velocity of 830 m/s and a maximum range of 29,000 m

Giat Industries cargo round projectile OGRE 155 G1 containing 63 bomblets with a maximum range of 28,500 m

The TR will also fire the more recent Giat Industries/Bofors Defence BONUS 155 mm artillery projectile which contains two top attack sub-munitions. This is now in production for France and Sweden.

The TR has been provided with a number of emergency systems to allow continued use in the event of damage. These are a battery for the hydraulics which makes it possible to fire a further six rounds, or take the weapon out of action if the thermal engine stops, connection of the gun to the tractor's hydraulic system powering all of the gun's firing operations and coming into and out of action (or two guns can be connected together), manual pumps for elevation and traverse and manual mechanisms for traversing, opening and closing the breech and firing.

Variants

155/45 gun

A prototype of this was completed and tested but marketing is now being concentrated on the 155 mm 52 calibre system.

155/52 gun

In 1990, Giat Industries announced that it had completed the prototype of a 155 mm towed gun TR with a 52 calibre ordnance that enables a range of 30,000 m to be achieved with basic projectiles and 42,000 m with ERFB-BB projectiles. This conforms to the new NATO ballistic MoU and was fired for the first time early in 1990. At the present time there is no formal French Army requirement for this system, although in the future it is possible that it may want to upgrade its current 155/39 calibre guns to the 155/52 configuration.

The main modifications include replacement of the barrel by the new 52 calibre ordnance, modified towbar, modified rammer and some modifications to the recoil system.

The 155/52 has similar specifications to the 155/39 except that it is slightly longer in the firing position and weighs 11,000 kg.

The complete upper part of the 155/52 calibre weapon forms the basis of the private venture Giat Industries CAESAR 155 mm (6 × 6) self-propelled artillery system which was first shown in 1994. Details of this are given in the Self-propelled guns and howitzers (wheeled) section. Late in 1999 the French Minister of Defence stated that the French Army had placed an order for five 155 mm CAESAR weapons fitted with a 155 mm/52 calibre ordnance.

Status

Production as required. In service with:

Country	Quantity
Cyprus	12
France	105
Saudi Arabia	28

Manufacturer

Giat Industries, Centre de Bourges.

UPDATED

Giat Industries 155 mm towed gun TR of the French Army (Rupert Pengelly)
2001/0105754

Giat Industries 155 mm towed gun TR of the French Army being prepared for towing (Rupert Pengelly)
2001/0105753

Giat Industries 155 mm howitzer M114

Development/Description

In 1987, Giat Industries showed the prototype of an upgraded 155 mm M114 towed howitzer. This was developed as a private venture and aimed specifically at the export market. As far as it is known, there have been no sales by Giat Industries of the 155 mm M114 upgrade package.

Giat Industries can either upgrade existing M114s in France, or supply kits enabling users to upgrade M114s in their own facilities. By 1990, development of the M114 was complete and it had successfully completed exhaustive firing trials in France. By early 2001, there were no known sales of this export package by Giat Industries.

The modification consists of the following:

(1) Replacement of the original barrel with the proven Giat Industries 39 calibre NATO barrel 155 TR (ordnance, muzzle brake, sleeve and breech), without modification of the original gun cradle
(2) Replacement of the equilibrators by new gas compensator equilibrators
(3) Addition of a loading tray (optional)
(4) Slight modification of the elastic couplings
(5) Advancing the firing jack by 20 cm
(6) Fitting of castor wheels to each trail leg to reduce time into and out of action (optional)
(7) One single-barrel handwheel for aiming in both elevation and traverse.

The original M114, developed over 50 years ago, has a crew of 11, whereas the upgraded M114 has a seven-man crew with time into action being 3 minutes compared to 6.

According to Giat Industries, the M114 has the following ranges:
18,500 m with US M107 HE projectile
24,000 m with Giat HE LU 211 HB projectile
30,400 m with Giat HE LU 211 BB projectile
32,900 m with ERFB-BB projectile.

The 155 mm M114 conforms to NATO 155 mm/39 calibre STANAG.

Specifications

Calibre: 155 mm
Barrel length: 6.025 m
Muzzle brake: double-baffle
Recoil: hydropneumatic
Breech mechanism: horizontal wedge
Shield: yes
Weight: 7,300 kg
Length:
(travelling) 10 m
Width:
(travelling) 2.44 m
Track: 2.08 m
Height:
(max, travelling with lunette 0.7 m from ground) 2.3 m
(trunnion) 1.42 m
Elevation/depression: +63/–2°
Traverse: 24° left/25° right
Rate of fire: 5 rds/min
Max range: 32,900 m (firing ERFB-BB projectile)
Crew: 7

Giat Industries M114 155 mm modernised howitzer in firing position (Rupert Pengelly) ***2000***/0085641

Status

Development complete. Ready for production on receipt of firm orders. As the French Army does not operate any 155 mm M114 systems this upgrade is aimed at the export market.

Manufacturer

Giat Industries, Centre de Bourges.

UPDATED

155 mm howitzer Model 50

Development

The 155 mm howitzer Model 50 was developed in France immediately after the Second World War and was also manufactured in Sweden by the now Bofors Defence for the Swedish Army, which called it the 15.5 cm Field Howitzer Fr. The French call the weapon the OB 155-50 BF.

The 155 mm Model 50 was used widely by the Israel Defence Force but has been withdrawn in favour of self-propelled artillery pieces such as the 155 mm M109. The Israel Defence Force also used a self-propelled

155 mm howitzer Model 50 being towed by Berliet GBU 15 (6 × 6) truck (Pierre Touzin)

155 mm howitzer Model 50 (Israeli Ministry of Defence)

Specifications

Calibre: 155 mm
Barrel length:
(from front of muzzle brake to rear part of breech ring) 4.41 m
Length of rifled bore: 2.727 m
Muzzle brake: multibaffle
Recoil: hydropneumatic
Breech mechanism: screw
Carriage: split trail
Shield: no
Weight:
(travelling order) 9,000 kg
(firing position) 8,100 kg
Length:
(travelling) 7.8 m
(firing) 7.15 m
Width:
(travelling) 2.75 m
(firing) 6.8 m
Height:
(travelling) 2.5 m
(firing) 1.65 m
Ground clearance: 0.25 m
Track: 2.31 m
Elevation/depression: +69/–4°
Traverse: (total) 80°
Rate of fire: (max) 3-4 rds/min
Max range:
(standard ammunition) 18,000 m
(with RAP) 23,300 m
Crew: 11
Towing vehicle: Berliet GBU 15 (6 × 6) truck (French Army service)

version of the Model 50 which is based on a modified Sherman chassis. It is possible that up to 120 of these may still be held in reserve. Details were given in *Jane's Armour and Artillery 1994-95,* page 546.

In the French Army the Model 50 has now been replaced by the 155 mm towed gun TR and has been supplemented in the Swedish Army by the now Bofors Defence FH-77A. A number of these weapons were captured by the Israel Defence Force from the PLO during the fighting in Lebanon in the summer of 1982.

More recently, some of these weapons have been observed to be operational in Ecuador, although their exact source is not known.

Until recently it was understood that the 155 mm howitzer Model 50 had been phased out of service with the French Army but recent information has indicated that a number of these systems are still in service, especially with artillery units supporting the French Foreign Legion overseas. According to the United Nations, between 1992 and 2000 France exported 18 of these systems to Morocco, all of these were delivered in 1997.

Description

The 155 mm barrel, with the breech ring, slides in the cradle during recoil and the breech ring is secured to the piston rod of the recoil buffer and to the recuperating rod. The recoil length varies automatically with elevation. The barrel, breech ring and cradle, which together form the elevating mass, are secured to the upper mounting by cradle trunnions. The elevating mass moves vertically in relation to the upper mounting when the elevating device is turned. The upper mounting rests on the lower carriage and turns horizontally in relation to it when the traversing device is turned. On the bottom of the lower mounting there is a pivot plate. The trails are pivoted on the sides of the lower carriage. Secured to each trail is a wheel bar which in turn holds two pneumatic puncture-proof tyres.

The weapon has a split trail carriage and, on the right trail, there is an unlimbering eye which during transport is hooked to the corresponding hook on the towing vehicle. During transport, the elevating mass is locked to the trails by a device in the rear part of the cradle. The weapon is fitted with a compressed air braking system which is operated from the towing vehicle.

In the firing position the 155 mm howitzer Model 50 rests on the pivot plate and the outer end of the trails. The trails, when opened out, are secured to the ground by pickets. The pivot plate is pivoted in two directions at right angles to each other and thus can be placed so that the carriage remains horizontal regardless of the ground angle.

The Model 50 fires an HE projectile weighing 43 kg with a maximum muzzle velocity of 650 m/s.

Status

Production complete. In service with:

Country	Quantity	Comment
Ecuador	n/a	
France	133	not all operational
Lebanon	20	estimate
Morocco	50	including 18 delivered in 1997
Senegal	6	
Sweden	114	local designation Type F

Manufacturer

Giat Industries.

UPDATED

Giat Industries 105 mm LG1 Mk II Light Gun

Development

The 105 mm LG1 Light Gun has been developed by Giat Industries, as a private venture, specifically for the export market.

By 1987, three prototypes had been built by the Etablissement d'Etudes et de Fabrications d'Armement de Bourges for extensive trials. These were followed by three preproduction weapons.

Late in 1990, following a competition between Giat Industries with the 105 mm LG1 Light Gun and Royal Ordnance of the UK with its 105 mm Light Gun, Singapore placed an order for 37 105 mm LG1 Light Guns. In Singapore the weapons have been issued to two artillery battalions each of which has three batteries with six guns. These were delivered between 1992 and 1993.

Early in 1994, Indonesia placed a firm order with Giat Industries for 20 105 mm LG1 Mk II Light Guns plus a

Belgian Army Giat Industries 105 mm LG II Light Gun with ordnance over closed trails (BP) ***2002***/0130006

Belgian Army Giat Industries 105 mm LG II Light Gun with ordnance traversed forwards (BP) ***2002***/0130005

significant quantity of ammunition and a training package. These were delivered in 1996.

In June 1994, the Public Works and Government Services Canada, on behalf of the Canadian Department of National Defense, placed an order for 28 Giat Industries 105 mm LG1 Mk II Light Guns worth C$18 million. These were delivered between 1996 and 1997.

Specifications

Mk II
Calibre: 105 mm
Muzzle brake: double baffle
Recoil: hydropneumatic
Breech mechanism: vertical sliding, automatic opening
Carriage: split trail
Shield: no
Weight: 1,520 kg
Length:
(barrel) 3.15 m
(barrel to rear) 5.32 m
(barrel deployed) 6.6 m
Width: 1.97 m
Height:
(travelling) 1.34 m
Ground clearance: 0.3 m
Track: 1.71 m
Elevation/depression: +70/–3°
Traverse: 36° (360° on firing platform)
Max rate of fire: 12 rds/min
Max range:
(HE M1) 11,400 m
(HE HB) 15,000 m
(HE BB) 18,500 m
Crew: 5

Giat Industries 105 mm LG1 Mk I Light Guns in service with Singapore and being towed by Mercedes-Benz MB 290 GD (4 × 4) vehicles (David Boey)

Status

Production as required. In service with the following countries:

Country	Order placed	Quantity	Model	Comment
Belgium	1996	14	Mk II	all delivered in 1997
Canada	1994	28	Mk II	8 delivered in 1996 and 20 in 1997
Indonesia	1994	20	Mk II	all delivered in 1996
Singapore	1990	37	Mk I	26 delivered in 1992 and 7 in 1993 which completed deliveries
Thailand	1996	24	Mk II	24, all delivered in 1996

In early 2000, the Canadian Army deployed its LG1 Mk II Light Guns for the first time when A Battery, 1st Regiment Royal Canadian Horse Artillery deployed six weapons to Bosnia.

Late in 1995, Giat Industries was awarded a contract from the Belgian Army for the supply of 14 105 mm LG1 Mk II guns valued at BFr 330 million (US$11.5 million).

The LG1 Mk II Light Guns were delivered from late 1996 through to early 1997 and have been issued to two batteries of the Belgian Army with each battery having six guns. The remaining two are used for training/reserve.

Early in 1996, it was announced that Giat Industries had been awarded a contract from Thailand for the supply of 24 LG1 Mk II guns for delivery from late 1996 through to early 1997.

The original LG1 Mk I gun has now been replaced in production by the LG1 Mk II weapon which differs from the earlier weapon in having a new autofrettaged barrel to enable it, in the future, to fire ammunition with a higher pressure.

The layout of the recoil mechanism has been improved to allow for easier maintenance and the shield has also been removed to reduce weight.

Giat Industries has developed a kit which enables existing users of the LG1 Mk I to upgrade these to the latest production LG1 Mk II standard in their own facilities.

Production of the Giat Industries 105 mm LG1 Mk II Light Gun is undertaken in Bourges where all French artillery and tank barrels are produced.

Description

The Giat Industries 105 mm LG1 Mk II Light Gun has a split trail carriage and a 30 calibre ordnance that is fitted with a double-baffle muzzle brake.

When deployed in the firing position, it is supported on its two trails and a circular firing platform that is lowered under the forward part of the carriage.

The 105 mm LG1 Light Gun Mk II can be brought into or taken out of action in just 30 seconds. A simple hydraulic system operated by a hand pump assists in opening out the trails.

According to Giat Industries, only two types of ammunition are required to cover ranges between 1.4 and 18.5 km, while the new Giat Industries base bleed projectiles cover the ranges from 7.3 to 18.5 km. Giat Industries has also proposed a hollow base projectile covering the ranges 2.5 to 15 km.

The Giat industries 105 mm LG1 Light Gun can be towed by a wide range of light vehicles including an ACMAT (4 × 4) truck, Peugeot P4 (4 × 4), Renault TRM 2000 (4 × 4), Land Rover (4 × 4) and Toyota (4 × 4).

It is fully airportable slung under a helicopter and four can be transported inside of a Lockheed Martin C-130 Hercules transport aircraft.

Manufacturer

Giat Industries, Centre de Bourges.

UPDATED

Giat Industries 105 mm M101A1 modernisation package

Development/Description

Giat Industries has developed, as a private venture, a modernisation package for the US 105 mm M101A1 towed howitzer which is used by many countries. The kit has been developed by Giat Industries at facilities in Bourges where all production of artillery and tank barrels in France is undertaken.

The modifications can be carried out by Giat Industries or the company can supply kits to enable the user to upgrade in his own facilities using his own personnel.

The modernisation includes replacing the original barrel with the more recent Giat Industries LG1 Mk I/II Light Gun already in service with Belgium, Canada, Indonesia, Singapore and Thailand. The breech ring and breech block of the original M101A1 is retained, with a new muzzle brake and balancer. Ballast on the trails is also installed.

The upgraded M101A1 is 233 kg heavier than the standard weapon but can fire all standard NATO ammunition including the US M1 projectile to a maximum range of 11,400 m. It can also fire a Giat 105 BB base bleed projectile to a range of 18,500 m or a

105 mm M101A1 towed howitzer modernised by Giat Industries and deployed in the firing position (Rupert Pengelly)
2001/0105750

Giat HB HE projectile to a maximum range of 15,000 m. It can also fire the new Giat Industries HE ER G2 round to a maximum range of 19,200 m after changing the breech ring.

In mid-1996, Giat Industries announced that it had been awarded a contract from the Philippines to upgrade a first batch of 12 old US-supplied 105 mm M101 towed artillery systems.

These were upgraded in a number of key areas including the replacement of the current barrel with the new barrel, similar to that fitted to the Giat Industries 105 mm LG1 Mk II light gun used by Belgium, Canada, Indonesia, Singapore and Thailand.

The Philippines is the second customer for the upgraded M101, the first being Thailand which ordered a total of 285 kits, all of which were delivered to Thailand by early 1997. Thailand has also taken delivery of significant quantities of 105 mm HE-BB (High Explosive Base Bleed) rounds from Giat Industries for use with these upgraded M101 weapons and the more recently acquired LG1 Mk II Light Guns covered in the previous entry.

Status
Production as required. In service with the Philippines (12) and Thailand (285).

Manufacturer
Giat Industries, Centre de Bourges.

VERIFIED

INDIA

Indian Ordnance Factories 105 mm Light Field Gun

Development
The Indian Ordnance Factories' 105 mm Light Field Gun (LFG) bears a resemblance to the now BAE Systems, RO Defence Light Gun. This is partly due to the fact that both were derived from the same 105 mm L13 gun used on the Vickers Defence Systems FV433 Abbot self-propelled gun. The Indian Army took delivery of a total of 68 Value Engineered Abbot 105 mm self-propelled guns (*Jane's Armour and Artillery 1995-96*, pages 586-587). The Abbot is understood to have been phased out of service with the Indian Army. The first 105 mm barrel, developed in India from the British original, was used on the 105 mm Field Gun Mk 1 (see next entry). This uses tubular bow trails similar to those of the British 105 mm Light Gun.

By early 1996, a total of 533 105 mm LFGs had been manufactured for the Indian Army. There are no known export sales of this weapon. It is believed that these weapons are mainly deployed with artillery units deployed in mountainous terrain.

As far as it is known there has been no recent production of the Indian Ordnance Factories 105 mm Light Field Gun. This has been confirmed by the United Nations who have stated that India did not export any artillery systems between 1992 through to the year 2000.

Description
The 105 mm LFG has been developed for use in numerous environments, ranging from mountains to deserts. To reduce weight the weapon has been constructed using light, high-strength alloy steels giving a weight in action of 2,380 kg.

The ordnance consists of a monobloc autofrettaged barrel with a vertical, sliding, hand-operated breech block. Firing is electrical. The cradle, saddle and undercarriage are lightweight fabricated structures. Other features include spring-type balancing gears, a telescopic type of hydropneumatic recuperator, a suspension system and bow-shaped tubular trails. The normal towing position is with the barrel folded back across the trails. A 360° traversing firing platform is provided. The LFG can be parachute dropped and carried by helicopter.

The 105 mm Light Field Gun fires separate ammunition produced by the Indian Ordnance Factories using a seven-charge system (1, 2, 3, 4, 4.5, 5 and Super). Projectiles include HE, smoke, base ejection, star and HESH (High Explosive Squash Head). Maximum range is 17,200 m. The normal rate of fire is 4 rds/min and the intense rate of fire is 6 rds/min for up to 10 minutes (charge 5). The sustained fire rate is one round every 2 minutes for up to 2 hours (charge Super).

The 105 mm LFG is fitted with a dial sight with a magnification of ×3 and a 13° field of view for use in the indirect fire role and a day telescope with a magnification of ×2 and a 16° field of view for use in the

Specifications
Calibre: 105 mm
Length:
(ordnance) 4.4 m
Barrel length: 3.8905 m
Max recoil length: 1.06 m
Nature of rifling: uniform 1 in 19 calibres right-handed
Number of rifling grooves: 28
Muzzle brake: double baffle
Breech mechanism: vertical sliding
Carriage: bow
Shield: yes
Weight:
(firing) 2,380 kg
Length:
(firing) 7.216 m
(travelling) 5.054 m
Width:
(travelling) 1.82 m
Height: 1.494 m
Ground clearance: 0.33 m
Track: 1.465 m
Elevation/depression: +73/−5°
Traverse:
(on carriage) 10°, 5° left and right
(on platform) 360°
Rate of fire:
(normal) 4 rds/min
(intense) 6 rds/min max up to 10 min (charge 5)
(sustained) 1 rd/2 min max up to 2 h
Max range: 17,200 m
Towing vehicle: Shaktiman Model 4200 (4 × 4) truck

Indian Ordnance Factories 105 mm Light Field Gun (Christopher F Foss)

Indian Ordnance Factories 105 mm Light Field Gun with ordnance at maximum elevation

direct fire role. The sighting system is situated on the left side of the mount.

Lighter version of 105 mm LFG

Early in 1993, the Indian Ordnance Factories stated that they were developing a lighter version of their 105 mm LFG with the following features:

(1) Subassemblies design optimised to reduce weight from 2,380 kg down to 2,100 kg

(2) Time into and out of action reduced by first underslinging the firing platform and second by modifying the wheel and saddle design to reverse the barrel over the ordnance without removing the right wheel.

As far as it is known this version has yet to enter production.

Status

Production complete. In service with the Indian Army.

Manufacturer

Gun Carriage Factory.

Enquiries to: The Secretary, Ordnance Board Factory.

UPDATED

Other Indian artillery developments

Development/Description

In addition to the Indian Ordnance Factories 105 mm Light Field Gun (see previous entry), at least two other Indian towed artillery equipments were developed and placed in production.

The lightest of these is the 75 mm Pack Gun-Howitzer Mk 1 which has a 24 calibre barrel and was designed for use in mountainous terrain. The carriage is of the split trail type and, if required, the wheels can be removed to allow the weapon to pivot on the trails and a firing platform under the carriage. The barrel has a single- or double-baffle muzzle brake. At least two slightly different models have been developed. The Mk 1A has a two-part flat shield with the top part of each half folding forwards. The Mk 1B has a one-piece shield with a flat lower half and a top half that slopes to the rear. This 75 mm Pack Gun-Howitzer is in service with the Indian Army.

Recent information has indicated that the requirement for a 75 mm gun to replace the old British 3.7 in howitzers was first indentified in 1959.

This was followed by limited production of a Canadian 75 mm mountain gun pending the purchase of the Yugoslav 76 mm Mountain Gun M48. It is believed that India has a total of 900 locally built 75 mm guns which serve alongside the 215 Yugoslav 76 mm M48 weapons. There is a possibility that the latter are now held in reserve.

The second weapon developed in India is the 105 mm Indian Field Gun. Development of this weapon began in the mid-1960s with a gun development team under the control of the Director General Ordnance Factories.

The 105 mm barrel was manufactured at the Ordnance Factory, Kanpur, the elevation and traversing mechanism at Machine Tool Prototype Factory, Ambarnath and the carriage and final equipment was produced by the Gun Carriage Factory , Jabalpur.

The 105 mm barrel of the Indian Field Gun was not autofrettaged. The carriage is of the split trail type with hydropneumatic fabricated recoil system and pneumatic balancing gear. The weapon has a maximum range of 17,400 m.

The 105 mm Indian Field Guns and the more recent 105 mm Light Field Gun have now replaced the British-supplied 25-pounder field guns. The 105 mm Indian Field Gun was replaced in production by the 105 mm Light Field Gun covered in detail in the previous entry.

It is believed that the total Indian fleet of 105 mm guns amounts to around 1,300 weapons of all types.

130 mm M-46 field gun

It is believed that the Indian Army has as many as 720 Russian-supplied 130 mm M-46 field guns in service with ammunition now being manufactured in India. A total of 36 regiments are equipped with the M-46 and this is expected to increase.

According to United Nations sources India took delivery of an additional batch of 120 M-46 weapons in 1994 and another batch of 202 weapons of an undisclosed type, probably also M-46 weapons, the following year. Some regiments equipped with the Indian 105 mm Field Gun may convert to the 130 mm M-46 in the future.

130 mm M-46 upgrade

Early in 2000, India awarded Soltam Systems of Israel a contract to upgrade 180 130 mm M-46 field guns to a 155 mm/39 and 155 mm/45 calibre standard. The first 180 weapons will be upgraded in Israel with the follow-on batch, expected to consist of 220 to 250 guns, being upgraded at the Indian Ordnance Factory Board facilities with kits provided by Soltam Systems. The work in India will be carried out at Kanpur and Jabalpur.

155 mm FH-77B

India has taken delivery of all the 410 Bofors Defence 155 mm FH-77B field howitzers together with associated ammunition, fire-control systems, tractors and training equipment. Under the second and third phases of the programme it was expected that local assembly and then local production would be undertaken in India.

From July 1999 India started to place contracts with Bofors Defence for spare parts for Indian 155 mm FH-77B.

In September 1999 India awarded a contract to Russia for the supply of 1,000 155 mm Krasnopol-M laser-guided artillery projectiles plus 10 associated tripod-mounted laser target designators and support equipment.

130 mm self-propelled gun

The Indian Army has modified a number of Vijayanta (Vickers Mk 1 MBTs) to mount the Russian 130 mm M-46 field gun. The modifications to the chassis have been extensive and include an additional roadwheel either side. Details of this system are given in the Self-propelled guns and howitzers (tracked) section under India.

155 mm self-propelled gun

In 1995, four 155 mm turret systems were evaluated in India on a locally manufactured T-72M1 MBT chassis to meet an Indian Army requirement for a 155 mm self-propelled artillery system.

These systems were the French GCT, Slovakian Zuzana, South African T6 and the UK AS90. Details of all these systems are given in the Self-propelled guns and howitzers (tracked) section.

In the end, none of these systems were selected and early in 1998, two systems, the South African LIW T6 and the UK's AS90, were shortlisted for a further series of trials. By January 2002 no production orders had been placed for the T6.

New 155 mm towed gun

The Indian Army also has a requirement for a new 155 mm/52 calibre towed artillery system with 400 expected to be procured, funding permitting. Seven companies are, in the future, expected to be invited to send their towed artillery systems to India for extensive trials. These are:

BAE Systems, RO Defence (UK)
Bofors Defence (Sweden)
Giat Industries (France)
LIW Division of Denel (South Africa)
Singapore Technologies Kinetics (Singapore)
Patria Vammas (Finland)
Santa Barbara (Spain).

India is also understood to be studying the possible acquisition of a 6 × 6 155 mm self-propelled artillery system with the contenders including the French Giat Industries CAESAR and the Swedish Bofors Defence FH-77BD, both of which are detailed in the Self-propelled guns and howitzer (wheeled) section. Another late contender could be the new Soltam Systems ATMOS 2000 SPG based on a Tatra (6 × 6) chassis. Details of this are given in the Self-propelled guns and howitzers (wheeled) section.

In both the self-propelled and towed artillery requirements, it is expected that some will be purchased off the shelf with progressive local production then being undertaken in India. Ammunition does not form part of the competition.

Status

Both the 75 mm Gun-Howitzer and 105 mm Field Gun are in service with the Indian Army.

UPDATED

105 mm Indian Field Gun (Indian Army)

Indian 75 mm Pack Gun-Howitzer Mk 1B (Indian Army)

INTERNATIONAL

NORICUM GH N-45 155 mm gun-howitzer

Development

From 1989, NORICUM manufactured an improved version of the 155 mm SRC International GC 45 155 mm gun-howitzer (see entry under Belgium in this section), under the designation GH N-45 (Gun-Howitzer, NORICUM, 45 calibre).

NORICUM assembled all but two of the 12 guns (two six-gun batteries) supplied to the Royal Thai Marines and then began deliveries of over 600 guns to various customers.

In June 1987, the Thai Cabinet approved a purchase of six GH N-45 weapons at a cost of US$7 million, all of which have now been delivered. These models are fitted with an auxiliary power unit.

In 1994, the Swiss-based company T & T Technology Trading Ltd purchased the design rights, weapons and spare parts of the Austrian NORICUM 155 mm GH N-45 towed artillery system.

No figures for Austrian production of the GH N-45 have been released but it is estimated that between 500 and 600 weapons were produced with production reaching 10 weapons per month.

Production by Voest-Alpine stopped in the late 1980s when it became apparent that the weapons were reaching the Middle East.

Known customers include Iran, Iraq and Thailand, but in some of these cases they were not supplied direct from Austria.

In a further effort to broaden its marketing activities, the NORICUM Division of T & T Technology Trading Ltd has held discussions with a number of NATO defence manufacturers under which these companies could manufacture some parts of the GH N-45 system as well as undertaking final assembly and delivery of the complete system to the customer.

Although parts of the GH N-45 could be manufactured in NATO countries, the NORICUM Division of T & T Technology Trading Ltd would still be prime contractor and be responsible for marketing.

In mid-1997, the Royal Thai Army placed an order for a second batch of 36 155 mm GH N-45 howitzers in the latest A1 towed configuration and these were delivered in 1998.

Specifications

Model	**Towed**	**APU**
Calibre:	155 mm	155 mm
Length:		
(barrel)	7.045 m	7.045 m
(rifling)	5.82 m	5.82 m
(travelling, barrel forward)	13.97 m	13.97 m
(travelling, barrel aft)	9.72 m	9.725 m
(firing)	11.4 m	11.4 m
Muzzle brake:	multibaffle	multibaffle
Carriage type:	split trail	split trail
Shield:	no	no
Weight:	10,100 kg	12,400 kg
(total)		
Width:		
(travelling)	2.5 m	2.75 m
(firing)	9.93 m	9.93 m
Height:		
(travelling)	2.05 m	2.05 m
(barrel aft)	2.26 m	2.26 m
Ground clearance:	0.31 m	0.315 m
Wheel track:		
(main)	2.22 m	2.47 m
(trail)	0.865 m	0.865 m
Tyres:		
(main)	9.00 × 20R	11.00 × 20R
(trail)	8.25 × 15R	8.25 × 15R
Elevation/depression:	+72/–5°	+72/–4°
Traverse:	30/40°	30/40°
(left/right)		
Range:		
(ERFB)	30,300 m	30,300 m
(ERFB BB)	39,600 m	39,600 m
Rate of fire:		
(normal)	2 rds/min unlimited	2 rds/min unlimited
(max)	7 rds/min	7 rds/min
(burst)	3 rounds/16 s	3 rounds/16 s
Crew:	6	6
Max towed speed:		
(road)	105 km/h	105 km/h
(track/cross-country)	50 km/h	50 km/h

NORICUM GH N-45 155 mm gun-howitzer fitted with an auxiliary propulsion system showing cross-country mobility

NORICUM GH N-45 155 mm gun-howitzer showing breech and rammer (Christopher F Foss)

The NORICUM 155 mm GH N-45 howitzer, firing an Extended Range Full Bore-Base Bleed (ERFB-BB) projectile, has a maximum range of 39.6 km so making it the longest range artillery system in service with the Royal Thai Army.

According to United Nations sources Thailand took delivery of 18 weapons in 1992, 18 in 1996 and 36 in 1998. All of these were delivered from Austria. There are no other known sales of the GH N-45 155 mm gun-howitzer in this period.

NORICUM GH N-45 155 mm gun-howitzer captured by Iran with ordnance over rear and trail partly open

Description

The GH N-45 is, in most respects, similar in appearance to the Belgian GC 45 but has been redesigned in certain key areas to improve both handling in the field and reliability and to make the weapon easier to manufacture.

The carriage uses a walking beam suspension for improved cross-country mobility and a secondary float is fitted for easy emplacement, displacement and general weapon handling. There is a quick release in the traversing gear which brings the barrel from the travelling jack-knife position to the firing position in a few seconds and there is a steering device that allows the weapon to travel in the tracks of the towing vehicle. The GH N-45 can be airlifted by a Lockheed Martin C-130 transport aircraft without preparation and the normal towing vehicle is a 5 or 10 tonne 6 × 6 truck, with the travelling speed on all types of terrain limited only by the capabilities of the towing truck.

The GH N-45 can be delivered in a basic towed form or with an APU; an APU kit can be supplied for fitting to towed models by the user. The main power source is a Porsche 236 6.9 four-cylinder, four-cycle air-cooled 2.3 litre engine delivering 120 hp. The APU is mounted on the forward part of the carriage with the driver on the right. The APU enables the weapon to propel itself at a maximum speed of 30 km/h and two 30 litre fuel tanks provide sufficient fuel for ranges of up to 130 km. With the APU the weapon can climb 40 per cent gradients at a speed of 5 km/h.

The GH N-45 has a 45 calibre barrel with a 22.94 m^3 chamber volume to enable base bleed ammunition to be fired to a maximum range of 39,600 m. The barrel has a semi-automatic breech mechanism with a pneumatic rammer. The overall length of the barrel, including muzzle brake and breech mechanism is 7,720 mm with the length of rifling being 5,830 mm and the length of chamber being 1,173 mm. The rifling has 48 grooves.

Extras available for the GH N-45 included a power elevating and traverse system, and an emergency ammunition bracket holding six standby 155 mm projectiles, both for the APU-powered version. A further option for the APU version was an ammunition handling device. The jib and power unit for this device are mounted on the top right of the barrel and are used to raise the ammunition tray from ground level to the loading position. Tracks could be fitted to the APU-powered version to increase traction over soft ground and enable gradients of up to 45 per cent to be climbed. Also available for the APU version was a remote-control system that enabled a 6 × 6 tractor to use the traction of the powered carriage over adverse terrain giving a 10 × 10 configuration.

In addition to standard NATO ammunition, the 155 mm GH N-45 can fire a complete family of 155 mm ammunition, including:

HE/BT, maximum range 30,300 m (charge N10)
HE/BB, maximum range 39,600 m (charge N10)
HC/smoke, maximum range 39,600 m (base bleed), 30,300 m (boat tail) (charge N10)
WP/smoke, maximum range 39,600 m (base bleed), 30,300 m (boat tail) (charge N10)
Illuminating, maximum range 39,600 m (charge N10)
HE, 24,700 m (charge N10)
HE (M107), 17,800 m (charge M119A2).

In addition to supplying the above types of 155 mm artillery projectile the company also claims to supply cargo munitions, terminally guided munitions and more advanced sensor fuzed munitions.

In addition, various types of artillery fuze can be supplied by NORICUM as well as various types of bagged and modular charge systems.

Status

Production. In service with a number of countries including Iran, Iraq and Thailand (army and marines).

Manufacturer

T & T Technology and Trading GmbH.

UPDATED

155 mm Field Howitzer 70 (FH-70)

Development

In the early 1960s, Germany, the UK and the USA agreed that they had a requirement for a new 155 mm towed field howitzer. Germany and the USA wanted to replace their 155 mm M114 howitzers and the UK wanted to replace its 5.5 in guns. In the end, the USA went on to develop the 155 mm M198 which has no APU. Germany and the UK agreed to develop a new weapon which became known as the FH-70.

The FH-70 MoU was signed in 1968 and the requirement was for a 155 mm howitzer to meet the following military and tactical requirements:

Continuous high rate of fire with a burst fire capability
High mobility with minimum effort for deployment
Increased range and lethality and a new family of ammunition.

The UK was the project leader for the towed FH-70 and Germany for the self-propelled version called the SP-70, development of which was suspended in 1986.

The first six FH-70 prototypes were completed between 1969-70 and, after accepting the Agreed Operational Characteristics, Italy joined the collaborative project in 1970 as a full partner. The second batch of eight prototypes was completed between 1971 and 1973 and the first international trials battery of six FH-70s was formed in 1975.

A total of 19 prototypes of the FH-70 were completed and after extensive trials the weapon was accepted for service in 1976, with the first production weapons being completed in 1978. Germany ordered 216, Italy 164 and the UK 71 FH-70s.

In May 1982, Saudi Arabia placed an order with International Military Services of the UK for the supply of 72 FH-70 howitzers plus ammunition, muzzle velocity measuring equipment, fire-control systems and training

155 mm FH-70 deployed in the firing position (PB) ***2001***/0105749

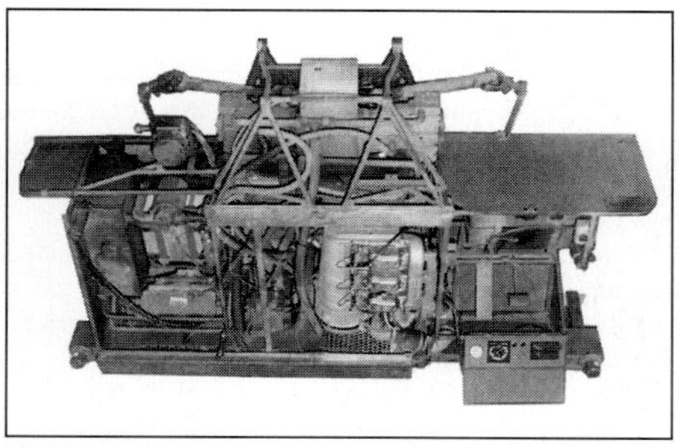

ARIS developed APU for installation on the 155 mm FH-70 artillery system **2001**/0034635

equipment. By early 1983, the first 18 weapons had been delivered.

The British designation for the FH-70 is the L121 while the Germans call it the FH 155-1. Production for all three countries was completed in 1982. Malaysia took delivery of nine FH-70 systems in 1989 with a further three being delivered in 1993. The FH-70 has also been manufactured under licence in Japan together with its ammunition.

The trilateral responsibilities for production were as follows:

United Kingdom: carriage, traversing gear, high-explosive shell and propelling charge (cartridges numbers 1 and 2)

Germany: ordnance, loading system, auxiliary propulsion unit, suspension, sighting equipment, smoke and illuminating ammunition and propelling charges (cartridge number 2)

Italy: cradle, recoil system, sights bracket, elevating gear and arc, high-explosive shell, smoke and illuminating ammunition and propelling charges (cartridges numbers 2 and 3).

Late in 1995, the German Army agreed to supply the Royal Netherlands Army with 15 155 mm FH-70 weapons. Morocco has taken delivery of an estimated 35 FH-70s which are believed to have come from Saudi Arabia.

Late in 1997, the Armaments Division of Vickers Shipbuilding and Engineering Limited (VSEL), (BAE Systems, RO Defence), the UK's prime contractor for the actual FH-70 weapon, took delivery of 24 155 mm FH-70 towed artillery systems which had become surplus to requirements of the British Army as a result of defence cutbacks.

BAE Systems, RO Defence is now marketing these 155 mm FH-70 systems to selected customers on a worldwide basis, particularly Europe and Asia. These FH-70s can be rapidly delivered as they are refurbished to an 'as new' condition.

This has been made possible as a marketing agreement has been reached between the now BAE Systems, RO Defence and the UK MoD Disposal Sales Agency, under which the company will store, maintain and sell the FH-70s on behalf of the UK MoD. The FH-70 has now been withdrawn from service with the British Army.

Specifications

Calibre: 155 mm
Barrel length: 6.022 m
Muzzle brake: double baffle
Breech mechanism: semi-automatic wedge
Carriage: split trail with APU
Shield: no
Weight:
(travelling) 9,300 kg
(firing) 9,300 kg
Length:
(travelling) 9.8 m
(firing) 12.43 m
Width:
(travelling) 2.58 m
(firing) 7.5 m
Height:
(travelling) 2.45 m
(firing) 2.192 m
Axis of bore:
(firing) 1.525 m
Ground clearance:
(travelling) 0.3 m
Track: 2.195 m
Wheelbase: 5.15 m
Fording:
(towed) 1.5 m
(APU in use) 0.75 m
Turning radius: 9.25 m
Tyres: 14.00 × 20
Elevation/depression: +70/–4.5°
Traverse: (total) 56°
Rate of fire:
(normal) 6 rds/min
(burst) 3 rds/13 s
(sustained) 2 rds/min for 1 h
Max range:
(standard projectile) 24,700 m
(base bleed projectile) 31,500 m
Crew: 7 or 8
Towing vehicles:
(Germany) MAN 6 × 6 7,000 kg truck
(Italy) FIAT 6605 TM 6 × 6 truck

155 mm FH-70s of the German Army coming into action (PB) **2002**/0130004

Description

The elevating mass of the FH-70 comprises the ordnance, cradle with integrated recoil system and the loading mechanism. The ordnance consists of a monobloc barrel fitted with a double-baffle muzzle brake, a breech mechanism incorporating a semi-automatic vertically sliding block and an automatic tube loader with a capacity of nine tubes, spent tube ejection and an emergency firing mechanism.

The cradle, which carries the ordnance on slides, includes the recoil system which consists of a buffer with high-angle cut-off gear and a recuperator. The mechanical main firing system, which is independent of elevation, is carried through one of the cradle trunnions.

The semi-automatic loading system operates at all elevations and traverses and consists of a loading tray which presents the round to the chamber. Together with an automatic tube loader, this system permits a normal rate of fire of 6 rds/min and a burst rate of 3 rounds in 13 seconds.

An automatic loading (flick ramming) system is available as an optional extra and provides a burst fire rate of 3 rounds in 8 seconds.

The carriage is of lightweight construction with saddle, split trails, self-digging spades, main and trail wheels which are operated hydraulically, a soleplate and a detachable APU. When travelling, the barrel is rotated through 180° and clamped between the trails.

The APU is housed in a space frame attached to the forward part of the carriage and consists of a commercial Volkswagen 1,800 cc engine with a gearbox, differential, batteries and a hydraulic pump. In the APU mode the FH-70 can attain speeds up to 16 km/h and is capable of negotiating slopes up to 34 per cent. When being towed, it can ford to a depth of 1.5 m and when using the APU to 0.75 m.

The functions of the APU include driving the main wheels for moving the howitzer, providing hydraulic power for steering and raising and lowering both the main and trail wheels. This facility allows for rapid unhooking, lowering the howitzer onto its soleplate, easy opening and lowering of the trails when going into action, as well as extracting the spades from the ground, lifting the equipment, rapid closing of the trails and limbering up. In an emergency, or when the APU has been removed, power is provided by a hand pump.

The main wheels are hydraulically suspended on swinging arms attached to the carriage body and are cushioned by hydraulic accumulators. Similar suspension elements are used for raising and lowering the trail wheels. Braking of both main wheels is possible from the towing vehicle, or from the APU driver's position on the left side of the carriage.

The trail wheels are attached to the trail end and can be steered, raised and lowered. The hydraulic units providing these functions are controlled from the driver's position and can operate with the trails open or closed. The steering is capable of 60° of movement either side of the centre and for spreading the trails the steering wheels can be turned through 90°.

The sighting system mounted on the left of the saddle consists of a support bracket, a dial sight carrier incorporating a quadrant elevation scale and levelling bubbles, a periscopic dial sight and a direct fire telescope. A direct fire night sight can also be fitted. All scales, graticules and bubbles are self-illuminating. The sight carrier incorporates transducers for azimuth and elevation and in the German and Italian armies these are connected to a Digital Display Unit (DDU) which also receives firing data (azimuth and elevation) from a fire-control computer. The layer lays the gun by nulling out the differences on the DDU and then laying the sight on the gun aiming point. The DDU was not used in UK service.

The FH-70 can be towed on a road at a maximum speed of 100 km/h. Germany uses the MAN (6 × 6) 7,000 kg truck to tow the FH-70, Italy uses the FIAT 6605 TM (TM69) and the UK used the Foden (6 × 6) medium-mobility vehicle. The FH-70 is airportable in a Lockheed Martin C-130 aircraft or underslung under the Boeing CH-47D helicopter.

A family of 155 mm ammunition was developed for the FH-70 which gives the weapon a significant increase in range and lethality for both direct and indirect firing against armoured targets. The FH-70 can also fire NATO standard ammunition including the US Copperhead CLGP. The range of ammunition consists of the following projectiles:

HE shell (L15A1) which weighs 43.5 kg and is thin walled, with high fragmentation effect and HE capacity of 11.3 kg

Base Ejection Smoke shell (DM105), where four smoke canisters fall to the ground giving a dense cloud of smoke in 30 seconds.

Illuminating (DM106) which provides illumination of 2.1 million candelas for 60 seconds.

The ERP (Extended Range Projectile) which will increase the range of the FH-70 to over 30,000 m is US M549A1 rocket-assisted projectile and RO30 base bleed.

The eight-zone charge system uses triple-based propellant and is divided into three separate cartridges, containing zones 1 and 2, 3 to 7 and 8 respectively. The charge system is ballistically balanced, has a clean combustion and is initiated by a percussion inner tube (DM 191A1). Muzzle velocity ranges from 213 to 827 m/s. Using the now BAE Systems, RO Defence BIS 14 propelling charge the FH-70 can fire an ERFB base bleed projectile to a range of 31,000 m.

Using the APU, the FH-70 can be unhooked, manoeuvred into position and put into action in less than 2 minutes and taken out of action in a similar period. If a target moves outside the traverse limits the trails can be raised hydraulically, thereby pulling the spades from the ground. The gun can then be wheeled about its soleplate to a new target. With the trails lowered the first shot will embed the spades.

Apart from the usual preliminary checking of the equipment, the gun detachment has only to open the breech by hand, load the first shell and propelling charge and place the second shell on the loading tray. The breech is closed and an ignition tube is automatically loaded from the nine-tube magazine. After this sequence, which takes less than 5 seconds, the FH-70 is ready for action. When the FH-70 is ready and laid correctly, all that is required for firing is for the layer to pull the firing handle. Thereafter, the sequence of operation becomes semi-automatic. The recoil system and muzzle brake jointly absorb the recoil forces as the gun recoils on the cradle slides to a maximum of 1.4 m. On running out to the firing position, the following operations take place simultaneously: The breech is opened automatically by a cam, the spent tube is ejected and the next 155 mm shell is lifted to the breech and aligned in the chamber before final hand ramming. The detachment then rams the shell and loads the propelling charge, places the next shell in the loading tray, closes the breech, if necessary relays the gun and fires again. A flick ramming device is available as an optional extra giving a burst rate of 3 rounds in 8 seconds and reducing crew fatigue.

To assist gun loading over prolonged periods of firing, the now BAE Systems, RO Defence developed an ammunition hoist device which effectively reduces the handling weight of a 155 mm projectile by about half. The device is suspended from a steel joist gantry mounted to the right-hand side of the main carriage, just forward of the breech. It is a spring-loaded bobbin which issues a steel wire cable with a projectile clamp at the end. When the clamp is secured over a projectile, the loader can lift the projectile with at least half the weight being taken up by the spring-wound bobbin. The gantry allows him to swing the projectile upwards and over to the loading tray where the projectile clamp is released. The device cuts loader fatigue over long periods and was troop-tested at the Royal School of Artillery, Larkhill, during mid-1982. It was offered as an optional extra.

Other options include a direct fire image intensification night sight, deflector boxes and mainwheel mudguards.

Variants

Japanese FH-70

The Japanese Ground Self-Defence Force selected the FH-70 to replace its 155 mm M114 towed howitzers and production is being undertaken under licence in Japan. Prime contractor in Japan was the Japan Steel Works. Rheinmetall was responsible for the Japanese contract.

Export version

For the world market, a simplified sighting system is available which utilises the now BAE Systems, RO Defence 105 mm Light Gun sighting system with a modified graticule in the direct fire telescope. It can be fitted to the standard gun via an adaptor on the left-hand trunnion.

FH-70 with 52 calibre ordnance

For trials purposes Rheinmetall has fitted an FH-70 with a 52 calibre ordnance that conforms with the new quadrilateral ballistic agreement. This enables a range of 30,000 m to be achieved with the L15A1 projectile, or 40,000 m with base bleed projectiles.

ARIS APU for FH-70

The Italian company of ARIS (Applicazioni Rielaborazioni Impianti Speciali) have completed development of a new diesel Auxiliary Power Unit (APU) for installation on the 155 mm FH-70 towed artillery system.

The original production FH-70 has the APU installed on the front of the carriage. This consists of a Volkswagen 1,800 cc engine with a gearbox, differential, batteries and hydraulic pump.

The installation of the APU allows the FH-70 to attain a maximum road speed of 16 km/h as well as climbing slopes of up to 34 per cent.

In addition to providing the power to move the FH-70 system over short distances, especially over terrain that cannot be reached by the 6 × 6 prime mover, the APU also provides hydraulic power for steering and raising and lowering the main and trail wheels. The installation of the APU reduces crew fatigue as well as speeding into and out of action times.

The new ARIS diesel powered APU can be installed on the 155 mm FH-70 in only 2 hours and uses standard proven Commercial-Off-The-Shelf (COTS) components to reduce overall life cycle costs.

The system is currently being evaluated by the Italian Army and is almost a direct replacement for the current APU and takes less than 2 hours to install. It gives the FH-70 a maximum road speed of 15 km/h and it can climb gradients of 40 per cent.

Manufacturers

BAE Systems, RO Defence.
Rheinmetall DeTec AG.
Otobreda SpA.

UPDATED

Status

Production complete in Europe but can be resumed if further orders are placed. Production continues under licence in Japan. In service with:

Country	Quantity	Comment
Germany	216	quantity delivered new, 15 transferred to Netherlands, total is now 192
Italy	164	quantity delivered new
Japan	463	local production
Malaysia	12	from UK, including three in 1993
Morocco	35	from Saudi Arabia
Netherlands	15	from Germany in 1995/96
Saudi Arabia	72	quantity delivered new, 35 transferred to Morocco
UK	71	quantity delivered new, 47 declared (under CFE Treaty) (but see development) (none in UK Army service)

IRAN

155 mm HM41 howitzer

Development

In 2001, it was revealed that in addition to building the 122 mm (HM51) and 155 mm (HM44) barrels used in the Thunder 1 and Thunder 2 self-propelled artillery systems (covered in detail in the Self-propelled guns and howitzers (tracked) section), the Hadid Armament Industries Group of the Iranian Defence Industries Organisation (DIO) was also producing a complete 155 mm/39 calibre towed artillery system. This is understood to be now fully operational with the Iranian Army.

Description

The 155 mm towed artillery system is designated the HM41 and is understood to be an upgraded version of the US 155 mm M114 towed howitzer, which was originally developed before the Second World War.

This fired a standard 155 mm M107 high-explosive projectile, which has been manufactured in Iran for many years by the Ammunition Division of the DIO, to a maximum range of 14,600 m.

The original M114 towed howitzer has been in service with the Iranian Army for many years. By upgrading these with a 155 mm/39 calibre barrel the range of the weapon is increased, allowing it to be deployed more to the rear, so increasing its survivability against counter battery fire.

According to Iranian sources, the 155 mm/39 calibre HM41 weighs 6,890 kg and firing a rocket-assisted projectile, a maximum range of 30,000 m can be achieved. Firing an unassisted projectile, a maximum range of 22,000 m can be obtained.

A maximum rate of fire of 4 rds/min can be achieved but there is no load assist device fitted. Elevation range is from 0 to plus 66° with traverse being 25° right and 23.5° left, all under manual control. Direct and indirect fire sights are fitted.

The new 155 mm/39 calibre ordnance is fitted with a double-baffle muzzle brake to reduce recoil forces, with the breech mechanism being of the screw type.

When deployed in the firing position the HM41 rests on the firing jack under the forward part of the carriage and on the spades attached to the two trails. The system is normally towed by a 6 × 6 10 ton truck which carries a quantity of ammunition (projectiles, charges and fuzes) plus the crew.

In appearance the Iranian HM41 system is very similar to the South Korean 155 mm/39 calibre KH179 howitzer developed by the WIA Corporation (previously the Kia Machine Tool Company) in the Republic of Korea (ROK).

Development of this system, which is based on the old US M114 towed artillery system, commenced in 1979, with deliveries to the ROK starting in 1983. Full details of the KH179 are given later in this section.

Other Iranian towed artillery systems

It is now known that Iran is also producing a version of the Russian 122 mm towed howitzer D-30, under the local designation of the HM40. This fires the same family of 122 mm separate loading ammunition as the Thunder 1 self-propelled artillery system.

Specifications

Calibre: 155 mm/39 calibre
Muzzle brake: double baffle
Carriage: split trail
Shield: no
Weight: 6,890 kg
Elevation: 0 to +66°
Traverse:
(left) 23.5°
(right) 23.5°
Rate of fire: 4 rpm
Maximum range:
(RAP) 30,000 m
(unassisted) 22,000 m

Status

Production. In service with Iran.

Manufacturer

Defence Industries Organisation, Hadid Armament Industries Group.

NEW ENTRY

IRAQ

122 mm Saddam (D-30) howitzer

Development/Description

The 122 mm Saddam (D-30) howitzer is essentially a locally produced copy of the Yugoslav version of the Russian 122 mm D-30 towed howitzer. The Saddam (D-30) weighs 3,350 kg in the firing position and 3,440 kg in the travelling configuration. Traverse is a full 360° with elevation being from –7 to +70°. Rate of fire is 6 to 8 rds/min with a maximum range of 17,133 m achieved. The HE projectile weighs 22 kg while the charge weighs 3.8 kg.

Status

Following the 1990-91 Gulf War, it was assumed that production of the 122 mm Saddam (D-30) howitzer had been suspended; it is in service with the Iraqi Army. It is understood that production of this 122 mm howitzer was undertaken at the Saad 5 (Saddam Engineering Complex) which also manufactured Ababil rockets, tanks and mortar sights. Recent information has indicated that production of the 122 mm Saddam (D-30) howitzer may possibly have started again, although, given that there is a UN embargo in force against Iraq, it is difficult to see where the raw materials for the production of this and other weapons are coming from.

130 mm M-46 field gun

NB: The Iraqi Army had a large inventory of Russian-supplied 130 mm M-46 towed field guns and it was intended that these would be upgraded to 155 mm/45 calibre configuration using the Yugoslav conversion kit described later in this section.

As far as is known this programme had not commenced at the time of the 1990-91 Gulf War. If it had gone ahead then Iraq would have had ammunition commonality with its South African LIW 155 mm/45 calibre G5 and Austrian-supplied NORICUM 155 mm GH N-45 towed artillery systems of which up to 500 were purchased.

Recent information has indicated that this 155 mm/45 upgrade package was developed by the former Space Research Corporation SA of Belgium, which no longer exists.

155 and 210 mm (6 × 6) SPGs

Iraq also developed to the prototype stage, with considerable assistance from companies in Western Europe, two 6 × 6 self-propelled artillery systems; one with a 155 mm ordnance and the other with a 210 mm ordnance; neither entered production.

UPDATED

Iraqi 130 mm M-46 field gun upgraded to 155 mm/45 calibre configuration

122 mm Saddam (D-30) howitzer deployed in firing position

ISRAEL

Soltam Systems 155 mm TIG gun-howitzer family

Development

The 155 mm TIG 2000 gun-howitzer family has been developed by Soltam Systems as a private venture and is a follow-on to the earlier Soltam Systems Model 839/839P and Model 945/945P gun-howitzers *(Jane's Armour and Artillery 1997-98,* pages 695 - 696). The P in this designation means that it is fitted with an auxiliary propulsion unit (APU).

In the designation TIG 2000, the TIG stands for Towed Independent Gun. According to Soltam Systems, the TIG 2000 is unique in that it can be fitted with a 155 mm 39, 45 or 52 calibre ordnance by the replacement of the barrel assembly and towing rod only. This can be accomplished at field level.

The TIG 2000 can be supplied as a conventional towed artillery system or fitted with an APU. In both cases, the system can be operated by a crew of four people.

It is fully air-transportable in aircraft such as the Lockheed Martin C-130 and the model with an APU can enter and leave the aircraft under its own power.

According to United Nations sources, between 1992 and 2000, Israel exported 155 mm towed artillery systems to just two countries, Cameroon and Slovenia. The former country took delivery of a total of 12 systems, four each in 1997, 1998 and 1999. Slovenia took delivery of a total of 18 systems, 10 in 1996 and eight the following year. The Slovenian company of RAVNE has also been involved in local production of the 155 mm Soltam Systems towed artillery systems. The Slovenian 155 mm system is not fitted with an APU.

Soltam Systems TIG 2000 artillery system with APU deployed in firing position 0058420

Description

The TIG 2000 has a conventional split carriage and when in the travelling position the 155 mm/52 calibre barrel is traversed through 180° and locked in position over the closed trails.

The 155 mm ordnance conforms to the Joint Ballistic Agreement and is of the monoblock autofrettaged type forged from high-strength steel. The ordnance is fitted with a muzzle brake and a fume extractor.

The breech ring mechanism consists of the breech ring assembly, breech block assembly and the automatic breech block operating mechanism.

The cradle, pivoted to the saddle by the trunnions, carries all parts of the gun which participate in recoil as well as in the vertical movement during elevation and depression.

The central part of the cradle carries the barrel assembly. On top are the recuperator cylinders and at the bottom, the buffer cylinder and elevating arc. At its lower ends are the trunnions, left and right and at its front end, the hinges for the equilibrators. The recoil length cut-off gear is located at the lower right front end of the cradle.

The recoil mechanism consists of a hydropneumatic recuperator, hydraulic buffer and recuperator and a mechanical recoil length cut-off gear.

Two equilibrators, energised by compressed gas, act as self-adjusting force, balancing the barrel, to allow for each laying and adjusting in elevation and depression.

The mount includes the saddle which carries the ordnance and enables laying in elevation and azimuth. The mounts main assemblies are saddle, elevating and traversing mechanisms.

The saddle is a welded construction which is bolted to the bearings rotating part in the bottom and carrying the cradle. The elevating and traversing gears are mounted on the saddle with the gunner's controls being on the left side.

The elevating mechanism comprises a number of gears incorporating worm and worm wheel with the complete mechanism being fitted on the left side of the saddle.

The traverse mechanism is mounted on the forward part of the saddle and serves to lay the gun in azimuth. A total deflection of 75° (1,334 mm) is possible without shifting the gun.

For fast deflection or when rotating the barrel from the firing position to the travelling position, or vice versa, the mechanism can be disengaged, using a quick release clutch allowing free lateral movement of the barrel.

The carriage serves as the weapon's main support in action and transport. It also carries all of the auxiliary systems that are required for operating the Soltam Systems 155 mm TIG gun-howitzer.

The main carriage assemblies are the central structure and main bearing, firing platform and jack, trails and towing rod assembly, running gears and walking beam, braking system, spade assemblies, support wheels and the electrical system and installation.

The central structure serves as the backbone of the whole weapon and is of all-welded steel construction carrying on its top the main bearing. On the underside is the firing jack with firing platform and on each side is a trail with running gear. On the front part of the structure is the electric battery and hydraulic controls compartment.

The main bearing is of the slewing wire, heavy-duty type. The lower and upper rings of the bearing are stationary and bolted to the central structure.

The central bearing ring carries the saddle and by means of a traversing gearbox, mounted on the saddle and being engaged with the bearing ring, allows traversing of the weapon in relation to the carriage.

The firing jack with its associated platform serves two purposes. First, to assure stability of the gun when it is firing with the gun resting on the platform at the front and spades at the rear. Second, to allow for the spreading and folding of the trails when the gun is brought into and out of action.

The jack is operated hydraulically by the APU or by a back-up electric pump powered by the batteries. Should the electric system fail then the jack will be operated by the manually operated hydraulic pump.

The trails also carry the APU (left trail) and associated systems including braking system, support wheels, spades, towing rod, travelling clamp and the rammer's pressure bottle.

Both trails are fitted with fixtures and mounts to carry the different accessories required to operate the gun. Handles are provided at the rear of each trail to facilitate movement of the weapon, this is in addition to sockets in each trail in which lifting rods can be inserted.

Standard size, four (two either side) running wheels that incorporate hydraulic motors in their hubs are mounted on walking beams for road and cross-country use. Each individual wheel incorporates a disc brake with two braking callipers.

The system is fitted with a hydro pneumatic braking booster which is energised by NATO standard dual-line pneumatic connections. The braking system, built into one of the howitzer's trails, incorporates an emergency pressure tank.

In ordinary operation, the howitzer's brakes are automatically applied simultaneously with the towing vehicles brakes by the drivers pedal. In case the howitzer and pneumatic lines become disconnected from the towing vehicle, the howitzers brakes are automatically activated, stopping the gun.

To support the gun against horizontal firing loads two self-digging spades are used. When travelling, the spades are carried on each trail at its rear and when required for action fold down. In this position, a special dual-purpose mechanism locks and secures the blade. The same lock is used for holding the blade in the travelling position.

One support wheel is located on each trail on the outboard sidewall. The wheels serve for supporting the gun while travelling as well as spreading and folding the trails when coming into and out of action.

Specifications

Model	**39 cal**	**45 cal**	**52 cal**
Calibre	155 mm	155 mm	155 mm
Barrel length			
(with breech)	6,670 mm	7,700 mm	8,745 mm
(without breech)	6,026 mm	7,035 mm	8,080 mm
Chamber entrance	164 mm	167 mm	167 mm
Number of grooves	48	48	48
Rifling inclination	Progressive, ended by 1 : 20	constant 1 : 20	constant 1 : 20
Breech block	wedge type, horizontal, all models		
Obturation	self-sealing metal rings, all models		
Firing mechanism	trip action	trip action	trip action
Weight			
(towed)	9,200 kg	9,400 kg	9,800 kg
(independent)	12,000 kg	12,200 kg	12,600 kg
Length			
(overall)	9.00 m	9.00 m	10.0 m
Width	2.60 m	2.60 m	2.60 m
Height	2.10 m	2.10 m	2.10 m
Ground clearance	0.35 m	0.35 m	0.35 m
Elevation	–3/+70°	–3/+70°	–3/+70°
Traverse			
(total)	78°	78°	78°
Wheels			
(driving)	4 × 11 × 20 in for all versions		
(support wheels)	2 × 8.5 × 10 in for all versions		
Low-speed range	0-8 km/h	0-8 km/h	0-8 km/h
High-speed range	0-17 km/h	0-17 km/h	0-17 km/h
APU	104 hp diesel in all versions		
Gradient	35%	35%	35%
Towing vehicle	5 t truck	5 t truck	5 t truck
Max towing speed	100 km/h	100 km/h	100 km/h
Towing hook load			
(towed)	750 kg	800 kg	900 kg
(independent)	1,000 kg	1,200 kg	1,400 kg

Soltam Systems TIG 2000 artillery system with APU deployed in travelling configuration 0058419

The electrical system comprises two 24 V, 120 Ah batteries as a power source, 24 V DC series motor driving a hydraulic pump, electronic controller for operating the pump motor and monitoring the battery charge state, lighting and signalling system and an electrical harness connection to the towing vehicle.

The TIG 2000 is fitted with a loading tray which incorporates a pneumatic ramming piston enabling uniform loading of the projectile into the barrel chamber at all angles of elevation without the need to lower the barrel.

According to Soltam Systems, the weapon imposes no limitations on the towing vehicles speed when being towed. Its low centre of gravity, overall length and the tandem wheels mounted on the walking beams assure smooth riding at all speeds, bends and rough terrain. The APU fitted to the towed independent version does not influence the towing properties in any way.

The automatic horizontal sliding breech mechanism, fume extractor and wide angle of traverse, together with the loader assist device, enable effective high rates of fire to be achieved. The load assist device transfers projectiles from the supply truck, or from the ground, to the loading tray of the howitzer. A manual back up is provided in case there is a power failure.

In the TIG 2000 fitted with the APU, the latter provides power not only for driving the howitzer but also for bringing it into and out of action. It also assists in preparing the howitzer for the transport mode and for ammunition hoisting.

The driver of the APU has a control lever of the joystick type which controls direction and speed, other controls include starter, speed range selector and parking brake.

By pushing the lever to the desired direction the system moves in that direction, as soon as it is released it returns to the neutral position and the weapon comes to a halt.

As this version can move in any direction up to 17 km/h, it can enter firing positions which could not be reached if the APU was not fitted.

The driver's seat and console are located on the left side of the howitzer and when deployed in the firing position, the driver's console is folded downwards. The system can be maneuvered from the driver's seat or from the ground with the operator walking alongside the gun.

The hydraulic power supplied from the APU powers the howitzer through hydrostatic motors incorporated into the four, (two either side) driving wheels. In addition, it lifts the howitzer off the wheels onto its firing platform, operates the auxiliary jack to hook or unhook the howitzer from the towing vehicle, spreads the trails and afterwards undigs the spades.

In the conventional towed model of the TIG 2000, an assistance hydroelectric power unit is installed to ease crew members tasks.

The TIG 2000 family can fire all natures of 155 mm ammunition, with the range depending on the type of projectile and charge used:

	Maximum range		
Projectile family	**39 cal**	**45 cal**	**52 cal**
M107/ M107A1/ M101	21,900 m	24,000 m	24,500 m
M56/M56A1	23,500 m	25,800 m	26,000 m
L15	24,700 m	27,500 m	30,000 m
M483A1	21,500 m	23,500 m	25,000 m
M397/401, LR-BB	30,000 m	33,000 m	34,500 m
ERFB-BB	28,000 m	30,000 m	41,000 m

The TIG 2000 can be used in the direct and indirect fire roles and to achieve these two objectives, separate sighting systems are installed. The main sighting system is intended for long-range fire with the secondary system serving for the direct fire role, especially self-defence. The sights are mounted on the left side of the TIG 2000.

The TIG 2000 can also be linked to any artillery command, control and communication system.

In place of the conventional indirect laying system, an Advanced Laying Instrument (ALI), based on ring laser gyro techniques, can be fitted. This also provides options for North Finder, GPS and navigation modules.

Soltam Systems 155 mm/52 calibre SPG

The upper part of the TIG 2000 is also used as the basis for the new Soltam Systems 155 mm/52 calibre 6 × 6 self-artillery system covered in an earlier section.

Status

Production as required. In service with undisclosed countries. The earlier Soltam Systems 155 mm Model 845 is known to be in service with Slovenia, which took delivery of 18 weapons between 1996 and 1998. Cameroon has taken delivery of 12 Soltam Systems 155 mm weapons of an undisclosed type. Botswana has taken delivery of a quantity of Soltam Systems 155 mm system of an unknown type.

Manufacturer

Soltam Systems Limited.

UPDATED

Soltam Systems 155 mm M-71 gun-howitzer

Development/Description

The Soltam Systems 155 mm M-71 gun-howitzer is a further development of the Soltam Systems 155 mm M-68 gun-howitzer and uses the same recoil system, breech and carriage. The first prototype was completed in 1974 and first production units the following year.

The main differences between the M-71 and the M-68 are that the former has a longer 39 calibre barrel and is fitted with a rammer driven by compressed air. This permits rapid loading at all angles of elevation. The rechargeable cylinder for the M-71 is mounted on the right trail. The barrel has 48 grooves, length of rifling is 5.038 m, width of groove 6.6 mm, width of land 3.5 mm and depth of groove is 1.2 mm.

Soltam Systems is now concentrating its marketing on the latest 155 mm TIG 2000 artillery system which is available with a 39, 45 or 52 calibre ordnance.

Specifications

Calibre: 155 mm
Barrel length: 6.045 m
(incl breech ring) 6.67 m
Muzzle brake: single baffle
Recoil: hydropneumatic
Breech mechanism: horizontal wedge
Carriage: split trail
Shield: no
Weight:
(travelling order) 9,200 kg
Length:
(travelling) 7.5 m
Width:
(travelling) 2.58 m
Height:
(travelling) 2.115 m
Ground clearance:
(travelling) 0.38 m
Track: 2.2 m
Elevation/depression: +52/–3°
Traverse:
(total) 84°
Rate of fire:
(short period) 5 rds/min
(sustained) 2 rds/min
Max range:
(standard projectile) 23,500 m
(base bleed projectile) 30,000 m
Crew: 8
Towing vehicle: 5 ton (6 × 6) truck
Max towing speed: 100 km/h

Variants

Soltam Systems developed to the prototype stage a self-propelled model of the M-71 gun-howitzer based on a Centurion MBT chassis. This never entered production or service.

Status

Production complete but can be resumed if additional orders are placed. In service with Israel (40 to 50), Singapore (38), South Africa (where it is known as the G4), Thailand (27) and three other undisclosed armed forces. According to United Nations sources, between 1992 and 2000, Israel exported 155 mm towed artillery systems to just two countries, Cameroon and Slovenia. The former country took delivery of 12 systems. Slovenia took delivery of 10 systems in 1996 and eight in 1998. South Africa has recently sold 24 of its G4 gun-howitzers to Chile which will be used in conventional role as well as the coast defence role.

Manufacturer

Soltam Systems Limited.

UPDATED

Soltam Systems 155 mm M-71 gun-howitzer in travelling position with barrel swung through 180° and locked over trails (Israeli Ministry of Defence)

Soltam Systems 155 mm M-71 gun-howitzer in firing position

Soltam Systems 155 mm/45 calibre M-46S field gun

Development

Based on its extensive experience in the design, development and production of towed and self-propelled artillery systems, Soltam Systems has completed and qualified the prototype of a 130 mm M-46 field gun (fully described under Russian Federation and Associated States) upgraded with a 155 mm/45 calibre ordnance.

Soltam Systems can either provide the kit, enabling the user to upgrade existing 130 mm M-46 weapons in its own facilities, or the company can undertake this work in its facilities in Israel.

The Israeli Defence Force is estimated to have 100 captured 130 mm M-46 guns in service but it is not known if this upgrade kit was aimed at the local requirement, for the export market or for both the home and export markets.

Early in 2000, India awarded Soltam Systems a contract to upgrade 180 130 mm M-46 field guns to a 155 mm/39 and 155 mm/45 calibre standard.

The first 180 weapons will be upgraded in Israel with the follow-on batch, expected to consist of 220 to 250 guns, being upgraded at the Indian Ordnance Factory Board facilities with kits provided by Israel.

Recent information has stated that the total value of the first contract to upgrade 180 weapons in Israel is about Rs2.07 billion (US$48.13 million). Upgrading of the follow on batch of weapons will be carried out in India at Kanpur and Jabalpur.

This upgrade is also applicable to the China North Industries Corporation (NORINCO) equivalent of the Russian M-46, the Type 59-1.

Description

The Soltam Systems upgraded M-46 is fitted with a monobloc autofrettaged 155 mm barrel 45 calibres long that incorporates a horizontal sliding breech block which is automatically operated. It uses standard M82 primers that are automatically ejected after firing.

The complete weapon system can be divided into two major assemblies, the ordnance and carriage.

Whereas the carriage, understructure and recoil mechanism remain unchanged, the ordnance is completely replaced by the new 155 mm 45 calibre barrel assembly that, together with the pneumatic ramming device, comprises the conversion kit.

The aiming and sighting systems can be brought up to Western standards as an option and a modern inertial aiming device can also be fitted.

The standard sighting system of the M-46 consists of optical sights for long range indirect fire and direct fire. The sight for indirect laying includes a sight instrument and a panoramic periscope. The sight for direct aiming includes a telescope mount and an elbow telescope.

The sight instrument is mounted on the left side of the saddle while the panoramic telescope, mounted on the top of the sight is used for azimuth laying. The elbow telescope is used for direct fire in case of defence against enemy armour or when flat trajectory fire is required.

The 155 mm conversion kit comprises the 45 calibre ordnance consisting of the barrel, breech ring assembly, breech block assembly, automatic breech block operating mechanism, new single-baffle muzzle brake, fume extractor and cradle adaptors with the loader assist device and its associated installation.

According to Soltam Systems, the upgraded weapon has a number of unique features:

(1) Firing cannot commence unless the breech block is completely closed and locked

(2) The horizontal breech mechanism, operated together with the pneumatic ramming device, enables a high rate of accurate fire to be achieved without the need to lower the ordnance for loading

(3) The gun crew is clear of the overpressure effects and clean of residual post-firing gases due to the efficient muzzle brake and bore evacuator.

The Soltam Systems upgraded M-46 gun can fire all the standard 155 mm types of ammunition in worldwide use starting with the M107 HE family of ammunition and ending with the latest 155 mm ERFB-BB family of long-range ammunition.

With unassisted ammunition, ranges of up to 30,000 m can be achieved while, with assisted ammunition, ranges of up to 39,000 m can be achieved.

Specifications

Calibre: 155 mm
Barrel length: 7,700 mm with breech ring
Muzzle brake: single baffle
Diameter of chamber: 172 mm
Volume of chamber: 25.7 litres
Rifling: 48 grooves
Rifling length: 5,851 mm
Rifling inclination: 1:20 constant right-hand twist
Breech block: horizontal wedge, automatically-operated
Carriage: split trail
Shield: yes
Weight:
(travelling) 8,850 kg
Length:
(travelling) 11.73 m
Width:
(travelling) 2.45 m
Height:
(travelling) 2.55 m
Elevation/depression: +45/–2.5°
Traverse: 25° left and right
Max range:
(standard ammunition) 25,800 m
(ERFB-BB) 39,000 m

Self-propelled 155 mm M-46

Soltam Systems have developed a new 155 mm/52 calibre self-propelled artillery system on a Tatra (6 × 6) truck chassis called the ATMOS 2000. Details of this are given in the Self-propelled guns and howitzers (wheeled) section. This system could also be fitted with the Soltam Systems 155 mm/45 calibre M-46S upgrade.

Status

Production as required. In early 2000 India ordered this upgrade package in 155 mm/39 calibre and 155 mm/45 calibre configurations.

Manufacturer

Soltam Systems Limited.

UPDATED

Soltam Systems 155 mm/45 calibre M-46S field gun in travelling configuration

Soltam Systems 155 mm/45 calibre M-46S field gun in firing configuration

Soltam Systems 155 mm M114S howitzer

Development

A number of manufacturers have developed modernisation packages that utilise a 155 mm/39 calibre ordnance, for the old US 155 mm M114 towed howitzer which was developed before the Second World War and is now obsolete.

The upgraded Soltam Systems M114S utilises a 155 mm/33 calibre barrel enabling a standard 155 mm M107 HE projectile to be fired to a maximum range of 18,300 m compared to the 14,600 m of a standard M114 and 18,100 m when fired from a 39 calibre barrel.

The conversion can be carried out in the user's own facilities using a kit supplied by Soltam Systems or Soltam Systems can carry out the upgrade work.

The Israel Defence Force has about 50 155 mm M114 towed howitzers held in reserve, as far as is known these have not been modified by Soltam Systems to the M114S standard. United Nations sources have shown that none of these weapons were exported between 1992 and 2000.

Soltam Systems 155 mm M114S upgraded howitzer in firing position

Specifications

Calibre: 155 mm
Barrel length: 33 calibres
Tube length: 4,990 mm
Muzzle brake: single baffle
Diameter of chamber: 172 mm
Volume of chamber: 17.5 litres
Rifling: 48 grooves
Rifling length: 4,200 mm
Rifling inclination: 1:20 constant, right-hand twist
Breech block: original M114
Carriage: split trail
Shield: yes
Weight: 5,900 kg
Length:
(travelling) 9.20 m
Width:
(travelling) 2.40 m
Height: 1.80 m
Elevation/depression: +63/–2°
Traverse: 25° left/24° right
Max range:
(M107 projectile) 18,300 m
(ERFB-BB) projectile) 22,500 m
Crew: 6

Description

The 155 mm/33 calibre version does not require such extensive modifications to the existing weapon when compared to the other 155 mm/39 calibre modernisation packages and, according to Soltam Systems, it is a more cost-effective solution.

The new 155 mm/33 calibre barrel has a high-efficiency muzzle brake and is fitted with the original breech ring and breech mechanism of the M114. The new ordnance has a constant 1:20 twist rifling with 48 lands and grooves. The muzzle brake is the same as that fitted to other Soltam Systems 155 mm towed artillery systems.

The cradle remains unchanged although a front bushing has to be added by inserting a new bushing into the original one without the need for extensive machining. The existing spring-loaded equilibrators remain untouched but the upper location is changed, using new brackets bolted to the original points on the saddle.

All other assemblies and components remain untouched, as well as all the integrated logistics aspects.

A loader assist device based on a pneumatic ramming piston is provided as an add-on kit. The loader assist device enhances the rate of fire and loading accuracy as well as easing crew workload by reducing the need to lower the gun for loading.

The conversion kit includes the following:

(1) 155 mm/33 calibre tube and muzzle brake
(2) Pneumatic ramming system and its installation
(3) Mounts and fixtures kit
(4) Direct firing telescope as an option
(5) Upgrading instruction and technical data pack.

The Soltam Systems upgraded M114S can fire all the standard types of 155 mm ammunition in worldwide use, starting with the M107 family of ammunition and ending with the latest ERFB - BB long-range ammunition family.

The following ranges are obtained:

M107 projectile	18,300 m
M56 projectile	19,100 m
ERFB-BB projectile	22,500 m

Due to improved gun balance, six crew members are enough for the deployment of the M114S instead of eight with the original M114.

Status

Development complete. Production as required. As far as is known this upgrade has not been adopted by the Israel Defence Force.

Manufacturer

Soltam Systems Limited.

UPDATED

ITALY

Otobreda 105 mm Model 56 Pack Howitzer

Development

The 105 mm Model 56 Pack Howitzer was developed by the Italian Army and produced by OTO Melara (now Otobreda) in the mid-1950s and entered production in 1957. Since then over 2,500 have been built and sales have been made to more than 30 countries.

The advantage of the Model 56 is that it can be dismantled into 11 subassemblies for transport across rough terrain, lifted by helicopters such as the Bell UH-1 and its profile can be lowered for anti-tank use.

United Nations sources state that between 1992 and 2000, Italy only exported this weapon to one customer, Brazil. This country took delivery of 18 weapons in 1995 and a further 10 in 1997. Canada exported surplus 105 mm Model 56 pack howitzers to Brazil with eight being delivered in 1998 and a further eight in 1999.

Early in 1997, it was revealed that China North Industries Corporation (NORINCO) had built a weapon almost identical to the Otobreda 105 mm Model 56 Pack Howitzer and details of this are given under China.

Specifications

[data in square brackets relate to anti-tank configuration]
Calibre: 105 mm
Barrel length: 1.478 m
Muzzle brake: multibaffle
Recoil: hydraulic buffer and helical recuperator
Breech mechanism: vertical sliding wedge
Carriage: split trail
Shield: optional
Weight:
(travelling order, section trails) 1,290 kg
(firing position, 3 section trails) 1,290 kg
Length:
(travelling) 3.65 m
(firing) 4.8 [4.7] m
Width:
(firing) 2.9 [4] m
(travelling) 1.5 m
Height:
(travelling) 1.93 m
(firing) 1.93 [1.55] m
Axis of bore:
(firing) 1.07 [0.693] m
Track: 1.14 [1.32] m
Elevation/depression: +65/–5° [+25/–5°]
Traverse:
(total) 36° [56°]
Rate of fire:
(for 30 min) 4 rds/min
(for 1 h) 3 rds/min
(anti-tank) 8 rds/min
Max range:
(normal HE projectile) 10,575 m
Crew: 7
Towing vehicle: LWB (4 × 4) Land Rover

Otobreda 105 mm Model 56 Pack Howitzer

Otobreda 105 mm Pack Howitzer in travelling configuration

As far as is known this has yet to enter volume production for the export market.

Description

The 105 mm barrel assembly is composed of the barrel, breech ring, slipper and muzzle brake. The weapon has a multibaffle muzzle brake. The vertical sliding breech block has a percussion firing mechanism which can be cocked automatically or by hand. Firing is either by hand lever or by lanyard.

The gun carriage comprises the undercarriage, saddle, cradle, upper and lower recoil systems, two balancing springs, two sights and the shield. The undercarriage consists of a cross head housing the saddle centre pivot pin, two stub-axles, spring equalising rocker arm and two articulated arms. Each of the stub-axle supporting arms has two bearings into which the stub-axle can be fitted, this carries the wheel assembly and its handbrake. Each of the articulating brackets carries a trail leg of two or three sections which can be folded up and interlocked. The end section of the trail may be anchored to the ground by a knife or rock spade.

The saddle, with its centre pivot mounting, is provided with two side members with rear-mounted lugs which carry the elevating and traversing gear. The recoil system is contained in two cradles, upper and lower, which slide on top of each other. The upper cradle contains the two recuperator-buffers and the lower cradle contains one buffer and two recuperators. The two balancing springs have helical springs operating in sliding sleeves and the cradle is mounted on two trunnion bearings.

The sighting equipment consists of an indirect sight on the left and an anti-tank direct sight on the right. The splinter-proof shield mounted on the saddle is in two halves, both of which can be removed to lighten the weapon.

As a field weapon the 105 mm Pack Howitzer wheels are overslung and the barrel has an elevation of +65°, a depression of −5° and a traverse of 18° left and right. As an anti-tank weapon the wheels are underslung and the barrel has an elevation of +25°, a depression of −5° and a traverse of 28° left and right.

The weapon can be dismantled into 11 subassemblies, the heaviest of which weighs 122 kg, in 3 minutes and can be reassembled in 4 minutes.

The 105 mm Pack Howitzer fires the same range of ammunition as the US 105 mm M101; HE(M1) with the complete round weighing 21.06 kg, a muzzle velocity of 472 m/s and the fixed HEAT (M67) round which weighs 16.7 kg, has a muzzle velocity of 387 m/s and will penetrate 102 mm of conventional steel armour.

Status

Production as required. In service with the following countries:

Country	Quantity	Comment
Argentina	84	used by army and marines
Bangladesh	30	
Botswana	4	
Brazil	100	including 26 recently supplied from Italy and 16 from Canada
Chile	36	
Ecuador	24	
Germany	19	
Greece	18	
India	50	
Iraq	n/avail	
Italy	267	
Kenya	8	
Malaysia	75	
Nigeria	200	
Nepal	14	
Pakistan	50	
Peru	20	
Philippines	n/avail	
Portugal	24	
Sierra Leone	n/avail	unconfirmed user
Somalia	90	
Spain	170 + 12	army + marines
Sudan	n/avail	
Venezuela	40	
Zambia	18	

Manufacturer

Otobreda SpA.

UPDATED

KOREA, SOUTH

WIA Corporation 155 mm KH179 howitzer

Description

Development of the KH179 155 mm howitzer by the Kia Machine Tool Company (which is today known as the WIA Corporation) began in 1979 and was completed in 1982. It has been in service with the Republic of Korea Army since early 1983. The conversion programme is believed to have been completed some years ago. It is estimated that South Korea has about 1,700 155 mm towed artillery systems but it is now known as to how many of these are M114s upgraded to the KH179 standard.

The 155 mm KH179 howitzer is a conversion of the American M114A1 howitzer and carriage upgraded to mount a new 155 mm/39 calibre barrel. Certain other parts of the carriage have been altered to suit the new barrel and new fire-control equipment has been fitted. The KH179 is light enough to be airlifted by a Boeing Vertol CH-47C/D helicopter and can also be carried inside a Lockheed Martin C-130 Hercules transport aircraft.

The new 155 mm/39 calibre barrel is of monobloc construction with an interrupted screw breech mechanism. The rifling has a constant 1:20 twist and there are 48 grooves. High-strength alloy steel is used throughout the barrel construction and heat treatment and autofrettaging are employed to provide maximum barrel life. Separate recoil and recuperator systems are used in what is stated to be a variable length, hydropneumatic, independent design.

On the carriage, the layer is on the left of the breech to operate the traverse handwheel on the left of the top carriage. Another member of the gun crew turns the elevating handwheel on the right of the breech. The layer is provided with a dial sight and telescope for aiming and another telescope for direct fire. All fire controls and some of the accessory equipment on the KH179 are equipped with radio-activated light sources.

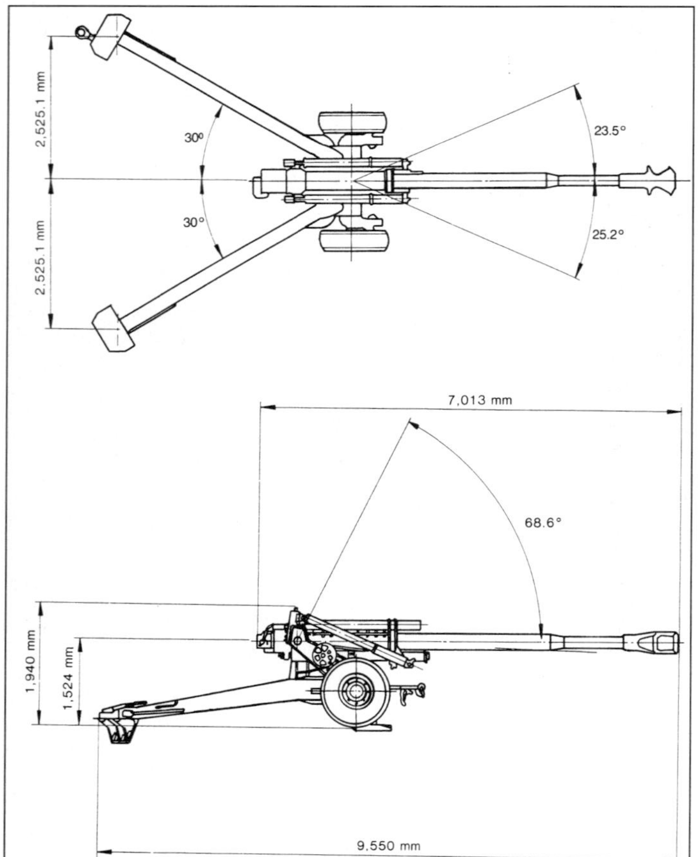

155 mm KH179 howitzer in travelling configuration **2002**/0130003

Top and side views of 155 mm KH179 howitzer in firing position

Specifications
Calibre: 155 mm
Barrel length:
(total) 7.013 m
Tube length: 6.096 m
Muzzle brake: double baffle
Breech mechanism: interrupted thread
Recoil mechanism: variable, hydropneumatic
Recoil length:
(long) 1.524 m
(short) 1.041 m
Carriage: split trail
Shield: no
Weight:
(total) 6,890 kg
Length:
(overall, travelling) 10.389 m
Width:
(overall) 2.438 m
Height:
(overall, travelling) 2.77 m
Ground clearance: 0.28 m
Elevation/depression: +68/0°
Traverse: 23.5° left, 25.2° right
Rate of fire:
(max) 4 rds/min
(sustained) 2 rds/min
Max range:
(standard HE) 22,000 m
(RAP) 30,000 m
Muzzle velocity: 826 m/s
Towing speed:
(road) 70 km/h
(cross-country) 16 km/h
Towing vehicle: 5 ton (6 × 6) truck

Pneumatic equilibrators are connected between the yoke of the recoil mechanism and the top carriage, to keep the handwheel loads within acceptable limits.

For indirect firing, a panoramic telescope with a x4 magnification is fitted on the left side. For direct firing, the elbow telescope with a x3.5 magnification and a 13.5° field of view is mounted on the right and has an effective performance up to 1,500 m.

The KH179 can fire existing NATO 155 mm ammunition and can also accommodate enhanced performance ammunition such as the ERFB projectiles. Range with RAP is stated to be 30,000 m and for normal HE projectiles the range is 22,000 m. Maximum rate of fire is 4 rds/min, dropping to 2 rds/min for sustained fire.

To reduce weight, the shield of the M114 has been removed on the upgraded KH179 weapon.

Iran 155 mm howitzer

The Iranian DIO has produced a very similar weapon to the K179 howitzer called the HM41. Details are given earlier in this section under Iran.

Status

Production complete. In service with the Republic of Korea Army. As far as is known this weapon has not been exported.

Manufacturer

WIA Corporation

UPDATED

WIA Corporation Tool Company 105 mm KH178 light howitzer

Development

In the preliminary stages of the design of a new 105 mm light howitzer for use by the Republic of Korea Army, the Kia Machine Tool Company (today the WIA Corporation) procured examples of the now BAE Systems, RO Defence 105 mm Light Gun and the German Rheinmetall conversion of the US 105 mm M101 howitzer.

Aspects of both have been incorporated in the KH178 105 mm light howitzer. Production of the KH178 started in 1984 and it is now in service with the Republic of Korea Army. This conversion programme is believed to have been completed some years ago.

It is believed that South Korea took delivery of about 1,700 M101s from the United States and that a significant portion of these have been upgraded to the 105 mm KH178 configuration.

The WIA Corporation 105 mm KH178 light howitzer and the 155 mm KH179 system, which features a 39 calibre ordnance, have both been offered on the export market but so far there have been no known sales.

105 mm KH178 howitzer in firing position

Description

The KH178 comprises the CN78 105 mm cannon, the RM78 recoil mechanism, the CG78 carriage and a new fire-control system. The barrel is 34 calibres long and is fitted with a double-baffle muzzle brake. A horizontal sliding breech block is used with a percussion firing mechanism. The barrel is autofrettaged and has a life of 7,500 EFC. Progressive twist rifling is used, changing from 1:35 to 1:18 at the muzzle.

The recoil mechanism is of the hydropneumatic type with a constant recoil length. The recoil and recuperator cylinders are mounted over and under the barrel. In order to reduce elevation handwheel loads, the equilibrators are of a new type using a hydropneumatic and spring system.

On the split trail carriage the trail legs are reinforced and a shield may be fitted. Run-flat tyres are used. All the fire-control equipment and some accessory devices are fitted with self-luminous sources.

The KH178 can fire all current standard 105 mm ammunition plus a new single-increment super propellant M200. This gives the weapon a range of 14,700 m. Firing a Rocket-Assisted Projectile (RAP) the range is 18,000 m.

Status

Production complete. In service with the Republic of Korea Army. As far as is known this weapon has not been exported.

Manufacturer

WIA Corporation.

UPDATED

Specifications
Calibre: 105 mm
Barrel length: 4.48 m
Tube length: 3.922 m
Muzzle brake: double baffle
Carriage: split trail
Shield: optional
Weight:
(total) 2,650 kg
Length: 7.56 m
Width: 2.146 m
Height: 1.575 m
Ground clearance: 0.386 m
Elevation/depression: +65/−5°
Traverse:
(total) 45.5°
Rate of fire:
(max) 15 rds/min
(sustained) 5 rds/min
Max range:
(HE) 14,700 m
(RAP) 18,000 m
Muzzle velocity: 662 m/s
Towing speed:
(road) 56 km/h
(cross-country) 16 km/h
Towing vehicle: 2½ ton (6 × 6) truck

NETHERLANDS

RDM Technology 155 mm M139 howitzer and M114/39 modified howitzer

Development

RDM Technology has developed a conversion kit which enables the old US 155 mm M114 towed howitzer to fire new ammunition including ERFB-BB. This kit is a further development of the SRC International 155 mm M114/39 modification kit. The company is no longer trading.

RDM Technology converted two Canadian M114s to the M114/39 configuration and these have both completed trials in Canada. At present there are no known plans for Canada to upgrade their M114 systems to this standard.

A Memorandum of Understanding (MoU) was signed late in 1986 between Denmark, the Netherlands and Norway for a total of 226 M114/39 conversion kits. RDM Technology was the prime contractor, the major subcontractors for offset work being DISA of Denmark and Norsk Forsvarsteknologi AS of Norway. The first M114/39 was delivered to the Royal Norwegian Army in August 1988 and the project was completed early in 1990.

RDM Technology can manufacture, in addition to the M114/39 modification kit, new guns under the designation M139. These are virtually identical to the M114/39 and their ballistic performance is the same.

RDM Technology handed over the first rebuilt systems in spring 1988. Each of the three countries was responsible for converting those weapons for its own army with each company feeding components to the others.

Following a competition, the now Bofors Defence of Sweden was selected to manufacture the barrels for the M114/39 conversion programme; the first sample barrel was delivered in 1984 for extensive tests and the first production barrels were delivered in 1987.

By early 2002, only sales of the upgraded 155 mm M114/39 had been made, there have been no sales of complete new guns.

In 1993, the Royal Netherlands Army stated that, as part of its overall defence cuts, only 51 155 mm M114/39s would be kept.

RDM Technology also offers upgrades for M109 self-propelled artillery systems with 39, 47 and 52 calibre barrels. Details of these are given in the Self-propelled guns and howitzers (Tracked) section.

Description

The modification is accomplished using the maximum number of existing gun parts and new components only where necessary. The major change involves a new 39 calibre barrel with a larger chamber, increased twist rifling and a muzzle brake which makes the weapon, with current projectiles and charges, ballistically similar to the US 155 mm M109A2 self-propelled and US 155 mm M198 towed howitzers.

The fitting of a new ordnance has required a number of changes to the carriage including moving the jack forwards and modifying the trails to compensate for the forward shift of the gun's centre of gravity, fitting larger capacity equilibrators with temperature compensating devices, a new metering rod in the recoil system with new seals, and the modification of the gear trains in both the elevating and traversing systems resulting in higher mechanical advantages.

The new 39 calibre 155 mm barrel has a constant 1:20 twist rifling with 48 lands and grooves. This provides a projectile travel of 5.08 m with a chamber volume of 19,020 cm^3. The barrel is autofrettaged to provide maximum fatigue life and is fitted with a screw-block breech mechanism, a conventional-type obturator pad and split ring seal obturator.

A three-port high-efficiency muzzle brake is fitted to reduce recoil. In order to accommodate the new barrel, the existing cradle has been replaced by a new structure which includes an anti-rotation device.

The original recoil and metering rod, which is in the recoil assembly, has been replaced by one which is optimised with respect to increased muzzle velocity, heavier weight of the ordnance and therefore longer recoil distance. The original seals have been replaced with modern components, in order to seal off the larger fluid pressure generated during recoil. This enables the weapon to be fired under all operational conditions without damage. The original front yoke has been replaced with a new structure allowing adaptation of the new ordnance and the provision of attachments for the new equilibrators. The original spring equilibrators were unable to compensate for the increased unbalanced moment and have been replaced by higher capacity pneumatic equilibrators.

The original equilibrator bracket is replaced by one which allows for temperature compensation compatible with the pneumatic equilibrators. The gear trains in the elevating and traversing mechanism have been slightly modified in order to increase the mechanical advantage, which has enabled the handwheel loads to be kept within acceptable operational limits.

In order to compensate for the shift in the centre of gravity of the weapon, the jack and the float assembly have been moved forward. This is accomplished by securing a spacer to the bottom carriage which in turn holds the jack. This provides a stable firing platform under all conditions.

Due to the forward shift of the centre of gravity, the original trails have been modified and this also allows the crew to manhandle the weapon easily when it is standing on its wheels. All other original components, including the sight and sight mount are retained, simplifying both training and logistic support.

A pneumatic power rammer can be incorporated as an optional extra to permit power loading at all angles of elevation. The rammer is powered by compressed air from an air cylinder mounted on the right trail leg.

RDM Technology has developed, to the prototype stage, a VW air-cooled diesel auxiliary power unit that can be installed on the forward part of the carriage. This develops 60 hp and enables the M139 and M114/39 to move around under their own power at 10 km/h. So far, no country is known to have adopted this auxiliary power unit.

RDM Technology can either convert existing guns at its own facilities or supply kits to enable users to undertake the conversion themselves.

Specifications

Calibre: 155 mm
Barrel length:
(muzzle face to rear face, not including muzzle brake) 6.016 m
Muzzle brake: multibaffle
Carriage: split trail
Shield: optional
Weight: 7,800 kg
Length:
(travelling) 10 m
Width:
(travelling) 2.44 m
Height:
(travelling with lunette 0.737 m from ground) 2.31 m
Trunnion height: 1.42 m
Track: 2.08 m
Elevation/depression: +63/-2°
Traverse: right 25°, left 24°
Rate of fire:
(max) 4 rds/min
Range:
(M107 projectile) 18,100 m
(ERFB) 24,600 m
(ERFB BB) 32,400 m
Crew: 11
Towing vehicle: 5 ton (6 × 6) truck

Status

Production as required. In service with the following countries:

Country	Quantity	Comment
Canada	2	trials only
Denmark	96	
Netherlands	82	31 for sale
Norway	48	

Manufacturer

RDM Technology bv.

UPDATED

Ballistic performance

Charge	Zone	M107 projectile Muzzle velocity (m/s)	Max range (m)	ERFB-HE projectile Muzzle velocity (m/s)	Max range (m)	ERFB + base bleed Muzzle velocity (m/s)	Max range (m)
M3A1	1	–	–	–	–	–	–
	2	237.7	5,000	*	*	*	*
	3	277.4	6,500	*	*	*	*
	4	318.5	8,300	*	*	*	*
	5	374.9	9,800	*	*	*	*
M4A2	3	292.6	7,200	*	*	*	*
	4	336.8	8,500	*	*	*	*
	5	393.2	10,300	*	*	*	*
	6	475.5	12,400	465	12,700	*	*
	7	565.4	14,800	552.7	15,700	*	*
M119A1	8	684.3	18,100	671.5	20,100	665	23,600
M9	9			785	24,600	785	30,400

* The usefulness of the ERFB projectile becomes apparent only with higher charges. ERFB can be used with smaller charges, but is not cost effective.

RDM Technology M114/39 155 mm howitzer in firing position (Directorate of Matériel, Royal Netherlands Army, Communication Office)

RDM Technology M114/39 155 mm howitzer in travelling configuration being towed by 6 × 6 truck
0006045

RDM Technology 155 mm M-46 field gun upgrade

Development/Description

In 1994, RDM Technology, as a private venture, completed a study to upgrade the older Russian 130 mm M-46 field gun which has a maximum range of 27,000 m. This upgrade is also applicable to the China North Industries Corporation (NORINCO) equivalent of the M-46, the Type 59-1.

The upgrade would include replacing the existing 130 mm ordnance with a new 155 mm 45 calibre ordnance which would enable a range of 39,000 m to be achieved firing an extended-range full-bore base bleed projectile.

Status

In early 2002 this remained at the study phase. There are no plans to build a prototype of this system yet.

Manufacturer

RDM Technology bv.

UPDATED

RDM Technology 105 mm M101/33 howitzer

Development

Building on its experience in the production of modernisation packages for the 155 mm M114 towed howitzer, RDM Technology has developed as a private venture, an upgrade package for the widely deployed US 105 mm M101 series of towed howitzer which was originally developed over 60 years ago.

When upgraded, the M101 can fire more modern 105 mm ammunition with enhanced range and lethality. The modernisation is accomplished by using a maximum of existing gun parts and only adopting new components where required.

In late 1990, an upgraded M101/33 howitzer was demonstrated to the Canadian Forces in Canada.

By late 1993, three prototypes of the upgraded RDM Technology 105 mm M101/33 howitzer had been completed. The third weapon underwent extensive trials in Indonesia, which were completed late in 1993.

In 1992, RDM Technology and the now BAE Systems, RO Defence of the UK reached an agreement whereby the two companies will co-operate in marketing their 105 mm artillery systems.

In March 1994, RDM Technology was awarded a contract by Canada to upgrade 96 M101 (plus an option on another batch of 36 which has yet to be exercised) to the M101/33 calibre configuration for the Canadian Forces with the complete programme to be carried out in the Netherlands.

First weapons were shipped back to Canada late in 1995 in batches of five and all work completed by 1998. The barrels for the Canadian weapons were manufactured by BAE Systems, RO Defence.

Chile awarded BAE Systems, RO Defence of the UK a contract to upgrade a substantial number of M101s and details of this are given later in this entry under the RDM Technology/BAE Systems, RO Defence teaming arrangement. RDM Technology completed this programme at the end of 2001.

RDM Technology 105 mm M101/33 howitzer deployed in firing position 0006046

US-built 105 mm towed howitzer upgraded to the 105 mm M101/37 configuration by BAE Systems, RO Defence/RDM Technology and deployed in the firing position

Description

The major modification includes the installation of a new 105 mm/33 calibre barrel with progressive rifling and a new muzzle brake. The M101 is also modified in a number of other areas including the incorporation of larger capacity equilibrators, new metering and recoil rod in the recoil system and new seals, replacement of the original elevating arc and reinforcement on the drawbar.

The new 105 mm/33 calibre barrel is ballistically similar to the latest generation of 105 mm artillery ammunition and suitable for high pressure, with the design being for 4,500 bars.

It has progressive rifling (1.35 to 1.18) over a length of 2,794 mm whereas rifling is constant (1.18) over the last 306 mm (muzzle end). The barrel has 36 lands and grooves and provides a shot travel of 3,100 mm with a chamber volume of 2,507 cm^3. This is identical to the standard M101 allowing the upgraded weapon to fire all existing types of 105 mm ammunition.

The modified breech mechanism is still of the horizontal sliding type but, due to the higher forces, the original breech ring is replaced by a stronger one which allows it to use all modern ammunition types.

The remaining parts of the original breech mechanism can still be used including the firing mechanism. In order to reduce the increased load on trunnions and recoil when using new ammunition, a new three-slotted muzzle brake is fitted to the threaded end of the barrel. This muzzle brake provides for a substantial reduction in the recoil forces while still maintaining blast overpressure within acceptable limits.

A modification to the recoil system was required due to the increase of the recoil momentum resulting from the use of the 33 calibre barrel and more modern ammunition. The increased recoil momentum is absorbed partly by the muzzle brake and partly by the recoil mechanism. The original control rod with the recoil cylinder is replaced and tailored with respect to the increased recoil momentum of the heavier cannon. To limit the recoil, the diameter of the recoil cylinder is increased. The recoil distances have not been changed and are still the same as the original. The original seals are replaced in order to retain the larger fluid pressure generated during the recoil.

As the original two-spring equilibrator is unable to compensate for the increased tipping movement, a new equilibrator has been mounted in such a way that it now houses three nested springs instead of the original two. The new equilibrator reduces the weight moment at all angles of elevation.

The cradle system does not need to be changed, although a small modification is required in order to connect the new equilibrator. No modifications are required to the carriage of the M101. A small reinforcement on the drawbar connection of the trails is required as well as reinforcement of the trails.

The original M101 howitzer provides for a maximum elevation of +65° but, in order to increase the overlap in ranges, the original elevation arc has been replaced which allows elevation to +70°.

All other components, including the sight and mount, are retained to simplify both training and logistic support.

RDM Technology/BAE Systems, RO Defence Teaming

Early in 1994, it was announced that the now BAE Systems, RO Defence of the UK had teamed with RDM Technology to offer a number of upgrade packages for the 105 mm M101/M101A1 towed howitzer.

Marketing lead for projects will be undertaken by either company according to perceived strength.

The original RDM Technology upgrade and the one selected by Canada, use a 105 mm/33 calibre barrel, but now an additional two upgrades are being offered. The first has a 30 calibre ordnance which is compatible with the BAE Systems, RO Defence L119 Light Gun and the second, a 37 calibre ordnance which is compatible with the BAE Systems, RO Defence L118 Light Gun.

The 105 mm/30 calibre barrel allows the standard US family of M1 ammunition to be fired to a maximum range of 14,200 m or, using the M913 rocket-assisted projectile, a range of 19,500 m can be achieved. Firing the Eurometaal ERM1 (Extended-Range M1) base bleed projectile a range of 18,500 m is achieved.

The 105 mm/37 calibre version allows a standard BAE Systems, RO Defence L31 high-explosive projectile (originally developed for the Abbot self-propelled gun) to be fired to a maximum range of 17,200 m or 20,500 m with an extended-range high-explosive projectile.

The L31 high-explosive projectile has 25 per cent greater lethality than the old M1 which was developed many years ago. The 37 calibre version has the larger 5 litre chamber which gives a stretch potential for new ammunition in the future.

The prototype of a Brazilian 105 mm M101 system upgraded to the M101/37 calibre system has now been successfully demonstrated with 100 rounds being fired, including Abbot charge super at high temperature.

Full inspection after the UK trials revealed no strength problems and the muzzle blast overpressure was acceptable.

Late in 1997, the now BAE Systems, RO Defence, was awarded a contract for an undisclosed amount by FAMAE Ordnance Limited of Chile for the supply of a large number of upgrade kits for the old US supplied 105 mm M101 towed artillery systems deployed by the Chilean Army.

This is the first upgrade contract for which BAE Systems, RO Defence has been selected as prime contractor since signing the co-operation agreement with RDM Technology of the Netherlands covering the joint marketing of the RDM designed upgrade.

Under the terms of the contract, BAE Systems, RO Defence supplied the 105 mm 33 calibre barrels to RDM Technology, the main subcontractor, which is supplying the other parts including new muzzle brake and breech ring, larger capacity equilibrators, new metering and recoil rod in the recoil system and new seals, replacement of the original elevating arc and reinforcements for the drawbar and trails.

Specifications

Model	**M101**	**M101/33 cal**	**M101/30 cal**	**M101/37 cal**
Calibre length	23	33	30	37
Max range	11,200 m	15,700 m	14,200 m	17,200 m
Max range (with RAP/BB round)	nil	19,500 m	19,500 m	20,500 m
Min range (high angle)	2,000 m	2,100 m	2,100 m	2,500 m
Length (overall)	6.00 m	7.40 m	7.08 m	7.85 m
Width (overall)	2.15 m	2.15 m	2.15 m	2.15 m
Height				
(M101)	1.52 m	1.55 m	1.52 m	1.52 m
(M101A1)	1.58 m	1.59 m	1.58 m	1.58 m
Total weight				
(M101)	2,030 kg	2,150 kg	2,130 kg	2,170 kg
(M101A1)	2,260 kg	2,380 kg	2,360 kg	2,400 kg
Max elevation	+65°	+70°	+70°	+70°
Min depression	–5°	–5°	–5°	–5°

Late in 1998, the first six complete kits were sent to FAMAE who upgraded the actual weapons under the supervision of staff from RDM Technology. FAMAE will also supply some components for this upgrade. Once the weapons have been assembled and tested, FAMAE will deliver them to the Chilean Army.

The upgraded M101 fitted with a 33 calibre barrel will fire a standard HE (high explosive) M1 projectile to a maximum range of 11, 900 m, or an ERM1 (extended range M1) projectile with R0663A2, zone 2 charge, to a maximum range of 19, 200 m.

105 mm MOBAT

Early in 1999, RDM Technology revealed their 105 mm MOBAT (Mobile Artillery) which has recently been tested in the United Arab Emirates. Details of this system, which as of early 2002 remained at the prototype stage, are given under the Netherlands in the Self-propelled guns and howitzers (wheeled) section.

Status

In production for Canada (96, plus option on 36) and Chile (ordered in 1997 with first deliveries in 1998).

Manufacturer

RDM Technology bv.

UPDATED

Specifications
(105 mm/33 calibre as selected by Canada)
Calibre: 105 mm
Barrel length: 3.513 m
Muzzle brake: multibaffle
Carriage: split trail
Shield: yes
Weight:
(M101) 2,150 kg
(M101A1) 2,380 kg
Length:
(travelling) 7.4 m
Width:
(travelling) 2.15 m
Height:
(travelling) 1.55 m
Trunnion height: 0.97 m
Track: 1.85 m
Elevation/depression: +70/–6°
Traverse: (total) 46°
Max range:
(M1 projectile, M67 charge) 11,900 m
(ERM1 with RO663A2, zone 1) 15,000 m
(ERM1 with RO663A2, zone 2) 19,200 m
(M760, M200 charge) 14,400 m
(M548 RAP, M176 charge) 15,500 m
(M913 RAP, M200 charge) 19,600 m

RDM Technology 105 mm M102 howitzer upgrade

Development/Description

In 1994, RDM Technology, as a private venture, completed a short study to upgrade the US 105 mm M102 towed howitzer, which has a maximum range of 11,500 m firing a standard M1 HE projectile.

The upgrade, which builds on experience obtained by RDM Technology in the upgrade of the 105 mm M101 towed howitzer (which has been produced for the export market) covered in the previous entry, includes replacing the current 105 mm ordnance with a new 105 mm/33 calibre ordnance.

This would allow the weapon to fire the new Eurometaal Extended Range M1 (ERM1) high-explosive projectile to a range of 19,500 m.

Development of the ERM1 was completed late in 1993.

The 105 mm M102 upgrade package was proposed to Malaysia early in 1994 but by 2001 no prototype had been built. As the Royal Netherlands Army does not use the 105 mm M102, this upgrade is aimed at the export market.

Status

Proposal. As of early 2001, no complete prototype of the upgraded M102 had been built.

Manufacturer

RDM Technology bv.

UPDATED

ROMANIA

152 mm Model M1985 gun-howitzer

Development/Description

The 152 mm M1985 gun-howitzer was developed in Romania. It is mounted on a split trail two-wheeled carriage, with each of the trails provided with a castor wheel to assist in bringing the weapon into and taking it out of action.

The carriage appears to be very similar to that used for the Russian 152 mm D-20 gun-howitzer although the Romanian weapon has a much longer 152 mm barrel and therefore a longer range. The D-20 has a maximum range, firing Russian standard ammunition, of 17,410 m.

Specifications
(Model M1985)
Calibre: 152.4 mm
Length:
(barrel, without muzzle brake) 7.105 m
(barrel, with muzzle brake) 8.025 m
(rifling) 6.102 m
(travelling) 11.17 m
Number of grooves: 48
Recoil: 910 mm
Carriage: split trail
Shield: yes
Weight:
(firing) 7,500 kg
(travelling) 7,550 kg
Width:
(travelling) 2.53 m
Ground clearance: 0.35 m
Elevation/depression: +57/–5°
Traverse: (total): 50°
Rate of fire: 2-4 rpm
Max range: 24,000 m

Romanian 152 mm M1985 gun-howitzer in travelling configuration

The 152 mm Model 1985 fires an HE projectile weighing 43.56 kg, maximum muzzle velocity of 825 m/s, with the propelling charge weighing 11.95 kg and a maximum range of 24,000 m.

According to the manufacturer, the 152 mm Model 1985 gun-howitzer fires a number of ammunition types including:

OF-540 high-explosive incendiary (maximum range of 24,000 m)

OF-550 high-explosive incendiary (maximum range of 22,000 m)

BR-540B armour piercing

When deployed in the firing position, the weapon is supported at three points, two spades at the rear and a circular firing jack under the forward part of the carriage. In the firing position the castor wheels are swung upwards and rest on top of the trails.

Romanian 152 mm Model 1985 gun-howitzer deployed in the firing position and with ordnance at maximum elevation. Note that the spades have not been bedded in. ***2002***/0130022

Variant

Romanian sources have stated that there is another 152 mm gun-howitzer called the Model 1984. This has an identical range and barrel length as the Model M1985 but elevation limits are from –5 to +57° with weight in firing position being 7,500 kg.

Types of ammunition fired are HE-FRAG PF-540 and OF-540 RM with a maximum muzzle velocity of 822 m/s and APC-T BR-540B.

Status

Production complete. In service with the Romanian Army. It is understood that Romania has a total of 111 of these weapons in service. According to United Nations sources, in 1993 Nigeria took delivery from Romania of four 152 mm M1985/M1981 systems.

Manufacturer

Arsenalul Armatei.

UPDATED

152 mm M1981 howitzer

Development/Description

The 152 mm M1981 howitzer was developed in Romania. It has split box-type trails, with a castor wheel provided to bring the weapon into action. When not required this is swung back and rests over the trail. The carriage of this weapon is similar in some respects to the Soviet 152 mm gun-howitzer D-20 and 122 mm field gun D-74, as a circular firing pedestal is fitted as standard and is inverted and secured just forward of the shield when travelling. When in the firing position, the firing pedestal is lowered allowing the weapon to be traversed fairly quickly through 360° using the castor wheels mounted on each trail leg.

The ordnance is fitted with a large double-baffle muzzle brake with a distinctive circular ring to its rear. The carriage used by this weapon is very similar to that used by the 130 mm gun Model 1982, which is covered in the following entry.

Types of separate loading ammunition (for example, projectile and charge) fired by the 152 mm M1981 howitzer include the following:

HE full variable charge, projectile weight 43.56 kg, maximum range 17,000 m.

HE full variable charge OF-550, projectile weight 43.56 kg, maximum range 24,000 m.

HE reduced variable charge, projectile weight 43.56 kg, maximum range 13,000 m.

HEAT, projectile weight 36.5 kg, maximum range 13,000 m.

HE full variable charge (APC-550), projectile weight 43.56 kg, maximum range 17,000 m.

HE full variable charge (CG-540), projectile weight 44.6 kg, maximum range 17,000 m.

Leaflet, projectile weight 43.56 kg, maximum range 17,000 m.

Illuminating, projectile weight 43.56 kg, maximum range 17,000 m.

Romanian 152 mm M1981 howitzer in travelling configuration (PB) ***2000***/0085644

Status

Production complete. In service with the Romanian Army. It is understood that the Romanian Army has 288 of these weapons in service. Recent information has indicated that Romania may be considering converting some of its 152 mm weapons into the NATO standard 155 mm. According to United Nations sources, in 1992 Nigeria took delivery from Romania of four 152 mm M1981/M1985 towed howitzers.

Manufacturer

Arsenalul Armatei.

UPDATED

130 mm gun Model 1982

Development/Description

This Romanian 130 mm gun Model 1982 is mounted on a split trail carriage which appears to be very similar to that used for the Romanian 152 mm Model 1985 gun-howitzer. It also utilises a similar top carriage, shield and recoil system.

Some Romanian sources have also referred to this as the 130 mm M-46 field gun as its ballistic characteristics are almost identical.

It has a long barrel with a double-baffle muzzle brake and no fume extractor is fitted. When deployed in the firing position, it rests on the two spades and a circular firing jack under the forward part of the carriage. When not required for use, the firing jack is swung upwards through 180° and is locked in position under the barrel.

The carriage and ordnance appear to have no relationship with the Russian 130 mm M-46 field gun. Each trail leg has a castor wheel to assist in bringing the weapon into action. According to the manufacturer, the 130 mm gun Model 1982 takes between 2 and 3 minutes to bring into action with a well-trained crew.

Types of separate loading ammunition (for example, projectile and charge), fired by the 130 mm gun Model 1982 include HE full charge and HE reduced charge, both of which can also be fired from the Russian 130 mm M-46 field gun.

Type	**HE** (full charge)	**HE** (reduced charge)
Fuze	V-429	V-429
Total weight	59 kg	52 kg
(cartridge)	25.6 kg	18.6 kg
(projectile)	33.4 kg	33.4 kg
Muzzle velocity	930 m/s	705 m/s
Max range	27,150 m	22,000 m

The manufacturer states that types of ammunition fired by this system include the following:
OF-482M high-explosive incendiary
BR-482B armour piercing
and newly created dedicated ammunition

Status

Production complete. In service with the Romanian Army. It is understood that the Romanian Army has 82 of these weapons in service. United Nations sources state that between 1992 and 2000 Romania exported the following quantities of 130 mm M-46 field guns:

Country	Quantity	Delivery
Bosnia Herzegovina	6	1998
Cameroon	12	1992
Guinea	12	2000
Nigeria	4 + 7	1992, 1994

Manufacturer

Arsenalul Armatei.

UPDATED

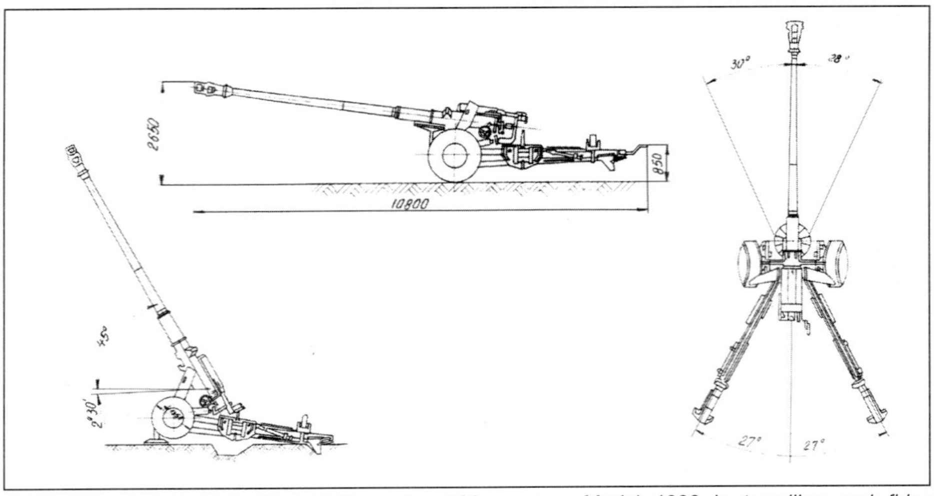

General arrangement drawing of Romanian 130 mm gun Model 1982 in travelling and firing configurations

Romanian 130 mm gun Model 1982 in travelling configuration

Specifications

Calibre: 130 mm
Length of barrel: 6,850 mm
Weight:
(firing) 6,150 kg
(travelling) 6,200 kg
Length:
(travelling) 10.80 m
Width:
(travelling) 2.59 m
Height:
(travelling) 2.65 m
Elevation/depression: +45/–2°
Traverse: 58° (30° left and 28° right)
Rate of fire: 7 rds/min
Max range: 27,150 m
Direct fire range: 1,100 m
Crew: 7
Maximum towing speed:
(road) 50 km/h
(off road) 20 km/h

100 mm towed anti-tank gun M1977

Development/Description

The 100 mm towed anti-tank gun M1977 was specifically developed by the Romanian defence industry to meet the requirements of the Romanian Army for a towed weapon to engage tanks and other armoured fighting vehicles under both day and night conditions, as well as being used in the fire support role. The Romanian Army nickname for the 100 mm towed anti-tank gun M1977 is the Misdrache.

It is understood that the Romanian Army has a total of 880 M1977 (or Gun 77 as they are also referred to) and 75 M1975 (or Gun 75 as they are referred to). It is believed that the M1975 is an earlier model of the M1977 with similar capabilities.

The carriage of the 100 mm towed anti-tank gun is a split carriage with one castor wheel to assist in bringing the weapon into action.

The 100 mm ammunition fired by the M1977 is identical to that fired by the 100 mm D-10 series guns installed in the T-54/T-55 MBTs, so the weapons may well be related.

Romanian sources have stated that this ammunition can also be fired from the Russian 100 mm field gun M1944 (BS-3) which was first fielded in 1944.

Romanian 100 mm towed anti-tank gun M1977 in travelling configuration with spades closed and castor wheel in position
2002/0130023

Romanian sources give the M1977 a maximum range of 4,000 m against tanks and armoured vehicles in the indirect fire mode and 1,100 m in the direct fire mode firing an armour-piercing projectile against a target height of 2 m. In the indirect fire mode using HE ammunition it has a maximum range of 20,700 m.

To bring the weapon into action from its travelling configuration takes 60 seconds.

Romanian sources provide the ammunition data in the table.

Type	**HE-FRAG**	**APC-T**	**HEAT**	**HEAT**
Designation	OF-412	BR-412D*	3BK-5M	BK-412R
Muzzle velocity				
(max)	900 m/s	887 m/s	900 m/s	950 m/s
(reduced)	600 m/s	n/a	n/a	n/a
Weight				
(propellant)	5.5 kg**	5.5 kg	4.5 kg	4.5 kg
(projectile)	15.6 kg	15.88 kg	12.2 kg	10.625 kg
Max range	20,700 m	4,000 m	2,200 m	2,200 m

* Also BR-412 and BR-412B with a muzzle velocity of 895 m/s
** Weight of reduced propellant charge is 4.5 kg.

Type	**HE** (full charge)	**HE** (reduced charge)	**AP-T**	**HEAT**
Fuze	V-429	V-429	MD-8	GPV-2R
Total weight	30.27 kg	26.74 kg	30.40 kg	21 kg
(cartridge)	11.5 kg	8.29 kg	11.4 kg	10.47 kg
(projectile)	15.6 kg	16.66 kg	15.88 kg	10.625 kg
Maximum range	16,000 m	11,000 m	4,000 m	3,000 m
Direct fire range	–	–	1,000 m	1,000 m
Muzzle velocity	900 m/s	600 m/s	887 m/s	950 m/s
Armour penetration	–	–	175 mm	80 mm

Romanian sources also state that it also fires a BM-412Sg APFSDS projectile with a muzzle velocity of 1,400 m/s that will penetrate 444 mm of armour at an inclination of 90° or at a range of 500 m or 328 mm at a range of 4,000 m.

In addition to the previously mentioned Russian rounds, the 100 mm M1977 can also fire these locally manufactured 100 mm projectiles (see tables above).

Optical equipment installed on the 100 mm towed anti-tank gun M1977 includes a mechanical sight, a panoramic sight with a magnification of ×4 and a 10° field of view, an optical sight with magnifications of ×4 and ×8 and a collimator.

Status

Production complete. In service with Romanian Army.

Manufacturer

Arsenalul Armatei.

Specifications
Calibre: 100 mm
Crew: 8
Length:
(ordnance) 5,350 mm
(rifling) 4,630 mm
Number of rifled grooves: 40
Pitch of rifling: 30 calibres
Width:
(land) 3.15 mm ± 0.3 mm
(groove) 4.7 mm ± 0.33 mm
Normal recoil: 750-830 mm
Max recoil: 840 mm
Time of recoil return cycle: 1-2 s
Weight: 3,200 kg
Axis of fire:
(at 0°) 0.9 m
Length:
(travelling) 9.37 m
Width: 2.24 m
Height:
(travelling) 1.9 m
Ground clearance: 0.36 m
Elevation/depression: +37/–5°
Elevation speed: 0.76°/rev
Traverse:
(total) 54° 60′ (27° 30′ in left and right)
Traverse speed: 1°/rev
Max range:
(OF-412 HE-FRAG projectile) 20,700 m
(APC-T projectile) 4,000 m
(HEAT projectile) 2,200 m

UPDATED

98 mm Model 93 mountain howitzer

Development/Description

For some years it has been known that the Romanian Army deployed a 76.2 mm mountain gun and a 100 mm mountain howitzer.

Under the Conventional Forces Europe Treaty each country has had to declare its tracked and self-propelled artillery systems, and mortars, with a calibre of more than 100 mm.

The 76.2 mm mountain gun did not have to be declared and the 100 mm mountain howitzer was never declared. The reason for this weapon not being disclosed is that its actual calibre is 98 mm and therefore it did not have to be declared in any case. Recent information has revealed that it is called the 98 mm Model 93 mountain howitzer.

Photographs now reveal that the 76.2 mm, which has no known designation and 100 mm Model 93 weapons are similar in some respects and it is possible that the 98 mm Model 93 weapon is a further development of the older 76.2 mm weapon.

The Romanian weapon is mounted on a conventional split trail carriage with the rear half of each carriage folding upwards and forwards through about 180° to rest on the front part of the trail when travelling.

The trails are then closed together and a locking pin with towing eye attached is fitted. Under this, a small rubber-tyred roadwheel is mounted to make deployment of the weapon easier.

When deployed in the firing position, spades help to absorb the recoil although stakes can be driven into the ground through the rear part of each trail if required.

The carriage has conventional rubber tyres, the wheels of which are provided with towing eyes. Positioned on the forward part of the carriage are two handbrakes which are used when positioning the 98 mm Model 93 mountain howitzer.

Elevation is manual from –5 to +65° with traverse being manual from 25° left and right. The elevation and traverse controls are located on the left side of the carriage where the direct and indirect sights are located. The weapon is provided with a small shield.

The 98 mm ordnance is fitted with a multibaffle muzzle brake and typical rate of fire is 6 rds/min.

This weapon fires ammunition of the separate loading type, for example projectile and charge which is in a brass cartridge case. Two natures of high-explosive ammunition are fired, the OF-402 and the OF-403.

Both of these are packed three rounds to a case and are fitted with a nose-mounted fuze which

Ammunition specifications

Type	**HE**	**HE**	**HEAT**
Designation	OF-402	OF-403	nil
Fuze designation	UP-2	UP-2	GPV-2
Length			
(total)	756 mm	759 mm	758 mm
(cartridge case)	241 mm	241 mm	241 mm
(projectile)	515 mm	518 mm	517 mm
Weight			
(total)	18 kg	18 kg	16 kg
(cartridge)	2.8 kg	2.8 kg	2.8 kg
(projectile)	13.2 kg	14.6 kg	12.2 kg
Muzzle velocity	235/470 m/s*	225/430 m/s*	470 m/s
Max range	10,800 m	10,500 m	10,800 m

* depends on charge

Romanian 98 mm Model 93 mountain howitzer in travelling configuration

Romanian 98 mm Model 93 mountain howitzer in firing configuration

can be set for impact, short delay and delayed functioning.

The only other known nature of ammunition is a fin-stabilised High-Explosive Anti-Tank (HEAT) type which is packed three to a case and fitted with a piezoelectric fuze.

If required, the weapon can be quickly disassembled into three loads each of which is carried on a carriage towed by a horse. Its normal towing vehicle is a DAC 444 (4 × 4) 2.5 tonne truck which carries the crew and ammunition.

Status

Production complete. In service with Romanian forces. There are no known exports of this weapon.

Specifications

Calibre: 98 mm
Muzzle brake: multibaffle
Carriage: split trail
Shield: no
Weight:
(travelling) 1,500 kg
(firing) 1,500 kg
Length: 4.25 m
Width: 1.65 m
Height: 1.70 m
Elevation/depression: +70/–5°
Traverse: 40° (23° right, 17° left)
Range:
(max) 10,800 m
(direct fire) 600 m
Rate of fire: 6 rds/min
Crew: 8
Towing vehicle: DAC 444 (4 × 4) truck and by horses

Manufacturer

Arsenalul Armatei.

UPDATED

76.2 mm mountain gun Model 1984

Development/Description

The Romanian-developed 76.2 mm Model 1984 mountain gun has been specifically developed to meet the requirements of Romanian mountain units and can be quickly disassembled into eight pack animal loads.

The crew of seven consists of commander, aimer, breech mechanism operator, loader, fuze handler, projectile handler and charge handler.

When being towed by a vehicle or horse team, the weapon can be brought into action in about 1 minute. When being carried by pack horse, the weapon takes between 6 and 8 minutes to come into action. This takes longer as the individual sections have to be assembled.

Types of ammunition fired include high explosive (maximum muzzle velocity 398 m/s), smoke and HEAT.

According to the manufacturer, the 76 mm mountain gun (or howitzer as at is also referred to), fires the following ammunition types as listed below. It should be noted that some of these with the M series designation are also used by the Yugoslav 76 mm mountain gun M48 which is closely related to the Romanian 76.2 mm mountain gun/howitzer Model 1984.

Charge round - HE OF-350M, fuze UP3
Charge round - M55, fuze M51A5
Charge round - smoke WP M60, fuze PD M48A3
Charge round - smoke D-350 TM, fuze UP3
Charge round - HEAT BK 354, piezo electric fuze CVP-312
Charge round - shell M50, fuze PD MK-451

Optical equipment fitted to the weapon consists of a day telescope with a magnification of ×3.

It has recently been revealed that the 76.2 mm mountain gun is related to the 98 mm Model 93 mountain howitzer covered earlier.

As far as is known the 76.2 mm and 98 mm Model 93 mountain howitzers have never been exported although they have been offered on the export market.

Status

Production as required. In service with Romanian Army.

Manufacturer

Arsenalul Armatei.

UPDATED

Romanian 76.2 mm M1984 mountain howitzer deployed in the firing position **2002**/0130024

Specifications

Calibre: 76.2 mm
Max recoil length: 830 mm
Weight:
(travelling, towed) 730 kg
(travelling, horse towed) 746 kg
(combat) 722 kg
Length:
(barrel, without breech mechanism) 1,178 mm
(travelling) 2.45 m
(in action) 3.10 m
Width:
(travelling) 1.33 m
(in action) 2.65 m
Height: 1.66 m
Ground clearance: 0.18 m
Track: 1.28 m
Elevation/depression:
(with lengthened trails) +45/–15°
(with shortened trails) +60/0°
(on slope with shortened trails) +60/0°
Traverse:
(total) 50°
Crew: 7
Max range:
(HE) 8,600 m
(HEAT) 1,000 m
Range in direct fire mode:
(for target with height of 2 m) 460 m
(for target with height of 3 m) 550 m
Rate of fire:
(first min) 25 rounds
(in 3 min) 40 rounds
(in 5 min) 50 rounds
(in 10 min) 60 rounds
(in 15 min) 70 rounds
(in 30 min) 90 rounds
(in 1 h) 120 rounds
(in 2 h) 210 rounds
(in each further hour) 80 rounds
Life of barrel: 15,000 rounds

RUSSIAN FEDERATION AND ASSOCIATED STATES (CIS)

152 mm howitzer 2A65 (M1987)

Development/Description

In addition to the 152 mm 2A36 gun described in the next entry, Russia also deployed a new 152 mm towed howitzer, designated the 2A65, which was allocated the NATO designation of the M1987. According to United Nations sources there were no exports of this weapon between 1992 and 2000.

The 152 mm 2A65 is also referred to as the MSTA-B and the weapon forms the main armament of the 152 mm self-propelled artillery system, the 2S19, which is also referred to as the MSTA-S. Full details of the latter are given in the Self-propelled guns and howitzers (tracked) section. In the designation MSTA-B the latter stands for Buksiruemyi, or towed. The MSTA-S has been offered on the export market but there are no known sales to date.

The 152 mm howitzer 2A65 (M1987) is mounted on a conventional split trail carriage and, when deployed in the firing position, rests on three points, the hydraulic circular firing jack under the forward part of the carriage and the two spades at the rear. Each of the box section trails has a castor wheel to assist the gun crew in bringing the weapon into action. When deployed in the firing position, these swing upwards through 180° and rest on top of each trail.

The 152 mm ordnance is fitted with a muzzlebrake and a semi-automatic breechmechanism, spring-operated ramming system, hydraulic counter-recoil device and a liquid-cooled recoil brake.

Elevation and traverse is manual, two-speed, with the direct and indirect sighting devices being located on the left side of the weapon. Pneumatic brakes are fitted as standard.

The 152 mm 2A65 fires the same ammunition types as the 152 mm 2S19 self-propelled artillery system and, more recently, a new family 152 mm has been introduced.

The OF45 high-explosive projectile weighs 43.56 kg, has a maximum muzzle velocity of 810 m/s and a maximum range of 24.7 km. The charges include OF72 (long range), OF58 (full charge) and OF73 (reduced charge).

The OF45 projectile can be fitted with different rear ends, for example various types of screw-on boat tails, or the OF61 base bleed projectile which weighs 42.86 kg, has a maximum muzzle velocity of 828 m/s and a maximum range of 29 km.

The OF23 cargo projectile weighs 42.8 kg, has a maximum range of 26,000 m and contains 42 HEAT bomblets, each of which can penetrate 100 mm of conventional steel armour. Other types of projectile include the HS30 jamming round and the Krasnopol laser-guided projectile which is covered in the entry for the 2S19 in the Self-propelled guns and howitzers (tracked) section.

In addition to these new 152 mm projectiles the 2A65 can fire all standard types of 152 mm ammunition fired by the older D-20 towed gun-howitzer and the 2S3 self-propelled gun-howitzer.

Variants

It is understood that a longer barrel version of the 2A65 is currently under development and there have been reports of a 155 mm model firing NATO standard ammunition. It is known that there is a 155 mm version of the 2S19 self-propelled artillery system.

Status

Production. Known to be in service with the following countries:

Country	Quantity	Comment
Belarus	136	
Georgia	11	
Kazakhstan	76	
Russia	600	estimate
Turkmenistan	72	
Ukraine	185	

Manufacturer

Motovilikha Plants Corporation.

UPDATED

Specifications

Calibre: 152.4 mm
Muzzle brake: multibaffle
Carriage: split trail
Shield: yes
Weight:
(firing) 7,000 kg
Elevation/depression: +70/–3.5°
Traverse: 27° left and right
Max rate of fire: 7 rds/min
Max range:
(standard ammunition) 24,000 m
Crew: 8
Max towing speed:
(roads) 80 km/h
(cross-country) 20 km/h
Towing vehicle: KrAZ-260 (6 × 6) or Ural 4320 (6 × 6)

152 mm towed howitzer 2A65 (M1987) deployed in the firing position

152 mm gun 2A36 (M1976)

Development

In the 1970s, Russia developed a new 152 mm towed gun which was allocated the industrial number of 2A36. This was first seen by NATO sources in 1976 and allocated the Western designation of 152 mm gun M1976.

The 152 mm gun 2A36 was developed from December 1968 at the Perm Machine Works following a requirement issued by the Russian Army for a new 152 mm weapon to replace the old 130 mm M-46 field gun. The main role of the 152 mm 2A36 system is counter battery fire.

Two different versions were developed, the towed 2A36 called the Giatsint-B (or Hyacinth) and the self-propelled Giatsint-S which has a 152 mm ordnance designated the 2A37.

The latter was eventually type classified as the 2S5 which is covered in the Self-propelled guns and howitzers (tracked section).

Both systems shared identical ballistics but fielded different types of ammunition.

Full development of the towed 152 mm 2A36 commenced in 1970 with the first two prototypes being completed in the Spring of 1971.

Full scale production commenced in 1976 and it was first observed by Western intelligence the same year and subsequently designated the 152 mm gun M1976 until its correct Russian designation became known.

The 152 mm 2A36 was first seen in public during a parade held in Moscow in May 1985 when it was towed by a KrAZ-260 (6 × 6) truck.

The 152 mm 2A36 is normally deployed in batteries of six or eight guns, with three batteries making up a battalion.

Although the 2A36 was first observed being towed by the KrAZ-260 (6 × 6) truck, it can also be towed by the KrAZ-255B (6 × 6) or lighter Ural 4320 (6 × 6) truck. In addition the 2A36 can also be towed by full tracked prime movers such as the AT-T, AT-S and ATS-59, although these are no longer used by the Russian Army in significant numbers.

According to United Nations sources, there were no exports of this weapon outside of the Russian Federation between 1992 and 2000.

Production of this system is understood to have ceased in the 1980s. Total production is understood to have amounted to at least 1,500 systems. It has been supplemented by the more recent 152 mm 2A65 weapon.

Description

The 152 mm 49 calibre barrel of the 2A36 is fitted with a multislotted muzzle brake weighing 141 kg. The recoil system consists of a buffer and a recuperator which are mounted above the ordnance towards the rear.

The horizontal sliding breech mechanism opens to the right automatically.

The ordnance pivots on the cradle which is of cast and welded steel construction. Elevation and traverse is

152 mm gun 2A36 (M1976) from the rear in travelling configuration

152 mm gun 2A36 (M1976) (PB) **2000**/0085645

manual with the former being from –2° 30' to +57° and traverse 25° left and right.

The direct and indirect sights are mounted on the left side of the carriage as are the elevation and traverse mechanisms.

Some of the eight person crew are provided with protection from shell splinters by the armoured shield that slopes to the rear and extends over the wheels.

Mounted below and to the rear of the breech is the load assist system which is referred to as a quick-firing loading system by the RFAS. This includes a hydraulic rammer operated from the hydropneumatic accumulator with controls provided for ramming and return.

A fuzed projectile is placed on the loading tray which is mounted to the right of the breech. When the breech is opened, this slides to the left until it lines up with the breech. The projectile is then rammed into the ordnance and the loading tray slides back to the right. The cartridge-cased charge is then located in a similar manner and, when the breech is closed, the rammer automatically returns to the side and the weapon is ready to fire. If the hydraulic rammer fails, loading takes place manually, although rate of fire is reduced.

The RFAS quotes a rate of fire of 6 rds/min for the 2A36 and that a battery can put almost 1 tonne of ammunition onto a target in 1 minute.

Types of separate-loading ammunition fired include High-Explosive Fragmentation (HE-FRAG) and Armour-Piercing Tracer (AP-T). The former weighs 46 kg and has a maximum muzzle velocity, using the top charge, of approximately 800 m/s. Before loading the projectile the fuzing mode is selected either for fragmentation or high-explosive action.

The AP-T round is used in the direct fire mode against tanks and other armoured fighting vehicles. The propelling charge, which weighs a maximum of 11 kg, is placed in a conventional cartridge case.

The standard HE-FRAG projectile has a maximum range of 27,000 m although a rocket-assisted HE-FRAG projectile, called the active-reactive projectile by Russia, can also be fired. This has a maximum range of 40,000 m.

Other types of projectile that can probably be fired by the 152 mm 2A36 include smoke, concrete-piercing and incendiary.

The ammunition system used by the 152 mm 2A36 is of a new design and is not interoperable with that of earlier artillery systems, such as the 152 mm 2S3 and may well be of a new streamlined design.

Specifications

Calibre: 152.4 mm
Barrel length: 8.197 m
Muzzle brake: 5-part multibaffle
Breech mechanism: semi-automatic horizontal sliding
Carriage: split trail
Shield: yes
Weight:
(travelling) 9,800 kg
(firing) 9,760 kg
Length:
(travelling) 12.92 m
(firing) 12.3 m
Width: 2.788 m
Height: 2.76 m
Ground clearance: 0.475 m
Track: 2.34 m
Tyres: 1,080 × 300 mm
Elevation/depression: +57/–2.5°
Traverse: 25° left and right
Rate of fire: 5-6 rds/min
Range:
(standard ammunition) 27,000 m
(with RAP) 40,000 m
Crew: 8
Towing vehicle: KrAZ-260 (6 × 6) or Ural 4320 (6 × 6) truck
Towing speed:
(road) 80 km/h
(cross-country) 30 km/h
Unit of fire: 60 rounds

The walking beam suspension gives the 152 mm gun 2A36 (M1976) improved cross-country mobility and has a total of four rubber-tyred braked roadwheels for high-speed towing and improved cross-country mobility. The tyres have a normal pressure of 4.8 kg/cm². Total wheel travel is 150 mm. Each of the front roadwheels has a hydraulic shock-absorber and a manual handbrake.

When in the firing position, the 152 mm 2A36 is supported on a circular jack located under the forward part of the carriage and on each of the box-type welded trails which is provided with a spade.

The winter spade is standard, but when firing in soft terrain much larger summer spades are used.

When coming out of action, the jack is raised clear of the ground and secured under the front, the spades removed and stowed on the trails, the trails brought together and clamped with a tie mechanism. The hinged towbar is fitted to the right trail and the weapon is then hitched up to the prime mover.

Improved Giatsint-B

Mention has also been made of an improved version of the 2A36 with a 53.8 calibre 152 mm ordnance compared to the 49 calibre of the standard production weapon.

Giatsint-BK

This did not pass the prototype stage.

Status

Production complete. In service with the following countries:

Country	Quantity	Comment
Armenia	26	
Azerbaijan	24	
Belarus	50	
Finland	24	Local designation 152 K 79
Georgia	3	
Iraq	n/k	
Kazakstan	48	
Moldova	21	
Russia	1,200	
Uzbekistan	36	
Ukraine	289	

Manufacturer

Perm Machine Works.

UPDATED

152 mm gun-howitzer D-20

Development

The 152 mm gun-howitzer D-20 was developed shortly after the end of the Second World War by the F F Petrov Artillery Design Bureau at Artillery Plant No 9 located in the town of Yekaterinburg.

The 152 mm gun-howitzer D-20 was first seen in public during the May Day parade in Moscow in 1955. It is understood that this weapon is no longer in front-line service with the Russian Army.

According to United Nations sources, Russia did not expect any surplus 152 mm gun-howitzer D-20 weapons between 1992 and 2000. Bulgaria did however export four D-20s to Angola in 1999, while China exported 18 of its equivalent Type 66 to Bolivia in 1992.

Description

The D-20 uses the same carriage and recoil system as the 122 mm field gun D-74 which was also shown for the first time in 1955. The D-20 can be distinguished from the D-74 by a much shorter and fatter stepped barrel and a larger double-baffle muzzle brake.

The shield has an irregular top with a sliding centre section. The top of the shield can be folded down to reduce the overall height of the weapon. A circular firing pedestal is fitted as standard on all D-20s and is inverted and secured just forward of the shield when travelling. In the firing position the firing pedestal is lowered, allowing the weapon to be traversed quickly through a full 360°. On each of the split box section trails is a castor wheel (on top) and a spade (underneath).

Gunlaying is carried out by the gunner placed on the left side. The gunner's sights are the indirect sight PG1M and the telescope OP4M for direct fire.

The ordnance of the 152 mm self-propelled gun-howitzer 2S3 is a development of that used on the towed D-20.

152 mm gun-howitzer D-20 in travelling configuration (Christopher F Foss)

152 mm gun-howitzer D-20 of the Sandinista Army (Julio Montes) 0021013

It fires the case-type, variable-charge, separate-loading ammunition listed in the table right.

Other types of ammunition include HE/RAP (range of 24,000 m), HEAT, illuminating (S-540), smoke (D-540) HEAT-SS (Spin-Stabilised), flechette, scatterable mines (anti-tank and anti-personnel) and semi-active laser guided. The last of these is called the Krasnopol by the RFAS and details of this are given in the entry for the 2S19 152 mm self-propelled gun.

Yugoslavia produced a long-barrelled version of the D-20 known as the 152 mm howitzer M84. See entry under Yugoslavia for illustrations. As far as is known there has been no production of this weapon in Yugoslavia in recent years.

Specifications

Calibre: 152.4 mm
Barrel length:
(overall) 5.195 m
Muzzle brake: double-baffle
Breech mechanism: semi-automatic vertical sliding wedge
Carriage: split trail with castor wheel on each
Shield: yes
Weight:
(travelling order) 5,700 kg
(firing position) 5,650 kg
Length:
(travelling) 8.69 m
Width:
(travelling) 2.40 m
Height:
(travelling) 1.925 m
Axis of bore:
(at 0°) 1.22 m
Ground clearance: 0.38 m
Track: 2.03 m
Tyres: 12.00 × 20
Elevation/depression: +63/–5°
Traverse: (total) 58°
Rate of fire: 5-6 rds/min
Range: 17,410 m
(with RAP) 24,000 m
Crew: 10
Towing vehicles:
(AT-S Medium-Tracked Artillery Tractor)
(Ural-375 (6 × 6) truck)
Max towing speed:
(road) 60 km/h
(cross-country) 15 km/h
Unit of fire: 65 rounds

Ammunition type	FRAG-HE	CP	AP-T
Projectile designation	OF-540*	G-545**	BR-540
Fuze model	RGM-2	KTD	MD-7
Weight of projectile	43.51 kg	56 kg	48.78 kg
Weight of bursting charge	6.25 kg	4.22 kg	1.2 kg
Type of bursting charge	TNT	TNT	n/a
Muzzle velocity	655 m/s	670 m/s	600 m/s
Armour penetration at 0°	n/a	n/a	124 mm/1,000 m

*other FRAG-HE projectiles also available
** other CP projectiles available

Status

Production complete. In service with the following countries:

Country	Quantity	Comment
Afghanistan	n/k	
Albania	75/90	Chinese Type 66
Angola	4	from Bulgaria in 1999
Armenia	34	
Azerbaijan	32	
Belarus	58	
Bolivia	18	from China in 1992
Bosnia	50	
Bulgaria	206	
China	1,400	Chinese Type 66
Congo	n/k	
Croatia	20	
Cuba	n/k	
Egypt	144	
Finland	126	local designation H55
Hungary	302	including reserves
Iran	30	
Iraq	n/k	
Kazakhstan	n/k	
Korea, North	n/k	
Libya	36	
Moldova	32	
Nicaragua	30	reserve
Poland	160	
Russia	1,000	estimate
Syria	20	
Ukraine	219	
Uzbekistan	49	
Vietnam	n/k	
Yemen	10	
Yugoslavia	25	plus 52 of local M84 variant (qv)

Manufacturers

Joint Stock Company Spetstehnika.
China North Industries Corporation (NORINCO).

UPDATED

152 mm 2A61 howitzer

Development/Description

Early in 1993, the RFAS released outline information of a new lightweight 152 mm towed howitzer designated the 2A61.

This has been developed by the famous F F Petrov artillery design bureau at Artillery Plant No.9, Yekaterinburg, which has designed many previous artillery weapons, including the 122 mm D-30 towed howitzer.

The 152 mm 2A61 light towed howitzer is based on the modified carriage of the late production 122 mm D-30A (2A18M) which has been modified to accept a larger calibre 152 mm ordnance. The latter is fitted with a large distinctive double 'T' muzzle brake with the towing eye located below.

The 152 mm 2A61 light towed howitzer is fitted with a ramming system which enables a rate of fire of 6 to 8 rds/min to be achieved.

When deployed in the firing position, the two pneumatic roadwheels are raised clear of the ground and the three trails are staked to the ground to provide a more stable firing platform. The weapon can be quickly traversed through a full 360° to be laid onto a new target.

When in the travelling position, the ordnance is over the closed trails. A small shield is fitted to provide some level of protection to the gun crew.

It is possible that this weapon was developed as a replacement for the 122 mm D-30 used by the RFAS air

152 mm 2A61 towed howitzer with ordnance horizontal

Specifications

Calibre: 152.4 mm
Muzzle brake: multibaffle
Shield: yes
Weight:
(travelling) 4,350 kg
Length:
(travelling) 6.36 m
Width:
(travelling) 2.2 m
Height:
(travelling) 1.97 m
Axis of bore: 0.93 m
Recoil length:
(up to +20°) 1.4 m
(over 20°) 0.98 m
Ground clearance:
(travelling) 0.35 m
Track: 1.84 m
Elevation/depression: +70/–5°
Traverse: 360°
Rate of fire: 6-8 rds/min
Range:
(max) 15,000 m
(min) 4,000 m
Crew: 7
Towing vehicle: ZIL-131 (6 × 6) truck or MT-LB multipurpose armoured vehicle
Max towing speed: 80 km/h

assault divisions. As far as is known, the 152 mm 2A61 towed howitzer has yet to enter volume production. It is expected that if and when placed in production, the 152 mm 2A61 will utilise existing D-30 carriages rather than brand new carriages. As of early 2002 there was no known production of the 152 mm 2A61 howitzer.

Variant

In 1997 the 2A61 Model 389 was revealed. This is essentially a 155 mm version chambered for NATO ammunition. The recoil system has been improved and the carriage and trunnions strengthened.

Status

Development complete. Ready for production.

Manufacturer

Former state factories.

The D-30 122 mm howitzer on which the 152 mm 2A61 howitzer is based is manufactured by the Joint Stock Company Spetstehnika which is also probably involved with the 152 mm 2A61 howitzer.

UPDATED

130 mm field gun M-46

Development

In April 1946, the Soviet Artillery Committee issued the technical requirements for 130 and 152 mm towed artillery systems, which were to use the same carriage.

Development was carried out by Factory No 172 and the Scientific Research Institute for Artillery Armaments with the 130 mm weapon being designated the M-46 and the 152 mm weapon the M-47.

The first prototypes of the 130 mm M-46 and 152 mm M-47 were completed in mid-1948. After trials, further work was carried out which resulted in the building of four additional 130 mm M-46 and 120 mm M-47 guns for troop trials that took place late in November 1950 after which they were accepted for service.

During 1951, production of both weapons commenced at Factory No 172 with 20 M-46 and four M-47 being built in that year. In the 1950s the following quantities of weapons were built:

	130 mm M-46	152 mm M-47
1951	20	4
1952	26	nil
1953	nil	nil
1954	16	16
1955	50	20
1956	62	32
1957	79	50

Production of both weapons continued after this date although the 152 mm M-47 was phased out of production.

The 130 mm M-46 was first seen in public during a parade held in Moscow in 1954 and was the replacement for the 122 mm M1931/37 (A-19) field gun.

The 130 mm M-46 has seen combat in Asia and the Middle East and has also been built in China as the Type 59-1 (qv) and in Egypt (qv). The weapon has now been phased out of front line service with the Russian Army although it is still used in many parts of the world.

Although production of the 130 mm field gun M-46 was completed many years ago, significant quantities of this weapon have been sold on the export market by various other countries in recent years.

According to the United Nations, between 1992 and 2000 the following quantities of weapons were exported:

From	To	Quantity	Date
Bulgaria	Eritrea	30	1999
China	Iran	106	1992
China	Pakistan	87	27 in 1998 and 60 in 2000
Germany	Finland	166	1994
Kazakhstan	Ethiopia	6	2000
Kazakhstan	Russia	24	1999
Romania	Bosnia and Herzegovina	8	1998
Romania	Cameroon	12	1992
Romania	Guinea	12	2000
Romania	Nigeria	13	4 in 1992 and 7 in 1993
Russia	India	120	1994

Description

When travelling, the 130 mm barrel is withdrawn by a mechanism on the right trail from battery to the rear to reduce the overall length of the weapon. The carriage is of the split trail type and is provided with a two-wheeled limber. When travelling, the spades are removed and carried on top of the two trails. The recoil system is mounted under the barrel and in front of the shield, which has been removed on some models, is an inverted U-shaped collar.

The M-46 field gun has the OP4M-35 direct fire sight with a field of view of 11° and a magnification of ×5.5, and an APN-3 active/passive night sight.

The M-46 fires case-type, variable-charge, separate loading ammunition as shown in the table.

Ammunition type	FRAG-HE	APC-T
Projectile designation	OF-482M/OF-33	BR-482B
Fuze model	RGM-2	DBR
Weight (of projectile)	33.4 kg	33.6 kg
Weight (of bursting charge)	4.63 kg	0.127 kg
Type of bursting charge	TNT	RDX/alum
Muzzle velocity	930 m/s	930 m/s
Armour penetration at 0°	n/a	230 mm/1,000 m

Other types of 130 mm projectile include:

PB-42, SAP-HE, weight 33.4 kg

SP-46, illuminating, weight 25.80 kg, muzzle velocity 687 m/s

OF-3S-42, semi-armour-piercing, weight 42 kg

DTs-1, target marker smoke, weight 32.80 kg, muzzle velocity 930 m/s

Israel Military Industries produces a 130 mm HE projectile identical to the original but fitted with a PD M739 fuze. This projectile is used by the Israel Defence Force.

NORINCO of China has developed a family of 10 130 mm projectiles and brief details of these are given in the entry for the Chinese 130 mm field gun Type 59-1.

Denel of South Africa also offers new types of 130 mm ammunition for this weapon.

Iranian 130 mm BB ammunition

The Iranian Army currently employs an estimated 1,000 130 mm M-46 field guns and its Chinese equivalent the NORINCO (China North Industries Corporation) Type 59-1.

The Ammunition Group of the Iranian Defence Industries Organisation (DIO) have revealed that they have developed and placed in production a new 130 mm HE fragmentation projectile with a composite base bleed (BB) unit.

This new Iranian 130 mm HE BB projectile weighs 35.5 kg and contains 3.7 kg of TNT, this TNT being smaller than the standard OF-482M due to the added weight of the base bleed unit.

The new round is fired like a conventional 130 mm HE projectile with a maximum muzzle velocity of 930 m/s. When in flight the base bleed unit is activated

130 mm M-46 of the Iraqi Army, deployed in firing position

130 mm M-46 field gun in the travelling configuration with limber attached (Christopher F Foss)

0006048

and enables the projectile to achieve a maximum range of 37,000 m, a significant increase over the 27,150 m of the standard OF-482M projectile.

This allows the 130 mm M-46 field gun to engage targets at a much longer range and also reduces the number of enemy weapons that can engage the system with counter battery fire.

Variants

Modified 130 mm M-46s with a longer barrel, recuperator and cradle appeared in Russian Army service in the mid-1970s. India has deployed a self-propelled model based on a lengthened Vijayanta MBT chassis. This is called the Catapult by the Indian Army. Brief details of this are given in the Self-propelled guns and howitzers (tracked) section under India.

155 mm conversions

The Chinese NORINCO concern is offering a 155 mm conversion known as the GM-45 (see entry under People's Republic of China for details) and Yugoslavia has offered conversion (see under Yugoslavia for details). Soltam of Israel is also offering a conversion pack for the 130 mm M-46 to 155 mm/45 calibre standard. Details of this are given under Israel. This was ordered by India early in 2000. RDM Technology of the Netherlands also offers an upgrade for this weapon, details of which are given under the Netherlands.

Specifications

Calibre: 130 mm
Barrel length:
(overall) 7.6 m
Muzzle brake: pepperpot
Recoil system: hydraulic buffer and hydropneumatic recuperator
Recoil length:
(long) 1,150-1,320 mm (–2° 30′ to +20°)
(short) 734-815 mm (+34° to +45°)
Breech mechanism: horizontal sliding wedge breech block
Carriage: split trail with limber
Shield: optional
Weight:
(travelling order) 8,450 kg
(firing position) 7,700 kg
Length:
(travelling) 11.73 m
Width:
(travelling) 2.45 m
Height:
(travelling) 2.55 m
Axis of bore:
(at 0°) 1.38 m
Ground clearance:
(carriage) 400 mm
(limber) 375 mm
Track: 2.06 m
Tyres: 13.50 × 38
Elevation/depression: +45/–2.5°
Traverse:
(total) 50° (25° left and 25° right)
Rate of fire: 5-6 rds/min
Range: 27,150 m
Crew: 8
Towing vehicles:
AT-S Medium-Tracked Artillery Tractor
ATS-59 Medium-Tracked Artillery Tractor
M1972 Medium-Tracked Artillery Tractor
Max towing speed: 50 km/h
Note: The Motovilkha Plants Corporation gives the maximum range of the 130 mm M-46 as 27,490 m with a maximum rate of fire of 8 rounds per minute.

130 mm field gun M-46 deployed in the firing position from the rear with ordnance elevated **2002**/0130002

Status

Production complete. In service with the following countries:

Country	Quantity	Comment
Afghanistan	n/avail	
Albania	n/a	Chinese Type 59-1
Algeria	10	
Angola	n/avail	
Bangladesh	40	Chinese Type 59-1
Bosnia-Herzegovina	20	12 from Egypt and 8 from Romania
Bosnian Serbs	24	
Bulgaria	72	
Cambodia	n/avail	Chinese Type 59-1
Cameroon	24	12 from China and 12 from Romania
China	n/avail	Chinese Type 59-1 (qv China)
Congo	5	
Congo, Democratic Republic of (Zaïre)	8	Chinese Type 59-1
Croatia	83	
Cuba	n/avail	
Egypt	420	local production (qv)
Eritrea	30	from Bulgaria in 1999
Ethiopia	6	
Finland	150	called K 54
Georgia	20	
Guinea	12	from Romania in 2000
Guyana	6	
India	750	also some self-propelled (qv). Including 120 from Russia in 1994. Some being upgraded to 155 mm
Iran	1,000	Russian M-46 and Chinese Type 59-1
Iraq	300	Russian M-46 and Chinese Type 59-1
Israel	100	
Kazakhstan	100	
Korea, North	n/avail	also some self-propelled versions
Lebanon	25	
Libya	330	
Laos	10	
Mali	6	
Mongolia	64	
Morocco	18	
Mozambique	24	
Myanmar	20	from North Korea
Nigeria	13	
Oman	24	12 Russian M-46 and 12 Chinese Type 59-1
Pakistan	400+	Chinese Type 59-1
Peru	30	
Russia	n/avail	reserve
Somalia	n/avail	status uncertain
Sri Lanka	12	Chinese Type 59-1
Sudan	40	M-46
	35	Type 59-1
Syria	800	
Tanzania	40	
Thailand	15	Chinese Type 59-1
UAE	20	Chinese Type 59-1 (reserve)
Uganda	12	
Vietnam	n/avail	
Yemen	90	
Yugoslavia	276	
Zambia	18	

Manufacturers

Motovilkha Plants Corporation.
China North Industries Corporation (qv).
Abu Zaabal Engineering Industries Company (qv).

UPDATED

125 mm 2A45M (Sprut-B) towed anti-tank gun

Development

The 125 mm 2A45M (Sprut-B) auxiliary-propelled towed anti-tank gun is also known as the Sprut-B (Russian for Octopus) and is the replacement for the 100 mm MT-12/T-12 towed anti-tank gun which entered service in 1955.

The weapon was developed in the late 1980s by the F F Petrov design bureau at the Artillery Plant No 9 at Yekaterinburg, which also designed the famous 122 mm D-30 artillery system as well as numerous tank guns.

According to Russian sources, the development of the 2A45M (Sprut-B) towed anti-tank gun can be traced back to 1967, when it became apparent that the 100 mm T-12 series towed ant-tank gun (covered in detail in a separate entry) could not defeat new generation tanks such as the UK's Chieftain and the MBT-70 (which in the end did not enter service.

In January 1968 OKB-9, which subsequently became part of the now Joint Stock Company Spetstehnika, was selected to develop a new 125 mm anti-tank gun. It had the same ballistics as the 125 mm D-81 smoothbore tank which subsequently became the standard Russian tank gun.

Originally two versions of the weapon were developed to the prototype stage, the D-13 towed and the SD-13 self-propelled. Further development of the latter resulted in the 2A45M (Sprut-B).

Description

The system is based on the well-known 122 mm D-30 towed howitzer carriage which has been in service with RFAS and many other countries for well over 30 years.

The 122 mm rifled ordnance of the D-30 has been replaced by a modified version of the 125 mm D-81 smoothbore ordnance used in the T-64/T-72/T-80/T-84/T-90 MBTs and features a thermal sleeve and a large single 'T'-type baffle muzzle brake with a towing eye mounted underneath like the earlier D-30. Ballistically, the 125 mm 2A45M and the 125 mm tank guns are identical. The cradle is also taken from the 125 mm D-81 tank gun.

The barrel of the 2A45M consists of a tube with a muzzle brake attached by a casing to the chamber part and a breech ring. The breech mechanism is of the semi-automatic mechanical vertical sliding wedge type. This is opened manually before the first round is loaded and after that opens automatically after the weapon has fired.

The recoil system is a counteracting piston-type hydraulic buffer with the counter recoil system being pneumatic, with both mounted above the ordnance.

The 2A45M has a mechanised system to convert from the firing position to the travelling position and this consists of a hydraulic jack and hydraulic cylinders. With the aid of the jack the carriage is raised to the required height for folding or unfolding the trails, then lowered onto the ground. The hydraulic cylinders raise the gun to the maximum clearance and also raise and lower the wheels.

When deployed in the firing position, the weapon rests on three trails which are normally staked to the ground to provide a more stable firing platform.

The system takes between 90 and 120 seconds to come into action and between 120 and 160 seconds to come out.

The system can be brought into action manually or using hydraulic assistance with the Auxiliary Propulsion Unit (APU) running.

The design of the carriage is such that the upper part of the weapon can be quickly traversed through a full 360° to be laid onto new targets.

A well-sloped shield provides the crew with limited protection from small arms fire and shell splinters. The 2A45M is fitted with an auxiliary propulsion unit consisting of an MeMZ-967A four-stroke petrol engine with hydraulic drive which allows it to propel itself around the battlefield at a maximum speed of 10/14 km/h. The engine is mounted on the right side with power transmitted to the roadwheels via cardan shafts. The driver's seat is mounted on the left side and is provided with the necessary controls for controlling the system when the APU is being used.

The 2A45M is normally towed by a 6 × 6 truck that also carries the crew and ammunition to as near the firing position as possible. A small wheel attached to one of the three trail legs is then lowered and the 2A45M moves under its own power to the firing position.

The driver is seated in front of the shield on the left side and steers using a conventional steering wheel which folds forward when not required.

The two large rubber-tyred roadwheels of the carriage are powered and claimed to be bulletproof with steering accomplished by altering the power to either roadwheel.

The 2A45M weighs 6,575 kg when deployed in the firing system compared to the standard 122 mm D-30 towed howitzer which weighs 3,210 kg.

The sights, elevation and traverse equipment are positioned on the left side. Optical systems fitted include the OP4M-48A direct fire day telescope and the 1PN53-1 night sight. The iron sight 2Ts33 is used together with the panoramic sight PG-1M for indirect firing.

The 2A45M fires the same family of separate-loading ammunition (projectile and semi-combustible cartridge case) as the T-64/T-72/T-80/T-84/T-90 MBTs and maximum rate of fire is quoted as 6 to 8 rds/min. A total of six ready rounds (projectile and charge) can be carried on the system for ready use with the remainder being carried in the towing vehicle.

The APFSDS-T projectile is used to engage MBTs and has an effective range of 2,100 m while the HE-FRAG projectile is used to engage general battlefield targets such as personnel. It contains 3.4 kg of A-IX-2 explosive and is fitted with a V-429E impact fuze.

Maximum range in the indirect fire role is 12,200 m compared to the 15,400 m (without using a rocket-assisted projectile) of the D-30, as the latter has a maximum elevation of +70°.

125 mm 2A45M (Sprut-B) towed anti-tank gun deployed in the firing position

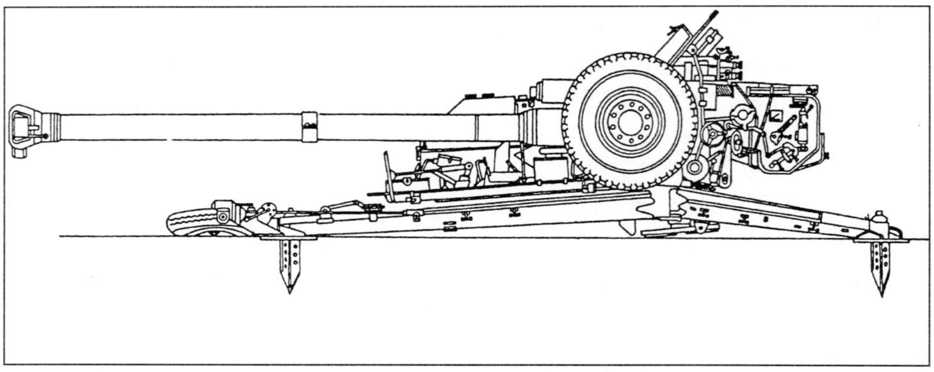

Provisional drawing of the 125 mm 2A45M (Sprut-B) towed anti-tank gun deployed in the firing position

Specifications

Calibre: 125 mm, 51 calibre
Muzzle brake: double 'T' type
Breech mechanism: semi-automatic, vertical wedge
Shield: yes
Weight:
(towing) 6,500 kg
(self-propelled) 6,800 kg
(firing) 6,575 kg
Length:
(towing) 7.120 m
(self-propelled) 6.790 m
Height:
(towing) 2.09 m
(self-propelled) 2.35 m
Width:
(towing) 2.66 m
(self-propelled) 2.66 m
Axis of bore: 0.925 m
Track: 2.20 m
Ground clearance: 0.36 m
Elevation/depression: +25/–6°
Traverse: 360°
Range
(APFSDS) 2,000 m
(missile) 5,000 m
(HE) 12,200 m
Traverse: 360°
Rate of fire: 6-8 rds/min
Fording: 0.9 m
Speed:
(in APU mode) 14 km/h
Range:
(in APU mode) 50 km
Crew: 7
Towing vehicle: Ural-4320 (6 × 6) truck, MT-LB multipurpose tracked vehicle
Towing speed: 80 km/h

Designation	Type	Muzzle velocity	Round weight	Projectile Weight	Projectile Length	Weight of filler
BM-17	APFSDS	1,700 m/s	20.7 kg	7.05 kg	620 mm	nil
BK-10	HEAT	905 m/s	29.58 kg	19.08 kg	675 mm	1.76 kg
OF-36	HE	850 m/s	32.5 kg	23.00 kg	675 mm	3.4 kg

The HEAT projectile has a V15 nose-mounted base activated fuze and contains 1.76 kg of A-IX-1 or OCFOL explosive which will penetrate first-generation MBTs such as the M60, M48 and Leopard 1 unless they are fitted with explosive reactive armour.

All projectiles use the same propelling charge which is 408 mm long and all that remains after firing is the stub case. Basic details of the key 125 mm ammunition types are given in the table above.

The 125 mm 2A54M anti-tank gun system can fire the 125 mm laser-guided projectiles fired by the latest RFAS MBTs, the 9K120 Refleks or Sniper. Both these projectiles have a range of 5,000 m and a single HEAT warhead which will penetrate 700 mm of steel armour. More recent versions have a tandem HEAT warhead to defeat explosive reactive armour (ERA).

In order to fire these 125 mm laser guided projectiles a 9S53 laser fire-control system has to be used.

Status

Development complete. As far as is known the 125 mm 2A45M anti-tank gun has not entered volume production. It is being offered for export. It is considered that if placed in volume production then existing 122 mm D-30 carriages may probably be used rather than brand new carriages. As of early 2002 there were no known sales of this 125 mm anti-tank gun.

Manufacturer

Joint Stock Company Spetstehnika.

UPDATED

122 mm field gun D-74

Development/Description

The 122 mm field gun D-74 was developed in the late 1940s by the F F Petrov design bureau and was seen for the first time in public during the 1955 May Day parade in Moscow. It appears that the 122 mm field gun D-74 and the 130 mm field gun M-46 were both designed to meet the Russian Army requirement for a gun with a longer range to replace the 122 mm corps gun M1931/37 (A-19).

After extensive troop testing the 130 mm field gun M-46 was selected in preference to the D-74 as it fired a heavier projectile to 27,150 m compared with the 24,000 m of the D-74.

The D-74 was, however, produced in some numbers, most of which were subsequently exported. It is now considered obsolete by the RFAS. The 122 mm field gun D-74 uses the same carriage as the 152 mm gun-howitzer D-20 but the D-20 can easily be distinguished from it by a much thicker barrel and a larger muzzle brake.

According to United Nations sources, there were no exports of the 122 mm field gun D-74 by any country between 1992 and 2000.

Table 1

Ammunition type	**FRAG-HE**	**APC-T**
Projectile designation	OF-472	BR-472
Fuze model	V-429	DBR
Weight		
(projectile)	27.3 kg	25 kg
(bursting charge)	2.95 kg	0.091 kg
Type of bursting charge	TNT	RDX/alum
Muzzle velocity	885 m/s	885 m/s
Armour penetration at 0°	n/a	185 mm/1,000 m

Other types of projectile include illuminating (m/v 885 m/s) and smoke (m/v 885 m/s with projectile weighing 21.7 kg)

The 122 mm field gun D-74 has also been manufactured for the home and export market by China North Industries Corporatiion (NORINCO) under the Chinese designation Type 60. This has not been built or marketed by the Chinese for some years.

The shield, which slopes slightly to the rear and extends over the road wheels, has an irregular top with a centre section which slides upwards so the ordnance can be elevated. A circular firing pedestal fitted as standard on all D-74s is inverted and secured just forward of the shield when travelling. In the firing position the firing pedestal is lowered to the ground, allowing the D-74 to be quickly traversed through a full 360°. On each of the split box section trails is a castor which folds up on to the top of the trail when travelling and a spade which folds up underneath the trail when travelling.

The D-74 fires variable charge, case-type, separate loading ammunition, details of which are in the table.

122 mm field gun D-74 on firing pedestal with spades lowered and support wheels raised (US Army)

Specifications

Calibre: 121.92 mm
Barrel length: (overall) 6.45 m
Muzzle brake: double-baffle
Recoil system: hydraulic buffer and hydropneumatic recuperator
Breech mechanism: semi-automatic vertical sliding wedge
Carriage: split trail with castor wheel on each
Shield: yes
Weight:
(travelling order) 5,550 kg
(firing position) 5,500 kg
Length: (travelling) 9.875 m
Width: (travelling) 2.35 m
Height: (travelling) 2.745 m
Axis of bore: (at 0°) 1.668 m
Ground clearance: 0.4 m
Track: 2.03 m
Tyres: 12.00 × 20
Elevation/depression: +45/–5°
Traverse:
(total) 58°
Rate of fire: 6-7 rds/min
Sustained rate of fire: (first hour) 75 rounds
Range:
(max) 24,000 m
(effective APHE) 1,070 m
Crew: 10
Towing vehicle: Ural-375D (6 × 6) truck
Unit of fire: 90 rounds

Status

Production complete. In service with the following countries:

Country	Quantity	Comment
Albania	150/280	Chinese Type 60
Algeria	28	
Cambodia	n/avail	Chinese Type 60
China	n/avail	local designation Type 60
Congo, Democratic Republic	14	Chinese Type 60
Egypt	144	Russian D-74
	48	Chinese Type 60
Iran	40	Chinese Type 60
Iraq	n/avail	
Korea, North	n/avail	possible local production
Libya	60	
Mauritania	20	
Nigeria	100	estimated figure
Pakistan	200	Chinese Type 60
Somalia	n/avail	status uncertain
Sudan	4/10	Chinese Type 60
Syria	n/avail	
Vietnam	n/avail	and Chinese Type 60
Zimbabwe	18/20	Type 60

Manufacturers

Joint Stock Company Spetstehnika.
China North Industries Corporation (NORINCO)

UPDATED

122 mm howitzer D-30

Development

The 122 mm howitzer D-30 entered service with the former Soviet Army in the early 1960s as the replacement for the 122 mm howitzer M1938 (M-30).

The 122 mm howitzer D-30 was designed by the F F Petrov design bureau at Artillery Plant No.9 in Sverdlovsk. The main improvements over the earlier weapon are increased range and the ability to rapidly traverse the weapon through 360°.

122 mm D-30 in firing position
2001/0105747

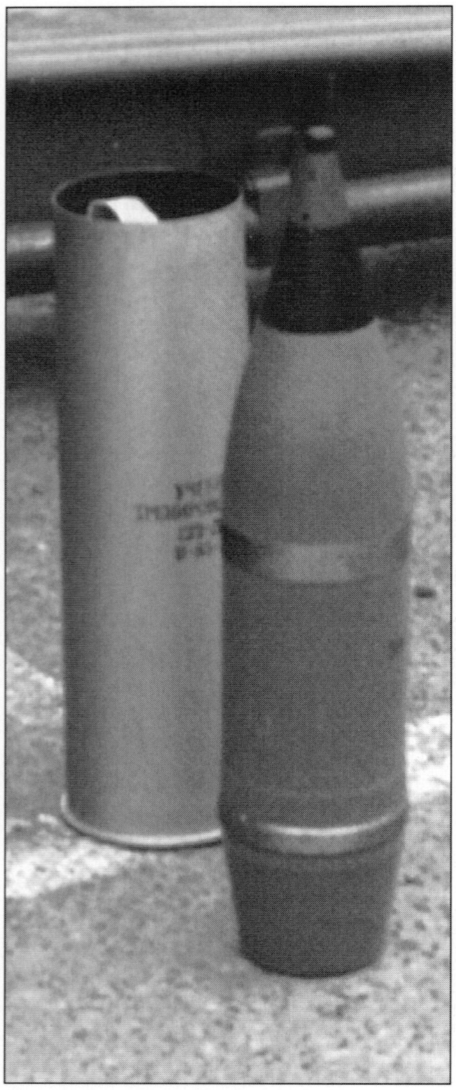

Ammunition fired by the 122 mm D-30 is of the separate loading type, for example projectile and charge (Paul Beaver)

Russian 122 mm D-30 howitzer on air drop platform without rigging attached (Richard Stickland)
0021014

122 mm D-30 in travelling configuration
2000/0085648

Late production models of the D-30 are designated the D-30M and have a number of modifications including a new double-baffle muzzle brake, a new central baseplate which is square rather than round and a towing lunette assembly. It also features modifications to its cradle, carriage and recoil system. As it is ballistically the same as the earlier D-30 it has the same range and capabilities.

According to the United Nations, between 1992 and 2000 the following quantities of D-30 122 mm howitzer weapons were exported:

From	To	Quantity	Date	Comment
Czech Republic	Peru	6	1998	
Egypt	Rwanda	6	1992	
Germany	Finland	218	1992	
Kazakstan	Angola	4	1998	
Kazakstan	Ethiopia	100	2000	
Romania	Moldova	18	1992	not confirmed as D-30
Slovakia	Bulgaria	7	1994	

Description

The 122 mm howitzer D-30 is towed muzzle first and on arrival at the battery position the crew first unlock the barrel travelling lock which then folds back onto the central trail. The firing jack under the carriage is then lowered, raising the wheels clear of the ground. The two outer trails are each spread through 120° and the firing jack is raised until all three trail ends are on the ground, when they are staked into position. The lunette is mounted either under the muzzle brake or on the barrel just to the rear of the muzzle brake. In either case it is normally swung through 180° so that it lies under the barrel before firing begins.

Fire-control instruments include a Type PG-1M panoramic sight, Type OP4M-45 telescopic sight and a Type K-1 collimator.

An unusual feature of the D-30 is that the recoil system is mounted over the barrel, which is designated the 2A18. Some of the projectiles fired by the D-30 are interchangeable with those fired by the 122 mm howitzer M1938 (M-30) but the D-30 also fires a HEAT (High Explosive Anti-Tank) projectile of the non-rotating fin-stabilised type.

The D-30 fires the following variable charge, case-type, separate loading ammunition:

Ammunition type	**FRAG-HE**	**HEAT-FS**
Projectile designation	OF-462*	BK-6M†
Fuze model	RGM-2	GPV-2
Weight		
(projectile)	21.76 kg	21.63 kg
(bursting)	3.675 kg	n/a
Type of bursting charge	TNT	RDX
Muzzle velocity	690 m/s	740 m/s
Armour penetration at 0°	n/a	460 mm/any range

* there are variants of projectile design giving different weights. Various other fuzes may also be used.
† rotating BP-463 may also be used.

Other types of projectiles include illuminating (S-463 weighing 22.4 kg), smoke (D-462 weighing 22.3 kg), leaflet, flechette and incendiary. Recently a Rocket-Assisted Projectile (RAP) has been introduced with a maximum range of 21,900 m.

NORINCO of China has developed a new family of ammunition which can be fired by the Russian-built 122 mm D-30 or the Chinese version, brief details of these are given in the entry for the D-30 under China.

In the Russian Army each tank division has an artillery regiment with two battalions each of 18 D-30s and one battalion of 18 152 mm (2S3) self-propelled gun-howitzers. Each motorised rifle division has 72 D-30s with each motorised rifle regiment having a battalion of 18 D-30s with the remaining two battalions of 18 weapons in the artillery regiment.

The main armament of the 122 mmself-propelled howitzer 2S1 is based on the ordnance of the D-30.

The D-30 is also produced in China (see separate entry), Croatia (made by RH ALAN under the designation of the 122 mm howitzer D-30 HR M94), Egypt (qv this section under Egypt), Iraq (called the Saddam, qv), Iran and Yugoslavia (122 mm howitzer D-30J, qv).

Upgrade package

Russia now has available an upgrade package for the 122 mm D-30 which enables it to be towed at higher

Specifications
D-30
Calibre: 121.92 mm (with muzzle brake)
Barrel length: 4.875 m
Muzzle brake: multibaffle
Breech mechanism: semi-automatic vertical sliding wedge breech block
Shield: yes
Weight:
(travelling order) 3,210 kg
(firing position) 3,150 kg
Length: (travelling) 5.4 m
Width: (travelling) 1.95 m
Height: (travelling) 1.66 m
Axis of bore: (at 0°) 0.9 m
Ground clearance: 0.325-0.345 m
Track: 1.85 m
Tyres: 9.00 × 20
Elevation/depression: +70/–7°
Traverse: 360°
Rate of fire: 7-8 rds/min
Sustained rate of fire: (first hour) 75 rounds
Range:
(max) 15,400 m
(with RAP) 21,900 m
Crew: 7
Towing vehicle:
MT-LB multipurpose tracked vehicle,
Ural-375D (6 × 6) truck
Unit of fire: 80 rounds

speeds, improving its reliability and enabling it to come into action more quickly.

122 mm D-30A

This is also referred to as the 2A18M and has a number of minor improvements including a new cradle and modified recoil system. It also features a new muzzle brake.

Brief specifications are:
Calibre: 121.92 mm
Max range: 15,300 m
Rate of fire: 6-8 rds/min
Axis of fire: 900 mm
Traverse: 360°
Elevation/depression: +70/–7°
Length:
(travelling) 5.4 m
Width:
(travelling) 2.2 m
Height:
(travelling) 1.8 m
Ground clearance: 0.325 m
Track: 1.85 m
Max towing speed: 80 km/h

SP122

This has been developed by United Defense LP of the USA and consists of a modified 155 mm M109 self-propelled howitzer/M992 Field Artillery Ammunition Support Vehicle (FAASV) chassis with a fully enclosed superstructure at the rear in which is mounted the complete upper carriage of the Egyptian-built D-30 towed howitzer. Details of this system, which is now in service with the Egyptian Army, are given in the Self-propelled guns and howitzers (tracked) section under Egypt. By early 2002 a total of 124 SP122 systems had been delivered to Egypt.

Status

Production complete. In service with the following countries:

Country	Quantity	Comment
Afghanistan	n/avail	
Algeria	198	
Angola	n/avail	
Armenia	59	
Azerbaijan	95	
Belarus	200	
Benin	n/avail	
Bosnia-Herzegovina	12	
Bosnian Serbs	60	
Bulgaria	380	
Burundi	18	
Cambodia	n/avail	
China	n/avail	local production (qv)
Congo	10	
Congo, Democratic Republic of (Zaïre)	0	estimate
Croatia	32	local production (qv)
Croatia HVO	48	
Czech Republic	148	
Cuba	n/avail	
Djibouti	6	
Egypt	156	local production (qv)
Eritrea	2	
Ethiopia	460	includes 122 mm M-30
Finland	486	local designation H-63
Georgia	57	
Ghana	6	
Guinea Bissau	9	estimate
India	550	
Iran	600	estimate, local production
Iraq	600	local production (qv)
Israel	100	
Korea, North	n/avail	
Kazakhstan	72	
Kyrgyzstan	72	
Laos	20	
Latvia	18	
Lebanon	10	
Libya	270	
Macedonia	142+	
Madagascar	12	
Mali	8	
Mauritania	20	
Moldova	18	not confirmed as D-30
Mongolia	120	
Morocco	50	
Mozambique	24	includes 122 mm M-30
Nicaragua	12	
Nigeria	100	estimate
Oman	25	
Peru	42	including 6 in 1998
Russia	2,000	estimate
Rwanda	6	from Egypt in 1998
Seychelles	3	
Slovakia	76	
Somalia	n/avail	
Sudan	25	
Syria	500	some were fitted onto T-34 chassis
Tajikistan	12	
Tanzania	20	
Turkmenistan	197	
Ukraine	435	
Uzbekistan	90	
Vietnam	n/avail	
Yemen	130	
Yugoslavia	310	local production (qv)
Zambia	25	

Manufacturer

Joint Stock Company Spetstehnika.

UPDATED

122 mm howitzer M1938 (M-30)

Development

The 122 mm howitzer M1938 (M-30) entered service with the Russian Army in 1939 after standardisation in September 1938. It was designed by the F F Petrov design bureau at Artillery Plant No.172 at Perm.

Until the introduction of the 122 mm howitzer D-30 it was the standard division howitzer of the Warsaw Pact and was issued on the scale of 36 per motorised rifle division (two battalions each consisting of 18 M1938s with three batteries each with six howitzers) and 54 per tank division (three battalions each consisting of 18 M1938s with three batteries each with six howitzers).

It is no longer in front-line service with the RFAS, although it can still be found in service in various parts of the world. By today's standards the 122 mm M1938 (M-30) is obsolete as it has poor mobility and a very short range.

According to United Nations sources, there were only two exports of the M-30 between 1992 and 2000. Bulgaria exported 108 systems to Macedonia while Kazakstan exported 24 systems to Russia.

Description

The carriage of the 122 mm M1938 (M-30) is of the riveted box section split trail type and is identical to that used for the 152 mm howitzer M1943 (D-1), which can

122 mm howitzer M1938 (M-30) with standard sponge rubber-filled tyres showing sliding vertical section in centre of shield

122 mm howitzer M1938 (M-30) in travelling configuration (IDF Spokesman)

Specifications
Calibre: 121.92 mm
Barrel length: (overall) 2.8 m
Muzzle brake: no
Recoil system: hydraulic buffer and hydropneumatic recuperator
Breech mechanism: hinged screw
Carriage: split trail
Shield: yes
Weight:
(travelling order) 2,450 kg
(firing position) 2,450 kg
Length: (travelling) 5.9 m
Width: (travelling) 1.975 m
Height: (travelling) 1.71 m
Axis of bore:
(at 0°) 1.2 m
Ground clearance: 0.33 m
Track: 1.6 m
Tyres: 11.80 × 15
Elevation/depression: +63.5/–3°
Traverse:
(total) 49°
Rate of fire: 5-6 rds/min
Sustained rate of fire:
(first hour) 75 rounds
Range:
(max) 11,800 m
(effective, HEAT) 630 m
Crew: 8
Towing vehicle:
Ural-375 (6 × 6) truck,
MT-LB multipurpose tracked vehicle
Unit of fire: 80 rounds

easily be distinguished from the 122 mm weapon by its longer and fatter barrel with a double-baffle muzzle brake. The top half of the shield slopes to the rear and in the centre of the shield is a section which slides upwards so the weapon can be elevated. The recoil system is under the 122 mm barrel and the counter-recoil system above. The elevating mechanism is on the right side of the carriage and the traverse mechanism is on the left. The 122 mm howitzer can be fired without spreading its trails, in which case traverse is limited to 1.5°. Maximum towing speed is 48 km/h. Each trail has two spades, one fixed for use on hard surfaces and one hinged for softer ground.

The 122 mm M1938 (M-30) fires the following variable charge, case-type, separate-loading ammunition which was developed many years ago and can also be fired by the more recent Russian 122 mm D-30 howitzer covered earlier in this section.

Ammunition type	FRAG-HE	HEAT
Projectile designation	OF-462*	BP-463†
Fuze model	RGM-2	GKB
Weight		
(projectile)	21.76 kg	14.8 kg
(bursting)	3.675 kg	n/a
Type of bursting charge	TNT	n/a
Muzzle velocity	515 m/s	570 m/s
Armour penetration at 0°	n/a	180 mm

* other FRAG and FRAG-HE projectiles available
† fired at full charge. Better design than earlier BP-460A (weight 14.8 kg) which was fired at charge 4 giving only 325 m/s
Other projectiles include illuminating (S-463) weighing 21 kg, smoke (D-462) weighing 22.4 kg and leaflet (A-462) weighing 21.5 kg.

Variants

The 122 mm howitzer M1938 (M-30) used by Bulgaria has new disc wheels with spoke-type impressions but retains the sponge rubber-filled tyres of the original model.

The M1938 has also been manufactured in China as the Type 54 and Type 54-1 and is issued on the scale of 12 per infantry division and 12 per armoured division. A self-propelled model of the Type 54-1 is essentially a Type 531 fully tracked APC with the Type 54-1 mounted on the rear of the vehicle firing forwards. Details of this appear in the Self-propelled guns and howitzers (tracked) section.

Polish upgraded 122 mm M1938 (M-30)

In the 1980s Poland upgraded some of their 122 mm M1938 (M-30) weapons to improve their direct fire capability. These are known as the wz 1938/1985.

Status

Production complete. In service with countries listed below. In some cases these weapons are held in reserve or used for training purposes.

Country	Quantity	Comment
Afghanistan	n/avail	
Albania	100	
Algeria	60	
Bangladesh	12/18	Chinese Type 54
Bosnia	n/avail	
Bulgaria	n/avail	108 sold to Macedonia in 1999
Cambodia	n/avail	Chinese Type 54
China	n/avail	made in China as Type 54 (qv)
Congo (Democratic Republic)	10	estimated figure
Croatia	45	
Cuba	n/avail	
Czech Republic	25	reserve
Egypt	359	
Ethiopia	360	D-30 and M-30
Guinea-Bissau	9	
Hungary	230	
Indonesia	28	
Iran	40	Chinese Type 54
Iraq	n/avail	
Korea, North	n/avail	
Kyrgystan	37	
Laos	20	estimate
Lebanon	33	
Macedonia	108	from Bulgaria in 1999
Moldova	17	
Mongolia	n/avail	
Mozambique	24	D-30 and M-30
Pakistan	30/200	Chinese Type 54
Poland	280	
Romania	204	
Russia	4,000	reserve
Slovakia	46	
Somalia	n/avail	
Sudan	24	M-30
	76	Type 54
Syria	150	estimate
Tanzania	85	
Uganda	18	
Ukraine	3	reserve
Vietnam	n/avail	
Yemen	50	
Yugoslavia	90	
Tanzania	100	Chinese Type 54
Uganda	20	
Zimbabwe	12	Type 54

Manufacturers

Joint Stock Company Spetstehnika.
China North Industries Corporation (NORINCO) (qv).

UPDATED

120 mm 2B16 (NONA-K) combination gun

Development

For some years the Russian Army has deployed two 120 mm gun/mortar systems, the 2B16 (NONA-K) towed system and the 2S9 full-tracked self-propelled system which is covered in detail in the Self-propelled mortar systems section. The 2S9 is based on the BMD-1 airborne combat vehicle chassis.

It is now known that these two weapons share the same 120 mm ordnance and fire the same family of ammunition with the obvious logistic and training advantages to the user.

These 120 mm weapons combine the features of a howitzer and mortar in one system. In many respects they are unique weapons as there is no similar equivalent weapon system in the West.

The 120 mm 2B16 (NONA-K) towed system was developed in the 1970s at the famous Perm artillery facility and the Central Scientific Research Institute for Precision Machinery Construction.

It was accepted for service with the Russian Army in 1986 with the main user being the Russian air assault forces.

More recently, Russia has developed a new full-tracked self-propelled artillery system called the 2S31 which fires the same family of ammunition as the 2B16, 2S9 and 2S23 as well as a complete new family of ammunition. The 2S31 remains at the prototype stage and is already being offered on the export market. Details of the 120 mm 2S31 are given in the Self-propelled mortar systems section.

The 120 mm 2S31 is based on a much modified BMP-3 infantry combat vehicle chassis.

United Nations sources have stated that Russia did not declare any export of the 120 mm 2B16 (NONA-K) for the years 1992 through to 2000.

120 mm 2B16 (NONA-K) combination gun deployed in the firing system and showing large muzzle brake and castor wheel on trail to assist in bringing the weapon into action ***2002***/0130025

Description

Within the designation NONA-K the K (or Kolesnaya in Russian) stands for wheeled, with the weapon being mounted on a split trail carriage and each trail being provided with a castor wheel to assist in bringing the weapon quickly into action.

When travelling, the trails are closed together and locked and the complete upper part of the weapon is traversed through 180° and locked in position over the closed trails.

The 2B16 is normally towed by the GAZ-66 (4 × 4) 2,000 kg light truck which also carries the five-man crew and a quantity of ready to use 120 mm ammunition. Maximum towed speed on a good road is quoted as 80 km/h. For short distances it can be towed by the smaller UAZ-469 (4 × 4) 695 kg light vehicle.

When deployed in the firing position, the NONA-K is supported on a small circular baseplate which is located under the forward part of the carriage. The rubber-tyred roadwheels are raised clear of the ground and the equipment rests on the two box-type spread trails, each of which is fitted with a spade.

The 120 mm/24.2 calibre ordnance, designated the 2A51, has 40 constant twist rifling grooves and is breech loaded. It can fire both Russian and Western natures of 120 mm rifled ammunition (for example that used by the French TDA 120 RT-61 towed mortar).

To reduce recoil forces on the barrel, a large box-type multibaffle muzzle brake is fitted which absorbs about 30 per cent of the recoil.

The hydraulic recoil system is mounted above the ordnance and extends about halfway along the barrel with the recuperator being of the hydropneumatic type. The breech mechanism is of the vertical sliding type with plastic gas obturator and a chamber indent device is fitted to retain the round in place when the weapon is fired at high elevation.

The thin shield either side of the ordnance provides the crew of the 2B16 with some protection from small arms fire and shell splinters, with the left side of the shield being slightly higher than the right to protect the direct and indirect sighting system. The manual elevation and traverse controls are also on the left side of the weapon.

Types of 120 mm projectiles fired by the 2B16 include HE-FRAG (high-explosive fragmentation), HEAT (high-explosive anti-tank) and smoke as well as 120 mm mortar bombs.

Details of the prerifled projectiles are given in the table.

The HE-FRAG projectile is called the OF-49, has a steel body and contains A-IX-2 explosive. There is also the OF-51 HE-FRAG projectile which also contains A-IX-2 explosive.

The rocket-assisted projectile is called the OF-50 with the motor cutting in 10 to 13 seconds after the projectile leaves the barrel. The HEAT projectile is fin-stabilised and will penetrate 650 mm of conventional steel armour at 90° to the vertical.

Specifications

2B16
Calibre: 120 mm
Calibre length: 24.2
Recoil: 400 mm
Muzzle brake: multibaffle
Carriage: split trail
Weight: 1,200 kg
Length: 5.9 m
Width: 1.79 m
Elevation/depression: +80/–10°
Traverse: 30° left and right
Rate of fire: 8-10 rds/min
Max range:
(HE-FRAG projectile) 8,700 m
(HE-FRAG mortar bomb) 7,100 m
Crew: 5

Type	HE-FRAG	HE-RAP	HEAT
Calibre	120 mm	120 mm	120 mm
Round weight	19.8 kg	19.8 kg	13.17 kg
HE content	4.9 kg	3.25 kg	n/a
Round length	820 mm	825 mm	960 mm
Muzzle velocity	367 m/s	367 m/s	560 m/s
Max pressure	100 MpA	100 MpA	110 MpA
Max range	8,850 m	12,850 m	1,000 m
Min range	1000 m	1000 m	40 m

Standard 120 mm mortar bombs fired include the OF-843B, OF-34 and OF-38 high-explosive fragmentation, S-843 and 3S9 illumination, 3Z2 incendiary and 3D5 smoke. These have muzzle velocities of between 119 and 331 m/s and ranges of between 430 and 7,150 m.

The 120 mm ammuntion for this as well as the 2S9, 2S23 and latest 2S31 has been designed and manufactured by the Bazalt State Research and Production Enterprise.

Russia has developed 122 and 120 mm laser-guided projectiles but there is no hard evidence that any of these have been designed for use with the 2B16.

The 2S9, also referred to as the NONA-S, has a similar ordnance to that used in the NONA-K but this is fitted with a system to purge the ordnance after the weapon has been fired. A pneumatically assisted rammer is fitted to increase the rate of fire and load the weapon at higher elevations. The ordnance is not fitted with a muzzle brake.

The latest 120 mm gun/mortar system to enter service is the 2S23, or NONA-SVK which is based on the chassis of the BTR-80 (8 × 8) APC which has been in service with the Russian Army for some years.

This uses a modified version of the 120 mm 2A51 ordnance which is called the 2A60, but ballistics and characteristics remain the same. Details are given in the Self-propelled mortar systems section.

Status

Production complete. In service with the following countries:

Country	Quantity
Azerbaijan	12
Russia	150
Ukraine	2

Manufacturer

Perm Artillery Factory.

UPDATED

100 mm anti-tank gun T-12 (2A19) and MT-12 (2A29)

Development

The 100 mm T-12 anti-tank gun was developed as the replacement for the older 85 mm D-48 towed anti-tank gun, which was covered in *Jane's Armour and Artillery 1995-96*. This may still be found in service in some parts of the world.

The 100 mm T-12 anti-tank gun has the industrial designation 2A19, was accepted for service in 1961 and was followed by the improved MT-12 which has the industrial designation 2A29. The MT-12 anti-tank gun entered production in 1970. The 100 mm T-12 was the first smoothbore anti-tank gun in the world to enter production.

According to Russian sources, the designers of the 100 mm T-12 smoothbore anti-tank gun at Factory No 75 decided on a smoothbore gun rather than a rifled gun for a number of reasons, this included:

(1) A spinning projectile tends to lose armour penetration from the jet of gases and hot metal that is produced when the HEAT (High Explosive Anti-Tank) projectile hits the target.
(2) As the smoothbore projectile is a tighter fit than a rifled projectile, the smoothbore gun can contain gas pressures that are higher when compared to a rifled tank gun. This means that the smoothbore gun has a higher muzzle velocity and therefore greater armour penetration characteristics than a comparable rifled gun.
(3) A rifled gun has lands and grooves which wear out when firing projectiles at high-muzzle velocities. Smoothbore guns have no lands or grooves and therefore have a longer life.

The T-12 anti-tank gun can be distinguished from the earlier 85 mm D-48 anti-tank gun by its large breech ring, slightly differently shaped shield with an opening in the left side for the sighting systems and its smaller diameter pepperpot muzzle brake which is also parallel to the muzzle.

The T-12/MT-12 is used by the anti-tank units of armoured and motorised rifled regiments, in order to protect open flanks against counterattack during rapid advances.

Although the T-12/MT-12 has been designed for use in the direct fire mode, both weapons also have indirect fire sights. Their range in this role, however, is limited as

100 mm anti-tank gun MT-12 with trails closed but spades still fitted and castor wheel lowered to ease field handling

100 mm anti-tank gun MT-12 being towed by an MT-LB multipurpose armoured tracked vehicle

maximum elevation is only +20°. For elevations above +15°, a pit has to be dug under and to the rear of the breech or the breech will hit the ground during recoil.

Description

The T-12/MT-12 is operated by a six-man crew consisting of detachment commander, driver of the towing vehicle, gunlayer on the left, loader on the right and two ammunition numbers.

The weapon comprises four main components, barrel and breech ring, semi-automatic breech, top carriage with sight and lower carriage. The actual mount used in the T-12 is the same as that used in the 85 mm D-48 rifled anti-tank gun.

The 100 mm smoothbore barrel is 6.30 m long with the front part strengthened and fitted with a perforated muzzle brake. When in the travelling position, the barrel is clamped tightly to the trails with a lug on the breech ring.

The breech ring contains the vertical breech block and some parts of the semi-automatic loading system. The breech block is manually opened and closed by the loader. Manually opening the breech is only required before the first round is fired and after that the semi-automatic loading system to the right of the breech ring on the upper carriage opens and closes the breech block. Consequently, the loader only has to reload.

The sights, elevating and traverse system are on the left. The aiming system consists of the S71-40 mount with the PG-1M panoramic sight for indirect aiming, the OP4M-4OU direct sight and the APN-6-40 (or older APN-5-40) for direct aiming at night. A range drum is fitted as standard.

The OP4M-4OU direct sight has a magnification of ×5.5 and an 11° field of view. The image intensification night sight has a magnification of ×6.8 and a 7° field of view.

The hydraulic buffer and the hydropneumatic recuperator are mounted on top of the ordnance forward of the vertical breech mechanism. Recoil is between 680 and 760 mm and the equilibrator of the MT-12 is mounted parallel and to the right of the recoil system.

The carriage of the MT-12 is of the split trail type with a castor wheel mounted on the left trail to assist bringing the weapon into action. The carriage is provided with a small shield with sides that slope to the rear. This provides protection for some of the crew from small arms fire and shell splinters.

The wheels of the T-12 can be removed and replaced by LO-7 type skis which allow the weapon to be fired on ice or snow at elevations of up to 16°.

The main difference between the T-12 and the MT-12 is the new carriage. The former was normally towed by a truck whereas the MT-12 is towed by a MT-LB tracked armoured vehicle. The original T-12 often turned over when being towed causing the MT-12 with the improved carriage to be introduced. The wheels are the same as those on the ZIL-150 (6 × 6) truck and have sponge rubber-filled tyres. The suspension of the MT-12 consists of one torsion bar and one shock-absorber per wheel. When the weapon is deployed in the firing position, the suspension system is disengaged to increase gun stability.

Specifications

Model	MT-12	T-12
Calibre	100 mm	100 mm
Muzzle brake	pepperpot	pepperpot
Breech mechanism	semi-automatic vertical sliding spring-actuated wedge type	
Carriage	split trail	split trail
Shield	yes	yes
Weight	3,050 kg	2,750 kg
Length	9.65 m	9.48 m
(travelling)	9.65 m	9.48 m
Width	2.31 m	1.795 m
(travelling)	2.31 m	1.795 m
Height	1.60 m	1.565 m
(travelling)	1.60 m	1.565 m
Ground clearance	380 mm	380 mm
Track	1.92 m	1.465 m
Tyres	9.00 × 20	6.50 × 20
Elevation/depression	+20/–6°	+20/–6°
Traverse:		
(total)	27°	27°
Max rate of fire	14 rds/min	14 rds/min
Crew	6	6
Towing speed		
(road)	70 km/h	60 km/h
(cross-country)	25 km/h	15 km/h
Towing vehicle	MT-LB	Ural-375D (6 × 6)

There are three types of fixed 100 mm ammunition fired by the T-12/MT-12 as detailed in the table.

Type	Muzzle velocity	Range	Weight of round
APFSDS	1,575 m/s	3,000 m (direct)	19.34 kg
HEAT	975 m/s	5,955 m (direct)	23.00 kg
HE	700 m/s	8,200 m (indirect)	18.6 kg

The APFSDS round will penetrate the following RHA at 90°:

Penetration	Range
230 mm	500 m
215 mm	1,000 m
190 mm	1,500 m
180 mm	2,000 m
165 mm	2,500 m
140 mm	3,000 m

The HEAT round will penetrate 350 mm of RHA at 90°. The T-12/MT-12 can also fire the 9M117 Kastet (Knuckleduster) (NATO designation Stabber) laser-guided projectile, which has a range of 100 to 4,000 m with its HEAT warhead penetrating between 550 and 600 mm of armour.

The laser designator is mounted on a tripod, normally to the left of the weapon and all the operator has to do is keep the cross-hairs of his sights on the target to ensure a hit.

The original version of the 9M117 missile only had a single HEAT warhead but more recently a version with a tandem HEAT warhead has been developed which will penetrate 550 mm of rolled homogenous armour protected by explosive reactive armour.

The laser-guided projectile version has been offered on the export market but there are no known sales as of early 2001.

Variants

Yugoslavia has placed the ordnance of the T-12 (designated the 2A19M) on a modified D-130 122 mm howitzer carriage. The complete system is designated the 100 mm anti-tank gun TOPAZ and details are given under Yugoslavia later in this section. As far as is known this never entered quantity production or service.

Status

Production complete. In service with the following countries:

Country	Quantity	Comment
Algeria	12	
Armenia	36	
Azerbaijan	30	
Belarus	40	
Bosnia/ Bosnia Serbs	90	
Bulgaria	200	
Cambodia	n/avail	
China	120	local designation Type 73
Croatia	133	
Cuba	n/avail	
Georgia	50	
Hungary	106	
Iraq	n/avail	
Kazakstan	125	
Kyrgystan	15	
Moldova	26	
Mongolia	24	
Russia	6,000	
Turkmenistan	48	
Ukraine	400	
Uzbekistan	39	
Yugoslavia	138	

Manufacturer

Factory Number 75.

UPDATED

85 mm divisional gun D-44

Development

The 85 mm divisional gun D-44 was developed by the F F Petrov design bureau during 1943-44 as the replacement for the 76 mm division gun M1942 (ZIS-3) but not issued until after the war. First production weapons were completed in 1945.

The barrel is a development of the 85 mm gun mounted in the T-34/85 medium tank and is basically identical to both this and the 85 mm M1939 (KS-12) anti-aircraft gun. Ammunition is interchangeable with that used in the Czech 85 mm M52 field gun and the auxiliary-propelled version of the D-44 called the SD-44.

Production of the 85 mm divisional gun commenced in 1945 and large numbers were built:

Year	Quantity
1945	7
1946	474
1947	1,803
1948	2,606
1949	2,500
1950	2,510
1951	1,508
1952	n/avail
1953	110

As far as is known, this weapon is no longer in front-line use with the RFAS, but may still be used for training. There is a separate entry under China for its version which is called the Type 56. As far as it is known, there has been no recent production of the Type 56 in China.

Poland has modified its D-44 and D-44ND with an electric system and these are now designated the D-44M and D-44MN.

Description

The shield of the 85 mm divisional gun D-44 has a wavy top and a small sliding panel in front of the main gun shield, moving in elevation with the barrel. The recoil system is mounted behind the shield and above the cradle and breech ring. A single castor mounted towards the spade on one of the trails assists in manoeuvring the weapon into or out of action. When fitted with an APN3-7 infra-red night device, which is mounted to the rear of the shield, the D-44 is known as the D-44-N.

The weapon is also fitted with gunner's telescope model OP1-7 and periscope PG-1M.

Ammunition is of the fixed type and the types available for this weapon include:

Type	Designation	Penetration*
AP-T	BR-357	120 mm
AT-T	BR-365	100 mm
AP-T	BR-365K	95 mm
HVAP-T	BR-367P	180 mm
HVAP-T	BR-365P	100 mm

* At a range of 1,000 m with conventional steel armour at an angle of 90°.

85 mm Auxiliary Propelled Field Gun SD-44

Details of the self-propelled version of the D-44, called the SD-44, were given in *Jane's Armour and Artillery 1994-95*. As far as is known this weapon is no longer in front-line service. This was developed mainly for use by Russian air assault divisions.

First production SD-44's were in fact conversions of earlier production D-44 weapons. A total of 88 were converted in 1954 with a further 250 the following year.

Status

Production complete. In service with the following countries:

Country	Quantity	Comment
Afghanistan	n/avail	Chinese Type 56
Albania	61	Chinese Type 56
Algeria	80	
Angola	n/avail	
Armenia	n/avail	
Azerbaijan	20	
Bulgaria	150	
Cameroon	n/avail	Chinese Type 56
China	n/avail	Chinese Type 56
Congo	n/avail	Chinese Type 56
Croatia	50	
Cuba	n/avail	
Democratic Republic of Congo	20	Chinese Type 56
Egypt	48	
Eritrea	12	
Ethiopia	n/avail	
Georgia	n/avail	
Guinea	6	
Guinea Bissau	8	
Hungary	162	reserve
Iran	100	
Iraq	n/avail	
Korea, North	n/avail	
Mali	6	
Mongolia	n/avail	
Morocco	36	
Mozambique	24	Russian Type D-44
	12	Chinese Type 56
Pakistan	180	Chinese Type 56
Poland	711	
Romania	18	estimate
Russia	n/avail	reserve
Tanzania	75/80	Chinese Type 56
Turkmenistan	6	
Somalia	100	
Sri Lanka	6	Chinese Type 56
Sudan	100	
Syria	n/avail	
Vietnam	n/avail	
Yemen	100	
Yugoslavia	45	

Manufacturers

Joint Stock Company Spetstehnika.
China North Industries Corporation (NORINCO).

UPDATED

Specifications
(D-44)
Calibre: 85 mm
Barrel length:
(overall) 4.693 m
Muzzle brake: double-baffle
Recoil system: hydraulic recoil buffer and hydropneumatic recuperator
Breech mechanism: semi-automatic vertical sliding wedge
Carriage: split trail
Shield: yes
Weight:
(travelling order) 1,725 kg
(firing position) 1,703 kg
Length:
(travelling) 8.34 m
Width:
(travelling) 1.78 m
Height:
(travelling) 1.42 m
Axis of bore: (at 0°) 0.825 m
Ground clearance: 0.35 m
Track: 1.434 m
Tyres: 6.50 × 20
Elevation/depression: +35/−7°
Traverse:
(total) 54°
Rate of fire: 15-20 rds/min
Range:
(max) 15,650 m*
(effective, HVAP) 1,150 m
Crew: 8
Towing vehicle: 6 × 6 truck

* With trails dug in to provide 45° elevation, a maximum range of 18,870 m can be achieved.

85 mm divisional gun D-44 of the Polish Army with trails spread (Jaroslaw Cislak) 0021015

82 mm Vasilyek (2B9) automatic mortar

Development/Description

The 82 mm Vasilyek (Cornflower) (2B9) automatic mortar was introduced into the former Soviet Army in the early 1970s and was used in Afghanistan. It has now been replaced in front-line service by the 120 mm 2B11 mortar system, but remains in service with RFAS air assault units. The 120 mm 2B11 is normally transported in the rear of a GAZ-66 truck (4 × 4) 2,000 kg.

The complete system is called the 2K21 and consists of the 82 mm 2B9 automatic mortar and the 2F54 transport vehicle based on the GAZ-66 (4 × 4) truck.

Compared to existing mortar systems the 2B9 has the added advantage that it can be used in the direct and indirect fire roles.

This flexibility allows a variable charge system to be used. Up to three charges can be used when the weapon is firing at high angles to provide the indirect fire support normally associated with conventional mortars.

A separate fixed charge can be used with the mortar when it is used to provide direct fire, similar to an anti-tank gun. According to Eastern sources, the weapon can produce consistently accurate fire in single shots, bursts, or at a sustained rate of fire of 120 rds/min. Its cyclic rate of fire is 170 rds/min.

When firing in the automatic mode the 82 mm mortar bomb has a maximum muzzle velocity of 270 m/s and

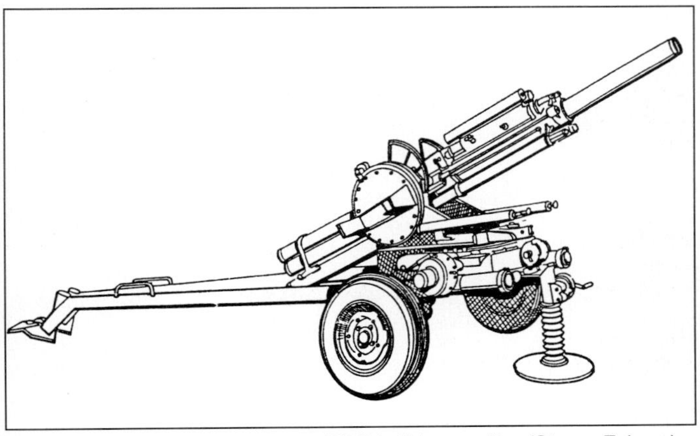

82 mm Vasilyek automatic mortar (2B9) in firing position (Steven Zaloga)

82 mm Vasilyek automatic mortar (2B9) mounted on the rear of an MT-LB multipurpose armoured vehicle

Polish 82 mm Vasilyek automatic mortar (2B9) in travelling configuration (Richard Stickland)
2002/0130017

when being used in the single-shot mode (muzzle loaded) it has a maximum muzzle velocity of 210 m/s. The primary charge weighs 7.5 g while the automatic loading charge weighs 75 g.

The HE-FRAG mortar bomb weighs 3.1 kg and is fitted with a direct action or graze fuze which is armed after 10 m. There is also an anti-armour projectile of the High-Explosive Anti-Tank (HEAT) type that will penetrate 100 mm of conventional steel armour.

Before loading, the moving parts are pulled back by the charger until they come to rest on the automatic sear. The lever of the disconnector mechanism is set at a position corresponding to the selected type of fire, automatic or single shot.

The mortar is then loaded and a clip holding four 82 mm mortar bombs inserted into the weapon through a cover on the right trunnion. The weapon is then fired by pressing the trigger lever.

As stated, the 82 mm Vasilyek (2B9) mortar can be fired singly or automatically and converts the energy of powder gases generated in firing the weapon to operate the charger.

The weapon fires during counter-recoil and as a result the recoil energy develops and is spent to decelerate and stop the moving parts and impart them with recoil energy.

The recoil energy is damped by the return springs with its excess being absorbed by the ring shock-absorbers of the piston rods.

After each 82 mm mortar bomb is fired, the feed mechanism automatically moves the clip by one pitch bringing the next mortar shell to the ramming line. When being used in the sustained fire role the loader links up during firing the first clip with the following clips so as to form a continuous belt.

Provision is made for a hydraulic buffer to absorb the counter-recoil energy of the moving parts in the event of a misfire or hangfire.

To enable sustained fire missions to be undertaken, the barrel is water cooled. This consists of a welded structure consisting of a jacket, steam collector and two valves, relief and check.

The screw breech block rams a mortar shell into the bore, indents the primer of the shell primary cartridge, obturates the powder gases and cocks the moving parts via the pressure of the powder gases.

To ensure an equal rate of counter-recoil at all angles of elevation, the mortar is provided with a booster mechanism. Interlocks exclude the triggering of the moving parts when the charger is engaged and there is no clip in the mortar as well as during the clip movement to bring the next round to the ramming line.

The 82 mm barrel is mounted on a towed, split trail, wheeled carriage and in action the weapon rests on the trail legs, each of which is provided with a spade and a forward extendable leg that is lowered by a handwheel, with the main carriage raised off the ground.

The ordnance then rests on a turning plate allowing 30° of traverse but the saddle is not fixed directly to the plate. Instead, it is supported by hydraulic elevators that can extend the ordnance into a high angle for indirect fire. For this the saddle, cradle, ammunition feed tray and ordnance are elevated together and moved forward so that at full elevation the trunnions are above the carriage axle.

Total traverse of the weapon is 60°, 30° left and right, while elevation is from −1 to +85°. The optical sight has magnifications of ×2.5 and ×3. The sight is designated the PAM-1 with the Lutch-PM2M illuminator enabling fire missions to be carried out in the dark. The sighting system is mounted on the left side of the mount.

The weapon can be towed but is normally transported under canvas covers on the cargo area of a modified GAZ-66 (4 × 4) 2,000 kg truck.

The vehicle is fitted with two ramps to assist in loading and offloading the 82 mm 2B9 automatic mortar.

Variants

Vasilyek on MT-LB

During the fighting in Afghanistan a number of MT-LB multipurpose tracked armoured vehicles were observed fitted with the 82 mm Vasilyek mortar with the wheels removed on their rear decking. This was probably a battlefield expedient and reappeared during the fighting in Chechnya.

Specifications

Calibre: 82 mm
Weight: 645 kg
Length:
(travelling) 4.115 m
Width:
(travelling) 1.576 m
Height: 1.18 m
Ground clearance: 0.26 m
Elevation/depression: +85/−1°
Traverse: 30° left and right
Max range: 4,270 m
Min range: 800 m
Rate of fire:
(maximum) 170 rds/min
(practical) up to 120 rds/min
Max towed speed:
(earth roads) 20 km/h
(asphalt roads) 60 km/h
Crew: 5 to 6

Vasilyek on ACRV

For trials purposes, Hungary has mounted the 82 mm Vasilyek automatic mortar system on a modified ACRV chassis. Full details of this were given in *Jane's Armour and Artillery 1999-2000* page 591. There are no known plans for this to enter production.

Vasilyek on BMP-1

Hungary has built the prototype of a self-propelled version of the Vasilyek mounted on a modified BMP-1 infantry combat vehicle chassis. Full details of this were given in *Jane's Armour and Artillery 1999-2000* page 592. There are no known plans for this to enter production.

Chinese Vasilyek

NORINCO (China North Industries Corporation) have developed a similar system called the 82 mm Type 82 mm Type W99 mortar and details of this are given earlier in this section under China.

Status

Production complete. In service with the following countries:

Country	Quantity	Comment
Bosnia-Herzegovina	n/avail	
Bosnian Serbs	n/avail	
Croatia and HVO	n/avail	
Hungary	n/avail	local production
Poland	n/avail	
Russia	n/avail	
Yugoslavia	n/avail	

Manufacturer

Russian state factories. Has also been manufactured under licence in Hungary.

UPDATED

76 mm mountain gun GP (M1966)

Development/Description

The 76 mm mountain gun GP is known in the West by the reporting designation M1966, but has been referred to as the M1969. The gun was first reported in 1966 and was used by Russian troops operating in Afghanistan.

The 76 mm mountain gun GP has a split trail and a carriage arrangement similar to that used by the Italian Otobreda Model 56 pack howitzer, including a split shield. The short barrel has no muzzle brake and reciprocates in a square cross-section slipper. The breech is a horizontal sliding wedge opening to the right. Panoramic and direct fire optical sights are located to the left of the barrel. The split trails can fold

Specifications

Calibre: 76.2 mm
Muzzle brake: no
Breech mechanism: horizontal sliding block
Carriage: split trail
Shield: yes
Weight:
(travelling) 780 kg
(firing) 780 kg
Length:
(travelling) 4.8 m
Width:
(travelling) 1.5 m
Height:
(travelling) 1.4 m
Elevation/depression: +65°/−5°
Traverse: (total) 50°
Max range: 10,500-11,500 m
Rate of fire:
(max) 15 rds/min
(sustained) 100 rds/h
Crew: 7
Towing vehicle: GAZ-66 (4 × 4) 2,000 kg truck or Ural-375D (6 × 6) 4,000 kg truck

sideways and the right-hand trail leg is provided with a metal dolly wheel for ease of handling with the trails folded. The rubber-tyred wheels are on stub axles that can be raised and lowered to alter the height of the 76 mm mountain gun GP (M1966) in action and to level the carriage on rough ground. It is assumed that the gun can be disassembled into loads for pack transport in a manner similar to the Otobreda Model 56.

It has been reported that the gun can fire any type of ammunition used by the 76 mm divisional gun M1942 (ZIS-3) including HE/FRAG, HEAT-FS, HEAT, HVAP and AP-T. The HE/FRAG projectile is fired at a muzzle velocity of 600 m/s, while the HEAT projectile can penetrate 300 mm of conventional steel armour set at right angles at any range. Maximum rate of fire is 15 rds/min with sustained fire being 100 rds/h.

Status

Production complete. In service with Afghanistan and the RFAS. This weapon has not so far been offered on the export market.

Manufacturer

Former Soviet state factories.

VERIFIED

76 mm mountain gun GP (M1966) being towed by a GAZ-66 (4 × 4) truck

76 mm mountain gun GP (M1966) in firing position

SINGAPORE

Singapore Technologies Kinetics 155 mm/52 calibre FH2000 artillery system

Development

The 155 mm FH2000 project was first conceived in January 1990 with the first prototype being completed the following year, the prime contractor being Ordnance Development & Engineering (ODE), a Division of Chartered Industries of Singapore. Early in 2000 this company became Singapore Technologies Kinetics.

This was followed by a preproduction example with a number of improvements. Final acceptance tests were completed in December 1993.

By mid-1995, a total of 18 155 mm FH2000 towed artillery systems had been deployed with the 23rd Battalion Singapore Artillery (23SA) in three batteries each of six guns. Since then production has continued.

The 155 mm FH2000 towed artillery system is a further development of the Singapore Technologies Kinetics 155 mm FH-88 gun howitzer which has a 39 calibre ordnance and entered service in 1987.

Some key parts of the earlier FH-88 are used in the FH2000 including the sighting system, APU and the 155 mm ammunition.

The 155 mm FH2000 was the first 155 mm/52 calibre artillery system, towed or self-propelled, to enter service with any army.

The 155 mm/52 calibre FH2000 artillery system is now being offered for export. Singapore Technologies Kinetics is also developing a self-propelled 155 mm/52 calibre system, which is understood to use the ordnance of the FH2000 installed in a fully enclosed turret on a full-tracked chassis.

Description

The most significant change compared to the earlier 155 mm FH-88 is the installation of the longer 155 mm/52 calibre ordnance with a 23 litre chamber which conforms to the Joint Ballistics Memorandum of Understanding.

The 155 mm/52 calibre ordnance is autofrettaged for durability and has a welded double-baffle muzzle brake.

The breech mechanism is semi-automatic with the breech opening automatically during counter-recoil. To allow space for a future automatic loading device and to enable easy charge loading, the breech opens

155 mm/52 calibre FH2000 gun-howitzer deployed in the firing position 0021019

Specifications

Calibre: 155 mm
Muzzle brake: double baffle
Breech mechanism: hinged, interrupted screw (opens automatically during counter recoil)
Carriage: split trail
Weight: (travelling) 13,500 kg
Length:
(travelling) 10.95 m
(firing) 12.038 m
Width:
(travelling) 2.80 m
(firing) 9.726 m
Height:
(with display unit) 2.55 m
(without display unit) 2.286 m
Ground clearance: 0.35 m
Fording: 0.76 m
Gradient: 45%
Angle of approach: 24°
Max speed:
(towed on paved roads) 80 km/h
(towed on unpaved roads) 50 km/h
(self-propelled paved roads) 16 km/h
(self-propelled cross-country) 8 km/h
(tandem drive) 16 km/h
Elevation/depression: +70/–3°
Traverse: 30° left/30° right
(on carriage) 360°
Rate of fire:
(sustained) 2 rds/min for 30 min
(max) 6 rds/min for 3 min
(burst rate) 3 rds in 15/20 s
Accuracy: at ⅔ maximum range
Accuracy: range
(ERFB BB) 0.5% at 27 km
(ERFB HB) 0.35% at 20 km
(M107) 0.25% at 16 km
Accuracy: azimuth
(ERFB BB) 1.0 mil at 27 km
(ERFB HB) 1.0 mil at 20 km
(M107) 0.56 mil at 16 km
Max range:
(M107 projectile) 19,000 m
(ERFB HB projectile) 30,000 m
(ERFB BB projectile) 40,000 m
Crew: 6
Towing vehicle: 6 × 6 truck

155 mm/52 calibre FH2000 gun-howitzer in self-propelled configuration 0021020

upwards. The breech can also be operated by remote control in the case of a misfire.

The primer feeding mechanism has been relocated on the cradle for built-in reliability. A new recoil system has been incorporated to tap the recoil energy for operation of the flick rammer, breech opening, primer feeding and thermal compensating equilibrators.

The FH2000 is fitted with a temperature indicator on the gun barrel which warns the gun crew of the temperature condition of the barrel before and during firing. This avoids cook offs.

The electronically controlled and hydraulically operated flick rammer has a safety device to ensure the safe loading of projectiles into the ordnance. For example, if there is no projectile in the ramming tray, it will not ram. If a projectile has been loaded into the chamber but not fired, it will not ram another round into the barrel.

The flick electronically controlled rammer has four modes of operation, automatic, automatic back up, electric manual and hydraulic manual. Therefore, if there is a failure, one of these alternatives is available to ensure successful accomplishment of the mission.

Traverse, elevation and firing of the 155 mm/52 calibre ordnance are purely mechanical. As a result, the weapon can still function when APU or hydraulic lines are damaged.

The four-wheeled split trail carriage is provided with an APU which is mounted on the forward part of the carriage. Each of the two trail legs has a retractable dolly wheel to assist in bringing the weapon into action. When in the travelling configuration, the 155 mm/52 calibre ordnance is swung through 180° and secured by the barrel clamps on the trail legs.

The APU consists of a 71 hp air-cooled turbo-charged diesel with a hydrostatic drive system which, in addition to providing the howitzer with a self-propelled capability, also powers the firing platform and the trail wheels. Steering is through the trail wheels.

If required the APU can be activated from the cab of a towing vehicle to provide additional traction in difficult terrain.

The weapon can be bought into action in 2 minutes in the following sequence:

(1) The weapon is unlocked from the tractor by lowering the trail wheels to the ground hydraulically which in turn lifts the trail end
(2) The trails are opened hydraulically
(3) Trails are lowered to the ground
(4) Spades are self-embedded
(5) The circular firing platform is lowered beneath the front end of the carriage.

The direction of fire can be changed in less than 2 minutes by raising the firing platform, lowering and steering the trail wheels, executing a pivot turn, steering and raising the trail wheels, embedding the spades and lowering the firing platform.

When deployed in the firing position the FH2000 rests on the firing platform under the carriage and the two trails, each of which is provided with a spade.

For improved reliability and for superior maintainability, all electrical relays in the APU have been replaced by a Programmable Logic Control (PLC) system.

The standard sighting system comprises a cant compensating mount, a peri-telescopic sight, a display unit and a direct aiming sight. The exact configuration selected can be tailored to meet particular battlefield conditions and to match contemporary military needs.

The technology of the sighting system ranges from a modern electro-optical sight to a sophisticated electronic sight.

The display unit provides an electronic visual aid for both layer and commander to fine-tune elevation and traverse (azimuth) settings for pinpoint accuracy. It is connected to the sighting system and receives firing data from the fire-control computer.

An onboard independent battery unit acts as a back-up system to ensure mission performance. Alternatively, the hydraulic system can be operated by a hand pump which gives the advantage of silent operation at night.

When designing the FH2000, the manufacturer incorporated a number of features for ease of maintenance and increased durability, especially when operating in a jungle environment.

All cables, wires and pipes are hidden away in the two trail legs while the batteries which power the gun's engine and electronic systems are housed in one of the two trail legs.

The number of parts has also been reduced where possible. For example, the breech of the FH2000 has 50 per cent less parts then the earlier FH88. The driver's instrument panel is made up of three modules so that the faulty module can be quickly replaced.

In addition to firing the standard 155 mm M107 HE projectile and the locally manufactured 155 mm ERFB-HB (Extended Range Full Bore – Hollow Base) and ERFB-BB (Extended Range Full Bore - Base Bleed) rounds, Singapore has recently developed a new 155 mm cargo projectile that is now in production for the Singapore Army.

Two versions of this have been developed, High Explosive – Base Bleed (HE-BB) and High Explosive – Hollow Base (HE-HB). Both of these carry dual purpose bomblets that have been designed to attack the vulnerable upper surfaces of armoured vehicles. Both carry 64 bomblets that are manufactured in Singapore and are fitted with a self-destruct mechanism.

Status

Production as required. In service with Singapore.

Manufacturer

Singapore Technologies Kinetics.

UPDATED

Singapore Technologies Kinetics 155 mm FH-88 gun-howitzer

Development

Ordnance Development and Engineering of Singapore (ODE), a member of Chartered Industries of Singapore, developed a 155 mm gun-howitzer with a 39 calibre barrel known as the FH-88. Early in 2000 the company became Singapore Technologies Kinetics.

There were five prototypes produced over a period of four years beginning in 1983. These were followed by a preproduction batch of six guns that incorporated a number of improvements as a result of trials with the prototype weapons.

First production FH-88s were completed in 1987, with the weapon becoming operational with the Singapore Army the following year.

It is believed that a total of 52 systems were built for the Singapore Army. These replaced Israeli older Soltam Systems 155 mm M71S (with the S standing for Singapore, as they have been modified) systems which have been placed in reserve. Indonesia took delivery of five 155 mm FH-88 systems in 1997.

The 155 mm/39 calibre FH-88 has been followed in production by the Singapore Technologies Kinetics 155 mm/52 calibre FH2000 which is covered in a separate entry.

The more recent FH-2000 artillery system uses some components of the FH-88 including the sighting

155 mm FH-88 gun-howitzer deployed in the firing position

Specifications

Calibre: 155 mm
Muzzle brake: double baffle
Breech mechanism: hinged, interrupted screw
Carriage: split trail
Shield: no
Weight: 12,800 kg
Length:
(barrel) 6.1 m
(travelling) 9.16 m
(firing) 9.88 m
Width:
(travelling) 2.8 m
(firing) 8.20 m
Height:
(with display unit) 2.57 m
(without display unit) 2.286 m
(towing eye) 1.32 m
Ground clearance: 0.35 m
Fording: 0.76 m
Gradient: 45%
Angle of approach: 24°
Max speed:
(towed on main roads) 80 km/h
(towed on secondary roads) 50 km/h
(self-propelled, roads) 16 km/h
(self-propelled, cross-country) 8 km/h
(tandem drive on tow) 8 km/h
(remote driving) 4 km/h
Elevation/depression: +70/–3°
Traverse: 60°
Rate of fire:
(sustained) 2 rds/min for 1 h
(max) 8 rds/min alternate minutes for 20 min
(burst) 3 rounds in 15 s
Max range:
(M107 HE projectile) 19,000 m
(ERFB HB projectile) 24,000 m
(ERFB BB projectile) 30,000 m
Accuracy:
(Probable error (range)) <0.22% of range at ⅔ max range
(Probable error (deflection)) <0.035% of range at ⅔ max range
Crew: 6
Towing vehicle: 6 × 6 truck

Breech of 155 mm FH-88 gun-howitzer in open position

155 mm FH-88 gun-howitzer in travelling configuration with ordnance in travel lock over closed spades

system, APU and the complete family of 155 mm ammunition.

Description

The monobloc 155 mm barrel is 39 calibres long and is autofrettaged. A double-baffle muzzle brake is fitted. The breech is of the hinged, interrupted screw type and automatically opens during the latter half of the return stroke of the recoil movement. The camplate lifting mechanism incorporates a safety lock to prevent accidental closing of the breech screw and the firing mechanism cannot be activated if the breech screw is not fully closed. Obturation is achieved using a resilient pad and split rings. The firing mechanism is a mechanical 'trip action' type which is actuated either by pulling the lanyard or operating the firing handle at the layer's seat. Primer feeding and extraction is automatic using a 12-round magazine. A flick rammer is used with power derived from an independent power pack.

The hydropneumatic recoil mechanism is of the independent-type single-recoil system. The recoil mechanism, connected to the recoiling and static parts of the gun, consists of a recoil brake and a recuperator/counter-recoil mechanism. The governor rod in the recoil brake varies the orifice for hydraulic oil to flow throughout the recoiling stroke, absorbs and ensures minimum and uniform recoil on the carriage. The pressurised nitrogen gas in the recuperator/counter-recoil mechanism is to return the barrel after recoil and also holds the barrel in its fully run-out position at all elevations.

There are two independent pneumatic equilibrators which balance the mount weight of the elevated mass throughout the full range of elevation. The equilibrator is made up of a cylinder, piston rod and a receiver using nitrogen gas as the working medium.

The cradle, which is made of high-strength alloy steel, supports and guides the recoiling parts. It houses the recoil mechanism and provides attachments for the equilibrators, camplate arm, elevating arc, flick rammer and the sighting system.

The saddle, which is made of high-strength alloy steel, comprises two side frames welded to a circular baseplate. It supports the elevating mass through the trunnion bearings and houses the elevating and traversing mechanism. It also provides attachments for the equilibrators and sighting equipment.

The elevating mechanism, with the aid of the equilibrators, enables the ordnance to be elevated or depressed with accuracy and ease. The traversing mechanism, mounted in front of the saddle, enables the ordnance to be swivelled horizontally.

The flick rammer is an electronically controlled hydraulic system which in turn operates the mechanical system. The latter basically consists of the swivel arm and the ramming unit. The swivel arm, with the loading tray attached, is pivoted at the rear end of the cradle.

The chassis serves as the base frame to join the upper carriage to the lower carriage. During firing it transmits all of the load on the upper carriage to the ground through the firing platform and trail legs.

The chassis is an all-welded structure of high-strength steel on a closed box design, with provision for the mounting of the main bearing, rocker beams, firing platform, trail legs and the APU. The firing platform is connected to the chassis assembly through jack arrangements and lifts the weapon off the ground for firing. The lowering of the platform is achieved through the tripod arrangement being operated by a telescopic cylinder, which is completely isolated from all firing loads when the platform is fully deployed. An indicator is incorporated to show when the platform is fully deployed.

The trails are made of high-strength steel and, together with the firing platform, serve as the main support for the weapon. In the closed position they serve as a base for locking the barrel in a fixed position for travelling. Hydraulically powered operations permit the trails to be spread or retracted within 10 seconds.

The two trail wheels, attached to the trail ends, can be steered, raised or lowered by hydraulically powered operations controlled by the driver.

The spades, an integral part of the trails, transmit the firing loads to the ground. As the spades are permanently hinged at the end of the trails, no carrying and mounting of spades is required during deployment.

The carriage is of the split trail type and is fitted with an APU. The 155 mm FH-88 can be towed at speeds up to 80 km/h and is self-propelled at speeds up to 16 km/h. The APU is a 96 hp Deutz air-cooled turbocharged diesel which powers a hydrostatic drive system on the front wheels. Back-up systems include a battery-operated pump and a manual pump. The APU can be operated in tandem with the towing vehicle for maximum traction and can be remote controlled. The APU has a 50 litre fuel tank providing an operating range of 60 km.

The diesel-powered APU consists of a hydrostatic drive system, braking system, operation hydraulic system and a back-up system. The basic functions of the APU can be summarised as follows:

(1) Self-propelled capability of the howitzer
(2) Mechanised hydraulic operation during deployment
(3) Ability to assist the prime mover in manoeuvring over rough terrain.

The rapid change in the zone of fire by the FH-88 to engage targets outside the primary zone is achieved by:

(1) Raising the firing platform
(2) Lowering and steering trail wheels
(3) Executing pivot turn
(4) Steering and raising trail wheels
(5) Embedding spades
(6) Lowering firing platform.

The basic sighting system of the FH-88 consists of a cant compensating mount, peri-telescopic sight (×4 magnification) and a direct aiming sight (×6 magnification). However, this can be upgraded to a sophisticated electronic sight and linked to battery fire-control computers.

The trail legs have wheels for ease of handling. They can be opened and closed by hydraulic power and the firing platform is raised and lowered by a double-stage hydraulic cylinder. The trail spades are self-embedding. With a crew of six the gun can be brought in or out of action in less than 1 minute.

The 155 mm FH-88 gun-howitzer can fire standard NATO ammunition such as the M107 HE projectile to a maximum range of 19,000 m. The ERFB HB projectile can be fired to 24,000 m while an ERFB BB can be fired to 30,000 m. The manufacturer claims a range accuracy of less than 0.22 per cent, with deflection less than 0.35 per cent at two-thirds of maximum range. The flick rammer, mounted on the left side of the mount, can be used to achieve a burst rate of three rounds in 15 seconds. The sustained fire rate is 2 rds/min for 1 hour.

In addition to firing the standard 155 mm M107 HE projectile and the locally manufactured 155 mm ERFB-HB (Extended Range Full Bore – Hollow Base) and ERFB-BB (Extended Range Full Bore - Base Bleed) projectiles, Singapore has recently developed a new 155 mm cargo projectile that is now in production for the Singapore Army.

Two versions of this have been developed, High Explosive - Base Bleed (HE-BB) and High Explosive Hollow Base (HE HB). Both of these carry dual purpose bomblets that have been designed to attack the vunerable upper surfaces of armoured vehicles. Both carry 64 bomblets that are manufactured in Singapore and are fitted with a self-destruct mechanism.

When fired with the locally developed C20 bagged propelling charge a maximum muzzle velocity of 710 m/s is obtained and a maximum range of 18 km.

When fired using the locally developed C30 combustible propelling charge a maximum muzzle velocity of 800 m/s is obtained together with a maximum range of 27.6 km.

Variants

Details of the latest Singapore Technologies Kinetics 155 mm/52 calibre FH 2000, are given in the previous entry. Main export marketing is now concentrating on the more recent 155 mm 52 calibre FH2000 system.

Status

Production complete. In service with Indonesia (5) and Singapore (52). Production can be resumed if further orders are placed

Manufacturer

Singapore Technologies Kinetics.

UPDATED

SOUTH AFRICA

LIW G5 155 mm gun-howitzer

Development

During operations carried out in 1975, the South African Defence Force (SADF) found its own artillery out-ranged and out-gunned by various forms of Russian artillery used by the Angolan forces.

Following this experience, the South African artillery initiated a staff requirement for a new artillery system that would encompass not only a new artillery piece but also an ammunition system, a new gun tractor, fire-control equipment, a fire-control computer system and other accessories.

The G5 became operational with the SADF during 1983. In mid-1996, it was stated that 72 155 mm G5 systems were in service with the South African Artillery.

The G5 155 mm gun-howitzer has seen combat service with the SADF in Southern Angola and South West Africa (now Namibia) and with the Iraqi Army during its war with Iran. Quantities of these weapons were captured by the Allied Coalition forces during Operation Desert Storm, the liberation of Kuwait, which took place early in 1991.

In November 2000, it was announced that an agreement worth about US$50 million had been signed between LIW and Malaysia for the supply of an artillery system.

Under the terms of this contract, the LIW Division of Denel will supply 22 G5 155 mm Mark 3 gun-howitzers to Malaysia together with ammunition, modular charges, fire-control systems, logistics support and training. The fire-control system selected is the African Defence Systems AS2000 which was originally developed to meet the operational requirements of the South African Defence Force.

First deliveries were made late in 2001 and all systems will be delivered by the end of 2002.

Description

The 155 mm G5 howitzer uses a 45 calibre barrel designed around the performance of the ERFB round originally developed by the Space Research Corporation of Canada.

The 155 mm G5 howitzer is produced and manufactured by LIW at Centurion, near Pretoria. The monobloc autofrettaged barrel has a single-baffle muzzle brake and uses a semi-automatic interrupted screw-type breech.

The 155 mm G5 incorporates a manually controlled, telescopic pneumatic cylinder rammer mechanism for ramming the projectiles in the chamber, at a uniform position whatever the elevation. Charges are loaded by hand.

The carriage is fitted with an APU to facilitate hydraulic powered deployment and self-propulsion. This APU comprises a 79 hp air-cooled diesel engine.

Hydraulic powered deployment includes the following actions:

(1) Aligning the trail wheels
(2) Lifting and lowering of the trail wheels
(3) Spreading and closing of the trail wheels
(4) Lifting and lowering of the firing platform

The G5 can be brought into action or taken out of action in 2 minutes. In case of APU failure, slave sockets and pipes are provided to couple the hydraulic system to that of another gun or tractor.

The G5 gun crew consists of five people, although in an emergency two can operate the weapon. Elevation and traverse controls are manual.

The G5 can be driven at a maximum speed of at least 16 km/h and has a gradient capability of 40 per cent. With the fuel tank capacity of 100 litres the G5 has an operating range of more than 100 km.

LIW 155 mm G5 gun-howitzer deployed in the firing position with ordnance elevated **2001**/0105799

LIW 155 mm G5 gun-howitzer deployed in the firing position **2000**/0085649

When being towed, the 155 mm ordnance can either be clamped over the trail or in the firing position. When travelling the ordnance will be clamped over the trail, while clamping in the firing position is intended for tactical movement.

In SANDF service the G5 is towed by a 6 × 6 tractor, based on a locally built SAMIL 100 (6 × 6) chassis. The gun tractor not only tows the G5 but also has accommodation for a crew of eight (normally five serve the gun in action). Storage is provided on the gun tractor for 32 155 mm projectiles and 48 charges.

To facilitate towing of the G5 system with virtually any prime mover, an extended towbar is fitted.

The ammunition system of the G5, which comprises the projectile, charges and fuzes, is identical to that of

Specifications

Calibre: 155 mm
Barrel length: 7.045 m
Muzzle brake: single baffle
Breech mechanism: interrupted thread
Carriage: split trail with APU
Shield: no
Weight: 13,750 kg (13,850 kg with extended tow bar)
Length:
(travelling) 9.5 m
(firing) 11 m
(on tow, barrel to rear) 12.2 m
(with extended tow bar) 13.1 m
Width:
(travelling) 2.53 m
(firing) 8.7 m
Height:
(travelling) 2.3 m
(barrel forward, clamped) 2.85 m
Height of towing eye:
(travelling) 0.95 m
Axis of bore:
(firing) 1.77 m
Ground clearance: 0.31 m
Fording: 0.6 m
Gradient: 40%
Axis of bore:
(firing) 1.77 m
Tyres:
(main wheels) 14.00 × 20
(trail wheels) 7.50 × 16 (Mk 2) (8.25 × 15) (Mk 3)
Elevation/depression: +75/–3°
Traverse:
(total, up to 15° elevation) 82°
(total, above 15° elevation) 65°
Rate of fire: (max) 3 rds/min for 15 min
Max range:
(sea level) standard 30,000 m
(BB) 39,000 m
(VLAP) 50,000 m
Min range (+75°): 4,500 m
Crew: 5
Towing vehicle: SAMIL 100 (6 × 6) gun tractor
Speed:
(in APU mode) 16 km/h

Improved G5 Mk 3 155 mm gun-howitzer with gun display unit on the left and muzzle velocity measuring unit over barrel rear 0021025

LIW 155 mm G5 MK 3 in production for Malaysia with carriage on left and weapon on right ***2002***/0130016

the LIW G6 (6 × 6) self-propelled artillery system covered in detail in an earlier section.

The layer's position on the left of the breech is provided with an upgraded metricated panoramic sight and a direct fire telescope for use up to 5,000 m.

The G5 features a gun management system which provides the layer with all relevant laying data.

A temperature warning device is available as an option. This warns the gun crew of high barrel-temperature conditions. Thus a high rate of fire can be maintained while avoiding a possible 'cook-off'.

The AS2000 Artillery Target Engagement System has been designed specifically for use with the G5 and G6 (6 × 6) artillery systems. This also includes a muzzle velocity meteorological ground station and a crew communication system.

Variants

Improved G5 Mk 3 155 mm towed gun

In co-operation with the South African National Defence Force, Denel (Pty) Ltd has further improved the G5. Designated the G5 Mk 3, a total of 35 modifications and improvements have been carried out to enhance the already proven reliability and operational efficiency of the gun.

User feedback based on many years of maintenance and operational experience has been incorporated in the G5 Mk 3, as well as results obtained from LIW's research and development centre.

On the main weapon the recoil/buffer cylinder and replenisher have been improved and a gas trap has been added which greatly improves the efficiency of the recoil system. The improved G6 breech and the G6 camplate have been adopted in the G5 Mk 3. A barrel temperature indicator and warning light has been installed.

Electrical power distribution to the top carriage has been improved to accommodate the state-of-the-art Artillery Communication System (ACS), muzzle velocity analyser and fire-control computer. The ACS has two dome antennas fitted on the gun which allow closed-circuit communication between the gun crew and gun commander and external communications between the gun commander and the battery control post.

Accessibility to the traverse bearing bolts has been improved. Improvements have been made to the traverse drive clutch, traverse drive shaft and elevation gearbox for easier maintenance and more efficient functioning. The advantages of these changes are improved gun alignment and greater durability. A greatly improved brake system has been introduced which automatically applies the brakes should the hydraulic drive system fail to stop the gun. When the APU is not running the brakes are engaged and only released if the air pressure provided by the newly fitted air compressor, is sufficient.

Attention to detail and to the users' requirements has led to minor modifications such as an easier to read diesel tank dipstick, additional bottom carriage drainholes, improved trail leg stops that improve reliability and ease of replacement, dog-clutch disengagement indication light, improved muzzle brake screw lock and improved hydraulic tank indicator, which all give the system better performance and reliability.

Trials with the upgraded G5, subsequently called the G5 Mk 3 were completed in 1996 and troop trials with a battery of six weapons took place in 1997 with conversions running through to 2000.

Extended range G5 and G6

Late in 1992, it was revealed that LIW, now part of the Denel Group, had built the prototype of a 155 mm ordnance with a calibre length of 52. This is being used in the development work and could be used to upgrade the G5 and G6 in the future, enabling the range advantage to be retained.

A Bi-Modular Charge System (M90) is available comprising modules of around 2.2 kg which can be used for 39, 45 and 52 calibre charge systems. The range with Denel's existing 155 mm ERFB-BB projectiles is 39 km with a 45 calibre barrel. When fired from a 155 mm/52 calibre ordnance, a range of 41 km is achieved.

With the new 155 mm ERFB-BB projectile under development, the respective ranges will increase to 41 km and 43 km respectively. The base bleed unit is of the screw on type which can be removed under field conditions if not required.

Velocity enhanced Long-range Artillery Projectile (VLAP)

The new Velocity enhanced Long-range Artillery Projectile (VLAP), first revealed early in 1997, is a rocket motor-assisted 155 mm projectile combined with the base bleed unit which extends the range of the 45 calibre gun to 50 km.

First production VLAP were completed early in 2001.

India is known to have recently ordered large quantities of 155 mm ammunition (projectiles and charges) from South Africa and could be the launch customer for the VLAP.

The VLAP was originally designed as a private venture by SOMCHEM to complement their existing family of 155 mm artillery projectiles which includes High Explosive - Extended Range Full Bore - Base Bleed (HE ERFB BB).

The current VLAP has a high explosive payload although other payloads are now under investigation by SOMCHEM.

SOMCHEM stress that the VLAP is complimentary to their existing family of projectiles and allows friendly artillery units to engage threat targets while positioned well to the rear and beyond range of enemy counter battery artillery fire. This will lead to higher survivability of friendly artillery systems.

To achieve this significant increase in range the VLAP incorporates base bleed and rocket assist. It can be fired from most existing towed and self-propelled 155 mm artillery systems including the South African LIW Division of Denel 45 calibre G5 (towed), 45 calibre G6 (6 × 6) self-propelled and the more recent 52 calibre T6 turret systems which has undergone extensive tests by the Indian Army.

Ranges of the VLAP compared to other types of 155 mm projectiles are given in the table above.

Weapon type	Boat tail	Base Bleed	VLAP
155 mm/39 calibre	23 km	30 km	39 km
155 mm/45 calibre	30 km	39 km	50 km
155 mm/52 calibre	32.5 km	41 km	52.5 km

The VLAP can be used in conjunction with a number of charge systems including the SOMCHEM M53, M90 Bi-Modular Charge System (BMCS) selected by the British Army after an international competition and the M64 BMCS.

PRO-RAM 155 mm projectile

Currently under development is a new 155 mm projectile called the PRO-RAM. This could have a range of up to 80 km, depending on which it is fired from.

Coastal defence G5

The LIW 155 mm/45 calibre G5 gun-howitzer is also being marketed for the coastal defence role. Additional details are given in the Coastal artillery guns and missiles section later in the volume.

Status

In production. In service with the following countries:

Country	Quantity	Comment
Iran	100	estimate, believed 200 delivered
Iraq	100	estimate, believed 200 delivered
Malaysia	22	Mark 3s ordered late 2000, first delivery 2001
Qatar	12	
South Africa	72	
Uganda	n/avail	unconfirmed user

Manufacturer

LIW, a division of Denel.

UPDATED

LIW 105 mm Light Experimental Ordnance

Development

The origins of the LIW Division of Denel 105 mm Light Experimental Ordnance, sometimes referred to as the LEO, can be traced back to 1995. This was when the South African Defence Force set the following design goal to LIW to 'Develop the technology for a light gun with a light logistic load that will equal or better the range and equal or better the terminal effect of current 155 mm light guns.'

A ballistic testbed was completed and test fired in 1997, with the first complete example of the 105 mm Light Experimental Ordnance completed in 2001. This has since undergone extensive firing trials at the Alkantpan range in South Africa, during which some 350 rounds were fired.

The LIW 105 mm Light Experimental Ordnance was first revealed late in 2000 and first exhibited overseas during the IDEX 2001 exhibition held in Abu Dhabi in March 2001.

LIW stress that this is a technology demonstrator and is not representative of a type classified and ready for production 105 mm artillery system. If adopted for service it will probably be given the designation of the GV7 or G7 for international marketing purposes. Its primary role is close support.

According to LIW, there is no other artillery system in the world with the capabilities of the 105 mm Light Experimental Weapon which includes a maximum range of 30,000 m.

This is similar to that of 155 mm/39 calibre artillery systems with the new 105 mm high-explosive artillery projectile, having a lethal area of 1,900 m^2 against targets. This is claimed to be better than a 155 mm artillery projectile.

A 105 mm Light Experimental Weapon together with 100 rounds of 105 mm ammunition currently weighs 6,000 kg which is being reduced still further. According to LIW this should be compared to more than 10,000 kg for a 155 mm system plus a similar number of rounds. Because of the lighter 105 mm ammunition, the crew requirement is reduced and it is also suitable for the 'lighter built' soldier.

The first example of the 105 mm Light Experimental Ordnance weighs 3,800 kg and the development team is now working to reduce this to 2,500 kg making it easier to deploy.

According to LIW the reduction in weight is especially important for fielding in peacekeeping operations or for early entry forces/light forces for rapid support of allies and for close fire support.

The 105 mm Light Experimental Weapon was originally developed using funding from the South African Defence Force. This has now expired and the programme is currently being funded by LIW, although it is reported that there is now strong overseas interest in the weapon.

The LIW division of Denel is still discussing with its own forces, the South African National Defence Force,

LIW 105 mm Light Experimental Ordnance (LEO) deployed in the firing position **2001**/0105978

Specifications

Calibre: 105 mm
Barrel length:
(with special muzzle brake) 57
(without muzzle brake) 52
Muzzle brake: pepperpot with special Leval shaped nozzles
Breech mechanism: semi-automatic, swing and slide
Carriage: split trail
Shield: no
Weight: 3,800 kg*
Length:
(travelling) 6.9 m
Width:
(travelling) 2.02 m
Height:
(travelling) 2.1 m
Height of towing eye: n/avail
Axis of bore: n/avail
Ground clearance: 0.3 m
Track: 1.82 m
Fording: n/avail
Gradient: n/avail
Tyres: n/avail
Elevation: –5 to +75°
Traverse: 40° (left and right)
Rate of fire: 6 rds/min
Max range:
(standard ammunition) 24,000 m
(base bleed) 30,000 m
Min range: 500 – 3,000 m (direct fire)
Crew: 5
Towing vehicle: 4 × 4 vehicle
* it is hoped to reduce this to around 2,500 kg

their requirements and possible further funding to continue development.

Description

The 105 mm Light Experimental Ordnance has a split trail carriage of high-strength steel and when deployed in the firing position is supported by a jack lowered under the forward part of the carriage and by the spade on the end of each trail.

The carriage is provided with an adjustable towing hook for towing by 4 × 4 vehicles and lifting hooks for transportation slung under a helicopter.

The 105 mm autofrettaged ordnance is 52 calibres long, or 57 calibres when fitted with the advanced muzzle brake, and has a rifling twist of 1:22. The high-efficiency pepperpot muzzle brake has special Lavel shaped nozzles and is rifled on the inside, which adds an effective five calibres to the barrel.

The ordnance has a fixed recoil length of 1 m with a gas type counter system being fitted. Counter recoil is also used to retract the gun using the hydraulics of the traverse and elevation system.

The breech mechanism is of the semi-automatic swing and slide type.

The traverse and elevation system of the 105 mm Light Experimental Ordnance consists of the following key parts:

(1) Hydraulic system fitted with proportional control valves
(2) Hydraulic elevation cylinders with integrated pneumatic equilibrators
(3) Traverse gearbox with hydraulic drive motor
(4) Hydraulic hand pump to supply hydraulic pressure manually
(5) Electrical driven hydraulic pump to supply the hydraulic pressure required
(6) Hydraulic hand pumps for manual elevation and traverse of the gun
(7) Traverse gearbox fitted with a clutch for rapid traverse of barrel from travelling to the firing position
(8) Joy stick control fitted for controlling elevation and traverse of the gun

The bottom carriage is of the split trail type made of high-strength steel. The two main wheels are mounted on a trailing arm suspension and are adjustable for height. The trails are provided with dig in spades designed for use in hard ground.

Mounted on the forward part of the carriage is a large manually operated stabilising platform that is lowered to the ground before firing commences. There is also a barrel clamp on the trail legs to clamp the barrel when the gun is towed in the travelling position.

The high-strength steel saddle is of the trough type with self adjustable type trunnion bearings. The gun control equipment is also mounted on the saddle on the left side.

An adjustable towing hook is provided to allow towing by a 4 × 4 or any 5-ton type of vehicle and lifting hooks can be fitted to allow transport of the gun by helicopter. The weapon can be towed on roads up to a maximum speed of 100 km/h with the ordnance in the firing position on locked over the closed trails.

To power the onboard electrical equipment, two 12 V maintenance-free batteries are fitted together with electrical harnesses, control boxes, tail lights, indicator lights and brake lights.

LIW 105 mm Light Experimental Ordnance (LEO) deployed in the firing position ***2001**/0105979*

The basic model is fitted with telescopic sight for direct fire up to a range of 3,000 m and a trunnion-mounted optical panoramic sight with compensating mechanism for trunnion cant for indirect fire. This is installed on the left side of the weapon. Aiming posts and beta lights are stored on the gun.

The advanced model will retain the telescopic sight for direct fire up to a range of 3,000 m but will also have a ring laser gyro, display unit and an automatic laying system.

The basic model is fitted with a manual loading system (hand ram) which enables the weapon to be loaded at all angles of elevation. The improved model will be fitted with a positive chain-push rammer which will be hydraulically operated and controlled. This will be able to ram both the projectile and charge at any elevation.

Maximum range with standard ammunition is 24,000 m or 30,000 m with a base bleed projectile which is still under development.

The standard 105 mm High Explosive Preformed Fragmented Projectile (HE-PFF) has a lethal area of 1,900 m^2. This can be compared to Denels 155 mm ERFB (Extended Range Full Bore) projectile with RDX/TNT fill which has a similar lethal area. A standard 155 mm M107 TNT filled projectile has a lethal area of only 1,000 m^2. These figures are for projectiles fitted with a proximity fuze.

Other 105 mm ammunition natures include a standard type high-explosive projectile and an APFSDS-T (armour piercing fin-stabilised discarding sabot – tracer) round for the engagement of armoured targets.

Other 105 mm projectiles to be developed include a long-range anti-armour projectile carrying 15 bomblets, smoke and illuminating plus a base bleed unit for all projectiles.

A five zone unimodular charge system has been developed by the Somchem division of Denel which is essentially a scaled-down, version of the M90 modular charge system, developed for the British Army for use with the AS90 155 mm self-propelled artillery system. Maximum muzzle velocity is from 950 to 1,050 m/s.

Ammunition is of the separate loading type, projectile and charge and the system has a maximum rate of fire of six rounds a minute. Various fuzes, including the new Fuchs electronics multi-option fuze have been tested with the 105 mm Light Experimental Ordnance. Fuze options include impact fuze M841, proximity fuze M8513 and time fuze M9220.

According to LIW, at 75 per cent of maximum range, dispersion is 0.3 per cent of range and 0.5 mil in deflection.

During the full scale development phase, LIW have stated that the following improvements will be addressed:

(1) Development of a base bleed unit to increase the range of the system to more than 30 km
(2) Further weight reduction making use of high-strength materials
(3) Methods of further improving accuracy and dispersion
(4) Adding an automatic laying system on models to be fitted with a laser inertial laying system
(5) A load-assist system
(6) Mobility and deployability improvements
(7) Changes to make it possible to fire other types of ammunition for training purposes.

Variants

It is possible that this could be mounted on the rear of a truck or a 6 × 6 or 8 × 8 armoured personnel chassis could be adopted for this weapon.

Status

Prototype. Not yet in production or service.

Manufacturer

LIW division of Denel.

UPDATED

SPAIN

Santa Barbara Sistemas 155 mm 155/52 APU SBT howitzer

Development

Based on the experience in the design and development of the 155 mm SB 155/39 towed howitzer, which never entered production (*Jane's Armour and Artillery 1997-98*), Santa Barbara Sistemas (and now owned by General Dynamics of the United States), in association with other defence equipment manufacturers, has developed a new 155 mm/52 calibre system called the 155/52 APU SBT howitzer.

The first prototype was completed in December 1997 and by May 1998 had fired over 170 rounds of 155 mm ammunition including Extended Range Full Bore and Extended Range Full Bore - Base Bleed.

In November 1999 Santa Barbara delivered six 155/52 APU SBT towed howitzers to the Spanish Army for an extensive series of troop trials.

These six preproduction 155/52 APU SBT towed howitzers, delivered to the Spanish Army, incorporate a number of improvements as a result of trials with the first prototype and have been built with no funding from the Spanish Army.

The Spanish Army currently operates a number of obsolete towed 105 mm and 155 mm artillery systems most of whose design is now 50 years old. The 155/52 APU SBT is expected to replace these systems as well as a number of obsolete coastal artillery weapons still used to defend key parts of the Spanish coast. The total Spanish requirement is for at least 100 units, funding permitting.

Early in 2001, it was stated that the Spanish Army had begun forming a Mobile Coastal Artillery Group with the first batch of six 155/52 APU SBT howitzers supplied by Santa Barbara.

The 155/52 APU SBT Coastal Howitzer C2 (command and control) system will co-ordinate data from coastal defence tracking radars to enhance target engagement capability.

Santa Barbara Sistemas 155 mm 155/52 APU SBT howitzer in firing configuration **2000**/0085652

Santa Barbara Sistemas 155 mm 155/52 APU SBT howitzer in self-propelled configuration **2001**/0067084

Description

The new Santa Barbara 155 mm SBT 155/52 APU SBT howitzer has a conventional split trail carriage and when deployed in the firing position is supported on a circular firing platform mounted under the carriage and on the two trails, each of which is provided with a spade and a small wheel. The latter is used to assist in bringing the weapon into action.

The carriage has four wheels and mounted on the front of the carriage is the 106 hp diesel auxiliary power unit (APU) with the driver being seated on the left side. The alternator and batteries are also located on the APU.

The suspension is of the walking beam type and the diesel APU supplies hydraulic power to the four drive wheels to enable the weapon to be driven across country under its own power. Maximum road speed is 18 km/h. In addition to providing power for the four main road wheels the APU also provides power for raising and lowering of the auxiliary wheels, opening of the trails and operation of the support foot and aiming systems.

When travelling, the upper part of the system, complete with 155 mm/52 calibre ordnance, is traversed through 180° and locked in position over the closed trails.

The 155 mm/52 calibre autofrettaged ordnance has a 23 litre chamber and is fitted with a three part slotted muzzle brake with an efficiency of 35 per cent. The screw breech mechanism is of the screw type with a plastic pad obturator and upon recoil opens automatically. The ordnance meets the NATO Joint Ballistic MoU. The firing system includes a detonation and primer magazine system with 10 M82 type primers in a drum type magazine.

The counter recoil system is of the oleo-pneumatic type. Recoil system is of the hydraulic rod type with an absorber. Elevation and traverse is hydraulic with the layer being seated on the left side of the carriage.

To enable a high rate of fire to be achieved and load the weapon at high angles of elevation, a loading tray and hydraulic projectile rammer are fitted as standard.

According to Santa Barbara Sistemas, the 155 mm 155/52 APU SBT howitzer has a maximum range of 31.7 km firing standard projectiles and 41 km firing base bleed projectiles. Standard bagged charges or modular artillery charge systems can be used with this equipment. Standard equipment includes a primer magazine, rounds counter, recoil counter and a chamber temperature detector.

Optional equipment includes a primer magazine, hydraulic rammer, automatic breech opening system,

Specifications

Calibre: 155 mm
Barrel length: 8.12 m
Length of rifling: 6,890 mm
Muzzle brake: slotted
Recoil:
(max, first stage) 1,525 mm
(max, second stage) 1,041 mm
Breech mechanism: screw type
Carriage: split trail with APU
Shield: no
Weight: 13,500 kg
Length:
(travelling) 10.35 m
(firing) 12.70 m (zero elevation)
Width: 2.82 m
Height: 2.20 m
Ground clearance: 0.35 m (under float)
Track: n/av
Max speed:
(towed road) 90 km/h
(trail) 50 km/h
(cross-country) 15 km/h
Max speed (APU):
(road) 18 km/h
(trail) 15 km/h
(off-road) 10 km/h
Gradient (APU): 30%
Fording: 0.70 m
Elevation/depression: +72/–3°
Traverse: 40° left/40° right
Rate of fire: n/av
Max range:
(standard projectiles) 31,700 m
(base bleed projectiles) 41,000 m
Crew: n/av
Towing vehicle: 4 × 4 or 6 × 6 truck

shot counter, recoil meter, chamber temperature measuring device, navigation system and various types of laying system. The following aiming devices are fitted:
Telescope M137
Telescope mount M171
Telescope square M17
Direct aiming sight M138
Sight mounting M172
Sight square M18
Muzzle centering device SKS155

The navigation system offered is inertial with odometer and GPS assistance. The ballistic computer is provided with ballistic patch calculation software with the muzzle velocity radar being the W-700 C. The latter is mounted over the 155 mm ordnance.

Variant

Santa Barbara Sistemas has also proposed that the ordnance used in the 155/52 APU SBT could be used to replace some of the 6 in weapons currently used in Spain's coastal defences.

Status

Six systems delivered to the Spanish Army for a series of trials from November 1999.

Manufacturer

Santa Barbara Sistemas (part of the General Dynamics Group).

UPDATED

SWEDEN

Bofors Defence 155 mm Field Howitzer 77B (FH-77B)

Specifications

Calibre: 155 mm
Muzzle brake: double baffle
Breech mechanism: screw
Carriage: split trail with APU
Shield: no
Weight:
(total) 12,000 kg
Length:
(travelling) 11.6 m
(firing) 11.16 m
(barrel) 6.045 m
Width:
(travelling) 2.65 m
(firing) 7.18 m
Height:
(travelling) 2.82 m
Axis of bore:
(firing) 1.7 m
Track:
(driving wheels) 2.2 m
(castor wheels) 2.2 m
Ground clearance: 0.42 m
Fording depth: 0.75 m
Max speed:
(towed) 70 km/h
(self-propelled) 8 km/h
Elevation/depression: +70/–3°
Total traverse: 60°
Rate of fire:
(max) 3 rds/12 s or 10 rds/min
Max range:
(standard) 24,000 m
(extended range) 30,000 m
Crew: 6
Engine: petrol or diesel
Towing vehicle: Saab-Scania SBAT IIIS (6 × 6) truck or equivalent vehicle

Development

The now Bofors Defence 155 mm FH-77B (now frequently referred to just as the FH-77) was originally developed as an export model of the FH-77A (see separate entry).

The main change was the ability to fire NATO standard ammunition, other improvements being made to its general handling, cross-country mobility and the introduction of a mechanised ammunition handling system.

Following an FH-77B order from Nigeria, the largest export order in Swedish history came when India placed an initial order for 410 FH-77B weapons and equipment. AB Bofors (today Bofors Defence) acted not only as manufacturer but as the major supply contractor for items such as Saab-Scania trucks, Barracuda camouflage netting, Marconi Radar and Control Systems (now Alenia Marconi Systems), Quickfire fire-control systems, Fairey Australia muzzle velocity radars, BEAB sighting equipment and surveying equipment from Wild of Switzerland and navigation equipment from the then Ferranti UK. First production FH-77Bs for India left the Bofors factory in August 1986.

The first Indian order for the FH-77B was for a total of 410 weapons with final systems delivered in 1990. The company delivered FH-77Bs to India in complete batteries to enable Indian artillery units to become fully operational as complete units.

The ammunition supplied by Bofors included smoke (FFV), illuminating (Bofors) and new Bofors High-Explosive Extended-Range (HEER) projectiles for which India was the first customer. There were two versions of the HEER supplied, one with boat tail and the other with base bleed. LIAB, a Bofors subsidiary, provided 155 mm fuzes.

When the original order was placed by India, it was envisaged that the total Indian requirement would consist of some 1,500 weapons.

The second batch of FH-77Bs would have been assembled in India from components supplied by Bofors with the third batch being made under licence in India. For a number of reasons, India did not place any further orders for the FH-77B.

In July 1999, Bofors Defence announced that it had received a small but important contract from India for spare parts for the 410 155 mm FH-77B field howitzers used by the Indian Army. This was the first order placed by India with the company since 1989.

The order covered the continuance of spare parts within the framework of the original FH-77B contract placed in 1986.

A second contract was placed in India with Bofors Defence in September 1999 for further spare parts worth a total of SKr 190 million.

In March 1990, Bofors announced that the Swedish Materiel Defence Administration had placed a US$162 million contract with Bofors for the supply of 48 FH-77B artillery systems, spares, software, ammunition and associated charges.

The Swedish Army FH-77Bs are similar to those produced for India but with some modifications to enable them to be operated at lower temperatures.

The Nigerian FH-77Bs and the original Indian FH-77Bs had a parallel pepperpot muzzle brake but, following trials with the more recent ammunition types, including HEER used with charge 9, a new and more efficient muzzle brake of the double-baffle type was developed. This was fitted to late production FH-77Bs and has now been retrofitted to earlier weapons.

Ammunition in the Swedish Army order includes the Bofors-developed HEER (High-Explosive Extended-Range) projectile, developed to meet the requirements of the Indian Army which, when used with charge 9, has a range of 30,000 m. HEER-BB (High Explosive Extended-Range - Base Bleed) was included in the order and the FH-77B will also be able to use ammunition already stockpiled by the Swedish Army.

Description

The FH-77B differs from the earlier FH-77A in three main areas; a slightly longer barrel, the introduction of a breech screw mechanism using bag charges, and a new hydraulic loading system. The FH-77B can be towed at speeds of up to 70 km/h, normally by the Saab-Scania SBAT 111S (6 × 6). For cross-country travel the driver of the tractor vehicle can remotely engage the gun roadwheels by starting the gun hydraulic motor making an 8 × 8 combination. Maximum speed in this configuration is 7 km/h and when this speed is exceeded the gun wheels are disengaged.

Steering is by regulating the drive to the independent roadwheels of the gun. The same hydraulic motor can also be used for several other functions. To get the gun into action the driver on the gun uses the hydraulic power to raise the gun castor wheels on the trail to disengage from the tractor towbar. In the self-propelled

Bofors Defence 155 mm FH-77B deployed in firing position. This weapon has the new muzzle brake

Bofors Defence 155 mm FH-77B from the front clearly showing new muzzle brake

mode the trails could be locked together or split, again using hydraulic power. The gun driver then reverses the gun to dig the trail spades into the ground. For making changes of fire direction the same process can be repeated. Using this system the driver on the gun can make rapid fire direction changes, get the gun in and out of action and move the gun from one firing location to another alone.

The ammunition loading system uses a hydraulic crane driven from the hydraulic system. The crane grab is designed to lift any 155 mm projectile in groups of three at a time, offered to the grab either horizontally or vertically and almost any projectile length can be handled, including the longer enhanced-range projectiles. From the crane the projectiles are loaded onto a loading table and are then guided onto the loading tray, while another loader places the propellant charge onto a separate loading trough/rammer. In two operations the projectile loading tray is swung over to line up with the chamber and the charge tray swings over behind it. The shell is then hydraulically rammed forward into the chamber with the charge pushed right forward behind the projectile. Once the projectile is seated in the driving bands the loading trough/rammer is withdrawn until the charge is stripped from the tray by a retaining catch. The rammer mechanism is fully withdrawn, the breech screw is closed and a primer cartridge automatically loaded into the primer seat from a magazine holding eight tubes. Using this system, projectiles and charges can be loaded into the gun at angles of elevation of up to +70°. The rate of fire, with three people on the gun, can be three rounds in 12 seconds. The subsequent rate of fire is governed only by the ammunition supply. An added bonus of the hydraulic ramming is that uniform ramming reduces the resultant projectile dispersion.

The full crew of the FH-77B is six: the gun commander, a layer and four loaders. One loader operates the mechanism for loading the projectiles from the loading table to the loading tray, one prepares the projectiles by inserting the fuzes, setting them and loading the projectiles in groups of three into the loading crane, one places the charges onto the loading tray and another loader operates the ammunition crane. If necessary, only the layer, crane operator and projectile loader combined, and a charge loader can operate the gun in action. There is a special internal communication system built into each gun, separate from the usual fire-control communication systems, with a 10 m range based on an induction loop with receivers built into the crew's ear protectors. Gunlaying is fully hydraulic. Two levers, one on the right for elevation and depression and one on the left for traverse allow barrel movement. Several types of sight may be fitted but the usual model fitted is the electronic BEAB BAAB RIA. The layer sits high on the left of the breech with a protective open cage over the seat and sight (the loading crane is on the right of the breech). A x4 telescope sight is fitted for direct fire. The BAAB NK24 night sight fitted in front of the ordinary telescopic sight permits direct fire under night conditions. As the FH-77B depends on hydraulic power for many of its systems, an electric motor for all operations using an internal or external power supply or two hydraulic hand pumps are provided in case the main system fails. One, situated just forward of the layer's position and in the power compartment itself, can provide manual hydraulic power for traverse, elevation and ramming. The other pump is on the other side of the gun and provides manual hydraulic power for operating the support wheels.

Unlike the FH-77A the ammunition system of the FH-77B is based on a bagged charge system which enables a wide range of 155 mm ammunition to be fired, including the following main types of projectile: Bofors HE M/77 B which is a high-capacity shell weighing 42.6 kg and containing 8 kg of TNT or Composition B. This has a maximum range of 24,000 m when fired from the FH-77B howitzer. The round is fully compatible with all Western-manufactured guns and internationally adopted fuzes.

HE M107, the standard NATO 155 mm howitzer projectile weighing 42.91 kg. It can be fired to a range of approximately 19,000 m.

Development of a base bleed round designated HEER (High-Explosive Extended-Range) has been completed by Bofors. HEER is compatible with all Western-manufactured 155 mm guns and international fuzes. Maximum range is 30,000 m and the weight is the same as conventional 155 mm rounds. This involves less stress on the barrel than firing other types of projectiles such as base bleed with a greater weight. Production is also made easier by omitting ERFB nubs.

Other rounds available are different types of smoke (including infra-red smoke) and the MIRA illuminating round in two versions, with maximum ranges of 17,000 m and 24,000 m respectively. For both, the time of illumination is 60 seconds, descending velocity 4 m/s and the mean illuminating intensity is 2.2 Mcd.

Standard NATO and other propellant charges can be used with the FH-77B but the company is now producing its own charges especially for the FH-77B. Charges available include M3 which provides a maximum muzzle velocity of 375 m/s to give a range of 10,300 m and M4 which provides a maximum muzzle velocity of 563 m/s to give a range of 15,500 m.

Charges include the Charge 8 which is roughly equivalent to M119 but gives more favourable pressure and burning properties at extreme temperatures. Charge 8 provides a muzzle velocity of 685 m/s to give a range of 19,000 m. Charge 9 is a new Bofors development to provide a muzzle velocity of 827 m/s and a range with standard projectiles of 24,000 m and with base bleed projectiles of 30,000 m.

The FH-77B can also fire the new 155 mm BONUS top attack artillery projectile developed to meet the operational requirements of the French and Swedish armies by Bofors Defence of Sweden and Giat Industries of France.

Variants

CD 77

In order to provide a version of the 120 mm KARIN/CD 80 coastal defence gun firing NATO standard ammunition, Bofors Defence has developed a version of the 155 mm FH-77B known as the CD 77. The CD 77 retains the same ordnance and carriage as the FH-77B and uses the same mechanical handling system but allied to the fire-control system used with the 120 mm KARIN/CD 80 or a similar system. Thus the CD 77 will be deployed as an individual gun and will use remote fire control coupled to an automatic sight. Weights and dimensions are the same as those for the FH-77B. There are no known plans for this to enter production or service.

Autonomous FH-77B

During late 1987, the first example was demonstrated of an FH-77B fitted with a combined land navigation gyro-stabilised system linked to the sight and fire-control equipment to permit autonomous positioning of the gun. In place of the normal sight a land navigation system (FIN 1150) from the then Ferranti Defence Systems of Edinburgh, Scotland, was used. The system gives the gun position and the attitude of the barrel (traversing, elevation and trunnion angle). This alleviates the need for surveying and alignment normally associated with bringing a gun into action and permits 'shoot and scoot' tactics to be employed more readily.

During one demonstration, a gun with a crew of four was deployed and four rounds were fired in 2 minutes 15 seconds. The gun was then redeployed approximately 50 m away and three further rounds were fired. Total time taken was 5 minutes and 10 seconds.

FH-77B with MAPS

The 48 155 mm FH-77Bs delivered to the Swedish Army by late 1992 were fitted with the Honeywell MAPS (Modular Azimuth Position System) land navigation system. The Honeywell MAPS system has also recently been fitted to the small number of Bofors Defence 155 mm Bandkanon 1A self-propelled artillery systems used by the Swedish Army. Details of this system are given in the Self-propelled guns and howitzers (tracked) section.

FH-77 (6 × 6)

Details of the Bofors Defence 155 mm FH-77 series artillery system fitted to a 6 × 6 chassis for trials purposes are given in the Self-propelled guns and howitzers (wheeled) section. As of early 2002, this system remained at the prototype stage.

Status

Production complete. In service with India (410), Nigeria (48) and Sweden (48).

Manufacturer

Bofors Defence AB (A United Defense company)

UPDATED

Bofors Defence 155 mm field howitzer 77A (FH-77A)

Development

Studies between 1965 and 1968 to establish the future requirements of Swedish Army artillery weapons led to the recommendation that a 155 mm howitzer with an Auxiliary Propulsion Unit (APU) should be developed. It was required to have a long range, high rate of fire, use an improved type of ammunition and have good mobility, especially across rough terrain and snow.

AB Bofors (which in 2000 became Bofors Defence and owned by United Defense of the USA) was subsequently awarded a development contract by the Army Materiel Department of the Defence Materiel Administration to develop a new weapon which became the FH-77A.

The first prototype was built in 1973 with the main production run being carried out from 1978 to 1984 for the Swedish Army. Further development resulted in the FH-77B covered in the previous entry.

It is estimated that total production of the FH-77A amounted to 200 guns before production switched to

Bofors Defence 155 mm Field Howitzer 77A in firing position

the FH-77B. It is understood that the Swedish FH-77A systems are now held in reserve.

The first customer for the latter was India which ordered 410 systems. It has also been adopted by Nigeria (48) and Sweden (48).

There has been no recent production of the Bofors Defence 155 mm FH-77A or FH-77B.

Description

The FH-77A is normally towed by a Saab-Scania SBAT 111S (6 × 6) truck at a maximum road speed of 70 km/h. When travelling across rough terrain the main wheels of the howitzer can be engaged making an 8 × 8 combination. Maximum speed in this configuration is 8 km/h and once this speed has been exceeded the wheels of the FH-77A are disengaged.

Mounted on the forward part of the carriage is a Volvo B20 engine which is connected to two variable hydraulic pumps. These are connected to hydraulic motors in the driving wheels and, when being used in the self-propelled mode, steering is by means of levers at the driver's position on the left side of the carriage. Both driving and steering are accomplished by the large roadwheels. The two support wheels, one on each trail, pivot freely and are also hydraulically operated and assist in limbering and unlimbering the weapon.

The elevating mass, which is supported in the cradle carrier, consists of a cradle which carries the barrel with the breech mechanism, recoil mechanism and loading tray and rammer for the loading system. The semi-automatic breech mechanism has a vertical sliding breech block which opens downwards. The firing device is electromechanical. Elevation and traverse are both hydraulic at a maximum speed of 3°/s.

On the right side of the FH-77A is the loading table on which the 155 mm projectiles are placed with the aid of an ammunition crane. The cartridge cases are placed manually on the loading tray. The loading device is operated by the loader who is seated on the right side of the howitzer.

Swedish Army weapons have the BEAB BAAB RIA sight which is operated by the driver/layer on the left side of the carriage.

The crew of six of the FH-77A consists of the commander, gunner who lays the gun and also acts as the driver, ammunition handler who places the cartridge case onto the loading table, loader who feeds the projectile from the loading table and operates the rammer, crane handler who lifts three projectiles at a time onto the loading table and the ammunition handler who screws the fuzes into the 155 mm projectile.

On arriving at the battery position the howitzer can be unlimbered by raising the castor wheels on the carriage trails with the aid of the hydraulics. The lunette is thereby raised and unhooked from the pintle. Once the howitzer has been unhooked the carriage trails are spread by releasing the locking device between the carriage trails. The howitzer is then swung first to one side and then the other, the carriage trails locking in their outer positions. The supporting wheels are turned up, and the howitzer, oriented in the firing direction ordered, is driven in reverse to dig in the trail spades. If the howitzer needs to be re-oriented to a new firing direction beyond its traverse limits it is driven forwards to free the trail spades and the supporting wheels are turned down. Once in the new firing position the supporting wheels are turned up again and the howitzer is again reversed to dig in the trail spades.

The FH-77A is also provided with hand pumps, one for laying and ramming and one for the castor wheels, which can be used to obtain the pressure required in the hydraulic systems if the auxiliary engine should be out of order. These hand pumps can also be used for quiet operation of the 155 mm FH-77A howitzer at night.

The unit loads of ammunition (both projectiles and charges) facilitate handling and can be stacked on each other. Each contains 12 projectiles or 12 cartridge cases and is kept unopened all the way to the ammunition supply at the gun. With the aid of the ammunition crane, three shells at a time can be taken directly from the unit load and transferred to the loading table of the gun. The cartridge cases are kept in special plastic tubes, with covers on one end. The tubes, which are inserted horizontally into the unit loads, can be opened and the cartridge case removed without having to break the unit load. If the situation should require manual moving of the cartridge cases, the unit load is divided into separate tubes in which the cartridge cases are carried.

The firing sequence is as follows. A cartridge case is placed on the loading tray and then the loader feeds a projectile from the loading table and actuates the ramming device when the projectile has slipped down into the cartridge case on the loading tray. The layer then fires the weapon. The cartridge case is ejected on recoil, after which the loading cycle is repeated.

The HE projectile is designated the M/77, weighs 42.4 kg complete with Bofors ZELAR fuze and contains 8 kg of TNT. The empty plastic cartridge case has a steel base and weighs 5.9 kg; cartridge case plus increments one to four weigh 11.9 kg complete and cartridge case plus increments five to six weigh 16.7 kg. The maximum muzzle velocity with the various charges is as follows: charge 1, 310 m/s; charge 2, 371 m/s, charge 3, 472 m/s, charge 4, 583 m/s, charge 5, 692 m/s and charge 6, 774 m/s.

Specifications

Calibre: 155 mm
Barrel length: 5.89 m
Muzzle brake: pepperpot
Breech mechanism: vertical sliding
Carriage: split trail with APU
Shield: no
Weight:
(travelling order) 11,500 kg
(on driving wheels) 9,500 kg
(on rear support wheels) 2,000 kg
Length:
(travelling) 11.6 m
(firing) 11.2 m
Width:
(travelling) 2.64 m
(firing) 7.18 m
Height:
(travelling) 2.75 m
(firing at 0° elevation) 2.75 m
(firing at 50° elevation) 6.75 m
Axis of bore: 1.6 m
Ground clearance: 0.445 m
Track:
(driving wheels) 2.2 m
(support wheels) 2.23 m
Elevation/depression: +50/–3°
Total traverse:
(elevation <+5°) 50°
(elevation >+5°) 60°
Rate of fire: 3 rounds in 6 to 8 s
6 rounds in 20 to 25 s
6 rounds every other min for 20 min
Max range:
(charge 6) 22,000 m
Crew: 6
Towing vehicle: Saab-Scania SBAT 111S (6 × 6) truck

Variants

The 120 mm KARIN mobile coastal defence gun uses the same carriage as the FH-77A. For details of KARIN see the separate entry in the Coastal artillery guns and missiles section. This is only used by the Swedish Navy.

FH-77 (6 × 6)

Details of the Bofors Defence 155 mm FH-77 series artillery system fitted to a 6 × 6 chassis for trials purposes are given in the Self-propelled guns and howitzers (wheeled) section. This remains at the prototype stage.

Tridon 40 mm

A similar chassis is used as the basis for the Bofors Defence Tridon 40 mm self-propelled anti-aircraft system. Details of this are given in *Jane's Land-Based Air Defence 2002-2003*. This remains at the prototype stage.

Status

Production complete. In service with the Swedish Army (held in reserve).

Manufacturer

Bofors Defence AB (A United Defense company).

UPDATED

TAIWAN

155 mm extended range gun

Development/Description

In the late 1970s, the Combined Service Forces (CSF) developed manufacturing capability for a version of the Israeli Soltam Systems 155 mm M-68/M-71 family of towed howitzers. It is believed that two prototypes were built in Taiwan. From the late 1970s, all effort was concentrated on the 155 mm towed extended range gun.

The 155 mm towed extended range gun uses the same ordnance and recoil system as the XT-69 self-propelled weapon. This is based on the chassis of the US supplied, now United Defense LP, M108/M109 and fully described in the Self-propelled guns and howitzers (tracked) section under Taiwan.

It is mounted on a two-wheeled split trail carriage similar to that of the US 155 mm M114 howitzer and incorporates some of the features found on the SRC International modernised M114/39 howitzer (*Jane's Armour and Artillery 1982-83*).

Further development of this by RDM Technology of the Netherlands resulted in M114/39, full details of which are given in this section under Netherlands.

Recent information has indicated that Taiwan has a total of 250 M114 towed artillery systems and most of these are believed to have been upgraded.

A handwheel is fitted at the elevation controls and the pneumatic equilibrators are of the high-capacity type. Mounted over the left trail are a pneumatic rammer, pressurised air bottle, swing-aside loading tray and a telescopic rammer mounted near the rear cradle. This enables a high rate of fire to be achieved and the weapon to be loaded at all angles of elevation.

The 155 mm 45 calibre ordnance is fitted with a multibaffle muzzle brake and a fume extractor. The weapon can fire all standard US/NATO ammunition as well as ERFB (with a maximum range of 30,000 m) which is now being manufactured in Taiwan. It is believed that 155 mm base bleed ammunition has now been developed which has extended the range of the weapon to around 39,000 m.

The original US M114 was also built in Taiwan under the designation of the T-65 and as far as it is known this is identical to the US model covered later in this section.

Status

Production complete. In service with Taiwanese Army.

Manufacturer

Combined Service Forces, Taiwan.

UPDATED

THAILAND

105 mm M425 howitzer

Development/Description

The Royal Thai Army has deployed a 105 mm towed howitzer designated the M425. This is believed to be the US-supplied 105 mm M101 upgraded in a number of key areas including the installation of a new and longer 105 mm barrel which has a life of 3,000 rounds.

In 1997, it was confirmed that Thailand had taken delivery from Giat Industries, France of a total of 285 kits to upgrade its old M101s with the complete barrel of the Giat Industries LG1 Light Gun. Details of this upgrade are given under France earlier in this section.

As well as the locally developed 105 mm M425, other sources have indicated that Thailand may have in service 32 locally built 105 mm guns under the designation of the M-618A2. At this stage it is not known if the M425 and the M-618A2 are in fact the same weapon.

The M425 has a conventional split carriage with a shield that extends either side over the wheels.

105 mm M425 towed howitzer with trails spread

Status

Believed to be in service with the Royal Thai Army. Production complete. Not offered on the export market.

Manufacturer

Upgrade carried out in facilities of the Royal Thai Army.

VERIFIED

Specifications
Calibre: 105 mm
Muzzle brake: double baffle
Recoil system: hydropneumatic
Breech mechanism: horizontal sliding wedge
Carriage: split trail
Shield: yes
Elevation/depression: +61.87/–5.34°
Traverse: 46° (total)
Rate of fire:
(max) 10 rds/min
(sustained) 4 rds/min
Max range: 14,500 m

UNITED KINGDOM

Lightweight Mobile Artillery Weapon System (LIMAWS)

Development/Description

Between April 1996 and March 1997, the UK MoD-funded Defence Evaluation and Research Agency (DERA) undertook a feasibility study for the British Army requirement for a Light Mobile Artillery Weapon System (LIMAWS) to provide indirect fire support for rapid deployment forces.

LIMAWS is to meet Staff Target (Land) 4095 with Staff Requirement (Land) due to be issued in the future.

It is expected that LIMAWS could be introduced into the Royal Artillery from 2006 onwards and replace most of the BAE Systems, RO Defence 105 mm Light Guns at present in service.

Late in 2001 it was stated that the British Army was considering a mix of 155 mm XM777 and 105 mm Light Guns, with the latter perhaps being used with enhanced ammunition with greater lethality.

Solutions being studied will include lightweight 155 mm towed artillery systems and lightweight multiple rocket launcher systems.

The 155 mm lightweight artilliary system expected to meet this requirement is the now BAE Systems, RO Defence, UFH (Ultralightweight Field Howitzer) designated the XM777 in the US.

For the lightweight multiple rocket requirement the British Army system will be called the LIMAWS (Rocket Platform). Additional details are given in the Multiple Rocket Launchers section. It is expected to be a towed mounted system.

The British Army requirement is expected to be for 90 155 mm guns and 30 rocket systems.

Status

Future British Army requirement.

Manufacturer

Not selected.

UPDATED

BAE Systems, RO Defence 155 mm Ultralightweight Field Howitzer (UFH)

Development

The Ultralightweight Field Howitzer (UFH) was originally developed as a private venture. Its origins can be traced back to the early 1980s when Vickers Shipbuilding and Engineering Limited (VSEL) (which in late 2000 became BAE Systems, RO Defence) originally perceived a potential market for a lightweight 155 mm towed howitzer.

In the spring of 1987 the project definition was completed. Its objective was to have a weapon with the same range as the US Army's M198 155 mm towed howitzer but weighing no more than 4,000 kg.

The current M198 weighs 7,163 kg which limits its air mobility; it can only be carried by two helicopters, the US Army CH-47 Boeing or the US Marine Corps Sikorsky CH-53.

The US Army was fully briefed on the system and agreed that if the company built a prototype of the system with its own money it would carry out a complete evaluation of the system.

In September 1987, the main board gave approval to build two prototypes of the system which subsequently became known as the UFH. Both were completed in late 1989.

The complete upper part of the UFH was test fired at Eskmeals in June 1989 with a total of 50 rounds being fired at all elevations, 12 of which were zone 8S (top charge).

Although the UFH was originally targeted at the US Army, the US Marine Corps took the initiative as it was looking for a lightweight 155 mm system to replace all current 105 mm and 155 mm towed artillery systems.

Following its unveiling at the 1989 Association of the United States Army Exhibition in Washington DC, one of the two prototypes went to the US for early evaluation.

This evaluation, under the supervision of the US Army Armament Research and Development Command on behalf of the US Marine Corps, took place in three phases through to 1990. Phase 1 was at Yuma Proving Ground where the system fired 437 rounds, 50 per cent at Charge 7W or higher (25 per cent at Charge M203 8S).

It also completed limited land mobility trials and airlift certification in single and split mode. At the end of Phase 1 the system was awarded limited live crew clearance for the USA.

Phase 2 was conducted at the US Marine Corps Base at Camp Lejeune, North Carolina and at the Naval Base at Little Creek, Virginia. During Phase 2 130 rounds were fired with a live crew and the system achieved a single lift with the UH-60L Black Hawk helicopter. Amphibious trials were carried out successfully at Little Creek.

The final phase took place at Aberdeen Proving Ground, Maryland where the system carried out successful climatic chamber firings at temperatures ranging from –25 to +145°C. These climatic firings were followed by air transportability (split lift) trials and 622 km of land mobility trials on test tracks ranging from trails to Belgian blocks and included wading to a depth of 1.5 m.

The US then had a competition which involved extensive tests with the UFH and the Light Towed Howitzer developed at the then Royal Ordnance facility at Nottingham. In the end the UFH was selected.

For the US programme, Textron Marine & Land Systems was selected to be the prime contractor with the then Vickers Shipbuilding and Engineering Limited being the main sub-contractor.

By 1998 it was clear that the US programme was running into problems and early in 1999 the now BAE

155 mm UFH deployed in firing position. The trails are of welded titanium alloy with the rear trails folding vertically for transport and forward trails folding horizontally alongside the weapon **2000**/0085639

155 mm UFH slung under a Sikorsky UH-60 helicopter 0006052

155 mm UFH deployed in the firing position during trials with a British Army crew 0021026

Systems, RO Defence, assumed the role of prime contractor of the troubled XM777 towed artillery system from its team member Textron Marine & Land Systems. This company no longer has any involvement with the XM777 programme.

Under the terms of the Engineering, Manufacturing and Development (EMD) contract awarded to Textron Marine & Land Systems early in 1997, a total of eight XM777s were to have been built and supplied between September 1998 and May 1999.

In September 2000, following an extensive competition, BAE Systems, RO Defence finally selected its core industrial supplier base for US production of the XM777 155 mm LTH.

Under current plans some 70 per cent of the XM777 will be manufactured in the US with final assembly, test and delivery taking place at United Defense LP facility at Pascagoula, Mississippi.

The body assembly will be manufactured by HydroMill Inc of Chatsworth, California, stabilisers, spades and trails will be supplied by Major Tool and Machining Inc of Indianapolis, Indiana, the breech operating load tray system will be provided by Rock Island Arsenal, Rock Island, Illinois with titanium being supplied by RTI International Metals Inc of Niles, Ohio.

BAE Systems, RO Defence have supplied eight XM777 155 mm LTH for the EMD (Engineering Manufacturing and Development) phase. The barrels for the XM777 are being supplied by Watervliet Arsenal.

Following extensive trials, which have already started at Yuma Proving Grounds, it is anticipated that orders will be placed for 413 weapons for the US Marine Corps and 273 for the US Army.

If all goes to plan then first production XM777's will be completed in 2003 with production to continue through to 2010.

The US weapons will be fitted with the General Dynamics Armament Systems Towed Artillery Digitisation (TAD) digital fire-control system under a separate contract.

In addition to the confirmation of the Marine Corp's commitment to complete development of the XM777, the US Army has also budgeted for completion of the Weapons Special Digital Fire-Control Suite, a Pre-Planned Product Improvement (P3I)

The US Army/Marines designation for the UFH is the XM777 Lightweight 155 mm Towed Howitzer and when accepted for service this will become the M777.

The US Marine Corps is funding the development of the XM777 while the US Army is funding development of the P3I part of the programme

According to the prime contractor, the US Army enjoyed extraordinary success in 1998 with its 2nd Generation P3I Advanced Technology Demonstrators (ATDs) at Fort Benning, Georgia in the Rapid Force Planning Initiative (RFPI) Field Experiment.

The Marine Corps' decision to also buy the final XM777 P31 systems also followed the successful field tests of ATD by the Marine Corp's Warfighting Lab.

Crew enhancements include an advanced direct fire sight, automatic primer feed mechanism and a powered rammer.

The US Army standard P3I will include a self-locating and orienting system, onboard ballistic computer and a self-contained power supply.

In the US Army, the 155 mm UFH will provide close and deep fire support and interdiction fire while for the US Marine Corps it will be used as a direct support weapon. The US Army airborne, air assault and light forces will use UFH as a general support weapon.

As stated in an earlier entry, the British Army has a requirement for a Lightweight Mobile Artillery Weapon System (LIMAWS) and the UK is now taking part in the XM777 programme and Italy has now joined the XM777 programme, although not for the complete weapon system.

Description

Although the UFH uses advanced materials in its construction, it is simple to operate and maintain under field conditions.

The 155 mm/39 calibre ordnance of the UFH is essentially that of the M284 barrel used by the US Army's M109A6 Paladin fitted with the M199 muzzle brake as used by the current towed M198 howitzer but modified to take a towing eye. The conventional screw breech is hydraulically operated and opens vertically. The complete ordnance for the UFH was built by Watervliet Arsenal for BAE Systems, RO Defence.

The LFH has elevation limits from –5 to +70° with traverse limits of 22.5° left and right. Zone 8S can be fired at all elevations above +20°.

To keep the weight down, extensive use has been made of extruded titanium alloy fabrications, designed by BAE Systems, RO Defence, but subcontracted to other companies, as well as efficient structural design. Some of the parts perform two functions, for example the Horstman Defence Systems designed hydropneumatic suspension system also operates as a hydraulic jack.

The titanium body of the UFH acts as the soleplate to which the two stabilisers and two short box-type trails with spades are attached; these open front and rear and are self-levelling. The trails lock up on firing. The rear two trails fold up vertically for travelling while the front stabilisers fold back alongside the body.

The leading arm suspension is of the hydropneumatic type with standard truck wheels and tyres being fitted; these have individual control and are fitted with vehicle towing disc brakes and handbrakes.

The saddle is also of titanium with worm-type elevating gear on the left. A simple traverse gearbox is mounted on the left side and moves with the saddle. A universal sight mount is fitted on the left side, but provision is made for a second sight on the right side for users requiring a two-man sighting system.

The cradle consists of four external, extruded titanium tubes with the trunnions at the rear. The two aluminium nitrogen cylinders are at the rear of the cradle. The hydropneumatic recoil system is a novel design with a maximum recoil length of 1.4 m.

A loading tray is provided to the rear of the breech which is opened automatically. BAE Systems, RO Defence is considering future options which could include a flick rammer, but the aim for thepresent is to keep the weapon as simple as possible.

The UFH can be brought into action in under 2 minutes and split into two loads in about 20 minutes.

When in the travelling configuration it is towed by its muzzle in a similar manner to the Russian 122 mm D-30 howitzer by any truck with a 2 tonne capacity (or greater) fitted with air-operated brakes.

It can be internally transported in a number of transport aircraft including the C-130, C-141, C-5 and C-160 Transall. It can also be transported slung under the UH-60L/UH-60M Black Hawk and CH-53E/CH-53D helicopters and the MV-22 Osprey.

In early 1996, the company completed the third (PT03) example of its 155 mm UFH which was used in the US Army/Marine Corps trials.

This has many detailed improvements incorporated as a the result of five years of trials with the first two examples of the 155 mm/39 calibre UFH (UFH001 and UFH002).

Improvements to the 155 mm/39 calibre UFH are mainly in the area of human factors to allow for easier operation by the gun crew as well as increasing the rate of fire, especially burst fire rate.

The breech now opens automatically after the gun has fired and a primer drum magazine has been fitted which holds a total of eight primers.

To increase the rate of fire a Bolt On Loading Tray System (BOLTS) has been fitted to the right side and like the breech opening mechanism is powered by the recoil system.

A thermal warning device has been fitted which warns the gun crew of high barrel temperatures and virtually eliminates the possibility of a cook-off. The saddle has been lightened and the balancer posts have been redesigned for ease of manufacture. The elevation and traverse mechanism has been modified and a traverse lock fitted.

Specifications

Calibre: 155 mm
Barrel length: 6.096 m
Muzzle brake: double baffle
Recoil: hydropneumatic
Max recoil: 1.4 m
Breech mechanism: screw
Shield: no
Weight:
(total) 3,745 kg
(elevating mass) 2,414 kg
(carriage weight) 1,331 kg
Length:
(towing) 9.527 m
(firing) 10.210 m
Width:
(towing) 2.770 m
(firing) 3.720 m
Height: (towing) 2.46 m
Trunnion height: (firing) 0.65 m
Ground clearance: (towing) 0.66 m (variable to suit terrain)
Track: 2.30 m
Elevation/depression: +70/–5°
Traverse: (on gun) 22.5° left and right
Rate of fire:
(intense) 5 rds/min
(sustained) 2 rds/min
Max range:
(unassisted) 24,700 m
(assisted) 30,000 m
Crew:
(normal) 7
(reduced detachment) 5
Towing vehicle: 2,500 kg (4 × 4) (airbraked)
Towing speed:
(road speed) 88 km/h
(cross-country) 50 km/h

Modifications and improvements have also been carried out to the firing lever, front-mounted stabilisers and the rear-mounted spades. For the US Army/Marine Corps trials an M137 direct fire sight and an M138 elbow sight are fitted. The 155 mm/39 calibre ordnance is from Watervliet Arsenal.

UFH003 obtained its live crew clearance from the UK Ordnance Board late in February 1996 and before being flown to the US to take part in the US Army/Marine Corps 155 mm competition, fired over 300 155 mm projectiles at various ranges in the UK.

In addition to firing all in-service 155 mm artillery projectiles, the XM777 will also fire more advanced 155 mm artillery projectiles such as SADARM and the XM982 Extended Range Projectile (ERP).

Status

Engineering and Manufacturing Development for US Army and Marine Corps. As of early 2002 the anticipated quantities of XM777 known to be required are:

Country	Quantity	Comment
Italy	65	anticipated requirement
UK	70	anticipated requirement
US	273	US Army
US	413	US Marine Corps

Manufacturer

BAE Systems, RO Defence.

UPDATED

BAE Systems, RO Defence 105 mm Light Gun

Development

In 1965, the British Army issued a requirement for a new 105 mm Light Gun to replace the Italian Otobreda 105 mm Pack Howitzer then in use with the light regiments of the Royal Artillery.

The requirement was for a gun with a longer range, a more stable firing platform, improved reliability and the ability to be towed across rough terrain at high speed. Design work began at the Royal Armament Research and Development Establishment (which is today part of the Defence Evaluation and Research Agency) at Fort Halstead the following year.

Trials with the prototypes were successful and in 1973 the 105 mm Light Gun (L118) was accepted for service with the British Army. First production guns were handed over at now BAE Systems, RO Defence, Nottingham in October 1974. The Nottingham facility closed down late in 2001 and so far no announcement has been made as to where future production of the Light Gun will be undertaken.

The 105 mm Light Gun, which uses the same range of 105 mm ammunition as the Vickers Defence Systems Abbot self-propelled gun, replaced all 105 mm Pack Howitzers in regular units of the Royal Artillery and has been issued to TA units. Each 105 mm Light Gun Regiment has three batteries, each with eight guns.

Royal Artillery 105 mm Light Gun regiments earmarked for use with the Royal Marines and airborne forces have three batteries each of six guns.

Late in 1981, Australia selected the 105 mm Light Gun, where it is known as the Hamel Gun, to meet its future requirements.

In early 1986, following an extensive series of trials and demonstrations, the Light Gun was type-classified by the US Army as the M119. The Light Gun version involved is the L119 with a barrel suitable for firing US 105 mm M1 ammunition. The type classification followed the evaluation of six Light Guns leased to the US Army in late 1984, followed in mid-1985 by a further 14 guns. These were evaluated by the 9th Infantry Division at Fort Lewis, Washington, as the XM119.

The US Army is the largest user of the BAE Systems, RO Defence 105 mm Light Gun having taken delivery of 147 direct from BAE Systems, RO Defence Nottingham with another 280 manufactured under licence in the USA by Watervliet Arsenal (ordnance) and Rock Island Arsenal (carriage). Licensed production of the 105 mm Light Gun has been completed in the USA.

In mid-1991, the Brazilian Army placed an order for four 105 mm Light Guns in the L118 configuration. These were delivered early in 1992.

In December 1994, Brazil placed an order for a further 36 105 mm L118 Light Guns and the first of these were delivered early in 1995. More recently the Brazilian Marines have placed an order for 18 L118 Light Guns.

Early in 1995, the Royal Netherlands Marine Corps received eight 105 mm Light Guns from the United Arab Emirates.

In July 1995, the Spanish Cabinet approved the purchase of 56 105 mm L118 Light Guns for use by the Spanish Army Rapid Deployment Forces (FAR).

These 105 mm Light Guns are used by three battalions each consisting of three batteries, each with six guns. The remaining two were for the artillery school.

Total value of the contract, which covered the supply of the guns, spares and ammunition is estimated to be Pta8,850 million (£48 million) over a five-year period ending in 1999.

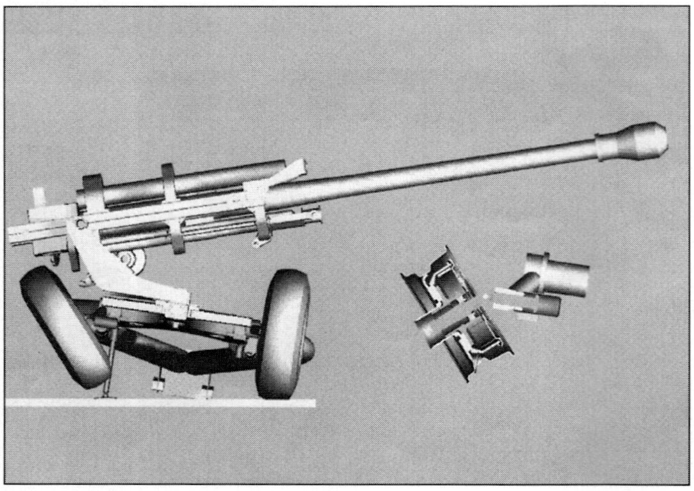

Close up of the new axle design of the BAE Systems, RO Defence 105 mm Light Gun with Mid-Life Upgrade which means the wheel no longer has to be removed to come into action **2000**/0048792

BAE Systems, RO Defence L118 105 mm Light Gun fitted with BAE Systems, Avionics APS 0036868

EXPAL and Santa Barbara are involved in the production of 105 mm ammunition in Spain. In addition, the contract includes 38 kits to convert the L118 to the L119 standard to allow standard 105 mm HE M1 ammunition to be fired for training purposes.

The first batch of L118 105 mm Light Guns, together with a quantity of ammunition, spare parts and kits to convert the weapons to the L119 configuration for training purposes, were shipped in January 1996.

By early 2002, over 1,000 105 mm Light Guns had been built, including those produced under licence in Australia and the USA.

Late in 1996, following an extensive series of user trials between the French Giat Industries 105 mm LG1 Mk II Light Gun and the UK's 105 mm Light Gun, the latter was selected to meet the requirements of the Portuguese Army.

The contract for 21 105 mm L119 Light Guns was signed late in December 1997 and the first batch of eight guns was shipped in February 1998.

Portugal selected the 105 mm L119 gun which fires standard 105 mm HE M1 semi-fixed high-explosive ammunition. The Portuguese order included a 100 per cent offset.

The Royal Malaysian Army is expected to issue a requirement for a new 105 mm light artillery system to replace its current 105 mm Otobreda Pack Howitzers.

To meet this future requirement, which could eventually be for more than 100 weapons, BAE Systems, RO Defence signed a Memorandum of Understanding (MoU) under which the Malaysian Naval Dockyard Sdn Bhd (NDSB), a subsidiary of Penang Shipbuilding and Construction Sdn Bhd, would collaborate in the promotion, manufacture and supply of the BAE Systems, RO Defence 105 mm Light Gun.

Malaysia would start off with the assembly of 105 mm Light Guns from components supplied by BAE Systems, RO Defence and then progress to shared manufacture. For the Malaysian market the 105 mm Light Gun is referred to as the SAKTI 105 mm.

The signing of the Confidentially Agreement between BAE Systems and Penang Shipbuilding and Construction/Naval Dockyard took place in April 2000 during which the first handover of the product drawings for the SAKTI 105 mm Light Gun took place. Malaysia has selected the L119 version of the Light Gun.

It has been confirmed that Bosnia-Herzegovina has taken delivery of 18 105 mm L119 Light Guns from the United Arab Emirates.

BAE Systems, RO Defence, 105 mm Light Gun being towed by Pinzgauer (6 × 6) vehicle (Michael Jerchel)
2001/0105797

BAE Systems, RO Defence, 105 mm Light Gun L118 deployed in the firing position with turntable in front (Michael Jerchel)
2002/0130015

Description

The 105 mm Light Gun consists of the following main components: elevating mass, saddle and controls, trail and spades, platform, suspension and sighting system.

The elevating mass consists of the ordnance, recoil system, cradle including trunnions, balancing gear and electric firing gear. The gun is fitted with a multibaffle muzzle brake which can be easily removed for cleaning. The vertical sliding breech block is actuated by a lever mounted on the top. The hydropneumatic recoil system has a separate recuperator and the buffer is fitted with a cut-off gear to shorten the recoil length as elevation is increased. The recoil system is mounted in a lightweight fabricated cradle which carries the elevating arc and a simple helical compression spring-balancing gear. The firing gear, which is mounted on the cradle, is fully waterproofed and is of the electric magneto type, specially designed for use with the Cap Conducting Composition primer of the 105 mm Abbot ammunition system.

The saddle is a lightweight fabrication on which the complete elevating mass is carried. It provides a top traverse of 11°. The elevating gear is a simple mechanical system providing constant handwheel effort throughout the range of elevation. A clutch is provided to reduce abnormal firing loads on the equipment.

The trail is fabricated in high-strength corrosion-resistant steel. It is bow shaped, enabling the breech operator and loader to remain within the trail and maintain a high rate of fire at all elevations. The platform is carried on top of the trail during travelling. The forward end carries the saddle, traversing gear, suspension and the layer's seat. At the rear are the non-rotating towing eye, overrun brake, barrel clamp and spades. The spade system, of which there are three types, caters for various firing and ground conditions. The first type is a combined rock and digger spade for use, normally with the platform, on rock or firm ground; the second is a field spade for use on very soft ground or when firing without a platform and the third is a snow spade.

The platform is a circular lightweight fabrication which gives a firm base and permanent gun stability under all conditions. The tyres run on the outer edge of the platform which is connected to the underside of the gun by four wire stays.

The wheels, which have large section 9.00 × 16 tyres, incorporate special hydraulically operated overrun brakes. This ensures safe towing at a high speed with a light vehicle such as a Land Rover (4 × 4). The brakes can also be operated during firing by a lever at the rear end of the trail. For manhandling purposes, an individual handbrake is fitted at each wheel. The trailing arm suspension has laminated torsion springs and shock-absorbers. The suspension system remains in operation during firing and assists in maintaining the stability of the equipment and reducing carriage stresses.

Indirect and direct sight system and controls are operated by the layer while seated. Indirect laying has been simplified by the abolition of a separate setting for angle of sight. The quadrant elevation is set on the elevation scale. A direct fire telescope is fitted which incorporates a moving illuminated graticule which is adjusted for lead to allow for target movement and range. A direct fire night sight is also available. Trilux activated light sources are fitted to illuminate all scales and graticules, eliminating the use of electric batteries on the equipment.

The L35 charge contains the first five charge zones in a single brass cartridge case. The propellant charge is ignited by the L10 electrical primer which is more reliable than a percussion primer. The Supercharge, L36, for maximum range, has a separate brass case and also has an L10 electrical primer. Polycarbonate spoiler rings can be clipped onto the nose of the projectile enabling targets to be engaged between 2,400 m and 3,600 m; these reduce the range by causing aerodynamic drag.

In mid-1986, it was revealed that BAE Systems, RO Defence was working on extended-range (base bleed) and cargo rounds as a private venture. The former has been fired with standard Supercharge L36 and has achieved a range of over 20,000 m compared to the 17,200 m of the standard projectile. The base bleed round is currently designated the X/RO 0381 ERHE which during trials in 1998 achieved a range of around 20.60 km.

The 105 mm Light Gun fires the following BAE Systems, RO Defence manufactured separate loading ammunition: HE (L31) with the projectile weighing 16.1 kg; HESH (L42) with the projectile weighing 10.49 kg (no longer produced by BAE Systems, RO Defence); illuminating (L43); SH/practice (L41 – no longer produced by BAE Systems, RO Defence); smoke (L45) with the projectile weighing 15.89 kg; smoke, coloured red (L37) and smoke, coloured orange (L38).

The charge system for the L118 Light Gun is as follows:

Charge 1	2,500 to 5,700 m (min range using shell spoilers)
Charge 2	2,700 to 7,200 m (min range using shell spoilers)
Charge 3	5,900 to 9,500 m
Charge 4	7,900 to 12,200 m
Charge 4½	8,700 to 13,600 m
Charge 5	15,300 m max
Charge Super	17,200 m max

It can also fire US 105 mm M1 semi-fixed ammunition by using a complete interchangeable percussion-fired ordnance (L119) which is available. The charge coverage is as follows:

Charge 1	1,800 to 3,400 m
Charge 2	2,100 to 4,100 m
Charge 3	2,600 to 4,800 m
Charge 4	4,100 to 6,300 m
Charge 5	4,200 to 7,900 m
Charge 6	5,000 to 9,500 m
Charge 7	6,200 to 11,500 m

A range of 14,300 m can be achieved by the L119 firing the M760 which is basically an M1 projectile with an M200 charge system. A range of 19,500 m can be achieved with the M913 HERA projectile and 19,000 m with the new ERM1 base bleed projectile.

The 105 mm Light Gun can be carried slung under a Puma helicopter as one load. The elevating mass can be removed to give two light helicopter (for example, Wessex) loads and the complete weapon can be reassembled with one simple tool in less than 30 minutes.

The 105 mm Light Gun is normally towed with the gun in the forward (firing) position. For long distances the gun is towed with the barrel rotated through 180° and clamped to the trail. This attitude is achieved by jacking up the equipment with a jack stowed on the trail, removing the quick-release wheel and traverse gear pin, clamping the barrel and replacing the wheel. The complete conversion takes less than 1 minute.

Firing US M1 105 mm ammunition and using the M200 charge system originally developed for the M204, a range of 14,000 m is achieved. Firing the M913 HERA cartridge, with the M229 charge 8 propelling charge, a range of 19,500 m is achieved. US changes include: modifications to the sight to permit use of the deflection method of laying, revision of the graticule pattern of the sight to incorporate 5 mil graduations instead of 10 mil, addition of attachment brackets for the battery computer, provision of attachment brackets for M90 muzzle velocity measurement system, labelling of all items containing radioactive elements (tritium sights), modification of tie down brackets, use of standard US paint type and US camouflage pattern, use of nitrogen in the recoil system instead of air, use of US tail-light patterns, incorporation of recent UK MoD minor modifications, elevation level guidance plate for M200 charge system, modified engraving and limited US thread forms in some areas of the weapon.

Variants

To meet the requirements of the Swiss Army the L127A1 Light Gun was developed by BAE Systems, RO Defence, with the ordnance being designated the L27A1. This is very similar to the British Army version but can fire Swiss ammunition. There were two guns delivered to Switzerland in 1979-80 with a further four following in 1981, but no production order was placed.

Enhanced ammunition for 105 mm Light Gun

The UK Defence Procurement Agency (DPA) has awarded a sole source contract to BAE Systems, RO Defence, to manage the assessment phase for enhanced ammunition for the 105 mm L118 Light Gun deployed by the Royal Artillery.

The British Army's future Light Mobile Artillery Weapons System (Gun) was originally expected to concentrate only on the BAE Systems, RO Defence 155 mm/39 calibre XM777 now under development for the US Army and Marines as the Light Towed Howitzer (LTH).

In some tactical situations the 155 mm/39 calibre XM777 may be difficult to deploy and the 105 mm L118 may well be retained in service with some units being dual capable.

BAE Systems, RO Defence, 105 mm Light Gun L118 being towed by the latest Alvis Vehicles SUPERCAT III (6 × 6) vehicle which is also towing a trailer with additional ammunition **2002**/0130014

Specifications (L118)

Calibre: 105 mm
Muzzle brake: double baffle
Recoil system: hydropneumatic
Recoil length:
(at 0° QE) 1.07 m
(at 70° QE) 0.33 m
Breech mechanism: vertical sliding
Carriage: bow
Shield: none
Weight:
(travelling order) 1,860 kg
(firing position) 1,860 kg
(load on towing eye) 168 kg
(elevating mass) 1,066 kg
(lower carriage and essential stores) 794 kg
Length:
(travelling, gun forward) 6.629 m
(travelling, gun over trail) 4.876 m
(firing, gun at 0°) 7.01 m
Width: 1.778 m
Height:
(travelling, ordnance forward) 2.63 m
(travelling, ordnance trails) 1.371 m
Ground clearance: 0.5 m
Track: 1.4 m
Tyres: 9.00 × 16
Elevation/depression: +70/–5.5°
Traverse:
(total) 11°
(on platform) 360°
Rate of fire:
(for 1 min) 12 rds/min
(for 3 min) 6 rds/min
(sustained) 3 rds/min
Max range: 17,200 m
Min range: 2,500 m

Firing a 105 mm L31 high-explosive projectile with Charge Super, a maximum range if 17,200 m can be achieved by the L118. Some years ago, RO tested a base bleed projectile (ERHE2) with a range of over 21,000 m, but at that time there was no market requirement for this projectile and development ceased.

It is expected that the DPA assessment contract will include a market survey of 105 mm ammunition (projectile and charge) which could be in service by 2004 with the main emphasis being on increased lethality, although enhanced range will also be taken into account. It may also include firepower, mobility and sustainability trials.

For several years BAE Systems, RO Defence, have as a privet venture, been working on a Mid Life Upgrade (MLU) for the Light Gun (see elsewhere in this entry). This mainly concentrates on improving the RAM-D (Reliability, Availability, Maintainability and Durability) of the system as well as reducing into action times. Range and lethality improvements are not part of the MLU, as this concentrated on the weapon.

It is considered that South Africa could be in a good position, as in 1999 the LIW Division of Denel revealed its 105 mm Light Experimental Ordnance (LEO).

This has a 105 mm/52 calibre ordnance and fires new ammunition with Naschem being responsible for the family of new projectiles and Somchem for the modular charge system. The new 105 mm High Explosive Pre-Formed Fragmentation (HE-PFF) projectile is more lethal than many current 155 mm projectiles.

105 mm Light Gun upgrade (Mid Life Upgrade)

BAE Systems, RO Defence, has developed an upgrade package for the L118 and L119 Light Gun which can be incorporated into new production weapons or retrofitted to existing weapons.

The main improvements can be summarised as:

(1) Currently one of the wheels has to be removed when rotating the gun to its stowed position. This is time consuming and adds to the burden of the gun crew. Instead of removing the wheel slide it partly out of the axle and let it hinge down. When the gun is rotated simply lift the wheel, slide it back into position and lock it in place

(2) The current platform must be handled by the crew when bringing the weapon into action. The platform is relatively heavy and weight saving would reduce crew fatigue and improve overall gun performance

A new platform has been constructed from aluminium and thus weight is significantly reduced

(3) The maintenance burden of the existing recoil system is relatively high and a maintenance free recoil system has been introduced with fewer parts and modern sealing technology

(4) When hosing the gun down with high pressure hose jets water can ingress into the gearbox. A thicker cover plate has been introduced with a modern sealing capacity

(5) Under some circumstances there is difficulty in attaching the 105 mm Light Gun to the towing vehicle and there is limited crew assistance for a large traverse movement. Handgrips have been introduced to the gun trails to assist the user

(6) The existing brake system is maintenance intensive and the handbrake lever can be difficult to operate. An updated brake system is introduced for easier user operation

(7) The existing A frame under the forward part of the weapon can become distorted and has difficult to engage locking pins. Into and out of action times are slower due to pin engagement difficulties. The existing A frame is replaced with a single strut and centre traverse lock

Differences between 105 mm Light Gun Ordnances

Weapon designation	Ordnance designation	Rifling length	Number of grooves	Twist	Firing mechanism	Muzzle brake	Max range
L118A1	L19A1	3.21 m	28	1 in 19 constant	electric	double baffle	17,200 m
L119A1	L20A1	2.779 m	36	1 in 18 to 1 in 35	percussion	single baffle	11,500 m
L127A1	L27A1	3.052 m	32	1 in 25.6	percussion	double baffle	n/av

(8) The clearance between the gun and towing vehicle can cause user problems with towing and manoeuvring. A new extended towing eye is introduced.

105 mm Light Gun with APS

Early in 1999, following an international competition, the UK MoD selected the now BAE Systems, Avionics, Ring Laser Gyro (RLG) based Automatic Pointing System (APS) to meet the requirements of the Royal Artillery

The contract, valued at over £10 million, was for the supply of 137 APS, or Laser Inertial Automatic Pointing Systems (LINAPS) as it is called by the company, for installation on the BAE Systems, RO Defence 105 mm L118 Light Guns used by the Royal Artillery.

First production LINAPS were delivered late in 1999 and deliveries ran to the year 2002. The contract also covers logistic support for an inertial period of five years.

The LINAPS consists of the proven FIN3110L integrated Inertial/Global Positioning System (GPS), a large screen Layers Display and Control Unit (LDCU), a wheel hub mounted odometer and a Power Management System (PMS). In addition there is a Navigation Display Unit (NDU) for use in the cab of the Steyr-Daimler-Puch Pinzgauer Turbo D 4 x 4 gun towing vehicle.

The mounting of the INS/GPS directly above the barrel permits the direct measurement of all elevation and azimuth and hence the rapid and accurate pointing of the gun onto the target. In addition, LINAPS provides the precise location of the 105 mm gun platform with a route navigation capability.

It provides the crew with orientation and location data without the requirement for much of the existing sighting and survey equipment and its associated training.

The installation of the APS on the BAE Systems, RO Defence 105 mm Light Gun L118 will make for a much reduced into action time as well as greater accuracy.

Status

Production as required. In service as follows:

Country	Quantity	Comment
Abu Dhabi	63	less eight supplied to Netherlands in 1995 and 18 to Bosnia-Herzegovina
Australia	111	of which 105 were built under licence
Bahrain	8	
Bosnia-Herzegovina	18	L118 from UAE
Botswana	36	Including 12 delivered in 1996
Brazil	58	40 army, 18 marines
Dubai	18	
Ireland	24	including 12 delivered in 1998
Kenya	58	
Malawi	9	
Morocco	36	refurbished
Nepal	8	delivered 1992
Netherlands	8	from Abu Dhabi in 1995
New Zealand	24	built in Australia under licence
Oman	42	
Portugal	21	deliveries from 1998
Spain	56	ordered 1995, first deliveries 1996 and all delivered by 1998
UK	168	including reserve and training
USA	147	manufactured by BAE Systems, RO Defence
	280	manufactured under licence in the USA

BAE Systems, RO Defence and RDM Technology upgrades

BAE Systems, RO Defence has teamed with RDM Technology of the Netherlands to offer two upgrades for the older US 105 mm M101 towed howitzer that uses the ordnance of the 105 mm Light Gun in its L118 and L119 versions. Details of these upgrades are given earlier in this section under the Netherlands.

BAE Systems, RO Defence has supplied the 105 mm barrels for the Canadian M101 upgrade. For the more recent Chilean M101 upgrade, BAE Systems, RO Defence is the prime contractor and is supplying the barrels, with RDM Technology being responsible for the other components.

Manufacturer

BAE Systems, RO Defence.

UPDATED

UNITED STATES OF AMERICA

8 in howitzer M115

Development

The Westervelt Board recommended the development of a howitzer with a calibre of approximately 8 in (203 mm) mounted on a carriage with an elevation of 0 to +65°, a traverse of 360° and a maximum range of 18,000 yards (16,800 m). It was also stated that it would be desirable for a carriage to be developed which could mount either an 8 in howitzer or a 155 mm gun. There were two 8 in howitzers subsequently developed, the M1920 and M1920M1, plus a carriage which could also be used for a 155 mm gun.

Owing to a shortage of funds the project was suspended until 1927. Between 1927 and 1929 the carriage was redesigned for high-speed transport and the breech ring of the 8 in howitzer modified. The howitzer was known as the T2 and was to have a centrifugally cast barrel, but as facilities for centrifugal casting were not then available, it was decided to design a barrel of cold-worked forged steel under the designation T3. Tests of the T3 howitzer began at Aberdeen Proving Ground, Maryland in 1939 and the following year the T3 howitzer was standardised as the 8 in Howitzer M1 and the 155 mm Gun/8 in Howitzer Carriage T2E1 with minor modifications was standardised as the M1.

After the Second World War the complete weapon was redesignated the Howitzer, Heavy, Towed: 8 in: M115. This consisted of Cannon Howitzer M2 or M2A1, Recoil Mechanisms M4 series (M4, M4A1, M4A2 or M4A3) and Carriage M1.

8 in howitzer M115 in travelling configuration and with limber attached (Richard Stickland) ***2002***/0130013

Description

The M2/M2A1 8 in cannon consists of the barrel, breech ring and breech mechanism. The M2A1 cannon is ballistically and physically interchangeable with the cannon M2, the major difference being that the M2A1 has been constructed with stronger steels. The carriage consists of equilibrator assemblies, elevating and traversing mechanisms, two single-wheel, single-axle heavy limber M5, two-axle bogie with eight tyres and two trails. Four spades, carried on the trails, are used to emplace the weapon. In recent years some armies have towed the weapon without the limber.

The 8 in howitzer M115 fires the following separate loading ammunition:

HE (M106) with the projectile weighing 92.53 kg, maximum muzzle velocity of 587 m/s to a maximum range (charge 7) of 16,800 m

HE (M404) (carries 104 M43A1 grenades) with the projectile weighing 90.72 kg, maximum muzzle velocity of 587 m/s to a maximum range (charge 7) of 16,800 m

HE (M509) (carries 195 M42 grenades) with the projectile weighing 93.66 kg, maximum muzzle velocity of 594.4 m/s to a maximum range of 16,000 m. It also fired a tactical nuclear projectile but these have now been phased out of service.

The carriage of the 8 in howitzer is also used to mount the 155 mm Gun M1 (Long Tom), which is still in use with some armies. In many countries the 8 in M115 has been replaced by a self-propelled version called the M110, details of which are given in the Self-propelled guns and howitzers (tracked) section. Versions of the M110 with the longer barrel are the M110A1 (no muzzle brake) and the M110A2 (fitted with muzzle brake). The 8 in M110 series of self-propelled howitzers were finally phased out of service with the US Army (Regular) in March 1995. The last unit was the 5th Battalion, 18th Field Artillery, based at Fort Sill, Oklahoma.

In most countries the M115 and the M110 have been phased out of service due to short range. In NATO service, their prime mission was to fire tactical nuclear rounds which have now been phased out of service.

Specifications
Calibre: 203.2 mm
Barrel length: 5.142 m
Muzzle brake: none
Recoil system: hydropneumatic
Breech mechanism: interrupted screw stepped thread
Carriage: split trail with limber
Shield: no
Weight:
(travelling order) 14,515 kg
(firing position) 13,471 kg
Length: (travelling) 10.972 m
Width:
(travelling) 2.844 m
(firing) 6.857 m
Height:
(travelling) 2.743 m
Ground clearance: 0.318 m
Track: 1.98 m
Tyres: 11.00 × 20
Elevation/depression: +65/–2°
Traverse:
(total) 60°
Rate of fire:
(max) 1 rd/min
(sustained) 1 rd/2 min
Max range: 16,800 m
Crew: 14
Towing vehicle: M125 (6 × 6) truck

Status
Production complete. In service with the following countries, in some cases these are held in reserve.

Country	Quantity	Comment
Denmark	12	
Iran	20	
Jordan	25	reserve
South Korea	50	estimate
Pakistan	26	
Saudi Arabia	8	reserve
Taiwan	70	
Turkey	162	

UPDATED

155 mm howitzer M198

Development
Development of a new 155 mm towed howitzer to replace the 155 mm towed howitzer M114/M114A1 began in 1968 under the direction of the Project Manager for Cannon Artillery Weapon Systems (CAWS) of the US Army Armament Command at Rock Island, Illinois.

A test rig was built to demonstrate how a lightweight towed 155 mm howitzer could fire the increased range of ammunition, with reasonable probability of stable dynamics and structural soundness. Design and fabrication of an advanced development prototype began in 1969 and firing trials commenced in 1970.

Rock Island was responsible for the carriage and recoil mechanism, Frankford Arsenal for the fire-control equipment, Watervliet Arsenal for the 155 mm ordnance and Picatinny Arsenal and the Harry Diamond Laboratory for the 155 mm ammunition.

The first two prototypes were delivered in April and May 1972 and were followed by a further eight prototypes.

After trials, the XM198 was standardised as the M198 and production commenced at Rock Island Arsenal in 1978 with first production weapons completed in July 1978.

The M198 replaced M114A1s in separate field artillery batteries assigned to corps/army and the general support field artillery battalions of the light divisions.

The first battalion of M198s became operational at Fort Bragg in April 1979. The first Marine Corps battalion to be equipped with the M198 was the 2nd Battalion, 10th Marines, at Camp Lejeune.

Production of the M198 was carried out at Rock Island Arsenal which also manufactured all components with the exception of the 155 mm ordnance (which came from Watervliet Arsenal), fire-control system and tyres.

In February 1992, the 1,672nd M198 rolled off the production line at Rock Island Arsenal. During FY91, a total of 120 were manufactured followed by 120 more in FY92.

In June 1992, Watervliet Arsenal was awarded a US$5.402 million increment as part of a US$5.509 million firm fixed-price contract for 292 product improvement kits for the M198 towed howitzer with all work completed by April 1995.

In December 1993, Watervliet Arsenal was awarded a US$5.525 million increment as part of a US$10.730 million firm fixed-price contract for 602 product improvement kits for the M198 towed howitzer (quantity increased from the original 292) with the work completed by April 1995.

Production of the M198 has now been completed although it could commence again if further orders were placed.

According to United Nations' sources, between 1992 and 2000 the United States exported M198 weapons to only two countries. These were Bahrain (19 delivered in 1992) and Pakistan (24 in 1996).

In US Army and Marine Corps service the 155 mm M198 will eventually be replaced by the (now BAE Systems, RO Defence) 155 mm Ultralightweight Field Howitzer which was selected in March 1997 following

Specifications
Calibre: 155 mm
Barrel length: 6.096 m
Muzzle brake: double baffle
Recoil: hydropneumatic
Breech mechanism: screw
Carriage: split trail
Shield: no
Weight:
(travelling order) 7,163 kg
(firing position) 7,163 kg
Length:
(travelling) 12.34 m
(firing) 11 m
(with ordnance over trails) 7.44 m
Width:
(travelling) 2.794 m
(firing) 8.534 m
Height:
(travelling) 2.9 m
(min firing) 1.803 m
(with ordnance over trails) 2.121 m
Ground clearance:
(travelling) 0.33 m
Track: 2.362 m
Tyres: 16.5 × 19.5
Elevation/depression: +72/–5°
Traverse: (total) 45°
Rate of fire: 4 rds/min
Max range:
(M107 projectile) 18,150 m (Z8 charge)
(M483A1 projectile) 22,000 m (Z8C charge)
(M549A1 RAP) 30,000 m (Z8S charge)
Towing speed:
(cross-country) 8 km/h
(improved roads) 72 km/h
(secondary roads) 40 to 48 km/h
Crew: 11
Towing vehicle: M939 5 ton (6 × 6) truck

155 mm howitzer M198 in firing position

155 mm howitzer M198 in service with Ecuador and with ordnance to the rear
2002/0130012

extensive trials with competing weapons. In US Army/ Marines service, the UFH has been designated the XM777 Lightweight 155 mm.Towed Howitzer and upon full type classification this will become the M777.

Description

The M198 weapon is mounted on a split trail carriage, fitted with a two-position rigid suspension system which can be rotated upwards to raise the wheels. This allows the weapon to rest on a non-anchored firing platform. When travelling the ordnance remains forward, but for storage it can be traversed through 180° and locked in position over the closed trails.

The carriage consists of the top carriage, elevating mechanism, equilibrators, traversing mechanism, bottom carriage, wheel suspension assembly, speed shift, trails and firing base. An auxiliary power unit, the MC-4000, has been developed for use on the M198 during amphibious operations, but no procurement has yet been made.

The recoil mechanism is of the hydropneumatic type with a variable recoil length. The autofrettaged ordnance has a screw-on double-baffle muzzle brake, and a screw breech mechanism with a conventional obturator pad and split ring seal obturator.

The fire-control equipment includes an M137 panoramic telescope with a magnification of x4 and a 10° field of view, two elevation quadrants (M17 and M18) and an M138 elbow telescope with a magnification of x8 and a 6° field of view.

The internal configuration of the Cannon Assembly M199 is in accordance with the German/British/ American/Italian quadrilateral agreement and will fire all current standard and developmental 155 m ammunition of the anti-material and anti-personnel types. The weapon can fire the following types of 155 mm ammunition:

Anti-tank (M718) (containing nine anti-tank mines) with the projectile weighing 46.72 kg, a muzzle velocity of 660 m/s and a maximum range of 17,740 m

Anti-tank (M741) (containing nine anti-tank mines) with the projectile weighing 46.72 kg, muzzle velocity of 660 m/s and a maximum range of 17,740 m

Copperhead Cannon Launch Guided Projectile (M712) to a range of 16,400 m

HE (M107) with the projectile weighing 42.91 kg, muzzle velocity of 684.3 m/s and, maximum range of 18,100 m

HE (M449) (containing 60 anti-personnel grenades) with the projectile weighing 43.09 kg and a maximum range of 18,100 m

HE (M483) (containing 88 dual-purpose grenades) with the projectile weighing 46.53 kg, muzzle velocity of 660 m/s and a maximum range of 17,500 m

HE (M692) (containing 36 anti-personnel mines) with the projectile weighing 46.5 kg, muzzle velocity of 660 m/s and a maximum range of 17,740 m

HE (M731) (containing 36 anti-personnel mines) with the projectile weighing 46.5 kg, muzzle velocity of 660 m/s and a maximum range of 17,740 m

HE (M795) registration projectile for M483 and maximum range of 22,400 m

HERA (M549 series) (RAP) with the projectile weighing 43.54 kg, maximum muzzle velocity of 826 m/s and amaximum range of 30,000 m

DPICM (Dual-Purpose Improved Conventional Munition) M864 (type-classified late in 1987)

Illuminating (M118 series) with the projectile weighing 46.26 kg, muzzle velocity of 684.3 m/s and a maximum range of 18,100 m

Illuminating (M485 series) with the projectile weighing 41.73 kg, muzzle velocity of 684 m/s and a maximum range of 17,500 m

REMBASS (XM694E1), under development

Smoke WP (M110 series) with the projectile weighing 44.4 kg, muzzle velocity of 684.3 m/s and a maximum range of 18,100 m

Smoke BE (M116 series) with the projectile weighing 42.22 kg, muzzle velocity of 684.3 m/s and a maximum range of 18,100 m

Smoke (M825), base ejection, with a muzzle velocity of 797 m/s and a maximum range of 22,600 m

It will also fire the latest 155 mm natures of ammunition such as the SADARM (Sense And Destroy Armor) and the future XM982 Extended Range Projectile now under development by Raytheon Systems. This is not expected to enter service for some years as it is still at the Engineering Manufacturing and Development (EMD) phase.

Improved M198

The US Army M198 improvement programme is a three tiered approach to extend the useful life of, increase the reliability and modernised the M198.

A Service Life Extension Programme (SLEP), consisting of grease fittings for the trunnion bearings and nitrogen gauges for the recoil mechanism, will lengthen the useful life of the M198.

A refurbishment programme will correct existing maintenance deficiencies and improve the readiness posture of the howitzer at significant cost savings over a depot overhaul.

This refurbishment will be a systematic process for inspecting and maintaining the M198 to identify and correct impending failures associated with normal usage and apply system modifications.

In addition, the 72 M198s used by the XVIII Airborne Corps will be rebuilt with major components that have passed a depot inspection. These M198s will receive more thorough checks to ensure that there are no catastrophic damages because of the robust operational environment experienced by XVIII Airborne Corps.

The M198 Product Improvement Plan (PIP) includes modifications and improvements to the M198 chassis, bringing the fleet up to Force XXI objectives.

The M198 PIP will include an automated fire-control system designed to interface with AFATDS and be interoperable with the new 155 mm XM982 projectile currently under development by Raytheon Systems Company.

It will also include the application of the Hydraulic Power Assist Kit (HYPAK) and a direct fire sight. A trail bogey wheel will aid the airloading of the M198 howitzer while a simple ammunition transfer mechanism is proposed to ease some of the traditional ammunition handling burden.

It is expected that a total of 275 M198 howitzers will be improved between FY00 and FY05 at a cost of US$276, 000 per gun.

Various other upgrades to the M198 have been proposed, including a new 155 mm/52 calibre ordnance, but none of these have passed the concept stage *(Jane's Armour and Artillery 1999-2000)*.

Status

Production complete but can be resumed if required. In service with following countries:

Country	Quantity	Comment
Australia	35	
Bahrain	28	including 19 delivered in 1992
Ecuador	12	
Honduras	4	
Lebanon	36	status uncertain
Morocco	26	
Pakistan	124	including 24 delivered in 1996
Saudi Arabia	66	
Somalia	18	status uncertain
Thailand	62	
Tunisia	57	
USA	478	Army
	599	Marines

Manufacturer

Rock Island Arsenal, Illinois.

UPDATED

155 mm howitzer M114

Development

In 1934, the development of a new split trail carriage was authorised under the designation T3. Development of the T3 was subsequently cancelled in favour of a new 155 mm howitzer (the T3) and a new carriage (the T2), which were standardised as the 155 mm Howitzer M1 and Carriage M1 in April 1941. The carriage was the same as that used for the 4.5 in field gun. By the time production of the M1 had been completed, just over 6,000 had been built.

After the Second World War the complete 155 mm weapon was redesignated the Howitzer, Medium, Towed: 155 mm, M114, consisting of the Cannon (M1 or M1A1), Carriage (M1A1) and Recoil System (M6, M6A1, M6B1 or M6B2). The M114A1 is almost identical to the M114 except that it has an M1A2 carriage.

According to United Nations' sources, between 1992 and 2000 the United States exported surplus M114s to Bosnia, which took delivery of 126 systems in 1997.

Description

The howitzer consists of the breech ring, breech block, breech mechanism and barrel. The carriage M1A1 or M1A2 consists of top and bottom carriages, wheels and brake mechanisms, trails, equilibrators and the firing jack. The carriage is of the unsprung two-wheel split trail type with the wheels fitted with combat tyres for high-speed towing. The carriage is equipped with airbrakes, which are operated from the prime mover, and there are handbrakes on each wheel for parking.

US Army M114 155 mm howitzer in firing position (US Army)

US Marine Corps M114 155 mm howitzer in travelling configuration without shield (Ray Young)

Specifications

Calibre: 155 mm
Barrel length:
(muzzle to rear end of breech mechanism) 3.778 m
(excluding breech) 3.626 m
Muzzle brake: no
Recoil system: hydropneumatic
Breech mechanism: stepped thread, interrupted screw
Carriage: split trail
Shield: optional
Weight:
(travelling order) 5,800 kg
(firing position) 5,760 kg
Length:
(travelling) 7.315 m
Width:
(travelling) 2.438 m
Height:
(travelling) 1.803 m
Axis of bore:
(at 0°) 1.676 m
Ground clearance: 0.229 m
Track: 2.07 m
Tyres: 14.00 × 20
Elevation/depression: +63/–2°
Traverse:
(right) 25°
(left) 24°
Rate of fire:
(first 30 s) 2 rounds
(first 4 min) 8 rounds
(first 10 min) 16 rounds
(per hour, sustained) 40 rounds
Max range: 14,600 m
Crew : 11
Towing vehicle: 5 ton (6 × 6) truck

155 mm M114A2 towed howitzer (US Army) 0008270

The barrel is balanced by spring-type equilibrators. The recoil mechanism, elevating and traversing mechanism, left and right shields, telescope mount and panoramic telescope are attached to the top carriage. The bottom carriage supports the weight of the top carriage and the barrel, and transmits operating stresses to the firing jack and trails.

The counter-recoil system consists of the recuperator cylinder, counter-recoil cylinder, and the counter-recoil and recuperator cylinder head stuffing box which links the two cylinders hydraulically. The recoil system consists of the recoil cylinder, replenisher and variable recoil cam assembly.

The M1 barrel is similar to the M1A1 but the M1A1 uses steels with increased physical properties. The M1A1 carriage has a rack-and-pinion-type firing jack while the M1A2 has a screw-type firing jack. The travelling lock of the M1A2 is equipped with a firing jack hanger whereas that of the M1A1 is not.

The M114 fires the following types of US 155 mm separate-loading ammunition:
HE (M107) with the projectile weighing 42.91 kg, maximum muzzle velocity of 563.9 m/s and a maximum range (charge 7) of 14,600 m
HE (M449) (carries 60 anti-personnel grenades) with the projectile weighing 43.09 kg, maximum muzzle velocity of 563.9 m/s and a maximum range (charge 7) of 14,600 m
Illuminating (M485) with the projectile weighing 41.73 kg, maximum muzzle velocity of 576 m/s and a maximum range (charge 7) of 13,600 m
Smoke (M116 series) with the projectile weighing 42.22 kg, maximum muzzle velocity of 563.9 m/s and a maximum range of 14,600 m
Smoke WP (M110 series) with the projectile weighing 44.4 kg, maximum muzzle velocity of 563.9 m/s to a maximum range (charge 7) of 14,600 m
M712 Copperhead Cannon Launched Guided Projectile
M804 practice with a maximum range of 14,600 m.
Note: Not all these types of 155 mm ammunition are in current use.

Variants

M114A2

It was originally understood that the M114A2 had the complete ordnance of the 155 mm M198 covered in detail in the previous entry.

It has now been confirmed that the barrel of the 155 mm M114A2 is only slightly longer than that of the M114A1 with the key identification feature being a groove that is cut around the outside of the barrel about 50 to 70 mm from the end.

The main tube difference is internal with the M114A2 tube posessing a 1 in 20 twist as opposed to the 1 in 12 twist used in earlier models. The greater twist extends the range of existing projectiles and other than the tube there are no additional changes in the M114A2 design.

In mid-1997, the US government announced that it was to supply the Bosnian Federation with 116 ex-US Army 155 mm M114A2 towed howitzers held in reserve and all of these were delivered by the end of 1997. An additional 145 systems were earmarked for spare parts.

Modified M114 (Netherlands)

Details of the M114/39 Modified Howitzer originally developed by SRC International and now produced by RDM Technology are given in this section under the Netherlands. Denmark, the Netherlands and Norway have now completed conversion of 230 M114s to the M114/39 configuration. There have been no recent sales of this weapon.

Modified M114 (Israel)

Full details of this conversion, developed as a private venture by Soltam Systems of Israel, are given under Israel.

155 mm howitzer M65

This was the Yugoslav version of the M114 but was built only in small numbers. Details were given in *Jane's Armour and Artillery 1997-98*, page 746.

M114 (France)

Giat Industries of France has developed a conversion package for the M114, details are given under France. This upgrade package remains at the prototype stage.

KH179

Details of the South Korean KH179, a conversion of the M114, are given under Korea, South.

M114 (Iran)

Iran is understood to have built an upgraded version of the M114 with an M114/39 calibre barrel under the designation of the HM41. Available details are given under Iran.

Status

Production complete. In service with the following countries:

Country	Quantity	Comment
Bosnian Federation	116	delivered 1997
Brazil	96	army
	6	marines
Chile	36	used by Marines
Denmark	96	upgraded to M114/39 standard
Ecuador	12	
Greece	267	
Iran	70	estimate, plus upgraded version
Iraq	n/avail	
Israel	50	reserve
Italy	54	reserve
Jordan	30	
Korea, South	n/avail	also local M114 versions called KH179
Laos	n/avail	
Lebanon	15	
Morocco	46	
Netherlands	82	upgraded to M114/39 standard with 31 for disposal
	27	not upgraded, for disposal
Norway	48	upgraded to M114/39 standard
Pakistan	60	
Peru	36	
Philippines	12	
Portugal	37	
Saudi Arabia	n/avail	reserve
Singapore	16	reserve
Somalia	6	status uncertain
Spain	84	
Sudan	12	status uncertain
Taiwan	250	also local self-propelled model (see Self-propelled guns and howitzers (tracked) section)
Thailand	56	
Tunisia	12	
Turkey	517	
Uruguay	5	
Vietnam	n/avail	
Yemen	12	
Yugoslavia	145	including small number of M65
Venezuela	12	

UPDATED

105 mm towed howitzer M119A1

Development/Description

Following the success of the BAE Systems, RO Defence 105 mm L118 Light Gun during the Falkland Islands campaign, the US Army purchased operational testing weapons to determine the feasibility of using this weapon in the US Army's new light division concept.

The key in the US Army's evaluation was the ability to airlift an entire division artillery within the light division air transport constraints.

Following extensive trials, the US Army negotiated both a licence agreement, for production within the US and a production contract with BAE Systems, RO Defence Nottingham in 1987.

Deliveries to the US Army commenced in December 1989 with the First Unit Equipped (FUE) being the 7th Infantry Division, Fort Ord, California.

The M119A1 105 mm towed howitzer is essentially the BAE Systems, RO Defence 105 mm L119 Light Gun, covered in detail under the United Kingdom, with some modifications. For example, the installation of a US-type sighting system, to meet the requirements of the US Army.

Production of the 105 mm towed howitzer M119A1 was undertaken at Rock Island in the US with the 105 mm barrels being provided by Watervliet Arsenal.

The US Army describes the M119A1 as a lightweight, air mobile, air droppable, low-altitude air extraction weapon system that provides direct and indirect fire support for use in selected field artillery battalions in light infantry, airborne and air assault divisions.

Using standard 105 mm ammunition, the M119A1 can reach targets at a range of 14,300 m. The M119A1 replaced the older US developed M101A1 and M102 towed howitzers with increased range and lethality.

There have recently been a number of new 105 mm artillery projectiles developed in the US which give the M119/M119A1 a significant range advantage.

The M915 rocket assisted projectile significantly increases lethality and improves range by 40 per cent over the standard US M1 HE projectile.

In contrast to the older M548 RAP, which uses the low end M176 propellant, the M913 uses the M229 maximum propelling charge (Charge 8, high end). The increased range is obtained by using the M229 propellant charge coupled with a modern rocket motor which will function about 1.5 seconds after leaving the barrel. With the rocket on the M913 it provides a range of more than 19.5 km.

The M916 is the same projectile as the M915 but will be fired with the M67 propellant (low end charge).

Both the M915 and M916 projectiles have a payload consisting of 42 M80 sub-munitions which incorporate a self-destruct mechanism.

The onboard fire-control system consists of an M137 panoramic telescope with a magnification of ×4, M90A2 direct fire telescope with a magnification of ×3, M186 direct fire mount and the M187 telescope mount.

The M119A1 is normally towed by the AM General M1097 (4 × 4) Heavy HMMWV and can be quickly moved and employed to provide maximum fire power with minimum of combat loaded weight.

M119A1 105 mm towed howitzer deployed in the firing position (US Army) 0021135

M119A1 105 mm towed howitzer in travelling configuration being towed by an M1097 (4 × 4) Heavy HMMWV (Steven Zaloga) 0021136

Light Artillery System Improvement Programme (LASIP)

As the M119A1 will remain in service well into the 21st Century the US Army is now embarking on a major modernisation and improvement programme for the M119A1 weapon to enhance its RAM-D (Reliability, Availability, Maintainability and Durability) characteristics under the Light Artillery System Improvement Programme (LASIP).

There are two major LASIP Blocks, I and II. Block I includes modifications to low temperature recuperator, replacement brakes, retrofit modified trunnion adapter, improve rear firing strays, improve travelling strays, trail lift handles, modify the cam follower, modify the M187 sight mount and modify the firing safety mechanism lock.

Block II components include a new rammer/extractor tool, improved elevating handwheel, firing platform clamps, improved direct fire scope, improved elevation clutch, improved buffer compensating tubes and a widened track for increased stability.

Under current plans Block I LASIP is fully funded. Block II is only partially funded and there may well be some changes to the final list of improvements. It is expected that 415 weapons will be upgraded under Block I LASIP.

Specifications
(from US Army)
Calibre: 105 mm
Muzzle brake: single baffle
Recoil system: hydropneumatic
Breech mechanism: vertical sliding
Carriage: bow
Shield: none
Weight:
(combat) 1,860 kg
Length: 4.87 m
Width: 1.78 m
Height: 1.37 m
Ground clearance: 0.50 m
Track: 1.40 m
Tyres: 9.00 × 16
Elevation/depression: +70/–5.5°
Traverse:
(total) 11°
(on turntable) 360°
Rate of fire: 6 rds/min
Range:
(unassisted) 11,500 m
(unassisted zone 8) 14,000 m
(rocket-assisted projectile) 19,000 m
Prime mover: M1097 (4 × 4) Heavy HMMWV

Status

Production complete. In service with US Army. A total of 147 were supplied from the UK production line, with another 280 being built under licence in the USA.

Manufacturers

Rock Island Arsenal (USA).
BAE Systems, RO Defence.

UPDATED

105 mm howitzer M102

Development

The requirement for a light towed 105 mm howitzer to replace the 105 mm M101 (which is covered in detail in a separate entry) in selected units was issued in 1960.

The first prototype was completed at Rock Island Arsenal in 1962 and was type-classified as Standard A in December 1963. First production M102s were completed at Rock Island in December 1965 and the howitzer was first used in combat in South Vietnam the following month. The weapon subsequently underwent major modifications to solve design problems.

Over 1,200 M102s were built by Rock Island Arsenal before production was finally completed in the early 1970s. A further 22 were built for foreign military sales

Specifications
Calibre: 105 mm
Barrel length: 3.382 m
Muzzle brake: none
Recoil: hydropneumatic
Breech: vertical sliding wedge
Carriage: box trail
Shield: no
Weight: 1,496 kg
Length:
(travelling) 5.182 m
Width:
(travelling) 1.964 m
Height:
(travelling) 1.594 m
Ground clearance:
(travelling) 0.33 m
Tyres: 7.00 × 16
Elevation/depression: +75/–5°
Traverse: 360°
Rate of fire: 10 rds/min
Max range:
(with HE (M1)) 11,500 m
(with HERA (M548)) 15,100 m
Crew: 8
Towing vehicle: 6 × 6 truck

in FY80. The main advantage of the M102 over the earlier M101 is that it only weighs 1,496 kg compared to 2,258 kg. It can also be traversed rapidly through 360° to engage targets in other sectors.

In 1985, the US Army had 526 M102 howitzers in its inventory. The M102 has been replaced by the British BAE Systems, RO Defence 105 mm Light Gun, designated the M119 (L119 in British Army service) in US Army service. The US Army holds quantities of the M102 for overseas sales to authorised countries.

Figures released by the United Nations for the years 1992 to 2000 show that there were no exports of the M102 in this period.

As the M119/M119A1 has some improvements over the British 105 mm Light Gun, there is a separate entry under the USA for this weapon which is now under going the Light Artillery System Improvement Program (LASIP).

105 mm howitzer M102 deployed in firing position (Mike Green)

105 mm M102 towed howitzer in service with El Salvador in travelling configuration (Julio Montes)
2001/0105795

Description
The 105 mm howitzer M102 consists of the M137 cannon, M37 recoil mechanism, M31 carriage and fire-control system.

The two-wheeled box carriage is constructed of aluminium. In the firing position, a circular baseplate is lowered to the ground under the forward part of the carriage and the complete carriage can be quickly swung through 360° by means of a roller located at the rear of the trail assembly. The baseplate is staked to the ground before firing begins. The elevating mechanism utilises a pair of elevating screw assemblies and permits an elevation of +75° and depression of –5°. The variable-length recoil system used on this weapon eliminates the need for a recoil pit.

The fire-control system consists of the M1A2 quadrant, M14A1 quadrant, M113A1 panoramic telescope and an M114A1 elbow telescope. Radioactive light sources are used to illuminate the fire-control system. The tritium light sources are a heavy radioactive isotope of hydrogen gas, sealed in a pyrex glass container, coated with phosphor. Weapons Command initiated a self-illumination development programme at Frankford Arsenal, Delaware in January 1969, in response to numerous reports of poor instrument light performance in Vietnam. The equipment successfully completed testing in February 1970. Production contracts for the modification kits were awarded in September 1973 and the first kits were issued in 1975.

The M102 fires the same range of ammunition as the M101 but with a slightly higher muzzle velocity and, therefore, a slightly longer range:
APERS-T (M546) with the complete round weighing 17.35 kg, maximum muzzle velocity of 549 m/s and a maximum range (charge 7) of 12,400 m
HE (M1) with the complete round weighing 21.06 kg, maximum muzzle velocity of 494 m/s and a maximum range (charge 7) of 11,500 m
HE (M413) (carries 18 M35 grenades) with the complete round weighing 19.06 kg, maximum muzzle velocity of 494 m/s and a maximum range (charge 7) of 11,500 m
HE (M444) (carries 18 M39 grenades) with the complete round weighing 19.06 kg, maximum muzzle velocity of 494 m/s and a maximum range (charge 7) of 11,500 m
HEP (M327) with the complete round weighing 15.27 kg, maximum muzzle velocity of 559 m/s and a maximum range of 8,685 m
HERA (M548) (RAP) with the complete round weighing 17.46 kg, maximum muzzle velocity of 548 m/s and a maximum range of 15,100 m
Illuminating (M314 series) with the complete round weighing 21.06 kg, maximum muzzle velocity of 494 m/s and a maximum range (charge 7) of 11,500 m
Smoke (M60 series) with the complete round weighing 19.46 kg, maximum muzzle velocity of 494 m/s and a maximum range (charge 7) of 11,500 m
Smoke (M84 series) with the complete round weighing 19.03 kg, maximum muzzle velocity of 494 m/s and a maximum range (charge 7) of 11,500 m
HE (M760) to a maximum range of 14,000 m.

Variants
Early in 1994, RDM Technology of the Netherlands announced that it was developing an upgrade for the 105 mm M102 towed howitzer. Details of this, which has yet to be completed in prototype form, are given earlier in this section under the Netherlands.

Status
Production complete. In service with the following countries:

Country	Quantity	Comment
Brazil	19	
El Salvador	36	
Guatemala	8	
Honduras	24	
Jordan	50	
Lebanon	10	
Malaysia	40	reserve
Philippines	n/av	
Thailand	12	
Saudi Arabia	40	National Guard
Uruguay	48	includes M101
USA	413	reserve

Manufacturer
Rock Island Arsenal.

UPDATED

105 mm howitzer M101

Development

The development of the 105 mm M101 towed howitzer can be traced back to 1919 and full details were given in *Jane's Armour and Artillery 2000 to 2001* pages 745 to 746. It entered service with the US Army in 1940. Shortly afterwards the 105 mm Howitzer Carriage M2A1 was standardised.

By the end of the Second World War, 8,536 105 mm towed howitzers had been built and post-war production continued at Rock Island Arsenal until 1953, by which time 10,202 had been built. In 1985 the US Army had six M101 and 494 M101A1 howitzers in its inventory. The M101 105 mm howitzer was replaced in US Army service by the M102 howitzer which in turn has now been replaced by the BAE Systems, RO Defence 105 mm Light Gun. According to United Nations' sources, between 1992 and 2000 there were only three declared exports of the 105 mm howitzer M101 and these were as follows:

From	To	Quantity	Comment
Netherlands	Brazil	6	3 in 1994 and 3 in 1997
USA	Colombia	78	delivered 1992
USA	Macedonia	18	delivered 1999

105 mm M101 series towed howitzer of the Canadian Forces deployed in the firing position (Canadian Forces)

105 mm M101 series towed howitzer in travelling configuration (PB) **2000**/0085640

Description

After the Second World War the complete weapon was redesignated the Howitzer, Light, Towed: 105 mm, M101 and M101A1. The M101 consists of cannon M2A1 or M2A2, carriage M2A1 and recoil system (M2A1, M2A2, M2A3, M2A4 or M2A5). The M101A1 is identical apart from an M2A2 carriage. The difference between the cannon M2A1 and the M2A2, is that the M2A1's muzzle end is straight and the M2A2's is bell-shaped. The M2A2 carriage is fitted with a main shield group composed of right and left upper and lower shields, right and left top flaps, a bottom flap, and left and right auxiliary shields; while the earlier M2A1 is equipped with left and right main shields and a top shield.

The barrel consists of the barrel assembly, breech ring and locking ring. The carriage is of the single axle split trail type with a drawbar for securing to the prime mover. The carriage consists of the equilibrator, shield, elevating mechanism, cradle gears, elevating arcs, traversing mechanisms, top carriage, wheels and trails. Fire-control equipment consists of a Telescope Elbow M16A1D with a magnification of ×3 and a 13° field of view, a Telescope Panoramic M12A7S with a magnification of ×4 and a 10° field of view and quadrant M4A1.

The M101 fires the following semi-fixed ammunition:
APERS-T (M546) with the complete round weighing 17.35 kg, maximum muzzle velocity of 504 m/s and a maximum range (charge 7) of 11,600 m
HE (M1) with the complete round weighing 21.06 kg, maximum muzzle velocity of 472.4 m/s and a maximum range (charge 7) of 11,270 m
HE (M444) (carries 18 M39 grenades) with the complete round weighing 19.05 kg, maximum muzzle velocity of 472.4 m/s and a maximum range (charge 7) of 11,270 m
HEP (M327) with the complete round weighing 15.17 kg, maximum muzzle velocity of 559 m/s and a maximum range of 8,685 m
HERA (M548) (RAP) with the complete round weighing 17.46 kg, maximum muzzle velocity of 548 m/s and a maximum range of 14,600 m
Illuminating (M314 series) with the complete round weighing 21.06 kg, maximum muzzle velocity of 472.4 m/s and a maximum range (charge 7) of 11,270 m
Smoke (M60 series) with the complete round weighing 19.46 kg, maximum muzzle velocity of 472.4 m/s and a maximum range (charge 7) of 11,270 m
Smoke (M84 series) with the projectile weighing 19.03 kg, maximum muzzle velocity of 472.4 m/s and a maximum range of 11,270 m
APERS-T (M546) with a maximum range of 11,600 m.

Variants

M101 modified (Germany)

The German Rheinmetall company has fitted all M101s used by the Federal German Army with a new longer barrel with a single-baffle muzzle brake and new sights. It fires an HE projectile weighing 21.06 kg, with a maximum muzzle velocity of 600 m/s and a maximum range of 14,100 m.

Upgraded M101 (France)

Details of the Giat Industries upgraded M101 are given in this section under France. The first customer for this was Thailand. In 1996 the Philippines became the second customer and ordered 12 upgrade packages from Giat Industries. The total number of M101 systems upgraded by Thailand was 285. There is a separate entry under Thailand for the upgraded M101 which has the local designation of the M425.

Upgraded M101 (Greece)

In 2000 it was stated that Greece intended to upgrade at least 10 M101's used by the 116th Medium Artillery Squadron. The two contenders for this programme are

Specifications

Calibre: 105 mm
Barrel length:
(muzzle to rear face of breech ring) 2.574 m
(bore) 2.363 m
Muzzle brake: none
Recoil system: hydropneumatic
Breech mechanism: horizontal sliding wedge
Carriage: split trail
Shield: yes
Weight:
(travelling order, M101) 2,030 kg
(travelling order, M101A1) 2,258 kg
(firing position, M101) 2,030 kg
(firing position, M101A1) 2,258 kg
Length: (travelling) 5.991 m
Width:
(travelling) 2.159 m
(firing) 3.657 m
Height:
(travelling, M101) 1.524 m
(travelling, M101A1) 1.574 m
(firing) 3.124 m
Axis of bore:
(at 0°) 1.3 m
Ground clearance: 0.356 m
Track: 1.778 m
Tyres: 9.00 × 20
Elevation/depression: +66/–5°
Traverse:
(total) 46°
Rate of fire:
(max) 10 rds/min
(sustained) 3 rds/min
Max range: 11,270 m
Crew: 8
Towing vehicle: 2½ ton (6 × 6) truck

105 mm M101 towed howitzer in service with El Salvador (Julio Montes) ***2001***/0105794

Giat Industries of France and RDM Technology of the Netherlands.

Upgraded M101 (Korea, South)

Details of the M101 upgrade package from the Kia Machine Tool Company are given under Korea, South. In addition to being in service with the South Korean Army it is believed that Chile (Marines) have taken delivery of 16 of these which have the designation of the K-178.

Upgraded M101 (Netherlands)

Details of the RDM Technology M101 are given in this section under the Netherlands. The first customer for this upgrade was Canada, followed by Chile.

M7 105 mm self-propelled howitzer

This is armed with an M2 or M2A1 howitzer with an elevation of +35°, a depression of –5°, and a traverse of 30° left and 15° right. Details were given in *Jane's Armour and Artillery 1983-84*. It is believed that Pakistan still has 50 of these systems held in reserve.

Manufacturer

Rock Island Arsenal.

UPDATED

Status

Production complete. In service with the following countries:

Country	Quantity	Comment
Argentina	15	marines
Australia	246	
Austria	108	
Bangladesh	50	
Belgium	5	reserve
Benin	4	
Bolivia	30	
Brazil	300	estimate, army and marines, some being replaced by 105 mm Light Gun
Burkina Faso	8	
Cameroon	16	
Canada	185	96 upgraded
Chile	91	army and marines, most being upgraded, including 78 for the US in 1992
Colombia	130	including 78 delivered in 1992
Croatia	90	
Chad	5	
Denmark	134	
Dominican Republic	22	
Ecuador	80	
El Salvador	24	
Gabon	4	
Germany	114	modified by Rheinmetall (reserve)
Greece	445	10 to be upgraded
Guatemala	12	
Indonesia	170	
Iran	130	estimate
Israel	60	
Japan	36	
Korea, South	1,700	including local K-178 model
Laos	25	
Lebanon	15	
Libya	60	
Macedonia	18	
Mauritiana	35	
Mexico	76	
Morocco	20	
Mozambique	n/avail	
Myanmar	96	
Netherlands	1	reserve
New Zealand	19	
Norway	36	
Pakistan	300	
Paraguay	15	
Peru	13	
Philippines	230	
Portugal	49	
Saudi Arabia	100	estimate
Senegal	6	
Slovenia	8	
Sudan	n/avail	status uncertain
Taiwan	650	including local T64
Thailand	280+	including local models, some upgraded by Giat Industries
Togo	4	
Tunesia	48	
Turkey	n/avail	
Uruguay	48	including M102
USA	355	marines
Venezuela	40	
Vietnam	n/avail	
Yemen	35	
Yugoslavia	54	

YUGOSLAVIA, FEDERAL REPUBLIC

155 mm converted gun M46/84

Development/Description

The 155 mm Gun M46/84 is a conversion of the Russian 130 mm Gun M-46 (or the Chinese North Industries Corporation Type 59-1) to accommodate a new 45 calibre barrel capable of firing Extended Range Full Bore – Base Bleed ammunition.

This conversion gives a range increase of 45 per cent over the M-46 when firing ERFB-BB ammunition. The efficiency of the 155 mm ERFB projectile is about three times that of the older Russian-designed 130 mm OF-482 projectile.

It is believed that this upgrade package was originally designed by the former Space Research Corporation headed by the late Dr Gerald Bull. The converted gun M46/84 was also sold to Iraq (qv) but it is understood that no conversions had been carried out prior to August 1990.

The M46/84 conversion involved replacing the existing 130 mm barrel with a new 155 mm barrel 45 calibres long. This is designed to make use of the original 130 mm sliding breech and cradle parts.

The barrel is fitted with a high-efficiency double-baffle muzzle brake of a new design. An updated equilibrator sealing system is introduced, involving a nitrogen pressure adjustment and an updated sealing system for the recuperator mechanism is also required.

The existing 130 mm sights are also revised to accommodate the new 155 mm ballistics with the optical sight graticule being changed. A new range quadrant drum with respective range scales is mounted.

The conversion allows existing 130 mm M-46 and the China North Industries Corporation (NORINCO) Type 59-1 users to fire standard 155 mm M107, ERFB and ERFB-BB ammunition to the ranges shown in the table right.

Variants

The converted gun could also be delivered in 152 mm calibre with this model being designated the cal. 152 M46/86. It uses the same system of propelling charges and Yugoslav projectiles M84 and M84-GG. Maximum range with the M84 projectile is 27,500 m and with the M84-GG base bleed it is 34,000 m. Details of the Yugoslav M84 weapon are given later in this section.

At the request of customers, both versions of the converted gun could be supplied for the use of ammunition with separate loading propelling charges.

Status

Development complete. Ready for production. As far as it is known, no production/upgrade weapons were made before the start of the Yugoslav civil war and this system is no longer being marketed by Yugoimport SDPR.

Contractor

Yugoimport SDPR.

UPDATED

PROJECTILES

Charge zone	**ERFB-BB**		**ERFB**		**M107**	
	m/v	**range**	**m/v**	**range**	**m/v**	**range**
3			266 m/s	5,900 m	292 m/s	7,100 m
4			302 m/s	7,300 m	334 m/s	8,800 m
M4A2 5			360 m/s	9,400 m	390 m/s	10,200 m
6			458 m/s	12,500 m	469 m/s	12,200 m
7			542 m/s	15,200 m	555 m/s	14,500 m
M2 8			658 m/s	19,300 m	675 m/s	17,850 m
9			796 m/s	25,100 m		
M11 10	897 m/s	38,600 m	897 m/s	30,300 m		

Specifications

Calibre: 155 mm
Barrel length: 6.975 m
Muzzle brake: double baffle
Breech mechanism: horizontal sliding wedge
Carriage: split trail with limber
Shield: yes
Weight:
(travelling) 8,428 kg
(in action) 7,680 kg
Mass:
(recoiling parts) 2,930 kg
(elevating parts) 3,860 kg
Length:
(travelling) 11.17 m
Width:
(travelling) 2.4 m
Height:
(travelling) 2.65 m
Axis of bore:
(at 0°) 1.38 m
Ground clearance: 0.35 m
Elevation/depression: +45/−2.5°
Traverse: 50°
Recoil length
(-2.5 to +20°) 1.28 m
(+36 to +45°) 0.775 m
Rate of fire:
(max) 6 rds/min
(normal) 4 rds/min
(in 5 min) 15 rounds
(in 10 min) 25 rounds
(in 30 min) 50 rounds
(per hour) 90 rounds
Range:
(ERFB) 30,300 m
(ERFB-BB) 39,000 m
(M107) 17,850 m
Crew: 8 + 1

155 mm gun M46/84 in travelling configuration

155 mm gun M46/84 in firing position

152 mm gun-howitzer M84, M84B1 and M84B2

Development/Description

The Yugoslav 152 mm gun-howitzer M84 is based on a similar carriage to the Soviet 152 mm gun-howitzer D-20 and can fire D-20 ammunition. The barrel is much longer than the D-20 and has a semi-automatic vertical wedge-type breech mechanism. The 152 mm ordnance is 39.73 calibre and is mounted on a carriage similar to the Soviet design. It is modified to improve ballistic performance and to enable easier and more rapid handling in the firing position. It is possible to fire at elevations of +45 to +63°. A counter-recoil system, consisting of a hydraulic brake, compensator and hydropneumatic recuperator, is located over the barrel and, when combined with the high-efficiency (approximately 50 per cent) multibaffle muzzle brake, the resultant recoil length is short even at high angles of elevation. The maximum rate of fire is 6 rds/min.

The welded steel carriage has a shield and split box section trails fitted with castor wheels to assist in handling. There are two hydraulic pumps provided to assist the nine-man crew in bringing the weapon from the travelling to the firing position. These set the weapon on a circular firing platform mounted on the forward part of the carriage and raise the wheels and suspension clear of the ground.

Sighting equipment includes a range quadrant, a direct fire ×5.5 optical sight, a ×3.7 panoramic sight, a gunner's quadrant, a collimator and a sight illuminating set.

152 mm gun-howitzer M84 in travelling configuration

The 152 mm gun-howitzer M84 takes between 3 and 5 minutes to be brought into and out of action. It is provided with a single and double line braking installation and a handbrake as well as a lighting system.

The 152 mm M84 can fire the complete range of D-20 ammunition, including the OF-540 FRAG-HE, to a range of 17,190 m with a muzzle velocity of 647 m/s. The M84 can also fire an M84 HE round weighing 43.56 kg, 7.677 kg of which is the cast TNT:RDX (50:50) explosive payload. When firing the M84 HE at a muzzle velocity of 810 m/s the maximum range is 24,160 m (minimum range 5,000 m). The propellant charge system uses one full charge and six increments. The M84B1 and M84B2 versions of the artillery weapon can fire the Russian OF-540 projectiles with Yugoslav propellant charges. The system of charges includes full charge and six zone charges.

Specifications
Calibre: 152.4 mm
Barrel length: 6,056 mm
Muzzle brake: multibaffle
Recoil system: hydropneumatic
Breech mechanism: semi-automatic vertical sliding wedge
Carriage: split trail
Shield: yes
Weight:
(firing position, M84) 7,080 kg
(firing position, M84B1/M84B2) 6,880 kg
Length:
(travelling) 11.21 m
(firing) 9.67 m
Width:
(travelling) 2.415 m
(firing) 5.73 m
Height: (travelling) 2.16 m
Axis of bore: 1.385 m
Ground clearance: 0.37 m
Tyres: 12.00 × 20
Elevation/depression: +63°/–5°
Traverse:
(M84 and M84B1) 50°
(M84B2, elevation –5 to +30°) 50°
(elevation +30 to +63°) 40°
Rate of fire:
(max) 6 rds/min
(normal) 4 rds/min
15 rounds in 3 min
40 rounds in 20 min
100 rounds in 1 h
Max range with HE projectile M84:
(M84) 24,160 m
(M84B1, M84B2) 24,000 m
Max range with HE base bleed projectile:
(M84-GG) 27,000 m
Max muzzle velocity: 810 m/s
Min range with HE projectile M84:
(M84) 5,000 m
(M84B1, M84B2) 5,400 m
Crew: 9
Towing vehicle: FAP 2026 BS/AV (6 × 6) truck
Max towing speed: 70 km/h

The M84 B2 version is fitted with a pneumatic loader which is operational at all gun elevations, with the capacity of over 120 work cycles from one standard compressed air tank mounted on one of the trails.

The Yugoslav illuminating projectile is called the M88 and falls to the ground at a speed of 5 m/s. It gives a maximum intensity of 1.3 million candela for a maximum period of 1 minute. There is also an HE/ICM (High-Explosive/Improved Conventional Munition) projectile which carries 63 KB-2 bomblets with a maximum range of 22,500 m. Each bomblet is 40 mm in diameter, 85 mm long and weighs 250 g, 35 g of which is FH-5 explosive.

The M84-GG base bleed projectile is in the final stages of development, the performance of which has already been proven with long-barrel systems. Maximum range of the M84B1 and M84B2 weapons is 27,000 m.

It is understood that the 152 mm M84 series is also referred to in Yugoslavia as the Nona.

152 mm gun-howitzer M84 in firing position

Variants

MB-160

This is used by the Bosnian Serbs (Minobaca 160 mm) and has the original ordnance replaced by a heavy 160 mm mortar without a muzzle break.

SP (8 × 8)

It is understood that an 8 × 8 152 mm or 155 mm self-propelled artillery system has been developed to the prototype stage.

Status

Production complete. No longer marketed. In service with:

Country	Quantity	Comment
Bosnia-Herzegovina	n/avail	unconfirmed user
Croatia	21	
Slovenia	n/avail	unconfirmed user
Yugoslavia	52	

Contractor

Yugoimport SDPR.

UPDATED

122 mm howitzer D-30J

Development/Description

The 122 mm howitzer D-30J is a Yugoslav version of the Russian 122 mm howitzer D-30 with minor differences. For full details refer to the entry under RFAS. Details provided are for the few changes from the original, using data supplied by Yugoslavia.

The D-30J is fitted with a more simple double-baffle muzzle brake.

Installation of a hydraulic firing jack (in place of the normal mechanical jack) facilitates and speeds up handling of the weapon in the firing position.

The maximum range when firing a standard HE projectile with a full charge is given as 17,300 m. The D-30J can fire the full family of ammunition developed for the Russian D-30. The Yugoslavs had under development an M82 HEAT (High Explosive Anti-Tank) projectile fitted with a UT PE M82 fuze and an M82 illuminating projectile fitted with an ETU 102 time fuze. The Yugoslav M76 projectile has a maximum range of 17,300 m.

Yugoslav 122 mm D-30J deployed in firing position with trails staked to the ground

Specifications
Calibre: 122 mm
Barrel length:
(with muzzle brake) 4,785 mm
(without muzzle brake) 4,270 mm
Length of rifling: 3,400 mm
Recoil: 740-930 mm
Muzzle brake: double baffle
Breech mechanism: semi-automatic vertical wedge
Shield: yes
Weight:
(travelling) 3,440 kg
(firing) 3,350 kg
Length:
(travelling) 5.33 m
Width:
(travelling) 1.95 m
Height:
(travelling) 1.66 m
Axis of bore: 0.9 m
Track: 1.855 m
Ground clearance: 325-345 mm
Tyres: 9.00 × 20
Elevation/depression: +70/–7°
Traverse: 360°
Rate of fire: 6-8 rds/min
Muzzle velocity: 735 m/s
Max powder charge pressure: 2,600 bar
Max range: 17,300 m
Time into action: 1.2-2.5 min
Max towing speed: 60 km/h

Iraq manufactured a version of the 122 mm D-30J under the local name of the Saddam. Available details are given under Iraq.

Variants

D-30 RH 94
RH ALAN of Croatia has built a further developement of the D-30 called the 122 mm howitzer D-30 RH M94 which, firing an extended range full bore-base bleed projectile, has a maximum range of 20,050 m. Full details of this system are given in this section under Croatia. The D-30 RH 94 is now being offered on the export market. There are however no known export sales of the D-30 RH 94 weapon.

100 mm coastal gun
This uses a modified 122 mm D-30J carriage and details are given in the Coastal artillery guns and missiles section.

100 mm M87 TOPAZ anti-tank gun
This uses a modified 122 mm D-30J carriage and details are given later in this section. As far as is known this system remains at the prototype stage.

Status
Production complete. In service with the following countries:

Country	Quantity	Comment
Bosnia-Herzegovina	12	
Croatia	32	local production (see Croatia)
Yugoslavia	310	

Contractor
Yugoimport SDPR.

UPDATED

105 mm howitzer M56

Development/Description
In addition to using the US 105 mm M101 (M2A1) howitzer and the German 105 mm M18, M18M and M18/40, Yugoslavia also uses a 105 mm howitzer of local design and manufacture called the M56. In many respects the M56 is similar to the German M18/40 but it fires semi-fixed ammunition of the US pattern, rather than the separate loading type used by the German 105 mm howitzers.

The semi-fixed ammunition fired by the M56 consists of three types: HE shell M1, weighing 15 kg and filled with 2.2 kg of explosive, fired at a muzzle velocity of 570 m/s and with a fuze that can be set for graze or 0.05 second delay; smoke shell WP M60 filled with white phosphorous and weighing 15.8 kg (this shell has the same fuze as the HE M1); the AP-T shell M67 HESH-T weighing 10 kg and fired at a muzzle velocity of 670 m/s to a maximum range of 12,000 m.

In use the M67 HESH range is normally limited to 870 m. The 2.2 kg of explosive filling can penetrate a conventional steel armour up to 100 mm thick, set at an angle of 30°. On such a plate, the HESH projectile will scab off a film of material opposite the strike of the projectile that is 25 mm thick, with the fragments weighing up to 5 kg and travelling at a velocity of 300 m/s. The 105 mm howitzer M56 may be fitted with a 20 mm subcalibre training device.

The 105 mm barrel is mounted on a tubular split trail carriage with a split shield sloping to the sides and rear. The most common model has pressed alloy wheels with solid rubber tyres, similar to those on the German M18 series weapons. However, some M56s are fitted with pneumatic-tyred wheels similar to those fitted to the US 105 mm M101 (M2A1). With the pneumatic wheels the M56 can be

Close up of rear of 105 mm M56 howitzer in service with El Salvador showing breech detail (Julio Montes) ***2001***/0105792

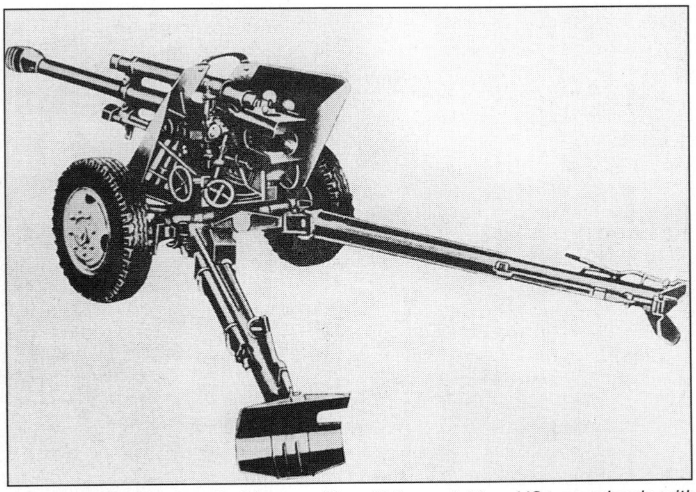

105 mm M56 howitzer in firing position. This model has US-type wheels with pneumatic tyres

105 mm M56 howitzer in service with Cyprus ***2002***/0130011

105 mm M56 howitzer in service with Guatemala and deployed in firing position

Specifications
Calibre: 105 mm
Barrel length:
(overall) 3.48 m
Muzzle brake: multibaffle
Recoil system: hydraulic recoil buffer and hydropneumatic recuperator
Breech mechanism: horizontal sliding wedge
Carriage: split trail
Shield: yes
Weight:
(travelling order) 2,100 kg
(firing position) 2,060 kg
Length:
(travelling) 6.17 m
(firing) 5.46 m
Width:
(travelling) 2.15 m
Height:
(travelling) 1.56 m
(at max elevation) 3.8 m
Axis of bore: (at 0°) 1.13 m
Ground clearance: 0.335 m
Track: 1.8 m
Elevation/depression: +68/–12°
Traverse:
(total) 52°
Rate of fire: 16 rds/min
Max range: 13,000 m
Crew: 7
Towing vehicle: TAM 1500 (4 × 4) truck

towed at speeds up to 70 km/h. Normally the M56 is supplied with a range quadrant, panoramic telescope (dial sight), direct firing anti-tank telescope and a gunner's quadrant, all equipped with lights for night firing. The direct fire telescope has a magnification of ×2.8. The panoramic telescope has a magnification of ×4 and weighs 6 kg.

While the carriage of the M56 is normally used for firing with the trail legs spread, in an emergency it can be fired from the travelling position. In this configuration the maximum elevation is limited to +16° and the total traverse to 16°.

There has been no production of the 105 mm howitzer M56 for some years.

Status

Production complete. In service with the following countries:

Country	Quantity	Comment
Bosnia-Herzegovina	n/av	
Croatia	n/av	
Cyprus	72	still operational
El Salvador	18	reserve
Guatemala	56	
Indonesia	10	
Mexico	24	army
	8	marines
Yugoslavia	256	

Contractor

Yugoimport SDPR.

UPDATED

100 mm M87 anti-tank gun TOPAZ

Development/Description

The 100 mm M87 anti-tank gun TOPAZ is essentially the Yugoslav-built 122 mm D-30J (qv) towed howitzer (which in turn is based on the Russian 122 mm D-30) fitted with the ordnance of the Russian 100 mm anti-tank gun T-12 (2A19M) (qv). As far as it is known the 100 mm M87 anti-tank gun has yet to enter production or service. It was marketed by Yugoslavia some years ago, but it is understood that there has been no recent marketing of the system.

Optical equipment installed on the system includes passive day/night sight ON 100-M78, panoramic sight P-M78, range-finder D100-M78 and collimator K-M78. The fire-control system has a laser range-finder with a range from 200 to 9,995 m. Ballistic data for four types of ammunition can be inserted. The fire-control system provides an automatic sighting mark in the field of view and takes into account the type of ammunition being used, meteorological conditions, charge temperature, target speed and deviation of the muzzle velocity of the projectile.

Compared with other eastern 100 mm anti-tank guns, for example the Russian T-12, the Yugoslav 100 mm M87 anti-tank gun is a much more flexible system and has a higher hit probability against stationary and moving targets.

For types of 100 mm fixed ammunition that can be fired by the system, see table right.

The main advantages of this system have been summarised by the manufacturer as follows:

(1) The weapon fires all types of 100 mm T-12 ammunition and has identical ballistics
(2) The use of the D-30J three trailed carriage enables the weapon to be quickly traversed through 360° to enable it to engage moving targets and its low silhouette gives it increased battlefield survivability
(3) It can be quickly taken into and out of action and has good mobility over rough terrain. Time into action is 150 seconds and time out of action is 180 seconds
(4) As the gunner's seat is mounted on the upper carriage, tracking of moving targets is simplified
(5) The auxiliary wheels fitted to the trails enable easy deployment and rapid positioning for combat
(6) An advanced opto-electronic day/night fire-control system that incorporates a laser range-finder enables targets to be engaged under both day and night conditions with a high first round hit probability. The fire-control system takes 75 seconds to install on the weapon and 50 seconds to be removed
(7) The installation of brakes and lights enable it to be deployed on public roads.

It is understood that volume production/upgrade work on this weapon had not started at the commencement of the Yugoslav civil war.

Specifications
Calibre: 100 mm
Barrel length: 6,300 mm
Shield: yes
Weight:
(total) 3,650 kg
Length:
(travelling) 7.670 m
Width:
(travelling) 1.95 m
Height:
(travelling) 1.66 m
Axis of fire: 0.90 m
Ground clearance: 0.33 m
Track: 1.856 m
Traverse: 360°
Elevation/depression: +18/–6°
Rate of fire:
(max) 14 rds/min
(normal) 8 rds/min
Max effective range:
(APFSDS-T) 2,230 m
(HEAT-T) 1,200 m
(HE) 7,800 m
Time into action: 1.2 - 2.5 min
Max towing speed: 60 km/h
Crew: 5-7

Designation	BM1/BM2	BK2	OF3
Type	APFSDS-T	HEAT-T	HE
Muzzle velocity	1,575 m/s	975 m/s	n/av
Max range			
(flat trajectory for 2 m high target)	1,880 m	1,020 m	n/av
(flat trajectory for 3 m high target)	2,230 m	1,200 m	n/av

100 mm M87 TOPAZ anti-tank gun deployed in the firing position with trails staked to the ground

Variants

M-87 100 mm gun

This was declared under the CFE Treaty by Bosnia-Herzegovina. This has the ordnance of the T-12 anti-tank gun from Russia but has a different carriage without wheels as it is used in the static defence role.

M-91 Monas

This is a Yugoslav weapon and is a D-30 carriage fitted with a 100 mm T-12 barrel.

Status

Development complete. Ready for production.

Contractor

Yugoimport SDPR.

UPDATED

76 mm mountain gun M48

Development/Description

The 76 mm mountain gun M48, which is often called the 'Tito Gun', was developed after the Second World War specifically to meet the requirements of Yugoslav mountain units, although it can also be used as a field gun. As far as it is known there has been no recent production of the 76 mm mountain gun M48.

There are at least four variants of the M48. The M48 (B-1) has pneumatic tyres and a maximum towing speed of 60 km/h. It can also be towed by animals in tandem or broken down into eight pack loads. The M48 (B-1A1-I) has the pneumatic tyres and wheels as fitted to the M48 (B-1), plus some of the features of the suspension of the M48 (B-1A2). The M48 (B-1A2) can also be used as a field piece but cannot be towed by animals or broken down for pack transport. The M48 (B-1A2) has light alloy wheels with solid rubber tyres and modified suspension, its maximum towing speed is 30 km/h. The final production model of the 76 mm mountain gun M48 was called the B-2 about which little is known at present.

Ammunition is of the semi-fixed type with four charges. It is based on that used for the obsolete Russian 76 mm Regimental Gun M1927 which fired fixed ammunition: HE M55 projectile weighing 6.2 kg with a muzzle velocity between 222 and 398 m/s; HEAT projectile weighing 5.1 kg which will penetrate 100 mm of conventional steel armour at a range of 450 m; and a smoke shell WP M60 weighing 6.2 kg.

The Yugoslav 76 mm mountain gun M48 is related to the Romanian 76 mm mountain howitzer M1984 (or Model 1984 as it is also referred to).

According to the Romanian manufacturer, the 76 mm mountain howitzer fires the following ammunition types as listed below. It should be noted that some of these with the M series designation are also used by the Yugoslav 76 mm mountain gun M48:

Charge round - HE OF-350M, fuze UP3
Charge round - M55, fuze M51A5
Charge round - smoke WP M60, fuze PD M48A3
Charge round - smoke D-350 TM, fuze UP3
Charge round - HEAT BK 354, piezo electric fuze CVP-312
Charge round - shell M50, fuze PD MK-451

All models of the M48 have split trails with each trail hinged in the centre and folding forwards through 180° for towing. At the end of each trail is a slot for the stake. The shield has winged sides and the centre of the shield has to be raised to obtain maximum elevation.

Specifications

M48 (B-1)
Calibre: 76.2 mm
Barrel length: 1.178 m
Muzzle brake: multibaffle
Recoil system: hydraulic
Breech mechanism: semi-automatic sliding wedge breech block
Carriage: split trail, folding
Shield: yes
Weight:
(travelling order) 720 kg
(firing position) 705 kg
Length:
(travelling) 2.42 m
Width:
(travelling) 1.46 m
(firing) 2.65 m
Height:
(travelling) 1.22 m
(elevation 45°) 1.67 m
Axis of bore:
(at 0°) 0.73 m
Ground clearance: 0.18 m
Track: 1.3 m
Tyres: 6.00 × 16
Elevation/depression: +45/−15°
Traverse:
(total) 50°
Rate of fire: 25 rds/min
Range:
(max) 8,750 m
(normal) 8,600 m
Crew: 6
Towing vehicle: Zastava (4 × 4)

76 mm Mountain Gun M48 with rectangular cross-section trail legs

Status

Production complete. In service with the following countries:

Country	Quantity	Comment
Bosnia-Herzegovina	n/av	unconfirmed user
Croatia	57	unconfirmed user
India	215	
Macedonia	55	
Myanmar	100	
Sri Lanka	14	
Yugoslavia	97	

Contractor

Yugoimport SDPR.

UPDATED

MULTIPLE ROCKET LAUNCHERS

MULTIPLE ROCKET LAUNCHERS

ARGENTINA

TAMSE VCLC series armoured rocket launchers

Development

During the early 1980s the Argentine TAMSE (Tanque Argentino Mediano Sociedad del Estado) concern,which had been formed to manufacture the German Thyssen Henschel (which in 2000 became Rheinmetall Land Systeme), designed the TAM medium tank and began to investigate a joint venture with Israel Military Industries Ltd to fit the company's modular rocket systems onto the TAM chassis.

In 1986, the first prototype vehicle, called the *Vehiculo de Combate Lanza Cohetes* (VCLC) underwent trials with the 160 mm LARS 160 rocket system. In early 1988, a second version was trialled with the 350 mm calibre MAR 350 system.

It is understood that a small quantity of VCLC-CAL 160 are now in service with the Argentine Army. Production of the TAM chassis on which the VCLC is based was completed some years ago. Under current plans, it is not expected that quantity production of the chassis will be undertaken in the immediate future.

Description

The major differences between the VCLC and the TAM medium tank chassis (described under Germany in the MBTs and medium tanks section) include modification of the chassis to incorporate additional hydraulic suspension, movement suppression systems to absorb the recoil forces during rocket launching and fitting of a new two-person fully enclosed armoured turret. The actual rocket launcher is mounted on top of this turret towards the rear.

Support arms for the rocket system modules are fitted at the rear of the turret on either side. They are traversed and elevated using the turret's electrohydraulic power unit. No modifications are needed if a change between the rocket types is required. For self-protection and limited air defence, a pintle-mounted 7.62 mm MAG or 0.30 in Browning MG is mounted above the front of the turret.

When the LARS 160 rocket system is carried the vehicle is designated the VCLC-CAL. This has two 18-round 160 CAL pod container-launchers which, once fired, are unloaded and replaced in a secure area by a supporting vehicle. In Argentine Army trials, the vehicle used was an United States supplied M809 Series (6 × 6) 5 ton flatbed truck fitted with a 12,000 kg capacity hydraulically operated extendable crane and carrying two of the 160 CAL pods. The whole reloading operation takes approximately 10 minutes.

If fitted with the MAR 350 system the vehicle is designated the VCLC-CAM and has two pairs of side-by-side cylindrical container-launchers for CAM 350 rockets. The resupply operation is similar to that used for the VCLC-CAL vehicle.

VCLC-CAL 160 multiple rocket launcher based on a modified TAM tank chassis

Specifications

Number of rockets: 2 18-round 160 CAL or 4 single-round 350 CAM
Crew: 3
Combat weight:
(with rocket system) 32,582 kg
Power-to-weight ratio: 22.1 hp/t
Ground pressure: 0.84 kg/cm²
Length:
(chassis) 6.75 m
Width:
(chassis) 3.12 m
Height:
(with launcher system) 3.05 m
Ground clearance: 0.45 m
Track: 2.62 m
Fuel capacity: 680 litres plus 400 litres in external auxiliary tanks
Engine: supercharged MTU MB 833 Ka-500 6-cylinder diesel developing 720 hp at 2,400 rpm
Secondary armament: 1 × 7.62 mm or 0.30 in MG
Transmission: Renk HSWL 204 planetary with 4 speeds both forward and reverse
Steering: double differential with hydrostatic steering control

In combat the basic six-launcher VCLC battery will have a truck-mounted fire-control radar system. Its job is to monitor the first ranging shot for between two-thirds to three-quarters of its flight. The round is deliberately aimed offset from the target so as not to alert it. The tracking data obtained is then fed into the vehicle's fire-control computer and used to calculate the launch information which is then transmitted to the battery launchers.

Status

A small quantity of the VCLC-CAL are in service with the Argentine Army but the VCLC-CAM is understood to be at the prototype stage.

Manufacturer (chassis)

TAMSE.

UPDATED

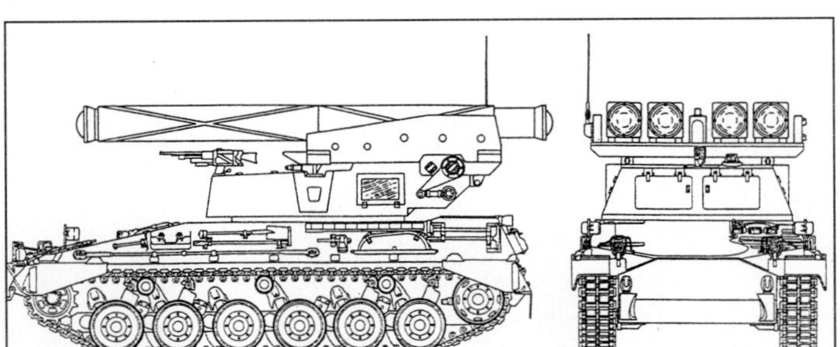

Argentine CAM 350 multiple rocket system mounted on a TAM tank chassis

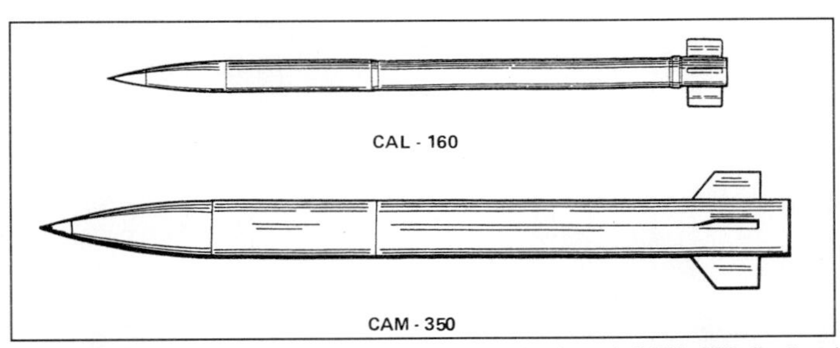

General arrangement drawings of CAL 160 (top) and CAM 350 (bottom) rockets

DGFM 127 mm (28-round) CP-30 multiple rocket launcher

Development/Description

Developed by the Research and Development Institute of the Argentine armed forces, the 127 mm CP-30 (28-round) system entered operational service in the 1980s. It is understood that a total of 30 systems are in service under the designation of the SLAM Pampero.

The 127 mm aluminium launch tube assembly can be configured in a variety of ways to suit various platforms, including the following: a 16-round (in two banks of eight 127 mm tubes) assembly on the rear bed of a modified Mercedes-Benz UNIMOG 416 (4 × 4) 1,500 kg light cross-country truck; a 36-round (in four banks of nine tubes) assembly on the rear bed of a French VLRA ACMAT (8 × 8) 7,000 kg cross-country truck; and a 28-round assembly also on the French

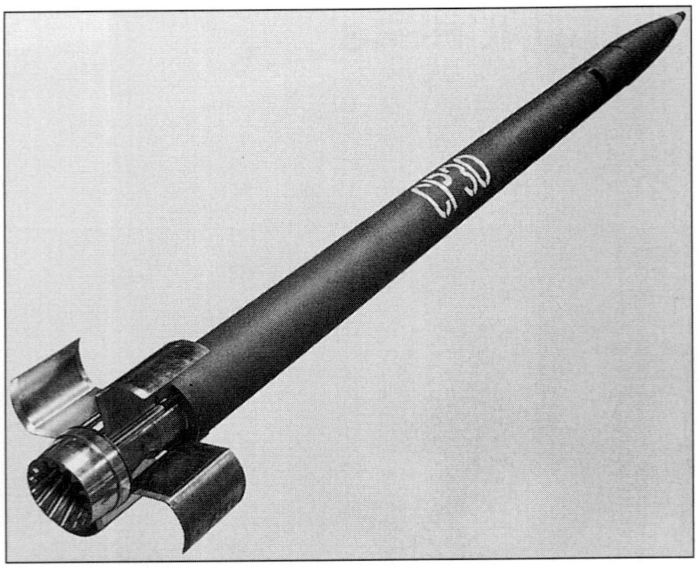

DGFM 127 mm (28-round) CP-30 multiple rocket launcher on Mercedes-Benz 416 (4 × 4) chassis with 16-round launcher elevated 0008303

DGFM 127 mm (28-round) CP-30 rocket showing wraparound fins at rear in unfolded position
0008304

VLRA ACMAT (8 × 8) truck chassis. As far as it is known, all production applications have utilised the German Mercedes-Benz UNIMOG 416 (4 × 4) 1,500 kg truck chassis.

To provide a more stable platform four hydraulically operated stabilisers are lowered. Traverse and elevation are also hydraulically operated systems. Sights are provided for direct and indirect fire.

Single-round and ripple firings up to the full salvo are possible. The 127 mm steel-bodied rocket uses a PHE-2 solid propellant filled rocket motor casing with four wraparound pop-up tail-mounted guidance fins. The warhead types available include: training, drill, HE anti-personnel, HE incendiary and anti-tank/anti-personnel submunition. Maximum rocket range is stated to be 30,000 m. The warshots have either contact or delayed action fuzing options.

Specifications

(28-round VLRA ACMAT (8 × 8) platform)
Calibre: 127 mm
Number of rocket tubes: 28
Elevation: +10 to +70°
Traverse: 90° left and right
Max range: 30,000 m
Max velocity: 1,000 m/s
Weight:
(fully laden) 14,000 kg
(empty) 11,760 kg
(rocket complete) 79 kg
(warhead) 20 kg
Length:
(travelling) 7.48 m
(rocket complete) 3.15 m
Width:
(travelling) 2.32 m
Height:
(travelling) 2.7 m
Engine: 6 cylinder petrol

Status

Production as required. In service with the Argentine Army. Offered for export. As far as it is known, there has been no recent production of the complete system.

Manufacturer

Direccion General de Fabricaciones Militares (DGFM).

UPDATED

DGFM 105 mm (16-round) Pampero multiple rocket launcher

Development/Description

Developed by the Research and Development Institute of the Argentine armed forces, the 105 mm (16-round) Pampero entered operational service in the mid-1980s.

The 105 mm 16-round launcher assembly, with two rows of eight tubes, is mounted on the rear bed of a modified Mercedes-Benz UNIMOG 416 (4 × 4) 1,500 kg light cross-country truck chassis. If required, the launcher can be fitted to other vehicle types with differing numbers of tubes.

This system is also referred to as the SLAM Pampero and about 40 systems are understood to have been built, all based on the Mercedes-Benz UNIMOG 416 1,500 kg (4 × 4) chassis with most being delivered to the army. There has been no recent production or marketing of this rocket system.

To provide a more stable firing platform four hydraulically operated stabilisers are lowered, two at the front of the vehicle and two at the rear. Traverse and elevation are manually set via handwheels. Sights are provided for both indirect and direct fire modes.

Single round, partial ripple or full salvo firings are possible. A full salvo takes 7.5 seconds and will fall within a 300 × 200 m area with a maximum round dispersion CEP of less than one per cent. The system has to be manually reloaded.

The single-stage 105 mm rocket utilises a PHE-2 solid propellant fuel filled rocket motor casing with four pop-up tail-mounted guidance fins. The warhead types include training, HE fragmentation, HE incendiary and HE anti-personnel. Maximum range is stated to be 10,100 m. The warshots have either contact or delayed action fuzing options.

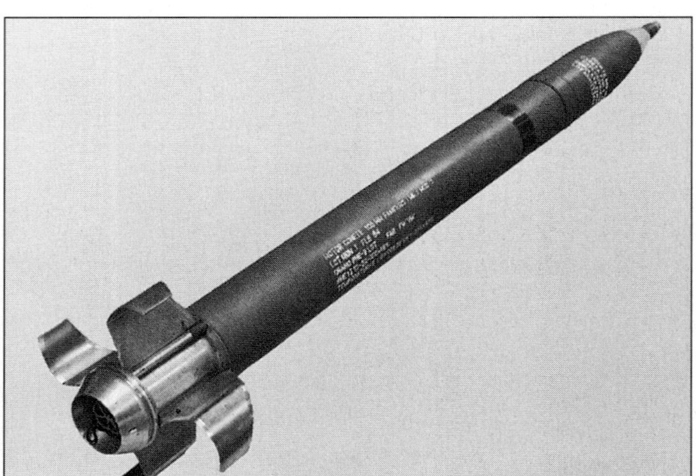

DGFM 105 mm Pampero rocket showing four unfolded fins at rear of rocket 0008306

DGFM 105 mm (16-round) Pampero rocket launcher deployed in the firing position on Mercedes-Benz 416 (4 × 4) truck chassis 0008305

Specifications
Calibre: 105 mm
Number of tubes: 16
Rate of fire: 16 rds/7.5 s
Elevation: +10 to +70°
Traverse: 90° left and right
Max range: 10,100 m
Max velocity: 530 m/s
Crew: 2-3
Weight:
(fully loaded) 6,490 kg
(empty) 6,035 kg
(rocket, complete) 28.5 kg
(rocket, warhead) 10.0 kg
Length:
(travelling) 5.22 m
(rocket) 1.476 m
Width: (travelling) 2.12 m
Height: (travelling) 2.38 m
Engine: 4-cylinder petrol

Status
Production as required. In service with the Argentine Army (36) and Marine Corps. There has been no recent production of this system.

Manufacturer
Direccion General de Fabricaciones Militares (DGFM).

UPDATED

BELGIUM

Forges de Zeebrugge FZ 70 mm (40-round) LAU97 multiple rocket launcher system

Development/Description
Developed specifically for the export market by Forges de Zeebrugge SA, the LAU97 is a light artillery rocket launcher system designed to provide saturation and holding fire at the regimental level. As far as is known this system has not been deployed by any country in NATO.

Recent information has stated that the Army of Shajah in the United Arab Emirates (UAE) has taken delivery of 18 of these systems, fitted on the rear of a French ACMAT VLRS 4.15 LRM (4 × 4) cross-country truck chassis.

It consists of a launcher box with five horizontal rows of eight 70 mm tubes and a 360° traversable base that allows elevation arcs of 0 to +55°. The system can be fitted on any military vehicle or trailer capable of carrying 1,200 kg payloads. The launcher's azimuth and elevation laying is performed by either electric motors or handwheels. A BR2 mortar sight or artillery panoramic sight is fitted for aiming the system with an electrically operated remote-control unit for firing.

The 70 mm unguided rockets can be launched either singly or in ripple sequences up to the full 40-round salvo, which takes less than 6 seconds. Using the FZ90 rocket motor and a 4.3 kg warhead, the maximum range is 9,000 m at a firing angle of 40° elevation. The rockets, using the standard FZ-68 rocket motor and a 4.3 kg warhead, have a maximum range of 7,900 m at a launcher elevation of +40°; if the elevation is set at +15° then the range is reduced to 6,000 m. If the Mk 40 rocket motor is used the maximum range at +40° elevation is reduced to 7,500 m.

A full complement of 70 mm unguided rockets can saturate an area measuring 200 × 300 m at maximum range. The FZ-100 6.2 kg cargo warhead with a payload of nine 0.48 kg anti-personnel/anti-tank bomblets is also available. The bomblets have a lethal radius of 10.5 m and are capable of penetrating up to 105 mm of conventional steel armour plate. The submunitions are released by a delay-adjustable time fuze over the target area.

Other warheads available include: the 4.3 kg FZ210 practice impact marking practice; the 3 kg FZ-49 anti-armour capable of piercing more than 350 mm of conventional steel armour plate; the 4.3 kg FZ-71 anti-personnel which forms more than 8,000 fragments on detonation and is lethal to between 18 and 21 m radius; and the 4.9 kg M257 illuminating type with a 120 second duration and one million candle lighting capacity. The launcher is operated by a crew of three, a layer, a commander and a gunner, who also reload the system by hand.

Variants
This 70 mm unguided rocket launcher can also be mounted on a variety of tracked and wheeled chassis and has been demonstrated on the rear of an Alvis Vehicles Streaker high-mobility load carrier in palletised form. The 70 mm (40-round) LAU97 multiple rocket system has been installed on an ACMAT VLRA Type TPK 4.15 LRM (4 × 4) truck, the Alvis Saladin armoured car (turret removed) and the Steyr-Daimler-Puch Pinzgauer (6 × 6) vehicle. There is also a towed version.

The FZ90 rocket motor with the flash warhead FZ181 is used in the BAE Systems, RO Defence Reduced Range Practice Rocket (RRPR) covered under the UK later in this section. This RRPR is in service with Norway and the UK.

LAU97 launcher installed on rear of ACMAT VLRS 4.15 LRM (4 × 4) truck chassis with stabilisers lowered and launcher traversed right
2002/0130029

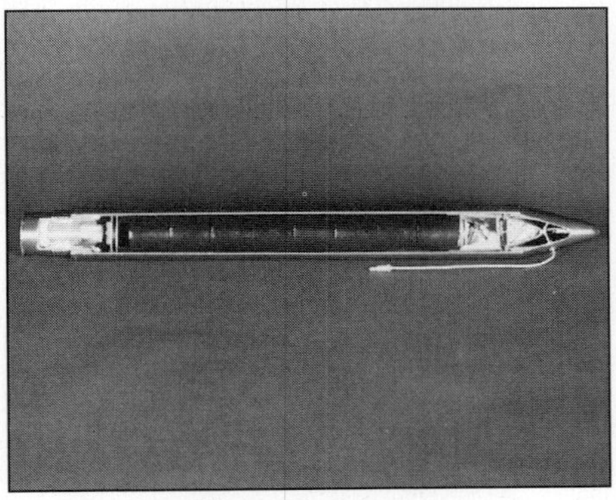

FZ-100 submunition warhead

Status
Production as required. In service with Sharjah in the United Arab Emirates (UAE) on ACMAT chassis. From 1985 IPTN of Indonesia started producing the 70 mm rocket system FFAR under licence from Forges de Zeebrugge SA. Indonesia has developed its own 70 mm launcher system (see entry this section) and also now manufactures the rockets under licence.

Specifications
(system installed on Alvis Vehicles Streaker tracked chassis)
Calibre: 70 mm
Number of barrels: 40
Number of rockets carried: 160
Max range: 9,000 m
Elevation/depression: +55/0°
Crew: 3
Weight:
(loaded) 5,350 kg
Length: 5.325 m
Width: 1.936 m
Height:
(travelling) 2.538 m
Ground clearance: 0.29 m
Max speed: 100 km/h
Range: 840 km
Fording: 0.9 m
Gradient: 65%
Engine: Perkins Engines Company T6.3544 6-cylinder diesel developing 120 hp at 2,800 rpm

Manufacturer
Forges de Zeebrugge FZ, a TDA subsidiary.

UPDATED

BRAZIL

AVIBRAS multiple rocket systems

AVIBRAS 108-R

This unguided rocket, which is manufactured by AVIBRAS, has the Brazilian Army designation FGT108-RA1 and is fitted with a single-stage solid propellant motor with a thrust of 1,250 kg. It is spin stabilised in flight by one central and six canted rocket nozzles. The motor burns for half a second to attain a maximum velocity of 409 m/s at burnout.

The rocket is launched from the X2A1 trailer launcher which has 16 1.1 m long tubes. The launcher, also manufactured by AVIBRAS, weighs 530 kg empty, 802 kg loaded and is 2.95 m long, 1.6 m wide and 1.37 m high. The wheelbase is 1.4 m and the tyre size 5.00 × 16. The launcher is normally towed by a ¾ ton 4 × 4 vehicle. It can also be airlifted by helicopter. The maximum rate of fire is 2 rds/s. The elevation is −1 to +50° and the traverse is ±12°. A 50 m cable is attached to the launcher for firing purposes. The maximum height of the rocket's trajectory is 4,000 m. The crew consists of four and reloads are carried in 16-round glass fibre containers. A battery will normally have four launchers each having at least 64 rockets available on its towing vehicle.

The complete family of AVIBRAS ASTROS II rockets, from top to bottom: SS-80, SS-60, SS-40 and SS-30

Specifications
Designation: 108-R
Calibre: 108 mm
Weight:
(complete round) 16.8 kg
(warhead) 7.8 kg
(explosive filler) 2.5 kg
Length: 0.931 m
Max speed: 440 m/s
Max range: 9,100 m
Number of rails/tubes: 16 tubes
Launcher: X2A1 trailer

AVIBRAS SBAT-70

This unguided rocket is essentially a standard AVIBRAS 70 mm SBAT-70 folding-fin air-to-ground rocket. For land application a 36 tube LM-07/36 (also referred to as the SS-07/36 system) launcher is used and this requires a crew of four. The maximum range is about 8,000 m. The following warhead types can be fitted: AVC-70/AC high-explosive anti-tank; AVC-70/AP high-explosive fragmentation (anti-personnel); AVC-70/AC/AP high-explosive anti-tank/anti-personnel; AVC-70/F flechette; AVC-70/FB white phosphorus smoke; AVC-70/EF practice smoke and AVC-70/E inert practice. The launcher weighs 700 kg empty and approximately 1,000 kg loaded. The length is 3.17 m, width 1.67 m, height 1.19 m and wheelbase 1.47 m. The elevation is 0 to +50° and the traverse ±12°. A 50 m cable is attached to the launcher for firing purposes. The system is not operational with the Brazilian Army but is offered for export. Brief specifications are:

Specifications
Designation: SBAT-70
Calibre: 70 mm
Weight:
(complete round) 8.5 kg
(warhead) 3.2 kg
Max range: 8,000 m
Launcher: LM-07/36 trailer

AVIBRAS SBAT-127

This unguided rocket, formerly manufactured by AVIBRAS, is an adaptation of the 127 mm SBAT-127 HVAR aircraft rocket. The 12-rail launcher was designed to be mounted either on a trailer or a vehicle. There are two types of high-explosive warhead used, these differ in weight. The heavier one produces a flight time of 70 seconds and the lighter one a flight time of 68 seconds for the rockets. Powered flight distances are 350 m and 275 m respectively. Maximum ranges are achieved at an elevation of +47° and the system was developed for export. Brief specifications are:

Specifications
Designation: SBAT-127
Calibre: 127 mm
Weight:
(complete round) 45/58 kg
(warhead) 19/32 kg
Max range: 15,400/13,700 m
Max speed: 690/680 m/s
Number of rails: 12
Launcher: trailer or vehicle

AVIBRAS Skyfire

A recent product developed by AVIBRAS is the Skyfire unguided family, a composite propellant 70 mm rocket with wraparound fins and larger warhead. This has greater accuracy, range and effectiveness in an air or land application. The warhead types are the same as for the SBAT-70, plus the Skyfire higher performance warheads which are: AVC-70/HE M2 (6.0 kg high-explosive anti-matériel/anti-personnel), AVC-70/MW (bomblets multiple warhead with electronic time fuze) and AVC-70/FE (anti-personnel flechette with electronic time fuze).

For the land application AVIBRAS has developed the 36 tube AV-LM-12/36 towed launcher. The launcher weighs 950 kg empty and 1,500 kg loaded. Brief specifications are:

Specifications
Designation: Skyfire
Calibre: 70 mm
Weight:
(complete round) 14.5 kg
(warhead) 6 kg
Max range: 12,000 m
Number of tubes: 36
Launcher: AV-LM-SS-12/36

AVIBRAS ASTROS II

AVIBRAS developed the ASTROS (Artillery SaTuration ROcket System) for the Brazilian Army and the export market. The first production units were completed in late 1983.

According to the United Nations report on Conventional Arms Transfers issued in 2001, AVIBRAS

X2A1 trailer launcher for 108-R rocket (Ronaldo S Olive)

LM-7/36 70 mm rocket launcher being carried slung under a helicopter
0058452

TECTRAN (6 × 6) AV-RMD resupply vehicle, left, loading new launcher pods onto TECTRAN (6 × 6) AV-LMU launcher, right

AVIBRAS AV-UCF radar fire-control vehicle deployed in operational configuration

exported none of its ASTROS systems between the years 1992 through to 2002.

In 2001 it was reported that Malaysia had ordered the ASTROS system to supplement its 155 mm towed artillery systems. First deliveries were made to Malaysia in early 2002.

The ASTROS II system consists of the launching vehicles (Universal Multiple Launcher: AV-LMU), ammunition resupply vehicles (AV-RMD with two complete reload sets for an AV-LMU) with onboard hydraulic cranes, a command and control vehicle (AV-VCC), a vehicle with radar for fire control (AV-UCF-optional) and several mobile workshop vehicles for electronic and mechanical field maintenance. All the vehicles are based on the same 6 × 6 10,000 kg cross-country truck (AV-VBA) designed and built by the AVIBRAS subsidiary company TECTRAN.

A typical battery comprises one AV-UCF, six AV-LMU and six AV-RMD. For a battalion-level headquarters there would be an AV-VCC and two mobile workshops. The AV-VCC provides co-ordination and fire-control direction for up to three ASTROS batteries. Other battery configurations possible include four or eight AV-LMU.

The launcher can fire four different types of unguided rocket, the AVIBRAS SS-30 (32 rounds per launcher), SS-40 (16 rounds per launcher), SS-60 (four rounds per launcher) and SS-80 (four rounds per launcher). The larger SS-40, SS-60 and SS-80 rounds have cluster munition warheads . The SS-40 warhead has 20 and the SS-60 warhead 65 dual-effect anti-armour/anti-personnel bomblets which improve the area saturation effectiveness. The weight of the SS-60 warhead is 212 kg. The SS-80 rocket has an increased range which has been achieved by carrying more propellant and reducing the weight of the warhead.

At the 1991 Paris Air Show AVIBRAS announced it had developed, to the production stage, three further submunition warheads for the SS-40/60/80 series rockets. These are:

(1) a high-explosive incendiary type utilising white phosphorus for anti-personnel, rapid smoke deployment and anti-materiel incendiary employment

(2) a mine deployment type capable of carrying three different mine types: anti-personnel with an effective radius of 30 m; anti-matériel with the ability to penetrate more than 100 mm of conventional steel armour on its main axis and up to 20 mm of mild steel at distances of more than 100 m; anti-armour with the ability to penetrate more than 120 mm of conventional steel armour plate

(3) an airfield denial type with a number of heavy TNT-filled delayed action fuzed submunitions capable of penetrating more than 400 mm of reinforced concrete. Typical dimension for a crater in concrete would be: depth of explosion centre 340 to 650 mm, depth of crater formed (from the actual concrete surface) 150 to 300 mm and radius of crater (from the centre of explosion to where crater breaches original concrete surface) 550 to 860 mm.

AVIBRAS AV-LMU firing SS-30 rocket ***2002***/0130030

Designation	**SS-30**	**SS-40**	**SS-60**	**SS-80**
Calibre	127 mm	180 mm	300 mm	300 mm
Weight				
(rocket)	68 kg	152 kg	595 kg	595 kg
Length	3.9 m	4.2 m	5.6 m	5.6 m
Range				
(min)	9 km	15 km	20 km	22 km
(max)	30 km	35 km	60 km	90 km
Number of tubes	32	16	4	4
Launcher	AVIBRAS/TECTRAN Universal Multiple Launcher (AV-VBA)			

The range of the unguided rockets vary from 9 to 90 km according to type. They all have wraparound fins for improved accuracy. The command and control vehicle is equipped with military computers for fire-control calculations and a radio set for data transfer and voice communication with the launchers, fire-control radar vehicle, ammunition resupply vehicles, forward observers and higher command echelons.

The ASTROS systems of the Iraqi Army saw extensive combat service in both the 1980-88 and

AVIBRAS Skyfire family, LM 70/7 SF launcher (background), Skyfire 70 M8/M9 and M10 folding fin rockets (left) and warheads (right) 0021035

Standard AVIBRAS 36 tube multiple launcher AV-LM-12/36 launching the older rocket and showing unfolding fins 0021036

1990-91 Gulf wars. Saudi Arabian systems saw extensive service in the 1990-91 Gulf War against Iraq.

Training

In 1997 AVIBRAS introduced a new concept for ASTROS II training means, which is the use of AVIBRAS Sub Calibre Training Ammunition 70 mm (AV-SS 09-TS).

This has been designed to help trainees to assimilate the operational, handling and storage procedures of the AVIBRAS ASTROS II rockets.

The AVIBRAS AV-SS 09-TS is an unguided rocket of 70 mm (2.75 in) in diameter using composite propellant, wraparound fins and spin stabilisation with high kinetic energy and superior accuracy.

The rockets are fired from their own reusable AVIBRAS Sub Calibre Training Container Launcher (AV-CL-SC70) and are fully compatible with the AV-LMU Universal Mulitple Launcher.

The AVIBRAS AV-CL-SC70 with eight tubes has been specifically designed to cope with high-energy rockets.

The ASTROS is also being marketed as the ASTROS II Coastal Defence System (see entry in the Coastal artillery guns and missiles section). This entered service with the Brazilian Army in 1998.

The Brazilian Army SS-40 ASTROS system has been filled with the Metsman meteorological sensor produced by Integrated Photomatrix Ltd of the UK.

Wing ASTROS rocket

Late in 2000, AVIBRAS confirmed that they had carried out the first launch of their latest Tactical Missile from their widely deployed ASTRO II (Artillery Saturation Rocket System).

The Tactical Missile is being developed as a private venture by the company enhancing the capability of the ASTROS II system to engage targets at a much longer range and with a significant increase in terminal accuracy over the current unguided system.

The first test, conducted at a Brazilian Army range, was to test the launch of the Tactical Missile from the ASTROS II launcher and the unfolding of the wings.

This was the first in a number of anticipated test launches, although AVIBRAS Aeroespecial is seeking external funding to complete full development.

An ASTROS II launcher would typically carry two Tactical Missiles which would be launched from the 450 mm launcher tubes. The solid propellant motor would boost the rocket out of the launcher after which the wings would unfold and the turbojet motor would cut in.

The Tactical Missile would have a maximum range of over 150 km and could be fitted with various types of warhead to meet different mission requirements.

These would range from a high-explosive warhead to one that would dispense submunitions out of either side of the rockets nose. It is believed that the Tactical Missile would have a guidance system that combined mid-course and terminal guidance.

Tactical Missile

In November 2001, it was stated that AVIBRAS was developing an autonomously guided version of the ASTROS II system called ASTORS TM (Tactical Missile).

This is understood to be an enhanced version of the previously mentioned Wing ASTROS rocket with flight tests starting in June 1999 at a Brazilian Army Range. In addition to its field and coastal defence roles a naval version is also being developed.

The ASTROS TM missile has a range of 300 km and uses an inertial guidance system which presumably incorporates a Global Positioning System (GPS) data for mid-course and possible terminal guidance.

No details of the payload of ASTROS TM have been released, although AVIBRAS have stated that its range and payload complies with the range and payload constraints of the Missile Technology Control Regime. This suggests a payload of less than 500 kg and probably includes high explosive and submunition types.

Status

The ASTROS II and SBAT-70 are in production for the Brazilian Army and export. An ASTROS II battery was ordered in early 1994 for the Brazilian Army. The first contract was for the supply of four launchers, two ammunition vehicles and one fire control vehicle.

Early in 1998, it was announced that the Brazilian Army had placed an order with AVIBRAS, for the supply of a further four batteries of the ASTROS II system.

In May 1998, another four launchers and two ammunition vehicles were delivered.

In 1999, another eight launchers, five ammunition vehicles and four meteorological vehicles were delivered to the Brazilian Army. Still to be delivered are four launchers, one ammunition vehicle, three fire control units, two maintenance vehicles and five meteorological post vehicles.

First photograph released of the AVIBRASTM Tactical Missile being launched from an AVIBRAS (6 × 6) launcher during trials ***2001***/0073666

The SBAT-70 airborne rocket system is in Brazilian Army service. The SBAT-70 is also in service with several unidentified countries and the ASTROS II is known to be in Brazilian, Iraqi (SS-30, SS-40 and SS-60, which are known locally under the names of Sijeel 30, Sijeel 40 and Sijeel 60), Qatari (believed three batteries), Malaysia and Saudi Arabian (believed 10 batteries) service (including the SS-30).

Manufacturer

Avibrás Indústria Aerospacial S/A.

UPDATED

CHINA, PEOPLE'S REPUBLIC

Beijing Bao-Long Science & Technology 300 mm (8-round) ANGEL-120 multiple rocket

Development/Description

The 300 mm calibre (8-round) ANGEL-120 multiple rocket launcher, understood to have been developed by the Beijing Bao-Long Science & Technology Developing Incorporation Ltd, is based on a heavyweight 8 × 8 truck chassis fitted with a launcher assembly of two by four rocket tubes on its rear decking.

The rocket is spin stabilised in flight by a spin imparted to it at launch by the rocket tube and is fitted with a composite propellant rocket motor. The warhead fitted is a cluster munition type and can carry either 40 anti-personnel bomblets or 40 dual anti-personnel/anti armour bomblets.

Other warhead types such as a minelet carrying one can be developed according to customer requirements.

The guided rocket comprises five major sub-systems, warhead, simplified guidance control system, propellant, stabilising system and the actual rocket structure. The rocket has wraparound fins which unfold as the rocket leaves the launcher.

The lethal radius of one rocket round is stated to be 100 m while the full 8-round salvo can cover a 300 m radius.

A typical ANGEL-120 battery would comprise four launcher vehicles, two rocket resupply vehicles with 32 reloads in total, a fire direction (control) vehicle, technical support facilities and a truck-mounted crane.

The manufacturer claims the following key advantages for the ANGEL-120 artillery rocket system: high mobility, long range, high accuracy, great lethality and high volume of fire.

It is also claimed that the ANGEL-120 has a number of advanced features including a simplified guidance system, cluster type warhead, composite solid propellant motor and rocket spin imparted by the launch tube when the rocket is launched.

Status

As of late 2001, the exact status of the ANGEL-120 system is uncertain. Offered for export. As far as it is known, the ANGEL-120 is not used by the People's Liberation Army.

Specifications

Calibre: 300 mm
Number of rounds: 8
Weight:
(complete round) 740 kg
(warhead) 155 kg
Length:
(complete round) 6.9 m
Rate of fire: 8 rds/30 s
Range:
(minimum) 50,000 m
(maximum) 120,000 m
Time into action: 10 min
Launcher: heavyweight cross-country truck chassis

Manufacturer

Beijing Bao-Long Science & Technology Developing Incorporation Ltd.

UPDATED

CPMIEC 320 mm (4-round) WS-1 artillery rocket system

Development/Description

China Precision Machinery Import and Export Corporation (CPMIEC) has developed a long-range unguided artillery rocket weapon system called the WS-1 (alternative designation believed to be M-1).

According to CPMIEC, the WS-1 system has been designed to bridge the gap between conventional towed and self-propelled artillery systems and tactical missile systems.

Mounted on the rear of a forward control Mercedes-Benz type 6 × 6 cross-country truck (which is manufactured in China by China North Industries Corporation) are four cylindrical launcher tubes for the WS-1 rocket. Each rocket is 4.52 m long, has a diameter of 320 mm and weighs 520 kg at launch, 150 kg of which is the warhead.

The WS-1 rocket has four fixed fins and reaches a maximum flight altitude of 30,000 m with a maximum speed of Mach 3.6.

Maximum range is quoted as 80,000 m with a minimum range of 20 to 30,000 m with probable deviation of 1 to 1.25 per cent of range. No details of the warhead type have been revealed although to be effective at maximum range it is logical that some type of cluster munition warhead would be carried.

The four-round launcher is mounted on a turntable at the rear of the vehicle and has powered traverse and elevation. Hydraulically operated stabilisers have to be lowered at the rear of the vehicle to provide a more stable firing platform.

Once the four rockets have been launched, new rockets are reloaded manually into the launcher tubes from a truck backed up to the rear of the firing platform.

It is stated that an onboard digital computer provides assistance to increase the accuracy of the rockets. Accuracy is stated to be within one per cent.

CPMIEC 320 mm (four-round) WS-1 artillery rocket system (right) with rockets in foreground

Variants

WS-1B Rocket System

A model of this was first shown in 1993. It is based on a Mercedes-Benz 6 × 6 truck chassis and has four or eight rocket launcher tubes each with a diameter of 302 mm. The rockets have a maximum range of 180 km and can be fitted with various types of warhead. This is currently being offered for export and details are given in a separate entry.

Status

Production as required. Understood to be in service with the People's Liberation Army. Early in 1997 it was reported that China and Turkey signed a contract to produce, in Turkey, five batteries of CPMIEC 320 mm (4-round) WS-1 artillery rocket systems for the Turkish Land Forces Command.

Specifications

Calibre: 320 mm
Number of rockets: 4
Max range: 80,000 m
Min range: 20,000-30,000 m
Length: 4.52 m
Weights:
(complete round) 520 kg
(warhead) 150 kg
Max speed: M 3.6

Manufacturer

China Precision Machinery Import and Export Corporation (CPMIEC).

UPDATED

CPMIEC 302 mm WS-1B (4-round) artillery rocket system

Development

The China National Precision Machinery Import and Export Corporation (CPMIEC) 302 mm WS-1B (4-round) Artillery Rocket System (ARS), is a further development of the 320 mm (4-round) WS-1 that is described in detail in an separate entry.

The more recent WS-1B is based on a similar Mercedes-Benz (6 × 6) chassis to the older WS-1. The most significant difference being the smaller diameter of the rocket (302 mm against 320 mm) and the significant increase in range. The older WS-1 has a maximum range of 80 km whereas the more recent WS-1B has a maximum range of 180 km.

The latest WS-1B ARS has a new high-performance solid-propellant rocket motor and can be supplied fitted with different advanced warheads to meet different users operational requirements.

Description

The WS-1B ARS is based on a Mercedes-Benz (6 × 6) cross-country truck chassis that has been manufactured under licence by China North Industries Corporation (NORINCO). This is used for a number of other Chinese systems, such as the NORINCO 122 mm (40-round) Type 90/90A multiple-rocket system.

The layout of the 302 mm WS-1B ARS is similar to the larger calibre 320 mm (4-round) WS-1 system with the two-door fully enclosed forward control cab at the front and the powered launcher, which is offered with four or eight launcher tubes, mounted at the very rear. To the immediate rear of the cab is a fully enclosed cabin for additional crew members.

When deployed in the firing position, four stabilisers are lowered to the ground hydraulically to provide a more suitable firing platform. One of these is mounted either side to the immediate rear of the cab with the other two stabilisers mounted at the very rear of the chassis.

Launcher elevation is powered from 0 to + 60° at a speed of between 0.1 and 3° a second while traverse is powered at a speed of 30° left and right at 0.1 to 4° per second. Manual controls are also provided.

The 302 mm unguided surface-to-surface rocket consists of the warhead and its associated fuze, FG-43 solid-propellant rocket motor and the tail section with its four fixed fins.

The FG-43 is a single-chamber solid-propellant rocket motor with an advanced Hydroxy-Terminated PolyButadine (HTPB) composition rocket propellant.

The rocket can be fitted with two types of warhead, the ZDB-2 blasting warhead or an SZB-1 submunition. The ZDB-2 is a blast type warhead high explosive, surrounded by steel balls and prefabricated fragments. The warhead weighs 150 kg of which 70 kg is high explosive. It explodes with 26,000 fragments and is said to have an effective lethal radius 70 m.

The SZB-1 submunition warhead is optimised for use against area targets such as concentrations of tanks and other armoured vehicles. When over the target area the warhead is activated and attacks the vulnerable upper surfaces of the target. This warhead is said to be 1,380 mm long and contains 466 'bullets' with a dispersion area of 28,000 m². Each of the 'bullets' will penetrate 70 mm of conventional steel armour.

According to the manufacture, the WS-1B takes about 20 minutes to prepare for action with information being transmitted from the DZ-88B firing command truck to the HF-4 rocket launcher.

The DZ-88B firing command truck contains the fire-control computer and simulation trajectory system, global positioning system, gyro theodolite direction system, infra-red ranging system, radio communication and data transmission system. A field meteorological detection system is also provided as in a communication control unit.

A typical WS-1B rocket battery consists of:
1 × DZ-88B firing command truck (6 × 6)
6 to 9 HF-4 rocket launcher truck (6 × 6)
6 to 9 QY-88B transport and loading truck

The QY-88B transport and loading truck has a similar chassis and carries rockets which can be loaded using mechanical means to the launcher when the two elements are back to back. The rockets are packed in individual boxes and are then removed and loaded onto the QY-88B transport and loading truck by a crane mounted on the vehicle.

The system can operate up to a maximum height of 3,000 m above sea level and in ambient temperatures from –30 to +45°C and with a relative humidity of up to 98 per cent (at 15°C). It can also operate with a wind velocity of up to 8 m/sec and in rain or snow of 3 mm per hour.

Specifications

Calibre: 302 mm
Number of rockets: 4 or 8
Max range: 180,000 m
Min range: 80,000 m
Max flight altitude: 60,000 m
Length
(complete rocket) 6.182 m
(rocket motor) 4.709 m
Weights
(complete rocket) 708 kg (at launch)
(rocket motor) 538 kg
(propellant) 370 kg
(warhead) 150 kg
Total impulse: 888 kN/s
Max thrust: 275 kN
Motor burn time: 5.3 s (maximum)
Max speed: M 5.00 approx
Error of probability: 1 to 1.5%
Max acceleration: 40 *g*
Elevation: 0 to +60°
Traverse: 30° left and right

Status

Development complete. Ready for production.

Manufacturer

China Precision Machinery Import and Export Corporation (CPMIEC)

NEW ENTRY

CPMIEC 300 mm (10-round) A100 multiple rocket system

Development

Early in 2000 it was revealed that China had developed a new 300 mm (10-round) Multiple Rocket System (MRS) to meet the operational requirements of the People's Liberation Army (PLA) called the A100.

As early as 1995, sources from Moscow stated that China was seeking a technology transfer of the latest Russian 300 mm BM 9A52 (12-round) Smerch MRS which has the longest range of all of the Russian unguided surface-to-surface rocket systems.

The 300 mm BM 9A52 (12-round) Smerch MRS was developed by the Splav Scientific Production Concern to meet the operational requirements of the Russian Army and was first fielded in 1987.

So far the only known customer is Kuwait who took delivery of 27 systems between 1995 and 1996.

The prime contractor for the new 300 mm (10-round) A100 MRS is now known to be China Precision Machinery Import and Export Corporation (CPMIEC).

Chinese A100 300 mm (10-round) MRS on a new 8 × 8 chassis shown launching a rocket (Arms Knowledge) ***2000***/0067705

Description

The new Chinese 300 mm (10-round) A100 system is very similar to the Russian 300 mm BM 9A52 (12-round) Smerch in missile diameter, range, rate of fire and overall appearance.

The standard production Russian 300 mm BM 9A52 (12-round) Smerch is based on the older MAZ-543M (8 × 8) chassis but the Chinese version is based on a different 8 × 8 chassis with a different stabiliser design.

Chinese sources state that the A100 system has a maximum road speed of 60 km/h with the 300 mm rockets having a minimum range of 40 km and a maximum range of 100 km.

Early in 2001, CPMIEC stated that the maximum range of the system would be 120 km and that it was fitted with a simplified guidance system to lower costs. Maximum range was stated to be 120 km with warhead types being bomblet, fragmentation and penetrator. The Russian system has 12 launch tubes while the Chinese system has only 10.

The Russian system was originally fielded capable of firing three different rockets with a minimum range of 20 km and a maximum range of 70 km. These were the 9M55K with 72 HE fragmentation sub-munitions, 9M55K1 with five top-attack sub-munitions and 9M55F with HE warhead. More recently two new warheads have been developed, the 9M55K4 with anti-tank mines and the 9M55S fuel air explosive.

Details have recently been released of a new series of Russian rockets with a minimum range of 25 km and a maximum range of 90 km with similar warheads to those of the first generation. Given the range quoted by China it is possible that the technology for these rockets has been transferred to China as well.

Chinese sources indicated that the new A100 MRS has been developed to bolster the PLA's surface-to-surface firepower capability against the outlying islands of Taiwan and the Penghu Archipelago.

The A100 is seen as a complimentary MRS to the previously fielded 320 mm (4-round) WS-1 with a maximum range of 80 km and the WS-1B with a maximum range of 180 km.

Status

Production. In service with the People's Liberation Army. There are no known exports of this system.

Manufacturer

China Precision Machinery Import and Export Corporation (CPMIEC).

UPDATED

NORINCO 425 mm Type 762 (2-round) mineclearing rocket system

Development/Description

This is the largest of the NORINCO (China North Industries Corporation) heavy mineclearing rocket systems. It is mounted on a cradle assembly carried on the rear of a tracked vehicle chassis which is very similar to that of the Type 83 152 mm self-propelled gun-howitzer. Full details of the latter are given in the Self-propelled guns and howitzers (tracked) section. This chassis is also used for the Type 1989 120 mm tank destroyer. Details of this system, which has never been offered on the export market, are given in the Tank destroyers section. There are no known exports of this system, which is believed to have been only deployed by the People's Liberation Army (PLA).

The driver is seated at the front of the hull on the left side with the powerpack, consisting of the engine, transmission and cooling system, to the right. The remainder of the crew are seated in the fully enclosed and protected compartment that is located at the rear. The warhead section weighs 621 kg and probably contains a sectionalised payload of fuel-air explosive containers, in order to create the blast overpressure required to clear a minefield. The range is from 800 to 1,000 m using a 129 kg, fin-equipped, solid propellant rocket motor section, to attain a maximum flight velocity of 98 m/s. Each rocket can clear a passage

Specifications

Calibre: 425 mm
Number of launch rails: 2
Weight:
(rocket at launch) 760 ± 5 kg
(warhead) 621 kg
Max rocket range: 800-1,000 m
Max velocity: 98 m/s
Min velocity: 31 m/s
Elevation/depression: +45/+5°
Traverse: ±90°
Crew: 4
Combat weight: 27,000 kg
Power-to-weight ratio: 19.3 bhp/t
Length:
(travelling) 7.162 m
(hull) 6.405 m
(rocket) 4.7 m
Width:
(travelling) 3.236 m
Height:
(travelling) 3.2 m
Ground pressure: 0.62 kg/cm²
Track: 2.62 m
Length of track on ground: 4.601 m
Max speed:
(road) 55 km/h
(cross-country) 30-32 km/h
Max road range: 450 km
Fording: 1.3 m
Gradient: 60%
Side slope: 40%
Vertical obstacle 0.7 m
Trench: 2.7 m
Secondary armament: 1 × 7.62 mm machine gun
Ammunition: 500 rounds

NORINCO 425 mm Type 762 mineclearing rocket system deployed in firing position
0058454

130 m long and between 12 and 22 m wide. The vehicle is equipped with a 7.62 mm machine gun for self-defence purposes which can be dismounted for use on a special anti-aircraft tripod. Firing ports for rifles are located on either side of the crew compartment and at the rear where a door is provided for access. The launcher system is hydraulically controlled. Other systems available for the crew include Type 803 intercom system, Type 889 radio set, panoramic sight Type WG002, 0.5 m stereo-optical range-finder system and a Type TYG-1 night vision set for the driver.

Status

Production probably complete. There are no known export sales of this system. In service with the People's Liberation Army.

Manufacturer

Chinese state arsenals, China North Industries Corporation (NORINCO).

UPDATED

NORINCO 305 mm (8-round) Type 79 minelaying rocket system

Development/Description

Mounted on the same CA-30A (6 × 6) truck chassis as the NORINCO 305 mm fuel air explosive minesweeping rocket system (qv) and NORINCO 284 mm Type 74 minelaying rocket system (qv), the NORINCO (China North Industries Corporation) 305 mm (8-round) Type 79 minelaying rocket system uses the same welded tubular steel launcher assembly frame of the former system instead of the rails of the Type 74.

The elevation and depression of the assembly is +4 to +50°, the traverse is 95° left and 45° right. The sighting system is mounted on the left side of the launcher.

The Type 79 minelaying rocket shell carries a payload of 10 Type 69 plastic anti-tank mines. A ring stabiliser is fitted to the tail fins. Maximum range is 2,600 m. Four vehicles can lay a 600 × 350 m minefield. Firing interval between the rockets is 1.75 seconds.

Ready for launch each Type 79 mine laying rocket shell weighs 156 kg to which the rocket motor adds an additional 77 kg to bring the total up to 233 kg.

Muzzle velocity is quoted as 45 m/s with maximum velocity being 185 m/s. It can be used in ambient temperature conditions from –40 to +40 °C.

Status

Production complete. In service with the People's Liberation Army. It is understood that this system is no longer being marketed.

Manufacturer

Chinese state factories, China North Industries Corporation (NORINCO).

VERIFIED

Specifications
Calibre: 305 mm
Number of launcher frames: 8
Number of rockets: 8
Max range: 2,600 m
Rate of fire: 1 rd/1.75 s
Rocket payload: 10 plastic anti-tank mines
Minelaying battery salvo zone: 600 × 350 m
Elevation/depression: +50/+4°
Traverse: 95° left, 45° right
Max speed: 65 km/h
Fording: 0.85 m
Weight:
(with 8 rockets) 8,730 kg
Length: 6.918 m
Width: 2.20 m
Height: 3.00 m
Length of rocket: 2.927 m
Weight of rocket: 156 kg
Muzzle velocity: 45 m/s
Maximum velocity: 185 m/s
Max chamber pressure: 220/260 kg/cm^2

NORINCO 305 mm (8-round) Type 79 minelaying rocket system with launcher elevated
0021039

NORINCO 305 mm (8-round) fuel air explosive minesweeping rocket system

Development/Description

Mounted on the same CA-30A (6 × 6) chassis as the NORINCO 284 mm Type 74 and NORINCO 305 mm Type 79 minelaying rocket systems, the NORINCO 305 mm (8-round) fuel air explosive minesweeping rocket system uses the same welded tubular steel launcher assembly frame of the latter system instead of the rails of the NORINCO Type 74.

When deployed in the firing position no stabilisers are used. The sighting system is fitted on the left side of the launcher.

The elevation and depression of the assembly is +4 to +50° with the traverse from 95° left to 45° right. Maximum speed of the vehicle on paved road is 65 km/h.

The Fuel Air Explosive (FAE) rocket is 2.927 m long, weighs 156 kg at launch and has a flight velocity of 185 m/s. A ring stabiliser is fitted to the tail fins. A single round can clear an area 48 m in diameter of anti-personnel mines and thick grass/undergrowth. When used as an anti-personnel weapon the round can kill and/or wound all exposed personnel within an area of 24 to 60 m in diameter. Maximum range is 2,700 m. For transportation, the rocket and its warhead are carried in separate boxes with the rocket box weighing 71 kg and warhead box weighing 152 kg. The system can be used in ambient temperatures of from –15 to +50°C.

Specifications
Calibre: 305 mm
Number of launcher frames: 8
Number of rockets: 8
Length:
(rocket) 2.927 m
Weight:
(rocket) 156 kg
Max range: 2,700 m
Max velocity: 185 m/s
Anti-personnel mine clearance zone: 48 m diameter circle
Anti-personnel lethal zone: 24-60 m diameter circle
Elevation/depression: +50/+4°
Traverse: 95° left, 45° right
Max speed: 65 km/h

Status

Production complete. In service with People's Liberation Army. There are no known export sales of this system. It is believed that the system is no longer being marketed.

Manufacturer

Chinese state factories, China North Industries Corporation (NORINCO).

VERIFIED

NORINCO 284 mm (10-round) Type 74 multiple rocket minelaying system

Development/Description

The Peoples Liberation Army have several minelaying heavy rocket systems in service, the NORINCO (China North Industries Corporation) 284 mm (10-round) Type 74 multiple rocket minelaying system is mounted on the rear platform of a CA-30A (6 × 6) truck.

This launcher comprises a 4 m long five-rail I-shaped frame system for 10 rockets with a team of four required to load each unguided rocket manually. The fin-stabilised Type 74 rockets each carry 10 plastic-cased Type 69 or Type 72 anti-tank mines in a bulbous warhead.

The Type 74 launchers are normally deployed in batteries of four which can lay a single 400 × 400 m minefield with one salvo. The maximum range of the rocket is 1,500 m and all 10 can be fired within 15 seconds.

Specifications
Calibre: 284 mm
Number of rails: 10
Length:
(rail) 4 m
(rocket) 2.47 m
Weight:
(rocket) 127 kg
(system unloaded) 6,954 kg
(system loaded) 8,780 kg
Max range: 1,500 m
Muzzle velocity: 34.6 m/s
Maximum velocity: 120 m/s
Elevation/depression: +7/+48°
Traverse: 90° left, 45° right
Crew: 6
Configuration: 6 × 6
Max speed: 40 km/h
Length travelling: 6.44 m
Width travelling: 2.27 m
Height travelling: 2.93 m

The usual crew totals six, four of whom sit behind the driver's cab on an open bench seat. For firing, the launcher has two stabiliser legs that are lowered at the rear of the chassis and steel shutters which are used to protect the cab windows.

A later rocket type used is the NORINCO 284 mm Type 85 mine-laying rocket shell. This has a maximum range of 3,500 m and contains 10 anti-tank mines. A complete battery salvo has a laying zone of 700 × 500 m. The rockets are 2.6 m long and weigh 156 kg at launch.

Status
Production complete. In service with the People's Liberation Army. There are no known export sales of this system. It is believed that NORINCO is no longer marketing this system.

Manufacturer
Chinese state factories, China North Industries Corporation (NORINCO).

VERIFIED

NORINCO 284 mm Type 74 (10-round) rocket minelayer on Chinese (6 × 6) CA-30A truck chassis

NORINCO 273 mm (8-round) WM-80 multiple rocket system

Development
This NORINCO (China North Industries Corporation) 273 mm (8-round) WM-80 multiple rocket system has been designed for use in brigade (regiment), battalion and battery units. According to the United Nations Report on Conventional Arms for 2001, between 1992 and 2000 there was only one export sale of the WM-80. This was to Armenia who took delivery of four units in 1999.

According to NORINCO, a complete WM-80 multiple rocket system would consist of three subsystems, the Firepower system, Fire Command and Control System and Technology Service System.

These are further broken down. The Firepower System covers the rocket launcher, ammunition carrier and the actual rockets. The Fire Command and Control System includes an optional LLX05 meteorological radar system, forward observation vehicle, command post vehicle and the optional ASN-206 UAV system. Finally, the Technology Service System includes the mechanical maintenance vehicle and the electronics maintenance vehicle.

Each brigade would have four battalions, each with three batteries. A total of six launcher vehicles and six resupply vehicles would be attached to each battery. The standard combat unit is considered to be the battalion with the customer choosing his own Table Of Equipment (TOE) if he so wishes.

Fire-control and command duties throughout the brigade are accomplished by use of an automatic distribution network. This consists of main computers located at the brigade, battalion and battery HQ command levels with various terminals and radio/wire communications links to automatically distribute the data as directed.

To undertake the command and control functions from brigade level downwards the following assets are attached:

(1) At the brigade headquarters level - a Brigade Fire-Control and Command Vehicle (to act as the Brigade Command Station) and a meteorological radar unit
(2) At each battalion headquarters level - a Battalion Fire-Control and Command Vehicle (to act as the Battalion Command Station) and a surveillance/spotting radar set
(3) At each battery headquarters level - a Battery Fire-Control and Command vehicle (to act as the Battery Command Station).

A 4 × 4 light vehicle is used as the forward observation vehicle for the unit's command and control system. This is fitted with a laser range-finder, a location and orientation navigation system, radio and target co-ordinate information transmission equipment.

It is understood that Armenia has recently taken delivery of eight of these WM-80 systems from NORINCO.

NORINCO 273 mm (8-round) WM-80 multiple rocket system deployed in the field **2002**/0130031

Specifications
Calibre: 273 mm
Number of launch tubes: 8
Length: 9.55 m
(launch tube) 4.885 m
(rocket) 4.582 m
Width: 3.06 m
Height: 3.3 m
Weight:
(rocket) 505 kg
(warhead) 150 kg
(complete system) 34,000 kg
Warhead types: HE blast fragmentation, submunition
Max range: 80,000+ m
Min range: 34,000 m
Rate of fire: 8 rds in 5 s
Max velocity: 1,140 m/s
Elevation/depression: +60/+20°
Traverse: ±20°
Emplacement/deplacement time: 3-5 min
Reload time: 5-8 min
Crew: 5
Configuration: 6 × 6
Max road speed: 70 km/h
Range: 400 km
Gradient: 35%
Engine: 525 hp air-cooled diesel

Description

The system is mounted on the rear of a TA-580 (8 × 8) cross-country truck chassis. An eight-round launcher assembly (comprising two four-round rail-type container-launcher box banks) is fitted on a hydraulically raised rear deck turntable platform. Attached to the assembly is the aiming system which comprises the elevating mechanism, traversing mechanism and sight.

Inside the five person fully enclosed forward control vehicle cab is the launcher firing display (to display any command vehicle data inputs and the launch vehicle status), the operating and aiming system to lay the launcher assembly and the system operating console and control panel. Optional wire-link or radio communications systems are also available to receive fire-control and targeting data and allow automatic weapon aiming and firing.

The operator can also fire the launcher himself or, if required, perform remote off-vehicle firing. Single round and salvo launches can be performed. For firing, four hydraulic jacks (two at the rear and one on either side section) are lowered to stabilise the vehicle.

The resupply vehicle is based on a TA-580 8 × 8 cross-country truck chassis and carries two four-round launcher-container boxes. These are transferred to the launch vehicle by means of a 10,000 kg capacity hydraulic crane fitted just behind the truck cab and carried over the centre line of the truck bed. It is locked in place on a raised frame at the truck rear for travel purposes. The resupply vehicle is called the WM-80DY and is fitted with four stabilisers which are lowered to the ground when the rocket launcher is being resupplied.

The rocket itself is 4.582 m long, 273 mm in diameter and weighs 505 kg in total. A small rotary motor is carried to impart rotational spin stabilisation to the rocket in flight. There are also four rear-mounted fins fitted.

The following types of warhead can be fitted:

(1) 150 kg HE blast fragmentation with 34 kg of TH (40/6) high explosive, that breaks the casing into 16,822 fragments and has an effective casualty diameter of 69.8 m

(2) 150 kg submunition dispensing warhead with a payload of 380 dual-purpose bomblets, each of which has a HEAT (High Explosive Anti-Tank) warhead which can pierce 80 to 100 mm of conventional steel armour plate.

There are two types of fuzes fitted: the WJ-6A mechanical impact fuze for direct contact; and the MD-23A electronic proximity fuze for airburst or submunition dispensing, height depending upon warhead type used. Both fuzes can be selected and preprogrammed with the appropriate parameters at any time during the engagement sequence by the fire-control system.

At launch the initial muzzle velocity is 40 m/s, which rapidly builds up to a maximum of 1,140 m/s, the rocket attaining a height of 31,000 m at its maximum range launch elevation. Flight duration is 165 seconds with a maximum range of 80,000 m plus. Minimum range is stated to be 34,000 m. The rocket motor section uses 205.5 kg of case-bonded central channel HPTB solid propellant fuel.

According to NORINCO, the dispersion circle probability error (CEP) is between one and two per cent. The system has an operating temperature range of between –40 and +50°C and an effective storage period of 10 years.

NORINCO 273 mm rocket out of its launcher

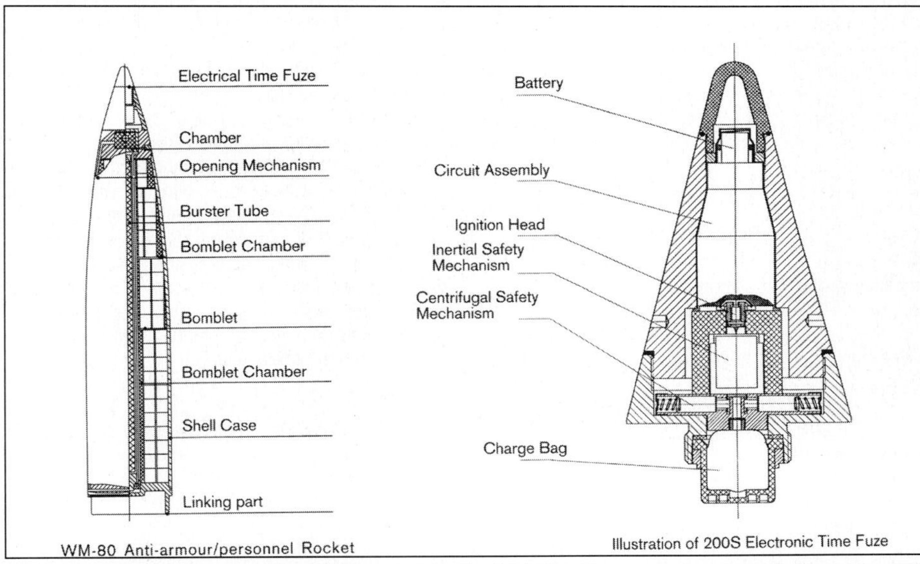

Two of the key subsystems of the AM-80 rocket are the WM-80 anti-personnel warhead (left) and the 200S electronic time fuze (right) **2002**/0130032

Status

Production as required. In service with Armenia (4 units). As far as it is known this system has yet to enter service with People's Liberation Army.

Manufacturer

Chinese state factories, China North Industries Corporation (NORINCO).

UPDATED

NORINCO 273 mm (4-round) Type 83 multiple rocket system

Development/Description

Developed by the No 123 and No 127 factories, the NORINCO (China North Industries Corporation) Type 83 273 mm (4-round) multiple rocket system was first begun in the 1960s. The design was changed many times before it gained its type classification in 1983. It is understood that there has been no recent production of this system by NORINCO. NORINCO is now concentrating its marketing efforts on other types of multiple rocket systems.

The system is based on the unarmoured Type 60-1 full-tracked artillery tractor with a five-man, fully enclosed cab at the front and a turntable with four 273 mm unguided rockets at the very rear. Before launching, the launch platform's two stabilisers are lowered at the rear of the chassis to provide a more stable firing platform.

Each rocket has a high-explosive warhead, although other types of warhead could easily be accommodated in a rocket of this calibre. The rocket has a maximum velocity of 811 m/s and all four rockets can be salvo-fired in 7.5 seconds with a time into action of 1 minute.

NORINCO 273 mm Type 83 MRS in firing position from rear with stabilisers lowered

Specifications

273 mm (4-round) Type 83
Calibre: 273 mm
Number of launch tubes: 4
Weight:
(rocket) 484 kg
(system, loaded) 17,542 kg
(system, empty) 15,134 kg
Length:
(travelling) 6.19 m
(rocket) 4.753 m
Max range: 40,000 m
Min range: 23,000 m
Max velocity: 811 m/s
Elevation/depression: +56/+5° 30'
Traverse: 10° left and right
Crew: 5
Width:
(travelling) 2.60 m
Height:
(travelling, over launcher) 3.18 m
(travelling, top of cab) 2.81 m
Ground clearance: 0.40 m
Track: 2.15 m
Length of track on ground: 3.248 m
Max road speed: 45 km/h
Range: 400 km
Min turning radius: 2.15 m
Fording: 1 m
Gradient: 60%
Side slope: 30%
Engine: 12150L-1 V-12 water-cooled 4-cycle direct injection diesel developing 300 hp at 1,600 rpm

The launcher is traversed and elevated manually, with the controls mounted on its left side. Fire-control equipment consists of one panoramic sight Type 85 and one indicator Type 85.

Details of the Type 60-1 tracked artillery tractor are given in *Jane's Military Vehicles and Logistics 2000-2001*. This is also used for towing field artillery and anti-aircraft guns.

Status

Production complete. In service with People's Liberation Army. There are no known export sales of this system. No longer being marketed.

Manufacturer

Chinese state factories, China North Industries Corporation (NORINCO).

UPDATED

NORINCO 253 mm (24-round) Type 87 minesweeping rocket system

Development/Description

Mounted on the hull of a modified NORINCO (China North Industries Corporation) Type 59 series MBT, the NORINCO 253 mm (24-round) Type 87 minesweeping rocket system uses a rectangular turret assembly fitted with 24 rocket launcher tubes in three layers of eight. It is also equipped with a laser range-finder to accurately measure minefield target distances. The system is used to clear anti-personnel minefields.

The system uses the following unguided minesweeping rocket types which are also used in the NORINCO 253 mm (10-round) Type 81 minesweeping rocket system covered in a separate entry:

Model	Type 81	Type 81-II
Calibre	253 mm	253 mm
Muzzle velocity	24 m/s	n/avail
Weight (rocket)	n/avail	74.5 kg
Max range	1,400 m	2,000 m (with jet efflux cone) 3,000 m (without jet efflux cone)
Mineclearing zone (salvo fire)	n/avail	10,000 m²

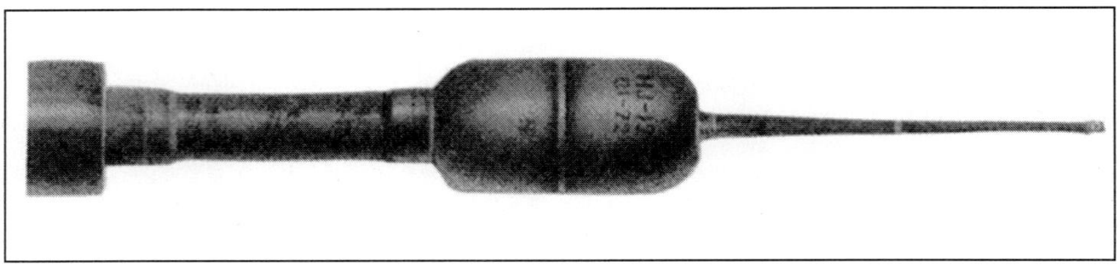

253 mm minesweeping rocket shell Type 81-II 0058455

The Type 81 round can be used in mountainous and jungle regions and apart from clearing anti-personnel mines, both rocket types may also be used for anti-personnel and fortification demolition roles.

The major difference between the two rounds is that the Type 81-II has a rigid-spike piezoelectric fuze fitted to its nose. This controls the height of the warhead's airburst, optimising its explosive effectiveness against semi-buried or ground-laid anti-personnel mines. A single Type 81-II round can also clear a 35 m diameter circle of anti-personnel mines. This type of system is also highly effective against infantry positions and other battlefield targets at close range.

Status

Production complete. In service with the People's Liberation Army. There are no known exports of this system.

Specifications

Calibre: 253 mm
Number of rocket tubes: 24
Number of rockets: 24
Max range: see text
Anti-personnel mine clearance zone: see text
Elevation/depression +45/0°
Traverse: 360°

Manufacturer

Chinese state factories, China North Industries Corporation (NORINCO).

VERIFIED

NORINCO 253 mm (10-round) Type 81 minesweeping rocket system

Development/Description

The NORINCO (China North Industries Corporation) 253 mm (10-round) Type 81 is a truck-mounted system for clearing anti-personnel mines for infantry assaults. It comprises 10 rocket launcher tubes (in two banks of five), fitted on a turntable assembly at the very rear of the truck's aft decking. Immediately to the fore of the assembly is a housing for 10 reload 253 mm rounds and a fuze container. The truck used is a 6 × 6 model and is fitted with an extended four door fully enclosed cab. When deployed in the firing position, shutters are raised to protect the windscreen and the windows in the upper part of the side door. To provide a more stabilised firing platform when the rockets are being fired, two stabilisers are lowered to the ground one either side at the rear of the chassis. A replacement wheel and tyre is carried in the horizontal position to the cab rear.

Specifications

Calibre: 253 mm
Number of rocket tubes: 10
Number of rockets: 10 + 10
Rate of fire:
(ripple) 12-14 s
Max range: 1,400 m
Anti-personnel mine clearance zone:
(full 10-round ripple) 60 × 10 m
Combat weight: 6,200 kg
Elevation/depression: +20/+50°
Traverse: 135°
Max road speed: 80 km/h
Fording: 0.850 m
Total weight of system: 8,300 kg
Length (travelling): 6.80 m
Width (travelling): 2.20 m
Height (travelling) 2.45 m

The NORINCO 253 mm (10-round) Type 81 Minesweeping Rocket System fires the same rockets as the Type 87 system covered in the previous entry.

Rate of fire is 1 round every 1.2 to 1.4 seconds. When deployed in the firing position shutters cover the front and side windows to protect them from blast.

Status

Production complete. In service with People's Liberation Army. There are no known exports of this system. It is understood that marketing of this system is now completed.

Manufacturer

Chinese state factories, China North Industries Corporation (NORINCO).

UPDATED

NORINCO 253 mm (10-round) Type 81 minesweeping rocket system in travelling configuration 0021040

NORINCO 253 mm (10-round) Type 81 minesweeping rocket system launching a rocket /021041

NORINCO 130 mm (30-round) Type 82 and Type 85 multiple rocket systems

Development/Description

The NORINCO (China North Industries Corporation) Type 82 multiple rocket system is an improved version of the NORINCO Type 63/70 family which is covered in detail in a separate entry.

The Type 82 consists of a 6 × 6 cross-country truck with a four-door, fully enclosed forward control cab. To the rear of this are 30 reserve rockets for immediate use and, at the very rear of the chassis is a 130 mm rocket launcher with three layers of 10 tubes each. Two salvos, each of 30 unguided rockets, can be launched in 5 minutes.

As the rocket launcher is mounted at the very rear of the chassis the foreward control cab is not fitted with any armoured shutters for its windows.

The launcher is also mounted on the top of the NORINCO Type 85 tracked APC to improve its cross-country capability. Full details of the NORINCO Type 85 APC are given in the Armoured personnel carriers (Tracked) section.

The 130 mm unguided rocket types available are shown in the table.

There is also a 1.05 m long single man-portable tripod-mounted launcher tube for a 130 mm rocket which is designed for use by commando, guerrilla or special forces units.

This can be disassembled into small loads for carriage – launch tube 10.5 kg, tripod 11.9 kg, sight 1.2 kg and remote firing device 1.25 kg. Its elevation range is +5 to +40° and its traverse range 7° to the left or right. Time to bring the weapon into action is about 1¾ minutes and to disengage between 1½ and 2 minutes.

Status

Production as required. In service with the People's Liberation Army (Type 82 and Type 85) and Thailand (Type 85). According to the United Nations Report on Conventional Arms issued in 2001, between 1992 and 2000 China exported the following multiple rocket launcher systems:

Country	System	Quantity
Armenia	WM-80	4
Sudan	not released	18
Bangladesh	not released	42

It is not known as to whether the latter two countries took delivery of 130 mm or 122 mm calibre systems.

Manufacturer

Chinese state factories, China North Industries Corporation (NORINCO).

UPDATED

Designation	Type 63 HE	Type 82 HE fragmentation	Type 82 HE-I	Extended Range HE
Diameter	130.43 mm	130.43 mm	130.43 mm	130.43 mm
Length	1.051 m	1.063 m	1.043 m	1.056 m
Weight (complete)	33 kg	32 kg	32.8 kg	32.5 kg
Max range	10,100 m	10,217 m	10,100 m	15,000 m
Warhead type	HE producing 1,916 fragments lethal out to 24 m radius	HE fragmentation with 2,600 6.35 mm diameter steel spheres. It produces 3,850 fragments lethal out to 42.1 m radius. Minimum rocket range is 3,000 m	HE with over 5,000 incendiary pellets producing over 2,367 fragments with a burning radius of 30 m	HE fragmentation with lethal radius of 32 m
Fuze type and settings	proximity or MJ-1 impact with instantaneous and delay actions			

Specifications

Model	Type 82	Type 85
Calibre	130.65 mm	130.65 mm
Number of barrels	30	30
Length of barrel	1.050 m	1.050 m
Number of rockets	60	60
Rate of fire	30 rds in 14.5-17.4 s	30 rds in 14.5-17.4 s
Carriage	6 × 6 truck	tracked Type 85 APC
Weight (in firing position)	9,000 kg	14,500 kg
Length	6.438 m	6.125 m
Width (travelling)	2.250 m	3.060 m
Height (travelling)	2.525 m	3.014 m
Elevation	0 to 50°	0 to 50°
Traverse	170° total	±15°
Crew	7	6
Gradient	30°	30°
Fording	0.85 m	amphibious
Speed (land)	80 km/h	60 km/h
Time (into action)	1 min	1 min
Time (out of action)	1.5-2 min	1.5-2 min

NORINCO Type 85 full-tracked APC fitted with 130 mm (30-round) Type 85 multiple rocket system **2000**/0085760

NORINCO 130 mm (30-round) Type 82 multiple rocket system in firing configuration **2000**/0085761

NORINCO 130 mm (19-round) Type 63 and Type 70 multiple rocket systems

Development/Description

The NORINCO (China North Industries Corporation) designed and manufactured 19-tube 130 mm calibre Type 63 rocket launcher was developed by the No 743 and No 247 factories and type classified in 1963.

It entered service with the People's Liberation Army mounted on the rear platform of the NJ-230 4 × 4 truck. There are two variants of the Type 63 mounted on this truck, the major difference being that the second variant, Type 63-1 has a covered crew cabin. In both cases the firing units are normally grouped into batteries of six vehicles.

The Type 70 version was developed from the Type 63 rocket launcher and mounted on the NORINCO Type 63 tracked APC chassis. It is assigned to rocket launcher units attached to the armoured and mechanised infantry divisions of the People's Liberation Army.

The launch tubes are arranged in two rows, the top one of 10 over another of nine.

The 130 mm unguided rocket types available are given in the previous entry. These were replaced in production by the 130 mm (30-round) Type 82 and Type 85 multiple rocket systems based on more modern tracked and wheeled chassis.

There is also a 1.05 m long single man-portable tripod-mounted launcher tube for 130 mm rockets which is designed for use by commando, guerrilla or special forces units. This can be disassembled into small loads for carriage – launch tube 10.5 kg, tripod 11.9 kg, sight 1.2 kg and remote firing device 1.25 kg. Its elevation range is +5 to +45° and its traverse range 7° to the left or right. Time to bring the weapon into action is about 1¾ minutes and to disengage between 1½ to 2 minutes.

Specifications

Model	Type 63-1	Type 70
Calibre	130.65 mm	130.65 mm
Length of barrel	1.050 m	1.050 m
Number of barrels	19	19
Number of rockets	38	39
Rate of fire	19 rds in 9.5-11.5 s	19 rds in 9.5-11.5 s
Carriage	4 × 4 cross-country truck	tracked YW 531 (Type 63) APC
Weight in firing position	6,887 kg	14,000 kg
Length		
(travelling)	6.659 m	5.746 m
Width		
(travelling)	2.140 m	2.798 m
Height		
(travelling)	2.435 m	2.625 m
Elevation	0-50°	0-50°
Traverse	-32° left to +148° right	180°
Crew	7	6
Gradient	30°	32°
Fording	0.85 m	1.232 m
Speed		
(road)	35-50 km/h	60 km/h
(water)	n/a	5.2 km/h
Time		
(into action)	2 min	1 min
(out of action)	2 min	2 min

Status

Production complete. In service with China, North Korea and Vietnam. This system is no longer marketed by NORINCO. According to the United Nations Report on Conventional Arms issued in 2001, between 1992 and 2000 China exported the following multiple rocket launcher systems:

Country	Systems	Quantity
Armenia	WM-80	4
Sudan	not released	18
Bangladesh	not released	42

It is not known as to whether the latter two countries took delivery of 130 mm or 122 mm calibre systems.

Manufacturer

Chinese state factories, China North Industries Corporation (NORINCO).

UPDATED

NORINCO 130 mm Type 63-1 MRS with covered crew cab on NJ-230 (4 × 4) truck chassis (Andrew Li)

NORINCO 130 mm Type 70 MRS on Type YW 531 APC **2001**/0085762

NORINCO 122 mm Type 89 (40-round) self-propelled multiple rocket system

Development/Description

The NORINCO (China North Industries Corporation) self-propelled 122 mm Type 89 (40-round) self-propelled rocket launcher was developed by No 674 factory, No 5137 factory and No 207 Research Institute and type classified in 1989.

It uses the same chassis as the Type 83 152 mm self-propelled gun-howitzer. This is covered in full detail in the Self-propelled guns and howitzers (tracked) section of this volume. As far as it is known, production of the Type 83 152 mm self-propelled gun-howitzer and the 122 mm Type 89 (40-round) self-propelled multiple rocket launcher has been completed. There are no known exports of either of these systems.

This chassis is also used as the basis for the NORINCO 425 mm Type 762 (2-round) mineclearing rocket system covered earlier in this section and the Type 1989 120 mm self-propelled tank destroyer. Details of the latter are given in the Tank destroyers section. It has not so far been exported.

This has been designed to accompany mechanised forces and, unlike many other self-propelled multiple rocket systems, has a fully armoured chassis.

The engine compartment is front right with the driver on the left-hand side. He is provided with a single-piece hatch cover that opens to the rear, a heated day periscope WG501 and a TYG-1 night driving periscope.

The other four crew members are seated in the fully enclosed armoured compartment at the back with bulletproof observation windows on either side and an entry door at the rear. It has torsion bar suspension and each side has six dual rubber-tyred roadwheels, a drive sprocket at the front, an idler at the rear and three track-return rollers. The first and last roadwheel stations have a hydraulic shock-absorber.

NORINCO 122 mm Type 89 (40-round) self-propelled multiple rocket system in travelling configuration

The vehicle commander has a Type WG501 heated periscope while the gunner has a Type 59 tank periscope, Type 84 panoramic telescope and a Type 66-1 direct fire sight.

The 122 mm (40-round) launcher is mounted on a turntable over the rear of the vehicle. It is almost identical to the 122 mm Russian BM-21 (40-round) system introduced in the early 1960s. The launcher can be traversed and elevated under full electric power and manual controls are provided for emergency use. The rockets can be fired singly or in a salvo with all 40 being launched in 20 seconds.

Mounted over the forward part of the vehicle is a reload pack with a further 40 122 mm rockets which can be reloaded into the launcher automatically. This reload feature is similar to that used on the Czech 122 mm RM-70 (40-round) multiple rocket system which made its first appearance in 1972.

Close-up of late production NORINCO 122 mm Type 89 (40-round) self-propelled multiple rocket system showing new armoured layer's position and different cover arrangement for forward pack of 40 rockets

Specifications

Calibre: 122.4 mm
Number of barrels: 40
Unit of fire: 40 + 40 rockets
Length:
(complete rocket) 2.87 m
(overall) 7.18 m
(chassis) 6.85 m
Weight:
(complete rocket) 66.88 kg
(propellant) 20.6 kg
Max range:
(rocket) 20,580 m
(vehicle, road) 450 km
Rate of fire: 40 rockets/18-20 s
Elevation/depression: +55/0°
Traverse: 102° left, 66° right
Crew: 5
Combat weight: 29,900 kg
Power-to-weight ratio: 17.39 hp/t
Ground pressure: 0.68 kg/cm²
Length: 7.18 m
Width: 3.145 m
(over tracks) 2.62 m
Height:
(overall) 3.18 m
(hull top) 1.47 m
Ground clearance: 0.45 m
Length of track on ground: 4.601 m
Track: 2.62 m
Max speed:
(road) 55 km/h
Average speed:
(road) 40-42 km/h
(cross-country) 30-32 km/h
Fording: 1.3 m
Gradient: 60%
Side slope: 30%
Vertical obstacle: 0.7 m
Trench: 2.6 m

Note: Extended-range (30,000 m) versions of the HE (lethal radius 28.2 m, 2.7 m long, 60 kg total round weight), HE cargo, HE fragmentation, HE incendiary and HE fragmentation incendiary (lethal radius 44.2 m and burning radius of about 30 m) rocket rounds are available. Details of these new rockets are given later in this section.

Late production NORINCO 122 mm Type 89 (40-round) self-propelled multiple rocket systems have a number of modifications including a different cover arrangement for the forward pack of 40 rockets and a fully enclosed armoured compartment for the layer on the left side of the launcher at the rear.

Status

Production complete. No longer marketed. In service with the People's Liberation Army. According to the United Nations Report on Conventional Arms issued in 2001, between 1992 and 2000 China exported the following multiple rocket launcher systems:

Country	Systems	Quantity
Armenia	WM-80	4
Sudan	not released	18
Bangladesh	not released	42

It is not known as to whether the latter two countries took delivery of 130 mm or 122 mm calibre systems.

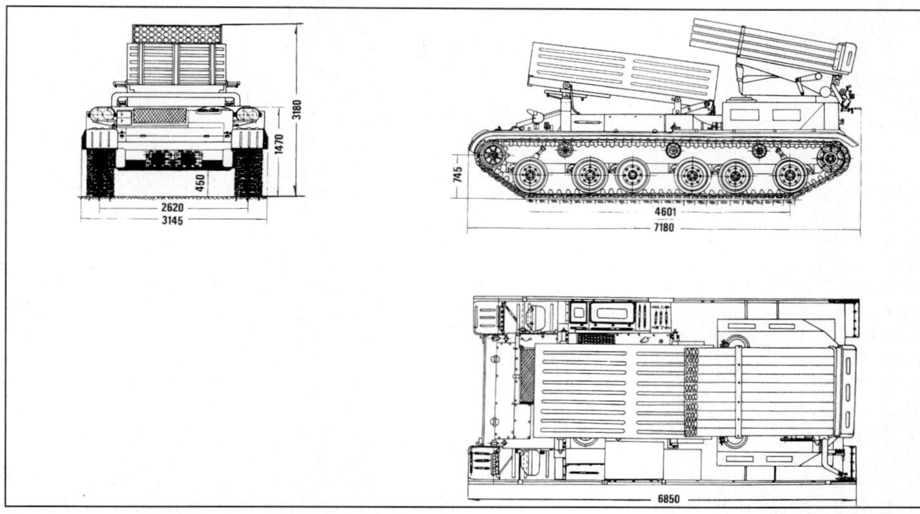

General arrangement drawing of the NORINCO 122 mm Type 89 (40-round) self-propelled multiple rocket system in travelling configuration

Manufacturer

Chinese state factories, China North Industries Corporation (NORINCO).

UPDATED

The 122 mm rocket types available are shown in the Table.

Type	HE	HE fragmentation	HE incendiary	HE cargo	HE fragmentation incendiary	Minelaying	Type 84 minelaying
Diameter	122 mm	122 mm	122 mm	122 mm	122 mm	122 mm	122 mm
Length	2.87 m	2.87 m	2.87 m	3.03 m	2.87 m	2.95 m	2.83 m
Weight							
(complete)	66.83 ± 0.66 kg	66.88 ± 0.66 kg	66.83 ± 0.66 kg	67 kg	66.8 ± 0.66 kg	58.2 kg	63 kg
(warhead)	18.3 kg	18.3 kg	18.3 kg	19.25 kg	18.3 kg	25.7 kg	33 kg
Max range	20,580 m	20,580 m	20,580 m	20,580 m	20,580 m	15,000 m	7,000 m
Warhead type	HE	6 kg HE with 4,100 steel balls	HE with 6,000 incendiary pellets and burning radius of 30 m	39 bomblets with 110 mm armour penetration and 7 m lethal radius	HE with 3,400 steel balls and burning radius of 30 m	6 AT mines or 96 anti-personnel mines (900 × 800 m full salvo laying zone)	8 AT mines or 128 anti-personnel mines (650 × 650 m full salvo laying zone)
Fuze type and settings	impact with instantaneous, short or long delay	proximity or impact with instantaneous, short or long delay	impact with instantaneous, short or long delay	proximity	impact with instantaneous, short or long delay	proximity	proximity

NORINCO 122 mm (40-round) Type 90 multiple rocket system

Development/Description

The NORINCO (China North Industries Corporation) Type 90 122 mm 40-round multiple rocket system is an indigenously designed and built system equipped with an automatic operating and laying system, an electric firing system and an automatically reloadable pack of 40 122 mm unguided rockets.

The chassis used is the Tiema SC2030 6 × 6 truck model with four rows of 10 tubes mounted above each other on a rotating cradle assembly. Elevation and traverse are performed by electrical systems.

A hydraulically operated collapsible awning is mounted over the rear deck for use as protection and camouflage. Conversion to the combat mode takes between 1½ and 2 minutes following arrival at the launch site.

The automatic operating and laying system consists of five main subsystems: an HJ-1 mini-computer; a launcher direction-finder (which takes less than 3 seconds to display the launcher's magnetic azimuth angle); longitudinal and transversal slope sensors; launcher assembly elevation and traverse position sensors; and a special vehicle power supply interface power pack unit.

The launcher is controlled by the computer through man/machine interaction. An interface is provided for connection to a battalion central command system, instructions from which can either be input directly via the interface or manually input by push-buttons. There are three modes of operating and laying the system: manual, electrical or automatic. Hydraulic jacks at the rear and sides of the vehicle are used to minimise the effect of uneven ground or vibration during firing. The computer automatically corrects the firing solution for any deviation of the firing angle caused by ground slope.

The electrical firing system fires the unguided rockets either singly or in a salvo with 0.5 seconds between rounds. The system comprises an electric igniter, remote firing device and ignition contacts and cables. Firing may be conducted both from the driver's cab or remotely outside the vehicle.

Specifications
Calibre: 122 mm
Number of tubes: 40
Max range:
(rocket) 30,000 m (40,000 m with long-range rocket)
(vehicle, road) 600 km
Unit of fire: 40 + 40 rockets
Rate of fire: 40 rds/18-20 s
Elevation/depression: +55/0°
Reload time:
(automatic) 3 min
Traverse: ±102° (left and right)
Combat weight: approx 20,000 kg
Configuration: 6 × 6
Length:
(travelling) 9.84 m
(tube) 3.0 m
Width:
(travelling) 2.5 m
Height:
(travelling) 3.245 m
Ground clearance: 0.412 m
Max road speed: 85 km/h
Max road range: 600 km
Fording: 0.9 m
Min turning radius: approx 12 m
Engine: air-cooled diesel developing 300 hp

NORINCO Type 90A (6 × 6) multiple rocket system deployed in the firing position
0058453

NORINCO Type 90A (6 × 6) multiple rocket system showing method of reloading 40 new 122 mm rockets
0058456

The fin-stabilised rockets used are of the 122 mm calibre extended-range composite solid propellant type and full details of these are given in the following entry. These rockets can be launched by other 122 mm multiple rocket systems such as those manufactured by Russia.

The automatic loading system consists of elevator, rack and feeder. Loading of the rockets can be performed either automatically, by push-button operation in the driver's cab, or manually by external operations. The loading mechanism can be rotated 90°. Automatic reloading of the launcher takes less than 3 minutes. A rocket-in-position and after-fire chamber display system in the driver's cab monitors the weapon status after loading or firing.

Variant

122 mm (40-round) Type 90A multiple rocket system

This is the latest version of the established NORINCO Type 90 and the four main improvements can be summerised as:

(a) ability to launch all types of unguided 122 mm rocket out to a maximum range of 40 km including high explosive and high explosive incendiary. Additional details of these rockets are given in the following entry
(b) installation of a new computerised fire control system which includes a Global Positioning System (GPS) and north finding device for greater accuracy. The panoramic sight is retained on the left side of the launcher for manual laying if required
(c) high level of automation for the operator
(d) command post vehicle can lay and control a number of Type 90A systems by remote control for maximum firepower.

The Type 90A has a modified chassis which is designated the Tienna XC 2200 (6 × 6) and this has a similar specification to the above apart from an overall length of 9.7 m, height of 3.20 m, fording depth of 0.7 m and a turning radius of 11 m.

A typical Type 90A battalion would consist of a battalion command vehicle based on a 6 × 6 chassis, three forward observation post vehicles mounted on a 4 × 4 light all terrain vehicle chassis and three batteries. Each of the latter has one battery command vehicle, six rocket launchers, six rocket resupply vehicles and one maintenance vehicle.

Status

Production. In service with the People's Liberation Army. According to the United Nations Report on Conventional Arms issued in 2001, between 1992 and 2000 China exported the following multiple rocket launcher systems:

Country	Systems	Quantity
Armenia	WM-80	4
Sudan	not released	18
Bangladesh	not released	42

It is not known as to whether the latter two countries took delivery of 130 mm or 122 mm calibre systems. Late in 2001, it was revealed that Oman had taken delivery of a quantity of Chinese 122 mm Type 90 series multiple rocket systems, which have been issued to at least one batallion.

Manufacturer

Chinese state factories, China North Industries Corporation (NORINCO).

UPDATED

NORINCO 122 mm calibre 30 km/40 km range rocket family

Development/Description

NORINCO (China North Industries Corporation) has developed a new family of 122 mm unguided calibre rockets for use with its range of 122 mm multiple rocket launchers and similar foreign made systems.

The rockets are called the 30 km range series with a further development being a 40 km range series.

In addition to being launched from Chinese North Industries Corporation 122 mm multiple rocket launchers, these new 122 mm rockets can also be launched by 122 mm systems manufactured by other countries.

Specifications

122 mm rocket specifications

Type	Length	Weight	Warhead weight	Max range	Min range
HE	2.87 m	67 kg	18.3 kg	20,000 m	9,600 m
HE*	2.87 m	67 kg	18.3 kg	20,000 m	9,600 m
HE*	2.87 m	67 kg	18.3 kg	20,000 m	12,000 m
HE	2.757 m	61 kg	18.3 kg	33,000 m	12,700 m
HE*	2.757 m	61 kg	18.3 kg	33,000 m	12,400 m
HE*	2.757 m	61 kg	18.3 kg	33,000 m	12,700 m
HE*	2.900 m	67 kg	22 kg	40,000 m	20,000 m
Cargo**	3.037 m	66 kg	18.3 kg	20,000 m	9,600 m
Cargo***	3.037 m	68 kg	28 kg	26,000 m	13,000 m
Cargo**	2.927 m	61 kg	18.3 kg	32,000 m	15,000 m
Cargo****	3.008 m	67 kg	22 kg	40,000 m	20,000 m
Mine*****	2.950 m	58 kg	26 kg	15,000 m	6,000 m

* contains high-explosive and steel balls for greater fragmentation effect
** contains 39 bomblets
*** contains 74 bomblets
**** contains 44 bomblets
***** contains six anti-tank mines
Note: In addition there are also steel high-explosive incendiary rockets that are designated SHEI

The 122 mm calibre unguided rocket is the most widely used system of its type in the world and is manufactured under licence or copied by many other countries.

Status

In production. In service with Chinese People's Liberation Army. Offered for export. It is considered probable that some of these 122 mm unguided rockets have been exported to undisclosed countries.

Manufacturer

Chinese state factories. China North Industries Corporation (NORINCO).

UPDATED

NORINCO 122 mm (40-round) Type 81 multiple rocket system

Development/Description

The NORINCO (China North Industries Corporation) 122 mm (40-round) Type 81 multiple rocket launcher system was developed by the No 743 and No 5137 factories and type classified in 1981.

The 40-tube launcher is carried on a CQ261 Hongyan 6 × 6 truck chassis. This is of the forward control type with a fully enclosed four door crew cab.

There is also a short 122 mm HE rocket which is used in conjunction with a 2.5 m long single-tube animal or man-portable tripod-mounted 122 mm rocket launcher. This is designed for use by guerrilla, commando or special forces units.

NORINCO 122 mm (40-round) Type 81 multiple rocket system

The unguided rocket weighs 46.3 kg complete and is 1.927 m long. Its maximum range is 13,400 m with the launcher having an elevation range of –3 to +50° and a traverse of +15°. The unguided rockets are fitted with a nose- mounted fuze with preselected instantaneous, short or long delay actions.

Total weight of the launcher when assembled for firing is 59 kg (which is made of the 25 kg tripod, the 30 kg launch tube and 4 kg for the sight and remote firing device with its 25 to 35 m long connector cable).

The launcher can also be fixed on a special mount for use on a light cross-country vehicle. In this application the launcher can either be fired from the vehicle or dismounted for ground use. A total of 10 to 15 reload rounds would normally be carried.

The Type 81 multiple rocket launcher system fires the same family of rockets as the 122 mm Type 83 already described. It can also fire the new NORINCO family of 122 mm long range rockets covered in a previous entry.

One part of this family has a maximum range of 30 km with the other having a maximum range of 40 km. There is also an anti-tank mine laying rocket with a maximum range of 15 km.

The standard indirect fire sighting system is on the left side of the launcher at the rear.

Status

Production as required. In service with the People's Liberation Army. It is considered that this system has now been replaced in production by the 122 mm Type 90/Type 90A system covered in the previous entry.

Manufacturer

Chinese state factories, China North Industries Corporation (NORINCO).

UPDATED

Specifications

Calibre: 122 mm
Number of rocket tubes: 40
Number of rockets: 40
Max range:
(rocket) 30,000 m (see text)
Elevation: 0 to +55°
Traverse: 102° left, 70° right
Rate of fire: 40 rds/18-20 s
Reload time: 8-10 min
Crew: 7
Combat weight: 15,160 kg
Configuration: 6 × 6
Length:
(travelling) 7.12 m
(rocket tube) 3.0 m
Width:
(travelling) 2.5 m
Height:
(travelling) 3.082 m
Max speed: n/avail

NORINCO 122 mm (24-round) Type 83 multiple rocket system and Type 84 minelaying rocket system

Development/Description

Development of the NORINCO (China North Industries Corporation) Type 84 minelaying rocket system began in 1977 and was jointly carried out by the People's Liberation Army Engineer Corps Technical Equipment Research Institute and factories Nos 743 and 5137.

The Type 84 consists of a 24-round 122 mm multiple rocket system mounted on the rear of a 6 × 6 cross-country chassis. The engine is at the front, the four-door, fully enclosed cab in the centre and the rocket launcher at the rear.

The sighting system is mounted on the left side of the launcher and consists of a panoramic sight Type 58 and a dependent yoke type sight collimator Type 58.

Speed of electro-operated laying is 5.7° in elevation and 9.3° in traverse. Speed of manual operated laying is 1.1 mil per turn of handle in elevation and 1.8 mil per turn of handle in traverse.

Reload rockets are carried by a dedicated resupply vehicle. The 122 mm minelaying cargo round carries

NORINCO 122 mm (24-round) Type 83 multiple rocket system

eight small anti-tank mines or 128 anti-personnel mines and has a maximum range of 7,000 m. A single vehicle salvo can lay a 650 m long by 400 m wide minefield at 6,000 m range. The rocket containing anti-tank mines weighs 62 kg and has a muzzle velocity of 350 m/s. The rocket is 2,770 mm long.

In addition to these there is also a short 122 mm HE rocket which is used in conjunction with a 2.5 m long single-tube, animal- or man-portable, tripod-mounted 122 mm rocket launcher. This is designed for use by guerrilla, commando or special forces units.

The rocket weighs 46.3 kg complete and is 1.927 m long. Its maximum range is 13,400 m with the launcher having an elevation range of -3 to +50° and a traverse of +15°. A contact fuze with preselected instantaneous, short or long delay actions is fitted.

Total weight of the launcher when assembled for firing is 59 kg (which is made up of the 25 kg tripod, the 30 kg launch tube and 4 kg for the sight and remote firing device with its 25 to 35 m long connector cable).

The launcher can also be fixed on a special mount for use on a light cross-country vehicle. In this application the launcher can either be fired from the vehicle or dismounted for ground use. A total of 10 to 15 reload rounds would normally be carried. The elevation/depression of the launcher is +55 to 0° and the traverse ±90° of the central axis. A full salvo takes between 15 and 20 seconds to fire.

The designation of the standard 122 mm (24-round) multiple rocket launcher system is Type 83.

The sighting system is on the left side of the launcher at the rear.

The multiple rocket launcher system fires the same family of rockets as the 122 mm Type 83 system already described. The Type 83 will also fire the latest generation of NORINCO 30 km and 40 km unguided rockets which are covered in a previous entry. These are fitted with a wide range of warheads enabling different battlefield targets to be engaged.

Specifications

Calibre: 122 mm
Number of barrels: 24
Max range:
(HE rocket) 20,580 m
(minelaying rocket) 7,000 m
(vehicle, road) 550 km
Elevation: 0/+55°
Traverse: 90° (left and right)
Crew: 6
Weight of system:
(loaded) 8,700 kg
Length:
(travelling) 6.51 m
(rocket tube) 3.0 m
Width:
(travelling) 2.2 m
Height:
(travelling) 2.7 m
Ground clearance: 0.285 m
Max road speed: 80 km/h
Fording: 0.85 m
Gradient: 60%
Engine: 135 hp diesel

Status

Production complete. In service with People's Liberation Army. This system is no longer marketed and as far as it is known, it was not exported.

Manufacturer

Chinese state factories, China North Industries Corporation (NORINCO).

UPDATED

NORINCO 107 mm (12-round) Type 63 and Type 81 multiple rocket systems

Development/Description

The China North Industries Corporation (NORINCO) 107 mm (12-round) Type 63 MRS was developed in the late 1950s as the replacement for the 102 mm (6-round) Type 427 and Type 488 multiple rocket systems which have now been withdrawn from service.

The Type 63 was developed by the No 803 and No 847 factories and type-classified in 1963. It was issued on the scale of 18 per Chinese infantry division. It is considered probable that this system has been replaced in most units of the People's Liberation Army by more recent NORINCO rocket launchers with a longer range and firing more effective rockets.

Large numbers of trailer and truck (for example GAZ-66) mounted 107 mm (12-round) multiple rocket launchers were observed during the fighting in Afghanistan late in 2001. At this stage it is not known as to the source of these systems, as the 107 mm rocket system has been made by many countries.

The launcher has three banks of four 107 mm barrels and is mounted on a rubber-tyred split pole-type carriage. In the firing position the wheels are removed and the launcher is supported by the two trails at the rear which are fitted with spades, which are carried on the top of the trails when travelling, and two short legs in the front of the carriage. The unguided rockets are spin stabilised.

An improved version of the Type 63 is mounted on a 4 × 4 truck, designated Type 81, with an enlarged cab to accommodate the launcher crew of four and 12 reload rounds. In transit the launcher is covered by a tarpaulin.

There is a seat for one crew member outside the cab and to the rear. When firing, the vehicle suspension is locked and the entire system can be removed from the vehicle if required and placed on a towing carriage.

Firing rate is 12 rockets in 10 to 12 seconds from the vehicle or 12 rockets is 6 to 9 seconds when deployed on the ground.

NORINCO 107 mm (12-round) Type 81 multiple rocket launcher on rear of 4 × 4 truck 0021044

The launcher can be fired from inside the cab or from a remote position. Maximum road speed is 100 km/h.

A pack model developed for use by airborne and mountain units weighs 281 kg in the firing position and can be dismantled into manpack loads. A lighter version of the Type 63 has been given the Western designation of Type 63-I. It is 136 kg lighter than the basic model and is distinguishable from it by its smaller spoked roadwheels and its four banks of three rockets.

The 107 mm rocket types available are given in the Table.

There is also a 0.9 m long Type 85 single man-portable tripod-mounted launcher tube for 107 mm rockets which is designed for use by commando, guerrilla or special forces units. This weighs 22.5 kg in total and can be disassembled into small loads for carriage. Its elevation range is +5 to +45° and its traverse range 7° to the left or right. Time to bring the weapon into action and to disengage is about 30 seconds. Typical rate of fire is 3 rds/min. A remote firing device is used to ignite the rocket.

The 107 mm rocket can also be fired by laying it on a suitable piece of sloped or raised terrain and pointing it in the direction of the target. In this application a set of small batteries with an electrical cable can be used as the firing mechanism. The accuracy is not very good against point targets unless the launch occurs very close to them but is more than sufficient for area targets such as population centres or large military installations.

Designation	Type 63-II HE	Type 63-III HE	Type 63-I HE incendiary	HE fragmentation	Type 63 HE Incendiary	Jamming Rocket
Calibre	107 mm	107 mm	107 mm	107 mm	107 mm	107 mm
Length	0.841 m	0.897 m (with mechanical fuze) 0.947 m (with electronic fuze)	0.915 m	n/avail	0.840 m	0.920 m
Weight (complete)	18.84 kg	n/avail	18.74 kg	18.9 kg	18.8 kg	18.3 kg
Max range	8,500 m	10,000 m	8,000 m	7,800 m	8,500 m	7,800 m
Warhead type	HE producing 1,214 fragments with lethal radius of 18.5 m	HE with 12.5 m lethal radius	HE incendiary with 40 s burning time	HE with some 1,600 6.35 mm diameter steel spheres	HE with some 1,600 incendiary pellets and a burning radius of over 21 m	parachute-deployed jammer unit with duration of up to 15 min
Fuze type and settings	radio frequency proximity or MJ-1 contact with instantaneous and delay actions					

Pedestal-mounted 107 mm (12-round) Type 63 multiple rocket launcher captured in Lebanon (Israeli Ministry of Defence)

NORINCO 130 mm single-tube rocket launcher deployed in the firing position 0021046

All of the rockets have a muzzle velocity of 34 m/s and can be used in ambient temperatures from –40 to +50°C.

Variants

During the 1982 invasion of Lebanon, a number of locally modified Type 63 MRSs were captured from the Palestinian guerrillas by the Israelis. These consisted of the basic 12-round launcher fitted to a crudely built pedestal and mounted on the rear of a light 4 × 4 truck chassis to increase mobility.

Iranian 107 mm MRS

The Defence Industries Organisation (DIO) of Iran are currently manufacturing a number of 107 mm (12-round) multiple artillery rocket systems including the FADGR and HASEB. Details of these are given later in this section under Iran.

Iraqi 107 mm MRS

Iraq has manufactured a copy of the Chinese Type 63 107 mm (12-round) Type 63 multiple artillery rocket launcher and additional details are given under Iraq.

North Korean 107 mm MRS

These are known to include a 107 mm (18-round) launcher on the VTT 323 (M1992) full tracked armoured personnel carrier and the 107 mm (24-round) M1992/2 self-propelled multiple rocket launcher.

South African 107 mm MRS

The South African company of Mechem Developments, part of the Denel Group, have reverse engineered the 107 mm (12-round) Type 63 system and now manufacture this under the designation of the RO 107 (12-round) Towed Multiple Rocket Launcher System. There is also a single 107 mm launcher called Inflict. Details are given later in this section under South Africa.

Turkish 107 mm MRS

MKEK of Turkey has manufactured a 107 mm (12-round) multiple rocket launcher and details of this are given under Turkey. Roketsan make the rockets for this system. Production of the launcher has been completed although Roketsan continues to manufacture 107 mm rockets for the home and export markets as and when required.

Specifications

Model	**Type 63**	**Type 81**
Calibre	107 mm	107 mm
Number of barrels	12	12
Number of rockets	12	24
Rate of fire	12 rds in 7-9 s	vehicle 12 rds/10-12 s ground 12 rds/6-9 s
Reload time	3 min	3 min
Carriage	split trail	4 × 4 light truck
Weight		
(firing position)	613 kg*	3,615 kg**
Length		
(travelling)	2.6 m	4.6 m
Width		
(travelling)	1.4 m	1.964 m
Height		
(travelling)	1.2 m	2.185 m
Elevation	0-60°	0-60°
Traverse	±15°	+45/–60°
Crew	5	4

Note:
* weight in travelling order without rockets is 385 kg.
** weight in travelling order with crew of four and one set of rockets is 3,709 kg.

107 mm (12-round) Type 63 multiple rocket launcher 0021045

Status

Production as required. In service with the following countries:

Country	**Quantity**	**Comment**
Albania	270	
Bosnia	n/avail	
Burkina Faso	n/avail	
Cambodia	n/avail	
China	n/avail	
Congo, Democratic Republic	10	
Iran	n/avail	local production
Korea, North	n/avail	local production
Libya	n/avail	
Myanmar	30	
Nicaragua	33	
Sudan	400	estimate
Syria	200	
Vietnam	360	
Zimbabwe	18	

They are also widely used by various guerilla forces in many parts of the world.

Manufacturer

Chinese state factories, China North Industries Corporation (NORINCO).

UPDATED

Other Chinese MRL systems

The People's Liberation Army is understood to have a number of other multiple rocket systems in service on which only limited information is available at present. It is understood that all of these use unguided rockets supplied by NORINCO (China North Industries Corporation).

180 mm (10-round) Type 71 multiple rocket launcher system

This is a truck-mounted system and has 10 cage-type launch rails for unguided 180 mm calibre rockets. These are 2.737 m long, weigh 134.8 kg at launch and carry a 45 kg HE warhead. Movable tail fins are fitted for stabilisation. Maximum velocity is around 620 m/s which gives a maximum range of 20,000 m.

20-round minelaying rocket system

This is believed to be truck-mounted. No other details are available.

10-round mineclearing rocket system

This is of unknown calibre and is carried on the rear of a 4 × 4 truck chassis. No other details are available.

Status

All systems are believed to be in service with the People's Liberation Army. As far as is known none of these multiple launch rocket systems has been exported outside of China. It is understood that there has been no recent production of any of these systems. None of the recent marketing material issued by NORINCO has listed any of these systems.

Manufacturer

Chinese state factories. It is probable that most of these systems have been developed by NORINCO (China North Industries Corporation).

UPDATED

CROATIA

RH ALAN 128 mm (24-round) LOV RAK 24/128 self-propelled rocket launcher

Development/Description

The LOV RAK 24/128 rocket launcher is based on the 4 × 4 LOV light armoured vehicle chassis with part of the hull modified to accept a 24-round 128 mm calibre rocket launcher assembly. Full details of the RH ALAN LOV family of 4 × 4 light armoured vehicles are given in the Armoured personnel carriers (wheeled) section. As far as it is known there have been no exports of the LOV (4 × 4) family.

The 128 mm (24-round) rocket launcher can fire two types of locally manufactured 128 mm rocket, designated the M91 and M93, both fitted with a nose-mounted impact fuze. Details of these unguided rockets are in the following Table:

Designation	**M91**	**M93**
Calibre	128 mm	128 mm
Weight		
(complete rocket)	23.5 kg	26 kg
(warhead)	8.5 kg	9 kg
(propellant)	4.5 kg	7.5 kg
Spinning speed	16,000	26,000
Max range	8,550 m	13,000 m

The launcher is fitted with an electronic elevation system of –5 to +45° and can be traversed a full 360°. An additional 24 rockets are stored in the rear section of the hull. The launcher system can be automatically levelled when at the firing position. Training the system and controlling the firing sequence is performed either on the vehicle or remote from it using a hand-held computer.

The official army designation for the system is the SVLR 128 M91 RAK-24 with SVLR standing for Samovozni Visecijevni Lanser Raketa.

Status

Production as required. In service with the Croatian and BiH armies. There are no known exports of this system.

Manufacturer

RH ALAN d.o.o.

UPDATED

RH ALAN 128 mm (24-round) LOV RAK 24/128 self-propelled rocket launcher in firing position
0008308

Specifications

Calibre: 128 mm
Number of barrels: 24
Length of barrel: 1.3 m
Sight system: PC-1 panoramic telescope
Fire-control computer: TT-2 hand-held unit
Traverse: 360°
Elevation: -5 to +45°
Rate of fire: single or ripple (all 24 rockets can be launched in 16.8 s)
Max range:
(standard rocket) 8,550 m
(extended range) 13,000 m
Basic load: 24 + 24 reload rounds
Crew: 3-4
Combat weight: 8,500 kg
Length: 5.89 m
Width: 2.36 m
Ground clearance: 0.315 m
Max speed: 100 km/h
Fuel capacity: 170 litres
Road range: 500-700 km
Fording: 1 m
Vertical obstacle: 0.5 m
Gradient: 65%
Side slope: 30%
Engine: Torpedo BF6L912S turbocharged diesel developing 130 hp at 2,650 rpm.
Transmission: manual gearbox giving 5 speeds, transfer case with 2 ranges, rigid axles with differential locks.
Steering: hydraulic, power-assisted
Tyres: 14.5R20 with run-flat inserts
Electrical system: 24 V
Armour protection: from 7.62 mm 51 API and HE shell fragments

RH ALAN 128 mm (12-round) VLR 128 M91A3 'RAK12' multiple rocket launcher

Development/Description

The RH ALAN 128 mm (12-round) VLR 128 M91A3 'RAK 12' is a mobile trailer rocket mounted system which fires the same family of unguided rockets as the RH ALAN 128 mm (24-round) LOV RAK

RH ALAN 128 mm (12-round) VLR 128 M91A3 RAK12 multiple rocket launcher deployed in firing position
0008309

24/128 unguided 128 mm self-propelled rocket launcher covered in a separate entry.

The 128 mm (12-round) rocket launcher can fire two types of locally manufactured 128 mm rocket, both fitted with a nose mounted impact fuze and details of these rockets are listed in the table:

Designation	**M91**	**M93**
Calibre	128 mm	128 mm
Weight		
(complete rocket)	23.5 kg	26 kg
(warhead)	8.5 kg	9 kg
(propellant)	4.5 kg	7.5 kg
Spinning speed	16,000	26,000
Max range	8,550 m	13,000 m

Specifications
Calibre: 128 mm
Number of barrels: 12
Length of barrel: 1.25 m
Weight:
(loaded) 850 kg
(unloaded) 580 kg
Max range:
(M91 rocket) 8,500 m
(M93 rocket) 13,000 m
Rate of fire: single round fire with 1 rocket/s, ripple fire with 1 rocket/0.75 s
Traverse ±15°
Elevation –1 to +46°

Status
Production as required. In service with the Croatian Army and Croatian Defence Council. There are no known exports of this rocket system.

Manufacturer
RH ALAN. d.o.o.

UPDATED

RH ALAN SVLR 122 mm (32-round) M96 Typhoon self-propelled multiple rocket launcher

Development/Description
The RH ALAN 122 mm (32-round) M96 Typhoon self-propelled multiple rocket launcher system is based on the former Czechoslovakia Tatra T813 (8 × 8) cross-country truck chassis. In the designation SLVR stands for Samovozni Visecijevni Lanser Raketa.

It has an unarmoured fully enclosed two-door cab of the forward control type with a pod of 32 rockets to the rear of the cab ready for rapid loading into the 32-round launcher assembly.

The launcher assembly is controlled by a microprocessor-based control panel system in the cab and mounted on the aft end of the truck's rear decking. Loading, under full power control, takes 2 minutes.

When travelling, a sliding tarpaulin cover extends right to the rear of the vehicle covering the launcher and the rockets so making the system difficult to distinguish from a conventional covered Tatra truck. With minor modifications the launcher system can be fitted to other truck types such as a MAN or Mercedes.

When deployed in the firing position four stabilisers are lowered to the ground. Standard equipment for the chassis includes a central tyre pressure regulation system that allows the driver to adjust the tyre pressure to suit the ground being crossed and a front-mounted winch.

The suspension system can also be locked when the truck is deployed in the firing system. The launcher has a powered traverse of 105° left and right with powered elevation from 0 to +50°. Manual back-up systems are fitted.

The 122 mm unguided rockets are normally fired in salvos with an interval of 0.6 seconds between each rocket being launched. If the operator selects single shots then one round is fired every second. The methods of firing are electronic and can be either from inside the cab itself or from a remote position up to approximately 30 m from the truck.

The latter also has a manual triggering option. The Typhoon is fitted with a Global Positioning System (GPS) and a gyroscope for accurate firing position and targeting purposes. Time to fire two complete salvos from the same launch position is 5 minutes.

The locally produced 122 mm RAK 122 M93 rocket has a maximum range of 20,400 m and is 2.878 m long. The rocket weighs 66 kg in total of which the warhead accounts for 18.5 kg and the propellant 20.5 kg. The maximum speed at motor burnout is 699 m/s. A point detonating fuzing system is fitted.

The full range of Russian Federation 122 mm calibre rockets and its clones can also be used. The Croatians are also using the 122 mm M21-OF 9M22U rocket with a point detonating fuze and similar 20,400 m maximum range to the M95.

RH ALAN SVLR 122 mm (32-round) M96 Typhoon multiple rocket launcher 0008310

Specifications
Calibre: 122 mm
Number of barrels: 32
Length of barrel: 3.00 m
Traverse: 105° left and right
Elevation: 0 to +50°
Rate of fire: single or salvo ripple
Max range: 20,400 m
Basic load: 32 + 32 reload rounds
Crew: n/avail
Combat weight: 23,500 kg
Empty weight: 19,000 kg
Length: 9.4 m
Width: 2.5 m
Height:
(travelling) 3.4 m
(at max elevation) 4.26 m
Levelling:
(longitudinal) ±6°
(traverse) ±10°
Track: 2.03 m
Wheelbase: 1.65 + 2.2 + 1.45 m
Max speed:
(road) 80 km/h
(cross country) 25 km/h
Fuel capacity: 520 litres
Road range: 500 km
Fording: 1.4 m
Vertical obstacle: 0.6 m
Gradient: 65%
Engine: Tatra T-930-3 12-cylinder air-cooled diesel developing 250 hp at 2,000 rpm
Transmission: dual-range 5-speed plus override providing a total of 20 forward and reverse gears
Tyres: 15.00 × 21
Brakes:
(main) air
(parking) mechanical
Electrical system: 24 V

In many respects, the RH ALAN 122 mm (32-round) M96 Typhoon self-propelled rocket launcher is a 122 mm version of the Yugoslav 128 mm (32-round) M-77 Oganj multiple rocket launcher developed in the early 1970s on a 6 × 6 chassis. Full details of this are given later in this section under Yugoslavia. As far as it is known there has been no recent production of this system. As production of the Tatra T813 truck chassis was completed some time ago, any new production would have to be on the more recent Tatra T815 (8 × 8) truck chassis.

Status
Production as required. In service with the Croatian Army. There are no known exports of this system although it is being offered for export.

Manufacturer
RH ALAN d.o.o.

UPDATED

RH ALAN 70 mm (40-round) VLR 70 M93A3 Heron mutiple rocket launcher

Development/Description
The RH ALAN 70 mm (40-round) VLR 70 M93A3 Heron multiple rocket launcher is a single-axle trailer-mounted all-weather cross-country assembly designed for the area fire support artillery mission.

The modular system has 40 launch tubes arranged in banks in two 20-round boxes. Each 1.6 m long tube is smooth, with a single stainless steel guide tube and a locking device to hold the rocket secure once loaded. When fully loaded the trailer is capable of being towed at speeds up to 15 km/h across country; on asphalt roads at speeds up to 80 km/h; and on tarmacadam roads at speeds up to 50 km/h. It is also air-portable as a helicopter load.

The Heron has the ability to launch unguided 70 mm calibre rockets either singularly at one per second or in ripples of two rockets per second up to the full capacity

Specifications
Calibre: 70 mm
Number of barrels: 40
Length of barrel: 1.6 m
Rocket weight: 8.6 kg
Warhead types: see text
Rocket length: 1.244 m
Max rocket speed at burnout: 440 m/s
Sight system: CN-4 mortar sight, M2HR quadrant
Time into action: 3 min max
Time out of action: 1.5 min
Weight:
(loaded) 1,250 kg
(empty) 920 kg
Length: 3.15 m
Width: 1.85 m
Height: 1.4 m
Traverse: ±15° (optional 360°)
Elevation: -1 to +46°
Rate of fire: single or salvo ripple
Maximum range: 8,100 m (M95 rocket system)
Basic load: 40 rounds
Launcher: M93A2 trailer system

RH ALAN 70 mm (40-round) VLR M93A3 Heron multiple rocket launcher deployed in firing configuration
0008311

of 40 rounds. A remote-control firing device can be used at a distance up to 25 m from the launcher.

When deployed in the firing position, the launcher is supported by a jack lowered under the tow bar and one either side on outriggers.

The current 70 mm rocket used is the RAK 70 mm TF M95, which has a launch weight of 8.6 kg with 1.7 kg of double base propellant and can be fitted with the following types of warhead:

(a) 3.7 kg M95 HE-fragmentation (with 1.45 kg HE and instantaneous action impact fuze)
(b) 3.7 kg M95-O illumination
(c) 3.7 kg M95-D smoke
(d) 3.7 kg M95-I incendiary
(e) 3.7 kg M95-T training

An improved 70 mm range rocket, which increases the maximum range to 10,000 m from 8,100 m, is in development. The rocket launch weight is increased to 10 kg with the additional weight accounted for by a larger rocket motor filled with 2.4 kg of double base propellant rather than the current 1.7 kg. Maximum velocity of the rocket is also increased to 550 m/s from 440 m/s.

Status
Production as required. In service with the Croatian Army.

Manufacturer
RH ALAN d.o.o.

UPDATED

RH ALAN 60 mm (24-round) VLR 60 M91 multiple rocket launcher

Development/Description
RH ALAN is manufacturing a mobile 24-round 60 mm calibre rocket launcher. The 24 rocket 60 mm launcher tube assembly is fitted to the trailer system used in the 12-round 128 mm RAK-12 system. The crew is three to five, with the 60 mm HE unguided rocket having a maximum range of 7,200 m. No other details are known at present.

Status
Production as required. In service with the Croatian Army.

Manufacturer
RH ALAN. d.o.o.

UPDATED

RH ALAN 60 mm (4-round) 4RL 60 M93A1 'Obad' self-propelled multiple rocket launcher

Development/Description
RH ALAN is manufacturing a small, man-portable 4-round 60 mm calibre rocket launcher. The launcher has four 60 mm rocket launcher barrels in a rectangular formation and is broken into two single-man loads for carriage. It weighs 49 kg in total. The crew is two to three with the 60 mm unguided HE rocket having a maximum range of 7,200 m. No other details are known at present.

Variant
The RH ALAN LOV-IZV reconnaissance vehicle version of the standard Torpedo LOV-OP 4 × 4 APC is equipped with a pedestal rocket launcher mount to the rear of its turret. The pedestal is equipped with a four-round M92A1 launcher assembly on either side.

Full details of the LOV family of 4 × 4 vehicles are given in the Armoured personnel carriers (wheeled) section earlier in this volume. As far as it is known all production of this system to date has been for the home market.

Status
Production as required. In service with the Croatian Army and Croatian Defence Council.

Manufacturer
RH ALAN. d.o.o.

UPDATED

LOV-IZV reconnaissance vehicle fitted with two four-round 60 mm M93A1 rocket launcher systems

CZECH REPUBLIC AND SLOVAKIA

122 mm RM-70 (40-round) multiple rocket system

Development/Description

The 122 mm RM-70 (40-round) MRS, which has the NATO designation M1972, was first seen in public during a parade in Czechoslovakia following the SHIELD-72 manoeuvres.

The system is essentially an armoured version of the Czech Tatra 813 (8 × 8) truck fitted with the same launcher as the Russian 122 mm BM-21 (40-round) MRS at the rear of the hull and an additional pack of 40 122 mm rockets to the rear of the cab for rapid loading.

Main advantages of the RM-70 over the Russian BM-21 are its rapid reloading capability, complete armour protection for the crew from small arms fire and shell splinters and much improved cross-country mobility.

The Tatra 813 chassis is fitted with a central tyre-pressure regulation system which enables the driver to adjust the tyre pressure to suit the type of ground being crossed.

Some vehicles have been fitted with a hydraulically operated BZ-T dozer blade at the front of the hull for preparing fire positions and clearing obstacles. It is also fitted with a winch with a capacity of 22,000 kg.

According to United Nations sources the following quantities of 122 mm RM-70 (40-round) series multiple rocket system were exported between 1992 and 2000:

Country	Quantity	Comment
Angola	40	from Slovakia in 1994
Ecuador	6	from Slovakia in 1995
Greece	150	from Germany in 1994
Rwanda	5	from Slovakia in 1997
Slovakia	24	from Czech Republic, 12 in 1999 and12 in 2000
Sri Lanker	8	from Czech Republic in 2000
Uruguay	2	from Czech Republic in 1996
	1	from Czech Republic in 1998
Zimbabwe	20	from Czech Republic in 1992

Russian designations	**M-21-OF, DB-1B**	**M-21-OF**	**DKZ-B**
Industrial index number	9M22U	9M22M	9M28
Diameter	122 mm	122 mm	122 mm
Length	3.226 m	2.870 m	1.905 m
Weight			
(complete)	77.5 kg	66 kg	45.8 kg
(HE fragmentation warhead)	19.4 kg	18.4 kg	19.4 kg
Range			
(max)	20,380 m	20,000 m	10,800 m
(min indirect)	1,500 m	1,500 m	2,500 m
(min direct)	500 m	500 m	400 m
Warhead types	HE fragmentation, smoke, incendiary		
Fuze settings	instant/single time delay/safe		
Fuze arming distance	150-400 m	150-400 m	150-400 m
Burnout velocity	690 m/s	700 m/s	450 m/s
Launchers	Grad RM-70 BM-21/SR-114 122 mm/DAC 665T Grad-1	Grad Grad-V RM-70 BM-21/SR-114 122 mm/DAC 665T Grad-1	Grad Grad-V RM-70 9P132 BM-21/SR-114 122 mm/ARO 122 mm/Trailer 122 mm/DAC 665T Grad-1

The original version of the RM-70 122 mm (40-round) multiple rocket system had a fully armoured cab (Willis A Bullard)

Specifications

RM-70
Calibre: 122.4 mm
Number of barrels: 40
Type of HE filler: 86% TNT, 14% Al
Unit of fire: 80 rockets
Reload time: 35 s
Elevation/depression: +55/0°
Traverse: 125° (left), 70° (right)
Crew: 6
Configuration: 8 × 8
Weight:
(system, loaded) 25,300 kg
(HE rocket fuze) 0.9 kg
(HE filler) 6.6 kg
Length:
(travelling) 8.8 m
Width:
(travelling) 2.55 m
Height:
(travelling) 2.96 m
Ground clearance: 0.4 m
Track: 2.03 m
Wheelbase: 1.65 + 2.2 + 1.45 m
Max speed: 75 km/h
Range: 1,100 km
Fuel capacity: 400 litres
Fording: 1.4 m
Gradient: 60%
Vertical obstacle: 0.6 m
Trench: 1.6 m
Engine: Tatra T-930-3 V-12 air-cooled diesel developing 270 hp at 2,700 rpm
Gearbox: dual-range 5-speed plus overdrive providing 20 forward and 4 reverse gears
Tyres: 15.00 × 20

RM-70/85
Crew: 4
Configuration: 8 × 8
Weight:
(with dozer blade) 25,400 kg
(without dozer blade) 24,300 kg
Length: (travelling) 8.70 m
Width: (travelling) 2.60 m
Height: (travelling) 3.03 m
Max speed: 85 km/h
Cruising range: 1,100 km
Vertical obstacle: 0.6 m
Trench: 2.00 m
Ground clearance: 0.41 m
Fording: 1.4 m

Latest version of the Mod 70/85 122 mm (40-round) multiple rocket system has an unarmoured cab and is shown here with launcher elevated
2000/0085763

Latest version of Mod 70/85 122 mm (40-round) multiple rocket launcher fitted with front-mounted dozer blade
0021048

The above figures do not always correspond exactly with those listed in the Status table as in many cases the RM-70 were delivered before the period 1992-1998.

122 mm Mod 70/85

This was introduced into service in the mid-1980s. It is the more recent Tatra T815 VNN (8 × 8) truck chassis fitted with the same 122 mm (40-round) multiple rocket system as the original RM-70 system. The major difference is that the Mod 70/85 is not fitted with a fully enclosed and armoured cab.

Standard equipment includes a front-mounted winch with a capacity of 12,000 kg and 85 m of cable and the cab is fitted with an NBC system of the collective and individual type.

The chassis has a central tyre-pressure regulation system that allows the ground pressure at each wheel station to be adjusted to suit the type of terrain being crossed. At the front of the vehicle is an SSP 1000 arrow-type snow-plough. A BZ-T815 dozer blade can also be quickly installed.

The fin-stabilised 122 mm unguided rocket types available are covered in the Table.

According to Czech Republic sources, the RM-70 122 mm launcher, when elevated at an angle of 50°, has a dispersion of 1/180 in range and 1/110 in traverse. A complete salvo of 40 122 mm rockets can be fired in 18 to 22 seconds. The system takes 150 seconds to come into action and 180 seconds to come out of action.

RM-70 upgrade

An international consortium headed by the German Diehl Group has developed an upgrade package for the RM-70 series 122 mm of rocket launchers.

The Slovak Army has a fully funded programme to upgrade 50 of its 87 systems to this new standard which is also applicable to the Russian BM-21 system and its many foreign derivatives.

Also working with the Diehl Group are Konstrukta-Defence (Slovakia), BGT (Germany), Junghans Geratetechnik (Germany) and Cerlag (France).

The existing 122 mm rocket has been modified with a new solid motor from Cerlag and a Honeywell/ Junghans DM42 fuzing system. These modifications increase the range of the rocket to 36 km.

For the launcher, the communications system, navigation system, fire-control system (including the computer), automatic aiming, accuracy enhancement and hydraulics re-evaluation have been improved.

Mod 70/85 122 mm (40-round) multiple rocket launcher from the rear (Michaell Jerchel) 0058457

In addition, the ammunition resupply and essential command and control systems used with these 122 mm rocket systems can also be enhanced.

122 mm M96 (32-round) Typhoon

Croatia has developed and placed in production the 122 mm M96 (32-round) Typhoon system on an unarmoured Tatra 8 × 8 chassis and details are given under Croatia earlier in this section. As far as is known this has yet to enter production for the export market.

Status

Production as required. In service with:

Country	Quantity	Comment
Angola	40	from Slovakia
Czech Republic	149	from Slovakia
Ecuador	6	
Finland	36	from Germany, called RaK H 89
Greece	150	from Germany
Libya	350	figure includes BM-21
Poland	30	
RFAS	n/avail	
Rwanda	5	from Slovakia
Slovakia	87	
Sri Lanka	8	from Czech Republic in 2000
Uruguay	3	from Czech Republic
Zimbabwe	52	some sources state 60

Manufacturer

ZTS Dubnica nad Váhom (Slovakia).

UPDATED

Military and Police Group Technopol 122 mm (40/80-round) VP 14 Krizan multiple rocket launcher system

Development/Description

The Slovak company Military and Police Group, Technopol has developed the 122 mm calibre 40-round VP 14 Krizan multiple rocket system for use with its 122 mm 40 Kus anti-personnel and Krizna minelet-carrying cargo rockets.

The launcher is mounted on the rear chassis of a Tatra T815VNN 8 × 8 truck as part of a combat engineer's minelaying system which comprises two vehicles, the minelaying vehicle itself and the ammunition resupply vehicle.

The minelaying vehicle has a variable superstructure onto which can be mounted either a mechanical minelaying system, or two 122 mm 40-round rocket minelaying systems or the mechanical minelaying system and a 122 mm 40-round rocket minelaying system. The actual rocket launcher system is understood to be the same as that used in the 122 mm Mod 70/85 122 mm (40-round) system covered in the previous entry and also based on a Tatra T815 series 8 × 8 cross-country truck.

The Kus rocket payloads of PPMi-S1 minelets can either be used alone to produce an anti-personnel minefield or be overlaid on a Krizna deployed anti-tank minelet field to hinder clearance.

This system has a fully enclosed and armoured cab to protect the crew from small arms fire and shell splinters. This is of a different design to that used on the RM-70 (8 × 8) 122 mm (40-round) multiple rocket system.

Status

Production as required. In service with armed forces (engineer units) of the Czech Republic and Slovakia. There are no other known exports of this system.

Manufacturer

Military and Police Group, Technopol, Slovakia.

UPDATED

Tatra T815 (8 × 8) chassis with VP 14 Krizan 122 mm (40-round) launcher in centre and mechanical minelaying system at the rear

Tatra T815 (8 × 8) chassis with two VP 14 Krizan launchers of 122 mm (40-round) rockets for laying anti-tank/anti-personnel minefields

Military and Police Group Technopol 122 mm MV-3 (3-round) rocket launcher system

Development/Description

The Slovak company Military and Police Group, Technopol has developed the 122 mm calibre MV-3 (3-round) launcher for use with its 122 mm unguided Krizna-S and Kus minelet cargo-carrying rockets.

The MV-3 consists of a lightweight wheeled trailer, which can either be pulled by two men or mounted on the rear of a light vehicle. The launcher is set up by extending the three legs (one of which also serves as the towing bar in the dismounted role), placing the unit on the ground pointing in the required direction and connecting the cable from the remote firing box.

Two 1.5 V batteries are then inserted in each minelet through openings in the rocket casing, following which the 122 mm unguided projectile is placed in a canister on the launch tray. The only adjustments are in launch angle, which is achieved using a small handwheel on the left of the launcher and in divergence (up to a maximum of 8° of the three projectiles). The remote firing box is also used to check the projectiles before launch, set the release timing fuzes according to the desired range and then carry out the actual launch. The maximum width of the minefield laid by one salvo is 100 × 100 m.

An optical sight is mounted on the left side of the launcher and the two rear tripod legs have adjustable feet.

122 mm MV-3 (three-round) rocket launcher deployed in the firing position with stabilisers in position, but with control box not attached **2002**/0130033

Specifications

Calibre: 122 mm
Number of tubes: 3
Length of tube: 1.35 m
Rocket types: Krizna-S, Kus
Crew: 2
Elevation: +15 to +40°
Traverse: 15° (total)
Length
(travelling) 1.60 m
Height
(travelling) 0.90 m
Diameter of wheels: 310 mm
Total weight: 103 kg
Weight of bag with launching apparatus: 12 kg
Launching facility: 2.55 kg
Weight (total): 130 kg (without rockets)
Dispersion: up to 8°
Rate of fire: 3 rounds in 4 s

Status

Production as required. In service with armed forces (engineer units) of the Czech Republic and Slovakia. There are no other known exports of this system.

Manufacturer

Military and Police Group, Technopol, Slovakia.

UPDATED

Military and Police Group Technopol 122 mm cargo rocket system

Development/Description

The Slovak company Military and Police Group, Technopol has developed a series of 122 mm calibre cargo rockets capable of laying remote anti-tank or anti-personnel mines. A mixture of the two can also be layed.

The 122 mm unguided rockets are:

(a) Krizna-R - used for long-range engagements (up to 17 km) and carrying four PTMi-D anti-tank minelets. It is fitted with a timing fuze to burst the casing open and deploy the minelets. The normal launching platforms are the heavier 122 mm calibre systems such as the Russian BM-21 (6 × 6) series and the former Czech RM-70/85 (8 × 8). Details of both of these systems are given elsewhere in this section.

The PTMi-D minelet is 116 mm in diameter, 124 mm high in armed position and weighs 2.7 kg. A 1.073 kg HE charge is used which enables it to destroy tank tracks and penetrate up to 110 mm of conventional steel rolled homogenous belly armour. When released from the cargo round's casing the minelet deploys a small parachute to orientate its downward trajectory and slow its descent. Upon impact with the ground, nine spring-loaded legs are released to ensure that it sits upright with the warhead uppermost.

The contact also primes the fuzing system and releases a short tilt rod, which detonates the mine when it is driven over by a heavy vehicle. The mine is also detonated if it is directly struck by a vehicle track and will self-destruct at one of three delay times that have been selected and set prior to launch. The warhead weighs 19.3 kg and the solid propellant rocket motor with fuel 46.9 kg.

(b) Krizna-S - used for short-range engagements (up to 3 km) and carrying the same fuzing system and number of PTMi-D anti-tank minelets as the Krizna-R. It is normally used with the MV-3 launcher assembly. The rocket motor weighs 14.7 kg and the warhead 19 kg. The solid propellant fuel load is 3.45 kg. Details of the MV-3 launcher are given in a separate entry. As far as it is known, there have been no exports of this system.

(c) Kus - used with either the MV-3 launcher assembly (covered in a previous entry) or the larger 40-round VP 14 Krizan system (covered in a previous entry) mounted on the rear of a Tatra T815 (8 × 8) truck chassis. The Kus carries five PPMi-S1 anti-personnel minelets. The minelets can either be used to form a separate anti-personnel minefield or be overlaid on a PTMi-D minefield to hamper clearance.

The PPMi-S1 minelet is 115.5 mm in diameter, 96 mm high and weighs 1.75 kg (with a 0.17 kg explosive charge). It contains some 1,900 fragments which are dispersed on detonation. The fragments have a minimum effective range of 8 m. The minelet also carries four decoys to mislead engineers trying to clear the minefield.

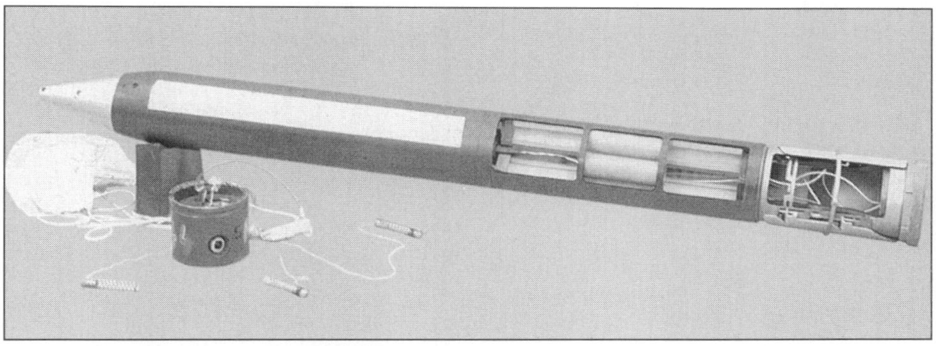

122 mm Kus cargo rocket which carries five PPMi-S1 anti-personnel mines

Specifications

Rocket	Krizna-R	Krizna-S	Kus
Calibre	122 mm	122 mm	122 mm
Length	3.187 m	1.56 m	1.4 m
Weight	67.5 kg	34.6 kg	32 kg
Cargo	4 × PTMi-D anti-tank mines	4 × PTMi-D anti-tank mines	5 × PTMi-S1 anti-personnel mines
Range	8 - 17 km	0.5 - 3 km	0.5 - 3 km
Launcher	RM-70/85	MV-3	MV-3
Type	wheeled	tripod	tripod

Status

Production as required. In service with armed forces (engineer units) of the Czech Republic and Slovakia. There are no other known exports of these rockets.

Manufacturer

Military and Police Group, Technopol, Slovakia.

UPDATED

Military and Police Group Technopol AGAT cargo warhead for 122 mm rockets

Development/Description

The Military and Police Group, Technopol of Slovakia is now marketing an improved cargo warhead, the AGAT, for use with 122 mm calibre unguided rockets of the type used with the RM-70, Mod 70/85, Russian BM-21 and equivalent 122 mm multiple rocket systems.

The AGAT can quickly replace the existing warhead or, alternatively, completely new rockets fitted with the warhead can be supplied. In the latter case the rear part of the rocket is designated JRM 122 mm thruster Type 70 with the complete rocket known as the JRKK-G. The weight of the complete round is 68 kg with range limits of 5,300 m minimum to 16,500 m maximum.

The AGAT warhead contains 56 dual-purpose anti-armour/anti-personnel bomblets which are ejected over the target area using a nose-mounted TM-120 time fuze. The bomblets are stabilised in flight by a rear-mounted ribbon. Each of the bomblets weighs 0.28 kg as is 38 mm in diameter. The cylindrical case is made from steel and contains a HEAT (High Explosive Anti-Tank) warhead that can penetrate between 110 and 130 mm of conventional steel armour. The casing also allows for a secondary fragmentation effect. An impact fuze and a self-destruct system is also fitted. The fuzing is designed to allow attacks on the vulnerable upper surfaces of armoured vehicles.

Specifications
(AGAT warhead)
Calibre: 122 mm
Number of bomblets: 56
Bomblet diameter: 0.038 m
Weight of bomblet: 0.28 kg
Warhead type: HEAT with steel bomblet body for fragmentation effect against personnel and matériel
Armour penetration: 110-130 mm steel plate

Status

Production as required. Believed to be in service with the armies of the Czech Republic and Slovakia. There are no other known customers of these rockets so far.

Manufacturer

Military and Police Group, Technopol, Slovakia.

VERIFIED

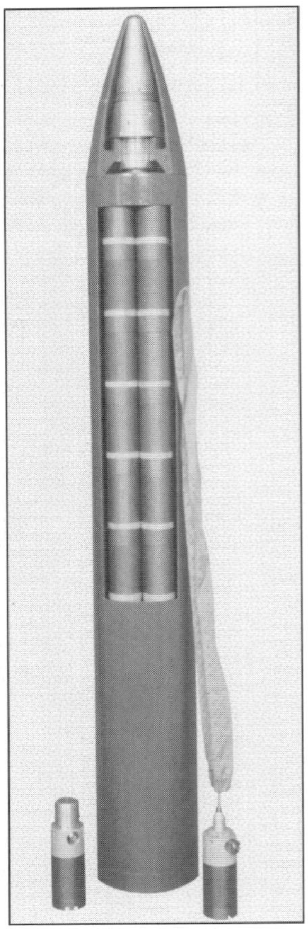

122 mm AGAT cargo projectile showing two of the submunitions
0058458

EGYPT

122 mm SAKR series multiple rocket launchers

SAKR-36 system

Following development work in the late 1970s, the SAKR Factory for Developed Industries, which comes under the control of the National Organisation of Military Production, successfully test-fired the SAKR-36 (or Hawk-36) 122 mm calibre unguided multiple rocket system in May 1981.

Based on the Russian 122 mm Grad BM-21 rocket, the weapon is fitted with interface systems allowing it to be fired from a number of rocket launcher types including the Helwan Machine Tools Company's 40-round launcher on a ZIL (6 × 6) truck chassis, the 30-round RL-21 launcher on a 3,500 kg Russian ZIL-131 6 × 6 or Japanese Isuzu 6 × 6 truck chassis, the 21-round launcher on a 4,000 kg modified Romanian 4 × 4 Bucegi SR-114 truck chassis, the lightweight 4/8-round light vehicle systems and the Russian BM-21 system. The RL 21 can also be mounted on other chassis such as the more recent Ural-4320 (6 × 6) and Mercedes-Benz 1222A/36.

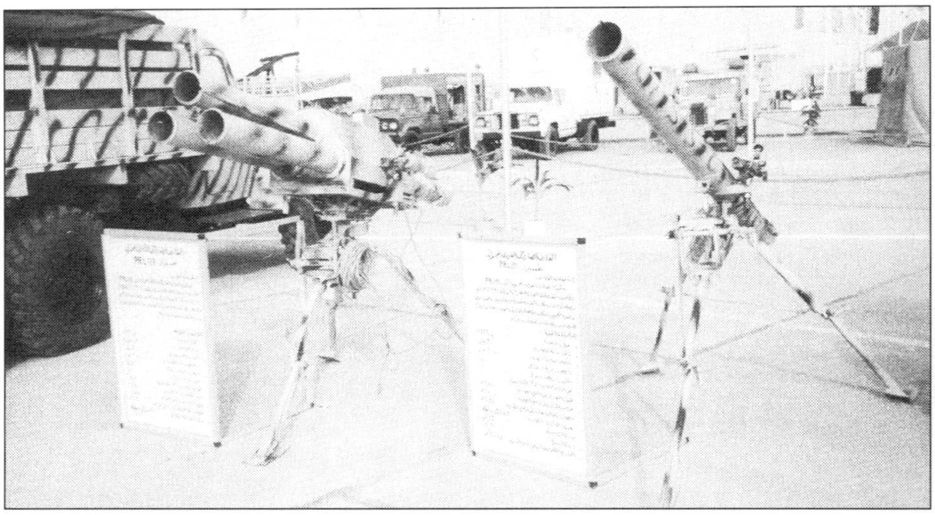

PRL-113 (three-round) and PRL-111 (single round) rocket launchers which have a maximum range of 10,800 m (Christopher F Foss)

Model RL 21 (30-round) 122 mm rocket launcher mounted on rear of ATS-59G full-tracked carrier

SAKR-36 rocket being launched during trials

Model PRL-111 122 mm single-tube rocket launcher deployed in firing position

Model PRL-81 107 mm single-tube rocket launcher deployed in firing position

Associated with the 122 mm MRL launchers is the Model RC 21 rocket carrier system. This is designed to hold 40 122 mm calibre 3 m long rockets in two 20-round protected stacks 3.5 m long, 1 m wide and 1 m high. There are also two protected boxes provided, one for rocket brake rings and the other for the rocket fuzes. Total weight of the carrier loaded is 3,150 kg and 450 kg empty.

The carrier is designed for mounting on any suitable military vehicle, typically a chassis of similar type to the launcher platform.

SAKR-18 system (bomblets)

To complement the SAKR-36 Egypt has also fielded the 122 mm calibre SAKR-18 system. This uses a 2 second burn rocket motor with 20.45 kg of double base propellant grain as used in the Egyptian copy of the Russian 122 mm Grad rocket. This gives a maximum range of 18,000 to 20,000 m at +48° launcher elevation. Rocket length is 3.25 m and weight complete is 67 kg.

The warhead contains 72 dual-purpose HEAT (High Explosive Anti-Tank) bomblets that can penetrate up to 110 mm of conventional steel armoured plate and has an effective anti-personnel lethal radius of 15 m. The fuze fitted is an electromechanical type which allows for dispensing altitudes between 2,500 and 3,000 m.

Specifications

30-tube RL-21 launcher
Calibre: 122.4 mm
Number of tubes: 30
Length: (tube) 3 m
Rate of fire: 30 rds/15 s
Elevation: 0 to +55°
Traverse: 102° left, 72° right

Rockets		
Type	**Long**	**Short**
Calibre	122 mm	122 mm
Weight		
(complete rocket)	66 kg	45.25 kg
(HE content)	6.4 kg	6.4 kg
Max range	20,400 m	11,000 m
Max velocity	690 m/s	450 m/s

Note: The above unguided rockets can be fitted with HE fragmentation, smoke and illuminating warheads

Platform: ZIL-131 (6 × 6)* Configuration	6 × 6
Weight	
(loaded)	11,250 kg
(unloaded)	8,500 kg
Length	
(travelling)	7.25 m
(firing)	7.25 m
Width	
(travelling)	2.6 m
(firing)	3.00 m
Height	
(travelling)	2.9 m
(firing)	4.2 m
Max speed	
(road)	60 km/h
(unpaved road)	40 km/h

* Other types of truck can be used including the Russian Ural-4320 (6 × 6), Mercedes-Benz Model 1222A/36 (4 × 4) and the ATS-59G full tracked prime mover. While the standard configuration is three rows of 10 tubes for some applications another row can be added making a total of 40 tubes

Lightweight man-portable and vehicle-mounted systems

In 1987, the Helwan Machine Tools Company Factory No 999, which manufactures the 30-round 122 mm rocket launcher system, exhibited two types of man-portable tripod-mounted 122 mm launchers for use by special forces and other mobile units. The single-round launcher is known as the PRL-111 while the three-round system has the designation PRL-113. Both have an optical mortar-type sight on their left sides which can be used both for direct and indirect fire roles. They are capable of being traversed 7° left and right and elevated from −10 to +40°. The rocket type used is the 122 mm SAKR-10 which gives the system a minimum range of 3,000 m and a maximum range of 10,800 m.

A 107 mm man-portable single-tube rocket launcher, the PRL-81 is also produced for the Egyptian armed forces and export. This is similar in many respects to the 107 mm Chinese Type 85 single-round man-portable system (see Chinese NORINCO 107 mm Type 63/81 entry this section).

More recently the Helwan Machine Tools Company has developed the Model RL812/TLC rocket launcher which is mounted on the rear of a 4 × 4 light vehicle. Details of this are given in a separate entry.

Specifications

PRL-111 launcher
Calibre: 122.4 mm
Rocket types: SAKR-10, RFAS 9M22M
Warhead types: HE fragmentation, illumination, smoke
Number of tubes: 1
Time into combat: 2.5 min
Rate of fire: 1 rd/min
Elevation: +10 to +40°
Traverse: 7° left and right
Min range: 3,000 m
Max range: 10,800 m
Length:
(deployed) 2.5 m
(rocket tube) 2.5 m
Width: (deployed) 1.5 m
Height: (max, deployed) 2.5 m
Weight:
(tube) 22 kg
(tripod) 32 kg
(fully deployed) 56±2 kg

PRL-113 launcher
Calibre: 122.4 m
Rocket types: SAKR-10, RFAS 9M22M
Warhead types: HE fragmentation, illumination, smoke
Number of tubes: 3
Time into combat: 3 min
Rate of fire: 3 rds/min
Elevation: +10 to +40°
Traverse: 7° left and right
Min range: 3,000 m
Max range: 10,800 m
Length:
(deployed) 2.5 m
(rocket tube) 2.5 m
Width: (deployed) 1.5 m
Height: (deployed) 1.79 m
Weight:
(single tube) 22 kg
(tripod) 74 kg
(fully deployed) 142 ±2 kg

PRL-81 launcher
Calibre: 107 mm
Rocket types: 107 mm models
Warhead types: HE fragmentation, illumination, smoke
Number of tubes: 1
Time into combat: 2 min
Rate of fire: 2 rds/min
Elevation: 0-45°
Traverse: 7° left and right
Min range: 3,500 m
Max range: 8,000 m
Length:
(deployed) 0.9 m
(rocket tube) 0.9 m
Width: (deployed) 0.78 m
Height: (max, deployed) 0.97 m
Weight:
(tube) 8.8 kg
(tripod) 13.3 kg
(fully deployed) 23 kg

Rocket
Calibre: 107 mm
Weight:
(complete rocket) 18.7 kg
(warhead) 8.33 kg
(HE content) 1.26 kg
Fuze: M491

Model RC 21 rocket carrier on ZIL (6 × 6) chassis which is designed to carry 40 122 mm rockets 0008313

Ural-4320-10 (6 × 6) truck chassis fitted with Egyptian-built RL-21 122 m (30 -round) multiple rocket system 0008312

Status

SAKR-36, production as required. In service with Egypt (with 21-round SAKR, 30-round SAKR and 40-round SAKR and BM-21 heavyweight rocket launchers) Iraq and Sudan.

SAKR-18, production as required. In service with Egypt (with 21-round SAKR, 30-round SAKR and 40-round SAKR and BM-21 heavyweight rocket launchers).

SAKR-10, production as required. In service with Egypt (with single-round PRL-111, three-round PRL-113).

PRL-81 and 107 mm rocket, in production. In service with Egypt and other countries.

According to the United Nations, Egypt did not declare any exports of any type of rocket launcher between 1992 and the year 2000.

Manufacturers

SAKR rocket systems and four/eight-round lightweight launchers: SAKR Factory for Developed Industries.

PRL-81/111/113, 21-, 30- and 40-round launchers: Helwan Machine Tools Company.

Enquiries to Helwan Machine Tools Company.

UPDATED

107 mm (12-round) RL812/TLC multiple rocket launcher

Development

For some years the Helwan Machinery and Equipment Factory marketed a 107 mm man-portable single-round type rocket launcher for the home and export markets called the PRL-81. Details of this are given in an earlier entry.

More recently the Helwan Machinery and Equipment Factory (or Factory 999 as it is also referred to) has developed the Model RL812/TLC rocket launcher system.

This is mounted on the rear of a 4 × 4 light cross-country vehicle, although it can be fitted to various other types of chassis, tracked and wheeled.

Specifications

Launcher
Calibre: 107 mm
Number of launcher tubes: 12
Length of launcher tube: 900 mm
Rate of fire: 12 rockets in 7 to 8 s
Max range: 8,200 m
Elevation: 0 to + 45°
Crew: 3
Configuration: 4 × 4
Weight
(with rockets) 3,108 kg
(without rockets) 2,900 kg
Height
(travelling) 2.08 m
(firing) 2.48 m (launcher at maximum elevation)

Rocket
Calibre: 107 mm
Maximum speed: 375 m/s
Weight:
(complete rocket) 18.7 kg
(warhead) 8.33 kg
(HE content) 1.26 kg
Fuze: M491

Description

The layout of the 107 mm (12-round) RL812/TLC multiple rocket launcher system is conventional with the engine at the front, fully enclosed two door cab in the centre and a flatbed type area at the rear. A replacement wheel and tyre is carried to the immediate rear of the cab and there is also additional unprotected seating at the rear.

Mounted on the rear of the chassis is a 12-round 107 mm rocket launcher with sighting system, manual traverse and elevation controls on the left side. This fires 107 mm unguided surface-to-surface rockets to a maximum range of 8,200 m.

These 107 mm rockets, which are manufactured by the SAKR Factory for Developed Industries, can be fitted with various warheads including high explosive, smoke and illuminating.

Status

Development complete. Production as required.

Manufacturer

Helwan Machine Tools Company

NEW ENTRY

107 mm (12-round) RL812/TLC rocket launcher mounted on 4 × 4 cross-country chassis
2002/0130018

GERMANY

Krauss-Maffei Wegmann/GUF 110 mm Light Artillery Rocket System (LARS)

Development

The 110 mm Light Artillery Rocket System (LARS) was developed in the 1960s and accepted for service with the Federal German Army in 1969, with first production systems being delivered the following year.

Prime contractor for the launcher was Wegmann and Company of Kassel (which merged with Krauss-Maffei in January 1999 with the new company being called Krauss-Maffei Wegmann GmbH & Co KG) and GUF was responsible for the rockets. The system was called the Artillerie Raketenwerfer 110 SF2 by the Federal German Army and was issued on the scale of one battery of eight launchers per division, with each battery having two (now Oerlikon Contraves) Fieldguard fire-control systems (on 4 × 4 truck chassis) and a resupply vehicle which carries 144 rockets.

The Federal German Army took delivery of 209 systems, but with the introduction of the longer-range 227 mm Lockheed Martin Missiles and Fire Control Multiple Launch Rocket System these 110 mm systems have now been phased out of service. It is possible that they may be passed onto another member of NATO.

The system underwent upgrading from LARS-1 to the LARS-2 in the mid-1970s. The programme has included a new fire-control system, more rocket types and mobility improvement by fitting the launcher to a new MAN (6 × 6) truck chassis. The MAN chassis is the standard 6 × 6 cross-country vehicle of the German Army and has improved mobility over the original chassis.

An air-transportable, towed, two-wheel, 15-round launcher was also developed by Dynamit Nobel but was not adopted by the Federal German Army. The same launcher was also tested on the rear of a Swedish Army Hägglunds Vehicle Bv 206 tracked vehicle, but by late 2001 this remained at the prototype stage.

Description

The 110 mm Light Artillery Rocket System system consists of a MAN (6 × 6) truck chassis fitted with a cab and a multiple rocket launcher at the rear. Mounted on the roof of the cab is a 7.62 mm MG3 machine gun with an elevation of +9 to +50°, a depression of 0° and a traverse of 360°. The crew of three, consisting of the commander, aimer and driver, all sit in the unarmoured cab.

The launcher at the rear consists of two banks each with 18 barrels and is balanced by a sprung equilibrator and elevated by a mechanical drive system. A drive is also provided for azimuth changes. The panoramic sight Model 59 is mounted between and above the two banks of 18 barrels and the aimer is provided with a seat which folds forwards for travelling. Before the rockets are fired from inside the cab two mechanical stabilisers are lowered at the rear of the vehicle and the covers at the end of each bank of launchers are folded forwards through 270° so that they rest on top of the launcher.

The unguided fin-stabilised 110 mm rockets can be fired singly, part rippled or in a complete ripple. When ripple firing there is a half-second gap between each rocket being launched and the complete load of 36 rockets can be launched in 17.5 seconds. The standard rocket has a solid propellant motor which burns for 2.2 seconds. It has a minimum range of 6,000 m and a maximum range of 14,000 m.

More recently two higher-performance rockets have been built using the same rocket motor casing but with improved propellants. These have maximum ranges of 19,000 m and 25,000 m respectively. Total length for all three rockets with their warheads attached is 2.263 m and their total weight is 35 kg. The interchangeable warheads are all 0.95 m long, 0.109 m in diameter and 17.3 kg in weight. There is a wide variety available which includes the following LARS-1 types: the Diehl training warhead, the Diehl practice warhead, the Diehl DM-15 smoke warhead (containing 6 kg of a smoke-generating compound for a 15 minute duration smoke screen), the Dynamit Nobel DM-11 fragmentation warhead with an impact fuze, the Dynamit Nobel DM-21 fragmentation warhead with a 20 m, altitude-actuated, ground proximity fuze and the Diehl DM-701 dispensing warhead containing eight AT1 anti-tank mines. For the LARS-2, Dynamit Nobel has produced the DM-28 practice warhead (converted from the older DM-15 smoke warhead), the DM-39 radar target warhead (fitted with a radar reflector to act as the pilot shot for the Fieldguard radar system to measure actual trajectory paths for salvo firing), the DM-711 dispensing warhead containing five parachute-retarded 2.2 kg AT2 anti-tank mines and a new smoke warhead (containing 8.4 kg of a smoke-generating compound for a 15 minute duration smoke screen). The company has also developed for the German MoD a bomblet warhead (containing 65 M42 or M77 0.21 to 0.23 kg multipurpose shaped-charge fragmentation submunitions, containing 0.03 kg of explosive each with an effective fragmentation radius of 3 m and an armour penetration capability of 65 to 100 mm) that has yet to be ordered for the German Army. Once the 36 rockets have been launched the vehicle normally moves off to be reloaded to avoid retaliatory fire. The launcher takes 15 minutes to be reloaded.

The rocket is also used in a subcalibre training role with the German Army's MLRS. Details of this are given in the entry for the MLRS under the US.

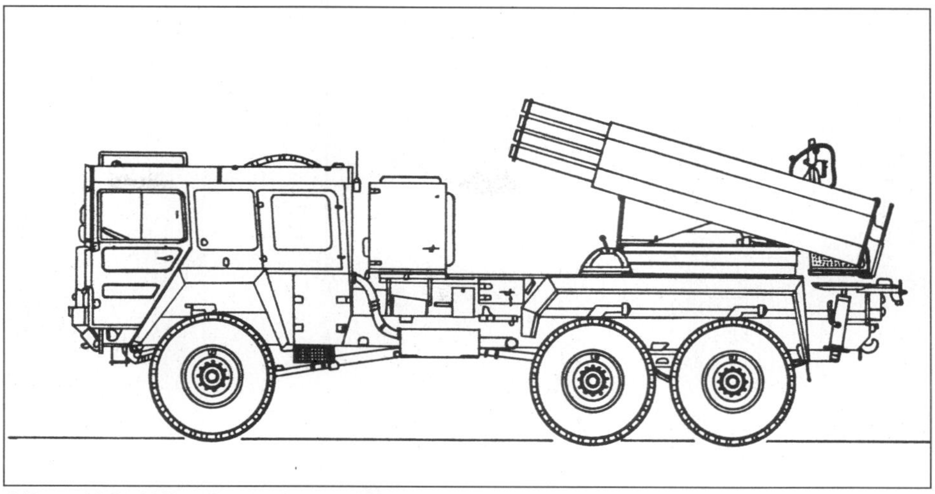

110 mm Light Artillery Rocket System 2 mounted on rear of MAN (6 × 6) 7,000 kg truck

Specifications

Calibre: 110 mm
Number of launcher tubes: 36
Elevation/depression: +55/0°
Traverse: 360°
Configuration: 6 × 6
Weight:
(system, loaded) 17,480 kg
Length: (travelling) 8.28 m
Width: (travelling) 2.5 m
Height: (travelling) 2.99 m

Status

Production complete. No longer in front line service with the German Army.

Manufacturers

Launcher: Krauss-Maffei Wegmann GmbH & Co KG.
Rockets: Dynamit Nobel Aktiengesellschaft.

UPDATED

Contraves-developed/produced Fieldguard fire-control system on MAN 5,000 kg (4 × 4) cross-country truck chassis. Radar operates in J-band with 300 to 20,000 m range (Michel C Klaver)

110 mm Light Artillery Rocket System on MAN (6 × 6) chassis with stabilisers lowered at rear and rocket being launched (Michael Jerchel)

INDIA

DRDO 214 mm (12-round) Pinacha multiple rocket launcher system

Development/Description

The Indian Defence Research and Development Organisation (DRDO) has developed a 214 mm calibre (12-round) multiple rocket launcher system. A battery of six launchers firing 72 unguided 214 mm rockets can neutralise a target area of 700 × 500 m.

The system has been designed for shoot-and-scoot fire missions with each launcher being provided with its own computerised fire-control system and automatic positioning system.

Mounted on the chassis of the Bharat Earth Movers Limited licence-built former Czechoslovakian Tatra 815 (8 × 8) Kolos cross-country truck chassis, the system comprises a fully enclosed five-man cab at the front with two pods each containing six rockets mounted at the rear of the chassis. Once the 214 mm rockets have been launched, replacement rockets have to be loaded one at a time.

The cab is fitted with an NBC system and standard equipment fit includes a central tyre regulation system,

DRDO 214 mm (12-round) Pinacha multiple rocket launcher system launching a rocket (DRDO) 0008314

that allows the driver to adjust the tyre pressure to suit the type of ground being crossed and passive night vision driving equipment. The maximum road speed of the vehicle is 80 km/h. When firing the unguided rockets, the cab windows are protected by shutters.

When deployed in the firing position, the system is stabilised by four hydraulically operated stabilisers which are lowered to the ground, two either side.

The 214 mm calibre rocket is 4.9 m long and weighs 275 kg at launch. The warhead weight is 100 kg. The warhead types available include anti-tank bomblet, anti-tank mine, incendiary and pilot shot. Maximum range is stated to be 45,000 m with all 12 rockets being ripple fired in 40 seconds.

The launcher assembly has electromechanical elevation and traverse, with traverse being 90° left and right of the centreline and elevation up to 55°.

Status

Believed to be in low rate production for the Indian Army. There are no known exports of this system.

Manufacturer

Designed by the DRDO. Produced by Indian state factories.

UPDATED

122 mm (40-round) multiple rocket launcher system

Development/Description

India is building an indigenously developed 40-round rocket launcher based on the Russian Splav Scientific Production Concern 122 mm BM-21 (6 × 6) design.

Mounted on the locally produced Shaktiman (6 × 6) 5,000 kg truck chassis, the launcher fires a new longer-range unguided 122 mm rocket called the Long Range Artillery Rocket (LRAR). This has been developed by the Armament Research and Development Establishment in Pune. The warhead contains 18 kg of TNT explosive and the rocket is fitted with a double-base solid propellant motor. The LRAR can also be fired from the Russian-designed and built 122 mm (40-round) BM-21. A full 40-round salvo can be fired within 20 seconds and reloading is thought to be manual.

Recent information has indicated that India has 150 122 mm systems in service. At this time it is not clear as to whether these are all Russian supplied BM-21 systems or a mixture of BM-21 and locally developed versions. These 122 mm (40-round) multple rocket systems have seen widespread use during the Indian/Pakistani border dispute.

Status

Production as required. In service with the Indian Army.

Manufacturer

Indian Ordnance Factories.

UPDATED

122 mm (40-round) BM-21 in service with the Indian Army (Indian Army)
2002/0130034

122 mm (40-round) BM-21 in service with the Indian Army (Indian Army)
2002/0130035

INDONESIA

IPTN 70 mm NDL-40 (20-round) multiple rocket launcher

Development/Description

IPTN has designed the 70 mm NDL-40 multiple rocket launcher for use in the ground support role. In the designation NDL-40, the former stands for Nusantara Dual Function Launcher. It comprises a launch assembly unit containing 20 launch tubes (in four rows of five) in a box mounted on a traversable base that is attached to a two-wheeled trailer fitted with four stabilisers, two at the front and two at the rear.

The complete unit weighs 754 kg and is aimed using a clip-on sight. The 70 mm unguided rockets may be fired singly or in salvos using a small remote-control device connected to the launcher by cable.

The standard production launcher has 20 launch tubes but according to the manufacturer this could be increased to 40 so doubling the firepower of the system. A vehicle as small as a jeep can tow the trailer. The 20-round launcher can also be mounted on a truck chassis or a fast patrol boat. If required, individual rocket tubes can be removed from the box and mounted on a tripod to give a fully man-portable system.

The 70 mm rocket used is the 2.75 in FFAR (Folding Fin Aerial Rocket) round fitted with the FZ-68 solid propellant motor and FZ-71 warhead. Maximum range in the ground role is 8,500 m, with a complete salvo covering a 200 × 300 m area.

The sighting system can be rotated over 6,500 mils and has a covered view of 200 mils and can be used for direct and indirect target engagement.

A typical battery would consist of six NDL-40 (20-round) rocket launchers which would be connected to a central fire control computer with forward observers providing target information for the computer prior to the targets being engaged.

The firing system is 265 mm long, 140 mm wide and 150 mm high, while the sighting system is 195 mm long, 145 mm wide and 200 mm high.

According to the manufacturer, the launcher tube has a life of 400 rounds before requiring replacement, while the detainer and contactor have a life of 4,000 rounds before requiring replacement.

Status
Production as required. In service with Indonesian Army (Special Forces). There are no known exports of this system.

Manufacturer
Industri Pesawat Terbang Nusantara (IPTN).
Enquiries to Divisi Sistem Hankam.

UPDATED

Specifications
Calibre: 70 mm
Number of rocket tubes: 20 (with growth to 40)
Elevation/depression: +65/-3°
Traverse: 180° (left and right)
Weight:
(launcher) 740 kg
(individual firing system) 10 kg
(sighting system) 4.5 kg
(firing control system commander) 2 kg
Length:
(launcher) 3.595 m
(launcher tube) 1.860 m
Width:
(launcher) 1.995 m
(launcher tube) 1.860 m
Height: (launcher) 1.600 m
Max range: 8,750 m
Towing vehicle: light truck (4 × 4)

IPTN 70 mm NDL-40 (20-round) multiple rocket launcher with stabilisers raised for travelling
0021050

Close-up of 70 mm NDL-40 (20-round) multiple rocket launcher being loaded with a 70 mm unguided rocket
2002/0130028

INTERNATIONAL

LAROM 160 multiple rocket launcher system

Development
The LAROM 160 Multiple Artillery Rocket System is a joint development between Aerostar of Romania and Israel Military Industries.

The system essentially consists of the Romanian Aerostar 122 mm (40-round) APRA multiple rocket launcher system, based on a DAC 15.215 (6 × 6) chassis upgraded and modified to accept the Israel Military Industries LAR 160 mm rockets. Full details of the LAR 160 rockets are given in a separate entry under Israel.

Israel Military Industries provides the 160 mm rockets and their associated pods and the Fire-Control System (FCS) computers while Aerostar and Israel Military Industries have jointly integrated the system.

It is understood that the Romanian Army hopes to upgrade its complete fleet of 160 existing 122 mm (40-round) APRA systems to the enhanced LAROM 160 configuration, with the first six systems already in service with another 12 systems following.

According to United Nations sources, the first three LAROM 160 systems were sent to Romania in 1999.

LAROM 160 artillery rocket system in travelling configuration
2001/0105967

160 mm rocket being launched from LAROM 160 artillery rocket system ***2001***/0073645

Description

The main improvement is that the system can use two Israel Military Industries 13-round pods of 160 mm LAR unguided rockets in place of the two 20-round pods that launch 122 mm Russian Grad type 9M22/9M22U rockets. The pods are loaded on to the launcher using a resupply vehicle fitted with a hydraulic crane.

To provide a more stable firing platform, four hydraulic stabilisers are lowered to the ground. These are positioned two either side. The first is positioned to the rear of the cab and the second at the very rear of the chassis.

By launching 160 mm LAR 160 rockets the range of the system is increased from just over 20 km to over 40 km. The 26 Mk 4 rockets can be launched in just over 46 seconds.

The LAR 160 rockets have a payload of 104 Israel Military Industries dual-effect AP/AM (anti-personnel/anti-material) bomblets, each of which has a self-destruct fuze. Two versions of the LAR 160 rocket are available, the standard unguided Mk 4 and the other fitted with the Accular guidance system which has a maximum range of 41 km.

For training and fire-adjustment missions LAR practice rockets can be used. These have a smoke signature charge which can be clearly seen by the forward observer.

In addition there is a training pod that has the same shape, dimensions and weight as the tactical pod launch container and is loaded with 13 inert 160 mm rockets that can be used in all operational modes. It is used to train the LAROM crew in launcher positioning, aiming, firing and reloading.

When deployed in the firing position the launcher is stabilised by four hydraulic jacks that are lowered to the ground by remote control. One of these is positioned either side of the chassis between the second and third road wheels with the other two at the rear of the chassis.

The LAROM 160 upgrade also adds a modern, automated FCS with Elbit computers being integrated into the launcher and in the Fire-Control Unit (FCU) which is installed in the cab of the vehicle.

The FCU provides an improved man machine interface and is operated from the cab of the vehicle or from a nearby shelter by a remote-control 50 m long cable. This indicates the number, type and status of the rockets in the pod, the physical and technical condition of the rockets and launchers.

It also allows for the input of all aiming sighting details, including meteorological conditions. The firing modes are manual or automatic and the operator can select either single shots, partial or complete salvos at four firing intervals, ranging from 0.5 seconds though to 1.8 seconds between rounds. At the slowest firing rate the complete load of 26 160 mm rockets can be launched in 46.8 seconds.

The new communication system is based on the Racal Panther VHF frequency hopping digital radio and allows status information to be transmitted back to higher command elements and firing requests to be sent direct to the fire units.

This information flow enables the system to come into action much quicker than the normal launcher and therefore gives a quicker target response time.

Variants

According to Aerostar, this upgrade is also applicable to the 122 mm (12 round) Awora multiple rocket launcher system covered in detail later in this section under Romania.

Status

Production. In service with the Romanian Army.

Manufacturers

Aerostar SA, Group Industrial Aeronautic.
Israel Milirary Industries.

UPDATED

FAMAE/RO Defence 160 mm Rayo multiple rocket system

Development

The 160 mm Rayo MRS has been developed jointly by FAMAE Ordnance of Chile and now BAE Systems, RO Defence, Rocket Motors Division of the UK. It is being designed to offer high lethality and high accuracy at ranges out to 45 km. The system is designed for use either as part of a rocket artillery battery or as an autonomous fire unit.

The 160 mm Rayo project, with its origins dating back to the mid-1980s is in three phases. The second phase already completed covered the development of the rocket and culminated in a qualification and proving programme. With the satisfactory conclusion of this programme Phase III was initiated.

Phase III, which commenced in October 1995, involves the completion of the full-scale qualification of the launcher and fire-control system and the integration of a complete 160 mm Rayo system onto a 6 × 6 wheeled vehicle chassis.

The programme also includes the development of a complementary training system comprising an embedded simulation package and a low-cost subcalibre reduced-range practice rocket with indicator warhead.

In August 1997, the now BAE Systems, RO Defence dispatched the German MAN SX 2000 (6 × 6) truck, platform and multimissile packs to FAMAE in Chile to integrate the first complete preproduction example of the RAYO 160 mm (24-round) multiple rocket launcher. This made its first public appearance in mid-September after which the Computing Devices Canada digital ballistic fire-control system was installed.

Early in 1999 the RAYO 160 mm (24 round) artillery rocket system completed a major milestone in its development following the successful completion of the first phase of its full Weapon Systems Integration Trials in Chile.

These trials took place in the extreme conditions of the Atacama desert and consisted of a series of rocket firings as part of a complete weapon system.

This was the first time that the 160 mm unguided rockets had been launched off the launch vehicle using the full vehicle mounted fire-control system. These confirmed the ballistic performance of the 160 mm rockets under a range of pre-conditions temperatures.

The most recent trials were part of a phased series of trials being carried out in Chile. During late 1998, an extensive series of sub-system proving trials took place under road and cross-country conditions.

During qualification trials, which took place in Chile in the first half of 2000, the 160 mm unguided rockets achieved a maximum range of 52 km against the required range of 45 km. The initial Chilean Army requirement is believed to be a batch of six systems, sufficient to form one battery.

Although the first customer for the 160 mm Rayo will be the Chilean Army, there is already considerable export interest in the system and a number of undisclosed potential customers have requested detailed briefings on the 160 mm Rayo.

In order to facilitate this, the construction of a rocket manufacturing plant was started in 1990 at FAMAE's facilities near Santiago. The plant was completed under Phase II by the now BAE Systems, RO Defence in 1992.

The 160 mm Rayo rocket has been developed to carry two types of warhead. In Phase III the rocket incorporates an HE warhead fitted with a proximity fuze designed to maximise fragmentation and the mean area of effect. The second type of warhead, using a bomblet dispensing system fitted with an airburst fuze, is expected to be developed later following successful qualification of the HE variant.

FAMAE and the now BAE Systems, RO Defence have established a joint venture company in Chile, FAMAE Ordnance Ltda to handle 160 mm Rayo sales.

Rayo 160 mm (24-round) multiple rocket launcher system on MAN SX2000 (6 × 6) chassis deployed in the firing position 0021052

Description

The 160 mm Rayo system is of modular design to facilitate mounting on a wide range of tracked and wheeled chassis types. The standard system comprises two removable rocket pod containers (RPC) each of 12 rockets, a launcher platform with power elevation and traverse and a ballistic fire-control system with an inertial navigation unit. This configuration enables all normal operations associated with deployment and firing of the system to be operated from the cab. The system is designed so that it can be operable by one person if needed, though a minimum crew of two would be the norm.

The Rayo rocket is approximately 3.2 m long, has a diameter of 160 mm and a normal sea-level range of 45 km. The rocket motor section case is of a strip laminate design and uses a cast HTPB composite propellant filling. The combination of propellant and case material has been demonstrated to meet the requirements for Insensitive Munitions, enhancing system survivability and crew safety. The HE warhead utilises a thin-wall high-strength case containing an RDX/TNT composition which maximises blast and fragmentation effects. The warhead is detonated by an electronic proximity fuze with preset burst height and back-up point-detonating mode to ensure maximum reliability of operation.

For ease of logistic handling and storage, rockets are supplied ready to fire in a disposable RPC. The rocket pack comprises a lightweight aluminium framework with 12 launch tubes made from glass fibre material each containing one rocket. The launch tubes are enviromentally sealed ensuring maximum life and reliability and allowing field storage in times of potential conflict. The rocket packs may be transported by air, land and sea, acting as a through life container.

In the standard system each launcher has its own independent digital fire-control system incorporating an inertial navigation system. This gives high first-round accuracy without the need for firing a sighting rocket, which could easily give away the launcher's position either visually or to counterbattery radar tracking. The integral INS provides both route navigation and obviates the need for ground survey giving added flexibility and minimising exposure times. The cab-mounted displays and controls enable the crew to rapidly compute the required ballistic solution and aim the system using either current or stored preplanned fire mission data. The system also allows a range of firing sequence options from single round to a full ripple launch.

Two RPCs are mounted on a launch platform which is hydraulically powered in traverse and elevation. For the last movement, a system called 'Ballscrew' has been developed to minimise friction and backlash. This enables the elevation movement to be accurately controlled. Firing platform stability is ensured by four hydraulic stabilisers mounted on the launch platform and deployed from within the cab. Redundancy is built in so the launcher can be traversed and fired either automatically or manually. Reload times are minimised by the use of the disposable RPCs. Reloading is carried out simply by lifting the fired RPC off the launcher with a jib crane on a support vehicle and replacing with a ready to fire RPC.

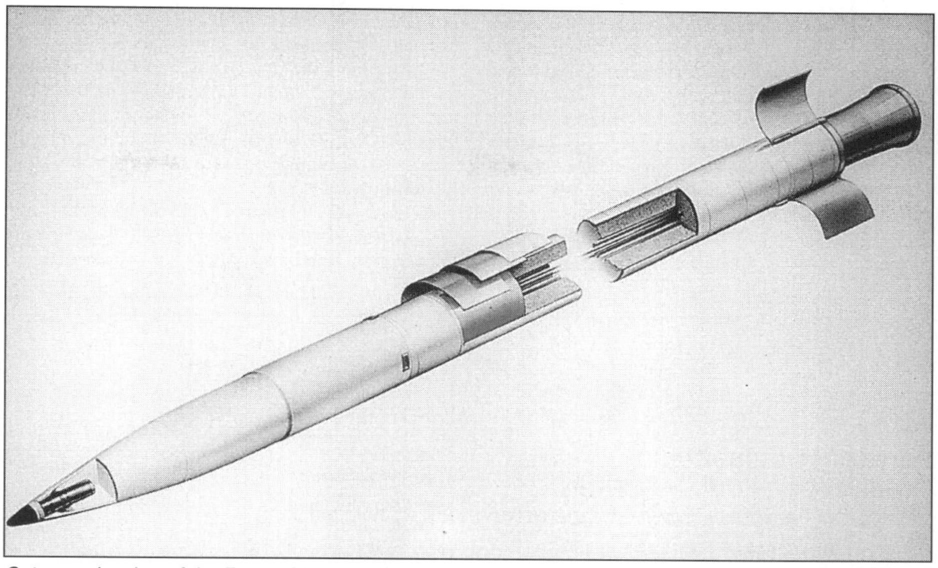

Cutaway drawing of the Rayo 160 mm rocket showing position of fuze, warhead and rocket motor

For Phase III the 160 mm Rayo weapon system has been integrated on a German MAN SX2000 (6 × 6) chassis providing high battlefield and tactical mobility. This launcher vehicle also offers Lockheed Martin C-130 air transportability with minimum preparation. The cab is provided with an air filtration/over-pressure system to prevent ingress of toxic fumes during firing, with environmental control as a standard option giving comfortable accommodation for up to three crew. An auxiliary storage module is mounted behind the cab providing additional stowage for crew and vehicle equipment. An auxiliary power unit (APU) has been incorporated to allow the platform and systems to be run with the main engine of the MAN truck to be switched off.

The 160 mm Rayo system also includes a complementary training system comprising an embedded simulation package and a low-cost subcalibre reduced-range practice rocket with indicator warhead. The simulation capability is inbuilt in the fire-control system, enabling the crew to carry out all aspects of planning and executing fire missions including simulated firings. A remote-control panel enables the instructor to introduce simulated faults and

Specifications

HE rocket
Calibre: 160 mm
Max range: 45,000 m
Min range: 10,000 m
Overall mass: 122 kg
Warhead: HE Frag, 11.3 kg RDX/TNT
Length: approx 3.2 m
Rate of fire: 24 rounds in 15 s

Training rocket
Calibre: subcalibre
Max range: 10 km
Warhead: Training flash/smoke

Launcher vehicle
MAN SX2000 6 × 6
Number of tubes: 2 × 12 rd packs
Elevation/depression: 60/0°
Traverse: ±105°
Max road speed: 90 km/h
Combat weight: 24,760 kg
Length: 8.60 m
Width: 2.65 m

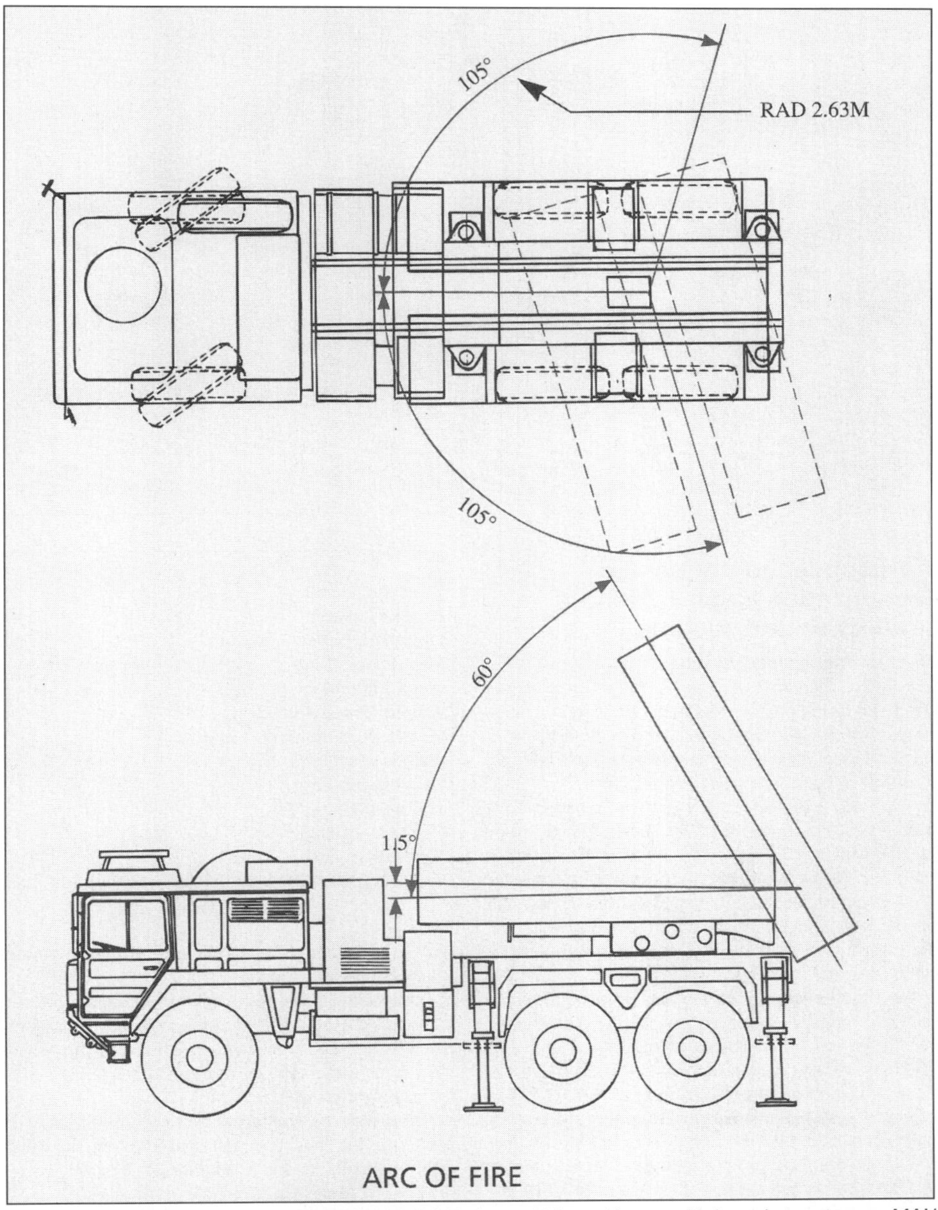

General arrangement drawing of the FAMAE/RO Defence 160 mm Rayo multiple rocket system on MAN (6 × 6) truck chassis
2002/0130027

munition failures (for example, hangfire or misfire) for instruction and training in operational safety procedures. For live fire training and exercises, reusable training rocket packs each containing 12 subcalibre reduced-range training rockets can be used in place of the standard weapon system rocket packs. This enables realistic live fire training to be carried out while reducing costs and giving increased flexibility in the use of peacetime firing range facilities.

Air-Log of the UK produced two rocket pod containers for RAYO. These comprise a tight tolerance steel structure coupled with a door closure system for the support and aiming of the launch tubes. The company has also designed a complex loom for communications between the fire-control system in the cab and the rocket sensing and firing circuits. Via this system, the crew can select single-round fire, through to a full ripple launch sequence of 24 rockets in under 60 seconds.

Air-Log has been associated with RAYO for some years and has also supplied other parts such as the adjustable firing stand and calibration rigs for test and trials use.

Status
Prototype qualification trials. As of late 2001, there were no stated plans for this system to enter quantity production for the Chilean Army.

Manufacturers
FAMAE Ordnance Ltda.
BA Systems, RO Defence, Rocket Motors Division.

UPDATED

IRAN

Aerospace Industries Organization Fadjr-5 333 mm (4- round) artillery rocket system

Development/Description
The Fadjr-3 333 mm (4-round) artillery rocket system has been developed by the Shahid Bagheri Industries division of the Aerospace Industries Organization of Iran and uses the same chassis as the Fadjr-3 240 mm (12-round) artillery rocket system. Details of the latter system are given in the following entry.

The Fadjr-5 is based on a modified Mercedes-Benz 2624 series (6 × 6) truck chassis with the engine and fully enclosed cab at the front. To the immediate rear of the cab is a small crew compartment. When launching rockets the cab windscreen would normally be protected by shutters.

Specifications
Calibre: 333 mm
Length of rocket: 6,485 mm
Max range: 75 km
Weight of rocket: 915 kg
Weight of warhead: 175 kg
Weight of HE content: 90 kg
Weight of system: 15,000 kg
Length travelling: 10.45 m
Width travelling: 2.54 m
Height travelling: 3.34 m
Max road speed: 60 km/h

Mounted on the rear of the chassis is the turntable with the manually operated elevating and traversing mechanism on the left side. The Fadjr-5 has a total of four 333 mm diameter rocket launcher tubes in one layer with launcher elevation being from 0 to +57° and with a traverse of 45° left and right.

Fadjr-5 333 mm (4-round) artillery rocket system in travelling configuration **2000**/0067779

It launches a solid propellant 333 mm unguided rocket with fins that unfold after launch and the rockets can be launched with an ambient temperature of between –40 to +50°C.

The Fadjr-5 system has a rate of fire of one rocket every four or eight seconds and replacement 333 mm Fadjr-5 rockets are packed one to a crate which weighs 1.21 tonnes.

To provide a more stable firing platform, four stabilisers are lowered to the ground before the rockets are launched, two of these are positioned at the rear of the vehicle with another one between the second and third axles on either side.

The rockets can be launched singly or in ripple fire and once the launcher has expended its rockets it would normally deploy to another position when new rockets would be loaded using a crane.

It is possible that some of the technology of these rocket systems could have come from North Korea.

Status
Production. In service with Iranian Army. There are no known exports of this system.

Manufacturer
Aerospace Industries Organization, Shahid Bagheri Industries.

UPDATED

Aerospace Industries Organization Fadjr-3 240 mm artillery rocket system

Development/Description
The Fadjr-3 240 mm (12-round) artillery rocket system has been developed by the Shahid Bagheri Industries division of the Aerospace Industries Organization of Iran and uses the same chassis as the Fadjr-5 333 mm (4-round) artillery rocket system (qv).

The Fadjr-3 artillery rocket system is based on a modified Mercedes-Benz 2624 series (6 × 6) truck chassis with the diesel engine and fully enclosed cab at the front. To the immediate rear of the cab is a small crew compartment. When launching rockets the cab windscreen would normally be protected by shutters. To the rear of the cab is a small compartment for additional crew members.

Mounted on the rear of the chassis is the turntable with the manually operated elevating and traversing mechanism on the left side. The Fadjr-3 has a total of 12 240 mm diameter rocket launcher tubes in two layers of six with launcher elevation being from 0 to +57° and with a traverse of 90° left and 100° right.

It launches a solid propellant rocket with fins that unfold after launch and the rockets can be launched with an ambient temperature of between –40 to +50°C.

The Fadjr-3 unguided artillery rocket system has a rate of fire of one rocket every four or eight seconds and replacement Fadjr-3 rockets are packed three to a crate which weighs 1.65 tonnes.

Specifications
Calibre: 240 mm
Length of rocket: 5,200 mm
Max range: 43 km
Weight of rocket: 407 kg
Weight of warhead: 90 kg
Weight of HE content: 45 kg
Weight of system: 15,000 kg
Length travelling: 10.45 m
Width travelling: 2.54 m
Height travelling: 3.34 m
Max road speed: 60 km/h

To provide a more stable firing platform four stabilisers are lowered to the ground before the rockets are launched, two of these are positioned at the rear of the vehicle with another one between the second and third axles on either side.

The rockets can be launched singly or in ripple fire and once the launcher has expended its rockets it would normally deploy to another position when new rockets would be loaded using a crane.

It is possible that some of the technology of these rocket systems could have come from North Korea.

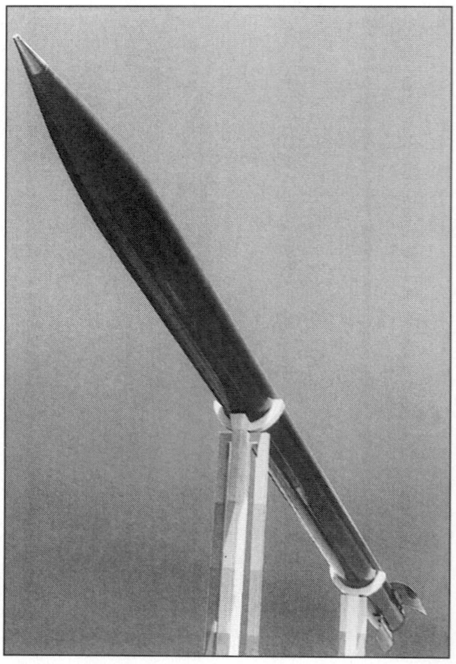
240 mm Fadjr 3 rocket out of launcher with fins unfolded 0008317

When first observed in 1996, this system was installed on a Japanese Isuzu (6 × 6) truck chassis and these were deployed with Hezbollah units operating in Southern Lebanon.

Status

In production. In service with the Iranian Army. Also used by Hezbollah.

Manufacturer

Aerospace Industries Organization, Shahid Bagheri Industries.

UPDATED

Fadjr-3 240 mm (12-round) artillery rocket system with launcher elevated
2000/0067780

Aerospace Industries Organization 610 mm Zelzal 2 heavy artillery rocket system

Development/Description

The Zelzal 2 unguided heavy artillery rocket is launched from a single rail launcher and is spin stablised in flight by small solid-propellant rocket motor nozzles located just behind the nose-mounted HE warhead and electromechanical fuzing section. All available details are given in the accompanying table.

Like all other Iranian rockets the 610 mm Zelzal 2 heavy artillery rocket system is produced by the Aerospace Industries Organization, Shahid Bagheri Industries facility.

The launcher is based on a modified Mercedes-Benz three-axle truck. When deployed in the firing position, four stabilisers are lowered to the ground. Of these four stabilisers, one is mounted on either side to the immediate rear of the cab and the other two are mounted at the very rear of the vehicle.

Status

In production. In service with Iranian Army. Offered for export. There are no known exports of this rocket system.

Manufacturer

Aerospace Industries Organization, Shahid Bagheri Industries.

UPDATED

Specifications
Calibre: 610 mm
Weight:
(complete round) 3,400 kg
(warhead) 600 kg
Length: 8.325 m
Warhead: HE with electromechanical-type fuzing system
Max range: 210,000 m

Zelzal 2 610 mm unguided artillery rocket on its launcher, based on a Mercedes-Benz truck
2001/0105902

Aerospace Industries Organization NAZEAT artillery rocket system family

Development/Description

The NAZEAT unguided heavy artillery rocket system family is an indigenous development by the Aerospace Industries Organization, Shahid Bagheri Industries facility.

The NAZEAT family is known to comprise the following versions: N5, N6, N7, N8, N9 and N10 (with a maximum range of 80,000 up to 150,000 m depending upon model type). All of these rockets use a solid propellant and all available model details follow.

Iranian 355.6 mm NAZEAT rocket

NAZEAT 6 355.6 mm unguided artillery rocket being launched 0021059

355.6 mm calibre NAZEAT N4 artillery rocket system

The unguided NAZEAT N4 carries an HE warhead fitted with an electromechanical-type fuze.

355.6 mm calibre NAZEAT N5 artillery rocket system

The unguided NAZEAT N5 heavy unguided artillery rocket system is carried either on a single rail elevating launcher fitted to the rear of a modified Mercedes-Benz 2624 (6 × 6) truck chassis to act as a Tractor-Erector-Launcher (TEL) or a single rail semi-trailer launcher. Different warhead types can be carried. The preparation time for a launch is approximately 1 hour. Maximum range is 120,000 m with a flight time of 180 seconds and maximum velocity of 1,800 m/s.

355.6 mm calibre NAZEAT N6 artillery rocket system

The unguided NAZEAT N6 carries an HE warhead fitted with an electromechanical-type fuze. According to Iranian sources, the rocket has a CEP of 5 per cent of range with the solid composite (PPG base) rocket motor having a maximum burn time of 9 seconds and specific impulse of 237 seconds. Total weight of propellant is stated to be 420 kg.

The NAZEAT 6 rocket is carried and launched from a Mercedes-Benz (6 × 6) truck chassis. When deployed in the firing position four stabilisers are lowered to the ground to provide a more stable firing platform.

450 mm calibre NAZEAT N10 artillery rocket system

The unguided NAZEAT N10 heavy unguided artillery rocket system comprises the following systems:

(a) Mercedes-Benz 911 Command and Control (4 × 4) truck with a rear-mounted cabin containing a Model 380DX computer processing system, communications equipment and other necessary apparatus.

(b) Mercedes-Benz 911 (4 × 4) meteorological truck with a rear-mounted cabin, containing the necessary equipment to measure the required meteorological conditions such as ambient air temperature, pressure and humidity, wind velocity and direction that can affect the rocket performance.

(c) Model WM 101 Global Positioning System console for surveying the launch platform site.

(d) Mercedes-Benz 2624 (6 × 6) transporter vehicle with four NAZEAT 10 reloads within its rear loading bed. It is fitted with a remote cable-operated hydraulic crane just behind the cab and two hydraulically operated stabilising jacks to facilitate loading of the launcher.

(e) A launcher platform which can be one of the following types:

  (i) A version of the Mercedes-Benz 2624 (6 × 6) truck chassis fitted with a single rail elevatable launcher assembly on its rear decking to act as a TEL vehicle. Maximum elevation angle is 80° and azimuth angle ±15°. Time to reload is 30 minutes. This is designated the NAZEAT 10 L1 Model Launcher.

NAZEAT 10 450 mm unguided artillery rocket being launched 0023375

NAZEAT 6 355.6 mm unguided artillery rocket ready to be launched from the back of its Mercedes-Benz (6 × 6) truck chassis **2002**/0130036

Specifications
NAZEAT N4
Calibre: 355.6 mm
Weight:
(rocket complete) 650 kg
(warhead) 150 kg
Length: (complete) 5.8 m
Max range: 80,000 m

Specifications
NAZEAT N5
Calibre: 355.6 mm
Weight:
(rocket complete) 850 kg
(warhead) 150 kg
Length:
(complete) 6 m
Max velocity: 1,800 m/s
Max range: 120,000 m
Warhead type: various, including HE

Specifications
NAZEAT N6
Calibre: 355.6 mm
Weight:
(rocket complete) 900 kg
(warhead) 130 kg
Length:
(complete) 6.29 m
Max range: 90,000 m
Warhead type: HE

Specifications
NAZEAT 10
Calibre: 450 mm
Weight:
(rocket complete) 1,850 kg
(warhead) 250 kg
(fuel) 1,000 kg
Length:
(complete) 8.3 m
Max velocity: Mach 4-5
Max range: 150,000 m

(ii) A four-wheel semi-trailer, with a different style single rail hydraulically operated elevatable launcher assembly from that fitted on the Mercedes-Benz truck version. The maximum elevation and traverse angles remain the same, as does the reload time. This is known as the NAZEAT 10 L2 Model Launcher.

(iii) The standard Russian BAZ-135L4 (8 × 8) Luna-M (FROG-7) artillery TEL vehicle. This is known as the NAZEAT 10 L3 Model Launcher. Details of this are given in the entry for the FROG-7 under the Russian Federation and Associated States.

The NAZEAT 10 is fitted with a single-stage solid propellant rocket motor with a maximum thrust of 280 kN, a propellant weight of 1,000 kg and specific impulse of 242 seconds. The fuze is of an electro-mechanical type.

Status

Production as required. In service with the Iranian Army and the IRGC. Not thought to have been exported.

Manufacturer

Aerospace Industries Organization, Shahid Bagheri Industries.

UPDATED

Aerospace Industries Organization 333 mm SHAHIN 1 artillery rocket system

Development/Description

The 333 mm SHAHIN 1 unguided heavy artillery rocket system is an indigenous development of the Aerospace Industries Organization.

The SHAHIN 1 has a solid propellant rocket motor with seven exhaust nozzles and is fitted with a 190 kg HE blast warhead. Maximum range is stated to be 13,000 m. It has four fixed fins at the rear.

The launcher is a four-wheel trailer-mounted elevating triple-rail assembly. Before firing, two pairs of stabiliser jacks have to be lowered to the ground manually, these provide a stable firing platform. The launcher platform is assigned to heavy artillery units.

Status

Production complete. In service with the Iranian Army and the IRGC. Not thought to have been exported.

Manufacturer

Aerospace Industries Organization, Shahid Bagheri Industries.

VERIFIED

Specifications
Calibre: 333 mm
Weight:
(rocket complete) 384 kg
(warhead) 190 kg
(fuel) 85±1 kg
Length:
(complete) 2.9 m
Max velocity: 450 m/s
Max range: 13,000 m
Warhead type: HE blast

Trailer-mounted Iranian SHAHIN rocket system with 333 mm SHAHIN 1 in centre position and 333 mm SHAHIN 2 on two outer rails

Aerospace Industries Organization 333 mm SHAHIN 2 artillery rocket system

Development/Description

The 333 mm SHAHIN 2 unguided heavy artillery rocket system is a further indigenous development of the 333 mm SHAHIN 1 system by the Aerospace Industries Organization, Shahid Bagheri Industries.

The SHAHIN 2 has a lengthened version of the solid propellant rocket motor that is fitted to the SHAHIN 1. It has the same seven-nozzle exhaust system and 190 kg HE blast warhead. Maximum range is stated to be 20,000 m. The SHAHIN 2 has four fixed fins at the rear of the rocket.

The launcher is a four-wheel trailer-mounted elevating triple-rail assembly. Before firing, two pairs of stabiliser jacks have to be lowered manually from the trailer to provide a stable firing platform. The launcher platform is assigned to heavy artillery units. The SHAHIN 2 can also be used as an unguided air-to-ground rocket. The SHAHIN 2 artillery rocket system has a storage temperature of –30 to +60°C.

Specifications
Calibre: 333 mm
Weight:
(rocket complete) 530 kg
(warhead) 190 kg
(fuel) 180 kg
Length:
(complete) 3.9 m
Max velocity: 660 m/s
Max range: 20,000 m
Warhead type: HE blast

Status

Production complete. In service with the Iranian Army and the IRGC. Not thought to have been exported.

Manufacturer

Aerospace Industries Organization, Shahid Bagheri Industries.

VERIFIED

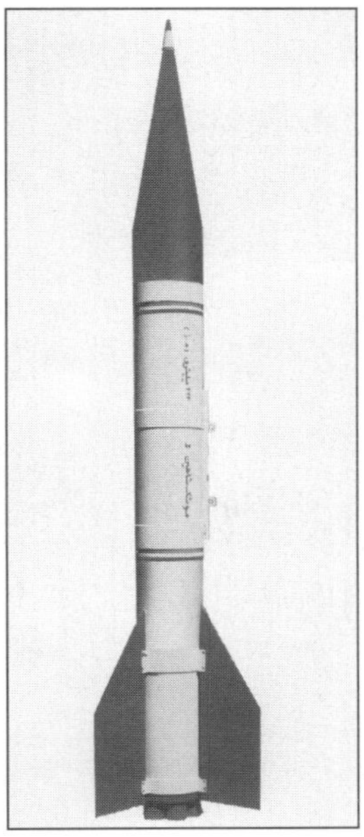

333 mm SHAHIN 2 artillery rocket
0021055

Aerospace Industries Organization 230 mm OGHAB artillery rocket system

Development/Description

The 230 mm OGHAB (Eagle) unguided artillery rocket system was developed indigenously during the 1980-88 Gulf War. The initial combat use of the rocket was in December 1986 when several rounds were fired at Basra.

From then to the end of the 1980-88 Gulf War approximately 330 rounds were fired by rocket units of both the regular army and the ground forces of the IRGC.

The 230 mm calibre OGHAB rocket has four fixed tail-mounted fins and is spin stabilised in flight. It carries a 70 kg HE fragmentation warhead. A contact fuze with instantaneous and delay actions is fitted. Propulsion is by a solid propellant rocket motor that develops a maximum thrust in the order of 76,480 kN.

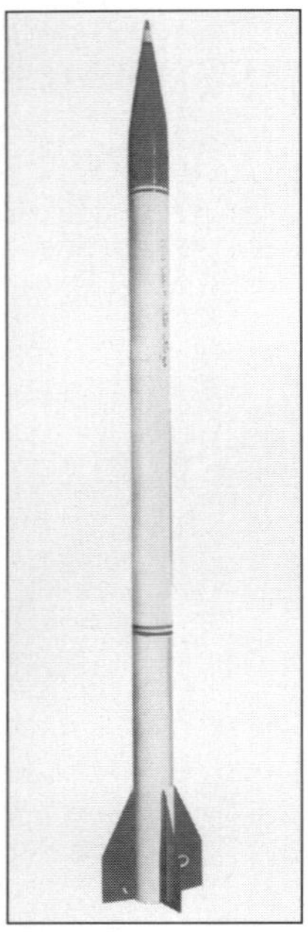

230 mm OGHAB unguided artillery rocket
0021057

The launcher is an elevatable triple-rail launcher assembly fitted to the rear of a Mercedes-Benz LA911B (4 × 4) 4,500 kg truck chassis. The truck is also known as the LA1113 and before 1963 as the LA328. Before firing, two pairs of stabiliser jacks have to be lowered to provide a stable firing platform. The launcher vehicle is assigned to heavy artillery units.

Like all other Iranian unguided artillery rockets, the 230 mm OGHAB has been developed and produced by the Aerospace Industries Organization, Shahid Bagheri Industries.

Status

Production complete. In service with the Iranian Army and the IRGC. Not thought to have been exported. Believed to have been supplemented by 240 mm Fadjr 3 system (qv entry this section).

Manufacturer

Aerospace Industries Organization, Shahid Bagheri Industries.

VERIFIED

Specifications
Calibre: 230 mm
Weight:
(rocket complete) 360 kg
(warhead) 70 kg
(fuel) 128 kg
Length:
(complete) 4.82 m
Max velocity: 750 m/s
Max thrust: 76,480 kN
Max range: 45,000 m
Warhead type: HE fragmentation

Iranian 230 mm OGHAB rocket

Aerospace Industries Organization 122 mm (40-round) HADID multiple rocket launcher system

Development/Description

The 122 mm HADID rocket launcher system is a 40-round launcher assembly system mounted on the rear of a Mercedes-Benz 2624 6 × 6 truck chassis. These chassis are also used for the Fadjr 333 mm (4-round) and Fadjr 240 mm (12-round) artillery rocket systems that are covered elsewhere in this section. The technology has certain operational characteristics that are similar to the Chinese 122 mm (40-round) Type 81 MRS and the Russian 122 mm (40-round) BM-21 systems, both of which are covered in detail elsewhere in this section.

The launcher assembly has two banks of 20 122 mm barrels in a side-by-side 4 × 5 row arrangement and is mounted on a rotating cradle fitted to the rear of the truck. The unguided rockets are spin stabilised and are of the 122 mm type (see the two 122 mm artillery rocket system entries following). The full 40-round load can be salvo-fired in 20 seconds, with the firing operations conducted either from the vehicle cab or from a ground position situated at a safe distance from the launcher. Reloading is performed manually by the launcher crew. The sighting system is mounted on the left side of the launcher at the rear.

These rocket systems have been developed by Aerospace Industries Organization Shahid Bagheri Industries.

The actual launchers have been developed by the Armament Industries Division of the Defence Industries Organisation. The launcher has also been referred to as the Hadid 122 mm (40-round) HM20 multiple rocket launcher system.

Iranian HADID 122 mm (30-round) MRS mounted on Mercedes-Benz LA911B (4 × 4) 4,500 kg truck chassis

Variants

122 mm (30-round) multiple rocket launcher system

A copy of the North Korean 30-round 122 mm calibre BM-11 rocket launcher system has also been produced. This comprises a single bank of three rows of 10 tubes mounted on a rotating cradle assembly. The chassis used is the Mercedes-Benz LA911B (4 × 4) truck chassis. Four stabilising jacks are lowered for firing and a hydraulic crane unit is also fitted immediately aft of the cab for use in reloading the rocket tubes.

The rear stabilisers are lowered to the ground manually while those to the immediate rear of the cab are lowered to the ground using an hydraulic system. It is possible that the launcher is fitted with an auxiliary power unit.

122 mm (16-round) multiple rocket launcher system

This has been designed for installation on boats and rafts and has two banks, each of eight 122 mm launcher tubes.

122 mm naval rocket launcher

An 8-round launcher assembly has been produced for use on naval vessels. The total launcher weight is 450 kg with the system having a maximum range of 14,000 m. Elevation range is 0-20°. The standard 122 mm rocket types are fired from the 3 m long, 23.4 kg weight barrels.

122 mm single barrel rocket launcher

A man-portable single-round 122 mm calibre rocket launcher has been produced. Mounted on a tripod the system has a maximum range of 20,400 m, an elevation range of 5-50° and a traverse of ±12°. The total assembly weighs 85-90 kg and uses the standard 3 m long, 23.4 kg weight barrel.

Specifications

Mercedes-Benz 2624 model
Calibre: 122.4 mm
Weight: 13,150 kg
Number of barrels: 40
Weight of barrel: 23.4 kg
Number of rockets: 40
Rate of fire: 40 rds in 20 s
Max range: 20,400 m
Length:
(rocket barrel) 3 m
Elevation: 1-55°
Traverse: 102° left, 72° right
Reloading time: approx 8-10 min
Chassis: Mercedes-Benz 2624 6 × 6 truck
Fording depth: 1.5 m
Crew: 6-7

Status

Production. In service with the Iranian Army and the IRGC. Not thought to have been exported.

Manufacturer

Aerospace Industries Organization, Shahid Bagheri Industries.

UPDATED

Aerospace Industries Organization 122 mm ARASH artillery rockets

Development/Description

The unguided 122 mm ARASH artillery rocket system is similar in appearance to the Russian and Chinese 122 mm calibre unguided artillery rockets in widespread use. It can be used in all the Aerospace Industries Organization, Shahid Begheri Industries such as the 122 mm (40-round) HADID multiple rocket launcher system and foreign-supplied 122 mm rocket launcher versions currently in Iranian use. It has a maximum range of 21,500 m.

The 122 mm calibre ARASH unguided rocket has four wraparound tail-mounted fins and is spin stabilised in flight. It carries an HE fragmentation warhead. A contact fuze with instantaneous and delay actions is fitted. It is probable that individual 122 mm rockets have been exported by Iran. The standard Russian Federation type 122 mm rocket is also manufactured. All of these rockets are of the solid-propellant type with wrap round fins at the rear.

Specifications

Model	Standard	Long range
Calibre	122 mm	122 mm
Weight		
(complete rocket)	65 kg	72 kg
(warhead)	18.38 kg	18 kg
Length of complete rocket	2.815 m	3.20 m
Max range	21,500 m	29,000 m
Max velocity	710 m/s	1,050 m/s
Thrust power	n/av	26,000 n
Time of flight	n/av	94 s

Variant

Extended-range 122 mm rocket

An extended-range 122 mm rocket is available. This rocket has a maximum range of 26,000 m.

Status

Production as required. In service with the Iranian Army and the IRGC.

Manufacturer

Aerospace Industries Organization, Shahid Begheri Industries.

UPDATED

Aerospace Industries Organization 122 mm NOOR artillery rockets

Development/Description

The 122 mm NOOR unguided artillery rocket system is similar in appearance to the Russian and Chinese 122 mm calibre short version artillery rockets in widespread use. It can be used in all of the Iranian Aerospace Industries Organization, Shahid Bagheri Industries facility, and foreign-supplied 122 mm rocket launcher versions currently in Iranian use.

The 122 mm calibre NOOR solid-propellant rocket has four wraparound tail-mounted fins and is spin stabilised in flight. It carries an HE fragmentation warhead. A contact fuze with instantaneous and delay actions is fitted. It is possible that individual 122 mm rockets have been exported by Iran.

Like other Iranian 122 mm unguided artillery rockets the 122 mm NOOR rocket can be stored in the temperature range of –30 to +60°C.

Specifications
Calibre: 122 mm
Weight:
(rocket complete) 45 kg
(warhead) 18.35 kg
Length: (complete) 2.050 m
Max velocity: 720 m/s
Max range: 18,000 m
Warhead type: HE fragmentation

Status

Production as required. In service with the Iranian Army and the IRGC. Not thought to have been exported.

Manufacturer

Aerospace Industries Organization, Shahid Bagheri Industries.

UPDATED

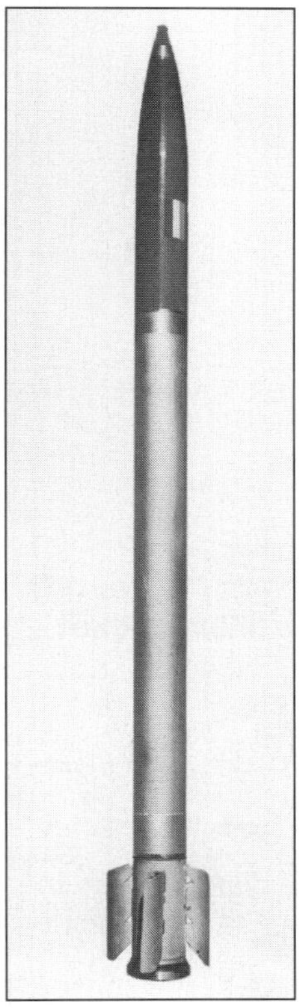

122 mm NOOR artillery rocket with fins unfolded
0021058

Aerospace Industries Organization 122 mm Fadjr 6 minelaying rocket

Development/Description

The Aerospace Industries Organization, Shahid Bagheri Industries, unguided 122 mm calibre Fadjr 6 minelaying rocket can be launched from standard 122 mm calibre multiple rocket launcher systems.

It carries a total of eight anti-tank/anti-personnel mines which are stabilised after release by parachute.

The pressure-activated mines are manufactured by the Ammunition Industries Division of the Defence Industries Organisation.

Status

In production. In service with Iranian Army. Offered for export.

Manufacturer

Aerospace Industries Organization, Shahid Bagheri Industries.

VERIFIED

Specifications
Calibre: 122 mm
Weight:
(complete round) 63 kg
(warhead) 32 kg
Length: 2.83 m
Payload: 8 anti-personnel/anti-tank mines
Range: 3,500-6,000 m

Aerospace Industries Organization 107 mm (12-round) FADGR 1 multiple rocket launcher system

Development/Description

The 107 mm FADGR 1 multiple rocket launcher system is similar in appearance to the Chinese 107 mm Type 63 MRS (see separate entry). In its standard two-wheeled trailer form it can be moved by vehicle, animals or personnel.

All of these launchers are manufactured by the Armament Industries Division of the Defence Industries Organisation with the rockets, which are unguided, being manufactured by the Aerospace Industries Organization, Shahid Bagheri Industries facility.

The launcher has three banks of four 107 mm barrels and in the trailer form is mounted on a rubber-tyred split pole-type carriage. In the firing position the wheels are normally removed and the launcher is supported by the two trails at the rear which are fitted with spades which are carried on the top of the trails when travelling, and two short legs in front of the carriage. It can also be fired with the two rubber tyred roadwheels still in position on their stub axles.

The unguided rockets are spin stabilised and are of the 107 mm HASEB type (see next entry). The full 12-round load can be salvo-fired in 7 to 8 seconds. Reloading is performed manually by the launcher crew.

The optical sight is fitted on the left side of the launcher.

Variants

107 mm (12-round) vehicle mount

The standard 107 mm (12-round) launcher assembly is available with a 50 kg weight, 1.4 × 0.585 × 0.22 m base plate for fitting to light 4 × 4 vehicles or 6 × 6 truck chassis. The optical sight is fitted on the left side of the launcher.

107 mm single-round rocket launcher

A single-round rocket launcher is available. Mounted on a tripod it weighs 23 kg in total and uses a generator or battery as the trigger system. Elevation range is 0-50° and traverse ±7°. The rocket tube is 0.8 m long. The optical sight is fitted on the left side of the launcher.

107 mm twin-round rocket launcher

This is similar to the single-round launcher but has two barrels. Total weight is 32 kg, elevation range is 0-50° and traverse ±10°.

107 mm (12-round) naval mounts

Two naval mounts exist, the Type ML-2 and Type ML-4. The former has 10 barrels in a 4:4: offset 2 arrangement and the latter the standard 12 barrel configuration.

Status

Production as required. In service with the Hezbollah, Iranian Army, Iranian Navy and the IRGC.

Manufacturer

Defence Industries Organisation (DIO), Armament Industries Division.

UPDATED

Specifications
Split trail version
Calibre: 107 mm
Number of barrels: 12
Number of rockets: 12
Rate of fire: 12 rds in 7-9 s
Max range: 8,500 m
Reloading time: 3 min (approx)
Carriage: split trail
Weight:
(launcher without rockets) 380 kg
(12-round load) 224 kg
(carriage) 195 kg
Length:
(travelling) 2.6 m
Width:
(travelling) 1.4 m
Height:
(travelling) 1.2 m
Elevation: 0-60°
Traverse: ±15°
Crew: 4-5

107 mm (12-round) FADGR 1 multiple rocket launcher deployed in firing position ***2001***/0105899

107 mm (12-round) FADGR 1 multiple rocket launcher being towed by a light 4 × 4 vehicle which has another system mounted in rear ***2001***/0105900

Aerospace Industries Organization 107 mm (12-round) HASEB multiple artillery rocket system

Development/Description

The 107 mm HASEB artillery rocket system is similar in appearance to the 107 mm Type 63 artillery rocket used in the Type 63 MRS (see entry this section). It can be used in all the NORINCO (China North Industries Corporation) 107 mm FADGR launcher versions. The 107 mm (12-round) launcher is also made in a number of other countries, including North Korea.

The 106.7 mm calibre rocket is spin stabilised and carries an HE fragmentation warhead. A contact fuze with instantaneous and delay actions is fitted, although an alternative proximity fuze is also believed to be available. These unguided rockets are manufactured by the Aerospace Industries Organization, Shahid Bagheri Industries.

Specifications
Calibre: 106.7 mm
Weight:
(rocket complete) 18.6 kg
(warhead) 6.39 kg
Length:
(complete) 0.838 m
Max velocity: 375 m/s
Max range: 8,500 m
Min range: 600 m
Warhead type: HE fragmentation

The rockets have a maximum rotating speed of 22,000 revolutions per minute and a lethal radius of 12.5 m. When stored in a warehouse they have a life of 13 years.

Status

Production as required. In service with Hezbollah, Iranian Army, Iranian Navy and the IRGC.

Manufacturer

Aerospace Industries Organization, Shahid Bagheri Industries.

VERIFIED

IRAQ

Iraqi multiple rocket launcher systems

Background

Apart from using large numbers of Brazilian, Egyptian and Russian unguided multiple rocket launchers, most of which were obtained in the 1980's, Iraq has developed a number of systems to meet its own operational requirements.

Many of these have been used in the war with Iran. Some of these are modifications of existing systems while others are surface-to-air missiles adopted for use in the ground-to-ground role. Following the 1990-91 Gulf conflict it is probable that none of these unguided rocket systems are now in quantity production. There have been no known exports of these systems in recent years.

550 mm Laith 90 artillery rocket system

This is a locally built cluster munition warhead version of the Russian Luna-M (Russian designation R-75, NATO designation FROG-7a) artillery rocket system. Maximum range has been increased from 70,000 to 90,000 m. Full details of the FROG-7 system are given later in this section under Russian Federation and Associated States.

Babeel 100 (4-round) launcher in ground emplaced configuration

107 mm (12-round) multiple rocket system (Christopher F Foss) 0021061

400 mm (4-round) Ababeel 100 multiple rocket system

Launcher 81 is a GAZ-66 (4 × 4) 2,000 kg truck fitted with an 81 mm multiple launcher mounted on rear

400 mm Ababeel 100 artillery rocket system

This is a 400 mm calibre 4-round truck-mounted rocket launcher with a maximum range of 100,000 m. A submunition dispensing warhead is fitted which carries either 300 anti-tank/fragmentation bomblets or 25 anti-tank minelets. The latter weigh 8 kg each and have a 2 m radius magnetic influence fuzing system that can fire the warheads even if they are lying on their sides.

262 mm Ababeel 50 multiple rocket launcher system

This is a co-produced version of the former Yugoslav M-87 (12-round) MRS developed with Iraqi assistance. A detailed description is given in the Yugoslav part of this section (qv). This has a maximum range of 50,000 m. A few preproduction series vehicles were delivered before the 1990-1991 Gulf War. Some of these have also been used in the recent Yugoslav conflict. As far as it is known, there has been no recent production of this system in Iraq or Yugoslavia. It is believed that some marketing of this system has recently been carried out in Asia by one part of the former Yugoslavia.

127 mm Sajeel 30, 180 mm Sajeel 40 and 300 mm Sajeel 60 multiple rocket launcher systems

These are respectively locally licence-built versions of the Brazilian AVIBRAS ASTROS II SS-30, SS-40 and SS-60 launchers and rockets. Details differ in some areas from those given by AVIBRAS (see Brazilian part of this section) and may reflect local variations.

The Sajeel 40 warhead contains 20 anti-armour/anti-personnel bomblets while the Sajeel 60 warhead contains 65 bomblets of the same type. Both are fitted with a variable time delay; they can be programmed before launch from 5.2 s onwards in order that the warhead detonates over the target area and not over friendly forces.

107 mm portable rocket launcher system

The Iraqis have designed and built a portable 12-round launcher system for use with an indigenous 107 mm calibre unguided rocket design. Aid may have come from the People's Republic of China in developing this system. This is almost identical to the Chinese North Industries Corporation (NORINCO) Type 63 107 mm

Specifications

Designation	**Sajeel 30**	**Sajeel 40**	**Sajeel 60**
Calibre	127 mm	180 mm	300 mm
Weight			
(launch)	68.1 kg	152 kg	587.65 kg
(warhead)	21 kg	n/avail	n/avail
Length			
(complete)	2.950 m	3.978 m	5.281 m
(warhead)	0.685 m	1.795 m	1.092 m
(rocket motor)	2.265 m	2.183 m	4.189 m
Thrust	1,650-2,000 kg	2,070-2,510 kg	4,600-6,400 kg
Burn time	2.25-2.75 s	4.1-5.1 s	6.2-8.8 s
Max range	30,000 m	35,000 m	60,000 m
Min range	9,000 m	15,000 m	20,000 m
Warhead type	HE	ICM	ICM

Sajeel multiple rocket launcher with stabilisers lowered and some of the rocket types in the foreground
0021060

(12-round) system covered in detail under China. The system is also used by the Iraqi-backed Iranian National Liberation Army as a light vehicle-mounted system, an MT-LB-mounted system (with two 12-round launcher assemblies mounted either side at the rear of the roof above the rear door entrances) and with the standard ground mount.

122 mm multiple rocket launcher systems

The Iraqis have taken locally built versions of the Egyptian Sakr-36 30-round 122 mm MRS launcher assembly and mounted them on the rear of the Russian MT-LB tracked APC and Swedish Scania SBA 111 series (4 × 4) truck.

30 mm al-Tawbiya multiple rocket launcher system

This weapon was used in quantity by the Iraqi Army during the latter stages of the 1980-88 Gulf War as an anti-personnel weapon to stem massed Iranian infantry attacks. No other details are available.

Launcher 81

GAZ-66 (4 × 4) 2,000 kg truck fitted with 35 round (five rows each of seven rockets) 81 mm rocket launcher at rear.

Specifications

107 mm portable rocket launcher system
Calibre: 107 mm
Number of barrels: 12
Max range: 8,000 m
Weight:
(system) 385 kg
(rocket complete) 19 kg
Length:
(firing) 3 m
Width:
(firing) 1.7 m
Height:
(firing) 1.1 m
Elevation: 0 to +60°
Traverse: 15° left and right
Rate of fire: 12 rockets/min

Status

Production as required. In service with Iraqi Army. No known exports.

VERIFIED

ISRAEL

Israel Military Industries Ltd artillery rocket systems

Background

In 1960, Israel Military Industries Ltd established the Rocket Systems Division, which started work in 1965 on the MAR-290 - a 290 mm artillery rocket with 25 km range. This entered Israel Defence Force (IDF) service mounted in frame racks, holding four rockets, carried on a heavily modified Sherman tank chassis.

Since these early beginnings, the company has developed an entire family of artillery rockets and multiple rocket launcher systems. Rockets are available in diameters from 160 mm to 350 mm, having ranges up to 100 km. The rockets are factory sealed into Launch Pod Containers (LPC) containing from 2 to 18 rockets depending on rocket size and launch vehicle type.

Tactical and logistics considerations learned through combat experience with the IDF produced a long list of improvements and system enhancements, many of which are unique to the Artillery Rocket family. These include longer effective range, fast reloading capability, adaptability to a wide range of carrier vehicles, modularity of ammunition packaging, regardless of size, easy adaptability of a variety of fire-control systems and flexibility in the use of battery support equipment.

United Nations sources state that between 1992 and 2000 Israel did not export any complete artillery rocket systems although it is possible that replacement rockets were exported to existing customers.

LAR multiple rocket launcher system (160 mm)

Development/Description

The Light Artillery Rocket system consists of a tracked or wheeled all-terrain vehicle or trailer on which is mounted a multiple rocket launcher holding two factory-sealed, expendable launch pod containers (LPC) with fuzed 160 mm diameter rockets in their launch tubes.

The LPCs, which also serve as storage and transportation containers for the rockets, can be rapidly replaced in the field after the rockets have been fired. When the rockets are fired, they burst frangible covers over the forward and aft ends of the launch tubes.

The Mk I system has been deployed on a French Giat Industries AMX-13 light tank chassis by the Venezuelan Army. The Mk II and Mk IV systems are currently in production for several other countries. The Mk IV system, having the same dimensions as the other rockets but using a modified propellant, has a range of 45 km.

The rockets are 160 mm in diameter and 3.4 m long, employing composite solid propellants. Wraparound stabilising fins deploy when the rocket exits the launcher. The Mk I version of the LAR rocket, weighing 100 kg, carries a 40 kg HE-COFRAM warhead activated by a super-quick impact fuze or proximity fuze. Maximum effective range is 30 km. Its launch pod containers hold 13 or 18 rockets each, with two pods on the launcher.

160 mm Rocket being fired from AMX-13 tank-mounted LAR system

The current production Mk II version rocket weighs 110 kg. Its 46 kg warhead can be an HE-COFRAM type or a cluster warhead containing 104 CL-3022-S4 AP/AM submunitions. A remotely set electronic time fuze opens the canister at the appropriate height to give area coverage of about 31,400 m^2 for each cluster warhead. Thirteen rockets are loaded in each launch pod container. Like the other versions of the LAR, the Mk II has a minimum range of 12 km. Its maximum range is 35 km.

In the current production, launchers are being mounted on heavy truck chassis, for example a German Mercedes Benz (6 × 6). Suitable chassis for mounting the modular launcher are light tanks (such as AMX-13), medium tanks (such as M47 and M48), heavy trucks (such as M809, MAN, Steyr and Mercedes), self-propelled howitzer (M109) and a trailer towed by a heavy truck. In its standard configuration, each launcher accommodates two 13-rocket launch pod containers. A light version with one launch pod is available for towing behind jeeps.

The LAR incorporates a modern C3I system called ACCS which has a total interface capability to all common artillery elements including meteorological unit, forward observers as well as sophisticated mapping, GPS and other items.

Elevation and traverse of the launchers are performed by an electrohydraulic system which is backed up by a manual system. When the system is fitted on a wheeled chassis two hydraulically operated stabilisers are lowered to the ground to provide a more stable firing platform.

160 mm ACCULAR being fired from Mercedes-Benz (6 × 6) truck-mounted LAR system 0021063

The launcher is operated by a computerised fire control system which is fully integrated with the ACCS so enabling each one to operate as a one gun battery.

All 26 rockets can be fired in less than 60 seconds and the launchers are reloaded in less than 5 minutes by the truck driver and launcher crew from conventional army trucks equipped with a 15 t/m hydraulic crane.

LAROM upgrade

Israel Military Industries teamed with Aerostar of Romania are upgrading the Romanian Aerostar 122 mm (40-round) APRA multiple rocket launcher system to a new configuration called LAROM. This integrates the Israel Military Industries 160 mm rocket system into the existing platform. Additional details are given under International. United Nations sources state that in 1999, three systems were sent to Romania following upgrade work in Israel.

Light LAR

A light launcher, which can be tailored for specific transport modes, has been specially supplied for several customers. It can be carried by helicopters and was designed to provide long-range, heavy fire support to light, highly mobile airborne units. This is an independent system which includes a launcher towed by a light vehicle such as a jeep or HMMWV. The towing vehicle is equipped with a light crane for loading the LPCs. A second vehicle, such as a light truck, carries the initial ammunition which includes at least two LPCs of eight or more rockets each. The launcher system is equipped with an optical aiming system, communications equipment and a land navigation system. The trailer mounted launcher is stabilised when deployed in the firing position by four hydraulically operated stabiliser legs.

The LAR launcher is not restricted to firing 160 mm rockets. Any launch pod container with a suitable interface can be mounted on the launcher. The pods are fully interchangeable with pods containing other size rockets, so a commander has a wide range of options without having to field completely separate systems. As new types of submunitions and mines are developed, they are being incorporated into submunition warheads.

Status

Mk I in service with Venezuela (20 systems on AMX-13 tracked chassis) and other armies. Mk II and Mk IV in service and production for undisclosed country.

MAR-350 medium artillery rocket system (350 mm)

Development/Description

Israel Military Industries Ltd has developed a 350 mm calibre artillery rocket system, the MAR-350, for use on the same range of chassis as the LAR. The initial firing trials took place in late 1986 and early 1988. The warhead is cluster munition type containing 770 CL-3022-S4 bomblets and a remotely set time fuze.

Fuzed rockets are factory sealed in their launch pod containers, two per container, which are loaded in the field for firing.

The basic mounting is the LAR hydraulically operated system with two launcher pod container-launchers in a side-by-side arrangement. Weight of the loaded two-round pod container is 2,000 kg, length 6.2 m, width 0.97 m and height 0.45 m. All four rockets can be fired within 30 seconds.

Status

Under development for undisclosed customers.

ACCULAR – trajectory corrected artillery rocket system

ACCULAR, a unique Israel Military Industries development, is an accurate artillery rocket system, based on the LAR family. With its unique trajectory correction system, accuracy of the rockets is said to be equal to that of common tube artillery.

The ACCULAR is a rocket weapon system developed by Israel Military Industries for the next generation of the LAR family, already operational for over a decade.

According to Israel Military Industries, it was only natural that this artillery rocket system will be significantly upgraded to better meet the requirements of the modern battlefield.

Fully demonstrated in field firings, ACCULAR is now under full scale development and production and will be deployed on several types of rocket system as well as the LAR.

As with other members of Israel Military Industries family of artillery rockets, the system can be used on various ground platforms including tanks, standard trucks, trailers and others.

In addition, for rapid deployment forces and their requirement for air transportability, special towed models are also available.

The modular design of the system allows varying the quantities of rockets (8, 10 or 13) in sealed launch pod containers on the launcher to meet the load capacity.

The ACCULAR concept can be adapted to different types and sizes of Israel Military Industries rockets and is valid through all artillery ranges.

The ACCULAR is claimed to represent only a moderate increase in cost. The major investment (technological and economic) is made on the exisiting command post while the cost increase in the rocket is relatively marginal.

By providing artillery rockets with increased accuracy to a level of that obtained by conventional tubed artillery systems, ACCULAR enables the optimal deployment of smart munitions (such as top attack type or scatterable mines) which require deployment in close vicinity to the target.

The Israel Military Industries ACCULAR concept may be adapted to different models and types of rockets and claimed to be a major breakthrough in this field as the upgrading of a free flight rocket has now become a reality.

This system is also being applied to the MLRS used by the Israel Defence Force under a contract awarded in mid-1998 to Israel Military Industries. Lockheed Martin Missiles and Fire Control – Dallas, prime contractor for the MLRS system are also involved in this programme.

It is expected that all 227 mm MLRS used by the Israel Defence Force will be fitted with ACCULAR by 2002. Israel has a total of 42 US-supplied MLRS.

The 160 mm ACCULAR is fitted with a Trajectory Correction System (TCS) which includes a steering unit mounted towards the nose that includes an electronic unit, gas generator and valve.

After launch, the ground control unit establishes a datalink with the rocket and calculates the current trajectory and corrections necessary to reach the target.

Specifications
Rocket
Calibre: 160 mm
Length: 3.70 m
Rocket weight: 120 kg
Payload weight: 60 kg
Payload: 104 bomblets with self-destruct fuze
Max range: 45 km

Pod
Number of rockets: 13
Total weight loaded: 2,050 kg
Length: 4,000 mm
Width: 980 mm
Height: 685 mm

The actual course correction phase is carried out as the rocket is accumulating environmental data that affects its flight course. Through this process, course correction is performed in a closed loop as the ground control unit commands the rocket steering to correct its trajectory.

The type of warhead depends on the user's operational requirement, for example a cargo warhead containing 104 HEAT bomblets fitted with a self-destruct fuze mechanism.

Specifications of the 160 mm ACCULAR rocket are given below as these are slightly different to those listed:

Status

A full Demonstration Programme has been completed. Full-scale development is under way for undisclosed customers.

Argentine rockets

Details of these systems, called the CAL 160 and CAL 350 and based on these Israel Military Industries rockets are given under Argentina earlier in this section. The Argentinian Army has at least two of these CAL 160 systems in service on a modified TAM tank chassis. These have the local designation of the VCLC-CAL 160.

Manufacturer

Israel Military Industries Ltd.

UPDATED

Specifications
Artillery rocket

Model	Mk I	Mk II	Mk IV	350
Rocket calibre	160 mm	160 mm	160 mm	350 mm
Rocket length	3.4 m	3.4 m	3.4 m	5.8 m
Rocket weight	100 kg	110 kg	117 kg	850 kg
Warhead weight	40 kg	46 kg	46 kg	300 kg
Range				
(min)	12 km	12 km	12 km	30 km
(max)	30 km	35 km	45 km	100 km

US 6 × 6 truck towing LAR 160 mm trailer-mounted rocket system with two pods each carrying 13 rockets
0021062

ITALY

BPD Difesa E Spazio SpA multiple rocket systems

BPD Difesa E Spazio SpA has developed, as a private venture, to the production stage two multiple unguided rocket systems called the FIROS 6 and FIROS 30. FIROS 6 is no longer being marketed by the company and it is understood that no complete FIROS 30 systems have been built in recent years. The primary role of these unguided rocket systems is area saturation.

According to United Nations sources, Italy did not export any multiple rocket systems between the years 1992 and 1998. It is possible that rockets have been supplied to replenish stocks used during training.

51 mm FIROS 6 multiple rocket system

Development/Description

The FIROS 6 is a 48-round MRS mounted on the rear of light 4 × 4 vehicles. The launcher is moved and aimed in traverse and elevation by handwheels and partial or full ripples can be fired at the rate of 10 rockets a second.

The crew can fire the rockets standing alongside the vehicle or from a remote position up to 30 m from the launcher via a cable. The launcher can be reloaded by hand within 5 minutes.

The 51 mm unguided rockets used in this system are the same solid propellant, with four folding fins, BPD Difesa E Spazio SpA 2 in rockets that are widely used in the air-to-ground role. Like the larger calibre 122 mm rockets manufactured by this company, these 51 mm rockets are unguided. Technical details of the rocket are as follows:

BPD Difesa E Spazio SpA 51 mm rocket

The anti-tank/anti-personnel rocket can penetrate over 250 mm RHA steel in addition to the secondary anti-personnel effect.

FIROS 6 multiple rocket system on an ACMAT 4 × 4 chassis in service with Mexican Navy (Julio Montes)

122 mm FIROS 30 multiple rocket systems

Development/Description

The FIROS 30 Field Rocket System is an area saturation weapon system and is based on the 122 mm BPD DIFESA E SPAZIO SpA unguided rockets. The system is intended for use either in battery or isolated missions and can cover large areas in the 8 to 34 km zone.

A typical battery would consist of a command post equipped with a fire direction system, six firing units and six to twelve escort units. The firing unit is a modular rocket launcher installed on a heavy (6 × 6) truck, for example IVECO (Italy) or Mercedes-Benz (Germany).

It is available in two different versions: a standard version, with a motorised/manual movement system simple to operate; and an automatic version, equipped with an inertial navigator, a servo-control system and a ballistic computer. This firing unit configuration enables the crew to perform the topographic and ballistic calculations and the aiming procedures through an Automatic Control Unit installed in the vehicle cab.

FIROS 30 multiple rocket launcher in action
2002/0130021

The launcher consists of two removable modules with 20 launching tubes each. The escort unit can carry four launching modules. The loading operations are carried out by removing the empty module with a jib crane fitted on the support vehicle and installing the filled one ready to fire.

The firing unit has a fire-control system and a link with the command post to acquire the necessary firing data in real time. The fire-control system has been designed to enable selection of the number (from 1 to 40) and type of rockets to be fired.

The FIROS 30 rockets differ in propellant types and maximum range and can carry a variety of different warheads.

Production versions of the FIROS 30 system for the Italian Army are mounted on an IVECO (6 × 6) 10,000 kg class chassis with a special armoured forward control cab. The United Arab Emirates use the FIROS system on Mercedes-Benz (6 × 6) truck chassis.

The 122 mm rockets launched by the 122 mm FIROS 30 system can also be launched by the widely used 122 mm Russian, Chinese or similar BM-21 systems.

Illustrations of the battery command post and escort (resupply unit) were given in *Jane's Armour and Artillery 2000-2001*.

Status

FIROS 6: Production complete. Only in service with Mexican Navy (on 4 × 4 light vehicle chassis). No longer marketed.

FIROS 30: Production complete. In service with a number of countries including Italy (20) and the United Arab Emirates. The latter country took delivery of 48 systems, but some of these are in reserve. Recent information has been that these are now up for disposal.

Manufacturer

BPD Difesa E Spazio SpA.

UPDATED

Specifications

Calibre: 51 mm
Length:
(without warhead) 744 mm
(with warhead) 1.05 m
Weight:
(without warhead) 2.6 kg
(with warhead) 4.8 kg
(propellant) 1.06 kg
Max range: 6,550 m
Time of flight to max range: 39 s
Propellant type: extruded double base
Total impulse: 210 kg/s
Max chamber pressure: 100 kg/cm^2
Average thrust: 200 kg
Acceleration: 60 g
Speed at burnout: 515 m/s
Motor case: steel, cold extruded
Nozzle fin: with thrust deflectors and folding free rotating fins
Warhead types: Anti-Tank/Anti-Personnel (AT/AP), High Explosive Incendiary (HEI), Illuminating (Ill), PreFormed Fragmentation (PFF), Smoke (SM), Target Practice (TP), Target Practice Flash (TP-FL), Target practice smoke (TP-SM), Target Practice Spotting (TP-SP)

Specifications

BPD 122 mm rocket
Calibre: 122 mm
Length of rocket.
(conventional warhead) 2.815 m
(submunition warhead) 3.321 m
Weight of rocket.
(conventional warhead) 65 kg
(submunition warhead) 71 kg
Max range.
(conventional warhead) 34,000 m
(submunition warhead) 30,000 m
Propellant weight: 24 kg
Total impulse: 6,000 kg/s
Chamber pressure: 210 kg/cm^2
Average thrust: 3,000 kg
Combustion time:
(at 50% max pressure) 2.05 s
Propellant type: composite

Conventional warheads
Weight: (total) 26 kg
Trajectory safety: 150 m
Types of warhead: High Explosive (HE), Target Practice (TP)
Fuze: Point Detonating (PD)

Submunition warheads
Weight: (total) 32 kg

Characteristics
Time into action: 5-10 min (1 min using automatic firing unit)
Ripple time: 16 s (40 rockets)
Disengagement time: 1 min (30 s using automatic firing unit)

Launcher Unit
Number of tubes: 40 (20 per module)
Elevation: 0 to +60°
Azimuth: ±105°
Crew: 3 (2 using automatic firing unit)

Loading Module
Length: 3.7 m
Width: 0.82 m
Height: 0.69 m
Weight empty: 400 kg
Weights loaded: 1,710 kg (conventional warhead), 1,810 kg (submunition warhead)

Specifications

FIROS 6 launcher module data
Number of tubes: 48
Length:
(module) 2.08 m
Width:
(module) 0.67 m
Height:
(module) 0.93 m
Weight:
(empty) 250 kg
(loaded) 480 kg
Elevation/traverse: manual
Elevation limits: –5 to +45°
Traverse: 360°
Power supply: 24 V DC
Rate of fire: 10 rockets/s

JAPAN

IHI AeroSpace 130 mm (30-round) Type 75 multiple rocket system

Development

Development of a 130 mm (30-round) self-propelled multiple rocket system by the Japanese Self-Defence Agency Technical Research and Development Institute began in 1965.

The first two prototypes were built in 1973 with Komatsu responsible for the chassis and the Aerospace Division of the Nissan Motor Company for the rocket. After operational tests the system was accepted for service with the Japanese Ground Self-Defence Force in 1975. The AeroSpace Division of Nissan Aerospace has recently been renamed IHI Aerospace.

The Japanese Ground Self-Defence Force took delivery of 66 Type 75 multiple rocket systems. The engine, suspension and tracks are the same as in the full-tracked Mitsubishi Heavy Industries Type 73 APC.

These 130 mm (30-round) Type 75 multiple rocket systems are supported by 15 Type 75 wind measurement vehicles to provide weather information to enhance the accuracy of the system. Recent figures issued by the Japanese Ground Self Defence Force have stated that at the end of 1999 a total of 64 systems remained in service alongside 14 Type 75 wind measurement vehicles.

130 mm (30-round) Type 75 MRS

Description

The crew compartment is at the front of the all-welded aluminium alloy hull with the driver sitting on the left and the commander on the right. The driver has three day periscopes to his front and a single-piece hatch cover that opens to the rear right. It contains a night vision periscope that can be traversed through a full 360°. The commander has a single-piece hatch cover that opens to the left rear containing six vision blocks. A 12.7 mm (0.50) M2 HB machine gun is provided at the front right commander's position for anti-aircraft defence. The operator is seated behind the commander and has a single-piece hatch cover that opens forward and to the right.

The engine compartment is on the left side of the hull behind the driver with the air inlet and air outlet louvres in the roof and the exhaust pipe running along the left side of the hull. The exhaust pipe is fitted with a silencer that is covered by a box-like mesh structure on the left middle of the hull top. The launcher is mounted on the top of the hull at the back and in the rear of the hull is a large door that opens to the right. The torsion bar suspension consists of five dual rubber-tyred roadwheels with the drive sprocket at the front and the idler at the rear. There are no track-return rollers. The Type 75 MRS has no amphibious capability but is fitted with infra-red driving lights.

The unguided 130 mm rockets have a single-stage solid propellant motor and are fin stabilised. The rockets can be fired singly or ripple fired with the complete unit of fire launched in 12 seconds. The launcher is electrically operated but can, in emergencies, be operated manually. Reload rockets are normally carried in accompanying cross-country trucks and reloaded manually.

Specifications

Calibre: 131.5 mm
Number of launcher frames: 30
Length:
(rocket) 1.856 m
(travelling) 5.78 m
Weight:
(complete rocket) 43 kg
(warhead) 15 kg
(launcher, empty) 2,000 kg
(launcher, loaded) 3,200 kg
(system, loaded) 16,500 kg
Max range: 15,000 m
Elevation/depression: +50/0°
Traverse: 50° left and right
Crew: 3
Width:
(travelling) 2.8 m
Height:
(travelling) 2.67 m
Ground clearance: 0.4 m
Track: 2.37 m
Track width: 400 mm
Length of track on ground: 3.3 m
Max road speed: 53 km/h
Range: 300 km
Ground pressure: 0.63 kg/cm^2
Gradient: 60%
Vertical obstacle: 0.7 m
Trench: 2.1 m
Engine: Mitsubishi 4ZF 2-stroke V4 air-cooled diesel developing 300 hp at 2,200 rpm
Suspension: torsion bar
Electrical system: 24 V
Armament: 1 × 12.7 mm M2 MG

The Type 75 is accompanied by a Type 75 self-propelled ground-wind measuring unit, which is a derivative of the Type 73 APC. It is also fitted with a wind-measuring mast that can be extended to 12.5 m and can measure wind velocities up to 30 m/s in two directions. Another variant is the Type 76 artillery location radar vehicle.

Status

Production complete. In service with the Japanese Ground Self-Defence Force. Japan is replacing the Type 75 with the American MLRS. The now IHI AeroSpace Company Limited is acting as the local system integrator. The first delivery of the MLRS to the Japanese Self-Defence Force was made by Nissan in March 1995. By the end of 1999 a total of 45 MLRS were in service with Japan. In 1996 the company was nominated by the Japan Defence Agency as the primary contractor for the MLRS Reduced Range Practice Rocket (RRPR). In May 1998 the IHI Aerospace Company Limited relocated to a new facility at Tomioka.

Manufacturer

IHI AeroSpace Company Limited.

UPDATED

KOREA, NORTH

North Korean multiple rocket launcher systems

Development/Description

Initial production of multiple rocket launchers began in North Korea during the early 1960s, with a locally built version of the 107 mm China North Industries Corporation (NORINCO) Type 63 system. Well over 5,000 were produced for the home and export markets.

With the Korean People's Army (KPA) the towed 107 mm system is found in the infantry division and brigade multiple rocket artillery battalions and batteries. In the mechanised and motorised division MRL battalions and in the specialised combined arms brigades battery the towed version is replaced by either a tracked version or a truck-mounted variant. The former has the launcher mounted on the licence-built Chinese full-tracked M1973 (Modified NORINCO Type YW 531 full tracked armoured personnel carrier). Apart from the 12-round type the self-propelled variant also exists in two other versions, the 18-round VTT-323 (M1992) system (which uses the launcher, in three layers of six tubes, mounted on the rear upper decking of the VTT-323 tracked APC) and the 24-round M1992/2 system (which uses the launcher, in three layers of eight tubes, mounted on the rear upper decking of the indigenously developed 4 × 4 armoured personnel carrier).

Considerable numbers have been exported and these have seen combat service with the Palestinian Liberation Organisation (PLO) and Syria in Lebanon and with Iran in the 1980-88 Gulf War.

Following delivery during the late 1960s of numbers of Russian 122 mm BM-21 MRL systems, the North Koreans began production of their own derivative in the mid-1970s. Known as the BM-11, this has two side-by-side 15-round banks of launcher tubes and is usually mounted on locally produced versions of the Russian Ural-375D (6 × 6) 4,000 kg truck, although for export purposes it can be mounted on other chassis. An example of this is the PLO's Isuzu (6 × 6) 2,500 kg systems which were captured by Israel during the 1982 war in Lebanon. Some of these were taken into service with the Israel Defence Force.

Specifications
(M1992/2)
Configuation: full tracked
Armament: 24 × 107 mm rocket launchers tubes
Weight (combat): 12,000 kg
Length (travelling): 6.10 m
Width (travelling): 2.80 m
Height (travelling): 2.60 m
Engine: 250 hp (unconfirmed)

Specifications
BM-11 provisional
Calibre: 122 mm
Number of tubes: 30
Rate of fire: 30 rounds in 15 s
Reload time: 8-9 min
Weight:
(complete rocket) 66 kg
(HE filler) 6.4 kg
Max range: 20,500 m
Min range: 2,500 m
Warhead types: HE fragmentation, smoke and chemical
Crew: 5
Launcher: locally built versions Ural-375D (6 × 6), ZIL-151 (6 × 6) or ZIL-157 (6 × 6) truck chassis

North Korean BM-11 version of Russian 122 mm MRS with two banks of 15 launcher tubes instead of single 40-round bank, mounted on rear of Japanese Isuzu (6 × 6) 2,500 kg truck chassis (Israeli Ministry of Defence)

In KPA service the BM-11 is used at army and corps level within MRL brigades (up to 72 systems). One of the roles assigned to these units is to provide chemical warfare delivery capability at the strategic level using chemical warheads that have been indigenously designed, manufactured and filled with agents including nerve gases. The BM-11 may also be found in the MRL battery of some of the combined arms brigades replacing the six 107 mm systems found in that unit.

There are two other 122 mm systems known. The 40-round M1977 version, which may be a direct copy of the BM-21 on a locally built chassis and the follow-on 40-round M1985, which is mounted on a Japanese Isuzu (6 × 6) truck chassis. The latter is believed to have a 40-round reload pack as part of the rear deck-mounted system.

The older systems are known to include varying numbers of truck-mounted Russian designed 132 mm BM-13-32, 140 mm BM-14-16, 200 mm BMD-20 and 240 mm BM-24 weapons plus small numbers of the towed 140 mm RPU-14 for airborne forces use. China is also known to have provided a number of its 130 mm Type 63 systems.

Status

107 mm system. Production as required. In service with Iran, Israel, Libya, North Korea (towed and self-propelled versions), the PLO and Syria.

122 mm BM-11. Production as required. In service with Algeria, Egypt, Iran, Iraq, Israel, North Korea, Lebanon, Libya, Nicaragua, Pakistan, PLO, Polisario Front, Syria, Uganda and Zimbabwe.

122 mm M1977. Production as required. In service with North Korea.

122 mm M1985. Production as required. In service with North Korea.

Specifications
(BM-11)
Configuration: 6 × 6
Crew: 5
Armament: 20 × 122 mm rocket launcher tubes
Range:
(max) 20,500 m
(min) 2,500 m
Weight of rocket: 66 kg
Reload time: 8-9 mins

Manufacturer

North Korean state factories.

UPDATED

240 mm (12-round) M1985 multiple rocket launcher

Development/Description

The 240 mm (12-round) M1985 MRL is mounted on a Japanese Isuzu 6 × 6 truck chassis and comprises 12 240 mm rocket tubes mounted in two banks of six on the rear decking. Each bank has two rows of three tubes. For firing, four manually operated stabilising jacks, two mounted at the rear of the vehicle and the other pair located just behind and either side of the cab, have to be lowered. All the traverse and elevation controls are manually operated.

The 240 mm calibre unguided rocket is spin stabilised in flight and fitted with a contact detonation fuzing system. Maximum range is said to be 43,000 m. The standard warhead used is a 90 kg HE fragmentation type although other types such as smoke, incendiary and chemical agent are believed to exist.

The launcher is supported by a transloader vehicle which carries at least one complete reload and a crane to assist in the manual reloading operation. The unguided 240 mm rockets are loaded one by one.

Variant

240 mm M1989 multiple rocket launcher

A 240 mm multiple rocket launcher, designated M1989 by the USA, is known to exist. No other details are known at present.

Status

Production as required. Both the M1985 and M1989 are in service with the North Korean Army. The designations M1985 and M1989 has been allocated by the US DoD as these were the years that they were first sighted.

Manufacturer

North Korean state arsenals.

It is probable that technology to build these 240 mm rockets has been transferred to Iran. The Iranian Defence Industries Organisation 240 mm Fadjr 3 (12-round) system (qv Iran) is very similar. Recent information gives this system a maximum range of 43,000 m with the 90 kg warhead containing 45 kg of high explosive. Recent information has indicated that at least one country in the Middle East, not Iran, has taken delivery of complete 240 mm multiple rocket launchers of an undisclosed type from North Korea.

UPDATED

240 mm (22-round) M1991 multiple rocket launcher

Development/Description

The 240 mm (22-round) M1991 MRL was first seen in 1991 and is mounted on a heavy 6 × 6 truck chassis with the cab located over the engine. Like the 240 mm M1985 system the M1991 is assigned to MRL battalions comprising three batteries, each of five 240 mm MRLs. The system fires the same 240 mm calibre unguided rockets as the M1985 and has similar HE, smoke, incendiary and chemical warhead capabilities. The designation M1991 has been allocated to this system by the US Department of Defense. In the designation M1991 the 1991 indicates the year that it was first observed by US intelligence.

Status

In production. In service with the North Korean Army. As far as is known this system has not been exported.

Manufacturer

North Korean state arsenals.

It is possible that technology to build these 240 mm rockets has been transferred to Iran. Recent information has indicated that at least one country in the Middle East, not Iran, has taken delivery of complete 240 mm multiple rocket launchers of an undisclosed type from North Korea.

UPDATED

KOREA, SOUTH

Daewoo Kooryong 130 mm (36-round) multiple rocket system

Development/Description

The Daewoo, Kooryong 130 mm 36-round, multiple rocket launcher has been designed and built in South Korea for corps and divisional level MRL units. Although fitted on the rear of the locally built 5 ton (6 × 6) KM809A1 truck chassis in South Korean Army service, it can be fitted to any wheeled or tracked vehicle with a 5 ton class load capacity.

The launcher uses a hydraulic motor to drive the azimuth, elevation and stabilising jack systems. If required all can be operated manually as well. The fire-control system consists of three solid-state components, the firing control box, the ignition distribution box and a circuit tester unit. A resupply KM813A1 6 × 6 5 ton truck is used to carry 72 reload rounds and to act as the loading platform. Reloading takes about 10 minutes.

The Kooryong 130 mm (36-round) multiple rocket launcher system fires at least two types of 130 mm unguided rocket which are manufactured by the Hanwha Corporation, these are called the standard rocket (K30) and the improved model (K33) and details are given in the Table following.

Single round, partial or full ripple salvo firings can be undertaken using the fire-control box from either the driver's seat, or off-vehicle from a site several metres away. A firing table or the fire direction calculator is used to provide the ballistic data. The latter provides a faster and more accurate method of calculation than the former.

The K30 and K33 solid propellant rocket motors are packed one round per wooden box with 12 boxes packed on a pallet for ease of transport in the field. The K37 and K38 warheads are packed two per wooden box and 16 of these boxes are then packed on a pallet.

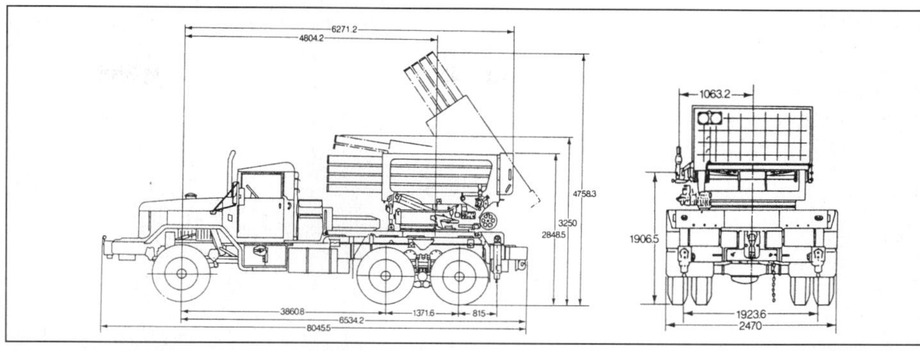

General configuration Kooryong 130 mm (36-round) Multiple Rocket Launcher System mounted on KM809A1 (6 × 6) truck chassis

Surface wind measurements are made using a 10.25 m mast installation mounted on the back of a 1¼ ton 4 × 4 KM450 light truck and connected to a microprocessor. This provides both digital average and instant windspeed data for the fire-control calculations.

It is believed that the South Korean Army has a total of 156 of these Kooryong 130 mm (36-round) multiple rocket launchers in service. They are now being supplemented by the US Lockheed Martin Missiles and Fire Control - Dallas 227 mm (12-round) Multiple Launch Rocket Systems.

Status

Production of the launcher is complete but can be resumed. Production of the rocket is still undertaken. In service with the South Korean Army. This system was never exported.

Manufacturers

Daewoo Heavy Industries Limited (launcher).
Hanwha Corporation (130 mm rockets).

UPDATED

Type	Standard	Improved
Designation	K30	K33
Calibre	130 mm	130 mm
Overall weight	55 kg	64 kg
Overall length	2,390 mm	2,528 mm
Max range	23,000 m	36,000 m
Max velocity	910 m/s	1,200 m/s
Weight of warhead	20.37 kg	20.37 kg
Rocket motor designation	Kooryong I	Kooryong II
Length of rocket motor	1,790 mm	1,924 mm
Propellant	composite (CTPB)	composite (HTPB)
Fuze designation	K505	K505
Fuze type	impact	impact
Warhead designation	K37	K38
Type	natural fragments	16,000 steel balls
Explosive weight	6.5 kg	2.8 kg
Explosive type	Composition B	Composition B

Specifications

Improved version KM809A1 truck chassis
Calibre: 130 mm
Number of tubes: 36
Length:
(tube) 3.4 m
(rocket) 2.55 m complete*
(rocket motor) 1.79 m*
Weight:
(complete rocket) 64 kg
(warhead) 21 kg (6.5 kg HE)
(rocket motor) 33.3 kg
Range: 12,000-36,000 m
CEP: 387 m
Max rocket velocity: 1,200 m/s
Rate of fire: 2 rockets/s
Elevation/depression: +55/0°
Traverse: 120° left and right
Reloading time: <10 min
Crew: 3
Configuration: 6 × 6
Combat weight: 17,100 kg
Length:
(travelling) 8.0455 m
Width:
(travelling) 2.47 m
Height:
(travelling) 2.8485 m
Max road speed: 80 km/h
Gradient: 55%
Power supply: 24 V DC
* This depends on the rocket used

Kooryong 130 mm (36-round) Multiple Rocket Launcher System mounted on KM809A1 (6 × 6) truck chassis

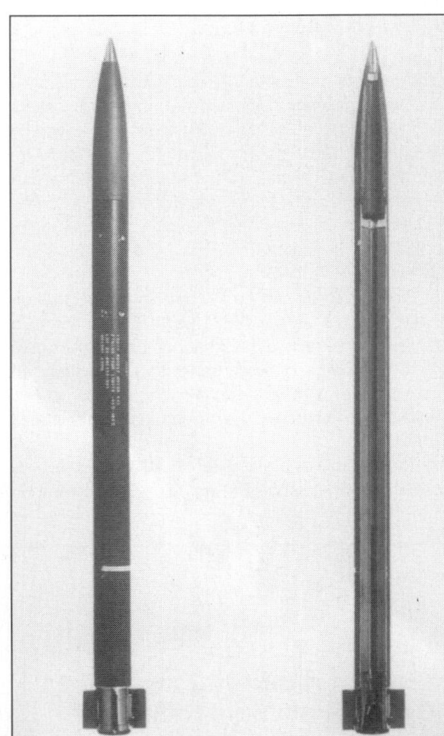

The two 130 mm rockets fired by the 130 mm (36 -round) Kooryong multiple rocket system are the K30 (left) and the K33 (right) 0021064

MYANMAR (BURMA)

Defence Products Industries 81 mm BA84 artillery rocket system

Development/Description

The Burmese Defence Products Industries is manufacturing a mobile 81 mm calibre unguided surface-to-surface rocket system known as the BA84. No other details are known at present. It is possible that this is an air to ground rocket modified for ground to ground use.

At this time it is not known as to whether this system is a single-round launcher or a truck or trailer-mounted multiple rocket launcher.

Status

Production. In service with the Burmese Army. There are no known exports of this artillery rocket system.

Manufacturer

Defence Products Industries.

VERIFIED

PAKISTAN

Dr A Q Khan Research Laboratories 122 mm (30-round) multiple rocket launcher system

Development/Description

In many respects the actual 122 mm (30-round) launcher, which is designed and built by the Dr A Q Khan Research Laboratories, is very similar to the widely deployed Russian 122 mm BM-21 (40-round) multiple rocket launcher.

The Pakistan system however consists of two banks of two launch tubes each of which consists of 15 launch tubes in three layers each of five tubes. The stainless steel tubes are 3.1 m long and have an actual calibre of 122.4 mm with the unguided rockets being manually loaded from the rear.

Each launch tube consists of a guide tube, projectile stopper and a contact device bracket. The guide tube has a spiral groove which has a uniform twist of 2-2.5° and at the rear end of the tube is a clamp which is fitted with a projectile stopper. The latter prevents the loaded 122 mm rocket from slipping out of the firing tube backwards and when activated, releases the guide pin when the propelling force of the rocket reaches 600 to 800 kg. This ensures the optimum muzzle velocity when fired.

The solid propellant rockets, which are manufactured by the Pakistan Ordnance Factories (POF) have wraparound fins at the rear and a nose mounted high explosive warhead and fuze system. The actual rockets are referred to as the Yarmuk by POF. The rocket warhead weighs 19.4 kg with fuze or 18.4 kg without fuze. The rocket is 2,875 mm long and weighs 66 kg of which 20.4 kg is the double base propellant and 6 kg the composition B bursting charge.

The rockets can be launched by an electric firing mechanism which is located in the vehicle cab, or, more usually, from a safe position up to 50 or 60 m from the launcher using a remote-control device which is connected to the launcher by a cable.

The rockets can be launched one at a time, or in salvo, with maximum range of the unguided rockets being 20,380 m. The launcher can be traversed through 180° (90° left and 90° right), with elevation being from zero to plus 55° . Launcher traverse and elevation is electric with manual controls being provided for emergency use. The latter are mounted at the rear of the launcher on the left side where the sighting system is also located.

In Pakistan Army service the system is mounted on the rear of a US supplied M35 series 6 × 6 truck or a Japanese 6 × 6 truck, although it can be installed on a wide range of other chassis, tracked and wheeled.

To provide a more stable firing platform two stabilisers are lowered to the ground, one either side at the rear.

In addition to manufacturing the 122 mm (30-round) rocket launcher, the Dr A Q Khan Research Laboratories are also involved in a number of other rocket and missile projects for the Pakistan armed forces.

These include the ANZA Mk 1, Mk 1A and Mk 11 manportable fire-and-forget surface-to-air missile systems, various rocket propelled mineclearing devices and the 3,000 m range Baktar Shikan anti-tank guided weapon which is very similar to the Chinese North Industries Corporation Red Arrow 8 system. The company also manufactures explosive reactive armour and rocket-propelled mine clearing systems.

Pakistan Army 122 mm (30-round) Dr A Q Khan Research Laboratories MRS mounted on a US supplied M35 series (6 × 6) truck

Pakistan Army 122 mm (30-round) Dr A Q Khan Research Laboratories MRS mounted on a Japanese 6 × 6 truck chassis
0021065

Status

Production as required. In service with the Pakistan Army. According to United Nations sources, Sri Lanka took delivery of six of these systems from Pakistan in 2000.

Manufacturer

Enquiries to Dr A Q Khan Research Laboratories, Kahuta.

UPDATED

POLAND

Polish upgraded 122 mm BM-21 (40-round) multiple rocket launcher

Development

The most widely used Multiple Rocket Launcher (MRL) in the world is the Russian 122 mm BM-21 (40-round) system, which is in service with well over 50 countries.

This system was developed from the late 1950s, with the Splav Scientific Production Concern being the overall system prime contractor and responsible for the 122 mm unguided rockets and the Motovilikha Plants Corporation being responsible for the launcher.

The original system entered service with the Russian Army in 1963 and has since seen combat in many parts of the world, especially Asia and the Middle East.

The baseline 122 mm BM-21 (40-round) system has also been copied or manufactured by many other countries, some of which use their local chassis, normally wheeled but some times tracked. These

Upgraded Polish BM-21 122 mm (40-round) MRL on Polish designed and built Star 1466 (6 × 6 chassis) with four door cab (Christopher F Foss)
2002/0096294

systems some times incorporate a number of improvements to meet local operational requirements.

The Polish Army has a large number of 122 mm BM-21 (40-round) systems installed on Russian Ural-375D (6 × 6) truck chassis. Some years ago a decision was taken to retain these systems in service by carrying out a major upgrade covering the chassis, 122 mm unguided rockets and fire-control system.

Description

For this upgrade the 122 mm (40-round) launcher has been removed from its Russian supplied Ural-375D (6 × 6) chassis and installed on a locally developed Star 1466 (6 × 6) truck chassis with a new fully enclosed forward control cab.

The conventional bonneted cab of the original Ural-375D only had seats for three people but the new Star cab has seats for six people including the driver. To allow rapid access to the powerpack located under the cab it can be tilted forwards hydraulically.

This chassis has excellent cross-country mobility and is fitted with a central tyre pressure regulation system that allows the driver to adjust the tyre pressure from within the cab to suit the terrain being crossed.

Mounted on the rear of the cab is the 122 mm (40-round) launcher, although the actual calibre of the launcher tubes is 122.4 mm. The launcher has elevation from 0 to +55° with traverse being 70° right and 102° left.

Elevation is powered at the rate of up to 7° a second with traverse being powered at the rate of up to 5° a second. In addition, there are manual back-up controls mounted on the left side of the launcher.

This part of the upgrade is carried out by HSW Stalowa Wola who is also prime contractor for the new Polish 155 mm/52 calibre Krab self-propelled artillery system.

This is essentially a locally designed and built chassis fitted with the complete turret of the UK BAE Systems, RO Defence AS90 Braveheart self-propelled artillery system. Details of this system are given in the Self-propelled guns and howitzers (tracked) section.

The standard Russian 122 mm BM-21 (40-round) MRL fires a 122 mm 9M22U unguided rocket to a maximum range of 20,380 m. By todays standards this was considered to be too short.

For many years the Polish company of Pressta Spotka Akcyjna (SA) has been engaged in the manufacture of a wide range of metal parts for ammunition as well as 122 mm rockets.

In fact of their military production, some 77 per cent of sales were for 122 mm rockets and their metal parts. In the early 1990s, development of a new and more effective 122 mm unguided artillery rocket commenced by the company with three new and more effective warheads being developed which could be backfitted to existing rockets. These were:

(1) High-explosive type with increased fragmentation effect called Spall which is designated the M-21FOB.
(2) Cargo type carrying five Polish developed scatterable anti-tank mines called Platan and which is designated the M-21FMK. The mines are designated the MN-121 and are made by BZU Belma.
(3) Cargo type carrying 42 Polish developed High Explosive Anti-tank Fragmentation (HEAT-FRAG) bomblets to attack the vulnerable upper surfaces of armoured vehicles. These will penetrate 120 mm of conventional steel armour. This is designated the M-21FK1.

The next step in the upgrade was to increase the range of the rocket and this part has been developed by the French company of CELERG SA who have extensive experience in rocket motor development. Some years ago CELERG were co-operating with Russia on a 122 mm rocket upgrade but in the end this came to nothing.

One of the key requirements was to increase the range of the system to 30 km. According to Pressta SA, not only has the new family of 122 mm rockets have increased range, but they are also more accurate and have longer shelf life.

CELERG developed a new composite solid-propellant rocket motor for the 122 mm rocket called Phenix (or Feniks). Modification of the airframe nozzle system and stabiliser unit was also carried out. The nose-mounted fuzes on all 40 rockets can be set in 3 minutes.

During acceptance trials in Poland in 1999, a maximum range of 41,000 m was achieved with the new rocket, with all rockets being launched in 20 seconds.

These unguided rockets are spin and fin stabilised with the latter unfolding after they have left the launch tube. It is understood that first production rockets will be complete in 2002.

Specifications
Calibre: 122 mm
Number of barrels: 40
Max range: 41,000 m (see text)
Elevation: 0 to +55°
Traverse
(left) 102°
(right) 70°
Crew: 3 (but can seat 6)
Configuration: 6 × 6
Combat weight: 14,000 kg
Length: 7.4 m
Width: 2.5 m
Height: 3.00 m
Max road speed: 86 km/h
Max range: 650 km
Fording: 1.2 m
Powerpack: MAN 6-cylinder diesel developing 162 kw at 2,400 rpm

Specifications of the Spall and Platan type unguided rockets are as follows:

Type	**HE**	**Cargo**
Max range	41 km	36 km
Length	2.67 m	2.82 m
Weight	61 kg	65 kg
Weight of warhead	19 kg	23 kg

In addition to being launched from the Russian 122 mm B-21 (40-round) MRL and its equivalent systems, the new rockets can also be launched from the RM-70 122 mm (40-round) system on a Tatra (8× 8) series cross-country truck, which is used by many countries.

In Polish Army service, the effectiveness of the upgraded BM-21 system has also been improved as it is provided with a new C4I (Command, Control, Communications, computer and Intelligence) system developed by the local company of W B Electronics.

For the Polish upgraded BM-21 system this includes a Battle Field Computer (BFC) 201 as a fire-control system terminal, RRC 9500 digital radio, FONET digital intercom and a navigation system.

Variants of this C4I system are also used on other Polish artillery mortar systems, including the Russian supplied 122 mm 2S1 self-propelled artillery system, Polish 155 mm/52 calibre Krab self-propelled artillery system and the 98 mm M-98 mortar.

Status

Development complete. Expected to soon enter production for the Polish Army. The first batch is understood to consist of 24 systems.

Manufacturer

HSW Stalowa Wola.

NEW ENTRY

The new 122 mm rockets for the upgraded Polish system have been developed by the Polish company of Pressta SA and CELERG of France (Christopher F Foss)
2002/0096295

ROMANIA

Aerostar 122 mm (40-round) APRA multiple rocket launcher system

Development/Description

There are two versions of the Aerostar 122 mm (40-round) APR-40 multiple rocket launcher system known to exist.

The basic model, covered below, is based on the locally built DAC 665T (6 × 6) truck chassis, has the Russian BM-21 launcher, as shown in the photograph, and has handrails mounted on the rear mudguards. This version is also referred to as the LRSV BM-21 (Croatia) and the SVLR 122 'Grad' (Bosnia).

The second and more recent model has a launcher with two removable modules as mentioned later in the text below, and the handrails have been replaced with sheet plate to protect the tyres. This is based on the DAC 15.215 DFAEG (6 × 6) chassis. In Croatian service, this is referred to as the LRSV M-91 'Vulcan', while the Bosnian Army calls the system the SVLR 122 APR-40.

The first model to enter service is called the APR-40 (Aruncator de Projectile Reactive) while the model with the two launcher modules is called 40 APRA 122 FMC.

The chassis is the forward control type with the two-door cab providing space for the complete crew of the system. A central tyre-pressure regulation system is fitted as standard and allows the driver to adjust the tyre pressure to suit the type of terrain being crossed.

The launcher consists of two removable modules each with 20 launching tubes that are banked in four rows of five. A resupply vehicle, the MITC, is based on the same chassis. This weighs 16,440 kg fully loaded, carries two reload modules and a rear-mounted MH 90 hydraulically operated crane with a lifting capacity of 6,000 kg at the minimum outreach of the arm and 2,000 kg at its maximum outreach. Maximum lifting height is 6 m.

Replacing the expended rocket modules takes about 10 minutes. The resupply vehicle can also tow a simple four-wheeled two-axle, Type RM13, trailer which carries four 20-round racks of 122 mm rockets that can be used to reload the modules by hand. To refill two 20-round launcher packs takes approximately 15 minutes. Total weight of a fully loaded trailer is 11,530 kg.

The fire-control unit is mounted in the forward control cab but if required the rockets can be launched by remote control via a cable up to 50 m from the cab.

The operator can select either single shots of a complete salvo at variable firing rates of up one, 0.7 second or 0.5 second.

The sighting system is mounted on the left side of the actual launcher where the elevation and traverse controls are located while the firing circuits are on the left side of the chassis, just above the diesel fuel tank.

Standard equipment includes a radio which is installed in the cab, this is used to communicate with other launchers or the next chain of command. When fitted with an antenna 1.2 m high the radio has a range of 10 to 12 km. The radio can also be operated up to 50 m away from the vehicle using a remote-control handset.

122 mm (40-round) APR 40 MRS based on DAC 665T (6 × 6) chassis (BA) ***2000***/0085764

The 122 mm unguided rocket type normally used is the standard Russian 122 mm M-21-OF, although other Russian 122 mm types and equivalent People's Republic of China and Yugoslav models can also be fired.

At maximum range the area covered by the 40-round salvo is 210,000 m². With small braking rings fitted to the rockets the area coverage at 15,000 m range is 100,000 m² while with large braking rings fitted the area coverage at 12,000 m is 75,000 m².

The Romanian Army has a total of 160 of these systems in operational use under the designation APR 40 with the latter indicating the number of barrels.

The Romanian company of RATMIL Regie Autonoma is now marketing 122 mm unguided rockets for the locally developed APRA system as well as the more widely deployed Russian 122 mm BM-21 series of truck mounted (6 × 6) multiple rocket systems.

Two 122 mm rockets are currently being marketed, the Type M21-OF-FP and the shorter range Type M21-OF-S and brief details of these are given in the Table:

Type	**M21-OF-FP**	**M21-OF-S**
Calibre	122 mm	122 mm
Length of rocket	2,870 mm	1,927 mm
Weight of rocket	65.4 kg	46.6 kg
Max range	20,400 m	12,700 m
Min range	5,000/6,000 m	1,000 m
Weight of explosive in warhead	6.35 kg	6.35 kg
Weight of propellant	20.65 kg	11.10 kg

Both of these rockets have a nose mounted fuze which is set before launch to give instantaneous, short or long delay functions.

Variants

Resupply vehicle

This is based on the same DAC 665T (6 × 6) truck chassis as the launcher and carries 40 122 mm rockets in two containers with an MH 90 hydraulic crane being fitted for reloading purposes. This has a maximum lifting height of 6 m and can lift a maximum load of 6 tonnes at minimum outreach and a maximum load of 2 tonnes at minimum out reach. This also tows a two-axle trailer with a further 80 122 mm rockets in four racks.

LAROM upgrade

This recent upgrade has been carried out by Israel Military Industries and Aerostar and is covered in detail in this section under International.

Coastal defence model

Special coastal and border defence versions of the 122 mm (40-round) rocket systems have also been proposed by Aerostar.

122 mm (21-round) APR 21 based on Bucegi SR-114 (4 × 4) chassis. This is no longer used by the Romanian Army (BA) ***2000***/0085765

Specifications
(DAC-665T chassis)
Calibre: 122 mm
Number of barrels: 40
Length:
(barrel) 3 m
(travelling) 7.39 m
Max range: 20,380 m
Elevation: 0 to +55°
Traverse: 110° left, 100° right
Crew: 1-5
Configuration: 6 × 6
Weight
(combat) 17,650 kg
(unloaded)15,010 kg
Length: 7.39 m
Width:
(travelling) 2.51 m
Height:
(travelling) 3.25 m
Ground clearance: 0.39 m
Fording: 1.2 m
Gradient: 60%
Side slope: 30%
Max speed:
(road) 80 km/h
(cross-country) 25 km/h
Engine: 6-cylinder, D2156 HMN8 water-cooled diesel developing 215 hp at 2,200 rpm
Range: 1,000 km

Typically these would be located underground together with their associated reloading mechanisms and raised into the operating position when required.

122 mm APRN

This is a 122 mm system for naval applications and consists of 2 × 40 122 mm rocket launchers which retract into the platform for reloading purposes.

APR-21 122 mm (21-round) multiple rocket launcher system

This is based on the locally developed Bucegi SR-114 (4 × 4) truck chassis and has three layers each of seven 122 mm launch tubes. This system is no longer employed by the Romanian Army.

Manufacturer

Aerostar SA, Group Industrial Aeronautic.

UPDATED

Status

Production as required. In service with:

Country	Quantity	Comment
Bosnia and Herzegovina	36	delivered 1997/98
Botswana	20	delivered 1996
Cameroon	20	delivered 1996
Croatian Army	n/avail	
Croatian Defence Council	n/avail	
Iranian National Liberation Army	n/avail	
Iraq	n/avail	
Liberia	n/avail	some being upgraded to LAROM configuration (qv)
Nigeria	18	
Romania	160	

Aerostar 122 mm (12-round) Awora multiple rocket launcher system

Development/Description

The 122 mm (12-round) Awora (or Aurora) multiple rocket launcher system was developed as a private venture by Aerostar SA which is also prime contractor for the 122 mm (40-round) APRA system on a 6 × 6 chassis and, the 122 mm (single round) rocket launcher covered elsewhere in this section. This has also been referred to as the 12 ARO 122 mm system. According to United Nations sources there were no exports of this system between the years 1992 to 2000.

The 122 mm (12-round) Awora is essentially a locally built ARO (4 × 4) light vehicle with the engine at the front, two door, two person fully enclosed cab centre and the 122 mm (12-round) rocket launcher carried on the rear decking.

The launcher has two banks of six 122 mm rocket launcher tubes within the restraining metal framework.

This comprises a complete pod which can be rapidly removed from the vehicle for reloading purposes. To provide a more stable firing platform, a hydraulically operated stabiliser is lowered to the ground either side at the rear of the vehicle.

While this 122 mm (12-round) multiple rocket launcher fires the same rockets as the 122 mm (40-round) system covered in the previous entry, it has a shorter range. This is due to the fact that the launcher rails are on 2.01 m long compared to 3.00 m of the 122 mm (40-round) system based on a 6 × 6 truck chassis.

The ADFM-01 fire-control unit is mounted in the cab but if required the rockets can be launched by remote control via a cable up to 100 m away from the cab. This is the same fire unit as used in the 122 mm (40-round) APRA system covered in the previous entry.

The operator can select either single shots of a complete salvo at variable firing rates of up one, 0.7 second or 0.5 second. When firing at the rate of 0.5 rockets per second a complete salvo can be fired in 6 seconds and, when firing at the rate of one rocket per second, a complete fire sequence takes 12 seconds.

The AO-AG 9M sighting system is mounted on the left side of the actual launcher where the elevation and traverse controls are located. This is also used on the APR 40 122 mm (40-round) multiple rocket launcher system.

Standard equipment includes a radio which is installed in the cab, this is used to communicate with other launchers or the next chain of command. When fitted with an antenna 1.2 m high the radio has a range of 10 to 12 km. The radio can also be operated up to 50 m away from the vehicle using a remote-control handset.

122 mm (12-round) Awora MRS mounted in rear of ASO (4 × 4) light chassis deployed in firing position with launcher traversed left
0058462

The 122 mm upguided rocket types normally used are the short version Russian DKZ-B and equivalent 122 mm Chinese round.

122 mm unguided rockets for this and other Romanian rocket launchers is now manufactured by the Romanian company of RATMIL Regie Autonoma.

Variants

Aurora 122 mm system

This is similar to the previous described system, but is mounted on a new chassis. It is provided with four hydraulic stabilisers that are lowered to the ground to provide a more stable firing platform.

Trailer system

A 122 mm (18-round) trailer-type multiple rocket launcher system was developed to the prototype stage. Details were given in *Jane's Armour and Artillery 1996-97.*

Status

Prototype. Not known to be in service.

Manufacturer

Aerostar SA, Group Industrial Aeronautic.

UPDATED

Specifications
Calibre: 122 mm
Number of barrels: 12
Length:
(barrel) 2.17 m
(travelling) 4.830 m
Max range: 13,000 m
Min range: 1,500 m
Rate of fire: 1 rd/0.5, 0.7 or 1 s
Elevation: 0 to +55°
Traverse: 90° left, 110° right
Width:
(travelling) 1.775 m
Height:
(travelling launcher at 35°) 2.815 m
(travelling launcher at 0°) 1.850 m
Ground clearance: 0.196 m
Wheelbase: 3.2 m
Max speed:
(road) 80 km/h
(cross-country) 25 km/h
Range: 1,000 km
Engine: Peugeot XD3TP-P158 indirect turbo injection diesel developing 95.1 hp
Crew: 2

Aerostar 122 mm (single-round) rocket launcher

Development/Description

In addition to the previous 122 mm truck mounted 6 × 6 and 4×4 Romanian multiple rocket launchers there is also a short 122 mm HE rocket which is used in conjunction with a 2.5 m long single-tube animal or man-portable tripod-mounted 122 mm rocket launcher. This is designed for use by guerrilla, commando and special forces units. Prime contractor for all of these Romanian 122 mm unguided rocket systems is Aerostar SA. This has also been referred to as the 122 mm Short-Type Rocket Launcher for Infantry and Paratroops.

The rocket weighs 46.3 kg complete and is 1.927 m long. Its maximum range is 13,400 m, the launcher has an elevation range of –3 to +50° and a traverse of +15°. A contact fuze with preselected instantaneous, short or long delay actions is fitted.

Total weight of the launcher when assembled for firing is 59 kg (which is made up of the 25 kg tripod, the

Specifications
Calibre: 122 mm
Weight
(with tripod) 63.2 kg
(without tripod) 23.2 kg
Length: 2,540 mm
Range: 11,400 m
Elevation: +1 to +50°
Traverse: 360°
Crew: 1

30 kg launch tube and 4 kg for the sight and remote firing device with 25 to 35 m connector cable).

The launcher can also be fixed on a special mount for use on a light cross-country vehicle. In this application the launcher can either be fired from the vehicle or dismounted for ground use. A total of 10 to 15 reload rounds would normally be carried. The number of reload rockets carried depends on the carrier platform.

122 mm unguided rockets for this and other Romanian rocket launchers is now manufactured by the Romanian company of RATMIL Regie Autonoma. These rockets can also be used in other countries' 122 mm artillery rocket systems.

Variant
This system can also be installed and fired from a vehicle.

Status
Production as required. In service with the Romanian Army. It is probable that this 122 mm launcher system has been exported.

Manufacturer
Aerostar SA, Group Industrial Aeronautic.

UPDATED

Romanian 122 mm (single round) rocket launcher in firing position

RUSSIAN FEDERATION AND ASSOCIATED STATES (CIS)

Free Rocket Over Ground (FROG) artillery rocket system

Development
The first Russian tactical rocket system was the 250 mm Korshun (Kite). It used the liquid propellant 3R7 tactical rocket which had a high explosive warhead. This was launched from an SM-44 launcher based on a YAZ-214/KrAZ-214 (6 × 6) cross-country truck chassis.

This entered production in 1957 and had six 250 mm tubes with the unguided rocket achieving a maximum range of 55 km. This was phased out of service many years ago.

This was followed by the first solid-propellant tactical rockets, the 3R1 Mars and the 3R2 Filin (Eagle Owl) with the latter entering service in 1955 and fitted with a tactical nuclear warhead.

The 3R1 Mars was a 324 mm diameter unguided rocket with a maximum range of 17.5 km while the 3R2 Filin was a 612 mm diameter unguided rocket with a maximum range of 25.7 km.

The single-rail Filin (NATO designation FROG-1) is carried on a modified IS-2 heavy tank chassis designated the 2P2 and the shorter-range single-rail Mars (NATO designation FROG-2) is carried on the 2P4 non-amphibious variant of the PT-76 light amphibious tank chassis.

The missiles were grouped together in special missile-launcher battalions of two FROG vehicles and attached to tank and mechanised (later motorised rifle) divisions.

By 1960, the cumbersome Filin system was being replaced with the similar performance 2K6 Luna family (3R-8 Luna, 3R-9 Luna-1 and 3R-10 Luna-2, respective NATO designations FROG-3, FROG-4 and FROG-5) based on a further modified version of the Mars non-amphibious PT-76 light tank chassis which was designated the 2P16. The major differences between the family variants which warranted the individual NATO designations were in the shape of the warheads.

It is now known that the 3R-9 was a high explosive fragmentation warhead whereas the 3R-10 was a tactical nuclear warhead.

FROG-7 artillery rocket system in travelling configuration and showing rocket reloading crane 0058463

More recently it has been stated that the 3R9 Luna rocket could be fitted with a tactical nuclear or high explosive warhead. A later high-explosive fragmentation warhead was designated the 3N15 while a new rocket, the 3R10, was developed for the tactical nuclear warhead designated the 3N14.

Complete with its rocket, the 2P16 system weighed a total of 18.8 tonnes and had a maximum road speed of 40 km/h. It took 11 minutes to come into action and launch a rocket and was normally operated by a crew of 11 men.

The launcher complex also included a ZIL-157 (6 × 6) tractor truck which towed a trailer carrying two replacement rockets and a crane truck which was required to lift the rockets from the trailer and load them onto the launcher.

Development of the new Luna-M (9M21) commenced in 1961 with the emphasis on mobility of the launcher and greater range and accuracy of the rocket.

The first test launch took place in December 1961 and it was accepted for service with the Russian Army in 1964 with production of the system being undertaken at the Barrikady facility. The Luna-M (9K52) system is also referred to in the West as the FROG-7a with the BAZ-135L4 (8 × 8) being the Transporter-Erecter-Launch (TEL) platform.

The Luna-M used the 9M21F rocket with the 9N18F warhead filled with 200 kg of explosive. Other rockets were the 9M21B with AA-21 nuclear warhead using a radio fuze and the 9M21B1 warhead with an AA-38 nuclear warhead.

The 9M21G rocket was equipped with the 9N18G chemical warhead. The 9M21A rocket was equipped with a 9N18A leaflet warhead. The 9M210F rocket has the 9N180F sub-munition warhead which carries

Specifications
FROG-7
Weight:
(at launch) 2,432-2,450 kg (depending upon warhead type)
(HE warhead) 450 kg
(chemical warhead) 436 kg
Length:
(FROG-7a) 8.96 m
(FROG-7b) 9.4 m
(travelling) 10.75 m
Diameter:
(body) 0.544 m
(HE warhead) 0.544 m
(chemical warhead) 0.544 m
Warhead types: nuclear, HE, chemical (and FROG-7b cluster)
Max range: 67,000-68,000 m
Min range: 12,000-15,000 m
CEP: 400 m
Crew: 7
Carriage: single-rail launcher on BAZ-135L4 (8 × 8) cross-country vehicle
Combat weight: 19,000 kg
Elevation: +15 to 65°
Traverse: -7 to +7°
Length: (travelling) 10.69 m
Width: (travelling) 2.8 m
Height: (travelling) 3.35 m
Height: (without rocket) 2.30 m
Track: 2.30 m
Engine: 360 hp
Ground clearance: 0.475 m
Max speed: 65 km/h
Cruising range: 650 km
Ground clearance: 0.5 m
Trench: 2.63 m
Vertical obstacle: 0.685 m
Fording: 1.2 m

Rear view of FROG-7 rocket clearly showing the small exhaust nozzles around the large exhaust venturi

42 sub-munitions. Training rockets were the 9M21Ye and 9M21Yel. All of these used the 3Kh18 solid propellant rocket motor.

There was also a special helicopter version called the 9K53 Luna-MV with the trailer launcher called the Br-257 (9P114) which could also propel itself up to a distance of 40 to 45 km. Prototypes of this system were built in 1964 and 1965. These were never placed in quantity production.

At the same time, the divisional artillery rocket battalion was doubled in size to a battalion of two firing batteries with two TELs each.

The FROG-7a was joined in service during 1968 by the FROG-7b variant (Russian designation R-70) which has a longer warhead section (total rocket length extended to 9.4 m from 8.95 m). This is fitted with a pair of air brakes on the rear of the rocket which can be locked open. The purpose of this is to lower the range of the rocket. The Luna-M rocket had a CEP of 1,200 to 2,000 m while the new Luna-Z (or R-70) was required to be much more accurate. This was launched from a new complex called the 9K52M which could fire either the Luna-M or Luna-Z rockets. Trials showed that the Luna-Z had even a larger CEP so this never entered quantity production. For this reason, work started on the Tochka (Point) guided missile system which has the NATO designation of the SS-21.

The FROG-1 and FROG-2 systems have been considered obsolete for many years but the FROG-3, FROG-4 and FROG-5 weapons can still be found in reducing numbers in front-line service with several armies.

The NATO-designated FROG-6 weapon is a non-operational dummy training round which is designated the PV-65 *(prakticheskaya vystrel-65).*

Early FROG types were exported to the former Warsaw Pact, Algeria, Cuba, Egypt, North Korea and Syria while the FROG-7 has been supplied to the former Warsaw Pact, Algeria, Cuba, Egypt, Iraq, North Korea, Kuwait, Libya, Syria and the Yemen.

Iraq has also manufactured its own variant of the FROG-7 called the Laith 90. This has a cluster munition warhead and an increased maximum range of 90,000 m. Iraq and North Korea are also known to have produced their own chemical warheads for FROG rockets. Iraq subsequently used its FROG-7 systems in the 1990-91 Gulf War against northern Saudi Arabia and against rebel positions in the ensuing civil war.

In 1985, the Soviet Army started using its FROG-7b systems in Afghanistan. Fitted with HE and submunition warheads the rockets were used against village targets as part of the effort to deny basic agricultural food supplies to the guerrilla forces.

In 1989, it was revealed that the Afghan Army had taken delivery of a number of FROG-7 launchers. These used life-expired FROG-7a/b rounds from Soviet stocks and took the place of the 1,000-odd Scud-B rounds delivered previously and fired against various guerrilla targets. The Serbian Army has also used FROG-7 weapons in the Yugoslav troubles.

The long-term replacement for the FROG-7 is the inertially guided US/NATO designated SS-21 Scarab (Russian name Tochka, meaning 'point') missile system. First deployed operationally in 1976, the weapon had, by 1989, replaced the majority of the FROG-7 systems in Eastern Europe and had been exported to Czechoslovakia, East Germany and Syria. Apart from various conventional warheads the SS-21 can carry the AA-60 nuclear warhead with 5, 10, 20 or 200 kT selectable yields available.

FROG-7 artillery rocket system deployed in the firing position and with launcher elevated

Description
(FROG-7)
A FROG-7 battalion consists of an headquarters battery and two firing batteries, each with two FROG-7 9P113 TELs, a GAZ-66 (4 × 4) fire direction vehicle and a trailer-mounted D-band RMS-1 (NATO designation End Tray) long-range meteorological radar. The TELs have no NBC protection for their crews.

Amongst its vehicles the headquarters battery contains four BAZ-135L4 9T29 TZMs *(Transportna-zaryazhyushcha Machina:* transporter-loader vehicle) which carry three reload FROG-7 rounds each. These are placed on the TEL 9P113 by using the vehicle's own onboard hydraulically operated goose-neck crane which has a capacity of 2.6 tonnes which is positioned on the right side of the launcher rail just behind and above the wheel arch of the centre pair of wheels. The reloading operation takes around 20 minutes in total.

Preparation time for firing can take 15 to 30 minutes depending upon what operations need to be performed and whether the unit has only just occupied the position. The Luna-M itself is aimed by raising the TEL's rail to the desired elevation and adjusting the weapon's speed brakes so as to operate during the flight to give the correct range. If needed, a limited traverse angle can also be set on the launch rail assembly to compensate for any last minute offset required.

For optimum usage the associated RMS-1 radar has to be used to establish the meteorological parameters that might affect the rocket's flight trajectory.

Offensive operations with FROG-7s require them to be positioned well forward, around 8 to 18 km from the line of contact in order to engage tactical targets in the enemy's rear. In defensive operations these distances may be up to 25 to 50 per cent further back.

The FROG-7 unguided is of conventional single-stage design with a cylindrical warhead of the same diameter as the rocket body. The combined booster-sustainer rocket motor uses solid propellant fuel and has a large exhaust venturi surrounded by 16 small booster nozzles. The four tail-mounted fins act to stabilise the rocket in flight. Flight time to maximum range is about 160 seconds. Burn time of the engine is between 6.8 and 10.8 seconds depending upon the propellant quality. Airbrakes can be fixed to the rear of the rocket to achieve ranges between 15 and 29 km.

The warheads available include nuclear, conventional HE, chemical and (in the case of the FROG-7b) cluster munition. Known rocket designations include 9M21 (Luna-M rocket), 9M21B (Luna-M rocket with one of three tactical nuclear warhead types), 9M21E (Luna-M rocket fitted with 9N18E dispenser warhead) and 9M21F (Luna-M rocket with 450 kg 9N18F unitary HE-fragmentation warhead). There is also a leaflet type warhead which is designated the 9M21D.

The 400 kg cluster munition warhead has the designation 9N18E and carries a payload of 42 9N22

HE bomblets. The 9N22 weighs 7.5 kg, is 540 mm long and 74 mm in diameter. A 9E237 impact fuze is fitted with a pyrotechnic back-up self-destruction function. Upon exploding, 690 prefragmented splinters are formed by the bomblet, each weighing from 1 to 4 g each. This results in a pure anti-personnel effect.

The three nuclear warhead types are the AA-22 initial version with 3, 10 and 20 kTs selectable yields, the AA-38, an improved type with same selectable yields as the AA-22, and the AA-52 with 5, 10, 20 and 200 kTs selectable yields.

The only official information available for a 9M21G chemical warhead is that for a 540 mm calibre type, weighing 436 kg and carrying a 216 kg payload of VX persistent nerve gas agent. This is dispersed as a dense aerosol/droplet cloud by a VT fuzed airburst at around 1,500 to 2,000 m altitude in order to disable unprotected personnel and contaminate material, terrain and/or engineering structures over approximately 6 km^2. Other bulk chemical warhead types are likely to be available for the FROG-7 including, according to French military sources, one filled with the blood gas agent hydrogen cyanide (HCN, agent designation AC).

During the Iran-Iraq War the Iraqis apparently used the vesicant (blister) agent, known as distilled mustard gas (sulphur mustard, agent designation HD), in FROG-7 warheads on several occasions against Iranian rear area troop concentrations. Although not as lethal as a nerve agent, mustard gas can penetrate normal clothing and rubber and is potentially very long-lived in a variety of environmental conditions. It can persist on the ground, on foliage, equipment and buildings; thus it can be used to seal off whole areas of a battlefield or provide flank protection during an offensive operation.

The conventional HE warhead weighs 450 kg but coupled with the FROG-7's CEP of 500 to 700 m is only really suited for bombardment tasks against area targets.

Variants

Export 9K52TS

In the late 1960s a special version of the 9K52 Luna-M was developed for use in hot climates, this was designated the 9K52TS. This used the 9P113TS launcher and 9T29TS transporter and was used only for rockets with conventional warheads.

LRSV M-1996 Orkan-2

This is based on the 9P113 TEL but is believed to be equipped with four barrels from the Yugoslav 262 mm M-87 Orkan covered in this section under Yugoslavia.

Status

Production complete. In service with the following countries, this includes launchers held in reserve:

Country	Version	Number of launchers
Algeria	FROG-7	18/24 number unconfirmed
Belarus	FROG-7	45
Bosnia	FROG-7	1 or 2
Bosnian Serbs	FROG-7	12
Bulgaria	FROG-7	28
Cuba	FROG-7	65-70 including FROG-4 and FROG-7
Czech Republic	FROG-7	16 estimate
Egypt	FROG-7	12
Iraq	FROG-7	n/avail
Korea, North	FROG-5	9
	FROG-7	18
Libya	FROG-7	40
Poland	FROG-7	35
Romania	FROG-3	9
Russian Federation	FROG-7	n/avail (reserve)
Slovakia	FROG-7	9
Syria	FROG-7	18
Turkmenistan	FROG-7	12
Ukraine	FROG-7	n/avail
Uzbekistan	FROG-7	6
Yemen	FROG-7	12, status uncertain
Yugoslavia (former)	FROG-7	8, used by Bosnian Serb Army and Yugoslav Army (Serbia and Montenegro)

Manufacturer

Former Soviet state factories.

UPDATED

GUP TOS-1 220 mm (30-round) rocket system

Development

It is understood that the TOS-1 'Fighting Vehicle' 220 mm (30-round) rocket system was developed by the GUP Design Bureau of Transport Machine Building based in Omsk in the late 1980s and subsequently fielded in small numbers in the early 1990s. It is reported that this system has the nickname of the Buratino.

Some sources have indicated that as few as 24 TOS-1 'Fighting Vehicle ' systems could have been manufactured with a small number of these having been deployed to Chechnya in 1999/2000 where they are claimed to have been very useful in urban fighting. Omsk is also where production of the T-80 series of MBTs is undertaken at a low rate.

The exact operational role of the TOS-1 'Fighting Vehicle' is still not clear but it is possible that the system could have been originally developed for use in Afghanistan where it would be highly effective against guerillas in mountainous terrain.

Another possible role could have been against dug in anti-armour teams in key defended areas, especially in urban terrain, which would be difficult to dislodge with conventional weapons.

The TOS-1 'Fighting Vehicle' may have been deployed at front level by the Russian Army but it has never appeared at any of the Russian military parades in Moscow, it was shown at a defence equipment exhibition in Russia in 1999. More recently a modified version of the TOS-1 has been observed and this has been offered on the export market.

Description

The TOS-1 'Fighting Vehicle' is essentially a T-72 MBT with its original 125 mm two person turret removed and replaced by a low profile two person armoured turret superstructure with armoured sights in its upper part.

On top of the turret is mounted a large box type structure with powered traverse and elevation. Within this structure are 30 × 220 mm rocket tubes in four layers, the top layer has six launcher tubes while the lower three layers have eight launcher tubes.

TOS-1 220 mm (30 round) rocket launcher based on a T-72 MBT chassis which has probably been designed for use in the defence suppression role ***2000***/0048633

The unguided 220 mm rockets have a maximum range of 3,500 m and a minimum range of 400 m with a full salvo of 30 rockets being launched in 7.5 seconds, although the operator can also select salvoes of two or more rockets.

The unguided rockets have a fuel air type explosive (or thermobaric) warhead which is highly effective against troops in the open or dug in, as well as being effective against soft-skinned and light armoured vehicles.

The fuel air explosive warhead creates a massive overpressure in the target area for which there is no practical defence.

In recent years Russia has developed thermobaric warheads for its anti-tank guided weapons such as the 9M133 Kornet (NATO AT-14) and 9M114 Kokon (AT-6).

Combat experience in Afghanistan and elsewhere showed that the normal HEAT (High Explosive Anti-Tank) warhead, used to defeat armoured targets, was not very effective against troops.

The complete TOS-1 system weighs 46 tonnes, has a maximum road speed of 60 km/h and an operational range of 550 km without external fuel tanks.

Standard equipment for the TOS-1 includes an NBC system, ability to lay a smoke screen by injecting diesel fuel into the exhaust outlet on the left side of the hull, unditching beam that is normally carried on the right side of the hull and a dozer blade which is mounted at the front of the hull.

The single photograph released to date shows the system on a T-72 MBT chassis but it could also be installed on other MBT chassis such as the T-80 or T-90.

Status

Production complete. In service with the Russian Army in small numbers.

Manufacturer

GUP Design Bureau of Transport Machine Building.

UPDATED

Specifications

Crew: 3
Combat weight: 46,000 kg
Max road speed: 60 km/h
Max road range: 550 km
Armament: 30 × 220 mm rocket launcher tubes
Range:
(max) 3,500 m
(min) 400 m

Splav 300 mm BM 9A52 (12- round) Smerch multiple rocket system

Development

The development of the 300 mm Smerch (Tornado) Front (Military District) level RSZO (*Reaktivaya Sistema Zalpovogo Ognya:* Rocket System for Salvo Fire) took place in the late 1970s/early 1980s under the guidance of the Splav Government Scientific Production Concern at Tula.

It was first seen by the West in 1983 and was given the STANAG designation M1983 by NATO. It entered Russian service in 1987.

While the overall prime contractor is the Splav Scientific Production Concern the actual launcher and transloader vehicles and associated equipment has been designed and built by the Motovilikha Plants Corporation based in Perm.

The main role of the Smerch system is to suppress missile, artillery and mortar batteries, destroy strongpoints and eliminate enemy nodes of resistance. It is assigned to Military Districts and deployed in brigades of four battalions each with 12 9A52 launchers.

The Russian industrial index number for the original launcher is 9A52 and the complete system has the designation 9K58. The latest launcher variant has the designator 9A52-2 and is known as Smerch-M.

In 1993, at the Idex-93 international weapons exhibition, this version was being marketed under the export name Smerch (Sandstorm). The system has also been referred to as the RS30 (Reaktivnaya Systema - Missile System).

The first export customer is Kuwait which took delivery of nine launchers in 1995 and 18 in 1996.

In July 1996, it was revealed that the United Arab Emirates had placed an order for a Smerch battery of six 9A52-2 launchers, six 9T234-2 loader-transporters and a Vivariy automated fire-control system.

Splav 300 mm BM 9A52-2 (12-round) MRS in travelling configuration and showing jack between third and fourth roadwheel station

Description

The 9A52-2 RSZO battalion comprises the following components:

(1) Battalion and battery 1K 123 Vivary fire-control system command and control units contained in KamAZ-4310 (6 × 6) truck chassis-based command vehicles fitted with a K4310 fully enclosed box body, each of which has its own power generator and a secure data and radio communications system that will support VHF communications to a range of 50 km and HF communications to a range of 350 km. The vehicles also contain the elements of the Vivariy automated fire-control system that can undertake the automated or non-automated command and control of a Smerch-equipped multiple rocket launcher brigade. The actual command and staff vehicle (CSV) used by the brigade commander and brigade staff is also the KamAZ-4310 (6 × 6) vehicle described previously. All command element vehicles carry onboard digital computers and associated displays and printers, in addition to the communications equipment mentioned. The latest version is designated the MP32M1 command and staff vehicle and includes workstations based on IBM-compatible computers with i80486 microprocessors interconnected via the Ethernet network with SVGA monitors for digital map visualisation and LCSs for alphanumeric data presentation, as well as a printer for information logging.

It also includes a set of data and voice HF/VHF radio communications sets, channel encryption and switching equipment, power supply, NBC and air conditioning system.

All brigade-level CSV are equipped with GLONASS/NAVSTAR receivers that indicate the precise time and location co-ordinates as well as training facilities.

Other elements include 1B44 meteorological station on ZIL-131 (6 × 6) unarmoured truck chassis, PM-2-70 MTO-V maintenance and repair unit also on ZIL-131 (6 × 6) truck chassis and the IT12M-2M topography vehicle on GAZ-66 (4 × 4) unarmoured truck chassis.

(2) Three batteries each of four launchers. The 9A52-2 launcher is based on a modified MAZ-543M (8 × 8) cross-country truck chassis and is fitted with a 12-round elevating launcher assembly on its rear decking. The 300 mm rocket tube arrangement is two separate banks of four with a connecting roof of the remaining four tubes overlying the inner tubes of the banks. Before firing, two stabilisers are lowered to the ground. These are positioned one either side between the rear two roadwheels. The MAZ-543M chassis is fitted with powered steering on the front two axles

300 mm 9T234-2 Splav-M transloader vehicle

and a central tyre pressure regulation system that allows the driver to adjust the tyre pressure to suit the terrain being crossed from within the cab.

The launch vehicle's cabin accommodates the launch preparation and firing equipment. Single-round or salvo firing is possible. The driver is seated at the front of the vehicle on the left side with the vehicle commander to his immediate rear. The fully enclosed cab is provided with an NBC system and two entry doors in the left side that open to the front.

Recent developments include an installation of an automated launch preparation equipment that has allowed for autonomous execution of fire

Rocket used in Splav 300 mm BM 9A52-2 (12-round) MRS the with tail fins folded (Christopher F Foss)
2000/0085766

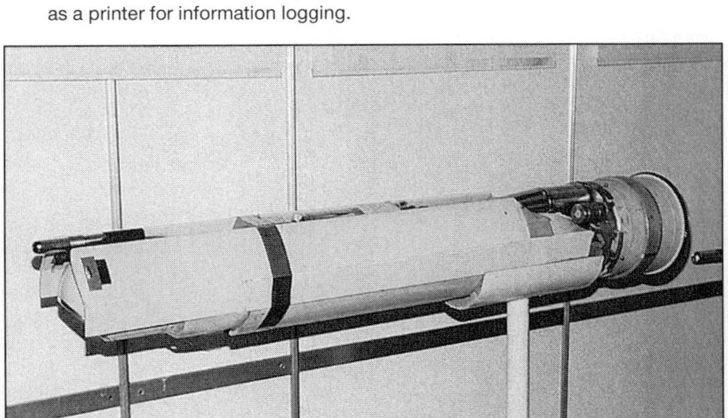

R-90 expendable UAV with wings and tails in the folded position 0008318

Splav 300 mm BM 9A52-2 (12-round) MRS deployed in firing position

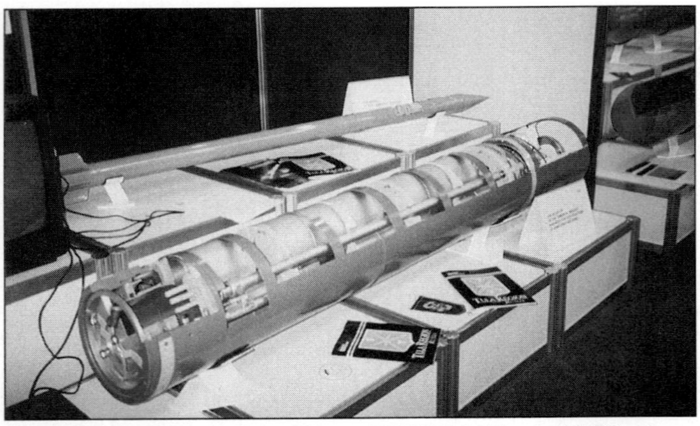

9M55K1 warhead for the Splav 300 mm BM 9A52-2 (12-round) MRS cutaway to show interior and MOTIV-3F submunitions (Christopher F Foss)

Splav 300 mm BM 9A52-2 (12-round) MRS with launcher traversed to right (Christopher F Foss) 0058464

missions. This allows for a sharp reduction in fire preparation time from unprepared positions.

This equipment provides for the topographical surveying and navigation of the launch vehicle, computation in the launch vehicle of fire settings and flight mission data of the basis of target and meteorological data provided by the Automated Fire-Control System, ramp laying on the target by the crew staying in the cab and using no aiming points, visual monitoring of rocket ramp laying, presentation on the display of the launch vehicle location, pre-launch testing on rocket projectile equipment, flight mission data input in rocket projectile onboard equipment and finally rocket launch with take off checks.

(3) Accompanying each 9A52-2 launcher is a 9T234-2 transloader vehicle based on the same MAZ-543M chassis. This carries 12 reload rounds and a hydraulic crane that can lift a maximum load of 850 kg on its rear decking to load and unload the 9A52-2. This vehicle is also referred to as the Oplot. The system takes a well trained crew 20 minutes to reload as the 300 mm rockets have to be loaded one at a time.

(4) The rocket used is of 300 mm calibre, fitted with a 'self-contained powered-flight phase control system which provides an increase in fire precision by a factor of two and in accuracy by a factor of three as compared to unguided rockets'. The accuracy is claimed as being 0.23 per cent of range. Minimum range is 20,000 m and maximum range 70,000 m. The rocket types available are:

(a) 9M55K which carries 72 fin-stabilised unguided fin-stabilised high-explosive fragmentation sub-munitions which are highly effective against soft skin and lightly armoured targets. The sub-munitions weigh 1.75 kg and are fitted with a self-destruct mechanism that operates after a period of 110 seconds. The rocket weighs 800 kg and has a minimum range of 20 km and a maximum range of 70 km.

(b) 9M55K1, which carries five Bazalt State Research and Production Enterprise parachute-retarded MOTIV-3F top-attack anti-armour submunitions. Each of the latter weigh 15 kg each, and are fitted with a two-spectrum infra-red seeker. The warhead will penetrate 70 mm of conventional steel armour. The seeker has a 30° field-of-view and when it is within some 150 m of the target the warhead is activated. The same munitions are used in the RBK-500 SPBE-D 500 kg cluster bomb. The sub-munition in this application is designated the MOTIV-3F. It has a minimum range of 20 km and a maximum range of 70 km.

(c) 9M55K4 carries 25 anti-tank mines each of which weighs 4.85 kg and carries 1.85 kg of high explosive. The mines are fitted with a self-destruct mechanism that operates after between 16 and 24 hours. The rocket weighs 800 kg and has a minimum range of 20 km and a maximum range of 70 km.

(d) 9M55F with a unitary warhead carrying 95.5 kg of high explosive which separates over the target and is slowed down by a parachute. This rocket weighs 810 kg and has a minimum range of 25 km and a maximum range of 70 km.

Specifications

9A52-2 launcher
Calibre: 300 mm
Number of tubes: 12
Rate of fire: 12 rds/38-40 s
Time into action: 3 min
Time out of action: 3 min
Elevation: 0-55°
Traverse: 30° left and right
Crew: 3
Configuration: 8 × 8
Weight:
(fully loaded) 43,700 kg
(empty) 33,700 kg
Length:
(travelling) 12.1 m
Width:
(travelling) 3.05 m
Height:
(travelling) 3.05 m
Ground clearance: 0.44 m
Engine: V-12, 4-stroke diesel developing 518 hp at 2,200 rpm
Max speed: (road) 60 km/h
Max range: 850 km
Fording: 1.1 m
Vertical obstacle: 0.78 m
Trench: 2.50 m

9T234-2 transloader
Number of reload rounds: 12
Crew: 3
Configuration: 8 × 8
Weight:
(fully loaded) 41,500 kg
(empty) 30,000 kg
Crane lifting capacity: 850 kg
Engine: V-12, 4 stroke diesel developing 518 hp at 2,200 rpm
Max speed: (roads) 60 km/h
Max range: 850 km
Fording: 1.1 m

9M55K rocket
Calibre: 300 mm
Length: 7.6 m
Weight: 800 kg
Warhead: see text
Min range: 20,000 m
Max range: 70,000 m

(e) 9M55S is the fuel air explosive member of the family. The rocket weighs 800 kg and has a minimum range of 25 km and a maximum range of 70 km. The actual rocket weighs 800 kg and contains 100 kg of fuel air explosive which is fitted with a self-destruct mechanism that operates after a period of 110 to 160 seconds.

(f) In addition there is a special RPV called the R-90 that can be launched from a 300 mm rocket tube. This was developed by the NITs ENIKS design bureau and could be used to acquire targets for the Smerch launcher or carry out battlefield reconnaissance after the target has been engaged. As far as it is known this remains at the prototype stage.

NEW LONG RANGE ROCKETS

More recently a new family of longer range rockets have been developed by Splav. These are all numbered in the 9M500 series and known rockets include:

(a) The 9M525 is an enhanced version of the 9M55K and carries 72 submunitions each of which weigh 7.75 kg and are now fitted with a self-destruct mechanism which operates after a period of 110 seconds. This rocket weighs 815 kg and has a minimum range of 25 km and a maximum range of 90 km.

(b) The 9M526 is an enhanced version of the 9M55K1 and carries five parachute-retarded, self-targeting submunitions that weigh 15 kg and will penetrate 70 mm of conventional steel armour. This rocket weighs 815 kg and has a minimum range of 25 km and a maximum range of 90 km.

(c) The 9M527 carries 25 anti-tank mines which are scattered over the target area. Each mine weighs 4.85 kg and carries 1.8 kg of high explosive and is fitted with a self-destruct fuze that operates after a period of 16 to 24 hours. The rocket weighs 815 kg and has a minimum range of 25 km and a maximum range of 90 km.

(d) The 9M528 is a new round that features a single, separable, high-explosive fragmentation warhead which is parachute retarded. This rocket weighs 815 kg and has a minimum range of 25 km and a maximum range of 90 km.

(e) The 9M529 is another new round and has a fuel air explosive-type warhead that carries 100 kg of explosive and is fitted with a self-destruct mechanism that operates after a period of between 110 and 160 seconds. The fuel air explosive is claimed to have an effective area of at least 25 m. This rocket weighs 815 kg and has a minimum range of 25 km and a maximum range of 90 km.

(f) The 90M530 is yet another new round that has been designed to penetrate key military infrastructure facilities such as airfields and fuel depots. It has a hardened high explosive warhead which is designed to penetrate the target before it explodes. The warhead contains 75 kg of high explosive. The rocket weighs 815 kg and has a maximum range of 90 km.

(g) The new 9M531 is of the high-explosive fragmentation type and contains 646 sub-munitions that are designed to attack troops in the open, light armoured vehicles or trucks. It carries a total of 646 submunitions, each of which is fitted with a self-destruction mechanism that operates after a period of 130 to 260 seconds. Each submunition is claimed to be able to penetrate 120 mm of conventional steel armour. The rocket weighs 815 kg and has a minimum range of 25 km and a maximum range of 90 km.

(h) In addition, there is an improved version of the original 9M55K designated the 9M55K5.

This carries 646 submunitions as carried by the 9M531 which are fitted with a self-destruct mechanism which operates after a period of 130 to 260 seconds. The rocket weighs 800 kg and has a minimum range of 25 km and a maximum range of 90 km.

(1) PM-2-70 MTO-V repair shop vehicles, 1I12-2M topographical vehicles and a 1B44 Ulybka meteorological complex vehicle.

Future developments

The current Smerch system is based on the MAZ-543A or MAZ-543M (8 × 8) truck chassis but it has been reported that consideration is being given to installing this system on a KZKT or MZKT chassis which are made in Russia rather than Belarus where MAZ trucks are built.

It is also believed that a new modular rocket pod system is being developed similar to that used for the US MLRS which has two pods each with six rockets.

The development of a pod-type rocket system for Smerch would considerably speed up the reload capability and therefore enhance the operational effectiveness of the system.

9A52-2T

This is the latest version of the Splav 300 mm BM 9A52 (12-round) Smerch multiple rocket system and is based on a new Tatra 10 × 10 cross-country truck chassis. This is aimed at the export market, especially the Middle East. Between the fourth and fifth roadwheel station is a hydraulic stabiliser that is lowered to the ground before firing commences, providing a more stable firing platform.

Chinese 300 mm (10-round) A100

China has recently revealed a 300 mm (10-round) A 100 multiple rocket system that is probably related to the Russian 300 mm BM 9A52 (12-round) Smerch system. Details are given under China.

Status

Production as required. In service with following countries:

Country	Quantity	Comment
Belarus	48	
India	n/av	may have placed an order
Kuwait	27	delivered between 1995 and 1996
Russia	300	estimate
Ukraine	94	
United Arab Emirates	6	

Manufacturers

Splav Scientific Production Concern (prime contractor and rockets).

Motovilikha Plants Corporation (launcher and transloader).

UPDATED

Splav 220 mm BM 9P140 (16-round) Uragan multiple rocket system

Development

Like all Russian multiple rocket launcher (MRL) systems, the 220 mm BM 9P140 (16-round) Uragan (Hurricane) was developed under the guidance of the Splav Government Scientific Production Concern at Tula, which is to the east of Moscow. While the overall prime contractor for the complete Urugan system is the Splav Scientific Production Concern, the actual launcher and transloader vehicles and associated equipment has been designed and built by the Motovilikha Plants Corporation based in Perm.

The system was designed in the early 1970s, with development being completed in 1975. It entered service with the Soviet Army in the same year and was allocated the STANAG designation of the M1977 by NATO, this being the year it was first identified. The 220 mm BM 9P140 (16-round) replaced the older 240 mm (12-round) BM-24 and 200 mm (4-round) BMD-20 systems.

The complete system has the industrial complex number of 9K57 which comprises the launcher, rockets and the essential rocket resupply vehicle. The system is also sometimes referred to as the BM-22 or BM-27.

Until the Splav 300 mm BM 9A52 (12-round) Smerch rocket system entered service with the Soviet Army in the early 1980s, the 220 mm 9P140 (16-round) was the largest system of its type in service.

The main advantage of the 220 mm BM 9P140 (16-round) is that it is mounted on an 8 × 8 chassis with improved tactical and strategic mobility. Its rockets have a much longer range and carry a variety of warheads, enabling a wider range of battlefield missions to be undertaken using the same launch platform.

The Uragan was used in action during the Soviet occupation of Afghanistan and in Chechnya. It is known to have been exported to Syria. Some years ago, Russia was offering the 9P140 launcher for around US$650,000, with each rocket costing an additional US$5,600.

Until the introduction of the larger 300 mm BM 9A52 (12-round) Smerch system, the 220 mm BM 9P140 (40-round) was normally deployed in rocket artillery regiment and brigade formations at Army or Front (Military District) level.

In the case of the former, the regiment Table of Organisation and Equipment (TOE) normally comprises three battalions of 12 launchers, which is normally increased to 18 launchers in time of war.

In a Military District, an independent Uragan rocket artillery regiment has the same TOE. But if the Uragan unit is assigned to an artillery division within the Military District, then the unit may be either a regiment or brigade with four battalions each with 12 launchers (which would be increased to 18 in time of war). If the unit were already a brigade then it would take on the extra six launchers per battalion to achieve its wartime TOE.

Typical roles of the Uragan system would be to suppress enemy missile, artillery and mortar batteries as well as destroying strongpoints and other areas of resistance. It would normally use 'shoot and scoot' techniques to avoid counterbattery fire.

220 mm BM 9P140 (16-round) Uragan (Hurricane) multiple rocket launcher system deployed in firing position with launcher elevated and stabilisers lowered **2001**/0056419

In some units it has replaced the shorter-ranged 122 mm BM-21 (40-round) rocket system mounted on a 6 × 6 truck chassis. The Uragan has been designed to operate in temperatures ranging from –40 to +50°C.

The Uragan battalion is controlled by the Kapustnik-B automated fire-control system. This has a Battalion commander's vehicle based on the BTR-80 (8 × 8) APC chassis, a Battalion Chief of Staff vehicle based on the Ural-4320 truck chassis, three Battery Commander's vehicles based on the BTR-80 APC chassis and three Battery Senior Officer's vehicles based on the Ural-4320 truck chassis.

The Kapustnik-B comprises subsystems for reconnaissance, initial battalion orientation and location fixing, weather reconnaissance and ballistic tracking information, interbattalion and higher echelon communication and data transfer. All the elements are connected to a single data processing computer.

If required, the system can automatically accept external reconnaissance data transfers and be connected to higher echelon automated control systems.

Description

The system is based on a modified ZIL-135LM 8 × 8 truck chassis which has excellent cross-country mobility and is used for a wide range of other battlefield missions. These include TEL (Transporter, Erector, Launcher) for the FROG-7 surface-to-surface rocket system (and its associated reload vehicle), Sepal cruise missile carrier, cargo truck and tractor truck, to name but a few.

The chassis was developed at the Likhachev Motor Vehicle Plant near Moscow but production was undertaken at the Bryansk Automobile Works, for which reason the family is sometimes referred to as the BAZ-135L4.

The fully enclosed unarmoured crew compartment is at the front with the two diesel engines to the rear. The turntable-mounted rocket launcher is mounted at the very rear of the chassis and is traversed to the front for travelling.

The rocket launcher pod consists of an upper layer of four tubes and two lower layers of six tubes so giving a total of 16 launch tubes. The sighting system is

9T452 transloader vehicle for 220 mm BM 9P140 (16-round) Uragan (Hurricane) multiple rocket system showing centrally mounted crane. The rockets are not being carried

mounted on the left side of the launcher and access to this is via a folding ladder.

The crew compartment accommodates the launch preparation and firing equipment and the operator can select single rockets or salvo firing. It is not fitted with an NBC system and as far as it is known does not have a land navigation system.

Steering is power assisted on the front and rear axles and a central tyre-pressure system is fitted as standard. This allows the driver to adjust the tyre pressure to suit the terrain being crossed without leaving his seat.

The 8 × 8 vehicle is powered by two diesel engines each of which drive the four wheels on one side of the vehicle.

When deployed in the firing position, steel shutters are normally raised over the front windscreens and two stabilisers are lowered at the very rear to provide a more stable firing platform.

The launcher is mounted on a turntable at the rear with powered elevation from 0 to +55° and powered traverse of 30° left and right.

There are at least five types of unguided artillery rocket launched by the BM 9P140 system, see Table 1.

The PGMDM scatterable anti-tank mine is also used in a number of other applications and consists of a liquid explosive charge contained in a thin flexible cover. It uses a pressure-activated fuze which is also used in the PFM-1 anti-personnel mine.

Unconfirmed reports indicate that there may well be a chemical rocket for this system and RFAS reports state that a Fuel-Air Explosive (FAE) submunition rocket is also available.

Each 9M27K1, 9M27K2, 9M27K3, 9M59 rocket is 220 mm in diameter, 5,178 mm long and weighs 270 kg at launch. The warhead weighs 100 kg. The 9M27F rocket is 220 mm in diameter, 4.832 m in length and weighs 280 kg. Minimum range is 10,000 m while maximum range is 35,000 m.

The 9M59 220 mm rocket carries nine anti-tank mines and is 5.178 m long and weighs 270 kg. It is fitted with a warhead that weighs 89.5 kg with the nine anti-tank mines capable of covering an area of 250 hectares.

The anti-tank mines weigh 5 kg and are fitted with a proximity fuze. They carry a 1.85 kg exposive direction charge which has a high armour-piercing capability. The mines have a self-destruct time of 16 to 24 hours.

The 9M27K2 rocket has a 89.5 kg cluster warhead carrying 24 anti-tank mines which cover an area of 150 hectares. The PTM-1 mines have been designed to destroy the tracks of tanks and other armoured vehicles, rather than to pierce the armour of the vehicle.

These mines weigh 1.5 kg and are fitted with a 1.1 kg PW-12S high-explosive charge. The self-destruct time of the PTM-1 anti-tank mine is 3 to 40 hours.

Close-up of the unguided 220 mm rocket out of the launcher showing four wraparound fins at the rear (Christopher F Foss)

Table 1

Rocket	**Warhead**	**Type**	**Payload**
9M27F	9N218F	unitary HE	100 kg warhead containing 51.7 kg of HE
9M27K1	9N218K1	ICM	30 × N9N210 anti-personnel anti-material bomblets
9M27K2	9N218K2	carrier	24 × PGMDM anti-tank mines
9M27K3	9N218K3	carrier	312 × PFM-1 anti-personnel minelets
9M59	n/avail	carrier	9 × bottom attack anti-tank mines with directional charge

The rockets are fin and spin stabilised and have wraparound fins at the rear that unfold after the rockets are launched.

According to RFAS sources, each salvo has a destructive area of 20,000 to 460,000 m². If the rockets are launched singly then one rocket is launched every 8.8 seconds. A complete salvo of 16 rockets takes 20 seconds to launch.

Maximum time quoted for the system to be prepared for action from the travelling configuration to ready to fire is 3 minutes with a similar time to come out of action.

Each BM 9P140 launcher is normally accompanied by at least one and sometimes two, 9T452 transloader vehicles based on the same ZIL-135LM 8 × 8 chassis.

The transloader vehicle carries 16 reload rounds arranged in two stacks positioned one either side of the vehicle. Mounted to the immediate rear of the cab is a hydraulic crane operated by one person which is used to resupply the rocket launcher. This crane has a maximum lifting weight of 300 kg with the operator being seated to the rear.

For reloading, the launcher has to be in the horizontal position and traversed to one side. The new rocket is then loaded into the tube via the crane and rammed home by the rammer. It is estimated that the reload time is 20 to 30 minutes, a long time when compared to the 227 mm US Lockheed Martin Missiles and Fire Control - Dallas Multiple Launch Rocket System (MLRS).

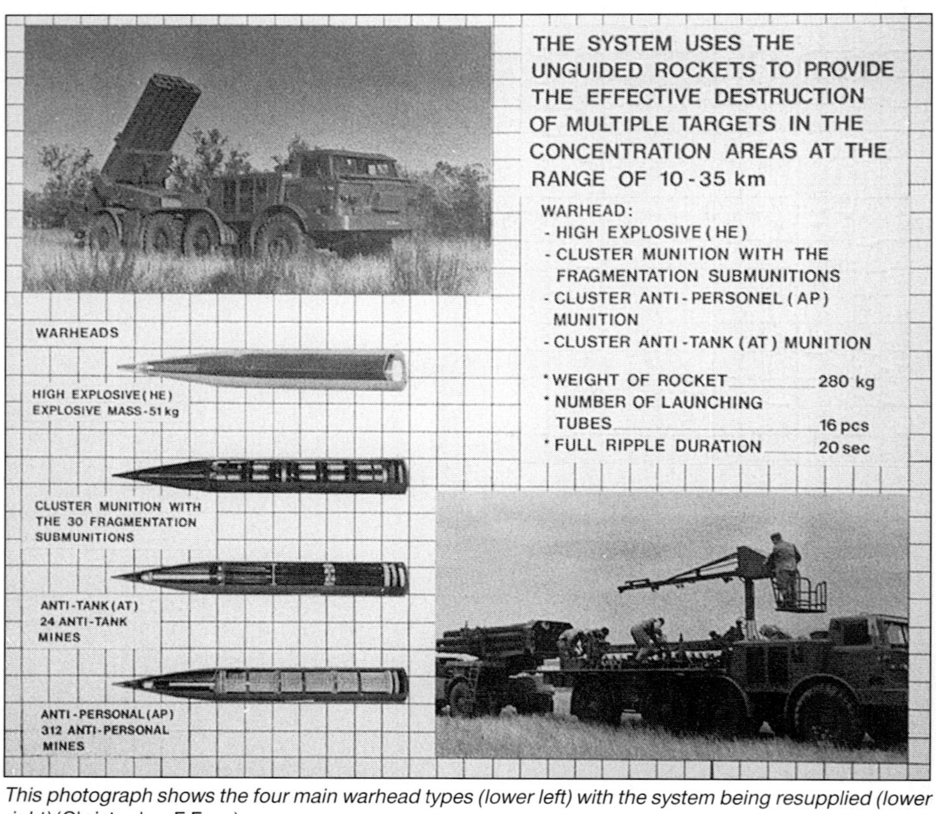

This photograph shows the four main warhead types (lower left) with the system being resupplied (lower right) (Christopher F Foss)

Specifications

Designation: 9P140
Calibre: 220 mm
Number of tubes: 16
Elevation: 0-55°
Traverse: ±30°
Range: 10,000-35,000 m
Rate of fire (full salvo of 16 rounds):
(constant) 8.8 s per rocket
(variable) 20 s for salvo
Time into action: 3 min
Time out of action: 3 min
Crew: 4
Configuration: 8 × 8
Weight:
(fully loaded) 20,000 kg
(empty) 15,100 kg
Length:
(travelling) 9.63 m
(firing) 10.83 m
Width:
(travelling) 2.8 m
(firing) 5.34 m
Height:
(travelling) 3.225 m
(firing, max elevation) 5.24 m
Ground clearance: 0.58 m (axles)
Track: 2.3 m
Wheelbase: 2.41 +1.5 +2.4 m
Engines: 2 × 8-cylinder 4-stroke petrol developing 177 hp at 3,200 rpm (each)
Gearbox: hydromechanical
Steering: power-assisted, 1st and 4th axles
Turning radius: 12.5 m
Max speed:
(road) 65 km/h
Max range: 570 km
Fuel capacity: 768 litres
Fording: 1.2 m
Gradient: 57%
Vertical obstacle: 0.6 m
Trench: 2.00 m

Note:
The 9T452 transloader vehicle has a similar performance and a crew of three. Fully loaded weight is 20,000 kg with 16 rounds.

Status
Production as required. In service with the following countries:

Country	Quantity	Comment
Afghanistan	18	
Belarus	84	
Guinea	3	delivered 1999
Kazakstan	50	
Moldova	15	
Russia	800+	estimate
Syria	n/av	unconfirmed user
Tadjikistan	12	
Turkmenistan	54	
Ukraine	139	
Uzbekistan	48	
Yemen	13	delivered 1994

Manufacturers
Splav Scientific Production Concern (prime contractor and rocket manufacturer).
Motovilikha Plants Corporation (launcher and transloader).

UPDATED

Splav 122 mm BM-21 (40 round) multiple rocket launcher upgrade

Development
The Splav State Unitary Enterprise, prime contractor for all Russian Multiple Rocket Launcher Systems (MRLS) has recently developed a package of enhancements for the 122 mm BM-21 (40 round) 'Grad' system. The system is used by many countries around the world, with an estimated 3,000 systems built by Russia alone.

Description
The improvements to the 122 mm BM-21 (40 round) system include the development of a fire-control system for more rapid and effective target engagement and the development of a new family of unguided 122 mm surface-to-surface rockets.

Not only do these 122 mm rockets have a longer range, due to the use of a new solid propellant, but also more effective warheads enabling the BM-21 to engage a wider range of battlefield targets at longer ranges with increased effectiveness.

The automated fire-control system is called the Kapustnik-BM. This includes the 1V153M fire control post mounted on a Ural-375 series (6 × 6) chassis, or the more recent Ural-4320 (6 × 6) truck. This receives target information from reconnaissance means such as aircraft, helicopters and battlefield radars or a command and observation post based on a BTR-80 (8 × 8) APC designated the 1V52M.

Target information is calculated and sent to each launcher that is fitted with a Automated Targeting and Fire Control System (ATFCS). The BM-21 is then rapidly laid onto the target and the rockets launched, with the crew remaining in the cab.

According to Splav, the time taken for a BM-21 to open fire from an unprepared firing position is between 25 and 35 minutes, with the ATFCS installed this is reduced to six minutes. The crew of the BM-21 normally consists of three people, with ATFCS it has been reduced to two.

The standard 122 mm unguided rocket is designated the 9M22U and is fitted with a high-explosive fragmentation warhead and has a maximum range of 20,380 m. The rocket weighs 66.6 kg and contains 18.4 kg of high explosive.

For engaging tanks and other armoured vehicles the 122 mm 9M217 rocket has been developed which has a maximum range of 30, 000 m and carries two self-targeting submunitions. Each of these has a special self-forging fragment warhead that will penetrate 60 to 70 mm of conventional steel armour. This would be used to attack the vulnerable upper surfaces of armoured vehicles.

The 122 mm 9M218 rocket contains 45 shaped charged (eg High Explosive Anti-Tank) submunitions that are highly effective against light armoured vehicles. This rocket has a maximum range of 30,000 m.

The 122 mm 9M521 high explosive fragmentation rocket is said to be twice as effective as the standard 9M22U rocket and has a maximum range of 40,000 m and carries 21 kg of high explosive plus spherical fragments.

The 122 mm 9M522 rocket has a maximum range of 37,500 m and is fitted with a high-explosive warhead that separates from the rocket over the target and is stabilised by a parachute. This explodes over the target for maximum effectiveness.

Other types of warhead are probably available including one carrying jammers, that is believed to be designated the 9M519 and possibly a fuel air explosive rocket.

These improvements will be offered on new production 122 mm BM-21 (40 round) MRLS as well as being capable of being back fitted to older systems.

Status
Development complete. Ready for production.

Manufacturer
Splav State Unitary Enterprise.

UPDATED

Specifications

Rocket	9M22U	9M217	9M218	9M521	9M522
Max range	20.1 km	30 km	30 km	40 km	37.5 km
Weigh RPt	66.6 kg	70 kg	70 kg	66 kg	70 kg
Weight of Explosive	18.4 kg	25 kg	25 kg	21 kg	25 kg

The 122 mm BM-21 (40 round) rocket system is now being offered with new longer range rockets and fire-control enhancements **2001**/0067869

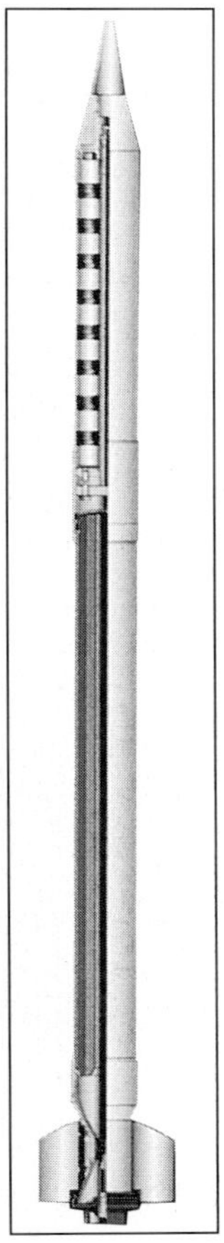

The 122 mm 9M218 rocket carries 45 shaped charge submunitions to attack the vulnerable upper surfaces of armoured vehicles **2001**/0067870

Splav 122 mm BM-21 multiple rocket launcher family

122 mm (40-round) BM-21 Grad

Development
The development of the 122 mm BM-21 Grad (Hail) divisional-level RSZO *(Reaktivaya Sistema Zalpovogo Ognya:* Rocket System for Salvo Fire) took place in the mid-1950s under the guidance of the Splav Scientific Production Concern at Tula.

The development phase was completed in 1958 and the BM-21 Grad entered operational service in 1963. The Grad is also known in the West by the US designation BM-21a.

The main role of the BM-21 Grad system is to support the division with suppressive fire to counter anti-tank missile, artillery and mortar positions, destroy strongpoints and eliminate enemy nodes of resistance on the immediate battlefield.

The BM-21 Grad is assigned to divisions as part of the divisional artillery structure in a battalion of 12 launchers (expanding to a Table of Organisation & Equipment (TOE) of 18 launchers in wartime). It may also be found at army or front (military district) level in independent regiments of three battalions each of 12 launchers (these expanding to a TOE of 18 launchers in wartime).

The BM-21 Grad system comprises the BM-21 launcher vehicle, a 9F37 ammunition resupply vehicle and the 9M22U index rocket (also designated M-21-OF).

While the prime contractor for the overall 122 mm BM-21 multiple rocket system is the Splav Scientific Production Concern based in Tula, the Motovilkha

Plants Corporation based in Perm is responsible for the actual launcher mechanism and in 1998 stated that it had manufactured over 2,000 BM-21 series systems for the home and export markets.

In addition to being manufactured in Russia, a number of other countries have designed and manufactured systems with similar capability, usually on local chassis.

According to United Nations Report on Conventional Arms for 2001, during the period 1992 through to 2000, Russia did not export any of these systems, although parts of the former USSR did:

From	To	Quantity	Delivery
Belarus	DR Congo	6	1997
Kazakstan	Angola	4	1998
Kazakstan	Congo	1	2000
Kazakhstan	Russia	2	1999

In addition, there were a number of unidentified transfers of undisclosed rocket systems, some of which could be 122 mm BM-21 type systems:

From	To	Quantity	Delivery
Russia	Angola	14	1993
Turkmenistan	Russia	54	1996

Description

The BM-21 Grad RSZO launcher is based on a Ural-375D (6 × 6) 4,000 kg cross-country truck chassis and is fitted with a 40-round elevatable launcher assembly on its rear decking. The rocket tube arrangement is a single rectangular bank of four layers each with 10 tubes. Before firing, two stabilisers are lowered to the ground; these are positioned one either side at the rear of the vehicle. The launch vehicle's cabin accommodates the launch preparation and firing equipment. Individual round, selective ripple or salvo firing is possible from either the cab or from a remote-control unit connected to the vehicle via a 60 m cable.

Specifications

Grad Ural-375D chassis
Calibre: 122.4 mm
Number of tubes: 40
Rate of fire: 40 rds/20 s
Reload time: 7 min
Elevation: 0-55°
Traverse: 120° left, 60° right
Time into action: 2.5 min
Time out of action: 0.5 min
Crew: 6
Configuration: 6 × 6
Weight:
(fully loaded) 13,700 kg
Length:
(travelling) 7.35 m
Width:
(travelling) 2.69 m
Height:
(travelling) 2.85 m
Engine: ZIL-375 7 litre V-8 water-cooled petrol developing 180 hp at 3,200 rpm
Gearbox: manual with 5 forward and 1 reverse gears
Clutch: twin dry disc
Transfer box: 2-speed
Steering: double-thread worm, hydraulic booster
Turning radius 10.5 m
Suspension:
(front) longitudinal semi-elliptical springs with hydraulic shock absorbers
(rear) bogie with longitudinal semi-elliptical springs
Tyres: 14.00 × 20
Brakes:
(main) air/hydraulic
(parking) mechanical
Electrical system: 12 V
Max speed: (road) 80 km/h
Max range: 1,000 km
Fuel capacity: 360 litres
Fording: 1.5 m
Gradient: 60%
Vertical obstacle: 0.65 m
Trench: 0.875 m

122 mm BM-21 (40-round) MRS of the Polish Army in travelling configuration (Jaaroslaw Cislak) 0021070

122 mm (12-round) Grad-V Multiple Rocket Launcher

Development

At the same time as the BM-21 Grad received its baptism of fire during the 1969 Sino-Soviet border skirmishes, the Splav concern was given a requirement from the VDV headquarters for a lightweight MRL system that could be transported by aircraft and dropped with troops via a parachute and carriage platform to make a soft landing. The result was the Grad-V or BM-21V (with the V standing for *vozdushnodesantii* or airborne) system. This first entered service with the Russian army in 1967.

Description

The Grad-V has the NATO STANAG designation BM-21 M1975 and is fitted with two tiers of six 122 mm tubes mounted on the lower chassis of the reduced weight airborne forces GAZ-66B (4 × 4) 2,000 kg truck.

The truck has a collapsible canvas cab, removable doors and windscreen, telescopic steering wheel and tie-down points for aiding parachute dropping and landing. The launcher assembly is traversed towards the cab for travelling and for firing requires the two stabilising jacks at the rear of the vehicle to be lowered.

The rockets can be either fired singly or in a salvo lasting six seconds. Manual reloading by the two-man crew takes approximately 5 minutes. The launcher can traverse a full 360°. The system is limited to firing HE fragmentation-type rockets only.

The BM-21V assembly was subsequently mounted on a tracked chassis, believed to be a variant of the BMD series airborne combat vehicle.

122 mm BM-21 (36-round) Grad-1

Development

At the same time as the BM-21 Grad was deployed as a divisional system a requirement was drawn up for a

122 mm BM-21 (40-round) MRS of the Polish Army deployed in the firing position with gunner using sight (Michael Jerchel) ***2001***/0105895

122 mm BM-21b (36-round) multiple rocket launcher on ZIL-131 (6 × 6) 3,500 kg truck chassis from the rear in travelling position ***2002***/0130020

Specifications
Grad-V GAZ-66B chassis – provisional
Calibre: 122.4 mm
Number of barrels: 12
Rate of fire: 12 rds/6 s
Time into action: 3.5 min
Time out of action: 0.5-1.0 min
Reload time: 5 min
Crew: 2
Configuration: 4 × 4
Weight:
(fully loaded) 6,000 kg
Length:
(travelling) 5.65 m
Width:
(travelling) 2.32 m
Height:
(travelling) 2.44 m
Engine: ZMZ-66 V-8 4.254 litre water-cooled petrol developing 115 hp at 3,200 rpm
Max speed:
(road) 85 km/h
Max range: 875 km
Fuel capacity: 210 litres

regimental MRL artillery system for issue to the independent six-launcher MRL batteries assigned to selected tank and motorised rifle regiments. The BM-21 Grad-1 launcher entered service in 1976 with the Russian military index number 9P138. It is also known by the US designation BM-21b and the NATO STANAG designation BM-21 M1976. The BM-21 Grad-1 system was subsequently produced on the MT-LB tracked chassis.

The BM-21 Grad-1 system comprises the 9P138 launcher, the 9T450 transport vehicle, 9F380 ammunition resupply vehicle and the 122 mm 9F28F rocket with HE warhead.

The associated 122 mm rockets were also modernised. The standard HE fragmentation round was produced with a prefragmented warhead while the incendiary round was modified to carry 180 incendiary elements as part of its payload.

122 mm BM-21M or BM-21b

Description

This is a modified version of the 122 mm BM-21 (40-round) multiple rocket launcher family based on a ZIL-131 (6 × 6) 3,500 kg truck chassis, with the complete system also being referred to as the BM-21b.

The main difference is that the BM-21b has had the two centre tubes in each of the lower banks removed to give a total of 36 rather than the normal 40 tubes.

This has allowed the installation of a new elevation mechanism. A quantity of these were supplied to the Iraqi Army and some of these were subsequently captured in 1991. These vehicles were fitted with a thermal shield over the rear part of the rocket launcher to prevent excessive solar heating of the rocket tubes.

122 mm (40-round) BM-21-1 Grad

Development/Description

During the 1990 exchange of data for the CFE Treaty it was noted that there was a BM-21-1 Grad divisional RSZO variant in service. This was subsequently identified as a modernised 40-tube Grad launcher assembly mounted on the rear of a modified Ural-4320 (6 × 6) 5,000 kg truck chassis. The Soviet industrial index number for the BM-21-1 is 9P137.

122 mm Grad modernisation

Development/Description

Splav is also developing a mondernisation package for the BM-21 launch vehicle. The basic Ural-375 or Ural-4320 chassis has its metal launch tube assembly replaced by two dispensable 20-round Launching Transporting Containers (LTC) made out of composite polymer materials. The LTCs are mounted on the launch vehicle via a special adaptor frame. The launcher is reloaded by replacing the LTC by crane. The containers are interfaced via a special electric connector unit. The BM-21 retrofit can be carried out as a field modification. An LTC weighs 1,770 kg with a 20-rocket load and 370 kg when empty. It takes 5 minutes to reload.

Specifications
Grad-1 ZIL-131 chassis – provisional
Calibre: 122.4 mm
Number of barrels: 36
Rate of fire: 36 rds/18 s
Reload time: 7-8 min
Time into action: 2.5 min
Time out of action: 0.5 min
Crew: 5
Configuration: 6 × 6
Weight:
(fully loaded) 10,500 kg
Length:
(travelling) 6.9 m
Width:
(travelling) 2.5 m
Height:
(travelling) 2.48 m
Engine: ZIL-131 6 litre V-8 water-cooled petrol developing 150 hp at 3,200 rpm
Max speed: (roads) 80 km/h
Max range: 525 km
Fuel capacity: 340 litres

122 mm BM-21-P (single round) launcher

Development/Description

For Spesnatz and other specialist users there is a single-round man-portable tripod launcher system known by the Russian industrial index designation 9P132 and service designation BM-21-P. This has a total weight of 49.54 kg, of which 21.82 kg is the 122 mm tube and 27.72 kg is the tripod. The launch tube is 2.4435 m long. The elevation limits are believed to be 0 to 35° and the traverse ±15°. Emplacing the system takes 3 minutes and displacing 30 seconds. The maximum rate of fire is 2 rds/min with manual reloading taking 30 seconds. Maximum range achieved during the Vietnam War was 10,973 m with the 122 mm calibre short-form DKZ-B rocket.

122 mm 9K510 portable rocket system

Development/Description

This comprises the 2P132 single-round 122 mm rocket launcher with the 9P611 launch control panel and fires the 9M42 flare rocket. The latter is 1.76 m long, has a calibre of 122 mm and weighs 27 kg at launch. Minimum and maximum ranges are 1,000 and 5,000 m. The flare component burns for 90 seconds and is a separable parachute-equipped flare. The launcher set up weighs 27 kg and is crewed by two men. Time to set up for launch is 2 minutes.

122 mm calibre rockets

Development/Description

The original 122 mm rocket produced for the Grad system was an HE fragmentation type fitted with four spring-out stabiliser vanes mounted at an angle of 1° to the longitudinal axis of the rocket at the tail. This was joined in 1971 by an incendiary variant. Maximum range of the rockets was 20,500 m. The fuzing system is not armed until the rocket has travelled between 150 and 400 m downrange.

With the advent of the BM-21 Grad-1 model the standard 122 mm rocket was modernised by producing it with a preformed HE fragmentation warhead. The incendiary round was also improved by being modified to carry 180 incendiary elements. Maximum range of the Grad-1 rocket family was 15,000 m. For the single-round 9P132 launcher a further short-form rocket was produced, the 9M28 or DKZ-B.

Available details of the rocket types are given in Table 1.

The fuze system has two armed settings, impact and delayed action and has to be set manually before launch.

In addition to the above rockets, other rocket types are known to have been developed.

The 9M53F has an enhanced blast fragmentation warhead which separates from the main body when a time fuzed setting is triggered.

The crew of the modified launcher remotely sets the time fuzes electronically from the cab of the launch vehicle. The actual fuze setters are housed in boxes on the upper right surface of the muzzle end of the tubes.

After separation, the parachute deploys at the rear end of the pre-formed HE fragmentation warhead causing it to descend vertically to approach the target at an angle of 85 to 90°. The round is then detonated above the ground by the proximity fuze.

Once detonated, the warhead forms some 2,500 fragments which, if the warhead is detonated as an airburst by the proximity fuze, are distributed in a circular pattern on the ground.

There are also rockets containing anti-tank mines and carrying two self-targeting top attack munitions that will penetrate 70 mm of conventional steel armour from above.

The 122 mm smoke round is designated the 9M43 which, upon hitting the ground, supplies a continuous smoke screen that is impenetrable to infra-red and laser devices operating in the 4 to 14μm waveband region. The smoke screen lasts for 320 seconds and, in a 10 rocket salvo, at up to 20,200 m range, covers a frontage of 1,000 m to a depth of 800 m.

The 122 m illumination round is designated the 9M42 and provides illumination to a diameter of 1,000 m at an altitude of 400 to 500 m for up to 90 seconds.

French/Russian BM-21 rocket developments

From 1994, the Splav Scientific Production Concern and the French company of CELERG started work on a new 122 mm solid propellant rocket motor. The two companies have now ceased this joint development programme and details were given in *Jane's Armour and Artillery 1999-2000* page 801.

122 mm (50-round) BM 9A41 Prima system

Details of this system, based on Ural-4320 (6 × 6) chassis is understood never to have entered production and is no longer being marketed. Details were given in *Jane's Armour and Artillery 1999-2000.*

Table 1

Original 122 mm calibre rockets Index number	**9M22U**	**9M22M**	**9M28**
Other designators	M-21-OF, DB-1B	M-21-OF	DKZ-B
Diameter	122 mm	122 mm	122 mm
Length	3.226 m	2.870 m	1.905 m
Weight			
(complete)	77.5 kg	66 kg	45.8 kg
(HE fragmentation warhead)	19.4 kg	18.4 kg	19.4 kg
(chemical warhead)	19.3 kg	n/avail	19.3 kg
Range			
(max)	20,380 m	20,000 m	10,800 m
(min indirect)	1,500 m	1,500 m	2,500 m
(min direct)	500 m	500 m	400 m
Warhead types	HE fragmentation, smoke, incendiary, HCN, GB and VX chemical agent*		
Fuze settings	instant/single time delay/safe		
Fuze arming distance	150-400 m	150-400 m	150-400 m
Burnout velocity	690 m/s	700 m/s	450 m/s

* Chemical payload rockets are being phased out of service in the RFAS

122 mm 9M519 jamming rocket

Splav has also jointly developed the 122 mm 9M519-1-7 (also known as Liliya-2) cargo round for use against enemy command and control systems with the Bulgarian company Electron Progress. Based on a set of seven rockets which cover the frequency range from 1.5 to 120 MHz the rockets can be fired in single or ripple mode. The basic rocket weighs 59 kg complete, is 3.026 m long and has a working time of 60 minutes for the jammer payload. Range limits are from 4,550 m to 18,300 m.

New long range 122 mm rockets

The Spav Scientific Production Concern has recently further enhanced the BM-21 (40-round) system as well as developing new rockets. Details of this upgrade are given in the following entry.

Belarussia upgrade BM-21 chassis

Government trials of the Russian designed BM-21 (6 ×6) 122 mm (40-round) multiple rocket launcher (MRL) installed on a MAZ-6317 (6 × 6) chassis have recently been undertaken by the Belarussian Army at the Baryowski proving ground.

The BM-21 (6 × 6) 122 mm MRL, originally designed in the late 1950s by the Spav Scientific Production Concern at Tula, entered service in the early 1960s based on the ZIL-131 (6 × 6) truck chassis. But later the improved Ural-375D (6 × 6) truck chassis was used.

During the recent trials, the BM-21 on the MAZ-6317 (6 × 6) chasis was tested alongside the older Ural-375D (6 × 6) chassis with the BM-21 and the former was found to have superior cross-country mobility. Some design faults were however noted, for example the special support struts which stabalise the launcher during firing will require some modification.

The MAZ-6317 (6 × 6) truck has been developed and manufactured by the Miinsk Wheel Tractor Plant in Minsk, Belarus who are well known for their extensive range of high mobility cross-country vehicles, especially tank transportter tractors and missile launcher crisis.

The MAZ-6317 (6 × 6) vehicle has now been selected to become the standard vehicle in its class in the Belarussian Army.

Variants

A number of countries have produced direct copies or modified versions of the BM-21 family and its 122 mm

Loading 12-round BM-21V (NATO designation M1975) MRSs based on GAZ-66 (4 × 4) truck chassis

rockets. These include China, Croatia, Egypt, India, Iran, Iraq, North Korea, Pakistan, Slovakia and Romania (see relevent entries).

Status

Production as required. In service with following countries:

Country	Quantity	Comment
Afghanistan	n/avail	
Algeria	48	
Angola	50	including 4 delivered in 1998
Armenia	47	
Azerbaijan	56	
Belarus	275	
Bosnia	18	
Bulgaria	222	
Burundi	12	
Cameroon	20	
Cambodia	8	
Congo, Democratic Republic	16	includes 6 delivered in 1997
Congo	8	
Croatia	40	
Cuba	150	estimate
Egypt	200	incl local models
Eritrea	6	
Ethiopia	n/avail	
Finland	24	called Rak H 76
Georgia	18	
Hungary	56	
India	120	
Iran	100	estimate
Iraq	n/avail	
Israel	40	
Kazakstan	250	
Korea, North	n/avail	
Kyrgistan	20	
Lebanon	25	
Libya	236	
Mali	2	
Mongolia	136	
Morocco	39	
Mozambique	30	
Namibia	5	
Nicaragua	18	
Pakistan	40	
Peru	14	
Poland	228	
Russia	3,000	also coastal defence
Seychelles	2	
Somalia	n/avail	status uncertain
Sudan	30	
Syria	280	
Tanzania	58	
Turkmenistan	60	
Uganda	n/avail	
Ukraine	373	
Uzbekistan	33	
Vietnam	350	
Yemen	185	
Yugoslavia	n/avail	
Zambia	50	

Manufacturers

Splav Scientific Production Concern (prime contractor and rockets).
Motovilikha Plants Corporation (rocket launcher).

UPDATED

SOUTH AFRICA

SOMCHEM 127 mm (40-round) Valkiri Mk II Multiple Artillery Rocket System (MARS)

Development/Description

The 127 mm (40-round) Valkiri MARS is a second-generation version of the SOMCHEM Valkiri Multiple Artillery Rocket System mounted on an all-terrain Samil 100 (6 × 6) truck chassis fitted with a five-man armoured (against small arms and shrapnel) and mine-protected crew cabin.

Development of the system began in 1985 under the leadership of SOMCHEM, a Division of Denel, with the first production units completed in 1989. Known by the name Bateleur to the South African National Defence Force, the first battery of eight launchers was delivered in March 1990 as the replacement for the 24-round Valkiri Mk I systems used by the mechanised brigades.

Details of this system, only used by the South African National Defence Force, are given in a separate entry.

The larger Samil 100 (6 × 6) chassis has sufficient space for all the launcher system equipment and rations, stores and water required for a 14 day operational cycle. The modular launch pack unit contains five rows of eight 127 mm rocket launch tubes and is fitted with electrohydraulically operated elevation and traverse gear. A manual back-up system is also fitted. The elevation and traverse controls are located on the left side of the launcher.

The Samil 100 chassis and suspension have been strengthened with the front axle upgraded from 6,500 to 7,500 kg. Improvements to the diesel engine and transmission give the vehicle 45 per cent better traction

SOMCHEM 127 mm (40-round) Valkiri Mk II MRS firing a rocket

and 90 per cent better gradient ability than a standard Samil 100.

A new option is a 5-speed automatic gearbox which has improved acceleration and top speed and reduced fuel consumption. Additional diesel fuel tanks extend the range to 800 km at a speed of 80 km/h.

The complete launcher unit can be mounted on other suitable 10,000 kg vehicle types if required. All production Valkiri II systems built to date have been fitted onto the locally built SAMIL 100 (6 × 6) chassis.

Sophisticated electronic navigation and north seeking equipment as well as computer-controlled

SOMCHEM 127 mm rockets showing folded fins and nose-mounted fuze

semi-automatic laying equipment has been developed for the weapon system. Together with a data communication radio, this brings the fire-control system of the weapon up to most modern standards.

Firing is usually carried out with the two hydraulically operated stabiliser jacks lowered at the rear and is performed from within the armoured cab or via an optional remote hand-held firing console.

The 127 mm unguided rockets may be fired either singly or in ripples with a separation interval of 0.5 second. Total salvo time for all 40 rockets is 20 seconds.

The 2.95 m long, 127 mm extended-range rockets weigh 63 kg overall and use a rocket motor with a high-energy double-base propellant. They can be fitted with a number of warhead types, the predominant one being the very effective HE prefragmented anti-personnel and light vehicle warhead with a thin-walled cast epoxy resin cylinder containing 9,700 steel balls. A proximity fuze assembly with an adjustable burst height between 3 and 8 m increases the kill probability. The proximity action is backed up by a super quick point detonation action.

The system can also fire the standard 54 kg, 22,000 m range 127 mm rocket.

Manual reloading in the field is undertaken by using an ergonomically designed loading platform. The option of mechanical reloading by replacement of the complete launch tube pack is also available.

SOMCHEM 127 mm (40-round) Valkiri Mk II MRS deployed in the field with launcher traversed left and stabilisers at rear lowered into position (T J Gander)

Specifications

Valkiri Mk II
Calibre: 127 mm
Number of tubes: 40
Length:
(tube) 3.0 m
(rocket) 2.95 m
Max range: 36,000 m
Min range: 8,000 m
Rate of fire: 2 rds/s
Launch time: 40 in 20 s
Reload time: 7-10 min
Elevation/depression: +50/0°
Traverse: 110°
Crew: 3
Configuration: 6 × 6
Combat weight: 21,500 kg
Length:
(travelling) 9.3 m
Width:
(travelling) 2.35 m
Height:
(travelling) 3.4 m
Max speed: 93 km/h
Fuel capacity: 400 litres
Max range: 800 km (at 80 km/h)
Ground clearance: 0.35 m
Fording depth: 1.2 m
Gradient: 35°
Vertical obstacle: 0.5 m
Tyres: 16.00 × R20 radial ply
(fitted with run flat)
Electrical system: 24 V DC
Engine: 4-stroke, V-10 air-cooled diesel developing 315 hp at 2,500 rpm
Transmission: 5-speed automatic

Status

Production as required. In service with the South African National Defence Force. It is believed that a total of 16 production systems were built. It has been demonstrated overseas but as of early 2002 there are no known sales of this system.

Manufacturer

SOMCHEM, a Division of Denel (Pty) Ltd.

UPDATED

SOMCHEM 127 mm (24-round) Valkiri Mk I 22 multiple rocket launcher system

Development

Development of the 127 mm (24-round) Valkiri Mk 1 22 multiple rocket launcher was undertaken by the SOMCHEM Division of Denel from November 1977 as a counter to the Russian BM-21 (40-round) 122 mm MRS in service with neighbouring African countries. It was completed in March 1981, with the system entering series production in April and first deliveries beginning later in 1982. It is understood that a total of 22 production systems were built. The system is no longer being marketed.

The 127 mm (24-round) Valkiri Mk I 22 is designed to be deployed either on its own in batteries of two troops with three launchers each, or in conjunction with more conventional artillery against area targets such as guerrilla camps, troop concentrations and soft-skinned vehicle convoys.

Due to its high mobility it is well suited to the 'shoot and scoot' technique whereby it would fire all its rockets at a target and then retire before counterbattery fire is initiated. The 127 mm (24-round) Valkiri Mk I 22 has seen extensive combat service in southern Angola since it entered the South African National Defence Force inventory.

The system was given the name *Chindungu* (the red pepper that bites) by UNITA. Within the South African National Defence Force this system is also referred to as the Visared.

Production of this system was completed some time ago and it was replaced in production by the more mobile and more well protected SOMCHEM 127 mm (40-round) Valkiri Mk II system on a SAMIL 100 (6 × 6) chassis. Details of this are in a separate entry. It is understood that this system is no longer in front line service in South Africa.

Specifications

Valkiri Mk I 22
Calibre: 127 mm
Number of tubes: 24
Length:
(tube) 3.0 m
(rocket) 2.68 m
(travelling) 5.35 m
Weight: (rocket) 53 kg
Max range: 22,000 m
Min range: 8,000 m
Unit of fire: 96 rockets
Rate of fire: 1 rd/s
Launch time: 24 in 24 s
Reload time: 10 min
Elevation/depression: +50/0°
Traverse: 110°
Crew: 2
Configuration: 4 × 4
Combat weight: 6,400 kg
Width:
(travelling) 2.3 m
Height:
(travelling) 2.32 m
Max speed: 90 km/h
Fuel capacity: 90 litres
Max range: 400 km
Ground clearance: 0.3 m
Fording: 0.6 m
Gradient: 30°
Vertical obstacle: 0.35 m
Tyres: 12.50 × R20
Electrical system: 24 V DC
Engine: 4-stroke, 6-cylinder water-cooled diesel developing 99 hp at 2,800 rpm
Transmission: 2 × 4 gear shift with differential lock

127 mm (24-round) Valkiri Mk I 22 MRLS in firing position (Terry Gander) 0021071

127 mm (24-round) Valkiri Mk I 22 MRLS firing 0021072

Description

The launcher unit consists of a 24-round pack of three rows of eight 127 mm launcher tubes mounted on the rear hull of an unarmoured Mercedes-Benz Unimog (4 × 4) truck chassis. Overhead canopy railings are fitted to disguise the vehicle as a normal truck. The vehicle carries hydraulically operated elevation and traversing gear for the launcher and is stabilised during firing by a pair of hydraulic supports mounted at the vehicle rear. A conventional artillery sight is mounted on the left side of the launcher and is stored in a special container on the vehicle when not in use.

Firing is either by a firing panel within the cab or via a 50 m plug-in cable connected to a remote firing unit. A test circuit is incorporated to check the rocket status and circuitry prior to firing. The unguided 127 mm rockets are launched either singly or in ripples of 2 to 24. The firing rate is one rocket per second.

The 127 mm standard 2.68 m, 53 kg weight rocket has a double-base propellant motor and a prefragmented anti-personnel warhead, with a thin-walled cast epoxy resin cylinder containing 8,500 steel balls to give a lethal area of 1,500 m². A contact or proximity fuze is screwed to the front of the warhead. The range varies from 8,000 to 22,000 m depending on whether spoiler rings are used on the rockets to slow them.

Reloading is manual and takes about 10 minutes. Upper atmosphere and surface meteorological data required for firing in order to achieve a high first hit probability is obtained by means of weather balloons, a vehicle-mounted telescopic mast with wind vane and anemometer close to the launchers and a wind gun that fires small calibre projectiles to estimate the prevailing ground wind conditions. The time taken for the launcher to deploy in a presurveyed site is 5 minutes; to come out of action takes only two minutes. A radio is fitted within the vehicle cab as standard with an extension to the remote fire unit.

Status

Production complete. In service with the South African National Defence Force. This system is no longer being marketed.

Manufacturer

SOMCHEM, a Division of DENEL (Pty) Ltd.

UPDATED

Mechem Developments 122 mm RO 122 artillery rocket system

Development/Description

Mechem Developments (a division of Denel) has developed a 122 mm unguided calibre rocket, the RO 122, for use with 122 mm multiple rocket launcher systems which are now manufactured by many countries including China, Croatia, Egypt, India, Iran, Iraq, North Korea, Pakistan, Romania, Russia, Slovakia, Turkey and Yugoslavia. The 122 mm unguided rocket has a new HE fragmentation warhead packed with some 9,800 steel balls and is fitted with a radio proximity fuze.

Mechem has developed a complete family of pre-fragmented warheads for the RO 122, RO 107 and RO 68 rockets covered in this section. In this designation the number after the RO indicates the actual calibre of the rocket in millimeters.

Status

Development complete. Production as required. This family of 122 mm rockets has been developed for the export market.

Manufacturer

Mechem Developments, a member of the Denel Group. Enquiries to Mechem Consultants.

VERIFIED

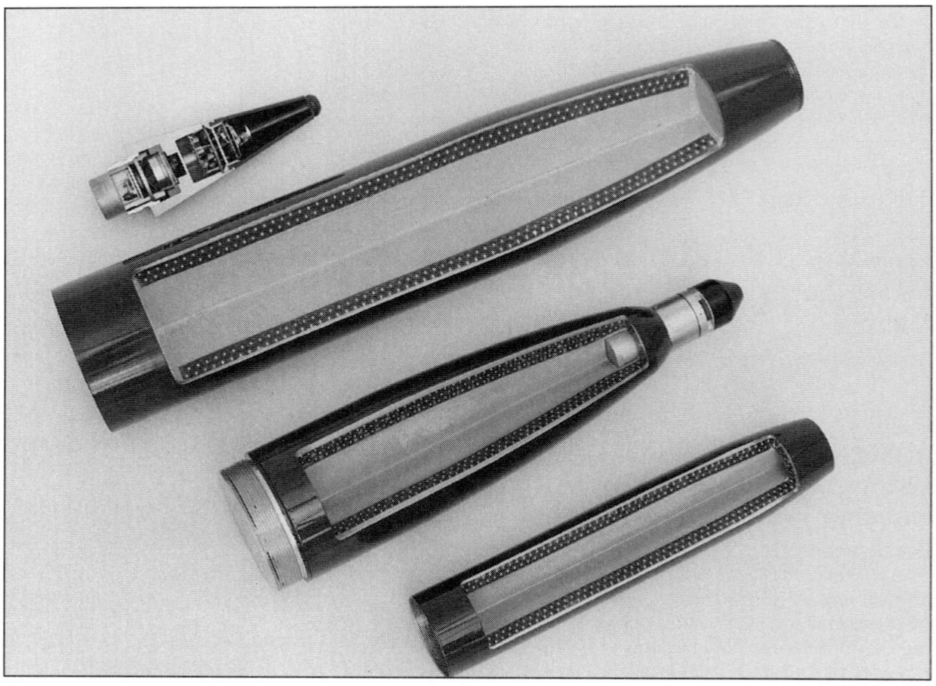

Mechem prefragmented rocket warheads, from top to bottom, fuze, RO 122, RO 107 and RO 68. The RO 107 is shown fitted with a nose-mounted Fuchs Electronics fuze

Mechem Developments 107 mm RO 107 (12-round) towed multiple rocket launcher system

Development/Description

The Mechem Developments 107 mm RO 107 (12-round) towed multiple rocket launcher MRS is designed for use as an anti-personnel weapon by light artillery, infantry and special forces units. It is basically a reverse engineered Chinese NORINCO (China North Industries Corporation) Type 63 107 mm trailer-mounted MRS (see entry in this section).

In addition to supplying its own version of the complete launcher, Mechem can supply its own version of the 107 mm unguided rocket. Mechem is a division of Denel.

The towed launcher has 12 barrels in three rows of four, weighs 380 kg and can be dismantled into units of 30 kg to allow manhandling across rough terrain. It takes about 5 minutes to dismantle and can be reassembled in 10 minutes. Elevation and traverse are set with handwheels mounted on the left. A simple mortar day sight is used for aiming.

All the standard Chinese 107 mm unguided rockets can be fired and because the upgraded Mechem RO 107 rounds use the same range tables as those prescribed for the standard ammunition, mixed salvos of standard and upgraded rockets can be fired using the same launch setting.

Mechem is prime contractor for the upgraded RO 107 rocket: the rocket motor is supplied by SOMCHEM; the rocket motor filling by Naschem; the warhead by Aserma and the RO 107 proximity fuze by Fuchs.

The improved 107 mm rocket is spin stabilised by seven canted venturis, and rotates at approximately 22,000 rpm. It develops an impulse of 6,000 N which gives it a maximum range of 8,500 m at sea level. Minimum firing angle can be as low as 4°. Approximately 80 per cent of the rounds fired to maximum range will fall within an area 150 × 200 m.

Mechem Developments 107 mm RO 107 (12-round) towed multiple rocket launcher system (Christopher F Foss) 0058467

Mechem Developments 107 mm RO 107 (12-round) towed multiple rocket launcher firing a rocket

The rocket can also be fired by equivalent Chinese 107 mm MRS types.

The rocket is fitted with a UHF radio band proximity fuze which arms after four seconds of flight time (approximately 1,400 m from the launcher). Frequency agility provides immunity to enemy jamming action. The proximity fuze causes a predictable airburst irrespective of the type of terrain. A point detonation back-up system is also provided.

The prefragmented warhead is packed with approximately 5,200 steel balls cast into epoxy resin. The shrapnel produced reaches an average velocity of 900 m/s which ensures the penetration of soft-skinned vehicles, support trucks, vehicle tyres and radiators as well as being lethal to personnel. Typical kill and maim areas are 120 m^2 and 504 m^2 respectively for an upgraded round, compared to 34 m^2 and 140 m^2 for a standard cast-iron warhead fitted with a point impact fuze.

Specifications

Calibre: 107 mm
Number of barrels: 12
Muzzle velocity: 30 m/s
Max rocket velocity: 370 m/s
Rate of fire: 12 rds/9 s
Min range: 1,500 m
Max range: 8,500 m
Weight:
(complete rocket) 19 kg
(warhead) 7.6 kg
(fuze) 0.28 kg
(system, empty) 380 kg
(system, loaded) 610 kg
Warhead type: HE fragmentation
Fuze type: radio band proximity with contact back-up
Carriage: two-wheel split trailer
Length:
(complete) 0.92 m
(travelling) 2.6 m
Width:
(travelling) 1.4 m
(deployed) 1.7 m
Height: 1.2 m
Elevation: 0-60°
Traverse: ±15°
Max towing speed:
(paved road) 60 km/h
(unpaved road) 15 km/h
(cross-country) 7 km/h

Variants

The RO 107 (12-round) 107 mm rocket system can be mounted on a variety of chassis, tracked and wheeled. For trials purposes it has already been installed on the Mechem Mamba light vehicle. As far as it is known, there have been no known customers for the vehicle mounted version of this system.

For vehicle applications, the 107 mm RO 107 (12-round) multiple rocket launcher system is palletised.

The upper part contains the complete rocket launcher with its elevation and traverse mechanism while under the lower part are tubes that hold reserve rockets.

Status

Production as required. In service with several undisclosed countries. It is believed that this system is in service with at least one country in Africa. A small quantity of these systems may be used by the South African airborne forces. In 1998 it was reported that Peru had taken delivery of four RO 107 rocket launchers and 5,000 rounds of 107 mm rockets from South Africa.

Manufacturer

Prime contractor: Mechem Developments, a member of the Denel Group.

VERIFIED

Mechem Developments 107 mm Inflict (single-tube) rocket launcher

Development/Description

The 107 mm Inflict single-tube lightweight rocket launcher was developed for the export market by Mecham Developments, a Division of Denel, for use as an anti-personnel weapon by specialist units.

It is fully compatible with the Mechem Developments 107 mm RO 107 12-round trailer-mounted MRS (see a separate entry in this section) with identical aiming and ranging tables.

The complete Inflict 107 mm launcher weighs 26 kg empty and can easily be carried by a two-man team into inaccessible terrain. The rocket tube is made of steel and is fitted inside a rectangular aluminium alloy holding frame with three foldable aluminium alloy legs. The fire-control unit consists of a hand-operated generator and a control box with two push-buttons, connected to the launcher by 15 m of cable.

The frame has a screw to adjust the traverse between −8° and +8°, a geared handwheel to change the elevation between 0° and +45° and a simple mortar sight attachment for aiming. The sight is mounted on the left side of the launcher.

All the standard NORINCO (China North Industries Corporation) 107 mm unguided rocket types can be fired as well as the upgraded RO 107 round. Mechem is prime contractor for the upgraded RO 107 rocket: the rocket motor is supplied by SOMCHEM; the rocket motor filling by Naschem; the warhead by Aserma and the RO 107 proximity fuze by Fuchs Electronics.

The improved unguided rocket is spin stabilised by seven canted venturis and rotates at approximately 22,000 rpm. It develops an impulse of 6,000 N which gives it a maximum range of 8,500 m at sea level.

Mechem Developments 107 mm Inflict single-tube rocket launcher deployed and with remote launcher control box below

Specifications
RO 107 rocket
Calibre: 107 mm
Number of barrels: 1
Muzzle velocity: 30 m/s
Max rocket velocity: 370 m/s
Min range: 1,500 m
Max range: 8,500 m
Weight:
(complete rocket) 19 kg
(warhead) 7.6 kg
(fuze) 0.28 kg
(system, empty) 26 kg
(system, loaded) 45 kg
Warhead type: HE fragmentation
Fuze type: radio band proximity with contact back-up
Length:
(complete rocket) 0.92 m
(folded) 1.4 m
Carriage: single-tube frame-type with three attached legs
Elevation: 0-45°
Traverse: ±8°

Minimum firing angle can be as low as 4° for a direct fire capability.

The rocket is fitted with a UHF radio band proximity fuze which arms after 4 seconds of flight time (approximately 1,400 m from the launcher). Frequency agility provides immunity to enemy jamming action. The proximity fuze causes a predictable airburst irrespective of the type of terrain. A point detonation back-up system is also provided.

The prefragmented warhead is packed with approximately 5,200 steel balls cast into epoxy resin. The shrapnel produced reaches an average velocity of 900 m/s which ensures the penetration of soft- skinned vehicles, support trucks, vehicle tyres and radiators as well as being lethal to personnel. Typical kill and maim areas are 120 m² and 504 m² respectively for an upgraded round, compared to 34 m² and 140 m² for a standard cast-iron warhead fitted with a point impact fuze. Photographs of the 107 mm unguided rockets appear in a previous entry.

An extended-range 107 mm calibre rocket with a range of 11,000 m is under development by Mechem.

Status

Production as required. In service with undisclosed countries. As far as is known the Mechem Developments 107 mm Inflict (single-tube) rocket launcher is not used within South Africa.

Manufacturer

Prime contractor: Mechem Developments, a member of the Denel Group.

VERIFIED

Mechem Developments 68 mm RO 68 (six-round) portable multiple rocket launcher system

Development/Description

The 68 mm RO 68 (six-round) portable rocket launcher is designed by Mechem Developments, a division of Denel, as a private venture, for use as an anti-personnel weapon by special forces units.

In the designation RO 68, the RO stands for rocket and 68 is the calibre of the rocket in millimetres. It is basically a 68 mm air-to-surface unguided rocket, which is normally fired from a pod carried under the wings, fitted with an upgraded HE fragmentation warhead for the surface-to-surface role and mated to a low-cost six-round portable lightweight launcher assembly.

The launcher has a single bank of six extruded aluminium rocket tubes in two vertical rows of three. The bank is mounted on a collapsible tripod unit and is fitted with a standard mortar sight for aiming purposes.

The launcher may be used in two modes:

(1) Conventionally, with the crew present and the rockets fired singly or in a rippled salvo by the control unit.
(2) Unconventionally, in the stand-off role when the six rockets can be fired in a salvo, without the crew being present, by use of a preset timer. The low cost of the launcher permits abandonment of it if necessary. The control unit may also be fitted with a self-destruct device as an option.

The 68 mm calibre RO 68 rocket has a prefragmented warhead containing over 3,000 steel balls. A turbine-powered UHF radio band proximity fuze is fitted causing the warhead to airburst at a consistent height of 3 m above ground. This creates a lethal radius in excess of 15 m against personnel and is also highly effective against soft- skinned vehicles. A back-up point detonation system is also fitted. Maximum rocket range is 6,500 m. The rocket has folding fins which unfold when the rocket leaves the launcher.

Specifications
Calibre: 68 mm
Number of barrels: 6
Max range: 6,500 m
Warhead type: HE fragmentation
Fuze type: radio band proximity with 3 m airburst setting and back-up contact
Carriage: collapsible tripod type
Weight:
(empty) 45 kg
Length:
(folded) 1.4 m
Elevation: 0-55°
Traverse: ±8°

Status

Production as required. In service with undisclosed countries. As far as is known the Mechem Developments 68 mm RO 68 (six-round) portable multiple rocket launcher system is not used within South Africa.

Manufacturer

Mechem Developments, a member of the Denel Group.

VERIFIED

Mechem Developments 68 mm RO 68 (6-round) portable multiple rocket launcher system (Christopher F Foss)

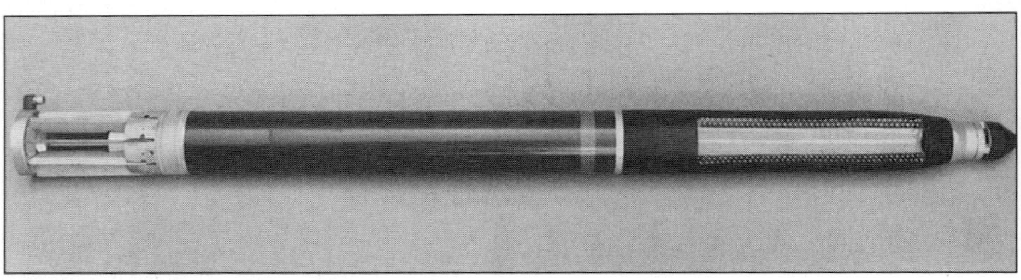
Mechem Developments 68 mm RO 68 rocket with warhead cutaway

SPAIN

Santa Barbara Sistemas 140 mm (40-round) Teruel multiple rocket launcher system

Development/Description

The 140 mm (40-round) Teruel MRS was developed by the Council for Rocket Research and Development of the Spanish Ministry of Defence to replace the older MRS in Spanish Army service.

The launcher has two 20-round packs arranged in five rows of four 140 mm launcher tubes each, mounted on the rear of a Pegaso 6 × 6 truck chassis. The vehicle is fitted with an armoured six-man crew cab that can mount a light 7.62 mm machine gun on the roof for local defence and anti-aircraft purposes.

As far as is known, the system has been sold to two countries, Spain (14 systems) and Gabon (8 systems).

Specifications
Calibre: 140.5 mm
Number of barrels: 40
Length:
(rocket) 3.23 m
Weight:
(high-explosive rocket) 76 kg
(high-explosive warhead) 18.6 kg
(submunition warhead) 21 kg
(smoke warhead) 21 kg
Max range: 28,000 m
Min range: 6,000 m
Max velocity: 687 m/s
Unit of fire: 80 rockets
Reload time: 5 min
Elevation/depression: +55/0°
Traverse: 240°
Crew: 5 (normal)

A longer rocket has also been developed and is fitted with a double-grain solid-fuel motor that, when fitted with aerodynamic airbrakes, allows three distinct trajectories to be flown using the same firing elevation.

The unguided 140 mm rockets also have pop-out fins for stabilisation in flight. The warhead can be either a 20 kg HE fragmentation or, in the case of the long rocket, a 21 kg submunition cargo warhead. This is available in four versions carrying 42 anti-personnel grenades each, with either 950 3.2 mm diameter steel balls, 60 g of Composition B explosive and a PO model percussion fuze, or 28 hollow charge grenades with contact fuzes and capable of penetrating 110 mm of conventional steel armour, six pressure-activated anti-tank mines fitted with anti-disturbance features or 14 smoke grenades each providing 4 minutes of smoke. The fuzes for the rockets can either be the PD M34A2 contact type, proximity or timed. All six rockets can be mixed within the launcher and can be selected according to the target to be engaged.

A Pegaso 6 × 6 resupply vehicle with four blocks of 20 long-rocket reload rounds or six blocks of 20 short-rocket reload rounds is assigned to each vehicle as part of the battery. Reloading is manual and takes 5 minutes while the time taken to emplace a launcher is 2 minutes, to fire a complete salvo takes 45 seconds and to come out of action another 2 minutes. The short HE rocket is 2.044 m in length, weighs 56.3 kg and has a maximum range of 18,000 m. The burn time of the motor is 1.6 seconds. The long rocket is 3.23 m in length, weighs 76 kg and has a burn time of 2.7 seconds, giving a maximum range of 28,000 m. Minimum range for both rockets is 6,000 m.

Production of this system was completed and halted some time ago and if production did commence again then a different chassis would have to be provided.

Status

Production as required. In service with Gabon (8) and the Spanish Army (14). As far as is known there has been no recent production of this complete system for the home or export markets.

Manufacturer

Santa Barbara Sistemas (a General Dynamics Group company).

UPDATED

140 mm (40-round) Teruel multiple rocket launcher system in the travelling configuration 0058469

140 mm (40-round) Teruel multiple rocket launcher in firing position with launcher elevated, stabiliser jacks lowered
0021077

TAIWAN

RT2000 Artillery Multiple Launch Rocket System (AMLRS)

Development

Early in 2000, the Missile & Rocket Systems Research Division of the Chung-Shan Institute of Science and Technology revealed that it had developed a new Artillery Multiple Launch Rocket System (AMLRS) called the RT2000 to meet the operational requirements of the Republic of China (ROC) Army. The RT in the designation means Ray Ting. It is understood that the ROC Army has a requirement for at least 150 RT2000 systems.

The RT2000 is also referred to as Thunder and late in 2001 it was stated that the ROC Army was to order an initial batch of 60 systems.

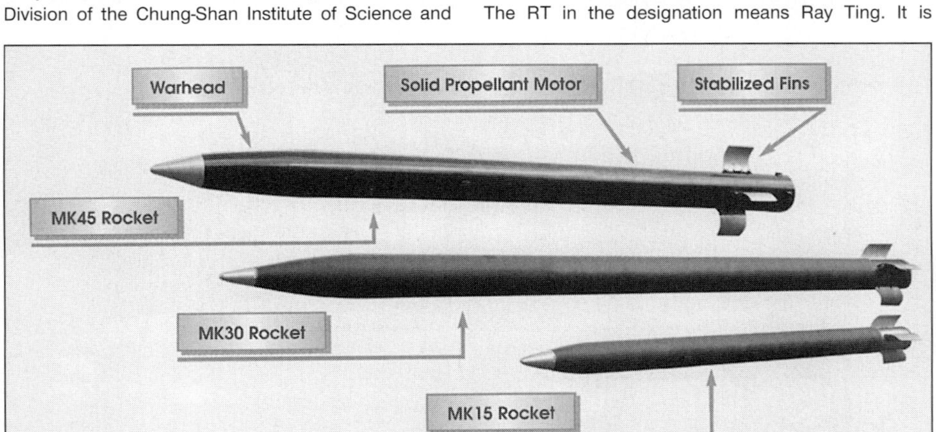

The RT2000 Artillery Multiple Launch Rocket System can launch three types of unguided rocket, the MK 15, MK 30 and MK 45, all have wrap round wrap round fins at the rear **2000**/0085769

Description

RT2000 is based on the chassis of the US Oshkosh Truck Corporation M977 series Heavy Expanded Mobility Tactical Truck (HEMTT) (8 × 8) which is used in large numbers by the ROC and has a nominal payload of 10 tons (US).

Mounted on the rear of the chassis is the powered launcher that can be fitted with pods of rockets that are aimed and fired with the crew remaining in the forward control cab. To provide a more stable firing platform two hydraulic stabilisers are lowered to the ground on either side of the truck.

The RT2000 fires three types of solid propellant, unguided rocket all of which have four wrap round fins at the rear that unfold after launch.

These rockets are designated the MK 15 (117 mm), MK 30 (180 mm) and MK 45 (230 mm) with the figure of 15, 30 and 45 indicating the range of the rocket in km.

Scale model of the RT2000 artillery multiple rocket system (AMLRS) which is based on the US HEMETT (8 × 8) truck chassis (Craig Hoyle/JDW) ***2000***/0067603

Specifications

Chassis: 8 × 8
Calibre:
(MK 15 rocket) 117 mm
(MK 30 rocket) 180 mm
(MK 45 rocket) 230 mm
Max range:
(MK 15 rocket) 15,000 m
(MK 30 rocket) 30,000 m
(MK 45 rocket) 45,000 m

Development of the MK 15 rocket is complete with that of the MK 30 and MK 45 almost complete. The rockets require no maintenance and are treated as 'wooden rounds'.

The rockets can be fitted with two types of warhead, high explosive and sub-munition carrying M77 dual-purpose bomblets that are highly effective against a variety of battlefield targets. The rocket warhead fuzes are set by remote control before the rockets are launched.

The rockets are transported and launched from pods that are rapidly loaded onto the launcher using a crane mounted above and to the rear of the second axle.

The MK 15 rockets come in pods of 20 rockets, MK 30 in pods of nine and MK 45 in pods of six. When using the MK 30 rockets the system carries a total of 27 rockets in the ready to launch position.

Mounted in the fully enclosed forward control cab is the fire-control panel which allows the crew to aim and fire the rockets with key elements of the system including the electronic unit, position and attitude determining system, controller box, servo motors, amplifier and the communications processor.

Primary role of the RT2000 system is the rapid neutralisation of amphibious craft before they are able to deposit troops and equipment on the beach.

Current ROC rocket systems, covered later in this section, include the 126 mm (40-round) Kung Feng III/IV and the 117 mm (45-round) Kung Feng VI mounted on a US-supplied, 5-ton (6 × 6) series truck. The latter rockets have a maximum range of 15 km and are similar to those used in the latest RT2000 AMLRS.

Status

Entering service with the Republic of China Army.

Manufacturer

Chung-Shan Institute of Science and Technology, Missiles & Rocket Systems Research Division.

UPDATED

126 mm (40-round) Kung Feng III/IV multiple rocket systems

Development/Description

The Kung Feng 126 mm (40-round) MRSs were designed in Taiwan by Sun Yat-sen Scientific Research Institute in the early 1970s.

The basic Kung Feng (Worker Bee) consists of two banks of 20 126 mm calibre rocket launcher tubes arranged in five rows of four. The tubes are over 1.6 m long and fire spin-stabilised unguided rockets. The rocket length is about 0.78 m and the weight approximately 24.94 kg. The maximum range is estimated to be 9,000 m, with the 40 electrically fired rockets ripple fired in 16 seconds.

The launcher can be mounted on a variety of platforms. In Taiwanese service the following are known.

Trailer-mounted version

A towed version for use by the infantry. The launcher is mounted on a two-wheeled carriage with a tubular frame and towed by a ¾ ton M37 (4 × 4) truck. This system is believed to be designated Kung Feng III.

M113A1 rocket launcher

A self-propelled version on the United States supplied United Defense LP M113A1 APC for mechanised and armoured units. The launcher is mounted just behind the troop compartment hatch pointing forward. Elevation and depression are hydraulically operated but there is no facility for traversing. A blast shield is fitted over the rear hull top to protect the dome ventilator. In order to fire, the crew must disembark. Use of the troop compartment is limited by the launcher's angle of elevation. This system is known as Kung Feng IV.

LVTP5 rocket launcher

A self-propelled version on the LVTP5 amphibious assault personnel carrier for the marine corps. The launcher is fitted just to the rear of the commander's hatch and can be fired while the vehicle is afloat. This is also known as the Kung Feng IV.

An ECM chaff-dispensing system known as the Kung Feng Sea is in use on Taiwanese naval destroyers, as far as it is known this rocket is not fired by the land based multiple rocket launchers.

Status

Production of all these 126 mm multiple rocket systems has been completed, although replacement rockets are built as and when required. In service with the Taiwanese armed forces. There are no known exports of this system. This system will soon start to be supplemented by the new locally developed RT2000 Artillery Multiple Launch Rocket System covered in the previous entry and based on the Oshkosh Truck Corporation M977 (8 × 8) chassis.

UPDATED

Kung Feng IV multiple rocket launcher on M113A1 APC chassis (L J Lamb)
0058470

Towed version of Kung Feng IV multiple rocket launcher on parade being towed by an M37 (4 × 4) truck

117 mm (45-round) Kung Feng VI multiple rocket launcher system

Development/Description

The 117 mm Kung Feng VI (45-round) multiple rocket launcher system was designed by the Sun Yat-sen Scientific Research Institute in the late 1970s. Intended only as a self-propelled MRS, the present model is designated Kung Feng VIA and is a 117 mm (45-round) launcher with five layers of nine tubes mounted on a traversable pedestal with hydraulic elevation equipment.

The US supplied carrier vehicle normally used is the M52A1 (6 × 6) 5 ton truck chassis with the launcher mounted on the rear. The cab is unarmoured so that the traverse is normally beyond 90° when firing. A support vehicle carrying reload rounds is required. Reloading is manual and takes up to 15 minutes for all 45 barrels.

The unguided rocket is 1.8 m long, weighs approximately 60 kg and is spin stabilised in flight by four fold-out fins at the tail. Maximum range is estimated to be 15,000 m, with a full salvo having a lethal area of some 30,000 m^2. The rockets are electrically fired with a 0.5 second delay between each, giving a total ripple firing time of 22 seconds. The warheads are of the high explosive blast fragmentation type. A computer-based fire-control capability is under development.

When deployed in the firing position, stabilisers are lowered to the ground to provide a more stable firing platform. The sighting system is mounted on the left side of the launcher.

This is expected to be replaced by the new RT2000 Artillery Multiple Launch Rocket System based on an Oshkosh M977 series truck.

117 mm Kung Feng VI 45-tube multiple rocket launcher mounted on rear of 6 × 6 cross-country truck (L J Lamb) 0058471

Details of this system, which fires MK 15 (117 mm), MK 30 (180 mm) and MK 45 (230 mm) rockets are given earlier in this section. The MK 15 rocket is similar to that used with the 117 mm (45-round) Kung Feng VI system.

Variant

An experimental version has been identified. A simple non-traversable launcher with hydraulic elevation controls was mounted in place of the twin 40 mm turret on an M42 tracked self-propelled anti-aircraft gun chassis. The current status is not known.

Status

Production of the complete system has been completed but production of the rockets is undertaken as and when required. In service with the Republic of China Army. There are no known exports of this system.

UPDATED

TURKEY

TOROS 230 mm and 260 mm Artillery Rocket System

Development

Turkey's largest ammunition manufacturer, MKEK (Makina ve Kimysa Endustrisi Kurmu) and the SAGE (Savunma Snayii Arastirma Gelistirme Enstitusu) Division of TUBITAK (Turkish Scientific Research Institute) are developing a new family of long-range unguided rocket systems to meet the future requirements of the Turkish Land Forces Command.

This is called TOROS (TOpcu ROketi Sistemi) artillery rocket system and is being developed in two versions, TOROS 230 A and TOROS 260 A, with the 230 and 260 indicating the calibre of the rocket in millimetres.

By early 2000, the first phase of development of the TOROS artillery rocket system had been completed, this being the development and testing of the sub-systems. The next phase is expected to be full-scale development of the overall system. New warheads and guidance systems are already said to be under development for the TOROS system.

Late in 2001 it was stated that the validation phase of TOROS had been completed.

In three series of firings of rockets fitted with smoke warheads, a total of 10 rockets have been successfully fired according to SAGE officials with a low dispersion being achieved. The last batch of firings took place with a basic launcher installed on the rear of a FAP (8 × 8) cross-country truck chassis.

SAGE is now seeking Turkish Land Force Command funding to allow all up trials to be completed. This would be followed by Phase II, during which full systems development would take place and some 168 rockets would be built and test fired. Phase III would be full scale production followed by further enhancements to the system.

Description

The TOROS ARS consists of five sub-systems, launcher vehicle, logistics vehicle and maintenance vehicle based on a 6 × 6 cross-country truck chassis, command and control vehicle on a 4 × 4 chassis and the TOROS unguided rockets.

Both of the TOROS rockets use a HTPB/AP solid composite propellant and feature four wrap around fins at the rear which unfold after launch with sabots towards the front which are discarded after launch. A fin detent system is fitted to reduce wind dispersion of the rockets.

TOROS 230 has a minimum range of 10 km and a maximum range of 65 km while TOROS 260 has a minimum range of 15 km and a maximum range of 100 km.

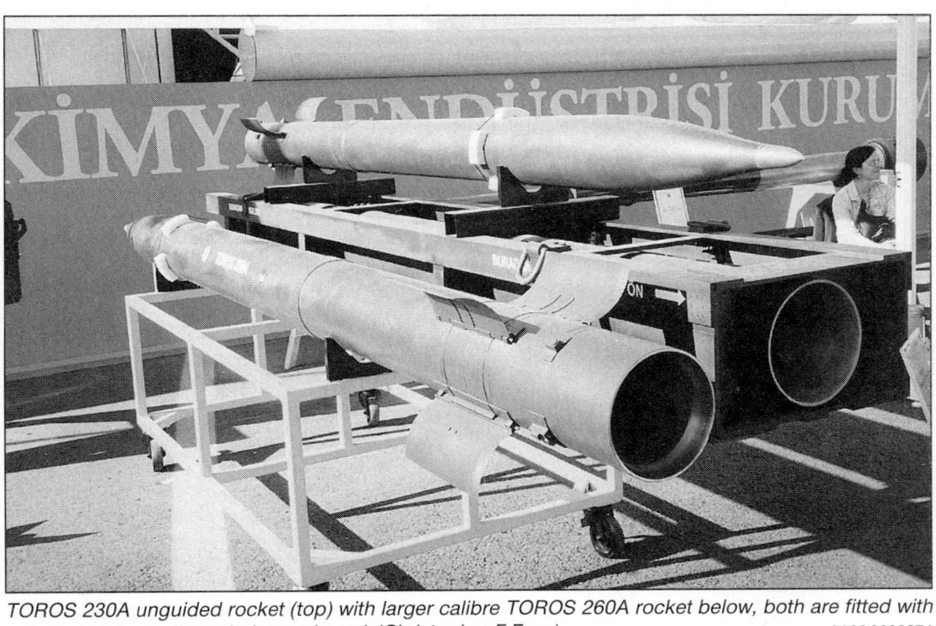

TOROS 230A unguided rocket (top) with larger calibre TOROS 260A rocket below, both are fitted with sabots which are discarded upon launch (Christopher F Foss) **2000**/0039274

First photograph released of the SAGE TOROS artillery system installed on an FAP (8 × 8) truck chassis for firing trials in Turkey. This has two pods of TOROS 260 rockets with a maximum stated range of 100 km in the ready to launch position **2002**/0096203

The unguided 230 and 260 mm rockets have been designed to have a shelf life of at least 10 years and to be used in a temperature range of from −30 to +60°C. Accuracy is being quoted as one or two per cent of range.

The basic rocket would have a high-explosive, preformed fragmentation warhead but other types of warhead could be fitted, for example cargo. The latter would carry sub-munitions to attack the vulnerable upper surfaces of armoured vehicles.

The high-explosive preformed fragmentation warhead of the TOROS 230 weighs 120 kg and contains 25,000 steel balls while the same type of warhead for the TOROS 260 weighs 145 kg and contains 30,000 steel balls. Both warheads have an inner high-explosive content. The rockets are activated by a nose-mounted fuze. Other warhead types are under active consideration and include a bunker buster and one carrying electronic warfare devices to lay behind enemy lines.

The rockets are transported and launched from pods with lightweight glass fibre launch tubes. The 230 mm pod contains six rockets and the 260 mm pod contains four rockets. The pods are mounted on a turntable installed on the rear of the launch platform with powered elevation and traverse. Each launcher would have two pods of TOROS rockets.

The TOROS launch platform would typically be a 6 × 6 all terrain truck with a forward control armour protected cab which would house the crew and the associated elevation, traverse and rocket launching equipment. A 7.62 mm or 12.7 mm machine gun could be installed on the cab roof for local and air defence purposes.

To provide a more stable platform while the rockets are launched, two hydraulically operated stabilisers would be lowered to the ground on either side.

A typical TOROS rocket launcher battery would consist of six launchers mounted on a 6 × 6 all terrain forward control cab and each of these would be supplied by an ammunition re-supply vehicle on a similar chassis. Each firing platform would have an advanced GPS and inertial navigation system to reduce into action times and improve target response times.

This would be provided with a hydraulic crane for re-supplying the rocket launcher with new pods of rockets. In addition there would be a battery command post and a reconnaissance vehicle such as the Otokar Akren (Scorpion) (4 × 4) or Cobra (4 × 4) light armoured vehicle which is already in service with the Turkish Army.

Status
Development. Not yet in production or service.

Manufacturers
MKEK-Cansas.
SAGE.

UPDATED

Specifications

Version	TOROS-230A	TOROS-260A
Configuration	6 × 6	6 × 6
Calibre	230 mm	260 mm
Length of rocket	4.1 m	4.8 m
Length of rocket motor	2.71 m	3.40 m
Min range	10 km	15 km
Max range	65 km	100 km
Total weight of rocket	326 kg	483 kg
Weight of rocket motor (including fins)	205 kg	329 kg
Weight of warhead	121 kg	144 kg
Length of warhead	1.45 m	1.45 m
Number of rockets in pod	6	4
Length of pod	4.2 m	4.9 m
Width of pod	1.2 m	1.1 m
Depth of pod	0.70 m	0.90 m
Weight of pod (empty)	508 kg	410 kg
Weight of pod (loaded)	2,468 kg	2,346 kg
Number of pods per launcher	2	2

MKEK 107 mm (12-round) multiple rocket launcher system

Development/Description
The Turkish company, MKEK-Çansas, has reverse-engineered the NORINCO (China North Industries Corporation) 107 mm (12-round) Type 63 multiple rocket launcher (see entry this section). Details of the system are believed to be similar to the NORINCO Type 63 specifications except where manufacturing techniques reflect local variations. It is understood that Turkish Land Forces Command facilities were also involved in production of this launcher.

The full range of 107 mm calibre rockets can be fired including the Roketsan TR-107 family. Details of these are given later in this section. This rocket is now in production for the Turkish Land Forces Command and has a maximum range of 10,000 m.

Specifications
Rocket
Weight of complete rocket: 18.80 kg
Length of rocket: 840 mm (with fuze)
Effective range: 8,500 m
Warhead type: TNT
Weight of warhead: 8.5 kg
Propellant type: N5 (double base)
Fuze: mechanical
Maximum rocket velocity: 370 m/s

Launcher
Calibre: 107 mm
Number of tubes: 12
Elevation: 0 to +55°
Weight
(empty) 435 kg
(loaded) 663 kg
Length: 2.525 m
Width: 1.525 m
Height: 1.60 m

Status
Recent information has indicated that the Turkish Army has deployed 48 of these systems. Production complete. No longer marketed.

Manufacturer
MKEK-Çansas.

UPDATED

MKEK 107 mm (12-round) multiple rocket launcher system deployed in the firing position (Christopher F Foss)

MKEK 70 mm (40-round) MAKSAM RA-7040 multiple rocket launcher

Development/Description
As part of the Turkish Land Forces Command weapons diversification programme, MKEK-Çansas has developed a 70 mm (2.75 in) multiple rocket launcher known as the RA-7040. In the latter designation, the 70 is for the calibre of the rockets in millimetres and 40 is the number of launcher tubes.

The RA-7040 multiple rocket launcher system is of a modular type and is mounted on a two-wheeled trailer with four hydraulically operated jacks for stabilisation purposes during firing. Any vehicle with a 600 kg towing capacity can tow the launcher. It can be adapted to fit on both land and amphibious forces vehicles if required. As far as it is known all examples built to date have been in the trailer-mounted configuration.

Specifications
Calibre: 70 mm
Number of tubes: 40
Carriage: two-wheel trailer
Weight:
(in firing position) 1,690 kg
(empty) 1,320 kg
Length:
(travelling) 3.470 m
Width:
(travelling) 2.060 m
Height:
(travelling) 2.135 m
Elevation: 0 to +55°
Traverse: 360°
Power supply: 24 V DC

The trailer-launcher weighs 1,320 kg when empty and can fire all 40 rockets in either single or ripple launch modes within 10 to 40 seconds. All types of 70 mm unguided rockets can be fired with a 40-round salvo typically covering an area of 200 × 300 m at a range of 7,400 m.

The actual rocket is also used in the air-to-surface roles with the double-base propellant giving a burn time of 1.42 seconds. Types of warhead that can be fitted include M151 (anti-personnel), M229 (anti-personnel), M427 (anti-personnel), flechette, Mk 5 and Mk 427 (anti-armour), M156 (smoke) and M245 (chemical).

The launcher has a full 360° traverse capability and is controlled by a remote-control device with 25 m of cable.

Status

Production complete. To date, 24 have been produced for the Turkish Land Forces Command. No longer marketed.

Manufacturer

Makina ve Kimya Endüstrisi Kurumu (MKEK-Çansas).

VERIFIED

MKEK-Çansas 70 mm (40-round) MAKSAM RA-7040 MRS deployed in firing position

Roketsan 122 mm Extended Range (+40 km) Multi-Barrel Rocket Launcher Ammunition, TR-122

Development/Description

The Turkish company Roketsan, has type classified a 122 mm calibre Extended Range Multi-Barrel Launcher Ammunition designated the TR-122. This new rocket extends the 122 mm calibre rocket launcher system's range to over 40,000 m.

The TR-122 unguided rocket can be fired from a Roketsan made 122 mm T-122 Multi-Barrel Rocket Launcher and its foreign equivalents, such as the widely deployed Russian 122 mm BM-21 Grad. Details of these are given earlier in this section with the T-122 rocket system, based on a 6 × 6 truck having been in service with Turkish Land Forces Command since 1996.

The Roketsan TR-122 Rocket Family is grouped into three as the TR-122, TRB-122 and TRK-122 having three kinds of warhead, high explosive, steel ball and submunition warheads respectively.

Their distinctive airframe/aerodynamics and high-energy composite propellant result in an excellent range and dispersion performance, according to Roketsan. The TR-122 is in service with the Turkish Armed Forces and is now offered to the export markets.

In addition to 122 mm extended range rockets, Roketsan also has the capacity for the production of classical 122 mm artillery rockets, namely SR-122 with a range of 20,000 m.

Roketsan 122 mm (40-round) T-22 Multi-Barrel Launcher System firing a Roketsan 122 mm TR-122 extended-range rocket **2002**/0021080

Specifications

TR-122 Rocket Family (TR-122 rocket and TRB-122 rocket)
Calibre: 122 mm
Max range: 40,000 m plus
Minimum range
21,000 m (without drag ring)
10,000 m (with drag ring)
Length: 2.93
Weight of complete round: 65.9 kg
Weight of warhead: 18.4 kg

TRK-122 rocket
Calibre: 122 m
Max range: 30,000 m plus
Minimum range: 16,000 m
Length: 3.24 m
Weight of complete round: 71.6 kg
Weight of warhead: 22.9 kg
Type of warhead: 50 APAM plus 6 incendiary submunitions

Status

In serial production. Multiyear contract placed by Turkish Land Forces Command. Offered for export.

Manufacturer

Roketsan.

NEW ENTRY

Roketsan 122 mm (40-round) T-122 multi-barrel rocket launcher system

Development/Description

The Turkish company, Roketsan, is the prime contractor for a 122 mm calibre multi-barrel rocket launcher system designated the T-122 which entered service with the Turkish Land Forces Command in 1996 and is now being offered on the export market.

According to Roketsan, the T-122 multi-barrel rocket launcher is an autonomous, artillery weapons system that is used for indirect fire-support missions under day and night conditions and in all weather conditions against area targets.

The T-122 has two banks of 20 122 mm rocket tubes mounted on the rear of a MAN (6 × 6) 26.281 cross-country truck chassis.

According to Rocketsan, the latest technology has been used in the design of the T-122 multi-barrel rocket launcher, including an advanced fire-control computer and digital secure communications system.

The launcher is hydraulically traversed and elevated with a panoramic M-115 sight mounted to the left of the launcher. The rockets can be fired singly or in salvo. A full salvo of 40 rockets can be launched in 80 seconds. Reloading is via a hydraulic loading crane which is mounted on the launcher.

When deployed in the firing position, four stabilisers are lowered to the ground under hydraulic control to provide a more stable firing platform.

The launcher can fire the full range of 122 mm unguided rockets available including Russian 122 mm GRAD rockets and Roketsan's 122 mm rockets including the SR-122. It also fires the extended range TR-122 artillery rocket family with a range of 40,000 m plus which is also in production and service with the Turkish Land Forces Command.

The TR-122 rocket family is grouped into three as the TR-122, TRB-122 and TRK-122 having three kinds of warhead, high explosive, steel ball and submunition respectively. Their distinctive airframe/aerodynamics and high-energy composite propellant results in an excellent range and dispersion performance. Additional details of the TR-122 family are given in the following entry.

Status

In serial production. Multiyear contract placed by Turkish Land Forces Command.

Manufacturer

Roketsan.

Roketsan 122 mm (40-round) T-122 multiple rocket launcher system deployed in firing position 0021078

Close-up of Roketsan 122 mm (40-round) T-122 multiple rocket launcher system (Christopher F Foss) 0021079

Specifications

T-122 launcher
Calibre: 122 mm
Number of rocket tubes: 40
Length of rocket tube: 3 m
Elevation: 0 to +55°
Traverse: ±110°
Rate of fire: 40 rockets/80 s
Max range:
40,000 m+ (TR-122 and TRB-122 rockets)
30,000 m+ (TRK-122 rocket with 56 submunitions)
20,000 m (SR-122 rocket)
Min range:
21,000 m (TR-122 and TRB-122 rockets without drag ring)
10,000 m (TR-122 and TRB-122 rockets with drag ring)
16,000 m (TRK-122 rocket with 56 submunitions)
3,000 m (SR-122 rocket)
Combat weight: 22,200 kg
Length: (travelling) 9.2 m
Width: (travelling) 2.5 m
Height: (travelling) 3.10 m
Max gradient: 60%
Max road speed: 75 km/h
Max road range: 970 km

UPDATED

Roketsan 107 mm TR-107 extended-range rocket

Development/Description

The Turkish company, Roketsan, has type classified a 107 mm calibre unguided rocket designated the TR-107 which extends 107 mm calibre rocket launcher system range to over 11,000 m.

The TR-107 can be fired from systems such as the NORINCO (China North Industries Corporation) Type 63 (qv) and its foreign equivalents such as the South African RO 107 (qv) and the MKEK 12-round launcher. It is understood that the Turkish Army has 48 of these 107 mm TR-107 systems in service. Details of the

Roketsan 107 mm calibre TR-107 extended-range rocket being launched from a MKEK 107 mm (12- round) launcher 0021081

MKEK 107 mm (12-round) rocket launcher are given earlier in this section.

The TR-107 is spin stabilised and has a reduced smoke solid propellant to provide the extended range. It can be fitted with one of three types of warhead: HE, HE fragmentation (containing steel spheres which are highly effective against soft targets) and incendiary. The warhead is activated by a nose-mounted impact fuze.

Status
Production as required. Multiyear contract with Turkish Land Forces Command.

Specifications
TR-107 rocket
Calibre: 107 mm
Max range: 11,000 m plus
Min range: 3,000 m
Length: 0.84 m
Weight of complete round: 19.5 kg
Weight of warhead: 8.4 kg

Manufacturer
Roketsan.

UPDATED

UKRAINE

State Scientific and Technical Centre of Artillery and Rifle Arms mobile 57 mm (32-round) and 80 mm (20-round) missile artillery system

Development/Description
The mobile missile artillery system, developed as a private venture by the State Scientific and Technical Centre of Artillery and Rifle Arms, consists of a standard Russian developed unguided air-to-surface 57 mm calibre 32-round, or 80 mm calibre 20-round, rocket pod mounted on a rotating pedestal assembly fitted to the rear of a light 4 × 4 truck chassis.

To provide a more stable firing platform, stabilisers are lowered to the ground manually before the unguided 57 or 80 mm rockets are launched.

Specifications
Calibre: 57 or 80 mm
Number of rounds: 32 × 57 mm or 20 × 80 mm
Rate of fire: single round or salvo at 4, 6, 8 or 16 rockets/min
Weight: (complete rocket pod)
(57 mm UB-5-32) 100 kg
(80 mm UB-8-20) 220 kg
Warhead: HE or HEAT
Max velocity: 700 m/s
Max range:
(57 mm rocket) 4,500 m
(80 mm rocket) 6,000 m
Launcher: 4 × 4 light truck chassis

It is designed for use against personnel, artillery positions, armour and lightly fortified positions. The rockets used are standard 57 mm S5M1 or 80 mm S-8M unguided rockets fitted with solid propellant rocket motors. The warhead types used can be either HE or HEAT. The rate of fire is stated to be single round or salvo at 4, 6, 8 or 16 rockets/min. Maximum rocket velocity at burnout is 700 m/s. All available details are given in the accompanying table. It is understood that the rocket pods are identical to those used on aircraft and helicopters.

A rocket fitted with a fire extinguishing composition is also now being marketed by the State Scientific and Technical Centre of Artillery and Rifle Arms.

Status
Ready for production. Offered for export.

Manufacturer
The State Scientific and Technical Centre of Artillery and Rifle Arms, Ukraine.

UPDATED

Light cross-country vehicle fitted with turntable mounted pod of unguided rockets in the rear 0023285

UNITED KINGDOM

Lightweight Mobile Artillery Weapons System (Rocket Platform)

Development/Description
The UK's Royal Artillery (RA) has a future requirement, Staff Target (Land) 4095, for a Lightweight Mobile Artillery Weapons System (LIMAWS).

LIMAWS comprises three elements, a 155 mm lightweight gun, LIMAWS-RP and a weapon locating radar (WLR) which is also referred to as MAMBA (Mobile Artillery Monitoring Battlefield Radar).

These weapons will be deployed by the Light Force Element of the UK's Rapid Reaction Force. The 155 mm Lightweight Gun is likely to be the now BAE Systems, RO Defence 155 mm/39 calibre XM777 Ultralightweight Field Howitzer (UFH) adopted by the US Marine Corps.

This British Army requirement is for about 65 155 mm XM777 systems with an in service date of around 2006.

The Lightweight Rocket Launcher System or LIMAWS (Rocket Platform) as it is referred to, is required to have a high level of commonality with the Lightweight Gun.

The current MLRS system used by the RA is based on a full tracked chassis with an armoured cab which, in turn, is based on automotive components of the United

Artist's impression of the INSYS Limited LIMAWS - RP deployed in the field and launching a rocket **2001**/0095103

Defense LP Bradley M2 infantry fighting vehicle. Fully loaded it has a combat weight of over 25 tonnes and has two pods, each of six 227 mm unguided rockets.

Currently under development for the US Army by Lockheed Martin Missile and Fire Control is the High-Mobility Artillery Rocket System (HIMARS) based on a Stewart & Stevenson 6 × 6 truck. This has one pod of six 227 mm MLRS rockets and weighs over 13 tonnes. Full details of this are given in a later entry under United States.

Early in 2000, the Defence Procurement Agency (DPA) issued Invitations To Tender (ITT) to six international companies for the Lightweight Mobile Artillery Weapon System (Rocket Platform) (LIMAWS – RP) Risk Reduction Demonstrator (RRD).

The LIMAWS – RP ITT, which was returned to the DPA towards the end of April, was issued to Alenia Marconi Systems (UK), BAE Systems, Land and Naval Systems (UK), Hunting Engineering (UK), LFK (Germany), Lockheed Martin Missiles and Fire Control (USA) and Matra BAe Dynamics (UK).

Early in 2001, the UK Defence Procurement Agency (DPA) awarded parallel risk reduction/simulation contracts worth £1.7 million to two UK consortiums for the LIMAWS – RP.

BAE Systems, RO Defence and Hunting Engineering Limited (today called INSYS Limited) have now each been awarded an 11 month contract for conceptual analysis of lightweight rocket launchers covering mobility, helicopter portability and the overall stability of such systems when firing the new generation of rocket munitions. Late in 2001, each contractor was awarded a three month extension to the original 11 month contract, this ran until March 2003.

LIMAWS – RP is required to launch the new Guided Multiple Launch Rocket System (MLRS) rocket currently under development or a Fibre Optic Missile (FOM). The latter could be the Polyphem under development by France, Germany and Italy.

It is also to be compatible with the MLRS replacement fire-control system that is due to be back fitted to European built MLRS in the future.

LIMAWS – RP will be a self-contained unit with an integral reload capability and carry one pod of six rockets that will be fully interchangeable with the current in-service MLRS system. Another alternative could be a prime mover carrying LIMAWS – RP that could be carried in a split load if required.

The INSYS LIMAWS – RP team includes Euro Rocket System (a joint venture company formed by Diehl Munitionssysteme and Lockheed Martin Missiles and Fire Control), HMT Supacat and the Defence Evaluation and Research Agency (DERA).

The artist's impression released by INSYS Limited of its proposal, shows that it comprises a prime mover vehicle and a Launcher Loader Module (LLM) that contains the Rocket Pod Container (RPC).

For most operational scenarios the integrated concept would be employed but if required the LLM and RPC can be detached from the prime mover vehicle and transported independently.

In addition to being fully air portable in the Lockheed Martin C-130 and Future Large Aircraft, LIMAWS – RP will be carried under a CH-47 Chinook and the projected Support, Amphibious and Battlefield Rotorcraft (SABR) and transported by a number ships such as LCU (Landing Craft Utility).

The stated aim of LIMAWS is to provide light forces, for example 16 Air Assault Brigade and 3 Commando Brigade, with the level of fire support normally associated with heavy armoured units and with an in-service date of 2008.

Status

Future requirement.

Manufacturer

Not selected. BAE Systems, RO Defence and INSYS were both contracted for the risk reduction/simulation phase.

UPDATED

BA Systems, RO Defence Reduced Range Practice Rocket (RRPR)

Development

Following an international competition, in mid-1995 the Rocket Motors Division of Royal Ordnance (BAE Systems, RO Defence, Rocket Motors) was selected by the UK Ministry of Defence to design, develop and produce the Reduced Range Practice Rocket (RRPR) for the Multiple Launch Rocket System (MLRS, covered in detail in the following entry) used by the Royal Artillery.

Total value of this contract was around £13 million. Major subcontractors to the Rocket Motors Division are Computing Devices Canada which wrote the software for the fire-control system, Air Log of the UK for the rocket pod containers and Forges de Zeebrugge of Belgium who supplied the unguided rockets.

This is the first contract for which the Rocket Motors Division of BAE Systems, RO Defence has been selected as prime contractor.

By August 1997, qualification trials of the RRPR had been completed as had the first production batch 'fly to buy' trials. It entered service with the Royal Artillery on completion of the contract in December 1997.

The first export customer for the RRPR was Norway with rocket and pod container deliveries completed in December 1997, acceptance and dynamic firings in February 1998 and in-service date of April 1998.

This contract covered the supply of 1,500 training rockets and six rocket pod containers plus training.

The introduction of the RRPR allows the Royal Artillery to carry out effective and realistic training on much smaller ranges than using the full-calibre rockets, but use the same operational drills from field reloading of the Rocket Pack Container (RPC) through to actual firing.

Description

The RRPR uses a subcalibre rocket fired from a reusable RPC which has a target service life of 10 years and interfaces, both mechanically and electrically, with the MLRS Self Propelled Loader Launcher (SPLL).

In this way training with the RRPR is identical to operating the full-calibre MLRS rockets. The key differences being reduced range, minimum environmental impact and significantly reduced training costs.

The RPC has a forward-opening cassette which facilitates loading/unloading of rockets and provides good access for maintenance. The launcher tubes are attached to the opening cassette with the igniter cables routeing from the launcher tube latching mechanism directly to the standard umbilical connector.

The FIS provides the appropriate responses to the launcher fire-control system when programming the fuze event time. Connection between the vehicle fire control and the RRPR RPC is retained via the vehicle umbilical connector.

Rear view of Multiple Launch Rocket System with BAE Systems, RO Defence Reduced Range Practice Rockets in Rocket Pod Container

BAE Systems, RO Defence Reduced Range Practice Rocket being launched from Multiple Launch Rocket System

0021082

The subcalibre rocket uses an extruded double-base propellant rocket motor instead of the HTPB composite motor as used for MLRS. The lower thrust of the subcalibre rocket combined with the acid free exhaust gas virtually eliminates blast damage to the launcher and obviates the need to decontaminate the launcher to neutralise the acidic corrosive film deposited from the MLRS propellant combustion.

According to BAE Systems, RO Defence, the life cycle costs of the launcher are therefore significantly reduced compared to training with full-calibre MLRS training rockets.

The range of the subcalibre rocket is less than 10 km and there are three ballistically matched warheads:

(1) Indicating pyrotechnic which gives a strong signal in the visible and IR spectrums
(2) High-explosive 1 kg RDX/TNT with an impact fuze
(3) Fully inert.

The actual preference for the warhead type is a matter of customer choice, taking into account the training process and available facilities and range safety considerations.

The comparison of ranges and determination of the fall of shot is achieved through the Range Predictive Software, run on an MLRS Aim Verification Device (MAVD) using a laptop computer.

The software relies upon the comparison of ranges through a common Quadrant Elevation (QE) which, when using the prevailing meteorological data, results in predictable impact points for either subcalibre or full-calibre rockets.

The software permits entry of the MLRS target data, as would normally be provided to the crew, or entry of the required RRPR impact point, which may be the optimum range available at the firing facility being used.

Assuming, for the purpose of this description, that for range safety the fall of RRPR shot must be ensured, then this will be the starting point. The MAVD will have been loaded with the same meteorological message as that given to the MLRS crew for loading into the MLRS fire-control system.

The MAVD calculates the appropriate RRPR Qe and Az, using the RRPR target data, launcher location and meteorological messages and also generates the MLRS full-calibre system target grid co-ordinates.

These latter target co-ordinates are those issued to the MLRS crew as MLRS targets. The MLRS crew can now enter the target co-ordinates into the MLRS fire-control computer and the respective Qe and Az calculated. The Qe and Az can then be compared and if within acceptable limits, for example no operator error has occured, the mission can proceed. Clearly, the comparison process is an additional range safety feature facilitating multi-use on crowded ranges.

The fielding of the RRPR offers the user a number of benefits. Realistic training can be carried out at significantly lower costs than would be possible using full-calibre rockets. Both the training ammunition and the reduced life cycle costs of the launcher contribute to lower training costs.

Additional savings result from the reduced range requirements resulting from the shorter range of the RRPR and also from the reduced safety traces resulting from inert or indicating warheads.

With the ability through the MAVD to double check the computed azimuth and elevation, not only is range safety enhanced, but it also opens up the possibility for co-operational use of ranges.

When compared to the full-calibre MLRS rocket, the RRPR generates minimum pollution and this, according to the BAE Systems, RO Defence, Rocket Motors Division, facilitates the use of the most environmentally sensitive ranges not possible with the full-calibre approach.

Status

Production as required. In service with Norway and the UK.

Manufacturer

BAE Systems, RO Defence, Rocket Motors Division.

UPDATED

UNITED STATES OF AMERICA

Lockheed Martin Missiles and Fire Control - Dallas 227 mm Multiple Launch Rocket System (MLRS)

Development

Early in 1976, the United States Army Missile Command at Redstone Arsenal, Alabama, initiated feasibility and concept formulation of a General Support Rocket System using a low-cost rocket that would be handled like a round of conventional ammunition and have a high rate of fire. This system would be used against troops and light equipment, air defence systems and command centres.

In March 1976, five companies, Boeing, Emerson Electric, Martin Marietta, Northrop and Vought were awarded contracts for the concept definition of the General Support Rocket System (GSRS).

After evaluating all five concept definition studies, in September 1977, Boeing Aerospace and the Vought Corporation (now Lockheed Martin Missiles and Fire Control - Dallas) were each awarded contracts for the 29-month validation phase of the GSRS competition.

Under the terms of the contract each company delivered three prototype Self-Propelled Launcher Loaders (SPLLs) and rockets for competitive testing at White Sands Missile Range, New Mexico.

Early in 1978, the US Army Missile Research and Development Command redirected the GSRS towards a standard NATO weapon that could be developed and produced in both the USA and Europe. The programme was renamed Multiple Launch Rocket System or MLRS.

In 1980, the then Ling Tempco Vought of Dallas, Texas (now Lockheed Martin Missiles and Fire Control - Dallas) was selected as prime US contractor for MLRS and was awarded a full-scale development contract. Teamed with the company were the Atlantic Research Corporation for the solid-propellant rocket motor, Bendix Guidance Systems Division for the stabilisation reference package/position determining system, the Brunswick Corporation for launch tubes, Norden Systems for the fire-control system and Vickers for the launcher drive system.

European participation was introduced by an MoU between the governments of the USA, Germany, the UK, France and Italy, allowing for the joint development, production and subsequent deployment of a multiple-launch, free flight rocket system.

In 1983, the first MLRS units entered service with the US Army. In the same year MLRS-Europaische Produktions Gesellschaft GmbH (MLRS-EPG), was established in Munich to undertake European production. Diehl of Germany, Hunting Engineering of the UK, Aerospatiale of France and SNIA BPD of Italy were invited to form this Prime Contract Consortium and sharing of manufacture was established in accordance with national offtake ratios.

The main European subcontractors to EPG were Aerospatiale for launcher integration, BPD for the rocket motor, Diehl for rocket integration and Thyssen Henschel (which has now become Rheinmetall Landsysteme) for the carrier vehicle.

The first European-built MLRS systems were handed over late in 1989 and production peaked at 9 to 10 launchers per month. The European MLRS programme involved the manufacture of:

284 launchers
128,000 M77 rockets and practice rockets
73,000 AT2 rockets.

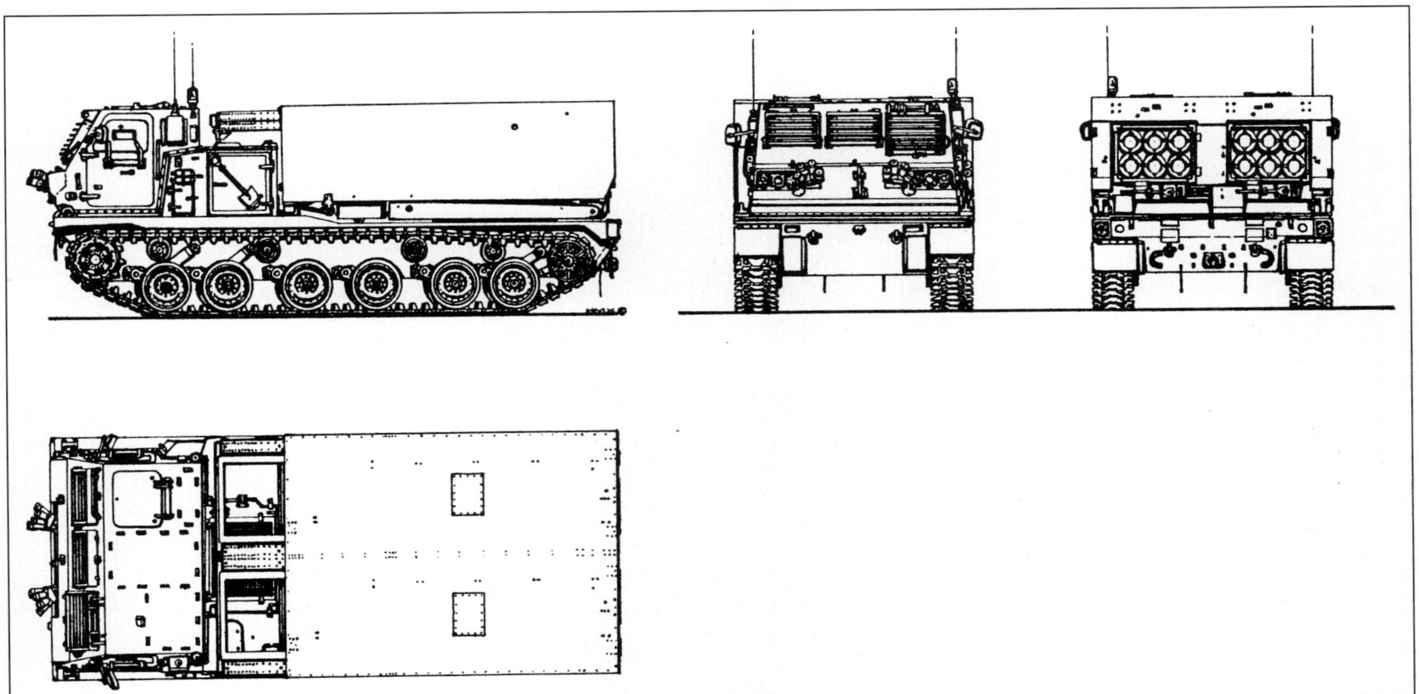

Four-view drawing of British Army M270 Armored Vehicle-Mounted Rocket Launcher (AVMRL) (Henry Morshead)

During Operation Desert Storm the USA deployed over 230 MLRS launchers while the British Army deployed 16 launchers.

Description

The development of the MLRS was planned in three distinct phases. Phase I covered the vehicle, launcher and an M26 anti-matériel/anti-personnel warhead containing M77 submunitions. Phase II addressed the AT2 scatterable anti-tank mine warhead. Phase III was to be the development of a Terminally Guided anti-armour Warhead (TGW). Development of this was eventually cancelled.

Phase I - M270 launcher

The lightly armoured M270 self-propelled launcher is a stretched version of the United Defense LP M2 Bradley infantry fighting vehicle. The Launcher Loader Module, mounted on the rear of the vehicle hull consists of a base, turret and cage. It contains the computerised fire-control system, a stabilisation reference package/ position determining system, a launcher drive system and a twin boom crane unit for self-loading and unloading.

The three-man crew consists of the driver, gunner and section chief, all of whom are seated in the fully enclosed cab at the front of the vehicle. This is protected against small arms fire and shell fragments by aluminium armour and louvred windows. The US Army has fitted some of its M270 launchers with a roof mounted 40 mm Mk 19 series automatic grenade launcher.

An overpressure ventilation system prevents rocket fumes from entering the cab and the vehicle is also equipped with a filtration unit to protect the crew from chemical, biological and radiological particles.

Equipment capabilities permit a reduced crew, or even one person, to accomplish a complete fire mission including the loading and unloading operations.

Upgraded M270A1 launcher

Since 1983 the US Army has taken delivery of over 850 227 mm (12-round) Multiple Launch Rocket Systems from the prime contractor, Lockheed Martin Missiles and Fire Control - Dallas.

The lightweight armoured M270 self-propelled launcher is manufactured by United Defence LP which then sends it to Lockheed Martin Missiles and Fire Control - Dallas for systems integration and final delivery to the customer.

The US Army has decided to upgrade these launchers to a new and enhanced configuration called the M270A1 with the first batch consisting of 35 units.

Of these, United Defense has remanufactured 18 launchers while the US Army Red River Army Depot has remanufactured the remaining 17.

In mid-1998, Lockheed Martin Missiles and Fire Control - Dallas was awarded another contract worth US$63 million to upgrade an additional 21 MLRS systems to the M270A1 standard with deliveries starting in April 2000.

As well as generally remanufacturing the launcher so that it is returned to the US Army in an almost new condition, the launcher is also being upgraded with the Improved Fire-Control System/Improved Launcher Mechanical System (IFCS/ILMS) which offers a number of significant advantages to the user including quicker into action times, faster reloading and reduced operating and support costs.

The IFCS involves a new fire-control panel with a mass storage device and the addition of a GPS augmented Positioning Determining/Navigation Unit (PDNU).

The ILMS upgrade involves electronic and hydraulic system modifications to decrease system response time. With current software improvements to the response time, this has been reduced from about 5 minutes to 2.5 minutes but with the ILMS the figure will be further reduced to 1.5 minutes.

Under the US Army programme, the M270A1 is a recapitalisation of the M270 legacy launcher designed to meet the US Army vision. Unlike the older M270 launcher, the M270A1 is capable of launching the Guided MLRS (GMLRS) rocket and Army TACMS (ATACMS) Block IA munitions variants.

As previously stated, MLRS systems for the French, German, Italian and UK armies were manufactured in Europe under the leadership of MLRS-EPG.

In May 1999, the M270A1 launcher successfully fired an ATACMS Block I missile and in November it conducted a successful launching of a rocket with six AT2 munitions.

New European FCS

The four members of the European Multiple Launch Rocket System (MLRS) consortium, France, Germany, Italy and the United Kingdom, are considering a programme to replace the current US designed computerised fire-control system installed in their European built 227 mm (12-round) MLRS systems.

The current system was designed in the US but built under licence in Europe under a build to print contract awarded to the then Marconi Command and Control Systems of the UK.

The components of the current European-built FCS are now becoming obsolete and the US Army is already upgrading its (now) Lockheed Martin Missiles and Fire Control - Dallas MLRS with a new fire-control system.

Europe has decided to develop and place in production its own fire-control system with the French Service des Programmes d'armement Terrestre (SPART) acting as the executive agency for the four European countries.

Expressions of interest by European companies with a proven record of FCS expertise were submitted to

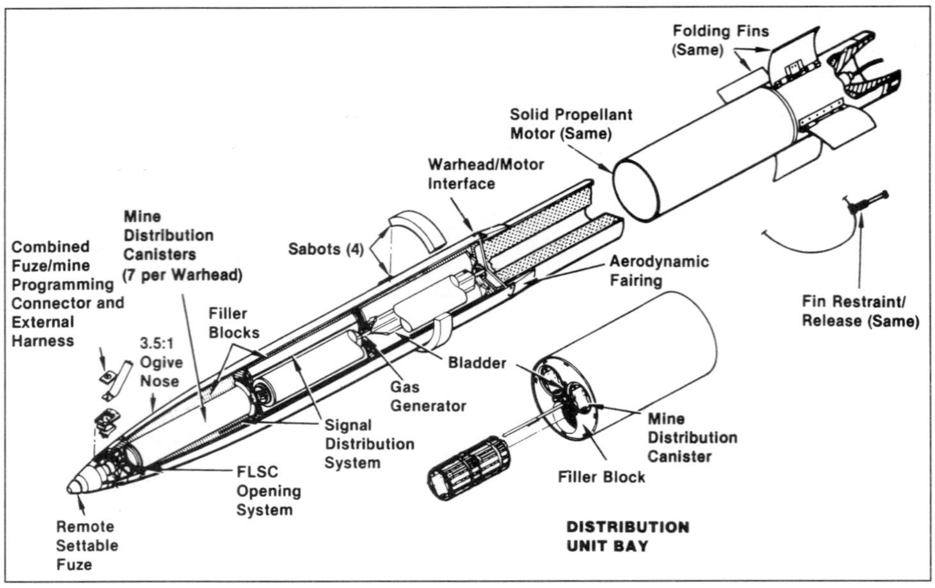

Major components of AT2 rocket (remote settable fuze is electronic time type)

US Army M270 AVMRL with launcher elevated and standard rocket being launched 0021084

M270 AVMRL with full-scale mockup of Army TACMS missile and standard rocket pod

M77 DPICM submunition, 644 of which are carried by each Phase I rocket

SPART in late November 1998. It is expected that a contract will be awarded to build a demonstrator FCS followed by production.

The new FCS will enable the MLRS to fire the existing unguided MLRS rockets as well as the new generation of longer range rockets now under development, including the guided rocket.

Euro Rocket

In June 1999, Lockheed Martin Missiles and Fire Control – Dallas and the German company Diehl, established a new joint venture company called Euro Rocket System GmbH.

The aim of this company is to provide European customers of the existing US-designed MLRS system with upgrades and new rockets to enable them to be effective well into the 21st Century.

MLRS rocket pod

The standard rocket pod, which is also the transportation and storage container, weighs 2,308 kg when loaded. It comprises six glass fibre tubes held rigidly in an aluminium case structure. Blow-off covers seal each rocket in its tube, so the rocket requires no maintenance, inspection or crew servicing from the factory to the launch site.

The rocket pods are loaded by the integral twin boom loader system of the launcher.

M26 warhead

The Phase I rocket has a minimum 20 year shelf-life, weighs 306 kg, is 3.937 m long and has a diameter of 0.227 m. The rocket is centred in the launch tube by a set of four forward sabots, two aft riders and two spin lugs that ride along helical rails. When the rocket is discharged, the sabots are discarded and four folding fins at the rear of the rocket motor are released by a cable-release initiated pyrotechnic delay device, attached to the launch tube. The warhead weighs 156 kg, has an aluminium skin and carries 644 M77 Dual-Purpose Improved Conventional Munition (DPICM) shaped-charge blast fragmentation bomblets weighing 213 g each, nestled with eight polyurethane foam supports. Armour penetration is 76 to 102 mm. An M445 remotely set electronic time fuze in the warhead initiates dispersal of the ribbon-stabilised bomblets in mid-air over the target. The range of this version is 31,600 m. At mid-range each warhead of a salvo will saturate a 200 × 100 m sized target. The actual ripple pattern size varies with the range and aimpoints plotted within the target zone. The Phase I rocket was used operationally with great success during Operation Desert Storm in 1991. The main drawback of this warhead is the submunitions are not fitted with a self-destruct mechanism.

German Army M270 AVMRL **2001**/0105901

Lockheed Martin Vought Systems Extended-Range Rocket (XR-M77)

As a result of the Iraqi use of superior ranged artillery, Lockheed Martin Missiles and Fire Control – Dallas began development work on an extended-range rocket for the MLRS in February 1991. The first prototype was tested in November 1991 and attained a range of 45,000 m. In April 1992, the US Army Missile Command announced its intention to buy the weapon from Lockheed Martin Missiles and Fire Control – Dallas. It is estimated that the development costs will be between US$30 million and US$33 million over three to four years. The warhead is shortened and contains only 518 bomblets instead of the standard 644. However, the reduced payload is compensated for by an increase in accuracy and a 274 mm stretch in rocket motor length to increase the range to 45,500 m. The extended-range rocket uses the current rocket pod container-launcher. A high-altitude fuze is fitted. Firing trials have also taken place at the French Centre D'Essais de la Mediterranée (CEM) range. The overall rocket weight is reduced slightly to 296 kg. The first export customer for this was Greece who ordered them in July 1999.

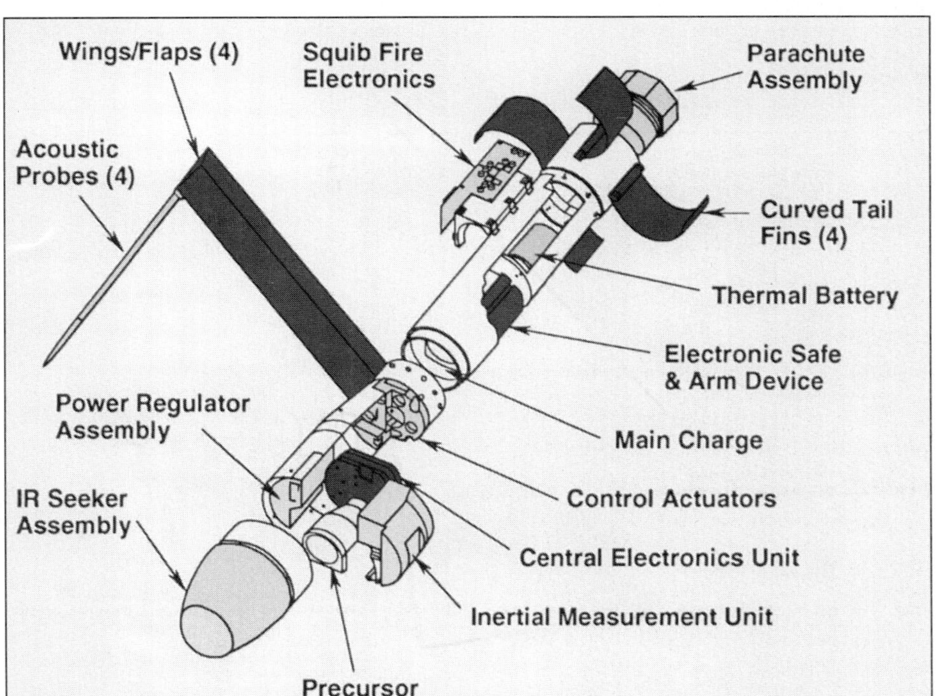

Main components of the Northrop Grumman Brilliant Submunition (Base BAT)

Guided MLRS

The five members (France, Germany, Italy, UK and USA) which signed the MLRS MoU, and have formulated their requirements for a new guided MLRS (GMLRS) which will have a greater range and accuracy.

In November 1998, Lockheed Martin Missiles and Fire Control – Dallas was awarded a 48-month Engineering and Manufacturing Development (EMD) contract worth US$121 million for the development of the Guided MLRS (GMLRS) rocket.

The GMLRS is an international co-operative programme with France, Germany, Italy, UK and the USA who are working together throughout the EMD phase of the rocket's life cycle.

The GMLRS rocket will include an Inertial Measurement Unit (IMU) aided by a Global Positioning System (GPS) integrated on a GMLRS rocket body. In addition, canards on the GMLRS nose will add basic manoeuvrability to further enhance the accuracy of the system.

Under the terms of this contract, the four European countries will contribute 12.5 percent each to the EMD phase while the US will contribute 50 percent.

In its basic form, the GMLRS is expected to have a warhead containing 416 to 462 M85 bomblets or Dual-Purpose Improved Conventional Munitions (DPICM), each of which will have self-destruct fuze.

The GMLRS will be 3.937 m long, have a diameter of 227 mm and weigh 302 kg.

If the programme runs on schedule then low rate initial production will commence early in FY02 and full rate production running through to FY12.

The European companies involved in the programme are Diehl, MBDA and Fiat Avio.

Under the US programme, the GMLRS rocket is a development effort to replace the M26 free flight tactical MLRS rocket and is designed to meet the US Army vision. It will increase accuracy and range while reducing the overall logistic burden to the force. It also provides MLRS with a guided rocket bus capability for the delivery of various future payloads. It can be launched from the upgraded M270A1 MLRS launcher and the HIMARS (6 × 6) wheeled variant of the MLRS.

GMLRS – unitary warhead

Late in 2000 it was stated that the US would develop a unitary warhead variant of the GMLRS to provide commanders with a precision strike capability in urban and complex terrain.

It is expected that this will have an improved guidance system to allow for precision strikes. It will incorporate a new differential global positioning system with an antenna that has greater resistance to jamming.

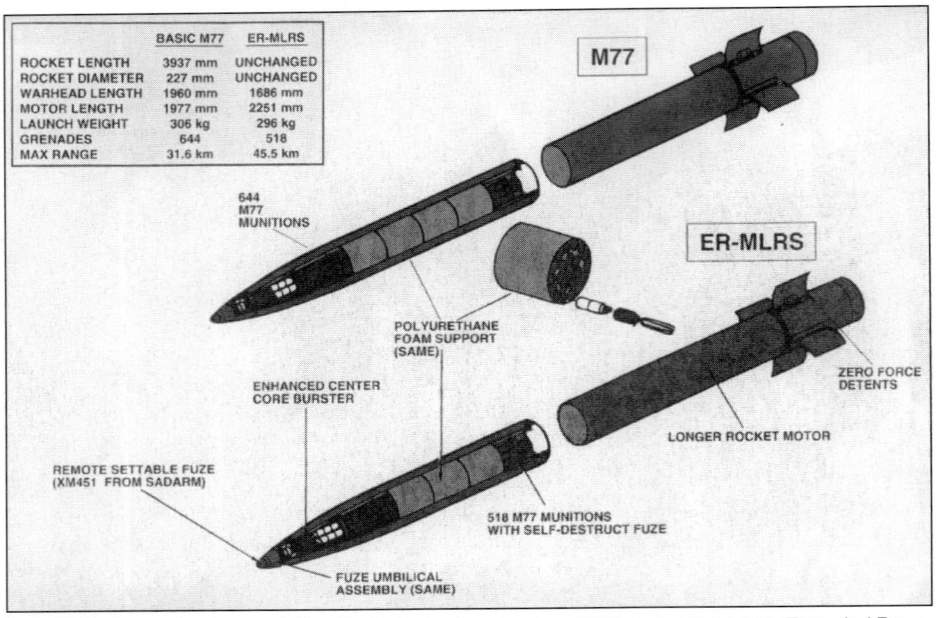

Lockheed Martin Missiles and Fire Control - Dallas standard M77 rocket (top) with Extended-Range Rocket (below) 0021089

These upgraded systems will feed mid-course updates to the separated warhead, allowing its four canards to make steering corrections that will improve the CEP from 10 m on the standard GMLRS to about 2 m on the unitary warhead variant.

In addition, the unitary warhead will also have the capability to selectively create an air burst, point detonation or delayed detonation for penetrating buildings and bunkers protected by 0.90 to 2.4 m of concrete. Range would be 70 km.

US Army Extended-Range Guided (ER-G) technology demonstration for MLRS

In late 1989- early 1999, the US Army Aviation and Missile Command's Research, Development and Engineering Centre, with industrial support, successfully demonstrated the ER-G technology for the GMLRS rocket. A low-cost GPS inertial guidance and control system was integrated into the extended range MLRS to allow ranges of between 15 to 60-70 km. The new guidance and control package consists of a GPS aided inertial measurement unit, a flight computer and canards housed in the nose section of the rocket. The payload is approximately 450 XM85 anti-personnel anti-light armour submunitions.

Guidance and control package for MLRS rockets

Lockheed Martin Missiles and Fire Control - Dallas has been awarded a US$7 million contract to design and test a low-cost guidance and control package that will be compatible with the basic MLRS rocket, extended-range rocket and future smart munition variants. The guidance package will occupy about 5 to 8 per cent of an extended-range rocket payload area and will further increase the accuracy of the MLRS rounds. See also the ER-G Technology Demonstration for MLRS in variants.

The programme is a joint advanced technology demonstration with the contracting agency, the US Army Missile Command. It is expected that the addition of the package to an extended-range rocket will provide a 3 mil accuracy throughout its range capabilities.

Multiple Launch Rocket System (MLRS) firing ATACMS missile during trials in the United States
2002/0130026

Phase II - AT2 warhead

The 1.959 m long AT2 anti-tank mine was developed by Germany and has been in service in the ground-scatterable version for some years. The mine warhead is 236.6 mm in diameter, weighs 108 kg and contains 28 AT2 DM1399 anti-tank mines. Due to its lighter 258 kg weight, the AT2 rocket has a range of 40 km, somewhat longer than the M77 rocket with the M26 warhead. The warhead is activated by a DM42 adjustable timer electronic fuze and is opened in flight by a small explosive charge. The seven mine dispensers, each with four mines, are pneumatically ejected and then retarded to release the mines which are parachute-stabilised in flight. The dispensers measure 105.7 mm in diametre, 548 mm in length and weigh 10.8 KG.

The 3.935 m long rocket usually dispenses the mines when overflying the target area at about 1,200 m. The 128.2 mm high AT2 anti-tank mine has a diameter of 103.5 mm and a deployed height of 165 mm with its ground orientation springs open it weighs 2.25 kg. Detonation of the shaped-charge full-width attack mine occurs either when an armoured vehicle passes over it or when it reaches 97 per cent of its useful lifetime, determined before launch from one of six available programmable settings. Armour penetration is in excess of 140 mm. An anti-handling/removal device is incorporated to disrupt mineclearing operations. The two launcher loads of AT2 rockets can create a minefield of 2,000 × 115 m. This has only been ordered by Germany and the UK, with production undertaken from 1993 to 1995.

Variants

US Army Tactical Missile System (Army TACMS)

Late in March 1986, the US Army announced that the now Lockheed Martin Missiles and Fire Control - Dallas had been selected to design and develop the Army Tactical Missile System (Army TACMS or ATACMS) and was subsequently awarded a US$180.3 million contract.

Army TACMS is fired from the MLRS M270 launcher using off-axis guidance techniques to prevent enemy radars from plotting the trajectory and initiating counterbattery fire against its launch point.

Each Block 1 missile is approximately 3.962 m long, 0.61 m in diameter and is powered by an advanced solid-propellant fuel to a range of over 165 km. Each 591 kg warhead contains 950 M74 anti-personnel/anti-materiel bomblets which weigh 0.59 kg each, are 0.059 m in diameter and are fitted with a tungsten

Specifications

Crew: 3
Weight:
(system, loaded) 25,191 kg
(system, empty) 20,189 kg
Length: (travelling) 6.972 m
Width: (travelling) 2.972 m
Height:
(travelling) 2.617 m
(elevated) 5.925 m
Ground clearance: 0.43 m
Track width: 533 mm
Length of track on ground: 4.33 m
Max speed: 64 km/h
Range: 483 km
Fuel capacity: 617 litres
Fording: 1.1 m
Gradient: 60%
Side slope: 40%
Vertical obstacle: 1 m
Trench: 2.29 m
Engine: Cummins VTA-903 turbocharged 8-cylinder diesel developing 500 hp at 2,400 rpm
Transmission: General Dynamics Land Systems HMPT-500 hydromechanical
Steering: hydrostatic
Suspension: torsion-bar-in-tube
Electrical system: 24 V

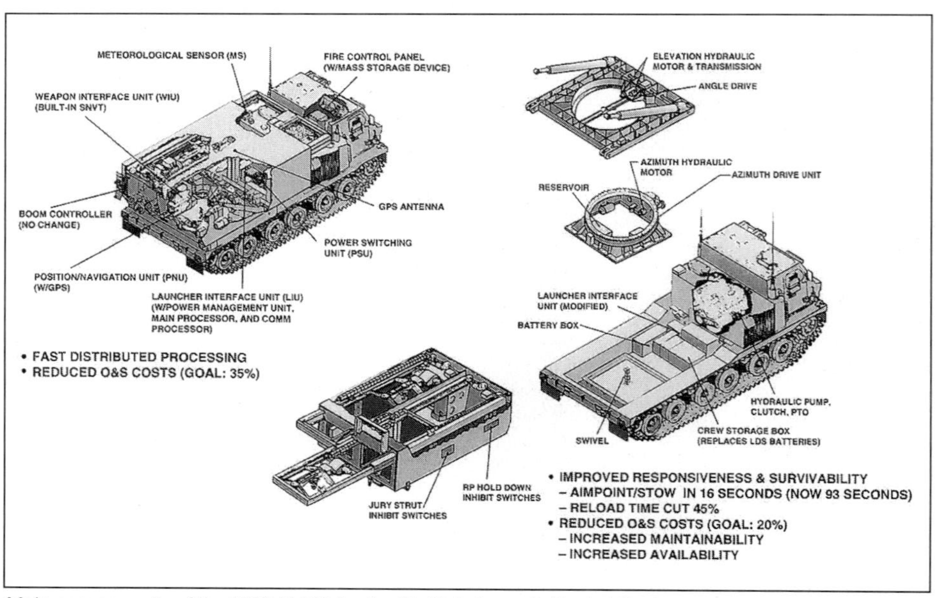

Main components of the IFCS/ILMS for the Multiple Launch Rocket System 0008322

fragmenting wall with steel overlay. They are dispersed over the target area to detonate upon impact.

Teamed with Lockheed Martin Missiles and Fire Control - Dallas are Atlantic Research Corporation for the solid rocket fuel propulsion and Honeywell Incorporated for the onboard inertial guidance system.

A follow-on warhead, currently under development would include the precision-guidance BAT sub-missiles for use against moving targets.

The first Army TACMS missile was fired in 1988 and the same year Lockheed Martin Missiles and Fire Control - Dallas was awarded an initial contract for 66 production missiles worth US$44 million. Early in 1990, the company was awarded a US$59.6 million fixed price contract for low-rate initial production of 104 missiles was delivered by late 1991. In December 1991, a US$114 million contract was awarded to Lockheed Martin Vought Systems for full-scale production. The first year procurement was for 300 missiles and launching assemblies. Further contracts have since been placed. First missiles were deployed to support Operation Desert Storm. A total of 32 out of 105 TACMS rounds taken were successfully used by the US Army in the 1991 Gulf War. Production of the Block I has been completed.

The US Army also awarded Lockheed Martin Missiles and Fire Control - Dallas a US$83 million contract to integrate Army TACMS into the existing M270 AVMRL. The changes are principally in the software of the system's fire-control computer and will allow the system to fire standard MLRS rockets or Army TACMS missiles.

Army TACMS allows battlefield commanders to strike well behind enemy lines, beyond the range of existing artillery, rockets or missiles.

In early 1994, Lockheed Martin Missiles and Fire Control - Dallas was awarded an US$18.7 million contract to design and test a long-range variant of ATACMS for the US Navy. Following this, the US Army approved plans for the development of the Block IA extended-range ATACMS with considerably improved accuracy. The Block IA round is the same size as the current weapon but has twice its range. The same solid fuel rocket motor is used but the range increase is due to a reduction in payload to 275 submunitions. An in-flight receiver for GPS satellite updates is fitted so that the missile ring laser gyro guidance system can make flight corrections. The first Block IA flight test was successfully performed in January 1995.

The US Army has ordered 573 Block IA missiles for delivery through to February 2003. Minimum range is 70 km and maximum range is 300 km.

Greece (71), South Korea and Turkey (72) have placed orders for ATACMS rockets. Greece has since placed additional orders for ATACMS with the last one being in July 1999 worth US$39 million for ATACMS and ER rockets. Late in 2001, Korea placed a contract for 111 ATACMS Block IA missiles, with deliveries to be completed by April 2004.

The latest ATACMS is the Block II and, in June 1999, the company was awarded a Low Rate Initial Production contract from the US Army for a total of 24 missiles at a cost of US$138.5 million. This has a range of more than 128 km and will carry 13 BAT submunitions (qv).

The US Army is also studying a Block IVA version with a unitary HE warhead.

US Army ATACMS Brilliant Submunition

Since 1984, the Northrop Grumman Corporation has been working on a US Army R & D programme to develop the Base BAT dual-sensor (acoustic and infra-red) brilliant submunition for use in the ATACMS Block II warhead.

The BAT programme was revealed in 1991 when the US Army awarded a US$344 million contract to the Northrop Grumman Corporation Electronics and Systems Integration Division to complete its development. Prototype fabrication go-ahead was given by the US Army in 1992. Tactical design flight tests began in June 1993 and met major performance test objectives. The EMD flight tests continued through March 1997 with Base BAT entering its Engineering and Manufacturing Development (EMD) phase later in the year.

The first phase is also referred to as Block II and will carry 13 BAT or P3I BAT with a maximum range of 140 km and a minimum range of 35 km.

In January 1995, the US Army awarded a US$72.4 million contract to Northrop Grumman for a preplanned production improvement programme (P3I) for BAT. This involves advanced seeker technologies that increase BAT's effectiveness against surface-to-surface missile launchers, such as the Scud, and other stationary vehicles. As system integrator, Northrop Grumman will manage the development of the dual-mode (millimetric-wave high-frequency radar and imaging infra-red high-resolution infra-red) seekers from Alliant Techsystems Inc and Westinghouse Electronics over the next three-and-a-half years. Northrop Grumman will then select one contractor to proceed to the engineering, manufacturing and development phase.

In July 1995, the US Army awarded the now Lockheed Martin Missiles and Fire Control - Dallas a US$180.2 million contract to modify the ATACMS to carry the BAT. The contract covers a US$155.2 million effort over four years with two options totalling US$24.98 million to extend the work to the year 2000. Block IIA will carry P3I BAT and have a minimum range of 100 km and a maximum range of 300 km. The P^{3}I BAT will be introduced to both variants at a later date.

In mid-1993, it was revealed that Lockheed Martin Missiles and Fire Control - Dallas is studying an MLRS-BAT round. This is part of a US Army programme called MSTAR (MLRS Smart Tactical Artillery Rocket) programme. This programme is a guided MLRS with either BAT P3I or SADARM PI on a smart rocket. The MSTAR programme was terminated in 2000.

The unpowered 20 kg Base BAT submunition is 914.4 mm long, 139.7 mm in diameter and is fitted with four foldable wings that each have a protruding acoustic sensor spike attached. A nose-mounted Raytheon Company infra-red sensor is also fitted together with a shaped-charge warhead. The four flip-out wraparound tail fins are used to stabilise the weapon in flight.

The projected Block III will be an earth penetrator but is not expected to enter production until FY09.

Resupply

The resupply vehicles used by the US Army are a Oshkosh Heavy Expanded Mobility Tactical Truck (HEMTT) (8 × 8), which carries four fully loaded rocket pods and its Heavy Expanded Mobility Ammunition Trailer (HEMAT), which carries a further four fully loaded rocket pods. The truck is air-transportable by C-130 or C-141 aircraft.

The European armies selected their own national resupply systems. The UK, for example uses its Demountable Rack Offloading and Pick-up System (DROPS). This is also a truck-trailer system consisting of the Leyland Medium Mobility Load Carrier (MMLC) and a trailer, both of which carry four rocket pods. Germany will utilise its MULTI system which is very similar to DROPS.

High-Mobility Artillery Rocket System (HIMARS)

Details of this system, based on a 6 × 6 truck chassis and still at the prototype stage, are given in the following entry.

Multiple Launch Rocket System (MLRS) of the Norwegian Army with one pod of six rockets and launcher traversed to left (Svein Wiiger Olsen)
2002/0130019

M28 practice rocket

The practice warhead was used as a training device for personnel. It uses steel pipe in place of the tactical munitions. Additionally, three smoke canisters are located in bay number 4. These canisters are made from plexiglass and contain 100 mg of titanium tetrachloride ($TiCl_4$).

On detonation, the canisters are broken by the central core burster and the $TiCl_4$ reacts with moisture in the air to produce smoke. This cloud is visible for a minimum of 10 seconds from 6,000 m. The M28 has been replaced by the M28A1 rocket.

Lockheed Martin Missiles and Fire Control - Dallas M28A1 Reduced Range Practice Rocket (RRPR)

This value engineered replacement for the M28 went into production in 1993. A blunt nose limits range to between 7.5 and 14.3 km so allowing its use on short artillery ranges. The M28A1 also deletes the time fuze and centre core burster in preference for a point detonating fuze and smoke charge on impact. The round uses a standard tactical rocket motor for training realism and launcher validation. Like the tactical rocket, no maintenance is required.

BAE Systems, RO Defence RRPR

Details of the now BAE Systems, RO Defence Reduced Range Practice Rocket, now in service with Norway and the UK are given under the UK.

Israel to upgrade MLRS

Late in 1998, Israel Military Industries awarded the now Lockheed Martin Missiles and Fire Control - Dallas of the US a contract worth US$60.2 million for development and production of a portion of the Multiple Launch Rocket System (MLRS) to be integrated with the Israel Military Industries-developed Trajectory Correction System (TCS).

Under the terms of this contract, Lockheed Martin Missiles and Fire Control - Dallas, prime contractor for the MLRS, will build the rockets and support flight testing in Israel during the development stage. The production phase runs concurrently.

Most of the work in the US will be carried out at the Lockheed Martin Missiles and Fire Control - Dallas headquarters in Dallas and at the MLRS rocket production facility at Camden, Arkansas.

The installation of the TCS onto the currently unguided MLRS rocket, provides in-flight control of the rocket's flight path and gives a much greater level of accuracy. This in turn means that fewer rockets are required to neutralise a given target.

Overall prime contractor for the MLRS TCS upgrade is Israel Military Industries who were recently awarded a US$170 million contract by the Israeli MoD. In addition to Lockheed Martin Missiles and Fire Control - Dallas there are a number of other sub-contractors including Elisra Electronic Systems in Israel.

The TCS is an add-on upgrading kit and can also be fitted to the Israeli Military Industries LAR 160 rocket which has a range of about 45 km and the MAR350 rocket with a range of about 80 km.

Dynamit Nobel subcalibre training system

This training system, developed by Dynamit Nobel to meet the requirements of the German Army, is based on the use of 110 mm Light Artillery Rockets (LAR) which have been in service with the German Army for many years and are covered in this section under Germany. These 110 m systems have almost been phased out of service.

These rockets meet the requirements concerning danger zones and low unit cost. They have a maximum range of 14,000 m and a minimum range of 6,000 m.

The use of the Dynamit Nobel subcalibre training system allows all MLRS training activities to be carried out including all tasks from the preparation of loaded rocket pods (RP) to rocket launching itself under realistic conditions even with the following constraints:

(1) Limitations/restrictions on national ranges (area, danger zones)
(2) Minimising the cost of practice fire missions.

Sensor pod on MLRS

Under contract to the German MoD, Krauss-Maffei Wegmann has fitted an MLRS carrier chassis with a mast-mounted sensor pod that can be elevated to 11 m.

Status

In production and service.

Bahrain – initial order for seven launchers increased to nine. All delivered in 1992.

Denmark – order for 8 launchers, 300 ER rocket pods, 25 practice rocket pods and spares placed. Total cost US$146 million. The first four MLRS launchers were delivered to Denmark in 1999.

Egypt - late in 2001, Egypt placed a contract for the supply of 26 MLRS launchers under the Foreign Military Sales (FMS) programme. Egypt also placed a contract worth US$72.2 million for the supply of more than 400 Extended Range MLRS rocket pods. These will be delivered to Egypt between late 2002 and 2003.

France – first launcher from USA with remainder built in Europe. Total of 55 launchers with final deliveries made in 1995. Two regiments, 12th and 74th, formed with 24 MLRS each. The 74th forms part of France's Rapid Reaction Force.

Germany – six launchers from the USA with remainder built in Europe. Total of 150 launchers required. First unit formed in 1991 with eight battalions, the 12th, 42nd, 52nd, 62nd, 122nd, 150th, 702nd, and 802nd, fielded. Batteries have nine launchers in platoons of four and five launchers respectively. Two battalions form part of Germany's Rapid Reaction Force.

Greece – order for nine launchers and 132 reduced range practice rockets placed in 1994. Greece took delivery of its first nine MLRS launchers in 1995 followed by 14 in 1999 and 18 in 2000. Greece has also ordered standard rockets, ER rockets and three lots of ATACMS.

Israel – initial order for six launchers, 726 tactical and 720 reduced range practice rockets placed in 1994 for delivery by December 1996. A further US$103.5 million package for 42 launchers and 1,500 rockets was placed in late 1995. According to the United Nations MLRS deliveries to Irael were as follows:

Year	Quantity
1995	6
1996	6
1997	16
1998	20
1999	16

Italy – one launcher from the USA with remainder built in Europe. A total of 24 launchers acquired. Single regiment, 3rd Volturno, equipped with 20 launchers in service.

Japan – fourth of four annual contracts each for nine launchers, with associated rockets and support equipment, signed in 1995 with Lockheed Martin Missiles and Fire Control – Dallas. The first contract was signed in late 1992. The first three launchers were shipped in May 1994, followed by three more in June and the final three of the first contract in August 1994. Upon delivery the now IHI AeroSpace and the official systems integrator for MLRS in Japan, makes a number of modifications primarily to the fire-control system before the launchers are handed over to the Japanese Ground Self-Defence Force. The last nine launchers have now been delivered. The initial order included 1,300 tactical rockets and 47 launcher trainer pods. By the end of 1999 Japan had taken delivery of a total of 45 MLRS systems. According to the United Nations MLRS deliveries were as follows:

Year	Quantity
1994	9
1995	9
1996	nil
1997	14
1998	12
1999	33
2000	nil

South Korea – As part of a US$624 million arms package from the USA, South Korea is receiving 29 MLRS launchers, 271 MLRS rocket pods, 168 reduced range practice rocket pods and nine MLRS fire-control proficiency trainers. Also included in the sale are 111 ATACMS rounds to counter hostile long-range artillery and rocket systems. 10 MLRS systems were delivered in 1998 and a further 19 in 1999.

Netherlands – 22 launchers and 16,000 rockets delivered by Lockheed Martin Missiles and Fire Control – Dallas. First European Army to deploy MLRS.

Norway – order placed for 12 launchers, 360 ER tactical rocket pods, 564 practice rocket pods and spares. Total cost US$199 million. Six systems were delivered to Norway in 1997 and a further six in 1998.

Turkey – inital nine launchers delivered in 1989 from USA with a further six in 1992. The latter order included more than 2,100 tactical rockets, 144 practice rockets and 12 training pods. Further orders for 24 launchers are expected. Has also ordered ATACMS.

United Kingdom – four from USA with remainder built in Europe. Total of 63 launchers required. Three artillery regiments equipped, the 5th, 32nd and 39th Heavy Regiments RA, with 18 launchers each.

USA – over 800 launchers and 450,000 tactical rockets are in service with the US Army (including national guard units).

Manufacturer and marketing

Lockheed Martin Missiles and Fire Control – Dallas.

UPDATED

Lockheed Martin Missiles and Fire Control – Dallas High-Mobility Artillery Rocket System (HIMARS)

Development

The High-Mobility Artillery Rocket System (HIMARS) was initially developed as a private venture by Lockheed Martin Missiles and Fire Control – Dallas to meet a US Army requirement for a lighter and more mobile version of MLRS that is capable of being carried by the C-130 transport aircraft.

According to Lockheed Martin Missiles and Fire Control – Dallas, the smaller HIMARS will require 30 per cent fewer airlifts to transport a battery.

Crew training is the same as the current MLRS system with the lightweight chassis offering higher road speeds and lower operating costs than its full-tracked counterpart.

The system was first unveiled in the US late in 1993 and was shown for the first time in Europe in September 1994.

Early in 1996, the US Army Missile Command (MICOM) awarded Lockheed Martin Missiles and Fire Control – Dallas a US$23.2 million contract to build four prototypes of the High-Mobility Artillery Rocket System (HIMARS) launcher.

Under the terms of the 53 month contract, the Lockheed Martin Missiles and Fire Control – Dallas has now built four systems which were subjected to a two year extended user evaluation. The fourth system was retained by the company for its testing and evaluation.

The contract awarded in 1996 was under the US Army Rapid Deployment Force Initiative (RFPI) Advanced Concept Technology Demonstration (ACTD) programme. Trials were successful and the US Army could order sufficient HIMARS for use by its light entry forces.

In July 1998 the US Army successfully conducted the first firing of an Army Tactical Missile Systems (ATACMS) from HIMARS.

Three of the four prototype systems built have been delivered to the 18th Airborne Corps Artillery for a two year extended user evaluation while the fourth has been retained by Lockheed Martin Missiles and Fire Control – Dallas for testing and evaluation.

In December 1999, the US Army Aviation & Missile Command awarded Lockheed Martin Missiles and Fire Control – Dallas, a US$65 million 36-month maturation/Engineering and Manufacturing Development (EMD) contract for the High Mobility Artillery Rocket System (HIMARS).

Under the terms of this contract, the company's Dallas, Texas and Camden, Arkansas facilities will build six HIMARS launchers for extensive US Army trials and evaluation before the award of a Low Rate Initial Production (LRIP) contract.

The first of six HIMARS launchers were delivered in the 4th quarter of 2001 with the sixth being delivered in the second quarter of 2002. The US Army has an acquisition objective of 363 HIMARS.

If all goes to plan, LRIP of HIMARS is expected to commence in FY03 with First Unit Equipment (FUE) scheduled for FY05. A milestone III decision for full rate

HIMARS in travelling configuration 0023286

HIMARS deployed in firing position 0023287

Specifications
Crew: 1 + 2
Configuration: 6 × 6
Weight:
(complete system) 13,696.36 kg
(carrier vehicle) 8,272.73 kg
(launcher/loader) 2,914.55 kg
(rocket pod) 2,281.81 kg
(crew) 227.27 kg
Length: 6.94 m
Width: 2.4 m
Height: 3.18 m
Ground clearance: 0.564 m
Wheelbase: 4.1 m
Angle approach/departure: 40/63°
Max speed: 89 km/h
Range: (loaded) 480 km
Max gradient: 60%
Fording:
(with kit) 1.538 m
(without kit) 0.923 m
Engine: Caterpillar 3116 ATAAC 6.6 litres 6-cylinder turbocharged and after-cooled diesel developing 290 hp at 2,600 rpm
Transmission: Allison MD-D7 automatic, 7-speed
Transfer case: Allison single-speed
Steering: power assisted
Suspension: parabolic leaf springs
Tyres: 395 R 20XML
Electrical system: 12/24 V
Alternator: 100 A

High-Mobility Artillery Rocket System (HIMARS) launching standard 227 mm series unguided rocket **2001**/0105898

production is scheduled for the second quarter of FY05.

During an exercise at Fort Bragg, North Carolina, a US Army platoon equipped with HIMARS was delivered to the base by a C-130 and conducted a live fire missions within 10 minutes of arrival.

In mid-2000, it was announced that an HIMARS had successfully fired a six-round ripple of standard MLRS rockets.

Under the US programme, HIMARS supports the US Army vision of a more deployable, lethal, survivable and tactically mobile force. It will launch all MLRS Family of Munitions (MFOM) rockets and Army TACMS (ATACMS) missiles.

The United States Marine Corps has placed an order for two HIMARS systems for trial purposes. Total requirement is 45 systems. When fielded, these would be issued to two battalions of the 14th Marine Regiment.

Description

HIMARS consists of a Stewart & Stevenson Family of Medium Tactical Vehicles (FMTV) 5 tonne (6 × 6) truck chassis, which is in full-scale production and service for the US Army, on the rear of which has been mounted a launcher which can accept a single pod of six MLRS rockets or a single ATACMS missile pod.

The system has the same operating procedures as the upgraded M270A1 launcher including fire-control, electronics and communications sub-units. It also retains the self-loading and autonomous features that have made the MLRS so successful. Although it has a three-man crew it can be operated by just one man.

HIMARS can fire to the same ranges as all the family of MLRS rockets in service and under development.

By late 1994, the HIMARS system had succesfully fired the standard Phase I rocket, the Reduced Range Practice Rocket and the Army Tactical Missile System (ATACMS).

Although the prototype system uses the 6 × 6 version of the FMTV chassis, the HIMARS launcher can be fitted to other 6 × 6 tactical truck chassis types.

The standard cab of the HIMARS provides the crew with no ballistic protection. Late in 2000, Stewart & Stevenson Tactical Vehicle Systems, revealed that they were developing two protected cabs for FMTV, Level 1 and Level 2.

Level 1 has been designed to protect the occupants against both missile blast and foreign object debris (FOD) during firings and to maintain cab overpressure.

For Level 2, Stewart & Stevenson has teamed with O'Gara-Hess and Eisenhardt to further expand FMTV armouring possibilities. This would include increasing the ballistic integrity to protect the occupants against 7.62 mm M80 ball ammunition throughout the cab.

In addition, mine blast protection could also be offered with a combination of special wheels and rims, deflective armour under the cab and special energy absorbing seats.

Utilising the modular systems and bulkhead connections of the HIMARS cab, a fully armoured cab could be installed on existing vehicles in about 4 hours.

Production HIMARS systems would also include an onboard munitions loading and reloading system.

In addition it would also include an Improved Position Determining System (IPDS) to allow operations with the Army Tactical Missile System Block 1A with global positioning update.

Status

Development. Not yet in production or service.

Manufacturer

Lockheed Martin Missiles and Fire Control - Dallas.

UPDATED

YUGOSLAVIA, FEDERAL REPUBLIC

262 mm (12-round) LRSV M-87 'Orkan' multiple rocket launcher system

Development/Description

Late in 1988, at a military exhibition in Iraq, the Iraqi army displayed a 262 mm 12-round multiple rocket launcher system known as the Ababeel 50. Examination of available evidence indicated that the system was Yugoslav in origin and mounted on the rear of the Yugoslav FAP 3235 heavy-duty (8 × 8) cross-country truck.

Since then further information, gained at the Iraqi's First Baghdad International Exhibition for Military Production in 1989, has shown that it is in fact a joint Iraqi-Yugoslav project with Iraq contributing to the development and producing some of the parts locally.

A typical battery of M-87 launchers would consist of:

four 8 × 8 launchers
four 8 × 8 resupply vehicles (each with 24 reload rockets)
one 8 × 8 command post vehicle
two 4 × 4 topographic survey light vehicles
two 4 × 4 observation post light vehicles
one 4 × 4 meteorological survey vehicle.

The command post vehicle is fitted with a German Teldix land navigation system. The resupply and command post vehicles are both on the same 8 × 8 chassis as the launcher vehicle and have a similar tarpaulin system to make identification of the launcher difficult.

Loading of the launcher is performed by a semi-automatic system with the preparations to fire taking 2 minutes. Firing of the rockets can be either singly or in a ripple mode by means of an electronic trigger in the vehicle cab. If needed, firing can also be done up to 20 to 50 m off-vehicle by means of a remote-control box with attached cable. All the traverse and elevation functions are automatically performed via control units with manual systems available as a back-up.

In Bosnian Serb and Yugoslav Army service the launcher is fitted with a 12.7 mm heavy AA machine gun on the cab roof and four 82 mm smoke dischargers in the front bumper.

The five-man crew has two 7.62 mm fully automatic rifles, two 7.62 mm semi-automatic rifles and a 90 mm RBR-90 hand-held anti-tank unguided rocket launcher for self-defence.

The 262 mm unguided rocket, known as the M-87, is 4.656 m long and uses a two-stage solid-propellant rocket motor. The booster stage contains 10 kg of fuel

262 mm (12-round) LRSV M-87 Orkan MRLS in travelling configuration with bows and tarpaulin cover stowed to immediate rear of cab

262 mm (12-round) LRSV M-87 Orkan MRLS deployed in firing position

Specifications
Calibre: 262 mm
Number of barrels: 12
Length:
(rocket) 4.656 m
(system) 9 m
Weight:
(APHE/bomblet rocket) 389 kg
(AT mine rocket) 381 kg
(APHE/bomblet warhead) 91 kg
(AT mine warhead) 83 kg
Max range: 50,000 m
Traverse: 110° left and right
Elevation: 0 to +56°
Dispersion at max range: 220 m range, 1-5 m azimuth
Max velocity: 1,200 m/s
Rate of fire: 1 rocket/2.3 to 4 s
Emplacement time: 2 min
Displacement time: 1 min
Crew: 5
Configuration: 8 × 8
Combat weight: 32,000 kg
Length: 9.00 m
Width: 2.64 m
Height: 3.84 m
Ground clearance: 0.38 m
Fording: 1.2 m
Turning radius: 13 m
Max road speed: 80 km/h
Max range: 600 km
Gradient: 60%
Side slope: 25%
Trench: 1.8 m
Engine: 354 hp OM422 turbocharged diesel
Power supply: 24 V DC

and burns for 200 ms to generate 8,000 kg of thrust to lift the rocket clear of the launcher tube. Once this happens the main sustainer motor cuts in with its 130 kg fuel load to give 18,000 kg of thrust during its 5 second burn time. This gives the rocket a maximum velocity of 1,200 m/s. Flight time to the maximum range of 50,000 m is 110 seconds with the highest possible altitude reached, depending upon launch angle, being 22,000 m. By opening and closing the four available aerodynamic brake assemblies one of four different ballistic range trajectories can be chosen: 24,000, 28,000, 37,000 or 50,000 m.

A TV camera made by Bosch of Germany is mounted on top of the left side of the launcher. It is used in a similar manner to the German 110 mm LARS in that a single rocket is launched and at some point in its flight is detonated so as not to alert the target area.

The Iraqi Ababeel 50 is essentially the Yugoslav M-87 Orkan MRS (Christopher F Foss) **2000**/0085772

The camera monitors this and the data collected is used to calculate the correct firing solution so that the remainder of the battery or vehicle salvo can be fired.

Several types of warhead can be fitted:

(1) 91 kg unitary APHE. An inertial fuze with instantaneous or delay action is fitted
(2) 91 kg cluster munition with a payload of 288 HEAT fragmentation bomblets. The shaped charge equipped bomblet also has some 420 steel spheres to enhance the fragmentation effect. Lethal radius is approximately 10 m and conventional armour plate penetration 60 mm plus. A typical dispersion pattern for a warhead at payload release height would be an ellipse 180 × 165 m. An alternative warhead with 420 anti-personnel submunitions has also been mentioned
(3) 83 kg cluster munition with a payload of 24 small cylindrical shaped-charge anti-tank KB-2 mines, each fitted with four flip-out curved vanes to stabilise their flight to the ground and ensure correct orientation. The mine can penetrate up to 40 mm plus of conventional steel armour plate.

Both cluster munition warheads are fitted with an electronic time fuze for bomblet/mine delivery.

The vehicle is fitted with a central tyre pressure regulation system which is operated by the driver from within the cab and to provide a more stable firing platform four stabilisers are lowered to the ground, by remote control, one either side to the rear of the second roadwheel and two at the very rear. When travelling, the launcher is traversed to the front and the whole launcher is covered by a tarpaulin cover with integral bows.

The 8 × 8 chassis of the 262 mm (12-round) LRSV M-87 'Orkan' multiple rocket launcher has also been used as a trials bed for the Turkish TOROS surface-to-surface rocket system covered in a separate entry and which remains at the prototype stage.

Variant

The LRSV M-1996 Orkan-2 is based on the Russian FROG-7 (8 × 8) launcher 9P113 but is believed to have four 262 mm barrels from the M-87 Orkan system.

Status

Production complete. In service with Bosnian Serb Army, Croatian Army (2), Iraq (2-3 systems delivered before Gulf War, as the Ababeel 50) and the Yugoslav Army. Recent information has indicated that limited marketing of this system has started again under the leadership of Bosnia-Herzegovina.

Contractor

Yugoimport SDPR.

UPDATED

128 mm (32-round) M-77 'Oganj' multiple rocket system

Development/Description

The M-77 128 mm (32-round) 'Oganj' (Fire) was developed in the early 1970s to meet the requirements of the Yugoslav Army. It is normally found in batteries of six launchers. It performs a similar role to the former Czech 122 mm RM-70 (40-round) MRS.

It does not, however, have an armoured cab or the excellent cross-country performance of the Czech vehicle which is based on the Tatra 813 (8 × 8) truck chassis.

The M-77 128 mm MRS basically consists of an FAP 2026 BDS/AV (6 × 6) series truck with a 32-barrel 128 mm rocket launcher mounted at the rear of the hull and an additional pack of 32 rockets to the rear of the cab for rapid reloading. The launcher fires a 128 mm calibre, 2.6 m long unguided rocket to a maximum range of 20,600 m.

Weight of the rocket is 67 kg. The 19.5 kg warhead contains 2,562 steel spheres 6.35 mm in diameter and 930 steel spheres of 10.32 mm diameter that are lethal out to a radius of over 30 m. The explosive content is 3.8 kg and a fuze with preselected super quick, inertia and delay action options is fitted as standard.

The rockets are either launched singly or in a ripple salvo, the latter taking a total of 25.6 seconds. The rockets can be launched either by manual, semi-automatic or fully automatic control with an optional 25 m off-vehicle remote-control box cable system.

First major production version of M-77 Oganj 128 mm MRL on rear of FAP 2026 BDS/AV (6 × 6) truck chassis, tarpaulin lowered showing bank of 32 128 mm rockets behind cab and rocket launcher at the rear. The stabilisers are mounted at the very rear of the vehicle and to the immediate rear of the front roadwheels

Once the 128 mm rockets have been launched, the launcher is traversed to the rear and depressed to the horizontal, the pack of 32 rockets to the rear of the cab is raised until it lines up with the launcher and the rockets are pushed into it. The reloading system then goes back onto the floor of the truck and the launcher is traversed forwards ready for firing again.

The complete reloading sequence takes 2 minutes with the second salvo capable of launch within 5 minutes of the first. Reloading can also be performed

M-77 Oganj 128 mm MRL on rear of FAP 2026 BDS/AV (6 × 6) truck chassis with tarpaulin covering launcher

128 mm lightweight single-tube rocket launcher in firing position, showing bipod and baseplate

Specifications
Calibre: 128 mm
Weight: (in combat position) 50 kg
Traverse: ±10°
Elevation: +7 to +48°
Max range: 20,600 m
Sight unit: NSB-4A
Firing of the launcher is performed using a remote hand-controller and cable from a distance of 25 m.
With FAP-2026 BDS/AV 6 × 6 chassis
Calibre: 128 mm
Number of barrels: 32
Max range: 20,600 m
Elevation/depression: +50/0°
Traverse: 185° left and right
Reload time:
(automatic) 2 min
Weight:
(system, loaded) 22,400 kg
Crew: 5
Configuration: 6 × 6
Length:
(travelling) 8.4 m
Width:
(travelling) 2.49 m
Height:
(travelling) 3.1 m
Track: 2.02 m
Wheelbase: 3.4 + 1.4 m
Max road speed: 80 km/h
Range: 600 km
Engine: V-8 direct injection water-cooled diesel developing 256 hp
Tyres: 15.00 × 21

manually. When travelling, the launcher is lowered and the launcher and reloading pack are covered by a tarpaulin which is folded up under the rear in the firing position. The time required to occupy or leave a firing position is less than 20 seconds, with the time taken for occupying a position, firing both rocket loads and leaving being 4 to 5 minutes in total.

It is believed that the original FAP-2220 BDS (6 × 4) chassis was used only for early production vehicles as it is not suitable for travelling across very rough country because of its low ground clearance. The chassis was then changed to the FAP 2020 BS (6 × 6) and FAP-2026 BDS/AV (6 × 6) trucks. These vehicles have large tyres with single rather than dual rear wheels and are fitted with a central tyre pressure regulation system that allows the driver to adjust the ground pressure to suit the type of ground being crossed.

M-77 Oganj Mod

This is a Bosnia version which, instead of having 32 × 128 mm rocket tubes, has 2 × 4 128 mm barrels and 2 × 2 122 mm barrels. The latter can launch standard 122 mm unguided rockets of Russian or equivalent types which are widely available.

128 mm single-tube launcher

To complement the M-77 there is a 128 mm lightweight single-tube rocket launcher for use by commando, special forces and territorial defence units.

In addition to being used by Yugoslavia, this 128 mm single-round launcher is known to be used by Macedonia (60 units) and Slovenia (56 units).

RH ALAN 122 mm MRS

The Croatian company of RH ALAN now manufactures a very similar system to the M-77 Oganj but based on a more mobile Tatra (8 × 8) chassis called the 122 mm (32-round) M96 Typhoon. Details of this system are given earlier in this section under Croatia. In addition to having greater mobility it has the added advantage of firing the more widely used Russian family of 122 mm unguided artillery rockets.

Available details of the five-man (two for the launcher, three for the rocket) bipod-mounted weapon are:

Status

It is believed that production of this system is now complete. In service with Bosnian Army, Bosnian Serb Army, Croatian Army (179, all types of 128 mm MRL), Croatian Defence Council, Iraqi Army and Yugoslav Army (64).

Contractor

Yugoimport SDPR.

VERIFIED

128 mm (32-round) M-85 multiple rocket launcher system

Development/Description

The first Baghdad International Exhibition for Military Production in Iraq revealed that the Yugoslav Army has a 128 mm rocket launcher, the M-85, based on the TAM truck chassis in service.

The 128 mm unguided rocket used weighs 25.5 kg and can be fired either singly or in a ripple salvo with set intervals of 0.2, 0.4 or 0.6 seconds.

It is understood that the rocket used by this system is the same Improved Rocket for the older M-63 covered in detail in the following entry.

Specifications
Calibre: 128 mm
Number of barrels: 32
Weight: (rocket) 25.5 kg
Max range: 12,500 m
Combat weight: 9,800 kg
Elevation/depression: +48/0°
Traverse: 360°

Status

Production complete. In service with the Bosnian Army, Bosnian Serb Army, Croatian Army, Croatian Defence Council and Yugoslav Army (Serbia and Montenegro). It is not expected that there will be any further production of this system.

Contractor

Yugoimport SDPR.

VERIFIED

128 mm (32-round) M-63 'Plamen' multiple rocket system

Development/Description

The 128 mm (32-round) M-63 'Plamen' (Flame) MRS was developed in Yugoslavia to meet the requirements of the then Yugoslav Army. It is normally found in MRL battalions of three batteries, each with four launchers. It is estimated that total production of this system amounted to about 500 units.

The M-63 is mounted on a split trail carriage with elevation, traverse, launch controls and sight on the left side of the launcher.

The spin-stabilised M-63 unguided rockets fired by the M-63 are 0.814 m long, weigh 23.1 kg, have a maximum velocity of 444 m/s and are fitted with a 7.55 kg warhead that is filled with 2.3 kg of high explosive.

A UTI M63(OV) point detonating fuze with super quick inertia action is used. A practice round of similar dimensions and weight for loading and unloading training is also used. The rockets are either fired singly or in a ripple salvo with set intervals of 0.2, 0.4 or 0.6 seconds.

An improved version of the M-63 with a wider carriage (2.212 m in travelling position) was also manufactured, and both 8- and 16-round launchers are available for export. If required, these can be mounted on a self-propelled system.

All versions of the M-63 launcher can be adapted for towing by animals.

Variants

128 mm M-71 'Partizan' single-tube rocket launcher

A 128 mm M-71 lightweight single-tube rocket launcher known as the 'Partizan' was also produced for use by commando, special forces and territorial defence units. It is known to be in service with Syria.

128 mm (32-round) M-63 rocket launcher in firing position with trails open

Close-up of rear of 128 mm (32-round) M-63 rocket launcher

Specifications
M-63 rocket used
Calibre: 128 mm
Number of barrels: 32
Length:
(barrel) 1.03 m
(rocket) 0.814 m
(travelling) 3.682 m
(firing) 3.65 m
Weight:
(rocket) 23.1 kg
(in firing position, unloaded) 1,395 kg
(in firing position, loaded) 2,134 kg
Max range: 8,545 m
Max velocity: 444 m/s
Reload time: 5 min
Crew: 7
Carriage: split trail
Width:
(travelling) 2.212 m
Height:
(travelling) 1.278 m
Ground clearance: 0.268 m
Track: 1.892 m
Tyres: 8.25 × 20
Elevation/depression: +48/0°
Traverse:
(total) 30° (15° left, 15° right)
Reload time: 3 min
Towing vehicle: TAM 1500 (4 × 4) truck

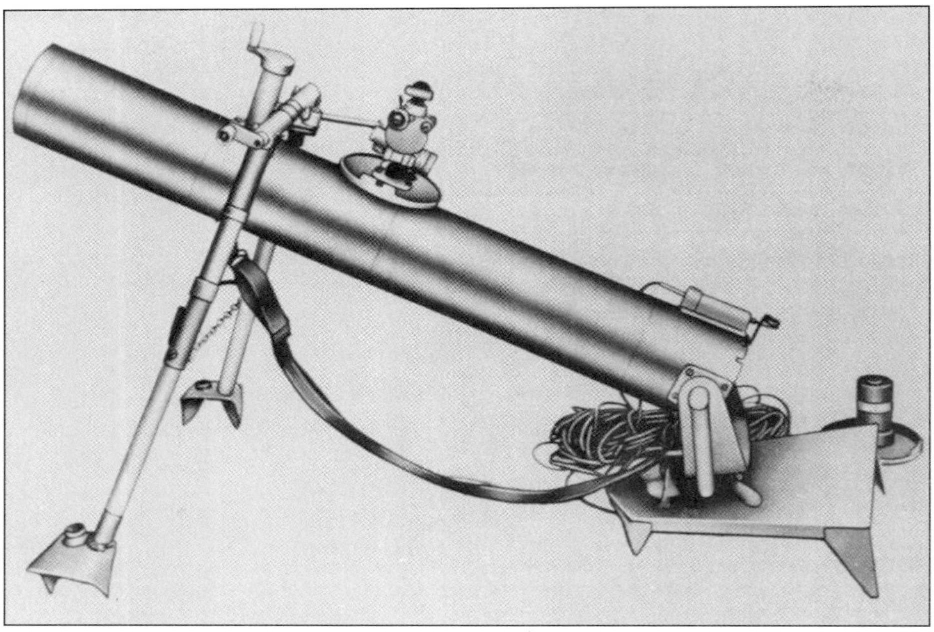

128 mm M-71 'Partizan' lightweight single-tube rocket launcher deployed in firing position and showing bipod and baseplate

Available details of the three-man bipod-mounted weapon are:

Time required for emplacement is 5 minutes and for displacement (in fired state) 3 minutes. The 128 mm unguided rocket type used is the M-63 type previously described.

Specifications
Calibre: 128 mm
Length:
(in travelling configuration) 1.133 m
(in combat configuration) 1.13 m
Weight:
(in travelling configuration) 21.5 kg
(in combat configuration, unloaded) 18.5 kg
(in combat configuration, loaded) 45 kg
Elevation: +5 to +45°
Max range: 8,564 m
Min range: 800 m
Sight unit: NBS-4A

It is understood that the correct designation of this system is the LRL 128 M-71 'Partizon'.

128 mm M-87 Improved Rocket

In 1980, Yugoslavia introduced the improved M-87 HE unguided rocket for use with both the M-63 and M-71 launchers. This weighs 25.5 kg in total, is 0.96 m long and uses an M-84 point detonating fuze with super quick inertia action. The warhead contains 3.3 kg of HE filler.

Maximum velocity is increased to 740 m/s with 7.8 kg of solid-propellent fuel carried giving a minimum range of 800 m and a maximum range of 12,800 m.

128 mm M-92 Improved version

This is an improved model used by Bosnia-Herzegovina and is fitted with a carrying handle.

128 mm (24-round) LOV RAK 24/128

Details of this system, developed by RH ALAN, are given under Croatia.

128 mm (12-round) VLR 128 M91A3 RAK 12

Details of this system, developed by RH ALAN, are given under Croatia.

Status

Production complete. In service with the following countries:

Country	Quantity	Comment
Bosnia-Herzegovina	n/avail	
Croatia	179	all types of 128 mm launchers
Cyprus	12	
Macedonia	15	
Slovenia	4	
Yugoslavia	103	

Contractor

Yugoimport SDPR.

UPDATED

COASTAL ARTILLERY GUNS AND MISSILES (INCLUDING FIXED FORTIFICATION WEAPONS)

COASTAL ARTILLERY GUNS AND MISSILES (INCLUDING FIXED FORTIFICATION WEAPONS)

BRAZIL

AVIBRAS ASTROS II coastal defence multiple rocket system

Development/Description

This is the Brazilian AVIBRAS ASTROS II multiple rocket system (see entry in the Multiple rocket launchers section) used in the coastal defence role.

The AVIBRAS ASTROS II cluster munition warhead-equipped rockets are fired to spread their submunitions in a top-attack mode over any approaching warship or amphibious assault vessel. The result, according to AVIBRAS, is an effective, low-cost coastal defence weapon.

The AVIBRAS ASTROS II coastal defence supporting artillery battery unit is identical to the surface-to-surface battery configuration. An additional feature of the coastal defence version is a truck-mounted long-range radar system to detect sea targets.

The Brazilian Army has taken delivery of SS-40 and SS-60 rockets for use with its coastal defence versions of the ASTROS II. As of early 2002, Brazil is the only known user of the AVIBRAS II coastal defence mutliple rocket system.

Key parts of the AVIBRAS ASTROS II Coastal Defence Multiple Rocket System. The AV-CBO appears to be the Swedish coastal Giraffe radar system

Status

Late in 1997, the Brazilian Army awarded AVIBRAS a contract for the ASTROS II coastal defence rocket system. In May 1998, four launching vehicles (6 × 6) and two ammunition vehicles (6 × 6) were delivered to the Brazilian Army and handed over to the 8th Motorised Coastal Artillery Group based in Rio de Janeiro. The Brazilian Army is the first customer for the coastal defence version of the AVIBRAS ASTROS II rocket system. After this first delivery, more ASTROS units were delivered to Brazilian Army to equip the 6th Motorised Coastal Artillery Group based in Praia Grande (São Paulo) and the 1st/10th Motorised Coastal Artillery Group based in Macaé (Rio de Janeiro). Each of those units received two launcher vehicles and one ammunition vehicle. The 8th Motorised Coastal Artillery Group was completed with the receipt of the fire-control unit.

Manufacturer

AVIBRAS Industria Aerospacial S/A.

UPDATED

AVIBRAS ASTROS II coastal defence systems leaving the AVIBRAS facility for delivery to the Brazilian Army 0014670

AVIBRAS ASTROS II coastal defence systems of the Brazilian Army launching SS-40 rockets during a training exercise **2002**/0130597

CHINA, PEOPLE'S REPUBLIC

CPMIEC Hai Ying-2 (Sea Eagle-2) coastal defence missile

Development/Description

The HY-2 (Sea Eagle-2) is a Chinese-developed product-improved version of the Russian P-15 (US/NATO designation SS-N-2 'Styx') anti-ship missile, assigned the Western designation CSS-N-3 'Seersucker'. The coastal defence version of the HY-2 is assigned the designation CSS-C-2 'Silkworm'.

In general configuration it is similar to the HY-1 (Sea Eagle-1) coastal defence missile system. As far as it is known, this is no longer in operational service and details were given in *Jane's Armour and Artillery 1999-2000*.

Guidance in the mid-course phase of the Hai Ying-2 attack profile is by gyro autopilot with the terminal

CPMIEC Hai Ying-2 (CSS-C-2 'Silkworm') on missile transloader being loaded onto launcher

CPMIEC Hai Ying-2 (CSS-C-2 'Silkworm') coastal defence system on its launcher in travelling configuration

CPMIEC Hai Ying-2 missile on ground trolley

phase depending upon which model HY-2 is used. Maximum effective range is between 20,000 and approximately 100,000 m.

The missile operates in conjunction with a 6 × 6 truck-mounted I/J-band target acquisition radar, which has a peak power output of 200 kW, a pulse-width of 0.2 µs and a range resolution of 60 m. The set can satisfactorily track a 7,500 m² radar cross-section target at 111,000 m distance.

Apart from its use with the Chinese armed forces, the HY-2 has been exported to North Korea (qv, where the missile in the past has been licence-built). It was also exported to both sides in the Iran/Iraq War, with the latter country using it and achieving mediocre results. Iraq used it during the 1991 Gulf War against Coalition naval units (qv Iraq) without success. The Iraqi versions are known as the FAW-70, FAW-150 and FAW-200 and additional details of these are given later in this section under Iraq. As far as is known, there has been no recent production of this missile in Iraq.

Variants

Basic active radar guided HY-2.

This uses an active I-band radar seeker for the terminal phase and cruises at an altitude of 200 m.

Specifications

Guidance: see text
Launch weight: 3,000 kg
Length: 7.36 m
Body diameter: 0.76 m
Wing span: 2.4 m
Warhead: 513 kg HE
Min range: 20,000 m
Max range: 95,000-100,000 m
Altitude of launch site: 0-400 m
Firing azimuth limits: ±85°
Propulsion: liquid-fuel rocket sustainer with jettisonable solid-fuel rocket motor booster
Speed: M0.9

Infra-red guided HY-2A

This uses a non-imaging infra-red seeker with a Cassegrain lens arrangement for the terminal homing phase. Cruise altitude is reduced to between 30 and 50 m.

Active radar guided HY-2G

This uses the same type of improved design altitude circuit with a high-precision altimeter as the HY-2A. The cruising altitude is 30 to 50 m.

Advanced active radar guided HY-2B

This uses an advanced centimetric wavelength monopulse active radar seeker (probably operating in the J-band) for the terminal homing phase.

A radio frequency altimeter is also fitted, allowing the missile to descend during the attack phase, from its cruise height of between 30 to 50 m to 8 m in order to present a more difficult target and air defence systems.

Status

Production complete. No longer marketed. In service with China, Iran, Iraq and North Korea.

Manufacturers

Chinese and North Korean state factories.
Enquiries to China Precision Machinery Import and Export Corporation (CPMIEC).

UPDATED

CPMIEC C-101 and C-301 Hai Ying-3 (Sea Eagle-3) coastal defence missile systems

Development/Description

The China Precision Machinery Import and Export Corporation (CPMIEC) C-301, alternative designation HY-3 (Hai Ying-3; Sea Eagle-3), is derived from the C-101 ramjet-powered missile developed in the late 1970s from the basic HY-1 family of airframes.

Specifications

Guidance: inertial with active radar terminal homing
Launch weight:
(C-101) 1,850 kg
(C-301) 4,900 kg
Length:
(C-101) 6.5 m
(C-301 with boosters) 9.85 m
(C-301 without boosters) 9.46 m
Body diameter:
(C-101) 0.54 m
(C-301) 0.76 m
Wing span:
(C-101) 1.62 m
(C-301) 2.24 m
Warhead:
(C-101) 300 kg HE SAP
(C-301) 500 kg HE
Range:
(C-101) 12,000-45,000 m
(C-301) 140,000 or 180,000 m
Propulsion:
(C-101) 2 ramjets with 2 jettisonable solid-fuel rocket motor boosters
(C-301) 2 ramjets with 4 jettisonable solid-fuel rocket motor boosters
Speed:
(C-101) M2.0
(C-301) M2.0 plus

The first showing of the C-101 (in model form) occurred in 1986. It has two side-mounted ramjet engines at the rear, two jettisonable solid-propellant boosters mounted beneath them, a V-shaped tail arrangement and a vertical fin with a control surface on its upper part.

Guidance is by a mid-course inertial system with an active radar terminal homing seeker. The warhead carried weighs 300 kg and is of the semi-armour-piercing HE type. Speed is said to be supersonic with the weapon cruising at around 50 m altitude.

The C-301 (US/NATO designation CSSC-6 'Sawhorse') is a heavier and longer range weapon and is similar in shape to the C-101. However, it has four solid-propellant booster motors with stabiliser fins mounted above and below the two ramjet sustainers.

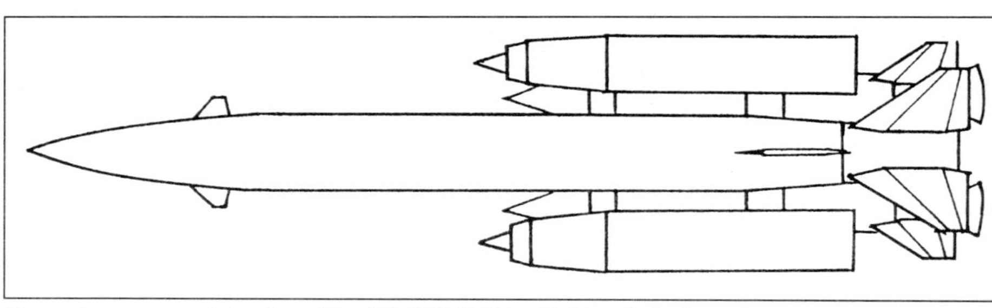

General arrangement drawing of the CPMIEC C-101 anti-ship missile from above 0008325

CPMIEC C-301 coastal defence missile on its launcher

The overall length of the C-301 with the four boosters attached is 9.85 m which reduces to the 9.46 m of the main missile body after they are jettisoned. Speed is M2.0 with a cruising height adjustable between 100 and 300 m or 6,000 m. The warhead weighs approximately 500 kg and is probably similar to the standard conventional HE configuration fitted to the other HY family members.

According to CPMIEC, a typical C-301 coastal defence battery would consist of a radar fire-control and command vehicle, three power supply vehicles, four single-rail launcher trailers and four tractor trucks.

Recent information has indicated that the radar command-post vehicle and power supply vehicles used with the C-101 coastal defence system are the same as that used for the more recent C-802 coastal defence missile system covered later in this section.

Status

As far as it is known, neither the C-101 or the C301 has entered production for the Navy of the People's Liberation Army in the coastal defence role. It has been offered on the export market for this application but there are no known sales to date.

Manufacturer

Chinese state factories.

Enquiries to China Precision Machinery Import and Export Corporation (CPMIEC).

VERIFIED

CPMIEC C-201W (Sea Eagle-4) coastal defence missile

Development/Description

The C-201W, alternative designation HY-4 (Sea Eagle-4), is a further development of the HY-2 series (also known as Sea Eagle 2 and covered in an earlier entry) to enhance the performance characteristics. The Western designation is CSSC-7 'Sadsack'. Prime contractor for all of these Chinese coastal defence missile systems is the China Precision Machinery Import and Export Corporation.

The general configuration remains essentially the same but the fuselage is lengthened and the liquid-fuel sustainer rocket motor is replaced by a small liquid-fuel turbojet propulsion unit with 200 kg of aviation kerosene, the air for which is drawn in through a ventral scoop intake located towards the rear of the underfuselage.

The initial launch and acceleration phases are still carried out by using a jettisonable solid-fuel rocket motor attached immediately behind the scoop.

The guidance system is believed to be similar to that of the older HY-2G, with an autopilot for the mid-course phase and an active monopulse radar seeker (probably operating in the J-band) for the terminal attack.

Cruise altitude can be varied between 70 and 200 m using the onboard radio frequency altimeter.

The C-201W missile is launched from a trailer type launcher rail which is very similar to that used for the older Sea Eagle-1 (covered in *Jane's Armour and Artillery 1999-2000* pages 826 to 827) coastal defence missile. The Sea Eagle-4 is transported on a trailer towed by a 6 × 6 tractor truck.

A typical HY-4 coastal defence battery would include four missile launchers, command post vehicle, surveillance radar vehicle plus a variety of support equipment including test trucks, fuel trucks and lifting equipment.

C-201W coastal defence missile on trolley

0021087

CPMIEC C-201W coastal defence missile fire-control centre (left) and tracking radar (right) both mounted on a 6 × 6 chassis

0021088

Status

Production as required. In service with the Chinese armed forces. There are no known exports of this missile system.

Manufacturer

Chinese state factories.

Enquiries to China Precision Machinery Import and Export Corporation (CPMIEC).

UPDATED

Specifications

Guidance: autopilot with active radar terminal homing
Launch weight: 1,950 kg (including 200 kg for the booster)
Length: 7.36 m
Body diameter: 0.76 m
Wing span: 2.4 m
Warhead: 500 kg HE
Range: 35,000-150,000 m
Propulsion: liquid-fuel turbojet sustainer with jettisonable solid-fuel rocket motor booster
Speed: M0.8

CPMIEC C-801 and C-802 coastal defence missiles

Development/Description

The China Precision Machinery Import and Export Corporation (CPMIEC) C-801 missile (alternative designations Ying Ji-1 Hawk Attack-1, YJ-1 and HY-5) is reminiscent of the MBDA (previously Aerospatiale Missiles) MM38 Exocet in appearance and is fired from a box-shaped container-launcher. The Western designation is CSS-N-4 'Sardine'.

A typical C-801 coastal defence battery comprises a radar command truck, a power supply truck and four launcher trucks each carrying two C-801 container-launchers. The radar command truck and power supply truck are both based on a 6 × 6 cross-country truck

CPMIEC C-801 coastal defence missile out of its launcher with container-launcher in background

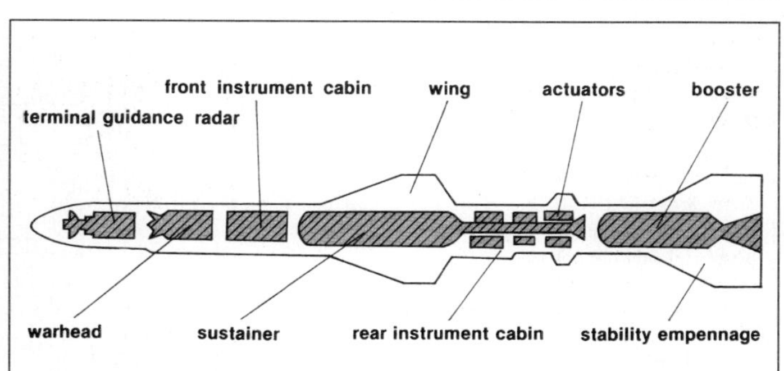

CPMIEC C-801 coastal defence missile showing position of main components

CPMIEC C-801 coastal defence missile out of its launcher

CPMIEC C-802 coastal defence missile with fins and wings unfolded

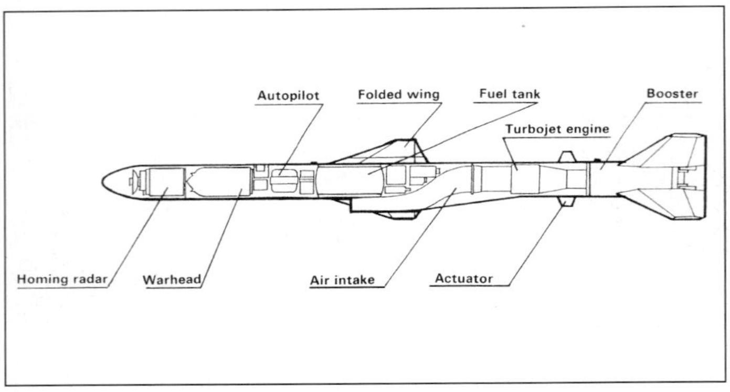

C-802 coastal defence missile showing main components 0008327

Model of CPMIEC C-802 coastal defence missile on 6 × 6 truck chassis showing three-round launcher in retracted position (Christopher F Foss) 0008328

chassis which is also used for the CPMIEC C-101 coastal defence missile system.

It uses a solid-fuel sustainer rocket motor with a jettisonable solid-fuel booster motor section for the propulsion system. Flight control is by four movable cruciform tailfins and four foldable mid-body wings.

Mid-course guidance is by autopilot with the terminal phase being controlled by a monopulse active radar seeker (probably operating in the J-band).

Cruise speed is M0.8 to 0.9, the cruise altitude being controlled by a radio altimeter at 20 m. For the terminal attack the missile is automatically commanded to descend to a height of 5 to 7 m.

In February 1989, the C-802 (alternative designation YJ-2) version was revealed. The Western designation is CSS-N-8 'Saccade'. The only external difference from the C-801 is that an air inlet scoop has been added beneath the main missile body. This is to serve the small Wopen air-breathing turbojet that has replaced the solid-fuel sustainer rocket motor. The jettisonable solid-fuel booster motor section is retained. Cruising altitude is controlled between 20 and 30 m with the terminal attack phase between 5 and 7 m. The C-802 missile can engage in a sector arc of 180°.

In its coastal defence role the C-802 is launched from a large truck (6 × 6) (approximately 10,000 kg), carrying three container-launchers on an elevating cradle fitted to its rear decking. To provide a more stable launch platform, two stabilisers are lowered to the ground before missile launch.

A C-802 coastal defence battery comprises four such trucks, a radar command truck and power supply truck. The radar command truck and power supply truck are also used in the C-801 and C-101 systems.

Air-launched versions, the C-801K and C-802K are known to exist. A typical helicopter-launched version would have one missile either side.

It is understood that Iran took delivery of up to 40 C-802 missiles by 1996. It is possible that others were transferred in 1996-98. In Iranian service the C-802 is also launched from 'Houdong' class patrol craft.

Specifications

Guidance: autopilot with active radar terminal homing
Launch weight:
(C-801) 815 kg
(C-802) 715 kg
Length:
(C-801) 5.814 m
(C-802) 6.392 m
Body diameter: 0.36 m
Wing span: 1.18 m
Warhead: 165 kg HE SAP self-forging fragment
Range:
(C-801) 8,000-42,000 m
(C-802) 15,000-120,000 m
Propulsion:
(C-801) solid-fuel rocket motor sustainer with jettisonable solid-fuel rocket motor booster
(C-802) liquid-fuel turbojet sustainer with jettisonable solid-fuel rocket motor booster
Speed:
(C-801) M0.8-0.9
(C-802) M0.8-0.9

Status

C-801. Production as required. In service with the Chinese armed forces and several Middle Eastern countries including Iran.

C-802 in production. In service with the Chinese armed forces and Iran. It is believed that neither of these two missiles are in service in China in the coastal defence role.

Manufacturer

Chinese state factories.

Enquiries to China Precision Machinery Import and Export Corporation (CPMIEC).

UPDATED

NORINCO 155 mm coastal defence system

Development/Description

The NORINCO (China North Industries Corporation) coastal defence system consists of a battery of six NORINCO Type WA 021 155 mm towed gun-howitzers, fully described in the Towed anti-tank guns, guns and howitzers section, a Fire-Control Computer (FCC), an advanced long-range Laser Range-Finder (LRF) and an advanced phased-array radar. All of the elements of the NORINCO 155 mm coastal defence systems are mounted on 4 × 4 or 6 × 6 trucks to provide good tactical and strategic mobility.

According to NORINCO, the system is capable of searching, intercepting, engaging and tracking surface targets in all weather conditions.

The NORINCO coastal defence system has two key parts; these are the battery of six Type WA 021 155 mm

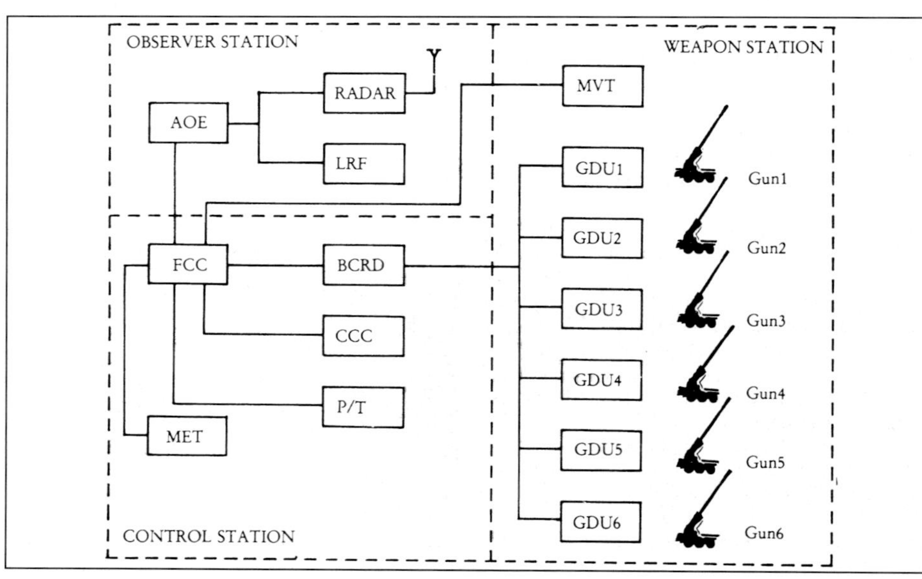

Block diagram of the proposed NORINCO 155 mm coastal defence system

45 calibre towed gun-howitzers, each of which is provided with a Gun Display Unit (GDU), and the fire-control system, with the latter normally being mounted in a cross-country vehicle.

The fire-control system utilises an advanced military 16-bit microcomputer in the FCC and an advanced communications system.

Interactive and automatic operating modes are used to input the various data into the fire-control computer.

Each fire-control computer can support six guns of its own battery and can also take over the control of a further 12 guns if their fire-control computers malfunction.

Information is fed into the central fire-control computer from the observer's station (radar and laser range-finder) and the meteorological station.

The Type WA 021 155 mm gun-howitzer has a maximum range of 30,000 m using ERFB projectiles and 39,000 m using ERFB-BB projectiles. Traverse is through a total of 70° (30° left and 40° right) and a rate of fire of 5 rds/min can be achieved.

In addition to firing high-explosive projectiles against sea targets, cargo projectiles would also be fired. These would be highly effective against small targets such as landing craft as well as damaging the upper parts of larger craft.

Status

Ready for production on receipt of orders. As far as is known a complete prototype of this system has yet to be demonstrated outside China. The only known user of the towed WAC 021 155 mm gun-howitzer outside China is Iran and these are understood to be deployed in the field role and not the coastal artillery role.

Manufacturer

NORINCO - China North Industries Corporation.

VERIFIED

CROATIA

RH ALAN MOL coastal defence system

Development/Description

Late in 1996, RH ALAN of Croatia released details of the MOL coastal defence missile system which is now operational with the Croatian Navy (*Hrvatska Ratna Mornarica*).

Before the break-up of Yugoslavia, orders were placed with the now SAAB Bofors Dynamics of Sweden for the supply of a number of RBS 15 Series anti-ship missiles for installation on surface craft being built in Yugoslavia.

The Croatian Navy has now deployed SAAB Bofors Dynamics RBS 15 on the Type 400 Corvette *Kralj Petar Kresimir IV*.

At least two batteries of the truck-mounted coastal defence version called MOL are understood to be in service with the Croatian Navy, with conversion work being carried out by the local company, RH ALAN.

The MOL system is mounted on the rear of a Slovak Tatra 815 (6 × 6) series cross-country truck which has a two-man forward control cab.

This chassis is also used as the basis for the Croatian RN ALAN 122 mm (32-round) M96 Typhoon self-propelled rocket launcher which is covered in detail in the Multiple rocket launchers section. So far this system has not been exported outside of Croatia.

To the immediate rear of the cab is the fully enclosed command and control cabin which contains the power supply system, missile fire-control system, simulation and test system for the training of operators, C^3I terminal, orientation and positioning system, air conditioning system and fire detection and suppression system. The system is normally operated by a two-man crew.

Mounted at the rear of the truck, in two layers of two, are four SAAB Bofors Dynamics RBS 15 anti-shipping missiles in sealed containers. When travelling, these are in the horizontal position and are elevated before launch. To provide a more stable firing platform two hydraulic jacks are lowered either side of the vehicle.

Croatian Navy MOL coastal defence system deployed in a typical fire position with stabilisers lowered and missile canisters elevated ready for launch

The RBS 15 anti-ship missiles have a range of over 70 km and are fitted with a high-explosive blast/fragmentation warhead weighing 179 kg. Target information would be provided by a radar system.

Full details of the complete SAAB Bofors Dynamics RBS 15 anti-ship missile system are given later in this section under Sweden. The missiles were designed to be air, ground or ship launched.

Status

In service with Croatia.

Missile specifications
(from RH ALAN sources)
Length: 4.33 m
Diameter: 0.50 m
Launch weight: 758 kg
Warhead weight: 179 kg
Speed: high-subsonic
Range: over 70 km

Manufacturer

RH ALAN.

UPDATED

FINLAND

Patria Vammas 130 mm coastal defence gun

Development/Description

This 130 mm coastal defence gun was developed to meet the requirements of the Finnish Army by the Tampella Defence Division, which has recently become Patria Vammas. It entered service in the mid-1980s. The 130 mm coastal defence gun has the local designation of the K-54 and some sources have indicated that up to 195 weapons may be in service.

The turret is of all-steel construction and is normally located on top of a hardened emplacement. Fire control is accomplished by a computerised fire-control system connected to a surveillance radar system.

A roof-mounted sight is provided and, if required, the crew can leave the turret via doors located in the rear. The 130 mm/53 calibre ordnance is fitted with a pepperpot-type muzzle brake.

The RAVAL automatic surveillance fire-control system for this 130 mm coastal defence gun was developed by the Finnish company, Elesco Oy.

The computer system has been designed to assist the operation of the 130 mm system and controls the

Patria Vammas 130 mm coastal defence gun

gunlaying, firing command logic and ammunition handling.

The computer is continually monitored by an independent safety logic, with specific displays being provided for the individual gun crew stations as well as a common system overview.

The system is adaptable to different weapon types and can also be used to modernise existing installations. The same basic system can be modified for use in mobile artillery applications.

In 1994, it was stated that a total of 70 of the gun control systems had been delivered with more than 30 still on order. It is not known if all of these are for the Patria Vammas 130 mm coastal defence gun.

It is possible that the 130 mm ammunition used by this weapon is similar to that used by the Russian 130 mm M-46 field gun which is used by Finland with the local designation of the K-54.

Variants

Although all production systems to date have been armed with a 130 mm ordnance, the turret system can also be fitted with 122, 152 or 155 mm barrels. It is known that Finland also deploys numbers of 100 mm armed T-54/T-55 tank turrets in the static defence role.

Status

Production as required. In service in Finland. There are no known export sales of this system. It is believed that production of this system was completed some years ago.

Specifications

Calibre: 130 mm
Length:
(barrel) 6,818 mm
Weight:
(turret) 16,000 kg
Elevation/depression: +50/–5°
Turret traverse: 360°
Max range: 24,000 m (standard ammunition)
Max rate of fire: 6 rds/min

Manufacturer

Patria Vammas.

UPDATED

FRANCE

Giat Industries 100 mm Compact multipurpose gun

Development/Description

Apart from its shipboard fitting, the Giat Industries 100 mm Compact gun can also be used for coastal defence in the anti-aircraft, anti-missile or anti-shipping roles. This turret was originally built for the export market by Creusot Loire Industrie (who were subsequently taken over by Giat Industries). As of early 2002, there are no known sales of this weapon for the coastal defence mission.

A proposed standard defence module or battery comprises one or two Compact 100 mm guns, a multisensor fire-control system (usually a search radar and optical surveillance system), the facilities necessary for the gun and the fire-control system, and the rooms required for the electric, hydraulic and compressed air subsystems. The equipment is normally buried and fully protected in concrete bunkers but can be produced in a mobile version on request.

The Compact turret houses a water-cooled 100 mm gun which can be remotely operated by one man. The maximum rate of fire is 90 rds/min but this can be selected according to the target type and ammunition required. A microprocessor-controlled fully reversible automatic ammunition replenishing system is fitted. This can select, by a single push-button control, any one of three ammunition types in the magazine. The normal ammunition load is 114 rounds but the system has the capability of being reloaded as firing takes place.

Production of the Giat Industries Compact Mk 2 naval turrets is now underway for the three air defence ships ordered by Saudi Arabia for its Sawari 2 programme.

Specifications

Crew:
(in turret) none
Weight: 13,750 kg
Armament: 1 × 100 mm water-cooled gun
Muzzle velocity: 867 m/s
Recoil force: 11,500 daN
Azimuth:
(limits) 350°
(velocity) 50°/s
(acceleration) 57°/s²
Elevation:
(limits) –15 to +80°
(velocity) 30°/s
(acceleration) 75°/s²
Rate of fire:
(high) 90 rds/min
(medium) 45 rds/min (adjustable)
(low) 20 rds/min (adjustable)
Selection of ammunition type: automatic - three types available
Fuze setting: yes
Ammunition: 114 (typically 90 Type 1 round + 12 Type 2 + 12 Type 3 with proximity and contact or time and contact fuzing)
Ammunition weights:
(total round) 25 kg
(projectile) 13.5 kg
Max range:
(surface target) 17,400 m
(airborne target) 6,000 m

Status

Ready for production. As far as is known there are no known customers for the coastal version of the 100 mm Compact multipurpose gun system.

Manufacturer

Giat Industries.

UPDATED

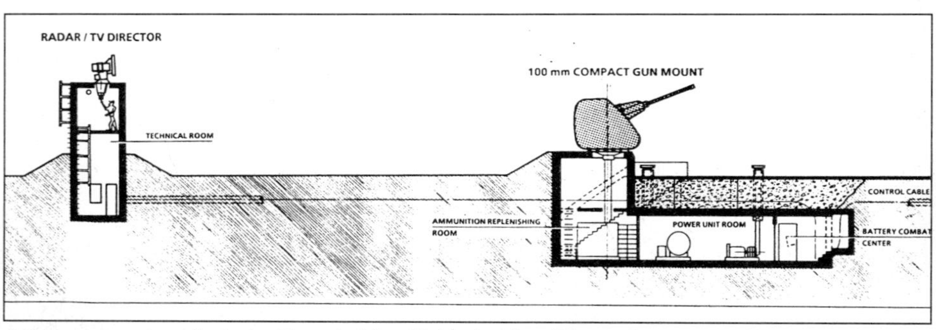

Artist's impression of a typical layout of the Giat Industries 100 mm Compact multipurpose gun in the coastal defence role

Although originally developed for installation on surface craft, the Giat Industries 100 mm Compact multipurpose gun has also been proposed for the coastal defence role ***2002***/0130598

MBDA MM40 Exocet coastal defence missile system

Development/Description

MBDA MM40 coastal battery is a derivative of the MM40 Exocet anti-ship missile system and in its latest form uses the MM40 Block 2 missile.

The MBDA Exocet MM40 coastal battery is designed to defend straits and choke points using an associated surveillance and control system to provide an enhanced protected zone over the designated defence area.

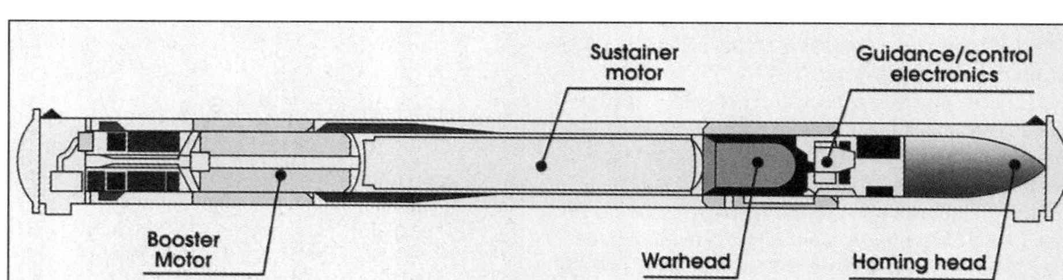

Main components of the MBDA *MM40 Exocet missile*

The basic MBDA Exocet MM40 coastal battery is a single fully autonomous firing unit loaded with four MM40 Exocet missiles and installed as a fixed, semi-mobile or fully mobile coastal battery integrated within the existing customer means: radio, radar, vehicles, logistic.

The MBDA MM40 Exocet coastal battery is also proposed as a complete system comprising a variable number of firing units ready for firing four MM40 missiles each, an ammunition supply unit with six missiles in their containers, a radar command post unit and one or two maintenance units.

Coastal defence version of MBDA *MM40 Exocet being launched*

All these units can operate over most terrain types and the operational units can deploy to either prepared or unprepared sites in less than 1 hour, the distance between the units depends on the type of communication system used, but can be up to 40 km from each other, the sites being datalinked for either a short-term or permanent mission.

Description

A complete MBDA Exocet MM40 coastal battery can comprise the following:

(1) A Radar Command Post unit (RCP). This is the command post for the whole coastal battery providing for all command, control and communication functions for the higher coastal defence command as well as the firing units and comprises:
- (a) A 4 × 4 truck equipped with a shelter which houses a two- to three-man command and control system with one or two surveillance radar consoles and the radio communication systems (HF voice with the higher command, VHF data and voice with the firing units)
- (b) A meteorological station
- (c) An air conditioned subsystem for both equipment and personnel on board
- (d) A towed surveillance and target acquisition radar with extending mast antenna.

(2) An RCP maintenance unit. This is a 4 × 4 truck mounting the operations room shelter with a two-axle trailer carrying the RCP's generator unit. It carries the first- and second-level spare parts giving to the RCP unit its complete autonomy far from the main base.

(3) A Firing Unit (FU). This unit insures the transportation and the setting of the missiles in firing position for the mobile version (6 × 6 truck) (the Renault TRM 10 000 is a typical platform) and comprises:
- (a) A shelter housing the Exocet coastal battery firing installation (ITBC) which computes automatically the missile's firing parameter settings according to the tactical situation surrounding the target and allows rapid fire of one to four missiles against one to four different targets on successive or salvo co-ordinate firings. In addition, this shelter is equipped with the radio communication means (VHF data and voice) with the RCP unit
- (b) Four MBDA MM40 Exocet missiles fixed onto the ramp and ready to be fired
- (c) A platform parameters set which allows the system to be fed with the firing unit heading and attitude
- (d) A hydraulic subsystem to stabilise horizontally the platform and to raise the missile's ramp to the firing elevation
- (e) An air conditioned subsystem for both equipment and personnel on board
- (f) A two-axle trailer carrying the FU's generator unit.

(4) An FU maintenance unit. This is a 4 × 4 truck mounting the operations room shelter. It carries the first- and second-level spare parts giving to the FU its complete autonomy far from the main base.

(5) An ammunition supply unit. This 6 × 6 truck usually stands in alert position ready to reload the firing units after the attacks have been completed. It is capable of carrying six MM40 missiles in their transportation containers. It is equipped with a hydraulic crane which allows four new missiles to be transferred to the firing unit within 1 hour.

The MM40 is stored in and fired like an ammunition round from a nitrogen pressurised glass fibre tubular container-launcher. At the firing, the jettisonable doors of the tube are discarded and as the missile leaves the launch container its folded wings and fins (centred inside the tube by pads) are deployed. The missile comprises the following subsystems:

- (a) a Block 2 active radar homing head
- (b) an equipment bay which contains the weapon's inertial navigation platform, digitial guidance computer, radio altimeter and thermal battery
- (c) a 165 kg HE shaped charged warhead
- (d) a two-stage solid propellant booster and sustainer rocket motor system
- (e) a rear-mounted compartment with auxiliary equipment, this includes a 400 Hz converter, thermal battery, guidance fin actuators and umbilical connectors.

The main characteristics are:

- (a) Fire and forget missile
- (b) Firing possible up to Sea State 7
- (c) Range 5 to more than 70 km
- (d) Offset angle of ±105°
- (e) Obstacle overflying capacity
- (f) True sea-skimming 1.5 to 3 m self-adapted flight altitude
- (g) Inertial mid-course guidance during the first large part of the flight
- (h) Terminal guidance with the Block 2 active radar homing head
- (i) Advanced ECCM against hard and soft kill
- (j) Improved target selectivity
- (k) Multitarget engagement capabilities
- (l) Warhead attack functions by impact and/or proximity able to neutralise any target size.

Specifications

Guidance: inertial with active radar terminal homing
Launch weight: 870 kg
Length: 5.8 m
Body diameter: 0.35 m
Wing span: 1.135 m
Warhead: 165 kg HE shaped charge
Range: 5,000 to more than 70,000 m
Propulsion: two-stage solid-fuel rocket motor sustainer with booster
Speed: M0.93

Status

In production. In service with several countries including the Qatari (at least one battery with three launchers) armed forces, Cyprus (one battery with three launchers) and one NATO navy which is believed to be Greece. A total of nine coastal batteries has been built to date with nearly 20 firing units. Cyprus took delivery of a total of 24 missiles and three launchers in 1994.

To meet the requirements of the Royal Navy, Vosper Thornycroft of the UK developed a coastal defence version of the Exocet MM38. These were deployed to Gibraltar but have now been withdrawn and sold to Chile. Details were given in *Jane's Armour and Artillery 1992-93.* Before being sent to Chile these were refurbished by MBDA. Chile took delivery of eight Exocet missiles of an undisclosed type in 1995.

Manufacturer

MBDA.

UPDATED

Coastal defence version of MBDA *MM40 Exocet deployed in the firing position with stabilisers lowered and launcher elevated*

INTERNATIONAL

MBDA OTOMAT Coastal Defence System (OCDS)

Development/Description

The OTOMAT Coastal Defence System (OCDS) uses the OTOMAT anti-ship missile developed by the now MBDA. The system was originally developed by Matra (France) and Otobreda (Italy).

By early 2002, total production of the OTOMAT Mk 1 and Mk 2 missiles amounted to over 1,000 units.

It has been sold to Egypt (which has three batteries in service and three of its Westland naval Sea King Mk 47 helicopters fitted for target updating of the system). Egypt uses the Berliet GBD (6 × 6) truck as the carrying platform.

A typical OCDS comprises: an Operational Centre Unit (OCU); one or more Coastal Defence Units. The Operational Centre Unit consists of a vehicle housing a tactical display and various communication systems. The tactical situation is established either by a surface search radar (typically the Thales Score coastal radar) or by other facilities (aircraft or helicopter) datalinked with the OCU.

A Coastal Defence Unit consists of:

(a) a Fire-Control Unit (FCU) in a vehicle with a control cabin housing a missile control console, a specific datalink transmitter for updating the target position during the missile flight and various communication systems

(b) two to four Firing Units (FU). Each firing unit consists of the launcher vehicle with two missiles in ready to fire container-launchers and an interface box which is linked by cables with the FCU within a range of several hundred metres. When travelling the missiles are horizontal. On arrival at the firing position two stabilisers are lowered hydraulically to the ground either side of the launcher to provide a more stable firing platform. The two-round launcher is elevated. Blast covers protect the forward control cab of the launch vehicle.

(c) two to four resupply vehicles with crane and two reload OTOMAT missiles in their container-launchers. Except the FCU and FUs all the other units of the OCDS can be separated for tactical and geographic reasons. They are linked together by protected VHF audio and datalinks.

The OTOMAT missile itself has a cylindrical body with a blunt nose and belled exhaust nozzle at the tail. There are delta-shaped folding wings at mid-body with small horizontal truncated triangular-shaped control fins at the rear.

Propulsion for the boost and acceleration phases is provided by two jettisonable TDA/SNPE 3,500 kg thrust, 4 second burn composite solid-propellant rocket booster motors, mounted either side of the missile body aft. The sustained rocket engine is a 400 kg thrust Turbomeca TR 281 Arbizon III turbojet. Cruise speed is M0.9.

Guidance is by a preprogrammed inertial autopilot and radio altimeter for the cruising phase. Terminal homing is by a two-axis Thales I/J-band active radar seeker.

Specifications
Guidance: inertial with active radar terminal homing
Launch weight: 770 kg
Length: 4.46 m
Body diameter: 0.46 m
Wing span: 1.35 m
Warhead: 210 kg HE SAP
Range: 6,000-100,000 m (6,000-180,000 m with fire-and-update facility)
Propulsion: turbojet sustainer with two jettisonable solid-fuel rocket motor boosters
Speed: M0.9

OTOMAT coastal defence missile being launched from its (6 × 6) launcher vehicle ***2001***/0105773

At launch, the weapon climbs to an initial height of 250 m at a distance of 600 m from the launcher. The autopilot/altimeter system then commands it to its cruising altitude for the cruising phase. For a long-range firing the last target position is sent to the missile by the specific updating datalink. At a predetermined point from the target, the radar seeker is activated in its elevation and 20° left/right azimuth scan search mode. When the target is found and the seeker locks on for continuous tracking, the missile is commanded to climb at a distance-to-target of 5,000 m to a height of 175 m whereupon, when it reaches a distance-to-target of 2,000 m it enters a 7° dive to impact the target.

Depending on the size of target and on the prevalent weather conditions, the terminal flight phase can also be conducted in a sea-skimming flight profile. The missile is commanded to descend to 2 m altitude. The choice of the missile attack profile is made before launch.

In the fire-and-forget role with only target designation at the launch phase, the effective range of the missile is around 100 km. With the use of the fire-and-update facility the effective range is increased to 180 km. In the Italian-made system an update can be obtained by means of helicopter-mounted PRT 405 tracking and datalink equipment. Salvo firing achieves a similar improvement to the effective range. Minimum engagement range is 6,000 m. The initial missile course can be up to 200° away from the launcher platform heading with the missile adopting an angled flight path if a helicopter is used for a mid-course guidance update.

Variant

An updated version of the OTOMAT missile, the Mk 3, is under development with a new engine and computer. In addition to the basic features of the Mk 1 and Mk 2 versions, the Mk 3 will have a number of improvements including obstacle avoidance, salvo co-ordination and target selection.

The guidance system will be the dual-mode homing head, involving an Officine Galileo (which is also part of Alenia Difesa) Ku-band radar and an Imaging Infra-Red (IIR) seeker, being developed for the Teseo Mk 3 stealth missile project.

Status

Production as required. Missile in service with Egypt, Italy, Kenya, Libya, Malaysia, Peru, Saudi Arabia and Venezuela. Of these, Egypt is known to operate the coastal defence version.

Manufacturer

MBDA.

UPDATED

OTOMAT coastal defence system on 6 × 6 truck chassis with launcher elevated

Polyphem coastal defence missile system

Development

In 1994, an international MoU was signed between France, Germany and Italy covering the exploratory development of the Polyphem optical fibre guided missile.

This international consortium comprised the then Aerospatiale Missiles of France, Dasa LFK of Germany and Italmissile of Italy and is also referred to as Trifom (Trinational Fibre Optic Missile).

The original intention was to develop Polyphem for land-based applications but it is now being developed for a much wider range of applications including air

Artist's impression of two Polyphem fibre optic guided missiles over the sea **2000**/0085804

Trial launch of a Polyphem fibre optic guided missile (DGL/CEL) **2000**/0069228

launched from a helicopter, ship and submarine and in a coastal defence application.

Polyphem is also being proposed for the British Army's LIMAWS (Rocket) programme, details of which are given in the Multiple rocket launchers section under the United Kingdom.

It is also possible that the British Army's current Multiple Launch Rocket System (MLRS) could have a dual role, for example, launch Polyphem and the current and future range of unguided and guided MLRS rockets.

The first firing of the air vehicle mock up was carried out at the Landes test centre in France in July 1995 and the following year an MBDA C22 target drone was used to test the in flight unwinding of a 60 km fibre optical cable to its full length.

Early in 1997, the first firing of a Polyphem missile from its launching container took place in Italy and this was followed by three further firings to validate the separation of the missile from the container.

In April 1997, the first complete flight was carried out at the Landes test centre. This experimental missile weighed 100 kg and used a reel of qualified cable and had a 600 kg plus solid propellant booster that fired for about 3.5 seconds after which the main engine cut in and the missile was guided to its target.

In March 1998, the governments of France, Germany and Italy issued a new MoU under which the manufacturers in the three countries would supply by 2001 a Polyphem missile system based on a light truck for trials purposes.

According to the consortium, the main features of the Polyphem in the coastal defence role can be summarised as:

(1) The ability to carry out precision strikes at ranges of up to 50 km plus

(2) The nose-mounted imaging infra-red camera assures precision targeting

(3) High ECM resistance and a low signature

(4) Advanced warhead technologies

Description

A typical Polyphem coastal defence battery would consist of a number of firing units, typically three and a fire direction centre where the missile operator would be seated.

Target information would come from radar or another sensor and all elements would be based on 6 × 6 trucks for improved cross-country mobility. All of these elements would be fully air portable in a Lockheed Martin C-130 transport aircraft.

The actual missile is stored in a sealed container which is also used for launch. It can be stored for more than 10 years without any preventative maintenance. The launcher platform aligns the missile for firing and ensures the connection of the mechanical, electrical and optical interfaces.

The missile is launched from its container by a solid propellant booster to an initial cruising speed of 180 m/s after which the main turbojet engine is ignited.

This powers the missile throughout the remainder of the flight that can be varied from 5 to 600 m above the surface, although normally it would fly well below the latter figure.

The use of a turbojet engine allows the guidance system and operator to control the speed of the missile. In addition to engaging surface targets it can also engage helicopters.

Specifications

Max range: 60 km
Speed: 120 to 200 m/s
Flight altitude: 20 - 400 m (variable)
Launch angle: 60°
Weight at launch: 145 kg
Length: 3.3 m
Weight: 20 kg shaped charge/fragmentation

The nose-mounted stabilised focal plane array seeker is of the 3.5 to 5.5 μm wavelength thermal imaging infra-red type and was successfully tested in 1994 on a captive missile carried under a Mystère trials aircraft.

The nose-mounted seeker can detect targets up to a range of 8,000 m which is reduced to 3,000 m under conditions of poor visibility. It has a field of view of 7.5 × 10° and can be panned plus/minus 30°. It is mounted on a gryostabilised platform which is claimed to provide excellent image sharpness for the processing system and operator display.

The heart of the navigation system is a laser inertial measurement unit supported by an altimeter and GPS receiver. This configuration ensures that the missile follows a predetermined trajectory into the designated target area.

Polyphem is designed to be fully automatic with the operator being involved to validate the automatic selection and, if required, adjust the final precise hit on the target.

The fibre optic datalink is capable of data transfer rates of several hundred megabytes per second thus

IVECO (4 ×4) with six round Polyphem fibre optic guided missile launcher elevated **2002**/0044134

Artists impression of the Transportpanzer 1 (6 × 6) armoured vehicle with pod of six Polyphem fibre optic guided missiles elevated and missile leaving launcher **2002**/0130599

allowing not only for real-time guidance but also for the acquisition and storage of operational intelligence for other applications.

The gunners firing post is normally installed in the fire direction centre and includes a console for the remote control of the actual missile and its nose-mounted seeker imager and a display unit for the images received.

It also includes the high-performance compact computer for guidance, control and image processing, mission planning and imbedded infra-red simulator for training purposes.

The fire-control computer fulfills all mission planning requirements before the actual target engagement including the display of waypoints, restricted areas and missile course during flight. Some applications will require a dual monitor solution.

The fire-control computer supports the trainer/simulator system for on the job training without any additional hardware.

The Polyphem is fitted with a modular warhead that is claimed to be highly effective against ships, helicopters, radar stations and bridges.

It consists of a multipurpose warhead, ignition electronics, impact contacts and safety devices. It contains at least 20 kg of explosive and the warhead serves both requirements of penetration and fragmentation.

Status
Development.

Manufacturer
See text

UPDATED

IRAQ

FAW coastal defence missile systems

Development/Description
At the First Baghdad International Exhibition for Military Production, held in 1989, Iraq showed a series of coastal defence missiles under the name FAW.

The smallest missile was designated FAW-70 and appears to be a locally built version of the Russian P-20/P-21 (US/NATO designations SS-N-2c 'Styx') weapon. It has a maximum range of 70,000 m and a minimum range, depending on a number of factors, of 5,400 m.

The general configuration of the FAW family of coastal defence missile systems is similar to that of a small aircraft, with delta planform foldable wings and a triple tail surface arrangement. A jettisonable solid-fuel booster motor is used for the launch and acceleration phases, after which an internal liquid-fuel sustainer rocket motor cuts in to provide a cruising speed of M0.9. Maximum length is 6.5 m, the body diameter is 0.78 m and launch weight (complete) is about 2,500 to 2,600 kg.

Guidance for the mid-course phase is by autopilot with the terminal phase guidance either active I-band radar or passive infra-red, depending upon the seeker head type fitted. The seeker is switched on at around 10,000 m from the predicted target position, with alternative wide and narrow gate facilities so that it can either attack any ship in a formation or one in particular.

The flight altitude can be preset in multiples of 50 between 50 and 300 m with the terminal trajectory being effectively a shallow dive. The warhead weighs 500 kg and contains conventional HE.

The FAW-150 is thought to be an improved version of the FAW-70 with increased length of 7.4 m and enlarged fuel tanks for a maximum range of 150 km, while the FAW-200 is an even longer (8.0 m), maximum 200 km range variant.

To achieve maximum effectiveness with all three missile types an over-the-horizon targeting platform, such as a ship, plane or helicopter equipped with radar and datalink to the battery launch position, is required.

Specifications
Provisional
Guidance: autopilot with active radar terminal homing
Launch weight:
(FAW-70) 2,500-2,600 kg
(FAW-150) 2,650 kg
(FAW-200) 2,740 kg
Length:
(FAW-70) 6.5 m
(FAW-150) 7.4 m
(FAW-200) 8.0 m
Body diameter: 0.78 m
Wing span: 2.75 m
Warhead: 513 kg HE
Range:
(FAW-70) 70,000 m
(FAW-150) 150,000 m
(FAW-200) 200,000 m
Propulsion: liquid-fuel rocket motor sustainer with jettisonable solid-fuel rocket motor booster
Speed: M0.9

Status
The FAW-70 entered operational service in 1990. The FAW-150 and FAW-200 are thought to have been developed to the preproduction stage.

As far as is known, none of the FAW family of coastal defence missiles are currently in producton and it is considered doubtful if any now remain in front line service. It is possible that some of the FAW 70 missiles are held in store or reserve. As far as it is known, none were ever exported by Iraq.

Manufacturer
Military Production Authority.

UPDATED

FAW-150 coastal defence missile (Christopher F Foss)

FAW-200 coastal defence missile (Christopher F Foss)

JAPAN

Mitsubishi Type 88 coastal defence missile

Development/Description
Development of the Type 88 (SSM-1) ground-launched coastal defence missile began in 1981 with the issue of the first proposal and it has been the joint responsibility of the Japanese Defence Agency's Technical Research and Development Institute and Mitsubishi Heavy Industries.

Like all Japanese missiles, the Type 88 coastal defence missile system has been developed only for the home market and has not been offered on the export market.

Based on the Type 80 ASM-1 air-launched anti-ship missile, the first prototype weapons were built in 1983 with the first trial launches occurring in 1985. The test programme was completed in 1987 and first production round deliveries were made in 1988.

Until recently, the Type 88 force structure of the Japanese Ground Self-Defence Force consisted of five artillery regiments. Of these, three were in the Northern Army and one each in the Northeastern and Western Armies.

Each of these regiments has 16 launchers (based on a 6 × 6 cross-country truck chassis) plus associated fire control and support vehicles. In addition, there is an eight launcher battalion in the Central Army.

Another batch of 24 launcher vehicles is now being delivered which will enable the Central Army battalion to be expanded to a regiment with 16 launchers. In addition, a regiment in the Eastern Army Area will be equipped with 16 launchers.

Other elements of a Type 88 regiment include a surveillance/targeting system on a Type 73 (4 × 4) light vehicle carrying Type 85/JTPS-P11 radars, a command and control vehicle, Type 73 (6 × 6) trucks fitted with a fire-control system, Type 73 data relay light trucks and crane equipped reload vehicles which use the same chassis as the Type 88 launchers.

The first six TELs and associated fire controls were delivered in mid-1988, for use by the Japanese Self-Defence Force's Fuji Instruction Brigade at the Fuji School, the Signal School and the Ordnance School.

The TEL has six hydraulically elevated cylindrical container-launcher canisters mounted in two layers of

Mitsubishi Type 88 coastal defence launcher deployed in firing position with resupply vehicle to rear (Paul Beaver)

Radar surveillance targeting vehicle based on the Type 74 (4 × 4) light vehicle (Paul Beaver)

Mitsubishi Type 88 coastal defence launcher in travelling configuration

three. The six canisters are reloaded as one unit direct from the reloader vehicle.

The length of the missile is 5.08 m with its jettisonable solid-propellant booster motor and 4 m without. The sustainer motor is an MHI TJM-2 microturbojet which gives a cruising speed of M0.9.

The Type 88 will be launched 50 km from the coast at targets up to 100 km out to sea. Once fired it adopts a terrain-avoiding flight profile using an inertial guidance package until it clears the coastline. It then descends to a low-altitude cruising height until the active radar seeker acquires a designated target. The missile finally adopts a sea-skimming flight profile for the terminal attack phase. The missile is fitted with a conventional high explosive warhead.

Specifications

Guidance: inertial with active radar terminal homing
Launch weight: 661 kg
Length: 5.08 m
Body diameter: 0.35 m
Wing span: 1.16 m
Warhead: 225 kg HE
Range: 150,000 m
Propulsion: turbojet sustainer with jettisonable solid-fuel rocket motor
Speed: M0.9

XSSM-2 missile

This is the follow on to the current Type 88 (SSM-1) and development is expected to be completed in the near future. It has been designed for the ship and coastal defence roles and will have a range of at least 250 km.

Status

Production complete. In service with Japanese Ground Self-Defence Force.

Manufacturer

Mitsubishi Heavy Industries.

UPDATED

KOREA, NORTH

130 mm M1992 self-propelled coastal gun

Development

The 130 mm M1992 self-propelled coastal gun is essentially the North Korean AT-S full tracked armoured chassis fitted with a turret at the rear. This is armed with the Russian 130 mm SM-4-1 coastal gun which, in its normal configuration, is based on a two-axle, four wheeled carriage and towed by a tracked prime mover or 6 × 6 truck. Full details of the 130 mm SM-4-1 coastal gun are given later in this section under Russian Federation and Associated States.

The driver and commander are normally seated at the front with the engine compartment located to their rear. The suspension either side consists of five road wheels with the drive sprocket at the front and the idler at the rear. There are no track return rollers.

When travelling, the 130 mm ordnance is held in a travel lock mounted on the top of the hull. To provide a more stable firing platform, spades are lowered at the rear of the chassis. It is understood that this is carried out manually. The 130 mm SM-4-1 coastal gun has a maximum range of 29,500 m and a maximum rate of fire of 5 rds/min.

Full details of the 130 mm SM-4-1 coastal gun are given later in this section under Russian Federation and Associated States. The standard towed 130 mm SM-4-1 is normally used in conjunction with an off carriage fire-control system. It is not known if this is used with the North Korean M1992 system. The M1992 designation is that allocated by the United States and indicates the first year that it was observed.

Specifications

Combat weight: 18,000 kg
Length: 8.3 m
Width: 2.78 m
Height: 2.9 m
Engine: 300 hp diesel
Armament: 1 × 130 mm gun
Max range: 29,500 m

Status

Production complete. In service with North Korea.

Manufacturer

North Korean state factories.

UPDATED

NORWAY

Kongsberg Defence & Aerospace 20 mm coastal defence stationary gun mount

Development/Description

Kongsberg Defence & Aerospace developed this 20 mm coastal defence stationary gun mount for the Royal Norwegian Navy. Production has now been completed and it is no longer being marketed. A 25 mm model armed with a German Mauser MK 25 Model E cannon was also proposed, but this is no longer being marketed.

Installed over a single-storey underground installation, the turret can be used as a self-contained

Specifications

Turret
Crew: 1 (turret gunner), other team members are within the underground installation
Weight: approx 1,000 kg
Turret bearing:
(outer diameter) 1,200 mm
Armament: 1 × Rheinmetall 20 mm MK Rh 202 automatic cannon
Rate of fire: 900 rds/min
Ammunition: as required in installation
Muzzle velocity: 20 mm Rh 202 1,050 m/s
Optics
Ground/sea target magnification: ×8
Air target:
(optical system) a combined day and night sight is an option
Elevation/depression: +50/−10°
Traverse: 360°
Drive: electrical-servo with variable traversing speeds from a few mrad/s to a max 90°/s or manual

Kongsberg Defence & Aerospace 20 mm coastal defence stationary gun mount

Artist's impression of Kongsberg Defence & Aerospace 20 mm coastal defence stationary gun mount and underground accommodation

fort unit operating for a predetermined period without being resupplied or used as a transportable fort unit for semi-permanent installation at prepared sites.

The German Rheinmetall 20 mm automatic cannon MK 20 Rh 202 turret armament can be used against ground, sea or air targets. In the latter role the 20 mm version is used by the Norwegian coastal artillery force to provide anti-aircraft defence of its coastal defence installations.

The turret is armoured but can accept additional armouring and camouflage coatings as required. It can also be fitted with ventilation and air conditioning systems for use in extreme environmental and NBC conditions. Optical sights for both ground and air targets are fitted.

Penguin coastal defence missile system

Details of the Kongsberg Defence & Aerospace coastal defence Penguin missile system were given *in Jane's Armour and Artillery 2001-2002.* The system never entered production in this application.

Status

Production complete. In service with Norwegian coastal artillery force (order placed January 1986 by Royal Norwegian Navy Matériel Command). No longer being marketed.

There are no known exports of either of this turret system. Kongsberg Defence & Aerospace is no longer marketing this system.

Manufacturer

Kongsberg Defence & Aerospace AS.

UPDATED

Norwegian mine and torpedo upgrade

Development/Description

In June 1990, the Royal Norwegian Navy Matériel Command placed a NKr500 million plus contract with Kongsberg Defence & Aerospace AS (at that time known as NFT), to undertake a comprehensive modification of existing underwater minefields and land-based torpedo batteries located at various coastal artillery fortresses along the Norwegian coast.

The contract involved the modernisation of three controllable minefields and three torpedo batteries, and included a three year development and test period before series production.

Following the completion of this programme, 14 batteries of elderly 105 and 150 mm guns will be discarded and the coastal defence force will essentially comprise nine batteries of modern 75 and 120 mm guns, 12 batteries of Hellfire missiles, a few batteries (out of the current seven) of 127 mm guns, the three modernised controllable minefields (out of a current eight minefields) and five torpedo batteries (with three modernised).

Artist's impression of Norwegian Mine and Torpedo Upgrade in operational use

Controllable minefields

The modernisation programme for the minefields involves the installation of a new weapon control system, based on Kongsberg Defence & Aerospace KMC 9000 tactical/fire-control console and KS 900 computer that have been used in Norway's 'Ula' class submarines, 'Oslo' class frigate modernisation and the NASAMS (Norwegian Advanced Surface-to-Air) programme. The computer is connected to a mine panel which receives regular status reports from the minefield and from individual mines.

The target detection is based on radar and electro-optic sensors which are connected to the weapon control via a fibre optical link. The electro-optical sensors include TV camera, laser range-finder and thermal camera, and are protected by a hydraulically operated, camouflaged, reinforced steel cupola. Both sensors are controlled from the operator's console.

During the development of the new controllable minefield, the highest priority has been peacetime safety. The Royal Norwegian Navy's moored and ground mines are upgraded with new safety and arming, new mine electronics and mine communication systems.

Horten T1 torpedo used in the Norwegian coastal defence role showing launcher rail

Torpedo systems

The torpedo battery modernisation included the same weapon control system as the controllable minefield. The Horten T1 torpedo (built at the Horten torpedo

Interior of the torpedo and minefield control system showing operator's controls and monitor screens

factory to the German Second World War G-7a torpedo plans) has been upgraded with a new guidance unit. The torpedo is launched from a ramp at an angle of 20°, emerging underwater. The weapon control concept and the loading and launching equipment will permit multiple torpedo guidance, autonomous torpedo operation and effective multisalvo capacity. The torpedo launching equipment is connected to the weapon control via a fibre opti link.

The old T1 torpedo is planned to be taken out of service and the Royal Norwegian Material Command will place an order with Kongsberg Defence & Aerospace to upgrade the torpedo batteries with a more recent Swedish torpedo.

Communications

Both the controllable minefields and the torpedo batteries will be equipped with new communications equipment for tactical and administrative use.

Training systems

Included in the contract is a comprehensive training package which includes the following components:

(1) Tactical trainer - Located at the Royal Norwegian Navy's training base, the trainer will comprise an instructor's console and two training consoles. The instructor will use the unit to simulate an operational scenario by projecting maps of threatened areas on a screen and tying in appropriate sensor inputs from associated databases. Confronted by the scenario, the trainees will have to react to the various threats and develop the appropriate combat response procedures.

(2) Mine simulator - This will provide trainees with an overview of all subsea activities and allow them to carry out simulated firings of all mine types, either individually or in groups against various types of target.

(3) Technical courses - A series of specialised technical courses will be undertaken at the appropriate equipment factory.

81 mm mortar systems

In a further programme to upgrade coastal defences, the Royal Norwegian Army awarded a £1.2 million contract in December 1991 to the now BAE Systems, RO Defence of the UK, for the supply of 110 81 mm L16 mortar tubes, baseplates and associated equipment. The mortars are operated by Royal Norwegian Navy personnel and are fitted in shore defence bunkers using existing elevating gear. Delivery of the weapons occurred in two batches, during September and December 1992.

Status

Production as required. Three controllable minefields and three torpedo batteries are now in service.

Prime contractor

Kongsberg Defence & Aerospace AS.

UPDATED

RUSSIAN FEDERATION AND ASSOCIATED STATES (CIS)

Barrikady 130 mm A-222 Bereg (Coastline) self-propelled coastal artillery system

Development/Description

Initial development of this system began by the Barrikady Production Association in Volgograd in 1976 with first prototype trials in January 1987. Government service testing took place between November 1992 and May 1993 in the Crimea. This system was retained by the Ukraine but has now been returned to Russia. For several years the system has been offered on the export market but as of early 2002 there are no known sales.

According to the manufacturer, the main advantages of the 130 mm Bereg coastal defence system can be summarised as:

(1) Ease of deployment, ease of concealment and the key parts of the system can be dispersed up to 2 km from each other in range and up to 300 m in height, this also helps to ensure the overall survivability of the system

(2) Effective engagement of sea targets travelling at a maximum speed of up to 200 kt out to a maximum range of 20,000 m, simultaneous data preparation on four targets and fire against two targets with any combination of weapons, in addition the system has a secondary role against ground targets

(3) Tracking of sea targets and correction of fire via the fire-control system of the central combat station

(4) Independent execution of combat missions by separate self-propelled 130 mm guns

(5) Independent autonomous duty for up to seven days.

The 130 mm medium-calibre Bereg self-propelled coastal artillery system is based on an all-terrain (8 × 8) MAZ-543M truck chassis. It is designed to engage small- and medium-sized surface ship and surface-

130 mm A-222S1 self-propelled coastal artillery mount with turret traversed right

BR-136 central station of the 130 mm A-222 self-propelled coastal battery deployed in firing position

Combat duty support vehicle of the 130 mm A-222 self-propelled coastal battery in travelling configuration

effect plane targets with speeds up to 200 kt and, if required, ground targets. The system takes between 5 and 30 minutes to come into action.

The basic battery comprises the following units on the MAZ-543M (8 × 8) chassis:

(1) A BR-136 central command and control station with a 360° elevatable surveillance and target acquisition radar, and ±135° scanning TV target detection and tracking system which also includes a laser range-finder. The BR-136 system provides all-weather fire control and ECCM for the battery. Up to four radar reflecting targets can be automatically tracked at any one time, with their movement parameters simultaneously computed for elevation and traverse firing solutions on all six battery guns. Any two of the four targets being tracked can be simultaneously engaged in independent actions by the battery, using the two tracking unit consoles and the two weapon fire-control consoles. The BR-136 can set the 130 mm guns to fire in single rounds or at a rate of 4, 6, 8 or 10 rds/min.

The van body also contains an auxiliary diesel power electric generator with sufficient fuel for seven days of continuous operation, a land navigation system and battery/higher echelon level communications and data transfer equipment. There are four hydraulic jacks used to stabilise the station at its operating site and ensure that the radar antenna is horizontal.

The maximum distance between the central station and the gun mounts is 1,000 m and in altitude 300 m. The system is designed to travel at a maximum road speed of 60 km/h and have a range of 650 km on internal fuel. Depending upon the situation the battery emplacement time can vary between 5 and 30 minutes. All three vehicles have NBC defence and warning equipment features fitted as standard

The central station can also be used with other coastal defence gun systems. The fire-control system is also referred to as the Podacha and is claimed to be highly effective even in a ECCM environment.

(2) Between four to six 130 mm A-222S1 self-propelled gun mounts comprise the battery's firepower element. The gun mount is on the rear of the 8 × 8 truck chassis and carried in the travel position with the barrel located forwards between a gap in the gun crew cab and its muzzle brake, overhanging the driver's cab. For firing, the gun tube is elevated and the turret traversed 180° to the rear. Before firing, four hydraulic stabilising jacks have to be lowered. The vehicle also has its own diesel-powered auxiliary electric generator system.

The gun uses components from the 152 mm 2S19 full tracked self-propelled artillery system that has been in service with the Russian Army for some years. The 130 mm gun is used in the A-222S1 and has identical ammunition and ballistics to the naval AK-130 ZIF-94 130 mm mount. It can operate in several modes:

(a) automatic, with the central station feeding digital codes into the gun's fire-control system

(b) semi-automatic, in which the gunner uses his own gunsight assembly

(c) manual, in which the system is independent of the central station and uses its onboard ballistic computer, target acquisition and tracking sight and range-finder assembly.

Types of 130 mm ammunition fired include high explosive, fragmentation, training and a nose fixed-air defence round. During firing, the laying of the gun is controlled by a correction input system which includes roll sensors and optical sight. The minimum basic load of 48 130 mm rounds is kept inside the turret in two ammunition stowage boxes. The shells are fed into the boxes by the aid of charging trays. Maximum effective range of the 130 mm gun is stated to be around 30 km.

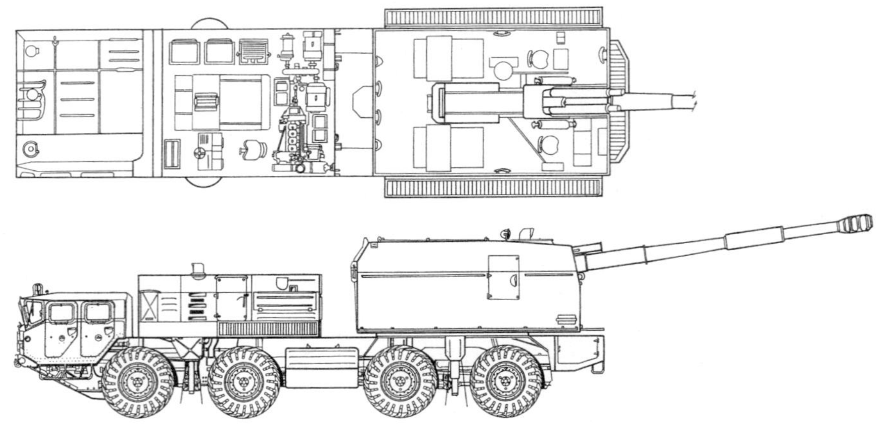

Bereg 130 mm A-222S1 Coastal Defence Artillery System. Top is plan view showing position of main components. Lower is side elevation with turret traversed to the rear (Steven Zaloga)

Specifications

BR-136 central station
Crew: 7
Configuration: 8 × 8
Combat weight: 43,700 kg
Weight:
(fire-control system) 5,000 kg
(radar antenna assembly) 470 kg
(TV/laser station) 235 kg
Length: 15.2 m
Width: 3.24 m
Height: 4.415 m
Max speed:
(road) 60 km/h
Max range:
(road) 650 km
Max radar target acquisition range: 35,000 m
Antenna speed: 15-25 rpm

Specifications

130 mm A-222S1 gun mount
Crew: 8
Configuration: 8 × 8
Armament: 1 × 130 mm gun
Elevation limits: –5 to +50°
Traverse: ±120°
Rate of fire: max 10 rds/min
Basic load: 48 rounds
Ammunition types: HE-Frag (with base fuze), AA (with nose fuze), Practice and Training
Combat weight: 43,700 kg
Length: 12.95 m
Width: 3.2 m
Height: 3.925 m
Max speed:
(road) 60 km/h
Max range:
(road) 650 km

Specifications

Combat duty support vehicle
Crew: 4
Passengers: 10
Configuration: 8 × 8
Fuel, food and water reserves: 7 days
Armament: 1 × 7.62 mm PKT MG
Combat weight: 43,500 kg
Length: 15.936 m
Width: 3.23 m
Height: 4.415 m
Auxiliary power generators: 2 × 380/220 V AC, 50 Hz, 30 kW
Speed:
(road) 60 km/h
Range:
(road) 650 km

Note: Details of the MAZ-543M (8 × 8) chassis, powered by a 525 hp diesel engine and manufactured by the Minsk Wheel Tractor Plant of Belarus, are given in *Jane's Military Vehicles and Logistics 2002-2003.*

(3) A combat duty support vehicle contains the main diesel-powered electric generator unit (with two 20 kW generators) supplying the central station at its command location. Sufficient fuel is carried for seven days of continuous operation. It also has two rest compartments, a kitchen and a dining room for the central station's crew. A small armoured turret is also fitted at the rear of the cab roof for limited anti-aircraft defence purposes. This is armed with a 7.62 mm PKT machine gun. The food reserves and kitchen equipment can feed 10 people for seven days.

Upgraded Bereg

The prime contractor has recently stated that if requested by the customer, Bereg can be fitted with a 127 mm gun rather than the normal Russian 130 mm gun.

In addition, the weapon could be provided with laser guided projectiles, which when used in conjunction with a laser designator, could give pin point accuracy to engage land and naval targets.

Other improvements suggested include a jam immune radar channel to exchange telecoded information between separate vehicles, an air defence capability and level of mechanisation and automation of operation during deployment to reduce into and out of action times.

Status
Development complete. Ready for production on receipt of orders.

Manufacturer
Barrikady Production Association.

UPDATED

130 mm coastal gun SM-4-1

Development/Description
Development of the 130 mm SM-4-1 towed coastal gun system began in May 1944 by the Naval Artillery Design Bureau (later known as TsKB-34). Factory and range testing took place in 1948-49. Government testing took place in 1950-51. Service acceptance was in October 1951. The 130 mm SM-4-1 system used the Moskva-TsN fire-control system and the ZALP-B radar.

As these did not meet all the requirements in terms of accuracy a new fire-control system, the Burya-MT-4, and radar, Burun, began development in 1950. Government testing and service acceptance took place in 1955. The new system had the designation SM-4-1B.

The carriage of the 130 mm SM-4-1 has removable front and rear axles each of which has four rubber-tyred roadwheels. In action, the axles are removed and the four stabilisers are extended and staked to the ground. The stabilisers are mounted two either side of the carriage and, when travelling, the stakes are mounted on the front of the shield, two either side. The shield has slightly curved sides and a removable centre section which allows the weapon to obtain a maximum elevation of +45°. The long-stepped 130 mm barrel is 58 calibres long and has an unusual muzzle brake and a travelling lock. Ammunition used with the 130 mm SM-4-1 is not interchangeable with the Russian 130 mm Field Gun M-46 or the 130 mm anti-aircraft gun KS-30.

Specifications
Calibre: 130 mm
Weight:
(travelling) 19,000 kg
(firing position) 16,000 kg
Length:
(barrel, overall) 7.6 m
(travelling) 12.8 m
Width:
(travelling) 2.85 m
Height:
(travelling) 3.05 m
Axis of bore:
(at 0°) 2.03 m
Ground clearance: 0.3 m
Tyres: 10.50 × 20
Elevation/depression: +45/–5°
Traverse: 360°
Rate of fire: 5 rds/min
Range: 29,500 m
Towing vehicle: AT-T Heavy Tracked Artillery Tractor

130 mm coastal gun SM-4-1 in travelling order

The 130 mm coastal gun SM-4-1 is known to fire an APHE projectile weighing 33.6 kg with a muzzle velocity of 1,050 m/s which will penetrate 250 mm of conventional steel armour at an incidence of 0° at a range of 1,000 m. It will also fire an HE projectile weighing 33.4 kg with a muzzle velocity of 1,050 m/s.

The SM-4-1 is normally used in four-gun batteries in conjunction with a radar fire-control system, which gives it the ability to engage targets in all weathers. Small quantities of these weapons were captured from Egypt when the Israel Defence Force captured Sharm el Sheikh in the 1967 Middle East War.

According to Russian sources, factory no 221 produced a total of 95 production SM-4-1 weapons (94 + 1 for training) between 1952 and 1956 with the yearly totals being 1952 (nil), 1953 (18), 1954 (36 + 1 training), 1955 (40), 1956 (40) and 1957 (8). Additional weapons were built by the SKMZ facility. In 1984, the Russian Navy had a total of 140 of these weapons of which 32 were deployed and 108 held in reserve.

Variant
North Korea has developed and deployed a full-tracked self-propelled version of the Russian-supplied 130 mm SM-4-1 towed coastal gun. This has the US designation of the North Korean M1992 130 mm SM-4-1 self-propelled gun.

The designation of M1992 is that assigned to the system by the US Army and indicates the year that it was first observed.

The chassis of the M1992 is based on an armoured North Korean AT-S chassis which is a further development of the Russian ATS-59 full-tracked chassis. Available details of this system, which has not been exported, are given earlier in this section under Korea, North.

Status
Production complete (SM-4-1 production 1952-56, SM-4-1B production 1955-56: total of 140 mounts produced of both types).

Known to be in service with Bulgaria (4), China (who also took delivery of the 100 mm KSM-65 weapon and its associated Moskva fire-control system), Egypt, North Korea (towed and self-propelled), Russia, Syria and the Yemen (36). In most of these countries the 130 mm SM-4-1/SM-4-1B coastal guns are understood to be now held in reserve.

Manufacturer
Russian state factories.

UPDATED

NPO Mashinostroyenie Bastion (SSC-X-5) coastal defence missile system

Development
Currently under development by the NPO Mashinostroyenie design bureau is a new supersonic anti-ship missile called Oniks (Onyx). Its Western designation is the SS-N-26.

This advanced missile has the Russian system designation of 3K55 with the export designation being the 3K55E Yakhont (Gem).

Although the first application of this missile is in the naval surface-to-surface role, it is also being marketed for other applications including submarine and air launched.

The coastal defence missile version, still at the design stage, is known in the West as the SSC-X-5 Bastion. Although development of the baseline missile started as far back as 1985, the first test launches did not take place until the early 1990s and by 1997 at least 10 all-up launches had taken place. As of early 2002, the Bastion coastal defence missile had not entered production or service.

Specifications
Missile
Guidance: inertial
Launch weight: 3,000 kg
Launch container weight: 900 kg
Length:
(missile) 8.9 m
(launch container) 9.0 m
Body diameter: 670 mm
Launch container diameter: 710 mm
Warhead: 250 kg anti-ship shaped charge
Min range: n/avail
Max range:
(hi-lo) 300 km
(lo-lo) 120 km
Max altitude: 14,000 m
Terminal approach: 5-15 m
Propulsion:
(boost) solid rocket
(sustainer) Plamya ramjet
Speed: Mach 2.00 to Mach 2.50

Description
The Bastion coastal defence missile version is based on the use of eight transporter-launcher vehicles each equipped with three integrated Transport Launching Containers (TLC), under the control of a single mobile command vehicle in a battery.

A complete battery salvo of 24 weapons can be fired simultaneously under the control of the command post.

The coastal defence version would require a system to locate the sea target before engagement. This could be in the form of a long range land radar system or an airborne radar system. The letter could be a Kamov Ka-31 helicopter with a NII-RT Yastreb search radar with a range of 250 km.

The system was to have been transported and launched from the MAZ-537 (8 × 8) heavy truck but it is understood that the launch platform will now be the more recent KZKT-74281 (8 × 8) cross-country vehicle.

The cruising speed of the missile is stated to be M2.0 to M2.5, using an initial solid-propellant rocket booster that is mounted integrally in the combustion chamber of the missile's ramjet main engine. It is ignition of the latter liquid-fuel engine that sustains the weapon's cruising speed.

The range varies from 120 to 300 km depending upon its mission profile. For the maximum range the missile cruises at 14,000 m altitude under inertial guidance to the pre-allocated target area where the active radar seeker is briefly activated to acquire the target.

It is understood that maximum acquisition ranges are 75 km for a cruiser, 50 km for a destroyer and 25 km for a missile craft. Maximum search angle is quoted as 45° either side.

The weapon then descends rapidly to between 5 to 10 m until impact. The homing seeker final switch on occurs beneath the radar horizon of the target, reducing its ECM defence period. For the 120 km short-range engagement the missile flies most of the profile at between 5 and 10 m with the seeker off. At approximately 25 km from the target the seeker is switched on.

Status

Development. Not yet in production or service.

Manufacturer

NPO Mashinostroyenie.

UPDATED

Raduga Rubezh-A (SSC-3 'Styx') coastal defence missile

Development/Description

In 1970, the MKB Raduga design bureau began development with TsNII Granit (fire-control system and overall integration) and the Barrikady Plant in Volgograd (for the launcher vehicle) of a coastal defence version of the P-15 (US/NATO designation SS-N-2) anti-ship missile. Known as the 4K40 Rubezh-A (Russian for Boundary), system trials were conducted in the mid-1970s with acceptance for service occurring in 1978. The first operational deployment occurring in 1979 when units were assigned to the Pacific Fleet.

The first exports were made in 1983-84 to East Germany (three batteries of six launchers each) and Yugoslavia. Since then, other countries to have accepted SSC-3 'Styx' shipments included Syria (one battalion with 18 launchers) and Libya.

In Russian service the SSC-3 was grouped at Fleet level into brigades of three battalions. Each battalion had three batteries with four Tractor-Erector-Launcher (TEL) vehicles. The SSC-3 supplemented rather than replaced the existing SSC-1b 'Sepal' (details of this system are given in a separate entry). The Russian Navy no longer has any of these SSC-3 'Styx' coastal defence missiles operational.

The TEL is a modified MAZ-543M (8 × 8) heavy wheeled transporter and has a large three-man fully enclosed engagement control compartment added behind the driver's station. Adjacent to this compartment, on the starboard side of the vehicle, is an elevatable I-band Garpun (Russian for Harpoon, NATO designation 'Plank Shave') surveillance radar for target acquisition and tracking. In the travelling configuration the radar antenna is telescoped down on to the roof of the driver's cab. The system also includes an IFF interrogation subsystem and secure radio communications.

To the rear of the control compartment is a tandem cylindrical container-launcher assembly for two KT-161 surface-to-surface missiles. For a missile launch it has to be rotated through 90° and elevated about +20° The front and rear canister covers are then folded open and the weapon fired.

The general configuration of the missile is like a small aircraft with a cigar-shaped body, a rounded nose and mid-body truncated delta planform wings which are foldable. The tail assembly is of a triple surface type and has the liquid fuel turbojet sustainer located beneath it. A jettisonable solid-fuel rocket booster motor for the launch and acceleration phases is attached beneath the missile body, just under the end of the wingroots.

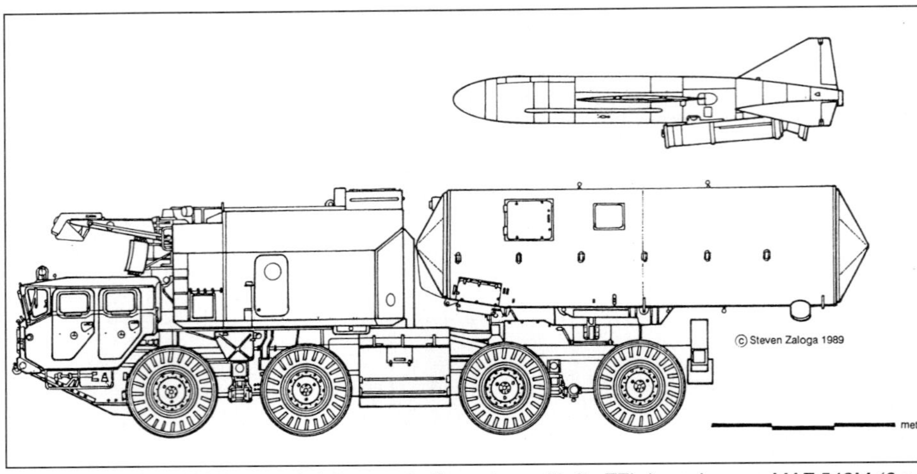

SSC-3 'Styx' (P-21/P-22) coastal defence missile system with its TEL based on an MAZ-543M (8 × 8) chassis (Steven Zaloga)

Guidance for the mid-course phase is by programmed autopilot with precision accelerometers. The terminal phase is handled, depending upon the missile type, either by an active I-band MS-2A radar with an auxiliary passive infra-red seeker in a bulge beneath the nose of the missile for the P-21 or an infra-red homing seeker for the P-22. The auxiliary infra-red seeker is switched on when the missile detects jamming and takes over terminal guidance when the hottest infra-red source has been detected and locked onto. The missile uses a Doppler radio altimeter to cruise at an intial preset cruise altitude of between 100 and 300 m over the sea.

The 'Plank Shave' radar, when elevated to its maximum height, allows engagements to a range of 35,000 m with the TEL's own facilities. The maximum effective range of these 'Styx' missiles, however, is about 80,000 m. This means that a forward observer system, such as an airborne platform, would have to be used to exploit fully the range capabilities.

For engagements, the missile climbs at an initial angle of +45° and then settles to its preset cruise level altitude. The terminal attack phase is effectively a shallow dive ending at approximately a 2 to 5 m sea-skimming height using the onboard radio altimeter to determine the altitude.

In all cases, the system operator manually sets a range gate around the designated target and feeds the derived target bearing and range data to the missile's autopilot/guidance system before launch.

The MAZ-543M (8 × 8) truck 8 × 8 chassis has good cross-country mobility and is fitted with a central tyre-pressure regulation system. This allows the driver to adjust the tyre pressure to suit the type of ground being crossed at any time from within the cab. When deployed in the firing position four stabilisers are lowered to the ground, two either side, to provide a more stable launch platform.

In addition to the TEL, other elements of the system include one SPTA-S set stored at base for up to four

Specifications

Guidance:

(P-21) autopilot with MS-2A active radar and passive infra-red terminal homing

(P-22) autopilot with passive infra-red terminal homing

Launch weight: 2,500 kg
Length: 6.55 m
Body diameter: 0.78 m
Wing span: 2.5 m
Warhead: 513 kg HE
Range: 8,000-80,000 m
Missile flight altitude: 25, 50 and 250 m
Propulsion: liquid amine/UDMH fuel turbojet sustainer with jettisonable solid-fuel rocket motor booster
Speed: M1.3

Launcher
Crew: 6
Configuration: 8 × 8
Weight: 40,900 kg
Length: 13.95 m
Width: 3.15 m
Height: 4.05 m
Max road speed: 60 km/h
Max range: 635 km/h

SSC-3 'STYX' coastal defence missile unit during a ceremonial parade. Front housing contains radar gear and operations cabin; rear housing is on a turntable and contains two 'Styx' anti-ship missiles on separate launch ramps. Note radar scanner in lowered position for travel

TEL and one SPTA-G set supplied for a group of up to 12 TEL. In addition there is also ground equipment, test equipment and a mobile test station for checking missiles before they are loaded in the TEL. Russia calls the TEL the SPL (Self-Propelled Launcher).

Status

Production complete. In service with Bulgaria, Libya (status uncertain), and Syria. It is believed that they are no longer used by Algeria, Romania, Russia, Yemen and Yugoslavia.

Manufacturer

Raduga design bureau.

UPDATED

Granit Central Institute Bal (SS-C-6) coastal defence missile system

Development/Description

The Bal (Russian for Whale) coastal defence missile system (Western designation SSC-6) is based on the same MAZ-543M all-terrain (8 × 8) vehicle chassis used for the Barrikady 130 mm Bereg self-propelled coastal artillery system. Details of this system, which remains at the prototype stage, are given in a separate entry earlier in this section.

The weapon used is the surface-to-surface version of the Zvezda Kh-35, NATO designation SS-N-25 'Switchblade'. This is very similar in appearance to The Boeing Company Harpoon, it is also known as 'Harpoonski'. The standard ship-launched version of this missile is in service with Algeria, India, Russia and Vietnam.

A typical battery comprises: two mobile command and communications vehicles, with 3Ts-24 elevatable surface surveillance radars; four launcher vehicles with eight ready to fire missiles in their steel container-launchers, elevatable assembly on the rear of the vehicle and four transloader vehicles each with eight reload container-launchers; hydraulic crane mounted at the rear for loading purposes. If required, the launcher units can operate in an autonomous mode.

The Kh-35 missile uses a solid-propellant rocket booster for the initial launch phase and an airbreathing turbojet sustainer for the main flight. Guidance is by inertial platform for the mid-course phase with a radar MMS ARGS-35 I-band coherent active radar seeker for the terminal sea-skimming (between 3 to 5 m according to sea conditions) homing phase. Acquisition range of the seeker is 20 km. The speed regime is transonic (more than 300 m/s).

Launch weight is 600 kg, missile length (without booster) is 4.4 m and body diameter 0.42 m. All the Kh-25's fins and control surfaces are folded against the weapon until launch when, as the missile leaves the canister, they unfold.

Artist's impression of the Granit Central Institute Bal coastal defence system deployed with the missile launchers in the background

According to Russian sources the Bal system only takes 10 minutes to come into action. A typical Bal battery would have a total of 46 officers and men.

AS-20 Kayak

This is the air launched version of the system which has been in service for some time and launched from aircraft and helicopters. This is in service with Russia and has been ordered by China.

Status

Offered for export as Bal-E system. As of early 2002, it is not certain as to whether a complete prototype of this coastal defence system had been built.

Manufacturer

Typhoon Joint Stock Company.

UPDATED

Specifications

Missile
Guidance: inertial mid-course with active radar terminal homing
Launch weight: 600 kg
Length:
(without booster) 4.4 m
Body diameter: 0.42 m
Warhead: 145 kg HE-SAP with fragmentation and incendiary effects
Min range: 7,000 m
Max range: 115,000-120,000 m
Propulsion: liquid-fuel turbojet with jettisonable solid-fuel rocket motor booster
Speed: transonic (300+ m/s)

Command, Control and Communications Post
Crew: 7
Configuration: 8 × 8
Weights:
(vehicle fully loaded) 38,200 kg
(onboard equipment) 16,500 kg
Length: 14.65 m
Width: 3.1 m
Height: 4.05 m
Max time for battery into action: 10 min
Max speed:
(road) 60 km/h
(cross-country) 20 km/h
Range: 850 km
Fuel capacity: 700 litres

Launcher vehicle
Crew: 6
Configuration: 8 × 8
Weights:
(vehicle fully loaded) 40,000 kg
(onboard equipment and missiles) 18,300 kg
Length: 13.5 m
Width: 3.1 m
Height: 4 m
Number of missiles: 8
Max speed:
(road) 60 km/h
(cross-country) 20 km/h
Range: 850 km

Transporter-reloader
Crew: 2
Configuration: 8 × 8
Weight:
(vehicle fully loaded) 42,500 kg
(without equipment) 21,700 kg
Length: 13.825 m
Width:
(with missile reloads) 3.62 m
(without missile reloads) 3.2 m
Height: 3.95 m
Hydraulic crane max lift capacity: 9,000 kg
Hydraulic crane max radius operation: 5.8 m
Time to reload launcher: 30 min
Speed:
(road) 60 km/h
(cross-country) 20 km/h
Range: 850 km

Chelomei SSC-1b 'Sepal' coastal defence missile

Development/Description

Development of the P-35 (US/NATO designation SSC-1b 'Sepal') coastal defence missile commenced in the late 1950s by the Chelomei design bureau, as a land-based variant of the standard P-5/P-5D/P-6/P-7/P-35 (respectively US/NATO designations SS-N-3c 'Shaddock'/SS-N-3c 'Shaddock'/SS-N-3a 'Shaddock'/SS-N-3c 'Shaddock'/SS-N-3b 'Sepal') family of anti-ship missiles.

The Russian designation for this system is the S-5 with the missile being referred to as the 4K95 and the launcher vehicle the 2P30. It was developed by the Chelomei Design Bureau (OKB-52) from 1954 with first tests being carried out in early 1957.

The original US/NATO designation for the system was the SSC-1a 'Shaddock', assigned in 1961 when the weapon was first spotted by Western intelligence in use as a naval coastal defence and an army tactical nuclear delivery system.

The first system was accepted for operational service in 1966 as the Redut. This was followed by an upgraded system, the Utes, in 1973 and finally by the definitive Progress system, which had a new onboard navigational system and a more automated fire-control system in 1982.

By the end of the 1970s, the Russian coastal missile artillery forces had approximately 100 launchers in

Specifications
Missile
Guidance: inertial with active radar terminal homing and mid-course update
Launch weight:
(launch) 5,400 kg
(flight) 4,600 kg
Length: 10.2 m
Body diameter: 1.0 m
Wing span: 2.6 m
Warhead: 1,000 kg HE or 350 kT yield nuclear (only fitted to Russian missiles)
Range:
(max) 270,000 m (see text)
(min) 25,000 m
Flight altitude: 100, 4,000 and 7,000 m
Propulsion: liquid-fuel turbojet sustainer with 2 jettisonable solid-fuel rocket motor boosters
Speed: M1.3

Launcher vehicle
Crew: 5
Configuration: 8 × 8
Length: 13.503 m
Width: 2.86 m
Height: 3.53 m
Max road speed: 40 km/h
Range: 500 km

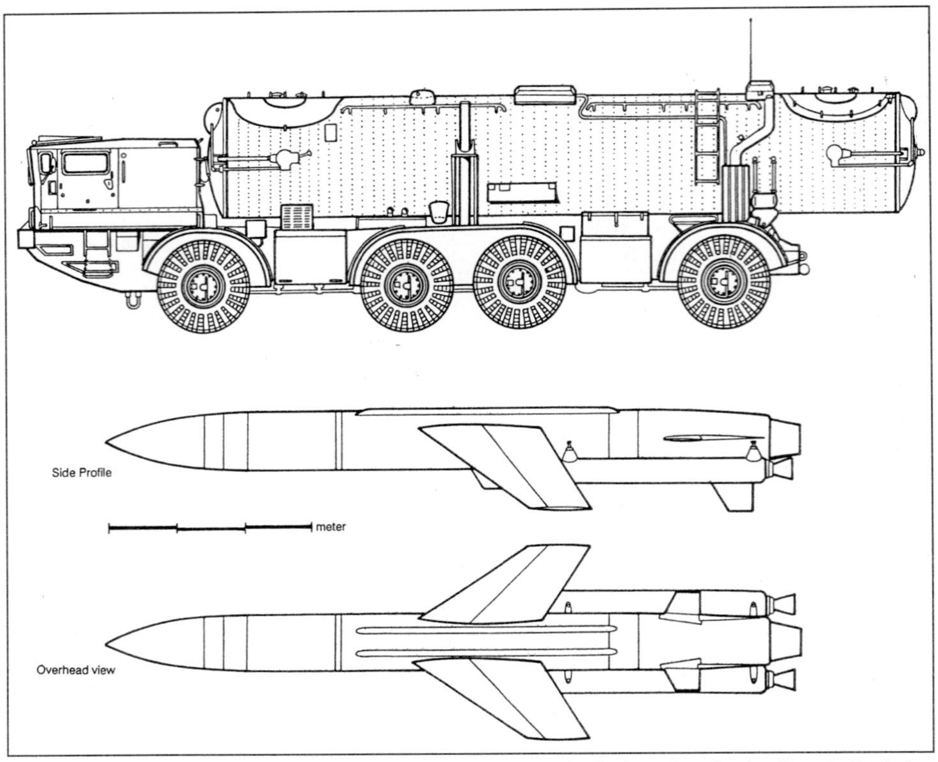

Latest version of SSC-1b 'Sepal' (P-35) coastal defence missile launcher (top) with missile below (Steven Zaloga)

independent coastal defence missile batteries of three launchers and coastal defence missile battalions, with five to six batteries totalling 15 to 18 launchers. The latter were grouped at Fleet level into brigades of three battalions with supporting units. In 1988, the battalion dispositions were: Northern Fleet, three; Baltic Fleet, six; Black Sea Fleet, five; Pacific Fleet, five. Recent information has indicated this system is no longer in front line service with the Russian Navy. The system still remains in service with a number of other countries in declining numbers.

As no missiles have been built for some years the overall combat effectiveness of the SSC-1b Sepal is decreasing year by year.

Initial target surveillance is by a target acquisition battery at brigade level, which hands over the contact to the battalion HQ target acquisition section. This monitors the contact by using its own target tracking radar and interrogates it by a 'Square Head' IFF system. If confirmed as hostile it is passed over to the battery's missile guidance platoon, whereby a modified truck-mounted NATO designation 'Front Door' D-band guidance radar accepts the target and a pair of missiles is launched from one of the battery's launcher sections.

The missile is carried in a sealed cylindrical container-launcher on the rear platform of the ZIL-135MB (SPU-35V) (8 × 8) cross-country vehicle with a hydraulically operated elevation mechanism. For firing, the launcher is elevated to +15 to 20°, the container's hemispherical front and rear ends are hinged upwards and the P-35 fired. The missile then uses a pair of jettisonable solid-fuel rocket booster motors, mounted either side of the sustainer, to eject itself clear for the launch and acceleration phases.

After booster burnout the cruise propulsion is taken over by the KRD-26 liquid-fuel turbojet sustainer with its air intake located, scoop fashion, beneath the belly of the missile.

Guidance for the initial stages of the flight is inertial, with mid-course corrections sent by datalink. If the target is located at the P-35's maximum effective range of 250 km, then the I-band seeker is activated in its search mode at a range-to-target of approximately 165 km. The range and speed of the missile depended on a number of factors including air temperature. For example, with an average speed of 345 m/s and a temperature of +20° C a range of 545 km was achieved.

With an ambient temperature of +40° C and a speed of 338 m/s a range of 650 km could be achieved but when the temperature was –24° C speed was 384 m/s and maximum range dropped to 431 km. The missile's cruising altitude is around 4,500 m. The data derived from the seeker is sent back via the video datalink to the command centre where it can be assessed and corrective guidance signals or target selection instructions retransmitted to the missile as appropriate. A change to the target acquisition mode, followed rapidly by the seeker lock on is normally performed for this target range at a range-to-target distance of 92 to 93 km. The missile then adopts a 20° dive angle to reach the target's vicinity for the final terminal-phase pursuit course run-in at a shallow angle.

For maximum effect in an over-the-horizon engagement, an airborne platform such as a Tupolev Tu-95 'Bear-D', Kamov Ka-25 'Hormone-B' or Tupolev Tu-16 'Badger D' fitted as a data relay platform to the command centre, has to be used, otherwise the missile's range is effectively limited to that of the radar horizon.

Variant

The Czechoslovakia army has deployed an RPV system called the VR-3 Rejs which is transported and launched from a very similar ZIL-135 series 8 × 8 platform to that used for the SSC-1b 'Sepel' coastal defence missile system. There is also a resupply vehicle on a similar ZIL-135 series 8 × 8 chassis.

Status

Production complete. In service with Angola, Bulgaria, Syria (one battalion of 18 launchers with conventional warhead-equipped S-35) and Yugoslavia (three batteries). As replacement missiles cannot be obtained, the long-term effectiveness of these systems must be in doubt. The Syrian missiles are operated by the army and the Yugoslav missiles are believed to be no longer operational.

Manufacturer

Chelomei design bureau.

Early version of SSC-1b 'Sepal' (P-35) coastal defence missile launcher which has a higher cab roof

UPDATED

Raduga 3M80 (SS-N-22 'Sunburn') coastal defence missile system

Development/Description

The 'Moskit' 3M80 missile is one of the weapons known by the NATO designation SS-N-22 'Sunburn' In its standard ship-launched version, the missile is in service with China, India and Russia.

The 3M80 'Moskit' (Russian designation P-270) is used on the 'Tarantul-III' class missile boat.

For coastal defence use it is fired from a cylindrical container-launcher. Although Russia is not believed to have any in coastal defence service, the Ukraine has sold eight launchers to Iran, which is believed to have employed the system in semi-mobile configuration to cover the Straits of Hormuz. These missiles are understood to be trailer mounted and are normally stored in underground bunkers until required for action.

Propulsion of the 3M80 missile is by an integrated solid-propellant rocket motor booster/airbreathing ramjet sustainer engine. The cruise speed is 2,800 km/h (around M3.0) with a missile range of 90,000 m.

Guidance is inertial in the cruise phase with an active/passive radar seeker system for the terminal flight profile. If very severe ECM conditions are met then the missile can automatically revert to a home-on-jam mode. The flight profile has a cruising altitude of 20 m with after-launch turns of 60° left or right of the launch bearing possible. The final attack phase is flown at a height of 7 m. The warhead section includes a 300 kg HE unit.

The number of missiles needed to incapacitate completely a destroyer-sized target is stated to be 1.2 and a 20,000 ton size transport ship 1.5. At M3.0 the 3M80 missile will cover the 90,000 m maximum range in only 2 minutes, thus giving the target only the minimum amount of time to react.

Variants

Improved 3M80

In 1998, the chief designer of the Moskit missile stated that it could be modified at minimal expense in order to increase its effectiveness by 70 per cent.

One option would be to increase the range of the missile to 150 km while the addition of a ventral fuel tank would increase the range to 200 km.

Altair 3M80E coastal defence system

The Altair missile bureau is currently working on a coastal defence system based on the 3M80E missile. There will be six to eight launcher vehicles linked to one command and control vehicle. Depending upon the vehicle used there will be two to three missiles per launcher. The extended-range 3M80E missile has minimum and maximum ranges of 10 to 120 km respectively.

Status

Production as required. The only user of this system in the coastal defence role is understood to be Iran. Ukraine sent Chelomei P-80 (SS-N-22 'Sunburn') missiles to Iran. These were first deployed on Project 956 EM destroyers of the Russian Navy in 1980. The Russian Navy does not deploy this missile in the coastal defence role.

Manufacturer

Raduga Design Bureau.

UPDATED

Specifications

Guidance: inertial mid-course with active/passive terminal homing and secondary home-on-jam capability.
Launch weight: 3,950 kg
Length: 9.385 m
Body diameter:
(with folded wings and fins) 1.3 m
Warhead: 300 kg HE
Max range: 90,000 m
Propulsion: cruise ramjet engine with built-in solid-propellant rocket motor booster
Speed: M3.0

Splav 122 mm (40-round) Damba coastal artillery multiple rocket system

Development/Description

The Splav Scientific Production Concern 40-round Damba coastal artillery rocket system is based on the Russian 122 mm (40-round) BM-21 (6 × 6) MRL series.

It is designed as a rapid deployment unit, for the engagement of detected small subsurface targets such as Swimmer Delivery Vehicles (SDVs) in the vicinity of naval base entrances/anchorages and coastal facilities, at ranges between 300 and 5,000 m.

The system comprises a 122 mm 40-round elevatable launcher assembly; four layers each of 10 tubes are carried on the rear of a Ural-375D (6 × 6) cross-country truck chassis. A resupply vehicle based on the same chassis carries 40 reload rockets. Either salvo or single round fire is possible. A full 40-round salvo can be fired in 20 seconds. The rockets may stay within their tubes for up to one year before having to be serviced. The sighting system is mounted on the left side of the launcher at the rear. Launcher traverse and elevation is manual.

The rocket used is a derivative of the standard 122 mm calibre system, but it is fitted with an ASW compatible fuzing system that can detonate the 20 kg HE warhead at any depth from 3 to 200 m.

Full details of the standard 122 mm BM-21 (40-round) multiple rocket launcher and its numerous variants, are given in the Multiple rocket launchers section.

While the overall prime contractor for this 122 mm (40-round) Damba system is the Splav Scientific Production Concern, the actual launcher system is manufactured by the Motovikha Plants Corporation which also supplies all other Russian multiple rocket launchers.

Status

Production as required. In service with Russian coastal defence forces. There are no known exports of this system.

122 mm (40-round) Damba coastal artillery multiple rocket system

Manufacturers

Splav Scientific Production Concern (rockets).
Motovikha Plants Corporation (launcher).

UPDATED

Specifications

Calibre: 122.4 mm
Number of barrels: 40
Length:
(rocket) 2.756 m
Weight:
(rocket) 75.3 kg
(system, Ural-375D launcher) 13,800 kg fully loaded
(system, Ural-375D reload vehicle) 9,900 kg fully loaded
Warhead type: 20 kg HE with ASW-type fuzing
Rate of fire: 40 rds/20 s
Min range: 300 m
Max range: 5,000 m
Min effective depth: 3 m
Max effective depth: 200 m
Engine: ZIL-375 7 litre V-8 water-cooled petrol developing 180 hp at 3,200 rpm
Gearbox: manual with 5 forward and 1 reverse gears
Clutch: twin dry disc
Transfer box: 2-speed
Steering: double-thread worm, hydraulic booster
Turning radius: 10.5 m
Suspension:
(front) longitudinal semi-elliptical springs with hydraulic shock absorbers
(rear) bogi with longitudinal semi-elliptical springs
Tyres: 14.00 × 20
Brakes:
(main) air/hydraulic
(parking) mechanical
Electrical system: 12 V

SOUTH AFRICA

LIW 155 mm G5 coastal defence gun

Development/Description

The standard towed LIW (a division of Denel) 155 mm G5 gun-howitzer (see entry in Towed anti-tank guns, guns and howitzers section) has been developed as a private venture or use in the coastal defence role.

Among the roles envisaged are the illumination of specific seaward areas, the laying of smoke screens to mask movement from seaward observers and to act as a deterrent against armed piracy and smuggling.

Firing from a fixed position, the 155 mm G5 coastal defence gun has an area coverage of almost 1,000 km^2. The firing areas of the 155 mm G5 gun are:

(a) at elevations of less than 15 to 82°
(b) at elevations greater than 15 to 65°.

The maximum range of the weapon is 39,000 m at sea level with ERFB base bleed projectiles. With 155 mm ERFB boat tail projectiles the maximum range is reduced to 30,000 m at sea level. The gun is also compatible with all NATO standard 155 mm ammunition.

The ammunition that can be fired for coastal defence applications includes: HE ERFB, HC M1 ERFB cluster with 42 dual-purpose anti-personnel/anti-armour bomblets, ERFB illuminating base ejection, ERFB radar echo, ERFB red phosphorus and ERFB screening smoke.

The accuracy of the weapon for the 50 per cent zone is a probable error of 0.48 per cent in range and 1.0 mil in breadth at 75 per cent of the maximum range of the standard base bleed projectiles being fired. This effectively equates to an area some 45 m wide by 216 m long.

The South African Institute for Maritime Technology was commissioned by Denel to undertake an operational research study, to examine the G5 for use in the coastal defence role. The study looked at large, medium and small maritime targets and the dispersion of projectiles.

The conclusion reached was that a G5 battery of three or six guns was an effective coastal defence system. Using cargo rounds with 42 bomblets per round, a 100 m^2 lethal area for each bomblet and a similar 39,000 m maximum range, the results showed that a single compensated salvo on a medium-sized target at 30,000 m had an 80 per cent hit probability. Even for small targets, the hit probability at 10,000 m was 60 per cent which reduced to 15 per cent at 35,000 m range.

Detection and identification of the target could be made by airborne, shipborne or coastal radar surveillance. For over-the-horizon detection even UAVs, such as Kentron's Seeker, can be used to provide accurate real-time information to the battery's Fire Direction Centre (FDC).

LIW 155 mm G5 gun-howitzer used in coastal defence role

This then provides accurate target data via the battery control post to the gun control computers. These then calculate a battery fire solution that ensures all the rounds fired by the battery fall simultaneously on the target for maximum effect.

The 155 mm G5 coastal defence gun can be mounted either in permanent gun sites or preprepared coastal sites in order to take advantage of the weapon's mobility.

VLAP projectile

Early in 1997, the SOMCHEM Division of Denel revealed that it had developed a new 155 mm Velocity enhanced Long-range Artillery Projectile (VLAP) which has a range of more than 50 km.

Production of the VLAP commenced in 2000 with first deliveries taking place in the first half of 2001.

India is known to have recently ordered large quantities of 155 mm ammunition (projectiles and charges) from South Africa and could be the launch customer for the VLAP.

So far most firings have utilised the latest 155 mm/52 calibre ordnance but additional qualification firings were planned with 155 mm/39 calibre and 155 mm/45 calibre weapons.

The VLAP was originally designed as a private venture by SOMCHEM to complement their existing family of 155 mm artillery projectiles which includes High Explosive - Extended Range Full Bore - Base Bleed (HE ERFB BB).

The current VLAP has a high-explosive payload although other payloads are now under investigation by SOMCHEM.

SOMCHEM stress that the VLAP is complimentary to their existing family of projectiles and allows friendly artillery units to engage threat targets while positioned well to the rear and beyond the range of enemy counter battery artillery fire. This will lead to higher survivability of friendly artillery systems.

To achieve this significant increase in range, the VLAP incorporates base bleed and rocket assist. It can be fired from most existing towed and self-propelled 155 mm artillery systems including the South African LIW Division of Denel 45 calibre G5 (towed), 45 calibre G6 (6 × 6) self-propelled and the more recent 52 calibre T6 turret systems which has undergone extensive tests by the Indian Army.

Ranges of the VLAP compared to other types of 155 mm projectile are as follows:

Weapon type	Boat tail	Base bleed	VLAP
155 mm/39 calibre	23 km	30 km	39 km
155 mm/45 calibre	30 km	39 km	50 km
155 mm/52 calibre	32.5 km	41 km	52.5 km

The VLAP can be used in conjunction with a number of charge systems including the SOMCHEM M53, M90 Bi-Modular Charge System (BMCS) selected by the British Army after an International competition and the M64 BMCS.

Status

Development complete. Ready for production on receipt of orders.

Manufacturer

LIW, a Division of Denel (Pty) Ltd.

UPDATED

SWEDEN

Bofors Defence 120 mm mobile coast defence gun KARIN/CD 80

Development

During the early 1970s, the now Bofors Defence (now part of United Defense LP of the United States) was asked to develop a new mobile coast defence gun by the Royal Swedish Coast Artillery and, as the FH-77A was under development by the now Bofors Defence at the same time, it was decided to develop the two weapons using the same split trail carriage.

Thus the FH-77A (see separate entry in Towed anti-tank guns, guns and howitzers section) and KARIN both use the same basic two-wheeled carriage fitted with an auxiliary power unit, carriage although KARIN uses a different ammunition handling system, an improved laying system which is fully remote controlled and enables moving naval targets to be engaged with high precision.

Specifications

Calibre: 120 mm
Muzzle brake: multibaffle
Breech mechanism: vertical sliding block
Carriage: split trail with APU
Shield: no
Weight: 12,500 kg
Length:
(barrel) 6.6 m
(travelling) 13.1 m
(firing) 12.15 m
Width:
(travelling) 2.65 m
(firing) 9.05 m
Height:
(travelling) 2.96 m
Axis of bore: 1.6 m
Ground clearance: 0.42 m
Track:
(driving wheels) 2.2 m
(castor wheels) 2.2 m
Fording: 0.75 m
Gradient: 50%
Elevation/depression: +50/–3°
Traverse:
(total) 60°
Rate of fire: up to 15 rds/min
Max range: 30,000 m
Towing vehicle: 6 × 6

The first complete KARIN was delivered to the Royal Swedish Coast Artillery in 1978 (to Coast Artillery Regiment (KA2) at Karlskrona). In 1981, a revised model, to operate with radar fire control, was delivered and, during 1985, the first complete training using the production version was carried out. KARIN is now often referred to as CD 80 as it is employed in special battalions known as KA-bataljon m/80.

These special mobile battalions are used to cover open stretches of coastline or reinforce fixed defences under attack. Each battalion has a battalion commander and staff, a staff and communications battery and two coastal artillery batteries. Each of the two coastal artillery batteries has four KARIN guns under the control of the battery commander and headquarters platoon. The rest of the battery is made up of an ammunition supply and transport platoon, a fire-control platoon, a self-defence infantry and traffic

Bofors Defence 120 mm mobile coast defence gun KARIN/CD 80 in travelling configuration

Bofors Defence 120 mm mobile coast defence gun KARIN/CD 80 firing

control platoon and an air defence platoon equipped with two Bofors Defence 40 mm L/70 anti-aircraft guns. A support platoon has repair, catering and emergency medical facilities.

Description

KARIN uses a 120 mm 55 calibre gun barrel and can fire at a rate of up to 16 rds/min. The ammunition feed system, sometimes referred to as a conveyor, is on the right trail leg of the carriage. The aiming system is known as MARIA and is operated by one man seated to the left of the gun.

The system receives digital fire-control data via wire or radio from a fire-control station housed in a container-type shelter. Supplied by Saab, this system combines a search and fire-control radar, computers, indicators and an electro-optic director. On the gun, the layer has a panoramic sight, a gun computer, a layer's panel and a computer panel. As most of the fire-control data is fed to the gun automatically, all the layer normally has to do is make slight corrections and monitor the process.

For long-distance moves KARIN is towed by a Scania 6 × 6 tractor at speeds up to 70 km/h. The auxiliary engine is used for local deployments at speeds up to 7 km/h and the engine is also used to provide hydraulic power to move the trail legs, operate the ammunition feed system and lay the gun.

KARIN fires the same projectiles as the Bofors Defence 120 mm ERSTA system (see following entry), a special naval target shell with a delay fuze to ensure that the charge detonates only after the shell has passed through the target armour, and an HE projectile. Both are fired at a maximum muzzle velocity of 830 m/s. When firing an enhanced range projectile known as HCER at a muzzle velocity of 900 m/s a range of more than 30,000 m is achieved; the normal range of KARIN using the naval target ammunition is 21,000 m. A complete round weighs 43 kg and the projectile 24 kg. An HE projectile is also available.

Variant

For trial purposes the upper part of the Bofors Defence 120 mm mobile coast defence gun KARIN/CD 80 has been installed on a highly mobile Volvo 6 × 6 cross-country chassis. There are no plans for this to enter service with the Swedish coastal artillery.

Trials have also been carried out with a Bofors Defence 40 mm L/70 anti-aircraft gun system installed on a similar chassis. This is called the TRIDON but has yet to pass the prototype stage.

By installing these 120 and 40 mm weapon systems on 6 × 6 cross-country chassis, not only will the systems have greater mobility but also faster into and out of action times and a significant reduction in manpower requirements.

Status

Deliveries to the Royal Swedish Coastal Artillery are complete. There are no known export sales of this weapon although it was offered on the export market.

Manufacturer

Bofors Defence AB (A United Defense company).

UPDATED

Bofors Defence 120 mm static coast defence gun ERSTA

Development

The ERSTA 120 mm static coast defence gun was developed by the now Bofors Defence in the early 1970s to meet Swedish defence force long-term requirements for high firing endurance, high-precision firing at long range, high protection levels for the gun and crew and high-level effects on target, all relating to coastal defence. ERSTA is thus an automatic turret gun with an automatic ammunition transport and loading system.

The installation of an ERSTA gun is a considerable civil engineering project and was undertaken by the Swedish ABV concern. Installation involved blasting positions in solid rock and the three-storied underground facilities include, quite apart from those of the gun and its associated mechanisms, a fire-control centre, kitchens, crew accommodation, washrooms, power plant (with two diesel generators - one acting as a spare - and several months' fuel supply for continuous running), ventilation compartment and desalination equipment to provide fresh water for drinking and other purposes.

The fire-control centre associated with ERSTA has armoured observation cupolas and armoured covers for folding radar antennas.

The Royal Swedish Coast Artillery is now placing emphasis on greater mobility for its artillery units and the 120 mm ERSTA will be the last weapon of its type to be developed by Bofors. There are no further improvements expected to be carried out on this weapon. It is considered unlikely that there will be any future production of the 120 mm ERSTA. The Norwegian systems are also to be placed in reserve.

Description

The 120 mm gun used with ERSTA is 62 calibres long and capable of rates of fire up to 25 rds/min. This rate of fire can be extended to the point where 200 rounds in 20 minutes can be fired. The 120 mm gun barrel is water cooled using a new system. The gun is protected by an armoured cupola which stands approximately 1.8 m above ground level and gives protection against both nuclear and conventional weapons. Extra protection is provided by lowering the cupola and the entire traversing portion of the gun by hydraulic power. This takes the cupola and the traverse portion off their normal spring-loaded roller bearings and locks them onto the foundation. The 120 mm 62 calibre barrel can be lowered into a concrete fork and the muzzle can be automatically sealed by a cover. Further protection is provided by special camouflage and possibly the use of dummy guns and installations nearby.

The cupola, which is covered in a neutron-absorbing material, sits over a long underground 'tube' which contains the ammunition transport system. The entire installation is set in concrete with the bulk of the

Specifications
Calibre: 120 mm
Length:
(gun) 7.44 m
Crew: 11
Elevation/depression: +50/–3°
Traverse: ±270°
Laying speed traverse: 10°/s slewing 0-3°/s tracking
Elevation: 5°/s slewing manual tracking
Rate of fire: 25 rds/min (200 rds in 20 min)
Weight:
(total) 70.5 t + 103.5 t concrete
Cupola armour thickness: 70 mm (steel armour)
Emplacement diameter: 4.04 m
Height of cupola above ground level: 1.8 m
Gunshaft below ground level: 18 m
Power requirements:
(AC) 3 × 380 V, 50 Hz, 6 kW
(DC) 24V, 0.1 kW

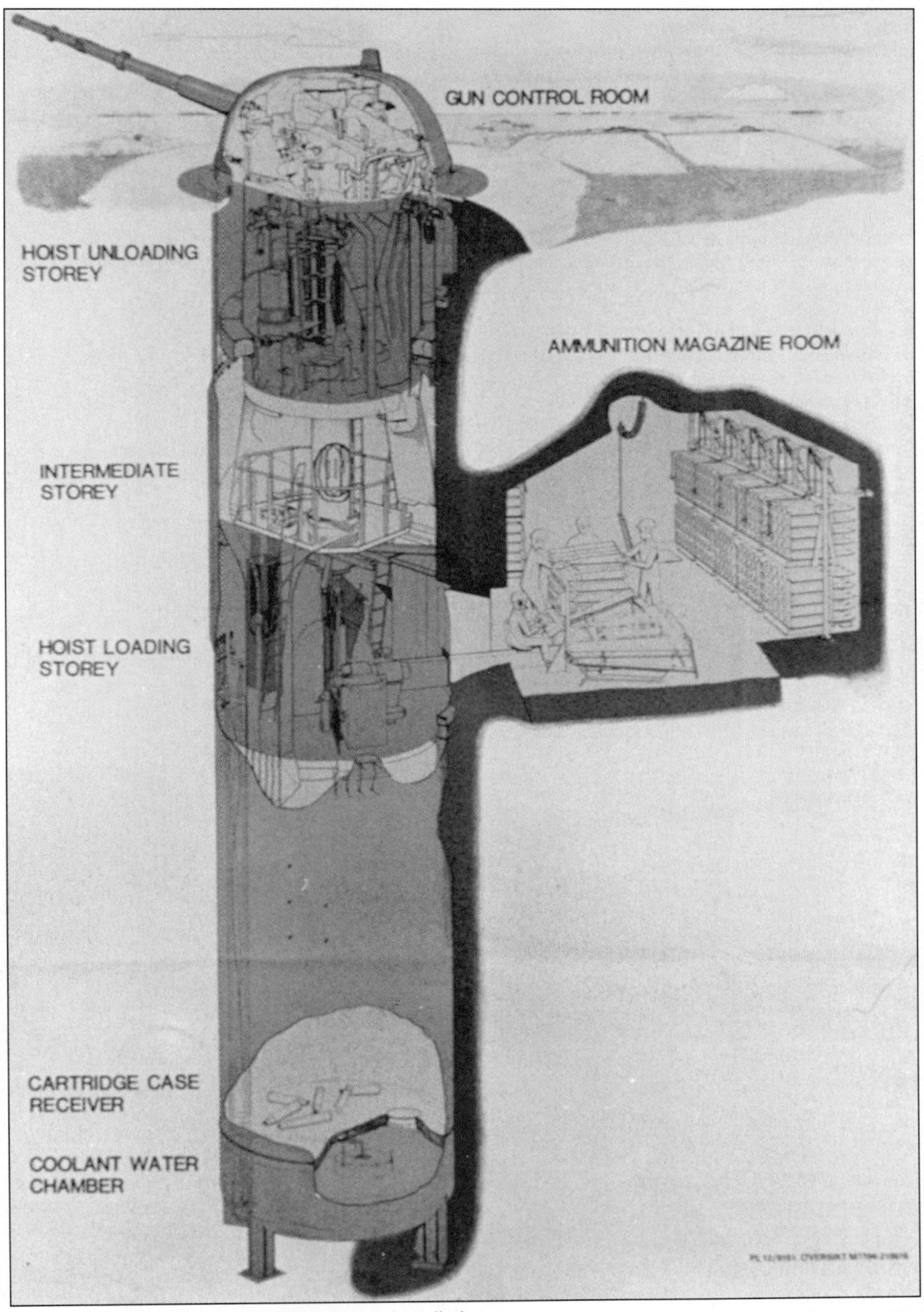

Layout of Bofors Defence ERSTA 120 mm installation

installation underground, including a chamber for the ammunition magazine and handling system. Inside this magazine, which is located to one side of the tube under about 5 m of overhead protection, the fixed rounds are stored in racks that can be moved by overhead handling equipment. In use these racks are moved from their storage positions around the magazine walls to the start of the transport system. The rounds are placed horizontally onto a handling table, then moved into a cross-shaped device that turns to point the round nose upwards for transport up a tube running through the centre of the cupola well. Halfway to the gun, the round enters a chamber where it is loaded onto a turntable device ready for further lifting to a position under the gun where it is turned to the appropriate angle of elevation for automatic loading. After firing, the spent 120 mm case is ejected to a chamber at the bottom of the well. Under this chamber is a water reservoir for the barrel cooling system.

ERSTA has a crew of 11 although for short periods the entire system can be operated by only three. Of these, two are in the cupola operating zero-setting instruments to lay the gun using data provided by a modern fire-control system using a Saab 400 (KONDOR) surveillance radar. This has the same protection level as the gun and can be rapidly covered by armour plates. Other sensors used include a TV tracker and laser range-finder.

The ammunition fired by ERSTA uses similar projectiles to those fired by the Bofors Defence KARIN/CD 80 120 mm towed system, covered in the previous entry in detail, allied to bigger cartridge cases to contain the larger propellant charges. The standard projectile is known as a naval target shell which is fuzed to burst only when it has penetrated a target and a standard HE projectile. Details of the existing 120 mm projectiles are as follows:

Type	**Naval target**	**HE**
Weight		
(shell)	24.5 kg	24.6 kg
(HE content)	2.6 kg	3.9 kg
Muzzle velocity	880 m/s	880 m/s
Max range	27,000 m	27,000 m

Status
Production complete. In service with Norway (8) and Sweden. With the increased emphasis being placed on more mobile coastal defence weapons, these 120 mm systems operated by Sweden are expected to be placed on a reserve status.

Manufacturer
Bofors Defence AB (A United Defense company).¼

UPDATED

Bofors Defence 75 mm TK75 coastal defence gun system

Description
The now Bofors Defence 75 mm TK75 semi-automatic coastal defence turret system was developed during the 1960s by the then Bofors AB to meet the requirements of the Royal Swedish Coastal Artillery.

The 75 mm gun is 60 calibres long and is mounted in a dome-like turret set into concrete with an underground ammunition department. The cast steel turret armour is 70 mm thick, provides protection against nuclear attack and is usually camouflaged. Ammunition is fed to the gun by an electrical hoist system that can have four or six steps according to the installation.

The gun has a crew of five to seven, traverse and elevation layers, a loader (in the turret) and two ammunition numbers (in the ammunition compartment). The ammunition numbers place the one-piece 75 mm ammunition onto a five-compartment loading table from which the rounds are moved under power to a hoist drum. Using a worm and bevel gear drive the round is fed up the hoist in steps, under microswitch control, to a pendulum where it is moved to a loading tube. The loading tube is supported on the pendulum which moves around the elevation trunnion centre. The raising of the pendulum and loading tube with the round is effected by a torsion spring placed in the trunnion centre. The torsion spring is cocked by the recoil. The loading tube delivers the round to a spring-operated rammer for loading into the breech which uses a vertical sliding block mechanism. The gun is fired by the traverse layer using a foot pedal. After firing, the spent cartridge case is ejected into a separate compartment. Using this semi-automatic ammunition hoist system the gun can fire at a rate of up to 25 rds/min. The hoist system uses three-phase 380 V AC supply and a mains-derived 24 V DC supply. In the event of an electrical failure the hoist system can be operated manually.

The 75 mm gun is hand layed in traverse and elevation by two hand cranks in the turret. For indirect laying follow-the-pointer dials are used. For direct aiming a periscopic sight with dual eyepieces, one for each layer, is used. The fire-control unit normally consists of one radar and one or two optical/laser range-finders which have the same level of protection as the gun installation.

The gun fires four types of fixed 75 mm ammunition, details of which are given in the above table.

Ammunition type	**Naval target shell**	**Naval target shell ER**	**HE shell**	**Practice shell**
Weight				
(projectile, with fuze)	5.73 kg	6.35 kg	5.52 kg	6.55 kg
(charge)	0.62 kg	0.81 kg	0.73 kg	n/a
(complete round)	11.5 kg	approx 12 kg	11.5 kg	12.2 kg
Muzzle velocity	850 m/s	830 m/s	870 m/s	850 m/s
Time of flight				
(to 3,000 m)	4.6 s	4.3 s	4.1 s	4.3 s
(to 6,000 m)	12.6 s	10.4 s	11.2 s	11.1 s
Max range				
(20° elevation)	11,400 m	13,300 m	12,000 m	12,200 m

Specifications
Calibre: 75 mm
Length:
(barrel) 4.245 m
Breech mechanism: vertical sliding block
Shield: 70 mm thick steel turret
Diameter:
(cupola) 2.3 m
Height:
(cupola, above ground level) 0.9 m
Total depth of gun shaft: 6 m (approx)
Weight:
(total) 17,500 kg
(4-step hoist) 16,750 kg
(6-step hoist) 17,500 kg
Elevation/depression: +20/–5°
Traverse: ±270°
Rate of fire:
(electrical hoist) 25 rds/min
(manual operation) 15 rds/min
(maximum) 200 rounds in 20 min
Max range: 11,400-13,300 m (depending upon ammunition type)
Crew: 5-7

Status
Production complete with over 100 systems built. In service with Norway, Sweden and several other undisclosed countries. There has been no production or marketing of this 75 mm turret system in recent years. With the increased emphasis being placed on more mobile coastal defence weapons, these 75 mm TK75 coastal defence turrets are to be phased out of service by Sweden in the near future. There are no known plans for the systems used by Norway to be upgraded in the future.

Manufacturer
Bofors Defence AB (A United Defense Company).

UPDATED

Bofors Defence 75 mm TK75 coastal defence gun before the addition of camouflage

Bofors Defence 75 mm TK75 coastal defence gun firing

SAAB Bofors Dynamics RBS 15K anti-ship missile system

Development/Description
The first development contract for the coastal defence version of the air- and sea-launched RBS 15 anti-ship missile, designated the RBS 15K by the Royal Swedish Coastal Artillery was awarded to the now SAAB Bofors Dynamics in 1984.

Users may have their own system configuration. The basic tactical unit is the missile battery. This consists of two missile platoons, each with one Missile Fire-Control Vehicle (MFCV) carrying a fire-control system in a cabin, and three launch vehicles carrying four missile canisters weighing 1,530 kg each, one service maintenance platoon and a guard platoon for local area security.

Target data is provided either from external sensors such as coastal defence radars, sensors on board naval vessels and aircraft or from a mobile radar (Ericsson Microwave Systems Giraffe CD) integrated into the launch platoon or battery.

Swedish RBS 15K anti-ship missile being launched from a Swedish Scania truck chassis

RBS 15K anti-ship missile system of Swedish Navy deployed in the firing position and with outriggers deployed and with command post in the foreground

Specifications

Guidance: autopilot with active radar terminal homing
Launch weight: 770 kg (the Mk 3 is quoted as having a total weight of 800 kg including boosters. Without boosters it weighs 630 kg)
Length: 4.35 m
Body diameter: 0.5 m
Wing span: 1.4 m
Warhead: 200 kg plus HE blast/fragmentation
Range: 110 km
Propulsion: liquid-fuel turbojet sustainer with 2 jettisonable solid-fuel rocket motor boosters
Speed: greater than M0.9

Each launch vehicle can be deployed up to 1,000 m from the MFCV. Target information is supplied over a datalink. Launch orders for a preprogrammed target can be issued within a few seconds. The vehicles will normally spend only a few minutes at a launch site and can be reloaded in the field very quickly.

The launcher vehicles may be located at the maximum distance of the missile from the coastline.

The coastal defence missile is identical to the ship-launched version. Launch weight (complete) is 770 kg with the two 85 kg weight jettisonable 3 second burn solid-fuel booster motors that boost the missile. The sustainer motor used is a 3,700 N thrust liquid-fuel Microturbo TRI 60 turbojet which provides a high-subsonic cruising speed.

Initial guidance is by preprogrammed autopilot with an FM-CW radar altimeter for height control. Once the missile has flown a certain distance, defined by the distance actually flown and the projected target location, the Saab AB Ku-band broad frequency-agile ECM-resistant radar seeker is activated in its search mode. The search pattern set depends upon the target data available. The final lock on mode can also be preset either as an active or active/passive operation.

Flight control is by cruciform wings with ailerons for combined pitch and roll plus canard fins at the nose of the weapon for yaw.

Maximum effective range is more than 110 km with a variable flight trajectory. Near to the target it adopts a sea-skimming terminal flight trajectory.

Sweden uses a Scania three-axle truck as the launch vehicle while Finland uses a Patria Vehicles three-axle launch truck. In both cases, two stabilisers are lowered to the ground hydraulically to provide a more stable platform before the RBS 15 missiles are launched.

Variants

Croatian RBS 15 coastal defence system

The Croatian company of RH ALAN have developed their own RBS 15 coastal defence system using missiles originally supplied to the Yugoslav Navy. The launch vehicle is a Slovak Tatra 6 × 6 truck chassis which is fitted with the four-round launcher-container assembly and associated electronics. A radar and fire-control unit are also mounted on other 6 × 6 chassis. Additional details of this system are given in this section under Croatia.

Status

Production. The RBS 15K is operational with the Swedish and Finnish coastal artillery forces. A locally derived version is in service with Croatia.

The Swedish Defence Matériel Administration has signed a contract with SAAB Bofors Dynamics for upgrading and lifetime extension of the Swedish Navy's RBS 15 system. Known as the RBS 15 Mk 2, both the shipborne and all coastal defence missiles will be identical. The upgrading involves extended-range capability, considerably increased tactical/operational flexibility, extended-coverage capability and increased capability to distinguish and select targets. The fire-control system uses a powerful graphical MMI (Man Machine Interface) and also has modes for planning and simulation. The contract value is approximately US$65 million.

In 2001, Saab Bofors Dynamics was awarded a contract to upgrade the Finnish Navy's inventory of RBS 15 missiles under a US$47 million contract. This version will be known as the RBS 15 SFIII. The work will be carried out in Sweden and extend the life of the missile for a further 15 years.

A Mk 3 version has also been in development since 1993. With a range of 200,000 m, the weapon has been designed to minimise its Radar Cross-Section (RCS) and also has a redesigned launch container that also has a lower RCS. It also uses all of the new features of the Mk 2 missile including the fire-control system, designated MEPS (Missile Engagement and Planning System). Mk 3 missiles will be identical for use of aircraft, ships and coastal defence roles.

In 2001, two additional customers for the ship-launched version of the Saab Bofors Dynamics RBS 15 Mk 3 missile were announced. The German Navy has selected the missile as the long-range surface-to-surface guided weapon for its new K 130 corvettes. Poland has ordered the RBS 15 Mk 3 for its three Type Orkan class Type 660 guided missile corvettes. Sweden has selected the RBS 15 Mk 3 for its new Visby class corvettes.

Mention has also been made of the RBS 15 Mk 3+ which could have a range of up to 400 km and have a low probability of intercept/multisensor seeker, a data link, terrain following reference navigation, automatic target recognition and alternative warheads.

Manufacturer

SAAB Bofors Dynamics AB.

UPDATED

Interior view of the Missile Fire-Control Vehicle (MFCV)

Artist's impression of the RBS 15K deployed to defend the coastline

Hellfire Shore Defence System (HSDS)

Development/Description

In late 1984, Rockwell International was awarded a US$8 million contract by the Swedish Defence Ministry to adapt its AGM-114 Hellfire anti-tank missile (which is normally launched from an attack helicopter such as an AH-64 Apache) to the all-weather rough terrain coastal defence role. This was to replace the obsolete French-supplied Aerospatiale Missiles SS-11 (Rb 52) wire-guided weapons then being used by the Swedish Coastal Defence Force.

Following the two year development programme, a US$65 million serial production contract was awarded in 1987 to Rockwell for single-rail launcher systems and

RBS 17 Hellfire shore defence missile with launcher 0021091

RBS 17 Hellfire shore defence missile being launched during trials 0021092

missiles. First deliveries were made in 1989 with final deliveries in 1991.

Final assembly of the Hellfire missile was conducted by the now Saab Bofors Dynamics in Sweden under the direction of Rockwell Tactical Systems Division. The adapted Hellfire missile is known as the RBS 17 in Swedish service.

In 1996, the Royal Norwegian Navy Matériel Command awarded a SKr400 million (US$58.73 million) contract to Saab Bofors Dynamics to supply the Royal Norwegian Coastal Defence Artillery with the HSDS. Deliveries began in 1997.

In January 1997, the Royal Norwegian Navy placed a US$12.5 million contract with the Laser Systems Division of Litton Industries for the supply of 54 Hellfire Shore Defence System - Designator Configuration (HSDS-DS) systems. The designators are used to observe, track and engage targets with HSDS under day and night conditions. Each system comprises a designator/range-finder, a French SAGEM 8 to 12 μm thermal camera and a British Instro-Precision Ltd tripod with traversing head.

The launcher system consists of the modified Hellfire missile, launch control equipment, launcher, a laser designator unit and missile container.

Each HSDS set for a site has two single-rail launchers, a launch control box, two cable reels, two safe and arming boxes and four power supply batteries.

The launcher is in two parts, the base with three folding legs weighing 21.8 kg and the trunnion with launch rail weighing 26.2 kg. The total weight is 48.0 kg with dimensions of 1.9 m long, 2.15 m wide and 1.05 m high. It takes less than 8 minutes to assemble on terrain with a 10° slope. The assembly is stabilised with stakes or weights at the foot pads. Alignment and levelling is accomplished by levels on the launcher itself.

The laser designator weighs 11 kg, has a length of 0.445 m, width of 0.27 m and height of 0.205 m. An interchangeable night sight weighs 7.3 kg, has a length of 0.489 m, a width of 0.235 m and a height of 0.205 m. The launch control box weighs 12.79 kg, is 0.36 m in length, 0.508 m in width and 0.195 m in height. A cable and reel weighs 8.52 kg, is 0.36 m in diameter, 0.21 m in width and 0.49 in height. The safe and arming box is 3 kg in weight, 0.37 m in length, 0.23 m in width and 0.07 m in height. A battery weighs 10.8 kg, is 0.1 m in length, 0.4 m in width and 0.399 m in height.

The missile transport and storage container can float with the missile inside. It is 1.772 m long, 0.336 m wide and 0.31 m high. The empty weight is 23 kg and with the missile inside weighs 71 kg.

Once set up, a two-man crew can load the launcher in less than 60 seconds. Launching is possible at elevations up to +10° for direct fire or +20° for indirect fire, with a traverse capability of ±30° without disturbing the tactical set up.

The missile itself has a long cylindrical body with a blunted rounded ogive and short limited-span fixed cruciform fins forward. Aft are the long short-span cruciform tail control surfaces.

Propulsion is by a single-stage solid-propellant smokeless rocket motor that gives the missile a speed of around M1.17. The effective range limits are from 300 to 10,000 m.

Guidance is by a semi-active terminal homing seeker that tracks the reflected laser energy from the target being illuminated by the laser designator. The CEP value is said to be less than 1.5 m at extreme range. The HSDS can operate with a variety of laser designators and night sights.

The missile can be fired in one of two modes: Lock On Before Launch (LOBL) or Lock On After Launch (LOAL) from 10 or 20° launcher elevation. The LOAL trajectory can be direct, low or high.

A full range of training equipment is available including a Special Training Missile (STM), dummy missile and both classroom and field configuration training aids.

The Royal Norwegian Navy has equipped 12 coastal batteries with the Hellfire missile and up to 36 Stridsbat (Combatboat) 90N vessels have been acquired to transport these units where the missiles would be deployed.

Specifications

Guidance: semi-active laser terminal homing
Launch weight: 48.25 kg
Length: 1.625 m
Body diameter: 0.1778 m
Wing span: 0.362 m
Warhead: 9 kg Bofors HE blast/fragmentation
Range: 300-10,000 m
Propulsion: solid-propellant rocket motor
Speed: M1.17

Status

Production complete for Norway and Sweden but can be resumed. In service with the Royal Norwegian Coastal Defence Artillery and Swedish Coastal Artillery Force. Finland is understood to have expressed an interest in purchasing this system for the coastal defence role.

Manufacturers

Hellfire Systems LLC (a Lockheed Martin and Boeing Company).
Saab Bofors Dynamics AB.

UPDATED

SWITZERLAND

SW Swiss Ordnance Enterprise 155 mm fortress gun 1993 L52 Bison

Development

SW Swiss Ordnance Enterprise Corp (part of RUAG Suisse) developed a 155 mm 52 calibre fortress gun system called the Bison, prototypes of which were constructed during 1988 to 1993.

The gun has been designed to have a range of 40,000 m and at the same time, for cost and logistical reasons, be able to use the existing family of 155 mm ammunition used by the 155 mm M109 series of self-propelled howitzers deployed by the Swiss Army in considerable numbers (see Self-propelled guns and howitzers section). Many of these are now being upgraded with a 155 mm/47 calibre barrel by SW Swiss Ordnance Enterprises Corp.

The first tranche of four batteries, each of which has two guns, was approved by the Swiss Parliament as part of the 1994 Equipment Programme.

Paired 155 mm Bison guns are installed in special reinforced concrete casemates providing a high degree of protection. The casemates are camouflaged to avoid detection by high-altitude photographic means and infra-red sensors.

The new 155 mm 52 calibre gun has a fully chromium-plated barrel to reduce barrel erosion.

The prototype of the Bison carried out its first ballistic firing in 1987. Range tests with Israel Military Industries cargo projectiles have taken place to confirm calculated ballistic values.

The Bison is fitted with a flick rammer loading system which is of modular design and is suitable for both automatic and semi-automatic loading on fortress guns, field howitzers or self-propelled artillery systems.

With the installation of the flick rammer, the rate of fire can be increased significantly and five rounds can be fired within 25 seconds in the burst fire mode.

Specifications

Calibre: 155 mm
Length:
(barrel) 8.1 m
Max range: 40,000 m

The SW Swiss Ordnance Enterprise 155 mm Bison fortress gun system from the front

Rear view of the SW Swiss Ordnance Enterprise 155 mm Bison fortress gun system showing the flick rammer with the charge on the left and the five-round magazine on the right

The version being produced is semi-automatic with an automatic feed and the charges being loaded manually. The primer charge feed is fully automated.

Early in 1998, the Swiss Ministry of Defence stated that with the exception of the new 155 mm L52 Bison guns, all Swiss Army fortress guns and mortars would be phased out by 1 January 2000. Additional details of the SW Swiss Ordnance 120 mm fortress mortar 59/83 were given in *Jane's Armour and Artillery 1998-99*, page 869.

Description

It has recently been stated that further thermal protection in the form of a water curtain is now being considered.

The use of terminally guided munitions could also be fired. The fire-control system is called Fargof and is similar to that used by Swiss field artillery.

Coastal role

The 152 mm fortress gun 1993 L52 is now being proposed in the coastal role with a typical battery of four weapons being controlled by a radar unit and fire-control unit.

Any individual gun or all four guns of the battery can be assigned a target. Each gun would fire a burst of five rounds in less than 25 seconds or a burst of 20 rounds in less than 25 seconds by the complete Bison battery.

The search radar unit (track while scan) locates and tracks sea surface targets. After firing, the surveyed impacts are used to optimise the effect, for example splash spotting. The radar can track up to 20 surface targets and four high precision tracks.

The fire-control unit allows system guiding and controlling. The tracked targets will be selected and transmitted to the fire-control system and integration in a higher level of command and control system is also possible.

Status

Production as required. In service with the Swiss Army.

Manufacturer

SW Swiss Ordnance Enterprise Corp.

UPDATED

TAIWAN

Hsiung Feng II (Proud Wind II) coastal defence missile

Development/Description

The Hsiung Feng II (Proud Wind II) anti-ship missile was revealed in March 1988 and is a longer range and faster missile than the Hsiung Feng I (Proud Wind I) which is based on the old Israeli Aircraft Industries Gabriel design and is covered in the following entry.

It has been produced in both a ship-mounted and a coastal defence version with the missiles mounted in containers carried on the rear of trucks or semi-trailers towed by cross-country vehicles.

Propulsion is by a Turbomeca Arbizon IV turbojet sustainer with the launch and acceleration phases carried out by a jettisonable solid-propellant booster rocket motor.

The practical guidance range using the inertial reference system with its mid-course correction capability, and the terminal attack phase active J-band radar seeker with its passive infra-red automatic target recognition image tracking module Mk 2 (mounted in a fairing above the nose-mounted seeker radome) is around 92,000 m.

Status

Production probably complete. In service with Taiwanese armed forces. These are understood to be manned by the navy. As far as is known Taiwan is the only country to use the Gabriel for the coastal defence role.

Manufacturer

Taiwanese state factories.

VERIFIED

Hsiung Feng II (Proud Wind II) missile out of its launcher for parade purposes (L J Lamb)

Specifications

Provisional
Guidance: inertial with active radar and passive infra-red terminal homing
Launch weight: 695 kg
Length: 4.845 m
Body diameter: 0.40 m
Wing span: 1.15 m
Warhead: 190 kg self-forming fragment
Range: 160,000 m
Propulsion: turbojet sustainer with jettisonable solid-fuel rocket motor booster
Speed: M0.9

Hsiung Feng I (Proud Wind I) coastal defence missile

Hsiung Feng I (Proud Wind I) coastal defence missile system during a parade in Taiwan with launcher covers open and radar vehicle in right background (L J Lamb)

Development/Description

The Hsiung Feng I (Proud Wind I, formerly Male Bee I) is a copy (or perhaps licenced-produced version) of the Israel Aircraft Industries Gabriel Mk II anti-ship missile fitted with a semi-active radar seeker head developed by Sun Yat-sen Institute. Neither the Proud Wind I or II were ever offered on the export market.

The missile is part of the Honeywell H930 Mod 3 mobile coast defence system which entered service in the early 1980s and is carried on semi-trailers towed by US 2.5 ton (US) M35 6 × 6 trucks. The system includes the Hsiung Feng triple-reinforced glass fibre container-launcher unit, a locally produced CS/SPG-24 fire-control radar (similar in many respects to the US RCA R-76 40 km range I-band monopulse naval fire-control radar system) for target tracking and illumination, a locally developed CS/UPS-60(X)C I/J-band surface search and target indication radar and a command and control centre (with three operators using two mini-computers and 11 microprocessors).

The Hsiung Feng I missile itself has a cylindrical body with a pointed nose. Flight control is by cruciform tailfins that are in line with larger rectangular mid-body wings. A 3 second burn solid-propellant booster and 200 second burn sustainer rocket motor is fitted. The semi-armour-piercing warhead weighs 180 kg and contains 75 kg of conventional HE. It is detonated by a delay action contact fuze.

Before launch, the missile's guidance system is programmed with the target data obtained from the unit's own radars. The missile is then fired and under control of its twin gyro system assumes an initial cruise height of approximately 100 m. At a range of 7,500 m from the launcher the onboard autopilot commands the missile to descend to 20 m altitude using a radar altimeter to maintain the height. At a predetermined distance between 1,000 and 2,000 m from the target the seeker is switched on, the target is acquired and locked on and the missile descends to one of its three possible preset attack altitudes for the final run-in. The actual set altitude varies between 1 and 3 m and depends upon the sea state encountered at the time.

If the primary semi-active terminal homing capability is being affected by countermeasures, then either the back-up manual/Honeywell optical tracking (with a locally produced passive infra-red night vision device) guidance or the command guidance with radar tracking option can be selected.

This missile system is now being supplemented by the more effective Hsiung Feng II (Proud Wind II) coastal defence missile system covered in the previous entry. It is possible that these have now been phased out of service.

Specifications

Guidance: autopilot with semi-active radar terminal homing (alternative manual guidance with optical tracking or command guidance with radar tracking options)
Launch weight: 537.5 kg
Length: 3.43 m
Body diameter: 0.34 m
Wing span: 1.36 m
Warhead: 180 kg HE SAP
Range: 36,000 m
Propulsion: combined solid-fuel sustainer and booster rocket motor
Speed: M0.7

Status

Production complete. In service with Taiwan. (Five Honeywell H930 Mod 3 systems were delivered from 1980 onwards.) It is understood that all coastal missile systems used by Taiwan are manned by the navy.

Manufacturer

Taiwanese state factories.

UPDATED

Hsiung Feng I (Proud Wind I) missiles in parade ground configuration (L J Lamb)

Fire-control radar system used in conjunction with the Hsiung Feng I (Proud Wind I) missile system (L J Lamb)

UNITED KINGDOM

MBDA Sea Skua coastal defence missile

Development/Description

In mid-1988, the then British Aerospace (which is today MBDA) announced the development of a coastal defence version of the air-launched Sea Skua anti-ship missile.

It is basically four missiles in their launch canister on a trainable base, a Seaspray 3000 radar on a hydraulic mast and a fire-control unit, all pallet mounted on a wheeled or tracked vehicle. The system uses proven elements of the contemporary ship- and helicopter-launched systems.

The missile body is cylindrical with a rounded nose and a taper to a smaller aft end which houses the propulsion motors. Electrically activated triangular-shaped cruciform control surfaces are located mid-body with interposed stabilising fins aft.

Propulsion is by separate solid-propellant booster and sustainer rocket motors. The sustainer has two angled exhaust nozzles. Cruise speed is about M0.85.

Guidance is by autopilot during the boost phase which stabilises the missile in roll and maintains it on a steady bearing towards the target. At boost burnout, the lateral guidance loop is closed and the I-band BAE Systems semi-active radar seeker head provides a proportional navigation course in the horizontal plane to the target. Shortly after this, the height loop is closed and the missile descends to the preselected mid-course height under radio altimeter control. There are four sea-skimming flight altitudes available, with the selection made according to the sea state and target size, before launch. Since the guidance systems are effectively independent, then several missiles can be ripple-fired at a target. The

Sea Skua anti-ship missile in flight **2000**/0085939

Coastal Battery Sea Skua shown fitted to the DAF YAZ 2000 (6 × 6) truck with a Seaspray 3000 radar. Mounted at the rear are the four ready to fire Sea Skuas in their container launchers

Specifications

Guidance: autopilot with semi-active radar terminal homing
Launch weight: 147 kg
Length: 2.5 m
Body diameter: 0.25 m
Wing span: 0.72 m
Warhead: 20 kg HE SAP
Range: 3,000-16,000 m
Propulsion: solid-fuel rocket motor sustainer with solid-fuel booster rocket motor
Speed: M0.85-0.9

complete weapon system can be operated by one person.

The coastal battery Sea Skua is self contained and provides multimissile flexibility with fast into and out of action capability. This means that the fire unit(s) can be rapidly redeployed to meet a changing threat.

MBDA has studied a number of improvements to the Sea Skua including a modified homing head and a longer burning sustainer motor to extend its range.

In mid-1997, the UK MoD awarded a Life Extension contract to MBDA to artificially age and then test life-expired rounds so as to increase the actual shelf life of the main stockpile. Production of the Sea Skua is undertaken as and when required.

Status

In production. The Sea Skua missile was operationally deployed during the Falklands and Gulf wars. In both cases it was launched from Westland Lynx helicopters of the Royal Navy. The coastal defence version has been proposed to a number of countries. By early 2002, no sales of the coastal defence version of the Sea Skua had been made. The basic Sea Skua missile is already in service with Brazil (air launched), Germany (air launched), Kuwait (sea launched), South Korea (air launched), Turkey (air launched) and the UK (air launched). The first customer for the ship-launched version was Kuwait who selected the Sea Skua for its Garoh patrol boat designed and built by Constructions Mécaniques de Normandie (CMN). In 2001, Malaysia ordered the Sea Skua for its Lynx helicopters.

Manufacturer

MBDA.

UPDATED

UNITED STATES OF AMERICA

The Boeing Company Harpoon Coastal Defense System

Development

The Boeing Company (previously McDonnell Douglas) Harpoon Coastal Defense System is a ground-based mobile anti-ship weapon system, providing area defence of high-value assets and control of shipping lanes from prepared and unprepared launch sites.

The basic Harpoon Coastal Defense System is a single battery comprised of one missile control unit and one launcher transporter unit. The battery can be configured with additional launcher transporter units and/or an optional launcher replenishment unit. The system is designed to operate over primary and secondary road systems with limited off-road capability, thus providing a highly survivable weapon system. The Harpoon Coastal Defense System is designed to achieve transition from the travelling configuration to launch ready status in less than 30 minutes.

According to the Boeing Company, the Harpoon Coastal Defence System has operating costs as low as 1 per cent of an equivalent ship based platform and is difficult to counter and destroy.

As far as is known, apart from Denmark, there have been no recent sales of the Harpoon in the coastal defence configuration.

In 1994 the Royal Danish Navy brought into service its two Harpoon Coastal Defence Systems batteries using Block IC missiles.

Both the Harpoon missiles and central sections of the launch control centre have been updated from the retired 'Peder Skram' class frigates.

Each battery consists of two launch vehicles and a mobile launch and command centre all based on Scania three axle trucks. Each of these also tows a

Artist's impression of Harpoon Coastal Defense System deployed

Harpoon missile in flight

support trailer/generator. Each launch vehicle has four Harpoon missiles giving the battery the capability to launch a total of 16 missiles.

Description

The individual system components comprise:

(1) Missile control unit - This is the only manned vehicle in the system during tactical operations. It provides for all command, control and communications functions for the Harpoon Coastal Defense System and comprises the following:
  (a) The shelter which houses the launch control and communications equipment and the Environmental Control Subsystem (ECS) which provides a controlled environment for both equipment and personnel within the shelter.
  (b) The Harpoon Ship Command Launch Control Set (HSCLCS) AN/SWG-1A which performs engagement planning, flight plan computations and initiation and launch of the Harpoon missiles.
  (c) Electrical Power Subsystem (EPS) which provides generation of electrical power, monitoring of power quality, conditioning and distribution of power, and circuit overload protection.
  (d) The Communications Subsystem (CS) which provides voice and optional datalinks for the transfer of target information, tactical unit co-ordination, and command and control direction for the battery.

(2) Launcher transporter unit – This provides the support structure and performs the transportation and launching functions for the canisterised Block 1C Harpoon missiles. The launcher transporter unit comprises the following:
  (a) The Launcher Erector Unit (LEU), which supports the four canisterised missiles during transportation and raises the missiles to the required launch angle at the launch site.
  (b) The Electrical Power Subsystem (EPS) which provides generation of electrical power, monitoring of power quality, conditioning and distribution of power, and circuit overload protection.
  (c) The Cable Reel Unit (CRU) which performs storage, deployment and retrieval functions for the cables that interconnect the battery vehicles.
  (d) The Launcher Parameters Set which determines the launcher heading and attitude, and interfaces this data to the HSCLCS for engagement planning and flight plan computations.

(3) Launcher replenishment unit – This is used to resupply the launcher transporter unit in the field. Typically, the launcher replenishment unit has a four-canister carried on a truck or trailer assembly which is transferred between the units by a hydraulically operated crane mounted on the launcher replenishment unit transporter.

(4) Missile – The Harpoon missiles used on the Harpoon Coastal Defense System are standard production RGM-84D Block 1C rounds in shock-resistant canisters.

Specifications

Guidance: autopilot assembly with active radar terminal homing
Launch weight:
(Block 1C) 722 kg
Length:
(Block 1C) 4.628 m
Body diameter: 0.343 m
Wing span: 0.914 m
Warhead: 227 kg HE blast/fragmentation
Range:
(Block 1C) 124,000 m
Propulsion: liquid-fuel turbojet sustainer with jettisonable solid-fuel rocket motor booster
Speed: M0.83

The missile has a cylindrical body of uniform diameter and a plastic-coated nose radome. Flight control is by tail control fins that are in line with mid-body cropped cruciform delta-shaped stabilising wings. A 2.9 second burn 5,600 kg thrust tandem-mounted solid-propellant booster rocket motor, is attached with in-line tailfins for the boost and acceleration phases.

The sustainer motor is a 300 kg thrust Teledyne CAE J402-CA-400 liquid-fuel turbojet with a flush ventral air intake. Cruising speed is M0.83.

The Harpoon is 4.628 m long, 0.343 m in body diameter and has a maximum wing span of 0.914 m. Launch weight (complete) is 722 kg, 227 kg of which is a penetration HE-blast/fragmentation warhead fitted with a contact time delay fuze assembly.

The missile's onboard data processor is programmed before launch. An attitude reference assembly and a digital computer form the Mid-course Guidance Unit (MGU) which acts as the autopilot controlling the mid-course phase of the flight. Altitude control is maintained by a short-pulse Honeywell or Kollsman AN/APN-194 radar altimeter.

The final active flight phase is controlled by a Raytheon Systems Company (originally Texas Instruments) J-band frequency-agile PR-53/DSQ-28 two-axis solid-state high-resolution active radar seeker.

The Range and Bearing Launch (RBL) mode is used against targets when known range and bearing information is available. A Harpoon-launched in this mode will wait until it approaches the target before seeker activation. The seeker is then commanded to search an area with limits equal to the target uncertainty boundary.

The latter is determined by the HSCLCS to include the initial target position uncertainty at launch, alignment and gyro drift errors, and random target motion after launch. After seeker acquisition and lock on the missile descends to a skim altitude that is adaptive to the sea state. The seeker then maintains target track until missile impact.

An alternative launch technique is used when the target bearing is known but range is uncertain. With this Bearing Only Launch (BOL) mode, the missile is directed along the target's known bearing line and the seeker scans an area either side of the directed missile bearing course. The missile limits the seeker minimum and maximum search ranges as determined before launch.

In either launch mode, once the seeker is locked on and in its tracking phase, the Block IC Harpoon can use a preselected terminal trajectory. This may be either a low-apogee 'pop-up' attack manoeuvre or a sea-skimming profile. The radar seeker and MGU continue to generate proportional navigation commands to the flight controls in order to ensure continued tracking and impact even on a high-speed manoeuvring target.

The Block IC can also be programmed to fly single or multiple 'doglegs' to the target (known as 'waypoints') to conceal the position of the battery and, if several Harpoons are being fired together, to allow the missiles to arrive simultaneously at the target from different directions. The missile launch angle can be up to 180° from the target bearing.

If required the Harpoon Coastal Defence System can be further improved by the addition of a Harpoon Coastal Defence System Radar Unit to provide additional target information.

Variants

Block 1G missile

The Block 1G missile entered production in 1995 and is similar to the Block 1C except it incorporates a re-attack capability originally intended to be introduced into service with the cancelled Block 1D variant.

There is a proposed Block II version of the Harpoon which will retain the airframe and propulsion systems of the Block 1C/1G, be compatible with the launch platforms but have a completely new repackaged Guidance Navigation Unit (GNU) in the space occupied by the original mechanical gyro. The GNU will integrate a Honeywell HG 1700 ring laser gyro inertial measurement unit, a GEM III GPS receiver and a 68040-based mission computer plus an associated input/output board. These components will allow the search pattern of the onboard radar to be tailored so as to minimise the effects of land clutter.

According to Boeing, the Harpoon Block II will be capable of executing both land-strike and anti-ship missions. The GPS aided inertial navigation system enables the missile to hit a designated target aimpoint.

The Harpoon incorporates key guidance technologies from two other Boeing weapon systems. The low-cost inertial measuring unit is from the Joint Direct Attack Munition (JDAM) while the software, mission computer, integrated Global Positioning System/Inertial Navigation System and GPS antenna and receiver are from the Standoff Land Attack Missile Expanded Response (SLAMER).

Status

In production (Block 1G since 1995). Deployed as a ship-, aircraft- and submarine-launched weapon with numerous countries. Coastal defence version – three systems deployed since 1988 (unidentified customer). Two systems produced for Denmark became operational in 1994. It is believed that Spain and South Korea deploy Harpoon in the coastal defence role.

Manufacturer

The Boeing Company.

UPDATED

YUGOSLAVIA, FEDERAL REPUBLIC

100 mm M-87 coastal defence gun system

Development

The 100 mm M-87 fixed coastal gun system was developed by merging the rifled 100 mm tan gun installed in the Russian designed T-54/T-55 MBT with a modified carriage of the towed 122 mm D-30J howitzer.

This is essentially the Russian 122 mm D-30 and details of both systems are given in the Towed anti-tank guns, guns and howitzer section.

Although designed primarily for deployment in coastal defence fortifications and dug-outs, its full circle traverse also allows it to provide fire support to army units, marine infantry and territorial defence units operating ashore.

This system was being marketed just before the start of the Yugoslav civil war but it is not known if it had actually entered volume production.

The 122 mm D-30J is the Yugoslav version of the Russian 122 mm D-30 towed weapon which is fully described in the towed anti-tank guns, guns and howitzers section.

RH ALAN of Croatia has developed and placed in production a further development of the 122 mm D-30J called the D-30 HR M94 which is also available with a 100 mm ordnance. Details of this are given under Croatia. As far as it is known, Croatia has not exported any of these weapons.

Description

The 100 mm M-87 coastal defence gun system is manned by a six-man crew: crew chief, gunner and four assistant gunners. The gun barrel is made from high-quality steel and is fitted with a muzzle brake (of 30 per cent efficiency) which does not generate high overpressures of propellant gases, an important feature when the weapon is emplaced in shelters.

The counter-recoil equipment and hydraulic brake (with compensator and hydropneumatic recuperator) are accommodated in the cradle where the barrel is hung by means of collars. The brake and recuperator cylinders are connected to the breech ring so that they recoil along with the barrel. The breech itself is of the vertical sliding wedge type.

100 mm M-87 coastal defence gun system from the rear with loader on the right

Type	**HE**	**HE**	**AP-T**	**HEAT-T**
Fuze	impact, super-quick inertia and delay action		BD, inertia action	impact, piezoelectric electric super-quick action
Length	1,095 mm	1,095 mm	1,020 mm	1,094 mm
Weight	30.2 kg	27.1 kg	30.5 kg	26 kg

The top carriage is of cast construction and carries all the main elevation mechanism assemblies as well as the stools for the gunner and assistant gunner. The elevation mechanism housing is connected to the carrier on the top left side of the carriage, allowing a –5 to +45° depression and elevation. The sighting system is mounted on the left side and direct and indirect fire capability is provided for.

The equilibrator is pneumatic and fulfils the role of balancing the gun elevation mechanism parts, thus reducing the force on the traverse mechanism wheel.

Specifications
Calibre: 100 mm
Length:
(barrel assembly with breech ring) 5.66 m
(bored rifle section) 4.63 m
(system) 7.6 m
(normal recoil) 790-930 mm
(max recoil) 940 mm
Muzzle brake: yes
Breech mechanism: vertical sliding wedge
Weight:
(gun mass) 3,200 kg
(barrel assembly) 1,255 kg
(recoil assembly) 1,435 kg
(elevation assembly) 1,900 kg
Width: 2.15 m
Height: 1.8 m
Elevation/depression: +45/–5°
Traverse: 360°
Max rate of fire: 15 rds/min
Max range: 20,000 m
Crew: 5

The bottom carriage is bell shaped with a worm wheel on top. This is coupled to the traversing mechanism worm and gear through which it is further connected to a bearing receiver via traverse transmission gear, universal shaft and multiplier. Since the traversing mechanism is designed for circular action, a switching limiter is fitted so that the weapon can be fully rotated only twice in one direction.

For semi-automatic loading at high rates of fire, an electromechanical loader, driven by two DC electric motors is fitted. This is attached to the cradle by rails so that loading is possible under any elevation conditions. Maximum rate of fire is 15 rds/min, with a sustained capability of 36 rounds in 4 minutes.

The ammunition types used are given in the table:

External fire control is achieved by use of the SUVOA-M70 fire-control system, which is capable of controlling mobile and stationary coastal gun or multiple rocket launcher batteries. The system allows both day and night engagements using both direct and indirect sighting and firing modes.

Variant

M-87 100 mm

Under the CFE Treaty, Bosnia-Herzegovina reported to have a gun in service called the M-87 100 mm. This has the ordnance of the 100 mm T-12 anti-tank gun but has a different carriage without wheels. This is used in the coastal defence role.

Status

Production as required. Believed to be in service with Croatia and Yugoslavia.

Contractor

Yugoimport SDPR.

UPDATED

ARMOUR AND ARTILLERY IN SERVICE

ARMOUR AND ARTILLERY IN SERVICE

INTRODUCTION

Listings by country of equipment in service, in reserve and due to enter service some time in 2002 (* denotes a later in-service date). These data are based on the latest available information at the time of going to press and updates that are given in the main text. This section excludes fixed fortifications and coastal defence weapons that have their own section. Details of mobile air defence systems, gun and missile, are given in *Jane's Land-Based Air Defence*.

AF: Air Force
ACV airborne combat vehicle
ARS artillery rocket system
ATG anti-tank gun
BG border guard
FG field gun
FH field howitzer
G gendarmerie
GH gun-howitzer
ICV infantry combat vehicle
IFV infantry fighting vehicle
IS internal security
Ma marines
MICV mechanised infantry combat vehicle
MRS multiple rocket system
MRV multirole vehicle
N navy
NG national guard
NI naval infantry
P phasing out
R reserve
Recce reconnaissance
SPG self-propelled gun
SPGH self-propelled gun-howitzer
SP/GM self-propelled gun mortar
SPH self-propelled howitzer
SPRR self-propelled recoilless rifle
T training
TD tank destroyer
UG upgraded
UN United Nations

AFGHANISTAN (current status of much of this equipment is uncertain)
T-62 MBT
T-54/T-55 MBT
PT-76 light tank
BRDM-1 recce
BRDM-2 recce
BMP-1 IFV
BMP-2 IFV
BTR-40 APC
BTR-50 APC
BTR-60 APC
BTR-70 APC
BTR-80 APC
BTR-152 APC
BRDM-2 Sagger TD
120 mm 2S9 SPM/H
76 mm M1938 mountain gun
76 mm M1966 mountain gun
76 mm M1942 FG
85 mm D-44 FG
85 mm D-48 ATG
100 mm M1944 FG
122 mm D-30 howitzer
122 mm M1938 howitzer
130 mm M-46 FG
152 mm D-1 howitzer
152 mm D-20 GH
122 mm BM-21 MRS
140 mm BM-14-17 MRS
220 mm BM 9P140 MRS

ALBANIA
Type 59 MBT
Type 62 light tank
BRDM-1 recce
Type 531 APC
SU-76 SPG
45 mm M1942 ATG
57 mm M1943 ATG
76 mm M1942 FG
85 mm D-44 FG
85 mm Type 56 FG
122 mm M1931/7 gun
122 mm M1938 howitzer
122 mm Type 60 howitzer
130 mm Type 59-1 field gun
152 mm Type 66 GH
107 mm Type 63 MRS

ALGERIA
T-72 MBT
T-62 MBT
T-54/T-55 MBT
BRDM-2 recce
AML-60 recce (G)
AM 100 ALG recce (G)
BMP-1 IFV
BMP-2 IFV
BTR-50 APC
BTR-60 APC
OT-64 APC
M3 APC
Fahd APC (IS)
SU-100 SPG
122 mm 2S1 SPH
152 mm 2S3 SPGH
BRDM-2 Sagger TD
57 mm M1943 ATG
85 mm D-44 FG
100 mm T-12 ATG
122 mm M1931/7 howitzer
122 mm M1938 howitzer
122 mm D-30 howitzer
122 mm D-74 FG
130 mm M-46 FG
152 mm M1931/7 GH
122 mm BM-21 MRS
140 mm BM-14-16 MRS
240 mm BM-24 MRS
FROG-7 ARS

ANGOLA
T-72 MBT
T-62 MBT
T-54/T-55 MBT
BRDM-2 recce
BMP-1 IFV
BMP-2 IFV
BMD-1 ACV
BTR-50 APC
BTR-60 APC
BTR-152 APC
EE-11 Urutu APC
Casspir APC
100 mm SU-100 SPG
122 mm 2S1 SPH
152 mm 2S3 SPH
76 mm M1942 FG
85 mm D-44 FG
122 mm D-30 howitzer
130 mm M-46 FG
152 mm D-20 GH
122 mm BM-21 MRS
140 mm BM-14-16 MRS
122 mm RM-70 MRS
240 mm BM-24 MRS

ARGENTINA
TAM MBT
Sherman tank (R)
Sherman Firefly (R)
AMX-13 light tank
Jagdpanzer SK 105
ERC 90 Lynx recce (Ma)
AML-90 recce
Shorland (IS)
VCTP
VCR-TT (Ma)
FIAT 6614 APC (AF)
AMX VCI APC
M113A1 APC
BDX APC
VAB APC
Condor APC
M3 half-track APC
Grenadier (G and Ma)
UR-416 APC (IS)
LVTP7 (Ma)
155 mm Mk F3 SPG
105 mm Model 56 P howitzer (and Ma)
105 mm M101 howitzer (and Ma)
155 mm M114 howitzer
155 mm CITEFA Mod 77 howitzer
155 mm CITEFA Mod 81 howitzer
155 mm CITEFA L 45 CALA 30/2 howitzer
127 mm CP3 MRS
127 mm SAPBA MRS
105 mm Pampero MRS (and Ma)
160 mm LAR-160 MRS

ARMENIA
BMP-1 IFV
BMP-2 IFV
BMD-1 ACV
BTR-60 APC
BTR-70 APC
BTR-80 APC
BTR-152 APC
MT-LB MRV
122 mm 2S1 SPH
152 mm 2S3 SPGH
85 mm D-44 gun
100 mm T-12 ATG
122 mm D-30 howitzer
152 mm D-1 howitzer
152 mm D-20 GH
152 mm 2A36 gun
122 mm BM-21 MRS
273 mm WM-80 ARS

AUSTRALIA
Leopard AS1 MBT
M113A1 APC
LAV-25 APC
Bison APC
Bushmaster*
81 mm M125A1 SPM
105 mm Hamel Light Gun
105 mm M101 howitzer
155 mm M198 howitzer

AUSTRIA
Leopard 2A5 MBT
M60A3 MBT (R)
Jagdpanzer SK 105
4K 7PA APC
4K 47A G-1 APC
4K 4FA G-2 APC
Pandur APC
Ulan IFV
155 mm M109/M109A2 SPH
155 mm M109A5Oe SPH
81 mm Gr W1 SPM
Jaguar 1 HOT TD
85 mm M52 FG
105 mm M101 howitzer

AZERBAIJAN
T-72 MBT
T-55 MBT
BMP-1 IFV
BMP-2 IFV
BMP-3 IFV
BMD-1 ACV
BTR-D APC
BTR-60 APC
BTR-70 APC
BTR-80 APC
MT-LB MRV
85 mm D-44 FG

100 mm T-12 ATG
120 mm 2S9 SPM/H
122 mm 2S1 SPH
122 mm D-30 howitzer
152 mm 2A36 gun
152 mm D-20 gun howitzer
122 mm BM-21 MRS

BAHRAIN
M60A3 MBT
Ferret recce
Saladin recce
AML-90 recce
AIFV
Saxon APC
M3 APC
Shorland APC
Hussar APC (IS)
203 mm M110A2 SPH
105 mm Light Gun
155 mm M198 howitzer
227 mm MLRS

BANGLADESH
T-54/T-55 MBT
Type 59 MBT
Type 62 light tank
BTR-80 APC
MT-LB MRV ?
6 pounder ATG
76 mm Type 54 FG
105 mm Model 56 P howitzer
105 mm M101 howitzer
122 mm Type 54 howitzer
130 mm Type 59-1 FG
Chinese MRS

BELARUS
T-72 MBT
T-62 MBT
T-55 MBT
BMP-1 IFV
BMP-2 IFV
BMD-1 ACV
BTR-60 APC
BTR-70 APC
BTR-80 APC
MT-LB MRV
100 mm T-12 ATG
120 mm 2S9 SPM/H
122 mm 2S1 SPH
152 mm 2S3 SPH
152 mm 2S5 SPG
152 mm 2S19 SPG
203 mm 2S7 SPG
BRDM-2 Spandrel TD
MT-LB Spiral TD
122 mm D-30 howitzer
152 mm D-20 GH
152 mm D-1 howitzer
152 mm 2A65 howitzer
152 mm 2A36 gun
122 mm BM-21 MRS
220 mm BM 9P140 MRS
300 mm BM 9A52 Smerch MRS
FROG-7 ARS

BELGIUM
Leopard 1A5 (BE) MBT
Leopard 1A1 MBT
Scimitar recce
AIFV
M113A1-B APC
Spartan APC
Pandur APC
S600 Shorland (G)*
M113A1-B MILAN TD
M109A2 SPH
105 mm M101 howitzer
105 mm LG 1 Mk II gun

BENIN
PT-76 light tank
M8/M20 recce
BRDM-2 recce
Buffalo APC
105 mm M2 howitzer
122 mm D-30 howitzer

BOLIVIA
Jagdpanzer SK 105
EE-9 Cascavel recce
M113/M113A1 APC
LAV-100 APC
Roland APC
EE-11 Urutu APC
4K 7FA APC
75 mm Bofors M1935 FG
75 mm M116 P howitzer
105 mm M101 howitzer
105 mm FH-18 howitzer
122 mm Type 54-1 howitzer

BOSNIA-HERZEGOVINA
M60A3 MBT
AMX-30 MBT
T-54/T-55 MBT
T-34/85 tank
AML recce
M113 APC
M80 APC
76 mm M48 mountain gun
100 mm T-12 ATG
105 mm Light Gun
105 mm M56 howitzer
120 mm 2S9 SPM/H
122 mm 2S1 SPH
122 mm D-30 howitzer (+ Bosnian Serbs)
122 mm D-30M howitzer
130 mm M-46 FG (+ Bosnian Serbs)
130 mm M-59-M FG
152 mm M-84 GH
152 mm D-20 GH
155 mm M114A2 howitzer
122 mm BM-21 MRS
122 mm (40-round) APRA MRS
128 mm M-63 MRS
262 mm M-87 Okran MRS
FROG-7 ARS

BOTSWANA
Scorpion 76 mm recce
Shorland (IS)
RAM-V2 APC
BTR-60 APC
LAV-150 APC
105 mm Light Gun
105 mm Model 56 P howitzer
155 mm Soltam howitzer
122 mm APRA MRS

BRAZIL
Leopard 1 MBT
M60A3 MBT
M41 light tank (and Ma)
Jagdpanzer SK 105 mm (Ma)
EE-9 Cascavel recce (and Ma)
M8 recce
M113A1 APC
EE-11 Urutu (and Ma)
LVTP7A1 (Ma)
Bernardini AM-IV (IS)
105 mm M108 SPH
105 mm M7 SPH
105 mm M101 howitzer (and Ma)
105 mm M102 howitzer
105 mm Model 56 P howitzer
105 mm Light Gun (and Ma)
155 mm M109A2 SPH
155 mm M114 howitzer
108m mm FGT 108R MRS (and Ma)
ASTROS II MRS

BRUNEI
Scorpion 76 mm recce
VAB APC (IS)

BULGARIA
T-72 MBT
T-54/T-55 MBT
BRM-23 recce
BRDM-1 recce
BRDM-2 recce
BTR-60 APC
BTR-152 APC (IS)
MT-LB MRV
BMP-23 ICV
BMP-23D ICV
BMP-30 ICV
BMP-1 ICV
BMP-2 ICV
BRDM-2 Sagger TD
SU-100 SPG
122 mm 2S1 SPH
120 mm MT-LB SPM
120 mm 2S12 SPM
85 mm D-44 FG
85 mm SD-44 FG
100 mm M1944 FG
100 mm T-12 ATG
122 mm D-30 howitzer
122 mm M1931/7 gun
122 mm M1938 howitzer
130 mm SM-4-1 gun (coastal)
152 mm D-20 GH
152 mm M1937 GH
122 mm BM-21 MRS
132 mm BM-13 MRS
FROG-7 ARS

BURKINA FASO
AML-60/90 recce
Ferret recce
M8/M80 recce
EE-9 Cascavel recce
M3 APC
105 mm M101 howitzer
107 mm Type 63 MRS

BURUNDI
AML-60/90 recce
BRDM-2 recce
Shorland (IS)
BTR-40 APC
M3 APC
Walid APC
122 mm D-30 howitzer
122 mm BM-21 MRS

CAMBODIA
T-54/T-55 MBT
T-55AM2 MBT
Type 59 MBT
PT-76 light tank
BTR-60 APC
BTR-152 APC
OT-64 APC
M113 APC
76 mm M1942 FG
100 mm T-12 ATG
122 mm D-30 howitzer
122 mm M1938 howitzer
130 mm Type 59-1 FG
122 mm Type 60 FG
107 mm Type 63 MRS
122 mm BM-21 MRS
132 mm BM-13 MRS
140 mm BM-14 MRS

CAMEROON
Ferret recce
VBL recce
RAM recce
LAV-150 APC (and G)
M3 half-track APC
81 mm LAV-150 SPM (and G)
75 mm M116 P howitzer
85 mm Type 56 FG
105 mm M101 howitzer
130 mm Type 59-1 FG
130 mm M-46 FG
122 mm BM-21 MRS
122 mm (40-round) APRA MRS

CANADA
Leopard C2 MBT
LAV-25 recce
Cougar APC
Grizzly APC
M113 APC
MTVL APC
LAV-25 APC
155 mm M109A1/M109A2 SPH
M113A2 TOW
105 mm LG1 Mk II

105 mm C1 howitzer
105 mm C1 howitzer (UG)

CAPE VERDE
BRDM-2 recce
75 mm gun
76 mm gun

CENTRAL AFRICAN REPUBLIC
T-55 MBT
Ferret recce
BTR-152 APC
VAB APC
ACMAT VBLA APC

CHAD
T-55 MBT
AML-60/90
ERC 90 Lynx recce
ERC 90 Sagaie recce
EE-9 Cascavel recce
LAV-150 S APC
M3 APC
105 mm M2 howitzer
130 mm BM-13 MRS

CHILE
AMX-30 MBT
Leopard 1 MBT
M41 light tank
EE-9 Cascavel recce
Scorpion 76 mm recce
Piranha APC (and Ma)
M113A1 APC
EE-11 Urutu APC
Roland APC (G and Ma)
LVTP5 (Ma)
Bernardini AM-IV (IS)
155 mm Mk F3 SPG
120 mm Piranha SPM
105 mm Model 56 P howitzer
105 mm M101 howitzer
155 mm M114 howitzer (Ma)
155 mm M71 GH
160 mm LARS MRS

CHINA, PEOPLE'S REPUBLIC
Type 59 MBT
Type 59-II MBT
Type 60 MBT
Type 69-I MBT
Type 69 II MBT
Type 79 MBT
Type 80/Type 80-II MBT
Type 85-II/Type 85-IIA
Type 88 MBT
Type 98 MBT
Type 62 light tank
Type 63 light tank
Type 99 light tank
Type 56 APC
Type 63 series APC
Type 55 APC
Type 77 series APC
WZ 501 series APC
Type 85 APC
YW 531 series APC
YW 534 APC
YW 307 IFV
YW 309 IFV
WZ 501 IFV
WZ 551 APC
Type 90 APC
120 mm 2S23 SP/GM
122 mm Type 54-1 SPH
122 mm Type 83 SPH
152 mm Type 83 SPH
120 mm Type 89 TD
YW 531 HJ-8 TD
WZ 501 HJ-73B TD
WZ 551 HJ-8 TD
WZ 551 (4 × 4) HJ-9 TD
155 mm PLZ45 SPH
122 mm Type 85 SPH
122 mm Type 85
122 mm YW 531C SPH
122 mm WZ 551 SPH
82 mm YW 304 SPM
120 mm YW 381 SPM
82 mm Type 85 SPM
120 mm Type 84 SPM
57 mm Type 56 ATG
76 mm Type 54 ATG
85 mm Type 56 FG
100 mm Type 73 ATG
100 mm Type 86 ATG
100 mm Type 59 FG
122 mm Type 60 FG
122 mm Type 54 howitzer
122 mm Type 83 howitzer
122 mm D-30 howitzer
122 mm M1931 gun
130 mm Type 59 FG
130 mm Type 59-1 FG
152 mm Type 66 GH
152 mmType 54 howitzer
152 mm Type 83 FG
152 mm Type 86 gun
155 mm Type WAC-21 GH
107 mm Type 63 MRS
107 mm Type 63-1 MRS
107 mm Type 81 MRS
122 mm Type 81 MRS
122 mm Type 84 MRS
122 mm Type 89 MRS
122 mm Type 90 MRS
130 mm Type 63/Type 63-1 MRS
130 mm Type 70 MRS
132 mm BM-13-16 MRS
140 mm BM-14-16 MRS
180 mm Type 71 MRS
252 mm Type 85 MRS
253 mm Type 81 MRS
253 mm Type 81-II MRS
273 mm Type 83 MRS
284 mm Type 74 MRS
300 mm A100 ARS
300 mm A120 ARS
305 mm Type 79 ARS
320 mm WS-1 ARS
350 mm M-1B MRS
425 mm Type 762 MRS

COLOMBIA
M3A1 Stuart light tank
M8/M20 recce modified
EE-9 Cascavel recce
EE-11 Urutu APC
M113A1 APC
Buffalo APC (IS)
RG-31 Nyala APC
RCV9 APC (IS)
M8 TOW TD
105 mm M101 howitzer

CONGO
T-54/T-55 MBT
Type 59 MBT
T-34/85 (R)
Type 62 light tank
PT-76 light tank
BRDM-1 recce
BRDM-2 recce
Eland recce
BTR-60 APC
BTR-152 APC
OT-62 APC
M113 APC
Mamba APC
57 mm M1943 ATG
75 mm M116 P howitzer
76 mm M1942 FG
85 mm Type 56 FG
100 mm M1944 FG
100 mm M1938 howitzer
122 mm D-30 howitzer
130 mm M-46 FG
152 mm D-20 GH
122 mm BM-21 MRS
140 mm BM-14 MRS

CONGO, DEMOCRATIC REPUBLIC
Type 59 MBT
Type 62 light tank
AML-60/90 recce
RAM recce
M3 APC
M113 APC
Type 531 APC
Fahd APC (IS)
90 mm M3 half-track TD
57 mm M1943 ATG
75 mm M116 P howitzer
85 mm Type 56 FG
122 mm M1938 howitzer
122 mm D-30 howitzer
122 mm Type 60 FG
130 mm Type 59-1 FG
107 mm Type 63 MRS
122 mm BM-21 MRS

COTE D'IVOIRE
AMX-13 light tank
AML-60/90 recce
ERC-90 Sagaie recce
M3 APC
VAB (G)
ACMAT VBLA APC (G)
Mamba RG-12 APC
105 mm M2 howitzer

CROATIA
M-84 MBT
M-84AB MBT
Degman MBT*
T-55 MBT
T-34/85 tank
PT-76 light tank
BRDM-2 recce
M-60P APC
M-80 APC
BTR-60 APC
LOV APC
BOV APC
122 mm 2S1 SPH
76 mm ZIS-3 FG
76 mm M48 mountain gun
85 mm D-44 FG
100 mm T-12 ATG
105 mm M-56 howitzer
105 mm M101 howitzer
120 m 2S9 SPM/H
122 mm M1938 howitzer
122 mm D-30 howitzer
130 mm M-46 FG
152 mm D-20 GH
152 mm M-84 GH
155 mm L33 CITEFA Mod 77 howitzer
70 mm Heron MRS
122 mm BM-21 MRS
122 mm LOV RAK 124/128 MRS
122 mm M96 Typhoon MRS
128 mm M-63 MRS
262 mm M-87 Okran MRS

CUBA
T-62 MBT
T-54/T-55 MBT
T-34/85 tank
PT-76 light tank
BRDM-1 recce
BRDM-2 recce
BMP-1 IFV
BTR-40 APC
BTR-60 APC
BTR-152 APC
SU-100 SPG
BRDM-2 Sagger TD
122 mm 2S1 SPH
122 mm 2S3 SPGH
76 mm M1942 FG
85 mm D-44 FG
85 mm SD44 FG
85 mm D-48 ATG
100 mm T-12 ATG
122 mm D-30 howitzer
122 mm M1938 howitzer
130 mm M-46 FG
152 mm D-20 GH
122 mm BM-21 MRS
140 mm BM-14 MRS
FROG-4 ARS
FROG-7 ARS

CYPRUS
T-80U/T-80UK MBT
AMX-30 B2 MBT
EE-3 Jararaca recce
EE-9 Cascavel recce
Shorland (IS)

BMP-3 ICV
VAB APC
EE-11 Urutu APC
Leonidas APC
AMX VCA carrier
EE-3 Jararaca MILAN TD
VAB HOT TD
155 mm Mk F3 SPH
155 mm/45 calibre Zuzana SPG
75 mm M116 P howitzer
25 pounder FG (R)
100 mm M1944 FG
105 mm M56 howitzer
128 mm M-63 MRS
155 mm TR gun

CZECH REPUBLIC
T-72 MBT
T-54/T-55 MBT
Sneza recce
BMP LOS recce
OT-65A recce
BRM-1 recce
VT-90 recce
BVP-1 IFV (and BG)
BVP-2 IFV
OT-62 APC (and BG)
OT-64 APC (and BG)
OT-90 IFV
MT-LB MRV
FUG D-944 APC
BRDM-2 Sagger TD
122 mm 2S1 SPH
152 mm Dana SPG
85 mm M52 FG
100 mm M53 FG
122 mm D-30 howitzer
122 mm M-30 FH (R)
122 mm RM-70 MRS
FROG-7 ARS

DENMARK
Leopard 1A5 MBT
Leopard 2 MBT
Eagle recce
M113/M113A1 APC
M113A2 25 mm APC
Piranha III APC
81 mm M125/M125A1 SPM
120 mm M106 SPM
155 mm M109A3 SPG
M113 TOW TD
105 mm M101 howitzer
155 mm M59 gun
155 mm M114/39 gun
203 mm M115 howitzer
227 mm MLRS

DOMINICAN REPUBLIC
M41 light tank
AMX-13 light tank
LAV-150 APC
M3A1 half-track APC
105 mm M101 howitzer

DJIBOUTI
AML-60/90 recce
VBL recce
BTR-60 APC
122 mm D-30 howitzer

ECUADOR
T-55 MBT
AMX-13 light tank
M3A1 Stuart light tank
AML-60/90 recce
EE-3 Jararaca recce
EE-9 Cascavel recce
AMX VCI APC
M113 APC
UR-416 APC
VAB APC
EE-11 Urutu APC
BMR-600 APC
BLR (IS)
155 mm MK F3 SPG
105 mm M101 howitzer
105 mm Model 56 P howitzer
155 mm M198 howitzer
155 mm M114 howitzer
122 mm RM-75 MRS

EGYPT
M1A1 MBT
M60A1/M60A3 MBT
T-54/T-55 MBT
T-62 MBT
BRDM-2 recce
Scout recce
M113A1/M113A2 series APC
BTR-50P APC (R)
AIFV
BMP-1 IFV (R)
OT-62 APC (R)
OT-64 APC (R)
BMR-600 APC
Fahd APC
Fahd 30 APC
Hussar (IS)
4984 (IS)
82 mm M125A2 SPM
120 mm M106A2 SPM
122 mm SP122 SPH
122 mm T-55 chassis SPH
155 mm M109A2 SPH
M901 ITV
AIFV TOW TD
BRDM-2 Sagger TD
85 mm D-44 FG
100 mm M1944 FG
122 mm D-74 FG
122 mm D-30 howitzer
122 mm M1938 howitzer
122 mm Type 60 FG
130 mm M-46 FG
130 mm Type 59-1M FG
152 mm D-20 GH
152 mm D-1 howitzer
155 mm EH52 (trials)
122 mm BM-21 MRS
122 mm Sakr-18 MRS
122 mm Sakr-36 MRS
227 mm MLRS*
FROG-7 ARS

EL SALVADOR
AML-90 recce
UR-416 APC
105 mm M56 howitzer
105 mm M101 howitzer
105 mm M102 howitzer

ERITREA
T-54/T-55 MBT
BRDM-2 recce
BMP-1 IFV
BTR-60 APC
85 mm D-44 FG
122 mm D-30 howitzer
130 mm M-46 FG
122 mm BM-21 MRS

ESTONIA
BRDM-2 recce
BTR-60 APC
BTR-70 APC
BTR-80 APC
105 mm m61/37 FH

ETHIOPIA (status of equipment is uncertain)
T-62 MBT
T-54/T-55 MBT
BRDM-1 recce
BRDM-2 recce
BMP-1 IFV
M113A1 APC
BTR-60 APC
BTR-152 APC
122 mm 2S1 SPH
85 mm D-44 FG
85 mm D-48 ATG
105 mm M101 howitzer
122 mm D-30 howitzer
122 mm M1938 howitzer
130 mm M-46 FG
152 mm D-1 howitzer
155 mm M114 howitzer
122 mm BM-21 MRS
200 mm BMD-20 MRS

EQUATORIAL GUINEA
BRDM-2 recce
BTR-152 APC

FINLAND
Leopard 2 (order expected 2002)
T-72 MBT
T-55M MBT
CV 9030 FIN*
BMP-1 IFV
BMP-2 IFV
MT-LB MRV
MT-LBV MRV
BTR-60PB APC
BTR-80 APC
XA-180 series APC
Sandringham APC
122 mm 2S1 SPH
152 mm 2S5 SPH
105 mm m61/37 FH
122 mm D-30 howitzer (H 63)
122 mm m/38 FH
130 mm m/54 FG (K 54)
152 mm 2A36 (15 K 89) gun
152 mm H 37 A GH
152 mm H 88-40GH
152 mm K 89 GH
155 mm 155 GH 52 APU
122 mm 2S1 SPH
152 mm 2S5 SPG
122 mm RM-70 MRS
122 mm BM-21 MRS

FRANCE
Leclerc MBT
AMX-30 B2 MBT
AMX-30 MBT
AMX-10RC recce
ERC 90 F4 Sagaie
VBL recce
AML-60 and 90 (G)
VBC-90 (G)
AMX-10P ICV
VAB APC (and G)
Bv 206S
VXB-170 (G)
155 mm GCT SPG
155 mm CEASAR SPG*
VAB HOT TD
155 mm Model 50 howitzer
155 mm TR gun
227 mm MRS

GABON
AML-60/90 recce
EE-3 Jararaca recce
EE-9 Cascavel recce
M8 recce
ERC 90 Sagaie
VBL recce
RAM recce
VXB-170 APC
LAV-150 APC
M3 APC
ACMAT VBLA APC
EE-11 Urutu APC
105 mm M2 howitzer
140 mm Teruel MRS

GAZA and JERICHO
BRDM-2 recce

GEORGIA
T-72 MBT
T-55 MBT
T-55AM2 MBT
BMP-1 IFV
BMP-2 IFV
BTR ACV
BTR-80 APC
152 mm 2S3 SPGH
152 mm 2S7 SPG
85 mm D-44 gun
100 mm T-12 ATG
122 mm D-30 howitzer
130 mm M-46 FG
152 mm 2A65 howitzer
122 mm BM-21 MRS

GERMANY
Leopard 2A5 MBT
Leopard 2A6 MBT
Leopard 2 MBT
Leopard 1 MBT
Spahpanzer Luchs recce
Fennek recce*
Wiesel recce/TD
Marder 1 ICV
Transportpanzer 1 APC
M113A1 APC series
UR-416 APC (IS)
TM 170 APC (IS)
APCV-2 (or Dingo)
155 mm M109A3G SPH
155 mm PzH 2000 SPG
Jagdpanzer Jaguar 1 HOT TD
Jagdpanzer Jaguar 2 TOW TD
120 mm M113 SPM
105 mm Model 56 P howitzer
105 mm M101 howitzer
155 mm FH-70
110 mm LARS MRS
227 mm MLRS

GHANA
EE-9 Cascavel recce
Piranha APC
Tactica APC (IS)

GREECE
Leopard 2 (selected March 2002)
Leopard 1A3 MBT
Leopard 1A5 MBT
M60A1/M60A3 MBT
M48A3/M48A5 MBT
AMX-30 MBT (R)
M8/M20 recce (R)
AMX-10P ICV (R)
M113/M113A1/M113A2 APC
Leonidas APC
BMP-1 IFV
Roland APC (G)
UR-416 APC (G)
105 mm M52A1 SPH
155 mm M109/M109A1B/
M109A2/M109A5 SPH
155 mm/45 calibre Zuzana SPG
155 mm/52 calibre PzH 2000 SPH
175 mm M107 SPH
203 mm M110 SPH
M113A2 TOW TD
M901 ITV TOW TD
81 mm M125A1 SPM
107 mm M106A1 SPM
105 mm M101 howitzer
105 mm Model 56 P howitzer
155 mm M114 howitzer
122 mm RM-70/85 MRS
227 mm MLRS

GUATEMALA
M41 light tank
M8 recce
RBY-1 recce
M113 APC
LAV-100 APC
Armadillo APC
75 mm M116 howitzer
105 mm M101 howitzer
105 mm M102 howitzer
105 mm M56 howitzer

GUINEA-BISSAU
T-34/85 tank
PT-76 light tank
BRDM-2 recce
BTR-40 APC
BTR-60 APC
BTR-152 APC
Type 56 APC
85 mm D-44 FG
122 mm M1938 howitzer
122 mm D-30 howitzer

GUINEA REPUBLIC
T-54 MBT
T-34/85 tank
PT-76 light tank
BRDM-2 recce
AML-90 recce
BTR-40 APC
BTR-50 APC
BTR-60 APC
BTR-152 APC
76 mm M1942 FG
85 mm D-44 FG
122 mm M1931/7 gun
130 mm M-46 FG
130 mm M1982 gun
220 mm BM 9P140 (16-round) MRS

GUYANA
Shorland (IS)
EE-11 Urutu APC
130 mm M-46 FG

HAITI (status of equipment is uncertain)
M3A1 Stuart light tank
M5A1 Stuart light tank
LAV-150 APC
37 mm M3A1 ATG
57 mm M1 ATG
75 mm M116 P howitzer
105 mm M101 howitzer

HONDURAS
Saladin recce
Scorpion 76 mm recce
Scimitar recce
RBY-1 recce
105 mm M101 howitzer
105 mm M102 howitzer
155 mm M198 howitzer

HONG KONG
Saxon (IS)

HUNGARY
T-72 MBT
T-55 MBT
FUG D-442 recce
BRDM-2 recce
BRDM-1 recce
BMP-1 IFV
BTR-80
BRT-80A APC
FUG D-944 APC
MT-LB MRV
120 mm 2S9 SPM/H
122 mm 2S1 SPH
BRDM-2 Sagger TD
85 mm D-44 FG
100 mm T-12 ATG
122 mm M1938 howitzer
152 mm D-20 GH
122 mm BM-21 MRS
FROG-7 ARS

INDIA
T-90S
T-72 MBT
T-54/T-55 MBT
Vijayanta MBT
Arjun* MBT
BRDM-2 recce
Ferret recce (IS)
BMP-1 IFV
BMP-2 IFV
Casspir APC
OT-62 (R)
OT-64 (R)
105 mm Abbot SPG (P)
130 mm Vijayanta SPG
76 mm mountain howitzer
76 mm M48 mountain gun (P)
105 mm FG
130 mm M-46 FG
155 mm M-46 FG (upgrade)*
5.5 in gun (P)
105 mm Model 56 P howitzer
155 mm FH-77B
122 mm D-30 howitzer
122 mm BM-21 MRS
122 mm LRAR MRS
214 mm Pinacha MRS

INDONESIA
AMX-13 light tank
PT-76 light tank (and Ma)
Ferret recce
Saladin recce
Scout recce
Scorpion 90 mm recce
VBL recce
BRDM-2 recce
AMX-10 PAC 90 (Ma)
AMX-10P APC (Ma)
AMX VCI APC
BTR-40 APC
BTR-50 APC (and Ma)
BMP-2 IFV
LAV-150 APC
Saracen APC
Ranger APC
Stormer APC
VAB APC
Tactica APC (IS)
ACMAT VBLA Yari APC (IS)
76 mm M48 mountain gun
105 mm Light Gun Mk II
105 mm M56 howitzer
105 mm M101 howitzer
122 mm M1938 howitzer (and Ma)
155 mm FH-88 FH
70 mm NDL-40 MRS

IRAN
Chieftain MBT
M48/M48A5 MBT
M60A1 MBT
M47/M47M tank
T-72 MBT
T-62 MBT
T-54/T-55 MBT
Type 59 MBT
Type 69-2 MBT
Type 72-Z MBT
Zulfiqar MBT
Safir-74 MBT
76 mm Scorpion recce
Tosan light tank
EE-9 Cascavel recce
BMP-1 IFV
BMP-2 IFV
BTR-50 APC
BTR-60 APC
M113A1 APC
Boragh APC
Cobra APC
122 mm Thunder 1 SPG
122 mm 2S1 SPH
155 mm M109/M109A1 SPH
155 mm Thunder 2 SPG
175 mm M107 SPG
170 mm M1978 SPG
203 mm M110 SPG
M113 TOW TD
85 mm D-44 FG
105 mm M101 howitzer
122 mm D-30 howitzer
122 mm HM40 howitzer
122 mm Type 60 FG
122 mm Type 54 howitzer
130 mm Type 59-1 FG
152 mm D-20 GH
155 mm M114 howitzer
155 mm HM41 howitzer
155 mm WAC-021 GH
155 mm G5 GH
155 mm GH N-45 GH
203 mm M115 howitzer
107 mm Type 63 MRS
122 mm Type 81 MRS
122 mm BM-21 MRS
NAZEAT MRS
SHAHIN 1 ARS
SHAHIN 2 ARS
OGHAB ARS
HADID ARS
ARASH ARS
NOOR ARS
FADGR ARS
HASEB ARS
ZELZAL ARS
FADJR 5 ARS
FADJR 3 ARS
FADJR 6 ARS

IRAQ (status of some equipment is now uncertain)
Asad Babl
T-72 MBT
T-62 MBT
T-55 MBT
T-55 mod MBT
T-54 MBT
Type 59 mod MBT
Type 69-2 MBT
Chieftain MBT
AML-60/90 recce
BRDM-1 recce
BRDM-2 recce
EE-9 Cascavel recce
PShZ-IV recce
BMP-2 IFV
BMP-1 IFV
BTR-50 APC
BTR-60 APC
BTR-152 APC
EE-11 Urutu APC
M3 APC
Walid APC
OT-62 APC
OT-64 APC
Roland (IS)
Type 531 APC
MT-LB MRV
BMD-1 ACV
AMX-10P ICV
M113 series APC
122 mm 2S1 SPH
152 mm 2S3 SPGH
155 mm Mk F3 SPH
155 mm M109A1 SPH
155 mm GCT SPG
240 mm 2S4 SPM
VCR/TR HOT TD
BRDM-2 Sagger TD
BRDM-2 Spandrel TD
76 mm M48 mountain gun
85 mm D-44 FG
100 mm T-12 ATG
105 mm Model 56 P howitzer
122 mm D-30 howitzer
122 mm Saddam howitzer
122 mm M1938 howitzer
122 mm D-74 FG
130 mm Type 59-1 FG
130 mm M-46 FG
152 mm D-1 howitzer
152 mm Type 83 gun
152 mm 2A36 gun
152 mm M1937 GH
152 mm Type 66 GH
155 mm M114 howitzer
155 mm GH N-45 GH
155 mm G5 GH
155 mm Type WA 021 GH
240 mm M-240 mortar
107 mm MRS
108 mm FGT 108R MRS
122 mm BM-21 MRS
122 mm M1975 MRS
122 mm BM-11 MRS
127 mm Sajeel 30 MRS
122 mm Scania MRS
132 mm BM-13 MRS
140 mm MRS
180 mm Sajeel 40 MRS
300 mm Sajeel 60 MRS
262 mm Ababeel MRS
40 mm Ababeel 100 ARS
FROG-7 ARS
Laith 90 ARS

IRELAND
Scorpion 76 mm recce
AML-60/90 recce
Panhard M3 APC
XA-180 APC (UN)
Piranha III (8 × 8)
25-pounder gun
105 mm Light Gun

ISRAEL
Merkava Mk 1/Mk 2/Mk 3/Mk 4
M48A5 MBT
M60/M60A1/M60A3 MBT
MAGACH-7 MBT
MAGACH-8 MBT
RBY-1 recce
BRDM-2 recce
Classical APC
M2/M3 half-track APC
M113A1/M113A2 APC
Puma APC
Achzarit APC
Nagmashot APC
Nagmachon APC
Nakpadon APC
155 mm M109/M109A1 SPH
155 mm L33 SPH
155 mm Sherman SPH (R)
175 mm M107 SPG
203 mm M110A2 SPH
81 mm M125A2 SPM
120 mm M106A2 SPM
160 mm Sherman SPM
M113 TOW TD
RBY Mk 1 TOW TD
RBY Mk 1 MAPATS TD
105 mm M101 howitzer
122 mm D-30 howitzer
130 mm M-46 field gun
155 mm M114 howitzer
155 mm M-71 GH
155 mm 845/839P GH
122 mm BM-21 MRS
240 mm BM-24 MRS
290 mm MAR ARS
350 mm MAR ARS
227 mm MLRS

ITALY
Ariete MBT
Leopard 1A5 MBT
Leopard 1A1 MBT
FIAT 6616 (IS)
Centauro 105 mm
Dardo IFV *
M113 APC (and Ma)
VCC-1 (and Ma)
LVTP-7 (Ma)
Bv 206S APC
FIAT 6614 APC (AF, Army and iS)
Puma (4 × 4)*
Puma (6 × 6)*
Boneschi MAV 5 APC
155 mm PzH 2000 SPG*
155 mm M109L SPH
203 mm M110A2 SPH
M113 TOW TD
105 mm Model 56 P howitzer
107 mm M106 SPM
155 mm FH-70
155 mm M114 howitzer
122 mm FIROS MRS
227 mm MLRS

JAMAICA
Ferret recce
LAV-150 APC

JAPAN
Type 90 MBT
Type 74 MBT
Type 61 MBT (P)
Type 82 recce
Type 87 recce
Type 89 MICV
Type 73 APC
Type 60 APC
Type 96 APC
105 mm Type 74 SPH
155 mm Type 75 SPH
155 mm Type 99 SPH
203 mm M110A2 SPH
81 mm Type SV SPM
107 mm Type SX SPM
106 mm Type 60 SPRR
105 mm M101 howitzer
155 mm FH-70
130 mm Type 75 MRL
227 mm MLRS

JORDAN
Centurion (Tariq) MBT
M48A5 MBT
M60A1/M60A3 MBT
Khalid MBT
Challenger 1 MBT
Saladin recce (G)
Ferret recce
Scorpion 76 mm recce
Saracen APC (G)
M113A1/M113A2 APC
BMP-2 IFV
EE-11 Urutu APC (G)
BTR-94 APC
105 mm M52 SPH
155 mm M109/M109A2 SPG
155 mm M44 SPH
203 mm M110A1/M110A2 SPH
81 mm M125 SPM
107 mm M106 SPM
M901 TOW TD
105 mm M101 howitzer
105 mm M102 howitzer
155 mm M59 gun
155 mm M114 howitzer
203 mm M115 howitzer

KAZAKHSTAN
T-72 MBT
BRDM-2 recce
BRDM-1 recce
BMP-1 IFV
BMP-2 IFV
BTR-80 APC
BTR-70 APC
MT-LB MRV
122 mm 2S1 SPG
152 mm 2S3 SPGH
100 mm T-12 ATG
122 mm D-30 howitzer
130 mm M-46 FG
152 mm D-20 GH
152 mm 2A65 howitzer
152 mm 2A36 gun
122 mm BM-21 MRS
220 mm BM 9P140 MRS

KENYA
Vickers Mk 3 MBT
AML-60/90 recce
Ferret recce
Shorland (IS)
M3 APC
UR-416 APC
25-pounder FG
105 mm Light Gun
105 mm Model 56 P howitzer

KOREA, NORTH
T-54/T-55 MBT
T-62 MBT
Type 59 MBT
Type 62 light tank
Type 63 light tank
M1985 light tank
T-34/85 tank
BMP-1 IFV
Type 531 APC
M1973 APC
BTR-40 APC
BTR-50 APC
BTR-60 APC
BTR-80 APC
BTR-152 APC
120 mm M1992 SPM
122 mm M1977 SPH
122 mm M1981 SPG
122 mm M1991 SPG
122 mm M1985 SPG
130 mm M1975 SPG
130 mm M1981 SPG
130 mm M1991 SPG
130 mm M1992 SPH
152 mm M1974 SPH
152 mm M1977 SPH
85 mm D-44 FG
85 mm Type 56 FG
85 mm D-48 ATG
100 mm M1944 FG
122 mm M1931/7 gun
122 mm M1938 howitzer
122 mm D-74 FG
122 mm Type 60 FG
122 mm Type 54 howitzer

122 mm D-30 howitzer
130 mm M-46 FG
130 mm Type 59 FG
152 mm M1937 GH
152 mm M1938 howitzer
152 mm Type 66 GH
107 mm Type 63 MRS
122 mm BM-11 MRS
122 mm BM-21 MRS
122 mm M1977 MRS
122 mm M1985 MRS
130 mm Type 63 MRS
140 mm RPU-14-16 MRS
200 mm BMD-20 MRS
240 mm BM-24 MRS
240 mm M1991 MRS
FROG-5 ARS
FROG-7 ARS

KOREA, SOUTH
K1/K1A1
T-80U MBT
M48A2/M48A3/M48A5
M47 (R)
KIFV
BMP-3 IFV
M113 series APC
FIAT 6614 APC
BTR-80 APC
LVTP7/LVTP7A1 (Ma)
155 mm M109A2 SPH
155 mm Thunder K9 SPG
175 mm M107 SPG
203 mm M110 SPH
KIFV SPM
105 mm M101 howitzer
105 mm KH-178 howitzer
155 mm KH-179 howitzer
155 mm M114 howitzer
155 mm M59 gun
203 mm M115 howitzer
130 mm KM809A1 MRS
227 mm MRS

KUWAIT
Chieftain MBT (R)
M-84/M-84AB MBT
M1A2 MBT
BMP-2 IFV
BMP-3 IFV
Desert Warrior IFV
M113A2 APC
Fahd APC
Pandur APC
VAB
S600 Shorland APC
TM-170 APC
Tactica (IS)
82 mm M125A2 SPM
155 mm Mk F3 SPG
155 mm M109A1B SPH
M901 ITV TD
M113 TOW TD
155 mm GCT SPG (R)
155 mm PLZ45 SPG
300 mm BM 9A52 Smerch
FROG 7 ARS

KYRGYZSTAN
T-72 MBT
BRDM-2 recce
BMP-1 IFV
BMP-2 IFV
BTR-80 APC
100 mm T-12 ATG
120 mm 2S9 SPM/H
100 mm M1944 ATG
122 mm D-30 howitzer
122 mm M198 howitzer
152 mm D-1 howitzer
122 mm BM-21 MRS

LAOS
T-54/T-55 MBT
T-34/85 tank
PT-76 light tank
BTR-40 APC
BTR-60 APC
BTR-152 APC
75 mm M116 P howitzer
85 mm D-44 FG
105 mm M101 howitzer
122 mm M1938 howitzer
122 mm D-30 howitzer
130 mm M-46 FG
155 mm M114 howitzer

LATVIA
T-55 MBT
BRDM-2 recce
SKPF m/42 APC
100 mm M53 FG
122 mm D-30 howitzer

LEBANON
AMX-13 FL-10 light tank
M48A5 (and R)
T-54/T-55 MBT
Saladin (R)
AML-90 recce
Ferret recce (R)
Staghound recce
M3 APC
M113A1/M113A2 APC
Saracen APC
VAB APC
AMX VCI APC
Chaimite (IS)
Shorland (IS)
81 mm M125A2 SPM
105 mm M101 howitzer
105 mm M102 howitzer
122 mm M1938 howitzer
122 mm D-30 howitzer
130 mm Type 59-1 FG
130 mm M-46 FG
155 mm Model 50 howitzer
155 mm M114 howitzer
155 mm M198 howitzer
122 mm BM-21 MRS
122 mm BM-11 MRS
FROG-7 ARS

LESOTHO
RBY 1 recce
AML-90 recce
Shorland (IS)

LIBERIA
M3A1 half-track APC
Piranha APC
75 mm M116 P howitzer

LIBYA
T-72 MBT
T-62 MBT
T-54/T-55
BRDM-2 recce
FIAT 6616 recce
EE-9 Cascavel recce
Shorland (IS)
BMP-1 IFV
BTR-50 APC
BTR-152 APC
OT-62 APC
OT-64 APC
Chaimite APC
M113A1 APC
FIAT 6614 APC
EE-11 Urutu APC
BRDM-2 Sagger TD
122 mm 2S1 SPH
152 mm 2S3 SPGH
152 mm Dana SPG
155 mm Palmaria SPG
155 mm M109 SPH
105 mm M101 howitzer
122 mm D-30 howitzer
122 mm D-74 FG
130 mm M-46 FG
152 mm D-20 GH
107 mm Type 63 MRS
122 mm BM-11 MRS
122 mm BM-21 MRS
122 mm RM-70 MRS
130 mm M51 MRS
FROG-7 ARS

LITHUANIA
BRDM-2 recce
SKPF m/42 APC
BTR-60 APC

LUXEMBOURG
Ranger APC

MACEDONIA
T-54/T-55 MBT
T-34/85 tank
BRDM-2 recce
BTR-70 APC
BTR-80 APC
M113 APC
TM-170 APC
76 mm M48 mountain gun
76 mm M1942 gun
105 mm M101 gun
122 mm D-30 howitzer
122 mm M1938 howitzer
122 mm M-63 MRS

MADAGASCAR
PT-76 light tank
Ferret recce
M8 recce
BRDM-2 recce
M3A1 half-track APC
76 mm M1942 FG
105 mm M2 howitzer
122 mm D-30 howitzer

MALAWI
Fox recce
Ferret recce
BRDM-2 recce
Eland recce
105 mm Light Gun

MALAYSIA
Ferret recce
Scorpion 90 mm recce
AML-60/90 recce
Shorland (IS)
Stormer APC
AIFV*
KIFV
SIBMAS APC
Condor APC
LAV-100/LAV-150 APC
M3 APC
SB 301 APC (IS)
Saxon APC (IS)
KIFV SPM
105 mm M102 howitzer (R)
105 mm Model 56 P howitzer
155 mm FH-70
155 mm G5 GH
ASTROSS II MRS

MALDIVES
Cobra (4 × 4) APC

MALI
T-54/T-55 MBT
T-34/85 tank
PT-76 light tank
BRDM-2 recce
BTR-40 APC
BTR-60 APC
BTR-152 APC
85 mm D-44 FG
100 mm M1944 FG
122 mm D-30 howitzer
130 mm M-46 FG
122 mm BM-21 MRS

MALTA
Humber FV1611 APC

MAURITANIA
T-54/T-55 MBT
AML-60/90 recce
Saladin recce
Saracen APC
122 mm D-30 howitzer
105 mm M101 howitzer
122 mm D-74 FG

MAURITIUS
VAB APC
Tactica APC

MEXICO
M8 recce
ERC-90 Lynx recce
VBL recce
DN-VII recce
MAC-1 APC
HWK 11 APC
BDX APC
M3 VTS APC
VCR APC
Sedana 100 APC
LAV-150 ST APC
AMX VCI APC
DN-IV APC
DN-V APC
Roland APC
75 mm M8 SPH
105 mm M7 SPH
VBL MILAN TD
75 mm M116 P howitzer
105 mm M56 howitzer
FIROS 6 MRS

MOLDOVA
BMP-1 IFV
BTR-60 APC
BTR-80 APC
BTR-D ACV
TAB-71 APC
MT-LB MRV
120 mm 2S9 SPM/H
BRDM-2 Spandrel TD
MT-LB Spiral TD
100 mm MT-12 ATG
122 mm D-30 howitzer
122 mm M1938 howitzer
152 mm 2A36 gun
152 mm D-20 GH
220 mm BM 9P140 MRS

MONGOLIA
T-62 MBT
T-54/T-55 MBT
BRDM-2 recce
BMP-1 IFV
BTR-60 APC
BTR-40 APC
BTR-152 APC
BRDM-2 Sagger TD
85 mm D-44 ATG
100 mm T-12 ATG
100 mm M1944 FG
122 mm D-30 howitzer
122 mm M1938 howitzer
130 mm M-46 FG
152 mm D-1 howitzer
122 mm BM-21 MRS

MOROCCO
M60A1/M60A3 MBT
M48A3/M48A5 MBT
T-72 MBT
T-54/T-55 MBT
Jagdpanzer SK 105
AML-20/60/90 recce
EBR-75 recce
AMX-10RC recce
RAM-V1 recce
Eland 20/60/90
M3 APC
M113A1 APC
OT-62 APC
OT-64 APC
VAB APC
UR-416 APC
Ratel 20/90 APC
EE-11 Urutu APC
BRDM-2 Sagger TD
UR-416 Cobra TD
SU-100 SPG
155 mm M109A1B SPH
155 mm Mk F3 SPG
203 mm M110A2 SPH
81 mm M125A1 SPM
M901 TOW TD
76 mm M1942 FG
85 mm D-44 FG
105 mm Model 56 P howitzer
105 mm M101 howitzer
105 mm Light Gun
122 mm D-30 howitzer
130 mm M-46 FG
155 mm M-50 howitzer
155 mm M198 howitzer
155 mm M114 howitzer
155 mm FH-70
122 mm BM-21 MRS

MOZAMBIQUE
T-54/T-55 MBT
T-34/85 tank
BRDM-2 recce
BRDM-1 recce
BTR-60 APC
BTR-152 APC
BMP-1 IFV
76 mm M1942 FG
85 mm D-44 FG
85 mm D-48 ATG
100 mm M1944 FG
122 mm M1938 howitzer
122 mm D-30 howitzer
130 mm M-46 FG
152 mm D-1 howitzer
122 mm BM-21 MRS

MYANMAR (BURMA)
Comet tank
Type 69-II MBT
Type 63 light tank
Ferret recce
Humber recce
Daimler recce
Type 85 APC
BAAC-83 APC
BAAC-84 APC
BAAC-85 APC
BAAC-86 APC
BAAC-87 APC
6-pounder ATG
76 mm M48 mountain gun
17-pounder ATG
25-pounder FG
105 mm M101 howitzer
130 mm M-46/Type 59 FG
5.5 in gun
107 mm Type 63 MRS

NAMIBIA
T-54/T-55 MBT
T-34/85 tank
BRDM-2 recce
BTR-152 APC
Casspir APC
Wolf APC
122 mm D-74 FG
122 mm BM-21 MRS

NEPAL
Ferret recce
3.7 in mountain howitzer
105 mm Model 56 P howitzer
105 mm Light Gun

NETHERLANDS
Leopard 2A5 MBT
Leopard 2 MBT
Fennek recce*
YPR-765 IFV
Transportpanzer 1 APC
UR-416 APC (IS)
155 mm PzH 2000 SPG*
105 mm M109A1/M109A2 SPH
107 mm M106 SPM
AIFV TOW TD
105 mm Light Gun
105 mm M101 howitzer
155 mm M114/39
155 mm FH-70
227 mm MLRS

NEW ZEALAND
Piranha LAV-III (8 × 8)*
M113A1 APC
105 mm Light Gun
105 mm M101 howitzer

NICARAGUA
T-55 MBT
PT-76 light tank
BRDM-2 recce
AMX VCI
BTR-60 APC
BTR-152 APC
BRDM-2 Sagger TD
76 mm ZIS-3 ATG
100 mm M1944 ATG
105 mm M101 howitzer
122 mm D-30 howitzer
152 mm D-20 GH
122 mm BM-21 MRS

NIGER
ERC 60/20 recce
AML-60/90 recce
VBL recce
M3 APC
UR-416 APC

NIGERIA
T-55 MBT
Vickers Mk 3 MBT
AML-60/90 recce
Fox recce
Scorpion 76 mm recce
Scimitar recce
EE-9 Cascavel recce
Sagaie recce
Hornet recce
Saracen APC
Steyr 4K 7FA APC
Piranha APC
M3 APC
Saxon APC
Casspir APC
EE-11 Urutu APC (?)
155 mm Palmaria SPG
105 mm Model 56 P howitzer
122 mm D-74 FG
122 mm D-30 howitzer
130 mm M-46 FG
122 mm BM-21 MRS
122 mm APR 40 MRS

NORWAY
Leopard 2A4 MBT
Leopard 1A5 NO
CV 9030N IFV
M113 APC
NM135 APC
XA-188 APC
XA-180 APC (UN)
VAB (UN)
155 mm M109A3GN SPH
NM142 TOW TD
105 mm M101 howitzer
155 mm M114/39
227 mm MLRS

OMAN
Challenger 2 MBT
M60A1/M60A3 MBT
Chieftain MBT (R)
Scorpion 76 mm recce
VBC 90 recce
VBL recce
OTO R 2.5
Piranha APC
Stormer APC
Fahd APC
Saxon APC (IS)
VAB (G)
VAB MILAN TD
155 mm G6 SP GH
105 mm Light Gun
122 mm D-30 howitzer
130 mm M-46 FG
130 mm Type 59-1M FG
122 mm Type 90 (40-round) MRL

PAKISTAN
Al Khalid MBT (trials)
T-54/T-55 MBT
T-80UD MBT
Type 59 MBT
Type 69 MBT
Type 85 MBT
M48A5 MBT
M47M tank
Ferret recce (IS)

M113A1 APC
Type 531 SPC
UR-416 APC (IS)
Transac APC (IS)
Shorland (IS)
BTR-80 APC (UN)
Akrep APC
105 mm M7 SPH
155 mm M109/M109A2 SPH
203 mm M110A2 SPH
M36 TD
M901 TOW TD
85 mm Type 56 FG
100 mm Type 59 FG
105 mm M101 howitzer
105 mm Model 56 P howitzer
122 mm Type 54 howitzer
122 mm Type 60 FG
130 mm Type 59-1 FG
130 mm M-46 FG
155 mm M59 gun
155 mm M114 howitzer
155 mm M198 howitzer
203 mm M115 howitzer
122 mm MRS
122 mm BM-11 MRL

PARAGUAY
M4A3 Sherman tank
Sherman Firefly tank
M3A1 light tank (Mod)
EE-9 Cascavel recce
EE-11 Urutu APC
75 mm Bofors M1935 FG
105 mm M101 howitzer

PERU
T-54/T-55 MBT
AMX-13 light tank
M8 recce
BRDM-2 recce (and AF)
FIAT 6616 recce
M113A1 APC
BTR-60 APC
BMR-600 APC
UR-416 APC
FIAT 6614 APC
Casspir APC
Chaimite APC (Ma)
Roland APC (and G)
81 mm FIAT 6614 SPM
107 mm M106A1 SPM
122 mm 2S1 SPH
155 mm Mk F3 SPG
155 mm M109A2 SPG ?
105 mm M101 howitzer
105 mm Model 56 P howitzer
122 mm D-30 howitzer
130 mm M-46 FG
155 mm M114 howitzer
107 mm RO 107 MRS
122 mm BM-21 MRS

PHILIPPINES
Scorpion 76 mm recce
AIFV
M113A1 APC
Chaimite APC
LAV-150 APC
LAV-300 APC
Simba APC
LVTP5 (Ma)
LVTP7A1 (Ma)
LVTH6 105 mm (Ma)
105 mm Model 56 P howitzer
105 mm M101 howitzer (& Ma)
155 mm M114 howitzer
155 mm M68 GH

POLAND
Leopard 2A4*
PT-91 MBT
T-72 MBT
T-55 MBT
BRDM-2 recce
BMP-1 IFV
OT-62 APC
OT-64 APC
BTR-60 APC
MT-LB MRV
BRDM-2 Sagger TD
120 mm 2S9 GPM/H
120 mm 2S12 SPM
122 mm 2S1 SPH
152 mm Dana SPG
152 mm D-20 GH
155 mm/52 calibre Krab SPG (trials)
203 mm 2S7 SPG
85 mm D-44 FG
122 mm M1938 howitzer
122 mm BM-21 MRS
122 mm RM-70 MRS
FROG-7 ARS

PORTUGAL
M60A3 MBT
M48A5 MBT
M47 tank
AML-60/90 recce
Saladin recce
VBL recce
Chaimite APC (and Ma)
Commando APC Mk 3
M3 APC
M113 APC
Condor APC
YP-408 APC
Shorland (IS)
155 mm M109A2 SPH
81 mm M125A1 SPM
107 mm M106A1/A2 SPM
M113A2 TOW TD
M901 TOW TD
105 mm M101 howitzer
105 mm Model 56 pack howitzer
155 mm M114 howitzer

QATAR
AMX-30S MBT
AMX-10RC recce
VBL recce
M1114 recce
AMX-10P IFV
AMX VCI APC
VAB APC
Piranha APC
LAV-150 APC
UR-416 APC (IS)
VCAC HOT TD
81 mm VAB SPM
155 mm Mk F3 SPG
155 mm G5
ASTROS MRS

ROMANIA
TR-55 MBT
TR-85 MBT
TR-580 MBT
BRDM-2 recce
TAB-C recce
MLI-84 IFV
BMP-1 IFV
MLVM MRV
ABAL support vehicle
TAB-71 APC
TAB-77 APC
B33 APC
122 mm 2S1 SPH
122 mm M1989 SPH
82 mm TAB-71AR SPM
82 mm TABC-79AR SPM
120 mm MLVM-AR SPM
BRDM-2 Sagger TD
57 mm M1943 ATG
76 mm M48 mountain gun
76.2 mm mountain gun
85 mm D-44 FG
85 mm D-48 ATG
98 mm mountain howitzer
100 mm Skoda gun (various)
105 mm Schneider
100 mm M1977 ATG
122 mm M1931/7 gun
122 mm M1938 howitzer
130 mm M1982 gun
150 mm Model 1934 howitzer
150 mm CEH M1937 gun
152 mm 1981 GH
152 mm M1984 howitzer
152 mm M1985 GH
152 mm M1938 howitzer
122 mm APR-40/MRS
160 mm LAROM 160 MRS
FROG-7 ARS

RUSSIA
T-90 MBT
T-80 MBT
T-72 MBT (and NI)
T-64 MBT
T-62 MBT
T-55 MBT (and NI)
T-54 MBT (and NI)
PT-76 light tank (and NI)
BRM/BRM-1/BRM-2 recce
BRDM-2 (and NI) recce
BRDM-1 recce (R)
BMP-1/BMP-1M IFV
BMP-2 IFV
BMP-3 IFV
MT-LB series MRV
BTR-50 APC (R)
BTR-60 APC (and NI)
BTR-70 APC
BTR-80 APC (and NI)
BTR-90 APC
BMD-1/BMD-1M ACV
BMD-2 ACV
BMD-3 ACV
BRDM-2 Sagger TD
BRDM-2 Swatter TD
BRDM-2 Spandrel TD
120 mm 2S9 SPM/H
120 mm 2S23 SP/GM
122 mm 2S1 SPH (and NI)
152 mm 2S3 SPH
152 mm 2S5 SPGH
152 mm 2S19 SPG
203 mm 2S7 SPG
82 mm BTR-60 SPM
120 mm MT-LB SPM
240 mm 2S4 SPM
76 mm M1966 mountain gun
85 mm D-44 field gun (R and NI)
100 mm T-12 ATG (and NI)
120 mm 2B16
122 mm D-30 howitzer
125 mm 2A45M ATG
130 mm M-46 FG (R)
152 mm D-1 howitzer
152 mm D-20 GH
152 mm 2A36
152mm 2A65
152 mm 2A61
122 mm BM-21 MRS (and NI)
122 mm M1975 MRS
122 mm M1976 MRS
220 mm Splav BM 9P 140 MRS
300 mm BM 9A52 Smerch MRS
FROG-7 ARS
TOS-1 MRS

RWANDA (status of equipment is uncertain)
T-54/T-55 MBT
AML-60/90 recce
VBL recce
Buffalo APC
Mamba Mk II APC
105 mm M101 howitzer
122 mm D-30 howitzer
122 mm RM-70 MRS

SAUDI ARABIA
M1A2 MBT
M60A3 MBT
AMX-30S MBT
AML-60/90 recce
OTO R 2.5 recce
M2 Bradley IFV
AMX-10P IFV
M3 APC
M113 APC
Al Faris 8-400
Transportpanzer 1 NBC
BMR-600 (N)
LAV-25 (and NG)
LAV-150 (NG)
UR-416 (IS)
Transac (IS)
Tactica (IS)

155 mm M109A1B/M109A2 SPH
155 mm GCT SPG
81 mm M125A1/M125A2
81 mm LAV-150 SPM (NG)
107 mm M106A1/M106A2 SPM
VCC-1 TOW TD
LAV-150 TOW (NG)
AMX-10 HOT TD
LAV HOT TD
LAV TOW TD (NG)
105 mm M101 howitzer
105 mm M102 howitzer (NG)
155 mm M114 howitzer
155 mm FH-70
155 mm M198 howitzer
ASTROS II MRS

SENEGAL
AML-60/90 recce
M8/M20 recce
M3 APC
VXB-170 (G)
M3 half-track APC
75 mm M116 P howitzer
105 mm M101 howitzer
155 mm Model 50 howitzer

SEYCHELLES
BRDM-2 recce
122 mm D-30 howitzer
122 mm BM-21 MRS

SIERRA LEONE
T-72 MBT
Ferret recce
Saladin recce
Piranha APC
Roland APC
OT-64 APC
25-pounder FG

SINGAPORE
Centurion MBT
AMX-13 SM-1 light tank
Bionix IFV
M113A2 APC
All-Terrain Tracked Carrier
AMX-10P 25 Dragar APC
AMX-10P PAC 90
LAV-150 APC
LAV-200 APC (AF)
81 mm M125A1/A2 SPM
120 mm M113 SPM
LAV-200 SPM (AF)
105 mm LG1 Light Gun
155 mm M114 howitzer (R)
155 mm M68 GH (R)
155 mm M71 GH (R)
155 mm FH-77 GH
155 mm FH-2000

SLOVAKIA
T-72 MBT
T-55 MBT
OT-65A recce
BRM-1 recce
VT-90 recce
BVP-1 IFV (and BG)
BVP-2 IFV (and BG)
OT-90 IFV
OT-64 APC (and BG)
MT-LB MRV
BRDM-2 Sagger TD
122 mm M-30 FH (R)
122 mm 2S1 SPG
152 mm Dana SPG
155 mm Zuzana SPG
100 mm M53 FG
122 mm D-30 howitzer
122 mm RM70 MRS
FROG-7 ARS

SLOVENIA
M-84 MBT
T-55 MBT
M-55 S1
T-34/85 tank
BRDM-2 recce
M1114 recce
M-80 APC
BOV-VP APC
BTR-50 APC
LOV Torpedo APC
Pandur APC
BOV Sagger TD
122 mm 2S1 SPH
105 mm M101 howitzer
155 mm Model 845 gun
128 mm M-63 MRS

SOMALIA (status of equipment is uncertain)
T-54/T-55 MBT
T-34/85 tank
M47 tank
BRDM-2 recce
AML-60/90 recce
FIAT 6616 recce
BTR-50 APC
BTR-60 APC
BTR-152 APC
LAV-150 APC
M113 APC
FIAT 6614 APC
M3 APC
M113A2 TOW TD
76 mm M1942 FG
85 mm D-44 FG
85 mm D-48 ATG
100 mm M1944 FG
105 mm Model 56 howitzer
122 mm Type 60 F
122 mm D-30 howitzer
122 mm M1938 howitzer
130 mm Type 59-1 FG
130 mm M-46 FG
152 mm D-1 howitzer
155 mm M114A1 howitzer
155 mm M198 howitzer
180 mm S-23 gun
122 mm BM-21 MRS
132 mm BM-13 MRS
140 mm BM-14-16 MRS
140 mm BM-14-17 MRS
240 mm BM-24 MRS

SOUTH AFRICA
Centurion MBT (Olifant)
Eland 60/90 (P)
Rooikat recce
Ferret recce
Swerwer (IS)
Ribbok (IS)
Ratel 20/90 APC
Buffel (and IS)
Bulldog (and IS)
Casspir (and IS)
Hippo (IS)
Rhino (and IS)
Wolf (IS)
Ingwe (IS)
Mamba
81 mm Ratel SPM
155 mm G6 SPG
Ratel ZT3 TD
155 mm G4 GH (R)
155 mm G5 GH
127 mm Valkiri MRS
127 mm Bataleur MRS

SPAIN
Leopard 2A5E* MBT
Leopard 2 MBT
M60A1/M60A3 MBT
M48/M48E/M48A5 MBT
AMX-30E MBT
VEC recce
Scorpion 76 mm (Ma)
Centauro 105 mm (8 × 8)
Pizarro IFV
BMR-600 APC
Piranha III*
Bv 206S APC
M3 APC
M113 APC
UR-416 APC (IS)
BLR APC
LVTP7 (Ma)
105 mm M108 SPH (P)
155 mm M109/M109A1/
M109A1B SPH
155 mm M44 SPH (R)
203 mm M110 SPH (R)
81 mm M125A1/A2 SPM
120 mm M125A1 SPM
105 mm Model 56 P howitzer (and Ma)
105 mm Light Gun
105 mm M-26
155 mm M114 howitzer
155 mm 155/52 APU SBT-1
140 mm Teruel MRS

Marines also operate M48E, M109A1B, BLR and M3

SRI LANKA
T-55 MBT
T-55AM2 MBT
Ferret recce
Saladin recce
Dingo recce
Daimler Mk II recce
BMP-1 IFV
BMP-2 IFV
Saracen APC
Buffel APC
BTR-80
BTR-152 APC
Sandringham APC
Type 85 APC
76 mm M48 mountain gun
85 mm Type 56 FG
25 pounder FG
130 mm Type 59-1 FG
122 mm RM-70 (40-round) MRL

SUDAN
M60A1/M60A3 MBT
T-54/T-55 MBT
Type 59 MBT
Type 62 light tank
Ferret recce
Saladin recce
BRDM-2 recce
AML-90 recce
BMP-2 IFV
BTR-50 APC
BTR-80 APC
BTR-152 APC
LAV-100/150/150S APC
OT-62 APC
OT-64 APC
M113 APC
Type 531 APC ?
Walid APC
M3 APC
AMX VTT APC
Fahd APC
81 mm M125A1/M125A2 SPM
155 mm Mk F3 SPG
76 mm Type 54 ATG
85 mm D-44 FG
85 mm D-48 ATG
85 mm M1944 FG
100 mm M1944 FG
105 mm Model 56 P howitzer
105 mm M101 howitzer
105 mm M1938 howitzer
122 mm D-30 howitzer
122 mm Type 54 howitzer
122 mm D-74 GH
122 mm Type 60 GH
130 mm Type 59-1 FG
130 mm M-46 FG
155 mm M114 howitzer
107 mm Type 63 MRS
122 mm Sakr MRS
122 mm BM-21 MRS

SURINAM
EE-9 Cascavel recce
EE-11 Urutu APC
YP-408 APC

SWEDEN
Leopard 2A5
Leopard 2
CV 9040 ICV
Pbv 302 APC
SKPF m/42 APC
BMP-1 IFV (Pbv 501)
MT-LB MRV (Pbv 401)

XA-180 APC
XA-203S APC
Mamba APC
155 mm Bandkanon 1A SPG
Pvrbv 551 TOW TD
155 mm FH-77A
155 mm FH-77B

SWAZILAND
RG-12 APC (IS)
RG-31 APC (IS)
RG-32 APC (IS)

SWITZERLAND
Pz 87 MBT
Pz 68 MBT
Eagle I/II recce
CV 9030CH*
M113 series APC
Piranha APC
120 mm MwPz 64 SPM
155 mm M109/M109A1 plus
M109 SPH (UG)
Piranha TOW TD
105 mm Model 46 FG (P)

SYRIA
T-72 MBT
T-62 MBT
T-54/T-55 MBT
T-34/85 tank
BRDM-2 recce
Shorland (IS)
BMP-1 IFV
BMP-2 IFV
BMP-3 IFV
BTR-40 APC
BTR-50 APC
BTR-60 APC
BTR-152 APC
OT-64 APC
BRDM-2 Sagger TD
100 mm M1944 T-34 SPG
122 mm 2S1 SPG
122 mm D-30 T-34 SPG
152 mm 2S3 SPGH
76 mm M48 mountain gun
85 mm D-44 FG
100 mm M1944 FG
122 mm M1938 howitzer
122 mm M1931/7 gun
122 mm D-74 FG
122 mm D-30 howitzer
130 mm M-46 FG
152 mm D-20 GH
152 mm M1937 GH
152 mm 2A36 gun
180 mm S-23 gun
240 mm M-240 mortar
107 mm Type 63 MRS
122 mm BM-21 MRS
122 mm BM-21 MRS
FROG-7 ARS

TAIWAN
M60A3 MBT
M48A2/M48A3/M48A5 MBT
M48H MBT
Type 64 light tank
M41D light tank
M41A1/M41A3 light tank
M24 light tank
M24 light tank flame thrower
M8 recce
AIFV
M113A1/M113A2 APC
LVT3C (Ma)
LVTP5A1 (Ma)
CM-21 APC
LAV-150 S APC
LVTH6 105 mm (Ma)
105 mm M108 SPH
155 mm M44 SPH
155 mm XT-69 SPH
155 mm XT-69 SPG
155 mm M109/M109A2 SPH (and Ma)
155 mm M109A5 SPH
203 mm M110A2 SPH (and Ma)
81 mm M125A2 SPM
81 mm LAV-150 SPM
107 mm M106A2 SPM
107 mm LAV-150 SPM
LAV-150 TOW TD
M42A2 TOW TD
M132 flame thrower
75 mm M116 P howitzer
105 mm M101 howitzer
105 mm T64 howitzer
155 mm T65 howitzer
155 mm M114 howitzer (and Ma)
155 mm M59 gun
203 mm M115 howitzer (and Ma)
240 mm howitzer
117 mm Kung Feng VI MRS
126 mm Kung Feng III MRS
126 mm Kung Feng IV MRS (and Ma)
RT2000 ARS

TAJIKISTAN
T-72 MBT
BMP-1 IFV
BMP-2 IFV
BTR-60 APC
BTR-70 APC
BTR-80 APC
122 mm D-30 howitzer
122 mm M1938 howitzer
220 mm BM 9P 140 ARS

TANZANIA
T-54 MBT
Type 59 MBT
Type 62 light tank
Scorpion 76 mm recce
BRDM-2 recce
BTR-40 APC
BTR-152 APC
Type 56 APC
Type 531 APC (?)
76 mm M1942 FG
85 mm Type 56 FG
122 mm D-30 howitzer
122 mm Type 54 howitzer
130 mm Type 59-1 FG
130 mm M-46 FG
122 mm BM-21 MRS

THAILAND
M60A3 MBT
M48A5 MBT
Type 69-2 MBT
M41 light tank
Stingray light tank
Scorpion 76 mm recce
Shorland Mk 3 recce
M113A1/M113A2/M113A3 APC
Type 85 APC
LAV-150 APC
Condor APC
81 mm M125A3 SPM
107 mm M106A1 SPM
120 mm M1064A3 SPM
M901A3 ITV TD
155 mm M109A5 SPH
105 mm LG1 Mk II
105 mm M425 howitzer
105 mm M101 howitzer (UG)
105 mm M101 howitzer
105 mm M102 howitzer
130 mm Type 59-1 FG
155 mm M71 GH
155 mm M114 howitzer
155 mm M68 GH (Ma)
155 mm GC 45 GH (Ma)
155 mm M198 howitzer
155 mm GH H-45 GH
70 mm MRS
130 mm MRS
122 mm Type 83 MRS
130 mm Type 85 MRS

TOGO
T-54/T-55 MBT
AML-60/90 recce
M8/M20 recce
EE-9 Cascavel recce
Scorpion 90 mm recce
VBL recce
BMP-2 IFV
M3 APC
UR-416 APC
M3A1 half-track APC
105 mm M102 howitzer
122 mm howitzer

TUNISIA
M60A1/M60A3 MBT
Jagdpanzer SK 105
AML-60 recce
Saladin recce
EE-9 Cascavel recce
M113A1 APC
FIAT 6614 (IS)
EE-11 Urutu (IS)
155 mm M109A2 SPH
81 mm M125A1 SPM
M113A2 TOW TD
107 mm M106A2 SPM
105 mm M101 howitzer
155 mm M114 howitzer
155 mm M198 howitzer

TURKEY
M60A1/M60A3 MBT
M48A5 MBT
M48A5T1/M48A5T2/M48A5T5
Leopard 1A3 MBT
Leopard 1A5 MBT
M8 recce (R)
Akrep recce
AIFV
M113 series APC
Condor APC
Dragoon APC (IS)
Shorland APC (IS)
UR-416 APC (IS)
BTR-60 (IS)
BTR-80 (IS)
LAV-150 (IS)
Transporterpanzer 1 APC (NBC)
105 mm M52/M52A1 SPH
155 mm M52T SPG
105 mm M108 SPH
155 mm M44T SPG
155 mm M109 SPH
175 mm M107 SPG
203 mm M110 SPH
155 mm K9 SPG*
203 mm M55 SPH (R)
107 mm M106/M106A1 SPM
M113 TOW TD
75 mm M116 P howitzer
105 mm M101 howitzer
155 mm M59 gun
155 mm M114 howitzer
203 mm M115 howitzer
107 mm MRS
122 mm T-122 MRS
227 mm MRS
320 mm WS-1 ARS

TURKMENISTAN
T-72 MBT
BRDM-2 recce
BMP-1 IFV
BMP-2 IFV
BTR-60 APC
BTR-70 APC
BTR-80 APC
120 mm 2S9 SPM/H
152 mm 2S3 SPGH
85 mm D-44 gun
100 mm MT-12 ATG
122 mm D-30 howitzer
152 mm D-1 howitzer
152 mm 2A65 howitzer
122 mm BM-21 MRS
220 mm BM 9P140 MRS
FROG-7 ARS

UGANDA
T-54/T-55 MBT
PT-76 light tank
BTR-60 APC
OT-64 APC
Casspir APC
Mamba Mk II APC
76 mm M1942 FG
122 mm M1938 howitzer
122 mm BM-21 MRS

UKRAINE
T-84 MBT
T-80 MBT
T-72 MBT
T-64 MBT
T-62 MBT
T-54/T-55 MBT
BRDM-2 recce
BMP-1 IFV
BMP-2 IFV
BMP-3 IFV
BMD-1 ACV
BMD-2 ACV
BTR-D APC
BTR-60 APC
BTR-70 APC
BTR-80 APC
MT-LB MRV
120 mm 2S9 SPM/H
122 mm 2S1 SPH
152 mm 2S3 SPGH
155 mm 2S5 SPG
152 mm 2S19 SPG
203 mm 2S7 SPG
BRDM-2 Spandrel TD
MT-LB Spiral TD
100 mm T-12 ATG
122 mm D-30 howitzer
152 mm D-20 GH
152 mm 2A65 howitzer
152 mm 2A36 gun
122 mm BM-21 MRS
132 mm BM-13 MRS
220 mm BM 9P140 MRS
300 mm BM 9A52 Smerch MRS
FROG-7 ARS

UNITED ARAB EMIRATES
Leclerc MBT
OF-40 Mk 2 MBT
AMX-30 MBT
AML-60/90 recce
Scorpion 76 mm recce
Saladin recce (R)
Shorland (IS)
Terrier APC (IS)
AIFV
BMP-3 IFV
AMX VCI APC
AMX-10P IFV
VAB APC
VCR/TT
M3 VTS
EE-11 Urutu APC
Saxon APC
155 mm M109L47
155 mm Mk F3 SPG
155 mm G6 SPG
EE-11 TOW TD
105 mm Light Gun
130 mm Type 59-1 FG (R)
70 mm (40-round) LAU(& RL (Sharjar)
122 mm FIROS-30 MRS
300 mm BM 9A52 Smerch MRS

UNITED KINGDOM
Challenger 2 MBT
Scimitar recce
Sabre recce
Scorpion (T)
Warrior IFV
FV432 APC
Spartan APC
Saxon APC
Alvis 8 APC
BvS 10 APC* (trials)
CAV-100 APC (IS)
Tactica APC
Tempest APC
Tavern APC
Transportpanzer 1 (M93) (NBC)
Striker Swingfire TD
81 mm FV432 SPM
155 mm AS90 SPG
105 mm Light Gun
155 mm FH-70
155 mm XM777 (trials)
227 mm MLRS

UNITED STATES OF AMERICA
M1A1/M1A2 MBT
M60/M60A1/M60A3 MBT (R)
M3 Bradley recce
M1114 recce
RG 31 Nyala APC
Pandur APC
Dragoon APC
M2 Bradley IFV
M113 series APC
M1117 APC
LAV-25 (Ma)
LVTP7A1 (Ma)
LAV-III APC
M1116 APC (AF)
Transportpanzer 1 (M93) (NBC)
81 mm M125A1 SPM
107 mm M106A1 SPM
120 mm M1064 SPM
155 mm M109 series SPH
203 mm M110A1/M110A2 SPG (R)
LAV TOW TD
M113 TOW TD
105 mm M101 howitzer (R)
105 mm M102 howitzer (R)
105 mm M119 Light Gun
155 mm M114 howitzer (R)
155 mm M198 howitzer
155 mm XM777 (trials)
227 mm MLRS

URUGUAY
T-55 MBT
M41 light tank
M24 light tank
M3A1 light tank
E-3 Jararaca recce
EE-9 Cascavel recce
BMP-1 IFV
M113 APC
EE-11 Urutu APC
OT-64 APC
Condor APC
75 mm Bofors M1935 FG
105 mm M101 howitzer
105 mm M102 howitzer
155 mm M114 howitzer
122 mm 2S1 SPH
122 mm RM-70 MRS

UZBEKISTAN
T-72 MBT
T-62 MBT
T-54 MBT
BRDM-2 recce
BMP-2 IFV
BTR-D ACV
BTR-60 APC
BTR-70 APC
120 mm 2S9 SPM/H
122 mm 2S1 SPH
152 mm 2S3 SPGH
203 mm 2S7 SPG
100 mm T-12 ATG
122 mm D-30 howitzer
152 mm D-20 GH
152 mm 2A36 gun
122 mm BM-21 MRS
FROG-7 ARS

VENEZUELA
AMX-30 MBT
AMX-13 light tank
AML-60/90 recce
M8 recce
Scorpion 90 mm recce
AMX VCI APC
EE-11 Urutu APC (and Ma)
LAV-100 APC
Dragoon APC
FIAT 6614 APC
Transportpanzer 1 APC (Ma)
UR-416 APC
LVTP7 (Ma)
Shorland (IS)
155 mm Mk F3 SPG
155 mm M109 SPH
75 mm M116 P howitzer
105 mm Model 56 P howitzer (and Ma)
105 mm M101 howitzer
155 mm M114 howitzer
160 mm LARS MRS

VIETNAM
T-54/T-55 MBT
T-62 MBT
Type 59 MBT
T-34/85 tank
Type 62 light tank
Type 63 light tank
PT-76 light tank
BRDM-1 recce
BRDM-2 recce
BMP-1 IFV
BTR-40 APC
BTR-50 APC
BTR-60 APC
BTR-152 APC
Type 531 APC
Type 56 APC
M113A1 APC
SU-100 SPG
120 mm 2S9 SPM/H
152 mm 2S3 SPG
175 mm M107 SPG ?
76 mm M1942 FG
85 mm Type 56 FG
85 mm D-44 FG
85 mm D-48 ATG
100 mm M1944 FG
105 mm M101 howitzer
122 mm D-30 howitzer
122 mm D-74 FG
122 mm M1938 howitzer
122 mm Type 54 howitzer
122 mm Type 60 FG
130 mm M-46 FG
130 mm Type 59 FG
130 mm Type 59-1 FG
152 mm Type 66 GH
152 mm D-20 GH
152 mm M1937 howitzer
152 mm D-1 howitzer
107 mm Type 63 MRS
122 mm BM-21 MRS
130 mm Type 63 MRS
140 mm BM-14-16 MRS

YEMEN
T-72 MBT
M60A1 MBT
T-62 MBT
T-54/T-55 MBT
T-55AM2 MBT
T-34/85 tank
BRDM-2 recce
AML-90 recce
BMP-1 IFV
BMP-2 IFV
BTR-40 APC
BTR-60 APC
BTR-152 APC
M113/M113A1 APC
Walid APC
SU-100 SPG
107 mm M106 SPM
122 mm 2S1 SPH
76 mm M1942 FG
85 mm D-44 FG
105 mm M101 howitzer
122 mm D-30 howitzer
122 mm M1938 howitzer
122 mm M1931/7 gun
122 mm M-46 FG
152 mm D-20 GH
152 mm M114 howitzer
122 mm BM-21 MRS
140 mm BM-14 MRS
220 mm BM 9140 (16-round) MRL
FROG-7 ARS

YUGOSLAVIA, FEDERAL REPUBLIC
M-84 MBT
T-72 MBT
T-54/T-55 MBT
T-34/85 (R)

PT-76 light tank
BRDM-2 recce
M60 APC
M80 IFV
BVP M80A IFV
MT-LB MRV
TAB-71 APC
TAB-72 APC
BOV-M APC
BOV-1 Sagger TD
M36B2 TD
M10 TD
M18 Hellcat TD
BRDM-2 Sagger TD
105 mm M7 SPH
122 mm 2S1 SPH
76 mm M48 mountain gun
76 mm M1942 FG
85 mm D-44 FG
100 mm T-12 ATG
100 mm M87 ATG
105 mm M56 howitzer
105 mm M101 howitzer
105 mm M18 FH
105 mm M18 (M) FH
105 mm M18/40 FH
120 mm 2S9 SPH/M
122 mm D-30 howitzer
122 mm M1938 howitzer
130 mm M-46 FG
152 mm M1937 GH
152 mm D-20 GH
152 mm M84 GH
155 mm M65 howitzer
155 mm M59 gun
155 mm M114 howitzer
122 mm BM-21 MRS
128 mm M-63 MRS
128 mm M-77 MRS
128 mm M-85 MRS
262 mm M-87 MRS
FROG-7 ARS

ZAMBIA

T-54/T-55 MBT
Type 59 MBT
PT-76 light tank
BRDM-1 recce
BRDM-2 recce
BTR-60 recce
76 mm M1942 FG
25-pounder FG
122 mm D-30 howitzer
130 mm M-46 FG
122 mm BM-21 MRS

ZIMBABWE

T-54/T-55 MBT
Type 59 MBT
Type 69 MBT
T-34/85 tank
Type 63 light tank
Eland 60/90 recce
EE-9 Cascavel recce
UR-416 APC
ACMAT TPK 420 BL (IS)
Type 531 APC
Hyena APC
Leopard APC
BTR-152 APC
Buffel APC
Hippo APC
Crocodile APC
122 mm Type 54 howitzer
122 mm Type 60 FG
107 mm Type 63 MRS
122 mm RM-70 MRS (Mod 70/85)

UPDATED

FREE ENTRY/CONTENT IN THIS PUBLICATION

Having your products and services represented in our titles means that they are being seen by the professionals who matter – both by those involved in procurement and those working for the companies that are likely to affect your business. We therefore feel that it is very much in the interests of your organisation, as well as Jane's, to ensure your data is current and accurate.

- **Don't forget –** You may be missing out on business if your entry in a Jane's book, CD-ROM or Online product is incorrect because you have not supplied the latest information to us.
- **Ask yourself –** Can you afford not to be represented in Jane's printed and electronic products? And if you are listed, can you afford for your information to be out of date?
- **And most importantly –** The best part of all is that your entries in Jane's products are TOTALLY FREE OF CHARGE.

Please provide (using a photocopy of this form) the information on the following categories where appropriate:

1. Organisation name: ______
2. Division name: ______
3. Location address: ______
4. Mailing address if different: ______
5. Telephone (please include switchboard and main department contact numbers, e.g. Public Relations, Sales, etc.): ______
6. Facsimile: ______
7. E-mail: ______
8. Web sites: ______
9. Contact name and job title: ______
10. A brief description of your organisation's activities, products and services: ______
11. Jane's publications in which you would like to be included: ______

Please send this information to:
Jacqui Beard, Information Collection, Jane's Information Group,
Sentinel House, 163 Brighton Road, Coulsdon, Surrey, CR5 2YH, UK
Tel: (+44 20) 87 00 38 08
Fax: (+44 20) 87 00 39 59
E-mail: yearbook@janes.co.uk

Copyright enquiries:
Contact: Keith Faulkner
Tel/Fax: (+44 1342) 30 50 32
E-mail: keith.faulkner@janes.co.uk

Please tick this box if you do not wish your organisation's staff to be included in Jane's mailing lists ☐

JAA

CONTRACTORS

Contractors

Abdallah Al-Faris Company for Heavy Industries
PO Box 5987
Dammam 31432
Saudi Arabia
Tel: (+966 3) 821 13 95
Fax: (+966 3) 821 15 40

Abu Zaabal Engineering Industries Co
PO Box 5888
Heliopolis West
Cairo
Egypt
Tel: (+20 2) 291 73 05; 70 33
Tx: 21487 hvrt un
Fax: (+20 2) 291 69 62

ACMAT, Ateliers de Construction Mécanique de l'Atlantique
Le Point du Jour
F-44600 Saint-Nazaire
France
Tel: (+33 2) 40 22 33 71
Tx: 700913
Fax: (+33 2) 40 66 30 96

ADI Limited
Military and Specialist Vehicles
Operations Group
Finn Street
Bendigo
Victoria 3550
Australia
Tel: (+61 3) 54 40 43 11
Fax: (+61 3) 54 41 67 64

Aerospace Industries Organisation
Shahid Bagheri Industries
Iran

Aerostar SA
9 Condorilor Street
R-5500 Bacau
Romania
Tel (+40 34) 17 50 70
Fax: (+40 34) 17 20 23

Alvis plc
34 Grosvenor Gardens
London SW1W OAL
Tel: (+44 020) 78 08 88 88
Fax: (+44 020) 78 08 88 89

Alvis Vehicles Limited
PO Box 106
Hadley Castle Works
Telford
Shropshire TF1 6QW
UK
Tel: (+44 1952) 22 45 00
Fax: (+44 1952) 24 39 10

ARIS SpA
Strada Cascina Bertola
I-10-10040 Lambardore
Torino
Italy
Tel: (+39 011) 995 61 50
Fax: (+39 011) 995 62 05

Arsenalul Armatei
State Owned Company
Blvd Timisoara Nr 58
Sector 6
cod 77309
Bucharest
Romania
Tel: (+40 1) 413 06 51

ARTEC GmbH
Krauss-Maffei Strasse
D-80997 Munich
Germany
Tel: (+49 89) 89 21 30; 0
Fax: (+49 89) 89 21 30 16

ASCOD
(See Santa Barbara of Spain and Steyr-Daimler-Puch of Austria)

Auverland SA
98, rue Point du Jour
F-92100 Boulogne
France
Tel: (+33 1) 46 21 21 09
Fax: (+33 1) 46 08 52 52

Avibras Industria Aeroespacial S/A
PO Box 278
12300-000
Jacarei
São Paulo
Brazil
Tel: (+55 12) 355 60 00
Fax: (+55 12) 351 62 77

AV Technology
A General Dynamics Company
14920 23 Mile Road
Shelby Township
Michigan 48315
USA
Tel: (+1 810) 532 10 00
Fax: (+1 810) 532 80 20

BAE Systems
RO Defence
Euxton Lane
Chorley
Lancashire PR7 6AD
UK
Tel: (+44 1257) 26 55 11
Fax: (+44 1257) 50 92 42

BAE Systems
RO Defence
Barrow-in-Furness
Cumbria LA14 1AF
UK
Tel: (+44 1229) 87 42 97
Fax: (+44 1229) 87 41 12

BAE Systems
RO Defence
Rocket Motors Division
Summerfield
Kidderminster
Worcestershire DY11 7RZ
UK
Tel: (+44 1562) 82 40 61
Fax: (+44 1562) 82 81 20

Barrikady State Production Association
400071 Volgograd
Russia
Tel: (+7 8442) 78 14 50
Fax: (+7 8442) 71 62 75

Beijing Bao-Long Science & Technology Developing Corporation Ltd
13 Bei San Haun Xi Lu
Beijing
People's Republic of China

Bernardini SA Indústria e Comércio
Rua Hipòlito Soares No 79
São Paulo 04201
Brazil
Tel: (+55 11) 274 80 33
Tx: 21605 bsaibr
Fax: (+55 11) 274 85 67

BETA PLC
BG-5980 Cherven Briag
Bulgaria
Tel: (+359 659) 24 70
Fax: (+359 659) 69 66

Boeing Company, The
PO Box 516
St Louis
Missouri 63166
USA

Bofors Defence AB
(A United Defense Company)
SE-691 80 Karlskoga
Sweden
Tel: (+46 586) 815 00
Fax: (+46 586) 850 30

BPD DIFESA E SPAZIO SpA
I-00034 Colleferro
Rome
Italy
Tel: (+39 06) 972 91
Fax: (+39 06) 97 29 22 05

BRAVIA sarl
Porto Alto
P-2135 Samora Correia
Portugal
Tx: 12715 BRAVIA P
Fax: (+351 1) 284 62 93

Bucher-Guyer AG
Vehicles Division
CH-8166 Niederweningen
Switzerland
Tel: (+41 1) 857 22 11
Fax: (+41 1) 857 22 49

Caliber Prague Ltd
Rejskova 7
CZ-120 00 Praha 2
Vinohrady
Czech Republic
Tel: (+420 2) 691 00 35
Fax: (+420 2) 691 03 18

Carrozzeria Boneschi Srl
Via G Boneschi 1
Viale Delle Industrie
I-20040 Cambiago
Milan
Italy
Tel: (+39 02) 95 30 80 45
Fax: (+39 02) 95 30 82 04

China Precision Machinery Import and Export Corporation (CPMIEC)
PO Box 845
Beijing
People's Republic of China
Tel: (+86 1) 89 50 12; 831 18 04
Tx: 22484 CPMC CN
Fax: (+86 1) 831 16 57

Chung-Shan Institute of Science and Technology
Missile & Rocket Systems Research Division
PO Box 90008-1
Lungtan
Taoyuan
Taiwan
Republic of China
Tel: (+886 2) 471 30 22
Fax: (+886 2) 471 41 83

Consorzio Iveco Oto
Via Tacito 90
I-00193 Rome
Italy
Tel: (+39 06) 32 01 46
Fax: (+39 06) 323 08 48

Daewoo Corporation
541 5-ga, Namdaemunno
Chung-gu
Seoul
South Korea
Tel: (+82 2) 771 91
Tx: K23341/4; K24295

Daewoo Heavy Industries Limited
23rd Floor Daewoo Center Building
541 5-ga, Namdaemunno
Chung-gu
Seoul
South Korea
Tel: (+82 2) 777 18 56
Tx: 23301

Defence Industries Organisation
Sangafzar Sazi
PO Box 17185
Tehran
Iran
Tel: (+98 21) 33 90 00
Fax: (+98 21) 34 50 21

Delco Defense
7410 Hollister Avenue
Goleta
California 93117
USA
Tel: (+1 805) 961 74 77

Denel (Pty) Ltd
PO Box 8322
Hennopsmeer 0046
South Africa
Tel: (+27 12) 428 06 09
Fax: (+27 12) 45 20 09

Dr A Q Khan Research Laboratories Kahuta
Director General Scientific and Technical
Co-operation Division
PO Box 502
Rawalpindi 46000
Pakistan
Tel: (+92 51) 928 05 41
Fax: (+92 51) 928 05 42;

DSD Mobiltec
CSL Armady 3/1697
036 57 Martin
Slovakia
Tel: (+421 842) 424 22 00
Fax: (+421 842) 424 23 77

Dynamit Nobel GmbH
Explosivstoff und Systemtechnik
PO Box 1261
Kaiserstrasse 1
D-53839 Troisdorf
Germany
Tel: (+49 2241) 89 13 52
Fax: (+49 2241) 89 16 69

E Falck Schmidt A/S
Tolclerlundsvej 106
Postbox 319
DK-5100 Odense C
Denmark
Tel: (+45) 66 13 11 00
Fax: (+45) 66 13 11 11

Egyptian Tank Plant (ETP)
Factory 200
5 Kassem Gouda Street
NASR City
Heliopolis 5952
Cairo
Egypt

ELBO, Hellenic Vehicle Industry SA
Industrial Area Sindos
GR-57022 Thessaloniki
Greece
Tel: (+30 31) 79 85 02; 79 86 02
Fax: (+30 31) 79 84 26

FAMAE
Avda Pedro Montt 1606
Santiago
Chile
Tel: (+56 2) 556 91 31
Tx: 242346
Fax: (+56 2) 551 18 94

Fighting Machines Command
ROC Army
Taichung
Taiwan 400

FNSS Defense Systems Inc
Tekox 37
TR-06830 Golbasi
Ankara
Turkey
Tel: (+90 312) 484 00 77
Fax: (+90 312) 484 00 73

Forges de Zeebrugge FZ
63 rue En Bois,
B-4040 Herstal
Belgium
Tel: (+32 4) 248 20 77
Fax: (+32 4) 248 25 10

General Dynamics Amphibious Operations
991 Annapolis Way
Woodbridge
Virginia 22191
USA
Tel: (+1 703) 492 32 20
Fax: (+1 703) 492 34 10

General Dynamics Land Systems
PO Box 2074
Warren
Michigan 48090-2074
USA
Tel: (+1 810) 825 40 00
Fax: (+1 810) 825 40 13

General Motors Defense
1660 L Street NW
Suite 401
Washington DC 20036
USA
Tel: (+1 202) 775 50 20
Fax: (+1 202) 775 50 46

General Motors Defense
General Motors of Canada Limited
PO Box 5160
London
Ontario N6A 4N5
Canada
Tel: (+1 519) 452 51 45
Fax: (+1 519) 452 54 88

Giat Industries
13 route de la Minière
F-78034 Versailles cedex
France
Tel: (+33 1) 30 97 37 37
Fax: (+33 1) 30 97 39 00

GLS GmbH
Krauss-Maffei Strasse 2
PO Box 500231
D-80997 Munich
Germany
Tel: (+49 89) 889 90
Fax: (+49 89) 812 63 80

Gorky Automobile Plant (Arzamas Machinery Plant)
603046 Nizhny Novgorod
Russia
Tel: (+7 8312) 56 41 45
Tx: 151127 OKA SU
Fax: (+7 8312) 56 98 42

Government of India
Ministry of Defence
Ordnance Factory Board
10A Auckland Road
Calcutta 700 001
India
Tel: (+91 33) 248 90 27
Fax: (+91 33) 248 17 48

Gun Carriage Factory
Jabalpur 482 011
India

GUP Design Bureau of Transport Machine Building
2 Krasny pereulok str
Omsk 644020
Russia
Tel: (+3812) 41 85 13
Fax: (+3812) 41 52 93

Hanwha Corporation
Hanwha Building
1 Changgyo–Dong
Chung-Ku
Seoul
Korea
Tel: (+82 2) 729 37 91
Fax: (+82 2) 729 35 35

Hägglunds Vehicle AB
SE-691 02 Örnskildsvik
Sweden
Tel: (+46 660) 805 00
Tx: 6051
Fax: (+46 660) 151 90

Heavy Industries Taxila (HIT)
Taxila Cantt
Pakistan
Tel: (+92 596) 927 05 71
Fax: (+92 596) 31 99

Heavy Vehicles Factory
Avadi
India

Hellfire Systems LLC
Lockheed Martin and Boeing Company
5798 South Semoran Boulevard
Orlando
Florida 32822
USA
Tel: (+1 407) 273 22 49
Fax: (+1 407) 282 47 50

Helwan Machine Tools Company
General Marketing Department
23 Talaat Harb Street
PO Box 1582
Cairo
Egypt
Tel: (+20 2) 55 59 30

Hotspur Armoured Products Limited
Division of Penman Engineering Limited
Heathall
Dumfries D61 3NY
UK
Tel: (+44 1387) 25 27 84
Tx: 779771
Fax: (+44 1387) 26 73 32

Hsing Hua Company Limited
PO Box 8746
Taipei
Taiwan

Huta Stalowa Wola SA (also referred to as HSW)
ul Kwiatkowskiego 1
37-450 Stalowa Wola
Poland
Tel: (+48 15) 843 42 12
Fax: (+48 15) 842 19 08

Hyundai Precision & Ind Co Ltd
140-2 Gye-dong
Chongro-Ku
Seoul
South Korea
Tel: (+82 2) 746 40 40
Fax: (+82 2) 741 22 48

IDEX
Foreign Trading, Contracting and Engineering Co Ltd
PO Box 24
H-1251 Budapest
Hungary
Tel: (+36 1) 115 92 90
Fax: (+36 1) 135 13 93

IHI Aerospace Company Limited
900 Fujiki
Tomioka-shi
Gunma 370-2398
Japan
Tel: (+81 274) 62 31 24

Instituto de Investigaciones Cientificas y Tecnicas de las Fuerzas Armadas (CITEFA)
Zufriategui y Varela
1603 Villa Martelli
Provincia de Buenos Aires
Argentina

INSYS Limited (previously Hunting Engineering Limited)
Reddings Wood
Ampthill
Bedford
Hertfordshire
MK45 2HD
United Kingdom
Tel: (+44 1525) 84 37 60
Fax: (+44 1525) 84 37 04

Israel Military Industries Ltd
64 Bialik Blvd
IL-47100 Ramat Hasharon
Israel
Tel: (+972 3) 548 56 17
Tx: 33179
Fax: (+972 3) 540 69 08

IVECO SpA
Defence Vehicles Division
Via A Volta 6
I-39100 Bolzano
Italy
Tel: (+39 0471) 90 51 11
Fax: (+39 0471) 90 54 44

Joint Stock Company Spetstehnika
Russia

Jung Jungenthal GmbH
D-57548 Kirchen
Germany
Tel: (+49 2741) 68 30
Tx: 875319
Fax: (+49 2741) 68 32 46

Kader Factory for Developed Industries
2 El Tayaran Street
Nasr City
Heliopolis
PO Box 287
Cairo
Egypt
Tel: (+20 2) 402 43 19
Fax: (+20 2) 260 87 18

KBM Engineering Design Bureau
42 Oksky Avenue
Kolomna
Moscow
Russia
Tel: (+7 9661) 332 77
Fax: (+7 9661) 330 64

Kentron Division of Denel (Pty) Ltd
PO Box 7412
Hennopsmeer 0046
South Africa
Tel: (+27 12) 671 19 11
Fax: (+27 12) 671 11 27

Kerametal Company Limited
Jasikova 2
SK-826 05 Bratislava
Slovakia
Tel: (+421 7) 299 11 11
Tx: 932 33 kmet sk
Fax: (+421 7) 522 58 04

Kharkov Morozov Design Bureau
126 Plekhanovskaya Street
Kharkov 310001
Ukraine
Tel: (+380 572) 28 31 90
Fax: (+380 572) 26 33 21

King Abdullah II Design and Development Bureau
Jordan
Tel: (+962) 625 60 24
Fax: (+962) 625 61 42

KINTEX Company
PO Box 209
66 James Baucher Blvd
BG-1407 Sofia
Bulgaria
Fax: (+359 2) 963 11 23; 963 16 12

Komatsu Ltd
3-3-1 Akasaka
Minato-ku
Tokyo
Japan

Kongsberg Defence & Aerospace AS
Kirkegardsveien 45
N-3600 Kongsberg
Norway
Tel: (+47 3273) 82 00
Fax: (+47 3273) 86 20

Krauss-Maffei Wegmann GmbH and Co KG (KMW)
Postfach 10 39 67
D-34112 Kassel
Germany
Tel: (+49) 561 10 50
Fax: (+49) 56 11 05 22 08

Krauss-Maffei Wegmann GmbH and Co KG (KMW)
Krauss-Maffei Strasse 2
D-80972 Munich
Germany
Tel: (+49 89) 81 40; 50
Fax: (+49 89) 81 40; 49

Kurgan Machine Construction Plant
17 Mashinostroitelej Avenue
640631 Kurgan
Russia
Tel: (+7 35222) 322 44
Fax: (+7 35222) 339 96; 314 97

Lockheed Martin Missiles and Fire Control
PO Box 650003
Dallas
Texas 75265-0003
USA
Tel: (+1 972) 603 10 00
Fax: (+1 972) 603 10 09

LOHR Industrie
29 rue du 14 Juillet
F-67980 Hangenbieten
France
Tel: (+33 3) 88 38 98 00
Fax: (+33 3) 88 96 06 36

Lyttelton Engineering Works (LIW)
PO Box 7710
Pretoria 0001
South Africa
Tel: (+27 12) 620 25 04
Fax: (+27 12) 620 22 83

Makina Ve Kimya Endustrisi Kurumu (MKEK)
Tandogan Meydani
Ankara
Turkey
Tel: (+90 312) 213 39 35
Fax: (+90 312) 222 22 41

Malyshev Plant
126 Plekhanovskaya Street
310001 Kharkov
Ukraine
Tel: (+380 0572) 27 97 82
Fax: (+380 0572) 27 75 04

MBDA France
20-22 rue Grange Dame Rose
F-78141 Velizy Villacoublay Cedex
France
Tel: (+33 1) 34 88 30 00
Fax: (+33 1) 34 88 22 88

MBDA UK
11 Strand
London WC2N 5RJ
United Kingdom
Tel: (+44 20) 74 51 60 00
Fax: (+44 20) 74 51 60 01

Mechem Consultants
PO Box 912454
Silverton 0127
South Africa
Tel: (+27 12) 803 72 90
Fax: (+27 12) 803 71 89

Military and Police Group
Technopol
Kutlikova 17
SK-852 50 Bratislava
Slovakia
Tel: (+421 7) 82 99 95
Fax: (+421 7) 83 30 13

Military Institute for Weapon and Ammunition Technology
Dlouha 300
CZ-763 21 Slavicin
Czech Republic
Tel: (+420 636) 712 53
Fax: (+420 636) 712 52

Military Production Authority
PO Box 5757
Baghdad
Iraq
Tx: 213670 SAFAN IK; 212461 SOTI IK
Fax: (+964 1) 538 46 34; 772 34 36

Minotor Service Company
38 Kuprevich Strasse
Minsk 220141
Republic of Belarus
Tel: (+375 17) 285 92 35
Fax: (+375 17) 263 86 87

Mitsubishi Heavy Industries, Sagamihara Ltd
Mitsubishi Heavy Industries
5-1 Marunouchi 2-chome
Chiyoda-ku
Tokyo
Japan
Tel: (+81 3) 212 31 11
Tx: J22282/J22443

Motovilikha Plants Corporation
614014 Perm
Russia
Tel: (+7 3422) 36 73 01, 36 79 01, 36 71 36
Tx: 134849 Stal SU
Fax: (+7 3422) 65 15 37, 65 38 50

MOWAG Motorwagenfabrik AG
Unterseestrasse 65
CH-8280 Kreuzlingen
Switzerland
Tel: (+41 71) 677 55 00
Fax: (+41 71) 672 28 86

MPP Vozila Limited
SI-2000 Maribor
Ptujska c 184
Slovenia
Tel: (+386 62) 450; 100
Fax: (+386 62) 412; 455

MTU GmbH
PO Box 2040
D-7990 Friedrichshafen
Germany
Tel: (+49 7541) 90-0
Tx: 734280-0 mt d
Fax: (+49 7541) 90 22 47

National Resources Development Company
PO Box 36
Amman 11941
Jordan
Tel: (+962 6) 533 79 18
Fax: (+962 6) 533 79 17

NORINCO
China North Industries Corporation
12A Guang'anmen Nanjie
Beijing
China
Tel: (+86 10) 63 52 99 88
Fax: (+86 10) 63 54 03 99

NP Aerospace Limited
473 Foleshill Road
Coventry
Warwickshire CV6 5AQ
UK
Tel: (+44 1203) 70 28 02
Fax: (+44 1203) 68 73 13

NPO Mashinostroyenie
Russia

Nurol Machinery and Industry Co Inc
Ankara Organize Sanayi Bolgesi
Avrupa Hun Bulvari No 6
TR-06395 Sincan
Ankara
Turkey
Tel: (+90 312) 267 05 30
Fax: (+90 312) 267 01 55

OBRUM
Osrodek Badawczo-Rozwojowy Urzadzen Mechanicznych
Ul Toszecka 102
PL-44-1-101 Gliwice
Skrytka Pocztowa 398
Poland
Tel: (+48 32) 231 72 14
Fax: (+48 32) 279 39 30

Oerlikon Contraves AG
Birchstrasse 155
CH-8050 Zurich
Switzerland
Tel: (+41 1) 316 22 11
Fax: (+41 1) 311 31 54

O'Gara-Hess & Eisenhardt Armoring Company
9113 LeSaint Drive
Fairfield
Ohio 45014
USA
Tel: (+1 513) 874 21 12
Fax: (+1 513) 874 25 58

Omsk Machine Construction Plant
2 Kransy perculok Street
Omsk 644020
RFAS
Tel: (+3812) 41 85 13
Fax: (+3812) 41 52 93

Ordnance Factory Medak
Yeddumailarm
Medak Dist 502 205
India
Tel: (+91 842) 24 10 21
Fax: (+91 842) 24 21 82

Otobreda
Via Valdilocchi 15
I-19136 La Spezia
Italy
Tel: (+39 0187) 58 11 11
Fax: (+39 0187) 58 26 69

OTOKAR AS
Ataturk Cad No 9
TR-54580 Arifiye
Adapazari
Turkey
Tel: (+90 264) 229 22 44
Fax: (+90 264) 229 22 42

Pakistan Ordnance Factories
Wah Cantonnement
Rawalpindi
Pakistan
Tel: (+92 586) 931 41 01
Fax: (+92 596) 931 40 58

Patria Vammas
PO Box 18
FIN-38201 Vammala
Finland
Tel: (+358 32) 19 71
Fax: (+358 35) 19 74 66

Patria Vehicles Oy
PO Box 186
FIN-13101 Hämeenlinna
Finland
Tel: (+358 3) 64 51
Fax: (+358 3) 619 67 10

Perm Machine Works (Perm Artillery Factory)
Russia

Peugeot Talbot España SA
Special Products Division
Apartado No 140
E-28080 Madrid
Spain
Tel: (+34 91) 347 29 31
Tx: 27590/42853/43603

Podpolianske strojarne as
Detva
Slovak Republic

PT Industri Pesawat Terbang Nusantara (PT IPTN)
Commercial Division for Helicopter, Defense System Aerospace and Software Systems (HDSSS)
Jalan Pajajaran 154 Bandung–40174
4th Floor
Indonesia
Tel: (+62 22) 600 27 02
Fax: (+62 22) 631 16 96

RAMTA Division
Israel Aircraft Industries
PO Box 323
IL- 84102 Beer Sheva
Israel
Tel: (+972 7) 627 09 50
Fax: (+972 7) 627 67 70

RDM Technology BV
PO Box 1039
NL-3000 BA Rotterdam
Netherlands
Tel: (+31 10) 487 91 11
Fax: (+31 10) 487 22 99

Renault VI
40 rue Pasteur
BP 302
F-92152 Suresnes cedex
France
Tel: (+33 1) 40 99 71 11

RH ALAN doo
Bosanka 26
HR-10000 Zagreb
Croatia
Tel: (+385 1) 378 08 00
Fax: (+385 1) 378 08 02

Rheinmetall Landsysteme GmbH
Postfach 101680
D-34016 Kassel
Germany
Tel: (+49 561) 801 64 04
Fax: (+49 561) 801 67 33

Rheinmetall Landsysteme GmbH
PO Box 9333
D-24159 Kiel
Germany
Tel: (+49 431) 39 99 22 92
Fax: (+49 431) 39 99 32 78

Rheinmetall Weapons & Munitions GmbH
Pempelfurstrstrasse/1
D-40880 Ratingen
Germany
Tel: (+49) 210 29 00
Fax: (+49) 21 02 47 35 53

Rocketsan
Ankara–Samsun Karayolu 40 km
P K 30 Elmadag
TR-06780 Ankara
Turkey
Tel: (+90 312) 863 42 00
Fax: (+90 312) 863 42 08

Rock Island Arsenal
Rock Island
Illinois 61299-5000
USA
Tel: (+1 309) 782 47 86
Fax: (+1 309) 782 67 82

Rockwell International
Tactical Systems Division
Defence Electronics Operations
1800 Satellite Boulevard
Duluth
Georgia 30136
USA

Rubtsovskiy Machine Building Plant
204 Prospekt Leninia
Rubtovsk 658215
Altai Territory
Russia
Tel: (+7 1) 37 55 36 75

Saab Bofors Dynamics AB
SE-581 88 Linkoping
Sweden
Tel: (+46 13) 18 60 00
Fax: (+46 13) 18 60 06

SAGE
The Scientific and Technical Research Council of Turkey
PO Box 216
Mamak
Ankara
Turkey
Tel: (+90 312) 399 03 38 45
Fax: (+90 312) 399 03 33

SAKR Factory for Developed Industries
PO Box 33
Heliopolis
Cairo
Egypt
Tel: (+20 2) 415 02 30; 415 02 50
Fax: (+20 2) 290 19 78

Samsung Techwin Ltd
Defense Program Division
14th Floor
Samsung Yok
647-9 Yoksam-Dong
Kangnam-Ku
Seoul
Korea 135–080
Tel: (+82 2) 34 67 78 32
Fax: (+82 2) 34 67 76 60

SANTA BARBARA SISTEMAS
Manuel Cortina, 2
E-28010 Madrid
Spain
Tel: (34 91) 585 01 00
Tx: 44466ENSB-E
Fax: (34 91) 585 02 44

Satory Military Vehicles
12 route de la Miniere
F-78036 Versailles Cedex
France
Tel: (+33 1) 39 49 29 00
Fax: (+33 1) 39 49 29 29

SIBAT–Foreign Defence Assistance and Defence Export
Ministry of Defence
8 David Elazar Street
IL-61909 Hakiryah
Tel Aviv
Israel
Tel: (+972 3) 697 55 15
Fax: (+972 3) 697 20 47

Sika International
BAE Systems
Chester House
Farnborough Aerospace Centre
Farnborough
Hampshire GU14 6YU
UK
Tel: (+44 1252) 38 46 11

Sika International
Lockheed Martin Inc
5600 Sand Lake Road
Orlando
Florida 32819-8907
USA
Tel: (+1 407) 356 22 07

Singapore Technologies Kinetics
5 Portsdown Road
Singapore 139296
Tel: (+65) 473 63 11
Fax: (+65) 473 06 62

Singapore Technologies Kinetics
15 Chin Bee Drive
Singapore 619863
Tel: (+65) 265 53 22
Fax: (+65) 261 40 53

S N ROMARM SA
Calea Victoriei (Pasajul Victoriei)
Nr 48-50, Sector 1
Cod R-70102
Bucharest
Romania
Tel: (+40 1) 313 43 66
Fax: (+40 1) 312 24 31

Société de Constructions
Mécaniques Panhard et Levassor
18 avenue d'Ivry
F-75013 Paris
France
Tel: (+33 1) 40 77 40 00
Fax: (+33 1) 45 82 73 26

SOFRAME
F-67980 Hangenbieten
Strasbourg
France
Tel: (+33 3) 88 48 53 00
Fax: (+33 3) 88 48 53 10

Soltam Systems Limited
PO Box 1371
IL-31013 Haifa
Israel
Tel: (+972 4) 989 62 82
Fax: (+972 4) 989 20 45

Somchem
Division of Denel (Pty) Ltd
PO Box 187
Somerset West
7129 South Africa
Tel: (+27 21) 850 25 44
Fax: (+27 21) 850 29 55

SP aerospace and vehicle systems bv
Eindhovenseweg 120
PO Box 436
NL-5661 AK Geldrop
Netherlands
Tel: (+ 31 40) 280 91 91
Fax: (+31 40) 280 91 23

Splav Scientific Production Concern
Splav State Research and Production Association
4 Shcheglovskaya Zaseka
Tula 300004
Russia
Tel: (+7 0872) 46 45 86

State Scientific and Technical Center of Artillery and
Rifle Arms
252057 Kiev 57
Letter Box 882/1
Ukraine
Tel: (+380 44) 446 24 59

Steyr-Daimler-Puch Fahrzeugtechnik AG and Co KG
Liebenauer Hauptstrasse 317
A-8041 Graz
Austria
Tel: (+43 316) 404 27 65
Fax: (+43 316) 404 30 93

Steyr-Daimler-Puch Spezialfahrzeug AG & Co KG
2 Haidequerstrasse
A-1111 Vienna
Austria
Tel: (+43 1) 760 64
Fax: (+43 1) 769 81 49

SW Swiss Ordnance Enterprise Corp
Allmendstrasse 86
CH-3602 Thun
Switzerland
Tel: (+41 33) 228 11 00
Tx: 921 256
Fax: (+41 33) 228 20 47

Systems & Electronics Inc
201 Evans Lane
St Louis
Missouri 63121
USA
Tel: (+1 314) 553 40 00
Fax: (+1 314) 553 49 49

SZ–Stroji in Tehnoloska Oprema doo
RAVNE na Koroskem
Koroska cesta 14
SI–2390
Slovenia
Tel: (+386) 60 22 11 31
Fax: (+386) 60 22 08 26

T & T Technology and Trading GmbH
Ocostrasse 20
CH-5330 Zurich
Switzerland
Tel: (+41 56) 249 08 55
Fax: (+41 56) 249 08 56

TAMSE
6615
Avenida San Martin
Buenos Aires
1419
Argentina
Tel: (+541) 45 72 13 14

TATRA as
Stefanikova 1163
CZ-742 21 Koprivnice
Czech Republic
Tel: (+420 656) 82 11 41
Fax: (+420 656) 82 11 48

TDA
Etablissement de Velizy
4 avenue Morane Saulnier
F-78140 Velizy-Villacoublay
France
Tel: (+33 1) 30 67 24 24
Fax: (+33 1) 30 67 24 44

Tenix Defence Systems Pty Ltd
Level 1
AMA Building
42 Macquarie Street
Barton ACT 2600
Australia
Tel: (+61 2) 62 73 42 99
Fax: (+61 2) 62 73 42 96

TERM Joint Stock Company
Bulgaria

Textron Marine & Land Systems
19401 Chef Menteur Highway
New Orleans
Louisiana 70129
USA
Tel: (+1 504) 245 66 00
Fax: (+1 504) 254 80 00

Typhoon Joint Stock Company
Kaluga 248009
Russia
Tel: (+7 0842) 22 78 83
Fax: (+7 0842) 52 23 57

United Defense LP
Armament Systems Division
4800 East River Road
Minneapolis
Minnesota 55421-1498
USA
Tel: (+1 612) 572 65 03
Fax: (+1 612) 572 25 67

United Defense LP
Defense Systems International Division
881 Martin Avenue
Box 58123
Santa Clara
California 05052
USA
Tel: (+1 408) 289 01 11
Tx: 6714210
Fax: (+1 408) 289 21 50

United Defense LP
Ground Systems Division
2830 De La Cruz Boulevard
Santa Clara
California 95052
USA
Tel: (+1 408) 289 01 11
Tx: 6714210
Fax: (+1 408) 289 21 50

United Defense LP
1525 Wilson Boulevard
Suite 700
Arlington
Virginia 22209
USA
Tel: (+703 312) 61 00
Fax: (+703 312) 61 48

United Defense LP
PO Box 15512
York
Pennsylvania 17405
USA
Tel: (+1 703) 312 61 00
Fax: (+1 703) 312 61 48

United States Army Tank-automotive and Armaments
Command (TACOM)
Warren
Michigan 48397-5000
USA
Tel: (+1 313) 574 57 88

Uraltransmash
Frontovikh Brigad Street 9
PO Box 620027
Ekaterinburg
Russia
Tel: (+3432) 34 45 43
Fax: (+3432) 34 46 61

Uralvagonzavod
622051
Vostochnoye Shosse 28
Nizhny Tagil
Sverdlovsk Region
Russia
Tel: (+7 3435) 23 17 74
Fax: (+7 3435) 23 34 92

Vickers Defence Systems
Armstrong Works
Scotswood Road
Newcastle Upon Tyne NE99 1BX
UK
Tel: (+44 191) 273 88 88
Fax: (+44 191) 273 23 24

Vickers OMC
12 Barnsley Road
Benoni 1500
Postbag X049
South Africa
Tel: (+27 11) 845 14 40
Fax: (+27 11) 845 13 79

Volgograd Tractor Plant
Joint Stock Company
1 Dzerzhinski Square
400041 Volgograd
Russia
Tel: (+7 8442) 77 14 44
Fax: (+7 8442) 71 14 55

Volsk Mechanical Plant (VMP)
Saratov Region
412680 Volsk
Russia
Tel: (+7 84) 59 33 26 77
Tx: 241622 CHAIKA

VOP 025 Novy Jicin sp
Dukelska 102
CZ-742 42 Senov u Noveho Jicina
Czech Republic
Tel: (+420 656) 78 33 16
Fax: (+420 656) 70 17 34

VOP 026 Sternberk sp
Olomoucka 175
CZ-785 01 Sternberk
Czech Republic
Tel: (+420 643) 48 31 11
Fax: (+420 643) 48 31 20

WIA Corporation
Fourth Floor
Kia Building
15 Yoido-dong
Youngdeungpo-gu
Seoul
South Korea
Tel: (+82 2) 786 94 10
Fax: (+82 2) 782 48 64

Wojskowe Zaklady Mechaniczne Siemianowice Slaskie
Ul Powstancow Slaskich 5/7
PL-41-100 Siemianowice Slaskie
Poland
Tel: (+48 32) 228 57 51
Fax: (+48 32) 228 12 52

Yugoimport SDPR
PO Box 89
Bulevar umetnosti 2
YU-11070 Novi Beograd
Yugoslavia
Tel: (+381 11) 311 27 43
Tx: 71000 JU SDPR
Fax: (+381 11) 130 26 31

Zaklady Mechaniczne Bumar-Labedy SA
ul Mechnaikow 9
PL-44-109 Gliwice
Poland
Tel: (+48) 734 51 11
Fax: (+48) 734 65 11

ZTS Dubnica
SI-01841 Dubnica nad Vahom
Slovakia
Tel: (+421 827) 227 05
Fax: (+421 827) 225 72

ZTS Martin AS
CSA 3
SI-036 57 Martin
Slovakia
Tel: (+42 1842) 323 71
Fax: (+42 1842) 328 22

INDEXES

Alphabetical index

To help users of this title evaluate the published data, *Jane's Information Group* has divided entries into three categories.
N NEW ENTRY Information on new equipment and/or systems appearing for the first time in the title.
V VERIFIED The editor has made a detailed examination of the entry's content and checked its relevancy and accuracy for publication in the new edition to the best of his ability.
U UPDATED During the verification process, significant changes to content have been made to reflect the latest position known to Jane's at the time of publication.
Items in italics refer to entries which have been deleted from this edition with the relevant page numbers from last year.

T

U

Weapon index by calibre

Manufacturers' index

Z